£37·15

HANDBOOK OF APPLIED HYDROLOGY

OTHER McGRAW-HILL HANDBOOKS OF INTE..

AMERICAN INSTITUTE OF PYHSICS · American Institute of Physics Handbook
AMERICAN SOCIETY OF MECHANICAL ENGINEERS · ASME Handbooks:
  Engineering Tables   Metals Engineering—Processes
  Metals Engineering—Design   Metals Properties
AMERICAN SOCIETY OF TOOL AND MANUFACTURING ENGINEERS:
  Die Design Handbook   Handbook of Fixture Design
  Manufacturing Planning and   Tool Engineers Handbook
    Estimating Handbook
ARCHITECTURAL RECORD · Time-Saver Standards
BEEMAN · Industrial Power Systems Handbook
BRADY · Materials Handbook
BURINGTON · Handbook of Mathematical Tables and Formulas
BURINGTON AND MAY · Handbook of Probability and Statistics with Tables
CARRIER AIR CONDITIONING COMPANY · Handbook of Air Conditioning System Design
CARROLL · Industrial Instrument Servicing Handbook
CONDON AND ODISHAW · Handbook of Physics
CONSIDINE · Process Instruments and Controls Handbook
CONSIDINE AND ROSS · Handbook of Applied Instrumentation
CROCKER · Piping Handbook
DAVIS · Handbook of Applied Hydraulics
DUDLEY · Gear Handbook
EMERICK · Heating Handbook
FACTORY MUTUAL ENGINEERING DIVISION · Handbook of Industrial Loss Prevention
FLÜGGE · Handbook of Engineering Mechanics
GARTMANN · De Laval Engineering Handbook
HARRIS · Handbook of Noise Control
HARRIS AND CREDE · Shock and Vibration Handbook
HEYEL · The Foreman's Handbook
HUSKEY AND KORN · Computer Handbook
JURAN · Quality Control Handbook
KALLEN · Handbook of Instrumentation and Controls
KING AND BRATER · Handbook of Hydraulics
KNOWLTON · Standard Handbook for Electrical Engineers
KOELLE · Handbook of Astronautical Engineering
KORN AND KORN · Mathematical Handbook for Scientists and Engineers
LASSER · Business Management Handbook
LAUGHNER AND HARGAN · Handbook of Fastening and Joining of Metal Parts
LEGRAND · The New American Machinists' Handbook
MACHOL · System Engineering Handbook
MAGILL, HOLDEN, AND ACKLEY · Air Pollution Handbook
MANAS · National Plumbing Code Handbook
MANTELL · Engineering Materials Handbook
MARKS AND BAUMEISTER · Mechanical Engineers' Handbook
MAYNARD · Industrial Engineering Handbook
MAYNARD · Top Management Handbook
MORROW · Maintenance Engineering Handbook
PERRY · Chemical Engineers' Handbook
PERRY · Engineering Manual
ROSSNAGEL · Handbook of Rigging
ROTHBART · Mechanical Design and Systems Handbook
SHAND · Glass Engineering Handbook
STANIAR · Plant Engineering Handbook
STREETER · Handbook of Fluid Dynamics
TOULOUKIAN · Retrieval Guide to Thermophysical Properties Research Literature

HANDBOOK OF APPLIED HYDROLOGY

A Compendium of Water-resources Technology

VEN TE CHOW, Ph.D., Editor-in-Chief

Professor of Hydraulic Engineering
University of Illinois

McGRAW-HILL BOOK COMPANY

New York   St. Louis   San Francisco   Düsseldorf   Johannesburg
Kuala Lumpur   London   Mexico   Montreal   New Delhi
Panama   Rio de Janeiro   Singapore   Sydney   Toronto

HANDBOOK OF APPLIED HYDROLOGY

Copyright © 1964 by McGraw-Hill, Inc. All Rights Reserved. Printed in the United States of America. No part of this publication may be reproduced, stored in a retrieval system, or transmitted, in any way or by any means, electronic, mechanical, photocopying, recording, or otherwise, without the prior written permission of the publisher. *Library of Congress Catalog Card Number* 63-13931.

ISBN 07-010774-2

9 – MAMM – 7 5

CONTRIBUTORS

Maurice L. Albertson, Ph.D., Sc.D.
Professor of Civil Engineering, Colorado State University, Fort Collins, Colorado
Section 7, Fluid Mechanics

John C. Ayers, Ph.D.
Oceanographer, Great Lakes Research Division, Institute of Science and Technology, and Professor of Oceanography, University of Michigan, Ann Arbor, Michigan
Section 23, Hydrology of Lakes and Swamps

Marion Clifford Boyer
Natural Resources Coordinator, Fresno County, California
Section 15, Streamflow Measurement

Ven Te Chow, Ph.D.
Professor of Hydraulic Engineering, University of Illinois, Urbana, Illinois
Section 1, Hydrology and Its Development; Section 2, Oceanography; Section 8-I, Frequency Analysis; Section 8-IV, Sequential Generation of Hydrologic Information; Section 14, Runoff; Section 26-II, Water-resources System Design by Operations Research; Section 29, Applications of Electronic Computers in Hydrology

S. V. Ciriacy-Wantrup, Dr.agr., Dr.habil.
Professor of Agricultural Economics, University of California, Berkeley, California
Section 28, Water Policy

Tate Dalrymple
Hydraulic Engineer, U.S. Geological Survey
Section 25-I, Flood Characteristics and Flow Determination

D. R. Dawdy
Hydraulic Engineer, U.S. Geological Survey
Section 8-III, Analysis of Variance, Covariance, and Time Series

J. W. Dixon
Water Resources Specialist, U.S. Area Redevelopment Administration
Section 26-I, Water-resources Planning and Development

Hans Albert Einstein, Dr.Tech.Scs.
Professor of Hydraulic Engineering, University of California, Berkeley, California
Section 17-II, River Sedimentation

CONTRIBUTORS

Le Roy Engstrom
Late Chief, River Control Branch, Tennessee Valley Authority
Section 25-III, Reservoir Regulation

Walter U. Garstka
Chief, Water Conservation Branch, U.S. Bureau of Reclamation
Section 10, Snow and Snow Survey

Charles S. Gilman, Sc.D.
Late Chief, Hydrometeorological Section, U.S. Weather Bureau
Section 9, Rainfall

Louis C. Gottschalk
Staff Geologist, Engineering Division, U.S. Soil Conservation Service
Section 17-I, Reservoir Sedimentation

Richard Hazen
Partner, Hazen and Sawyer, Engineers, New York
Section 18, Droughts and Low Streamflow

Robert L. Hobba
Hydrologist, U.S. Forest Service
Section 22, Hydrology of Forest Lands and Rangelands

H. N. Holtan
Director, U.S. Hydrograph Laboratory, U.S. Agricultural Research Service
Section 12, Infiltration

H. E. Hudson, Jr.
Partner, Hazen and Sawyer, Engineers, New York
Section 18, Droughts and Low Streamflow

Stifel W. Jens
Consulting Engineer, St. Louis, Missouri
Section 20, Hydrology of Urban Areas

Don Kirkham, Ph.D., D.H.C.
Charles F. Curtiss Distinguished Professor in Agriculture, and Professor of Soils and Physics, Iowa State University, Ames, Iowa
Section 5, Soil Physics

Edward A. Lawler
Chief, Hydrology and Reservoir Regulation Section, Ohio River Division, Corps of Engineers, U.S. Army
Section 25-II, Flood Routing

Howard W. Lull, Ph.D.
Chief, Division of Watershed Management Research, Northwestern Forest Experiment Station, U.S. Forest Service
Section 6, Ecological and Silvicultural Aspects

CONTRIBUTORS

J. A. Mabbutt
Geomorphologist, Division of Land Research and Regional Survey, Commonwealth Scientific and Industrial Research Organization, Australia
Section 24, Hydrology of Arid and Semiarid Regions

N. C. Matalas, Ph.D.
Hydraulic Engineer, U.S. Geological Survey
Section 8-III, Analysis of Variance, Covariance, and Time Series

George B. Maxey, Ph.D.
Research Professor of Geology and Hydrology, Desert Research Institute, University of Nevada, Reno, Nevada
Section 4-I, Hydrogeology

M. B. McPherson
Professor of Hydraulic Engineering, University of Illinois, Urbana, Illinois
Section 20, Hydrology of Urban Areas

Mark F. Meier, Ph.D.
Project Hydrologist, U.S. Geological Survey
Section 16, Ice and Glaciers

Victor Mockus
Hydraulic Engineer, U.S. Soil Conservation Service
Section 21, Hydrology of Agricultural Lands

G. W. Musgrave
Consultant in Hydrology, St. Petersburg, Florida
Section 12, Infiltration

T. J. Nordenson
Chief, Hydrologic Investigations Section, U.S. Weather Bureau
Section 25-IV, River Forecasting

Harold O. Ogrosky
Chief, Hydrology Branch, U.S. Soil Conservation Service
Section 21, Hydrology of Agricultural Lands

Sverre Petterssen, Ph.D.
Professor of Meteorology and Chairman, Department of the Geophysical Sciences, The University of Chicago, Chicago, Illinois
Section 3, Meteorology

Sheppard T. Powell
Consulting Engineer, Baltimore, Maryland
Section 19, Quality of Water

M. M. Richards
Special Assistant, Hydrologic Services Division, U.S. Weather Bureau
Section 25-IV, River Forecasting

CONTRIBUTORS

J. Marvin Rosa
Hydraulic Engineer, U.S. Agricultural Research Service
Section 22, Hydrology of Forest Lands and Rangelands

Edward J. Rutter
Chief (retired), Flood Control Branch, Tennessee Valley Authority
Section 25-III, Reservoir Regulation

Daryl B. Simons, Ph.D.
Acting Chief, Civil Engineering Research Center, and Professor of Civil Engineering, Colorado State University, Fort Collins, Colorado
Section 7, Fluid Mechanics

R. O. Slatyer, Sc.D.
Climatologist, Division of Land Research and Regional Survey, Commonwealth Scientific and Industrial Research Organization, Australia
Section 24, Hydrology of Arid and Semiarid Regions

Herbert C. Storey
Director, Division of Watershed Management and Recreation Research, U.S. Forest Service
Section 22, Hydrology of Forest Lands and Rangelands

Arthur N. Strahler, Ph.D.
Professor of Geomorphology, Columbia University, New York
Section 4-II, Quantitative Geomorphology of Drainage Basins and Channel Networks

David Keith Todd, Ph.D.
Professor of Civil Engineering, University of California, Berkeley, California
Section 13, Groundwater

Frank J. Trelease
Dean and Professor of Law, College of Law, The University of Wyoming, Laramie, Wyoming
Section 27, Water Law

Frank J. Veihmeyer, Ph.D.
Professor of Irrigation, Emeritus, University of California, Davis, California
Section 11, Evapotranspiration

Gilbert F. White, Ph.D.
Professor of Geography, The University of Chicago, Chicago, Illinois
Section 25-V, Floodplain Adjustments and Regulations

Vujica M. Yevdjevich, Eng.D.
Professor of Civil Engineering, Colorado State University, Fort Collins, Colorado
Section 8-II, Regression and Correlation Analysis

James H. Zumberge, Ph.D.
President, Grand Valley State College, Allendale, Michigan
Section 23, Hydrology of Lakes and Swamps

PREFACE

Water not only serves as a vital substance for human existence but also plays an important role in advancing civilization. Owing to the rapid growth in world economy and civilization, the need for the development of water resources has become more urgent than ever before. Many governments and private organizations have already taken positive actions to cope with this situation. The United Nations (UNESCO) has scheduled an International Hydrological Decade beginning in 1965 for promoting and coordinating long-term international cooperation of programs in hydrology.

Along with all the intensive activities in promoting hydrology and water resources, there has been a tremendous increase and expansion in the scientific and technological knowledge about water, and there is a great need for an authoritative compilation of such knowledge. Since water is related to so many things in nature as well as in human society, this knowledge is extremely broad and interdisciplinary.

This handbook contains a wealth of information on hydrology and water-resources technology that has not been collected in a single volume, and it provides an interdisciplinary coverage of the up-to-date information on the subject that has not been so treated elsewhere. The handbook is intended to serve practicing scientists, engineers, consulting engineering firms, administrators, planners, designers, college faculties, and undergraduate and graduate students in various fields relating to water, including agriculture, agronomy, biology, chemistry, city planning, climatology, economics, forestry, geography, geology, horticulture, law, meteorology, oceanography, physics, recreation, and many branches of engineering such as agricultural, civil, drainage, hydraulic, irrigation, flood-control, hydropower, sanitary, and water-resources engineering To the students, the handbook is not merely a general reference; it is equivalent to several textbooks and various portions of it can also be selected as text material for many subjects which are usually presented in science and engineering.

As hydrology is not an exact science, application of hydrologic knowledge to practical problems requires a great deal of rich experience and sound judgment of the hydrologist. For this reason, one of the basic requirements to assure the success of this handbook project is that the contributors should have been very active in the field and that they should have long reached their senior positions in their profession. It is indeed most heartening that a panel of outstanding authorities, who are actively contributing to advances in the field of hydrology, have willingly joined in producing this work. The present status, the latest theories, and the most promising methods of analysis are presented for various fields of the hydrologic profession. Of the 45 contributors, 23 are government experts, 17 are university educators, and 5 are consulting engineers. This distribution of the contributors' background provides a broad spectrum of knowledge, experience, and viewpoints in the field of hydrology.

Because handbook size is limited, careful selection of topics is necessary. An attempt was made to obtain a comprehensive treatment covering various aspects in applied hydrology. The sections in general are interrelated yet stand by themselves with but minimum dependence on other sections. Some duplication from section to section was permitted to reduce the necessity for continual cross-reference. Ample references are supplied at the end of each section to assist the reader in pursuing the subject further.

Various sections of the handbook can be roughly divided into four groups. The first group deals with closely related sciences upon which hydrology depends heavily, including oceanography, meteorology, hydrogeology, geomorphology, soil physics, plant ecology, silviculture, fluid mechanics, statistics, probability, operations research, and electronic computers. The second group covers various phases of the hydrologic cycle and phenomena, including rainfall, snow, evapotranspiration, infiltration, groundwater, runoff, ice and glaciers, reservoir and river sedimentation, droughts and low streamflow, and quality of water. The third group is devoted to practice and application of hydrology in such various fields as flow determination, flood routing, streamflow measurement, reservoir regulation, river forecasting, urban hydrology, agricultural lands, forest lands and rangelands, lakes and swamps, and arid and semiarid regions. The fourth group is related to some socio-economic aspects of hydrology, including water-resources planning and development, floodplain adjustment and regulation, water law, and water policy. In modern society, water is not only a necessity for daily living but also an important commodity which has socio-economic value. Although the fourth group is not directly considered in the definition of hydrology as a physical science, it is of great importance to practicing hydrologists who apply the knowledge of hydrology to the planning and development of water resources as the ultimate objective of the science.

The editor-in-chief is deeply indebted to all those who have generously given advice, information, and encouragement on the preparation of this handbook. He is especially grateful to the contributors for their willingness and cooperation, to the publishers for their support and assistance, and to the members of his family for their understanding that almost all of his spare time in the past several years had to be spent in burning midnight oil in order to bring this project to its completion.

Ven Te Chow

CONTENTS

Contributors v
Preface ix

Section 1.	Hydrology and Its Development	1-1
Section 2.	Oceanography	2-1
Section 3.	Meteorology	3-1
Section 4.	Geology	4-1
PART I.	*Hydrogeology*	4-1
PART II.	*Quantitative Geomorphology of Drainage Basins and Channel Networks*	4-39
Section 5.	Soil Physics	5-1
Section 6.	Ecological and Silvicultural Aspects	6-1
Section 7.	Fluid Mechanics	7-1
Section 8.	Statistical and Probability Analysis of Hydrologic Data	8-1
PART I.	*Frequency Analysis*	8-1
PART II.	*Regression and Correlation Analysis*	8-43
PART III.	*Analysis of Variance, Covariance, and Time Series* . .	8-68
PART IV.	*Sequential Generation of Hydrologic Information* . .	8-91
Section 9.	Rainfall	9-1
Section 10.	Snow and Snow Survey	10-1
Section 11.	Evapotranspiration	11-1
Section 12.	Infiltration	12-1
Section 13.	Groundwater	13-1
Section 14.	Runoff	14-1
Section 15.	Streamflow Measurement	15-1
Section 16.	Ice and Glaciers	16-1
Section 17.	Sedimentation	17-1
PART I.	*Reservoir Sedimentation*	17-1
PART II.	*Channel Sedimentation*	17-35
Section 18.	Droughts and Low Streamflow	18-1
Section 19.	Quality of Water	19-1
Section 20.	Hydrology of Urban Areas	20-1
Section 21.	Hydrology of Agricultural Lands	21-1
Section 22.	Hydrology of Forest Lands and Rangelands . . .	22-1
Section 23.	Hydrology of Lakes and Swamps	23-1
Section 24.	Hydrology of Arid and Semiarid Regions	24-1
Section 25.	Hydrology of Flow Control	25-1
PART I.	*Flood Characteristics and Flow Determination* . .	25-1
PART II.	*Flood Routing*	25-34
PART III.	*Reservoir Regulation*	25-60
PART IV.	*River Forecasting*	25-98
PART V.	*Floodplain Adjustments and Regulations*	25-112

Section 26.	**Water-resources Planning and Development**	26-1
PART I.	*Planning and Development*	26-1
PART II.	*System Design by Operations Research*	26-30
Section 27.	**Water Law**	27-1
Section 28.	**Water Policy**	28-1
Section 29.	**Applications of Electronic Computers in Hydrology**	29-1

Index follows Section 29.

Section 1

HYDROLOGY AND ITS DEVELOPMENT

VEN TE CHOW, *Professor of Hydraulic Engineering, University of Illinois.*

I. Definitions and Scope................................... 1-1
   A. Hydrology as a Science............................. 1-1
   B. Hydrologic Cycle................................... 1-2
   C. Scope of Hydrology................................. 1-5
II. Historical Development................................ 1-7
   A. Period of Speculation (Ancient–1400)............... 1-7
   B. Period of Observation (1400–1600).................. 1-7
   C. Period of Measurement (1600–1700).................. 1-7
   D. Period of Experimentation (1700–1800).............. 1-7
   E. Period of Modernization (1800–1900)................ 1-8
   F. Period of Empiricism (1900–1930)................... 1-8
   G. Period of Rationalization (1930–1950).............. 1-9
   H. Period of Theorization (1950–date)................. 1-10
III. Professional Status.................................. 1-11
   A. Hydrologic Organizations........................... 1-11
      1. International Organizations.................... 1-11
      2. Organizations in the United States............. 1-12
      3. Organizations in the U.S.S.R................... 1-15
   B. Hydrologic Publications............................ 1-16
      1. Books.. 1-16
      2. Periodicals.................................... 1-16
   C. Hydrologic Education............................... 1-17
IV. References... 1-17

I. DEFINITIONS AND SCOPE

A. Hydrology as a Science

Hydrology in its broadest sense is the science that relates to water. Since it deals with water primarily on earth, it is an earth science. For practical reasons, however, hydrology is sometimes limited in various respects; for example, it may not cover all studies of oceans (oceanography), and it is not concerned with the medical uses of water (medical hydrology).

The author is indebted for helpful comments and assistance to his colleagues and friends: W. C. Ackermann, W. B. Langbein, R. K. Linsley, Jr., and Abel Wolman.

In both Europe and the United States, "hydrology" has been used to denote the study of water below the land surface, while other terms such as "hydrography" and "hydrometry" denoted the study of the surface water. However, these terms now have their specific meanings. *Hydrology* refers to the general science of water or it may have other specific definitions to be given later. *Hydrography* is the science that describes the physical features and conditions of all the waters of the earth's surface, particularly, charting the bodies of water for navigational purposes. For example, the Hydrographic Department of the British Admiralty, established in 1795, is under the Hydrographer to the Admiralty, and the U.S. Hydrographic Office, established in 1866, is under the Chief of Naval Operations in the U.S. Department of Navy. *Hydrometry* is the science of the measurement of water.

Hydrology is not entirely a pure science, for it has many practical applications. To emphasize its practical importance, the term "applied hydrology" is commonly used. Since numerous applications of hydrologic knowledge are in the field of hydraulic, sanitary, agricultural, water resources, and other branches of engineering, the name "engineering hydrology" is also employed. Also, the expression "scientific hydrology" has been originated in Europe in order to distinguish its field from medical hydrology or from hydrology of spas. However, the use of these expressions is rather to identify their purposes than to draw any lines of demarcation which are knowingly impossible.

Various elaborate definitions of hydrology have been proposed. Webster's Third New International Dictionary (Merriam-Webster, 1961 [1]) describes hydrology as "a science dealing with the properties, distribution, and circulation of water; specifically, the study of water on the surface of the land, in the soil and underlying rocks, and in the atmosphere, particularly with respect to evaporation and precipitation."

To define hydrology as a science, as distinguished from the application of science-in-general to problems of water assessment and management, the Ad Hoc Panel on Hydrology of the Federal Council for Science and Technology (established by the President of the United States in 1959) recommended the following definition [2]: "*Hydrology is the science that treats of the waters of the Earth, their occurrence, circulation, and distribution, their chemical and physical properties, and their reaction with their environment, including their relation to living things.*" The domain of hydrology embraces the full life history of water on the Earth."

Among a number of definitions to emphasize the practical importance of hydrology concerning water resources on the land, Wisler and Brater [3] offer the following: "*Hydrology is the science that deals with the processes governing the depletion and replenishment of the water resources of the land areas of the earth.* It is concerned with the transportation of water through the air, over the ground surface, and through the strata of the earth. It is the science that treats of the various phases of the hydrologic cycle."

B. Hydrologic Cycle

Hydrology can be seen as the scientific examination and appraisal of the whole continuum of a *hydrologic*, or *water, cycle*. The hydrologic cycle is by no means a simple link but a group of numerous arcs which represent the different paths through which the water in nature circulates and is transformed. These arcs penetrate three parts of the total earth system: atmosphere, hydrosphere, and lithosphere. The *atmosphere* is a gaseous envelope above the hydrosphere; the *hydrosphere* is the bodies of water that cover the surface of the earth; and the *lithosphere* is the solid rock below the hydrosphere. The activities of water extend through these three parts of the earth system from an average depth of about a half mile in the lithosphere to a height of about 10 miles in the atmosphere. They create a gigantic system of great complexity and intricacy.

The hydrologic cycle has no beginning or end, as water evaporates from the oceans and the land and becomes a part of the atmosphere. The evaporated moisture is lifted and carried in the atmosphere until it finally precipitates to the earth, either on land or in the oceans. The precipitated water may be intercepted or transpired

DEFINITIONS AND SCOPE 1-3

by plants, may run over the ground surface and into streams, or may infiltrate into the ground. Much of the intercepted and transpired water and the surface runoff returns to the air through evaporation. The infiltrated water may percolate to deeper zones to be stored as groundwater which may later flow out as springs or seep into streams as runoff and finally evaporate into the atmosphere to complete the hydrologic cycle. Thus, the hydrologic cycle undergoes various complicated processes of evaporation, precipitation, interception, transpiration, infiltration, percolation, storage, and runoff. Many diagrams have been designed to illustrate the hydrologic

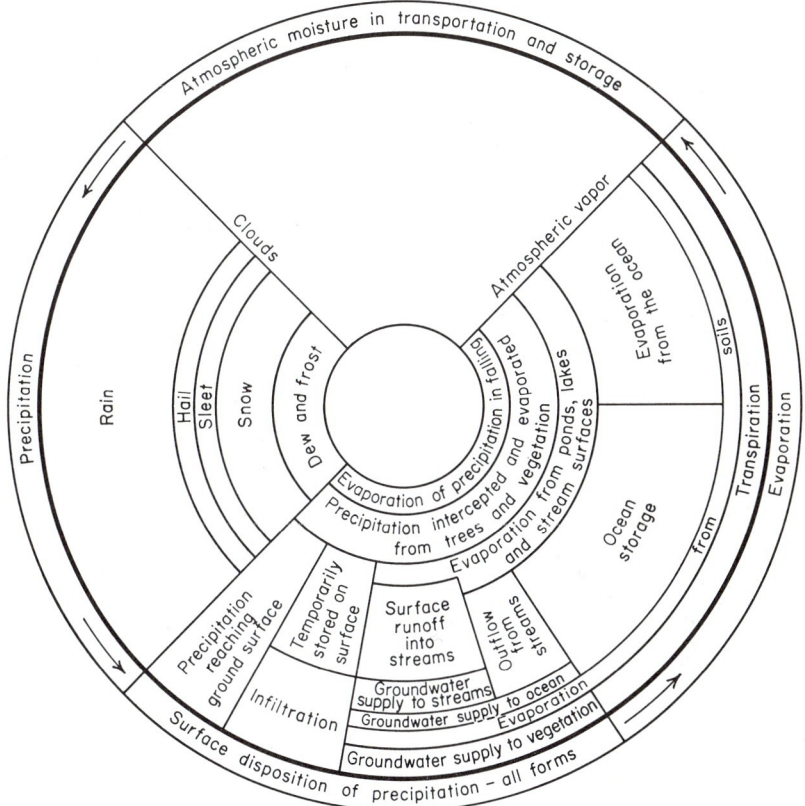

FIG. 1-1. The hydrologic cycle—a qualitative representation. (*Horton* [4].)

cycle; some are qualitative (Fig. 1-1), some descriptive (Fig. 1-2), and some quantitative (Fig. 1-3).

Wolman [7] wrote that 97 per cent of all the water in the world, or one quadrillion (10^{15}) acre-ft, is contained in the oceans. If the world were a uniform sphere, this quantity would be sufficient to cover it to a depth of 800 ft. The total amount of fresh water is estimated at about 33 trillion acre-ft. It is distributed roughly as follows: 75 per cent in polar ice and glaciers; 14 per cent in groundwater between depths of 2,500 and 12,500 ft; 11 per cent in groundwater at depths less than 2,500 ft; 0.3 per cent in lakes; 0.06 per cent as soil moisture; 0.035 per cent in atmosphere; and 0.03 per cent in rivers. These are, however, stationary estimates of distribution. While the water content of the atmosphere is relatively small at any given moment, immense quantities of water pass through it annually. The annual precipitation on land surfaces alone is 7.7 times as great as the moisture contained in the entire atmosphere at any one time, that is, about 30 times as great as the moisture in the air over the land. According to Reichel [8], the mean annual precipitation for the entire earth is about 86 cm/year (34 in./year) which under stationary conditions is balanced by an equally large evaporation amount. Thus, the average evaporation for the whole earth amounts to 2.37 mm (about 0.1 in.) of water per day.

The amounts of evaporation, precipitation, runoff, and other hydrologic quantities are not evenly distributed on the earth, either geographically or momently. The overall distribution of precipitation can be shown schematically in Fig. 1-4, in which the numerical values in average correspond to estimates only for the continental

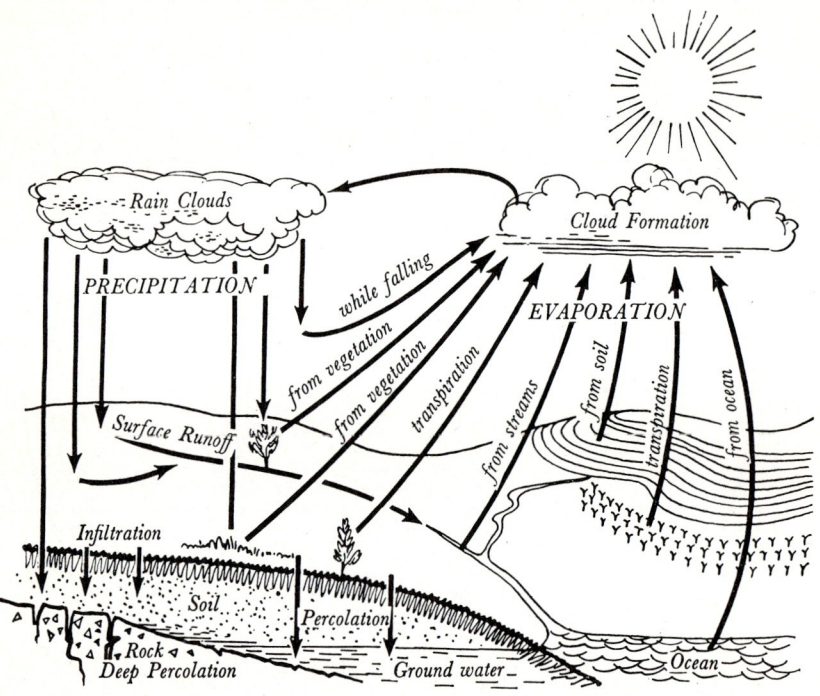

Fig. 1-2. The hydrologic cycle—a descriptive representation. (*Ackermann, Colman, and Ogrosky* [5].)

United States. About 70 to 75 per cent of the precipitation is returned to the atmosphere by evapotranspiration and direct evaporation, while the remaining 30 per cent becomes runoff. About one-fourth of the runoff is diverted. About two-thirds of that diverted is fed back into the stream and eventually goes to oceans for storage and evaporation, and the remaining one-third is consumed and returns to the atmosphere directly.

The quantities of water going through any arc or arcs of the hydrologic cycle can be evaluated usually by the so-called *hydrologic equation*, which simply states as

$$I - O = \Delta S \tag{1-1}$$

where I is the inflow during a given period to a problem area including, for instance, the total inflow of the channel and overland runoff to the area above the ground surface and of the groundwater across the boundaries of the area plus the total precipitation over the area during the period; O is the outflow during the given period to the area including, for instance, total evaporation, transpiration, and outflow of surface runoff in channel and on overland from the area above the surface plus the overflow of groundwater across the boundaries of the area; and ΔS is the change in storage in various forms of retention, depression, and interception. Equation (1-1) is essentially a form of continuity equation.

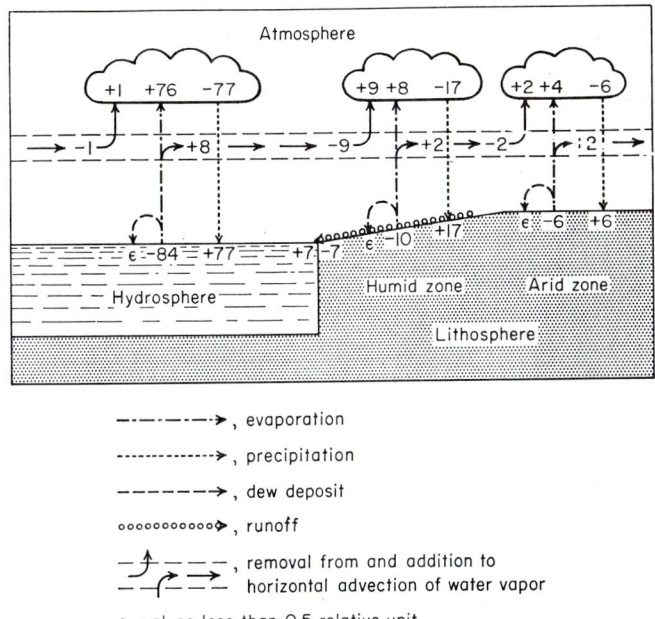

FIG. 1-3. The hydrologic cycle—a quantitative representation. 100 relative units = 85.7 g/cm²/year or 857 mm (33.8 in.) global annual mean of precipitation. (Lettau [6]).

C. Scope of Hydrology

In view of the extensiveness of the hydrologic cycle to be dealt with, hydrology is a very broad science. It is an interdisciplinary science because it borrows heavily from many other branches of science and integrates them for its own interpretation and uses. The supporting sciences required for hydrologic investigations are physics, chemistry, biology, geology, fluid mechanics, mathematics, statistics, and operations research. As the arc of the hydrologic cycle reaches into the atmosphere, hydrology traverses the domains of *hydrometeorology*, *meteorology*, and *climatology*. In the hydrosphere, hydrology crosses, or embodies as it may be so defined, the domains of *potamology* (surface streams), *limnology* (lakes), *cryology* (snow and ice), *glaciology*, and *oceanology*. In the lithosphere, hydrology relates to *agronomy*, *hydrogeology* (emphasizing hydrologic aspects), *geohydrology* (emphasizing geologic aspects), and *geomorphology*. As water affects plants as well as animals, hydrology extends itself into *plant ecology*, *silviculture*, *biohydrology* (emphasizing hydrologic aspects) and *hydrobiology* (emphasizing biologic aspects). Since hydrology is a science that underlies the development and control of water resources, it has its important influence in agriculture, forestry, geography, watershed management, political science (water law and policy), economics (*hydroeconomics*), and sociology; and it has practical applications in structural design, water supply, waste-water disposal and treatment, irrigation, drainage, hydropower, flood control, navigation, erosion and sediment control, salinity control, pollution abatement, recreational use of water, fish and wildlife preservation, insect control, and coastal works.

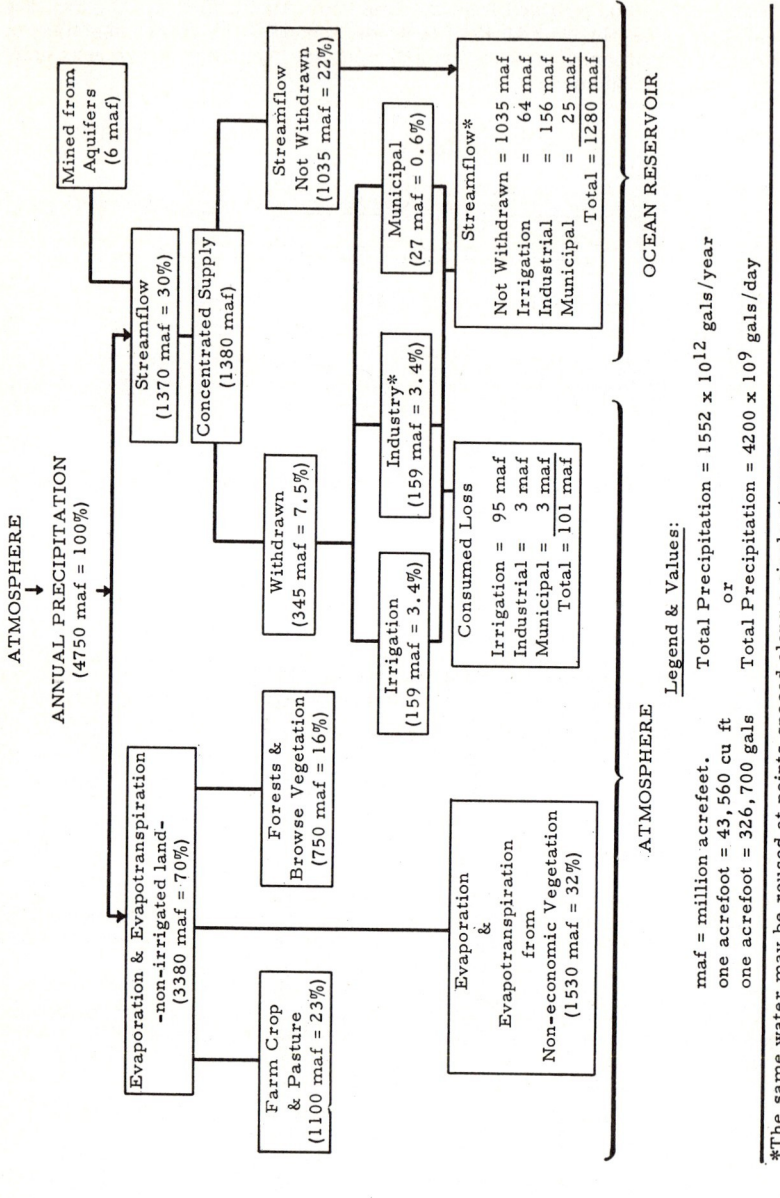

Fig. 1-4. Average distribution of precipitation in the continental United States. (*Wolman* [7].)

II. HISTORICAL DEVELOPMENT

Many words have been written on the history of hydrology [9–22]. Among these writings the work by Meinzer [14, pp. 8–30; 10] is most extensive. In general, the historical development of hydrology can be viewed through a series of periods. Since these periods may overlap, their time division should not be considered exact.

A. Period of Speculation (Ancient–1400)

From ancient times to about 1400 A.D., the concept of the hydrologic cycle was speculated upon by many philosophers, including Homer (about 1000 B.C.), Thales, Plato, and Aristotle in Greece; Lucretius, Seneca, and Pliny in Rome; and many Bible scholars during that time [14, 19, 22]. While most of these philosophical concepts were erroneous, Marcus Vitruvius, who lived about the time of Christ, conceived a theory which is now generally accepted, since he stated that the groundwater is for the most part derived from rain and snow by infiltration from the surface. Thus, the theory of Vitruvius may be considered the forerunner of modern concepts of the hydrologic cycle.

During this period, however, men may have learned a great deal of practical knowledge of hydrology through the constructions of great hydraulic works known in history, such as the ancient Arabian wells, the Persian kanats, the Egyptian and Mesopotamian irrigation projects, the Roman aqueducts, the water-supply and drainage projects in the Indus Valley, and the Chinese irrigation systems, canals, and flood-control works.

B. Period of Observation (1400–1600)

During the period known as the Renaissance, a gradual change was perceptible from the pure philosophical concepts of hydrology toward the observational science of the present day. For example, based on observations, Leonardo da Vinci and Bernard Palissy [23] achieved a correct understanding of the hydrologic cycle, especially the infiltration of rain and return of the water through springs.

C. Period of Measurement (1600–1700)

The modern science of hydrology may be considered to have begun in the 17th century with measurements. For example, Pierre Perrault [24] measured rainfall, evaporation, and capillarity in the drainage basin of the Seine; Edmé Mariotte [25] computed river discharges in the Seine at Paris after measuring the channel cross section and the flow velocity; and Edmond Halley [11] measured the rate of evaporation and stream discharge for the study of the Mediterranean Sea. From these measurements, they were able to draw correct conclusions on the observed hydrologic phenomena.

During the same period, the study of artesian water was pursued even before the emergence of the science of geology. Pioneers in such studies included Giovanni Cassini, Bernardini Ramazzini, and Antonio Vallisnieri [16].

D. Period of Experimentation (1700–1800)

During the 18th century, hydraulic experimental studies of hydrology were flourishing. As a result, much in the way of new discoveries and understanding of hydraulic principles was obtained. Notable examples are Bernoulli's piezometer, the Pitot tube, Woltman's current meter, Smeaton's scale models, the Borda tube, D'Alembert's principle, the Bernoulli theorem, and Chézy's formula [17, 18]. All such developments have accelerated greatly the beginning of hydrologic studies on a quantitative basis.

In the same period, important contributions to hydrology include the siphon theory of ebbing and flowing springs by J. T. Desaguliers [14], publication of a book on rivers and torrents by Paul Frisi [26], and extension of Mariotte's work on infiltration by La Metherie [9].

E. Period of Modernization (1800–1900)

The 19th century was in many ways the grand era of experimental hydrology that had been started in the preceding period of experimentation and was now greatly modernized such that the foundation of the modern science of hydrology was firmly laid. The sign of modernization can be seen from many significant contributions to modern hydrology. Most contributions, however, were in groundwater hydrology and in surface-water measurement.

In the field of groundwater, the knowledge of geology was for the first time applied to hydrologic problems by William Smith [27], and also many basic discoveries were made, including Hagen-Poiseuille's equation of capillary flow (1839, 1840) [28, 29]; Darcy's law of groundwater flow (1856) [30]; Dupuit-Thiem's well formula (1863, 1906) [31, 32]; and Ghyben-Herzberg's principle of salt-water balance (1889, 1901) [33, 34]. Other significant contributions were also made, notably by Eugène Belgrand, Abbé Paramelle, Jean Dumas, Gustave Dumont, G. A. Daubrée, Édouard Imbeaux, Léon Pochet, Edmond Maillet, and Édouard Martel in France; O. E. Meyer, K. G. Bischof, Adolph Thiem, and Konrad Keilhack in Germany; William Whittaker and H. B. Woodwood in England; Eugene Dubois, J. Pennink, and J. J. Versluys in Holland; René D'Andrimont in Belgium; J. G. Richert in Sweden; P. Ototzky and Alexander Lebedeff in Russia; Albert Heim and Arnold Engler in Switzerland; D. Spataro, G. Cuppari, and M. Canavari in Italy; G. Garcia in Spain; Eduard Suess and Phillip Forchheimer in Austria; and T. C. Chamberlin, J. W. Powell, R. T. Hill, WJ McGee, I. C. Russell, N. H. Darton, Robert Hay, G. K. Gilbert, Frank Leverett, Warren Upham, G. H. Eldridge, W. H. Norton, T. W. Vaughn, Edward Orton, S. W. McCallie, W. D. Johnson, Allen Hazen, F. H. King, C. S. Slichter, E. W. Hilgard, and L. J. Briggs in the United States.

In the field of surface water, hydrometry was greatly advanced, including the development of many flow formulas and measuring instruments and the beginning of the systematic stream gaging. Outstanding contributions are Humphreys and Abbot's discharge measurement in the Mississippi River (1861) [35]; publication of Francis' weir-discharge formula (1855) [36]; Ganguillet and Kutter's determination of Chézy's coefficient (1869) [37]; proposal of the Manning's flow formula (1889) [38]; and development of the Ellis and Price current meters (1870, 1885) [39–40]. Significant flow measurements were made in many rivers [20, pp. 13–14]. In 1850 the Congress of the United States authorized a topographical and hydrographical survey of the delta of the Mississippi River. In November, 1867, the first international discharge measurement was organized on the Rhine River at Basle [41].

In the field of evaporation, John Dalton (Ref. 4 in Sec. 11) first recognized the relationship between evaporation and vapor pressure (1802) and thus established Dalton's law. In the field of rainfall, J. F. Miller (1849) [42] made a basic attempt to correlate rainfall and altitude and Lorin Blodget (1857) [43] published an important book describing the rainfall distribution in the United States.

During this period, several governmental hydrologic agencies were founded in the United States, including the U.S. Army Corps of Engineers (1802), the Geological Survey (1879), the Weather Bureau (1891), and the Mississippi River Commission (1893). Important hydrologic studies on engineering works in the United States during the time were the planning of the Erie Canal by DeWitt Clinton [44] and the water supply investigation for the Croton Aqueduct in New York.

F. Period of Empiricism (1900–1930)

Although much work on the modernization of hydrology had been started during the 19th century, the development of quantitative hydrology was still immature.

The science of hydrology was largely empirical, as the physical basis for most quantitative hydrologic determinations was not well known, nor were many research programs of any consequence conducted to produce quantitative information for use by hydrologists and engineers in solving practical problems. During the last part of the 19th century and the following 30 years or so, empiricism in hydrology became more evident; for example, hundreds of empirical formulas were proposed [45] and the selection of their coefficients and parameters had to depend mainly upon judgment and experience.

As most empirical approaches to the solution of practical hydrologic problems were soon found to be unsatisfactory, many governmental agencies increased their effort in hydrologic investigations, and many technical societies were organized either primarily or secondarily for the advancement of the science of hydrology.

Major governmental agencies founded in the United States during this period that are interested in hydrology as part of their functions include the Bureau of Reclamation (1902), the Forest Service (1906), the Pittsburgh Flood Commission (1908), the Miami Conservancy District (1914), the Los Angeles County Flood Control District (1915), the Franklin County Conservancy District (1915), and the U.S. Army Engineers Waterways Experiment Station (1928).

In 1922, the Section of Scientific Hydrology, later called the International Association of Scientific Hydrology (IASH), was organized in the International Union of Geodesy and Geophysics (IUGG). The American Geophysical Union (AGU) was organized in the National Research Council of the National Academy of Sciences in 1919 as a member of IUGG, and the Section of Hydrology was created in 1930 as a constituent unit of AGU and as the American representation to IASH. The first All-Russian Hydrologic Congress was held in Leningrad in 1924. The first Baltic Hydrologic Conference took place in Riga in 1926.

During this period, several other international organizations that related in part to hydrology were also founded, including the International Association of Physical Oceanography (1919), the International Society of Soil Science (1924), the World Power Conference (1924), and the International Congress on Large Dams (1928).

G. Period of Rationalization (1930–1950)

During this period there emerged many great hydrologists who used rational analysis instead of empiricism to solve hydrologic problems. In 1932, Sherman [46] made a distinct advance in hydrologic thought by demonstrating the use of the unit hydrograph for translating rainfall excess into runoff hydrograph. In 1933, Horton [47] initiated the most successful approach to date in the problem of determining rainfall excess on the basis of infiltration theory. In 1935, Theis [48] introduced the nonequilibrium theory which revolutionized the concept of hydraulics of wells. In 1941, Gumbel [49] proposed the use of the extreme-value distribution for frequency analysis of hydrologic data. He and many others revitalized the use of statistics in hydrology advocated earlier by Hazen [50]. In 1944, Bernard [51] discussed the role of meteorology, thus marking the beginning of the science of hydrometeorology. In 1950, Einstein (Ref. 17 in Sec. 17-II) developed the bed-load function which introduced theoretical analysis to sedimentation studies.

A notable development in this period was the establishment of many hydraulic and hydrologic laboratories throughout the world, including the U.S. National Hydraulics Laboratory (1932). In the United States, more governmental agencies were organized, reorganized, or simply changed their names to strengthen their activities relating to water studies, such as the Tennessee Valley Authority (1933), the Muskingum Valley Conservancy District (1933), the Soil Conservation Service (1935), the Weather Bureau (1940), and the Bureau of Public Roads (1949). During the economic depression of the 1930s, many committees were set up to investigate the national water resources, including the President's Committee on Water Flow (1934), the Mississippi Valley Committee of the Public Works Commission (1934), the National Resources Board (1934), and the National Resources Committee (1935).

H. Period of Theorization (1950–date)

Since about 1950, theoretical approaches have been used extensively in hydrologic problems. As many rational hydrologic principles have been proposed, they can now be subjected readily to mathematical analysis. As sophisticated instruments and high-speed computers are being developed, they can now be employed to measure delicate hydrologic phenomena and to solve complicated mathematical equations involved in the application of hydrologic theories. Furthermore, the emergence of modern fluid mechanics from the traditional hydraulics has helped greatly in promoting the development of theoretical hydrology. Examples of theoretical hydrologic studies are the linear and nonlinear analysis of hydrologic systems, the adoption of transient and statistical concepts in groundwater hydrodynamics, the application of heat- and mass-transfer theories to evaporation analysis, the study of energetics and dynamics of soil moisture, the sequential generation of hydrologic data, and the use of operations research in water-resources system design.

With the increase of world population and the improvement of economic conditions after World War II, there has been a rapidly growing need for solving all kinds of water problems, and thus great interest and concern have been generally developed in basic research and education in hydrology as well as in water resources. This can be seen through many commission and committee activities in the United States. In 1950, the President's Water Resources Policy Commission (Cooke Commission) was established to investigate and make recommendations on the development, utilization, and conservation of water resources in the United States [52]. In the same year, the Water Policy Panel of the Engineers Joint Council was organized [53]. In 1953, a Task Force on Water Resources and Power was set up in the Commission on Organization of the Executive Branch of the Government (Second Hoover Commission) to issue a report on water resources and power [54]. In 1954, a Presidential Advisory Committee on Water Resources Policy was created [55]. In 1959, a Senate Select Committee on National Water Resources (Kerr Committee) was formed to study water-resources problems in the United States for the future up to 1980 (Ref. 20 in Sec. 26-I). As a result of the investigations by the Kerr Committee, Senate Bill S-2 (Anderson Bill or Water Resources Research Act) was passed in 1964 to establish water-resources research centers in colleges and universities in all states to conduct basic research in water resources and to promote a national program of water research. Also, reports on water resources and hydrology were issued by the Federal Council for Science and Technology [27] and the National Academy of Sciences [7, 56]. To investigate the status and needs in hydrology, a Committee was formed in the American Geophysical Union [57]. Concerning education in hydrology, a Universities Council on Hydrology (UCOH) was organized in 1963 [58].

International activities in water resources and hydrology were also developed. In 1959, a Water Resources Development Center (WRDC) [59] was established in the United Nations to promote coordinated efforts for the development of water resources in member nations. Hydrologic studies of specific problems were encouraged by many other organizations, including the United Nations Educational, Scientific and Cultural Organization (UNESCO) through its Arid Zone Program, the World Meteorological Organization (WMO), the Food and Agriculture Organization (FAO), the World Health Organization (WHO), and the International Atomic Energy Agency (IAEA). In October, 1961, an International Hydrologic Decade (IHD) to coordinate international research and training programs in hydrology was proposed [60, 61] and approved in principle by the Executive Committee of the IASH. UNESCO adopted and rectified the IHD proposal in September, 1962 [62] and made a resolution in November-December of the same year to promote a long-term program of international cooperation in hydrology after the framework of IHD, to begin its operation in 1965. In the United States, a Conference of American Hydrologists [57] was created in 1962 in the Ad Hoc Committee on International Programs in Atmospheric Sciences and Hydrology (CIPASH), which operates under the Geophysics Research Board of the National Academy of Sciences. In the following year, the Conference developed an outline of international programs in hydrology [63].

III. PROFESSIONAL STATUS

A. Hydrologic Organizations

1. International Organizations. The *International Association of Scientific Hydrology* (IASH) is the only international nongovernmental organization concerned exclusively with scientific hydrology.* It was organized in 1922 in Rome during the General Assembly of the International Union of Geodesy and Geophysics (IUGG) as an autonomous constituent association of the Union for the purpose of encouraging and developing the study of hydrology and coordinating research into hydrological problems, which necessitates international participation. Regular meetings of IASH have been held as a part of the triennial assemblies of the Union in Madrid (1924), Prague (1927), Stockholm (1930), Lisbon (1933), Edinburgh (1936), Washington, D.C. (1939), Oslo (1948), Brussels (1951), Rome (1954), Toronto (1957), Helsinki (1960), and Berkeley (1963). In addition to these meetings, various symposia devoted to a particular problem have been arranged, including the Symposia Darcy at Dijon (1956), the Symposium at Chamonix-Mont Blanc on the physics of ice movement (1958), the Symposium at Hannoversch-Munden on water-and-woodlands and lysimeters (1959), the Symposium in Athens on groundwater resources in arid zones in collaboration with UNESCO (1961), the Symposium at Obergurgle (Tyrol) on variations of existing glaciers (1962), and the Symposium at Bari on land erosion (1962). The headquarters of the association is at rue des Ronces 61, Gentbrugge, Belgium.

There are many other international organizations relating to hydrology. The important ones are listed alphabetically as follows:

Food and Agricultural Organization of the United Nations (FAO)—founded in 1945, with headquarters at Viale delle Terme di Caracalla, Rome, Italy.

International Association for Hydraulic Research (IAHR)—founded in 1935, with headquarters at Raam 61, Delft, Netherlands.

International Association of Hydrogeologists—founded in 1960, with headquarters at 77, rue de la Fédération, Paris 15ᵉ, France.

International Association of Meteorology and Atmospheric Physics (IAMAP)—founded in 1919, with headquarters at Meteorological Branch, Air Services, Department of Transport, 315 Floor St. West, Toronto 5, Ontario, Canada.

International Association of Physical Oceanography (IAPO)—founded in 1919, with headquarters at Oceanografiska Institutet, Box 1038, Göteborg 4, Sweden.

International Association of Theoretical and Applied Limnology (IAL)—founded in 1922, with headquarters at Freshwater Biological Association, Ferry House, Far Sawrey, Ambleside, Westmoreland, England.

International Atomic Energy Agency (IAEA)—founded in 1957, with headquarters at Kaerntnerring 11, Vienna 1, Austria. It sponsored a symposium on the application of radioisotopes in hydrology in 1963 in Tokyo.

International Commission of Agricultural Engineering (ICAE)—founded in 1930, with headquarters at 15, avenue du Maine, Paris 15ᵉ, France.

International Commission on Irrigation and Drainage (ICID)—founded in 1950, with headquarters at Central Office, 184 Golf Links Area, New Delhi 3, India.

International Commission on Large Dams of the World Power Conferences (ICOLD-WPC)—founded in 1928, with headquarters at 91, rue Saint Lazare, Paris 9ᵉ, France.

International Grassland Congress (IGC)—founded in 1927, with headquarters at the Grassland Research Institute, Harley (near Maidenhead), Berkshire, England.

International Hydrographic Bureau (IHB)—founded in 1921, with headquarters at Quai des États-Unis, Monte Carlo, Principality of Monaco.

International Institute for Land Reclamation and Improvement (IILC for Inter-

---

* The international nongovernmental organization concerned with medical hydrology is the International Society of Medical Hydrology and Climatology, which was also founded in 1922. Its aim is to encourage the clinical and experimental study of hydrology, particularly the therapeutic effects of medicinal waters, and the study of climatology and its effects on health. The headquarters is at Villa Joal, Allée Dr. Percepied, 6, Mont Dare, Puy de Dôme, France.

national Instituut voor Landaanwinning en Cultuurtechniek)—founded in 1956, with headquarters at Prinses Marijkeweg 15, P. O. Box 45, Wageningen, Netherlands.

International Water Supply Association (IWSA)—founded in 1947, with headquarters at 34 Park St., London W. 1, England.

Permanent International Association of Navigation Congresses (PIANC)—founded in 1900, with headquarters at 60, rue Juste Lipse, Brussels 4, Belgium.

United Nations Economic Commission for Asia and the Far East (ECAFE)—founded in 1945, with headquarters at Sala Santitham, Bangkok, Thailand. It sponsors Interregional Hydrologic Seminars (on hydrologic networks and methods, 1959, in collaboration with WMO; on field methods and equipment used in hydrology and hydrometeorology, 1961, with WMO and U. N. Bureau of Technical Assistance Operations) and Regional Technical Conferences on Water Resources Development.

United Nations Educational, Scientific and Cultural Organization (UNESCO)—founded in 1945, with headquarters at Place de Fontenoy, Paris 7^e, France. It conducts Arid Zone Research and Humid Tropics Research and supports hydrologic programs of other organizations.

Water Resources Development Center (WRDC)—established in 1959 in the United Nations, with headquarters in the United Nations, New York.

World Health Organization (WHO)—founded in 1948, with headquarters at Palais des Nations, Geneva, Switzerland.

World Meteorological Organization (WMO)—founded in 1950, with headquarters at 41 Avenue Giuseppe Motta, Geneva, Switzerland. It sponsored a Commission for Hydrological Meteorology (changed to Hydrometeorology in 1963), having its first session in Washington, D.C. in 1961.

2. Organizations in the United States. In the United States numerous organizations involved in hydrology may be classified into several groups, including federal executive departments, independent federal agencies, interboundary agencies, state and local agencies, private agencies, educational institutions, and technical societies. Some discussions on Federal and state organizations are given in Subsec. 28-IV. The major organizations of these groups and some of their primary functions relating to hydrology are shown as follows:

  a. Federal Executive Departments
    (1) Executive Office of the President
      (*a*) White House Office. The Standing Committee of the Federal Council for Science and Technology in this Office established in 1960 an *Ad Hoc Panel on Hydrology* to report on ways of stimulating research in hydrology [2, 56]. The Committee and the Panel were terminated in 1962.
      (*b*) Office of Science and Technology. A *Committee on Water Resources Research* was recommended [64] and established in 1963 in this Office to coordinate hydrologic research and other researches relating to water resources among Federal agencies.
    (2) Department of Agriculture
      (*a*) Agricultural Research Service (ARS, 1953). Research in watershed engineering and water management is conducted in the *Soil and Water Conservation Research Division* at numerous locations throughout the United States and Puerto Rico as well as at the *Agricultural Research Center* of the Service in Beltsville, Maryland.
      (*b*) Forest Service (1906). Research in flood control and watershed management is conducted in *The Watershed Management Research Division* and *The Range, Wildlife Habitat and Recreation Research Division* at 10 regional forest and range experiment stations and in research units in Alaska and Puerto Rico.
      (*c*) Soil Conservation Service (SCS, 1935). Engineering phases of water conservation, flood control, and water-supply forecasting are investigated at many state and territorial offices and at the *Central Technical Unit* in Beltsville, Maryland.
    (3) Department of Commerce
      (*a*) National Bureau of Standards (NBS, 1901). The *National Hydraulics Laboratory* was founded in the Bureau in 1932 to conduct hydraulic research and experiments.

(b) Coast and Geodetic Survey (CGS, 1878; Coast Survey, 1807). It provides charts and related information for the safe navigation of marine vessels and also basic data through research for engineering and scientific progress of the development of oceanic resources through its *Marine Data and Operations Division* (Office of Oceanography) and the *Nautical Chart Division* (Office of Cartography).
(c) Weather Bureau (USWB, 1940; under Department of Agriculture, 1890). It conducts hydrologic studies, designed to investigate problems in river stage forecasting, flood warning, and application of hydrometeorological information to water-resources planning, through the *Hydrologic Services Division* of the Bureau. Its *Office of Climatology* and *Office of Meteorological Research* also contribute knowledge to hydrology.
(d) Bureau of Public Roads (BPR, 1949; Public Roads Administration, under Federal Works Agency, 1939; Bureau of Public Roads, 1918, and Office of Public Roads and Rural Engineering, 1916, under Department of Agriculture; Office of Road Inquiry, 1894). It conducts research on highway drainage in its *Division of Hydraulic Research.*

(4) Department of the Army
   (a) Corps of Engineers (1802)
      1. Office of the Chief of Engineers and Offices of 10 Division Engineers and Offices of many District Engineers. They administer all matters in hydrology related to the planning, construction, operation, and maintenance of works for improvement of rivers, harbors, and waterways for navigation, flood control, and related purposes, including shore protection projects.
      2. Beach Erosion Board (1930). It conducts hydrologic investigations of natural phenomena related to erosion of the shore by waves and currents of the sea. In 1963, it was transferred to the Board of Engineers for River and Harbors and reorganized as U.S. Army Coastal Engineering Research Board and U.S. Army Coastal Engineering Research Center.
      3. The Board of Engineers for Rivers and Harbors (1902). It is concerned with hydrology in connection with the review of projects for river and harbor, flood control, and multiple-purpose improvements by the Corps of Engineers.
      4. California Debris Commission (1893). It is concerned with hydrology in regulating hydraulic mining and flood control in most rivers in California.
      5. United States Lake Survey (1841). It conducts hydrographic and hydrologic investigations in the Great Lakes.
      6. Mississippi River Commission (1893). It is concerned with hydrology in the control of floods in the Mississippi alluvial valley.
      7. The U.S. Army Waterways Experiment Station, Vicksburg, Mississippi (WES, 1929). It conducts research and development in military hydrology and hydraulics.
      8. The U.S. Army Cold Regions Research and Engineering Laboratory, Hanover, New Hampshire (1961). It conducts research and development in cryological phenomena pertaining to snow, ice, and frozen ground on and beneath the earth's surface. See also Ref. 4 in Sec. 10.

(5) Department of the Navy. In the Office of the Chief of Naval Operation, the *Hydrographic Division* provides information on hydrography and oceanography. It maintains the *Hydrographic Office* and the *National Oceanographic Data Center.* The Office of Naval Research, Bureau of Ships, Bureau of Naval Weapons, and Bureau of Yards and Docks also conduct research and development in hydrodynamics.

(6) Department of Health, Education, and Welfare
   (a) Public Health Service (PHS, 1953; under Federal Security Agency, 1939; under Department of the Treasury, 1798).
      1. National Institutes of Health (NIH). It sponsors health-related research in hydrology and water resources.
      2. Bureau of State Services. It assists state and local government on health-related hydrologic programs through its *Division of Water Supply and Pollution Control.*

(7) Department of the Interior

(a) Geological Survey (USGS, 1879). Through the *Branches of Surface Water, Ground Water, Quality of Water,* and *General Hydrology* of its *Water Resources Division*, it determines and describes the source, occurrence, quantity, quality, and availability of surface water and groundwater in the United States, its territories and possessions; and it conducts research in hydrologic principles and processes to better understand water in the hydrologic cycle and for the development and application of investigative techniques to water problems and appraisal. The *Branch of Waterpower Classification* of the *Conservation Division* gathers basic hydrologic data for investigation of waterpower potentials.

(b) Office of Saline Water (1952). It conducts hydrologic research and development in desalination of sea water.

(c) Bureau of Reclamation (USBR, 1903; Reclamation Service, 1902). It conducts hydrologic and hydraulic research and planning for the storage, diversion, and development of waters for the reclamation of arid and semiarid lands in the Western states in the *Engineering Laboratories Division* of the *Reclamation Engineering Center*, Denver, Colorado.

b. Independent Federal Agencies

(1) Atomic Energy Commission (AEC, 1946). It conducts and sponsors research on the use of radioisotopes in hydrology and on the hydrologic aspects of atomic waste disposals.

(2) Federal Power Commission (FPC, 1920). It conducts hydrologic investigation of river basins for hydropower development.

(3) National Science Foundation (NSF, 1950). It sponsors basic research and education in hydrology.

(4) Tennessee Valley Authority (TVA, 1933). It conducts research and development in hydrology, hydraulics, and water resources in the Tennessee River basin.

c. Interboundary Agencies. There are many interstate and international agencies and interagencies that are concerned with hydrology. For example, the *International Boundary and Water Commission, United States and Mexico* (1889) conducts surveys and investigations between the two countries relating to flood control, irrigation, water supply, water resources, conservation and utilization of water, river pollution, and channel stabilization. The *Inter-Agency Committee on Water Resources* coordinates hydrologic operations through its *Subcommittee on Hydrology*.

d. State and Local Agencies. There are numerous state and local agencies that are concerned with hydrology. These agencies include many public corporations, such as the Pittsburgh Flood Commission (1908), the Miami Conservancy District (1914), the Los Angeles County Flood Control District (1915), the Franklin County Conservation District (1915), and the Muskingum Valley Conservancy District (1933). In the State of Illinois, for example, state agencies include the Illinois State Water Survey (1895), the Illinois State Geological Survey (1905), the Illinois Division of Waterways (1917), and the Illinois Division of Highways (1917).

e. Private Agencies. There are many private engineering offices which are concerned with hydrology in the design and operation of hydraulic and water-resources projects. Hydrologic research is also conducted by private research institutes such as the Travelers Research Center, Inc.; Arthur D. Little, Inc.; the California Research Corporation; the Southwest Research Institute; the Schlumberger Well Surveying Corporation; The Rand Corporation; and the Sperry Rand Research Center, Inc.

f. Educational Institutions. Research on various hydrologic problems are conducted at universities and colleges. Many universities have established Water Resources Centers or Institutes to develop hydrologic research and education.

g. Technical Societies. Numerous scientific and engineering societies are concerned with hydrology. The most important one is the American Geophysical Union (1919), which promotes the science of hydrology through its Sections on Hydrology, Meteorology, and Oceanography. The American Society of Civil Engineers (1852) also contributes to hydrology through its many Divisions, particularly of Hydraulics (1938), Irrigation and Drainage, Sanitary Engineering, and Waterways and Harbors. Some other important technical societies include American Society of Agricultural

Engineers, American Meteorological Society, American Society of Agronomy, American Society of Limnology and Oceanography, American Water Works Association, National Water Well Association, and Geological Society of America (organized in 1888 with a hydrogeology group established in 1959).

3. Organizations in the U.S.S.R. In the U.S.S.R. "hydrology" has a broad connotation, referring to all the waters of the earth both on the land and in the oceans. According to Sweet [65], hydrologic service in the U.S.S.R. is administered on an All-Union basis by the Main Administration of the Hydrometeorological Service (GUGMS), with headquarters in Moscow. This body was established in 1929, with a responsibility for the development of hydrologic networks, coordination of operations, and improvement and expansion of a program for the investigation of water resources. There are other establishments and institutions which are involved in hydrology, mostly of an educational and research nature. These and GUGMS are administered directly under the Council (Soviet) of Ministers of the U.S.S.R. The following list shows these establishments and institutes:

    a. Academy of Sciences of the U.S.S.R.
        (1) Academy Establishments
            (*a*) Water Conservation Council
            (*b*) 12 Academy Research Institutes: Geographic Institute, Moscow; Hydrogeological Laboratory, Moscow; Applied Geophysics Institute, Moscow; Soil Science Institute, Moscow; Forestry Research Institute, Uspenskoye; Transportation Institute, Moscow; Lake Research Institute, Leningrad; Oceanographic Institute, Moscow; Marine Hydrophysical Institute, Liublino; Hydrochemical Institute, Novocherkassk; Reservoir Biological Research Institute, Borok; Baikal Limnological Station, Irkutsk
            (*c*) 11 Provincial Academy Departments: Irkutsk, Yakutsk, S. Sakhalinsk, Kazan, Kirovsk, Kishinev, Novosibirsk, Petrozavodsk, Syktyvkar, Sverdlovsk, Vladivostok
        (2) Academies of Sciences of 11 Republics: Alma-Ata, Baku, Frunze, Yerevan, Kiev, Minsk, Riga, Stalinabad, Tashkent, Tbilisi, Vilnius; maintaining 16 research institutes of these Academies

    b. GUGMS Moscow
        (1) Central Establishments
            (*a*) State Hydrologic Institute (GGI) Leningrad, with 3 field laboratories: Valday, Dubrovskaya, Zelenogorye
            (*b*) Central Forecasting Institute, Moscow (TsIP)
            (*c*) Main Geophysical Observatory, Leningrad (GGO)
            (*d*) State Oceanographic Institute, Moscow, branch in Leningrad
            (*e*) Central Aerological Observatory, Moscow
            (*f*) Aeroclimatological Scientific Research Institute, Moscow
            (*g*) Hydrometeorological Instrument Development Research Institute, Moscow
            (*h*) State Hydrometeorological Publishing Agency, Leningrad (Gidrometeoizdat) with a branch in Moscow
        (2) Regional Establishments
            (*a*) 34 Regional Hydrometeorological Offices: Alma-Ata, Arkhangelsk, Ashkhabad, Baku, Chita, Frunze, Gorki, Khabarovsk, Irkutsk, Yakutsk, Yerevan, S. Sakhalinsk, Kiev, Kishinev, Krasnoyarsk, Kuibyshev, Kursk, Leningrad, Magadan, Minsk, Moscow, Murmansk, Novosibirsk, Omsk, Petropavlovsk, Riga, Rostov, Stalinabad, Sverdlovsk, Tallinn, Tashkent, Tbilisi, Vilnius, Vladivostok
            (*b*) 5 Regional Hydrometeorological Research Institutes: Alma-Ata, Kiev, Tashkent, Tbilisi, Vladivostok
            (*c*) 5 Hydrometeorological Technical Schools: Kharkov, Kherson, Moscow, Rostov, Tashkent

    c. Ministry of Higher Education
        (1) Universities and Technical Colleges
            (*a*) 5 Schools of Geography of State Universities

(b) 2 Hydrometeorological Schools: Leningrad and Odessa
(c) 2 Water Transportation Engineering Colleges: Leningrad and Novosibirsk
(d) Agricultural Engineering Colleges: Moscow, Kiev, and Novocherkassk
(e) 9 other technical schools, 7 schools of Agriculture, Forestry, Conservation, etc.
(2) Research Institutes and Planning Agencies of Regional Offices of GUGMS
(a) 8 All-Union and 11 Regional Hydraulic Construction, Soil Improvement, Hydrogeological, Water Supply, and similar research institutes
(b) 10 All-Union Water Conservation Planning Agencies
(c) 45 Institute Extensions and Regional Agencies

B. Hydrologic Publications

1. Books. Numerous books have been written on subjects related to hydrology, but only a few books treat hydrology in general scope.* The oldest British book on hydrology is believed to have been prepared by the English civil engineer Nathaniel Beardmore (1862) [67]. The first American book on hydrology is by Mead (1904) [68]. Other well-known works in English are by Meyer (1917) [69], Mead (1919) [70], Meinzer (1942) [14], Barrows (1948) [71], Foster (1948) [72], Johnstone and Cross (1949) [73], Wisler and Brater (1949) [3], Linsley, Kohler, and Paulhus (1949) [74], (1958) [75], Williams (1950) [76], Kuenen (1956) [77], and Butler (1957) [78]. Handbooks and manuals on hydrology have been prepared by the American Society of Civil Engineers (1949) [79] and by the U.S. Bureau of Reclamation (1947) [80], U.S. Soil Conservation Service (1956) [81], U.S. Forest Service (1959) [82], U.S. Geological Survey (1960) [83], and U.S. Agricultural Research Service (1962) [84].

Other books on hydrology have been written in many languages. There are hydrologic textbooks in Russian by Shestopal (1923) [85], Sovetov (1933) [86], Ogievskiĭ (1933) [87], Basin (1936) [88], Blizniak and Nikol'skiĭ (1946) [89], Poliakov (1946) [90], Ioganson and Ioganson (1947) [91], Velikanov (1948) [92], Gavrilov and Bogomazova (1948) [93], Leĭvikov (1949) [94], Luchsheva (1950) [95], Chebotarëv (1950) [96], Domanevskiĭ (1951) [97], Lebedev (1952) [98], Sokolovskiĭ (1952) [99], Chebotarëv and Klibashëv (1956) [100], and Chebotarëv (1960) [101]; in German in "Handbuch der Ingenieurwissenschaften" (since 1876) [102] and by Thiem (1906) [32], Brauer (1907) [103], Gravelius (1914) [104], Prinz (1919) [105], Drenkhahn (1927) [106], Schaffernak (1935) [107], Streck (1953) [108], and Wundt (1953) [109]; in French by Pardé (1933) [110], (1943) [111], Réméniéras (1959) [112], (1960) [113], and Roche (1963) [114]; in Italian by Giandotti (1937) [115], Arredi (1947) [116], and Tonini (1963) [117]; in Hungarian by Németh (1954) [118]; in Croatian by Yevdjevich (1956) [119]; in Czech by Novotný (1952) [120]; in Slovak by Dub (1957) [121]; in Polish by Pomianowski, Rybczyński, and Wóycicki (1933, 1934, 1939, 1947) [122], Debski (1948, 1955, 1959) [123], and Czetwertyński (1955) [124].

Since 1933, bibliographies on hydrology in many countries have been prepared annually by IASH through cooperative efforts of those countries. Since 1941, the United States has published independently 3 annotated bibliographies on hydrology for 1941–50 [125], 1951–54 [126], and 1955–58 [127]. The State Hydrologic Institute of the U.S.S.R. has established a Central Bureau of hydrologic bibliography under G. Iu. Vereschagin, which prepared several annual indexes.

2. Periodicals. Hydrologic papers and articles are currently published in hundreds of scientific and engineering periodicals. The earliest hydrologic periodical is probably the *Zeitschrift für Gewässerkunde* in Germany, which began in 1898 and ended during the first World War. Since its organization, the International Association of Scientific Hydrology publishes *Publications*, which contain papers presented during meetings and symposia, and since 1956, a quarterly *Bulletin*, which carries news of the Association as well as scientific papers. Since 1963, a journal of *Hydrology* is

---

* The first book under the title of hydrology is probably the "Hydrologia" by Melchior (1694) [66], which deals with mineral springs of Wiesbaden in connection with their significance to health.

published by the North-Holland Publishing Company, P. O. Box 103, Amsterdam, The Netherlands, which is intended to be an international journal. Also started in 1963 is *Hydraulic Research (Recherches Hydrauliques)*, the Journal of the International Association for Hydraulic Research, which publishes some papers relating to hydrology.

In the United States, the most important periodicals are the *Transactions* (since 1921), *Journal of Geophysical Research* (since 1959), *Soviet Hydrology: Selected Papers* (since 1962), and *Reviews of Geophysics* (since 1963) of the American Geophysical Union and the *Transactions, Proceedings*, and *Journals of Divisions* of the American Society of Civil Engineers. Since 1963, a journal of *Ground Water* is being published by the National Water Well Association, P. O. Box 222, Urbana, Illinois. Since 1964, a series of *Advances in Hydroscience*, edited by Ven Te Chow, is being published annually by Academic Press Inc. of New York, which carries articles on hydrology as well as on other scientific fields related to water. For many publications issued by government agencies, see Subsec. 26-I-IV.

In the U.S.S.R., important periodicals are *Meteorologiia i Gidrologiia*, a monthly journal and the *Trudy* (Transactions) and *Izvestiia* (Bulletins) of the State Hydrologic Institute (GGI), the Ukrainian Scientific Investigations Hydrologic Institute (UN-IGI), the Central Forecasting Institute (TsIP), the Voeĭkov Main Geophysical Observatory (GGO), and other institutions Three important journals available in English translations are *Doklady* (Proceedings) of the Academy of Sciences of the U.S.S.R., Earth Science Section, Geochemistry Section, Hydrogeology Section; *Geochemistry* (U.S.S.R.); and *Geophysics Series*, a monthly journal of the Academy of Sciences of the U.S.S.R., translated by AGU.

Current hydrologic research projects in the United States and in Canada are published in *Hydraulic Research in the United States*, an annual publication by the U.S. National Bureau of Standards, while those in other countries are in *Recherches Hydrauliques*, an annual publication by the International Association for Hydraulic Research.

C. Hydrologic Education

As pointed out by many authors [128–137, 56], hydrologic education becomes extremely important in training man power for research and planning in the vast programs to develop and control water resources.

The first university course in hydrology given in the United States is believed to have been started at the University of Wisconsin in 1904 by the late Professor Daniel W. Mead [131]. Throughout the past 60 years, educational programs in hydrology have been gradually taking shape. While most programs are at the graduate level, many universities are introducing hydrology in their undergraduate curricula. At first, hydrologic programs were developed mostly in the departments of civil engineering. Since hydrology is a borderline science, it not only integrates many other sciences for its own interpretation and uses but also serves as a basic course to support many other disciplines. Consequently, the programs are becoming interdisciplinary in nature. Many schools now offer degrees in civil engineering with hydraulic option specializing in hydrology or in geology with groundwater option specializing in hydrology, mostly designed on an interdisciplinary basis. In 1961, the University of Arizona became the first American university to offer the degrees of Bachelor of Science, Master of Science, and Doctor of Philosophy in Hydrology.

In view of the importance of hydrologic education, a Universities Council on Hydrology (UCOH) was established in 1963 by 29 universities interested in hydrology for promoting hydrologic research and education [58]. In order to broaden the scope and to include the social-science aspects of water, UCOH was transformed to UCOWR (Universities Council on Water Resources) in 1964.

IV. REFERENCES

1. Merriam-Webster editorial staff: "Webster's Third New International Dictionary of The English Language, Unabridged," G. and C. Merriam Company, Springfield, Mass., 1961.

2. "Scientific Hydrology," Ad Hoc Panel on Hydrology, U.S. Federal Council for Science and Technology, Washington, D.C., June, 1962.
3. Wisler, C. O., and E. F. Brater: "Hydrology," John Wiley & Sons, Inc., New York, 1949; 2d ed., 1959.
4. Horton, R. E.: The field, scope and status of the science of hydrology, *Trans. Am. Geophys. Union*, vol. 12, pp. 189–202, 1931.
5. Ackermann, W. C., E. A. Colman, and H. O. Ogrosky: From ocean to sky to land to ocean, in "Water," The Yearbook of Agriculture 1955, U.S. Dept. Agr., pp. 41–51.
6. Lettau, H.: A study of the mass, momentum and energy budget of the atmosphere, *Arch. Meteorol. Geophys. Bioklimatol.*, Vienna, Ser. A, vol. 7, pp. 131–153, 1954.
7. Wolman, Abel: Water resources, a report to the Committee on Natural Resources of the National Academy of Sciences—National Research Council, *Pub.* 1000-B, Washington, D.C., 1962.
8. Reichel, E.: Der Stand des Verdunstungsproblems (The status of the evaporation problem), *Ber. Deut. Wetterdienst*, Bad Kissingen, vol. 35, p. 155, 1952.
9. Keilhack, Konrad: "Grundwasser und Quellenkunde" (Groundwater and the Hydrology of Springs), Berlin, 1912.
10. Meinzer, O. E.: The history and development of groundwater hydrology, *Washington Acad. Sci. J.*, vol. 24, no. 1, pp. 6–32, 1934.
11. Baker, M. N., and R. E. Horton: Historical development of ideas regarding the origin of springs and ground water, *Trans. Am. Geophys. Union*, vol. 17, pt. II, pp. 395–400, 1936.
12. Tolman, C. F.: "Ground Water," McGraw-Hill Book Company, Inc., New York, 1937.
13. Follansbee, Robert: "A History of the Water Resources Branch of the United States Geological Survey to June 30, 1919," privately printed, 1938.
14. Meinzer, O. E.: "Hydrology," vol. IX of "Physics of the Earth," McGraw-Hill Book Company, Inc., New York, 1942; reprinted by Dover Publications, Inc., New York, 1949.
15. Hackett, J. E.: The birth and development of groundwater hydrology—A historical summary, *Trans. Wisconsin Acad. Sci.*, vol. 41, pp. 201–206, 1952.
16. Adams, F. D.: "The Birth and Development of Geological Sciences," Baltimore, 1938; reprinted by Dover Publications, Inc., New York, 1954.
17. Rouse, Hunter, and Simon Ince: "History of Hydraulics," Iowa Institute of Hydraulic Research, State University of Iowa, 1957.
18. Kolupaila, Steponas: Early history of hydrometry in the United States, *Proc. Am. Soc. Civil Engrs., J. Hydraulics Div.*, vol. 86, no. HY1, pp. 1–51, January, 1960.
19. Krynine, P. D.: On the antiquity of "sedimentation" and hydrology (with some moral conclusions), *Bull. Geol. Soc. Am.*, vol. 71, pp. 1721–1726, 1960.
20. Kolupaila, Steponas: "Bibliography of Hydrometry," University of Notre Dame Press, Notre Dame, Ind., 1961. It contains an account on historical development of hydrometry, pp. 1–18.
21. Jones, P. B., G. D. Walker, R. W. Harden, and L. L. McDaniels: The development of the science of hydrology, *Circular* 63-03, Texas Water Commission, April, 1963.
22. Parizek, R. R.: Development of the hydrologic cycle concept and our challenge in the 20th century, *Mineral Ind., Penn. State Univ.*, vol. 32, no. 7, pp. 1–8, April, 1963.
23. Palissy, Bernard: "Discours admirable de la nature des eaux et fontaines tant naturelles qu'artificielles" (Admirable Discourse on the Nature of the Waters of Both Natural and Artificial Fountains), Paris, 1580. Also, Henry Morley: "The Life of Bernard Palissy of Saintes," 2 vols., Boston, 1853; and translation by Aurele La Rocque: "The Admirable Discourses of Bernard Palissy," The University of Illinois Press, Urbana, Ill., 1957.
24. Perrault, Pierre: "De l'origine des fontaines" (On the Origin of Fountains), Pierre le Petit, Paris, 1674; 2d. ed., 1678.
25. Mariotte, Edmé: "Traité du mouvement des eaux et des autres corps fluides" (Treatise on the Movement of Waters and Other Fluids), E. Michallet, Paris, 1686; 2d ed., J. Jombert, Paris, 1700.
26. Frisi, Paul: "A Treatise on Rivers and Torrents," 1762; translated from the 2d Italian edition by John Garstin of the Bengal Engineers, John Weale, London, 1860.
27. Sheppard, Thomas: "William Smith, His Maps and Memoirs," A. Brown & Sons, Ltd., Hull, England, 1920. Also, *Proc. Yorkshire Geol. Soc.* (n.s.), vol. 19, pt. 3, pp. 75–253, March, 1917.
28. Hagen, G. H. L.: Ueber die Bewegung des Wassers in engen cylindrischen Röhren (On flow of water through small cylindrical pipes), *Poggendorffs Ann. Physik Chem.*, vol. 16, pp. 423–442, 1839.
29. Poiseuille, J. L.: Recherches expérimentales sur le mouvement des liquides dans les tubes de très petits diameters (Experimental studies on the movement of fluids in

REFERENCES 1-19

tubes of very small diameters), *Compt. Rend.*, vol. 11, pp. 961 and 1041, 1840; vol. 12, p. 112, 1841; *Mémoires des Savants Étrangers*, vol. 9, pp. 433–543, 1846.
30. Darcy, Henri: "Les fontaines publiques de la ville de Dijon" (The Public Fountains of the City of Dijon), V. Dalmont, Paris, 1856.
31. Dupuit, A. J.: "Études théoriques et pratiques sur le mouvement des eaux dans les canaux découverts et à travers les terrains perméables" (Theoretical and Practical Studies on the Flow of Water in Open Channels and through Permeable Terrains), 2d ed., Dunod, Paris, 1863.
32. Thiem, Günther: "Hydrologische Methoden" (Hydrologic Methods), J. M. Gebhart, Leipzig, 1906. This deals mainly with groundwater.
33. Badon Ghyben, W.: Nota in verband met de voorgenomen putboring nabij Amsterdam (An account in connection with the proposed well boring near Amsterdam), *Tijdschr. Koninkl. Inst. Ingrs.*, The Hague, pp. 8–22, 1888–1889.
34. Herzberg, Baurat: Die Wasserversorgung einiger Nordseebäder (Water supply of North Sea sea-side resorts), *J. Gasbeleucht. Wasserversorg.*, vol. 44, pp. 815–819, 842–844, Munich, 1901.
35. Humphreys, A. A., and H. L. Abbot: "Report Upon the Physics and Hydraulics of the Mississippi River," *Profess. Papers Corps Topograph. Engrs.*, Philadelphia, 1861; 2d ed., Washington, D.C., 1876.
36. Francis, J. B.: "Lowell Hydraulic Experiments," Little, Brown and Company, Boston, 1855; 2d ed., D. Van Nostrand Company, Inc., New York, 1868; 3d ed., 1871; 4th ed., 1883; 5th ed., 1909.
37. Ganguillet, Emil, and W. R. Kutter: Versuch zur Aufstellung einer neuen allgemeinen Formel für die gleichförmige Bewegung des Wassers in Canälen und Flüssen (An attempt to design a new general formula for the uniform flow of water in canals and rivers), *Z. Oesterr. Ingr.—Architekten-Vereins*, Vienna, vol. 21, no. 1, pp. 6–25, no. 2–3, pp. 46–59, 1869; "Bewegung des Wassers in Kanälen und Flüssen" (Flow of Water in Canals and Rivers), 2d ed., Lang and Co., Bern, 1877.
38. Manning, Robert: On the flow of water in open channels and pipes, *Trans. Inst. Civil Engrs. Ireland*, Dublin, vol. 20, pp. 161–207, 1891; supplement, vol. 24, pp. 179–207, 1895. The formula was first presented on Dec. 4, 1889 at a meeting of the Institution.
39. Ellis, T. G.: Surveys and examinations of Connecticut River, 45th Congress, 3d Session, *House Document* 1, pt. 2, *Report of the Secretary of War*, vol. 2, p. 1; *Report of the Corps of Engineers for* 1878, App. B14, pp. 248–391, Washington, D.C., 1878. The cup-type current meter was actually first adopted for rivers by D. F. Henry in 1867, and later used by Ellis in 1870.
40. Price, W. G.: Current meter, U.S. Patent Office, specification forming part of Letters Patent No. 325,011, dated Aug. 25, 1885.
41. Grebenau, Heinrich: "Die internationale Rheinstrom-Messung bei Basel vorgenommen am 6–12. November 1867" (The International Rhine River Measurement at Basle Undertaken Nov. 6–12, 1867), J. Lindauer, Munich, 1873.
42. Miller, J. F.: On the meteorology of the lake district of Cumberland and Westmoreland; including the results of experiments on the fall of rain at various heights above the earth's surface, up to 3166 feet above the mean sea level, *Phil. Trans. Roy. Soc. London*, pt. 1, pp. 73–89, pt. 2, pp. 319–329, 1849.
43. Blodget, Lorin: "Climatology of the United States," Philadelphia, 1857.
44. Clinton, DeWitt: "Letters on the Natural History and Internal Resources of the State of New York [By Hibernicus (pseudonym)]," New York, 1822.
45. Chow, V. T.: Hydrologic determination of waterway areas for the design of drainage structures in small drainage basins, *Univ. Illinois Eng. Exp. Sta., Bull.* 462, 1962. See App. 1: A compilation of formulas for waterway area and determination, pp. 66–91.
46. Sherman, L. K.: Streamflow from rainfall by the unit-graph method, *Eng. News-Rec.*, vol. 108, pp. 501–505, Apr. 7, 1932.
47. Horton, R. E.: The role of infiltration in the hydrologic cycle, *Trans. Am. Geophys. Union*, vol. 14, pp. 446–460, 1933.
48. Theis, C. V.: The relation between the lowering of the piezometric surface and the rate and duration of a well using ground-water recharge, *Trans. Am. Geophys. Union*, vol. 16, pp. 519–524, 1935.
49. Gumbel, E. J.: The return of flood flows, *Ann. Math. Statis.*, vol. XII, no. 2, pp. 163-190, June, 1941.
50. Hazen, Allen: "Flood Flows," John Wiley & Sons, Inc., New York, 1930.
51. Bernard, Merrill: The primary role of meteorology in flood flow estimating, *Trans. Am. Soc. Civil Engrs.*, vol. 109, pp. 311–382, 1944.
52. "A Policy for the American People," The Report of the President's Water Resources Policy Commission 1950; vol. 1: General report; vol. 2; ten rivers in America's future; vol. 3: water resources law, U.S. Government Printing Office, Washington, D.C., 1950.

53. "Principles of a Sound National Water Policy," prepared by the National Joint Council of Engineers Joint Council, Edwards Brothers, Inc., Ann Arbor, Mich., July, 1951.
54. "Report on Water Resources," prepared for the Commission on Organization of the Executive Branch of the Government by the Task Force on Water Resources and Power, 3 vols., U.S. Government Printing Office, Washington, D.C., June, 1955.
55. "Water Resources Policy," Presidential Advisory Committee on Water Resources Policy, U.S. Government Printing Office, Washington, D.C., 1955.
56. "Report of the Ad Hoc Committee on Education in Hydrology," National Science Foundation, Washington, D.C., 1962.
57. Ackermann, W. C.: Conference of American Hydrologists (CIPASH) and Committee on Status and Needs in Hydrology (AGU), *Trans. Am. Geophys. Union*, vol. 44, no. 3, pp. 705–708, September, 1963.
58. Todd, D. K.: Inter-University Conference on Hydrology, *Trans. Am. Geophys. Union*, vol. 43, no. 4, pp. 491–495, December, 1962.
59. "Water Resources Development Centre," E/3319, United Nations, New York, 1960.
60. A proposal for international cooperation in hydrology, by a Panel of Hydrologists (USA), *Bull. Intern. Assoc. Sci. Hydrol.*, VIe Année, no. 4, pp. 5–9, December, 1961.
61. Nace, R. L.: A plan for international cooperation in hydrology—Panel on Hydrology (USA), *Bull. Intern. Assoc. Sci. Hydrol.*, VIe Année, no. 4, pp. 10–26, December, 1961; also *Intern. Union Geodesy Geophys. Chron.*, January, 1962.
62. "Proposal for a Programme in Hydrology," UNESCO General Conference, Paris, October, 1962.
63. An outline of international programs in hydrology, *Pub.* 1131, National Academy of Sciences—National Research Council, Washington, D.C., 1963.
64. Water resources research, Memorandum of the Chairman, Report of the U.S. Senate Committee on Interior and Insular Affairs, 87th Congress, 2d Session, U.S. Government Printing Office, September, 1962.
65. Sweet, J. S.: The science and services of hydrology in the Soviet Union, *Trans. Am. Geophys. Union*, vol. 43, no. 1, pp. 20–33, March, 1962.
66. Melchior, Eberhard: "Hydrologia" (Hydrology), J. D. Zunnern, Frankfort, 1694.
67. Beardmore, Nathaniel: "Manual of Hydrology," Waterlow and Sons, London, 1862. This is a 3d ed. of "Hydraulic Tables," 1850; 2d ed., 1851.
68. Mead, D. W.: "Notes on Hydrology," Press of S. Smith and Co., Chicago, 1904.
69. Meyer, A. F.: "The Elements of Hydrology," John Wiley & Sons, Inc., New York. 1917; 2d ed., 1928.
70. Mead, D. W.: "Hydrology," 1st ed., McGraw-Hill Book Company, Inc., New York, 1919; 2d ed., revised by H. W. Mead and H. J. Hunt, Mead and Hunt, Inc., Madison, Wisc., 1950.
71. Barrows, H. K.: "Floods, Their Hydrology and Control," McGraw-Hill Book Company, Inc., New York, 1948.
72. Foster, E. E.: "Rainfall and Runoff," The Macmillan Company, New York, 1948.
73. Johnstone, Don, and W. P. Cross: "Elements of Applied Hydrology," The Ronald Press Company, New York, 1949.
74. Linsley, R. K., Jr., M. A. Kohler, and J. L. H. Paulhus: "Applied Hydrology," McGraw-Hill Book Company, Inc., New York, 1949.
75. Linsley, R. K., Jr., M. A. Kohler, and J. L. H. Paulhus: "Hydrology for Engineers," McGraw-Hill Book Company, Inc., New York, 1958.
76. Williams, G. R.: Hydrology, chap. 14 in Hunter Rouse (ed.): "Engineering Hydraulics," John Wiley & Sons, Inc., New York, 1950, pp. 229–320.
77. Kuenen, P. H.: "Realms of Water," John Wiley & Sons, Inc., New York, 1956.
78. Butler, S. S.: "Engineering Hydrology," Prentice-Hall, Inc., Englewood Cliffs, N.J., 1957.
79. "Hydrology Handbook," American Society of Civil Engineers, Manuals of Engineering Practice, no. 28, 1949.
80. Flood hydrology, pt. 6 of vol. IV on Water Studies, "Bureau of Reclamation Manual," U.S. Department of Interior, Bureau of Reclamation, 1947; rev., 1951.
81. Hydrology, sec. 4 of "Engineering Handbook," U.S. Department of Agriculture, Soil Conservation Service, 1956; Supplement A, 1957. This is a revision of "Hydrology Guide for Use in Watershed Planning."
82. Forest and range hydrology handbook, "Forest Service Category 2 Handbook," 2518, U.S. Department of Agriculture, Forest Service, April, 1959.
83. "Manual of Hydrology," published in many parts as *U.S. Geol. Surv. Water-Supply Papers*, since 1960. This was formerly published in parts as "Handbook for Hydrologists," since 1946.
84. Field manual for research in agricultural hydrology, *Agricultural Handbook* 224, U.S.

REFERENCES 1-21

Department of Agriculture, Agricultural Research Service, Soil and Water Conservation Research Division, June, 1962.
85. Shestopal, O. S.: "Gidrologiia i gidrometriia" (Hydrology and Hydrometry), Gostransizdat, Moscow, 1923; 2d ed., 1935.
86. Sovetov, S. A.: "Osnovy gidrologii" (Elements of Hydrology), ONTI NKTP SSSR, Novosibirsk, 1933.
87. Ogievskii, A. V.: "Gidrologiia sushi" (Continental Hydrology), ONTI, Moscow-Leningrad, 1933; 2d ed., Moscow-Leningrad, 1941; 3d ed., Moscow, 1952.
88. Basin, M. M.: "Gidrometriia i osnovy gidrologii" (Hydrometry and Elements of Hydrology), ONTI, Leningrad-Moscow, 1936.
89. Blizniak, E. V., and V. M. Nikol'skii: "Gidrologiia i vodnyie issledovaniia" (Hydrology and water investigations), Izd. MRF, Moscow-Leningrad, 1946.
90. Poliakov, B. V.: "Gidrologicheskii analiz i raschëty" (Hydrologic Analysis and Computations), Gidrometeoizdat, Leningrad, 1946.
91. Ioganson, E. I., and V. E. Ioganson: "Osnovy gidrologii i gidrometrii" (Elements of Hydrology and Hydrometry), Gosenergoizdat, Moscow-Leningrad, 1947.
92. Velikanov, M. A.: "Gidrologiia sushi" (Continental Hydrology), Gidrometeoizdat, Moscow, 1925; 2d ed., 1932; Leningrad, 4th ed., 1948.
93. Gavrilov, A. M., and Z. P. Bogomazova: "Prakticheskaia gidrologiia" (Practical Hydrology), Gidrometeoizdat, Leningrad, 1948.
94. Leivikov, M. L.: "Meteorologiia, gidrologiia i gidrometriia" (Meteorology, Hydrology and Hydrometry), Sel'khozgiz, Moscow, 1949; 2d ed., 1955.
95. Luchshev, A. A.: "Prakticheskaia gidrologiia" (Practical Hydrology), Gidrometeoizdat, Leningrad, 1950; 2d ed., 1959.
96. Chebotarëv, A. I.: "Gidrologiia sushi i rechnoi stok" (Continental Hydrology and River Runoff), Gidrometeoizdat, Leningrad, 1950; 2d ed., 1953; 3d ed., 1955.
97. Domanevskii, N. A.: "Gidrologiia i gidrometriia" (Hydrology and Hydrometry), Rechizdat, Moscow, 1951.
98. Lebedev, V. V.: "Gidrologiia i gidrometriia v zadachakh" (Hydrology and Hydrometry in Problems), Gidrometeoizdat, Leningrad, 1952; 2d ed., 1955.
99. Sokolovskii, D. L.: "Rechnoi stok" (River Runoff), Gidrometeoizdat, Leningrad, 1952; 2d ed., 1959.
100. Chebotarëv, A. I., and K. P. Klibashëv: "Gidrologicheskie raschëty" (Hydrologic Computations), Gidrometeoizdat, Leningrad, 1956.
101. Chebotarëv, A. I.: "Obshchaia gidrologiia" (General Hydrology), Gidrometeoizdat, Leningrad, 1960.
102. Die Gewässerkunde (Hydrology), vol. 1, part 3 of Der Wasserbau in "Handbuch der Ingenieurwissenschaften" (Handbook of Engineering Science), W. Engelmann, Leipzig, 1876, 1877; 3d ed., 1892; 4th ed., 1905, 1911; 5th ed., 1923. Various subjects prepared by Edward Schmitt, Paul Gerhardt, Robert Jasmund, and J. F. Bubendey.
103. Brauer, Richard: "Die Grundzüge der praktischen Hydrographie" (Outlines of the Practical Hydrography), Bibliothek der gesamten Technik No. 53, Dr. Max Jänecke, Hannover, 1907.
104. Gravelius, Harry: "Flusskunde" (River Hydrology) vol. 1 in "Grundriss der Gesamten Gewässerkunde" (Outlines of General Hydrology), G. J. Göschen, Berlin-Leipzig, 1914. Other volumes were never completed.
105. Prinz, E.: "Handbuch der Hydrologie" (Handbook of Hydrology), Springer-Verlag, Berlin, 1919; 2d ed., 1923. This book deals mainly with the hydrology of groundwater.
106. Drenkhahn, Rudolf: "Kreislauf des Wassers und Gewässerkunde" (Water Cycle and Hydrology), Walter de Gruyter and Co., Berlin-Leipzig, 1927.
107. Schaffernak, Friedrich: "Hydrographie" (Hydrography; or, actually, Hydrology), Julius Springer, Vienna, 1935; reprinted by Akademische Druck- und Verlagsanstalt, Graz, Austria, 1959.
108. Streck, Otto: "Grundlagen der Wasserwirtschaft und Gewässerkunde" (Fundamentals of Water Control and Hydrology), Springer-Verlag, Berlin-Göttingen-Heidelberg, 1953.
109. Wundt, Walter: "Gewässerkunde" (Hydrology), Springer-Verlag, Berlin-Göttingen-Heidelberg, 1953.
110. Pardé, M. E.: "Fleuves et rivières" (Streams and Rivers), Armand Colin, Paris, 1933; 2d ed., 1947; 3d ed., 1955.
111. Pardé, M. E.: "Cours de potamologie" (Studies of Potamology), 2 vols.: vol. 1, Hydrologie fluviale (River hydrology); vol. 2, Dynamique fluviale (River dynamics), École des Ingénieurs Hydrauliciens, Grenoble, 1943; 2d ed., 1949.
112. Réméniéras, Gaston: "Éléments d'hydrologie appliquée" (Elements of Applied Hydrology), Librairie Armand Colin, Paris, 1959.

113. Réméniéras, Gaston: "L'Hydrologie de l'ingénieur" (The Engineer's Hydrology), Les Éditions Eyrolles, Paris, 1960.
114. Roche, M.: "Hydrologie de surface" (Hydrology of Surface Runoff), Gauthier-Villars, Paris, 1963.
115. Giandotti, Mario: "Idrologia generale" (General Hydrology), Florence, 1937.
116. Arredi, Filippo: Idrologia (Hydrology), vol. 1 of "Construzioni e impianti idraulici," G. Principato, Milan, 1947.
117. Tonini, Dino: "Elementi di idrografia ed idrologia" (Elements of Hydrography and Hydrology), Libreria universitaria, Venice, 1959.
118. Németh, Endre: "Hydrológia és hidrometria" (Hydrology and Hydrometry), Tankönyvkiadó, Budapest, 1954.
119. Jevdević, Vujica (V. M. Yevdjevich): "Hidrologija," vol. 1, Hidrotechnicki Institut, Belgrade, 1956.
120. Novotný, J.: "Hydrologia" (Hydrology), CMT, Prague, 1925.
121. Dub, Oto: "Hydrologia, hydrografia, hydrometria" (Hydrology, Hydrography, Hydrometry), Slovenské Vydavatel'stvo technickej literatúry, Bratislava, 1957.
122. Pomianowski, Karol, Mieczysław Rybczyński, and Kazimierz Wóycicki: "Hydrologia" (Hydrology); vol. 1, Opad-odpływ (Precipitation-runoff), Warsaw, 1933; vol. 2, Wody gruntowe (Groundwater), Warsaw, 1934; vol. 3, Hydrografia i hydrometria wód powierzchniościowych (Hydrography and Hydraulics of Surface Water), Warsaw, 1939; vol. 4, Hydrostatyka, hydraulika teoretyczna i stosowana, pochodzenie i ruch rumowiska rzecznego (Hydrostatics, Theoretical and Applied Hydraulics, Origin and Movement of Sediment), Danzig, 1947.
123. Dębski, Kazimierz: "Hydrologia i hydraulika w zakresie średnim" (Hydrology and Hydraulics for Technicians), Seria A, Instrukcje i podręczniki, no. 7, Państwowy Instytut Hydrologiczno-Meteorologiczny, Warsaw, 1948; and "Hydrologia Kontynentalna" (Continental Hydrology), nos. 31 and 39, Wydawnictwa Komunikacyjne, Warsaw, 1955 and 1959.
124. Czetwertyński, E.: "Hydrologia" (Hydrology), Państwowy Wydawnictwa Naukowy, Warsaw-Poznań, 1955.
125. Annotated bibliography on hydrology, 1941–1950 (United States and Canada), compiled under the auspices of the Subcommittee on Hydrology, Federal Inter-Agency River Basin Committee, by American Geophysical Union, *Notes on Hydrologic Activities, Bull.* 5, U.S. Government Printing Office, June, 1952.
126. Annotated bibliography on hydrology, 1951–54 and sedimentation, 1950–54 (United States and Canada), prepared under the auspices of Subcommittee on Hydrology and Sedimentation—Inter-Agency Committee on Water Resources and compiled and edited by American Geophysical Union, *Joint Hydrology-Sedimentation Bull.* 7, U.S. Government Printing Office, December, 1955.
127. Annotated bibliography on hydrology and sedimentation, 1955–58 (United States and Canada), *U.S. Geol. Survey Water-Supply Paper* 1546, 1962.
128. Wood, H. W., Jr.: Hydrology in the civil engineering curriculum, *J. Eng. Educ.*, vol. 36, pp. 519–521, April, 1946.
129. Dubois, R. H.: A course in agricultural hydrology, *Agr. Eng.*, vol. 29, pp. 165–166, April, 1948.
130. The role of hydrology in engineering, a paper prepared by the Committee on Hydrology of the Hydraulics Division, American Society of Civil Engineers, October, 1950.
131. Lenz, A. T.: Educational facilities for hydrology, unpublished paper presented at the first ASCE Hydraulics Division Conference in Vicksburg, Mississippi, November 1–3, 1950.
132. Wilm, H. G., and others: The training of men in forest hydrology and forest management, *Rept. Comm. Soc. Am. Foresters*, 1956.
133. Linsley, R. K., Jr.: The role of hydrology in engineering, *Civil Eng., Bull. Am. Soc. Eng. Educ.*, vol. 21, pp. 7–9, 1956.
134. Rousseau, C. A.: Algunas consideraciones sobre la hidrogeologia moderna. Su enseñanza y applicación en los Estados Unidos (Some considerations about modern hydrogeology. Its education and application in the United States), *Ciencia y Tecnologia*, vol. 1, no. 24, Pan-American Union, 1957.
135. Cosens, K. W.: Hydrology in the sanitary engineering curriculum, *J. Eng. Educ.*, vol. 47, pp. 421–423, January, 1957.
136. Langbein, W. B.: Hydrologic education, *Intern. Assoc. Sci. Hydrol., Bull.* 11, pp. 27–30, September, 1958.
137. Hackett, J. E., and W. C. Walton: Educational and academic research facilities in ground-water geology and hydrology in the United States and Canada, Technical Division, National Water Well Association, 1961.

Section 2

OCEANOGRAPHY

VEN TE CHOW, *Professor of Hydraulic Engineering, University of Illinois.*

I. Introduction	2-2
II. Oceans and Sea Water	2-2
A. Characteristics of Oceans	2-2
B. Properties of Sea Water	2-2
1. Composition	2-3
2. Density	2-4
3. Viscosity	2-4
4. Compressibility	2-4
5. Vapor Pressure	2-4
6. Freezing Point	2-4
7. Temperature	2-4
8. Ice Formation	2-6
III. Ocean Currents	2-6
A. Types of Currents	2-6
B. Hydrodynamics of Ocean Currents	2-8
C. Drift Current	2-9
D. Slope Current	2-11
E. Density Current	2-11
F. Measurement of Currents	2-13
1. The Eulerian Method	2-13
2. The Langrangian Method	2-13
3. The Geostrophic Method	2-13
4. The Electromagnetic Method	2-14
IV. Ocean Waves and Tides	2-14
A. Variation of Sea Level	2-14
B. Theories and Classification of Surface Waves	2-15
1. Capillary Waves	2-16
2. Gravity Waves	2-16
3. Tides	2-17
4. Trans-tidal Waves	2-17
C. Gravity Waves	2-17
D. Wind Waves	2-17
E. Tides	2-18
F. Internal Waves	2-19
G. Wave Pressure	2-19
V. Thermal Interactions between Ocean and Atmosphere	2-20
A. Advective and Convective Processes	2-20
B. Heat Budget of the Ocean	2-21
C. Evaporation from Sea Surface	2-21
D. Ocean Clouds and Fogs	2-22
VI. References	2-22

I. INTRODUCTION

The oceans, covering about 71 per cent of the globe surface and containing about 98 per cent of all water on the earth, act as the source and sink for the circulation of water in the hydrosphere. Although the major scientific studies of oceans are in the domain of oceanography, many aspects of oceans directly affect hydrologic phenomena, especially the movement of moisture-carrying air masses between oceans and land, the actions of ocean waves on coasts and shores, and the direct effects of the oceans on continental waters in estuaries and in aquifers.

Literally speaking, the entire science of the study of oceans should be called *oceanology*, while its part dealing with hydrographical description of seas and oceans be called *oceanography*. However, in most literature written in the English language, the term "oceanography" is generally used in lieu of "oceanology."

There are many branches of oceanography such as physical oceanography, chemical oceanography, and biological oceanography. This section presents very briefly certain features in physical oceanography that are considered important to hydrologists. As the name implies, *physical oceanography* deals with the physical properties and processes of seas and oceans. For a broad study of this subject, the reader should refer to Refs. 1 to 12.

II. OCEANS AND SEA WATER

A. Characteristics of Oceans

Oceans are large bodies of salt water that cover 361,254,000 km² (about 139,500,000 sq mi) or about 71 per cent of the earth surface (510,100,000 km² or about 196,950,000 sq mi). The oceans have an average depth equal to approximately 3.79 km (2.35 miles), lying in a space of about 5 km (3 miles) between the *continental platform*, which embraces sea level and includes the continental shelf and most land, and the *oceanic platform*.

Of the ocean area, 85 per cent lies between 500 and 3,500 fathoms (1 fathom = 6 ft = 1.8288 m) while only 0.1 per cent lies below 3,500 fathoms forming the *trenches*. The area of the earth's solid surface above any given level of elevation or depth can be shown by a *hypsographic curve* (Fig. 2-1). The frequency distributions of elevations and depths are shown at the left of Fig. 2-1. The hypsographic curve represents the summation of areas between certain levels without respect to their location or to the relation of elevations and depressions, and therefore it should not be interpreted as an average profile of the land surface and sea bottom.

The oceans play a vital role in the hydrologic cycle (Sec. 1). Persistent wind systems transport huge air masses across the oceans and in and out of the land areas. More or less stable currents carry their temperature characteristics to far-distant parts of the oceans affecting the climatic conditions in bordering lands. The inland evaporation of about 15,000 cu mi yearly occurring in dry air masses is transported as water vapor out to sea and is precipitated. Of the 81,000 cu mi of water evaporated yearly from the ocean, a considerable portion is carried by maritime air masses to the land, where total precipitation amounts to about 24,000 cu mi annually. To compensate for the excess of precipitation over land evaporation, 9,000 cu mi of water returns to the sea by runoff. In addition to this large-scale phenomenon, there are numerous interactions between the contacting areas of the oceans and land which affect synoptic meteorology (Sec. 3) and cause salt-water intrusion problems (Sec. 13).

B. Properties of Sea Water

The physical properties of sea water are different from those of fresh water, because they vary not only with temperature and pressure but also with the concentration of salt. Some important properties are listed below, while detailed information may be found in Refs. 11 and 12.

1. Composition. Sea water is generally composed of water, solids, and gases. The solids contain suspended particles of inorganic matters, living organisms, and organic detritus as well as salts. The salts, constituting about 99.975 per cent of the total solids in typical sea water, are almost completely ionized. The proportions of the salts in per cent by weight are [2]: Na^+, 30.61; Mg^{++}, 3.69; Ca^{++}, 1.16; K^+, 1.10; Sr^{++}, 0.04; Cl^-, 55.04; SO_4^{--}, 7.68; HCO_3^-, 0.41; Br^-, 0.19; and H_3BO_3, 0.07.

The ratio of total solids to total sample of sea water is approximately what is known as the *salinity*. The ratio of the halide ions (Cl^-, Br^-, and I^- measured

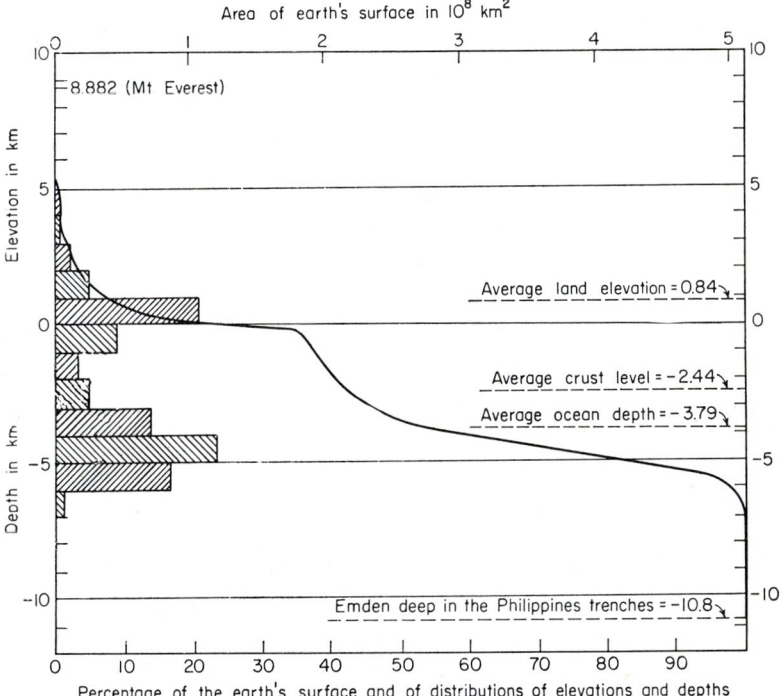

FIG. 2-1. Hypsographic curve for the earth's surface and distribution diagram of elevations and depths. (*Modified from Sverdrup, Johnson, and Fleming* [2].)

by chemical titration) to total sample of sea water is approximately what is known as the *chlorinity*. The relation between salinity and chlorinity can be expressed by the *Knudsen formula* [14]:

$$\text{Salinity} = 0.03 + 1.805 \times \text{chlorinity} \qquad (2\text{-}1)$$

Both salinity and chlorinity are expressed in parts per thousands or per mille, denoted by ‰. The salinity of most sea water varies between 33 and 37‰ and the average 35‰ is usually chosen as standard.

The precise definitions of salinity and chlorinity are as follows: Salinity is the total amount of solid material in grams contained in one kilogram of sea water when all the carbonate has been converted to oxide, the bromine and iodine replaced by chlorine, and all organic matter completely oxidized. Chlorinity is the total amount of chlorine, bromine, and iodine in grams contained in one kilogram of sea water, assuming that the bromine and the iodine had been replaced by chlorine. Since this definition of chlorinity gives numerical values which change with each refinement of the values of atomic weights, a new definition independent of such changes has been given as: chlorinity in grams per kilogram of sea water is identical with the mass in grams of "atomic-weight silver" just needed to precipitate the halogens in a 0.3285233-kg sample of sea water. The determination of salinity and chlorinity according to their precise definitions is difficult and slow [2, pp. 50–54]. For practical purposes, the approximate definitions given in the previous paragraph may be used.

Because sea water has more or less a uniform composition, its density and some other physical properties depend almost entirely on temperature, salinity, and pressure. Figure 2-2 shows some of the properties at a pressure of 1 atmosphere.

2. Density. The density of sea water depends mainly on temperature, salinity, and pressure. Many tables are available for computation of density and specific volume of sea water [13–15]. The average density of surface sea water varies between 1.022 and 1.028 g/cm³. In oceanography, the density of sea water is often referred to as the density of a parcel after the pressure has been reduced to 1 atmosphere, that is, brought to the sea surface. This transformation eliminates adiabatic effects. Furthermore, the density ρ is often expressed in terms of a quantity known as "sigma-tee" or

$$\sigma_t = (\rho - 1)1{,}000 \tag{2-2}$$

This conversion magnifies the variation in the numerical value of the density. Thus, σ_t of the surface sea water varies between the limits of 22.0 and 28.0.

The maximum density of sea water is reached at a temperature which depends on salinity and pressure. Under 1 atmosphere, this temperature in °C is approximately equal to $4.00 - 0.215 \times$ salinity in ‰.

When the salinity of a sea water sample is plotted against the corresponding temperature on rectangular coordinates, a so-called *temperature-salinity diagram*, or *TS-diagram*, can be constructed (Fig. 2-2). When water in a certain region of the sea possesses a definite temperature and salinity for a wide variety of conditions, it forms a continuous *water mass*. If the water mass is homogeneous, then the oceanographic factors in it are constant and it is plotted as a single point on the TS-diagram. The position of the single point will not change if this homogeneous water mass is moved in any direction. Due to certain processes such as mixing, radiation, or evaporation, however, the water mass may lose its homogeneity and its position on the diagram will change. Such changes usually occur in the surface layer down to about 500 ft where the water is subject to climatic influence. On the TS-diagram, lines of equal density σ_t, or ρ, or any other properties can be added to supply certain instructive description of the water mass. For example, addition of lines of equal σ_t, or *isopycnals*, will serve to indicate stability of vertical stratification. If the plotted curve on the diagram is parallel to the isopycnals, only the surface layer is stable; but if the curve intersects the isopycnals at a wide angle, the stability is larger.

The use of TS-diagrams was first proposed by Helland-Hansen [16]. One practical application is to enable the oceanographer to delete errors and correct them in the preparation of oceanographic data. If the data do not plot consistently on the diagram, usually an observational error or a faulty computation is indicated. Corrections can be made therefore by comparison with similar plots for neighboring stations.

3. Viscosity. The viscosity of sea water changes linearly with salinity. With a salinity of 35‰, the dynamic viscosity in g/cm/sec is 0.01877 at 0°C and 0.01075 at 20°C, and the kinematic viscosity in cm²/sec is 0.01826 at 0°C and 0.01049 at 20°C.

4. Compressibility. Water that is less saline or cooler is more compressible.

5. Vapor Pressure. The vapor pressure over sea water is somewhat lower than that over fresh water, depending on salinity. It can be computed approximately by multiplying the vapor pressure of distilled water at the same temperature by $1 - 0.000537 \times$ salinity in ‰.

6. Freezing Point. The freezing point of sea water is lower than that of fresh water because of its salinity. The freezing point in °C of water is approximately equal to $-0.053 \times$ salinity in ‰.

7. Temperature. Since the specific heat of sea water, about 0.95, is much higher than that of the land, about 0.50 on average, the temperature variation in sea water is

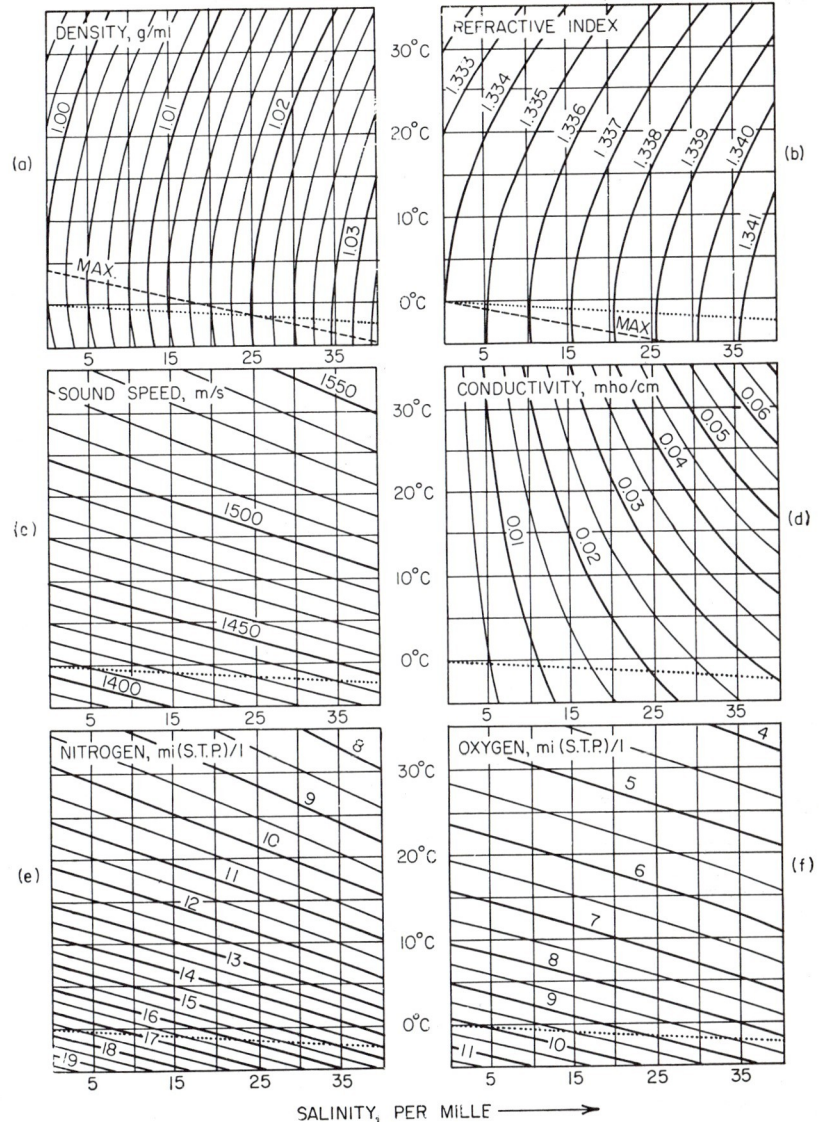

FIG. 2-2. Temperature-salinity diagrams for sea water at 1 atm pressure: (a) density, (b) refractive index for sodium light (0.5893 micron) relative to air, (c) sound speed, (d) conductivity, (e) concentration of nitrogen in equilibrium with 1 atm (1,013.25 mb) of air saturated with aqueous vapor, (f) concentration of oxygen under same equilibrium conditions. Freezing point is shown by dotted line; values below it pertain to undercooled water. (After R. B. Montgomery [11].)

much slower than that of the land. The temperature in the open ocean varies between about 32°C and the freezing point, −2°C, and more than half the ocean surface is warmer than 20°C. About 93 per cent of the ocean volume is colder than 10°C and 76 per cent colder than 4°C.

The horizontal variation of the surface temperature of sea water depends on many factors, such as solar radiation, ocean currents, flow from rivers, evaporation, wind, and precipitation.

The vertical variation in temperature may divide the depth of ocean into several layers. The shallow *surface layer* usually maintains a high close-to-surface temperature which decreases very gradually with increase in depth. This layer is followed by a *thermocline layer* which characterizes rapidly decreasing temperature with depth. Below the thermocline layer is the *deep-water layer* in which the temperature decreases gradually again with the depth and at its lower end the temperature scale becomes asymptotic to the constant low temperature of the *bottom-water layer*. The thermocline layer has a mean thickness of 200 meters (650 ft). It may be permanent and extend to more than a mile deep. It may also be shallow and vary with the seasons, thus becoming known as *seasonal thermocline*.

8. Ice Formation. Two types of ice can be formed in the oceans: *sea ice* which is frozen sea water, and *icebergs* which are broken-off pieces of glaciers (see Sec. 16).

From the known data on maximum density and freezing point of sea water, it can be computed that the temperature of maximum density coincides with the freezing point at a salinity of 24.7‰, a temperature equal to −1.3°C. Since sea water has a uniform salinity higher than 24.7‰, the maximum density is reached only if the water is undercooled. In general, the salinity is uniform in a surface layer of about 10 to 20 fathoms thick, and this whole layer has to be cooled to freezing point before ice can form. When ice begins to form, elongated crystals of pure ice are produced and interwoven with brine cells of high salt concentration. The process of ice formation in sea water is therefore different from that in fresh water (Sec. 16).

Sea ice forming a solid immobile cover is called *fast ice*. When broken up by winds and currents, sea ice is called *drift ice*. Drift ice becoming jammed together to form a continuous cover or rugged mass is often described as *pack ice*. In the Antarctic, the icebergs originate from the polar icecaps. In the Northern Hemisphere, the icebergs are derived mainly from the glaciers on Greenland. Icebergs are generally flat and may be 10 to 20 miles wide, up to 50 miles long, and more than 2,400 ft thick, thus having a corresponding height of 250 ft or so above the water surface.

Limits indicating extent of the presence of sea ice can be found on the charts of mean monthly sea-surface temperatures issued by the U.S. Navy Hydrographic Office.

III. OCEAN CURRENTS

A. Types of Currents

Ocean currents are of many types. *Drift*, or *wind*, *currents* are those developed from friction exerted on sea surface by steadily blowing wind. *Slope currents* are caused by inclination of the water surface developed by wind, barometric differentials, or inflow from rivers. *Density currents* are caused by difference in density of the water. *Tidal currents* are caused by tidal actions. To satisfy the law of continuity of flow, *compensation currents* are developed to fill up spaces left by activating currents. Currents may also be generated by internal waves propagated along surfaces of large density discontinuity within the body of the sea.

The average surface ocean currents are composed of large-scale circulation of water due to the distribution of mass in the sea which is developed by heating at the tropics and cooling at the poles. This heating and cooling process induces a general surface motion toward the poles and a depth motion toward the equator. The motions are further modified by the effect of earth rotation, gravity, and irregular solid boundaries of the sea, and by wind effects. In contrast to other small-scale, more or less local, currents, the large-scale circulating currents transport great amounts of water.

Figure 2-3 shows the prevailing surface ocean currents in the Northern Hemisphere

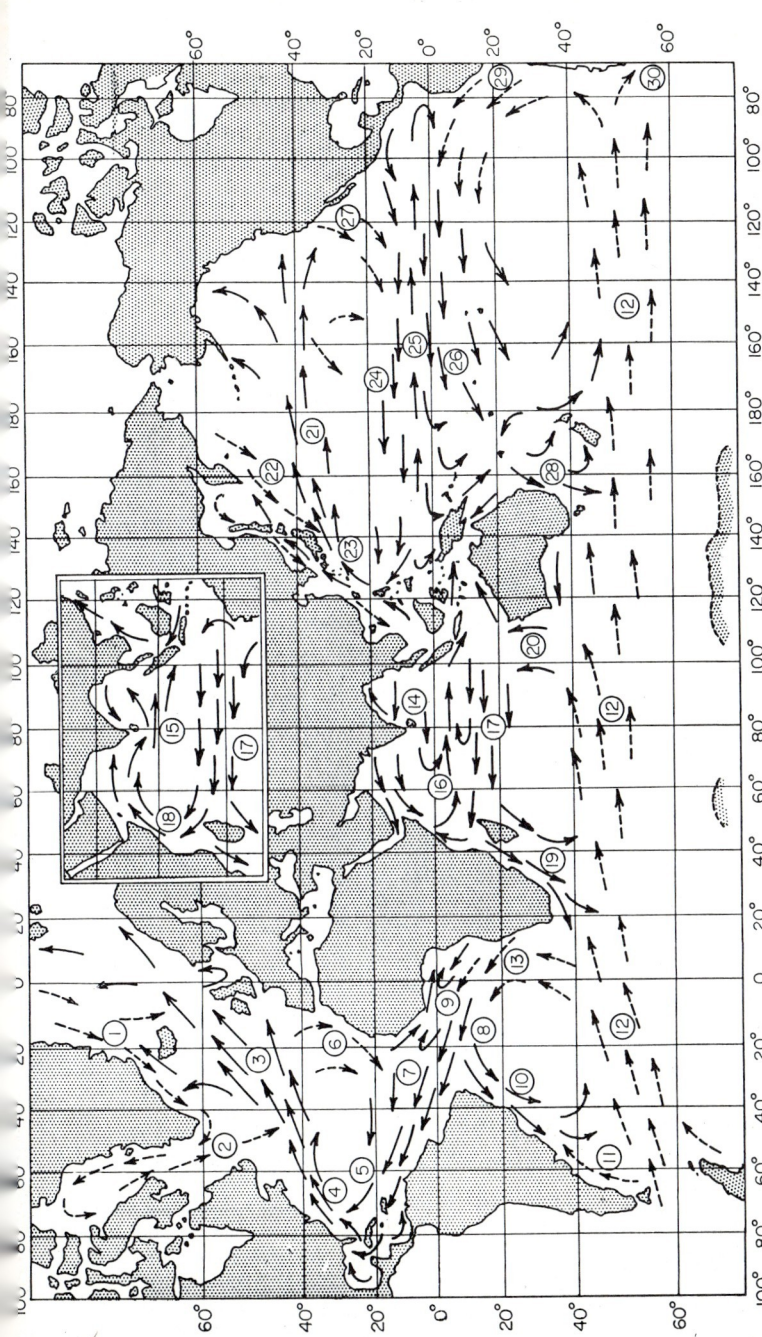

FIG. 2-3. Prevailing surface ocean currents in Northern Hemisphere in winter. The top figure represents the condition in summer in Indian Ocean. (1) East Greenland Current, (2) Labrador Current, (3) North Atlantic Current, (4) Gulf Stream, (5) Antilles Current, (6) Canaries Current, (7) North Equatorial Current, (8) South Equatorial Current, (9) Guiana Current, (10) Brazil Current, (11) Falkland Current, (12) Circumpolar Current, (13) Benquela Current, (14) North-east Monsoon Current, (15) South-west Monsoon Current, (16) Equatorial Countercurrent, (17) South Equatorial Current, (18) Somali Current, (19) Agulhas Current, (20) West Australian Current, (21) North Pacific Current, (22) Oyashio Current, (23) Kuroshio Current, (24) North Equatorial Current, (25) Equatorial Countercurrent, (26) South Equatorial Current, (27) Californian Current, (28) East Australian Current, (29) Peru Current, (30) Cape Horn Current.

in winter and the reversed currents in the Indian Ocean in summer. These surface currents are composed mainly of drift currents and their compensation currents. Equatorial currents are caused by trade winds while the equatorial countercurrents are compensation currents. The Antarctic circumpolar currents are caused by the west wind drift. Their countercurrents and the density currents constitute the polar currents. Figure 2-4 shows the meridional vertical circulation of ocean currents.

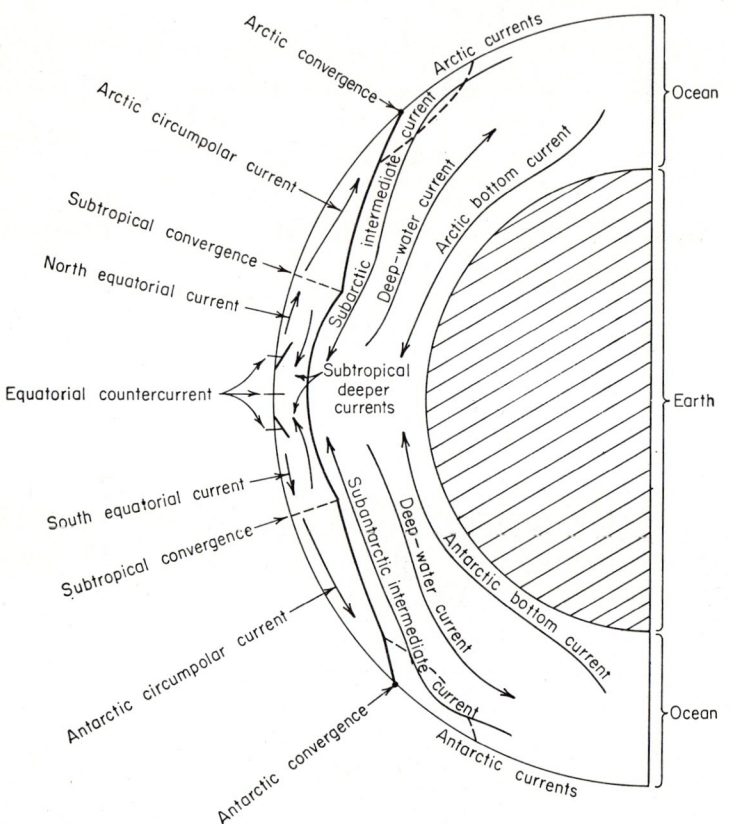

FIG. 2-4. Meridional vertical circulation of ocean currents.

B. Hydrodynamics of Ocean Currents

The hydrodynamics of ocean currents can be described by the dynamic equation of incompressible fluid [17].

Let x, y, and z be the three coordinates in space; u, v, and w, the component velocities in x, y, and z directions, respectively; and X, Y, and Z, the external body forces in x, y, and z directions, respectively; p, the pressure; p_o, the pressure on sea surface; ρ, the density; Ω, the angular velocity of the earth ($= 7.292 \times 10^{-5}$); φ, the latitude; t, the time; g, the gravitational acceleration; R_x and R_y, resistance forces in x and y directions, respectively. The z axis is assumed to be perpendicular to the surface of the sea. Then, the motion of the sea water can be described approximately by the following dynamic equations:

$$\frac{\partial u}{\partial t} = X + 2\Omega v \sin \varphi - \frac{1}{\rho}\frac{\partial p}{\partial x} + \frac{1}{\rho}R_x \qquad (2\text{-}3a)$$

$$\frac{\partial v}{\partial t} = Y - 2\Omega u \sin \varphi - \frac{1}{\rho}\frac{\partial p}{\partial y} + \frac{1}{\rho}R_y \qquad (2\text{-}3b)$$

$$p = p_o(x,y) + \rho g z \qquad (2\text{-}3c)$$

which should be solved along with the following continuity equation:

$$\frac{\partial u}{\partial x} + \frac{\partial v}{\partial y} + \frac{\partial w}{\partial z} = 0 \qquad (2\text{-}4)$$

In Eqs. (2-3), the terms including $\sin \varphi$ are due to the deflecting force of the earth's rotation or *Coriolis force*. If horizontal turbulence is neglected, the frictional forces can be evaluated by Prandtl's theory of momentum transport [18] as:

$$R_x = \frac{\partial}{\partial z}\left(\epsilon \frac{\partial u}{\partial z}\right) \quad \text{and} \quad R_y = \frac{\partial}{\partial z}\left(\epsilon \frac{\partial v}{\partial z}\right) \qquad (2\text{-}5)$$

where ϵ is the eddy viscosity. By Taylor's theory of vorticity transport [19], the frictional forces are

$$R_x = \epsilon \frac{\partial^2 u}{\partial z^2} \quad \text{and} \quad R_y = \epsilon \frac{\partial^2 v}{\partial z^2} \qquad (2\text{-}6)$$

although ϵ may be variable. It appears that the theory of momentum transport or Eqs. (2-5) give better agreement with observed conditions. The observed value of ϵ varies broadly depending on individual determinations. For sea water, the average value varies from 1 to 1,000 g/cm/sec. The value for the surface layer (50 to 500) is far greater than that for the deep water layer (10). The value for the surface layer increases with the increase of wind velocity.

C. Drift Current

The tractive force τ_o on sea surface due to a wind velocity of w may be expressed by

$$\tau_o = \rho k^2 w \qquad (2\text{-}7)$$

where ρ is the density of air or about 0.0013 and k is a coefficient equal to about 0.0025 all in metric units. This value may be decreased to 0.001 if the wind velocity is less than 6.6 m per sec or 20 ft per sec.

The velocity V_o of the surface current can be estimated by many formulas. The simplest one is

$$V_o = k_o w \qquad (2\text{-}8)$$

where k_o is a *wind coefficient* which ranges between 0.025 and 0.035 approximately.

Because of the effect of both the Coriolis force and the eddy viscosity, the drift current in deep water decreases in velocity and changes in direction as depth increases. This phenomenon can be best illustrated by the mathematical model developed by Ekman [20]. The schematic representation of this model is shown in Fig. 2-5. The arrows represent the velocities at depths of equal intervals. Together they form a spiral staircase known as the *Ekman spiral*, the steps of which rapidly decrease in width as they proceed downward. The horizontal projection of the arrows forms a logarithmic spiral. The average deflection of the drift current from the direction of the wind has been observed to be about 45° to the right of the wind direction, independent of latitude. This is in agreement with Ekman's theory.

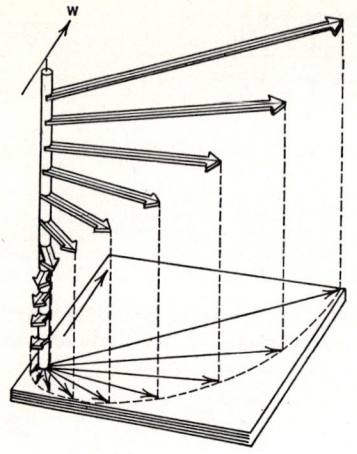 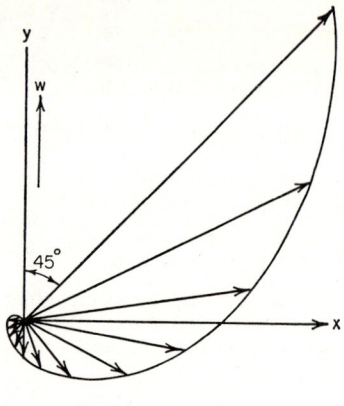

Fig. 2-5. The Ekman model for drift current. w indicates wind direction.

Under steady conditions with constant ϵ, Eqs. (2-3) may be written as

$$\frac{\epsilon}{\rho}\frac{d^2u}{dz^2} = -2\Omega v \sin\varphi \tag{2-9a}$$

and

$$\frac{\epsilon}{\rho}\frac{d^2v}{dz^2} = 2\Omega u \sin\varphi \tag{2-9b}$$

If the wind is in the direction of the y axis and h is the depth of sea water, solution of the above equations gives

$$u = A \sinh a(h-z) \cos a(h-z) - B \cosh a(h-z) \sin a(h-z) \tag{2-10a}$$

and

$$v = A \cosh a(h-z) \sin a(h-z) + B \sinh a(h-z) \cos a(h-z) \tag{2-10b}$$

where

$$a = \sqrt{\frac{\rho\Omega \sin\varphi}{\epsilon}}$$

$$A = \frac{\tau_o D}{\pi \epsilon} \frac{\cosh ah \cos ah + \sinh ah \sin ah}{\cosh 2ah + \cos 2ah}$$

and

$$B = \frac{\tau_o D}{\pi \epsilon} \frac{\cosh ah \cos ah - \sinh ah \sin ah}{\cosh 2ah + \cos 2ah}$$

in which $D = \pi/a$. The angle θ between the directions of the wind and the surface current is

$$\tan\theta = \frac{\sinh 2ah - \sin 2ah}{\sinh 2ah + \sin 2ah} \tag{2-11}$$

When the depth h becomes very large, Eqs. (2-10) become

$$u = V_o e^{-az} \cos(45° - az) \tag{2-12a}$$

and

$$v = V_o e^{-az} \sin(45° - az) \tag{2-12b}$$

where $V_o = \tau_o/\sqrt{2\epsilon\rho\Omega \sin\varphi}$ is the velocity of surface current. At depth $z = \pi/a$, the current is directed opposite to the surface current and its velocity is only $e^{-\pi} = \frac{1}{23}$ of V_o. This depth is known as the *depth of frictional resistance, D*.

D. Slope Current

The slope current is caused by wind, barometric differentials, inflow of fresh river water, or other reasons. The slope of the sea surface caused by wind may be expressed by

$$\tan \beta = \frac{3}{2} \frac{\tau_o}{g\rho h} \tag{2-13}$$

where β is the slope angle with the horizontal plane, τ_o is the tractive force of wind on sea surface, and h is the depth of sea water. In the Baltic Sea, A. Colding [2, p. 490] observed that the slope due to wind of surface velocity w is

$$\tan \beta = 4.8 \times 10^{-9} \frac{w^2}{h} \tag{2-14}$$

where w is in meters per second and h is in meters.

Considering β as the angle between the sea surface and the y axis, $X = 0$, and $Y = g \sin \beta$, Eqs. (2-3) give the following equations for steady-state conditions:

$$\frac{d^2u}{dz^2} + 2a^2 v = 0 \tag{2-15a}$$

and

$$\frac{d^2v}{dz^2} - 2a^2 u + \frac{\rho g \sin \beta}{\epsilon} = 0 \tag{2-15b}$$

Solution of these equations gives

$$u = \frac{g \sin \beta}{2\Omega \sin \varphi} \left[1 - \frac{\cosh a(h+z) \cos a(h-z) + \cosh a(h-z) \cos a(h+z)}{\cosh 2ah + \cos 2ah} \right] \tag{2-16a}$$

and

$$v = \frac{g \sin \beta}{2\Omega \sin \varphi} \left[\frac{\sinh a(h+z) \sin a(h-z) + \sinh a(h-z) \sin a(h+z)}{\cosh 2ah + \cos 2ah} \right] \tag{2.16b}$$

For $h > D$ or π/a, the velocities in the surface layer are $u_o \approx g \sin \beta / 2\Omega \sin \varphi$ and $v_o \approx 0$, and the direction of the current is perpendicular to the direction of surface slope.

E. Density Current

Density currents are caused by variation in density of sea water which may be developed due to temperature, sediment, or salinity differentials. For steady-state conditions, Eqs. (2-3) give

$$\frac{D^2}{2\pi} \frac{d^2u}{dz^2} + v = \frac{1}{2\epsilon a^2} \frac{\partial p}{\partial x} \tag{2-17a}$$

and

$$\frac{D^2}{2\pi} \frac{d^2v}{dz^2} - u = \frac{1}{2\epsilon a^2} \frac{\partial p}{\partial x} \tag{2-17b}$$

where the notations have been defined previously. For given conditions, a simple solution of the above equations may be made by Bjerknes' theorem of circulation [21–22], but a complete solution is difficult due to nonhomogeneity of the fluid under consideration.

Density current in estuaries is of specific importance to hydraulic engineers and hydrologists. According to Pritchard [23], a sequence of estuarine types of distinct density stratification and circulation pattern can occur in various ways depending

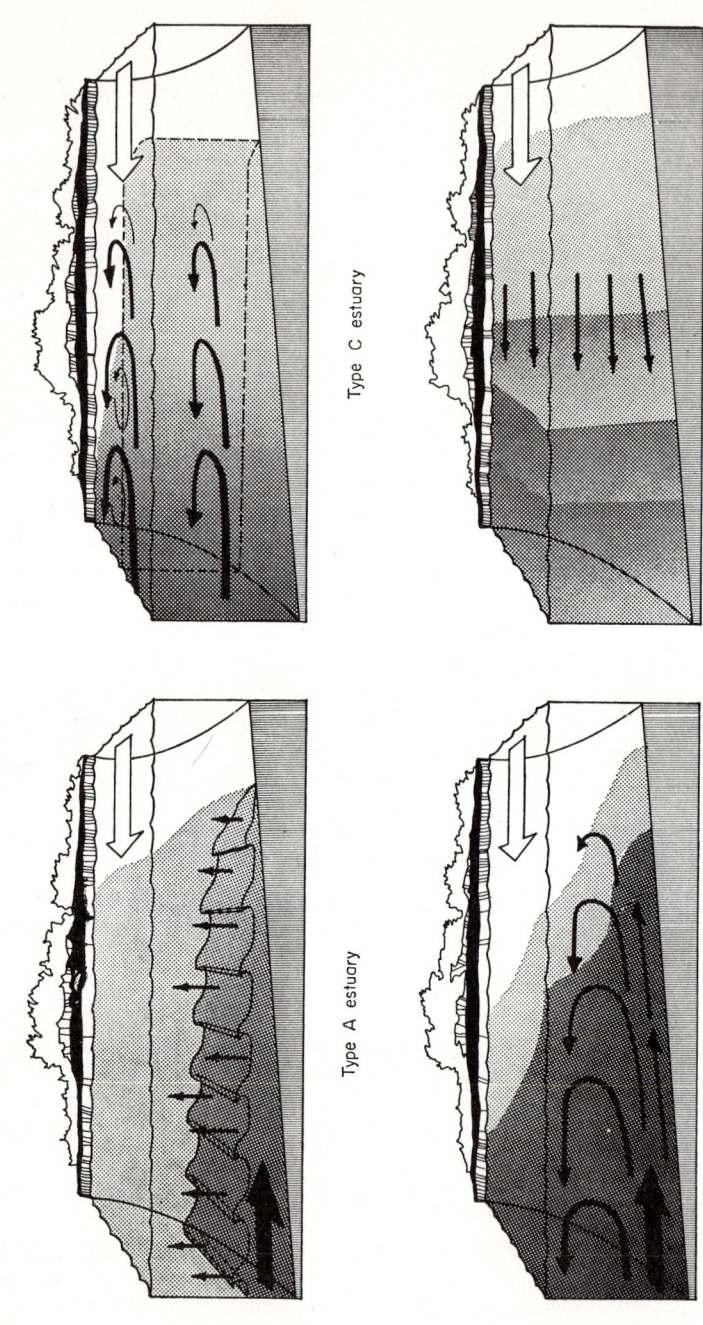

Fig. 2-6. Circulation patterns in estuaries. (*After Pritchard* [23].)

on river flow, tidal flow, width, and depth. Four sequential types of estuaries are shown in Fig. 2-6.

Type A is a salt-wedge estuary which occurs when the ratio of river flow to tidal flow is relatively large and the ratio of width to depth is relatively small. The circulation pattern in this type is dominated by the river flow, which overlays a salt-water wedge and the latter can be pushed back to the sea when the river flow increases. Salt water is advected into the upper layer from the wedge, but there is little or no mixing of fresh water down into the wedge. As salt water is added from below, the volume of upper-layer flow increases and a relatively small return flow must occur in the wedge to compensate for this loss of water from the wedge. Tidal mixing is slight in this type of estuary.

In type B estuary, tidal mixing becomes important. The mixing occurs between the upper, seaward flowing layer, and the dense salt water flowing upstream along the bottom. The volume of compensation current involved in this circulation pattern is far greater than the fresh-water outflow.

In type C estuary, the vertical mixing is so intense that the estuary becomes vertically homogeneous. In order to maintain salt balance, a lateral flow pattern exists in which lateral advection and eddy mixing of salt flux take place. The outflow occurs along the right side of the estuary, while the higher salinity return flow is found along the left side.

In type D estuary, the width is sufficiently narrow so that lateral mixing cannot occur and no upstream advection can develop when tidal mixing is sufficient to produce vertical homogeneity.

While other factors, such as wind velocity, air temperature, solar radiation, and bottom roughness, are kept constant, an estuary tends to shift from type A through type B to type C or D with decreasing river flow, increasing tidal velocities, increasing width, and decreasing depth.

F. Measurement of Currents

There are numerous methods to measure ocean currents. According to von Arx [1], they can be divided into direct and indirect categories. The direct methods can be subdivided into *Eulerian* and *Langrangian* methods, while the indirect methods can be subdivided into *geostrophic* and *electromagnetic* methods.

1. The Eulerian Method. This method consists of mechanical or dynamical measurements of the flow past a geographically fixed point, such as an anchored ship, sea bed, or any fixed man-made structures in shallow waters. The velocity of flow is determined as a function of depth and time. The flow of water can be measured by *current meters* or counting against time the rotations of a free-turning propeller; by measuring the torque of an arrested propeller or rotor; by measuring the ram pressure on a plate, membrane, sphere, or Pitot orifice; by measuring the deflection of a suspended cable supporting a known drag; by measuring the change of the velocity of sound between two points of a known distance apart; or by measuring the motional emf of the flow through a known magnetic field.

2. The Langrangian Method. In this method, the trajectories of water parcels are tracked and plotted in space and time with the aid of tracers. The tracers may consist of the following devices and practices: drift bottles, radio buoys, contaminants or dye stuffs, current poles, deep drogues (devices having a large drag at the level of measurement and hanging on a small floating buoy with a fine wire), neutrally buoyant floats, and ship drift.

3. The Geostrophic Method. In this method a very useful approximation to geostrophic motion of fluid is made. When fluid flow is both unaccelerated and frictionless, its motions are known as "geostrophic," that is, "earth-turned." In this case the pressure difference, Coriolis force, and gravity balance one another, and the currents thus developed are known as *geostrophic currents*. In oceanic regions remote from solid boundaries and the free surface of the sea, such geostrophic currents have been found to be good approximations to the actual currents.

For geostrophic currents, Eqs. (2-3) can be written as

$$\frac{1}{\rho}\frac{\partial p}{\partial x} = 2\Omega v \sin \varphi \qquad (2\text{-}18a)$$

$$\frac{1}{\rho}\frac{\partial p}{\partial y} = 2\Omega u \sin \varphi \qquad (2\text{-}18b)$$

$$\frac{1}{\rho}\frac{\partial p}{\partial z} = g \qquad (2\text{-}18c)$$

The third equation is the hydrostatic equation. The first two equations can be reduced to

$$\frac{1}{\rho}\frac{\partial p}{\partial n} = 2\Omega c \sin \varphi \qquad (2\text{-}19)$$

or

$$c = \frac{1}{2\rho\Omega \sin \varphi}\frac{\partial p}{\partial n} \qquad (2\text{-}20)$$

where $c = \sqrt{u^2 + v^2}$ and $\partial n = \sqrt{(\partial x)^2 + (\partial y)^2}$. Equation (2-20) can be used to compute the horizontal velocity component c of the geostrophic current from the measurements of the horizontal gradients in the field of pressure, or $\partial p/\partial n$. This component c is at right angles to the pressure difference ∂p along a line segment ∂n also normal to c. This geostrophic equation may be called the *Mohn-Sandström-Helland-Hansen formula* due to its original contributors [24–25]. The geostrophic velocity so computed, however, is only relative in the velocity field; it must be calibrated with known boundary conditions in order to develop an approximation of the actual velocity field. The practical procedure is to apply the geostrophic method to measurements at successive observation stations, and this is usually known as the *method of dynamic sections*. This method essentially incorporates the following assumptions: the flow is unaccelerated; the flow is frictionless; the pressure field is hydrostatic; and water properties do not change at all observations.

4. The Electromagnetic Method. The sea water, containing an abundance of highly ionized salts, is an electrolyte of relatively high conductivity. The motion of this electrolyte through the earth's magnetic field produces electromagnetic effects By this principle, ocean currents can be determined from measurements of the electromagnetic effects by means of electrodes being towed by a ship. The modern instrument used for such purposes is known as a *geomagnetic electrokinetograph* or usually abbreviated as GEK.

IV. OCEAN WAVES AND TIDES

A. Variation of Sea Level

The position of the sea surface is represented by *sea level* which is the height of the boundary between sea and air. The sea level is measured relative to a fixed reference point on land, and hence is susceptible to changes in time with respect to the following major causes: the distribution of oceanographical factors, such as temperature and salinity of sea water; the distribution of climatological factors, such as wind and barometric pressure; changes in the distance of the reference point from the center of the earth due to reasons such as sudden movements in the earth's crust, glaciation, and tide-generating forces; and the long-period astronomical and pole tides.

Mean sea level is the plane about which the tide oscillates. It is determined from tidal observations by averaging the recorded hourly heights of the tide over a period of several years. Mean sea level varies with locality of observation and method of computation, and hence it has no absolute constant value. In order to simplify the computation, mean tide level is sometimes used as a substitute for mean sea level. *Mean tide level* is simply equal to the average of the observed high and low waters, and it is considered a poor substitute in many cases because of the appreciable influence of short-period tides.

Fluctuations of sea level range from short-period oscillations of surface waves and tides to long-period secular variations. Figure 2-7 is a schematic diagram suggested by Stommel [26] to depict various components of the spectral distribution of sea level. In this diagram, one horizontal ordinate represents the logarithm of the period of fluctuation P of the sea level in seconds, the other horizontal ordinate represents the logarithm of the wavelength L of the fluctuation in centimeters, and the vertical ordinate represents the height of sea level above mean sea level in meters. This diagram shows a whole spectrum of phenomena, on both the periodic and the geometric scale, associated with variations in sea level, for example, gravity

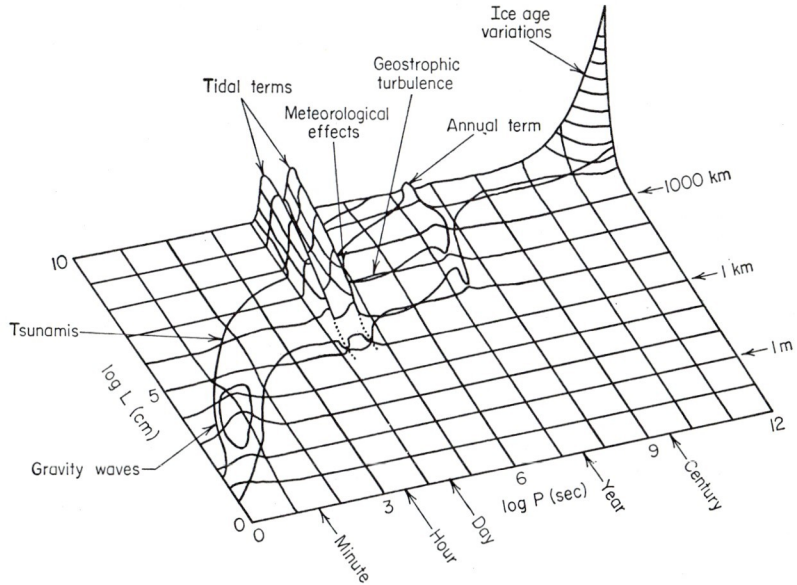

FIG. 2-7. Schematic diagram of the spectral distribution of sea level. (*After Stommel* [26].)

waves and tides. The largest range is in the ice-age variations. There are also variations of sea level associated with geostrophic turbulence, and with synoptic meteorological events.

B. Theories and Classification of Surface Waves

Theoretical studies of surface waves were made as early as 1802 by Gerstner [27] and 1845 by Airy [28]. In later years, classic theories of surface waves were extensively treated by many others including Lamb [29], Coulson [30], Stoker [31], and Eckart [17]. Classic theories deal with periodic motions of waves by assigning average values of wavelengths, directions, etc. Only in recent years, the randomness of sea waves has been studied by statistical analysis and measured by modern recording instruments. In such modern approaches, the variables are treated not as analytical functions but as stochastic processes, definable only in terms of probabilities. The modern approaches not only have explained some observed features, such as tidal waves, that cannot be analyzed satisfactorily by classic theories, but also have already been successfully applied to important engineering problems connected with the sea, such as the motions of ships and beach erosion.

If the rise and fall of a sea surface is recorded at a fixed point in the sea, it can be observed that the surface rises to a peak in a certain time and falls to a depression in another time. The recorded data would exhibit a very complicated wave pattern. If a photograph of the sea surface is taken from the sky at a certain time, it will show peaks and valleys of the surface at various places. This means that the wave pattern is formed not only with respect to time but also with respect to space. From the complicated wave pattern, basic surface waves of simple patterns can be derived. A combination of various basic surface waves forms the random geometry of an ordinary sea surface.

In classic theories, only basic surface waves are treated. It is assumed that by superposition such waves can reproduce an observed sea surface. In modern approaches, the sea surface is treated as a whole statistically in terms of both time and space. The modern approaches, however, have yet to be further developed and their discussions are scattered in literature only of the last decade or so (e.g., [32-34]).

In trains of sinusoidal waves, the *wavelength L* is the horizontal distance between successive points in phase on the waves, and the *period t* is the time elapsed for the wave to travel a distance equal to its wavelength. Thus, the velocity of a wave

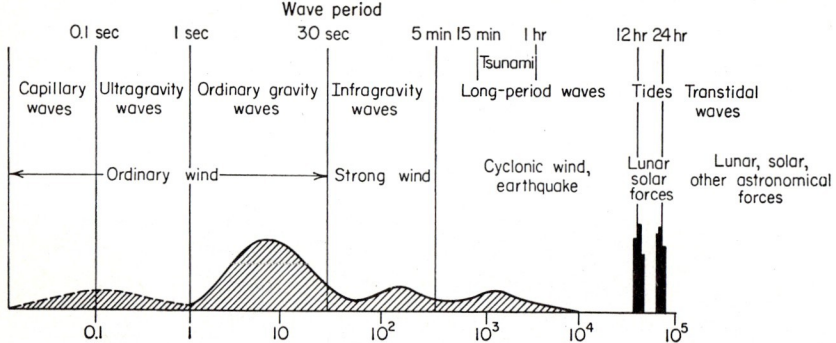

FIG. 2-8. Schematic diagram of the ocean-wave spectrum. (*After Munk* [35].)

measured relative to the liquid, known as *celerity c*, is equal to the wavelength divided by the period, or $c = L/t$. The wave frequency n is the number of the waves passing a point in the liquid per unit time. Hence, it is the reciprocal of the period, t, or $n = 1/t$. Consequently, $c = nL$.

In the modern concept the wave spectrum represents the distribution of energy, or mean-square oscillation, over a scale of frequency or period. Figure 2-8 shows a schematic representation of the ocean-wave spectrum. On the basis of the period, ocean waves can be classified as follows:

1. Capillary Waves. Capillary waves on the sea surface are small ripples produced by wind having velocities greater than 1 meter per second (3 ft/sec) and controlled by forces associated with surface tension and gravity. The wave period is less than 0.1 sec, and the wavelength is less than 1.73 cm, the celerity is therefore less than 17.3 cm/sec.

2. Gravity Waves. These waves are controlled by gravitational and inertial forces. Their wavelength is greater than 1.73 cm.

a. Ultra-gravity Waves. These are wind-generated waves with a period of 0.1 to 1 sec. They usually accompany well-developed waves in the storm area.

b. Ordinary Gravity Waves. These are generated by winds having velocities greater than 6.5 to 7 m/sec. Having periods between 1 and 30 sec, these wind waves are commonly observed on the sea, composed mainly of swell and sea (Subsec. IV-D).

c. Infragravity Waves. These are degenerated wind waves which cause obvious oscillation of the sea surface. The period is between 30 sec and 5 min.

d. Long-period Waves These waves are mostly generated by strong wind, cyclonic wind, and earthquakes. Their period is between 5 min and 12 hr. When caused by weather influence, the period is between 15 to 20 min and the amplitude is largely below 5 cm.

3. Tides. These waves are represented by line-spectra with well-defined phases. They are generated by lunar and solar tidal forces. Their periods are between 12 and 24 hr. They also include various transient phenomena, such as storm surges or waves from explosions or eruptions.

4. Trans-tidal Waves. These waves are controlled by lunar, solar, and other astronomical tidal forces. Their period is greater than 24 hr.

C. Gravity Waves

Gravity waves in oceans are mostly wind waves. Their properties can be considered in terms of a train of long crested sinusoidal or trochoidal waves of small amplitude (ratio of height to length is $1/100$ or less) moving on an ideal fluid of uniform depth d. The celerity of the waves can be mathematically represented by

$$c = \sqrt{\frac{gL}{2\pi} \tanh 2\pi \frac{d}{L}} \qquad (2\text{-}21)$$

where L is the wavelength and g is the gravitational acceleration. For *deep-water waves*, $d > L/2$, the above equation approaches $c = \sqrt{gL/2\pi}$. For *shallow-water waves*, $d < L/20$, the celerity becomes $c = \sqrt{gd}$. The group speed is half the phase speed for deep-water waves, but it is independent of wavelength and equals the phase speed for shallow-water waves of small amplitude.

In deep-water waves of small amplitude, the water particles move in circles. The radius r of the circles decreases with increasing depth by the following relationship

$$r = ae^{-2\pi \frac{z}{L}} \qquad (2\text{-}22)$$

where a is the amplitude or half the wave height, L is the wavelength, and z is the depth below the undisturbed water surface. The particles complete one revolution in the time T and hence their velocity is $2\pi r/T$. In shallow-water waves, the vertical motion of particles is restricted and the particles move back and forth horizontally and independent of depth.

The energy contained in a wave is the average energy over one wavelength and is a combination of kinetic and potential energies. In surface waves, the total average energy per square foot is $\frac{1}{2}g\rho a^2$, equally divided between kinetic and potential forms. Since g and the density, ρ, of water are constant, this total energy is proportional to the square of the amplitude and hence to the square of the wave height. In deep-water waves, only half the energy advances with wave velocity, whereas, in shallow-water waves, all the energy advances with the velocity.

Waves of large amplitude (ratio of height to length is greater than $1/100$ or $a/L > 1/200$) do not exhibit sinusoidal forms but assume a trochoid for a/L greater than $1/200$ but less than $1/50$. When $a/L > 1/50$, the wave form deviates from the trochoid to forms of wider and flatter troughs and narrower and steeper crests. When a/L approaches a critical value of $1/14$, the wave form becomes unstable. The value of a/L for most ocean waves remains less than this critical value, but relatively short waves may grow so rapidly that the critical value is reached and thus break up and form *white caps* in the sea. Increasing value of a/L increases wave velocity but the increase in velocity never exceeds 12 per cent.

D. Wind Waves

Wind waves are mostly ordinary gravity waves that grow in height under the effect of wind. The ordinary gravity waves normally observed at sea are composed of *swell* and *sea*. The *swell* is produced by winds and storms at some distance from the point of observation. As it advances to the region of weaker wind, it decreases in its height. They are long and relatively symmetrical waves having a period in the order of 10 sec. On the other hand, the *sea* is the local wind-generated wave motion usually of shorter period with asymmetrical slopes and steep or white-capped crests. It causes the irregular broken appearance of the sea surface known as *sea state*.

In addition to ordinary seas and swell, there are infragravity waves. They may be produced by interference of two different wind wave trains which also results in *surf beat* having the beat frequency and an amplitude of a few centimeters. Also long-period waves generally arise due to wind shifts and pressure changes associated with meteorological disturbances.

Various empirical relationships between wind and waves have been observed by a number of workers as follows [36]: For a given wind velocity, the wave velocity increases with increasing *fetch* (the distance on sea surface over which the wind blows without appreciable change of direction) and the wave height becomes greater the longer the fetch. For fetches longer than 10 nautical miles, the maximum probable wave height in feet for very strong wind is about $1.5 \sqrt{F}$, where F is the fetch in nautical miles. At low wind velocities, the maximum wave height is about $0.8v$ and the average height over the entire range of wind velocities is about $0.026v^2$, where v is the wind velocity in knots. The ratio of wave velocity to wind velocity varies from 0.1 to nearly 2.0. The average maximum wave velocity slightly exceeds the wind velocity at wind velocities less than about 25 knots but it is somewhat less at higher wind speeds. For a given fetch and wind velocity, the wave velocity increases rapidly with time. The time to develop waves of maximum height increases with increasing wind velocity and takes less than 12 hr in case of strong winds.

E. Tides

Tides are very long waves that are caused by either periodic or aperiodic action of certain forces. If the tide-producing forces are of astronomical origin, such as the periodic gravitational attraction of the sun and moon, the tides are known as *astronomical tides*. If the forces are of meteorological origin, such as forces due to strong winds, barometric changes, and temperature differences, the tides are called *meteorological tides*. Earthquakes and volcanic eruptions also cause *tidal waves*, known as *tsunami*. The currents accompanying all tides are tidal currents.

Astronomical tides can be considered as composed mainly of a series of harmonic oscillations, known as *partial tides, tide species,* or *tidal harmonics*, having the periods of the tide-producing forces. When the tidal record at a given place has been analyzed for the amplitudes of each of these partial tides and the phase angles for each of the harmonic terms are known for a given initial time, it is possible to reconstruct the tide at any time in the past or future relative to the initial moment. Discounting the aperiodic influences of wind and atmospheric pressure on local sea level, the reconstruction or predictions of tides by the tidal harmonic analysis are usually good, to within 0.1 ft over the period of the ensuing year. Table 2-1 lists some selected partial tides. According to their period, they can be grouped as semidiurnal, diurnal, and long-period tides.

Many terms for describing tides have been developed. Some important ones are as follows:

Spring tides are those of greatest amplitude which are produced by the combined forces of the sun and moon at the time of the new and full moons, i.e., when the sun, moon, and earth are in line.

Neap tides are those of least amplitude which are developed when the moon is in the first or third quarter, i.e., 90° away from the sun.

Diurnal tide is a tide which has only one high and one low water daily.

Mixed tide is a tide which has two high and two low waters daily with a marked diurnal inequality in the two high and two low tides.

Tropic tides are those that display the greatest diurnal inequalities, which occur when the moon is at its maximum or minimum declination or nearly above the Tropic of Cancer or of Capricorn.

Mean range is the difference in height of average high water and average low water.

OCEAN WAVES AND TIDES   2-19

Spring range is the mean range of spring tides.
Neap range is the mean range of neap tides.
Diurnal range is the difference between the elevations of mean higher high water and mean lower low water.

Table 2-1. Selected Partial Tides

Name	Symbol	Period	Speed, °/hr
Semidiurnal:			
Lunisolar declinational	K_2	11.97 hr	30.0821
Principal solar	S_2	12.00 hr	30.0000
Principal lunar	M_2	12.42 hr	28.9841
Larger lunar elliptic	N_2	12.66 hr	28.4397
Diurnal:			
Lunisolar declinational	K_1	23.93 hr	15.0411
Principal solar declinational	P_1	24.07 hr	14.9589
Principal lunar declinational	O_1	25.82 hr	13.9430
Lunar declinational	Q_1	26.87 hr	13.3987
Long-period:			
Lunar fortnightly	M_f	13.66 day	1.0980
Lunar monthly	M_m	27.55 day	0.5444
Solar semiannual	S_{sa}	0.5 year	0.0821
Solar annual	S_a	1.0 year	0.0411

F. Internal Waves

Internal waves are undulating swells which form between subsurface water layers of varying density in a nonhomogeneous element such as the sea. Such waves exist in all oceans, probably also in most bays and lakes, and vary widely in amplitude, period, and depth. Because of the buoyant forces acting, they may have a much greater amplitude than surface waves. They also progress very slowly and have periods of many minutes.

G. Wave Pressure

When a moving wave encounters a solid vertical wall, a negative wave is reflected with the resultant absolute velocity at the wall reduced to zero. The heights of the incident and negative waves are equal and superposed at the wall, resulting in a standing wave known as *clapotis* with its height equal to twice that of the incident wave. The dynamic pressure on the wall due to the clapotis may be represented approximately by $ABCD$ in Fig. 2-9. Of many methods, the Sainflou theory [37] has been used most commonly for computing the pressure in the design of sea walls, breakwaters, pile-supported shore structures, and other coastal engineering works. The simplified Sainflou formulas [38] are as follows:

$$p_1 = (p_2 + \gamma h) \frac{H + h}{h + H + h_o} \tag{2-23a}$$

$$p_2 = \frac{\gamma H}{\cosh(2\pi h/L)} \tag{2-23b}$$

$$h_o = \frac{\pi H^2}{L} \coth\left(\frac{2\pi h}{L}\right) \tag{2-23c}$$

where L is the incident wavelength, γ is the unit weight of liquid, H is the height of the original incident wave, h_o is the rise of the original water level, and other notations are shown in Fig. 2-9.

A clapotis is usually established when the ratio of wave height to bottom depth (h in Fig. 2-9 or D in Fig. 2-10) is less than about 0.75. Otherwise, a breaking wave will develop.

Pressures due to breaking waves, however, may reach substantially higher magnitudes than for the clapotis condition, apparently owing to the particles of air trapped between the wall and the wave. The dynamic and static pressures due to breaking waves may be approximately represented in Fig. 2-10. A common

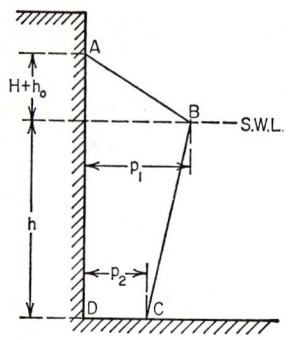

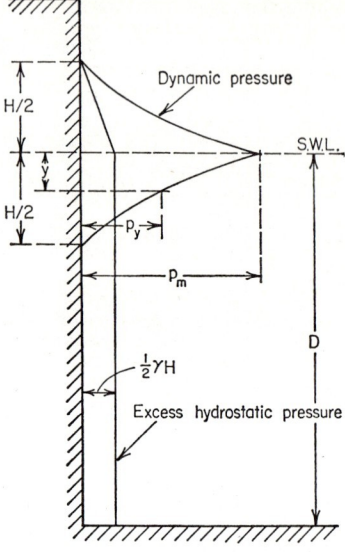

Fig. 2-9. Pressure distribution on vertical wall due to clapotis.

Fig. 2-10. Pressure distribution on vertical wall due to breaking waves.

method for computing the dynamic pressure is the use of Minikin formulas [39]:

$$p_m = 2\pi g \gamma D \frac{H}{L} \tag{2-24a}$$

and

$$p_y = p_m - \left\{ p_m \left[1 - \left(\frac{H - 2y}{H} \right)^2 \right] \right\} \tag{2-24b}$$

The total pressure on the wall is equal to the sum of the static and dynamic pressures.

V. THERMAL INTERACTIONS BETWEEN OCEAN AND ATMOSPHERE

A. Advective and Convective Processes

Important thermal processes involved in the interactions between ocean and atmosphere are advective and convective processes. The *advective process* is the transference of heat by horizontal currents of air. The *convective process* is the transmission of heat by the mass movement of air or water.

The large-scale advective circulations of the oceans and atmosphere are controlled by solar radiation and earth rotation. In equatorial regions the amount of radiation received from the sun and the sky exceeds the loss by nocturnal radiation. At higher latitudes, however, the contrary is true. The temperature differences that arise between high and low latitude are the major cause of the flow of heat from the equator to the poles. This heat transport takes place in the main thermocline in the oceans and below the tropopause of the atmosphere.

In the atmosphere, the heat supply is mainly from the sea surface and land, and thus atmospheric circulations depend on heating from below. In the oceans, on the

THERMAL INTERACTIONS BETWEEN OCEAN AND ATMOSPHERE

other hand, heat supply is mainly from above. Since heating and cooling take place at nearly the same level, these processes cannot maintain large-scale circulations in the oceans but they are of importance to the development of convection currents in the surface layer, to the local exchange of energy with the atmosphere, and to the slow deep-water circulation. Large-scale circulations in the oceans are mainly due to distribution of mass in the sea and the superimposing effect of the prevailing winds.

Although some new theories have been advanced in recent years [40-42], the convective circulations in the oceans are not yet as well understood as in the atmosphere. It is known, however, that the heat exchange between the ocean and the atmosphere follows a pattern similar to evaporation. Where the sea surface is warmer, heat is transferred from the sea to the air and is transported to great heights by eddy conductivity and by convection currents in the air. Such heat transfer is small, however, in the tropics where radiation surplus is mainly used for evaporation.

As the ocean has an enormous heat-storage capacity, it tends to stabilize the atmospheric circulations and characteristics and thus to reduce the range of weather extremes closer to the mean than is the case on land.

In general, thermal processes control oceanic climate greatly. Ocean climates are largely determined by the response of the ocean to atmospheric circulation. In turn, both ocean climate and atmospheric circulation respond to the distribution of solar heat over the earth's surface.

B. Heat Budget of the Ocean

The heat content in the ocean does not vary appreciably from year to year. In this quasi-stationary state all the supply in energy is balanced by equally large losses of energy.

The major energy sources are (1) radiation absorbed from the sun and sky, (2) convection of sensible heat from atmosphere to sea, (3) conduction from the sea floor, (4) conversion of mechanical energy into heat, (5) heat produced by chemical and biological processes, (6) condensation of water vapor on the sea surface, and (7) radioactive disintegration in the sea water.

The major energy losses are (1) radiation from the sea surface to space, (2) convection of sensible heat from sea to atmosphere, (3) evaporation from the sea surface, and (4) conduction to the sea floor.

In the above items, the radioactive and conductive exchanges with the atmosphere and space and the evaporative processes are of greatest importance. The conversion of energy of winds and ocean currents is relatively very small, but locally it may be somewhat greater where tides occur in shallow seas of great bottom friction.

C. Evaporation from Sea Surface

Evaporation from the sea surface has been estimated by either direct measurement or indirect computation. Schmidt [43] proposed the use of the method of heat budget, i.e., the energy-balance method (Sec. 11), for evaluating the evaporation on the oceans on the assumption that the annual change of temperature at the surface and the advection are negligible. By heat budget, all gains and losses of heat must balance from day to day, or by Eq. (11-10),

$$Q_s = Q_r + Q_e + Q_h \tag{2-25}$$

where Q_s is the short-wave radiation received by the oceans from the sun and sky, Q_r is the energy lost by the oceans in long-wave radiation, Q_e is the energy lost by the oceans in evaporation at the surface, and Q_h is the heat lost to the atmosphere by conduction. The mean annual evaporation E is defined by Q_e/L, where L is the heat of evaporation of water. Thus, from Eq. (2-25)

$$E = \frac{Q_s - Q_r}{L(1 + R)} \tag{2-26}$$

where R is the Bowen ratio equal to Q_h/Q_e (see Subsec. 11-II-D-3). The values of Q_s and Q_r in the above equation are known approximately from measurements of solar radiation at sea and from climatological data on cloudiness over oceans The ratio R has been estimated at about 0.1. Using Schmidt's method, Mosby [44] found the average evaporation from the sea surface to be 106 cm/year (4.18 in./year), a figure that is widely quoted.

For short-term estimates of evaporation, a practical formula based on Dalton's law may be used (Subsec. 11-II-A), that is

$$E = C(e_w - e_a)w \qquad (2\text{-}27)$$

where E is the evaporation rate in inches per day or month, C is a coefficient equal to 3.5×10^{-3} for daily estimates and 3×10^{-3} for monthly estimates, e_w is the vapor pressure in mb at the sea surface, e_a is the vapor pressure in mb in the air at a height of 10 to 15 meters (20 to 30 ft), and w is the wind velocity in knots at a height of 10 to 15 meters.

D. Ocean Clouds and Fogs

In general, the air in contact with the sea surface has a high relative humidity, about 80 per cent. Where the sea surface is warmer than the air, convection currents and clouds exist, whereas where the sea surface is cooler, the sky may be clear or fog may form.

In trade-wind areas, the sea surface is constantly warmer than the air and convection currents develop down to the trade-wind inversion. Below the inversion, cumulus or stratocumulus clouds exist regularly.

In middle and higher latitudes, the sea surface is much warmer than the air in the winter but often colder in the summer. In winter, therefore, convection clouds will occur in stratus, stratocumulus, or fractostratus forms below high polar air masses, and in cumulonimbus form with showers if instability develops in air masses. In summer, a clear sky is frequently observed but fog prevails over the sea surface when the temperature difference becomes great.

When moist air flows over colder water, *advection fog* will be formed within the temperature inversion established at the sea surface. This is the commonest type of fog that is usually observed over the open oceans. If the temperature difference is great, *steam fog* is formed but it occurs only in protected coastal waters. If a layer of moist air underlies very dry air in a calm sea, *radiation fog* may be formed because of loss of heat from the sea surface by nocturnal radiation and by radiation from the moist air.

Formation of clouds and fogs depends largely on the temperature difference between the air and the surface. Ocean weather maps have been prepared to show such differences for forecasting purposes.

VI. REFERENCES

1. von Arx, W. S.: "An Introduction to Physical Oceanography," Addison-Wesley Publishing Company, Inc., Reading, Mass., 1962.
2. Sverdrup, H. U., M. W. Johnson, and R. H. Fleming: "The Oceans: Their Physics, Chemistry, and General Biology," Prentice-Hall, Inc., Englewood Cliffs, N.J., 1942.
3. Defant, Albert: "Physical Oceanography," 2 vols., Pergamon Press, Ltd., London, and The Macmillan Company, New York, 1961.
4. Bruns, Erich: "Ozeanologie," vols. 1 and 2, VEB Verlag der Wissenschaften, Berlin, 1958.
5. Proudman, Joseph: "Dynamical Oceanography," John Wiley & Sons, Inc., New York, 1953.
6. King, C. A. M.: "Oceanography for Geographers," Edward Arnold (Publishers) Ltd., London, 1962.
7. Hill, M. N., editor: The sea; ideas and observations on progress in the study of the seas, vol. 1 of "Physical Oceanography," Interscience Publishers, Inc., New York, 1962.
8. Sears, Mary (ed.): "Oceanography; Invited Lectures Presented at the International

REFERENCES

Oceanographic Congress," American Association for the Advancement of Science, Washington, D.C., 1961.
9. Sverdrup, H. U.: Oceanography, Sec. XIV of "Handbook of Meteorology," ed. by F. A. Berry, Jr., E. Bollay, and N. R. Beers, McGraw-Hill Book Company, Inc., New York, 1945, pp. 1029–1056.
10. Sverdrup, H. U.: Oceanography, in "Handbuch der Physik," S. Flügge, ed., Bd. XLVIII, Geophysik II, Springer-Verlag, Berlin, 1957, pp. 608–670.
11. Montgomery, R. B.: Oceanographic data, Art. 2k in "American Institute of Physics Handbook," ed. by Dwight E. Gray, 2d ed., McGraw-Hill Book Co., Inc., New York, 1964, pp. 2-123 to 2-132.
12. Bibliography of oceanographic publications, ICO Pamphlet No. 9, Interagency Committee on Oceanography of the Federal Council for Science and Technology, U.S.A., Washington, D.C., April, 1963.
13. LaFond, E. C.: Processing oceanographic data, U.S. Navy Hydrographic Office Publication 614, 1951.
14. Knudsen, Martin: "Hydrographical Tables," G. E. C. GAD, Copenhagen, 1901; Williams & Norgate, London, 2d ed., 1931.
15. Tables for sea water density, U.S. Navy Hydrographic Office Publication 615, 1952.
16. Helland-Hansen, Bjørn: Nogen hydrografiske metoder, *Forh. skand. naturf. Møte*, vol. 16, pp. 357–359, 1916.
17. Eckart, C. H.: "Hydrodynamics of Oceans and Atmospheres," Pergamon Press, London and New York, 1960.
18. Prandtl, Ludwig: Bericht über Untersuchungen zur ausgebildeten Turbulenz, *Z. angew. Math. Mech.*, vol. 5, no. 2, p. 136, 1925. Also, Goldstein, Sydney: "Modern Developments in Fluid Mechanics," vol. 1, p. 205, Oxford University Press, London, 1938.
19. Taylor, G. I.: The transport of vorticity and heat through fluids in turbulent motion, *Proc., Royal Soc. London*, Series A, vol. 135, 1932, pp. 685–701.
20. Ekman, V. W.: On the influence of the earth's rotation on ocean currents, *Ark. f. Mat. Astr. och Fysik. K. Sv. Vet. Ak.*, Stockholm, 1905–06, vol. 2, no. 11, pp. 1–53, 1905.
21. Bjerknes, V. F. K.: Über einen hydrodynamischen Fundamentalsatz und seine Anwendung, besonders auf die Mekanik der Atmosphäre und des Weltmeers, *Kgl. Svenska Vetenskapsakad. Handl.*, vol. 31, no. 4, pp. 1–35, 1898.
22. Bjerknes, V. F. K., J. Bjerknes, H. Solberg, and T. Bergeron: "Physikalische Hydrodynamik," Julius Springer, Berlin, 1933.
23. Pritchard, D. W.: Estuarine circulation patterns, *Proc. Am. Soc. Civil Engr.*, vol. 81, no. 717, pp. 1–11, June, 1955.
24. Mohn, Henrik: Die Strömungen des Europäischen Nordmeeres, *Petermannes Geogr. Mitt.* 17, 1885.
25. Sandström, J. W., and B. Helland-Hansen: Über die Berechung von Meeresströmungen, *Rep. on Norw. Fishery and Mar. Investigations*, Bergen, vol. 2, no. 4, pp. 1–43, 1903.
26. Stommel, Henry: Varieties of oceanographic experience, *Science*, vol. 139, pp. 572–576, 15 February 1963.
27. von Gerstner, F. J.: "Theorie der Wellen," Abhandlungen der königlichen böhmischen Gesellschaft der Wissenschaften, Prague, 1804.
28. Airy, G. B.: "Tides and Waves," Encyclopaedia Metropolitana, London, 1845.
29. Lamb, Sir Horace: "Hydrodynamics," 6th ed., Dover Publications, New York, 1932.
30. Coulson, C. A.: "Waves," 3d ed., Oliver and Boyd, Edinburgh, 1944.
31. Stoker, J. J.: "Water Waves," vol. IV of "Pure and Applied Mathematics," Interscience Publishers, Inc., New York, 1957.
32. Pierson, W. J. Jr., Gerhard Neumann, and R. W. James: Practical methods for observing and forecasting ocean waves by means of wave spectra and statistics, *H. O. Pub. No.* 603, U.S. Navy Hydrographic Office, Washington, D.C., 1955.
33. Longuet-Higgins, M. S.: The statistical analysis of a random moving surface, *Phil. Trans. Roy. Soc. London*, A 249, pp. 321–387, 1957.
34. Munk, W. H., M. J. Tucker, and F. E. Snodgrass: Remarks on the ocean wave spectrum, chap. III of "Naval Hydrodynamics," *Pub.* 515, National Academy of Sciences, Washington, D.C., pp. 45–60, 1957.
35. Munk, W. H.: Origin and generation of waves, chap. 1, pt. 1, *Proc. 1st Conf. Coastal Eng.*, Council on Wave Research, The Engineering Foundation, pp. 1–4, 1951.
36. Techniques for forecasting wind waves and swell, *H. O. Pub. No.* 604, U.S. Navy Hydrographic Office, Washington, D.C., 1951.
37. Sainflou, George: Essai sur les diques maritimes verticales, *Ann. des Ponts et Chaussées*, Paris, vol. 98, no. 4, pp. 5–48, 1928.
38. Hudson, R. Y.: Wave forces on breakwaters, *Trans. Am. Soc. Civil Engrs.*, vol. 118, pp. 653–674, 1953.

39. Minikin, R. R.: "Winds, Waves and Maritime Structures," Chas. Griffin and Co., Ltd., London, 1950.
40. Stommel, Henry, and A. B. Arons: On the abyssal circulation of the world ocean, *Deep-Sea Res.*, vol. 6, pt. I, pp. 140–154, pt. II, pp. 217–233, 1960.
41. Lineikin, P. S.: On the determination of the thickness of the baroclinic layer in the sea, *Doklady Akad. Nauk S. S. S. R.*, vol. 101, pp. 461–464, 1955.
42. Ichiye, T.: On convective circulation and density distribution in a zonally uniform ocean, *Oceanog. Mag.*, vol. 10, pp. 97–135, 1958.
43. Schmidt, Wilhelm: Strahlung und Verdunstung an freien Wasserflächen; ein Beitrag zum Wärmehaushalt des Weltmeers und zum Wasserhaushalt der Erde, *Ann. d. Hydrogr. u. Mar. Meteor.*, Berlin, vol. 43, pp. 111–124, 169–178, 1915.
44. Mosby, Håkon: Verdunstung und Strahlung auf dem Meere, *Ann. d. Hydrogr. u. Mar. Meteor.*, Berlin, vol. 64, pp. 281–286, 1936.

Section 3

METEOROLOGY

SVERRE PETTERSSEN, *Professor of Meteorology and Chairman, Department of the Geophysical Sciences, The University of Chicago.*

I. Introduction	3-2
II. The Atmosphere	3-2
A. Dry Air	3-2
B. Ozone	3-3
C. Carbon Dioxide	3-4
D. Water Vapor	3-4
E. The Precipitable Water	3-6
F. The Mass of the Atmosphere	3-6
G. Adiabatic Rates of Cooling	3-6
H. Potential Temperatures	3-8
I. Stability and Instability	3-8
J. Barometry	3-9
K. The Standard Atmosphere	3-9
L. The Spheres	3-9
III. Radiation, Heat, and Temperature	3-11
A. Radiation Balance	3-11
B. Variation with Latitude	3-12
C. Conduction Properties	3-13
D. Surface Influences on Temperature	3-13
E. Role of the Water Cycle	3-15
F. Annual Variation of Temperature	3-16
G. Diurnal Variation of Temperature	3-17
IV. Condensation and Precipitation	3-18
A. Condensation Nuclei	3-18
B. Growth of Cloud Droplets	3-18
C. Precipitation Processes	3-19
1. The Coalescence Process	3-19
2. The Ice-crystal Process	3-19
3. Artificial Stimulation	3-20
D. Classification of Clouds	3-20
E. Classification of Precipitation	3-22
V. Semipermanent Wind Systems	3-23
A. The Equations of Motion	3-23
B. The Geostrophic Wind	3-24
C. The Thermal Wind	3-24
D. Global Circulations	3-24
E. Monsoons	3-27

	F. Land and Sea Breezes................................... 3-28
	G. Mountain and Valley Winds........................... 3-28
VI.	Weather Systems and Regimes............................. 3-29
	A. Terminology.. 3-29
	1. Air Mass.. 3-29
	2. Front... 3-29
	3. Squall Line... 3-29
	4. Cyclone... 3-29
	5. Anticyclone... 3-30
	6. Heat Lows and High-level Highs...................... 3-30
	7. Hurricane... 3-30
	8. Tornado... 3-30
	9. Convection.. 3-30
	B. Extratropical Cyclones................................. 3-30
	C. Convective Weather..................................... 3-32
	D. Tropical Disturbances and Storms....................... 3-33
	1. Easterly Waves...................................... 3-33
	2. Tropical Depressions............................... 3-34
	3. Tropical Storms..................................... 3-34
	4. Hurricanes.. 3-34
	E. Weather Regimes.. 3-35
VII.	References.. 3-38

I. INTRODUCTION

The word *atmosphere* is derived from the Greek words *atmos* (vapor, or breath) and *spheria* (sphere, or ball), and it is now taken to refer to the gaseous envelope of any heavenly body, and especially that of the earth. Similarly, the word *meteorology* is derived from the words *meteoros* (lofty) and *logos* (discourse). In its widest sense meteorology, as a science, is concerned with the states and the chemical, physical, and dynamical processes of the entire gaseous envelope of the earth. In the past however, the meteorologists have been concerned mainly with the atmosphere below about 20 miles (33 km), while the physicists have interested themselves in the processes of the uppermost layers. Recently the word *aeronomy* has been used to name the science of the high atmospheric layers. This distinction, which is quite arbitrary, is likely to vanish as more observations from the outer regions become available. However, the present account, because of its relation to hydrology, will deal almost exclusively with the lower layers of the atmosphere, or the layers within which the water substances play important roles.

The purpose of this section is to give an introductory outline of meteorology. Readers who require further details should consult the publications listed at the end of this section.

II. THE ATMOSPHERE

That invisible and odorless something which we breathe, which sustains life and fire and produces an infinite variety of scenic nuances, is what we call *air*. Though its composition is exceedingly complex, it is useful here to simplify and consider natural air as consisting of three main parts, namely, dry air, water vapor, and such impurities as are important in connection with the water cycle.

A. Dry Air

The dry component of the atmosphere is a mechanical mixture of a number of individual gases. In order to compare the amounts of the several constituents, it is convenient to measure them in terms of mole fractions, or volume percentages. This measure refers to the volume that each gas would occupy if the component gases were separated and brought to the same temperature and pressure. The relative amounts of the four principal gases are shown in Table 3-1. Actually, these four gases account for about 99.997 per cent of the whole. The remainder is made up of minute traces of such gases as neon, helium, krypton, hydrogen, xenon, ozone, radon, etc. Of these, only ozone is of any particular importance in meteorology.

Table 3-1. The Four Principal Gases of the Dry Atmosphere Below about 15 Miles (25 Km)

Gas	Mole fractions (or volume per cent)
Nitrogen, N_2	78.09
Oxygen, O_2	20.95
Argon, Ar	0.93
Carbon dioxide, CO_2	0.03
Total	100.00 (approx.)

Observation shows that, apart from small variations in the amount of carbon dioxide, the composition of the dry atmosphere is constant all over the world up to about 15 miles (25 km) above sea level, indicating that a state of complete mixing exists. At great heights chemical processes maintain a variable composition.

B. Ozone

The overwhelming part of the oxygen of the atmosphere is diatomic (O_2). A small portion (a little less than 1 part in 400,000 by weight) consists of triatomic oxygen (O_3), or *ozone*, and is found mainly at great heights. Though present only in minute amounts, the ozone is important in that it prevents harmful ultraviolet radiation from reaching down to levels where biological processes are prominent.

The formation and destruction of ozone may be accounted for as follows. In the first place, an ordinary oxygen molecule (O_2) dissociates completely when it absorbs a quantum of energy corresponding to a free wavelength of less than 0.240 micron. Thus, with customary symbols,

$$O_2 + h\nu(\lambda < 0.240\mu) \rightarrow 2O$$

where $h\nu$ is the energy of a photon of light.

Now, when an oxygen atom (O) collides with an oxygen molecule (O_2) and with any third neutral molecule (say, M), O_2 and O combine to form ozone. Thus

$$O_2 + O + M \rightarrow O_3 + M$$

A three-body collision is necessary to stabilize the process.

Next, ozone is very unstable in the presence of sunlight, and when it absorbs energy at wavelengths less than 1.100 microns, it reverts to diatomic and monatomic forms, so that

$$O_3 + h\nu(\lambda < 1.100\mu) \rightarrow O_2 + O$$

whereafter

$$O_3 + O \rightarrow 2O_2$$

The productive and destructive processes are active mainly in the layer from 15 to 40 miles (20 to 70 km) above the ground, but the maximum concentration of ozone is normally found in the lower part of this layer.

Minute traces of ozone are sometimes formed in the lower atmosphere as a result of electric discharges, but these are negligible as compared with the amounts found in the upper layers.

C. Carbon Dioxide

The amount of carbon dioxide in the air is not quite constant. It is continuously consumed by the vegetable world, and it is produced by the animal world, by the burning of fuels, by volcanic actions, and by various processes of decay in the soil. Although these processes are not always balanced, the oceans absorb carbon dioxide so readily that only a small (and slightly variable) portion of it remains in the air. In a manner of speaking, the carbon dioxide behaves like an iceberg: the bulk of it goes into the ocean.

The carbon dioxide in the air is important for the heat budget of the atmosphere, for, together with the water vapor, it causes part of the long-wave radiation from the earth to be absorbed by the air. There is some indication that the amount of carbon dioxide has increased by about 10 per cent since the turn of this century [1, 2], and it is believed that this trend is due to increased burning of fuels. Also, since the beginning of this century, there has been a marked rise in the air temperature, particularly in high and middle latitudes. The cause of this rise is not well understood, but it is possible that absorption by carbon dioxide of long-wave radiation has been a contributing factor.

D. Water Vapor

In many respects water vapor is the most important constituent of the atmosphere. While the temperature is very much higher than the critical temperatures at which the component gases of the dry air liquefy, condensation of water vapor and freezing of water occur well within the atmospheric range.

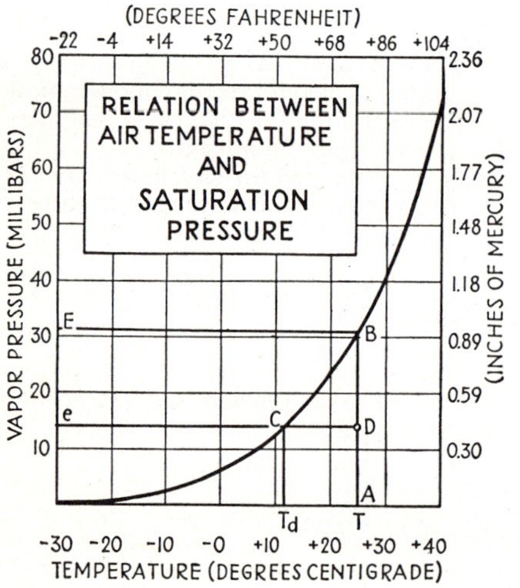

The amount of water vapor present in the air may be expressed as the pressure that the vapor would exert if the other gases were absent. This is called the *vapor pressure;* it is usually denoted by the symbol e and expressed in millibars or inches of mercury. The maximum amount of vapor E in the air depends upon the temperature; the higher the temperature, the more vapor the air can hold (Fig. 3-1), and the air is said to be saturated when this amount is reached. In some connections it is necessary to define the term saturation more precisely. Thus *saturation* is a state in which water vapor (or air containing water vapor) can exist in equilibrium with a plane surface of pure liquid water at the same temperature. If the water is frozen or impure, if the surface is curved, or if the temperatures are different, other equilibria may exist or equilibrium may not be attainable. Such conditions are important and will be discussed in connection with the formation of clouds and precipitation.

FIG. 3-1. Heavy curve shows relation between temperature and saturation vapor pressure. A plot D has a temperature T and a vapor pressure e. The relative humidity is $100DA/BA$, and the dew point is T_d.

Apart from the vapor pressure, a number of other humidity measures are used, and these may be defined as follows.

The *relative humidity* U is the percentage ratio of the actual vapor pressure e to the saturation vapor pressure E at the observed temperature. Thus

$$U = 100\frac{e}{E} \tag{3-1}$$

THE ATMOSPHERE

The *vapor density* ρ_v is the mass of water vapor contained in a unit volume. Thus, from the equation of state for water vapor,

$$\rho_v = \frac{e}{R_v T} \tag{3-2}$$

where R_v is the specific gas constant of water vapor, and T is the Kelvin temperature. If R is the specific gas constant of dry air, we have with satisfactory accuracy $R_v = \tfrac{8}{5} R$.

The *mixing ratio* r is defined as the ratio of the mass of water vapor to the mass of the dry air with which the vapor is associated.

The term *specific humidity* s is defined in like manner, except that the term moist air is substituted for dry air. From the foregoing definitions and the appropriate equations of state, one readily obtains for r and s

$$r = \frac{\rho_v}{\rho_a} = \frac{5}{8}\frac{e}{p-e} \quad \text{and} \quad s = \frac{\rho_v}{\rho_a + \rho_v} = \frac{5}{8}\frac{e}{p - \tfrac{3}{8}e} \tag{3-3}$$

where ρ_a is the density, and $p - e$ is the partial pressure of the dry air.

In practice, the measure of r and s is augmented by a factor of 1,000, so that these symbols express the number of grams of water vapor per kilogram of air. Since p is normally about 100 times larger than e, the difference between r and s is very small, and the two may be used interchangeably unless great accuracy is required.

The *dew-point temperature* T_d is the temperature to which the air must be cooled at constant pressure and with constant water-vapor constant in order to reach saturation (Fig. 3-1) with respect to water. The *frost-point temperature* is defined in like manner except that saturation refers to ice. From experimental data on water substances it is found that the following formula holds with satisfactory accuracy for temperatures t between 40 and $-40°C$.

$$e_s = 6.11 \times 10^{at/(t+b)} \tag{3-4}$$

where $a = 7.5$ and $b = 277.3$ for water
and $a = 9.5$ and $b = 265.5$ for ice

The *wet-bulb temperature* T_w is the lowest temperature to which the air can be cooled at constant pressure by evaporating water into the air (until saturation is reached). From thermodynamical considerations one finds the following approximation:

$$T_w = T - \frac{L}{c_p}(r_s - r) \tag{3-5}$$

where r is the mixing ratio of the air, and r_s is the saturation value corresponding to T_w. Furthermore, L is the latent heat of vaporization and c_p is the specific heat of air at constant pressure.

In some connections it is useful to compute the increment in the air temperature that would result if all water vapor were condensed and the liberated latent heat used to heat the air at constant pressure. With the symbols given above, the increment is $dr\, L/c_p$ and the so-called *isobaric equivalent temperature* T_e is defined thus:

$$T_e = T + \frac{L}{c_p} r \tag{3-6}$$

which, together with Eq. (3-5), gives

$$T_e = T_w + \frac{L}{c_p} r_s \tag{3-7}$$

For moist unsaturated air $T_e > T > T_w > T_d$.

The various humidity measures are most readily obtained by converting the observed quantities with the aid of tables (e.g., Smithsonian Meteorological Tables, The Smithsonian Institution, Washington, D.C., 1951).

E. The Precipitable Water

This term refers to the amount of liquid water that would form in an air column of unit cross section if all water vapor in it were condensed. If dz is an element of height, the amount of precipitable water, P, in the column between z_0 and z is expressed by

$$P = \int_{z_0}^{z} \rho_v \, dz \tag{3-8}$$

Since the atmosphere is very nearly in hydrostatic equilibrium, the increments of height and pressure, dp, are related so that

$$-dp = \rho g \, dz \tag{3-9}$$

where g is the acceleration of gravity and ρ is the density of natural air (that is, $\rho = \rho_a + \rho_v$). Elimination of dz gives

$$P = \frac{1}{g} \int_{p}^{p_0} s \, dp = \frac{1}{g} (p_0 - p) \bar{s} \tag{3-10}$$

where p and p_0 are the pressures at the top and the base of the column, s is the specific humidity, and the bar signifies the mean value.

For an air column extending to the top of the atmosphere ($p = 0$),

$$P = \frac{1}{g} p_0 \bar{s} \tag{3-11}$$

The overwhelming part of the vapor content of the atmosphere is present in the lower layers, and unless great accuracy is required, it is satisfactory to ignore the amounts present above the 18,000-ft (or 500-mb) level.

F. The Mass of the Atmosphere

The total weight of the atmosphere is enormous; it amounts to 5.6×10^{15} metric tons (1 metric ton = 1.102 short tons = 0.984 long ton). In comparison, the weight of the water vapor is 1.5×10^{13} tons, and that of the ozone is 3.3×10^{9} tons. The weight of the atmosphere is equivalent to the weight of a layer of water about 33 ft (10 m) deep and spread evenly over the face of the earth; it is also equivalent to the weight of a layer of mercury 76 cm (29.92 in.) deep. The pressure exerted by the atmosphere at sea level under standard conditions is called one *atmosphere*. Converted to pressure units, 1 atm = 1,013.25 mb.

G. Adiabatic Rates of Cooling

If dH denotes the amount of heat given to a unit mass of air, the first law of thermodynamics may be written

$$dH = c_v \, dT + p \, d\alpha \tag{3-12}$$

where c_v is the specific heat of air at constant volume, and $\alpha \, (= 1/\rho)$ is the volume occupied by a unit mass, or the specific volume.

Using the equation of state ($\alpha p = RT$), Eq. (3-12) may be written

$$dH = c_p \, dT - \alpha \, dp \tag{3-13}$$

If the pressure change is due to vertical displacement, dp may be replaced by dz according to Eq. (3-9). Thus

$$dH = c_p \, dT + g \, dz \tag{3-14}$$

An adiabatic process is one in which no heat is supplied to or withdrawn from the substance taking part in the process. Consider an adiabatic process in which the air remains unsaturated. Thus, with $dH = 0$, Eq. (3-14) gives, for the rate of cooling γ_d of a parcel of air ascending along the vertical,

$$\gamma_d = -\left(\frac{dT}{dz}\right)_{\text{dry}} = \frac{g}{c_p} \qquad (3\text{-}15)$$

This is very nearly equal to 0.01°C/m, or 1°C per 100 ascent, or 5.4°F per 1,000 ft.

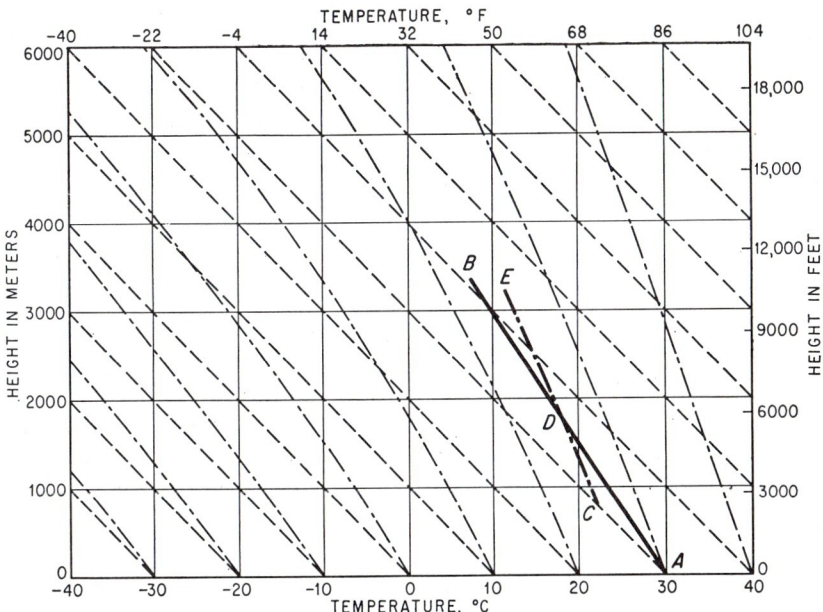

Fig. 3-2. Temperature-height diagram with dry (– – –) and wet (– · – · –) adiabatics.

If saturation occurs while no heat is supplied from external sources, the amount of liberated latent heat is $-L\,dr$, where L is the latent heat of vaporization and dr is the change in mixing ratio ($dr < 0$) resulting from condensation. With a high degree of approximation one may ignore the heat of the condensed water. Thus, putting $dH = -L\,dr$, Eq. (3-14) gives, for the saturated adiabatic rate of cooling,

$$\gamma_s = -\left(\frac{dT}{dz}\right)_{\text{sat}} = \gamma_d + \frac{L}{c_p}\frac{dr}{dz} \qquad (3\text{-}16)$$

Since dr is negative, the saturated rate is smaller than the dry one. While γ_d is constant, the last term on the right of Eq. (3-16) varies with the air temperatures as shown in Fig. 3-2. At such high temperatures as occur in the equatorial belt, γ_s is about 35 per cent of γ_d; at very low temperatures, the two rates differ but little.

It should be noted that γ_d and γ_s, as defined above, refer to the successive states of individual parcels of air. Normally, the temperature at any given time decreases along the vertical, and the rate of decrease, γ, is called the *lapse rate*. Thus $\gamma = -dT/dz$. The differences $\gamma_d - \gamma$ and $\gamma_s - \gamma$ are important in considerations of the static stability of air columns.

H. Potential Temperatures

If Eq. (3-13) is divided by T, one obtains

$$dS = \frac{dH}{T} = c_p d(\ln T) - R d(\ln p) \tag{3-17}$$

Similar expressions are obtained from Eqs. (3-12) and (3-14). The quantity dS is an element of entropy. Since $dS = 0$ when $dH = 0$, an adiabatic process is also isentropic.

Upon integration of Eq. (3-16) between the pressure levels p_0 and p, one obtains

$$\frac{T_0}{T} = \left(\frac{p_0}{p}\right)^\kappa \quad \text{with } \kappa = \frac{R}{c_p} \tag{3-18}$$

It will be seen that, in an adiabatic process, the temperature is a function of pressure only. The foregoing equation yields an expression for the *potential temperature* θ, which is the temperature that a parcel of air, with temperature T and pressure p, would have if it were brought dry-adiabatically to a standard pressure of 1,000 mb. Thus, with $p_0 = 1{,}000$ mb and $T_0 = \theta$, we have

$$\theta = T\left(\frac{1{,}000}{p}\right)^\kappa \tag{3-19}$$

Since θ depends upon the initial conditions only, the potential temperature of a parcel of air is conserved in a dry-adiabatic process. Formulas similar to Eq. (3-19) hold for the potential wet-bulb temperature θ_w and the potential equivalent temperature θ_e. Thus

$$\theta_w = T_w\left(\frac{1{,}000}{p}\right)^\kappa \quad \text{and} \quad \theta_e = T_e\left(\frac{1{,}000}{p}\right)^\kappa \tag{3-20}$$

These quantities are quasi-conservative in adiabatic processes whether or not condensation takes place.

I. Stability and Instability

Though the hydrostatic equation [Eq. (3-9)] holds with a high degree of accuracy, small imbalances between the vertical component of the pressure force and the force of gravity may, in certain cases, result in vertical overturnings. The stratification is said to be unstable if a parcel of air that is displaced adiabatically a small distance Δz along the vertical obtains a buoyancy acceleration in the direction of the displacement. If the acceleration is opposed to the direction of the displacement, the stratification is said to be stable. Approximate criteria for stability and instability are readily obtained by considering a buoyant parcel displaced in an environment which remains in hydrostatic equilibrium. Referring the forces to a unit mass, the acceleration a of the parcel is

$$a = -\alpha \frac{dp}{dz} - g$$

and for the environment

$$0 = -\alpha_1 \frac{dp}{dz} - g$$

since dp/dz is the same for both. Eliminating the pressure gradient and substituting from the equation of state, one obtains

$$a = g\frac{T - T_1}{T_1}$$

THE ATMOSPHERE

Now, if T_0 is the temperature before the displacement, one has $T = T_0 - \gamma_d \Delta z$ and $T_1 = T_0 - \gamma \Delta z$. Thus, for a parcel of nonsaturated air,

$$a = \frac{\gamma - \gamma_d}{T} g \Delta z \qquad (3\text{-}21)$$

If the parcel is saturated, γ_d should be replaced by γ_s. It will be seen that the state is stable or unstable according as the actual lapse rate γ is smaller or larger than the appropriate adiabatic rate γ_d or γ_s. When the critical lapse rate is exceeded, convective currents set in, and if the unstable layer is sufficiently deep, cumulus or cumulonimbus clouds develop. In pronounced cases, thunderstorms, hailstorms, squalls, and even tornadoes may form.

J. Barometry

The equation of state and the hydrostatic equation may be combined to give a useful formula for computation of heights. Thus

$$-dp = \rho g \, dz \qquad \text{and} \qquad p = \rho R T \qquad (3\text{-}22)$$

give

$$g \, dz = -RT d(\ln p)$$

or, when integrated between the levels z_0 and z_1,

$$z_1 - z_0 = \frac{R}{g} \int_{p_1}^{p_0} T d(\ln p) = \frac{R}{g} \bar{T} \ln \frac{p_0}{p_1} \qquad (3\text{-}23)$$

If heights are expressed in meters, $R = 287$ kilo-joules per ton per °C is the specific gas constant for dry air, g is expressed in meters per second squared, and $\bar{T}$ is the mean of the air temperature after adjustment for the effect on the density of the moisture content. This adjusted temperature is called the *virtual temperature* T_v. Thus

$$T_v = \frac{T}{1 - 0.378 e/p} \qquad (3\text{-}24)$$

Since e is very much smaller than p, the adjustment may be omitted unless great accuracy is required.

K. The Standard Atmosphere

Equation (3-23) may be used to obtain a standard relation between pressure and height by specifying the temperature as a function of height. In the *standard atmosphere*, sea-level temperature and pressure are taken to be 15°C and 1,013.25 mb, respectively. The lapse rate of temperature is 6.5°C/km below 10,769 km and then zero up to 20 km. It is this standard-atmosphere relationship that is used in constructing the customary altimeters for aircraft. The corresponding values of pressure and height, height and density, and height and temperature are readily obtained from tables. For heights above 20 km a tentative standard has been established, but it is likely to be revised as more observations become available.

L. The Spheres

The average state of the atmosphere below about 100,000 ft has been fairly well explored with the aid of customary meteorological instruments. Some knowledge of the structure of the uppermost atmosphere has been gained from studies of ozone, auroras, meteors, and propagation of radio waves. In recent years rockets and satellites have added direct observations, and it is now possible to outline the predominant features of the atmosphere up to 1,000 miles or more. On the whole, one may distinguish between five more or less concentric shells or spheres. These are shown schematically in Fig. 3-3.

The lowest layer is called the *troposphere;* it contains about 75 per cent of the mass and almost all the moisture and dust of the atmosphere. It is the abode of all the phenomena which are commonly called *weather*. The top of the troposphere is called the *tropopause;* its height varies from about 5 miles (8 km) over the poles to a little over 10 miles near the equator. In individual cases it varies with the changing weather conditions; it is normally high over warm high-pressure areas and low over cold low-pressure areas.

The most striking feature of the troposphere is the relatively steep lapse rate of temperature (Fig. 3-4). On the whole, the degree of static stability is low and the air is often unstable in places. The air is subject to frequent overturnings, and these

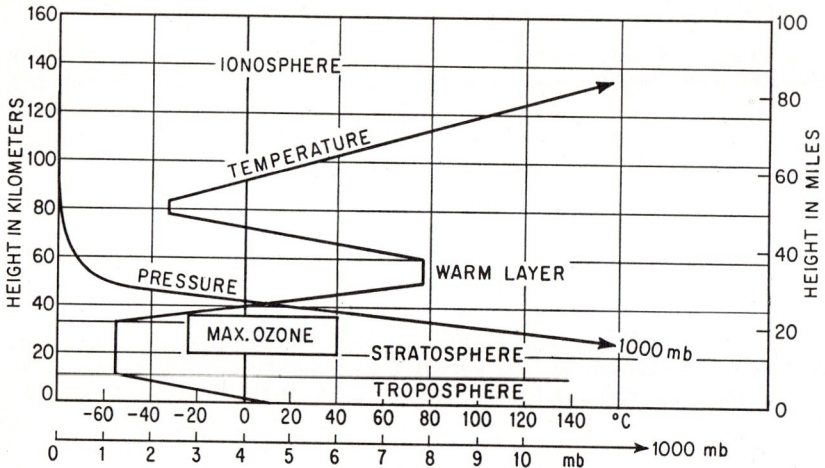

FIG. 3-3. Idealized structure of the atmosphere below the exosphere.

occur on a variety of scales. At the lower end one meets with random turbulence and gusts with horizontal dimensions of a few yards. The heap-shaped clouds (called *cumulus*) occur in connection with vertical currents with horizontal dimensions of a few miles; the larger variety of such clouds (called *cumulonimbus*) are accompanied by showers or thunderstorms and hailstorms; in extreme cases, destructive squalls and tornadoes may occur. These small-scale circulations often occur in groups, such as clusters of thunderstorms, squall lines, etc., with horizontal dimensions of 100 miles or more. Such groups, or clusters, are often referred to as *mesoscale systems*. Their lives are relatively short, and the vertical velocities in them may be as large as 30 ft/sec, or 10 m/sec. At the other end of the scale one has the large migratory storms (commonly called *cyclones*) with horizontal dimensions of 2,000 miles or more. Their lives span over several days, and the vertical velocities in them rarely exceed 4 in./sec, or 10 cm/sec.

Above the tropopause is the *stratosphere*. Here the temperature is more or less constant with elevation and the stratification is very stable. Broadly speaking, the tropopause is the upper limit of the overturnings that result from daily and yearly heating of the earth's surface. The stratosphere is, by and large, separated from these direct influences, and for this reason it contains very little moisture and dust except such as may be brought to high levels by major volcanic eruptions. Another characteristic of the stratosphere is that it contains much of the ozone that was discussed in Subsec. II-B.

Above the stratosphere is a warm layer which has variously been called the *meso-*

sphere, or the *chemosphere*. The highest temperature within this layer is higher than the temperature at the ground, even on warm days. The high temperature is mainly due to selective absorption of ultraviolet radiation from the sun.

The fourth major layer is the *ionosphere*, the base of which normally is about 45 to 50 miles (70 to 80 km) above the earth. At such low pressures as prevail at these heights ionization processes are lively, resulting in a layer with a high concentration of free electrons which reflects ordinary radio waves while letting visible electromagnetic waves through without noticeable distortion. The ionosphere is divided into a number of layers with different electrical characteristics and varying intensities.

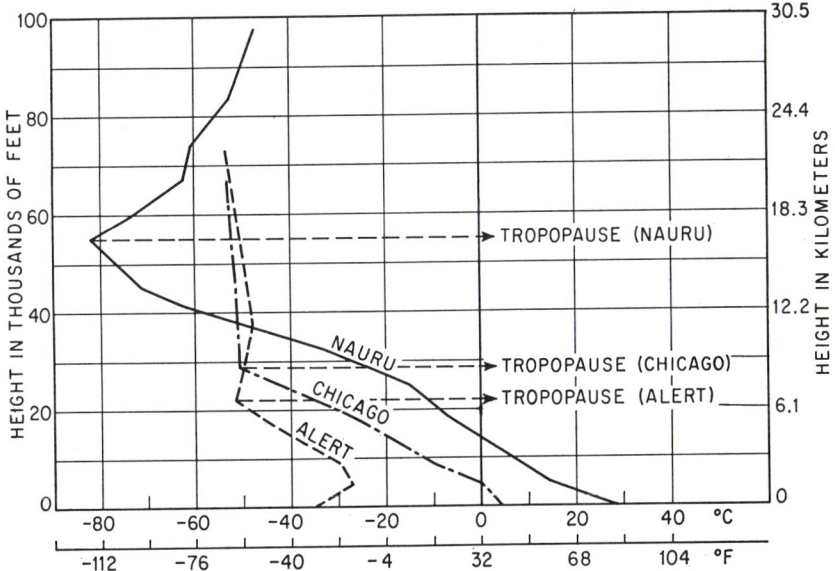

FIG. 3-4. Typical soundings of temperature showing variation with latitude.

The ionosphere, which is the abode of the auroras, merges gradually into the outermost shell, the *exosphere*. Here the mean free path is very large and the atmosphere has lost the property of a continuum.

III. RADIATION, HEAT, AND TEMPERATURE

While the sun is the source of all the energy that maintains the motion of the atmosphere, the physical properties of the earth's surface play an important part in converting solar radiation into sensible heat and in apportioning the heat between the atmosphere and the surface substances.

A. Radiation Balance

The portion of the total incoming radiation that is reflected back to space is called the *albedo*. Under average conditions the albedo is about 40 per cent. Of the remaining 60 per cent, a small portion is absorbed in the atmosphere, but the bulk is absorbed at the earth-atmosphere interface and there shared between the atmosphere and the surface substances.

The average value of the energy available to the earth-atmosphere systems amounts to about 0.30 cal/cm^2/min of the earth's surface. However, the distribution varies greatly with latitude and season, and it varies locally with the changing weather conditions. If all this heat were used to warm the atmosphere, the temperature would rise about 1.5°C (3°F) a day. Since the climate does not change perceptibly, the earth-atmosphere as a whole must lose as much as it gains. This loss is ultimately brought about by long-wave radiation back to space. Since the earth's surface and the clouds of the atmosphere radiate very nearly as a black body, this radiation is proportional to the fourth power of the temperature (*Stefan's law*). Some of this radiation passes directly back to space, but some of it is absorbed by the water vapor and the carbon dioxide of the air and reradiated at the temperatures prevailing, in accordance with *Kirchhoff's law*. This law states that a body which is a good absorber in a certain wavelength is also a good emitter in the same wavelength.

For the system earth-atmosphere as a whole, the average outgoing radiation must be equal to the incoming, which is 0.30 cal/cm²/min, and corresponding to this is the radiation of a black body having a temperature of 240°K (−33°C, or −27°F) radiating into vacuum. This, then, must be the temperature at some significant level within the atmosphere, and it explains why the observed temperatures are what they are and not, say, 100° higher or lower. The general level of temperature being thus determined from the radiation balance, the variation with height, above and below this level, is governed by the principle of lapse rate.

B. Variation with Latitude

The mean annual amount of incoming radiation varies with latitude, and some average values for zonal bands are given in Table 3-2. It will be seen that the absorbed radiation decreases rapidly with increasing latitude, while the loss through

Table 3-2. Annual Heat Balance, by Latitude Zones

Zones of latitude, deg	Fraction of total area	Short-wave radiation absorbed, cal/cm²/min	Long-wave radiation emitted, cal/cm²/min	Poleward transport of heat across latitude parallels, cal/min
0–20	0.34	0.39	0.30	
				57×10^{15} (20°)
20–40	0.30	0.34	0.30	
				77×10^{15} (40°)
40–60	0.22	0.23	0.30	
				50×10^{15} (60°)
60–90	0.14	0.13	0.30	
Weighted mean.........	....	0.30	0.30	

long-wave emission is sensibly constant. To maintain a steady state the air and ocean currents must transport the amounts of heat given in the last column of the table. The bulk of this heat is transferred in the atmosphere, mainly by the large-scale motion systems. As a result of this poleward export of heat from the low-latitude belt, the temperature difference between low and high latitudes is greatly reduced.

Some information on the seasonal variation of the incoming radiation is given in Table 3-3. According to Stefan's law, the temperature is proportional to the fourth root of the radiation, and the values of these roots are given in the parentheses. It will be seen from the last column that the mean annual variation of temperature tends to increase poleward. This is true for the system earth-atmosphere as a whole. In any particular place the annual variation of the air temperature depends upon how the heat is apportioned between the air and the surface substance.

The diurnal variation of temperature follows closely the diurnal change in the altitude of the sun. In general, it decreases with increasing latitude, but here too the amounts are greatly influenced by the properties of the underlying surface.

Table 3-3. Incoming Radiation in Latitude Zones*

Latitude zones	Short-wave radiation absorbed, cal/cm²/min		
	Summer	Winter	Difference
0–20°	0.42 (0.805)	0 36 (0.775)	0.06 (0.030)
20–40°	0.42 (0.805)	0.26 (0.714)	0 16 (0.091)
40–60°	0.35 (0.769)	0.12 (0.589)	0.23 (0.180)
60–90°	0.24 (0.700)	0.02 (0.376)	0 22 (0.324)

* Figures in parentheses are fourth roots of radiation.

C. Conduction Properties

The incoming radiation that is absorbed at the earth's surface is used to warm the air and the surface substance, and part of the heat is conducted away from the interface. One defines the distance of penetration at either side of the interface at a given time as the distance at which the temperature rise is a small fraction, say, 5 per cent, of the rise at the interface. The manner in which the two adjoining media share the heat depends upon the density ρ, the specific heat c, and the temperature conductivity K of the substances involved. In the case of solids, K refers to molecular processes, and in the case of fluids, it refers to eddy transfer of heat.

From the theory of conduction it is found that the distance of penetration at any given time is proportional to $\sqrt{K}$. The heat required to raise the temperature of a unit volume by the amount ΔT is $c\rho \Delta T$. Since the volume affected is proportional to $\sqrt{K}$, the rate of heating must be proportional to $c\rho \Delta T \sqrt{K}$, so that the temperature rise is inversely proportional to $c\rho \sqrt{K}$.

The quantity $c\rho \sqrt{K}$ is called the *conductive capacity;* it is a measure of the ability of a substance to conduct heat away from the place where it is supplied. Now if heat is supplied at the earth-atmosphere interface, the temperature rise on either side will be inversely proportional to the conductive capacities, and the distance of penetration, into the air and into the underlying medium, will be proportional to $\sqrt{K}$ of the substance involved. The range of temperature at the interface must be the same for both media, and it will be determined approximately by the sum of the two conductive capacities.

D. Surface Influences on Temperature

The conduction properties of air and typical surface substances are given in Table 3-4. Disregarding freshly fallen snow (the life of which is quite short), one may distinguish between three rather distinct types of surface substances, namely (1) snow and dry sand, (2) ice and all kinds of customary soils, and (3) stirred water. Still water need hardly be considered here since natural water is more or less in a stirred state. However, if the waters of the earth were still, the heat conduction would be effected through molecular processes and there would be no appreciable difference between the oceans and the continents. The vast differences that are observed are mainly due to the mobility of the water along the vertical. The range of mobility of air is much larger than that of water; it is mainly determined by the lapse rate of temperature such that it increases as the stability decreases.

From the data in Table 3-4 some general conclusions may be drawn. (1) Since the heat capacity of water is very much larger than that of the air, the oceans are effective reservoirs of heat. (2) Since the sum of the conductive capacities of water and air is about 7 while the corresponding sum over customary land substances is

Table 3-4. Conduction Properties of Air and Typical Surface Substances, CGS-Cal Units

Substance	Heat capacity per unit volume, $c\rho$	Conductive capacity, $c\rho \sqrt{K}$	Surface-stirred air ratio
Snow, fresh	0.03	0.002	0.02
Snow, aged	0.022	0.012	0.12
Sand, dry	0.3	0.011	0.11
Sand, wet	0.4	0.04	0.4
Sandy clay*	0.6	0.037	0.37
Soil, organic	0.57	0.04	0.4
Soil, wet	0.7	0.038	0.38
Ice	0.45	0.05	0.5
Water, still	1.00	0.039	0.39
Water, stirred	1.00	7†	70
Air, stirred	0.0003	0.1‡	

* 15 per cent water.
† Varies from about 0.3 to 17, depending upon stability.
‡ Varies from about 0.01 to 1.0, depending upon stability.

about 0.14, the temperature range over land will be very much larger than at sea. (3) Since the temperature conductivity of customary land masses is 0.01 or less while the values for stirred water and air are 50 and 100,000, respectively, the depth of penetration of the annual and diurnal cycles will vary accordingly. A few typical values are given in Table 3-5.

Table 3-5. Penetration of Diurnal and Annual Heating

Surface substance	Penetration, ft	
	Diurnal	Annual
Sand, dry	0.6	11
Customary soils	1.6	30
Ice	1.8	34
Water, still	0.6	12
Water, stirred*	100	Variable
Air, stirred	5,000	Tropopause

* The annual penetration depends greatly upon the stability, which usually increases downward.

If radiation and conduction were the only processes at work, the annual and diurnal ranges of temperature would be considerably larger than what is observed. The difference is due partly to heat being expanded in evaporating water and partly to the moderating influences of horizontal air movements.

E. Role of the Water Cycle

When the atmosphere and the earth are considered as separate entities, it is found that radiation and conduction do not provide balanced heat budgets, for the earth's surface comes out with a gain, and the free atmosphere with a loss, of heat. The link between the gain and loss is the water cycle: evaporation of terrestrial water, condensation of water vapor in the air, and precipitation of water back to the earth.

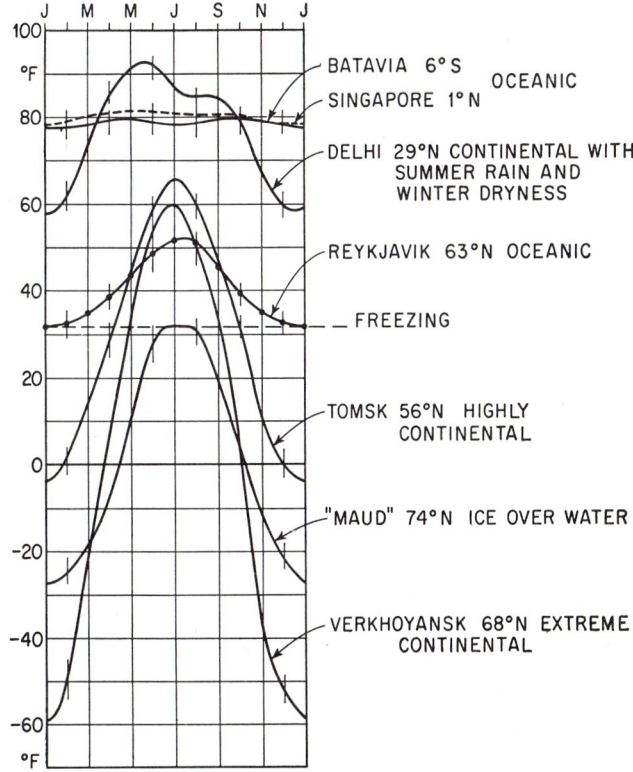

Fig. 3-5. Examples of annual variations of temperature showing variations with latitude over continents and oceans.

Part of the heat absorbed by the earth's surface is expended in evaporation and thus transferred to latent heat, which later is realized as sensible heat and given to the air when the vapor condenses to clouds. The evaporation is particularly intense where relatively cool air sweeps over warmer oceans. In the Northern Hemisphere the highest values are found in the Atlantic and Pacific trade-wind belts, south of about 30°N. Large values are found also over the northwestern parts of the North Pacific and North Atlantic Oceans during the colder part of the year when cold continental air masses sweep over the warmer waters.

The average life of the water-vapor molecules in the air may vary within wide limits, from an hour to several days. Latent heat will normally be liberated far from the regions where the evaporation took place. In particular, this is true of the evaporation in the trade-wind belts, which supply much of the vapor that results in clouds and precipitation in middle and high latitudes. Similarly, the rivers of the continents are

3-16   METEOROLOGY

the counterparts to the transport of water vapor from oceanic to continental areas. On the whole, the circulation of water substances is an important part of the transfer of heat from low to high latitudes and from oceans to continents.

F. Annual Variation of Temperature

The general principles discussed in the foregoing sections suffice to explain the broad aspects of the observed annual range. Examples of the dependence of the

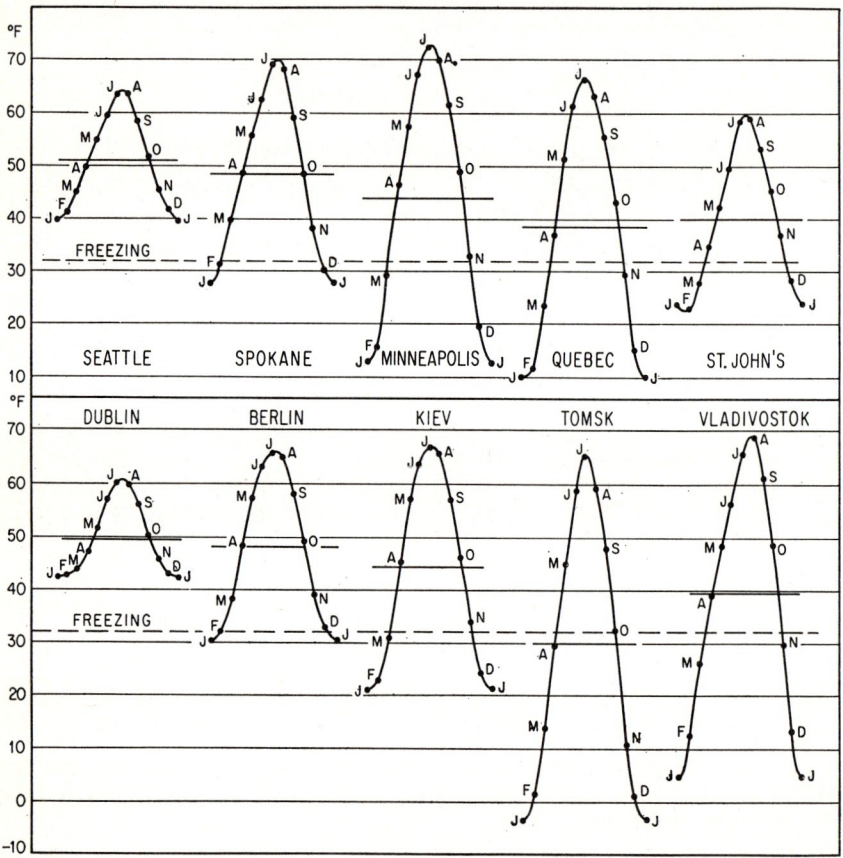

FIG. 3-6. Examples of annual variation of temperature showing variations with degree of continentality.

annual range upon latitude is shown in Fig. 3-5. Near the equator (Batavia and Singapore) the range is very small; at Delhi it is large, with a disruption of the normal trend during the monsoon season when the incoming radiation is much reduced because of increased albedo. The difference between oceans and continents is clearly shown in the curves for Reykjavik, Tomsk, and Verkhoyansk. Of special interest is the curve for "Maud," where the summer maximum is flat and coincides with the melting temperature of ice, while the winter minimum is relatively high as a result of heat conduction from the water under the ice.

Examples of the influences of oceans and continents and the effects of horizontal transports of sensible and latent heat are shown in Fig. 3-6. At Dublin and Seattle the annual range is small because of the moderating influence of the oceans. In Eurasia, where no mountain barrier obstructs the heat and moisture-carrying westerly winds from the Atlantic, the lowest winter temperatures are found far downwind from the west coast. In North America, on the other hand, where the Rocky Mountains form an effective barrier, the transition is far more abrupt.

G. Diurnal Variation of Temperature

The processes are the same as those in the annual cycle, but there are differences in degree. Since the daily cycle is short, the penetration is shallow and the range of

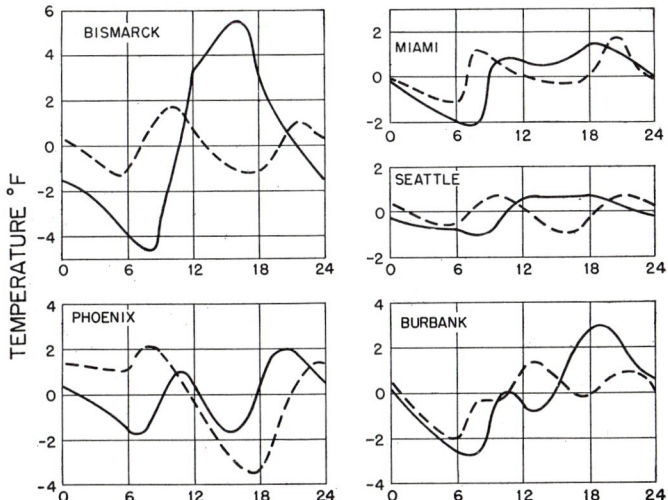

Fig. 3-7. Examples of daily variation of dew-point temperature.

temperature is relatively large. Furthermore, on account of the shortness of the cycle, the horizontal transport of heat and vapor is relatively unimportant, except in the sea-breeze zone along the coasts and in hilly terrain where drainage winds may influence the temperature, particularly during the night. Because of the high conductive capacity of water, the diurnal range is very small over oceans (about 1°F); over continents, the range is generally large, and it decreases with increasing latitude. On windy days the mobility along the vertical is enhanced and the daily range of temperature is decreased.

The mobility of the air has a marked influence on the moisture content. As the sun rises in the morning, water begins to evaporate from the ground and the moisture content (and the dew-point temperature) increases at first. Before noon the lapse rate normally becomes so steep that turbulent eddies carry vapor upward faster than the dried-out surface can supply water for evaporation; as a result, the moisture content decreases and reaches a minimum in the late afternoon. At this time, the moisture gradient is reversed and the eddy motion returns moisture to the ground layer, with the result that the dew point rises. With further cooling dew forms on the ground and equilibrium is reached in the late evening, whereafter the dew point falls again until the sun again rises. Thus, though the diurnal variation of temperature is a single wave, that of the dew point is normally a double wave. Some examples are shown in Fig. 3-7. Irregularities may occur along coasts where the sea breeze causes alternation between dry and humid air (e.g., Burbank) and also over snow-covered ground (e.g., Bismarck).

IV. CONDENSATION AND PRECIPITATION

It is commonly observed that some clouds may exist without yielding precipitation while other clouds release large elements that fall to the ground or evaporate below the cloud base. It is necessary, therefore, to consider condensation and precipitation as separate processes.

A. Condensation Nuclei

Both observation and theory show that condensation of water vapor into cloud droplets takes place on certain hygroscopic particles which are commonly called *condensation nuclei*. Without such nuclei condensation could be initiated on certain large ions, but the relative humidity would then have to be several hundred per cent. In natural air the degree of supersaturation (when such occurs) is usually a fraction of one per cent. The most active nuclei are particles of sea salt and such products of combustion as those which contain sulfurous and nitrous acids. The customary nuclei are generally less than 1 micron in diameter, but on occasion a few giant nuclei (5μ or so) may be present. The number of salt nuclei varies from 10 to 1,000 per cubic centimeter. The combustion nuclei are generally small, and their number varies greatly with the industrial activity.

B. Growth of Cloud Droplets

From physical chemistry it is known that the equilibrium vapor pressure (Subsec. II-D) is reduced when salt and similar substances are dissolved in water, the reduction being expressed by

$$\frac{e_1}{e_s} = 1 - CM$$

where e_s is the saturation vapor pressure corresponding to pure water, e_1 is the equilibrium pressure of the molar aqueous salt solution containing M moles of solute per liter, and C is a temperature-variable factor which depends upon the particular substance. Now, if vapor condenses on a nucleus, M decreases, and so does the solute effect; when the radius has increased to about 2 microns, the solute effect is negligible.

From the theory of surface tension it is known that the equilibrium vapor pressure e_2 over a curved droplet is larger than that over a plain water surface. The relation is well represented by the approximate formula

$$\frac{e_2}{e_s} = \left(1 - \frac{K}{r}\right)^{-1}$$

where r is the radius of the droplet, and K is constant at any given temperature. K is a very small quantity, and the curvature effect is generally negligible for values of r greater than about 2 microns. Thus the solute effect and the curvature effect are of consequence only while the cloud droplets have radii less than about 2 microns. In natural air the process follows a path (Fig. 3-8) which is a compromise between the effects of solute and curvature. It is readily seen that the larger the original nucleus, the smaller is the supersaturation needed for the droplet to grow while the 100 per cent level of relative humidity is approached.

Computations show that, under normal conditions, it takes a nucleus a few seconds to grow to a droplet of 10 microns, a few minutes to reach 100 microns, 3 hr to grow to 1,000 microns (1 mm), and about 24 hr to become a small raindrop (diameter of 3 mm). It follows then that condensation is incapable of producing raindrops though it may well produce oversize cloud droplets.

C. Precipitation Processes

In the main there are two processes that will cause clouds to release precipitation; one of these is called the coalescence process, and the other may be referred to as the ice-crystal process.

1. The Coalescence Process. A water drop in the air is exposed to the force of gravity and the frictional drag. After a relatively short while the two forces balance, and the drop has then acquired a maximum velocity called the *terminal velocity*. This velocity is proportional to the square of the radius, with the result that

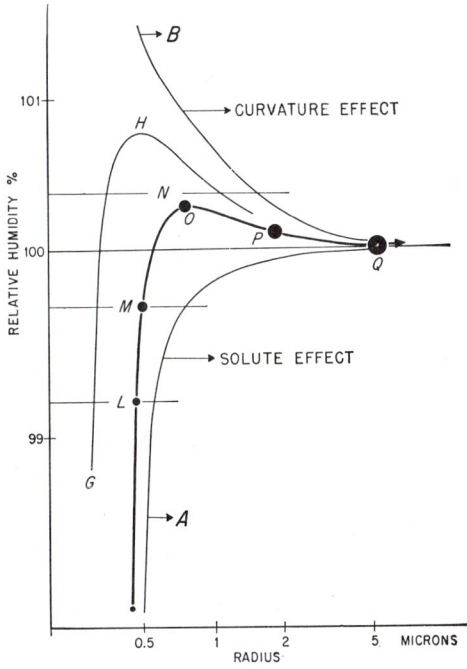

FIG. 3-8. The combination of the effects of solute and curvature. The curve $LMOPQ$ corresponds to a moderately large nucleus, and GHQ to a smaller one.

large drops fall faster and may collect smaller drops on their forward sides. Furthermore, a falling drop leaves behind it a turbulent wake which allows somewhat smaller drops to fall faster and overtake the leading drop. Hence the large drops will grow as a result of (1) direct capture of small drops on their forward side, and (2) wake capture of drops of almost equal size.

As a drop grows to a diameter of 7 mm, the fall velocity increases to about 10 m/sec, or 30 ft/sec. At such high speeds the drop flattens out and breaks up into drops of the size of small rain or drizzle. Each of these will fall faster than the cloud droplets, with the result that the capture processes are repeated. In clouds with strong updrafts (e.g., thunderclouds) the drops may fall and grow, splinter and rise, several times, thus multiplying in number as if a chain reaction were in operation.

2. The Ice-crystal Process. Observations show that water droplets may exist in clouds at subfreezing temperatures down to about $-40°C$ ($-40°F$). Within this range solidification occurs in connection with certain particles which are now called *freezing nuclei.* A large number of such materials have been identified, but most prominent are clay minerals (kaolin, montmorillonite, illite, pyrite, etc.) and organic and common sea salts. When ice elements form in an undercooled cloud, an imbalance is created, for the equilibrium vapor pressure over the water drops is higher than over the ice elements (Fig. 3-9). As a result, the water tends to evaporate and condense on the ice, thus bringing about a growth of some cloud elements and shrinking of others. The uneven size distribution resulting therefrom favors further growth through collision and coalescence. In warm clouds, i.e., clouds which do not reach up to layers with subfreezing temperatures, giant salt nuclei (when present) serve to create some unusually large drops which, through fragmentation, may develop the precipitation process.

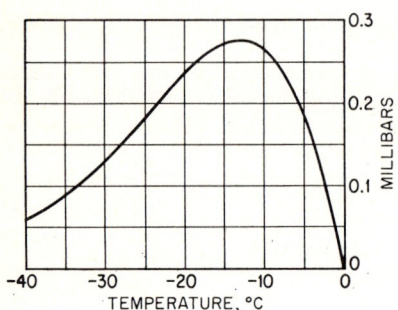

FIG. 3-9. Difference between the saturation vapor pressure over water and ice.

3. Artificial Stimulation. Recent attempts at artificial stimulation of rainfall have been based upon the general principles outlined above. Silver iodide has been supplied to the air to imitate the effects of ice crystals; frozen carbon dioxide has been used to cool the air to stimulate formation of natural ice crystals; and water has been sprayed in warm clouds to produce some large drops which could foster chain reactions. Experiments to determine the economic value of such techniques are in progress.

D. Classification of Clouds

As seen from below, the clouds are referred to three main groups, namely (1) *cirrus*, or feathery streaks of clouds; (2) *stratus*, or *layer clouds;* and (3) *cumulus*, or *heap-shaped clouds*. These basic forms may be present simultaneously in various combinations and at different heights. Depending upon their basic forms and their height above the ground, the cloud forms may be divided into 10 principal types as shown in Table 3-6.

Table 3-6. Principal Types of Clouds

Name	Symbol	Height (approximate)
Cirrus.............	Ci	
Cirrostratus.........	Cs	High (20,000–40,000 ft)
Cirrocumulus........	Cc	
Altostratus..........	As	
Altocumulus.........	Ac	Medium (8,000–20,000 ft)
Stratus.............	St	
Stratocumulus.......	Sc	Low (below 8,000 ft)
Nimbostratus........	Ns	
Cumulus............	Cu	
Cumulonimbus.......	Cb	Vertical development

Cirrus clouds occur in the upper troposphere; they consist of fine ice crystals and have a delicate silky appearance without shadows. A streaky type is shown in Fig. 3-10. Other forms are *cirrostratus*, which is a thin white sheet, and *cirrocumulus*, which consists of small white flakes arranged in a pattern resembling lamb's wool; they are often called *mackerel clouds*.

Altocumulus differs from cirrocumulus in that the cloud sheet is lower and the flakes are larger and often show light shadows.

Altostratus (Fig. 3-11) is a dense sheet of gray or bluish-gray color, often showing a slight fibrous structure. Clouds of this type are present in the middle troposphere, and they are usually followed by precipitation of a continuous type.

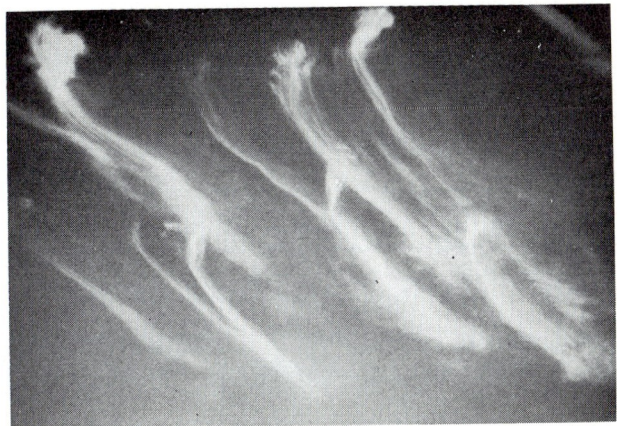

Fig. 3-10. Feathery streaks of cirrus clouds [3].

Fig. 3-11. A typical sheet of altostratus through which the sun is barely visible [3].

Stratocumulus is a low cloud layer consisting of large lumpy masses or rolls of dull gray color with brighter interstices.

Nimbostratus is a dense, shapeless, and ragged layer of clouds from which precipitation usually falls. It is often connected with altostratus at higher levels.

Cumulus is a thick cloud whose upper surface is dome-shaped, often of a cauliflower structure, and whose base is more or less horizontal. Cumulus clouds are often divided into three groups, namely, *cumulus humulis*, which are flat fair-weather clouds; *cumulus congestus*, or towering cumulus, which reach to considerable heights without producing precipitation; and *cumulonimbus* (Fig. 3-12), which are great masses of cloud rising like mountains, towers, or giant anvils and having a base that looks like nimbostratus. The cumulonimbus clouds are accompanied by showers, squalls, or thunderstorms, and sometimes hail. In pronounced cases destructive winds or tornadoes may occur. A severe variety is the line-squall cloud, which extends like a dark low arch across the sky.

Stratus is a uniform layer of low foglike cloud which does not touch the ground.

Fig. 3-12. A cumulonimbus, or thundercloud, of the squall-line type [3].

Fog and *smog* are low clouds that touch the ground; in the former the visibility is reduced mainly by cloud droplets, and in the latter mainly by impurities which have absorbed water.

Haze consists of dry dust and soots which reduce the visibility and affect the coloring of distant objects.

E. Classification of Precipitation

The number of different forms of precipitation is very large, and the description here must be limited to the more common types.

Drizzle is a fine sprinkle of very small and rather uniform waterdrops, with diameter less than 0.02 in. (0.5 mm). The drops are so small that they seem to float in the air and follow the irregularities in the air motion. To qualify as a drizzle the drops must not only be small, but they must also be very numerous.

Rain is precipitation of liquid water in which the drops, as a rule, are larger than in drizzle. On occasion, the drops may be of drizzle size, but they are then few and far between, and this distinguishes them from drizzle.

Snow is precipitation of solid water, mainly in the form of branched hexagonal crystals, or stars. Even at temperatures well below freezing, the crystals carry a thin coating of liquid water, and when they collide, they become matter together as large flakes. At very low temperatures the crystals are dry and large flakes are not seen.

Sleet (British) is melting snow or a mixture of snow and rain. Rain which freezes to pellets of ice when falling through a layer of cold air is called sleet in North America.

Glaze, or *freezing rain*, is reported when rain falls into a cold layer of air and freezes when it strikes objects on the ground.

Ice needles, or *diamond dust*, are thin shafts or small thin plates of ice which are so light that they seem to float in the air. They occur only at very low temperatures and are therefore rarely seen except in high mountains and in the polar regions.

Granular snow is opaque small grains falling from stratus clouds. It is the frozen counterpart of drizzle.

Hail is usually referred to three groups. *Soft hail* is round and opaque grains of snowlike structure with diameters from 0.08 to 0.2 in. (2 to 5 mm). The grains are crisp, and they rebound and disintegrate easily when falling on a hard surface. *Small hail* is semitransparent round grains of about the same size as soft hail. The grains have a core of soft hail surrounded by a crust of ice, which gives it a glazed appearance. Small hail falls from cumulonimbus clouds together with rain. The word *hail* is used to indicate large balls or lumps of ice with average diameter from 0.2 to 2 in. or more. Such hail falls almost exclusively from violent thunderstorms in which the vertical currents are very strong. Often the hailstones show concentric layers of clear ice alternating with layers of snow.

Dew forms directly by condensation on the ground, mainly during the night when the surface has been cooled by outgoing radiation.

Hoar frost consists of ice crystals in the form of scales, needles, feathers, and fans. It forms in the same manner as dew, except that the water vapor of the air is transformed directly into ice crystals.

Rime consists of white layers of ice crystals and forms when the droplets of undercooled clouds and fogs strike obstacles.

V. SEMIPERMANENT WIND SYSTEMS

The atmosphere, being an open system consisting of a thermally active substance, exhibits motions of extraordinary complexity and on a variety of scales. However, in the space available here only a brief outline of the most common types can be given.

A. The Equations of Motion

Observation shows that the motion associated with the large-scale systems is overwhelmingly horizontal, so that, as far as the balance of the forces is concerned, the influences due to the vertical motion may be neglected. Furthermore, the effect of the frictional force is unimportant except below about 2,000 ft (600 m) above the ground. In a rectangular system of coordinates with the z axis along the vertical, the simplified equations of motion may be written

$$\dot{u} = -\alpha \frac{\partial p}{\partial x} + v 2\Omega \sin \varphi$$

$$\dot{v} = -\alpha \frac{\partial p}{\partial y} - u 2\Omega \sin \varphi$$

$$0 = -\alpha \frac{\partial p}{\partial z} - g$$

where u, v are the horizontal components of the wind along the x and y axes, respectively, and $\dot{u}$, $\dot{v}$ are the corresponding accelerations. Furthermore, α is specific volume, and the terms containing α are the components of the pressure force per unit mass; g is the acceleration of gravity; Ω is the angular velocity of the earth's rotation; φ is latitude; and the terms containing Ω represent an apparent force due to the circumstances that the motion is considered relative to a system of coordinates that rotates with the earth. This latter force is commonly called the *Coriolis force*.

If the specific volume is eliminated with the aid of the third of the above equations, one obtains

$$\dot{u} = -g \frac{\partial Z}{\partial x} + v 2\Omega \sin \varphi$$
$$\dot{v} = -g \frac{\partial Z}{\partial y} - u 2\Omega \sin \varphi$$
(3-25)

where Z is the height of the isobaric surface.

B. The Geostrophic Wind

Observation shows that in the large-scale motions in middle and high latitudes the pressure force and the Coriolis force are very nearly balanced, so that the acceleration is very small as compared with either of the force terms. The wind corresponding to this balance is called the *geostrophic wind* (u_g, v_g). Thus, putting $f = 2\Omega \sin \varphi$,

$$u_g = -\frac{g}{f}\frac{\partial Z}{\partial y} \quad \text{and} \quad v_g = \frac{g}{f}\frac{\partial Z}{\partial x} \qquad (3\text{-}26)$$

It will be seen that the geostrophic wind blows along the contour lines of the isobaric surface and its speed is inversely proportional to the distance apart of the contours. The geostrophic wind blows along the isobars in a level surface, and since α varies but little in the horizontal, its speed is very nearly inversely proportional to the distance apart of the isobars. In the Northern Hemisphere ($\varphi > 0$) the geostrophic wind blows along the contours (isobars) with high pressure to the right; in the Southern Hemisphere ($\varphi < 0$) the low pressure is to the right; at the equator the Coriolis force vanishes, and Eqs. (3-26) have no meaning.

C. The Thermal Wind

If Eqs. (3-26) are written down for two different levels (1 and 2) and the difference taken, one obtains

$$\Delta u_g = -\frac{g}{f}\frac{\partial (Z_2 - Z_1)}{\partial y} \quad \text{and} \quad \Delta v_g = \frac{g}{f}\frac{\partial (Z_2 - Z_1)}{\partial x} \qquad (3\text{-}27)$$

where $Z_2 - Z_1$ signifies the thickness of the layer between the isobaric surfaces p_2 and p_1. From Eq. (3-23) it will be seen that the thickness is proportional to the mean temperature of the layer. The shear along the vertical of the geostrophic wind is called the *thermal wind*; it is related to the thickness lines in the same manner as the geostrophic wind is related to the contour lines.

D. Global Circulations

Wind or pressure charts for different levels show rather complex patterns near the earth's surface, a maximum of simplicity in the upper troposphere (say, at 30,000 ft,

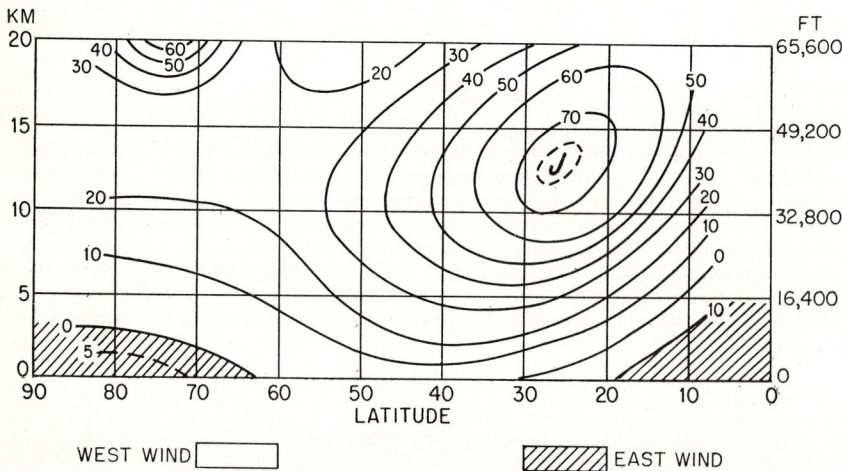

FIG. 3-13. Zonal wind speed, miles per hour, in winter, averaged around the world.

SEMIPERMANENT WIND SYSTEMS 3-25

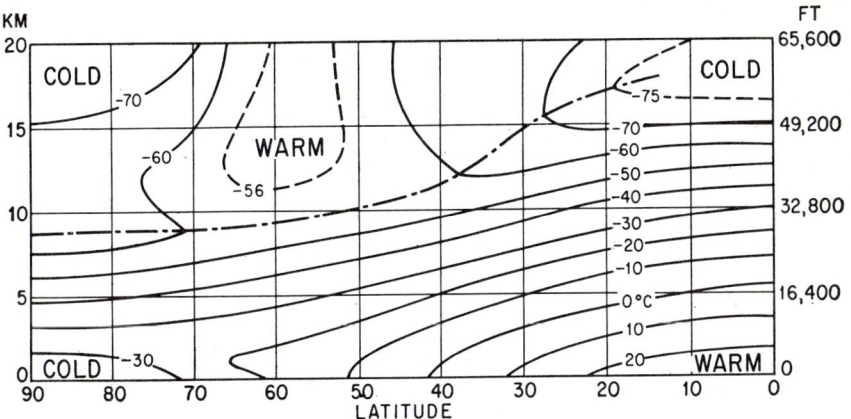

FIG. 3-14. Temperature, degrees centigrade, in winter, averaged around the world.

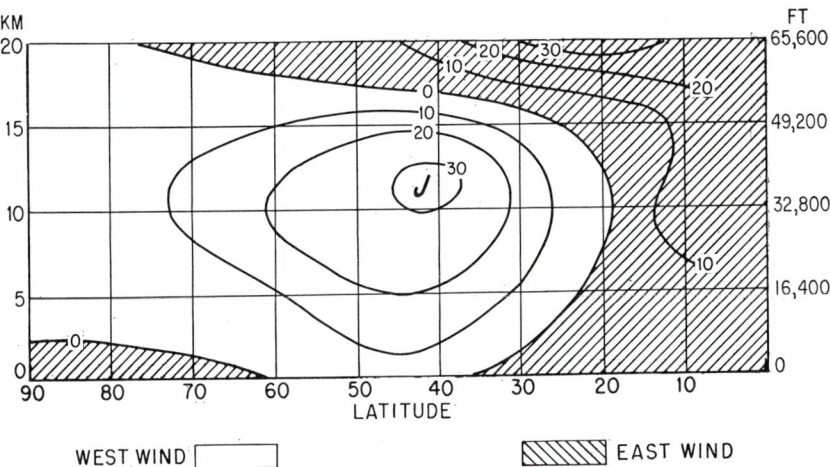

WEST WIND   EAST WIND

FIG. 3-15. Zonal wind speed, miles per hour, in summer, averaged around the world.

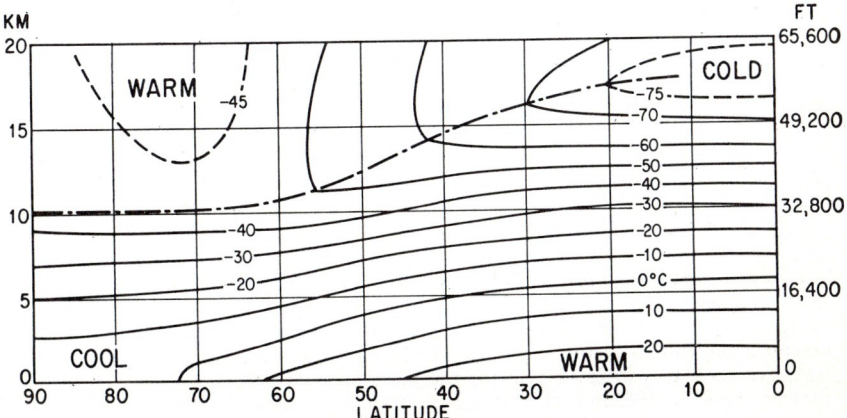

FIG. 3-16. Temperature, degrees centigrade, in summer, averaged around the world.

10 km, or 300 mb), and then increasing complexity up to great heights. In the upper troposphere and lower stratosphere (Figs. 3-13 and 3-15) the predominant features are (1) an equatorial belt of easterlies which is strongly developed in the summer hemisphere; (2) a strong circumpolar jet stream between 10 and 15 km in middle latitudes; (3) relatively light winds in the polar regions, except in winter when a

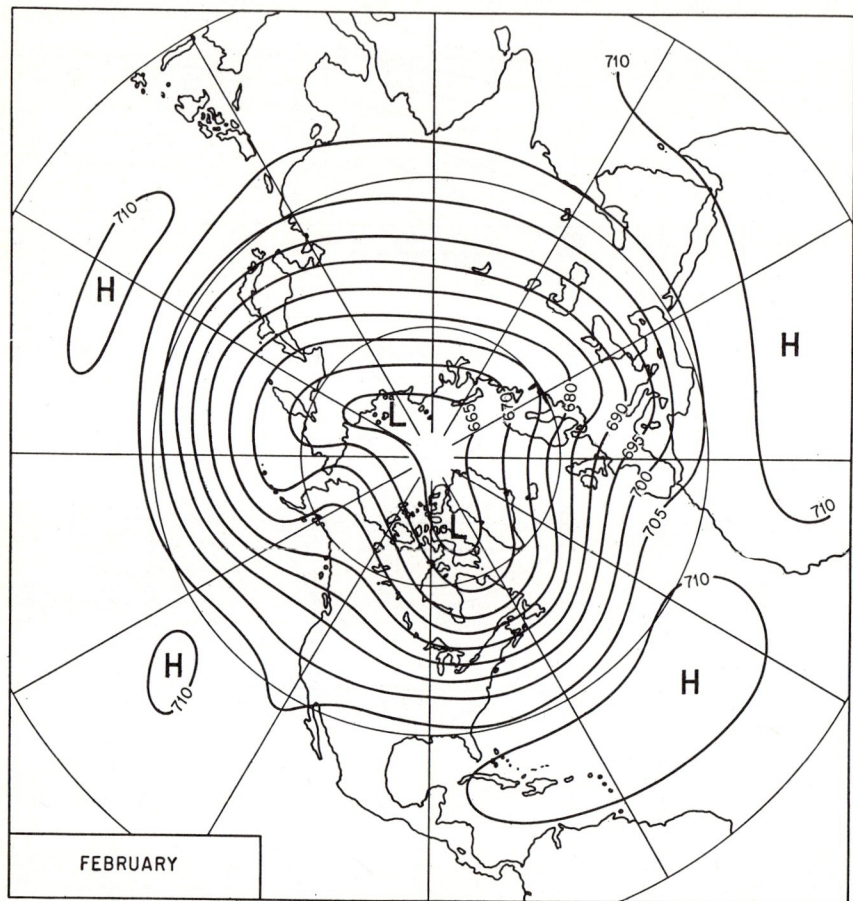

Fig. 3-17. Mean pressure at the 10,000-ft level in winter. The winds may be assumed to be very nearly geostrophic. (*After U.S. Weather Bureau.*)

polar jet stream is present at and above about 20 km. These upper winds reflect the meridional differences in temperature (Figs. 3-14 and 3-16), in accordance with the thermal-wind equation [Eq. (3-27)].

In the lower part of the troposphere (Fig. 3-17) the circumpolar vortex is prominent and so is a broad intertropical belt of relatively light winds. In summer the conditions are essentially the same. The main differences are that the mid-latitude westerlies are weaker and the subtropical belts of high pressure are more pronounced, except in the Indian sector, where the summer monsoon is active.

At sea level in winter (Fig. 3-18) the most prominent features are (1) vast low-pressure areas with counterclockwise wind circulations over the northern parts of the oceans (the Icelandic and the Aleutian lows); (2) a belt of high pressure in the lower-middle latitudes; (3) high-pressure areas over the northern continents; and (4) a belt of relatively low pressure near the equator. In summer (Fig. 3-19) the subtropical high-pressure areas over the oceans are well developed and displaced poleward; the prevailing westerlies in middle latitudes are relatively weak; the pressure over the

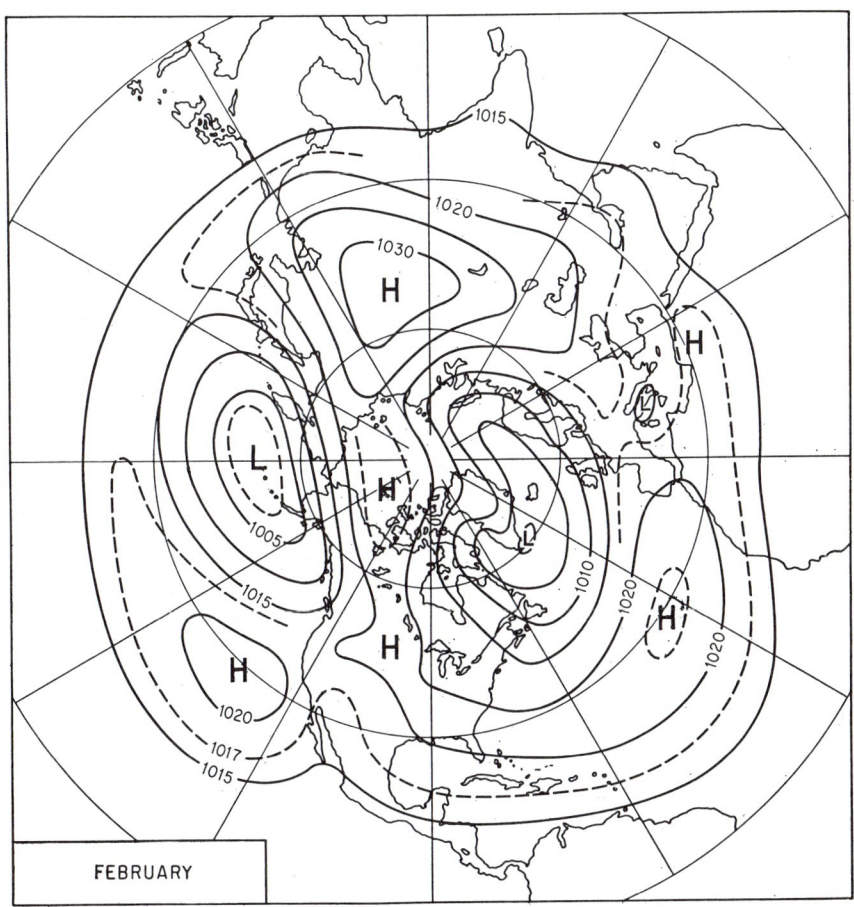

Fig. 3-18. Mean pressure at sea level in winter. (*After U.S. Weather Bureau.*)

continents is relatively low, and monsoon systems are prominent in South Asia, Africa, and the northern part of South America.

E. Monsoons

Monsoons are wind systems with an annual oscillation, blowing from oceans to continents in summer and in the reverse direction in winter. These oscillations, which are in response to the annual heating and cooling of the underlying surface, are quite general, though the Indian monsoon is most widely known, mainly because of the excessive dryness during the winter and the equally excessive rainfall during the summer season. The monsoon rains in South Asia begin in April or early May in Burma.

During a 3-week period in late May and early June, the monsoon sweeps northward into central India; it is accompanied by heavy rain, squalls, and occasional violent storms of the hurricane type. With the cooling of the Eurasian continent in the autumn, the warm-season monsoon dies out in October and is replaced by off-land winds with marked dryness.

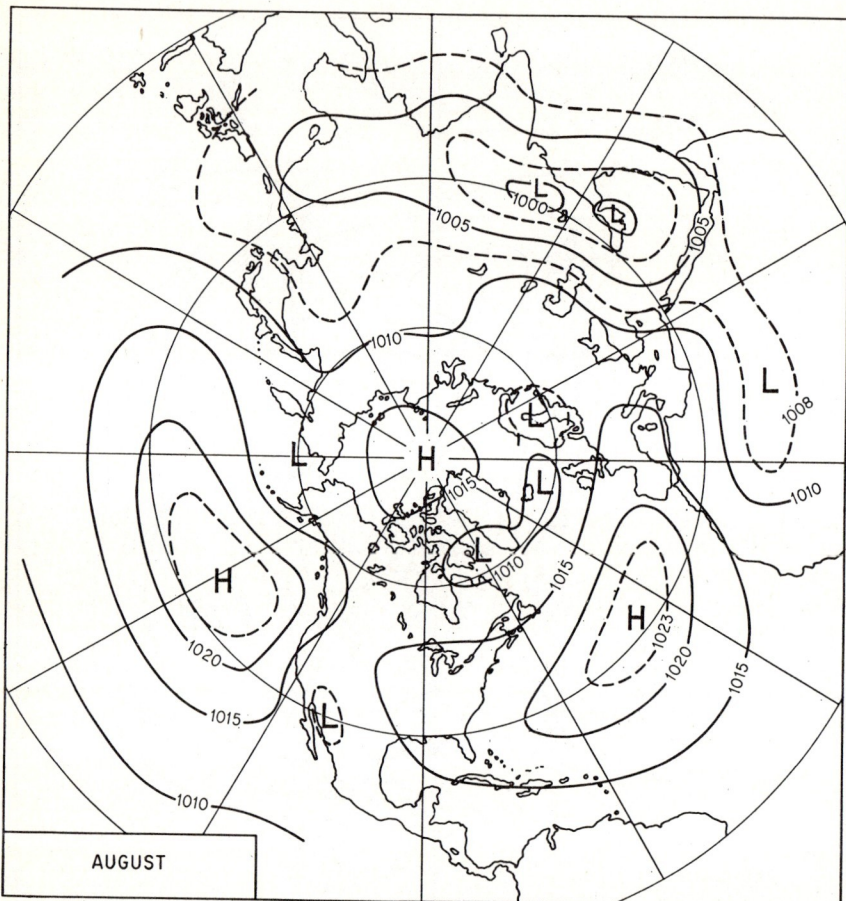

Fig. 3-19. Mean pressure at sea level in summer. (*After U.S. Weather Bureau.*)

F. Land and Sea Breezes

These are winds which develop in response to the diurnal heating of land masses adjacent to oceans and blow landward during the day and seaward during the night. The landward wind, which is called the *sea breeze*, may sometimes be of considerable strength (20 mph or more) in places, while the land breeze is almost invariably weak. The sea-breeze system is very shallow except along mountainous coasts.

G. Mountain and Valley Winds

These are currents which develop in response to the diurnal heating of mountain-sides. On sunny days a gentle breeze is often observed to blow upslope, and on clear nights when the mountainsides cool off, the wind reverses its direction, the cool air seeping down through valleys and creeks to the lowlands. The night wind may be strong where valleys converge.

VI. WEATHER SYSTEMS AND REGIMES

Almost all weather phenomena are related, directly or indirectly, to the motion of larger or smaller bodies of air, and many of the general features of the weather systems are most conveniently considered in connection with air motion. Furthermore, on account of the close relation that exists between motion and pressure (Subsec. V-B), it is often convenient to describe weather systems in relation to pressure configurations.

A. Terminology

Some frequently used terms may be defined as follows.
1. Air Mass. An *air mass* is a vast body of air whose physical properties are more or less uniform in the horizontal while abrupt changes occur along the border between neighboring masses. The physical properties of an air mass reflect the climate of the geographical region where the air sojourned sufficiently long to acquire quasi equilibrium with the underlying surface. Such regions are called *air-mass-source regions*, and in the Northern Hemisphere the most prominent sources are associated with semipermanent high-pressure areas, notably those over (1) the arctic fields of snow and ice, (2) the subpolar continents, (3) the subtropical ocean areas, and (4) the arid subtropical lands.

When an air mass moves out of its source region it is usually referred to as a cold or a warm mass. A warm air mass is warmer than the surface over which it moves; such masses develop a stable stratification, and the typical cloud forms are of the stratus type. On the other hand, a cold mass is colder than the surface over which it moves. The air is often unstable, the clouds are generally of the cumulus family, and when precipitation occurs it is of the showery or squally type.

2. Front. A *front* is the border region between adjacent air masses. The colder air forms a wedge under the warmer air, and in normal cases the layer of transition has a slope of about $1/100$. In general, there is a tendency for the warm air to ascend over the cold wedge, and as a result of adiabatic cooling, clouds are often present in the ascending air. On occasion fronts at the earth surface may be so sharp that they appear as a discontinuity in temperature or a sudden shift of wind; at other times the frontal zone may be wide and indistinct. In the upper air the fronts almost always appear as broad zones of transition.

The most prominent frontal zone in the Northern Hemisphere is the *polar front*, which is the zone separating polar air from tropical air. An *Arctic front* is sometimes observed where arctic air is brought into juxtaposition with milder air from the subpolar regions. In winter the *Mediterranean front* separates cold European air from the much milder air farther to the south.

The fronts are classified according to their motion: a *cold front* ▲▲▲) is a front along which colder air replaces warmer air, and a *warm front* (●●●) is similarly defined. On occasions a cold front may overtake a warm front in such a manner that the warm air between them becomes displaced upward; such systems are said to be *occluded* (▲●▲) .

3. Squall Line. A *squall line*, or *instability line*, is a line or narrow band of thunderstorms accompanied by wind shift and squalls. The air in the rear of the line is cooled by falling rain or hail; the squall line therefore resembles a sharp cold front.

4. Cyclone. A *cyclone* in middle latitudes is a large low-pressure area with the attendant winds. Normally, a cyclone has a frontal structure the characteristics of which reflect its stage of development. Cyclones are usually accompanied by extensive cloud systems and precipitation. On occasions strong and even destructive winds may be present. The qualification *extratropical* is often used to distinguish mid-latitude cyclones from those which develop in the intertropical belt.

5. Anticyclone. An *anticyclone* is a large area of high pressure with the attendant wind system. In general, anticyclones are accompanied by fair weather, though clouds may be present at low levels, particularly at sea and along coasts. The winds are generally moderate, except toward the border of well-developed cyclones.

6. Heat Lows and High-level Highs. *Heat lows* are low-pressure systems which occur in subtropical latitudes during the warmer part of the year. They are brought about by excessive heating of the air over arid land masses, and they differ from extratropical cyclones in that they are stationary and cloudless. The heat lows are shallow formations and are overlain by high-pressure systems (high-level highs, or

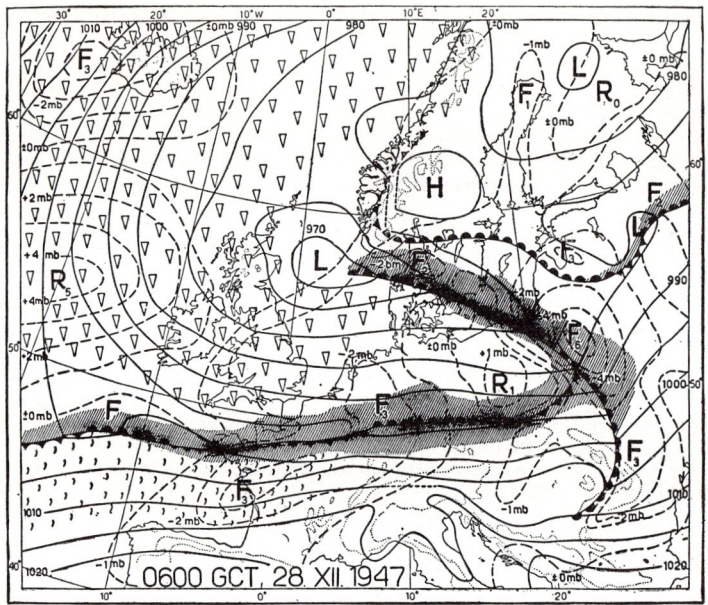

FIG. 3-20. An example of mobile weather systems. (*After Berggren* [4].)

anticyclones) in the middle and upper troposphere. It will be seen from Fig. 3-19 that the preferred regions of heat lows coincide broadly with the deserts.

7. Hurricane. A *hurricane* is an intense cyclone of tropical origin and of relatively small horizontal dimensions (100 to 300 miles). Storms of this type are called *cyclones* in India, *typhoons* in the Far East, and more generally, *tropical revolving storms*, or *tropical cyclones*.

8. Tornado. A *tornado* is a violent whirl of destructive winds which nearly always is accompanied by a funnel cloud hanging down from a cumulonimbus. The diameter of a tornado is usually a few hundred yards, and the maximum wind is in excess of 100 mph. In North America tornadoes are popularly called *cyclones*.

9. Convection. This term is loosely used in meteorology as a name for vertical overturnings resulting from static instability (Subsec. II-I). *Convective currents* are air motions with appreciable updrafts and downdrafts; *convective clouds* are clouds of the cumulus family; and *convective weather* includes showers, squalls, thunderstorms, etc.

B. Extratropical Cyclones

Figure 3-20 shows a highly simplified, but nevertheless typical, chart of mobile weather systems at sea level. Noteworthy features are (1) a large mass of moisture-laden tropical air with low stratus and drizzle (෨ ෨ ෨ ෨) ; (2) a vast mass of polar air streaming southward while developing instability and producing numerous showers (∇ ∇ ∇) ; and (3) a polar front separating the two masses and maintaining an extensive band of continuous precipitation (///////) . The center of low pressure, L, over the North Sea is a decaying cyclone, and the trailing part of the front, across Germany and France, exhibits minor disturbances in the form of frontal waves.

Observation shows that cyclones undergo stages of development, broadly as shown in Fig. 3-21. Diagram A shows a quasi-stationary front between a warm

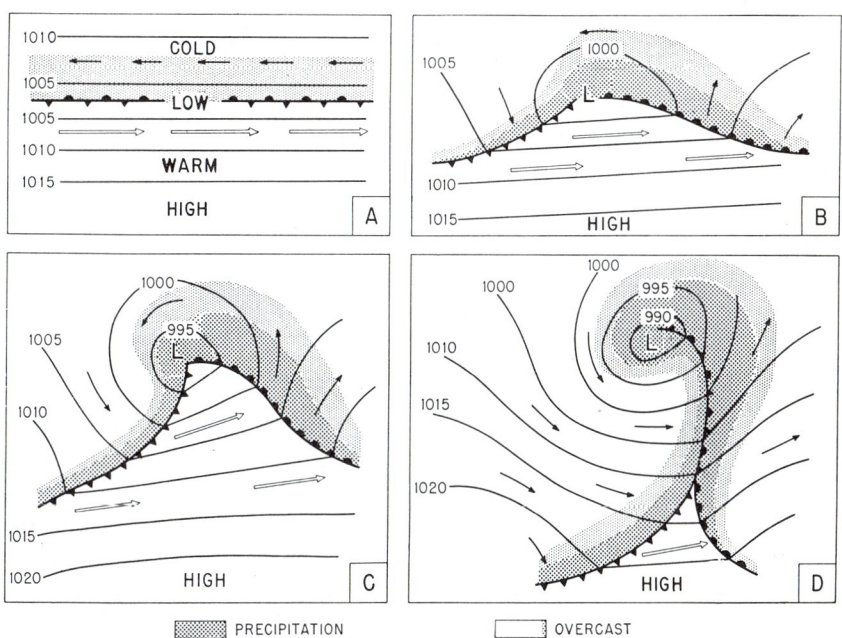

Fig. 3-21. Stages in the development of extratropical cyclones.

and a cold air mass. In diagram B, a wave disturbance has formed, showing growing amplitude and the development of a system of continuous clouds and precipitation, mainly on the cold side of the front. During the further development the amplitude grows; the cold front overtakes the warm front; the cyclone continues to intensify and occlude; and in the final stage (not shown) the fronts dissolve and the cyclone gradually dissipates its energy.

Horizontal and vertical sections through a warm-sector cyclone are shown in Fig. 3-22. As the warm air ascends over the cold wedges, adiabatic cooling leads to condensation and precipitation. Furthermore, as the cyclone occludes, cold air replaces warm air at low levels; the center of gravity of the system is lowered, and potential energy becomes converted into kinetic energy of the winds.

Fronts and cyclones of the type here described dominate the weather conditions in middle and high latitudes, especially during the cold season. Areas with particularly high frequency of cyclones in winter are (1) a broad belt from the east coast of the Carolinas and Virginia to northern Norway, with a maximum of frequency near Iceland; (2) a broad belt from the Philippines to the west coast of Canada, with a maximum in the Gulf of Alaska; and (3) the Mediterranean region. Moderate frequencies during the cold season are found over the Great Lakes, the Baltic, the

Black Sea, and the Caspian Sea. In all seasons, the leeside of mountain ranges are favored regions for cyclone development.

In the warm season the mid-latitude cyclones are generally weak; the highest frequency is found in a broad belt along the 55th-parallel circle, with some preference for the oceanic sections. The Mediterranean maximum vanishes in spring, and at

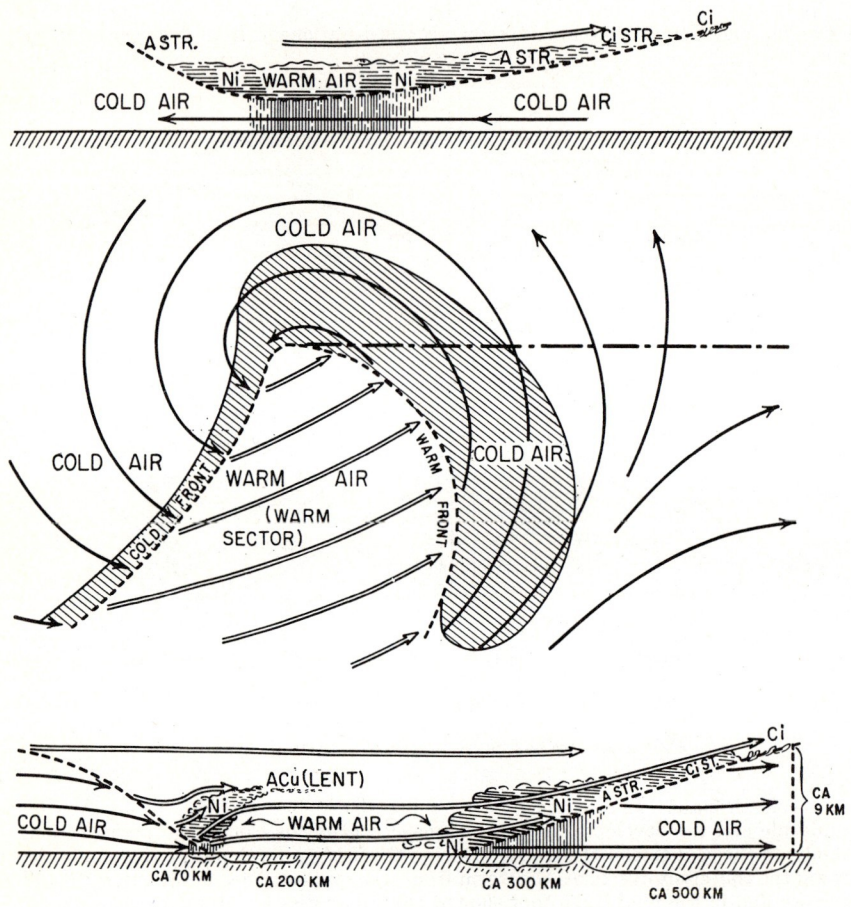

Fig. 3-22. Bjerknes cyclone model.

this time wet monsoon systems or dry-heat lows begin to establish themselves in subtropical latitudes.

C. Convective Weather

Much of the precipitation that falls on the earth comes from convective clouds, or clouds of the cumulonimbus type. These clouds, which form as a result of instability, are characterized by strong updrafts and downdrafts. Over the oceans convective weather is usually widespread in the cyclonic parts of cold air currents moving over warmer water (Fig. 3-20). Over the continents, where the moisture supply is less plentiful, convection is often limited to relatively small areas or narrow bands along cold fronts and squall lines.

A typical example of the structure of a convective system is shown in Fig. 3-23, where updrafts (UPD) and downdrafts (DWD) are superimposed upon a general current. In the incipient stage liquid and solid water accumulate in the cloud until the weight can no longer be supported by the updraft. A heavy downpour then commences. The falling precipitation, partly through cooling of the air and partly by frictional drag, creates a downdraft which spreads out under the cumulonimbus.

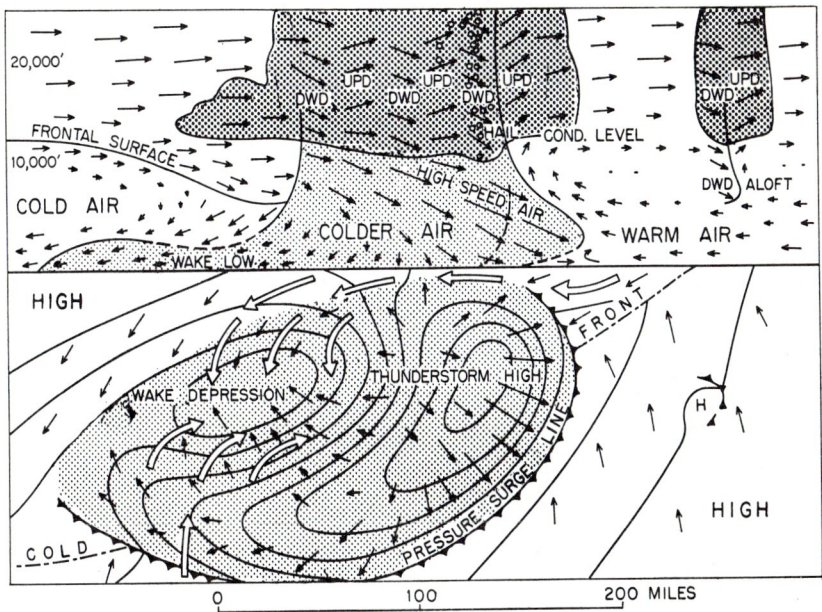

Fig. 3-23. Typical structure of a convective system. Upper part: vertical section; lower part: chart near the ground. (*After Fujita* [5].)

Normally, this spreading downdraft is much colder than the air that was present before the downdraft reached the ground. The excess weight due to the rain-cooled air is reflected in a small high-pressure area under the convective cloud system, and this is often referred to as the *thunderstorm high*. In pronounced cases the thunderstorm high is followed by a small low, called the *wake depression*. Usually the thunderstorm high reflects the presence of a cluster or band of individual storms. On occasion the thunderstorm high may be elongated and extended over a few hundred miles ahead of an advancing cold front, but normally the dimensions of the high are of the order of 50 to 200 miles. Since the dimensions of a typical extratropical cyclone are about 1,000 to 2,000 miles, the convective systems are often referred to as *mesoscale phenomena*. In systems as small as these the winds may deviate much from the geostrophic wind.

D. Tropical Disturbances and Storms

The vast areas of the subtropical belt have weather systems which differ essentially from those of the mid-latitude zone. It is customary to distinguish between the following types.

1. Easterly Waves. These are relatively small wave-shaped disturbances superimposed upon the easterly winds (trade winds) in the zone between the equator and the subtropical high-pressure belts. The wave troughs are almost invariably accompanied by convective clouds, showers, and squalls, and the wedges have fair weather. The regions with frequent occurrence of such waves are shown in Fig. 3-24. In the Indian sector the predominant weather systems are superimposed upon the monsoon.

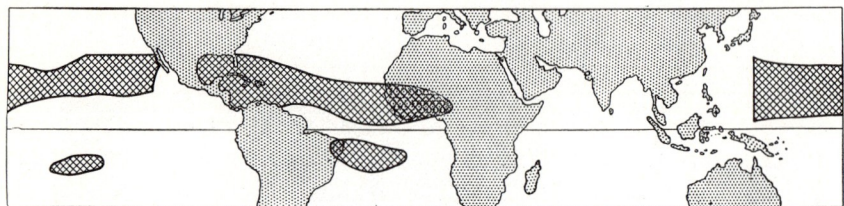

Fig. 3-24. Regions where easterly waves are frequent. (*After Berry, Bollay, and Beers* [6].)

2. Tropical Depressions. These are centers of low pressure which form in the troughs of easterly waves or in the convergence zone where the trade winds from the two hemispheres meet. Though these depressions are small and often indistinct, they produce deep clouds and much precipitation, mainly of the convective type.

3. Tropical Storms. These are well-developed low-pressure systems surrounded by strong winds and much rain. By convention, a system qualifies as a tropical storm if the winds are in the range from 25 to 75 mph.

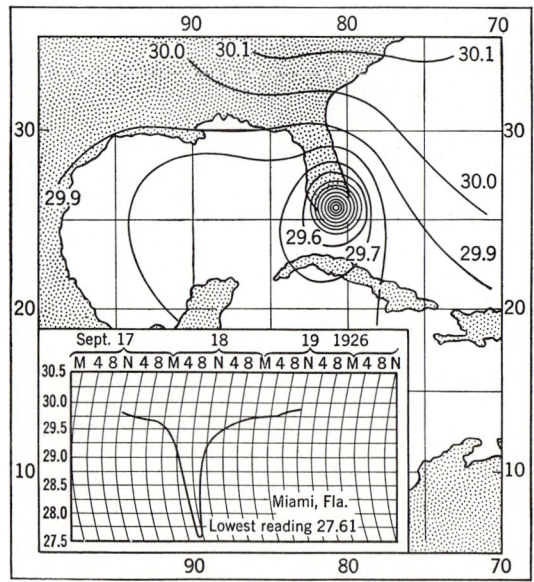

Fig. 3-25. Pressure distribution associated with a strong hurricane. (*After Trewartha* [7].)

4. Hurricanes. These are very intense low-pressure systems with winds in excess of 75 mph. A typical example of the pressure field surrounding a hurricane is shown in Fig. 3-25, and the regions which are frequently visited by hurricanes are shown in Fig. 3-26. On the whole, hurricanes form where the sea-surface temperature is above 27°C (80°F).

In general, hurricanes are accompanied by torrential rains. The highest rain intensity is found at some distance from the center, in the band where the winds are strongest. Rainfalls as high as 6 to 10 in. in 24 hr are common over level land, and on slopes of hills and mountains much higher values occur. In an extreme case, 75 in. in 24 hr was measured at Baguio, in the Philippines.

On the north Atlantic and on the north Pacific to the west of Mexico, the hurricane season begins in June, ends in November, and reaches its peak in September. A similar trend is present in the Far East, but here the frequency is higher and no

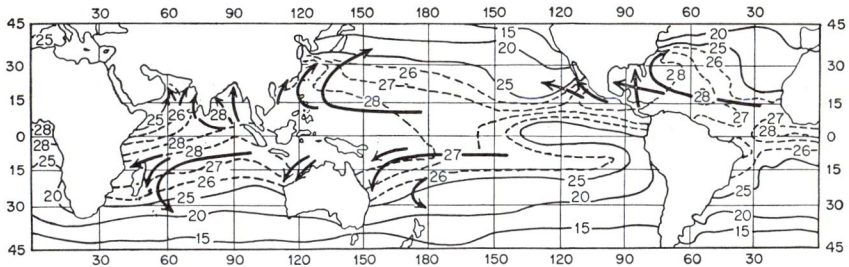

FIG. 3-26. Principal hurricane tracks and isotherms of the sea surface. (*After Palmén* [8].)

month is quite free of typhoons. In the Indian Ocean the frequency is generally lower, no month is excluded, and there is a tendency for maxima in the early part of the summer monsoon (May to July) and during the change from summer to winter monsoon (October to November).

In the Southern Hemisphere storms of the hurricane type reach their highest frequency in late summer, except on the south Atlantic, where none exists.

E. Weather Regimes

On a uniform wet earth the average weather conditions would exhibit very simple patterns and there would be no variation dependent upon longitude. The temperature would decrease monotomically from the equator to the poles, and the rainfall would have three maxima, one in the equatorial belt and one in each of the mid-latitude zones. The equatorial rainfall regime would be maintained by ascending currents resulting from the convergence of the hemispheric winds and the superimposed mobile disturbances discussed in the foregoing section. The mid-latitude zones of wetness would be maintained by the fronts, cyclones, and air-mass transformations described earlier. The equatorial and the mid-latitude rainfall regimes would be separated by a zone of relative dryness coinciding with the subtropical belt of high pressure. Finally, the polar caps would have little precipitation, one reason being that the temperature is so low that the air can hold but little water.

Figure 3-27 shows the rainfall averages around the world. The existence of the ideal zonal pattern is clearly evident, but there are large deviations, particularly in subtropical latitudes. In these belts we find extensive areas of deserts as well as tropical rain forests. In general, the wetness in the subtropical belt is limited to regions near east coasts (e.g., Mobile, Ala., and Asunción, Paraguay) while deserts or semiarid regions occupy the interior of the continents (e.g., Sahara, Arabia, Pakistan, etc.) and extend westward to the coasts (San Diego, Calif., and Iquique, Chile). These and other deviations from the ideal zonal arrangement are due partly to the land masses, which distort the distributions of heat and moisture supply, and partly to hills and mountain barriers, which cause the air currents to ascend and shed water on their windward sides.

The broad aspects of the major weather regions are shown in Figs. 3-28 and 3-29 for winter and summer, respectively. During the northern winter the equatorial rainfall regime is present mainly in the Southern Hemisphere and the subtropical high-pressure belt is displaced southward. During the northern spring these belts migrate northward, causing an annual rhythm. In general, regions near the equator have a tendency toward a double rainfall maximum, while the northern zone of the equatorial regime has summer rain and winter dryness. With increasing distance from the equator, dryness becomes increasingly prominent. With the exceptions mentioned below, the normal sequence of zones, as one goes from the equator northward, would be (1) plentiful rain in all seasons and tropical rain-forest climate; (2) summer rain and winter dryness, with climate of the Savanna type; (3) slight summer rain and long winter dryness, with steppe climate; and (4) dryness in all seasons, with desert climate.

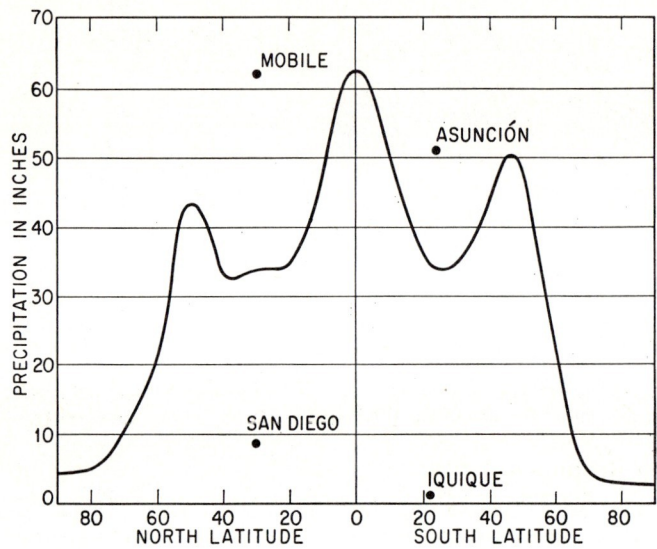

Fig. 3-27. Average rainfall around the world.

Along the west coast of continents the winds (around the oceanic subtropical highs) are equatorward, with the result that the equatorial rainfall regime is relatively narrow. On the other hand, along the east coast of America, and also in the Far East, the winds, particularly during the warm season, are poleward, with the result that the equatorial rainfall regime is wide and tends to merge with that of the midlatitude belt. These conditions are reflected in the data shown in Fig. 3-27.

In the South Asian sector the normal arrangement is disturbed by the monsoons, so that summer wetness alternates with winter dryness over much of the area to the east of West Pakistan.

In the Mediterranean area cyclones are frequent in winter, and some of these move northeastward while others drift across the Near East, causing the Mediterranean winter rain belt to extend, with diminishing amounts, as far east as Kashmir. As summer approaches, the cyclone activity fades out and is followed by dryness as the high-level anticyclone (Fig. 3-29) becomes established.

During the cold season the cyclone activity on the north Pacific and north Atlantic is intense, and though the frequency and intensity are greatest in the Gulf of Alaska and near Iceland, some of the disturbances penetrate far into North America and Eurasia. The oceanic cyclone belts occupy their southernmost positions in midwinter, and winter rainfall is typical along the west coasts north of lower California and to the north of the Canary Islands. As the warm season approaches, the oceanic subtropical anticyclones and the cyclone belts move northward and the activity diminishes; as a result, relative dryness spreads northward along the west coasts in the middle latitudes. In summer the cold continental anticyclones are absent and the cyclone belt is more or less continuous around the globe. In general, the oceans and the coastal areas affected by the mid-latitude cyclones have most rainfall during the colder part of the year, while the continental areas experience most rainfall in

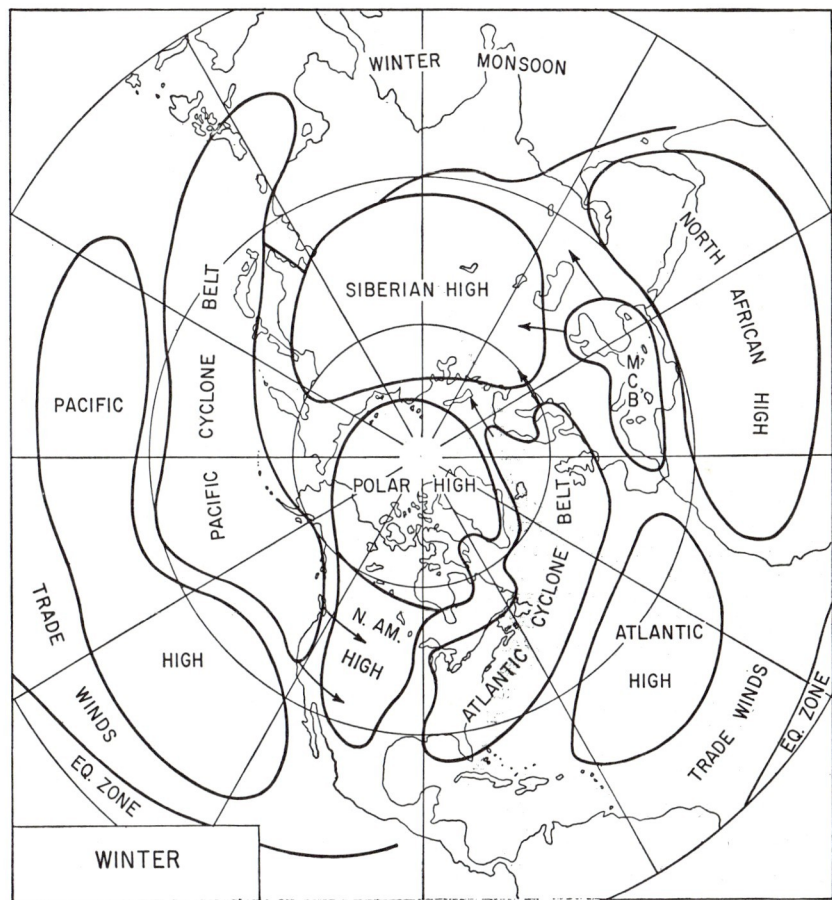

Fig. 3-28. Predominant weather regimes in winter.

summer. This latter maximum is due mainly to increased shower activity resulting from heating of the land masses.

The arctic fields of tundra, snow, and ice have little precipitation in all months, and the annual sum is generally less than 10 in. The largest amounts fall in winter as a result of penetrations by mid-latitude cyclones.

In all precipitation regimes, orographic effects are important. Much water is shed on the windward side of hills and mountains, while dryness is typical of the lee side. Orographic effects are not only important in connection with the large-scale patterns associated with major mountain barriers; very large differences are sometimes observed over distances of only a few tens of miles.

In addition to the mean daily, monthly, or annual amounts of precipitation, it is of interest to know the frequency with which precipitation falls and the regularity of its recurrence. The variability, or the percentage departure from the established normal, is commonly taken to be a measure of the reliability, such that low variability indicates high reliability. Only very small regions have a variability less than 10 per cent. A variability of 10 to 15 per cent indicates high reliability, and a variability of 40 per cent or more indicates very unreliable conditions.

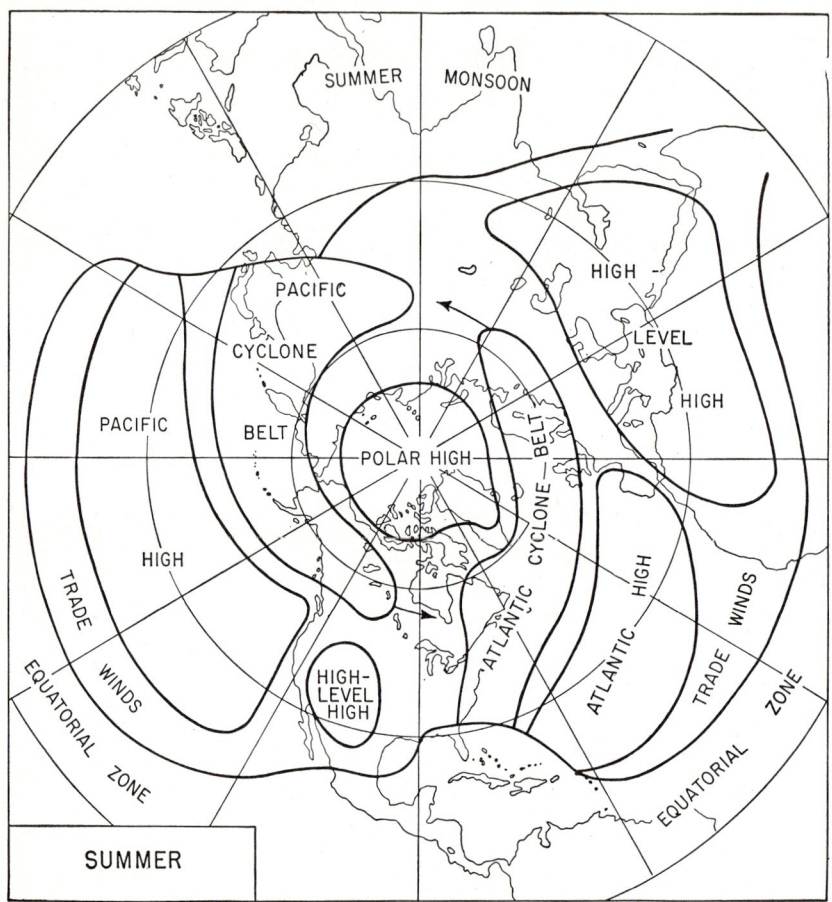

FIG. 3-29. Predominant weather regimes in summer.

In general, it is found that the equatorial regime and the mid-latitude cyclone belts are characterized by high reliability while the deserts are highly unreliable. As a rough rule it may be said that the larger the mean monthly, seasonal, or annual rainfall is, the greater is the reliability of that rainfall.

VII. REFERENCES

1. Callendar, G. S.: The artificial production of carbon dioxide and its influence on temperature, *Quart. J. Roy. Meteorol. Soc.*, vol. 64, pp. 223–240, 1938.
2. Bolin, Bert. and Erik Eriksson: "Changes in the Carbon Dioxide Content of the Atmosphere and Sea Due to Fossil Fuel Combustion," Rossby Memorial Volume,

American Meteorological Society, Boston, and Rockefeller Institute Press, New York, 1959, pp. 130–142.
3. "International Cloud Atlas," World Meteorological Organization, Geneva, 1956, vols. I and II.
4. Berggren, Roy: On the frontal analysis in the higher troposphere and the lower stratosphere, *Arkiv Geofysik* (Stockholm), vol. 2, no. 2, pp. 13–58, 1953.
5. Fujita, Tetsuya: Results of detailed synoptic studies of squall lines, *Tellus*, vol. 7, no. 4, pp. 405–436, 1955.
6. Berry, F. A., Jr., E. Bollay, and N. R. Beers (eds.): "Handbook of Meteorology," McGraw-Hill Book Company, Inc., New York, 1945.
7. Trewartha, G. T.: "An Introduction to Climate," 3d ed., McGraw-Hill Book Company, Inc., New York, 1954.
8. Palmén, E. H.: On the formation and structure of tropical hurricanes, *Geophysica* (Helsinki), vol. 3, no. 26, pp. 26–38, 1948.

General References

9. Battan, L. J.: "Radar Meteorology," The University of Chicago Press, Chicago, 1959.
10. Brooks, C. P. E.: "Climate in Everyday Life," Philosophical Library, Inc., New York, 1951.
11. Byers, H. R.: "General Meteorology," 3d ed., McGraw-Hill Book Company, Inc., New York, 1959.
12. Eckart, Carl: "Hydrodynamics of Oceans and Atmospheres," Pergamon Press, New York, 1960.
13. Geiger, Rudolf: "The Climate near the Ground," Harvard University Press, Cambridge, Mass., rev. ed., 1957 (transl. by M. N. Stewart).
14. Johnson, J. C.: "Physical Meteorology," John Wiley & Sons, Inc., New York, 1954.
15. Kendrew. W. G.: "The Climates of the Continents," Oxford University Press, London, 1942.
16. Landsberg, H.: "Physical Climatology," Pennsylvania State College, State College, Pa., 1958.
17. Petterssen, Sverre: "Introduction to Meteorology," 2d ed., McGraw-Hill Book Company, Inc., New York, 1958.
18. Petterssen, Sverre: "Weather Analysis and Forecasting," 2d ed., McGraw-Hill Book Company, Inc., New York, 1956, vols. I and II.
19. Priestley, C. H. B.: "Turbulent Transfer in the Lower Atmosphere," The University of Chicago Press, Chicago, 1959.
20. Riehl, Herbert: "Tropical Meteorology," McGraw-Hill Book Company, Inc., New York, 1954.
21. Sutton, O. G.: "Micrometeorology," McGraw-Hill Book Company, Inc., New York, 1953.

Section 4-I

GEOLOGY

PART I. HYDROGEOLOGY

GEORGE B. MAXEY, *Research Professor of Geology and Hydrology, Desert Research Institute, University of Nevada.*

  I. Introduction.. 4-1
 II. Geologic Factors in Surface-water Studies..................... 4-2
     A. Geologic Activity of Broad Effects........................ 4-2
     B. Geologic Effects of Small Scale........................... 4-2
     C. Soil and Topographic Effects.............................. 4-3
III. Geologic Factors in Studies of Soil Moisture and Vadose Water.... 4-6
     A. Geologic Factors in Soil Development...................... 4-6
     B. Soil Profiles and Their Effect on Water Movement.......... 4-6
     C. Secondary Cementation and Water Movement.............. 4-8
     D. Summary.. 4-9
 IV. Geologic Factors in Groundwater Studies...................... 4-9
     A. General... 4-9
     B. Nature and Occurrence of Aquifers........................ 4-10
     C. Permeability and Porosity................................ 4-12
     D. Movement and Storage of Groundwater.................... 4-22
     E. Exploration for Groundwater.............................. 4-26
        1. Surficial Geologic Studies............................ 4-27
        2. Surficial Geophysical Methods........................ 4-28
        3. Geochemical Methods................................. 4-31
        4. Subsurface Geologic Methods......................... 4-31
        5. Subsurface Geophysical Methods...................... 4-32
        6. Subsurface Geochemical Methods...................... 4-33
     F. Wells and Springs.. 4-33
        1. Wells... 4-34
        2. Springs... 4-34
  V. References... 4-36

I. INTRODUCTION

In the hydrologic cycle, water modifies rocks by erosion, solution, and deposition. The rocks contribute mineral constituents to, and alter the heat content of, the water and thus materially change the character of the latter. The study of this complicated and widespread interaction between the geologic framework and water constitutes the science of *hydrogeology*.

Applied hydrology is concerned primarily with the small fraction (2 to 3 per cent) of terrestrial water that is fresh and easily available for human use. Therefore, the scope of this treatment of hydrogeology is essentially limited to the interaction between the geologic framework and the fresh water. Various phases of hydrogeology that relate to water are described in many other sections of this handbook (Secs. 5, 12, 13, and 24). The following discussion is therefore somewhat general and directed toward describing the basic geologic factors involved in the studies of surface-, soil-, and groundwater reservoirs. Geologic methodology and field practice are also emphasized as much as possible.

That part of the hydrologic cycle which involves fresh water and operates within the geologic framework takes place in the surface-, soil-, and groundwater reservoirs. Water is in contact with earth materials from the moment it strikes the surface of the earth as precipitation until it is evaporated or transpired. The relative amounts of water that enter, are stored in, and leave each of the three reservoirs, depend largely upon the nature of the soil and rocks. The nature and topographic expression of these materials also govern in part the *evaporation opportunity*, which is the amount of water made available for discharge into the atmosphere by evaporation and transpiration. Diagrammatic representation of the movement, storage, occurrence, and interrelationship of surface, ground, and soil water appears in Figs. 13-1 and 13-3. Figure 13-1 demonstrates the close relationship of the three reservoirs and clearly shows how critical is the effect of shallow earth materials upon the distribution of water between them. The water profile, as shown in Figs. 13-1 and 13-3, clearly illustrates the general differences in storage and movement of water within these reservoirs as well as the effect of geologic controls reponsible for the differences. Geohydrologic explanations to supplement these figures are given in Sec. 13.

II. GEOLOGIC FACTORS IN SURFACE-WATER STUDIES

Geologic factors that act as controls on surface-water phenomena may be classified broadly, as (1) *lithologic*, depending primarily on the composition, texture, and sequence of rock types, and (2) *structural*, including chiefly faults and folds that interrupt the continuity or uniformity of occurrence of a rock type or sequence of rock types. Often structures such as beds and joints also materially influence movement of water and development of drainage patterns. These factors in combination with hydroclimatic processes control the development of soils and topography, which in turn profoundly affect distribution and movement of water. Together the geologic, pedologic, and topographic features influence regimen, stability, form, and distribution of streams and other surface-water bodies. The magnitude of effects resulting from geologic controls alone or in combination with other controls ranges from very broad, such as the outlines, relief, and distribution of continents and ocean basins, to very small, affecting the character of the smallest creek.

A. Geologic Activity of Broad Effects

The broad regional effects result primarily from activity along and near great tectonic structures of continental dimensions over long periods of time. This activity and the resulting relief and changes in erosional and depositional patterns not only control the placement, size, and other characters of surface-water bodies but even influence regional climate.

B. Geologic Effects of Small Scale

Geologically controlled effects of a smaller scale also influence patterns of drainage and distribution of surface-water bodies. Thus one of the commonest manifestations of the effect of varied and folded strata is the distinctive *trellis* drainage pattern (Fig.

4-I-1a). Where the bedrock strata are more or less horizontal and homogeneous, the drainage pattern is usually *dendritic* (Fig. 4-I-1b). Structural controls are often manifested in the development of sharp turns and extraordinary straight reaches of streams that have selectively eroded easily cut rocks along joints and faults. Tectonic movements have been known to deflect streams or to dam them to form lakes of considerable size. Two notable examples in historical times are the formation of Reelfoot Lake along the Mississippi River in Kentucky and Tennessee in 1811 and 1812 and of Earthquake Lake along the Madison River in Montana in 1959 [1, 2].

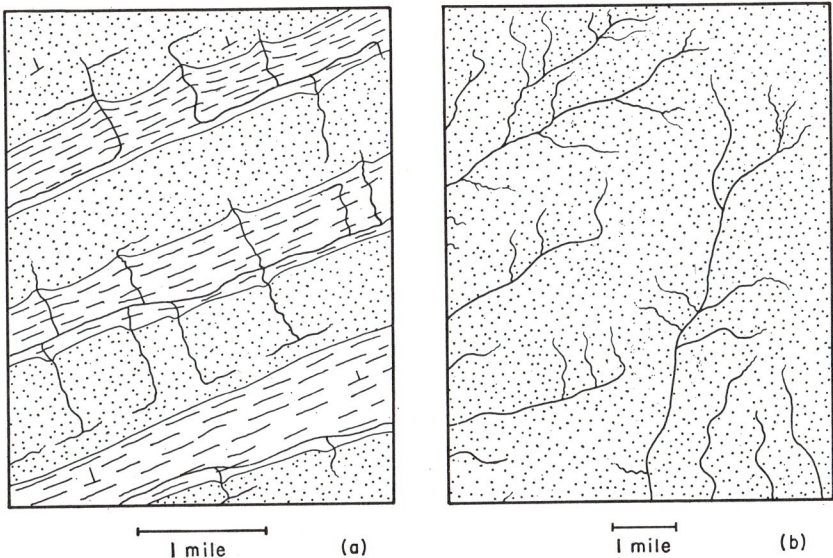

FIG. 4-I-1. (a) Trellis drainage pattern; (b) dendritic drainage pattern.

Igneous activity, such as lava flows, has also materially altered stream courses and formed lakes. Ice sheets and glaciers have affected stream and lake regimen, not only by damming and deflection, but also by supplying abundant meltwaters to the streams.

C. Soil and Topographic Effects

Soil and topographic development also affect stream runoff directly or indirectly. The most commonly noted effects on streamflow behavior are related to the lithologic character of the bedrock in a given area. For example, in some limestone and lava terranes, especially where the bedrock is cut by open fractures and solution channels or lava tubes and where the soil cover is permeable or poorly developed, a large fraction of the precipitation moves directly into the groundwater reservoir and eventually reappears as *base flow* (*sustained* or *fair-weather runoff*), and a very small fraction, if any, runs off. This results in a slightly fluctuating runoff pattern because down gradient the streams are fed from springs and seepage from the groundwater reservoir, and only a small fraction of the flow is flood or melt runoff. Figure 4-I-2 illustrates this effect in the Deschutes River Valley in Oregon, which drains a large area of basalt terrane and is said to be more stable than any other stream of its size in the United States [3, p. 11]. The hydrograph of the Deschutes River may be compared with that of the Yellowstone River in Montana (Fig. 4-I-2), a stream of comparable size in a somewhat comparable climate but flowing over a terrane of varied lithology and complicated structure. The high flood peaks and the relatively low base flow of the Yellowstone are characteristic of streams in such a terrane.

Another common geologically controlled characteristic of streams in limestone and lava terranes is that, for given basins, they may exhibit extremely low runoff relative to the precipitation in those basins or stream reaches, whereas in other basins or reaches the runoff may be far higher than is expectable from the precipitation. This condition often reaches an extreme where streams flowing strongly in the upper reaches of the drainage basin disappear entirely for long reaches, only to appear again as springs. Not infrequently these streams do not appear again within the

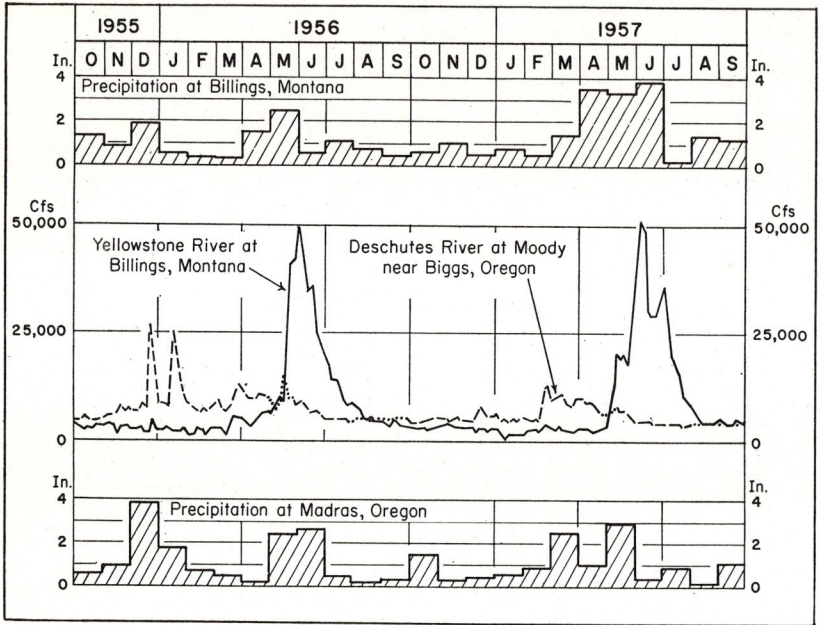

FIG. 4-I-2. Hydrographs of the Deschutes and Yellowstone Rivers, showing effect of terrane on distribution of annual discharge.

same or even adjacent surface-water drainage basins. These are "lost" rivers so commonly encountered in such areas as the Columbia Lava Plateau in Idaho, Oregon, and Washington and in the karst areas of Kentucky and Central Europe. Many studies in these and similar areas demonstrate that this anomalous distribution of runoff from basin to basin occurs because the surface and underground drainage basins do not match. In these regions the groundwater divides are much farther apart than topographic divides, and where most of the discharge from the area is groundwater, the runoff is low. In basins where the groundwater comes to the surface, streamflow is greatly increased and thus exceeds the expectable, or "normal," runoff.

Topographic basins containing large outcrop areas of confined aquifers of broad regional extent also may have "abnormally" low runoff. Again the cause may be related to disparity between topographic and groundwater divides. In this case water from precipitation either enters an aquifer directly or infiltrates from streams. It is then "piped" by the aquifer to long distances away from the basin in which it fell. Examples of streams that lose water in this manner are numerous. Typical ones are the eastward-flowing rivers in the Black Hills region of South Dakota [4] and many streams issuing from the mountains onto highly permeable alluvial aprons in the valleys of the Basin and Range Province in western United States.

Such gross effects on runoff as have been described are nearly all related to specific geologic conditions in more or less monolithologic terranes. More common and widespread effects of geologic and topographic controls on annual runoff may be found in the interplay between them and the climatic controls in regions underlain by heterogeneous lithologies of varying permeability and with varying depths and types of soil and vegetal cover.

In general, the amount of water from precipitation which may be lost by evapotranspiration depends in part on the amount that infiltrates deeply enough to be out of reach of the atmosphere and plants. In areas where highly permeable materials crop out, a large fraction of the precipitation infiltrates quickly, percolates to the groundwater reservoir, and finally discharges into streams. This process modulates and eventually increases the amount of water available for runoff from the area by diverting water that might otherwise be discharged by evapotranspiration. Detailed studies have been made in several areas, including the sand hills of Nebraska [5], some dune-sand regions in Kansas, beach and dune sands in the vicinity of Coos Bay in Oregon, and glacial-drift sands and gravels on Long Island, N.Y. These studies demonstrate this phenomenon, as do the examples of volcanic and limestone terrane previously cited.

Runoff may also be augmented by the effect of highly impermeable surface materials, especially in mountainous terrane. Here the combination of steep slopes and low permeability results in rapid runoff of a large fraction of water from precipitation before enough time has elapsed to allow much water to be lost by evapotranspiration. The runoff pattern is erratic, the flow being strong during flood periods and dropping off sharply, or even ceasing, when precipitation does not occur. Troxell's studies in southern California [6, pp. 104–109] demonstrate that, as expected, the least permeable terranes yield higher runoff for given amounts of precipitation than the most permeable terranes. In most instances augmentation of runoff in this manner is probably never as beneficial as runoff augmented as a result of permeable terranes, because most of the runoff occurs as floods immediately following the storms. Such floods are difficult to control and cause considerable damage. Further, it is difficult to put appreciable amounts of the floodwater to beneficial use. In the permeable terranes, temporary storage of storm waters in the groundwater reservoirs allows for slower release and increased base flow of streams with only little to moderate flooding, a condition subject to beneficial use and efficient management. However, in permeable terranes where evapotranspiration opportunity is great, a large fraction of the water from precipitation infiltrates, but is diverted from runoff by evapotranspiration. Troxell found this to be true in his studies.

Extremely high runoff is well illustrated in mountainous areas where relief is great and the terrane is composed of highly indurated rocks or other impermeable cover. Extremely low runoff occurs in broad, relatively flat areas of poorly developed drainage such as the High Plains region of Texas, New Mexico, and Oklahoma. Here shallow depressions in the relatively permeable alluvial cover collect the rainfall, which then mostly evapotranspires. Only a fraction of an inch of runoff and groundwater recharge occurs, and most of the latter is discharged by evapotranspiration at the edges of the plateau. In one large part of the region nearly 21,000 sq mi is considered as contributing no surface outflow [3, p. 11]. Langbein [3, plate 1] shows that the annual runoff from all the High Plains south of the North Platte River is less than 0.25 in., perhaps about one-fourth of what might be expected from the climatic characteristics of the area.

Generalization regarding the interplay of geologic and climatic controls that influence runoff and regimen is difficult since only a few detailed quantitative studies, mostly on small drainage basins, are available. Much quantitative and extended geologic, pedologic, and hydrologic research is needed before the broad variations in runoff that are known to exist in the various terranes can be adequately explained and interpreted for efficient application. In any case, analysis and prediction of runoff regimen and other characteristics of streams should consider the effects of geologic controls as fully as possible.

III. GEOLOGIC FACTORS IN STUDIES OF SOIL MOISTURE AND VADOSE WATER

The soil surface is the interface between the atmosphere and the earth's crust, and the future course of water falling on it depends primarily upon its physical nature. Thus the fraction of the precipitation that goes into streams, that enters the soil, and that passes through the soil (vadose water, Figs. 13-1 and 13-3) into the groundwater reservoir depends in large part upon the topographic configuration, composition, permeability, moisture content, and plant cover of the soil. The chemical nature of water is also conditioned by the soil and parent material, usually by the addition of humic acids and by solution of various more or less easily soluble mineral salts, especially calcium carbonate. Storage of water in the soil (soil moisture) depends primarily upon porosity and is of tremendous importance to man since the soil-water reservoir supplies the largest fraction of fresh water that benefits man, that is, the water used in the production of nonirrigated crops.

Some of the effects of the soil and underlying parent material on the movement, chemical character, and storage of water are mentioned in, or are apparent from, the following discussion. Geologic factors in soil development and water movement in the zone of aeration are also discussed.

A. Geologic Factors in Soil Development

The geologic factor most influential in soil formation is the nature of the parent materials. For many years before the recognition of the climatic effects in soil development, the geologic nature of parent materials was the sole basis used for classification of soils. The reason for this is self-evident if one examines a commonly accepted developmental series such as

$$\text{Rock} \rightarrow \text{weathered rock} \rightarrow \text{immature soil} \rightarrow \text{mature soil}$$

or if one considers soil formation in a monolithologic terrane where relatively uniform climate and vegetation prevail. With few exceptions it appears that immature soils and soils developed in arid or frigid zones will most closely reflect the character of the parent material. Under conditions of severe weathering and leaching and in mature soils the effects of the parent material will be least evident. However, many exceptions to this occur in nature. For example, in terranes underlain by rocks of high feldspar content, the clay content in mature overlying soils is also high and varies about as the feldspar content of the rocks. In contrast, soils developed on predominantly sandy terranes contain little clay and remain sandy even under the most rigorous climatic development. Further, the constituents of a soil may not be identical with those of its parent material, but the nature of the latter may still strongly and characteristically influence the type of soil that develops on it. For example, soils formed in carbonate rock terranes are closely related to the impurities in the parent rock because the limestone is more or less readily dissolved and carried away. Thus the sand, silt, and clay fractions, approximating less than 10 per cent in most limestones, make up the "parent material" for the soil. Further, the presence of acid-neutralizing, alkaline-earth carbonates retards soil-forming processes. Until the carbonates are removed the soil profiles develop slowly, if at all. As a result of these and possibly other factors, limestones tend to produce immature soils that are surprisingly alike from region to region. Also, topographic conditions in most karst areas are not conducive to widespread thick-soil accumulation. Thus, in many carbonate terranes the soil is thin, immature, and infertile.

B. Soil Profiles and Their Effect on Water Movement

The term *soil* is used here in the genetic sense [7, pp. 1-20; 8, pp. 887-896] to mean the near-surface layer of rock material, ranging in thickness from a few inches to

20 or even 30 ft, that has been modified to its present form by the interaction of weather, plants, and animals on existing rock materials. The process of soil formation is a dynamically evolving system wherein the shallow rocks of the earth are converted by disintegration, decomposition, eluviation, deposition, compaction, and alteration. Soil evolution continues until the surface materials are buried or eroded away. The active processes occur at the surface, and their effect is downward. Thus more or less strongly developed horizontal layering results in what is commonly referred to as the *soil profile*. Since the effect of continuing processes is progressively downward, "normal" soil zones, or *horizons*, as they are called by pedologists, are extended progressively deeper, unless surface erosion keeps pace with the development. (See also Sec. 5.)

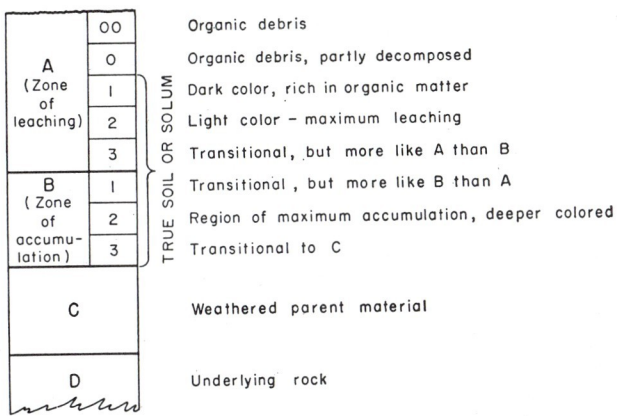

Fig. 4-I-3. Normal or zonal soil profile. (*Modified from Rice and Alexander* [8].)

Figure 4-I-3 is a diagram of a normal, or *zonal*, soil profile, showing the commonly developed horizons. The *A-horizon*, being at the surface, is the most intensely weathered, with the more soluble minerals leached away and most other minerals altered. Usually, the A-horizon is darker because the mineral matter is mixed with organic substances. Structurally the A-horizon is friable, granular, or platy. When friable or granular, the A-horizon is permeable, usually much more so than the underlying *B-horizon*, which is, most frequently, a zone of clay accumulation. Although some clay is produced by weathering of minerals in place, part of it is moved downward from the A-horizon by vadose water. The B-horizon generally displays vertical structure in widely or closely spaced joints that may bound columnar blocks. Most commonly, the B-horizon exerts the greatest influence on the movement of water in the soil profile. When the clay is dry, the vertical joints allow relatively rapid downward movement, but when it is wetted, the clay expands, closing the joints so that this zone may become essentially impervious and water may move only as a result of capillary forces, if at all. The B-horizon may become such an effective barrier to downward movement that the A-horizon becomes saturated, at least in part, for appreciable periods.

The horizon of only slightly altered original, or "parent," material subjacent to the B-horizon has been called the *C-horizon*. In some places carbonates are known to have accumulated in this zone, and the rocks may be more or less oxidized. Usually it contains less carbonate and is more permeable than the B-horizon. Therefore movement of vadose water reaching the C-horizon from above is not impeded, and the water passes downward in the section.

When calcium carbonate accumulates in the lower part of a soil profile, the soil is commonly called a *pedocal*. Although such accumulation may take place to a degree that it impedes gravity movement of water, usually it may be interpreted that the quantity of water moving downward was not sufficient to carry its load of mineral material to the zone of saturation. Alumina and iron-rich soils that lack calcium carbonate enrichment are often called *pedalfers*.

In addition to zonal soils, two other major categories are *intrazonal* and *azonal soils*. *Intrazonal soils* develop as a result of predominance of the influence of one genetic element which has distorted the profile and prevented the development of normal horizons. Often this predominating element is climatic; frequently it is the amount and movement of water; and sometimes it is the nature of the parent material. Where the downward and lateral drainage of water has been retarded or prevented, *hydromorphic*, or *humic-gley*, *soils* develop. When the parent material is predominantly clay under semiarid or arid conditions, the resulting soil is often classed as intrazonal. *Azonal soils* are usually thin, without distinct layering, and result from special climatic conditions. Included with them are the *lithosols*, which consist of a thin, granular, incomplete cover, barely thick enough to support growth of some plants, overlying indurated or semi-indurated rocks. Most intrazonal soils are more or less clay-rich and possess low permeability. As a result they retard or stop water movement downward but may store considerable quantities, much as the B-horizon does. The thin granular azonal soils have little effect on the movement or storage of vadose water.

C. Secondary Cementation and Water Movement

In some areas, especially in arid and semiarid regions, zones of cementation in the lower part of the B-horizon or upper part of the C-horizon have a strong influence on downward water movement. When these zones are cemented by calcium carbonate, they have been called *hardpan, caliche, mortar beds,* and locally by other names. Such a secondary cementation of soils and other unconsolidated materials range in appearance from isolated or disconnected nodules, to *stringers* and *pipes* through networks of interconnected streaks or zones (honeycomb or boxwork), and to more or less uniform, densely cemented zones. The content of secondary cement may range from a few per cent to over 50 per cent of the deposit.

Genetically, these concentrations of calcium carbonate at or near the land surface in arid and semiarid regions can be placed in three categories:

1. Those developed as a result of evaporation of groundwater in the shallow soil or at the surface. Usually this occurs in deserts where capillary action may draw groundwater from the water table or from perched water bodies to, or near to, the surface, where it evaporates, leaving a secondary deposit of cement. Deposits of this type are probably of only local importance, but where present, they do impede movement of water. Although the calcium carbonate may be derived from the soil, these are not wholly soil phenomena, since precipitation of the cement depends upon the position of the water table, varying permeabilities, and other factors. This category is difficult to define and may include materials of different origins, but in any case it includes the so-called *groundwater*, or *water-table, cements*.

2. *Soil caliches*, that is, calcium carbonate concentrations associated with the lowermost part of the B-horizon and the upper part of the C-horizon in a developed, zonal soil profile. These caliches may range from a few inches up to several feet thick, occur as continuous zones, and be densely cemented in the upper part. Typically, they are sharply limited at the top by a crenulate plane separating the zone of calcium carbonate accumulation from the leached material above. They are gradational at the base. Such zones may retard downward movement of vadose water, but are properly regarded as a reflection of insufficient quantities of water moving downward to remove the mineral matter during the development of the soil profile. The carbonates are thus taken into solution near the surface. Water losses by evapotranspiration cause reprecipitation of most of the carbonate at a deeper position.

3. Derivatives of first-cycle soil caliches described above. During climatic changes, either wetter or drier than a former interval of equilibrium conditions, the friable A-horizon may be removed, exposing a zone of normally developed caliche. When

solution and reprecipitation of the exposed calcium carbonate occur, the physical character of a former soil caliche is radically modified, generally in the direction of increased density and decreased permeability. Modified caliches of this type may pass through several cycles of climatic change and result in dense limestones of very low permeability. They may be near-surface or exposed units, such as the Ogallala *caprock* of western Texas, or buried in a sequence of alluvial deposits, such as commonly found in the Great Basin (Utah, Arizona, and Nevada).

In humid climates, calcium carbonate does not normally accumulate in the soil profile because the quantity of vadose water moving downward is adequate to carry dissolved mineral matter to the zone of saturation. However, in some pedalfer soil profiles secondary accumulations of mineral matter, usually iron compounds, do occur, usually low in the profile and commonly within the C-horizon. Common names applied to these accumulations include *Ferrito zone*, *hardpan*, and *iron zone*. Observers in the United States report such accumulations more often from buried soils than in actively developing profiles and indicate that they are related to zones of higher permeability in the parent material. These zones of iron accumulation do not materially affect water movement except in limited areas. They are undoubtedly related to the occurrence of bothersome amounts of iron in groundwater in some places.

D. Summary

The effects of the soil and surficial geologic formations on water movement and distribution include the following:

1. The amount of vadose water infiltrating the land surface and percolating to the water table varies directly with the permeability, composition, and structure of the soil. In general, strongly profiled and thick soils allow much less water to percolate deeply. In humid climates much, and sometimes all, of the water entering such soil is lost to evapotranspiration and runoff, at least during the growing season.

2. In contrast, granular soils allow more rapid and deep percolation and usually store less water for plant use; thus runoff and evapotranspiration are reduced and groundwater recharge is increased.

3. Especially in arid climates, caliche zones may form and impede deep percolation of large quantities of water. These may be more or less widespread or only of local nature. In humid climates, *ironstone* hardpans may have the same effect locally.

Quantitative methods to determine the effect of soil in hydrogeologic studies have not been widely developed and are available for only a few specially studied areas. In general, the effect of soil is determined indirectly by other quantitative methods used to determine the amount and distribution of stream losses, infiltration, and groundwater recharge (Secs. 5, 12, and 13).

IV. GEOLOGIC FACTORS IN GROUNDWATER STUDIES

A. General

Groundwater occurs in the large reservoir beneath the water table. It more or less saturates the earth materials through which it is moving and in which it is stored. Movement of groundwater is from areas of greater to areas of lesser hydrostatic head. The prime moving force is gravity, that is, the weight of the water column, but other forces, such as natural-gas pressure, are known to contribute locally to hydrostatic head. By far the largest fraction of fresh water available for water supply is stored in and transmitted through the groundwater reservoir, although only about one-seventh of the total fresh water used in the United States is withdrawn directly from it.

Occurrence, movement, and storage of groundwater are influenced by the sequence, lithology, thickness, and structure of rock formations. Movement and storage capacity are chiefly controlled by permeability and porosity. An *aquifer* is a lithologic unit (or combination of such units) which has an appreciably greater transmissibility than adjacent units and which stores and transmits water commonly recoverable in economically usable quantities.

The lithologic units of low permeability which bound the aquifer are commonly called *confining beds*, or *aquicludes*. Appreciable quantities of water may move through confining beds, and when this occurs, these beds have been referred to as *aquitards*. An aquifer or a combination of aquifers and confining beds that comprise

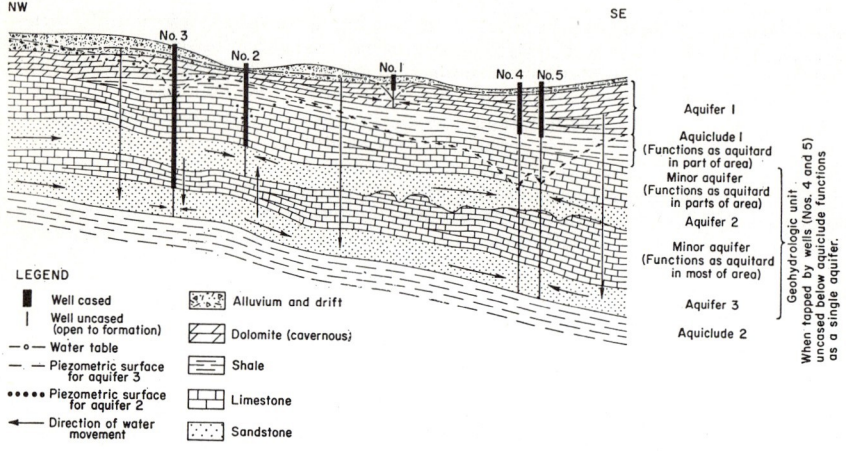

FIG. 4-I-4. Aquifers, aquicludes, and geohydrologic units.

a framework for a reasonably distinct hydraulic system may be considered as a *geohydrologic unit* (Fig. 4-I-4). An impermeable formation neither containing nor transmitting water, such as solid granite, is known as an *aquifuge*.

B. Nature and Occurrence of Aquifers

The chief function of a hydrogeologist is to locate, describe, and analyze aquifers and geohydrologic units. He should understand the character, distribution, and extent of aquifers and the stratigraphic and geographic occurrence of groundwater.

The most valuable aquifers have long been recognized, and most of them have been utilized as sources of water supply for many years. Meinzer [9, 10] was the first hydrologist to describe the occurrence of groundwater and water-bearing formations in the United States. He outlined 21 groundwater provinces in a comprehensive classification based upon aquifers composed of glacial drift, valley fill of the western basins, Tertiary lava, Miocene and Pliocene formations of the interior, Tertiary formations of the coastal plains, Cretaceous formations, Paleozoic formations, and pre-Cambrian and other crystalline rocks. The descriptions of the water-bearing formations arranged according to geologic age as well as to stratigraphic and geographic position have rendered Meinzer's work the most comprehensive and detailed discussion of the subject available in modern scientific literature. More recent contributions include a map of groundwater areas in the United States by Thomas [11, p. 32] on which he delineates aquifers according to whether they are associated with water courses in which groundwater areas can be replenished by perennial streams (primarily unconsolidated alluvial channel deposits), consolidated rock aquifers, and unconsolidated or semiconsolidated rock aquifers—all of which can be expected to yield 50 gal/min or more of water containing less than 2,000 parts per million of dissolved solids. Thomas also [12, pp. 14–73] published maps of descriptions of 10 groundwater regions, which he delineated by modifying Meinzer's 21 provinces.

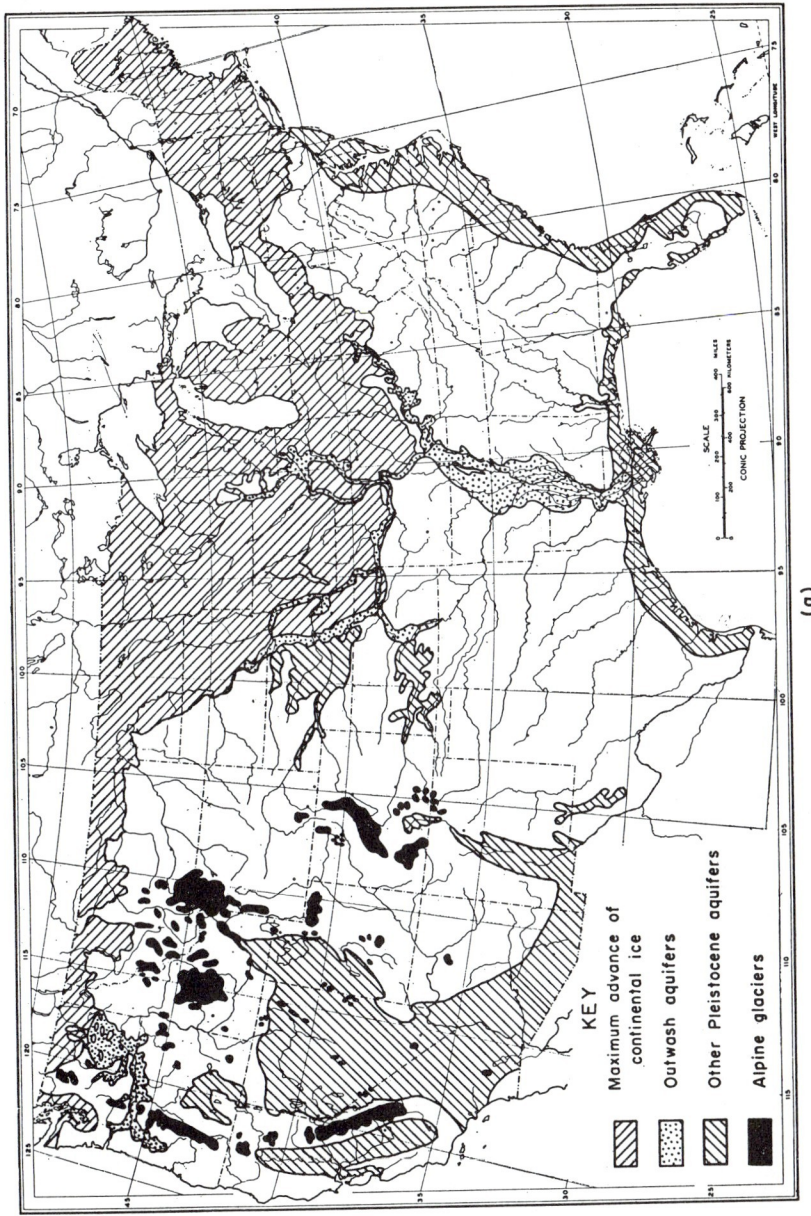

FIG. 4-1-5. (a) Quaternary aquifers in the United States. (*Modified from Meinzer* [9].)

Figures 4-I-5a and b and Table 4-I-1 show the occurrence of various aquifers in the United States. This information is derived mainly from the works by Meinzer and Thomas.

Most of the groundwater withdrawn by pumping wells in the United States comes from the following four distinctive geologic terranes:

1. Quaternary and Pliocene deposits of unconsolidated or poorly consolidated sand and gravel commonly interbedded with finer clastic rocks and some, but very few, carbonate units.

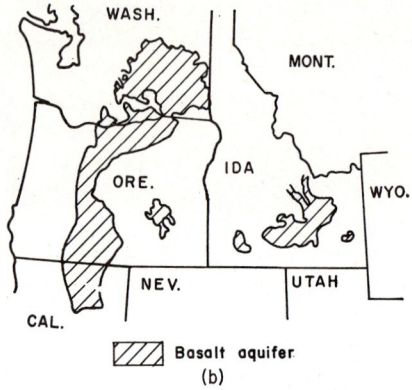

Fig. 4-I-5. (b) Outcrops of basaltic aquifer in the northwest part of the United States. (*Modified from Meinzer* [9].)

2. Semiconsolidated and consolidated conglomerate and sandstone formations which are often jointed and therefore possess both primary (intergranular) and secondary (fracture) porosity and permeability.

3. Limestone and dolomite formations possessing joints and other fractures, which have nearly always been enlarged by solution of the rock, and possessing little intergranular porosity and permeability.

4. Volcanic rocks, mostly fractured and jointed basalts, possessing little intergranular porosity and permeability. Some porosity and permeability result from primary structures such as vesicles and tubes, but often porosity and permeability are secondary.

Figures 4-I-5a and b show in a general way the distribution of two of these terranes in the United States and indicates their geologic age. More detailed discussions of the various rock types are given below.

C. Permeability and Porosity

The open spaces, voids, or interstices in earth materials comprising geohydrologic units are the receptacles that store and transmit ground water. The size, type, shape, and arrangement of the voids are the chief factors controlling the storage capacity and transmissibility of the earth materials. Several types of rock interstices and the relation of rock texture to porosity are shown in Fig. 4-I-6. They may be classified into two general categories:

1. Intergranular pores, exemplified by A to D, which are characteristic of the clastic rocks, both consolidated and unconsolidated.

2. Pores resulting from joints and other secondary openings (Fig. 4-I-6, E and F) in rocks. These may be subdivided into two groups:

    a. Crystalline, igneous and metamorphic rocks, some relatively impermeable carbonate rocks, indurated clastic rocks and some fine-grained clastic rocks such as shales, all of which may transmit and store water in joints and other open fractures.

b. Jointed and bedded rocks that are subject to solution and therefore possess solution channels developed primarily along fractures and bedding planes. These include the carbonates and evaporites.

Figure 13-2 illustrates empirically the porosity and Fig. 13-8 the permeability for various unconsolidated clastic materials. These diagrams show that the better aquifers among aggregates consist of clean coarse sands and mixtures of sand and gravel and fine-grained clastics (fine sands, silts, clay). Various clastic mixtures containing considerable amounts of silt and clay comprise the poorer aquifers. Thus, even though fine-grained clastics frequently possess high porosity, they transmit little water and therefore are poor aquifers. In many instances interbedded fine-grained

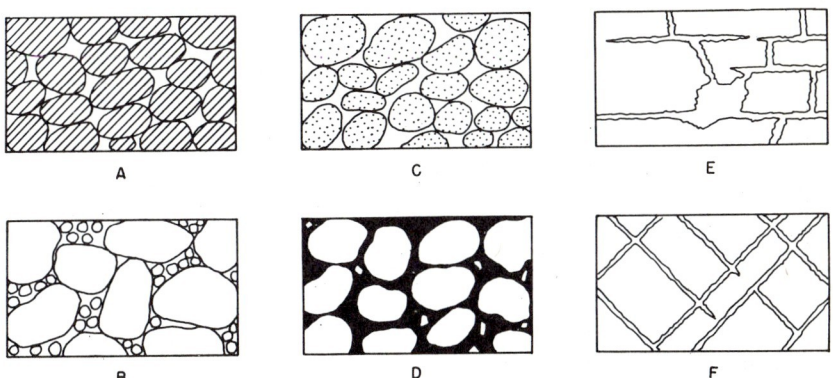

Fig. 4-I-6. Types of rock interstices. (*A*) Openings between relatively impermeable particles; (*B*) openings between relatively impermeable particles of different sizes; (*C*) openings between permeable particles made up of vastly smaller grains; (*D*) openings between grains reduced by cementation; (*E*) openings formed by solution along joints and bedding planes in carbonate rocks; (*F*) fractures in crystalline rocks. (*Modified from Meinzer* [9].)

and coarse-grained layers, as might occur in glacial terranes, form valuable aquifer systems, because requisite permeability is supplied by the coarse sediments, and storage is available in the highly porous fine-grained sediments which may drain into the coarser beds when the head is lowered.

For consolidated and semiconsolidated rock aquifers, porosity and permeability data that are reasonably applicable to any aquifer as a whole are difficult to find. However, many determinations have been made of the transmissibility and storage coefficients of such aquifers and of the specific capacity of wells penetrating them. The *coefficient of transmissibility* (Sec. 13) is a measure of the permeability. If expressed in English units (gallons per square foot per day), it may be converted to permeability in Meinzer's units (the *field coefficient of permeability* defined as the flow of water in gallons per day through a cross section of aquifer one foot thick and one mile wide under a hydraulic gradient of one foot per mile at field temperature) by dividing it by the thickness of the aquifer in feet. The *coefficient of storage* (Sec. 13) is defined as the volume of water, measured as a fraction of a cubic foot, released from storage in each column of the aquifer having a base of one square foot and a height equal to the full thickness of the aquifer, when the head is lowered one foot. Thus, the coefficient of storage is a measure of the porosity when the aquifer is unconfined. In any case, it gives an approximation of the available storage in the aquifer.

Another measure of permeability commonly used is the *specific capacity* of a well. This is expressed in gallons of yield per foot of drawdown when the well is being pumped. Although many factors other than permeability enter into these measurements, specific capacity is often indicative of the relative permeability of the aquifer from which the well withdraws water.

Table 4-I-1. Occurrence of Aquifers in the United States

(1) Geologic age and rock type	(2) Western Mountain Ranges	(3) Arid Basins	(4) Columbia Lava Plateau	(5) Colorado Plateau	(6) High Plains	(7) Unglaciated Central Region	(8) Glaciated Central Region	(9) Unglaciated Appalachian Region	(10) Glaciated Appalachian Region	(11) Atlantic and Gulf Coastal Plain	(12) Special comments
Cenozoic											
Quaternary											
Alluvium and related deposits (primarily Recent and Pleistocene sediments and may include some of Pliocene age)	S and G deposits in valleys and along stream courses. Highly productive but not greatly developed—P to M.	S and G deposits in valleys and along stream courses. Highly developed with local depletion. Storage large but perennial recharge limited—P.	S and G deposits along streams, interbedded with basalt—I to M.	U	S and G along water courses. Sand dune deposits—P (in part).	S and G along water courses and in terrace deposits—I (limited).	S and G along water courses—M.	S and G along water courses and in terrace deposits. Not developed.	S and G along water courses and in terrace deposits. Not developed.	S and G along water courses and in terrace and littoral deposits, especially in the Mississippi and tributary valleys. Not highly developed in East and South. Some depletion in Gulf Coast—I.	The most widespread and important aquifers in the United States. Well over one-half of all groundwater pumped in United States is withdrawn from these aquifers. Many are easily available for artificial recharge and induced infiltration. Subject to salt-water contamination in coastal areas.
Glacial drift, especially outwash (Pleistocene)	S and G deposits in northern part of region—I.	S and G deposits especially in northern part of region and in some valleys—I.	S and G outwash, especially in Spokane area—I.	U	S and G outwash, much of it reworked (see above)—I.	S and G outwash especially along northern boundary of region—I.	S and G outwash, terrace deposits and lenses in till throughout region—P (in part).	S and G outwash in northern part. Not highly developed—M.	S and G outwash, terrace deposits, and lenses in till. Locally highly developed—I.	S and G outwash in Mississippi Valley (see above)—I.	
Other Pleistocene sediments	Alluvial Fm and other basin deposits in the southern part—M to P (see Alluvium above).		U	U	Alluviated plains and valley fills—M to I.		U	U	U	Coquina, limestone, sand, and marl Fms in Florida—M.	

Era											
Tertiary Sediments Pliocene	S and G in valley fill and terrace deposits. Not highly developed—M.	Some S and G in valley fill—M.	U	U	Ogalalla Fm in High Plains. Extensive S and G with huge storage but little recharge locally. Much depletion—P (in part).	U	U	Absent	Absent	Dewitt Ss in Texas. Citronelle and LaFayette Fms in Gulf States—I.	Aquifers in coastal areas subject to salt-water encroachment and contamination.
Miocene	Ellensburg Fm in Washington—I; elsewhere—U.	U	Ellensburg Fm in Washington—I; elsewhere—U.	U	Arikaree Fm—M.	Arikaree Fm—M.	Flaxville and other terrace deposits. S and G in northwestern part—M.	Absent	Absent	New Jersey, Maryland, Delaware, Virginia—Cohansey and Calvert Fms—I. Delaware to North Carolina—St. Marys and Calvert Fms—I. Georgia and Florida—Tampa Ls, Alluvium Bluff Gp, and Tamiami Fm—I. Eastern Texas—Oakville and Catahoula Ss—I.	
Oligocene	U	U	U	U	Brule clay, locally—I; elsewhere—U.		U	Absent	Absent	Suwannee Fm, Byram Ls, and Vicksburg Gp—I.	

Table 4-1-1. Occurrence of Aquifers in the United States (Continued)

(1) Geologic age and rock type	(2) Western Mountain Ranges	(3) Arid Basins	(4) Columbia Lava Plateau	(5) Colorado Plateau	(6) High Plains	(7) Unglaciated Central Region	(8) Glaciated Central Region	(9) Unglaciated Appalachian Region	(10) Glaciated Appalachian Region	(11) Atlantic and Gulf Coastal Plain	(12) Special comments
Eocene	Knight and Almy Fm in southwest Wyoming —M.	U	U	Knight and Almy Fm in southwest Wyoming, Chuska Ss, and Tohatchi Sh in northwest Arizona and northeast New Mexico —M.	U	Claibourne and Wilcox Gp in southern Illinois (?), Kentucky, and Missouri—M; elsewhere—U.	Absent	Absent	Absent	New Jersey, Maryland, Delaware, Virginia—Pamunkey Gp—I. North Carolina to Florida—Ocala Ls and Castle Hayne Marl —P (in part). Florida—Avon Park Ls., South Carolina to Mexican border, Claibourne Gp, Wilcox Gp—I.	Includes the principal formations (Ocala Ls, especially) of the great "Floridian" aquifer. Subject to salt-water contamination in coastal areas but source of largest groundwater supply in southeastern United States.
Paleocene	U	U	U	U	Ft. Union GP —M.	Ft. Union Gp —M.	Ft. Union Gp —M.	Absent	Absent	Clayton Fm in Georgia —I.	
Volcanic rocks, primarily basalt	U	Local flows— M.	Many interbedded basalt flows from Eocene to Pliocene —P.	Local flows— M.	Absent	Absent	Absent	Absent	Absent	Absent	

GEOLOGIC FACTORS IN GROUNDWATER STUDIES 4–17

Mesozoic						
Upper Cretaceous	U	Ss lenses in southern California—M; elsewhere—U.	U	Dakota Ss and other not clearly distinguishable Ss a notable source of water from Minnesota and Iowa to the Rocky Mountains and south into New Mexico; also in Utah and Arizona—I. In northwestern part of region Fox Hills and related Ss (Lennep, Colgate, etc.) locally valuable as water sources—M. Ss of Montana Gp—M. Ss members of Mesaverde Gp in Wyoming, Colorado, Utah, New Mexico, and Arizona—M. In Texas aquifers listed under col. 11—I.	U	U	New Jersey, Maryland, Delaware—Magothy and Raritan Fm—I. North and South Carolina—Peedee and Black Creek Fms—I. Alabama and Georgia—Ripley and Eutaw Fms—I. Tennessee, Kentucky, Illinois—McNairy Ss—I. Arkansas to Texas—Navarro Gp and Taylor Fm—I.	In coastal areas subject to saltwater encroachment and contamination. Ss aquifers of the Central Regions and the West primarily valuable when water from other sources is unavailable.
Lower Cretaceous	U		U	In northern part of these regions Lakota, Cloverly, and Kootenai Ss—M. In southern part Purgatoire and Dakota (?) Ss—M. Texas aquifers listed in col. 11—I.	U	U	Texas—Woodbine Ss—I. New Jersey, Maryland, Delaware—Patapsco and Patuxent Fms—I. West of Mississippi River, especially in Texas—Edwards Ls and Ss in Trinity Gp—I.	

Table 4-I-1. Occurrence of Aquifers in the United States (Continued)

(1) Geologic age and rock type	(2) Western Mountain Ranges	(3) Arid Basins	(4) Columbia Lava Plateau	(5) Colorado Plateau	(6) High Plains	(7) Unglaciated Central Region	(8) Glaciated Central Region	(9) Unglaciated Appalachian Region	(10) Glaciated Appalachian Region	(11) Atlantic and Gulf Coastal Plain	(12) Special comments
Jurassic	Locally—Ss Fm—M.	Locally—Ss Fm—M.	U	Ss Fms. Some may not be developed—I.	U	U	Absent	Absent	Absent	U	
Triassic	Locally—Ss and C Fms—M.	U	U	Ss and C Fms used locally. Shinarump C and correlatives give rise to springs —I.	U	U	Absent	Ss, C, jointed shale, and basalt beds of Neward Gp in Massachusetts, Connecticut, New Jersey, Pennsylvania, Maryland, Virginia, and North Carolina—M.	Absent	U	Water from Ss, C, and Ls Fms west of Mississippi River, especially valuable when water from other sources is unavailable.
Paleozoic Permian	U	U	U	DeChelly Ss —I. Kaibab Ls— M.	U	San Andres Ls in Roswell Basin—P. Quartermaster Gp gives rise to many springs—M. Other Ss and Ls in Kansas, Oklahoma, and Texas—M.	U	U	Absent	U	
Pennsylvanian	Tensleep Ss in Wyoming and other Ss elsewhere —M.	U	U	U	U	Ss and C beds from the Appalachians to Iowa and eastern Kansas—M to I.			Jointed and weathered Sh, Ss, and C in Rhode Island and Massachusetts —M.	U	

GEOLOGIC FACTORS IN GROUNDWATER STUDIES 4–19

Mississippian	Ls locally but little developed; springs arise from Ls in Rocky Mountains—M.	A few springs arise from Ls locally—U.	U	Some springs arise from Ls locally—U.	U	In Illinois, Iowa, Missouri, and Kentucky the Burlington, Keokuk, and St. Louis Ls—I. Some Ss (primarily Chester)—M. In Alabama and Tennessee—the Ft. Payne chert, Gaspar Fm, and St. Genevieve and Tuscumbia Ls—I. In Kentucky many springs arise in Ls.	U	Do.
Devonian	U	U	U	U	U	U, except locally in Michigan (Traverse Fm), Illinois, Missouri, Ohio (Columbia Ls), and Kentucky—M.	U	Jointed Ls, Ss, and Sh, some highly metamorphosed. M locally and little used.
Silurian	U	U	U	U	U	Ls and dolomite Fms in New York, Kentucky, Tennessee, Ohio, Illinois, and Iowa. Better-known aquifers include Monroe dolomite and related carbonate Fms in Ohio—I; "Niagaran" dolomite in Illincis—P (in part).	U	
Ordovician	U	U	U	U	U	In Arkansas, Missouri, Iowa, Illinois, eastern Indiana, southern Wisconsin, southeastern Minnesota, the St. Peter Ss—I. Overlying and subjacent Ls and Ss where present in above states and in Kansas, Oklahoma, and New York—M to I. In Kentucky and Tennessee—Ls Fm—M to I.	U	Locally Ls and Ss Fms; not highly developed—M.
Cambrian	U	U	U	U	U	Ss beds in Wisconsin, Minnesota, Iowa, and Illinois include Jordan Ss, "Dresbach Fm" (Galesville Ss, Eau Claire Fm, Mt. Simon Ss)—P (in part). Ls and Ss Fms in Missouri and Arkansas give rise to many large springs and yield water to many wells—P.	U	Ls Fms give rise to large springs in southern Appalachians. Otherwise—U. Eastern New York and New England Ss Fms—M; otherwise—U.

Table 4-I-1. Occurrence of Aquifers in the United States (Continued)

(1) Geologic age and rock type	(2) Western Mountain Ranges	(3) Arid Basins	(4) Columbia Lava Plateau	(5) Colorado Plateau	(6) High Plains	(7) Unglaciated Central Region	(8) Glaciated Central Region	(9) Unglaciated Appalachian Region	(10) Glaciated Appalachian Region	(11) Atlantic and Gulf Coastal Plain	(12) Special comments
Precambrian (including crystalline rocks which may be younger)	Weathered and jointed rocks. Locally—M.	Weathered and jointed rocks.	U	U	U	U	Weathered and jointed rocks locally in Minnesota, Wisconsin, northern Michigan, Piedmont Plateau, New England—M to I. Some Ss in North Central States.			U	Do.

Abbreviations: (1) aquifers: P, principal aquifer in region; I, important aquifer in region; M, minor aquifer in region; U, unimportant as an aquifer in region. (2) Rock terms: S, sand; Ss, sandstone; G, gravel; C, conglomerate; Sh, shale; Ls, limestone; Fm, formation; Gp, group.

In the following tables and discussion, values of the coefficients of transmissibility and storage and of specific capacity are given to indicate broadly the permeability and porosity or storage capacity of common aquifers. Table 4-I-2 gives the values for various aquifers composed of aggregates, for the most part unconsolidated sand or sand and gravel.

Table 4-I-2. Characteristics of Aquifers Composed of Aggregates

Material	Thickness, ft	Coefficient of transmissibility, gpd/ft	Coefficient of storage	Reference
1. Glacio-fluvistile deposits, Hanford, Wash.	45	3,000,000	0.20	[13]
2. Glacio-fluvistile deposits, Hanford, Wash.	30	380,000	0.06	[13]
3. Ringold Conglomerate	85	34,000	0.00007	[13]
4. Alluvial sand and gravel, Gallatin Valley, Mont.	26	100,000	0.006	[14]
5. Alluvial fan deposits, Gallatin Valley, Mont.	63	36,000	0.06	[14]
6. Sand and gravel outwash, Mattoon, Ill.	16	25,600	0.0015	[15]
7. Sand and gravel outwash, Barry, Ill.	36	119,000	0.003	[15]
8. Valley train sand and gravel, Fairborn, Ohio	80	280,000	0.0008	[16]
9. Glacial outwash, Bristol Co., R.I.	60	350,000	0.007	[17]
10. Glacial outwash, Providence, R.I.	55	150,000	0.20	[17]

Aquifer characteristics of some well-known indurated or semiconsolidated clastic-rock aquifers are shown in Table 4-I-3. The permeability and porosity of the semiconsolidated aquifers (items 3 to 7) result from intergranular openings in the rocks and are closely comparable with these characteristics to the aggregates. In other aquifers (items 1 and 2), the permeability and porosity result not only from intergranular openings, but also from fractures in the rocks. This is a common characteristic that has been demonstrated to exist in other well-known indurated-rock aquifers, such as the St. Peter sandstone.

Table 4-I-3. Characteristics of Aquifers Composed of Semiconsolidated and Consolidated Clastic Rocks

Aquifer	Thickness, ft	Coefficient of transmissibility, gpd/ft	Coefficient of storage	Reference
1. Cambrian-Ordovician sandstone (Wis.)	?–850	23,800 (av.)	0.00039 (av.)	[18]
2. Cambrian-Ordovician aquifer (Ill.)	?	17,400	0.00035	[19]
3. Carrizo sandstone	120	32,200	0.000138	[20]
4. Aquia greensand	20	10,000–20,000	0.00023	[21]
5. Patuxent formation	70	4,210	0.00037	[21]
6. Patuxent formation	14	450	0.00013	[21]
7. Patapsco formation	25	35,600 (av.)	0.0002 (av.)	[21]
8. Wissahickon formation (weathered material above the schist)	65–100(?)	3,000–10,000	0.002–0.01	[22]

Values for the porosity and permeability of fractured and cavernous rocks are somewhat more difficult to determine, primarily because they depend upon the gross nature and distribution of the interstices, both of which are extremely variable and sporadic in comparison with the relatively orderly distribution of pore spaces in the clastic rocks. Such rock types as metamorphic and igneous crystalline rocks and carbonate rocks are usually dense and nonporous. Since the volume of the fractures and caverns is very small in comparison with the mass of such rocks, the storage capacity is very small, even though they transmit large quantities of water.

Basalts and rhyolites often form highly productive aquifers when they are much jointed or cut by other fractures. These rocks usually have very low storage capacity, but frequently they are vesicular or cut by lava tubes and thus may store relatively large quantities of water. For example [23], in 11 aquifer tests in the Snake River basalt near Minidoka, in Mud Lake Basin; and at the National Reactor Test Site near Arco, Idaho, the coefficient of transmissibility ranged from 1×10^5 to 1.8×10^7 and averaged about 4×10^6 gpd/ft. In the same tests the coefficients of storage ranged between 0.02 and 0.06. Specific capacity data for 238 production wells ranged from 6 to 22,000 and averaged about 2,100 gpm/ft of drawdown. The average penetration of the wells was about 100 ft below the regional water table. However, not all basalt aquifers are so productive. For example, specific capacities of 128 wells in Washington and Oregon studied by W. C. Walton (personal communication) show a wide range from 0.1 to 4,200 gpm/ft of drawdown, but average only 121 gpm/ft, much less than the Idaho values.

Frequently, fine-grained clastic rocks (especially shales) which possess little intergranular permeability are jointed sufficiently to be more or less permeable and form locally important aquifers. Notable examples include the Brunswick shale in New Jersey, the Brule clay in parts of the Great Plains, and shales of Pennsylvanian age in the Central States.

Table 4-I-4 shows values of coefficients of transmissibility and storage determined for various carbonate aquifers in the United States. Although carbonate aquifers may have some primary porosity, the highly transmissible ones are usually jointed and the joints have been enlarged by solution. Thus transmissivity ranges widely from practically nothing in dense, poorly jointed, essentially insoluble carbonates to very large values in porous, much jointed, and cavernous formations.

Table 4-I-4. Aquifer Characteristics of Carbonate Rocks

Formation	Thickness, ft	Coefficient of transmissibility, gpd/ft	Coefficient of storage	Reference
1. Fort Payne chert (limestone)	20–40 to 95–360	4,800–1,360,000	0.0289–000454, 0.005 (av.)	[24]
2. Renault-St. Genevieve	125–175	126,000 (av.)	0.00029 (av.)	[25]
3. Tymochtee dolomite	225	8,000 (av.)	0.002 (av.)	[26]
4. Silurian dolomite	250	61,000	0.00035	[27]
5. Cockneysville marble (weathered to sand)	8	35,000		[22]

D. Movement and Storage of Groundwater

Detailed discussions of water storage and movement, methods of determining directions and rates of flow, and other quantitative information are given in Sec. 13. These discussions indicate that the common methods of determining the various quantities depend upon the use of only a few basic formulas and techniques, which include:

1. The Darcy formula and its derivatives
2. The Theis formula and its derivatives

3. Flow-net analyses based on graphical solutions or laboratory-model studies
4. Laboratory tests of porosity and permeability
5. Calculations based on water-budget studies

Frequently, various combinations of these methods are used either to complement or supplement one another or to check the results.

In order to apply the above methods effectively and accurately, an intimate knowledge of geologic factors is necessary. Thus the lithology, structure, sequence, thickness, and lateral extent of both aquifers and aquicludes must be known. Further, values of permeability and porosity must be determined (often by application of the same formulas and methods) or estimated before rates of flow and storage capacity can be ascertained.

The physical characteristics of rocks, including mineralogical and chemical composition, grain or crystal size, sorting, packing, and primary structures that make up the overall lithology have been previously mentioned in regard to their obvious geohydrologic significance. These are the features that most frequently determine the permeability and porosity of the aquifers, and hence the rate of movement and storage of water. Geologists have not yet developed quantitative methods of determination of aquifer characteristics based on these features, but useful approximations of permeability and porosity can be and are frequently made by visual examination; by geophysical methods such as calculations from potential, resistivity, radioactive, and sonic logs of boreholes; and by mechanical analysis of the sediments. A geologic factor of great importance in the determination of aquifer potential that is often overlooked or otherwise unaccounted for is the gradual lateral change in lithology. This change often occurs in aquifers and produces "boundaries" which critically affect the movement and storage of water as well as the yield from wells.

Structure of the rocks also has a strong bearing on the movement and storage of water. The function of secondary structures such as joints, bedding planes, and faults has already been discussed in connection with the occurrence of water in consolidated clastic rocks and in crystalline igneous, metamorphic, and carbonate rocks. Boundaries occur in aquifers as a result of faulting, bedding, nonconformity, and unconformity. The location, extent, and nature of such boundaries must be known in order to apply successfully the quantitative methods of determining aquifer characteristics and potentials.

Sequence of aquifers and aquicludes often critically determines the potential storage of the groundwater and the direction and rate of water movement in a given region. For example, in an area of heterogeneous lithology, several aquifers, separated by aquicludes of varying permeability and porosity, may be present. Although most of the water may move through the aquifers, some of it also permeates the confining layers and moves in the direction of the hydraulic gradient from one aquifer to another. Thus, depending upon the hydraulic gradient and the permeability of both the aquifers and the aquicludes, water may move in various directions (Fig. 4-I-4). Under some conditions (Fig. 4-I-4) water may not only recharge an aquifer from the land surface, but may also move from one aquifer to another. Further, the storage capacity of a sequence of aquifers may be considerably increased by the occurrence of water in the usually highly porous, though relatively impermeable, confining layers.

Both the lateral and vertical extent (thickness) of aquifers must be accurately known in the application of most quantitative methods of determining the storage and rate of movement of groundwater. Thickness is especially important in the determination of permeability of both aquifers and aquicludes. Lateral extent is of prime importance in ascertaining boundaries and extent and location of recharge areas.

Accurate knowledge of the geologic parameters is also necessary in the determination of permeability and porosity. By analysis of field pumping tests, the Darcy law, or formulas derived from it such as the Thiem (equilibrium) formula, is commonly used to determine the coefficient of permeability. Basic assumptions underlying these formulas are:

1. The aquifer is homogeneous and isotropic.
2. Flow in the aquifer is laminar.
3. The discharging well completely penetrates the aquifer and is infinitesimal in diameter.

In addition to these assumptions, the following must be made when using the Theis formula:
1. The aquifer is of infinite areal extent.
2. The aquifer is bounded by impermeable strata above and below.
3. The coefficients of transmissibility and storage are constant in all directions at all times.
4. Water is released instantaneously with decline in head.

Obviously, none of these assumptions are completely satisfied in natural earth materials. However, many of the resulting deviations from true values are negligible or fall within reasonable limits of error in ordinary geohydrologic studies. Most aquifers are known not to be homogeneous and isotropic, yet analysis of many pumping tests in many different aquifers have yielded values of both coefficients which have been checked agreeably by various means. On the other hand, numerous instances are known where inadequate values have been determined. Such values are inaccurate, probably because of lithologic heterogeneity within the aquifer. Laminar flow seems to, and may be expected to, occur under most conditions of groundwater flow. However, in jointed and cavernous rocks and in coarse aggregates turbulence undoubtedly occurs, especially under conditions of high hydraulic gradient as in the near vicinity of a well. Complete penetration of the aquifer by pumping wells is often difficult to attain under many prevailing field conditions. When only partially penetrating wells must be tested, corrections can be applied as explained in Sec. 13, provided the correct boundaries of the aquifers are known.

Mathematical corrections of values which are inadequate because of the heterogeneous lithology of the aquifer or because of turbulent flow have not come into common use. Knowing the geologic nature of the aquifer may allow recognition of the cause of the inaccuracy, but allows little help in compensating for it.

Lack of uniformity in the coefficients of transmissibility and storage in all directions in the aquifer results chiefly from anisotropic conditions and lack of homogeneity. Treatment of problems resulting from such nonuniformity has not been successfully developed. The resulting inaccuracy is often within the limits of error in practical problems, but inadequate values in some instances undoubtedly result from the nonuniformity. Again, accurate and detailed geologic data may allow recognition of the cause of the inaccuracy.

Most aquifers contain lenses or particles of silt and clay interspersed with coarser-grained particles. Further, the confining beds of the aquifer are made up predominantly of such materials. When the head declines in a confined aquifer as a result of pumping, water is released from storage primarily from these fine-grained sediments, which are far more compressible than the coarser aquifer materials. Since the silts and clays are not very permeable, the water moves slowly through them, not instantaneously as is demanded by the theoretic conditions of the Theis formula. With a decline in head, there is a slight expansion of the water, an instantaneous effect which is also expressed as an element of the coefficient of storage. If the aquifer is elastic and expands as an effect of head decline, a further instantaneous increment to the coefficient of storage would result. Thus, in pumping tests in very clean sandstone and in jointed limestone aquifers bounded by impermeable strata above and below, exceedingly small values of the coefficient of storage are determined, but they are relatively reliable. In aquifers where a relatively large fraction of the coefficient of storage results from the movement of water from fine sediments in the aquifer or in the aquiclude, the determined value of the coefficient may be in considerable error unless the pumping test continues for a long time. The accuracy of determination of the coefficient is further affected by the movement of water from one aquifer to another through the confining beds. When this occurs the coefficient of storage does not apply to the aquifer, but to the total thickness of the beds and to the contained water, which are affected by the head decline.

The assumption that aquifers are bounded by impermeable strata above and below is violated by most known instances in nature. Actually, the aquifers from which are withdrawn over half of the groundwater in the United States consist of interbedded sands and gravels. Most of these aquifers contain considerable clay and silt, either as layers or lenses, and are more or less confined by aquicludes of predominantly finer materials but with a wide range in permeability. Thus the methods of determination of the aquifer characteristics must allow for slow drainage from these finer-grained elements and for the movement of water from one aquifer to another. Relatively accurate methods have been described to compensate for these factors (Sec. 13), but they require the aid of more geologic information in greater detail, especially in regard to the nature of the confining beds and their thickness.

The withdrawal of large quantities of groundwater from storage in poorly consolidated and unconsolidated aquifers is known to result in land subsidence. This subsidence is believed to occur because decline in head has removed hydraulic support of overlying deposits, thus increasing the effective load and causing compaction of the more compressible materials in the section.

Subsidence studies have been conducted in considerable detail in several areas, the most notable being the San Joaquin Valley, California, and Mexico City, D.F., where economic damage at the surface has occurred and further damage is expected. In the San Joaquin Valley, subsidence on the order of 20 ft and affecting over 1,100 sq mi has been measured [28]. In Mexico City subsidence of about the same order of magnitude has been observed [29]. Subsidence is not limited to these areas and has been observed in many other places, including Las Vegas Valley, Nev. [30], Charleston, S.C., and London, England. Further, it is not confined to water-well fields, but occurs in oil fields, as at Houston and Goose Lake, Tex. [31], and at Long Beach, Calif. [32].

Much consideration has been given to the possibility that compaction of the sediments in and around the aquifers may damage them. In general, it seems that little material damage has occurred and that primarily a loss in storage capacity, as yet unevaluated, has resulted. Although the mechanism and effects of compaction are not yet fully understood, it seems doubtful that even in extreme cases the capacity of the aquifers to transmit water is materially affected, except in instances of collapse around individual wells.

The limitations and uses of geologic data as discussed above are not only applicable in studies involving the Thiem and Theis formulas, but also apply to other quantitative methods now in use or in experimental stages. For example, the application of flow nets, analogs, and statistical and numerical analyses requires the use of, and justifies the acquisition of, far more detailed and accurate geologic data than are commonly collected.

In nearly all groundwater investigations, contour maps of the piezometric surface of the water in one or more aquifers are constructed and analyzed to determine the general direction of flow and, when necessary data are available or can be estimated, the amount of water available under a given set of conditions. Normally, these maps are constructed by measuring water levels in wells and boreholes and plotting the values, adjusted to some datum, on maps of proper scale. Contour lines of equal elevation, that is, equipotential lines, are then drawn to reproduce the water surface under water-table conditions and the pressure surface under artesian conditions. Figure 4-I-7 illustrates two very simple water-table contour maps showing general conditions under different climatic environments. In both plan views the solid equipotential lines are crossed at right angles by arrows indicating the general direction of flow to be expected at the point of intersection.

When the thickness and permeability of the aquifer are known, the rate of movement and quantity of water moving through a given cross section can be determined by using Darcy's formula or one of several derivations from it (Sec. 13). The accuracy of values determined by this method is directly related to the accuracy and density of the control upon which the piezometric map is based, the accuracy of the permeability determinations, and the fundamental assumptions which must be made regarding Darcy's formula. If the rate of movement or quantity of water moving through a given cross section of the aquifer is known, apparent values of the permeability can be determined. Use of this method obviously requires accurate knowledge of the extent, thickness, and lithology of the geologic section. Also, if quantities of water are to be determined, the permeability of the aquifer must be known.

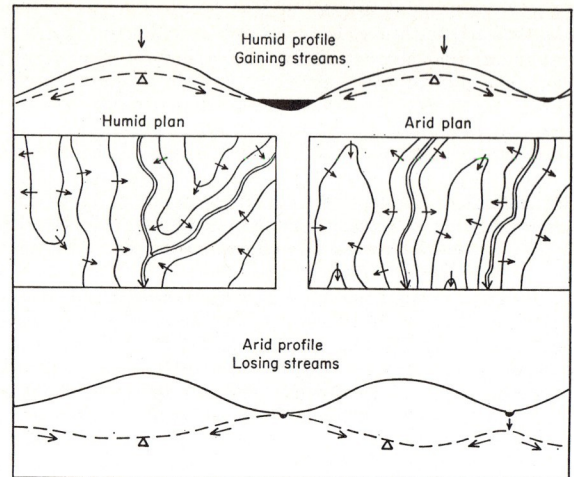

Fig. 4-I-7. Contour maps of water tables in arid and humid zones.

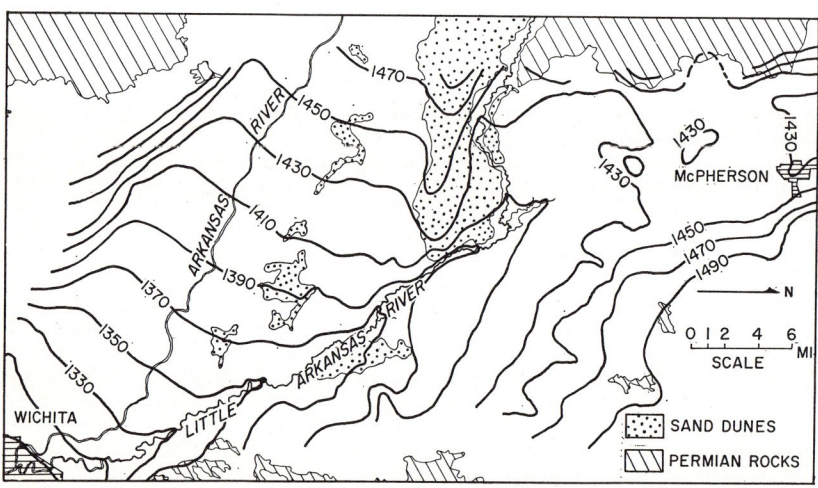

Fig. 4-I-8. Contour map of the water table in the vicinity of Wichita, Kans. (*After Williams and Lohman* [33].)

Frequently, piezometric contour maps show the effect of changes in lithology and other geologic features. For example, the permeable sand dunes in the north central part of Fig. 4-I-8 are clearly reflected by the water-table contours forming a groundwater ridge in that area. Also, the closely spaced contours in the southwest corner of the map reflect the lower permeability of the aquifer there.

E. Exploration for Groundwater

The usual objectives of exploratory groundwater investigations are to locate and evaluate new sources of groundwater or to enlarge withdrawals from known sources. Such studies require the location and delineation of aquifers, the determination of structural characteristics and lithology, and the determination of the quantity and quality of water that may be yielded by wells penetrating them. Often a knowledge of the hydrologic relationships between the water in the aquifer and the water on the land surface and in soils is also needed.

Various exploration methods have been devised and are commonly used to determine any one or more of these factors, but no one method is commonly adequate to evaluate or predict accurately all of them. Ordinarily the demands of a given investigation dictate which methods or method will be required for successful results. Thus, location and development of domestic or other low-yield wells might be confidently undertaken by a relatively simple geologic examination of the area in question

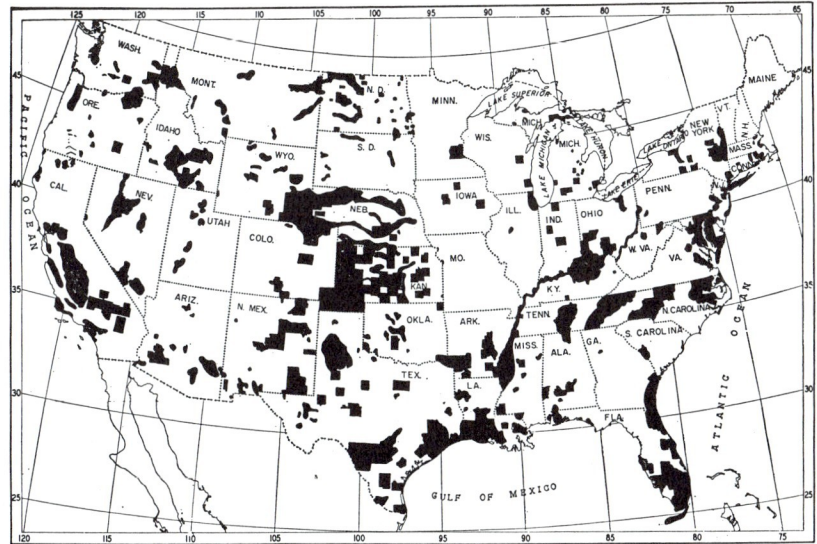

Fig. 4-I-9. Areas of substantial groundwater information in the United States, 1960. (*After LeMoreaux and Maxey* [34].)

or even by a study of the available geologic literature of the area if geologic and hydrologic conditions are favorable. In contrast, location and development of even low-yield wells might tax the efficiency of any or all known methods in poorly watered or more or less impermeable terranes.

The various methods may be most conveniently classified as to (1) whether they are conducted on the land surface and involve primarily surface phenomena or (2) whether they are primarily subsurface studies. Each class may then be subdivided into geologic, geophysical, and geochemical operations. Hydrologic methods by pumping tests and other analyses are discussed in Sec. 13.

1. Surficial Geologic Studies. *Surficial geologic study* is nearly always required for groundwater studies and is most logically initiated first because the results will indicate the need for the employment of other methods. The first step in geologic studies consists of the collection, analysis, compilation, and hydrogeologic interpretation of existing geologic maps, aerial photography, and other pertinent records. Some recorded geologic data are available for nearly any point in the United States, but the amount and detail of such data are highly variable and their value for groundwater studies varies accordingly. Figure 4-I-9 [34] shows in a general way the availability of groundwater information in the United States. Sources of geologic information may be obtained from publications, records, and personnel of the Federal Geological Survey and the various state geological surveys and universities and from publications of many scientific societies. Consulting geologists and engineers and well drillers frequently possess much valuable geologic information.

After existing photographs, maps, and geological data have been studied, further needed work can be evaluated. Depending on the adequacy of available information, the hydrogeologist may be required to conduct further geologic studies in the field and to make tests in the laboratory, including mapping, stratigraphic and structural work, and collection, examination, and analysis of samples. Concurrently with the office and field study it is usually possible and necessary to collect and incorporate into the study considerable hydrologic information including water-level and stream-flow measurements, observations and records of springs and wells, water analyses, and other geochemical and biological data. Frequently such information is lacking and facilities to measure and collect hydrologic records must be provided. If this is the case, such activities should take place early in the study so that records can accumulate while the geologic work is progressing.

Commonly, results of the geologic study will indicate within some limits the areas to be further explored and the most applicable methods of exploration to be used. Thus, areas in which subsurface data are lacking may be designated for test-drilling. Sometimes the nature of the terrane will be adaptable to application of geophysical methods which in turn might narrow down the area to be test-drilled.

2. Surficial Geophysical Methods. *Geophysical methods* conducted on the land surface that are most useful in groundwater exploration include electrical resistivity, seismic, and gravity surveys.

a. Resistivity Methods. Electrical-resistivity surveys have proved valuable to solve problems involving shallow aquifers. Normally, this method does not allow recognition of fresh-water bodies but it does indicate the presence and approximate thickness of formations that may be water-bearing and therefore should be test-drilled to prove the presence of usable water supplies. Where layers of fresh and brackish water or brackish or saline-water bodies occur, the method has been used successfully to delineate their thickness and lateral extent.

The principles underlying the use of electrical resistivity are relatively simple. Voltages between two electrodes in the earth are measured. The differences in electrical response between one pair of points and another reflect the differences in resistivity of the earth materials beneath them. The differences in resistivity must be due to some geologic cause. The values thus determined are not strictly the actual resistivity of the material being tested since under natural conditions many factors would influence the determinations. For this reason most workers refer to the values obtained in the field as the *apparent resistivity* (R_a). Three fundamental factors primarily affecting the value of apparent resistivity are:

1. Water saturation. In unsaturated materials the presence of hydroscopic, pellicullar, and other vadose water and in the saturated zone the presence of groundwater result in increased conductance of electric current. In general, the more water present in a formation, the lower the apparent resistivity.

2. Water quality. Many studies have shown that, as the ionic content of water increases, the apparent earth resistivity decreases.

3. Geologic factors. These include the amount and nature of the porosity, sorting, packing, and shape of particles, and presence of conductive solids such as certain clay minerals.

In field operations some variation of the Wenner electrode configuration is commonly used (Fig. 4-I-10). This scheme consists of four equally spaced electrodes in a straight line. Current is applied to the outside electrodes (C_1, C_2). The voltage drop between the inside electrodes (P_1, P_2) is then measured. The instrument most used to measure the voltage drop is some modification of one developed by Gish and Rooney [36].

With the Wenner configuration it is assumed that the horizontal distance between the electrodes is roughly equal to the depth of penetration, an assumption that is satisfactory for most practical applications. For a more exact determination of depth to a resistivity boundary, theoretical interpretations such as curve matching or Tagg's method [37] should be applied.

Two procedures may be followed in the use of the Wenner configuration:

1. A series of measurements may be taken using different distances between electrodes (expanding electrode-separation method), thus yielding a series of values of apparent resistivity with depth. From these data a *depth profile* can be drawn which may be used as an indication of vertical geologic changes.

2. A series of measurements at a constant electrode spacing may be taken at closely spaced intervals along the line of a traverse, thus giving indications of lateral changes in geology at a constant depth.

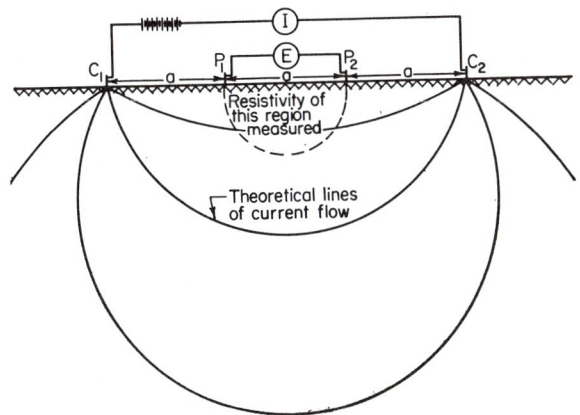

FIG. 4-I-10. The Wenner four-electrode configuration. (*After Hackett* [40].)

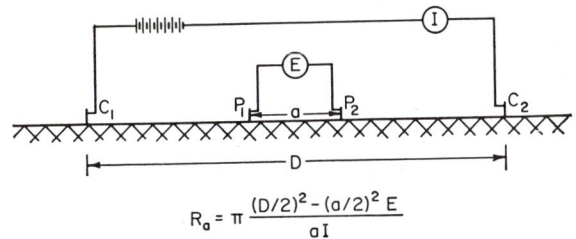

$$R_a = \pi \frac{(D/2)^2 - (a/2)^2}{aI} E$$

FIG. 4-I-11. The Schlumberger configuration. (*La Compagnie Générale de Géophysique* [48].)

Usually a resistivity survey combines these two procedures by taking several depth profiles at properly spaced stations so that both vertical and horizontal variations in apparent earth resistivity can be recorded. Resistivity data are most useful in conjunction with a sound geological knowledge of the given area. Often a nearby logged well is used for control. The value of the resistivity method is limited when test-hole data are not available or when the geology is poorly known. Resistivity-spacing curves for a two-, three-, and four-layered earth and curve-matching techniques have been published by Mooney and Wetzel [38]. However, these curves are theoretically derived and have not been commonly used in practical problems. Excellent descriptions of the use of the resistivity method are available in the discussions by Buhle [39], Hackett [40], Roman [41], Wollard and Hanson [42], and Moore [43]. The resistivity method has been used also to determine salt-water bodies [44–47]. Such an application in Holland [47] is based upon the Schlumberger configuration, which differs from Wenner's in that the potential electrodes are spaced close together. The Schlumberger configuration (Fig. 4-I-11) has been used infrequently in the United States. Mooney and Wetzel [38] have published tables of potentials about a point electrode from which resistivity-spacing curves can be computed for the Schlumberger configuration. La Compagnie Générale de Géophysique [48] has also published such curves.

b. Seismic Methods. Seismic methods are also used in exploration for groundwater sources. As with the resistivity methods, seismic studies are most helpful when some geologic control is available. Usually they may not prove to be particularly helpful without nearby logged wells or other sources of geologic knowledge. Also, the seismic method has not been found especially successful in the location of water directly, but it is of much value in the location and delineation of potentially water-bearing rocks. Two seismic techniques commonly used in exploratory operations are the reflection and refraction methods.

Principles upon which the *reflection method* is based involve the passage of an elastic wave, outward in all directions through earth materials. This wave originates from an exploded charge at a selected point. Portions of the energy in the wave are reflected from lithologic or structural interfaces and return to the land surface. These reflected waves generate earth motion, causing seismometers to move. The movements are recorded on an oscillograph, which shows a trace of the wave and its relative arrival time. From the arrival time of a particular reflection and a knowledge of the velocities in the various strata above the reflecting horizon, the depth of the reflecting interfaces can be computed and plotted to show the geologic structure. The reflection method has long been used successfully in petroleum exploration. An example showing the use of the reflection method in groundwater studies is given by Warrick and Winslow [49].

The *refraction method* is based upon Fermat's principle; that is, seismic waves travel along least-time paths. Velocities are greatest in solid crystalline rocks and least in poorly indurated and unconsolidated rocks. Changes in wave velocities are influenced primarily by the elastic properties of the rock formation. Also, porosity decreases wave velocity, and in porous formations such as glacial drift, water saturation increases it. The common field procedure is to set off a charge or to cause an impact at a selected distance from an array of "in-line" seismometers or detectors. The elastic wave propagates outward from the shot point, until it encounters high-velocity layers along which it is refracted, eventually returning to the surface, where its arrival is picked up by the detectors and recorded by an oscillograph. This refracted wave overtakes the direct wave, which travels more or less horizontally through the surficial material. The first arrival time at each seismometer is plotted against the distance from shot to detector, producing a seismic travel-time curve from which depths and thicknesses of layers can be computed. Where low-velocity layers underlie high-velocity material, erroneous values may result. Therefore the best determination with the refraction method is obtained when velocities increase with depth.

Typical case histories for seismic methods have been published by Linehan [50], Linehan and Keith [51], Shepard and Wood [52], Johnson [53], and Warrick and Winslow [49]. An excellent review of various methods of seismic prospecting has been written by Dix [54].

Frequently, the resistivities of adjacent layers of rocks are similar, but other physical characteristics differ, so that the velocity of seismic waves may differ considerably. Also, under natural conditions high-velocity layers frequently overlie low-velocity material. Under such conditions the electrical-resistivity method used in conjunction with the refraction method may yield information not available from the separate use of either method. McGinnis and Kempton [55] have recently reviewed several case histories which illustrate the advantage of using the two methods simultaneously in groundwater exploration.

c. The Gravity Method. This method involves measurement of the gravitational field of the earth at various points on the surface. A contour map based on these measurements will show fluctuations in the force of gravity at any assumed datum. These fluctuations are caused by variations in rock density. The method is chiefly valuable in groundwater studies as a means of determining the location and configuration of certain deposits which may be water-bearing or otherwise affect the occurrence of groundwater. For example, Woollard and Hanson [42] have mapped valleys in igneous rocks underlying drift.

Romberg [56] has written an excellent up-to-date review of exploration geophysics which briefly covers theoretical and practical considerations of the various methods. General work on geophysical methods has been variously published [57–60].

3. Geochemical Methods. The chemical quality of water frequently reflects the composition of the rocks making up the aquifer. The temperature of water may also more or less indicate the source and depth of the aquifer. Therefore a study of water qualities and temperatures may assist materially in groundwater exploration. Such studies probably are of greatest help when conducted in conjunction with surficial geologic and geophysical work, although sometimes the quality or temperature of the water may be a deciding factor in the exploration program, especially when the intended use of water may require a specific limitation on these water characteristics.

Various methods of obtaining and plotting geochemical data may be utilized to assist in exploration. Already mentioned is the use of electrical resistivity in delineation of salt-water bodies. A closely related method is the use of potential and resistivity logs to determine salinity of water. Contour maps may also be drawn by plotting data from water analyses and connecting points with lines of equal concentration (isocons). Examples of useful results from such studies include the work in Holland [47], studies in northeastern Illinois [19], and numerous studies of salt-water intrusion in Hawaii, California, Texas, Florida, and New Jersey. Similar techniques may also be used to determine and delineate both chemical and bacteriological contamination in aquifers.

4. Subsurface Geologic Methods. The selection and use of *subsurface-exploration methods* depend heavily upon the completeness of information derived from surface methods, the nature of the terrane being studied, and the standard of information needed to satisfy the requirements of any given project. Since most subsurface exploration depends primarily upon test-well drilling, a relatively costly procedure, the requirements must normally justify considerable expenditure.

The geologic phases of subsurface exploration include the location and drilling of the test holes, the collection and examination of the samples, and the interpretation and correlation of the acquired data with geohydrologic and hydrometeorologic data. Location of test sites and holes is normally decided by consideration of the results of surface exploration and the demands placed upon the project in terms of proposed geohydrologic testing, land ownership, placement of the proposed user's facilities, distribution system, and other needs. Almost always the chosen location of test holes is a compromise between the demands of the aforementioned factors. Drilling of the test holes should be a carefully planned and highly technical operation aimed at getting the most desirable information in terms of overall long-term potential of the area, and not just at determination of yield of individual wells from an aquifer or a combination of aquifers. Thus, as the test holes are drilled, a careful sampling procedure should be followed and provision should be made for proper study of samples. Normally, samples of at least every 5 ft of formation and at every change in formation should be collected. Samples of the aquifer materials, at least, should be as close to the undisturbed as possible. To accomplish this, core samples of consolidated formations and, if possible, split-spoon or Shelby-tube samples of the unconsolidated formations should be collected. Such samples allow for more realistic estimates of the nature of the formations and much more reliable laboratory determination of aquifer characteristics than ordinary driller's samples (cuttings). The latter inevitably are disturbed, mixed, and washed free of much fine material as a result of the drilling operations. Also, the structures in the rocks are preserved in cores and split-spoon samples. Thus the samples will allow for longer-range and more detailed interpretation, especially in regard to extent of the various elements of the aquifer and other beds. The sample-study program should not only include visual examination of the materials, but should allow for careful sedimentological analysis, including microscopic examination, mechanical analysis, and, if possible determination of values of porosity and permeability of the aquifers. All geologic subsurface studies should be carefully correlated with geophysical studies to be described in a following paragraph. Observations during drilling and the results of the sampling program should indicate the aquifers which might profitably be pump-tested in order to determine the coefficients of transmissibility and storage, movement of water between aquifers, recharge and discharge boundaries, possible future drawdown rates, and aquifer capacity and potentiality for use. Methods for determining these factors are described in detail in Sec. 13. Pumping tests normally involve the use of test holes both for pumping and observation to obtain the best results.

Probably all well-drilling methods have been used for test drilling (see Sec. 13 for descriptions of the various methods), but *cable-tool* and *rotary methods* are most common. In general, cable-tool methods are somewhat slower than rotary, but the samples recovered are usually considered to be more representative of the formation. For shallow testing in unconsolidated materials a combination of the hydraulic jetting method and a split-spoon sampler yields by far the best samples. In deep drilling and in consolidated rocks, better and more accurate samples may be preserved most economically with a core-barrel attachment on a rotary rig. Ordinary drill cuttings from either method are usually very poor and not wholly adequate for analysis and understanding of the earth materials. Drill cuttings taken under expert supervision from either cable-tool or rotary operations do, of course, yield useful information. In any case, since test drilling is a relatively expensive operation and the best collection methods are relatively inexpensive, it is a waste of money and time to test-drill but not to obtain the best possible information. Frequently, test holes closely adjacent to holes selected for pumping tests are needed for geohydrologic observations. Such holes need not be so large in diameter nor sampled so accurately and therefore may be drilled more cheaply. For example, in shallow unconsolidated sediments the methods of jetting and driving such wells may be much more economical and just as satisfactory.

Certain other drilling records and operations are highly valuable to the hydrogeologist and may be easily collected during a test-drilling program. For example, a log of drilling time and the strata penetrated should be kept, along with notations of where water has been encountered and with any observations as to changes in water levels and chemical quality. A caliper log of the hole diameter is also useful in most instances and is necessary in order to make computations from certain geophysical logging operations. Caliper logs are now normally obtained in conjunction with geophysical logs.

Some subsurface study can be accomplished without test drilling and is usually carried out during the course of the surface-geology phase of the program. Thus all available well logs and records, geophysical logging records, and geochemical records should be studied at that time.

5. Subsurface Geophysical Methods. *Subsurface geophysical methods* include primarily down-the-hole logging procedures which measure certain electrical, sonic, and radioactive characteristics of the earth materials. Space limitations allow only a brief discussion of these methods. Detailed descriptions and methods of computation may be found elsewhere [61–63], and basic theory is also discussed in the literature [59, 60].

a. The Electric Methods. Electric logging normally includes measurement of the resistivity and spontaneous potential of earth materials in the vicinity of the borehole. A sonde containing four electrodes, two for emitting current and two for potential measurements, properly spaced to fit both the nature of the formations and the recording equipment, is lowered into the test hole. The electric log resulting from a series of measurements consists of a series of traces representing a *normal* and a *lateral* curve, both usually on the right side of the log, and a *spontaneous potential curve* on the left side of the log. Various configurations can be used, depending upon the information desired and the nature of the liquids in the hole and of the formations surrounding it. In general, formation boundaries can best be located by short electrode spacing. Characteristics of the fluids in the formations can best be studied with long spacings. The information yielded by a resistivity log may include the lithology and sequence of rock formation and formation boundaries; the presence, approximate ionic concentration, and location of fresh- and salt-water bodies present; the amount of casing in an old well; and an estimate of porosity. The spontaneous potential log indicates permeable zones (in terms of relative, but not absolute, permeability) and may be used to compute groundwater resistivity. Usually the resistivity and potential logs supplement one another.

   b. *The Radiation Method.* The commonest methods of radiation logging include the *gamma-ray* and *neutron* methods. The gamma-ray sonde may contain an ionization chamber and a Geiger counter or a scintillation counter. When lowered into a hole, it will record the emission of gamma rays by elements in the rock formations. The greatest concentrations of radioactive salts seem to occur in shales; therefore the gamma-ray method essentially distinguishes shales from other formations. Gamma-ray logs are especially valuable because they may be recorded in cased holes containing practically any fluid or in empty holes. The neutron sonde contains a source of neutrons and a counter spaced about 18 in. apart. In operation, neutrons from the source are released at high velocities into the formation, bombarding the atomic nuclei of the materials in the formation and thus losing energy expressed as "capture" gamma rays, which return to the counter or detector, where a count of them is recorded. The counting rate increases with decreasing hydrogen concentration in the formation, since hydrogen atoms have a small nucleus of almost the same mass as the neutron and therefore are most effective in slowing down the latter process. Since liquids filling the pores of the formations (commonly water and oil) abound in hydrogen, the porosity of the formation is reflected by the gamma-ray count. As with the gamma log, the neutron log may be used in cased or empty holes and therefore is very useful in surveying old wells. It provides the best information in regard to porosity and supplements information from other methods of logging.

   c. *The Sonic Method.* Sonic logging is a relatively new development utilizing the velocity of sound waves traveling through a formation. The travel time of the waves is recorded versus depth as a sonic sonde is pulled out of the borehole. Sonic velocities in consolidated rocks range widely (between 6,000 and 25,000 fps), and the speed depends upon elastic properties, porosity, fluid content, and pressure in the formations. The sonic log reflects changes in lithology and gives reliable indications of the porosity of formations, especially the low porosities. The sonic sonde consists of a sound generator and a receiver or two receivers positioned at a fixed distance beneath it. The sonic log is usually run with a self-potential log or with a gamma-ray log. The most common use for it is in the correlation and the porosity determination of consolidated formations.

   All down-the-hole logging surveys and interpretations of the results are highly technical operations and should be conducted by specialists. Certain relatively simple operations can be and are performed by less experienced operators, such as determining the proper location and length of well screens opposite aquifers, some other lithologic determinations, and the location of casing and liners in old wells. An excellent discussion of logging methods in groundwater work has been prepared by Jones and Skibitzke [64].

   6. Subsurface Geochemical Methods. *Geochemical subsurface methods* are closely tied in with the electrical-logging programs since the ionic concentration of groundwater may be estimated from its resistivity and also computed from the spontaneous-potential log data. Examples of methods of determination of amounts of dissolved solids in water are given in Refs. 65 to 69. Temperature logs may also be run in conjunction with other down-the-hole logging and have been found to be useful in identification of water from different aquifers. The data obtained from subsurface studies can be handled in the same way as has been described in the discussion of surface geochemical studies. Used in conjunction with geologic, geohydrologic, and geophysical data, they are very useful in groundwater exploration.

F. Wells and Springs

Groundwater is commonly recovered for ordinary use by means of wells and springs. A *spring* is a natural discharge point where groundwater issues from soil or rocks in concentrated flow. *Wells* are artificial excavations, usually dug, bored, or drilled holes or tunnels, which penetrate the water-yielding beds and allow water to flow or to be pumped to the land surface. Use of groundwater must have started with diversion from springs, and the latter are still an important source of supply. However, by far the greatest amount of groundwater used for ordinary purposes is now withdrawn by means of wells.

1. Wells. The most efficient water well is developed so as to yield the greatest quantity of water with the least drawdown (reduction in head) and the lowest velocity in the vicinity of the well. Usually, this condition is approached by constructing a well the wall of which is as permeable as, or is more permeable than, the formation materials immediately adjacent to the well. Thus properly finished wells are often complex structures, and successful completion of them depends, in some degree, on accurate knowledge of the formation materials. Methods of drilling, construction, and finishing wells are described in Sec. 13. Proper methods of sampling the formations, analyzing samples, and determining the thickness of the aquifers are discussed in the paragraphs on test-well drilling and down-the-hole geophysical studies in this section.

Large production wells are drilled into chiefly unconsolidated and semiconsolidated sand and gravel formations, consolidated sandstones and conglomerates, cavernous limestones, or highly fractured volcanic rocks. Methods of well construction and development differ in these different aquifer types. Normally, large wells in the unconsolidated and semiconsolidated deposits are drilled by the rotary and reverse-rotary method, although cable tools may be used. Almost always some method of gravel packing is used, and the efficiency of the well will depend upon the selection of well screen and the adequacy of development of the gravel pack and the formation. Both the selection of the screen and successful well development in turn depend in large part upon the analysis and interpretation of the nature and thickness of the formation and the gravel pack (Sec. 13). In consolidated rock wells, well screens and gravel packs are not commonly used and the development of the well consists primarily of surging it in some manner. In some areas where incoherent layers of sand are interbedded with consolidated rocks, screening and gravel packing has been used and could be successfully used in many more instances than it has been.

Evidently considerable geologic and geophysical experience and knowledge can be profitably used in well construction and development.

2. Springs. Conditions necessary to produce springs are numerous and are related to many and various combinations of geologic, hydrologic, hydraulic, pedologic, climatic, and even biologic controls. Therefore numerous specific descriptive terms for spring have arisen, and no single basis of classification of springs has satisfied the needs of general usage. Among the various bases of classification that have been used, the most common include:

1. Character of openings from which water issues
2. Character of the water-bearing formations
3. The rock structure and resulting force that brings the water to the surface
4. Quantity of water discharged
5. Uniformity and periodicity of the rate of discharge
6. Chemical quality of the water discharged
7. Temperature of the water
8. Deposits and other features produced by the springs
9. Source of the water, whether shallow or deep-seated
10. Direction of movement of the water

Various suggested classifications based on one or more of the above items may be found in Refs. 70 to 73. Classifications that best meet practical needs and that include the common types of springs are given in Table 4-I-5.

Springs are not only used as sources for water supply but have also been exploited for recreational and therapeutic purposes and for the production of power.

One of the more recent developments in power production has been utilization of steam from wells drilled in hot-spring areas. Although hot springs had been used to provide heat for buildings and baths for many centuries, it was not until 1904 that wells to produce steam for the generation of electrical power were drilled. In that

Table 4-I-5. Classifications of Springs

A. Classification based upon the type of water-bearing formation or type of opening (*after Tolman* [71])
   I. Springs issuing from permeable-veneer formations overlying relatively impermeable bedrock with irregular surface; bedrock outcrop or near approach to surface is controlling feature; includes *contact* and so-called *gravity* springs, *perched, talus, pocket, mesa* or *cuesta*, and *barrier* springs. Usually are small, but some are large, and discharge may vary periodically.
   II. Springs issuing from thick permeable formations; water movement unaffected by deeply lying impermeable bedrock. Controlling feature is intersection of water table with the land surface. Include all *water-table* springs, among special types of which are *channel, cliff, valley, dimple,* and *boundary,* or *alluvial-slope* springs. Discharge is usually small.
   III. Springs issuing from interstratified permeable and impermeable formations; aquifers are stratiform and may be structurally deformed. Control of springs is outcrop of aquifer. Springs may draw on confined water (*artesian springs*), or water may be unconfined. Include *contact, monoclinal, synclinal, anticlinal,* and *unconformity* springs. Discharge of springs usually small, but some may be large.
   IV. Springs issuing from solution openings formed primarily along fractures and bedding planes in carbonate rocks. Discharge of these springs is often large and fluctuates considerably.
   V. Springs issuing from fractures and tubes in lava and from thin interbedded porous strata. Discharge is often large and usually steady.
   VI. Springs issuing from fractures intersecting permeable materials and impermeable materials and fractures supplied in part by waters of deep-seated unknown origin. For the most part, discharge of such springs is small, but some may be large.

B. Classification of springs according to magnitude of discharge (*after Meinzer* [70])

Magnitude	Discharge
First	100 cfs or more
Second	10 to 100 cfs
Third	1 to 10 cfs
Fourth	100 gpm to 1 cfs (448.8 gpm)
Fifth	10 to 100 gpm
Sixth	1 to 10 gpm
Seventh	1 pt to 1 gpm
Eighth	Less than 1 pt/min

C. Classification according to variability and permanence of discharge (*after Meinzer* [70])
   I. Perennial springs (springs that discharge continuously)
      a. Constant—springs with a variability of not more than 25 per cent
      b. Subvariable—springs with a variability of more than 25 but not more than 100 per cent
      c. Variable—springs with a variability of more than 100 per cent
N.B.: Variability of a spring is defined by Meinzer as the ratio of the discharge fluctuation to its average discharge within a given period of record. Thus $V = 100(a - b)/c$, where V is the variability in per cent, a the maximum, b the minimum, and c the average, discharge.
   II. Intermittent springs (all are variable since they discharge only during certain periods and are dry at other times; geysers are a special type)

D. Classification according to temperature (*after Meinzer* [70])
   I. Nonthermal springs
      a. Ordinary springs whose waters have temperatures closely approximating the local mean annual temperature of the atmosphere
      b. Cold springs whose waters have temperatures appreciably below the local mean annual temperature
   II. Thermal springs
      a. Hot springs whose waters have temperatures higher than 98°F
      b. Warm springs whose waters have temperatures higher than the local mean annual temperature of the atmosphere but lower than 98°F.

year, at Larderello, about 40 miles from Florence, the first such well was drilled [74]. Present production from several wells at Larderello now exceeds 300,000 kw. In 1950 the New Zealand government initiated a steam-power project at Wairakei which, by 1958, produced about 65,000 kw from 39 wells. In 1960, the first power plant in the United States utilizing steam from wells went into operation at The Geysers, Sonoma County, Calif. This plant is rated to produce 12,500 kw. Two other countries, Russia and Mexico, are known to be producing power from natural steam, and several, including Iceland, Chile, Fiji Islands, Nicaragua, El Salvador, Guatemala, and Japan, have started well steam-power development programs [74]. At least 13 areas in the western United States (chiefly in California, Nevada, and Oregon) have been tested by private companies for natural steam for power.

The rocks in nearly all the areas investigated and developed are much faulted and fractured and are chiefly volcanic in origin. They are most likely underlain by large quantities of molten rock, closer to the land surface than in "normal" areas where hot springs are not found. This molten rock is the heat source, and heat flow through adjacent solid rocks serves to raise the temperature of water in contact with them. Very little of this water is magmatic in origin, certainly less than 5 per cent [75], and most of it is meteoric, having circulated from the surface to considerable depths (as much as 10,000 ft) in fractures and permeable beds. The permeability and porosity of the solid rocks control, for the most part, the amount of water that circulates deeply. In the less permeable rocks less water circulates, and thus is heated more rapidly than large quantities of water which may circulate in more permeable areas. The balance of heat supply and water supply will thus determine whether an area contains dry steam or wet steam and hot water.

The waters in hot-spring areas contain many minerals and gases. For example, at Larderello about 5 per cent of the natural vapor consists of carbon dioxide, hydrogen, hydrogen sulfide, nitrogen, boric acid, ammonia, helium, and argon. Such constituents may cause critical problems as a result of deposition and corrosion of equipment, and special treatment may be required in order to utilize the steam. At Larderello the noncondensable gases at first proved to be a hindrance, but by special treatment they are now collected and are marketed as a profitable by-product.

V. REFERENCES

1. Hadley, J. B.: The Madison Canyon landslide, *Geotimes*, vol. 4, no. 3, pp. 14–17, 1959.
2. Witkind, I. J.: The Hebgen Lake earthquake, *Geotimes*, vol. 4, no. 3, pp. 13–14, 1959.
3. Langbein, W. B.: Annual runoff in the United States, *U.S. Geol. Surv. Circ.* 52, 1949.
4. Brown, C. B.: Report on an investigation of water losses in streams flowing east out of the Black Hills, South Dakota, *U.S. Dept. Agr., Soil Conserv. Serv., Sedimentation Sec., Spec. Rept.* 8, 1944.
5. Lohman, S. W.: Sand Hills area, Nebraska, chap. 5 in "Sub-surface Facilities of Water Management and Patterns of Supply: Type Area Studies," U.S. Congress, House of Representatives, Interior and Insular Affairs Committee, 1953, pp. 79–91.
6. Troxell, H. C.: "Hydrology of Western Riverside County, California," Riverside County Flood Control and Water Conservation District, 1948.
7. Jenny, Hans: "Factors of Soil Formation," McGraw-Hill Book Company, Inc., New York, 1955.
8. Rice, T. D., and L. B. Alexander: The physical nature of soil in "Soils and Men," U.S. Department of Agriculture Yearbook, 1938.
9. Meinzer, O. E.: Occurrence of ground water in the United States, *U.S. Geol. Surv. Water-Supply Paper* 489, pp. 1–321, 1923.
10. Meinzer, O. E.: Ground water in the United States: a summary, *U.S. Geol. Surv. Water-Supply Paper* 836-D, pp. 157–229, 1939.
11. Thomas, H. E.: "The Conservation of Ground Water," McGraw-Hill Book Company, Inc., New York, 1951.
12. Thomas, H. E.: "Ground-water Regions of the United States: Their Storage Facilities," U.S. Congress, House of Representatives, Interior and Insular Affairs Committee, 1952.
13. Bierschenk, W. H.: Hydraulic characteristics of Hanford aquifers, *Atomic Energy Comm. Rept.* HW-48916, 1957.
14. Hackett, O. M., F. N. Visher, R. G. McMurtrey, and W. L. Steinhilber: Geology and ground-water resources of the Gallatin Valley, Gallatin County, Montana, *U.S. Geol. Surv. Water-Supply Paper* 1482, 1960.
15. Walton, W. C.: Leaky artesian aquifer conditions in Illinois, *Illinois State Water Surv. Div. Rept. Invest.* 39, 1960.
16. Walton, W. C., and G. D. Scudder: Ground-water resources of the valley train deposits in the Fairborn area, Ohio, *Ohio Div. Water Tech. Rept.* 3, 1960.
17. Lang, W. M., W. H. Bierschenk, and W. B. Allen: Hydraulic characteristics of glacial outwash in Rhode Island, *Rhode Island Water Resources Coordinating Board Hydrol. Bull.* 3, 1960.

REFERENCES 4-37

18. Foley, F. C., W. C. Walton, and W. J. Drescher: Ground-water conditions in the Milwaukee-Waukesha area, Wisconsin, *U.S. Geol. Surv. Water-Supply Paper* 1229, 1953.
19. Suter, Max, R. E. Bergstrom, H. F. Smith, G. H. Emrich, W. C. Walton, and T. E. Larson: Preliminary report on ground-water resources of the Chicago region, Illinois, *Illinois State Geol. Surv. and Illinois State Water Surv. Cooperative Ground Water Rept.* 1, 1959.
20. Guyton, W. F.: Results of pumping tests of the Carrizo sand in the Lufkin area, Texas, *Trans. Am. Geophys. Union*, vol. 22, pt. 1, pp. 40–48, 1942.
21. Otten, E. G.: Ground-water resources of the southern Maryland coastal plain, *Maryland Dept. Geol., Mines, Water Resources Bull.* 15, 1955.
22. Dingman, R. J., and H. F. Ferguson: The water resources of Baltimore and Harforal counties, *Maryland Dept. Geol., Mines, Water Resources Bull.* 17, 1956.
23. Walton, W. C., and J. W. Stewart: Aquifer tests in the Snake River Basalt, *Proc. Am. Soc. Civil Engrs., J. Irrigation and Drainage Div.*, vol. 85, no. IR3, September, 1959.
24. Malmberg, G. T., and H. T. Downing: Geology and ground-water resources of Madison County, Alabama, *Geol. Surv. Alabama Co. Rept.* 3, 1957.
25. Walker, E. H.: Ground-water resources of the Hopkinsville quadrangle, Kentucky, *U.S. Geol. Surv. Water-Supply Paper* 1328, 1956.
26. Walton, W. C.: The hydraulic properties of a dolomite aquifer underlying the village of Ada, Ohio, *Ohio Div. Water Tech. Rept.* 1, 1953.
27. Zeizel, A. J., W. C. Walton, R. T. Sasman, and T. A. Prickett: Ground-water resources of DuPage County, Illinois, *Illinois State Geol. Surv. and Illinois State Water Surv. Coop. Rept.* 2, 1963.
28. Poland, J. F., and others: "Progress Report on Land-subsidence Investigations, San Joaquin Valley, California through 1957," Inter-Agency Committee on Land Subsidence in the San Joaquin Valley, Sacramento, Calif., 1958.
29. Loehnberg, Alfred: Aspects of the sinking of Mexico City and proposed countermeasures, *J. Am. Water Works Assoc.*, vol. 50, no. 3, pp. 432–440, 1958.
30. Maxey, G. B., and C. H. Jameson: Geology and water resources of Las Vegas, Pahrump, and Indian Spring Valleys, Clark and Nye Counties, Nevada, *Nevada State Engineer's Office Water Res. Bull.* 5, pp. 69–71, 1948.
31. Winslow, A. G., and L. A. Wood: Relation of land subsidence to ground-water withdrawals in the upper Gulf Coast region, Texas (abstract), *Mining Engineering*, vol. 10, no. 12, p. 1243, 1958.
32. Gilluly, James, and U. S. Grant: Subsidence in the Long Beach harbor area, California, *Geol. Soc. Am. Bull.*, vol. 60, pp. 461–530, 1949.
33. Williams C. C., and S. W. Lohman: Geology and ground-water resources of a part of south-central Kansas, *State Geol. Surv. Kansas Bull.* 79, plate 1, 1949.
34. LeMoreaux, P. E., and G. B. Maxey: Ground water in the United States, National Report to the Twelfth General Assembly of the International Union of Geodesy and Geophysics, *Trans. Am. Geophys. Union*, vol. 41, no. 2, pp. 311–314, 1960.
35. Weener, Frank: A method of measuring resistivity in the earth, *U.S. Bur. Stand. Bull.* 12, pp. 469–478, 1916.
36. Gish, O. H., and W. J. Rooney: Measurement of resistivity of large masses of undisturbed earth, *Terrestrial Magnetism*, vol. 30, no. 4, pp. 161–188, 1925.
37. Tagg, G. F.: Interpretation of resistivity measurements, *Trans. Am. Inst. Mining Met. Eng., Geophys. Prospecting*, 1934, pp. 135–145.
38. Mooney, H. M., and W. W. Wetzel: "The Potentials about a Point Electrode and Apparent Resistivity Curves for a Two-, Three-, and Four-layered Earth," Univ. Minn. Press, Minneapolis, 1956.
39. Buhle, M. B.: Earth resistivity in ground-water studies in Illinois, *Trans. Am. Inst. Mining Engrs. Tech. Paper* 3496L, *Mining Eng.*, 1953, pp. 395–399.
40. Hackett, J. E.: Relation between earth resistivity and glacial deposits near Shelbyville, Illinois, *Illinois State Geol. Surv. Circ.* 223, 1956.
41. Roman, Irwin: Resistivity reconnaissance, *Am. Soc. Testing Mater. Spec. Tech. Publ.* 122, 1952.
42. Woollard, G. P., and G. F. Hanson: Geophysical methods applied to geologic problems in Wisconsin, *Wisconsin Geol. Surv. Bull.* 78, Sci. Ser. 15, 1954.
43. Moore, R. W.: Applications of electrical resistivity measurements to subsurface investigations, *Public Roads*, vol. 29, no. 7, 1957.
44. Sayre, A. N., and E. L. Stephenson: The use of resistivity methods in the location of salt-water bodies in the El Paso, Texas area, *Trans. Am. Geophys. Union*, vol. 18, pp. 393–398, 1937.
45. Volker, Adrian, and J. C. van Dam: Geo-elektrisch onderzoek bij nitvuering van waterbouwkuddige werken, *Serv. for Water Resources Develop.* "Rijkwaterstaat," The Netherlands, 1954.

46. van Dam, J. C.: Geo-electrical investigations in the Delta area of the Netherlands, *Serv. for Water Resources Develop.* "*Rijkwaterstaat,*" The Netherlands, 1955.
47. Volker, Adrian, and J. Dijkstra: Détermination des salinités des eaux dans le sous-sol du Zuiderzee par prospection géophysique, *Geophys. Prospecting*, vol. 3, pp. 111–125, 1955.
48. La Compagnie Générale de Géophysique: Abaques de sondage électrique, *Geophys. Prospecting*, vol. 3, suppl. 3, pp. 21–46, 1955.
49. Warrick, R. E., and J. D. Winslow: Application of seismic methods to a ground-water problem in northeastern Ohio, *Geophysics*, vol. 25, no. 2, pp. 505–519, 1960.
50. Linehan, Daniel: Seismology applied to shallow zone research, *Am. Soc. Testing Mater. Spec. Tech. Publ.* 122, pp. 156–170, 1951.
51. Linehan, Daniel, and Scott Keith: Seismic reconnaissance for ground-water development, *J. New England Water Works Assoc.*, vol. 63, no. 1, pp. 76–92, 1949.
52. Shepard, E. R., and A. E. Wood: Application of the seismic refraction method of subsurface exploration to flood-control projects, *Trans. Am. Inst. Mining Met. Engrs.*, vol. 138, pp. 312–325, 1940.
53. Johnson, R. B.: Use of the refraction seismic method for differentiating Pleistocene deposits in the Arcola and Tuscola quadrangles, Illinois, *Illinois State Geol. Survey Rept. Invest.* 176, 1954.
54. Dix, C. H.: "Seismic Prospecting for Oil," Harper & Row, Publishers, Incorporated, New York, 1952.
55. McGinnis, L. D., and J. P. Kempton: Integrated seismic, resistivity, and geologic studies of glacial deposits, *Illinois State Geol. Surv. Circ.* 323, 1961.
56. Romberg, F. E.: Exploration geophysics: a review, *Geol. Soc. Am. Bull.*, vol. 72, pp. 883–932, 1961.
57. Nettleton, L. L.: "Geophysical Prospecting for Oil," McGraw-Hill Book Company, Inc., New York, 1940.
58. Dobrin, M. B.: "Introduction to Geophysical Prospecting," 2d ed., McGraw-Hill Book Company, Inc., New York, 1960.
59. Heiland, C. A.: "Geophysical Exploration," Prentice-Hall, Inc., Englewood Cliffs, N.J., 1940.
60. Jakosky, J. J.: "Exploration Geophysics," 3d ed., Times-Minnor Press, Los Angeles, 1960.
61. Wyllie, M. R. J.: "The Fundamentals of Electric Log Interpretations," Academic Press Inc., New York, 1954.
62. LeRoy, L. W.: "Subsurface Geologic Methods," 2d ed., Colorado School of Mines, Golden, Colo., 1951.
63. "Introduction to Schlumberger Well Logging," *Schlumberger Well Logging Corp. Doc.* 8, Houston, Tex., 1958.
64. Jones, P. H., and H. E. Skibitzke: Subsurface geophysical methods in ground-water hydrology, *Advan. in Geophys.*, vol. 3, pp. 241–300, 1956.
65. Jones, P. H., and T. B. Buford: Electric logging applied to ground-water exploration, *Geophysics*, vol. 16, pp. 115–139, 1951.
66. Poland, J. F., and R. B. Morrison: An electrical resistivity apparatus for testing well waters, *Trans. Am. Geophys. Union*, vol. 21, pp. 35–46, 1940.
67. Bryan, F. L.: Application of electric logging to water-well problems, *Water Well J.*, vol. 4, no. 1, pp. 3–7, 1950.
68. Pryor, W. A.: Quality of water estimated from electric resistivity logs, *Illinois State Geol. Surv. Circ.* 215, 1956.
69. Priddy, R. R.: Fresh-water strata of Mississippi as revealed by electric studies, *Mississippi Geol. Surv. Bull.* 83, 1955.
70. Meinzer, O. E.: Outline of ground-water hydrology, *U.S. Geol. Surv. Water-Supply Paper* 494, pp. 50–55, 1923.
71. Tolman, C. F.: "Ground Water," McGraw-Hill Book Company, Inc., New York, pp. 435–466, 1937.
72. Bryan, Kirk: Classification of springs, *J. Geol.*, vol. 27, pp. 522–561, 1919.
73. Keilhack, K.: "Grundwasser und Quellenkunde," 3d ed., Gebrüder Borntraeger, Berlin, 1935, pp. 257–359.
74. McNitt, J. R.: Geothermal power, *Calif. Div. Mines Inform. Serv.*, vol. 13, no. 3, 1960.
75. White, D. E., and Harmon Craig: Isotope geology of the Steamboat Springs area, Nevada (abstract), *Econ. Geol.*, vol. 54, no. 7, pp. 1343–1344, 1959.

Section 4-II

GEOLOGY

PART II. QUANTITATIVE GEOMORPHOLOGY OF DRAINAGE BASINS AND CHANNEL NETWORKS

ARTHUR N. STRAHLER, *Professor of Geomorphology, Columbia University.*

I. Introduction	4-40
II. Basic Concepts	4-40
A. Open Systems and Steady States	4-40
B. Dimensional Analysis	4-41
C. Statistical Analysis	4-42
D. Plan of Morphometric Analysis	4-43
III. Linear Aspects of the Channel System	4-43
A. Stream Orders	4-43
B. Stream Lengths	4-45
C. Length of Overland Flow	4-47
IV. Areal Aspects of Drainage Basins	4-48
A. Arrangement of Areal Elements	4-48
B. Frequency Distribution of Basin Areas	4-48
C. Law of Stream Areas	4-48
D. Relation of Area to Length	4-49
E. Relation of Area to Discharge	4-50
F. Basin Shape (Outline Form)	4-51
G. Drainage Density	4-52
H. Constant of Channel Maintenance	4-54
I. Stream Frequency	4-55
V. Relief (Gradient) Aspect of Drainage Basins and Channel Networks	4-56
A. Channel Gradients	4-56
1. Single-channel Profiles	4-56
2. Cause of Profile Upconcavity	4-57
3. Fitted Regression Functions	4-57
4. Derivative Functions	4-59
5. Profile Segmentation	4-60
6. Composite Profiles	4-60
7. Channel Slope as a Function of Order	4-60
8. Main-stream-slope Factor	4-61
B. Channel-cross-section Geometry	4-61

  C. Ground-surface Gradients............................... 4-61
    1. Relationship of Ground and Channel Slopes........... 4-61
    2. Maximum Valleyside Slopes.......................... 4-62
    3. Total Surface-slope Distribution..................... 4-63
  D. Relief Measures.. 4-65
    1. Relief... 4-65
    2. Relief Ratios...................................... 4-66
  E. Ruggedness and Geometry Numbers....................... 4-67
  F. Hypsometric (Area-Altitude) Analysis..................... 4-68
 VI. Theory of Drainage-basin Dynamics........................ 4-69
  A. Statement of Variables................................. 4-69
  B. Solution by Pi Theorem................................. 4-70
  C. Steady-state Relationships.............................. 4-71
  D. Upsets of Steady State................................. 4-71
 VII. Observed Complex Relations among Hydrologic and Geometric
    Properties.. 4-72
  A. Control of Basin Geometry by Climatic Factors........... 4-72
  B. Relation of Basin Geometry to Stream Flow.............. 4-72
 VIII. Notations... 4-73
 IX. References.. 4-74

I. INTRODUCTION

 Under the impetus supplied by Horton [1, 2], the description of drainage basins and channel networks was transformed from a purely qualitative and deductive study [3] to a rigorous quantitative science capable of providing hydrologists with numerical data of practical value. Horton's work was supplemented by Langbein [4], then developed in detail by Strahler [5–11] and his Columbia University associates [14–18, 21, 22, 25, 26, 39, 66, 68].
 This section treats quantitative land-form analysis as it applies to normally developed watersheds in which running water and associated mass gravity movements, acting over long periods of time, are the chief agents in developing surface geometry. Emphasis is upon the geometry itself, rather than upon the dynamic processes of erosion and transportation which shape the forms (see Sec. 17-I for erosion and transportation).

II. BASIC CONCEPTS

A. Open Systems and Steady States

 Of fundamental importance is the concept of a drainage basin as an open system tending to achieve a steady state of operation. Strahler [5, p. 676] applied open-system biologic concepts [12] to a graded drainage system. An open system imports and exports matter and energy through system boundaries and must transform energy uniformly to maintain operation. In a drainage basin the land surface within the limits of the basin perimeter constitutes a system boundary through which precipitation is imported. Mineral matter supplied from within the system and excess precipitation leave the system through the basin mouth. In a graded drainage basin the steady state manifests itself in the development of certain topographic characteristics which achieve a time-independent state. Erosional and transportational processes meanwhile produce a steady flow (averaged over periods of years or tens of years) of water and waste from the basin. Potential energy of position is transformed into kinetic energy of water and debris motion or into heat. Considered over a very long span of time, however, continual readjustment of components in the steady state is required as relief lowers and available energy diminishes. The topographic forms will correspondingly show a slow evolution.

BASIC CONCEPTS

Where geologic events have brought into being a new land mass not previously acted upon by running water, the steady state is preceded by a transient state in which a new channel system grows and deepens rapidly as the ground slopes are transformed to contribute most efficiently to the drainage network. In the terminology of the earlier, classical descriptive geomorphology, the transient state was referred to as the stage of youth in the cycle of erosion; the steady state, as the stage of maturity [3].

Validity of the Horton system of fluvial morphometry depends upon the theory that, for a given intensity of erosion process, acting upon a mass of given physical properties, the conditions of surface relief, slope, and channel configuration reach a time-independent steady state in which morphology is adjusted to transmit through the system just the quantity of debris and excess water characteristically produced under the controlling regimen of climate. Should controlling factors of climate or geologic material be changed, the steady state will be upset. Through a relatively rapid series of adjustments, serving to reestablish a steady state, appropriate new values of basin geometry are developed [11, pp. 295–296]. In brief, steady state manifests itself by invariant geometry; transient state, by rapid changes in geometry in which new sets of forms replace the old.

B. Dimensional Analysis

Dimensional analysis forms a sound basis for study of the geometrical and mechanical aspects of drainage basins [11]. The fundamental dimensions of length, mass, and time, whether used singly or combined as products, suffice to define all geometrical and mechanical properties of a drainage basin. Many of the form elements have the simple dimension of length, for example, stream length, basin perimeter, or basin

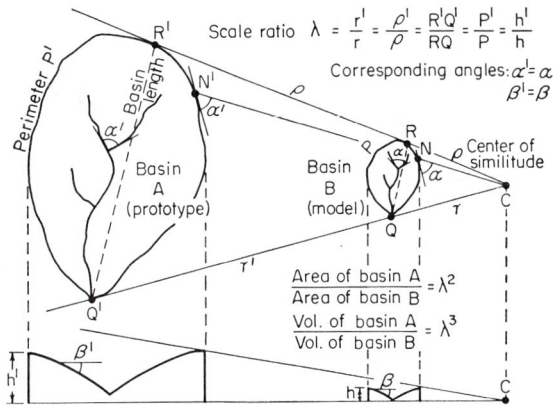

FIG. 4-II-1 Geometrical similarity of two drainage basins. (*After Strahler* [11, p. 291; 10, p. 914].)

relief. Measures of area have the dimension of length squared; volumes, the dimension of length cubed. Another class of geometrical properties consists of the dimensionless ratios of one length property to another. Dimensionless ratios describe pure shape, or form, irrespective of absolute size. For tables of land-form parameters and their dimensional properties, see Ref. 11, pp. 282–283.

Geometrical similarity is an important concept of dimensional analysis applied to drainage basins. Systems of land forms evolving from the same geologic processes and materials possess a high degree of geometrical similarity, an attribute that makes possible the recognition and classification of land forms. Figure 4-II-1 shows the meaning of geometrical similarity as applied to two drainage basins which differ in size but not in shape. Basins A and B are said to be *homothetic* [13, pp. 14–20], because any two corresponding points in the basins lie on the same radius vector from a center of similitude, i.e., are *collinear*.

The points Q' and Q mark the mouths of the basins, at corresponding distances r' and r from the center of similitude. Two other corresponding points are R' and R, collinear and located at distances ρ' and ρ, respectively, from the center of similitude. For reasons that are self-evident from considerations of Euclidian geometry of similiar triangles,

$$r' = \lambda r \quad \text{and} \quad \rho' = \lambda \rho \tag{4-II-1a}$$

where λ is the linear scale ratio. Consequently,

$$\lambda = \frac{r'}{r} = \frac{\rho'}{\rho} \tag{4-II-1b}$$

which is to say that the radius vectors of any two collinear points are always in the ratio λ. Hence, in geometrically similar drainage basins, the distances between corresponding points in the system have the same scale ratio. In geometrical similarity, the tangents to corresponding points on curved lines in the two systems are always equal. In Fig. 4-II-1 the tangents of α' and α at the points N' and N are equal, whereas the degrees of curvature at corresponding points on the two figures are inversely related to the linear scale ratio.

In summary, two drainage basins are geometrically similar when all corresponding land-form elements having the dimension of length are in the same ratio λ, when all corresponding measures having the dimension of inverse of length are in the scale ratio $1/\lambda$, and when those of dimension length squared are in the same ratio, λ^2. Furthermore, all dimensionless properties must have identical values in the corresponding parts of both systems. Although perfect similarity is not to be expected in drainage basins, a high degree of similarity has been found over a great size range when planimetric (horizontal) aspects are considered [11, pp. 292–294]. Lack of similarity among drainage basins may result from strong geologic inhomogeneities which distort the basin shapes.

As applied to scale-model studies, the ratio λ is taken as the ratio of length in prototype to length in model. Mechanical (kinematic and dynamical) similarity between basins is not treated here, but would be essential aspects of model studies.

C. Statistical Analysis

Application of principles of mathematical statistics to quantitative geomorphology is essential if meaningful conclusions are to be achieved (see Sec. 8 for statistical methods). Groups of measurements of drainage-basin characteristics constitute samples drawn from vastly greater populations [8]. At the outset of any investigation, a sampling procedure must be designed to assure the highest possible degree of objectivity in selection of observations. Use of grids or randomly selected orthogonal coordinates may provide a means of sampling the surface characteristics of a watershed where operator bias must be avoided. Tables of random numbers can provide unbiased selection of sample elements from among many individual items, or can provide random azimuths and distances for unbiased field traverses.

Mathematical statistics is concerned with the making of inferences from a small sample about the characteristics of a vast population whose absolute parameters can never be known. Tests are concerned with ascertaining the probability of being right or wrong in stating some hypothesis concerning the relation of one or more samples to the population or populations from which they have been drawn.

In practice, a particular geometric property of a drainage basin, for example, the length of stream segments, is sampled by measuring from maps or air photographs or by direct field surveys. When a sample of, say, 50 or 100 measurements is thus obtained, the standard methods of frequency-distribution analysis are used [8, pp. 2–6]. The individual measurements, termed variates, are grouped into classes, and the nature of the distribution examined. If strongly skewed, a logarithmic transformation of variates may be required. The mean $\bar{x}$, variance s^2, and standard deviation s of the population, as estimated from the sample, are next computed, and serve to describe the geomorphic property in objective and useful terms. Next, the sample frequency distribution is compared with the normal, or Gaussian, distribution, and a test performed to assure that the normal distribution can be assumed [8, pp. 8–10]. Many geometric properties of drainage basins, particularly those having the dimensions of length, area, or volume, are characteristically log-normal in distribution, whereas other properties, particularly dimensionless ratios and angular values, tend to be arithmetically normal in distribution [18]. Melton [14] collected an extensive body of morphometric data on drainage basins and discussed the sample-size requirements for use in statistical tests.

Two or more samples can be compared by statistical tests to reach a decision as to whether the samples are likely to have been drawn from the same or from different populations [8, pp. 10–17]. Such tests are essential to avoid unwarranted assumptions as to the significance of the observed differences in means and variances of the samples themselves.

Relationship of a dependent variable to an independent variable, as, for example, the influence of infiltration capacity upon drainage density, is treated by regression analysis [8, pp. 18–24]. Linear and nonlinear equations may be used to obtain the best fit of data. Significance of the observed relationship can be evaluated by rigorous tests. Multiple regression, in which the combined effect of several independent variables upon one dependent variable can be considered, has been extensively used in drainage-basin analysis [15–17]. Machine methods of multiple-correlation-and-regression analysis have been introduced [18].

D. Plan of Morphometric Analysis

Systematic description of the geometry of a drainage basin and its stream-channel system requires measurement of linear aspects of the drainage network, areal aspects of the drainage basin, and relief (gradient) aspects of channel network and contributing ground slopes. Whereas the first two categories of measurement are planimetric (i.e., treat properties projected upon a horizontal datum plane), the third category treats the vertical inequalities of the drainage-basin forms. Although not free from inconsistencies, the above plan of morphometric analysis is useful operationally and is followed throughout the remainder of this section.

III. LINEAR ASPECTS OF THE CHANNEL SYSTEM

A. Stream Orders

The first step in drainage-basin analysis is designation of stream orders, following a system introduced into the United States by Horton [2, pp. 281–282] and slightly modified by Strahler [7, p. 1120]. Melton [19, pp. 345–346] has explained the mathematical concepts involved. Assuming that one has available a channel-network map including all intermittent and permanent flow lines located in clearly defined valleys, the smallest fingertip tributaries are designated order 1 (Fig. 4-II-2). Where two first-order channels join, a channel segment of order 2 is formed; where two of order 2 join, a segment of order 3 is formed; and so forth. The trunk stream through which all discharge of water and sediment passes is therefore the stream segment of highest order.

Usefulness of the stream-order system depends on the premise that, on the average, if a sufficiently large sample is treated, order number is directly proportional to size of the contributing watershed, to channel dimensions, and to stream discharge at that place in the system. Because order number is dimensionless, two drainage networks differing greatly in linear scale can be compared with respect to corresponding points in their geometry through use of order number. After the drainage-network

QUANTITATIVE GEOMORPHOLOGY

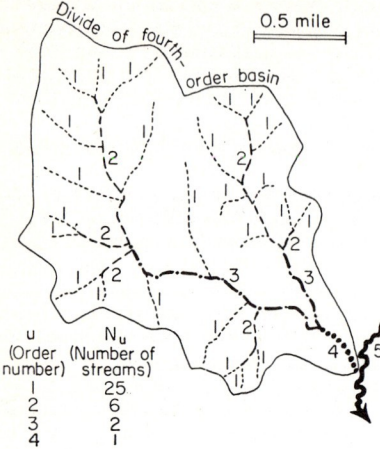

FIG. 4-II-2. Designation of stream orders. (*After Strahler* [10, p. 914].)

elements have been assigned their order numbers, the segments of each order are counted to yield the number N_u of segments of the given order u (Fig. 4-II-2).

It is obvious that the number of stream segments of any given order will be fewer than for the next lower order but more numerous than for the next higher order. The ratio of number of segments of a given order N_u to the number of segments of the higher order N_{u+1} is termed the *bifurcation ratio* R_b:

$$R_b = \frac{N_u}{N_{u+1}} \quad (4\text{-}II\text{-}2)$$

The bifurcation ratio will not be precisely the same from one order to the next, because of chance variations in watershed geometry, but will tend to be a constant throughout the series. This observation is the basis of Horton's [2, p. 291] *law of stream numbers*, which states that the numbers of stream segments of each order form an inverse geometric sequence with order number, or

$$N_u = R_b{}^{k-u} \quad (4\text{-}II\text{-}3)$$

where k is the order of the trunk segment, and the other terms are as previously defined. The law has received verification by accumulated data from many localities [7, p. 1137; 20, p. 18; 21, p. 603; 22, p. 1002; 16, p. 48]. When logarithm of number

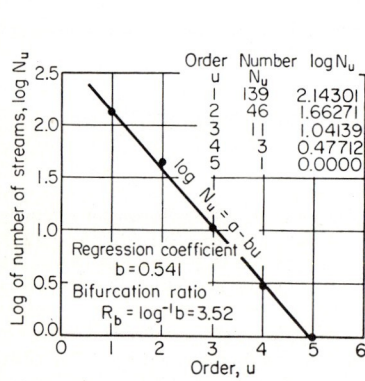

FIG. 4-II-3. Regression of number of stream segments on order. (*After Strahler* [10, p. 915], *based on data by Smith* [22, p. 1003].)

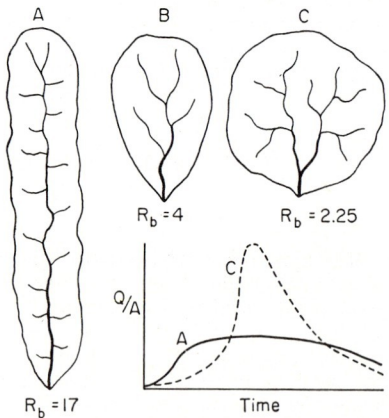

FIG. 4-II-4. Hypothetical basins of extreme and moderate bifurcation ratios, with schematic hydrographs.

of streams is plotted against order, most drainage networks show a linear relationship, with small deviation from a straight line (Fig. 4-II-3).

Calculation on an average value of R_b for a given channel network can be made by determining the slope of the fitted regression of logarithm of numbers (ordinate) on order (abscissa) [23]. The regression coefficient b is identical with the logarithm of R_b. In Fig. 4-II-3 the bifurcation ratio is estimated to be 3.52, which means that, on the average, there are $3\frac{1}{2}$ times as many channel segments of any given order as of the next higher order.

Bifurcation ratios characteristically range between 3.0 and 5.0 for watersheds in which the geologic structures do not distort the drainage pattern. The theoretical minimum possible value of 2.0 is rarely approached under natural conditions. Because the bifurcation ratio is a dimensionless property, and because drainage systems in homogeneous materials tend to display geometrical similarity, it is not surprising that the ratio shows only a small variation from region to region.

Abnormally high bifurcation ratios might be expected in regions of steeply dipping rock strata where narrow strike valleys are confined between hogback ridges. Basin A in Fig. 4-II-4 shows such an elongate basin compared with a normal basin (basin B) and one approaching the theoretical minimum value of 2.0 (basin C). The effects of such distortions upon maximum flood discharges, assuming precipitation and other controls to be the same throughout, are suggested by hydrographs (Fig. 4-II-4). Whereas the elongate basin with high R_b would yield a low but extended peak flow, the rotund basin with low R_b would produce a sharp peak. Basin B would lie somewhere between these two extremes.

Horton [2, p. 286] shows that the total number of streams of all orders in a network can be computed if the bifurcation ratio R_b, and trunk order k are known:

$$\sum_{i=1}^{k} N_u = \frac{R_b{}^k - 1}{R_b - 1} \qquad \text{(4-II-4)}$$

B. Stream Lengths

Mean length $\bar{L}_u$ of a stream-channel segment of order u is a dimensional property revealing the characteristic size of components of a drainage network and its contributing basin surfaces. Channel length is measured with the chartometer (map measurer) directly from the map and therefore represents the true length somewhat shortened by projection upon a horizontal plane [24]. Because in practice all segments of a given order within the specified drainage network are measured successively without pause for recording, the cumulative length appears on the dial of the chartometer. To obtain the mean length of channel $\bar{L}_u$ of order u, the total length is divided by the number of segments N_u of that order, thus:

$$\bar{L}_u = \frac{\sum_{i=1}^{N} L_u}{N_u} \qquad \text{(4-II-5)}$$

Treating each channel segment as a statistical variate, the frequency distribution of segment lengths of a given order was studied by Miller [25], who observed that the distribution of first- and second-order segments was strongly skewed to the right. Schumm [21, p. 607] corrected this skewness by use of logarithm of length. It is recommended that logarithmic transformation of the raw data be made before grouping into classes. Computation of mean, variance, and standard deviation can then be made on a logarithmic basis as in Fig. 4-II-5 [8, pp. 7–8].

The first-order stream channel with its contributing first-order drainage-basin surface area should be regarded as the unit cell, or building block, of any watershed. Because first-order drainage basins tend to be geometrically similar over a wide range of sizes, it matters little what length property is chosen to provide the characteristic measurement of size by which systems are compared from region to region. Thus, while length of first-order channel is a convenient and easily obtained length measure, it might be equally valid to select basin perimeter, basin length, drainage density, or square root of basin area as alternative indices of scale of the unit basin.

As expected, mean length of channel segments of a given order is greater than that of the next lower order but less than that of the next higher order. Horton [2, p. 291] postulated that the length ratio R_L (which is the ratio of mean length $\bar{L}_u$ of segments of order u to mean length of segments of the next lower order $\bar{L}_{u-1}$) tends to be constant throughout the successive orders of a watershed. He was therefore able to state the *law of stream lengths*, that the mean lengths of stream segments of each of the successive orders of a basin tend to approximate a direct geometric sequence in which the first term is the average length of segments of the first order:

$$\bar{L}_u = \bar{L}_1 R_L^{u-1} \qquad (4\text{-II-}6)$$

If the law of stream lengths is valid, a plot of logarithm of stream length (ordinate)

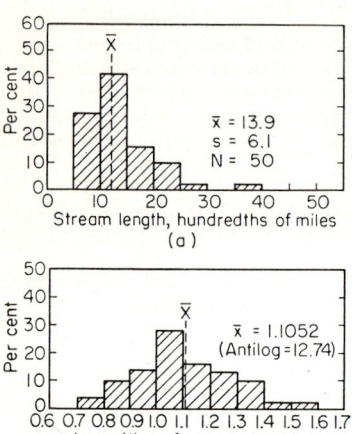

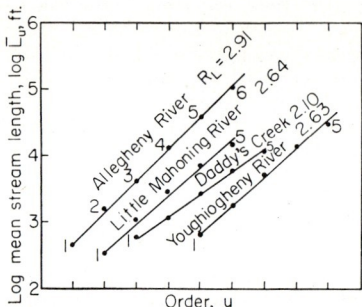

Fig. 4-II-5. Histogram showing (*a*) arithmetic distribution of lengths of first-order streams developed on Copper Ridge dolomite, Virginia; (*b*) distribution of logarithms of same stream lengths. Frequencies computed and grouped from logs of variates. (*After Strahler* [8, pp. 7–8], *based on data from Miller* [25, table 2].)

Fig. 4-II-6. Regression of logarithm of stream-segment length on order for four drainage basins in the Appalachian Plateau Province. (*After Morisawa* [17, p. 48].)

as a function of order (abscissa) should yield a set of points lying essentially along a straight line (Fig. 4-II-6). Confirmation of the law seems amply demonstrated by data from many watersheds [21, pp. 604–605; 20, p. 13; 17, pp. 49–50; 26, p. 5].

As with the law of stream numbers, the law of stream lengths is essentially an exponential function defined only for integer values of the independent variable. The length ratio R_L is therefore obtained as the antilogarithm of the regression coefficient b of a line fitted by inspection or by least-squares method to the plot of logarithm of stream length on order (Fig. 4-II-6).

Verification of Horton's laws of stream numbers and lengths supports the theory that geometrical similarity is preserved generally in basins of increasing order. In other words, a basin of the third order would tend to be geometrically similar to the second-order basins which lie within it, and these in turn would be similar to the first-order basins within them. Hack [27, pp. 63–64] casts doubt on this theory by finding that stream length (measured cumulatively from the stream head) varies as the 0.6 power of area in basins spanning nearly four orders (0.01 to 100 sq mi). An exponent of 0.5 is required if geometrical similarity is to be perfectly preserved, whereas the observed value of 0.6 requires that basins become somewhat longer and narrower as their size increases.

Of interest in the estimation of channel storage capacity for an entire watershed is Horton's [2, p. 291] observation that the laws of stream numbers and lengths can be combined as a product to yield an equation for the total length of channels of a given order u, knowing only the bifurcation and length ratios, the mean length $\bar{L}_1$ of the first-order segments, and the order of the trunk segment, thus:

$$\sum_{i=1}^{N} L_u = \bar{L}_1 R_b{}^{k-u} R_L{}^{u-1} \tag{4-II-7}$$

Furthermore, Horton [2, p. 293] shows that the total length of channels for all orders of a watershed of order k can be estimated as

$$\sum_{i=1}^{k}\sum_{i=1}^{N} L_u = \bar{L}_1 R_b{}^{k-1} \frac{R_{Lb}{}^k - 1}{R_{Lb} - 1} \tag{4-II-8}$$

where
$$R_{Lb} = \frac{R_L}{R_b}$$

A somewhat different approach to a measure of stream length representative of a given drainage basin does not use the Horton stream-order concept, but instead measures length L_{ca} from basin mouth (or other reference point on a stream such as a gage) to a point on the main stream channel opposite the computed center of gravity (centroid) of the total drainage area [28; 29; 30, p. 456; 31].

Snyder [28, p. 450] found from study of a large number of basins that the lag in time between center of mass of surface-producing runoff and resulting peak discharge at a given station varied as the 0.6 power of distance in miles from station to center of area. Taylor and Schwartz [29, p. 235] found distance L_{ca} from gage to computed center of gravity (centroid) of drainage area to be a significant factor in unit-hydrograph lag (Sec. 14).

C. Length of Overland Flow

Surface runoff follows a system of downslope flow paths from the drainage divide (basin perimeter) to the nearest channel. This flow net, comprising a family of orthogonal curves with respect to the topographic contours, locally converges or diverges from parallelism, depending upon position in the basin. Horton [2, p. 284] defined *length of overland flow* L_g as the length of flow path, projected to the horizontal, of nonchannel flow from a point on the drainage divide to a point on the adjacent stream channel. He noted [2, p. 284] that "length of overland flow is one of the most important independent variables affecting both the hydrologic and physiographic development of drainage basins." (See also Sec. 14.)

During evolution of the drainage system, L_g is adjusted to a magnitude appropriate to the scale of the first-order drainage basins and is approximately equal to one-half the reciprocal of the drainage density [2, p. 284].

Because the number of starting points on a basin perimeter is infinite, the choice of flow path to represent length of overland flow must be specified. An average length can be computed from measurements of a number of paths emanating from points uniformly spaced around the entire basin perimeter or extended upward from uniformly spaced points along the channel. A maximum length can be obtained for any given first-order basin by taking the longest possible flow path contributing to the tip of the first-order channel [26, pp. 6, 39].

A particular case of the length of overland flow is that used to describe the length of a triangular element of ground surface lying between two adjacent tributary basins and the larger stream they join. The relationships of these surfaces, which drain directly into channels of order higher than first, without themselves being included in any lower-order basin, are described below in the discussion of basin areas. The maximum horizontal length of one such element of surface, from its apex to the adjacent channel, is here designated as *interbasin length* L_0.

IV. AREAL ASPECTS OF DRAINAGE BASINS

A. Arrangement of Areal Elements

The perimeters of all first, second, and higher orders of basins may be drawn on the topographic map of a watershed. The area A_u of a basin of a given order u is defined as the total area projected upon a horizontal plane, contributing overland flow to the channel segment of the given order and including all tributaries of lower order (Fig. 4-II-2). For example, the area of a basin of the fourth order, A_4, would cumulate the areas of all first-, second-, and third-order basins, plus all additional surface elements, known as *interbasin areas*, contributing directly to a channel of order higher than first [21, p. 608].

In Fig. 4-II-7, A_2, the area of the second-order basin, consists of the sum of the two first-order basins plus the areas of two interbasin areas contributing directly to the second-order channel segment, and may be written as

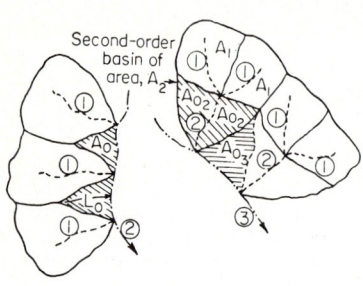

Fig. 4-II-7. Interbasin areas contributing runoff directly to second- and third-order channel segments.

$$A_2 = \sum_{i=1}^{N} A_1 + \sum_{i=1}^{N} A_{O_2} \qquad (4\text{-II-}9)$$

where A_{O_2} is the area of an interbasin area contributing to a second-order segment.

In the general case, the total area A_u of a basin of the order u may be written as

$$A_u = \left(\sum_{i=1}^{N} A_1 + \sum_{i=1}^{N} A_2 + \cdots + \sum_{i=1}^{N} A_{u-1} \right)$$
$$+ \left(\sum_{i=1}^{N} A_{O_1} + \sum_{i=1}^{N} A_{O_2} + \cdots + \sum_{i=1}^{N} A_{O_u} \right) \qquad (4\text{-II-}10)$$

B. Frequency Distribution of Basin Areas

The areas of drainage basins of a given order can be measured by planimeter from a map on which the perimeters have been outlined for each order. Frequency distribution of areas has been studied by Miller [25, p. 14] and Schumm [21, p. 607], who found that a strong right skewness in the distributions could be largely corrected by using the logarithm of area (Fig. 4-II-8). For a given order, area characteristics can be described in terms of mean, variance, and standard deviation computed from the sample. Although individual basin areas deviate widely from the mean, the means themselves show a progressive increase with order.

C. Law of Stream Areas

Horton [2, p. 294] inferred that mean drainage-basin areas of progressively higher orders should increase in a geometric sequence, as do stream lengths. Schumm [21, p. 606] expressed this relationship in a *law of stream areas*: the mean basin areas of stream of each order tend closely to approximate a direct geometric sequence in which the first term is the mean area of the first-order basin. This law may be written as

$$\bar{A}_u = \bar{A}_1 R_a^{u-1} \qquad (4\text{-II-}11)$$

where $\bar{A}_u$ is mean area of basins of order u, $\bar{A}_1$ is mean area of the first-order basins, and R_a is an area ratio analogous to the length ratio R_L.

AREAL ASPECTS OF DRAINAGE BASINS   4-49

As with stream length, the regression of logarithm of basin area on order is linear (Fig. 4-II-9). Morisawa [17, p. 51] confirmed this relationship in representative watersheds of the Appalachian Plateau Province. Leopold and Miller [20, pp. 19–20] found the law to apply to basins of ephemeral streams in central New Mexico.

D. Relation of Area to Length

Assuming the validity of the laws of stream lengths and basin areas, in which both properties are related in an exponential function with order, length should be related to area by a power function. Morisawa [17, pp. 12, 58–61] plotted both logarithm of mean stream length and logarithm of cumulative length against logarithm of basin area for each order of representative basins of the Appalachian Plateau Province, obtaining highly linear relationships (Fig. 4-II-10).

Absolute stream length, measured headward to the divide from a given point on a stream, plotted against area of watershed contributing to the stream above the given point, also shows a strongly linear relationship when the logarithms of both variables

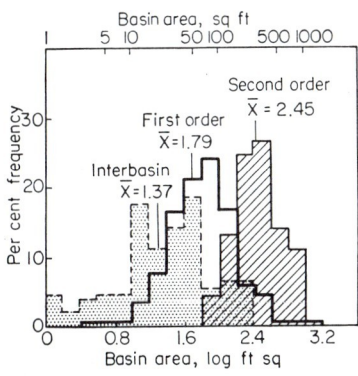

FIG. 4-II-8. Frequency distributions of logarithm of basin area for interbasin, first-order, and second-order areas at Perth Amboy badlands. (*After Schumm* [21, p. 607].)

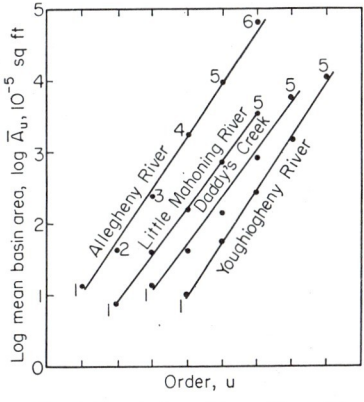

FIG. 4-II-9. Regression of logarithm of basin area on order for four drainage basins of the Appalachian Plateau Province. (*After Morisawa* [17, p. 51].)

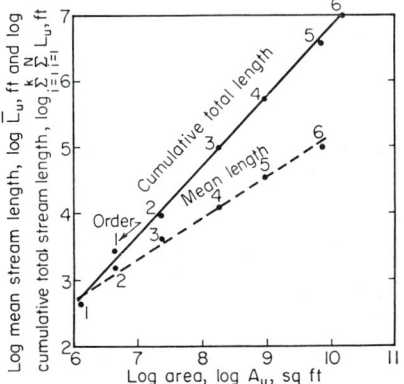

FIG. 4-II-10. Relation of stream length to basin area, order for order, for Allegheny River. (*After Morisawa* [17, p. 58].)

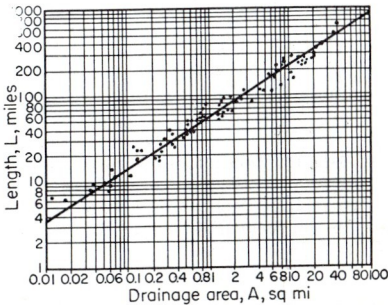

FIG. 4-II-11. Relation of stream length to drainage area for seven areas in Virginia and Maryland. (*After Hack* [27, p. 64].)

are used. Hack [27, p. 64] demonstrated the applicability of the power function relating length and area as thus defined for streams in seven areas of Virginia and Maryland (Fig. 4-II-11). He used the equation

$$L = 1.4 A^{0.6} \qquad (4\text{-II-}12)$$

where L is stream length in miles measured to a point on the drainage divide, and A is area in square miles. He plotted 400 similar measurements made by Langbein [4, p. 145] at gaging stations in the northeastern United States and found the same equation to apply. Hack noted [27, p. 64] that if geometrical similarity is to be preserved as a drainage basin increases in area downstream, the exponent in the above equation should be 0.5. An observed exponent larger than 0.5 requires that drainage basins change their overall shape in a downstream direction, becoming longer and narrower as they enlarge. Hack [27, pp. 65–67] further examined the relationships between area and length in terms of Horton's laws of drainage-network composition, showing that the area A_u of a basin of order u can be derived as

$$A_u = \bar{A}_1 R_b{}^{u-1} \frac{R_{Lb}{}^u - 1}{R_{Lb} - 1} \qquad (4\text{-II-}13)$$

where $\bar{A}_1$ is the mean area of first-order basins, R_b is the bifurcation ratio, and R_{Lb} is Horton's term for the ratio of length ratio to bifurcation ratio. This equation gives close agreement with the actual plot of logarithm of length as a function of logarithm of area for the watershed of Christian's Creek, Virginia [27, p. 67].

E. Relation of Area to Discharge

Empirical equations relating stream discharge to basin area have long been in general use in the form

$$Q = jA^m \qquad (4\text{-II-}14)$$

where Q is some measure of discharge in cfs, such as the mean annual flood; A is the watershed area in suitable areal units; and the constants j and m are derived by fitting a regression line to the available data. The exponent m generally falls in the range 0.5 to 1.0 [32, p. 329].

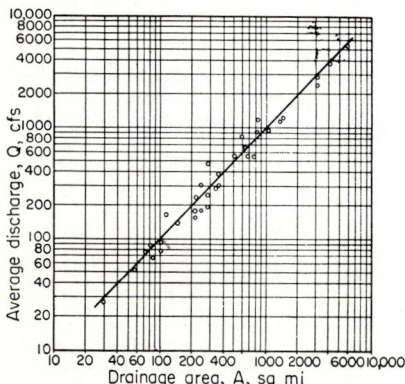

FIG. 4-II-12. Relation of discharge to drainage area for all gaging stations in the Potomac River basin, fitted by a regression line with coefficient of 1. (*After Hack* [27, p. 54].)

For example, Hack [27, p. 54] plotted average discharge (cubic feet per second) against drainage area (square miles) on logarithmic paper for all gaging stations in the Potomac River basin and fitted a regression line with an exponent of 1.0 (Fig. 4-II-12). From this he concluded that studies of relationship of basin area with respect to other variables, such as order, channel, slope, channel width, and stream length, would apply, by direct proportionality to average annual discharge as well.

Leopold and Miller [20, pp. 23–24] found that for 12 streams of central New Mexico the discharge-area relationship can best be described by the equation

$$Q_{2.3} = 12 A^{0.79} \qquad (4\text{-II-}15)$$

where $Q_{2.3}$ is flood discharge in cfs equaled or exceeded in 2.3 years, and A is drainage area in sq mi. They were then able to combine the discharge-area graph with an order-area graph to show the relationship of discharge to stream order.

F. Basin Shape (Outline Form)

The shape, or outline form, of a drainage basin, as it is projected upon the horizontal datum plane of a map, may conceivably affect stream-discharge characteristics. As explained above, long narrow basins with high bifurcation ratios would be expected to have attenuated flood-discharge periods, whereas rotund basins of low bifurcation ratio would be expected to have sharply peaked flood discharges.

Horton [33, pp. 303–304] described the outline of a normal drainage basin as a pear-shaped ovoid. He depicted the average outline of six great rivers of the world by a composite of their perimeters reduced to equal basin area and superimposed on one another. He regarded the pear shape as one proof that drainage basins are formed largely by sheet-erosion processes acting upon an initially inclined surface.

Quantitative expression of drainage-basin outline form was made by Horton [1, p. 351] through a form factor R_f, which is the dimensionless ratio of basin area A_u to the square of basin length L_b, thus:

$$R_f = \frac{A_u}{L_b^2} \tag{4-II-16}$$

In its inverted form, L_b^2/A_u, this ratio was used in unit hydrograph applications by the U.S. Army Corps of Engineers [34].

Miller [25, p. 8] used a dimensionless *circularity ratio* R_c, defined as the ratio of basin area A_u to the area of a circle A_c having the same perimeter as the basin. He found that circularity ratio remained remarkably uniform in the range 0.6 to 0.7 for first- and second-order basins in homogeneous shales and dolomites, indicating the tendency of small drainage basins in homogeneous geologic materials to preserve geometrical similarity. By contrast, first- and second-order basins situated on the flanks of moderately dipping quartzite strata of Clinch Mountain, Virginia, were strongly elongated and had circularity ratios of between 0.4 and 0.5 generally.

Schumm [21, p. 612] used an *elongation ratio* R_e, defined as the ratio of diameter of a circle of the same area as the basin to the maximum basin length. This ratio runs between 0.6 and 1.0 over a wide variety of climatic and geologic types. Values near to 1.0 are typical of regions of very low relief, whereas values in the range 0.6 to 0.8 are generally associated with strong relief and steep ground slopes.

The inappropriateness of a circle as the standard figure of reference in comparison with a pear-shaped drainage basin, which has a sharply defined point at the mouth, led Chorley, Malm, and Pogorzelski [35, pp. 138–141] to use as a model the lemniscate function

$$\rho = L_b \cos p\theta \tag{4-II-17}$$

where ρ and θ are radius and angle, respectively, in polar coordinates. L_b is basin length measured from mouth to most distant point on the perimeter, and p is a coefficient which determines the rotundity of the basin. When $p = 1$, the basin outline is a circle. Basin area A_u is obtained by integration of Eq. (4-II-17) between the limits $-\pi/2p$ and $+\pi/2p$, giving

$$A_u = \frac{\pi L_b^2}{4p} \tag{4-II-18}$$

and

$$p = \frac{\pi L_b^2}{4 A_u} \tag{4-II-19}$$

Thus the coefficient p, which expresses rotundity, is readily obtained by substitution of measurements of basin length and basin area. The degree of approach of actual basin form to the pure lemniscate form is measured by a *lemniscate ratio*, the ratio of perimeter of the lemniscate to actual perimeter of the basin.

Morisawa [36, pp. 587–591] tested the effectiveness of the above measures of basin outline form as factors in the hydrology of a watershed. For 25 watersheds of the Appalachian Plateau, regressions of the runoff-rainfall ratio on five measures of form showed a significant regression coefficient at the 5 per cent level only with elongation ratio R_e and circularity ratio R_c, but the standard error of estimate is relatively high in both cases. From this it is concluded that controls other than drainage-basin outline form dominate the hydrologic characteristics of a basin.

G. Drainage Density

An important indicator of the linear scale of land-form elements in stream-eroded topography is *drainage density D*, introduced into the American hydrologic literature by Horton [1, p. 357; 2, p. 283]:

$$D = \frac{\sum_{i=1}^{k} \sum_{i=1}^{N} L_u}{A_u} \qquad (4\text{-II-}20)$$

Thus D is simply the ratio of total channel-segment lengths cumulated for all orders within a basin to the basin area (projected to the horizontal). Dimensionally, this ratio reduces to the inverse of length, L^{-1}. Horton used units of miles per square mile, and most later workers followed suit. Drainage density may be thought of as an expression of the closeness of spacing of channels. If geometrical similarity exists between two drainage systems, their drainage densities will be related in the same ratio as the inverse of the linear scale ratio (Fig. 4-II-1). Thus, broadly considered, drainage density is simply one of several linear measures by which the scale of features of the topography can be compared.

Measurement of drainage density is made from a map with the planimeter and chartometer. If a complete morphometric analysis is being made, the necessary stream lengths and basin areas will have been measured in the course of the analysis. An average drainage-density value for each order can be computed and designated $D_1, D_2, \ldots, D_k$ through the highest order k. Extreme care must be taken to include all permanent stream channels to their upper ends. Checking in the field and on air photographs is an essential step in verification of topographic maps. A rapid approximation method of drainage-density determination is described by Carlston and Langbein [37].

Drainage-density measurements have been made over a wide range of geologic and climatic types of the United States. The lowest values, between 3.0 and 4.0 miles/sq mi, are observed in resistant sandstone strata of the Appalachian Plateau Province [38, pp. 658–659; 17, pp. 84–86] (Fig. 4-II-13A). Values in the range 8 to 16 are typical of large areas of the humid central and eastern United States on rocks of moderate resistance under a deciduous forest cover [7, p. 1135; 39, pp. 19, 59] (Fig. 4-II-13B). Comparable values are found in parts of the Rocky Mountain region [15, table 2], but in the drier areas of that region values range from 50 to 100.

Coast ranges of southern California, where strongly fractured and deeply weathered igneous and metamorphic rocks have evolved under a dry-summer subtropical climate, show drainage-density values in the range 20 to 30 (Fig. 4-II-13C) [38, p. 659; 7, p. 1135; 18, appendix I], but where weak Pleistocene sediments are exposed, values of D rise to 30 to 40. A still higher order of magnitude of drainage density is observed in badlands, developed on weak clays barren of vegetation. Smith [22, p. 999] measured values of 200 to 400 in Badlands National Monument, S.D. (Fig. 4-II-13D). Schumm [21, p. 616] measured values as high as 1,100 to 1,300 in badlands developed on weak clay at Perth Amboy, N.J.

Factors controlling drainage density are the same as those that control the characteristic length dimension of any group of first-order basins. A complete discussion is not appropriate here. In general, low drainage density is favored in regions of highly resistant or highly permeable subsoil materials, under dense vegetative cover, and where relief is low. High drainage density is favored in regions of weak or impermeable subsurface materials, sparse vegetation, and mountainous relief. A

AREAL ASPECTS OF DRAINAGE BASINS

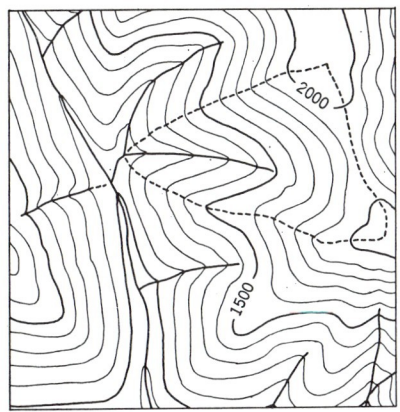

A. Low drainage density or coarse texture, Driftwood, Pennsylvania, Quadrangle.

B. Medium drainage density or medium texture, Nashville, Indiana, Quadrangle.

C. High drainage density or fine texture, Little Tujunga, California, Quadrangle.

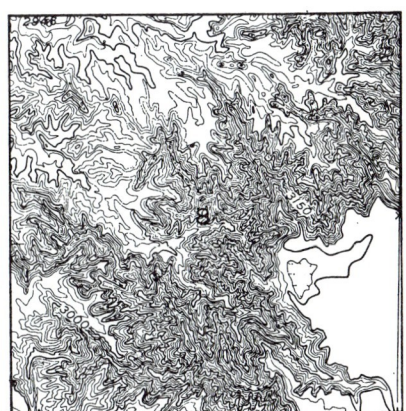

D. Extremely high drainage density or ultrafine texture, Cuny Table West, South Dakota, Quadrangle.

Fig. 4-II-13. Topographic maps of 1 sq mi each, illustrating natural range in drainage density. (*From maps of the U.S. Geological Survey. Reproduced by permission from Strahler, Physical Geography, copyright by John Wiley & Sons, Inc., New York, 1960.*)

comprehensive study of drainage-density controls by Melton [15, pp. 33–35] used multiple-regression-and-correlation analysis in which drainage density is the dependent variable with respect to Thornthwaite's precipitation-effectiveness index[1] and infiltration capacity, vegetative cover, surface roughness, and a runoff-intensity index as independent variables. Of these, only surface roughness has no significant correlation with drainage density. The strongest related factor appears to be

[1] PE index $= 10 \sum_{1}^{12} (P/E)$, where P is the average precipitation for each month, and E the average evaporation for each month. It is a measure of the availability of moisture to vegetation. (See also Subsec. 11-V-B.)

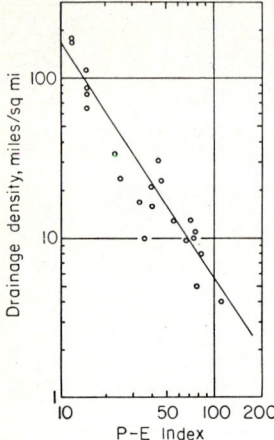

FIG. 4-II-14. Regression of drainage density on Thornthwaite's precipitation-effectiveness index for 22 small drainage basins of Colorado, Arizona, New Mexico, and Utah. (*After Melton* [15, p. 73].)

Thornthwaite's *PE* index, with a simple correlation coefficient of -0.943; a two-variable regression of this relationship is shown in Fig. 4-II-14.

Horton [2, p. 293] combined the laws of stream numbers and lengths with his definition of drainage density to yield

$$D_u = \frac{\bar{L}_1 R_b{}^{u-1}}{A_u} \frac{R_{Lb}{}^u - 1}{R_{Lb} - 1} \quad (4\text{-II-}21)$$

where D_u is the drainage density of an entire basin of order u. Horton noted that this equation combined all the geometric factors which determine the composition of the drainage net of a stream system into one expression. This can be regarded as the quantitative statement of a major part of *Playfair's classic law of streams*[1] stated in 1802.

The average length of overland flow $\bar{L}_g$ is approximately half the average distance between stream channels and is therefore approximately equal to half the reciprocal of drainage density [2, p. 284]:

$$\bar{L}_g = \frac{1}{2D} \quad (4\text{-II-}22)$$

In order to take into account the effect of slope of the stream channels and valleysides, Horton [2, p. 285] refined this generalization to read

$$\bar{L}_g = \frac{1}{2D \sqrt{1 - (\theta_c/\theta_g)}} \quad (4\text{-II-}23)$$

where θ_c is channel slope, and θ_g is average ground slope in the area.

H. Constant of Channel Maintenance

Schumm [21, p. 607] used the inverse of drainage density as a property termed *constant of channel maintenace C*. Thus

$$C = \frac{1}{D} = \frac{A_u}{\sum_{i=1}^{k} \sum_{i=1}^{N} L_u} \quad (4\text{-II-}24)$$

This constant, in units of square feet per foot, has the dimension of length and therefore increases in magnitude as the scale of the land-form units increases. Specifically, the constant C tells the number of square feet of watershed surface required to sustain one linear foot of channel. The relation of C to stream order is shown in Fig. 4-II-15 [10, p. 917], in which logarithm of basin area (ordinate) is plotted against logarithm of cumulative stream length (abscissa). Both lengths and areas are those measured when projected to a horizontal plane. For three basins shown, the series of points

---

[1] The law of streams originally stated by John Playfair reads as follows: "Every river appears to consist of a main trunk, fed from a variety of branches, each running in a valley proportioned to its size, and all of them together forming a system of valleys connecting with one another, and having such a nice adjustment of their declivities that none of them join the principal valley either on too high or too low a level; a circumstance which would be infinitely improbable if each of these valleys were not the work of the stream which flows in it."

numbered 1 to 5, represent the data for successive orders within each basin. In each basin the points fall close to a straight line of 45° slope, indicating that a linear relationship may be used. If the logarithm of the intercept is read at log stream

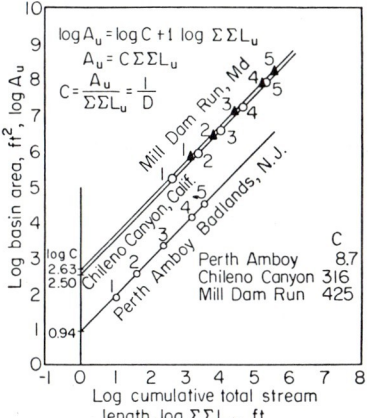

Fig. 4-II-15. Constant of channel maintenance throughout successive orders of three drainage basins. (*Data by Schumm* [21, p. 606], *modified by Strahler* [10, p. 917].)

length = 0 and the antilog of this intercept is taken, the constant of channel maintenance C for the whole basin is obtained. The value of $C = 8.7$ in the Perth Amboy badlands means that, on the average, 8.7 ft² of surface is needed to support each linear foot of channel. By contrast, 316 ft² of surface is required to support one foot of channel in the Chileno Canyon watershed of the San Gabriel Mountains.

I. Stream Frequency

Horton [1, p. 357; 2, p. 285] introduced *stream frequency* (or *channel frequency*) F as the number of stream segments per unit area, or

$$F = \frac{\sum_{i=1}^{k} N_u}{A_k} \tag{4-II-25}$$

where $\sum_{i=1}^{k} N_u$ is the total number of segments of all orders within the given basin of

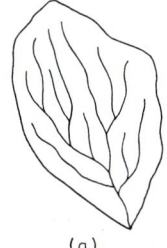  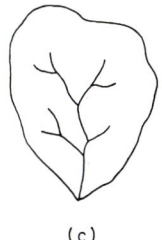 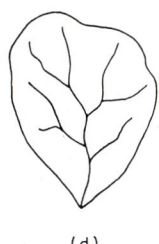

(a) (b) (c) (d)

Fig. 4-II-16. Hypothetical basins a and b have the same drainage densities but different stream frequencies; basins c and d have the same stream frequencies but different drainage densities.

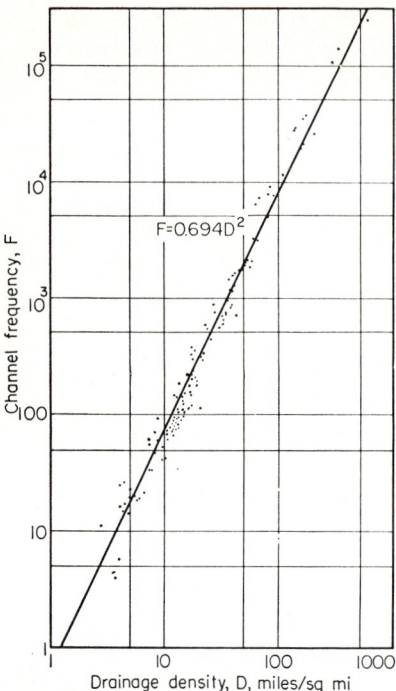

Fig. 4-II-17. Regression of channel frequency on drainage density for 156 drainage basins. (*After Melton* [16, p. 37].)

order k, and A_k is the area of that basin in square miles. Stream frequency has the dimensions L^{-2}.

Melton [16, pp. 35–54] analyzed in detail the relationships between drainage density and stream frequency, both of which measure the texture of the drainage net, but each of which treats a distinct aspect. As shown in Fig. 4-II-16, it is possible to construct two hypothetical drainage basins having the same drainage density but different stream frequency, and on the other hand, it is possible to have two basins of the same frequency but different density. Melton tested this possible range of variation in nature by plotting F versus D for 156 drainage basins covering a vast range in scale, climate, relief, surface cover, and geologic type (Fig. 4-II-17). Remarkably small scatter exists, showing that the relationship of density to frequency tends to be conserved as a constant in nature. The slope of a least-squares line was not significantly different from 2.0, from which Melton [16, p. 36] derived the dimensionally correct equation

$$F = 0.694 D^2 \qquad (4\text{-II-}26)$$

and from this the dimensionless number F/D^2, which tends to approach the constant value 0.694, despite vast variations in linear scale.

V. RELIEF (GRADIENT) ASPECT OF DRAINAGE BASINS AND CHANNEL NETWORKS

A. Channel Gradients

1. Single-channel Profiles. The longitudinal profile of a stream channel may be shown graphically by a plot of altitude (ordinate) as a function of horizontal distance (abscissa) (Fig. 4-II-18A). Altitude is commonly stated in feet or meters above the sea-level datum; distance, in miles or kilometers from stream head, stream mouth, or some other convenient reference point. For streams of large discharge and high order, a considerable factor of vertical exaggeration is used, whereas for streams of low order in regions of strong relief, none may be required.

A single-channel profile follows one channel continuously despite the junction of tributaries of equal or lower stream order [26, p. 5]. Within a given basin that part of the profile following the trunk stream of highest order is unambiguous, whereas to continue the profile headward into channels of lower order requires choice of one of two alternatives at the head of each segment of a given order. Choice may be governed by which branch falls most directly in line with the higher-order segment or which branch leads eventually to the longest total stream length in the entire basin.

Where profiles are plotted from topographic maps, stream elevation is estimated from the contour crossings, the distinction between stream bed and stream surface in streams of low order being undetermined or neglected as trivial. Where, however, the profile of a perennial stream of large discharge is plotted, the elevation of the stream surface may be defined as the elevation of mean low water, or actual lowest low water [40, pp. 624–626; 41, p. 650], or as the elevation at some rigorously defined stage, such as the mean or median annual discharge. Actual elevation of the stream bed along the thalweg, or line of maximum depth in channel cross section, is used instead of stream-surface elevation where attention is focused upon processes of scour and aggradation [40, p. 622] and where profiles are made from direct instrumental field surveys in streams of small discharge [42, p. 11].

2. Cause of Profile Upconcavity. Single-channel profiles of almost all streams, under a wide range of climatic and geologic conditions, show *upconcavity*, i.e., a persistent downstream decrease in gradient. Causes of upconcavity cannot be treated in detail here. Gilbert [43, pp. 103–104, 107–108] explained upconcavity as an effect of increasing stream discharge. His *law of declivities* states that declivity (gradient) bears an inverse relation to discharge because, as discharge increases, channel cross-section increases, reducing proportionately the frictional losses of the stream and enabling it to carry its bedload on a lesser slope.

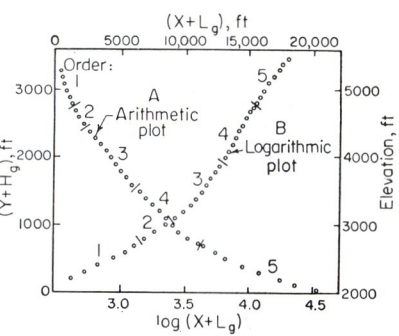

FIG. 4-II-18. Single-channel longitudinal profile of Chileno Canyon, California. (A) Arithmetic plot; vertical exaggeration ×5. (B) Logarithmic plot. Terms defined in Fig. 4-II-19. (*After Broscoe* [26, p. 72].)

Others have attributed upconcavity to decreasing caliber (diameter) of bed-load particles downstream [40; 44, pp. 660–662], using the reasoning that a lesser gradient suffices for the transport of finer bed materials.

3. Fitted Regression Functions. The longitudinal stream profile may be fitted by an equation expressing the statistical regression of elevation Y as the dependent variable on distance X as the independent variable. Four simple regression equations may be considered [26, pp. 5–6]:

1. Simple linear form, in which both altitude and distance are plotted on arithmetic scales. A straight line on such a plot is represented by the regression equation of basic form

$$Y = a - bX \qquad (4\text{-II-27})$$

Although useful in providing a visual impression of the longitudinal profile (Fig. 4-II-18), the arithmetic plot typically yields a strong upconcave-profile line to which the linear equation is poorly fitted.

2. Exponential form, in which altitude is on a logarithmic scale while horizontal distance is on an arithmetic scale. A straight line on such a plot is represented by the basic regression equation

$$\log Y = a - bX \qquad (4\text{-II-28})$$

3. Logarithmic form, in which altitude is plotted on an arithmetic scale on the ordinate against distance scaled logarithmically on the abscissa (Fig. 4-II-18). A straight line on such a plot is represented by the regression equation of the basic form

$$Y = a - b \log X \qquad (4\text{-II-29})$$

4. Power form (log-log form), represented by the basic regression equation

$$\log Y = \log a - b \log X \qquad (4\text{-II-30})$$

With appropriate definitions of Y and X, the exponential, logarithmic, and power functions are capable of making upconcave profiles more nearly approximate straight lines and therefore minimizing the deviations from the ideal mathematical expression selected for description and prediction.

A serious problem in plotting the stream profiles in exponential, logarithmic, and power forms is the selection of meaningful reference points from which the arbitrary constants of these equations are derived [26, p. 6]. Assignment of arbitrary constants does not affect the linear equation. In Eq. (4-II-29), where X is defined as distance downstream from the profile head, the equation cannot be solved for $X = 0$, because logarithm of zero is not defined. If an arbitrary constant C is added to X, the stream head may be plotted, for when $X = 0$, $Y = a - b \log C$. An element of horizontal distance has thus been added to the stream head to define the origin, or reference point,

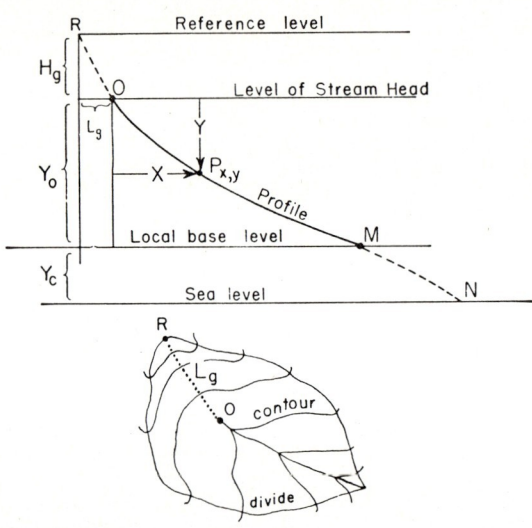

FIG. 4-II-19. Reference points in profile plotting.

Points: R reference point
           O stream head
           M end of segment, gaging station, or local base level
           N mouth at sea level

Constants: L_g horizontal distance of overland flow on orthogonal from reference point R to head O
            H_g vertical drop from R to O
            Y_o vertical distance from stream head O to local base level M
            Y_c vertical distance between local base level and sea level

Variables: Y drop in elevation from stream head O to any point P on stream profile
            X horizontal distance from stream head O to any point P on stream profile

(*After Broscoe* [26, p. 39].)

for the measurement of downstream distance. Degree of curvature of the plotted profile now depends upon the arbitrary constant selected, enabling the equation to be so adjusted as to fit closely a given stream profile. In so doing, however, the application of the logarithmic function as a general case is greatly impaired.

Broscoe [26, p. 6] proposed that the arbitrary constant in the logarithmic function be unambiguously defined by selecting a reference point R on the drainage divide of the watershed, where it is intersected by a profile line projected in the upstream direction along the longest path of overland flow leading to the channel head. The length of this line is equivalent to the term L_g, defined previously as length of overland flow. Justification for use of maximum L_g is that, during the evolution of the drainage basin, L_g has been adjusted in magnitude to the scale of the first-order drainage basins and is approximately equal to one-half of the reciprocal of drainage density.

Figure 4-II-19 shows a complete definition of reference point and constants in stream-profile plotting. In the logarithmic equation, the horizontal distance X is now

ASPECT OF DRAINAGE BASINS AND CHANNEL NETWORKS   4–59

replaced by the term $X + L_g$. Correspondingly, a constant of vertical distance H_g is defined as the elevation difference between the point R and the head of the stream channel O. The variable Y is then defined as vertical drop from O to any point P on the stream. The variable X is defined as the horizontal distance in the downstream direction from O to any point P. In the regression equations, then, the origin of numerical values shifts from point O to point R; the dependent variable becomes $Y + H_g$; the independent variable becomes $X + L_g$.

Using the new definitions, Eqs. (4-II-27) to (4-II-30) can be rewritten, respectively, as:

Linear: $Y + H_g = a + b(X + L_g)$ (4-II-31)
Logarithmic: $Y + H_g = a + b \log (X + L_g)$ (4-II-32)
Exponential: $\log (Y_c + Y_0 - Y) = -b(X + L_g)$ (4-II-33)
Power: $\log (Y_c + Y_0 - Y) = \log a - b \log (X + L_g)$ (4-II-34)

4. Derivative Functions. A vexing problem of arbitrary constants arises in Eq. (4-II-33). In most exponential plots of stream profiles [40, 44–46], sea level is

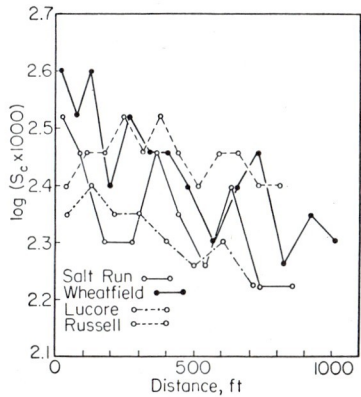

FIG. 4-II-20. Exponential relation of channel slope to downstream distance from stream head for four first-order stream-channel segments of the Appalachian Plateau in north-central Pennsylvania. (*After Broscoe* [26, p. 45]).

taken as the arbitrary reference level. Although sea level is a natural, readily defined geologic feature, related to stream development, there are good reasons to think that sea level does not relate dynamically to the control of stream slope in the upper reaches of the stream. Rubey [46, p. 134] has stated succinctly the reasons for concluding that the level of water body into which a stream flows controls the vertical position of the profile, but not its shape.

A solution to the problem of an arbitrary base of reference is to examine the relation of stream slope S_c to distance downstream, where S_c is defined as dY/dX. Differentiation removes reference to sea level, which is in effect a constant of integration in Eqs. (4-II-31) to (4-II-34). Differentiation yields:

Linear: $S_c = b$ (4-II-35)
Logarithmic: $S_c = \dfrac{b}{X + L_g}$ (4-II-36)
Exponential: $\log S_c = \log b - b(X + L_g)$ (4-II-37)
Power: $\log S_c = \log ab - (b + 1) \log (X + L_g)$ (4-II-38a)

or, letting $a' = \log ab$, and $b' = b + 1$,

$\log S_c = a' - b' \log (X + L_g)$ (4-II-38b)

An example of the exponential slope plot [Eq. (4-II-37)] of four stream segments of the first order is shown in Fig. 4-II-20. To the extent that the observed points fall on a straight, sloping line, the exponential profile can be used. If a fitted regression line has a slope not different from zero, the linear equation is suggested (Russell Hollow, in Fig. 4-II-20). Shulits [40, pp. 624–626] and Yatsu [44, pp. 656–659] use exponential plots of slope as a function of distance and seem to have accepted the exponential form as the best description of stream profiles. Hack [27, pp. 69–70] fitted the power-slope function [Eq. (4-II-38b)], to several streams of Virginia and Maryland. Broscoe [26, pp. 45, 72] observed that the logarithmic plot [Eq. (4-II-36)] gave a good approach to a straight line for several single-channel profiles (Fig. 4-II-18). No generalization as to the best description of the single-channel longitudinal profile seems yet warranted in view of such divergent observations.

5. Profile Segmentation. That the longitudinal profile of a stream channel consists of series of connected segments, "each differing from those that adjoin it, but all closely related parts of one system," has been pointed out by Mackin [47, p. 491], who states further that "each segment has the slope that will provide the velocity required for transportation of all of the load supplied to it from above, and this slope is maintained without change as long as controlling conditions remain the same." To describe a single-channel profile by one continuous mathematical function is therefore unrealistic in failing to take segmentation into account, but may nevertheless be useful in certain applications.

Abrupt changes in gradient marking the discontinuities between adjoining channel segments may result from changes in discharge-load ratios, in caliber of load, or in channel characteristics [44, p. 657; 47, p. 491; 48, p. 819].

Considering the convergence of a drainage network into channel segments of increasing order, it is obvious that the formation of a segment of a given order by the junction of two segments of the next lower order will normally mark an abrupt reduction in gradient, for reasons explained by Gilbert's law of declivities, discussed above. Between tributary junctions the profile may be expected to approach a straight line of uniform slope, discounting the slight upconcavity to be expected from gradual increases in discharge and load from directly contributing valleyside slopes. Actual plots of single-channel profiles do not show obvious segmentation associated with changes from one order to the next [26, pp. 44, 72], but the principle is strongly displayed in the composite profiles described below.

6. Composite Profiles. A composite stream profile combines the segments of a given order within the watershed into a single average segment whose vertical drop is the mean drop of all the segments and whose horizontal distance is the mean horizontal length of all the segments (Fig. 4-II-21). The average slope of the channel segments of a given order is thus the slope of the hypotenuse of a right triangle defined by the average vertical drop and the average horizontal distance. Triangles for each order are connected in sequence to produce the composite profile shown in Fig. 4-II-21.

The succession of order segments of the composite profile may be fitted, if desired, by a continuous mathematical function, using one of the four forms [Eqs. (4-II-31) to (4-II-34)]. Similarly, the derivative equations of slope [Eqs. (4-II-35) to (4-II-38b)] may be fitted to the composite profile by plotting slope of the segment against horizontal distance $X + L_g$, measured to the mid-point of the segment. Details and examples are given by Broscoe [26, pp. 8, 16, 29–30, 46, 73], who found that the composite profiles are best fitted by a logarithmic equation.

7. Channel Slope as a Function of Order. The average slope of segments of a given order in a drainage net, measured as described above, will obviously be less than the average slope for the next lower order but greater than that for the next higher order. Horton [2, p. 295] expressed this relationship in a *law of stream slopes*, an inverse-geometric-series law, which is analogous to the law of stream numbers.

$$\bar{S}_u = \bar{S}_1 R_s{}^{k-u} \qquad (4\text{-II-}39)$$

where $\bar{S}_u$ is average slope of segments of order u; $\bar{S}_1$ is average slope of first-order segments; R_s is a constant slope ratio, analogous to bifurcation ratio; and k is the order of the highest-order segment.

The law of stream slopes has been applied to many watersheds [17, pp. 9–10, 52–53;

ASPECT OF DRAINAGE BASINS AND CHANNEL NETWORKS 4–61

20, p. 21; 21, p. 605; 26, p. 15] and appears to be generally valid, provided that the geologic materials in which the channels are carved are free of strong inhomogeneities.

8. Main-stream-slope Factor. Use of a single numerical value representing slope of the main stream of a drainage basin has been made by Taylor and Schwartz [29, pp. 235–238, 244]. Their slope factor, S_{st}, termed equivalent main-stream slope, is the slope, in feet per foot, of a uniform channel having the same length as the

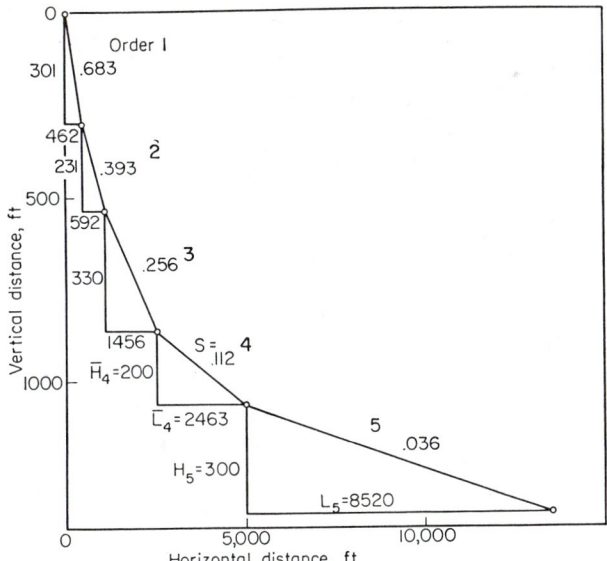

Fig. 4-II-21. Composite profile of stream segments of five orders in the Chileno Canyon watershed, California. (*After Broscoe* [26, p. 72].)

longest watercourse and an equal time of travel. This slope factor proved to be a significant drainage-basin characteristic influencing unit-hydrograph lag and peak flow of 20 basins ranging from 20 to 1,600 sq mi in the North and Middle Atlantic States. For further discussions of a main-stream-slope factor, see Refs. 31, 34, and 49.

B. Channel-cross-section Geometry

In this section the channel network has been treated only as a system of branching lines, without consideration of the fact that channels have finite depths and widths and that these parameters change systematically as the channels are followed downstream or with fluctuations of discharge. The field of hydraulic geometry of streams, which relates width, depth, cross-sectional area, and form to such factors as distance, gradient, discharge, and load, is treated in Sec. 17-II of this handbook. Detailed treatment given in Refs. 20, 27, 42, and 50 to 53 constitutes an integral part of the field of quantitative geomorphology of drainage basins and channel systems.

C. Ground-surface Gradients

1. Relationship of Ground and Channel Slopes. The inclinations, or gradients, of the ground-surface elements of a watershed are closely tied in with its channel gradients and relief (elevation differences). In mountainous regions, where relief is great, erosion intensity is correspondingly high. Steep slopes contribute large quantities of relatively coarse textured detritus to channels, which must have steep gradients to enable stream flow to transport the debris as bed load through the channel system. In regions of low relief, slopes are gentle and shed relatively small quantities of fine-textured detritus, which in turn requires correspondingly low channel gradients for its transport. Strahler [5, p. 689] has observed a close quantitative relationship of

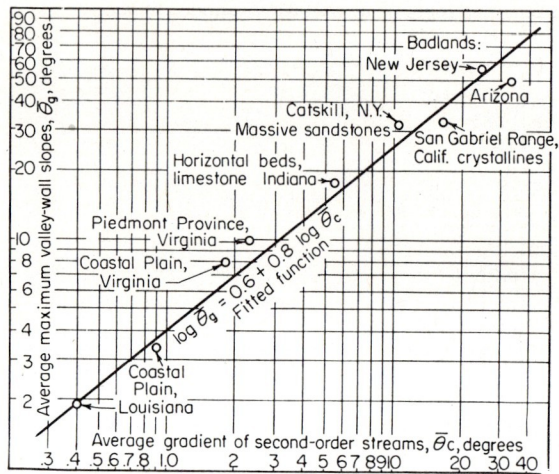

FIG. 4-II-22. Relation of valleyside slope to slope of immediately adjacent stream channel in nine maturely dissected regions of widely different relief, climate, and rock type, fitted by least squares with a power-regression equation. (*After Strahler* [5, p. 689].)

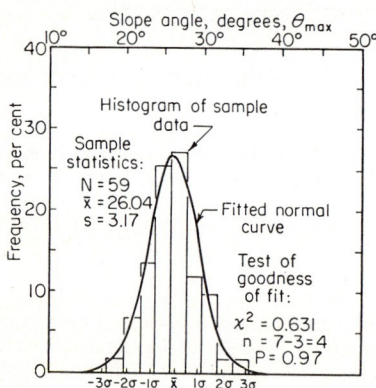

FIG. 4-II-23. Histogram of frequency distribution of maximum valleyside-slope angles fitted by a normal curve and tested for goodness of fit. Field measurements from dissected Santa Fe formation near Bernalillo, N.M. (*After Strahler* [5, p. 683].)

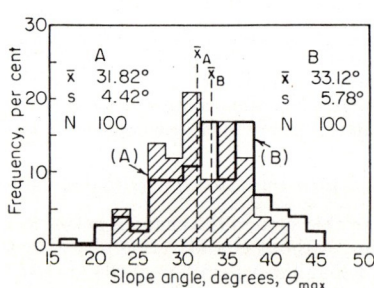

FIG. 4-II-24. Comparison of valleyside-slope samples in homogeneous clastic rocks in western Virginia. (*A*) Athens formation, shale; (*B*) Pennington formation, interbedded sandstone and shale. (*After Strahler* [8, p. 15]; *data from Miller* [25, p. 18].)

power form between channel slope θ_c and valleyside slope θ_g over a wide range of geographical regions (Fig. 4-II-22).

2. **Maximum Valleyside Slopes.** A significant indicator of overall steepness of ground slopes in a watershed is the maximum valleyside slope θ_{max}, measured at intervals along the valley walls on the steepest parts of the contour orthogonals running from divides to adjacent stream channels.

ASPECT OF DRAINAGE BASINS AND CHANNEL NETWORKS   4-63

Maximum valleyside slope has been sampled over a wide variety of geologic and climatic environments [5, 15, 16, 21, 22, 25, 39, 54, 55]. A sample of 50 to 100 or more slope readings, taken according to a plan of uniformly spaced sample points, may be measured directly in the field with the Abney level or with dividers and scale from topographic maps of suitably large scale and high degree of accuracy. The variates of the sample may then be grouped into classes and treated by standard procedures of frequency-distribution analysis, including calculation of arithmetic mean, variance, standard deviation, skewness, and goodness of fit to the normal curve (Fig. 4-II-23) (see Sec. 8-I). These statistics not only serve to describe the slope characteristics of a region, but they may be used in rigorous statistical tests of differences in means and variances between two regions or among several regions. For example, valleyside-slope samples collected by Miller [25] from adjacent localities of somewhat different lithologic composition prove to have significantly different variances when tested by the F ratio of sample variances [8, p. 15] (Fig. 4-II-24).

Based on field and map data from a large number of small watersheds in Arizona, Colorado, and New Mexico, Melton [16, p. 46] found by regression analysis that valleyside slope θ_{max} may be estimated by the equation

$$\theta_{max} = 27.53 \frac{\Sigma L_u^{0.25} H^{0.5}}{(\sqrt{A_u})^{0.75}} \qquad (4\text{-II-}40)$$

where ΣL_u is total channel length in the basin in miles, H is total basin relief in ft, and A_u is basin area in sq mi. These exponents show that area and relief have the greatest effect on valleyside slope, whereas channel length is less important.

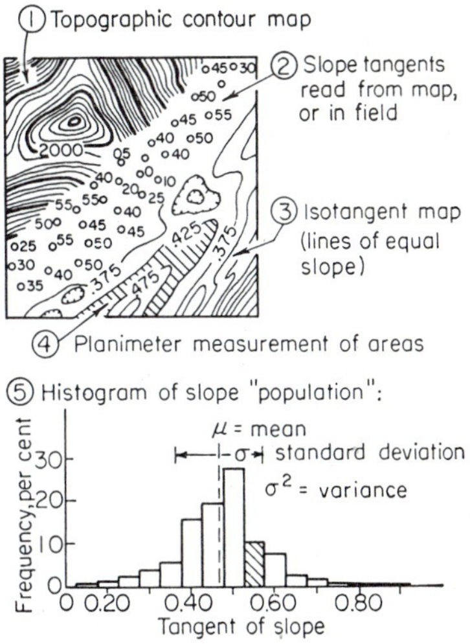

FIG. 4-II-25. Construction of isotangent slope map and slope-frequency histogram. (*After Strahler* [9, p. 575].)

3. Total Surface-slope Distribution. Slope conditions over an entire watershed may be shown by means of a slope map [9], which shows distribution of the degree of surface inclination. Procedure is as follows (Fig. 4-II-25): (1) A good contour topographic map is obtained. (2) Slopes of short segments of line normal to the contours are determined at many points over the map. Tangent or sine of slope angle may be recorded, depending upon the function desired. (3) The readings are contoured with lines of equal slope (isotangents or isosines). (4) The areas between successive slope contours are measured with planimeter and summed for each slope class. (5) This summation yields a slope frequency-percentage distribution from which mean, variance, and standard deviation can be computed.

Figure 4-II-26 shows topographic and isotangent maps compared for two small drainage basins in the Appalachian Plateau of north-central Pennsylvania. Figure 4-II-27 shows the resulting slope histogram for the area within the square. Remarkable similarity in both means and variances is typical of basins throughout a geologically uniform region.

Construction of slope maps and their areal measurement is extremely time-consuming. Essentially, the same information can be obtained by random-coordinate and grid sampling [9, pp. 589-594]. In the *random-coordinate method* a sample square is scaled in 100 units of length on a side. From a table of random numbers the coordinates of sample points are drawn for whatever size sample is desired (Fig. 4-II-28). The *grid-square method* achieves similar results, but is not flexible as to sample size. Point samples, easily and quickly taken, compare favorably in frequency-distribution properties with samples obtained by planimetry of a slope map.

4-64 QUANTITATIVE GEOMORPHOLOGY

A method of estimating the average slope of the ground surface within a drainage basin was used by Horton [56] and is described in detail by Wisler and Brater [32, pp. 46–47]. Average slope of each contour belt is computed, after which the average slope of the entire basin is computed, weighting each contour belt according to proportionate area. Strahler [7, pp. 1125–1128] describes a similar procedure and depicts a mean-slope profile for the complete basin.

Chapman [57] developed a method of analyzing both azimuth and angle of slope from contour topographic maps. Although based on petrofabric methods and designed largely for use in geologic analysis of terrain, his method might be applied to a watershed as a means of assessing both slope steepness and orientation simultaneously. For discussions of more general methods of slope analysis, see Refs. 31, 34, and 58 to 63.

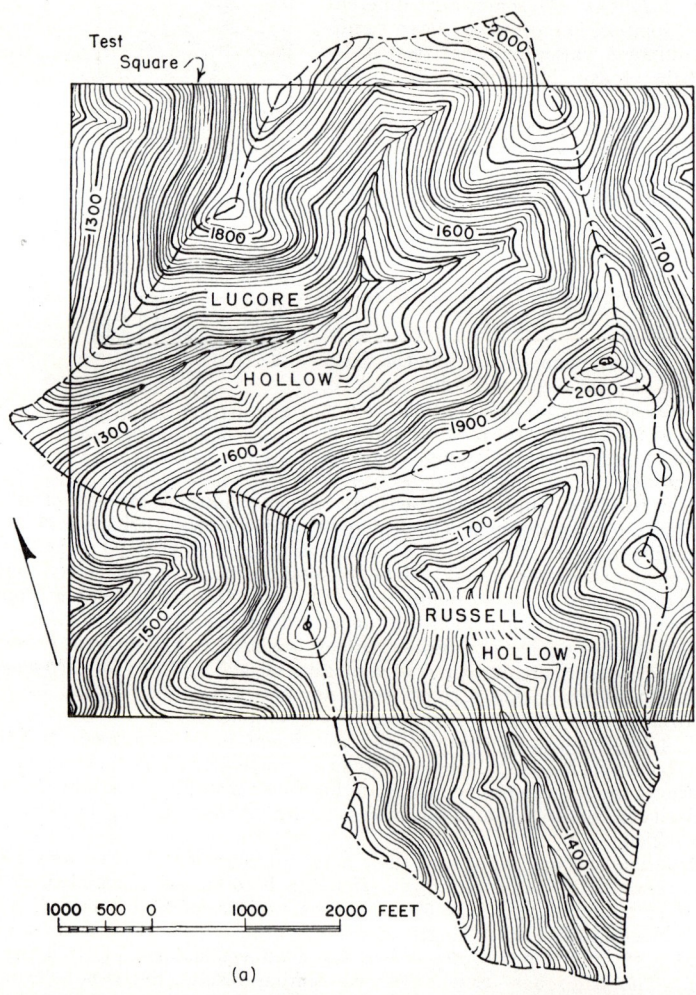

FIG. 4-II-26. Topographic contour map (a) and isotangent slope map (b) compared for two (*Topography by U.S. Geological*

D. Relief Measures

1. Relief. *Relief H* is the elevation difference between reference points defined in any one of several ways. *Maximum relief* within a region of given boundary is simply

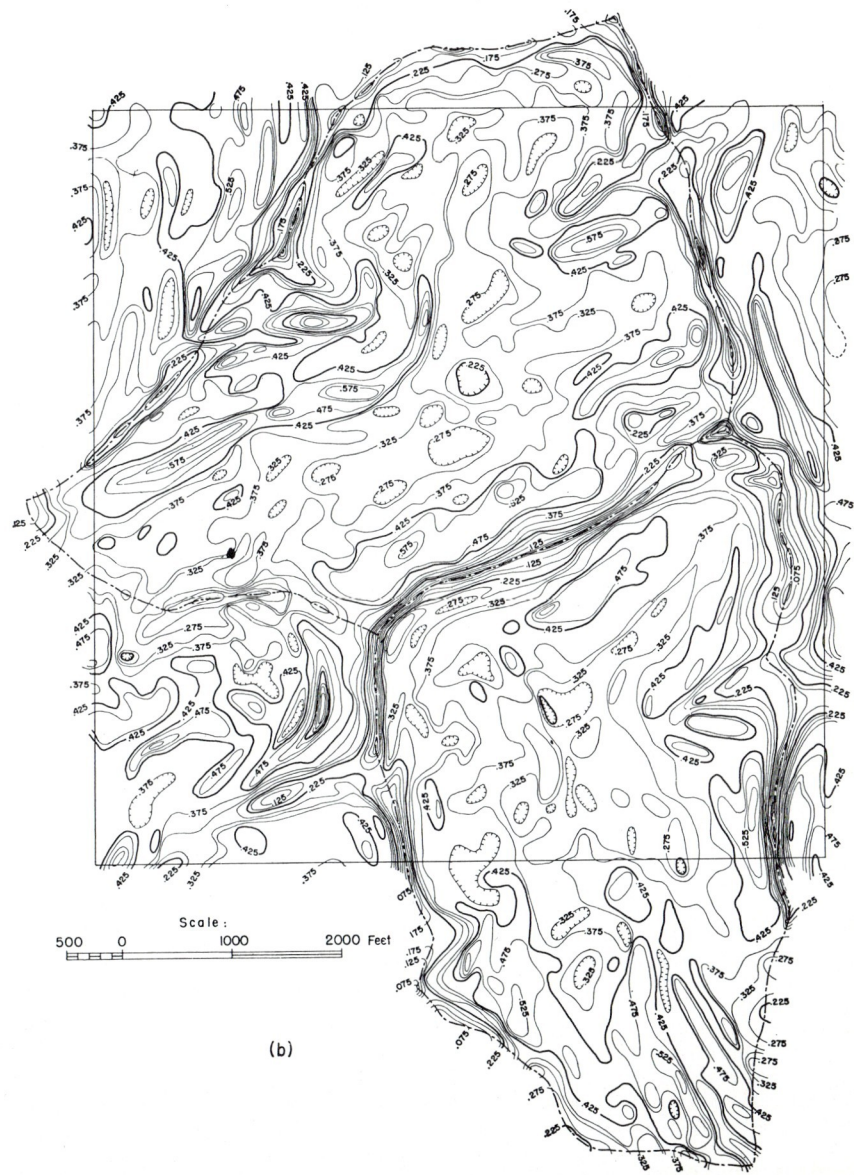

(b)

small drainage basins and an overlapping sampling square. Emporium, Pa., Quadrangle. Survey. After Strahler [9, pp. 586, 590].)

the elevation difference between highest and lowest points. *Maximum basin relief* is the elevation difference between basin mouth and the highest point on the basin perimeter, usually stated in units of feet or meters. Schumm [21, p. 612] measured basin relief along "the longest dimension of the basin parallel to the principal drainage line." Maxwell [18] measured relief along the basin diameter, an axial line found by use of rigorously defined criteria. Still another basin-relief measure may be obtained by determining the mean height of the entire basin perimeter above the mouth, thus minimizing the spurious effects of sharply pointed summits. Whatever criteria are

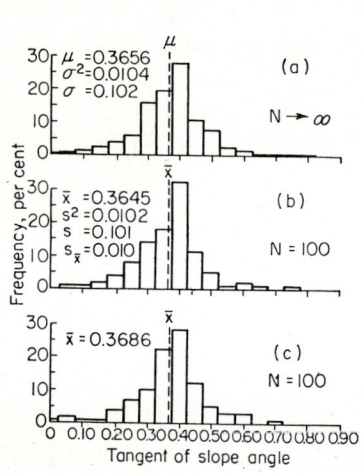

Fig. 4-II-27. Frequency-distribution histograms compared for sample square outlined in Fig. 4-II-26. (*a*) Total, or population, distribution measured by planimeter from isotangent map; (*b*) random-coordinate sample; (*c*) grid sample. (*After Strahler* [9, p. 592].)

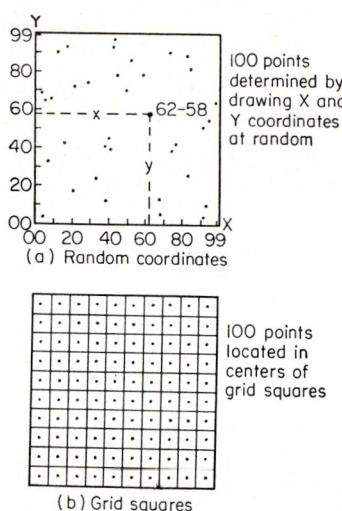

Fig. 4-II-28. Sampling of slope by random-coordinate and grid-square methods. (*After Strahler* [9, p. 591].)

selected for measuring basin relief, they may be applied to basins of a given order to yield a mean basin relief $\bar{H}_u$ of basins of order u.

Still another means of measuring relief is to take the elevation difference between a point on drainage divide and another on the nearest adjacent stream channel, where both points lie at the ends of a line orthogonal to the contours (e.g., the surface-flow path) [11, p. 295].

Relief measures are indicative of the potential energy of a drainage system present by virtue of elevation above a given datum. In the absence of information as to the horizontal distances over which the relief measurement applies, however, one cannot directly relate relief to ground and channel slopes.

2. Relief Ratios. When basin relief H is divided by the horizontal distance on which it is measured, there results a dimensionless *relief ratio* R_h [21, p. 112]. Taking vertical and horizontal distances as legs of a right triangle, relief ratio is equal to the tangent of the lower acute angle and is identical with the tangent of the angle of slope of the hypotenuse with respect to the horizontal. The relief ratio thus measures the overall steepness of a drainage basin and is an indicator of the intensity of erosion processes operating on slopes of the basin.

Schumm [21, p. 112] measured relief ratio R_h as the ratio of maximum basin relief to horizontal distance along the longest dimension of the basin parallel to the principal

ASPECT OF DRAINAGE BASINS AND CHANNEL NETWORKS 4-67

drainage line. Melton [15, p. 5] used relative relief R_{hp}, expressed in per cent, as

$$R_{hp} = \frac{100H}{5,280P} \qquad (4\text{-}II\text{-}41)$$

where H is maximum basin relief in ft, and P is basin perimeter in miles.

Use of the perimeter as a horizontal-length dimension does away with difficulties of locating a suitable axial line in the basin but has the shortcoming that minor crenulations of the perimeter may greatly increase its length without representing any actual change in characteristic areal dimensions of the basin. Maxwell [18] used basin diameter as the horizontal distance for calculation of a relief ratio.

Possibility of a close correlation between relief ratio and hydrologic characteristics of a basin is suggested by Schumm [64], who found that sediment loss per unit area is closely correlated with relief ratio (Fig. 4-II-29). The significant regression with small scatter suggests that relief ratio may prove useful in estimating sediment yield if the appropriate parameters for a given climatic province are once established.

Maner [65] used a relief-length ratio in correlation with sediment-delivery rates of watersheds in the Red Hills area of southern Kansas, western Oklahoma, and western Texas. This ratio yielded a higher correlation with sediment delivery rate than did relief and length treated together as variables. Moreover, it gave a much closer correlation than did other individually treated geometrical factors of length-width ratio of basin, sediment-contributing area, basin relief alone, or average land slope.

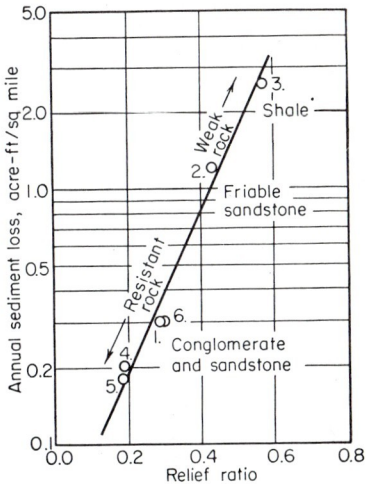

Fig. 4-II-29. Relation of sediment loss to relief ratio for small drainage basins in six localities of the Colorado Plateau Province. (*After Schumm* [64, p. 218].)

E. Ruggedness and Geometry Numbers

To combine the qualities of slope steepness and length, a dimensionless *ruggedness number* HD is formed of the product of relief H and drainage density D, where both terms are in the same units. If D should be increased while H remains constant, the average horizontal distance from divides to adjacent channels is reduced, with an accompanying increase in slope steepness. If H is increased while D remains constant, the elevation difference between divides and adjacent channels will also increase, so that slope steepness increases. Extremely high values of the ruggedness number occur where both variables are large, that is, when slopes are not only steep but long as well [11, p. 289]. Observed values of the ruggedness number range from as low as 0.06 in the subdued relief of the Louisiana coastal plain to over 1.0 in coast ranges of California or in badlands on weak clays.

The dimensionless property of slope can be introduced into the ruggedness number in the following way. Consider that the horizontal distance between a drainage divide and the adjacent stream channel is equal to about one-half the reciprocal of the drainage density D [2, p. 284] and that local relief H is measured as the vertical drop from divide to adjacent channel. Thus the slope S_g of the ground surface from divide to stream will be related to H and D by the equation

$$S_g = H \times 2D \qquad (4\text{-}II\text{-}42a)$$

where S_g is the tangent of the ground slope θ_g in degrees. Then

$$\frac{HD}{S_g} = \frac{1}{2} \qquad (4\text{-II-}42b)$$

Because the geometrical relations of H, D, and S_g will not be those of a perfect right triangle, the constant $\frac{1}{2}$ should be replaced by some dimensionless constant, determined empirically, that will differ little from unity, despite a wide range in the ruggedness number (numerator). Strahler [11, p. 296] computed values of HD/S_g,

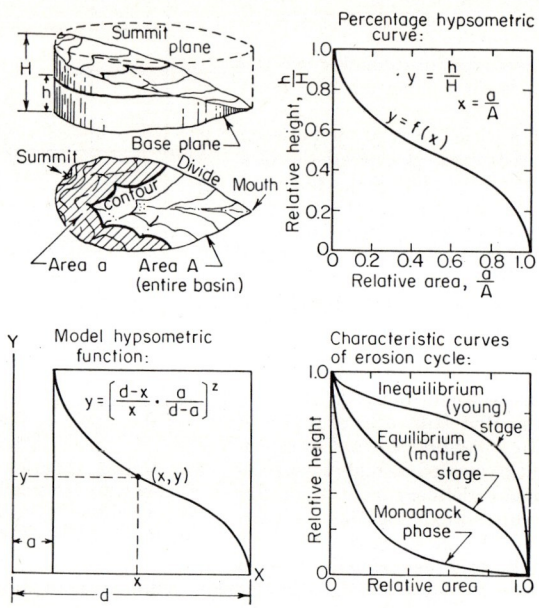

Fig. 4-II-30. Definitions and functions in hypsometric analysis of small drainage basins. (*After Strahler* [10, p. 919].)

named the *geometry number*, and found them to fall in the range 0.4 to 1.0 for six regions differing greatly in the individual components of D, H, and S_g. From this it was concluded that the geometry number tends to be conserved about a common value and that a change in any one of the three components is compensated for by changes in one or both of the other two, thus tending to keep the product constant.

F. Hypsometric (Area-Altitude) Analysis

Hypsometric analysis, or the relation of horizontal cross-sectional drainage-basin area to elevation, was developed in its modern dimensionless form by Langbein [4], who applied it to large watersheds. Application to small drainage basins of low order has been made by Strahler [7], Miller [25], Schumm [21], and Coates [39]. Similar methods have been described by Golding and Low [31].

Figure 4-II-30 illustrates the definition of the two dimensionless variables involved in hypsometric analysis. Taking the drainage basin to be bounded by vertical sides and a horizontal base plane passing through the mouth, the relative height y is the ratio of height of a given contour h to total basin height (relief) H. Relative area x is the ratio of horizontal cross-sectional area a to entire basin area A. The percentage hypsometric curve is a plot of the continuous function relating relative height y to relative area x. As shown in Fig. 4-II-30 (lower right), the shape of the hypsometric curve varies in early geologic stages of development of the drainage basin, but once a steady state is attained (mature stage), tends to vary little thereafter, despite lowering relief [7, pp. 1128–1132]. Isolated bodies of resistant rock may form prominent hills (monadnocks) rising above a generally subdued surface; the result is a distorted hypsometric curve, termed the *monadnock phase*.

Certain dimensionless attributes of the hypsometric curve, useful for comparative purposes, include the integral or relative area lying below the curve, the slope of the curve at its inflection point, and the degree of sinuosity of the curve. Many hypsometric curves seem to be closely fitted by the model function shown in Fig. 4-II-30 (lower left), although no rational basis is known for using this function. Hypsometric curves plotted for hundreds of small basins in a wide variety of regions and conditions show generally stable curve properties where the rock masses are homogeneous and the erosion stage is conventionally described as mature. Small but distinct differences in curve form from region to region appear to exist.

In hydrologic applications the hypsometric curve can be of use where some hydrologic factor, such as precipitation or evaporation, varies with altitude, or where the vegetative cover shows an altitude stratification. Langbein [4, p. 141] states: "For example, snow surveys generally show an increase in depth of cover and water equivalent with increase in altitude; the area-altitude relation provides a means for estimating the mean depth of snow or its water equivalent over a drainage basin."

VI. THEORY OF DRAINAGE-BASIN DYNAMICS

A. Statement of Variables

In quantitative studies of geomorphic processes and forms, the relationships between form elements, described above, and causative factors need to be expressed by dimensionally correct rational equations. It has already been noted that drainage basins developed in homogeneous bedrock materials under a given set of climatic conditions tend to develop a characteristic linear-scale dimension. Because of the tendency to geometrical similarity of the horizontal, or planimetric, aspects of such systems, one is free to select any property having the dimension of length or a product of length (inverse of length, length squared) to serve as the indicator of characteristic size of the elements in the system. Thus one might select mean length of first-order stream channels, or the mean perimeter of second-order basins, or the mean area of first-order basins. One of the most extensively known scale measures is drainage density D, the length of channels per unit area of watershed. Drainage density has the dimension of inverse of length, L^{-1}, and varies from values as high as 500 to 1,000, where first-order basins are only a few feet across, to values as low as 2 to 3, where the first-order basins are about a half mile wide. Drainage density is therefore used as the dependent variable in developing an equation relating the horizontal scale of the land-form units to a series of independent or controlling variables:

$$D = f(Q_r, K, H, \rho, \mu, g) \qquad (4\text{-II-}43)$$

All terms of this equation, together with their definitions and dimensional properties, are explained in Table 4-II-1.

The first independent variable, runoff intensity Q_r, combines rainfall intensity and infiltration capacity in a single term. Rainfall intensity represents a major climatic control; infiltration capacity, a major physical factor, expresses state of the ground surface and subsoil. Both components have the dimensions of velocity LT^{-1}; runoff intensity is simply the excess of rainfall intensity over infiltration capacity.

The second independent variable is an *erosion proportionality factor* K, defined by Horton [2, p. 324] as the ratio of erosion intensity to eroding force. Erosion intensity has the dimensions of mass rate of removal per unit area; eroding force, the dimensions of force per unit area. Thus K has the dimensions $L^{-1}T$, the inverse of velocity, and may be thought of as a measure of the erodibility of the ground surface.

Relief H, the third independent variable, represents the vertical dimension of the basin geometry and may vary independently of the horizontal scale. Relief represents potential energy of the system and is directly related to its total erosion intensity. Relief may be measured in various ways, described above, but is most meaningful in the analysis when defined as local, or basin, relief.

Table 4-II-1. Factors Controlling Drainage Density

Symbol	Term	Dimensional quality	Dimensional symbol
D	Drainage density	Length divided by area	$\dfrac{L}{L^2} = L^{-1}$
Q_r	Runoff intensity	Volume rate of flow per unit area of surface	$\dfrac{L^3 T^{-1}}{L^2} = LT^{-1}$
K	Erosion-proportionality factor	Mass rate of removal per unit area divided by force per unit area	$\dfrac{ML^{-2}T^{-1}}{ML^{-1}T^{-2}} = L^{-1}T$
H	Relief	Length	L
ρ	Density of fluid	Mass per unit volume	ML^{-3}
μ	Dynamic viscosity of fluid	Mass per unit length per unit time	$ML^{-1}T^{-1}$
g	Acceleration of gravity	Distance per unit time per unit time	LT^{-2}

The remaining variables—density ρ, viscosity μ, and acceleration of gravity g—are significant properties of a fluid system, introduced here because the drainage system is developed by water erosion on slopes and in channels, acting in a force field of gravity. Note that the first four variables involve no mass dimension; hence an analysis limited to these four would include only geometric and kinematic factors of time and length. Introduction of mass through density and viscosity brings force into the analysis and makes possible scale-model comparisons.

B. Solution by Pi Theorem

The variables of Eq. (4-II-43) may be grouped into the functional relationship

$$f'(D, Q_r, K, H, \rho, \mu, g) = 0 \qquad (4\text{-II-}44)$$

The seven variables in this function may be reduced to four through application of the pi theorem (Sec. 7, Subsec. II-B). Solution of the four pi terms, described in detail by Strahler [11, p. 290], yields a function of four dimensionless groups:

$$\phi\left(HD,\ QK,\ \frac{Q_r \rho H}{\mu},\ \frac{Q_r^2}{Hg}\right) = 0 \qquad (4\text{-II-}45)$$

Solving for drainage density gives

$$D = \frac{1}{H} f\left(Q_r K,\ \frac{Q_r \rho H}{\mu},\ \frac{Q_r^2}{Hg}\right) \qquad (4\text{-II-}46)$$

The term HD is the ruggedness number, previously described as expressing essential geometrical characteristics of the drainage system. It may be replaced by the dimensionless geometry number HD/S_g explained above, without affecting the dimensionless nature of the group. The second term, $Q_r K$, is the *Horton number*, expressing the relative intensity of the erosion process in the drainage basin. The third term, $Q_r \rho H/\mu$, is a form of the Reynolds number, in which Q takes the place of the velocity term and H the characteristic length. The fourth term, Q_r^2/Hg, is a form of Froude number. Reduction of the seven variables into four dimensionless groups focuses attention upon dynamic relationships, simplifies the design of controlled empirical observations, and establishes conditions essential to the validity of comparisons of models with prototypes.

C. Steady-state Relationships

Conditions for a steady state within a drainage basin are such that, for a given Horton number (i.e., for a given intensity of erosion process), values of local relief, slope, and drainage density reach a time-independent steady state in which basin geometry is so adjusted as to transmit through the system just that quantity of runoff and debris characteristically produced under the controlling climatic regime.

D. Upsets of Steady State

Consider possible upsets and readjustments of the steady state. If a forested land surface is denuded of its vegetative cover and intensively cultivated, the Horton

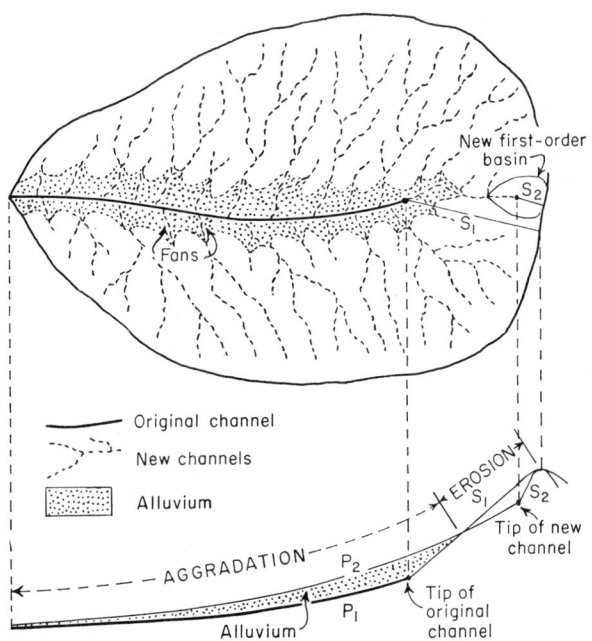

Fig. 4-II-31. Drainage-density transformation accompanying severe, accelerated slope erosion and development of badlands. (*After Strahler* [11, p. 297].)

number will undergo a sharp increase, either through increase in runoff intensity or surface susceptibility to erosion or through simultaneous increase in both. In compensation, basin geometry is altered by gully development to increase greatly the drainage density, to increase channel and ground-surface gradients, but to decrease the local relief [11, p. 296].

When transformation has been completed, a new set of drainage-basin forms, very much smaller in characteristic-length dimension, has replaced the original basins. A new steady state of erosion is achieved on a much higher level of intensity. Thus, badlands, from which sediment is produced at a rapid rate, replace the former long, gentle, smooth slopes of the land. Many new channel segments of first, second, and third orders have come into existence where formerly a single first-order basin had been (Fig. 4-II-31). Steepening of main-stream channel gradients through aggradation is also a characteristic result of the transformation. The various manifestations of accelerated soil erosion are thus seen to be related to a general theory of drainage-basin dynamics.

VII. OBSERVED COMPLEX RELATIONS AMONG HYDROLOGIC AND GEOMETRIC PROPERTIES

A. Control of Basin Geometry by Climatic Factors

Students of quantitative geomorphology of drainage basins have attempted to relate empirically one or more of the geometric elements of drainage basins to several independent variables, including climatic, vegetative, and hydrologic factors. Statistical methods of correlation and regression have been applied to observational data on basin geometry, precipitation and runoff, vegetative cover, geologic type, and soils.

Melton [15] used multiple-regression-and-correlation analysis upon 23 small drainage basins of Arizona, Colorado, New Mexico, and Utah. Multiple-correlation analyses showed that valleyside slopes and drainage density are related to climate and to properties of mantle and vegetal cover. Slopes were found to be higher with greater values of infiltration capacity and Thornthwaite's precipitation-effectiveness index (see Subsec. IV-G) but to vary inversely with soil strength and runoff intensity-frequency. Melton found that drainage density varies directly with per cent of bare area and runoff intensity-frequency, but inversely with precipitation-effectiveness index and infiltration capacity, confirming Horton's infiltration theory of erosion. Melton [54] carried his analysis further by examining the correlation structure of morphometric properties of drainage systems and their controlling agents. He classified variable systems by the presence and direction of feedback among the variables.

Chorley [66] attempted to establish the relationship of basin geometry to climate by a comparative study of three areas of similar gross geology but greatly different climate. A climate-vegetation index, combining mean annual rainfall, mean monthly precipitation in 24 hr, and Thornthwaite's precipitation-effectiveness index, was computed for each region. The climate-vegetation index was found to be closely correlated with ogarithms of stream length, basin area, and drainage density.

B. Relation of Basin Geometry to Stream Flow

Effect of drainage-basin characteristics upon unit-hydrograph lag and peak flow has been reported by Taylor and Schwartz [29], using the data of 20 basins ranging in area from 20 to 1,600 sq mi, and located in the North and Middle Atlantic States. Drainage area, length of longest watercourse, main-stream length to centroid of area, and equivalent main-stream slope were judged the most significant geometrical variables.

A regression of peak stream discharge upon factors of topography, basin area, and rainfall was determined empirically by Potter [67, p. 69] for 51 basins in the Appalachian Plateau. Potter's T factor, representing basin geometry, is the ratio of longest length of principal stream to square root of average channel slope from head to mouth. The T factor was judged to be significant in multiple regression with basin area and measures of rainfall intensity and frequency.

Morisawa [17, pp. 16–17] substituted other geomorphic properties for Potter's T factor in an effort to explain still more of the observed variance. When relief ratio, circularity ratio, and frequency of first-order streams were combined as a product to yield a new T factor, multiple regression for 10 of Potter's basins on rainfall intensity and frequency yielded an equation in which the standard error of estimate was considerably reduced and a high correlation was established with peak intensity of runoff. In another study, Morisawa [68] established significant regressions for average runoff and peak runoff on stream length, relief ratio, and shape ratio within subdivisions of a single small watershed.

Maxwell [18] used digital computers to relate stream-discharge characteristics to several elements of drainage-basin geometry in the San Dimas Experimental Forest of southern California. He computed multiple correlations between peak discharge and storm rainfall, cover density, antecedent rainfall, and nine geomorphic properties taken five at a time. The geomorphic variables considered were fifth-order area and diameter; means of second-order area, diameter, relief, drainage density, channel frequency, and relief ratio; and watershed bifurcation, length, diameter, and area ratios. It was concluded that fifth- and second-order areas or diameters, together with second-order drainage density and relief ratio, provide a good estimate of the variability in peak discharge which can be explained by geomorphic variation between watersheds.

VIII. NOTATIONS

A	watershed area above gage or other reference point, sq mi
A_c	area of a circle of same perimeter as basin, sq mi
A_u	area of basin of order u, sq mi
$\bar{A}_u$	mean area of basins of order u, sq mi
A_0	area of interbasin area, ft^2, sq mi
A_1	area of a first-order basin, sq mi
a	cross-sectional area of basin at a given contour level, sq mi; a numerical constant
b	regression coefficient, dimensionless
C	constant of channel maintenance, ft^2/ft; a constant of integration
D	drainage density, mi/sq mi
D_u	drainage density of entire basin of order u, mi/sq mi
D_1	drainage density of first-order basins, mi/sq mi
F	stream frequency; channel frequency, no./ sq mi
g	acceleration of gravity, ft/sec^2
H	basin relief, ft
H_g	elevation difference; stream head to divide reference point, **ft**
$\bar{H}_u$	mean relief of basins of order u, ft
h	height of given contour above basin mouth, ft
i	item number in summation
j	a constant, dimensionless
K	erosion proportionality factor, sec/ft
k	highest stream order in a given basin, no.
L	stream length from gage to point on divide, mi
L_b	basin length, mi
L_{ca}	main-stream length from gage to centroid, mi
L_g	length of overland flow, ft, mi
$\bar{L}_g$	mean length of overland flow, ft, mi
$\bar{L}_u$	mean length of stream segments of order u, mi
$\sum_{i=1}^{N} L_u$	total length of stream segments of order u, mi
L_0	interbasin length, ft, mi
$\bar{L}_1$	mean length of first-order stream segments, mi
m	an exponent, dimensionless
N_u	number of stream segments of order u, no.
P	basin perimeter, ft, mi
p	constant in lemniscate model of basin shape, dimensionless
Q	discharge, cfs
Q_r	runoff intensity, fps
$Q_{2.3}$	discharge equaled or exceeded in 2.3 years, cfs
R_a	basin-area ratio, dimensionless
R_b	bifurcation ratio of stream segments, dimensionless
R_c	circularity ratio, dimensionless
R_e	elongation ratio, dimensionless
R_f	form ratio of basin, dimensionless
R_h	relief ratio, dimensionless

R_{hp} relative relief, dimensionless
R_L length ratio of stream segments, dimensionless
R_{Lb} ratio of R_L to R_b, dimensionless
R_s slope ratio of stream segments, dimensionless
S_c channel slope, ft/ft, %
S_g ground-surface slope, ft/ft, %
S_{st} equivalent main-stream slope, ft/ft, %
$\bar{S}_u$ mean slope of stream segments of order u, ft/ft, %
$\bar{S}_1$ mean slope of first-order stream segments, ft/ft, %
s standard deviation, estimated from sample
s^2 variance, estimated from sample
u a given order of stream segments, no.
X horizontal distance downstream from stream head, ft, mi
x relative area of horizontal cross section to basin area, dimensionless
$\bar{x}$ arithmetic mean of a sample
Y vertical distance downward from stream head, ft
Y_c elevation difference between local base level and sea-level datum, ft
Y_0 elevation difference between stream head and local base level, ft
y relative height of given contour above basin mouth, dimensionless
λ linear scale ratio in model analysis, dimensionless
μ viscosity (absolute or dynamic) of a fluid; population mean (statistical)
ρ density of a fluid; radius vector in polar coordinates
$\sum_{i=1}^{N}$ summation of terms from 1st to nth
σ population standard deviation
σ^2 population variance
θ angle in polar coordinates
θ_c gradient of stream channel, deg
θ_g gradient of ground surface, deg
θ_{max} maximum angle of valleyside slopes, deg

IX. REFERENCES

1. Horton, R. E.: Drainage basin characteristics, *Trans. Am. Geophys. Union*, vol. 13, pp. 350–361, 1932.
2. Horton, R. E.: Erosional development of streams and their drainage basins: hydrophysical approach to quantitative morphology, *Bull. Geol. Soc. Am.*, vol. 56, pp. 275–370, 1945.
3. Davis, W. M.: "Geographical Essays," Ginn and Company, Boston, 1909. (Reprinted 1954 by Dover Publications, Inc., New York.)
4. Langbein, W. B., and others: Topographic characteristics of drainage basins, *U.S. Geol. Surv. Water-Supply Paper* 968-C, 1947.
5. Strahler, A. N.: Equilibrium theory of erosional slopes approached by frequency distribution analysis, *Am. J. Sci.*, vol. 248, pp. 673–696, 800–814, 1950.
6. Strahler, A. N.: Dynamic basis of geomorphology, *Bull. Geol. Soc. Am.*, vol. 63, pp. 923–938, 1952.
7. Strahler, A. N.: Hypsometric (area-altitude) analysis of erosional topography, *Bull. Geol. Soc. Am.*, vol. 63, pp. 1117–1142, 1952.
8. Strahler, A. N.: Statistical analysis in geomorphic research, *J. Geol.*, vol. 62, pp. 1–25, 1954.
9. Strahler, A. N.: Quantitative slope analysis, *Bull. Geol. Soc. Am.*, vol. 67, pp. 571–596, 1956.
10. Strahler, A. N.: Quantitative analysis of watershed geomorphology, *Trans. Am. Geophys. Union*, vol. 38, pp. 913–920, 1957.
11. Strahler, A. N.: Dimensional analysis applied to fluvially eroded landforms, *Bull. Geol. Soc. Am.*, vol. 69, pp. 279–300, 1958.
12. Von Bertalanffy, Ludwig: The theory of open systems in physics and biology, *Science*, vol. 111, pp. 23–28, 1950.
13. Duncan, W. J.: "Physical Similarity and Dimensional Analysis," Edward Arnold (Publishers) Ltd., London, 1953.
14. Melton, M. A.: List of sample parameters of quantitative properties of landforms: their use in determining the size of geomorphic experiments, *Project* NR 389-042, *Tech. Rept.* 16, Columbia University, Department of Geology, ONR, Geography Branch, New York, 1958.

REFERENCES

15. Melton, M. A.: An analysis of the relations among elements of climate, surface properties, and geomorphology, *Project* NR 389-042, *Tech. Rept.* 11, Columbia University, Department of Geology, ONR, Geography Branch, New York, 1957.
16. Melton, M. A.: Geometric properties of mature drainage systems and their representation in an E_4 phase space, *J. Geol.*, vol. 66, pp. 35–54, 1958.
17. Morisawa, M. E.: Relation of quantitative geomorphology to stream flow in representative watersheds of the Appalachian Plateau Province, *Project* NR 389-042, *Tech. Rept.* 20, Columbia University, Department of Geology, ONR, Geography Branch, New York, 1959.
18. Maxwell, J. C.: Quantitative geomorphology of the San Dimas Experimental Forest, California, *Project* NR 389-042, *Tech. Rept.* 19, Columbia University, Department of Geology, ONR, Geography Branch, New York, 1960.
19. Melton, M. A.: A derivation of Strahler's channel-ordering system, *J. Geol.*, vol. 67, pp. 345–346, 1959.
20. Leopold, L. B., and J. P. Miller: Ephemeral streams: hydraulic factors and their relation to the drainage net, *U.S. Geol. Surv. Profess. Paper* 282-A, 1956.
21. Schumm, S. A.: Evolution of drainage systems and slopes in badlands at Perth Amboy, New Jersey, *Bull. Geol. Soc. Am.*, vol. 67, pp. 597–646, 1956.
22. Smith, K. G.: Erosional processes and landforms in Badlands National Monument, South Dakota, *Bull. Geol. Soc. Am.*, vol. 69, pp. 975–1008, 1958.
23. Maxwell, J. C.: The bifurcation ratio in Horton's law of stream numbers (abstract), *Trans. Am. Geophys. Union*, vol. 36, p. 520, 1955.
24. Morisawa, M. E.: Accuracy of determination of stream lengths from topographic maps, *Trans. Am. Geophys. Union*, vol. 38, pp. 86–88, 1957.
25. Miller, V. C.: A quantitative geomorphic study of drainage basin characteristics in the Clinch Mountain area, Virginia and Tennessee, *Project* NR 389-042, *Tech. Rept.* 3, Columbia University, Department of Geology, ONR, Geography Branch, New York, 1953.
26. Broscoe, A. J.: Quantitative analysis of longitudinal stream profiles of small watersheds, *Project* NR 389-042, *Tech. Rept.* 18, Columbia University, Department of Geology, ONR, Geography Branch, New York, 1959.
27. Hack, J. T.: Studies of longitudinal stream profiles in Virginia and Maryland, *U.S. Geol. Surv. Profess. Paper* 294-B, 1957.
28. Snyder, F. F.: Synthetic unit-graphs, *Trans. Am. Geophys. Union*, vol. 19, pp. 447–454, 1938.
29. Taylor, A. B., and H. E. Schwartz: Unit-hydrograph lag and peak flow related to basin characteristics, *Trans. Am. Geophys. Union*, vol. 33, pp. 235–246, 1952.
30. Linsley, R. K., M. A. Kohler, and J. L. H. Paulhus: "Applied Hydrology," McGraw-Hill Book Company, Inc., New York, 1949.
31. Golding, B. L., and D. E. Low: Physical characteristics of drainage basins, *Proc. Am. Soc. Civil Engrs., J. Hydraulics Div.*, vol. 86, no. HY3, pp. 1–11, 1950.
32. Wisler, C. O., and E. F. Brater: "Hydrology," John Wiley & Sons, Inc., New York, 1959.
33. Horton, R. E.: Sheet erosion: present and past, *Trans. Am. Geophys. Union*, vol. 22, pp. 299–305, 1941.
34. "Unit Hydrograph Compilations," U.S. Corps of Engineers, Department of the Army, Washington District, Civil Works Invest., Project CW 153 (three volumes 1949; one volume 1954).
35. Chorley, R. J., Donald E. G. Malm, and H. A. Pogorzelski: A new standard for estimating drainage basin shape, *Am. J. Sci.*, vol. 255, pp. 138–141, 1957.
36. Morisawa, M. E.: Measurement of drainage-basin outline form, *J. Geol.*, vol. 66, pp. 587–591, 1958.
37. Carlston, C. W., and W. B. Langbein: Rapid approximation of drainage density: line intersection method, *U.S. Geol. Surv. Water Resources Div., Bull.*, p. 11, Feb. 10, 1960.
38. Smith, K. G.: Standards for grading texture of erosional topography, *Am. J. Sci.*, vol. 248, pp. 655–668, 1950.
39. Coates, D. R.: Quantitative geomorphology of small drainage basins of southern Indiana, *Project* NR 389-042, *Tech. Rept.* 10, Columbia University, Department of Geology, ONR, Geography Branch, New York, 1958.
40. Shulits, Samuel: Rational equation of river-bed profile, *Trans. Am. Geophys. Union*, vol. 22, pp. 622–631, 1941.
41. Shulits, Samuel: Graphical analysis of trend profile of a shortened section of river, *Trans. Am. Geophys. Union*, vol. 36, pp. 649–654, 1955.
42. Miller, J. P.: High mountain streams: effects of geology on channel characteristics and bed material, *New Mexico Bur. Mines & Mineral Resources*, Mem. 4, 1958.

43. Gilbert, G. K.: Report on the geology of the Henry Mountains, U.S. Geographical and Geological Survey of the Rocky Mountain Region, Washington, D.C., 1877.
44. Yatsu, Eiju: On the longitudinal profile of the graded river, *Trans. Am. Geophys. Union*, vol. 36, pp. 655–663, 1955.
45. Krumbein, W. O.: Sediments and the exponential function, *J. Geol.*, vol. 45, pp. 577–601, 1937.
46. Rubey, W. W.: Geology and mineral resources of the Hardin and Brussels quadrangles (Illinois), *U.S. Geol. Surv. Profess. Paper* 218, 1952.
47. Mackin, J. H.: Concept of the graded river, *Bull. Geol. Soc. Am.*, vol. 59, pp. 463–512, 1948.
48. Woodford, A. O.: Stream gradients and Monterey Sea Valley, *Bull. Geol. Soc. Am.*, vol. 62, pp. 799–852, 1951.
49. Benson, M. S.: Channel-slope factor in flood frequency analyses, *Proc. Am. Soc. Civil Engrs., J. Hydraulics Div.*, vol. 85, no. HY4, pp. 1–9, 1959.
50. Lane, E. W.: Stable channels in erodible material, *Trans. Am. Soc. Civil Engrs.*, vol. 102, pp. 123–194, 1937.
51. Leopold, L. B., and Thomas Maddock, Jr.: The hydraulic geometry of stream channels and some physiographic implications, *U.S. Geol. Surv. Profess. Paper* 252, 1953.
52. Leopold, L. B.: Downstream change of velocity in rivers, *Am. J. Sci.*, vol. 251, pp. 606–624, 1953.
53. Wolman, M. G.: The natural channel of Brandywine Creek, Pennsylvania, *U.S. Geol. Surv. Profess. Paper* 271, 1955.
54. Melton, M. A.: Correlation structure of morphometric properties of drainage systems and their controlling agents, *J. Geol.*, vol. 66, pp. 442–460, 1958.
55. Melton, M. A.: Intravalley variation in slope angles related to microclimate and erosional environment, *Bull. Geol. Soc. Am.*, vol. 71, pp. 133–144, 1960.
56. Horton, R. E.: Derivation of runoff from rainfall data, Discussion, *Trans. Am. Soc. Civil Engrs.*, vol. 77, pp. 369–375, 1914.
57. Chapman, C. A.: A new quantitative method of topographic analysis, *Am. J. Sci.*, vol. 250, pp. 428–452, 1952.
58. Wentworth, C. K.: A simplified method of determining the average slope of land surfaces, *Am. J. Sci.*, 5th ser., vol. 20, pp. 184–194, 1930.
59. Smith, Guy-Harold: The relative relief of Ohio, *Geograph. Rev.*, vol. 25, pp. 272–284, 1935.
60. Raisz, Irwin, and Joyce Henry: An average slope map of New England, *Geograph. Rev.*, vol. 27, pp. 467–472, 1937.
61. Calef, Wesley: Slope studies of northern Illinois, *Trans. Illinois Acad. Sci.*, vol. 43, pp. 110–115, 1950.
62. Calef, Wesley, and Robert Newcomb: An average slope map of Illinois, *Ann. Assoc. Am. Geographers*, vol. 43, pp. 305–316, 1953.
63. Ruhe, R. V.: Graphic analysis of drift topographies, *Am. J. Sci.*, vol. 248, pp. 435–443, 1950.
64. Schumm, Stanley: The relation of drainage basin relief to sediment loss, *Intern. Union Geodesy Geophys., Tenth Gen. Assembly* (Rome), *Intern. Assoc. Sci. Hydrol. Publ.* 36, vol. 1, pp. 216–219, 1954.
65. Maner, S. B.: Factors affecting sediment delivery rates in the Red Hills physiographic area, *Trans. Am. Geophys. Union*, vol. 39, pp. 669–675, 1958.
66. Chorley, R. J.: Climate and morphometry, *J. Geol.*, vol. 65, pp. 628–638, 1957.
67. Potter, W. D.: Rainfall and topographic factors that affect runoff, *Trans. Am. Geophys. Union*, vol. 34, pp. 67–73, 1953.
68. Morisawa, M. E.: Relation of morphometric properties to runoff in the Little Mill Creek, Ohio, drainage basin, *Project* NR 389-042, *Tech. Rept.* 17, Columbia University, Department of Geology, ONR, Geography Branch, New York, 1959.

Section 5

SOIL PHYSICS

DON KIRKHAM, *Charles F. Curtiss Distinguished Professor in Agriculture, and Professor of Soils and Physics, Iowa State University.*

I. Introduction	5-2
A. Some Soil Terms	5-3
II. Formation and Mechanical Composition of Soils	5-3
A. Formation of Soils	5-3
B. Mechanical Composition of Soils	5-3
III. Soil Classification	5-4
A. Pedological System	5-4
1. Higher Categories	5-4
2. Lower Categories	5-6
3. Soil Surveys and Maps	5-6
B. Engineering Unified Soil Classification System	5-6
C. Other Engineering Classification Systems	5-7
D. Land Capability Classification System	5-8
E. Comprehensive Pedological System	5-8
F. European and Forest Soils Classifications	5-9
IV. Soil Water	5-9
A. Water Retention in Soil	5-9
1. Surface Tension, Capillarity, Surface Energy	5-9
2. Measurement of Soil-moisture Tension	5-13
3. Measurement of Soil-moisture Tension above 0.85 Atm	5-14
4. Hysteresis	5-14
5. Retention Forces Other Than Those of Capillarity	5-14
6. Osmotic Forces and Soil-water Retention	5-15
7. Soil-moisture Retention and the Outflow Law	5-15
8. Soil-moisture-retention Terminology	5-15
9. Measurement of Water Retained in Soil	5-15
10. The Neutron Soil-moisture Meter—Some Soil-moisture Profiles	5-16
B. Water Movements in Saturated Soils	5-16
1. Darcy's Law for Water-saturated Soils; Hydraulic Conductivity; Permeability	5-16
2. Laplace's Equation; Dupuit-Forchheimer (DF) Theory	5-17
3. A Drain-spacing Equation and Nomograph	5-17
4. Soil Factors Governing Water Entry to Tile Drains	5-17
5. Leaching	5-19
C. Water Movement in Unsaturated Soils; Ice Lenses	5-19

SOIL PHYSICS

V. Soil Air, Soil Temperature, Soil Structure.................... 5-19
   A. Soil Air... 5-19
      1. Renewal of Soil Air 5-20
      2. Diffusion of Soil Oxygen; Air Permeability............... 5-20
   B. Soil Temperature....................................... 5-20
      1. Soil Thermal Constants............................. 5-20
   C. Soil Structure.. 5-21
      1. Definition of Soil Structure......................... 5-21
      2. Measurement of Soil Structure....................... 5-22
      3. Soil Compaction.................................. 5-22
VI. References... 5-22

I. INTRODUCTION

In this section the physical properties of soil will be discussed. There is currently only one textbook [1] available on the subject. Therefore a rather long list of original references will be cited. The main topics to be discussed, in order, are formation and mechanical composition of soil, soil classification, soil water, soil air, soil temperature, and soil structure; emphasis is on soil water. A number of these topics, in connection with plant growth, are discussed extensively in a treatise edited by Shaw [2]. The subject soil classification is not ordinarily found in a treatment on soil physics, but it is involved in soil physics and is included here for completeness.

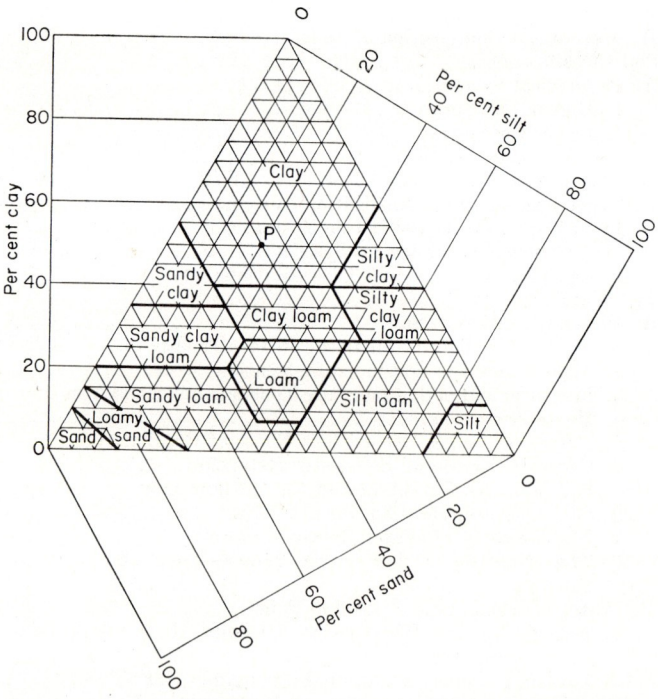

Fig. 5-1. U.S. Department of Agriculture textural classification triangle [3, p. 209] with axes added. The point P represents a clay (soil) containing 50 per cent clay, 20 per cent silt, and 30 per cent sand.

FORMATION AND MECHANICAL COMPOSITION OF SOILS

Symbols in equations: Equations are not used until Subsec. IV and will be defined where they are used. The same symbols may have different meanings in different equations.

A. Some Soil Terms

Sand, silt, and *clay* of respective sizes [3, p. 207] 2 to 0.05 mm, 0.05 to 0.002 mm, and less than 0.002 mm are the *primary particles* of soil which form its *texture*. Sand, silt, and clay do not always refer to primary soil particles. Clay (soil) may, for example, be a soil with 50 per cent clay-size particles (Fig. 5-1). To estimate the texture of a soil see Refs. 3, p. 212, and 4, p. 68.

Soil is generally divided into *A-, B-,* and *C-horizons,* a procedure stemming from Russian soil scientists [5, p. 32]. The A-horizon is the surface layer of soil, usually darker than the others, because it contains decayed vegetable matter. The B-horizon, below the A-horizon, usually contains more clay than, and differs greatly in color from, the A. The A- and B-horizons together are called the *solum*. The C-horizon is below the B-horizon and consists usually of the soil *parent material,* from which the A- and B-horizons were formed. If the A- and B-horizons are not formed from the horizon below them, then the horizon below the A and B is called the *D-horizon*. The A-, B-, and C-horizons, when exposed in a vertical cut, constitute the soil *profile*. The profile may have subhorizons in the A, B or C, which may be indicated by a subscript or added number as A_2 or A-2.

II. FORMATION AND MECHANICAL COMPOSITION OF SOILS

Two recent reference books on this topic are those of Thompson [6] and Lyon and others [7].

A. Formation of Soils

Soils are formed from exposed masses of partially weathered rocks of the earth's crust. A weathering process of tremendous influence is glaciation, especially active in latitudes north of 40°N. Through radiocarbon dating of trees in buried soil, it is now known that glaciations have occurred much more recently than was formerly believed [8]. The most recent substage of the Wisconsin glaciation occurred in Iowa only 13,000 years ago. The sheetlike deposits of glaciers are called *till,* which is a heterogeneous, unsorted, nonstratified mixture of clay, silt, sand, gravel, and boulders. Plant deposits may also occur in till. Water erosion, with or following glaciations, has laid down *alluvium* in flood plains or bottomlands of streams. Alluvium includes clays and silts and occasional gravel deposits. After the retreat of glaciers, winds raised clouds of particles which had been ground by the sheets of glacial ice. These particles were transported, generally southward, to form *loess*. Thickness of loess has been found to follow an exponential-decay law with distance from the source [9].

Soil is often considered to be formed by five factors: climate, organisms, topography, parent material, and time. For a discussion of these see Ref. 10. Man may have an influence on soil formation. His influence has been noticed particularly in Europe [11, p. 382; 12, p. 216], where the soils have been tilled for centuries. In the Netherlands the use of uprooted heather, brought in from a distance in the fall for sheep bedding, and the scattering, in the spring, of this heather, along with the mineral matter brought with the roots the previous fall, on the farmstead field have created man-made soils of about 1-m depth [13]. Clearing of forests and addition of lime and fertilizer through the centuries has made, from acid forest soils, nonacid agricultural soils, called by the Germans and Austrians *Ackerpodsole* [11, p. 309].

B. Mechanical Composition of Soils

A unit volume of an "average" soil, a silt-loam surface soil, has about 50 per cent solids space and 50 per cent air space. This 50-50 division is based on an average oven-dry *bulk density* (sometimes called *volume weight*) of 1.3 g/cm³ bulk soil (solids space plus voids space) [14, esp. p. 412] and a particle density of 2.6 g/cm³ [1, p. 57]. For humid-region soils, about 45 per cent of the solid space is occupied by clay minerals and 5 per cent by organic matter. Under normal growing conditions, after *gravitational water* has been drained from the soil following a rainfall, the water held in the soil will vary from about 30 per cent of the bulk volume to 15 per cent; the air space will vary accordingly from about 20 to 35 per cent. The air space in soil may go as low as about 10 per cent and still support plant growth [15]. The clay minerals may be in the form of sand, silt, or clay. The clay may be of a swelling type, as montmorillonite, or of nonswelling type, as kaolinite. Much of the clay may be of colloidal size 0.5 to 0.01 micron ($=10^{-6}$ cm), the 0.01-micron size being about 100 times larger than a "normal" molecule (10^{-8} cm). Brownian movement—the movement of very small particles due to the bombardment of molecules—keeps colloidal clay from settling in a water suspension. To make clay in a suspension settle, a *flocculating agent* must be added, as calcium chloride.

The clay fraction is of particular interest because of its large specific surface. A piece of gravel 1 cm on an edge (1 cm³), if divided into cubes of length 0.01 micron, or 10^{-6} cm, on an edge, will yield $(10^6)^3$ cubes, each having $6 \times (10^{-6})^2$ cm² of surface area, making the total surface for the 1 cm³ of gravel equal to 6×10^6 m², or 600 m². For 0.1 micron clay, the corresponding specific surface is 60 m²; for 1 micron clay, 6 m². Such large surface areas, when acted on by water, put soil minerals in solution for plant food. Clay minerals which do not break down readily to the smaller sizes are not active suppliers of plant nutrients.

The ease of breakdown (weathering) of minerals to produce a large specific surface is known. The following list [6, p. 116] gives the more easily weathered mineral first: (1) gypsum, (2) calcite, (3) hornblende, (4) biotite, (5) albite, (6) quartz, (7) muscovite, (8) vermiculite, (9) montmorillonite, (10) kaolinite, (11) gibbsite, (12) hematite, and (13) anatose. As for individual mineral components, the order of weathering of those of particular interest for plant growth is $Ca > Mg > Na > K$. The chemical composition of a number of clay minerals may be found in Grim [16, esp. pp. 370–373]. A treatise on the surface chemistry of soil colloids, based mainly on Russian work, has been written by Tschapek [17].

III. SOIL CLASSIFICATION

There are a number of soil classification systems. They are based on the intended use of the system. If the use is for engineering works, the classification will be different from one for agronomic purposes. Also, schemes vary from country to country. Most systems have a number of points in common with classifications used in the United States. Some of these United States systems will be briefly described.

A. Pedological System

This system, stemming largely from Russian soil scientists [5, pp. 6–7], depends on observation of the soil profile down into the C-horizon. The system has been developed primarily with agriculture in mind, but it is not intended to serve agriculture alone. This system has the two broad divisions, *higher categories* and *lower categories*.

1. Higher Categories. The higher categories of the pedological system (Table 5-1) are *order*, *suborder*, and *great soil groups*. Under order belong the *zonal*, *intrazonal*, and *azonal* soils. Zonal soils depend primarily on the climatic zone in which they have developed. Intrazonal soils depend on climate and also on some local conditions, such as poor drainage. They cross zonal boundaries; hence the term intrazonal. Azonal soils do not depend on zones. They include rocky soils (*lithosols*), dry sands (*regosols*), and alluvial sediment.

The zonal soils have six suborders, with the suborders running from cold to tropical zones. In suborder 5 there occur *podzolized* soils (*podsols*, or *podzols*) of the timbered regions. These soils are of ashy-gray color in the lower part of the A-horizon because acid which has leached from fallen tree leaves has dissolved and carried downward with it dark-colored material as iron compounds. In suborder 6 there occur the tropical *lateritic soils*, red in color and having a high content of iron oxide and hydroxide of aluminum. The intrazonal soils have three suborders: (1) salt and alkali soils, (2) soils of wet areas, and (3) soils rich in calcium. The azonal soils have no suborders.

Table 5-1. Soil Classification in the Higher Categories*

Order	Suborder	Great soil groups
Zonal soils	1. Soils of the cold zone	Tundra soils
	2. Light-colored soils of arid regions	Desert soils Red desert soils Sierozem Brown soils Reddish-brown soils
	3. Dark-colored soils of semiarid subhumid and humid grasslands	Chestnut soils Reddish chestnut soils Chernozem soils Prairie soils Reddish prairie soils
	4. Soils of the forest-grassland transition	Degraded chernozem Noncalcic brown, or Shantung brown soils
	5. Light-colored podzolized soils of the timbered regions	Podzol soils Gray wooded, or gray podzolic soils Brown podzolic soils Gray-brown podzolic soils Red-yellow podzolic soils
	6. Lateritic soils of forested warm-temperate and tropical regions	Reddish-brown lateritic soils Yellowish-brown lateritic soils Laterite soils
Intrazonal soils	1. Halomorphic (saline and alkali) soils of imperfectly drained arid regions and littoral deposits	Solonchak, or saline soils Solonetz soils Soloth soils
	2. Hydromorphic soils of marshes, swamps, seep areas, and flats	Humic-glei soils (includes Wiesenboden) Alpine meadow soils Bog soils Half-bog soils Low-humic-glei soils Planosols Groundwater packed soils Goundwater laterite soils
	3. Calcimorphic soils	Brown forest soils (Braunerde) Rendzina soils
Azonal soils		Lithosols Regosols (includes dry sands) Alluvial soils

* After Thorp and Smith [18].

The names of the great soil groups in Table 5-1 define the soils in a general way. Some uncommon names, mainly Russian, appear. *Sierozems* are pale-grayish soils found in temperate to cool arid regions. *Chernozems* are black fertile soils formed under prairie grass and found in temperate to cool subhumid climates. *Shantung* soils are noncalcic brown soils found mainly under deciduous forest in warm climates. *Solonchak* soil has a gray salty surface crust and is salty throughout the profile. *Solonetz* soils result from improvement in drainage of solonchak soils. *Soloth* soils have a thin friable brown surface layer over a gray horizon. *Wiesenboden* soils are wet marsh soils. *Glei* refers to a mottled horizon in a poorly drained mineral soil subject to a rising and falling water table. The mottled colorations are red, yellow, or brown due to oxidized iron and manganese which form when the water table is low and blue and gray due to unoxidized iron and manganese which form when the water table is high. *Rendzina* soils have developed from cretaceous chalk; the blacklands of Texas are an example. *Brunizem*, a term not listed in the table, is a prairie soil developed under tall grass. Podzols, the soils with an ashy-gray A-2 horizon, have been mentioned; in podzols the layer beneath the A-2 is rich in *sesquioxides* (Al_2O_3 and Fe_2O_3).

Some names not mentioned in Table 5-1 and not necessarily used at the classification levels of Table 5-1 may be added. *Ando* soils [18] have developed from volcanic ash in very humid tropical climates. *Planosols* have a surface layer from which much clay has been leached and which rests on compact or cemented subsoil. *Grumusols* [19] is a grouping term for several soils, namely, rendzina, the black cotton soils of India also known as *regur*, and other black earths of warm regions. *Pedocals* are calcium-rich soils; *pedalfers* are iron-rich. A *catena* (chain) is a group of soil series (see below) within a particular climatic zone developed under similar parent material but differing in characteristics due to drainage or relief. A *complex* consists of soils so intricately associated in small areas that they are mapped together.

2. Lower Categories. The first subdivision of a great soil group of Table 5-1 is called a *soil series*. Series are then divided into *types*, and types in turn are divided into *phases* [20]. The soils of any one series have, by definition, similar profile characteristics, except for the texture of the surface layer. More precisely, a series is a group of soils developed from the same type of parent material. Type is determined by the texture of the A-horizon. Phase is determined by some deviation from the normal, as erosion, slope, stoniness, or soluble-salt content. In the United States each series is given a name, usually from some city, village, river, or county or the like. In naming a soil at the series level, the series type and phase are all given. An example is Houston clay, stony phase. For a list of soil types which have been classified in the United States and their locations, see the Federal Housing Administration report [21].

3. Soil Surveys and Maps. The pedological system has been used by the U.S. Department of Agriculture and state soil scientists to provide *soil surveys* with maps for many United States counties.[1] A recent example is that of Jefferson County, Iowa [22], which describes 35 soil series and 2 soil complexes and gives 12 soil-survey maps on a 1:31680 basis (one inch for one-half mile). For information on how to make soil surveys, see the "Soil Survey Manual" [3].

Some recent county soil surveys include soil-classification information pertinent to the *Unified System* and to the *AASHO System*, engineering classification systems described below.

B. Engineering Unified Soil Classification System

This system, often designated as the Unified System, is described and used by several public agencies [21, 23, 24]. The system is not applicable to agronomic soil classification, but is intended for soil classification for foundations and hydraulic structures.

The system (Table 5-2) has only two main categories: *coarse-grained soils* and *fine-grained soils*. The coarse-grained soils, those with particles larger than the No. 200 U.S. Standard Series sieve size (0.074 mm), are divided further into gravels and sands. The fine-grained soils, or *fines*, those with particles smaller than 0.074 mm, are also divided further into two silt-clay combinations, depending on values of the *liquid limit*. The liquid limit (LL) is the moisture content (per cent referred to oven-dry weight) at which a soil-water mixture will just flow. Silt and clay are both included in the fines, but the fines are not subclassified into silt and clay. Ultimately, for the coarse-grained soils, as GW, GP, etc. (last column of Table 5-2), the classification depends on how well or how poorly graded the soil material is, the *degree of gradation*

---

[1] Obtainable from the Superintendent of Documents, U.S. Government Printing Office, Washington 25, D.C.

depending on a *coefficient of uniformity* and a *coefficient of curvature*, each obtained from a particle-size-analysis curve of the soil material [23, pp. 5–6]. Ultimately, for the fine-grained soils, as for ML, CL, etc., of the table, the classification depends on the plasticity constants, these being the liquid limit (LL) already defined, the *plastic limit* (PL), which is the soil-moisture content when the soil can just be molded, and the *plasticity index* (PI), which is given by the equation PI = LL − PL.

Table 5-2. Categories and Group Symbols of Engineering Unified Soil Classification System*

Category				Group symbols
Coarse-grained soils	Gravels	Clean gravel	Gravel, well graded Gravel, poorly graded	GW GP
		Gravels with fines	Gravels, mixed non-pl.† fines Gravels, clayey-pl. fines	GM GC
	Sands	Clean sands	Sands, well graded Sands, poorly graded	SW SP
		Sands with fines	Sands, mixed non-pl. fines Sands, clayey-pl. fines	SM SC
Fine-grained soils	Silts and clays	LL‡ less than 50	Mineral silts, low pl. Clays (mineral), low pl. Organic silts, low pl.	ML CL OL
		LL greater than 50	Mineral silts, high pl. Clays (mineral), high pl. Organic clays, high pl.	MH CH OH
Highly organic soils			Organic soils as peat	Pt

* Adapted from "Earth Manual" [23, pp. 1–23].
† pl. = plastic, or plasticity.
‡ The liquid limit (LL) is the ratio of weight of water to weight of dry soil, expressed as a per cent, for a soil-water mixture that just flows under the pull of gravity.

The Unified System does not seem to give an exact definition of peat. In pedology, organic soils, as peat, contain about 80 per cent or more by weight of organic matter; mineral content is thus about 20 per cent or less [7, p. 7]. Once a soil has been classified into GP, GW, etc., of Table 5-2, its suitability can be determined for various engineering purposes from charts, reproduction of which space does not permit here. Such charts are in the "Earth Manual" [23, pp. 22–23] and the Federal Housing Administration manual [21, pp. 32–33]. Recent county soil surveys, as has been implied, give engineering uses of the different soils listed (see, for example, Mathews [25, pp. 95–97]).

C. Other Engineering Classification Systems

Two other important engineering soil-classification systems [26] are that of the American Association of State Highway Officials (AASHO) [27] and that of the U.S. Civil Aeronautics Administration [28]. They are similar to the Unified System. Soil texture and plasticity are stressed. The U.S. Federal Housing Administration uses, as has been noted, the Unified System. For an engineering description of the soils of North America, see Woods and Lovell [29].

D. Land Capability Classification System

A soil classification system, widely used in farm soil surveys and other surveys by the U.S. Soil Conservation Service, is one based on land capability [30]. The system provides a practical grouping based on the needs and limitations of the soils, on the risks of damaging them (as by exposing them to erosion), and on their response to management. There are eight main classes [20, pp. 5–6; 25, pp. 11–12; 31], which may be described briefly as follows. Class I soils have the largest range of use. Class II soils can be cultivated regularly. Class III soils can be cropped regularly. Class IV soils should be cultivated only occasionally. Class V soils are flat and subject to overflow and should not be cultivated for annual crops. Class VI soils are steep and should not be cultivated for annual crops. Class VII soils can provide fair yields of forage or timber. Class VIII soils provide wildlife habitats or scenery.

E. Comprehensive Pedological System

This classification system [5], sometimes denoted here for brevity by "the comprehensive system," has been under development by the Soil Survey staff of the U.S. Department of Agriculture and other interested soil scientists since 1951, for the following reasons: (1) Earlier pedological systems have stressed the profiles of virgin soils. Few soils now are virgin. (2) Earlier systems have been developed primarily for application to Russian and Western European soils and to soils of the U.S.; little provision has been given to tropical soils, for which many data are now accumulating. (3) In the earlier pedological systems, soils were classified primarily into soil series, with few relations, if any, shown between the series, even when the series was only a few hundred miles apart. (4) In earlier classifications much of the work was done as an art. Physical and chemical measurements data were not widely taken or used. (5) Earlier systems have not had sufficiently descriptive and logical nomenclature for identifying soils and showing their interrelations. (The comprehensive system has radically and completely new nomenclature.) (6) Earlier pedological systems often stressed genetic factors of soil formation instead of giving emphasis to the soil properties as found. (The comprehensive system places emphasis on soil properties rather than on genetic factors.)

The categories of the new system are orders, suborders, subgroups, families, and series. The category soil type, as loam, sandy loam, etc., is not included in the nomenclature. There are 10 orders, the names all ending in -*sols* (soils) as follows.

Entisols are often *recent* soils; *vertisols* are often *inverted* soils, in the sense that surface soil has sloughed into cracks and subsoil has been pushed, by swelling action, to the surface. Vertisols crack markedly when dry, so that fence posts—even trees—may be tilted [5, p. 124]. *Inceptisols* are often young (*inception*) soils, as the ando soils, which are formed of young volcanic ash. *Aridisols* are often *arid* soils, as desert soils. *Mollisols* often have a crumbly (soft) (L. *mollis*, soft, mollify) surface layer, as do the chernozems and prairie soils. *Spodosols* (Gr. *spodos*, wood ash) often have a horizon at least 15 cm (6 in.) below the soil surface, containing free sesquioxides (Al_2O_3, Fe_2O_3) and organic carbon which have leached from the surface layer; a podsol (with its ashy-gray layer which overlies the sesquioxide layer) is an example. *Alfisols* (*alfi*: al, aluminum, and f(e), iron) do not, as the name may suggest, contain a subsurface layer rich in the sesquioxides; they have a clay-enriched subsoil. Like the spodosols, they often have an ashy-gray subsurface horizon. The former gray-brown podzolic soils are classed with alfisols. *Ultisols* (L. *ultimus*, last) are often very old and hence are found only in humid (never postglacial) climates. Many of the former red and yellow podzolic soils are classed as ultisols. *Oxisols* contain horizons rich in *oxides* of silica of iron and of aluminum; the minerals of these horizons are all strongly weathered. The oxisols are restricted to tropical and subtropical regions and have earlier been called *latosols*. *Histosols* (Gr. *histo*, tissue) contain to a large extent, or actually are, residues of plant tissues. They are organic soils such as peat and muck.

This completes the orders. The suborders, subgroups, families, and series cannot be described here, for lack of space. This new system has overcome so many of the difficulties of earlier pedological systems that it is being accepted enthusiastically.

F. European and Forest Soils Classifications

For classifications and descriptions of European soils and for extensive references to the European soils literature, see Franz [11, pp. 232–282]. An earlier book is that of Scheffer and Schachtschabel [12]. The European classification is much as in Table 5-1. For forest soils, see Wilde [32, esp. pp. 19–174]. The forest soils classification has only two main groups: the *upland forest soils* and the *bottomland forest soils* (hydromorphic).

IV. SOIL WATER

For a literature review on the soil physics of soil water up to 1930, see Zunker [33]. Two recent books, neither of them extant in the English language, are those of Rode [34], translated from Russian into German, and of Tschapek [35], in Spanish. For a review on water and its relation to soils and crops, see Russell [36].

Water retention and water movement in soil are of primary interest in soil-water physics. What happens can be brought out with diagrams, compressed, for space saving, into multiple figures.

A. Water Retention in Soil

1. Surface Tension, Capillarity, Surface Energy. Water is retained in soil largely by surface-tension forces. Surface tension has been explained by Laplace, who considers that water molecules attract each other over a small distance r (Fig. 5-2a). A molecule a in the body of the water is, on the average, attracted equally on all sides. But a molecule b in the surface of the water is attracted by the molecules only below the surface. One neglects the attraction of the few vapor molecules in the air. The molecule b at the surface, and other similar surface molecules, thus exert a pressure at the air-water interface, as if the surface were covered by a (Laplace's equivalent) membrane under tension. To attach a numerical value to surface tension, imagine a strip of Laplace's equivalent membrane (Fig. 5-2b) 1 cm wide and very thin in the direction perpendicular to the plane of the paper and containing an imaginary line PQ shown. The surface tension developed along the 1-cm long line PQ is, by definition, the *surface-tension coefficient*, symbol σ, of water with respect to air. The Greek letter γ is also often used. The value of σ for an air-water interface is $\sigma = 73$ dynes/cm at 20°C.

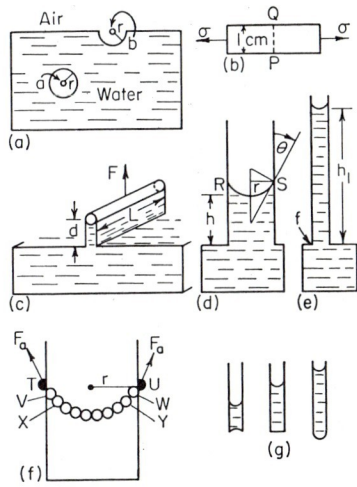

Fig. 5-2. Diagrams illustrating surface tension.

The surface-tension coefficient σ has an energy definition. Figure 5-2c represents a long thin horizontal rigid wire being lifted through an air-water interface. Neglecting end effects and the pull of gravity, one sees that the upward pull F, to offset the surface tension, is given by

$$F = 2\sigma L \quad \text{dynes} \tag{5-1}$$

where the factor 2 enters because surface tension acts on each side of the wire. The work W done in raising the wire a height d is $W = Fd$; that is

$$W = 2\sigma L d \qquad \text{dyne-cm, or ergs} \tag{5-2}$$

While this work W is being done, the surface area of the water increases an amount $2Ld$; that is, the work done per unit increase in surface area of the water is

$$\frac{W}{2Ld} = \frac{2\sigma L d}{2Ld} = \sigma \qquad \text{ergs/cm}^2 \tag{5-3}$$

which gives the energy definition of σ as the energy stored in a surface by increasing its area by unit area.

Surface tension in water and adhesive forces in the wall of a capillary tube cause capillary rise in soil pores. Such pores might be cylindrical in cross section (Fig. 5-2d), as from a decayed root channel. At equilibrium, in a pore, a meniscus as RS is formed, making an angle θ, the *wetting angle*, with the pore wall. To obtain the height of rise h, one equates the vertical component $2\pi r \sigma \cos \theta$ of the surface-tension force at the circumference of the pore against the weight $\pi r^2 h \rho g$ of the water (ρ = density of water, g = 981 in cgs system) and solves for h. The small weight of water in the meniscus is neglected. The result for h is

$$h = \frac{2\sigma \cos \theta}{r \rho g} \tag{5-4}$$

This equation shows (cf. Fig. 5-2d and e) that, for a large-diameter pore, the height of rise is smaller than for a small pore. But the equation does not show how the adhesive forces in the wall of the capillary are involved.

Figure 5-2f shows how adhesive forces are involved. The molecules in the surface layer of the meniscus of Fig. 5-2d are represented by open circles, and a cross section of a ring of molecules TU in the wall is represented by darkened circles. The ring TU exerts adhesive forces F_a on the ring of water molecules VW. The ring VW simultaneously exerts a surface-tension force $2\pi r \sigma$ on a contiguous ring XY of water molecules. Thus the force of adhesion F_a acting through the ring of water molecules VW is just equal to the surface-tension force $2\pi r \sigma$, the upward component of the adhesive force being equal to $2\pi r \sigma \cos \theta$. It is now clear that the adhesive forces hold up the column of water. The above picture is approximate, and one may ask, what about the ring of molecules in the tube wall just below TU? The answer is, this band would pull the ring VW outward with essentially no effective vertical pull. The rings TU, VW, and XY are supposed, of course, to contain many molecules.

When water is poured or sprinkled on the soil surface, the water will enter the soil and move downward, but it may not move very far because it may be held in soil pores, as demonstrated by the three capillary tubes of Fig. 5-2g. In the right-hand tube the addition of more water could cause the air-water interface at the base to break with the escape of a drop of water.

Consider again Fig. 5-2e. The work w required to raise a small mass m of water from a free-water level f to a height h_1 just below the meniscus is

$$w = mgh_1 \tag{5-5}$$

so the potential energy E per gram of water just under the meniscus is

$$E = \frac{w}{m} = gh_1 \qquad \text{ergs/g} \tag{5-6}$$

Although, in Eq. (5-6), the quantity gh_1 represents the work to raise 1 g of water through a height h_1, the energy due to surface tension does not necessarily arise as a consequence of gravity. On the contrary, one sees (Fig. 5-3a) that the tension or capillary energy per gram E_P, E_Q, and E_R at the points P, Q, and R under the surfaces of menisci of vertical and horizontal capillary tubes of the same diameter must be the same; that is,

$$E_P = E_Q = E_R = gh_1 \quad \text{ergs/g} \tag{5-7}$$

so that h_1 can be taken as a measure of the capillary energy at the point R of the horizontal capillary, as well as at the point F of the vertical one. In Fig. 5-3a it is observed that, although the capillary energy at points Q and R are equal, the gravitational energy is greater for Q than for R, since the pore containing the point R is at a lower level. In solving problems of water flow, both the capillary and gravitational energy must be considered unless the flow is in a horizontal direction.

It is often more useful to have the energy expressed in terms of the pore radius rather than the height h_1 as in Fig. 5-3a. To get the appropriate expression one substitutes h_1 as given by the capillary-rise formula [Eq. (5-4)] into Eq. (5-7) and obtains

$$E_P = \frac{2\sigma \cos \theta}{r\rho} \tag{5-8}$$

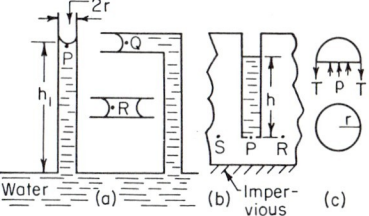

Fig. 5-3. Diagrams for soil-water phenomena.

One can also get the energy in terms of the tension T_P, the pull per unit area at the point P under the meniscus. To do this one equates the total tension force $T_P \pi r^2$ at the level of the point P in Fig. 5-3a (the same tension force would act across vertical planes perpendicular to the plane of the paper through points P and Q) to the total upward pull $2\pi r \sigma \cos \theta$ and solves for T_P to obtain

$$T_P = \frac{2\sigma \cos \theta}{r} \tag{5-9}$$

Now compare this result with the preceding equation, observing there that, in the cgs system, ρ is unity. Hence observe that T_P and E_P are numerically equal in the cgs system, both expressions giving ergs per gram of capillary energy under a meniscus. The various ways commonly used in soil physics for expressing energy, etc., of soil-pore water, for a wetting angle of zero degrees, are given in Table 5-3, for tensions up to 15 atm.

Table 5-3 raises a question. The values given imply that tensions up to 15 atm may be measured. They cannot be. In fact, *tensiometers*, which are devices for measuring soil-water tension, operate only up to about 0.85 atm tension [37, 38]. The tensiometers fail because dissolved air and the impurities in the soil water reduce the water's tensile strength [39].

There is another important observation about Table 5-3. The table is based on the wetting angle between the water and the capillary tube used being zero. This condition, although ordinarily assumed for soil pores, is not found. Organic matter makes the angle greater than 0 [42], and the angle may be even greater than 90°, in which case the soil will not wet [43]. Linford [44] stated that the soil minerals—calcite, orthoclose quartz, mica, and limestone—when strictly clean, as on a newly broken crystal cleavage plane, gave zero wetting angles. "Old faces and cleaned faces, after they had been handled, showed finite and crudely varying angles of contact." Oleic acid on mica resulted in a contact angle of 18° and on galena, a contact angle of 86°.

So far tension forces have been mentioned as holding water in the soil. The water may also be held under pressure. At the point P in Fig. 5-3b, which represents a test well (auger hole) drilled into soil below its water table, one sees that the pressure is $\rho g h$ dynes/cm². The pressure will also be $\rho g h$ at points S and R if the water is static, as due to the shown impervious layer. For nonstatic conditions the pressures at points S, P, and R will not be equal. If water is moving outward from the hole, the pressure will be less at points R and S than the pressure ρgh at P and will be more at points R and S than ρgh if the water is moving toward the hole.

Water in unsaturated soil can also be under pressure. In Fig. 5-3c, lower portion, a spherical soil particle of radius r surrounded by a film of water is shown. Above the particle, for purposes of force equilibration, there has been "isolated" one-half of the sphere and its surrounding water film. The water film exerts a downward force

Table 5-3. Energy and Tension Value* to Remove Water from a Circular Pore of a Given Diameter When the Wetting Angle of the Water against the Pore Wall Is Zero

Tension, atm	Tension, cm of water column	Log, cm of water column†	Energy, ergs/g	Tension, dynes/cm^2	Tension, mb	Diameter of pore, μ
0	0	$-\infty$	0	0	0	0
0.058	60	1.78	5.89×10^4	5.89×10^4	58.9	49.44
0.100	103	2.01	1.01×10^5	1.01×10^5	101.0	28.80
0.333	344	2.54	3.38×10^5	3.38×10^5	337.5	8.62
0.500	516	2.71	5.06×10^5	5.06×10^5	506	5.75
1	1,033	3.01	1.01×10^6	1.01×10^6	1,013	2.87
2	2,066	3.32	2.03×10^6	2.03×10^6	2,027	1.44
3	3,099	3.49	3.04×10^6	3.04×10^6	3,040	0.96
5	5,165	3.71	5.07×10^6	5.07×10^6	5,067	0.57
10	10,330	4.01	1.01×10^7	1.01×10^7	10,134	0.29
15	15,495	4.19	1.52×10^7	1.52×10^7	15,201	0.19

* Values are based on (1) 1 atm = 1,033 cm of water column of density 1 g/cm^3; (2) surface tension of water σ = 72.75 dynes/cm at 20°C; (3) g = 981 dynes/g.

† The entries in this column are sometimes, especially in Europe, designated as "pF values" [40]. Such designation should be used with care [41].

$2\pi r\sigma$, which is offset by an upward force $p\pi r^2$ across the base of the hemisphere, p being gage pressure. Thus, for equilibrium,

$$2\pi r\sigma = p\pi r^2 \tag{5-10}$$

so that the gage pressure is

$$p = \frac{2\sigma}{r} \tag{5-11}$$

This pressure will be transmitted to points just inside the surface of the water film and also to points farther inside the water film (but not through the air-water surface). Thus this film water is under a positive gage pressure p. If the sphere of Fig. 5-3c is of the size of colloidal clay, 0.2 micron (radius 0.1×10^{-4} cm), the pressure in excess of atmospheric is

$$p = \frac{2 \times 73 \text{ dynes/cm}}{0.1 \times 10^{-4} \text{ cm}} = 14{,}600{,}000 \text{ dynes/cm}^2$$

or (Table 5-3) about 15 atm. Because of such high pressure water molecules escape by evaporation with difficulty through such a film of water.

In all the examples above, spherical surfaces have been considered. For soil, this is not realistic, since in soil the curvature of pore water is more complex than spherical. Even so, the situation can be analyzed because it is known that at any point on a curved surface the curvature can be obtained by passing, in imagination, two mutually perpendicular planes through the point. Let the trace of the surface with one such plane have a radius of curvature r_1, and let the trace of the surface with the other plane have a radius of curvature r_2. Then, if this surface is an air-water interface, the gage pressure beneath the surface is [39]

$$p = \sigma \left(\frac{1}{r_1} + \frac{1}{r_2} \right) \quad (5\text{-}12)$$

where σ is the surface-tension coefficient. If $r_1 = r_2 = r$, as for the round soil particle enclosed by water of Fig. 5-3c, then one has $p = 2\sigma/r$, as before. If $r_1 = r_2 = -r$, as for a meniscus in a capillary tube of zero wetting angle, then one has $p = -2\sigma/r$; that is, one has, for the tension T, $T = 2\sigma/r$, as before.

2. Measurement of Soil-moisture Tension. To understand how a tensiometer works, see Fig. 5-4a, which shows several soil particles cemented together at A, B, C, D by, say, colloidal clay, to form a water-filled soil pore, the pore being shown in

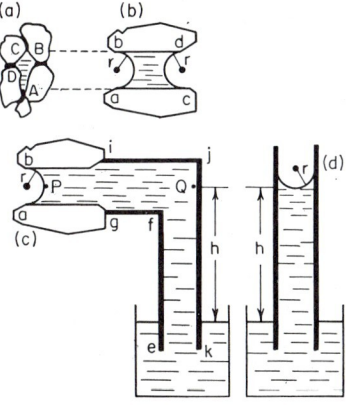

FIG. 5-4. Measurement of water tension in soil pores.

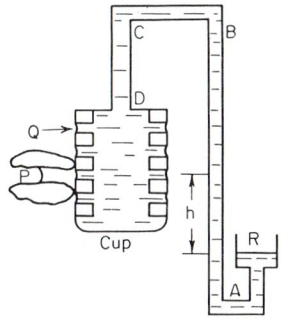

FIG. 5-5. A porous cup tensiometer; in practice, the water column AB is replaced by mercury so that the reservoir R is above the cup.

longitudinal cross section in Fig. 5-4b. Menisci ab and cd each have the same radius of curvature r, since the water in the pore is everywhere under the same tension. For simplicity, circular cross sections for the menisci are shown and the wetting angle is shown as zero.

Now imagine, in Fig. 5-4c, that a bent water-filled tube $efgijk$ makes sealed contact with the pore at the end of meniscus cd with water connection with the pore water. And suppose further that it is found that a water column of height h is supported by the meniscus. The tension then under the meniscus at a point P and along a line PQ is just equal to ρgh dynes/cm². The water-filled tube with the water container shown below it may be called a *tensiometer*. Figure 5-4d shows the familiar tension column of height h equal to that of Fig. 5-4c.

The actual commercial tensiometer is inserted into the soil at a desired depth to measure soil-water tension. The unit consists of a porous ceramic cup (Fig. 5-5) having many fine openings as at Q in its walls which make contact with soil particles and their pore water; one soil pore P is shown. A tension h corresponding to the tension under the pore water P may be created by means of water in tubing $ABCD$ and an associated water reservoir R. Now, in practice, tensiometer cups are inserted 6 to 36 in. in the ground. Furthermore, the reservoir R would have to be, as is seen in the figure, at a still lower depth. To keep the reservoir R and some of the associated tubing aboveground, one may replace each 13.6 cm of water column AB by 1 cm of mercury column or equivalent vacuum gage tension. The fact that mercury does not stick to water in the tension column does not matter since the tensions are *gage* tensions, and these, as has been stated, never reach more than about 0.85 atm. In Fig. 5-5 the pores in the wall of the cup which do not make contact with soil water develop the same curvature as that in the soil pores. These cups have pores such that air will not enter until about 0.85 atm tension is reached. Tensiometers are commercially available. Those used in irrigation work [45] are simpler than the one of Fig. 5-5, because, for them, the reservoir R and tubing ABC are replaced by a mechanical vacuum gage at C. For details on tensiometer construction, see Richards [38].

3. Measurement of Soil-moisture Tension above 0.85 Atm. To measure tension greater than 0.85 atm one applies a pressure to the top surface of soil pores whose lower ends are next to a porous plate or porous membrane of much smaller pore size than that of the soil pores, the pore water in the porous plate or membrane to be at atmospheric pressure. If the pressure applied to the soil does not drive water from the soil pores, then the upward capillary force, at the soil-water interface, is just holding up the applied pressure plus the weight of the pore water below the meniscus. Since this latter weight is negligible, one can take the upward capillary force to be just equal to the applied pressure. Thus the applied pressure gives a measure of the capillary force, and hence of the pore radius or diameter. In fact, one has for the pore diameter $2r$, when one neglects the weight of the pore water, the relation

$$2\pi r \sigma = p \pi r^2 \quad (5\text{-}10)$$

that is, one has

$$p = \frac{2\sigma}{r} \quad (5\text{-}11)$$

which is a result previously found.

As a numerical example consider equilibrium conditions for 5 atm ($= 5.07 \times 10^6$ dynes/cm^2) of applied pressure. The last equation then yields (Table 5-3)

$$5.07 \times 10^6 = \frac{2 \times 72.75}{r}$$

or

$$r = 0.285 \text{ micron}$$

that is, the pore diameter $2r$ is (as in Table 5-3)

$$2r = 0.57 \text{ micron}$$

If the pressure in the apparatus is released, as for removal of the soil, then menisci will form at the bottom of the soil pores to bring about a force balance with the menisci at the top of the pores where the air pressure was previously applied.

4. Hysteresis. Because soil pores lose more water from a water-saturated condition than they take up from a dry or partially dry condition, the moisture content of a soil sample can contain different amounts of water for a same soil-moisture tension, depending on whether the soil took up or drained water to reach its moisture status.

The phenomenon is called *hysteresis*. For an example of hysteresis when a groundwater table rises and falls, see Ref. 46.

5. Retention Forces Other Than Those of Capillarity. Consider two soil particles, each covered with a film of water. Let the particles be considered as discus-shaped disks, and let them come in coaxial contact with each other. A ring of water will form near the point of contact, the point of contact being the center of the ring. The pressure in this water ring will be negative because the radius of the ring as observed in a plane containing the axis of the particles is much smaller than the radius as observed in a plane perpendicular to the latter plane and passing through the axis of the disks. But the water covering the disks, on the sides of the disks removed from their sides of contact, will be under positive pressure because of surface-tension effects, as is clear from Subsec. IV-A-2. Thus it is apparent that forces other than those of surface tension must exist under the water of these soil particles if the water is to be in equilibrium; otherwise, all the water would accumulate in the space near the point of contact of the disks. These other forces are primarily electrical in nature and have been considered by Derjaguin and Melnikova [47].

6. Osmotic Forces and Soil-water Retention. Salts in soil water tend to hold the water in the soil against removal forces. Richards [48] refers to the suction due to salt forces as *solute suction*. He shows how this suction may be separated out, by measurement, from the *total soil suction*. The difference between the total suction and the solute suction he terms *matric suction*, which would be numerically equal to the capillary suction referred to in previous subsections. The solute suction is numerically equal to the osmotic pressure of the soil solution. Independently of toxic effects, plants cannot, because of osmotic forces, pull as much water from soils with large salt contents as they can from soils with smaller salt contents.

7. Soil-moisture Retention and the Outflow Law. Sometimes auger holes or trenches are dug into soil to collect seepage water to serve as evidence that irrigation or rainfall has penetrated into the soil. But water will not enter such holes or trenches even if the water has seeped into the soil, unless the soil water is at a pressure greater than atmospheric. Here the *outflow law* governs, which is [49]: "Outflow of free water from soil occurs only if the pressure in the soil water exceeds atmospheric pressure."

8. Soil-moisture-retention Terminology. In many soils not having a water table and having a homogeneous texture with depth, the moisture content will reach a quasi-equilibrium condition in about 1 to 3 days after the occurrence of a heavy rainfall or heavy application of irrigation water. This moisture content, per unit of oven-dry weight or per unit bulk soil volume, is designated as the *field capacity* (FC). It is the moisture retained largely by capillary forces. It is the moisture content of the soil after the so-called *gravitational water* has been removed by deep seepage. When plants extract soil water to the extent that the plants wilt, the moisture content is said to be at the *wilting point* (WP) [50]. The difference between FC and WP is called the plant *available water* (AW). The concepts FC and WP, and hence AW, must be used with care. As for the care in using WP, it has been hypothesized [51, 52], and experimentally verified [53, 54], that plant wilting depends importantly on (1) the *capillary conductivity* of the soil (a term to be considered below) and (2) weather conditions. Plants growing in soils of low capillary conductivity and under hot, dry meteorological conditions may wilt when the soil contains more water than for the WP determined under cool, humid meteorological conditions. A WP determined for a soil when the plant top is in a humid, not too hot atmosphere may be called a *permanent-wilting point* (PWP). As for use of the concept FC, one should observe that, if this concept FC is applied to layered soils or to soils with water tables, several (nonunique) FCs for the same soil and sampling location may be found [55, 56]. Another soil-moisture term which has had much use is the *moisture equivalent* (ME). The moisture equivalent is the water retained by a soil when it has been centrifuged at 1,000 times the force of gravity in a special centrifuge. The ME has been used as a measure of the FC [57].

9. Measurement of Water Retained in Soil. The most common and direct way to determine the amount of water in a soil sample is to weigh it before and after oven-drying it at 105°C. The difference in weights divided by the oven-dry weight and multiplied by 100 is the *moisture percentage on oven-dry-weight basis*. If the volume of moisture removed by the oven drying is divided by the initial volume of the bulk soil, then the result, when multiplied by 100, is the *moisture percentage by volume*. The soil volume used in the last calculation is usually that of the soil sample as found at the FC. Multiply the moisture percentage on an oven-dry-weight basis by the *bulk density* of the oven-dry soil to obtain the moisture percentage by volume. The bulk density, unless otherwise specified, is ordinarily defined as the grams per cubic centimeter, or the pounds per cubic foot, of the bulk soil when it is oven-dry—not its bulk volume at the field capacity when the soil may be swollen.

Some indirect methods for measuring soil moisture are listed in Baver [1]. Tensiometers [45] and plaster of paris [58] and fiberglas and nylon [59] units are used rather extensively. The tensiometers have been described. In the plaster of paris, fiberglas, and nylon units, two separated electrodes are embedded in the porous material. The units are buried in the soil and come to moisture equilibrium with soil. The resistance between the electrodes when the units are in the soil is governed by the soil moisture. Measurement of this resistance and calibration of the units for each soil type gives the moisture content.

10. The Neutron Soil-moisture Meter—Some Soil-moisture Profiles. The neutron-scattering soil-moisture meter [60] has come into wide use. For description of a commercial meter and a long list of references, see Kuranz [61]. The neutron meter has the advantage that a single calibration serves for a wide range of mineral soils; also, the meter gives the moisture percentage on a volume basis, which is often the most useful result for hydrology work. One neutron-meter measurement is as good as seven conventional oven-drying measurements [62]. Some moisture profiles, as determined by the neutron meter, are available [63]; also field capacities (FCs) [56].

B. Water Movements in Saturated Soils

1. Darcy's Law for Water-saturated Soils; Hydraulic Conductivity; Permeability. Consider a column of uniform soil of length L and cross section A, through which Q units of water move per unit time when the hydraulic head at the top of the column is h_t and at the bottom h_b, where h_t and h_b are both to be measured from the same reference level. Then one finds experimentally the relation (*Darcy's law*)

$$Q = \frac{k(h_t - h_b)A}{L} \tag{5-13}$$

where k is the *hydraulic conductivity*, related to the soil *permeability* k' [64] by the expression

$$k = \frac{k'\rho g}{\eta} \tag{5-14}$$

ρ being the density of the water, g the acceleration of gravity, and η the viscosity. Since Q/A is the *velocity* v (of equivalent surface water) and $(h_t - h_b)/L$ is the hydraulic gradient i, the law is often given in the simpler form

$$v = ki \tag{5-15}$$

It is important, in using Darcy's law applied to the soil column, as the one considered above, to recognize that $(h_t - h_b)/L$ is a hydraulic gradient, not a pressure gradient. This is seen as follows. Let the base of the soil column be the reference level. Then, if y_t is the piezometric height for points in the top surface of the soil column (here piezometric height means height above the top soil surface to which water stands in piezometers whose bottom ends just touch the top of the soil column), one has

$$h_t = L + y_t \tag{5-16}$$

so that one may write

$$Q = \frac{k(L + y_t - h_b)A}{L} \tag{5-17}$$

where, if h_b and y_t are both zero, there is zero pressure gradient across the sample, but nonzero flow. In fact, one has for the zero pressure gradient the relation

$$Q = \frac{kLA}{L} \tag{5-18}$$

or one has

$$\frac{Q}{A} = v = k \tag{5-19}$$

The last expression, aside from showing that zero pressure gradient need not result in zero flow, shows that, when a thin layer of water is maintained on the surface of a homogeneous soil in which vertical seepage occurs, the bottom of the column being at zero gage pressure, the velocity of in-seeping surface water is equal to the hydraulic conductivity of the soil. The factor k'—the permeability—noted above is a measure of the tortuosity and size of the soil pores in a porous medium, while, as was seen, k, the hydraulic conductivity, includes properties of the fluid (ρ and η) and the value g of gravity. If gravity were zero (as in a space experiment) for the soil column discussed under Darcy's law, there would be for zero pressure gradient also zero hydraulic gradient and zero flow. For measuring hydraulic conductivity in the laboratory, see Wenzel [65]; for measuring hydraulic conductivity in the field, see Kirkham [66] and Bouwer [67]; for measuring direction of water movement in soil, see Donnan [68, pp. 455–458]; for measuring direction of water movement in gravel, see Andreae [69].

2. **Laplace's Equation; Dupuit-Forchheimer Theory.** Imagine a rectangular x, y, z system of coordinates to be established in a homogeneous porous medium of constant hydraulic conductivity, and let h be the hydraulic head referred to an arbitrary reference level for a point (x, y, z); then, from the *equation of continuity* [70] and Darcy's law, one may, for incompressible steady-state flow in a porous medium where k is a constant, derive the expression

$$\frac{\partial^2 h}{\partial x^2} + \frac{\partial^2 h}{\partial y^2} + \frac{\partial^2 h}{\partial z^2} = 0 \tag{5-20}$$

as the expression governing the groundwater flow. Many solutions of this equation, especially when the surface of the groundwater is horizontal, are available [71–74]. Since Laplace's equation is difficult to solve when there are curved groundwater surfaces, the *Dupuit-Forchheimer* (DF) theory has been developed [72, 75]. In the DF theory one supposes that all the groundwater moves in horizontal sheets and thus neglects convergence loss of hydraulic head as groundwater enters into, say, a tile drain or ditch drain. Therefore drain spacings as computed by this theory will be on the unsafe side. But DF theory, although inexact, has enabled approximate solutions of a number of non-steady-state groundwater problems. Some recent examples are those of Maasland [76] and of Glover and Bittinger [77]. Some earlier examples are in the paper by Van Schilfgaarde et al. [75]. These non-steady-state groundwater problems involve the water table as a moving boundary. Exact solutions for the problems will be difficult to obtain, as is evident from a paper by Gibson [78]. The falling-water-table problem can, however, independently of DF theory, be studied by models [79].

3. **A Drain-spacing Equation and Nomograph.** A drain-spacing equation based on an exact solution of Laplace's equation, field-tested and theoretically valid on the safe side for the removal of steady rainfall or steady excess irrigation water, has been put in convenient nomographic form by Toksöz and Kirkham [80] and is presented herewith as Fig. 5-6. In the figure, R is rainfall rate, k is hydraulic conductivity, and the other symbols are defined in the inset of the figure, which also contains a clarifying example. Where the family of curves tend to merge into a single line, the drain spacing becomes independent of the size and shape of the drains; in other words, the DF theory then begins to apply. For the nomograph to be valid, observe that k/R must be greater than 1. If k/R is less than 1, the soil cannot take the rain as fast as it falls, and the soil will become waterlogged to the surface, a point discussed in the next subsection. For simultaneous drainage of surface and artesian water, see Hinesly [81].

4. **Soil Factors Governing Water Entry to Tile Drains.** Water will not enter tile drains unless its pressure exceeds atmospheric pressure (see outflow law of Subsec. IV-A-7). Therefore there must be either (1) a soil layer below the drains which is less permeable than the layer where the drains are located (to build up a water table) or (2) upward artesian pressure below the drains. If, near the soil surface, there is a tight layer as a plow sole or a layer tramped and puddled by cattle, and if there is no source of groundwater other than from surface water, then tiles under these tight surface layers will not intercept water, since any water penetrating through the surface

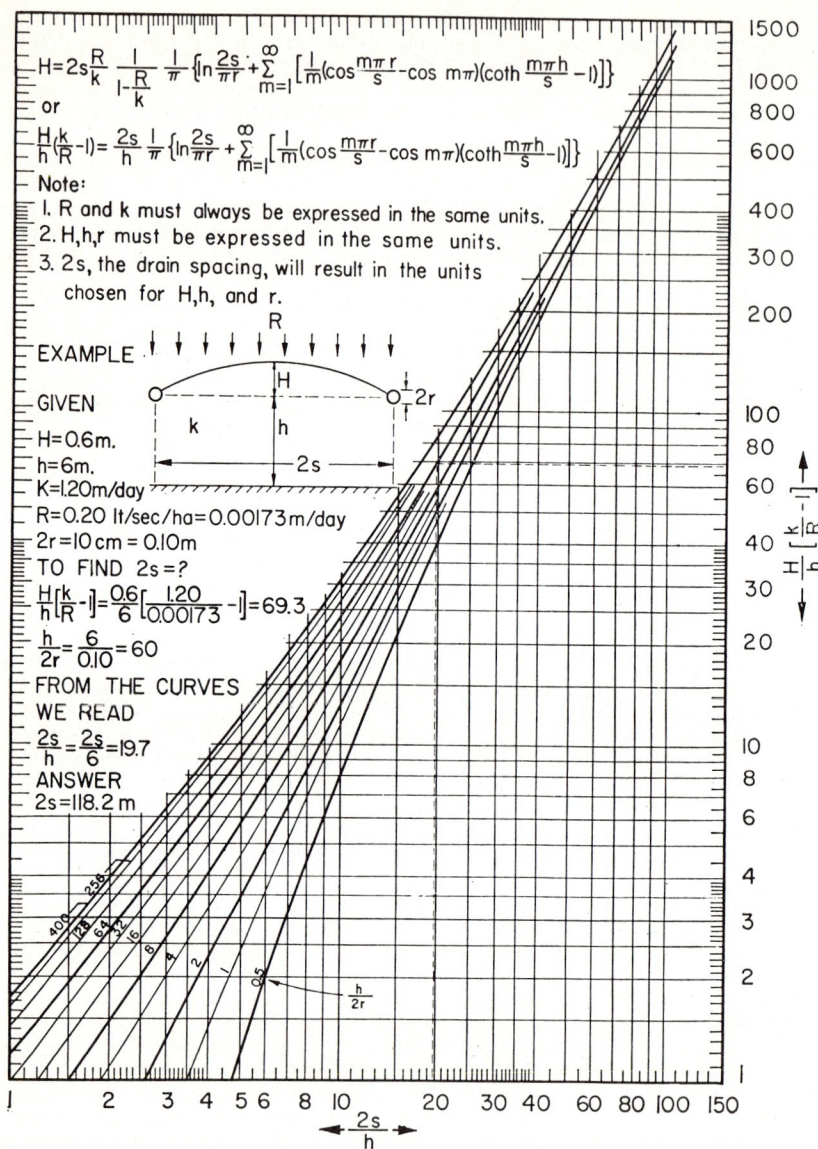

Fig. 5-6. Dimensionless chart of $(H/h)(k/R-1)$ versus $2s/h$ for various values of $h/2r$, for obtaining drain spacings $2s$ for the removal of steady rainfall or excess irrigation water (*Toksöz and Kirkham* [80].)

layer will be moving vertically downward at less than atmospheric pressure, none of this water having a lateral component of hydraulic force to bring it to the drains. If, however, below the tiles there is a tighter layer than any layer above the tiles, the water which moves through the surface layer can build up a water table, and movement to the tiles would then depend on the hydraulic conductivity of the soil between the very tight deep layer and the less tight surface layer. From these considerations

SOIL AIR, SOIL TEMPERATURE, SOIL STRUCTURE 5–19

it is clear that, if the average rainfall rate, or average rate of addition of excess irrigation water, exceeds the hydraulic conductivity of the tight surface layer, the drains will not control the water table in the surface soil. Instead, a *perched* water table will form, and it will rise to the surface; ponded surface water will result.

5. Leaching. To remove excess salts from soil, applications of excess, relatively salt-free irrigation water may be used [82–84]. Such leaching is generally needed only in arid areas. In humid regions excess rainfall may leach desired nutrients from the soil [85]. For properties of water solutions that influence water movement into soil, see Fletcher [86].

C. Water Movement in Unsaturated Soils; Ice Lenses

Using the equation of continuity and assuming that Darcy's law holds for unsaturated moisture flow, one may derive the following equation [70]:

$$\frac{\partial}{\partial x} k \frac{\partial h}{\partial x} + \frac{\partial}{\partial y} k \frac{\partial h}{\partial y} + \frac{\partial}{\partial z} k \frac{\partial h}{\partial z} = \frac{\partial \theta}{\partial t} \qquad (5\text{-}21)$$

where k is now called the *capillary conductivity* rather than the hydraulic conductivity, h is the hydraulic head, and θ is the fraction of the soil bulk volume occupied by water. If the soil is saturated, the equation reduces to Laplace's equation because k and θ become constants. The unsaturated-flow equation has not yet been solved for the general case, but it has been solved (1) for one-dimensional x-direction (horizontal) flow [87, 88] and (2) for z-direction (vertical) flow [51, 52, 89, 90]. For each solved case the result has depended upon (1) replacing h by

$$h = \psi + z \qquad (5\text{-}22)$$

where $\psi = p/\gamma g$ is called the *capillary potential* (rather than pressure potential, since the soil is now not saturated); (2) introducing a term called the *diffusivity D*, defined by

$$D = k \frac{\partial \psi}{\partial \theta} \qquad (5\text{-}23)$$

and (3) making a so-called mathematical transformation, called the *Boltzmann transformation*, which involves the time t to the one-half power. Recent careful experimental work [91] shows that the above theory fails at moisture contents where the soil-water tension is greater than 2 mb. See also Hallaire and Hénin [92].

Ice lenses form when the soil is in an unsaturated condition and when freezing occurs slowly. With slow freezing the water has time to flow to the seed of the incipient lens. In a rapid freeze the soil water freezes in place; small crystals rather than lenses form. Agriculturally, ice-lens formation is damaging because it tears plants from the soil; it is useful in that it mellows fall-plowed land. Ice heaving, a consequence of formation of large ice lenses, is troublesome to engineers because of its damage to roads. Heaving may be prevented by (1) keeping the soil well drained or (2) putting a barrier to capillary flow below the zone of freezing danger to prevent water from moving upward into the freezing zone. Coarse sand or gravel can act as such a barrier. For references on ice heaving as related to roads and engineering works, see two Highway Research Board reports [93]. Literature on ice-heaving damage to plants is rare [94, 95]. Literature on ice damage to plants, aside from ice heaving, is plentiful [96].

V. SOIL AIR, SOIL TEMPERATURE, SOIL STRUCTURE

A. Soil Air

Air is needed in the soil for micro and macro soil organisms and in particular for the growth of plant roots [1 and 97].

1. Renewal of Soil Air. Soil air contains about seven times as much CO_2 as does normal air, and the amount may be much greater. The removal of the excess CO_2 and its replacement primarily by O_2 is referred to as *soil aeration*. Several ways in which soil aeration occurs have been considered by Romell [1, pp. 209–211]. Daily variations in soil temperature account for less than $1/800$ of normal aeration. Temperature differences between the soil and the atmosphere provide $1/240$ to $1/480$ of normal aeration. Changes in barometric pressure yield not more than $1/100$. Wind action produces about $1/1,000$. The carrying of oxygen into soil by rain water gives $1/16$. The remainder, which is about $15/16$ of the total, is due to *diffusion*.

2. Diffusion of Soil Oxygen; Air Permeability. Diffusion of oxygen results from a concentration gradient. If c is the oxygen concentration, dc/dx the gradient in an x direction, and k the *diffusivity*, then the quantity Q of oxygen moving across a unit area per unit time in the x direction at a certain instant is given by

$$Q = -k \frac{\partial c}{\partial x} \tag{5-24}$$

In dry soil k is found to be proportional to the porosity (pore space divided by bulk volume). In soil which is moist, k is found to vary approximately linearly with the non-water-filled pore space; k is not directly proportional to this pore space, because at zero diffusion flow there may be as much as 10 per cent air in the soil. Ordinarily, a soil with about 10 per cent of its bulk volume occupied by air will be adequately aerated [15]. The diffusivity k in the last equation is quite different from the diffusivity D defined previously for unsaturated-moisture flow. Diffusion transfer of oxygen in soil may be measured by platinum microelectrodes [98, 99].

Air permeability refers to the ease with which air moves through the soil due to *mass flow*, that is, due to (total) pressure difference. In fact, the mass moved is proportional to the total pressure gradient [100].

To measure air permeability one may use an air permeameter, like the ones described by Grover [101] and Tanner and Wengel [102]. To correct an error, multiply the ordinates of Grover's figure 4 by 0.229.

B. Soil Temperature

Soil temperature varies with soil depth, amount and type of vegetative cover, color of the soil, amount of soil moisture, amount of large, small, and total pores, angle of surface exposure to insulation, and amount of insulation. Plant growth may be highly sensitive to soil temperature. Willis et al. [103, esp. fig. 2] found in the greenhouse that corn plants (harvested when 38 cm tall) had one-seventh more dry matter when grown at 70°F than when grown at 60°F, and in the field that leaves of corn plants 84 cm tall grew at a rate of 1.1 mm/hr at 67°F and 1.3 mm/hr at 70°F. The same authors report that corn seedlings will not sprout below 49°F, sprout fastest at 95°F, and fail to sprout above 115°F.

1. Soil Thermal Constants. One finds, as for water movement and air movement in soil, that heat follows a linear-flow law; that is, one finds that the number of calories of heat Q transmitted per unit area per unit time through a soil in the x direction where the temperature is T and the temperature gradient is dT/dx is given by the relation (*Fourier's law*)

$$Q = -K \frac{dT}{dx} \tag{5-25}$$

where K is defined as the *thermal conductivity*. The differential equation for heat flow in an x, y, z system of coordinates is [104]

$$K \left(\frac{\partial^2 T}{\partial x^2} + \frac{\partial^2 T}{\partial y^2} + \frac{\partial^2 T}{\partial z^2} \right) = \rho c \frac{\partial T}{\partial t} \tag{5-26}$$

where ρ is the soil bulk density (water and solids included), and c is the specific heat,

SOIL AIR, SOIL TEMPERATURE, SOIL STRUCTURE

that is, the calories of heat required to raise the temperature of one gram of the soil one degree centigrade. One-dimensional vertical heat flow is often pertinent in soils work. Then one writes the above equation, taking x instead of z as the vertical coordinate, as

$$k \frac{\partial^2 T}{\partial x^2} = \frac{\partial T}{\partial t} \tag{5-27}$$

where $k = K/\rho c$ is defined as the (thermal) *diffusivity*. Some values of K, ρ, c, and k are given in Table 5-4. The values stem from the International Critical Tables

Table 5-4. Thermal Constants, in the CGS System of Units, for Some Soil Substances

Substance	Density ρ	Specific heat c	Conductivity K	Diffusivity $k = K/\rho c$
Air....................	0.00129	0.240	0.000058	0.187
Average rock............			0.0042	0.0118
Ice.....................	0.092	0.502	0.0053	0.0115
Fresh snow..............	0.1	0.5	0.00025	0.0050
Average soil.............	2.5	0.2	0.0023	0.0046
Sandy dry soil...........	1.65	0.19	0.00063	0.0020
Sandy soil, 8% moist.....	1.75	0.24	0.0014	0.0033
Water..................	1.0	1.0	0.00144	0.00144
Ground cork............	0.15	0.48	0.0001	0.0014

(McGraw-Hill Book Company, Inc., New York, 1926–1930, 1933). For values of some of these constants as they depend on moisture content, see Jackson and Kirkham [105]. For values of the constants for a number of soil types, as well as for much other valuable material on soil temperature and plant growth, see [106, esp. pp. 324–325]. For mathematical solutions of the heat-flow equation, many of which are applicable to soil problems, see Carslaw and Jaeger [104]. See also a book by van Wijk [107]. For some non-steady-state solutions, see Hanks and Bowers [108] and van Duin [109].

An important thermal constant, not given in Table 5-4, is the latent heat of vaporization of water which, at 10°C, is 590 cal/g. Thus, in undrained land, as in the spring, one sees that about 590 cal is needed to evaporate 1 g (1 cm³) of water from the soil. Suppose that the undrained soil in question were the moist sandy soil of Table 5-4. Then this same 590 cal could raise the temperature of a volume V of this soil an amount of 10°C, where

$$V = \frac{590}{1.75 \times 0.24 \times 10} = 140.5 \text{ cm}^3$$

and the numbers 1.75 and 0.24 in the calculation come from Table 5-4.

Another important constant is the latent heat of fusion of water, which is 80 cal/g at 0°C. This constant implies that the water in wet soil will give up heat ("absorbs cold") when freezing temperatures occur. Thus the temperature in a soaked soil may not fall as low as in an unsoaked one. But frost heaving may result when soils are wetted heavily before a freeze occurs. Mulches—the ground cork of Table 5-4 would be an excellent one—may be used to prevent soils from freezing or from getting too hot. Straw mulches are found to reduce the soil temperatures in citrus groves [110]. But mulches may prevent soil from warming in the spring, and this lack of warming may and does reduce plant growth [111].

C. Soil Structure

1. Definition of Soil Structure. The term *soil structure* has two principal connotations. The term may refer to soil units found in the soil profile, units perhaps

1 cm or more on an edge, which the soil surveyor, when he surveys the soil profile, may designate as blocky, prismatic, platy (squamose), nuciform (nutlike), etc. [3, p. 228]. Or the term may refer to soil crumbs found in the plow layer after a cultivation. If the crumbs are friable and resistive to the slaking action of water and the beating action of rain, the soil of this plow layer is said to be in good structure. A soil in good structure will not contain much dust and will contain many large pores for water entry and aeration. Such a soil may or may not be erosive. If there is a tight layer of soil at the plow-layer depth, the soil may be erosive because the plow layer may easily become water-saturated, so that runoff and washoff of the soil crumbs, whether stable or not, may occur. Organic matter and/or colloidal clay, especially the sesquioxides Al_2O_3 and Fe_2O_3, tend to form stable soil crumbs. Soil crumbs removed from a field long in meadow will not break up when allowed to fall through a column of ordinary tap water. Soil from an adjacent plot of intertilled crops, as corn, will break up and muddy the water. A simple method for measuring decrease of aggregate sizes by the slaking action of water and also the hydraulic conductivity of the slaked soil has been presented by Williams and Cooke [112]. A review on soil-aggregate formation, etc., has been prepared by Martin et al. [113]. See also articles by De Leenheer [114].

2. Measurement of Soil Structure. The expression *measurement of soil structure* does not necessarily imply that the size of the soil crumbs, or *aggregates*, is to be measured. The term implies that any one of a number of soil physical properties associated with the concept soil structure may be measured. Such properties are bulk density, pore-size distribution, stability to water breakdown, dry strength, and others. For methods of measuring soil structure, see Baver [1] and the *Proceedings* of an international soil-structure symposium held at Ghent, Belgium [115]. Although many clever ways of measuring soil structure have been devised—even the depth of rifle-bullet penetration into the soil [116]—one has found it difficult, with few exceptions [117], to correlate crop yields with the measurements. Because of this, visual ratings of soil structure for plant yield have been made [118]. The difficulty of correlating measured structure values with yields probably lies in the failure to relate the physically measured values to moisture, oxygen, and nutrient supply to plants. Then too, De Boodt et al. [117] found that structure as measured, for example, by water stability may be nearly optimum over a wide range of soil conditions. Finally, highly variable weather effects (as too much or too little rain) may obscure any structure influences to which a soil may be subject on yields.

3. Soil Compaction. Soil compaction often restricts plant growth as a consequence of poor aeration, because in a compacted soil most of the pores will be water-saturated much of the time. But sufficiently compacted soils, as at bulk density 1.75 g/cm³ for sands and 1.46 to 1.63 g/cm³ for clay soils, may also reduce plant growth because of decreased pore size, independent of oxygen supply [119, 120]. The data just cited are from greenhouse experiments. Field experiments have also shown decreases in plant yields due to compaction when aeration was apparently adequate. Corn yields on nonfertilized plots dropped from 85 to 65 bu/acre as the plow-layer bulk density increased from 1.15 to 1.40 g/cm³, and on fertilized plots, dropped from 100 to 90 bu/acre for the same increase in bulk density [121].

For European and other recent literature on soil compaction see "International Treatise on Soils" [122].

VI. REFERENCES

1. Baver, L. D.: "Soil Physics," 3d ed., John Wiley & Sons, Inc., New York, 1956.
2. Shaw, B. T. (ed.): Soil physical conditions and plant growth, *Agronomy*, vol. 2, Academic Press Inc., New York, 1952.
3. "Soil Survey Manual," U.S. Department of Agriculture Handbook 18, 1951.
4. Clark, G. R.: "The Study of the Soil in the Field," 4th ed., Oxford University Press, London, 1957.
5. "Soil Classification: A Comprehensive System, 7th Approximation," U.S. Department of Agriculture, Soil Survey Staff, August, 1960.
6. Thompson, L. M.: "Soils and Soil Fertility," 2d ed., McGraw-Hill Book Company, Inc., New York, 1957.

REFERENCES

7. Lyon, T. L., H. O. Buckman, and N. C. Brady: "The Nature and Prpoerites of Soils," 5th ed., The Macmillan Company, New York, 1952.
8. Scholtes, W. H., and Don Kirkham: The use of radiocarbon for dating soil strata, (Belgian Journal of) *Pédologie*, vol. 7, pp. 316–323, 1957. Also see there work cited of Ruhe and Scholtes.
9. Hutton, C. E.: Studies of loess-derived soils in southwestern Iowa, *Soil Sci. Soc. Am. Proc.*, vol. 12, pp. 424–431, 1948.
10. Jenny, Hans: "Factors of Soil Formation," McGraw-Hill Book Company, Inc., New York, 1941.
11. Franz, Herbert: "Feldbodenkunde," Verlag Georg. Fromme and Co., Vienna and Munich, 1960 (reviewed in *Soil Sci. Soc. Am. Proc.*, August, 1961, p. iv).
12. Scheffer, Fritz, and Paul Schachtschabel: "Bodenkunde," Ferdinand Enke Verlagsbuchhandlung, Stuttgart, 1952.
13. Edelman, C. H.: "Soils of the Netherlands," North Holland Publishing Company, Amsterdam, 1950, pp. 23–24.
14. Shaw, R. H., D. R. Nielsen, and J. R. Runkles: Evaluation of some soil moisture characteristics of Iowa soils, *Iowa Agr. Expt. Sta. Res. Bull.* 465, pp. 411–420, February, 1959.
15. Wesseling, Jans, and W. R. van Wijk: Soil physical conditions in relation to soils and crops, in Drainage of agricultural lands, *Agronomy*, vol. 7, J. N. Luthin (ed.), American Society of Agronomy, pp. 461–504, especially pp. 468 and 472, 1957.
16. Grim, R. E.: "Clay Mineralogy," McGraw-Hill Book Company, Inc., New York, 1953.
17. Tschapek, M. W.: Quimica coloidal del suelo. I. Fenomenos de superficie, *Commun. Inst. Nat. Invest. Ciencias Nat.*, Publ. de Extension Cultural Didactica, no. 3, Buenos Aires, 1949.
18. Thorp, James, and G. D. Smith: Higher categories of soil classification: order, suborder, and great soil groups, *Soil Sci.*, vol. 67, pp. 117–126, February, 1949.
19. Oakes, Harvey, and James Thorp: Dark-clay soils of warm regions variously called rendzina, black cotton soils, regur, tirs, *Soil Sci. Soc. Am. Proc.*, vol. 15, pp. 347–354, 1950.
20. Riecken, F. F., and G. D. Smith: Lower categories of soil classification: family series, type, and phase, *Soil Sci.*, vol. 67, pp. 107–115, 1949.
21. "Engineering Soil Classification for Residential Developments," Federal Housing Administration, F.H.A. 373, August, 1959.
22. Smith, S. M., and F. F. Riecken: "Soil Survey of Jefferson County, Iowa," U.S. Department of Agriculture and Iowa Agricultural Experiment Station, March, 1960.
23. "Earth Manual," U.S. Department of the Interior, Bureau of Reclamation, 1960.
24. "Unified Soil Classification System," U.S. Army Corps of Engineers Waterways Experiment Station, Technical Manual 3-357, Vicksburg, Miss., March, 1953.
25. Mathews, E. D.: "Soil Survey of Frederick County, Maryland," U.S. Department of Agriculture and Maryland Agricultural Experiment Station, September, 1960.
26. Spangler, M. G.: Engineering soil classification, in K. B. Woods (ed.), "Highway Engineering Handbook," McGraw-Hill Book Company, Inc., New York, 1960.
27. The classification of soils and soil-aggregate mixture for highway construction purposes, Designation M145-49, in "Highway Materials, Part I, Specifications," 7th ed., American Association of State Highway Officials, Washington, D.C., 1955, pp. 15–51.
28. "Airport Paving," U.S. Civil Aeronautics Administration, 1956.
29. Woods, K. B., and C. W. Levell, Jr.: Engineering description of soils of North America, in K. B. Woods (ed.), "Highway Engineering Handbook," McGraw-Hill Book Company, Inc., New York, 1960.
30. Hockensmith, R. D., and J. G. Steele: Recent trends in the use of the land-capability classification, *Soil Sci. Soc. Am. Proc.*, vol. 14, pp. 383–388, 1949.
31. Eikelbury, R. W., and E. H. Templin: "Soil Survey of Brown County, Kansas," U.S. Dept. of Agriculture and Kansas Agricultural Experiment Station, September, 1960.
32. Wilde, S. A.: "Forest Soils," The Ronald Press Company, New York, 1958.
33. Zunker, Ferdinand: Das Verhalten des Bodens zum Wasser, in Edwin Blanck (ed.), "Handbuch der Bodenlehre," Springer-Verlag OHG, Berlin, 1930, pt. 6, pp. 66–220.
34. Rode, A. A.: "Das Wasser im Boden," Akademie-Verlag GmbH. Berlin, 1959 (transl. from the Russian by M. Trénel).
35. Tschapek, M. W.: "El Aqua en el Suelo," El Instituto Nacional de Technologic Agropecuaria, Buenos Aires, 1959.
36. Russell, M. B. (coordinator): Water and its relation to soils and crops, *Advan. in Agron.*, A. G. Norman (ed.), Academic Press Inc., New York, 1959, pp. 1–122.
37. Richards, L. A.: The usefulness of capillary potential to soil moisture and plant

investigators, *J. Agr. Res.*, vol. 37, pp. 719–742, 1928 (p. 738 for dissolved air or impurities in a water column).
38. Richards, L. A.: Methods for measuring soil moisture tension, *Soil Sci.*, vol. 68, pp. 95–112, July, 1949.
39. Poynting, J. H., and J. J. Thomson: "A Textbook of Physics," 5th ed., Charles Griffin & Company, Ltd., London, 1909, vol. 1, pp. 143–145. [For the law $p = \sigma(1/r_1 + 1/r_2)$; for tensile strength of water, p. 174.]
40. Schofield, R. K.: The pF of water in soil, *Trans. Intern. Congr. Soil Sci.*, vol. 3, no. 2, pp. 37–48, 1935.
41. Richards, L. A., and L. R. Weaver: Moisture retention by some irrigated soils as related to soil-moisture tension, *J. Agr. Res.*, vol. 69, esp. pp. 232–234, 1944.
42. Swartzendruber, Dale, M. F. DeBoodt, and Don Kirkham: Capillary intake rate of water and soil structure, *Soil Sci. Soc. Am. Proc.*, vol. 18, pp. 1–7, 1954.
43. Wander, I. W.: Interpretation of the cause of water-repellent sandy soils found in citrus groves of central Florida, *Science*, vol. 110, pp. 229–301, Sept. 23, 1949.
44. Linford, L. B.: Soil moisture phenomena in a saturated atmosphere, *Soil Sci.*, vol. 29, pp. 227–237, 1930.
45. Richards, S. J., and A. W. Marsh: Irrigation based on soil suction measurements, *Soil Sci. Soc. Am. Proc.*, vol. 25, pp. 65–69, 1961.
46. Tolman, C. F.: "Ground Water," McGraw-Hill Book Company, Inc., New York, 1937, p. 157.
47. Derjaguin, B. V., and N. K. Melnikova: Mechanism of moisture equilibrium and migration in soils, in H. F. Winterkorn (ed.), "Water and Its Conduction in Soils: An International Symposium," Highway Research Board Special Report 40, pp. 43–54, 1958.
48. Richards, L. A.: Advances in soil physics, *Trans. Seventh Intern. Congr. Soil Sci.* (Madison, Wis., August, 1960), vol. 1, pp. 67–79, Elsevier Publishing Company, Amsterdam, 1961.
49. Richards, L. A.: Laws of soil moisture, *Trans. Am. Geophys. Union*, vol. 31, pp. 750–756, 1950.
50. Veihmeyer, F. J., and A. H. Hendrickson: The permanent wilting percentage as a reference for the measurement of soil moisture, *Trans. Am. Geophys. Union*, vol. 29, pp. 887–896, 1948.
51. Gardner, Wilford: Dynamic aspects of water availability to plants, *Soil Sci.*, vol. 89, pp. 63–73, 1960.
52. Philip, J. R.: The physical principles of water movement during the irrigation cycle, *Third Congr. Intern. Comm. on Irrigation and Drainage*, question 8, pp. 8-125 to 8-154, 1957.
53. Letey, John, and G. B. Blank: Influence of environment on the vegetative growth of plants watered at various soil moisture suctions, *Agron. J.*, vol. 53, pp. 151–153, 1961.
54. Denmead, O. T., and R. H. Shaw: Availability of soil water to plants as affected by soil moisture content and by meteorological conditions, *Agron. J.*, vol. 54, no. 5, pp. 385–390, September-October, 1962.
55. Nielsen, D. R., Don Kirkham, and W. R. van Wijk: Measuring water stored temporarily above the field moisture capacity, *Soil Sci. Am. Proc.*, vol. 23, pp. 408–412, 1959.
56. Burrows, W. C., and Don Kirkham: Measurement of field capacity with a neutron meter, *Soil Sci. Soc. Am. Proc.*, vol. 22, pp. 104–105, 1958.
57. Kirkham, Don: Soil physical properties, in C. B. Richey, Paul Jacobson, and C. W. Hall (eds.), "Agricultural Engineers' Handbook," McGraw-Hill Book Company, Inc., New York, 1961, pp. 793–801.
58. Bouyoucos, G. J.: More durable plaster of paris moisture blocks, *Soil Sci.*, vol. 76, pp. 447–451, December, 1953.
59. Weaver, H. A., and V. C. Jamison: Limitations in the use of electrical resistance soil moisture units, *Agron. J.*, vol. 43, pp. 602–605, 1951.
60. Stone, J. F., Don Kirkham, and A. A. Reed: Soil moisture determination by a portable neutron scattering moisture meter, *Soil Sci. Soc. Am. Proc.*, vol. 19, pp. 419–423, 1955.
61. Kuranz, J. L.: Measurement of moisture and density in soils by the nuclear method, in Symposium on Applied Radiation and Radioisotope Test Methods, *Am. Soc. Testing Mater. Spec. Tech. Publ.* 268, pp. 40–54, 1960.
62. Stone, J. F., R. H. Shaw, and Don Kirkham: Statistical parameters and reproducibility of the neutron method of measuring soil moisture, *Soil Sci. Soc. Am. Proc.*, vol. 25, pp. 435–438, 1961.
63. Nielsen, D. R., Don Kirkham, and W. C. Burrows: Solids–water–air space relations of some Iowa soils, *Iowa State Coll. J. Sci.*, vol. 33, pp. 111–116, 1958.
64. Richards, L. A.: Report of the Sub-committee on Permeability and Infiltration, Committee on Terminology, Soil Science Society of America, *Soil Sci. Soc. Am. Proc.*, vol. 16, pp. 85–88, 1952.

REFERENCES

65. Wenzel, L. K.: Methods for determining permeability of water-bearing materials, *U.S. Geol. Surv. Water-Supply Paper* 887, pp. 1–192, 1942.
66. Kirkham, Don: Measurement of the hydraulic conductivity of soil in place, in Symposium on Permeability of Soils, *Am. Soc. Testing Mater. Spec. Tech. Publ.* 163, pp. 80–97, 1955.
67. Bouwer, Herman: A double tube method for measuring hydraulic conductivity of soil *in situ* above a water table, *Soil Sci. Soc. Am. Proc.*, vol. 25, pp. 334–339, 1961.
68. Donnan, W. W.: Soil drainage investigation methods, in Drainage of agricultural lands, *Agronomy*, vol. 7, J. N. Luthin (ed.), American Society of Agronomy, 1957.
69. Andreae, Horst:"Grundwassermessungen," VEB Verlag Technik, Berlin, 1959, esp. pp. 38–42.
70. Childs, E. C.: The physics of land drainage, in Drainage of agricultural lands, *Agronomy*, vol. 7, J. N. Luthin (ed.), American Society of Agronomy, esp. p. 75, 1957.
71. Scheidegger, A. E.: "The Physics of Flow through Porous Media," rev. ed., The Macmillan Company, New York, 1960.
72. Muskat, Morris: "Flow of Homogeneous Fluids through Porous Media," McGraw-Hill Book Company, Inc., New York, 1937, or J. W. Edwards, Publisher, Incorporated, Ann Arbor, Mich., 1946.
73. Luthin, J. N. (ed.): Drainage of agricultural lands, *Agronomy*, vol. 7, American Society of Agronomy, 1957.
74. Kirkham, Don: The ponded water case, in Drainage of agricultural lands, *Agronomy*, vol. 7, J. N. Luthin (ed.), American Society of Agronomy, 1957.
75. Van Schilfgaarde, Jan, Don Kirkham, and R. K. Frevert: Physical and mathematical theories of tile and ditch drainage and their usefulness in design, *Iowa Agr. Expt. Sta. Res. Bull.* 436, 1956.
76. Maasland, Marinus: Water table fluctuations induced by intermittent recharge, *J. Geophys. Res.*, vol. 64, pp. 549–559, 1959.
77. Glover, R. E., and M. W. Bittinger: Drawdown due to pumping from an unconfined aquifer, *Proc. Am. Soc. Civil Eng., J. Irrigation and Drainage Div.*, vol. 86, no. IR3, pp. 64–70, September, 1960.
78. Gibson, R. E.: A one-dimensional consolidation problem with a moving boundary, *Quart. Appl. Math.*, vol. 18, pp. 123–129, 1960.
79. Grover, B. L., and Don Kirkham: A glass-bead glycerol model for non-steady state tile drainage, *Soil Sci. Soc. Am. Proc.*, vol. 25, pp. 91–94, 1961.
80. Toksöz, Sadik, and Don Kirkham: Graphical solution and interpretation of a new drain-spacing formula, *J. Geophys. Res.*, vol. 66, pp. 509–516, 1961.
81. Hinesly, T. D.: Seepage of surface and artesian water into soil drains, Ph.D. thesis, Iowa State University, Ames, Iowa, 1961. (Also obtainable from University Microfilms, Ann Arbor, Mich.)
82. Reeve, R. C., L. E. Allison, and D. F. Peterson: Reclamation of saline-alkali soils by leaching, delta area, Utah, *Utah Agr. Expt. Sta. Bull.* 335, 1948.
83. Luthin, J. N.: Proposed method of leaching tile-drained land, *Soil Sci. Soc. Am. Proc.*, vol. 15, pp. 61–68, 1951.
84. Richards, L. A. (ed.): "Diagnosis and Improvement of Saline and Alkaline Soils," U.S. Department of Agriculture Handbook 60, 1954.
85. Massey, H. F., and M. L. Jackson: Selective erosion of soil fertility constituents, *Soil Sci. Soc. Am. Proc.*, vol. 16, pp. 353–356, 1952.
86. Fletcher, J. E.: Some properties of water solutions that influence infiltration, *Trans., Am. Geophys. Union*, vol. 30, pp. 548–554, 1949.
87. Klute, Arnold: A numerical method for solving the flow equation for water in unsaturated materials, *Soil Sci.*, vol. 73, pp. 105–116, 1952.
88. Philip, J. R.: The theory of infiltration: the influence of the initial moisture content, *Soil Sci.*, vol. 84, pp. 329–339, 1957 (see also four earlier articles of series).
89. Youngs, E. G.: Moisture profiles during vertical infiltration, *Soil Sci.*, vol. 84, pp. 283–290, 1957.
90. Nielsen, D. R., Don Kirkham, and W. R. van Wijk: Diffusion equation calculations of field soil, *Soil Sci. Soc. Am. Proc.*, vol. 25, pp. 165–168, 1961.
91. Nielsen, D. R., J. W. Biggar, and J. M. Davidson: Experimental consideration of diffusion analysis in unsaturated flow problems, *Soil Sci. Soc. Am. Proc.*, vol. 26, pp. 107–111, March-April, 1962.
92. Hallaire, Marc, and Stéphane Hénin: Sur la non-validité de l'équation de conductivité pour exprimer le mouvement de l'eau non saturante dans le sol, *Compt. Rend.* (Paris), vol. 246, pp. 1720–1722, 1958.
93. Highway Research Board, Highway Research Council: Frost action in roads and airfields: a review of the literature, 1765–1951, by A. W. Johnson, *Spec. Rept.* 1; also, Frost action in soils, a symposium presented at the Thirtieth Annual Meeting, Jan. 9–12, 1951, *Spec. Rept.* 2, Washington, D.C., 1952.

94. Johnson, W. C.: Frost heaving, soil type, and soil water tension, Ph.D. thesis, Iowa State University, Ames, Iowa, 1955, esp. pp. 33–35.
95. Edwards, R. S.: Frost heaving of the soil and its effect on the overwinter survival of winter oats, *J. Agr. Soc. Univ. Coll. Wales*, vol. 38, pp. 21–28, 1957.
96. Levitt, Jacob: The hardiness of plants, *Agronomy*, vol. 6, Academic Press Inc., New York, 1956.
97. Russell, M. B.: Soil aeration and plant growth, in Soil physical conditions and plant growth, *Agronomy*, vol. 2, B. T. Shaw (ed.), Academic Press Inc., New York, 1952.
98. Lemon, E. R., and A. E. Erickson: The measurement of oxygen diffusion in the soil with a platinum microelectrode, *Soil Sci. Soc. Am. Proc.*, vol. 16, pp. 160–163, 1952.
99. Erickson, A. E., and D. M. van Doren: The relation of plant growth and yield to soil oxygen availability, *Trans. Seventh Intern. Congr. Soil Sci.* (Madison, Wis., August, 1960), vol. 3, pp. 428–432, Elsevier Publishing Company, Amsterdam, 1961.
100. Kirkham Don: Field method for determination of air permeability in its undisturbed state, *Soil Sci. Soc. Am. Proc.*, vol. 11, pp. 93–99, 1947.
101. Grover, B. L.: Simplified air permeameters for soil in place, *Soil Sci. Soc. Am. Proc.*, vol. 19, pp. 414–418, 1955.
102. Tanner, C. B., and R. W. Wengel: An air permeameter for field and laboratory use, *Soil Sci. Soc. Am. Proc.*, vol. 21, pp. 663–664, 1957.
103. Willis, W. O., W. E. Larson, and Don Kirkham: Corn growth as affected by soil temperature and mulch, *Agron. J.*, vol. 49, pp. 323–328, 1957.
104. Carslaw, H. S., and J. C. Jaeger: "Conduction of Heat in Solids," 2d ed., Clarendon Press, Oxford, 1959.
105. Jackson, R. D., and Don Kirkham: Method of measurement of the real thermal diffusivity of moist soil, *Soil Sci. Soc. Am. Proc.*, vol. 22, pp. 479–482, 1958.
106. Richards, S. J., R. M. Hagan, and T. M. McCalla: Soil temperature and plant growth, in Soil physical conditions and plant growth, *Agronomy*, vol. 2, B. T. Shaw (ed.), Academic Press Inc., New York, pp. 303–491, 1952.
107. van Wijk, W. R.: "Physics of Plant Environment," North Holland Publishing Company, Amsterdam, 1963.
108. Hanks, R. J., and S. A. Bowers: Non-steady moisture, temperature, and soil air pressure approximation with an electric simulator, *Soil Sci. Soc. Am. Proc.*, vol. 24, pp. 247–252, 1960.
109. van Duin, R. H. A.: Influence of tilth on soil and air temperature, *Neth. J. Agr. Sci.*, vol. 2, pp. 229–242, 1954.
110. Smith, G. E. P.: Control of high soil temperatures, *Agr. Eng.*, vol. 17, pp. 383–385, September, 1936.
111. van Wijk, W. R., W. E. Larson, and W. C. Burrows: Soil temperature and the early growth of corn from mulched and unmulched soil, *Soil Sci. Soc. Am. Proc.*, vol. 23, pp. 428–434, 1959.
112. Williams, R. J. B., and G. W. Cooke: Some effects of farmyard manure and of grass residues on soil structure, *Soil Sci.*, vol. 92, pp. 30–39, 1961.
113. Martin, J. P., and others: Soil aggregation, *Advan. in Agron.*, vol. 7, A. G. Norman (ed.), Academic Press Inc., New York, pp. 1–35, 1955.
114. De Leenheer, Louis: Articles on soil compaction, in International treatise on soils, in preparation, Springer-Verlag, Vienna.
115. De Leenheer, Louis (ed.): International Symposium on Soil Structure Proceedings, *Mededel. Landbouwhogeschool Opzoekingsta. Staat Ghent*, vol. 24, pp. 1–434, 1959.
116. Culpin, Claude: Studies on the relation between cultivation implements, soil structure and the crop I (use of rifle bullets), *J. Agr. Sci.*, vol. 26, pp. 22–35, esp. p. 30, 1936.
117. De Boodt, M. F., Louis De Leenheer, and Don Kirkham: Soil aggregate stability indexes and crop yields, *Soil Sci.*, vol. 91, pp. 138–146, 1961.
118. Peerlkamp, P. K.: A visual method of soil structure evaluation, in International Symposium on Soil Structure Proceedings, *Mededel. Landbouwhogeschool Opzoekingsta. Staat Ghent*, vol. 24, pp. 216–221, 1959.
119. Wiersum, L. K.: The relationship of the size and structural rigidity of pores to their penetration by roots, *Plant and Soil*, vol. 9, pp. 75–85, 1957.
120. Viehmeyer, F. J., and A. H. Hendrickson: Soil density and root penetration, *Soil Sci.*, vol. 65, pp. 487–493, 1948.
121. Phillips, R. E., and Don Kirkham: Soil compaction in the field and corn growth, *Agron. J.*, vol. 54, no. 1, pp. 29–34, January-February, 1962.
122. International treatise on soils (or revised title), articles by Richards, Schuffelen, Muckenhausen, Frese, Flaig, De Leenheer, and others, in preparation, Springer-Verlag, Vienna.

Section 6

ECOLOGICAL AND SILVICULTURAL ASPECTS

HOWARD W. LULL, *Chief, Division of Watershed Management Research, Northwestern Forest Experimental Station, U.S. Forest Service.*

```
     I. Introduction................................................  6-2
    II. Importance of Water Relations.............................  6-2
   III. Effects of Water on Vegetation............................  6-3
        A. Functions of Water.....................................  6-3
        B. Movement of Water......................................  6-4
        C. Water and Plant Habitats...............................  6-5
           1. Hydrophytes.........................................  6-5
           2. Xerophytes..........................................  6-5
           3. Mesophytes..........................................  6-6
    IV. Interception...............................................  6-6
        A. Interception of Rainfall...............................  6-6
           1. Stemflow............................................  6-7
           2. Effect on Rainfall Intensity........................  6-7
        B. Dew Formation and Fog..................................  6-7
        C. Snow Interception......................................  6-8
        D. Factors Influencing Interception.......................  6-8
           1. Size of Storm.......................................  6-8
        E. Interception Effect on Transpiration...................  6-9
        F. Interception by Herbaceous Vegetation..................  6-9
        G. Interception by Forest Vegetation......................  6-10
        H. Ecological Aspects.....................................  6-11
        I. Silvicultural Aspects..................................  6-11
     V. Infiltration..............................................  6-13
        A. Factors Influencing Infiltration.......................  6-13
           1. Soil Texture and Structure..........................  6-13
           2. Soil-moisture Content...............................  6-14
           3. Vegetation..........................................  6-14
           4. Compaction..........................................  6-14
           5. Frost...............................................  6-15
        B. Ecological Aspects.....................................  6-15
        C. Silvicultural Aspects..................................  6-16
    VI. Evapotranspiration........................................  6-17
        A. Transpiration..........................................  6-17
        B. Evaporation............................................  6-18
        C. Factors Influencing Transpiration......................  6-19
        D. Evapotranspiration by Herbaceous Vegetation............  6-21
        E. Evapotranspiration by Forest Vegetation................  6-21
```

F. Ecological Aspects.................................... 6-23
G. Silvicultural Aspects................................. 6-23
   1. Clear-cutting...................................... 6-24
   2. Seed-tree Method................................... 6-24
   3. Shelterwood, Diameter-limit, and Selection Systems..... 6-24
   4. Cleanings and Liberation Cuttings................... 6-26
   5. Reproduction....................................... 6-26
VII. Glossary.. 6-27
VIII. References... 6-28

I. INTRODUCTION

The purpose of this section is to describe the water relations of vegetation and the effects on them of plant succession and silvicultural practices.

This section deals mainly with native (i.e., United States wildland) vegetation and largely with forest vegetation. The water relations to be described and interpreted in respect to ecology and silviculture are interception, evapotranspiration, and infiltration.

The first two, interception and evapotranspiration, concern the plant or tree itself. *Interception* deals with the amount of moisture that is caught and stored on leaves and stems. *Evapotranspiration* deals with the movement of water through the plant to the atmosphere (transpiration) and the evaporation of water from neighboring soil or litter surfaces. *Infiltration* concerns the movement of water into the soil; though this does not involve vegetation directly, it is governed in part by the amount and kind of vegetation covering the soil surface.

The amount of water that is intercepted by vegetation, the amount that is transpired into the atmosphere and evaporated from the soil and vegetal surfaces, and the rate of movement of water into the soil depend to various degrees on the type of vegetation and its density. For the native vegetation, these characteristics in turn depend upon the point of *ecological succession* (the development from one type of vegetation to another)—the point the vegetation occupies between the pioneer and climax stages.

Water relations also depend upon the condition of vegetation in respect to its use. Regarding forest cover, this is a concern of *silviculture*, that branch of forestry that deals with the establishment, development, care, and reproduction of stands of timber and whose aim is the continuous production of wood. In achieving this aim the forest stand can be subjected to many different kinds of treatment, each of which affects water relations. The silvicultural status of the forest stand determines whether it is in the seedling, sapling, pole, or sawlog stages or its condition in regard to cutting practice—whether it has been clear-cut or subjected to shelterwood, seed-tree, diameter-limit, or more intensive selection types of cutting.

The ecological and silvicultural effects on water relations for the most part can be interpreted only in general terms from the little information available. Few studies have been made of these effects, and these few studies have dealt more with recording the growth and change of vegetation in various environments, or after specific treatments, than with the consequent changes in water relations.

However, a growing source of information on this subject has developed within the past few years in the field of forest-watershed-management research, with its objective of determining the effects of forest type, condition, and treatment on stream flow. To meet this objective, small experimental forested watersheds are gaged, calibrated, and treated. Consequent effects on water yield can be interpreted, with additional plot studies, in respect to infiltration, interception, and evapotranspiration. Results of some of this research are described here.

II. IMPORTANCE OF WATER RELATIONS

Interception, evapotranspiration, and infiltration are somewhat interrelated. In this interrelation water that is intercepted is a subtraction from water that otherwise could be evaporated from the ground surface, run off to the stream, or infiltrated into the soil. The water that is infiltrated and stored in the soil becomes the water that is later evapotranspired or becomes subsurface runoff. And, to complete this circuit, the evapotranspiration process is limited during periods of interception.

The water relations of vegetation and the effects on them of ecological succession and silviculture are subjects of some practical importance. Consider, for example, the usefulness of interception data in interpreting precipitation data. Rainfall is commonly measured by gages placed in openings free of obstructions. Readings from these gages require corrections to estimate the amount of rainfall that reaches the soil surface in the extensive areas covered by native vegetation. To do so requires some knowledge of the effects of interception of precipitation by vegetation.

The importance of interception to rainfall disposition has been long recognized; studies date back to the late nineteenth century. In the United States Horton [1] made the first major study of interception; he caught samples of water passing through the crowns of single trees of various species and then derived equations by which interception could be predicted. But his results, as Wicht [2] has pointed out, cannot be used satisfactorily to estimate the interception of water by stands of trees.

Evapotranspiration also has its importance. Water losses from the soil to the air are increased greatly by plant cover. The aggregate leaf surface of vegetation may be 20 times greater than the area of the soil surface it occupies. Plants can withdraw water from considerable depths, whereas surface evaporation commonly affects only the upper 6 to 15 in. As plants withdraw water from the soil by transpiration, storage opportunity is created for succeeding rainfalls. During the growing season only the heaviest rainfall can satisfy this storage and still supply some water to groundwater storage or to streamflow. Plants, and particularly forest vegetation, are extravagant users of water; research has indicated the possibilities of reducing this use and increasing water supplies by controlling vegetation.

And infiltration has its importance. In those situations where the rate of infiltration is less than the rate of rainfall intensity, overland flow results. If it continues, soil erosion may start. Overland flow can be the principal contributor of peak flows and of sediment. Continued overland flow on any one area washes away seed and prevents establishment of vegetation. Rehabilitation of such areas consists primarily in seeding, mulching to absorb rainfall impact and hold the seed in place, and diversion of overland flow from the treated areas.

III. EFFECTS OF WATER ON VEGETATION

The processes described before deal, then, with the effects of vegetation on water. Before examining in some detail the nature of these processes and their ecological and silvicultural aspects, some attention will be paid to the effects of water on vegetation, first, as to the functions of water in plants, next, to the way in which water moves through plants, and finally, to the response of plants to habitats containing relatively different amounts of water.

A. Functions of Water

In the physiology of plants, water is important in many ways. It is the most abundant compound: plant tissues actively growing have a water content equal to as much as 85 to 95 per cent of their fresh weight. As a near-universal solvent, water dissolves all minerals contained in the soil, and yet is not too active chemically. Water is the connecting link between the soil and the plant. It is the medium by which solutes enter the plant and move about through its tissues. And it is a raw material of photosynthesis.

Water is responsible for the turgidity of the cells; without water, cells cannot function actively. Protoplasm, to exist, requires water; very few tissues survive if water content is reduced as low as 10 per cent. Also, water creates uniform temperature conditions for biochemical reactions: to a remarkable degree, water can absorb much heat from warm surroundings with relatively little change in temperature and thereby slow the rate of temperature change in protoplasm.

In plant reproduction, water often assumes a transporting function. It carries pollen for aquatic plants and seeds and spores for land plants. Seed plants on remote islands may owe their existence to water-borne seeds that survive salt-water travel.

Water may also deter reproduction. Heavy rainfalls during the pollinating season can have a very detrimental effect, washing wind-borne pollen out of the air and grounding pollinating insects. Also, seeds of many plants, when wetted, are easily killed by lack of aeration.

Water may be necessary even before germination. The seeds of some plants (for example, willows and cottonwoods) must make contact with moist soil within a few days after maturation or otherwise perish. After germination and early growth, moisture supplies can direct the form of a plant: root-shoot ratios increase as soil moisture decreases, the roots extending into the dry soil if another part of the root system has access to growth water. A high water table enforces shallow rooting. And finally, the drier the soil, the earlier a plant may mature.

Water is a necessary element for plant succession on dry land. Plant pioneers are usually simple organisms that can live on bare rock and the rainfall that sticks to it. The most successful of these are the lichens, composed of two organisms, a fungus and alga. The alga is parasitized by the fungus, which is able to store water in its cells when it rains and hold it through dry periods. The alga, then, may draw water from the fungus mutualistically and photosynthesize food to supply both organisms with needed energy.

Lichens fasten on rocks, and in the process of extracting from them necessary mineral constituents, small quantities of soil are formed. Mosses ordinarily follow the lichens, moss spores finding lodging in the soil and water built up by the lichens. Mosses, larger and more capable of disintegrating rock, collect dust and create deeper soils on the rock. Then spores of ferns or seed plants become established in the soil.

Thus the successful invasion of dry land depends upon obtaining and retaining sufficient water. As plant successions proceed, the soil becomes deeper, more water is stored upon the land surface, more water become available for plant growth, more plant growth is produced, and finally, in some areas, a heavy forest cover climaxes the succession.

B. Movement of Water

Water in the plant moves upward from the root to stem to petiole and thence to the leaf surface from which it escapes into the atmosphere (Fig. 6-1).

It enters the plant through epidermal cells and roothairs (filamentous outgrowths of epidermal cells that greatly increase the absorbing surface of roots) and leaves mainly through epidermal openings in leaves, the stomata, controlled by guard cells lying on either side.

From roots to leaves, practically all the water movement occurs in the specialized water-conducting vessels, the *xylem*. Before the water can enter the xylem of roots, however, it must pass through several layers of cells, the *cortex*, and after it passes out of the xylem of the smallest veins of the leaves, it again passes through several cells, the *mesophyll*, before reaching those which are losing water by evaporation. The cells between the xylem and root or leaf surfaces transmit water more slowly than the xylem.

The motivating forces for water movement are several. The most important is the *diffusion-pressure deficit*, or tension that is set up in the spongy cells of the leaf when water evaporates through the stomata. This tension, maintained through the cohe-

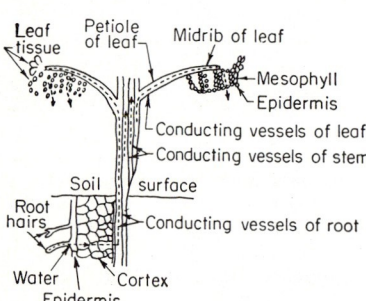

FIG. 6-1. The pathway of water from the soil to the leaves. (*From Harry J. Fuller and Oswald Tippo: "College Botany,"* rev. ed., Holt, Rinehart and Winston, Inc. Copyright 1949, 1954.)

sive attraction of water molecules, is passed on down through the xylem to the root tips. When a greater diffusion-pressure deficit exists in the root cells than in the soil, the roots absorb water. Osmotic and imbibing forces also play some part in the absorption of water by roots.

The high cohesiveness of the water molecules and the high diffusion-pressure deficits created by transpiration are sufficiently great to move water to the top of the tallest trees.

From the tree trunk to roots and branches the conducting system divides and subdivides repeatedly until it consists of millions of individual elements in the smallest veins of the leaves and the youngest roots. Within the tree trunk, sapwood is most actively involved in conduction, since xylem in the heartwood is blocked by accumulation of gums, resins, and other substances. In some species of ring-porous trees, including chestnut, oak, and red oak, water movement is largely confined to the outermost annual ring. In diffuse-porous trees a considerable number of annual rings are involved in conduction.

The *specific conductivity* (the volume of water moved per unit of time under a given pressure through a segment of given length and cross section) of hardwoods is about 3 to 6 times greater than that of conifers, and the conductivity of vines is about 2 to 10 times greater than that of hardwoods. Roots have even greater conductivity than vines. The specific conductivity of branches and twigs is lower than that of the trunk, and the specific conductivity of the trunk decreases from bottom to top. Rates of water movement may vary from about 5 ft/hr when the transpiration rate is high to 2 or 3 in./hr when the rate is low.

Many studies have shown that the absorbing and conducting systems of terrestrial plants are relatively inefficient in supplying water to meet the demands of transpiration, even at times when growth water is plentiful. In conifers, especially, the conductive elements consist entirely of small vessels which would be quite incapable of conducting water rapidly enough to meet the demands of leaves with a high transpiration rate. As a result of inefficient conduction, most plants, even if rooted in moist soil, lose more water during the daylight hours than they can absorb and conduct to the leaves. Should the water deficit attain considerable magnitude, leaves and stems may wilt. At night the transpiration rate declines sometimes to no more than 3 to 5 per cent of the diurnal rate, and this permits a reverse in the direction of the water-balance trend, so that the deficit due to transpiration is gradually changed to a surplus which may even result in guttation.

C. Water and Plant Habitats

Plants have been classified in respect to the relative wetness of their habitats into hydrophytes, xerophytes, and mesophytes, or plants living in wet, dry, and moist sites, respectively.

1. Hydrophytes. Hydrophytes normally grow in water or in swamps and bogs. Some, like water hyacinth, float in the water, with a portion of the plant at the water surface; some, like phytoplankton, are suspended beneath the surface; another group, such as pondweeds, grow suspended in the water but are attached to the bottom; another group, also anchored to the bottom, has its leaves floating on the surface, for instance, water lilies; and one group, that includes rice and cattail, grows in shallow water with shoots extended well above the surface.

Hydrophytes are characterized by spongy tissues which provide gas storage and buoyancy in the water, and the organs that normally grow below the water lack an outer bark. Roots are shorter and less branched than those of mesophytes or xerophytes.

2. Xerophytes. Plants that grow in dry situations and regularly endure and survive droughts have developed several quite different means of survival. One such measure is the short life cycle typical of the small annuals that in desert regions flourish briefly during a brief rainy season and are at seed stage in the dry season, thus avoiding its rigors.

A quite different adaptation is the development of water-storage tissue by some plants, such as cacti, that furnishes water reserves during the dry seasons. Cacti are also water conservers in that their stomata are closed during the day when transpiration stress is greatest and open at night when it is least.

Other adaptations include rapid elongation of taproots for the pursuit of soil moisture, the development of extensive root systems to draw water from greater volumes of soil, the development of high osmotic pressures which may delay or prevent irreversible changes in protoplasmic colloids under extreme drying, the drying back or shedding of leaves during the dry season or the changing of their position to reduce the direct amount of radiation received, and the reduction of leaf blade to reduce exposure to strong solar radiation.

No one of these adaptations is common to all plants. Each species has solved its own water problem and in so doing has produced the great variety in form that characterizes desert vegetation.

3. Mesophytes. Plants that do not inhabit water or wet soil or habitats where water is in short supply do not possess the structural and physiologic adaptation of the hydrophyte and xerophyte. They survive both short periods of wetness and drought.

IV. INTERCEPTION

Rainfall, snowfall, and fog are intercepted by vegetation to a degree that affects their distribution and the use of water by plants.

A. Interception of Rainfall

Rainfall is caught by the vegetative canopy and redistributed as throughfall, stemflow, and evaporation from the vegetation. Horton [1] admirably described the process as follows:

When rain begins, drops striking leaves are mostly retained, spreading over the leaf surfaces in a thin layer or collecting in drops or blotches at points, edges, or in depressions of the leaf surface. Only a meager spattered fall reaches the ground, until the leaf surfaces have retained a certain volume of water, dependent on the position of the leaf surface, whether horizontal or inclined, on the form of the leaf, and on the surface tension relations between the water and the leaf surface, on the wind velocity, the intensity of the rainfall, and the size and impact of the falling drops. When the maximum surface storage capacity for a given leaf is reached, added water striking the leaf causes one after another of the drops to accumulate on the leaf edges at the lower points. Each drop grows in size (the air being still) until the weight of the drop overbalances the surface tension between the drop and the leaf film, when it falls, perhaps to the ground, perhaps to a lower leaf hitherto more sheltered. These drops may also be shaken off by wind or by impact of rain on the leaf. The leaf system temporarily stores the precipitation, transforming the original rain drops usually into larger drops. In the meantime the films and drops on the leaves are freely exposed to evaporation.

It is evident that the amount of interception in a given shower comprises two elements. The first may be called interception storage. If the shower continues, and its volume is sufficient, the leaves and branches will reach a state where no more water can be stored on their surfaces. Thereafter, if there is no wind, the rain would drop off as fast as it fell, were it not for the fact that even during rain there is a considerable evaporation loss from the enormous wet surface exposed by the tree and its foliage. As long as this evaporation loss continues and after the interception storage is filled, the amount of rain reaching the ground is measured by the difference between the rate of rainfall and the evaporation loss. When the rain ceases, the interception storage still remains on the tree and is subsequently lost by evaporation. If there is wind accompanying the rain, then, owing to motion of the leaves and branches, it is probable that the maximum interception storage capacity for the given tree is materially reduced, as compared with still air conditions. Furthermore, in such a case, after the rain has ceased, a part of the interception storage remaining on the tree may be shaken off by the wind, and the storage loss in such a case is measured only by the portion of the interception storage which is lost by evaporation and is not shaken off the tree after the rain has ceased. One effect of wind is, therefore, to reduce materially the interception storage. As regards evaporation loss during rain, the effect of wind is, of course, to increase it materially.

Hamilton and Rowe [4] have used a number of terms to describe the various components of interception. *Interception loss* is the part of rainfall that is retained by the aerial portion of the vegetation and is either absorbed by it or is returned to the atmosphere through evaporation. *Throughfall* is the part of rainfall that reaches the ground directly through the vegetative canopy, through intershrub spaces in the canopy, and as drip from the leaves, twigs, and stems. Generally, between 0.02 and 0.10 in. of rain is held on foliage before appreciable drip takes place. *Stemflow* is that part of rainfall that, having been intercepted by the canopy, reaches the ground by running down the stems. *Gross rainfall* is the total amount of rainfall as measured in the open or above the vegetative canopy. And *net rainfall* is the quantity that actually reaches the ground; it is the sum of throughfall and stemflow.

1. Stemflow. Water flowing down the stems concentrates at their bases, where the soil is apt to be most highly receptive to water.

The amount of stemflow depends largely upon the roughness of the bark. The amount of rain at which stemflow starts may be as low as 0.01 in. for smooth-bark species like beech or as high as 0.07 in. for rough-bark species. For smooth-bark brush species in California, Rowe [5] found that stemflow was equal to 0.15 to $0.01P$ (P is storm rainfall in inches depth). Kittredge [6] reported that stemflow for rough-barked Canary pine was equal to 0.03 to $0.02P$. Generally, the decreasing order of stemflow is beech, maple, ash, elm, pine, basswood, hemlock, oak, and shagbark hickory.

Stemflow can produce striking differences in soil-moisture content over very small distances. Specht [7] has noted that the soil-moisture content around heath stems could be at field capacity, whereas a foot away it was at wilting point. Hoover [8] has noted that when 1 in. of rainfall fell in the open, about $\frac{1}{2}$ gal/ft^2 was received in the space between loblolly pine trees, but 8 gal of water was poured into the ground at the base of each tree. Because of this, the soil at the tree base often receives a thorough soaking when the space between the trees is only dampened. Voigt [9] found that, during the growing season, the soil at the base of red pine received about one-fifth of the amount that fell in the open, soil at the base of hemlock received the same amount of rainfall as the open areas, and soil around the beech stem received about $2\frac{1}{2}$ times the amount falling on the area. These differences in stemflow were related to bark texture.

2. Effect on Rainfall Intensity. Rainfall intensity beneath the forest stand may be made more uniform by interception, for the small drops of fine misty rain are combined into larger ones that drip from the leaves, and large raindrops are broken by the foliage, reducing their size and velocity. Drops that drip from the leaves are generally larger than raindrops, and their terminal velocity is reached by the time they have fallen 25 ft. Drops falling from canopy heights of more than 25 ft exceed the striking force of rainfall when they reach the ground.

Trimble and Weitzman [10] found that the maximum reduction in high-intensity rainfall under even-aged mixed hardwood stands about 50 years old was less than 20 per cent. They concluded that a high tree canopy has only limited value in reducing erosion potential of rainfall intensity, but a heavy stand of reproduction or a forest with a canopy that reaches close to the ground could reduce intensity.

B. Dew Formation and Fog

Whenever loss of heat by radiation cools a surface below the dew point, water vapor from the air will condense on it as a film of dew. Turbulent air prevents the required temperature gradient, but gentle air currents thicken dew films by bringing fresh supplies of air into contact with the surface. Such dew falling on leaves may be absorbed through the cuticle of normal epidermal cells or through specialized cells.

Fog accelerates this process, depositing minute water droplets as it passes through the canopy. Such precipitated moisture may be absorbed directly, and on the rainless coast of Peru it serves as the sole source of water for plants. In less arid regions the water may be deposited so heavily that drops form on foliage and fall to the ground, materially augmenting the supply of soil moisture. Fog drip is important in determining plant distribution along the west central coast of North America. Along the southeast coast of Hokkaido in Japan, special strips of forest are maintained along the seashore to comb a large part of the water out of the fogs that move landward [11].

C. Snow Interception

The interception of falling snow by conifer forests is one of the most significant effects of forests on supplies of snow water. Conifers offer a canopy on which a sizable fraction of the falling snow becomes lodged and is kept from immediately reaching the ground or snowpack surface. This snow is exposed on all sides to evaporation losses. Snow depth under conifers mirrors the interception effect, being deeper between the trees and shallower toward the trunk.

Snow interception in hardwood stands is negligible. Sapling and pole stands of hardwoods make ideal areas for snow-depth and moisture-content measurement because drifts are not formed in them and measurements tend to be rather uniform.

Shelter belts cause an unequal distribution of snow when drifts form to their lee. There the moisture content of the soil is raised at the expense of other areas. The contrasting influence of windbreaks in a northern and southern part of the central prairies of North America illustrate this point. Northward, where much of the precipitation falls as snow, windbreaks accumulate snow and build up moisture at the expense of adjacent fields, so that tree growth remains vigorous during all but severe droughts. Southward, there is little snow, and the trees intercept only rain, decreasing the amount of moisture the soil beneath the trees would otherwise receive. In this region, windbreaks suffer frequently from drought [11].

D. Factors Influencing Interception

Interception varies with species composition, age, and density of stands, with season of the year, and with regional differences in the way rainfall is received.

Of the total precipitation that falls during a growing season, 10 to 20 per cent is intercepted and evaporated back into the air. There is surprisingly little difference among trees, shrubs, and grass vegetation with respect to the amount of water they intercept seasonally, despite the seemingly large difference in foliage area. Foliage area per unit of ground area varies considerably between species. For instance, values from 1.7 to 9.2 have been reported for chaparrall, 4.2 for ponderosa pine, and 5.4 for white fir [6].

Variations in crown density between crowns of different kinds of trees and even under different parts of the crown of the same trees are almost infinite, and consequently the amount of precipitation that reaches the floor may be quite varied. For summer throughfall in hardwoods, Leonard [12] has reported variations ranging from 1.05 to 1.73 in. for a gross rainfall of 1.47 in. He found no consistency, storm by storm, in the distribution of throughfall beneath the stand.

In respect to single trees, interception is greatest close to the boles, becomes less under the central and outer parts of the crown, and is still less under the edge of the crown where there is some concentration of dripping, like that from a peaked roof.

On a leaf-by-leaf basis, it has been estimated that a leaf of oak or aspen may retain over 100 drops of water. If a tree of 500,000 leaves averaged 20 drops per leaf, each drop having a diameter of $\frac{1}{8}$ in., the tree would contain 5.92 ft^3 of intercepted storage water. If the crown diameter were 40 ft with a projection area of 1,256 ft^2, the interception storage would amount to 0.06 in. [6].

1. Size of Storm. Interception is essentially satisfied out of the first part of the rain, and since most storms yield only small amounts of precipitation, interception by forests or other dense cover commonly amounts to 25 per cent of the annual precipitation. If the yearly precipitation on any areas made up of small storms is separated by periods of clear weather, evaporation from crowns is high, as much as 35 to 50 per cent of the yearly total. Where storms are larger and much cloudy weather occurs, the relative amount of interception loss is smaller.

The relationship of interception, expressed as a percentage of the precipitation to the amount of storm rainfall, is shown in Fig. 6-2. These curves, corrected for stemflow, indicate interceptions of 40 to 100 per cent in showers less than 0.01 in., and 10 to 40 per cent in showers over 0.04 in. Stemflow (Fig. 6-3), for five of the six hardwood species shown, begins after precipitation of 0.01 to 0.10 in. For 1 in. of rainfall, stemflow amounts to less than 5 per cent for hardwood species, except beech.

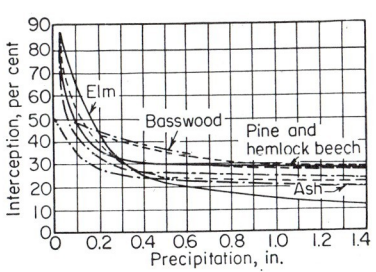
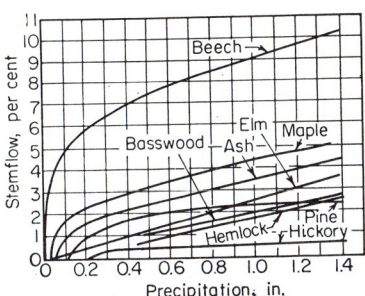

FIG. 6-2. Percentage interception for different species in relation to precipitation per storm. (*After Kittredge* [6].)

FIG. 6-3. Percentage of precipitation stemflow for different species in relation to precipitation per storm. (*After Kittredge* [6].)

Some time is required to saturate the foliage of a large tree. Precipitation at an intensity of 0.7 in./hr requires about 15 min to saturate the foliage, at which point dripping and evaporation equal rainfall application. When the foliage is saturated, the weight of the intercepted water can exceed the dry weight of the entire plant [6].

Even during rain there is considerable evaporation from the wet foliage. The evaporation rate per unit of surface area is normally very small under these conditions. However, when multiplied by the ratio of foliage area to the projected area of the tree, it becomes significant.

For any one storm the total interception is the sum of the interception storage S, a value that varies from 0.01 to 0.05 in., and the proportion of the rainfall evaporated before it reaches the ground—the product of the ratio of the evaporating leaf surface to the projectional area K, evaporation depth per hour during the rain E, and the duration of the storm in hours T. Or, interception $= S + KET$. Depth of interception loss increases as each of the other factors increases. S and K are functions of the amount of foliage and size of crowns [6].

E. Interception Effect on Transpiration

Rainfall and dew deposited on leaf surfaces are equivalent to moisture brought up through the plant system since heat required for their evaporation is thus not available for transpiration that would otherwise occur. Burgy and Pomeroy [13] found that, when grasses were transpiring, rainfall interception losses were accompanied by an equal reduction of moisture used from the soil. Lysimeter measurements have shown that, when there was little or no dew, larger quantities of water were used in the evapotranspiration process. Thus it matters little whether water evaporating from a plant comes from the soil by the root system or is merely intercepted rain. Both processes, Thornthwaite [14] points out, require the same quantity of energy; both constitute evapotranspiration. Unless soil moisture is at the wilting point, the energy consumed in evaporating the intercepting water would have been used to evaporate transpired water. Thus intercepted rainfall, while not contributing directly to soil-moisture supplies available to plants, does reduce the drain on these supplies.

F. Interception by Herbaceous Vegetation

Grasses and herbs intercept surprisingly large amounts of rainfall. Their leaf-area to ground-area ratio approaches that of forest vegetation. For instance, a square meter of big bluestem had a total leaf surface of 6 m², including both sides of the leaves. In mixtures with other grasses a similar quadrat had a leaf surface of 3 m³. A pure stand of blue grama grass exposed 2.5 m² of leaf surface. A square foot of little bluestem had 20 ft² of leaf surface; slough grass, 9 ft²; and big bluestem, 12.5 ft², exclusive of other species in the area [15].

Clark [16] found that the amounts of rainfall that were intercepted when applied at rates of ½ in. in 30 min were as follows: little bluestem, 50 to 60 per cent; big bluestem and tall panic grass, 57 per cent; bindweed, 17 per cent; and buffalo grass, 31 per cent. Little water appeared to run down the stems.

Beard [17] determined water-retention capacity of natural grass of various types by cutting all the grass from exactly 1 ft², standing it upright in a wired basket, sprinkling, shaking, and weighing. An average capacity of 0.07 in. was found, which, if applied to the number of rain days, gave total interception of 14 to 19 per cent. The height of the grass was about 3 ft. Burgy and Pomeroy [13] found storage values of 0.040 to 0.048 in. for a mixture of tall fescue and soft chess about 10 in. high.

In California, undisturbed grass of *Avena, Stipa, Lolium,* and *Bromus* species was found to intercept 26 per cent of a 33-in. seasonal rainfall; stemflow was not determined. In Missouri, bluegrass intercepted 17 per cent of the rainfall during the month prior to harvest. Musgrave [18] also noted that interception increased as crops matured and as the density of the cover increased. Haynes [19] found the interceptions for alfalfa, corn, soybean, and oats as given in Table 6-1.

Table 6-1. Interceptions for Various Crops

Description	Alfalfa	Corn	Soybean	Oats
During growing season:				
Rainfall, in.	10.81	7.12	6.25	6.77
Canopy penetration, in.	6.18	4.84	4.06	6.30
Stemflow, in.	0.76	1.18	1.28	
Interception, in.	3.87	1.10	0.91	0.47
Interception, %	35.8	15.5	14.6	6.9
During low-vegetation development, %	21.9	3.4	9.1	3.1

Thus, for dense grasses and herbs approaching full growth, interception loss is nearly as great as for deciduous trees in full leaf. However, their season is short, and as a result, their total annual interception is considerably less than that of deciduous trees [20].

G. Interception by Forest Vegetation

Coniferous trees intercept more rainfall than deciduous trees in full leaf. The deciduous canopy tends to be more dense, but raindrops remain hanging on the separate conifer needles, whereas in hardwoods they flow together and drip off the leaves or run down the trunks. The large number of small linear leaves in conifers hinders the free flow of raindrops and presents numerous cavities in which water can be trapped.

For any given depth of precipitation, winter and summer losses appear to be about the same for conifers, but for deciduous trees summer losses are 2 or 3 times greater than winter losses [20].

According to Trimble [21], a dense even-aged conifer stand will usually intercept about one-fourth to more than one-third of the precipitation that falls, the more tolerant (thicker crowns) and stiffer-needled conifers intercepting a larger amount. Stemflow is less than 5 per cent. Stemflow on hardwoods varies greatly, depending on position of the tree in the canopy, smoothness of bark, branching habit, and season of the year. Approximate values for seasonal rainfall interception by northeastern forest types for dense stands at age of greatest interception are given in Table 6-2.

Table 6-2. Interceptions for Various Forest Types

Forest type	Gross interception		Stemflow		Net interception	
	With leaves, %	Without leaves, %	With leaves, %	Without leaves, %	With leaves, %	Without leaves, %
Northern hardwood	20	17	5	10	15	7
Aspen-birch	15	12	5	8	10	4
Spruce–spruce-fir	35	...	3	...	32	...
White pine	30	...	4	...	26	...
Hemlock	30	...	2	...	28	...
Red pine	32	...	3	...	29	...

Approximate snow interception for conditions in the White Mountains in the Northeast was estimated as follows:

Types	Net interception, %
Northern hardwoods	10
Aspen-birch	7
Spruce–spruce-fir	35
White pine	25
Hemlock	25
Red pine	30

Note that snow-interception values are within 1 to 5 per cent of interception values estimated for rainfall.

The difference in interception between conifers and hardwoods is of some significance in those regions where conifers have been cut and replaced by hardwoods. In the Northeast, for instance, where the spruce-fir is largely cut out, rainfall interception probably has been reduced from a maximum of 30 per cent at the time of settlement to 10 per cent at the present. In snowfall interception the virgin forest probably held back and evaporated 25 per cent of the precipitation. At present it is estimated that the forest evaporates 10 per cent [22].

H. Ecological Aspects

Interception loss varies with species and forest types because of differences in thickness and density of foliage in crowns. In general, intolerant species have less dense crowns than tolerant species. As succession approaches the climax, the growing space is more completely utilized because of the greater variety of species and the stratification of their roots and tops. In earlier stages the species are less effective in utilizing growing space; they are limited in number and usually about the same size. Hence tolerant species intercept more than intolerant, climax more than preclimax, and mesophytic more than xerophytic [6].

Kittredge [6] pieced together seasonal interception and stemflow figures to illustrate a successional series though taken from different localities: jack pine in Wisconsin, an intolerant pioneer species, intercepted 21 per cent; white and red pine in Ontario, 37 per cent; maple-beech climax in New York, 43 per cent; and tolerant climax hemlock in Connecticut, the maximum interception of 48 per cent.

I. Silvicultural Aspects

The effects of various cutting practices on rainfall interception would appear to be in proportion to the amount of canopy removed. Two factors mitigate against this:

(1) branches and leaves of slash continue to intercept rainfall for some time; and (2) understory growth generally increases as the stand is opened by cutting. Thus silvicultural treatments would be expected to have only a short-term and minor effect on rainfall interception.

It would be expected that forest regeneration by natural means or by plantations would lead to a gradual increase in interception as the stand developed. This particularly would be the case during the period of closure. In New England, for instance, crowns of white pine, jack pine, and Norway spruce plantations close at 14 to 15 years on the average, although the time may vary from 8 to 25 years. Red pine crowns close at 11 years, with a range of 7 to 18 years with 6- by 6-ft spacing and 4 or 5 years later with 8- by 8-ft spacing [6]. After plantations close or a dense stand of natural reproduction develops, rainfall interception increases but slowly.

Little is known about the effects of cutting practices on dew formation, though Stone [23] has noted that European foresters since 1912 have assumed that dew was

Table 6-3. Effect of Stand Densities of Lodgepole Pine on Snow and Rainfall Interception, Snow Evaporation, and Evapotranspiration

	Merchantable reserve stand per acre, bd ft				
	11,900 (uncut)	6,000	4,000	2,000	0
Precipitation reaching the ground:					
Snowfall...	7.60	8.41	8.61	9.09	9.59
Rainfall...	9.17	10.63	11.28	11.24	12.19
Evaporation from snow...	0.80	1.40	1.60	1.80	2.00
Evapotranspiration...	5.63	6.26	5.97	6.09	6.26

important to tree growth and have devised cutting systems to encourage dew formation and retention.

The effects of cutting practices on snow interception by conifers have been studied in relation to snow accumulation and water available for streamflow. Weitzman and Bay [24] have reported on the water content of snow at the period of greatest accumulation for differently cut stands of black spruce: clear-cut strip, 4.0 in.; single-tree selection, 2.6 in.; uncut, 2.5 in.; shelterwood, 3.2 in.; and open patch, 4.5 in. The clear-cut strip and the open patch contained 60 and 80 per cent more snow water than the uncut stand.

Most of the research on snow interception has been conducted in lodgepole pine in the Rocky Mountains. The effects of thinning dense second-growth lodgepole pine stands on snow and rainfall accumulation and interception have been reported by Goodell [25]. Two methods of thinning were applied. One consisted of reserving better trees spaced about 8.5 ft apart over the plot and cutting all other trees. This left about 600 trees per acre. The other method consisted of cutting openings 8 ft in radius around each of 100 selected crop trees per acre. All trees outside the lined circle were left standing. This left about 2,000 stems per acre.

Over 3 years of measurement the average water equivalent increase in net snowfall reaching the ground surface due to crop-tree thinning was 1.69 in., while the increase caused by single-tree treatment was 2.31 in.—increases of 17 and 23 per cent, respectively. For 2 years of measurement the average increase in net summer rainfall caused by crop-tree thinning was 0.49 in., or 14 per cent of the summer rainfall received under the untreated plots. The single-tree thinning produced an increase of 0.65 in., or 18 per cent. Soil-moisture sampling indicated that losses of moisture from the soil through evaporation and transpiration were unaffected by the thinning treatments.

A second, and earlier, study in the lodgepole pine type concerned the effect of different stand densities, as measured in merchantable volumes, on snow and rainfall interception, evaporation, and evapotranspiration [26]. A merchantable stand of lodgepole pine was cut so as to leave volumes of 0, 2,000, 4,000, and 6,000 bd ft/acre. The heaviest cut stand contained, on the average, 147 trees larger than 3.5 in. in diameter, or about 38 per cent of the average total number on the uncut plots (382 trees). The other treatments left an average of 206, 181, and 223 trees per acre for the 2,000, 4,000, 6,000 bd ft stands, respectively. Inches of snow and rainfall that reached the ground, estimated evaporation from snow, and evapotranspiration during the growing season were found as shown in Table 6-3.

When the figures on net snowfall and rainfall are combined, an average of 16.77 in. of water passed through the forest canopy on the uncut plots and 21.78 in. on the heaviest-cut plots, a difference of 30 per cent. Total evapotranspiration for the extremes amounted to 6.43 in. for the uncut stand and 8.26 in. for the most heavily cut.

V. INFILTRATION

Infiltration concerns the movement of water into the soil (see Sec. 12 for details). By some, it has been considered as a surface phenomenon, governed entirely by the surface conditions, particularly the noncapillary porosity of the soil surface. Others have used infiltration to describe the movement of water into the soil as governed by the permeability of the entire profile.

Thousands of infiltration measurements have been made with a great variety of instruments, ranging from a simple ring infiltrometer, into which water is poured and its rate of disappearance measured, to rainfall simulators, intricate in design, with which known amounts and intensities of water can be applied. Results of some of these measurements have been published; most of them have not.

Infiltration measurements have regularly shown an initial high rate, at the beginning of the test, decreasing rapidly and then more slowly, until it approaches a constant rate in a period of $\frac{1}{2}$ to $1\frac{1}{2}$ hr or longer. This typical trend is the response to a decreasing hydraulic gradient as the water moves deeper into the soil, to filling of pores with water, and to changes in the soil such as dispersion of aggregates, puddling of the surface layer by the impact of raindrops, and swelling of colloids and closing of soil cracks.

When turbid water is applied to the soil, noncapillary pores are quickly choked by the sediment that settles out and infiltration rate is sharply reduced. The same result is obtained when raindrops strike the unprotected soil, detaching soil particles which block the pores.

A. Factors Influencing Infiltration

From the foregoing, it is obvious that infiltration is strongly influenced by soil texture and structure, which govern noncapillary porosity, soil wetness, and the amount of protection from rainfall impact offered by vegetation.

1. Soil Texture and Structure. Water may infiltrate into very coarse texture soils or well-aggregated soils so readily that none is lost in runoff even in the heaviest downpour. By contrast, the surface layer of a bare clay soil may soak up the first moisture that falls; then it may swell and become a dense waterproof layer that sheds the remainder of the water [11]. In this situation the effect of texture is jointly correlated with soil-moisture content.

The influence of texture and structure on infiltration, as described through noncapillary porosity, clay content, and organic-matter content, was shown in a study in which infiltration rates of 68 soils were related to various soil properties. Noncapillary porosity of the subsoil gave a correlation coefficient of 0.54; surface organic matter, 0.50; clay content of the subsoil, 0.42; and organic matter in the subsoil, 0.40. When the factors were combined in multiple correlations, the highest multiple coefficient of 0.71 was obtained with noncapillary porosity, organic matter of both surface and subsoil, and clay content of the subsoil [27].

2. Soil-moisture Content. The rate of infiltration into a soil is at a maximum when a soil is fairly dry, for after water is added, the pore space becomes full. Then colloids swell, and the rate of entry of additional water declines to a low but uniform level [11]. To avoid the effect of different soil-moisture content on infiltration, infiltrometer runs are usually made after prewetting the soil.

3. Vegetation. Infiltration is maintained at a maximum by undisturbed natural forest canopy and floor. Rarely under forest condition are evidences of overland flow visible except in those areas that have been disturbed and compacted by roads or logging.

The forest floor particularly affects infiltration. Trimble and Weitzman [10] have pointed out that the rainfall intensity under a high forest is very similar to that in the open, and therefore the protective function of the forest, in respect to rainfall impact, is provided by the forest floor only. In a recharge area in New Jersey where the forest vegetation had been killed by spraying of water in amounts up to 600 in./year, high infiltration rates have been maintained because the forest floor has not been disturbed [28].

When the forest floor is removed, infiltration rate is reduced. For instance, Arend [29] reported the following average infiltration rates from a study in Minnesota:

	Inches per hour
Forest floor, undisturbed	2.36
L- and F-layers* removed mechanically	1.94
Forest floor, burned annually	1.58
Unimproved pasture	0.95

* See Subsec. VIII, Glossary.

Johnson [30], in Colorado, found that removing the forest floor reduced infiltration capacity from 1.52 to 0.92 in./hr, a reduction of about 40 per cent.

About ½ in. of litter is the minimum depth to provide adequate protection. Rowe [31] found that surface runoff was less for soils with a forest floor ½ in. deep than for soils with a shallower covering, but remained about the same for forest floors more than ½ in. in depth.

Penetration and subsequent decay of tree roots in forest soil are also responsible for their higher rates of infiltration as compared with cultivated and pasturelands. Gaiser [32] found at least 4,000 highly permeable root cavities per acre in a 51-year-old white oak forest in Ohio. These cavities penetrated into and through a clay subsoil of very low permeability.

The density of herbaceous vegetation is also closely related to infiltration, as has been attested by several studies on the western range. Packer [33], for instance, found that the per cent of the soil covered by living or dead plant parts was closely related to runoff, and therefore to infiltration. As cover density increased to about 70 per cent on wheatgrass and cheatgrass areas, overland flow decreased. At densities above 70 per cent, there was little further decrease.

Fibrous-rooted vegetation such as wheatgrass has been found to be much more effective in controlling runoff than taprooted annual weeds.

4. Compaction. A major factor influencing infiltration is the degree the surface soil is compacted. Under wet conditions, one pass of a tractor has been known to reduce noncapillary pore space by half and infiltration rate by 80 per cent [34]. In another instance, two passes with a tractor reduced infiltration rate from 1.4 to 0.6 in./hr [35].

When a soil is compacted, its total porosity is decreased, the major reduction being in noncapillary porosity. Compaction is one reason why cultivated fields, compacted by farm vehicles, have much lower infiltration rates than nearby woodlands. At the recharge area in New Jersey mentioned above, the spraying of only 2 in. of water a day resulted in saturation and surface runoff from a cultivated field, whereas 400 ft away, in a wooded area, over 150 in. of water was applied during a 10-day period without noticeable saturation [36]. Compaction of rangeland and pasture by grazing has similar effects on infiltration.

INFILTRATION

5. Frost. Where frost is discontinuous, as in the honeycomb and stalactite types, a soil may be as permeable as, or even more permeable than, frost-free soil. Concrete frost, on the other hand, is almost impermeable. In the Northeast, infiltration was zero on concrete frost in the open and forest areas. However, where the forest soil was traversed by large holes in which water had not frozen, infiltration was not affected [37]. Infiltration tests on concrete frost in northern Minnesota forest and grassland gave 0.09 in. of infiltration per hour in silt-loam soils and 0.47 in. on sands [38]. Concrete frost in the forest is apt to be less widely distributed than in open fields, so that any water running off concretely frozen forest soil may go but a short distance before reaching and infiltrating unfrozen soil.

B. Ecological Aspects

When primary succession begins on a rock, with no soil covering, there is no infiltration. As soon as soil forms under the protective influence of vegetation, its existence is evidence that infiltration is not a limiting factor to its stability and development. Thus shallow soil stabilized on parent-rock material must have an infiltration rate as great as, or greater than, maximum rainfall intensities.

This is not the case in secondary succession where vegetation has been removed and the soil bared, leaving it unprotected from rainfall impact. Infiltration rates may then be very low, particularly if the soil is fine-textured, and the soil will be unstable. As vegetation invades this area, infiltration rates will increase and will continue increasing until a stable situation exists.

One of the few studies designed particularly to determine the effect of various stages of plant succession upon infiltration capacity was made in central Pennsylvania by Alderfer and Bramble [39]. Infiltration was measured with a type F infiltrometer. Infiltration rates, in inches per hour, for a rainfall intensity of 3 in./hr were as follows:

```
Old poverty–oat grass pasture........ 1.68
Virginia pine forest:
    30 years old..................... 2.95
    60 years old..................... 2.47
Oak-hickory forest................... 3.00
```

In this study, the main improvement in infiltration developed between the old pasture and the Virginia pine stand.

Infiltration does not necessarily increase with development of forest vegetation. For instance, Coile [40] reported the following comparative milliliters of water absorbed in 5 min from infiltrometer-ring tests:

```
Broom sedge field..................... 60
Loblolly pine, age class:
    10 years.......................... 15
    20 years.......................... 53
    70 years.......................... 77
White oak, black oak, red oak......... 40
```

The difference between pine and oak was attributed to the relatively high activity of soil animals and considerable mixing of humus and mineral soil in the pine stands, whereas in the oak type the humus mat, bound together by roots and fungous mycelia, had fewer water passageways. The presence of deeply incorporated organic matter in soil with broom sedge cover was considered responsible for its greater absorption rate than that in the young pine stand. However, broom sedge disappears under well-stocked pine stands.

Percolation rates for the surface 6 in. of soil under old-field pine–hardwood conditions in South Carolina were reported by Metz and Douglass [41] to be as shown in Table 6-4.

Sharp reduction in percolation rate below the surface 2 in. is illustrated in Table 6-4. Again, in this study, rates under broom sedge were greater than rates under pine.

Table 6-4. Percolation Rates under Pine–hardwood Conditions in South Carolina

Cover type	0–2 in. (in./hr)	2–4 in. (in./hr)	4–6 in. (in./hr)
Barren	0.06	0.06	0.06
Broom sedge field	44.79	19.86	12.26
Loblolly pine plantation	23.17	9.58	1.88
Shortleaf pine	4.98	4.61	1.43
Shortleaf pine–hardwood	90.00	13.24	11.34

Rothacher [42], in a study in eastern Tennessee, found the greatest difference in infiltration between cultivated fields and broom sedge. The number of seconds for successive $\frac{1}{2}$-in. increments of water to infiltrate into four cover conditions was as shown in Table 6-5.

Table 6-5. Infiltration Time in Seconds under Various Cover Conditions

Description	First	Second	Third	Fourth	Ratio of sum
Cultivated preceding year	290	472	571	546	93.3:1
Broom sedge, abandoned 5 years	33	67	77	92	1.4:1
Old-field stand	39	52	62	73	1.2:1
Hardwood stand, second growth	36	46	53	62	1

How retrogression of herbaceous cover through grazing affects factors that influence infiltration has been described by Weaver [43]. Under continued grazing the true prairie grasses were gradually replaced by bluegrass, by the short grama grass, by buffalo grass, or by a mixture of the three. Not only was forage yield greatly decreased, but the underground plant parts were reduced a third or more. However, the fibrous root system still anchored the soil. Under long-continued grazing and trampling, the sod was broken and poor-rooted annual weeds such as wire grass, poverty grass, crab grass, and others became dominant. The weight of underground plant material dropped from 4.12 for big bluestem to 0.95 ton/acre for weedy annuals. Runoff increased manyfold.

C. Silvicultural Aspects

As a general relationship, for any one species or type, the greater the amount of forest cover, the higher the infiltration rate. Infiltrometer measurements in plantations of black locust and sassafras of different ages have indicated progressive improvement of infiltration as the trees became older. Under plantations more than 25 years old, the rate may approach that of the natural forest. Likewise, second-growth stands may have infiltration rates as high as old growth.

According to Kittredge [6], infiltration is greater in dense than in open plantations, greater in old than in young plantations, and greater in unthinned or lightly thinned than in heavily thinned stands.

As long as a protective forest floor is maintained, silvicultural methods may have little effect on infiltration rates. However, once the forest floor is disturbed, infiltration rates may be sharply reduced.

Principal disturbance is through logging. In the usual type of logging in the West, at least 10 per cent of the forest land being logged is used for tractor and truck roads and landings, and the proportion of land put to this use may go as high as 20 per cent. In the East, the area disturbed varies with logging methods and volume and type of timber removed. Under certain conditions it may be less than 10 per cent, and with careful selection of road location, under 5 per cent [44].

The type of equipment used in logging makes considerable difference in soil disturbance and exposure. In a mixed-conifer stand in the Cascade range in Washington, for instance, tractor logging left exposed mineral soil on 22.2 per cent of the area as compared with 5.4 per cent on an area logged with the Skyline Crane [45].

Trimble and Weitzman [46] found that it took 619 times longer for a given quantity of water to enter the soil of a skid road than to enter the A-horizon of an undisturbed forest soil, and 20 times longer than to enter the B-horizon. In the Douglas-fir region Steinbrenner and Gessel [47] found that silty-clay and clay-loam soils from tractor-logged areas had a 35 per cent loss in permeability as compared with soils in undisturbed timber. Tractor roads showed a 92 per cent loss in permeability.

Changes in infiltration induced by logging may last for some time. In the Pacific Northwest the percolation rate of the upper 3 in. of soil on second-growth Douglas-fir stands tends to increase at a slow rate after logging and burning. Most improvement appeared to be gained during the first 20 to 25 years.

VI. EVAPOTRANSPIRATION

Evapotranspiration is the process by which water moves from the soil to the atmosphere (see Sec. 11 for details). It consists of transpiration, the movement of water through the plant to the atmosphere, and evaporation, the movement of water vapor from soil and vegetative surfaces. For practical purposes, these two processes are considered together, their sum representing that part of precipitation that is returned to the atmosphere as opposed to that part which goes into runoff. As to amount of water involved, transpiration is by far the more important of the two processes during the growing season. During the dormant season, evaporation is responsible for most moisture loss.

The proportion of moisture transpired or evaporated depends in part on the distribution of rainfall. Where rains are small and infrequent, a high percentage of annual precipitation is lost by surface evaporation. Conversely, the proportion of larger rainfalls that is evaporated is smaller.

The influences of vegetation in shading the soil, reducing wind velocity, and giving off water vapor tend to reduce direct evaporation from soil or snow cover. However, the vigorous absorption of soil moisture by roots, together with the losses due to interception, usually more than offset the effects of vegetation in retarding evaporation from the soil. Thus the soil in forest openings tends to have more moisture than soil beneath the trees.

A. Transpiration

Transpiration differs from evaporation from a water surface in that it is subject to the effects of structural and functional features peculiar to the plant and is strongly influenced by light. Evaporation rates cannot be assumed to indicate transpiration rates, although the two curves do correspond closely under certain conditions. When the stomata are open, the transpiration curve tends to follow the evaporation curve until a water deficit develops and the stomata begin to close. With the stomata nearly closed, there is no apparent relationship between the rate of transpiration and evaporation, transpiration being controlled chiefly by the diameter of the stomata openings. The effect of stomata closure on transpiration is clearly shown by the sharp reduction in transpiration when a leaf wilts or when nightfall closes the stomata. Transpiration continues slowly through the cuticle after the stomata close, but at a much lower rate than the evaporative power of the air [11].

Transpiration is beneficial in that it transports nutrients to the upper part of the plant and, more important, cools the leaves. Leaves exposed to the sun can become much warmer than air when transpiration is stopped. The temperature of a leaf undergoing rapid transpiration may be as much as 27°C below the temperature of the surrounding air [11].

Transpiration beyond the amount useful in cooling leaves can damage the plant. Excessive transpiration may desiccate protoplasm below the minimum water content (30 to 50 per cent) which permits it to remain alive. Warm weather or spring days may result in excessive transpiration because water cannot be absorbed rapidly from the cold or frozen soil. Leaves then dry and become discolored. In the Adirondacks entire mountainsides of conifers so discolored have been noted [48].

Excessive transpiration is also common during the growing season, for the absorbing and conducting systems of terrestrial plants are relatively inefficient in supplying water to meet transpiration demand, even when soil water is plentiful. As a result, most plants, even if rooted in moist soil, lose more water during daylight hours than they can absorb and conduct to the leaves. At night the transpiration rate declines to no more than 3 to 5 per cent of the daylight rate, so that the deficit to the transpiration is gradually changed to a surplus and guttation occurs. It has been said that, if night failed just once to come to the rescue of plants, many of them would perish [11].

Maximum rates of evapotranspiration during the growing season (and thus largely transpiration) appear to be about 0.2 in./day, though higher rates have been reported. Hendrickson and Veihmeyer [49], for instance, found an average daily loss of 0.293 in. as computed from soil-moisture measurements under 10-year-old almond trees, taken to a depth of 12 ft. Broadfoot [50] reported a forest water use of 0.35 to 0.50 in./day in July and August from a 50- to 60-year-old stand of hardwoods in a river-bottom area in Mississippi in which water had been impounded to a depth of 2 ft.

Only a small proportion of the water that the plant absorbs from the soil is retained in the plant. Roughly, there is about 5 to 10 lb of water in a plant per pound of dry matter. But for each pound of dry matter produced, the plant must absorb and transpire several hundred pounds of water. In relation to pine production in the mid-South, Zahner [51] estimated that it requires about $\frac{3}{4}$ million gallons of water to grow 1 cord/acre/year, or, on a dry-weight basis, 2,000 tons of water to grow 1 ton of wood. Even desert vegetation is not economical in use of water; rather, it uses water extravagantly when it is available. Desert plants owe their existence to an ability to grow rapidly when moisture is present and to remain alive when moisture is absent.

B. Evaporation

When a soil is wet, it loses water rapidly by evaporation. However, as the surface layer dries out, the rate of water loss diminishes rapidly, even though the subsoil remains moist. Surface evaporation can desiccate a normal soil to a depth of 8 to 12 in.; rainfall that is inadequate to penetrate deeper than this can be drawn back into the atmosphere directly [11]. Some observations indicate losses to greater depths in areas with long periods of no rain—6 months or more.

Water evaporates more rapidly from compact than from loose soils and from dark- than from light-colored soils. Soils with medium-size particles permit greatest evaporation. The formation of crumb structure reduces evaporation.

The opportunity for soil evaporation is considerably less in forest cover than in grass and bare areas. Hursh [52] found that monthly evaporation in a denuded zone in Tennessee was 5 times greater than evaporation in the forest during the summer months. During winter months evaporation in the bare zone was about twice that in the forest. Evaporation in a grass zone was similar to, but less than, that in the bare zone. During June, the month of greatest evaporation in the open, 7.62 in. were evaporated in the bare zone, 6.00 in the grass, and 1.58 in the forest. Rowe [31] reported that, although annual evaporation from the forest floor increased as the depth of the floor increased, the sum of evaporation from the floor and soil decreased. This decrease was very small for litter accumulations greater than $\frac{1}{2}$ in. in depth.

C. Factors Influencing Transpiration

The rate of transpiration depends on the evaporative power of the air as determined by air temperature, wind, saturation deficit, the amount of light, which partly controls opening of the stomata, and the availability of moisture in the leaf tissues, which in turn depends on soil-moisture availability.

If the evaporative power of the air is suddenly increased during the period when soil moisture is being depleted to the permanent-wilting percentage, plants will wilt before this point is reached. Thus drought effects may be produced just as readily by increasing the transpiring power of the air as by a sudden reduction in moisture availability [11].

Air temperature and light, two of the most influential factors, are controlled by solar radiation. This explains the striking difference between nocturnal and diurnal transpiration in most plants. Because the diurnal period of greatest saturation deficit coincides with the period of highest temperature and brightest light, transpiration stress is greatest, and these factors attain their greatest combined intensity at midday or in early afternoon. Light increases transpiration more than it does evaporation, whereas wind movement increases evaporation more than it increases transpiration [11].

A part of the solar radiation absorbed by the earth's surface heats the air, soil, and vegetation, but the major portion is used in converting water in the soil or in vegetation from liquid to vapor state. About 70 per cent of the incoming radiation is absorbed by the leaf. Of this amount, perhaps 20 per cent is used for heating and about 1 per cent is used for photosynthesis. The remaining 49 per cent is used for evaporation and transpiration, mostly the latter. Evaporation from forest soil is necessarily small, as a complete forest cover intercepts almost all the insolation.

During a growing-season day of adequate soil-moisture supply, transpiration will be occurring most rapidly where sunlight directly illuminates foliage. Leaves inside the canopy and vegetation growing on the forest floor will transpire but a fraction of the amount lost from the lighted canopy, a fraction equivalent to the fraction of energy received.

At night, transpiration continues, utilizing in part the heat stored during the day in air, vegetation, and soil. For crops, surface transpiration is about 5 to 10 per cent of the daytime value [53].

As the soil dries and water supply becomes less available, less heat energy is used in evaporation and transpiration. Consequently, the surface temperature rises and there is a greater transfer of heat from the hot surface to the air. At this point moisture supply becomes the limiting factor to transpiration, rather than solar radiation.

Evaporation from a free-water surface, on the other hand, is limited only by the heat available. Evaporation from a very wet soil has approached 0.9 that from a large free-water surface [54]. Likewise, when moisture at the plant surface is plentiful, there is a high correlation between transpiration and available heat.

The distribution of plants often reveals the influence of solar radiation and air temperature on water loss. Algae and other nonvascular plants are characteristically more abundant on the north side of tree trunks because higher evaporation due to insolation effectively retards growth on the south side. On mountain slopes vegetative zones are found at different altitudes according to the exposure to the heat of the sun. In the Northern Hemisphere, moisture-demanding species are restricted to higher levels on south-facing slopes than they are on the cooler opposite side. The variation in the amount of rains required for a good growth of short grass furnishes another illustration. In Montana a precipitation of about 14 in. is sufficient, but in northwestern Texas a rainfall of 21 in. is needed to produce the same amount of grass; evaporation in the six summer months in Montana amounts to 33 in., whereas in Texas it averages 54 in. [55].

At Coshocton, Ohio, lysimeter measurements indicated that evapotranspiration in the warm months was from 3 to 4 times greater than for the colder months [56].

June and July were the months of highest water demand. Average daily evapotranspiration by months was as follows:

	Inches of water
January	0.06
February	0.08
March	0.10
April	0.13
May	0.16
June	0.18
July	0.19
August	0.15
September	0.12
October	0.09
November	0.06
December	0.06

During the warm summer months, soil moisture is evaporated and transpired so rapidly that the upper foot of a wet soil may approach wilting point in a few days. With intense radiation, high air temperature, high saturation deficit, and active plant growth, soil-moisture losses are high regardless of the kind of vegetation. Under these conditions, the rate of loss is controlled almost entirely by the amount of moisture in the soil. Once the soil dries to the permanent-wilting point, moisture losses practically cease.

Wind velocity also exerts a major effect on the loss of water. At 5 mph the transpiration of plants is increased 20 per cent over that in still air; at 10 mph, 35 per cent; and at 15 mph, 50 per cent. Desiccating action of warm dry winds sometimes prevents the invasion of vegetation on windward slopes, even though the same vegetation grows perfectly well on the leeward sides of the same slopes.

Saturation deficit is better correlated with evapotranspiration than relative humidity because it expresses the absolute capacity of the air for additional moisture, whereas relative humidity indicates no more than the degree of saturation. For example, the evaporative power of the air at 60 per cent relative humidity is not the same at different temperature levels, as the following comparison indicates [11]:

Relative humidity, %	Temperature, °F	Saturation deficit, mm
60	80	10.38
60	70	7.44
60	60	5.24

At 80°F the capacity of air for water vapor is twice that at 60°F.

Each locality has a certain potential evapotranspiration, set primarily by the energy received as solar radiation, as this raises temperature and increases saturation deficit. So long as soil moisture is not limiting, actual evapotranspiration equals potential evapotranspiration. In this situation, evapotranspiration is very similar for closed vegetation as diverse as grassland and forest, provided the surfaces of the canopies are relatively level, the foliage has similar reflectivity, and the vegetative activities of the plants are near maximum [11].

The more of the incoming solar radiation that is reflected back into the sky, the less that remains for heating the vegetation. Rye and wheat fields reflect 10 to 25 per cent of solar radiation; deciduous forest, 15 to 20 per cent; coniferous, 10 to 15 per cent; semidesert, 25 per cent; desert, 30 per cent; and rock, 12 to 15 per cent.

Potential evapotranspiration is least from the grassland and greatest from the pine forest. It ranges from less than 18 in. in the high mountains of the West to more than 60 in. in Arizona and southern California. It is less than 21 in. along the Canadian border and more than 48 in. in Florida and southern Texas. It is greatest in July, ranging from 5 in. along the Canadian border to 7 in. in the Gulf Coast [57].

D. Evapotranspiration by Herbaceous Vegetation

Considered as separate individuals, different species and types of plants make widely different demands on soil moisture even under the same environmental conditions. For example, a corn plant may transpire 2 qt of water a day and an oak tree 170, the oak roots drawing their water from a much greater volume of soil.

Root growth depends on the physical characteristics of the soil as well as on species. Shallow soils and those with high water tables restrict root depth. The effect of root depth on water removal has been illustrated by a study by Croft, in Utah [58]. Available storage space of 11, 8, and 3 in. was found under adjacent aspen, herbaceous, and bare plots, respectively, at the end of the growing season. Aspen roots penetrated 2 to 3 ft deeper than roots from the herbaceous cover. Moisture losses were least in the bare plots where roots were absent. In the Piedmont of South Carolina, forest types depleted soils of moisture to depths of at least 66 in., whereas, under old-field herbaceous vegetation and bare soil, moisture was lost primarily from the surface 30 in. [41].

Different species also start and stop transpiration at different times. For example, broom sedge in the Southeast begins moisture extraction almost a month later in the spring than do pine trees. Lemon [59] has reported that the rapid drop in transpiration rate of cotton in late August coincided with evident maturation of the early bolls. Despite the fact that growth continued until December frost and most of the leaves on the plants were retained in a healthy green condition, the plants' transpiration rate was reduced.

Treatment of vegetation can also affect water use. Frequent cutting of grass results in less water use but with a higher proportion of the water coming from the surface layer. Cutting the top growth reduces root growth and produces a shallow-rooted plant that secures most of its water from the surface layers.

The position of the plant within the plant community also affects transpiration. As a barren area becomes clothed with vegetation, the shade, reduced air temperature, interference with wind movement, and the water vapor given off by the leaves, all work to reduce the evaporation rate. The rate just above the tips of the leaves is several times greater than that just below them in certain plant communities [11]. In swampy woodland, for example, evaporation midway between the ground and treetops was 30 per cent less than that near the tops. Lower down, near the damp soil, plants were subject to an evaporation rate of only 7 per cent of that near the treetops [55].

E. Evapotranspiration by Forest Vegetation

Different trees are very differently equipped to grow in dry areas. For example, in the central Rocky Mountain region, on moist sites, white fir and aspen grow in closed stands and reach a similar mature size. On dry sites in the same region the aspen is reduced to dense thickets of trees only 8 to 10 ft tall, while white fir maintains a height of 60 to 70 ft, but the trees are widely spaced in parklike stands. The aspen has maintained itself mainly by reducing its transpiring area, the fir by increasing its absorbing area per tree [60].

There are major differences in rooting habits among the tree species. However, some trees, such as yellow birch, produce shallow roots even when growing on deep soils. Red maple, on the other hand, can adapt its root system to a range of soil depth. Upland hickories, red cedar, and oaks generally require deep soils for their deeply penetrating taproots.

Annual use of water by forests has been estimated to range from 3 in. in boreal forests to 125 in. in tropical forests. In the United States, it may range from 10 to 55 in.

Annual water losses by forest and desert types have been computed by Kittredge [6] from isohyetal maps of rainfall and runoff, as shown in Table 6-6.

Table 6-6. Annual Water Losses by Various Forest and Desert Types

Eastern regions	Inches	Western regions	Inches
Longleaf-loblolly-slash pine	30–40	Pacific Douglas-fir	25–60
River-bottom hardwoods and cypress	30–40	Redwood	25–55
		Sugar and ponderosa pine	15–40
Oak-pine	25–35	Western larch-western white pine	15–20
Oak-chestnut-yellow poplar	20–30		
Oak-hickory	20–30	Spruce-fir	10–20
Tall grass	20–30	Ponderosa pine	10–20
Birch-beech-maple-hemlock	15–20	Short grass	10–20
White-red-jack pine	15–20	Lodgepole pine	10–15
Spruce-fir	10–20	Piñon-juniper	5–15
		Chaparral	5–10
		Sagebrush	5–10
		Desert shrub	4–10

In the East, losses are at a maximum in the southern part, where the temperatures are highest and the growing season longest, and decrease progressively toward a minimum in the North. In the West the progression from south to north is disturbed by the low precipitation which prevails over much of the region and limits losses.

In 1960, estimates of evapotranspiration for western types were as shown in Table 6-7 [61].

Table 6-7. Estimated Evapotranspiration for Western Types of Vegetation

Vegetation type	Precipitation	Water yield		Evapotranspiration
		Range	Average	
	Inches			
Forest:				
Lodgepole pine	20–45	6–30	14	19
Engelmann spruce-fir	20–45	8–40	18	15
White pine-larch-fir	24–60	5–40	20	22
Mixed conifer	17–70	1–50	22	22
True fir	40–100	20–60	36	24
Aspen	20–45	1–20	10	23
Pacific Douglas-fir-hemlock-redwood	20–100	10–80	45	30
Interior ponderosa pine	20–30	0.5–15	4	17
Interior Douglas-fir	22–35	1–20	7	21
Chaparral and woodland:				
Southern California chaparral	12–40	0–20	5	20
California woodland-grass	12–40	0–25	7	18
Arizona chaparral	12–25	0–5	1.5	17.5
Piñon-juniper	10–20	0–3	0.5	14.5
Semiarid grass and shrub	5–20	0.1–1.0	0.4	10.6
Alpine	25–80	15–65	32	20

Ranges in precipitation and water yield give some conception of the variation in evapotranspiration within vegetation types.

F. Ecological Aspects

Susceptibility to an unfavorable water balance varies continuously throughout the life cycle of the average plant. Seeds are often capable of enduring extreme drought and, indeed, may require this condition to maintain their viability. Some seedlings maintain a high degree of drought resistance until they are several days old, but after their first leaves appear, they become very sensitive to desiccation. When plants approach maturity, they lose some of the sensitivity to drought that characterizes the seedling stages [11].

Annual crops require little water while they are still young and small. At the same time, however, the roots are limited to a small volume of soil. The plants are especially sensitive to shortage of water at this stage of growth. When plants are larger and have a more extensive root system, their water requirement is higher, but the roots can reach a larger reservoir of moisture and the plant is less sensitive, so that it can withstand short periods of wilting, during the heat of the day, without damage.

Perennial crops require water earlier in the growing season than do the annual crops. Their root systems are well established in the spring, so that as soon as conditions are right for growth, there will be a relatively high demand for water. The requirement will be somewhat less if the leafy portion does not cover the ground completely, because less transpiring surface is exposed to the sun's radiation [62].

Moisture relationships usually control the ability of pioneer plants to establish themselves on barren areas. Invasion of vegetation results in mesic conditions. Xeric habitats become moister and hydric habitats become drier as succession progresses.

Hydric succession begins in open water wherever vegetation can become established. It progresses in response to any environmental change that reduces the water (depth or saturation) and improves aeration in the soil. The direction of change is from aquatic toward terrestrial habitat. Peat-filled bogs are a result, then lowland or bog trees. The trees lower the water tables, with consequently improved soil aeration.

Xeric succession on rock follows a definite pattern whose progress is controlled by the rate at which soil forms and accumulates. Pioneer lichens and mosses build up the soil sufficiently to provide necessary anchorage and water-retaining ability for seed plants. Early stages in both hydric and xeric habitats are apt to be extremely slow. Later stages speed up as reaction of the vegetation becomes more effective. The final stage, as when trees become dominant, is again very slow [63].

Xeric succession increases evapotranspiration. From a bare rock, evapotranspiration is no more than the depression storage. As soil is formed, developed, and deepened, more of the rainfall remains on the site and more is evapotranspired. The development of deeper-rooted vegetation gives access to greater soil-water volumes; the development of heavier canopies leads to greater interception of precipitation. Succession stops when growth is in equilibrium with the environment. For upland soils in temperate latitudes, the climax vegetation generally evapotranspires most of the growing-season rainfall.

Hydric succession decreases evapotranspiration, the progression being the reverse of that described above. As the water surface is invaded by vegetation, it is shielded from solar radiation, dropping its temperature and evaporation rate.

Secondary succession, the invasion of vegetation on a denuded area, would, like xeric succession, tend to increase evapotranspiration.

G. Silvicultural Aspects

The different techniques of silvicultural practices create various kinds of openings in the growing space of forest stands. Most intermediate cuttings, thinnings, and cleanings are for the sole purpose of allotting more growing space to desirable trees; vacancies involved are relatively small and temporary. Natural regeneration is obtained by creating larger vacancies through harvest cuttings ranging from clear-cutting of all trees to a selection system in which only a few trees may be removed during any one operation. Each type of cutting has certain effects on evapotranspiration.

1. Clear-cutting. Clear-cutting can eliminate all transpiration, though in most forest operations understory vegetation of varying densities remains. It causes changes in microclimate, which may last for years. In swampy places, particularly where the climate is cool, the reduction of transpiration caused by the removal of the trees may result in a rise of the water table. Clear-cutting also tends to reduce protection against erosion, landslides, snowslides, and overland flow. It should be avoided on slopes where control of floods is an objective of management [64].

Perhaps the best-known study of the effects of clear-cutting on water yield—and thereby, evapotranspiration—was conducted at the Coweeta Hydrologic Laboratory of the U.S. Forest Service in western North Carolina. An old-growth hardwood forest on a 33-acre watershed was clear-cut without disturbing the ground cover (trees and shrubs were left where they fell). The first year streamflow increased 17 in., or 60 per cent, an approximation of the annual transpiration for the area. After several years, during which sprouts were cut annually, the increase fell off to 11 in., an average of about 30 per cent above streamflow in previous years. Another watershed was clear-cut, but allowed to regrow. Runoff increased 15 in. the first year, 11 in. after 3 years, 7 in. after 7 years, and 5 in. after 12 years. The increase, it is estimated, will be negligible after 35 years [65].

In Colorado, clear-cutting in alternate strips of a lodgepole pine–spruce–fir watershed increased water yield [66]. About 30 per cent of the watershed was cleared. Results were as follows:

	1956, in.	1957, in.
Recorded yield..........	15.6	23.0
Predicted yield..........	11.2	19.6
Increased yield..........	4.2	3.4

Part of the increase was due to reduced snow interception, and part to reduced transpiration, during the growing season.

A number of soil-moisture studies have shown higher soil-moisture contents, or lower soil-moisture losses, under bare or herbaceous areas as compared with forested areas. These reflect the difference in magnitude of water loss by evaporation from bare areas as compared with evapotranspiration from forested plots [41, 67, 68].

2. Seed-tree Method. In the seed-tree method 2 to 20 mature trees are left per acre to reseed the cutover area. Next to clearcutting, it has the greatest effect on transpiration. It also affords relatively little protection against erosion, landslides, snowslides, and overland flow.

3. Shelterwood, Diameter-limit, and Selection Systems. The effects of these harvesting systems on evapotranspiration will depend on the amount of material removed. Some conception of their effect on evapotranspiration is evident from a watershed study at the Fernow Experimental Forest in West Virginia [69]. Four watersheds, originally supporting a 50-year-old hardwood forest cover ranging from 11,000 to 13,000 bd ft/acre, were subjected to four harvesting treatments. Streamflow during the first growing season after treatment, as compared with predicted values developed from the control watershed, was found as shown in Table 6-8.

The heavier the cut, the greater the increase, or the less the evapotranspiration. Streamflow from the commercially clear-cut area was more than doubled for the May to October period; when less than half as much volume was removed, the increase was about one-half; and when lighter cuts were made, the increase was still less.

Thinnings may also reduce evapotranspiration, the reduction depending on the amount of cover removed. If the stand is young, the reduction in transpiration may be somewhat limited, because full root development has not yet taken place. Only recently have the effects of thinning on soil-moisture content and water losses been described. For instance, before thinning a 16-year-old loblolly pine plantation, soil moisture was found to be distributed fairly evenly between trees at the beginning of the growing season. After thinning, at the end of the growing season, soil moisture

Table 6-8. Effect of Harvesting Systems on Evapotranspiration

	Volume of timber removed, 1,000 bd ft/acre	Streamflow, May to October		
		Actual, in.	Predicted, in.	Increase, in.
Commercial clear-cut............	9.0	5.67	2.59	3.08
Diameter limit.................	4.3	5.23	3.39	1.84
Extensive selection.............	3.5	5.64	4.22	1.42
Intensive selection.............	1.8	3.72	3.41	0.31

was highest midway between the trees and lowest adjacent to the trees. The differences in moisture level between trees and under trees averaged 3 in. in 1957 and 2 in. in 1958, which could indicate that the effect of the thinning is decreasing [70].

Soon after thinning, the roots of harvested trees die. This creates growing space for the remaining trees and also lessens the number of roots striving to remove soil moisture. Later the dead roots decay, leaving root channels that serve as conduits for gravitational water. Some years after thinning and depending upon its severity, the soil again becomes occupied by living roots. It is during the interim period when the soil is partially freed of active roots that increases in soil moisture and reduction in evapotranspiration can be expected [71].

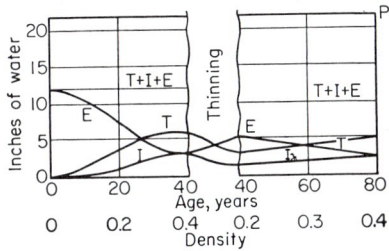

FIG. 6-4. Effect of thinning on transpiration T, interception I, and evaporation E. Assumed precipitation $P = 22$ in. (*After Kittredge* [6].)

The effects of thinning a stand from a crown closure of 0.4 to one of 0.2 on evaporation, transpiration, and interception have been diagrammatically described by Kittredge as follows (Fig. 6-4):

By the thinning the transpiration and interception might be approximately halved and the evaporation something less than doubled, because the forest floor would still protect the soil from evaporation after the thinning as before. After thinning the stand might regain its former crown closure of 0.4 at an age of 80 years, and, during that period, transpiration and interception would again increase and evaporation decrease. Immediately after the thinning the summation of the three losses would be less than before, and the trends have been drawn so that the summation remains constant during the 40 years at a value of 10 in. as compared with 12 in. before the thinning. The stream flow following the thinning would consequently be 2 in. more than before.

In this example an average rainfall of 22 in. was assumed. The difference between $T + I + E$ and the rainfall value is equal to a streamflow value of 10 in. before thinning and 12 in. thereafter.

4. Cleanings and Liberation Cuttings. These cuttings are done for the purpose of removing valueless species in order to make dominant the reproduction of commercial species. A wide variety of tree sizes may be removed ranging from dominant wolf trees to water-consuming underbrush. The effects on evapotranspiration will be equally variable.

At Coweeta, cutting out a dense stand of laurel and rhododendron from beneath a hardwood forest increased runoff (or decreased evapotranspiration) an average of 2 in. during the first 6 years of treatment [65].

In Arkansas, midsummer water-loss rates were about 25 per cent faster in loblolly pine plots with the hardwood understory left in place than in those where it had been

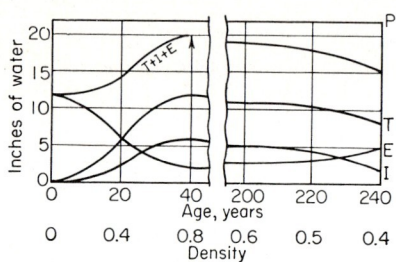

Fig. 6-5. Transpiration T, interception I, evaporation E, and runoff in relation to age and density. Assumed precipitation $P = 22$ in. (*After Kittredge* [6].)

eradicated. When the soil was moist, evapotranspiration from both types was nearly equal. By July, as the soil dried, moisture level in plots with no hardwood understory was more than 50 per cent greater than in plots with the understory present [72].

5. Reproduction. The effect of plantations or of natural reproduction on evapotranspiration depends on the nature of the vegetation previously occupying the area. Changes in water yield or evapotranspiration from the area will also depend on the proportion of the area planted. Rothacker [42] reported no significant change in total water yield after 15 years' growth of planted pines and yellow poplar at the White Hollow watershed in Tennessee. He inferred that transpiration increase had been balanced by a decrease in evaporation. Thirty-four per cent of the watershed had been planted. In a study in New York State of the effects of reforestation on streamflow, no significant results on water yield were observed, partly because the land planted had been well covered with native growth [73].

Kittredge [6] has described the effects of reproduction growth on evaporation, transpiration, and interception as follows (Fig. 6-5):

At zero age, transpiration and interception would obviously be zero. Both would increase slowly at first and most rapidly during the period of most rapid growth to a maximum at the culmination of the annual growth, which is assumed at 40 years. Thereafter the trends would be slightly downward, becoming more definitely so with the mortality and opening of the stand in old age. Transpiration is likely to be higher than interception in a forest stand. Evaporation would start at a maximum at zero age and would decrease to a minimum at the age of maximum development of canopy at 40 years, when it would be less than interception. Thereafter there would be a gradual increase somewhat accelerated in old age.

The values for the three forms of water loss are summed at different ages and plotted as a summation curve for transpiration, interception, and evaporation. Because transpiration and interception after 20 years are decidedly greater than evaporation, the trend of the summation curve starting from the value of evaporation at zero age follows somewhat the trend of the transpiration and interception curves, reaching a peak at the 40-year age. An assumed value of 22 inches for average precipitation has also been plotted as a straight line.

Again, differences between the rainfall value and the sums of the water losses, $T + I + E$, equal streamflow. In this example streamflow decreased from 10 in. at age zero to 2 in. at 40 years and then slowly increased to about 6 in. at 240 years.

VII. GLOSSARY[1]

Clear-cut: Removal of the entire stand in one cut.
Climax: The final stage of a succession which continues to occupy an area as long as climatic or soil conditions remain unchanged.
Crown: The branches and foliage of a tree; the upper portion of a tree.
Diameter-limit cut: Removal of mature timber to a specified diameter limit.
F-layer: A layer of humus consisting of partly decomposed organic matter, below the litter and above the H-layer.
H-layer: The lower part of the humus layer, consisting principally of amorphous organic matter.
Hydrophyte: A plant which grows under wet or moist conditions or which requires a large amount of moisture.
Intolerance: The incapacity of a tree to develop and grow in the shade of, and in competition with, other trees.
Landing: A place where logs are assembled for transportation.
Litter: The uppermost layer of the organic debris, composed of freshly fallen or slightly decomposed organic materials.
Lysimeter: A vessel or container placed below the ground surface to intercept and collect water moving downward through the soil.
Mesophyte: A plant that grows under intermediate moisture conditions.
Permanent-wilting percentage: The moisture content of the soil at the time when the leaves of plants growing in that soil first become permanently wilted.
Phreatophyte: Water-loving plants that grow mainly along stream courses, where their roots reach into the capillary fringe overlying the water table.
Pioneer species: Species capable of invading bare areas, often in large numbers and over considerable areas, and of persisting until displaced as succession proceeds.
Plant ecology: The science which deals with the relation of plants to their environment and to the site factors that operate in controlling their distribution and growth.
Pole: A young tree 4 in. or more in diameter breast-high. The maximum size is usually placed between 8 and 12 in.
Preclimax: A successional stage before the climax.
Sapling: A young tree less than 4 in. in diameter breast-high. The minimum diameter is usually placed at 2 in.
Sawlog: A log large enough to produce lumber or other products that can be sawed. Its size and quality vary with utilization practices of the region.
Secondary succession: Plant succession which is subsequent to the destruction of part or all of the original vegetation.
Seed-tree cut: Removal of the mature timber in one cut, except for a small number of seed trees left singly or in small groups.
Selection cut: Removal of mature timber, usually the oldest or largest trees, either as single scattered trees or in small groups at relatively short intervals, commonly 5 to 20 years, repeated indefinitely, by means of which the continuous establishment of natural reproduction is encouraged and an uneven-aged stand is maintained.
Shelter belt: A wind barrier of living trees and shrubs maintained for the purpose of protecting farm fields.
Shelterwood cut: Removal of the mature timber in a series of cuttings which extend over a period of years, by means of which the establishment of natural reproduction under the partial shelter of seed trees is encouraged.
Silviculture: The art of producing and tending a forest; the theory and practice of controlling forest establishment, composition, and growth.
Stand: An aggregation of trees or other growth occupying a specific area and sufficiently uniform in composition (species), age arrangement, and condition as to be distinguishable from the forest or other growth on adjoining areas.
Stomata: Minute openings bordered by guard cells, in the epidermis of leaves and stems through which gases pass.
Tolerance: The capacity of a tree to develop and grow in the shade of, and in competition with, other trees.

[1] Most of these definitions are from "Forest Terminology: A Glossary of Technical Terms Used in Forestry," 3d ed., Society of American Foresters, Washington, D.C., 1958.

Xerophyte: A plant that grows in dry situations.
Xylem: The lignified water-conducting, strengthening, and storage tissues of branches, stems, and roots.

VIII. REFERENCES

1. Horton, R. E.: Rainfall interception, *U.S. Monthly Weather Rev.*, vol. 47, pp. 603–623, 1919.
2. Wicht, C. L.: An approach to the study of rainfall interception by forest canopies, *J. S. Africa Forestry Assoc.*, no. 6, pp. 54–70, 1941.
3. Fuller, H. J., and Oswald Tippo: "College Botany," rev. ed., Holt, Rinehart and Winston, Inc., New York, 1954.
4. Hamilton, E. L., and P. B. Rowe: "Rainfall Interception by Chaparral in California," California Department of Natural Resources, Division of Forestry, 1949.
5. Rowe, P. B.: Some factors of the hydrology of the Sierra Nevada foothills, *Trans. Am. Geophys. Union*, pt. I, pp. 90–100, 1941.
6. Kittredge, Joseph: "Forest Influences," McGraw-Hill Book Company, Inc., New York, 1948.
7. Specht, R. L.: Dark Island heath (Ninety-mile Plain, South Australia). IV. Soil moisture patterns produced by rainfall interception and stem-flow, *Australian J. Botany*, vol. 5, no. 2, pp. 137–150, 1957.
8. Hoover, M. D.: Interception of rainfall in a young loblolly pine plantation, *U.S. Forest Serv., Southeast. Forest Expt. Sta., Sta. Paper* 21, 1953.
9. Voigt, G. K.: Distribution of rainfall under forest stands, *Forest Sci.*, vol. 6, no. 1, pp. 2–10, 1960.
10. Trimble, G. R., Jr., and Sidney Weitzman: Effect of a hardwood forest canopy on rainfall intensities, *Trans. Am. Geophys. Union*, vol. 35, pp. 226–234, 1954.
11. Daubenmire, R. F.: "Plants and Environment," 2d ed., John Wiley & Sons, Inc., New York, 1959.
12. Leonard, R. E.: Net precipitation in a northern hardwood forest, *J. Geophys. Res.*, vol. 66, no. 8, pp. 2417–2421, August, 1961.
13. Burgy, R. H., and C. R. Pomeroy: Interception losses in grassy vegetation, *Trans. Am. Geophys. Union*, vol. 39, pp. 1095–1100, 1958.
14. Thornthwaite, C. W., and F. K. Hare: Climatic classification in forestry, *Unasylva*, vol. 9, no. 2, pp. 51–59, 1955.
15. Clark, O. R.: Interception of rainfall by prairie weeds, grasses, and certain crop plants, *Ecol. Monographs*, vol. 10, pp. 243–277, 1940.
16. Clark, O. R.: Interception of rainfall by herbaceous vegetation, *Science*, vol. 86, ns. 2243, pp. 591–592, 1937.
17. Beard, J. S.: Results of the Mountain Home Rainfall Interception and Infiltration Project on Black Wattle, 1953–1954, *J. S. Africa Forestry Assoc.*, vol. 27, pp. 72–85, 1956.
18. Musgrave, G. W.: Field research offers significant new findings, *Soil Conserv.*, vol. 3, pp. 210–214, 1938.
19. Haynes, J. L.: Interception of rainfall by vegetative canopy, U.S. Department of Agriculture Soil Conservation Service (American Society of Agronomy Annual Meeting), mimeo., 1937.
20. Wisler, C. O., and E. F. Brater: "Hydrology," John Wiley & Sons, Inc., New York, 1959.
21. Trimble, G. R., Jr.: A problem analysis and program for watershed-management research, *U.S. Forest Serv., Northeast. Forest Expt. Sta., Sta. Paper* 116, 1959.
22. Baldwin, H. I., and C. F. Brooks: Forests and floods in New Hampshire, *New England Region Planning Comm. Publ.* 47, 1936.
23. Stone, E. C.: Dew as an ecological factor. I. A review of the literature, *Ecology*, vol. 38, no. 3, pp. 407–413, 1957.
24. Weitzman, Sidney, and R. R. Bay: Snow behavior in forests of Northern Minnesota and its management implications, *U.S. Forest Serv., Lake States Forest Expt. Sta., Sta. Paper* 69, 1959.
25. Goodell, B. C.: Watershed-management aspects of thinned young lodgepole pine stands, *J. Forestry*, vol. 50, no. 5, pp. 374–378, 1952.
26. Wilm, H. G., and E. G. Dunford: Effect of timber cutting on water available for stream flow from a lodgepole pine forest, *U.S. Dept. Agr. Tech. Bull.* 968, 1948.
27. Free, G. R., G. M. Browning, and G. W. Musgrave: Relative infiltration and related physical characteristics of certain soils, *U.S. Dept. Agr. Tech. Bull.* 729, 1940.

REFERENCES

28. Little, Silas, H. W. Lull, and Irwin Remson: Changes in woodland vegetation and soils after spraying large amounts of waste water, *Forest Sci.*, vol. 5, pp. 18–27, 1959.
29. Arend, J. L.: Infiltration as affected by the forest floor, *Soil Sci. Soc. Am. Proc.*, vol. 6, pp. 430–435, 1942.
30. Johnson, W. M.: Infiltration capacity of forest soil as influenced by litter, *J. Forestry*, vol. 38, p. 520, 1940.
31. Rowe, P. B.: Effects of the forest floor on disposition of rainfall in pine stands, *J. Forestry*, vol. 53, no. 5, pp. 342–348, 1955.
32. Gaiser, R. N.: Root channels and roots in forest soils, *Soil Sci. Soc. Am. Proc.*, vol. 16, pp. 62–65, 1952.
33. Packer, P. E.: An approach to watershed protection criteria, *J. Forestry*, vol. 49, no. 9, pp. 639–644, 1951.
34. Steinbrenner, E. C.: The effect of repeated tractor trips on the physical properties of forest soils, *Northwest Sci.*, vol. 29, pp. 155–159, 1955.
35. Doneen, L. D., and D. W. Henderson: Compaction of irrigated soils by tractors, *Agr. Eng.*, vol. 34, pp. 94–95, 102, 1953.
36. Mather, J. R.: The disposal of industrial effluent by woods irrigation, *Trans. Am. Geophys. Union*, vol. 34, pp. 227–239, 1953.
37. Trimble, G. R., Jr., R. S. Sartz, and R. S. Pierce: How type of frost affects infiltration, *J. Soil Water Conserv.*, vol. 13, pp. 81–82, 1958.
38. Stoeckeler, J. H., and Sidney Weitzman: Infiltration rates in frozen soils in Northern Minnesota, *Soil Sci. Soc. Am. Proc.*, vol. 24, no. 2, pp. 137–139, 1960.
39. Alderfer, R. B., and W. C. Bramble: The effect of plant succession on infiltration of rainfall into gilpin soil in Central Pennsylvania, *Penn. State Forest School Res. Paper* 5, 1942.
40. Coile, T. S.: Soil changes associated with loblolly pine succession on abandoned agricultural land of the piedmont plateau, *Duke Univ. School Forestry Bull.* 5, 1940.
41. Metz, L. J., and J. E. Douglass: Soil moisture depletion under several piedmont cover types, *U.S. Dept. Agr. Tech. Bull.* 1207, 1959.
42. Rothacker, J. S.: White hollow watershed management: fifteen years of progress in character of forest, runoff, and streamflow, *J. Forestry*, vol. 51, no. 10, pp. 731–738, 1953.
43. Weaver, J. E.: Effects of roots of vegetation in erosion control, U.S. Department of Agriculture, Soil Conservation Service (American Society of Agronomy Annual Meeting), 1937.
44. Mitchell, W. C., and G. R. Trimble, Jr.: How much land is needed for the logging transport system?, *J. Forestry*, vol. 57, no. 1, pp. 10–12, 1959.
45. Wooldridge, D. D.: Watershed disturbance from tractor and skyline crane logging, *J. Forestry*, vol. 58, no. 5, pp. 369–372, 1960.
46. Trimble, G. R., Jr., and Sidney Weitzman: Soil erosion on logging roads, *Soil Sci. Soc. Am. Proc.*, vol. 17, no. 2, pp. 152–154, 1953.
47. Steinbrenner, E. C., and S. P. Gessel: The effect of tractor logging on physical properties of some forest soils in southwestern Washington, *Soil Sci. Soc. Am. Proc.*, vol. 19, pp. 372–376, 1955.
48. Curry, J. R., and T. W. Church, Jr.: Observations on winter drying of conifers in the Adirondacks, *J. Forestry*, vol. 50, no. 2, pp. 114–116, 1952.
49. Hendrickson, A. H., and F. J. Veihmeyer: Daily use of water and depth of rooting of almond trees, *Am. Soc. Hort. Sci. Proc.*, vol. 65, pp. 133–138, 1955.
50. Broadfoot, W. M.: A method of measuring water use by forests on slowly permeable soils, *J. Forestry*, vol. 56, no. 5, p. 351, 1958.
51. Zahner, Robert: Takes water to make wood, *U.S. Forest Serv., South. Forest Expt. Sta., Sta. Notes* 104, 1956.
52. Hursh, C. R.: Local climate in the copper basin of Tennessee as modified by the removal of vegetation, *U.S. Dept. Agr. Circ.* 774, 1948.
53. Tanner, C. B.: Factors affecting evaporation from plants and soils, *J. Soil Water Conserv.*, vol. 12, pp. 221–227, 1957.
54. Penman, H. L.: Natural evaporation from open water, bare soil, and grass, *Proc. Roy. Soc. London*, Ser. B, vol. 193, pp. 120–145, 1948.
55. Clarke, G. L.: "Elements of Ecology," John Wiley & Sons, Inc., New York, 1954.
56. Harrold, L. L., and F. R. Dreibelbis: Agricultural hydrology as evaluated by monolith lysimeters, *U.S. Dept. Agr. Tech. Bull.* 1050, 1951.
57. Thornthwaite, C. W.: An approach toward a rational classification of climate, *Geograph. Rev.*, vol. 38, no. 1, pp. 55–94, 1948.
58. Croft, A. R.: A water cost of runoff control, *J. Soil Water Conserv.*, vol. 5, pp. 13–15, 1950.

59. Lemon, E. R., A. H. Glasser, and L. E. Satterwhite: Some aspects of the relationship of soil, plant, and meteorological factors to evapotranspiration, *Soil Sci. Soc. Am. Proc.*, vol. 21, no. 5, pp. 464–468, 1957.
60. Baker, F. S.: "Theory and Practice of Silviculture," McGraw-Hill Book Company, Inc., New York, 1934.
61. Evapo-transpiration reduction, in "Water Resources Activities in the United States," Committee Print No. 21, Select Committee on National Water Resources, United States Senate, 86th Congress, 2d Session, U.S. Government Printing Office, Washington, D.C., February, 1960.
62. Taylor, Sterling: Use of moisture by plants, in "Soil," U.S. Department of Agriculture Yearbook 1957, pp. 61–66.
63. Oosting, H. J.: "The Study of Plant Communities: An Introduction to Plant Ecology," 2d ed., W. H. Freeman and Company, San Francisco, 1956.
64. Hawley, R. C., and D. M. Smith: "The Practice of Silviculture," John Wiley & Sons, Inc., New York, 1954.
65. Dils, R. E.: "A guide to the Coweeta Hydrologic Laboratory," U.S. Forest Service, Southeastern Forest Experiment Station, 1957.
66. Goodell, B. C.: A preliminary report on the first year's effects of timber harvesting on water yield from a Colorado watershed, *U.S. Forest Serv., Rocky Mt. Forest and Range Expt. Sta., Paper* 36, 1958.
67. Lull, H. W., and John H. A.: Forest soil-moisture relations in the coastal plain sands of southern New Jersey, *Forest Sci.*, vol. 4, no. 1, pp. 2–19, 1958.
68. Koshi, P. T.: Soil-moisture trends under varying densities of oak overstory, *U.S. Forest Serv., South. Forest Expt. Sta., Occas. Paper* 167, 1959.
69. Northeastern Forest Experiment Station: Annual Report 1959, U.S. Forest Service, 1960.
70. Douglass, J. E.: Soil moisture distribution between trees in a thinned loblolly pine plantation, *J. Forestry*, vol. 58, no. 3, pp. 221–222, 1960.
71. Della-Bianca, Lino, and R. E. Dils: Some effects of stand density in a red pine plantation on soil moisture, soil temperature, and radial growth, *J. Forestry*, vol. 58, no. 5, 1960.
72. Zahner, Robert: Hardwood understory depletes soil water in pine stands, *Forest Sci.*, vol. 4, no. 3, pp. 178–184, 1958.
73. Ayer, G. R.: A progress report on an investigation of the influence of reforestation on streamflow in state forests in central New York, U.S. Geological Survey and N.Y. Conservation Department, Albany, N.Y., 1949.

Section 7

FLUID MECHANICS

MAURICE L. ALBERTSON, *Professor of Civil Engineering, Colorado State University.*

DARYL B. SIMONS, *Acting Chief, Civil Engineering Research Center, and Professor of Civil Engineering, Colorado State University.*

I. Introduction	7-2
II. Physical Properties of Fluids	7-2
A. Individual Properties	7-2
1. Density	7-2
2. Specific Weight	7-3
3. Compressibility	7-3
4. Viscosity	7-3
5. Vapor Pressure	7-4
6. Surface Energy	7-4
B. Dimensionless Parameters	7-5
III. Fluid Statics	7-6
A. Fluid Pressure	7-6
B. Forces on Submerged Surfaces	7-8
C. Buoyancy, Flotation, and Stability	7-8
D. Forces in Pipes and Conduits	7-9
IV. Fluid Dynamics	7-9
A. Definitions	7-9
B. Continuity Equation	7-10
C. Momentum Equation	7-10
D. Energy Equation	7-12
V. Fluid Viscosity and Turbulence	7-12
A. Laminar Flow	7-12
B. Turbulent Flow	7-13
VI. Resistance to Flow	7-14
A. Resistance and Drag	7-14
B. Boundary Layer	7-14
VII. Flow in Closed Conduits	7-15
A. Flow Development	7-15
B. Steady Uniform Flow	7-16
C. Solution of Problems	7-18
D. Steady Nonuniform Flow	7-19
E. Compound Pipelines	7-20
F. Looping Pipes	7-21
G. Branching Pipes	7-21
H. Pipe Networks	7-22

VIII. Flow in Open Channels.. 7-22
    A. Flow Classifications.. 7-22
    B. Uniform-flow Equations... 7-23
        1. The Chézy Equation... 7-23
        2. The Manning Equation.. 7-24
    C. Natural Channels.. 7-25
    D. Bed Roughness in Alluvial Channels................................ 7-26
    E. Design of Stable Channels in Alluvial Material..................... 7-26
    F. Meanders.. 7-26
    G. Sediment Transport... 7-29
        1. Suspended Load.. 7-31
        2. Bed Load... 7-31
        3. Total Sediment Load... 7-31
    H. Simple Waves and Surges.. 7-32
    I. Hydraulic Jump.. 7-34
    J. Transitions in Open Channels....................................... 7-36
    K. Specific-head Diagram... 7-37
    L. Discharge Diagram... 7-38
    M. Gradually Varied Flow.. 7-38
    N. Classification of Flow Profiles..................................... 7-39
    O. Computation of Backwater Curves................................. 7-42
IX. Flow around Submerged Objects.. 7-42
    A. Shear Drag and Pressure Drag...................................... 7-42
    B. Fall Velocity.. 7-43
X. Flow Measurement and Control... 7-44
    A. Orifices.. 7-44
    B. Weirs.. 7-45
        1. Sharp-crested Weirs... 7-45
        2. Broad-crested Weirs... 7-46
    C. Gates.. 7-46
    D. Overflow Spillways.. 7-47
    E. Venturi Meter... 7-47
    F. The Parshall Flume... 7-47
    G. Contracted Opening... 7-48
XI. References... 7-48

I. INTRODUCTION

The information presented in this section includes basic concepts and fundamental principles of fluid mechanics with widespread application to various aspects of hydrology—which is concerned primarily with water and, to a certain extent, air. Detailed treatment of the subject is available in Refs. 1 to 9 and other standard books on fluid mechanics. The English system of units is used in this text except in certain specified cases. Where units are not given, a consistent system must be applied.

II. PHYSICAL PROPERTIES OF FLUIDS

A. Individual Properties

Fluid properties are dependent upon the molecular characteristics of the fluid, including weight, spacing, attraction, and activity of the molecules. Individual properties commonly encountered in hydrology are as follows.

1. Density. The *density* ρ (in lb-sec^2/ft^4 or slugs/ft^3) of a fluid is the mass which it possesses per unit volume. Since a molecule has a definite mass regardless of its state (solid, liquid, or gas), the density is proportional to the number of molecules in a

PHYSICAL PROPERTIES OF FLUIDS

unit volume of the fluid. Air and water in the gaseous state have a molecular spacing greater than that in the liquid or the solid state. Steam, therefore, has much smaller density than water or ice, for which the molecular spacing is nearly the same.

2. Specific Weight. The *specific weight* γ (in lb/ft³) is the weight per unit volume. It is related to the density by

$$\gamma = \rho g \tag{7-1}$$

where g is the gravitational acceleration in ft/sec². The specific weight (or density) of a fluid or solid relative to the specific weight (or density) of water at a standard temperature (usually 38°F, or 4°C) is a dimensionless ratio known as the *specific gravity* (sp. gr.).

The specific weight of a gas, in contrast to that of a liquid, changes greatly with a variation of either temperature or pressure or both. This variation is attributable to the molecular structure of the gas in which the molecules have so much kinetic energy that their spacing is relatively sparse. The molecular spacing and activity depend on temperature and pressure. The basic relationships between specific weight, pressure, and temperature may be stated as a combination of *Boyle's law*, p/γ = constant for constant temperature, and *Charles' law*, p/T = constant for constant specific weight, to yield the *ideal-gas equation*, or the *equation of state*,

$$\gamma = \frac{p}{RT} \tag{7-2}$$

where γ is specific weight in lb/ft³, p is absolute pressure in lb/ft², T is absolute temperature in degrees (°F + 460°), and R is a gas constant in length per degrees absolute.

For an ideal gas, $mR = 1{,}545$ in the foot-pound-second system of units, but for real gases, the value of mR ranges between 1,512 and 1,546, in which m is molecular weight. Air and other real gases that are far removed from the liquid state may usually be considered as ideal gases. In the region of liquefaction, where a gas becomes a vapor, the above equation does not apply.

3. Compressibility. *Compressibility* of a fluid is a measure of the change in volume of the fluid when it is subject to external forces. It is expressed quantitatively by means of its bulk modulus of elasticity E in pounds per square inch:

$$E = \rho \frac{dp}{d\rho} \tag{7-3}$$

where p is pressure in psi, and ρ is density in slugs/ft³. The bulk modulus of elasticity for water is approximately 300,000 psi, and for air it is approximately 15 psi.

The compressibility of water may be considered negligible except in certain cases involving study of water hammer and transmission of elastic sound waves in bodies of water. For a gas in an *isothermal* process, that is, expansion and compression under constant temperature, $E = p$, while for an *adiabatic* process, $E = np$, where n is the ratio of the specific heat of the gas at constant pressure to that at constant volume. For air at normal temperature $n = 1.4$. For an adiabatic process, the pressure–specific weight relation is p/γ^n = constant.

4. Viscosity. *Viscosity* is the property of a fluid which resists relative motion and deformation in the fluid and causes internal shear. Therefore viscosity is a property exhibited only under dynamic conditions of motion. It is due to the cohesiveness of the molecules in liquids and the active and repeated impingement of the molecules upon each other in a gas. According to Newton, the shear τ at a point within a fluid is proportional to the velocity gradient dv/dy at that point, or

$$\tau = \mu \frac{dv}{dy} \tag{7-4}$$

where μ, in lb-sec/ft^2, is a proportionality constant known as *dynamic viscosity*. When divided by the density ρ in slugs/ft^3, it becomes *kinematic viscosity* ν in ft^2/sec, or $\nu = \mu/\rho$. Under ordinary conditions of pressure, viscosity has been found to vary only with temperature.

5. Vapor Pressure. The activity of the molecules at the surface of a liquid creates a *vapor pressure* which is a measure of the rate at which the molecules leave the surface. When the vapor pressure of the liquid is equal to the partial pressure of the molecules from the liquid which are in the gas above the surface, the number of molecules leaving is equal to the number entering. At this equilibrium condition,

Table 7-1. Properties of Water*

Temperature, °F	Specific weight, γ, lb/ft^3	Mass density, ρ, lb-sec^2/ft^4	Dynamic viscosity, $\mu \times 10^5$, lb-sec/ft^2	Kinematic viscosity, $\nu \times 10^5$, ft^2/sec	Surface energy,† $\sigma \times 10^3$, lb/ft	Vapor pressure, p_v, lb/in.^{2}	Bulk modulus, $E \times 10^{-3}$, lb/in.^{2}
32	62.42	1.940	3.746	1.931	5.18	0.09	290
40	62.43	1.940	3.229	1.664	5.14	0.12	295
50	62.41	1.940	2.735	1.410	5.09	0.18	300
60	62.37	1.938	2.359	1.217	5.04	0.26	312
70	62.30	1.936	2.050	1.059	5.00	0.36	320
80	62.22	1.934	1.799	0.930	4.92	0.51	323
90	62.11	1.931	1.595	0.826	4.86	0.70	326
100	62.00	1.927	1.424	0.739	4.80	0.95	329
110	61.86	1.923	1.284	0.667	4.73	1.24	331
120	61.71	1.918	1.168	0.609	4.65	1.69	333
130	61.55	1.913	1.069	0.558	4.60	2.22	332
140	61.38	1.908	0.981	0.514	4.54	2.89	330
150	61.20	1.902	0.905	0.476	4.47	3.72	328
160	61.00	1.896	0.838	0.442	4.41	4.74	326
170	60.80	1.890	0.780	0.413	4.33	5.99	322
180	60.58	1.883	0.726	0.385	4.26	7.51	318
190	60.36	1.876	0.678	0.362	4.19	9.34	313
200	60.12	1.868	0.637	0.341	4.12	11.52	308
212	59.83	1.860	0.593	0.319	4.04	14.7	300

* Adapted from Ref. 1.
† In contact with air.

the vapor pressure is known as the *saturation pressure*. If more molecules leave the surface than enter it, *evaporation* is occurring, whereas if more molecules are entering the surface than are leaving it, *condensation* is occurring.

The vapor pressure depends upon the temperature, because molecular activity depends upon heat content. As the temperature increases, the vapor pressure increases until the *boiling point* is reached for the particular ambient atmospheric pressure.

6. Surface Energy. *Surface energy*, in pounds per foot or pounds per inch, also inadequately known as *surface tension*, of a liquid, is caused by cohesion and adhesion of molecules. *Cohesion* is the force of attraction between molecules of the same kind, and *adhesion* is the force of attraction between molecules of different kinds. Surface energy causes the spherical shape of water droplets and their great internal pressure, as well as certain types of movement of water and other fluids through porous media. The difference in pressure across the surface of a droplet or bubble may be expressed as

$$\Delta p = \frac{2\sigma}{r} \tag{7-5}$$

where Δp is the pressure difference in psi, σ is the surface energy per unit area in lb/ft, and r is the radius of the droplet in ft. Note that, for convenience, inches can be substituted in place of feet as long as it is done consistently.

Surface energy also causes the rise of liquid in a capillary tube. The rise h in feet is

$$h = \frac{2\sigma}{\gamma r_0} \tag{7-6}$$

where γ is the specific weight of the liquid in lb/ft^3, and r_0 is the radius of the tube in ft.

Table 7-1 gives several properties of water at different temperatures.

B. Dimensionless Parameters

By relating various fluid properties as variables, significant dimensionless parameters have been developed. Determination of these parameters can be made by a special procedure of *dimensional analysis* involving the use of the *pi theorem*, which was introduced by Buckingham in 1915. If any variable a_1 depends upon the independent variables $a_2, a_3, \ldots, a_n$, and upon no others, the relationship may be written as

$$a_1 = f(a_2, a_3, \ldots, a_n) \tag{7-7}$$

The pi theorem states that, if all these n variables may be described with m fundamental dimensional units, they may then be grouped into $n-m$ dimensionless parameters, or π terms:

$$f(\pi_1, \pi_2, \pi_3, \ldots, \pi_{n-m}) = 0 \tag{7-8}$$

In each term there will be $m + 1$ variables, only one of which need be changed from term to term. When the force-length-time system is used, there are only three fundamental dimensional units, and then $m = 3$, and Eq. (7-8) will contain $n - 3$ dimensionless parameters. For a detailed treatment of dimensional analysis and its applications see Refs. 1 or 10.

The dimensionless parameters most likely to be encountered in hydrology include Reynolds number, Froude number, and Weber number.

The *Reynolds number*

$$\text{Re} = \frac{\rho V L}{\mu} = \frac{V L}{\nu} \tag{7-9}$$

relates the inertia forces to the viscous forces and is usually involved wherever viscosity is important, such as in slow movement of fluid in small passages or around small objects.

The *Froude number*

$$\text{Fr} = \frac{V}{\sqrt{\Delta \gamma L/\rho}} = \frac{V}{c} \tag{7-10}$$

relates the inertia forces to the gravitational effects and is important wherever the gravity effect is dominating, such as in waves, flow in open channels, gravity currents of sediment in reservoirs, salt-water intrusions, and mixing of air masses of different specific weights.

The *Weber number*

$$\text{We} = \frac{\rho V^2 L}{\sigma} \tag{7-11}$$

relates the inertia forces to the forces of surface energy and is important especially in connection with formation of water droplets and movement of water in capillaries and interstices of porous media such as soil.

In the above equations, V is the velocity of flow, L a characteristic length, ρ the density, μ the dynamic viscosity, ν the kinematic viscosity, $\Delta \gamma$ the difference in specific weight across the interface, c the celerity of a small wave in quiet fluid, and σ the surface energy. The units of these variables and parameters must be consistent.

III. FLUID STATICS

A. Fluid Pressure

Pressure of a fluid is the force per unit area which acts at a point. Within the fluid it acts in all directions, but against a boundary it acts perpendicularly. In foot-pound-second units, pressure is usually expressed as pounds per square inch (psi), pounds per square foot (psf), inches of mercury, or feet of water.

Pressure is commonly expressed relative to either of two reference data: *absolute zero*, giving absolute pressure in psia or psfa, and *atmospheric pressure*, giving *gage pressure* indicated by a gage. A negative pressure, or *vacuum*, is less than the atmospheric reference datum, but is positive when read as absolute pressure. At mean sea level under normal conditions, 1 atmospheric pressure = 14.7 psia = zero gage pressure = 30 in. Hg = 34 ft of water.

Hydrostatic pressure within a liquid is composed of the weight of the column of liquid of unit cross section plus the ambient pressure above the column. For atmospheric pressure acting over the liquid surface, the hydrostatic pressure p in terms of gage pressure (psf) is

$$p = \gamma h \tag{7-12}$$

where γ is the specific weight in lb/ft³, and h is the vertical depth in ft from the surface of the liquid to the point in question. It is convenient to express the pressure in terms of feet of fluid. For this purpose, the foregoing equation can be written as

$$\frac{p}{\gamma} = h \tag{7-13}$$

where p/γ is known as the *pressure head*, or simply *head*.

In incompressible fluids, such as most liquids, the specific weight does not vary significantly with pressure. This same relationship can be applied also to compressible fluids, such as gases, provided the change in elevation is not great. If the change in elevation is great, then the variation of γ with p must be considered. Such is the case of variation of pressure in the atmosphere, which depends also on temperature. Within the atmosphere the pressure variation with elevation is expressed in terms of the temperature, or the pressure as

$$\frac{p}{p_0} = \left(1 - \frac{n-1}{n} \frac{\Delta z}{RT_0}\right)^{n/(n-1)} = \left(1 - \frac{n-1}{n} \frac{\Delta z}{p_0/\gamma_0}\right)^{n/(n-1)} \tag{7-14}$$

where p = pressure at ground level, psia
p_0 = initial pressure at ground level, psia
R = a gas constant [Eq. (7-2)], or 53.3 for air
Δz = change in elevation, ft
T_0 = initial absolute temperature, deg, at ground level
γ_0 = initial specific weight of air at ground level, lb/ft³
n = ratio of specific heat of air at constant temperature to that at constant volume

Equation (7-14) is applicable for any value of n, except for $n = 1$, for which the following applies:

$$\frac{p}{p_0} = e^{-\Delta z/RT_0} = e^{-\Delta z/(p_0/\gamma_0)} \tag{7-15}$$

As mentioned previously (Subsec. II-A-3), $n = 1$ for isothermal conditions and $n = 1.4$ for dry-adiabatic conditions. In the atmosphere the value of n varies between $n = 1.2$ for wet-adiabatic, and $n = 1.4$ for dry-adiabatic, conditions. This is shown in Fig. 7-1, which gives a graphical representation of the above equations.

FLUID STATICS

Measurement of pressure is made at some point on the boundary of a container or of a submerged object such as a probe. When motion of the fluid is involved, the shape of the opening of the measuring device is of paramount importance because projections, depressions, or unsymmetrical shape of the opening will cause an inaccurate reading. The pressure-sensing element for pressure measurement is usually made of the following [1, p. 450; 11, p. 186]:

1. A mechanism with an element which changes shape or location with change in pressure, such as the *Bourdon gage, recording barometer*, and *barograph*. Each mechanism indicates the gage pressure.

2. A column of fluid, which is used to balance the pressure by connecting to the fluid at one end and to some reference pressure at the other end of the column. Examples

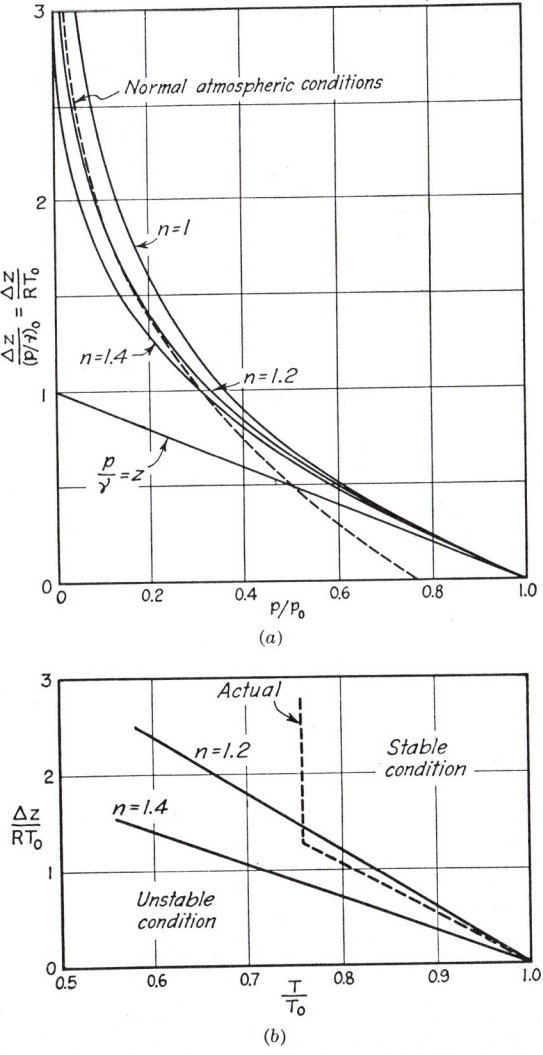

FIG. 7-1. Variations of (a) pressure and (b) temperature with elevation in the atmosphere.

are *mercury barometer, piezometer, open manometer, differential manometer, single-tube manometer,* and *micromanometer.*

3. A *pressure cell* such as a *piezoelectric cell* which changes in its electrical properties with pressure fluctuations. A transducer may be used to transform the mechanical pressure impulses into electrical signals for recording.

B. Forces on Submerged Surfaces

Basically, the pressure at a point on a submerged surface is given by Eq. (7-12). The total force which results from the integration of the pressure on the submerged plane surface is

$$F = \gamma \bar{h} A \qquad (7\text{-}16)$$

where F is the force in lb, γ is the specific weight of the liquid in lb/ft^3, A is the area of the submerged surface in ft^2, and $\bar{h}$ is the vertical distance in ft from the liquid surface to the center of gravity (the centroid) of the submerged surface. The total force acts perpendicularly to the plane surface and at the *center of pressure,* which is located a distance e below the centroid of the plane surface (Fig. 7-2):

$$e = S_0 - \bar{S} = \frac{\bar{I}}{\bar{S}A} = \frac{k^2}{\bar{S}} \qquad (7\text{-}17)$$

where S_0 and $\bar{S}$ are the moment arms as shown in Fig. 7-2, $\bar{I}$ is the moment of inertia of the area about its centroidal axis parallel to the axis 0-0 on the liquid surface, and k is the radius of gyration of the area.

For the determination of the total force on a submerged curved, or inclined plane, surface, the procedure is as follows:

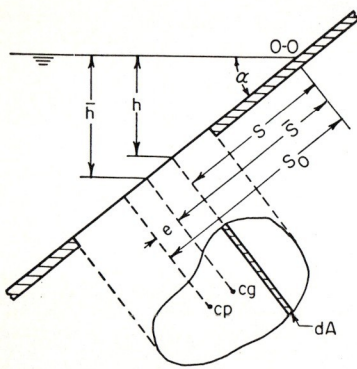

Fig. 7-2. General diagram for submerged plane surfaces.

1. The horizontal component F_H of the force is computed by projecting the surface horizontally onto a vertical plane and treating the projected area as a submerged vertical plane surface. By Eq. (7-16), $F_H = \gamma \bar{h}_p A_p$, where the subscript p indicates the projected area.

2. The vertical component F_V of the force is equal to the weight of the liquid, either real or imaginary as the case may be, vertically above the submerged surface. If the space above the surface is occupied by the liquid, the weight is real and the force acts vertically downward. If the space above the surface is not occupied by the liquid, the weight is imaginary and the force acts vertically upward.

3. The total force is given by $F = \sqrt{F_H^2 + F_V^2}$, and the angle which the force makes with the horizontal is $\theta = \tan^{-1}(F_V/F_H)$.

C. Buoyancy, Flotation, and Stability

A submerged or a floating object is buoyed up with a force equal to the weight of the fluid displaced by the object. This is the *Archimedes principle.* The *center of buoyancy* is at the center of gravity of the displaced volume of fluid.

Stability of a floating or submerged body is the tendency of the body to maintain a given position and to return to this position if rotated through a small angle. When a floating body has been rotated through an angle, the *metacenter* can be determined as the intersection of the vertical line through the center of buoyancy with the vertical line through the center of gravity before rotation while the body was in equilibrium. Stability or instability may be determined by consideration of the relative locations of the center of gravity, the weight of the body, the center of buoyancy, and the metacenter. The stability criteria are as follows:

	Submerged body	Floating body
Stable...............	Center of gravity *below* center of buoyancy	Center of gravity *below* metacenter
Unstable..............	Center of gravity *above* center of buoyancy	Center of gravity *above* metacenter
Neutral..............	Center of gravity *at* center of buoyancy	Center of gravity *at* metacenter

A given floating body rides higher in a heavy liquid than in a light one. The density, and hence the specific gravity, of a liquid may be determined by noting the depth to which a specially calibrated floating body sinks. Such a device is known as a *hydrometer*.

D. Forces in Pipes and Conduits

Static forces are created in pipes and other conduits because of the relative internal and external pressures of liquids or gases. The force F in pounds acting along the circumference of a pipe is

$$F = dL(p_i - p_e) \qquad (7\text{-}18)$$

where d is the diameter of the pipe in ft, L is the length of the pipe in ft, and p_i and p_e are, respectively, the internal and external pressures in psf. Such a force F per unit length of the pipe is known as the *hoop tension*.

Other factors which must be considered in design and construction of pipelines are the *external pressure*, which may cause the pipe to collapse if this is too great relative to the internal pressure, and the *flotation* forces, if the pipe under water becomes filled with air and the anchoring forces are not large enough to hold the pipe in place. Forces related to acceleration, and changes in pressure and temperature, may require special anchoring of the pipe to hold it in place.

IV. FLUID DYNAMICS

A. Definitions

Fluid dynamics treats fluids in motion. Many problems encountered in fluid motion are solved by application of the principles of continuity, momentum, and energy. In order to use such principles certain terms must be defined first as follows:

Velocity is the linear rate of movement or displacement of a point with respect to time (fps or mph).

Discharge is the volume rate of movement or flux of a quantity of fluid past a given point with respect to time (ft³/sec, or cfs, or sec-ft).

A *streamline* is an imaginary line within the flow for which the tangent at any point is the time average of the direction of motion at that point. A *streamtube* is a tube of fluid bounded by a group of streamlines which enclose the flow.

Steady flow exists if the velocity at a point remains constant with respect to time. Conversely, *unsteady flow* exists if the velocity changes either in magnitude or in direction with respect to time. Steady flow is usually much easier to analyze and solve than unsteady flow. In fact, a strictly rigorous solution for unsteady flow is sometimes impossible. Many hydrologic phenomena involve unsteady flow, so that only approximate solutions are possible. Such solutions are usually based on statistical averages or a step analysis for which there is only a small change in velocity over a short increment of time, and for this small time increment the flow is assumed steady. Most methods of flood routing (see Sec. 25-II) are of this step analysis.

If at a given instant the velocity remains constant with respect to distance along a streamline, the flow is *uniform*. If there is a change either in magnitude or in direction along the streamline, the flow is *nonuniform*. In flow around a bend of a pipe or channel, the *direction* changes with distance, and in flow with changing cross section, the *magnitude* changes with distance; hence the flow is nonuniform. In open channels where the velocity changes slowly along the channel, the flow is assumed uniform for an increment of length and is known as *gradually varied flow* (Subsec. VIII-M).

A *flow net* is a system of streamlines and orthogonal lines which show the idealized flow pattern for a given system of boundaries. The flow net is an important tool for the analysis of groundwater flow in porous media, as the latter is largely a potential flow, to be defined next.

Irrotational flow, also known as *potential flow*, exists if each fluid element in a flow system has no net rotation about its own mass center. In other words, if any part of a fluid element rotates in one direction, another part of the element rotates in the opposite direction, so that the net rotation is zero, and only distortion, rather than rotation, of the fluid element occurs. The flow net is based on the assumption of irrotational flow and therefore is not applicable to *rotational flow*.

Separation is a phenomenon which occurs in a fluid at changes in boundary shape because of either the inertia of the fluid or the reduced velocity of the flow near the boundary. Careful study of the separation zones is important in analyzing flow patterns and pressure distributions. Within the zone of separation the flow-net analysis cannot be applied. However, the flow net can be constructed with reasonable accuracy if the boundary is assumed at the edge of the separation zone.

B. Continuity Equation

The continuity of flow arises from the basic law of conservation of mass, which states that the mass rate of flow is the same at all sections of the flow in a streamtube. At a given section, the mass rate G in slugs per sec is

$$G = \rho Q = \int \rho v \, dA \tag{7-19}$$

where Q is the discharge in cfs, ρ is the density in slugs/ft^3, v is the local velocity in fps, and dA is the increment of cross-sectional area over which the local velocity applies. If the density is constant from section to section and the average velocity is used for each section, the above equation can be written as

$$Q = A_1 V_1 = A_2 V_2 = A_3 V_3 \tag{7-20}$$

where A is the cross-sectional area in ft^2, V is the average velocity in fps, and the subscripts refer to different sections. This equation is known as the *continuity equation for steady flow*.

C. Momentum Equation

The momentum equation is based on Newton's second law of motion, and it states that, in order to change the momentum flux of the flow $M_x = \int \rho v_x \, dQ = \int \rho v_x v \, dA$, a force must be applied in the x direction so that

$$\Sigma F_x = (K_m \rho Q V_x)_2 - (K_m \rho Q V_x)_1 \tag{7-21}$$

This is the *momentum equation* written in the x direction. Similar equations can also be written in the y and z directions. In Eq. (7-21), K_m is a *momentum-flux correction coefficient*, ρ is the density of the fluid, Q is the discharge, V_x is the average velocity of flow in the x direction, and the subscripts represent different sections of the flow. The velocity distribution across a section of flow is generally nonuniform. Since the average velocity is used in Eq. (7-21), the coefficient K_m is used to correct the effect of nonuniform distribution of velocity. If A is the cross-sectional area, v the local

FLUID DYNAMICS

velocity on the area, and V the mean velocity, then the correction coefficient is

$$K_m = \frac{1}{A} \int_A \frac{v_x}{V_x} \frac{v}{V} dA \qquad (7\text{-}22)$$

In practice, the velocity distribution is usually assumed to be uniform as a first approximation, which means that $v \approx V$ and K_m can be taken as 1.00. In reality, however, this assumption may be markedly in error because of the difference between the square of the average velocity and the average of the squares of the local velocity in Eq. (7-22). Under ordinary conditions of flow in a pipe, $K_m = 1.33$ for laminar flow, and K_m varies from 1.01 to 1.07, with an average value of about 1.03 for turbulent flow.

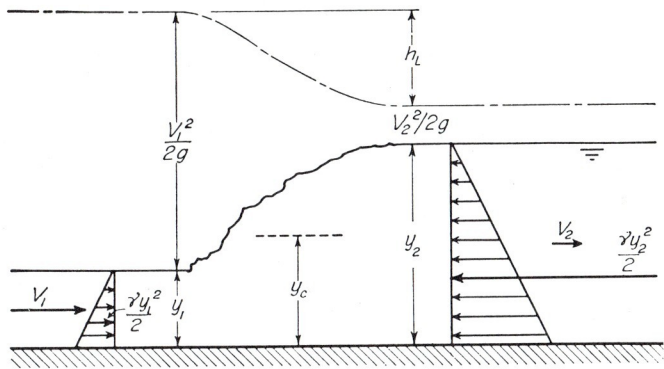

FIG. 7-3. Hydraulic jump.

The momentum equation, which contains vector quantities, can be used for the solution of such problems as the forces on a pipe elbow, the thrust of a propeller, the force of a jet of fluid impinging on a boundary, the hydraulic jump, and jet propulsion. Sometimes the first three of these phenomena can be used as a means of measuring the velocity and/or discharge of the flow.

Taking the hydraulic jump (Fig. 7-3) as an example and assuming $K_m = 1$, Eq. (7-21) gives the following expression for the flow per unit width:

$$\frac{1}{2} \gamma y_1^2 - \frac{1}{2} \gamma y_2^2 = \rho q V_2 - \rho q V_1 \qquad (7\text{-}23)$$

where γ is the specific gravity, y_1 and y_2 are depths of flow before and after the jump, ρ is the density, q is the discharge per unit width, and V_1 and V_2 are average velocities of flow before and after the jump. The continuity equation for the discharge per unit width is

$$q = V_1 y_1 = V_2 y_2 \qquad (7\text{-}24)$$

Equations (7-23) and (7-24) yield

$$\frac{y_2}{y_1} = \frac{1}{2} (\sqrt{1 + 8 Fr_1^2} - 1) \qquad (7\text{-}25)$$

where $Fr = V_1/\sqrt{gy_1}$ as defined by Eq. (7-10). Equation (7-25) gives the relationship between the depths of flow before and after the jump, called *initial depth* and *sequent depth*, respectively, and the Froude number of the incoming flow. Further discussion of hydraulic jump and of this equation is given later in flow in open channels.

D. Energy Equation

The energy equation can be derived from Newton's second law of motion. For one-dimensional, or irrotational, steady flow, the following *energy equation* can be written for any two cross sections 1 and 2 of the flow:

$$K_{e1}\frac{V_1^2}{2g} + \frac{p_1}{\gamma} + z_1 + E_M + E_H = K_{e2}\frac{V_2^2}{2g} + \frac{p_2}{\gamma} + z_2 + H_L \qquad (7\text{-}26)$$

where $K_{e1}(V_1^2/2g)$ and $K_{e2}(V_2^2/2g)$ are *velocity heads* at sections 1 and 2, respectively; p_1/γ and p_2/γ are the *pressure heads* at the two sections; z_1 and z_2 are *elevation heads*, or the elevation of the two sections above a certain datum plane; E_M is the mechanical energy added between the sections; E_H is the heat energy added between the sections; K_{e1} and K_{e2} are *energy-flux correction factors* for the two sections; V_1 and V_2 are average velocities at the two sections; p_1 and p_2 are pressures at the two sections; g is the gravitational acceleration; and γ is the specific weight. The sum of the pressure head and the elevation head is termed the *piezometric head*, $h = (p/\gamma) + z$.

The velocity distribution across the flow is usually nonuniform. Since the average velocities are used in Eq. (7-26), the effect of nonuniform distribution of velocity is corrected by K_e, which is defined as

$$K_e = \frac{1}{A}\int_A \left(\frac{v}{V}\right)^3 dA \qquad (7\text{-}27)$$

where A is the cross-sectional area of flow, v is the local velocity for the increment of area dA, and V is the average velocity over the area A.

In practice, K_e is usually taken as 1.00. In reality, for laminar flow in a pipe $K_e = 2.0$, and for turbulent flow in a pipe K_e varies from 1.02 to 1.15, with an average value of about 1.06.

The energy equation, which contains scalar quantities, can be applied to the solution of such problems as jets issuing from an orifice, jet trajectory, flow under a gate, flow over a weir, siphons, transition flow in pipes and open channels, and flow associated with pumps and turbines. Furthermore, such phenomena frequently exist in flow systems and can sometimes be used as a means of measuring velocity, pressure, or discharge of the flow.

V. FLUID VISCOSITY AND TURBULENCE

Because of the effect of viscosity, the flow in a system can be either laminar or turbulent or a combination of both.

A. Laminar Flow

Laminar flow exists if the momentum transfer within a fluid flowing past a boundary, or another separate body of fluid acting as a fixed or moving boundary, is by molecular action only. In such a flow the forces of viscosity dominate other forces such as inertia forces. Consequently, the fluid particles appear to move in definite smooth paths and there is no significant transverse mixing as the fluid moves from point to point along the flow.

For the analysis of laminar flow, Eq. (7-4), or

$$\tau = \mu \frac{dv}{dy} = (\mu/\rho)\frac{d(\rho v)}{dy} = \nu \frac{d(\rho v)}{dy}$$

since $\mu = \rho \nu$, is used. In Eq. (7-4) the *shear* is the internal stress within a fluid which resists deformation or change of shape of fluid masses during motion.

FLUID VISCOSITY AND TURBULENCE 7-13

Representing the rate of transfer of momentum flux per unit area, the shear exists only under dynamic conditions of motion and is proportional to the rate of deformation as indicated by Eq. (7-4).

Based on Eq. (7-4), many problems of laminar flow can be solved. Table 7-2 gives the solutions for certain cases of laminar flow.

Table 7-2. Solutions for Laminar Flow*

Flow case	Head loss h_f or drag h_D	Mean velocity V	Local velocity v
Parallel boundaries at distance B apart	$h_f = 12\dfrac{\mu VL}{\gamma B^2}$	$\dfrac{\gamma B^2}{12\mu}\dfrac{h_f}{L}$	$-\dfrac{\gamma}{2\mu}\dfrac{dh}{dx}(By - y^2)$
Free surface of flow depth y_0 and slope S	$h_f = 3\dfrac{\mu VL}{\gamma y_0^2}$	$\dfrac{\gamma y_0^2}{3\mu}\dfrac{h_f}{L}$	$\dfrac{\gamma S}{2\mu}(2yy_0 - y^2)$
Circular conduits of radius r_0	$h_f = 8\dfrac{\mu VL}{\gamma r_0^2}$	$\dfrac{\gamma r_0^2}{8\mu}\dfrac{h_f}{L}$	$-\dfrac{\gamma}{4\mu}\dfrac{dh}{dx}(r_0^2 - r^2)$
Porous media of grain size d and permeability coefficient k	$h_f = k\dfrac{\mu VL}{\gamma d^2}$	$\dfrac{\gamma d^2}{k\mu}\dfrac{h_f}{L}$	
Submerged sphere of diameter d	$h_D = 3\dfrac{\mu Vd}{\gamma d^2}$	$\dfrac{\gamma d^2}{3\mu}\dfrac{h_D}{L}$	

* Use consistent units for all formulas. Other notation: dh/dx is gradient of head in direction of flow; y is distance from boundary to point in flow; L is distance considered in direction of flow; r is radius at a point where velocity is v; γ is specific weight of flowing fluid; and μ is dynamic viscosity of flowing fluid.

B. Turbulent Flow

The Reynolds number defined by Eq. (7-9) relates the inertia forces to the viscous forces, and it is generally used to identify laminar flow and turbulent flow. When the number is small so that the viscous forces are predominant, the flow is laminar. As the number is increased, the forces of inertia become increasingly large, until finally instability develops and the flow becomes turbulent. When flow is *turbulent*, the fluid particles move in irregular paths which are neither smooth nor fixed, but which in the aggregate will represent the forward motion of the entire stream.

In laminar flow the mixing or diffusion of momentum is by the action of *molecular viscosity*. In turbulent flow, however, diffusion takes place much more rapidly through the action of eddies which are finite groups of molecules and are referred to as *eddy (molar) viscosity*. Consequently, for the analysis of turbulent flow, Eq. (7-4) must be modified as

$$\tau = (\mu + \eta)\frac{dv}{dy} = (\nu + \epsilon)\frac{d(\rho v)}{dy} \qquad (7\text{-}28)$$

where $\epsilon = \eta/\rho$ is the *kinematic eddy viscosity*, and η is the *dynamic eddy viscosity*. Turbulence is commonly described by an *intensity* of $\overline{v'^2}$ and a *scale* of l, where v' is the velocity deviation from the mean velocity at a point. The eddy viscosity can be related to the intensity and scale of the turbulence by

$$\epsilon = cl\sqrt{\overline{v'^2}} \qquad (7\text{-}29)$$

where c is a proportionality constant. The eddy viscosity varies in magnitude with distance from the boundary or from the zone of greatest shear, whereas the molecular viscosity is a constant throughout the flow if the temperature remains constant.

7-14 FLUID MECHANICS

VI. RESISTANCE TO FLOW

A. Resistance and Drag

Resistance is the force transmitted from a boundary to the fluid, and *drag* is the force of the fluid on the boundary.

Resistance or drag can be divided into one or a combination of two kinds: (1) *shear resistance*, or *shear drag*, a tangential stress caused by the fluid viscosity and taking place along a boundary of the flow in the tangential direction of local motion, and

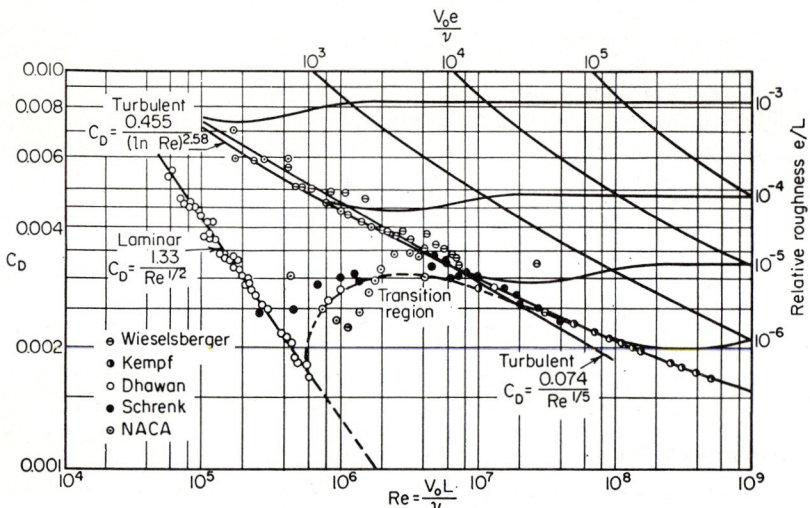

FIG. 7-4. Variation of drag coefficient with Reynolds number for a flat plate parallel to the flow.

(2) *pressure resistance*, or *pressure drag*, a normal stress caused by acceleration of the fluid, which results in a decrease in pressure from the upstream to the downstream side of an object, and acting perpendicularly to the boundary. Each kind can be integrated over the whole boundary to obtain the total shear or pressure and hence the resulting total resistance or drag.

The general *drag equation* is

$$F_D = \frac{C_D A \rho V_0^2}{2} \tag{7-30}$$

where C_D is a drag coefficient, and F_D is the drag on any boundary having the area A, the ambient velocity V_0 of the flow, and the density ρ of the fluid. The drag coefficient depends upon the viscous effects relative to the inertial forces, as contained in the Reynolds number, the shape of the boundary, and the boundary roughness in terms of *relative roughness*, e/L, the ratio of the roughness size to a characteristic length. Figure 7-4 shows the variation of drag coefficient with Reynolds number for a flat plate of length L parallel to the flow.

B. Boundary Layer

Near the boundary of a flow system a layer of fluid is decelerated because of the resistance to flow which is offered by the shear at the boundary. This relatively thin layer is called the *boundary layer*, which can develop into either laminar flow or turbulent flow. For practical purposes, the outer edge of a boundary layer is defined arbitrarily as being where the local velocity is equal to 0.99 of the ambient velocity.

The thickness δ of a *laminar boundary layer* over a flat plate at a distance x from the upstream edge of the plate is

$$\delta = \frac{5.2x}{\sqrt{V_0 x/\nu}} = \frac{5.2x}{\text{Re}^{0.5}} \tag{7-31}$$

The corresponding drag coefficient is

$$C_D = \frac{1.33}{\text{Re}^{0.5}} \tag{7-32}$$

As the laminar boundary layer continues downstream from the leading edge of a flat plate, it expands in thickness until it becomes unstable and a *turbulent boundary layer* is formed. Immediately adjacent to a smooth boundary associated with a turbulent boundary layer, there is a thin film of essentially laminar flow known as the *laminar sublayer*.

For a turbulent boundary layer, there are two limiting types of boundaries: (1) a *smooth boundary*, for which the roughness elements are covered with the laminar sublayer so that the roughness has no influence on the flow within the boundary layer, and (2) a *rough boundary*, for which the laminar sublayer is disturbed by the roughness elements, which are therefore contributing directly to the turbulence.

For the turbulent boundary layer over a smooth boundary, the thickness δ of the layer over a flat plate at a distance x from the leading edge can be derived on the basis of the *one-seventh power law* of velocity distribution, which has application over the range of $2 \times 10^5 < \text{Re} < 2 \times 10^7$, as

$$\delta = \frac{0.38x}{\text{Re}^{0.2}} \tag{7-33}$$

The corresponding drag coefficient is

$$C_D = \frac{0.074}{\text{Re}^{0.2}} \tag{7-34}$$

On the basis of the *logarithmic law* of velocity distribution, which has application over the range of $2 \times 10^6 < \text{Re} < 2 \times 10^9$, the drag coefficient becomes

$$C_D = \frac{0.455}{(\ln \text{Re})^{2.58}} \tag{7-35}$$

For turbulent flow along a rough boundary the drag coefficient varies only with relative roughness but not with Reynolds number.

VII. FLOW IN CLOSED CONDUITS

A. Flow Development

Flow in closed conduits involves a combination of steady or unsteady flow, uniform or nonuniform flow, laminar or turbulent flow, and flow over smooth or rough boundaries.

At the upstream end of a pipe there is a region of flow development in which the boundary layer is developing and the flow is technically nonuniform because the velocity varies from point to point along the streamline. Therefore the velocity distribution changes from section to section, which means that, in this region of flow development, a standard calibration of measuring devices and the standard loss coefficients for pipe fittings (minor losses) are not applicable. In other words, calibrations must be made and coefficients must be determined in place if reasonable accuracy is to be achieved. Furthermore, in this region the greater boundary shear and any initial separation cause a greater amount of energy loss per unit length of pipe than in the region of fully developed flow downstream. The length of the region of flow development in a pipe of diameter D is approximately 0.06 Re D for laminar flow and 0.7 $Re^{0.25}$ D for turbulent flow throughout the length from the pipe entrance.

B. Steady Uniform Flow

Problems involving steady uniform flow in closed conduits may be solved by the energy equation [Eq. (7-26)], which is written for two sections 1 and 2 as follows, assuming $K_e = 1.00$:

$$\frac{V_1^2}{2g} + \frac{p_1}{\gamma} + z_1 = \frac{V_2^2}{2g} + \frac{p_2}{\gamma} + z_2 + H_L \qquad (7\text{-}36)$$

where H_L is the sum of the losses caused by both the shear resistance h_f and the pressure resistance h_L; that is, $H_L = h_f + h_L$.

The shear resistance can be evaluated by the *Darcy-Weisbach equation*,

$$h_f = f \frac{L}{D} \frac{V^2}{2g} \qquad (7\text{-}37)$$

where f is a *resistance coefficient*, L is the length of the pipe in ft, D is the diameter in ft, V is the mean velocity of flow in fps, and g is the gravitational acceleration, or 32.2 ft/sec². The resistance coefficient depends upon the Reynolds number of flow and the relative roughness e/D, where e is the average size of the roughness element. For laminar flow or for turbulent flow with a smooth surface, the relative roughness is unimportant, and hence f depends on Re alone. For a rough boundary, Re is unimportant, and then f depends on e/D alone. The relationship between f, Re, and e/D is shown in Fig. 7-5 by the so-called *Moody resistance diagram*. In this diagram the roughness e for various pipe materials and inside coatings is given. The average value of the range of e should be used unless additional information gives reason to use the smaller or larger values of the range. However, it may be seen from the diagram that a rather large error in the estimate of e would result in a smaller error in f.

In Fig. 7-5 it is seen that, when Re is less than 2,000, the flow is laminar and $f = 64/\text{Re}$. When Re increases, the laminar sublayer is penetrated by roughness elements and the flow becomes turbulent. The region between Re = 2,000 to approximately Re = 3,500 indicates an indefinite transition for flow to change from laminar to turbulent. For turbulent flow, the resistance coefficient can be estimated from the following equations:

For turbulent boundary layer over smooth boundary,

$$\frac{1}{\sqrt{f}} = 2 \log (\text{Re} \sqrt{f}) - 0.8 \qquad (7\text{-}38)$$

For turbulent boundary layer over rough boundary,

$$\frac{1}{\sqrt{f}} = 2 \log \frac{D}{e} + 1.14 \qquad (7\text{-}39)$$

For the transition from smooth to rough boundary, the above two equations can be combined to produce the following semiempirical form:

$$\frac{1}{\sqrt{f}} = 1.14 - 2 \log \left(\frac{e}{D} + \frac{9.35}{\text{Re} \sqrt{f}} \right) \qquad (7\text{-}40)$$

which is known as the *Colebrook-White equation*. This equation reduces to Eq. (7-38) for flow in smooth pipes and to Eq. (7-39) for flow in rough pipes.

FLOW IN CLOSED CONDUITS 7-17

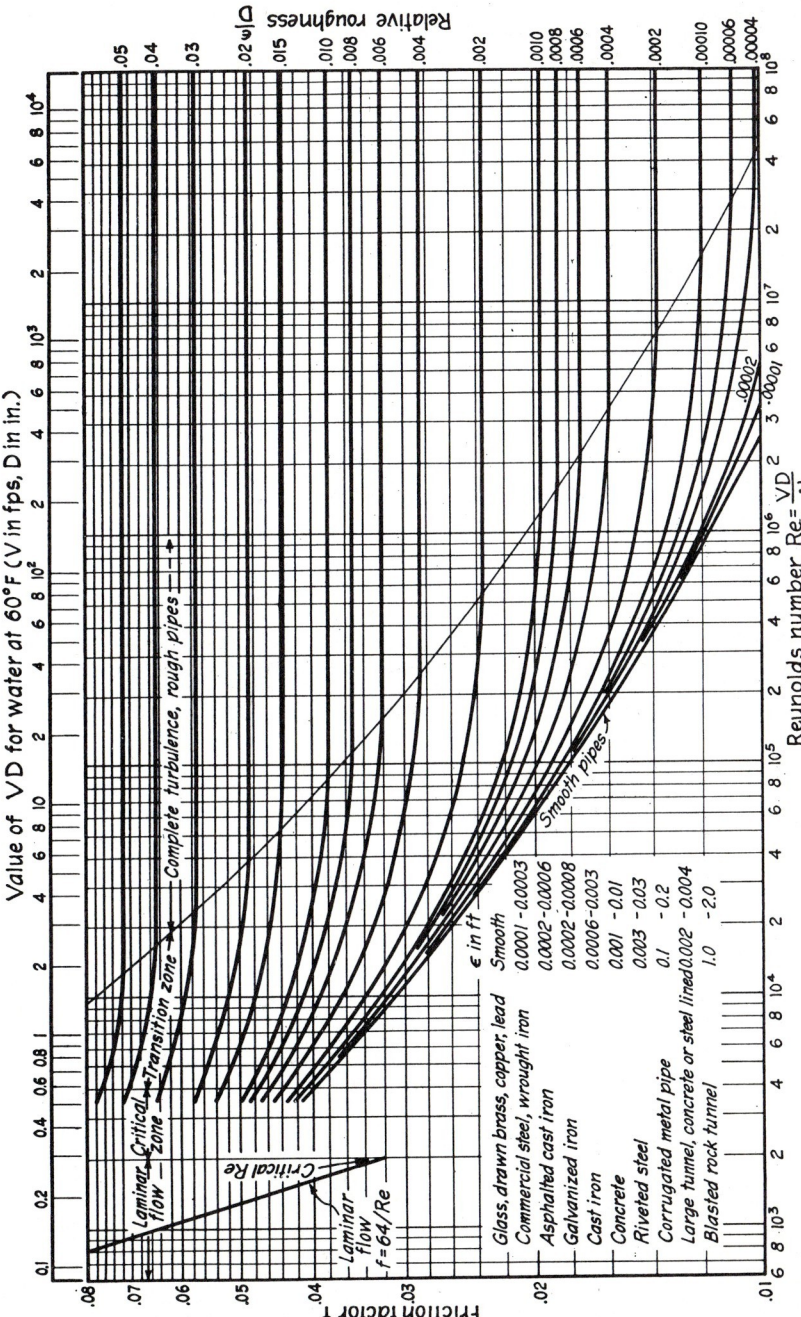

Fig. 7-5. Moody diagram.

Pipes having a noncircular cross section but a simple geometrical shape, such as a rectangle, a trapezoid, or an ellipse which does not differ markedly from circular, can be solved by Fig. 7-5 if the hydraulic radius equivalent to that of a circular pipe is used. Thus $R = D/4$, or $D = 4R$, and Eq. (7-37) becomes $h_f = f(L/4R)(V^2/2g)$. For turbulent flow, this use of hydraulic radius gives reasonably accurate results. For laminar flow, however, it gives increasingly inaccurate results as the shape of conduit differs more and more from circular.

C. Solution of Problems

The following three types of problems occur most frequently in the design of pipelines and pipe systems:

1. Usually, the head loss h_f is needed for a given length of pipe L, mean velocity V or discharge Q, pipe diameter D, roughness e, fluid viscosity ν. For this problem the solution is direct. From Reynolds number Re, relative roughness e/D, the resistance coefficient f can be found from Fig. 7-5 and the head loss solved from Eq. (7-37).

2. When V or Q instead of head loss is needed in the above problem, however, Re cannot be computed directly. Hence another parameter Re $\sqrt{f}$ is used, which does not involve either V or Q. By use of Re $\sqrt{f}$ and e/D, the value of f can be determined directly from Eq. (7-40), and V can be computed by Eq. (7-37) and Q by the continuity equation $Q = VA$.

By assuming the pipe being rough with a given value of e/D, f can be determined from Fig. 7-5. After computing V as above, the value of f can be checked by calculating Re to see if the pipe has the roughness assumed. If the pipe is not rough as assumed but is in the transition zone, adjust the answer by using a new value of f from the Re versus f curves.

3. When the diameter D is unknown, a trial-and-error method can be used. The f value is estimated (between 0.02 and 0.04 in most problems) and a trial D value is determined from Eq. (7-37), and also from $Q = VA$ if V is unknown but Q is known. The trial D is used to compute both Re and e/D; then f is found from Fig. 7-5. This new f value is now used as the new estimate of f, and the process is repeated until the estimated f is in good agreement with the computed f.

Although the Darcy-Weisbach equation and the Moody resistance diagram give the rational solution of pipe flow problems, various empirical formulas for flow of water in pipes have been developed from data taken from the laboratory or in the field. Perhaps the best known and most extensively used of these is the *Hazen-Williams formula*,

$$V = 1.32 C_1 R^{0.63} S^{0.54} \tag{7-41}$$

where V is the velocity in cfs, C_1 is a discharge coefficient, R is the hydraulic radius in ft, which is equal to the cross-sectional area A of the pipe divided by the wetted perimeter, and S is the slope of the energy line or hydraulic gradient h_f/L, where h_f is head loss in pipe of length L.

Table 7-3. Hazen-Williams Coefficients C_1 for Flow in Pipes

Description of pipe	C_1 value
Extremely smooth and straight	140
Very smooth	130
Smooth wood and wood stave	120
New riveted steel	110
Vitrified	110
Old riveted steel	100
Old cast iron	95
Old pipes in bad condition	60–80
Small pipes badly tuberculated	40–50

The Hazen-Williams formula is widely used in waterworks design and is most applicable for pipes of 2 in. or larger and velocities less than 10 fps. The principal advantage of this formula is that the coefficient C_1 does not involve Reynolds number,

FLOW IN CLOSED CONDUITS 7-19

and hence all problems have direct solutions. This is also a disadvantage, however, since temperature and viscosity variations are ignored. These are at times very important, and ignoring them can cause serious errors.

D. Steady Nonuniform Flow

Nonuniform flow occurs when either the magnitude or direction of the velocity of flow varies with distance along a streamline. Tangential acceleration occurs if the

Table 7-4. Form-loss Coefficient

Sudden expansion, $C_L = (1 - A_1/A_2)^2$

A_1/A_2	0	0.2	0.4	0.6	0.8	1.0
C_L	1	0.64	0.36	0.16	0.04	0

Sudden contraction, $C_L = (1/C_c - 1)^2$

A_2/A_1	0	0.2	0.4	0.6	0.8	1.0
C_L	0.5	0.45	0.36	0.21	0.07	0

90° smooth-pipe bend

r/D	1	2	4	6	8	10
C_L	0.25	0.14	0.10	0.085	0.08	0.08

Gradual expansion with $\theta = 15°$, Re $= 1.5 \times 10^5$

A_2/A_1	2	4	6	8	10	20
C_L	0.11	0.21	0.25	0.28	0.29	0.32

Commerical-pipe fitting

Globe valve, fully open.............................	10
Angle valve, fully open.............................	5
Swing check valve, fully open......................	2.5
Closed return bend.................................	2.2
Tee, through side outlet............................	1.8
Short-radius elbow.................................	0.9
Medium-radius elbow..............................	0.8
Long-radius elbow.................................	0.6
45° elbow...	0.4
Gate valve, fully open..............................	0.2
Three-quarters open............................	1
One-half open..................................	5.6
One-quarter open...............................	24

NOTATION: C_L = form-loss coefficient; A_1 and A_2 = respective areas upstream and downstream of source of loss; r = radius of pipe bend; D = pipe diameter; C_c = coefficient of contraction = A_c/A_2, where A_c = area of contracted jet at vena contracta; θ = total angle of expansion; Re = Reynolds number.

velocity is changed in magnitude, and normal acceleration occurs if the velocity is changed in direction. These changes in velocity result in a change in momentum flux, which is a vector quantity. The latter change is accomplished only by pressures against the fluid in addition to the pressures which would be associated with uniform flow. When such changes in velocity occur, zones of separation and secondary flow frequently result, and this consequently increases the shear and the turbulence at the expense of the piezometric head. Hence head losses h_L result. Since the foregoing changes in velocity and the resulting head losses are caused by nonuniform distribution of pressures on the boundary, the losses are termed *form losses* because of pressure resistance and the associated changes (usually increases) in shear resistance.

The form losses can be expressed as

$$h_L = C_L \frac{V^2}{2g} \tag{7-42}$$

where C_L is called the *form-loss coefficient*, and V is the mean velocity of flow. Table 7-4 lists the average values of C_L for various kinds of form losses. These form losses are sometimes called *minor* losses. Such a term represents the true situation literally when the pipeline is relatively long and the shear-loss coefficient $f(L/D)$ in Eq. (7-37) is large by comparison with C_L. For shorter pipelines, however, the form losses caused by pressure resistance may be of *major* importance.

E. Compound Pipelines

The principles presented in all the foregoing discussion can be used in combination to solve problems involving compound pipelines. Figure 7-6 is an example of a compound pipeline which consists of an entrance, a sudden expansion, a sudden contraction, a valve, a bend, a gradual expansion, an outlet, and pipes of different diameters. Each of these items involves a head loss. The straight pipe involves shear

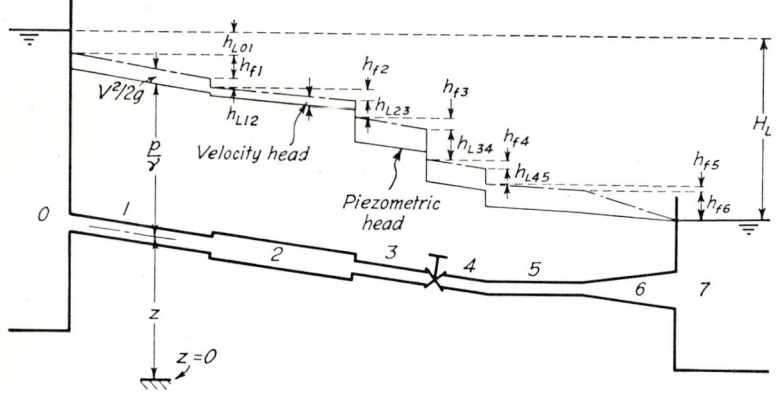

FIG. 7-6. Energy diagram for compound pipelines.

resistance, and each of the others involves both shear and pressure resistance to make up the form losses. The energy equation [Eq. (7-26)] may be written for any reach of pipe between sections a and b:

$$\frac{V_a^2}{2g} + \frac{p_a}{\gamma} + z_a = \frac{V_b^2}{2g} + \frac{p_b}{\gamma} + z_b + H_L \tag{7-43}$$

If the upstream reservoir is chosen as section a and the downstream reservoir as b, then H_L is the sum of all the losses indicated in Fig. 7-6, or

$$\begin{aligned}
H_L &= h_{L01} + h_{f1} + h_{L12} + h_{f2} \\
\text{total} & \quad \text{entrance} \quad \text{pipe} \quad \text{expansion} \quad \text{pipe} \\
\text{loss} & \quad \text{loss} \quad \text{loss} \quad \text{loss} \quad \text{loss} \\
&+ h_{L23} + h_{f3} + h_{L34} + h_{f4} + h_{L45} + h_{f5} + h_{L57} + h_{L67} \\
& \quad \text{contraction} \quad \text{pipe} \quad \text{valve} \quad \text{pipe} \quad \text{bend} \quad \text{pipe} \quad \text{gradual} \quad \text{exit} \\
& \quad \text{loss} \quad \text{loss} \quad \text{loss} \quad \text{loss} \quad \text{loss} \quad \text{loss} \quad \text{expansion} \quad \text{loss} \\
& \quad\quad\quad\quad\quad\quad\quad\quad\quad\quad\quad\quad\quad\quad\quad\quad\quad\quad\quad\quad \text{loss}
\end{aligned} \tag{7-44}$$

Each of the losses must be determined by the methods already discussed, and then added together to get H_L. Losses due to pumps and turbines may also be added to the system if they are involved.

F. Looping Pipes

Looping pipes are a simple and common problem which frequently arises in connection with increasing the capacity of a pipeline. Figure 7-7 shows a schematic

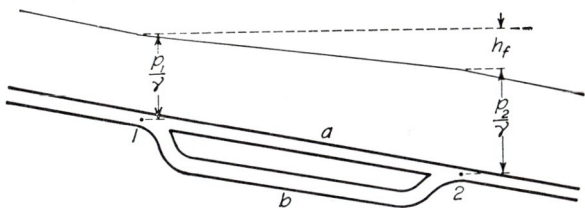

Fig. 7-7. Looping pipes.

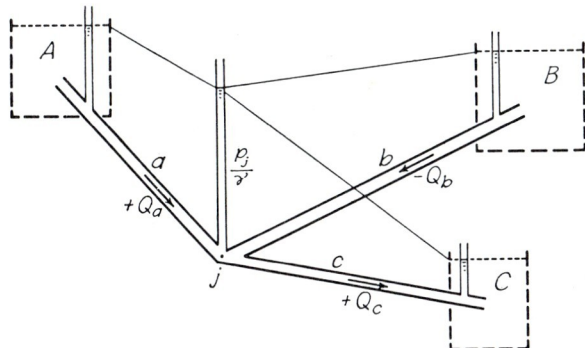

Fig. 7-8. Branching pipes.

representation of a looping-pipe system. The continuity equation for this system is

$$Q_1 = Q_a + Q_b = Q_2 \qquad (7\text{-}45)$$

The energy equation written from point 1 to point 2 is

$$\frac{V_1^2}{2g} + \frac{p_1}{\gamma} + z_1 = \frac{V_2^2}{2g} + \frac{p_2}{\gamma} + z_2 + h_f \qquad (7\text{-}46)$$

Since the energy loss is the same between the ends of either of paths a and b, Eq. (7-37) gives $h_f = f_a \dfrac{L_a}{D_a} \dfrac{V_a^2}{2g} = f_b \dfrac{L_b}{D_b} \dfrac{V_b^2}{2g}$, and either pipe can be used to determine the loss.

G. Branching Pipes

Pipe systems which branch but do not rejoin as the looping pipes do differ from the looping pipes in that the piezometric heads at the downstream ends of the branches may not be equal. Figure 7-8 illustrates a branching-pipe system and indicates that the flow into the junction must equal the flow out of the junction. Furthermore, the piezometric head at the junction is common for all three pipes. The three piezometric readings at A, B, and C can be considered as the water-surface elevations in three reservoirs, as shown by broken lines, since the velocity head is considered as insignificant in these problems when compared with the head losses due to boundary resistance.

There are three different flow conditions for the continuity equation, any one of which may be applicable for a given problem. Each flow condition depends upon the slope of the hydraulic gradient, as follows: (1) flow from pipe a into pipes b and c, so that the piezometric head line for pipe b slopes downward to the right and

$$Q_a = Q_b + Q_c$$

(2) flow from pipes a and b into pipe c, so that the piezometric-head line from pipe b slopes downward to the left and $Q_a + Q_b = Q_c$; (3) flow from pipe a into pipe c, with no flow in pipe b, so that the piezometric head line for pipe b is horizontal and $Q_a = Q_c$ while $Q_b = 0$.

The solution of a specific problem depends further on other known and unknown variables: (1) pipe sizes and lengths, (2) the piezometric head at the ends of each of the pipes, which includes the elevation of the piezometric head lines at the junction and the water-surface elevations in each reservoir; and (3) the discharges Q_a, Q_b, and Q_c in the three lines.

If the pipe sizes and lengths, the flow in one line, and the water-surface elevations in two reservoirs are known, the solution to the problem is direct, by the use of the Moody resistance diagram to determine the piezometric head at the junction. The solution of most other problems is indirect, by assuming values for the discharge or the piezometric head at the junction. When the water-surface elevation in each of the three reservoirs is given, together with the pipe sizes and lengths, the piezometric head at the junction can be assumed as equal to the water surface in reservoir B to determine whether flow is into or out of reservoir B. New assumptions can then be made to complete the solution.

H. Pipe Networks

Pipelines are frequently branching and looping in a complex network fashion. This is particularly true in connection with city water-distribution systems. Because of the complexity of these problems, the controlled trial-and-error solutions proposed for a simple looping or branching system are difficult and time-consuming to use. One of the most widely used solutions for balancing flow distribution in pipe networks is the *Cross-Doland method* [1, 12, 13]. This method requires that the flow in each pipe be assumed so that the principle of continuity is satisfied at each junction, and the flow is then balanced by step approximations. For fast solutions of complex problems, the *McIlroy network analyzer* has been developed for use. This device is based on the principle of electrical analogy [14, 15] between the pressure and flow distribution in pipe networks and the voltage and current in electrical systems.

VIII. FLOW IN OPEN CHANNELS

A. Flow Classifications

Flow in open channels has been nature's way of conveying water on the surface of the earth through rivers and streams. Furthermore, these streams have constantly been the subject of study by man, as he has been alternately blessed by the life-giving quality of streams under control and plagued by the destructive power of streams out of control, such as in time of flood.

Open channels include not only those which are completely open overhead, but also closed conduits which are flowing partly full. Examples of such closed conduits are tunnels, storm sewers, sanitary sewers, and various types of pipelines.

Flow in open channels involves a *free surface*, which is actually an *interface* between two fluids having different specific weights, such as air and water. Because of the presence of the free surface, flow in open channels has an additional classification to the three that have been given previously to the flow in pipes. The four classifications are

(1) uniform or nonuniform, (2) steady or unsteady, (3) laminar or turbulent, and (4) tranquil, rapid, or critical.

Uniform flow in open channels has no change with distance in either the magnitude or the direction of the velocity along a streamline. Otherwise the flow is *nonuniform*, or *varied*. Strictly uniform flow rarely exists. For practical purposes, flow in an open channel is generally considered as uniform if the depth of flow is approximately constant in the direction of flow. The depth of uniform flow is called *normal depth*.

Steady flow occurs when the velocity at a point does not change with time. Otherwise, the flow is *unsteady*, such as traveling surges and flood waves in an open channel.

Whether *laminar flow* or *turbulent flow* exists in an open channel depends upon the Reynolds number of the flow, just as it does in pipes. Like the flow in pipes, turbulent flow may be over either a smooth boundary or a rough boundary, depending on the relative size of the roughness elements as compared with the thickness of the laminar sublayer.

Unlike laminar and turbulent flow, *tranquil flow* and *rapid flow* occur only with a free surface or interface. The criterion for this classification of flow is the Froude number $Fr = V/\sqrt{gy}$, as defined by Eq. (7-10). When $Fr = 1.0$, the flow is *critical*; when $Fr < 1$, the flow is *tranquil*; and when $Fr > 1$, the flow is *rapid*.

Uniform flow in an open channel occurs with either a *mild*, a *critical*, or a *steep slope*, depending on whether the flow is tranquil, critical, or rapid, respectively.

B. Uniform-flow Equations

Two most common equations for uniform flow in open channels are the Chézy and the Manning equations.

1. The Chézy Equation. This equation, proposed by Chézy in 1769, may be written as

$$V = C\sqrt{RS} \qquad (7\text{-}47)$$

where V is the mean velocity of flow in cfs, C is the *Chézy discharge coefficient*, R is the hydraulic radius in feet, and S is the slope of the channel or the sine of the slope angle. The evaluation of the coefficient C can be made as follows.

For laminar flow in a wide channel, assuming a parabolic distribution of velocity, the value of C can be determined by the following equation:

$$\frac{C}{\sqrt{g}} = \sqrt{\frac{\text{Re}}{8}} \qquad (7\text{-}48)$$

where $\text{Re} = 4VR/\nu$.

For turbulent flow in a wide channel, the velocity distribution may be assumed to be logarithmic, as

$$\frac{(v-V)C}{V\sqrt{8g}} = 2\log\frac{y}{y_0} + 0.88 \qquad (7\text{-}49)$$

where v is the local velocity at a depth y, and y_0 is the total depth. This equation, however, does not apply near the bed or near the surface of the flow.

In alluvial channels the magnitude of C depends upon the form of the boundary roughness.

Expressed in terms of the Darcy-Weisbach resistance coefficient f, the coefficient C is

$$C = \sqrt{\frac{8g}{f}} \qquad (7\text{-}50)$$

2. The Manning Equation.

In an effort to correlate and systematize existing data from natural and artificial channels, Manning in 1889 proposed an equation which was developed into

$$V = \frac{1.5}{n} R^{2/3} S^{1/2} \qquad (7\text{-}51)$$

where n is the *Manning roughness coefficient*. By comparing this equation with the Chézy equation, the following relationship can be written:

$$C = 1.5 \frac{R^{1/6}}{n} \qquad (7\text{-}52)$$

This relationship indicates that the Chézy discharge coefficient is a function of the Manning coefficient and the hydraulic radius.

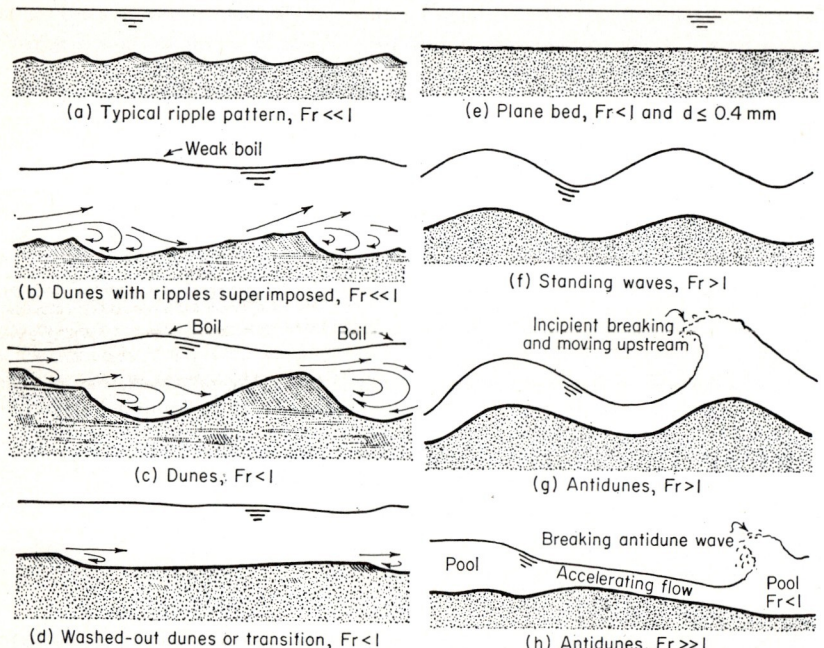

FIG. 7-9. Forms of bed roughness in alluvial channels. (*After Richardson, Simons, and Haushild* [16, 17].)

The Manning n was developed empirically as a coefficient which remained approximately a constant for a given boundary condition, regardless of slope of channel, size of channel, or depth of flow. As a matter of fact, however, each of these factors causes n to vary to some extent. In other words, the Reynolds number, the shape of the channel, and the relative roughness have an influence on the magnitude of Manning's n. Furthermore, for a given alluvial bed of an open channel, the size, pattern, and spacing of sand waves (Fig. 7-9) vary with slope, discharge, channel shape, and character of bed material, so that n also varies. Despite the shortcomings of the Manning roughness coefficient, however, it is used extensively in Europe, India, Egypt, and the United States.

The approximate magnitude of Manning roughness is given in Table 7-5 for rigid channels and alluvial channels. Considering alluvial channels, note that, as the form

FLOW IN OPEN CHANNELS 7-25

of bed roughness (Fig. 7-9) changes from dunes through transition to plane bed or standing waves, the magnitude of Manning n decreases by approximately 50 per cent.

Table 7-5. Manning Roughness Coefficients for Various Boundaries

Boundary	Manning roughness n, ft$^{1/6}$
Very smooth surfaces such as glass, plastic, or brass	0.010
Very smooth concrete and planed timber	0.011
Smooth concrete	0.012
Ordinary concrete lining	0.013
Good wood	0.014
Vitrified clay	0.015
Shot concrete, untroweled, and earth channels in best condition	0.017
Straight unlined earth canals in good condition	0.020
Rivers and earth canals in fair condition—some growth	0.025
Winding natural streams and canals in poor condition—considerable moss growth	0.035
Mountain streams with rocky beds and rivers with variable sections and some vegetation along banks	0.040–0.050
Alluvial channels, sand bed, no vegetation	
1. Lower regime	
Ripples	0.017–0.028
Dunes	0.018–0.035
2. Washed-out dunes or transition	0.014–0.024
3. Upper regime	
Plane bed	0.011–0.015
Standing waves	0.012–0.016
Antidunes	0.012–0.020

C. Natural Channels

The natural shape of an open channel may be markedly different from simple geometric shapes. However, it is usually possible to break down the complex shape of a natural open channel into simple elementary shapes for analysis. For example, consider Fig. 7-10, in which flow occurs not only in the main channel, but also in the

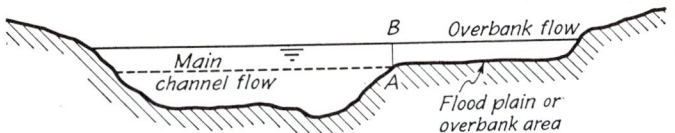

FIG. 7-10. Cross section of a natural stream channel.

overbank or flood-plain area. In this case, the hydraulic radius R, which would be obtained by using the area and the wetted perimeter for the entire section, would not be truly representative of the flow. Furthermore, the grain roughness, the form roughness, and hence the roughness coefficient in the overbank area are usually different from the coefficient in the main channel. Therefore such a section should be divided along AB and treated as two separate sections. The plane AB, however, is not considered as a part of the wetted perimeter, since there is no appreciable shear in this plane.

Along a natural channel, there are frequently pools with a flatter slope, and riffles or rapids with a steeper slope, than the average slope of the channel taken over an appreciable distance. Therefore care must be taken in studies of natural streams to consider the correct slope for the particular discharge and particular reach of the stream in question. In computing the flow by a uniform-flow equation, the slope of the water surface, or more precisely the slope of the energy grade line of the flow, is generally used for the slope in the equation. Unless there is considerable variation in channel shape with distance, it is then necessary to work with the energy grade line.

D. Bed Roughness in Alluvial Channels

In open channels with rigid bed, the roughness is independent of the flow. In alluvial channels with movable sand or gravel bed, however, the roughness is dependent upon the flow and the forms of the bed roughness. The forms of bed roughness in alluvial channels are illustrated in Fig. 7-9. (See some definitions in Sec. 17-II.) As the Froude number of the flow increases in a given channel, the bed-roughness form changes in the order of *ripples, dunes, washout dunes, plane bed, standing waves,* and *antidunes*. For shallow depths, $D < 1.0$ ft, when the median size of bed material d is less than approximately 0.4 mm, the dunes wash out at Fr ≈ 0.5, and a plane bed and plane water surface persists until Fr ≈ 1, at which time antidunes develop. If the median diameter is larger than about 0.4 mm, Fr ≈ 1 when the dunes are completely washed out. With further increase in shear and Froude number, standing sand and water waves, which are in phase, develop and persist until Fr = 1.2, at which time antidunes develop. Considering coarser alluvial materials, ripples will not form when the median diameter $d \geq 0.6$ mm; that is, when sufficient shear is developed on the bed to cause bed-material movement, conditions yielding ripples have been exceeded and thus dunes form. On the other hand, with large depths, dunes apparently develop, even for coarse bed materials such as gravel, if the flow can exert a tractive force on the stream bed of sufficient magnitude to move the bed material within the tranquil flow regime, Fr < 1.

E. Design of Stable Channels in Alluvial Material

One of the oldest methods for the design of stable channels in alluvial material is the *method of permissible velocity*, in which the channel is designed for the greatest mean velocity that will not cause erosion of the channel body [18]. This method is now largely replaced by other methods because the permissible velocities are very uncertain and variable.

For erodible channels which scour but do not silt, the *method of tractive force* is applicable [18, 19]. The *tractive force* is the drag or shear that is developed on the wetted area of the channel bed and acts in the direction of the flow. This force per unit wetted area is called the *unit tractive force* τ_0 and can be expressed as

$$\tau_0 = \gamma RS \qquad (7\text{-}53)$$

where γ is the specific weight of water, R is the hydraulic radius, and S is the slope of the channel bed. To design a channel by the tractive-force method is to assure that the unit tractive force of the bed material will not exceed a permissible value over the entire wetted area. For noncohesive material, the permissible unit tractive force depends largely on the size and angularity of the material. For cohesive material, the permissible value depends largely on compactness or voids ratio of the material.

For alluvial channels encountering both scouring and silting, the *method of regime theory* has been developed for design purposes. The theory is largely empirical but recognizes that width, depth, and slope of the channel are variables and that three or more independent equations can be written to determine them. Some of the most common and useful regime equations are summarized in Table 7-6. Those developed by Lacey, Bose, and Inglis are probably the most widely accepted. In general, these equations are applicable to the design of stable channels which have mobile beds and carry a relatively small bed-material load—less than 500 ppm [20]. Alluvial channels designed to carry more than 500 ppm of bed load usually require bank stabilization.

F. Meanders

Winding of natural channels is known as *meandering*. Although the mechanics of meandering is not completely understood, at least four variables: valley slope, sediment

Table 7-6. Summary of Regime Equations*

Engineer	Date	Velocity V, ft	Slope S	Silt factor	Area A, ft²	Wetted perimeter P, ft	Hydraulic radius R, ft	Width B, ft	Depth D, ft
Kennedy	1895	CD^m or $0.84D^{0.64}$		$C = \phi(d)$					
Lindley	1919	$0.95D^{0.57}$ or $0.57B^{0.36}$						$3.80D^{1.61}$	$0.44B^{0.62}$
	1929	$1.17f^{1/2}R^{1/2}$	$\dfrac{f^{3/2}}{2{,}587Q^{1/6}}$	$\phi(d)$ and $\dfrac{0.75V^2}{R}$	$\dfrac{3.8V^5}{f^2}$	$2.67Q^{1/2}$			
	1929†	$\dfrac{1.346R^{3/4}S^{1/2}}{N_a}$ $16.12R^{2/3}S^{1/2}$	$\dfrac{f^{3/2}}{2{,}614R^{1/4}}$	$\left(\dfrac{N_a}{0.0225}\right)^4$	$\dfrac{4.0V^5}{f^2}$				
	1934	$1.155f^{1/2}R^{1/2}$ $16.0R^{2/3}S^{1/2}$ $0.79Qf^{1/6}f^{1/3}$	$\dfrac{5.47 \times 10^{-4}f^{5/3}}{Q^{1/6}}$	$f_{VR} = \dfrac{0.75V^2}{R}$	$\dfrac{1.26Q^{5/6}}{f^{1/3}}$	$8\tfrac{3}{4}Q^{1/2}$	$\dfrac{0.47Q^{1/3}}{f^{1/3}}$		
Lacey	1939		$\dfrac{3.5 \times 10^{-4}}{R^{1/2}}$	$f_r = Kd^{1/2} = 1.76d^{1/2}$ $f \propto \dfrac{d^{1/2}g^{3/2}}{(vg)^{1/2}}$					
	1946	$\dfrac{Kv}{S}$ $\dfrac{K(R^{1/2}S)^n}{S}$ $1.60(R^{1/2}S)^{1/3}$		$f_{SV} = 488S^{1/2}V^{1/2}$					
Bose	1936	$1.12R^{1/2}$	$\dfrac{2.09 \times 10^{-3}d^{0.86}}{Q^{0.21}}$		PR	$2.8Q^{1/2}$	$0.47Q^{1/3}$		
Malhotra	1939	$18.18R^{0.63}S^{0.34}$							
White	1939	$\dfrac{0.7d^{1/4}g^{2/5}R^{1/2}}{Q^{1/20}}$		$f_{VR} = \dfrac{0.37g^{4/5}\omega^{1/2}}{Q^{1/10}}$					
	1941	$12.0(R^2S)^{2/7}$	$\dfrac{f_m^{1.53}}{2{,}110Q^{0.145}}$						
Inglis	1947	$\dfrac{\alpha_3 g^{7/18}Q^{1/6}(C\omega d)^{1/2}}{v^{1/36}}$	$\dfrac{\alpha_5(C\omega d)^{5/12}}{v^{5/36}g^{7/18}Q^{1/6}}$	$\dfrac{\alpha(C\omega)^{1/2}}{gv^{1/6}}$	$\dfrac{\alpha_2 v^{1/36}Q^{5/6}}{g^{7/18}(C\omega d)^{1/12}}$			$\dfrac{\alpha_1 Q^{1/2}(C\omega)^{1/4}}{g^{1/8}v^{1/2}d^{1/4}}$	$\dfrac{\alpha_4 v^{1/6}Q^{1/2}d^{1/6}}{g^{1/18}(C\omega)^{1/3}}$

Table 7-6. Summary of Regime Equations* (Continued)

Engineer	Date	Velocity V, ft	Slope S	Silt factor	Area A, ft²	Wetted perimeter P, ft	Hydraulic radius R, ft	Width B, ft	Depth D, ft
Blench	1939 1941 1946 1952 1957	$(F_bF_sQ)^{1/6}$	$\dfrac{F_b^{5/6}F_s^{1/2}\nu^{1/4}}{3.63(1+aC)gQ^{1/6}}$	$F_b = \dfrac{V^2}{D} \propto d$ $F_s = \dfrac{V^3}{B} \propto d$	BD			$\left(\dfrac{F_bQ}{F_s}\right)^{1/2}$	$\left(\dfrac{F_sQ}{F_b^2}\right)^{1/3}$

* Prepared by D. B. Simons and E. V. Richardson for the American Society of Civil Engineers Task Force on Resistance to Flow in Alluvial Channels.
† Equations developed in answer to original discussions.

NOTATION: a = about $1/400$ for uniform sands originally used in the experiment and probably about $1/233$ for natural river-bed sands; C = silt factor in the Kennedy equation, sediment charge or sediment transport divided by water discharge in the Inglis equations, or bed-load charge in hundred-thousandths by weight; d = representative diameter of sediment particles, usually d_{50} or the size smaller than 50 per cent of the weight; f, f_{VR}, and f_{SV} = silt factor relative to the sediment; f_m = a constant; F_b = bed factor; F_s = side factor; K = a constant = $D^{1/2}S$; m = an exponent; N_a = rugosity factor = $0.0225f^{1/4}$; Q = discharge of water; w = width of channel; ω = fall velocity of the particle of size d; α, α_1, α_2, α_3, α_4, and α_5 = coefficients; and ν = kinematic viscosity. All dimensions are in English units except d, which is in millimeters.

load, discharge, and the characteristics of the bed and bank material, are known to control the process of meandering. The process of meandering can be triggered by an increase or a decrease in bed-material load. In a stable channel when the sediment load exceeds that required for stability, the slope of the stream steepens because of deposition. Then the boundary shear is increased, and if the banks of the channel are not sufficiently resistant, the banks begin to erode. As the banks erode unevenly, only a slight misalignment of flow is necessary to shift the thalweg of the stream toward one side or the other, thus increasing the rate of attack by water on the bank and initiating meandering.

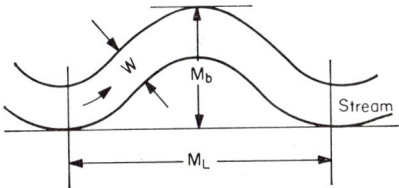

FIG. 7-11. Characteristic dimensions of a meander.

The shape of meanders as illustrated in Fig. 7-11 have been studied under both laboratory and field conditions. As a part of the regime theory, the length M_L, the breadth M_b, and the width W, all in feet, for rivers in flood plains have been expressed empirically in terms of the discharge Q in cubic feet per second as follows [21]:

$$M_L = 29.6 \sqrt{Q} \tag{7-54}$$
$$M_b = 84.7 \sqrt{Q} \tag{7-55}$$
$$W = 4.88 \sqrt{Q} \tag{7-56}$$

Similar equations with different coefficients have been prepared for incised rivers.

Experiments reported by the U.S. Army Engineers Waterways Experiment Station [22] indicate the following: Meandering is caused by local bank erosion, increased sediment load, and deposition of some of the coarser material; the radius of curvature of the bends increases with discharge and/or slope; the meanders move downstream; and many irregularities result because of differences in bank and bed material.

If the slope of a stream is excessive or if the discharge is increased to a relatively large magnitude, the local rate of bank scour and deposition may be of sufficient magnitude to cause the stream to braid. In general, *braiding* is associated with steeper slopes and larger sediment loads than meandering [23].

G. Sediment Transport

The transport of sediment in natural channels may be classified in three ways: (1) suspension, (2) contact, and (3) saltation. Generally speaking, the median diameter of the suspended load is smaller than that of the bed load or contact load, and since saltation is a transitional type of movement between suspension and contact, its median diameter is intermediate.

There are two types of sediment transported in streams: (1) the *bed-material load*, which includes all sizes of sediment found in appreciable quantities in the bed material, and (2) the *fine-sediment load*, or *wash load*, which is not found in appreciable quantities in the bed material, usually supplied from bank erosion or some external upstream source such as overland flow.

The magnitude of suspended, contact, and bed load which comprise the bed-material load varies with the physical size, fall velocity, and gradation of the bed material, as well as with other fluid, flow, and geometric variables, many of which are interrelated. For example, the magnitude of the bed-material load is intimately related to the form roughness (such as ripples, dunes, and antidunes). The form roughness is in turn related to the fall velocity of the bed material. The fall velocity varies with the apparent viscosity of the sediment-water mixture, and the apparent viscosity varies with such variables as the temperature of the sediment-water mixture, the concentration of bed-material load, and the geological properties of the sediments [24]. It is possible to change the fall velocity of the bed material by changing either the concentration of fine sediment or the temperature of the liquid, or both,

Table 7-7. Variation of Factors with Regimes of Flow and Forms of Bed Roughness*

Regime	Forms of bed roughness	0.28 mm sand					0.45 mm sand				
		Total load concentration, ppm	f	n, ft$^{1/6}$	Fr	$S \times 10^2$	Total load concentration, ppm	f	n, ft$^{1/6}$	Fr	$S \times 10^2$
Lower flow regime	Plane	0	0.0301	0.016	0.17	0.011	0	0.0359	0.016	0.18	0.015
	Ripples	1–150	0.0635–0.1025	0.02–0.027	0.17–0.37	0.023–0.11	1–100	0.0521–0.1330	0.020–0.028	0.14–0.28	0.016–0.11
	Dunes	150–800	0.0612–0.0791	0.021–0.026	0.32–0.44	0.09–0.16	100–1,200	0.0489–0.1490	0.019–0.033	0.28–0.65	0.06–0.30
Transition		1,000–2,400	0.0250–0.0344	0.014–0.017	0.55–0.67	0.13–0.17	1,400–4,000	0.0415–0.0798	0.016–0.022	0.61–0.92	0.37–0.49
Upper flow regime	Plane	1,500–3,100	0.0244–0.0262	0.013–0.014	0.71–0.92	0.15–0.28	4,000–7,000	0.0200–0.0406	0.011–0.015	1.0–1.6	0.36–0.62
	Standing waves	....	....	....	1.0–1.3	....	6,000–15,000	0.0247–0.0292	0.012–0.014	1.4–1.7	0.66–1.0
	Antidunes	5,000–42,000	0.0281–0.0672	0.014–0.022	....	0.33–1.0					

* After Simons and Richardson [25].

NOTATION: f = Darcy-Weisbach's resistance coefficient, n = Manning's roughness coefficient, Fr = Froude number, and S = slope.

to such an extent that the form roughness may change, for example, from dunes to plane bed, or vice versa [24]. Such changes significantly affect bed-material transport.

The importance of the size of bed material and the form of bed roughness on bed-material discharge is clearly illustrated in Table 7-7, which is based upon laboratory flume data.

1. Suspended Load. Lane and Kalinske [26] proposed a practical method of computing suspended load in alluvial streams. By assuming a Prandtl-Kármán logarithmic velocity distribution this method gives the following equation for the total suspended load q_s carried through the stream section per unit of time per unit of width:

$$q_s = qC_{sa}\xi e^{15\chi a/D} \qquad (7\text{-}57)$$

where q is the water discharge per unit of width; C_{sa} is the average concentration of suspended sediment at distance a above the bed; χ is the ratio of sediment-settling velocity to the value of $\sqrt{gDS}$, in which D is the depth of flow and S is the slope of the bed; and ξ is a function of χ and the relative roughness $n/D^{1/6}$, in which n is the

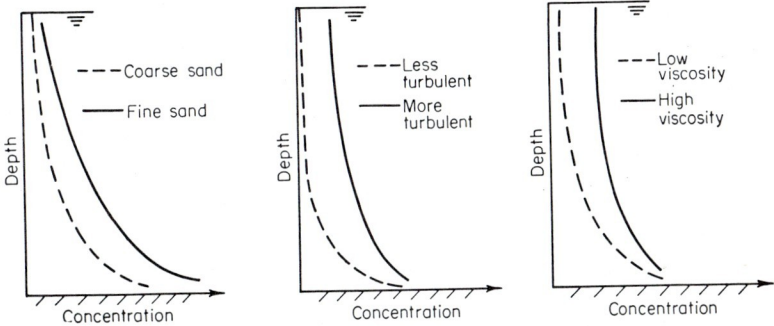

FIG. 7-12. Qualitative effects of size of suspended sediment, turbulence, and viscosity on the vertical distribution of suspended sediment.

Manning roughness coefficient. More fundamentally, Eq. (17-II-18) provides a basis for computing the distribution of sediment concentration in the vertical and the suspended sediment load. To apply this theory it is essential to know the concentration of suspended sediment C_a at distance a above the bed and the exponent

$$z = \frac{v_s}{\kappa u_*} \qquad (7\text{-}58)$$

where v_s is the fall velocity of the particles, κ is the von Kármán universal constant, and u_* is the shear velocity.

The qualitative effects of size of suspended material, turbulence, and viscosity on the vertical distribution of the suspended sediment load are illustrated in Fig. 7-12.

2. Bed Load. Various empirical formulas such as those proposed by Chang, Schoklitsch, MacDougall, O'Brien, and Meyer-Peter (see list of formulas in [27]) were developed along the lines of the classical formula of Du Boys. Currently, such bed-load equations as those of Kalinske [28], Einstein, Meyer-Peter, and Müller (Sec. 17-II) are widely used. The Kalinske and Einstein equations are best suited to sand-channel streams, whereas the Meyer-Peter and Müller equations are more specifically applicable to streams with bed material coarser than sand, such as in semi-Alpine and Alpine streams.

3. Total Sediment Load. The total sediment load includes both the bed-material load and the wash load. The total bed-material load can be estimated by applying the so-called Einstein's *bed-load function*, which is fully discussed in Sec. 17-II. The total sediment load including wash load can be estimated by the modified

Einstein procedure [29]. The major disadvantage of the modified Einstein method is that the suspended load should be measured, which excludes its direct application to design problems.

Other important bed-material-load relations have been proposed by Bagnold [30] and Colby [31]. Bagnold's concepts stress the relation of stream energy to bed-material transport, whereas Colby related velocity of flow to bed-material discharge

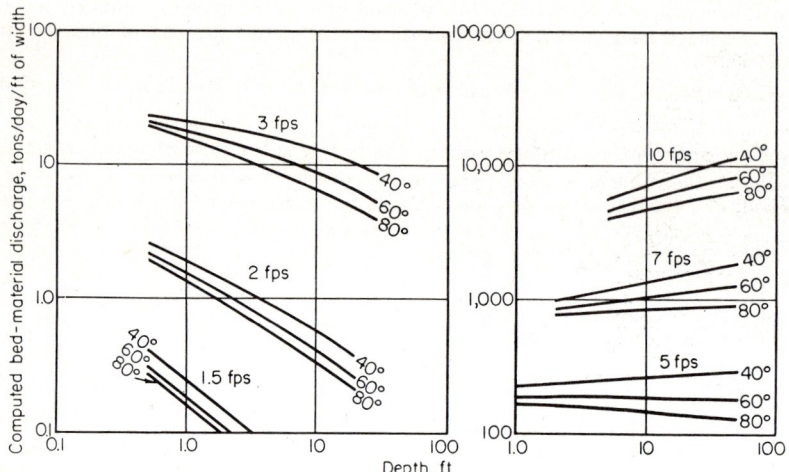

FIG. 7-13. Qualitative effects of depth on the relation between mean velocity and discharge of bed material (0.4-mm median diameter). (*After Colby* [31].)

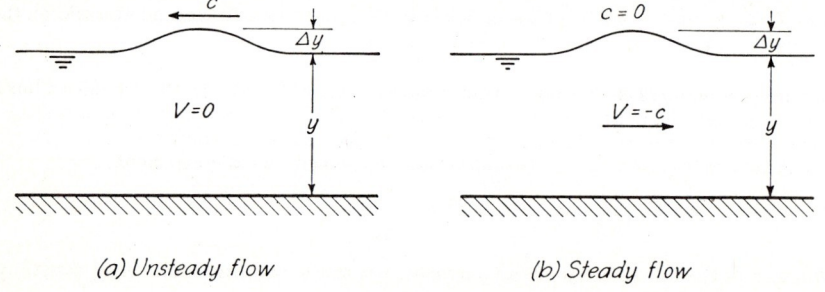

(a) Unsteady flow (b) Steady flow

FIG. 7-14. Small gravity waves.

and shows the significant effect of depth of flow on transport of bed material (Fig. 7-13).

H. Simple Waves and Surges

Various types of waves and surges may occur in open channels and cause a locally unsteady flow. The simplest is the small surface wave, which is observed to progress radially outward from the point of a small disturbance. The rate at which this wave progresses in *still* water is called *celerity c*, as indicated in Fig. 7-14a. The unsteady flow in Fig. 7-14a can be transformed into steady flow simply by superposing on the system a velocity equal in magnitude but opposite in direction to the celerity, as shown in Fig. 7-14b.

FLOW IN OPEN CHANNELS

Assuming the energy loss in the wave to be negligible, the energy equation, together with the continuity equation, can be applied to the steady flow in Fig. 7-14b to develop a solution for the rate of propagation of a solitary wave in terms of the depth of flow y and the height of the wave Δy:

$$V = c = \sqrt{g} \left[\frac{2(y + \Delta y)}{1 + y/(y + \Delta y)} \right]^{1/2} \tag{7-59}$$

When the wave height Δy is small compared with the depth y, this equation reduces to

$$c = \sqrt{gy} \left(1 + \frac{3\Delta y}{2y} \right)^{1/2} \tag{7-60}$$

Furthermore, as the ratio of wave height Δy to depth y approximates zero, Eq. (7-60) becomes

$$c = \sqrt{gy} \tag{7-61}$$

which is the celerity for *shallow-water waves*. The celerity for *deep-water waves* is, however, independent of the depth y, or approximately equal to

$$c = \sqrt{\frac{g\lambda}{2\pi}} \tag{7-62}$$

where λ is the length of the wave and $\lambda < y/2$. As the amplitude or height of the deep-water wave increases, this equation must be modified until the limiting condition $2\Delta y/\lambda = \frac{1}{7}$ is reached, where the wave becomes unstable and breaks, as commonly illustrated by ocean or lake waves generated by the wind. (See also Subsec. 2-IV-C.)

In Eqs. (7-59) to (7-61), it can be seen that, if the velocity of a stream is equal to or greater than the celerity of the wave, the wave cannot move upstream; that is, when $V > c$, the wave moves downstream, and when $V = c$, the wave is stationary. For *very small stationary waves*

$$V = \sqrt{gy} \tag{7-63}$$

which may be rearranged as

$$\frac{V}{\sqrt{gy}} = 1 \tag{7-64}$$

This equation is the principal definition of critical flow; that is, critical flow occurs when the velocity of flow is just equal to the celerity of a small wave in quiet water at the same depth. Equation (7-64) is the Froude number Fr for critical-flow conditions. The general equation of the Froude number for a liquid-gas interface is

$$\text{Fr} = \frac{V}{\sqrt{gy}} \tag{7-65}$$

which shows that the Froude number is the ratio of the velocity of flow to the celerity of a very small gravity wave. When Fr < 1.0, the wave can move upstream, and when Fr > 1.0, the wave is carried downstream.

As the magnitude of the wave height Δy is increased, Eq. (7-60) (for a small wave) becomes less and less applicable and, at $\Delta y \approx y$, the wave becomes unstable and breaks. This may be understood by considering a large wave or surge as a series of small incremental surges of height Δy, as shown in Fig. 7-15a. Each of these small individual surges has a celerity corresponding to the depth of flow over which it is moving; that is, at the left side the depth is y_1, so that the celerity c is

$$c = \sqrt{gy_1} \tag{7-66}$$

while at the right side the depth is $y_2 = y_1 + n \, \Delta y$ and the celerity c is

$$c = \sqrt{gy_2} \tag{7-67}$$

Since $y_2 > y_1$, the incremental surges on the right have a celerity to the left which is greater than those surges on the left. Hence the surges on the right will overtake those on the left, and a very steep surge will develop. As the height of this surge

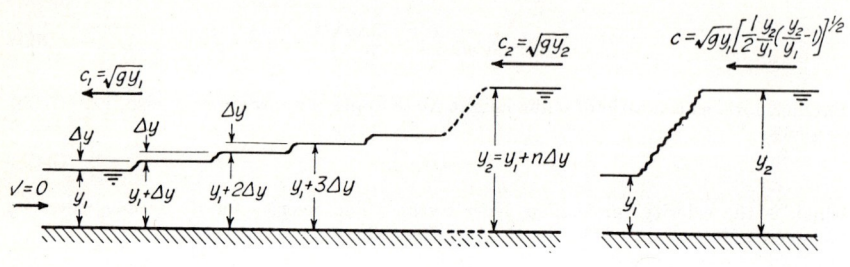

FIG. 7-15. Analysis of the breaking surge. (a) A series of surges; (b) breaking surge, or bore.

$n \, \Delta y$ increases, it becomes undulatory, and as it approaches y_1, the flow at the face of the surge becomes unstable and breaks, as shown in Fig. 7-15b.

I. Hydraulic Jump

The breaking surge can be changed to steady flow by superposing on it a velocity of flow to the right which is equal in magnitude and opposite in direction to the net celerity of the breaking surge. Consequently, a *standing surge*, or *hydraulic jump*, will result (Fig. 7-3). The relationship between the depths of flow before and after the jump (respectively known as *initial depth* and *sequent depth*) and the Froude number of the incoming flow can be expressed by Eq. (7-25) as derived previously. When $Fr_1 = 1$ (critical flow), this equation gives $y_1 = y_2$. Equation (7-25), however, is not simple to apply, and therefore, for ease of computation, the following approximate equation may be used:

$$\frac{y_2}{y_1} = 1.4 \, Fr_1 - 0.4 \tag{7-68}$$

By application of the momentum and continuity equations, it can be shown that the depths y_1 and y_2 are related to the velocity of approach V_1 as

$$V_1 = \sqrt{gy_1} \left[\frac{1}{2} \frac{y_2}{y_1} \left(\frac{y_2}{y_1} + 1 \right) \right]^{\frac{1}{2}} \tag{7-69}$$

Figure 7-16 is a dimensionless representation of the characteristics of the hydraulic jump. The following significant facts may be observed from a study of this representation:
 1. When $Fr_1 < 1.0$, no hydraulic jump can exist, because the celerity of a wave is greater than the velocity of flow.
 2. When $1.0 < Fr_1 < 1.7$ and $y_2/y_1 < 2.0$, there is very little head loss, because the depth ratio y_2/y_1 is less than 2, and hence the surge is undulatory and not breaking.
 3. When $Fr_1 > 2$ and $y_2/y_1 > 2.4$, the surge is breaking and the head loss h_L increases very rapidly. This shows that the hydraulic jump is an excellent energy dissipator. The head loss can be easily determined from the energy equation.

FLOW IN OPEN CHANNELS

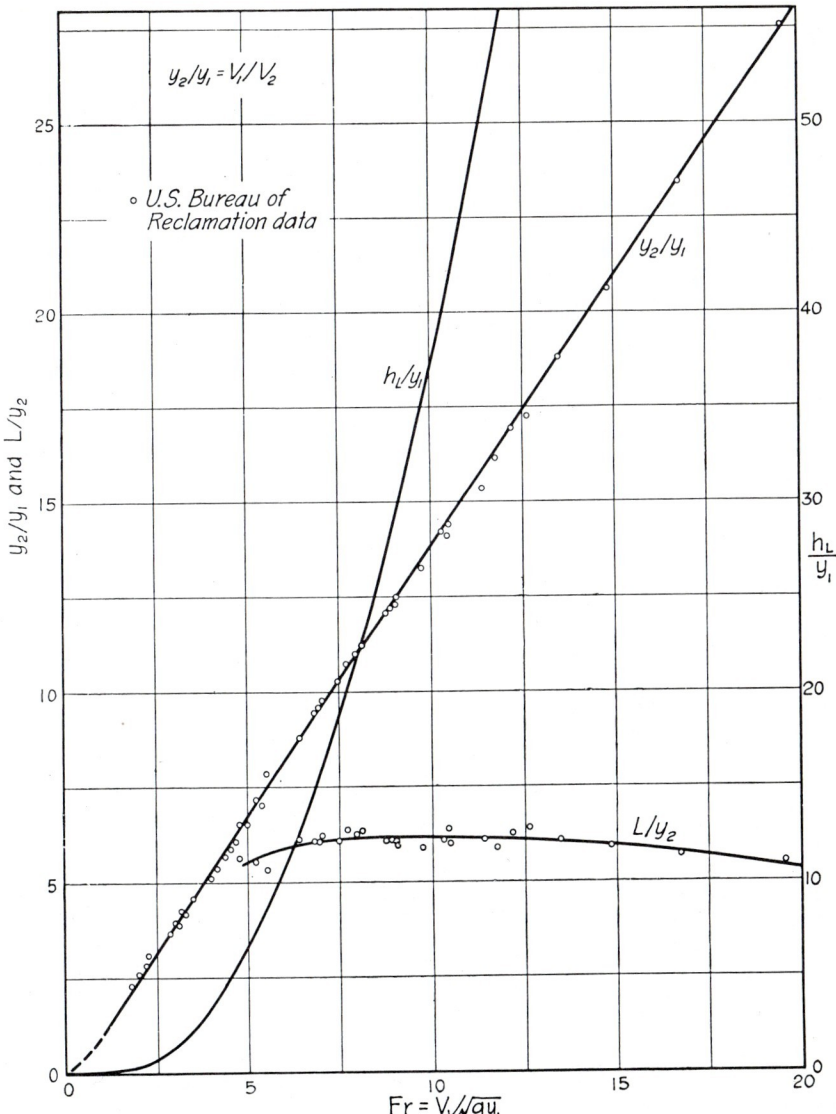

FIG. 7-16. Dimensionless characteristics of hydraulic jump.
$$y_2/y_1 = V_1/V_2$$
y_1 = initial depth, y_2 = sequent depth, V_1 = initial velocity, V_2 = sequent velocity, L = length of jump, h_L = head loss. (*U.S. Bureau of Reclamation data.*)

4. The relative length of the hydraulic jump L/y_2 is nearly constant at about 5.5 to 6.0 beyond Fr = 5.

The hydraulic jump may occur as a *moving surge* which is traveling either upstream if $V < c$ or downstream if $V > c$. To analyze this phenomenon, the flow system is first changed to steady flow by superposing on the entire system a velocity which is equal in magnitude and opposite in direction to the absolute velocity of the surge. Once the flow is steady, the problem can be analyzed by the usual procedure for the hydraulic jump. By doing so it can be shown that the celerity c of a moving surge is equal to V_1 represented by Eq. (7-69).

A wave caused by a sudden release of additional water upstream in the channel may develop into a moving surge, and Eq. (7-69) is applicable. If the channel is dry, a sudden release of flow may be represented approximately by

$$c \approx 1.5 \sqrt[3]{gq} \approx 2 \sqrt{g\,\Delta y} \tag{7-70}$$

where c is the velocity (celerity) of the wavefront, q is the steady discharge introduced into the channel, and Δy is the height of the wave.

J. Transitions in Open Channels

Flow through relatively short transitions in open channels can be described by the energy equation, since the resistance forces over these short distances are relatively small. As shown in Fig. 7-17, an increase in velocity caused by a reduction in the

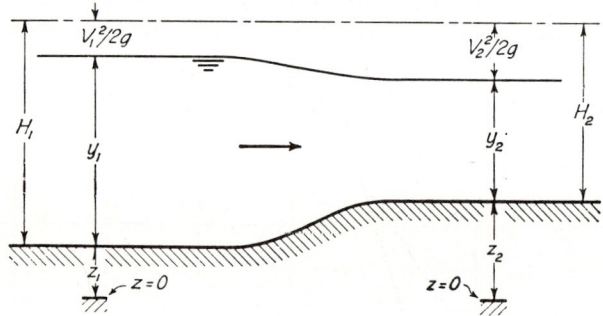

Fig. 7-17. Flow in transitions.

area of cross section, where either the floor is raised or the sides contracted, results in an increase in velocity head, and hence a decrease in water-surface elevation. The energy equation

$$\frac{V_1^2}{2g} + y_1 + z_1 = \frac{V_2^2}{2g} + y_2 + z_2 \tag{7-71}$$

is applicable to this flow pattern and may be used to determine the drop in water-surface elevation.

In connection with a detailed study of Fig. 7-17 and various other transitions, Eq. (7-71) can be modified as

$$H_1 + z_1 = H_2 + z_2 \tag{7-72}$$

in which, ignoring the subscripts and expressing in general terms,

$$H = \frac{V^2}{2g} + y \tag{7-73}$$

The new term H is known as the *specific head* (also known as *specific energy*), or the total head above the floor of the channel. As will be seen in the following discussion, the specific head is a very useful tool in analyzing the flow in open channels. By use of the continuity equation, Eq. (7-73), flow in a rectangular channel can be expressed in terms of q, the discharge per unit width of the channel, as

$$H = \frac{q^2}{2gy^2} + y \qquad (7\text{-}74)$$

This equation can be analyzed from two viewpoints:
1. Variation of y with H when q is held constant, which yields a so-called *specific-head diagram* (Fig. 7-18).
2. Variation of y and q when H is held constant, which yields a *discharge diagram* (Fig. 7-19).

K. Specific-head Diagram

The specific-head equation [Eq. (7-74)] can be plotted as shown in Fig. 7-18 to show how the specific head H varies with the depth of flow y for progressively increasing values of discharge per unit width: q_1, q_2, q_3, etc. This diagram shows that two different depths can exist with a given specific head H and discharge q. In Fig. 7-18, for example, the depth at A is small where the velocity is great, and the other depth at B is great where the velocity is small. These depths are termed *alternate depths*, because they can occur at the same specific head, but independently of each other, depending only upon the boundary conditions of the channel. Also of significance is the fact that there is a minimum value of specific head for a given discharge, such as at C_1, C_2, C_3 in Fig. 7-18.

It can be shown that this minimum specific head corresponds to the condition of a critical flow. Thus the depth of flow for

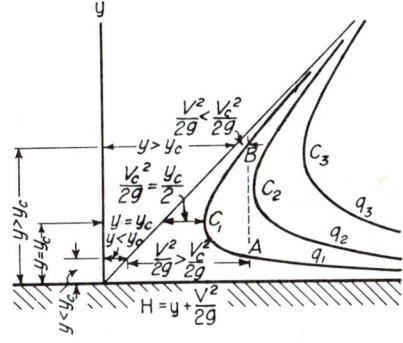

Fig. 7-18. Specific-head diagrams.

the minimum value of the specific head H is equal to the critical depth y_c. In a rectangular channel, the critical depth can be evaluated by differentiating Eq. (7-74) with respect to y and setting it equal to zero and rearranging to yield

$$q = \sqrt{gy_c^3} \qquad (7\text{-}75)$$

From Eqs. (7-75) and (7-73),

$$y_c = \sqrt[3]{\frac{q^2}{g}} = 2\frac{V_c^2}{2g} = \tfrac{2}{3}H \qquad (7\text{-}76)$$

where V_c is the critical velocity.

For nonrectangular channels the equation for critical velocity V_c is

$$V_c = \sqrt{\frac{gA_c}{K_eB_c}} \qquad (7\text{-}77)$$

where A_c and B_c are, respectively, the cross section and top width of the critical flow, and K_e is the energy-flux correction coefficient. The critical depth y_c must be determined as that depth which corresponds to the critical area A_c in a plot of A versus y. The ratio A_c/B_c is an average depth. For a rectangular cross section, this average depth is identical with the critical depth. From Eq. (7-77), the velocity head $K_eV_c^2/2g$ for critical flow can be shown to be equal to half the average depth A_c/B_c. In a rectangular channel with $K_e = 1$, it follows that $V_c^2/2g = y_c/2$ (Fig. 7-18).

L. Discharge Diagram

When the discharge q in Eq. (7-74) is plotted as a function of the depth of flow y for a constant specific head H, the resulting curve as shown in Fig. 7-19 forms a *discharge diagram*. This curve indicates a maximum discharge q_{max}. By differentiating q in Eq. (7-74) with respect to y and setting $dq/dy = 0$, it can be shown that this maximum discharge occurs at the critical-flow condition and is equal to

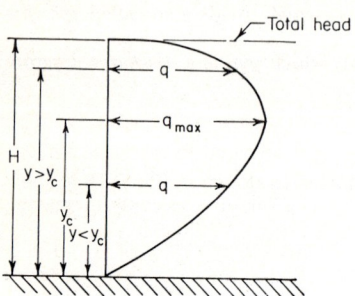

FIG. 7-19. Discharge diagram.

$$q_{max} = \sqrt{g(\tfrac{2}{3}H)^3} = \sqrt{gy_c^3} \qquad (7\text{-}78)$$

M. Gradually Varied Flow

When the cross section of flow in an open channel varies gradually along the channel so that the resulting changes in velocity take place very slowly, and thus the accelerative effects are negligible, the flow is known as *gradually varied flow*. The water surface of a gradually varied flow is called the *flow profile*, or commonly known as the *backwater curve*.

Changes in cross section of the flow may result either from a change in geometry of the channel, such as a change in slope or cross-sectional shape, or an obstruction; or from an unbalance between the forces of resistance tending to retard the flow and the forces of gravity tending to accelerate the flow.

There are several types of flow profiles. In order to analyze these profiles, the total head H_T at a channel section can be expressed as

$$H_T = K_e \frac{V^2}{2g} + y + z = K_e \frac{Q^2}{2gA^2} + y + z \qquad (7\text{-}79)$$

where K_e is the energy-flux correction coefficient, y is the depth of flow, z is the elevation of the channel bed above some arbitrary datum, Q is the discharge, and A is the cross section of the flow. Since the variation of these terms with distance x along the channel is desired, assuming $K_e = 1$, Eq. (7-79) can be differentiated with respect to x to obtain

$$\frac{dH_T}{dx} = -\frac{Q^2}{gA^3}\frac{dA}{dx} + \frac{dy}{dx} + \frac{dz}{dx} \qquad (7\text{-}80)$$

Let $dA = B\,dy$, where B is the top width of the cross section of flow. Then

$$\frac{dH_T}{dx} = -\frac{Q^2 B}{gA^3}\frac{dy}{dx} + \frac{dy}{dx} + \frac{dz}{dx} \qquad (7\text{-}81)$$

The gradient of total head dH_T/dx can be set equal to the negative of the slope obtained from the Chézy equation, or $S = (Q/A)^2/C^2R$, and the bed slope is equal to $dz/dx = -(Q/A_0)^2/C_0^2R_0 = -S_0$ for uniform-flow conditions. The subscript 0 represents the uniform-flow condition. For simplicity, however, a wide rectangular channel can be assumed, so that $Q/B = q$ equal to the discharge per unit width and the hydraulic radius $R = A/B = y$. Equation (7-81) then becomes

$$-\frac{q^2}{C^2 y^3} = \frac{dy}{dx}\left(1 - \frac{q^2}{gy^3}\right) - \frac{q^2}{C_0^2 y_0^3} \qquad (7\text{-}82)$$

Furthermore, $q^2/g = y_c^3$, so that Eq. (7-82) can be rearranged to solve explicitly for dy/dx, which is the rate of change of the depth of flow with respect to the distance along the channel. Thus

$$\frac{dy}{dx} = \frac{q^2/C_0^2 y_0^3 - q^2/C^2 y^3}{1 - (y_c/y)^3} \qquad (7\text{-}83)$$

which simplifies to

$$\frac{dy}{dx} = S_0 \frac{1 - (C_0/C)^2 (y_0/y)^3}{1 - (y_c/y)^3} \qquad (7\text{-}84)$$

If the change in the Chézy C is not great from one point to another along the channel, the ratio C_0/C can be considered equal to 1.0. However, the Manning n is usually more nearly constant from section to section. Hence Eq. (7-51) can be used in Eq. (7-84) to yield

$$\frac{dy}{dx} = S_0 \frac{1 - (n/n_0)^2 (y_0/y)^{19/3}}{1 - (y_c/y)^3} \qquad (7\text{-}85)$$

Using Eq. (7-85), it is possible to classify the various flow profiles which may occur in open channels.

N. Classification of Flow Profiles

The analysis of flow profiles depends first upon the sign of dy/dx. If dy/dx is positive, the depth is increasing downstream, and if it is negative, the depth is decreasing downstream. From Eq. (7-85), it can be seen that the slope dy/dx depends upon S_0, n/n_0, y_0/y, and y_c/y. In the following analysis, it is assumed that $n/n_0 = 1.0$. Although this assumption is not justified for all conditions, it may be taken as sufficiently accurate for the purpose of this analysis. Hence

$$\frac{dy}{dx} = S_0 \frac{1 - (y_0/y)^{19/3}}{1 - (y_c/y)^3} \qquad (7\text{-}86)$$

The slope of the channel serves as the primary means of classification. If the bed slope S_0 is negative, the bed rises in the direction of flow. This slope is called an *adverse slope*, and the flow profiles over it are known as *A profiles*. If $S_0 = 0$, the bed slope is horizontal and the profiles over it are *H profiles*. When $S_0 > 0$, the bed slope may be mild, steep, or critical, and the corresponding flow profiles are *M profiles*, *S profiles*, or *C profiles*, depending upon the ratio y_0/y_c. When $y_0/y_c > 1.0$, an *M* profile exists; when $y_0/y_c = 1.0$, a *C* profile exists; and when $y_0/y_c < 1.0$, an *S* profile exists.

A further classification of flow profiles depends upon the ratios y_c/y and y_0/y. If both y_c/y and y_0/y are less than 1.0, then the profile is designated as type 1, for example, M_1, S_1, and C_1. If the depth y is between the normal depth y_0 and the critical depth y_c, then it is type 2, such as M_2, S_2, H_2, and A_2. If both y_c/y and y_0/y are greater than 1.0, then the profile is type 3, such as M_3, C_3, S_3, H_3, and A_3.

Various flow profiles are shown in Fig. 7-20, where the longitudinal distance has been shortened and the slopes have been exaggerated for the sake of clarity. The general characteristics of the flow profiles are summarized in Table 7-8.

Figure 7-21 illustrates actual conditions where various flow profiles can occur. The vertical scales in the illustrations are again very much exaggerated. A sluice gate could be placed where the dam is located in each case, and the same profiles would exist. It should be observed that a smooth, although relatively sharp, profile is created when it crosses the critical-depth line from a greater to a lesser depth, such as in the A_2 profile and the S_2 profile in Fig. 7-21b. As the profile crosses from a lesser to a greater depth, however, there is a hydraulic jump which causes a sudden change in elevation such as in the downstream sections of Fig. 7-21a, d, and e.

The location of a hydraulic jump is a problem which frequently must be solved. This is accomplished by computing the flow profiles M_2 and M_3, H_3 and H_2, or A_3 and A_2 in the direction of increasing depth until the computed downstream depth reaches the point where it is the sequent depth for the computed upstream depth.

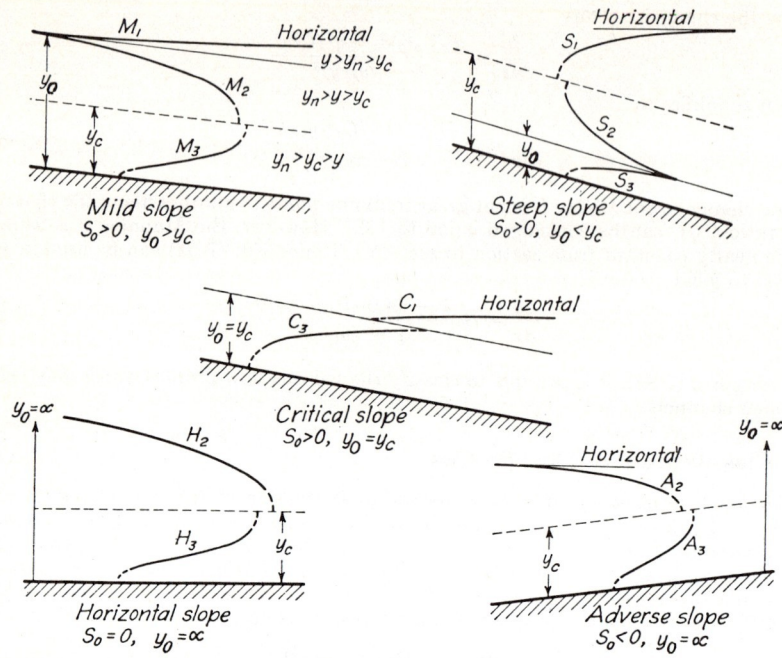

Fig. 7-20. Classification of flow profiles.

In other words, the surge which causes the jump is propagated upstream or downstream until the approaching velocity of flow is just equal to the celerity of the surge in quiet water at the same depth.

In summary, the following statements can be made about the flow profiles:
1. The type 1 profiles are all above both y_c and y_0.
2. The type 2 profiles are all between y_c and y_0.
3. The type 3 profiles are all below both y_c and y_0.
4. The type 3 profiles all approach the bottom of the channel at a certain angle.

Table 7-8. Characteristics of Flow Profiles

Class	Bed slope	$y : y_0 : y_c$	Type	Symbol
Mild............	$S_0 > 0$	$y > y_0 > y_c$	1	M_1
	$S_0 > 0$	$y_0 > y > y_c$	2	M_2
	$S_0 > 0$	$y_0 > y_c > y$	3	M_3
Critical.........	$S_0 > 0$	$y > y_0 = y_c$	1	C_1
	$S_0 > 0$	$y < y_0 = y_c$	3	C_3
Steep...........	$S_0 > 0$	$y > y_c > y_0$	1	S_1
	$S_0 > 0$	$y_c > y > y_0$	2	S_2
	$S_0 > 0$	$y_c > y_0 > y$	3	S_3
Horizontal......	$S_0 = 0$	$y > y_c$	2	H_2
	$S_0 = 0$	$y_c > y$	3	H_3
Adverse.........	$S_0 < 0$	$y > y_c$	2	A_2
	$S_0 < 0$	$y_c > y$	3	A_3

FLOW IN OPEN CHANNELS   7-41

5. The profiles which approach the y_0 depth line approach it asymptotically except for C profiles, where $y_0 = y_c$.

6. The profiles which approach the y_c depth line approach it vertically, except for C profiles, where $y_0 = y_c$.

7. The profiles which approach the horizontal line approach it asymptotically, except for the C_3 profile.

8. The profiles over steep slopes, where $y_0 < y_c$, are unaffected upstream from any disturbance because the wave of the disturbance is swept downstream by the rapid flow.

9. The profiles over mild slopes, where $y_0 > y_c$, are influenced by any downstream disturbance, because the celerity of the wave of the disturbance is so great that it travels upstream against the oncoming flow.

The foregoing discussion of the characteristics of flow profiles in open channels is applicable in general not only to wide uniform channels, but also to natural open

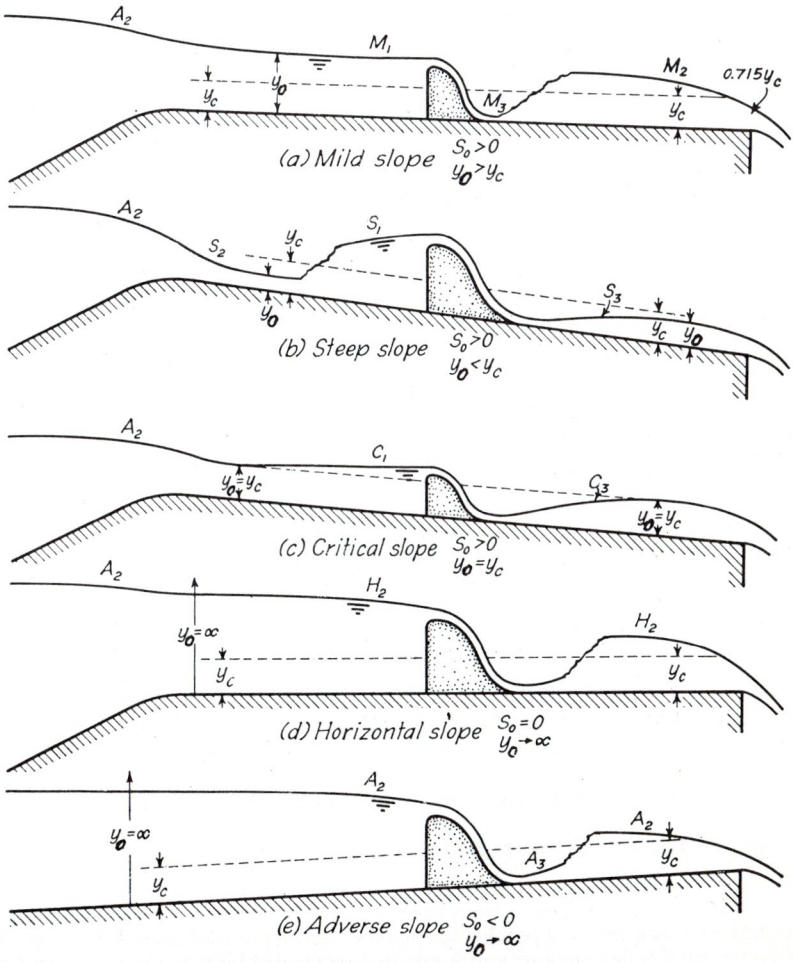

Fig. 7-21. Examples of flow profiles.

channels, provided the changes in shape, roughness, and slope are taken into consideration properly. For detailed discussion see Ref. 18.

O. Computation of Backwater Curves

By integrating Eq. (7-86), a mathematical equation can be obtained to represent the surface profile of a gradually varied flow. For practical purposes, however, a *step method*, described below, is widely used.

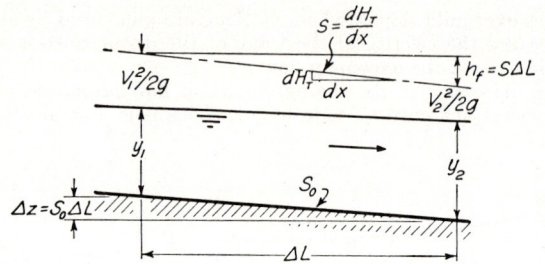

FIG. 7-22. Definition sketch for computation of backwater curves.

Figure 7-22 illustrates a channel of length ΔL which is sufficiently short so that the water surface can be approximated by a straight line. By geometry or from Eq. (7-80), it can be shown that $\Delta H/\Delta L = S_0 - S$, or

$$\Delta L = \frac{\Delta H}{S_0 - S} \qquad (7\text{-}87)$$

where H is the specific head; $S = -dH_T/dx$, the average energy gradient; and

$$S_0 = -\frac{dz}{dx}$$

the slope of the channel bed. The energy gradient at a channel section can be computed by the Manning equation as $S = Q^2 n^2/(1.5 A R^{2/3})^2$. The average of the energy gradients at the two end sections of the reach is used for S in Eq. (7-87).

The step method is characterized by dividing the channel into short reaches and applying Eq. (7-87) by steps from one end of a reach to the other. To start the computation, the depth of flow at the beginning section should be given or assumed. From a given discharge and channel conditions, the specific heads at the two end sections and their difference ΔH and the energy slope at the two end sections and their average are computed. Substituting these quantities and the channel slope S_0 in Eq. (7-87), the length of the reach is computed. By repeating the computation for the subsequent reaches, the entire flow profile or backwater curve can be determined. It should be noted that the step computation should be carried upstream if the flow is tranquil and downstream if the flow is rapid. If carried in the wrong direction, the computation tends inevitably to make the result diverge from the correct flow profile. For a comprehensive treatment of the computation of flow profiles, see Ref. 18.

IX. FLOW AROUND SUBMERGED OBJECTS

A. Shear Drag and Pressure Drag

Flow around submerged objects will develop two basic types of resistance or drag: *shear drag* and *pressure drag*. Pure shear drag is developed by the flow around a flat plate or a disk oriented parallel to the flow. Pure pressure drag is developed by the flow around a flat plate or a disk oriented perpendicularly to the flow. In most problems, however, both types of drag occur.

FLOW AROUND SUBMERGED OBJECTS

The general drag equation for flow around submerged objects is

$$F_D = \frac{C_D A \rho V_0^2}{2} \tag{7-88}$$

where F_D is the drag, C_D is a drag coefficient, A is the projected cross-sectional area of the object in the direction of flow, ρ is the density of the fluid, and V_0 is the velocity of the ambient fluid.

When the Reynolds number is very small, say, Re < 0.5, the flow about a submerged object is laminar and the shape of the object is of secondary importance in regard to the drag as compared with the size of the object, the viscosity of the fluid, and the velocity of flow. The drag coefficients for various objects in laminar flow are shown in Table 7-9.

For purely laminar flow around a sphere, Stokes developed a theory which has been proved by experiment to be accurate. The theory involves the following: (1) The shear drag is two-thirds, and the pressure is one-third, of the total drag; (2) at all points on a sphere the longitudinal components of shear drag and pressure drag are combined to produce the same value of unit total drag over the entire surface of the sphere; and (3) the total drag on the sphere is equal to the product of the surface area of the sphere and the unit total drag, or

$$F_D = \pi d^2 \frac{3\mu V_0}{d} = 3\pi\, d\mu V_0 \tag{7-89}$$

where F_D is the drag, d is the diameter of the sphere, μ is the dynamic viscosity, and V_0 is the terminal velocity of the sphere. When combined with Eq. (7-88), this equation will produce the drag coefficient for a sphere as listed in Table 7-9.

Table 7-9. Drag Coefficients for Laminar Flow

Object	Range of Re	Value of C_D
Sphere	<0.5	24/Re
Disk perpendicular to flow	<0.5	20.4/Re
Disk parallel to flow	<0.1	13.6/Re
Circular cylinder	<0.1	8π/Re (2.0 − ln Re)
Flat plate perpendicular to flow	<0.1	8π/Re (2.2 − ln Re)
Flat plate parallel to flow	<0.01	4.12/Re
	<6 × 10^5	$1.33/\sqrt{\text{Re}} + 4.12/\text{Re}$

NOTATION: Re = $V_0 d/\nu$, where V_0 = ambient velocity; d = diameter or length; ν = kinematic viscosity.

At small Reynolds numbers the influence of inertia is insignificant compared with the influence of viscosity. As the Reynolds number is increased, the influence of inertia becomes increasingly pronounced, until eventually, at large Reynolds numbers, the situation is completely reversed and the influence of viscosity becomes small compared with the influence of inertia. The variations of the drag coefficient with Reynolds number for several submerged objects are shown in Fig. 7-23. It can be seen that the change from one condition to another usually takes place gradually. The sudden decrease in C_D near Re = 2 × 10^5 for rounded objects is caused by a change from a laminar boundary layer to a turbulent boundary layer and by the resulting change in location of the point of separation of flow.

B. Fall Velocity

Frequently, in the analysis of sediment and other falling bodies, the size and weight of particles are known and it is desirable to determine the velocity of fall in a fluid.

Equation (7-89) can be used to determine the fall velocity V_0 of a spherical or nearly spherical particle in a fluid since the drag is the weight of the particle minus the buoyancy force.

At high Reynolds numbers, the velocity must be determined by trial and error from the plot of C_D versus Re in Fig. 7-23. However, a direct solution can be made by

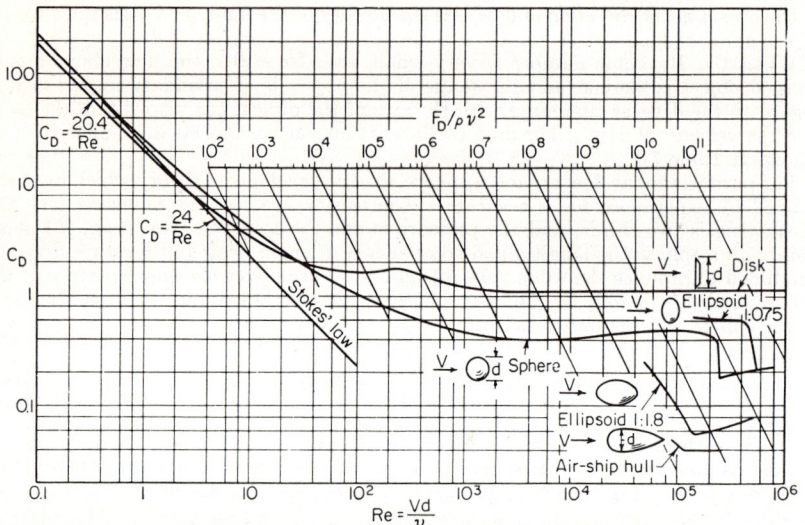

FIG. 7-23. Drag coefficients for spheres and other bodies of revolution.

plotting, with either C_D or Re, a parameter which does not contain the velocity. This has been done in Fig. 7-23 by the scale of $F_D/\rho\nu^2$, which can be obtained by dimensional analysis or by the following relationship:

$$C_D (\text{Re})^2 = \frac{F_D}{A\rho V^2/2}\left(\frac{Vd}{\nu}\right)^2 \sim \frac{F_D}{\rho\nu^2} \qquad (7\text{-}90)$$

At any point on the diagonal broken lines for constant values of $F_D/\rho\nu^2$, the corresponding values of C_D and Re are those required to satisfy the particular values of F_D, ρ, and ν. In fact, the $F_D/\rho\nu^2$ scale can be employed to determine the fall velocity of any object by using the submerged weight of the object as F_D and following down the proper $F_D/\rho\nu^2$ line to the curve for the shape of object involved, from which either C_D or Re can be determined to solve for the fall velocity.

X. FLOW MEASUREMENT AND CONTROL

There are many devices and methods for measuring the discharge of a fluid, and most of them can or do function also as a flow control [1, pp. 439–481, and 11]. Only those most frequently used, however, are discussed below.

A. Orifices

An orifice is an opening with a closed perimeter through which a fluid flows. The upstream edge of an orifice may be rounded or sharp, and the cross section may be any shape, such as circular, square, or rectangular. An orifice with prolonged sides in the direction of flow is called a *tube*, or *nozzle*, which may have a constant or variable cross section. An orifice in a flat plate can be inserted in a pipeline to form an *orifice meter* to measure the flow in the line.

The general equation for the discharge Q through an orifice or nozzle is

$$Q = C_d A \sqrt{2gH} \tag{7-91}$$

where C_d is a discharge coefficient (which varies with relative size, shape of opening, and Reynolds number), A is the cross-sectional area at the smallest section, and H is the head, or the vertical distance from the center of the orifice to the upstream free-water surface (piezometric head), provided the height of the orifice is small compared with the head. Equation (7-91) applies also to submerged orifices where H is the difference in elevation of the upstream and downstream liquid surfaces. Because of simplicity of design and construction, sharp-edged circular orifices are most common for fluid measurement. The value of C_d for such orifices is usually about 0.61 to 0.65, although, for relatively large orifices and small Reynolds numbers, it may be 0.8 [1, p. 466].

B. Weirs

A weir is an overflow structure built across an open channel for the purpose of measuring the flow. With reference to the shape of the notch through which the liquid flows, weirs may be classified as rectangular, triangular, trapezoidal, circular,

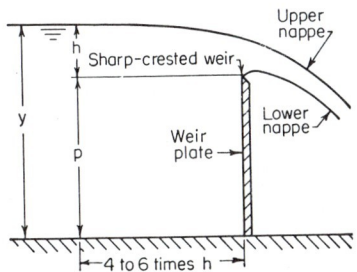

Fig. 7-24. Sharp-crested weir.

parabolic, or of any other geometrical form. With reference to the form of the crest, weirs may be classified as *sharp-crested* or *broad-crested*. The notch of a weir may have one, two, or no *end contractions*, the effect of which is to reduce the flow below that which would occur under the same head over an uncontracted weir of the same length. The flow over a weir may be either *free* or *submerged*, depending upon whether the water surface downstream from the weir is lower or higher than the crest. This submergence does not significantly reduce the discharge of flow over the weir for a small submergence. When the submergence is $0.2h$, the reduction in discharge is about 2 per cent and for $0.5h$, the reduction is about 15 per cent.

1. Sharp-crested Weirs. The *uncontracted*, or *suppressed*, sharp-crested weir (Fig. 7-24) is a standard weir which has the following basic principles of design: (1) The weir plate is vertical, and the upstream face essentially smooth; (2) the crest is horizontal and normal to the direction of flow, and it must be sharp so that the water springs free from the edge in a predictable and reproducible manner; (3) the pressure along the upper and the lower nappe is atmospheric; (4) the approach channel is uniform in cross section, and the water surface is free of surface waves; and (5) the sides of the channel are vertical and smooth, and they extend downstream from the crest of the weir. If the velocity of approach is negligible, the discharge Q in cubic feet per second is expressed by the well-known *Francis formula*

$$Q = 3.33 L h^{3/2} \tag{7-92}$$

where L is the length of the weir in ft, and h is the head on the weir in ft. However, the equation suggested for such weirs [31] is

$$Q = C_d L_e h_e^{3/2} \tag{7-93}$$

where $C_d = 3.22 + 0.40h/P$, in which P is the height of the weir in ft, and

$$L_e = L + K_L$$

is the effective width. K_L is a width-adjustment factor given in Fig. 7-25a; $h_e = h + 0.003$ is the effective head; and h is the piezometric head measured above the crest of the weir. When the weir has side contractions (a weir notch) the magnitude of C_d, which is a function of the contraction ratio b/B and h/P, can be obtained from Fig. 7-25b, and L_e and h_e are determined as for the uncontracted (suppressed) weir.

The *trapezoidal weir* can be considered a combination of a triangular weir and a rectangular suppressed weir. Because there are no calibration data for various side slopes, this weir must be calibrated in place, except for the *Cipolletti weir*, which has side slopes of four vertical to one horizontal. For the Cipolletti weir, the suppressed weir equation [Eq. (7-92)] applies.

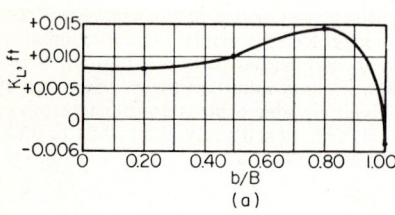
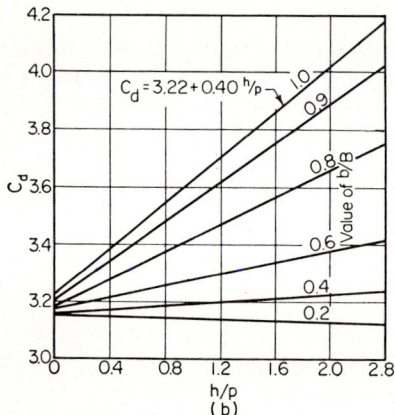

Fig. 7-25. K_L and C_d for contracted sharp-crested weirs. (*After Kindsvater and Carter* [32].)

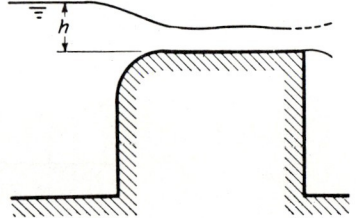

Fig. 7-26. Broad-crested weir.

For a *triangular weir* with a notch angle θ, the discharge can be expressed by

$$Q = C_d \times \tfrac{8}{15} \sqrt{2g} \left(\tan \frac{\theta}{2}\right) h^{2.5} \qquad (7\text{-}94)$$

where the coefficient C_d is determined through experiment. Experiments for $\theta = 90°$ have shown that $Q = 2.5h^{2.5}$ is an accurate representation of the flow.

2. Broad-crested Weirs. When the breadth of crest is less than about $\tfrac{2}{3}h$, the nappe will ordinarily spring clear and the weir is in effect sharp-edged. Greater crest breadth changes the form of the nappe, and sharp-crested weir coefficients no longer apply. The discharge q per unit width of such broad-crested weirs (Fig. 7-26) is $q = C_d h^{1.5}$, where C_d is a discharge coefficient (approximately 3), which can be obtained from tables [32] or curves [33], and h is the height of the upstream water surface above the crest of the weir when the velocity of approach is negligible. See [34].

C. Gates

A gate is a metering device in a pipeline or open channel or an opening in a dam or other hydraulic structure to control the flow. Gates have the hydraulic properties of orifices. In open channels, the flow through a gate may be either free or submerged. The coefficient of discharge varies widely, depending on the design, the points at which the head is measured, and the conditions of flow.

FLOW MEASUREMENT AND CONTROL

There are various types of gates, such as a radial gate or roller gate on the crest of a spillway or in an open channel and a slide gate or butterfly valve in a pipeline. A gate which has its lower edge in or near the bed of a channel is a *sluice gate*. The discharge Q through a sluice gate is

$$Q = C_d A \sqrt{2g\,\Delta y} \tag{7-95}$$

where C_d is a discharge coefficient, depending on the size of the opening, the upstream depth y_1, and the design of the gate; A is the cross-sectional area of the opening under the gate; and Δy is the difference between y_1 and the downstream depth y_2. Equation (7-95) applies also to other types of gate, where Δy represents the difference between upstream and downstream piezometric heads. The value of C_d can be determined by field calibration or by model testing in a laboratory.

D. Overflow Spillways

Overflow spillways in open channels behave as weirs. If the spillway is not submerged, the depth of flow over the crest is the critical depth. A spillway having a rounded-crest shape, similar to the underside of the nappe springing from a sharp-crested weir, is called an *ogee spillway*. Its discharge per unit width is approximately $q = 3.7h^{3/2}$, where h is the height of the upstream water surface above the crest, while the velocity of approaching flow is negligible.

E. Venturi Meter

Venturi meter is a device to measure flow in a pipe. This device consists of a nozzle-like reducer followed by a more gradual enlargement to the original size of the pipe. The reducer reduces the pipe cross section to a constricted "throat" and thus produces an increased velocity accompanied by a reduction in pressure. In the enlargement the velocity is transformed back into pressure with slight friction loss. By applying the energy equation to a section immediately upstream from the entrance section and the throat section, the discharge Q can be shown as

$$Q = \frac{C_v A_2}{[1 - (D_2/D_1)^4]^{1/2}} \sqrt{2g\,\Delta h} \tag{7-96}$$

where C_v is a velocity coefficient, A_2 is the cross-sectional area of the throat, D_1 is the diameter of the upstream section, D_2 is the throat diameter, and Δh is the difference in piezometric head between the upstream section and the throat. The coefficient C_v depends on the Reynolds number at the throat and the diameter ratio D_2/D_1. The loss of energy through a Venturi meter is slight, depending on the Reynolds number and the diameter ratio, but it is approximately 0.1 of the velocity head at the throat. For accuracy in use, the Venturi meter should be preceded by a straight pipe of at least 5 to 10 pipe diameters in length.

F. The Parshall Flume

By applying to open-channel flow a principle similar to that for the Venturi meter, a *Venturi flume*, or *Parshall flume* (Fig. 7-27), has been developed. Under free-flowing (unsubmerged) conditions, the Parshall flume is a critical-depth meter, somewhat like the weir or free overfall over a spillway, which creates a critical depth to establish a relationship between the stage and discharge.

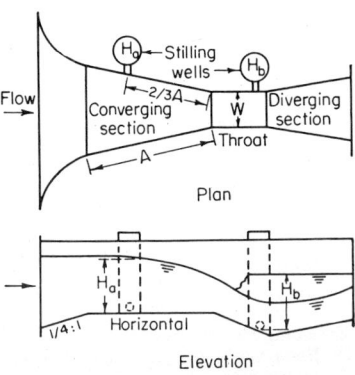

Fig. 7-27. Parshall flume.

Under free-flowing conditions, the discharge in cubic feet per second measured by the Parshall flume of standard dimensions has been calibrated as

$$Q = 4WH_a{}^{1.522W^{0.026}} \quad \text{for } W \text{ from 1 to 8 ft} \tag{7-97}$$

and
$$Q = (3.688W + 2.5)H_a{}^{1.6} \quad \text{for } W \text{ from 8 to 50 ft} \tag{7-98}$$

In these two equations, W is the throat width in ft and H_a is the upstream head in ft (Fig. 7-27). When the flume is operating under submerged conditions with $H_b/H_a >$ 0.75, it is necessary to measure the head H_b downstream as well as H_a upstream for the determination of the discharge.

A recent study of the Parshall flume [35] shows that the relation

$$y_0 + \frac{Q^2}{2y_0{}^2(1 + 0.4x_0)^2} = 1.351Q^{0.645} \tag{7-99}$$

closely fits available Parshall-flume data, which indicates its applicability for determining discharge in flumes of all sizes. In this relation b is the channel width where H_a is measured, x_1 is the distance from the throat to the point where H_a is measured, y_1 is the depth at the measuring section H_a, x_0 is the nondimensional distance x_1/b, and y_0 is the nondimensional distance y_1/b.

Equation (7-99) has the advantage that it permits use of the pressure opening for the stilling well to be set wherever desirable. Despite the fact that this equation is somewhat awkward to use, it is more general than the equations previously presented.

G. Contracted Opening

Flow through a contraction, such as a bridge or culvert, in a stream can provide a means of measuring the discharge of a flood. The same principles are involved as for the Venturi flume. Although a head loss is usually considered to take place from upstream to the section where the area is measured, this loss can usually be ignored, because it is very small compared with the errors which are bound to occur in even the best of the rough field measurements which are possible. The best measurements are obtained if they are made while the water is flowing through the contracted opening. In this case, care should be taken to discount any water which is in a separation zone immediately adjacent to the boundary of the contracted opening. In other words, it is necessary to measure the cross-sectional opening of the water which is flowing at a high velocity through the contracted opening. If it is necessary to make the measurements after the flow has passed, each of the variables is extremely difficult to obtain accurately.

XI. REFERENCES

1. Albertson, M. L., J. R. Barton, and D. B. Simons: "Fluid Mechanics for Engineers," Prentice-Hall, Inc., Englewood Cliffs, N.J., 1960.
2. Streeter, V. L. (ed.): "Handbook of Fluid Dynamics," McGraw-Hill Book Company, Inc., New York, 1961.
3. Vennard, J. K.: "Elementary Fluid Mechanics," 4th ed., John Wiley & Sons, Inc., 1961.
4. Streeter, V. L.: "Fluid Mechanics," 3d ed., McGraw-Hill Book Company, Inc., New York, 1962.
5. Jaeger, Charles: "Engineering Fluid Mechanics," Blackie & Son, Ltd., Glasgow, 1956 (transl. by P. O. Wolf).
6. Rouse, Hunter (ed.): "Engineering Hydraulics," John Wiley & Sons, Inc., New York, 1950.
7. Davis, C. V.: "Handbook of Applied Hydraulics," 2d ed., McGraw-Hill Book Company, Inc., New York, 1952.
8. King, H. W., and E. F. Brater: "Handbook of Hydraulics," 5th ed., McGraw-Hill Book Company, Inc., New York, 1963.
9. Daugherty, R. L., and A. C. Ingersoll: "Fluid Mechanics," 5th ed., McGraw-Hill Book Company, Inc., New York, 1954.

REFERENCES   7-49

10. Langhaar, H. L.: "Dimensional Analysis and Theory of Models," John Wiley & Sons, Inc., New York, 1951.
11. Howe, J. W.: Flow measurement, chap. 3, in Hunter Rouse (ed.), "Engineering Hydraulics," John Wiley & Sons, Inc., New York, 1950.
12. Steel, E. W.: "Water Supply and Sewerage," 4th ed., McGraw-Hill Book Company, Inc., New York, 1960.
13. Babbitt, H. E., J. J. Doland, and J. L. Cleasby: "Water Supply Engineering," 6th ed., McGraw-Hill Book Company, Inc., New York, 1962.
14. McIlroy, M. S.: Pipe networks studied by nonlinear resistors, *Trans. Am. Soc. Civil Engrs.*, vol. 118, pp. 1055–1067, 1953.
15. McPherson, M. B.: Water distribution design and the McIlroy network analyzer, *Proc. Am. Soc. Civil. Engrs., J. Hydraulics Div.*, vol. 84, no. HY2, pp. 1–15, April, 1958.
16. Richardson, E. V., D. B. Simons, and W. L. Haushild: Boundary form and resistance to flow in alluvial channels, *U.S. Geol. Surv. Profess. Paper*, 1961.
17. Simons, D. B., and E. V. Richardson: Forms of bed roughness in alluvial channels, *Proc. Am. Soc. Civil Engrs., J. Hydraulics Div.*, vol. 87, no. HY3, pt. 1, pp. 87–105, May, 1961.
18. Chow, V. T.: "Open-channel Hydraulics," McGraw-Hill Book Company, Inc., New York, 1959.
19. Lane, E. W.: Design of stable channels, *Trans. Am. Soc. Civil Engrs.*, vol. 120, pp. 1234–1260, 1955.
20. Simons, D. B., and M. L. Albertson: Uniform water conveyance channels in alluvial material, *Proc. Am. Soc. Civil Engrs., J. Hydraulics Div.*, vol. 86, no. HY5, pp. 33–71, May, 1960.
21. Inglis, C. C.: "The Relationship between Meandering Belts, Distance between Meanders on Axis of Stream, Width and Discharge of Rivers in Flood Plains and Incised Rivers," Government of India, Central Board of Irrigation and Power, Annual Report, 1938–1939, New Delhi.
22. Friedkin, J. F.: "Laboratory Study of the Meandering of Alluvial Rivers (Studies Conducted from 1942 to 1944)," U.S. Army Corps of Engineers Waterways Experiment Station, Vicksburg, Miss., May, 1945.
23. Leopold, L. B., and H. G. Wolman: River channel patterns: brained, meandering and straight, *U.S. Geol. Surv. Profess. Paper* 282-B, 1951.
24. Simons, D. B., E. V. Richardson, and W. L. Haushild: Some effects of fine sediment on flow phenomena, *U.S. Geol. Surv. Water-Supply Paper* 1498-G, 1963.
25. Simons, D. B., and E. V. Richardson: Forms of bed roughness in alluvial channels, *Proc. Am. Soc. Civil Engrs., J. Hydraulics Div.*, vol. 87, no. HY3, pp. 87–105, May, 1961.
26. Lane, E. W., and A. A. Kalinske: Engineering calculations of suspended sediment, *Trans. Am. Geophys. Union*, vol. 22, pp. 603–607, 1941.
27. Brown, C. B.: Sediment transportation, chap. 12, in Hunter Rouse (ed.), "Engineering Hydraulics," John Wiley & Sons, Inc., New York, 1950.
28. Kalinske, A. A.: Movement of sediment as bed load in rivers, *Trans. Am. Geophys. Union.*, vol. 28, pp. 615–620, August, 1947.
29. Colby, B. R., and C. H. Hembree: Computations of total sediment discharge, Niobrara River, near Cody, Nebraska, *U.S. Geol. Surv. Water-Supply Paper* 1357, 1955.
30. Bagnold, R. A.: The flow of cohesionless grains in fluids, *Phil. Trans. Roy. Soc. London*, vol. 249, no. 964, pp. 235–297, 1956.
31. Colby, B. R.: Effect of depth of flow on discharge of bed material, *U.S. Geol. Surv. Water-Supply Paper* 1498-D, 1961.
32. Kindsvater, C. E., and R. W. Carter: Discharge characteristics of rectangular thin-plate weirs, *Trans. Am. Soc. Civil Engrs.*, vol. 124, pp. 772–801, 1959.
33. King, H. W., and E. F. Brater: "Handbook of Hydraulics," 4th ed., McGraw-Hill Book Company, Inc., New York, 1954.
34. Tracy, H. J.: Discharge characteristics of broad-crested weirs, *U.S. Geol. Surv. Circ.* 397, 1957.
35. Davis, Sidney: Unification of Parshall flume data, *Proc. Am. Soc. Civil Engrs., J. Irrigation and Drainage Div.*, vol. 87, no. IR4, December, 1961.

Section 8-I

STATISTICAL AND PROBABILITY ANALYSIS OF HYDROLOGIC DATA

PART I. FREQUENCY ANALYSIS

VEN TE CHOW, *Professor of Hydraulic Engineering, University of Illinois.*

I. Introduction...	8-2
A. Importance of Statistical and Probability Analysis..........	8-2
B. Hydrologic Frequency Studies............................	8-3
1. Flood and Streamflow Studies.........................	8-3
2. Rainfall Studies.....................................	8-4
3. Drought and Low Streamflow Studies...................	8-4
4. Water Quality Studies................................	8-4
5. Water Wave Studies..................................	8-4
II. Fundamentals...	8-4
A. Statistical Variables.....................................	8-4
B. Frequency, Probability, and Statistical Distributions........	8-5
1. For Discrete Random Variables........................	8-5
2. For Continuous Random Variables.....................	8-6
C. Statistical Parameters...................................	8-6
1. Measures of Central Tendency........................	8-6
2. Measures of Variability..............................	8-7
3. Measures of Skewness...............................	8-8
D. Statistical Moments.....................................	8-8
E. Hydrologic Models, Processes, and Systems................	8-9
F. Statistical Homogeneity..................................	8-10
1. Time Homogeneity—Trend, Periodicity, and Persistence..	8-10
2. Space Homogeneity..................................	8-13
III. Probability Distributions.....................................	8-13
A. Rectangular Distribution.................................	8-13
B. Binomial Distribution....................................	8-13
C. Poisson Distribution.....................................	8-14
D. Normal Distribution.....................................	8-14
E. Gamma Distribution.....................................	8-14
F. Pearson Distributions....................................	8-14
1. Type I Distribution...................................	8-15
2. Type III Distribution.................................	8-15
G. Extremal Distributions..................................	8-16
1. Type I Distribution...................................	8-18

2. Type II Distribution ... 8-16
3. Type III Distribution ... 8-16
H. Logarithmically Transformed Distributions ... 8-17
1. Lognormal Distribution ... 8-17
2. Logextremal Distributions ... 8-17
3. Truncated Lognormal Distributions ... 8-17
IV. Procedure of Analysis ... 8-17
A. Treatment of Raw Data ... 8-18
1. Data Sampling ... 8-18
2. Observation Errors ... 8-18
3. Inherent Defectiveness ... 8-18
B. Selection of Data Series ... 8-19
C. Recurrence Interval ... 8-22
D. Frequency Analysis Using Frequency Factors ... 8-23
1. General Equation for Hydrologic Frequency Analysis ... 8-23
2. The K-T Relationship ... 8-23
E. Probability Paper ... 8-27
F. Plotting of Data ... 8-28
G. Curve Fitting ... 8-29
1. Method of Moments ... 8-30
2. Method of Least Squares ... 8-31
3. Method of Maximum Likelihood ... 8-31
H. Reliability of Analysis ... 8-31
1. Sampling Reliability ... 8-31
2. Prediction Reliability ... 8-34
I. Theoretical Justifications ... 8-34
1. Type I Extremal Distribution ... 8-35
2. Lognormal Distribution ... 8-35
3. Exponential Distribution ... 8-35
4. Logextremal Distribution ... 8-35
J. Regional Analysis ... 8-36
V. References ... 8-37

I. INTRODUCTION

A. Importance of Statistical and Probability Analysis

Quantitative scientific data may be classified into two kinds: experimental data and historical data. The *experimental data* are measured through experiments and usually can be obtained repeatedly by experiments. The *historical data*, on the other hand, are collected from natural phenomena that can be observed only once and then will not occur again. Most hydrologic data are historical data which were observed from natural hydrologic phenomena.

Since hydrologic data are the only source of information upon which quantitative hydrologic investigations are generally based, their measurements have been continuously expanding and resulting in ever-increasingly large amounts of sampled data. *Statistics* deals with the computation of sampled data, and *probability* deals with the measure of chance or likelihood based on the sampled data. The mounting quantities of hydrologic data can suitably be expressed in statistical terms and be treated with probability theories. Furthermore, natural hydrologic phenomena are highly erratic and commonly stochastic in nature, and therefore are amenable to statistical interpretation and probability analysis. Section 8 covers some fundamental principles and methods of statistics and probability that are useful in the solution of hydrologic problems.

One of the important problems in hydrology deals with interpreting a past record of hydrologic events in terms of future probabilities of occurrence. This problem arises

INTRODUCTION   8–3

in the estimates of frequencies of floods, droughts, storages, rainfalls, water qualities, waves, etc; the procedure involved is known as *frequency analysis*. The methods of frequency analysis and some fundamentals in statistics and probability are discussed in this Part I of Section 8.

For general discussions on statistical and probability analysis of hydrologic data, reference may be made to Refs. 1-4. For frequency analysis of hydrologic data in particular, Refs. 5-8 may be found to be useful.

B. Hydrologic Frequency Studies

1. Flood and Streamflow Studies. The frequency analysis of streamflow data is believed to have been first applied to flood studies by Herschel and Freeman (see [9]) in 1880 to 1890 by means of a graphical procedure of using flow-duration curves (Subsec. 14-V-A). According to Fuller [10], the use of probability methods in runoff studies had been suggested to him in 1896 by George W. Rafter. Owing to the dearth of long-period records on American rivers at that time, the use of probability methods for flood frequency analysis was apparently hindered until later years.

The Gaussian law of probability, or the normal law of errors, is the basic and simplest tool for frequency analysis. It was therefore used for flood studies in the very early days. For such studies, Horton [11] discussed briefly in 1913 the earlier applications of the Gaussian law, and in 1914 Fuller [10] gave a full account of the first really comprehensive study of statistical methods applied to floods in the United States.

However, Hazen [12] soon discovered that if the logarithms representing the annual floods are used instead of the numbers themselves, the agreement with the normal law of errors is closer. This is true because the frequency distributions of annual floods are usually skewed or asymmetrical and the distribution can be suitably represented by such frequency distribution laws as the Galton, or lognormal-probability, law. He proposed the use of lognormal-probability paper [13] and developed a procedure of analysis [14]. Hazen's method requires a table of factors for computing theoretical frequency curves by means of the coefficients of variation and skewness. The table [13, p. 219; 14, pp. 49, 188] was originally obtained by empirical methods and hence has been found to be inaccurate. A corresponding table of exact factors based on a mathematical procedure was later prepared by Chow [15] (Table 8-I-1). For the study of streamflow variability, Lane and Lei [16] made use of the lognormal-probability plotting of flood flows to determine the variability index (Subsec. 14-V-A).

Other laws of frequency distribution and methods of frequency analysis of floods were also proposed by many hydrologists. Type 1 and Type 3 of Karl Pearson's curves of frequency distribution were put in a form convenient for use in flood studies by Foster [17]. A table of frequency factors similar to Hazen's table was given by Foster and extended by Switzer and Miller [18]. Hall [19] proposed a special "hydraulic probability paper" in which the probability scale was obtained empirically from flow-duration curves of 35 California streams. Goodrich [20] proposed a special skew-frequency paper which was later tested and refined by Harris [21]. Up to 1934, Slade [22] derived various skew probability functions to which was introduced an ultimate limiting magnitude of flood flow or the limiting flood potentialities of the drainage basin.

In 1941, Gumbel [23] published the first of a great number of papers (e.g., Refs. 24–29) on the application of the Fisher-Tippett theory of extreme values to flood frequency analysis. The use of extreme-value theory has been further extended by other hydrologists. Powell [30] derived an extremal probability paper for graphical application of the method. Cross [31] soon applied it to the study of flood frequencies in Ohio. As the extremal distribution assumes a constant skewness, the variate of a given recurrence interval should theoretically depend on the coefficient of variation and the mean. Potter [32] applied this assumption to 370 extremal probability curves and derived practical graphic relationships between variate, mean, and coefficient of variation. Benson [33] developed a synthetic "1,000-year record" of peak floods based on a straight-line plotting on the extremal probability paper.

Both the lognormal-probability law and the extreme-value law have been used extensively in recent years. From a theoretical point of view, Chow [15] has shown that the extreme-value law is practically a special case of the lognormal-probability law, or it is practically identical with the latter for a skewness coefficient of 1.139 and a coefficient of variation equal to 0.364. He also proposed a flexible straight-line fitting of flood data based on the merits of both methods. See Sections 14 and 25-I and Refs. 4–8 and 34–49 for other discussions of flood frequencies.

2. Rainfall Studies. Many frequency studies on rainfalls and other meteorological events have been made. An extensive rainfall frequency study in the United States was first made by Yarnell [50] in 1935 (Subsec. 9-VI). This study produced the well-known *Yarnell rainfall frequency data*, which are a set of 56 isohyetal maps of the continental United States, covering the range of durations of 5, 10, 30, 60, and 120 min for 2-year frequency and of the same durations plus 4, 8, 16, and 24 hr for 5, 10, 25, 50, and 100-year frequencies. For longer durations of 1, 2, 3, 4, 5, and 6 days, the Miami Conservancy District [51] published the data for 15, 25, 50, and 100-year frequencies in 1936, covering that part of the continental United States east of the 103d meridian.

Since Gumbel proposed the Type I extremal distribution for flood frequency analysis, Chow [52] applied it to the study of the rainfall-intensity frequency in Chicago, Illinois in 1953, and also published a design chart [53] for approximate determination of rainfall frequencies in the continental United States.

As more rainfall data were collected, extended and detailed rainfall frequency analyses were made by the U.S. Weather Bureau. In 1961, a rainfall frequency atlas of the United States was published by the Weather Bureau, which completely revises and supersedes the Yarnell data ([54]; Subsec. 9-VI).

Comparisons of several methods of rainfall frequency analysis have been made by Huff and Neill [55] and by Hershfield [56].

3. Drought and Low Streamflow Studies. Type III extremal distribution was first proposed by Gumbel [57] for drought frequency analysis in 1954. The method was later applied to actual problems [29, 58], including graphical applications to Michigan streams [59] and to streams in eastern United States [60].

Other frequency studies of droughts and low streamflows are discussed in Section 18 and Refs. 61–63.

4. Water Quality Studies. Frequency analysis has been applied to virus, bacteria, alkalinity, salinity, chlorides, sulfates, and other dissolved and undissolved materials in water. Some recent studies are discussed in Ref. 64.

5. Water Wave Studies. Frequency analysis of water waves constitutes its own unique field, as it has its special purposes in oceanography. Important developments in this field may be found in Refs. 65–69. The Type I extremal distribution also has been applied to wave frequency analysis first by Gumbel [70] and Jasper [71], and later by Bennet [72] and others.

II. FUNDAMENTALS

A. Statistical Variables

Hydrologic data can be treated as statistical variables. In statistics, the whole collection of objects under consideration is called a *population*, or *universe*. A segment of a population may have one or more characteristics associated with them. Their characteristics are called *variables*, usually designated as X. An individual observation or the value x of any variable X is known as a *variate*. In hydrologic phenomena, for example, the variable X may be the depth of rainfall, and it may have a value, say, $x = 1.45$ in.

Variables may be obtained by an experiment consisting of random operations known as *trials*. The result of an unspecified trial is called a *random variable*. The collection of all possible values for the random variables associated with an experiment is called a *sample space*. In hydrologic phenomena, the observations for a certain period may be considered as a trial. By this trial, the rainfall depth, for example, is obtained as a random variable. Since the value of the rainfall depth can have all possible nonnegative values, the sample space is infinite.

FUNDAMENTALS

Random variables are of two kinds: *discrete* and *continuous*. The discrete random variable has finite sample space, whereas the sample space of a continuous variable has an interval of real numbers or a union of such intervals. For example, the number of rainy days is a discrete random variable, while the depth of rainfall is a continuous random variable. For practical purposes, however, it is sometimes necessary to treat arbitrarily the discrete variables as continuous variables by fitting a continuous function to the variates, or vice versa by breaking down the continuous variable into intervals and then grouping them as discrete numbers.

B. Frequency, Probability, and Statistical Distributions

1. For Discrete Random Variables. For discrete random variables, the number of occurrences of a variate is generally called *frequency*. When the number of occurrences, or the frequency, is plotted against the variate as the abscissa, a pattern of distribution is obtained. This pattern is called *frequency distribution* (Fig. 8-I-1a).

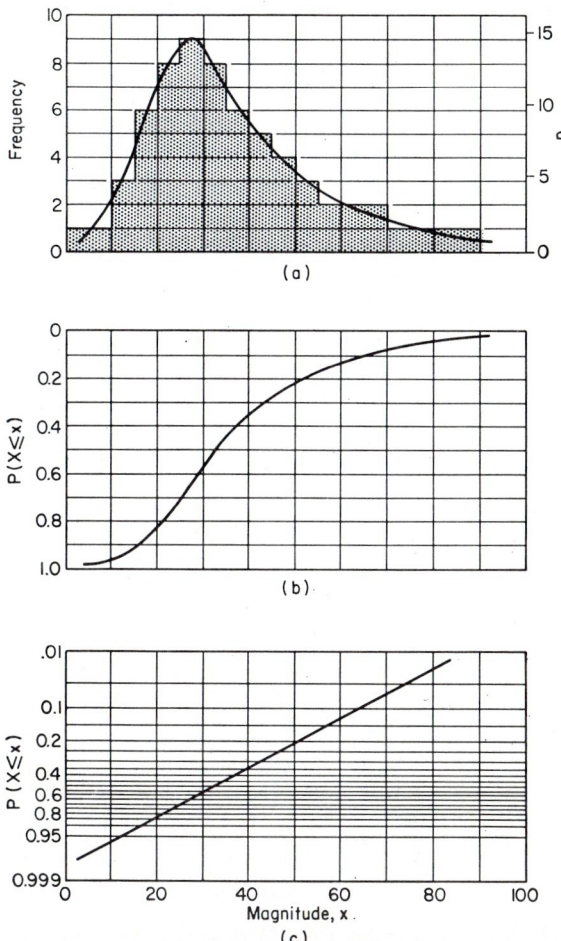

Fig. 8-I-1. Linearization of a statistical distribution. (a) Frequency or probability distribution curve; (b) cumulative probability curve plotted on rectangular coordinates; (c) cumulative probability curve linearized on probability paper.

When the number of occurrences of a discrete variate is divided by the total number of occurrences, the result is a probability p of the variate. The total probability for all variates should be equal to unity, or $\Sigma p = 1$. Distribution of the probabilities of all variates, instead of their frequencies, is called *probability distribution* (Fig. 8-I-1a).

The ordinates of the frequency distribution and its corresponding probability distribution are obviously proportional to each other. Both distributions may be called *statistical distributions*.

The *cumulative probability* of a variate (Fig. 8-I-1b) is the probability that the random variable has a value equal to or less than certain assigned value, say x. This cumulative probability may be designated as $P(X \leq x)$. Thus, the probability of being equal to or greater than x is equal to $1\text{-}P(X \leq x)$ or designated by $P(X \geq x)$.*

2. For Continuous Random Variables. For continuous random variables, the probability of a variate can be considered as the probability $p(x)$ of a discrete value grouped in the range from x to $x + \Delta x$. As x is a continuous value or Δx becomes dx, the probability $p(x)$ becomes a continuous function called *probability density*. The cumulative probability $P(X \leq x)$ is an integral function of the probability density (Fig. 8-I-1b), or

$$P(X \leq x) = \int_{-\infty}^{x} p(x)\, dx \tag{8-I-1}$$

where the probability distribution is considered unlimited. When the upper limit of $x = \infty$, $P(X \leq x) = 1$. If the probability distribution is limited, or the probability density $p(x)$ is defined only for a range of $a \leq x \leq b$, the above equation is also valid by assuming that $p(x) = 0$ for values outside the range of x.

C. Statistical Parameters

The characteristics of a statistical distribution may be described by *statistical parameters*. While such parameters are many, only the important ones are defined below:

1. Measures of Central Tendency. The parameters generally representing measures of the central tendency of a statistical distribution are the *averages*, including mean, median, and mode.

a. The Mean. There are three kinds of mean: arithmetic, geometric, and harmonic.

The familiar *arithmetic mean* is usually referred to simply as the *mean* and is designated by

$$\bar{x} = \frac{\Sigma x}{N} \tag{8-I-2}$$

where x is the variate and N is the total number of observations. The above equation gives the sample mean, while the population mean is usually denoted by μ. It may be noted that an unbiased estimate of the population mean is equal to the sample mean.

The *geometric mean* is the Nth root of the product of N terms or can be designated by

$$\bar{x}_g = (x_1 \cdot x_2 \cdot x_3 \cdots x_N)^{1/N} \tag{8-I-3}$$

The logarithm of the geometric mean is obviously the mean of the logarithms of the individual values. In a lognormal distribution, the geometric mean has the properties analogous to those of the arithmetic mean of a normal distribution.

The *harmonic mean* is the reciprocal of the mean value of the reciprocals of individual values. It can be expressed as

$$\bar{x}_h = \frac{N}{\Sigma(1/x)} \tag{8-I-4}$$

---

* Theoretically, for discrete variables, $1 - P(X \leq x) = P(X > x)$ or $1 - P(X < x) = P(X \geq x)$, since $P(X < x) + P(X = x) + P(X > x) = 1$; and for continuous variables, $1 - P(X < x) = P(X > x)$, since $P(X = x)$ is infinitesimal. For practical purposes, $P(X \leq x) + P(X \geq x) = 1$ is acceptable for both discrete and continuous variables.

FUNDAMENTALS
8–7

b. The Median. The *median* is the middle value or the variate which divides the frequencies in a distribution into two equal portions.

The arithmetic mean is more commonly used than other measures of central tendency as a "normal" or "standard" on account of its computational simplicity and, in general, its greater sampling stability. For example, the U.S. Weather Bureau has a long-established practice of using the mean as the precipitation normal. However, in extremely skew distributions the mean may be misleading. In such cases, the median will provide a better indication, particularly for a continuous variable because all variates greater or less than the median always occur half the time. Also, the use of median makes it easy to locate an internal estimate by adding and subtracting a specified amount from this central value so that the portions of the distribution outside the interval have the same probability.

c. The Mode. In a distribution of discrete variables, the *mode* is the variate which occurs most frequently. In a distribution of continuous variables, this is the variate which has a maximum probability density, i.e., $dp/dx = 0$ and $d^2p/dx^2 < 0$.

2. Measures of Variability. The important parameters representing variability or dispersion of a distribution are mean deviation, standard deviation, variance, range, and coefficient of variation.

a. The Mean Deviation. The mean of the absolute deviations of values from their mean is called *mean deviation*, or

$$\text{M.D.} = \frac{\Sigma |x - \bar{x}|}{N} \qquad (8\text{-I-}5)$$

This parameter was used frequently to describe meteorological data, but it has been now superseded largely by the standard deviation.

b. The Standard Deviation. This parameter as a measure of variability is most adaptable to statistical analysis. It is the square root of the mean-squared deviation of individual measurements from their mean and is designated by

$$\sigma = \sqrt{\frac{\Sigma (x - \mu)^2}{N}} \qquad (8\text{-I-}6)$$

This equation represents the *standard deviation* of the population. An unbiased estimate of this parameter from the sample is denoted by s and computed by

$$s = \sqrt{\frac{\Sigma (x - \bar{x})^2}{N - 1}} = \sqrt{\frac{N}{N - 1} (\overline{x^2} - \bar{x}^2)} \qquad (8\text{-I-}7)$$

where $\overline{x^2} = (\Sigma x^2)/N$.

The standard deviation of the sampling distribution of a statistical parameter is known as the *standard error* of that parameter. It can be shown that the standard error of the mean is $\sigma/\sqrt{N}$, the standard error of the standard deviation is $\sigma/\sqrt{2N}$, and the standard error of the difference between the means of samples from two independent populations is $\sqrt{\sigma_{\bar{x}}^2 + \sigma_{\bar{y}}^2}$ where $\sigma_{\bar{x}} = \sigma_x/\sqrt{N_1}$ and $\sigma_{\bar{y}} = \sigma_y/\sqrt{N_2}$ with σ_x and σ_y equal to the standard deviations from the two populations and N_1 and N_2 equal to the numbers of variates sampled from the respective populations.

c. The Variance. The square of the standard deviation is called *variance*, which is denoted by σ^2 for the population. The unbiased estimate of the population variance is s^2.

d. The Range. The difference between the largest and the smallest values is the *range*.

e. The Coefficient of Variation. The standard deviation divided by the mean is called the *coefficient of variation*, or

$$C_v = \frac{\sigma}{\mu} \approx \frac{s}{\bar{x}} \qquad (8\text{-I-}8)$$

FREQUENCY ANALYSIS

3. Measures of Skewness. The lack of symmetry of a distribution is called *skewness* or *asymmetry*. The statistical parameter to measure this property is the *skewness* defined as

$$\alpha = \frac{1}{N} \sum (x - \mu)^3 \qquad (8\text{-I-}9)$$

This equation represents the skewness for the population. An unbiased estimate of this parameter from the sample is

$$a = \frac{N}{(N-1)(N-2)} \sum (x - \bar{x})^3$$

$$= \frac{N^2}{(N-1)(N-2)} (\overline{x^3} - 3\overline{x^2}\bar{x} + 2\bar{x}^3) \qquad (8\text{-I-}10)$$

where $\overline{x^3} = (\Sigma x^3)/N$ and other notations have been defined previously.

One commonly used measure of skewness is the *coefficient of skewness* represented by

$$C_s = \frac{\alpha}{\sigma^3} \approx \frac{a}{s^3} \qquad (8\text{-I-}11)$$

For a symmetrical distribution, $C_s = 0$. A distribution with $C_s > 0$ is said to be skewed to the right (with a long tail on the right side), while a distribution with $C_s < 0$ is said to be skewed to the left.

Another measure of skewness often used in practice is *Pearson's skewness*, or

$$S_k = \frac{\mu - \text{mode}}{\sigma} \approx \frac{\bar{x} - \text{mode}}{s} \qquad (8\text{-I-}12)$$

D. Statistical Moments

In a statistical distribution, the rth moment about the origin $x = 0$ of the variates $x_1, x_2, \ldots, x_k$, having a weighted mean $\bar{x}$, is

$$\nu_r = \frac{1}{N} \sum_{i=1}^{k} p_i x_i^r \qquad (8\text{-I-}13)$$

where p_i is the frequency or probability of x_i and $N = \Sigma p_i$ with $i = 1, \ldots, k$.

The rth (central) moment about the weighted mean $\bar{x}$ of the variates $x_1, x_2, \ldots, x_k$, is

$$\mu_r = \frac{1}{N} \sum_{i=1}^{k} p_i (x_i - \mu)^r \qquad (8\text{-I-}14)$$

For the first three moments, with $r = 1, 2,$ and 3, it can be shown that

$$\begin{aligned}\mu_1 &= 0 \\ \mu_2 &= \nu_2 - \nu_1^2 = \sigma^2 \\ \mu_3 &= \nu_3 - 3\nu_2\nu_1 + 2\nu_1^3 = \sigma^3 \alpha\end{aligned} \qquad (8\text{-I-}15)$$

and

$$\begin{aligned}\nu_1 &= \bar{x} \\ \nu_2 &= \mu_2 + \nu_1^2 \\ \nu_3 &= \mu_3 + 3\mu_2\nu_1 + \nu_1^3\end{aligned} \qquad (8\text{-I-}16)$$

The above equations show that the mean is equal to the first moment about the origin, the standard deviation is the square root of the second moment about the mean, and the skewness is the third moment about the mean divided by the cube of the standard deviation. For a detailed discussion of the statistical moments, see Refs. 8, 73, and 74.

Moments of order higher than three are not commonly used in the statistical analysis of hydrologic data because most hydrologic data do not have sufficiently long length of record and thus cannot warrant reliable estimates of the moments of higher order.

E. Hydrologic Models, Processes, and Systems

Hydrologic models considered here are mathematical formulations to simulate natural hydrologic phenomena which are considered as processes or as systems.

Any phenomenon which undergoes continuous changes particularly with respect to time may be called a *process*. As practically all hydrologic phenomena change with time, they are *hydrologic processes*. If the chance of occurrence of the variables involved in such a process is ignored and the model is considered to follow a definite law of certainty but not any law of probability, the process and its model are described as *deterministic*. On the other hand, if the chance of occurrence of the variables is taken into consideration and the concept of probability is introduced in formulating the model, the process and its model are described as *stochastic* or *probabilistic*. For example, the conventional routing of flood flow through a reservoir is a deterministic process (Sec. 25-II), and the mathematical formulation of the unit-hydrograph theory (Subsec. 14-III) is a deterministic model. As the probability of the flow is taken into account in the probability routing (Subsec. 14-V-C), the process and the queuing model employed to simulate the process are considered as stochastic or probabilistic.

Strictly speaking, a stochastic process is different from a probabilistic process, as the former is generally considered as time-dependent and the latter as time-independent. For the time-independent *probabilistic process*, the sequence of occurrence of the variates involved in the process is ignored and the chance of their occurrence is assumed to follow a definite probability distribution in which variables are considered pure-random. For the time-dependent *stochastic process*, the sequence of occurrence of the variates is observed and the variables may be either pure-random or non-pure-random, but the probability distribution of the variables may or may not vary with time. If *pure-random*, the members of the time series are independent among themselves and thus constitute a random sequence. If *non-pure-random*, the members of the time series are dependent among themselves, are composed of a deterministic component and a pure-random component, and thus constitute a nonrandom sequence. For example, the flow-duration-curve procedure (Subsec. 14-V-A) is probabilistic, whereas the probability routing mentioned above is stochastic.

In reality, all hydrologic processes are more or less stochastic. They have been assumed deterministic or probabilistic only to simplify their analysis. Mathematically speaking, a *stochastic process* is a family of random variables $X(t)$ which is a function of time (or other parameters) and whose variate x_t is running along in time t within a range T. Quantitatively, the stochastic process, which may be discrete or continuous, can be sampled continuously or at discrete or uniform intervals of $t = 1, 2, \ldots$, and the values of the sample form a sequence of $x_1, x_2, \ldots$, starting from a certain time and extending for a period of T. This sequence of sampled values is known as a *time series*, which may be discrete or continuous. For example, a hydrograph is a continuous time series. Daily, monthly, and annual discharges represent a discrete time series.

The random variable $X(t)$ has a certain probability distribution. If this distribution remains constant throughout the process, the process and the time series are said to be *stationary*. Otherwise, they are *nonstationary*. For example, a virgin flow (Subsec. 14-I) with no significant change in river-basin characteristics or climatic conditions for the period of record is considered as a stationary time series. If it is affected by man's activities in the river basin or nature's large accidental or slow modifications of the rainfall and runoff conditions, the recorded or historical flow is a nonstationary time series. Since a nonstationary process is very complicated mathematically, hydrologic processes are generally treated as stationary.

For clarity, the classification of hydrologic processes may be shown below. It may be noted that actual hydrologic processes are processes following the path of the heavy line while the processes following the thin lines are only approximations which may be assumed in order to simplify the analysis.

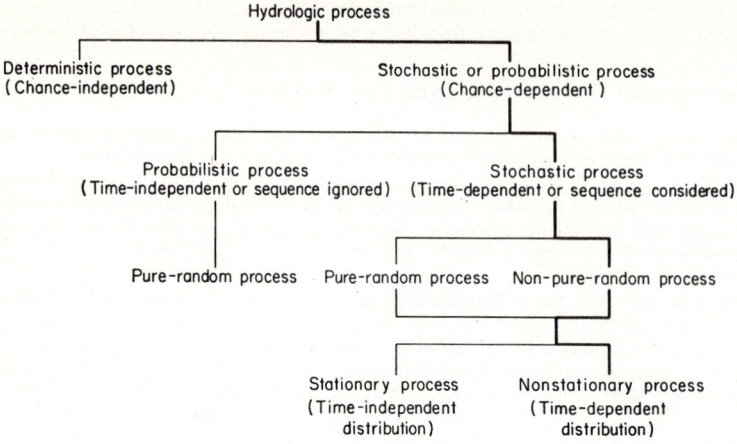

In this section, the probabilistic models and the frequency analysis of the probabilistic process are mainly discussed, since the stochastic process and various stochastic models are covered in several other sections, including Section 8-III (moving averages, sum of harmonics, autoregression, correlograms), Section 8-IV (Markov process), Section 18 (queuing theory), Section 14 (ranges, queuing theory, theory of storage), and Section 26-II (stochastic programming models).

A *system* is an aggregation or assemblage of objects united by some form of regular interaction or independence. The system is said to be *dynamic* if there is a process taking place in it. If the process is considered probabilistic or stochastic, the system is said to be *stochastic*. Otherwise, it is a *deterministic system*. Furthermore, the system is called *sequential* if it consists of *input*, *output*, and some working fluid (matter, energy, or information) known as *throughput* passing through the system. The hydrologic cycle or a drainage basin is a sequential, dynamic system in which water is a major throughput. Since a stochastic system is very complicated analytically, the hydrologic system has been generally treated as deterministic and its formulation by deterministic models, such as instantaneous unit hydrographs, has been proposed (Subsec. 14-III).

For detailed mathematical discussions on stochastic processes and on systems, see Refs. 75-82.

F. Statistical Homogeneity

The nature of homogeneity in hydrologic processes can be examined statistically with respect to time and space.

1. Time Homogeneity—Trend, Periodicity, and Persistence. A process or time series may be considered *time-homogeneous* if the identical events under consideration in the series are equally likely to occur at all times. Thus, purely random and stationary processes or time series are time-homogeneous. In hydrology, strictly time-homogeneous data are practically nonexistent because various kinds of variations of natural or artificial causes exist in most hydrologic phenomena. However, such variations, if appreciable, may be analyzed by various techniques.

Types of departure from true time homogeneity in hydrologic data may be roughly classified as trend, periodicity, and persistence.

a. Trend. This is a unidirectional diminishing or increasing change in the average value of a hydrologic variable, such as the trend of annual precipitation that is often plainly visible on a plotted graph. A number of statistical techniques may be used to determine the trend. A commonly used method is to analyze the trend by the method of moving averages (Subsec. 8-III-II-B).

FUNDAMENTALS

b. Periodicity. This represents a regular or oscillatory form of variations, such as diurnal, seasonal, and secular changes that exist frequently in hydrologic phenomena. Such variations are of nearly constant length and they may be assumed sinusoidal and determined by harmonic analysis.

In the *harmonic analysis* a Fourier series is used to represent the time series $x_1, x_2, \ldots, x_N$ of a total period of length T:

$$x_t = \tfrac{1}{2} A_0 + \sum_{j=1}^{T/2} \left(A_j \cos \frac{360 jt}{T} + B_j \sin \frac{360 jt}{T} \right) \quad (8\text{-}I\text{-}17)$$

where A_0 is a constant, t is the time, and the coefficients A_j and B_j are *amplitudes* being expressed by

$$A_j = \frac{2}{N} \sum_{t=1}^{N} y_t \cos \frac{360 jt}{T} \quad (8\text{-}I\text{-}17a)$$

$$B_j = \frac{2}{N} \sum_{t=1}^{N} y_t \sin \frac{360 jt}{T} \quad (8\text{-}I\text{-}17b)$$

where y_t is the deviation of x_t from the arithmetic straight-line trend for the period selected, with $j = 1, 2, \ldots,$ and N being the number of years of record used in the analysis. The sum of the squared amplitudes is

$$R_j^2 = A_j^2 + B_j^2 \quad (8\text{-}I\text{-}17c)$$

If no periodic fluctuations are present in the series; that is, if the series is a pure-random (nonautocorrelated) series of N terms having a normal distribution, the mean-squared amplitude of the series is

$$R_m^2 = \frac{4\sigma^2}{N} \quad (8\text{-}I\text{-}17d)$$

where σ^2 is the variance of the series y_t.

If the series has periodic fluctuations, three tests for periodicity are available [83–85]:

(1) Schuster Test. According to Schuster [86], the probability P_s in per cent that the squared amplitude R_j^2 is k times the mean-squared amplitude R_m^2 is

$$P_s = e^{-k} \quad (8\text{-}I\text{-}18)$$

where
$$k = \frac{R_j^2}{R_m^2} = -\ln P_s \quad (8\text{-}I\text{-}18a)$$

The value of R_j^2 for a given series can be tested to see if it differs from R_m^2 derived from a pure-random series. It is apparent that the higher the probability P_s the more likely the series is pure-random since the hypothesis being tested is that the series is pure-random. Generally $P_s = 10$ per cent may be taken as the level of significance. The corresponding value of $k = 2.303$. Thus, $R_j^2 = 2.303 R_m^2 = 9.212\sigma^2/N$. Computing this value and substituting it in Eqs. (8-I-17a to c), the value of j can be computed. The possible hidden periodicity is equal to T/j.

(2) Walker Test. According to Walker [87], the probability that *at least* one squared amplitude R_j^2 will be k times R_m^2 is

$$P_w = 1 - (1 - e^{-k})^{N/2} \quad (8\text{-}I\text{-}19)$$

which may be used for a periodicity test as in the Schuster test.

(3) Fisher Test. Let R_J^2 be the largest of the squared amplitudes R_j^2. According to Fisher [88], the probability P_f that $R_J^2/2s^2$, where s^2 is the unbiased estimate of

σ^2, is greater than a given value g is

$$P_f = \sum_{i=0}^{m} (-1)^i \binom{j}{i} (1 - ig)^{j-1} \qquad (8\text{-}I\text{-}20)$$

where m is the greatest integer less than $1/g$, and $j = 1, 2, \ldots$ is the number of periods. This probability may be used for a periodicity test as in the Schuster test.

It must be noted that the above tests are based on normal distribution of the deviations and they apply only to strict periodicity and to nonautocorrelated data. Therefore, before using these tests, the effect of persistence or autocorrelation should be eliminated and the deviations be known as reasonably normally distributed.

Periodicity of secular nature is a matter of controversy [89]. Although the 11-year sunspot cycle is generally believed to have effect of various degrees upon hydrologic phenomena through the corresponding variations in solar radiant energy, statistical tests of possible astronomical effects on hydrologic phenomena have failed to show any statistical significance. Huntington [90] believed that very long secular variations are really not truly cyclic and therefore described them as *pulsations*.

Because of the seasonal effect on most hydrologic phenomena, *water year* instead of calendar year is usually adopted for the analysis of annual data. The water year will vary somewhat materially with the climatic conditions in various parts of the world. Water year usually starts when the ground and surface storage are both reduced to a minimum. The U.S. Geological Survey arranges the runoff data for a water year from October 1 to September 30. In England, the water year from September 1 to August 31 is sometimes used. Brakensiek [91] suggested that an optimum water year for tabulating water yield data can be determined by correlation analysis.

c. Persistence. This means that the successive members of a time series are linked among themselves in some persistent manner, resulting in non-pure-randomness. Due to meteorological and climatic causes, it has been found that wet years tend to occur in groups and dry years to occur together likewise. This tendency in grouping having the carryover effect of the immediate antecedent hydrologic conditions is the indication of the presence of persistence in hydrologic phenomena.

Since the carryover effect plays a significant part in hydrologic phenomena, it and hence the persistence are inversely related to the time interval between observations of such effects. When the time interval is shorter, the carryover effect or the persistence becomes more pronounced. As the effect of persistence exists, the degree of pure-randomness of the hydrologic data reduces. The magnitude of persistence may be determined by serial correlation analysis and correlograms (Subsec. 8-IV-II). It has been found that the magnitude depends on the type of hydrologic data; for example, it is higher in streamflows than in rainfalls.

Leopold [92] has described the nature of persistence with reference to probability analysis applied to a water-supply problem. He pointed out that Hurst [93] analyzed the longest record of river stage in the world (1,050 years of recorded stage of the Nile at the Roda gage) and obtained the evidence that the tendency for wet years to occur together and dry years together increased variability of means of various periods. In other words, the variability of groups of streamflows in their natural order of occurrence is actually larger than if the same flows occurred in random sequence. To illustrate this point, Leopold prepared Fig. 8-I-2 to show the variability of mean values of streamflow for records of various lengths. The dashed curve was plotted with grouped data taken from some longest streamflow records in the United States and Europe. If the annual streamflows were to occur in random sequence, the variability of means of groups would decrease inversely as the square root of the number of years comprising the group. Thus, the means of 100-year groups would be $1/\sqrt{100}$ or 1/10 as variable as 1-year values. The solid curve represents this random-sequence data. The difference between the dashed and solid curves represents the effect of persistence.

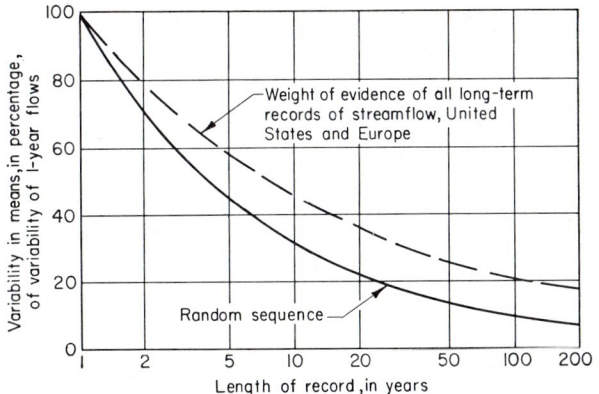

FIG. 8-I-2. Variability of mean values of streamflow for various lengths. (*Leopold* [92].)

2. Space Homogeneity. *Statistical meteorological homogeneity*, or *statistical hydrologic homogeneity, in space* implies that the occurrences of a particular meteorological, or hydrologic, event at all places within a so-called *statistically homogeneous area* are equally likely within a tolerable statistical difference. Because of the changes in geographical environment, statistically homogeneous areas are limited and can be delineated by statistical *regional analysis* (Subsec. IV-J).

III. PROBABILITY DISTRIBUTIONS

There are many probability distributions that have been found to be useful for hydrologic frequency analysis. Theoretical derivations and detailed discussions of such distributions can be found in many standard textbooks on statistics [73–75, 94–95].

A. Rectangular Distribution

The rectangular distribution is a uniform distribution of a continuous variable X between two constants a and b. The probability density of this distribution is

$$p(x) = 0 \quad \text{for } x < a$$
$$p(x) = \frac{1}{b-a} \quad \text{for } a \leq x \leq b \quad \quad (8\text{-}I\text{-}21)$$
and
$$p(x) = 0 \quad \text{for } b < x$$

The statistical parameters are: Mean $= (b + a)/2$; and variance $= (b - a)^2/12$.

B. Binomial Distribution

This is one of the most commonly used discrete distributions. It represents the distribution of probabilities in Bernoulli trials, say tossing a coin. The probability density is

$$p(x) = C_x^N p^x q^{N-x} \quad \quad (8\text{-}I\text{-}22)$$

where p is the probability of occurrence of an event, for example, a success in tossing a coin; C_x^N is the number of combinations of N things taken x at a time; q is the probability of failure or $1 - p$; N is the total number of trials; and x is the variate or the number of successful trials.

The statistical parameters are: Mean $= pN$; standard deviation, $\sigma = \sqrt{pqN}$; and skewness, $\alpha = \mu_3/\sigma^3 = (q-p)/\sqrt{pqN}$, where μ_3 is the third moment about the mean. When $p = q$, the distribution is symmetrical.

In a binomial distribution, the events or trials can be classified into only two categories: success and failure, yes and no, rainy and clear, etc. The probabilities p and q remain constant from one trial to another, i.e., the events are independent to each other.

C. Poisson Distribution

If N is very large and p is very small so that $pN = m$ is a positive number, then

$$p(x) = \frac{m^x e^{-m}}{x!} \tag{8-I-23}$$

gives a close approximation to binomial probabilities when m is small. A distribution with this probability density is called the *Poisson distribution* and is generally referred to as the *law of small numbers*. It is most useful when neither N nor p is known but their product pN is given or can be estimated.

The statistical parameters are: Mean = m; standard deviation = m; and skewness = $1/\sqrt{m}$.

D. Normal Distribution

This is a symmetrical, bell-shaped, continuous distribution, theoretically representing the distribution of accidental errors about their mean, or the so-called *Gaussian law of errors*. The probability density is

$$p(x) = \frac{1}{\sigma\sqrt{2\pi}} e^{-(x-\mu)^2/2\sigma^2} \tag{8-I-24}$$

where x is the variate, μ is the mean value of the variate, and σ is the standard deviation. In this distribution, the mean, mode, and median are the same. The total area under the distribution is equal to 1.0. The cumulative probability of a value being equal to or less than x is

$$P(X \leq x) = \frac{1}{\sigma\sqrt{2\pi}} \int_{-\infty}^{x} e^{-(x-\mu)^2/2\sigma^2} \, dx \tag{8-I-25}$$

This represents the area under the curve between the variates of $-\infty$ and x. Areas for various values of x have been calculated by statisticians, and tables for such areas are available in many textbooks and handbooks on statistics.

E. Gamma Distribution

The probability density of this distribution is

$$p(x) = \frac{x^a e^{-x/b}}{b^{a+1}\,\Gamma(a+1)} \tag{8-I-26}$$

with $b > 0, a > -1$ for $x = 0$
and $p(x) = 0$ for $x \leq 0$

where a and b are constant and $\Gamma(a+1) = a!$ is a *gamma function*. The cumulative probability being equal to or less than x ($< \infty$) is known as the *incomplete gamma function*.

The statistical parameters are: Mean = $b(a+1)$; and variance = $b^2(a+1)$.

F. Pearson Distributions

Karl Pearson [96] has derived a series of probability functions to fit virtually any distribution. Although these functions have only slight theoretical basis, they have

PROBABILITY DISTRIBUTIONS   8–15

been used widely in practical statistical works to define the shape of many distribution curves. The general and basic equation to define the probability density of a Pearson distribution is

$$p(x) = e^{\int_{-\infty}^{x} (a+x)/(b_0 + b_1 x + b_2 x^2) dx} \tag{8-I-27}$$

where a, b_0, b_1, and b_2 are constants. The criteria for determining types of distribution are β_1, β_2, and κ being defined as follows:

$$\beta_1 = \frac{\mu_3^2}{\mu_2^3} \tag{8-I-28}$$

$$\beta_2 = \frac{\mu_4}{\mu_2^2} \tag{8-I-29}$$

and

$$\kappa = \frac{\beta_1(\beta_2 + 3)^2}{4(4\beta_2 - 3\beta_1)(2\beta_2 - 3\beta_1 - 6)} \tag{8-I-30}$$

where μ_2, μ_3, and μ_4 are the second, third, and fourth moments about the mean (Sheppard's corrections may be made if necessary).

With $\beta_1 = 0$, $\beta_2 = 3$, and $\kappa = 0$, the resulting Pearson distribution is identical with the normal distribution. Types I and III distributions are often used in the hydrologic frequency analysis.

1. Type I Distribution. For Type I, $\kappa < 0$. This is a skew distribution with limited range in both directions, usually bell-shaped but may be J-shaped or V-shaped. Its probability density is

$$p(x) = p_0 \left(1 + \frac{x}{a_1}\right)^{m_1} \left(1 - \frac{x}{a_2}\right)^{m_2} \tag{8-I-31}$$

with $m_1/a_1 = m_2/a_2$ and the origin at the mode. The values of m_1 and m_2 are given by

$$m_1 \text{ or } m_2 = \frac{1}{2}\left[r - 2 \pm r(r + 2)\frac{\sqrt{\mu_2 \beta_1}}{2(a_1 + a_2)}\right] \tag{8-I-31a}$$

When μ_3 is positive, m_2 is the positive root and m_1 is the negative root; and vice versa in signs. The other values are

$$r = \frac{6(\beta_2 - \beta_1 - 1)}{6 + 3\beta_1 - 2\beta_2} \tag{8-I-31b}$$

$$a_1 + a_2 = \tfrac{1}{2} \sqrt{\mu_2[\beta_1(r+2)^2 + 16(r+1)]} \tag{8-I-31c}$$

and

$$p_0 = \frac{N}{a_1 + a_2} \frac{m_1^{m_1} m_2^{m_2}}{(m_1 + m_2)^{m_1 + m_2}} \frac{\Gamma(m_1 + m_2 + 2)}{\Gamma(m_1 + 1)\Gamma(m_2 + 1)} \tag{8-I-31d}$$

where N is the total frequency.

The statistical parameters are: Mean = mode $- (\mu_3/2\mu_2)[(r+2)/(r-2)]$; standard deviation = $\sqrt{\mu_2}$; and Pearson's skewness = $(\sqrt{\beta_1}/2)[(r+2)/(r-2)]$.

2. Type III Distribution. For Type III, $\kappa = \infty$ or $2\beta_2 = 3\beta_1 + 6$. This is a skew distribution with limited range in the left direction, usually bell-shaped but may be J-shaped. Its probability density with the origin at the mode is

$$p(x) = p_0 \left(1 + \frac{x}{a}\right)^c e^{-cx/a} \tag{8-I-32}$$

where

$$c = \frac{4}{\beta_1} - 1 \tag{8-I-32a}$$

$$a = \frac{c}{2}\frac{\mu_3}{\mu_2} \tag{8-I-32b}$$

$$p_0 = \frac{N}{a}\frac{c^{c+1}}{e^c \Gamma(c+1)} \tag{8-I-32c}$$

The statistical parameters are: Mean = mode $- \mu_3/2\mu_2$; standard deviation = $\sqrt{\mu_2}$; and Pearson's skewness = $\sqrt{\beta_1}/2$.

G. Extremal Distributions

Fréchet (on Type II) in 1927 [97] and Fisher and Tippett (on Types I and III) in 1928 [98] independently studied the distribution of extreme values and found that the distribution of the N largest (or the N smallest) values, each of which values is selected from one of m values contained in each of N samples, approaching a limiting (asymptotic) form as m is increased indefinitely. The type of the limiting form depends on the type of the initial distribution of the Nm values. For three different types of initial distribution, three asymptotic *extremal distributions* can be derived. A systematic study of the three asymptotes to the corresponding types of initial distributions was made by von Mises [99]. For detailed discussions on these distributions, see Refs. 100–102.

1. Type I Distribution. This distribution results from any initial distribution of *exponential type* which converges to an exponential function as x increases. Examples of such initial distributions are the normal, the chi-square, and the lognormal distributions. The probability density of Type I distribution is

$$p(x) = \frac{1}{c} e^{-(a+x)/c - e^{-(a+x)/c}} \qquad (8\text{-I-}33)$$

with $-\infty < x < \infty$, where x is the variate, and a and c are parameters. The cumulative probability is

$$P(X \leq x) = e^{-e^{-(a+x)/c}} \qquad (8\text{-I-}34)$$

By the method of moments, the parameters have been evaluated as

$$a = \gamma c - \mu \qquad (8\text{-I-}34a)$$

and

$$c = \frac{\sqrt{6}}{\pi} \sigma \qquad (8\text{-I-}34b)$$

where $\gamma = 0.57721 \ldots$ a Euler's constant, μ is the mean, and σ is the standard deviation. The distribution has a constant coefficient of skewness equal to $C_s = 1.139$.

2. Type II Distribution. This distribution results from an initial distribution of *Cauchy type* which has no moments from a certain order and higher. The cumulative probability is

$$P(X \leq x) = e^{-(\theta/x)^k} \qquad (8\text{-I-}35)$$

with $0 \leq x < \infty$, where the parameter θ is the expected largest value defined for a sample of size n and increases with n, and k is an order of moments and independent of n.

3. Type III Distribution. This distribution results from a type of initial distribution in which x is limited by $x \leq \epsilon$. The cumulative probability is

$$P(X \leq x) = e^{-[(x-\epsilon)/(\theta-\epsilon)]^k} \qquad (8\text{-I-}36)$$

with $-\infty < x \leq \epsilon$. The parameter k is the order of the lowest derivative of the probability function that does not vanish at $x = \epsilon$, and θ is the expected largest value.

In application, Type I distribution is sometimes known as *Gumbel distribution* since Gumbel [23] first applied it to flood frequency analysis. Type III is known as *Weibull distribution* since Weibull [103–104] first applied it to the description of the strength of brittle materials although Gumbel [57] also applied it later to drought frequency analysis.

H. Logarithmically Transformed Distributions

Many probability distributions can be transformed by replacing the variate with its logarithmic value. Three transformed distributions commonly used in hydrologic studies are as follows:

1. Lognormal Distribution. This is a transformed normal distribution in which the variate is replaced by its logarithmic value. This distribution represents the so-called *law of Galton* because it was first studied by Galton [105] as early as 1875. Its probability density is

$$p(x) = \frac{1}{\sigma_y e^y \sqrt{2\pi}} e^{-(y-\mu_y)^2/2\sigma_y^2} \qquad (8\text{-I-}37)$$

where $y = \ln x$, x is the variate, μ_y is the mean of y, and σ_y is the standard deviation of y. This is a skew distribution of unlimited range in both directions.

Chow [15] has derived the statistical parameters for x as

$$\mu = e^{\mu_y + \sigma_y^2/2} \qquad (8\text{-I-}37a)$$

$$\sigma = \mu(e^{\sigma_y^2} - 1)^{1/2} \qquad (8\text{-I-}37b)$$

$$\alpha = (e^{3\sigma_y^2} - 3e^{\sigma_y^2} + 2)C_v^3 \qquad (8\text{-I-}37c)$$

$$M = e^{\mu_y} \qquad (8\text{-I-}37d)$$

$$\frac{\mu}{M} = e^{\sigma_y^2/2} \qquad (8\text{-I-}37e)$$

$$C_v = (e^{\sigma_y^2} - 1)^{1/2} \qquad (8\text{-I-}37f)$$

$$C_s = 3C_v + C_v^3 \qquad (8\text{-I-}37g)$$

where μ is the mean, σ is the standard deviation, C_s is the coefficient of skewness, M is the median, and C_v is the coefficient of variation. Chow [15] has also shown that the Type I extremal distribution is essentially a special case of the lognormal distribution when $C_v = 0.364$ and $C_s = 1.139$. For other discussions, see Refs. 106–109.

2. Logextremal Distributions. Let x be replaced by y in Eq. (8-I-34) and then equate Eq. (8-I-34) to Eq. (8-I-35) and to Eq. (8-I-36). It can be found that for Type II extremal distribution, y is a linear function of $\ln x$, and for Type III extremal distribution, y is a linear function of $\ln (x - \epsilon)$. In other words, if the variate x in Type I distribution is replaced by a linear function of the logarithm of x and $x - \epsilon$, the resulting logarithmically transformed distributions become Type II and Type III distributions respectively.

3. Truncated Lognormal Distributions. Slade [22] introduced two truncated and shifted logarithmically transformed normal distributions for hydrologic frequency analysis. One is called the *partly bounded distribution* which has an unlimited range only in the positive direction of the variate. Its probability density is

$$p(x) = ae^{-c^2[\ln d(x+b)]^2} \qquad (8\text{-I-}38)$$

with $-b \leq x < \infty$, where a, b, c, and d are parameters which can be derived from the first three statistical moments.

The other is called the *totally bounded distribution* which has the maximum and minimum limits of fluctuations from the mean. Its probability density is

$$p(x) = ae^{-p^2c^2\{\ln d[(x+b)/(g-x)]\}^2} \qquad (8\text{-I-}39)$$

with $-b < x < g$, where the parameters a, p, c, and d are determined empirically.

IV. PROCEDURE OF ANALYSIS

Frequency analysis of hydrologic data starts with the treatment of raw hydrologic data and finally determines the frequency or probability of a hydrologic design value. Since the time sequence of hydrologic phenomena is not considered primarily in this section, probabilistic frequency analysis is mainly discussed here. In such analyses, a probability distribution is assumed as a mathematical model to which the hydrologic frequency data are to be fitted without considering the sequence of occurrence of the data. See Subsec. II-E.

A. Treatment of Raw Data

1. Data Sampling. Large masses of hydrologic data are unwieldy and uneconomical to analyze and their population is infinite or nearly infinite in size. For the use in analysis, the data must be sampled. For probabilistic frequency analysis, it is required that samples be pure-random. In other words, they should be unbiased, independent, and homogeneous.

A sample for which the sampling procedure is entirely by chance is called an *unbiased*, or a *random*, sample. In order to prevent the sample's being biased, the sample must be as representative as possible of the total population. In the collection of rainfall data in a drainage basin, for example, the stations should be so located that a large part of the basin would be covered and that various types of basin conditions would be represented. Such a representative sample is called a *stratified* sample, which is the opposite of a *spot* sample that is taken only from one small area or class of population.

Dependence of data may be referred to on either time or space. *Time dependence* is the major cause for non-pure-randomness of the data. For example, two successive floods occurring very closely may result in a high degree of dependence as the storm producing the first flood may effectively affect the meteorological condition that produces the second flood. *Space dependence* may be a major reason to produce unstratified data. For example, two rainfall stations placed closely together will produce practically identical data and should be considered only as one station in computing mean rainfall.

Lack of *homogeneity* means that the samples are taken from two different populations. For example, temperatures taken under the sun should not be averaged with those taken in the shade if the two conditions are considered as constituting different populations in the analysis.

2. Observation Errors. Nowadays vast amounts of hydrologic data are being collected. The basic form of such data is generally a continuous record in time, which is too bulky for publication. Usually, only selected or processed data are published.

Measurement and publication often involve instrumental and human errors. Such errors may be considered of two kinds, namely accidental and systematic errors, although it is sometimes difficult to distinguish between them and many errors are a combination of the two kinds. *Accidental errors* are usually due to the observer and sometimes due to the uncertain nature of the measuring instrument. Such errors may be considered random errors; they are disordered in their incidence and variable in magnitude, positive and negative values occurring in repeated measurements in no ascertainable sequence. On the other hand, *systematic errors* may arise from the observer or the instrument. Such errors are not random; they may be constant and create a trend, or vary in some regular way and produce periodicity.

3. Inherent Defectiveness. Major defectiveness of hydrologic data, such as non-pure-randomness, nonstationarity, missing data, etc., should be investigated. If they affect measurably the basic assumptions required in probabilistic frequency analysis, the raw data should be adjusted accordingly by various methods, such as serial correlation analysis for persistence, moving averages for trend, and Fourier or harmonic analysis for periodicity (Subsec. II-F-1). Missing data may sometimes be estimated by regional analysis by correlation with other hydrologic data in the neighborhood (Subsec. 9-V).

Statistical properties of hydrologic phenomena may also depend on inferences derived from long-term nonhydrologic natural data. Examples of such data which may possibly be used for this purpose include widths of tree rings, pattern of fossil pollen, distribution of clay varves, fluctuations in levels of closed lakes, glacial movement, and very long-range historical records of extraterrestrial phenomena. These nonhydrologic phenomena contain the intrinsic record of the nature of non-pure-randomness, nonstationarity, and other characteristics of time-series events. They may be used to improve the quality of hydrologic data through statistical inferences. However, they are available only in limited number. Furthermore, their use in statistical inferences requires the understanding of the processes involved in these natural phenomena and the correct interpretation of the results obtained from the inferences.

It may be noted that hydrologic phenomena seldom completely satisfy the requirements of the statistical theory. Before the raw data are used for frequency analysis, they should be examined for possible observation errors and inherent defectiveness. If such errors and defectiveness are appreciable, they should be analyzed and corrected before the frequency analysis can be suitably applied.

B. Selection of Data Series

The available hydrologic data are generally presented in chronological order. Figure 8-I-3 exhibits a hypothetical set of such data for a certain period of observation, say 20 years as shown in the figure. The magnitude of data is expressed in an arbitrary unit. Since all available data are shown, they constitute a *complete-duration series*.

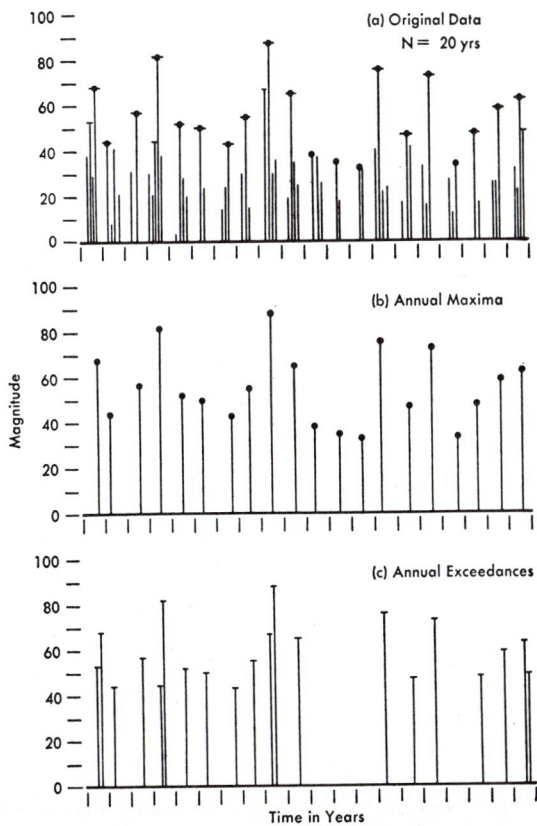

Fig. 8-I-3. Hydrologic data arranged in the order of occurrence.

Experience has shown that many of the original data have practically no significance in the analysis because the hydrologic design of a project is usually governed by a few critical conditions only. Thus, except sometimes in a few cases such as in the analysis by duration curves and mass curves, the complete-duration series is not always used. In order to save labor and time in the publication and analysis of the data, the data of insignificant magnitude should be excluded. For this purpose, two types of data are generally selected from the complete-duration series: the partial-duration series and the extreme-value series.

The *partial-duration series*, or *partial series*, is a series of data which are so selected that their magnitude is greater than a certain *base value*. If the base value is selected so that the number of values in the series is equal to the number of the record, the series is called *annual exceedance series* as shown in Fig. 8-I-3c.

The *extreme-value series* includes the largest or smallest values with each value selected from an equal time interval in the record. The time interval is usually taken as one water year and the series so selected is the *annual series*. For largest annual values, it is an *annual maximum series* as shown in Fig. 8-I-3b. For smallest annual values, it is an *annual minimum series*. When the time interval decreases, the dependence between observations and the number of selected values increase. If the time interval is less than one year, the seasonal variation will further introduce nonhomogeneity to the data. However, homogeneity of the data may be maintained at least for practical purposes if the data are selected only from a particular season, month, or other definite duration within a year [48]. For example, summer storms and spring floods can be selected to form their own independent series.

For clarity, the classification of hydrologic data series may be shown as follows:

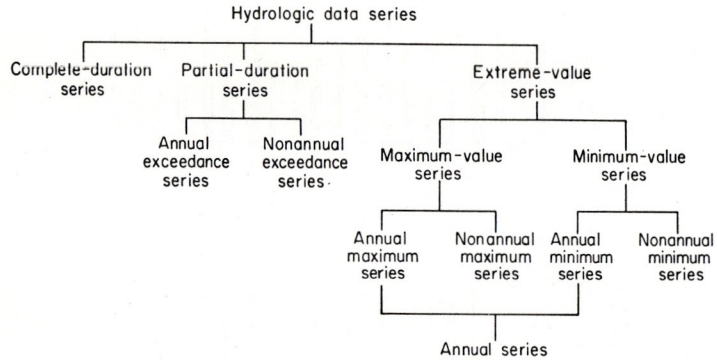

The annual maximum values and the annual exceedance values of the hypothetical data in Fig. 8-I-3a are arranged graphically in Fig. 8-I-4 in the order of magnitude. The figure shows that many annual exceedances surpass the annual maxima in magnitude. In this particular example, only five of the twenty values are the same in both annual exceedance and annual maximum series. Figure 8-I-3a also shows that the second largest value in several years outranks many annual maxima in magnitude. Thus, in the annual maximum series for which only the annual maxima are selected these second largest values are omitted, resulting in the neglect of their effect in the analysis. On the other hand, in the annual exceedance series where several values which occurred close together may be included in the same year, the selected data may be less independent than the annual maxima because one hydrologic event can affect another which follows very closely after, e.g., one flood and one storm may influence the meteorological condition for the subsequent ones.

From the logical point of view, the selection of hydrologic data in designing a structure may be judged by the type of structure or project. The annual exceedances or the partial-duration series should be used if the second largest values in the year would affect the design. For instance, the damage to bridge foundations caused by flooding sometimes results from the repetition of flood occurrence rather than from a single peak flow. A culvert subject to flood damage or destruction may be rapidly and economically repaired or restored and then soon again exposed to future damage. Similarly, in highway drainage, the loss due to traffic interruption as a result of flooding will be weighed by the number of flood peaks and the extent of flooding which are largely caused by associated peak flows. In other cases where the design is governed by the most critical condition, such as the design of a spillway, the annual maxima

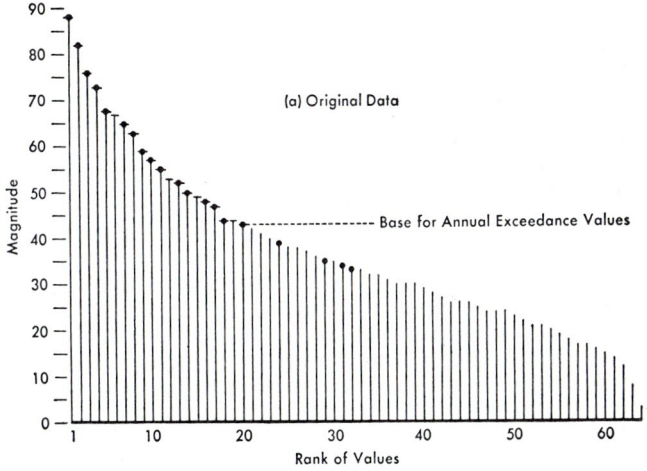

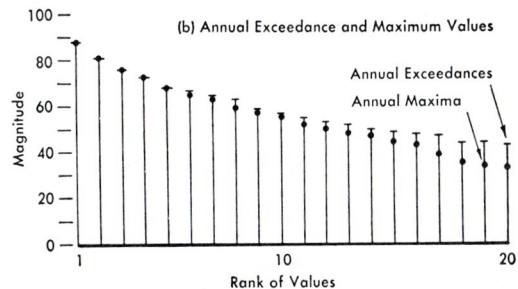

FIG. 8-I-4. Hydrologic data arranged in the order of magnitude.

should be used. For practical purposes, the partial-duration series and the annual maximum series do not differ much except in the values of low magnitude. Usually, both series are used in an analysis for comparison purposes.

The relationship between the probabilities of the partial-duration series (or the annual exceedance series) and the annual maximum series has been investigated by Langbein [37] and a corresponding theoretical relationship was derived by Chow [52, 110] as follows:

Let P_E be the probability of a variate in the partial-duration series (or the annual exceedance series) being equal to or greater than x, and let m be the average number of events per year or mN be the total number of events in N years of record. Then P_E/m is the probability of an event being equal to x or greater, and $1 - P_E/m$ is the probability of an event being less than x. Thus the probability of an event of magnitude x becoming a maximum of the m events in a year is $(1 - P_E/m)^m$. This probability approaches e^{-P_E} when P_E is small compared with m, which is true for most cases. Therefore, the probability P_M of an annual maximum of magnitude x being equaled or exceeded is equal to

$$P_M = 1 - e^{-P_E} \quad (8\text{-}I\text{-}40)$$
or
$$P_E = - \ln (1 - P_M) \quad (8\text{-}I\text{-}41)$$

It can be shown that $P_M \approx P_E$ as both P_M and P_E become large.

C. Recurrence Interval

The primary objective of the frequency analysis of hydrologic data is to determine the recurrence interval of the hydrologic event of a given magnitude x. The *average*

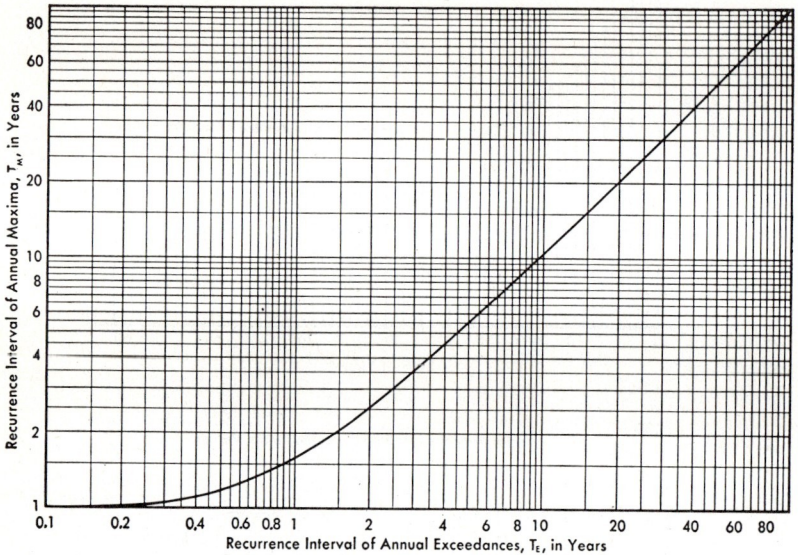

FIG. 8-I-5. Relationship between recurrence intervals T_M and T_E.

interval of time within which the magnitude y of the event will be equaled or exceeded once is known as *recurrence interval, return period,* or simple *frequency,* to be designated by T.

If a hydrologic event equal to or greater than x occurs once in T years, the probability $P(X \geq x)$ is equal to 1 in T cases, or

$$P(X \geq x) = \frac{1}{T} \quad (8\text{-}I\text{-}42)$$

Hence,
$$T = \frac{1}{P(X \geq x)} = \frac{1}{1 - P(X \leq x)} \quad (8\text{-}I\text{-}43)$$

If T_M and T_E are the recurrence intervals of the annual maximum series and the partial-duration series (or the annual exceedance series) respectively, and P_M and P_E are their corresponding probabilities of being equal to and greater than the magnitude x, Eq. (8-I-42) gives $P_M = 1/T_M$ and $P_E = 1/T_E$. Substituting these expressions of P_M and P_E in Eq. (8-I-40) and simplifying, the relationship between the two recurrence intervals is

$$T_E = \frac{1}{\ln T_M - \ln (T_M - 1)} \quad (8\text{-}I\text{-}44)$$

This relationship is plotted as shown in Fig. 8-I-5. Langbein [37] has plotted a number of actual cases which were found to be very close to this theoretical curve. It can be shown from this curve that a difference between P_M and P_E is equal to about 10 per cent when P_E is 5 years, and the difference becomes about 5 per cent when P_E is 10 years. In ordinary hydrologic analysis, a 5 per cent difference may be considered tolerable. Therefore, it may be concluded that the two recurrence intervals are practically identical for recurrence intervals greater than 10 years.

It should be noted that the recurrence interval as usually defined is a mean time interval based on the distribution of the variate x; it is not the actual time interval between two events of equal magnitude x. Thom [111] has studied the distribution of the actual time interval (assuming a Poisson distribution) instead of the distribution of the magnitude x. He found that the relationship between the recurrence interval T (i.e., the mean time interval) and the actual time interval distribution can be expressed as

$$T = \frac{\tau}{-\ln P(\tau)} \qquad (8\text{-I-}45)$$

where $P(\tau)$ is the probability that the actual time interval τ is to be equaled or exceeded.

D. Frequency Analysis Using Frequency Factors

1. General Equation for Hydrologic Frequency Analysis. The variate x of a random hydrologic series may be represented by the mean $\bar{x} \approx \mu$ plus the departure Δx of the variate from the mean, or

$$x = \bar{x} + \Delta x \qquad (8\text{-I-}46)$$

The departure Δx depends on the dispersion characteristic of the distribution of x and on the recurrence interval T and other statistical parameters defining the distribution. Thus, the departure may be assumed equal to the product of the standard deviation σ and a *frequency factor* K, i.e., $\Delta x = \sigma K$. The frequency factor is a function of the recurrence interval and the type of probability distribution to be used in the analysis. Equation (8-I-46) may therefore be expressed as

$$x = \bar{x} + \sigma K \qquad (8\text{-I-}47)$$

or

$$\frac{x}{\bar{x}} = 1 + C_v K \qquad (8\text{-I-}48)$$

where $C_v = \sigma/\bar{x}$. The above equation was proposed by Chow [112] as the *general equation for hydrologic frequency analysis*. This equation is applicable to many probability distributions proposed for use in hydrologic frequency analysis. For a proposed distribution a relationship can be determined between the frequency factor and the corresponding recurrence interval. This relationship can be expressed in mathematical terms, by tables, or by curves called *K-T curves*. In applying the general equation, the statistical parameters required in the proposed distribution are first computed from the random hydrologic data series. For a given recurrence interval, the frequency factor can be determined from the K-T relationship for the proposed distribution and the magnitude x for the recurrence interval can be computed by Eq. (8-I-47) or (8-I-48), using the corresponding frequency factor and the computed statistical parameters.

2. The K-T Relationship. Based on the observations of many streams, Fuller [10] derived the earliest empirical formula for the frequency analysis of annual maximum daily flow as

$$x = \bar{x}(1 + 0.8 \log T) \qquad (8\text{-I-}49)$$

Comparing this formula with Eq. (8-I-48), the frequency factor can be easily found as

$$K = \frac{0.8}{C_v} \log T \qquad (8\text{-I-}50)$$

which is a function of C_v and T. The value of C_v varies from 0.1 to 2.0, having an average of 0.50. The Fuller formula is actually based on an empirical statistical distribution and thus the K-T relationship so derived is also empirical. Since Fuller's

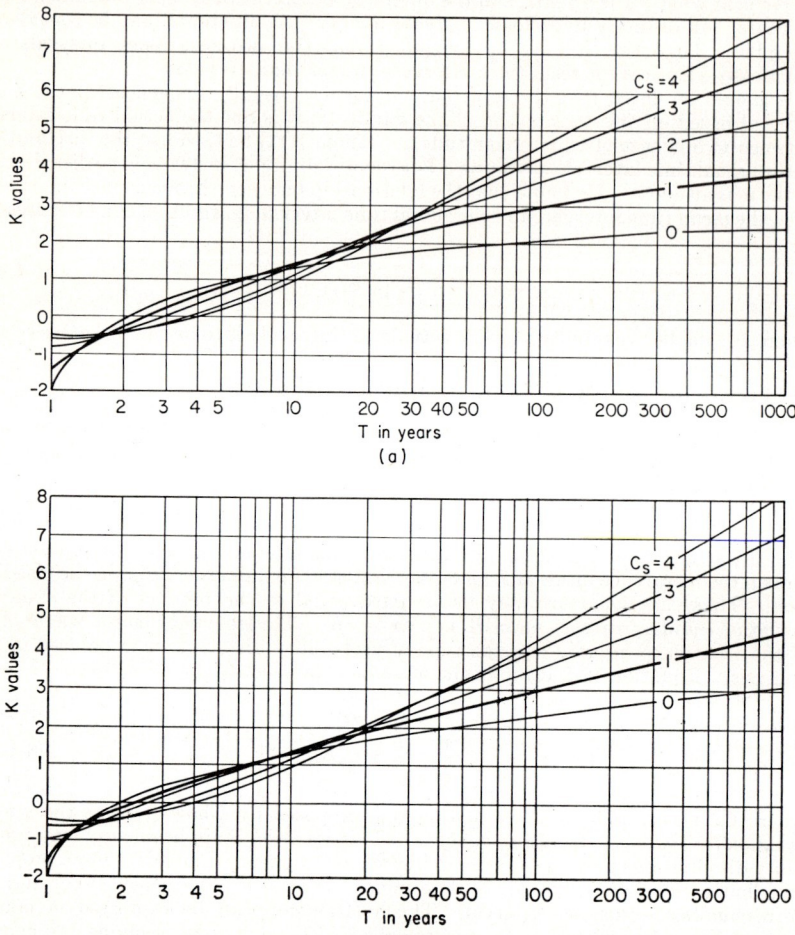

Fig. 8-I-6. K-T curves for Pearson distributions. (a) Pearson Type I distribution. (b) Pearson Type III distribution.

time, many methods using theoretical distributions have been proposed. The K-T relationships for some important theoretical distributions are discussed below:

a. Normal Distribution. Taking $\bar{x} = \mu$, the frequency factor can be expressed from Eq. (8-I-47) as

$$K = \frac{x - \mu}{\sigma} \tag{8-I-51}$$

Substituting this expression in Eq. (8-I-25),

$$P(X \leq x) = \frac{1}{\sqrt{2\pi}} \int_{-\infty}^{K} e^{-K^2/2}\, dK \tag{8-I-52}$$

Values of $P(X \leq x)$ for various values of K can be found from normal probability tables in many textbooks and handbooks on statistics. The corresponding values of T can be obtained from Eq. (8-I-43).

  b. *Pearson Distributions.* Foster [17] proposed a method in which the Pearson Type I and Type III distributions are used. From Foster's derivation, the frequency factor of these distributions can be shown by K-T curves in Fig. 8-I-6. Foster suggested that the coefficient of skewness computed by Eq. (8-I-11) should be multiplied

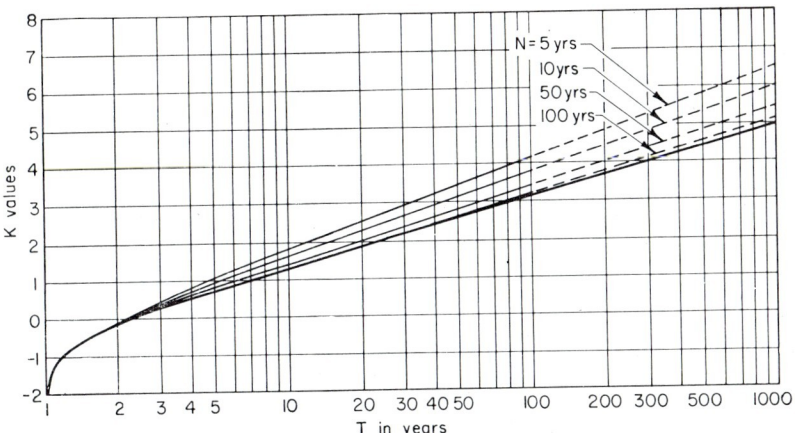

Fig. 8-I-7. K-T curves for Type I extremal distributions.

by $1 + 8.5/N$ for Type I distribution and by $1 + 6/N$ for Type III distribution in order to adjust the influence due to the length of hydrologic records.

  c. *Extremal Distribution.* The frequency factor for the Type I extremal distribution can be derived from Eq. (8-I-33) in a manner similar to that described above for the normal distribution. It has been given by Chow [52, 112] as

$$K = -\frac{\sqrt{6}}{\pi}\left[\gamma + \ln\ln\left(\frac{T}{T-1}\right)\right] \quad (8\text{-}I\text{-}53)$$

which is plotted as the heavy line in Fig. 8-I-7. Potter [32] studied 370 Type I extremal distributions for maximum rainfall intensities of various durations, monthly and annual rainfall amounts, and peak rates of surface runoff. The result of his study can be plotted in K-T curves as shown in Fig. 8-I-7. It can be seen that the K-T relationship so obtained depends on the number of years of record, N. These curves are shown by thin lines with the dashed portions extrapolated. As N increases, the K-T relationship approaches the theoretical relationship which is derived for the population. The curve for $N = 100$ years is practically identical with the theoretical curve.

  When $x = \bar{x}$, Eq. (8-I-48) gives $K = 0$ and thus Eq. (8-I-53) results in $T = 2.33$ years. This is the recurrence interval of the mean of the Type I extremal distribution. It is taken by U.S. Geological Survey as the *recurrence interval of a mean annual flood.*

  d. *Lognormal Distribution.* Substituting $x = e^y$, $\bar{x} \approx \mu$ by Eq. (8-I-37a), and σ by Eq. (8-I-37b) in Eq. (8-I-47), Chow [15, 106] has derived the frequency factor for the lognormal distribution as

$$K = \frac{e^{\sigma_y K_y - \sigma_y^2/2} - 1}{(e^{\sigma_y^2} - 1)^{1/2}} \quad (8\text{-}I\text{-}54)$$

where $K_y = (y - \bar{y})/\sigma_y$ and can be expressed in a form similar to Eq. (8-I-47) as

$$y = \bar{y} + \sigma_y K_y \quad (8\text{-}I\text{-}55)$$

8-26   FREQUENCY ANALYSIS

Equation (8-I-37) shows that y is normally distributed while its antilogarithm x is lognormally distributed. For a given recurrence interval T, or probability $P(X \geq x)$ or $P(X \leq x)$, the value of K_y can be computed in a manner similar to that described above for the normal distribution. When K_y is known, the value of K can be computed by Eq. (8-I-54) for any given value of σ_y. The value of σ_y and the corresponding

Table 8-I-1. Frequency Factors for Lognormal Distribution

C_s	Probability at mean	Probability in per cent equal to or greater than the given variate									Corresponding C_v
		99 −	95 −	80 −	50 −	20 +	5 +	1 +	0.1 +	0.01 +	
0	50.0	2.33	1.65	0.84	0	0.84	1.64	2.33	3.09	3.72	0
0.1	49.3	2.25	1.62	0.85	0.02	0.84	1.67	2.40	3.22	3.95	0.033
0.2	48.7	2.18	1.59	0.85	0.04	0.83	1.70	2.47	3.39	4.18	0.067
0.3	48.0	2.11	1.56	0.85	0.06	0.82	1.72	2.55	3.56	4.42	0.100
0.4	47.3	2.04	1.53	0.85	0.07	0.81	1.75	2.62	3.72	4.70	0.136
0.5	46.7	1.98	1.49	0.86	0.09	0.80	1.77	2.70	3.88	4.96	0.166
0.6	46.1	1.91	1.46	0.85	0.10	0.79	1.79	2.77	4.05	5.24	0.197
0.7	45.5	1.85	1.43	0.85	0.11	0.78	1.81	2.84	4.21	5.52	0.230
0.8	44.9	1.79	1.40	0.84	0.13	0.77	1.82	2.90	4.37	5.81	0.262
0.9	44.2	1.74	1.37	0.84	0.14	0.76	1.84	2.97	4.55	6.11	0.292
1.0	43.7	1.68	1.34	0.84	0.15	0.75	1.85	3.03	4.72	6.40	0.324
1.1	43.2	1.63	1.31	0.83	0.16	0.73	1.86	3.09	4.87	6.71	0.351
1.2	42.7	1.58	1.29	0.82	0.17	0.72	1.87	3.15	5.04	7.02	0.381
1.3	42.2	1.54	1.26	0.82	0.18	0.71	1.88	3.21	5.19	7.31	0.409
1.4	41.7	1.49	1.23	0.81	0.19	0.69	1.88	3.26	5.35	7.62	0.436
1.5	41.3	1.45	1.21	0.81	0.20	0.68	1.89	3.31	5.51	7.92	0.462
1.6	40.8	1.41	1.18	0.80	0.21	0.67	1.89	3.36	5.66	8.26	0.490
1.7	40.4	1.38	1.16	0.79	0.22	0.65	1.89	3.40	5.80	8.58	0.517
1.8	40.0	1.34	1.14	0.78	0.22	0.64	1.89	3.44	5.96	8.88	0.544
1.9	39.6	1.31	1.12	0.78	0.23	0.63	1.89	3.48	6.10	9.20	0.570
2.0	39.2	1.28	1.10	0.77	0.24	0.61	1.89	3.52	6.25	9.51	0.596
2.1	38.8	1.25	1.08	0.76	0.24	0.60	1.89	3.55	6.39	9.79	0.620
2.2	38.4	1.22	1.06	0.76	0.25	0.59	1.89	3.59	6.51	10.12	0.643
2.3	38.1	1.20	1.04	0.75	0.25	0.58	1.88	3.62	6.65	10.43	0.667
2.4	37.7	1.17	1.02	0.74	0.26	0.57	1.88	3.65	6.77	10.72	0.691
2.5	37.4	1.15	1.00	0.74	0.26	0.56	1.88	3.67	6.90	10.95	0.713
2.6	37.1	1.12	0.99	0.73	0.26	0.55	1.87	3.70	7.02	11.25	0.734
2.7	36.8	1.10	0.97	0.72	0.27	0.54	1.87	3.72	7.13	11.55	0.755
2.8	36.6	1.08	0.96	0.72	0.27	0.53	1.86	3.74	7.25	11.80	0.776
2.9	36.3	1.06	0.95	0.71	0.27	0.52	1.86	3.76	7.36	12.10	0.796
3.0	36.0	1.04	0.93	0.71	0.28	0.51	1.85	3.78	7.47	12.36	0.818
3.2	35.5	1.01	0.90	0.69	0.28	0.49	1.84	3.81	7.65	12.85	0.857
3.4	35.1	0.98	0.88	0.68	0.29	0.47	1.83	3.84	7.84	13.36	0.895
3.6	34.7	0.95	0.86	0.67	0.29	0.46	1.81	3.87	8.00	13.83	0.930
3.8	34.2	0.92	0.84	0.66	0.29	0.44	1.80	3.89	8.16	14.23	0.966
4.0	33.9	0.90	0.82	0.65	0.29	0.42	1.78	3.91	8.30	14.70	1.000
4.5	33.0	0.84	0.78	0.63	0.30	0.39	1.75	3.93	8.60	15.62	1.081
5.0	32.3	0.80	0.74	0.62	0.30	0.37	1.71	3.95	8.86	16.45	1.155

value of C_v for an assigned value of C_s can be computed by means of Eqs. (8-I-37b) and (8-I-37g). Table 8-I-1 is a list of frequency factors so computed for assigned values of C_s and of various probabilities $P(X \geq x)$. The table also lists the probabilities at mean, which occur when $K = 0$ or $K_y = \sigma_y/2$. The recurrence interval T is related to $P(X \geq x)$ by Eq. (8-I-43). Thus, this table can be used to plot K-T curves using C_s as the parameter.

E. Probability Paper

The cumulative probability of a distribution may be represented graphically on a *probability paper* which is designed for the distribution. On such paper the ordinate usually represents the value of x in certain scale and the abscissa represents the probability $P(X \geq x)$ or $P(X \leq x)$, or the recurrence interval T. The ordinate and abscissa scales are so designed that the distribution plots as a straight line and the

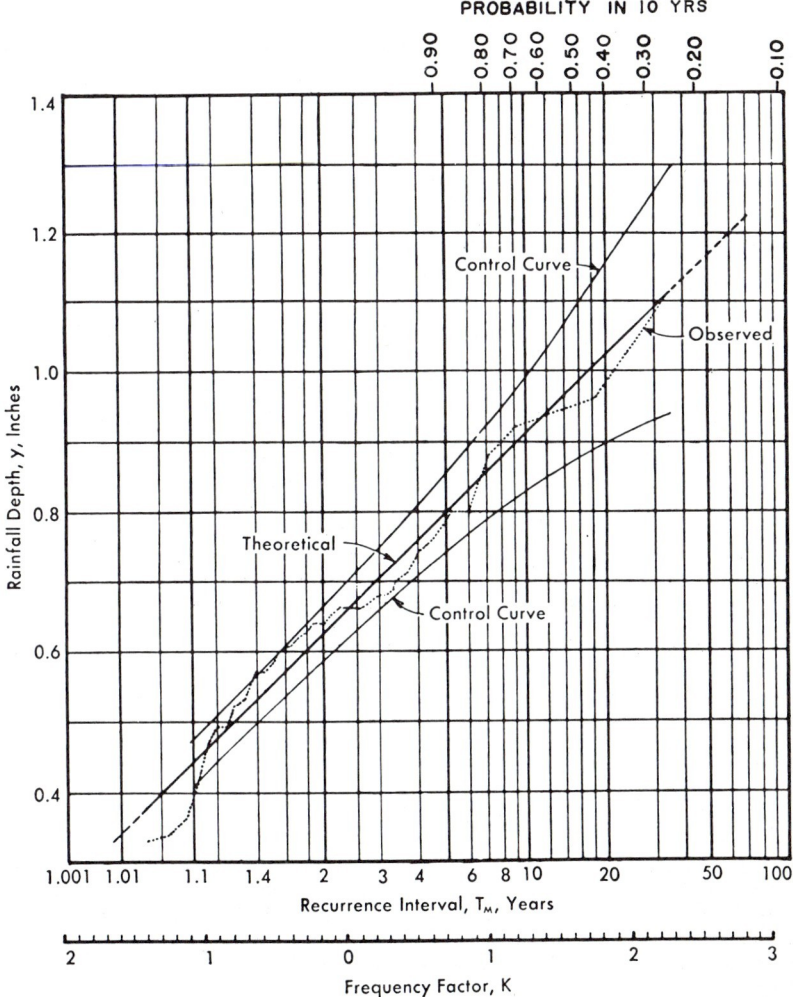

FIG. 8-I-8. Probability plotting of annual maxima of 10-min rainfall depth.

data to be fitted appear close to the straight line. The objective of using the probability paper is to linearize the distribution so that the plotted data can be easily analyzed for extrapolation or comparison purposes. In case of extrapolation, however, the effect of sampling errors is often magnified. Therefore, hydrologists should be warned against such practice if no consideration is paid to this effect in interpreting the extrapolated information.

Figure 8-I-1 shows the linearization of a distribution and the formation of a probability paper. The top diagram represents a frequency distribution and its probability distribution of the data is shown in Fig. 8-I-3a. The center diagram represents the cumulative probability curve plotted on a paper with rectangular scales. The lower diagram shows the straight-line plotting of the cumulative probability curve on a probability paper having a special scale for the probability designed for the given distribution. It should be noted that in practice the probability is plotted as abscissa rather than as ordinate as shown in the diagram.

Linearization of probability plotting has been proposed by many hydrologists since Hazen [113] first suggested graphical linearization of the normal distribution in 1914. By linearization, Powell [30] constructed the probability paper for Type I extremal distribution proposed by Gumbel [23] for flood frequency analysis. Such paper may be called *extremal probability paper* or *Gumbel-Powell probability paper*. A further modification of the paper was made by Court [2]. Type III extremal distribution was first used by Weibull [103–104] for stress analysis and later by Gumbel [57] for drought frequency studies. Linearization for this distribution can be made on *logextremal probability paper* or *Weibull probability paper*. On this paper, the ordinate has a logarithmic scale for the variate and the abscissa has a special scale for Type I extremal probability since Type III is essentially a logarithmically transformed Type I distribution. For linearization of the lognormal distribution, a *lognormal probability paper* may be used. On this paper, the ordinate has the logarithmic scale for the variate and the abscissa has a normal probability scale. On a probability paper it is frequently necessary to provide a scale for recurrence interval along with the probability scale or sometimes to replace the latter.

Probability paper is generally designed for plotting hydrologic data of annual and complete series, which possess a complete frequency distribution. For partial-duration series and annual exceedance series, which constitute only a "tail" distribution, the data are plotted on semilog paper where the ordinate has the rectangular scale for the variate and the abscissa has the logarithmic scale for the recurrence interval.

In case a probability paper is not available, the probability scale may be constructed by use of Eq. (8-I-47). For example, Fig. 8-I-8 shows an external probability paper in which the scale for recurrence interval is constructed with reference to a rectangular scale for frequency factor K. The recurrence intervals corresponding to various K values on the rectangular scale can be computed by Eq. (8-I-53) and used to construct the required scale. By means of Eq. (8-I-43), the probability scales can also be constructed. Similarly, probability paper of any distribution can be constructed if the K-T relationship is known.

When a probability paper is available, it is possible to determine graphically the frequency factor of the distribution for which the paper is constructed. Chow [107] has proposed such a graphical method for the determination of frequency factors of a lognormal distribution.

F. Plotting of Data

When a probability paper is chosen for use, the plotting of data on the paper requires the knowledge of *plotting positions*. Numerous methods have been proposed for the determination of plotting positions. Most of them are empirical. Table 8-I-2 lists some important plotting-position formulas but the list by no means shows all the methods, since there are many other methods [26, 121–125] which cannot be expressed by simple formulas.

Equation (8-I-56a) is believed to be the earliest formula for computing plotting positions. Use of this formula is known as the *California method*, since it was first employed to plot flow data of the California streams. Chow [52] has demonstrated theoretically that this method is suitable for plotting annual exceedance series or partial-duration series. However, this simple formula plots data at the edges of group intervals and produces a probability of 100 per cent which cannot be plotted on a probability paper. Thus, it was gradually replaced by the *Hazen formula*, Eq. (8-I-56b), which plots data at the centers of group intervals. As the extremal distribution was later introduced to frequency analysis, the *Weibull formula*, Eq. (8-I-56c), was soon found to be very satisfactory. Chow [52] has shown that this formula is theoretically suitable for plotting the annual maximum series. A comparative study of the Beard, Hazen, and Weibull methods by Benson [126] has also revealed that, on the basis of theoretical sampling from extreme values and normal distributions, the Weibull formula provides the estimates that are consistent with experience. The *Chegodayev formula*, Eq. (8-I-56e), is an empirical formula commonly used in the U.S.S.R., but Eq. (8-I-56c) has been recommended as the All-Union Standard 3999-48 in 1948 [116]. Equation (8-I-56e) is a mathematical approximation of Eq. (8-I-56d). In order to simplify the visual inspection of a plotted set of ordered observations on extremal probability paper, Gringorten [120] further recommended Eq. (8-I-56h) for computing plotting positions.

Table 8-I-2. Plotting-position Formulas

Name	Date	Formula* for T or $1/P(X \geq x)$	Equation
California [114]	1923	$\dfrac{N}{m}$	(8-I-56a)
Hazen [14]	1930	$\dfrac{2N}{2m-1}$	(8-I-56b)
Weibull [103–104]	1939	$\dfrac{N+1}{m}$	(8-I-56c)
Beard [35]	1943	$\dfrac{1}{1-0.5^{1/N}}$	(8-I-56d)†
Chegodayev [115–117]	1955	$\dfrac{N+0.4}{m-0.3}$	(8-I-56e)
Blom [118]	1958	$\dfrac{N+\frac{1}{4}}{m-\frac{3}{8}}$	(8-I-56f)
Tukey [119]	1962	$\dfrac{3N+1}{3m-1}$	(8-I-56g)
Gringorten [120]	1963	$\dfrac{N+0.12}{m-0.44}$	(8-I-56h)

* N = total number of items; m = order number of the items arranged in descending magnitude, thus $m = 1$ for the largest item.

† This formula applies only to $m = 1$; other plotting positions are interpolated linearly between this and the value of 0.5 for the median event.

It may be noted that all methods of determining plotting positions give practically the same results in the middle of a distribution but produce different positions near the "tails" of the distribution. Thus, the choice of a plotting-position formula becomes important. According to Benson [26], it is believed that only by use of a method that gives the mathematically expected value of the probability does the expected recurrence equal that experienced over a long period of time, and that commonly used methods may overestimate the benefit-cost ratios of proposed projects if the methods do not furnish the mathematically expected value. Therefore, a refined choice of a method depends on the acceptance of certain statistical principles and on the aim of the analysis.

G. Curve Fitting

After the hydrologic data are plotted on a probability paper, a curve may be fitted to the plotted points. The curve is a straight line if linearization of the distribution is attempted. The straight line can be essentially represented by Eq. (8-I-47). Curve fitting may be done either mathematically or graphically. In general, a mathematical curve fitting can be achieved by three methods: the method of moments, the method of least squares, and the method of likelihood. Of course, the mathematical fitting does not necessarily require data plotting on a probability paper. By graphical fitting, a straight line is simply drawn to fit the plotted data by eye-fit, and this method is the simplest but involves human error.

Table 8-I-3. Frequency Analysis of Annual Maximum Values of 10-min Duration Rainfall Depth at Chicago, Illinois

m	x	x^2	T_M	$y = K$	y^2	xy	Δx
(1)	(2)	(3)	(4)	(5)	(6)	(7)	(8)
1	1.11	1.2321	36.000	2.332	5.4382	2.5885	0.177
2	0.96	0.9216	18.000	1.783	3.1791	1.7117	0.124
3	0.94	0.8836	12.000	1.455	2.1170	1.3677	0.091
4	0.92	0.8464	9.000	1.218	1.4835	1.1206	0.079
5	0.88	0.7744	7.200	1.033	1.0671	0.9090	0.070
6	0.80	0.6400	6.000	0.878	0.7709	0.7024	0.064
7	0.80	0.6400	5.143	0.745	0.5550	0.5960	0.059
8	0.76	0.5776	4.500	0.627	0.3931	0.4765	0.056
9	0.74	0.5476	4.000	0.522	0.2725	0.3863	0.052
10	0.71	0.5041	3.600	0.425	0.1806	0.3018	0.050
11	0.70	0.4900	3.272	0.337	0.1136	0.2359	0.047
12	0.68	0.4624	3.000	0.255	0.0650	0.1734	0.045
13	0.68	0.4624	2.769	0.177	0.0313	0.1204	0.044
14	0.66	0.4356	2.571	0.102	0.0104	0.0673	0.042
15	0.66	0.4356	2.400	0.032	0.0010	0.0211	0.041
16	0.66	0.4356	2.250	−0.035	0.0012	−0.0231	0.039
17	0.65	0.4225	2.118	−0.100	0.0100	−0.0650	0.038
18	0.64	0.4096	2.000	−0.164	0.0269	−0.1050	0.037
19	0.64	0.4096	1.895	−0.225	0.0506	−0.1440	0.037
20	0.63	0.3969	1.800	−0.286	0.0818	−0.1802	0.036
21	0.62	0.3844	1.715	−0.346	0.1197	−0.2145	0.035
22	0.61	0.3721	1.636	−0.405	0.1640	−0.2471	0.035
23	0.60	0.3600	1.565	−0.464	0.2153	−0.2784	0.034
24	0.58	0.3364	1.500	−0.523	0.2735	−0.3033	0.034
25	0.57	0.3249	1.440	−0.582	0.3387	−0.3317	0.033
26	0.57	0.3249	1.385	−0.643	0.4134	−0.3665	0.033
27	0.53	0.2809	1.333	−0.704	0.4956	−0.3731	0.033
28	0.52	0.2704	1.285	−0.768	0.5898	−0.3994	0.033
29	0.49	0.1401	1.242	−0.834	0.6956	−0.4087	0.033
30	0.49	0.2401	1.200	−0.904	0.8172	−0.4430	0.033
31	0.47	0.2209	1.162	−0.980	0.9604	−0.4606	0.033
32	0.41	0.1681	1.125	−1.064	1.1321	−0.4362	0.034
33	0.36	0.1296	1.092	−1.159	1.3433	−0.4172	0.035
34	0.34	0.1156	1.058	−1.277	1.6307	−0.4342	0.037
35	0.33	0.1089	1.029	−1.445	2.0880	−0.4769	0.042
$\Sigma =$ 22.71		15.8049		−0.987	27.1261	4.6705	

$\bar{x} = 0.6489 \quad \bar{x}^2 = 0.4516 \quad \bar{y} = -0.0282 \quad \bar{y}^2 = 0.7750 \quad \overline{xy} = 0.1334$
$B = 0.1960 \quad A = 0.6544 \quad \sigma = 0.1775$
Line of best-fit: $x = 0.1960K + 0.6544$

1. Method of Moments. By this method, the statistical parameters or moments are computed from the data and then substituted in the probability function of the given distribution. This method gives a theoretically exact fitting but the accuracy can be substantially affected by any errors involved in the data at the tails of the distribution where the moment arms are long and the errors are thus magnified. The method originally proposed by Gumbel [23] to fit Type I extremal distribution is a method of moments. Lieblein [121] modified this method by order statistics and developed a procedure which maintains the original time order of the extreme-value series, divides the values into subgroups, and then weighs each observation according to its ordered rank in the subgroup which in turn is a function of the sample size. Hershfield [56] made a comparison of the two procedures and concluded that the Gumbel procedure gives a better estimate beyond the range of data for the areally independent data tests, but overestimates the longer recurrence-intervals in the dependent data tests.

  2. **Method of Least Squares.** By this method, a regression line is computed to fit the plotted data (Sec. 8-II). The curve so obtained may not represent the exact theoretical distribution but it gives a better overall fit than the method of moments. For extremal distributions, Gumbel [101] introduced a modified least-squares method by minimizing both vertical and horizontal deviations and taking the geometric mean of the parameters obtained from the two minimizations. Based on the general equation for hydrologic frequency analysis, Eq. (8-I-47), proposed by Chow [112], a least-squares procedure for fitting a normal, lognormal, or extremal distribution was developed by Brakensiek [127].

  Table 8-I-3 shows the computation for fitting annual maximum values plotted on an extremal probability paper in Fig. 8-I-8. In this table, m is the rank number, x is the variate or the rainfall depth, and y is the frequency factor K. The recurrence interval T is computed by Eq. (8-I-56c), and the frequency factor by Eq. (8-I-53). The coefficients A and B of the least-squares equation are computed by Eqs. (8-II-8) and (8-II-9).

  3. **Method of Maximum Likelihood.** By this method, the value of a parameter is determined to make the probability of obtaining the observed outcome as high as possible. Mathematically, $\partial \log p(x)/\partial u = 0$, where $p(x)$ is probability density and u is a statistical parameter. This method provides the best estimate of the parameters but it is usually very complicated for practical application. Kimball [128–129] has suggested this method for fitting extremal distributions, and a practical procedure was later developed by Panchang and Aggarwal [130].

H. Reliability of Analysis

  1. **Sampling Reliability.** The fact that observed data may exhibit a straight-line trend on a suitable probability paper but do not exactly follow the theoretical curve to be fitted leads to the belief that singular sampled events cannot be represented with perfect confidence by the theory of probability. It is therefore important to know the reliability of results obtained by the frequency analysis; i.e., to know how well the individual event agrees with the theoretical prediction derived from the sampled data.

  The curve or distribution function fitted to the hydrologic data can be considered only to represent either the mean or sometimes the mode of the data at a given cumulative probability or recurrence interval. The distribution of the data for the given cumulative probability or recurrence interval can be described by the so-called *confidence limits* established on both sides of the fitted curve. Such confidence limits define the probability density areas on both sides of the mean, or of the mode of an assumed distribution of the data for the given cumulative probability or recurrence interval. *Control curves*, which join the equal confidence limits, can then be drawn to show the *confidence bands*. The reliability of any plotted point lying within the confidence band is thus indicated by the probability on which the confidence limits are based.

  Gumbel [28] has proposed a method for establishing the confidence limits in the plotting of annual maximum values. This method is based on the principle that the theoretical value of rank m situated on the straight line and corresponding to a given recurrence interval is the approximation to the most probable mth value. Considering a probability equal to 68.269 per cent (i.e., the probability for a deviation of $\pm\sigma$ from the predicted value in a normal distribution), the control curves can be constructed, respectively, above and below the fitted theoretical straight line on the extremal probability paper with a vertical distance of Δx from the line. In other words, the mth

observation is contained in the confidence band defined by $x - \Delta x < x < x + \Delta x$
It is expected that 68.269 per cent of all possible values would fall within the band
Gringorten [131] has developed a set of graphs for use in contructing such confidence
limits. For practical purposes, the method can be simplified by the approximate

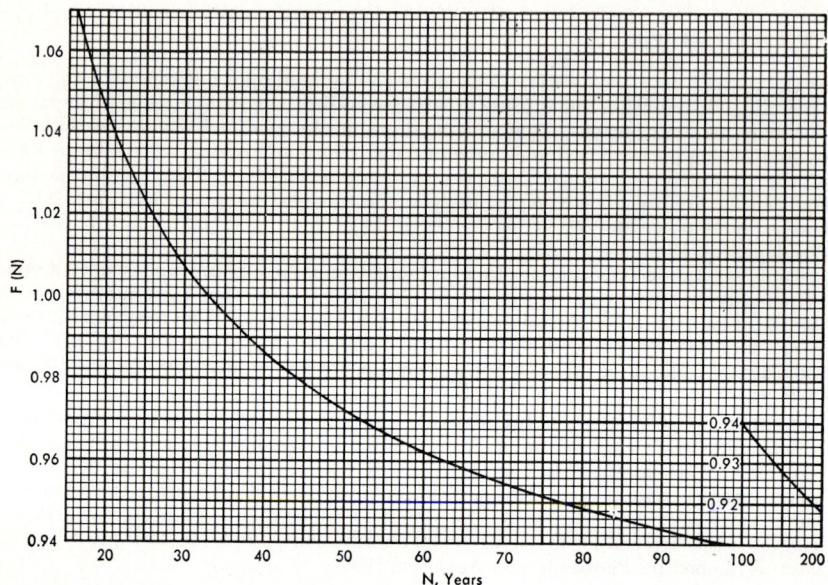

FIG. 8-I-9. Relation between N and $F(N)$.

procedure given below [52]. In Fig. 8-I-8, the control curves are constructed by this
method and the computation is given in Table 8-I-3.

a. For the largest value with $m = 1$, the half vertical width of the confidence band is

$$\Delta x_1 = sF(N) \tag{8-I-57}$$

where s is the standard deviation of the observed data and $F(N)$ is a function of the N
years of record as expressed graphically in Fig. 8-I-9.

b. For the second largest value with $m = 2$,

$$\Delta x_2 = \frac{0.661(N+1)}{N-1} \Delta x_1 \tag{8-I-58}$$

c. For intermediate values of rank m,

$$\Delta x_m = \frac{0.877}{\sqrt{N}} \Delta x_1 F(T_M) \tag{8-I-59}$$

where $F(T_M)$ is a function of the recurrence interval T_M as expressed graphically in
Fig. 8-I-10. When T_M is greater than 10 years, $F(T_M) = \sqrt{T_M}$.

d. For very small values, control curves are generally not necessary. For extrapolation beyond the largest value, Gumbel suggested that the control curves be drawn
as two lines parallel to the extrapolated straight line. This suggestion, however, will
result in sudden breaks on the control curves and in narrowing down the growing
width of the confidence band. Therefore, it is generally not followed. Kimball [12]

has suggested a procedure for constructing the control curves by a method of order statistics, which avoids such difficulties in extrapolation.

In the above discussion, the confidence limits, which represent the probable errors of estimate, are computed on the basis of an assumed distribution of errors, since the actual distribution is unknown. In the case of regional analysis (Subsec. IV-J), an approximate actual distribution of the sampling errors may be established from the data obtained from a group of stations in the region of statistical homogeneity. The

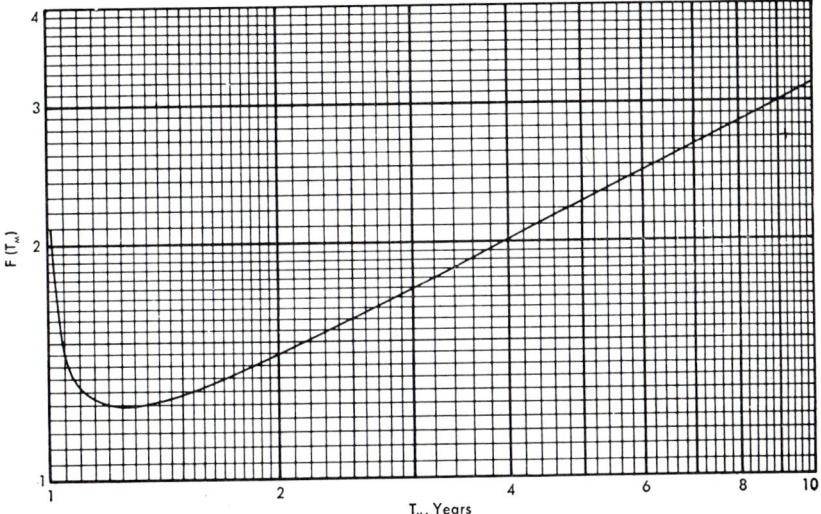

Fig. 8-I-10. Relation between T_M and $F(T_M)$.

sampling errors from the station mean for a given recurrence interval can be computed and analyzed to develop a distribution function. From this function, the per cent of errors to be equaled or exceeded among a given per cent, say 68.269 per cent, of the sampled stations can be determined. This per cent of errors multiplied by the variate x of the given recurrence interval will give the range Δx which can be used to construct the confidence limits. For other recurrence intervals, the confidence limits can be computed similarly.

Occasionally, the plotted point or points near the two ends of the distribution depart markedly from the general path indicated by the curve of best-fit. The departure may be so great that the plotted points fall far outside the confidence band, becoming the so-called *off-control data*. It has been explained that such off-control data follow some other type of distribution which applies to events that may occur at very long recurrence intervals. In other words, such events are statistically incompatible with the events with which they are associated in a given sample. Therefore, it is logical to recognize that these data may have an actual recurrence interval many times greater than the length of record. Whether or not these data follow the usual pattern or some other pattern of distribution than that represented by the rest of the data, it would seem obvious that they cannot be assumed to have a recurrence interval equal to the value computed by the plotting-position formula for which the available length of record is used. Consequently, inclusion of the off-control data in the computation of a theoretical probability curve would produce a different result from the one the sample should indicate. Since the off-control data are, in a sense, nonhomogeneous with the rest of the sample, they should be excluded from the fitting of the data. If the theoretical curve first computed using all data produces off-control data, the curve should be recomputed by excluding the latter unless the off-control data are *assumed* to be homogeneous with the total sample.

2. Prediction Reliability.

In the above discussion on sampling reliability, the probable error in the estimate of the variate by frequency analysis for a given recurrence interval is considered. Such errors which would affect the reliability of the result are largely due to sampling defects. There is another problem which relates to the probability of an event of a given average recurrence interval occurring during a given period of time. This probability would affect the reliability of prediction on the basis of the recurrence interval obtained from data fitting.

As described in Subsection IV-C, the recurrence interval is considered as the mean time interval but not as the actual time interval between two events of the given magnitude. Once the variate of a given recurrence interval is obtained from the fitted data, it is desirable to know the probability that this variate will occur in the forthcoming period of n years. This probability or the variation of recurrence interval of an event having a given magnitude about the mean recurrence intervals has been studied by many hydrologists using nonparametric probabilities [36, 42, 132–133].

From a fitted cumulative probability curve, the probability $P(X \leq x)$ corresponding to a variate x represents the probability that the value x will not be equaled or exceeded during a certain time interval. By the multiplicative law of probability, the probability of not exceeding the value of x in n years for an independent series of events is

$$P(X \leq x)_n = [P(X \leq x)]^n \qquad (8\text{-I-}60)$$

or

$$P(X \geq x)_n = 1 - [1 - P(X \geq x)]^n \qquad (8\text{-I-}61)$$

Since the recurrence interval $T = 1/P(X \geq x)$,

$$P(X \geq x)_n = 1 - \left(1 - \frac{1}{T}\right)^n \qquad (8\text{-I-}62)$$

Thus,

$$n = \frac{\ln P(X \leq x)_n}{\ln P(X \leq x)} = \frac{\ln [1 - P(X \geq x)_n]}{\ln [1 - P(X \geq x)]}$$

$$= \frac{\ln [1 - P(X \geq x)_n]}{\ln [(T-1)/T]} \qquad (8\text{-I-}63)$$

The value of $P(X \leq x)$, $P(X \geq x)$, or T of a given variate x can be obtained from the fitted data. The probability that this variate will occur in a period of n years can be computed by Eq. (8-I-61) or (8-I-62). If this probability $P(X \geq x)_n$ is given according to a design policy, the value of n, known as a *design period*, can be computed by Eq. (8-I-63). As an illustration, an additional scale is shown on top of the diagram in Fig. 8-I-8 for $P(X \geq x)_n$ in $n = 10$ years. The scale is computed by means of Eq. (8-I-63). Thus, the probability that a 10-min rainfall depth of 0.91 in. with a recurrence interval of 10 years will have a chance of 65 per cent to occur in the next 10 years. If a chance of 50 per cent occurrence in the next 10 years is considered in the design, the design rainfall should be 0.97 in. and have a recurrence interval of 15 years. Similarly, another scale for 20 years can be drawn and it can be shown that this rainfall of 0.91 in. will have a chance of 88 per cent to occur in the next 20 years.

I. Theoretical Justifications

From a practical point of view, the frequency analysis is only a procedure to fit the hydrologic data to a mathematical model of distribution. It is only experience and verification of data that decide the use of a certain distribution. However, there are several theoretical interpretations or reasonings for the preference of one distribution to another. Such interpretations would describe a physical process of the hydrologic phenomena and thus help to understand the procedure of frequency analysis and the significance of the results, but they are usually based on a number of assumptions which may not be readily satisfied in the real world. Most theoretical distributions recommended for hydrologic frequency analysis are asymptotic. The asymptotic condition is valid only when the number of variates becomes indefinitely large. Most of these distributions also assume independent variables. In actual hydrologic phenomena, however, the number of variates is always limited and mostly of small size, and the variables are likely to be interdependent to a certain extent.

1. Type I Extremal Distribution. This distribution was first proposed by Gumbel [23] for the analysis of flood frequencies. Gumbel considered the daily flow as a statistical variable unlimited to the positive end of the distribution, and defined a flood as being the largest value of the 365 daily flows. The flood flows are therefore the largest values of flows. According to the theory of extreme values, the annual largest values of a number of years of record will approach a definite pattern of frequency distribution when the number of observations in each year becomes large. Thus, the annual maximum floods constitute a series which can be fitted in the theoretical extremal distribution of Type I. Although it has been questioned whether the number of observations in a year is large enough for the asymptotic distribution to be approached, practical applications have shown satisfaction with the use of this theory to many problems.

In applying this theory to some meteorological data, Barricelli [134] and Brooks and Carruthers [1] have found some defects and therefore modified the theory in their use. They found that for temperatures, the Gumbel approximation overestimates, and for rainfall, underestimates, the maximum values reached in long periods. As a further improvement, Jenkinson [135] derived a general solution of the functional equation which should satisfy the extreme values of all types of distributions applicable to meteorological data. Borgman [136] proposed a distribution of near extremes which can be applied to limited and small-size samples.

2. Lognormal Distribution. This distribution has been used empirically for hydrologic frequency analysis since Hazen [12] first proposed it in 1914 for flood studies. In 1955, Chow [15] offered a theoretical interpretation to justify its use. Chow considered that the occurrence of a hydrologic event is a result of the joint action of many causative factors. Thus, a variate x is equal to the product of a large number of r independent magnitudes $x_1, x_2, \ldots x_r$ which are respectively due to the r causative factors. The logarithm of x is therefore equal to the sum of logarithms of a very large number of independent variates. By the central limit theorem, it can be shown that the logarithm of x is normally distributed when r becomes infinitely large.

The lognormal distribution contains three interdependent parameters. When hydrologic data are plotted on a lognormal probability paper, a straight-line trend is possible only for one value of C_s when C_v is given. Figure 8-I-11 shows that the plot is theoretically a straight line only for $C_s = 1.139$ when $C_v = 0.364$. For other values of C_s, the plots are curved for $C_v = 0.364$. Since the value of C_s computed from ordinary hydrologic data is not so reliable, it has been suggested that if the plot shows curvature, the value of C_s should be modified so that a straight line is obtained. Otherwise, a special probability paper may be constructed if a straight-line plot for the original value of C_s is desired [15].

3. Exponential Distribution. The exponential distribution has been applied empirically to partial-duration series. However, Chow [52] reasoned that the probability $p(x)$ of occurrence of a variate is the product of the probabilities of r number of causative factors. Thus, $p(x) = p^r$ where p is the geometric mean probability of all causative factors. When r is infinitely large and x is of high magnitude, it can be shown mathematically that the distribution of x is exponential.

4. Logextremal Distribution. The Type III extremal distribution was first proposed by Gumbel [57] for drought frequency analysis. Gumbel defined the drought as the smallest annual values of the mean daily discharges of a river. Since there is always a limit to the drought with a minimum of zero, Type III extremal distribution is assumed to be suitable. In this distribution, Eq. (8-I-36), three parameters are involved. The parameter ϵ is the lower limit called the *minimum drought*. This is the drought for which the probability of a value equal to or greater than it is unity and the recurrence interval is infinite. The parameter θ is called the *drought characteristic*, which has a recurrence interval of 1.58198 years. The parameter k has no particular name with reference to the drought, but its reciprocal is a scale parameter which defines skewness.

In applying Gumbel's method to drought solution, the droughts can be plotted on logextremal probability paper. If $\epsilon = 0$, the plot should have a straight-line trend. If $\epsilon > 0$, the plot will appear curved at the side of the long recurrence interval.

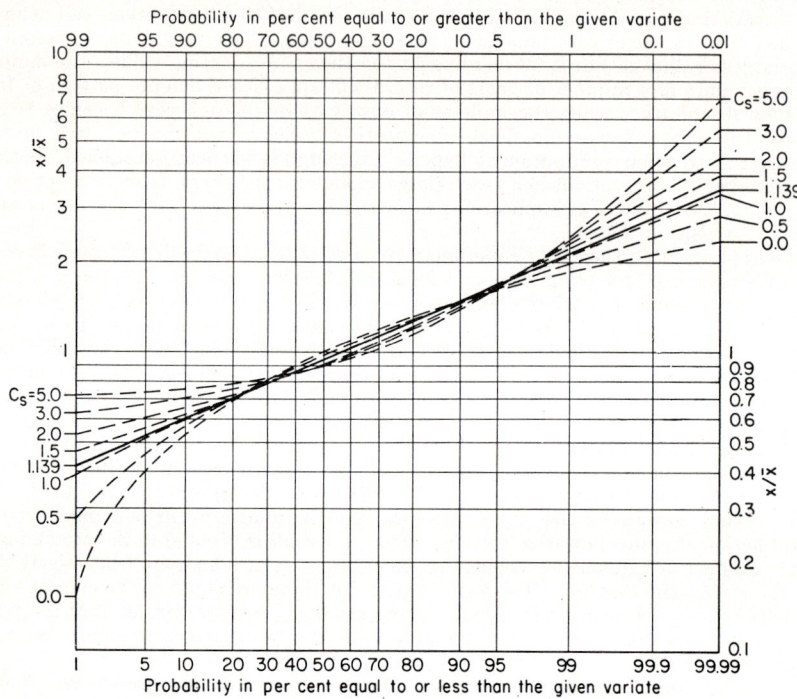

FIG. 8-I-11. Lognormal probability plotting.

The parameters of the distribution may be computed by statistical fitting of the distribution function. For extrapolation, the computed distribution function can be used since graphical extension of the curve may not be easy.

J. Regional Analysis

The observations at a geographical point, such as at the site of a rain gage or a stream gaging station, are *point data*. Extension of the results of the frequency analysis of the point data to an area requires *regional analysis*. Various methods of regional analysis have been developed, including the station-year method for rainfall analysis (Sec. 9) and the regional methods for flood analysis [38] (Subsec. 25-I-III-B-I). For all these methods, a statistically homogeneous region is defined. Within such regions, the results of point-data analysis can be averaged to best represent the frequency characteristics of the whole region. Usually, an average probability curve so obtained is applicable throughout the region.

In order to define a homogeneous region, a test has been developed by Langbein [7, 137] for regional flood-frequency analysis practiced by the U.S. Geological Survey (Subsec. 25-I-III-B-1). This *homogeneity test* requires a study of the 10-year flood as estimated from the probability curve at each station within a region. These 10-year floods expressed as ratios to mean annual floods (which have a recurrence interval of 2.33 years according to the extremal distribution) are averaged to obtain the mean 10-year ratio for the area. The recurrence interval corresponding to the mean annual flood times the averaged 10-year ratio is determined from the probability curve of each station and plotted against the number of years of record on a test graph (Fig. 8-I-12). If the points for all of the stations lie between the two control curves (indicating 95 per cent reliability) on the graph, they are considered homogeneous.

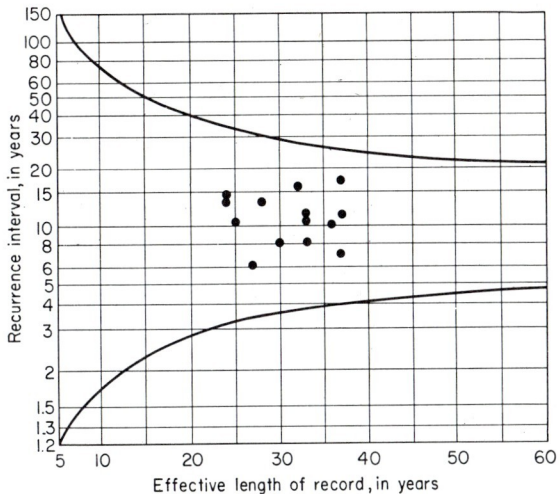

FIG. 8-I-12. Homogeneity test graph. (*U.S. Geological Survey.*)

Points outside the curves indicate that the region should be subdivided for homogeneity. Repeat the procedure until all the subdivided areas pass the homogeneity test.

The principle of the above test is to determine in a statistical sense whether the records in a group differ from one another by amounts that cannot reasonably be expected by chance. If the differences are found to be no more than those due to the operations of chance, they can be considered to represent merely different aspects of the same event and thus can be grouped. The test graph is constructed on the basis of the extremal distribution. The control curves represent a range of variation equal to two standard deviations of the reduced variate [i.e., $(a + x)/c$ in Eq. (8-I-33)] on the 10-year flood. This means that 95 per cent of the estimate will lie within 2σ of the most probable value of a recurrence interval of 10 years. The 10-year flood is used in this test because this is the longest recurrence interval for which most flood records will give dependable estimates.

V. REFERENCES

1. Brooks, C. E. P., and N. Carruthers: "Handbook of Statistical Methods in Meteorology," Great Britain Meteorological Office, H.M.S.O., London, 1953.
2. Court, Arnold: Some new statistical techniques in geophysics, in "Advances in Geophysics," ed. by H. E. Landsberg, vol. 1, Academic Press Inc., New York, 1952, pp. 45–85.
3. Johnstone, Don, and W. P. Cross: "Elements of Applied Hydrology," The Ronald Press Company, New York, 1949, pp. 236–264.
4. Foster, E. E.: "Rainfall and Runoff," The Macmillan Company, New York, 1948.
5. Beard, L. R.: Statistical methods in hydrology, U.S. Army Engineer District, Corps of Engineers, Sacramento, Calif., January, 1962.
6. Benson, M. A.: Evolution of methods for evaluating the occurrence of floods, *U.S. Geol. Surv. Water-Supply Paper* 1580-A, 1962.

7. Dalrymple, Tate (ed.): Flood-frequency analysis, Manual of Hydrology, pt. 3, Flood-flow techniques, *U.S. Geol. Surv. Water-Supply Paper* 1543-A, 1960.
8. Foster, H. A.: "Methods of Analyzing Hydrological Records," privately published, 1955.
9. Foster, H. A.: Duration curves, *Trans. Am. Soc. Civil Engrs.*, vol. 99, pp. 1213–1235, 1934.
10. Fuller, W. E.: Flood flows, *Trans. Am. Soc. Civil Engrs.*, vol. 77, pp. 564–617, 1914.
11. Horton, R. E.: Frequency of recurrence of Hudson River floods, *U.S. Weather Bur. Bull. Z*, pp. 109–112, 1913.
12. Hazen, Allen: Discussion on Flood flows by W. E. Fuller, *Trans. Am. Soc. Civil Engrs.*, vol. 77, p. 628, 1914.
13. Hazen, Allen: Discussion of Probable variations in yearly runoff by L. S. Hall, *Trans. Am. Soc. Civil Engrs.*, vol. 84, pp. 214–224, 1921.
14. Hazen, Allen: "Flood Flows, A Study of Frequencies and Magnitudes," John Wiley & Sons, Inc., New York, 1930.
15. Chow, V. T.: The log-probability law and its engineering applications, *Proc. Am. Soc. Civil Engrs.*, vol. 80, paper no. 536, pp. 1–25, November, 1954.
16. Lane, E. W., and Kai Lei: Stream flow variability, *Trans. Am. Soc. Civil Engrs.*, vol. 115, pp. 1084–1098, 1950.
17. Foster, H. A.: Theoretical frequency curves and their application to engineering problems, *Trans. Am. Soc. Civil Engrs.*, vol. 87, pp. 142–173, 1924.
18. Switzer, F. G., and H. G. Miller: Floods, *Cornell Univ. Eng. Exp. Sta. Bull.* 13, December 15, 1929.
19. Hall, L. S.: The probable variations in yearly runoff as determined from a study of California streams, *Trans. Am. Soc. Civil Engrs.*, vol. 84, pp. 191–213, 1921.
20. Goodrich, R. D.: Straight-line plotting of skew-frequency data, *Trans. Am. Soc. Civil Engrs.*, vol. 91, pp. 1–43, 1927.
21. Harris, R. M.: Straight-line treatment of hydraulic duration curves, *Univ. Wash. Eng. Exp. Sta. Bull.* 65, May, 1932.
22. Slade, J. J., Jr.: An asymmetric probability function, *Proc. Am. Soc. Civil Engrs.*, vol. 60, pp. 1007–1023, 1934; also *Trans. Am. Soc. Civil Engrs.*, vol. 101, pp. 35–61, 1936.
23. Gumbel, E. J.: The return period of flood flows, *Ann. Math. Statist.*, vol. XII, no. 2, pp. 163–190, June, 1941.
24. Gumbel, E. J.: Probability interpretation of the observed return periods of floods, *Trans. Am. Geophys. Union*, vol. 21, pp. 836–850, 1941.
25. Gumbel, E. J.: Statistical control-curves for flood-discharges, *Trans. Am. Geophys. Union*, vol. 23, pp. 489–500, 1942.
26. Gumbel, E. J.: On the plotting of flood discharges, *Trans. Am. Geophys. Union*, vol. 24, pp. 699–719, 1943.
27. Gumbel, E. J.: Floods estimated by probability methods, *Eng. News-Rec.*, vol. 134, no. 24, pp. 97–101, June 14, 1945.
28. Gumbel, E. J.: The statistical forecast of floods, *Bull.* 15, The Ohio Water Resources Board, Columbus, Ohio, 1949.
29. Gumbel, E. J.: Statistical theory of floods and droughts, *J. Inst. Water Engrs.*, vol. 12, no. 3, pp. 157–184, May, 1958.
30. Powell, R. W.: A simple method of estimating flood frequencies, *Civil Eng.*, vol 13, pp. 105–106, February, 1943.
31. Cross, W. P.: Floods in Ohio, magnitude and frequency, *Bull.* 7, The Ohio Water Resources Board, Columbus, Ohio, 1946.
32. Potter, W. D.: Simplifications of the Gumbel method for computing probability curves, *U.S. Dept. Agr. Soil Conserv. Serv.* SCS-TP-78, May, 1949.
33. Benson, M. A.: Characteristics of frequency curves based on a theoretical 1,000-year record, in Ref. 7, pp. 57–74; also *U.S. Geol. Surv. Open-file Report*, 1952.
34. Jarvis, C. S., and others: Floods in the United States, *U.S. Geol. Surv. Water-Supply Paper* 771, 1936.
35. Beard, L. R.: Statistical analysis in hydrology, *Trans. Am. Soc. Civil Engrs.*, vol. 108, pp. 1110–1160, 1943.
36. Thomas, H. A., Jr.: Frequency of minor floods, *J. Boston Soc. Civil Engrs.*, vol. 34, pp. 425–442, October, 1948.
37. Langbein, W. B.: Annual floods and the partial-duration flood series, *Trans. Am. Geophys. Union*, vol. 30, pp. 879–881, 1949.
38. Dalrymple, Tate: Regional flood frequency, in "Surface Drainage," *Highway Res. Board Res. Rept.* 11-B, pp. 4–20, December, 1950.
39. Bodhaine, G. S., and W. H. Robinson: Floods in Western Washington—frequency and magnitude in relation to drainage basin characteristics, *U.S. Geol. Surv. Cir.* 191, 1952.

REFERENCES

40. Review of flood frequency methods, Final Report of the Subcommittee of the Joint Division Committee on Floods, *Trans. Am. Soc. Civil Engrs.*, vol. 118, pp. 1220–1230, 1953.
41. Mitchell, W. D.: Floods in Illinois—magnitude and frequency, Illinois Division of Waterways in cooperation with U.S. Geological Survey, 1954.
42. Gumbel, E. J.: The calculated risk in flood control, *Appl. Sci. Res.* (The Hague), sec. A, vol. 5, pp. 273–280, 1955.
43. "Studies of Floods and Flood Damages 1952–1955," American Insurance Association, New York, May, 1956.
44. Chow, V. T.: Hydrologic studies of floods in the United States, in "Symposia Darcy," *Intern. Assoc. Sci. Hydrology Pub.* 42, pp. 134–170, 1956.
45. Rowe, R. R., G. L. Long, and T. C. Royce: Flood frequency by regional analysis, *Trans. Am. Geophys. Union*, vol. 38, pp. 879–884, 1957.
46. Moran, P. A. P.: The statistical treatment of flood flows, *Trans. Am. Geophys. Union*, vol. 38, pp. 519–523, 1957.
47. Chow, V. T.: Frequency analysis in small watershed hydrology, *Agr. Eng.*, vol. 39, pp. 222–225, and 231, April, 1958.
48. Rangarajan, R.: A new approach to peak flow estimation, *J. Geophys. Res.*, vol. 65, no. 2, pp. 643–650, February, 1960.
49. Hall, W. A., and D. T. Howell: Estimating flood probabilities within specific time intervals, *J. Hydrology*, vol. 1, no. 3, pp. 265–271, 1963.
50. Yarnell, D. L.: Rainfall intensity-frequency data, *U.S. Dept. Agr. Misc. Publ.* 204, 1936.
51. Engineering Staff of Miami Conservancy District: Storm rainfall of eastern United States, *Tech. Rept.*, pt. V, Miami Conservancy District, Dayton, Ohio, 1917.
52. Chow, V. T.: Frequency analysis of hydrologic data with special application to rainfall intensities, *Univ. Illinois Eng. Exp. Sta. Bull.* 414, July, 1953.
53. Chow, V. T.: Design charts for finding rainfall intensity frequency, *Water and Sewage Works*, vol. 99, no. 2, pp. 86–88, February, 1952; also *Concrete Pipe News*, vol. 4, no. 6, pp. 8–10, June, 1952.
54. Hershfield, D. M.: Rainfall frequency atlas of the United States for durations from 30 minutes to 24 hours and return periods from 1 to 100 years, *U.S. Weather Bur. Tech. Rept.* 40, May, 1961.
55. Huff, F. A., and J. C. Neill: Comparison of several methods for rainfall frequency analysis, *J. Geophys. Res.*, vol. 64, no. 5, pp. 541–547, May, 1959.
56. Hershfield, D. H.: An empirical comparison of the predictive value of three extreme-value procedures, *J. Geophys. Res.*, vol. 67, no. 5, pp. 1535–1542, April, 1962.
57. Gumbel, E. J.: Statistical theory of droughts, *Proc. Am. Soc. Civil Engrs.*, vol. 80, sep. no. 439, pp. 1–19, May, 1954.
58. Gumbel, E. J.: Statistical forecast of droughts, *Bull. Intern. Assoc. Sci. Hydrology*, 8th year, no. 1, pp. 5–23, April, 1963.
59. Velz, C. J., and J. J. Gannon: Drought flow characteristics of Michigan streams, Department of Environmental Health, School of Public Health, University of Michigan, in cooperation with Michigan Water Resources Commission, 1960.
60. Hardison, C. H., and R. O. R. Martin: Low-flow frequency curves for selected long-term stream-gaging stations in eastern United States, *U.S. Geol. Surv. Water-Supply Paper* 1669-G, 1963.
61. Hudson, H. E., Jr., and W. J. Roberts: 1952–55 Illinois drought with special reference to impounding reservoir designs, *Illinois State Water Surv. Bull.* 43, 1955.
62. Stall, J. B., and J. C. Neill: A partial duration series for low-flow analysis, *J. Geophys. Res.*, vol. 66, no. 12, pp. 4219–4225, 1961.
63. Matalas, N. C.: Probability distribution of low flows, *U.S. Geol. Surv. Profess. Paper* 434-A, 1963.
64. Ledbetter, J. O., and E. F. Gloyna. Predictive techniques for water quality inorganics, *Proc. Am. Soc. Civil Engrs.*, *J. Sanitary Eng. Div.*, vol. 90, no. SA1, pp. 127–151, February, 1964.
65. Seaway, chap. 1 in B. V. Korvin-Kroukovsky (ed.): "Theory of Seakeeping," The Society of Naval Architects and Marine Engineers, 1961.
66. "Ocean Wave Spectra," Proceedings of a Conference arranged by the National Academy of Sciences, Prentice-Hall, Inc., Englewood Cliffs, N.J., 1963.
67. Longuet-Higgins, M. S.: On the statistical distribution of sea waves, *Sears Foundation: J. Marine Res.*, vol. XI, no. 3, pp. 245–266, December, 1952.
68. Putz, R. R.: Ocean wave record analysis, ordinate distribution and wave heights, *Univ. Calif. Inst. Eng. Res. Tech. Rept.* ser. 3, issue 351, 1953; also summarized as Measurement and analysis of ocean waves, in chap. 5 of J. W. Johnson (ed.): "Ships

and Waves," Council on Wave Research and Society of Naval Architects and Marine Engineers, 1955, pp. 63–72.
69. Cartwright, D. E., and M. S. Longuet-Higgins: The statistical distribution of the maxima of a random function, *Proc. Royal Soc. ser. A*, vol. 237, pp. 212–232, 1956.
70. Gumbel, E. J.: Statistical distribution patterns of ocean waves, *Trans. Soc. Naval Arch. Marine Engrs.*, vol. 8, p. 427, 1956.
71. Jasper, N. H.: Statistical distribution patterns of ocean waves and of wave induced ship stresses and motions, with engineering applications, a doctorate dissertation, The Catholic University of America Press, Washington, D.C., 1956; also summarized as Distribution patterns of wave heights, and ship motions and hull stress, chap. 34 in J. W. Johnson (ed.): "Ships and Waves," Council on Wave Research and Society of Naval Architects and Marine Engineers, 1955, pp. 489–503.
72. Bennet, Rutger: Stress and motion measurements on ships at sea, The Swedish Shipbuilding Research Foundation, *Rept.* 13, Goteborg, Sweden, 1963.
73. Elderton, W. P.: "Frequency Curves and Correlation," Cambridge University Press, London, 4th ed., 1953.
74. Burington, R. S., and D. C. May: "Handbook of Probability and Statistics with Tables," Handbook Publishers, Inc., Sandusky, Ohio, 1953; reprinted, McGraw-Hill Book Company, Inc., New York, 1958.
75. Feller, William: "An Introduction to Probability Theory and Its Applications," vol. 1, John Wiley & Sons, Inc., 2d ed., 1957.
76. Doob, J. L.: "Stochastic Processes," John Wiley & Sons, Inc., New York, 1953.
77. Takács, Lajos: "Stochastic Processes," Methuen and Co., Ltd., London, 1960.
78. Goode, H. H., and R. E. Machol: "System Engineering," McGraw-Hill Book Company, Inc., New York, 1957.
79. Riordan, John: "Stochastic Service Systems," John Wiley & Sons, Inc., New York, 1962.
80. Brown, R. G.: "Smoothing, Forecasting and Prediction of Discrete Time Series," Prentice-Hall, Inc., Englewood Cliffs, N.J., 1963.
81. Rosenblatt, Murray (ed.): "Time Series Analysis," John Wiley & Sons, Inc., New York, 1963.
82. Hannan, E. J.: "Time Series Analysis," Methuen and Co., Ltd., London, 1960.
83. Davis, H. T.: "The Analysis of Economic Time Series," The Principia Press, Inc., Bloomington, Ind., 1941.
84. Tinter, Gerhard: "Econometrics," John Wiley & Sons, Inc., New York, 1952.
85. Kendall, M. G.: "The Advanced Theory of Statistics," vol. 2, Charles Griffen and Co., Ltd., London, 1948.
86. Schuster, A.: On the investigation of hidden periodicities with application to a supposed 26-day period meteorological phenomena, *Terrestrial Magnetism*, vol. 3, no. 1, pp. 13–41, March, 1898.
87. Walker, Sir Gilbert: On periodicity, *Roy. Meteorol. Soc. J.*, vol. 51, pp. 337–345; disc. 345–346, October, 1925.
88. Fisher, R. A.: Tests of significance in harmonic analysis, *Royal Soc. London Proc.*, ser. A, vol. 125, no. 796, pp. 54–59, 1929.
89. Chow, V. T.: Do climatic variations follow definite cycles? *Civil Eng.*, vol. 20, no. 7, p. 470, July, 1950.
90. Huntington, Ellsworth: "Civilization and Climate," Yale University Press, 3d ed., 1924.
91. Brakensiek, D. L.: Selecting the water year for small agricultural watersheds, *Trans. Am. Soc. Agr. Engrs.*, vol. 2, no. 1, pp. 5–8 and 10, 1959.
92. Leopold, L. A.: Probability analysis applied to a water supply problem, *U.S. Geol. Surv. Cir.* 410, 1959.
93. Hurst, H. E.: Long-term storage capacity of reservoirs, *Trans. Am. Soc. Civil Engrs.*, vol. 116, pp. 770–799, 1951.
94. Dixon, W. J., and F. J. Massey, Jr.: "Introduction to Statistical Analysis," McGraw-Hill Book Company, Inc., New York, 2d ed., 1957.
95. Mood, A. M., and F. A. Graybill: "Introduction to the Theory of Statistics," McGraw-Hill Book Company, Inc., New York, 2d ed., 1963.
96. Pearson, Karl: "Tables for Statisticians and Biometricians," Part I, The Biometric Laboratory, University College, London; printed by Cambridge University Press, London, 3d ed., 1930.
97. Fréchet, Maurice: Sur la loi de probabilité de l'écart maximum (On the probability law of maximum error), *Ann. Soc. Polonaise Math.* (Cracow), vol. 6, pp. 93–116, 1927.
98. Fisher, R. A., and L. H. C. Tippett: Limiting forms of the frequency distribution of the smallest and largest member of a sample, *Proc. Cambridge Phil. Soc.*, vol. 24, pp. 180–190, 1928.

REFERENCES

99. von Mises, R.: La distribution de la plus grande de n valeurs (The distribution of the largest of n values), *Revue Math. l'Union Interbalcanique* (Athens), vol. 1, pp. 1–20, 1936.
100. Gumbel, E. J.: "Statistics of Extreme Values," Columbia University Press, New York, 1958.
101. Statistical theory of extreme values and some practical applications, *U.S. Nat. Bur. Stds. Appl. Math. Ser.* 33, 1954.
102. Probability tables for analysis of extreme-value data, *U.S. Nat. Bur. Stds. Appl. Math. Ser.* 22, 1953.
103. Weibull, W.: A statistical theory of the strength of materials, *Ing. Vetenskaps Akad. Handl.* (Stockholm), vol. 151, p. 15, 1939.
104. Weibull, W.: The phenomenon or rupture in solids, *Ing. Vetenskaps Akad. Handl.* (Stockholm), vol. 153, p. 17, 1939.
105. Galton, Francis: Statistics by intercomparison with remarks on the law of frequency of error, *The London, Edinburgh, Dublin Phil. Mag. J. Sci.*, 4th ser., vol. XLIX, p. 33, January–June, 1875.
106. Chow, V. T.: On the determination of frequency factor in log-probability plotting, *Trans. Am. Geophys. Union*, vol. 36, pp. 481–486, 1955.
107. Chow, V. T.: Determination of hydrologic frequency factor, *Proc. Am. Soc. Civil Engrs., J. Hydraulics Div.*, vol. 85, no. HY7, pp. 93–98, July, 1959.
108. Aitchison, J., and J. A. C. Brown: "The Lognormal Distribution, with Special Reference to the Uses in Economics," Cambridge University Press, London, 1957.
109. Alekseyev, G. A.: Determination of standard parameters of a logarithmically normal distribution curve by three reference ordinates, *Soviet Hydrology: Selected Papers*, American Geophysical Union, no. 6, pp. 637–684, 1962.
110. Chow, V. T.: Discussion on Annual floods and the partial duration flood series, by W. B. Langbein, *Trans. Am. Geophys. Union*, vol. 31, pp. 939–941, 1950.
111. Thom, H. C. S.: Time interval distribution for excessive rainfalls, *Proc. Am. Soc. Civil Engrs., J. Hydraulics Div.*, vol. 85, no. HY7, pp. 83–91, July, 1959.
112. Chow, V. T.: A general formula for hydrologic frequency analysis, *Trans. Am. Geophys. Union*, vol. 32, pp. 231–237, 1951.
113. Hazen, Allen: Storage to be provided in impounding reservoirs for municipal water supply, *Trans. Am. Soc. Civil Engrs.*, vol. 77, pp. 1539–1659, 1914.
114. Flow in California streams, *Calif. State Dept. Pub. Works Bull.* 5, chap. 5, 1923.
115. Alekseyev, G. A.: O formule dlya vychisleniya obespechenosti gidrologicheskikh velichin (Formulas for the calculation of the confidence of hydrological quantities), *Metrologiia i Gidrologiia* (Leningrad), no. 6, pp. 40–43, November–December, 1955.
116. Leivikov, M. L.: "Meterologiia, gidrologiia i gidrometriia" (Meteorology, Hydrology and Hydrometry), Sel'khozgiz, Moscow, 2d ed., 1955.
117. Benard, A., and E. C. Bos-Levenbach: The plotting of observations on probability paper (Dutch), *Statistica* (Rijkswijk), vol. 7, pp. 163–173, 1953.
118. Blom, Gunnar: "Statistical Estimates and Transformed Beta-Variables," John Wiley & Sons, Inc., New York, 1958.
119. Tukey, J. W.: The future of data analysis, *Ann. Math. Statist.*, vol. 33, no. 1, pp. 1–67, 1962.
120. Gringorten, I. I.: A plotting rule for extreme probability paper, *J. Geophys. Res.*, vol. 68, no. 3, pp. 813–814, 1963.
121. Lieblein, Julius: A new method of analyzing extreme-value data, *Natl. Advs. Comm. Aero. Tech. Note* 3053, Washington, D.C., 1954.
122. Chernoff, Herman, and G. J. Lieberman: Use of normal probability paper, *J. Am. Statist. Assoc.*, vol. 49, pp. 778–785, 1954.
123. Kimball, B. F.: The bias in certain estimates of the parameters of the extreme-value distribution, *Ann. Math. Statist.*, vol. 27, pp. 758–767, 1956.
124. Kimball, B. F.: On the choice of plotting positions on probability paper, *J. Am. Statist. Assoc.*, vol. 55, no. 291, pp. 546–560, 1960.
125. Ferrell, E. B.: Plotting experimental data on normal or lognormal probability paper, *Indus. Qual. Control*, vol. 15, no. 1, pp. 12–15, July, 1958.
126. Benson, M. A.: Plotting positions and economics of engineering planning, *Proc. Am. Soc. Civil Engrs., J. Hydraulics Div.*, vol. 88, no. HY6, pt. 1, November, 1962.
127. Brakensiek, D. L.: Fitting generalized lognormal distribution to hydrologic data, *Trans. Am. Geophys. Union*, vol. 39, pp. 469–473, 1958.
128. Kimball, B. F.: Sufficient statistical estimation functions for the parameters of the distribution of maximum values, *Ann. Math. Statist.*, vol. 17, no. 3, pp. 299–309, September, 1946.
129. Kimball, B. F.: An approximation to the sampling variation of an estimated maximum

value of a given frequency based on fit of doubly exponential distribution of maximum values, *Ann. Math. Statist.*, vol. 20, no. 1, pp. 110–113, March, 1949.
130. Panchang, G. M., and V. P. Aggarwal: Peak flow estimation by method of maximum likelihood, *Tech. Memo.* HLO2, Government of India, Central Water and Power Research Station, Poona, India, March, 1962.
131. Gringorten, I. I.: Envelopes for ordered observations applied to meteorological extremes, *J. Geophys. Res.*, vol. 68, no. 3, pp. 815–826, 1963.
132. Kendell, G. R.: Statistical analysis of extreme values, First Canadian Hydrology Symposium, National Research Council of Canada, November 4 and 5, 1959.
133. Riggs, H. C.: Frequency of natural events, *Proc. Am. Soc. Civil Engrs., J. Hydraulics Div.*, vol. 87, no. HY1, pt. 1, pp. 15–26, January, 1961.
134. Barricelli, N. A.: Les plus grands et les plus petits maxima ou minima annuels d'une variable climatique (The largest and smallest of annual maxima or minima of a climatic variable), *Archiv for Mathematick og Naturvidenskab* (Oslo), vol. XLVI, no. 6, 1943.
135. Jenkinson, A. F.: The frequency distribution of the annual maximum (or minimum) values of meteorological elements, *Quart. J. Roy. Meteorol. Soc.*, vol. 81, no. 348, pp. 158–171, April, 1955.
136. Borgman, L. E.: The frequency distribution of near extremes, *J. Geophys. Res.*, vol. 66, no. 10, pp. 3295–3307, 1961.
137. Langbein, W. B., and others: Topographic characteristics of drainage basins, *U.S. Geol. Surv. Profess. Paper* 968-C, 1947, pp. 125–157.

Section 8-II

STATISTICAL AND PROBABILITY ANALYSIS OF HYDROLOGIC DATA

PART II. REGRESSION AND CORRELATION ANALYSIS

VUJICA M. YEVDJEVICH, *Professor of Civil Engineering, Colorado State University.*

I. Introduction..	8-44
II. Basic Definitions and Tools.................................	8-44
A. Hydrologic Variables and Series.........................	8-44
B. Definitions of Regression and Correlation.................	8-45
C. Curve Fitting..	8-46
D. General Model for Regression and Correlation Analyses.....	8-47
E. Transformation of Variables..............................	8-48
F. Use of Ungrouped and Grouped Data.....................	8-48
G. Statistical Inference in Regression and Correlation Analyses.	8-50
III. Simple Linear Regression and Correlation.....................	8-50
A. Regression Lines...	8-50
1. Analytical Method of Determining Parameters..........	8-50
2. Graphical Method of Determining Parameters..........	8-51
3. Grouped Data in the Bivariate Distribution............	8-52
B. Measures of Linear Correlative Association................	8-52
1. Correlation Coefficient and Coefficient of Determination..	8-52
2. Standard Deviation of Residuals......................	8-54
3. Interpretation of Parameters.........................	8-55
C. Statistical Inference.....................................	8-55
1. Correlation Coefficient..............................	8-55
2. Regression Coefficient...............................	8-56
3. Intercept...	8-57
4. Inference in the Case of Nonnormal Distributions of Variables Which Are Internally Dependent..................	8-57
IV. Simple Curvilinear Regression and Correlation.................	8-58
A. Curvilinear Regression...................................	8-58
B. Correlation Index..	8-58
C. Regression and Correlation Inference.....................	8-59

- V. Multiple Linear Regression and Correlation.................. 8-59
  - A. General... 8-59
  - B. Linear Regression with Three Variables................... 8-60
  - C. Linear Regression with Several Variables................. 8-60
  - D. Measures of Multiple Linear Correlative Association........ 8-61
    1. The Standard Deviation of Residuals................... 8-61
    2. Multiple Correlation Coefficient........................ 8-62
    3. Coefficient of Determination............................ 8-62
    4. Partial Correlation Coefficients........................ 8-62
    5. Beta Coefficients.. 8-64
  - E. Statistical Inference of Regression Coefficients.............. 8-64
- VI. Multiple Curvilinear Regression and Correlation................ 8-65
  - A. Analytical Method... 8-65
  - B. Graphical Method... 8-65
- VII. Multivariate Analysis... 8-66
- VIII. References.. 8-67

I. INTRODUCTION

The regression and correlation analysis is one of the oldest statistical tools used in hydrology. It was first used for filling missing data and extending short records at one hydrologic station by relating the available data at this station with those at adjacent stations. Now its application has been broadened to cover the study of the relationship between two or more hydrologic variables and also the investigation of dependence between the successive values of a series of hydrologic data.

This section deals with the basic definitions and presents discussion necessary for the understanding of the concepts of regression and correlation and the methods of their applications.

The simple linear regression and correlation are presented according to classical treatments, with special emphasis placed on the statistical inference for computing the parameters. For both grouped data in bivariate distribution and ungrouped data, the methods of computing the parameters are discussed. The simple curvilinear regression and correlation, mostly employing the polynomials as the regression function, is only briefly presented.

The multiple linear and curvilinear regression and correlation are discussed in some detail, especially with respect to the various ways of measuring the multiple and partial correlation and to the statistical inference of regression coefficients. The multiple regression and correlation analysis is used a great deal nowadays because such complicated analyses can be made practicable through numerical computations and thus can be economically executed with the aid of electronic digital computers.

II. BASIC DEFINITIONS AND TOOLS

A. Hydrologic Variables and Series

A *variable* in hydrology can be represented by either a *continuous series* (such as a recorded hydrograph) or a *discontinuous series* (such as annual flow values, flood peaks, etc.). The measuring interval may be selected in terms of time (hour, day, month, year), length (foot, mile), or surface area (acre, square mile). By use of the mean or total value of the variable in such intervals, a continuous series can be transformed into a discontinuous series. Generally, only the discontinuous series of data is statistically analyzed in hydrology.

Inasmuch as the hydrologic data are obtained by observations and by further appraisal of observed values, the hydrologic series are subject to human errors (random and systematic) and are often *nonhomogeneous*. *Random errors* are always present because of the inaccuracy in measurements and observations. *Systematic errors,* or *errors of inconsistency,* refer to errors occurring in one direction, such as trends or jumps in the series. Nonhomogeneity of the data results from changes due either to natural catastrophes, such as fires, landslides, etc., or to man-made developments. It is advisable to appraise the data in terms of their probable errors and nonhomogeneity before using them for statistical analysis and to consider the validity of the data in drawing conclusions concerning the reliability of the statistical parameters and relationships determined from them. This is especially important when the available data series represents a small sample, that is, where the sample size is smaller than about fifty items.

The relationships of variables in hydrology may either show the cause and effect at one hydrologic site (such as the precipitation-runoff relationship at one river basin) or correlate the effects only at neighboring sites (correlating precipitations at two stations or runoffs of two adjacent river basins). The most common case attempted is that of showing the relationship of an effect to many causes, of which a small number of the causes exert greater influence than do all others. In such a case, when neglected variables and inherent errors and nonhomogeneity of data have relatively small effects, the relationship between the remaining limited number of variables would indicate a narrow spread around a basic function. This is the form of relationship generally required and used, since pure, functional relationships in hydrology are rare.

B. Definitions of Regression and Correlation

If two variables, given as a series with concurrent values (x_i, y_i), show a concentration around an imaginary curve when plotted on a graph (Fig. 8-II-1), then for a large

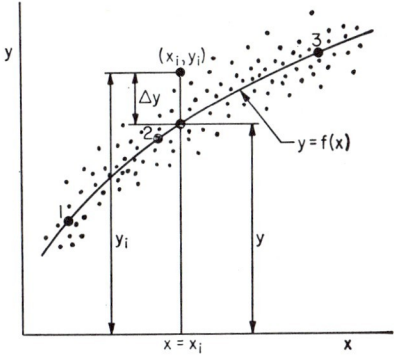

Fig. 8-II-1. Schematic representation of a simple regression and correlation analysis.

series there will always be a distribution of y values for a given value of x_i, or more precisely, a distribution of y values for a given interval Δx around x_i. The mean value y_0 of all y values for this given interval Δx around x_i is the expected value of y for the given $x = x_i$. A curve fitted to all mean values, y_0, is called the *regression line of y versus x.* On the other hand, the curve fitted to all expected (mean) values, x_0, for the given $y = y_i$, defines the *regression line of x versus y.* These two lines do not coincide, but have different parameters, showing the regressional relationships between the variables.

A pure functional relationship between variables assumes that all points would follow a curve, without spread. Inasmuch as the spread of points around the regression lines may actually be great or small, the degree of association of the variables involved is generally called *correlation* and is defined by the parameters of correlation. The correlation is greater when the points are closer to the lines.

C. Curve Fitting

The methods of curve fitting may be graphical or analytical. If any curve of the type

$$y = f(x, a, b, c, \ldots) \tag{8-II-1}$$

is fitted to the plotted points, the graphical procedure is simply that of tracing a curve by eyes through the mean of the spread (Fig. 8-II-1). This procedure is often subject to large errors, but is practical and expedient.

The simplest analytical procedure is to fit a function employing as many parameters $a, b, c, \ldots$, in Eq. (8-II-1), as the number of selected points. These points can be selected from the sample on the basis of their positions within the sample. Thus, for three parameters, the use of points 1, 2, and 3 is shown in Fig. 8-II-1. The values (x_i, y_i) for each point must satisfy Eq. (8-II-1), so that there are m equations for the determination of m parameters. Better yet, they may be points which represent a weighted average of the groups of points in the sample. Thus, for three parameters, the points would be divided into three adjacent groups, the weighted mean of each group would be determined, and the equation fitted to pass through these three means. These points may be fitted graphically to produce curves. The parameters are then determined from the curves. This procedure is called the *three-point method*.

The currently approved analytical method of fitting curves to scattered points is to minimize the sum of squares of departures $\Delta_{yi} = y_i - y$, where, for a given x_i, the value y is determined from the fitted curve, and y_i is from the observed point (Fig. 8-II-1). This is known as the *least-squares method*. In order that the line of Eq. (8-II-1) may have the minimum $\sum_{1}^{N} (\Delta_{yi})^2$, where N is the number of points, all partial derivatives of this sum with respect to the parameters $a, b, c, \ldots$ should be zero, so that

$$\frac{\partial \sum_{i=1}^{N} (y_i - y)^2}{\partial a} = 0; \quad \frac{\partial \sum_{i=1}^{N} (y_i - y)^2}{\partial b} = 0; \ldots \tag{8-II-2}$$

These partial derivatives give m equations for the determination of m parameters. The simplest functions, given by Eq. (8-II-1), are those which are linear with respect to the parameters. Among such functions the polynomial function for x is a general case, since other functions can be approximated by polynomials if they are developed in power-series form. There must be more points than parameters, or $N > m$. Generally, N should be much greater than m, if the derived line is to be used for prediction purposes.

For example, the fitting of a quadratic parabola of the form

$$y = a + bx + cx^2 \tag{8-II-3}$$

gives the following three equations by using Eq. (8-II-2):

$$\begin{aligned} aN + b\Sigma x_i + c\Sigma x_i^2 &= \Sigma y_i \\ a\Sigma x_i + b\Sigma x_i^2 + c\Sigma x_i^3 &= \Sigma x_i y_i \\ a\Sigma x_i^2 + b\Sigma x_i^3 + c\Sigma x_i^4 &= \Sigma x_i^2 y_i \end{aligned} \tag{8-II-4}$$

with the summations taken from $i = 1$ to N. The solution of these three equations gives a, b, and c.

BASIC DEFINITIONS AND TOOLS   8-47

Table 8-II-1 gives the results of fitting a parabolic rating curve to measured discharges Q and stages H. The sums in Eq. (8-II-4) are given in the table. The computation must be carried out to the same decimal point for all sums, because some of the sums are of the same order of magnitude as the differences involved in the use of much larger numbers. The rating curve by the least-squares method is found to be $Q_1 = 29.9 + 0.567H + 0.00720H^2$. The parabola through three selected points (6, 11, and 13 in Table 8-II-1) is $Q_2 = 38.6 + 0.396H + 0.007597H^2$.

Table 8-II-1. Fitting of Parabola to Rating Curve

	x_i (stage H), cm*	y_i (discharge Q), m³/sec†	Estimates of Q		χ_1^2	χ_2^2
			Least-squares Q_1	Three-point Q_2		
1	−23	15.55	20.7	35.5	1.25	9.55
2	−22	15.46	20.9	33.6	1.39	9.78
3	−16	20.07	23.6	34.2	0.26	5.00
4	−16	21.99	23.6	34.2	0.11	4.33
5	14	36.11	39.3	45.6	0.26	1.98
6	33	59.82	56.5	59.9	0.19	0.00
7	46	86.58	70.9	72.9	3.50	2.57
8	69	110.96	103.3	102.1	0.57	0.77
9	88	136.52	135.6	132.3	0.01	0.13
10	120	204.40	200.7	190.5	0.07	1.00
11	136	232.87	240.2	233.0	0.22	0.00
12	220	492.50	503.2	493.4	0.23	0.00
13	400	1,412.48	1,409.0	1,412.5	0.01	0.00
					8.07	35.12

* 1 cm = 0.3937 in.
† 1 m³/sec = 35.314 cfs.
$N = 13$; $\Sigma x_i = 1,049.00$; $\Sigma x_i^2 = 258,727.00$; $\Sigma x_i^3 = 80,006,477.00$;
$\Sigma x_i^4 = 28,580,940,099.00$; $\Sigma y_i = 2,846.31$; $\Sigma(x_i y_i) = 754,258.87$;
$\Sigma(x_i^2 y_i) = 258,951,971.73$.

The *goodness of fit* of a function, fitted to the observed points by any procedure, can be measured and tested approximately by the χ^2 *parameter (fitting parameter)*. This test is known as the *chi-square test*. The parameter is defined as [1, p. 203]

$$\chi^2 = \sum_{i=1}^{N} \frac{(y_i - y)^2}{y} = \sum_{i=1}^{N} \frac{(\Delta_y i)^2}{y} \qquad (8\text{-II-}5)$$

For the above example of fitting the parabolic rating curve, $\chi_1^2 = 8.07$ for the least-squares method and $\chi_2^2 = 35.12$ for the three-point method (Table 8-II-1), indicating that the least-squares method gives a χ^2 value four times smaller than that given by the three-point method.

It is assumed for the chi-square test that values of y_i are normally distributed around y_0 for any given interval Δx for x_i. This assumption is used as a basis either for comparison of the different procedures of fitting the functions to observed data or for statistical inference.

A similar procedure is used when the sum of $(x_i - x)^2$ is minimized for the regression of x versus y.

D. General Model for Regression and Correlation Analyses

A variable that occurs as a sequence of consecutive values may be composed of values that are *internally independent* of each other (*random variable*) or *internally dependent* (*autocorrelated*, or *serially correlated, variable*). These conditions are called here *internal dependence* and *internal independence of a variable*. A series of monthly streamflow values is usually internally dependent, because of the carryover of stored water and the influence of the annual cycle of climate.

The regressional or correlative relationship of m variables has two extreme cases. When one variable is dependent on all other $m - 1$ variables, the latter may be either mutually independent, known as *externally independent*, or interdependent, known as *externally dependent*. Runoff data depend on such variables as precipitation, evapotranspiration, topography, geology, and water retention. These are, in turn generally interdependent i.e., externally dependent.

The frequency distribution of each individual variable (called the *marginal distribution* in multivariate distributions) is usually skewed in hydrology. The normal distribution is generally used as a standard for comparison of such distributions.

The basic relationship for the external or internal dependence of variables can be expressed arbitrarily by any desired function.

When a statistical model is being set up for regression and correlation analyses, the statistical relationships should comply with the true hydrologic behavior. Several of the most important hydrologic variables probably are related among themselves. They show a residual scatter of points around the curve of a selected functional relationship because of the effect of other neglected variables and because of inherent errors. Most hydrologic variables are mutually dependent externally. Nearly all hydrologic variables are skew-distributed in their marginal distributions. And many hydrologic variables are internally dependent (autocorrelated).

Difficulties in statistical treatment of the complex model are usually encountered in hydrology, especially in appraising the reliability (or errors) of the computed statistical parameters defining the regression or correlation. Such difficulties may be overcome by many simplifications in the use of a general hypothetical model.

When only two variables are related, the analysis is one of a *simple regression or correlation*. When three or more variables are involved, the analysis is one of a *multiple regression or correlation*. The approach to simplification is to limit the number of variables to the most significant ones and thereby reduce the labor of computation. Current hydrologic practice generally restricts the analysis to that of simple regression or to multiple regression involving variables limited to a maximum of 5 to 10. Generally, one variable is considered dependent, and all others are considered to be independent. It is also assumed that the marginal distributions of all variables can be approximated by a normal distribution or that, by transformation, they can approach to a normal distribution.

Internal dependence is important only in making statistical inference.

The complexity of the mathematical functions involved can be avoided either by transformation to linearize the relation or by replacing the complex function with a polynomial, provided the function can be developed to a power series with a limited number of terms. Currently, the linear relationship is most used in regression and correlation analyses in hydrology.

E. Transformation of Variables

The transformation of variables has three basic advantages, which may or may not all occur in a given transformation: (1) A simple linear relationship is obtained among the transformed variables. (2) The marginal distributions of the transformed variables approach more closely to the normal distribution than do those of the untransformed variables. (3) The variation of the points along the regression line is more homogeneous. In other words, the variance, or standard deviation, is stabilized.

Table 8-II-2 gives the transformations for linearizing a two-variable relationship. A logarithmic transformation is generally used in hydrology more frequently than others.

F. Use of Ungrouped and Grouped Data

In the regression and correlation analysis of a small number of variables with a large sample size, the use of ungrouped data can be cumbersome and time-consuming if

Table 8-II-2. Transformations for Linearization of Different Functions

	Type of function	Straight-line coordinate Abscissa	Straight-line coordinate Ordinate	Equation in linear form
1	$y = a + bx$	x	y	$[y] = a + b[x]$
2	$y = be^{ax}$	x	$\log y$	$[\log y] = \log b + (a \log e)[x]$
3	$y = ax^b$	$\log x$	$\log y$	$[\log y] = \log a + b[\log x]$
4	$y = a_0 + a_1 x + a_2 x^2$	$x - x_0$	$\dfrac{y - y_0}{x - x_0}$	$\left[\dfrac{y - y_0}{x - x_0}\right] = a_1 + 2a_1 x_0 + a_2[(x - x_0)]$
5	$y = a + b/x$	$1/x$	y	$[y] = a + b[1/x]$
6	$y = x/(a + bx)$	x	x/y	$[x/y] = a + b[x]$
7	$y = a/(b + cx)$	x	$1/y$	$[1/y] = (b/a) + (c/a)[x]$
8	$y = c + be^{ax}$	x	$\log \dfrac{\Delta y}{\Delta x}$	$\left[\log \dfrac{dy}{dx}\right] = \log(ab) + (a \log e)[x]$
9	$y = c + ax^b$	$\log x$	$\log \dfrac{\Delta y}{\Delta x}$	$\left[\log \dfrac{dy}{dx}\right] = \log(ab) + (b - 1)[\log x]$
10	$y = c + \dfrac{b}{x - a}$	$x - x_0$	$\dfrac{x - x_0}{y - y_0}$	$\left[\dfrac{x - x_0}{y - y_0}\right] = -\dfrac{a - x_0}{c - y_0} + \dfrac{1}{c - y_0}[x - x_0]$
11	$y = c + \dfrac{x}{a + bx}$	x	$\dfrac{x - x_0}{y - y_0}$	$\left[\dfrac{x - x_0}{y - y_0}\right] = (a + bx_0) + \dfrac{b(a + bx_0)}{a}[x]$
12	$y = d + cx + be^{ax}$	x	$\log \dfrac{\Delta^2 y}{\Delta x^2}$	$\left[\log \dfrac{d^2 y}{dx^2}\right] = \log(a^2 b) + (a \log e)[x]$
13	$y = dc^x b^m$, where $m = a^x$	x	$\log \dfrac{\Delta^2(\log y)}{\Delta x^2}$	$\left[\log \dfrac{d^2(\log y)}{dx^2}\right] = \log\left[\dfrac{(\log b)(\log a)^2}{(\log e)^2}\right] + (\log a)[x]$
14	$y = de^{cx} + be^{ax}$	$\dfrac{y_{k+1}}{y_k}$	$\dfrac{y_{k+2}}{y_k}$	$\left[\dfrac{y_{k+2}}{y_k}\right] = -e^{(a+c)\Delta x} + (e^{a \Delta x} + e^{c \Delta x})\left[\dfrac{y_{k+1}}{y_k}\right]$
15	$y = e^{ax}(d \cos bx + c \sin bx)$	$\dfrac{y_{k+1}}{y_k}$	$\dfrac{y_{k+2}}{y_k}$	$\left[\dfrac{y_{k+2}}{y_k}\right] = -e^{2a \Delta x} = (2e^{a \Delta x} \cos b\, \Delta x)\left[\dfrac{y_{k+1}}{y_k}\right]$

[Row for type 12 also shows: $[y - be^{ax}] = d + c[x]$]

[Row for type 14 also shows: $[\log y - a^x \log b] = \log d + (\log e)[x]$ and $[ye^{-cx}] = d + b[e^{(a-c)x}]$]

[Row for type 15 also shows: $\left[\dfrac{yc^{-ax}}{\cos bx}\right] = d + c[\tan bx]$]

REMARK: In types 14 and 15, y_k, y_{k+1}, and y_{k+2} are consecutive values for an increment Δx.

ordinary numerical methods of computation are used. The grouping of the data by deriving a *bivariate distribution* (for simple regression and correlation) or a *multivariate distribution* (for three or four variables) may save much of the computational work. However, the use of grouped data gives somewhat greater correlation parameters and decreases the measure of spread of points around the regression line in comparison with the case for ungrouped data. This bias must be considered when statistical inference gives results close to the limits of the prescribed intervals. An example of analysis for grouped data will be given later.

G. Statistical Inference in Regression and Correlation Analyses

As the regression lines are defined by parameters and the correlation is measured by parameters, the reliability of these computed parameters is important.

The sample of the data available for an analysis is considered as a small part of an infinite *universe* of values. The general case, treated in hydrology, is that the sample is a part of a homogeneous universe, assuming thus that the same cause-effect relation applies to all other data from the universe. A change in homogeneity in either time or space, such as the change of climate, the change of river-basin factors, etc., must be corrected for nonhomogeneity of the data. In other words, any trend of sample changes must be corrected.

If a large number of samples of the same size as the analyzed sample were available, then any parameter describing the samples would have its own distribution. The *standard deviation* of this distribution, which generally is unknown in exact form because only a small sample is usually available in hydrology, is considered as the measure of reliability of the determined parameter. The general rule is that the larger the sample size, the smaller the standard deviation of the computed parameter, and thus the more reliable the parameter, all other conditions being the same.

The *confidence interval* is defined as an interval around the computed parameter within which a given percentage of parameters of a large number of samples is expected to be found. This given percentage is the *level of confidence*. The confidence interval at the 95 per cent level means that, out of 100 samples of equal size, it is expected that 95 values of a parameter would be inside that interval. Sometimes the confidence interval is chosen as the expected percentage of parameters to fall outside the confidence interval; in this case the level is 5 per cent. The *confidence limits* are numerical values describing the boundaries of the confidence interval.

If the computed parameter falls inside the selected confidence limits, it is considered either as *significant* or as *nonsignificant*, depending on the problem at hand. If it falls outside, it is considered either as *not significant*, or *unreliable* for use, or as *significant* as the case may be. The selection of level is made either by convention or by judgment. In hydrology the limits are generally chosen at between 80 and 95 per cent.

If a regression line with m parameters is fitted to a set of data points and if N is the sample size, the number of degrees of freedom is $N - m$. Thus every parameter takes away one degree of freedom in the analysis of the data. When $m = N$, the line passes through all points, so that there are no degrees of freedom. Since the errors of the parameter are inversely related to the degrees of freedom, this line cannot be used for prediction; otherwise, the errors would become infinite. For small-size samples, the sample size N used in related formulas should be replaced by the degrees of freedom $(N - m)$. This correction is necessary for avoiding a bias in the measure of reliability of the computed statistical parameters. If no such correction is made, the computed standard deviations of the parameters are significantly smaller in very small samples than their true values.

III. SIMPLE LINEAR REGRESSION AND CORRELATION

A. Regression Lines

1. Analytical Method of Determining Parameters. The straight-line regression for variable y versus variable x is defined by a straight line which gives the *best* *estimate* of y for a given value of x. Similarly, the best estimate of x for a given y is given by the regression line of x versus y.

The line connecting the mean values y_0 of all y_i for given x_i, or more precisely, for the interval Δx around x_i, is for all practical purposes the regression line. In the simple linear regression analysis (Fig. 8-II-2) a straight line is fitted through these values.

The straight regression line is generally fitted analytically by the method of least squares of the departures from the line. The individual mean values y_0 for given x_i, or x_0 for given y_i, however, are not usually computed. For the regression line y versus x, the departures are Δ_{yi} along the ordinate. For the regression line x versus y the departures are Δ_{xi} along the abscissa. The equations of the two lines are

$$y = A_y + B_y x \quad (8\text{-II-}6)$$
$$x = A_x + B_x y \quad (8\text{-II-}7)$$

where B_y is the *regression coefficient* of y versus x, B_x is the regression coefficient of x versus y, and A_y and A_x are the respective *intercepts* for the regression lines. The regression line of y versus x is not the same as that of x versus y.

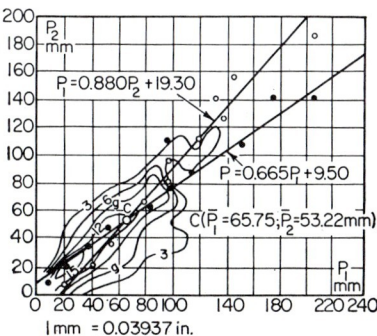

Fig. 8-II-2. Simple linear regression analysis for monthly precipitations (P_1, P_2), using the grouped data in a bivariate distribution. Small circles represent mean values of P_1 in a given class interval ΔP_2; small dots represent mean values of P_2 in ΔP_1.

The sum of $(\Delta_{yi})^2 = (y_i - y)^2$ is minimized, where y_i is the observed value and y is the value from the straight regression line for a given x_i. If $\bar{x}$ and $\bar{y}$ are the mean values of x and y for the sample size N, then

$$B_y = \frac{\Sigma x_i y_i - N\bar{x}\bar{y}}{\Sigma x_i^2 - N\bar{x}^2} = \frac{\Sigma \Delta x_i \, \Delta y_i}{\Sigma (\Delta x_i)^2} \quad (8\text{-II-}8)$$

where the summations are taken from 1 to N, with $\Delta x_i = x_i - \bar{x}$ and $\Delta y_i = y_i - \bar{y}$.

The intercept is

$$A_y = \bar{y} - B_y \bar{x} = \bar{y} - \bar{x}\frac{\Sigma \Delta x_i \, \Delta y_i}{\Sigma (\Delta x_i)^2} \quad (8\text{-II-}9)$$

The parameters B_x and A_x are obtained in a similar way by interchanging Δx_i with Δy_i, and x with y, in Eqs. (8-II-8) and (8-II-9).

To determine both regression lines the following five summations of the sample should be performed: Σx_i, Σy_i, $\Sigma(\Delta x_i)^2$, $\Sigma(\Delta y_i)^2$, and $\Sigma(\Delta x_i \, \Delta y_i)$.

2. Graphical Method of Determining Parameters. The graphical method is based on the principle that the graphically fitted straight line, minimizing deviations from the line, leaves on both sides the same amount of scattered points and that these points are nearly homogeneously distributed along both sides of the line. Graphical construction of a regression line usually does not give an accurate determination of the parameters, which are obtained from the line as the intercept at the ordinate A_y and the slope of the line $B_y = \tan \alpha$, α being the slope angle. The computed χ^2 of goodness of fit is generally much greater than that by the analytical least-squares method, because the fit is less accurate.

3. Grouped Data in the Bivariate Distribution. When a large sample necessitates the use of grouped data and the bivariate distribution (x_i, y_j) is used for the determination of parameters (disregarding the small bias obtained by using grouped data), the regression coefficient for the grouped data [2] is computed by the formula

$$B_y = \frac{\sum_{i=1}^{p} \sum_{j=1}^{q} f_{ij} \Delta x_i \, \Delta y_j}{\sum_{i=1}^{p} f_{xi}(\Delta x_i)^2} \tag{8-II-10}$$

where $\Delta x_i = x_i - \bar{x}$, x_i being the middle of each class interval for x
$\Delta y_j = y_j - \bar{y}$, y_j being the middle of each class interval for y
f_{xi} = frequency, or number of occurrences, of class interval x_i, f_{yj} being the same for class y_j (both are frequencies, or numbers of occurrences, of marginal distributions)
f_{ij} = bivariate frequency, or number of occurrences, for (x_i, y_j) class intervals
p, q = number of class intervals in x and y, respectively
B_y = regression coefficient of the straight-line regression y versus x

The parameter A_y is computed by Eq. (8-II-9), with $N\bar{x} = \sum_{i=1}^{p} f_{xi} x_i$ and $N\bar{y} = \sum_{j=1}^{q} f_{yj} y_j$, where N is the total number of occurrences. For the regression line x versus y, the parameters A_x and B_x are determined by interchanging x and y in Eqs. (8-II-9) and (8-II-10).

Table 8-II-3 gives the monthly precipitations ($P_1 = x_i$ and $P_2 = y_j$) for two rainfall stations, with their bivariate distribution f_{ij} and marginal distributions f_{xi} and f_{yj}. It also shows a practical procedure for computing the sums necessary to be used in Eqs. (8-II-9) and (8-II-10). In order to avoid computation with large absolute values of P_1 and P_2, the intervals are coded. For *coding* or *indexing* class intervals, new dimensionless variables u and v are introduced: $u = (113 - P_1)/\Delta P_1$ and $v = (83 - P_2)/\Delta P_2$, where the length of class intervals is $\Delta P_1 = \Delta P_2 = 15$ mm (1 mm = 0.03937 in.). The new variables u and v are selected, with an arbitrary central class interval introduced, such that the central point is equal to zero. The variables u and v have positive and negative integers as marks of successive class intervals. The means of u and v are $\bar{u} = 3.150$ and $\bar{v} = 1.985$, respectively. By decoding class intervals, $\bar{P}_1 = 65.75$ mm and $\bar{P}_2 = 53.22$ mm (1 mm = 0.03937 in.). The regression coefficients are $B_y = 0.665$ and $B_x = 0.880$, and the two regression lines are $P_2 = 9.50 + 0.665 P_1$ and $P_1 = 19.30 + 0.880 P_2$.

Figure 8-II-2 shows the relation between monthly precipitations P_1 and P_2 for the two stations, with the isolines of equal bivariate frequencies; the points of the mean values y_0 of all y_j for given x_i and the mean values x_0 of all x_i for given y_j of class intervals; and the two regression lines, with means $\bar{P}_1$ and $\bar{P}_2$.

B. Measures of Linear Correlative Association

1. Correlation Coefficient and Coefficient of Determination. The *correlation coefficient* is the most commonly used statistical parameter for measuring the degree of association of two linearly dependent variables. It is defined [2] as

$$r = \frac{\Sigma(\Delta x_i \, \Delta y_i)}{\sqrt{\Sigma(\Delta x_i)^2 \Sigma(\Delta y_i)^2}} = \frac{\Sigma x_i y_i - N\bar{x}\bar{y}}{s_x s_y (N - 1)} \tag{8-II-11}$$

Table 8-II-3. Computation of Sums for Determination of Parameters in Linear Regression and Correlation Analysis of Precipitations

							Monthly precipitation, P_1 (mm) = x_i												Monthly precipitation, P_2 (mm) = y_j
$v\Sigma f_{ij}u$	$\Sigma f_{ij}u$	$f_{yj}u^2$	$f_{yj}v$	v	f_{vj}	8	23	38	53	68	83	98	113	128	143	158	173	188	203	
30	−6	75	−15	−5	3														1	158
16	−4	64	−16	−4	4									1			1			143
15	−5	27	−9	−3	3								1	1					1	128
20	−10	48	−24	−2	12							1	1	5			2			113
−16	16	14	−14	−1	14			1	2	1	2	3	2	2	2					98
0	28	0	0	0	27				1	3	12	3	4	2	2		1	1		83
90	90	39	39	1	39		3	2	7	9	9	2	2	2	1					68
312	156	200	100	2	50		6	7	14	12	6	9	1	2						53
534	178	414	138	3	46		20	12	12	7	4	3	2							38
930	240	800	200	4	50		11	14	7	6	2	5	1							23
960	192	775	155	5	31	14		5		1										8
2921		2456	554		279	14	40	41	43	39	35	28	14	15	3	0	4	1	2	f_{xi}
						7	6	5	4	3	2	1	0	−1	−2	−3	−4	−5	−6	u
					876	98	240	205	172	117	70	28	0	−15	−6	0	−16	−5	−12	$f_{xi}u$
					4546	686	1440	1025	688	351	140	28	0	15	12	0	64	25	72	$f_{xi}u^2$
						70	159	133	97	82	39	9	−6	−15	2	0	−8	0	−8	$\Sigma f_{ij}v$
					2921	490	954	665	388	246	78	9	0	15	−4	0	32	0	48	$u\Sigma f_{ij}v$

NOTATION
$u\Sigma f_{ij}v = v\Sigma f_{ij}u$
1 mm = 0.03937 in.
Entries corresponding to x_i and y_j in the table are f_{ij} values.

where s_x and s_y are the standard deviations of x_i and y_i, respectively. As s_x and s_y are positive, the sign of r depends on the sum of the cross products $\Delta x_i \, \Delta y_i$. Since this sum can vary between $+s_x s_y$ and $-s_x s_y$, the correlation coefficient varies from $+1$ to -1. If the sum of the cross products $\Delta x_i \, \Delta y_i$ is zero, the variables x and y are linearly independent and the correlation coefficient is zero. However, it does not follow that the variables are independent if the sum of the cross products is zero. If the points (x_i, y_i) fall symmetrically around and all along a circle, r is zero, or the linear dependence is zero. However, in this case there is a high correlation for the function which has a circle as the regression line.

The correlation coefficient is unity only if all points fall on a straight line. A positive value of r means that y increases with an increase of x. A negative value of r means that y decreases with an increase of x.

If there is no linear relationship, $r = 0$. If there is a functional linear relationship, $r = \pm 1$. All values of r between these limits describe the various degrees of correlative association. The greater the absolute value of r, the greater is the linear correlation.

By geometrical interpretation, the more the bivariate-frequency isolines (Fig. 8-II-2) are concentrated along a straight line, the greater is the correlation coefficient. The correlation coefficient will be unity for the case where all isolines coincide with the straight line. When the isolines are concentrated approximately in circles, the correlation coefficient is around zero, and there is no significant linear dependence.

In the case of grouped data in a bivariate distribution, the correlation coefficient [2] can be computed from the expression

$$r = \frac{\sum_{i=1}^{p} \sum_{j=1}^{q} f_{ij} \Delta x_i \, \Delta y_j}{\sqrt{\sum_{i=1}^{p} f_{xi}(\Delta x_i)^2 \sum_{j=1}^{q} f_{yj}(\Delta y_j)^2}} \qquad (8\text{-II-12})$$

Using Eqs. (8-II-8) and (8-II-11), or Eqs. (8-II-10) and (8-II-12), the regression coefficients may be expressed in terms of the correlation coefficient r and the standard deviations s_x and s_y:

$$B_y = r \frac{s_y}{s_x} \qquad \text{or} \qquad B_x = r \frac{s_x}{s_y} \qquad (8\text{-II-13})$$

Furthermore,
$$D = r^2 = B_x B_y \qquad (8\text{-II-14})$$

where D is defined as the *coefficient of determination*. It is a measure of the degree to which the variance or square of the standard deviation, $s_y{}^2$ and $s_x{}^2$, is explained or accounted for by the linear regression. In other words, it is a measure of the difference between the variance of the observed (actual) values y_i and the variance of the values determined for given values of x_i by the use of the linear-regression line. The greater D is, the smaller is this difference.

For the example of grouped data in Table 8-II-3, $D = r^2 = 0.581$ and $r = 0.762$. The coefficient D tells that 58.1 per cent of the variance is accounted for by the linear-regression relationship in this example.

2. Standard Deviation of Residuals. The *residuals* of the straight-line regression y versus x are $\Delta_{yi} = y_i - y$, where y_i is the observed value and y is the value determined from the straight regression line for a given $x = x_i$. The *standard deviation of the residuals* for the straight regression line y versus x is defined [2, p. 72] as

$$S_y = \sqrt{\frac{\sum_{i=1}^{N}(\Delta_{yi})^2}{N}} = s_y \sqrt{1 - r^2} \qquad (8\text{-II-15})$$

SIMPLE LINEAR REGRESSION AND CORRELATION 8–55

A similar expression for S_x is obtained if Δ_{xi} and s_x replace Δ_{yi} and s_y, respectively.

Taking into account the number of degrees of freedom in the case of small samples, the unbiased standard deviation of residuals, $\hat{S}_y$, is given as

$$\hat{S}_y = \sqrt{\frac{N-1}{N-2}} s_y \sqrt{1 - r^2} \qquad (8\text{-II-}16)$$

The greater $\hat{S}_y$ or $\hat{S}_x$, the wider is the spread of the points around the regression line and the less accurate are the values determined from the regression lines. Referring to the example in Table 8-II-3, the standard deviations for the variables u and v are $s_u = 2.52$ and $s_v = 2.20$. By transforming u and v to P_1 and P_2, respectively, the standard deviations for the variables P_1 and P_2 are $s_x = 37.80$ mm and $s_y = 33.00$ mm. Since N is very large, Eq. (8-II-15) gives $S_y = 21.3$ mm and $S_x = 24.4$ mm.

3. Interpretation of Parameters. To interpret properly the practical aspects of linear regression, all three parameters B_y, r, and S_y (or B_x, r, and S_x) are useful and necessary. B_y tells how fast the dependent variable increases or decreases with a change of the other variable. r measures the degree of the association. S_y measures the spread of points around the straight line.

In the example of Table 8-II-3, $B_y = 0.665$, $r = 0.762$, and $S_y = 21.3$ mm. Thus P_2 increases very fast with a corresponding increase of P_1, the association is very high, and despite the high association, the spread of points is also large.

C. Statistical Inference

1. Correlation Coefficient. To make a statistical inference about the computed correlation coefficient r, the type of distribution of r should be known for an infinite number of samples of size N, all drawn from the same universe of data. The general assumption is that in all samples both x_i and y_i are normally distributed and are random in sequence; that is, they are internally independent variables. The correlation coefficient r of the sample available is only an estimate of ρ, the *correlation coefficient for the universe*.

The frequency distribution of r is bounded at both sides ($r = +1$, $r = -1$), but depends only on ρ and N. For $\rho = 0$, it is a bounded symmetrical function, but for large absolute value of ρ, it is a highly skewed distribution. The degree of skewness is a function of ρ.

For ρ close to zero, an approximate method for testing the hypothesis that r is different from zero is based on the assumption that r is normally distributed. It uses the *standard deviation of the correlation coefficient* S_r as a criterion.

$$S_r = \frac{1 - r^2}{\sqrt{N}} \qquad (8\text{-II-}17)$$

where ρ should have been used but is replaced by its sample value r [2, p. 143].

Instead of using the confidence interval (such as 90 per cent, 95 per cent, etc.), the limits for r are sometimes given either as three standard deviations on both sides of r (that is, $r \pm 3S_r$) or as four standard deviations for a stronger test of significance (that is, $r \pm 4S_r$). If both limits ($r + 3S_r$ and $r - 3S_r$, or $r + 4S_r$ and $r - 4S_r$) are of the same sign as r, it is considered by this test that r is *significantly different from zero*. If, however, these limits have different signs, then r is not considered as significantly different from zero.

For the example in Table 8-II-3, $S_r = 0.025$, so that $r + 3S_r = 0.837$ and $r - 3S_r = 0.687$, or $r + 4S_r = 0.862$ and $r - 4S_r = 0.662$. All are positive, so that $r = 0.762$ is significantly different from zero with respect to either of these empirical confidence limits.

If a test of significance of r for ρ close to zero is to be made on the basis of confidence interval and if $N \geq 5$, the distribution of r follows approximately the Pearson type II function, which has the shape of a symmetrical bell-shaped curve. As N becomes large, or $N \geq 50$, the function approaches the normal distribution and the standardized variable t (with mean equal to zero and standard deviation equal to unity) is

$$t = r\sqrt{N-1} \qquad (8\text{-II-}18)$$

For any prescribed level of significance, the significant departure of r from zero may be tested by using this function. The table for the Student t distribution should be used for the prescribed level of significance if Eq. (8-II-18) is applied. Such a table is generally available in standard statistical textbooks.

If ρ is different from zero, Fisher's transformation [2, p. 200] with a new variable

$$z = \tfrac{1}{2} \ln \frac{1+r}{1-r} \qquad (8\text{-II-}19)$$

should be used, with ln denoting the natural logarithm. For normally distributed x and y, z is normally distributed, with the mean

$$\bar{z} = \tfrac{1}{2} \ln \frac{1+\rho}{1-\rho} \qquad (8\text{-II-}20)$$

and the standard deviation

$$s_z = \frac{1}{\sqrt{N-3}} \qquad (8\text{-II-}21)$$

From this the standardized variable, or argument t (with mean equal to zero and standard deviation equal to unity), can be determined by

$$t = \frac{z - \bar{z}}{s_z} \qquad (8\text{-II-}22)$$

For the example in Table 8-II-3, with $N = 279$, the values of t-distribution function for the 95 per cent confidence level are $t = \pm 1.97$. Replacing ρ by $r = 0.762$ in Eq. (8-II-20) as an estimate of ρ results in $\bar{z} = 1.0$, $s_z = 0.063$ from Eq. (8-II-21), $z_1 = 1.119$, and $z_2 = 0.881$. Finally, from Eq. (8-II-19), $r_1 = 0.807$ and $r_2 = 0.706$. By this test $r = 0.762$ is highly significant, because, on the average, of 95 out of 100 samples of size $N = 279$ and $\rho = 0.762$, r would be between 0.706 and 0.807, a rather narrow confidence interval for the correlation coefficient. This assumes that P_1 and P_2 are internally independent; i.e., they are not serially correlated. This is not the case, however.

It is generally considered in hydrology that a linear-regression approximation is good if r by these tests is significantly different from zero, and if the absolute value of r is greater than 0.6; that is, $r > 0.6$ and $r < -0.6$. However, if the physical relationship can be well justified, the linear regression can be accepted even if the absolute value of r is smaller than 0.6. If there is no physical reason to justify a computed absolute correlation coefficient greater than 0.6, especially for small N, the linear regression must be used discreetly, in spite of the high degree of correlation.

2. Regression Coefficient. Assuming an infinite number of values B_y and an infinite number of samples (x_i, y_i) of size N, the unbiased estimate of the standard deviation s_b of the distribution of B_y is [3, p. 337]

$$\hat{s}_b = \sqrt{\frac{B_y^2(1-r^2)}{(N-2)r^2}} \qquad (8\text{-II-}23)$$

The standardized variable of the B_y distribution is

$$t = \frac{r(B_y - \beta_y)}{B_y}\sqrt{\frac{N-2}{1-r^2}} \qquad (8\text{-II-}24)$$

where β_y is the true value of B_y. For $\beta_y = 0$, the standardized variable t is

$$t = r\sqrt{\frac{N-2}{1-r^2}} \qquad (8\text{-II-}25)$$

which is dependent on B_y only through the relation of r and B_y.

Using Eq. (8-II-25) and the table of the t distribution for the selected level of significance, the test that B_y is not significantly different from zero can be made. If the t value computed from Eq. (8-II-25) is much greater than that which corresponds to the selected level of significance, then B_y is significantly different from zero. For the example from Table 8-II-3, Eq. (8-II-25) gives $t = 19.5$, for $r = 0.762$ and $N = 279$. The t table shows that the level of significance is very close to 100 per cent for $t = 19.5$, so that $B_y = 0.665$ is significantly different from zero at either the 90 or the 95 per cent level of significance. For other values of B_y, the test that B_y does not depart from B_y significantly on the selected level is made by using Eq. (8-II-24). If $\beta_y = 0.50$ in the above example and the level of significance is selected to be 95 per cent, Eq. (8-II-24) for $t = \pm 1.97$ gives the values for the limits of confidence interval: $B_1 = 0.555$ and $B_2 = 0.455$. So, at this level, $B_y = 0.665$ is significantly different from $\beta_y = 0.50$, because it is outside the confidence interval. The variables P_1 and P_2 are assumed internally independent.

3. Intercept. The standardized variable of the distribution of A_y is

$$t = \frac{r(A_y - \alpha_y)}{B_y}\sqrt{\frac{N-2}{(1-r^2)(s_x^2 + \bar{x}^2)}} \qquad (8\text{-II-}26)$$

where α_y is the true value of A_y.

The standard test is to determine whether or not A_y is significantly different from zero, that is, whether the regression line may be assumed to pass through the coordinate origin. If the regression does pass through the origin, α_y is taken to be zero. Strictly, B_y, s_x, and $\bar{x}$ should be replaced by their true values, or values of the universe, β_y, σ_x, and μ, but the estimates B_y, s_x, and $\bar{x}$ would give results sufficiently accurate for practical purposes, since they are the only values available. Also, the true value of r is unknown; hence the estimate of r is the only value that can be used.

For the above example, with $A_y = 9.50$ mm, $B_y = 0.665$, $r = 0.762$, $N = 279$, $s_x^2 = 1428$, $P_1^2 = \bar{x}^2 = 4{,}330$, and for the 95 per cent level of significance, the limits of the confidence interval are $A_1 = 5.08$ mm and $A_2 = -5.08$ mm. At that level, the value $A_y = 9.50$ mm is significantly different from zero. At the 98 per cent level, $A_{1,2} = \pm 6.04$ mm and $A_y = 9.50$ mm is still significantly different from zero.

4. Inference in the Case of Nonnormal Distributions of Variables Which Are Internally Dependent. If the distributions of x_i and y_i are not highly asymmetrical, the above formulas and procedures can be used as approximations for statistical inference. However, when the distributions are highly asymmetrical, transformations of the variables x and y may give more symmetrical distributions, and thus provide more reliable statistical inference for the regression and correlation parameters.

When x and y are internally dependent (serially correlated), an approximate method of statistical inference can be accomplished through the computation of a sample of reduced size. The standard deviations of the parameters and the confidence intervals are larger in the case of a serially correlated series than in the case of the same series when internally independent.

If the variables x and y follow a cyclic trend, this cyclic trend should be removed. If they do not, but are internally dependent, the sample size N may be reduced to [4]

$$N' = \frac{N}{1 + 2r_1r_1' + 2r_2r_2' + \cdots} \qquad (8\text{-II-}27)$$

where r_i and r_i' are the serial correlation coefficients of order i for the variables x and y, respectively.

For example, consider the annual flows of the St. Lawrence River at Ogdensburg and the Mississippi River at St. Louis, for which the records for 100 years of concurrent observations are available. The first-order serial correlation coefficients are $r_1 = 0.705$ for St. Lawrence and $r_1' = 0.294$ for Mississippi. All higher-order values for the Mississippi River are not significantly different from zero. Therefore the reduced sample size $N' = N/(1 + 2r_1 r_1') = 70.7$, or about 71, a 29 per cent reduction in the size of the sample resulting from the effect of serial correlation of the data.

IV. SIMPLE CURVILINEAR REGRESSION AND CORRELATION

A. Curvilinear Regression

Linear regression is a special case of *curvilinear regression*, where either the variables x and y are originally linearly related or, by a transformation, a curvilinear relationship is linearized. All nonlinear functions which describe the relation between the variables are easier to determine if they can be reduced by transformation to a linear relationship.

Theoretically, any function $y = f(x)$ can be fitted to data by computing its parameters by the method of least squares of the departures $\Delta_{yi} = y_i - y = y_i - f(x_i)$. The other regression line of the same family, given as $x = g(y)$, can be determined by minimizing the squares of departures $\Delta_{xi} = x_i - x = x_i - g(y_i)$.

All functions which cannot be easily transformed to a linear relationship or fitted linearly to the data may be developed in the form of a power series, by neglecting the higher-order terms and reducing the function to a polynomial. In that case, the polynomial can be fitted to the points (x_i, y_i). Linear regression is then a special case, where all terms of the power series of order higher than first are neglected.

The general form of a polynomial regression of y versus x is represented by

$$y = A_0 + A_1 x + A_2 x^2 + \cdots + A_m x^m \tag{8-II-28}$$

or for x versus y by

$$x = B_0 + B_1 y + B_2 y^2 + \cdots + B_m y^m \tag{8-II-29}$$

where the coefficients A_i and B_i are the parameters to be determined by the least-squares method. The number m is so chosen as to minimize the sum of squares of departures from the line. It must be stressed, however, that m should be much lower than the number of points N, in order to have enough degrees of freedom $(N - m)$ for making a reliable estimate of the standard deviation. Also, in order to keep the arithmetical work reasonably simple, m must be small. Generally, m is in the order of 2 to 4.

To select the number m, either a graphical plot will suggest the order of magnitude of m or two or three different m values may be selected, and the standard deviation of residuals from the line for each case is taken as a measure to decide the most suitable m for fitting.

When a polynomial is selected for curvilinear regression, two methods of treatment are used. They are (1) to compute the parameters by the least-squares method directly, as was explained earlier by the example in Table 8-II-1, and (2) to consider x, x^2, x^3, ... as the variables u, v, ... z and apply multiple-linear-regression procedures (described in Subsec. V). The least-squares method in treatment 1 involves the solution of $m + 1$ linear equations involving $m + 1$ moments of x and y and the cross products of powers of x and y, with the highest sum of power $m + 1$.

B. Correlation Index

As was shown earlier in the case of the correlation coefficient for linear association, $r^2 = 1 - S_y^2/s_y^2 = 1 - S_x^2/s_x^2$. In that case $S_y^2/s_y^2 = S_x^2/s_x^2$, where S_x, S_y, s_x, and s_y are standard deviations of residuals and standard deviations of marginal distributions for x and y, respectively. In the case of curvilinear association, these two ratios are not equal, so that the measure of correlation has to depend on the regression curve of either y versus x or x versus y.

For the regression curve of y versus x, the parameter to be used is

$$R_y = \sqrt{1 - \frac{S_y^2}{s_y^2}} \qquad (8\text{-II-30})$$

which is called the *correlation index* for y versus x.
For the regression curve of x versus y,

$$R_x = \sqrt{1 - \frac{S_x^2}{s_x^2}} \qquad (8\text{-II-31})$$

is the correlation index for x versus y. These two indices indicate the closeness with which the points (x_i, y_i) approximate the regression lines. As s_x and s_y are different from zero, the absolute values of R_y and R_x can be unity only when $S_y = S_x = 0$, or when all points are on the curve. Also, R_x and R_y are zero if $S_y = s_y$ and $S_x = s_x$, respectively.

C. Regression and Correlation Inference

When the polynomial regression lines are used, the test of significance for the computed regression coefficients is less simple than it is in the case of multiple linear regression. The general hypothesis to be tested is whether a selected power m of the polynomial represents the relationship or a term of higher power contributes significantly to the relationship. In applying polynomial regression, it is not permitted to omit any intermediate term between the intercept A_0 or B_0 and the regression coefficient of the highest-power term.

Two tests are generally applicable. The first is the test of whether the intercept differs from zero. The second is the test of the highest significant term. Thus the tests are made at the two extremes. The procedure is first to select a polynomial with m power terms. If by test the regression coefficient of the highest-power term is significantly different from zero, then the polynomial with $m + 1$ power terms should be selected and the test repeated. If the regression coefficient of the highest-power term $m + 1$ is not significantly different from zero, then the polynomial with m power terms is finally selected. If the regression coefficient of the highest-power term m is insignificant, the regression coefficients must be computed for the polynomial of $m - 1$ power terms, and the regression coefficient of the highest-power term $m - 1$ tested first, etc. [5, pp. 40–42].

V. MULTIPLE LINEAR REGRESSION AND CORRELATION

A. General

The association of three or more variables can be investigated by the multiple regression and correlation analysis.
The multiple-regression relation may be expressed in the form

$$x_1 = f(x_2, x_3, \ldots, x_m) \qquad (8\text{-II-32})$$

where $x_1, x_2, \ldots, x_m$ are m variables. This equation gives the estimate of x_1 for given values of all other variables.
If Eq. (8-II-32) is linear, the regression is referred to as *multiple linear regression* and the association is *multiple linear correlation*.
Because linear equations are easier to treat than nonlinear multiple relations, variables of nonlinear relations in hydrology are often transformed to linear relations for multiple-regression analysis.

B. Linear Regression with Three Variables

The linear-regression relation for three variables is

$$x_1 = B_1 + B_2 x_2 + B_3 x_3 \tag{8-II-33}$$

where x_1 is the dependent variable and x_2 and x_3 are externally independent variables. In this equation there are three parameters to be determined. The parameter B_1 is the *intercept*. The parameter B_2 is the *multiple regression coefficient* of x_1 on x_2 when x_3 is kept constant. The parameter B_3 is the multiple regression coefficient of x_1 on x_3 when x_2 is kept constant.

The residuals of x_1, or the difference between the observed value x_1' and the estimated value x_1 from Eq. (8-II-33), are $\Delta_1 = x_1' - x_1 = x_1' - B_1 - B_2 x_2 - B_3 x_3$. By use of the least-squares method to minimize the sum of Δ_1^2, the three partial differential equations in B_1, B_2, and B_3 give three linear equations in B_1, B_2, and B_3 for the fitting of curves:

$$\begin{aligned}B_2 \Sigma (\Delta x_2)^2 + B_3 \Sigma (\Delta x_2 \, \Delta x_3) &= \Sigma (\Delta x_1 \, \Delta x_2) \\ B_2 \Sigma (\Delta x_2 \, \Delta x_3) + B_3 \Sigma (\Delta x_3)^2 &= \Sigma (\Delta x_1 \, \Delta x_3) \\ B_1 &= \bar{x}_1 - B_2 \bar{x}_2 - B_3 \bar{x}_3 \end{aligned} \tag{8-II-34}$$

where $\bar{x}_i = \Sigma x_i / N$ and $\Delta x_i = x_i - \bar{x}_i$, with i taken from 1 to 3, and N is the sample size. The sums in the equations may be expressed and computed as $\Sigma (\Delta x_i)^2 = \Sigma x_i^2 - N \bar{x}_i^2$; and $\Sigma (\Delta x_i \, \Delta x_j) = \Sigma x_i x_j - N \bar{x}_i \bar{x}_j$, with i and j having the corresponding values 1, 2, and 3 in the equations. As the sum $\Sigma (\Delta x_1)^2$ is necessary for the computation of other parameters in the multiple-correlation analysis, nine sums of the sample must be computed altogether for the three variables—for the variables themselves, their squares, and their three different cross products. The solution of these linear equations gives the three coefficients B_1, B_2, and B_3.

C. Linear Regression with Several Variables

If there are m variables to correlate, including one dependent and $m - 1$ externally independent, the general equation for multiple linear regression is

$$x_1 = B_1 + B_2 x_2 + \cdots + B_i x_i + \cdots + B_m x_m \tag{8-II-35}$$

where B_1 is the intercept and B_i is the multiple regression coefficient of the dependent variable x_1 on the independent variable x_i, with all other variables kept constant.

Applying the least-squares method for the sum of residuals $\Delta_1 = x_1' - B_1 - B_2 x_2 - \cdots - B_m x_m$, m partial differential equations in B_1, B_2, $\ldots$, B_m give m linear equations:

$$\begin{aligned} B_2 \Sigma (\Delta x_2)^2 + B_3 \Sigma (\Delta x_2 \, \Delta x_3) + \cdots + B_m \Sigma (\Delta x_2 \, \Delta x_m) &= \Sigma (\Delta x_1 \, \Delta x_2) \\ B_2 \Sigma (\Delta x_2 \, \Delta x_3) + B_3 \Sigma (\Delta x_3)^2 + \cdots + B_m \Sigma (\Delta x_3 \, \Delta x_m) &= \Sigma (\Delta x_1 \, \Delta x_3) \\ \cdots \\ B_2 \Sigma (\Delta x_2 \, \Delta x_m) + B_3 \Sigma (\Delta x_3 \, \Delta x_m) + \cdots + B_m \Sigma (\Delta x_m)^2 &= \Sigma (\Delta x_1 \, \Delta x_m) \\ B_1 &= \bar{x}_1 - B_2 \bar{x}_2 - B_3 \bar{x}_3 - \cdots - B_m \bar{x}_m \end{aligned} \tag{8-II-36}$$

where $\Delta x_i = x_i - \bar{x}_i$, with $i = 1$ to m. These equations enable the determination of m parameters. The computational work increases with the increase of either the number m of the variables or the sample size N.

When there are a small number of variables (say, three or four), either ungrouped or grouped data may be used for a multiple regression and correlation analysis. For more than four variables the use of multivariate distributions in the form of grouped data will become cumbersome, because of the difficulties in obtaining a suitable representation of the multivariate distributions.

The selection of a feasible method of solving the linear equations depends on the number of variables and the available computational devices. When the number of variables and the sample size are relatively small, desk computers can be used for an analytical solution, or a graphical approach can be taken. The practice in the past has been to restrict the number of variables. Nowadays, card tabulators and digital computers make possible the use of any number of variables which are statistically significant.

The IBM-type tabulators, designed to solve multiple-regression problems [6], can speed up the computational process tremendously, especially if they are equipped with automatic multiplying devices. Electronic computers can perform any operation in solving the multiple-linear-regression problems by following a suitable set of instructions given to the machines.

In a study of floods in New England [7], a multiple linear regression and correlation analysis was made. By this analysis, the annual peak discharge in cubic feet per second for a recurrence interval of T years was found to be

$$Q_T = aA^b S^c S_t^d I^e t^f O^g \qquad (8\text{-II-}37)$$

where A = drainage area, sq mi
S = slope of main channel, ft/mile
S_t = per cent of surface storage area plus 0.5 per cent
I = 24-hr rainfall, in in., of a recurrence interval of T years
t = average temperature in January, °F below freezing
O = orographic factor
b, c, d, e, f, g = multiple linear regression coefficients when a logarithmic transformation is applied to all variables

By a logarithmic transformation of all seven variables, the general relationship expressed by the above equation can be transformed to a multiple-linear relationship of the type represented by Eq. (8-II-35). Table 8-II-4 shows the computation of the coefficients $a, b, c, \ldots$ for three recurrence intervals. The computation work was made by a desk computer. The results given in the table show how the additional variables affect the values of the multiple regression coefficients computed previously.

D. Measures of Multiple Linear Correlative Association

The degree of correlation of a dependent variable x_1 to many externally independent variables in a multiple-linear association is measured by any of the five parameters, the standard deviation of residuals, the multiple correlation coefficient, the coefficient of multiple determination, the partial correlation coefficients, and the beta coefficients.

1. The Standard Deviation of Residuals. This is determined as the standard deviation of the differences between the observed values x_1' and the values of x_1 estimated from Eq. (8-II-35). This parameter is a measure of the closeness with which the observed values approach the estimated values. It will be designated by S_1, to be distinguished from the standard deviation s_1 of the marginal distribution of x_1.

If the total number of points $(x_1, x_2, x_3, \ldots, x_m)$ is small in comparison with the number m of the variables involved, the standard deviation of residuals computed by the usual procedure is biased, since the degrees of freedom $N - m$ are much smaller than the sample size N.

Taking into account m parameters and $N - m$ degrees of freedom,

$$\hat{S}_1 = \sqrt{\frac{NS_1^2}{N-m}} = \sqrt{\frac{\Sigma \Delta_1^2}{N-m}}$$
$$= \sqrt{\frac{1}{N-m}[\Sigma(\Delta x_1)^2 - B_2 \Sigma \Delta x_1 \Delta x_2 - \cdots - B_m \Sigma \Delta x_1 \Delta x_m]} \qquad (8\text{-II-}38)$$

where S_1 is the *biased*, and $\hat{S}_1$ the *unbiased, standard deviation of residuals*. If, however, $\hat{S}_1$ exceeds $\hat{s}_1$ (the unbiased standard deviation of the marginal distribution of x_1), $\hat{s}_1$ would be used for the standard deviation of residuals instead of $\hat{S}_1$. This can occur only for a very small number of degrees of freedom.

The addition of a new independent variable x_{m+1} in a multiple linear regression will decrease the standard deviation of residuals if this variable influences the dependent variable x_1. Whether this variable should be included in the multiple regression depends on how much the standard deviation of residuals is decreased.

For the study of floods in New England (Subsec. V-C), Table 8-II-4 shows how the standard deviation of residuals in percentage of the mean flood discharge changes with a change of the number and selection of independent variables.

2. Multiple Correlation Coefficient. The correlation coefficient is a measure of the linear association of two variables. In this case, the two variables are the estimated and actual values. Thus the correlation coefficient is the ratio of the standard deviation of estimated values to that of actual values, or

$$r = \frac{s_{e1}}{s_1} = \sqrt{1 - \frac{S_1^2}{s_1^2}} \tag{8-II-39}$$

where s_{e1} is the standard deviation of estimated values, s_1 is the standard deviation of actual values of x_1, and S_1 is the standard deviation of residuals.

In a similar manner, the measure of linear association of many variables, including one dependent and the others externally independent of one another, is expressed by the ratio of the standard deviation of estimated values divided by the standard deviation of the actual values of the dependent variable. This is the *multiple correlation coefficient*, or the *coefficient of multiple correlation*. It measures the combined influence of all externally independent variables in terms of the difference between the actual and the estimated values of the dependent variable. The coefficient is expressed as

$$R_1 = \frac{s_{e1}}{s_1} = \sqrt{1 - \frac{S_1^2}{s_1^2}} = \sqrt{\frac{B_2 \Sigma \, \Delta x_1 \, \Delta x_2 + B_3 \Sigma \, \Delta x_1 \, \Delta x_3 + \cdots + B_m \Sigma \, \Delta x_1 \, \Delta x_m}{\Sigma (\Delta x_1)^2}} \tag{8-II-40}$$

The unbiased value $\hat{R}_1$ is obtained by putting in Eq. (8-II-40) the unbiased values $\hat{S}_1$ and $\hat{s}_1$ for S_1 and s_1, respectively. Table 8-II-4 gives the values of R_1 for the example of floods in New England (Subsec. V-C), showing the change of R_1 with the change of number and selection of variables.

The multiple correlation coefficient is a dimensionless measure of linear association while the standard deviation of residuals is a measure having the dimension of the variable x_1. As s_1 is always the same, regardless of the number of other variables, a decrease of the standard deviation of residuals will cause an increase of the multiple correlation coefficient.

The multiple correlation coefficient can be defined as the correlation coefficient of linear association of two variables, one being the actual values of the dependent variable and the other the estimated values of the dependent variable, based upon all other externally independent variables.

3. Coefficient of Determination. The square of the multiple correlation coefficient, R_1^2, is referred to as the *coefficient of multiple determination D_1*.

The square of the multiple correlation coefficient indicates the part of the variance in the dependent variable s_1^2 which has been mathematically accounted for, whereas $1 - D_1 = 1 - R_1^2$ indicates the part of the variance which has not been accounted for by the multiple linear correlation of the variables $x_1, x_2, \ldots, x_m$.

Table 8-II-4 gives the values of D_1 in percentages for different numbers and selections of variables in the example of floods in New England (Subsec. V-C).

4. Partial Correlation Coefficients. The *partial correlation coefficients* measure the association of the dependent variable x_1 with any given independent variable x_i. Thus they are measures of the variation of x_1, which is explained by, and only by, the given x_i. A simple method of estimating the partial correlation coefficient r_{1-i} involves the determination of (1) the multiple correlation coefficient R_1 between x_1 and all the independent variables, and (2) the multiple correlation coefficient R_{1-i}, between x_1 and all the independent variables except the chosen x_i. The partial correlation coefficient r_{1-i} is then determined from [8, p. 193]

$$r^2_{1-i} = \frac{(1 - R^2_{1-i}) - (1 - R_1^2)}{1 - R^2_{1-i}} = 1 - \frac{1 - R_1^2}{1 - R^2_{1-i}} \tag{8-II-41}$$

Table 8-II-4. Multiple Linear Regression and Correlation Analysis of Floods in New England*

Recurrence interval T	No. stations N	Independent variables included	a	b	c	d	e	f	g	R_1	Standard deviation, % of mean Q	D_1
10	164	A	113.06							.883	$\bar{Q} = 164.4$ cfs: 63.6	.780
		A, S	5.9808	.7858						.942	44.4	.888
		A, S, O	3.4623	1.0182	.5720					.965	34.5	.932
		A, S, O, S_t	6.1854	1.0804	.6119	.2645			.8482	.972	30.4	.945
		A, S, O, S_t, t	5.1680	1.0508	.5153	.2524	.1915	.3549	.8543	.978	27.2	.959
		A, S, O, S_t, t, I	3.6365	.9707	.4306	.2592		.3963	1.0815	.978	27.2	.959
				.9726	.4255				1.0668			
50	116	A	342.32							.875	$\bar{Q} = 145.4$ cfs: 59.9	.765
		A, S	17.892	.6764						.931	44.6	.865
		A, S, O	9.7302	.9197	.5704					.962	33.0	.925
		A, S, O, S_t	17.115	1.0006	.5727	.2520			.9894	.969	29.9	.940
		A, S, O, S_t, t	13.433	.9675	.4978	.2554	.8773	.2165	.9288	.971	29.0	.940
		A, S, O, S_t, t, I	1.8401	.9354	.4602	.2757		.4504	1.1089	.975	26.8	.951
				.9458	.4349				1.0866			
100	100	A	395.75							.862	$\bar{Q} = 134.6$ cfs: 62.8	.743
		A, S	25.504	.6922						.910	50.6	.827
		A, S, O	13.937	.9095	.5438					.950	37.7	.902
		A, S, O, t	10.964	.9922	.5332		.9507	.2837	1.0627	.952	36.8	.905
		A, S, O, t, I	1.1194	.9480	.4667		1.0685	.4409	1.2930	.958	34.6	.920
		A, S, O, t, I, S_t	1.4203	.9524	.4409	−.2308		.5632	1.2730	.965	32.2	.935
				.9226	.3823				1.2129			

*From Benson [7].

Since R^2_{1-i} is smaller than R_1^2, then $r_{1-i} \leq 1.0$. The greater the difference between R_1 and R_{1-i}, the smaller is the ratio on the right side of the equation and the greater is r_{1-i}. It is seen from Eq. (8-II-41) that a considerable amount of computation is necessary in the case of many variables if all partial correlation coefficients are to be determined.

It is important to stress that the partial correlation coefficients should have the sign of their corresponding regression coefficients. If B_i is negative, r_{1-i} is also negative. For a given number of variables and their multiple correlation coefficient R_1, the numerator of the fraction in the right end of Eq. (8-II-41) is a constant, and only the denominator changes from one partial correlation coefficient to another. This agrees with the definition of the coefficient, which measures the reduction of variation of the dependent variable when all other variables except the considered one have been taken into account.

If a new variable x_{m+1} is added, all the partial correlation coefficients computed for the m variables will change.

The partial correlation coefficients are important parameters, since they measure the association of each independent variable with the dependent one after the influence of certain related variables has been accounted for. It is possible that the simple correlation coefficient between the dependent and the new independent variable is very small, and this variable may still have a significant influence on the variation of the dependent variable. It may be reminded that the converse can also occur. The partial correlation coefficient is the measure of the association, after the relation of the dependent variable to other independent variables has been considered and included in the analysis. Therefore the simple correlation coefficient is not as true a measure of association between two variables in multiple linear correlation as is the partial correlation coefficient.

An independent variable may be highly correlated with another independent variable, when simple linear correlation is applied between the two variables, but it may not be highly correlated with the dependent variable in a multiple correlation. The partial correlation coefficient may give a reliable answer as to the real effect of an independent variable on the variation of the dependent variable. In hydrology one may find many examples of this type (e.g., the example given in Table 8-II-4).

5. Beta Coefficients. The effect of individual independent variables on the dependent variable may be measured by a dimensionless form of the regression coefficient. Expressing each variable in ratio to its standard deviation, Eq. (8-II-35) may be written as

$$\frac{x_1}{s_1} = \beta_1 + \beta_2 \frac{x_2}{s_2} + \beta_3 \frac{x_3}{s_3} + \cdots + \beta_m \frac{x_m}{s_m} \qquad (8\text{-II-}42)$$

where

$$\beta_1 = \frac{B_1}{s_1}; \ \beta_2 = B_2 \frac{s_2}{s_1}; \ \beta_3 = B_3 \frac{s_3}{s_1}; \ \cdots \ ; \ \beta_m = B_m \frac{s_m}{s_1} \qquad (8\text{-II-}43)$$

These are *beta coefficients*, the dimensionless parameters which measure the effect of the individual independent variables on the variation of the dependent variable.

The comparison of beta coefficients with partial correlation coefficients shows, in general, that the rank or order of magnitude for the effect of the independent variables is the same for both these measures. This does not always hold true, since they are two different sets of measures. It is easier to compute beta coefficients than partial correlation coefficients, since B_i values and s_i values in Eq. (8-II-43) must always be computed for regression analysis. For this reason, there is a general preference for beta coefficients, although the partial correlation coefficients are better measures of the influence of individual independent variables.

E. Statistical Inference of Regression Coefficients

The same general conclusions for the correlation coefficient in simple linear association can be applied to the measures of association in multiple linear regression.

When only the extreme low and extreme high values of variables in multiple linear regression are available, the multiple correlation coefficient, the partial correlation coefficients, and the beta coefficients are likely to be higher. Thus, if the sample values of variables are available only for a narrow range, these coefficients are likely to be smaller than the true coefficients for the universe.

The unbiased standard deviation of any multiple regression coefficient B_i is [8, p. 283]

$$\hat{S}_{bi} = \sqrt{\frac{\hat{S}_1^2}{N s_i^2 (1 - R_i^2)}} = \frac{S_1}{s_i} \sqrt{\frac{1}{(N - m)(1 - R_i^2)}} \qquad (8\text{-II-}44)$$

where s_i is the standard deviation of marginal distribution of variable x_i, $\hat{S}_1$ is the unbiased standard deviation of residuals for variable x_1, S_1 is the biased standard deviation of residuals for variable x_1, $N - m$ is the number of degrees of freedom, and R_i is the multiple correlation coefficient of x_i with respect to all variables except the variable x_1, with i varying from 2 to m. If the variables $x_2, x_3, \ldots, x_m$ are externally independent among themselves, $R_i = 0$, and Eq. (8-II-44) becomes

$$\hat{S}_{bi} = \frac{S_1}{s_i \sqrt{N - m}} \qquad (8\text{-II-}45)$$

It may be noted that the more the variables $x_2, x_3, \ldots, x_m$ are related among themselves, the greater is R_i. Consequently, $\hat{S}_{bi}$ is larger, and the regression coefficient is less reliable.

The standard deviation of individual values of B_i can be very high. Some of the values are more accurate and some are less accurate than the others. If x_1 is estimated using all B_i values, the error in one value of B_i may be compensated by an error in another value of B_i, so that the error of estimated x_1 may be smaller than that which may appear from the errors in B_i. If the extreme values of x_1 are estimated by the multiple regression linear equation, the error may be large, because B_i values have much greater error for the extremes.

VI. MULTIPLE CURVILINEAR REGRESSION AND CORRELATION

The use of any function in multiple regression and correlation analysis other than a linear function is referred to as *curvilinear regression and correlation*. Either analytical or graphical methods of curvilinear regression are used in hydrologic analyses.

A. Analytical Method

The analytical method applies the least-squares fitting of a regression function. The computational work increases greatly and the determination of the parameters becomes more difficult as a more complex mathematical function is used or as the number of variables increases. The use of digital computers enables the application of different functions and of many variables in a multiple curvilinear regression analysis. The method is similar to that used for multiple linear regression, except that the derived equations for determining the regression coefficients are more complicated.

B. Graphical Method

The *coaxial graphical method* [8] has been widely used in hydrology [9].

An example, taken from Ref. 9 and represented in Fig. 8-II-3, can be used to explain this method. This example relates the basin recharge as the dependent variable to the antecedent precipitation, date (or week number), the rainfall amount, and the rainfall duration as the independent variables. A three-variable relationship is developed (chart A) by plotting antecedent precipitation vs. recharge, labeling the points with week numbers and fitting a family of curves, with one curve for each week. The second step (chart B) relates the observed recharge to the recharge computed from chart A, and the points are labeled with rainfall duration. A family of curves is fitted to represent the effect of different durations on recharge. The third step (chart C) relates the observed recharge to the recharge computed from charts A and B, labeling the points with rainfall amount. A family of curves is fitted to represent the effect of rainfall amounts on recharge. Charts A, B, and C constitute the first approximation [9, pp. 173-175]. Chart D indicates the accuracy of the derived charts. A process of successive approximations must be used to converge upon the correct graphical solution to the relation. This process becomes more necessary as the interdependence of the independent variables increases. The successive approximation may be accomplished by a reverse process, beginning with the last independent variable used in each approximation and working back to the dependent variable, refining

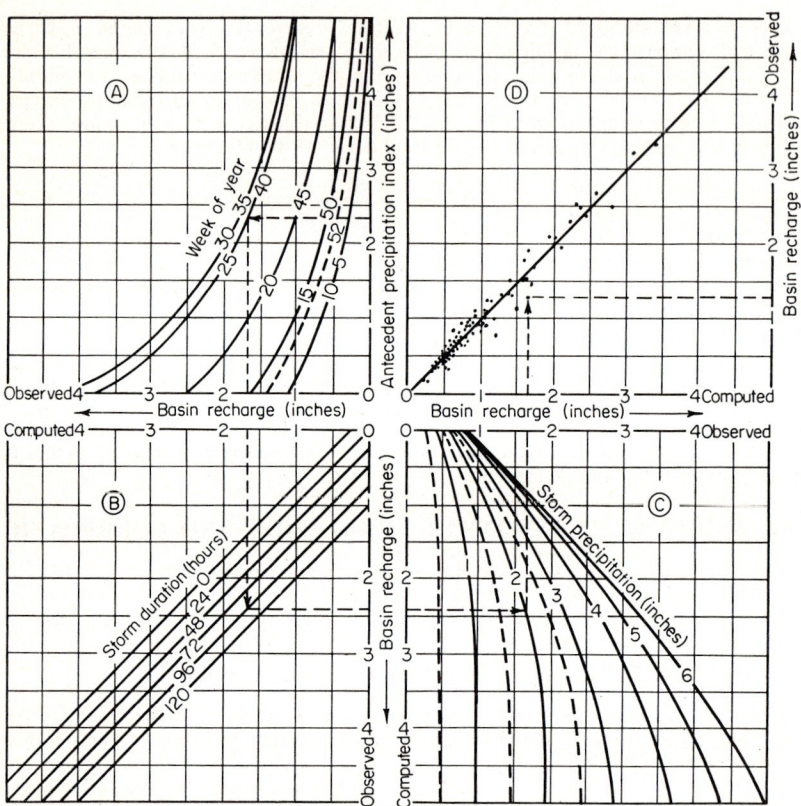

Fig. 8-II-3. Basin-recharge relation for the Monocacy River at Jug Bridge, Md., determined by the coaxial graphical method. (*After U.S. Weather Bureau and Linsley, Kohler, and Paulhus* [9].)

all families of curves along the way. Further approximations then follow the same procedure in both directions alternatively.

The advantage of the graphical coaxial method lies in the fact that the families of curves are independent of any analytical function and its parameters. The disadvantages are the laboriousness of successive approximations and the difficulties in making a reliable statistical inference.

VII. MULTIVARIATE ANALYSIS

Multiple regression is sometimes used to evaluate numerically the assumed model of a hydrologic process with the assumption that independent variables are not correlated. Snyder [10] pointed out that in hydrologic analyses the independent variables may not be uncorrelated. For example, a multiple regression solution does not usually provide a satisfactory estimate of the independent contributions of the rainfall and its duration to the runoff because a positive correlation may exist between the amount and duration of rainfall. In such cases, multivariate analysis appears to offer a workable and mathematically oriented solution in fitting prediction equations to the observed hydrologic data.

Multivariate analysis is the study of arrays of variables. In such analyses, a vector of means and a matrix of covariances of several variables are used instead of the simple mean and the variance of a single variable. This concept allows the association of errors with more than one variable, thus making possible the determination of all the truly independent components of variation in an array of variables. Detailed discussion of the analysis is beyond the present scope and the reader may refer to Refs. 11 to 14.

VIII. REFERENCES

1. Elderton, W. P.: "Frequency Curves and Correlation," 4th ed., Cambridge University Press, Cambridge, and Harren Press, Washington, D.C., 1953.
2. Weatherburn, C. E.: "A First Course in Mathematical Statistics," Cambridge University Press, London, 1952.
3. Kendall, N. G.: "The Advanced Theory of Statistics," 5th ed., Hafner Publishing Company, Inc., New York, 1952, vol. 1.
4. Bartlett, M. S.: Some aspects of the time-correlation problem in regard to tests of significance, *J. Roy. Statist. Soc.*, vol. 98, pp. 536–543, 1935.
5. Williams, E. F.: "Regression Analysis," John Wiley & Sons, Inc., New York, 1959.
6. Allan, D. H. W., and R. F. Attridge: The applications of an IBM calculating punch to solve multiple regression problems, *Proc. Seventh Ann. Conf. Am. Soc. Quality Control*, pp. 521–533, 1953.
7. Benson, M. A.: Factors influencing the occurrence of floods in a humid region of diverse terrain, *U.S. Geol. Surv. Water-Supply Paper* 1580-B, 1962.
8. Ezekiel, Mordecai, and K. A. Fox: "Methods of Correlation and Regression Analysis: Linear and Curvilinear," 3d ed., John Wiley & Sons, Inc., New York, 1959.
9. Linsley, R. K., Jr., M. A. Kohler, and J. L. H. Paulhus: "Hydrology for Engineers," McGraw-Hill Book Company, Inc., New York, 1958.
10. Snyder, W. M.: Some possibilities for multivariate analysis in hydrologic studies, *J. Geophys. Res.* vol. 67, no. 2, pp. 721–729, February, 1962.
11. Kendall, M. G.: "A Course in Multivariate Analysis," Hafner Publishing Co., Inc., New York, 1957.
12. Rao, C. R.: "Advanced Statistical Methods in Biometric Research," John Wiley & Sons, Inc., New York, 1952.
13. Anderson, T. W.: "An Introduction to Multivariate Statistical Analysis," John Wiley & Sons, Inc., New York, 1958.
14. Roy, S. N.: "Some Aspects of Multivariate Analysis," John Wiley & Sons, Inc., New York, 1958.

Section 8-III

STATISTICAL AND PROBABILITY ANALYSIS OF HYDROLOGIC DATA

PART III. ANALYSIS OF VARIANCE, COVARIANCE, AND TIME SERIES

D. R. DAWDY and **N. C. MATALAS**, *Hydraulic Engineers, U.S. Geological Survey.*

I. Analysis of Variance and Covariance........................ 8-69
   A. Introduction... 8-69
      1. Definition... 8-69
      2. Chi-square Distribution.............................. 8-69
      3. F Distribution..................................... 8-70
   B. Analysis-of-variance Models............................. 8-72
      1. One-way Classification............................... 8-72
      2. Two-way Classification............................... 8-73
      3. Linearity of Regression.............................. 8-74
   C. Analysis-of-covariance Models............................ 8-75
      1. One-way Classification............................... 8-75
      2. Study of Regression Effect........................... 8-76
II. Analysis of Time Series.................................... 8-78
   A. Introduction... 8-78
      1. Definition of Time Series............................ 8-78
      2. Characteristics of Time Series....................... 8-79
      3. Properties of the Nonrandom Element.................. 8-79
   B. Trend Analyses.. 8-81
      1. Use of Moving Averages............................... 8-81
      2. Slutzky-Yule Effect.................................. 8-82
   C. Tests for Serial Dependence.............................. 8-82
      1. Parametric Test of Significance...................... 8-82
      2. Nonparametric Tests of Significance.................. 8-83
   D. Generating Processes.................................... 8-84
      1. Definition... 8-84
      2. Moving-average Process............................... 8-84
      3. Sum-of-harmonics Process............................. 8-84
      4. Autoregression Process............................... 8-85
      5. Correlograms... 8-85

ANALYSIS OF VARIANCE AND COVARIANCE

E. Effect of Serial Correlation.................................. 8-85
   1. Estimation of the Variance............................ 8-85
   2. Correlation and Regression Analyses................... 8-86
III. References.. 8-89

I. ANALYSIS OF VARIANCE AND COVARIANCE

A. Introduction

1. Definition. Analysis of variance is a statistical technique which tests mean values by a partitioning of the total variance of a sample into component parts, each of which can be assigned to a particular cause. Thus, in a study of point rainfall, several rain gages may be used and several storms measured by each. Within the total variance of all measured values of precipitation, there is a portion of the variation which is due to the variation of the mean values recorded for each gage and there is a portion which is the result of the variation of individual values about these mean values. A comparison of the two variances can aid in determining whether any measured difference in average rainfall is due to chance. The partitioning of the variance in an analysis of variance is determined by the test to be made.

The partitioning of the variance can include the covariance of the variable being studied with another independent variable. Thus the difference in mean values can be studied after they have been corrected for the effect of the correlated independent variable.

Before discussing the analysis of variance, it is well to consider two probability distributions which play an important role in the analysis. These are the chi-square distribution and the F distribution.

2. Chi-square Distribution. The distribution of the variance is fundamental to many tests of statistical inference. First, consider the distribution of the sum of squares of a variable. Let $x_1, x_2, \ldots, x_n$ be normally and independently distributed variables, each with mean 0 and variance 1. Then

$$\chi^2 = x_1^2 + x_2^2 + \cdots + x_n^2 \qquad (8\text{-III-1})$$

is called *chi-square* and has the probability density function

$$p(\chi^2) = \frac{1}{2^{\frac{1}{2}\nu}\Gamma\left(\frac{\nu}{2}\right)} (\chi^2)^{\frac{1}{2}(\nu-2)} \exp\left(-\frac{1}{2}\chi^2\right) \qquad (8\text{-III-2})$$

where ν is called the *number of degrees of freedom,* and Γ represents the gamma function. Figure 8-III-1 shows this distribution, which is tabulated in most standard statistics books for $\nu \leq 30$ (i.e., Hoel [1]). For larger values of ν, the quantity $(2\chi^2)^{\frac{1}{2}} - (2\nu - 1)^{\frac{1}{2}}$ is approximately normally distributed with mean 0 and variance 1. Since, from Eq. (8-III-1), it can be shown that ns^2/σ^2 is distributed like χ^2 with $\nu = n - 1$ degrees of freedom, where s^2 and σ^2 are the sample and population variances, respectively, then the sample variance of the x_i's is distributed as χ^2 with $n - 1$ degrees of freedom [2].

Equation (8-III-2) is the basis for a test of whether or not a sample variance is significantly different from a presumed population variance. If, for instance, a regionalized flood-frequency curve gives a variance for annual floods of σ_0^2, and if n floods for a given station which was not used to determine σ_0^2, and is therefore independent of it, have a variance of s^2, then the ratio of the sample variance to the "true," or regional, variance can be tested as $s^2/\sigma_0^2 = \chi^2/n$. The critical value of χ^2

for rejecting the null hypothesis $H_0:s^2 = \sigma_0^2$ is taken from tables of the chi-square distribution at a chosen confidence level with $\nu = n - 1$ degrees of freedom.

Often it is desirable to test the hypothesis that $\sigma_1^2 = \sigma_2^2 = \cdots = \sigma_k^2$. Consider that a record of annual precipitation is available for n years and that during this period of time the location of the rain gage has been moved k times. Thus the homogeneity of the rainfall record is questioned. In order to test the hypothesis of the homogeneity of the variance, the rainfall record is divided into $k + 1$ parts, each being that part of the record during which time the rain gage was at a particular location. Let n_i denote the number of years in the ith segment of the record and s_i^2 the variance of the rainfall data in the ith segment, where $i = 1, 2, \ldots, k + 1$. Bartlett [3] has

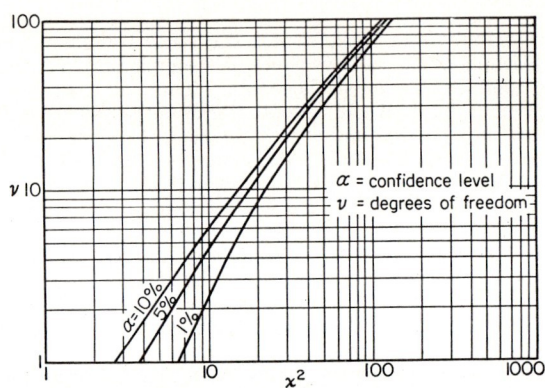

FIG. 8-III-1. Critical values of χ^2.

shown that the hypothesis of homogeneous variance can be tested by means of the chi-square distribution, where χ^2 for k degrees of freedom is defined as

$$\chi_k^2 = \frac{2.3026}{c}\left[\nu \log\left(\frac{1}{\nu}\sum_{i=1}^{k+1}\nu_i s_i^2\right) - \sum_{i=1}^{k+1}\nu_i \log s_i^2\right] \quad (8\text{-III-3})$$

where

$$c = 1 + \frac{1}{3k}\left(\sum_{i=1}^{k+1}\frac{1}{\nu_i} - \frac{1}{\nu}\right) \quad (8\text{-III-4})$$

and

$$\nu = \sum_{i=1}^{k+1}\nu_i = \sum_{i=1}^{k+1}(n_i - 1) = n - (k + 1) \quad (8\text{-III-5})$$

If the value of χ^2 computed by Eq. (8-III-3) exceeds the tabulated value of χ^2 for k degrees of freedom at a chosen confidence level, then the hypothesis of homogeneity of the variance is rejected. It should be noted that the value of c is greater than unity and tends to unity as the number of degrees of freedom increases. Hence, the value of c may be set equal to unity in Eq. (8-III-3) unless there is some doubt about the significance of χ_k^2. An insignificant value of χ_k^2 cannot be made significant by using the actual value of c computed by Eq. (8-III-4).

3. F Distribution. Assume that x_i $(i = 1, \ldots, n_1)$ and y_j $(j = 1, \ldots, n_2)$ are two independent random samples and that s_1^2 and s_2^2 are unbiased estimates of the variance for each sample. The question arises whether or not the two samples have been drawn from the same normal population having variance σ^2. Thus it is necessary to test the hypothesis $\sigma_1^2 = \sigma_2^2$.

ANALYSIS OF VARIANCE AND COVARIANCE 8-71

The basis for testing the hypothesis $\sigma_1^2 = \sigma_2^2$ is the ratio of the two sample variances, which is defined as

$$F = \frac{s_1^2}{s_2^2} = \frac{\nu_1 s_1^2/\nu_1\sigma^2}{\nu_2 s_2^2/\nu_2\sigma^2} \qquad (8\text{-III-}6)$$

whereby

$$\frac{\nu_1 F}{\nu_2} = \frac{\nu_1 s_1^2/\sigma^2}{\nu_2 s_2^2/\sigma^2} \qquad (8\text{-III-}7)$$

From the previous discussion of the chi-square distribution, it is seen that the numerator and denominator of the right-hand side of Eq. (8-III-7) are distributed inde-

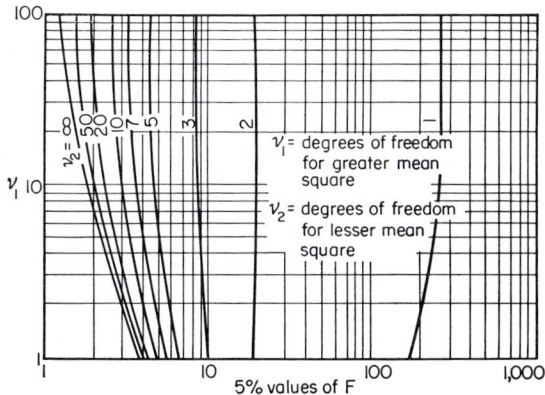

FIG. 8-III-2. Critical values of F at the 5 per cent level.

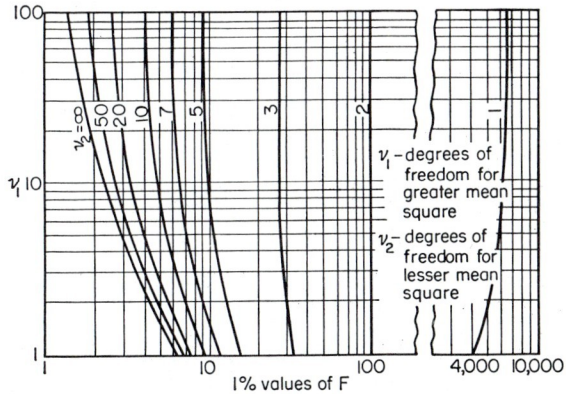

FIG. 8-III-3. Critical values of F at the 1 per cent level.

pendently as χ^2, with ν_1 and ν_2 degrees of freedom, respectively. It can be shown that the probability density function of F is

$$p(F) = \frac{\nu_1^{\frac{1}{2}\nu_1}\nu_2^{\frac{1}{2}\nu_2} F^{\frac{1}{2}(\nu_1-2)}}{B(\frac{1}{2}\nu_1, \frac{1}{2}\nu_2)(\nu_1 F + \nu_2)^{\frac{1}{2}(\nu_1+\nu_2)}} \qquad (8\text{-III-}8)$$

where B denotes the beta function. Figures 8-III-2 and 8-III-3 show values of F corresponding to cumulative probabilities of 0.01 and 0.05 for various values of ν_1 and ν_2. These values, which are available in tabular form (i.e., Hoel [1]), facilitate the

B. Analysis-of-variance Models

1. One-way Classification. Quite often it is necessary to determine whether or not an abrupt change in the mean value of some hydrologic statistic, e.g., measured precipitation or streamflow, has occurred or whether an appreciable difference in rainfall or runoff exists among several experimental watersheds. This test can be made through an analysis of variance.

Assume that there are k drainage areas in a given region and that it is desired to determine if the regional runoff can be considered as homogeneous. Let x_{ij} denote the mean annual discharge per square mile for the ith drainage area ($i = 1, \ldots, k$) during the jth year of record ($j = 1, \ldots, n_i$). Thus the data may be considered as divided into k classes, with n_i items in the ith class. The total number of values within all the classes is $N = \sum_{i=1}^{k} n_i$, the mean for the values in the ith class is

$$\bar{x}_i = \frac{1}{n_i} \sum_{j=1}^{n_i} x_{ij}$$

and the mean for the values in all the classes is

$$\bar{x} = \frac{1}{N} \sum_{i=1}^{k} \sum_{j=1}^{n_i} x_{ij}$$

The total sum of squares of the departures of x_{ij} from $\bar{x}$ may be divided into two parts. The first is due to the variability which cannot be explained by differences of regional effects between classes. The second is due to the variability between classes averaged over the individual items in each class. Thus

$$\sum_{i=1}^{k} \sum_{j=1}^{n_i} (x_{ij} - \bar{x})^2 = \sum_{i=1}^{k} \sum_{j=1}^{n_i} (x_{ij} - \bar{x}_i)^2 + \sum_{i=1}^{k} n_i (\bar{x}_i - \bar{x})^2 \qquad \text{(8-III-9)}$$

By taking the expectation of both sides of Eq. (8-III-9), it is seen that

$$(N - 1)\sigma^2 = (N - k)\sigma^2 + (k - 1)\sigma^2 \qquad \text{(8-III-10)}$$

Thus the three sums of squares given in Eq. (8-III-9) give unbiased estimates of the variance when they are divided by the appropriate number of degrees of freedom given in Eq. (8-III-10).

The total response x_{ij} of the jth individual in the ith class is assumed to be made up of an overall effect μ, a part β_i characteristic of the ith drainage area, and a part ϵ_{ij} which can be regarded as error. These parts are assumed to be additive, so that

$$x_{ij} = \mu' + \beta_i' + \epsilon_{ij} \qquad \text{(8-III-11)}$$

where μ' is adjusted ($\mu' = \mu + \bar{\beta}_i$) so that $\sum_{i=1}^{k} \beta_i' = 0$. It is also assumed that each ϵ_{ij} is an independent random normal variate with expectation 0 and variance σ^2, independent not only of the other ϵ's, but also of the β's.

The hypothesis which is to be tested is that the β's are all zero, in which case the regional runoff may be considered as homogeneous. The testing of this hypothesis may be summarized in an analysis-of-variance table, such as Table 8-III-1.

ANALYSIS OF VARIANCE AND COVARIANCE 8-73

Table 8-III-1. Test of Hypothesis for One-way Classification

Source of variation	Degrees of freedom	Sum of squares	Mean square	F
Between class means.....	$k - 1$	$A = \sum_{i=1}^{k} n_i(\bar{x}_i - \bar{x})^2$	$\dfrac{A}{k-1}$	$\dfrac{A(N-k)}{B(k-1)}$
Within classes..........	$N - k$	$B = \sum_{i=1}^{k} \sum_{j=1}^{n_i} (x_{ij} - \bar{x}_i)^2$	$\dfrac{B}{N-k}$	
Total.................	$N - 1$	$C = \sum_{i=1}^{k} \sum_{j=1}^{n_i} (x_{ij} - \bar{x})^2$	$\dfrac{C}{N-1}$	

If the computed value of F exceeds the value of F with $N - k$ and $k - 1$ degrees of freedom for a chosen confidence level, the hypothesis that the region is homogeneous is rejected. It should be noted that B is used instead of C for determining F. On the basis of the null hypothesis, C is the most reliable, since it is based on the largest number of degrees of freedom; however, B is generally used since it is valid even when the null hypothesis is not true.

A one-way-classification analysis of variance may be used for studying the coefficient of skewness of low-flow data based on various durations in order to determine if the variation of the skewness between different durations is significant [4].

2. Two-way Classification. In the analysis based on a one-way classification, the values in each class are considered as replicates of one another, subject only to random variation. However, there may be a possible significant variation between the individual values in each of the k classes. In order to investigate this possible source of variation, assume that the number of items in each class is a constant equal to n. Each item in each class corresponds to a given year, where each year is referred to as a group. It should be noted that in each of the k classes there is one value from each group, and in each of the n groups there is one value from each class. This arrangement of the values of the variable being studied holds for all two-way-classification analyses. Thus the data may be considered as divided into k classes and n groups. In addition to class means and total mean defined for the one-way-classification analysis, the group means are given by

$$\bar{x}_j = \frac{1}{k} \sum_{i=1}^{k} x_{ij}$$

The total sum of squares of the departures of x_{ij} from $\bar{x}$ may be divided into three parts, the first of which is due to the variability between the classes, the second to the variability between groups, and the third to error or residual variability. Thus

$$\sum_{i=1}^{k} \sum_{j=1}^{n} (x_{ij} - \bar{x})^2 = \sum_{i=1}^{k} n(\bar{x}_i - \bar{x})^2 + \sum_{j=1}^{n} k(\bar{x}_j - \bar{x})^2 + \sum_{i=1}^{k} \sum_{j=1}^{n} (x_{ij} - \bar{x}_i - \bar{x}_j + \bar{x})^2 \quad (8\text{-III-}12)$$

By taking the expectation of both sides of Eq. (8-III-12), it is seen that

$$(nk - 1)\sigma^2 = (k - 1)\sigma^2 + (n - 1)\sigma^2 + (nk - n - k + 1)\sigma^2 \quad (8\text{-III-}13)$$

whereby the four sums of squares given in Eq. (8-III-12) provide unbiased estimates of the variance when they are divided by their appropriate number of degrees of freedom given in Eq. (8-III-13).

The mathematical model is now

$$x_{ij} = \mu + \alpha_i + \beta_j + \epsilon_{ij} \qquad (8\text{-III-}14)$$

where μ is the overall effect, α_i is the characteristic of the ith drainage area, β_j is the characteristic of the jth year of record, and ϵ_{ij} is the error. The overall effect μ is considered to be adjusted so that $\sum_{i=1}^{k} \alpha_i = 0$ and $\sum_{j=1}^{n} \beta_j = 0$. It is assumed that the ϵ_{ij}'s are all normally and independently distributed about zero with the same variance σ^2.

The hypothesis which is to be tested is that all the α_i's and β_j's are zero, in which case the region may be considered as homogeneous with respect to space and time. The testing of this hypothesis is summarized in Table 8-III-2.

Table 8-III-2. Test of Hypothesis for Two-way Classification

Source of variation	Degrees of freedom	Sum of squares	Mean square	F
Between classes...	$k - 1$	$A = \sum_{i=1}^{k} n(\bar{x}_i - \bar{x})^2$	$\dfrac{A}{k - 1}$	$\dfrac{A(n - 1)}{C}$
Between groups...	$n - 1$	$B = \sum_{j=1}^{n} k(\bar{x}_j - \bar{x})^2$	$\dfrac{B}{n - 1}$	$\dfrac{B(k - 1)}{C}$
Error...	$(k - 1)(n - 1)$	$C = \sum_{i=1}^{k} \sum_{j=1}^{n} (x_{ij} - \bar{x}_i - \bar{x}_j + \bar{x})^2$	$\dfrac{C}{(k - 1)(n - 1)}$	
Total...	$nk - 1$	$D = \sum_{i=1}^{k} \sum_{j=1}^{n} (x_{ij} - \bar{x})^2$	$\dfrac{D}{nk - 1}$	

If the first value of F is found to exceed the value of F with $(k - 1)(n - 1)$ and $k - 1$ degrees of freedom, then the hypothesis is rejected that all the α's are zero, and if the second value of F exceeds the value of F with $(k - 1)(n - 1)$ and $n - 1$ degrees of freedom, then the hypothesis is rejected that all the β's are zero. If both computed values of F are found to be significant, then the entire hypothesis of homogeneity is rejected.

3. Linearity of Regression. Many hydrologic studies are based on regression analyses where generally linear functions are fitted to the data. However, in some studies, a linear function may be questionable and it may be necessary to test for nonlinearity of the regression line. This test may be made by an appropriate analysis of variance.

The data should be partitioned into k arrays according to the values of the independent variable. The range of values of the independent variable within each array should be narrow enough so that the range in the values of the dependent variable in each array approximates the spread of the values of the dependent variable about the regression line within each array. Let Y_i be the estimate of the dependent variable from the regression line for the mean value of the independent variable in the ith array, let y_{ij} be the jth value ($j = 1, \ldots, n_i$) of the dependent variable in the ith array, let $\bar{y}_i$ be the mean of the values of the dependent variable in the ith array, and let $\bar{y}$ be the mean for all the values of the dependent variable.

ANALYSIS OF VARIANCE AND COVARIANCE

The sum of squares of $(y_{ij} - \bar{y})^2$, which is proportional to the variance of the dependent variable, may be divided into three parts. The first part is due to the regression function itself; the second part is due to the deviations of the means from the regression line; and the third part is due to the variation within the arrays. Thus

$$\sum_{i=1}^{k} \sum_{j=1}^{n_i} (y_{ij} - \bar{y})^2 = \sum_{i=1}^{k} n_i (Y_i - \bar{y})^2 + \sum_{i=1}^{k} n_i (\bar{y}_i - Y_i)^2 + \sum_{i=1}^{k} \sum_{j=1}^{n_i} (y_{ij} - \bar{y}_i)^2 \quad (8\text{-III-}15)$$

By taking the expectation of Eq. (8-III-15), it is seen that

$$(N - 1)\sigma^2 = \sigma^2 + (k - 2)\sigma^2 + (N - k)\sigma^2 \quad (8\text{-III-}16)$$

where $N = \sum_{i=1}^{k} n_i$. Thus the four sums of squares given in Eq. (8-III-15) provide unbiased estimates of the variance when they are divided by their appropriate number of degrees of freedom given in Eq. (8-III-16). The test for linearity is summarized in Table 8-III-3.

Table 8-III-3. Test for Linearity

Source of variation	Degrees of freedom	Sum of squares	Mean square	F
Linear regression	1	$A = \sum_{i=1}^{k} n_i (Y_i - \bar{y})^2$	$\dfrac{A}{1}$	
Deviation of means from regression line	$k - 2$	$B = \sum_{i=1}^{k} n_i (\bar{y}_i - Y_i)^2$	$\dfrac{B}{k - 2}$	$\dfrac{B(N - k)}{C(k - 2)}$
Within arrays	$N - k$	$C = \sum_{i=1}^{k} \sum_{j=1}^{n_i} (y_{ij} - \bar{y}_i)^2$	$\dfrac{C}{N - k}$	
Total	$N - 1$	$D = \sum_{i=1}^{k} \sum_{j=1}^{n_i} (y_{ij} - \bar{y})^2$	$\dfrac{D}{N - 1}$	

On the assumption of linearity of regression, the sum of squares denoted by B is due to sampling errors and the estimate of σ^2 obtained from B should not be greater than that derived from the sum of squares within arrays, which is denoted by C. If the computed value of F is found to be significant, then the hypothesis of linearity of regression is rejected.

C. Analysis-of-covariance Models

1. One-way Classification. At times it is desirable to test the significance of the difference in mean values of a variable after these means have been corrected for the effect of another correlated variable. Thus a study of the effect of deforestation or reforestation upon streamflow must first eliminate the effect of varying precipitation through the test period. If the data in this example were classified by years, the covariance of streamflow with precipitation might be removed and the remainder would be variance between years.

The covariance first is partitioned into

$$\sum_i \sum_j (x_{ij} - \bar{x})(y_{ij} - \bar{y}) = \sum_i \sum_j (x_{ij} - \bar{x}_i)(y_{ij} - \bar{y}_i)$$
$$+ \sum_i n_i(\bar{x}_i - \bar{x})(\bar{y}_i - \bar{y}) \quad (8\text{-III-17})$$

where the first term on the right represents covariance within classes (years), and the second term that between classes (years). All three terms give an estimate of the covariance, assuming there is no difference between years. The analysis of covariance, as shown in Table 8-III-4, is intended to test the null hypothesis that there is no difference in covariance between years and in the population as a whole. Thus our null hypothesis would be that reforestation had no effect on streamflow. This type of analysis can also be used for studying the significance of changes in the slopes of double-mass curves [5].

Table 8-III-4. Analysis of Covariance

Source of covariance	Degrees of freedom	Sum of cross products	F
Between classes (years)	$k - 1$	$A = \sum_i n_i(\bar{x}_i - \bar{x})(\bar{y}_i - \bar{y})$	
Within classes (years)	$N - k$	$B = \sum_i \sum_j (x_{ij} - \bar{x}_i)(y_{ij} - \bar{y}_i)$	$(k - 1)B/(N - k)A$
Total...............	$N - 1$	$C = \sum_i \sum_j (x_{ij} - \bar{x})(y_{ij} - \bar{y})$	

A more exact method of testing the same hypothesis would be to determine the significance of the coefficient of regression of streamflow with precipitation within years. If this is significant, then the yearly means can be corrected for the regression effect, and the differences between the corrected yearly means then are tested for significance. In addition, the significance of the difference between the two regression coefficients for between and within years can be tested to determine whether the classification by years has an effect upon the degree of association of the variables.

The coefficients of regression are given in Table 8-III-5 with the t statistic

$$t_{N-k} = A \sqrt{\frac{(N - k) \sum_i \sum_j (x_{ij} - \bar{x}_i)^2}{\sum_i \sum_j (y_{ij} - \bar{y}_i)^2}} \quad (8\text{-III-18})$$

used to test the significance of the within-years coefficient.

2. Study of Regression Effect. In order to correct the yearly means for the regression effect, the variance must be partitioned. First, it is partitioned into that due to the regression of streamflow with precipitation and that due to deviations from the regression line. The latter part, then, is further partitioned into that within years and that between years.

The variation due to regression is

$$A = \frac{\left[\sum_i \sum_j (x_{ij} - \bar{x})(y_{ij} - \bar{y})\right]^2}{\sum_i \sum_j (x_{ij} - \bar{x})^2} \quad (8\text{-III-19})$$

ANALYSIS OF VARIANCE AND COVARIANCE

Table 8-III-5. Test of Coefficients of Regression

Source of covariance	Degrees of freedom	Coefficient of regression
Between classes (years)..........	$k - 1$	$A = \dfrac{\sum_i n_i(\bar{x}_i - \bar{x})(\bar{y}_i - \bar{y})}{\sum_i n_i(\bar{x}_i - \bar{x})^2}$
Within classes (years)..........	$N - k$	$B = \dfrac{\sum_i \sum_j (x_{ij} - \bar{x}_i)(y_{ij} - \bar{y}_i)}{\sum_i \sum_j (x_{ij} - \bar{x}_i)^2}$
Total.......................	$N - 1$	$C = \dfrac{\sum_i \sum_j (x_{ij} - \bar{x})(y_{ij} - \bar{y})}{\sum_i \sum_j (x_{ij} - \bar{x})^2}$

The variation between years can be determined in two ways, however, depending upon whether the between- or within-years regression coefficient is used as an adjusting factor. The total variation due to deviations from the regression line is

$$B = \sum_i \sum_j (y_{ij} - \bar{y})^2 - \frac{\left[\sum_i \sum_j (x_{ij} - \bar{x})(y_{ij} - \bar{y})\right]^2}{\sum_i \sum_j (x_{ij} - \bar{x})^2} \qquad (8\text{-III-20})$$

That owing to within-years variation corrected for its regression coefficient is

$$C = \sum_i \sum_j (\bar{y}_{ij} - \bar{y}_i)^2 - \frac{\left[\sum_i \sum_j (x_{ij} - \bar{x}_i)(y_{ij} - \bar{y}_i)\right]^2}{\sum_i \sum_j (x_{ij} - \bar{x}_i)^2} \qquad (8\text{-III-21})$$

and that due to between-years variation corrected for its regression coefficient is

$$D = \sum_i n_i(\bar{y}_i - \bar{y})^2 - \frac{\left[\sum_i n_i(\bar{x}_i - \bar{x})(\bar{y}_i - \bar{y})\right]^2}{\sum_i n_i(\bar{x}_i - \bar{x})^2} \qquad (8\text{-III-22})$$

whereas the difference between Eqs. (8-III-20) and (8-III-21) gives the variation due to between-years variation corrected for the within-classes regression coefficient. The analysis of covariance is given in Table 8-III-6.

The within-classes sum of squares divided by its degrees of freedom is the estimate of the variance used for testing. An F test may be applied first to the regression effect

$$F_{1, N-k-1} = \frac{(N - k - 1)A}{C} \qquad (8\text{-III-23})$$

Table 8-III-6. Analysis of Covariance

Source of variation	Degrees of freedom	Sum of squares
Regression．．．	1	A
Deviations about regression line．．．．．．．．．．．．．．．．．．	$N - 2$	B
Within classes．．	$N - k - 1$	C
Between classes based on between-classes regression．．	$k - 2$	D
Between classes based on within-classes regression．．	$k - 1$	$B - C$

then to either or both measures of the between-years variation,

$$F_{k-2, N-k-1} = \frac{(N - k - 1)D}{(k - 2)C} \tag{8-III-24}$$

and

$$F_{k-1, N-k-1} = \frac{(N - k - 1)(B - C)}{(k - 1)C} \tag{8-III-25}$$

although the second test is more meaningful, if a class effect exists. To test whether the regression coefficients based on within- and between-years variation are significantly different, the difference of the two estimates of between-years variation may be used. This variation, which is the result of the difference in regression coefficients, may be used with one degree of freedom to compute an estimate of the variance, and this tested against the within-groups variance. The resulting F test

$$F_{1, N-k-1} = \frac{(N - k - 1)(B - C - D)}{C} \tag{8-III-26}$$

may be used to test whether the two regression coefficients are significantly different. The null hypothesis tested in this example is that there is no time variation of the relation of streamflow to precipitation, the assumption being that any time variation which does exist is the effect of reforestation.

II. ANALYSIS OF TIME SERIES

A. Introduction

1. Definition of Time Series. A *time series* is a sequence of values arrayed in order of their occurrence which can be characterized by statistical properties. The sequence of values is represented by $x(t_1), x(t_2), x(t_3), \ldots$, where $t_1 < t_2 < t_3 \cdots$. The daily hydrograph is a graphical representation of a time series of daily discharges. Other examples of hydrologic time series are the annual sequences of floods, low flows, and mean discharges. A time series may be a function of time explicitly or a function of any single variable which takes the place of time. Examples of sequences ordered by distance rather than time are the width and roughness of a stream channel as a function of distance.

Generally, it is possible to classify time series as being either of two types: *stationary* or *nonstationary*. Assume that a time series is divided into several segments and that a statistical parameter such as the mean is used to characterize the data within each section. If the expected value of the statistical parameter is the same for each section, the time series is said to be stationary. If the expected values are not the same, the time series is nonstationary. In stationary time series, absolute time is not important, and the series may be assumed to have started somewhere in the infinite past. However, in nonstationary time series, it is necessary to consider absolute time since the series cannot be assumed to have begun prior to the time of the initial observation.

2. Characteristics of Time Series. Most of the statistical methods used in hydrologic studies are based on the assumption that the observations are independently distributed in time. The occurrence of an event is assumed to be independent of all previous events. This assumption is not always valid for hydrologic time series. Observations of daily discharges do not change appreciably from one day to the next. There is a tendency for the values to cluster, in the sense that high values tend to follow high values and low values tend to follow low values. Thus the daily discharges are not independently distributed in time. The dependence between monthly discharges is less than that between daily discharges, and the dependence between annual discharges is less than that between monthly discharges. Thus the dependence between hydrologic observations decreases with an increase in the time base.

Hydrologic time series may be considered as composed of the sum of two components: a *random element* and a *nonrandom element*. A nonrandom element is said to exist when observations separated by k time units are dependent. If the values of x_i are linearly dependent upon the values of x_{i+k}, then the correlation between x_i and x_{i+k} may be taken as the measure of dependence. This correlation is referred to as the *k*th-order *serial correlation*.

The *serial correlation coefficient* is analogous to the product-moment correlation coefficient for two sets of data. If x_i and x_{i+k} are considered as two sets of data then the kth-order serial correlation coefficient is defined as

$$r_k = \frac{\frac{1}{N-k}\sum_{i=1}^{N-k} x_i x_{i+k} - \frac{1}{(N-k)^2}\left(\sum_{i=1}^{N-k} x_i\right)\left(\sum_{i=1}^{N-k} x_{i+k}\right)}{\left[\frac{1}{N-k}\sum_{i=1}^{N-k} x_i^2 - \frac{1}{(N-k)^2}\left(\sum_{i=1}^{N-k} x_i\right)^2\right]^{1/2} \left[\frac{1}{N-k}\sum_{i=1}^{N-k} x_{i+k}^2 - \frac{1}{(N-k)^2}\left(\sum_{i=1}^{N-k} x_{i+k}\right)^2\right]^{1/2}} \quad (8\text{-III-27})$$

where N is the length of the time series. For $k = 0$, it follows that $r_0 = 1$, and for $k \geq 1$, $-1 \leq r_k \leq 1$.

If a time series is random, $r_k = 0$ for all values of $k \geq 1$. However, for a sample of finite size, computed values of r_k may differ from zero because of sampling errors. Since N is small for most hydrologic sequences, the sampling errors are very large, so that it is necessary to test the values of r_k to determine if they are significantly different from zero. A test of significance for r_k is given below.

An example of the computation of the first-order serial correlation coefficient r_1 for the low flows (annual minimum daily discharges) for Middle Branch Westfield River near Goss Heights, Mass., is shown in Table 8-III-7. The period of record is from 1913 to 1950 ($N = 38$). In the table, the columns headed x_i and x_{i+1} give the low-flow values from 1913 to 1949 and from 1914 to 1950, respectively.

In order to determine r_2, it is necessary that x_i and x_{i+2} denote the low-flow values from 1910 to 1953 and from 1912 to 1955, respectively. Similarly, by forming two sets of data, the values of r_3, r_4, etc., can be determined.

3. Properties of the Nonrandom Element. The nonrandom element may be composed of both a trend, or a long-term movement, and an oscillation about the trend. Both of these parts need not be present in a particular time series. The first step in analyzing a time series is to separate the nonrandom element from the random element.

Trend is usually thought of as a smooth motion of the series over a long period of time. For any given time series, the sequence of values will follow an oscillatory pattern. If this pattern indicates a more or less steady rise or fall, it is defined as a *trend*. However, no matter what the length of a time series is, it can never be stated

ANALYSIS OF VARIANCE, COVARIANCE, AND TIME SERIES

Table 8-III-7. Computation of the First-order Serial Correlation Coefficient

	x_i	x_{i+1}	x_i^2	x^2_{i+1}	$x_i x_{i+1}$
	1.6	0.4	2.56	0.16	0.64
	0.4	0.4	0.16	0.16	0.16
	0.4	2.9	0.16	8.41	1.16
	2.9	5.4	8.41	29.16	15.66
	5.4	5.0	29.16	25.00	27.00
	5.0	7.5	25.00	56.25	37.50
	7.5	5.0	56.25	25.00	37.50
	5.0	14	25.00	196	70.00
	14	15	196	225	210.00
	15	2.5	225	6.25	37.50
	2.5	3.0	6.25	9.00	7.50
	3.0	9.1	9.00	82.81	27.30
	9.1	4.0	82.81	16.00	36.40
	4.0	6.8	16.00	46.24	27.20
	6.8	14	46.24	196	95.20
	14	4.0	196	16.00	56.00
	4.0	4.7	16.00	22.09	18.80
	4.7	4.8	22.09	23.04	22.56
	4.8	2.1	23.04	4.41	10.08
	2.1	4.6	4.41	21.16	9.66
	4.6	6.0	21.16	36.00	27.60
	6.0	5.5	36.00	30.25	33.00
	5.5	2.5	30.25	6.25	13.75
	2.5	6.9	6.25	47.61	17.25
	6.9	10	47.61	100	69.00
	10	2.6	100	6.76	26.00
	2.6	4.6	6.76	21.16	11.96
	4.6	2.5	21.16	6.25	11.50
	2.5	4.4	6.25	19.36	11.00
	4.4	4.5	19.36	20.25	19.80
	4.5	4.8	20.25	23.04	21.60
	4.8	11	23.04	121	52.80
	11	3.5	121	12.25	38.50
	3.5	3.6	12.25	12.96	12.60
	3.6	2.6	12.96	6.76	9.36
	2.6	1.8	6.76	3.24	4.68
	1.8	3.6	3.24	12.96	6.48
Σ	193.6	195.6	1,483.84	1,494.24	1,134.70
$\Sigma/(N-1)$	5.23	5.29	40.10	40.38	30.67

$$r = \frac{33.37 - (5.23)(5.29)}{[40.10 - (5.23)^2]^{\frac{1}{2}}[40.38 - (5.29)^2]^{\frac{1}{2}}} = 0.24$$

with certainty that an apparent trend is not part of a slow oscillation, unless the series ends.

An oscillatory pattern is often confused with a cyclical pattern. For a *cyclical time series*, the maximum and minimum values occur at equal intervals of time with constant amplitude. The random element, if present, tends to distort this pattern. In an oscillatory time series, the amplitude and the interval of time between maximum and minimum values are distributed about mean values. A cyclical time series is oscillatory, but an oscillatory time series is not necessarily cyclical.

B. Trend Analysis

1. Use of Moving Averages. Various methods of removing trend are available. All the methods, however, are not fully understood as to how they affect the time series. The most general method involves the *fitting of a polynomial* to the data. This method has two principal objections: (1) the coefficients of the polynomial must be defined by high-order moments which are unreliable because of their large sampling errors since N is small, and (2) the coefficients of the polynomial must be recomputed each time a new value is added to the time series because they are based on the available data of the time series.

An alternative method of trend elimination is that of *moving averages*, which consists of finding a polynomial which will fit part of the record and using different polynomials for different parts of the record. This method permits the addition of new values without altering the previously fitted polynomials.

In order to remove the trend, it is necessary to smooth out irregularities in the time series. Assume that the observations $x_i, x_2, \ldots, x_N$ are taken at equal intervals of time. The method of moving averages consists of determining overlapping means of m successive weighted values. An example of moving averages of $m = 3$ is

$$y_2 = \frac{b_1 x_1 + b_2 x_2 + b_3 x_3}{3}$$

$$y_3 = \frac{b_1 x_2 + b_2 x_3 + b_3 x_4}{3}$$

$$\ldots \ldots \ldots \ldots$$

$$y_{N-1} = \frac{b_1 x_{N-2} + b_2 x_{N-1} + b_3 x_N}{3}$$

(8-III-28)

The weights of the moving average, b_1, b_2, and b_3, are such that their sum equals 3. In general, for moving averages of m,

$$\sum_{i=1}^{m} b_i = m \qquad (8\text{-III-}29)$$

The weights may be either positive or positive and negative. A simple moving average refers to the case where each of the weights equals 1. Although a simple moving average tends to smooth out the data, it does not preserve the main features of the time series as well as a weighted moving average.

It is convenient to use odd values of m so that the computed values of y correspond in time to the middle value of the x's being averaged. A moving average of m applied to a sequence of N terms yields a sequence of $N - 2n$ terms, where $n = (m - 1)/2$. Thus, if $m = 3$, $n = 1$, so that one term is lost at the beginning and end of the time series. Although it is possible to use moving averages of $m = 2, 3, \ldots, N - 1$, it is necessary that m be small relative to N.

Generally, even a smooth trend obtained by the method of moving averages cannot be represented conveniently by a mathematical equation. If a mathematical trend is fitted to the data, a simple relation should be used unless logic indicates otherwise. The simplest mathematical expression is a straight line. However, a time series is apt to be such that a single linear trend cannot be used throughout the time of observation. In such cases, it is possible to use linear trends for portions of the time series.

After a trend has been established, it is possible to remove the trend from the data in one of several ways. One way is to take as a new variable the deviations about the trend line. It is necessary that these deviations constitute a stationary time series. This procedure of trend removal is widely used in hydrologic studies. With some time series, such as tree-ring series, the deviations of the data about the trend line decrease with time [6]. Hence, in this case, dividing the deviations by their corresponding trend values yields a series which may be considered as stationary.

As an example of trend analysis, simple moving averages of $m = 3$ and $m = 5$ are applied to the low-flow data for the Middle Branch Westfield River near Goss Heights, Mass. These results are shown graphically in Fig. 8-III-4. Both moving averages indicate an apparent trend. This apparent trend may, however, be part of an oscillatory movement. With such a short series, it is difficult to prove that the apparent trend is significant, and not part of the oscillatory movement of the series.

2. Slutzky-Yule Effect. Assume that a time series has an oscillatory component about a trend. If a moving average is used to determine the trend, a long-period oscillation tends to be included as part of the trend. Oscillations which are comparable in period with the length of the moving average m, or even shorter, are damped out. The moving average also introduces an oscillatory movement into the random element. These consequences of the moving-average method are referred to as the *Slutzky-Yule effect* [7, 8].

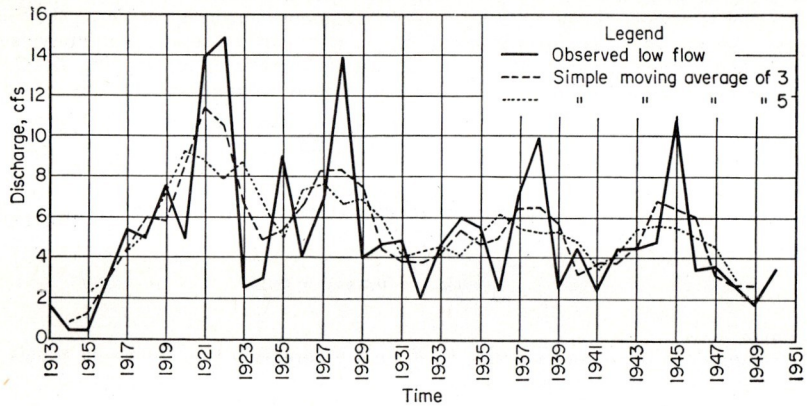

Fig. 8-III-4. Annual daily low flow for Middle Branch Westfield River near Goss Heights, Mass.

If a simple moving average is used, the variance of the induced oscillation is $1/m$ times the variance of the random element, and the average length of this induced oscillatory movement is $360°/\cos[(m - 1)/(m + 1)]$. If a weighted, instead of a simple, moving average is used, the Slutzky-Yule effect is magnified. Because of the Slutzky-Yule effect, care must be exercised in discussing the oscillatory character of a time series if its trend has been removed by means of the moving-average method.

C. Tests for Serial Dependence

1. Parametric Test of Significance. The variables x_i and x_{i+k} used to determine the serial correlation coefficients are actually parts of the same time series. The serial correlation coefficients cannot be tested for significance by means of the test for the ordinary product-moment correlation between two random series unless N is very large. A reliable test of significance must be based on small-sample theory.

Anderson [9] developed a test of significance based on a normal random time series which is circular. A *circular time series* is defined as a time series where the last value is followed by the first value so that the series repeats itself. As N tends to infinity, the confidence limits based on a circular time series converge to those based on an open time series. If N is small, only the low-order (k small) serial correlation coefficients may be tested for significance. Blackman and Tukey [10] recommend that k/N should not exceed 0.10. This rule appears to be satisfactory for deciding upon the highest-order serial correlation which may be tested for significance by Anderson's method.

With respect to the first-order serial correlation coefficient r_1, Anderson showed that, for a normal random time series of N values, the expected value and the variance of r_1 are $-1/(N-1)$ and $(N-2)/(N-1)^2$, respectively. Since r_1 is nearly normally distributed, the confidence limits (CL) for a computed value of r_1 are given by

$$\operatorname{CL}(r_1) = -\frac{1}{N-1} \pm t_\alpha \frac{\sqrt{N-2}}{N-1} \qquad (8\text{-III-}30)$$

where t_α is the standardized normal variate corresponding to the probability level $1 - \alpha$.

If the computed value of r_1 lies within the confidence limits, then r_1 is considered to be insignificantly different from zero at the probability level $1 - \alpha$. An insignificant r_1 is a necessary, but not a sufficient, condition for deciding that a time series is random. In order that a time series be regarded as random, it is necessary that r_k for $k \geq 1$ be insignificant. Because of sampling errors, the serial correlation coefficients for some values of k will be found to be significant even if the observed time series is a sample from a random time series. However, the number of significant serial correlation coefficients should not be greater than that expected by chance from the total number of serial correlation coefficients tested. The reader is referred to Anderson's paper [9] for tests of significance of r_k's where $k \geq 2$.

For the low-flow data for Middle Branch Westfield River near Goss Heights, Mass., r_1 is 0.24. At the 95 per cent level, $t_\alpha = 1.96$. Thus the 95 per cent confidence limits are 0.30 and -0.34. Since r_1 is below the upper confidence limit, it may be regarded as insignificant at the 95 per cent level.

2. Nonparametric Tests of Significance. A nonrandom time series has an oscillatory component. The observed values in a purely random time series fluctuate erratically about some mean value. The fact that a time series exhibits more or less erratic fluctuations suggests that the number of times that the values are above or below a given value is indicative of the randomness or nonrandomness of the time series.

A nonparametric method of determining if a time series is random is that of the *median cross*. For a time series of N values, the median is determined. From the sequence of N values, the number of times that the series crosses the median is determined. Let this number be denoted by n. The expected value and variance of n are $(N-1)/2$ and $(N-1)/4$, respectively. Since n is nearly normally distributed, it is possible to test if n is significantly different from $(N-1)/2$ by

$$t = \frac{n - (N-1)/2}{\sqrt{(N-1)/4}} \qquad (8\text{-III-}31)$$

If the absolute value of t is greater than the absolute value of t_α, the normal deviate at the probability level $1 - \alpha$, the time series is regarded as nonrandom. If the contrary is true, the time series is considered to be random.

Another nonparametric method for determining if a time series is random is the *turning-point test*. A turning point is associated with a value x_i, where either $x_{i+1} > x_i < x_{i-1}$ or $x_{i+1} < x_i > x_{i+1}$. Let m denote the number of turning points. The expected value and the variance of m are $2(N-2)/3$ and $(16N-29)/90$, respectively. Since m is nearly normally distributed,

$$t = \frac{m - 2(N-2)/3}{\sqrt{(16N-29)/90}} \qquad (8\text{-III-}32)$$

The significance or nonsignificance of m is determined in the same manner described for n in the median-cross test. By using the median-cross and the turning-point tests to determine if the Middle Branch of Westfield River data are random, the t values are

0.82 and 0, respectively. Since both t values are less than 1.64, the data are assumed to be random. It is interesting to note that the nonparametric tests indicate randomness and the parametric test indicates nonrandomness. The parametric test is stronger and more reliable than the nonparametric tests.

D. Generating Processes

1. Definition. A *generating process* is the manner by which the causal forces act to produce a time series. Some processes can be expressed mathematically, in which case it is possible to determine directly the various statistical characteristics of the time series. Often a time series is approximated by a certain process. The choice of the process is based upon how well the mathematical structure of the process conforms to the physical characteristics of the time series. The processes which have been used in hydrologic studies are (1) the moving average, (2) the sum of harmonics, and (3) the autoregression.

2. Moving-average Process. The moving-average process may be expressed as

$$x_i = b_0 + b_1 y_i + b_2 y_{i-1} + \cdots + b_m y_{i-(m-1)} \tag{8-III-33}$$

where y is a random variable and m is the extent of the moving average. Equation (8-III-33) may be taken as the model representing the relation between annual runoff x and annual effective precipitation y, where m is the extent of the carryover due to the water-retardation characteristics of the river basin. For such a model, the weights $b_0, b_1, \ldots, b_m$ must all be positive and sum to unity. By virtue of the moving average on the y's, the generated series x is not random. The serial correlation coefficients for the x's are given by Wold [11]:

$$r_k = \frac{\sum_{i=0}^{m} b_i b_{i+k}}{\sum_{i=0}^{m} b_i^2} \qquad 0 \leq k \leq m - 1 \tag{8-III-34}$$

where

$$r_k = 0 \qquad k \geq m \tag{8-III-35}$$

It should be noted in Eqs. (8-III-34) and (8-III-35) that dependence between values does not extend throughout the time series. Values separated by m or more time units are independent.

3. Sum-of-harmonics Process. A simple model of the generating process of the sum of harmonics is

$$x_i = A \sin \theta i + y_i \tag{8-III-36}$$

where A and θ are the amplitude and period of cyclicity, respectively, and y is a random component. Equation (8-III-36) may be taken as a model representing seasonal discharges. For example, if the x's denote monthly discharges and if there is a distinct period of high flow and of low flow, then $\theta = \pi/6$, and i would represent the months from 1 to 12. The generated x's are nonrandomly distributed in time. The serial correlation coefficients are defined by

$$r_k = \frac{A^2}{2 \operatorname{Var}(x)} \cos \theta k \tag{8-III-37}$$

where the variance of x, Var (x), is defined by

$$\operatorname{Var}(x) = \frac{A^2}{2} + \operatorname{Var}(y) \tag{8-III-38}$$

Equation (8-III-36) is a special case, where only one harmonic is involved. It is often argued that there are hidden periodicities in hydrologic data of annual sequences.

ANALYSIS OF TIME SERIES

The periodicities are called hidden since the superposing of many series of different harmonics yields a series which is seemingly random. A recent study involving the search for hidden periodicities in rainfall has been made by Abbott [12].

4. Autoregression Process. An *autoregression process* is used in hydrologic studies for representing sequences whose nonrandomness is due to storage in the basin (groundwater, lake, or channel storage). There are many autoregressive models; however, the first-order process is defined as

$$x_{i+1} = r_1 x_i + \epsilon_{i+1} \qquad (8\text{-III-}39)$$

where r_1 is the first-order serial correlation coefficient for the x's and ϵ is a random component. This process is often referred to as the *first-order Markov process*. For this process, the serial correlation coefficients are given by

$$r_k = r_1^k \qquad (8\text{-III-}40)$$

If r_1 is positive, then all values of r_k are positive and $r_1 > r_2 > \cdots$. If r_1 is negative, then r_k is positive for even values of k and negative for odd values of k. The absolute value of r_k decreases as k increases.

5. Correlograms. A *correlogram* is a graphical representation of the r_k's as a function of k where the values of r_k are plotted as ordinates against their respective values of k as abscissas. In order to reveal the features of the correlogram better, the plotted points are joined each to the next by a straight line.

From Eqs. (8-III-34) and (8-III-35), it is seen that the correlogram for a moving average may oscillate, depending upon the b's, but it will vanish for all values of $k > m$. It is seen from Eq. (8-III-37) that the correlogram for a harmonic process will oscillate with period θ and amplitude $A^2/2 \text{ Var } (x)$. The period of oscillation of the correlogram is the same as that for the time series itself. For the autoregression process, it is seen by Eq. (8-III-40) that, if r_1 is positive, the correlogram will decrease monotonically from $r_0 = 1$ to $r_\infty = 0$. If r_1 is negative, the correlogram will oscillate with period unity above the abscissa with a decreasing but nonvanishing amplitude.

The correlogram provides a theoretical basis for distinguishing among the three types of oscillatory time series. From a set of data, the serial correlation coefficients can be determined and the correlogram can be constructed. The shape of the correlogram is indicative of the generating process in the manner described above. In practice, the number of observations forming a sequence is small, so that observed correlograms always show less damping than theoretical correlograms because the observed serial correlation coefficients are inflated by sampling errors. Thus one cannot easily discern what the generating process is simply by observing the correlogram.

At present there is no adequate small-sample test for distinguishing among the generating processes. Quenouille [13] has developed tests of significance of the correlograms for various autoregressive models. However, these tests are based on the length of sequence being very large.

E. Effect of Serial Correlation

1. Estimation of the Variance. Serial correlation represents a tendency for fluctuations about the mean to perpetuate themselves. In nonrandom hydrologic time series, r usually is positive, so that high values tend to follow high values and low values tend to follow low values. Thus values near x_i yield little new information concerning the true fluctuation of the events about the mean. The amount of information which is furnished varies inversely with r_1. If $r_1 = 0$, then each successive event furnishes new information. If $r_1 = 1$, then each event contains all the available information, so that each successive event furnishes no new information. Thus, for a given nonrandom time series of length N, the information given by the N values is equal to that given by a random time series of length N', where $N' < N$. N' is often referred to as the *effective length of record*. With respect to a given value of N, the larger r_1 is, the smaller N' is.

For a sequence of N' events, taken from a random time series, the unbiased estimator of the variance is given by

$$\hat{\sigma}^2 = \frac{N'}{N'-1}\left[\sum_{i=1}^{N'} \frac{(x_i - \bar{x})^2}{N'}\right] = \frac{N'}{N'-1} S^2 \qquad \text{(8-III-41)}$$

so that the expected value of $\hat{\sigma}^2$ is σ^2. For a sequence of N events, from a time series generated by a first-order Markov process, the unbiased estimator of the variance is given by

$$\hat{\sigma}^2 = \left[1 - \frac{1 - r_1^2}{N(1-r_1)^2} + \frac{2r_1(1 - r_1^N)}{N^2(1-r_1)^2}\right]^{-1} S^2 \qquad \text{(8-III-42)}$$

If $r_1 = 0$ and $N = N'$, Eq. (8-III-42) reduces to Eq. (8-III-41). By equating Eqs. (8-III-41) and (8-III-42), it is possible to determine N' for given values of r_1 and N. A graphical procedure facilitates the determination of N'. In Fig. 8-III-5 a family of curves is shown for $S^2/\hat{\sigma}^2$ versus N as a function of r_1. As N tends to infinity,

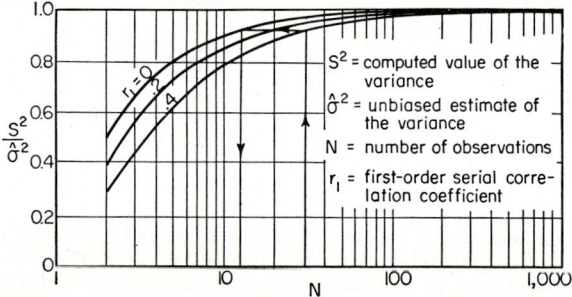

Fig. 8-III-5. Relation between $s^2/\hat{\sigma}^2$ and N as a function of r_1.

$S^2/\hat{\sigma}^2$ tends to unity for all values of r_1. The larger r_1 is, the slower is the rate of convergence. For a given sequence, N is known and r_1 can be determined by Eq. (8-III-27). Thus, starting with the value of N on the abscissa, a vertical line is drawn upward until it intersects the curve corresponding to r_1. From this point of intersection, a horizontal line is drawn to the left until it intersects the curve for $r_1 = 0$. From this point of intersection, a vertical line is drawn downward to the abscissa scale to determine N'. An example is shown in Fig. 8-III-5. It is assumed that $N = 30$ and $r_1 = 0.4$, so that $N' = 12$.

2. Correlation and Regression Analyses. In hydrologic studies, one is often interested in whether or not two or more variables are associated (see also Sec. 8-II). Extensive theory has been developed for determining the degree of association by means of the correlation coefficient when each variable is randomly distributed. The correlation between two nonrandom time series can be determined, but cannot be tested for significance in the same manner as the correlation between two random variables. If N pairs of observations are available, each observation cannot be considered as contributing new information about the correlation if the two time series are nonrandom.

The test of significance for the correlation between two random variables is based on the t test [2], where

$$t = r\sqrt{\frac{n}{1-r^2}} \qquad \text{(8-III-43)}$$

Where r denotes the correlation between the two variables, and n, which is equal to $N - 2$, where N is the number of pairs of observations, denotes the number of degrees of freedom.

ANALYSIS OF TIME SERIES

In order to test the correlation between two nonrandom time series for significance, it is necessary to replace n by the effective number n' of degrees of freedom. From Bartlett's work [14] it can be shown that, for very large sample sizes,

$$n' = \frac{n}{1 + 2r_1r_1' + 2r_2r_2' + \cdots} \qquad (8\text{-III-}44)[1]$$

where $r_1, r_2, \ldots$ are the serial correlation coefficients for one of the time series and $r_1', r_2', \ldots$ are the serial correlation coefficients for the other time series. To determine n', it is necessary to compute the serial correlation coefficients for many orders. This is very laborious, and because of the sampling errors associated with the serial correlation coefficients, it is not possible to determine n' accurately.

A useful formula for n' is obtained by considering each of the time series to be generated by a first-order Markov process. For this process, $r_k = r_1^k$ and $r_k' = (r_1')^k$. By using these relations in Eq. (8-III-44), n' becomes

$$n' = n \left(\frac{1 - r_1 r_1'}{1 + r_1 r_1'} \right) \qquad (8\text{-III-}45)$$

By Eqs. (8-III-44) and (8-III-45), it can be seen that if either time series is random, $n' = n$. This is consistent with the fact that each observation of the random time series contributes completely new information on the value of the correlation between the random and nonrandom time series.

Regression analysis is used in hydrologic studies to establish the relation between a given variable (referred to as the dependent variable) and one or more variables (referred to as the independent variables). The classical theory of regression analysis is based on the assumption that each variable is randomly distributed. If both the dependent and independent variables are time series, it is necessary to determine if the variables are random or not. An ordinary regression analysis involving time series is valid under two conditions: (1) if either the dependent or the independent variables are random, and (2) if the deviations from the line of regression are serially independent.

The serial dependence of the deviations from the line of regression may be determined by means of serial correlation coefficients. However, in testing the serial correlations of the deviations for significance, it is necessary to allow for the fitting of the regression line. No exact test of significance is available. An approximate test of significance is given by Durbin and Watson [15]. This test, summarized in Table 8-III-8, is based on giving correction terms for determining the effective number of deviations.

Table 8-III-8. Corrections to Number of Observations

Number of independent variables	Level of significance	
	$P = 0.05$	$P = 0.02$
1	(-1) (20)	(-1) (16)
2	(-5) (35)	(-5) (30)
3	(-10) (60)	(-10) (50)
4	(-15) (100)	(-15) (75)

Table 8-III-8 is used in the following manner. Assume that a regression analysis involving two time series is based on $N = 20$ pairs of observations. The serial correlation of the deviations from the line of regression may be tested for significance by

[1] See also Eq. (8-II-27).

Eq. (8-III-43), where $N = 20 - 1 = 19$ and $N = 20 + 20 = 40$. At the 95 per cent confidence level t is approximately 2. Thus, for $N = 19$, $r = 0.46$, and for $N = 40$, $r = 0.31$. If the computed serial correlation is greater than $r = 0.46$, then the serial correlation is significantly greater than 0 at the 95 per cent level. A computed serial

Table 8-III-9. Data for Studying the Effect of Serial Correlation on Correlation and Regression Analyses

Year	x	y	ϵ_x	ϵ_y	y'
1913	21	1.6	.....	......	−3.03
1914	22	0.4	13.69	−0.32	−4.43
1915	20	0.4	12.34	0.22	−4.03
1916	45	2.9	38.04	2.72	−6.53
1917	30	5.4	29.34	4.09	−1.03
1918	32	5.0	21.56	2.55	−1.83
1919	24	7.5	12.86	5.23	2.27
1920	25	5.0	16.65	1.60	−0.43
1921	31	14	22.30	11.73	7.37
1922	33	15	22.21	8.66	7.97
1923	20	2.5	8.52	−4.30	−1.93
1924	18	3.0	11.04	1.87	−1.03
1925	16	9.1	9.74	7.74	5.47
1926	14	4.0	8.43	−0.12	0.77
1927	20	6.8	15.13	4.99	2.37
1928	43	14	36.04	10.92	4.97
1929	20	4.0	5.04	−2.34	−0.43
1930	15	4.7	8.04	2.89	1.27
1931	14	4.8	8.78	2.67	1.57
1932	15	2.1	10.13	−0.07	−1.33
1933	21	4.6	15.78	3.65	−0.03
1934	23	6.0	15.69	3.92	0.97
1935	29	5.5	21.00	2.78	−0.73
1936	20	2.5	9.91	0.01	−1.93
1937	26	6.9	19.04	5.77	1.27
1938	24	10	14.95	6.87	4.77
1939	22	2.6	13.65	−1.93	−2.23
1940	25	4.6	17.34	3.42	−0.83
1941	14	2.5	5.30	0.42	−0.73
1942	14	4.4	9.13	3.27	1.17
1943	19	4.5	14.13	2.51	0.27
1944	21	4.8	14.39	2.76	0.17
1945	41	11	33.69	8.83	2.37
1946	39	3.5	24.73	−1.48	−4.73
1947	30	3.6	16.43	2.01	−2.83
1948	19	2.6	8.56	0.97	−1.63
1949	18	1.8	11.39	0.62	−2.23
1950	21	3.6	14.74	2.78	−1.03

correlation is nonsignificant if it is less than $r = 0.31$. If a computed serial correlation lies between these two values, then there is doubt about the significance at the 95 per cent level, since it is not certain if the 95 per cent level is reached [13].

If the residuals are serially uncorrelated, an ordinary regression analysis is valid. However, if the residuals are serially correlated, it is necessary to take this fact into account in the regression analysis. Quenouille [13] suggests that this may be done in one of two ways. The first way consists in calculating the deviations from a serial regression of the dependent variable upon previous values of itself and using these deviations in a regression analysis on the independent variables and their previous values. The second way is to make a regression analysis of the dependent variable upon the independent variables and upon previous values of the dependent variables and itself. The first method may be used to predict the random variation in the dependent variable from the independent variables. By the second method, values of the dependent variable may be estimates from the independent variables and previous observations.

In order to clarify the above discussions an example is given. In Table 8-III-9, under the columns headed by x and y, respectively, the correlation between x and y, $r(xy)$, is 0.47, and the first-order serial correlations for the x and y series are $r_1(x) = 0.35$ and $r_1(y) = 0.24$, respectively. By assuming that both series are generated by a first-order Markov process, then the effective number of degrees of freedom is, according to Eq. (8-III-40), 32. By using Eq. (8-III-43), $t = 2.91$. Since the value of t at the 95 per cent level, 2.056, is less than 2.91, the correlation between x and y is significant.

The equation for the regression of y on x is

$$y = 0.31 + 0.20x \qquad (8\text{-III-}46)$$

The deviations from this regression are given in Table 8-III-9 under the column headed by y'. The first-order serial correlation of the deviations, $r_1(y')$, is 0.333. By using the corrections, given in Table 8-III-8, to the number of observations and applying Eq. (8-III-43), it is seen that $r_1(y')$ is significant at the 95 per cent level. Since the x, y, and y' series are nonrandom, an ordinary regression analysis cannot be made.

The deviations from the serial regression of the independent variable upon previous values of itself are given by

$$x_{i+1} - 0.453x_i = (\epsilon_x)_{i+1} \qquad (8\text{-III-}47)$$

These deviations are given in Table 8-III-9 under the column headed by ϵ_x. Similarly, the deviations from the serial regression of the dependent variable upon previous values of itself can be determined. These deviations are given in Table 8-III-9 under the column headed by ϵ_y. The first-order serial correlation coefficients for these two sets of deviations are $r_1(\epsilon_x) = 0.226$ and $r_1(\epsilon_y) = -0.176$. Both of these coefficients are insignificant at the 95 per cent level. Thus the ϵ_x and ϵ_y series may be considered as random.

If the first method of accounting for the serial correlation is used, it is necessary to determine the regression of ϵ_y on ϵ_x. This regression gives

$$(\epsilon_y)_{i+1} = -0.782 + 0.232(\epsilon_x)_{i+1} \qquad (8\text{-III-}48)$$

so that y_{i+1} might be predicted using

$$y_{i+1} = -0.782 + 0.348y_i + 0.232x_{i+1} - 0.105x_i \qquad (8\text{-III-}49)$$

If the second method is used, a multiple regression of y_{i+1} on y_i, x_{i+1}, and x_i must be carried out. This regression gives

$$y_{i+1} = 1.210 + 0.602y_i + 0.244x_{i+1} - 0.205x_i \qquad (8\text{-III-}50)$$

In order to use either Eq. (8-III-49) or Eq. (8-III-50), it is necessary that the deviations from the line of regression be serially independent.

III. REFERENCES

1. Hoel, P. G.: "Introduction to Mathematical Statistics," John Wiley & Sons, New York, 1954.

2. Weatherburn, C. E.: "A First Course in Mathematical Statistics," Cambridge University Press, London, 1952.
3. Bartlett, M. S.: Properties of sufficiency and statistical tests, *Proc. Roy. Soc., London,* ser. A, vol. 160, pp. 268–282, 1937.
4. Characteristics of low flow volume-duration-frequency statistics, *U.S. Army Corps Engrs. Tech. Bull.* 1, 1960.
5. Double-mass curves, *U.S. Geol. Surv. Water-Supply Paper* 1541-B, 1960.
6. Schulman, Edmund: "Dendroclimatic Changes in Semiarid America," University of Arizona Press, Tucson, Ariz., 1954, pp. 29–30.
7. Slutzky, Eugen: The summation of random causes as the source of cyclic processes, *Econometrika,* vol. 5, pp. 105–146, 1937.
8. Yule, G. U.: On the time series problem, *J. Roy. Statist. Soc.,* vol. 84, pp. 497–526, 1921.
9. Anderson, R. L.: Distribution of the serial correlation coefficient, *Ann. Math. Statist.,* vol. 8, no. 1, pp. 1–13, 1941.
10. Blackman, R. B., and J. W. Tukey: "The Measurement of Power Spectra," Dover Publications, Inc., New York, 1959.
11. Wold, Herman: "A Study in the Analysis of Stationary Time Series," Almqvist and Wiksell, Stockholm, 1954.
12. Abbott, C. G.: A long-range forecast of United States precipitation, *Smithsonian Inst. Misc. Collections,* vol. 139, no. 9, 1960.
13. Quenouille, M. H.: "Associated Measurements," Academic Press Inc., New York, 1952, pp. 165–187.
14. Bartlett, M. S.: Some aspects of the time-correlation problem in regard to tests of significance, *J. Roy. Statist. Soc.,* vol. 98, pp. 536–543, 1935.
15. Durbin, James, and G. S. Watson: Testing for serial correlation in least squares regression, *Biometrika,* vol. 37, pp. 409–428, 1950.

Section 8-IV

STATISTICAL AND PROBABILITY ANALYSIS OF HYDROLOGIC DATA

PART IV. SEQUENTIAL GENERATION OF HYDROLOGIC INFORMATION

VEN TE CHOW, *Professor of Hydraulic Engineering, University of Illinois.*

I. Introduction	8-91
II. Generating Techniques	8-92
A. Sampling of Cards	8-92
B. The Table of Random Numbers	8-92
C. The Markov Process	8-93
D. Other Mathematical Models	8-95
III. Evaluation of the Approach	8-95
IV. References	8-96

I. INTRODUCTION

Sequential generation of hydrologic information is a statistical process using Monte Carlo methods for generating sequentially synthetic hydrologic records. The so-called *Monte Carlo method* refers to a process by which "experience" data are produced synthetically by a sampling technique or some form of random-number generator. For a general discussion of such methods, see Refs. 1 and 2.

The application of the method of sequential generation to generate streamflows [3, 4] for the analysis of river-basin systems ([5]; see also Subsec. 26-II-III-D) has been referred to as *synthetic hydrology*. However, this term has met with much criticism because it tends to produce some confusion among hydrologists, and perhaps total perplexity among nonspecialists, since "synthetic hydrology" can be easily interpreted in a broad sense as any type of hydrologic method that involves synthesis and simulation, thus including synthetic unit hydrographs or hydrologic simulation by field plots, laboratory models, or electronic analogs. Another term *stochastic hydrology*, as a distinction from *parametric hydrology*, has been also proposed.

The concept of sequential generation of hydrologic data is not new. In 1914, Hazen [6] described generation of a runoff sequence of 300 years by combining the annual-mean-flow series for 14 streams. This is similar to the station-year method for multiplying rainfall data (see Subsec. 9-V-A-2). Sudler [7] obtained an artificial runoff record of 1,000 years by dealing 20 times a deck of 50 cards, on each of which was printed a representative annual streamflow. Barnes [8] used a similar method to synthesize a 1,000-year sequence of streamflows by means of a table of random variables, but assigned the synthetic flows as normal variates with the same mean and standard deviation as the historical flows.

In recent years, the subject has received renewed emphasis because of theoretical work done in the field of mathematical statistics and probability. By using random numbers of extreme-value distribution, Benson [9, 10] developed 1,000 synthetic flood peaks which correspond to 1,000 numbers in random order, representing annual peak flows that fit an extreme-value distribution. Using the mathematical model of a circular random walk, Thomas and Fiering [5] synthesized 510 years of monthly flows for each of the five inflow sites of a simplified river-basin system by combining serial and cross correlations. By means of a simple autoregressive model, Julian [3, pp. 134–143] generated synthetic hydrologic data of yearly flows in the Colorado River at Lees Ferry, Ariz., from the annual precipitations. Brittan [4] developed synthetic hydrologic records at Lees Ferry by two probability approaches: one by determination of the probability distribution of mean flows in relation to the range and the other by use of a Markov-chain model.

The development of the subject is still in progress and applications of its techniques have been made only by few who were developing them. The discussion in this section is therefore brief, but factual enough to stimulate attention and interest.

II. GENERATING TECHNIQUES

A. Sampling of Cards

Sampling of cards is the simplest method of generating hydrologic data. By this method, the historical data, such as annual flows, are first written on cards, one for each value. The cards are then shuffled and a card is drawn at random, and its value is taken as the first value of the generated data. Another card is drawn at random. The value is taken as the next value of the generated data. This procedure is repeated until all values are drawn.

In following the prescribed procedure, the card being drawn may or may not be replaced by the same card after each drawing. If the card is not replaced, the pack of cards will be exhausted when the number of drawings equals the number of cards. Then the cards are reshuffled and drawn; another, usually different, series of data is created. This is the method used by Sudler [7]. Its drawback is that, once a card is drawn, the population of the deck is changed and the probability of drawing another of the same rank in any successive trial becomes zero. Also, all series of generated data have the same mean, standard deviation, and range. Accordingly, a better simulation of the method to the natural hydrologic system would be to replace the card after each drawing.

In order to provide a realistic distribution of hydrologic data, the individual cards may be labeled in accordance with a desirable probability distribution. Thus Barnes [8] used a normal distribution approximating annual flows of a stream and Benson [9, 10] fitted an extreme-value distribution of theoretical flows to the historical record.

By applying the method of sampling cards, it is assumed that the hydrologic data are purely random and further that the magnitudes of the generated data will be the same as those of the historical data. These assumptions are not realistic, and therefore better methods are now being used.

B. The Table of Random Numbers

The procedure of sampling by shuffling cards, or the like, can be simplified by use of random-number tables. Such tables have been constructed by more efficient methods, for example, by applying mathematical theories of numbers and by operating a specially designed chance device. All published tables have been subject to the standard statistical tests for randomness and are thereby considered acceptable for general sampling use. Large tables of random numbers are available and may be found in most mathematics libraries and statistics laboratories and offices. Ordinarily, the tables present random decimal digits uniformly distributed over the real line (0,1), i.e., having a rectangular distribution. Their quantities vary from a few thousands to a million, like the one constructed by the Rand Corporation. For special purposes, the random numbers have also been developed to fit a known law of distribution, such as the normal-distribution law. However, it is possible to develop random variables of any given distribution from an ordinary random-number table for rectangular distribution. A direct method is the use of the probability-integral transformation and its inverse [2, p. 250]. Consider the cumulative probability function for any distribution of variables. It can be shown that the distribution of probabilities of this function has a rectangular distribution over line (0,1). From the ordinary table, random numbers can be selected to represent the random sample of the cumulative probabilities. Since the variable of the given distribution is uniquely related to its cumulative probability, by inverse transformation a random sample of the variable can be obtained.

To use the table, one closes his eyes and places his finger or a pencil point on any page, and the number so indicated is a random number. Usually, the other random numbers are taken consecutively from the table directly succeeding this first number. In order to eliminate the need for extensive use of such tables in complicated problems, mathematical programs for generating *pseudo-random numbers* have been developed and recorded on tapes or IBM cards, which can be used as input to high-speed computers.

By means of a table of random numbers, Brittan [4] simulated streamflows in the Colorado River by selecting from the table 100 random samples of 5 each corresponding to a 5-year runoff sequence. Since there are many different combinations of five which will yield the same mean, the samples were chosen subject to the following constraints: (1) the annual runoff should have a range between the upper and lower limits set by the historical record; and (2) the 5-year sequences of runoff should be distributed according to the distribution of the mean and the ratio of the range to the mean of the historical data. Then 100 samples of 30 inflows (6 samples of 5-year sequence) were chosen at random from the 100 samples. These simulated flows should exhibit the same statistical characteristics as the historical flows, as required by the constraints.

C. The Markov Process

The Russian mathematician A. A. Markov (1856–1922) introduced the assumption that the outcome of any trial depends only on the outcome of the directly preceding trial. This assumption led to the formulation of the classical concept of a stochastic process known as the *Markov process*, or *Markov chain* [11]. In a Markov process, the probability at any time of a system being in a given state depends only on the knowledge of the state of the system at the immediately preceding time.

For generating annual flows in the Colorado River at Lees Ferry, Julian [3, 12] used a simple autoregressive model, or a first-order Markov process. By Eq. (8-III-39), the following may be written:

$$x_t = rx_{t-1} + \epsilon(y)_t \qquad (8\text{-IV-1})$$

where x_t is the annual runoff at year t; x_{t-1} is the runoff at the preceding, or the $(t-1)$st, year; r is the first-order serial correlation coefficient for the runoff, or a *Markov-chain coefficient;* and $\epsilon(y)_t$ is a random uncorrelated component due to annual rainfall. This equation indicates that the runoff at a given year is equal to a constant times the runoff of the preceding year plus a random component. Julian used $r = 0.2$ in one case [3] and $r = 0.25$ in the other [12]. He used Eq. (8-IV-1) to generate annual runoff from runoff of the preceding year and the random component due to rainfall, or actually, he computed the power spectrum of x_t. The *power spectrum* is the distribution of the variance of x_t on a frequency scale [13]. A chi-square test of fit between the actual spectrum of the Lees Ferry runoff and the generated Markov spectrum did not show any significant difference on the 5 per cent level. However, this test was not considered conclusive because the length of record was too short.

From the historical record of runoff at Lees Ferry, Brittan [4] generated 20 flow sequences of 50 years each by means of the following Markov-chain model of a type similar to Eq. (8-IV-1):

$$x_t = rx_{t-1} + (1 - r)\bar{x} + s_x(1 - r^2)^{\frac{1}{2}}\epsilon \qquad (8\text{-IV-2})$$

where x_t is the annual runoff at year t; x_{t-1} is the annual runoff at the preceding, or the $(t - 1)$st, year; $\bar{x}$ is the mean annual flow computed from the historical record; s_1 is the standard deviation of the historical runoff; r is the Markov-chain coefficient taken as 0.25; and ϵ is the random variate assumed normally distributed with mean = 0 and standard deviation = 1. Brittan found that the generated flow contains negative values, and thus explained that this may be due to the incorrect assumption of a normal distribution for the random component ϵ.

Thomas and Fiering [5, 14] used essentially the same Markov-chain model represented by Eq. (8-IV-2) in which the variate at the tth time is comprised of a component linearly related to that at the $(t - 1)$st time and a random additive component. In applying this model to generating monthly flows at a diversion dam by serial correlation of monthly flows, the following recursion equation was used:

$$Q_{i+1} = \bar{Q}_{j+1} + B_j(Q_i - \bar{Q}_j) + s_{j+1}(1 - r_j^2)^{\frac{1}{2}}\epsilon_i \qquad (8\text{-IV-3})$$

where Q_i and Q_{i+1} are the discharges during the ith and $(i + 1)$st month, respectively, counted from the start of the generated sequence; $\bar{Q}_j$ and $\bar{Q}_{j+1}$ are the mean monthly discharges during the jth and $(j + 1)$st month, respectively, within a repetitive annual cycle of 12 months; B_j is the regression coefficient for estimating flow in the $(j + 1)$st month from the flow in the jth month; s_{j+1} is the standard deviation of flows in the $(j + 1)$st month; r_j is the correlation coefficient between the flows of the jth and $(j + 1)$st month; and ϵ_i is a random normal and independent variate with zero mean and unit variance.

Similarly, by cross correlations between flows at different pairs of gaging stations, the recursion equation is

$$Q_{iY} = \bar{Q}_{jY} + B_j(Q_{iX} - \bar{Q}_{jX}) + s_{jY}(1 - r_j^2)^{\frac{1}{2}}\epsilon_i \qquad (8\text{-IV-4})$$

where Q_{iY} and Q_{iX} are the discharges at stations Y and X, respectively, during the ith month; $\bar{Q}_{jY}$ and $\bar{Q}_{jX}$ are the mean discharges at stations Y and X, respectively, during the jth month within a repetitive annual cycle of 12 months; B_j is the regression coefficient for Y on X during the jth month; s_{jY} is the standard deviation of discharges at station Y during the jth month; r_j is the correlation coefficient between flows at stations X and Y during the jth month; and ϵ_i is as defined in Eq. (8-IV-3).

For the analysis of flood damages, several types of monthly floods were selected from records and represented by dimensionless hydrographs in the form of histograms whose abscissas were expressed in 5 to 12 fractional periods of a month and whose ordinates were expressed in ratio of the total monthly volumes of flow of specific floods of record. With the total volumes of monthly flows known from the generated flow sequences, it was possible to reproduce synthetic hydrographs for the various types. The base of the hydrographs was subdivided into 120 six-hour intervals, and the flows during these intervals were used in flood routing. The flood-peak intervals were separated from the remaining, or residual, monthly flood volumes. The peaks were further correlated with the corresponding total monthly flows by linear regression. To the expected peak flow from this regression was then added a random component based on the appropriate standard error of estimate and a table of normal random numbers. These synthesized flood peaks were used to evaluate flood-control benefits in the simulation study of a river basin.

D. Other Mathematical Models

Many other mathematical models have been proposed for sequential generation of hydrologic data, including the moving average [12], autoregression of high order [3, 12], and correlograms ([15]; also see Subsec. 8-III-II-D).

III. EVALUATION OF THE APPROACH

Observed hydrologic records are usually short. Unless the record is too meagre to be considered as a representative sample, the statistical parameters derived from it should enable the hydrologist to construct a stochastic model that will generate hydrologic information for as long a period of time as desired. Since the statistical parameters of the population of the generated data are necessarily the same as those estimated from the historical data, the new information is limited by errors of measurement and sampling that are inherent in the observed record. As far as the quality of the information is concerned, the new data are no better than the data from which the new data were generated. If a stochastic model of hydrologic sequences is suitable in all respects to the problem under consideration, the historical and generated data cannot be theoretically distinguished significantly by the usual statistical tests.

The major advantage of sequential generation is to create synthetic records longer than the historical. Consequently, the sequential-generation approach makes possible to produce as many combinations of hydrologic sequences as desired for use in hydrologic analyses. This possibility is particularly valuable in the study of reservoir operations and in the design of complex water-resources systems (Subsec. 26-II-III-D). In the design of a system of hydraulic structures or water-resources projects, the generated information helps to overcome the paucity of possible patterns of extreme cases by providing a large number of new sets of data that could be obtained from the given hydrologic record. Thus it provides flexibility, and hence the possibility of examining the broad spectrum or extent to which a specific design may be overloaded or underloaded by different sets of statistically compatible generated sequences of hydrologic events. As a result, a fairly well balanced design can be evolved.

For the planner of hydraulic or water-resources systems, the generated information enables him to make more alternative designs for comparison or optimization in economic analysis than from the short historical data. The variation in the results thus obtained should be of help to him in identifying and developing an ultimate design with further consideration of other nontechnical factors.

According to Bower [16], the rationale for the utilization of sequentially generated hydrologic data in water-resources system planning seems clear as follows. The problem is to decide what water-resources system to build. Solving this problem involves determining the response of each proposed water-resources system to a sequence of hydrologic events. The questions which the system planner seeks to answer are (1) whether the system can produce the desired outputs when subjected to future hydrologic events, and (2) how much risk and uncertainty are associated with the proposed investment. In answering the first question, the system planner has traditionally relied on the analysis of system response to only one (the historical), or a portion of one, of the many sequences of hydrologic events which are possible. Sequential generation can be used to produce additional probable sequences of hydrologic events with which system response can be tested. In answering the second question, which is rarely attempted, some estimate of the degree of variation in outputs, and hence of the risk and uncertainty of the investment, can be made by the use of multiple sequences of generated data.

Bower also pointed out the importance of the fact that the tangible outputs involved in the water-resources investment-decision problem are defined in terms of dollars. There is no intrinsic value to a particular physical output as such. The same physical output is worth different amounts, in dollar terms, at different points in time over the period of analysis—seasonally as well as yearly. Except for the simplest case of a single reservoir with a single demand which is constant over time, benefits produced by a water-resources system depend on the sequence of hydrologic events. Consequently, no qualified statement can be made about the expected performance of such a system based on one sequence only. Using a number of sequences, an estimate can be made of the expected system performance and of its variation. It appears to Bower that in simple cases in water-resources system planning, that is, with one reservoir, one purpose, no holdover storage, and little regulation of streamflow at the given reservoir site, there is little justification for the utilization of the generated information. However, when either or both of the following two conditions occur, the use of sequentially generated hydrologic data appears logical and necessary: (1) when the system under study is relatively complex, because of economic, social, and/or physical factors, and (2) when the time period over which maximization is taking place is relatively long compared with the total length of the historic record available.

It appears that, properly done, sequential generation yields hydrologic information in a form of great practical use in water-resources development and management [17, 18]. However, the techniques themselves must be further refined; suitable stochastic models, better than the proposed ones, must be developed; and the generated information must be rigorously tested statistically for its precision and validity. Despite the practical value of sequential generation, this field of study requires further research and investigation, but potentially it has a promising future.

IV. REFERENCES

1. Householder, A. S., G. E. Forsythe, and H. H. Germond (eds.): Monte Carlo method, *U.S. Natl. Bur. Standards, Appl. Math. Ser.*, vol. 12, 1951.
2. Meyer, H. A. (ed.): "Symposium on Monte Carlo Methods," John Wiley & Sons, Inc., New York, 1956.
3. Fishman, Leslie, and P. R. Julian: A synthetic hydrology for the Colorado River, in H. L. Amoss (ed.), "Water: Measuring and Meeting Future Requirements," Western Resources Papers 1960, University of Colorado Press, Boulder, Colo., 1961, pp. 125–145.
4. Brittan, M. R.: Probability analysis applied to the development of synthetic hydrology for Colorado River, pt. IV of "Past and Probable Future Variations in Stream Flow in the Upper Colorado River," University of Colorado, Bureau of Economic Research, Boulder, Colo., October, 1961.
5. Thomas, H. A., Jr., and M. B. Fiering: Mathematical synthesis of streamflow sequences for the analysis of river basins by simulation, chap. 12 in Arthur Maass, M. M. Hufschmidt, Robert Dorfman, H. A. Thomas, Jr., S. A. Marglin, and G. M. Fair (eds.), "Design of Water-resource Systems," Harvard University Press, Cambridge, Mass., 1962, pp. 459–493.
6. Hazen, Allen: Storage to be provided in impounding reservoirs for municipal water supply, *Trans. Am. Soc. Civil Engrs.*, vol. 77, pp. 1539–1669, 1914.
7. Sudler, C. E.: Storage required for the regulation of stream flow, *Trans. Am. Soc. Civil Engrs.*, vol. 91, pp. 622–660, 1927.
8. Barnes, F. B.: Storage required for a city water supply, *J. Inst. Engrs., Australia*, vol. 26, pp. 198–203, 1955.
9. Benson, M. A.: Characteristics of frequency curves based on a theoretical 1,000-year record, in Tate Dalrymple (ed.), Flood-frequency analyses, Manual of Hydrology, pt. 3, Flood-flow technique, *U.S. Geol. Surv. Water-Supply Paper* 1543-A, pp. 57–74, 1960; also, U.S.G.S. open-file report, 1952.
10. Benson, M. A.: Plotting positions and economics of engineering planning, *Proc. Am. Soc. Civil Engrs., J. Hydraulics Div.*, vol. 88, no. HY6, pp. 57–71, November, 1962.
11. Bharucha-Reid, A. T.: "Elements of the Theory of Markov Processes and Their Applications," McGraw-Hill Book Company, Inc., New York, 1960.
12. Julian, P. R.: A study of the statistical predictability of stream-runoff in the Upper Colorado River basin, pt. II of "Past and Probable Future Variations in Stream Flow in the Upper Colorado River," University of Colorado, Bureau of Economic Research, Boulder, Colo., October, 1961.
13. Blackman, R. B., and J. W. Tukey: "The Measurement of Power Spectra," Dover Publications, Inc., New York, 1958.
14. Fiering, M. B.: Queuing theory and simulation in reservoir design, *Trans. Am. Soc. Civil Engrs.*, vol. 127, pt. I, pp. 1114–1144, 1962.
15. Yevdjevich, V. M.: Some general aspects of fluctuations of annual runoff in the Upper Colorado River basin, pt. III of "Past and Probable Future Variations in Stream Flow

in the Upper Colorado River," University of Colorado, Bureau of Economic Research, October, 1961.
16. Bower, B. T.: Synthetic hydrology: "Where are we?," notes stemming from synthetic hydrology discussion at Colorado State University, Fort Collins, Colo., Aug. 10, 1961, private communication.
17. Clough, D. J., Jr.: Measures of value and statistical models in the economic analysis of flood control and water conservation schemes, *ASME-EIC Hydraulics Conf. Paper* no. 61-EIC-5, Engineering Institute of Canada, 2050 Mansfield Street, Montreal, Canada, 1961.
18. Maughan, W. D., and R. Y. Kawano: Project yields by a probability method, *Proc. Am. Soc. Civil Engrs., J. Hydraulics Div.*, vol. 89, no. HY3, pt. 1, pp. 41–60, May, 1963.

Section 9

RAINFALL

CHARLES S. GILMAN, *Late Chief, Hydrometeorological Section, U.S. Weather Bureau.*

I. Introduction	9-2
II. The Measurement of Rainfall	9-2
A. Types of Gages	9-2
1. Nonrecording Gages	9-4
2. Recording Gages	9-5
3. Storage Gages	9-5
B. Radar Measurement of Rainfall	9-5
C. Rain-gage Networks	9-7
1. Bucket Surveys	9-8
D. Records of Rainfall	9-8
E. Sources of Rainfall Data	9-9
III. Physics and Hydrodynamics of Rain	9-9
A. Formation of Rain	9-9
1. Mechanism for Cooling—Adiabatic Reduction of Pressure Associated with Upward Motion	9-10
2. Mechanism for Condensation—the Nuclei and Molecular Diffusion	9-10
3. Mechanisms for Droplet Growth—Collision and the Coexistence of Ice Crystals and Water Droplets	9-11
4. Role of the Freezing Nuclei	9-13
5. Mechanism for Accumulation of Moisture—Convergence	9-13
B. Artificial Production of Precipitation	9-14
1. Cloud Seeding	9-15
C. Net Inward Transport of Water Vapor	9-15
1. Storage in a Fixed Volume in Space	9-15
2. Storage in a Fixed Earth-space Volume	9-16
3. Storage in a Fixed Mass of Air	9-16
4. Effect of Local Evaporation on Local Precipitation	9-16
IV. Synoptic Meteorology of Rain	9-18
A. Rainfall with Extratropical Systems	9-18
B. Local Convective Rains and Thunderstorms	9-21

NOTE: After completing the manuscript for this section, Dr. Gilman was involved in a fatal aircraft accident on Jan. 22, 1962, on Lake O'Higgins in the south of Chile while he was heading a surveying mission for the United Nations to help development of water resources in Chile. Publication of this section is dedicated to him for his most distinguished services and outstanding contributions to the science and engineering of hydrology and meteorology.—Editor-in-chief.

C. Tropical Storms.. 9-23
D. Orographic Rainfall... 9-24
E. Quantitative Precipitation Forecasting...................... 9-25
V. Space-Time Characteristics of Rainfall........................ 9-26
   A. Techniques of Analysis....................................... 9-26
      1. Double-mass Curve....................................... 9-26
      2. The Station-year Method................................ 9-27
      3. Interpolation of Rainfall Records....................... 9-28
      4. Average Depth of Rainfall over Area.................... 9-28
      5. Effect of Network Density on Areal Rainfall Average...... 9-29
      6. Techniques for Orographic Precipitation................ 9-29
      7. Depth-Area-Duration Analysis........................... 9-32
      8. Excessive Precipitation................................. 9-35
   B. Results of Space-Time Analyses of Rainfall................. 9-41
VI. Design Applications of Rainfall Data......................... 9-49
   A. Frequency Analysis... 9-49
      1. Area-Depth Relationship................................. 9-57
      2. Seasonal Variation....................................... 9-58
      3. Calculated Risk... 9-59
      4. Synthesis of Frequency Regimes......................... 9-60
   B. Frequency Formulas... 9-60
   C. Storm Transportation....................................... 9-60
   D. Probable Maximum Precipitation............................ 9-62
   E. Standard Project Storms................................... 9-64
VII. References... 9-65

I. INTRODUCTION

This section proposes to give enough of the meteorology of rainfall to aid the hydrologist in understanding the problems of importance to his field, to enable him to appreciate the overlap between the two sciences, and to inform him where and how to seek further knowledge if necessary. Many concepts of the science of meteorology carried over from past years are definitely out of date, and a general orientation toward the present state of knowledge of the physics and synoptic meteorology of rainfall production is necessary for the hydrologist to realize the impact of current work in meteorology in his field. The first subsection deals in broad view with the measurement of rainfall, the basis for all hydrologic and hydrometeorological work in the field. The second treats the physics and hydrodynamics, including the artificial inducement of precipitation. This aspect has been the subject of much work in the past decade, and further advances may be expected in the near future. Consequently, an attempt has been made to concentrate on those aspects that are likely to be of permanent interest. It is in the third subsubsection, on the synoptic meteorology of rain, that most deviation from the older textbooks is likely to be noted. In this field practice has, in some respects, outpaced reporting, making it necessary to evaluate critically some of the concepts that found their way into textbooks two or more decades ago. Subsections V, on the space-time characteristics of rainfall, and VI, on design applications of rainfall data, are of direct application to many hydrologic problems. An attempt has been made to use the latest information in these applications. (See also Sec. 3.)

II. THE MEASUREMENT OF RAINFALL

A. Types of Gages

Rainfall and other forms of precipitation are measured in terms of depth, the values being expressed in inches in the United States and Canada and in millimeters in other

(a)

(b)

Fig. 9-1. Standard 8-in. precipitation gage. (a) Unassembled, showing overflow can, measuring tube, funnel receiver, and measuring stick; (b) gage in stand ready for service. (*U.S. Weather Bureau.*)

Fig. 9-2. Weighing-type precipitation gage. (*Friez Instrument Co.*)

countries, including the British Isles. (Australia uses inches for some internal purposes and millimeters for international purposes.)

Rain gages are based on the simple idea of exposing in the open a hollow cylindrical vessel with a bottom but no top. Rain (or other forms of precipitation) falls into the vessel, and its depth (or volume or weight) is measured, snow or other frozen forms being melted before the measurement. The principal difficulties are: (1) The presence of the gage may disturb the wind field so that the free fall of precipitation is interfered with. (2) Piled-up snow or ice on the opening of the cylinder also may interfere. (3) Trees, buildings, a roof serving as a catchment area, or other objects may make the exposure site unrepresentative. (4) Some of the precipitation may be lost by evaporation or by wetting the sides of the gage or the measuring tube. (5) Various other factors such as dents in the rim of the receiver or the measuring volume may give false answers. (6) In extreme cases, splash into or out of the objects may modify the true value of rainfall. Most of these causes may be minimized with proper care.

1. **Nonrecording Gages.** The 8-in. gage (Fig. 9-1) is standard in the United States. It consists of a copper vessel 8 in. in diameter, a receiver (or funnel), and a measuring tube, whose cross-sectional area is one-tenth that of the gage. During warm weather the measuring tube is placed in the gage and the receiver is fitted over the top of it. If the depth of precipitation is less than one-tenth of the height of the gage, it is simply measured in the tube by dividing the actual depth in the tube by 10. For greater depths the measuring tube overflows, and the water must be poured back from the gage into the tube in successive fillings until the total depth is measured. During cold weather the receiver and tube

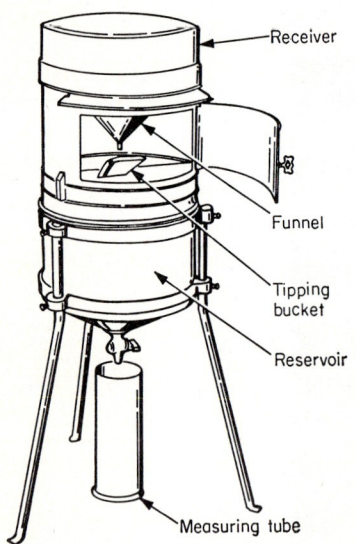

Fig. 9-3. Tipping-bucket rain gage. (*U.S. Weather Bureau.*)

are removed from the gage so that they will not interfere with the accumulation of snow.

2. **Recording Gages.** To know the intensity of rainfall or the amounts for shorter durations than can be obtained by manual measurements at a regular rain gage, recording gages are used that give a continuous pen trace on a clock-driven drum. Several mechanisms are provided for moving the pen: (1) floats, (2) weighing devices, or (3) the tipping bucket. It is usually desirable that the total rainfall be retained so that manual measurements can provide a check on the total rainfall and a means of calibrating the rates of fall. The general principles of the first two are self-evident. The tipping bucket consists of a pair of small containers designed so that when a certain amount of rainfall (0.01 in. in the U.S. Weather Bureau type) falls in one of the containers, it tips, brings the other container into position to receive the next rainfall, empties into a storage container, and closes an electrical contact that causes a mark by a pen on the recorder chart. *Float-type gages* are used in the British Isles. *Weighing types* (Fig. 9-2) are preferred for cold climates where it is desired to record snow as well as rainfall. The *tipping-bucket type* (Fig. 9-3) is especially adapted to remote recording and has been in use at the U.S. Weather Bureau first-order stations since the early 1890s. Most recording stations in the United States are equipped with the weighing type.

3. **Storage Gages.** Storage gages are used for remote locations that require servicing only every 2 to 6 months. Those located in areas of heavy snowfall must be high enough for the orifice to be above the greatest snow depth (Fig. 9-4).

B. Radar Measurement of Rainfall

Though radar was initially designed to detect aircraft, it is an excellent detector of all types of hydrometeors in the air (Fig. 9-5). The ability to determine the areal distribution of precipitation intensities depends on the type of radar employed, the best all-purpose weather-search radar being one that would have the properties: (1) wavelength such that the effect of precipitation attenuation is minimized, (2) power and pulse length selected to ensure that the lowest significant amounts of precipitation are detected at maximum range, (3) correction for range, (4) beam width as narrow as possible, (5) antenna large enough to receive weakest possible reflected energy. Of the several types of radars in use today for weather search, the Weather Bureau WSR-57 meets the needs of the meteorologist and hydrologist.

Fig. 9-4. Tower to lift gage above maximum snow cover. (*U.S. Weather Bureau.*)

Experience has shown that the *hydrologic range* of the WSR-57 radar, that is, the maximum range to which it can detect virtually all rainfall of significance to flood forecasting and reservoir control, is about 125 miles. Beyond this range the beam is at such a high altitude and is so broad that results become distorted in azimuth. Intense rainfall centers, however, can be detected beyond this range.

Photographic procedures for integrating radar echoes and correlating them with precipitation consist of exposures at 5-min intervals over a period of from 1 to 3 hr. The resulting integrated echo intensities are calibrated with reported precipitation. An experienced radar operator can then make a fairly good quantitative analysis of the rainfall distribution. It will be necessary for some time to come to depend upon

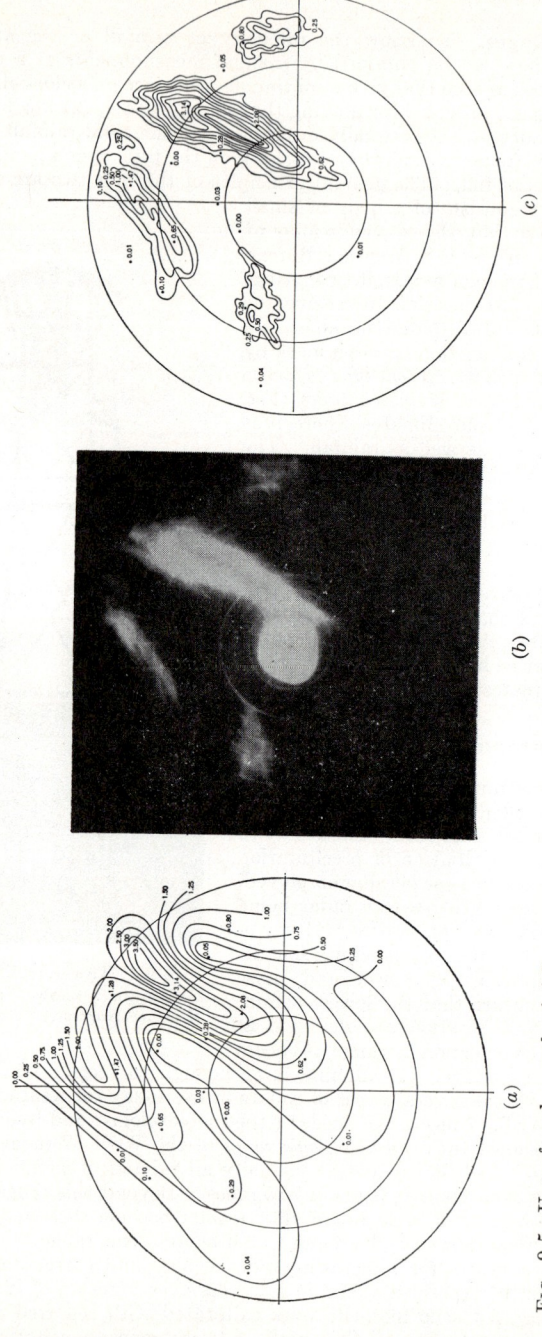

Fig. 9-5. Use of radar and rainfall measurements to define the rainfall pattern. (a) Rainfall map based on gage measurements only; (b) 6-hr multiple-exposure radar-echo integration; (c) rainfall map using integrated radar echoes with observed precipitation. (*U.S. Weather Bureau.*)

rain gages for calibration of the radar echoes, but experience and the development of improved electronic storage devices will probably overcome this limitation in the not-too-distant future.

C. Rain-gage Networks

The density of rain gages varies greatly from country to country and even within the same country (Fig. 9-6). Over vast areas of the world there is actually very little information on the amount of rainfall. Even the total amount of precipitation over

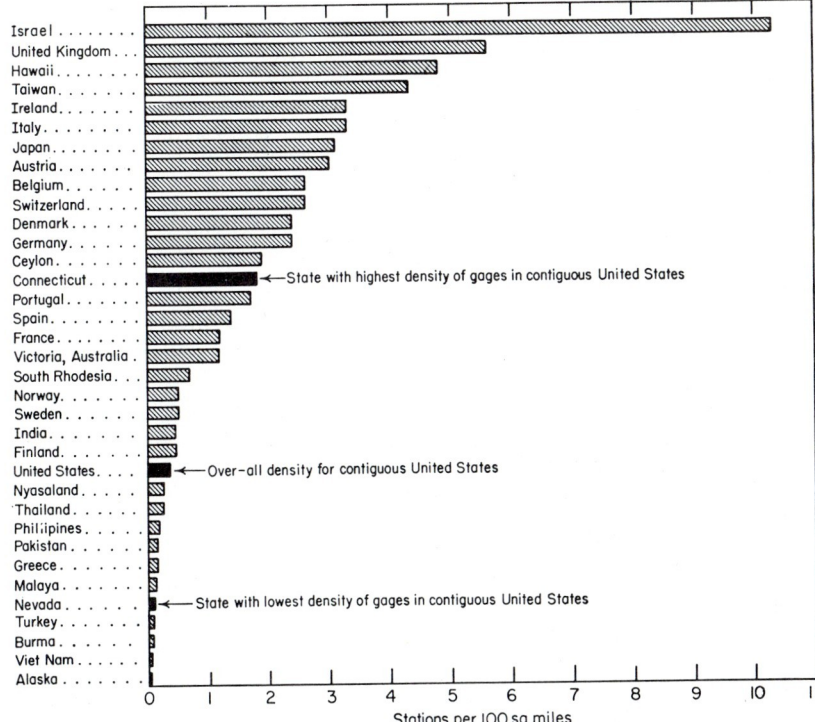

FIG. 9-6. Comparative rain-gage density (the data are accurate as of mid-1960). (*U.S. Weather Bureau.*)

the globe is only approximately known. In the more developed and densely populated countries, there have been fairly good networks for 70 to 100 years. The most dense networks have been established, as might be expected, in areas where there is great economic significance in local variations of rainfall. For example, in the Hawaiian Islands, the intense agriculture of pineapples and sugar is closely related to the rainfall. Also, in California, the rapid population growth has caused a great demand for long-range planning of the water resources. There are great variations in topography closely correlated with variations in rainfall. These facts have created a need for a very dense network of rain gages, especially since the end of World War II. Unfortunately, it will be several decades before stable normals for such a network can be determined.

In the United States there are several basic networks of rain gages: (1) The *synoptic network* at first-order stations takes rainfall measurements as part of regular weather observations. These data are used in weather forecasting and are usually taken and transmitted to collecting points every 6 hr. (2) The *special reporting network* takes rainfall measurements every 24 hr but transmits the values only when they exceed a certain amount, say, 1 in. in the 24-hr period. Such values are frequently taken in connection with river-stage reports. Similar in intent are *radio rain gages* that transmit automatically the amount of precipitation received in a container. *Radar beacons* have been devised by means of which such information can be received from transmitting points on a regular radar set being used for other purposes. (3) The *recording network*, also called the *hydrologic network*, consists of about 3,500 weighing gages in the United States. The records from this network are worked up regularly. (4) The *nonrecording network* consists of about 11,000 gages, which are read every 24 hr by volunteer observers. The data for this network are transmitted only at the end of the month and are published during the succeeding month.

There are probably several times as many observations as the number included in all the groups mentioned above that are maintained with some degree of regularity by persons having a personal or economic interest in the amount of rainfall. For example, many farmer-supply companies furnish rain gages as an advertisement. On occasions of very heavy rainfall some of the amounts may be reported on the national teletype circuits and radio and television networks.

In addition to the basic networks, many networks have been maintained for special purposes. A number of these have been high-density networks for studying the effect of rainfall on certain specialized characteristics such as runoff and soil erosion. For example, one may include those of the French Broad Basin in the Tennessee Valley with one gage per 40 sq mi, the Muskingum Basin in Ohio with one gage per 18 sq mi, the Mexican Spring area in New Mexico with one gage per 3 sq mi, and the San Dimas Experimental Forest in southern California with 14 gages per square mile. The Illinois State Water Survey also maintains several large-density networks.

1. Bucket Surveys. In cases of suspected record rainfall special reconnaissance teams are sent into the field to collect all possible information from rain gages and from other vessels that may fortuitously have caught some of the heavier rainfall amounts. Such vessels may include stock-watering troughs, old bathtubs, milk buckets, etc. Obviously, much care must be exercised to ensure that values are not exaggerated. Instructions covering the wording of questions in these surveys would do credit to a cross-questioning attorney. In spite of all the limitations of such surveys, most of our knowledge of the larger amounts over smaller-size areas must continue to come from them because any other method of obtaining such measurements is several times as expensive.

The optimum density of rain gages depends on many hard-to-measure considerations. The existing networks world-wide seem to be related to the density of population, the economic development of the region, and the importance of water resources in the economy. For any quantitative idea of the density required, the network must be related to some purpose, and the purposes can vary from time to time and from region to region. For general climatological purposes, the present plan of the U.S. Weather Bureau calls for one gage per 625 sq mi, of which some four-fifths are in operation. Many of the remaining sites required are in uninhabited areas. A still smaller network is to consist of benchmark stations, those that can be maintained for many years without environmental change for the primary purpose of studying climatic trends. For hydrologic purposes, Linsley [1] suggests that each study basin (preferably less than 500 sq mi in size) should have a first-order hydrometeorological station and three or four precipitation stations within it, or about one station per 100 sq mi. For analysis of thunderstorm rainfall or self-help flash-flood forecasting systems, a density of one gage per square mile has been advocated. (See also Subsec. 26-IV-A-2.)

D. Records of Rainfall

Nearly all countries make some effort to collect and publish the measurements of rainfall. After a few years the manuscript records become voluminous and their management becomes a problem. A summary of the disposal of the records in the

United States gives some indication of the procedures followed. Before 1870, meteorological observations in the United States were made by the surgeons at army posts, and these records are preserved by the National Archives in the reports of the Surgeon General. Also, many records collected from the network established by the Board of Regents of New York University, the Smithsonian Institution, and other observers are preserved in the National Archives. From 1870 to 1891, the weather service was operated by the Signal Corps, and records for those years are similarly preserved by the National Archives under material of the Chief of the Corps. The Weather Bureau was established in 1891, and records since that time are maintained by the Bureau. Naturally, the methods of handling the data and its publication have varied greatly from time to time during the period since 1891. Now the reports are sent by the observers to three weather-records processing centers, where they are summarized and published by machine methods. Then the records are sent to the National Weather Records Center at Asheville, N.C., where they are preserved in manuscript or on microfilm. The Library of the U.S. Weather Bureau has many records of foreign stations and knowledge of the records kept by other countries.

E. Sources of Rainfall Data

At present, precipitation data are published for the United States in state and national summaries which give 24-hr amounts for all stations and hourly amounts for recording rain gages. The place of publication has changed quite a bit through the years. Before the early 1890s several of the sections (roughly equivalent to states) published their own records of 24-hr precipitation. From 1909 to 1914 the data were published in the *Monthly Weather Review*. Then they appeared in climatological data up until the late 1940s. Hourly data appeared in the *Hydrologic Bulletins* from the time of the establishment of the hydrologic network about 1940 until the publishing was converted to machine methods—1948 or 1949, in different parts of the country. From that time through July, 1951, they were published in *Climatological Data*, and since then in *Hourly Precipitation Data*. Weekly and monthly maps of precipitation over the United States appear in *Weekly Weather and Crop Bulletin* and the *Monthly Weather Review*. The *Climatic Summary for the United States* summarized monthly and annual data up to 1950.

Several states and agencies publish data from rain gages maintained by them. Notable are the *Water Bulletin* of the International Boundary and Water Commission for the Rio Grande Basin and *Precipitation in the Tennessee River Basin* of the Tennessee Valley Authority.

III. PHYSICS AND HYDRODYNAMICS OF RAIN

A. Formation of Rain

According to the best of present knowledge, four conditions are necessary for the production of the observed amounts of rainfall: (1) a mechanism to produce cooling of the air, (2) a mechanism to produce condensation, (3) a mechanism to produce growth of cloud droplets, and (4) a mechanism to produce accumulation of moisture of sufficient intensity to account for the observed rates of rainfall. So far as is known, the simultaneous occurrence of all these mechanisms is a sufficient condition for heavy rainfall. It is, of course, possible that some of these conditions are always fulfilled in the atmosphere and that rainfall occurs when events bring about a change from nonfulfillment to fulfillment of the others. For example, there seem to be no cases, or at most very few, where cooling does not result in condensation. Also, it is possible that fulfillment of some one of the conditions implies necessary fulfillment of the others. Many synoptic meteorologists feel that, if the net moisture inflow is strong enough, then heavy rain will fall, and therefore the other three conditions must necessarily be fulfilled. Some cloud physicists believe that the nonfulfillment of condition 3 when the other conditions are fulfilled may occur, and a change from nonfulfillment to fulfillment will produce rainfall.

1. Mechanism for Cooling—Adiabatic Reduction of Pressure Associated with Upward Motion. Physical studies show that the following methods of cooling are too small to account for precipitation, except possibly light drizzle or fog: (1) adiabatic cooling by horizontal motion toward lower pressure, (2) radiational cooling, (3) cooling by contact with a colder land or sea surface, and (4) mixing of two air masses.

The pressure reduction when air ascends from near the surface to upper levels in the atmosphere is the only known mechanism capable of producing large enough lowering of temperature to account for all precipitation rates except the very lowest. The cooling lowers the capacity of a given volume of space to contain water vapor. The rate at which the vapor content necessary for saturation lowers during ascending motion may be called the *rate of production of moisture excess over saturation*.

Thermodynamics gives this rate in terms of the vertical speed of the air and the mean temperature of a saturated layer of air. The results show that if the rate of precipitation is equal to the rate of production of moisture excess, appreciable upward speeds of the order of miles per hour at 10,000- to 20,000-ft altitude must accompany heavy rainfall rates.

Alternative partial explanations of high rainfall rates are (1) that horizontal convergence of the falling raindrop causes the rainfall rate to be greater than the rate of production of moisture excess and (2) that large amounts of liquid water, supported temporarily by a vertical current, fall rapidly when the current slackens or ends.

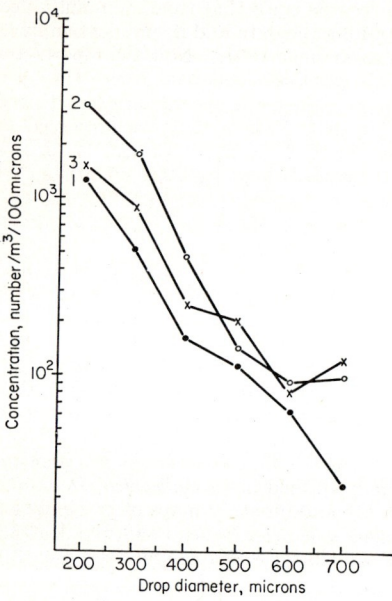

Fig. 9-7. Drop-size (150 to 650 microns in diameter) distributions in tropical cumuli. Distribution 1 (46 clouds): cloud tops estimated 6,000 to 9,000 ft; flight altitude 5,500 to 7,000 ft; concentration 2,131 per cubic meter. Distribution 2 (35 clouds) cloud tops estimated 9,000 to 12,000 ft; flight altitude 7,000 to 9,000 ft; concentration 5,979 per cubic meter. Distribution 3 (12 clouds): cloud tops estimated greater than 12,000 ft; flight altitude 7,000 to 9,000 ft. (*After Brown and Braham* [2].)

2. Mechanism for Condensation—the Nuclei and Molecular Diffusion. Under certain controlled conditions in the laboratory, lowering of the temperature of air does not produce condensation. Investigators have found that air from which all foreign particles have been removed can be cooled until the relative humidity is as much as 1,000 per cent before droplets of liquid water form—presumably on aggregates of molecules. In clean air exposed to ionization, droplets begin to form when the relative humidity is 400 to 500 per cent. Condensation in the atmosphere takes place on *hygroscopic nuclei*, small particles of substances that have an affinity for water, even when the air is not saturated. The best examples of such substances are sodium chloride and sulfur trioxide. In ordinary air each drop forms around a small particle of foreign substance, known as a *condensation nucleus*, much smaller than the dust particles seen in a beam of light in a dark room. The size of the nuclei is of the order of 10^{-3} to 10 microns. Observations before and after a duststorm in the Sahara Desert showed that the number of nuclei active was the same. Similar results are obtained before and after beating a carpet in a room. The number of nuclei varies greatly from region to region—from several million per cubic centimeter in heavily polluted industrial air to only a few per cubic centimeter in mountainous regions or at high levels. The source of the nuclei is thought to be mostly terrestrial—the breaking of surf or waves, turbulence in the lower layers as exemplified by duststorms, or combustion processes. Chemical analysis of rainwater has shown that the proportion of sodium to magnesium and potassium chloride is roughly the same as in sea water. Evidently there are always sufficient hygroscopic nuclei in the atmosphere to accommodate condensation processes. If air in the lower atmosphere is cooled to saturation, condensation will always occur—in many cases before the saturation point is reached. The growth of water droplets has been studied in detail in its theoretical aspects. Energy considerations show that the saturation vapor pressure of a surface increases as the curvature increases. Therefore condensation occurs on small droplets whose surfaces have large curvature only if the air is supersaturated with respect to a plane surface. During the time that a droplet is very small, the effects of curvature and hygroscopicity counteract each other. As water is attracted to the particle, the latter goes into solution and its attraction for water decreases. On the other hand, the curvature of the droplets decreases, so that less supersaturation is required for condensation. The effects of curvature and hygroscopicity become negligible after the droplet has attained a radius of about 10^{-4} cm. Thereafter the growth of the droplets must be studied by consideration of the process of molecular diffusion, by which the water passes from the vapor state in the surrounding air to the liquid state as part of the drop. This process has been studied mathematically. An important consequence of such studies is that the droplets tend to become uniform in size as they age. This theoretically derived result is supported by observations. A further consequence is that it is impossible for condensation alone to produce droplets larger in diameter than about 200 microns.

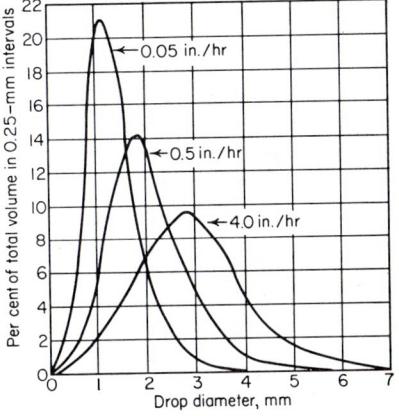

Fig. 9-8. Per cent of total volume of liquid water in drops of different sizes for three rainfall ratios. (*After Laws and Parsons* [3].)

Droplet sizes in clouds have been measured by (1) collecting them on a glass plate coated with carbon black and magnesium oxide or with two oils of different density, then observing with a microscope or photographing, (2) making theoretical calculations based on optical laws and the observed diameters of optical phenomena such as halos, (3) observing the rates of accumulation of ice on rotating cylinders of various sizes, and (4) observing the falling speeds. Figure 9-7 shows for a typical cloud the per cent of liquid water contained in droplets of various sizes.

Raindrop sizes have been measured by (1) observing the diameters of blots produced as the drops hit paper covered with a water-soluble dye, and (2) weighing the dough pellets produced as drops fall into freshly ground flour. Both methods are calibrated by use of artificial droplets of known sizes. Results of measurements for rains of different intensity are shown in Fig. 9-8. Obviously, raindrops are of much larger size than cloud drops and are far too large to be produced by the condensation process alone.

3. Mechanisms for Droplet Growth—Collision and the Coexistence of Ice Crystals and Water Droplets. Clouds can be regarded as colloidal-like suspensions, or, more properly, *aerosols*. A tendency for the droplets to remain small and therefore not to fall is called *colloidal stability*. On the other hand, if the droplets tend to coalesce, thereby becoming large enough to overcome the frictional resistance to falling, the cloud is said to be *colloidally unstable*.

Falling speed of droplets as a function of their size has been investigated by (1) measuring the speed of an air stream required to support them, (2) using a shutter mechanism as a velocity selector, (3) using stroboscopic methods and, (4) inducing an electric charge on the droplet and then measuring with an oscillograph the difference in time between pulses produced as the droplet falls through two inducing rings a known vertical distance apart. Results obtained by method 4 in 1949 are given in Figs. 9-9 and 9-10.

Current evidence is that coalescence of cloud droplets to form raindrops cannot be accounted for by the following processes: (1) electric charges on the drops, (2) hydrodynamic forces, (3) evaporation of smaller drops and condensations on larger ones due to the radius-of-curvature effect, and (4) turbulence. Theoretical studies have shown that each of these processes is too weak. The two processes regarded as most effective are (1) the difference in speeds between large droplets and small droplets and (2) the coexistence of ice crystals and water droplets. In the second explanation, the difference in saturation pressure over water and over ice accounts for the evaporation of waterdrops and the resulting condensation of much of the water thus vaporized onto the ice crystals.

In earlier stages of droplet growth, the

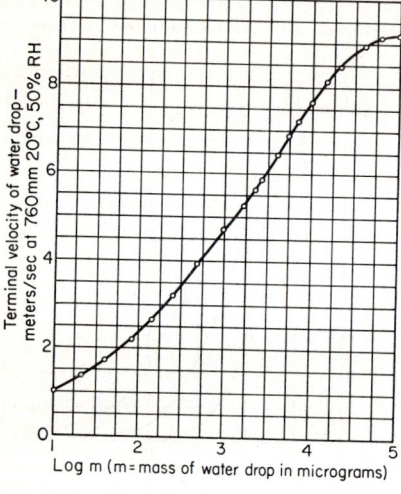

FIG. 9-9. Terminal velocity of distilled-water droplets in stagnant air at 76 cm pressure, 20°C, and 50 per cent relative humidity. (*After Gunn and Kincer* [4].)

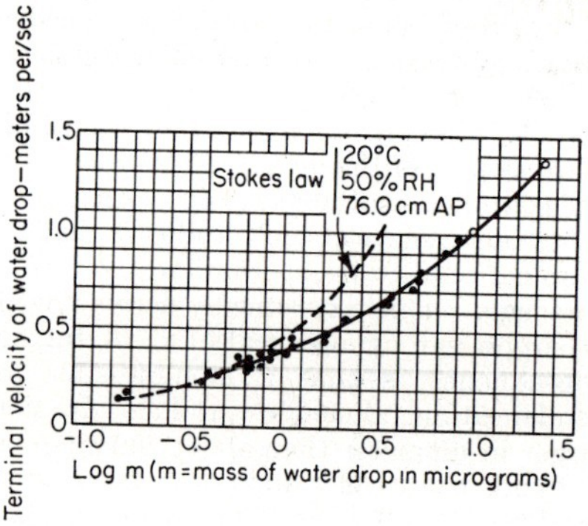

FIG. 9-10. Terminal velocity of very small droplets. (*After Gunn and Kincer* [4].)

ice-crystal effect may be more important, and in later stages, the collision effect, according to Houghton [5]. He states:

Quantitative estimates show that the ice-crystal effect is much more rapid than the collision effect for growth of droplets from cloud element size to crystals with mass equivalent to that of drizzle drops. On the other hand, the collision process is more rapid than the ice-crystal effect for elements of mass comparable to raindrops. It is also shown that the collision effect alone can start precipitation in a "warm" cloud if the liquid water content is high, the cloud is deep and the cloud drop-size distribution is large.

It has also been suggested that the difference in saturation pressure between drops having different temperatures may have an effect similar to the ice-water difference.

For several years the evidence for the ice-crystal effect was regarded by many meteorologists as very convincing. Recently, the position of most scientists has become less certain on this point. The principal points of evidence in favor of the ice-crystal theory are the following: (1) Most cumulonimbus clouds in temperate latitudes begin to form precipitation at about the time their tops are observed to become ice clouds. Also, other clouds from which precipitation is falling in temperate latitudes usually have subfreezing temperatures somewhere within them. (2) In artificial clouds such as are produced by blowing the breath into a deep freeze, the introduction of a very cold object causes a few ice crystals to form; then the whole fog is converted to ice crystals, which settle out. If the breath is again blown into the freezer a few seconds after all crystals have disappeared from view, the new water droplets will disappear and be replaced by ice crystals. If, however, the blowing of the breath is delayed for a longer period, the new fog of water droplets will persist. (3) The dissipation of supercooled natural clouds as dry ice is dropped into them may be regarded as evidence in favor of the theory. In refutation to each of these arguments critics of the theory offer counterarguments and observational evidence: Against argument 1, in the tropics, heavy rainfall has been observed from clouds whose temperatures were entirely above freezing; and against arguments 2 and 3, the disappearance of the water droplets does not necessarily mean that they have grown to the size of raindrops or that they would if the process continued longer.

A cold rod at a temperature near $-40°C$ has the same effect as dry ice in a cloud chamber. Many substances, notably silver iodide, are able to produce ice crystals at a much higher temperature.

4. Role of the Freezing Nuclei. There is abundant evidence that supercooled water clouds are a common occurrence in the atmosphere, especially between 0 and $-15°C$. Water fogs have been produced in filtered air in a cloud chamber at $-50°C$. A case has been reported where waterdrops were kept in the liquid state at $-70°C$ for several hours. On the other hand, it has been found impossible to supercool water droplets below $-33°C$ in the presence of solid, insoluble, wettable particles. These and many other observations have indicated that ice crystals form on foreign particles. If the crystals form by sublimation, the particles are called *sublimation nuclei*, and if the crystals form by freezing of liquid water, they are called *freezing nuclei*. Evidence indicates that true sublimation nuclei are probably extremely rare in the atmosphere. Silver iodide crystals become active as particularly effective insoluble freezing nuclei, their crystal structure being fairly close to that of ice. Ice crystals in the presence of silver iodide appear only at the dew point and at temperatures as high as $-4°C$.

The results from experiments in dropping dry ice and silver iodide into supercooled clouds demonstrate conclusively that this is one phase of the precipitation-forming process that can be influenced greatly by artificial means. Its efficacy in causing appreciable precipitation to reach the earth's surface is another question.

5. Mechanism for Accumulation of Moisture—Convergence. Regardless of whether or not the other conditions for precipitation are fulfilled, simple continuity considerations demand that there must be a good amount of moisture present in order that the evaporation losses between cloud and ground be overcompensated if there is to be appreciable rain. Also, and more important, heavy rainfall amounts exceed by far the amount of water vapor in a vertical column at the beginning of the rainfall, and the amount of water vapor in a vertical column over the rainfall area remains the same or actually increases during the rainfall process—at least up to a certain time. For these reasons there must be a large net horizontal inflow of water vapor into the column above the rain area. This process is called *convergence*, which is defined as the net horizontal influx of air per unit area.

A consequence of the definition is that the horizontal area of a column moving with the air flow must decrease with time. Thus convergence may also be shown to be equal to the percentual time rate of decrease of such an area. Other meteorological considerations show that divergence must be present in the upper atmosphere over convergence in the lower. Figure 9-11 shows the change in shape of an original area to give the amounts of precipitation over the original area as indicated. The change of shape takes place during the rain regardless of the time during which it occurs. It will be seen that the vertical motion is large, the change of shape is enormous when high rates of rainfall occur, and large amounts of air are transported from the lower atmosphere to the upper atmosphere.

The amount of liquid water in clouds has been measured (1) by noting the increase of weight of phosphorous pentoxide or other drying agent when a known amount of air collected in a cloud is passed through it, (2) by noting the amount of vapor increase when the air is heated by taking initial and final psychrometer readings, (3) from visibility measurements, (4) by actually capturing the water from an airplane flying through the clouds, and (5) from radar measurements. For most clouds these results are of the order of magnitude 0.1 to 0.5 g/m³. Even a cloud 20,000 ft thick may contain only a few hundredths of an inch of liquid water. Some observations in cumulus congestus and rain clouds have indicated values of 1 to 10 g/m³. In the cases with higher values, precipitation was actually falling, or fell shortly afterward.

The conclusion to be drawn from such considerations is that, even if all the liquid water in most clouds at a given instant can be induced to fall to the earth as precipitation, the amounts would, in most cases, be negligibly small. If active lifting and convergence are present, then the amounts of moisture may be supplied for appreciable precipitation. Whether or not such lifting and convergence can be speeded up or increased by artificial means is controversial, the best evidence at present indicating considerable doubt.

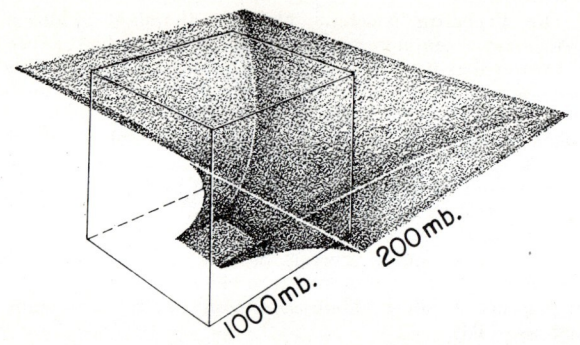

Fig. 9-11. Change in shape of an air mass during precipitation of 2 in. or over. The isometric line drawing shows the original shape, and the shaded volume, the final shape. The 1,000-mb surface is near the ground, and 200 mb is at about 50,000 ft. The essential features are contraction in the lower layer, upward motion, and expansion aloft. Of course, there must be compensating sinking of another mass to take the place of the contraction in lower levels and make way for the expansion at high levels.

B. Artificial Production of Precipitation

In the most primitive tribes some individuals have claimed the ability to make rain. Throughout our civilization there have been individuals claiming this ability. Several patents were granted to "rainmakers" and other "weather controllers" during the nineteenth century. In the 1890s Congress appropriated $10,000 for experiments to produce rain by cannonading. The man who conducted the experiments concluded that they had been moderately successful. A case is on record where an association agreed to pay a rainmaker $6,000 if a normal amount of rain fell. Many other cases can be cited. No doubt many of the practitioners have believed that they influenced the weather, but serious scientists have shown fairly conclusively that the older claims are unsubstantiated.

Several considerations should be borne in mind when one attempts to evaluate both sides in a subject as controversial as the artificial production of precipitation. In the first place, scientific progress can be made only by starting off with a hypothesis as to how something might be done and then testing this hypothesis by experiment. It is natural that many hypotheses will be proposed that will be disproved by later experience. Therefore scientists are inclined to doubt any hypothesis until it has been tested and retested. Second, in dry areas the economic consequences of successful rainmaking would be so great that businessmen and farmers might well consider it justifiable to invest considerable sums for even a small possibility of success.

Third, since the present knowledge of the causes of precipitation are still imperfect, it is very difficult to tell whether any particular amount is the result of natural or artificial causes. Finally, if statistical arguments are to be used, the experiments must be carefully planned by a person familar with experimental-design considerations.

1. Cloud Seeding. The theory proposed by proponents of cloud seeding postulates that there exist colloidally stable supercooled clouds containing appreciable amounts of liquid water, that this water or a large part of it would not fall as precipitation under natural conditions, and that the cloud can be made colloidally unstable by the addition of dry ice, silver iodide, or other chemical agents so that a certain part of this otherwise unavailable water will reach the ground as precipitation. The uncertain parts of the theory are whether this *Bergeron-Findeisen condition* is absolutely necessary and whether sufficient liquid water is present in colloidally stable clouds to give precipitation amounts of economic importance.

Authoritative statement of the present position of these ideas has been prepared by the American Meteorological Society, the World Meteorological Organization (Subsec. 24-II), and the U.S. Advisory Committee on Weather Control. The statement of the latter [8] is, in part:

> On the basis of its statistical evaluation of wintertime cloud seeding using silver iodide as the seeding agent, the Committee concluded that:
> 1. The statistical procedures employed indicated that the seeding of winter-type storm clouds in mountainous areas in western United States produced an average increase in precipitation of 10 to 15 per cent from seeded storms with heavy odds that this increase was not the result of natural variations in the amount of rain.
> 2. In non-mountainous areas, the same statistical procedures did not detect any increase in precipitation that could be attributed to cloud seeding. This does not mean that effects may not have been produced The greater variability of rainfall patterns in non-mountainous areas made the techniques less sensitive for picking up small changes which might have occurred there than when applied to the mountainous regions.
> 3. No evidence was found in the evaluation of any project which was intended to increase precipitation that cloud seeding has produced a detectable negative effect on precipitation.
> 4. Available hail frequency data were completely inadequate for evaluation purposes and no conclusions as to the effectiveness of hail suppression projects could be reached.

C. Net Inward Transport of Water Vapor

An equation of the storage of water may be written

$$P = E + R + \Delta S_g \tag{9-1}$$

where P is the volume of precipitation falling over a fixed area of the earth's surface during a certain time interval, E is the volume of evapotranspiration into the atmosphere from the surface of this fixed area, R is the volume of runoff from this same area, and ΔS_g is the change in the amount of water in the ground under the area during the same time interval. On the average, ΔS_g may be neglected and the equation becomes

$$P = E + R \tag{9-2}$$

Comparison of streamflow records with precipitation records for the continental area of the earth as a whole has shown that R is about $\frac{1}{4}P$; therefore E for the continental area of the whole earth is about $\frac{3}{4}P$.

1. Storage in a Fixed Volume in Space. A second equation takes the form

$$I + E = O + P + \Delta S_a \tag{9-3}$$

where I and O, respectively, are the amounts of moisture inflow to and outflow from a fixed volume in space during a certain time interval; E and P are, respectively, the evaporation into and precipitation from the volume during the interval; and ΔS_a is the change in moisture storage in the volume during the same interval. As before, ΔS_a may be neglected on the average and the equation becomes

$$I + E = O + P \tag{9-4}$$

2. Storage in a Fixed Earth-space Volume. If a volume of space exactly overlying an area on the earth's surface is chosen, the evaporation into the volume is equal to the evaporation off the surface, and the precipitation out of the volume is equal to the precipitation onto the surface. Then the two preceding equations may be combined:

$$R = I - O \qquad (9\text{-}5)$$

Recent studies have indicated that R is much smaller than I and O. For the Mississippi Valley, Benton and Estoque [9] have estimated that R is 6.6 in. and that I is 146 in. for a period of a year.

3. Storage in a Fixed Mass of Air. Another form of the storage equation for the water vapor in the atmosphere is

$$E_m - P_m = \Delta S_m \qquad (9\text{-}6)$$

where E_m is the amount of evapotranspiration into a fixed air mass during a certain time, P_m is the precipitation falling from that air during the same time, and ΔS_m is the moisture change in the air mass during the same interval.

Holzman [10] used reasoning based on the last equation, together with a synoptic study of the moisture changes in air masses as they proceed across the country, to infer that evaporation in the central United States takes place mostly into polar continental air and that most of the volume of precipitation falls from tropical maritime air. It has been estimated that in the Mississippi Valley the per cent of precipitation from cP (continental polar) air is 10 and from mT (maritime tropical) air, 90. However, nearly equal amounts of water are evaporated into the two types of air mass.

4. Effect of Local Evaporation on Local Precipitation. It was formerly believed that land-use practice that increased evaporation could cause precipitation on nearby areas to increase appreciably. This view has been almost entirely abandoned by meteorologists.

When it became known that runoff measurements showed that evaporation over the continents is capable of supplying three-fourths of the moisture needed for the observed precipitation, many meteorologists jumped to the conclusion that this evaporation must be the principal source of moisture in the atmosphere over continents. Actually, they were neglecting Eq. (9-4), which shows that the relative importance of sources of moisture for any area can be judged only from estimates of the moisture transport by the winds into and out of the volume overlying that area.

Holzman [10] showed that evaporation over a continent could make but a negligible contribution to precipitation over the continent. His argument may be understood by the use of Eq. (9-6). From synoptic studies he concluded that most evaporation in the eastern United States (north of about 35° according to Benton and Estoque [7]) takes place into continental air, which then travels across the country without losing much, if any, of the moisture thus gained. Most precipitation, on the other hand (at least east of the Great Plains), falls from maritime tropical air, which usually has its greatest moisture content just before moving inland. He emphasized that the cycle of air masses in which continental polar air flows southward from the continent to the ocean there gains moisture by evaporation and also heat, becoming a maritime tropical mass; then flows northward over the continent, losing moisture by precipitation and heat by radiation and conduction, until it eventually becomes a continental polar mass again. As for the effect of increased evapotranspiration on precipitation in the same locality, it should be remembered that the air into which the water is evaporated is usually some hundreds of miles away even a few hours later.

The source of the moisture is, of course, at the surface—moisture being added by evaporation and transpiration. For the year as a whole and the earth as a whole, by far the largest amount of moisture is added over the oceans. However, warm, moist areas such as the tropical rain-forest belts or the southeastern United States in summer may also serve as important sources.

The moisture thus added to the atmosphere is ordinarily transported very large distances before it falls again to the earth as precipitation. No exact figures are available, but they are probably of the order of several hundreds of miles in summer

PHYSICS AND HYDRODYNAMICS OF RAIN 9–17

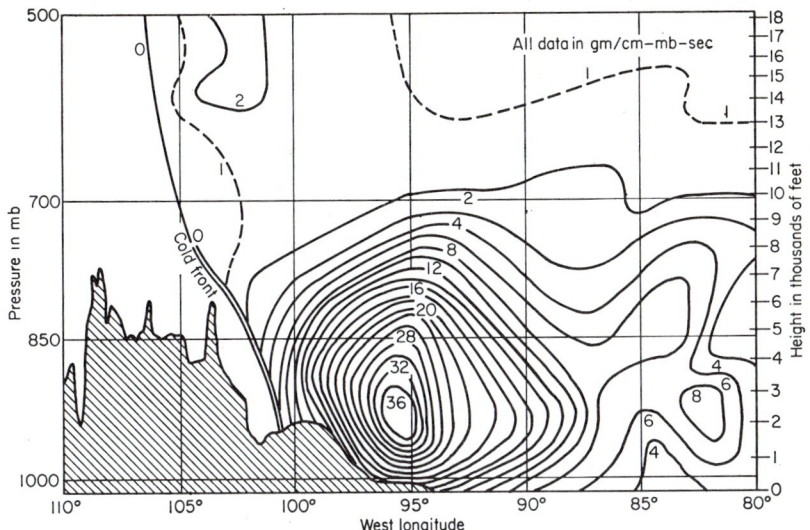

Fig. 9-12. Transport of water vapor during a storm situation northward across 30°N latitude at 9 p.m. EST, May 8, 1943. (*After Lott and Myers* [6].)

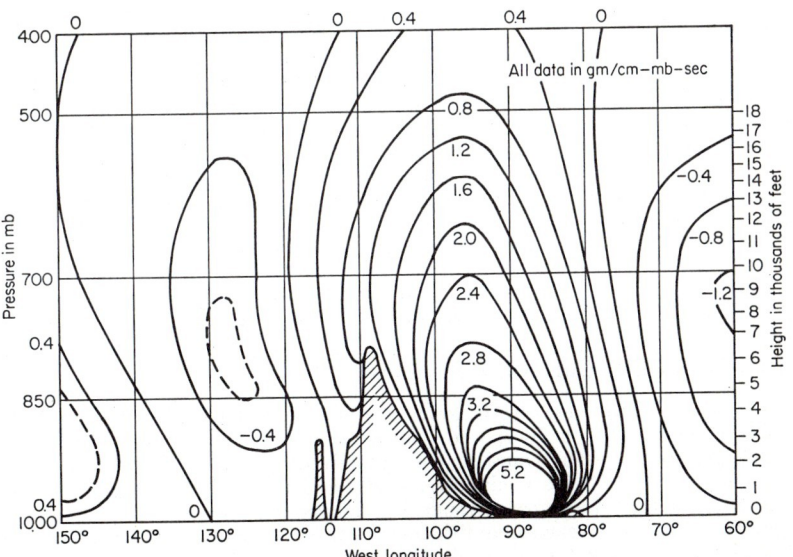

Fig. 9-13. Long-time transport of water vapor northward across 30°N latitude averaged for the winter, 1949. (*After Benton and Estoque* [7].) Note that in both Figs. 9-12 and 9-13 the maximum transport is in the lower 2,000 or 3,000 ft of the atmosphere and that the major fraction is below 5,000 ft.

and a thousand miles or so in winter. The transport of this moisture takes place, for the most part, in the very lowest layers of the atmosphere. Figures 9-12 and 9-13 show typical distributions of water-vapor transport with height. Then, when the moist current reaches a region of active vertical motion, it rises many thousands of feet in a few hours.

The moisture-storage capacity of the atmosphere is small. Here again exact figures are not available, but a good approximation is that in the whole atmosphere at any one time, there is enough water vapor to account for about 2 weeks of the average precipitation over the whole earth.

IV. SYNOPTIC METEOROLOGY OF RAIN

This subsection proposes (1) to summarize the empirical knowledge acquired in synoptic meteorology as to the relation of rainfall—especially heavy rainfall—to other observed meteorological elements, such as temperature, pressure, and wind; (2) to summarize the tendency of heavy rainfall to occur in definite patterns with respect to time and area; and (3) to relate this knowledge in a general way to the energy-transformation processes of the atmosphere.

Potential and thermal energy in the atmosphere are represented by the contrast between warm air masses and cold air masses. They are converted to the kinetic energy of the winds by the upward movement of the warmer air simultaneously with the downward movement of colder air. Two types of initial distributions can lead to development of winds: (1) a primarily horizontal contrast between cold and warm air and (2) a vertical distribution characterized by the presence of cold air in the upper layers of the atmosphere over warm, and possibly moist, air in the lower layers. If the line of demarcation between cold and warm air is sharp, it is called a *front*. The vertical distribution alluded to is called a *condition of instability*. Part of the warmth of the rising air in instability cases comes from the release of heat of condensation. Since upward motion of air is necessary for rainfall and since upward motion of warm air is necessary for kinetic-energy production, there is a general correlation between the two processes. However, the correlation is not exact. Many heavy rains occur in connection with the dying stages of atmospheric circulation systems. This is especially true of the dying stages of tropical cyclones.

A possible confusion as to the use of the word "storm" should be borne in mind. The synoptic meteorologist usually means by the word any area of low pressure. It is almost synonomous with the word "cyclone." On the average, two or three storms a week pass over the United States during winter. The hydrologist or hydrometeorologist, on the other hand, usually applies the word to a period of heavy rainfall over a specific drainage basin, which may be large or small, or a number of basins. Several cyclones may pass over the basin during the period covered by the "storm" in the hydrometeorological sense. Shorter periods, when the rainfall rate is heavy, are referred to by the hydrometeorologist as "bursts." These bursts are often, but not always, associated with the passage of the "storms" of the synoptic meteorologist. In the rest of this section the word will be used in the hydrometeorological sense.

A. Rainfall with Extratropical Systems

The excess of surface heating in lower latitudes and of low-level cold in higher latitudes is compensated for by sporadic outbreaks of warm air moving poleward and cold air equatorward. The kinetic energy is derived by sinking of the cold air and ascent of the warm. The warm air currents are frequently moist; they are very much concentrated in time and area in tonguelike intrusions; they are most frequently located east of surface cyclones; and their vertical magnitude may be of the order of tens of thousands of feet. Usually, all the air in such a tongue is eventually lifted to very high levels, and it thus yields practically all its moisture as precipitation sooner or later. The depths of precipitation depend on the degree of concentration of the lift and on the total amount of moisture originally in the tongue. In many cases the air-mass contrast is between very cold arctic air and moderately warm, not very moist middle-latitude air. Such cases yield a much smaller volume of precipitation than one in which the warm air tongue consists of air that originated with a strong, long-continued flow from the tropics.

Depth in atmospheric systems has been emphasized by meteorological studies in recent years. The disturbance observed on surface weather maps is associated with very pronounced disturbances throughout the atmosphere up to heights of 80,000 ft or higher. Deep masses of cold and warm air frequently lie very close together and are associated with a strong belt of winds at levels of about 40,000 ft. This belt is called a *jet stream*, and the winds in it frequently reach speeds of 100 to 200 mph. The degree of concentration of upward motion is related to certain characteristics of the jet streams.

At heights of about 10,000 ft many rainfall situations are characterized, in the eastern United States, by southerly flow ahead of a trough of lower pressure. It has been found that such locations are associated with horizontal convergence, upward vertical motion, abundant moisture, and convective instability, which characterize southerly flow from the Gulf of Mexico.

In the lower layers, heavy rainfall is usually associated with temperature contrast and low pressure. However, the relation is not simple. In the case of pressure, the heaviest precipitation does not usually occur at the lowest pressure. More important in locating the precipitation exactly are the curvature of the isobars and contrasts between weak and strong pressure gradients.

In the case of temperature contrasts, heavy rainfall is usually located near fronts. The *frontal model* of an extratropical cyclone is often used to classify the occurrences of precipitation. However, it fails, as any model must, at a certain point to describe exactly the real world. And it fails to account for the amounts and the locations of heavy precipitation. According to the model, precipitation should be on the cold-air side of the warm front and the cold front. The precipitation to the north of the warm front should be of a uniform continuous type; that associated with the cold front should be showery, and it should be back of the front. In agreement with the model, continuous drizzle or light rainfall is frequently observed ahead of the warm front, and cellular precipitation frequently occurs near the time of passage of the cold front. Also, heavy rainfall or snow frequently occurs near the center and in the cold air of a slowly moving, deepening cyclone. Many other facts, however, are not in agreement with the model: (1) The vertical speeds associated with simple glide up frontal surfaces (horizontal wind times slope of front) are enough to account only for drizzle or very light rain. (2) Frequently the rainfall associated with a warm front is not uniform but increases very markedly about 50 to 100 miles north of the front and in an order of magnitude greater than that which would be accounted for by simple upglide. (3) The very greatest amounts of precipitation occur overwhelmingly in the warm sector at a distance of from 100 to 200 miles from either front, in locations where no precipitation at all would be called for by the frontal theory. (4) Most cold fronts in the United States have very little or no precipitation associated with them. The rainfall that does occur with them, while it may be very intense for a short period, seldom lasts long enough in one place to give very large amounts. Needless to say, the failure of the cyclone model to account for the facts of heavy rainfall in no way decreases its value for other problems in meteorology.

Some meteorologists argue that the frontal theory can be modified to take care of some of the above-mentioned objections. For example, it is argued that north of the warm front there would sometimes be a very shallow layer of cold air in the surface layers and that the real slope of the cold wedge begins some distance to the north where the heavy precipitation occurs. Apart from the *ad hoc* nature of this explanation, it still does not explain why the heavy precipitation is so much larger than would be accounted for by any reasonable horizontal wind multiplied by the slope of the front. Again, the unsatisfactory explanation of the phenomena at the cold front is attributed to the fact that the cold air aloft outruns and overruns the front in the lower layers, thus setting a condition of very pronounced thermodynamic instability. The heavy precipitation well within the warm sector is attributed to the presence of a frontal-like formation called a *squall line*, or an *instability line*.

The rainfall pattern associated with a cyclone is not always the same even for the same cyclone at different periods in its life history. For example, many cyclones start in the Great Plains of the United States as very dry ones. They may give little or no precipitation as they move eastward across the country until they reach the Eastern Seaboard, when they may pick up moisture and then give very heavy rainfall.

The *cyclone* is of the nature of a moving wave, through which new air continuously moves, passes through, and is replaced by other air. The air at heights of 10,000 ft or higher in a cyclone today may have been hundreds of miles away and in the lower 2,000 ft of the atmosphere 24 hr earlier. While the storm was moving eastward, this

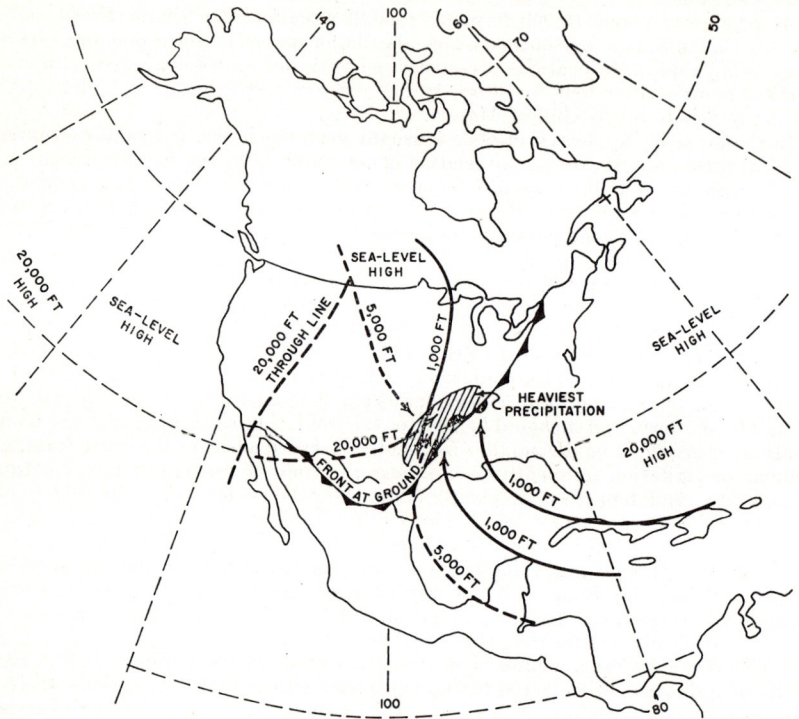

Fig. 9-14. Conditions associated with heavy precipitation in the Central Mississippi Valley. The lines with arrowheads represent airflow. (*After Lott and Myers* [6].)

air may have moved westward and then northward to meet the wavelike cyclone. The air in the lower layers west of the cyclone, to the contrary, probably has come into the cyclone system from several hundred miles to the north or northeast. Also, a cyclone is a continuously changing structure, which develops, intensifies, and then dies during the course of several days.

A fast-moving cyclone tends to give moderate rainfall over a large area, while a stationary cyclone tends to give heavy depths over a small area. Widespread light precipitation frequently occurs with old, nearly stationary lows after the warm air has been lifted to great heights. An ideal situation for heavy rainfall is one in which an active equatorward flow of cold air a few hundred miles ahead of a cyclone slows its movement and adds to its energy.

Figure 9-14 illustrates schematically some of the conditions occurring together with heavy rains in the Mississippi Valley.

B. Local Convective Rains and Thunderstorms

The vertical distribution of potential and thermal energy (instability) is more capable of concentrating kinetic-energy release, rainfall, and other atmospheric elements than is the horizontal. *Tornadoes,* strong vertical turbulence and hailstorms, for example, require the prior presence of instability.

Both vertical and horizontal energy-releasing distributions are present for the more severe tornado outbreaks and for the heavy local rainstorms. For example, the Holt, Mo., storm of June 22, 1947, when 11 in. of rain fell in 42 min, was an example of such vertical instability released as a front approached (Fig. 9-15). Over Florida it has

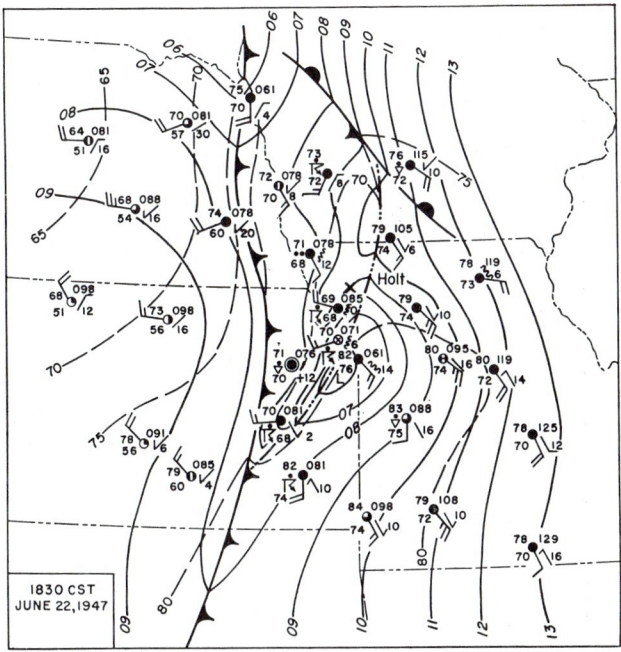

FIG. 9-15. Weather map 2 hr before Holt, Mo., storm. The cold front was located about 100 miles west of the storm site. The heavy rainfall occurred near the small-scale low-pressure area which had formed during the last 2 hr. (*After Lott* [11].)

been found that the occurrence of scattered thunderstorms (which are associated with the vertical contrast) varies as the convergence of the sea breezes around the peninsula, the latter depending on the contrast between the heating over the land and the lack of heating over the sea during the day, a horizontal effect. In tornado forecasting, the instability is considered to be released by a trigger effect, which is associated with larger-scale atmospheric processes, such as the movement of low-pressure areas, the approach of middle-level troughs, or the expected characteristics of the jet stream.

Rainfall occurrences cover a double spectrum: (1) from large amounts to small amounts, and (2) from very large areas to areas of only a square mile or so. At one corner of the square represented by these two spectra are local heavy rains. With respect to destructive effects other than rainfall, this corner might be represented by the occurrence of tornadoes. There is a common conception, not without some validity, that the storms thus represented are of a different generic type from more general rains. It will be understood from the foregoing discussion that the distinction is not exact. Many large-areal mid-latitude systems are primarily of horizontal contrast, but may have instability effects added during a part of their existence. *Thunder* is not necessarily associated closely with heavy local rainfall. Often thunder is heard over extensive areas, with little or no rainfall—the dry thunderstorms. It has been heard in widespread snow situations in winter. The intensity of thunder associated with heavy rain in the subtropics is very weak as compared with midwest thunderstorms that give less rainfall. (The meteorological services define a *thunderstorm* as the sound of thunder.)

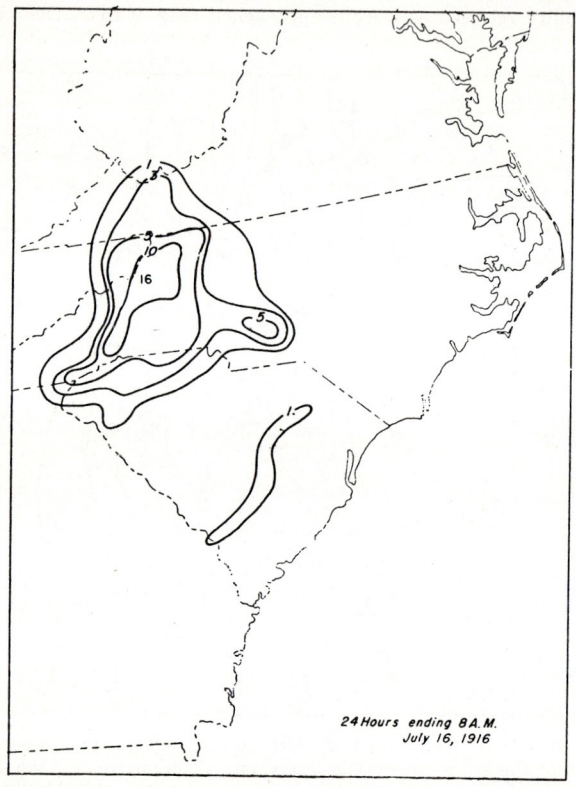

FIG. 9-16. Rainfall distribution (in inches) over Carolina and Virginia during a hurricane situation. Twenty-four hours ending 8 A.M. July 16, 1916. (*After Schoner and Molansky* [12].)

Air-mass showers is the name given to scattered rainfall amounts that are associated with no synoptic pressure system. They occur relatively frequently in the southeastern part of the United States in summer (the West Coast is practically rainless in summer). Usually they are light in amount, being of much more practical importance for agriculture than for setting the standards for storm-sewer design, for example. Their intensity and areal coverage varies considerably from day to day. There is no clear-cut line between air-mass showers and *prefrontal showers*. *Instability showers* occur at times when cold air moves over a warm surface, for example, off the Atlantic Coast when a cold air mass moves over it. Many very heavy snowstorms occur on the shores of the Great Lakes in winter when cold air remains over the relatively warm Lakes.

C. Tropical Storms

The vertical cold-over-warm-and-moist energy distribution prevails generally over oceans in lower latitudes during the latter part of summer and during fall. At other seasons of the year it is present from time to time. The destructive winds of the hurricane almost certainly owe their kinetic energy to the transformation processes associated with the rising of this warm moist air. The question as to how the rising is initiated and maintained is, among others, now being vigorously attacked by a research project. During the life of a tropical storm an enormous quantity of air is

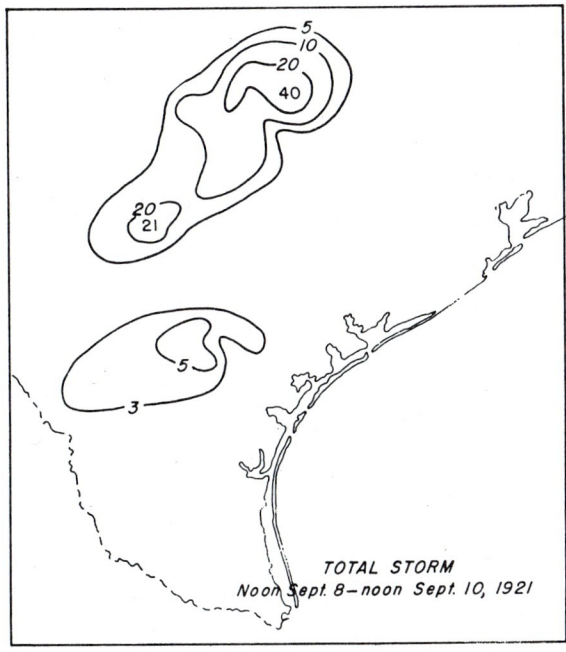

FIG. 9-17. Rainfall distribution (in inches) over Texas during a decadent hurricane situation. Total storm noon Sept. 8 to noon Sept. 10, 1921. (*After Schoner and Molansky* [12]).

pulled into the system in the lower layers, has its moisture condensed as rainfall, and is discharged from the storm circulation in the upper layers of the atmosphere. The resulting rain falls mostly in the forward quadrants and mostly near the path of the trajectory of the storm center, perhaps more to the left than to the right, in the Northern Hemisphere. There are, however, many exceptions to these generalities.

Many rainstorms in association with tropical storms have occurred in the states that border on the Gulf of Mexico. Others have occurred in many other parts of the United States. Directly associated with such storms have been the record flood-producing rains in Pennsylvania and New England in August, 1955; in western New York and southern Ontario in October, 1955; in North Carolina in 1916 and 1940 (Fig. 9-16); in Louisiana in August, 1940; and many others. In addition to such direct associations, there have been many heavy rainfall occurrences with the decadent stages of tropical storms, sometimes 2 or 3 days after the heavy winds of the storm have died down. A number of examples may be cited: The Thrall, Texas, storm of Sept. 8 to 10, 1921 (Fig. 9-17), occurred along the projected trajectory of a hurricane which entered the Mexican coast near Tampico and whose wind pattern died out very rapidly. The Baldwin, Maine, storm of Aug. 21, 1939, followed by 2 days a hurricane that went off the New England Coast. The Covesville, Va., storm of Sept. 15 to 20, 1944, followed by 2 days the passage of the Great Atlantic storm off the Coast on Sept. 16. During September, 1939, three tropical hurricanes moved up the Gulf of California, and subsequently heavy local rainstorms occurred in Arizona and in the Imperial Valley and adjacent areas of California. A heavy rainstorm in the Great Valley of California at Red Bluff in September, 1918, has been associated with a decadent hurricane which moved from off the West Coast of Mexico. Apparently the vigorous circulation of a hurricane disturbs atmospheric conditions with respect to temperature, wind, and moisture over a region to such an extent that successive rainfall can occur even in otherwise unlikely places. For example, the Covesville storm was near the center of a high-pressure area.

Many tropical systems with weak winds also give copious rainfall. Some are associated with weak, rather diffuse low pressures or with troughs not having a definite center.

D. Orographic Rainfall

The windward sides of most of the higher mountain ranges experience heavy precipitation; the leeward slopes and regions downwind are usually very dry, being said to be in the *rain shadow* of the mountain range. A good example is afforded by the state of Washington. Along the Pacific Coast, where there is little topographic variation, the rainfall averages about 60 in. annually. Where there is even a little topography, the annual total may increase by 20 in. to give a total of 80 in. On the western slopes of the Olympic Mountains, amounts up to 150 in. are recorded. However, just to the east in the Puget Sound–Willamette Valley, amounts average around 40 in. Further to the east, the western slopes of the Cascades receive amounts of 60 to 70 in. Across the Cascade Divide there is a sharp transition from rain forest to sage brush. Amounts in the valleys east of the ranges are very low; Yakima receives 7 in. annually. Farther east amounts gradually increase again, reaching about 20 in. around Spokane.

In some areas the direction of moisture-bearing winds is not so unique as in Washington. The eastern slopes of the Appalachians, for example, may be the windward ones in some situations and the lee slopes in others. However, moisture usually approaches from a southerly direction.

The amount of precipitation in an orographic situation is roughly proportional to the wind speed up the slope and to the amount of moisture in the air. Mountains are not so efficient as cyclonic systems in removing the water from the air. It was found in a number of orographic situations of strong wind flow across the Coast Range and Sierras in California that about 30 per cent of the moisture flowing into the area fell as precipitation. Thus orography is less effective than other causes in removing moisture from a given current of air. Presumably this is because the total amount of lifting is much less in the orographic cases than in the others. However, orography is very effective in causing precipitation to reach the ground at the same place time after time or even continuously for extended periods. The highest ratios of 24-hr-point precipitation to total moisture in a column are found in orographic areas.

Precipitation produced by convergence often falls in association with pure orographic precipitation. Many of the storms that give heaviest precipitation in California have characteristics similar to storms that give heavy winter precipitation elsewhere in temperate latitudes. These characteristics include strong temperature contrast, the presence of low-pressure troughs, and most precipitation in the warmer air. In some cases there are relatively heavier amounts in the valleys as compared with the slopes and the ridges than in other cases. In fact, there have been storms around Los Angeles where the amounts of rainfall in the flat area near the city have been larger than those in the mountains to the north.

The relation of storm precipitation to the mean annual pattern depends largely on the relative importance of precipitation amount and of frequency in determining the annual pattern. For example, the area in the vicinity of the California-Oregon border has storms much more frequently than does southern California, and the mean annual totals are much larger, although the highest annual 24-hr totals are no larger. In the whole West Coast area, the mean annual patterns may be regarded as the accumulation of many storms of similar patterns. To a certain extent the rainfall pattern of each major winter storm that occurs in the region reproduces features of the mean annual pattern. Most of the rainfall occurs on the windward slopes; the leeward slopes and adjacent valleys are very dry. However, the degree of reproduction of the pattern is never perfect. The maximum percentages in individual storms of the mean annual totals is usually confined to a band perpendicular to the ridge, with smaller percentages on each side of this band. When convergence precipitation is associated with the orographic, the pattern may be modified markedly.

Some evidence exists that quite small topography has some influence on the frequency of light or moderate rainfall. Also, it has been noticed that several large storms have occurred along or near the Balcones Escarpment in eastern Texas. The general opinion of meteorologists who have studied the synoptic characteristics of these storms is that the storms will occur more frequently near the escarpment than in other locations because of tripper influences brought about by the escarpment itself, but that occasionally storms just as heavy will occur in nearby areas. Thus the escarpment is barely noticeable in the probable maximum values over the area. In many other cases slight orography is thought to be more effective as a trigger effect on the frequency of rainfall than in producing heavy amounts. This is especially true of hills a few hundred feet in height. Popular local belief usually ascribes far more significance to these hills than meteorologists do. In fact, most hydrometeorologists probably feel that general meteorologists overemphasize the effects of slight orography, especially in connection with the heavier storm amounts.

E. Quantitative Precipitation Forecasting

Since the early 1940s forecasts of the depth of precipitation in inches expected over areas of about 10,000 sq mi have been made by the U.S. Weather Bureau for hydrologic users. Most of the skill in the early part of this period was in forecasting amounts of less than an inch. Recently, a certain amount of work has been devoted to forecasting the heavier amounts. Two general approaches have been used. One uses flow of moisture plus an approximation of vertical speeds obtained as a by-product of the mathematical models used in numerical weather forecasting. These speeds show a considerable amount of correlation with areas of cloudiness and widespread areas of light rainfall. Their order of magnitude is too small to account for most of the hydrologically important rainfall amounts. There is undoubtedly a large field of work in developing more sophisticated models that will give closer approximations to true conditions.

The methods that have yielded more immediate practical results have utilized close study of moisture inflow into convergence-producing isobaric configurations, the relations of the pressure field to precipitation volume, the study of the interrelated movements of cold air and warm air at different levels and their effects on vertical speeds, and the development of pressure systems.

In addition, several studies have been made that show considerable skill in relating future moderate rainfall amounts over a particular locality or basin to very simple meteorological parameters using statistical procedures. A very effective procedure has been devised by Thompson (Fig. 9-18) for estimating probabilities for rainfall amounts at Los Angeles.

Forecasts of tercile frequency of 5-day totals are made for all areas in the United States three times a week; that is, the forecast states whether the amount over an area is expected to be as large as the upper third, middle third, or lower third of the totals that have occurred for the like period in previous years. Verifications show a significant skill over chance. It seems quite likely that there will be a substantial improvement in quantitative precipitation forecasting in the next few years. Engineers charged with operations that could utilize such forecasts should plan for this development and should let their needs be known so that the aims will not be set too low.

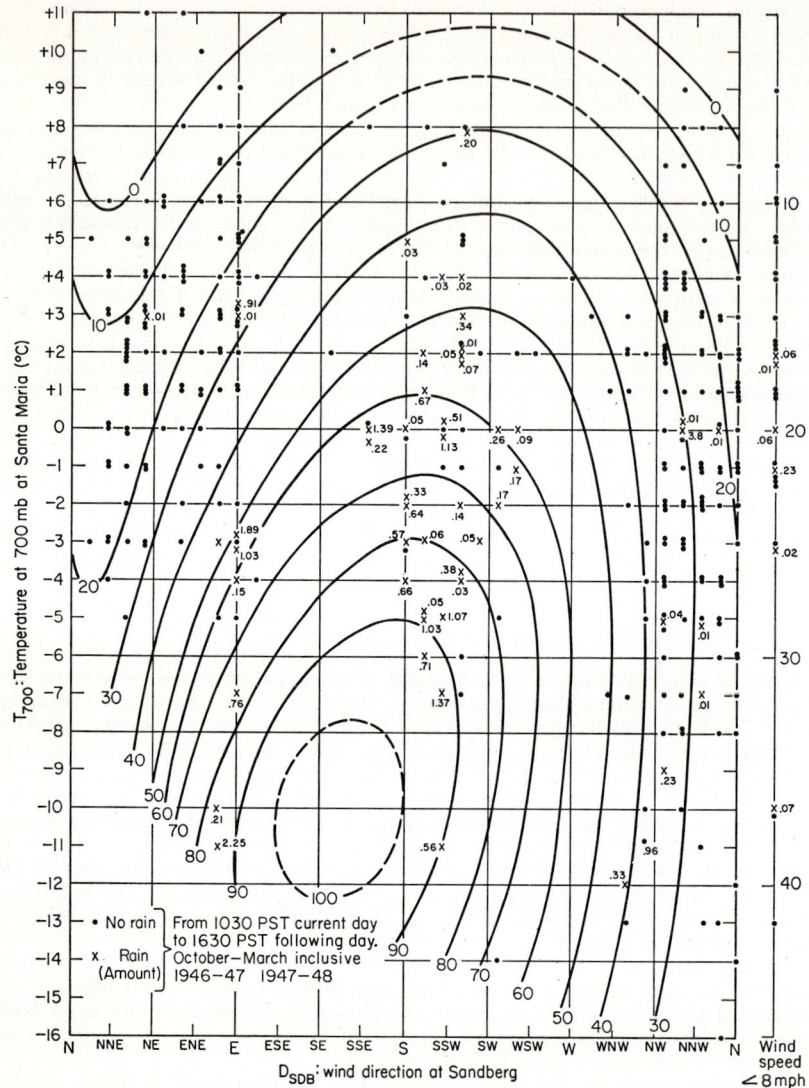

Fig. 9-18. Relationship of rainfall at Los Angeles to simple meteorological parameters. Three similar diagrams are used to give an estimation of the probability of rainfall of various amounts. (*After Thompson* [13].)

V. SPACE-TIME CHARACTERISTICS OF RAINFALL

A. Techniques of Analysis

1. The Double-mass Curve. This technique is a method of adjusting precipitation records to take account of nonrepresentative factors, such as change in location or exposure of the rain gage [15–19]. It accomplishes this by comparison of the accumulated totals of the gage in question with the corresponding totals for a representative group of nearby gages. Obviously, the more homogeneous the records of the nearby gages, the more accurate the procedure will be. Usually at least ten stations are used as the basic network for the comparison. It has been found that decided changes in slope of the curve frequently appear on such analyses and that these can frequently be related to changes in location of the gage or other changes. For example, in Fig. 9-19 the change in the slope of the curve can be explained by the fact that the location of the gage was moved several miles in 1923. The records for the earlier years should be decreased by the factor of $237/278$ to make them consistent with the more recent period. For the maximum degree of accuracy, the records of each of the stations in the basic group should be tested by the double-mass-curve method. Those stations showing appreciable changes of slope should be dropped from the basic group

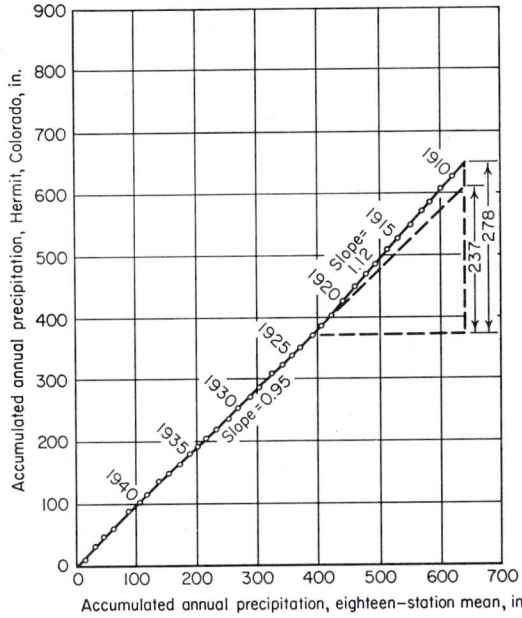

Fig. 9-19. Double-mass curve. (*After Linsley, Kohler, and Paulhus* [14].)

and their records adjusted as above. Seasonal accumulative totals are more representative than annual totals for this procedure in regions where there is a pronounced seasonal variation in precipitation.

There is a tendency on the part of some analysts to overdo the double-mass-curve technique. A change of slope should not be considered significant unless it persists for at least 5 years and there is collaborating evidence of a change in location or exposure or other change. If one particular year shows an abnormal accumulation, two parallel segments should be used. Occasionally one station will have a larger amount during one particular year than the nearby stations. This may be especially true where a large percentage of the annual precipitation falls in one or two large storms. The technique is not recommended for storm or daily rainfall amounts.

2. The Station-year Method. This is a method of extending the length of record for a frequency curve at a station, based on the assumption that the records for the same or different periods of records at a number of stations may be considered as a composite record for a single station for a period equal to the total number of years involved [20–21]. For example, if 50 years of record was available for 100 stations, it might be assumed that this was the equivalent of 5,000 years at a single station. Two important objections can be offered to the method. In the first place, it cannot be applied very rigorously unless one expects that over very long periods of time, say, thousands of years, the stations will have very nearly the same frequency distribution. In that case the stations are said to be *meteorologically homogeneous*. This requirement of homogeneity practically precludes use of the method in mountainous areas. Again, if the maximum amounts at different stations tend to occur in the same storms or even in the same group of storms or in any other nonrandom way, the method is not strictly applicable. In the extreme case, where the maximum amounts always occur together, the method may be no more significant than the record of a single station. In areas where the maximum amounts are associated with geographically separated intense local storms, the method may be considered more applicable. It has been suggested that the estimated return periods are roughly N times that from a single-station record, where N is the number of stations sampling each storm.

In spite of the limitations of the station-year method, it is probably as reliable a method as can be developed for estimating the amounts of rainfall for a station for return periods that are several times the length of record at any of the stations. It may yield reliable approximations when used by someone who understands thoroughly the statistical pitfalls involved and who has studied the records in detail, including tests for dependence between maximum amounts of rainfall at the various stations.

3. Interpolation of Rainfall Records. Frequently, records of rainfall for a certain station are missing for a day or several days, especially for cooperative stations. In order not to lose valuable information, it is desirable to have techniques for estimating the amounts for such days in calculating monthly and annual totals. The U.S. Weather Bureau [22] uses two procedures for these estimations, both based on simultaneous records for three stations as close to and as evenly spaced around the station with missing records as possible: (1) If the normal annual precipitation at each of these stations is within 10 per cent of that for the station with missing records, a simple arithmetic average of the precipitation at the three stations is used for the estimated amount. (2) If the normal annual precipitation at any of the three stations differs from that of the station with missing records by more than 10 per cent, the *normal-ratio method* is used. This method consists of weighing, by the ratios of the *normal-annual-precipitation* values,

$$P_x = \frac{1}{3}\left(\frac{N_x}{N_A}P_A + \frac{N_x}{N_B}P_B + \frac{N_x}{N_C}P_C\right) \tag{9-7}$$

where N is the normal annual precipitation. It is readily seen that the second method is adaptable to regions where there is large orographic variation in the precipitation. The two procedures have been adapted to machine methods and are used routinely by the Weather Bureau.

4. Average Depth of Rainfall over Area. Several methods are used to give average depth of rainfall over area. The simplest and the one most often used, especially in relatively flat country, is the simple taking of the arithmetic mean of a number of stations. For many years the U.S. Weather Bureau published the arithmetic mean of the monthly rainfall of all the stations in each state (or section). This practice was discontinued in 1956 because it was thought that it was unrepresentative in many instances. Instead, arithmetic means are now published for climatological divisions of states.

The *Thiessen method* [23] assumes that the amount at any station can be applied halfway to the next station in any direction. It is applied by constructing a Thiessen polygon network, the polygons being formed by the perpendicular bisectors of the lines joining nearby stations (Fig. 9-20). The area of each polygon is determined and is used to weight the rainfall amount of the station in the center of the polygon. The polygons must be changed each time a station is added to or taken from the network or each time the amount for any station is missing.

The *isohyetal method* consists of drawing lines of equal rainfall amount (*isohyets*) using observed amounts at stations and any additional factors available to adjust or interpolate between observation stations [24]. The average depth is then determined by computing the incremental volume between each pair of isohyets, adding these incremental amounts, and dividing by the total area.

In the *percentage-of-mean-annual method,* much used for storm averages in orographic regions, the ratio of the storm amount to the mean seasonal or annual precipitation is plotted for each observation station. Isolines of these ratios are plotted. The means of the ratios are determined for incremental area; these are multiplied by the mean annual amounts to give storm depths for the incremental areas; multiplication by the size of the incremental areas, addition of these incremental volumes, and division by the total area give the average depth.

For the *abbreviated isopercentual method,* one calculates the storm total to mean annual ratio for each station, weights these ratios by Thiessen polygons, and multiplies this weighted ratio by the basin mean average annual precipitation. The abbreviated method is especially useful when several storms or storm periods are to be worked up for the same basin.

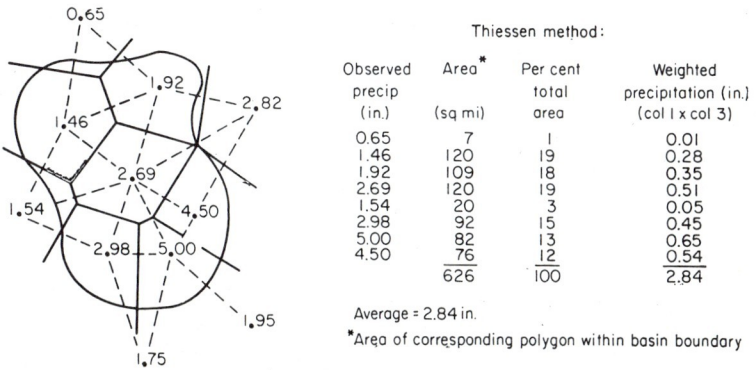

Thiessen method:

Observed precip (in.)	Area* (sq mi)	Per cent total area	Weighted precipitation (in.) (col I x col 3)
0.65	7	1	0.01
1.46	120	19	0.28
1.92	109	18	0.35
2.69	120	19	0.51
1.54	20	3	0.05
2.98	92	15	0.45
5.00	82	13	0.65
4.50	76	12	0.54
	626	100	2.84

Average = 2.84 in.
*Area of corresponding polygon within basin boundary

Fig. 9-20. Thiessen diagram. (*After Linsley, Kohler, and Paulhus* [14].)

5. Effect of Network Density on Areal Rainfall Average. Because of the concentrated and spotty areal distribution of many heavy rainfall centers (Fig. 9-21), the effect of increasing the density of gages is more often than not to increase the apparent average areal rainfall. Several investigators have studied the effect of density on this apparent average. Figure 9-22 is one example.

6. Techniques for Orographic Precipitation. In most orographic regions it is difficult to install and maintain gages over the parts of the area where precipitation is heaviest. Gages, usually located in the valley, give an unrepresentative idea of the annual and storm precipitation over the whole area. Two techniques have been developed for estimating mean annual precipitation, and one for estimating storm precipitation. (1) The *Parsons,* or *Sacramento, method,* developed at the Corps of Engineers District Office in Sacramento [25], is used to determine mean annual precipitation in orographic areas. It uses measurements of runoff and the analyst's knowledge of the soil and vegetation, as well as measurements of mean annual precipitation. The method proposes to construct a mean-annual-precipitation map that will be in agreement with the observed precipitation values and such that the differences between the average precipitation value for any area and the measured runoff from that area will form a consistent pattern that agrees with the character of the soil, vegetation, and the analyst's judgment. It has been most used in the western United States in areas where most of the precipitation stations are located in valleys, but where there are many stream-gaging stations that measure runoff from higher elevations. (2) The *Spreen method* [26] consists of a statistical analysis of the effect of various topographic parameters on the annual or seasonal totals at the gages (Fig. 9-23). The relations are then applied to estimate amounts in areas between gages.

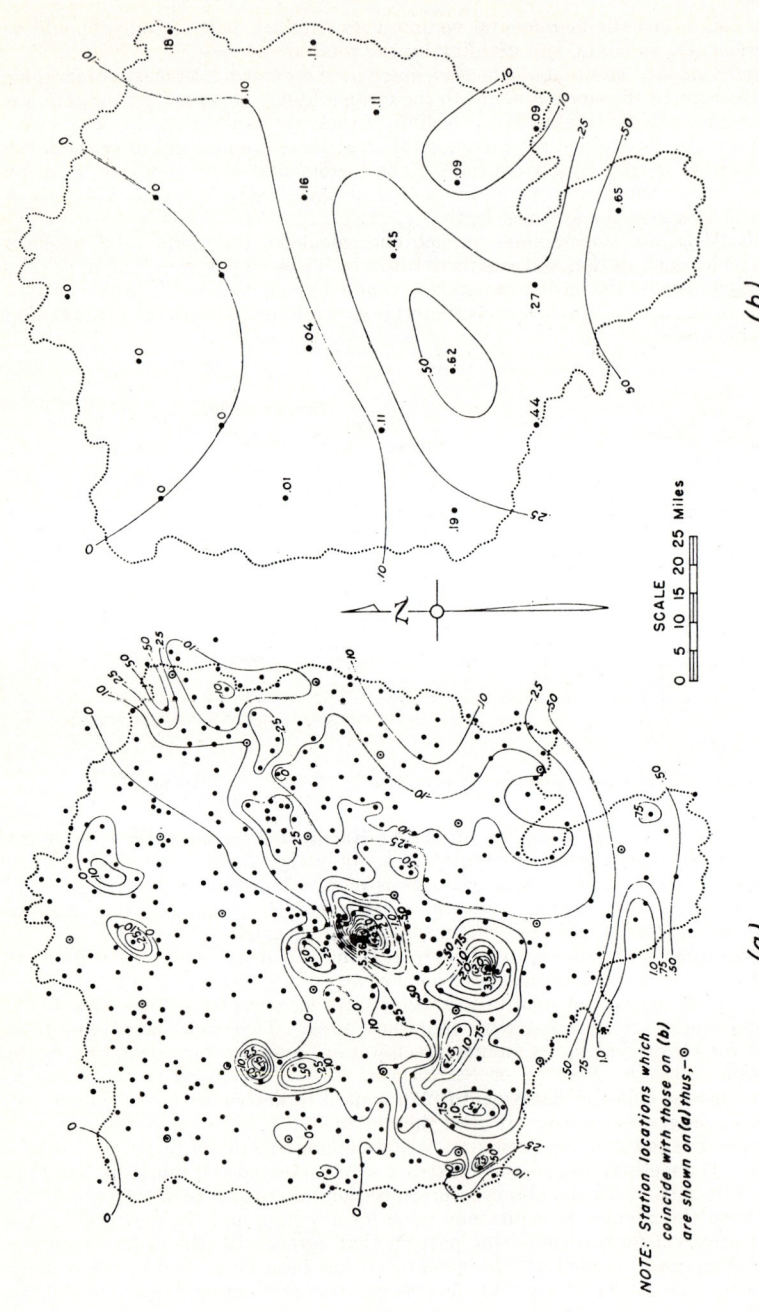

Fig. 9-21. Effect of rain-gage density on analysis of rainfall pattern. Isohyetal maps of Aug. 3, 1939 storm, Muskingum Basin, Ohio. (a) 449 rain gages (1 gage per 18 sq mi); (b) 22 rain gages (1 gage per 375 sq mi). (*U.S. Weather Bureau.*)

The parameters of the relation are (1) *elevation* of the precipitation station (in 1,000 ft); (2) *rise*, difference in elevation (in 1,000 ft) between station and highest point within a 5-mile radius; (3) *exposure*, the sum (in degrees) of those sectors of a 20-mile-radius circle about the station not containing a barrier of 1,000 ft or more above the station elevation; and (4) *orientation*, the direction to eight points of the compass of the greatest exposure defined above. Other parameters have been used by other investigators. Perhaps the two most commonly found significant have been *rise*, as defined above or similarly, and *barrier height* in the direction from which the moisture-bearing winds flow toward the station. Figure 9-24 shows a result of applying Spreen's method.

For orographic storm rainfall, the *isopercentual method* of analysis is most frequently

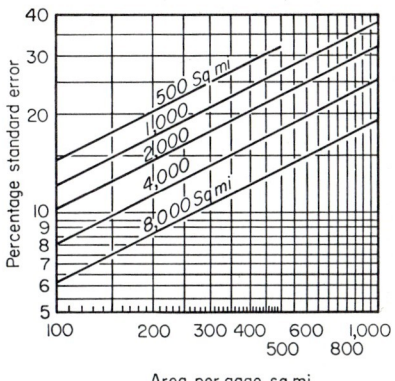

Fig. 9-22. Standard error of average precipitation as a function of network density and drainage area for the Muskingum Basin. (*U.S. Weather Bureau*.)

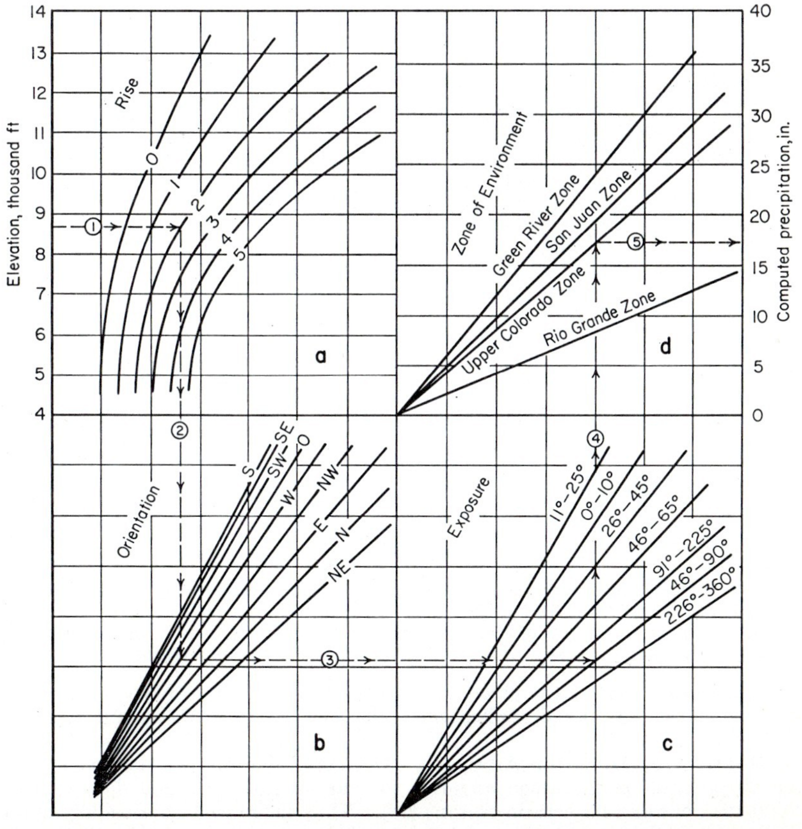

Fig. 9-23. Relation between seasonal precipitation and topographic parameter for western Colorado. (*After Spreen* [26].)

9-32 RAINFALL

used. This consists of calculating the percentage of the storm-rainfall total to the mean-annual or seasonal amounts at all gages, drawing isolines of these percentages on a map, and multiplying the mean-annual or seasonal totals at all points by the percentages thus obtained to obtain the storm totals. Of course, the method may be

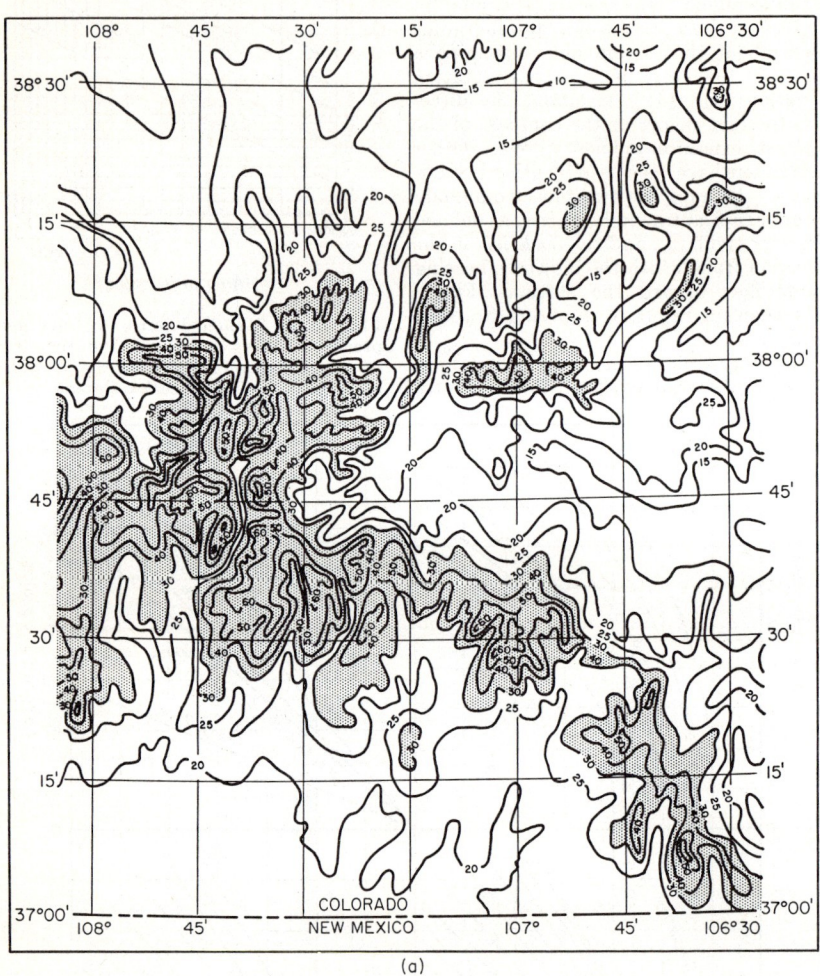

(a)

FIG. 9-24. Influence of topographic adjustment on mean annual precipitation pattern for western Colorado. (*After Spreen* [26].) Normal annual precipitation for southwestern Colorado. (a) Adjusted (26 million acre-feet). Based on adjusted climatological data and values computed by use of topographic parameters (5-in. isohyetal intervals for values

used in the same way for the precipitation values for a shorter duration rather than the whole storm. A special application is the *normal-ratio method* for interpolating missing records.

7. Depth-Area-Duration Analysis. The DAD analysis was devised to determine the greatest precipitation amounts for various-size areas and durations over different regions and for certain seasons [28]. Thus the procedure is usually applied only to a storm that conceivably might give at least one such maximum depth. It is designed (1) to select the several areas or centers within the storm that have greater

DAD values than the areas immediately adjacent to each of them and (2) to compare the values from these several areas to determine which is the greatest for each size of area and each duration. A detailed time analysis, called Part I, is the first step, since most of the data consist of 24-hr amounts and hydrologic considerations require

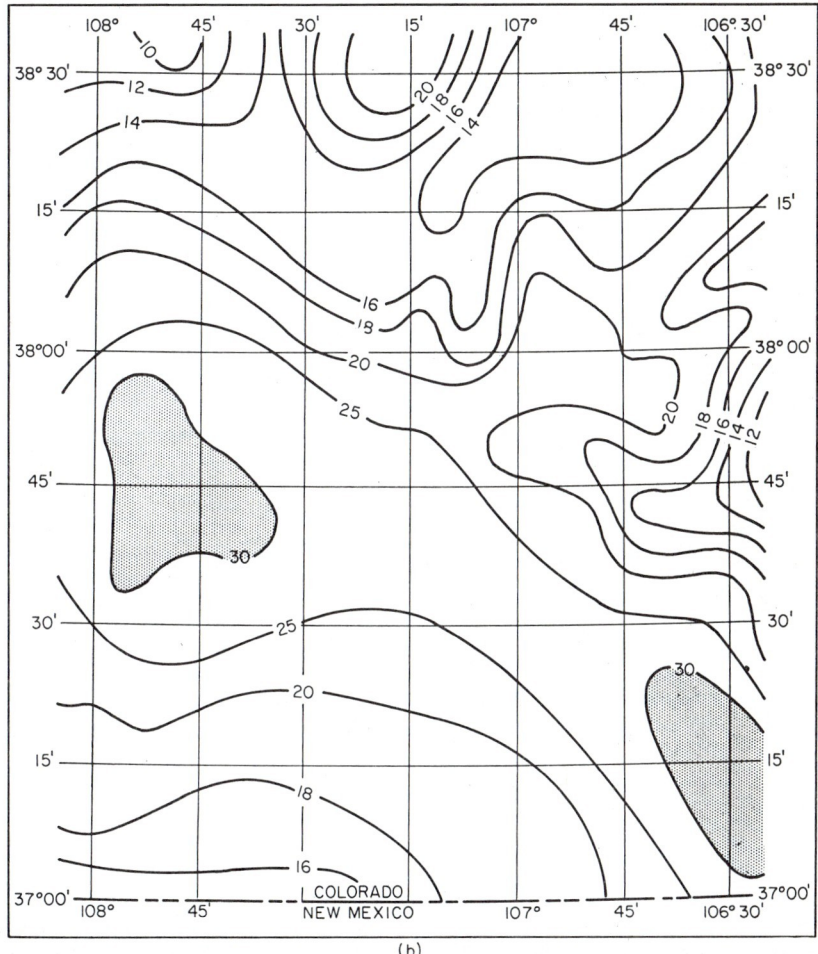

(b)

up to 30 in.; 10-in. intervals for greater values). (b) Unadjusted (20 million acre-feet). Based on unadjusted climatological data, from "Climate and Man," U.S. Department of Agriculture Yearbook, 1941 (2-in. isohyetal intervals for values up to 20 in.; 5-in. intervals for greater values).

6-hr amounts. Full use of recording gages and synoptic meteorology is required to give the time breakdown, the final result being the preparation of a mass curve for each station. Part II consists of the following:

1. Preparation of several total-storm depth-area curves by
    a. Dividing the total-storm rainfall map into its major rainfall centers
    b. Determining the total size of area and the average depth within the area enclosed by successive total-storm isohyets around each center

9-34　　　　　　　　　　　　RAINFALL

   2. A time breakdown of each of these depth-area curves by
      a. Weighting the amount of each station within each of the total-storm isohyets by the ratio of its Thiessen area within that isohyet to the total area within that isohyet, except that, if there are at least six stations fairly well distributed within the isohyet, each station is given equal weight
      b. Distributing timewise the total-storm average depth within each isohyet according to the time distribution of the total-storm precipitation at these stations, weighted as indicated
   3. Comparison of the values thus obtained by plotting all of them on a diagram and drawing enveloping curves

The procedure is best illustrated by an example. Figure 9-25 presents a set of mass curves. Figure 9-26 shows the total-storm isohyetal map with division into two zones

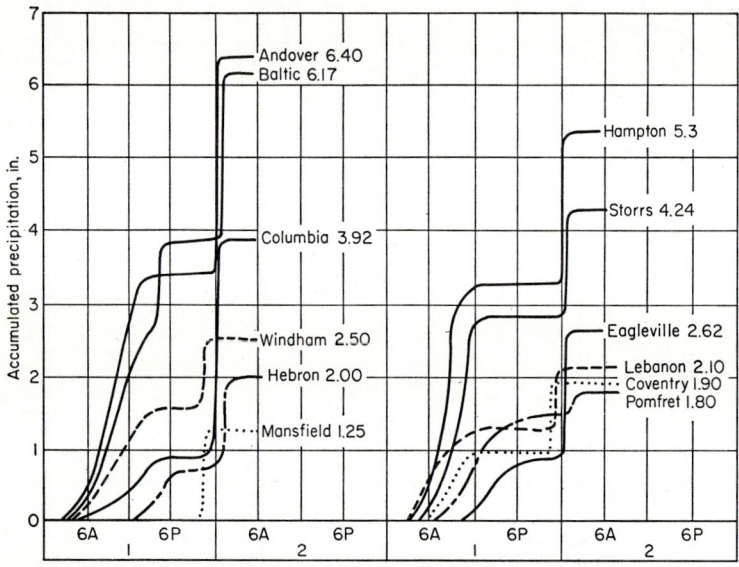

Fig. 9-25. Mass curves of rainfall. (*After Linsley, Kohler, and Paulhus* [14].)

and the Thiessen polygons. The results of planimetering the isohyetral lines appear in Fig. 9-27, columns 5 and 10, containing the data for the total-storm area-depth curve. The values in column 7 are estimated by eye, and represent the average depths between successive isohyets or within a central isohyet. In Fig. 9-28, the breakdown into 6-hr increments of the station totals is carried out. The values under "absolute max precipitation" are the highest that can be found by trial anywhere on the mass curve for the particular duration and station. They are selected only near intense centers and are usually arbitrarily assigned to an area of 10 sq mi. The values under "contemporaneous accumulated precipitation" are read directly from the mass curves (Fig. 9-25). The procedure for distributing a real rainfall with respect to duration by means of the station amounts is illustrated in Fig. 9-29. The data under "encompassing isohyet" are copied directly from Fig. 9-27. The data under "effective area controlled by station" are obtained by planimetering, the object being to determine the per cent of the total area within the encompassing isohyet that also lies within the Thiessen polygon for each station. The numbers on station lines in the last four columns are obtained by multiplying these weighting factors by the contemporaneous accumulated amounts from Fig. 9-28. The values on lines *a* result from the addition of these amounts for each area enclosed by an encompassing isohyet. The total-storm average value so obtained in lines a, last column, will not ordinarily be the same as that obtained in Fig. 9-27. The latter is entered on line b, last column, and each of the other values in line a is multiplied by the ratio of the value in line b, last column, to that in line a, last column. The adjusted mass-curve values thus obtained appear in line b. Line c results from subtraction of successive values in line b. The largest

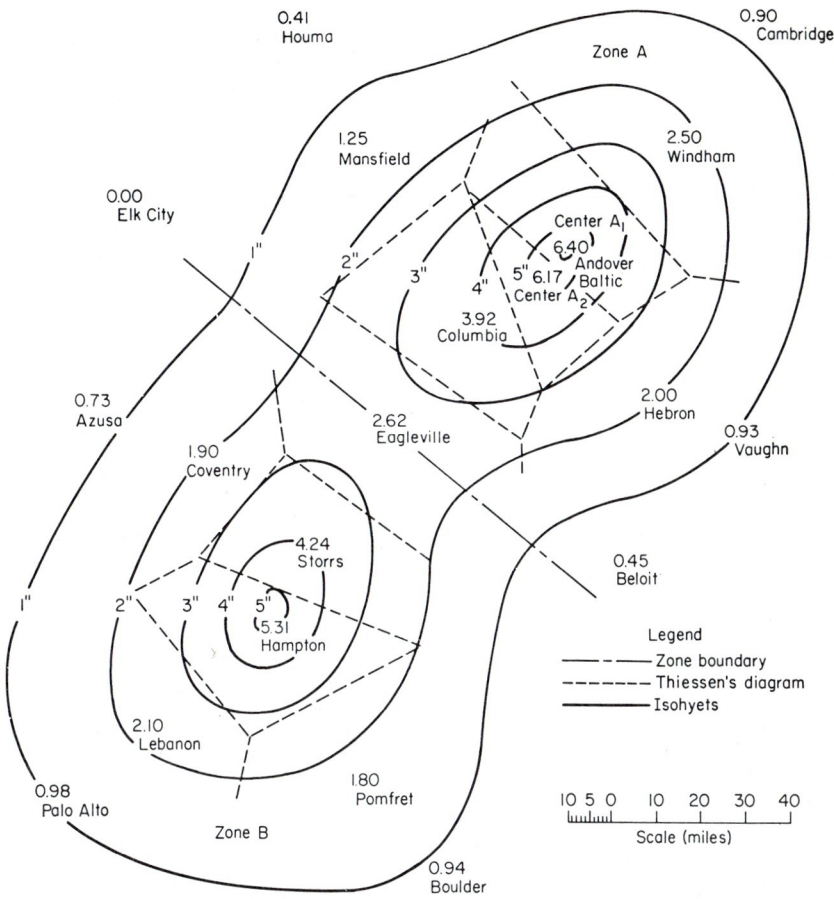

Fig. 9-26. Total-storm isohyetal map (hypothetical storm) for 24 hr, from 3 A.M., Sept. 1, 1947, to 3 A.M., Sept. 2, 1947. (*After Linsley, Kohler, and Paulhus* [14].)

of the values in line c is entered under 6-hr in line d; the largest total of two successive values is entered under 12 hr, etc. The values from lines d are plotted against the area encompassed by the corresponding isohyet in the depth-area-duration diagram (Fig. 9-30). Drawing of enveloping lines on this diagram and selection from these lines of depth values for standard sizes of area and durations (Fig. 9-31) completes the procedure.

8. Excessive Precipitation. As defined by the U.S. Weather Bureau, *excessive precipitation* is any precipitation that falls at a rate equaling or exceeding that indicated by the following formula for durations of 5, 10, 15, 20, 30, 45, 60, 80, 100, 120, 150,

9-36 RAINFALL

Line No.	Rainfall center or zone	isohyet inches	Area enclosed		Net area in sq mi	Average depth of rainfall in inches	Volume of rainfall in inches/sq mi		Average depth of rainfall in inches (Col.9÷Col.5)
			Planimeter reading	Area in sq mi			Increment (Col. 6 x Col.7)	Accumulative	
1	2	3	4	5	6	7	8	9	10
1	A	6.4	–	STATION	–	–	–	–	6.4
2	Center A_1	6	9	14	14	6.2	87	87	6.2
3									
4	Center A_2	6	7	11	11	6.1	67	67	6.1
5									
6	Center A_1+A_2	6	–	25	–	–	–	154	–
7		5	123	189	164	5.5	902	1056	5.6
8		4	531	815	626	4.5	2817	3873	4.8
9		3	1580	2424	1609	3.5	5632	9505	3.9
10		2*	3652	5602	3178	2.5	7945	17450	3.1
11		1*	6704	10284	4682	1.5	7023	24473	2.4
12									
13	B	5	36	55	55	5.3	292	292	5.3
14		4	319	489	434	4.5	1953	2245	4.6
15		3	1171	1796	1307	3.5	4574	6819	3.8
16		2*	3122	4789	2993	2.5	7482	14301	3.0
17		1*	6749	10353	5564	1.5	8346	22647	2.2
18									
19									
20	A+B	2	–	10391	–	–	–	31751	3.1
21		1	–	20637	–	–	–	47120	2.3
22									
23									
24	* Isohyet does not close within this zone								
25									

Fig. 9-27. Computation of depth-area data from isohyetal map. (*After Linsley, Kohler, and Paulhus* [14].)

Line No.	Zone or center	Station	Time of observation	Absolute max. precip. Duration in hours				Contemporaneous acc. precip. in inches Time in hours at end of period					
				6	12	18	24	6 9A1	12 3P1	18 9P1	24 3A2	30	36
1	A	Andover		3.0	3.4	5.6	6.4	1.3	3.4	3.4	6.4		
2		Baltic		2.4	3.7	5.4	6.2	0.9	2.6	3.8	6.2		
3		Columbia		3.0	3.3	3.8	3.9	0.2	0.8	0.9	3.9		
4		Windham						0.8	1.5	1.6	2.5		
5		Hebron						0.0	0.2	0.7	2.0		
6		Mansfield						0.0	0.0	0.0	1.2		
7													
8	B	Hampton		2.8	3.2	4.4	5.3	2.0	3.2	3.2	5.3		
9		Storrs		2.3	2.8	3.7	4.2	0.9	2.8	2.8	4.2		
10		Eagleville						0.0	0.5	0.8	2.6		
11		Lebanon						0.9	1.3	1.3	2.1		
12		Coventry						0.5	1.0	1.0	1.9		
13		Pomfret						0.2	1.1	1.5	1.8		
14													
15													
16													
17													
18													
19													
20													

Fig. 9-28. Tabulation of data from mass curves of rainfall. (*After Linsley, Kohler, and Paulhus* [14].)

SPACE-TIME CHARACTERISTICS OF RAINFALL 9-37

Line No.	Center or zone	Station or item description	Encompassing isohyet			Effective area controlled by sta.		Product (1) Time in hours (2)			
			Inches	Average P in Inches	Area inclosed sq mi	Planim. reading K=1.534	Station weight in %	6 9a1	12 3p1	18 9p1	24 3a2
1	A	Andover (Absolute max. station precip.)						3.0	3.4	5.6	6.4
2		(e) End of period of max. precip.						2a2	3p1	1a2	3a2
3											
4	A	Baltic (Absolute max. station precip.)						2.4	3.7	5.4	6.2
5		(e) End of period of max. precip.						3a2	2a2	2a2	3a2
6											
7											
8	A	Andover	6	6.2	14	—	100	1.3	3.4	3.4	6.4
9	(Center A₁)	(b) Adjusted mass curve						1.3	3.3	3.3	6.2
10		(c) Adjusted increment						1.3	2.0	0.0	2.9
11		(d) Max. depth-duration						2.9	3.3	4.9	6.2
12		(e) End of period for (d)						3a2	3p1	3a2	3a2
13											
14											
15	A	Baltic	6	6.1	11	—	100	0.9	2.6	3.8	6.2
16	(Center A₂)	(b) Adjusted mass curve						0.9	2.6	3.7	6.1
17		(c) Adjusted increment						0.9	1.7	1.1	2.4
18		(d) Max. depth-duration						2.4	3.5	5.2	6.1
19		(e) End of period for (d)						3a2	3a2	3a2	3a2
20											
21											
22	A	Andover	5	5.6	189	67	54	0.70	1.84	1.84	3.46
148	B		2	3.0	4789	—	46	0.43	0.89	0.94	1.59
149	A+B		2	3.1	10,391	—	100	0.71	1.66	1.85	3.64
150								0.6	0.8	0.2	1.5
151								1.5	1.7	2.5	3.1
152								3a2	3a2	3a2	3a2
153											
154		Computations based on unweighted av. of mass curves for various stations inclosed by									
155									various isohyets		
156											
157	A	Sum of precip. at 6 stat	1	2.4	10,284			3.2	8.5	10.4	22.2
158	B	Sum of precip. at 6 stat	1	2.2	10,353			4.5	9.9	10.6	17.9
159	A+B	Sum of precip. at 12 stat	1	2.3	20,637			7.7	18.4	21.0	40.1
160		(a) Unweighted mass curve (Average of 12 stations)						0.64	1.53	1.75	3.34
161		(b) Adjusted mass curve						0.4	1.1	1.2	2.3
162		(c) Adjusted increment						0.4	0.7	0.1	1.1
163		(d) Max. depth-duration						1.1	1.2	1.9	2.3
164		(e) End of period for (d)						3a2	3a2	3a2	3a2
165											
166											
167											
168											
169											
170											

REMARKS: (1) Accumulative station rainfall x station weight, except as otherwise noted.
(2) At end of period (or duration of maximum precipitation)

FIG. 9-29. Duration breakdown of average precipitation over area.

9-38 RAINFALL

and 180 min:

$$P = t + 20 \qquad (9\text{-}8)$$

where P is the precipitation in hundredths of an inch, and t is the time in min.

Prior to 1936, the published data generally present the accumulated precipitation amounts during storms in which the rate equaled or exceeded $P = t + 20$, the tabulation beginning with the 5-min period where the rate of 0.25 in. in 5 min began and

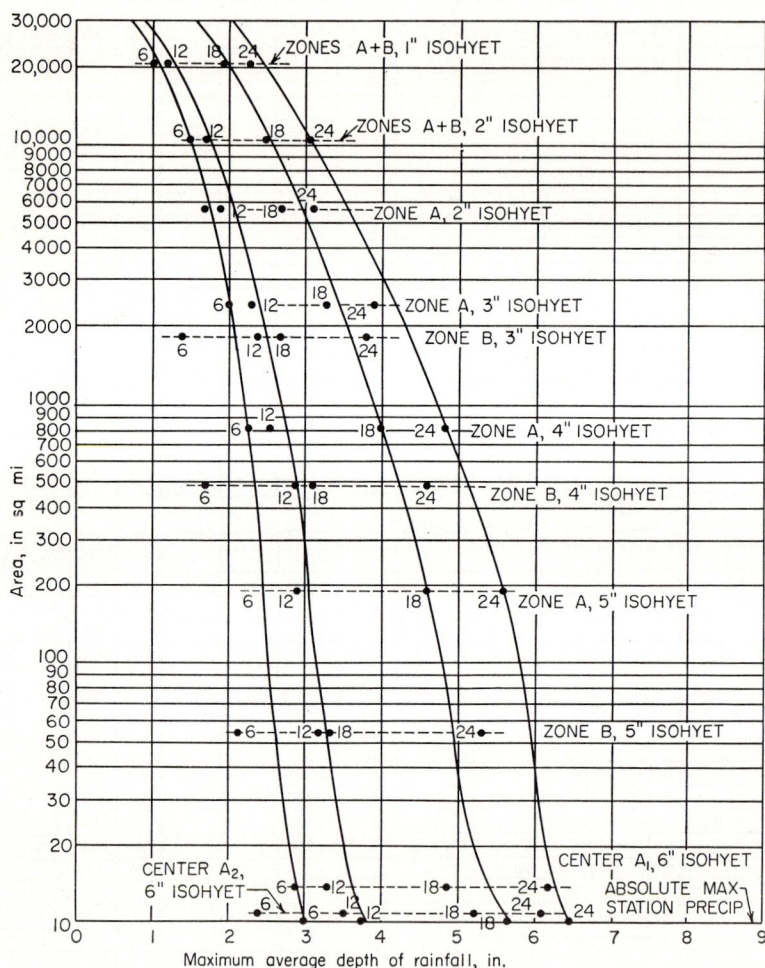

Fig. 9-30. Maximum depth-area-duration curves. The plotted points represent the maximum average depth over the area designated, within the number of hours indicated by the figures beside the points.

continuing by 5-, 10-, or 20-min intervals for as long as the excessive rate prevailed. However, the tabulations for 1933, 1934, and 1935 were continued for a total period of 120 min, even though the excessive rate terminated sooner. It can be seen that the amount listed for a given duration under the method of tabulation used until 1936 is not necessarily the maximum for that duration.

Beginning with 1936, tabulations of excessive precipitation give maximum precipitation for periods from 5 to 180 min, the amounts being for the periods in which the fall was actually the greatest for the particular duration. Thus, for example, the amount listed for a particular duration, say, 5 min, is the greatest amount for any 5-min interval of the 180-min period. From 1936 to 1948, the lower limit of excessive precipitation for the states of North Carolina, South Carolina, Georgia, Florida, Alabama, Mississippi, Tennessee, Arkansas, Louisiana, Texas, and Oklahoma was defined by the formula $P = 2t + 30$, but the use of the formula $P = t + 20$ for the entire United States was resumed in 1949.

Area in sq mi	Duration of rainfall in hours									
	6	12	18	24	30	36	48	60	72	96	120
10	3.0	3.7	5.6	6.4							
100	2.5	3.2	4.8	5.8							
200	2.4	3.0	4.5	5.5							
500	2.3	2.9	4.2	5.1							
1000	2.2	2.7	3.9	4.8							
2000	2.0	2.4	3.5	4.3							
5000	1.8	2.1	3.0	3.6							
10000	1.5	1.8	2.4	3.1							
20000	1.1	1.3	2.0	2.5							

Fig. 9-31. Maximum average depth of rainfall (in inches).

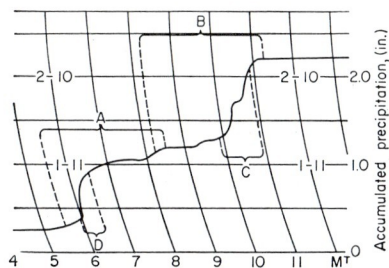

Fig. 9-32. Excessive-precipitation data. A indicates the time of occurrence of the maximum amount of precipitation associated with the period of excessive precipitation D for an interval of 180 min. Similarly, B indicates the maximum amount for 180 min associated with the period C. The tabulation lists the excessive precipitation data for the two periods of excessive precipitation. Note that C and D are separate periods of excessive precipitation, since they are separated by 180 min and each does not equal or exceed 1.80 in. Also note that since A and B overlap, only the greater amount is tabulated. (*U.S. Weather Bureau.*)

Interval	5	10	15	20	30	45	60	80	100	120	150	180
Amount for period D	0.20	0.35	0.43	0.46	0.52	0.58	0.63	0.69	0.73	0.76	0.80	
Amount for period C	0.18	0.25	0.33	0.38	0.45	0.70	0.82	0.90	0.95	1.00	1.00	1.06

Figure 9-32 illustrates the tabulation of excessive precipitation from an actual record.

Summaries of excessive-precipitation data have appeared in the Annual Report of the Chief of the Weather Bureau from 1895 to 1934, in the United States Meteorological Yearbook from 1935 to 1949, and since that time in the annual issues of *Climatological Data, National Summary.* The techniques used throughout the years are described in detail in *Key to Meteorological Records Documentation* 3081 prepared by the Weather Bureau and available from the U.S. Superintendent of Documents.

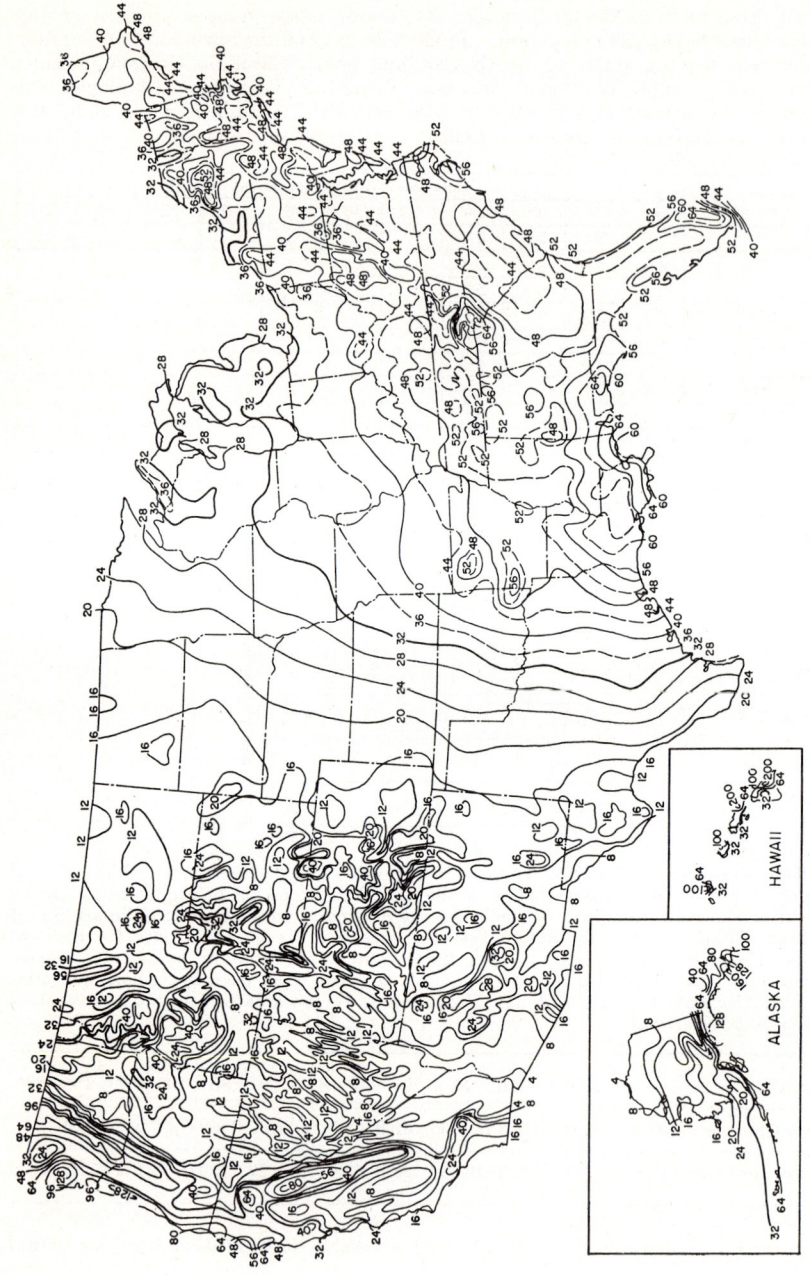

Fig. 9-33. Mean annual total precipitation in the United States (in inches). (*U.S. Weather Bureau*.)

B. Results of Space-Time Analyses of Rainfall

A great number of analyses have been prepared and are generally available. Here an attempt will be made to present a sample chosen to show the most useful ones and to suggest what others can be obtained if needed.

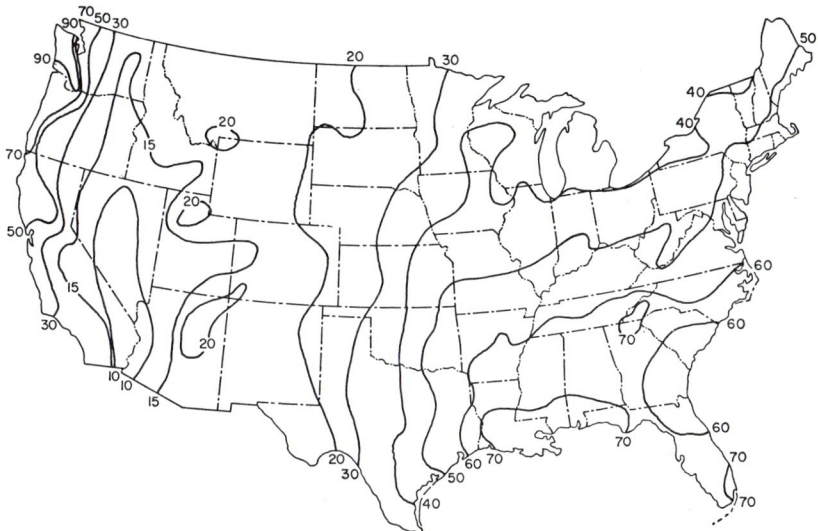

FIG. 9-34. Total precipitation in a year so wet that only one-eighth of the years receive more. (*After Visher* [29].)

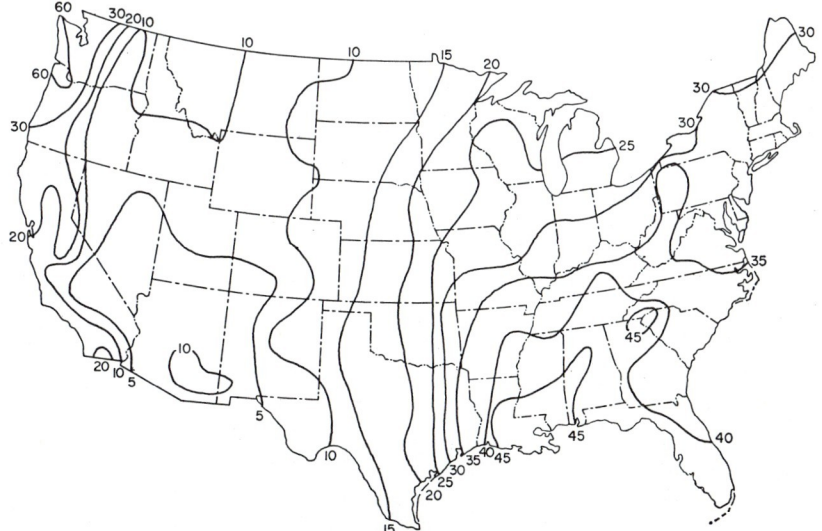

FIG. 9-35. Total precipitation in a year so dry that only one-eighth of the years receive less. (*After Visher* [29].)

9-42   RAINFALL

Mean-annual-precipitation (Fig. 9-33) maps are generally available, as are also mean seasonal, mean monthly, mean biweekly, and mean weekly amounts. Reliability of the annual amounts has received a great deal of study. Maps are available of median, quartile, quintile, octile (Figs. 9-34 and 9-35), and the greatest (Fig. 9-36) and least (Fig. 9-37) observed amounts.

Seasonal variations of monthly totals are quite different in different parts of the country (Fig. 9-38). Notable are the dry summers on the West Coast, the evenness of

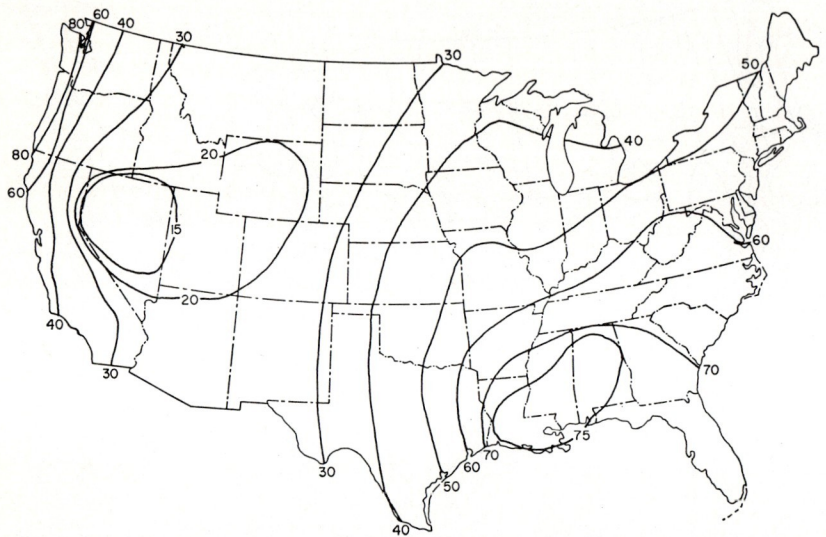

FIG. 9-36. Maximum annual precipitation.   (*After Lackey* [30].)

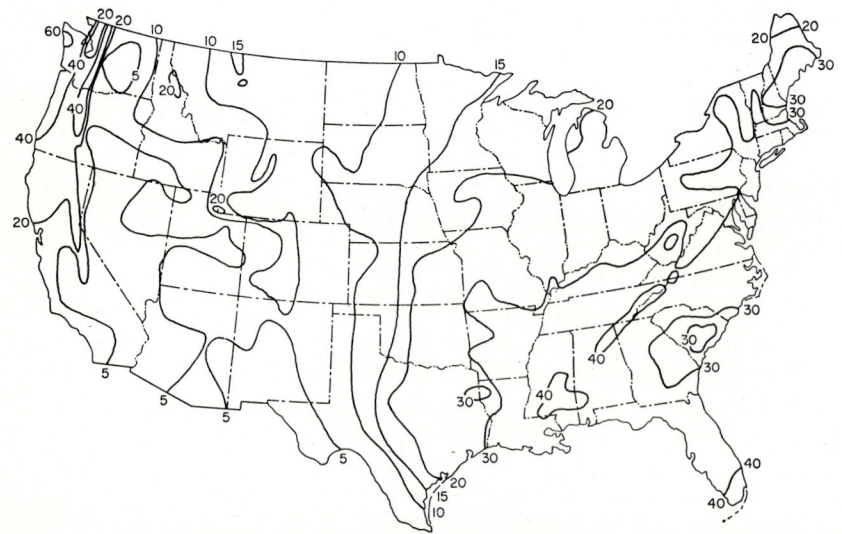

FIG. 9-37. Least annual precipitation.   (*After Lackey* [30].)

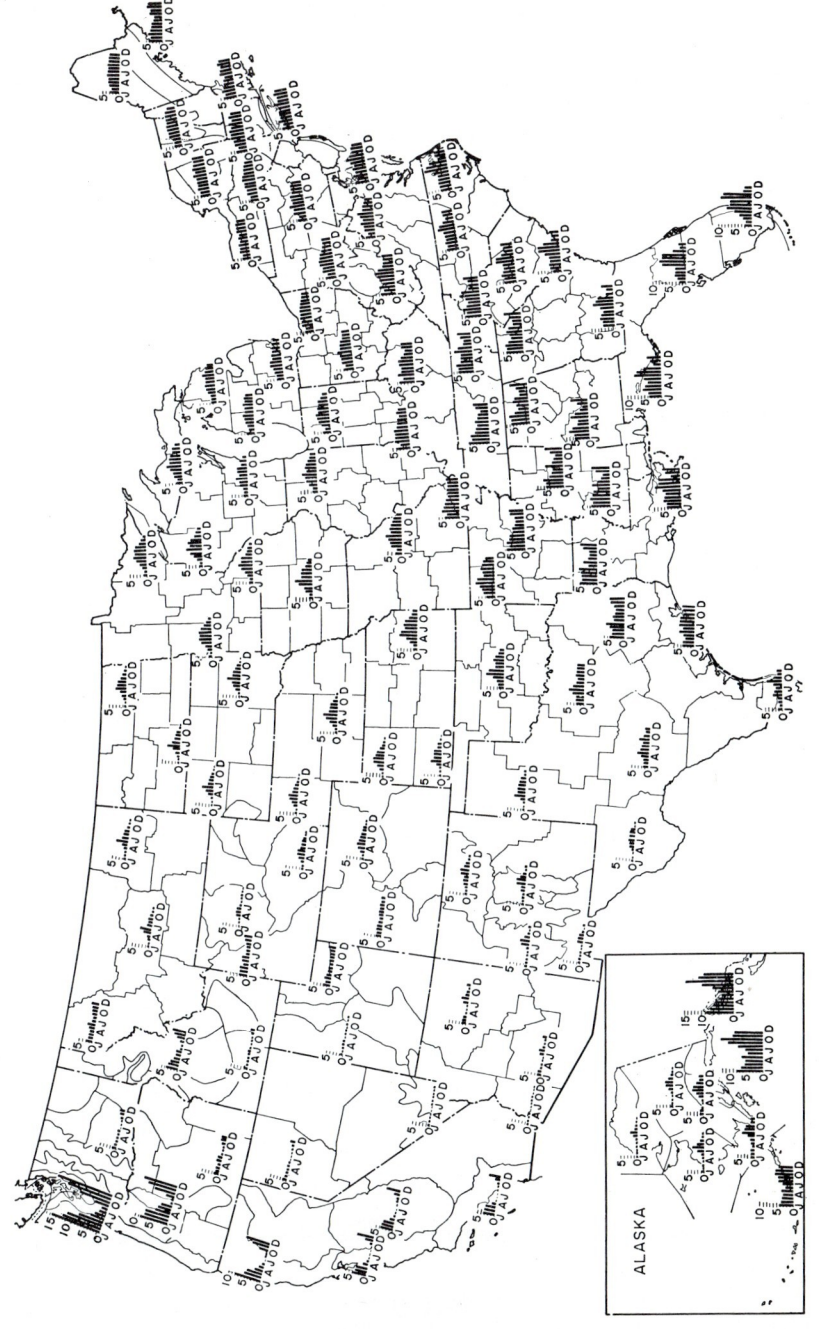

Fig. 9-38. Month-to-month variation of precipitation in the United States. (*U.S. Weather Bureau.*)

9-43

RAINFALL

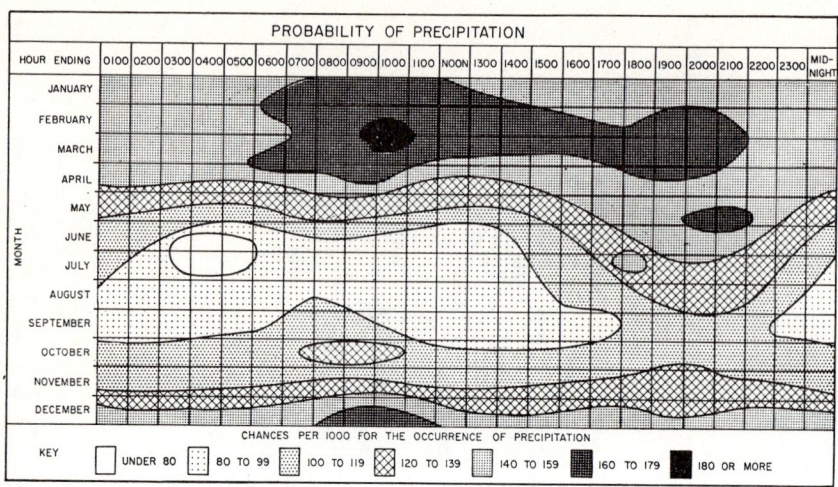

Fig. 9-39. Probability of precipitation occurrence by hours at Washington, D.C. (*U.S Weather Bureau.*)

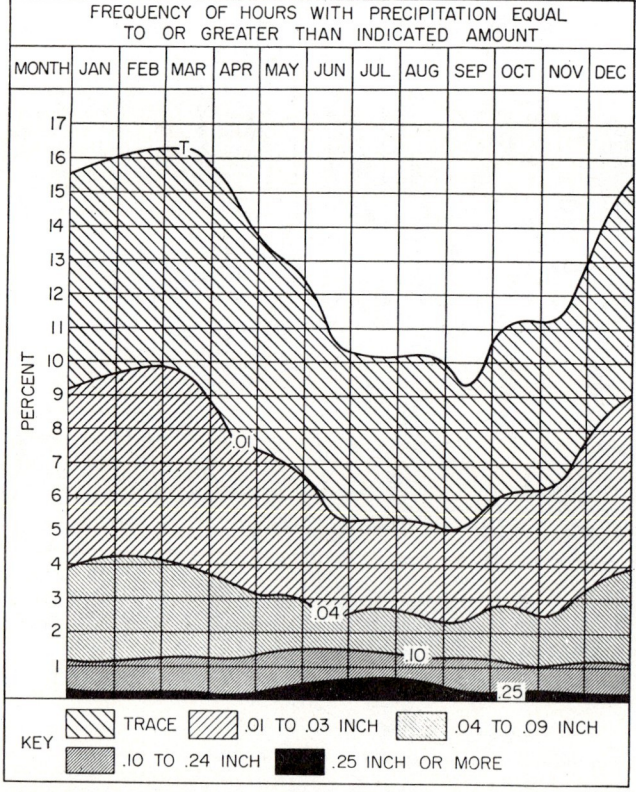

Fig. 9-40. Frequency of hourly amounts of precipitation at Washington, D.C. (*U.S. Weather Bureau.*)

SPACE-TIME CHARACTERISTICS OF RAINFALL 9–45

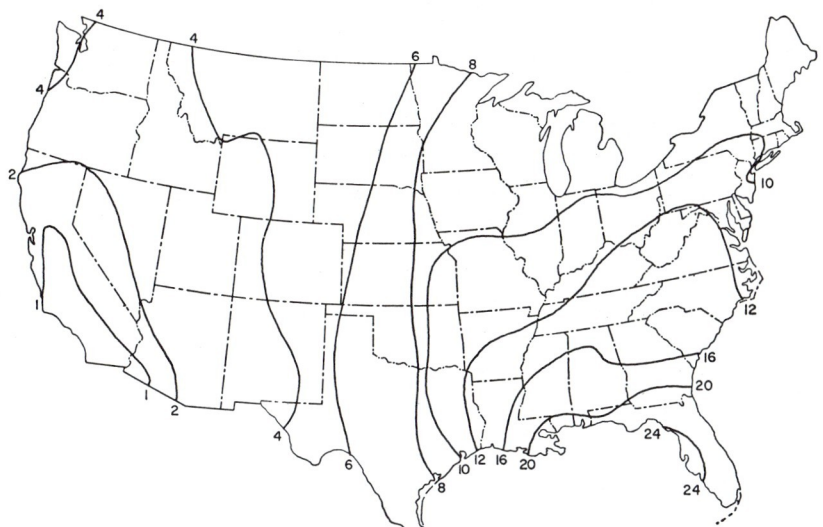

Fig. 9-41. Normal warm-season daytime precipitation. (*After Kincer and Visher* [29].)

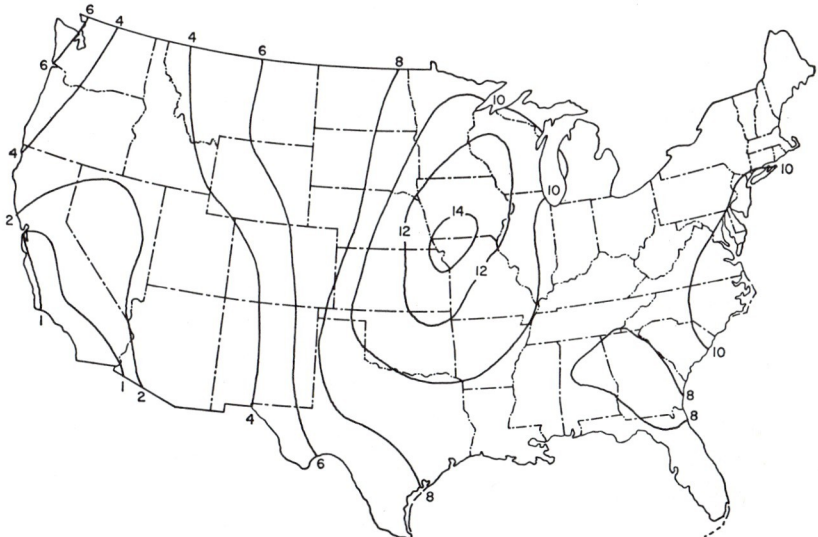

Fig. 9-42. Normal warm-season nighttime precipitation. (*After Kincer and Visher* [29].)

monthly totals of precipitation throughout the year in New England, the winter and spring maxima from the Central Gulf to the Ohio Valley states, and the sudden transition from June drought to July thunderstorms in Arizona. Seasonal variation by amounts has been worked up for only a few stations (Figs. 9-39 and 9-40). Many places are similar to Washington, D.C., in that they experience larger totals of rainfall in summer but have many more hours of light precipitation in winter.

Twenty-four-hour maxima for each month are worked up by first-order Weather

Bureau stations routinely and published in *Local Climatological Summary*. The calendar-day maxima for each of the 12 months during the period of record at several thousand cooperative stations over the United States are published in *Weather Bureau Technical Paper* 16 [31].

Diurnal variations in precipitation occurrence and precipitation amount (Figs. 9-41 and 9-42) have been intensively studied. The nocturnal maximum in the Midwest as compared with the late-afternoon maximum in the eastern part of the country has invoked several explanations.

The amount of precipitation per rainy day shows a significant variation from one part of the country to another (Fig. 9-43).

The maximum amounts that have occurred in each of the 12 months for 1, 2, 3, 6, 12, and 24 hr at about 2,500 of the recording rain gages in the United States is being prepared in *Weather Bureau Technical Paper* 15 [33]. The results are mostly completed for the eastern half of the country and for certain states in the West.

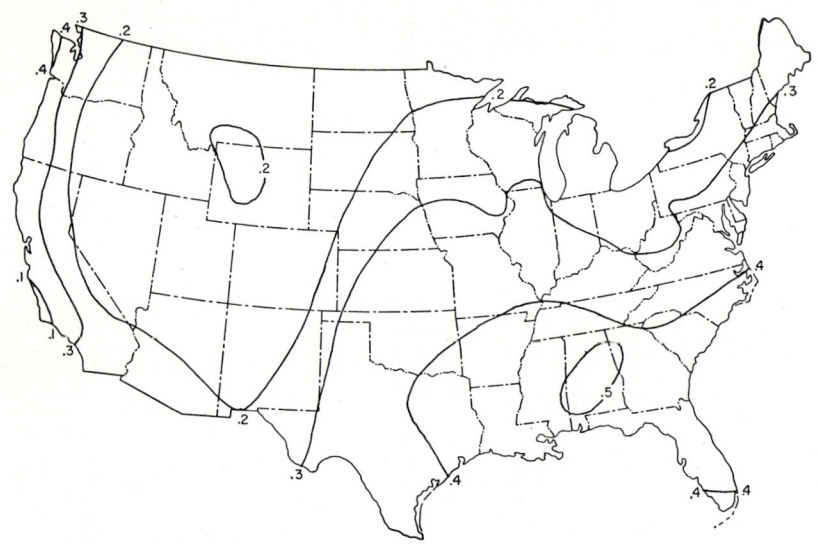

Fig. 9-43. Normal precipitation per rainy day. (*After Miller* [32].)

Maximum recorded values for periods of less than an hour are available for many stations. Typical values appear in Table 9-1. The maximum point values observed anywhere in the world are summarized in Fig. 9-44. Maximum depth-area-duration values observed anywhere in the United States are presented in Table 9-2. These values are the greatest of those compiled by the Corps of Engineers [36] and the Bureau of Reclamation.

Cycles in precipitation data have received much study. There seems to be a human tendency to see cycles in any sort of data. Experiments have been made wherein a set of random numbers was given to a series of technically trained persons, many of whom thought they detected regular cycles in the data. If cycles of less than a geologic age are present, their total magnitude in precipitation amount is probably quite small. The ones with the most supporting evidence, other than 24-hr and 365-day, are those related to the sunspot cycle of about 11 years.

Trends have also received much attention. It is certain that there are years with marked flooding over large areas and other years with much dryness. A definite trend of winter temperatures at middle- and high-latitude stations in the past half century has been demonstrated. However, there is no reliable means at present of projecting any trend very far into the future. Therefore, for practical planning,

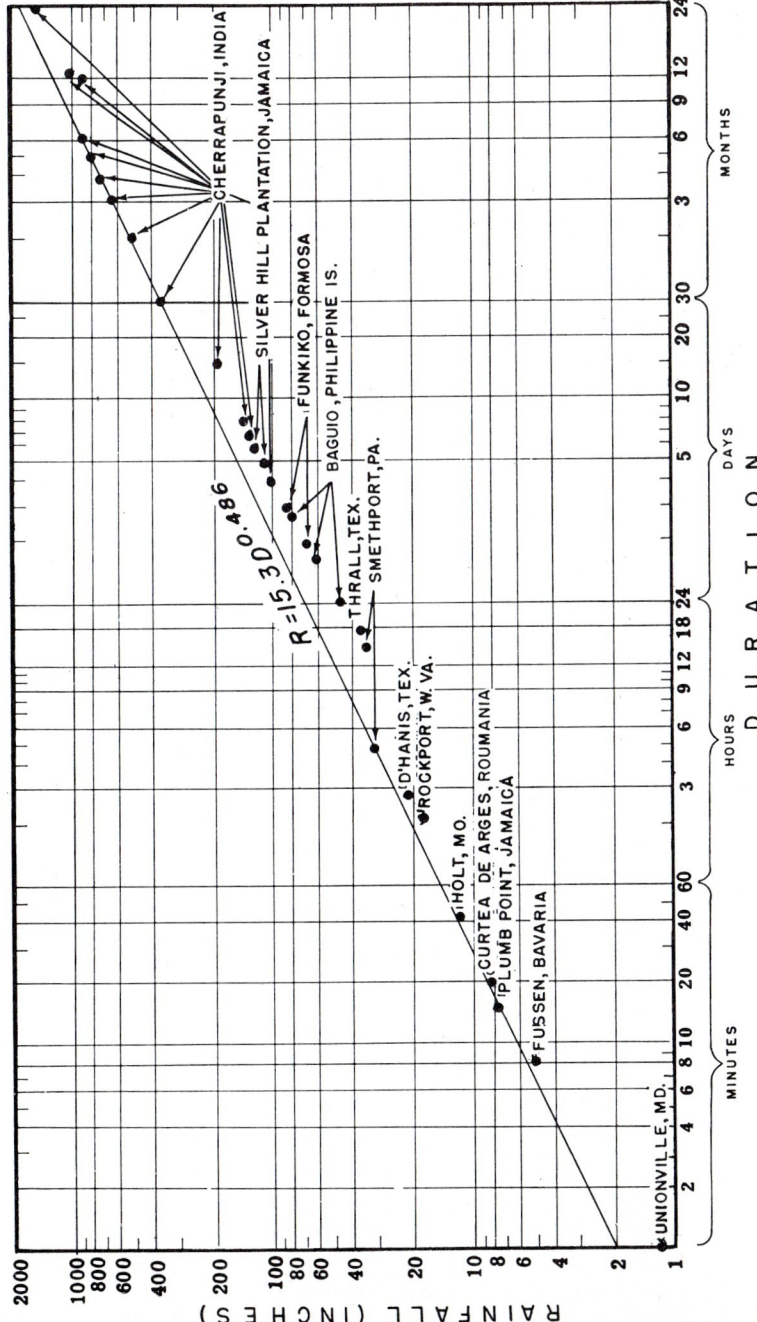

Fig. 9-44. The world's record point rainfall values. (*After Jennings* [35].)

Table 9-1. Maximum Recorded Rainfall*

Station	Duration, min			
	5	15	30	60
Boston, Mass............	0.56 8/7/08	1.12 8/7/08	1.45 8/7/08	1.80 8/24/01
Chicago, Ill..............	0.64 7/15/06	1.31 9/13/36	2.03 7/7/21	2.81 7/6/43
Atlanta, Ga..............	0.88 6/10/33	1.64 6/10/33	2.43 8/11/26	3.23 8/20/14
Denver, Colo.............	0.91 7/14/12	1.54 7/14/12	1.72 7/14/12	2.20 8/23/21
Seattle, Wash............	0.29 8/24/21	0.52 6/23/17	0.61 9/11/07	0.80 9/11/07
Los Angeles, Calif........	0.44 1/14/08	0.81 2/18/14	1.12 2/18/14	1.51 2/18/14

* After Shands and Ammerman [34].

Table 9-2. Maximum Observed United States Rainfall
(Revised June, 1960)

Area, sq mi	Duration, hr						
	6	12	18	24	36	48	72
10	24.7a	29.8b	36.3c	38.7c	41.8c	43.1c	45.2c
100	19.6b	26.3c	32.5c	35.2c	37.9c	38.9c	40.6c
200	17.9b	25.6c	31.4c	34.2c	36.7c	37.7c	39.2c
500	15.4b	24.6c	29.7c	32.7c	35.0c	36.0c	37.3c
1000	13.4b	22.6c	27.4c	30.2c	32.9c	33.7c	34.9c
2000	11.2b	17.7c	22.5c	24.8c	27.3c	28.4c	29.7c
5000	8.1bd	11.1b	14.1b	15.5c	18.7e	20.7e	24.4e
10,000	5.7d	7.9f	10.1g	12.1g	15.1e	17.4e	21.3e
20,000	4.0d	6.0f	7.9g	9.6g	11.6e	13.8e	17.6e
50,000	2.5gh	4.2i	5.3g	6.3g	7.9g	8.9g	11.5j
100,000	1.7h	2.5hk	3.5g	4.3g	5.6g	6.6j	8.9j

Storm	Date	Location of center	Assignment no.
a	July 17–18, 1942	Smethport, Pa.	OR 9-23
b	Sept. 8–10, 1921	Thrall, Tex.	GM 4-12
c	Sept. 3–7, 1950	Yankeetown, Fla.	SA 5-8
d	June 27–July 4, 1936	Bebe, Tex.	GM 5-6
e	June 27–July 1, 1899	Hearne, Tex.	GM 3-4
f	Apr. 12–16, 1927	Jefferson, Parish La.	LMV 4-8
g	Mar. 13–15, 1929	Elba, Ala.	LMV 2-20
h	May 22–26, 1908	Chattanooga, Okla.	SW 1-10
i	Apr. 15–18, 1900	Eutaw, Ala.	LMV 2-5
j	July 5–10, 1916	Bonifay, Fla.	GM 1-19
k	Nov. 19–22, 1934	Millry, Ala.	LMV 1-18

trends should be largely neglected. More reliance should be placed on frequency studies of the past and on the extremes at either end of a spectrum.

VI. DESIGN APPLICATIONS OF RAINFALL DATA

The degree of safety optimum for any structure depends, of course, on the economics associated with the cost of increased safety and the loss occasioned by failure. The flooding once a year of a small airfield may be acceptable as compared with the cost of an expensive drainage system. On the other extreme, it would be uneconomical and unethical to take a calculated risk of the failure of a large dam above a large city. Various design criteria are needed for different parts of this scale.

A. Frequency Analysis

Frequency analysis of point rainfall values are used for many purposes: design of storm sewers in urban developments, culvert design on highways, drainage design on airfields, and others. For many years such analyses were made on an *ad hoc* basis, each particular need being served by a specialized study. The first extended rainfall-frequency study in the United States was made by Yarnell [37] in the early 1930s and was presented in the form of a series of maps for several combinations of return periods and durations. The U.S. Weather Bureau has brought this work up to date in their recently prepared *Rainfall Frequency Atlas* [38]. Approximately 4,000 stations, or 20 times the number of stations available to Yarnell, were used to define the positions of the isolines. Recent statistical innovations include procedures for determining the frequencies on a seasonal basis and for extending the point frequency values to areal frequency values.

The Yarnell-type rainfall-frequency data cover durations from 5 min to 24 hr and return periods from 2 to 100 yr. For longer durations, the Miami Conservancy District [39] developed data for durations from 1 to 6 days and return periods from 15 to 100 yr. In this study records of deficient length were analyzed by the station-year method (Subsec. V-A-2) and the area covered is limited to that part of the continental United States east of the 103d meridian.

One method of selecting extreme-rainfall data, the *annual series*, involves selecting the maximum value for each calendar year of record. The other, *partial-duration series*, involves selecting the maximum N or more values from the N years of record, it being possible that more than one value would come from certain years. The annual series is more readily available since it is laborious to compile the partial-duration series. Table 9-3 gives empirical factors for converting the partial-duration series to the annual series. For example, if the 2-, 5-, and 10-year partial-duration-series values estimated from the maps (presented here) are 1.00, 2.00, and 3.00 in., respectively, the annual-series values are 0.88, 1.92, and 2.97 in., after multiplying by the conversion factors in Table 9-3. The *extreme-value distribution* used to analyze the rainfall data for the atlas is partly empirical and partly theoretical. From 1 to 10 years it is entirely empirical, based on the partial-duration series. For the 20-year and longer return periods, reliance was placed on Gumbel's analysis of annual-series data [40].

Table 9-3. Empirical Factors for Converting Partial-duration Series to Annual Series

Return period	Conversion factor
2-year	0.88
5-year	0.96
10-year	0.99

Individual stations are analyzed by the method described immediately above. Typical results of such analyses appear in Fig. 9-45.

The set of 12 maps presented in Figs. 9-45 to 9-57 on the partial-duration series represent data from more than 4,000 stations. The durations are for the true maxima

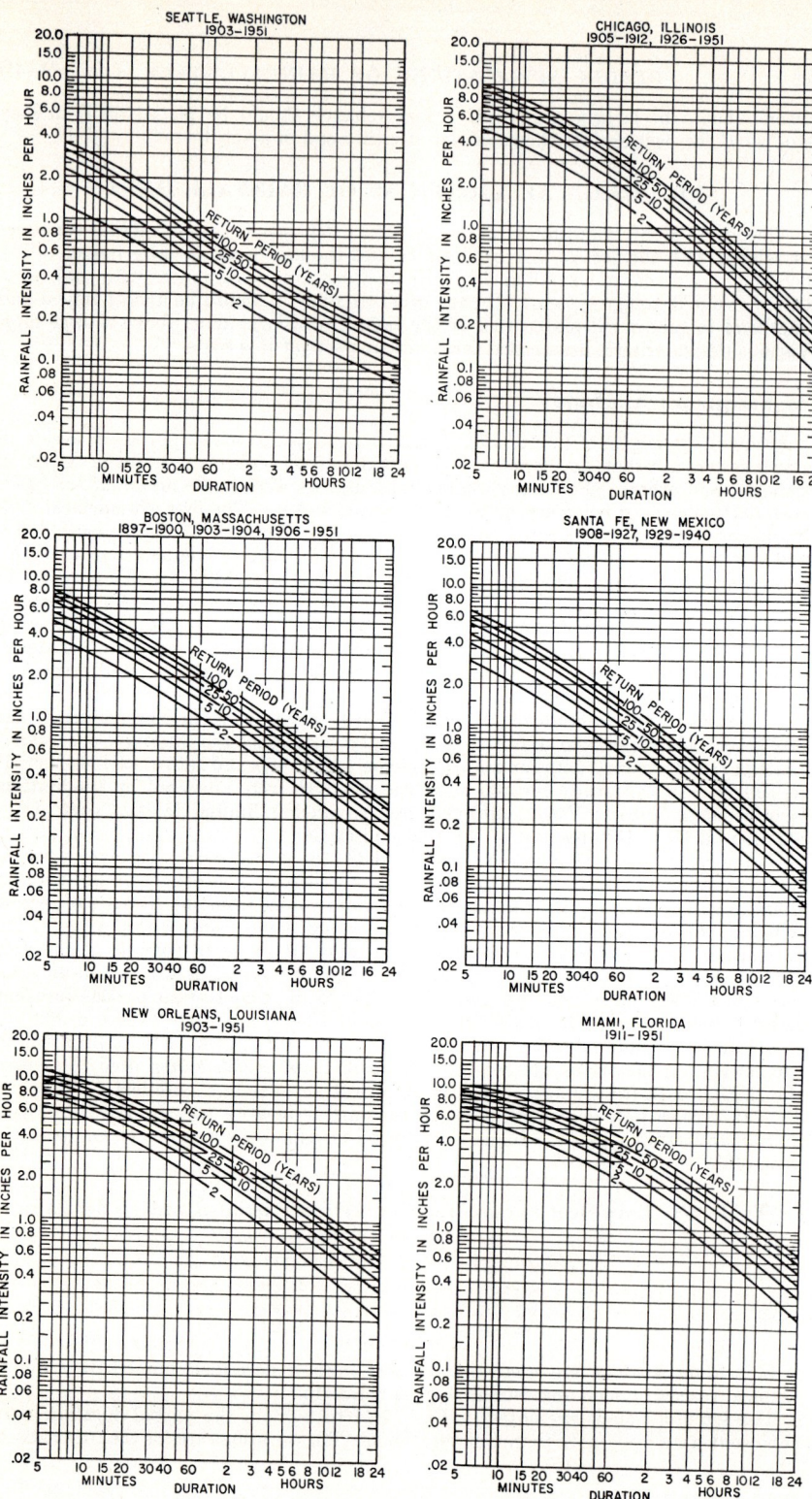

Fig. 9-45. Typical point rate-duration-frequency curves. (*U.S. Weather Bureau* [38].)

DESIGN APPLICATIONS OF RAINFALL DATA 9-51

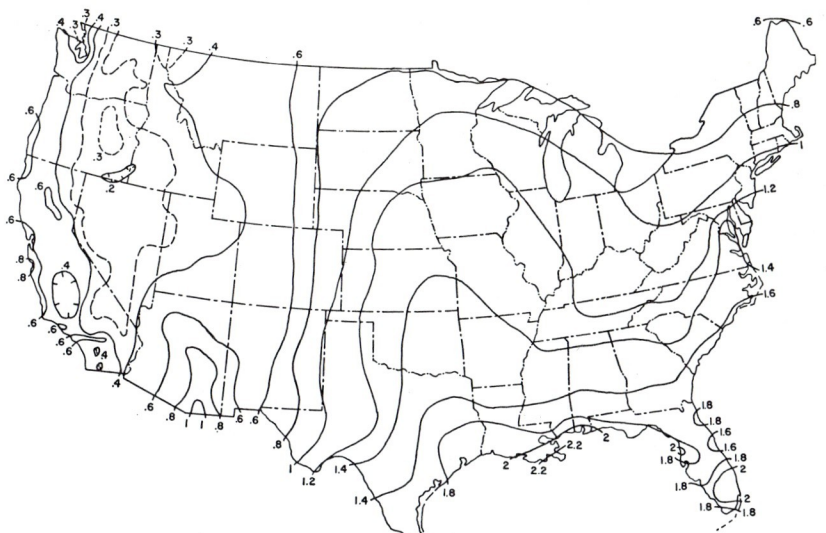

FIG. 9-46. Two-year 30-min rainfall (inches). (*U.S. Weather Bureau* [38].)

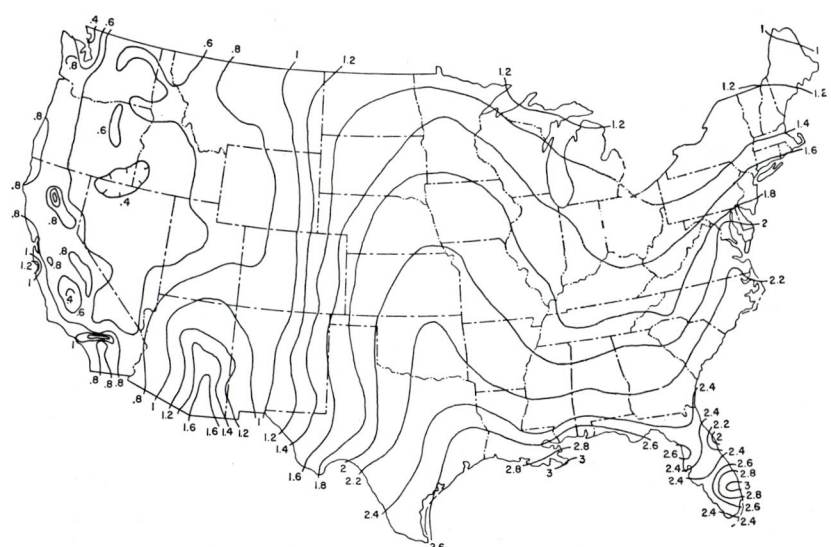

FIG. 9-47. Ten-year 30-min rainfall (inches). (*U.S. Weather Bureau* [38].)

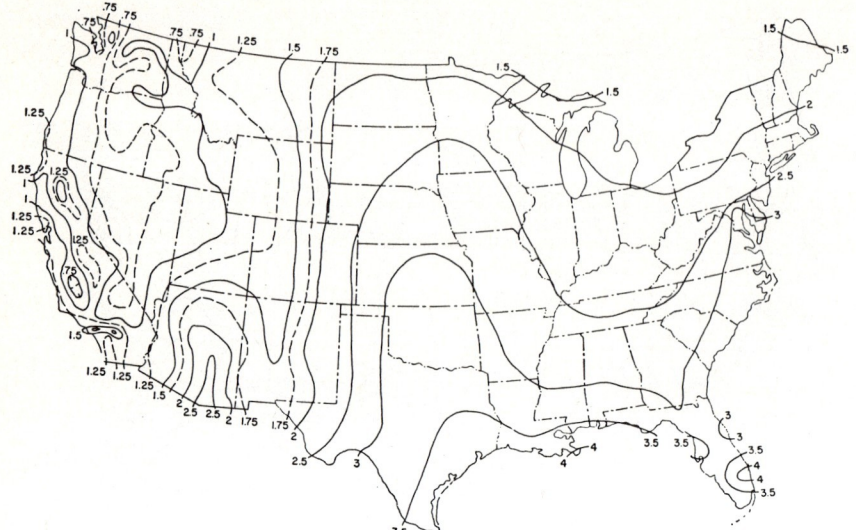

Fig. 9-48. 100-year 30-minute rainfall (inches). (*U.S. Weather Bureau* [38].)

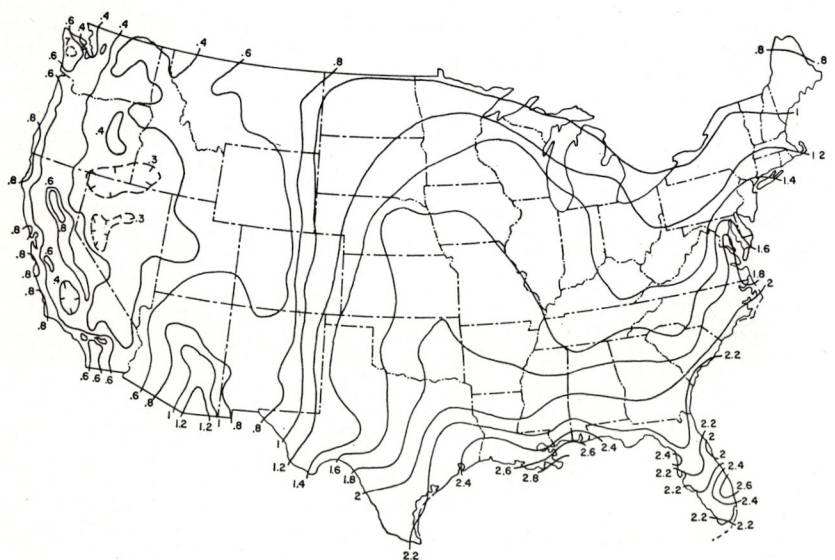

Fig. 9-49. Two-year 1-hr rainfall (inches). (*U.S. Weather Bureau* [38].)

DESIGN APPLICATIONS OF RAINFALL DATA 9–53

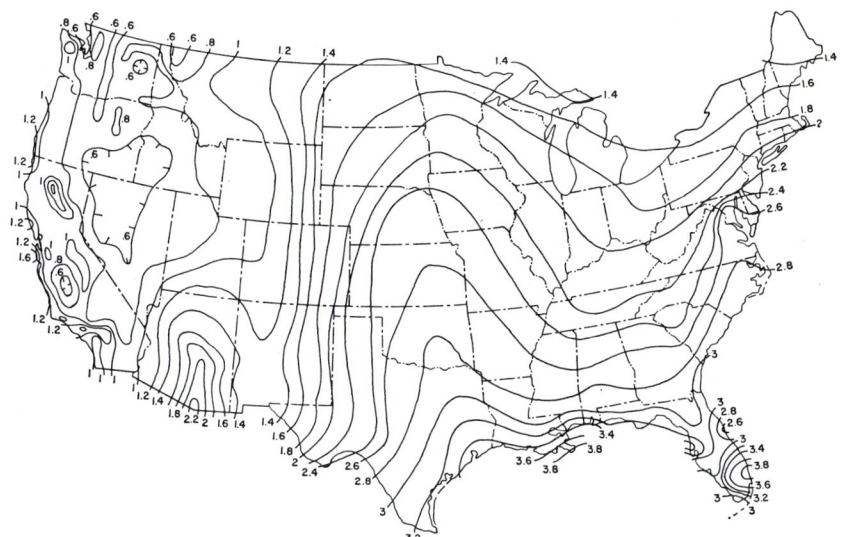

Fig. 9-50. Ten-year 1-hr rainfall (inches). (*U.S. Weather Bureau* [38].)

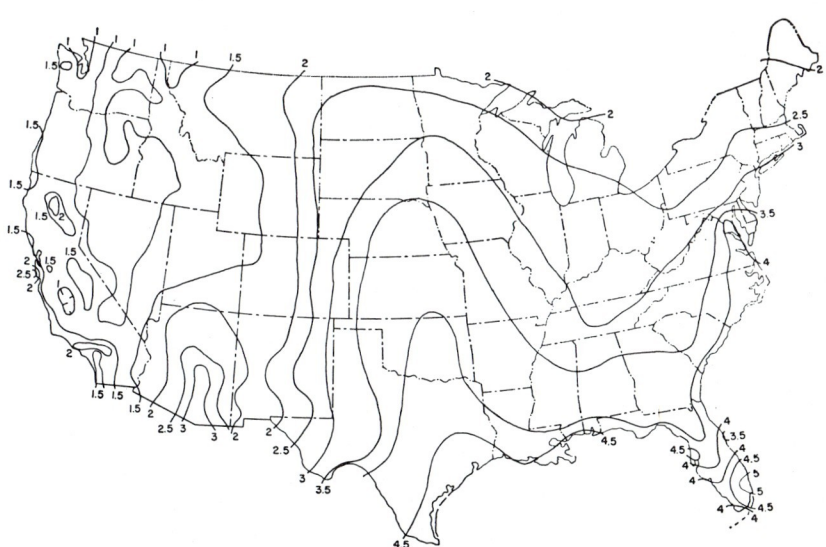

Fig. 9-51. 100-year 1-hr rainfall (inches). (*U.S. Weather Bureau* [38].)

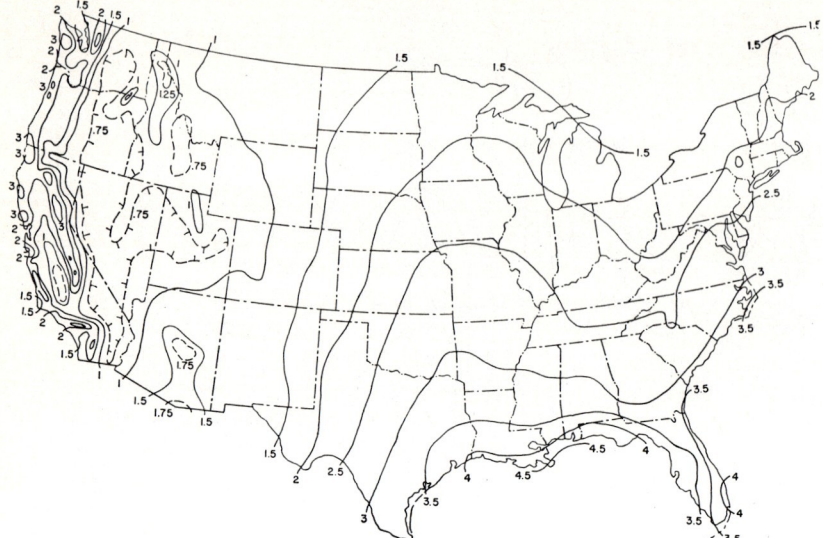

Fig. 9-52. Two-year 6-hr rainfall (inches). (*U.S. Weather Bureau* [38].)

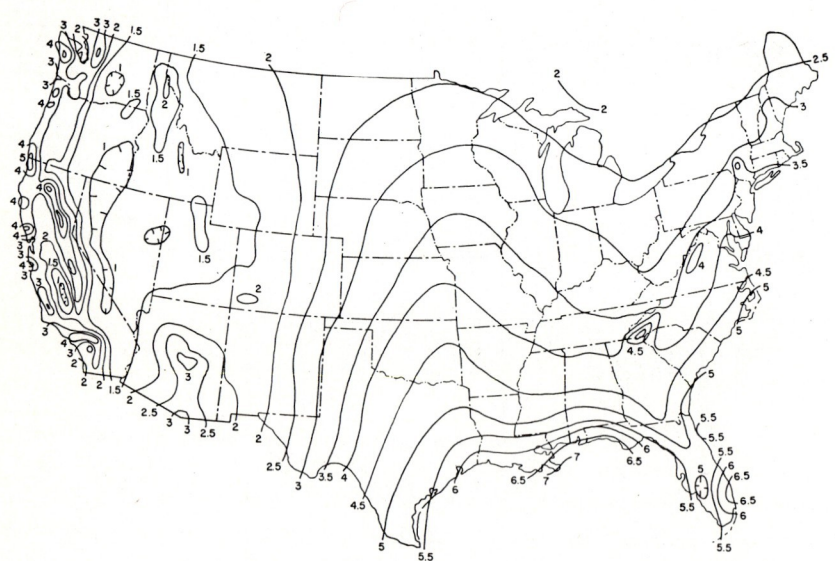

Fig. 9-53. Ten-year 6-hr rainfall (inches). (*U.S. Weather Bureau* [38].)

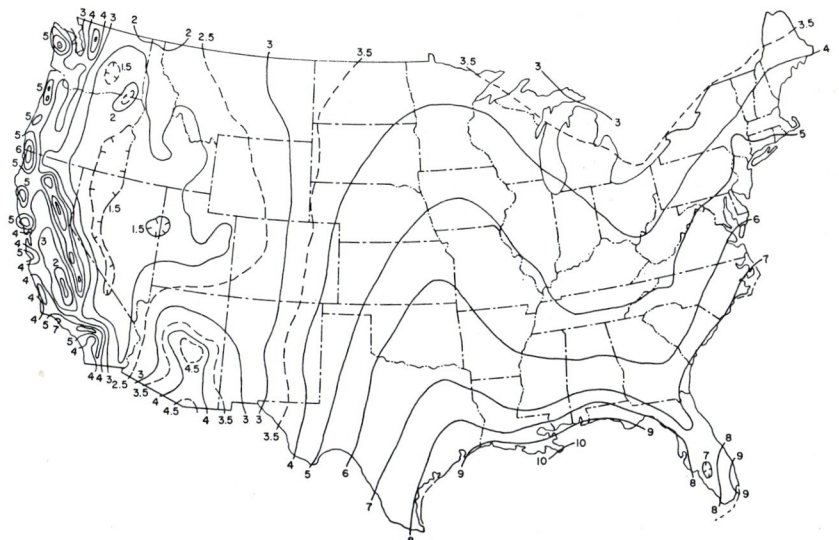

Fig. 9-54. 100-year 6-hr rainfall (inches). (*U.S. Weather Bureau* [38].)

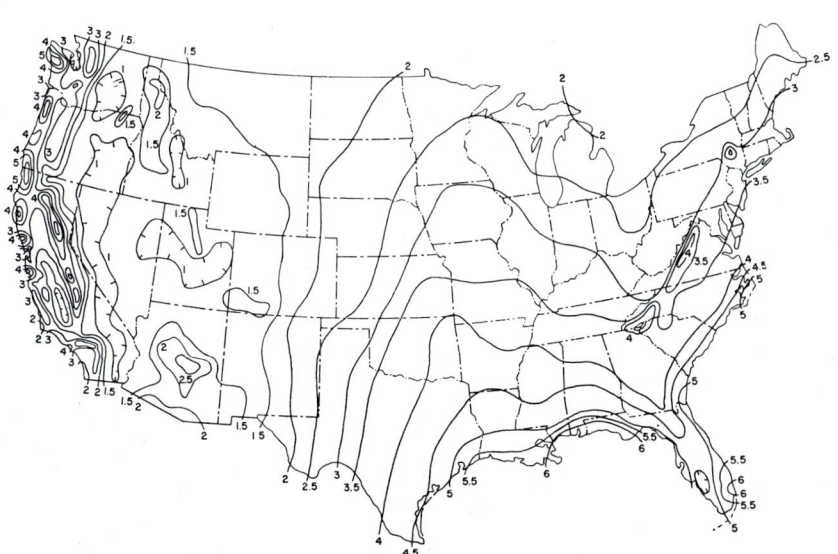

Fig. 9-55. Two-year 24-hr rainfall (inches). (*U.S. Weather Bureau* [38].)

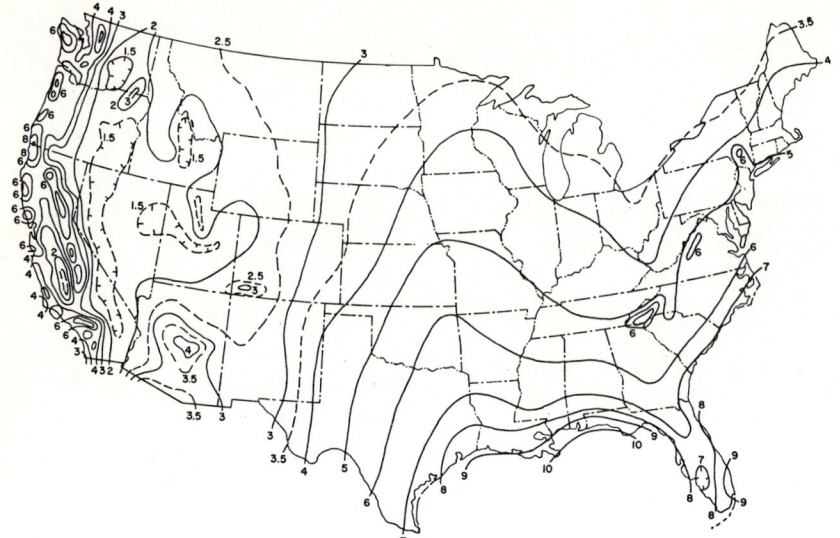

Fig. 9-56. Ten-year 24-hr rainfall (inches). (*U.S. Weather Bureau* [38].)

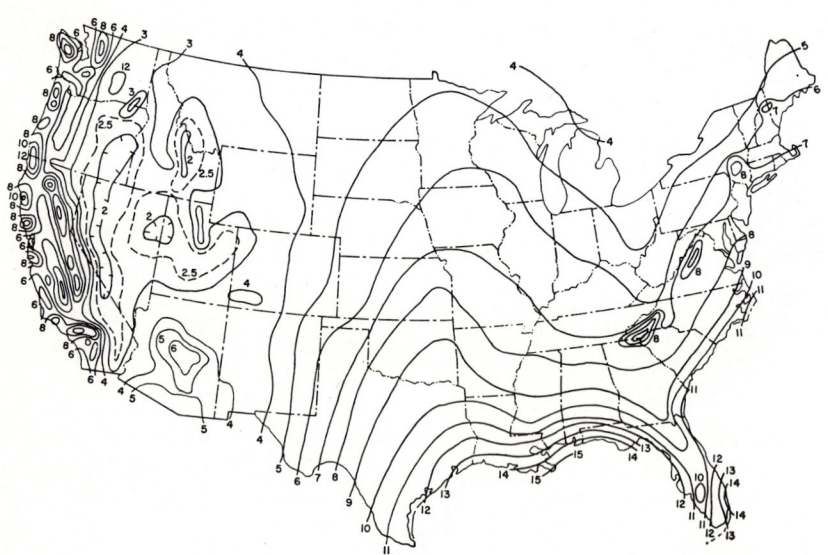

Fig. 9-57. 100-year 24-hr rainfall (inches). (*U.S. Weather Bureau* [38].)

periods, that is, for example, the 1,440-consecutive-minute period containing the maximum rainfall, not 24-clock-hour or calendar-day rainfall.

For interpolation among the durations and return periods given on these maps, it is suggested that a series of 12 values—one from each map—be read from a particular geographical location. Smooth curves drawn through these points on logarithmic paper provide a basis for estimating additional return periods and/or durations. The set of curves in Fig. 9-58 illustrates this procedure for the point at 41°N and 91°W.

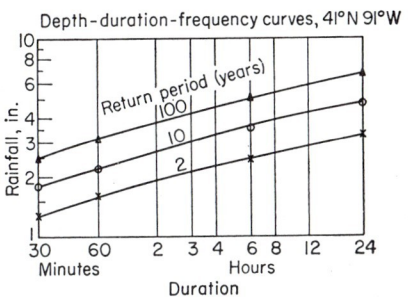

Fig. 9-58. Interpolation of depth-duration-frequency values.

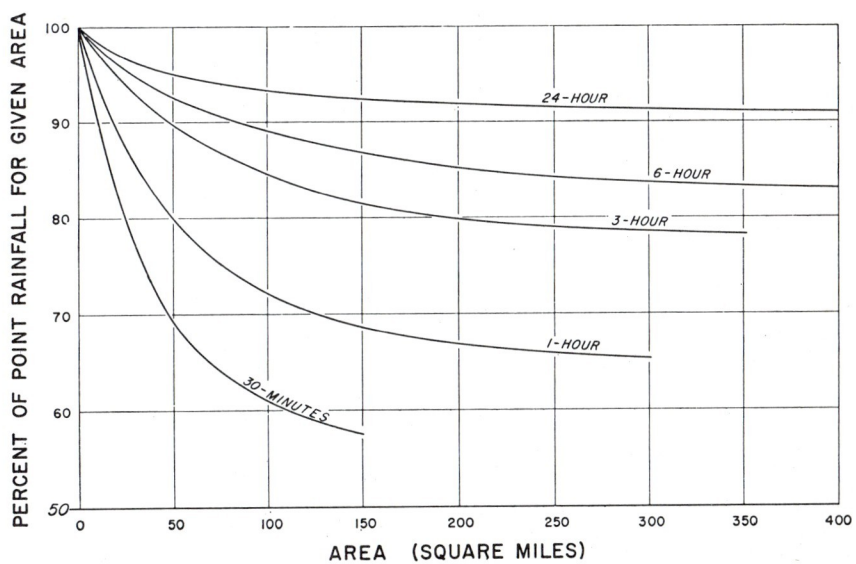

Fig. 9-59. Area-depth curves for use with duration-frequency values. (*U.S. Weather Bureau.*)

1. Area-Depth Relationship. The average area-depth relationship, as a per cent of the point values, can be obtained from the diagram of Fig. 9-59 for areas up to 400 sq mi. For example, the average value of 10-year, 6-hr rainfall in the vicinity of 40°N and 91°W is 3.45 in., but the average 6-hr depth over the drainage area would be less than 3.45 in. for the 10-year return period. Referring to Fig. 9-59, it is seen that the 6-hr curve intersects the area scale at 85 per cent. Accordingly, the 10-year, 6-hr average depth over 200 sq mi in the example would be 3.45 times 0.85, or 2.9 in.

2. Seasonal Variation. The frequency analysis presented thus far is based on the partial-duration series, that is, the first N (or more) maximum events for N years of record. Obviously, some months of the year contribute more events to this series than others, and in fact, some months might not contribute at all to the series. The main reason for concern with seasonal variation may be illustrated by the fact that

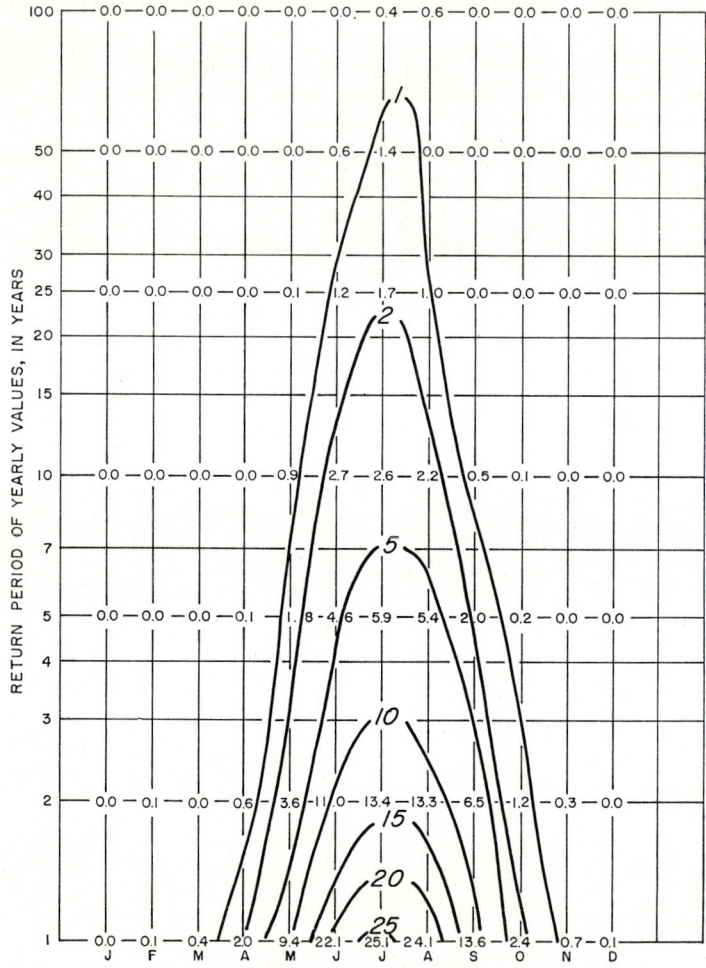

Fig. 9-60. Seasonal probability of intense rainfall of 1-hr duration (Great Lakes region). Probability in per cent of obtaining a rainfall in any month of a particular year equal to or exceeding the yearly return-period values taken from the isopluvial maps and diagrams. (*U.S. Weather Bureau.*)

the 10-year, 1-hr rain may be associated with a summer thunderstorm, with considerable infiltration, whereas the 10-year flood may come from a lesser storm occurring on frozen or snow-covered ground in the late winter or early spring. Determination of the season a particular rainfall will occur then becomes an essential preliminary for design purposes.

Seasonal-variation diagrams, similar to the one presented in Fig. 9-60, are available for several durations for five subregions of the United States east of 90°W. Suppose it is required to determine the probability of a 5-year, 1-hr rainfall for the months of June to August. From Fig. 9-60 the probabilities of each month are interpolated to be 5, 7, and 6 per cent, respectively. In other words, the probability of occurrence of a 5-year, 1-hr rainfall in June of any particular year is 5 per cent; for July, 7 per cent; and for August, 6 per cent.

3. Calculated Risk. Use of a *return period* (i.e., *recurrence interval*), regardless of how large (small probability), means that the design engineer is willing to take a

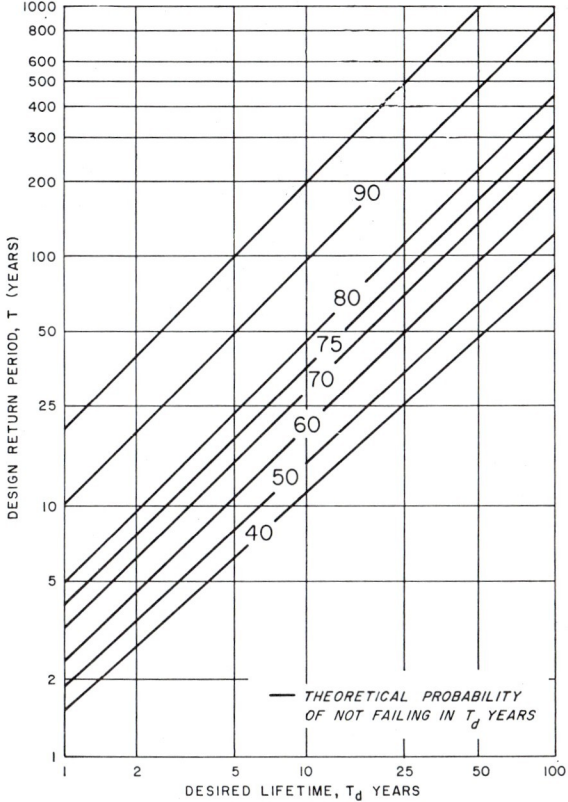

Fig. 9-61. Calculated-risk diagram. (*U.S. Weather Bureau.*)

calculated risk. However, there is a good chance that the large return period of rainfall will be exceeded at least once in N years. The probability of a rainfall having a given return period occurring at least once in N years is given by the formula

$$P = 1 - q^N \tag{9-9}$$

where q is equal to the probability of not occurring in a particular year. As an example, if an event has a probability of ⅕ for each trial (5-year return period), what is the probability that it will occur at least once in 3 years? Here, q is $1 - \frac{1}{5}$, N is 3, and P is $1 - (\frac{4}{5})^3$, or $^{61}\!\!/_{125}$. This means that there is approximately 1 chance out of 2 that the 5-year value will be exceeded once in the next 3 years.

The relationship between the calculated risk and the design return period, prepared from theoretical computations, is illustrated in graphical form in Fig. 9-61. If, for example, the design engineer wants to be approximately 90 per cent certain that a 10-year or greater rainfall does not occur in the next 10 years, he should design for the 100-year rainfall.

The U.S. Weather Bureau Rainfall Frequency Atlas [38] contains 49 maps for the contiguous United States for return periods from 1 to 100 years and durations from 30 min to 24 hr.

4. Synthesis of Frequency Regimes. A certain amount of work has been done on relating various meteorological parameters to the frequency regimes. Such relationships can be used, for example, for estimating the 2-year, 1-hr value of rainfall in a region where there are no observations taken oftener than once a day (Fig. 9-62).

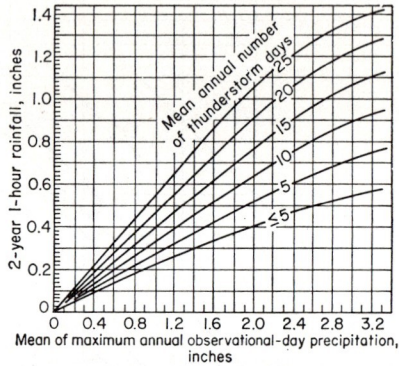

FIG. 9-62. Synthetic depth-frequency relation. Diagram for estimating 2-year, 1-hr rainfall. (*U.S. Weather Bureau.*)

B. Frequency Formulas

Several formulas have been proposed for expressing rainfall-frequency relations. Most of the early ones were simple in form; they considered only one or two of the causal factors and were derived from scanty localized data. It has been suggested that graphic portrayal is as accurate as the data warrant and that the development of a formula is often a mere exercise in curve fitting. At times a formula may be necessary for use with electronic computers. Table 9-4 presents a summarization of the principal formulas and curves that have been suggested.

C. Storm Transposition

Storm transposition consists of the bodily movement of an observed *depth-area-duration pattern* from the basin or area in which it occurred to another basin or area. Transposition was developed rather intensively by the Miami Conservancy District during the 1920s and early 1930s. Patterns for transposition were developed by that agency, but most examples nowadays are taken from storm rainfall patterns of the U.S. Army Corps of Engineers, Bureau of Reclamation, and Soil Conservation Service. Also, the procedure has largely been adopted for Probable Maximum and Standard Project studies and is infrequently used by itself. Its greatest usefulness is in setting a definite absolute lower limit for probable maximum precipitation.

Limits of transposition are usually determined by a hydrometeorologist. Topography, sources of moisture-bearing winds for the area of the observed storm as compared with the basin under investigation, storm tracks, etc., are factors considered in setting the limits of transposition.

Table 9-4. Summary of Rainfall Intensity-Duration-Frequency Formulas, Curves, and Charts*

(1) Item	(1) Locality to which formula, curve, or chart applies	(2) Formula or chart proposed by:	(3) Date	(4) Background date	(5) Formulas: Amount = I Intensity = i	(6) Limits of t, min	(7) Frequency, once in no. of years	(8) Remarks
1	United States east of Rockies	A. F. Meyer [42]	1917	Excessive precipitation, 1,962 storms at 43 stations 1896–1914	$I = \dfrac{At}{t+b}$ $i = \dfrac{A't}{t+b'}$	5 to 120	1, 2, 5, 10, 25, 50, 100	Formulas are given for five geographic groups and for the frequencies noted in col. 7.
2	All of contiguous United States	D. L. Yarnell [37]	1935	Excessive precipitation from 211 recording gages—all data up to and including 1933	62 isohyetal charts	5, 10, 15, 30, 60, 120, 240, 480, 960, 1,440	2, 5, 10, 25, 50, 100	Values can be read directly from charts for any geographic location. Values for Far West to be used with considerable caution.
3	United States east of 101st meridian	Merrill Bernard [43]	1932	Data of Meyer and of Morgan	$i = \dfrac{KF^x}{t^n}$	120 to 6,000–9,000 (12 hr to 6 days)	5, 10, 15, 25, 50, 100	Values of K, x, and n are given for any location east of 101st meridian.
4	All of contiguous United States	U.S. Army Corps of Engineers [44]	1945	Yarnell's data and other data	Intensity-duration curves	0–240 min	Values along any one curve have same frequency	Requires determination of 60-min average intensity for the desired frequencies at the point under design. Maximum 60-min rate on curves: 4 iph.
5	United States east of 103d meridian	A. E. Morgan, Miami Conservancy District [39] (Ohio 1936 rev.)	1917	Revised 1936 edition based on all available rain data through 1934	Graphs of time-area-depth curves for northern and southern states	1, 2, 3, 4, 5, 6 days	12, 25, 50, 100	Isopluvial charts given maximum 1-, 2-, 3-, 4-, 5-, or 6-day rainfall for frequency in col. 7. report also gives precipitation vs. area for every storm.
6	All of contiguous United States	D. M. Hershfield, U.S. Weather Bureau [38]	1961	Based on all available data	49 isopluvial maps	30, 60, 120, 180, 360, 720, 1,440 min	1, 2, 5, 10, 25, 50, 100 years	Return periods based on partial-duration series. Durations are for in-consecutive minutes containing maximum rainfall.

* Suggested by Jens [44].

D. Probable Maximum Precipitation

For design of spillways on large dams it is necessary to use a much higher precipitation estimate than is the case for almost any other structure. In the first place, the structure itself is often very expensive, so that its loss would involve a large economic expense. Also, the construction creates a risk of great loss of life and property downstream that otherwise would not exist. Often the downstream interests involved receive little or no direct benefit from the construction of the project. Another important consideration is that failure of such a structure with a dramatic, publicized event might adversely affect public opinion so as to jeopardize a whole public or private program. For these reasons, design engineers have asked meteorologists for estimates of the probable maximum precipitation as the basis for design of such spillways.

The estimates represent the best judgment of the meteorologists of the realistic upper limit of precipitation that can occur. Many meteorologists have thought that there is no upper limit on precipitation amount—that any given amount can conceivably occur. Such a view is not realistic mathematically or physically speaking, since it is certainly possible to put an upper bound on the amount of precipitation that can occur. And if it is possible to fix an upper bound, there must exist a least upper bound, which might be called the *possible maximum*, or *probable maximum*, *precipitation* (PMP). (See Sec. 25-I.)

A procedure to determine PMP for regions of little to moderate topographic variation has been developed and widely applied to give one approximation [45–46]. The procedure involves two steps: (1) the preparation of probable maximum depth-area-duration curves representative of the region of which the basin under study is a part, and (2) the selection by means of these curves of a pattern storm for use in the basin. The probable maximum depth-area-duration curves are prepared by simple enveloping of the moisture-adjusted, depth-area-duration values for all storms considered transposable to the region under study. The *moisture-adjustment factor* is the ratio of the maximum total moisture in an atmospheric column of unit cross section in the region to the total moisture in a similar column that occurred during the storm. Obviously, the maximum values for different-size areas and for different durations may and probably do come from different storms. The pattern storm depends on the size of area and, to a lesser extent, on the duration deemed hydrologically critical. For areas up to a few hundred square miles in size not too peculiar in shape, the pattern is not usually very critical and a single pattern may be applicable to many basins. For areas of tens of thousands of square miles or more in size and areas of unusual shape, close examination of pattern may be necessary to determine one that is realistic. In ordinary cases the pattern will probably be based on one of the storms that gave the enveloping depth-area-duration values near the area size and duration of hydrologic importance. All depth-area-duration values for the storm will be multiplied by a factor, which will be the factor that brings the values of certain points to tangency with the enveloping curves. In cases of unusual basin shape or of large topographic variation, the pattern storm may be based on combinations of different parts of actual storms.

Advantages of the procedure used are several. It provides empirical or statistical controls. The values are directly related to the largest that have occurred. The experience of a basin is extended through transposition. The use of actual storms for patterns ensures realism in that nature's integrations are used rather than hard-to-justify synthetic ones. Thus overcompounding of probabilities is minimized.

Several features of the results obtained are worthy of note. The highest estimates of probable maximum precipitation often exceed the greatest value of observed precipitation in certain basins by only a small per cent. In other basins they may be several times as great as the maximum observed. The greatest maximizing process for a given basin is *storm transposition*. If a precipitation value several times as large as any over a given problem basin has been observed over a nearby basin, then it is considered that the observed isohyets in the actual storm can be transferred, or

DESIGN APPLICATIONS OF RAINFALL DATA   9–63

transposed, so as to indicate the maximum amount over the problem basin. Obviously, there are geographical, topographical, and synoptic limits to how far this transposition can be extended. A difficulty arises in areas that are small in geographical extent but are obviously not homogeneous meteorologically with other areas. Then simple transposition is obviously impossible, but a lack of any substitute for transposition will certainly result on the average in values that are very much

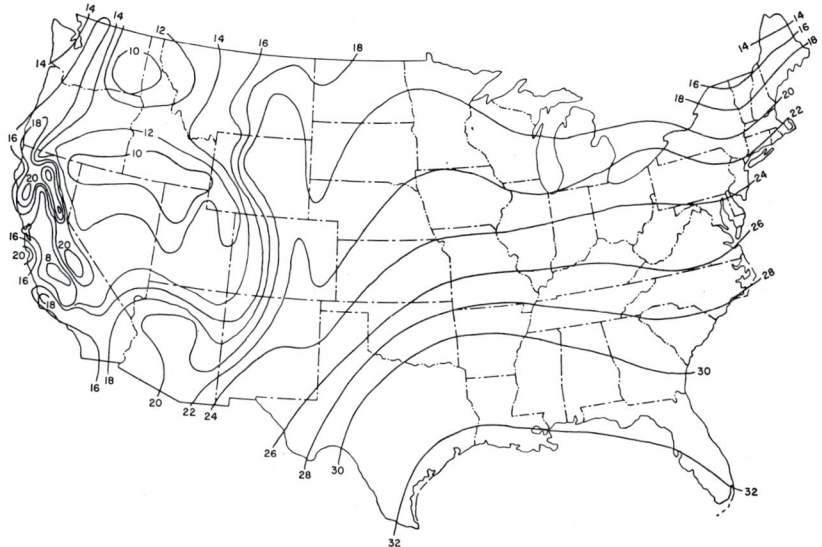

FIG. 9-63. Probable maximum precipitation for 24 hr and 200 sq mi. (*U.S. Weather Bureau.*)

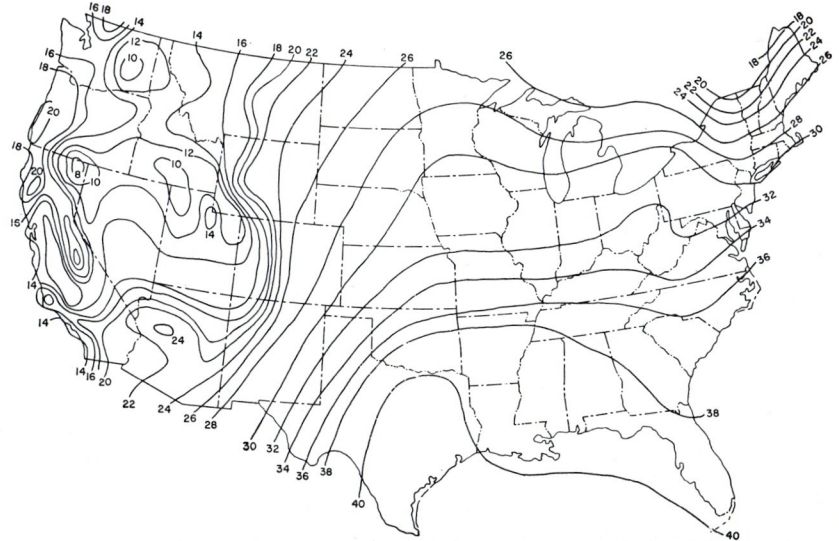

FIG. 9-64. Probable maximum precipitation for 24 hr and 10 sq mi. (*U.S. Weather Bureau.*)

lower, comparatively, than would be obtained from transposition in a large homogeneous area. The principal place this difficulty appears is in rugged mountainous areas such as the West Coast area of the United States. The estimation of probable maximum precipitation in mountainous areas is usually considered to be much more uncertain than that in large homogeneous areas.

In regions of pronounced orography, such as the West Coast states of the United States, the procedures described above cannot apply. As noted, transposition is severely limited by the lack in similarity of orography in nearby areas. Patterns are likely to be based on observed storms in the basin under study. Some estimates have been based on the transposition of percentages of the mean annual pattern or various statistical parameters derived from analyses of station records. Recently, an analysis

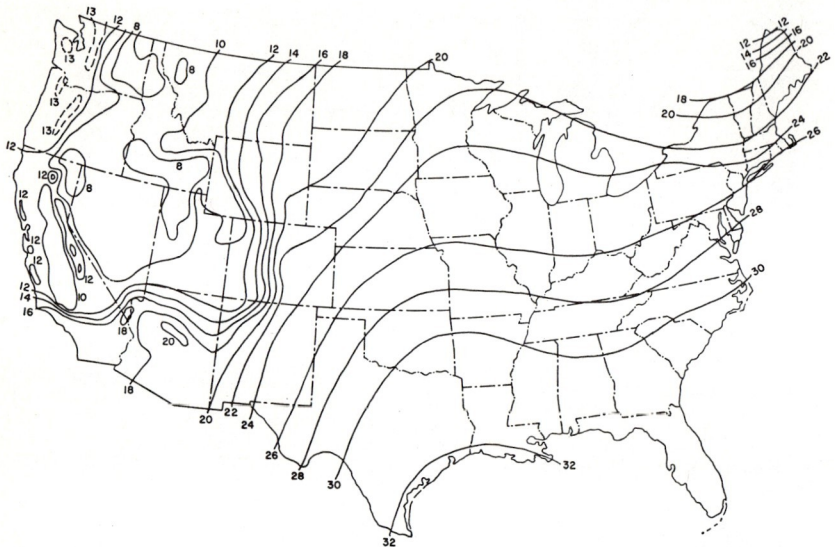

FIG. 9-65. Probable maximum precipitation for 6 hr and 6 sq mi. (*U.S. Weather Bureau.*)

has been used that combined separate maxima of estimates of orographic and convergence precipitation, the former being based on the strongest continued upslope winds that have occurred and the latter on an analysis similar to that used in regions of little topography. Controls based on empirical and statistical considerations are still quite desirable.

Generalized charts of probable maximum precipitation, together with its seasonal variations, have been prepared for smaller areas for the United States (Figs. 9-63 to 9-65).

E. Standard Project Storms

The standard project storm has been developed as a guide to the design of projects not requiring the large degree of safety that the probable maximum precipitation represents. Generalized procedures are almost always used. An attempt is made to underdevelop a certain number of storm values as uniformly as possible with respect to geographical location and season of the year. Since the requirements are frequently closely allied with agency policy, the particular offices involved should usually be consulted as to criteria [47–49].

As a reference, the specifications of a standard project storm (SPS) recommended by the U.S. Army Corps of Engineers [44] are quoted as follows (see Sec. 25-I):

The "standard project storm" estimate for a particular drainage area and season of year in which snow-melt is not a major consideration should represent the most severe flood-producing rainfall depth-area-duration relationship and isohyetal pattern of any storm that is considered reasonably characteristic of the region in which the drainage basin is located, giving consideration to the runoff characteristics and existence of water regulation structures in the basin. In deriving standard project storm rainfall estimates applicable to seasons and areas in which melting snow may contribute a substantial volume of runoff to the standard project flood (SPF) hydrograph, appropriate allowances for snow-melt are included with and considered as a part of the standard project storm rainfall quantities in computing the SPF hydrograph. Where floods are predominately the result of melting snow, the SPF estimate is based on estimates of the most critical combinations of snow, temperature and water losses considered reasonably characteristic of the region.

The term "snow" is used in a broad sense to mean any period or sequence of rainfall events that may contribute to critical flood events in the particular drainage basin under study.

The term "region" as used above is construed to include the area surrounding the given basin in which storm producing factors are substantially comparable; i.e., the general area within which meteorological influences and topography are sufficiently alike to permit adjustment of storm data to a common basis of comparison with practical degree of reliability. Such a "region" includes a very large area in the eastern half of the United States where relief is generally moderate, and relatively small areas in certain sections of the western United States where extreme topography is encountered.

A general comparison of maximum storms of record in the region, supplemented by meteorological investigations, serve as a basis in selecting rainfall criteria to represent the most severe storm that is considered "reasonably characteristic" of a region. Certain storms of extraordinary severity may be eliminated as too unusual and extreme to warrant adoption as the standard project storm.

VII. REFERENCES

1. Linsley, R. K., Jr.: Techniques for surveying surface water resources, *World Meteorol. Organization Tech. Note* 26, Geneva, 1958.
2. Brown, E. N., and R. R. Braham, Jr.: Precipitation-particle measurements in trade-wind cumuli, *J. Meteorol.*, vol. 16, no. 6, pp. 609–616, December, 1959.
3. Laws, J. O., and D. A. Parsons: The relation of raindrop size to intensity, *Trans. Am. Geophys. Union*, vol. 24, pt. II, pp. 452–460, 1943.
4. Gunn, Ross, and G. D. Kincer: The terminal velocity of fall for water droplets in stagnant air, *J. Meteorol.*, vol. 6, no. 4, pp. 243–248, August, 1949.
5. Houghton, H. G.: Preliminary quantitative analysis of precipitation mechanisms, *J. Meteorol.*, vol. 7, no. 6, pp. 363–369, December, 1950.
6. Lott, G. A., and V. A. Myers: Meteorology at flood-producing storms in the Mississippi River basin, *U.S. Weather Bur. Hydrometeorol. Rept.* 34, 1956.
7. Benton, G. S., and M. A. Estoque: Water-vapor transfer over the North American Continent, *J. Meteorol.*, vol. 11, no. 6, pp. 462–477, December, 1954.
8. Final report of the Advisory Committee on Weather Control, *Am. Meteorol. Soc. Bull.*, vol. 39, pp. 583–598, 1958.
9. Benton, G. S., R. T. Blackburn, and V. O. Snead: The role of the atmosphere in the hydrologic cycle, *Trans. Am. Geophys. Union*, vol. 31, pp. 61–73, 1950.
10. Holzman, Benjamin: Sources of moisture for precipitation in the United States, *U.S. Dept. Agr. Tech. Bull.* 589, 1937.
11. Lott, G. A.: The world-record 42-minute Holt, Missouri, rainstorm, *Monthly Weather Rev.*, vol. 82, no. 2, pp. 50–59, February, 1954.
12. Schoner, R. W., and S. Molansky: Rainfall associated with hurricanes, *U.S. Weather Bur. Hurricane Res. Project Rept.* 3, 1956.
13. Thompson, J. C.: A numerical method of forecasting rainfall in the Los Angeles area, *Monthly Weather Rev.*, vol. 78, no. 7, pp. 113–124, July, 1950.
14. Linsley, R. K., Jr., M. A. Kohler, and J. L. H. Paulhus: "Applied Hydrology," McGraw-Hill Book Company, Inc., New York, 1949.
15. Kincer, J. B.: Determination of dependability of rainfall records by comparison with nearby records, *Trans. Am. Geophys. Union*, vol. 19, pt. 1, pp. 533–538, 1938.
16. Merriam, C. F.: Progress report on the analysis of rainfall data, *Trans. Am. Geophys. Union*, vol. 19, pt. 1, pp. 529–532, 1938.
17. Kohler, M. A.: Double-mass analysis for testing the consistency of records and for making required adjustments, *Am. Meteorol. Soc. Bull.*, vol. 30, pp. 188–189, May, 1949.

18. Weiss, L. L., and W. T. Wilson: Evaluation of significance of slope changes in double-mass curves, *Trans. Am. Geophys. Union*, vol. 34, pp. 893–896, December, 1953.
19. Searcy, J. K., and C. H. Hardison: Double-mass curve, Manual of Hydrology, Part 1, General Surface-water Techniques, *U.S. Geol. Surv. Water-Supply Paper* 1541-B, 1960.
20. Clarke-Hafstad, Katherine: A statistical method for estimating the reliability of a station-year rainfall record, *Trans. Am. Geophys. Union*, vol. 19, pt. II, pp. 526–529, 1938.
21. Clarke-Hafstad, Katherine: Reliability of station-year rainfall frequency determinations, *Trans. Am. Soc. Civil Engrs.*, vol. 107, pp. 633–683, 1942.
22. Paulhus, J. L. H., and M. A. Kohler: Interpretation of missing precipitation records, *Monthly Weather Rev.*, vol. 80, pp. 129–133, August, 1952.
23. Thiessen, A. H.: Precipitation for large areas, *Monthly Weather Rev.*, vol. 39, pp. 1082–1084, July, 1911.
24. Reed, W. G., and J. B. Kincer: The preparation of precipitation charts, *Monthly Weather Rev.*, vol. 45, pp. 233–235, May, 1917.
25. Sacramento method of correlating storm precipitations with normal seasonal precipitations and runoff, *U.S. Army Corps Engrs. Rept.*, Sacramento District Office, 1941.
26. Spreen, W. C.: A determination of the effect of topography upon precipitation, *Trans. Am. Geophys. Union*, vol. 28, pp. 285–290, 1947.
27. "Climate and Man," *U.S. Dept. Agr. Yearbook*, 1941.
28. Manual for depth-area-duration analyses of storm precipitation, *U.S. Weather Bur. Coop. Studies Tech. Paper* 1, 1946.
29. Visher, S. S.: "Climatic Atlas of the United States," Harvard University Press, Cambridge, Mass., 1954.
30. Lackey, E. E.: Annual rainfall variability maps of the United States, *Monthly Weather Rev.*, vol. 67, p. 201, 1939.
31. Jennings, A. H.: Maximum 24-hour precipitation in the United States, *U.S. Weather Bur. Tech. Paper* 16, 1952.
32. Miller, E. R.: Raininess charts of the United States, *Monthly Weather Rev.*, vol. 61, pp. 44–45, 1933.
33. Maximum station precipitation for 1, 2, 3, 6, 12 and 24 hours, *U.S. Weather Bur. Tech. Paper* 15. Pt. I: Utah, pt. II: Idaho (1951); pt. III: Florida (1952); pt. IV: Maryland, Delaware, and District of Columbia, pt. V: New Jersey, pt. VI: New England, pt. VII: South Carolina (1953); pt. VIII: Virginia, pt. IX: Georgia, pt. X: New York (1954); pt. XI: North Carolina, pt. XII: Oregon, pt. XIII: Kentucky, pt. XIV: Louisiana, pt. XV: Alabama (1955); pt. XVI: Pennsylvania, pt. XVII: Mississippi, pt. XVIII: West Virginia, pt. XIX: Tennessee, pt. XX: Indiana (1956); pt. XXI: Illinois, pt. XXII, Ohio (1958); pt. XXIII, California, pt. XXIV: Texas (1959); pt. XXV: Arkansas (1960); pt. XXVI: Oklahoma (1961).
34. Shands, A. L., and D. Ammerman: Maximum recorded United States point rainfall for 5 minutes to 24 hours at 207 first order stations, *U.S. Weather Bur. Tech. Paper* 2, 1947.
35. Jennings, A. H.: World's greatest observed point rainfall, *Monthly Weather Rev.*, vol. 78, pp. 4–5, 1950.
36. "Storm Rainfall in the United States: Depth-Area-Duration Data," U.S. Army Corps of Engineers, since 1946.
37. Yarnell, D. L.: Rainfall intensity-frequency data, *U.S. Dept. Agr. Misc. Publ.* 204, 1935.
38. Hershfield, D. M.: Rainfall frequency atlas of the United States, for durations from 30 minutes to 24 hours and return periods from 1 to 100 years, *U.S. Weather Bur. Tech. Rept.* 40, May, 1961.
39. Engineering Staff of Miami Conservancy District: Storm rainfall of eastern United States, *Tech. Repts.*, pt. V, Dayton, Ohio, 1917.
40. Gumbel, E. J.: Statistical theory of extreme values and some practical applications, *U.S. Bur. Std. Appl. Math. Ser.* 33, Feb. 12, 1954.
41. Jens, S. W.: Engineering meteorology, in Proceedings of the Conference on Water Resources, October 1, 2, 3, 1951, *Illinois State Water Surv. Bull.* 41, pp. 99–116, 1952.
42. Meyer, A. F.: "The Elements of Hydrology," John Wiley & Sons, Inc., New York, 1st ed., 1917; 2d ed., 1928.
43. Bernard, Merrill: Formulas for rainfall intensities of long duration, *Trans. Am. Soc. Civil Engrs.*, vol. 96, pp. 592–624, 1932.
44. Hathaway, G. A.: Military airfields—a symposium: Design of drainage facilities, *Trans. Am. Soc. Civil Engrs.*, vol. 110, pp. 697–733, 1945.
45. Bernard, Merrill: The primary role of meteorology in flood flow estimating, *Trans. Am. Soc. Civil Engrs.*, vol. 109, pp. 311–382, 1944.

REFERENCES

46. Generalized estimates of maximum possible precipitation over the United States east of the 105th meridian, *U.S. Weather Bur. Hydrometeorol. Rept.* 23, June, 1947.
47. Generalized estimates of probable maximum precipitation for the United States west of the 105th meridian for areas to 400 square miles and durations to 24 hours, *U.S. Weather Bur. Tech. Paper* 38, 1960.
48. Standard project flood determinations, *U.S. Army Corps Engrs. Civil Eng. Bull.* 5-2-8, Mar. 26, 1952, unpublished.
49. "Standard Project Rain-flood Criteria, Sacramento-San Joaquin Valley, California," U.S. Army Corps of Engineers, Sacramento District, 1957.

Other Important References

On Instruments and Observations

50. Middleton, W. E. K., and A. F. Spilhaus: "Meteorological Instruments," 3d ed., University of Toronto Press, Toronto, Canada, 1953.
51. Instructions for climatological observers, *U.S. Weather Bur. Cir.* B, 10th ed., October, 1955.
52. Kadel, B. C.: Measurement of precipitation, *U.S. Weather Bur. Circ.* E, 4th ed., 1936.
53. Kurtyka, J. C.: Precipitation measurements study, *Illinois State Water Surv. Rept. Invest.* 20, 1953.
54. Battan, L. J.: "Radar Meteorology," University of Chicago Press, Chicago, 1959.
55. Stout, G. E.: Radar for rainfall measurements and storm tracking, *Proc. Am. Soc. Civil Engrs., J. Hydraulics Div.*, vol. 85, no. HY1, pp. 1–16, January, 1959.
56. Kohler, M. A.: Design of hydrological networks, *World Meteorol. Organization Tech Note* 25, Geneva, 1958.

On Precipitation Formation and Artificial Induction

57. MacDonald, J. E.: The physics of cloud modification, *Advan. in Geophys.*, vol. 5, pp. 223–298, 1958.
58. Mason, B. J.: "The Physics of Clouds," Oxford University Press, London, 1957.
59. Dufour, L., F. Hall, F. H. Ludlam, and E. J. Smith: Artificial control of clouds and hydrometeors, *World Meteorol. Organization Tech. Note* 13, 1955.
60. Weickmann, Helmut, and Waldo Smith (eds.): "Artificial Stimulation of Rain," Pergamon Press, Ltd., London, 1957.
61. Weather Modification—a symposium: Seeding of west coast winter storms, by R. D. Elliot, and Seeding of clouds in tropical climates, by W. E. Howell, *Trans. Am. Soc. Civil Engrs.*, vol. 127, pt. III, pp. 327–371, 1962.

On Depth-Area-Duration Relations

62. Thunderstorm rainfall, *U.S. Weather Bur. and U.S. Corps Engrs. Hydrometeorol. Rept.* 5, Waterways Experiment Station, Vicksburg, Miss.,.1947.
63. Huff, A. F., and G. E. Stout: Area-depth studies for thunderstorm rainfall in Illinois, *Trans. Am. Geophys. Union*, vol. 33, pp. 495–498, August, 1952.
64. Huff, F. A., and J. C. Neill: Areal representativeness of point rainfall, *Trans. Am. Geophys. Union*, vol. 38, no. 3, pp. 341–345, June, 1957.
65. Fletcher, R. D.: A relation between maximum observed point and areal rainfall values, *Trans. Am. Geophys. Union*, vol. 31, no. 3, pp. 344–348, June, 1950.

On Frequency of Rainfall Intensity

66. Chow, V. T.: Frequency analysis of hydrologic data with special application to rainfall intensities, *Univ. Illinois Eng. Expt. Sta. Bull. Ser.* 414, July, 1953.
67. Rainfall intensities for local drainage design in the United States, for durations of 5 to 240 minutes and 2-, 5-, and 10-year return periods, *U.S. Weather Bur. Tech. Paper* 24: pt. I, West of 115th meridian, 1953; pt. II, between 105 and 115°W, 1954.
68. Rainfall intensity-duration-frequency curves, for selected stations in the United States, Alaska, Hawaiian Islands, and Puerto Rico, *U.S. Weather Bur. Tech. Paper* 25, 1955.
69. Rainfall intensities for local drainage design in western United States, for durations of 20 minutes to 24 hours and 1- to 100-year return periods, *U.S. Weather Bur. Tech. Paper* 28, 1956.
70. Rainfall intensity-frequency regime, *U.S. Weather Bur. Tech. Paper* 29: Pt. 1, Ohio Valley, 1957; pt. 2, Southeastern United States, 1958; pt. 3, Middle Atlantic Region, 1958, pt. 4, Northeastern United States, 1959; pt. 5, Great Lakes Region, 1960.
71. Rainfall-frequency atlas of the Hawaiian Islands for areas to 200 square miles, durations to 24 hours, and return periods from 1 to 100 years, *U.S. Weather Bur. Tech. Paper* 43, 1962.

72. Probable maximum precipitation and rainfall-frequency data for Alaska for areas to 400 square miles, durations to 24 hours and return periods from 1 to 100 years, *U.S. Weather Bur. Tech. Paper* 47, 1963.

On Probable Maximum Precipitation

73. Commonwealth of Australia, Bureau of Meteorology: Conference on Estimation of Extreme Precipitation, Melbourne, 1958.
74. "Design of Small Dams," U.S. Bureau of Reclamation, 1960.
75. Showalter, A. K., and S. B. Solot: Computation of maximum possible precipitation, *Trans. Am. Geophys. Union*, vol. 23, pp. 258–274, 1942.
76. Seasonal variation of the probable maximum precipitation east of the 105th meridian for areas from 10 to 1,000 square miles and duration of 6, 12, 24 and 48 hours, *U.S. Weather Bur. Hydrometeorol. Rept.* 33, April, 1956.
77. Hershfield, D. M.: Estimating the probable maximum precipitation, *Proc. Am. Soc. Civil Engrs., J. Hydraulics Div.*, vol. 87, no. HY5, pt. 1, pp. 99–116, September, 1961.
78. Knox, J. B.: Procedures for estimating maximum possible precipitation, *Calif. State Dept. Water Resources Bull.* 88, May, 1960.
79. Generalized estimates of probable maximum precipitation and rainfall frequency data for Puerto Rico and Virgin Islands, *U.S. Weather Bur. Tech. Paper* 42, 1961.

Section 10

SNOW AND SNOW SURVEY

WALTER U. GARSTKA, *Chief, Water Conservation Branch, U.S. Bureau of Reclamation.*

I. Introduction	10-2
II. Snow and Its Classification	10-2
III. Distribution of Snow	10-6
IV. Ripening of Snow	10-8
V. Measurement of Snow at Time of Fall	10-9
A. Precipitation Gages	10-11
1. Standard and Recording Rain Gages	10-11
2. Seasonal Storage Precipitation Gages	10-12
B. Snow Boards	10-13
C. Snow Stakes	10-14
VI. Snow Surveying	10-14
A. Snow-course Layout on the Ground	10-14
B. Snow-surveying Instruments	10-15
C. Snow Surveying on the Ground	10-17
D. Snow Observations from the Air	10-19
E. Radioisotope Snow Gage	10-20
F. Sources of Snow-survey Data	10-21
VII. Objectives of Runoff from Snowmelt Computations	10-23
A. Seasonal Water-yield Forecasting	10-23
B. Runoff Forecasting for River Regulation	10-27
C. Design Floods Due to Snow	10-27
VIII. Factors Affecting Runoff from Snowmelt	10-27
A. Sources of Energy for Snowmelt	10-27
1. Radiation	10-28
2. Sublimation	10-29
3. Heat Exchange	10-30
B. Snowpack Characteristics	10-30
C. Site Conditions	10-31
D. Antecedent Conditions	10-32
E. Rainfall	10-33
IX. Techniques of Analysis of Snowmelt for Forecasting Runoff from Snowmelt	10-33
A. Degree-day Correlations	10-33
B. Basin Indexes	10-34
C. Recession Analysis	10-35
D. Correlation Analyses	10-38

   E. Physical Equations.................................... 10-38
     1. The Light Equation............................... 10-38
     2. U.S. Army Corps of Engineers Equations............ 10-39
  X. Snow Compaction... 10-41
  XI. Snow Avalanches.. 10-41
  XII. Snow Loads.. 10-42
   A. On Buildings... 10-42
   B. On Power Lines and Transmission Towers............... 10-43
  XIII. Snow Produced by Artificial Means...................... 10-45
  XIV. Miscellaneous Problems Related to Snow.................. 10-45
   A. Ice in Streams....................................... 10-46
   B. Frost in the Soil.................................... 10-47
   C. Permafrost.. 10-48
   D. Induced Melting of Snow.............................. 10-49
  XV. References... 10-49

I. INTRODUCTION

  Snow is of great importance to the hydrologist practicing in the Northern and Southern Hemispheres where the winter precipitation is apt to fall as snow, whether or not it stays on the ground for any appreciable time.

  In many parts of the world streamflow consists mainly of water released by the melting of snow. The coming of spring exposes the snow to heat, causing a rapid melt and a short period of runoff. Since the waters yielded by the melting of snow appear in the natural stream channels out of phase with the demand for water for human endeavors, extensive water-resources engineering systems have been developed to store snowmelt runoff and make it available throughout the remainder of the year.

  In many parts of the world snow acquires a great importance in the occurrence of floods in the springtime. The presence of a snow cover on a drainage basin influences very greatly the runoff characteristics of the area under certain basic conditions of air temperature and rainfall.

  Snow is of great importance to the structural engineer in dealing with snow loads on buildings and to the electrical engineer in connection with electrical-energy transmission lines. The effect of snow on surface transportation systems is widely appreciated.

  Snow can be looked upon as a soil which changes its texture with time and with progression of melting and which disappears when the temperatures persist above the melting point of ice. A study of the physical characteristics of snow and ice has received special attention in recent years, and the general subject of the formation, the ripening, the melting, and the disappearance of snow is an entrancing subject of great complexity, concerning which there is a voluminous and expanding literature.

  This section should be looked upon as a condensed assembly of the presentation of techniques supported by pertinent references rather than as a monograph on any one of the many phases of snow and snow hydrology.

II. SNOW AND ITS CLASSIFICATION

  Schaefer [1] defines snow as follows: "Snow is the solid form of water which grows while floating, rising, or falling in the free air of the atmosphere."

  Snow crystals are in the hexagonal system. However, there are many variations of the hexagonal system. The Commission on Snow and Ice of the International Association of Scientific Hydrology made an extensive study [2], from which an abstract of the classification of snow at the time of fall is presented in Fig. 10-1. The definitions used in English [2] follow closely those given by Seligman [3].

  In the International Snow Classification F is the symbol used for the form of the crystal, D (dimension) the size, K (kohäsion) the cohesion, and W the wetness.

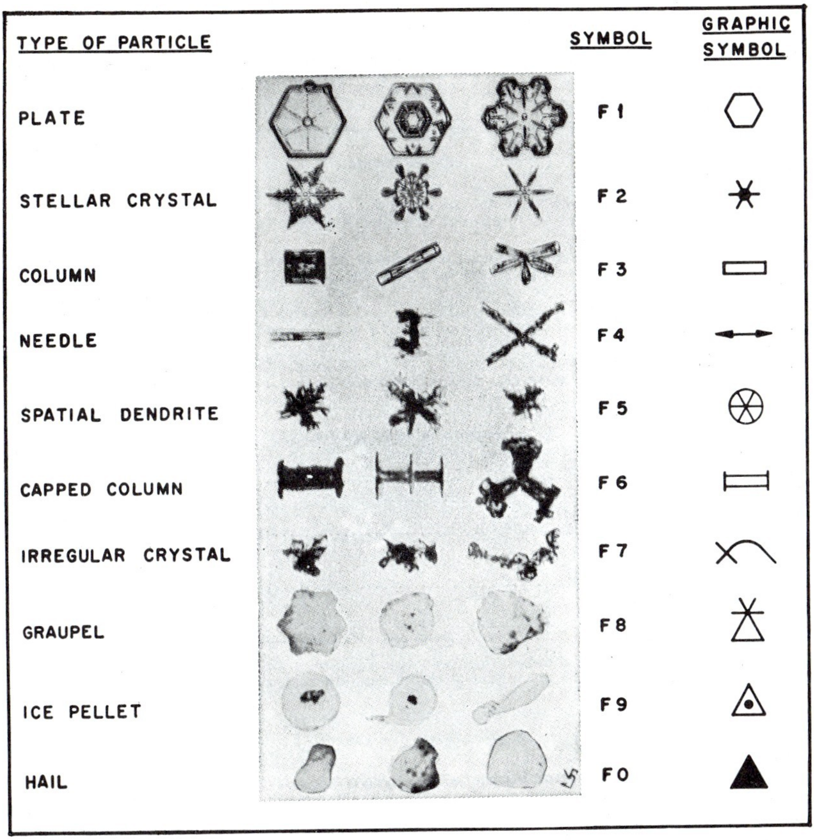

SNOW AND ITS CLASSIFICATION 10–3

TYPE OF PARTICLE		SYMBOL	GRAPHIC SYMBOL
PLATE		F 1	⬡
STELLAR CRYSTAL		F 2	✻
COLUMN		F 3	▭
NEEDLE		F 4	↔
SPATIAL DENDRITE		F 5	⊛
CAPPED COLUMN		F 6	⊨
IRREGULAR CRYSTAL		F 7	⌒
GRAUPEL		F 8	⧖
ICE PELLET		F 9	⊙
HAIL		F 0	▲

MODIFYING FEATURE	BROKEN CRYSTALS	RIME COATED CRYSTALS	CLUSTERS	WET
SYMBOL SUBSCRIPT	p	r	f	w

SIZE OF PARTICLE D MEASURED IN MILLIMETERS.

FIG. 10-1. Classification of solid precipitation. (*Schaefer* [2].)

Each of the properties is classified in five degrees designated by a, b, c, d, and e, referring to the order of magnitude or the state of metamorphism. A complete description of snow includes not only the use of the above terms, but also the description of depth, surface conditions, surface features, density, hardness, and temperature as they may pertain to each profile of a snow layer. The U.S. Army Corps of Engineers [4] has developed instructions for making and recording snow observations based upon Ref. 2. Depending on the engineering purpose of the snow classification, measurements may include compressive yield strength, tensile strength, shear strength at zero normal stress, and hardness, all of which require highly specialized instrumentation.

Since the delicate and complex equipment necessary for a full description of snow characteristics by layers may not be available, the Corps of Engineers developed a

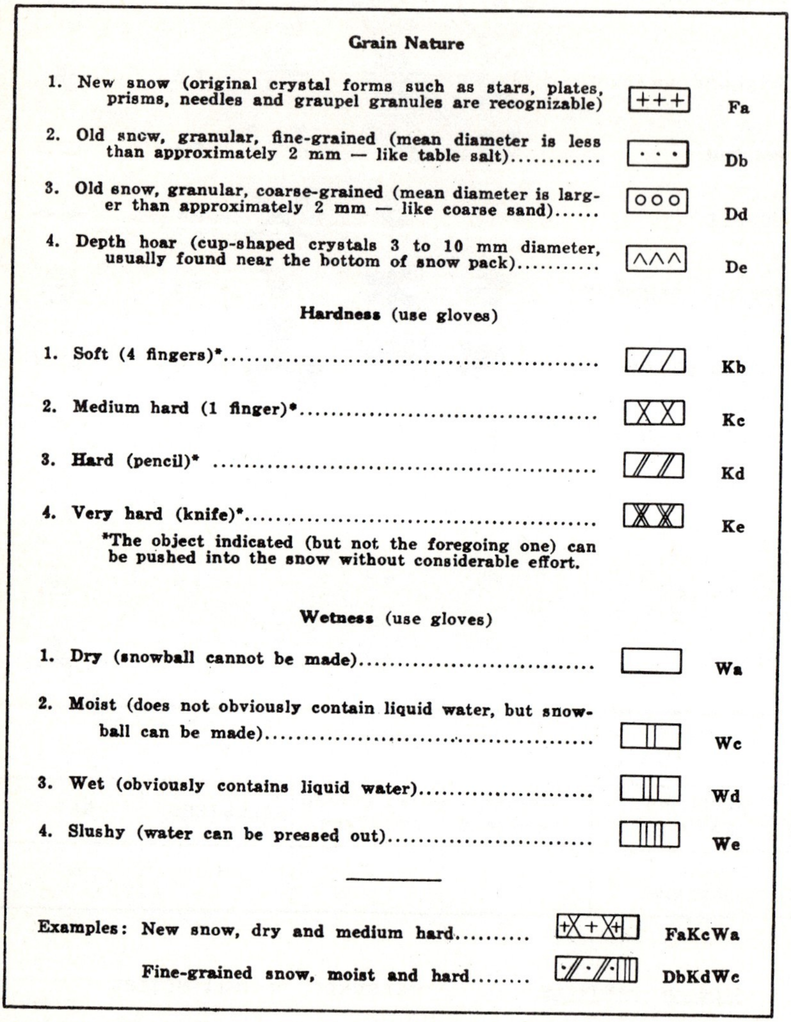

Grain Nature

1. New snow (original crystal forms such as stars, plates, prisms, needles and graupel granules are recognizable) [+++] Fa
2. Old snow, granular, fine-grained (mean diameter is less than approximately 2 mm — like table salt).......... [...] Db
3. Old snow, granular, coarse-grained (mean diameter is larger than approximately 2 mm — like coarse sand)...... [ooo] Dd
4. Depth hoar (cup-shaped crystals 3 to 10 mm diameter, usually found near the bottom of snow pack).......... [∧∧∧] De

Hardness (use gloves)

1. Soft (4 fingers)*.. [//] Kb
2. Medium hard (1 finger)*..................................... [XX] Kc
3. Hard (pencil)*.. [//] Kd
4. Very hard (knife)*... [XX] Ke

*The object indicated (but not the foregoing one) can be pushed into the snow without considerable effort.

Wetness (use gloves)

1. Dry (snowball cannot be made)............................ [ ] Wa
2. Moist (does not obviously contain liquid water, but snowball can be made).. [|] Wc
3. Wet (obviously contains liquid water)..................... [||] Wd
4. Slushy (water can be pressed out)......................... [|||] We

---

Examples: New snow, dry and medium hard.......... [+X+X+] FaKcWa
Fine-grained snow, moist and hard........ [/·/·|||] DbKdWc

Fig. 10-2. Simplified field classification of natural snow type for engineering purposes. Gloves should be of single-thickness fabric or leather finger gloves only. (*U.S. Army Corps of Engineers* [5].)

simplified field classification of natural snow type for engineering purposes [5]. This field classification, presented in Fig. 10-2, can be supplemented by whatever measurements may be made with instruments available, such as snow density, water equivalent, and temperature.

There is an extensive literature concerning the properties of water in a solid state. A monumental work is Dorsey's "Properties of Ordinary Water Substance" [6]. An extension of this work, with special emphasis on the mechanical properties of snow and ice, is presented in a very comprehensive review edited by Mantis [7].

An outstanding discussion of ice engineering with special reference to the St. Lawrence River in Canada is given by Barnes [8].

Still another field of knowledge is concerned with water in a solid state in glaciers and icebergs and polar icecaps (Sec. 16). The existence of water substance on the earth in its three phases, vapor, liquid, and solid, with their thermodynamic transfers of energy, exerts a profound effect on all human endeavors. The heat of fusion of ice, commonly taken as 79.7 cal [9], has at 0°C, according to the most careful observations [6, p. 615], a value of 79.69 cal$_{15}$/g. The calories$_{15}$ is the amount of heat required to raise one gram of water from 14.5°C to 15.5°C with the water to be under an air pressure of one atmosphere [6, p. 254].

Table 10-1. Comparison of Latent Heat of Fusion and Specific Heat of Water with Those of Other Substances [10]

Substance	Latent heat of fusion		Specific heat	
	Heat required, cal/g	Ratio of water to substance	Heat required, cal/g/1°C	Ratio of water to substance
Water.........	79.7	1.0	1.000	1.0
Aluminum.......	38.3	2.1	0.224	4.5
Copper.........	50.6	1.6	0.09	11.1
Iron...........	65.5	1.2	0.117	8.5
Lead...........	6.3	12.7	0.031	32.1
Magnesium......	90	0.9	0.285	3.5
Nickel..........	73.8	1.1	0.11	9.1

The latent heat of vaporization of water at 0°C is 596 cal. It has been assumed that the latent heat of sublimation of ice-I (ordinary ice) at 0°C is the sum of the latent heat of fusion of ice-I and the latent heat of vaporization of water at the same temperature [6, p. 614]. However, Barnes [8, pp. 31-33] reported that he and Vipond observed the latent heat of sublimation of ice to be the same as that of water at the instant of evaporation, which would indicate that the molecules of water were leaving the ice as a polymeric vapor without having passed through the liquid phase and that the polymer broke down into ordinary water vapor after leaving the ice, this breakdown requiring exactly the latent heat of fusion of ice. Thus the sublimation of ice-I at 0°C to form ordinary water vapor requires 676 cal$_{15}$/g [6, using JF value for 0°C in Table 272].

The storage of water in the form of ice not only withdraws vast quantities of water from participation in the evaporation and condensation portions of the hydrologic cycle, but also has a profound effect on the climate of the earth because of the heat energy required to effect any changes of state.

The latent heat of fusion is the amount of heat required to change a solid to a liquid state with no change in temperature. Specific heat is the amount of heat required to raise a substance, without change in state, over a temperature interval. Water possesses among the largest latent heats of fusion and specific heats; for example, it requires about 13 times as much heat to melt a gram of ice as it does to melt a gram of lead, and about twice as much heat to melt a gram of ice as it does a gram of aluminum. The specific heat of water is over 8 times as great as that of iron and over 32 times as great as that of lead. Table 10-1 is a comparison of these characteristics for some common substances.

A continuing problem with aircraft is that of icing. The Wright Development Center of the U.S. Air Force has prepared an extensive bibliography [11] on ice and frost control in which are contained many abstracts relating to the physics of water at the freezing point.

As the Arctic and Antarctic regions are the last remaining frontiers on the surface of the earth, there is an intense international interest in these regions. In order to provide a key to scientific publications relating to the polar regions and to areas of low temperature, such as found in the higher altitudes of mountain regions, the Arctic Institute of North America has published a series of bibliographies, an example of which is Ref. 12, which alone contains 7,192 abstracts.

The U.S. Army Corps of Engineers has prepared annotated bibliographies on snow, ice, and permafrost, of which Ref. 13, issued as of January, 1959, is an example.

III. DISTRIBUTION OF SNOW

The usual climatological records refer to the total depth of snow as measured at the time of fall. Figure 10-3 presents the mean annual total snowfall for the United States, including Hawaii and Alaska, based upon a study performed by the U.S. Weather Bureau using data from 3,515 stations for the period 1931 to 1952.

Kincer [15] presents data based upon the period 1895 to 1914, including the average annual snowfall in inches unmelted, the average data of the first snowfall in the autumn, the average number of days with snow cover, and the average number of days with snowfall of 0.01 in. or more as melted [15].

The 1941 Yearbook of Agriculture [16, p. 728] presents the average annual number of days with snow cover of 1 in. or more. The U.S. Weather Bureau's series Climatic Summary of the United States [17] summarizes the snowfall totals for the stations published in the monthly *Climatological Data* bulletin for the various states.

For many hydrologic interpretations, the depth of snow as measured at the time of fall is of but limited usefulness. The depth of snowfall and of the new snow reaching the ground may vary greatly within short distances in mountainous forested areas such as in the western United States, where only a few miles horizontally may separate bare ground from snow depths of many feet. The significance of metamorphosis, density changes, and water equivalent is discussed below.

Snow at the time of fall may have a density as low as 0.01 to as high as 0.15; snowfall in the form of dry snow, not sleet or graupel, may vary in density between 0.07 and 0.15. The average for the United States is taken to be 0.10. It is common, therefore, to interpret depths of fall by assuming that they have a density of 0.10. Thus a 10-in. snowfall is assumed to have 1 in. of water.

A distinction is to be made between snow density at the time of fall, which is a ratio between the volume of meltwater derived from a sample of snow, the initial volume of the sample, and the water equivalent of the snowpack, which is the depth of water that would result from melting of the snowpack. Uniform density in a snowpack even shortly after the time of fall is rarely to be found. Furthermore, unless exceptionally cold weather prevails, metamorphic changes which occur in the snow often rearrange the distribution of water within a snowpack. The water equivalent is the depth of water which would result from the melting of the snow without regard to the density distribution or to the amount of liquid water retained by the snow by capillary retention.

The U.S. Weather Bureau has summarized for the United States extremes of snowfall [18]. Unfortunately, authentic accounts or official records are very difficult to secure for many of the great snowstorms which reportedly have occurred around the world. However, taking into account orographic features of the areas and applying knowledge of storm paths, one can conjecture what the possibilities may be in view of the following recorded snowfalls:

1. The greatest seasonal snowfall in the Weather Bureau's records is 1,000.3 in. (more than 83 ft), observed at the Paradise Ranger Station in the state of Washington during the winter of 1955–1956.

2. Eighty-seven inches of snow fell at Silver Lake, Colo., in 27.5 hr, Apr. 14–15, 1921.

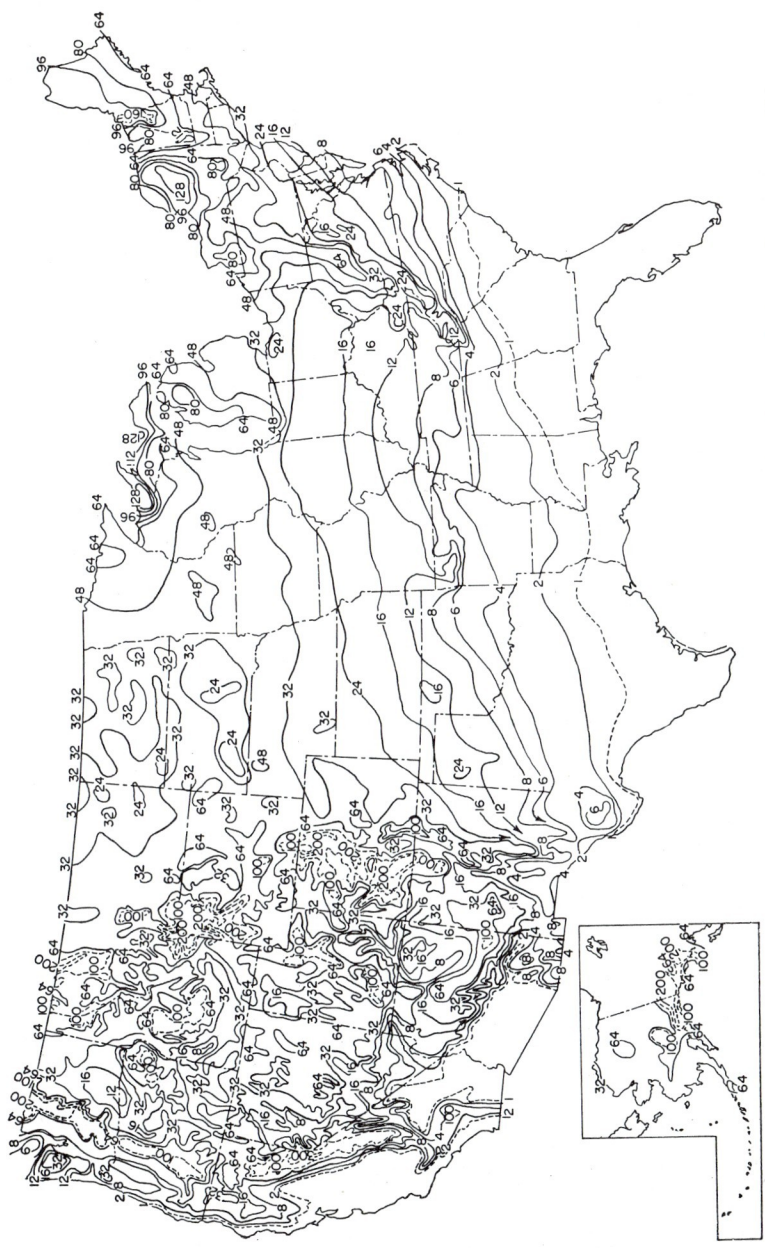

Fig. 10-3. Mean annual total snowfall in inches for the United States (1931–1952). (*U.S. Weather Bureau* [14].)

3. Sixty inches of snow fell in one day at Giant Forest, Calif.
4. Forty-eight inches of snow fell at Idaho Springs, Idaho, in 26.5 hr on Apr. 14–15, 1921.
5. Eighty-seven inches of snow fell in three days at Giant Forest, Calif., Feb. 12–14, 1926.
6. One hundred and eight inches of snow fell in 4 days at Tahoe, Calif.
7. Forty-two inches of snow fell in two days at Angola, N.Y.
8. Fifty-four inches of snow fell in three days at The Dalles, Ore.
9. Twenty-six inches of snow, weighing nearly 100 million tons over the the city, fell on New York City, Dec. 26–27, 1947.

The effects of the paths of air masses as influenced by orographic features such as mountain ranges have been studied extensively in relation to precipitation (Sec. 9). It is known that, as elevation above sea level increases, so does total precipitation. Furthermore, as precipitation increases, so usually does the total amount of snowfall, since in mountainous areas even in the vicinity of the equator precipitation at the higher altitudes is apt to fall in the form of snow, depending upon the moisture-bearing characteristics of the air masses. There is, however, an upper limit to the maximum precipitation, after which the amount of precipitation tends to decrease with altitude. Very few mountain ranges exhibit this reduction because they are usually not sufficiently high for discernment of this trend.

Garstka [19] discusses the snow-water equivalent in relation to elevation in the drainage basin of the Snake River above Jackson Lake, Wyoming, and also for the Snake River downstream from Jackson Lake for that portion of the drainage basin of Jackson Lake, Wyoming, to Heise, Idaho.

Although the fraction of the area of a drainage basin at higher altitudes is smaller in comparison with the area at lower elevations, the very great increase in snow-water equivalent accumulating over the winter at the higher elevations makes those elevations very important in yielding large volumes of spring snowmelt runoff.

IV. RIPENING OF SNOW

Snow cover and soil have much in common. Both possess a solid and a nonsolid phase. Many of the concepts and methods of mathematical analysis of soil mechanics apply to snow mechanics as set forth by Haefeli [20, chap. 2]. Whereas the weathering of minerals to form soil results in the production of finer and finer particle sizes, the metamorphism of snow tends to increase particle size. The ripening and metamorphism of a snowpack might be looked upon as paralleling changes in a "soil" which began as a silt loam and ended as fine gravel. The stellar crystal snowflake, in all its beautiful hexagonal forms in an infinite variety of arrangements, is very evanescent under natural conditions.

The term metamorphism applies to the transformation which snow undergoes in the period after deposition in the snowpack until its disappearance by melting or evaporation. The points, sharp edges, and abrupt angles, characteristic of stellar ice crystals, are unstable and tend to disappear, through a process involving a supersaturation of the atmosphere immediately in the vicinity of the sharp edges, as part of the common tendency of crystals to reduce their overall surface area. This supersaturation, a function of surface tension, results in the stellar snowflakes disappearing, often in a matter of hours after the individual snowflakes have fallen. Metamorphism takes place unceasingly as long as ice crystals remain in air spaces, even at low temperatures. At the end of winter a snowpack usually consists of very uniform, coarse, large ice crystals. A comprehensive discussion of metamorphism is given by Bader [20, chap. 1].

The interrelationships of liquid water, ice, and water vapor are very complex. The *triple point* of water at which solid, liquid, and vapor phases are all in equilibrium can be expected to be present some of the time during the active snowmelt period at the end of winter. The triple-point temperature is not the same as the melting point. The triple point for water is not measured at atmospheric pressure (760 mm), but is determined under the pressure of the water vapor alone, which at the triple point is about 4.6 mm. The triple point of water is $+0.0075°C$ [7, p. 87].

Dorsey [6, pp. 603–604] defines the *melting point* as:

By definition, the melting-point of ice-I (ordinary ice) in contact with water saturated with air at a pressure of one normal atmosphere, but otherwise pure, the entire system being subjected to a uniform hydrostatic pressure of one normal atmosphere, is 0°C. This is often called the normal melting point of ice; also, the ice point.

Of the seven known crystalline ices, only the ordinary form, ice-I, expands upon freezing and thus floats in the liquid phase of water. The six other forms [6, p. 608] are denser than water under the conditions of temperature and pressure at which they exist. Dorsey [6, p. 397] states that, as a consequence, the pressure exerted by the freezing of water in a confined space cannot exceed about 30,000 psi since at or above that pressure the bulky ice-I cannot exist.

The metamorphism of snow produces density changes with time. The density changes, according to Work [21], bear little relationship to the depth of snow or to the weights of the various layers. Density is directly related to the size and arrangement of the ice crystals of the snow in a particular layer. A diminishing snow depth therefore may be a very poor indicator of possible changes in its water content. The differences in metamorphism of snow deposited by separate storms, as influenced by the weather between storms, are so complex as to render practically valueless to the hydrologist the estimates of either snow density or of water equivalent from depth-time relationships, alone, of a snow cover [21]. This has a bearing upon snow surveying (Subsec. VI).

Rapid changes in temperature during the winter in the Northern Hemisphere are not uncommon. Figure 10-4 presents the weather extremes around the world.

In addition to the changes brought about by vapor-phase metamorphism, *regelation*, the freezing together of two pieces of ice [6, p. 413], takes place especially in the presence of free water in a snowpack. Vapor-phase metamorphism and regelation result in producing a uniform and coarse structure of the snow. The process of the formation of coarse crystals is commonly called *ripening*. Ripe snow has a remarkably uniform density, as observed throughout the world, of from 45 to 50 per cent.

Unconsolidated snow, when subjected to melting, will absorb, in the ripening process, the liquid phase released by the initial melt. Rainfall upon snow would likewise be withheld in the unripe snowpack by the ripening process. The similarity of snow and soil is again evident, the new snow being capable of withholding far greater percentages of water than the fully ripe snow. In a fully ripe snow the ice-crystal size is so great and the capillary retentivity of ripe snow during the active water-yielding portion of the snowmelt season is so low that very little water is withheld by the snow.

A melting snowpack consists of a mixture of ice crystals and the small amount of free water. Thermal quality of snowpack [23, chap. 2] containing no free water would have a value of 1.0 since heat would have to be supplied to melt all the water equivalent of the snowpack. A melting snowpack of fully ripe snow normally averages from 3 to 5 per cent of liquid water, and the thermal quality would normally range from 0.95 to 0.97.

Metamorphism, ripening, and thermal quality as they affect density, water equivalent, and liquid-phase retentivity of a snowpack require different engineering interpretations, especially with regard to the hydrographs which may be yielded by snow-covered drainage basins subjected to flood-producing rainstorms [23, chap. 3].

V. MEASUREMENT OF SNOW AT TIME OF FALL

As snow is a form of precipitation, climatological records [17] commonly report snowfall depth as measured at time of fall, and the water equivalent of the snow is included in precipitation totals. The U.S. Weather Bureau's *Circular* B [24] gives detailed instructions for climatological observers. Among the many approaches used in the attempt to attain a consistent measurement of water equivalent of snow at time of fall are (*a*) precipitation gage with designs of various types, including seasonal storage gages, (*b*) snow boards, and (*c*) snow stakes.

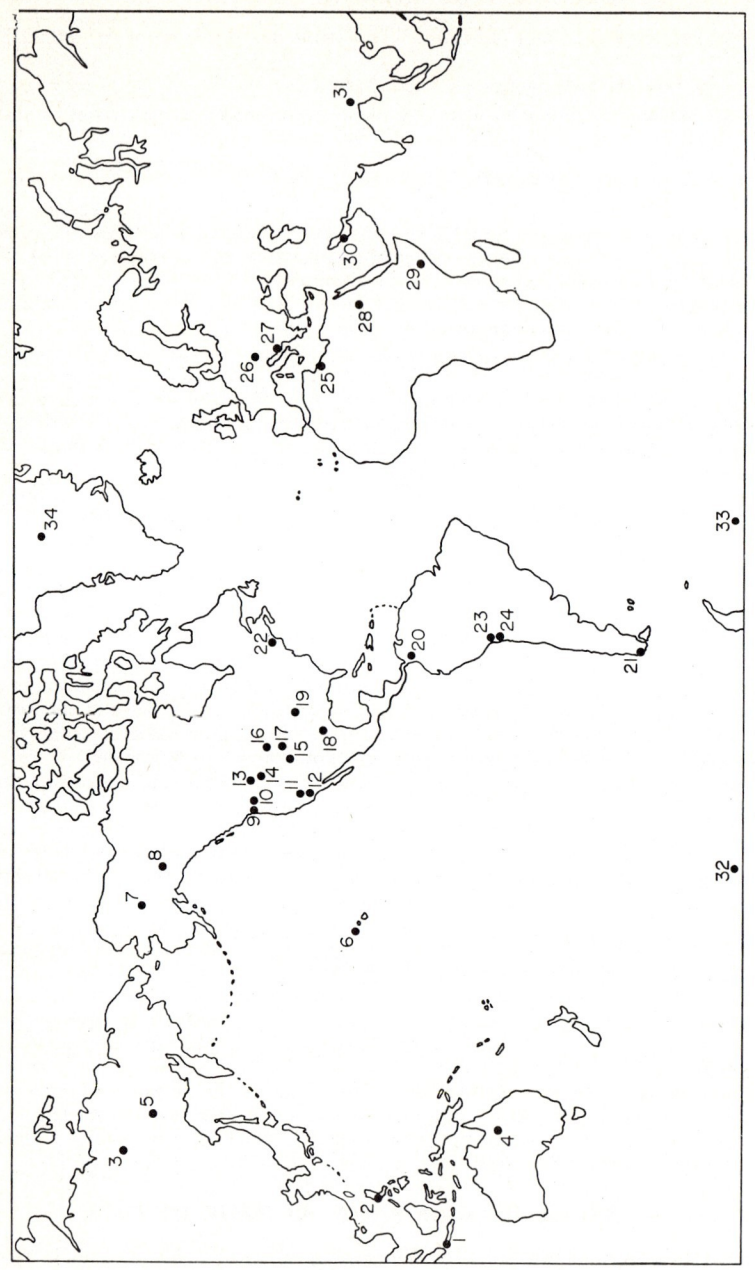

Fig. 10-4 See opposite page for descriptive legend.

A. Precipitation Gages

1. Standard and Recording Rain Gages. The 8-in. nonrecording rain gage, which is standard in the United States (Sec. 9 and [24, p. 20]) is also used for the measurement of snow. The rainfall receiver, a short, sharp-edged metal cylinder, exactly 8 in. inside diameter, provided with a funnel-shaped bottom, is ordinarily removed

FIG. 10-4. Weather extremes around the world. Undoubtedly, more extreme values have occurred than have been recorded. Some recorded values may not have been accepted as official because of incomplete information or nature of exposure. Some of the official values are under question. (*U.S. Army Quartermaster* [22], revised *April*, 1959.)

1. Average annual thunderstorm days, 322, Buitenzorg, Java.
2. World's greatest 24-hour rainfall, 46 in., July 14–15, 1911, Baguio, Luzon.
3. Northern Hemisphere's lowest temperature, −90°F, Verkhoyansk (1892), Siberia.
4. Australia's highest temperature, 127°F, January, Cloncurry, Queensland.
5. Northern Hemisphere's lowest temperature, −90°F, 1933, and Northern Hemisphere's unofficial lowest temperature, −108°F, Oimekon, Siberia.
6. World's greatest average annual precipitation, 472 in., Mt. Waialeale, Kauai, Hawaii.
7. Alaska's lowest temperature, −76°F, Tanana, January, 1886.
8. North America's lowest temperature, −81°F, Feb. 3, 1947, Snag, Yukon.
9. United States greatest average annual precipitation, 156 in., Wynoochee, Wash.
10. United States greatest single season snowfall, 1,000.3 in., Paradise Ranger Station, Washington, 1955–1956.
11. United States highest temperature, 134°F, July 10, 1913, Death Valley, Calif.
12. United States longest dry period. 767 days, October, 1912–November, 1914, Bagdad, Calif.
13. United States greatest 24-hour temperature fall, 100°F, from 44 to −56°F, Jan. 23–24, 1916, Browning, Mont.
14. United States lowest temperature (excluding Alaska), −70°F, Jan. 20, 1954, Rogers Pass, Mont.
15. United States greatest 24-hour snowfall, 76 in., Apr. 14–15, 1921, Silver Lake, Colo.
16. United States greatest 2-min temperature rise, 49°F, from −4° to 45°F, Jan. 22, 1943, Spearfish, S.D.
17. Largest officially recorded hailstone, 5.41-in.-diameter, Potter, Nebr., July 7, 1928.
18. United States unofficial greatest 12-hour rainfall, 32 in., Sept. 9, 1921, Thrall, Tex.
19. World's greatest 42-min rainfall, 12 in., June 22, 1947, Holt, Mo.
20. South America's greatest average annual rainfall, 342 in., Buena Vista, Colombia.
21. Bahia Felix, Chile, has an average of 325 days/year with rain.
22. World's highest surface wind speed, 231 mph, Apr. 12, 1934, Mt. Washington, N.H.
23. World's lowest average annual rainfall, 0.02 in., Arica, Chile.
24. Iquique, Chile, had no rainfall for 14 years.
25. World's highest temperature, 136°F, El Azizia, Libya, Sept. 13, 1922.
26. Average daily total solar radiation, 770 g cal/cm^2, June, Davos, Switzerland.
27. Europe's greatest average annual precipitation, 183 in., Crkvice, Yugoslavia.
28. Wadi Halfa, Sudan, had no rain in a 19-year record of observations.
29. World's highest average annual temperature, 88°F, Lugh Ferrandi, Somalia.
30. World's highest mean monthly dew point, 83°F, August, Bahrein Island.
31. Cherrapunji, India, had 12.5 feet rain in one 5-day period, August, 1841; had world's greatest rainfall in one month, 366 in., July, 1861; had world's greatest rainfall for one year, 1,042 in., August, 1860–July, 1861.
32. Antarctica's lowest annual average temperature, −71°F, Sovietskaya, at 12,200 ft.
33. World's lowest temperature, −125°F, Vostok, Antarctica, Aug. 25, 1958.
34. Greenland's lowest temperature, −87°F, at 9,820 ft, Dec. 6, 1949.

REFERENCES:

New York Times, Jan. 2, 1959.
Weekly Weather and Crop Bulletin, USWB, vol. 43, no. 5, Jan. 30, 1956; Feb. 27, 1956.
Monthly Weather Review, USWB, vol. 78, no. 1, January, 1950; vol. 81, no. 2, February, 1953, no. 7, July, 1953.
Naval Air Pilot, East Central Africa, prepared by USWB for USND, 1943.
Journal of Glaciology (Loewe), vol. 2, no. 19, 1956.
Weatherwise, vol. 9, no. 6, December, 1956.
F. Prohaska, *Gerlands Beitr Geophys.*, vol. 59, no. 3–4, 1943.
USWB Letter, May 4. 1956.

for winter operation. The measuring tube is likewise removed from the overflow can for winter operation. The overflow can is used to gage the snow. The funnel-bottomed receiver can be used in conjunction with snow boards for securing samples of freshly fallen snow. Observers commonly bring the overflow can into a warm interior and measure the water from the melted snow in the usual manner. Samples may be taken with the overflow can.

Weighing-type recording rain gages have likewise been operated with the additional provision of antifreeze compounds in the catchment bucket for measuring snowfall.

The influence of wind velocity and of shape of precipitation gages upon the catchment of precipitation long has been recognized. In 1878 Nipher [25] wrote about the desirability of shielding of rain gages. Snow is very sensitive to air-velocity differences. The inability of rain gages to measure snow has led to numerous attempts to control the wind flow at the orifice of the gage. Alter [26] described in 1937 a shield for use with precipitation gages which, with minor modifications, is widely used.

Warnick [27] conducted extensive investigations both in the wind tunnel using sawdust to simulate snow and in the forests and mountains in northern Idaho, aimed at the development of improved shielding of snow gages.

2. Seasonal Storage Precipitation Gages. Since the heavy snowfalls of economic and hydrologic importance often fall in sparsely inhabited areas, numerous designs of storage gages have been worked out. Storage gages may have the capacity of up to several hundred inches of water equivalent as expressed in terms of their intake orifice area [28].

Depending upon the frequency of attendance and upon the expected total precipitation, the U.S. Weather Bureau [29] has placed in use three types of storage gages:

1. The 8-in. cylindrical rain gage 24, 36, or 42 in. in height, made at installations which may be measured at intervals of from 7 to 30 days by weighing the gage and computing the increment of precipitation.

2. The Sacramento-type, conical, seasonal storage precipitation gage with capacities of 60, 100, 200, and 300 in. The seasonal catchment is determined by weighing once or twice each year, at which times withdrawal of the liquid may be necessary. Intermediate observations are made with a calibrated stick.

3. The standpipe seasonal precipitation gage made up of a vertical steel smokestacklike structure 12 in. in diameter which converges to the standard 8-in.-diameter intake orifice.

The three-section standpipe gage has a capacity of 250 in.; the four-section, 475 in. The seasonal catch is determined once or twice each year by withdrawing the liquid and weighing it. Intermediate measurements are made with a dipstick or tape.

All three gages are provided with Alter shields [26] with baffles not rigidly restrained at the lower ends (Fig. 10-11).

For the 8-in.-diameter intake orifice, 1 lb of snow equals 0.55 in. of water equivalent. To facilitate observations the U.S. Weather Bureau [24] has developed spring scales calibrated to read, when used with an 8-in. gage, directly in inches of water equivalent of depth of precipitation. Observations are recorded on a special form [29, WB Form 612-41].

Storage gages are protected against freezing usually through the use of either calcium chloride solutions or ethylene glycol solutions. Care should be exercised in the placing of the calcium chloride antifreeze solution since this substance may form several hydrates differing greatly in their effectiveness in depressing the freezing point of the solution [30]. The greatest depression of the freezing point is attained with calcium chloride at $-59.8°F$ by a concentration of 29.6 per cent of the salt by weight. An increase in the amount of the calcium chloride results in a rise in the freezing point. Increase in the strength of the calcium chloride solution which has been diluted by precipitation will be effective only if an aqueous solution of calcium chloride of lesser concentration than mentioned above is used. Addition of the solid calcium chloride may result in the formation of a hydrate and cause loss of protection of the storage gage against damage by freezing, with possible resultant loss of the record. Freezing-point diagrams [31] for calcium chloride and ethylene glycol solutions are given in Fig. 10-5.

MEASUREMENT OF SNOW AT TIME OF FALL

Transformer oil [32] has been found to be superior to other substances for use in storage gages to prevent loss of catchment by evaporation. Such loss can be significant since storage gages may be prepared for the winter as early as mid-September and not be observed until the following March or April.

It is not unusual for high rates of snowfall as expressed in water equivalent to take place exactly at 32°F under microclimatological conditions conducive to the generation of a cohesive and sticky snow. Under such conditions, because of the arch action which snow exhibits to a remarkable degree, storage gages are often very adversely affected and at times completely incapacitated by the snow bridging over the intake orifice. It has been demonstrated that the supplying of relatively small amounts of heat to the orifice of a precipitation gage can very successfully prevent capping over of the intake. Allen and others [32, 33] describe the development of the liquid-phase heat-transfer system used in a radio-reporting rain and snow gage system [34] operated on call by the U.S. Bureau of Reclamation in the Central Valley of California [35].

Warnick [36, 37] describes the development and initial field observations of an electrically heated orifice for a storage precipitation gage which incorporates upper- and lower-limit switches to reduce electrical-energy requirements supplied by storage batteries.

There is a need in river-regulation activities for automatic, unattended, radio-reporting instrumentation capable of describing at the time of report the depth, water equivalent, thermal quality, and other characteristics of a snowpack. Recent successful developments in radio reporting of water equivalent are described in Subsec. VI-E. Maxwell and others [38] describe the development, as yet incomplete, of a radio-reporting complex of instrumentation, as part of which snow depth would be sensed by a chain-mounted, motor-driven, photoelectric-cell system.

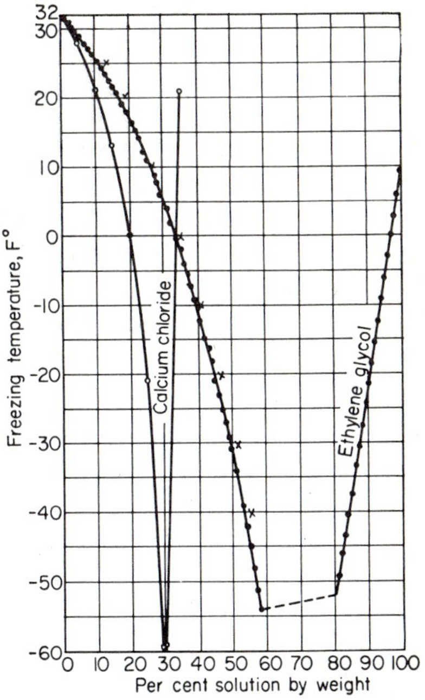

FIG. 10-5. Freezing-point diagrams for calcium chloride and ethylene glycol solutions. (*Garstka, Love, Goodell, and Bertle* [31, p. 157].) • *From C. S. Cragoe: "Properties of Ethylene Glycol and Its Aqueous Solutions," Cooperative Research Council, New York, 1943, p. 13.* × *From Hodgman and Holmes: "Handbook of Chemistry and Physics," 26th ed., Chemical Rubber Publishing Co., Cleveland, Ohio, 1942, p. 1713.* ° *From U.S. Weather Bureau, University of Nevada Cooperating, Snow Studies at the Cooperative Snow Research Project, Soda Springs, Calif., Annual Report 1943–1944, p. 5.* □ *Lowest plot, from U.S. Weather Bureau: "Operation and Maintenance of Storage Precipitation Gages," 1951, p. 3.*

B. Snow Boards

The U.S. Weather Bureau *Circular* B [24] describes the use of snow boards or snow markers at least 16 in. square which are laid on the previous accumulations of snow. The freshly fallen snow can thus be readily identified, and snow samples can be cut from it (Subsec. V-A-1). Snow boards can be placed only in locations where wind-flow and snow-accumulation patterns are known to be such that the observations made using the snow boards would be acceptable as being indicative of snowfall at that site. Several snow boards should be used to attain more representative data.

C. Snow Stakes

Wooden stakes, 1¾ in. square in cross section, provided with angle-iron supports and calibrated in inches, in a design resembling the markings of a stadia rod, are used [24] to indicate the vertical depth of snow on the ground. Depth of snow can also be measured by plunging a stick or metal rod to the ground surface.

Snow stakes or markers of altogether different design are used in facilitating observations of snow from aircraft, as discussed in Subsec. VI-D.

VI. SNOW SURVEYING

To the hydrologist, many of the characteristics of the snowpack, such as the water equivalent, density, and depth, are determined at the desired time and place by snow surveys.

A. Snow-course Layout on the Ground

Snow surveying is performed by taking samples of the snowpack with suitable core-cutting equipment. Since determination of water equivalent by melting is not practical in the snow fields, the water equivalent is customarily determined by weight of the snow at the time of sampling. A *snow course* consists of a series of sampling points, usually not fewer than 10. The points are located along a predetermined geometric pattern at a spacing of 50 to 100 ft. The ends and pivots of the pattern are permanently marked to make certain that the snow is surveyed at the same locations year after year.

The snow courses should be so located as to yield, year after year, data on snow which will reflect differences in seasonal accumulations uninfluenced by encroachments of forest growth or cultural disturbances. It is very important that snow surveys be performed at the established course. Unless this is done, the commonly observed great range of variability in very short distances in the accumulated snow conditions could result in yielding data not indicative of differences in the snowfall from year to year. Figure 10-6 shows a standard snow-course marker.

Locations for the establishment at snow courses should be inspected, first, in the winter, when the snowpack is present, but the snow courses should be established when the area is free of snow, preferably in late spring or early summer when any sidehill seepage can be seen. The markers should be placed high enough aboveground so that they will remain visible even after the maximum depth of snow has accumulated. Only by laying out the snow course in the summer is it possible to choose spacing for sampling points so that they will not be located on down timber, rock ledges, stumps, or intermittent or flowing stream channels.

If over-snow vehicles are to be used, advance preparation can be made in the summer for improving the access trail to the snow course. In remote uninhabited country it may be necessary to build and stock a shelter cabin which, in areas of great depth of snow, may need a special access provided by a "Santa Claus chimney."

The performance of a snow survey requires experienced and rugged individuals, confident of their ability to survive under extremely hazardous weather conditions. The Snow Survey Safety Guide [39] is recommended not only to snow surveyors, but also to anyone working outdoors in the winter.

In most parts of the West a snow course for each 100 sq mi of drainage basin is considered to be an indicative network, and snow-course systems of one course to 300 or 400 sq mi are not unusual. In designing a new snow-survey system for a drainage basin, it is wise to establish a considerably greater number of snow courses than experience in other basins would indicate, which may ultimately be used in the forecasting system.

Since it is desirable to accumulate at least 10 years of data before correlation analyses are made of the snow-survey data in streamflow forecasting, a failure to establish a sufficient number of courses at the very beginning introduces, later on, difficulties in statistical and correlation analysis. Should the hydrologist decide that more snow courses are needed, the expanded snow-survey system would yield data from snow courses of unequal lengths of record. The degree of confidence which can be placed in the forecasts from such a system would be reduced until a sufficient number of years has elapsed to establish the value of a particular snow course.

FIG. 10-6. Standard Federal-state cooperative snow-survey snow-course marker. The snow surveyor is holding a snow sampler. (*U.S. Soil Conservation Service*.)

Marr [40] and Work [41] describe snow-survey operations in further detail. The layout of the Brush Creek Summit, Montana, snow course is shown in Fig. 10-7. Detailed instructions for sampling of both shallow snows and of snow depth exceeding the length of the sampling equipment are given in Ref. 42.

B. Snow-surveying Instruments

The funnel of a standard 8-in. rain gage can be used to cut cores of shallow snow for subsequent determination of water equivalent by melting (Subsec. V). In an emergency, any cylinder, such as a stovepipe, can be used to take samples of snow. Variations in snow density, and especially the occurrence of layers of ice in a deep snow (Subsec. IV), have required the development of highly specialized equipment for snow surveys. The Mount Rose snow-sampling equipment was originally developed by Church [43, p. 97, footnote 2]. Following several improvements, there evolved what is now known as the *Federal snow sampler* [44]. The configuration of the hardened tool-steel cutter teeth is such that a core of snow and ice can be extracted without incorporating in the sample the chips of ice scraped from the layers being drilled as the snow sampler is twisted through a deep snow bank. It has been established that the Mount Rose cutter having an inside diameter of 1.485 in. (which makes one ounce weight equivalent to one-inch depth of water) is just as accurate as, and much easier to handle than, larger-diameter cutters [45].

10-16 SNOW AND SNOW SURVEY

As it is not unusual to find the snow (in deep accumulations at some distance below the surface) to be at exactly 32°F when the surface air temperature may be considerably below freezing, special surface treatment of the sampler with hard wax or paraffin is required to forestall plugging of the tubes when they are inserted into the snow.

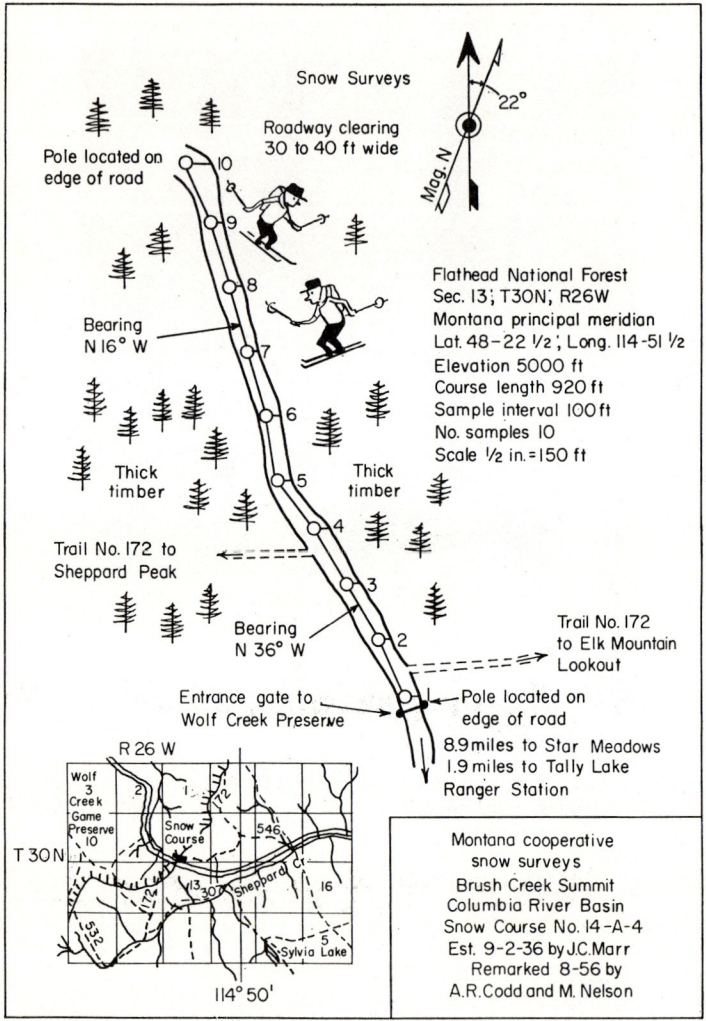

FIG. 10-7. Description of Brush Creek Summit, Montana, snow course. (*U.S. Soil Conservation Service* [42, p. 4].)

The capacity of the equipment depends upon the total number of 30-in.-long sampling tubes constituting the set. In the Pacific Northwest snow depths of 150 in., with water equivalents of 50 in. on April 1 or May 1, are not uncommon. Snow-survey instruments are commercially available from meteorologic- and hydrologic-instrument manufacturing companies. Snow surveyors have designed a wide variety of both firm and canvas type of carrying case and back packs to hold the snow-survey equipment. Figure 10-8 shows, schematically, a snow-sampling kit assembled for wrapping in canvas [42].

Figures 10-9 and 10-10 are examples [42] of data recorded when the snow course of Fig. 10-7 was surveyed. The Soil Conservation Service developed the form based upon extensive field experience. There are, as of 1962, about 1,400 snow courses in the West. These are usually surveyed once a month, in February, March, and April and in some instances on May 1.

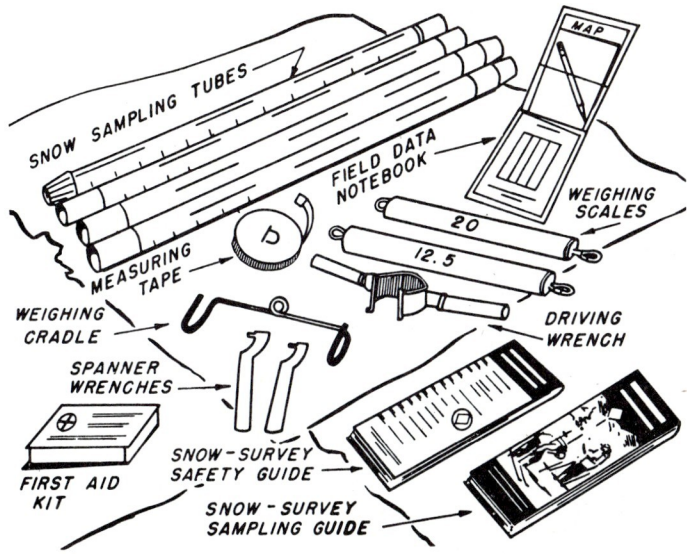

Fig. 10-8. Snow-sampling kit assembled for wrapping in canvas. (*U.S. Soil Conservation Service* [42].)

C. Snow Surveying on the Ground

Access on foot to snow courses usually requires snowshoes or skis. On-foot cross-country travel in the mountains is hazardous, laborious, and time-consuming. Therefore extensive effort has been expended on the development of an over-snow vehicle. As travel over snow takes on many of the aspects of navigation over water, the requirements of satisfactory over-snow travel differ markedly from those which are acceptable for travel on solid terrain. Over-snow vehicles may apply their propelling force in various ways. There are (1) air-driven vehicles, (2) track-laying vehicles, and (3) sliding vehicles.

The air-driven vehicles are chiefly very lightly built sleds propelled by an airscrew. They can be used only in open country where easy maneuverability and relatively flat terrain make possible the travel at high speeds which characterizes this type of over-snow vehicle.

The track-laying vehicles are similar to the caterpillar tractor in their manner of propulsion. The track-laying vehicle moves on wheels upon track which rests, ideally, without motion on the snow surface. Any movement of the track may result in digging in and sinking of the vehicle. The U.S. Army Weasel and M-7 over-snow vehicles are examples of this system of propulsion.

The sliding vehicle is propelled by a ladderlike movable endless belt which is laid on the snow beneath pontoons. Ideally, the ladder does not move, and the vehicle slides forward on the pontoons over the snow like a ski. Propulsion requires that the shear resistance of the snow exceed the forces needed to propel the vehicle.

According to Work [46], an acceptable over-snow vehicle should have mechanical dependability, ability to travel steadily and consistently in loose, deep snow or on soft, sticky snow, on sidehills, and on bare ground. It should have hill-climbing ability, maneuverability, including ability to surmount obstacles such as fallen trees, economy, speed, ability to cross upon frozen streams, portability, safety, and comfort. No one vehicle as yet possesses, to the extent desired, all these features. Bekker [47] reviewed the concepts pertinent to vehicle design.

SCS-708 (5-56)		UNITED STATES DEPARTMENT OF AGRICULTURE SOIL CONSERVATION SERVICE FEDERAL AND STATE COOPERATIVE SNOW SURVEYS						
	State	MONTANA						
	Drainage Basin	Columbia — Kootenai						
	Snow Course	BRUSH CREEK SUMMIT						
	Party	R. Funke — T. Triplett						
	Date	Feb. 27, 1958						

*Description or Number of Course	†Sample Number	Depth of Snow Inches	Length of Core Inches	Weight of Empty Tube	Weight of Tube and Core	Water Content Inches	Density Per-cent	Remarks (see reverse)
	1	38½	32	23	33½	10½	27	G.F. Damp
	2	42½	37		35	12	28	" "
	3	40	37		33½	10½	26	" "
	4	43	36½	23	35	12	28	G.N.F. Damp
	5	41½	37		34½	11½	28	" "
	6	45	42		36	13	29	Dry Dirt
	7	44	40	23	35½	12½	28	Damp Dirt
	8	38½	34½		34	11	29	Damp Needles
	9	44	39	*44	57½	13½	31	Wet Dirt
	10	38½	34½	44	55	11	29	Water
Total	⑩	415½				117½	28	
Average	⑩	41.6				11.8		
			*Driving wrench added.					

*Show number or description as given on sketch map, i. e., "Course No. 1," or "Major Course," or "N 5° E," etc.
†Always start measurements for sampling from the *initial* point as shown by the sketch map of the course and follow the spacing for samples as indicated. Particular care should be taken to note any *irregular* spacing between samples.

No. ..1.. of ..1.. sheets. Comp. by ..R—F.. Checked by ....

Fig. 10-9. Front of Federal and state cooperative snow-surveys record form. See Figs. 10-7 and 10-10. (*U.S. Soil Conservation Service* [42, p. 25].)

Most concepts relating to vehicular travel over soil apply to snow, except that the mechanical characteristics of snow may vary greatly throughout the day, depending on temperature and thermal quantity. Most over-snow vehicles exert a very light pressure on the bare surface, usually in the range of about ½ to 1 psi. Diamond [48] and Lanyon [49] describe extensive tests of several vehicles.

Over-snow vehicles have reached such a degree of development that they are now in common usage, not only by snow surveyors, but also by mining interests, power-line maintenance crews, naturalists, foresters, and winter-resort operators. A detailed description, profusely illustrated, of over-snow travel along the spine of the Oregon Cascade Mountains is given by Brown [50].

Another manner of travel to snow courses is by air. This form of travel is very successful where adequate landing areas, experienced pilots, and suitable aircraft are available, according to Codd and Nelson [51]. The greatest disadvantage of air travel to snow courses is that the weather at the date scheduled for the snow survey may not be suitable for flying in mountain terrain. Proximity of the landing field to the snow course is important, since travel of several miles on foot after landing can be so time-consuming and costly in terms of aircraft rental as to reduce greatly the economic advantages of traveling by air. Aircraft, to be suitable for travel to snow fields, should be able to take off and land on either bare ground or snow; completely satisfactory landing equipment, capable of service on snows of different degrees of ripeness, for the small but powerful aircraft now available, has not as yet been developed.

```
NOTE.—Please fill in while in the field.
                              a. m.              a. m.
DATE OF SURVEY: Began 10:20 p.m.  Ended 11:10 p.m.
                    SAMPLING CONDITIONS
          (Please check items descriptive of present conditions.)
                  Weather at Time of Sampling
...... Clear, ......... Partly cloudy, ......... Overcast, ......... Raining,
..✓.. Snowing, ......... Blowing, ......... Freezing, ......... Thawing.
                Snow Conditions at Snow Course
Snow samples obtained with ..✓.. ease, .......... moderate difficulty.*
Snow samples obtained with .......... extreme difficulty.*
Ground under snow: ......✓...... frozen, .................. not frozen,
.............. dry, .....✓..... damp, .................. wet (saturated).
Ice layer on ground None.. How thick? .......... inches.
                  General Snow Conditions
1. What elevation is snow line generally? .......... ft.
2. Is snow melting on north and east slopes? ..No..........
3. Is snow melting on south and west slopes? ..No..........
4. How many inches of new snow at snow course? ..........2...... in.
5. Is there evidence of snowslides? .None....
              General Stream-Flow Conditions
1. Are very small streams running? Yes ......... No ..✓........
2. Are small streams bridged over by snow? Yes ..✓... No ..........
3. Are streams clear or muddy? (Check one) Clear ..✓... Muddy ......
*Explain fully under remarks.
                    PRECIPITATION DATA
```

Month	Day	Year	Precipitation	Readings	Dipstick	Weight
			Current		Made by (check)	
			Previous			
Station name			Catch, inches		Scale number	
			After recharge			

```
REMARKS: ......................................................................
    .....Truck..Roundtrip...............58.Mi.....
    ..........Sno.Cat....".................12.Mi.....
    ..........Foot.Travel.................6.Mi.....
    ........Used..3..sections..tubing.................
    ........Used..driver..at..Samples..9.and..10....
```

FIG. 10-10. Back of Federal and state snow surveys record form. See Figs. 10-7 and 10-9 (*U.S. Soil Conservation Service* [42, p. 26].)

D. Snow Observations from the Air

An increasing demand for greater precision in water-supply forecasting has led to an endeavor aimed at securing more data on the snow in the headwaters of drainage basins where the terrain may be rugged and where access on foot or by over-snow-vehicle travel may be time-consuming and expensive. For water-supply forecast computations the ideal would be to have all the snow courses in a drainage basin sampled at the same time. This is seldom possible, and, for most purposes, statistical analysis can be made if all the courses have been surveyed during the period of uninterrupted clear weather. The occurrence of a major snowstorm after some of the snow courses have been surveyed either requires a resurvey or results in complicated statistical computations aimed at converting the snow-survey data to an assumed equally indicative basis. These considerations have led to the development of snow surveying from the air [52].

The advantages of aerial snow markers are that they permit extensive observations within very short time of the snow depth prevailing over the drainage basin. Hannaford [53] mentions two important disadvantages of aerial snow markers: one is that they are single-point measurements, and the other is that they yield only depth, and not water content or density data. The requirement for weather suitable for flying and for observation may set a date which may not coincide with the established dates of surveys of snow courses.

Dean [54] reports that helicopters obtain maximum efficiency by distributing power between forward motion and lift, vertical lift being inefficient at all altitudes and practically impossible at higher elevations when a helicopter is carrying two persons and the supplies and equipment needed for snow surveying. However, with continuing improvement of helicopters, their future widespread usage in snow surveying appears assured. Since late 1961, landing and takeoff have been made at elevations of 12,000 ft.

Washichek [55] found, working in Colorado with markers in snowpacks varying from 100 to 150 in. in depth, that the computed water equivalent agreed within 1 to 3 in. of water equivalent as determined by snow surveys. A greater agreement existed before the spring melt actively began. Agreement was poor on the May snow surveys.

Aerial photographic snow-depth measurements for a large number of markers in California are reported in a basic-data supplement to the State of California water resources series of bulletins entitled *Water Conditions in California* (Subsec. VI-F).

E. Radioisotope Snow Gage

The value of information on the current condition of the snowpack in remote and inaccessible drainage basins has long been recognized. The recent availability of radioactive isotopes of elements capable of yielding known intensities of electromagnetic radiation has made possible the development by Gerdel and others [56] of a snow-water equivalent gage based upon the absorption of gamma radiation by water substance. The radioisotope snow gage, however, does not discriminate between the various forms of water substances, whether in the liquid, solid, or vapor phase. The loss of intensity of radiation as measured by a Geiger-Müller tube is in proportion to the water equivalent of the snowpack, regardless of the density and its variations or thermal quality.

The radioisotope snow gage as developed by the U.S. Army Corps of Engineers and the U.S. Weather Bureau has a capacity of 55 in. water equivalent and is capable of determining the water equivalent of a snowpack without disturbance within a 2 to 5 per cent deviation [57]. This Corps of Engineers–Weather Bureau radioisotope-snow-gage configuration places the source of gamma radiation below ground level, with the Geiger-Müller tube suspended in the air. The opposite of this configuration is used by Itagaki [58]. In this Japanese system the source of radiation is suspended in the air. Itagaki's gage has a capacity of 60 cm of water substance.

Both radioisotope gages are provided with automatic radio-reporting telemetering facilities which are described, for Gerdel's system by Doremus [59] and by Itagaki [58]. The capacity for a radioisotope snow gage requires a judicious choice of the intensity of gamma radiation with due regard to public safety. Whatever the configuration of the system may be, it is essential that the intermittent radiation as picked up by the Geiger-Müller tube be of such strength as to be indicative of snow-water equivalent, the natural background radiation at the site notwithstanding.

The U.S. Army Corps of Engineers is operating a system of radio-reporting radioisotope snow gages in the King's River drainage basin in California. Figure 10-11 shows a Sacramento-type seasonal storage precipitation gage and the radioisotope snow gage as installed at Mitchell Meadow in the King's River Basin. Hildebrand [60] gives a comprehensive report on the theoretical physical background and the development of equipment, including power supply, of the installations in California.

Close agreement has been observed between water equivalents as surveyed at snow courses with those reported by radio as determined by the radioactive snow gages in King's River, California [61].

SNOW SURVEYING

Robinson [62] describes a Corps of Engineers radioisotope radiotelemetering system of three snow gages in the Clearwater River watershed in Idaho. The gamma radiation emitted, by initially 80 millicuries of cobalt 60 in two energy bands 1.33 and 1.17 Mev, is detected in the air, 15 ft above the buried lead-shielded collimator, by a 1½-in.-diameter thallium-activated sodium iodide scintillation crystal and a photomultiplier tube. Capacity of this snow gage is 50 in. water equivalent. Signals transmitted from these snow gages are automatically received on a modified adding-machine mechanism.

FIG. 10-11. Sacramento-type seasonal-storage-precipitation gage with Alter shield mounted on tower, on the left, and radioisotope radio-telemetering snow gage on the right, as installed at Mitchell Meadow, King's River drainage basin, California, as seen from the air on Sept. 27, 1958. Depth of snow on the ground was about 8 ft. (*Courtesy of U.S. Army Corps of Engineers, Sacramento District.*)

F. Sources of Snow-survey Data

More than 1,400 snow courses are surveyed each year in the West. Table 10-2 lists information on the sources of data for the Federal-State-Private Cooperative Snow Survey System which is coordinated by the U.S. Soil Conservation Service.

The State of California maintains its own snow-survey system. Reports are issued monthly, February through May, by the California State Department of Water Resources, Sacramento, in two series, one of which presents forecasts and the other supplements them with basic data. The latter gives data not only on snow surveys performed on the ground, but also on snow depths as observed from the air (Subsec. VI-D) and also on water equivalents as reported by the radioisotope radiotelemetering snow gages (Subsec. VI-E).

British Columbia snow courses are reported monthly, February through June, by the Comptroller, Water Rights Branch, Department of Lands and Forests, Vancouver, B. C., Canada.

There is no central coordinating office for snow surveys for the Lake States or for the eastern United States or eastern Canada. It is known that, in the East, various Federal, state, and private agencies perform snow surveys, not only for hydrologic purposes, but also in connection with fish and wildlife, recreational, and highway-maintenance activities. It is suggested that those interested in snow data for the central or eastern parts of the continental United States refer to a nearby office of the

U.S. Weather Bureau or of the Water Resources Division of the U.S. Geological Survey (Subsec. V).

The U.S. Soil Conservation Service issues summaries of snow-survey data for the same areas listed in Table 10-2. These summaries are brought up to date at approximately 5-year intervals.

Table 10-2. Reports for That Portion of the Western Federal-State-Private Cooperative Snow Survey System Coordinated by the U.S. Soil Conservation System*

Report	Frequency of issue	Location	Cooperating with:
River basins:			
Colorado and state of Utah	Monthly (Jan.–May)	Salt Lake City, Utah	Utah State Engineer and other agencies
Columbia	Monthly (Jan.–May)	Boise, Idaho	Idaho State Reclamation Engineer
Upper Missouri and state of Montana	Monthly (Feb.–May)	Bozeman, Mont.	Montana Agricultural Experiment Station
West-wide	Oct. 1, Apr. 1, May 1	Portland, Ore.	All cooperators
States:			
Alaska	Monthly (Mar.–May)	Palmer, Alaska	Alaska Conservation District
Arizona	Semimonthly (Jan. 15–Apr. 1)	Phoenix, Ariz.	Salt River Valley Water Users Association, Arizona Agricultural Experiment Station
Colorado and New Mexico	Monthly (Feb.–May)	Fort Collins, Colo.	Colorado Agricultural Experiment Station, Colorado State Engineer, New Mexico State Engineer
Idaho	Monthly (Feb.–May)	Boise, Idaho	Idaho State Reclamation Engineer
Nevada	Monthly (Feb.–Apr.)	Reno, Nev.	Nevada Department of Conservation and Natural Resources, Division of Water Resources
Oregon	Monthly (Jan.–May)	Portland, Ore.	Oregon Agricultural Experiment Station, Oregon State Engineer
Washington	Monthly (Feb.–May)	Spokane, Wash.	Washington State Department of Conservation
Wyoming	Monthly (Feb.–June)	Casper, Wyo.	Wyoming State Engineer

* Copies of these various reports may be secured from Soil Conservation Service, Water Supply Forecasting Section, Portland, Ore.

The Western Snow Conference, organized on Feb. 18, 1933, by G. D. Clyde, H. M. Stafford, and J. E. Church [63], has issued, with a few interruptions, a series of *Proceedings* of its meetings. Prior to 1949, the *Proceedings* were published in the *Transactions of the American Geophysical Union*. Beginning with 1949, the *Proceedings* have been published separately by the Western Snow Conference, being printed by the Colorado State University, Fort Collins, Colo.

The Central Snow Conference held one meeting at the Michigan State University, East Lansing, Mich., in December, 1941 [64], and has been inactive since that meeting.

OBJECTIVES OF RUNOFF FROM SNOWMELT COMPUTATIONS   10–23

Although organized prior to December, 1941, the Eastern Snow Conference held its second meeting in 1949. To date six volumes of *Proceedings* have been published as follows:
Volume 1, Ninth Annual Meeting, 1952
Volume 2, Tenth and Eleventh Annual Meetings, 1953, 1954
Volume 3, Twelfth Annual Meeting, 1955
Volume 4, Thirteenth and Fourteenth Annual Meetings, 1956, 1957
Volume 5, Fifteenth Annual Meeting, 1958
Volume 6, Sixteenth and Seventeenth Annual Meetings, 1959, 1960

A complete set of *Proceedings* is on file at the Library of Congress, at Syracuse University, and at the Yale University Forestry Library. The secretaryship of the Eastern Snow Conference reposes with personnel of the U.S. Department of the Interior, Geological Survey, Water Resources Division, Branch of Surface Water, Albany, N.Y., and any inquiries concerning the activities of the Eastern Snow Conference should be forwarded to that address.

Norwich University, Northfield, Vt., is microfilming all available papers which have not been published in any of the Eastern Snow Conference volumes enumerated above, with the intention that microfilm prints should be available upon request to interested persons.

Literature dealing with snow surveying is to be found incorporated in publications of practically all the meteorological, geophysical, agricultural, and civil-engineering periodicals and publications.

VII. OBJECTIVES OF RUNOFF FROM SNOWMELT COMPUTATIONS

Computations and forecasts of runoff from snowmelt may be performed for a number of objectives, among which the most important hydrologic usages are seasonal water-yield forecasting and rate-of-runoff forecasting. Both of these objectives may require snowmelt computations either for the operation of existing projects or for the use of designing future projects. Snow- and icemelt computations are pertinent, likewise, to the hydrologic operations in drainage basins partly fed by water yielded from glaciers. The computation of snowmelt is inherent in the clearance of highways and walkways by induced melting (Subsec. XIV) and to snow and ice loads on power lines (Subsec. XII). The microclimatological and physical factors underlying the complex processes by which runoff is yielded by melting snow are discussed in Subsecs. VIII and IX.

It should be kept in mind that a great many of the computations relating to runoff from snowmelt make no use whatever of a knowledge of the processes of the melting of snow. Seasonal water-yield computations are essentially correlation analyses, ranging from simple graphic correlations to complex curvilinear multiple correlations. Expressed in its simplest terms, runoff from snowmelt is the result of heating an accumulation of ice crystals. A great number of procedures have been developed. The choice of a particular one depends upon the data available, lengths of records, analytical procedures used, and especially upon the objective for which the runoff from snowmelt computation is being made.

A. Seasonal Water-yield Forecasting

As precipitation and snowfall vary from season to season and from year to year, wide variations in the water supply yielded by drainage basins have been experienced. Seasonal water-yield forecasting has been developed to provide information, in advance, on the amount of water which may be available for the remainder of the water year. Such information is invaluable in guiding the planning of operations of reservoirs and of reservoir systems. Irrigators are dependent on seasonal water-yield forecasting for planning their crop programs as discussed, with references to specific instances, by Fredericksen [65]. Hydroelectric generating systems are concerned with seasonal water-yield forecasting to assist in securing the most efficient management with the highest overall income as discussed in detail, with examples, by Blanchard [66].

Flood control, navigation, municipal and industrial water supply, and practically every other human endeavor make use of seasonal water-yield forecasting. Confidence in the accuracy of seasonal water-yield forecasting runoff from snowmelt has made possible the lowering of the contents of multiple-purpose reservoirs to have space available for flood control and, still, to result in a reservoir filled to its allotted capacity for irrigation at the beginning of the irrigation season.

The diversity of usages of seasonal water-yield forecasts has resulted in the development of techniques of analysis of data and of computations of forecasts related to the usage. For example, irrigation-project operators are interested in a forecast of the highest possible accuracy of the total volume of flow, while peak rates of runoff and their distribution throughout the season may be a very minor interest, provided that the snowmelt waters are conveyed in channels of sufficient capacity and are impounded in reservoirs for release later on as needed. Hydroelectric-power-system operators are interested in having the greatest possible accuracy of the minimum inflow which may be expected during the forecasted period, since their commitments for the delivery of electrical energy require a conservative estimate for profitable economic management of the system. The demand for the highest accuracy in a flood-control system is aimed at a maximum flow which may occur during a given flood season.

It is not surprising, therefore, that a number of forecasts, appearing to differ very widely in the amounts forecasted, may be computed by various interests from identical snow-water equivalent and related data. In their endeavors to arrive at the most accurate seasonal water-yield forecast for a particular objective, hydrologists have not only made use of snow-water equivalent data yielded by snow surveys (Subsec. VI), but also have introduced other factors, such as antecedent precipitation, temperature, and any factor for which data may be available and for which there is logical reason for incorporation.

Probably one of the earliest, if not the first, usage of snow surveying may have been that referred to by Ayer [67] in describing, in his history of snow surveying in the East, the report submitted on Jan. 14, 1836, by John B. Jervis, Chief Engineer of the Chenango Canal, New York. Jervis [68, p. 58], without giving details of his methods, concluded that snow on the ground which fell in November and December, 1834, on the drainage basin of Madison Brook, having an area of 6,000 acres, amounted to 87,120,000 ft^3 of water. In the western United States, Church performed the first snow surveys on the slopes of Mount Rose in Nevada on the shores of Lake Tahoe in 1909 [43, 69]. A general discussion of forecasting seasonal-water yield, summarizing the state of knowledge up to that time, is presented in Church [70, in the chapter on Snow and Snow Survey: Ice].

Work [41, pp. 8–9] discusses a simple correlation between the peak accumulation of the snow-water equivalent of the Diamond Lake, Ore., snow course and the April-through-September volume of seasonal water yield of the North Fork of the Rogue River above Prospect, Ore. Garstka [19, p. 417, fig. 8] presents a simple correlation between the arithmetic average of snow courses above Jackson Lake, Wyo., and the April-through-July flow of the Snake River above Jackson Lake. When the water-equivalent data as observed from the snow-course network above Jackson Lake, Wyo., is adjusted in accordance with the area-elevation-weighting concept, considerably improved simple forecasting correlation is derived, as illustrated in Ref. 19.

Although it is evident that the water equivalent in storage in the snow is obviously a source of the spring runoff released by snowmelt, experience of many hydrologists indicated that factors other than snow in storage were influencing seasonal water yield. The antecedent precipitation and temperature conditions were found to be the cause of departures of the actual from the forecasted water yields. High temperatures during the preceding summer season, coupled with deficient rainfall, would result in a dry soil mantle in the water-yielding drainage basins. The deficit in soil moisture would have to be satisfied by the initial water yielded from snow before the snowmelt could reach the stream channels. Unusual snow distribution such as heavy depositions of snow at low altitudes or total absence of snow at lower elevations further complicated the snow storage-runoff relationship. Warm spells during the winter would melt much, if not all, of the snow. A considerable portion, if not all, of the waters released would be retained as soil moisture without being accounted for in the spring-season snow survey. Such warm spells also would introduce reasons for departures from the expected yields. All these factors are referred to as antecedent conditions since they occur before the dates of the forecast computations.

The importance of soil moisture has long been recognized. Stockwell [71] describes the use of electrical-resistance soil-moisture units as developed by Coleman [72] and Bouyoucos [73]. The U.S. Soil Conservation Service has established, as of late 1961, about 110 stations in the western United States for determining soil moisture beneath the snowpack. Fredericksen [65, pp. 33–34] describes a saving of approximately one-third of a million dollars in one year alone to the Twin Falls Soil Conservation Service District. This resulted from the realistic acceptance of a deficient water yield resulting from excessive moisture deficiency in the soil as indicated by the system developed by Stockwell.

The intensiveness of the efforts expended by hydrologists in their endeavor to reduce costs without sacrificing accuracy of forecast is indicated by the work performed by Court [74]. In the early days of snow-course layout it was not unusual for a course to require measurements every hundred feet over a distance of $\frac{1}{2}$ to 1 mile (Subsec. VI). When travel to snow courses on foot required several days' time in getting to and from the snow course, a few additional measurements did not matter. Snow courses which require 3 or 4 days' time by a crew of two men can now be reached in a day by over-snow vehicles and in an hour by helicopter, provided that the elevation of the course permits the helicopter to land and take off. With mechanized travel an excessive number of sampling points can become very expensive. Court [74] concludes that having at least a dozen years of record, by statistical analyses, it should be possible to reduce the number of sampling points very markedly. For the Onion Creek course in the South Yuba River, California, Court concludes that the accuracy of the forecasts would not be adversely affected if the number of sampling points at this course were reduced from 19 to only 5, with the added thought that the median of the 5 points would give as good a forecast as the mean of 19 points.

An appreciation of the importance of the antecedent conditions led to the development of the seasonal storage precipitation gages (Subsec. V).

Departures of the actual from the expected seasonal water yields may also be caused by events subsequent to the date of the computations of the forecast. Excessively high temperatures may cause evapotranspiration losses. Marked departures from the precipitation considered normal subsequent to the date of the forecast can cause various errors in the forecast. Deficient precipitation tends to increase the waters subtracted from the forecasted season yield by requiring greater amounts for soil-moisture replenishment to satisfy evapotranspiration loss, whereas excessively abundant precipitation tends to reduce the relative importance of the factors operating to deplete the seasonal water yield.

Alter [75] published in 1940 an article setting forth his estimates of future trends in developments of seasonal water-yield forecasting, which stimulated much thought. Alter thought that snow surveys would become obsolete.

Extreme positions could be taken in the approach to the computation of seasonal water-yield forecasts: one would be that of using snow-water equivalent alone, the other that of using precipitation as measured in rain gages in the valleys. It may easily be that, for a particular drainage basin depending upon its soils, vegetation, elevation above sea level, and geography, one extreme position or the other can be shown to be far superior. Nevertheless, the realistic hydrologist engaged in seasonal water-yield forecasting has developed techniques making use to the fullest extent of all the data available to him for a particular drainage basin for computing a seasonal water-yield forecast for a specific objective.

Methods of application of statistical mathematical multiple-correlation analyses in forecasting a seasonal runoff are discussed in Subsec. IX-D.

The U.S. Weather Bureau's Hydrologic Services Division issues a series of water-supply forecasts based upon statistical analyses, including certain assumptions of probabilities applying to precipitation subsequent to the date of the forecast. An example of this forecast is Ref. 76. This series of water-supply forecasts presents water-year forecasts ending, for the most part, through September for approximately 380 points.

Table 10-2 lists the reports published by the Federal, state, and private cooperative snow-survey systems coordinated by the U.S. Soil Conservation Service. The bulletins listed in the table include not only information on the snowpack, but also on reservoir storage and soil-moisture data, in addition to seasonal water-yield forecast as computed by the Soil Conservation Service and its collaborators.

In addition to the seasonal water-yield forecasts issued publicly by the U.S. Weather Bureau and the U.S. Soil Conservation Service, a great many agencies, Federal, State, and private, responsible for the operation of specific projects, compute their own forecasts, tailored to suit a particular objective. An example of such forecasting is that described by Blanchard [66]. Such operational forecasts are seldom released to the public, although the techniques used in their derivation often appear in the hydrologic literature.

Taking into account the complexity of statistical tools available to the hydrologist, the use of various concepts in averaging snow-water equivalent and precipitation data, the great range of possibilities of mathematical procedures and of judgment in the transposition of data from one drainage basin to another, and especially the requirements to be met to attain the objective for which seasonal water yields have been computed, it is not surprising that wide differences are produced in the numerical values of seasonal water-yield forecasts computed from identical basic data, even when the forecasts are made for similar purposes. In view of these differences it is to be expected that comparisons would be made of the results attained by various schools of thought. Examples of such comparisons are not new. Kohler [77] presents an excellent summary, including an annotated bibliography of 22 references of water-supply forecasting developments during the period 1951 to 1956.

Work and Beaumont [78] discuss several forecast methods, including methods of expressing the accuracy of forecasts. With regard to the Rio Grande and Arkansas River basins, they describe the need for suitably designed precipitation gages at water-producing elevations to be of practical help in reducing forecast errors.

Kohler [79] discusses Work's and Beaumont's paper [78] in considerable detail, especially with reference to the method of expressing the accuracy of the forecast. This is difficult. Kohler also discusses a comparison of 1,211 forecast cases in which both the Soil Conservation Service and the Weather Bureau had 22 errors which exceeded 150 per cent. A comparison of forecasts by Kohler [79] is summarized as follows:

Date of forecast	Number of cases	Average error (% observed, flow)	
		SCS	WB
February 1	41	49.1	43.5
March 1	173	39.9	39.3
April 1	513	34.2	36.4

In his conclusions Kohler [79, p. 32], speaking for the Weather Bureau, says:

The Weather Bureau has used and will continue to use such snow survey and other hydrologic data as are available to us at the time the forecasts are prepared. I am certain, on the other hand, that the forecasters in the Soil Conservation Service feel that winter precipitation at so-called "valley stations" is of some value, since these data are included in their forecast bulletins.

FACTORS AFFECTING RUNOFF FROM SNOWMELT 10-27

It is evident that the differences in the accuracy of the forecasts as determined by the divergent philosophies delineated by Alter [75] are so minor as to appear to be of little practical significance when reviewed in the light of 20 years' experience. The comparison of the averages of the results of the two methods would conclude that the accuracies of both methods have attained a plateau, possibly set by the indicativeness of the basic data. It should be kept in mind that the most useful forecast for a given objective must be derived for each particular drainage basin, and thus, except for a specific case, a comparison of West-wide averages of accuracy is meaningless. New sources of data such as aerial observations of snow depth (Subsec. VI), radioisotope radiotelemetering gages, storage gages, better methods of averaging the water equivalent, determination of distribution and of the rate of disappearance of snow in storage in regard to aspect and elevation, the use of radar, and the utilization of automatic data-processing machinery are all relatively recent developments. The possibility of introducing a low-flow index of water yielded by a drainage basin to account for time trends in precipitation-runoff relationship is discussed by Peck [80]. The effect of forest and range management and of agricultural irrigation practices with regard to the utilization of water and the recent increasing use of thermal power to be associated with hydroelectric power will all introduce new and challenging problems which are bound to have a profound effect on the future development on the science and art of seasonal water-yield forecasting. However, a very great improvement in seasonal water-yield forecasting could take place as the result of increased accuracy in long-range weather forecasting when such is attained even for a period of 6 months in advance of the date of the forecast.

B. Runoff Forecasting for River Regulation

The waters released from the melting of snow are just one component of runoff. Since snow is a form of precipitation in storage, forecasting of runoff from snowmelt requires, first, consideration of sources of heat for melting a specific amount of a particular snowpack in a certain drainage basin and, second, the application of hydrologic techniques to determine the water yield from the melting of snow as runoff. Factors affecting runoff from snowmelt are discussed in detail in Subsec. VIII. Techniques of analysis of snowmelt and their application to forecasting of runoff from snowmelt for river regulation and design-flood application are discussed in Subsec IX.

Brief discussions of the general subject of runoff from snowmelt are presented by Linsley, Kohler and Paulhus [81, pp. 141–143, 428–432], Linsley and Franzini [82, pp. 44–48], Butler [83, chap. 3], and Linsley, Kohler, and Paulhus [84, pp. 181–190]. Kittredge [85, chap. 14] treats snowmelt as part of a broad discussion of the influence of forests on snow.

C. Design Floods Due to Snow

Design floods are extensively discussed in Sec. 25-I. Depending upon the location and elevation of the engineering structure, the design flood may be due to runoff from snowmelt alone or to the result of combination of rain and melting snow. The hydrometeorologic implications for flood runoff due to snow are very complex and require special treatment for each site. For the design of multiple-purpose reservoirs, Riesbol [86] discusses the role of snow hydrology. Another detailed consideration of design floods in snow hydrology is treated in Ref. 87. Summersett [88] describes a method of reproduction of snowmelt floods in the Boise River above Diversion Dam in Idaho, which may be used to determine design floods.

VIII. FACTORS AFFECTING RUNOFF FROM SNOWMELT

A. Sources of Energy for Snowmelt

The sources of heat which are necessary to melt snow may be classified into three broad categories: (1) radiant heat from the sun, (2) latent heat of vaporization released by the condensation of water vapor, and (3) heat by conduction from the environment in contact with the snow, such as from the ground, the rainfall, and the air. It is appreciated that snowpack is not always the recipient of heat; it may lose heat to its environment by radiation, by sublimation, and by conductivity. It is quite possible for a snowpack to be gaining heat through one process and losing heat through another. The interactions of the various phenomena of heat exchange make the problem of the melting of snow one of the most complex in the field of hydrology. The above-mentioned three sources of heat energy for snowmelt should be kept in mind throughout further discussions and in all problems relating to the accumulation and melting of snow and ice.

1. Radiation. The sun is the source, practically, of all energy at the earth's surface. The heat equivalent of solar energy above the earth's atmosphere and nor-

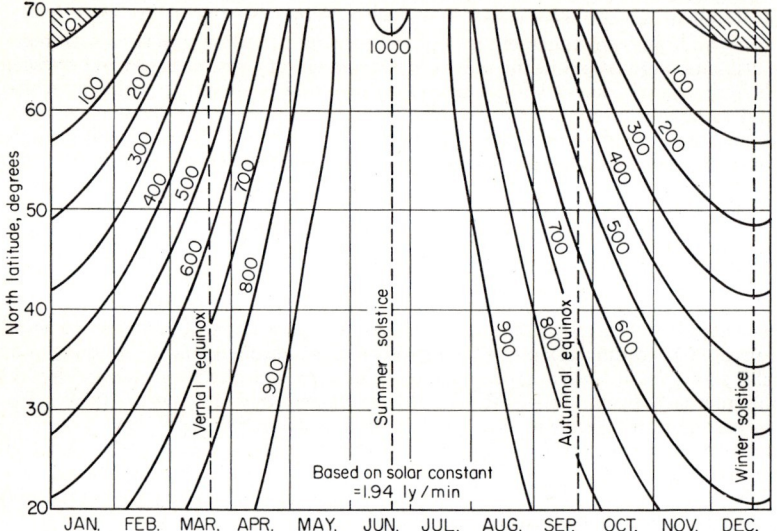

FIG. 10-12. Seasonal and latitudinal variation of daily solar radiation outside the earth's atmosphere, in langleys. (*U.S. Army Corps of Engineers* [23].)

mal to the path of radiation is taken to be 1.94 langleys/min (1 langley = 1 g cal/cm^2; 1 small, or gram, calorie = 0.0039685 btu). Figure 10-12 gives the seasonal and latitudinal variation of daily solar radiation outside of the earth's atmosphere in langleys. Methods of estimating solar radiation are summarized in Ref. 87, chaps. 5 and 6, plate 6-1, drawing PD-20-25/36. The radiation incident upon a particular snowpack varies greatly not only with the prevailing sky-cloudiness condition, but also with slope, aspect, and forest cover of the land.

The influence of forest cover on solar radiation and light is discussed in general terms by Kittredge [85, chap. 6]. An excellent discussion of the technique background of radiation as it applies to snow is given in Ref. 87, secs. 5-02 and 5-03. Several basic physical laws apply to solar radiation. *Planck's law* states that the spectral distribution of the energy of a radiating black body is a function of the temperature of the radiating body. Generally speaking, the higher the temperature, the greater the intensity of the total radiation and the shorter the wavelength of the maximum intensity. *Stefan's law*, which is derived by integrating Planck's law for all wavelengths, states that the total energy emitted by a black body in all wavelengths, per unit time and area, is expressed as

$$E = \sigma T^4$$

where E is the total radiation, σ is Stefan's constant having the value of 0.826×10^{-10} langley/min/°K^{-4}, and T is the absolute temperature in °K.

In so far as the radiation characteristics of the snowpack are involved, the radiation lies in the range of 0.15 to 80 μ (microns). The solar short-wave radiation is in the range of 0.15 to 4 μ. This includes the visible spectrum of 0.4 to 0.7 μ, with the maximum intensity of the visible spectrum occurring at about 0.5 μ. Radiation involved in snowmelt extends into both the ultraviolet and the infrared. Terrestrial, or long-wave, radiation is generally in the range of 3 to 80 μ, with a maximum intensity in the infrared at about 11 μ. Since snow radiates as a black body according to Planck's law and since the total energy radiated is according to Stefan's law, both laws are operative in a very complex manner where clouds and tree cover are concerned in snowmelt.

Albedo is the reflectivity of short-wave radiation of a snowpack. Since short-wave solar radiation, when reflected, is lost and not effective as sensible heat, the albedo of snow is important in the snowmelt phenomena. The especially high albedo of snow covers a range of from about 40 per cent for melting late-season snow to about 80 per cent for freshly fallen snow [23, p. 5] and in a few instances as high as nine-tenths [7, p. 64]. This explains why the snow persists and air temperatures remain cool in the winter under intense clear-sky solar radiation. At an albedo of 80 per cent, four-fifths of the incoming short-wave solar radiation is reflected without having any warming effect on the air or without being effective in melting the snow. Garstka [89] offers a possible explanation for the severe winter of 1948–1949 in the Great Plains, on the basis of the albedo of snow.

Of the solar radiation not reflected in short-wave form as albedo, a portion is absorbed by the snowpack and converted to heat energy, which may be utilized either in melting snow or radiated as long-wave heat or conducted through the snowpack. In the case of frozen ground, some of the heat may be absorbed by the ground without any melting. The ability of solar radiation to penetrate snow varies with the density and the state of metamorphism. The radiation-absorbing characteristics of snow are very markedly influenced by foreign substances such as dust, bark fragments, twigs, and practically any substance other than snow. The importance of foreign substances on energy conversion of radiation is discussed in Subsec. XIV.

The ability of light to penetrate snowpacks is important to the growth and survival of plants under snowpack. Work [21, p. 545] reports that light sufficient to permit formation of chlorophyll in newly sprouted grass penetrated a 35- to 40-in. depth of dense snow. A comprehensive discussion of heat exchange and melt of late-season snow patches in heavy forest is given by Boyer [90], with discussion by Garstka [91]. Other valuable information on snowmelt due to radiation is given in Refs. 87 and 23.

2. Sublimation. Changes of state of water substances were discussed previously in Subsec. II. The condensation of one gram of water releases sufficient heat to melt 7½ g of ice, a process which will yield 8½ g of snowmelt, with the condensate included. Evaporation from snow can take place only when the vapor-pressure at dew point is not in equilibrium with the vapor pressure of snow at the prevailing air temperature and when the air is unsaturated with water vapor. When the dew point is above 32°F, condensation occurs with the release of latent heat of vaporization. When the dew point is lower than the snow surface temperature, evaporation occurs.

The sublimation of snow either by evaporation or condensation is so intimately related to the mass-transfer and turbulent-exchange phenomena of air that further discussion of this source of heat will be included under the discussion of heat exchange. A comprehensive discussion of evaporation theory is given by Anderson, Anderson, and Marciano [92].

The structure of the boundary layer and mass-transfer equations are treated in detail by Marciano and Harbeck [93]. A review of evaporation from snow is given in Ref. 31, sec. 10. Six graphs relating to convection-condensation melt are given in Ref. 87, of which two relate to melt determined for 1-mph wind speed. Hildebrand [94], in reporting upon studies performed in a snow lysimeter in the central Sierra snow laboratory at Soda Springs, California, compares convection and condensation coefficients with those derived by de Quervain, Sverdrup, and Light.

The determination of the portion of the snowpack lost by evaporation is very difficult. Based upon the physical considerations, which are supported to some extent by field observations, it is to be expected that the portion of the snowpack lost by evaporation would probably be very small during the active snowmelt period when dew points are above 32°F. In many mountainous areas during the winter, even though temperatures are low, dew points likewise are low and the total evaporation loss from the snowpack in areas of low snow accumulations may be significant, especially when the snowpack is shallow. In the middle latitudes in the winter and early spring, before snowmelt runoff begins, evaporation loss from the snow surface cover can be assumed, for hydrologic computation purposes, to be about ½ in. of water equivalent per month [23, p. 25].

3. Heat Exchange. The heat secured by a snowpack by contact with its environment is the third major source available for snowmelt. Heat exchange may take place by conduction from the earth, by contact with warm rain, and by contact with the atmosphere.

a. Heat from the Earth. Wilson [95] concludes that after the thirtieth day of snow cover, the rate of heat transfer from the earth to the snow is so small as to be insignificant in supplying heat during the active snowmelt season. It may be assumed [23, p. 10] that snowmelt from ground heat during the snowmelt season is in the order of 0.02 in./day.

Work [21, p. 545] observed that frost will penetrate a snow cover at least 24 in. thick to freeze the earth beneath. It is a common observation by snow surveyors that the soil under the deep snowpack in the mountains is rarely found to be frozen at the beginning of the snowmelt season. It is not unusual for the soil to be frozen in the autumn at the time of the first snowfall. Unless the snow cover is very thin, there is normally sufficient residual heat in the soil remaining from the previous summer to thaw out the soil before the spring snowmelt season. Extensive observations and literature are in support of the fact that practically all the melting of snow takes place at the snow surface exposed to the air.

b. Warm Rain. Since rainfall during the spring, when snow is on the ground, is apt to fall at low temperatures, a relatively small quantity of heat is available for transfer from rain to the snow. Wilson [95, p. 186] presents a chart for determining, graphically, the water melted from snow in inches for wet-bulb temperatures observed during the rain in the range of 32 to 70°F for depths of rain up to 10 in. Wilson's chart indicates that, at a 50°F wet-bulb temperature, 4 in. of rain will melt only ½ in. of water. A hydrologically important process that takes place when rain falls on snow is that of metamorphism, which is discussed in detail in Subsec. IV.

The daily snowmelt due to rain may be estimated by the following equation [23]:

$$M_p = 0.007 P_r (T_a - 32)$$

where M_p is the daily snowmelt from rain in in., P_r is the daily rainfall in in., and T_a is the air temperature in °F, mean for the day of saturated air at the 10-ft level.

c. Contact with the Atmosphere. Some of the theoretical considerations underlying turbulent exchange and mass transfer were discussed in Subsec. VIII-A-2. Boyer [90], with discussion by Garstka [91], found that the melting of the last remnants of the seasonal snow cover, at the U.S. Army Corps of Engineers Willamette Basin Snow Laboratory in Oregon, was accounted for by sources of heat as follows: 15 per cent short-wave radiation, 50 per cent long-wave radiation, 20 per cent convective heat, and 15 per cent latent heat of condensation. Boyer [90] concluded that 85 per cent of the melt at the snow surface was accounted for by air temperature and relative humidity and 15 per cent by short-wave radiation.

The use of degree-day correlation, basin indexes, and equations including wind as a factor is discussed in detail in Subsec. IX.

B. Snowpack Characteristics

Work [21] presents a very thorough discussion of snow-layer density changes as the snowfall accumulates and the melt season proceeds. The ripening of snow was discussed in detail in Subsec. IV. The significance of snowpack characteristics as influenced by heat and by climatological factors such as wind is discussed in relation to avalanche in Subsec. XI. Since snow acts as a soil, the capillary retentivity must be satisfied before gravitationally free water can depart from the snowpack. In general [23, p. 21], the capillary retentivity of snowpack is usually within a range of 2 to 5 per cent of its total water equivalent. This retentivity varies throughout the day and also throughout the season, depending upon the metamorphosis of the snow.

During the melt season, especially at high altitudes on clear nights, the snow may freeze to a depth of as much as 10 in. because of loss of heat by long-wave radiation to the sky, since under such conditions the snow behaves as a practically perfect black body. The nocturnal crust and the temperature of the snow below the freezing point act in effect to delay the yielding of water when sufficient heat becomes available from whatever sources may be operative to cause snowmelt. Clyde [96, pp. 19, 27] reported in 1931 on his extensive snowmelt investigations, in which his use of fluorescein dye established a fact which has been substantiated extensively by all workers since then, that the melting of snow occurred at the surface exposed to the air.

In the mountainous areas with slopes sufficiently steep for ready drainage, the net storage effect of water draining through snowpack is about 3 to 4 hr for moderately deep packs with the snow fully ripened at a density of about 50 per cent [23, p. 21]. The retentivity of fully ripened snowpack is very low, since it consists largely of coarse ice crystals. Thus, for operational and flood-design processes, the water in transit either resulting from rainfall or from the melting can be assumed to be about 6 per cent of the snowpack water equivalent [23, p. 23].

C. Site Conditions

The influences of the antecedent summer and fall seasons' precipitation on soil moisture, as a factor in the amount of water available from runoff due to snowmelt, were discussed in Subsec. VII-A. In the western United States 3 to 5 in. of water may be required to satisfy the soil-moisture deficiency before any marked runoff appears in the stream channels. Since the melting of snow usually takes place at rates considerably below the infiltration capacity of the soil, overland flow from snowmelt is rare. Overland flow has been observed in only a few instances, when the soil mantle is very shallow or when disappearance of snow on south-facing slopes exposes the soil to frost penetration in the late spring.

It is not unusual in the Great Plains to have snow fall upon frozen ground. The evanescent character of such snow covers in a region characterized by the occurrence of foehn winds results in a very complex snowmelt-infiltration situation. Field observations concerning such details are rare. McKay and Blackwell [97], in reporting on snow observations in the prairies of Canada, conclude that estimates of snowpack water equivalent may be attainable by an accounting procedure of losses and gains. For collecting information for the estimates they describe [97] a large-diameter "prairie" snow sampler.

The effect of forest management upon runoff from snowmelt has been the subject of intensive study. The Fraser Experimental Forest in Colorado is the site of a long-term investigation for such studies [98].

A depredation of pine and spruce stands by bark beetles in the White River drainage basin in Colorado presented Love [99] with an opportunity to ascertain the effect of the resultant thinning of the forest upon water yield. Love found 22 per cent increase in the streamflow.

Martinelli [100], working in the Alpine snow fields above 12,000-ft elevation in Colorado, found that each acre of snow present during July and August released an average of 60,000 gal, or 0.19 acre-foot, of water per day for streamflow. Martinelli suggested the installation of artificial barriers above the timber line to induce accumulation of snow fields so that the latter would persist into the summer. Kittredge [85] gives a comprehensive review of the influences of the forest on snow.

Goodell [101] reports that the water yield from areas under study in the Fraser Forest, Colorado, increased 24 per cent over a period of 3 years, according to records available since the timber was harvested. The possibilities of increasing water yield through snowpack management are discussed in detail by Anderson [102] in California, by Hoover [103] in the southern Rocky Mountain area, by Lull and Pierce [104] in the northeastern United States, and by Packer [105] in the western pine forests of the northern Rocky Mountain region.

As was discussed in Sec. VII-A, it is important that the locations at which snow gages and snow courses are located be indicative of differences in precipitation from year to year. Since much of the water supply is yielded from forested drainage basins, the influences of logging on snow accumulation, soil moisture, and water loss could have an important bearing on the indicativeness of the data when used for seasonal water-yield and rate-of-runoff forecasting. Anderson and Gleason [106], in discussing the effects of logging, conclude that the maximum snowpack was increased noticeably, the rate of snowmelt was greater in commercially logged areas, the soil-moisture content was increased on logged-over areas, and the actual yield of water was increased by about 20 per cent on strip-cut logged areas, by about 14 per cent on block-cut areas, and by about 7 per cent on commercially logged areas.

Schneider and Ayer [107] and Schneider [108] conclude that reforestation of abandoned farm lands in central New York State resulted, in a period of about 25 years, in modifying the distribution of volumes of the seasonal runoff from snowmelt and in reducing the runoff to about three-fourths of that yielded by the drainage basins before forest conditions were established.

D. Antecedent Conditions

All precipitation since the preceding snowmelt season can be properly considered as antecedent precipitation. In developing an estimating equation of April-through-July runoff of the Colorado River at Cameo, Ford [109] uses the July-through-September antecedent precipitation and the October-through-January antecedent precipitation separately. The antecedent precipitation is also used in the multiple-correlation forecasting equations developed by the U.S. Weather Bureau for computing water supply in the western United States [76].

Antecedent precipitation, provided it were sufficient to satisfy the capillary retentivity of the soil to field capacity and no depletion of that soil moisture would take place, need not be considered in seasonal water-yield forecasting and design-flood synthesis. The soil moisture from that portion of the antecedent precipitation which would not depart from the drainage basin as streamflow is available for loss through the process of evapotranspiration.

The multiple-correlation analysis as used in seasonal water-yield forecasting makes no attempt to assign numerical values to the evapotranspiration loss. However, in maximum probable flood-design synthesis based upon a point snowmelt approach, such as that used in Refs. 23 and 87, the loss of water can be estimated through methods [23] developed by Thornthwaite [110], Penman [111], and Halstead [112].

Ford [109] found that the temperature, during the snow-accumulation period in the Swan River in Montana, had a significant effect on variations in flood-season runoff. The coefficient of multiple regression of temperature, during the snow-accumulation period, was found to be negative, meaning that the higher the winter temperature, the less runoff during the following melting season. The Swan River basin in Montana is frequently subjected to warm winds during the winter. The statistical process of analysis in this case does not discern between loss of snow by evaporation or departure of water equivalent from the snow storage during short periods of midwinter runoff.

Another indirect approach to determining antecedent precipitation conditions and the evapotranspiration losses (the interrelationships of which are influenced by the climatological and physiographic characteristics of the drainage basin) is the use of base flow. Nelson, McDonald, and Barton [113], with discussion by Dean [114], found the base flow to be a very significant factor in the forecasting of April-June runoff of a number of streams in the Pacific Northwest. The base flow accounted for 18 per cent of the variance of the Columbia River at Castlegar, B.C., and for $4\frac{1}{2}$ per cent of the variance of the Clearwater River at Spalding, Idaho. The use of base flow as a parameter in forecasting runoff from snowmelt has resulted in substantial improvement in the accuracy of the forecasts. Dean [114] points out that the flow of the upper Columbia River above Kootenay consists on the average during the winter of 80 per cent of the runoff volume from the ground-water storage, whereas the Clearwater River at Spalding receives only 30 per cent of its winter runoff from the base flow.

Peck [80] describes the use of low winter streamflow in water-supply forecasting of the Blue River at Dillon, Colo. Linsley, Kohler, and Paulhus [81, pp. 434–439] present a comprehensive discussion of the influence of antecedent precipitation on water yield. Further considerations of an antecedent-precipitation condition are given in Subsec. IX.

E. Rainfall

The influence of water flowing through the snow was discussed in detail in Subsec. IV. Rain as a source of heat is discussed in Subsec. VIII-A. Rainfall as a factor in affecting runoff from snowmelt can be treated also in any general terms. Riesbol [86] classifies the combination of rain-and-snowmelt floods into two categories, namely, "rain-on-snow" and "snowmelt-followed-by-rain."

It should not be assumed that a combination rain-and-snowmelt flood will be more severe than a snowmelt flood alone. An example may be cited with reference to the procedure used by the Corps of Engineers in deriving the maximum probable snowmelt flood for the Salmon River basin, which drains 14,100 sq mi of a rugged and mountainous region in central Idaho [23, chap. 11]. Runoff in this basin is chiefly from snowmelt, with 72 per cent of the annual flow occurring in the 4-month period April to July. The winter snowpack accumulates to as much as 50 in. of water equivalent in the higher elevations, whereas at the lower elevations the annual precipitation may be as little as 6 in. It was discovered that the peak discharges for the maximum probable flood, with the superimposition of the maximum observed rainstorm, yielded a flood crest slightly less than that for the snowmelt flood alone. This was due to the hydrometeorologic considerations, which resulted in less heat being available for snowmelt when a low-temperature rainfall-producing air mass was considered as operative in the area than the amount computed from the procedures, set forth in Ref. 23, under clear-weather snowmelt conditions.

IX. TECHNIQUES OF ANALYSIS OF SNOWMELT FOR FORECASTING RUNOFF FROM SNOWMELT

A distinction should be made between the melting of snow at a point and the yielding of runoff as streamflow from a drainage basin. The melting of snow, as discussed in Subsec. VIII, is a physical process involving a change of state requiring heat, whereas the yielding of water from a drainage basin is a much more complex hydrologic process.

A. Degree-day Correlations

As used in the United States, a *degree-day*, in its broadest sense, is a unit expressing the amount of heat in terms of the persistence of a temperature for a 24-hr period of one-degree-Fahrenheit departure from a reference temperature. As often applied in snowmelt studies, the degree-day is computed by subtracting the average of the daily maximum and the daily minimum temperatures from 32°F. For example, if a daily mean thus computed were 32°F, there would be 0 degree-day; a daily mean of 37°F would yield 5 degree-days. There are many ways of computing the mean temperature; the method used above is just one example. From observed daily maximum temperature and the average daily minimum temperature of the preceding and following days, Cavadias [115] gives two nomographs for computing degree-days above 32°F for the St. Maurice River drainage basin, in Quebec, Canada. One nomograph is for daily minimum temperatures above 32°F, and the other for below 32°F.

Degree-days below a reference temperature of 65°F is found to be a useful indicator of a heating demand. The cumulative normals of such degree-days since September 1 during the winter months and the degree-days by months are generally published, for example, in Ref. 116.

Stevens [117], in his analysis of river-ice formation in the Upper Missouri River basin, used degree-day departures both above and below a reference temperature of 32°F in deriving his ice-forming factor. He plotted mean daily temperatures for the months of November through March. The winter season was considered to begin on the first day in the autumn on which the mean daily temperature fell below 32°F and ended on the first day in the spring on which the mean temperature exceeded 32°F. Stevens defined the *ice-forming factor* as the excess of degree-days below freezing over those above freezing for a winter season.

Degree-days above 32°F and degree-days above other reference temperatures have been used in point-snowmelt and in runoff-snowmelt computations. An accurate computation of degree-days could be made from hourly observations or from a chart recording temperatures. Air-temperature thermograph records useful for such computations are relatively rare, especially for the areas on which the winter snowpack accumulates. Computations of degree-days from the daily mean may be misleading. In many parts of the West the drop of temperature to the minimum may be so great as to yield daily means below 32°F, indicating no degree-days, whereas snowmelt conditions may have prevailed during part of the day when air temperatures were much above the freezing point. Accordingly, Garstka, Love, Goodell, and Bertle [31] used the commonly available maximum and minimum temperatures to estimate the degree-days in a detailed interpretive study.

According to a study by the U.S. Army Corps of Engineers [23], the daily springtime snowmelt M in inches may be estimated by the following correlation equations as a function of the mean daily temperature T_{mean}, the maximum daily temperature T_{max}, and the relative forest cover:

1. For open sites,

$$M = 0.06(T_{mean} - 24)$$
$$M = 0.04(T_{max} - 27)$$

2. For forest sites,

$$M = 0.05(T_{mean} - 32)$$
$$M = 0.04(T_{max} - 42)$$

These equations are applicable for T_{mean} in the range of 34 to 66°F and for T_{max} in the range of 44 to 76°F.

Correlations of snowmelt with air temperature are much poorer in the open than in the forest, probably because of reradiation of heat from the forest canopy. Since heat supplied by the air is only one of the sources of heat available for snowmelt at a point, considerably superior correlations exist when several sources of heat are taken into account.

The point melt rate in inches per degree-day above 32°F may vary from as little as 0.015 to as much as 0.20 in./degree-day. For a spring snowmelt period, an average of 0.05 in./degree-day may be used, but with discretion. Such point melt values for inches melt per day must be distinguished from the basin-index melt values, discussed below.

B. Basin Indexes

Endeavors to establish a workable relationship between degree-days and runoff were not very successful for short periods, although such a relationship between degree-days and snowmelt was shown to be possible [118, 119]. Wilson [118] showed an S-shaped double-mass curve of seasonal runoff at the Gardiner River at Mammoth, Yellowstone River Basin, Wyo., plotted against the accumulated degree-days above 32°F at Mammoth, both for the period subsequent to Mar. 1. Linsley [119] also obtained an S-shaped curve showing the variation in the degree-day number, i.e., runoff per degree-day, against the calendar date, for the Stanislaus and Tuolumne Rivers in California. Both endeavors were admittedly only attempts to derive a single-index snowmelt-temperature relationship for a complex system. The reason for the S shape of the curve will be discussed in Subsec. IX-C.

The average rate of snowmelt D, in inches per day, is generally given [120–122] as

$$D = K(T - 32)$$

where T is the air temperature in °F, and K is the melting constant, or degree-day, factor. The value of K lies between 0.02 and 0.11, with a maximum of 0.30 [120]. It is believed that the above basin-melt equation is derived from the following equation given by Wilson [95]:

$$D = KV(T - 32°)$$

where D is the depth of water melted from snow in 6 hr, V is wind velocity in mph, T is dry-bulb temperature in °F, and K is a constant involving the latent heat of ice, exposures of instruments, air density, conversion units, and certain considerations involved in the theory of turbulence.

Basin-index equations can be useful if the basin index K has been established from long hydrologic records. An example of the application of degree-days as a basin index is given by Martin [123].

C. Recession Analysis

To forecast runoff from snowmelt, an analysis may be made of snowmelt hydrographs [31]. This analysis is based upon field experience in observing watershed runoff from snowmelt. In forested areas there is, in effect, no basin-wide surface runoff from snowmelt. Practically all snowmelt runoff enters stream channels as subsurface or groundwater flow, or usually as a combination of both. It may be reasoned, therefore, that the hydrograph recession toward the end of the snowmelt season should apply to each day's snowmelt contribution to the total flow. The total flow is essentially the summation of the daily contributions of flow through porous media according to the Darcy law rather than the summation of overland hydraulic flows.

Daily average recession discharges in St. Louis Creek near Fraser, Colo. (basin area = 32.8 sq mi) are plotted against the daily average discharges on the preceding day for the 1947 and 1948 snowmelt seasons (Fig. 10-13a) and for the same data with the addition of 1949 and 1950 data (Fig. 10-13b). Recession factors can therefore be derived from the slopes of the straight lines being fitted to the plotted points.

The recession factor K_r may also be derived by plotting of the hydrograph for the 1949 snowmelt season on semilogarithmic paper [124] as shown in Fig. 10-14.

From Fig. 10-13 or 10-14, the recession factor is shown to be 0.933 for discharges above 30 cfs and 0.981 for discharges between 8 and 30 cfs. A winter flow of 8 cfs is considered to be groundwater base flow, which, for each day's contribution, cannot be easily segregated.

The recession factor can be used to derive the recession portion of the hydrograph as shown in Fig. 10-15.

The volumes constituting a given day's contribution to the snowmelt hydrograph are delineated in Fig. 10-16. The recession portions of the hydrograph are extended by lines having slopes which are computed from the recession factor for the corresponding discharge at the trough of the hydrograph. These straight lines can be extended to the base flow of 8 cfs for St. Louis Creek. A given day's contribution therefore consists of two areas: area 1, which is the volume of the day's snowmelt runoff appearing between troughs of two successive days, and area 2, which is the recession flow. These areas can be computed mathematically [31] on the basis of the following equation originally proposed by Barnes [125]:

$$Q = Q_0 K_r{}^t$$

where Q is the flow in cfs at time t in days, Q_0 is the flow in cfs at the beginning of the computation period or at $t = 0$, and K_r is the daily-runoff recession coefficient.

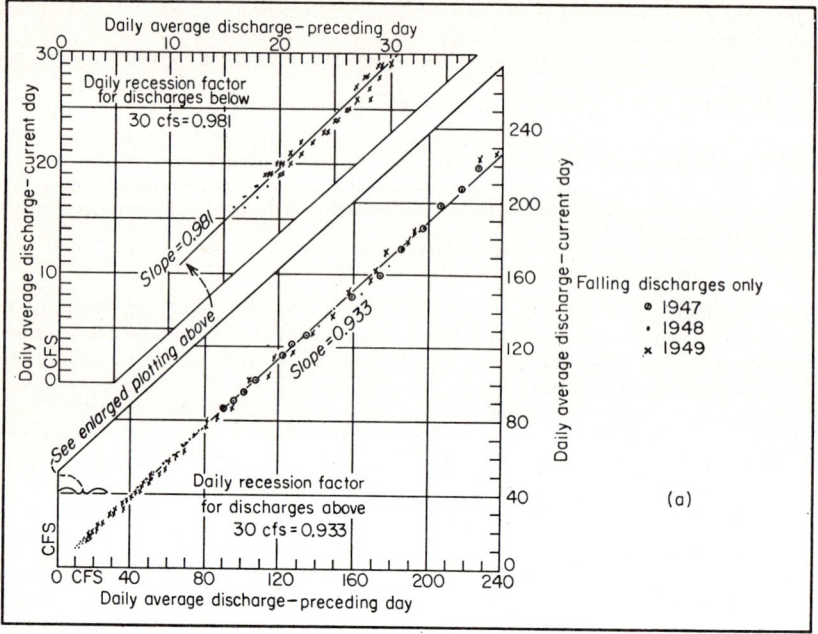

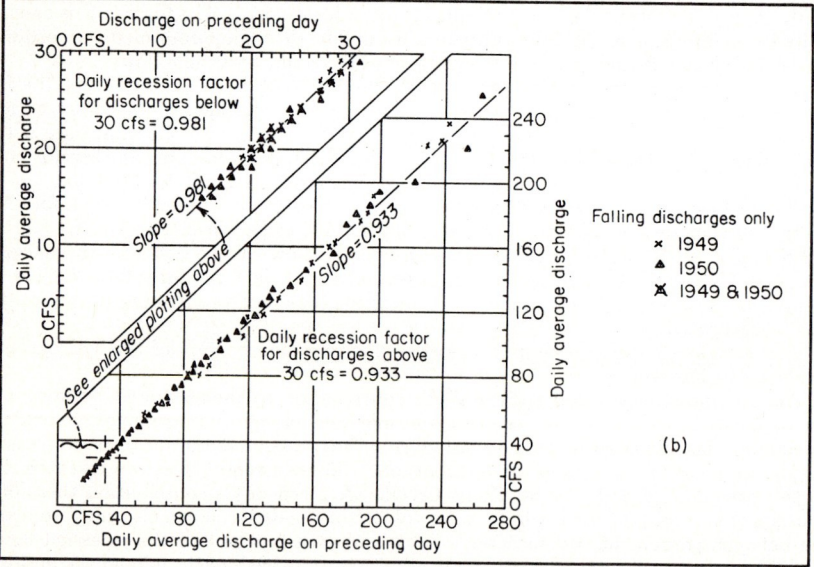

Fig. 10-13. Daily recession analysis for St. Louis Creek near Fraser, Colo. Slopes were derived from 1947 and 1948 points only. 1949 and 1950 points were added later. These recession factors fit the yearly snowmelt period of the hydrographs for all years 1935 to 1953. (a) Derivation, using data from 1947 and 1948; (b) check of recession slopes derived from 1947–1948 data with observed 1949–1950 flows. (*Garstka, Love, Goodell, and Bertle* [31].)

TECHNIQUES OF ANALYSIS OF SNOWMELT   10-37

The above equation has been restated by W. T. Moody [31] as

$$Q = Q_0 e^{-kt}$$

where e is the base of natural logarithms, and k is a factor related to Barnes's K_r by $k = -\ln K_r$. Both equations yield identical results, but may have individual advantages in different methods of computation by either analytical procedure or data-processing machinery.

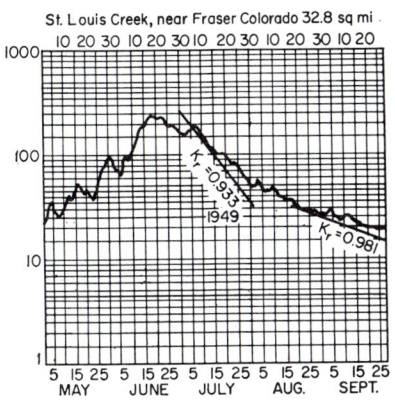

FIG. 10-14. Application of recession slopes to a semilogarithmic plot of the 1949 hydrograph, St. Louis Creek near Fraser, Colo. (*Garstka, Love, Goodell, and Bertle* [31].)

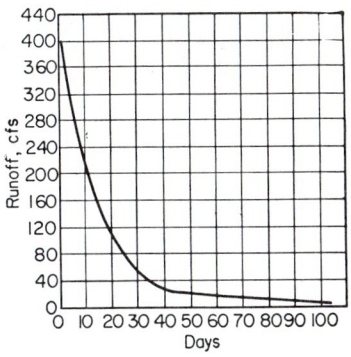

FIG. 10-15. Snowmelt-runoff recession curve for St. Louis Creek near Fraser, Colo. $K_r = 0.933$ above 30 cfs, $K_r = 0.981$ between 30 and 8 cfs. Flows below 8 cfs are considered to be inseparable as to individual day's contributions. (*Garstka, Love, Goodell, and Bertle* [31].)

In the St. Louis Creek studies [31] it was found that about one-eighth of the runoff released from snowmelt on a given day (area 1 of Fig. 10-16) appeared in the hydrograph for that day, and the remaining departed from the drainage basin as the recession flow (area 2). The concept of the assignment of different volumes to a given day's melt is for the analysis of snowmelt runoff. This is independent of the method used to arrive at the heat units which were utilized to melt the snow. In St. Louis Creek, it was therefore possible to have a peak flow of 52 cfs on May 14, 1948, with 14.9 degree-days, but a much higher peak flow of 168 cfs on May 23, 1948, with 14.4 degree-days, as shown in Fig. 10-17.

In Fig. 10-16, AB is the height to peak and CD is the height to trough, both above the preceding day's recession. These dimensions, together with the base flow, have been utilized to make possible the mathematical forecasting of snowmelt runoff either on a short-term basis for operational purposes or on an extended-period basis for design floods [31].

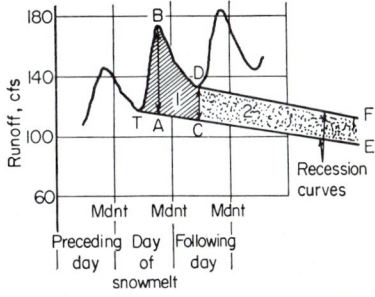

FIG. 10-16. Separation of snowmelt hydrograph showing contribution from one day's melt. Area 1 = volume of a day's snowmelt appearing in the first 24-hr period (first-day volume). Area 2 = volume of a day's snowmelt in recession flow. Total snowmelt for the day = area 1 + area 2. (*Garstka, Love, Goodell, and Bertle* [31].)

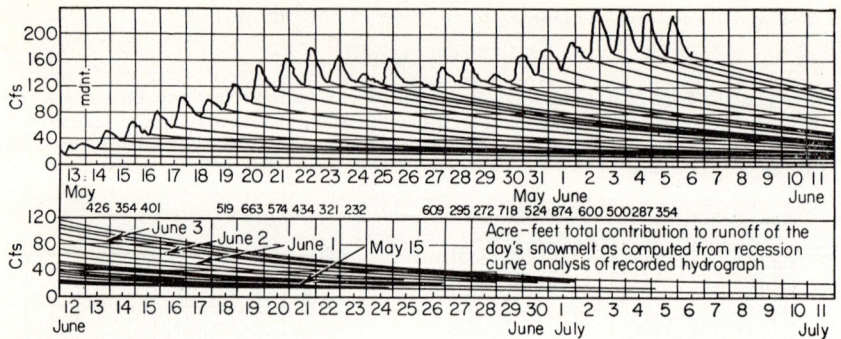

Fig. 10-17. Snowmelt runoff, 1948, St. Louis Creek, near Fraser, Colo. (*Garstka, Love, Goodell, and Bertle* [31].)

D. Correlation Analyses

Correlation analyses and its applications to hydrologic data in general are discussed in Sec. 8-II. Such analyses have been applied to correlation of the factors affecting snowmelt runoff from a drainage basin in rugged mountains [31] and to forecasting spring runoff from snowmelt [109, 115].

E. Physical Equations

Equations based on physical principles for estimating snowmelts have been variously derived [23, 95, 126].

1. The Light Equation. A widely used physical equation is the Light equation [126], which is derived from an eddy-conductivity equation based on a theory of atmospheric turbulence [127]. The general statement of the Light equation in cgs units is

$$D = \frac{\rho k_0^2}{80 \ln (a/z_0) \ln (b/z_0)} U \left[c_p T + (e - 6.11) \frac{423}{p} \right]$$

where D = effective snowmelt, cm per sec
  ρ = density of air
  k_0 = von Kármán's coefficient = 0.38
  a = elevation of anemometer in mb at hygrothermograph level
  z_0 = roughness parameter = 0.25
  b = elevation of hygrothermograph
  U = wind velocity of anemometer level
  c_p = specific heat of air at constant pressure = 0.24
  T = air temperature, °C, at hygrothermograph level
  e = vapor pressure of air
  p = atmospheric pressure, mb

The wind velocity may be assumed to vary with the logarithm of height. Consequently, the wind velocity used in this equation may be converted from one height to another on the basis of a logarithmic relationship. For practical applications, the Light equation has been reduced to the following simple form in English units for a = 50 ft and b = 10 ft:

$$D = U_m[0.00184(T_f - 32)10^{-0.0000156h} + 0.00578(e - 6.11)]$$

where D = effective snowmelt, in. per 6 hr
  U_m = average wind velocity, mph, at 50-ft height
  T_f = air temperature, °F, at 10-ft height
  e = vapor pressure, mb, at 10-ft height
  h = station elevation, ft above mean sea level

In using the Light equation, it is essential that the vapor pressure and air temperature should refer to the same period of time under consideration. Also, the use of average values of air temperature and dew point for periods longer than 6 hr may yield different results from those obtained by summing up the computed values for 6 hr periods.

The Light equation presented above gives only theoretical point values of snowmelt. In applying such values to an actual drainage basin, consideration must be given to two modifying factors, namely, the surface roughness and the forest cover ratio. These two factors can be included in a basin constant K, which is to modify the theoretical point value. Thus the basin snowmelt $= KD$.

The basin constant for a drainage basin is the ratio of the observed snowmelt to the theoretical snowmelt computed by the Light equation. The value of K can be obtained through correlation of the two snowmelt values. Theoretically, $K = 1.0$ for a flat basin uniformly covered with snow and without surface obstructions.

Two different concepts can be applied to the computation of basin constants. One concept computes the value from runoff volumes represented by the snowmelt hydrographs up to the day of peak runoff, including the recession flows as shown in Fig. 10-16. The other concept computes the value from the actual daily discharges up to the day of peak flow. For the 1948, 1949, and 1950 snowmelt seasons at the St. Louis Creek Basin above Fraser, Colo., the first concept yields basin constants of 0.901, 0.993, and 0.954, respectively, with an average of 0.951. By the second concept, the corresponding values are 0.495, 0.470, and 0.511, respectively, with an average of 0.494, which is nearly half the value obtained by the first concept.

2. U.S. Army Corps of Engineers Equations. By evaluating snowmelt on the theoretical basis of heat transfer involving radiation, convection, and conduction, the U.S. Army Corps of Engineers [23] has derived physical equations for basin snowmelt during rain as well as during rain-free (clear-weather) periods.

a. Basin Snowmelt during Rain. During rainstorms, solar-radiation melt is relatively small, and snowmelt resulting from long-wave radiation is easily evaluated from known theoretical considerations. Heat transfer by convection and condensation represents the major source of energy for snowmelt, and the snowmelt equation depends somewhat upon the type of area. Basin coefficients must therefore be assumed which expresses the wind effect over the basin for the convection-condensation melt.

The total basin melt during rain is equal to the sum of various components of melt, respectively, due to short-wave and long-wave radiation, to convection and condensation, to heat from rain, and to ground heat. For open (<10 per cent cover) or partly forested (10 to 60 per cent cover) basin areas, the total snowmelt is expressed as

$$M = (0.029 + 0.0084kv + 0.007P_r)(T_a - 32) + 0.09$$

For heavily forested (>80 per cent cover) areas, the total snowmelt is

$$M = (0.074 + 0.007P_r)(T_a - 32) + 0.05$$

In these equations, M is the total daily snowmelt in in., T_a is the mean temperature in °F of saturated air at the 10-ft level, v is the mean wind velocity in mph at the 50-ft level, P_r is the rate of precipitation in in./day in open portions of the basin, and k is a basin constant which represents the mean exposure of the basin or segment thereof to wind, considering topographic and forest effects [23, p. 10]. The value of k would be 1.0 for unforested plains, but for forests it could be as low as 0.3, depending upon stand density. The values of wind and temperature in the equations should be representative of average conditions over the snow-covered area of the basin. Where measurements of air temperature and wind are at different levels from those specified in the equations, it is essential to adjust them to the standard levels. The adjustment may be made by multiplying $T_a - 32$ and v, respectively, by correction factors $1.47Z_a^{-1/6}$ and $1.92Z_b^{-1/6}$, where Z_a and Z_b are the heights of measurement, for air temperature and wind speed, respectively, in feet above the snow surface [23, p. 11].

b. Basin Snowmelt during Clear-weather Periods. During rain-free periods, solar and terrestial radiations both become important variables in the balance of heat exchange to the snowpack. They may require direct evaluation for specific meteorological conditions, depending upon basin forest cover. Convection and condensation are generally less important heat sources than radiation. By assuming the loss by forest transpiration to be equal to short-wave radiation melt, which is not computed for forested areas, the total snowmelt for a ripe snowpack (isothermal at 32°F and with 3 per cent free-water content) may be expressed by the following equations [23, p. 15]:

For heavily forested area (>80 per cent cover),

$$M = 0.074(0.53T_a' + 0.47T_d')$$

For forested area (60 to 80 per cent cover),

$$M = k(0.0084v)(0.22T_a' + 0.78T_d') + 0.029T_a'$$

For partly forested area (10 to 60 per cent cover),

$$M = k'(1 - F)(0.0040I_i)(1 - a) + k(0.0084v)(0.22T_a' + 0.78T_d') + F(0.029T_a')$$

For open area (<10 per cent cover),

$$M = k'(0.00508I_i)(1 - a) + (1 - N)(0.0212T_a' - 0.84) \\ + N(0.029T_c') + k(0.0084v)(0.22T_a' + 0.78T_d')$$

where M = snowmelt, in./day

T_a' = difference between air temperature at 10-ft level and snow surface temperature, °F

T_d' = difference between dew-point temperature at 10-ft level and snow surface temperature, °F

v = wind speed at 50-ft level, mph

I_i = observed or estimated solar radiation on horizontal surface, langleys

a = observed or estimated average snow surface albedo

k' = basin short-wave radiation factor (between 0.9 and 1.1), depending on average exposure of open areas to short-wave radiation in comparison with an unshielded horizontal surface

F = estimated average basin forest canopy cover expressed as a decimal fraction

T_c' = difference between cloud base temperature and snow surface temperature, °F (estimated from upper-air temperatures or by lapse rates from surface station, preferably on a snow-free site)

N = estimated cloud cover expressed as a decimal fraction

k = basin convection-condensation melt factor, depending on relative exposure of area to wind

In some cases, it may be necessary to compute snowmelt for subperiods of less than a day. This may be done by applying mean values of air temperature, wind, and precipitation for the particular subperiod and dividing the computed melt by the number of subperiods per day. However, this approach may not be applicable to open areas where solar radiation is involved directly.

A critical item of information of great importance to all snowmelt computations is that of the location and extent. In all snowmelt computations it is important to know that, although snow accumulates by elevations, except in very general terms, snow does not disappear by elevations, but rather by aspects. South-facing slopes are denuded of snow cover perhaps weeks earlier than north-facing slopes. At present no adequate technique has been developed to determine with desired accuracy the location and areal extent of the snow cover. This deficiency in knowledge tends to impair the accuracy expected of the results obtained by the physical equations.

X. SNOW COMPACTION

Compaction of snow by means of rollers, rut cutters, or water-spray devices was a common practice in the United States to prepare stable road surface for the movement of sleds over snow-covered lands and for logging operations in forest. With intensive explorations of the polar regions in recent years, new techniques for snow compaction have been devloped.

By the method developed by the U.S. Navy, snow can be compacted so effectively as to permit landing and take-off of aircraft, weighing up to 27,000 lb with provision for either skis or usual wheels, on two rubber-tired wheels inflated to 60 psi of air pressure [128].

The U.S. Navy constructed a 15-acre compacted-snow public parking lot on a meadow at Squaw Valley, California, to accommodate up to 2,000 cars at the North American Olympic Trials during February, 1959 [129]. It is reported [129] that a compressive compaction, using four-roller-pass techniques, can be used to build a traffic-bearing snow mat provided the initial old snowpack is not over 24 in. and the successive compacted layers do not exceed 6 in. If the old snowpack exceeds 24 in., two passes by a Pulvi-Mixer plus three passes of a roller are required for initial snow-mat construction.

The compacted snow should be protected against any inflow of water from adjacent areas. Sawdust must be used to protect the compacted snow from high ambient temperatures, and especially from the intense solar radiation expected in the spring. A sawdust layer of about $\frac{1}{2}$ in. in thickness spread uniformly is necessary, not only to protect the snow from melting because of ambient heat, but also to provide traction for vehicles. All debris, especially dark-colored soil and gravel, should be removed daily, because it will absorb radiation heat easily, and thus cause localized high rate of melting.

For successful use of a snow-compacted surface, continuous maintenance and repairing are necessary. Also, it is essential that the temperature during the day does not exceed 35°F for a prolonged time, unless the snow is heavily insulated with sawdust. In any case, the temperature should fall below the freezing point during the night.

Recent snow-compaction equipment and snow drags used by the U.S. Navy are described by Camm [130]. It is also reported that snow-leveling and snow-finishing drags are necessary for the construction and maintenance of compacted snow using cold-processing techniques.

XI. SNOW AVALANCHES

The instability of snow in mountainous regions may lead to sudden downward movements of the snow known as *snow avalanches*. The avalanche is a natural phenomenon that releases great natural forces suddenly, like lightning, tornadoes, floods, and earthquakes. The mechanics of snow avalanches is discussed by Bucher [131].

Snow avalanches are classified as either *loose snow* or *slab avalanches*, each of which may be further classified as being dry, damp, or wet, according to the amount of liquid water present in the snow. Ice avalanches may occur at glaciers or at continental ice shelves. A comprehensive discussion of snow avalanches with special reference to forecasting and control measures is given in Ref. 132.

Dry-powder snow avalanches are fastest-moving, whereas wet-snow avalanches are slowest. Church [70] refers to Coaz as having given a mean speed of 350 k/hr (217 mph) for the great Glärnisch powder-snow avalanche of Mar. 6, 1898, on an average slope of 44°. He refers to Mougin's and Bernard's estimate of the speed of wet-snow avalanche as being 17 mph on a slope of 45°. Since it is not unusual for avalanches to cross a valley and climb partly up the opposite slope, it may be dangerous for one to stay on the projection of the apparent path of avalanches.

On Jan. 10, 1962, an ice avalanche broke loose from Mt. Huascarán, the highest mountain in the Peruvian Andes, having an elevation of 22,205 ft. This ice avalanche turned into a mud flow 40 ft deep, demolishing a number of villages, with great loss of life. According to Acoca [133], this avalanche descended with an average speed of 65 mph.

Published references to avalanche velocities in North America are rare. The following two reports were given to the author in personal communication with Richard M. Stillman, an avalanche-hazard forecaster for the Arapahoe National Forest of the U.S. Forest Service. Stillman timed with a stop watch the flow of an avalanche at the Stanley Slide, which is on the eastern slope of Berthoud Pass, Colorado, and computed the peak velocity of the avalanche to be 130 mph. He also reported upon the avalanche of Jan. 22, 1962, at the Twin Lakes, Colorado. This was a wind-slab avalanche descended from the slopes of Mt. Elbert, the highest peak in Colorado, having an elevation of 14,431 ft. The peak velocity was estimated between 150 and 175 mph.

There are various approaches to control avalanches and prevent damage [134]. These may consist of reforestation of the slopes at the site of the avalanches and construction of control structures such as masonry walls; ridges of wood, steel, or concrete; vertical fences; heavy-wire steel supports; wind baffles; snow fences; earthen or masonry bunkers; mounds; snow sheds; and deflection walls, bunkers, or ramps to split avalanches. A forest cover may be used for avalanche prevention, but it must be dense and be extended to the points of avalanche release.

For the protection of railroads against snow avalanches, fixed and portable snow fences are commonly used to control drifting snow, and snow sheds can be built to protect the track against slides and excessive level fall and to prevent filling of cuts [135]. For heavy sheds, reinforced-concrete structures may be used to reduce maintenance cost and to avoid fire risk.

For the protection of highways against snow avalanches, precast-concrete roof beams have been used to construct snow sheds in order to save time of construction [136]. Other methods, including snow sheds, earth dams, mounds, and use of explosives, were also reported and discussed [137, 138].

Avalanche hazards may be reduced by preventing the accumulation of potentially avalanche-producing deposits of snow at critical spots. This may be done by deliberately starting small avalanches by skilled skiers or by the use of explosives or artillery fire. The use of 75-mm field artillery has been eminently successful in preventing destructive avalanches in the Rocky Mountains of Colorado [139].

Favorable conditions for snow-avalanche hazards in the western United States are in general as follows [39]: accumulation of 10 in. or more of new snow at the rate of 1 in. or more per hour; wind at 12 mph or more; fair weather with rapidly rising temperature after a heavy storm; persistence of above-freezing temperatures for 36 hr over a heavy snowpack in the spring; and rain. Avalanche danger is greatest after 10 in. or more of snow has accumulated in one storm, and the danger persists for 24 to 48 hr following that storm.

XII. SNOW LOADS

A. On Buildings

The design criteria for snow loads are apt to vary geographically. The probable snow load varies with latitudes and elevations above sea level. For roof trusses three possible combinations of loads may produce the maximum stresses, namely, dead load plus snow load, dead load plus wind load, and dead load plus one-half snow load plus wind load. In the middle latitudes the last combination is usually most critical.

In Canada, an extensive analysis of hydrologic data has been made to develop design criteria for snow loads [140]. The density of snow at time of fall is known to vary, and the metamorphosis of snow further changes the density as the snow ripens. By assuming that one inch of snow will exert a pressure of one pound per square foot, corresponding to an accepted density of 0.192 for snow, a map showing the maximum depth of snow in inches on ground for a given recurrence interval, taken as 30 years in Canada, can be used to derive another map showing the maximum snow load in pounds per square foot for the given recurrence interval. In the analysis, the time occurrence of the maximum depth and the effect of rainfall were also considered. If the rain was totally absorbed by the snow, the snow load is increased by an amount of 5.2 psf for each inch of rainfall. The type of roofs was also specified. On flat or low-pitch roofs, the maximum snow load is assumed to be 80 per cent of that on the ground.

For the continental United States, the Weather Bureau and the Housing and Home Finance Agency have produced maps showing estimated weight of seasonal snowpack (Fig. 10-18) [141]. The minimum recommended snow loads on roofs are given in Table 10-3. In choosing the design-load values, however, it is always desirable to consult also the building code pertinent to the area under consideration.

B. On Power Lines and Transmission Towers

The snow load is usually considered together with sleet and ice loads in the design and operation of power lines and transmission towers. The consolidation of a snowpack as a result of metamorphosis can result in very destructive effects. As the snow

Table 10-3. Minimum Uniformly Distributed Design Vertical Live Roof Loads*

	Pounds per square foot of horizontal projection of roof slopes			
	3 in/ft or less†	6 in/ft	9 in/ft	12 in/ft or more
Southern states..................	20	15	12	10
Central states...................	25	20	15	10
Northern states..................	30	25	17	10
Great Lakes, New England, and Mountain areas‡.............	40	30	20	10

* From *Home and Housing Finance Agency Housing Research Paper* 19, May, 1952 [141].

† For flat roofs used for sun decks or promenades, a 60-psf minimum load is specified by most building codes, which is governed by operational loads, such as a crowd of people, rather than by snow load.

‡ Great Lakes and New England areas include Minnesota, Wisconsin, Michigan, New York, Massachusetts, Vermont, New Hampshire, and Maine. Mountain areas include the Appalachians above 2,000-ft elevation, Pacific coastal ranges above 1,000-ft elevation, and Rocky Mountains above 4,000-ft elevation

density increases with the approach of spring, the reduction in depth of the snowpack may cause very heavy loadings on horizontal structural members due to an arch action developed in the snowpack.

In British Columbia, where winds are strong and snow may be as great as 25 ft in depth, transmission towers were designed with five or six aluminum legs of 38 in. in diameter [142]. Both aluminum and steel were used for all components of the transmission system to withstand extremely heavy ice loads up to 40 lb/foot of length at 0°F. On Jan. 26, 1955, a dry-snow avalanche destroyed one aluminum and two steel towers of the transmission line. Because such towers cannot be rebuilt economically to withstand future snow avalanches, the towers were replaced by a suspension system to carry the power line across the avalanche area [143].

There are many problems specifically related to the suspended conductors under winter conditions. The density of hoar frost and ice was found to vary from a few pounds to 57 lb/ft^3 in studies at nine locations in the northwestern United States [144]. Visual observations alone were found to be exaggerating the seriousness of the ice conditions. For example, 2.5 in. of radial ice observed in the form of hoar frost may be actually only ½ in. of solid radial ice. Severe icing conditions in the Pacific Northwest were found to occur commonly in 3- to 7-year cycles.

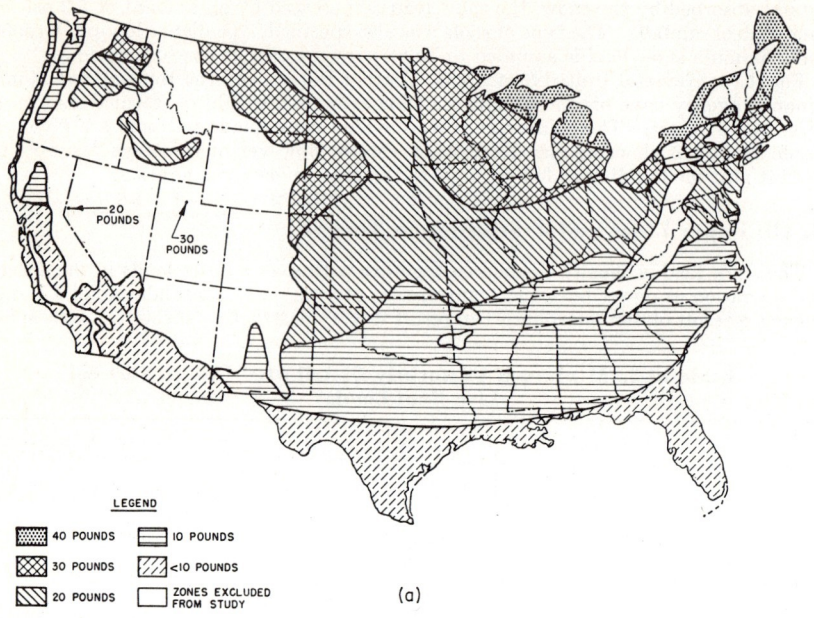

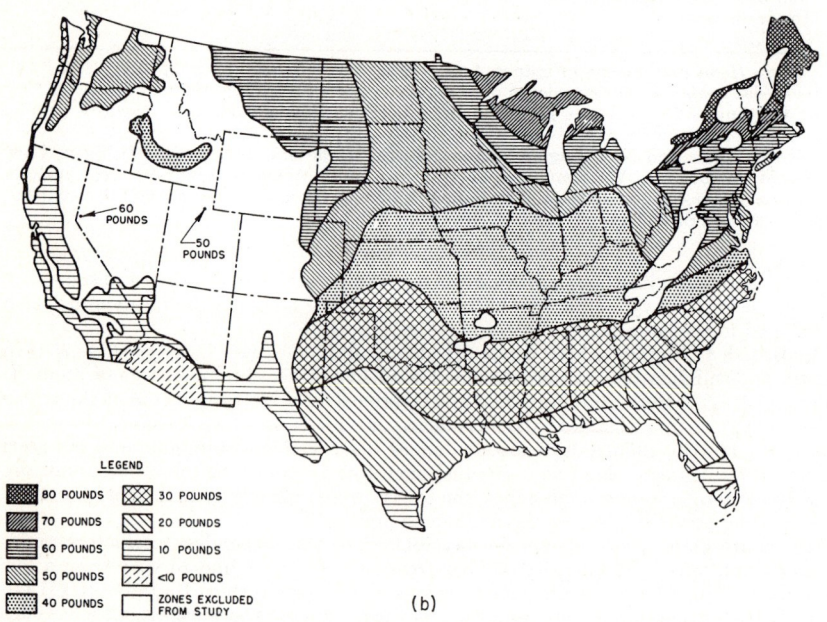

Fig. 10-18. Estimated weight of seasonal snowpack in pounds per square foot (*U.S. Home and Housing Finance Agency* [141]). (a) Equaled or exceeded 1 year in 10; (b) maximum accumulation on the ground plus weight of maximum probable snowstorm.

The accumulation of sleet on suspended conductors tends to develop a cross section converting the circular section of the cable to a section approximating an airfoil [145]. When wind, with a certain distribution of gustiness, impinges on suspended cables having an airfoil section, strong uplift may be generated aerodynamically sufficient to cause the glazed conductor to rise [146]. The mid-point of the conductor may rise practically as far above the line of supports as it is below when at rest.

The vibration of ice-encrusted cables may be caused also by a sudden change in the center of gravity of the cables due to breaking off of the brittle sleet caused by flexure of the cables [147]. This may result in setting up a very complex oscillatory system, causing each individual cable in a transmission line to vibrate in an individual manner.

The vibration of power lines or telephone wires due to such natural causes is generally described as "dancing," or "galloping" [147, 148]. When this occurs in cables conducting high-voltage electrical current, short circuits or flashovers may occur, causing at times extensive damage to equipment and interruption of service. In order to avoid such difficulties, the bracing, sags, and dimensions of the transmission lines should be adequately designed and proportioned [143]. Various attempts have been made to arrive at mathematical solutions to the problem [149, 150]. However, the almost infinite combinations of various factors involved have not yet produced exact solutions.

Transmission-line outages due to vibration in the vertical plane of spans following sudden unsymmetrical releases of ice loads have been greatly reduced, especially for periods free of winds, by horizontally offsetting the conductors from 1.5 to 3 ft [149]. For design purposes, the horizontal wind loads on transmission-line components, including conductors and hangers, may be taken as 8 psf acting on two-thirds of the projected area of each component [149].

Recent research on the mechanism of galloping may lead to new ways of preventing or reducing the damages [151, 152]. Despite many practical limitations, new techniques are also being studied to heat the conductors by high amperages. A rise of 9°C above ambient temperature has been recommended to prevent the formation of ice on a conductor [153].

XIII. SNOW PRODUCED BY ARTIFICIAL MEANS

Although many attempts were made in the past to cover bare areas on otherwise snow-covered slopes by hauling snow or fine ice chips and depositing them on bare spots, the patchings thus made were not desirable for skiing. Through serendipity, Philip L. Tropeano of the Larchmont Engineering Company, Lexington, Mass., while working on a specialized sprayer being developed for the protection of fruit trees against early frost, found that his sprayer was producing snowflakes [154]. This led to the development of specialized apparatus for producing snow.

Pierce [155] has patented a snow-making machine in which separate pipe systems are used for bringing compressed air and water to the nozzle, which is so designed as to cause the two fluids to impinge outside the nozzle. The sudden decompression of the air causes a sufficient drop in the temperature to produce crystallization of very fine droplets of water, with the resultant formation of true snowflakes. A big installation for making snow was used at Hot Springs, Virginia [156]. This installation consumed 500 gal of water per minute, 360 gal being used to make snow and the remainder being recirculated to prevent freezing. Each snowmaker required 12 gal/min of water at a pressure of 100 to 150 psi and 100 ft^3/min of air at 100 psi.

For successful snow production, the air temperature must be 28°F or colder at night, the preferred time for snowmaking, and 20°F or colder when the sun is shining. Twelve nozzles can produce 4 to 8 in. of well-packed snow in a day, covering an area of 1,000 by 250 ft.

XIV. MISCELLANEOUS PROBLEMS RELATED TO SNOW

Several problems that are closely akin to snow may be briefly discussed in this section for the benefit of hydrologic engineers and practicing hydrologists.

A. Ice in Streams

Although the physical behavior of ice is thoroughly discussed in Sec. 16, the problem of ice in streams as it affects engineering works may be discussed here.

A major problem in river regulation, forecasting, and operations is the river ice. Under turbulent-flow conditions, needlelike *frazil ice* may be formed because of supercooling (Sec. 16). Deposition of frazil ice on banks, rocks, or engineering structures may take place at surprisingly rapid rates and possess sufficient mechanical strength to stop the flow of water through trash racks completely, as reported by Schaefer [157, p. 887]. An increase in temperature of only a few thousandths of degrees centigrade of a flow containing frazil ice was found to be enough to prevent adherence of the ice, and thus to obviate operational difficulties at hydropower installations [158, 159]. When a flowing stream becomes loaded with a sufficiently heavy amount of frazil particles, it is not uncommon to freeze the flow rapidly and solidly to depths of several feet. In the St. Lawrence River drainage basin, such freezing has been observed to occur at depths up to 70 ft.

Anchor ice is another source of trouble in river regulation and power-plant operation [160]. This is a type of ice formed on underwater rocks or engineering structures. The formation is due to the cooling down below the freezing point by back radiation through the clear water. Formation of anchor ice has not been observed under ice cover or under bridges or culverts, since back radiation from the covering over the water surface effectively cuts down the supercooling of the underwater surfaces.

Freezing of water surface to form an ice sheet in streams, lakes, and reservoirs is an important concern in engineering design and hydrologic practice. The rate of thickening a sheet of ice by conduction alone to a cold atmosphere can be calculated by a formula given by Barnes [160, 8]:

$$t = \frac{LSE}{K\theta}\left(1 + \frac{E}{2}\right)$$

where t is the time in sec for the ice sheet to attain a thickness E in cm; L is the heat of fusion, or 79.7 cal/grain at 0°C; S is the density of ice, or 0.9166; K is the conductivity of ice, or 0.0057 cal/°C difference of temperature per square centimeter per second; and θ is the difference in temperature in °C between the underside of the ice sheet (taken as 0°C) and the air temperature. This formula is based on an analysis of the actual observed data, assuming that the surface is clear of snow and water at 0°C. Although the formula does not consider other atmospheric factors such as wind, humidity, and radiation, it has been found to hold quite accurately for fairly large temperature differences.

Expansion of an ice sheet, as the result of a temperature rise after a cold wave, can develop considerable thrust against abutting structures. Many studies have been made to determine such ice pressures [161, 162]. Monfore [162] gives the curves, in Fig. 10-19, which can be used to estimate the maximum pressure and the time of temperature rise for a given rate of ice-temperature rise.

The formation of ice in streams will affect the flow to various degrees. Where ice covers the stream throughout the winter, a rather definite annual cycle of streamflow can be observed [163]. In general, large quantities of frazil ice will cause slight rise in stage due to the resulting expansion of volume and increase in viscosity of the water. Anchor ice will do likewise, but its effect is never large. An ice sheet has a rough undersurface which will reduce the discharge of the flow. As the spring comes, the ice sheet will melt and become honeycombed to form small reservoirs to hold water. With the rise of temperature, the first breakup of such reservoirs will hasten to move the whole pack of ice downstream.

Accumulation of large volumes of frazil ice and anchor ice will cause ice jams and gorges and thus may result in serious flooding. When the ice is deposited on flood plains to form a broad sheet, the river may flow in a braided channel over the ice to form overflow gorges [164]. Ice jams may be broken up by various ways of induced melting, to be discussed later. Ice gorges may be prevented by building reservoirs or check structures to intercept the frazil or anchor ice being brought downstream from headwaters.

Many publications on ice problems in hydraulic structures are given in Refs. 165 and 166.

B. Frost in the Soil

Frost in the ground is important to hydrologists because it affects runoff and infiltration, and to engineers because it may damage ground surface structures such as highways, airfield pavements, and canals.

The occurrence of frost in the ground depends on complex interrelations between many factors, including air temperature, soil condition, moisture content, vegetal cover, and snow on the ground.

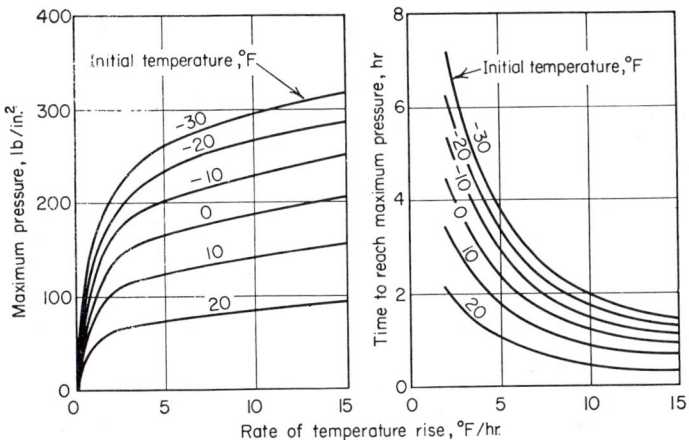

Fig. 10-19. Maximum ice pressure and time of temperature rise related to rate of ice-temperature rise. (*Monfore* [162].)

For ice segregation to occur in soil, freezing temperatures must penetrate the soil for a sufficient period of time. To measure the combined duration and magnitude of below-freezing temperatures occurring during any given freezing season, a *freezing index* is generally used by the U.S. Army Corps of Engineers [167] and U.S. Bureau of Public Roads [168]. This index is defined as the number of degree-days between highest and lowest points on a curve of cumulative degree-days vs. time for one freezing season. The so-called *air freezing index* is the freezing index determined for air temperatures at 4.5 ft above the ground surface. Penetration of freezing temperatures below a pavement kept cleared of snow may be correlated approximately with the air freezing index and with water content and dry unit weight of the base and subgrade materials lying in the annual frost zone. The *annual frost zone* is the top layer of ground subject to annual freezing and thawing.

The type and condition of soils mainly influence the structure of the frost formation that occurs. The moisture in soil affects the rate of freezing because of its latent heat of fusion and its effect on the specific heat of the soil. According to Casagrande [170], inorganic soils containing 3 per cent or more of grains finer than 0.02 mm in diameter, by weight, are generally frost-susceptible. Based on a combination of laboratory tests and field experience, this criterion is not precise, but can be taken as an engineering approximation.

The role played by vegetal cover depends on the humus content, root channels, transpiration, nocturnal radiation, and many other factors. In general, a good vegetal cover will slow frost formation and reduce the depth of penetration.

A snow cover of sufficient depth forms a good insulating blanket against freezing air temperatures, which affects greatly the frost penetration in the soil. Atkinson and Bay [171] report tests showing faster thawing of the frozen ground under thicker snow cover. By theoretical computations Berggren [172] shows that a 4-in. snow cover would reduce frost penetration to one-sixth that in a bare ground. Other studies on this subject are variously reported [173–175].

Maps showing approximate annual depths of frost penetration have been prepared for the United States [17, 176]. Such information is offered as a guide only because it is based on limited observations and hence does not attempt to show local variations, which may be substantial, particularly in mountainous areas.

Freezing and thawing of soil and their resulting effects on contacting materials and structures are called *frost action*. The raising of the ground surface due to the formation of ice in the underlying soil is known as *frost heave*. Both are complicated phenomena. For details the reader should also consult Refs. 177 to 186.

C. Permafrost

The ground perennially below the freezing temperature is known as *permafrost*. An irregular surface representing the upper limit of permafrost is the *permafrost table*. The entire layer of ground above the permafrost table is *suprapermafrost*. In arctic

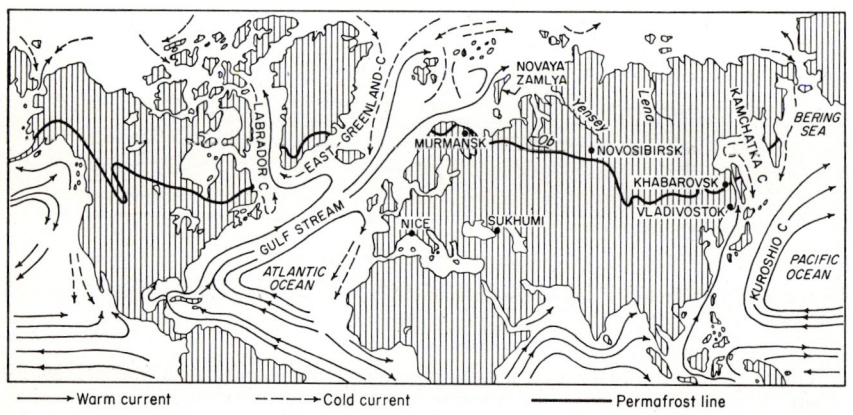

Fig. 10-20. Permafrost line in Arctic region. (*Prepared by Ven Te Chow.*)

and subarctic regions where annual freezing penetrates to the permafrost table, suprapermafrost is identical with the annual frost zone. Figure 10-20 shows the *permafrost line*, which represents the approximate border of the arctic permafrost. The thickness of the permafrost varies roughly with annual mean temperature. It varies from a maximum of about 2,000 ft reported at Nordvik in Siberia to about 920 and 170 ft at Umiat and Northway, respectively, in Alaska.

Permafrost may or may not contain segregated ice. Appreciable bodies of ground ice in various forms are often found in permafrost consisting of fine-grained soils under poorly drained positions. Their presence generally creates serious engineering problems.

Experience of building foundations in permafrost regions has shown the necessity of keeping the permafrost in a perpetually frozen condition. Any local disturbance may cause instability, such as sudden outpours of water, with subsequent freezing. Presence of subsurface groundwater channels may cause thawing, with resulting progressive differential settlement in summer and of excessive heave in winter. As for construction of dams on permafrost, one should recognize the risk of leakage and possible failure of the dam, because the impounding water may cause thawing of the foundation soil [187].

It is reported [188] that the presence and character of permafrost may be indicated by tree species. Spruce indicates a wet site with the permafrost close to the surface; birch, unfrozen ground; willow, presence of underground water; and pine and fir, well-drained soils, with the permafrost, if present, at some depth below the soil surface. Possible permafrost problem areas are frequently indicated by various ground configurations such as polygons, pingos, stone rings, and stone stripes.

Muller [189] offers an extensive discussion on engineering problems in permafrost regions. For other details, the reader is referred also to Sec. 16 and Refs. 190 to 192.

D. Induced Melting of Snow

Induced melting provides a way to clear snow. This may be done by heating or by employing chemicals. Both methods in various forms are commonly used to clear snow on highways and walkways, in cities and airfields, and even over large areas.

Heat used for induced melting may be obtained from solar energy, from burning of combustible substances, and from a surface heated by a centralized source.

Solar energy can be absorbed by a coating of dark materials over the snow and thereby converted into heat for melting. The dark materials used for such purposes may include soot, finely powdered coal, industrial slag, and finely powdered organic matter. Lang [193] obtained the best results on highways at elevations over 9,000 ft in the Sierra Nevadas, California, by a spread of $\frac{1}{4}$ lb of soot per 100 ft^2 of snow or about 100 lb/mile of a 7-ft spread over the highway. Landsberg [194] reported that coal dust would absorb about 90 per cent of the incident radiation. However, caution should be taken not to overuse the heat-absorbing material, which may serve as a layer of heat insulation and consequently may reduce its value for induced melting [195].

Burning of combustible substances provides a direct application of heat to melt snow. For this purpose a snow melter has been patented [196] which consists of a snow-loading conveyor, a combustion chamber, and a controlled-discharge system. It can clear about 50 tons of snow per hour.

Surface or panel heating may be achieved by installing a system of hot-water pipes [197] or of electric-resistance wires underneath the surface to be heated. When hot water is used, salt or ethylene glycol may be added to the water to prevent possible freezing.

The use of chemicals to induce melting of snow is based on the principle that the lowering of the vapor pressure of the solution formed by adding the chemical to the snow lowers the freezing point of water [198, 199]. The ready availability and low costs of many salts, including rock salt, sodium chloride, and calcium chloride, make them conveniently useful for this purpose. Calcium chloride is found to be particularly effective since it has greater solubility, positive heat of solution, and more operative ability at low temperatures and humidities than other salts [200]. For practical purposes, a mixture of two salts with addition of cinders, sand, or final gravel is commonly used.

XV. REFERENCES

1. Schaefer, V. J.: Snow, in "Encyclopaedia Britannica," vol. 20, p. 854, 1956.
2. Schaefer, V. J., G. J. Klein, and M. R. de Quervain: The International classification for snow (with special reference to snow on the ground), issued by the International Association of Scientific Hydrology, Commission on Snow and Ice, published as *Tech. Mem.* 1 by National Research Council, Associate Committee on Soil and Snow Mechanics, Ottawa, Canada, August, 1954.
3. Seligman, Gerald: "Snow Structure and Ski Fields," Macmillan & Co., Ltd., London, 1936.
4. "Instructions for Making and Recording Snow Observations," U.S. Army Corps of Engineers, Snow Ice and Permafrost Research Establishment, Instruction Manual 1, May, 1954. (Effective July 15, 1961, U.S. Army Corps of Engineers, Snow, Ice and Permafrost Research Establishment, Wilmette, Ill., and U.S. Army Engineering Divi-

sion, New England, Arctic Construction and Frost Research Laboratory, Waltham, Mass., were combined into one organization with new headquarters at U.S. Army Cold Regions Research and Engineering Laboratory, Hanover, N.H.)

5. Simplified field classification of natural snow type for engineering purposes, U.S. Army Corps of Engineers, Snow, Ice and Permafrost Research Establishment, GPO 801882-1, Wilmette, Ill., December, 1952.
6. Dorsey, N. E.: "Properties of Ordinary Water Substance," Reinhold Publishing Corporation, New York, 1940.
7. Mantis, H. T., and others: Review of the properties of snow and ice, U.S. Army Corps of Engineers, Snow, Ice, and Permafrost Research Establishment, *SIPRE Rept.* 4, July, 1951.
8. Barnes, H. T.: "Ice Engineering," Renouf Publishing Company, Montreal, Canada, 1928.
9. "Handbook of Chemistry and Physics," 42d ed., Chemical Rubber Publishing Company, Cleveland, 1960–1961.
10. "American Society for Metals Handbook," 8th ed., vol. I, American Society for Metals, Metals Park, Ohio, 1961.
11. McConica, T. H., III: Bibliography of ice and frost control, WADE, *Tech. Rept.* 56-338, pt. 1, ASTIA, Document AD142317, Wright Air Development Center, Wright-Patterson Air Force Base, Ohio, January, 1958.
12. Tremaine, Marie: "Arctic Bibliography," U.S. Department of Defense, 1960, vol. IX; also preceding volumes.
13. Bibliography on snow, ice and permafrost with abstracts, Snow, Ice and Permafrost Research Establishment, U.S. Army Corps of Engineers, *SIPRE Rept.* 12, vol. XIII, January, 1959.
14. "Mean Annual Total Snowfall (Inches)," U.S. Weather Bureau, Office of Climatology, 1960, sheet of the National Atlas of the United States.
15. Kincer, J. B.: Precipitation and humidity, in "Atlas of American Agriculture," U.S. Department of Agriculture, separate issue, Mar. 15, 1922.
16. "Climate and Man," U.S. Department of Agriculture Yearbook of Agriculture, 1941.
17. *Climatological Data*, a series published monthly by the U.S. Weather Bureau for individual states and, in a few instances, a combination of states.
18. "Some Extremes of Snowfall," U.S. Weather Bureau publication on reverse side of the Daily Weather Map for Dec. 23, 1958.
19. Garstka, W. U.: Interpretation of snow surveys, *Trans. Am. Geophys. Union*, vol. 30, no. 3, pp. 412–420, June, 1949.
20. Bader, H., R. Haefeli, E. Bucher, J. Neher, O. Eckel, and C. Thams, with an introduction by P. Niggli: Snow and its metamorphism (Der Schnee und seine Metamorphose), *Beitr. Geol. Schweiz, Geotechn. Ser.*, *Hydrologie*, No. 3, Bern, 1939, U.S. Army Corps of Engineers, Snow, Ice and Permafrost Research Establishment, Transl. 14, Wilmette, Ill., January, 1954.
21. Work, R. A.: Snow-layer density changes, *Trans. Am. Geophys. Union*, vol. 29, pp. 525–546, 1948.
22. Weather extremes around the world, U.S. Army Quartermaster, Research and Engineering Command, Environmental Protection Research Division, Natick, Mass., rev., April, 1959.
23. "Runoff from Snowmelt," U.S. Army Corps of Engineers, Engineering and Design Manuals, EM 1110-2-1406, Jan. 5, 1960.
24. Instructions for climatological observers, *U.S. Weather Bur. Circ.* B, 11th ed., rev. January, 1962.
25. Nipher, F. E.: On the determination of the true rainfall by elevated gages, *Proc. Am. Assoc. Advan. Sci.*, St. Louis, Mo., vol. 27, pp. 103–108, 1878.
26. Alter, J. C.: Shielded storage precipitation gages, *Monthly Weather Rev.*, vol. 65, no. 7, pp. 262–265, July, 1937.
27. Warnick, C. C.: Influence of wind on precipitation measurements at high altitudes, *Univ. Idaho Eng. Expt. Sta. Bull.* 10, Moscow, Idaho, April, 1956.
28. Codd, A. R.: Seasonal storage precipitation gages, *Trans. Am. Geophys. Union*, vol. 28, no. 6, pp. 899–900, December, 1947.
29. "Operation and Maintenance of Storage Precipitation Gages," U.S. Weather Bureau, 1951.
30. "Instructions for the Operation and Maintenance of the Sacramento-type Storage Precipitation Gage," U.S. Weather Bureau, Division of Climatological and Hydrologic Services, 1946.
31. Garstka, W. U., L. D. Love, B. C. Goodell, and F. A. Bertle: "Factors Affecting Snowmelt and Streamflow," U.S. Bureau of Reclamation and U.S. Forest Service, 1958.

32. Hamilton, E. L., and L. A. Andrews: Control of evaporation from rain gages by oil, *Am. Meteorol. Soc. Bull.*, vol. 34, no. 5, pp. 202–204, May, 1953.
33. Allen, F. D., R. E. Glover, W. U. Garstka, and H. M. Posz: Heated precipitation gage intake tube, U.S. Patent 2,701,472, Feb. 8, 1955, with certificate of correction, Mar. 22, 1955.
34. Daum, C. R., R. H. Kuemmich, and W. U. Garstka: Electronic coding device, U.S. Patent 2,748,376, May 29, 1956.
35. Push-button flood control, *Reclamation Era*, vol. 40, no. 1, pp. 1–3, February, 1954.
36. Warnick, C. C.: Rime ice and snow capping on high altitude precipitation gage, *Proc. 25th Annual Meeting Western Snow Conf.*, April, 1957, Colorado State University, Fort Collins, Colo., pp. 24–34, November, 1957.
37. Warnick, C. C.: A study of rime ice and snow capping on high altitude precipitation gages, *Univ. Idaho, Eng. Expt. Sta., Spec. Res. Project* 17A, *Progr. Rept.* 3, Moscow, Idaho, June, 1958.
38. Maxwell, L. M., C. C. Warnick, L. A. Beattie, and G. G. Hespelt: Automatic measurement of hydrologic parameters at remote locations, *Proc. Western Snow Conf.*, Santa Fe, N.M., April, 1960, Colorado State University, Fort Collins, Colo., pp. 25–31, September, 1960.
39. Snow survey safety guide, *U.S. Soil Conserv. Serv. Agr. Handbook* 137, January, 1958.
40. Marr, J. C.: Snow surveying, *U.S. Dept. Agr. Misc. Publ.* 380, June, 1940.
41. Work, R. A.: Stream-flow forecasting from snow surveys, *U.S. Soil Conserv. Serv. Circ.* 914, March, 1953.
42. Snow survey sampling guide, *U.S. Soil Conserv. Serv. Agr. Handbook* 169, December, 1953.
43. Church, J. E.: Principles of snow surveying as applied to forecasting stream flow, *J. Agr. Res.*, vol. 51, no. 2, pp. 97–130, July 15, 1935.
44. Specifications for snow sampling equipment, U.S. Soil Conservation Service, Water Supply Forecasting Section, Portland, Ore., 1956.
45. Goodell, B. C., and K. L. Roberts: Test of snow-sampling tubes of large and small diameter, *Trans. Am. Geophys. Union*, pt. I, pp. 151–152, 1941.
46. Work, R. A.: General requirements of a satisfactory over-snow machine, *Proc. Western Snow Conf.*, April, 1951, Fort Collins, Colo., pp. 69–74, July, 1951.
47. Bekker, M. G.: Soil-vehicle concepts found impeding design, *Soc. Automotive Engrs. J.*, vol. 58, no. 5, pp. 20–24, May, 1950.
48. Diamond, Marvin: Studies on vehicular trafficability of snow, U.S. Army Corps of Engineers, Snow, Ice and Permafrost Research Establishment, *SIPRE Rept.* 35, pt. 1, April, 1956.
49. Lanyon, J. L.: Studies on vehicular trafficability of snow, U.S. Army Corps of Engineers, Snow, Ice and Permafrost Research Establishment, *SIPRE Rept.* 35, pt. II, July, 1959.
50. Brown, A. H.: Sno-cats mechanize Oregon snow surveys, *Natl. Geograph. Mag.*, vol. 96, pp. 691–710, November, 1949.
51. Codd, A. R., and M. W. Nelson: Use of aircraft in snow surveying, *Proc. 22d Annual Meeting Western Snow Conf.*, Salt Lake City, Utah, April, 1954, Fort Collins, Colo., pp. 83–85, October, 1954.
52. Henderson, T. J.: The use of aerial photographs of snow-depth markers in water supply forecasting, *Proc. Western Snow Conf.*, April, 1953, Fort Collins, Colo., pp. 40–47, September, 1953.
53. Hannaford, J. F.: A proposal for incorporating aerial snow-depth markers into water supply forecasts, *Proc. 28th Annual Meeting, Western Snow Conf.*, Santa Fe, N.Mex., April, 1960, Colorado State University, Fort Collins, Colo., pp. 9–14, September, 1960.
54. Dean, W. W.: Helicopters, from the fieldman's viewpoint, *Proc. Western Snow Conf.*, April, 1957, Colorado State University, Fort Collins, Colo., pp. 71–72, November, 1957.
55. Washichek, J. N.: Snow survey by air, paper presented at Snow Surveyors' School at McCall, Idaho, Jan. 19, 1954, mimeographed by the U.S. Soil Conservation Service, Fort Collins, Colo.
56. Gerdel, R. W., B. L. Hansen, and W. C. Cassidy: The use of radioisotopes for the measurement of the water equivalent of a snow pack, *Trans. Am. Geophys. Union*, vol. 31, no. 3, p. 449, June, 1950.
57. Gerdel, R. W., and C. W. Mansfield: The use of radioisotopes in research on snow melt and runoff, *Proc., Western Snow Conf.* Boulder City, Nev., pp. 5–17, April, 1950.
58. Itagaki, K.: An improved radio snow gage for practical use, *J. Geophys. Res.*, vol. 64, no. 3, pp. 375–383, March, 1959.
59. Doremus, J. A.: Telemetering system for radioactive snow gage, *Electronics*, vol 25, pp. 88–91, 1951.

60. Hildebrand, C. E.: Development and test performance of radioisotope-radiotelemetering snow gage equipment, U.S. Army Corps of Engineers, South Pacific Division, Civil Works Investigation Project CWI-170, Los Angeles, Calif., April, 1956.
61. Radio transmitted snowpack measurements, King's River Watershed, 1958–1959 season, Water Conditions in California Basic Data Supplement, as of May 1, 1959, State of California Department of Water Resources, Division of Resources Planning, Sacramento, Calif., 1959.
62. Robinson, C.: Gamma radiation gauges snow pack water content, *Elec. World*, vol. 154, no. 17, cover photo and pp. 82–83, Oct. 24, 1960.
63. *Proc. Western Interstate Snow Surv. Conf.*, Feb. 18, 1933, and June 28, 1933, University of Nevada, Agricultural Experiment Station, Reno, Nev., 1934.
64. *Proc. Central Snow Conf.*, Dec. 11–12, 1941, Michigan State College, East Lansing, Mich., 1942, vol. 1.
65. Fredericksen, D. G.: Use of snow survey data by soil conservation districts, *Proc. Western Snow Conf.* Bozeman, Mont., April, 1958, Colorado State University, Fort Collins, Colo., pp. 32–35, October, 1958.
66. Blanchard, F. B.: Operational economy through applied hydrology, *Proc. Western Snow Conf.*, April, 1955, Colorado State University, Fort Collins, Colo., pp. 35–48, August, 1955.
67. Ayer, G. R.: History of snow surveying in the East, *Proc. Western Snow Conf.* Reno, Nev., April, 1959, Colorado State University, Fort Collins, Colo., pp. 12–16, September, 1959.
68. Documents of the Assembly of the State of New York, 59th Sess., 1836: Report of John B. Jervis, vol. II, E. Croswell, Albany, N.Y., 1836, pp. 55–60.
69. Stafford, H. M.: History of snow surveying in the West, *Proc. Western Snow Conf.*, Reno, Nev., April, 1959, Colorado State University, Fort Collins, Colo., pp. 1–9, September, 1959.
70. Church, J. E.: Snow and snow survey: ice, chap. 4, in Oscar E. Meinzer (ed.), "Physics of the Earth," vol. IX, "Hydrology," McGraw-Hill Book Company, New York, Inc., 1942, and Dover Publications, Inc., New York, 1949.
71. Stockwell, H. J.: Use of soil moisture resistance units in water supply forecasting, *Proc. Western Snow Conf.*, Reno, Nev., April, 1959, Colorado State University, Fort Collins, Colo., pp. 35–39, September, 1959.
72. Coleman, E. A.: Manual of instructions for use of fiber glass soil moisture instrument, California Forest and Range Experiment Station, Berkeley, Calif., rev. ed., April, 1950.
73. Bouyoucos, G. J.: Improved soil moisture meter, *Agr. Eng.*, vol. 37, no. 4, pp. 261–262, April, 1956.
74. Court, Arnold: Selection of "best" snow course points, *Proc. Western Snow Conf.*, Bozeman, Mont., April, 1958, Colorado State University, Fort Collins, Colo., pp. 1–11, October, 1958.
75. Alter, J. C.: The mountain survey: its genesis, exodus and revelations, *Trans. Am. Geophys. Union*, pt. III, pp. 892–899, 1940.
76. Water supply forecasts for the western United States including forecasts prepared by U.S. Weather Bureau and the State of California Department of Water Resources, vol. 13, 1960–1961 water year, nos. 1–5, issued as of Jan. 1, Feb. 1, Mar. 1, Apr. 1, and May 1, 1961, respectively. On request from U.S. Weather Bureau, Hydrologic Services Division.
77. Kohler, M. A.: Water supply forecasting developments, 1951–1956, *Proc. Western Snow Conf.*, Santa Barbara, Calif., April, 1957, Colorado State University, Fort Collins, Colo., pp. 62–68, November, 1957.
78. Work, R. A., and R. T. Beaumont: Basic data characteristics in relation to runoff forecast accuracy, *Proc. Western Snow Conf.*, Bozeman, Mont., April, 1958, Colorado State University, Fort Collins, Colo., pp. 45–53, October, 1958.
79. Kohler, M. A.: Preliminary report on evaluating the utility of water supply forecasts, *Proc. Western Snow Conf.*, Reno, Nev., April, 1959, Colorado State University, Fort Collins, Colo., pp. 26–33, September, 1959.
80. Peck, E. L.: Low winter streamflow as an index to the short- and long-term carry-over effects in water supply forecasting, *Proc. Western Snow Conf.*, Salt Lake City, Utah, April, 1954, State University, Fort Collins, Colo., pp. 41–47, October, 1954.
81. Linsley, R. K., Jr., M. A. Kohler, and J. L. H. Paulhus: "Applied Hydrology," McGraw-Hill Book Company, Inc., New York, 1949.
82. Linsley, R. K., Jr., and J. B. Franzini: "Elements of Hydraulic Engineering," McGraw-Hill Book Company, Inc., New York, 1955.
83. Butler, S. S.: "Engineering Hydrology," Prentice-Hall, Inc., Englewood Cliffs, N.J., 1957.

REFERENCES

84. Linsley, R. K., Jr., M. A. Kohler, and J. L. H. Paulhus: "Hydrology for Engineers," McGraw-Hill Book Company, Inc., New York, 1958.
85. Kittredge, Joseph: "Forest Influences," McGraw-Hill Book Company, Inc., New York, 1948.
86. Riesbol, H. S.: Snow hydrology for multiple purpose reservoirs, *Trans. Am. Soc. Civil Engrs.*, vol. 119, pp. 595–627, 1954.
87. "Snow Hydrology," Summary report of the snow investigations, U.S. Army Corps of Engineers, North Pacific Division, Portland, Ore., June 30, 1956.
88. Summersett, John: Reproduction of snowmelt floods in the Boise River, *Proc. Western Snow Conf.*, Boise, Idaho, April, 1953, Colorado State University, Fort Collins, Colo., pp. 36–43, September, 1953.
89. Garstka, W. U.: Why the snow?, *Reclamation Era*, vol. 35, no. 6, pp. 126, 127, 137, June, 1949.
90. Boyer, P. B.: Heat exchange and melt of late season snow patches in heavy forest, *Proc. Western Snow Conf.*, April, 1954, Colorado State University, Fort Collins, Colo., pp. 54–68, October, 1954.
91. Garstka, W. U.: Discussion of Heat exchange and melt of late season snow patches in heavy forest, by Peter B. Boyer, *Proc. Western Snow Conf.* April, 1954, Colorado State University, Fort Collins, Colo., pp. 68–70, October, 1954.
92. Anderson, E. R., L. J. Anderson, and J. J. Marciano: A review of evaporation theory and development of instrumentation, *U.S. Navy Electronics Lab. Rept.* 159, San Diego, Calif., Feb. 1, 1950.
93. Marciano, J. J., and G. E. Harbeck, Jr.: Mass-transfer studies, in Water loss investigations, Lake Hefner Studies, Technical Report, vol. 1, *U.S. Geol. Surv. Profess. Paper* 269, 1954.
94. Hildebrand, C. E.: Lysimeter studies of snowmelt, *Proc. Western Snow Conf.*, Santa Barbara, Calif., April, 1957, Colorado State University, Fort Collins, Colo., pp. 94–105, November, 1957.
95. Wilson, W. T.: An outline of the thermodynamics of snowmelt, *Trans. Am. Geophys. Union*, pt. I, pp. 182–195, July, 1941.
96. Clyde, G. D.: Snow-melting characteristics, *Utah State Agr. Expt. Sta. Tech. Bull.* 231, Logan, Utah, October, 1931.
97. McKay, G. A., and S. R. Blackwell: Plains snowpack water equivalent from climatological records, *Proc. Western Snow Conf.*, April, 1961, Colorado State University, Fort Collins, Colo., pp. 27–43, September, 1961.
98. Wilm, H. G., and E. G. Dunford: Effect of timber cutting on water available for streamflow from a lodgepole pine forest, *U.S. Dept. Agr. Tech. Bull.* 968, 1948.
99. Love, L. D.: The effect on streamflow of the killing of spruce and pine by the Englemann spruce beetle, *Trans. Am. Geophys. Union*, vol. 36, no. 1, pp. 113–118, February, 1955.
100. Martinelli, M., Jr.: Some hydrologic aspects of alpine snow fields under summer conditions, *J. Geophys. Res.*, vol. 64, no. 4, pp. 451–455, April, 1959.
101. Goodell, B. C.: Management of forest stands in western United States to influence the flow of snow-fed streams, Symposium of Hannoversch-Münden, *Intern. Assoc. Sci. Hydrol. Publ.* 48, pp. 49–58, September, 1959.
102. Anderson, H. W.: Prospects for affecting the quantity and timing of water yield through snowpack management in California, *Proc. Western Snow Conf.*, Santa Fe, N.M., April, 1960, Colorado State University, Fort Collins, Colo., pp. 44–50, September, 1960.
103. Hoover, M. D.: Prospects for affecting the quantity and timing of water yield through snowpack management in southern Rocky Mountain area, *Proc. Western Snow Conf.*, Santa Fe, N.M., April, 1960, Colorado State University, Fort Collins, Colo., pp. 51–53, September, 1960.
104. Lull, W., and R. S. Pierce: Prospects in the northeast for affecting the quantity and timing of water yield through snowpack management, *Proc. Western Snow Conf.*, Santa Fe, N.M., April, 1960, Colorado State University, Fort Collins, Colo., pp. 54–62, September, 1960.
105. Packer, P. E.: Some terrain and forest effects on maximum snow accumulation in a western white pine forest, *Proc. Western Snow Conf.*, Santa Fe, M.N., April, 1960, Colorado State University, Fort Collins, Colo., pp. 63–66, September, 1960.
106. Anderson, H. W., and C. H. Gleason: Logging effects on snow, soil moisture, and water losses, *Proc. Western Snow Conf.*, Reno, Nev., April, 1959, Colorado State University, Fort Collins, Colo., pp. 57–65, September, 1959.
107. Schneider, W. J., and G. R. Ayer: Effect of reforestation on streamflow in central New York, *U.S. Geol. Surv. Water-Supply Paper* 1602, 1961.

108. Schneider, W. J.: Changes in snowmelt runoff caused by reforestation, *Proc. Western Snow Conf.* April, 1961, Colorado State University, Fort Collins, Colo., pp. 44–50, September, 1961.
109. Ford, P. M.: Multiple correlation in forecasting seasonal runoff, *U.S. Bur. Reclamation Eng. Monograph* 2, Denver, Colo., 2d ed., rev., June, 1959.
110. Thornthwaite, C. W.: An approach toward a rational classification of climate, *Geograph. Rev.*, vol. 38, no. 1, pp. 55–94, 1948.
111. Penman, H. L.: Estimating evaporation, *Trans. Am. Geophys. Union*, vol. 37, no. 1, pp. 43–46, February, 1956.
112. Halstead, M. H.: Theoretical derivation of an equation for potential evaporation, *Johns Hopkins Univ. Lab. Climatol., Micrometeor. Surface Layer of Atmosphere, Interim Rept.* 16, pp. 10–12, 1951.
113. Nelson, M. W., C. C. McDonald, and M. Barton: Base flow as a parameter in forecasting the April-June runoff, *Proc. Western Snow Conf.*, Boise, Idaho, April, 1953, Colorado State University, Fort Collins, Colo., pp. 61–66, September, 1953.
114. Dean, W. W.: Discussion of Base flow as a parameter in forecasting the April-June runoff by M. W. Nelson, C. C. McDonald, and M. Barton, *Proc. Western Snow Conf.*, Boise, Idaho, April, 1953, Colorado State University, Fort Collins, Colo., pp. 67–68, September, 1953.
115. Cavadias, G. S.: An approach to forecasting the spring runoff in Quebec, *Proc. Western Snow Conf.*, Bozeman, Mont., April, 1958, Colorado State University, Fort Collins, Colo., pp. 35–45, October, 1958.
116. Degree-days for December, 1960, *Air Conditioning, Heating and Ventilating*, vol. 58, no. 2, pp. 140–141, February, 1961.
117. Stevens, J. C.: Winter overflow from ice-gorging on shallow streams, *Trans. Am. Geophys. Union*, pt. III, pp. 973–978, September, 1940.
118. Wilson, W. J.: Some factors in relating the melting of snow to its causes, *Proc. Central Snow Conf.*, Michigan State College, East Lansing, Mich., pp. 33–41, 1942.
119. Linsley, R. K., Jr.: A simple procedure for the day-to-day forecasting of runoff from snowmelt, *Trans. Am. Geophys. Union*, pt. III, pp. 62–67, November, 1943.
120. Babbitt, H. E.: Water Supply and Treatment, sec. 10 in "Civil Engineering Handbook," L. C. Urquhart (ed.), 4th ed. McGraw-Hill Book Company, Inc., New York, 1959.
121. Abbett, R. W. (ed.): "American Civil Engineering Practice," John Wiley & Sons, Inc., New York, 1956.
122. "Hydrology Handbook," Hydraulics Division, Committee on Hydrology, American Society of Civil Engineers, Manuals of Engineering Practice, no. 28, Jan. 17, 1949.
123. Martin, J. T.: Use of snowmelt forecasts by the Corps of Engineers for flood control operations on the Rio Grande, *Proc. Western Snow Conf.*, Santa Fe, N.M., April, 1960, Colorado State University, Fort Collins, Colo., pp. 5–8, September, 1960.
124. Barnes, B. S.: The structure of discharge-recession curves, *Trans. Am. Geophys. Union*, pt. IV, pp. 721–725, 1939.
125. Barnes, B. S.: Discussion of Method of predicting the runoff from rainfall by R. K. Linsley, Jr., and W. C. Ackerman, *Trans. Am. Soc. Civil Engrs.*, vol. 107, pp 836–841, 1942.
126. Light, Phillip: Analysis of high rates of snowmelting, *Trans. Am. Geophys. Union*, pt. I, pp. 195–205, 1941.
127. Sverdrup, H. U.: The eddy-conductivity of the air over a smooth snow field, Results of the Norwegian-Swedish Spitsbergen Expedition in 1934, *Gofys. Publ.*, Oslo, vol. 11, no. 7, 1936.
128. Snow compaction, pt. IV in Technical Report on Experimental Arctic Operation Hard Top II, 1954, *U.S. Naval Civil Eng. Lab. Tech. Rept.* R-007, Port Hueneme, Calif., Dec. 29, 1955.
129. Coffin, R. C., Jr.: Squaw Valley winter trials, 1958–1959, Compacted-snow parking lot study on meadow land, *U.S. Naval Civil Eng. Lab. Tech. Rept.* 051, Point Hueneme, Calif., Nov. 16, 1959.
130. Camm, J. B.: Snow compaction equipment: snow drags, *U.S. Naval Civil Eng. Lab. Tech. Rept.* 109, Point Hueneme, Calif., Oct. 20, 1960.
131. Bucher, Edwin: Contributions to the theoretical foundations of avalanche defense construction, Zurich, October, 1948, U.S. Corps of Engineers, Snow, Ice and Permafrost Research Establishment, Transl. 18, Vicksburg, Miss., February, 1956 (transl. by Jan C. Van Tienhoven).
132. "Snow Avalanches," U.S. Forest Service January, 1961. A handbook of forecasting and control measures.

REFERENCES

133. Acoca, Miguel: "Scar on Huascarān . . . on a worn face, a tear," *Life Mag.*, vol. 52, no. 4, pp. 26–35, Jan. 26, 1962.
134. Martinelli, M., Jr.: A look at avalanche control structures in Europe, *Proc. 28th Annual Meeting Western Snow Conf.*, Santa Fe, N.M., April, 1960, Colorado State University, Fort Collins, Colo., pp. 67–70, September, 1960.
135. Hay, W. W.: Railroad engineering, in R. W. Abbett (ed.), "American Civil Engineering Practice," vol. I, John Wiley & Sons, Inc., New York, 1956.
136. Blomberg, Ray: Snow shed protects cascade highway, *Public Works*, vol. 82, no. 10, p. 54, October, 1951.
137. Schaerer, Peter: The avalanche defense on the Trans-Canada Highway at Roger's Pass, *Proc. 28th Annual Meeting Western Snow Conf.*, Colorado State University, Fort Collins, Colo., pp. 71–78, September, 1960.
138. Borland, W. M.: Discussion of the avalanche defense on the Trans-Canada Highway at Roger's Pass by Paul Schaerer, *Proc. 28th Annual Meeting Western Snow Conf.*, Colorado State University, Fort Collins, Colo., pp. 79–80, September, 1960.
139. Avalanche, *Contractors and Engrs.*, p. 13, October, 1959.
140. Boyd, D. W.: Maximum snow depths and snow loads in Canada, *Proc. Western Snow Conf.*, April, 1961, Colorado State University, Fort Collins, Colo., pp. 6–16, September, 1961.
141. Snow load studies, *U.S. Home and Housing Finance Agency Housing Research Paper* 19, May, 1952.
142. Huber, W. G.: Kemano-Kitimat transmission line, built to resist heavy ice and wind loads, *Civil Eng.*, vol. 25, no. 1, pp. 52–56, January, 1955.
143. White, H. B.: Cross suspension system, Kemano-Kitimat transmission line, *J. Eng. Inst. Canada*, July, 1956.
144. Winkelman, P. F.: Transmission spans provide icing data for transmission design, *Elec. Light and Power*, vol. 34, no. 22, pp. 148–150, Oct. 15, 1956.
145. Davison, A. E.: Dancing conductors, *Trans. Am. Inst. Elec. Engrs.*, pp. 1444–1449, October, 1930.
146. Davison, A. E.: Ice-coated electrical conductors, *Bull. Hydro-electric Power Comm. Ontario*, vol. 26, no. 9, pp. 271–280, September, 1939.
147. Archbold, W. K.: Dancing cables affect service, *Elec. World*, pp. 199–202, Jan. 26, 1929.
148. "Summary of Reports on Galloping of Transmission Line Conductors (compiled from data supplied to Utilities Research Commission Committee 104 for Investigation of Galloping Conductors, July 15, 1949)," Utilities Research Commission, Chicago, 1949.
149. Knowlton, A. E. (ed.): "Standard Handbook for Electrical Engineers," 9th ed., McGraw-Hill Book Company, Inc., New York, 1957.
150. Tornquist, E. L., and C. Becker: Galloping conductors and a method of studying them, *Trans. Am. Inst. Elec. Engs.* vol. 66, pp. 1154–1164, 1947.
151. Richardson, A. S., Jr., J. R. Martuccelli, and M. Wohltmann: Research study on galloping of electric power transmission lines, *MIT Aeroelastic and Structures Res. Lab. Prog. Rept.* 5, May, 1961.
152. Foss, K. A.: Methods for computing the mechanical dynamic response of electric power transmission lines: research study on galloping of electric power transmission lines, *MIT Aeroelastic and Structures Res. Lab. Prog. Rept.* 4, Aug. 1, 1960.
153. Clem, J. E.: Currents required to remove conductor "sleet," *Elec. World*, vol. 96, pp. 1053–1056, Dec. 6, 1930.
154. Lawrence, L.: The story of man-made snow for skiers, *Reader's Digest*, vol. 77, no. 464, pp. 121–123, December, 1960.
155. Pierce, W. M., Jr.: Method for making and distributing snow, U.S. Patent 2,676,471, Apr. 27, 1954.
156. Allan, Roger: Design for artificial snowfall, *Air Conditioning, Heating, and Ventilating*, vol. 58, no. 2, pp. 87–90, February, 1961.
157. Schaefer, V. J.: The formation of frazil and anchor ice in cold water, *Trans. Am Geophys. Union*, vol. 31, no. 6, pp. 885–893, December, 1950.
158. Gisiger, P. E.: Safeguarding hydro plants against the ice menace, *Civil Eng.*, vol. 17, no. 1, pp. 24–27, January, 1947.
159. Granbois, K. J.: Combatting frazil ice in hydroelectric stations, *Am. Inst. Elec. Engrs. Tech. Paper* 53-8, October, 1952.
160. Barnes, H. T.: "Ice Formation with Special Reference to Anchor Ice and Frazil," John Wiley & Sons, Inc., New York, 1906.
161. Rose, Edwin: Thrust exerted by expanding ice sheet, *Trans. Am. Soc Civil Engrs.*, vol. 111, pp. 871–885, 1947.

162. Ice pressure against dams: a symposium, *Trans. Am. Soc. Civil Engrs.*, vol. 119, pp. 1–42, 1954. Includes the following papers with their discussions: Bertil Löfquist, Studies of the effects of temperature variations; A. D. Hogg, Some investigations in Canada; and G. E. Monfore, Experimental investigations by the Bureau of Reclamation.
163. Parsons, W. J.: Ice in the northern streams of the United States, *Trans. Am. Geophys. Union*, vol. 21, pp. 970–973, 1940.
164. Parsons, W. J., Jr.: The evolution of ice in streams, in O. R. Meinzer (ed.): "Hydrology, Physics of the Earth, IX," McGraw-Hill Book Company, Inc., New York, 1942; reprinted by Dover Publications, Inc., New York, 1949, pp. 137–142.
165. Selected references on ice as it affects hydraulic structures and channels; its prevention and removal, U.S. Bureau of Reclamation, Technical Reference Section, February, 1949.
166. Ice problems in hydraulic structures, International Association for Hydraulic Research, Seminar 1, Montreal, Canada, Aug. 27, 1959.
167. Addendum 1, 1945–1947, to Report on frost investigation, 1944–1945, U.S. Army Corps of Engineers, Frost Effects Laboratory, Boston, 1949.
168. Linell, K. A.: Frost design criteria for pavements, in Soil temperature and ground freezing, *Highway Res. Board Bull.* 71, pp. 18–32, 1953.
169. Aldrich, H. P., Jr.: Frost penetration below highway and airfield pavements, in Factors influencing ground freezing, *Highway Res. Board Bull.* 135, pp. 124–144, 1956.
170. Casagrande, Arthur: Discussion on frost heaving, *Proc., Highway Res. Board*, vol. 11, pt. 1, pp. 168–172, 1931.
171. Atkinson, H. B., and C. E. Bay: Some factors affecting frost-penetration, *Trans. Am. Geophys. Union*, vol. 21, pp. 935–951, 1940.
172. Berggren, W. P.: Prediction of temperature distribution in frozen soils, *Trans. Am. Geophys. Union*, vol. 24, pp. 71–77, 1943.
173. Crabb, G. A., Jr., and J. L. Smith: Soil-temperature comparisons under varying covers, in Soil temperature and ground freezing, *Highway Res. Board Bull.* 71, pp. 32–80, 1953.
174. Willis, W. O., C. W. Carlson, J. Alessi, and H. J. Haas: Depth of freezing and spring runoff as related to fall soil-moisture level, *Canadian J. Soil Sci.*, vol. 41, pp. 115–123, February, 1961.
175. Garstka, W. U.: Hydrology of small watersheds under winter conditions of snow-cover and frozen soil, *Trans. Am. Geophys. Union*, vol. 25, pp. 838–871, 1944.
176. Siple, P. A.: Climatic aspects of frost heave and related ground frost phenomena, in Frost action in soils, *Highway Res. Board Spec. Rept.* 2, pp. 10–16, 1952.
177. Beskow, Gunnar: Soil freezing and frost heaving with special application to roads and railroads, *Swedish Geol. Soc.*, *26th Yearbook*, no. 3, 1935, ser. C, no. 375, with a special supplement for the English translation of progress from 1935 to 1946, Northwestern University, Technological Institute, Evanston, Ill., November, 1947 (transl. by J. O. Osterberg).
178. Jumikis, A. R.: Theoretical basis for frost-action research, *Highway Res. Board Bull.* 111, pp. 76–84, 1955.
179. Jumikis, A. R.: The soil freezing experiment, *Highway Res. Board Bull.* 135, pp. 150–165, 1956.
180. Bouyoucos, G. J.: Movement of soil moisture from capillaries to the large capillaries of the soil upon freezing, *J. Agr. Res.*, vol. 24, no. 5, pp. 480–491, 1923.
181. Taber, Stephen: Frost heaving, *J. Geol.*, vol. 37, no. 5, pp. 428–461, 1929.
182. Taber, Stephen: Freezing and thawing of soils as factors in the destruction of road pavements, *Public Roads Bull.* 11, no. 6, pp. 113–132, 1930.
183. Taber, Stephen: The mechanics of frost heaving, *J. Geol.*, vol. 38, pp. 303–317, 1930.
184. Johnson, A. W.: Frost action in roads and airfields: a review of the literature, *Highway Res. Board Spec. Rept.* 1, 1952.
185. Lovell, S. W., Jr., and Moreland Herrin: Review of certain properties and problems of frozen ground including permafrost, *Snow, Ice and Permafrost Res. Estab. Rept.* 9, Purdue University Engineering Experiment Station, Lafayette, Ind., March, 1953.
186. Linell, K. A.: Frost action and permafrost, sec. 13 in K. B. Woods (ed.), "Highway Engineering Handbook," McGraw-Hill Book Company, Inc., New York, 1960, pp. 13–1 to 13–35.
187. Savarenskii, F. P.: Dams in permafrost regions, *Izbrannye Sochinenniia*, 1950, pp. 370–371, U.S. Army Corps of Engineers, Arctic Construction and Frost Effects Laboratory, Transl. 29, Waltham, Mass., June, 1960 (transl. by Orest Popovych).

REFERENCES

188. A test study of foundation design for permafrost conditions, *Eng. News-Rec.*, vol. 139, no. 12, pp. 104–107, Sept. 18, 1947.
189. Muller, S. W.: "Permafrost or Permanently Frozen Ground and Related Engineering Problems," J. W. Edwards, Publisher, Inc., Ann Arbor, Mich., 1947.
190. Black, R. F.: Permafrost: a review, *Bull. Geol. Soc. Am.*, September, 1954.
191. Terzaghi, Karl: Permafrost, *J. Boston Soc. Civil Engrs.*, vol. 39, no. 1, January, 1952.
192. "Arctic and Subarctic Construction," Engineering Manual for Military Construction, U.S. Army Corps of Engineers, pt. XV, chaps. 1–7, EM-1110-345-370 to 376.
193. Lang, W. A.: The use of soot for snow removal purposes, *Proc. Western Snow Conf.* Sacramento, Calif., April, 1952, Colorado State University, Fort Collins, Colo., pp. 29–36, October, 1952.
194. Landsberg, H.: The use of solar energy for the melting of ice, *Bull. Am. Meterol. Soc.*, vol. 21, pp. 102–107, March, 1940.
195. Martinelli, M., Jr.: Alpine snow research, *J. Forestry*, vol. 58, no. 4, pp. 278–281, April, 1960.
196. Altenburg, W. M.: Snow-melting machine and method, U.S. Patent 2,977,955, Apr. 4, 1961.
197. Harris, W. S.: Residential snowmelting systems, *Air Conditioning, Heating, and Ventilating*, vol. 58, no. 1, pp. 51–54, January, 1961.
198. Kersten, M. S., L. P. Peterson, and A. J. Toddie, Jr.: A laboratory study of ice removal by various chloride salt mixtures, *Highway Res. Board Bull.* 220, pp. 1–13, July, 1959.
199. Dickinson, W. E.: Ice-melting properties and storage characteristics of chemical mixtures for winter maintenance, *Highway Res. Board Bull.* 220, pp. 14–22, July, 1959.
200. Lang, C. H., and W. E. Dickinson: Chemical mixture test program in snow and ice control, *Highway Res. Board Bull.* 252, pp. 1–8, June, 1960.

Section 11

EVAPOTRANSPIRATION

FRANK J. VEIHMEYER, *Professor of Irrigation, Emeritus, University of California.*

   I. Introduction 11-2
      A. Definitions 11-2
  II. Evaporation from Free-water Surfaces 11-3
      A. Nature of Process 11-3
      B. Factors Affecting Evaporation 11-4
         1. Temperature 11-4
         2. Wind 11-5
         3. Atmospheric Pressure 11-5
         4. Soluble Solids 11-6
         5. Nature and Shape of Surface 11-6
      C. Measurement of Evaporation 11-6
         1. Evaporation Pans 11-6
         2. Atmometers 11-9
      D. Estimation of Evaporation 11-10
         1. Empirical Evaporation Equations 11-10
         2. Water-balance Method 11-11
         3. Energy-balance Method 11-11
         4. Mass-transfer Method 11-13
      E. Evaporation Reduction 11-14
         1. Surface-area Reduction 11-14
         2. Mechanical Covers 11-14
         3. Surface Films 11-14
 III. Evaporation from Soil Surfaces 11-15
      A. Energetics of Soil Moisture 11-16
      B. Movement of Moisture in Soil 11-17
      C. Evaporation from Unsaturated Soil 11-17
      D. Evaporation from Saturated Soils 11-19
      E. Controlling Evaporation Losses from Soil 11-19
         1. Dust Mulch 11-19
         2. Paper Mulch 11-19
         3. Chemical Alteration 11-19
         4. Pebble Mulch 11-19
      F. Evaporation from Bare Soils versus Transpiration 11-19
 IV. Transpiration 11-20
      A. Nature of the Process 11-20
      B. Factors Affecting Transpiration 11-21
         1. Temperature 11-21

  2. Solar Radiation.................................... 11-22
  3. Wind... 11-22
  4. Soil Moisture..................................... 11-22
 C. Determination of Transpiration..................... 11-22
V. Evapotranspiration... 11-23
 A. Methods of Estimation.............................. 11-23
  1. Soil-moisture Sampling........................... 11-23
  2. Lysimeter Measurements........................... 11-24
  3. Inflow-outflow Measurements...................... 11-24
  4. Integration Method............................... 11-25
  5. Energy Balance................................... 11-25
  6. Vapor Transfer................................... 11-25
  7. Groundwater Fluctuations......................... 11-25
 B. Evapotranspiration Equations....................... 11-25
 C. Evaporation as an Index of Evapotranspiration...... 11-31
 D. Consumptive-use Requirements....................... 11-32
VI. References... 11-33

I. INTRODUCTION

 This section deals with basic principles of evapotranspiration that are important to the hydrologist for an understanding of this phase of the field of hydrology. The first subsection deals with evaporation from free-water surfaces. Several recent texts on hydrology deal adequately with methods of studying, measuring, and analyzing the results of evaporation measurements. Therefore the subject is treated here in a broad way (see also Sec. 6.)

 Other subsections—evaporation from soil surfaces and transpiration—receive more detailed treatment because there is lack of agreement, in some instances, as to methods of studying and conclusions drawn from the data reported. Transpiration receives fuller treatment also because engineers are not always as familiar with this field as with evaporation from water surfaces.

 Some data are given on evapotranspiration, but it is not the intention to compile all that have been obtained. Rather, the objective is to supply the reader with information that will help him judge what to expect in a given locality on the basis of climatic, soil, and crop conditions.

A. Definitions

 The following specialized terms are commonly used in connection with evapotranspiration:

 Evaporation: The process by which water is changed from the liquid or solid state into the gaseous state through the transfer of heat energy [1].

 Transpiration: The evaporation of water absorbed by the crop and transpired and used directly in the building of plant tissue, in a specified time. It does not include soil evaporation [2].

 Evapotranspiration: The process by which water is evaporated from wet surfaces and transpired by plants.

 Consumptive use: The quantity of water per annum used by either cropped or natural vegetation in transpiration or in the building of plant tissue, together with water evaporated from the adjacent soil, snow, or from intercepted precipitation. It is sometimes termed "evapotranspiration" [3].

 Duty of water: The quantity of irrigation water applied to a given area for the purpose of maturing its crop [2].

 Irrigation requirement: The quantity of water, exclusive of precipitation, that is required for crop production. It includes surface evaporation and other economically unavoidable wastes [2].

Water requirement: The quantity of water, regardless of its source, required by a crop, in a given period of time, for its normal growth under field conditions. It includes surface evaporation and other economically unavoidable wastes [2].

From the foregoing definitions it may be noted that evaporation and transpiration are essentially the same processes, but differing in the kind of surface from which the water vapor escapes.

The terms *evapotranspiration* and *consumptive use* are frequently used interchangeably. However, they are synonymous only if evapotranspiration is used to indicate the amount of water consumed (used up and lost by change from a liquid state to a vapor state) in evaporation and transpiration in raising plants. Engineers are usually interested in valley or basin use of water, while agriculturalists are mostly concerned with the water consumed in growing crops. Diversion of water from one watershed to another is considered to be consumptively used even though it is not evaporated or transpired. From the second watershed the water may flow to the ocean, but as far as the first watershed is concerned, it is still consumptive use. Thus the two terms do not always have the same meaning.

The terms *duty of water* and *consumptive use* are also used interchangeably. If all the water applied to the irrigated area is consumed, then the duty of water will be equal to the consumptive use, but there is usually considerable waste of water incident to applying it to the land, so the terms may not be synonymous.

II. EVAPORATION FROM FREE-WATER SURFACES

A. Nature of Process

By evaporation, water in liquid state is changed to vapor state. This change occurs when some molecules in the water mass have attained enough kinetic energy to eject themselves from the water surface. Escaping molecules, of course, are attracted by other molecules which tend to hold them together within the water. Only those molecules possessing greater kinetic energy than the average within the liquid escape from the surface. The temperature of the liquid is lowered by such escape, so evaporation results in cooling. Molecules may leave a solid surface in the same way. This change without passing through the usual intermediate liquid state is called *sublimation*.

The motion of the molecules through the water surface produces a pressure. This pressure of the aqueous vapor is called *vapor pressure*. More correctly, it is only the partial pressure of the water vapor in the atmosphere, because in a mixture of gases each gas exerts a partial pressure which is independent of the other gases in the mixture.

Escaping molecules collide with those in the air, and some of the former will drop back into the water. When the number of molecules that escape equals the number of those that fall back into the water, an equilibrium is reached between the pressure exerted by the escaping molecules and the pressure of the surrounding atmosphere. This equilibrium condition is known as *saturation*. Further, some of the molecules in the gaseous phase have kinetic energy sufficient to cause them to penetrate the liquid, and others will condense from a vapor to a liquid state. Thus evaporation from and condensation onto the liquid surface are continuous processes. Evaporation is faster than condensation if the space above the water surface is not saturated.

The rate of evaporation, then, will be determined by the difference between the vapor pressure of the body of water and that of the air above the water surface. Under given conditions, evaporation is proportional to the deficit in vapor pressure, which is the difference between the pressure of saturated vapor at the temperature of the water and the aqueous vapor pressure of the air. This fact has been known since 1802, when Dalton [4] first recognized it as a law. *Dalton's law* has the general form shown in Table 11-1 by Eq. (11-1). This is the basis of many other equations. Some popular ones are listed in the table.

Table 11-1. Evaporation Equations Based on Dalton's Law

Name	Date	Equation		
Dalton [4]	1802	$E = C(e_w - e_a)$		(11-1)
Fitzgerald [5]	1886	$E = \psi(e_w - e_a)$	$\psi = 0.4 + 0.199w$	(11-2)
Meyer [6]	1915	$E = C(e_w - e_a)\psi$	$\psi = 1 + 0.1w$	(11-3)
Horton [7]	1917	$E = 0.4(\psi e_w - e_a)$	$\psi = 2 - e^{-0.2w}$	(11-4)
		For large areas, E is multiplied by $(1 - P) + P\dfrac{\psi - 1}{\psi - h}$		
Rohwer [8]	1931	$E = 0.771(1.465 - 0.0186B)\psi(e_w - e_a)$	$\psi = 0.44 + 0.118w$	(11-5)
Lake Hefner [9]	1954	$E = 0.00177w(e_w - e_a)$		(11-6)
Lake Mead [10]	1958	$E = 0.001813w(e_w - e_a)t[1 - 0.03(T_a - T_w)]$		(11-7)

NOTATIONS

B = mean barometric reading, in in. Hg, at 32°F
C = coefficient depending on various uncounted factors affecting evaporation or $C = 15$ for small, shallow water and $C = 11$ for large, deep water in Meyer equation
e = base of natural logarithms
e_a = actual vapor pressure in air based on monthly mean air temperature and relative humidity at nearby stations for small bodies of shallow water, or based on information about 30 ft above water surface for large bodies of deep water, in Meyer equation; or mean vapor pressure of saturated air at temperature of dew point, in mb, for Lake Hefner and Lake Mead equations, or in Hg in other equations
e_w = maximum vapor pressure, in Hg, corresponding to monthly mean air temperature observed at nearby stations for small bodies of shallow water, or corresponding to water temperature instead of air temperature for large bodies of deep water, in Meyer equation; or mean vapor pressure at water-surface temperature in mb in Lake Hefner and Lake Mead equations or in in. Hg in other equations
E = rate of evaporation, in in. per 30-day month in Meyer equation, in inches per t days in Lake Mead equation, or in in. per 24 hr in other equations
h = relative humidity
P = fraction of time during which wind is turbulent
t = number of days in period for evaporation
T_a = average air temperature, °C + 1.9°C
T_w = average water-surface temperature, °C
w = monthly mean wind velocity, in mph, at about 30 ft aboveground in Meyer equation, or mean wind velocity near surface of ground or water in knots in Lake Mead equation or in mph in other equations
ψ = wind factor

B. Factors Affecting Evaporation

The factors controlling evaporation have been known for a long time, but their evaluation is difficult because of their interdependent effects. As pointed out previously, the rate of evaporation depends on the vapor pressure of the body of water and that of the air. These vapor pressures depend on temperatures of the water and air, wind, atmospheric pressure, quality of the water, and the nature and shape of the surface.

1. Temperature. The vapor pressure of a body of water increases with temperature because the kinetic energy of the water molecules is raised with increasing temperature. An increase in the temperature of the ambient air, when the increase is commensurate with the increase of the temperature of water surface, will also increase the aqueous vapor pressure of both air and water. Since evaporation is proportional to the vapor-pressure difference between the water and the air, equal temperature increases may not increase the rate of evaporation. For evaporation to continue, heat must be applied to the water as it is cooled by evaporation. Otherwise, when air and water temperatures become equal, evaporation ceases.

Since air-temperature records are readily available for most localities, evaporation measurements are frequently compared with air temperatures. However, evaporation is not dependent on air temperature alone. A typical record of the evaporation from a U.S. Weather Bureau pan and the air temperature is shown in Fig. 11-1. It indicates that the mean monthly evaporation for months having the same mean monthly air temperatures is higher during the spring and summer months than during the last summer and fall.

In shallow bodies of water the temperature of the water may lag behind air temperature. When water temperatures are averaged for monthly periods, they may not be greatly different from air temperatures. In deep water the temperature lag may be great and heat may be stored to considerable depths. As the surface water is cooled or heated, vertical density currents will be set up.

The cooling of the surface water in the fall causes the water to sink and the warmer water below to rise. This process continues until the temperature reaches 39.2°F, the temperature of maximum density, throughout the entire body of the water. Further cooling of the surface water will decrease its density, and the water will remain at the surface. Thus, in deep water, the relatively low temperature of the surface water in the spring will decrease evaporation, while the higher temperature in the fall tends to increase evaporation. The utilization of the solar radiation to heat the deep layers of water lessens the amount of energy available for evaporation. Thus a high correlation between the air temperature and evaporation cannot be expected.

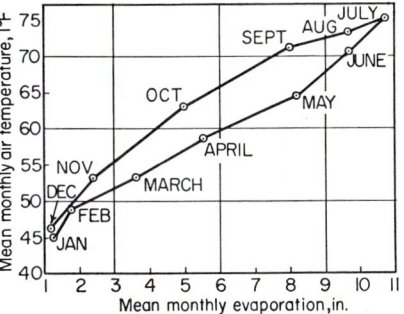

Fig. 11-1. Evaporation from U.S. Weather Bureau Class A pan at Davis, Calif., averaged for 1926 to 1959.

2. Wind. Water molecules escaping from the water surface may collide with others already present in the air. Wind is effective in removing water molecules in the air and bringing in air capable of holding more water vapor. When the wind velocity is great enough to remove all the water molecules escaping from the water surface, a further increase in velocity will not increase evaporation appreciably. If the incoming air is preheated by flowing over heated surfaces, it will supply additional energy for evaporation. Conversely, a cool air mass may decrease evaporation and even surrender heat to the underlying surface and cause condensation. This phenomenon is called *advection*.

The effect of wind on evaporation may be more pronounced over large bodies of water than over small areas. In the case of pans, a slight increase in wind may be sufficient to remove the water vapor as fast as it appears. Large water areas may require high velocity and turbulent air movement for water to evaporate at its maximum rate. Winds up to 25 mph may be needed before evaporation is increased. The effect of wind is considered in many equations, as shown in Table 11-1. Horton suggested a wind factor ψ which decreases with increased wind velocity, becoming negligible at velocities greater than about 20 mph.

3. Atmospheric Pressure. The decrease in barometric pressure with increased altitude should increase the rate of escaping water molecules from a free-water surface because there are fewer molecules in the atmosphere above the evaporating surface, hence less interference. Decreasing evaporation with increasing altitude would occur only if all other climatic factors affecting the aqueous vapor pressure of the air remained the same.

Meyer [6] believes that the effect of atmospheric pressure is compensated for by the change in vapor pressure with change in altitude.

Rohwer [8] measured the daily rate of evaporation at several places to detect the effect of atmospheric-pressure changes. Such effect is expressed by the factor $1.465 - 0.0186B$ in Eq. (11-5).

4. Soluble Solids. The vapor pressure of pure water under given conditions is determined by its temperature. When a solute is dissolved in water, the vapor pressure of the latter is reduced. Since the rate of evaporation is proportional to the difference in vapor pressure between the water and atmosphere, lowering that of the former will reduce the rate of evaporation. The lowering of the vapor pressure of a solution, according to *Raoult's law*, is equal to the mole function of the solute in the solution, or

$$\frac{P_0 - P}{P_0} = \frac{n_2}{n_1 + n_2} \tag{11-8}$$

where P_0 is the vapor pressure of the pure solvent at a given temperature, P is the vapor pressure of the solution at the same temperature, n_1 is the moles of solvent, and n_2 is the moles of solute.

Solutions that obey Raoult's law exactly are called *ideal solutions*. Actually, very few solutions behave ideally, and some deviation from the law is to be anticipated. For dilute solutions, however, these deviations are small and can usually be ignored. Within the limits of usual concentrations, the effect of soluble solids on rate of evaporation will be determined by the amount of material in solution.

Evaporation from salt solutions was studied by Rohwer [11], Lee [12], and Adams [13]. It was found that the evaporation rate decreases as specific gravity increases. The evaporation rate decreases about 1 per cent for each 1 per cent increase in specific gravity until crusting takes place, usually at a specific gravity of 1.30. Evaporation is about 2 to 3 per cent less from sea water than from fresh water. Rohwer [11] states that if the vapor pressure is corrected for the effect of the sodium chloride, the computed rates of evaporation agree well with the observed rates, but the agreement is closer for the weaker than for the more concentrated solutions.

5. Nature and Shape of Surface. A body of water with a flat surface has greater vapor pressure than one with a concave surface, but less than one with a convex surface under the same conditions.

Many attempts have been made to find a correct equation for evaporation from surfaces of various kinds. In 1881, Stefan [14] investigated evaporation from flush circular and elliptical water surfaces at constant temperature and in still air. According to Humphreys [15], Stefan was able to show that evaporation under the restricted conditions is proportional to the diameter (or other linear dimension) of the evaporating surface, but not to the evaporating area. Brown and Escombe [16] reported later that diffusion of vapor through small openings is more nearly proportional to their diameter or perimeter than to their area. They stated that, by imposing a thin septum pierced with a circular aperture at any point in the line of flow of vapor or gas, the rate of flow across the unit area of the aperture is greater than it would be across an equal area of the unobstructed column at this point. Similar conclusions were also reported by other investigators. Plant physiologists have accepted these conclusions as being applicable to diffusion of water vapor through stomata in plants.

C. Measurement of Evaporation

1. Evaporation Pans. The *evaporation pan* is the most widely used instrument for evaporation measurement today. Several types of pans commonly used in the United States are as follows:

a. U.S. Weather Bureau Class A Land Pan. As shown in Fig. 11-2a, this instrument [17, 18] is made of unpainted galvanized iron 4 ft in diameter and 10 in. deep. The bottom, supported on a wooden frame, is raised 6 in. above the ground surface. The water surface, maintained between 2 and 3 in. below the rim of the pan, is measured daily with a hook gage in a stilling well. The daily evaporation is computed as the difference between observed levels, corrected for any precipitation measured in an adjacent or nearby standard rain gage. The rate of evaporation from small areas is greater than that from large areas. Therefore a *pan coefficient* of 0.7 has been recommended for converting the observed value to an estimated value for lakes or reservoirs. However, true values of the coefficient have been reported to vary from 0.6 to 0.8. This instrument is used at all official U.S. Weather Bureau stations and at the cooperative stations in the United States. The advantages of this pan are (1) more data available from this type of pan, (2) stable pan coefficient, (3) easy access for observations, (4) more stability than a floating pan, (5) relative freedom from drifting snow, dirt, and debris, and (6) reasonable cost of installation [19].

(a)

(b)

FIG. 11-2. Pans for evaporation measurements. (a) U.S. Weather Bureau Class A land pan; (b) U.S. Bureau of Plant Industry sunken pan.

b. *U.S. Bureau of Plant Industry Sunken Pan.* As shown in Fig. 11-2b, this pan is 6 ft in diameter and 2 ft deep, buried in the ground within 4 in. of the top. The water surface is supposed to be not more than ½ in. above the ground level or below it. Because of its size, this pan provides by far the best index to lake evaporation. The pan coefficient is normally about 0.95, but smaller in spring and larger in fall [19, 20].

c. *Colorado Sunken Pan.* This pan is made of unpainted galvanized iron, buried in the ground to within 4 in. of its rim. It is 3 ft square, with a depth from 18 in. to 3 ft. The pan coefficient ranges from 0.75 to 0.86, with a mean of 0.78 [19, 20].

d. *U.S. Geological Survey Floating Pan.* This pan is 3 ft square and 18 in. deep, supported by drum floats in the center of a raft 14 by 16 ft. The water level in the pan is the same as that of the surrounding water. Diagonal baffles in the pan are used to reduce wave action and wash. The pan coefficient ranges from 0.70 to 0.82, but is generally recommended as 0.80 [19].

For evaporation studies, the U.S. Weather Bureau land pan is generally recommended as the first choice and the Colorado sunken pan as the second choice [19].

Based on the records of 108 Weather Bureau Class A land pans and 42 Bureau of Plant Industry sunken pans, Horton [21] was able to prepare maps showing the normal evaporation in the United States in terms of values by Class A pans for a year and for periods from May to October and from November to April. Such maps have been brought up to date by the U.S. Weather Bureau in recent years (Fig. 11-3).

According to T. W. Robinson of the U.S. Geological Survey, the annual evaporation ending May 1, 1959, at the National Park Survey Headquarters on Cow Creek in Death Valley, California, was reported as 155.05 in., slightly less than 1 in. under 13 ft. This is probably the highest value ever recorded in the United States.

The measurement by evaporation pans is affected by many variables, including vapor-pressure difference, wind movement, pan diameter, water temperature, air pressure, rim height, color of pan, and depth of pan. These variables were correlated by Hickox [23] with the experimental results which he obtained from a small pan under controlled conditions.

Many studies were also made to determine reliable relations between pan evaporation and meteorological factors. Some results so obtained have been incorporated in various methods of evaporation determination. Some of the relations so developed involve the substitution of air temperature for water temperature, with a resultant seasonal and geographic bias. Thus the need for water-temperature observations can be eliminated, and the use of air temperature is more practical for design purposes. Penman [24] proposed such a method through simultaneous solution of an empirical mass-transfer equation and an energy-balance equation.

Observations indicate that pan evaporation is also affected by the sensible-heat transfer across the pan walls in either direction and by advection to and by the energy storage in the reservoir or lake. Such effects have been analyzed through the use of the Bowen ratio and energy-balance concept [25]. When such effects are negligible, the pan coefficient of a U.S. Weather Bureau Class A pan is, for practical purposes

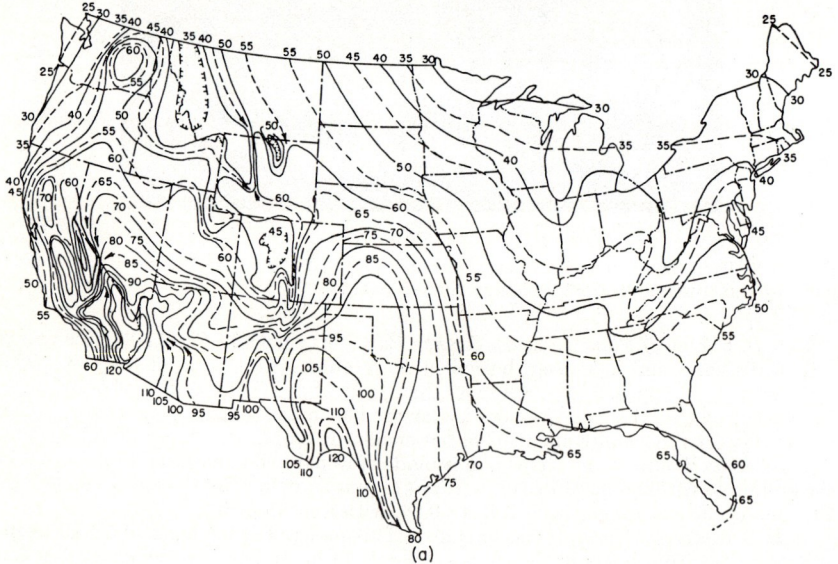

FIG. 11-3. Evaporation maps. (a) Average annual Class A pan evaporation in inches (1946–1955); (b) average annual lake evaporation in inches (1946–1955); (c) average May–October evaporation in per cent of annual. (NOTE: Seasonal per cent is based primarily on pan data, but limited testing indicates that the map is equally applicable to lake evaporation, assuming no change in heat storage.) (*U.S. Weather Bureau* [22].)

EVAPORATION FROM FREE-WATER SURFACES 11–9

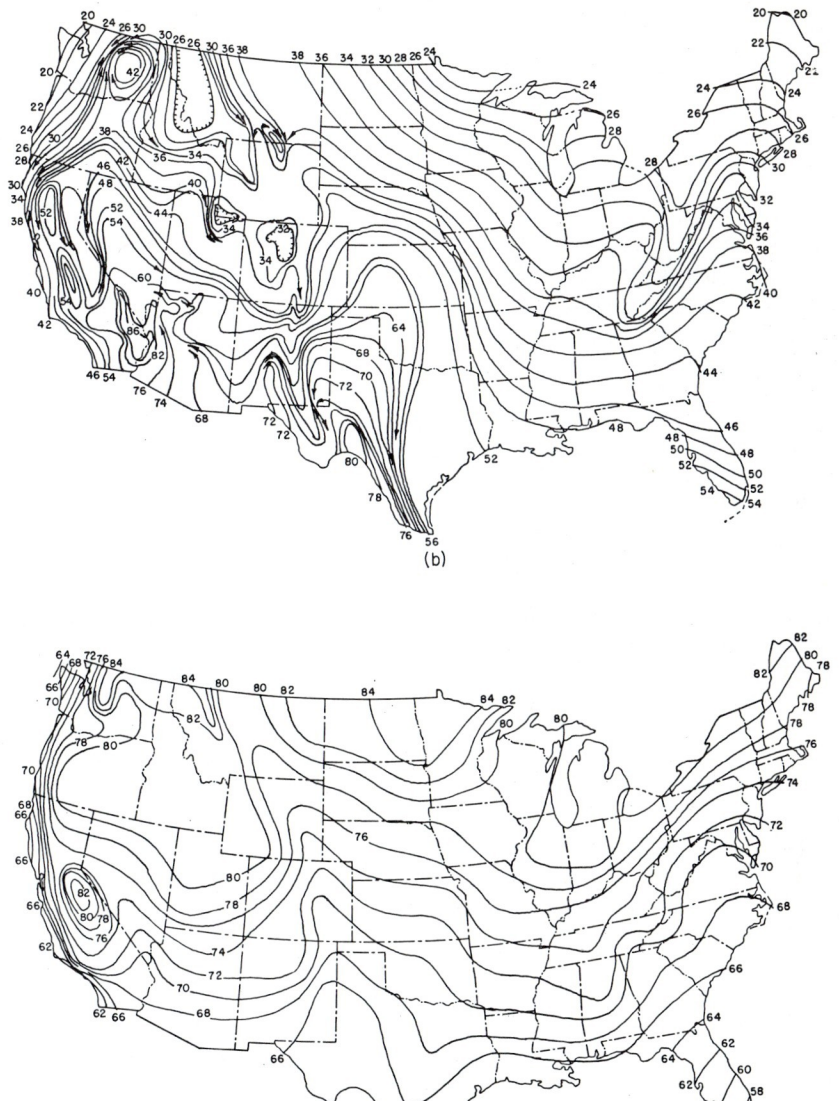

Fig. 11-3 (*Continued*)

0.70. The pan under this condition then corresponds to a "theoretical" pan, which is essentially independent of climatic variation.

2. Atmometers. According to Livingston [26], the *atmometer* is "any instrument of whatever form, for measuring or estimating different intensities of evaporativity, which is frequently called evaporation." Several types of atmometers are as follows *a. Livingston Atmometer.* This is a hollow porous porcelain sphere 5 cm in diameter with walls about 3 mm thick. When in operation, the moist surface of the sphere, without any superficial film of water, assumes a constant and uniform exposure in all directions except below, where the narrow cylindrical neck is connected to a supply reservoir of distilled water. Loss of water due to evaporation on the surface can be measured. Because of the variation in size and in the characteristics of the porous material, the reading must be compared with a standard and multiplied by a calibration coefficient supplied by the manufacturer of the sphere.

Fig. 11-4. A set of black and white Livingston spherical porcelain atmometers used at a field station. Evaporation is measured by refilling the jar, to the mark on the neck, with distilled water from a burette. The measurement mark is thickened here so that it is made visible; actually, it is only a thin scratch.

The spherical bulb of Livingston atmometers may be white or coated with a black film (Fig. 11-4). The difference in evaporation arising from this difference in color should be highly correlated with the intensity of solar-energy radiation. Halkias and others [27] reported that the mean monthly difference between black and white atmometers, in cubic centimeters, is equal to 0.028 times the mean monthly radiation measured by an Eppley pyrheliometer, in gram calories per square centimeter. This relation has a coefficient of correlation equal to 0.94. O'Connor [28] found that the black and white atmometers are feasible in measuring global radiation for climatological and ecological purposes. For accuracy of such measurements, the bulbs, particularly the white ones, should always be kept clean; only distilled water should be used, an accumulation of air bubbles should be avoided, and careful handling of the instrument is necessary.

Measurement of evaporation in very cold weather is difficult by Livingston atmometers because the bulb will burst on freezing [29].

The Livingston instrument is commonly used by plant physiologists to study loss of moisture from plants. The readings cannot be directly applicable to the evaporation from a water surface. However, Halkias and others [27] found that a very high coefficient of correlation equal to 0.98 exists between the measurements by a spherical Livingston white atmometer and a U.S. Weather Bureau Class A pan. The relation shows that the pan reading in inches is equal to 0.0051 times the atmometer reading in cubic centimeters.

b. Bellani Atmometer. This is similar to the Livingston atmometer, except that porous porcelain flat plates are used instead of spheres.

c. Piche Atmometer. This consists of a graduated tube, about 9 in. in length and 0.4 in. inside diameter, with one end closed and with the other flat end covered by a circular piece of filter paper pressed against it by a disk. When in operation, the tube is filled with distilled water. After the paper and the disk are put on, the instrument is inverted. This atmometer is very sensitive to winds. The diameter of the paper is limited by the rate at which water can move outward radially through the paper. If the paper is too large, its edge tends to dry out during rapid evaporation, and the measurement will not represent the condition for a constant surface of evaporation.

D. Estimation of Evaporation

1. Empirical Evaporation Equations. Many empirical or semiempirical equations have been developed to estimate evaporation from free-water surfaces. Most

EVAPORATION FROM FREE-WATER SURFACES 11-11

of them are based on Dalton's law, with modifications for factors affecting evaporation. The equations given in Table 11-1 are typical of many that have been proposed. They all have the common feature that the energy factor is the difference in vapor pressure between water and air. Their application is difficult because it may not be possible to obtain the information needed for their solution. For example, the temperature of the water surface over a considerable period may not be known. Most of the quantities used are average values, whereas, in fact, evaporation depends upon the total quantity of incoming energy, and the average may not represent the total. The relative humidity, or vapor-pressure deficit, measured in early morning and another measurement in late afternoon may give averages that do not indicate the total quantities needed to determine evaporation. This is also true in the case of mean temperature as a basis for the measure of the total amount of incoming energy.

2. Water-balance Method. The *water-balance*, or *water-budget*, *method* is a measurement of continuity of flow of water. This holds true for any time interval and applies to any drainage basin and to the earth as a whole.

According to Horton [30], the water-balance equation may be written as

$$E = I - O - S \tag{11-9}$$

where E is the evaporation; I is the inflow, or precipitation; O is the outflow, or total runoff; and S is the change in reservoir contents.

Theoretically, it is possible to use the water-balance method for the determination of evaporation from any lake or reservoir. Practically, however, it is difficult to do so because of the effects of errors in measuring the various items. Evaporation as determined by this method is a residual and therefore may be subject to considerable error if it is small compared with other items.

The outflow in Eq. (11-9) may contain outflow due to subsurface seepage. Seepage is usually the most difficult item to evaluate since it must be estimated indirectly from measurements of groundwater levels, permeability, etc. When seepage nearly equals or exceeds evaporation, the water-balance method cannot be very reliable. If a stage-seepage relation can be derived, the water balance becomes applicable on a continuing basis. Recent studies show that a simultaneous solution of the water-balance and heat-transfer equations can produce estimates of evaporation and seepage [31].

When the method was applied to Lake Hefner [9], evaporation was found to be a major item in the daily water balance. During a 16-month period of observation, outflow was only 10 per cent greater than evaporation. Inflow was generally much less because of the sporadic nature of reservoir replenishment. The error in computed monthly evaporation was less than 5 per cent. Daily figures were subject to much larger percentage errors. This is particularly so during the spring and fall months when the thermal expansion of water in the reservoir introduces appreciable systematic, but unavoidable, error in daily figures.

3. Energy-balance Method. The *energy-balance*, or *energy-budget*, *method* is similar to the water-balance method except that it deals with the continuity of flow of energy instead of water. This method was applied to obtain estimates of the annual evaporation from the oceans, first by Schmidt [32] and then by Sverdrup [33]. Ångström [9, pp. 71–119], Richardson [34], and Cummings [35] applied this concept successively to lakes.

Like the water-balance method, the energy-balance method is complicated by the difficulties of evaluating the needed items for the solution of the energy-balance equation, including such items as atmospheric radiation, long-wave radiation from the body of water, and energy storage [36, 37]. The conduction of sensible heat to or from the body of water is also a difficult item to evaluate. With the energy-balance equation, it is possible to obtain the sum of energy conducted as sensible heat and energy utilized by evaporation. The ratio of these two terms is known as *Bowen's ratio* [38].

According to Anderson [9], the energy-balance equation applied to a body of water with free-water surfaces may be expressed as

$$Q_s - Q_r - Q_b - Q_h - Q_e = Q_\theta - Q_v \tag{11-10}$$

where Q_s = solar radiation incident to water surface
Q_r = reflected solar radiation
Q_b = net energy lost by body of water through the exchange of long-wave radiation between atmosphere and body of water
Q_h = energy conducted from body of water to atmosphere as sensible heat
Q_e = energy utilized for evaporation
Q_θ = increase in energy stored in body of water
Q_v = net energy advected into body of water

This equation assumes the principle of conservation of energy but neglects items of small magnitude such as heat transformed from kinetic energy, heating due to chemical and biological processes, and conduction of heat through the bottom.

From Eq. (11-10), the evaporation can be expressed as

$$E = \frac{Q_s - Q_r - Q_b + Q_v - Q_\theta}{\rho L(1 + R)} \tag{11-11}$$

where E is the evaporation in cm, ρ is the density of water, L is the latent heat of evaporation, and R is the Bowen ratio expressed as

$$R = \frac{0.61P}{1,000} \frac{T_w - T_a}{e_w - e_a} \tag{11-12}$$

where P is the atmospheric pressure in mb, T_w is the water-surface temperature in °C, T_a is the air temperature in °C, e_w is the saturated vapor pressure in mb at T_w, and e_a is the vapor pressure of air in mb.

Since the sensible-heat transfer to the atmosphere, Q_h, cannot be readily measured, this term is excluded from the energy-balance equation through the use of the Bowen ratio.

The short-wave radiation Q_s incident to the water surface can be evaluated by either indirect computation based on easily measured quantities or direct measurement by *radiometers* [39]. There are many methods of indirect measurement. The following two methods are examples which are also widely used for oceanographic and limnologic studies.

a. The Mosby Formula. This is an empirical formula for computing the incoming radiation on a horizontal surface, in terms of the average altitude of the sun and the average cloudiness [40]:

$$Q_s = k(1 - 0.071C)\bar{h} \text{ cal/cm}^2/\text{min} \tag{11-13}$$

where Q_s is the solar radiation, k is a constant that is a function of latitude, C is the average cloud cover in tenths of sky covered, and $\bar{h}$ is the average altitude of the sun in degrees. The value of $\bar{h}$ should be less than 60°. Otherwise, the result is not valid unless the value is replaced by reduced altitudes. This formula is designed to provide evaluation for long time intervals, such as monthly or annual values of solar radiation.

b. The Kennedy Method. This method makes use of a pyrheliometer station, similar in latitude and altitude to the place where the radiation data are desired [41]. Data are extrapolated from the station to the place under consideration. The solar radiation Q_s is evaluated by the equation

$$Q_s = L_0 a^m \tag{11-14}$$

where L_0 is the solar radiation received on a horizontal surface at the exterior of the earth's atmosphere, a is the atmospheric transmission coefficient, and m is the solar air mass or the ratio of the length of the actual path of the solar beam to the path through zenith. By Eq. (11-14) the daily value of a can be computed from known values of Q_s, L_0, and m at a pyrheliometer station. The values of a thus computed can be plotted against cloudiness in tenths to produce a smooth curve of fitting. This curve is then used to estimate the value of a at the unknown station for a given degree of cloudiness. Utilizing this value of a and tables giving L_0 and m as functions of latitude and the declination of the sun, the solar radiation at the unknown station can be computed by Eq. (11-14).

The Kennedy method represents an extrapolation designed to provide evaluation for short time intervals, usually the daily solar-radiation values.

In the Lake Hefner studies [9], it was concluded that the indirect methods are inadequate, and for requisite accuracy, direct measurement of solar radiation is necessary.

The reflected solar radiation Q_r is always positive and varies from slightly more than zero to about 10 cal/hr/cm². It may thus decrease the amount of solar energy available for evaporation by about 13 per cent.

The net energy loss Q_b through exchange of long-wave radiation between atmosphere and water is generally expressed by

$$Q_b = Q_a - e\sigma T_s^4 \tag{11-15}$$

where Q_a is the atmospheric radiation, principally from water vapor, cloud droplets, carbon dioxide, and ozone; T_s is the water-surface temperature in °K; σ is the *Stefan-Boltzmann constant*, or 1.72×10^{-9} Btu/hr/ft²/°K⁴; and e is the emissivity of the water surface, variously taken from 0.9 to 1. The reflection of atmospheric radiation has usually been neglected.

The algebraic sum of incident and reflected sun and sky short-wave radiation, incident and reflected atmospheric long-wave radiation, and long-wave radiation emitted by the water body is known as *net radiation*. The net radiation can be measured also by radiometers, or by the *Cummings radiation integrator* (CRI) [9, pp. 120–126]. The CRI is a thermally insulated pan to measure certain net-radiation items by maintaining an energy balance. It is assumed that incident and reflected solar and long-wave radiation are the same for the pan as for an adjacent body of water. Having the sum of these terms for the pan and computing the long-wave radiation from the water on the basis of its surface temperature, net radiation for the water body can be evaluated.

The stored energy in water, Q_θ, may vary from about -10 to about 10 cal/hr/cm², while the advected energy Q_v may be about 10 cal/hr/cm² for a pan. The terms $Q_v - Q_\theta$ in Eq. (11-11) can be computed from an approximate water balance [Eq. (11-9)] and temperatures of the respective water volumes. Thus

$$Q_v - Q_\theta = \frac{1}{A}(\Sigma I T_I - \Sigma O T_O - E T_E - \Sigma S T_s) \tag{11-16}$$

where A is the surface area of the water body, and T_I, T_O, T_E, and T_s are the temperatures, respectively, of I, O, E, and S defined in Eq. (11-9) for the considered period.

The Bowen ratio R in Eq. (11-12) may vary, for a short period, from -1 to 1, but for a 24-hr interval, it is seldom greater than 0.30 and frequently less than 0.20.

The constant in Eq. (11-12) has received much discussion [42, pp. 43–48]. Bowen found the limiting values of 0.58 and 0.66, depending on atmospheric stability, and considered 0.61 for normal atmospheric conditions.

In the Lake Hefner studies [9], it was concluded that the energy-balance method is probably adequate when applied to periods of less than 7 days. For greater periods, an accurate evaluation of evaporation by the method will result in a maximum accuracy approaching ±5 per cent of the mean energy-balance evaporation.

4. Mass-transfer Method. Based on the concepts of discontinuous and continuous mixing applied to mass transfer in the boundary layer, a mass-transfer theory has been developed to derive evaporation equations, notably those of Thornthwaite and Holzman [43], Sverdrup [44], and Sutton [45]. A detailed description of the mass-transfer theory and the derived equations is beyond the scope of this discussion.

From the studies at Lake Hefner [9] it was found that the Sverdrup and Sutton equations gave good results for field use and that the Thornthwaite-Holzman equation [Eq. (11-17)] would give satisfactory results with proper instrumentation, but the instrument requirements are exacting.

The Sverdrup and Sutton equations gave good results in the Lake Hefner study [9] but were considered inadequate in the Lake Mead study [10]. From these studies, the Thornthwaite-Holzman equation was believed to give satisfactory results with instrumentation meeting the exacting requirements. Assuming an adiabatic atmospheric condition and logarithmic distribution of wind speed and moisture in the vertical, this equation may be expressed as

$$E = \frac{133.3(e_1 - e_2)(w_2 - w_1)}{(T - 459.4) \ln (h_2/h_1)^2} \qquad (11\text{-}17)$$

where E is the evaporation in in./hr; e_1 and e_2 are vapor pressures in in. Hg at the lower height h_1 and upper height h_2, respectively, above the water surface; w_1 and w_2 are wind velocities in mph at h_1 and h_2, respectively; and T is the mean temperature in °F of the air between h_1 and h_2.

E. Evaporation Reduction

The reduction of evaporation by controlling the rate at which water vapor escapes from water surfaces is of great economic importance. Many discussions and methods have been developed for such evaporation suppression from free-water surfaces [46–50].

The methods of evaporation reduction may be classified into three groups: those that reduce the surface area, those that mechanically cover the area, and those that cover the surface with a film.

1. Surface-area Reduction. The surface area may be reduced by constructing reservoirs with minimum ratio of area to storage [51–53], by storing water below ground [51–55] and in one large reservoir instead of several small ones [51, 53] by proper selection of reservoir sites [52, 54, 55], and by straightening stream channels and thus reducing meandering of surface areas of water.

2. Mechanical Covers. This method applies mostly to small reservoirs. The covers include roofs [51, 56], floating rafts [51], and windbreaks [51]. Removal of phreatophytes near the water body may also save loss of water through transpiration [51, 52, 54, 56, 57].

3. Surface Films. This method appears to be the most economical and offers the greatest potential for evaporation reduction. Miss Pockels [58] first discovered the formation of thin surface films of spreading oils on water, which were found to be of monomolecular thickness by Lord Rayleigh [59]. Du Nouy [60] believed that he published the first paper [61] on the influence of a monomolecular layer on evaporation of water. The monomolecular film of fatty acid was found by Hedestrand [62] to reduce evaporation very little but others reported contrary findings [63, 64].

Hexadeconal has received much study as an effective film-forming agent to reduce evaporation from water surfaces. This material is a white, waxy, crystalline solid, generally available in flakes or powder form. It is relatively tasteless and odorless. It is derived from tallow, sperm oil, or coconut oil, and is also called *cetyl alcohol*.

The use of hexadeconal for evaporation control was begun by Mansfield [65, 66], of the Commonwealth Scientific and Industrial Research Organization in Melbourne, Australia.

The monomolecular film formed by hexadeconal on water surfaces is only about six ten-millionths of an inch in thickness. This film offers a barrier to prevent water molecules from escaping from the water body. The film is penetrable by raindrops, but will close again after being broken. It is flexible and moves with the motion of the water surface. It does not easily break up, as do thick films formed, for example, by heavy oils.

Hexadeconal is a polar compound in which one end of the molecule has a great affinity for water (hydrophylic) and the other end repels water (hydrophobic). When in contact with water, the molecules literally stand on end. The film thus formed is so tight that water molecules cannot penetrate and escape from it. However, the film is pervious to oxygen and carbon dioxide. Normal gas exchange between the atmosphere and the body of water is unhindered.

The use of hexadeconal in the amount and manner proposed to reduce a monomolecular film on water surfaces in evaporation reduction seems to be no hazard to public health [49]. No apparent toxicity of hexadeconal to wildlife and fish life has been observed, although long-term ill effects are yet to be determined and may be produced [47]. Moran and Garstka [67] stated that "toxicity of hexadeconal itself offers no deterrent to its use in reservoirs. However, there was a reasonable question as to the possible collateral effects of degradation products when hexadeconal is exposed on a sizable body of water under natural conditions."

By studying 152 compounds as potential evaporation retardants, Cruse and Harbeck [49] considered that the homologous straight-chain fatty alkanols are the best materials for retardants although they are susceptible to biochemical oxidation. Of all the chemicals tested, Roberts [68] stated that combination of two fatty alcohols, octadeconal and hexadeconal, had been found to have superior film-forming properties.

The effectiveness of the monofilm depends on many factors, such as solubility, wind action, method of application, and oxidation or degradation by microorganisms. Since most fatty alcohols are practically insoluble, their effectiveness is determined principally by their persistence controlled by other factors. Wind may sweep off the film and pile up the material on shore. This becomes a major problem, especially on large reservoirs. Effective application of the material is also a big technical problem to a large reservoir. The material may be applied in a solvent such as kerosene or any flammable solvent, as a liquid, as an emulsion, or as a solid in the form of powder, pellets, or cakes. In a solvent, or as a liquid or emulsion, it may be applied by especially designed dispensers, by an aerosol-type sprayer, by a low-pressure application system, by cannisters with perforated bottoms, or by drippers. As a solid, it may be distributed by rafts or by projection. According to Cruse and Harbeck [49], the wick-type drippers for the application of liquids and cage rafts for the application of solids appear to be the most promising from an economic viewpoint, although both methods have their disadvantages. The biochemical oxidation may be reduced by addition of bactericides (such as $CuSO_4 \cdot 5H_2O$) or by the use of bacteria-resistant materials (such as $RSiH_2OH$, which is being proposed), but most of such materials pose a potential hazard because of their determined or undetermined toxicity.

The reduction of evaporation by use of monofilms varies with the kind of material and the condition under which it is used. One report states that normal evaporation may be reduced by as much as one-third by efficient use of monofilms [69]. Theoretically, 0.02 lb of hexadeconal forms a compact monolayer film on an acre of water surface. In practice, considerably more chemical is required because of the removal of the material by wind, birds, insects, and aquatic life and because of biologic attrition of the layer. For example, 6 lb of powdered hexadeconal was used in 7 days in the month of August on a pond of 2.8 acres [70]. Much work is in progress which attempts to develop a practical, safe, and effective method of evaporation reduction by monofilms.

III. EVAPORATION FROM SOIL SURFACES

Atmospheric environmental conditions that affect evaporation from a free-water surface also affect that from soils. The difference, however, is in the nature of the surface from which evaporation takes place. Furthermore, water molecules have to overcome greater resistance to escape from soils than from a free-water surface. The factors affecting the escape of water molecules from a free-water surface, as previously discussed, are temperature, wind, material in solution in the water, and, to a slight degree, barometric pressure. In addition, the escaping molecules of water from soil must overcome the resistance due to the attraction of the soil particles for the water, that is, the body or field forces existing between the soil and water. When the moisture content of the surface soil becomes relatively low, the loss of moisture by surface evaporation practically ceases.

A. Energetics of Soil Moisture

In unsaturated soils the water surfaces are not flat, as in a free-water surface. They are concave with respect to the air, and consequently the vapor pressure is lower than that from a flat surface. The vapor pressure of the water in the soil in relation to that of pure free water at the same temperature can be used to determine the *free energy*, or *thermodynamic potential*, of the water. Usually, a free flat surface of pure water is taken as a zero potential level. The absolute free energy of this datum is not equal to zero, but it serves as a reference point. The total free energy is composed of several components [71, 72], including those due to hydrostatic pressure, to osmotic value given by dissolved material, to field force such as adsorption, and to surface tension if the water is in an interface. The field force may be important. At high soil-moisture content, the adsorptive field force probably plays a minor role, but as the soil moisture decreases, it becomes increasingly important. If the moisture in the whole system is

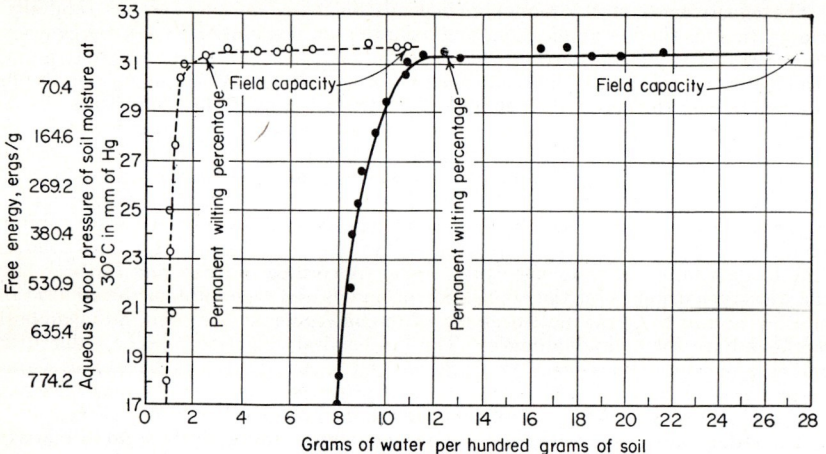

FIG. 11-5. Relations between free energy and soil moisture. The dotted line is for sand, and the solid line for a clay soil. (*After Edlefsen* [71].)

in equilibrium, the total free energy is a constant, but the magnitude of each component generally varies from point to point in the system.

Since the energy required to remove a unit quantity of water from the soil can be measured, it is logical to consider soil-moisture relations from the energetics of the system. Figure 11-5 shows two typical relations between free energy and soil moisture. No points of sudden change can be seen on the curves to indicate that the water at one level is in a condition different from that at another. At the permanent wilting percentage, however, there is a rapid change in the energy and soil-moisture relations. This lends support to the selection of free energy as a reference for the measurement of soil moisture [73]. Gross measurements of energy changes therefore seem to be the most reliable basis for studying soil moisture.

It is sometimes reasoned that, since resistance to removal by the plant increases as the soil-moisture content decreases, transpiration and growth of plants must likewise decrease. This is not necessarily so when the soil-water-plant system is considered as a whole. The total free energy of the soil moisture surrounding the roots at permanent wilting percentage is about -0.16×10^8 ergs per gram, and if leaves are exposed to the atmosphere with a 40 per cent relative humidity at 30°C, the free energy of the air is -12.8×10^8 ergs per gram (Subsec. IV-B). There is, then, an overall drop in free energy of -12.6×10^8 ergs per gram from soil at permanent wilting percentage to the air. If this drop occurred through a tree 10 meters high there would be an average force tending to pull water up the tree of 12.6×10^5 ergs per gram, a force of about 1,300 times gravity. The total energy required to take water from a soil at the permanent wilting percentage is only 0.07 per cent more than at saturation.

There are various methods of measuring free energy and soil moisture, as well as many ways of expressing the moisture content, such as in terms of field capacity[1] and permanent wilting percentage.[2] Detailed discussions on these subjects are beyond the scope of this section, and the reader should refer to Sec. 5, to Ref. 74, and to standard textbooks on soil physics and plant physiology [75–77].

B. Movement of Moisture in Soil

To reach the surface of the soil and dissipate into the atmosphere, moisture must move from the lower depths to the surface. If evaporation is to be a continuous process, movement will have to take place through considerable distances in unsaturated soils.

Soil investigators have spent much effort in studying soil-moisture movement, especially in unsaturated soils. The literature dealing with the subject is too voluminous to permit reference to more than a few papers.

For water to move, there must be a driving force. If the free energy of the water is known as well at one point in the soil as at another point, the direction of flow can be determined. The space rate of change of free energy represents a force per unit mass that causes water to move in the soil or plants. If the change in the free energy in different parts of the soil or in the plant is known, it is possible to determine where the water will move, and if the conditions are specified, the magnitude of the flow may also be calculated. Unfortunately, the conditions that may affect the rate of flow of water in soils and plants are usually unknown. In other words, it may be possible to know the driving force, but not the magnitude of the resistance to flow.

The relation between the mass of water passing in unit time through unit cross section normal to the direction of flow, V, and the change in total free energy or potential ϕ per unit distance, grad ϕ, is

$$V = -K \text{ grad } \phi \qquad (11\text{-}18)$$

where the negative sign is due to the selection of a free-water surface as the datum. Moisture above this datum will have a negative free energy or potential (see also Sec. 13).

The coefficient K in Eq. (11-18), called the *specific conductivity*, is really a proportionality factor, generally determined empirically. It is related to the nature of the soil and its moisture content and therefore is a variable that depends on the amount of soil moisture. Texture and structure have decided effects upon the conductivity. The number and size of the pores in the soil greatly influence the soil-moisture movement. The viscosity of the water will also influence flow. Equation (11-18), known as *Darcy's law* (Secs. 5 and 13), is believed to hold for both saturated and unsaturated flow of water in soils.

The flow through a smooth capillary tube of circular section for conditions of streamline flow can be expressed by the Poiseuille equation (Sec. 7).

C. Evaporation from Unsaturated Soil

The upward movement of soil moisture in maintaining evaporation was first stressed by King [78]. Hilgard and Loughridge [79] and Widtsoe [80] held that water moved

---

[1] *Field capacity* is the amount of water held in the soil after the excess gravitational water has drained away and the rate of downward movement of water has materially decreased.
[2] *Permanent wilting percentage*, or *permanent wilting point*, is the moisture content of soil at which soil cannot supply water at a sufficient rate to maintain turgor and the plant wilts.

upward through great distances and, consequently, evaporation would continue in significant quantities for long periods, but others [81–84] disagreed.

Veihmeyer [85, 86] showed that the movement of moisture from moist to dry soil was slight in extent and amount when the soil was not in contact with a free-water surface. His studies were made with soils in containers and in the field. Figure 11-6 shows the distribution of moisture under an irrigation furrow just after the water is distributed and again after a relatively long period, 56 days. The ground was covered to prevent evaporation, and no plants were allowed to grow during this time. The upward, downward, and lateral movements of moisture were too slight to be measured.

The success of the practice of summer fallow, whereby the land is left without plants for one season to store rainfall for use by crops in the following season, is based on the fact that there is only slight upward movement of water to the surface, where it can evaporate.

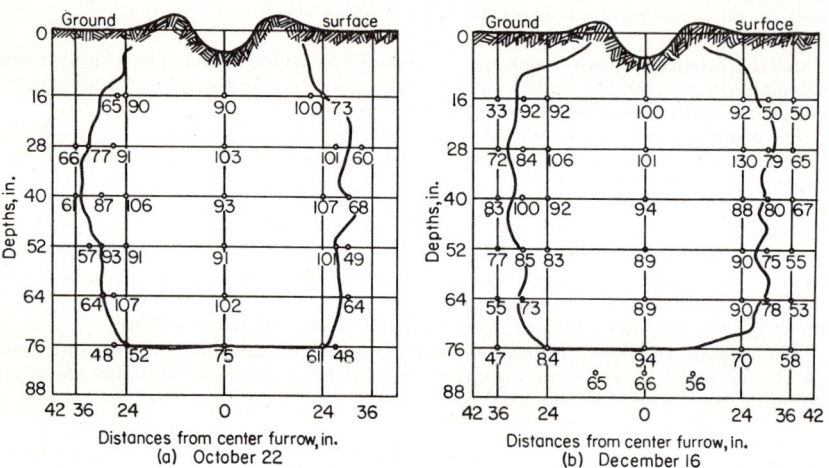

Fig. 11-6. Distribution of water from a furrow just after irrigating on October 22 and 55 days later. Ground surface is covered to prevent evaporation. Heavy lines indicate limit of penetration of water. Numbers are relative wetness in per cent, or ratio of moisture contents to moisture equivalents. (*After Veihmeyer* [85].)

Veihmeyer [85] showed that soil in large tanks could be irrigated early in the summer and exposed to evaporation until late fall (167 days) before seed was planted. No water was applied to the soil during the interim. The successful crop indicated the small loss of water by evaporation from unsaturated soil.

Evaporation from the soil surface will continue as long as the shallow surface layer, about 4 in. for clays and about 8 in. for sands, remains moist. When the time between applications of water to the soil is long, evaporation losses will be small during the growing season. Where rain or irrigation is frequent, however, evaporation may be great. If only the shallow surface layer is wet each time water is applied, evaporation may approach 100 per cent of the water added to the soil; otherwise direct evaporation from the surface of the soil when not in contact with a free-water surface will be very small.

The rate of drying is as rapid from dry soil (at permanent wilting percentage) as from wet soil. Some investigators [87–91] found that wet soils maintain a practically constant rate of evaporation at a certain range of moisture content until a low moisture content is reached. Veihmeyer and Hendrickson [92] showed that this rate, under constant evaporation conditions, is constant until the permanent wilting percentage is reached.

D. Evaporation from Saturated Soils

The preceding discussion concerned soil-moisture movement when the soil was not saturated nor in contact with a free-water surface. In the presence of a free-water surface, moisture movement may be appreciable. Upward movement of water from a water table with the deposition of salts at the soil surface is a common occurrence.

Lee [93] reviewed the results of work on evaporation from bare soil moistened from a water table. Veihmeyer and Brooks [94] reported the results of measurements on evaporation from soils in contact with free water. They showed that evaporation rates do not bear a linear relation to the depth of the water table from the soil surface, a condition that has been held true by others. If a water table is within one foot of the bare surface, the soil evaporation loss is comparable with the transpiration loss for an irrigated crop. In unsaturated soils, on the contrary, the loss is only a very small portion of the total loss of water from the soil.

E. Controlling Evaporation Losses from Soil

There are various methods of controlling evaporation losses from soil.

1. Dust Mulch. This is an age-old practice in cultivation of soil to keep it loose on the surface. It is based on the theory that loosening the surface will permit rapid drying and remove the soil particles from contact with each other. Rapid drying will develop dry soil to act as a blanket to suppress evaporation. Reducing points of contact between soil particles will lessen capillary rise.

Sewell [95] concluded that soil cultivation by tillage may be necessary only to kill weeds and to keep the soil in a receptive condition to absorb water. Chilcott and Cole [96] found that deep tillage is futile as a means to overcome drought or to increase yield. Call and Sewell [97] showed that mulching not only decreased the amount of water in the soil, but also caused loss of more moisture than in the bare, undisturbed soils. In tank and field trials, Veihmeyer [86] also found that mulching by thorough cultivation at weekly intervals failed to save soil moisture, but the surface shallow layer, by drying quickly, acted as a deterrent to further loss of moisture.

Since these early investigations, the results of many others have been published. Many agricultural experiment stations have studied this problem, resulting in conclusions similar to those mentioned. Shaw [98], however, concluded that the soil mulch can reduce moisture loss only when the water table, perched or permanent, is within the capillary rise of the surface.

2. Paper Mulch. Covering the soil with paper to reduce evaporation was in considerable use in the late 1920s, but is now rarely done. Smith [99] reported that the effect of paper mulch was confined to the surface 4 in. of soil and was due to condensation of water beneath the paper.

3. Chemical Alteration. Lemon [100] found that chemical alteration of the soil-moisture characteristics may decrease evaporation. Hedrick and Mowry [101] state that addition of polyelectrolytes to soils decreases the rate of evaporation and increases the water available to plants.

4. Pebble Mulch. This was reported in use for partial control of evaporation in some dry sections of China [102].

F. Evaporation from Bare Soils versus Transpiration

As pointed out earlier, unless the soil surface is kept moist, either by rainfall or irrigation or by rise of water from the water table, it will dry rapidly and evaporation will practically cease.

In a study [103] of the moisture regimen in a soil kept free of vegetation, the soils were sterilized to kill plants but the surface was undisturbed. The rainfall during the summer was negligible. Evaporation from the surface through an entire season was found not to reduce the moisture content to the permanent wilting percentage in the top foot of the soil. The loss of moisture in the lower depth was insignificant.

On the other hand, in a study [103] on areas covered with vegetation and adjacent to the bared land, during dry summer, soil moisture was reduced to the permanent wilting percentage throughout the entire depth of the soil by early July.

Under semiarid and arid conditions, transpiration is the predominant cause of loss of water from soils.

IV. TRANSPIRATION

Water in a plant may be lost in various ways by transpiration. By *stomatal transpiration*, water escapes through numerous pores of *stomata* in the leaves. By *cuticular transpiration*, water evaporates from moist membranes into the atmosphere through the cuticle. By *guttation*, water is forced out of the plant through special organs called *hydathodes* when the transpiration is at a low rate. Under certain conditions, water may also exudate from cut surfaces of the plant. By far the greatest source of water loss in a plant is the stomatal transpiration [104]; other losses are too small to warrant discussion here (see also Sec. 6).

A. Nature of the Process

Transpiration is essentially the same as evaporation except that the surface from which the water molecules escape is not a free-water surface. The surface for transpiration is largely in leaves. Leaves are composed mainly of thin-walled cells (*mesophyll*). A layer of cells (*epidermis*), generally one cell thick, covers the entire leaf. It is relatively impervious to moisture and gases but contains numerous stomata. The intercellular spaces in the mesophyll join and unite with large air spaces beneath each stoma. The moisture on the surface of the mesophyll cells vaporizes through these spaces and escapes from the leaf through stomata. The area of moist cell surfaces is many times greater than the external leaf surface. Water escaping from the stomata behaves as does water vapor passing through perforated membrane, where the rate of diffusion is proportional to the diameter but not to the area of the pores. The transpiration process is greatly affected by the length of passage of the vapor from the mesophyll cell walls to the stoma opening on the leaf surface.

Stomata may be sunk below the leaf surface as in conifers and the rubber plant. Consequently, water vapor must diffuse through a relatively long passage, which may be tortuous, before reaching the leaf surface and dissipating in the atmosphere. Transpiration thus becomes very slow.

Many plants have leaves that are covered with hairs growing out of the epidermal cells. The hairs sometimes form a rather dense cover. It may influence the wind movement over the epidermis, but no evidence has been found to indicate any transpiration reduction by its presence.

Stomata are most common on green aerial parts of the plant, particularly the leaves. They exist on either one or both sides of the leaves, but more frequently on the underside. The number of stomata on leaves varies from 50,000 to nearly 800,000 per square inch, depending on plant species and environmental conditions such as humidity, light, and soil moisture.

The stomata are tiny long openings through epidermis. Each stoma lies between two *guard cells*, which in turn are bordered by subsidiary cells. As water moves into the guard cells, there is a tendency to stretch them, causing the opening of the stoma. Thus the opening and closing of the stoma is produced by changes in turgor of the guard cells and results from unequal thickening of the guard-cell walls. The important factors affecting such an operation are light intensity, moisture supply of leaves, temperature of air, humidity, and chemical changes. Stomata usually open in the light and close in the dark. They close with reduced moisture, which causes the guard cells to lose turgor. Temperature affects speed of opening.

The guard cells contain chloroplasts, which possess chlorophyll. Within the chloroplasts, carbon dioxide and water react to form sugar. Starch is usually found in the protoplasm of guard cells. The change in turgor of these cells is associated with changes in their starch content. High turgor and open stomata are associated with little or no starch, and low turgor and closed stomata with abundant starch.

When the sugar content of the guard cells increases, as it usually does in daylight hours, the osmotic pressure of the cell sap increases, water is drawn in from adjacent cells, the turgor pressure of the guard cells increases, and the stomata open. When the sugar content decreases, changing to starch, as usually occurs in the dark, the osmotic, and consequently the turgor pressure of the guard cells decreases, and the stomata close.

Contrary to general belief, the stomata exercise a very limited control on the transpiration rate. They close after the wilting or darkness begins, and not in anticipation of it. When the stomata are fully opened, the transpiration rate is determined by the same factors that control evaporation alone. The stomata exert a slight regulatory influence only when they are almost closed.

In a living plant, there is a continuous column of water which extends from the root in the soil, through the stem and leaves, to the walls of mesophyll cells. The forces controlling the movement of water in the continuous column are not well understood. Of the many proposed theories, Bonner [105] believes that the tension-cohesion theory of water transport remains unchallenged. By this theory, a column of water in a thin capillary tube can transmit a considerable tension because of cohesion of water molecules. This theory explains why, in large trees, the height of the water column may exceed 300 ft and the total flow through the xylem (wood) may amount to many hundred gallons per day.

B. Factors Affecting Transpiration

The factors affecting transpiration may be physiological or environmental. Important physiological factors are density and behavior of stomata, extent and character of protective coverings, leaf structure, and plant diseases. The kind of plant, however, may not materially influence water use when extensive areas of crops are considered. The amount of ground coverages may be important [27]. The essential environmental factors include temperature, solar radiation, wind, and soil moisture when the permanent wilting percentage is reached. The physiological factors have been generally discussed previously. The environmental factors will therefore be discussed here.

Since the loss of water from a plant is governed by the difference in vapor pressure in the space below the stomata and that of the atmosphere, the aqueous vapor-pressure deficit of the air is the principal cause of transpiration. For example, if the relative humidity in the leaf at 30°C is 100 per cent, the vapor pressure is 31.86 mm Hg, and if the relative humidity in the air at 30°C is 80 per cent, the aqueous vapor pressure is 25.49 mm Hg; the difference in vapor pressure is 6.37 mm Hg. If the relative humidity in the air becomes 40 per cent with a vapor pressure of 12.74 mm Hg, the difference is 19.12 mm Hg. This is equivalent to a gradient in free energy of 12.80×10^8 ergs/g.

The difference in vapor pressure between the space in the leaf and the outside air is, of course, a measure of the energy required to move the water from the leaf to the air (Subsec. III-A). This is, therefore, the basic factor influencing transpiration. The gradient of vapor pressure between the leaf and the air will also be influenced by the morphological features of the leaf, the position of the leaf, and the relation of the leaf to neighboring leaves.

1. Temperature. The foregoing calculations are based on the assumption that the temperature of the water in the leaf is the same as that of the surrounding air. This is rarely the case while the leaf is exposed to the sun. The temperature of the leaf will be higher than that of the air. This is true even though evaporative cooling from transpiration lowers the temperature of the leaf, but it is impossible to lower the temperature of a leaf below that if the air when the leaf is in direct sunlight.

The aqueous vapor pressure, of course, is influenced by temperature. If an increase in air temperature increases the temperature of the water in the leaf by the same amount, the vapor pressure of both leaf and air will be raised equally. There would be no increase in transpiration from this cause alone. Because the leaf temperature in sunlight is usually above that of the air, with a consequent greater increase in its vapor pressure over that of the air, transpiration will usually result from an increase in air temperature. The relative humidity within an unwilted leaf is always close to 100 per cent.

2. Solar Radiation. The primary source of energy, of course, is the sun. Absorption of this energy by the leaf raises its temperature and its aqueous vapor pressure. The result is that transpiration increases along with insolation. Briggs and Shantz [106] found good correlation between radiation and transpiration. Others have reported similar results [27].

Many methods of reducing transpiration are based on reduction of solar radiation. Spraying leaves with solutions which change their color, and hence may affect the absorption of solar radiation, has been tried to reduce transpiration. Bordeaux mixture (an insecticide containing copper sulfate and hydrated lime) has been known to increase the transpiration rate of most plants to which it is applied. Adding lampblack to Bordeaux mixture was found to be more effective in increasing transpiration in full sunlight than in shade. Spraying leaves with hexadecanol has also been suggested. Preliminary experiments indicate that this method may be a practical means of reducing transpiration.

3. Wind. The removal of water vapor next to the leaf surface by wind might increase transpiration. In most leaves the stomata are on the lower side. There is a tendency for moist air to form under the leaf. Wind might carry away this layer of moist air and replace it with drier air, thus increasing the vapor-pressure difference between inside and outside of the leaf, with an increase in transpiration.

A gentle wind may be as effective as a strong wind in removing the outside molecules. Hence the wind's effect on transpiration may be limited to relatively light winds. In fact, a gentle breeze is relatively much more effective in increasing transpiration rate than are winds of greater velocity [107]. Giddings [108] found that the activity of leaves increased with wind velocity up to 8 mph. Thereafter the rate of transpiration became less. Brown [109] found that gentle air currents stimulated transpiration but strong air currents might check it.

4. Soil Moisture. Transpiration is definitely affected by the soil-moisture content when reduced to the permanent-wilting percentage. However, the effect of varying amount of soil moisture upon transpiration and plant growth is still a matter of controversy and conflicting evidence. Some investigators believe that transpiration and plant growth are affected by the decreasing soil moisture in rough proportion to that which is available to the plant. Others hold that transpiration and plant growth are unlimited by the reduced soil moisture that is available to the plant, because water is readily available when the moisture content is between the field capacity and the permanent-wilting percentage. This is an important question. Evapotranspiration equations which are said to apply only if soil moisture is adequate, without defining adequacy, are useless if it is true that the level of soil moisture does affect transpiration markedly; that is, soil moisture should be a factor in the equation. The only constant level of soil moisture in a drained soil is field capacity, and for most crops this level of moisture would be injurious if maintained.

It is believed that the controversy and apparent discrepancies are due to different interpretation of the supporting data and to lack of exhaustive experimental data. A discussion on the unsettled questions is beyond the scope of this section. For such discussion the reader may refer to Stanhill [110]; whose report contains a partial list of 80 papers dealing with soil moisture and plant growth but not transpiration; to Veihmeyer and Hendrickson [74, 104, 111–112] and Richards and Wadleigh [113], for a review of divergent views on plant-soil-water relations; and Penman [114, 115], on rationalizing the differences in opinion on the subject.

C. Determination of Transpiration

For a small plant the transpiration may be measured for short periods by placing the plant in a closed container and computing the changes in humidity in the container. The excessive humidity thus produced can be reduced by use of a drying agent placed inside the container, but the computed transpiration must be corrected for the moisture absorbed by the drying agent.

A *phytometer* provides a practical method for measuring transpiration. This is a large vessel filled with soil in which one or more plants are rooted. The soil surface is sealed to prevent evaporation so that the only escape of moisture is by transpiration, which can be determined from the loss in weight of plant and vessel. This method gives satisfactory results provided the simulated testing condition is comparable with the natural environment under investigation. A small phytometer containing only water instead of soil is called a *potometer*.

Transpiration may also be determined by watershed studies. This method studies the effect of removing the vegetative cover from the watershed and thus computes by statistical analysis the apparent amount of reduction in runoff measurements resulting from such effects.

For a given kind of plant the ratio of the weight of water transpired to the weight of dry matter produced, exclusive of roots, is called its *transpiration ratio*. This ratio varies widely, depending largely on soil type, relative humidity, and other climatic factors. It ranges from 200 to 500 for crops in humid areas. Many experiments have been made for the determination of transpiration ratio, including those by Briggs and Shantz [116–118] at Akron, Colo., during 1911 to 1917 for agricultural crops, by Franz von Höhnel (reported by Horton [119]) at Maraibrunn, Australia, from 1878 to 1880 for forest trees, and by others [93, 120]. Similarly, transpiration is also expressed in terms of depth of water consumed annually by the plant, and such depths for various plants were evaluated. Unfortunately, such data, i.e., transpiration ratio and transpiration depth, are of no practical value, because they do not usually represent a true measure of the field conditions and they may be misleading unless the exact conditions under which the measurements were made are fully specified.

The amount of transpiration depends on many variables. Its precise determination cannot be easily obtained, and no accurate single value can be assigned to any crop without specifying all the variables. Therefore the estimated transpiration may be extremely unreliable, serving only as a measure of the relative water use by plants under similar conditions. This is the reason why so much effort has been made to correlate climatic factors to evapotranspiration. Thus the measurements in one locality under known evapotranspiration conditions may be applied to another in which the environmental conditions affecting the use of water can be measured.

V. EVAPOTRANSPIRATION

A. Methods of Estimation

Evapotranspiration, or consumptive use, as defined previously, is the evaporation from all water, soil, snow, ice, vegetative, and other surfaces, plus transpiration. Since Thornthwaite [121] thought soil moisture may have an effect upon evapotranspiration, he has suggested a term *potential evapotranspiration* to define the evapotranspiration that would occur were there an adequate soil-moisture supply at all times. He believed that this requires a very high soil-moisture content.

There are many methods of estimating evapotranspiration and potential evapotranspiration, but no one method can be applied generally for all purposes. Most methods for estimating evapotranspiration apply also to estimation of potential evapotranspiration, provided the area under observation has sufficient water at all times. All methods, however, may fall into three general categories: theoretical approaches based on the physics of the evapotranspiration process, analytical approaches based on the balance of energy or water amounts, and empirical approaches based on the regional relation between the measured evapotranspiration and the climatic conditions. Several methods widely used in engineering investigations are given as follows.

1. Soil-moisture Sampling. This method is usually suitable for irrigated field plots where soil is fairly uniform and the depth to ground water is such that it will not influence soil-moisture fluctuations within the root zone. Soil samples are taken in the area before and after each irrigation, and their moisture contents are determined by standard laboratory practices. A standard soil tube is generally used to take samples in 6-in. sections for the first foot, and thereafter in 1-ft sections in the major root zone. The samples are weighed and dried, and the dry weights determined. From the moisture percentage thus obtained, the quantity of water in acre-inches per acre (inches) removed by evapotranspiration from each foot of soil is computed by

$$D = \frac{PVd}{100} \qquad (11\text{-}19)$$

where P is the moisture percentage of the soil by weight, V is the apparent specific gravity or volume weight of the soil, d is the depth of soil in in., and D is the equivalent depth of water in in. lost by the soil. If d is expressed in centimeters then, of course, D will be in centimeters. Acre-inches of water extracted from the soil for any desired period, say, 30 days, is computed. The losses may be plotted against time and a use-of-water curve for the season obtained. This method usually requires a large number of measurements covering representative locations in order to obtain desired accuracy [122, 123].

Various devices exist to measure soil moisture without the drudgery of soil sampling. In general, these instruments have the advantage of immediate readings. But they also have serious disadvantages. Some are not always reliable at low moisture contents, others are not accurate at high moisture contents. Their calibrations may not remain correct after these instruments have been in place for some time.

Certain precautions must be taken with any method of determining changes in soil-moisture contents in the field. The soil must be deep enough to accommodate the root system of the plant to the full depth to which it would normally penetrate the soil.

Soil sampling should, of course, represent the full depth of soil occupied by the roots. The area sampled should be large, so as to avoid border effects. The place of sampling should be some distance from the outside rows of plants, to prevent undue influence from advected energy. A free-water surface within the reach of the roots precludes use of the soil-sampling method in the field.

Some plants have poor root systems because their roots do not thoroughly penetrate the soil. This is generally not the case for permanently rooted crops, such as alfalfa and trees. Because portions of the soil are not occupied by roots, sampling will include some soil devoid of roots. In this case, the average moisture content of the sample will not give the true value for the soil that is in contact with the absorbing portion of the roots.

Some annual plants have sparse root development for only the first part of their life cycle; later, they develop good root systems. Others have poor root systems that persist throughout the entire season. Soil conditions sometimes restrict the development of roots. For example, the soil may be too dense to permit penetration.

2. Lysimeter Measurements. This method is commonly used to determine evapotranspiration of individual crops and natural vegetation, by growing the plants in tanks, or *lysimeters*, and then measuring the losses of water necessary to maintain the growth satisfactorily. The tanks used are generally about 2 to 3 ft in diameter and 6 ft deep, although large sizes up to 10 ft in diameter and 10 ft deep have been used [122, 124]. In general, the overall condition in the tank may not closely simulate the field conditions, and hence the result thus obtained may not be converted reliably to an acreage basis for a much larger area under consideration [125].

If the condition in the tank is maintained almost the same as the field conditions by using a large-size tank and keeping the surface of the soil in the tank at the same level as that in the surrounding field, the result may become relatively satisfactory [126].

3. Inflow-outflow Measurements. This method involves the application of the water-balance principle to large land areas, which may be up to 400,000 acres [127–129]. The amount of water entering a known area of land during a certain period is measured and compared with the recorded precipitation on the area for the same period. The difference between these two items and the amount flowing out of the area, adjusted by the change in groundwater storage, during the same period will be a measure of losses by evapotranspiration for the period. The difference between storage of water at the beginning and end of the period is usually considered to be negligible. It is assumed that stream measurements are made on bedrock controls [122, 130] and that the subsurface inflow is about the same as subsurface outflow. It is apparent that this method usually presents difficulties in determining the flow quantities to a desired accuracy.

4. Integration Method. This method determines evapotranspiration by the summation of the products of evapotranspiration for each crop times its area, plus the evapotranspiration of natural vegetation times its area, plus water-surface evaporation times water-surface area, plus evaporation from bare land times its area [127, 128]. In applying this method, of course, it is necessary to know unit evapotranspiration and the areas of various classes of agricultural crops, natural vegetation, bare land, and water surfaces.

5. Energy Balance. This method assumes that the energy received by a surface through radiation equals the energy used for evaporation and heating the air and the soil, plus any extraneous or advective energy. For short periods, such as daily and monthly balances, the energy for heating the soil and the advective energy may be neglected. The method is based on the simple principle of energy balance discussed in Subsec. II-D-3. Its application to cropped land and the instrumentation necessary for measuring the energy items are discussed by Tanner [131] and Levine [132].

6. Vapor Transfer. This method uses the Thornthwaite-Holzman equation [Eq. (11-17)] for evapotranspiration estimation [43, 121] by modifications suggested by Pasquill [133-135]. The method requires strict adherence to boundary conditions and to the limitations imposed by the sensitivity of the instrumentation. These requirements will usually put the method beyond most facilities available for the measurement of evapotranspiration.

7. Groundwater Fluctuations. Daily rise and fall of the water table give an indication of evapotranspiration losses. The evapotranspiration by overlying vegetation can therefore be computed on the basis of diurnal measurements of the groundwater-table fluctuations in observation wells. This method has been used by the U.S. Geological Survey with great success [130, 136, 137].

B. Evapotranspiration Equations

The lack of basic data and the difficulties in measurement required in the field methods have accounted for the great efforts made to develop evapotranspiration equations that can relate the evapotranspiration with some readily available climatic data. A number of equations have been suggested for this purpose [138, 139]. Some typical ones are given in Table 11-2.

The *Hedke equation* [Eq. (11-20)] is based on a method by which the evapotranspiration, or water use, is estimated by summing for the growing season the values of available heat, expressed in degree-days above the germinating or minimum growing temperature. Some minimum growing temperatures in degrees Fahrenheit are 33° for alfalfa, 38° for pasture, 40° for vegetables, 44° for forage and grains, 46° for fruits, and 48° for cotton. Considerable judgment is required in selecting the coefficient k since only limited data are available. A value of about 0.0004 has been determined for Cache la Poudre Valley in Colorado, where there is a relatively high standard of agriculture [140].

The *Lowry-Johnson equation* [Eq. (11-21)] assumes a linear relationship between the effective heat and evapotranspiration. The effective heat is defined as the accumulated degree-days of maximum daily temperature above 32°F for the growing season. This equation is designed to estimate valley consumptive use for agricultural lands on an annual basis.

The *Blaney-Criddle equation* [Eq. (11-25)] and the *Blaney-Morin equation* [Eq. (11-22)] are similar, except that the latter has an additional factor for relative humidity. The coefficient k varies with the type of vegetation (Table 11-3). It also varies with the month and locality. For accurate results it must be known for

Table 11-2. Evapotranspiration Equations

Name	Date	Period for U	Unit for U	Equation	
Hedke [140]	1930	Annual	Feet	$U = kH$	(11-20)
Lowry-Johnson [129]	1942	Annual	Feet	$U = 0.000156H + 0.8$	(11-21)
Blaney-Morin [141]	1942	m months	Inches	$U = k \sum_{1}^{m} pt(114-h)/100$	(11-22)
Thornthwaite [142]	1944	Monthly	Centimeters	$U = 1.6 \left(\dfrac{10t}{TE}\right)^a$	(11-23)
Penman [24]	1948	Daily	Millimeters	where $a = 0.000000675(TE)^3 - 0.0000771(TE)^2 + 0.01792TE + 0.49239$ $U = \dfrac{AH + 0.27E}{A + 0.27}$ where $E = 0.35(e_a - e_d)(1 + 0.0098w_2)$ $H = R(1-r)(0.18 + 0.55S) - B(0.56 - 0.092e_d^{0.5})(0.10 + 0.90S)$	(11-24)
Blaney-Criddle [143]	1950	m months	Inches	$U = k \sum_{1}^{m} pt = kF$ where $F = \sum_{1}^{m} pt$	(11-25)
Halkias-Veihmeyer-Hendrickson [27]	1955	Monthly	Inches	$U = SD$	(11-27)
Hargreaves [144]	1956	m months	Inches	$U = \sum_{1}^{m} kd(0.38 - 0.0038h)(t - 32)$	(11-26)

NOTATION

A = slope of saturated-vapor-pressure curve of air at absolute temperature in °F, or de_a/dt in mm Hg/°F (Fig. 11-8)
B = a coefficient depending on temperature (Table 11-7)
D = difference in evaporation between white and black atmometers in cm³
d = monthly daytime coefficient dependent upon latitude (Table 11-9)
e_a = saturation vapor pressure at mean air temperature in mm Hg (Fig. 11-7)
e_d = saturation vapor pressure at mean dew point (i.e., actual vapor pressure in the air) in mm Hg, being equal to e_a multiplied by relative humidity in per cent
E = daily evaporation in mm
h = mean monthly relative humidity at noon, in Eq. (11-26), or annual mean relative humidity, in per cent, in Eq. (11-22)
H = accumulated degree-days above minimum growing temperature for growing season, in Eq. (11-20); or accumulated degree-days of maximum daily temperature above 32°F for growing season, in Eq. (11-21); or daily heat budget at surface in mm of water, in Eq. (11-24)
k = annual, seasonal, or monthly consumptive-use coefficient
p = per cent of daytime hours of the year, occurring during the period, divided by 100 (Table 11-4)
r = estimated percentage of reflecting surface
R = mean monthly extraterrestrial radiation in mm of water evaporated per day (Table 11-6)
S = estimated ratio of actual duration of bright sunshine to maximum possible duration of bright sunshine; or slope of regression line between D and U in Eq. (11-27)
TE = Thornthwaite's *temperature-efficiency index*, being equal to the sum of 12 monthly values of *heat index* $i = (t/5)^{1.514}$, where t is mean monthly temperature in °C
t = mean monthly temperature in °F, in Eqs. (11-22), (11-25), and (11-26), or in °C in Eq. (11-23)
U = evapotranspiration or consumptive use for given period
w_2 = mean wind velocity at 2 m above the ground in miles/day, or equal to w_1 (log 6.6/log h), where w_1 is measured wind velocity in miles/day at height h in ft

EVAPOTRANSPIRATION   **11-27**

Table 11-3. Seasonal Consumptive-use Coefficients k in Blaney-Criddle and Blaney-Morin Equations, for Irrigated Crops in Western United States*

Crop	Length of growing season or period	k
Alfalfa	Between frosts	0.80–0.85
Beans	3 months	0.60–0.70
Corn	4 months	0.75–0.85
Cotton	7 months	0.65–0.75
Flax	7–8 months	0.80
Grains, small	3 months	0.75–0.85
Sorghums	4–5 months	0.70
Orchard, citrus	7 months	0.50–0.65
Walnuts	Between frosts	0.70
Deciduous	Between frosts	0.60–0.70
Pasture, grass	Between frosts	0.75
Ladino clover	Between frosts	0.80–0.85
Potatoes	3½ months	0.65–0.75
Rice	3–5 months	1.00–1.20
Sugar beets	6 months	0.65–0.75
Tomatoes	4 months	0.70
Vegetables, small	3 months	0.60

* After Blaney [145] and Criddle [138].
NOTE: The lower values of k are for coastal areas; the higher values, for areas with an arid climate. See Table 21-1 for maximum monthly values and Table 21-2 for monthly values.

the time and place where the estimate of evapotranspiration is to be made. The value of p can be computed from the U.S. Weather Bureau's "sunshine tables" [146], as shown in Table 11-4.

The *Thornthwaite equation* [Eq. (11-23)] is based on an exponential relationship between mean monthly temperature and mean monthly consumptive use, the same factors elements used by Blaney and Criddle. The relationship is based largely on experience in the central and eastern United States. The temperature-efficiency

Table 11-4. Daytime-hours Percentages, or $100p$, in Blaney-Criddle and Blaney-Morin Equations
(Annual value of $p = 1.00$)

Latitude, deg	J	F	M	A	M	J	J	A	S	O	N	D
North:												
60	4.67	5.65	8.08	9.65	11.74	12.39	12.31	10.70	8.57	6.98	5.04	4.22
50	5.98	6.30	8.24	9.24	10.68	10.91	10.99	10.00	8.46	7.45	6.10	5.65
40	6.76	6.72	8.33	8.95	10.02	10.08	10.22	9.54	8.39	7.75	6.72	6.52
35	7.05	6.88	8.35	8.83	9.76	9.77	9.93	9.37	8.36	7.87	6.97	6.86
30	7.30	7.03	8.38	8.72	9.53	9.49	9.67	9.22	8.33	7.99	7.19	7.15
25	7.53	7.14	8.39	8.61	9.33	9.23	9.45	9.09	8.32	8.09	7.40	7.42
20	7.74	7.25	8.41	8.52	9.15	9.00	9.25	8.96	8.30	8.18	7.58	7.66
15	7.94	7.36	8.43	8.44	8.98	8.80	9.05	8.83	8.28	8.26	7.75	7.88
10	8.13	7.47	8.45	8.37	8.81	8.60	8.86	8.71	8.25	8.34	7.91	8.10
0	8.50	7.66	8.49	8.21	8.50	8.22	8.50	8.49	8.21	8.50	8.22	8.50
South:												
10	8.86	7.87	8.53	8.09	8.18	7.86	8.14	8.27	8.17	8.62	8.53	8.88
20	9.24	8.09	8.57	7.94	7.85	7.43	7.76	8.03	8.13	8.76	8.87	9.33
30	9.70	8.33	8.62	7.73	7.45	6.96	7.31	7.76	8.07	8.97	9.24	9.85
40	10.27	8.63	8.67	7.49	6.97	6.37	6.76	7.41	8.02	9.21	9.71	10.49

index, TE, an integral element of Thornthwaite's classification of climate [147], is the sum of 12 monthly values of the *heat index* $i = (t/5)^{1.514}$, where t is the mean monthly temperature in degrees centigrade. Equation (11-23), however, gives only unadjusted rates of potential evapotranspiration. Since the number of days in a month varies from 28 to 31 and since the number of hours in the day between the onset of evapotranspiration in the morning and its termination in the evening varies with the season and with latitude, it becomes necessary to reduce or increase the unadjusted rates by a factor which varies with the month and the latitude. These factors are given in Table 11-5. Hargreaves [144] found that computation of consumptive use by sugar cane requires factors other than mean temperatures and daylight hours.

Table 11-5. Mean Possible Duration Expressed in Units of 30 Days of 12 Hr Each, or the Adjusting Factor for Potential Evapotranspiration Computed by the Thornthwaite Equation*

Latitude, deg	J	F	M	A	M	J	J	A	S	O	N	D
0	1.04	0.94	1.04	1.01	1.04	1.01	1.04	1.04	1.01	1.04	1.01	1.04
10	1.00	0.91	1.03	1.03	1.08	1.06	1.08	1.07	1.02	1.02	0.98	0.99
20	0.95	0.90	1.03	1.05	1.13	1.11	1.14	1.11	1.02	1.00	0.93	0.94
30	0.90	0.87	1.03	1.08	1.18	1.17	1.20	1.14	1.03	0.98	0.89	0.88
35	0.87	0.85	1.03	1.09	1.21	1.21	1.23	1.16	1.03	0.97	0.86	0.85
40	0.84	0.83	1.03	1.11	1.24	1.25	1.27	1.18	1.04	0.96	0.83	0.81
45	0.80	0.81	1.02	1.13	1.28	1.29	1.31	1.21	1.04	0.94	0.79	0.75
50	0.74	0.78	1.02	1.15	1.33	1.36	1.37	1.25	1.06	0.92	0.76	0.70

* After Criddle [138].

The *Penman equation* [Eq. (11-24)] is based on the most complete theoretical approach, showing that evapotranspiration is inseparably connected to the amount of radiative energy gained by the surface. This equation has been widely used in England and to some extent in Australia and the eastern part of the United States. The values of e_a and A can be obtained from Figs. 11-7 and 11-8, respectively. The value of R and B can be obtained from Tables 11-6 and 11-7, respectively.

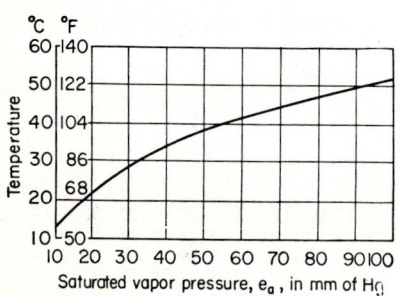

Fig. 11-7. Temperature vs. saturated vapor pressure. (*Criddle* [138].)

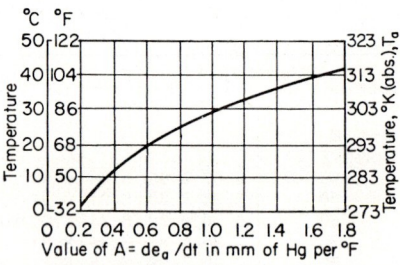

Fig. 11-8. Temperature versus A for use in the Penman equation [Eq. (11-24)]. (*Criddle* [138].)

The *Hargreaves equation* [Eq. (11-26)] is based on the assumptions that evaporation is a physical process and can be computed and that an empirical relationship exists between computed evaporation and consumptive use by various crops. The consumptive-use coefficient k depends on the individual crops grown. The values are given in Table 11-8. The monthly daytime coefficient d depends on latitude. The values are given in Table 11-9.

EVAPOTRANSPIRATION

Table 11-6. Midmonthly Intensity of Solar Radiation on a Horizontal Surface, in Millimeters of Water Evaporated per Day, or the Value R in the Penman Equation*

Latitude, deg	J	F	M	A	M	J	J	A	S	O	N	D
North												
90	....	....	....	7.9	14.9	18.1	16.8	11.2	2.6	....	....	....
80	....	....	1.8	7.8	14.6	17.8	16.5	10.6	4.0	0.2	....	....
70	....	1.1	4.3	9.1	13.6	17.0	15.8	11.4	6.8	2.4	0.1	....
60	1.3	3.5	6.8	11.1	14.6	16.5	15.7	12.7	8.5	4.7	1.9	0.9
50	3.6	5.9	9.1	12.7	15.4	16.7	16.1	13.9	10.5	7.1	4.3	3.0
40	6.0	8.3	11.0	13.9	15.9	16.7	16.3	14.8	12.2	9.3	6.7	5.5
30	8.5	10.5	12.7	14.8	16.0	16.5	16.2	15.3	13.5	11.3	9.1	7.9
20	10.8	12.3	13.9	15.2	15.7	15.8	15.7	15.3	14.4	12.9	11.2	10.3
10	12.8	13.9	14.8	15.2	15.0	14.8	14.8	15.0	14.9	14.1	13.1	12.4
0	14.5	15.0	15.2	14.7	13.9	13.4	13.5	14.2	14.9	15.0	14.6	14.3
South												
10	15.8	15.7	15.1	13.8	12.4	11.6	11.9	13.0	14.4	15.3	15.7	15.8
20	16.8	16.0	14.6	12.5	10.7	9.6	10.0	11.5	13.5	15.3	16.4	16.9
30	17.3	15.8	13.6	10.8	8.7	7.4	7.8	9.6	12.1	14.8	16.7	17.6
40	17.3	15.2	12.2	8.8	6.4	5.1	5.6	7.5	10.5	13.8	16.5	17.8
50	17.1	14.1	10.5	6.6	4.1	2.8	3.3	5.2	8.5	12.5	16.0	17.8
60	16.6	12.7	8.4	4.3	1.9	0.8	1.2	2.9	6.2	10.7	15.2	17.5
70	16.5	11.2	6.1	1.9	0.1	....	....	0.8	3.8	8.8	14.5	18.1
80	17.3	10.5	3.6	....	....	....	....	....	1.3	7.1	15.0	18.9
90	17.6	10.7	1.9	....	....	....	....	....	....	7.0	15.3	19.3

* After Criddle [138].
NOTE: Values from a table by Shaw [148] multiplied by 0.86 and divided by 59 give the radiation in millimeters of water per day.

Table 11-7. Values of B in the Penman Equation*

T_a, °abs.	B, mm water/day	T_a, °F	B, mm water/day
270	10.73	35	11.48
275	11.51	40	11.96
280	12.40	45	12.45
285	13.20	50	12.94
290	14.26	55	13.45
295	15.30	60	13.96
300	16.34	65	14.52
305	17.46	70	15.10
310	18.60	75	15.65
315	19.85	80	16.25
320	21.15	85	16.85
325	22.50	90	17.46
		95	18.10
		100	18.80

* After Criddle [138].
NOTE: $B = \sigma T_a^4$, where σ is Boltzmann constant, or 2.01×10^{-9} mm/day, and T_a is the air temperature. Heat of evaporation was assumed to be constant at 500 ca/g of water.

Table 11-8. Consumptive-use Coefficients k in the Hargreaves Equation*

Crop	Mar.	Apr.	May	June	July	Aug.	Sept.	Oct.	Nov.	Seasonal
At Davis, Calif.										
Alfalfa	0.41	0.70	0.64	0.67	0.74	0.67	0.64	0.40	0.41	0.59
Almonds	0.16	0.36	0.34	0.52	0.48	0.34	0.29	0.48	0.21	0.36
Asparagus	0.16	0.11	0.12	0.18	0.46	0.81	0.84	0.99	0.51	0.46
Beans, Lima	....	....	....	0.41	0.51	0.61	0.32	....	....	0.46
Beans	....	....	....	0.15	0.28	0.66	0.51	....	....	0.40
Cantaloupes	....	....	0.24	0.31	0.37	0.61	0.38	....	....	0.48
Carrots	0.16	0.18	0.19	0.52	0.64	0.28	....	....	....	0.33
Celery	....	....	....	0.15	0.14	0.25	0.45	0.70	0.85	0.42
Citrus	0.41	0.36	0.44	0.43	0.44	0.41	0.41	0.64	0.41	0.44
Corn	....	....	....	0.12	0.38	0.42	0.26	0.10	....	0.26
Fruit, deciduous	0.14	0.45	0.49	0.74	0.71	0.55	0.43	0.36	....	0.48
Grain, sorghums	....	....	0.07	0.30	0.39	0.30	0.15	....	....	0.24
Grain and hay	0.50	0.75	0.58	0.12	....	....	....	....	....	0.49
Grapes, Muscat	....	0.13	0.24	0.26	0.31	0.26	0.26	0.18	....	0.23
Hops	....	0.07	0.12	0.31	0.61	0.61	0.38	....	....	0.35
Ladino clover	0.50	0.81	0.55	0.77	0.83	0.76	0.70	0.44	....	0.67
Onions, early	0.28	0.45	0.30	....	....	....	....	....	....	0.34
late	0.28	0.45	0.30	0.31	0.28	....	....	....	....	0.32
Pasture	0.11	0.25	0.29	0.33	0.31	0.32	0.32	0.22	0.14	0.25
Peaches	0.22	0.45	0.43	0.46	0.51	0.51	0.38	0.60	0.41	0.44
Peas	0.28	0.36	0.49	0.31	....	....	....	....	....	0.36
Potatoes, early	0.55	0.72	0.73	0.62	....	....	....	....	....	0.66
Prunes	0.17	0.34	0.34	0.50	0.48	0.32	0.42	0.48	0.24	0.37
Rice	....	0.32	1.34	1.42	1.40	1.44	0.51	....	....	1.07
Sudan grass	....	....	0.24	0.33	0.37	0.35	0.28	0.24	....	0.30
Sugar beets	0.19	0.27	0.55	0.87	0.69	0.36	0.15	0.10	0.03	0.36
Tomatoes	....	....	....	0.32	0.41	0.71	0.67	0.81	....	0.58
Walnuts	....	....	0.36	0.43	0.57	0.67	0.63	0.26	0.36	0.24	0.44
Watermelons	....	....	....	0.15	0.18	0.25	0.51	....	....	0.27

At Caribbean Area

Crop	Jan.	Feb.	Mar.	Apr.	May	June	July	Aug.	Sept.	Oct.	Nov.	Dec.	Seasonal
Sugar cane (planted in March)	0.77	0.69	0.49	0.52	0.53	0.56	0.59	0.73	0.85	0.84	0.91	0.86	0.70
Bananas	0.86	0.85	0.73	0.88	0.85	0.86	0.85	0.78	0.88	0.86	0.76	0.78	0.83

* After Criddle [138].

Table 11-9. Monthly Daytime Coefficients d in the Hargreaves Equation*

N. latitude, deg	J	F	M	A	M	J	J	A	S	O	N	D
5	1.01	0.91	1.02	0.99	1.03	1.00	1.03	1.03	0.98	1.02	0.98	1.00
10	0.98	0.89	1.02	1.01	1.05	1.03	1.06	1.05	0.99	1.00	0.95	0.97
15	0.96	0.88	1.01	1.01	1.08	1.06	1.08	1.06	0.99	0.99	0.93	0.95
20	0.93	0.87	1.01	1.02	1.10	1.08	1.11	1.08	0.99	0.98	0.91	0.92
25	0.91	0.86	1.01	1.03	1.12	1.11	1.13	1.09	1.00	0.97	0.89	0.89
30	0.88	0.84	1.00	1.05	1.14	1.14	1.16	1.11	1.00	0.96	0.86	0.86
35	0.85	0.83	1.00	1.06	1.17	1.17	1.19	1.12	1.00	0.94	0.84	0.82
40	0.81	0.81	1.00	1.08	1.20	1.21	1.23	1.14	1.01	0.93	0.81	0.78
45	0.77	0.79	0.99	1.09	1.24	1.26	1.27	1.17	1.01	0.91	0.77	0.74
50	0.72	0.76	0.99	1.11	1.28	1.32	1.32	1.20	1.01	0.89	0.73	0.68

* After Criddle [138].

C. Evaporation as an Index of Evapotranspiration

Observations have shown that evaporation is a satisfactory index of evapotranspiration [149]. In fact, many evapotranspiration equations were evolved, either knowingly or not, from evaporation equations.

The pan evaporation has been found to be a good index to evapotranspiration [150–152]. The black Bellani plate atmometer is an instrument for measuring the drying ability of the air, or the *latent evaporation*. Robertson and Holmes [152] suggested a maximum factor of about 0.0034 in./cm³ to change the latent evaporation to evapotranspiration.

The difference in evaporation D between black and white atmometers gives a measure of the intensity of radiation. Since solar radiation is an important factor affecting transpiration, Halkias and others [27] found correlation coefficients of 0.95 or better for the correlation between the evaporation difference D and the monthly evapotranspiration U. Hence the following correlation equation was proposed:

$$U = SD \tag{11-27}$$

where S represents the slope of the regression line between U in in. and D in cm³. The value of S seemed to be governed by the ground coverage afforded by the crop, rather than by the kind and height of the plants. Similar conclusions have also been reached by others [153, 154]. Table 11-10 lists the values of S and the correlation coefficients

Table 11-10. Coefficients S in the Equation $U = SD$ and the Correlation Coefficients*

Kind of crop	Coefficient S	Correlation coefficient
Alfalfa	0.0134	0.99
Walnuts	0.0135	0.97
Apricots	0.0120	0.95
Peaches	0.0110	0.98
Prunes	0.0108	0.98
Cotton	0.0105	0.97
Sugar beets	0.0096	0.99
Grapes	0.0086	0.98
Tomatoes	0.0082	0.98
Artichokes	0.0073	0.98

* After Halkias and others [27].

for several kinds of crop. The correlation coefficients were determined from the equation derived for small samples. The value of S holds for any time and any place if the soil moisture is above the permanent wilting percentage. Since D is highly correlated to solar-energy radiation [27], it can be expressed in terms of the Eppley pyrheliometer readings Q_s in gram-calories per square centimeters. Thus, the above equation becomes $U = 0.028 S Q_s$.

Analysis of the data on consumptive use indicated a high degree of correlation between U.S. Weather Bureau pan evaporation and consumptive use [155]. The equation of correlation developed by the method of least squares is

$$U = k(E + 2.70) \tag{11-28}$$

where U is the monthly consumptive use of water by the crop, k is a monthly consumptive-use coefficient, depending primarily upon the extent of ground coverage by the crop, and E is the monthly evaporation in inches.

D. Consumptive-use Requirements

Based on studies of weather and crop data from irrigated areas over the western United States, Munson [139] found that the following monthly P/E ratios (precipitation-evaporation ratios) hold adequately for normal plant growth:

	J	F	M	A	M	J	J	A	S	O	N	D
P/E	1.0	1.8	3.2	4.4	5.8	6.0	6.8	6.1	4.6	3.5	2.3	1.5

Table 11-11. Examples of Measured Monthly Consumptive Use (Evapotranspiration) by Irrigated Crops during Irrigation Season in Western United States*

Location	Crop	Consumptive use (evapotranspiration), in.							Authority
		Apr.	May	June	July	Aug.	Sept.	Oct.	Total	
Arizona, mesa...	Alfalfa	5.0	6.5	9.0	12.0	10.0	6.0	4.0	52.5	Harris
	Dates	6.2	7.6	8.3	9.2	8.4	7.2	5.7	52.6	Harris
California:										
Los Angeles†..	Lemons	2.1	2.6	3.3	3.9	3.7	3.4	2.8	21.8	Blaney
	Oranges	2.2	2.2	3.1	3.4	3.7	3.1	2.9	20.6	Blaney
	Walnuts	3.8	5.0	5.9	6.1	5.0	2.8	2.0	30.6	Blaney
	Alfalfa	3.3	6.7	5.4	7.8	4.2	5.6	4.4	37.4	Blaney
Coastal.......	Alfalfa	4.9	4.9	4.3	5.2	5.9	5.5	4.7	35.4	
Ontario.......	Peaches	1.0	3.5	6.7	8.0	6.5	2.7	1.4	29.8	Blaney
Shafter.......	Cotton	0.5	1.0	4.0	8.5	9.7	5.8	3.2	32.7	Beckett
Firebaugh....	Cotton	...	0.8	1.1	7.3	7.8	3.6	2.0	22.6	Adams
	Cotton	...	0.4	0.7	8.4	9.5	3.0	2.5	24.5	Adams
Delta‡.......	Alfalfa	3.6	4.8	6.0	7.8	6.6	6.0	1.2	36.0	Mathew
	Potatoes	...	1.8	4.6	6.2	3.6	1.8	...	18.0	Mathew
	Truck	1.2	3.0	6.0	5.4	5.4	3.6	1.8	26.4	Mathew
	Sugar beets	1.6	3.8	6.1	7.3	6.4	2.4	...	27.6	Mathew
	Beans	1.9	2.4	1.7	2.9	6.9	4.4	...	20.2	Mathew
	Fruit	2.2	3.8	6.0	6.8	4.8	2.8	0.8	27.2	Mathew
	Onions	1.6	3.2	5.9	5.2	2.4	1.9	...	19.8	Mathew
Davis........	Sugar beets	...	5.2	5.7	7.1	5.8	...	...	23.8	Veihmeyer
	Tomatoes	...	...	3.2	6.2	4.9	4.7	...	22.3	Veihmeyer
	Alfalfa	...	6.8	7.9	8.3	7.1	4.3	...	....	Veihmeyer
	Prunes	...	5.8	6.0	7.6	6.5	5.0	...	....	Veihmeyer
	Peaches	...	5.4	6.4	7.9	7.2	5.0	...	....	Veihmeyer
	Walnuts	...	6.6	6.7	8.4	7.2	4.8	...	....	Veihmeyer
	Grapes	...	4.6	4.9	6.2	5.3	4.3	...	....	Veihmeyer
Winters......	Apricots	...	...	5.6	6.8	6.5	4.9	...	....	Veihmeyer
Nebraska, Scottsbluff....	Alfalfa	1.4	4.0	7.0	7.1	6.4	3.0	...	28.9	Bowen
	Beets	1.9	3.3	5.2	6.9	5.8	1.1	...	24.2	Bowen
	Potatoes	...	...	...	3.4	5.8	4.4	...	....	Bowen
	Oats	...	3.0	6.1	5.1	....	....	...	14.2	Bowen

* After H. F. Blaney [147].
† In San Fernando Valley, City of Los Angeles, Calif.
‡ In Sacramento–San Joaquin Delta, Calif.

For a given month, the corresponding value of the P/E ratio and the average monthly temperature t in °F can be substituted in the following equation to solve P:

$$P = 0.014(t-10) \left(\frac{P}{E}\right)^{0.9} \tag{11-29}$$

This equation is based on a formula derived by Thornthwaite [147], in which P is the required monthly precipitation, or actually the consumptive-use requirement, in inches, for crop production on an area basis. The sum of the 12 P/E ratios is known as the PE index. The method is therefore called the PE-index method. It gives the monthly, hence annual, consumptive-use requirements directly. As is the case with any theoretical approach, however, it is necessary to exercise judgment in the use of this method.

The monthly water use by several crops as measured by soil-moisture sampling at Davis, Calif., was compared by tabulation with the result obtained by the PE-index method and with those by other methods, including Blaney-Criddle, Lowry-Johnson, Thornthwaite, and atmometers [139]. Such information should be of use to those who are interested in comparing the various methods.

Data for consumptive-use requirements have been developed in many studies as reported particularly in Refs. 2, 3, 93, and 156 to 158. Table 11-11 provides some examples of measured consumptive use by irrigated crops during the irrigation season in western United States. Although such information usually refers to a certain specific condition, it is useful as a guide to practicing hydrologists and engineers in planning and designing water-use projects.

VI. REFERENCES

1. "Hydrology Handbook," American Society of Civil Engineers, Manuals of Engineering Practice, no. 28, 1949.
2. Consumptive use of water: a symposium, *Trans. Am. Soc. Civil Engrs.*, vol. 117, pp. 948–1023, 1952. Includes H. F. Blaney, Definitions, methods, and research data, pp. 949–967; L. R. Rich, Forest and range vegetation, pp. 974–987; W. D. Criddle, Irrigated crops, pp. 991–1000; G. B. Gleason, Municipal and industrial areas, pp. 1004–1009; R. L. Lowry, Special case in Rio Grande Basin, pp. 1014–1020; and discussions.
3. Lee, C. H., and others: Report of the Committee on Absorption and Transpiration, 1933–34, *Trans. Am. Geophys. Union*, vol. 15, pt. II, pp. 295–296, 1934.
4. Dalton, John: Experimental essays on the constitution of mixed gases; on the force of steam or vapor from water and other liquids in different temperatures, both in a Torricellian vacuum and in air; on evaporation; and on the expansion of gases by heat, *Manchester Lit. Phil. Soc. Mem. Proc.*, vol. 5, pp. 536–602, 1802.
5. FitzGerald, Desmond: Evaporation, *Trans. Am. Soc. Civil Engrs.*, vol. 15, pp. 581–646, 1886.
6. Meyer, A. F.: Computing run-off from rainfall and other physical data, *Trans. Am. Soc. Civil Engrs.*, vol. 79, pp. 1056–1155, 1915.
7. Horton, R. E.: A new evaporation formula developed, *Eng. News-Rec.*, vol. 78, no. 4, pp. 196–199, Apr. 26, 1917.
8. Rohwer, Carl: Evaporation from free water surfaces, *U.S. Dept. Agr. Tech. Bull.* 271, 1931.
9. Harbeck, G. E., Jr., and others: Water-loss investigations, vol. 1, Lake Hefner Studies, Technical Report, *U.S. Geol. Surv. Paper* 269, 1954. Previously published as *U.S. Geol. Surv. Circ.* 229, 1952, as part of *U.S. Navy Electron. Lab. Rept.* 237.
10. Harbeck, G. E., Jr., M. A. Kohler, G. E. Koberg, and others: Water-loss investigations, Lake Mead Studies, *U.S. Geol. Surv. Profess. Paper* 298, 1958.
11. Rohwer, Carl: Evaporation from salt solutions and from oil-covered water surfaces, *J. Agr. Res.*, vol. 46, pp. 715–729, 1933.
12. Lee, C. H.: Discussion of Evaporation on United States reclamation projects by I. E. Houk, *Trans. Am. Soc. Civil Engrs.*, vol. 90, pp. 330–343, 1927.
13. Adams, T. C.: Evaporation from Great Salt Lake, *Am. Meteorol. Soc. Bull.*, vol. 15, no. 2, pp. 35–39, 1934.
14. Stefan, J.: Über die Verdampfung aus einer Kreisförmig oder elliptisch begrenzten Becken, *Sitzber. K. Akad. Wiss. Wien*, vol. 83, pp. 943–954, 1881.

15. Humphreys, W. J.: "Physics of the Air," 3d ed., McGraw-Hill Book Company, Inc., New York, 1940.
16. Brown, H. T., and F. Escombe: Static diffusion of gases and liquids in relation to the assimilation of carbon and translocation in plants, *Phil. Trans. Roy. Soc. London*, ser. B., vol. 193, pp. 223–291, 1900.
17. Instructions for the installation and operation of Class A evaporation stations, *U.S. Weather Bur. Circ.* L, 1919.
18. Instructions for climatological observers, *U.S. Weather Bur., Circ.* B, 10th ed., November, 1952.
19. Evaporation from water surfaces: a symposium, *Trans. Am. Soc. Civil Engrs.*, vol. 99, pp. 671–718, 1934. Includes Carl Rohwer, Evaporation from different types of pans, pp. 673–703; Robert Follansbee, Evaporation from reservoir surfaces, pp. 704–715, 719–747; F. C. Scobey, I. E. Houk, and R. L. Parshall, Standard equipment for evaporation stations, final report of Subcommittee on Evaporation of the Special Committee on Irrigation Hydraulics, pp. 716–718.
20. Rohwer, Carl: Evaporation from free water surfaces, *U.S. Dept. Agr. Tech. Bull. Relating to Irrigation* 271, 1931.
21. Horton, R. E.: Evaporation maps of the United States, *Trans. Am. Geophys. Union*, vol. 24, pt. II, pp. 743–753, 1943.
22. Kohler, M. A., T. J. Nordenson, and D. R. Baker: Evaporation maps for the United States, *U.S. Weather Bur. Tech. Paper* 37, 1959.
23. Hickox, G. H.: Evaporation from a free water surface, *Trans. Am. Soc. Civil Engrs.*, vol. III, pp. 1–33, 1946. Appendix I contains a good bibliography on evaporation. Discussions by Carl Rohwer, C. W. Thornthwaite, H. F. Blaney, A. A. Young, A. A. Kalinske, A. F. Meyer, M. L. Albertson, and G. H. Hickox, pp. 34–66.
24. Penman, H. L.: Natural evaporation from open water, bare soil, and grass, *Proc. Roy. Soc. London*, ser. A, vol. 193, pp. 120–145, 1948.
25. Kohler, M. A., T. J. Nordenson, and W. E. Fox: Evaporation from pans and lakes, *U.S. Weather Bur., Res. Paper* 38, May, 1955.
26. Livingston, B. E.: Atmometers of porous porcelain and paper: their use in physiological ecology, *Ecology*, vol. 16, pp. 438–472, 1935.
27. Halkias, N. A., F. J. Veihmeyer, and A. H. Hendrickson: Determining water needs for crops from climatic data, *Hilgardia*, vol. 24, pp. 207–233, December, 1955.
28. O'Connor, T. C.: On the measurement of global radiation using black and white atmometers, *Geofisica Pura Appl.*, Milan, vol. 30, pp. 130–136, 1955.
29. Livingston, B. E., and F. Haasis: The measurement of evaporation in freezing weather, *J. Ecol.*, London, vol. 17, pp. 315–328, 1929.
30. Horton, R. E.: Hydrologic interrelations between lands and oceans, *Trans. Am. Geophys. Union*, vol. 24, pt. II, pp. 753–764, 1943.
31. Langbein, W. B., C. E. Hains, and R. C. Culler: Hydrology of stock-water reservoir in Arizona, *U.S. Geol. Surv. Circ.* 110, 1951.
32. Schmidt, W.: Strahlung und Verdunstung in freien Wasserflächen: ein Beitrage zum Wärmehaushalt des Weltmeers und zum Wasserhaushalt der Erde, *Ann. Hydrographie Maritimem Meteorol.*, vol. 43, pp. 111–124, 169–178, 1915.
33. Sverdrup, H. U.: On the annual and diurnal variation of the evaporation from the oceans, *J. Marine Res.*, vol. 3, no. 2, pp. 93–104, 1940.
34. Richardson, Burt: Evaporation as a function of insolation, *Trans. Am. Soc. Civil Engrs.*, vol. 95, pp. 996–1011, 1931.
35. Cummings, N. W.: Evaporation from water surfaces: status of present knowledge and need for further investigations, *Trans. Am. Geophys. Union*, vol. 16, pt. 2, pp. 507–509, 1935.
36. Holzman, Benjamin: The heat-balance method for the determination of evaporation from water surfaces, *Trans. Am. Geophys. Union*, vol. 22, pt. 3, pp. 655–659, 1941.
37. Penman, H. L.: Evaporation in nature, *Phys. Soc. London Repts. on Progr. in Phys.*, vol. 11, pp. 366–388, 1946–1947.
38. Bowen, I. S.: The ratio of heat losses by conduction and by evaporation from any water surface, *Phys. Rev.*, ser. 2, vol. 27, pp. 779–787, June, 1926.
39. Dunkle, R. V., and others: Non-selective radiometers for hemispherical irradiation and net radiation interchange measurements, *Univ. California, Eng. Dept., Thermal Radiation Project Rept.* 9, Oct. 1, 1949.
40. Mosby, H.: Verdunstung und Strahlung auf dem Meere, *Ann. Hydrographie Maritimen Meteorol.*, vol. 64, p. 281, 1936.
41. Kennedy, R. E.: Computation of daily insolation energy, *Am. Meteorol. Soc. Bull.*, vol. 30, no. 6, pp. 208–213, June, 1949.
42. Anderson, E. R., J. L. Anderson, and J. J. Marciano: A review of evaporation theory

and development of instrumentation, *U.S. Navy Electron. Lab. Rept.* 159, February, 1950.
43. Thornthwaite, C. W., and Benjamin Holzman: The determination of evaporation from land and water surfaces, *Monthly Weather Rev.*, vol. 67, no. 1, pp. 4–11, 1939.
44. Sverdrup, H. U.: The humidity gradient over the sea surface, *J. Meteorol.*, vol. 3, no. 1, pp. 1–8, March, 1946.
45. Sutton, O. G.: The application to micrometeorology of the theory of turbulent flow over rough surfaces, *Roy. Meteorol. Soc. Quart. J.*, vol. 75, no. 326, pp. 335–350, October, 1949.
46. Magin, G. B., and L. E. Randall: Review of literature on evaporation suppression, *U.S. Geol. Surv. Profess. Paper* 272-C, 1960.
47. "Water-loss Investigations: Lake Hefner 1958 Evaporation Reduction Investigations," U.S. Bureau of Reclamation, Denver, Colo., June, 1959. Report by the collaborators: City of Oklahoma City, Okla., Oklahoma State Department of Health, U.S. Public Health Service, U.S. Weather Bureau, U.S. Geological Survey, and U.S. Bureau of Reclamation.
48. 1960 evaporation reduction studies at Sahuaro Lake, Arizona, and 1959 monolayer behavior studies at Lake Mead, Arizona-Nevada, and Sahuaro Lake, Arizona, *U.S. Bur. Reclamation Chem. Eng. Lab. Rept.* SI-32, Aug. 15, 1961.
49. Cruse, R. R., and G. E. Harbeck, Jr.: Evaporation control research, 1955–58, *U.S. Geol. Surv. Water-Supply Paper* 1480, 1960.
50. La Mer, V. K. (ed.): "Retardation of Evaporation by Monolayers: Transport Processes," Academic Press Inc., New York, 1962.
51. Beadle, B. W., and R. R. Cruse: Water conservation through control of evaporation, *Am. Water Works Assoc. J.*, vol. 49, pp. 397–404, 1957.
52. Eaton, E. D.: Control of evaporation losses, interim report, Senate Committee on Interior and Insular Affairs, Print no. 1, 85th Congr., 2d Sess., 1958.
53. Freese, S. W.: Reservoir evaporation control by other techniques, *Proc. First Intern. Conf. on Reservoir Evaporation Control*, Apr. 14, 1956, Southwest Research Institute, San Antonio, Tex.
54. Arthur, G. B.: The water we do not use, *Public Works*, vol. 88, no. 9, pp. 134–136, 176, 178, 180, 182, 184, 185, 1957.
55. Dressler, R. G.: The southwest cooperative research project on reservoir evaporation control, *Proc. First Intern. Conf. on Reservoir Evaporation Control*, Apr. 14, 1956, Southwest Research Institute, San Antonio, Tex.
56. Fleming, Roscoe: The problem of water, in "Britannica Book of the Year," Encyclopaedia Britannica, Inc., Chicago, 1957, pp. 1–32.
57. Robinson, T. W.: Phreatophytes, *U.S. Geol. Surv. Water-Supply Paper* 1423, 1958.
58. Pockels, Agnes: Surface tension, *Nature*, vol. 43, pp. 437–439, 1891.
59. Rayleigh, Lord: Investigation in capillarity, *Phil. Mag.*, 5th ser., vol. 48, pp. 321–337, 1899.
60. Du Nouy, P. L.: Concerning the rate of evaporation of water through oriented monolayers on water, *Science*, vol. 99, no. 2575, p. 365, 1944.
61. Du Nouy, P. L.: Further evidence indicating the existence of a superficial polarized layer of molecules at certain dilutions, *J. Exptl. Med.*, vol. 39, p. 717, 1924.
62. Hedestrand, G.: The influence of thin surface films on the evaporation of water, *J. Phys. Chem.*, vol. 28, pp. 1245–1252, 1924.
63. Rideal, E. K.: The influence of thin surface films on the evaporation of water, *J. Phys. Chem.*, vol. 29, pp. 1585–1588, 1925.
64. Langmuir, Irving, and D. B. Langmuir: The effect of monomolecular films on the evaporation of ether solutions, *J. Phys. Chem.*, vol. 31, pp. 1719–1731, 1927.
65. Mansfield, W. W.: Effect of surface films on the evaporation of water, *Nature*, vol. 172, p. 247, 1953.
66. Mansfield, W. W.: Summary of field trials on the use of cetyl alcohol to restrict evaporation from open storages during the season 1954–55, *Commonwealth Sci. Ind. Res. Org., Div. Ind. Chem.*, serial 74, Melbourne, 1955.
67. Moran, W. T., and W. U. Garstka: The reduction of evaporation through the use of monomolecular films, *Third Intern. Comm. on Irrigation and Drainage Rept.* 29, San Francisco, Calif., 1957, pp. 9–468 to 9–481.
68. Roberts, W. J.: Reduction of transpiration, *J. Geophys. Res.*, vol. 66, no. 10, pp. 3309–3312, October, 1961.
69. Roberts, W. J.: Evaporation suppression from water surfaces, *Trans. Am. Geophys. Union*, vol. 38, no. 5, pp. 740–744, October, 1957.
70. Roberts, W. J.: Reducing lake evaporation in the Midwest, *J. Geophys. Res.*, vol. 64, no. 10, pp. 1605–1610, October, 1959.

71. Edlefsen, N. E.: A new method of measuring the aqueous vapor pressure of soils, *Soil Sci.*, vol. 38, pp. 29–35, 1934.
72. Edlefsen, N. E., and A. B. C. Anderson: Thermodynamics of soil moisture, *Hilgardia*, vol. 15, no. 2, pp. 31–298, 1943.
73. Veihmeyer, F. J., and A. H. Hendrickson: The permanent wilting percentage as a reference for the measurement of soil moisture, *Trans. Am. Geophys. Union*, vol. 29, pp. 887–896, 1948.
74. Veihmeyer, F. J., and A. H. Hendrickson: Methods of measuring field capacity and permanent wilting percentage of soils, *Soil. Sci.*, vol. 68, pp. 75–94, 1949.
75. Baver, L. D.: "Soil Physics," 3d ed., John Wiley & Sons, Inc., New York, 1956.
76. Shaw, B. T. (ed.): "Soil Physical Conditions and Plant Growth," *Agronomy*, vol. II, 1952.
77. Kramer, P. J.: "Plant and Soil Water Relationships," McGraw-Hill Book Company, Inc., New York, 1949.
78. King, F. H.: "Soil Management," Orange Judd Co., New York, 1914.
79. Hilgard, E. W., and R. H. Loughridge: Endurance of drought in soils of the arid regions, *Calif. Agr. Expt. Sta. Rept.* 1897–1898, pp. 40–46, 1898.
80. Widtsoe, J. A.: "Principle of Irrigation Practice," The Macmillan Company, New York, 1914.
81. Alway, F. J.: Some studies in the dry land regions, *U.S. Dept. Agr. Bur. Plant Ind. Bull.* 130, 1918.
82. Burr, W. W.: The storage and use of soil moisture, *Nebraska Agr. Expt. Sta. Res. Bull.* 5, 1914.
83. Keen, B. A.: Roots search for water, *Monthly Sci. News*, no. 24, p. 3, 1943.
84. Keen, B. A.: What happens to the rain? *Quart. J. Roy. Meteorol. Soc.*, vol. 65, pp. 123–137, 1939.
85. Veihmeyer, F. J.: Evaporation from soils and transpiration, *Trans. Am. Geophys. Union*, vol. 19, pt. 2, pp. 612–619, 1938.
86. Veihmeyer, F. J.: Some factors affecting the irrigation requirements of deciduous orchards, *Hilgardia*, vol. 2, pp. 125–291, 1927.
87. Fisher, E. A.: Some factors affecting the evaporation of water from soil, II, *J. Agr. Sci.*, vol. 17, pp. 407–419, 1927.
88. Fisher, E. A.: Some fundamental principles of drying, *J. Soc. Chem. Ind. London*, vol. 54, pp. 343–348, 1935.
89. Sherwood, T. K.: Application of theoretical diffusion equations to the drying of solids, *Trans. Am. Inst. Chem. Engrs.*, vol. 27, pp. 190–200, 1932.
90. Keen, B. H., E. M. Crowther, and J. R. H. Coutts: The evaporation of water from soil. III. A critical study of the technique, *J. Agr. Sci.*, vol. 16, pp. 105–122, 1926.
91. Patten, H. E., and F. E. Gallagher: Absorption of vapors and gases by soils, *U.S. Dept. Agr. Bur. Soils Bull.* 51, 1908.
92. Veihmeyer, F. J., and A. H. Hendrickson: Rates of evaporation from wet and dry soils and their significance, *Soil Sci.*, vol. 80, pp. 61–67, 1955.
93. Lee, C. H.: Transpiration and total evaporation, chap. 8 in O. E. Meinzer (ed.): "Hydrology," vol. IX of "Physics of the Earth," McGraw-Hill Company, Inc., New York, 1942; reprinted by Dover Publications, Inc., New York, 1949.
94. Veihmeyer, F. J., and F. A. Brooks: Measurements of cumulative evaporation from bare soil, *Trans. Am. Geophys. Union*, vol. 35, pp. 601–607, 1954.
95. Sewell, M. C.: Tillage: a review of literature, *J. Am. Soc. Agron.*, vol. 11, pp. 269–290, 1919.
96. Chilcott, E. C., and J. S. Cole: Subsoiling, deep tilling, and soil dynamiting in the Great Plains, *J. Agr. Res.*, vol. 14, pp. 481–521, 1918.
97. Call, L. E., and M. C. Sewell: The soil mulch, *J. Am. Soc. Agron.*, vol. 9, pp. 49–61, 1917.
98. Shaw, C. F.: When the soil mulch conserves moisture, *J. Am. Soc. Agron.*, vol. 21, pp. 1165–1171, 1929.
99. Smith, A.: Effect of paper mulches on soil temperature, soil moisture, and yields of certain crops, *Hilgardia*, vol. 6, pp. 159–201, 1931.
100. Lemon, E. R.: The potentialities of decreasing soil moisture evaporation loss, *Proc. Soil Sci. Soc. Am.*, vol. 20, pp. 120–125, 1956.
101. Hendrick, R. M., and D. T. Mowry: Effect of synthetic polyelectrolytes on aggregation, aeration, and water relations of soil, *Soil Sci.*, vol. 73, pp. 427–441, 1952.
102. Hide, J. C.: Observations on factors influencing the evaporation of soil moisture, *Proc. Soil Sci. Soc. Am.*, vol. 18, pp. 234–239, 1954.
103. Veihmeyer, F. J.: Use of water by native vegetation versus grasses and forbes on watersheds, *Trans. Am. Geophys. Union*, vol. 34, pp. 201–212, 1953.

REFERENCES

104. Veihmeyer, F. J., and A. H. Hendrickson: Does transpiration decrease as the soil moisture decreases? *Trans. Am. Geophys. Union*, vol. 36, pp. 425–428, 1955.
105. Bonner, J.: Water-transport: This classical problem in plant physiology is becoming amenable to mathematical analysis, *Science*, vol. 129, pp. 447–450, 1958.
106. Briggs, L. J., and H. L. Shantz: Comparison of the hourly evaporation rate of atmometers and free water surfaces with the transpiration rate of *Medicago sativa, J. Agr. Res.*, vol. 9, pp. 277–292, 1917.
107. Martin, E. V., and F. E. Clements: Studies of the effect of artificial wind on growth and transpiration in *Helianthus annuus, Plant Physiol.*, vol. 10, pp. 613–660, 1935.
108. Giddings, L. A.: Transpiration of *Silphium lacinniatum, Plant World*, vol. 17, pp. 309–328, 1914.
109. Brown, M. A.: The influence of air currents on transpiration, *Proc. Iowa Acad. Sci.*, vol. 17, pp. 13–15, 1910.
110. Stanhill, G.: The effect of differences in soil moisture status on plant growth: a review of analysis of soil moisture regime experiments, *Soil Sci.*, vol. 84, pp. 205–214, 1957.
111. Veihmeyer, F. J., and A. H. Hendrickson: Soil moisture in relation to plant growth, *Ann. Rev. Plant Physiol.*, vol. 1, pp. 285–304, 1950.
112. Veihmeyer, F. J., and A. H. Hendrickson: Some plant and soil moisture relations, *Am. Soil Surv. Assoc. Bull.*, vol. 15, pp. 76–80, 1934.
113. Richards, L. A., and C. H. Wadleigh: Soil water and plant growth, chap. 3 in B. T. Shaw (ed.): "Soil Physical Conditions and Plant Growth," *Agronomy*, vol. II, pp. 75–251, 1952.
114. Penman, H. L.: Evaporation: an introductory survey, *Netherlands J. Agr. Sci.*, vol. 4, pp. 9–29, 1956.
115. Penman, H. L.: The movement and availability of soil waters, *Soils and Fertilizers*, vol. 19, pp. 221–225, 1956.
116. Briggs, L. J., and H. L. Shantz: The water requirement of plants, pt. 2A, Review of the literature, *U.S. Dept. Agr., Bur. Plant Ind. Bull.* 285, 1913.
117. Briggs, L. J., and H. L. Shantz: Relative water requirements of plants, *J. Agr. Res.*, vol. 3, no. 1, pp. 1–63, Oct. 15, 1914.
118. Briggs, L. J., and L. N. Piemeisel: The water requirement of plants at Akron, Colo., *J. Agr. Res.*, vol. 34, no. 12, pp. 1093-1190, 1927.
119. Horton, R. E.: Transpiration by forest trees, *Monthly Weather Rev.*, vol. 51, pp. 571–581, November, 1923.
120. Lyon, T. L., H. O. Buckman, and N. C. Brady: "The Nature and Properties of Soils," 5th ed., The Macmillan Company, New York, 1952, pp. 221–222.
121. Thornthwaite, C. W.: An approach toward a rational classification of climate, *Geograph. Rev.*, vol. 38, pp. 55–94, 1948.
122. Blaney, H. F., C. A. Taylor, and A. A. Young: Rainfall penetration and consumptive use of water in Santa Ana River valley and coastal plain, *California State Div. Water Resources Bull.* 33, Pomona, Calif., 1930.
123. Beckett, S. H., H. F. Blaney, and C. A. Taylor: Irrigation water requirement studies of citrus and avocado trees in San Diego County, California, 1926 and 1927, *Univ. Calif. Agr. Expt. Sta. Bull.* 489, 1930.
124. Young, A. A., and H. F. Blaney: Use of water by native vegetation, *California State Div. Water Resources, Bull.* 50, Sacramento, Calif., 1942.
125. Kittredge, J.: Report of Committee on transpiration and evaporation, 1940-41, *Trans. Am. Geophys. Union*, vol. 22, pp. 906–915, 1941.
126. Pruitt, W. O., and D. E. Angus: Large weighing lysimeter for measuring evapotranspiration, *Trans. Am. Soc. Agr. Engrs.*, vol. 3, pp. 13–18, 1960.
127. Blaney, H. F., P. A. Ewing, O. W. Israelsen, Carl Rohwer, and F. C. Scokey: Water utilization, Upper Rio Grande Basin, pt. III, U.S. National Resources Committee, February, 1938.
128. Blaney, H. F., P. A. Ewing, K. V. Morin, and W. D. Criddle: Consumptive water use and requirements, report of the participating agencies, Pecos River, Joint Investigation of the National Resources Planning Board, Washington, D.C., June, 1942.
129. Lowry, R. L., Jr., and A. F. Johnson: Consumptive use of water for agriculture, *Trans. Am. Soc. Civil Engrs.*, vol. 107, pp. 1243–1266, 1942.
130. Blaney, H. F., C. A. Taylor, M. G. Nickle, and A. A. Young: Water losses under natural conditions from wet areas in Southern California, pt. I, *California State Div. Water Resources Bull.* 44, Pomona, Calif., 1933. Troxell, Harold: *Ibid.*, pt. II.
131. Tanner, C. B.: Energy balance approach to evapotranspiration from crops, *Proc. Soil Sci. Am.*, vol. 24, pp. 1–9, 1960.
132. Levine, G.: Methods of estimating evaporation, *Trans. Am. Soc. Agr. Engrs.*, vol. 2, pp. 32–34, 1959.

133. Rider, N. E.: Water loss from various land surfaces, *Quart. J. Roy. Meteorol. Soc.*, vol. 83, pp. 181–193, 1957.
134. Pasquill, F.: Some estimates of the amount and diurnal variation of evaporation from pasture in fair spring weather, *Proc. Roy. Meteorol. Soc.*, ser. A, vol. 198, pp. 116–140, 1949.
135. Pasquill, F.: Some further considerations on the measurement and direct evaluation of natural evaporation, *Quart. J. Roy. Meteorol. Soc.*, vol. 76, pp. 287–301, 1950.
136. White, W. N.: A method of estimating ground-water supplies based on discharge by plants and evaporation from soil: results of investigation in Escalante Valley, Utah, *U.S. Geol. Surv. Water-Supply Paper* 659-A, 1932.
137. Gatewood, J. S., T. W. Robinson, B. R. Colby, J. D. Hem, and L. C. Halpenny: Use of water by bottom-land vegetation in Lower Safford Valley, Arizona, *U.S. Geol. Surv. Water-Supply Paper* 1103, 1950.
138. Criddle, W. D.: Methods of computing consumptive use of water, *Proc. Am. Soc. Civil Engrs., J. Div. Irrigation and Drainage*, vol. 84, no. IR1, pp. 1–27, January, 1958.
139. Munson, W. C.: Method for estimating consumptive use of water for agriculture, *Trans. Am. Soc. Civil Engrs.*, vol. 127, pt. III, pp. 200–212, 1962.
140. Harding, S. T., and others: Consumptive use of water in irrigation: progress report of the Duty of Water Committee of the Irrigation Division, *Trans. Am. Soc. Civil Engrs.*, vol. 94, pp. 1349–1399, 1930.
141. Blaney, H. F., and K. V. Morin: Evaporation and consumptive use of water empirical formulas, *Trans. Am. Geophys. Union*, vol. 23, pt. I, pp. 76–83, 1942.
142. Thornthwaite, C. W., H. G. Wilm, and others: Report of the Committee on Transpiration and Evaporation, 1943–44, *Trans. Am. Geophys. Union*, vol. 25, pt. V, pp. 683–693, 1944.
143. Blaney, H. F., and W. D. Criddle: Determining water requirements in irrigated areas from climatological and irrigation data, *U.S. Dept. Agr., Div. Irrigation and Water Conserv.*, SCS TP-96, August, 1950.
144. Hargreaves, G. H.: Irrigation requirements based on climatic data, *Proc. Am. Soc. Civil Engrs., J. Div. Irrigation and Drainage*, vol. 82, no. IR3, pp. 1–10, November, 1956.
145. Blaney, H. F.: Monthly consumptive use requirements for irrigated crops, *Proc. Am. Soc. Civil Engrs., J. Div. Irrigation and Drainage*, vol. 85, no. IR1, pt. 1, pp. 1–12, March, 1959.
146. Sunshine tables, *U.S. Weather Bur. Bull.* 805, 1905.
147. Thornthwaite, C. W.: The climates of North America according to a new classification, *Geograph. Rev.*, vol. 21, pp. 633–655, 1931.
148. Shaw, Napier: Comparative meteorology, in "Manual of Meteorology," vol. II, 2d ed., Cambridge University Press, London, 1936.
149. Mortenson, E., and L. R. Hawthorn: The use of evaporation records in irrigation experiments with truck crops, *Am. Soc. Horticultural Sci.*, vol. 30, pp. 466–469, 1934.
150. Pruitt, W. O.: Relation of consumptive use of water to climate, *Trans. Am. Soc. Agr. Engrs.*, vol. 3, pp. 9–17, 1960.
151. Pruitt, W. O., and M. C. Jensen: Determining when to irrigate, *Agr. Eng.*, vol. 36, pp. 389–393, 1955.
152. Robertson, C. W., and R. H. Holmes: Estimating irrigation water requirements, *Canada Dept. Agr. Exptl. Farm Ser.*, 1956.
153. Scofield, R. K., and H. L. Penman: The principles governing transpiration by vegetation, *Proc. Biol. Civil Eng.*, pp. 75–98, 1948.
154. Scofield, R. K.: Soil moisture and evaporation, *Trans. Fourth Intern. Congr. Soil Sci.*, vol. 2, pp. 20–28, 1950.
155. Hargreaves, G. H.: Closing discussion on Irrigation requirements based on climatic data, *Proc. Am. Soc. Civil Engrs., J. Irrigation and Drainage Div.*, vol. 84, no. IR1, pp. 7, 8, January, 1958.
156. Houk, I. E.: Agricultural and hydrological phases, in "Irrigation Engineering," vol. I, John Wiley & Sons, Inc., New York, 1951.
157. Israelsen, O. W., and V. E. Hansen: "Irrigation Principles and Practices," 3d ed., John Wiley & Sons, Inc., New York, 1962.
158. Woodward, G. O.: "Sprinkler Irrigation," 2d ed., Sprinkler Irrigation Association, Washington, D.C., 1959.

Section 12

INFILTRATION

G. W. MUSGRAVE, *Consultant in Hydrology, St. Petersburg, Florida.*

H. N. HOLTAN, *Hydraulic Engineer, U.S. Agricultural Research Service.*

    I. Introduction.. 12-2
   II. Early Concepts of Infiltration in America.................. 12-2
  III. Factors Affecting Infiltration............................ 12-2
      A. Surface Entry.. 12-2
      B. Transmission through the Soil........................ 12-2
      C. Depletion of Available Storage Capacity in the Soil... 12-3
      D. Characteristics of the Permeable Medium.............. 12-3
      E. Characteristics of the Fluid.......................... 12-4
   IV. Measurement of Infiltration............................. 12-6
      A. Infiltrometers.. 12-6
        1. Flooding-type Infiltrometers...................... 12-7
        2. Rainfall Simulators............................... 12-7
        3. Reflection of Factors by Infiltrometer Data....... 12-12
      B. Hydrograph Analyses to Estimate Infiltration......... 12-13
        1. The Detention-flow-relationship Method........... 12-14
        2. The Time-condensation Method.................... 12-16
        3. The Block Method................................ 12-17
    V. Relation of Infiltration to Runoff........................ 12-19
   VI. Infiltration in Computations of Runoff................... 12-22
      A. On Small Areas...................................... 12-22
        1. Estimating Storm Volumes of Infiltration......... 12-22
        2. Rate and Volume of Infiltration throughout the Storm. 12-22
      B. On Larger Watersheds............................... 12-24
        1. The Standard Curve.............................. 12-27
        2. Soil and Vegetative Arrays and Soil Moisture..... 12-27
      C. Infiltration Indexes.................................. 12-28
        1. The Φ Index..................................... 12-29
        2. The W Index..................................... 12-29
        3. The W_{min} Index................................. 12-29
        4. Initial Abstractions.............................. 12-29
        5. Rainfall Distribution............................. 12-29
  VII. A Glance Ahead....................................... 12-30
 VIII. References... 12-30

I. INTRODUCTION

The flow of water through the soil surface is called *infiltration*. Generally, infiltration has a high initial rate that diminishes during continued rainfall toward a nearly constant lower rate.

II. EARLY CONCEPTS OF INFILTRATION IN AMERICA

Infiltration is a relatively recent concept. The rain and other forms of precipitation that fall upon the earth, their evaporation, and the surface waters of the earth—the lakes, the seas, the groundwater beneath the surface, the runoff of surface water, the movement of water through plants (called *transpiration*)—all had been observed, measured, and studied for some years before attention was given to the water that enters the earth. This, perhaps, was natural because measurement of the amount and the rate at which precipitation moves into the soil is difficult, and even today is arrived at mostly by indirect means.

The term *infiltration* was used in the United States by George P. Marsh, a world traveler and careful observer of natural phenomena, who in 1864 published "Man and Nature." Ivan Houk in 1921, in connection with the studies leading to the development of the Miami Conservancy District in Ohio, under Arthur E. Morgan, reported the results of measurements of "soil absorption" for several conditions in the vicinity of the Miami River. This term was used in reference to the difference between rainfall and runoff on small areas and, together with other studies in the same area, in formulating the plans for the Miami flood-control project.

In the mid-1930s, when the American Geophysical Union established a research committee on infiltration, there were already in that organization active committees on precipitation, runoff, evaporation, and other phases of hydrology as they are known today. In spite of technological advances in all these fields, there is still a great need for more knowledge of infiltration.

Infiltration affects many aspects of hydrology. Infiltration influences runoff—so apparent in the differences in streamflow patterns in permeable and impermeable areas. Infiltration determines the moisture in the soil. Infiltration is related to the transpiration of plants and the evaporation of soil moisture. Infiltration is of interest to the conservationist and the farmer, for the farmer may modify the infiltration of his fields through management.

III. FACTORS AFFECTING INFILTRATION

Infiltration may be considered as a three-step sequence: surface entry, transmission through the soil, and depletion of storage capacity in the soil. These are important factors affecting infiltration, in addition to the characteristics of the permeable medium and percolating fluid.

A. Surface Entry

The surface of the soil may be sealed by the inwash of fines or other arrangements of particles that prevent or retard the entry of water into the soil. Soil having excellent underdrainage may be sealed at the surface and thereby have a low infiltration rate.

B. Transmission through the Soil

Water cannot continue to enter the soil more rapidly than it is transmitted downward. Conditions at the surface cannot increase infiltration unless the transmission capacity of the soil profile is adequate.

Transmission rates may vary for successive horizons of the soil profile. A surface horizon may become compacted by wet-weather traffic, or it may be naturally impermeable because of its texture and structure. A plow sole may have formed or a tight layer may have been created by fines moving downward from repeated cultivation of the surface. In the same manner, clay or caliche pans that restrict transmission may have developed naturally during the processes involved in soil formation, or restricting bedrock may exist at shallow depths.

After saturation, the rate of infiltration is limited to the lowest transmittal rate encountered by the infiltrating water up to that time. This concept may be visualized by assuming a theoretical soil profile in which three master *horizons*, like layers of a cake, are identified by the letters A, B, and C, in the order of the depth (Sec. 5-I-A). The B-horizon has a rate of transmission lower than that of the A- and C-horizons and a surface-entry rate higher than the transmission rate of any horizon. Infiltration will then equal the transmission rate of the A-horizon until the available storage in the A-horizon is exhausted. Thereafter infiltration will be limited to the transmission rate of the B-horizon. The transmission rate of the C-horizon is not at its capacity, since the B-horizon of the theoretical soil profile is the least permeable layer.

If the surface-entry rate is slower than the transmission capacity of any horizon, infiltration is limited to the surface-entry rate throughout an entire storm.

C. Depletion of Available Storage Capacity in the Soil

The storage available in any horizon depends on porosity, thickness of the horizon, and the amount of moisture already present. Texture, structure, organic-matter content, biologic activity, root penetration, colloidal swelling, and many other factors determine the nature and magnitude of the porosity of the soil horizon. Total porosity, as well as the size and arrangement of pores, has a significant bearing upon the availability of storage. The infiltration that occurs in the early part of the storm will largely be controlled by the volume, sizes, and continuity of noncapillary or large pores, because such pores provide easy paths for the movement of percolating water.

Storage capacity may directly affect infiltration rates during the storm. When infiltration rates are controlled by transmission rates through soil strata, the infiltration rates will diminish as storage above a restricting stratum is exhausted. The infiltration rate will then be equal to the transmission rate until a second, more restricting stratum is encountered.

D. Characteristics of the Permeable Medium

Factors that affect infiltration basically are the characteristics of the permeable medium—in the present case, soil—and the characteristics of the percolating fluid. In soil the problem concerns itself largely with pore size and pore-size distribution, that is, the proportion of different sizes present, as well as their relative stability during storms, irrigations, or other applications of water. In sands, the pores are relatively stable, since the sand particles that form them are not readily disintegrated, nor do they swell upon wetting. During a storm or irrigation, they may rearrange themselves into a more dense mix than formerly. However, this change in condition of the sand is relatively slow when compared with changes that occur in silts or clays.

Soils with appreciable amounts of silt or clay are subject during a storm to the disintegration of the crumbs or aggregates which in their dry state may provide relatively large pores. These soils also normally contain more or less colloidal material, which in most cases swells appreciably when wet. Thus a deterioration in permeability of the mass is much more readily accomplished than in sands. The impact of raindrops breaks down soil crumbs, there is a melting of aggregates, and the very small particles of silt and clay float across the surface and penetrate previously existing pores, thus clogging them and greatly reducing infiltration.

The degree of swelling that occurs in soil also is affected by the content and kind of clay minerals present. In some clays of the tropics, termed *laterites*, very little swelling occurs and storms are followed by rapid drainage in the soil. In a practical way, the amount of swelling common to a given soil may be estimated by the amount of shrinkage observable when it dries. The soil cracks that form on drying are relatively large and deep in the Houston black clay. It is indicative of the high swelling per cent characteristic of this soil when wetted.

Modification of pore sizes and pore-size distribution occurs commonly in the field. A simple example is that of wet-weather traffic producing a compacted surface layer. The converse is found in tillage, when, at least for a short time, the soil is opened up and large pores are provided.

Many studies have shown that vegetation is one of the most significant factors affecting surface entry of water. Vegetation, or mulches, protect the soil surface from rainfall impact. Massive root systems such as sods perforate the soil, keeping it unconsolidated and porous. The organic matter from crops promotes a crumb structure and improves permeability. On the other hand, vegetation, such as a row crop, gives less protection from impact, depending upon the stage of growth, and the root systems perforate only small portions of the soil profile and the accompanying tillage reduces permeability.

Forest litter, crop residues, and other humus materials also protect the soil surface. High biotic activity in and beneath such layers opens up the soil, resulting in high entrance capacities. Nonerosive pavements similarly protect soil surfaces from raindrop impact, but do not stimulate biotic activity to the same extent as vegetative litter, and therefore do not promote high intake capacities. Some types of litter may have an opposite effect if their chemical constituents tend to make soils resistant to weather. Only recently has it been realized that the compaction of soil surface by farm tractors, grazing livestock, and other movement upon the soil, particularly when it is wet, has a deleterious effect on the surface-entry rates.

Many experiments have shown that pore sizes and pore-size distribution are greatly affected by the content of soil organic matter, since both the sizes of soil aggregates and their stability in water are related to the amount of soil organic matter. This is not only true in silts and clays, but also in most soils containing colloidal material. The addition of organic matter or its removal, as by intensive cultivation and oxidation, changes the prevailing permeability. While these effects are more pronounced in some silts and clays than in others, they are nevertheless widely observed and often of large magnitude.

Frozen ground affects infiltration. If the soil is frozen while saturated, a very dense, nearly impermeable mass often results. However, if frozen when very dry, some soils, such as the Marshall silt loam, are fluffed up and may pass through the winter beneath a snow blanket in a physical condition not unlike sand. Wet and dry seasons, of course, have their effects upon soil moisture, and this affects infiltration.

Inherent soil characteristics are often masked by prior land use. Likewise, the potential effects of a good protective cover may be nullified by a nearly impermeable soil profile. The degree to which such combined effects reduce infiltration obviously cannot be predicted, except in a very general way. In the same manner, the effects of the duration of rainfall, of the depth of water on the surface of the land, and of a given soil-moisture increase upon the amount of swelling in different clays and colloids, all are highly complex and generally not quantitatively predictable.

Improved infiltration is well demonstrated in many studies of forest hydrology also, since most forest soils are richer in organic matter than cultivated ones. A soil with an old established grass cover, like pasture land or prairie, is likewise more permeable, all other things being equal. In Subsec. VI-B-2-b on the land-use array, more detail is given on the forms of land use and their relation to infiltration.

In brief, the characteristics of the permeable medium are affected primarily by the kind of soil, its texture, its structure, the amount and kind of clay and colloid that it contains, the depth and thickness of its more permeable layers, and its prior history of land use. On occasion, some of these elements have very large effects upon infiltration, and at other times, relatively small effects. The history of land use has particular interest because soil management has important and enduring effects on the infiltration capacity of a soil. Land uses, being subject to management, provide a useful tool for modifying infiltration within considerable limits.

E. Characteristics of the Fluid

Another group of factors that affect infiltration, though usually to a lesser degree, are those that modify the physical characteristics of the fluid, namely, water. In practice, it is seldom, if ever, that pure rainwater enters the soil without some prior contamination. One of the most potent of these changes in the infiltrating fluid is its pollution by the fine clays and colloids common in greater or lesser degree to almost all soils. Dr. W. C. Lowdermilk, while in China, showed that, when clear and turbid water supplies were alternately applied to the soil, runoff was much greater with the turbid water.[1] There have been many experiments, and there is a large body of experience, showing runoff to be less from areas in grass or forest, where the water is relatively low in suspended material, than is the runoff in cultivated areas, where the water is relatively high in suspended materials. The effect, of course, of such suspended materials in the infiltrating fluid is that of blocking the fine pores through which it must pass if appreciable rates of infiltration are to be maintained.

The infiltrating fluid is often contaminated by the salts, particularly in alkali soils and to some extent in many other soils. These salts may affect the viscosity of the fluid and form complexes with the soil colloids which affect the swelling rate when wet. Irrigation waters very often contain residues of fertilizer, particularly when they are re-used. Water in farm ponds may contain impurities that modify infiltration. Although difficult to handle, these items must be recognized by the practicing hydrologist, especially in the design of irrigation or drainage systems.

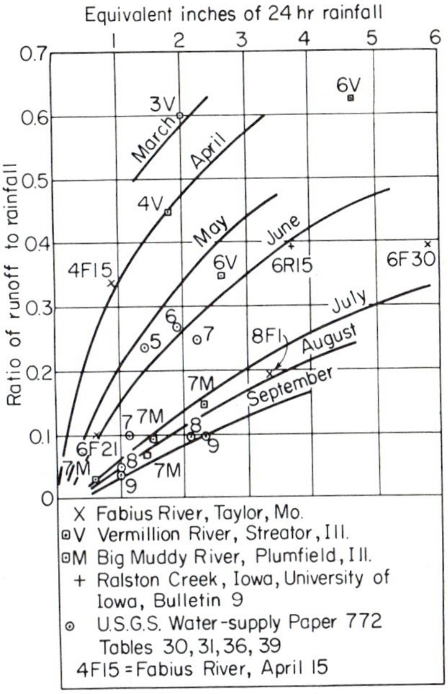

The temperature and viscosity of the fluid obviously affect the rate at which it moves through the soil. Little attention has been given in practice to this effect, though it is not uncommon to find higher rates of runoff in the cool months of the year than in the summer months. In many cases where attempts have been made to isolate and measure the temperature factor in the field, results have been negative, possibly because temperature may also affect the soil. In at least one instance, however, with temperature variations of 25 to 30° diurnally, relationships that could be accounted for by viscosity changes were observed.

Fig. 12-1. Relation of runoff to rainfall, indicating relatively less infiltration during the cool spring months than during the warm summer periods. (*Courtesy of Dover Publications, Inc.* [2].)

Mavis and Wilsey [1], in their study of sands for filtration beds, found direct and positive relations between temperature and rate of filtration. The practicing hydrologist will find plenty of records of rainfall and runoff where rates of runoff were higher in spring and fall from less rain than occurred on the same watersheds in summer months.

In Fig. 12-1, Sherman has shown the ratio of runoff to rainfall for different months of the year for a number of midwestern streams. It will be noted that the highest per cent of runoff occurred in March and the next in April. These are the coolest months of the year in the record and bear out the point that a greater proportion of rainfall occurs as runoff in the cool months.

The depth of water in an area also affects the rate and amount of infiltration. With an increasing head of water, some increase in infiltration is to be expected, though carefully performed experiments show that the magnitude of increase is extremely small—indeed, often unobservable—for differences in head of water of less than ½ in. Of greater magnitude and more frequent occurrence is the increase in the infiltration

---

[1] Personal communication.

when more rain falls and more ground is covered. When the rainfall is extremely light and the depth of water very slight, parts of the land surface may really be above the water level. Figure 12-2 shows what happens when the depth of water is increased on the miniature hummocks that occur in almost any field.

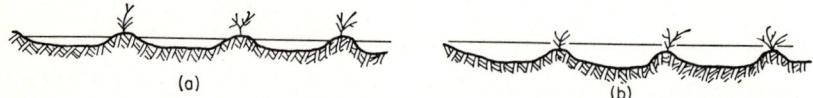

FIG. 12-2. The depth of water on a field may affect the rate and amount of infiltration. (a) When the depth of water on a field is very slight, many parts of the land surface may be above the water level, thus limiting infiltration; (b) when the depth of water is increased, miniature hummocks on a field are covered and infiltration opportunity is increased.

In brief, the characteristics of the fluid that affect infiltration are primarily (1) its turbidity, particularly its content of clay and colloid, (2) the kind and amount of salts it contains, whether from alkali lands, fertilizer residues, or other sources, and (3) its temperature and viscosity.

IV. MEASUREMENT OF INFILTRATION

The three-step sequence surface entry, transmission, and exhaustion of storage presents difficulties in the measurement of infiltration. The many factors affecting each phase may occur in a multitude of combinations. The early work of Schiff [3] and the more recent work of Philips [4] made definite progress in routing water into and through the soil. However, the techniques have not as yet been adapted for general application in hydrology. For the most part, hydrologists determine the rate and the amount of in-soak and attempt to correlate this with various combinations of soil, vegetation, and antecedent soil moisture.

There are two general approaches to the determination of infiltration capacity of a soil-cover and soil-moisture complex. One of these is the analysis of hydrographs of runoff from natural rainfall on plots and watersheds. The other is the use of infiltrometers with artificial application of water to enclosed sample areas.

Concerning hydrograph analysis, the derived estimate of infiltration can ultimately be no more accurate than the precision of measurement of rainfall and runoff. In practice, both are subject to some error, and the rainfall variation on different parts of a watershed is often completely unknown. However, improvements in the methods of measurement and the increase in the detail of rainfall records, particularly through the use of radar, are now coming into vogue. Future difficulties may be much less than those of the past. Present practice in hydrograph analysis is described in Subsec. IV-B.

A. Infiltrometers

Infiltrometers are often used on small watersheds or on experimental or sample areas within large watersheds. Where there is a wide variation in either soils or vegetation on the area, the watershed is subdivided into relatively uniform subareas, each being a single soil-cover complex. By making a number of repetitions of the test runs, reliable data for each subarea can be obtained.

Infiltrometers may be considered in two general groups: (1) *rainfall simulators*, with the water applied in the form and at the rate comparable with natural rainfall, and (2) *flooding type*, with the water applied in a thin sheet upon an enclosed area and usually in a manner to obtain a constant head. Various kinds of equipment are in use for both types. They vary in size, in the quantity of water that is required, and in methods of measuring the water. In a rainfall simulator, the drops are of raindrop size, having an energy of impact similar to that of the natural rain with which it is to be compared. In the flooding type, the essentials include a good control of the head of water on the enclosed area and a depth of water that is comparable with field applications. On irrigated areas, the depth may be much greater than that provided by rain on a hillside. An essential in the operation of either type of infiltrometer is that of replication sufficient to obtain results fully representative of the variable local conditions. A further need is that of installation on each site with a minimum of disturbance of soil structure.

1. Flooding-type Infiltrometers. The most commonly used flooding types of infiltrometers include tubes and concentric rings. The *tubes* are cylinders usually about 9 in. in diameter and 18 to 24 in. long. These tubes are jacked into the soil to depths of 15 or 21 in. Water is then applied from graduated burettes to maintain a constant head of water sufficiently deep to submerge plant crowns. Reading of the burettes at successive time intervals permits direct determination of rates and amounts of infiltration. In effect, this device measures the rate of water movement downward through a confined column of soil 9 in. in diameter and 15 or 20 in. long. The surface of the soil is generally protected by a perforated disk of sheet metal, so that turbidity of the surface water is at a minimum.

Concentric rings, as the name implies, consist of two concentric rings, the inner one, usually about 9 in. in diameter, being used to determine the infiltration rate, while the concentric 14-in.-diameter ring is flooded to the same depths to decrease border effects on the inside ring. Rings, as opposed to tubes, are inserted into the soil at the minimum depth necessary to prevent leakage from the rings. Water may be applied at a constant depth or head and measured as in the case of the tubes. Reading of the burettes at intervals indicates directly the rates and amounts of infiltration.

2. Rainfall Simulators. Infiltrometers of this general class include many modifications. The most common types in use are the type F and the type FA. The type F generally utilizes a plot 6 ft wide and 12 ft long, although any reasonably longer length of plot can be used. Simulated rainfall in rather large drops is applied to the plot and surrounding area from two rows of special type F nozzles mounted along each long side of the plot. These special nozzles direct their spray upward and slightly inward to cover the plot and surrounding area with relatively uniform rainfall intensities of about 1.75, 3.50, or 5.25 in./hr, depending on how many sets of nozzles are used. The drops normally reach a height of 6 or 7 ft above the plot surface, and hence do not quite attain full terminal velocity, which would require 16 to 20 ft of fall. The impact, however, is great enough to produce erosion and surface conditions resembling those of natural rain.

The type FA infiltrometer employs the same nozzle as the type F, but it is normally operated at a lower water pressure; hence the drops do not rise as high or fall as far. The normal size of plot is 1 ft wide and 2½ ft long. Many modifications of plot size have been used, however. A border area around the plot about 1½ ft wide is also sprinkled by the nozzle. Simulated rainfall intensities are varied by turning nozzles on or off. Intensities may be varied, therefore, by multiples of about 1.5 in./hr.

A typical test run and analytical run with the type F infiltrometer are shown schematically in Fig. 12-3. The data, also schematic, are represented in Table 12-1.

A plastic sheet or a metal pan is placed over the plot during rainfall-calibration runs prior to the infiltration test runs and following the analytical run. The average rate of rainfall is computed from these runs.

After the plastic sheet or metal pan has been removed, the test run is started and continued until the rate of runoff becomes constant.

The analytical run is started just as the free water disappears from the soil surface, before any recovery of infiltration capacity can occur. Therefore, during the analytical run, the rate of infiltration is a constant, as indicated by the broken line in Fig. 12-3a, and the difference between the quantity $(p - f)$ and q is due to detentions of flow, D_a, and storage in depressions, V_d. Since detentions of flow ultimately occur as recession flow after rainfall stops, depression storage is computed by the equation

$$V_d = P - F - Q - D_a \tag{12-1}$$

In Table 12-1, $V_d = 0.285$ in. over the plot surface.

The relationship between D_a and the rate of runoff q is derived by summing, in

Table 12-1. Computations in Analyses of Hydrographs from Type F Infiltrometer (Schematic)
(Test-run data)

Time, min	q, in./hr	ΔQ, in.	Q, in.	D_a,* in.	P_e,† in.	P, in.	P-P_e, in.	Remarks
0	0	0	0	0	0	0	0	Rainfall started
5	0	0	0	0	0	0	0	Runoff started
10	0.04	0.002	0.002	0.005	0.007	0.333	0.326	
20	0.16	0.016	0.018	0.019	0.037	0.667	0.630	
30	0.36	0.043	0.061	0.052	0.113	1.000	0.887	
40	0.68	0.086	0.147	0.117	0.264	1.333	1.069	
50	0.98	0.138	0.285	0.180	0.465	1.667	1.202	
60	1.20	0.182	0.467	0.235	0.702	2.000	1.298	
70	1.40	0.217	0.684	0.285	0.969	2.333	1.364	
80	1.52	0.243	0.927	0.310	1.237	2.667	1.430	
90	1.60	0.260	1.187	0.316	1.503	3.000	1.497	
100	1.60	0.267	1.454	0.316	1.770	3.333	1.563	
110	1.60	0.267	1.721	0.316	2.037	3.667	1.630	
120	1.60	0.267	1.988	0.316	2.304	4.000	1.696	Rainfall stopped

Recession accumulated in reverse

Time, min	q, in./hr	ΔQ, in.	Q, in.	D_a, in.				Remarks
120	1.60	0	0	0.316				
130	0.70	0.192	0.316	0.124				
140	0.30	0.083	0.124	0.041				
150	0.10	0.033	0.041	0.008				
160	0	0.008	0.008	0				Runoff stopped
180	0	0	0	0				Pocket storage disappeared, rainfall started
190	0.26	0.022	0.022					
200	0.68	0.078	0.100					
210	1.19	0.156	0.256					
220	1.48	0.223	0.479					
230	1.58	0.255	0.734					
240	1.60	0.265	0.999	0.316				Runoff constant, rainfall stopped

Recession accumulated in reverse

Time, min	q, in./hr	ΔQ, in.	Q, in.	D_a, in.				
240	1.60			0.316	$P - F = 2.00 - 0.40$		$= 1.600$ in.	
250	0.70	0.192	0.316	0.124	$Q + D_a = 0.999 + 0.316$		$= 1.315$	
260	0.30	0.083	0.124	0.041	$V_d = P - F - Q - D_a$		$= \overline{0.285}$	
270	0.10	0.033	0.041	0.008				
280	0	0.008	0.008					

* D_a, uncorrected for infiltration during recession, read from Fig. 12-4.
† P_e is rainfall excess $= Q + D_a$.

Symbols used as column heads and in text are defined as follows:

Term	Rate, or increment	Accumulation, or volume
Rainfall	p	P
Runoff	q	Q
Infiltration	f	F
Volume of depressions	...	V_d
Detentions of flow (average)	...	D_a
Initial abstractions (interceptions $+ V_d$)	...	I_a
Retentions ($I_a + F$)	...	R
Time	t	T

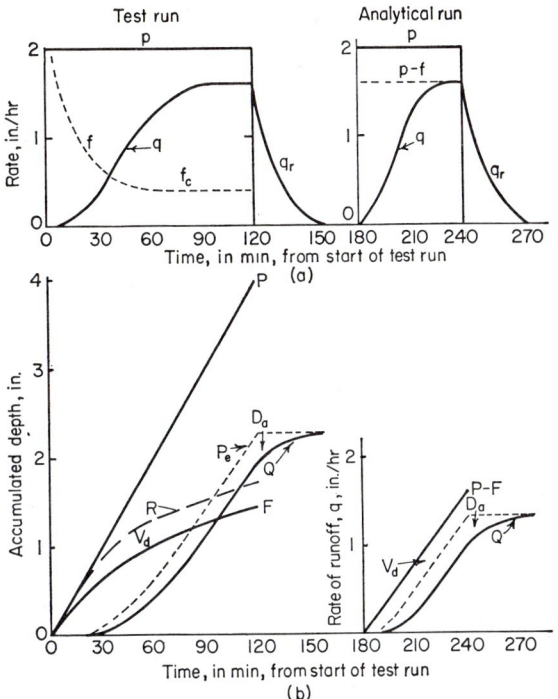

Fig. 12-3. Analysis of hydrograph typical of type F infiltrometer data (schematic).

reverse, the volumes of recession flow (Table 12-1) and plotting these volumes vs. instantaneous rates on logarithmic scales (Fig. 12-4). Although not illustrated in Table 12-1 or Fig. 12-3, a diminishing rate of infiltration will be found during this recession. These rates are estimated using the equation

$$f_r = \frac{f_c}{q_c} q_r \qquad (12\text{-}2)$$

where the subscript r denotes recession and the subscript c denotes the final constant rate.

Infiltration during the recession, f_r, is added to observed q_r to obtain revised estimates of D_a for each rate.

Detentions are estimated from Fig. 12-4 and added to the accumulated runoff during the test run to derive rainfall excess P_e in Table 12-1. The difference between rainfall and rainfall excess is retention R and is equal to infiltration plus pocket storage, $F + V_d$ as interception is nil.

Retention R is plotted against the time in Fig. 12-3b. It is assumed that V_d must be essentially satisfied before high rates of runoff can occur. It is further assumed that the mass curve of infiltration starts at zero rainfall and is a smooth curve of diminishing slope. Therefore the mass

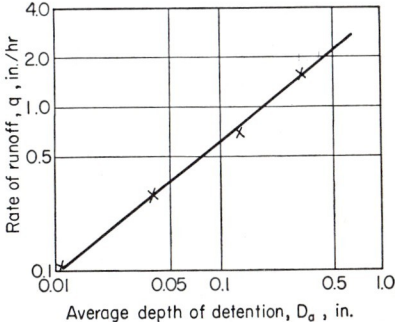

Fig. 12-4. Detention-flow relationship from analysis-of-recession curve (schematic infiltration run).

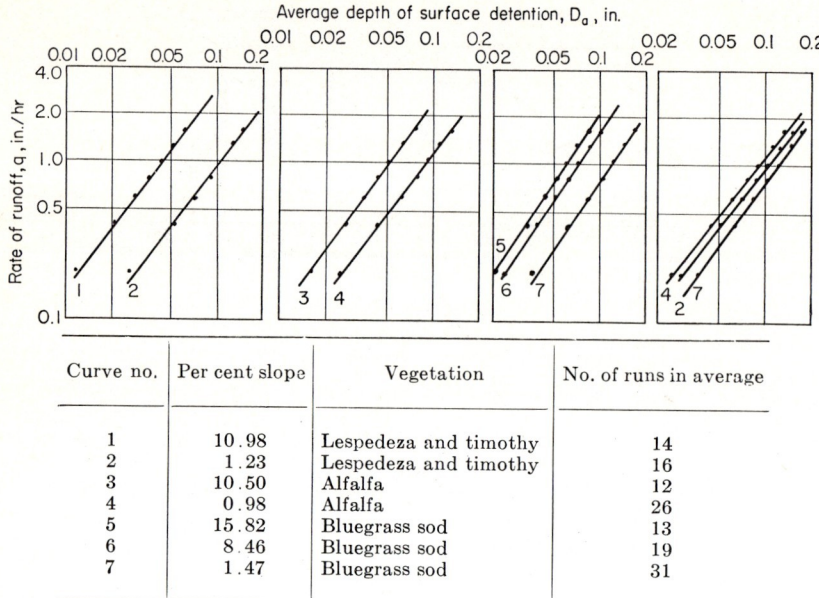

FIG. 12-5. Detention-flow relationships of overland flow, from type F infiltrometer runs, 6- by 12-ft plot.

Curve no.	Per cent slope	Vegetation	No. of runs in average
1	10.98	Lespedeza and timothy	14
2	1.23	Lespedeza and timothy	16
3	10.50	Alfalfa	12
4	0.98	Alfalfa	26
5	15.82	Bluegrass sod	13
6	8.46	Bluegrass sod	19
7	1.47	Bluegrass sod	31

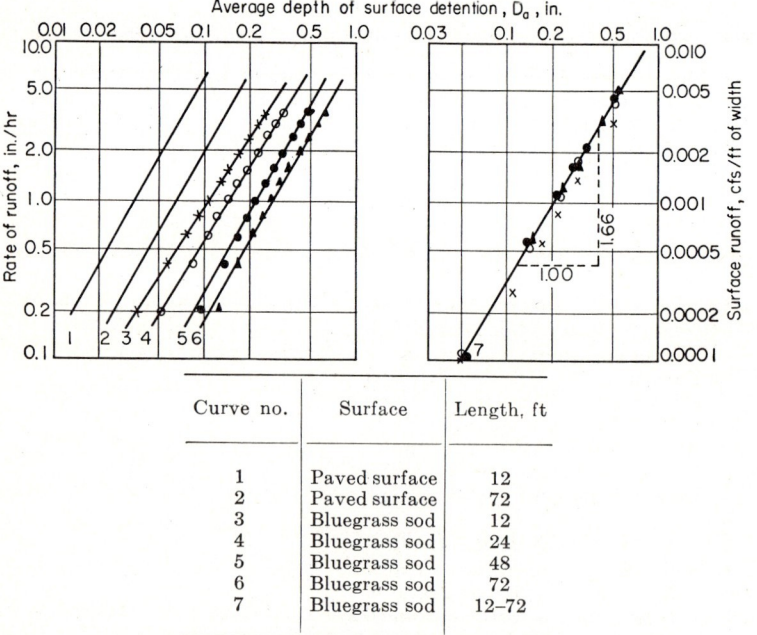

Curve no.	Surface	Length, ft
1	Paved surface	12
2	Paved surface	72
3	Bluegrass sod	12
4	Bluegrass sod	24
5	Bluegrass sod	48
6	Bluegrass sod	72
7	Bluegrass sod	12–72

FIG. 12-6. Detention-flow relationships of overland flow. Infiltrometer data from 12-ft plots by H. N. Holtan; other lengths by C. F. Izzard.

curve of infiltration must start to rise at the beginning of rainfall, must have a diminishing slope, and has a ceiling of $R - V_d$ during periods of significant runoff rate. The analytical run provides estimates of depression storage and detention-flow relationships which are very useful in applications of the infiltration data obtained during test runs. The detention-flow relationships for various types of vegetation and land slopes in Fig. 12-5 are useful for routing computed rainfall excess through overland flows to derive inflow to channels or gutters.

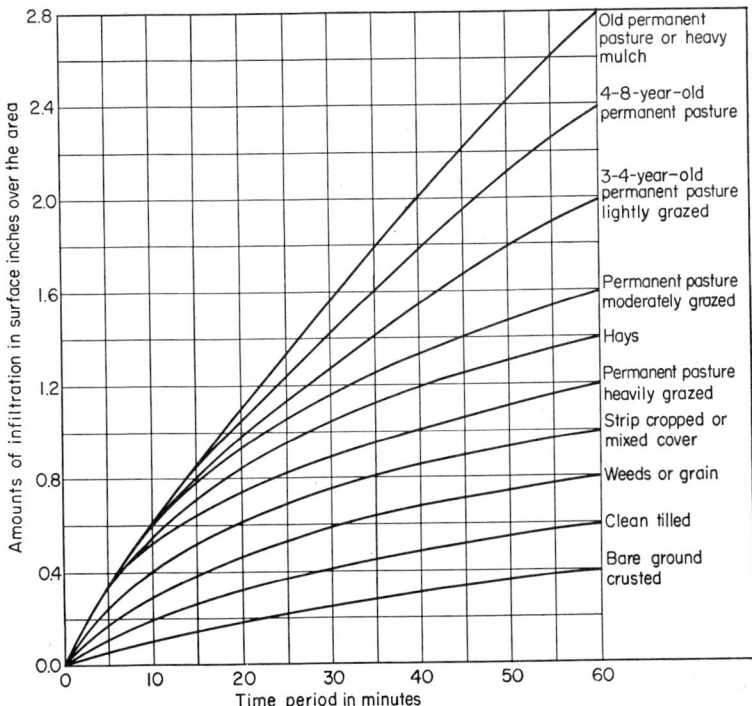

FIG. 12-7. Typical mass-infiltration curves based on data from Cecil, Madison, and Durham soils. (*Holtan and Kirkpatrick* [5].)

In units of inches per hour, the curves of Fig. 12-5 are applicable only to the 12-ft lengths of slope on which they were obtained. However, if they are converted to units of cubic feet per second, as illustrated in Fig. 12-6, they are pertinent to any length of slope.

The family of curves in Fig. 12-7 are typical of mass infiltration measured for a group of related piedmont soils planted to various types of vegetation. Mass curves are a convenient form for subtracting infiltration from mass rainfall.

Horton [6, 7] developed a mathematical equation for defining the rate curve of infiltration capacity:

$$f = f_c + (f_0 - f_c)e^{-kt} \qquad (12\text{-}3)$$

where f_0 is initial rate of infiltration capacity, e is base of natural logarithms, k is a constant depending primarily upon soils and vegetation, t is time from start of rainfall, and other symbols are as previously given.

Values of f_0, f_c, and k are associated with soil-cover complexes and provide a convenient means of expression. The effect of variations in the value of k is evident in Fig. 12-8.

The use of infiltrometers on watersheds has the inherent advantage that it permits the mapping of *isopotal areas*, i.e., areas of equal infiltration capacities. These, when studied in comparison with isohyetal maps, give more precision because the variation in rainfall and the variation in infiltration on different parts of the watershed may be considered in proper relation to each other.

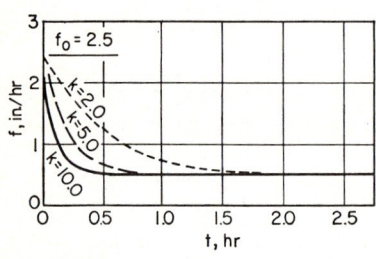

Fig. 12-8. Effects of variations in k in Horton's equation for the infiltration-rate curve:

$$f = f_c + (f_0 - f_c)e^{-kt}$$

Each curve asymptotic to $f_c = 0.5$.

3. Reflection of Factors by Infiltrometer Data. That infiltrometers provide relative, rather than absolute, values is demonstrated by the fact that, when used on a series of soil-cover complexes, different types of instruments give different magnitudes of F although the various conditions are ranked in the same relative order. For example, Table 12-2 gives the mean results from replicated tests on adjoining areas by two different, though commonly used, types of instruments. The concentric rings supply water to the soil surface without impact, whereas the rainfall simulator provides impact similar to that of natural rain. In both types there is an exterior wetted border about the experimental portion, though somewhat larger in the case of the type F infiltrometer. Without the rain impact, that is, with the concentric rings, the results for protective cover are about twice those for nonprotective. Where the effects of rain impact are measured as with the type F infiltrometer, the ratio of protected to nonprotected is about 8:1.

Table 12-2. Comparative Data from Two Types of Infiltrometers

Infiltrometer	Infiltration f_c, in./hr.	
	Overgrazed, poor cover	Grazed, good cover
Rainfall simulator, type F..................................	0.14	1.13
Flooding type, concentric rings, inner-ring area...........	1.11	2.35

The example given in the table is typical of other comparisons between flooding and rainfall simulator types of infiltrometers though the ratios may differ on different soils and covers. Generally, the type F gives lower magnitudes of infiltration, and the results more nearly approach those of natural rains on watersheds than do either smaller rainfall simulators or flooding types.

An example of the differences in infiltration on soils varying from deep to shallow, each having contrasting covers or land-use conditions, is shown in Table 12-3. The soils are all silt loams, differing mainly in depth and content of organic matter. The Muscatine is a deep, very dark colored soil rich in organic matter. The Viola is a relatively shallow soil, comparatively low in content of organic matter. Tama, Berwick, and Clinton are listed in the table in approximate order of depth and organic-matter content between Muscatine and Viola.

The bluegrass, of course, provides a dense surface cover highly protective against rain impact. The tests were made on farms where the grass was under practical grazing conditions. The corn is not noteworthy for any great protective effects, and intertillage tends to break down soil aggregation or crumb structure.

The data of Table 12-3 are means of replicated wet runs of the type F infiltrometer where the simulated rainfall was at 1.80 in./hr, with a large drop size and an energy of impact similar to that of natural storms of this size. The results show (1) the consistent and wide differences between the two kinds of land use on each of the soils, and (2) the steadily decreasing infiltration under the protective cover from the 5.38-in. total on Muscatine—the deep soil with high content of organic matter—to the 1.63-in. total on the shallow Viola. But this close relationship to soil-profile characteristics is not found under the less protective corn, where surface conditions rather than profile characteristics tend to govern intake. Or to look at the matter another way, the

Table 12-3. Infiltration on Soils of Varying Depth and Organic Matter, with Contrasting Covers*

Silt loam soils	Total infiltration in 5 hr, in.		Difference due to land use, in.
	Bluegrass pasture	Cornland	
Muscatine........	5.38	1.34	4.04
Tama............	5.03	1.51	3.52
Berwick..........	3.48	1.21	2.27
Clinton..........	2.77	2.17	0.60
Viola............	1.63	1.28	0.35

*From Holtan and Musgrave [8].

differences in soil characteristics—even where they are rather large as in this case—have relatively little effect under adverse cover conditions. Here, without protection against rain impact, the mud slurry and partial seal of the soil surface largely overshadowed the inherent soil characteristics. In brief, the example shows the more protective land use to be superior for each of the soils without exception. But the better soil does not necessarily rank best under adverse land use.

The practicing hydrologist is probably interested in the results found during the last hour of this 5-hr storm during which a total of 9 in. of water was applied. These are shown in Table 12-4.

Table 12-4. Rates of Infiltration during Fifth Hour

Silt loam soils	Bluegrass, in./hr	Cornland, in./hr
Muscatine..........	0.61	0.11
Tama.............	0.77	0.14
Berwick............	0.34	0.12
Clinton...........	0.29	0.18
Viola.............	0.16	0.08

As expected, the rates during this fifth hour are less than the average for the entire period. Of real interest is the fact that, without exception, the more protective cover on each soil is producing greater infiltration. Again it is seen that the soils tend to arrange themselves under the bluegrass in order of their depth and content of organic matter. And again, under the less protective cover, soil differences tend to be overshadowed by what has obviously happened on the surface, namely, a clogging of pores.

B. Hydrograph Analyses to Estimate Infiltration

The curve of infiltration, as it progresses with time, can be estimated for a plot or watershed through analyses of the runoff hydrograph. Such analyses have distinct advantages over infiltrometer techniques in that they consider the vagaries of rainfall and incorporate the influences of factors such as length of overland flow, slope, soil and vegetation difference, depression storage, and surface detention as they occur in actual application. Numerous techniques have been developed which present distinct advantages for given situations. These methods utilize one or more of three basic principles, i.e., detention-flow relationships to derive rainfall excess, time condensations to eliminate periods of inadequate rainfall, and "block" methods of dividing the

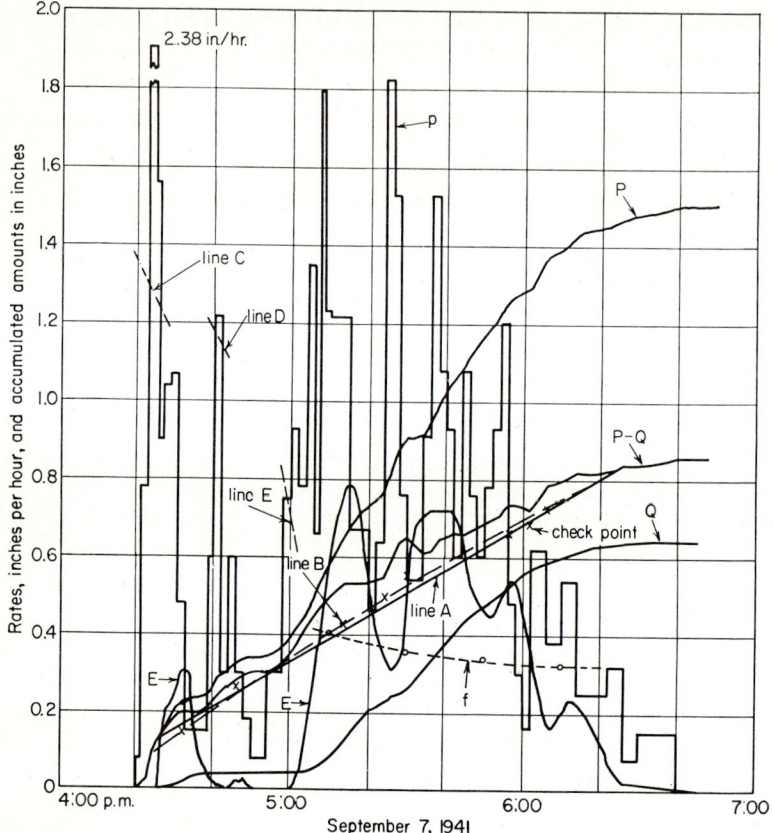

FIG. 12-9. Infiltration f derived from analyses of storm hydrograph, control plot number 6, La Crosse, Wis.

storm into a series of blocks, or showers, to obtain a succession of average rates of infiltration throughout the storm.

1. The Detention-flow-relationship Method. On plots and small homogeneous watersheds (those up to a few acres in size) the method used to analyze hydrographs of infiltrometer plots may be adapted for analyzing natural hydrographs and deriving infiltration curves. The method in this case is given more fully elsewhere [9, 10]. Only the essentials of the analysis of one type of storm will be presented here. A storm suitable for analysis must be of sufficient duration and intensity so that the infiltration rate f will have become relatively constant. The rainfall is intermittent; hence there will be rises and falls in the rate of runoff. Such a storm and hydrograph is shown in Fig. 12-9.

Steps in analysis are as follows:
1. Subtract the Q curve from the P curve, using a scale or pair of dividers, and plot the $P - Q$ curve as shown in Fig. 12-9. When $Q = 0$, the $P - Q$ curve is equal to $F + I_a + D_a$ plus negligible amounts of evaporation and transpiration.
2. Separate the $P - Q$ curve by trial and error into two parts, D_a and $F + I_a$. When D_a is negligible, $F + I_a$ coincides with $P - Q$. In the first trial draw $F + I_a$ as a straight line from the point on the P curve where the $P - Q$ curve begins to some

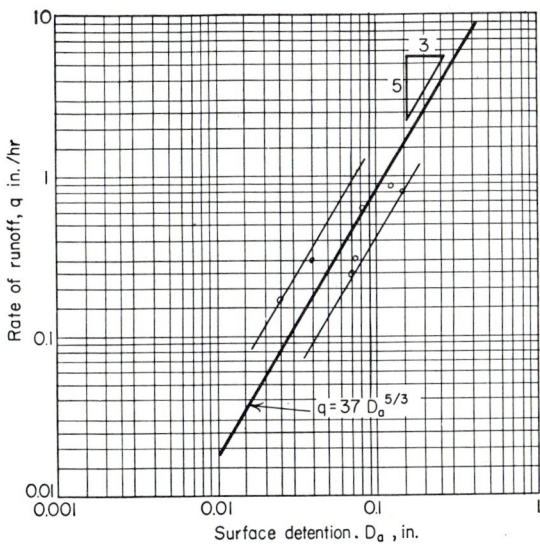

Fig. 12-10. Steps in developing a q versus D_a relation:
1. Plot points.
2. When only a few points are plotted and no very definite trend is apparent, as in this example, draw upper and lower envelope lines as shown, using a slope of 3:5.
3. Draw the line of relation for q and D_a, using a slope of 3:5, at half the distance between the envelopes.
NOTE: The slope of 3:5 is based on the assumption that Manning's formula applied to overland flow. R in the formula becomes $D_a/12$ for overland flow, so that

$$v = \frac{1.5}{n}\left(\frac{D_a}{12}\right)^{\frac{2}{3}} S^{\frac{1}{2}} \quad \text{fps}$$

Since $q_w = vD_a/12$, where q_w is the discharge in cubic feet per second per unit width, then $q = CD_a^{\frac{5}{3}}$, where C is a coefficient containing the conversion constants and the n and S factors. C is the intercept on the q axis, where $D_a = 1$. In this example, $C = 37$.

convenient point on the $P - Q$ curve near the end of runoff. This is line A in Fig. 12-9.
3. Plot selected peak and trough values of D_a versus corresponding values of q as shown on Fig. 12-10. D_a values are differences between the $P - Q$ and $F + I_a$ lines. Peaks and troughs of q will lag behind corresponding D_a values.
4. Draw the line of relation between q and D_a as shown on Fig. 12-10.
5. Check the trial $F + I_a$ line, line A on Fig. 12-9.

    a. Select a peak or trough q on Fig. 12-9 that was used to make the line of relation on Fig. 12-10.
    b. On Fig. 12-10, using the line of relation, find the corresponding D_a.
    c. Subtract the D_a value from the $P - Q$ line of Fig. 12-9, working at a time ahead of the given q value by the lag for that pair of values.

12-16 INFILTRATION

    d. Plot the D_a value. These are shown as crosses on Fig. 12-9.
    e. Plotted points will fall close to the trial $F + I_a$ line if this line is adequate.

    6. If a second trial line of $F + I_a$ is needed, draw a curve as shown by line B of Fig. 12-9. Steps 3, 4, and 5 may be repeated when a large change is made in drawing the second trial line. They are not repeated for this example.
    7. Compute points for the f curve, using the second (or later) trial line of $F + I_a$. Since I_a is assumed to be a constant, the slopes of $F + I_a$ will give the same values regardless of the magnitude of I_a. For example, at 5:20 and 5:40 P.M. the $F + I_a$ values are 0.465 and 0.585 in., respectively. The average rate of f during this 20-min time interval is

$$\frac{0.585 - 0.465}{20/60} = \frac{0.120}{0.333} = 0.36 \text{ in./hr}$$

where 20/60 converts the 20 min into 0.333 hr. The average rate is plotted at the mid-point of the time interval—in this example, at 5:30 P.M. Other f points are similarly computed and plotted.
    8. Draw the f curve as shown on Fig. 12-9. An inspection of the early part of the storm shows that the f curve must come through locations marked by lines C and D. Assuming the rainfall data are correct, there was some recovery of the infiltration capacity after 4:52 P.M. The shape of the f curve for the early part of the storm may be approximated if actual I_a values are known. Generally they are not, and the f curve is either left incomplete or the curve is roughly sketched, starting where line E is shown.

    2. The Time-condensation Method. This method attempts to eliminate variations in rainfall by theoretically condensing time [11]. With rainfall at a constant rate, the principles of analyses derived under the controlled conditions of the sprinkler-type infiltrometer can be applied. A storm which occurred on a 27-acre watershed in Illinois is analyzed by this procedure in Fig. 12-11. Rainfall and runoff are depicted as volumes only. Rates may be readily discerned as the slope of the volume curve in any particular increment.
    Time is theoretically *condensed* in such a manner as to cause the mass rainfall curve to become a straight line. Identical time condensations applied to the mass curve of runoff result in a path of points indicated as Q_{tc} in Fig. 12-11. The theory applied is that, if the rainfall rate were truly constant, the curve of runoff plus detention Q_{tc} would be either a straight line or show a smooth upward curvature approaching the slope of rainfall as an upper limit. Therefore a curve (straight line or upward curvature) fitted to those points where detention can be assumed negligible (derived from flat portions of the Q_c curve) represents the curve of runoff plus detentions of flow. This curve is labeled $(P - R)_{tc}$ in Fig. 12-11.
    By expanding the time scale as indicated by rainfall, the $(P - R)_{tc}$ curve becomes the $P - R$ curve as it occurred during the storm. Differences between the $P - R$ curve and the Q_c curve (runoff at the weir) are due to surface detention.
    Retention R, including interception, pocket storage, evaporation, and infiltration, is used here in reference to those quantities or volumes of rainfall which do not occur as surface runoff at any point on the watershed above the point of consideration. In Fig. 12-11 retention is derived as the difference between the P curve and the $P - R$ curve.
    Infiltration F is estimated from the retention curve R under the following premises:
    1. Initial abstractions such as interception and volume of depressions V_d must be essentially satisfied at the time the runoff starts. Depression storage will approach the same constant for each successive period of rainfall excess.
    2. The F curve must be essentially parallel to the R curve during extended periods of rainfall excess and runoff.
    3. During periods of recession runoff, or no runoff, the F curve will continue with little or no change in curvature until it intersects with the R curve. Rate of infiltration f is derived as the slope of F-curve segments.

3. The Block Method. Estimates of infiltration on larger watersheds may be made either by adapting methods for plots and small watersheds or by a method similar to the determination of F_a, as developed by Horner and Lloyd [12] and Sherman [13]. The *block*, or *average infiltration method*, is generally adequate since the distribution of rainfall on larger watersheds is seldom known closely enough to warrant the use of more laborious methods. An application of the method to Johnson Creek was made by A. L. Sharp in Ref. 9.

Johnson Creek (Fig. 12-12), with a drainage area of 28.2 sq mi above the U.S. Geological Survey gaging station, drains an area of rolling mountain foothills in cut-over timber, pasture, and cropland. Soils are residual from basalt and on the heavy

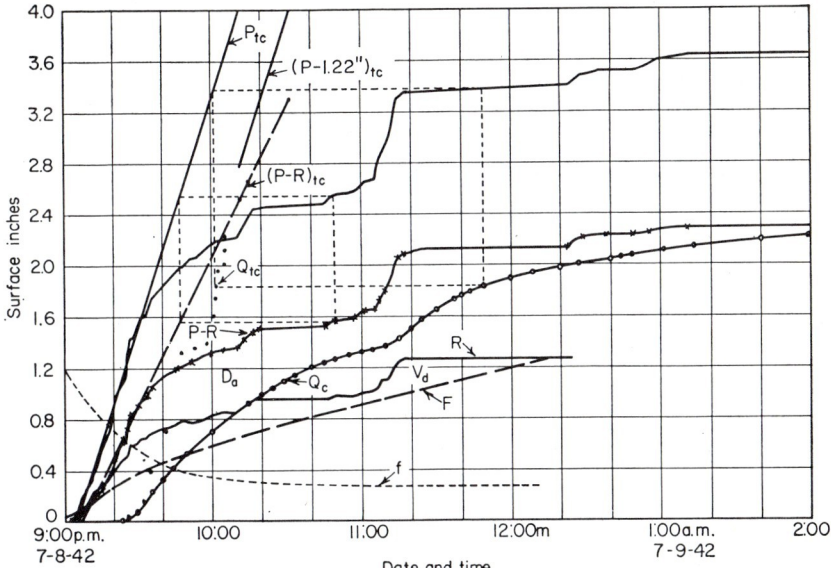

Fig. 12-11. Infiltration estimated by the time-condensation method of hydrograph analysis. P = accumulated rainfall; R = retention, including infiltration; D_a = detentions of flow; Q_c = runoff as measured at weir; tc indicates that time is condensed; V_d = pocket storage and interception; F = accumulated infiltration; f = rate of infiltration in inches per hour.

side. Stream gradients are relatively low just above the gaging station and become progressively steeper in the headwaters. Total fall in the watershed is about 600 ft. The watershed is about 14 miles long and averages about 2 miles in width.

One recording gage is located at Gresham, Ore., near the center of the watershed. Other gages, mostly standard, surround the watershed at distances of 10 to 30 miles away. In this general area, there is a close relationship between altitude and precipitation. Watershed rainfall was estimated using the isohyetal method (Sec. 9).

Figure 12-13a and b shows hydrographs on Johnson Creek during the period Feb. 14 to 24, 1949. These figures also show the hourly rainfall at the Gresham gage P_G; an estimated base flow curve; watershed precipitation mass curve P_w; cumulative surface or storm runoff Q_s; infiltration F; time of excess precipitation T_e; and the computation of average infiltration rate f_a.

A series of three rains and associated streamflow rises, Feb. 14 to 22, are shown in Fig. 12-13a. An isohyetal map was prepared for each storm and used with the Gresham gage record to obtain P_w. Each hydrograph was separated from the succeeding ones by transposing recession curves. Base flow was deducted from total flow, and the storm runoff was obtained for each separate rise (the small rise of Feb. 19 and 20 was omitted). Durations of precipitation excess T_e were estimated by inspection of the Gresham record. Average infiltration rates f_a were computed for each storm as shown in Fig. 12-13a and b.

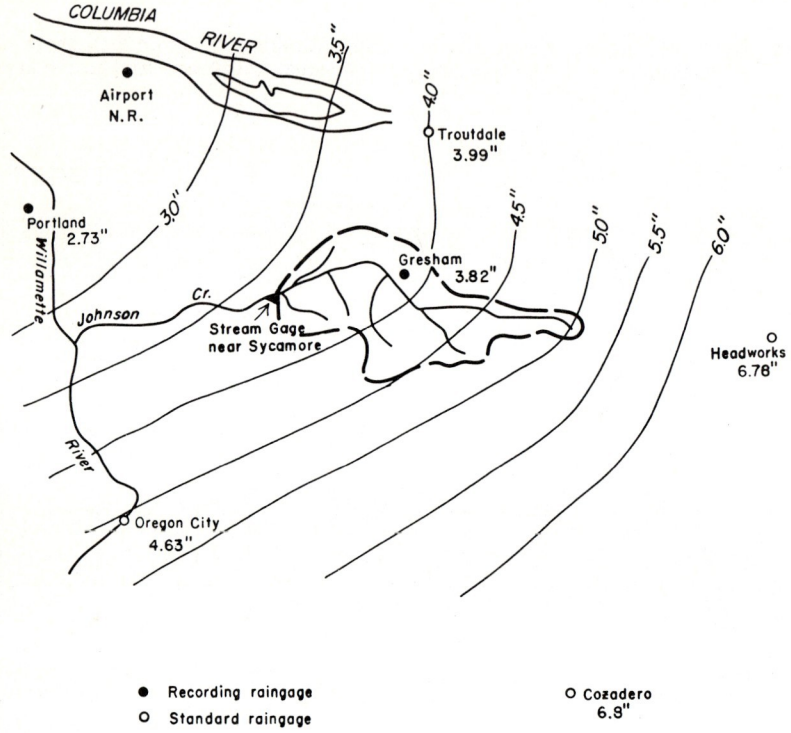

Per cent of watershed	Average precipitation, in.	Weighted watershed precipitation, in.
48	3.80	1.82
32	4.23	1.36
20	4.75	0.95
Total.......	....	4.13

FIG. 12-12. Storm isohyets, Johnson Creek, Oregon, watershed. Storm of Feb. 17–19, 1949.

Intermittent showers occurred on Feb. 19, and steady rain began after noon of Feb. 20, continuing after a 1-hr break until the morning of Feb. 22. The isohyetal map for this storm showed that rainfall was relatively uniform over the watershed. The Gresham gage record was used without an areal adjustment. Estimated f_a was only 0.007 in./hr for a 42-hr period.

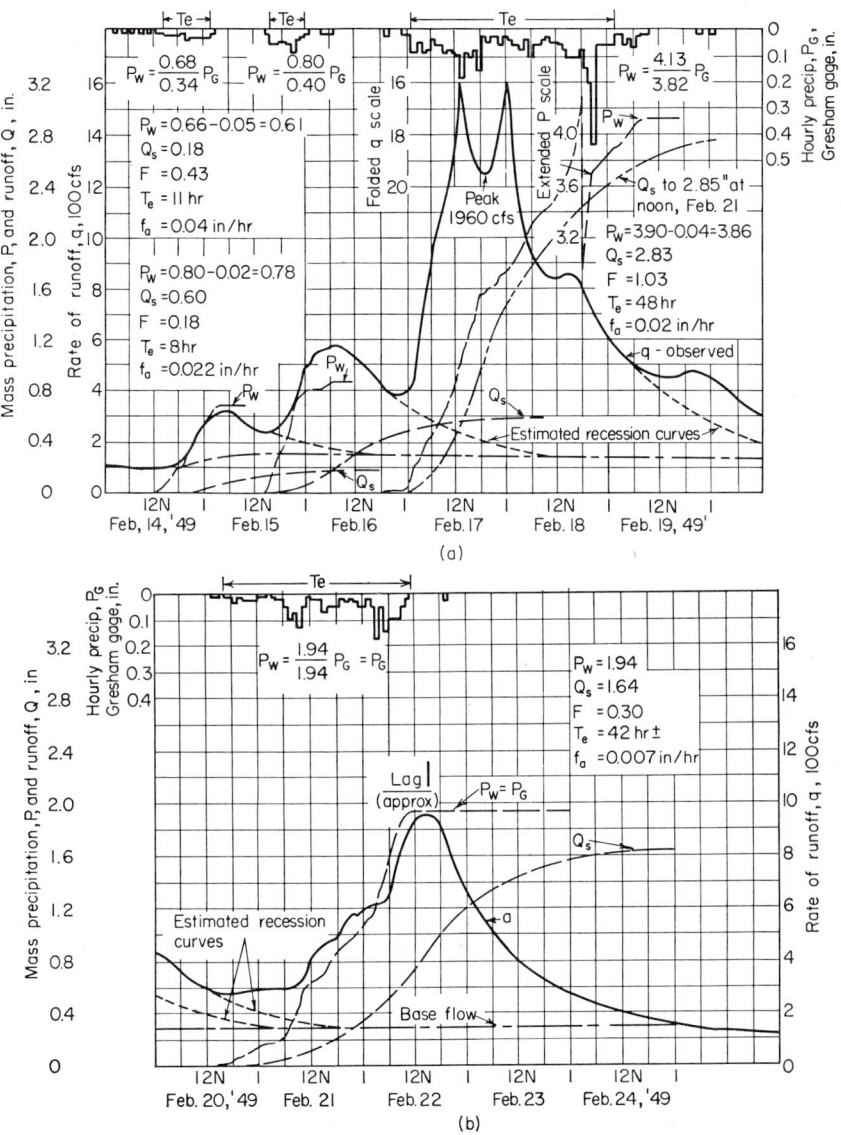

Fig. 12-13. Infiltration estimated by the block method of hydrograph analysis.

V. RELATION OF INFILTRATION TO RUNOFF

An intensive study of the role of infiltration in the rainfall-runoff relationship is evident in comparisons of the hydrograph of runoff from an agricultural watershed with those observed on small 6- by 12-ft rectangular plots located within the watershed for a given storm.

Two watersheds, one of 27 acres all in alfalfa and the other 50 acres in various stands and species of pasture, both instrumented for measurement of rainfall and runoff, were selected as typical of the clay-pan soils of Illinois. These two watersheds were surveyed by use of a type F infiltrometer (1.78 in./hr sprinkled on 6- by 12-ft rectangular plots) to delineate isopotal areas (Subsec. IV-A-2) in Fig. 12-14. Thereafter one 6- by 12-ft rectangular plot equipped for measurement of rainfall and runoff was established in each isopotal area. The nine plots, shown on Fig. 12-14, represent various percentages of the watershed area as indicated. These percentages were used

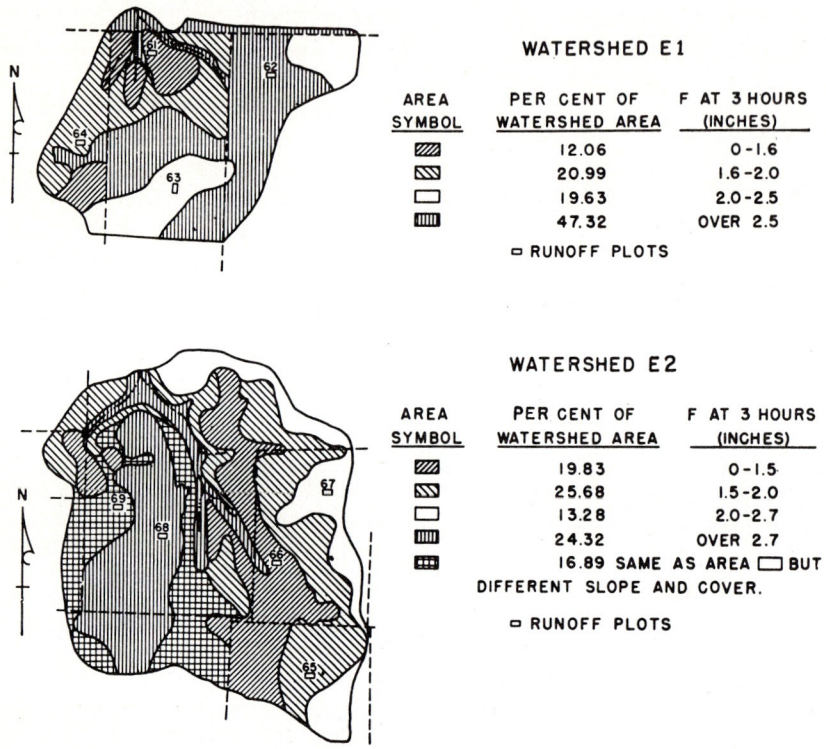

FIG. 12-14. Isopotal areas delineated by survey with type F infiltrometer.

as factors in deriving weighted averages of plot runoff for comparison with observed watershed runoff.

The storm of July 8 and 9, 1942, provided a good opportunity to compare the weighted average of plot runoff with the observed watershed runoff throughout the storm. In Figs. 12-15 and 12-16 the weighted average runoff from small plots (labeled ΣQ) precedes watershed runoff labeled Q_{E1} or Q_{E2} in both time and volume by amounts which could sensibly be attributed to detentions of channel flow. Detentions of overland flow and the curve of infiltration F in these figures were derived by the hydrograph analysis illustrated in Fig. 12-11.

Hydrograph analysis is of necessity a reversal of the actual order of events. Synthesis of the hydrograph, however, considers each event in the order of its occurrence. Rainfall minus interceptions being the causal factor, infiltration proceeds at capacity rate during all periods when the rate of rainfall reaching the ground is adequate and for continued periods to the extent of depression storage or detentions of overland and channel flow; i.e., rainfall excess must be routed through overland flow and, subsequently, through channel flow.

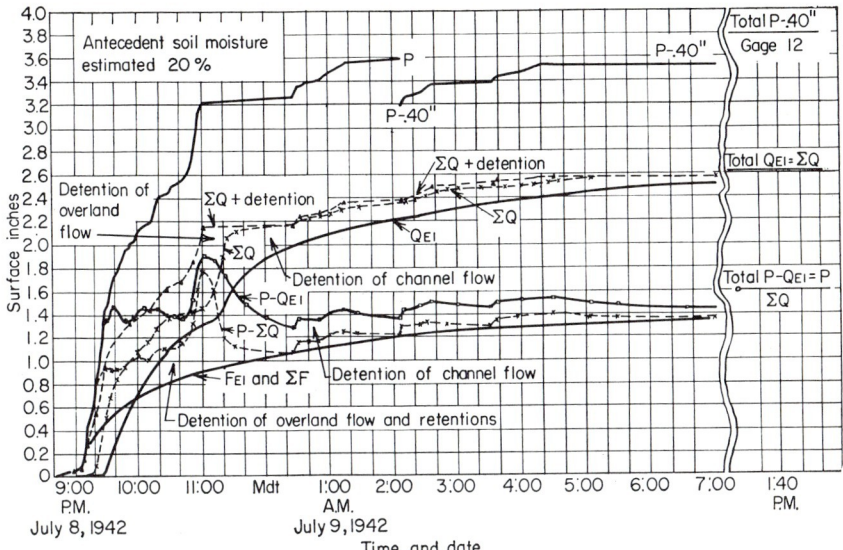

FIG. 12-15. Rainfall, runoff, and infiltration as indicated by plot averages and by measurements on watershed $E1$, Edwardsville, Ill. P = mass rainfall, Q_{E1} = mass runoff from watershed, ΣQ = mass runoff from plots, F_{E1} = mass infiltration on watershed, ΣF = mass infiltration on plots. Detentions are head causing flow given in units of watershed "surface inches." Retentions include pocket storage, interception, evaporation, etc.

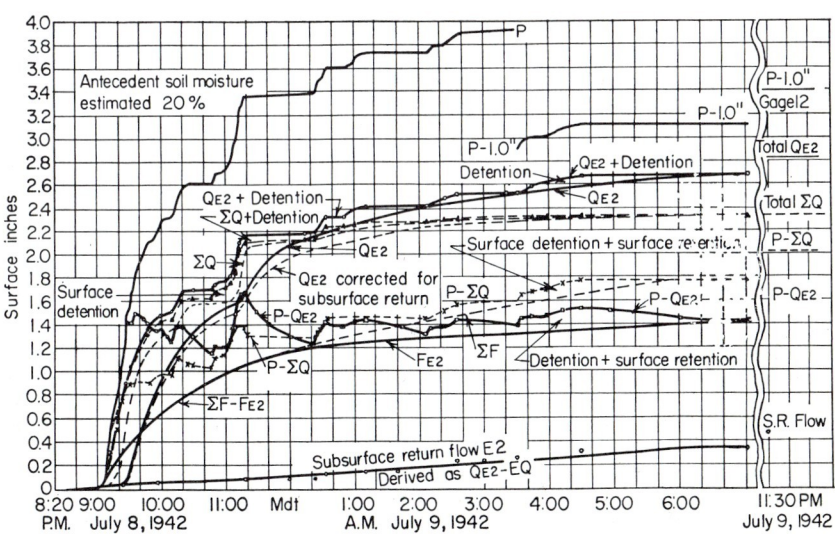

FIG. 12-16. Rainfall, runoff, and infiltration as indicated by plot averages and by measurements on watershed $E2$, Edwardsville, Ill. P = mass rainfall, Q_{E2} = mass runoff from watershed, ΣQ = mass runoff from plots, F_{E2} = mass infiltration on watershed, ΣF = mass infiltration on plots. Detentions are head causing flow given in units of watershed "surface inches." Retentions include pocket storage, interception, evaporation, etc.

Return flow, as illustrated in Fig. 12-16, is not reflected by the hydrologic performance of small plots in the uplands. This is perhaps the greatest hazard in the application of small-plot data to watersheds. Information on the factors influencing return flow is one of our greatest current needs.

Similar unreported attempts using data from a companion study at Colorado Springs, Colo., revealed additional ramifications to be considered. In this drier climate a cycle occurred; prolonged opportunity for infiltration in the draws (i.e., upper reaches of the stream channel in which no defined channel is evident) greatly improved the vegetation, which in turn increased the infiltration capacity of the deep colluvial fills (i.e., sediment deposition from the hill or hills surrounding the draws). Channel losses occurred which reduced 2 and 3 in. of sidehill runoff by two-thirds before it reached the watershed gaging station. These losses were essentially accounted for by applying the infiltration-capacity rate, as measured with the type F infiltrometer, to the area of the draw for the duration of runoff.

VI. INFILTRATION IN COMPUTATIONS OF RUNOFF

Economics of the project and availability of hydrologic information usually dictate the fineness of runoff computations. Projects involving runoff from small areas are quite often high-finance projects such as urban drainage, airports, etc., which can justify a greater expenditure, per unit drainage area, for design. Also, information needed for detailed hydrologic computations can be readily obtained on small areas, whereas larger areas present problems, not only of financing a detailed consideration, but also of attainable accuracy. Rainfall, for example, is a rather straightforward measurement at a given point, but detailed measurement and accurate representation of rainfall over a large area is almost a physical impossibility. Other factors on infiltration such as soils, land use, and soil moisture present similar problems of true representation over large areas. Consequently, the approaches to hydrologic problems on small areas differ materially from the approaches used on larger areas.

A. On Small Areas

Data obtained with the infiltrometer, during isopotal surveys, revealed three major factors affecting the infiltration capacity of the Illinois clay-pan soils, i.e., depth of topsoil, soil moisture, and vegetation. The influence of these factors is evident in Fig. 12-17.

1. Estimating Storm Volumes of Infiltration. The graphs of Fig. 12-17 provide a basis for quickly estimating the volume of infiltration, and subsequently the volume of runoff, to be expected from a given amount of rainfall on these soil-cover complexes. Rainfall minus the related infiltration, corrected for canopy interception and depression storage, is the estimated volume of runoff. Principles of the unit hydrograph are then applied to the runoff volume to give a distribution in time.

2. Rate and Volume of Infiltration throughout the Storm. Because of a short time of concentration, runoff from small areas is very sensitive to changes in rainfall intensity. In projects such as urban drainage where instantaneous rates of runoff are critical to design, it is often desirable to determine the hydrograph of runoff from a specific storm pattern. The curves of infiltration capacity in Fig. 12-17 are typical of the information needed.

The infiltration-capacity curve for a given soil-cover–moisture complex is subtracted from the curve of storm-rainfall pattern as indicated in Fig. 12-18 to derive excess rainfall (labeled on-site runoff in the legend of this figure).

The accumulative curve of infiltration capacity is positioned on the same graph with the curve of accumulated rainfall, base lines coincident, so that the rainfall curve is tangent to it at some point. Thereafter infiltration proceeds at capacity rate whenever the rainfall curve is steeper than the infiltration-capacity curve and will continue beyond the cessation of rainfall to the extent of depression storage. When rainfall resumes, both depression storage and the rate of infiltration must be satisfied before runoff can occur. If the period of cessation is significant, percolation may remove

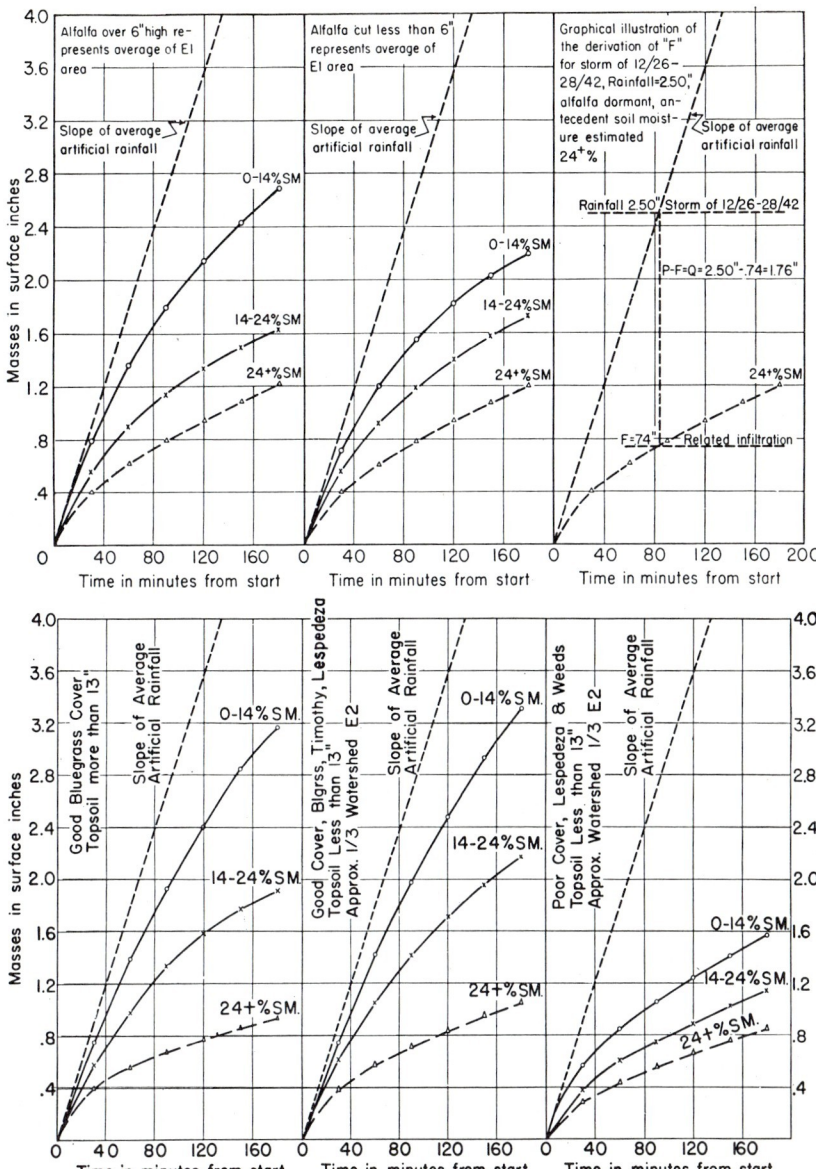

FIG. 12-17. Average infiltration curves as determined by the type F infiltrometer and a graphical illustration of the derivation of F for a natural storm. (SM = soil moisture.)

some of the gravitational water to the extent that porosity storage is available. This porosity storage and depression storage cause what is commonly termed an *infiltration recovery*. Estimates of porosity storage thus made available would be very difficult and are not usually attempted. Estimates of depression storage are given for a few conditions in Table 12-5.

The curve of infiltration plus storage is subtracted from the curve of rainfall to derive the curve of rainfall excess, labeled $Q = P - F - Sd$ in Fig. 12-18. Rainfall excess is then routed through overland flow by use of detention-flow relationships such as those presented in Fig. 12-6 to obtain the hydrograph of inflow to gutters or channels as the case may be.

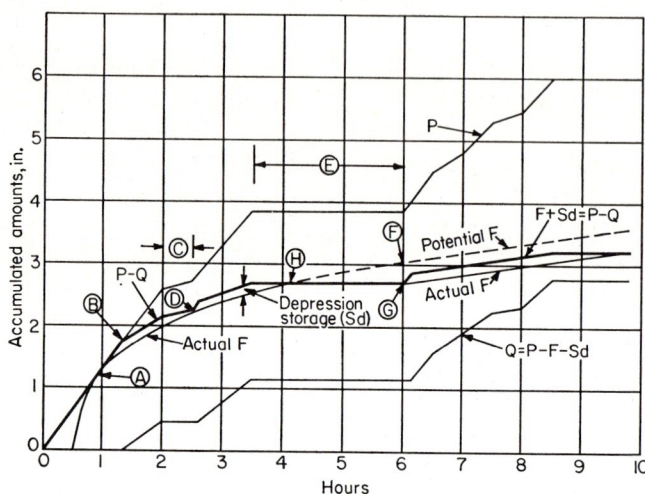

Fig. 12-18. Application of an infiltration curve in estimating runoff. (*U.S. Soil Conservation Service* [9].)

1. Plot accumulative rainfall P.
2. Plot accumulative infiltration F. Match F to P curve where rates (slopes of lines) are equal as at A.
3. Estimate initial abstractions to be used. In this example, only a depression storage of 0.15 in. is used. At B, where $F + Sd = P$, the line of $P - Q$ diverges from and remains below the P line.
4. During time C, rainfall is less than infiltration, and Sd is reduced. Since $Q = 0$ when $\Delta F \geq \Delta P$, then $\Delta P - \Delta F = \Delta Sd$. For the period 2:00 to 2:30, $\Delta Sd = 0.12 - 0.23 = -0.11$.
5. Sd begins to rise at D. When Sd again equals 0.15 in., the Q curve begins to rise. Other forms of initial abstraction, such as interception, are sometimes used only once, at the beginning of the storm.
6. During the time E, potential F continues to rise. When the slope of P again exceeds the F slope, the infiltration curve is brought down parallel, F to G, and the work continues. (Theoretically, the line should be brought across from H to G. Often, as in this case, there is little difference in the results.)
7. Q is on-site runoff. For downstream locations, either route Q or use unit hydrographs

If the plots used for infiltrometer tests or rainfall-runoff studies are of sufficient area, good estimates of depression storage and of the detention-flow relationship (Fig. 12-6) can be derived by hydrograph analysis [14]. The 6- by 12-ft rectangular plot used in the isopotal survey proved adequate, but smaller plots, 1 by 2 ft rectangular, were not suited for this purpose. Musgrave and Norton [15] derived mathematical relationships between storages in mechanically induced depressions and the slope of the land. These are given in Fig. 12-19.

B. On Larger Watersheds

The application of small-plot and infiltrometer-test data to predict the hydrologic performance of larger complex watersheds is extremely difficult and usually limited by

INFILTRATION IN COMPUTATIONS OF RUNOFF

Table 12-5. Estimated Limits of Depression Storage*

Land slope, %	Alfalfa,† in.	Pasture,† in.	Contour furrows,‡ in.
0	0.10–0.30	0.04–0.26	2.0–3.0
2	0.07–0.15	0.02–0.17	
4		0.02–0.13	
6	0.05	0.02–0.12	1.0–2.1
8	0.03–0.10	0.02–0.11	
10	0.02–0.08	0.02–0.10	
12	0.02–0.07	0.02–0.09	0.5–1.3
16	0.02–0.06	0.02–0.07	0.2–0.8
20			0–0.6
24			0–0.3

* Large depressional areas such as potholes, sinks, etc., are not included.
† Data obtained from analyses of rainfall and runoff on 6- and 12-ft plots.
‡ Computed for unbroken furrows. Average represents ordinary lister furrows on the contour. Some breakage of furrows on cultivated land may be expected during large storms.

a paucity of detailed information needed for such computations. Perhaps the greatest contributions resulting from efforts along this line are the concepts thus derived. Generally, greater reliance is placed upon less detailed analyses of data from watersheds which are not strictly homogeneous and applying the results to other watersheds of similar heterogeneity.

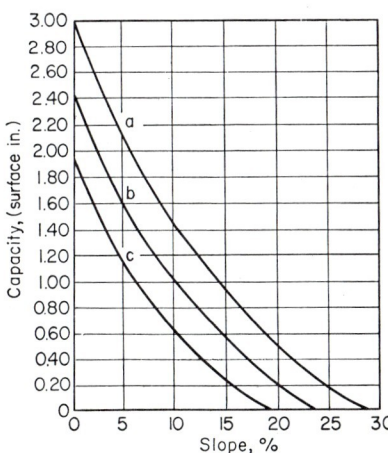

FIG. 12-19. Relation of degree of land slope to theoretical capacities for impounding surface water on listed corn. (*Musgrave and Norton* [15].) The amount of water which may be impounded by a given treatment depends largely on the land slope to which the treatment is applied. *a, b, c* denote various treatments; for example, *a* for 3.0 surface inches at 0% slope.

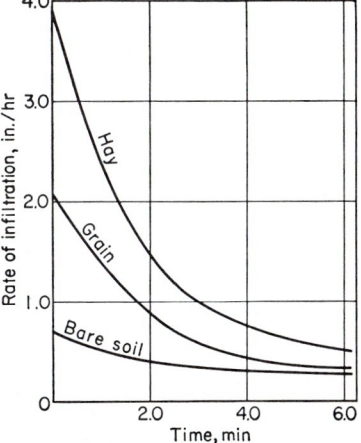

FIG. 12-20. Typical standard infiltration curves for hay, grain, and bare soil. (Computed from Fig. 12-7.)

Table 12-6. Array of Soils in Order of Minimum Infiltration Rate*

D—Lowest Group

Minimum infiltration rate: 0 to 0.05 in./hr.
Includes soils of high swelling per cent, heavy plastic clays, and certain saline soils.
Examples (from low to high continuing from first to second and third columns):

Houston	Trinity	Lufkin
Austin	Susquehanna	Some gumbos

C—Below-average Group

Minimum infiltration rate: 0.05 to 0.15 in./hr.
Includes many clay loams, shallow sandy loams, soils low in organic matter, and soils usually high in clay.
Examples (from low to high):

Bellmont	Ava	Bates
Bluford	Berwick	Shelby
Cisne	Bogota	Iredell
Eylar	Del Rey	Elkton
Jacob	Atterbury	Vernon
Okaw	Batavia	Cecil clay loam
Racoon	Clarksdale	Dunkirk
Rushville	Elliott	Miami
Weir	Shiloh	Fillmore
Breese	Upshur	Butler
Cowden	Putnam	Kirkland
Ebbert	Muskingum	Rosebud
Clarence	Westmoreland	Myatt
Patton	Parsons	Kalmia
Rantoul	Volusia	Appling
Swygert	Viola	Seneca
Wabash	Crown heavy clay	

B—Above-average Group

Minimum infiltration rate: 0.15 to 0.30 in./hr.
Includes shallow loess and sandy loams.
Examples (from low to high):

Arenzville	Melbourne	Carrington
Camden	Sylvan-Blair	Hopi
Youthful Ava	Athena	Ruston
Walla Walla	Davidson	Aiken
Sharpsburg	Monona-Marshall	Hagerstown
Selah	Ida	Hamburg
Buell	Tama	Muscatine
Badger	Marshall	Saybrook
Clinton	Fremont	Harpster
Colby	Webster-Clarion	Ellison
Greenville	Fayette	Kincaid
Boone	Seaton	Waukesha
Red Bay	Sylvan	Judson
Cecil fine sandy loam	Flanagan	Honeoye
Palouse	Huntsville	Madison
Dubuque	Tama	Durham
Kirkland	Orangeburg	

A—Highest Group

Minimum infiltration rate: 0.30 to 0.45 in./hr.
Includes deep sand, deep loess, aggregated silts.
Examples (from low to high):

Knox	Nebraska sand hills	Southeast sand hills
Other deep loess		

* The soils listed have had some measurements of infiltration, on which the tentative array here presented is based. It is recognized that some shifting of order is still necessary. Other soils may be added on the basis of the judgment of soils technicians. The names of some of the soils have been changed through recorrelation. The entire list, therefore, is tentative.

INFILTRATION IN COMPUTATIONS OF RUNOFF   12-27

1. The Standard Curve. The use of *standard infiltration-rate curves* is becoming quite a common practice. Curves derived from analyses of a number of storms on single-practice watersheds are used in arriving at a standard curve of infiltration-capacity rates. The standard infiltration-rate curve for a given soil–land-use complex is derived as the average curve of recession in the infiltration-capacity rate with time, under conditions of continued adequate supply.

Various segments of the standard curve are then associated with antecedent conditions of soil moisture. The three curves presented in Fig. 12-20 are typical standard infiltration curves derived from hay, grain, and bare soil, respectively, on certain soils of the piedmont. For most purposes, these curves would be determined for time durations greater than 1 hr.

Some readily identifiable point on the standard infiltration-rate curve such as the rate at the end of 1 hr (f_1) is selected for association with rather broad groups of soils. Thus the standard curve provides a basis for classifying or grouping soils in accordance with their infiltration capacity.

Similarly, the ratio of f_1 for a soil with a given vegetation to the f_1 of the same soil in a bare condition provides a means of hydrologically evaluating various types and conditions of vegetation.

The segment of the standard infiltration-rate curve selected to represent soil, vegetation, and antecedent moisture conditions, is positioned on the same graph with the histogram of rainfall intensities with base lines coincident. The area between the two curves is rainfall excess to be routed through overland and channel flows.

2. Soil and Vegetative Arrays and Soil Moisture. On a production basis, the engineer usually cannot afford to apply detailed computational methods in evaluating the hydrologic performance of large drainage areas. Generally, it is more feasible to estimate the volume of runoff by applying a coefficient to the volume of precipitation and to distribute this volume in time by applying principles of the unit hydrograph. The relationship of storm runoff volume to storm rainfall is greatly affected by soils, vegetation, and antecedent soil moisture, as shown for infiltration in Figs. 12-7 and 12-17. These three factors are used as bases for grouping data to determine the rainfall-runoff relationship within each soil-cover–moisture complex. The result is a family of curves representing rainfall-runoff relationships for the various complexes.

a. The Soil Array. Rainfall and runoff records are available for a number of the major soils in the United States. Hydrograph analyses have provided some very good estimates of the infiltration capacities of these soils. Although the infiltration capacity of a given soil varies considerably, because of differences in vegetation and soil moisture, the relatively constant rate of infiltration after a period of prolonged wetting provides a fairly stable value for a given land use.

The soils listed in Table 12-6 are grouped in accordance with their infiltration capacity, after a period of prolonged wetting, when planted to clean tilled crops [15]. The array of these soils, both intergroup and introgroup, was derived mostly from analyses of runoff hydrographs. Infiltrometer data and intimate knowledge were used to establish the juxtaposition of others. The order of listing in Table 12-6 and the values indicated in the graphical array of Fig. 12-21 are tentative, subject to revision or verification by further testing. In their present state, Table 12-6 and Fig. 12-21 are used primarily as a basis for relative grouping in derivations and applications of rainfall-runoff relationships.

b. The Land-use Array. The hydrologic classification of vegetation does not necessarily agree with its economic value. Massiveness of root system, protection afforded the ground surface, effects upon soil structure, and the rate of moisture extraction are more pertinent to the infiltration capacity than is the forage production or market value of a crop. For example, alfalfa is a better producer of forage than bluegrass, yet the massive root system of the latter increases the infiltration capacity of a soil considerably beyond that of alfalfa. Table 12-7 lists various land-use practices in the estimated order of their influence upon the infiltration capacities of various soils. The order is that indicated by analyses of hydrographs from plots and single-practice watersheds and by infiltrometer tests.

12-28 INFILTRATION

  c. *Antecedent Soil Moisture.* Of the three major factors affecting infiltration, the least factual data are available on soil-moisture conditions at the start of the storm. Recently developed techniques using neutron scatter appear to offer good possibilities for determining the zone of hydrologic activity and for quantitatively representing the soil-moisture status of a watershed at some future time. For the present, the four

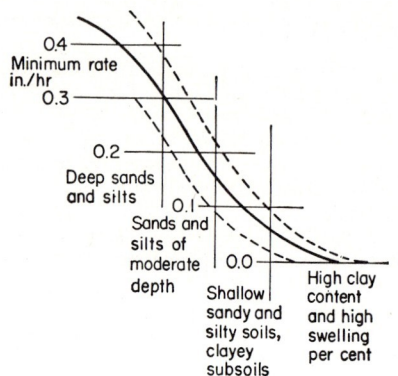

FIG. 12-21. Range of minimum infiltration rates with row crops on wet soils (the variation due to past treatment shown by dashed lines about the mean). (*Musgrave* [16].)

points of equilibrium, i.e., saturation, field capacity, the wilting point, and hygroscopic moisture, are used for moisture classification in determining the rainfall-runoff relationship. Many schemes have been devised for estimating the antecedent moisture status relative to these points of equilibrium. Included are:

  1. Indices based upon summations of rainfall during a fixed number of days previous to the storm producing runoff, or based upon the summation of the quotients obtained by dividing each preceding rainfall amount by the number of days previous to the runoff-producing storm
  2. Indices based upon the status of base flow in the area
  3. Indices derived by soil-moisture accounting

  None of these indices are suited for universal application, yet each of them is credited with applicability in some area. Progress in this particular phase of hydrology is

Table 12-7. Land Uses in Order of Associated Infiltration*

1. Fallow
2. Row crops, poor rotation†
3. Row crops, good rotation‡
4. Pasture, poor
5. Legumes after row crops
6. Small grains, poor rotation
7. Small grains, good rotation
8. Pasture, fair
9. Woods, poor
10. Pasture, good
11. Woods, fair
12. Meadows
13. Woods, good

* From U.S. Soil Conservation Service [9].
† One-fourth or less in hay or sod.
‡ More than one-fourth of rotation in hay or sod.

dependent upon the development of techniques for obtaining continuous records of soil moisture and a means of representing the watershed condition of soil moisture rather than moisture at a single site.

C. Infiltration Indexes

  Infiltration indexes, in general, express infiltration as an average rate throughout the storm. Since the infiltration capacity actually decreases with prolonged rainfall, the use of an average value assumes too little infiltration during the first part of the storm and too much near the end of it. For this reason, infiltration indexes are best suited for derivation and use pertinent to major flood-producing storms occurring on wet soils, or storms of such intensity and duration that the rate of infiltration might be assumed to have reached a final constant rate prior to or early in the storm. The application of such indexes to moderate storms is entirely an empirical procedure, and strict attention must be given to matching conditions of antecedent soil moisture and storm durations [17].

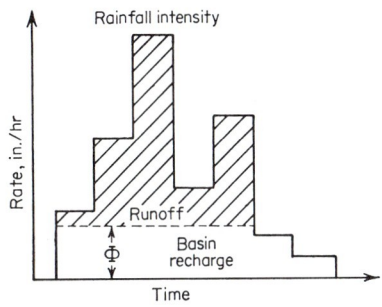

FIG. 12-22. Schematic diagram illustrating the derivation and meaning of the Φ index. (*Linsley, Kohler, and Paulhus* [18].)

1. **The Φ Index.** The Φ index is an average rate of infiltration derived from a time-intensity graph of rainfall, as illustrated in Fig. 12-22, in such a manner that the volume of rainfall in excess of this rate will equal the volume of storm runoff.

2. **The W Index.** This index is a refinement of the Φ index in that it excludes surface storage and retentions. The equation is as follows:

$$W = \frac{P - Q - S}{t_f} \qquad (12\text{-}4)$$

where W is the average rate of infiltration, P is the total storm rainfall, Q is the total storm runoff, S is the volume of depression storage plus canopy interception, and t_f is the total time during which rainfall intensity is greater than W.

3. **The W_{min} Index.** As indicated by the subscript, the W_{min} index is obtained under very wet conditions when the infiltration capacity has reached a final, constant, minimum rate.

4. **Initial Abstractions.** Initial abstractions, sometimes referred to as initial losses, are defined as the maximum amounts of rainfall that can be absorbed under specific conditions without producing runoff. Estimates of initial losses for the humid areas of the United States may range from a few tenths of an inch in wet seasons to approximately 2 in. during the dry summer and fall months [19]. For conditions usually preceding major floods in humid regions, initial losses approximate 0.2 to 0.5 in. and are relatively small in comparison with the flood-runoff volumes [19]. Consequently, they may be neglected in these instances without appreciable error being introduced. The initial abstraction is an important estimate for storms occurring on dry antecedent conditions since it tends to offset underestimation of initial infiltration capacity resulting from the use of an average rate.

5. **Rainfall Distribution.** Unequal distribution of rainfall over the drainage area might affect the derivation and application of the infiltration index. If significant portions of the area receive rainfall at rates appreciably less than the infiltration capacity while rainfall in other portions is producing runoff, the difference between average rainfall and runoff would be a poor estimate of the infiltration capacity.

The usual practice is to subdivide the drainage area by Thiessen polygons or by isohyetal bands and to prepare average curves of rainfall vs. time for each subarea.

Rainfall is then tabulated for successive time increments for each polygon or isohyetal area. Standard increments of time are selected to isolate, as closely as possible, periods of rainfall in excess of infiltration from periods of lesser intensities. Infiltration cannot exceed rainfall, except as briefly supplied by surface flow or surface storage; hence rainfall excess is determined for each polygon or isohyetal area for each intensity period. The summation of rainfall excess volumes, weighted by area, should equal the total volume of storm runoff from the basin.

VII. A GLANCE AHEAD

Despite the fact that infiltration is a relatively youthful branch of the science of hydrology, much progress in its understanding has been made in recent years. The research of the future may be expected to make still greater progress. The fact that it is subject to modification by man may readily lead to its greater utilization by coming generations. As infiltration is a primary source of water used by plants, animals, and man, future research quite likely will lead to greater efficiency in such use, a matter which rapidly growing populations make vitally necessary.

VIII. REFERENCES

1. Mavis, F. T., and E. F. Wilsey: A study of the permeability of sand, *Univ. Iowa Studies in Eng. Bull.* 7, 1936.
2. Sherman, L. K.: The unit hydrograph method, in O. E. Meinzer (ed.), "Physics of the Earth," vol. IX, "Hydrology," Dover Publications, Inc., 1949, p. 520.
3. Schiff, Leonard: Surface runoff supply estimates based on soil-water movements and precipitation patterns, *U.S. Dept. Agr., Soil Conserv. Serv.*, TP-86, October, 1949.
4. Philips, J. R.: The theory of infiltration, pts. I–VII, *Soil Sci.*, vol. 83, pp. 345–375, 435–448, 1957; vol. 84, pp. 163–178, 257–264, 329–346, 19J7; vol. 85, pp. 278–286, 333–337, 1958.
5. Holtan, H. N., and M. H. Kirkpatrick, Jr.: Rainfall, infiltration, and hydraulics of flow in run-off computation, *Trans. Am. Geophys. Union*, vol. 31, pp. 771–779, 1950.
6. Horton, R. E.: Analysis of runoff-plat experiments with varying infiltration-capacity, *Trans. Am. Geophys. Union*, vol. 20, pp. 693–711, 1939.
7. Horton, R. E.: An approach toward a physical interpretation of infiltration capacity, *Proc. Soil Sci. Soc. Am.*, vol. 5, pp. 399–417, 1940.
8. Holtan, H. N., and G. W. Musgrave: Soil water and its disposal under corn and under bluegrass, *U.S. Dept. Agr., Soil Conserv. Serv.*, TP-68, July, 1947.
9. Hydrology, sec. 4 of "Engineering Handbook," U.S. Department of Agriculture, Soil Conservation Service, 1956.
10. Sharp, A. L., and H. N. Holtan: Extension of graphic methods of analysis of sprinkled-plot hydrographs to the analysis of hydrographs of control plots and small homogeneous watersheds, *Trans. Am. Geophys. Union*, vol. 23, pp. 578–593, 1942.
11. Holtan, H. N.: Time condensation in hydrograph analysis, *Trans. Am. Geophys. Union*, vol. 26, pp. 407–413, 1945.
12. Horner, W. W., and C. L. Lloyd: Infiltration capacity values, *Trans. Am. Geophys. Union*, vol. 21, pp. 522–541, 1940.
13. Sherman, L. K.: Derivation of infiltration capacity from average loss rates, *Trans. Am. Geophys. Union*, vol. 21, pp. 541–550, 1940.
14. Sharp, A. L., and H. N. Holtan: A graphical method of analysis of sprinkled-plot hydrographs, *Trans. Am. Geophys. Union*, vol. 21, pp. 558–570, 1940.
15. Musgrave, G. W., and R. A. Norton: Soil and water conservation investigations at the Soil Conservation Experiment Station, Missouri Valley loess region, Clarinda, Iowa, *Sta. Prog. Rept.* 1931–1935 and *U.S. Dept. Agr., Tech. Bull.* 558, pp. 58–60, February, 1937.
16. Musgrave, G. W.: How much of the rain enters the soil? in "Water," U.S. Department of Agriculture Yearbook, 1955, pp. 151–159.
17. "Hydrology Handbook," American Society of Civil Engineers, Manuals of Engineering Practice, no. 28, pp. 57–60, 1949.
18. Linsley, R. K., Jr., M. A. Kohler, and J. L. H. Paulhus: "Applied Hydrology," McGraw-Hill Book Company, Inc., New York, 1949, pp. 417–427.
19. "Flood-hydrograph Analysis and Computations," U.S. Army Corps of Engineers, Engineering and Design Manuals, EM 1110-2-1405, Aug. 31, 1959.

Section 13

GROUNDWATER

DAVID KEITH TODD, *Professor of Civil Engineering, University of California.*

I. Introduction	13-2
II. Groundwater Resources	13-3
A. Groundwater in the Hydrologic Cycle	13-3
B. Occurrence of Groundwater	13-3
C. Groundwater Use in the United States	13-5
III. Groundwater Movement	13-7
A. Darcy's Law	13-7
B. Coefficient of Permeability	13-8
C. Permeability Measurement	13-10
1. Formulas	13-10
2. Laboratory Measurements	13-10
3. Field Measurements	13-11
D. Tracing Groundwater Movement	13-12
E. Basic Flow Equations	13-13
1. Steady Flow	13-13
2. Unsteady Flow	13-14
F. Flow-line Analysis	13-14
1. Flow Nets	13-14
2. Refraction at Permeable Boundaries	13-15
IV. Well Hydraulics	13-15
A. Steady Radial Flow	13-16
1. Confined Aquifer	13-16
2. Unconfined Aquifer	13-17
B. Well in a Uniform Flow	13-17
C. Unsteady Radial Flow—Confined Aquifer	13-18
D. Unsteady Radial Flow—Unconfined Aquifer	13-20
E. Unsteady Radial Flow—Leaky Aquifer	13-21
F. Well Flow near Aquifer Boundaries	13-22
1. Well Flow near a Stream	13-22
2. Location of an Aquifer Boundary	13-23
3. Multiple Boundaries	13-25
G. Multiple Well Systems	13-25
H. Partially Penetrating Wells	13-26
I. Well Losses	13-27
V. Water Wells	13-28
A. Shallow Wells	13-28
B. Deep Wells	13-28

C. Well Completion, Development, and Testing............ 13-30
  D. Sanitary Protection of Wells......................... 13-31
  E. Maintenance and Repair of Wells..................... 13-31
  F. Collector Wells and Galleries........................ 13-31
VI. Groundwater Fluctuations.............................. 13-33
  A. Secular and Seasonal Effects........................ 13-33
  B. Streamflow Effects.................................. 13-34
  C. Evapotranspiration Effects.......................... 13-35
  D. Atmospheric-pressure Effects........................ 13-36
  E. Tidal Effects....................................... 13-37
VII. Groundwater Management............................... 13-38
  A. Safe Yield.. 13-38
  B. Equation of Hydrologic Equilibrium.................. 13-39
  C. Basic-data Collection............................... 13-39
    1. Surface Inflow and Outflow; Imported and Exported Water... 13-39
    2. Precipitation..................................... 13-40
    3. Consumptive Use................................... 13-40
    4. Changes in Surface Storage........................ 13-40
    5. Changes in Groundwater Storage.................... 13-40
    6. Subsurface Inflow and Outflow..................... 13-40
  D. Determination of Safe Yield......................... 13-40
  E. Conjunctive Use of Water Resources.................. 13-41
VIII. Artificial Recharge of Groundwater................... 13-41
  A. Artificial-recharge Practice........................ 13-41
  B. Water Spreading..................................... 13-42
    1. Basin Method...................................... 13-43
    2. Modified-stream-bed Method........................ 13-43
    3. Ditch or Furrow Method............................ 13-43
    4. Flooding Method................................... 13-44
  C. Recharge through Pits............................... 13-44
  D. Recharge through Wells.............................. 13-44
  E. Sewage and Waste-water Recharge..................... 13-45
  F. Induced Recharge.................................... 13-46
IX. Salt-water Intrusion of Coastal Aquifers................ 13-46
  A. Salt-water Sources.................................. 13-46
  B. Intrusion in the United States...................... 13-47
  C. The Ghyben-Herzberg Concept......................... 13-48
  D. The Dynamic Concept................................. 13-49
  E. Location of the Interface........................... 13-49
  F. Structure of the Interface.......................... 13-50
  G. Control of Intrusion................................ 13-51
  H. Development of Groundwater in Intrusion Areas....... 13-52
X. Notation.. 13-52
XI. References... 13-53

I. INTRODUCTION

Groundwater represents the largest available source of fresh water in the hydrologic cycle. Emphasis in this section is on quantitative aspects of groundwater as a water-supply source. The fundamentals of groundwater flow and well hydraulics are presented. Water wells are described, followed by discussion of groundwater fluctuations and management of groundwater. Special problems of artificial recharge and salt-water intrusion are included.

II. GROUNDWATER RESOURCES

Groundwater refers to water in a saturated zone of a geologic stratum. Water also occurs underground in unsaturated zones where voids are filled with water and air. Underground water constitutes the largest available source of fresh water—far greater than all the lakes, reservoirs, and streams combined [1]. As such, the proper development and utilization of this major renewable natural resource is of interest for all water-supply requirements.

A. Groundwater in the Hydrologic Cycle

The use of groundwater dates from ancient times, but its origin, occurrence, and movement were not clearly understood until the end of the seventeenth century [2].

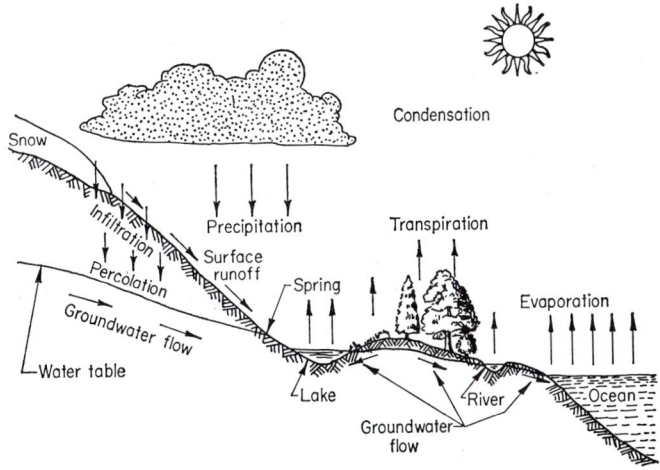

FIG. 13-1. The hydrologic cycle. (*After Todd* [1].)

Groundwater is one portion of the earth's hydrologic cycle. Figure 13-1 indicates the role of groundwater in the cycle. Permeable formations in the earth's crust serve as conduits for transmission and as reservoirs for storage of water. Essentially, all groundwater is in motion; velocities range from a few feet per day to only a few feet per year. As such, groundwater provides large, widely distributed sources of water supply. Groundwater discharging into streams maintains streamflow during periods of no surface runoff. In arid zones, pumped water is the only available supply for large portions of each year.

Groundwater originates for all practical purposes as surface water. Water infiltrates into the ground from natural recharge of precipitation, streamflow, lakes, and reservoirs. In addition, efforts by man constitute artificial recharge. Once underground, the water moves downward, under the action of gravity. When a zone of saturation is reached, the water flows in a direction controlled by the hydraulic boundary conditions. Discharge of groundwater represents a return of the water to the earth's surface. Most discharge is into surface-water bodies. Spring flow, evaporation, and transpiration are other modes of discharge. Pumping of wells is the primary artificial discharge method.

B. Occurrence of Groundwater

Usable groundwater occurs in permeable geologic formations known as *aquifers*. These permit appreciable water to move through them under usual field conditions.

Types of aquifers and the geologic conditions under which they occur are described in Sec. 4-I.

Groundwater occurs in the voids, or pores, of geologic formations. *Porosity* measures this void space and is defined by

$$\alpha = \frac{w}{V} \tag{13-1}$$

where w is the volume of water required to saturate all voids, and V is the total volume of the rock. It is usually expressed as a percentage. Representative values are listed in Table 13-1 for sedimentary materials.

Table 13-1. Representative Porosity Ranges for Sedimentary Materials*

Material	Porosity, %
Soils	50–60
Clay	45–55
Silt	40–50
Medium to coarse mixed sand	35–40
Uniform sand	30–40
Fine to medium mixed sand	30–35
Gravel	30–40
Gravel and sand	20–35
Sandstone	10–20
Shale	1–10
Limestone	1–10

* From Todd [1].

When groundwater is pumped or drained, some water is retained by molecular and surface-tension forces. *Specific retention* is the ratio (usually expressed as a percentage),

$$S_r = \frac{w_r}{V} \tag{13-2}$$

where w_r is the volume occupied by retained water. The water removed by the force of gravity is the *specific yield*, or *effective porosity*. It is defined by

$$S_y = \frac{w_y}{V} \tag{13-3}$$

where w_y is the volume of water drained and is usually expressed as a percentage. It follows that

$$\alpha = S_r + S_y \tag{13-4}$$

Specific yield depends upon grain size, shape and distribution of pores, and compaction of the formation. In alluvial aquifers values in the range of 10 to 20 per cent are common. Data in Fig. 13-2 from an alluvial basin in southern California show the relation of α, S_r, and S_y as a function of grain size.

Various zones of vertical distribution of groundwater are recognized. Under homogeneous subsurface conditions the distribution of zones shown in Fig. 13-3 would result. Detailed analysis of the zones above the water table is presented in Sec. 5; here discussion is limited to the zone of saturation. This zone is of major importance as a water-supply source.

Water added to or removed from an aquifer represents a change in the storage volume. The *storage coefficient* of an aquifer is the volume of water it releases from or takes into storage per unit surface area of aquifer per unit change in the component of head normal to that surface. Figure 13-4a shows a vertical column of unit area extending through a *confined, or artesian, aquifer*. The storage coefficient S equals the

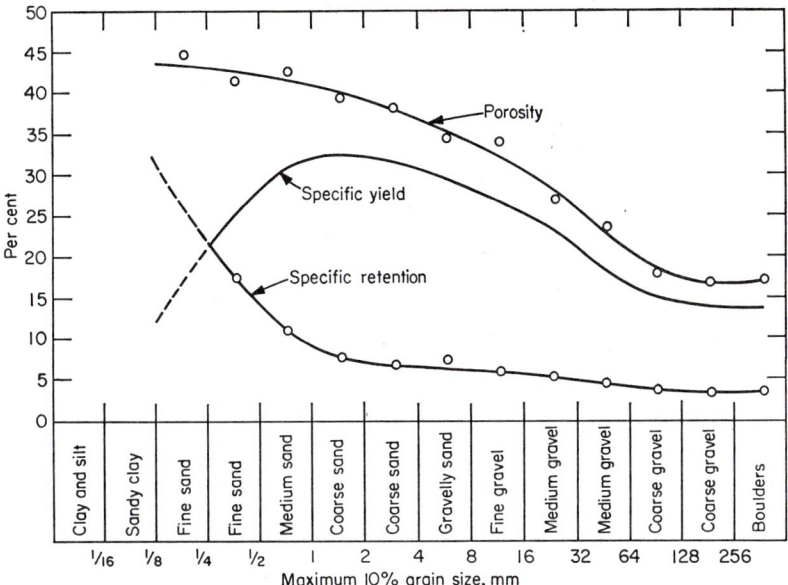

FIG. 13-2. Porosity, specific-yield, and specific-retention variations with grain size, South Coastal Basin, California. The grain size in which the cumulative total, beginning with the coarsest material, reaches 10 per cent of the total sample. (*After Eckis* [3].)

volume of water released when the piezometric head declines a unit distance. Typical values are in the range $0.00005 \leq S \leq 0.005$, indicating that changes in hydrostatic loading cause only very small yields of water from a confined aquifer. For an *unconfined aquifer* (Fig. 13-4b), the storage coefficient equals the specific yield.

C. Groundwater Use in the United States

The rate of use of groundwater in the United States as of 1960 approximated 50 billion gallons per day. This equals about 20 per cent of the total national water use and represents a fivefold increase since 1935. A 50 per cent increase by 1975 is anticipated. Groundwater is pumped from some 14 million wells, the large majority of which are single-family domestic wells [4]. Each year a total of 400,000 new wells are drilled by the more than 9,000 licensed well-drilling contractors.

Data on the relative demand for groundwater by different users are summarized in Table 13-2. California is the leading user of groundwater in the United States, followed by Texas and Arizona. The demand in these three states constitutes more than

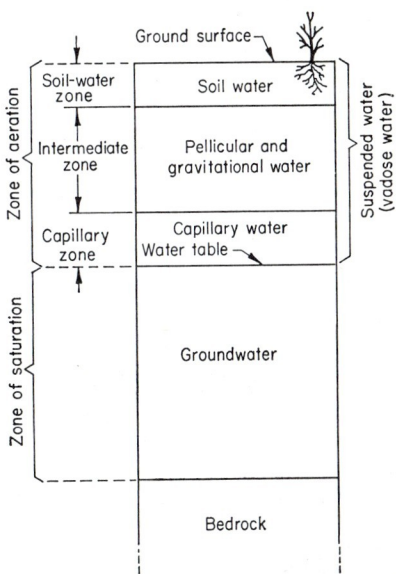

FIG. 13-3. Divisions of subsurface water. (*After Todd* [1].)

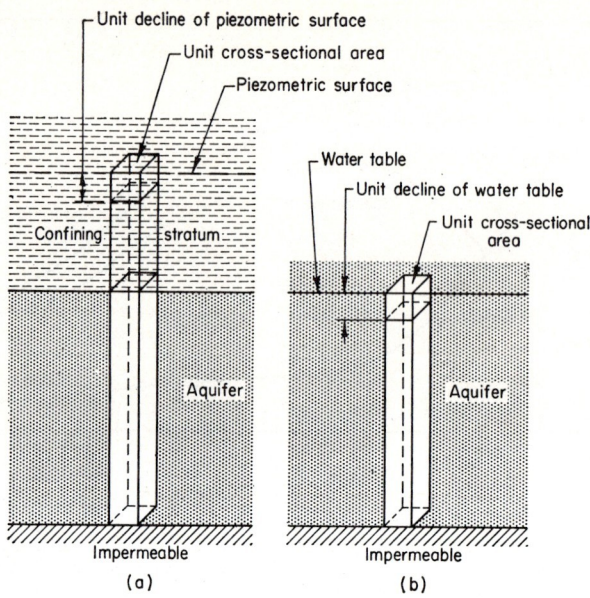

Fig. 13-4. Illustrations defining storage coefficient of (a) confined and (b) unconfined aquifers. (*After Todd* [1].)

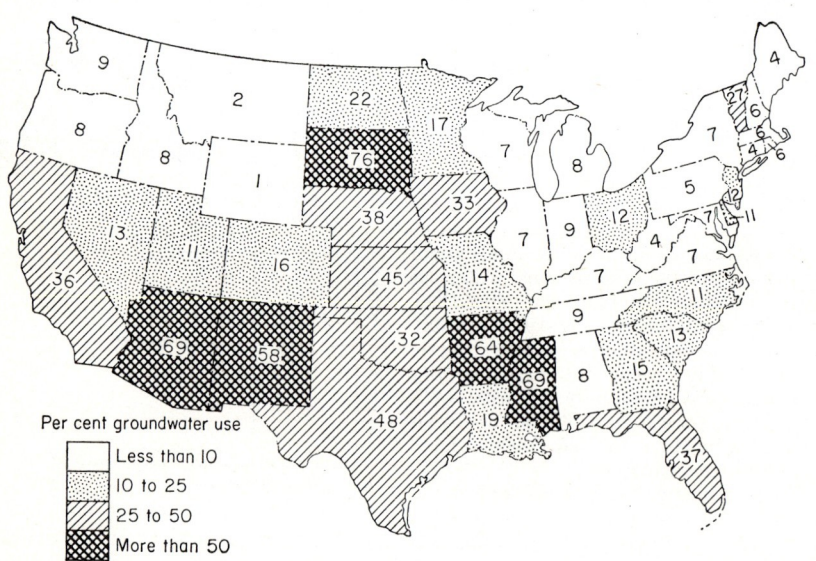

Fig. 13-5. Groundwater use relative to total water use in the United States. (*After MacKichan* [5].)

GROUNDWATER MOVEMENT

half of all groundwater used in the United States. Relative use of groundwater by states is shown in Fig. 13-5. Extensive data on groundwater in the United States are available in publications of the U.S. Geological Survey [6, 7] and in Thomas [8].

Table 13-2. Distribution of Groundwater Use by Type of Consumer*

Consumer	Per cent of total groundwater use
Irrigation	65
Self-supplied industrial water supplies	21
Public water supplies	10
Rural water supplies	4
Total	100

* From MacKichan [5].

In the future, as the demand for groundwater continues to increase, the creation or modification of aquifers for maximum beneficial use is within the realm of possibility. To this end the Plowshare program of the U.S. Atomic Energy Commission is studying uses of nuclear explosives for creating underground reservoirs for water storage or permeable zones permitting recharge of aquifers. Figure 13-6 gives an example of what could be accomplished.

FIG. 13-6. Schematic diagram of creation of underground reservoir by nuclear explosives for storage and recharge of groundwater. (*After Johnson and Brown* [9].)

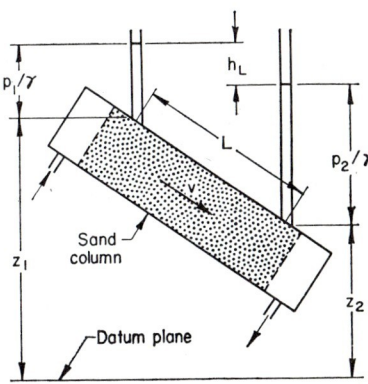

FIG. 13-7. Pressure distribution and head loss in flow through a sand column. (*After Todd* [1].)

III. GROUNDWATER MOVEMENT

Most groundwater is in motion as a result of natural and man-made influences; therefore an understanding of hydraulic principles governing flows through aquifers is essential for quantitative evaluation of groundwater supplies (see also Sec. 5).

A. Darcy's Law

In 1856 Henry Darcy, in a treatise on water supply, reported on experiments of the flow of water through sands. He found that flows were proportional to the head loss and inversely proportional to the thickness of sand traversed by the water. Considering the generalized sand column shown in Fig. 13-7 with a flow rate Q through a cylinder of cross-sectional area A, Darcy's law can be expressed as

$$Q = KA\frac{hL}{L} \qquad (13\text{-}5)$$

More generally, the velocity

$$v = \frac{Q}{A} = K\frac{dh}{dL} \qquad (13\text{-}6)$$

where dh/dL is the hydraulic gradient. The quantity K is a proportionality constant known as the *coefficient of permeability*, or *hydraulic conductivity*. The velocity in Eq. (13-6) is an apparent one, defined in terms of the discharge and the gross cross-sectional area of the porous medium. The actual velocity varies from point to point throughout the column.

Darcy's law is applicable only within the laminar range of flow where resistive forces govern flow [10]. As velocities increase, inertial forces, and ultimately turbulent flows, cause deviations [1] from the linear relation of Eq. (13-6). Fortunately, for most natural groundwater motion, Darcy's law can be applied.

B. Coefficient of Permeability

Rearranging Eq. (13-6),

$$K = \frac{Q}{A(dh/dL)} \qquad (13\text{-}7)$$

illustrating that K has the dimensions of $[L/T]$, or velocity. Units such as centimeters per second and meters per day are commonly used in the metric system. In the United States the U.S. Geological Survey has adopted two definitions of the coefficient of permeability:

Symbol	Name	Definition
K_s	Laboratory (or standard) coefficient of permeability	Flow of water at 60°F in gal/day through a medium having a cross-sectional area of 1 ft² under a hydraulic gradient of 1 ft/ft
K_f	Field coefficient of permeability	Flow of water in gal/day through a cross section of aquifer 1 ft thick and 1 mile wide under a hydraulic gradient of 1 ft/mile at field temperature

For most natural aquifers values of K_s fall in the range of 10 to 5,000.

Analysis of units of K_s and K_f reveals that the only difference is that of temperature. Neglecting the temperature effect on density of water, only viscosity need be considered. Therefore

$$\frac{K_s}{K_f} = \frac{\mu_f}{\mu_{60}} \qquad (13\text{-}8)$$

where μ_f is the viscosity at field temperature, and μ_{60} that at 60°F.

A convenient quantity in calculations of groundwater flow is the *coefficient of transmissibility* T, defined by

$$T = K_f b \qquad (13\text{-}9)$$

where b is the aquifer thickness in ft.

The permeability of a porous medium expresses the ease with which a fluid will pass through it. A rational analysis of permeability requires that it be defined in terms of porous-media properties and independent of fluid properties such as viscosity μ and specific weight γ. Introducing d, a representative pore diameter, then, by dimensional analysis,

$$K = \frac{Cd^2\gamma}{\mu} \tag{13-10}$$

where C is a dimensionless constant. A *specific* (or *intrinsic*) *permeability* k may now be defined by

$$k = Cd^2 \tag{13-11}$$

As the dimensions of k are $[L^2]$, or area, the quantity can be interpreted as a unique pore area governing the flow. From a practical standpoint it is usually necessary to assume d to be proportional to some representative grain diameter. The quantity C depends upon other properties of the medium affecting flow, including porosity, packing, and grain-size distribution and shape. Experimental evidence has shown that the same value of k can be obtained for different liquids and gases flowing through a given medium [11].

Substituting Eqs. (13-10) and (13-11) into Eq. (13-7) leads to

$$k = \frac{\mu Q/A}{\gamma(dh/dL)} \tag{13-12}$$

As the value of k in ordinary length units is extremely small, the *darcy* has been adopted as a more practical unit. From Eq. (13-12),

$$1 \text{ darcy} = \frac{1 \text{ centipoise} \times 1 \text{ cm}^3/\text{sec}/1 \text{ cm}^2}{1 \text{ atm}/1 \text{ cm}}$$

Inserting compatible viscosity and atmosphere units reduces the darcy to units of area. Similarly, with

$$K = \frac{k\gamma}{\mu} \tag{13-13}$$

from the above and appropriate values of γ and μ for water at 60°F, an equivalency between a darcy and K_s can be obtained. Table 13-3 summarizes definitions and

Table 13-3. Permeability Units and Conversions

Permeability units

Laboratory (standard) coefficient of permeability:

$$1K_s = \frac{1 \text{ gal of water at } 60°\text{F}/\text{day}}{(1 \text{ ft}^2)(1 \text{ ft}/\text{ft})}$$

Field coefficient of permeability:

$$1K_f = \frac{1 \text{ gal of water at field temperature per day}}{(1 \text{ ft} \times 1 \text{ mile})(1 \text{ ft}/\text{mile})}$$

Specific (intrinsic) permeability:

$$1 \text{ darcy} = \frac{(1 \text{ centipoise} \times 1 \text{ cm}/\text{sec})/1 \text{ cm}}{1 \text{ atm}/\text{cm}}$$

Conversion factors

$1K_s = 4.72 \times 10^{-5}$ cm/sec
      $= 0.0408$ m/day
      $= 0.134$ ft/day
1 darcy $= 0.987 \times 10^{-8}$ cm^2
        $= 1.062 \times 10^{-11}$ ft^2
        $= 18.2K_s$ for water at 60°F
        $= 0.966 \times 10^{-3}$ cm/sec   for water at 20°C
1 cm/sec $= 1.02 \times 10^{-5}$ cm^2 for water at 20°C

conversion factors for permeability. Figure 13-8 indicates relative magnitudes of k and K_s for different soil classes.

C. Permeability Measurement

Quantitative permeability values can be obtained from formulas based on physical properties of porous media, from laboratory measurements of aquifer samples, and from various field tests.

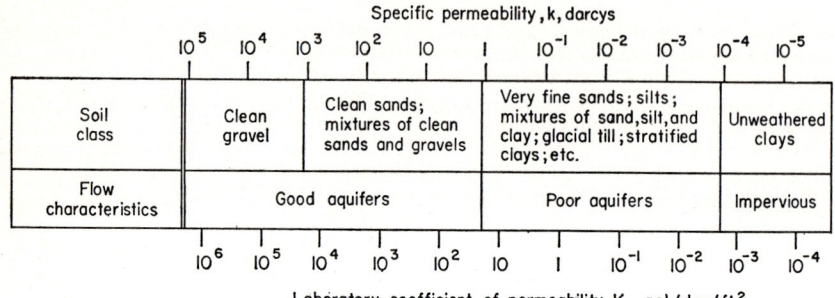

FIG. 13-8. Magnitude of specific permeability and laboratory coefficient of permeability for different classes of soils. (*After Todd* [1].)

1. Formulas. No general expression has been developed for relating permeability to the properties of porous media, although many formulas of limited application appear in the literature. One of the best-known expressions, developed from dimensional considerations and verified experimentally on sands, is that of Fair and Hatch [12]. Specific permeability

$$k = \frac{1}{m\left[\frac{(1-\alpha)^2}{\alpha^3}\left(\frac{\theta}{100}\sum \frac{P}{d_m}\right)^2\right]} \qquad (13\text{-}14)$$

where α is porosity, m is a packing factor equal to about 5, θ is a sand-shape factor varying from 6.0 for spherical grains to 7.7 for angular grains, P is the percentage of sand held between adjacent sieves, and d_m is the geometric mean of rated sizes of adjacent sieves. As the equation is dimensionally correct, any consistent system of units may be used.

2. Laboratory Measurements. Two types of *permeameters* for laboratory measurement of permeability are shown in Fig. 13-9. These can be used for consolidated or unconsolidated formations. The permeability with the constant-head permeameter is given by

$$K = \frac{VL}{Ath} \qquad (13\text{-}15)$$

where V is the flow volume in time t, and the other quantities are as shown in Fig. 13-9a. For the falling-head permeameter, the applicable equation is

$$K = \frac{d_t^2 L}{d_c^2 t} \ln \frac{h_0}{h} \qquad (13\text{-}16)$$

where all quantities are as defined in Fig. 13-9b.

A laboratory measurement of permeability gives only a point result, whereas large variations within an aquifer may exist in the field. Difficulties are often encountered in working with unconsolidated formations. If samples are disturbed and repacked, permeabilities may be significantly modified.

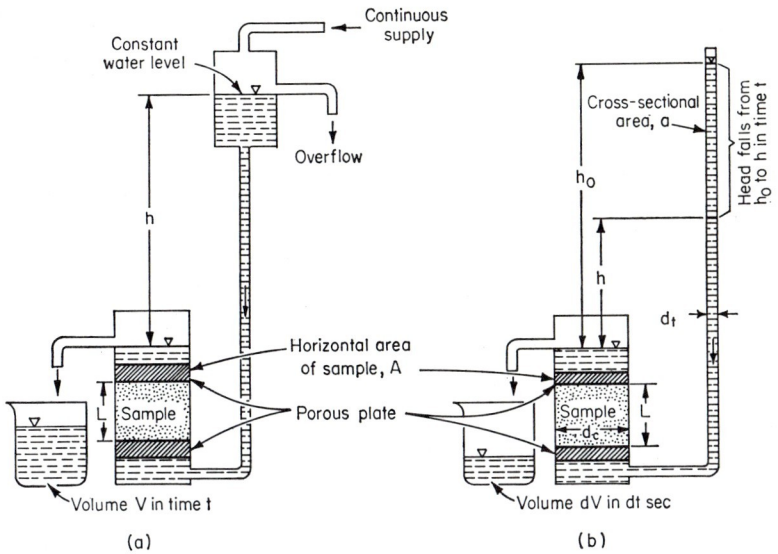

Fig. 13-9. Permeameters: (a) Constant-head; (b) falling-head. (*After Todd* [1].)

3. Field Measurements. Permeabilities at shallow depths can be determined from a single auger hole penetrating the water table (Fig. 13-10). After determining the elevation of the water table in the hole, water is pumped out until a new level is reached and the rate of rise of water is measured. For a homogeneous soil the empirical equation of Ernst (see [13, p. 424]) is applicable. Here

$$K = \frac{4{,}000}{(20 + h/d)(2 - y/h)} \frac{a}{y} \frac{\Delta y}{\Delta t} \quad (13\text{-}17)$$

where all quantities are as defined in Fig. 13-10. The coefficient of permeability is expressed in meters per day, and all other quantities are in centimeters or seconds. According to Ernst, K is accurate to within 20 per cent if $3 < a < 7$, $20 < d < 200$, $0.2 < h/d < 1$, and $s > d$. Several similar techniques, developed for drainage design, are available for other subsurface conditions [13].

A second method of field measurement is by tracers. A tracer substance is introduced into groundwater at an upstream point; the time for it to appear at a downstream point and the hydraulic gradient are measured. As flow occurs only within the pores of the formation, the average velocity v_a measured by the tracer

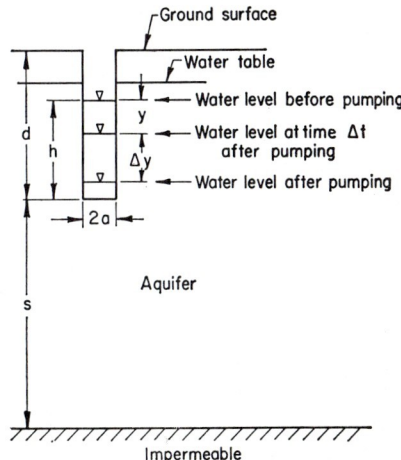

Fig. 13-10. Sketch of an auger hole for measurement of permeability in a homogeneous formation.

$$v_a = \frac{\Delta L}{\Delta t} = \frac{Q}{A\alpha} \quad (13\text{-}18)$$

where ΔL is the distance between observation points, Δt is the travel time, and α is porosity. Substituting in the Darcy equation [Eq. (13-6)] and solving for the coefficient of permeability,

$$K = \frac{\alpha v_a \, \Delta L}{\Delta h} \qquad (13\text{-}19)$$

where $\Delta h/\Delta L$ is the hydraulic gradient. In practice, the method is seldom applicable to distances of more than several feet. Tracers and their movement in porous media are discussed below.

A third method for evaluating permeability is by pumping tests of wells. Because an integrated value is obtained over a relatively large area of aquifer, the method is more reliable than point methods. Pumping-test methods are described in Subsec. IV.

D. Tracing Groundwater Movement

Ideally, a groundwater tracer should move in an identical manner as the groundwater. Although many substances have been employed as tracers, only a few merit serious consideration [1]. Tracers may be grouped by method of detection; for example, dyes are detected by colorimetry, electrolytes by electrical conductivity, and radioisotopes by their nuclear radiations.

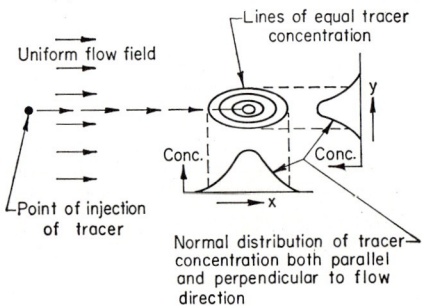

FIG. 13-11. Dispersion of a point tracer with distance. (*After de Josselin de Jong* [16].)

A common dye is sodium fluorescein, which can be detected in very low concentrations. The chloride ion is a satisfactory tracer in low concentrations and can be detected by a simple titration technique. In recent years a variety of radioisotopes have been investigated to determine their feasibility as tracers. Problems of length of half-life, base exchange, and adsorption eliminate all but a relatively few. Of these, tritium (H^3) seems to have the greatest potentialities as a tracer. It has a half-life of about 12.5 years and a low activity level. The fact that a water molecule containing tritium (often designated HTO) differs only slightly from a normal water molecule makes it useful as a water tracer. Laboratory tests have demonstrated its feasibility, although, where clay fractions are present, tritium moves more slowly than other tracers because it exchanges with bound water in clay lattices [14]. Data are as yet inadequate to show its capabilities under field conditions.

If a tracer is introduced at a point into a flow through a porous medium, it disperses longitudinally and laterally as it travels downstream. Dispersion is a hydrodynamic phenomenon resulting from microvelocity variations inherent in laminar flow and diffusion. Considerable analytic work, treating the phenomenon on a probability basis [15], and experimental data indicate that the tracer tends toward an expanding normal distribution with distance. As the longitudinal dispersion is usually several times the lateral [16], an expanding ellipsoid appears downstream from a point injection, as shown in Fig. 13-11.

Tritium is produced in the atmosphere by cosmic radiation (and thermonuclear explosions) and appears at the earth's surface in rain water. For rainfall infiltrating underground, the tritium radioactivity decreases in a predictable exponential manner. Dr. W. F. Libby suggested that analyses of tritium content of groundwater samples might provide a means of dating groundwater. Limited field studies indicate that age differentiations of even a few years are possible by the technique [17].

E. Basic Flow Equations

1. Steady Flow. The general equations governing groundwater flow can be derived, starting from Darcy's law, in the general form

$$v = K \frac{\partial h}{\partial s} \tag{13-20}$$

where v, K, and h are as previously defined, and s is distance along the average direction of flow. In a homogeneous and isotropic medium, it follows that the directional velocity components are

$$v_x = K \frac{\partial h}{\partial x} \qquad v_y = K \frac{\partial h}{\partial y} \qquad v_z = K \frac{\partial h}{\partial z} \tag{13-21}$$

Defining a velocity potential $\varphi = -Kh$ and substituting in Eq. (13-21),

$$v_x = -\frac{\partial \varphi}{\partial x} \qquad v_y = -\frac{\partial \varphi}{\partial y} \qquad v_z = -\frac{\partial \varphi}{\partial z} \tag{13-22}$$

which reveal that a velocity potential exists for groundwater flow.

Assuming that water is incompressible, its density may be taken as constant. The equation of continuity for this case is

$$\frac{\partial v_x}{\partial x} + \frac{\partial v_y}{\partial y} + \frac{\partial v_z}{\partial z} = 0 \tag{13-23}$$

Substituting Eq. (13-22) and replacing φ by $-Kh$ leads to

$$\frac{\partial^2 h}{\partial x^2} + \frac{\partial^2 h}{\partial y^2} + \frac{\partial^2 h}{\partial z^2} = 0 \tag{13-24}$$

This is the general equation for steady flow of water in homogeneous and isotropic media.

For one-directional flow in a confined aquifer, Eq. (13-24) reduces to

$$\frac{\partial^2 h}{\partial x^2} = 0 \tag{13-25}$$

Integration and selection of boundary conditions gives

$$h = \frac{vx}{K} \tag{13-26}$$

showing that the head decreases linearly in the direction of flow with flow in the negative x-direction.

For one-directional flow in an unconfined aquifer no direct analytic solution is possible because the water table forms the upper boundary but also is a flow line governed by the flow distribution. Making the Dupuit assumptions that (1) the velocity is proportional to the tangent of the hydraulic gradient, and (2) the flow is everywhere horizontal and uniform in a vertical section, an approximate solution can be obtained.

From Fig. 13-12, the discharge per unit width q is

$$q = Kh \frac{dh}{dx} \tag{13-27}$$

which, after integration, leads to

$$q = \frac{K}{2x}(h^2 - h_0^2) \tag{13-28}$$

The water table thus has a parabolic form. The actual water table for the same boundaries is also shown in Fig. 13-12. The difference between the two surfaces increases as the flow deviates increasingly from the Dupuit assumptions. At the downstream boundary, a flow discontinuity results in a seepage face with discharge above the outflow water boundary.

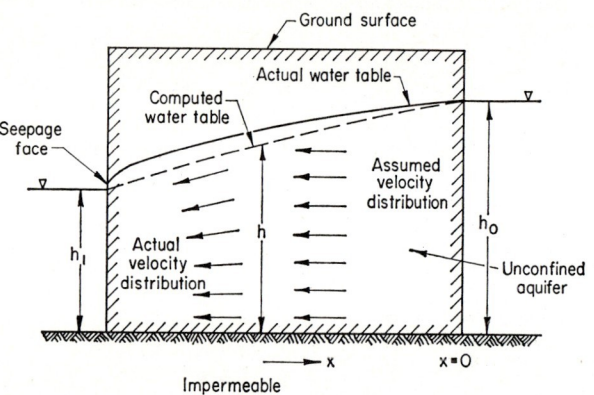

FIG. 13-12. Steady flow in an unconfined aquifer between two water bodies with vertical boundaries. (*After Todd* [1].)

2. Unsteady Flow. Derivation of a general unsteady flow equation requires consideration of the aquifer storage coefficient S. For an unconfined aquifer this equals the specific yield; for a confined aquifer it is a measure of the compressibility of the aquifer and water. Taking this into account, it can be shown [1] that the general equation has the form

$$\frac{\partial^2 h}{\partial x^2} + \frac{\partial^2 h}{\partial y^2} + \frac{\partial^2 h}{\partial z^2} = \frac{S}{Kb}\frac{\partial h}{\partial t} \tag{13-29}$$

for a confined aquifer of uniform thickness b. The comparable equation for an unconfined aquifer is nonlinear [18]; however, Eq. (13-29) can be applied where variations in saturated thickness are relatively small.

F. Flow-line Analysis

1. Flow Nets. Because groundwater flow is a potential flow system, for specified boundary conditions flow lines and equipotential lines can be mapped for a two-dimensional cross section. These form an orthogonal flow net. Generally, graphical methods based on trial-and-error sketching [19] or laboratory-model studies [1] are necessary to define flow-net configurations.

Once a flow net has been established, the groundwater flow can be computed directly from the geometry of the net. Thus

$$Q = \frac{Kmh}{n} \tag{13-30}$$

where Q is the discharge, K is permeability, m is the number of channels formed by the flow lines, n is the number of squares between two adjacent flow lines, and h is the total head loss.

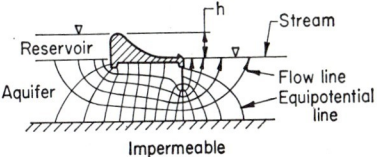

FIG. 13-13. Flow net under a dam on a permeable foundation. (*After Casagrande* [20].)

Example. In Fig. 13-13 $m = 5$ and $n = 15$. If $h = 20$ ft and $K = 100$ ft/day, then the total underflow

$$Q = \frac{(100/7.48)(5)(20)}{15} = 89 \text{ gal/day/ft width}$$

2. Refraction at Permeable Boundaries. Flow lines parallel a water table if no flow crosses the free surface. For water percolating to a water table, a refraction of flow lines occurs. The angle of refraction ϵ is given by [1]

$$\epsilon = \tan^{-1}\left(\frac{K}{v_u} \tan \delta\right) - \delta \tag{13-31}$$

where K is permeability, and other quantities are as defined in Fig. 13-14a.

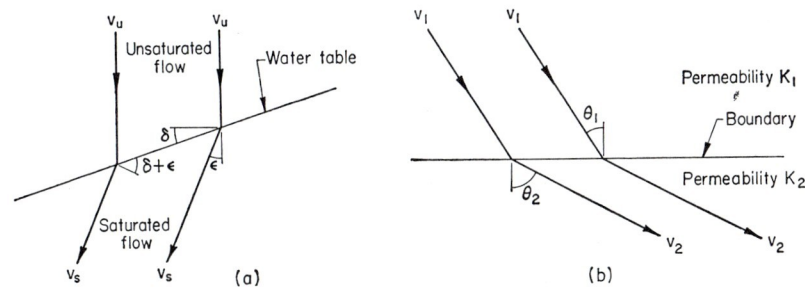

FIG. 13-14. Refraction of flow lines (a) across a water table and (b) across a boundary of different permeabilities. (*After Todd* [1].)

At a boundary between media of different permeabilities, a refraction of flow lines also occurs. The change in direction can be expressed by Fig. 13-14b

$$\frac{K_1}{K_2} = \frac{\tan \theta_1}{\tan \theta_2} \tag{13-32}$$

IV. WELL HYDRAULICS

The effect of a pumping well on groundwater flow and distribution in an aquifer depends upon the well construction and operation, aquifer conditions,[1] and aquifer boundaries. Well hydraulics assists in evaluating aquifer properties, defining boundaries, and predicting yields and future effects of pumping.

[1] The well is described as *artesian* or *water-table* depending on whether it is in a confined or an unconfined aquifer.

A. Steady Radial Flow

As water is removed from an aquifer by a pumping well, the groundwater level—water table or piezometric surface—is lowered. *Drawdown* is the distance the level is lowered; distance-drawdown and time-drawdown curves reveal the space and time variations, respectively. The drawdown curve around a well (Fig. 13-15) in three dimensions forms a cone of depression, the outer boundary of which defines the area of influence of the well.

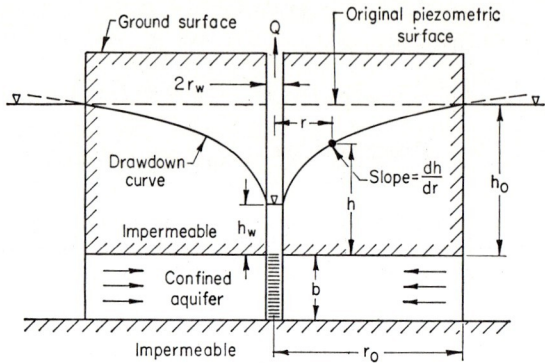

Fig. 13-15. Steady radial flow to a well penetrating a confined aquifer on an island. [*After Todd* [1]].

1. Confined Aquifer. Consider a well pumping from a confined aquifer on an island (Fig. 13-15). For homogeneous and isotropic conditions, the Dupuit assumptions apply. The well discharge

$$Q = Av = 2\pi r b K \frac{dh}{dr} \tag{13-33}$$

for steady flow at any distance r from the well. Integration leads to

$$Q = 2\pi K b \frac{h_0 - h_w}{\ln(r_0/r_w)} \tag{13-34}$$

where all symbols are as defined in Fig. 13-15.

More generally, where an extensive confined aquifer is involved, there is no limit on r. Thus, from Eq. (13-34),

$$Q = 2\pi K b \frac{h - h_w}{\ln(r/r_w)} \tag{13-35}$$

which shows that h increases indefinitely with r. Yet the maximum h is h_0; hence, theoretically, steady radial flow cannot exist under these conditions. As an approximation, a value of $r = r_0$ can be assumed where h approaches h_0.

Equation (13-35) is known as the *equilibrium*, or *Thiem*, *equation*. It can be employed to evaluate aquifer permeability from measurements around a pumping well [21]. Noting that the drawdown curve is a linear function of the distance from the well, measurements of drawdown in two observation wells at different distances from a pumped well are sufficient to define the curve. Permeability is given by

$$K = \frac{Q}{2\pi b(h_2 - h_1)} \ln \frac{r_2}{r_1} \tag{13-36}$$

where r_1 and r_2 are the distances, and h_1 and h_2 are the heads of the respective observation wells. Pumping should continue at a uniform rate until the drawdown changes negligibly with time, thereby approaching a steady-state condition.

2. Unconfined Aquifer. With the Dupuit assumptions the discharge of a well penetrating an unconfined aquifer can be determined. From Fig. 13-16 the flow toward the well at a distance r is

$$Q = 2\pi r K \frac{dh}{dr} \qquad (13\text{-}37)$$

which, after integration, yields

$$Q = \pi K \frac{h_0^2 - h_w^2}{\ln(r_0/r_w)} \qquad (13\text{-}38)$$

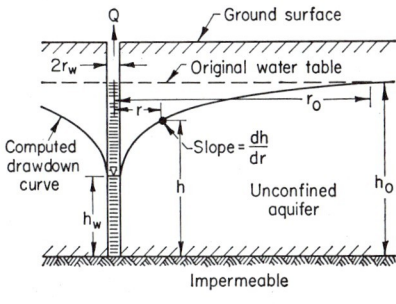

Fig. 13-16. Radial flow to a well penetrating an unconfined aquifer. (*After Todd* [1].)

The drawdown curve near the well is inaccurate; however, estimates of Q for given heads are satisfactory. For practical application, values of r_0 in the range of 500 to 1,000 ft may be selected. Fortunately, Q varies by only a small amount for large variations in r_0. Pairs of distances and heads can be substituted into Eq. (13-38), similar to Eq. (13-36).

B. Well in a Uniform Flow

For a well pumping from an aquifer with a uniform-flow field, the circular area of influence for the radial-flow case is modified. However, for most relatively flat natural slopes, Eqs. (13-35) and (13-38) can be applied without appreciable error.

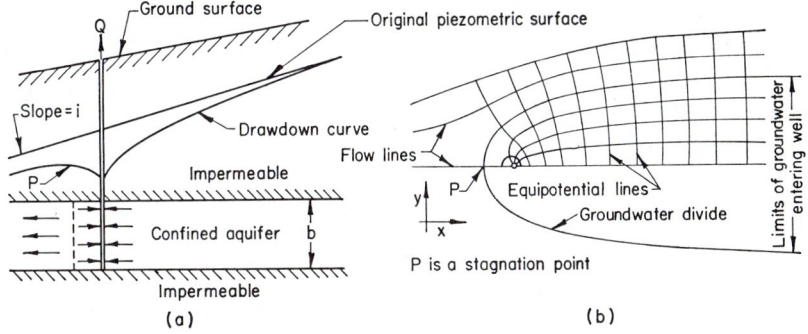

Fig. 13-17. Flow to a well penetrating a confined aquifer having a sloping plane piezometric surface. (*a*) Vertical section; (*b*) plan view. (*After Todd* [1].)

In cases of significant uniform flow, a groundwater divide identifies the area of aquifer contributing to the well (Fig. 13-17). For a very long pumping period, the area would extend upstream to the aquifer boundary. With the Dupuit assumptions and a confined aquifer, the divide can be described by

$$-\frac{y}{x} = \tan\left(\frac{2\pi K b i}{Q} y\right) \qquad (13\text{-}39)$$

where x and y are coordinates measured from the well. The inflow area has a width of

$$y = \pm \frac{Q}{2Kbi} \qquad (13\text{-}40)$$

as $x \to \infty$ and extends downstream to

$$x = -\frac{Q}{2\pi Kbi} \tag{13-41}$$

Unconfined aquifers can be treated similarly by replacing b in Eqs. (13-39) to (13-41) by h_0, the uniform saturated thickness. The drawdown should be small in relation to the aquifer thickness.

C. Unsteady Radial Flow—Confined Aquifer

A well pumping at a constant rate from an extensive confined aquifer produces an area of influence which expands with time. Water is taken from storage within the aquifer as the piezometric head is reduced; therefore no steady-state flow develops. Rewriting Eq. (13-29) in plane polar coordinates,

$$\frac{\partial^2 h}{\partial r^2} + \frac{1}{r}\frac{\partial h}{\partial r} = \frac{S}{T}\frac{\partial h}{\partial t} \tag{13-42}$$

This equation was solved by Theis [22] based on the analogy between groundwater flow and heat conduction. For conditions of $h = h_0$ at $t = 0$ and of $h \to h_0$ as $r \to \infty$ for $t \geq 0$ (after pumping begins),

$$h_0 - h = \frac{Q}{4\pi T}\int_u^\infty \frac{e^{-u}\,du}{u} \tag{13-43}$$

where

$$u = \frac{r^2 S}{4Tt} \tag{13-44}$$

Equation (13-43) is the *nonequilibrium*, or *Theis, equation*.

The nonequilibrium equation permits evaluation of S and T from pumping tests. Field measurements consist of recording drawdowns in an observation well as a function of time. Rapid approximate solutions are necessary for field applications. A method by Theis [22] is described below. Other convenient methods developed by Jacob [23] and Chow [24] are described in detail elsewhere [1].

Equation (13-43) becomes

$$h_0 - h = \frac{114.6Q}{T} W(u) \tag{13-45}$$

when $h_0 - h$ is the drawdown in ft, Q is the well discharge in gal/min, T is the coefficient of transmissibility in gpd/ft, and $W(u)$ is the *well function*, which is equivalent to the exponential integral of Eq. (13-43). The argument

$$u = \frac{1.87 r^2 S}{Tt} \tag{13-46}$$

where r is distance from observation well to the pumping well in ft, and t is the pumping time in days. The Theis method is a graphical one based on superposition of curves. On logarithmic paper $W(u)$ is plotted against u from data in Table 13-4. A curve, known as a *type curve*, is drawn through the points. Using paper of the same scale, values of $h_0 - h$ from the observation well are plotted against values of r^2/t. The two sheets are superimposed with the coordinate axes parallel. When a position is found with most of the observed points falling on a portion of the type curve, any set of coincident values of $W(u)$, u, $h_0 - h$, and r^2/t are noted. Substituting these in Eqs. (13-45) and (13-46), S and T are obtained.

WELL HYDRAULICS

Table 13-4. Values of $W(u)$ for Values of u*

u	1.0	2.0	3.0	4.0	5.0	6.0	7.0	8.0	9.0
× 1	0.219	0.049	0.013	0.0038	0.0011	0.00036	0.00012	0.000038	0.000012
× 10^{-1}	1.82	1.22	0.91	0.70	0.56	0.45	0.37	0.31	0.26
× 10^{-2}	4.04	3.35	2.96	2.68	2.47	2.30	2.15	2.03	1.92
× 10^{-3}	6.33	5.64	5.23	4.95	4.73	4.54	4.39	4.26	4.14
× 10^{-4}	8.63	7.94	7.53	7.25	7.02	6.84	6.69	6.55	6.44
× 10^{-5}	10.94	10.24	9.84	9.55	9.33	9.14	8.99	8.86	8.74
× 10^{-6}	13.24	12.55	12.14	11.85	11.63	11.45	11.29	11.16	11.04
× 10^{-7}	15.54	14.85	14.44	14.15	13.93	13.75	13.60	13.46	13.34
× 10^{-8}	17.84	17.15	16.74	16.46	16.23	16.05	15.90	15.76	15.65
× 10^{-9}	20.15	19.45	19.05	18.76	18.54	18.35	18.20	18.07	17.95
× 10^{-10}	22.45	21.76	21.35	21.06	20.84	20.66	20.50	20.37	20.25
× 10^{-11}	24.75	24.06	23.65	23.36	23.14	22.96	22.81	22.67	22.55
× 10^{-12}	27.05	26.36	25.96	25.67	25.44	25.26	25.11	24.97	24.86
× 10^{-13}	29.36	28.66	28.26	27.97	27.75	27.56	27.41	27.28	27.16
× 10^{-14}	31.66	30.97	30.56	30.27	30.05	29.87	29.71	29.58	29.46
× 10^{-15}	33.96	33.27	32.86	32.58	32.35	32.17	32.02	31.88	31.76

* After Wenzel [21].

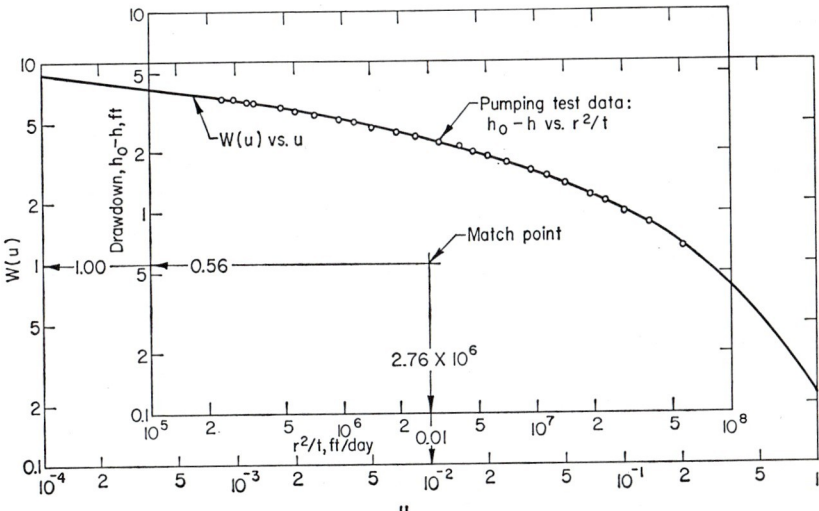

FIG. 13-18. Theis method of superposition for solution of the nonequilibrium equation.

Example. Drawdowns in an observation well 200 ft from a well penetrating a confined aquifer and pumping 500 gpm are listed in Table 13-5. Values of r^2/t are computed and plotted against $h_0 - h$ (Fig. 13-18). On a second sheet values $W(u)$ and u from Table 13-4 are plotted. The sheets are superposed, as shown in Fig. 13-18, and the coincident values at a convenient match point are noted. From Eq. (13-45),

$$T = \frac{114.6Q}{h_0 - h} W(u) = \frac{(114.6)(500)(1.00)}{0.56} = 102{,}000 \text{ gpd/ft}$$

and from Eq. (13-46),

$$S = \frac{uT}{1.87 r^2/t} = \frac{(0.01)(102{,}000)}{1.87(2.76 \times 10^6)} = 0.000198$$

Table 13-5. Pumping-test Data*
($r = 200$ ft)

Time since pumping began, t		Drawdown in observation well, $h_0 - h$, ft	r^2/t, ft^2/day
Minutes	Days		
0	0	0.00	∞
1.0	6.96×10^{-4}	0.66	5.76×10^7
1.5	1.02×10^{-3}	0.87	3.84×10^7
2.0	1.39×10^{-3}	0.99	2.88×10^7
2.5	1.74×10^{-3}	1.11	2.30×10^7
3.0	2.09×10^{-3}	1.21	1.92×10^7
4	2.78×10^{-3}	1.36	1.44×10^7
5	3.48×10^{-3}	1.49	1.15×10^7
6	4.17×10^{-3}	1.59	9.6×10^6
8	5.57×10^{-3}	1.75	7.2×10^6
10	6.96×10^{-3}	1.86	5.76×10^6
12	8.33×10^{-3}	1.97	4.80×10^6
14	9.72×10^{-3}	2.08	4.1×10^6
18	1.25×10^{-2}	2.20	3.2×10^6
24	1.67×10^{-2}	2.36	2.4×10^6
30	2.09×10^{-2}	2.49	1.92×10^6
40	2.78×10^{-2}	2.65	1.44×10^6
50	3.48×10^{-2}	2.78	1.15×10^6
60	4.17×10^{-2}	2.88	9.6×10^5
80	5.57×10^{-2}	3.04	7.2×10^5
100	6.96×10^{-2}	3.16	5.76×10^5
120	8.33×10^{-2}	3.28	4.8×10^5
150	1.02×10^{-1}	3.42	3.84×10^5
180	1.25×10^{-1}	3.51	3.2×10^5
210	1.46×10^{-1}	3.61	2.74×10^5
240	1.67×10^{-1}	3.67	2.4×10^5

* From U.S. Geological Survey.

D. Unsteady Radial Flow—Unconfined Aquifer

The derivation of the nonequilibrium equation assumes that water is released instantaneously from storage by lowering the head. In a confined aquifer this is essentially true since the water released comes from expansion of the water and compression of the aquifer. In an unconfined aquifer, however, the water comes chiefly from gravity drainage of the void space within the cone of depression. During a pumping test the storage coefficient varies with time, increasing at a diminishing rate. Ultimately it is equivalent to the specific yield.

The nonequilibrium equation can be applied to unconfined aquifers if the following limitations are observed:

1. The drawdown should be small in relation to the saturated thickness.
2. The nonequilibrium equation should not be applied to water-table conditions until after a minimum pumping time which is governed by aquifer properties. This

can be expressed [25] as

$$t_{min} = \frac{37.4 S h_0}{K} \qquad (13\text{-}47)$$

where t_{min} is the time after pumping began in days, S is the coefficient of storage, h_0 is the saturated aquifer thickness in ft, and K is the coefficient of permeability in gpd/ft^2. For example, if $S = 0.15$, $h_0 = 100$ ft, and $K = 2{,}000$ gpd/ft^2,

$$t_{min} = \frac{37.4(0.15)(100)}{2{,}000} = 0.28 \text{ day} = 6.7 \text{ hr}$$

Equation (13-47) is valid only when the observation well is greater than $0.2 h_0$ from the pumping well.

E. Unsteady Radial Flow—Leaky Aquifer

Aquifers which are overlain or underlain by semipermeable strata are referred to as *leaky aquifers.* Such aquifers are confined in the sense that pumping does not dewater the aquifers, but a significant portion of the yield may be derived by vertical leakage through the confining formations into the aquifers.

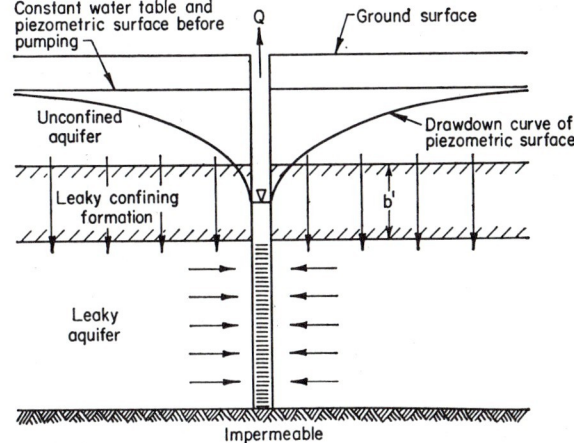

FIG. 13-19. Radial flow to a well penetrating a leaky aquifer.

A solution to the unsteady-radial-flow situation posed by a well pumping from a leaky aquifer was obtained by Hantush and Jacob [26]. The method is similar to that by Theis for a confined aquifer. The leakage is assumed to occur vertically through the confining bed and is proportional to the drawdown. The water level in the formation supplying leakage is assumed constant (Fig. 13-19).

The basic equation may be expressed simply as

$$h_0 - h = \frac{114.6 Q}{T} W\left(u, \frac{r}{B}\right) \qquad (13\text{-}48)$$

where $W(u, r/B)$ is the well function for leaky aquifers. The quantity

$$u = \frac{1.87 r^2 S}{Tt} \qquad (13\text{-}49)$$

and

$$\frac{r}{B} = \frac{r}{\sqrt{T/(K'/b')}} \qquad (13\text{-}50)$$

For convenience, the symbols and their units are as follows (Fig. 13-19):

$h_0 - h$ = drawdown in observation well, ft
r = distance from pumped well to observation well, ft
Q = well discharge, gpm
t = time of pumping, days
T = coefficient of transmissibility of aquifer, gpd/ft
S = coefficient of storage of aquifer
K' = vertical permeability of confining bed, gpd/ft^2
b' = thickness of leaky confining bed, ft

Values of $W(u, r/B)$ are plotted against values of $1/u$ for various values of r/B in Fig. 13-20. The relation of $W(u, r/B)$ to $1/u$ is the same as of $h_0 - h$ to t. A solution is obtained by plotting drawdown vs. time on logarithmic paper. This plot is superposed on curves of $W(u, r/B)$ versus $1/u$ on the same scale of paper. The respective

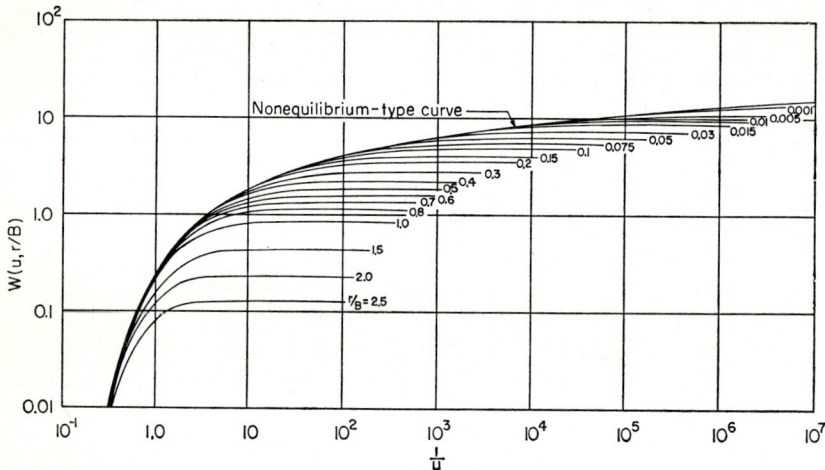

Fig. 13-20. Type curves for unsteady radial flow to a well in a leaky aquifer. (*After* Walton [27].)

axes are kept parallel. From any match point, values of $W(u, r/B)$, $1/u$, $h_0 - h$, and t are noted. The value of T is computed from Eq. (13-48), and S is found from Eq. (13-49). By noting the value of r/B of the curve which gave the best fit to the field data, K' can be computed from Eq. (13-50), knowing T and b'.

Note that as $K' \to 0$, $r/B \to 0$. Thus, for an impermeable confining layer, Eq. (13-48) reduces to Eq. (13-45) for a confined aquifer.

F. Well Flow near Aquifer Boundaries

Where a well is located close to an aquifer boundary, such as an impermeable formation or a surface-water body, marked deviations from a radial-flow system can occur. Solutions are readily obtained by application of the *method of images*. An image, which may be an imaginary discharging or recharging well, creates a hydraulic system which is equivalent to the effects of a known physical boundary on the flow system. In essence, images enable an aquifer of finite extent to be transformed to one of infinite extent. This enables radial-flow equations to be applied to the modified system.

1. Well Flow near a Stream. Consider the situation of a well pumping near a stream intersecting an aquifer, as shown in Fig. 13-21a. This can be transformed into an infinite aquifer by replacing the stream by an imaginary recharge well opposite to and at the same distance from the stream as the pumping well (Fig. 13-21b). The image well is assumed to operate simultaneously and at the same rate as the pumping well. The build-up of head around the recharge well equals the drawdown around the pumping well. Along the line of the stream, these cancel, giving a constant-head equivalent to the stream surface. The summation of the heads resulting from each well at any point on the pumping-well side of the stream gives the actual resultant drawdown.

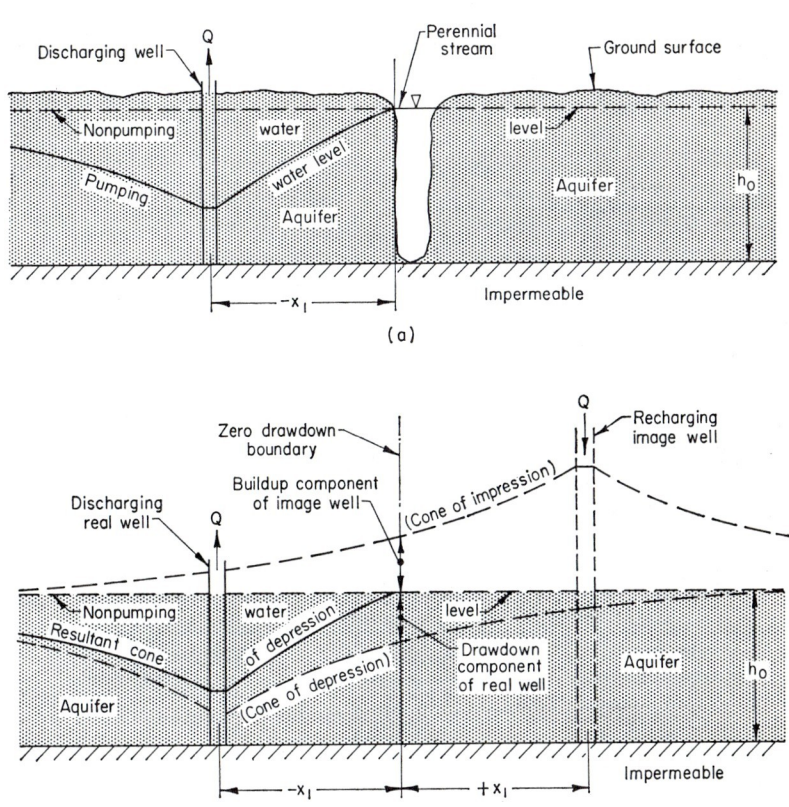

Fig. 13-21. Sectional views of (a) a discharging well near a perennial stream, and (b) the equivalent hydraulic system in an aquifer of infinite areal extent. Aquifer thickness h_0 should be very large compared with resultant drawdown near real well. (*After U.S. Geological Survey.*)

Analysis using the nonequilibrium equation enables the fraction Q_s/Q of stream contribution to total well discharge to be evaluated. For a confined aquifer the solution can be expressed graphically by Fig. 13-22. Here x_1 is the distance from the well to the stream, b is the aquifer thickness, and K, S, and t are as previously defined. The solution is also applicable to an unconfined aquifer if the drawdown is small in relation to the saturated thickness.

2. Location of an Aquifer Boundary. A common field problem is that of identifying and locating aquifer boundaries. Pumping-test data can provide quantitative answers after application of the method of images and the nonequilibrium equation. Let a pumping well and an observation well be located near an unknown impermeable aquifer boundary as in Fig. 13-23. An image pumping well will furnish

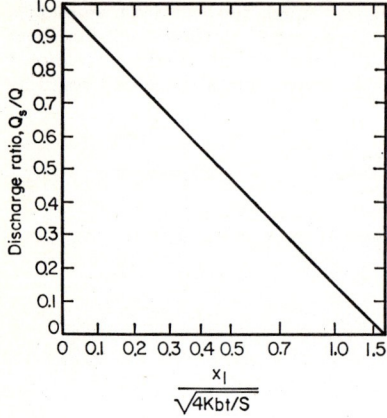

FIG. 13-22. Graph for determining the portion of well discharge furnished by a nearby stream. (*After Glover and Balmer* [28].)

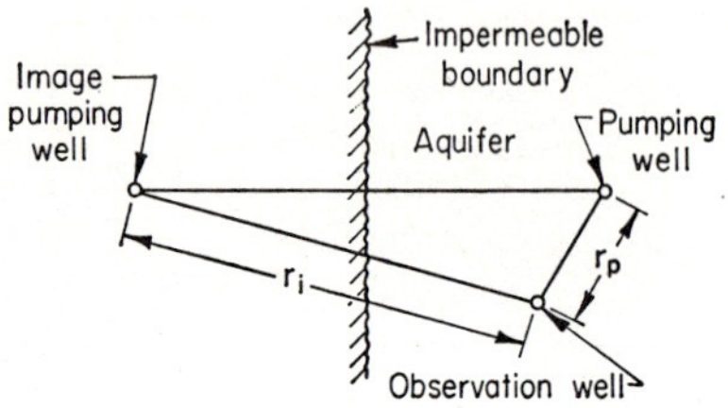

FIG. 13-23. Definition sketch showing pumping, observation, and image wells near an impermeable aquifer boundary.

the equivalent hydraulic flow system. From the nonequilibrium equation it can be shown that [29]

$$\frac{r_p^2}{t_p} = \frac{r_i^2}{t_i} \tag{13-51}$$

where r_p and r_i are as shown in Fig. 13-23. The value t_p is the time since pumping began to any selected drawdown before the boundary affects the drawdown. The value t_i is the time since pumping began when the divergence of the drawdown curve

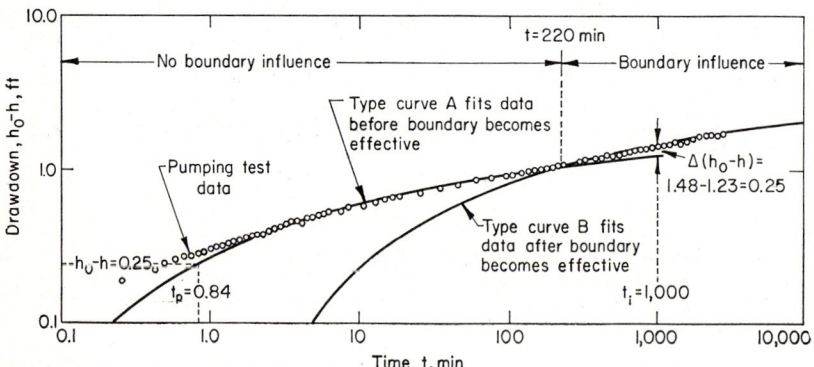

FIG. 13-24. Graph of drawdown vs. time with fitted-type curves showing effect of impermeable boundary. (*After Walton* [30].)

from the type curve equals the selected drawdown. The unknown distance r_i can be found from the known value of r_p, and the values t_p and t_i determined from plotting of drawdown data.

Example. In Fig. 13-24 pumping-test data show the effect of an impermeable boundary. The observation well is 125 ft from the pumping well. Type curve A provides a good fit to the observed data until 220 min. This is the drawdown curve of the observation well before it is affected by the boundary. After 220 min the data fit type curve B, which is produced by the image pumping well. Selecting $h_0 - h = 0.25$ ft, $t_p = 0.84$ min. A divergence of 0.25 ft between the two curves occurs at 1,000 min; this is the value of t_i. Hence

$$r_i = \sqrt{\frac{t_i}{t_p}} r_p = \sqrt{\frac{1,000}{0.84}} (125) = 4,320 \text{ ft}$$

Where no information is available about a boundary, three observation wells are necessary to pinpoint the orientation and location of the boundary. This is accomplished by finding the common intersection of three arcs having radii of the computed values of r_i (Fig. 13-25). The boundary lies along the perpendicular bisector of the line connecting the pumping well and the intersection point.

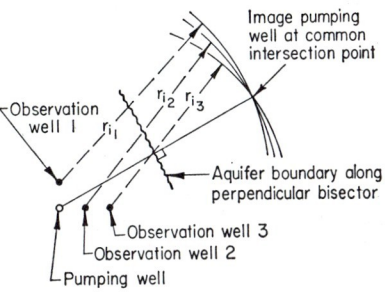

Fig. 13-25. Location of image well and aquifer boundary from intersection of radii r_i from observation wells.

3. Multiple Boundaries. The above procedures can be extended to analyze pumping effects and locate multiple aquifer boundaries [1, 31]. The complexity of solution increases with the number, irregularity, and different types of boundaries present.

G. Multiple Well Systems

Where two or more wells are pumping from the same aquifer and near each other, the areas of influence can overlap. This is known as *interference of wells*. Wells close

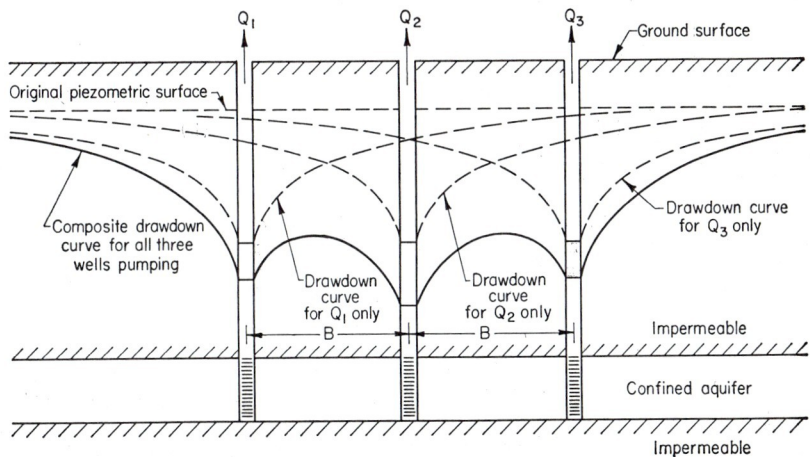

Fig. 13-26. Individual and composite drawdown curves for three wells in a line. (*After* Todd [1].)

together seriously interfere with one another in that drawdowns are greater, leading to increased pumping lifts and costs. Proper well field design attempts to provide the most economical pumping system, taking into account effects of interference.

The drawdown at any point in an area near several pumping wells is equal to the sum of the individual well drawdowns at that point. At a given location, total drawdown

$$D_t = D_a + D_b + D_c + \cdots + D_n \qquad (13\text{-}52)$$

where $D_a, D_b, D_c, \ldots, D_n$ are the drawdowns at that location resulting from the discharge of wells $a, b, c, \ldots, n$, respectively. Figure 13-26 shows the resulting drawdown curve from three wells in a line, all discharging at the same rate.

Computations of well field drawdown can be made by the equilibrium or nonequilibrium equation. For an approximate steady-state condition in a confined aquifer with n wells,

$$h_0 - h = \sum_i^n \frac{Q_i}{2\pi Kb} \ln \frac{R_i}{r_i} \tag{13-53}$$

where $h_0 - h$ is the drawdown at a given point, R_i is the distance from the ith well to a point at which the drawdown becomes negligible, and r_i is the distance from the ith well to the given point.

H. Partially Penetrating Wells

A partially penetrating well is one whose length of water entry is less than the aquifer thickness. Typically, this occurs when drilling intersects a satisfactory aquifer

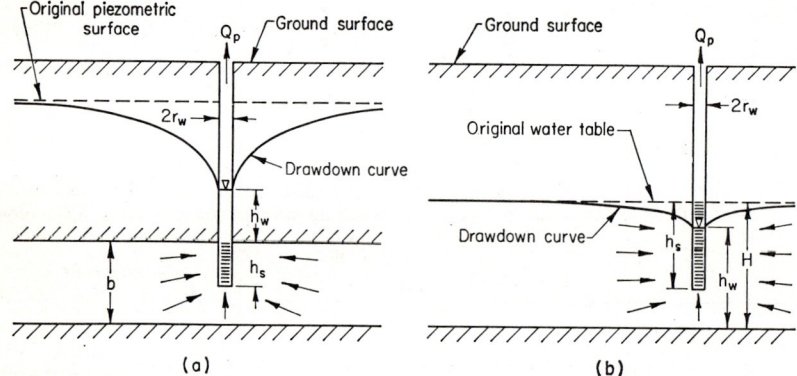

FIG. 13-27. Partially penetrating wells in (a) confined and (b) unconfined aquifers. (*After Todd* [1].)

and no effort is made to extend the well down to the bottom of the formation. Examples of such wells in confined and unconfined aquifers appear in Fig. 13-27. The radial flow into fully penetrating wells (implicitly assumed heretofore) is modified by vertical convergence near partially penetrating wells. As a result,

$$(\Delta h)_p > \Delta h \quad \text{for } Q_p = Q$$
and
$$Q_p < Q \quad \text{for } (\Delta h)_p = \Delta h$$

where Q is well discharge, Δh is drawdown, and the subscript p refers to a partially penetrating well.

The drawdown of a partially penetrating well in a confined aquifer can be expressed by [1]

$$h_0 - h_w = \frac{Q_p}{2\pi K}\left(\frac{1}{h_s}\ln\frac{\pi h_s}{2r_w} + \frac{0.10}{b} + \frac{1}{b}\ln\frac{r_0}{2b}\right) \tag{13-54}$$

where h_0 is the head at radius of influence r_0, and other quantities are as identified in Fig. 13-27a. The equation is valid for $h_s/b \geq 0.77$ and for $h_s/2r_w \geq 5$. If Eq. (13-54) is divided by the drawdown equation for a fully penetrating well,

$$\frac{Q_p}{Q} = \frac{\ln(r_0/r_w)}{(b/h_s)\ln(\pi h_s/2r_w) + 0.10 + \ln(r_0/2b)} \tag{13-55}$$

This gives the ratio of yields for the same drawdown. A graph of Eq. (13-55) is given in Fig. 13-28 for selected values of r_0/r_w and $h_s/2r_w$.

Equation (13-55) and Fig. 13-28 can be applied to unconfined aquifers by replacing b by H (Fig. 13-27b) and if the drawdown is small in relation to the saturated thickness.

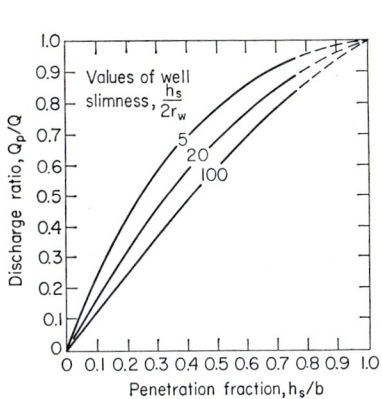

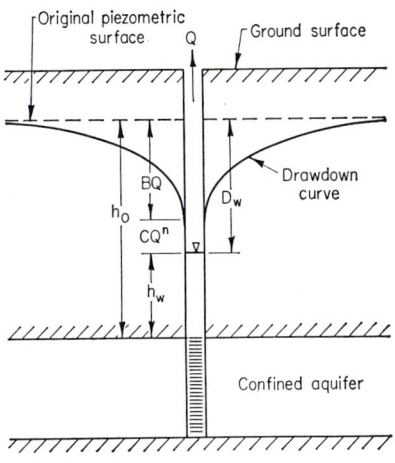

FIG. 13-28. Discharge of a partially penetrating well Q_p to that of a fully penetrating well Q for the same drawdown as a function of the penetration fraction h_s/b and well slimness $h_s/2r_w$. $r_0/r_w = 1,000$. (*After Todd* [1].)

FIG. 13-29. Relation of well loss CQ^n to drawdown for a well penetrating a confined aquifer. (*After Todd* [1].)

I. Well Losses

The drawdown in the immediate vicinity of a well is augmented by a *well loss*. This results from flow through the well screen and from flow inside the well to the pump intake. As these are turbulent flows, the loss is proportional to an nth power of the discharge, where n is about 2. Modifying Eq. (13-34), the drawdown D_w at a well may be written

$$D_w = h_0 - h_w = \frac{Q}{2\pi Kb} \ln \frac{r_0}{r_w} + CQ^n \quad (13\text{-}56)$$

where C is a constant dependent upon the radius, construction, and condition of the well. Letting

$$B = \frac{\ln (r_0/r_w)}{2\pi Kb} \quad (13\text{-}57)$$

then

$$D_w = BQ + CQ^n \quad (13\text{-}58)$$

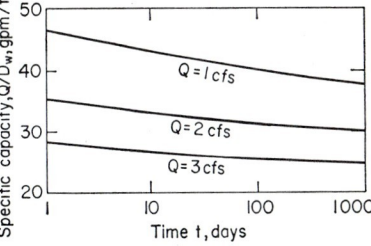

FIG. 13-30. Variation in specific capacity of a pumping well with discharge and time. (*After Jacob* [18].)

Thus, as shown in Fig. 13-29, the drawdown at a well consists of an *aquifer loss* BQ and a *well loss* CQ^n. Well losses are minimized by using large-diameter wells and by maintaining low pumping rates.

The *specific capacity* of a well is the ratio of its discharge to drawdown. A high specific capacity indicates an efficient, good-yielding well. The quantity is not constant, however, but varies with discharge and time [1]. Well data in Fig. 13-30 illustrate these variations for one well.

V. WATER WELLS

Water wells serve primarily as means of access so that groundwater can be brought to the surface for use. Wells are constructed by a variety of methods; factors of cost, depth, formations to be penetrated, and purpose of the well have important bearing on the type of well required.

A. Shallow Wells

Shallow wells, generally but not necessarily less than 50 ft in depth, are constructed by digging, boring, driving, or jetting [32].

Dug wells are usually constructed with a large diameter and to a depth a few feet below the water table. Such wells are most common for individual domestic water supplies since their size permits storage of considerable quantities of water. Construction is usually by hand, and linings of a variety of materials are employed. Yields considerably less than 100 gpm are typical.

Bored wells are constructed by hand-operated or power-driven earth augers. Hand-bored wells are usually less than 8 in. in diameter; power-driven augers may form holes up to 36 in. in diameter. Augers operate most successfully in noncaving formations and are especially useful in sticky clays. Casings of concrete, tile, or metal are placed after the hole is constructed.

Driven wells are created by driving a series of connected lengths of pipe into the ground. Typical diameters are $1\frac{1}{4}$ to 4 in. Water enters the well through a special screened cylindrical section, known as a *drive point*, at the lower end of the well. For use with suction-type pumps, water tables should be within 10 to 15 ft of ground surface.

Jetted wells are constructed by a high-velocity stream of water which is directed downward into the earth. By attaching the well pipe to a self-jetting well point, consisting of a screened section ending in a nozzle, the well pipe is jetted directly into place. Diameters are usually only a few inches. Both jetted and driven wells are used in well-point systems for dewatering excavation sites.

B. Deep Wells

Large, deep wells of high capacity for industrial, irrigation, or municipal use are constructed by drilling [33]. Methods of construction are by cable tool or by various forms of rotary drilling [32, 34–36].

A *cable-tool well* is constructed by the regular lifting and dropping of a string of tools, at the bottom of which a bit breaks the rock by impact. A string of tools (Fig. 13-31) consists of a rope socket, a set of jars (to aid in loosening tools stuck in a hole), a drill stem (for weight and length), and a drilling bit. The tools together may weigh several thousand pounds. Cuttings are removed from the well by a bailer (Fig. 13-31). Tools are lifted by a mast and a multiline hoist and engine, forming a drilling rig, which is usually truck-mounted.

The *cable-tool method* is adapted to drilling deep holes of from 3 to 24 in. in diameter through consolidated rock materials. To protect the hole from caving, casing is usually driven down to near the bottom of the hole. In consolidated water-bearing formations at the lower portion of a well, casing can be omitted entirely.

The other major type of deep-well drilling is by the rotary process. Most common is the *hydraulic rotary method*. A hollow rotating bit loosens the rock, while a mixture of clay and water (drilling mud) forced down through the drill rod carries the cuttings upward in the rising mud. The mud serves the additional purpose of forming a clay lining on the wall of the well, thereby preventing caving and making casing unnecessary during drilling. The drill rod is turned by a rotating table which permits the rod to slide downward as the hole deepens (Fig. 13-32). Drilling mud emerging from the hole is conducted to a pit where cuttings settle out and from where the mud can be pumped back down the drill rod again. After drilling, perforated casing is lowered

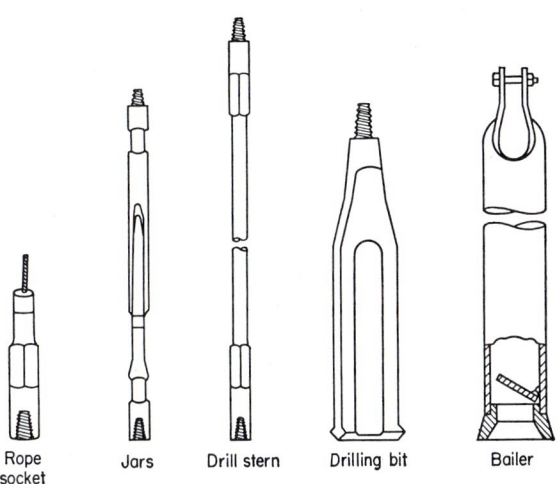

FIG. 13-31. Basic well-drilling tools for the cable-tool method. (*After Todd* [1].)

FIG. 13-32. A truck-mounted hydraulic rotary well-drilling rig. (*Bucyrus-Erie Co.*)

into the hole and the clay lining is washed from the wall by injecting water down the drill rod. Wells up to 18 in. in diameter are constructed by this method; with reamers, even larger diameters are possible.

The *reverse rotary method* uses water instead of mud and operates as a suction dredging method in which cuttings are removed by a rotating suction pipe [37]. Hydrostatic pressure acting against fine-grained-material deposits supports the walls. The method is useful for large-diameter wells in unconsolidated materials. An adequate supply of water is necessary for effective operation. Injection of air into the water column may speed circulation and drilling rates.

Direct rotary drilling by air is rapid and convenient for small-diameter holes in consolidated formations [38]. The fastest drilling in hardrock formations is possible by a recently developed *rotary-percussion procedure* using air. Directly above the rotating bit an air hammer delivers 600 to 1,000 blows per minute to the bottom of the hole. Penetration rates of up to 1 ft/min have been reported [39]. Air rotary techniques do not operate satisfactorily in water-bearing or caving formations; where these are encountered, a change to conventional rotary drilling with mud is usually necessary.

C. Well Completion, Development, and Testing

Completion of a well in an unconsolidated formation requires that it be cased to support the surrounding material and to provide free entry to water. Perforations can

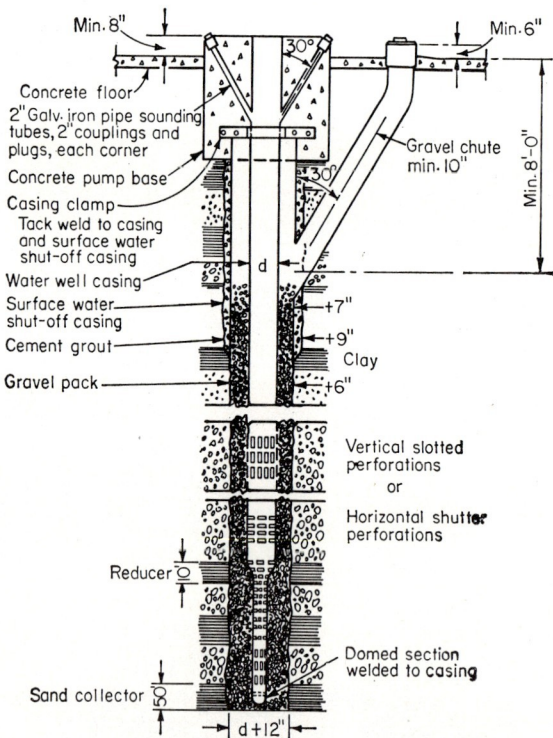

Fig. 13-33. Rotary-drilled gravel-packed well. (*American Water Works Association.*)

be made in the field, or machine-perforated casing is available. Openings should permit 50 to 80 per cent of the surrounding grains to pass into the well. Well screens available in various designs, diameters, slot sizes, and metals, can be inserted as casing sections. When selected on the basis of grain-size distributions of aquifer samples, these provide proper control of material entering the well. Aquifers should be sealed off from other contaminating waters by blank casings and grouting around the casing.

A *gravel-packed well* contains a gravel envelope around the perforated casing (Fig. 13-33). The gravel increases the effective well diameter, keeps fine material out of the well, and prevents caving of surrounding formations. Gravel packing is especially important for wells in sandy aquifers and for recharge wells. Gravel size should be related to grain sizes of the aquifer and to the opening size of the casing.

A new well should be developed to increase its specific capacity, prevent sanding, and to obtain maximum economic well life. This requires that the finer material in the formations surrounding the perforated casing should be removed. Development may be accomplished by a variety of procedures, including pumping, surging, injection of compressed air, backwashing, and addition of solid carbon dioxide [1, 32].

After development a new well should be tested to determine its yield and drawdown. This information is helpful in selecting the type and size of pump for the well.

D. Sanitary Protection of Wells

Wells furnishing drinking water should be properly sealed to prevent contamination from surface or subsurface sources. To keep surface water from entering a well, the annular space outside of the casing should be filled with cement grout for the first few feet below ground surface. The top of the well should contain a watertight seal. Surface covers around the well should be made of concrete and should slope away from the well [40].

To avoid subsurface contamination care should be exercised in locating the well so that it is not near or on the downhill sides of obvious contamination sources. Particular care is important in consolidated formations since they do not have the filtration capacity of unconsolidated formations. New wells should, of course, be chlorinated before use.

E. Maintenance and Repair of Wells

Most wells need some periodic maintenance. A failing well is one yielding less water with time. Many factors can contribute to a failing well, including a pump in need of repair, casing breaks, or corrosion or incrustation of perforated casing. Corrosion can be minimized by installation of corrosion-resistant metal screens and by providing cathodic protection. Incrustation results from reactions taking place at the well leading to precipitation of soluble matter in the groundwater. Deposits of calcium carbonate and ferric hydroxide are most common. Cleaning with hydrochloric acid or Calgon, followed by agitation and surging, is the best remedy. Most well-development methods can be employed to stimulate yields of old wells [41].

F. Collector Wells and Galleries

A *collector well*, illustrated by Fig. 13-34, consists of a cylindrical concrete caisson from which perforated pipes extend radially. The cylinder is sunk down into the aquifer, after which it is sealed at the bottom by concrete. The radial pipes are jacked hydraulically into the formation. Fine material around the pipes is removed by washing during construction. Collector wells are most often constructed in alluvial formations adjoining rivers. The radial pipes extend toward and under the river, thereby inducing movement of water downward through the stream bed to the pipes. Yields average about 5,000 gpm.

An *infiltration gallery* is a horizontal conduit for intercepting groundwater. Galleries are often placed parallel to rivers so that a perennial supply of water is available. Construction may be of vitrified clay, brick, concrete, or cast iron. Diameters of 2 to 5 ft are common; yields are in the range of 700 to 3,500 gpm per 1,000 ft of gallery. Galleries are suited to high-water-table conditions and give a minimum drawdown.

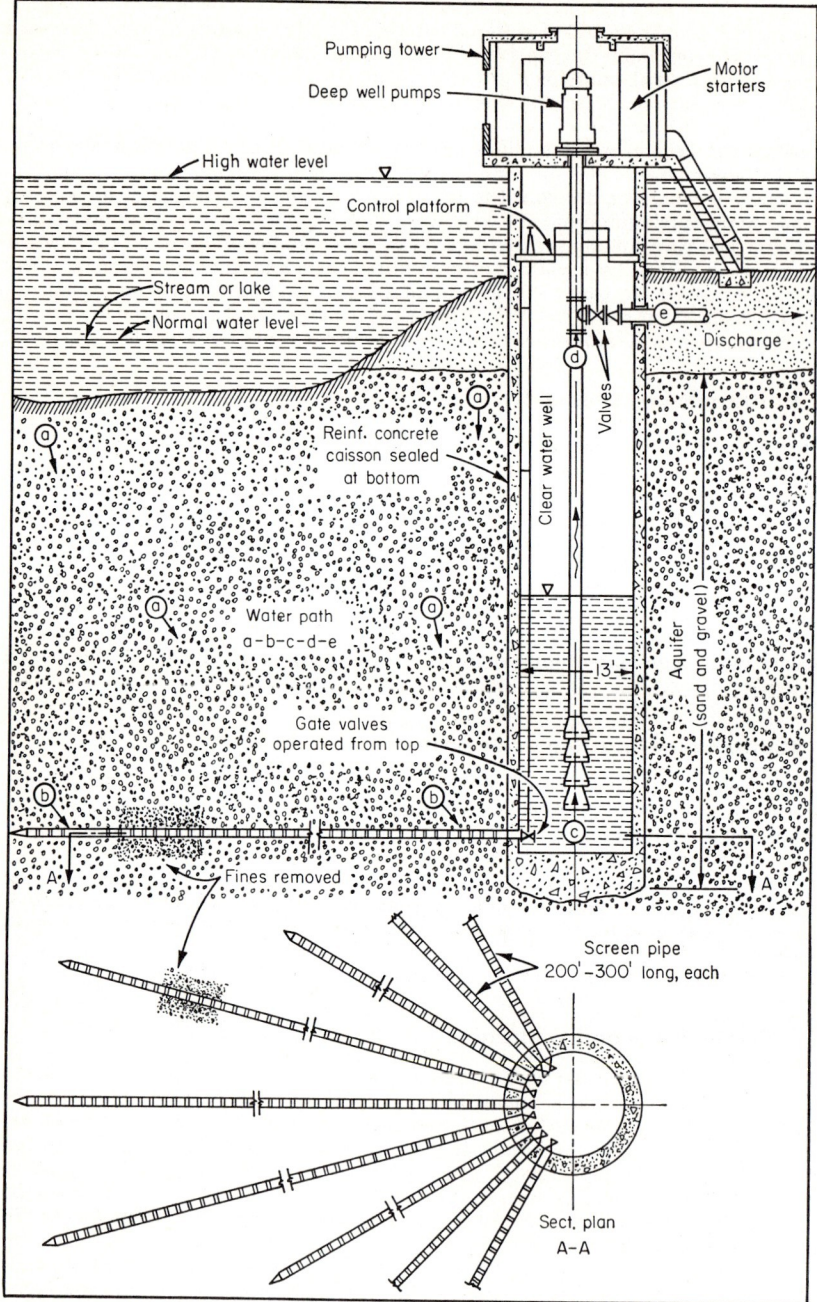

Fig. 13-34. A collector well located near a surface-water body. (*Ranney Method Water Supplies, Inc.*)

VI. GROUNDWATER FLUCTUATIONS

Groundwater levels, including water tables and piezometric surfaces, indicate the elevation of atmospheric pressure of aquifers. Changes of groundwater in storage by natural or artificial means cause changes in water levels. Similarly, changes in external loads will induce variations in groundwater levels.

A. Secular and Seasonal Effects

Secular variations are those extending over periods of several years or more. These are commonly produced by alternating series of wet and dry years in which rainfall is above or below the mean [42]. The rainfall- and groundwater-level records from San Bernardino Valley, California, shown in Fig. 13-35, provide a good illustration. As rainfall is the primary source of recharge in many aquifers, variations of rainfall

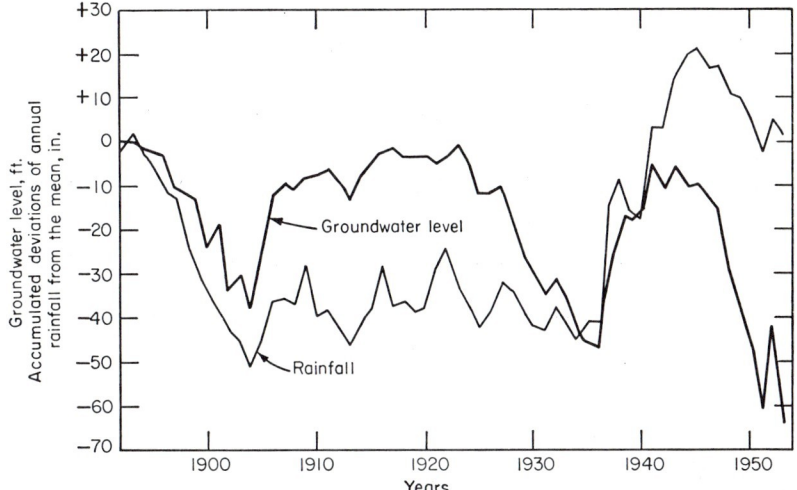

FIG. 13-35. Secular variations of maximum annual groundwater level and annual rainfall in San Bernardino Valley, California. (*After Todd* [1].)

and groundwater levels are closely correlated. The correlation is imperfect, however, because differences in rainfall intensity and distribution produce different amounts of recharge from the same total annual rainfall.

Where overdraft is a continuing phenomenon in a groundwater basin, a downward trend in water levels over a period of years is apparent (Fig. 13-36).

Seasonal fluctuations are often observed in the western United States. Here the alternating influences of recharge from rainfall and discharge by pumping for irrigation become apparent. These annual cycles can be seen in Fig. 13-36.

Groundwater-level changes can also produce changes in elevation of the land surface. Three distinct effects, all resulting in land subsidence, are recognized. In peat lands initially having water tables near ground surface, land subsidence occurs when water tables are lowered for agricultural purposes by drainage. Oxidation and wind erosion of the exposed peat are responsible. In the peat soils of the Sacramento–San Joaquin Delta in California, a steady subsidence averaging 3 in./year has been measured for more than 30 years.

A second form of land subsidence can occur when excessive pumping causes marked declines of water levels in confined aquifers. Benchmark and piezometric surface elevations in Fig. 13-36 indicate the relation involved. The subsidence is attributed to the reduction of hydrostatic pressure in the aquifer, which increases the stress on confining clay layers, causing them to be compressed. This form of subsidence has been observed in several areas in California [44].

The third type of land subsidence is that occurring when water is applied to certain soils in arid regions. Studies by Lofgren [45] in the San Joaquin Valley of California showed a surface subsidence of 10 ft within 18 months after application of water on a

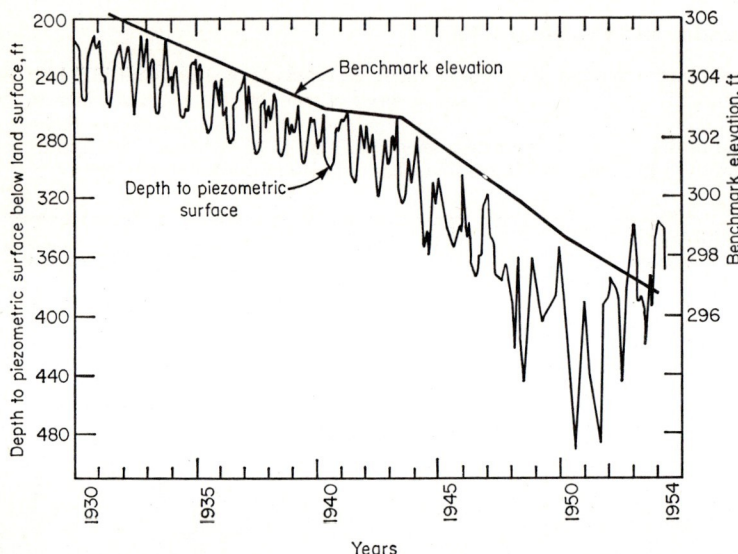

Fig. 13-36. Changes in ground-surface and piezometric-surface elevations near Delano, California. Annual fluctuations result from seasonal pumping for irrigation. (Land subsidence)/(water-level change) = $\frac{1}{23}$ for period 1931–1951. (*After Poland and Davis* [43].)

test plot. The subsidence is attributed to a compaction of the upper soil layers produced by wetting from above. Alluvial deposits of low field density and of extreme dryness appear to be most easily affected.

B. Streamflow Effects

Where a stream channel is in contact with an unconfined aquifer, water may flow from the stream into the ground, or the reverse, depending upon the relative water levels. An influent stream supplies water to aquifers; an effluent stream receives water from the aquifer. A stream may be influent in one location and effluent in another; also, changes can occur with time as stream stages relative to nearby groundwater levels shift.

During a flood period groundwater levels may be temporarily raised near a channel by inflow from the stream. This water is known as *bank storage*. Under idealized conditions, or where complete geologic and hydrologic data are available, bank storage can be evaluated quantitatively. Data from a model study [46], shown in Fig. 13-37, indicate the volume and flows to and from bank storage as function of time in flood periods.

Groundwater discharging into a stream forms the base flow of the stream. This may vary from total flow during periods of no surface runoff to a negligible fraction of the total flow during periods of high surface runoff. Empirical methods, described in Sec. 14, have been developed for estimating base flow during floods and for segregating surface runoff from base flow.

C. Evapotranspiration Effects

Water tables near ground surface frequently exhibit diurnal fluctuations resulting from evaporation and/or transpiration. Both processes cause the release of groundwater into the atmosphere, and both are highly correlated with temperature.

Evaporation from groundwater can be neglected unless the capillary zone above the water table approaches ground surface. Measurements of groundwater evaporation in tanks filled with different soils have been compared with evaporation from pans on the ground surface [47]. Results, presented in Fig. 13-38, indicate that evaporation is insignificant for water tables 3 ft or more below ground surface.

Where the root zone of vegetation approaches the water table, the uptake of water by roots equals for practical purposes the transpiration rate. Figure 13-39 shows the effect of transpiration on water levels

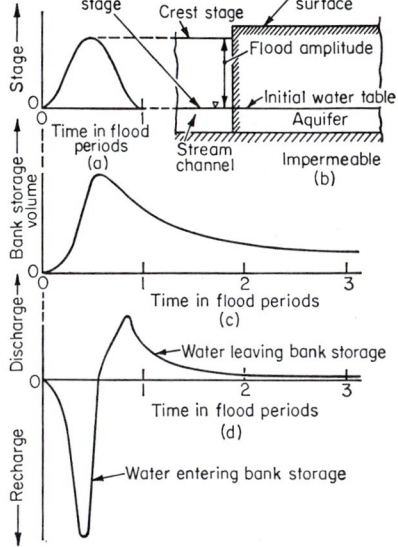

FIG. 13-37. Groundwater flow in relation to a flooding stream determined by a laboratory-model investigation for an idealized situation. (a) Flood hydrograph; (b) vertical cross section of field conditions simulated by model; (c) volume of bank storage as a function of time; (d) groundwater flow to and from bank storage. (*After Todd* [46].)

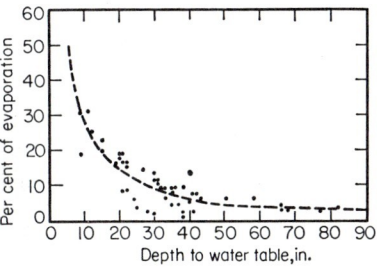

FIG. 13-38. Groundwater evaporation, expressed as a percentage of pan evaporation, as a function of depth to the water table. (*After White* [47].)

in a thicket of willows during the growing season and after heavy frosts. Magnitude of transpiration fluctuations depends upon type of vegetation, season, and weather.

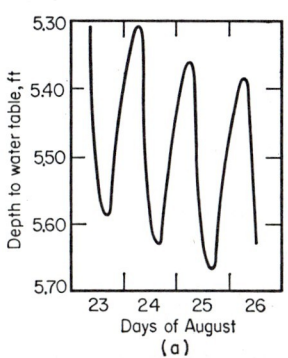
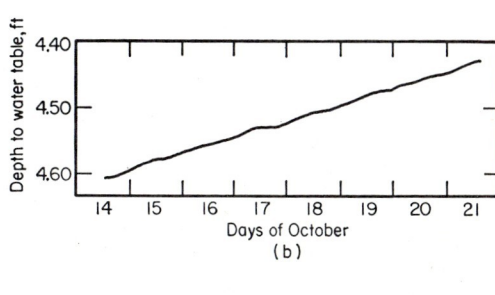

FIG. 13-39. Effect of transpiration discharge on groundwater levels in (a) summer and (b) after frost, near Milford, Utah. (*After White* [47].)

A typical pattern of diurnal fluctuation of a water table by evapotranspiration is given by Fig. 13-40. The quantity of groundwater withdrawn during a day can be computed [47] by assuming that evapotranspiration is negligible between midnight and 4 A.M. and that the water-table level during this interval approximates the daily mean. Letting h equal the hourly rise from midnight to 4 A.M. and letting s be net fall or rise of the water table in one day, then the daily volume discharged can be approximated by

$$V_{ET} = S_y A (24h \pm s) \qquad (13\text{-}59)$$

where S_y is specific yield and A is the area of vegetation.

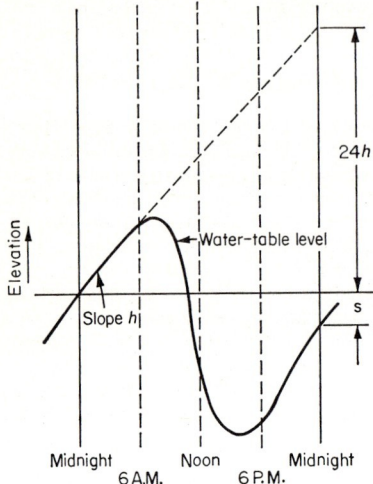

FIG. 13-40. Diurnal fluctuation of a water table as a result of transpiration. (*After Troxell* [48].)

D. Atmospheric-pressure Effects

Changes in atmospheric pressure have no effect on water tables (neglecting entrapped air below the water table), but do produce sizable fluctuations in wells penetrating confined aquifers. Increases in atmospheric pressure cause decreases in water levels, and conversely (Fig. 13-41). The ratio of water-level change to pressure change, both in terms of the same units, defines the *barometric efficiency* of an aquifer. Most observations fall in the range of 20 to 75 per cent.

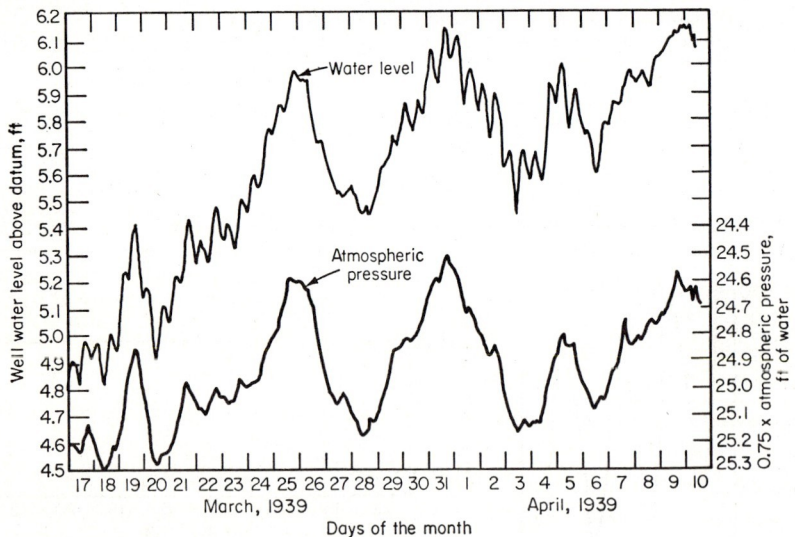

FIG. 13-41. Response of water level in a well penetrating a confined aquifer to atmospheric-pressure changes, showing a barometric efficiency of 75 per cent. (*After Robinson* [49].)

The phenomenon is a result of the elasticity of aquifers. It can be shown [50] that the barometric efficiency

$$B = \frac{\alpha E_s}{\alpha E_s + E_w} \qquad (13\text{-}60)$$

where α is porosity, E_s is the modulus of elasticity of structure of the aquifer, and E_w is the bulk modulus of compression of water (approximately 300,000 psi). Thick impermeable confining strata resist pressure changes and are associated with high barometric efficiencies; thinly confined aquifer show low efficiencies.

Because the barometric efficiency is related to the compressibility of a confined aquifer, it can be used in defining the storage coefficient S of an aquifer. The relation has the form

$$S = \frac{\alpha \gamma b}{E_w B} \tag{13-61}$$

where γ is the specific weight of water, and b is the aquifer thickness.

E. Tidal Effects

Groundwater levels in coastal aquifers fluctuate in response to ocean tides. Three situations, illustrated by Fig. 13-42, will be considered. For convenience, assume the tides to have a simple harmonic form.

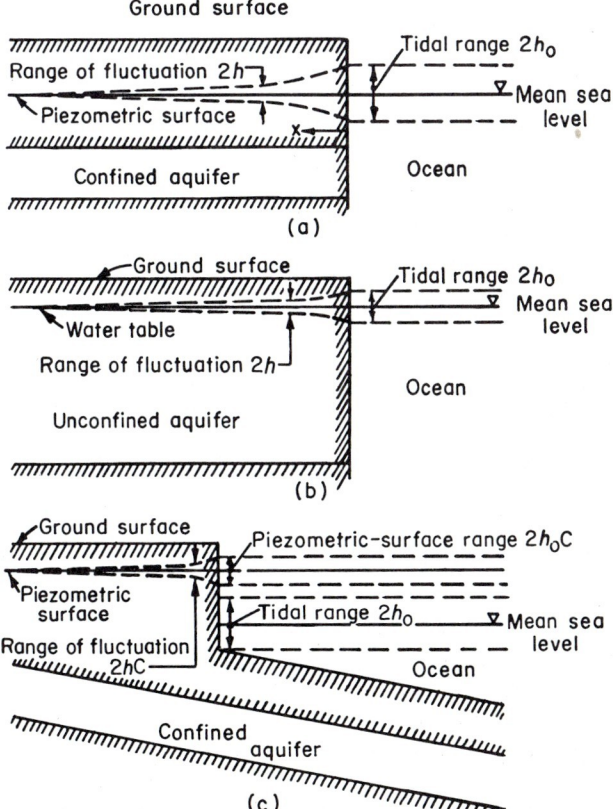

FIG. 13-42. Groundwater-level fluctuations produced by ocean tides. (a) Confined aquifer; (b) unconfined aquifer; (c) loading of a confined aquifer. (*After Todd* [1].)

In a confined aquifer outcropping to the sea, one-dimensional flow is involved. Flow is governed by

$$\frac{\partial^2 h}{\partial x^2} = \frac{S}{T} \frac{\partial h}{\partial t} \tag{13-62}$$

where h and x are as defined in Fig. 13-42a, S is storage coefficient, T is the coefficient of transmissibility, and t is time. For a tidal amplitude h_0, the boundary conditions of $h = h_0 \sin \omega t$ at $x = 0$ and $h = 0$ at $x = \infty$ are applicable. With a tidal period t_0, the solution of Eq. (13-62) is [51]

$$h = h_0 e^{-x\sqrt{\pi S/t_0 T}} \sin\left(\frac{2\pi t}{t_0} - x\sqrt{\frac{\pi S}{t_0 T}}\right) \qquad (13\text{-}63)$$

It follows that the amplitude decreases exponentially inland, so that, at a point x,

$$h_x = h_0 e^{-x\sqrt{\pi S/t_0 T}} \qquad (13\text{-}64)$$

The time lag t_L between any given maximum or minimum in the ocean and its subsequent occurrence inland is given by

$$t_L = x\sqrt{\frac{t_0 S}{4\pi T}} \qquad (13\text{-}65)$$

From this, the wave velocity

$$v_w = \frac{x}{t_L} = \sqrt{\frac{4\pi T}{t_0 S}} \qquad (13\text{-}66)$$

and the wavelength

$$L_w = v_w t_0 = \sqrt{\frac{4\pi t_0 T}{S}} \qquad (13\text{-}67)$$

Water flows into the aquifer for half of each tidal period and out for the other half. The volume of water involved in the interval $t_0/2$ per foot of coast amounts to

$$V = h_0 \sqrt{\frac{2 t_0 S T}{\pi}} \qquad (13\text{-}68)$$

For an unconfined aquifer connecting with the ocean (Fig. 13-42b), the above analysis is applicable if the fluctuations are small in relation to the saturated thickness.

Confined aquifers may be subjected to loading variations of tides (as shown in Fig. 13-42c) without being directly in contact with the ocean. The loading effect is direct, opposite to that of atmospheric pressure. The ratio of piezometric-level amplitude to tidal amplitude measures the tidal efficiency of the aquifer. The *tidal efficiency C* is related to the barometric efficiency B by [52]

$$C = 1 - B \qquad (13\text{-}69)$$

The aquifer response to loading is given by Eq. (13-63) when multiplied by C.

VII. GROUNDWATER MANAGEMENT

Development of groundwater resources for maximum utilization requires recognition that groundwater occurs in underground reservoirs, or basins. These hydrologic units must be managed to obtain desired benefits in a similar manner as in the operation of a surface-water reservoir. By planning and by coordinating groundwater and surface-water resources, maximum beneficial use can be achieved [53].

A. Safe Yield

Safe yield may be defined as the amount of water which can be withdrawn annually from a groundwater basin without producing an undesired result. Draft in excess of safe yield is *overdraft*. The concept of safe yield is based on the principle that groundwater is a renewable natural resource. If groundwater extractions tend to exhaust or harm the supply in some manner, then the resource is being impaired with time to the extent that the safe yield is exceeded.

Several undesired results of excess pumpage are recognized [54]. One is that of quantity of water available. If the annual pumpage consistently exceeds the inflow to the basin, depletion, or mining, of groundwater will occur. Either the physical size of the basin or the rate at which water enters the basin may govern the available supply [8]. If lowering of groundwater levels to a point where the cost of pumping becomes excessive, a limit defining safe yield may be reached. Similarly, if pumping beyond some rate produces water of inferior quality as a result of inflow of degrading water, safe yield has been exceeded. And finally, if pumping interferes with prior water rights of others, a legal limitation for safe yield can be determined.

It should be recognized that safe yield of a groundwater basin is subject to considerable variation with time. As its determination is based upon specified conditions, either existing or assumed, changes in these produce changes in the safe yield. Thus changes in pumping rates and locations will affect groundwater levels, which in turn modify inflow to the basin. Changes in land use could affect recharge rates; changes in water use could make large differences in defining an economic pumping lift. In practice, safe-yield calculations should be reexamined from time to time in terms of changing conditions within a basin and of more comprehensive hydrologic and geologic data being available.

B. Equation of Hydrologic Equilibrium

The available water limit for safe yield is most amenable to quantitative evaluation; in many instances one or more of the other results may be incurred when this limit is exceeded. Determination of safe yield is based upon solution of the equation of hydrologic equilibrium. This is a statement that all waters entering and leaving a basin must form a balance. In general form this may be expressed as

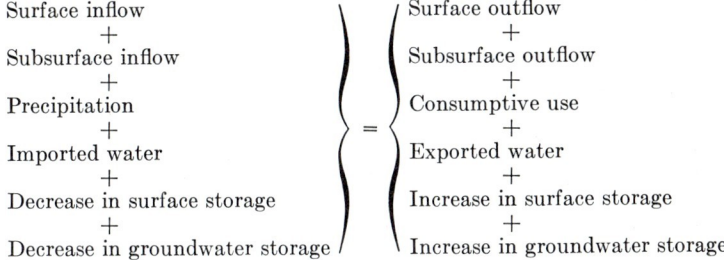

Because of the interrelations of surface and groundwaters, all waters must be included in the equation. Often certain terms can be omitted because they are negligible. Each item represents a discharge of water. Any consistent units of volume and time can be employed; acre-feet per water year is common. Inaccuracies in basic data and approximations rarely yield an exact balance. Adjustments require good judgment and careful analysis of the area under study.

C. Basic-data Collection

Solution of the equation of hydrologic equilibrium requires comprehensive collection of basic data within the basin of interest. Following is an outline of the data required and the methods of analysis necessary [55].

1. Surface Inflow and Outflow; Imported and Exported Water. Surface-water flows are measurable by standard hydrographic procedures. If all flows to and from the basin are not measured by existing gaging stations, additional gages should be installed to obtain at least a short record of all flows.

2. Precipitation. Data on precipitation are generally available from the Weather Bureau. Additional gages should be installed where necessary to provide a reasonable areal coverage. Average annual precipitation over the basin is obtained from point measurements by the isohyetal or Thiessen-polygon methods.

3. Consumptive Use. Consumptive use, or evapotranspiration, can be evaluated for an area by determining the various types of land use and estimating the value of consumptive use for each. Standard methods are available for crop and native vegetation lands; sampling may be necessary for urban areas. The sum of the products for each land use of area times consumptive use yields the total consumptive use.

4. Changes in Surface Storage. These can be evaluated directly from areas and water-level changes of surface-water bodies.

5. Changes in Groundwater Storage. By selecting time intervals in which the amount of water in unsaturated storage is nearly equal, only changes in storage within the zone of saturation need be considered. From measurements of water levels in wells, maps of groundwater changes can be prepared. The product of change in water level times specific yield (or storage coefficient) times area gives change in underground storage.

6. Subsurface Inflow and Outflow. These quantities are difficult to evaluate in practice. Investigation may show one to be absent, or the two to be nearly equal, and therefore they cancel from the equation. Where subsurface data are available, flows can be computed directly by Darcy's law.

D. Determination of Safe Yield

Several convenient methods for evaluating safe yield are available where the quantity of water governs. These are simplifications of the equation of hydrologic equilibrium. Three are described below; other more specialized methods can be found elsewhere [1, 53].

By plotting annual changes in elevations of groundwater levels against annual draft, R. A. Hill [56] measured safe yield as the draft corresponding to zero change in elevation. For the safe yield to be representative, the supply during the period of

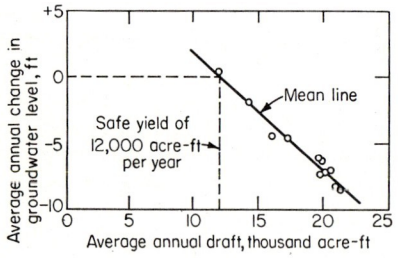

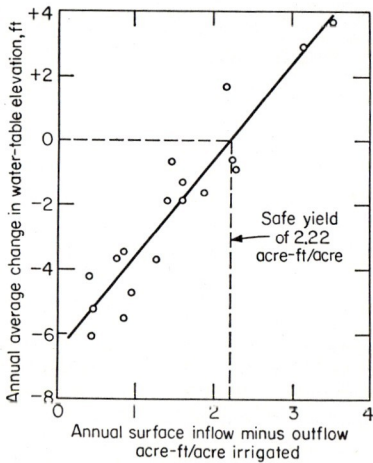

Fig. 13-43. Determination of safe yield by the Hill method for the Pasadena Basin, Los Angeles County, California. (*After Conkling* [56].)

Fig. 13-44. Determination of safe yield by the Harding method for the Tule River–Deer Creek area, San Joaquin Valley, California. (*After Ingerson* [58].)

record should approximate the long-time mean supply. Figure 13-43 is an example of the application of the method.

In a similar manner Harding [57] evaluated safe yield by plotting retained inflow (total inflow minus total outflow) vs. annual changes in water-table elevation. The annual draft should be reasonably constant, and the supply should approximate the long-time mean. An example is shown in Fig. 13-44; here subsurface flows could be neglected.

If the mean groundwater elevation of a basin is the same as at some time several years earlier, the average annual net draft on the basin is a measure of safe yield. The draft before and after the period should approach overdraft conditions, and the annual supply should approximate the long-time mean. Figure 13-45 shows an example of the application of the method.

E. Conjunctive Use of Water Resources

Future demand for water requires planning for the maximum utilization of all existing supplies. This can most economically be attained by conjunctive use of surface and groundwater reservoirs. In operation, surface reservoirs impound streamflow for transfer at an optimum rate to underground reservoirs. During normal and wet years, surface storage will meet most needs, while groundwater storage can be retained for use during years of subnormal precipitation. Groundwater levels will be raised during wet periods and lowered during dry periods. Artificial recharge (Subsec. VIII) is necessary to store a maximum volume of water underground when it is available at the surface.

Conjunctive use provides a greater firm yield than separate surface and groundwater sources could yield individually. Other than provision for artificial recharge, the method is largely a matter of coordinated use of the available water resources. Advantages and disadvantages, described elsewhere [1], need to be considered in applying the method in a given basin. Examples of the procedure have been published [53, 60].

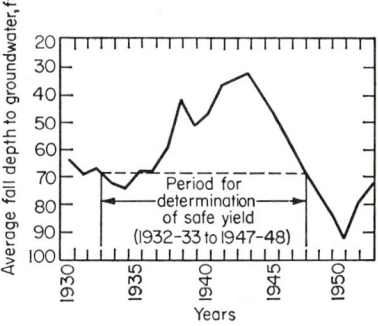

FIG. 13-45. Determination of safe yield based on zero net groundwater fluctuation for South Santa Clara Valley, California. (*After Haley and others* [59].)

VIII. ARTIFICIAL RECHARGE OF GROUNDWATER

Artificial recharge may be defined as augmenting the natural replenishment of groundwater storage by some method of construction, spreading of water, or by artificially changing natural conditions. It is useful for reducing overdraft, conserving surface runoff, and increasing available groundwater supplies [61]. Recharge may be incidental or deliberate, depending on whether or not it is a by-product of normal water utilization [62]. Only deliberate recharge will be considered here, although locally incidental recharge, such as by excess irrigation, may be important.

A. Artificial-recharge Practice

In the United States recharging began near the end of the nineteenth century, and the practice has increased steadily since then. Quantities of water artificially

Table 13-6. Artificial Recharge of Groundwater in the United States, 1955*

Source of water	Quantity, million gpd
Air-conditioning return	41
Industrial wastes	49
Surface water	540
Public water supplies	71
Total	701

* From MacKichan [5].

recharged during 1955 are listed in Table 13-6. The total amounts to 1.5 per cent of the total national groundwater use [1].

More than half of the total recharge in Table 13-6 was conducted in California. Data for the state agencies conducting recharging, type and numbers of recharge projects, and quantities of water recharged during the 1957–1958 season are summarized in Table 13-7.

Table 13-7. Summary of Artificial Recharge in California, 1957–1958*

Agencies conducting recharging		Recharge quantities and projects		
Type	No.	Method of recharge	Quantity, acre-ft	No. of projects
Irrigation districts....................	9			
Water-conservation districts.............	7	Basin	367,700	149
Flood-control districts.................	7	Modified stream bed	185,900	42
Other water districts...................	7	Ditch or furrow	59,400	22
Municipalities.........................	4	Pit	8,500	19
Mutual water companies................	10	Well	6,100	33
Other water companies and associations...	6	Flooding	2,400	11
Miscellaneous.........................	4			
Total..............................	54	....	630,000	276

* From Richter and Chun [61].

In Europe artificial recharge is widely practiced. Many Swedish cities recharge river or lake water for purification, storage, and temperature-control purposes. In Germany polluted river waters are recharged to help meet industrial and municipal water demands. Recharge basins in coastal sand dunes are an integral part of the water-supply systems for Amsterdam, Leiden, and The Hague, in the Netherlands.

B. Water Spreading

A variety of methods have been developed for artificially recharging groundwater; the choice may be governed by topographic, geologic, hydrologic, economic, or

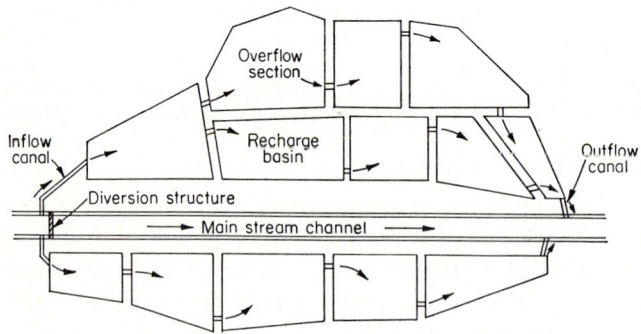

Fig. 13-46. Layout of a series of connected recharge basins adjoining a stream channel.

climatic conditions. The most widely practiced methods come under the heading of *water spreading*. This involves the release of water over the ground surface, thereby increasing the wetted area over which infiltration into the ground can occur. Four types of spreading are recognized: basin, modified stream bed, ditch or furrow, and flooding.

1. Basin Method. In the basin method water is contained in a series of basins formed by a network of dikes or levees generally constructed to make maximum advantage of the local topography. Basins are often interconnected so that overflow from one basin will be released into the next basin at a slightly lower elevation. Figure 13-46 shows the layout of a basin system. As indicated by California data in Table 13-7, the basin method is a common means of recharge. The efficient use of space, adaptability to irregular terrain, and ease of construction by bulldozers are important advantages. Silt deposition, particularly in the upper basins of a series, should be removed periodically as necessary. Representative recharge rates are listed in Table 13-8.

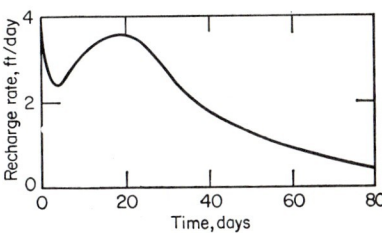

FIG. 13-47. Typical recharge-rate variation with time for water spreading on undisturbed soil. (*After Muckel* [64].)

Water spreading in basins is most economical with high infiltration rates. A typical rate curve, shown in Fig. 13-47, shows a pronounced tendency to decrease with time. Where subsurface strata or water-table conditions do not impede the downward movement of water, the gradual reduction can be attributed to microbial growths clogging the soil pores. Research programs have studied soil treatments and operational methods as means to maintain high infiltration rates [64]. Although progress has been made, particularly in understanding the causes of clogging, no general solution to the problem has yet been found.

Table 13-8. Spreading-basin Recharge Rates*

Location	Rate, ft/day
Santa Cruz River, Ariz.	1.1–3.8
Los Angeles County, Calif.	2.2–6.2
Madera, Calif.	1.0–4.1
San Gabriel River, Calif.	1.9–5.4
Santa Ana River, Calif.	1.8–9.6
Santa Clara Valley, Calif.	1.4–7.3
Tulare County, Calif.	0.4
Ventura County, Calif.	1.2–1.8
Des Moines, Iowa	1.5
Newton, Mass.	4.3
East Orange, N.J.	0.4
Princeton, N.J.	0.1
Long Island, N.Y.	3.1
Richland, Wash.	7.7

* After Todd [63].

2. Modified-stream-bed Method. The objective of this method is to extend the time and area over which water is recharged from a naturally influent channel. In southern California releases from upstream flood-control reservoirs are often regulated so as not to exceed the absorptive capacity of downstream channels. Often works of a temporary nature such as widening, leveling, or scarifying a bed are effective. Low check dams, dikes, or ditches to distribute flows may be constructed. Examples are sketched in Fig. 13-48. Dams have been constructed of earth, brush, and rock and wire. In some instances permanent dams or weirs of concrete have been constructed.

3. Ditch or Furrow Method. In irregular terrain, water can be spread through a series of ditches or furrows. These should be shallow, flat-bottomed, and closely spaced to obtain maximum water-contact area. Gradients of major ditches should be sufficient to carry suspended material through the system since deposition of fine-grained material clogs surface openings. The design of a ditch system is governed by the topography and size of area. Various types have been devised; common ones are the contour, lateral, and tree-shaped types [61]. On steep slopes checks give better water distribution. A collecting ditch is necessary at the lower end of the area to return excess water to the main channel.

4. Flooding Method. The flooding method involves diversion of water to form a thin sheet flowing over relatively flat land. Although little in the way of control works is required, small canals and gullies aid in releasing the water uniformly over an

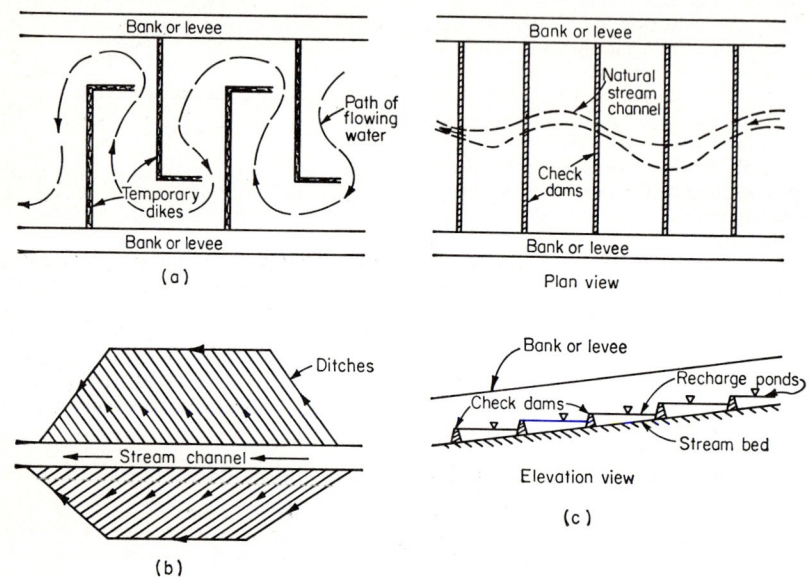

Fig. 13-48. Examples of channel spreading. (a) Temporary earth dikes; (b) ditch network; (c) check dams and ponds.

area. Flow rates should be minimum so as not to disturb vegetation and soil covering; highest infiltration rates can thus be obtained. Peripheral dikes or ditches should surround the entire flooding area.

C. Recharge through Pits

Where shallow subsurface strata restrict the downward passage of water, water spreading is ineffective. However, if the relatively impermeable material can be removed by excavation, recharging in pits may be feasible. The high cost of construction makes the procedure economical only under special circumstances. Abandoned gravel pits and pits formed by sale of the excavated material are useful situations. At Peoria, Ill., a pit excavated adjacent to the Illinois River proved to be a highly beneficial means of recharging river water into locally overdrawn aquifers [65]. Sides of pits should be steep enough so that silt settles to the bottom, leaving the sides relatively free for infiltration. Provision must be made to remove the silt accumulation periodically.

D. Recharge through Wells

Water can be recharged underground through wells where deep confined aquifers or space limitations preclude application of other methods. From a hydraulic standpoint, pumping and recharge rates should be equal for wells having the same head conditions (Fig. 13-49); however, in practice, recharge rates seldom equal pumping rates. Differences can be attributed to clogging in the immediate vicinity of the perforated casing. Clogging can result from entrapment of fine aquifer particles, from filtration of suspended material in the recharge water, from air binding by dissolved air, from bacteria, and from chemical reactions between recharge and natural waters. Considerable care must be exercised in the construction and operation of recharge wells to minimize these effects. Table 13-9 lists average well-recharge rates.

Wells that are gravel-packed can be recharged more efficiently than those without gravel envelopes. Concrete seals at the top of the input aquifer prevent upward movement of water along the outside edge of the casing. Complete development to remove fines from around the well is important. Recharge water should be clear, should not have a high sodium content, and should be chlorinated. Recharging should be done at a relatively constant pressure, and water should not fall freely in the well because of the resulting aeration.

Recharge wells have been extensively employed on Long Island, New York, to recharge uncontaminated cooling and air-conditioning water [66, 67]. At Manhattan Beach, California, a coastal line of recharge wells has served to control seawater intrusion [68, 69]. In a few localities recharge wells dispose of storm runoff; however, the practice should be limited to formations where danger of pollution is low.

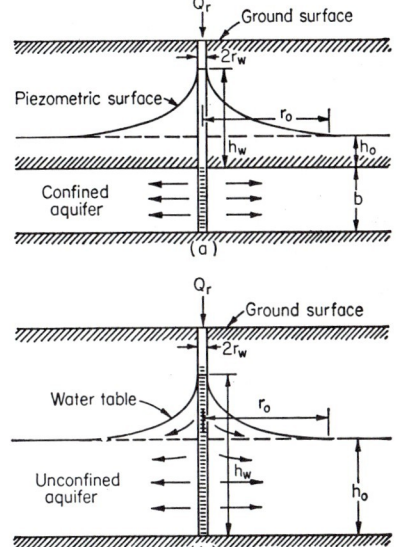

FIG. 13-49. Radial flow from recharge wells penetrating (a) $Q_r = \dfrac{2\pi K b (h_w - h_0)}{\ln (r_0/r_w)}$, confined, and (b) $Q_r = \dfrac{\pi K (h_w^2 - h_0^2)}{\ln (r_0/r_w)}$, unconfined aquifers. (*After Todd* [1].)

Table 13-9. Well-recharge Rates*

Location	Rate, cfs
Fresno, Calif.	0.2–0.9
Los Angeles, Calif.	1.2
Manhattan Beach, Calif.	0.4–1.0
Orange Cove, Calif.	0.7–0.9
San Fernando Valley, Calif.	0.3
Tulare County, Calif.	0.12
Orlando, Fla.	0.2–21
Mud Lake, Idaho	0.2–1.0
Jackson County, Mich.	0.1
Newark, N.J.	0.6
Long Island, N.Y.	0.2–2.2
El Paso, Tex.	2.3
Williamsburg, Va.	0.3

* From Todd [63].

E. Sewage and Waste-water Recharge

Sewage and waste waters constitute valuable sources of recharge water. Because of higher concentrations of suspended matter and bacteria, recharge rates for sewage effluents are lower than for fresh water. Investigations of sewage spreading for water reclamation in California yielded recharges of from 0.2 to 1.2 ft/day [63]. Alternate wetting and drying periods of 7 to 14 days, with cultivation during the dry cycle, gives maximum spreading rates. Where spreading is conducted on fine-grained materials, natural filtration makes the water potable after only a few feet of percolation.

Treated sewage can also be recharged through wells. Tests [70] have shown clogging rates to be proportional to the suspended solids of the sewage water. Continuous operation is possible with regular chlorine injection and redevelopment by pumping about 4 per cent of the recharged water.

F. Induced Recharge

Induced recharge is water entering the ground from a surface-water source as a result of withdrawal of ground water adjacent to the source. Figure 13-50 illustrates how a well pumping near a stream induces flow into the ground. Thus wells, infiltration galleries, and collector wells located directly adjacent to and fed largely by surface water cause surface water to be recharged underground. The method is useful in locations where surface waters may not be of high quality and large groundwater reservoirs are not available. By natural filtration the pumped water is of higher

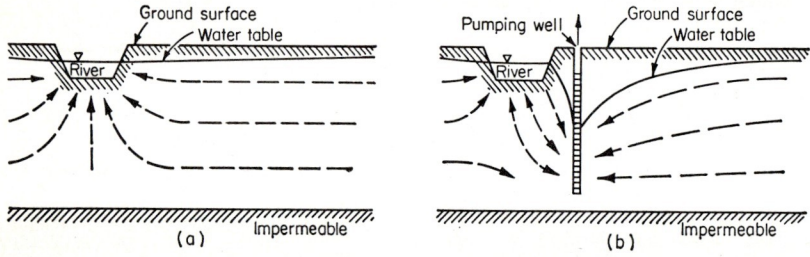

Fig. 13-50. Induced recharge resulting from a well pumping near a river. (a) Natural flow pattern; (b) flow pattern with pumping well. (*After Todd* [1].)

quality than the source stream [71] and no depletion of groundwater reserves is produced. Pumpage for induced recharge is common in the Mississippi Valley of this country and along rivers in Germany.

IX. SALT-WATER INTRUSION OF COASTAL AQUIFERS

Salt-water intrusion may be defined as the increase in salinity of groundwater at a given location and depth produced by acts of man. Intrusion can result from a variety of degrading influences; however, a unique groundwater problem of worldwide significance is that of intrusion in coastal aquifers. Here excessive pumping may cause lateral or upward movement of saline water into wells.

A. Salt-water Sources

Salt water from coastal aquifers originates from two primary sources, connate water and sea water. Connate waters are those remaining in sedimentary rocks from the time of deposition. Most of such deposition took place under submarine conditions, but the water is often considerably modified from its original composition. Increased salinities up to brine concentrations can result from solution of rock minerals or saline invasions by shifting sea levels. Decreased salinities can be produced from dilution or incomplete flushing by meteoric water. In layered sedimentary formations large variations in salt content from one stratum to another can be found. These result from differences in circulation, affected by rock structure and permeability. An example from the Texas coast appears in Fig. 13-51.

Sea water enters aquifers most commonly through submarine outcrops. Surface sources, which may be of local importance, include leakage downward around wells, tidal waves, hurricane-blown sea spray, and tidal marshes. Special surface situations

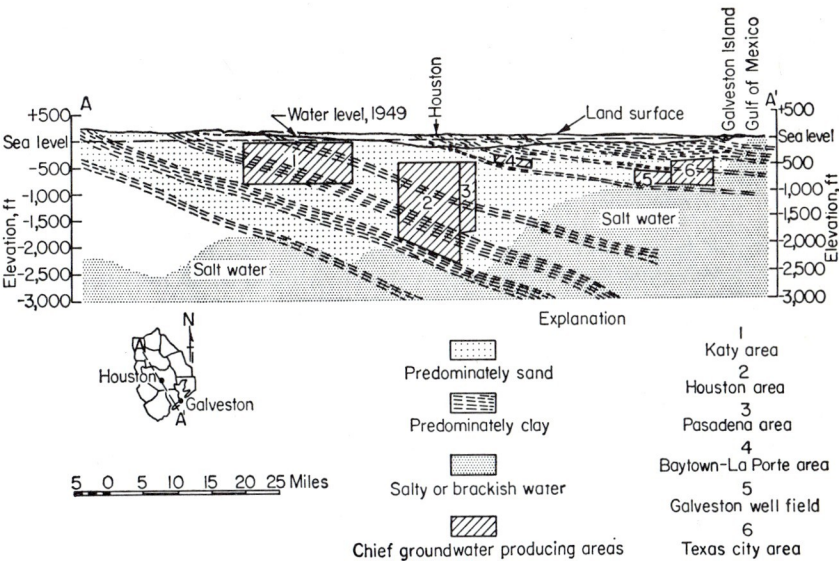

Fig. 13-51. Cross section showing distribution of salt water in Coastal Plain deposits in the Houston Gulf Coast region. (*After Goines, Winslow, and Barnes* [72]*.*)

Fig. 13-52. Locations of salt-water intrusion of coastal aquifers in the United States. (*After Todd* [73]*.*)

may be conducive to intrusion, such as sea-level canals and regulated streamflows which permit sea water to advance inland.

B. Intrusion in the United States

Locations where intrusion in coastal aquifers of the United States is known to be occurring are shown in Fig. 13-52. It can be seen that intrusion is common to all coasts; invariably it occurs where a coastal metropolitan area is largely dependent upon an underground water supply. The states of Florida, California, Texas, and New York, in that order, are most seriously affected [73].

In Florida, a combination of circumstances has caused extensive intrusion. The state is a lowland, is underlain by saline waters of connate origin in permeable limestone aquifers, and has a lengthy coast line which is noted for its salubrious climate. As a consequence, salt-water intrusion is occurring in 28 specific locations, and some 18 municipal water supplies have been adversely affected. Interior drainage canals in the Miami area permitted invasion of sea water by tidal action during droughts. Although installation of control dams has stabilized the advance since 1946, three municipal well fields have been affected since 1925.

Again, in California, the coastal concentration of population has been responsible for several intrusion areas. By 1957 nine groundwater basins had been critically affected, while 71 others were areas of suspected intrusion. Almost all were contaminated by the lateral advance of sea water into confined aquifers.

Residual saline waters in the confined Coastal Plain sediments have contributed to intrusion in southern Texas (Fig. 13-51). Areas at Galveston, Texas City, Houston, and Beaumont–Port Arthur have been adversely affected. In New York intrusion at the western end of Long Island has existed for many years.

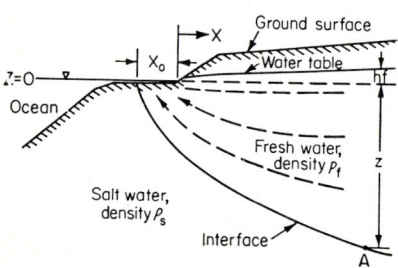

Fig. 13-53. Sketch of a fresh-water flow in an unconfined coastal aquifer.

C. The Ghyben-Herzberg Concept

Consider an unconfined coastal aquifer, as shown in Fig. 13-53, containing fresh water and salt water. The fluids have densities ρ_f and ρ_s, respectively, and have a common boundary, or interface. The shape and movement of the interface depend upon a hydrodynamic balance between the two fluids.

An early simplified picture of the distribution of the fluids was based on a hydrostatic equilibrium. In Fig. 13-53 the hydrostatic pressure on each side of the interface at point A gives

$$\rho_s g z = \rho_f g h_f + \rho_f g z \tag{13-70}$$

where g is the acceleration of gravity, and z and h_f are as shown in Fig. 13-53. This reduces to

$$z = \frac{\rho_f}{\rho_s - \rho_f} h_f \tag{13-71}$$

which is the *Ghyben-Herzberg relation*, named after its European discoverers. If $\rho_s = 1.025$ g/cm³ and $\rho_f = 1.000$ g/cm³, then

$$z = 40 h_f \tag{13-72}$$

Although the Ghyben-Herzberg relation implies no flow, groundwater is invariably moving in coastal areas. Without flow, a horizontal interface would develop with fresh water floating above salt water. A more correct picture of intrusion is that shown in Fig. 13-53. Where the flow lines are sloping upward, the Ghyben-Herzberg relation gives too small a depth to salt water. Further inland, where the flow lines are nearly horizontal, the error is negligible.

The Ghyben-Herzberg relation is also applicable to confined aquifers, the piezometric surface replacing the water table. It should be noted that a fresh-water flow to the sea is necessary to stabilize the interface.

D. The Dynamic Concept

In many intrusion areas the depth to salt water computed from the Ghyben-Herzberg relation differs markedly from observations. Often these discrepancies can be attributed to assumptions that the head in the salt water is at mean sea level and that the fresh and salt waters are static. An expression taking into account flows in each fluid can be derived by considering the fluid heads [74, 75].

If h_f and h_s are heads in regions occupied by fluids of densities ρ_f and ρ_s, respectively, then

$$h_f = z + \frac{p}{\rho_f}$$
$$h_s = z + \frac{p}{\rho_s}$$
(13-73)

where z is elevation, and p is pressure. At a point on the interface, Eq. (13-73) reduces to

$$z = \frac{\rho_s}{\rho_s - \rho_f} h_s - \frac{\rho_f}{\rho_s - \rho_f} h_f$$
(13-74)

Thus the depth to salt water can be defined by the heads and densities across the interface. Taking mean sea level as datum and letting $h_s = 0$, Eq. (13-74) reduces to Eq. (13-71), the Ghyben-Herzberg relation.

Example. Given the situation sketched in Fig. 13-54. Substituting values into Eq. (13-74),

$$z = \frac{1.025}{1.025-1.000}(-5.0) - \frac{1.000}{1.025-1.000}(2.5) = -305 \text{ ft}$$

Neglecting the salt-water head and computing the depth to salt water,

$$z = -40(2.5) = -100 \text{ ft}$$

This indicates the invalidity of the Ghyben-Herzberg relation for general application in intrusion areas.

E. Location of the Interface

For a homogeneous and isotropic aquifer, the fresh–salt water interface can be closely approximated by a flow-net analysis [77]. Assume a fresh-water flow discharging into a shallow sea along the line $z = 0$, $x_0 < x < 0$, as shown in Fig. 13-53. The small fraction of fresh-water flow above sea level is neglected; this is equivalent to a confining layer along the line $z = 0$, $x > 0$. The interface can be located from

$$z^2 = \frac{2\rho_f q}{K(\rho_s - \rho_f)}\left(x + \frac{\rho_f q}{2K(\rho_s - \rho_f)}\right)$$
(13-75)

where q is the fresh-water flow per unit width, K is permeability, and ρ_f and ρ_s are the fresh- and salt-water densities, respectively. The length of the submarine outflow section x_0 is obtained by setting $z = 0$ in Eq. (13-75), so that

$$x = -\frac{\rho_f q}{2K(\rho_s - \rho_f)}$$
(13-76)

The head along the line $z = 0$ is given by

$$h_f = \sqrt{\frac{2qx(\rho_s - \rho_f)}{\rho_f K}}$$
(13-77)

where h_f is the height of the water table, or the piezometric head for a confined case, at a distance x from the shore line. Note in Eq. (13-76) that the outflow section is proportional to the magnitude of the fresh-water flow.

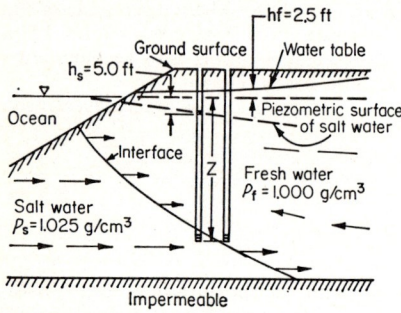
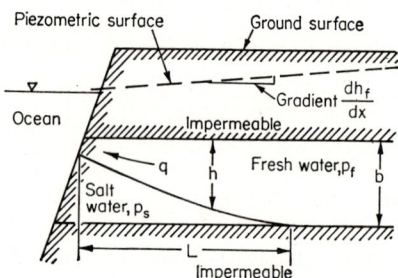

FIG. 13-54. Example of salt-water intrusion with the salt-water head below mean sea level.

FIG. 13-55. Salt-water wedge in a confined aquifer.

For a confined aquifer having horizontal fresh-water flow in equilibrium with salt water (Fig. 13-55), an expression for the position of the interface can be derived directly from Darcy's law. Thus

$$q = Kh \frac{dh_f}{dx} \qquad (13\text{-}78)$$

where q and K are as previously defined, and h and dh_f/dx are as shown in Fig. 13-55. Substituting the Ghyben-Herzberg relation and integrating yields

$$q = \frac{1}{2} \frac{\rho_s - \rho_f}{\rho_f} \frac{Kb^2}{L} \qquad (13\text{-}79)$$

where b and L are as identified in Fig. 13-55. For uniform aquifer and fluid conditions, Eq. (13-79) shows that the length of the intruded wedge is inversely proportional to the fresh-water flow.

F. Structure of the Interface

Field measurements of interfaces have revealed a mixing zone ranging in thickness from a few feet to several hundred feet. The interface is not, therefore, an ideal flow line of zero thickness, but instead is a zone resulting from various external influences. These consist of natural recharge and discharge of the fresh-water, pumping, and tidal action. These influences cause the interface to shift continually toward a new equilibrium position. Each movement, however, causes dispersion to occur, so that a transition zone marked by a well-defined salinity gradient is established. Dispersion depends upon the coefficient of dispersion of the aquifer and the distance traversed by groundwater (Subsec. III-D). The thickness of the transition zone at any location depends upon the coefficient of dispersion, the unsteady fresh-water-flow field, the permeability, and the tidal pattern. Current research is expected to elucidate some of the complexities of the problem.

As salt water disperses upward within the interfacial transition zone by externally caused fluctuations, a portion of it is swept seaward by the fresh-water flow. This creates a small density gradient and induces a slow net landward flow of the salt-water body. A sketch of the circulation in an idealized aquifer appears in Fig. 13-56. Although several investigators have postulated this salt-water-flow system [73], only recently have studies at Miami, Fla., by the U.S. Geological Survey provided field verification [77].

G. Control of Intrusion

Once intrusion develops in a coastal aquifer, it is not easy to control. The slow rates of groundwater flow, the density differences between fresh and salt waters, and the flushing required usually mean that contamination, once established, may require years to remove under natural conditions.

To control intrusion several methods have been suggested [78]. Reduction of pumping, or modification of pumping, eliminates the overdraft which causes intrusion and allows formation of a stable position of the interface. Water rights of landowners and the cost of supplemental water or of a surface distribution system are practical difficulties. Artificial recharging, a second method, would in many instances be economically infeasible in terms of capital and annual costs of a large artificial recharge system. The pumping-trough method, shown in Fig. 13-57a, would be created by a line of wells paralleling the coast. By maintaining a minimum gradient on the landward side, most water pumped would be sea water.

The pressure-ridge method (Fig. 13-57b) is the reverse of the pumping trough. Fresh water is recharged into a line of wells (or basins in unconfined aquifers) to main-

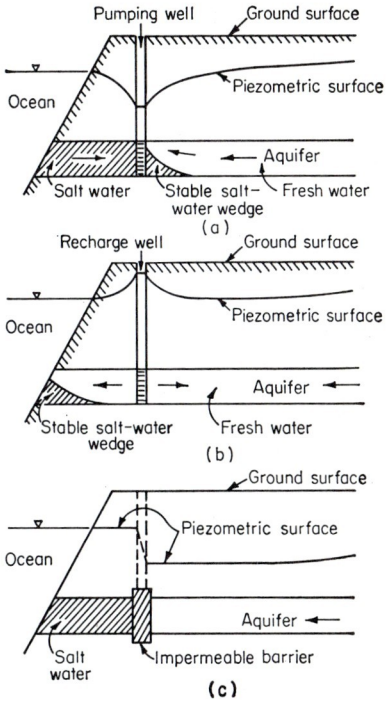

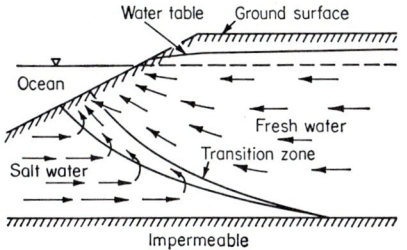

FIG. 13-56. Idealized sketch of mean flow pattern in an unconfined coastal aquifer showing effects of tides and dispersion.

FIG. 13-57. Methods for control of salt-water intrusion. (a) Pumping trough paralleling the coast; (b) pressure ridge paralleling the coast; (c) impermeable subsurface barrier.

tain a ridge above sea level. Proper operation would allow a minimum discharge of fresh water into the sea. The method has been tested on a field basis by a 4,000-ft line of wells at Manhattan Beach, California [69]. Using high-quality chlorinated water, the method has proved technically feasible after several years of operation. Plans are under way to extend the line to provide complete protection to the West Coastal Basin of Los Angeles. The final method involves construction of an impermeable subsurface barrier (Fig. 13-57c) to prevent inflow of sea water. The method has many technical and economical limitations; it would be useful only for closing a shallow, narrow opening to the sea.

None of the last three methods solves the basic problem of overdraft, which is responsible for intrusion. Each provides only a separation between fresh and salt waters. The aquifer is thus protected until such time as supplemental water can be provided to replenish the aquifer or to reduce the pumping demand.

H. Development of Groundwater in Intrusion Areas

Pumpage in areas of known intrusion should be designed, if possible, to minimize contamination of water supplies. The total rate of pumping should be reduced to the safe yield of the basin in order to stabilize the salt-water interface. Well fields should be dispersed so as to reduce excessive lowering of groundwater levels at any location, particularly near the coast line. Where the intrusion has the form of a wedge lying at the bottom of the aquifer (Fig. 13-56), a number of shallow wells rather than a few deep ones will avoid coning of salt water upward into the bottom of wells.

There are several instances along the coasts of the United States where, as a remedy for intrusion, a new well field was constructed a short distance inland. With the same pumping rate resumed, the new well field in a short time showed a sharp increase in salinity. The procedure can be repeated with the same result, each well field pulling the salt-water tongue farther inland. To limit the intrusion, pumping must be reduced or dispersed or one of the control methods must be established.

In coastal-dune areas natural and artificial recharging can be integrated with groundwater extractions to stabilize salt water. Along the Dutch coast the equilibrium has been calculated for certain conditions [79]. Here pumping from shallow wells and horizontal drains is counterbalanced by natural recharge from rainfall and artificial recharge through spreading basins.

In permeable oceanic islands groundwater derived solely from rainfall floats as a lens on underlying saline water. The interface for an idealized circular island can be defined [1] approximately as

$$z^2 = \frac{\rho_f^2 W}{2\rho_s(\rho_s - \rho_f)K}(R^2 - r^2) \tag{13-80}$$

where R is the island radius, W is the recharge rate from rainfall, K is permeability, z is the depth to the interface below sea level at radius r, and ρ_f and ρ_s are the fresh- and salt-water densities, respectively. To avoid upward movement of saline water, groundwater extractions must produce minimum drawdowns. On the Hawaiian Islands [80] and elsewhere, infiltration galleries and horizontal collecting tunnels have proved successful.

X. NOTATION

A	area	S	coefficient of storage
a	length	S_r	specific retention
B	barometric efficiency, constant	S_y	specific yield
b	aquifer thickness	s	length
C	tidal efficiency, constant	T	coefficient of transmissibility
D	drawdown	t	time
d	grain, or pore, diameter	u	argument of well function
E	modulus of elasticity	V	electric potential, volume
e	base of Naperian logarithms	v	velocity
g	acceleration of gravity	W	recharge rate
H	height	w	volume
h	head, height	x	coordinate length
I	electric current	y	coordinate length
K	coefficient of permeability	z	coordinate length, elevation
k	specific permeability	α	porosity
L	length	δ	angle
m	factor, number	γ	specific weight
n	number	ϵ	angle
P	percentage	θ	angle, factor
p	pressure	μ	dynamic viscosity
Q	discharge	π	3.1416
q	discharge per unit width	ρ	density, resistivity
R	radius, electric resistance	φ	velocity potential
r	radius		

XI. REFERENCES

1. Todd, D. K.: "Ground Water Hydrology," John Wiley & Sons, Inc., New York, 1959.
2. Meinzer, O. E.: The history and development of ground-water hydrology, *J. Washington Acad. Sci.*, vol. 24, pp. 6–24, 1934.
3. Eckis, R. P.: South Coastal Basin investigation, geology and ground-water storage capacity of valley fill, *California Div. Water Resources Bull.* 45, Sacramento, Calif., 1934.
4. Picton, W. L.: The water picture today, *Water Well J.*, vol. 10, no. 4, pp. 10–11, 25, 26, 29, 1956.
5. MacKichan, K. A.: Estimated use of water in the United States, 1955, *J. Am. Water Works Assoc.*, vol. 49, pp. 369–391, 1957.
6. Waring, G. A., and O. E. Meinzer: Bibliography and index of publications relating to ground water prepared by the Geological Survey and cooperating agencies, *U.S. Geol. Surv. Water-Supply Paper* 992, 1947.
7. Vorhis, R. C.: Bibliography of publications relating to ground water prepared by the Geological Survey and cooperating agencies, 1946–1955, *U.S. Geol. Surv. Water-Supply Paper* 1492, 1957.
8. Thomas, H. E.: "The Conservation of Ground Water," McGraw-Hill Book Company, Inc., New York, 1951.
9. Johnson, G. W., and Harold Brown: "Non-military uses of nuclear explosives, *Sci. Am.*, vol. 199, no. 6, December, 1958.
10. Hubbert, M. K.: Darcy's law and the field equations of the flow of underground fluids, *Trans. Am. Inst. Mining Met. Engrs.*, vol. 207, pp. 222–239, 1956.
11. Muskat, Morris: "The Flow of Homogeneous Fluids Through Porous Media," McGraw-Hill Book Company, Inc., New York, 1937.
12. Fair, G. M., and L. P. Hatch: Fundamental factors governing the streamline flow of water through sand, *J. Am. Water Works Assoc.*, vol. 25, pp. 1551–1565, 1933.
13. Luthin, J. N. (ed.): "Drainage of Agricultural Lands," American Society of Agronomy, Madison, Wis., 1957.
14. Kaufman, W. J., and G. T. Orlob: Measuring ground water movement with radioactive and chemical tracers, *J. Am. Water Works Assoc.*, vol. 48, pp. 559–572, 1956.
15. Bear, Jacob: On the tensor form of dispersion in porous media, *J. Geophys. Res.*, vol. 66, no. 4, pp. 1185–1197, April, 1961.
16. de Josselin de Jong, G.: Longitudinal and transverse diffusion in granular deposits, *Trans. Am. Geophys. Union*, vol. 39, pp. 67–74, 1958.
17. Carlston, C. W., L. L. Thatcher, and E. C. Rhodehamel: Tritium as a hydrologic tool: the Wharton Tract Study, *Assemblée Gén. Helsinki, Assoc. Intern. Hydrol. Sci. Publ.* 52, pp. 503–512, 1960.
18. Jacob, C. E.: Flow of ground water, in Hunter Rouse (ed.), "Engineering Hydraulics," John Wiley & Sons, Inc., New York, 1950, pp. 321–386.
19. Casagrande, Arthur: Seepage through dams, *J. New England Water Works Assoc.*, vol. 51, pp. 131–172, 1937.
20. Casagrande, Arthur: Discussion on Security from under-seepage masonry dams on earth foundations by E. W. Lane, *Trans. Am. Soc. Civil Engrs.*, vol. 100, pp. 1289–1294, 1935.
21. Wenzel, L. K.: Methods for determining permeability of water-bearing materials with special reference to discharging-well methods, *U.S. Geol. Surv. Water-Supply Paper* 887, 1942.
22. Theis, C. V.: The relation between the lowering of the piezometric surface and the rate and duration of discharge of a well using ground-water storage, *Trans. Am. Geophys. Union*, vol. 16, pp. 519–524, 1935.
23. Jacob, C. E.: Drawdown test to determine effective radius of artesian wells, *Trans. Am. Soc. Civil Engrs.*, vol. 112, pp. 1047–1070, 1947.
24. Chow, V. T.: On the determination of transmissibility and storage coefficients from pumping test data, *Trans. Am. Geophys. Union*, vol. 33, pp. 397–404, 1952.
25. Boulton, N. S.: The drawdown of the water-table under non-steady conditions near a pumped well in an unconfined formation, *Proc. Inst. Civil Engrs.*, vol. 3, pt. 3, pp. 564–579, 1954.
26. Hantush, M. S., and C. E. Jacob: Non-steady radial flow in an infinite leaky aquifer, *Trans. Am. Geophys. Union*, vol. 36, pp. 95–100, 1955.
27. Walton, W. C.: Leaky artesian aquifer conditions in Illinois, *Illinois State Water Surv. Rept. Invest.* 39, 1960.
28. Glover, R. E., and G. G. Balmer: River depletion resulting from pumping a well near a river, *Trans. Am. Geophys. Union*, vol. 35, pp. 468–470, 1954.

29. Brown, R. H.: Selected procedure for analyzing aquifer test data, *J. Am. Water Works Assoc.*, vol. 45, pp. 844–866, 1953.
30. Walton, W. C.: The hydraulic properties of a dolomite aquifer underlying the Village of Ada, Ohio, *State of Ohio, Div. Water, Tech. Rept.* 1, 1953.
31. Lang, S. M.: Interpretation of boundary effects from pumping test data, *J. Am. Water Works Assoc.*, vol. 60, pp. 356–364, 1960.
32. "Wells," U.S. Department of the Army, Technical Manual TM5-297, 1957.
33. "AWWA Standard for Deep Wells," American Water Works Association, New York, 1958.
34. Gordon, R. W., "Water Well Drilling with Cable Tools," Bucyrus-Erie Co., South Milwaukee, Wis., 1958.
35. Moss, Roscoe, Jr.: Water well construction in formations characteristic of the Southwest, *J. Am. Water Works Assoc.*, vol. 50, pp. 777–788, 1958.
36. Guardino, S. T.: Developments in the design and drilling of water wells, *J. Am. Water Works Assoc.*, vol. 50, pp. 769–776, 1958.
37. Gossett, O. C.: Reverse-circulation rotary drilling, *Water Well J.*, vol. 12, no. 3, pp. 6, 7, 22, 48, 1958.
38. Yoeman, R. A.: Direct rotary air drilling, *Water Well J.*, vol. 12, no. 7, pp. 12, 32, 34, 1958.
39. Yellig, E. J.: Down-the-hole air percussion drilling, *Water Well J.*, vol. 12, no. 5, pp. 8, 22, 24, 27, 28, 30, 1958.
40. Joint Committee on Rural Sanitation: Individual water supply systems, *U.S. Public Health Serv. Publ.* 24, 1950.
41. Koenig, Louis: Survey and analysis of well stimulation performance, *J. Am. Water Works Assoc.*, vol. 52, pp. 333–350, 1960.
42. Fishel, V. C.: Long-term trends of ground-water levels in the United States, *Trans. Am. Geophys. Union*, vol. 37, pp. 429–435, 1956.
43. Poland, J. F., and G. H. Davis: Subsidence of the land surface in the Tulare-Wasco (Delano) and Los Banos-Kettleman City Area, San Joaquin Valley, California, *Trans. Am. Geophys. Union*, vol. 37, pp. 287–296, 1956.
44. Poland, J. F.: Land subsidence due to ground-water development, *Proc. Am. Soc. Civil Engrs., J. Irrigation and Drainage Div.*, vol. 84, no. IR3, 1958.
45. Lofgren, B. E.: Near-surface land subsidence in Western San Joaquin Valley, California, *J. Geophys. Res.*, vol. 65, pp. 1053–1062, 1960.
46. Todd, D. K.: Ground-water flow in relation to a flooding stream, *Proc. Am. Soc. Civil Engrs.*, vol. 81, separate issue 628, 1955.
47. White, W. N.: A method of estimating ground-water supplies based on discharge by plants and evaporation from soil, *U.S. Geol. Surv. Water-Supply Paper* 659, pp. 1–105, 1932.
48. Troxell, H. C.: The diurnal fluctuation in the ground-water and flow of the Santa Ana River and its meaning, *Trans. Am. Geophys. Union*, vol. 17, pp. 496–504, 1936.
49. Robinson, T. W.: Earth-tides shown by fluctuations of water-levels in wells in New Mexico and Iowa, *Trans. Am. Geophys. Union*, vol. 20, pp. 656–666, 1939.
50. Tuinzaad, H.: Influence of the atmospheric pressure on the head of artesian water and phreatic water, *Assemblée Gén. Rome, Assoc. Intern. Hydrol. Sci., Publ.* 37, pp. 32–37, 1954.
51. Ferris, J. G.: Cyclic fluctuations of water level as a basis for determining aquifer transmissibility, *Assemblée Gén. Bruxelles, Assoc. Intern. Hydrol. Sci., Publ.* 33, pp. 148–155, 1951.
52. Jacob, C. E.: On the flow of water in an elastic aquifer, *Trans. Am. Geophys. Union*, vol. 21, pp. 574–586, 1940.
53. "Ground Water Basin Management," American Society of Civil Engineers, Manual of Engineering Practice no. 40, 1961.
54. Banks, H. O.: Utilization of underground storage reservoirs, *Trans. Am. Soc. Civil Engrs.*, vol. 118, pp. 220–234, 1953.
55. Simpson, T. R.: Utilization of ground water in California, *Trans. Am. Soc. Civil Engrs.*, vol. 117, pp. 923–934, 1952.
56. Conkling, Harold: Utilization of ground-water storage in stream system development, *Trans. Am. Soc. Civil Engrs.*, vol. 111, pp. 275–354, 1946.
57. Harding, S. T.: Ground water resources of Southern San Joaquin Valley, *Calif. Div. Eng. Irrigation Bull.* 11, Sacramento, 1927.
58. Ingerson, I. M.: The hydrology of Southern San Joaquin Valley, California, and its relation to imported water supplies, *Trans. Am. Geophys. Union*, vol. 22, pp. 20–45, 1941.

59. Haley, J. M., and others: Santa Clara Valley investigation, *Calif. State Water Resources Board Bull.* 7, Sacramento, 1955.
60. Thomas, R. O., and others: Ground-water development: a symposium, *Trans. Am. Soc. Civil Engrs.*, vol. 122, pp. 421–517, 1957.
61. Richter, R. C., and R. Y. D. Chun: Artificial recharge of ground water reservoirs in California, *Proc. Am. Soc. Civil Engrs., J. Irrigation and Drainage Div.*, vol. 85, no. IR4, pp. 1–27, 1959.
62. Schiff, Leonard (ed.): *Proc. Bien. Conf. on Ground Water Recharge*, 1959, University of California, Berkeley, Calif., and U.S. Department of Agriculture, Fort Collins, Colo.
63. Todd, D. K.: Annotated bibliography on artificial recharge of ground water through 1954, *U.S. Geol. Surv. Water-Supply Paper* 1477, 1959.
64. Muckel, D. C.: Replenishment of ground water supplies by artificial means, *U.S. Dept. Agr. Tech. Bull.* 1195, 1959.
65. Suter, Max: The Peoria recharge pit: its development and results, *Proc. Am. Soc. Civil Engrs., J. Irrigation and Drainage Div.*, vol. 82, no. IR3, pp. 1–17, 1956.
66. Brashears, M. L., Jr.: Artificial recharge of ground water on Long Island, New York, *Econ. Geol.*, vol. 41, pp. 503–516, 1946.
67. Johnson, A. H.: Ground-water recharge on Long Island, *J. Am. Water Works Assoc.*, vol. 40, pp. 1159–1166, 1948.
68. Baumann, Paul: Basin recharge, ground-water development: a symposium, *Trans. Am. Soc. Civil Engrs.*, vol. 122, pp. 458–473, 1957.
69. Laverty, F. B., and H. A. van der Goot: Development of a fresh-water barrier in southern California for the prevention of sea-water intrusion, *J. Am. Water Works Assoc.*, vol. 47, pp. 886–908, 1955.
70. Krone, R. B., P. H. McGauhey, and H. B. Gotaas: Direct recharge of grround wate with sewage effluents, *Proc. Am. Soc. Civil Engrs., J. Sanitary Eng. Div.*, vol. 83, no. SA4, pp. 1–25, 1957.
71. Klaer, F. H., Jr.: Providing large industrial water supplies by induced infiltration, *Mining Eng.*, vol. 5, pp. 620–624, 1953.
72. Goines, W. H., A. G. Winslow, and J. R. Barnes: Water supply of the Houston Gulf Coast region, *Texas Board of Water Engrs. Bull.* 5101, Austin, Tex., 1951.
73. Todd, D. K.: Salt water intrusion of coastal aquifers in the United States, *Assemblée Gén. Helsinki, Assoc. Intern. Hydrol. Sci. Publ.* 52, pp. 452–461, 1960.
74. Hubbert, M. K.: The theory of ground-water motion, *J. Geol.*, vol. 48, pp. 785–944, 1940.
75. Perlmutter, N. M., J. J. Geraghty, and J. E. Upson: The relation between fresh and salty ground water in Southern Nassau and Southeastern Queens Counties, Long Island, New York, *Econ. Geol.*, vol. 54, pp. 416–435, 1959.
76. Glover, R. E.: The pattern of fresh-water flow in a coastal aquifer, *J. Geophys. Res.*, vol. 64, pp. 457–459, 1959.
77. Kohout, F. A.: Cyclic flow of salt water in Biscayne aquifer of Southeastern Florida, *J. Geophys. Res.*, vol. 65, pp. 2133–2141, 1960.
78. James, L. B., and others: Sea-water intrusion in California, *Calif. Dept. Water Resources Bull.* 63, Sacramento, 1958.
79. Todd, D. K., and L. Huisman: Ground water flow in the Netherlands coastal dunes, *Proc. Am. Soc. Civil Engrs., J. Hydraulics Div.*, vol. 85, no. HY7, pp. 63–81, 1959.
80. Morgan, E. J.: Honolulu water supply, *J. Am. Water Works Assoc.*, vol. 49, pp. 1403–1413, 1957.

Section 14

RUNOFF

VEN TE CHOW, *Professor of Hydraulic Engineering, University of Illinois.*

I. Introduction	14-2
II. Runoff Phenomena	14-3
A. Runoff Process	14-3
1. Runoff Cycle	14-3
2. Factors Affecting Runoff	14-4
B. Relations between Precipitation and Runoff	14-5
1. Precipitation and Runoff Correlation	14-5
2. The Rational Formula	14-6
III. Time Distribution of Runoff	14-8
A. Hydrograph Analysis	14-8
1. Features of a Hydrograph	14-8
2. Components of a Hydrograph	14-10
3. Base-flow Separation	14-11
4. Relation between Hyetograph and Hydrograph	14-12
5. Methods of Analysis	14-13
B. Unit Hydrographs	14-13
1. Unit-hydrograph Theory	14-13
2. Unit-hydrograph Analysis	14-15
3. Modifications of Unit Hydrograph	14-22
C. Instantaneous Unit Hydrograph	14-24
1. IUH and Its Development	14-24
2. Methods of Determining IUH	14-25
D. Conceptual Models of IUH	14-27
1. Components of Models	14-27
2. Formulation of Models	14-28
E. Nonlinearity of Runoff Distribution	14-34
IV. Space Distribution of Runoff	14-35
A. Overland Flow	14-35
B. Streamflow	14-37
V. Variability of Runoff	14-42
A. Flow-duration Curve	14-42
B. Flow-mass Curve	14-44
C. Time-series Analysis	14-46
1. Analysis of Ranges	14-47
2. Queueing Theory	14-48
VI. References	14-50

I. INTRODUCTION

Runoff is that part of the precipitation, as well as any other flow contributions, which appears in surface streams of either perennial or intermittent form. This is the flow collected from a drainage basin or watershed, and it appears at an outlet of the basin. Specifically, it is the *virgin flow*, which is the streamflow unaffected by artificial diversions, storage, or other works of man in or on the stream channels, or in the drainage basin or watershed. It may be noted that the terms "drainage basin" and "watershed" used here are often considered synonymously. Strictly speaking, however, a watershed is the divide separating one drainage basin from another. In British literature, the drainage basin is called the *catchment*.

According to the source from which the flow is derived, runoff may consist of surface runoff, subsurface runoff, and groundwater runoff. The *surface runoff* is that part of the runoff which travels over the ground surface and through channels to reach the basin outlet. The part of the surface runoff that flows over the land surface toward stream channels is called *overland flow*. After the flow enters a stream, it joins with other components of flow to form *total runoff*. In contrast to overland flow, the total runoff confined in stream channels is called *streamflow*.

The *subsurface runoff*, also known as *subsurface flow, interflow, subsurface storm flow*, or *storm seepage*, is the runoff due to that part of the precipitation which infiltrates the surface soil and moves laterally through the upper soil horizons toward the streams as ephemeral, shallow, perched groundwater above the main groundwater level. A part of the subsurface runoff may enter the stream promptly, while the remaining part may take a long time before joining the streamflow.

The *groundwater runoff*, or *groundwater flow*, is that part of the runoff due to deep percolation of the infiltrated water which has passed into the ground, has become groundwater, and has been discharged into the stream.

For the practical purpose of runoff analysis, total runoff in stream channels is generally classified as direct runoff and base flow. The *direct runoff*, *direct surface runoff*, or *storm runoff* is that part of runoff which enters the stream promptly after the rainfall or snow melting. It is equal to the sum of the surface runoff and the prompt subsurface runoff, plus channel precipitation. *Channel precipitation* is the precipitation that falls directly on the water surfaces of lakes and streams. Owing to its relatively small amount, it is usually treated as a part of surface runoff. It must be noted that the practice of hydrograph analysis varies. In certain procedures, the direct runoff is assumed either to include or to exclude the entire subsurface runoff, not just a portion of it.

The *base flow*, or *base runoff*, is defined as the sustained or fair-weather runoff. It is composed of groundwater runoff and delayed subsurface runoff. However, in certain procedures of hydrograph analysis, the base flow is assumed either to include or exclude the total subsurface runoff but not any portion of it.

During a runoff-producing storm, the total precipitation may be considered to consist of precipitation excess and abstractions. The *precipitation excess* is that part of the total precipitation that contributes directly to the surface runoff. When the precipitation is rainfall, the precipitation excess is known as *rainfall excess*. The abstractions are the remaining parts which do not eventually become surface runoff, such as interception, evaporation, transpiration, depression storage, and infiltration. They are also called *losses*, which is not an appropriate name since water cannot be really lost in the hydrologic cycle.

The part of precipitation that contributes entirely to the direct runoff may be called the *effective precipitation* or *effective rainfall* if only rainfall is involved. The effective precipitation consists of the precipitation excess and that part of the precipitation which becomes prompt surface runoff. Thus, the difference between the total precipitation and the effective precipitation may be called the *effective abstractions* as far as the direct runoff is considered. It must be noted that in some procedures the subsurface runoff is entirely excluded from the direct runoff and then the effective precipitation (or rainfall) is equivalent to the precipitation (or rainfall) excess.

The above are some important terms that will be used in this section. From total precipitation to total runoff, the various terms can be related as follows:

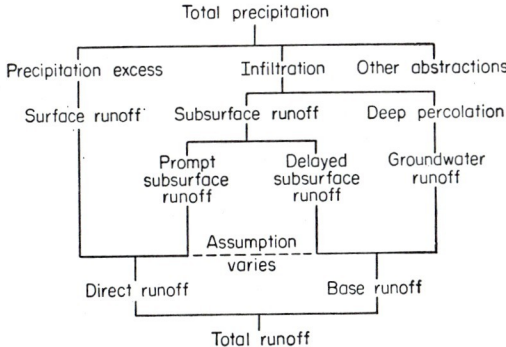

Other sections of this handbook also cover various aspects of runoff, including groundwater runoff (Secs. 4-I and 13); drainage basins and channel networks (Sec. 4-II); mechanics of flow (Sec. 7); statistical analysis of runoff (Sec. 8); snowmelt runoff (Sec. 10); streamflow measurement (Sec. 15); glacier runoff (Sec. 16); low streamflow (Sec. 18); urban runoff (Sec. 20); runoff from agricultural lands (Sec. 21); runoff from forest lands and rangelands (Sec. 22); runoff in arid and semiarid regions (Sec. 24); flood and runoff control (Sec. 25); and legal aspects of runoff (Sec. 27). The present section, therefore, covers only the basic aspects of runoff, including terminology, phenomena, time and space distributions, variability, and some other aspects that are not fully discussed elsewhere in this handbook.

II. RUNOFF PHENOMENA

A. Runoff Process

1. Runoff Cycle. The phenomenon of runoff may be visualized as a cycle dependent on the nature of the supply. Hoyt [1, 2] has described such a cycle by considering it to be comprised of five phases as follows:

The first phase relates to rainless periods just prior to the beginning of rainfall and after an extended dry period. During this phase, the groundwater table is low and its elevation continues to decrease gradually. In mountainous areas or under steep slopes, there may be no continuous water table and the streamflow is maintained by draining such areas of gravity water in perched layers or in rock interstices. In arid regions, where there may be no water table or perched groundwater contributing flow to streams, the channels are dry. In addition to disposal of water by streamflow, water is also lost through evaporation on land and water surface and by transpiration from plants. If snow, ice, or frost is present and temperatures are below freezing, the phenomena described remain almost unchanged. If temperatures are above freezing, the snow, ice, or frost will melt as runoff, thus entering the second phase of the cycle.

The second phase relates to the initial period of rain. As the rain starts, its amount is divided among channel precipitation, interception by vegetation, infiltration into the soil, and temporary retention in surface depressions. The infiltrated water results in a gradual increase of water in the zone of aeration after the natural storage or field moisture capacity is satisfied. During this phase there is little overland flow except on impervious surfaces, while evaporation and transpiration are slight. Groundwater runoff to the stream channels may or may not continue, depending on whether the first phase continued until streamflow ceased. If snow is present, it will absorb part of the falling rain and its storage effect will lengthen the period of this phase. If frost is present, infiltration will be reduced when moisture content is high, but increased when moisture content is low and the ground is frozen. The runoff will be augmented only when thawing releases stored water in the snow, ice, and frost.

The third phase relates to a continuation of rain at variable intensity. As rain continues, the capacities of vegetable interception and retention of surface depressions are reached, and the excess rain becomes a source of runoff and detention storage on land surfaces and in channels. Overland flow occurs when the net rate of rain exceeds the infiltration rate; but it may or may not reach the stream channels, depending on retention and detention capacities of the land surface over which it travels. The infiltrated water will saturate the upper part of the zone of aeration which has been depleted in the previous phases and will then move down to the water table. If rain continues, the water table will rise and the groundwater contribution to streamflow will increase. As the zone of aeration is saturated, subsurface runoff may contribute also to the streamflow. If the stage of flow in the channels rises rapidly and becomes higher than the relatively slowly rising groundwater table, the streams will change from *effluent streams* to *influent streams*, contributing to the groundwater and developing bank storage of water. During this phase, evaporation and transpiration are slow. If snow, ice, and frost are present, their effects will be about the same as in the previous phase. When frost is gone, the groundwater recharge will momentarily increase rapidly and the overland flow and subsurface runoff will diminish accordingly. Through groundwater recharge, the subsurface runoff will increase in later periods.

The fourth phase relates to a continuation of rainfall until all natural storage has been satisfied. The infiltration rate will approach the rate of water transmission through the zone of aeration to both groundwater table and subsurface runoff. The amount of subsurface runoff, which will join the streamflow almost as promptly as the overland runoff, apparently depends on the porosity of the material through which it is transmitted. As the rain continues, the water table rises constantly until the groundwater runoff balances the maximum rate of recharge possible, and all additional rain results in direct increment to runoff. Although this ultimate condition may rarely be reached, it is approximately attained in flat swampy areas after periods of heavy and prolonged rainfall. The effects of snow, ice, and frost are similar to those in the third phase.

The fifth phase relates to the period between the termination of rain and the time when the first stage is to be reached. This usually involves a relatively long time for channel storage and surface retention to become depleted. Evaporation and transpiration are active and infiltration continues. Water in the zone of aeration is reaching the water table or the stream channels. Streamflow is sustained by releasing stored water from stream channels, subsurface flow, and groundwater flow. The water table is rising and then falling when its peak stage is over and the stored water is diminishing. When temperature is below freezing, the presence of snow and ice has little effect; if the temperature rises above the freezing, their presence prolongs this phase.

It should be noted that the above is an oversimplified description of the runoff phenomenon. The actual process is more complicated and variable since it is affected by numerous factors.

2. Factors Affecting Runoff. From the hydrologic point of view, the runoff from a drainage basin may be considered as a product in the hydrologic cycle, which is influenced by two major groups of factors: climatic factors and physiographic factors. Climatic factors include mainly the effects of various forms and types of precipitation, interception, evaporation, and transpiration, all of which exhibit seasonal variations in accordance with the climatic environment. Physiographic factors may be further classified into two kinds: basin characteristics and channel characteristics. Basin characteristics include such factors as size, shape, and slope of drainage area, permeability and capacity of groundwater formations, presence of lakes and swamps, and land use. Channel characteristics are related mostly to hydraulic properties of the channel which govern the movement of streamflows and determine channel storage capacity. It should be noted, however, that the above classification of factors is by no means exact because many factors are interdependent to a certain extent. For clarity, the following is a list of the major factors:

RUNOFF PHENOMENA   14-5

a. Climatic factors

Precipitation: Form (rain, snow, frost, etc.), type, intensity, duration, time distribution, areal distribution, frequency of occurrence, direction of storm movement, antecedent precipitation, and soil moisture (see Secs. 3, 9, and 10).

Interception: Vegetation species, composition, age, and density of stands; season of the year; size of storm (see Sec. 6).

Evaporation: Temperature, wind, atmospheric pressure, soluble solids, nature and shape of evaporative surface (see Sec. 11).

Transpiration: Temperature, solar radiation, wind, humidity, soil moisture, kinds of vegetation (see Secs. 6 and 11).

b. Physiographic factors

1. Basin characteristics (see Sec. 4).
Geometric factors: Size, shape, slope, orientation, elevation, stream density.
Physical factors: Land use and cover, surface infiltration condition, soil type, geological conditions such as the permeability and capacity of groundwater formations, topographical conditions such as the presence of lakes and swamps, artificial drainage.

2. Channel characteristics (see Sec. 7).
Carrying capacity: Size and shape of cross section, slope, roughness, length, tributaries.
Storage capacity: Backwater effect.

The factors affecting runoff generally tend to cause most large drainage areas to behave differently from most small drainage areas on the basis of hydrologic behavior. Consequently, drainage basins may be classified as large and small, not on the basis of the size alone, but on the effects of certain dominating factors. Frequently two basins of nearly the same size may behave entirely differently in runoff phenomena. One drainage basin may show prominent channel storage effects, like most large basins, while the other may manifest strong influence of the land use, like most small basins. A distinct characteristic of small basins is that the effect of overland flow rather than the effect of channel flow is a dominating factor affecting the peak runoff. Also, small basins are very sensitive both to high-intensity rainfalls of short duration and to land use. On large basins, the effect of channel storage is so pronounced that such sensitivities are greatly suppressed. Therefore, a *small drainage basin* may be defined as one that is so small that its sensitivity to high-intensity rainfalls of short durations and to land use is not suppressed by the channel characteristics. According to this definition, the size of a small basin may be from a few acres to 1,000 acres, or even up to 100 sq mi. The upper limit depends on the condition at which the abovementioned sensitivity becomes minimized due to the overwhelming channel-storage effect.

B. Relations between Precipitation and Runoff

In hydrologic analysis and design, it is often necessary to develop relations between precipitation and runoff, possibly using some of the factors affecting runoff as parameters. Such relations are also useful for extrapolation or interpolation of runoff records from generally longer records of precipitation. The relations between precipitation and runoff differ with the type of precipitation, the consideration of the volume or peak of runoff, or the time distribution of runoff. In this section only the relation between rainfall and runoff is discussed. The runoff due to snowmelt is covered in Sec. 10. Since the time distribution of runoff will be discussed in Subsec. III, this subsection deals only with runoff volume and the relation given by the rational formula. An extended discussion of peak runoff is given in Sec. 25-I.

1. Precipitation and Runoff Correlation. Because the various factors affecting runoff usually vary considerably on a drainage basin during different storms, a direct plotting of rainfall against runoff for individual storms does not usually produce a satisfactory correlation. The result is likely to be a wide scatter of plotted points. However, the relation may be greatly improved by taking into account as parameters some influencing factors, such as storm frequency, initial soil-moisture condition, storm duration, and time of year. Using mean annual temperature as a parameter, for example, Langbein and others [3] have shown a satisfactory relationship between mean annual precipitation and runoff.

The initial soil-moisture condition, or *antecedent moisture condition*, is a parameter which cannot be determined directly and used reliably. For practical purposes, this parameter is usually expressed by an index which is so defined that it can be roughly representative of the initial soil-moisture condition and also can be easily measured. The following are some such indexes that have been proposed: groundwater flow at the beginning of the storm, basin evaporation, and antecedent precipitation.

When groundwater flow at the beginning of the storm is used as an index, it should be supplemented by a weighted index of the rainfall for several days preceding because recent rains affect current moisture content.

Accumulated evaporation from a standard evaporation pan has been used as an index of field-moisture deficiency. Linsley and Ackermann [4] found that the deficiency at any time was equal to 0.9 times the accumulated pan evaporation since the ground was last saturated minus any additions due to intervening rains.

The antecedent precipitation can be used as an index because it affects the soil-moisture condition. For the correlation of annual runoff with rainfall, Butler [5] showed an *antecedent precipitation index* (API), P_a, which can be expressed essentially as

$$P_a = aP_0 + bP_1 + cP_2 \tag{14-1}$$

where P_0, P_1 and P_2 are the annual rainfalls for the current year, the antecedent year, and the second antecedent year, respectively. The weighting coefficients a, b, and c, with their sum equal to unity, are determned by trial and error in order to obtain a best correlation between the runoff and the weighted API. For individual storms, Kohler and Linsley [6] proposed a similar API:

$$P_a = b_1P_1 + b_2P_2 + \cdots + b_tP_t \tag{14-2}$$

where b_t is a constant less than unity, and P_t is the amount of precipitation which occurred t days prior to the storm under consideration. The constant b_t is commonly assumed as a function of t. If a day-by-day value of the index is required, b_t may be assumed to decrease exponentially with t, or $b_t = k^t$, where k is a *recession constant*. Thus,

$$P_{at} = P_{a0}k^t \tag{14-3}$$

where P_{a0} is the initial value of API and P_{at} is the reduced value t days after. The index after any day is related to the index of the day before as $P_{a1} = kP_{a0}$ since $t = 1$. The value of k normally ranges between 0.85 and 0.98.

By using API, week of year, and storm duration as parameters, Kohler and Linsley [6] developed a relationship between the storm runoff and precipitation by a graphical method of coaxial correlation (see Figs. 8-II-3 and 25-IV-2). Hopkins and Hackett [7] extended such analyses by including an index of antecedent temperature (ATI) and the average annual basin temperature as additional parameters.

2. The Rational Formula. The relation between rainfall and peak runoff has been represented by many empirical or semiempirical formulas [8]. The *rational formula* can be taken as a representative of such formulas (see also Secs. 20, 21, and 25-I). Although this formula is based on a number of assumptions which cannot be readily satisfied under actual circumstances, its simplicity has won it popularity. The origin of this formula is somewhat obscure. In American literature, the formula was first mentioned in 1889 by Kuichling [9] for a determination of peak runoff for sewer design in Rochester, New York, during the period from 1877 to 1888. Some authors believe that the principles of the formula were explicit in the work of Mulvaney [10] in 1851. In England, the method using the rational formula is often referred to as the *Lloyd-Davis method* owing to the implication ascribed to a paper of 1906 [11].

RUNOFF PHENOMENA 14-7

The rational formula is
$$Q = CIA \tag{14-4}$$

where Q is the peak discharge in cfs, C is a runoff coefficient depending on characteristics of the drainage basin, I is the rainfall intensity in in. per hr, and A is the drainage area in acres. The formula is called rational because the units of the quantities involved are numerically consistent approximately.

When using the rational formula, one must assume that the maximum rate of flow, owing to a certain rainfall intensity over the drainage area, is produced by that rainfall which is maintained for a time equal to the period of concentration of flow at the point under consideration. Theoretically, this is the *time of concentration*, which is the time required for the surface runoff from the remotest part of the drainage basin to reach the point being considered. For uniform rainfall intensity, this would be the *time of equilibrium* at which the rate of runoff is equal to the rate of rainfall supply. For natural drainage basins of large size and complex drainage pattern, runoff water originating in the most remote portion may arrive at the outlet too late to contribute to the peak flow. Accordingly, the time of concentration is generally greater than the lag time of the peak flow. For small drainage basins with simple drainage patterns, the time of concentration may be very close to the lag time of the peak flow. For small agricultural drainage basins, Ramser [12] has determined the time of concentration by noting the time required for the water in the channel at the gaging station to rise from the low to the maximum stage as recorded by the water-stage recorder. An empirical formula for the time of concentration in hours thus determined by Kirpich [13] is

$$t_c = 0.00013 \frac{L^{0.77}}{S^{0.385}} \tag{14-5}$$

where L is the length of the basin area in feet, measured along the watercourse from the gaging station and in a direct line from the upper end of the watercourse to the farthest point on the drainage basin; and S is the ratio in feet to L of the fall of the basin from the farthest point on the basin to the outlet of runoff, or approximately the average slope of the basin in dimensionless ratio.

Some values of the runoff coefficient C are reported by a joint committee of the American Society of Civil Engineers and the Water Pollution Control Federation [14] as given in Table 14-1. These values are applicable for storms of 5 to 10-year frequencies. Less frequent higher-intensity storms will require the use of higher coefficients because infiltration and other abstractions have a proportionally smaller effect on peak runoff. Average values of C for agricultural lands are also given in Sec. 21.

According to Krimgold [15], the assumptions involved in the rational formula are:

(1) The rate of runoff resulting from any rainfall intensity is a maximum when this rainfall intensity lasts as long or longer than the time of concentration.
(2) The maximum runoff resulting from a rainfall intensity, with a duration equal to or greater than the time of concentration, is a simple fraction of such rainfall intensity; that is, it assumes a straight line relation between Q and I, and $Q = 0$ when $I = 0$.
(3) The frequency of peak discharges is the same as that of the rainfall intensity for the given time of concentration.
(4) The relationship between peak discharges and size of drainage area is the same as the relationship between duration and intensity of rainfall.
(5) The coefficient of runoff is the same for storms of various frequencies.
(6) The coefficient of runoff is the same for all storms on a given watershed.

It is believed that these assumptions might nearly hold for paved areas with gutters and sewers of fixed dimensions and hydraulic characteristics. The formula has thus been rather popular for the design of drainage systems in urban areas and airports. The exactness and satisfaction of these assumptions in application to other drainage basins, however, have been questioned. In fact, many hydrologists have called attention to the inadequacy of this method.

Table 14-1. Values of Runoff Coefficient C

Type of drainage area	Runoff coefficient, C
Lawns:	
Sandy soil, flat, 2%	0.05–0.10
Sandy soil, average, 2–7%	0.10–0.15
Sandy soil, steep, 7%	0.15–0.20
Heavy soil, flat, 2%	0.13–0.17
Heavy soil, average, 2–7%	0.18–0.22
Heavy soil, steep, 7%	0.25–0.35
Business:	
Downtown areas	0.70–0.95
Neighborhood areas	0.50–0.70
Residential:	
Single-family areas	0.30–0.50
Multi units, detached	0.40–0.60
Multi units, attached	0.60–0.75
Suburban	0.25–0.40
Apartment dwelling areas	0.50–0.70
Industrial:	
Light areas	0.50–0.80
Heavy areas	0.60–0.90
Parks, cemeteries	0.10–0.25
Playgrounds	0.20–0.35
Railroad yard areas	0.20–0.40
Unimproved areas	0.10–0.30
Streets:	
Asphaltic	0.70–0.95
Concrete	0.80–0.95
Brick	0.70–0.85
Drives and walks	0.75–0.85
Roofs	0.75–0.95

III. TIME DISTRIBUTION OF RUNOFF

A. Hydrograph Analysis

1. Features of a Hydrograph. A graph showing stage, discharge, velocity, or other properties of water flow with respect to time is known as a *hydrograph*. When the stage is plotted against time, the graph is a *stage-time graph* or *stage hydrograph*, which is usually shown on the recorder chart from a recording-gage station (Subsec. 15-VIII-B). When the discharge is shown against time, the graph is a *discharge hydrograph*, or commonly called simply a "hydrograph." By use of the stage-discharge relation at a gaging station, the discharge hydrograph can be obtained by conversion from a given stage hydrograph (Subsec. 15-XI).

The hydrograph can be regarded as an integral expression of the physiographic and climatic characteristics that govern the relations between rainfall and runoff of a particular drainage basin. It shows the time distribution of runoff at the point of measurement, defining the complexities of the basin characteristics by a single empirical curve. A typical hydrograph produced by a concentrated storm rainfall is a single-peaked skew distribution curve; multiple peaks may appear on a hydrograph indicating abrupt variation in rainfall intensity, a succession of storm rainfalls, abnormal groundwater recession, or other causes. In hydrograph analysis, multiple-peaked *complex hydrographs* may be separated into a number of single-peaked hydrographs by methods to be discussed later.

A typical single-peaked *simple hydrograph* (Fig. 14-1) consists of three parts: the *approach segment (limb* or *curve) AB*, the *rising* (or *concentration) segment (limb* or *curve) BD*, and the *recession (falling* or *lowering) segment (limb* or *curve) DH*. The lower portion of the recession segment is a *groundwater recession* (or *depletion*) *curve* which shows the decreasing rate of groundwater inflow. On these segments are shown the *point of rise B*, two *points of inflection C* and *E*, the *peak point D*, and two other characteristic points F and G. The segment CE is the *crest* (or *peak*) *segment*. The time at point B is the *time of rise*, at D is the *time of peak flow*, and from center of mass (or beginning) of rainfall to center of mass (or peak) of runoff is variously defined as the *lag time*.

The shape of the rising segment depends on the duration and intensity distribution of rainfall, and the antecedent condition and time-area diagram (Subsec. III-D-1) of the drainage basin. For a dry antecedent ground condition, a uniform rainfall will lose most of its water through abstractions. Thus, the effective rainfall will increase steadily, causing a steadily increasing rate of runoff or a steadily rising segment. For most basins, the time-area diagram is close to a pear shape with the area between isochrones being largest in the middle or upstream part of the basin. This tends to cause the rising segment to be concave upward, rising gradually and then, rapidly toward the end of the rise.

The peak of a hydrograph represents the highest concentration of the runoff from a drainage basin. It occurs usually at a certain time after the rain has ended, and this time depends on the areal distribution of the rainfall. However, if the storm pattern is of an advanced type with a diminishing rainfall intensity, the peak may occur before the end of the rainfall. The multiple peaks of a hydrograph may occur in any basin as the result of multiple storms being developed close to each other. If a hydrograph shows double or triple peaks, fairly regularly, the reason may be due

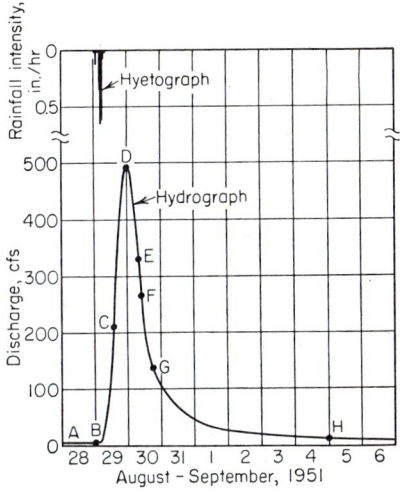

FIG. 14-1. A typical hydrograph with corresponding hyetograph. (Panther Creek at El Paso, Illinois.)

to either nonsynchronization of the runoff contributions from several tributaries to the main stream, or a kidney-like shape of the drainage basin.

The recession segment represents withdrawal of water from storage after all inflow to the channel has ceased. Therefore, it is more or less independent of the time variations in rainfall and infiltration; it may be slightly dependent, however, on areal rainfall distribution and heavily dependent on ground conditions. By matching the recession segments of several hydrographs observed on a drainage basin, an envelope curve derived from the segments may be developed to represent the *groundwater recession curve*. Since ground conditions vary with season, generally two different groundwater recession curves may be obtained respectively for growing and dormant seasons. The physical process of releasing water from storage in the ground is a phenomenon which can be described by the law of bacteria decay or an exponential law. This process can also be simulated by a linear reservoir whose outflow is directly proportional to storage (Subsec. III-D-1) and can be derived from Eq. (14-32) in the following form:

$$Q_t = Q_0 K_r{}^t \tag{14-6}$$

where Q_t is the flow at any time t after Q_0 and K_r is a *recession constant* which is less than unity. Integrating the above equation and noting that $Q_t\,dt = -dS_t$ will show that the storage S_t remaining in the basin at time t is $S_t = -Q_t/\ln K_r$. The preceding equation will plot as a straight line on semilogarithmic paper with the discharge on the logarithmic scale. Barnes [16] found from actual hydrographs that this is nearly true for the lower end of the recession segment of a hydrograph. The upper part of the recession segment contains surface flow and subsurface flow, each having different lag characteristics; therefore, it does not show as an exact straight line but as a curve with gradually decreasing slope, thus indicating an increasing value of K_r instead of a constant. Langbein [17] suggested that values of Q_0 be plotted against Q_t after some fixed time, say 24 hr. If Eq. (14-6) holds true, the plot will show a straight line trend and thus a constant value of K_r. When deviation from a straight line is indicated, the flow may include component flows of different recession constants.

2. **Components of a Hydrograph.** The hydrograph represents the distribution of total runoff in a stream at the given gaging station. It is practically impossible to draw exact division lines to disclose the hydrograph's flow components. However, approximate empirical procedures have been proposed to separate these flow components for the purpose of hydrograph analysis.

As described above, an exponential law of flow may be disclosed by plotting on semilogarithmic paper the discharge against time with the former on logarithmic scale. Barnes [16] proposed that the flow components may be separated by plotting

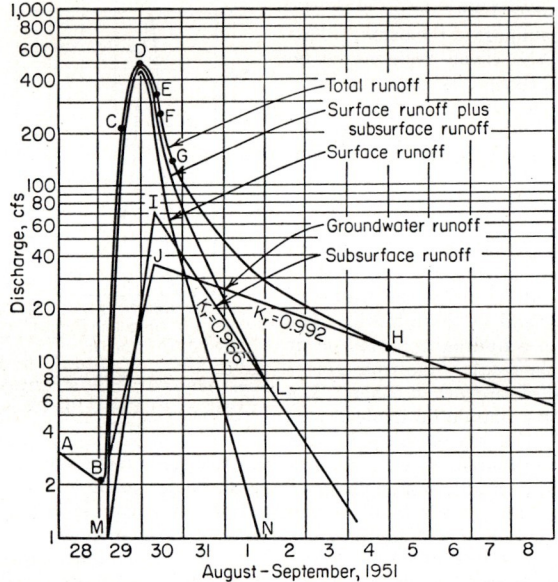

FIG. 14-2. Semilogarithmic plotting of a hydrograph, showing separation of runoff components. (Panther Creek at El Paso, Illinois.)

a hydrograph on semilogarithmic paper. Figure 14-2 shows the semilogarithmic plot of the actual hydrograph in Fig. 14-1. In this plot the groundwater recession can be approximated by a straight line with $K_r = 0.992$. This straight line can be extended back under the hydrograph to a point J which is arbitrarily located directly below the point of inflection E on the hydrograph. The approach segment AB is plotted also as an approximate straight line. The points B and J are connected arbitrarily by a straight line. Consequently, the area under the hydrograph and above BJH is considered roughly to represent the combined surface and subsurface runoff. This combined runoff is replotted and a straight line IL with $K_r = 0.966$ can be fitted to the recession of the runoff and extended back to I under E. A straight line is then connected between I and the beginning point M. Thus, the broken line MIL divides the replotted hydrograph into the surface runoff on top and the subsurface runoff below.

A hydrograph can be also studied by a so-called *hydrophase diagram*, which is a curve plotted by the rate of change of discharge dQ/dt against the discharge Q. The hydrophase diagram of the hydrograph of Fig. 14-1 is shown in Fig. 14-3. Such a diagram appears to be closed in a rectangle having its sides defined by the point of rise B, the points of inflection C and E, and the peak point D. The recession segment appears to consist of a straight line FG and a flat curve (approximately a straight line) GH. The hydrographs of the combined subsurface and groundwater runoff and of surface runoff as separated by the method shown in Fig. 14-2 are also plotted. It appears that the straight-line portion on the surface runoff recession curve corresponding to GF on the total runoff hydrograph is greatly lengthened. It may be interesting to note that plotting of the separated hydrographs for subsurface and groundwater runoff would produce triangular hydrophase diagrams with one side perpendicular to the discharge axis.

3. Base-flow Separation. If the base flow is assumed either to include or to exclude the entire portion of the subsurface flow, the method described above for separating the runoff components can be used for base-flow determination and for base-flow separation from the direct runoff on a hydrograph. However, as defined in Subsec. I,

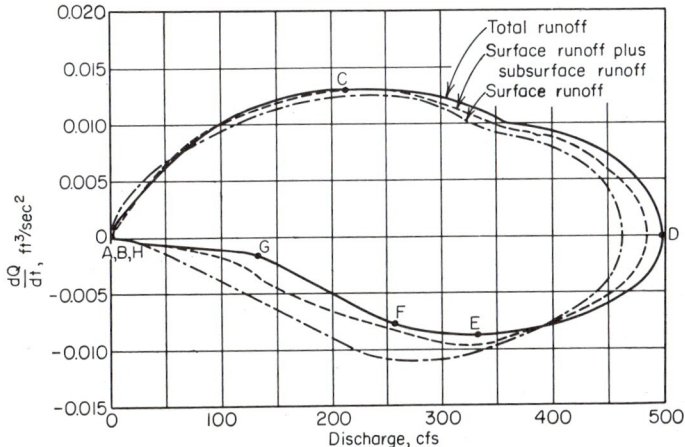

Fig. 14-3. Hydrophase diagram of a hydrograph. (Panther Creek at El Paso, Illinois, for storm of August 29, 1951.)

the total runoff may be divided into only two parts: direct runoff and base flow, and both parts may contain a certain amount of the subsurface runoff. In practical hydrograph analysis, the base-flow separation is made usually in an arbitrary manner, and it is not significant even to consider the exact amount (which is unknown anyway) to be included in or excluded from the base flow.

The simple way to make a base-flow separation is to draw a straight line from the point of rise to an arbitrary point on the lower portion of the recession segment of the hydrograph. This arbitrary point may be so chosen that the base-flow separation line should not be too long and, on the other hand, the base flow should not rise too high. This point may also be chosen as the beginning point of the fitted part of a groundwater recession curve being matched to the recession segment. According to Linsley, Kohler, and Paulhus [18], the point may be approximately taken as $A^{0.2}$ days after the time at peak flow, where A is the drainage area in square miles.

Another simple procedure is to extend the approach segment forward to a point directly below, or a little beyond, the peak flow, and then to connect this point to the arbitrary point on the recession segment with a straight line. The extended line of the approach segment usually represents the decreasing groundwater contribution during the rise of the stage in the stream. If this reasoning is correct, it is possible to have a negative base flow since the stage in the stream may become so high that the stream water flows into the banks.

A slightly elaborate procedure for base-flow separation will be described in an example given in Subsec. III-B-2.

It should be noted that any procedure for base-flow separation is arbitrary unless the exact amount of the base flow can be determined. Fortunately, for most hydrograph analysis in which the base flow is a small percentage of the critical peak flow, any errors involved in base-flow separation are not significant.

4. Relation between Hyetograph and Hydrograph. In hydrograph analysis involving storms of highly nonuniform rainfall distribution, it may be necessary to separate the effective rainfall from abstractions on a hyetograph in a way similar to the base-flow separation on a hydrograph. This procedure will be discussed in an example given in Subsec. III-B-2. Again, any such procedure is only approximate. If the effective rainfall is considered equivalent to rainfall excess, the separation may

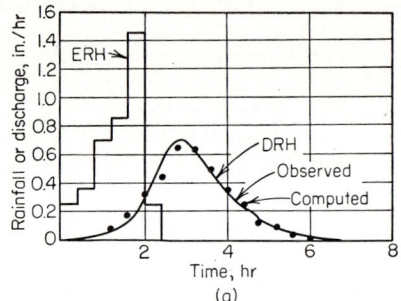

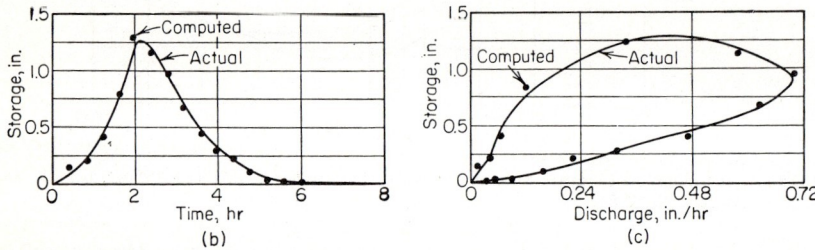

Fig. 14-4. ERH, DRH, and storage curves. (Watershed 97, Coshocton, Ohio; storm of June 28, 1957.)

be done by constructing an infiltration capacity curve (Sec. 11) or by using infiltration indexes. When the infiltration index is used, the U.S. Army Corps of Engineers [19] recommends the following procedure: An initial loss is first estimated and then a trial value of the infiltration index in inches per hour is assumed. The equivalent infiltration loss, in inch-square miles, for each station polygon (i.e., Thiessen polygon in Sec. 9) is computed. The rainfall quantities of all station polygons in excess of the trial value of the infiltration index, after the initial loss has been satisfied, are added and compared with the quantity of the computed direct runoff. This procedure is repeated until the value of the infiltration index necessary to give rainfall excess equal to the correct volume of direct runoff is determined.

The hyetograph minus the abstractions may be replotted as the *effective-rainfall hyetograph* (ERH). The corresponding *direct-runoff hydrograph* is abbreviated as DRH. The ordinates of the ERH and DRH may both be expressed in same units, say inch per hour. They can be plotted on the same scales as shown in Fig. 14-4a, where the ERH can be considered as an input to the basin system and the DRH as the output from the basin system. By the concept of continuity, the accumulative differences between the ordinates of the ERH and DRH gives the storage in the basin system as shown in Fig. 14-4b. A plot of the storage in average depth against the outflow will usually appear as the loop in Fig. 14-4c. Kulandaiswamy [20] analyzed such loops and represented them by a nonlinear storage equation expressed by Eq. (14-48). The dotted points represent the computed values made on the basis of the storage equation. The loop indicates that the storage is not a linear single-valued function of the discharge, reflecting the unsteady effect of the flow. The basin system is further said to be *nonlinear* if the parameters defining the loop function are functions of inflow and/or outflow.

5. Methods of Analysis. Numerous methods of hydrograph analysis have been proposed and developed [21-43], including Ross' time-contour plan [21], Sherman's unit hydrograph [23], Zoch's theoretical analysis [24], Bernard's distribution graph [25], Snyder's synthetic unit hydrograph [26], Horton's virtual channel-inflow graph [28], Commons' dimensionless hydrograph [30], Edson's mathematical model [31], Appleby's runoff dynamics [32, 33], Iwagaki's method of characteristics [34], Snyder's least-squares method [35, 36], Sugawara's and Maruyama's rainfall model [37], Cuénod's characteristic hydrograph [38], and Linsley's and Crawford's Stanford Watershed Models [39-41].

The unit-hydrograph method developed by Sherman and extended by many others will be thoroughly discussed in the remaining portion of this subsection.

The Stanford Watershed Models (Mark I, II, and III) utilize daily, hourly, or smaller time increments of rainfall as an input to generate continuous mean daily or hourly hydrographs for a drainage basin. The development of the model parameters consists principally of adjusting the infiltration capacities, groundwater recession rates, and evapotranspiration losses in a general digital-computer model to match those characteristics of the particular basin. The general model is constructed so that the model parameters are closely related to the corresponding physical parameters. The model parameters discovered in this way have been found to exhibit considerable regional stability. If they could be correlated successfully with physical parameters, the hydrologic regime of an ungaged area could be predicted.

B. Unit Hydrographs

1. Unit-hydrograph Theory. The relationship between rainfall and runoff distribution for individual storms appears to have been studied as early as 1929 by Folse [22], who presented the ideas of base-flow separation, rainfall reduction by variable infiltration losses, and derivation of physical constants. These are, in effect, the successive ordinates of a 24-hr unit hydrograph to be proposed by Sherman three years later. Two years later, a committee of the Boston Society of Civil Engineers [44] reported that flood hydrographs afford the best basis for the study of drainage areas and concluded that "the base of the flood hydrograph appears to be approximately constant for different floods" and "peak flows tend to vary directly with the total volume of flood runoff."

In 1932, Sherman [23] proposed the well-known theory of unit hydrographs. The *unit hydrograph* (originally named *unit-graph*) of a drainage basin is defined as a hydrograph of direct runoff resulting from 1 in. of effective rainfall generated uniformly over the basin area at a uniform rate during a specified period of time or duration. Sherman originally used the word "unit" to denote the specified period of time or a "unit of time" of the effective rainfall. Later, however, the word "unit" was often misinterpreted to denote 1 in. or "unit depth" of the effective rainfall. Sherman classified the runoff only into surface runoff and groundwater runoff since subsurface runoff was not recognized during his time. Consequently, he defined the unit hydrograph only for the use of surface runoff.

The unit hydrograph defined above can be used to derive the hydrograph of runoff due to any amount of effective rainfall. The above definition and the following basic assumptions constitute the so-called *unit-hydrograph theory:*

a. The effective rainfall is uniformly distributed within its duration or specified period of time.

b. The effective rainfall is uniformly distributed throughout the whole area of the drainage basin.

c. The base or time duration of the hydrograph of direct runoff due to an effective rainfall of unit duration is constant.

d. The ordinates of the direct-runoff hydrographs of a common base time are directly proportional to the total amount of direct runoff represented by each hydrograph.

e. For a given drainage basin, the hydrograph of runoff due to a given period of rainfall reflects all the combined physical characteristics of the basin.

Under the natural condition of rainfall and drainage basins, the above assumptions cannot be satisfied perfectly. However, when the hydrologic data used for unit-hydrograph analysis are carefully selected so that they meet the above assumptions closely, the results obtained by the unit-hydrograph theory have been found acceptable for practical purposes. Although the unit-hydrograph method was originally devised for large drainage basins, Brater [45] later showed that it was also applicable to small drainage basins varying in size from about 4 acres to 10 sq mi. It is known that there are exceptional cases which do not support the use of unit hydrographs for small drainage basins. In such cases, one or more of the assumptions may not have been well satisfied. For similar reasons, Sherman ruled out the application of the unit-hydrograph theory to runoff originating from snow or ice and to conditions having the duration of effective rainfall greater than the time of concentration (or the time of rise as suggested later by Wisler and Brater [46]).

Regarding assumption a, the storms selected for analysis should be of short duration since they would most likely produce an intense and nearly uniform effective rainfall, yielding a well-defined single-peaked hydrograph of short time base. Such a storm may be called a *unit storm*. Assumption a, however, becomes irrelevant when the theory of instantaneous unit hydrographs is used (Subsec. III-C).

Regarding assumption b, the unit hydrograph may become inapplicable when the drainage area is too large to be covered by nearly uniform distribution of rainfall. In such cases, unless the selected storm covers the entire drainage basin, the basin has to be divided into small subbasins and each subbasin is subjected to analysis for storms covering the whole subbasin.

Regarding assumption c, the base of the hydrograph of direct runoff is unknown but depends on the method of base-flow separation. The base is usually short if the direct runoff is considered to include only surface runoff; it is long if the direct runoff also includes subsurface runoff. Theoretically speaking, the recession curve of a hydrograph decreases exponentially with time and should have an infinite time base. This assumption of constant hydrograph base becomes invalid when conceptual models of hydrographs are used since these models usually consider an exponential decrease of the recession of runoff (Subsec. III-D).

Regarding assumption d, the principle involved is variously known as the *principle of linearity*, *principle of superposition*, or *principle of proportionality*, since the ordinates of the direct-runoff hydrographs are also mutually proportional and thus can be added or superimposed numerically in proportion to the total amount of direct runoff. The conventional theory of unit hydrographs based on this assumption is specifically known as the *linear unit-hydrograph theory* in order to distinguish it from the *nonlinear unit-hydrograph theory* (Subsec. III-E) which treats the unit hydrograph independently of this assumption.

For a given drainage basin, the ordinate at time t of the unit hydrograph having a unit duration of Δt_0 is represented by $u(\Delta t_0, t)$, where t is any time after the beginning of effective rainfall. The effective rainfall for a given storm may be considered to consist of n blocks of different intensity I_i and of same duration equal to Δt_0, where the subscript i denotes the number representing a block. By the principle of linearity, the ordinate of the direct-runoff hydrograph for the given storm may be expressed as

$$Q(t) = \sum_{i=1}^{n} u[\Delta t_0, t - (i-1)\Delta t] I_i \, \Delta t \tag{14-7}$$

This is a summation form of the convolution integral [Eq. (14-17)] to be discussed later (Subsec. III-C).

Regarding assumption e, the hydrograph of direct runoff from a drainage basin due to a given pattern of effective rainfall at whatever time it may occur is invariable. This is known as the *principle of time invariance*. It is known that basin characteristics change with seasons, man-made adjustments on the basin, conditions of the flow, etc. Apparently, this principle is valid only when the time and condition of the drainage basin are fixed or specified.

The linearity and time invariance are the two fundamental principles of the unit-hydrograph theory. Other assumptions are often made for practical convenience, but they are not very essential from a fundamental and theoretical point of view.

2. Unit-hydrograph Analysis. A unit hydrograph can be derived from an observed hydrograph or hydrographs. In selecting a hydrograph for such an analysis, it is desirable to consider carefully that the assumptions involved in the unit-hydrograph theory are satisfied as closely as possible. A hydrograph resulting from an isolated, intense, short-duration storm of nearly uniform distribution in space and time is the most desirable. If such well-defined single-peaked hydrographs are unavailable, it is necessary to derive the unit hydrograph from a complex hydrograph.

Equation (14-7) may be used to separate a complex hydrograph into a number of simple hydrographs. After the base flow is separated from the observed hydrograph, the ordinates of the direct-runoff hydrograph at convenient equal time intervals Δt can be expressed by Eq. (14-7) in terms of the ordinates of the unit hydrograph. Consequently, a number of equations containing the ordinates of the unit hydrograph $u[\Delta t_0, t - (i - 1) \Delta t]$ can be written. A simultaneous solution of these equations will give the ordinates of the unit hydrograph. In these equations, the intensity of effective rainfall I_i must be known. Of course, the computation will be simplified if I_i can be assumed uniform or constant. Because of computational errors or other causes, such as nonlinearity of the data (Subsec. III-E), the computed unit-hydrograph ordinates may show negative values or erratic variations. If this occurs, a smooth curve may be fitted to these ordinates to produce an approximation of the unit hydrograph. For a more elaborate analysis, the unit hydrograph can be derived as an average from the unit hydrographs of several observed hydrographs and the fitting can be done by the method of least squares [35, 47]. With the aid of electronic computers, more precise unit hydrographs can be obtained by the theory of instantaneous unit hydrograph and the conceptual models (Subsecs. III-C and D). Finally, as a check of the theory, the area covered by the computed unit hydrograph should be equal to unity. Also, the computed unit hydrograph, or just an assumed unit hydrograph without using the above methods of computation, can be used to reconstruct the observed direct-runoff hydrograph. If the reconstructed hydrograph does not agree with the observed hydrograph, the computed unit hydrograph is modified. This checking procedure is repeated until a satisfactory agreement is obtained.

The derivation of the unit hydrograph from a simple hydrograph is straightforward. By the unit-hydrograph theory, the ordinates of the required unit hydrograph are simply equal to the corresponding ordinates of the given direct-runoff hydrograph divided by the total amount of direct runoff in inches. In this procedure, the effective rainfall is assumed uniformly distributed. If it is not, the methods for complex hydrographs as described above or a *method of successive approximation* by Collins [48, 49] can be used. The procedure by Collins includes the following four steps: (1) assume a unit hydrograph, and apply it to all effective-rainfall blocks of the hyetograph except the largest; (2) subtract the resulting hydrograph from the actual direct-runoff hydrograph, and reduce the residual to unit hydrograph terms; (3) compute a weighted average of the assumed unit hydrograph and the residual unit hydrograph, and use it as the revised approximation for the next trial; (4) repeat the previous three steps until the residual unit hydrograph does not differ by more than a permissible amount from the assumed hydrograph.

When a unit hydrograph of a given effective-rainfall duration is available, the unit hydrographs of other durations can be derived. If other durations are integral multiples of the given duration, the new unit hydrographs can be easily computed by the application of the principle of superposition. However, a general method of derivation applicable to unit hydrographs of any required duration may be used on the basis of the principle of superposition. This method is known as the *S-hydrograph method*, first suggested by Morgan and Hullinghors [50].

The theoretical *S-hydrograph* is a hydrograph produced by a continuous effective rainfall at a constant rate for an indefinite period (Fig. 14-5a). The curve assumes a deformed S-shape and its ordinates ultimately approach the rate of effective rainfall either as a limit or at a time of equilibrium. The S-hydrograph can be constructed graphically by summing up a series of identical unit hydrographs spaced at intervals

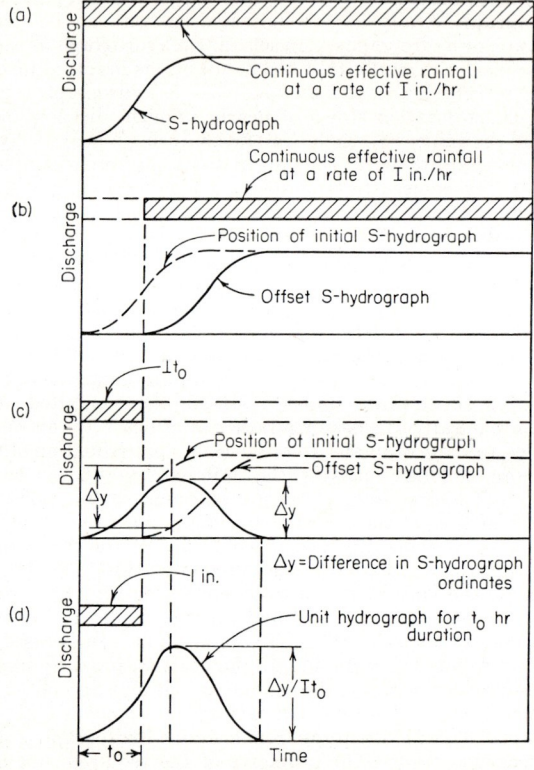

Fig. 14-5. Derivation of a unit hydrograph by the S-hydrograph method.

equal to the duration of the effective rainfall from which they were derived. After the S-hydrograph is constructed, the unit hydrograph of a given duration can be derived as follows (Fig. 14-5): Assume that the S-hydrograph is produced by a continuous effective rainfall at a constant rate of I in. per hr. Then advance or offset the position of the S-hydrograph for a period equal to the desired duration of t_0 hours and call this S-hydrograph an *offset* S-hydrograph. The difference between the ordinates of the original S-hydrograph and the offset S-hydrograph, divided by It_0, should result in the desired unit hydrograph.

It can be shown that the discharge of the S-hydrograph at the time of equilibrium is equal to $1.008AI$, where A is the drainage area in acres. If C is a runoff coefficient and I is replaced by CI, then this discharge is $1.008CIA$ or approximately $Q = CID$, which is the well-known rational formula (Subsec. II-B-2).

The unit-hydrograph analysis can be illustrated by a practical example [51] as shown in Fig. 14-6. This example consists of three parts: analysis of rainfall data, derivation of the unit hydrograph, and derivation of unit hydrographs for durations other than the duration of the original effective rainfall.

a. Analysis of Rainfall Data. This part of the analysis may not always be necessary unless a comprehensive study of the problem is required. For this analysis, the average depth of rainfall can be estimated in a number of ways (see Sec. 9). In this example, the isohyetal method is used. Figure 14-6a is the isohyetal depth of the given storm. By planimetering the areas within the isohyetals and weighting the indicated rainfall depths, the average depth of rainfall is found to be 2.49 in. The mass rainfall curves obtained from the recording charts of all gages are plotted as shown in Fig. 14-16b. From these curves an average mass curve is constructed. Consideration should be given to the weights of individual curves. For example, the curve for station SF deviates greatly from the other curves. As this station is located at a far distance outside the drainage basin, the corresponding curve should be given less weight in computing the average curve.

From the average mass curve of rainfall, a hyetograph as shown in Fig. 14-6c is constructed. In this example, a convenient time interval of $\frac{1}{4}$ hr is adopted. As a check, the area under the hyetograph may be measured and it should be equal to the total average rainfall depth or 2.49 in.

b. Derivation of the Unit Hydrograph. The observed hydrograph, usually converted from a stage hydrograph at the gaging station, is plotted in Fig. 14-6d. As described previously, there are several empirical methods for the base-flow separation. In this example, a normal groundwater depletion curve is used and fitted in the recession part of the observed hydrograph. At the beginning of the hydrograph, the base flow is represented by the continuation of the groundwater recession curve prior to the time of rise. As shown in Fig. 14-6d, the base flow is indicated by the extension of the previous groundwater recession curve from the rising point A of the hydrograph, to a point B which is arbitrarily taken (about $\frac{1}{10}$ of the base of the hydrograph, say 1 hr) beyond the time of peak flow. The curve AB is inclined toward the right at a slowly decreasing rate. At the recession end of the hydrograph, the normal recession curve ED and the groundwater recession curve CD are traced in smoothly. Point C on the curve CD is arbitrarily located at the middle of points B and D. Between points B and C, a smooth curve with a curvature convex upward is introduced. Then, the entire base flow of the hydrograph is represented by curve $ABCD$. Since the procedure of base-flow separation is arbitrary anyway, a much simpler separation could be made by just drawing a straight line between A and D or between B and D. In most flood hydrographs for analysis, the base flow usually constitutes only a small percentage of the peak flow, and thus any errors introduced in base-flow separation would be very small.

The ordinate of direct-runoff hydrograph is equal to the ordinate of the observed hydrograph minus base flow. Near the recession end, the observed hydrograph is modified to follow the normal recession curve, which is essentially a fitted smooth exponential curve designed to separate any bumps on the recession part of the hydrograph. In Fig. 14-6d, the direct-runoff hydrograph is shown with a dashed line. The area below the direct-runoff hydrograph is planimetered as 1,190 cfs-hr or 4,280,000 cu ft. As the drainage area is 2,850 acres, the total direct runoff is equal to 0.42 in. This is 16.9 per cent of the total rainfall. The unit hydrograph as shown in Fig. 14-6d is obtained by dividing the ordinates of the direct-runoff hydrograph by 0.42. It should be noted that two different scales for discharge are used in order to accommodate the hydrographs in the same figure for this particular example.

The duration of the effective rainfall is determined by drawing a horizontal line on the hyetograph (Fig. 14-6c) in such a way that the area of the hyetograph above the horizontal line is equal to the volume of the direct runoff, or 0.42 in. The area below the horizontal line represents the abstractions. The duration of the effective rainfall is the time elapsed between the beginning and end of the effective rainfall, which is indicated on the horizontal line as $\frac{3}{4}$ hr. The centroid of the mass of effective rainfall is found to be 1 hr and 10 min ahead of the peak flow.

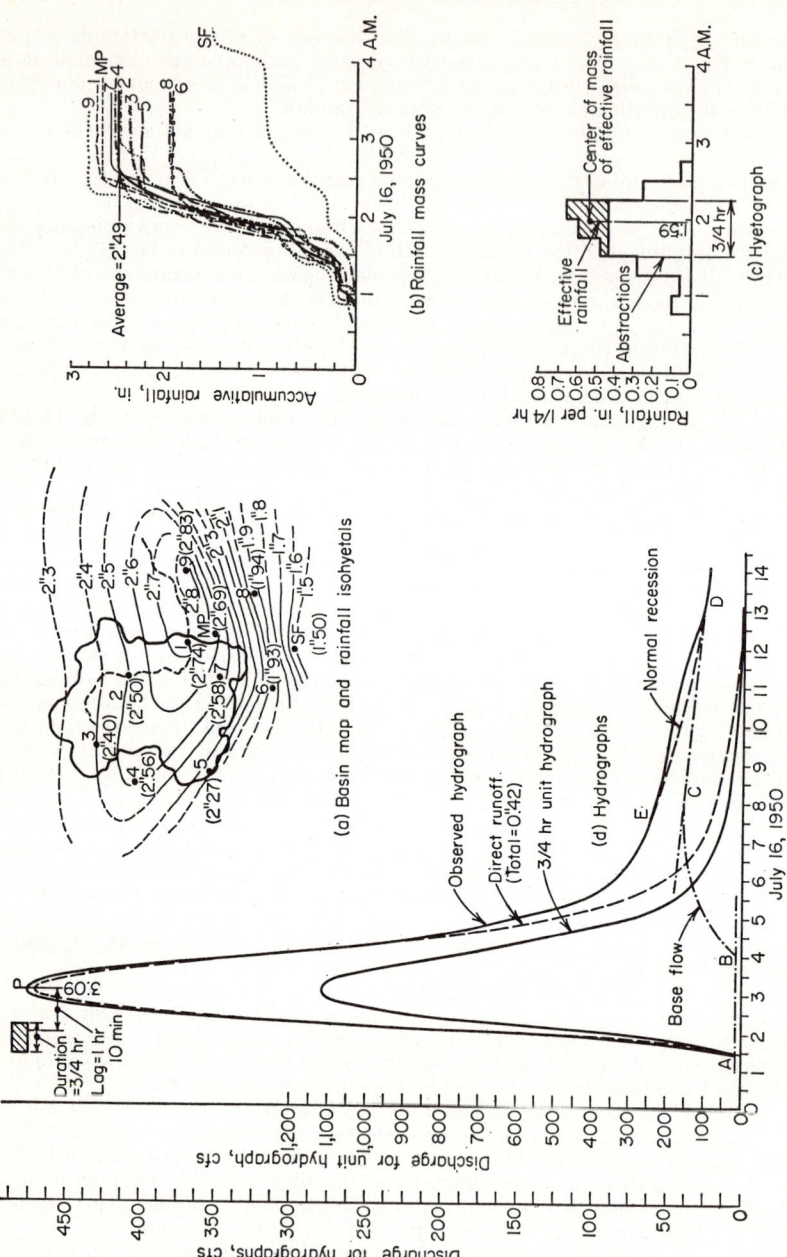

FIG. 14-6. An example of unit-hydrograph analysis. (Boneyard Creek, Champaign-Urbana, Illinois.)

TIME DISTRIBUTION OF RUNOFF

It must be noted that for refined analysis more elaborate methods, such as infiltration indexes and curves (Sec. 11), than constructing a horizontal line can be used to separate the effective rainfall from the abstractions. However, any such method is based on certain assumptions, and more research is needed for the development of a most reliable and correct procedure.

c. Derivation of Unit Hydrographs for Other Durations. After the ¾-hr unit hydrograph is derived, unit hydrographs for other durations can be computed by the S-hydrograph method. Summing up the ¾-hr unit hydrographs spaced at intervals of ¾ hr will produce an S-hydrograph, which is a hydrograph caused by effective rainfall at a rate of 1⅓ in. per hr. For simplification of the procedure, the computation is illustrated in Table 14-2, which is self-explanatory. A graphical computation of this procedure is shown in Fig. 14-7. Table 14-3 gives the computation of an S-hydrograph caused by a rate of effective rainfall equal to 1 in. per hr.

Table 14-2. Computation Form for an S-hydrograph

Time	Unit hydrograph ordinates	Offsets	S-hydrograph ordinates (for 1 in. per unit duration)	S-hydrograph ordinates (for 1 in./hr)
(1)	(2)	(3)	(4)	(5)
↑ Time interval = duration of effective rainfall ↓	a b c d e	0 0 0 0 0	Col. 2 + Col. 3 Col. 2 + Col. 3 Col. 2 + Col. 3 Col. 2 + Col. 3 Col. 2 + Col. 3	Col. 4 × unit duration Col. 4 × unit duration Col. 4 × unit duration Col. 4 × unit duration Col. 4 × unit duration
	f g h i j	a + 0 b + 0 c + 0 d + 0 e + 0	Col. 2 + Col. 3 Col. 2 + Col. 3 Col. 2 + Col. 3 Col. 2 + Col. 3 Col. 2 + Col. 3	Col. 4 × unit duration Col. 4 × unit duration Col. 4 × unit duration Col. 4 × unit duration Col. 4 × unit duration
	k l m n o	a + f + 0 b + g + 0 c + h + 0 d + i + 0 e + j + 0	Col. 2 + Col. 3 Col. 2 + Col. 3 Col. 2 + Col. 3 Col. 2 + Col. 3 Col. 2 + Col. 3	Col. 4 × unit duration Col. 4 × unit duration Col. 4 × unit duration Col. 4 × unit duration Col. 4 × unit duration
	etc.	etc.	etc.	etc.

In general, the computed ordinates of an S-hydrograph will not fall on a smooth curve, and it must be smoothed off in the analysis. In fact, the computed ordinates may show a periodic up-and-down variation known as *hunting* or *waving* when nonuniform or abrupt distribution actually occurs in the effective rainfall, whereas it is, of course, assumed uniform in the computation.

Two S-hydrographs for effective rainfall of 1 in. per hr are spaced at a horizontal offset distance equal to the duration of the required unit hydrograph. The unit hydrograph of the given duration is equal to the difference between the two S-hydrographs divided by the given duration. By this procedure, the unit hydrographs for durations of 1, 2, 4, and 6 hr are constructed in Fig. 14-8, in which the offset S-hydrograph for deriving the 4-hr unit hydrograph is shown. These unit hydrographs may also be derived from the S-hydrograph by a numerical procedure using the same principle of linearity. Table 14-4 shows the computation in which the unit-hydrograph ordinates are obtained by subtracting the offset S-hydrograph ordinates from the ordinates of the original S-hydrograph, and then dividing the differences by the duration of the required unit hydrograph.

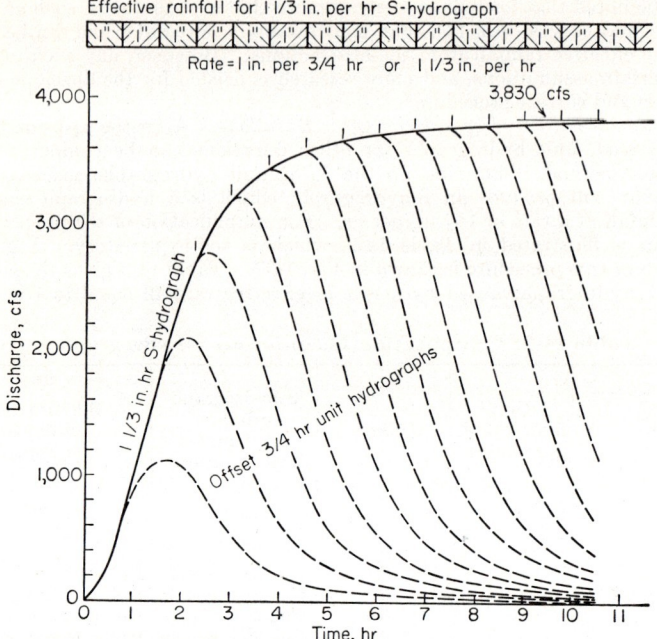

Fig. 14-7. Derivation of S-hydrograph.

Table 14-3. Computation of 1-in./hr S-hydrograph

Time (interval = ¾ hr) hr	Unit hydrograph ordinates, cfs	Offset, cfs	1⅓-in./hr S-hydrograph, cfs	1-in./hr S-hydrograph, cfs
(1)	(2)	(3)	(4)	(5)
0.00	0	.....	0	0
0.75	580	0	580	440
1.50	1,100	580	1,680	1,260
2.25	960	1,680	2,640	1,980
3.00	530	2,640	3,170	2,380
3.75	260	3,170	3,430	2,570
4.50	140	3,430	3,570	2,680
5.25	80	3,570	3,650	2,740
6.00	50	3,650	3,700	2,770
6.75	40	3,700	3,740	2,800
7.50	30	3,740	3,770	2,830
8.25	20	3,770	3,790	2,840
9.00	20	3,790	3,810	2,860
9.75	15	3,810	3,825	2,870
10.50	5	3,825	3,830	2,880
11.25	0	3,830	3,830	2,880
12.00	0	3,830	3,830	2,880
etc.	do.	do.	do.	do.

TIME DISTRIBUTION OF RUNOFF   14-21

When a number of unit hydrographs of various durations are obtained by the S-hydrograph method, the computed unit-hydrograph peak flow can be plotted against duration for interpolation purposes. Experience has shown that such a curve

Table 14-4. Derivation of Unit Hydrographs from S-hydrograph

Time, hr	1-in./hr S-hydrograph, cfs	1-hr offset, cfs	1-hr u.h., cfs	2-hr offset, cfs	2-hr u.h., cfs	4-hr offset, cfs	4-hr u.h., cfs	6-hr offset, cfs	6-hr u.h., cfs
(1)	(2)	(3)	(4)	(5)	(6)	(7)	(8)	(9)	(10)
0	0	.....	0	.....	0	.....	0	.....	0
0.50	225	.....	225	.....	113	.....	56	.....	37
1.00	714	0	714	.....	357	.....	179	.....	119
1.50	1,260	225	1,035	.....	630	.....	315	.....	210
2.00	1,790	714	1,076	0	895	.....	448	.....	299
2.50	2,140	1,260	880	225	958	.....	535	.....	356
3.00	2,380	1,790	590	714	833	.....	595	.....	397
3.50	2,530	2,140	390	1,260	635	.....	632	.....	421
4.00	2,640	2,380	260	1,790	425	0	660	.....	440
4.50	2,690	2,530	160	2,140	275	225	616	.....	449
5.00	2,730	2,640	90	2,380	175	714	504	.....	455
5.50	2,760	2,690	70	2,530	115	1,260	375	.....	460
6.00	2,790	2,730	60	2,640	75	1,790	250	0	465
6.50	2,800	2,760	40	2,690	55	2,140	165	225	430
7.00	2,810	2,790	20	2,730	40	2,380	108	714	349
7.50	2,820	2,800	20	2,760	30	2,530	72	1,260	260
8.00	2,830	2,810	20	2,790	20	2,640	47	1,790	175
8.50	2,840	2,820	20	2,800	20	2,690	38	2,140	117
9.00	2,850	2,830	20	2,810	20	2,730	30	2,380	78
9.50	2,860	2,840	20	2,820	20	2,760	25	2,530	55
10.00	2,865	2,850	15	2,830	18	2,790	19	2,640	37
10.50	2,870	2,860	10	2,840	15	2,800	18	2,690	30
11.00	2,875	2,865	10	2,850	13	2,810	16	2,730	24
11.50	2,878	2,870	8	2,860	9	2,820	14	2,760	20
12.00	2,880	2,875	5	2,865	8	2,830	13	2,790	15
12.50	2,880	2,878	2	2,870	5	2,840	10	2,800	13
13.00	2,880	2,880	0	2,873	3	2,850	8	2,810	12
13.50	2,880	2,880		2,878	1	2,860	5	2,820	10
14.00	2,880	2,880		2,880	0	2,865	4	2,830	8
14.50	2,880	2,880		2,880		2,870	2	2,840	7
15.00	2,880	2,880		2,880		2,875	1	2,850	5
15.50	2,880	2,880		2,880		2,878	1	2,860	3
16.00	2,880	2,880		2,880		2,880	0	2,865	2
16.50	2,880	2,880		2,880		2,880		2,870	2
17.00	2,880	2,880		2,880		2,880		2,875	1
17.50	2,880	2,880		2,880		2,880		2,878	0
18.00	2,880	2,880		2,880		2,880		2,880	0
etc.	do.	do.		do.		do.		do.	

is better plotted with the peak flows computed from unit-hydrographs derived from individual observed hydrographs, thus avoiding the use of the S-hydrograph. This is because the S-hydrograph method assumes the principle of superposition, whereas the actual hydrographs for a drainage basin may be somewhat nonlinear in nature [8, p. 42].

3. Modifications of the Unit Hydrograph. Since the formulation of the unit-hydrograph theory, a number of modifications of the unit hydrograph have been proposed. In 1935, Bernard [25] proposed a *distribution graph* which is a unit hydrograph whose ordinates are expressed in per cent of the total direct runoff. Such a graph can be derived by dividing the ordinates of a unit hydrograph by the sum of all the ordinates, thus producing a curve under which the area is equal to 100 per cent.

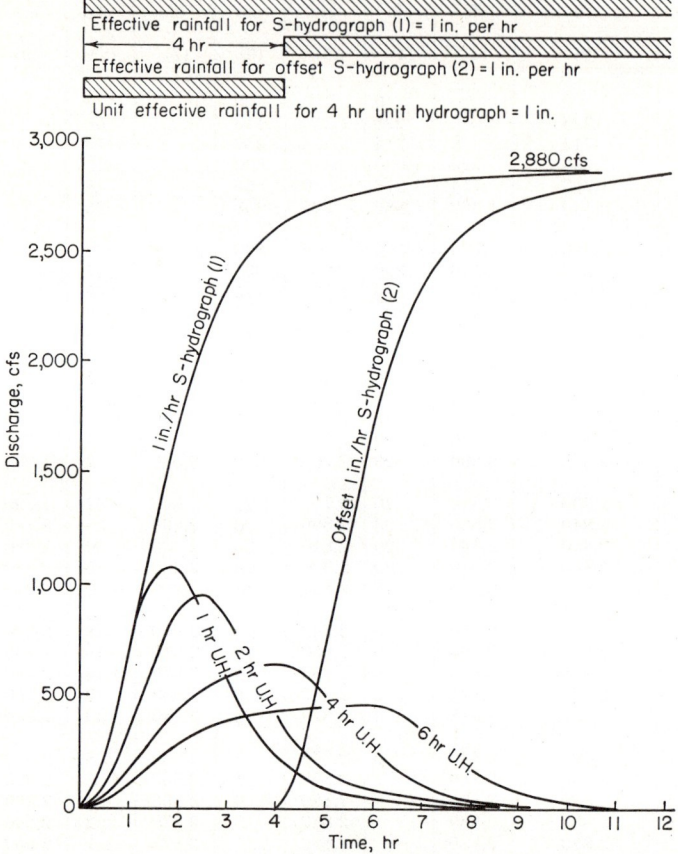

FIG. 14-8. Derivation of unit hydrographs by S-hydrograph.

In 1938, McCarthy [52] proposed a method of *synthesis of unit hydrographs*. In this method, a correlation analysis was made between three unit-hydrograph parameters (peak discharge, lag time from beginning of rain to peak flow, and total base time) and three basin characteristics (size of basin area, slope of basin area-elevation curve, and number of major streams in the basin). From the resulting correlation curves, it is possible to estimate the three unit-hydrograph parameters for an ungaged area when the three basin characteristics are known.

In the same year that McCarthy developed his hydrograph synthesis, a better-known method of *synthetic unit hydrographs* was developed by Snyder [26]. Snyder analyzed a large number of hydrographs from drainage basins in the Appalachian Mountain region in the United States and developed the following group of equations:

TIME DISTRIBUTION OF RUNOFF

$$t_p = C_t(LL_c)^{0.3} \tag{14-8}$$

$$t_r = \frac{t_p}{5.5} \tag{14-9}$$

$$q_p = \frac{640 C_p}{t_p} \tag{14-10}$$

$$t_{pR} = t_p + 0.25(t_R - t_r) \tag{14-11}$$

$$q_{pR} = \frac{640 C_p}{t_{pR}} \tag{14-12}$$

$$q_{pR} = q_p \frac{t_p}{t_{pR}} \tag{14-13}$$

The notations in the above equations are explained as follows:

t_p = lag time from midpoint of effective-rainfall duration t_r to peak of a unit hydrograph, hr
t_r = duration of effective rainfall equal to $t_p/5.5$
t_R = duration of effective rainfall other than standard duration t_r adopted in specific study, hr
t_{pR} = lag time from midpoint of duration t_R to peak of unit hydrograph, hr
q_p = peak discharge per unit drainage area of unit hydrograph for standard duration t_r, cfs/mi^2
q_{pR} = peak discharge per unit drainage area of unit hydrograph for duration t_R, cfs/mi^2
L_c = river mileage from the station to center of gravity of the drainage area
L = river mileage from the given station to the upstream limits of the drainage area
C_t and C_p = coefficients depending upon units and basin characteristics

The average values of C_p and C_t have been found to be respectively 0.63 and 2.0 in the fairly mountainous Appalachian Highlands; and an extreme range of the values are from 0.94 and 0.4 in southern California to 0.61 and 8.0 in sections of states bordering in the eastern Gulf of Mexico [19]. Although Snyder indicated that the coefficient C_t is affected by basin slope, Taylor and Schwarz [53] found, from an analysis of 20 basins in the North and Middle Atlantic States, that $C_t = 0.6/\sqrt{s}$, where s is the basin slope. Linsley, Kohler, and Paulhus [54] showed an expression for the lag time containing the slope as

$$t_p = C_t \left(\frac{LL_c}{\sqrt{s}}\right)^n \tag{14-14}$$

where $n = 0.38$ and $C_t = 1.2$ for mountain drainage areas, 0.72 for foothill areas, and 0.35 for valley areas. This expression, of course, resembles the type of the equation for the time of concentration [Eq. (14-5)].

From a given hydrograph of a drainage basin, the corresponding unit hydrograph of duration t_R, lag time t_{pR}, and peak discharge per unit drainage area q_{pR} can be computed. The values of L and L_c can be obtained from the basin map. Assume t_{pR} equal to t_p and compute t_r by Eq. (14-9). If the computed t_r happens to be equal to or close to t_R, further assume $q_{pR} = q_p$ and compute C_t and C_p by Eqs. (14-8) and (14-10). If the computed t_r does not equal t_R, use t_R, t_{pR}, and Eq. (14-9) for t_r in Eq. (14-11) and compute t_p. Then the value of C_t is computed by Eq. (14-8) and C_p by Eq. (14-12).

From a unit hydrograph of a gaged drainage basin, the basin coefficients C_t and C_p can be computed by the method described above. If the ungaged basin and the gaged basin are located within a region of homogeneous hydrologic condition, the coefficients so obtained can be used to derive a unit hydrograph in the ungaged basin of duration t_R. First, L and L_c can be measured from the map of the drainage basin. By means of Eqs. (14-8), (14-9), (14-11), and (14-12), t_{pR} and q_{pR} can be computed, and the computed q_{pR} multiplied by the drainage area in square miles will give the peak discharge of the required unit hydrograph. The width W_{75} of the unit hydrograph at discharge equal to 75 per cent of the peak discharge, in hours, and the width W_{50} at discharge equal to 50 per cent of the peak discharge, in hours, can be estimated from the following empirical formulas developed by the U.S. Army Corps of Engineers [19]:

$$W_{75} = \frac{440}{q_{pR}^{1.08}} \quad \text{and} \quad W_{50} = \frac{770}{q_{pR}^{1.08}} \qquad (14\text{-}15)$$

The base length of the unit hydrograph, in days, may be estimated by the following empirical formula [26]:

$$t_b = 3 + \frac{t_p}{8} \qquad (14\text{-}16)$$

The base length depends largely on the method of base-flow separation. This formula usually gives long base length for small basin areas because it may include the effect of subsurface runoff.

By synthesis of unit hydrographs, a number of dimensionless hydrographs have been developed by Commons [30]; Williams [55]; Mitchell [56]; U.S. Soil Conservation Service [57]; Hickok, Keppel, and Rafferty [58]; Bender and Roberson [59]; and Gray [60, 61]. The *dimensionless hydrograph* is essentially a unit hydrograph for which the discharge is expressed by the ratio of discharge to peak discharge, and the time by the ratio of time to lag time, thus eliminating the effect of basin size and much of the basin shape. Knowing the peak discharge and lag time for the duration of effective rainfall, the unit hydrograph can be estimated from a synthesized dimensionless hydrograph for the given basin region [62]. The peak discharge q_p may be estimated by use of a plot relating q_p to t_p or by the Snyder method; the lag time can be computed by a correlation formula, such as Eqs. (14-8) and (14-14).

C. Instantaneous Unit Hydrograph

1. IUH and Its Development. If the duration of the effective precipitation becomes infinitesimally small, the resulting unit hydrograph is called an *instantaneous unit hydrograph* (IUH) which is, expressed by $u(0,t)$ or simply $u(t)$. In other words, for an IUH the effective precipitation is applied to the drainage basin in zero time. Of course, this is only a fictitious situation and a concept to be used in hydrograph analysis. By the principle of superposition in the linear unit-hydrograph theory (Fig. 14-9), when an effective rainfall of function $I(\tau)$ of duration t_0 is applied, each infinitesimal element of this ERH will produce a DRH equal to the product of $I(\tau)$ and the IUH expressed by $u(0,t - \tau)$ or $u(t - \tau)$. Thus, the ordinate of the DRH at time t is

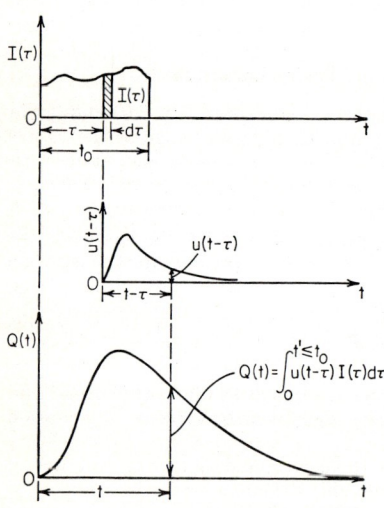

Fig. 14-9. Convolution of $I(\tau)$ and IUH.

$$Q(t) = \int_0^{t' \leq t_0} u(t - \tau) I(\tau) \, d\tau \qquad (14\text{-}17)$$

which is called the *convolution integral*, also known as the *Duhamel integral*, in which $u(t - \tau)$ is a *kernel function*, $I(\tau)$ is the input function, and $t' = t$ when $t \leq t_0$ and $t' = t_0$ when $t > t_0$.

The shape of an IUH resembles a single-peaked hydrograph (Fig. 14-1). If the rainfall and runoff in the convolution integral are measured in the same units, the ordinates of the IUH must have a dimension of [time^{-1}]. The properties of the IUH are as follows:

$0 \leq u(t) \leq$ a positive peak value for $t > 0$ (14-18)
$u(t) = 0$ for $t \leq 0$ (14-19)
$u(t) \to 0$ for $t \to \infty$ (14-20)

$$\int_0^\infty u(t)\, dt = 1.0 \qquad (14\text{-}21)$$

$$\int_0^\infty u(t) t\, dt = t_L \qquad (14\text{-}22)$$

where t_L is the lag time of the IUH. Since the convolution integral is a linear process, it can be shown that t_L is also equal to the time interval between the centroid of the effective rainfall and that of the direct runoff.

The major advantage of the IUH in comparison with a unit hydrograph is that the IUH is independent of the duration of effective precipitation, thereby eliminating one variable in hydrograph analysis. Furthermore, the use of IUH is better suited for the needs of theoretical investigations on the rainfall and runoff relationship in drainage basins. The determination of IUH is numerically more tedious than that of unit hydrographs, but it can be simplified through the use of electronic computers.

The idea that the hydrograph produced by an instantaneous storm could provide an indication of drainage-basin characteristics was first expressed in a committee report by the Boston Society of Civil Engineers [44] in 1930. It was Clark [63] in 1945 who adapted this idea to the unit-hydrograph theory, thus introducing the use of IUH for the first time in hydrograph analysis. In 1951, Edson [31] derived a two-parameter equation for unit hydrographs, unknowingly creating a conceptual model essentially identical to one developed later by Nash (if the input to the equation were considered instantaneous—see next subsection). In 1952, Paynter [64, 65] proposed an analog computer for flood routing for which he compared the drainage network to an admittance network where an *admittance function* is the output for a unit step input. The derivative of the admittance function can be shown to be an IUH. In 1956, Cuénod [38] called such an output or outflow hydrograph produced by a unit step inflow as the *characteristic hydrograph* (l'hydrogramme indiciel). He used a reversed process of numerical convolution to compute the ordinates of the outflow hydrograph by a composite long division procedure of a time series. Although Paynter and Cuénod did not mention IUH, their approaches implied the concept of IUH and the convolution integral. Since 1955, several investigations directly related to IUH have been made by O'Kelly (1944, [66]), Sugawara and Maruyama (1956, [37]), Nash (1957 [67]), Dooge (1959, [68]), Boersma (1959, [69]), O'Donnell (1960, [70]), Singh (1962, [71]), Chow (1962, [8]), Henderson (1963, [72]), Wu (1963, [73]), Diskin (1964, [74]), and Kulandaiswamy (1964, [20]).

2. Methods of Determining IUH. There are various methods for the determination of an IUH from the given ERH and DRH. Chow [8] used a procedure for an approximate determination of IUH. In this procedure, the IUH ordinate at time t is simply equal to the slope of an S-hydrograph at time t. This S-hydrograph can be derived from a given set of rainfall and runoff data as described before and its ordinates are adjusted such that the maximum ordinate at the time of equilibrium is made equal to a discharge corresponding to a direct runoff or unity. The procedure is based on the fact that this S-hydrograph is an integral curve of the IUH or its ordinate at time t is equal to the integration of the area under the IUH from time 0 to time t. Since such an S-hydrograph derived from the actual data cannot be too exact, the IUH so obtained is only approximate.

A procedure for deriving an IUH by harmonic analysis was proposed by O'Donnell [70]. In this procedure, the ERH, DRH, and IUH are represented by three harmonic series. Since the harmonic coefficients of these series can be related by simple equations, the harmonic coefficients of an IUH can be derived from those of the ERH and DRH. The ordinates of IUH can then be calculated.

A theoretically possible procedure for determining an IUH is by the Laplace-transform method, which was first mentioned by Paynter [64] in 1952 and later suggested by Amorocho and Orlob [75] in 1961. In 1964, Diskin [74] used the method to investigate the nonlinear properties of hydrographs. The Laplace-transform function (LTF) of the IUH $u(t)$ is expressed as

$$\bar{u}(p) = \int_0^\infty e^{-pt} u(t)\, dt \tag{14-23}$$

where p is the variable of the Laplace transform. The dimension of p is [time^{-1}] and the LTF itself is dimensionless. The properties of the LTF can be derived from those of the IUH. As $u(t)$ is a bounded function of exponential order, $\bar{u}(p)$ is finite for all positive values of p. Since both $u(t)$ and e^{-pt} are positive functions in the range of integration, $\bar{u}(p)$ must be nonnegative. Since $u(t)$ is of exponential order, $\bar{u}(p)$ approaches zero when p increases. For $p = 0$, the LTF is reduced to $\bar{u}(p) = 1$ as a result of the unit-area characteristic of the IUH curve. Also, at $p = 1$, the first derivative of $\bar{u}(p)$ with respect to p, or $d\bar{u}(p)/dt$, can be shown equal to $-t_L$ as a result of the characteristic of the IUH given by Eq. (14-22). It can also be shown that $d\bar{u}(p)/dt$ is always negative and increases to zero as p increases to infinity, which means

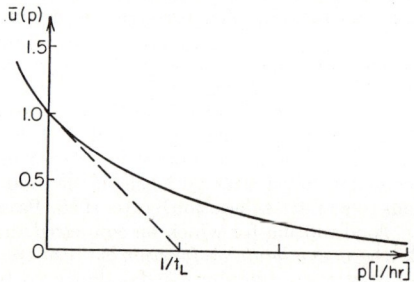

Fig. 14-10. Laplace transform function of an IUH.

that the slope of the LTF curve increases from its initial value of $-t_L$ but always remains negative, and the curve approaches the p-axis asymptotically. Considering the properties of the LTF, the curve of this function is shown in Fig. 14-10.

A feature of the Laplace transform which makes it particularly suitable for hydrograph analysis is the fact that the Laplace transform of the convolution integral by Eq. (14-17) is equal to the product of the Laplace transforms of two functions involved in the integral [76–80], or

$$\bar{Q}(p) = \bar{u}(p) \cdot \bar{I}(p) \tag{14-24}$$

where $\bar{Q}(p)$ and $\bar{I}(p)$ are the LTF's respectively of the DRH $Q(t)$ and ERH $I(t)$, or

$$\bar{Q}(p) = \int_0^\alpha e^{-pt} Q(t)\, dt \tag{14-25}$$

and

$$\bar{I}(p) = \int_0^\alpha e^{-pt} I(t)\, dt \tag{14-26}$$

From Eq. (14-24), the Laplace transform of the IUH can be written as

$$\bar{u}(p) = \frac{\bar{Q}(p)}{\bar{I}(p)} \tag{14-27}$$

TIME DISTRIBUTION OF RUNOFF

For a numerical evaluation of the LTF, the integrals of the functions may be replaced by summations, and the ERH and DRH are divided into elements for small time intervals. When $\bar{u}(p)$ is computed by Eq. (14-27), it can be transformed back to $u(t)$. In order to do so, an empirical equation or an equation representing a conceptual model (see next subsection) may be fitted to the computed $\bar{u}(p)$.

Other methods of determining IUH involve the use of various conceptual models. By fitting a conceptual model to the given hydrologic data, the IUH can be computed.

D. Conceptual Models of IUH

There are various conceptual models that have been proposed to delineate an IUH. These models may be of physical analogy or of mathematical simulation, all being composed of simulated components such as linear reservoirs, linear channels, or time-area diagrams.

1. Components of Models. A *linear reservoir* is a fictitious reservoir in which the storage S is directly proportional to the outflow Q, or

$$S = KQ \qquad (14\text{-}28)$$

where K is a reservoir constant, called *storage coefficient*. Since the difference between inflow I to the reservoir and the outflow Q is the rate of change of storage, the continuity equation is

$$I - Q = \frac{dS}{dt} \qquad (14\text{-}29)$$

Substituting Eq. (14-28) in Eq. (14-29) and considering the condition that $Q = 0$ when $t = 0$, the following equation for outflow can be derived:

$$Q = I(1 - e^{-t/K}) \qquad (14\text{-}30)$$

When $t = \infty$, the above equation gives $Q = I$, which means that the outflow approaches an equilibrium condition, becoming equal to inflow. If the inflow terminates at time t_0 since outflow began, a similar derivation gives the outflow at t in terms of discharge Q_0 at t_0 as

$$Q = Q_0 e^{-\tau/K} \qquad (14\text{-}31)$$

where $\tau = t - t_0$, being equal to the time since inflow terminated.

For an instantaneous inflow which fills the reservoir of storage S_0 in $t_0 = 0$, Eq. (14-28) shows $Q_0 = S_0/K$ and Eq. (14-31) gives the outflow as

$$Q = Ie^{-t/K} = \frac{S_0}{K} e^{-t/K} \qquad (14\text{-}32)$$

For a unit input or $S_0 = 1$, the IUH of the linear reservoir is therefore

$$u(t) = \frac{1}{K} e^{-t/K} \qquad (14\text{-}33)$$

which is represented by the hydrograph for the outflow from the first reservoir as shown in Fig. 14-13.

A *linear channel* is a fictitious channel in which the time T required to translate a discharge Q of any magnitude through a given channel reach of length x is constant. Thus, when an inflow hydrograph is routed through the channel, its shape will not be changed. If the inflow function is $I = f(t)$, the outflow function is identical with the inflow function except that a change in time is effected equal to the time of translation, or $Q = f(t - T)$. Accordingly, the velocity is constant for all discharges at any section of the channel, but may vary from section to section along the channel. At a given section, the relation between the water area A and the discharge Q is linear, or

$$A = CQ \qquad (14\text{-}34)$$

where $C = f(T)$, a function of the translation time called *translation coefficient*, which is constant at the given section.

If a segment of inflow of duration Δt and volume S is routed through a linear channel (Fig. 14-11), the outflow is

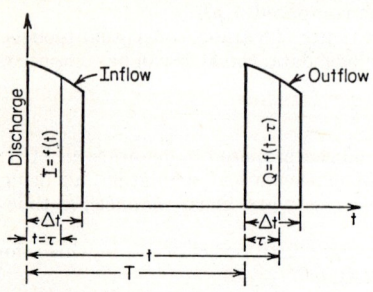

FIG. 14-11. Routing a segment of hydrograph through a linear channel.

$$Q = S\delta(t,\Delta t) \qquad (14\text{-}35)$$

where

$$\delta(t,\Delta t) = \frac{1}{\Delta t} \qquad (14\text{-}36)$$

for $0 \leq \tau \leq \Delta t$ and $t = \tau + T$; it is zero otherwise, where τ is the time measured from the beginning of the segment.

Equation (14-36) is a *pulse function*. When Δt approaches zero, this equation becomes an *impulse function* $\delta(t)$, known as a *Dirac-delta function*, which represents the IUH for the linear channel.

A drainage area can be considered analogous to a linear channel containing spatially varied flow (Fig. 14-12a). The total area a is divided into n subareas of size Δa_j by isochrones (contours of equal travel time Δt) with $j = 1, 2, \ldots, n$ to denote the order of the isochrones or subareas counted upstream from the outlet. The flow from

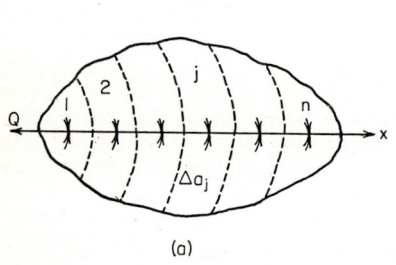

(a)

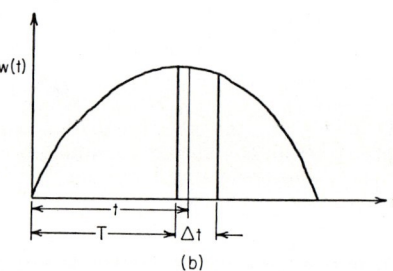

(b)

FIG. 14-12. Development of area-time diagram. (a) Drainage basin simulated by a linear channel, (b) area-time diagram.

subarea j is equal to $i_j \Delta a_j$ where i_j is the effective rainfall. The outflow at the outlet for this flow is therefore

$$Q = i_j \Delta a_j \, \delta(t - T, \Delta t) \qquad (14\text{-}37)$$

where $T = (j - 1) \Delta t$. When Q is divided by a and plotted against t, the diagram shown in Fig. 14-12b is produced, being represented by

$$w(t) = \sum_{j=1}^{n} \frac{i_j \Delta a_j}{a} \delta(t - T, \Delta t) \qquad (14\text{-}38)$$

When Δt approaches zero and i_j is constant, the diagram is represented by a curve with its ordinates directly proportional to the shape of the drainage area projected onto the channel and may be called a *time-area*, or *time-area-concentration*, *diagram*.

2. **Formulation of Models.** The concept of a linear reservoir was implied in an analysis of rainfall and runoff relationship by Zoch [24], published in 1934 to 1937, in which he introduced the use of Eq. (14-28). In 1941, Turner and Burdoin [27] considered the runoff hydrograph as the product of routing a time-area diagram through a

reservoir having a linear storage coefficient derived from the recession curve of the hydrograph. Later, Clark [63] in 1945 and O'Kelly [66] in 1955 considered the IUH as the result of the same process in their proposed methods of hydrograph analysis. In 1956, Sugawara and Maruyama [37] introduced the conceptual use of a simple mechanism, consisting essentially of linear reservoirs, as analogous physical models to analyze river runoff. In all these studies, however, the concept of linear reservoirs is implied, but not mentioned explicitly; nor are the models for IUH expressed mathematically.

Since 1957 and the following years, Nash [67, 81–86] proposed a conceptual model by considering a drainage basin as n identical linear reservoirs in series (Fig. 14-13). By routing a unit inflow through the reservoirs, a mathematical equation for IUH can be derived. The outflow from the first reservoir is obviously expressed by Eq. (14-33). In routing the flow, this outflow is considered as the inflow to the second reservoir. Using Eq. (14-33) as the input function with τ being the variable and using the kernel

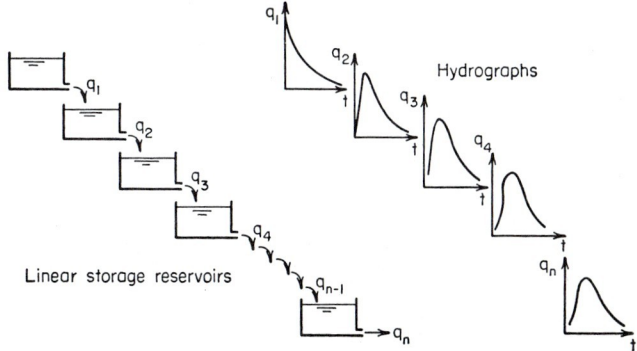

FIG. 14-13. Routing of instantaneous inflow through a series of linear storage reservoirs (Nash's model).

function with $t - \tau$ being the variable, the convolution integral of Eq. (14-17) gives the outflow from the second reservoir as

$$q_2 = \int_0^t \frac{1}{K} e^{-\tau/K} \frac{1}{K} e^{-(t-\tau)/K} \, d\tau = \frac{t}{K^2} e^{-t/K} \qquad (14\text{-}39)$$

This outflow is then used as the inflow to the third reservoir and routed through the latter. Continuing this routing procedure gives the outflow q_n from the nth reservoir as

$$u(t) = \frac{1}{K(n-1)!} \left(\frac{t}{K}\right)^{n-1} e^{-t/K} \qquad (14\text{-}40)$$

which is the IUH of the simulated drainage basin, mathematically representing a gamma distribution.

The values of K and n in Nash's model can be evaluated by the method of moments. It can be shown that the first and second moments of IUH about the origin $t = 0$ are respectively

$$M_1 = nK \qquad (14\text{-}41)$$

and

$$M_2 = n(n+1)K^2 \qquad (14\text{-}42)$$

The first moment M_1 represents the lag time of the centroid of IUH. By applying IUH in the convolution integral to relate ERH with DRH, the principle of linearity requires each infinitesimal element of the ERH to yield its corresponding DRH with the same lag time. In other words, the time difference between the centroids of ERH and DRH should be equal to M_1. If M_{ERH1} and M_{DRH1} are the first moment arms respectively of ERH and DRH about the time origin, then

$$M_{DRH1} - M_{ERH1} = nK \tag{14-43}$$

It can be further proved that

$$M_{DRH2} - M_{ERH2} = n(n+1)K^2 + 2nKM_{ERH1} \tag{14-44}$$

where M_{DRH2} and M_{ERH2} are the second moment arms respectively of DRH and ERH about the time origin. As the first and second moment arms of ERH and DRH can be computed from the given ERH and DRH, the values of n and K defining the IUH can be found by Eqs. (14-43) and (14-44).

The Nash model is linear since the storage coefficient K is constant. It may become nonlinear if K is made a variable or a function of Q. Rosa [87] has built an electronic analog to represent a series of five linear reservoirs with variable storage

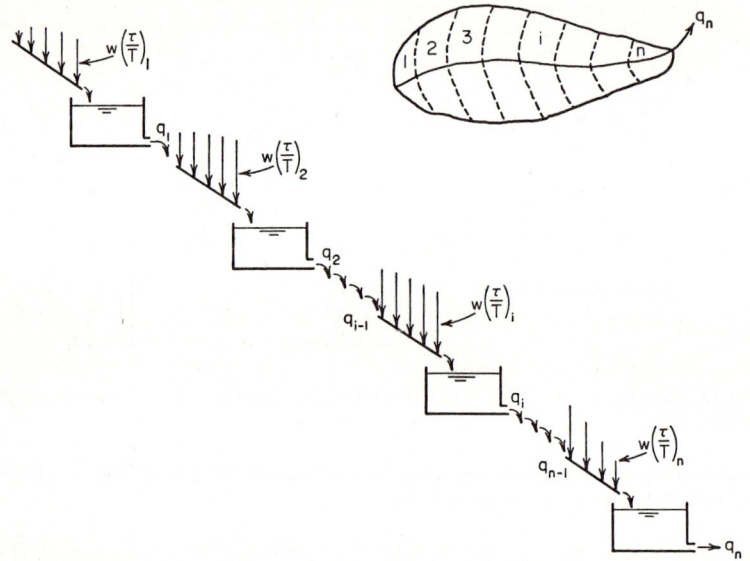

Fig. 14-14. Routing flow through linear channels and reservoirs (Dooge's model).

coefficients (see Fig. 29-3). By means of this analog it is possible to fit various DRH's on a nonlinear basis, although no derivation of the IUH has been made.

The Nash model does not involve the concept of translation of flow. In order to incorporate this concept into the analysis, Dooge [68] proposed to use the concept of a linear channel for the first time and to represent the basin system by a series of alternating linear channels and linear reservoirs (Fig. 14-14). In Dooge's model, which is linear, the drainage area is divided by isochrones into n number of subareas. (The idea of isochrones is believed to have been first proposed by Ross [21] in 1921.) Each subarea is represented by a linear channel in series with a linear reservoir. The outflow from the linear channel is represented by a time-area diagram which, together with outflow from the preceding subarea, serves as the inflow to the linear reservoir. The IUH of the simulated drainage area can be shown as

$$u(t) = \frac{S}{T} \int_0^{t' \leq T} \left[\frac{\delta(t-\tau)}{\prod_{i=1}^{i(\tau)} (1+K_i D)} \right] w\left(\frac{\tau}{T}\right) d\tau \tag{14-45}$$

where S is the input volume taken as unity; T is the total translation time of the basin (equal to LC with L as the total channel length of the basin and C as a constant translation coefficient for all linear channels) from which the isochrones are constructed; i is the order of reservoirs equal to 1, 2, ... counted downstream to the basin outlet; $i(\tau)$ is a function of τ representing an integer equal to the order number of the subarea where τ is considered; K_i is a storage coefficient; D is the differential operator d/dt; $\delta(t - \tau)$ is the Dirac-delta function where t is the time elapsed and τ is the translation time between the elements in the subarea and the outlet; and $w(\tau/T)$ is the ordinate of a dimensionless time-area diagram. Although the Dooge model takes into account the translation effect of flow in a drainage basin, the equation for IUH cannot be easily solved for practical applications.

Under the direction of the author, improved models have been developed from 1961 to 1964 at the University of Illinois by Singh [71], Diskin [74], and Kulandaiswamy [20]. To overcome the difficulties of the Dooge model for practical applications, Singh developed a model which consists of a linear channel of translation coefficient C and two linear reservoirs of different storage coefficients K_1 and K_2 in series. The IUH for this model is

$$u(t) = \frac{1}{K_2 - K_1} \int_0^{t' \leq T} [e^{-(t-\tau)/K_2} - e^{-(t-\tau)/K_1}] w(\tau) \, d\tau \qquad (14\text{-}46)$$

The linear channel is used to produce a time-area diagram for the whole drainage basin with variable areal distribution of the instantaneous effective rainfall; that is, i_j in Eq. (14-38) is not a constant. In applying this model to actual data, K_1 is assumed to have a constant average value of about 0.25 and the time-area diagram to be one of a number of basic geometric forms such as a rectangle, triangle, trapezoid, and sine curve, instead of the actual diagrams.

Diskin proposed a model which consists of two branches of linear reservoirs in parallel, one branch consisting of n_1 identical linear reservoirs of storage coefficient K_1 in series and the other of n_2 identical linear reservoirs of storage coefficient K_2 in series. The input to the first branch is α and to the second branch is β. For a total unit input, $\alpha + \beta = 1$. The lag time of the centroid of the IUH can be shown to be $t_L = \alpha n_1 K_1 + \beta n_2 K_2$. The equation of IUH of this model can be derived from Eq. (14-40) as

$$u(t) = \frac{\alpha}{k_1(n_1 - 1)!} \left(\frac{t}{K_1}\right)^{n_1-1} e^{-t/K_1} + \frac{\beta}{K_2(n_2 - 1)!} \left(\frac{t}{K_2}\right)^{n_2-1} e^{-t/K_2} \qquad (14\text{-}47)$$

If an instantaneous unit input is subtracted from an IUH at its centroid, the resulting function is called a *special IUH*. The Laplace transform of this special IUH is called the *special transfer function of IUH*, whose maximum ordinate is denoted by X_2 and ordinate at $pt_L = 5$ by X_3, where p is the variable of the Laplace transform. Diskin made a comparison of his model with the models of Nash and Singh as shown in Fig. 14-15 in which $X_5 = X_2/X_3$ is plotted against X_2. It can be shown that Diskin's and Singh's models cover a broader region than the curve representing the locus of the Nash models for different numbers of reservoirs and are therefore more flexible to fit any given hydrograph data on such a plane.

Kulandaiswamy [20] considered the rainfall and runoff relationship by system analysis and proposed a general equation of storage for nonlinear reservoir as follows:

$$S = \sum_{n=0}^{N} a_n(Q,I) \frac{d^n Q}{dt^n} + \sum_{m=0}^{M} b_m(Q,I) \frac{d^m I}{dt^m} \qquad (14\text{-}48)$$

By combining this equation with the continuity equation, the resulting general differential equation can be shown to be that of many mechanical and electrical dynamic systems. As a simplification and approximation, the coefficients are assumed to be functions of the average values $\bar{Q}$ and $\bar{I}$ instead of Q and I. Substituting the above modified equation in the continuity equation [Eq. (14-29)] and dropping the insignificant terms in the differential equation after being tested by actual data, the resulting differential equation proposed for analysis is

$$a_2 \frac{d^3Q}{dt^3} + a_1 \frac{d^2Q}{dt^2} + a_0 \frac{dQ}{dt} + Q = -b_1 \frac{d^2I}{dt^2} - b_0 \frac{dI}{dt} + 1 \qquad (14\text{-}49)$$

The outflow by this equation can be written as

$$Q(t) = \frac{-b_1 D^2 - b_0 D + 1}{a_2 D^3 + a_1 D^2 + a_0 D + 1} I(t) \qquad (14\text{-}50)$$

where D is the differential operator d/dt. To compensate for the terms being dropped from the general equation, which would take the translation effect into consideration,

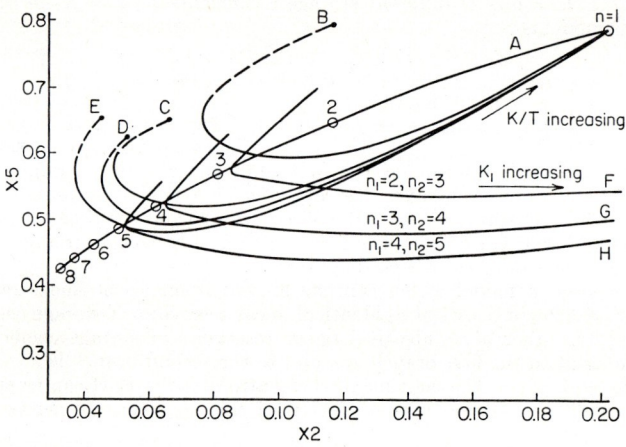

FIG. 14-15. Comparison of some conceptual models. Nash's model: n reservoirs in series (A). Singh's model with $K_2 = K$ and $K_1 = 0$: rectangular (B), parabolic (C), isosceles (D), and end-peaked triangular (E) time-area diagrams routed through a reservoir. Diskin's model: n_1 reservoirs in parallel with n_2 reservoirs having $\alpha = 0.1$ and K_1 varying from 0.1 to about 2.0.

it is desirable to replace $I(t)$ by $I(t - \tau_0)$ where τ_0 is the translation time which is equal to the difference in time between the beginning of the effective rainfall and direct runoff and may be determined from the actual data. When $I(t)$ or $I(t - \tau_0)$ is a unit instantaneous input or $\delta(t - \tau_0)$, Eq. (14-50) represents the function of an IUH.

Equation (14-50) is a polynomial of third degree and has three roots. Assuming that the system is stable, the following four cases describing different mathematical models are possible, depending on the nature of the roots.

Case 1. When all the roots are real and unequal, Eq. (14-50) reduces to

$$Q(t) = \left[\frac{C_1}{1 + K_1 D} + \frac{C_2}{1 + K_2 D} + \frac{C_3}{1 + K_3 D} \right] I(t - \tau_0) \qquad (14\text{-}51)$$

where C_1, C_2, C_3 are constants in terms of a_2, a_1, a_0, b_1, and b_0.

Case 2. When all the roots are real and two of them are equal, Eq. (14-50) reduces to

$$Q(t) = \left[\frac{C_1}{1 + K_1 D} + \frac{C_2}{1 + K_2 D} + \frac{C_3}{(1 + K_2 D)^2} \right] I(t - \tau_0) \qquad (14\text{-}52)$$

Case 3. When all the roots are real and equal, Eq. (14-50) reduces to

$$Q(t) = \left[\frac{C_1}{1 + K_1D} + \frac{C_2}{(1 + K_1D)^2} + \frac{C_3}{(1 + K_1D)^3} \right] I(t - \tau_0) \qquad (14\text{-}53)$$

Case 4. When one root is real and two are complex conjugates, Eq. (14-50) reduces to

$$Q(t) = \left[\frac{C_1}{1 + ZD} + \frac{\bar{C}_1}{1 + \bar{Z}D} + \frac{C_2}{1 + KD} \right] I(t - \tau_0) \qquad (14\text{-}54)$$

where C_1, $\bar{C}_1$, Z, and $\bar{Z}$ are complex numbers in terms of a_2, a_1, a_0, b_1, and b_0.

The above cases can be represented by block diagrams as in Fig. 14-16. Each block, or so-called *operational block*, represents the mathematical model of a linear

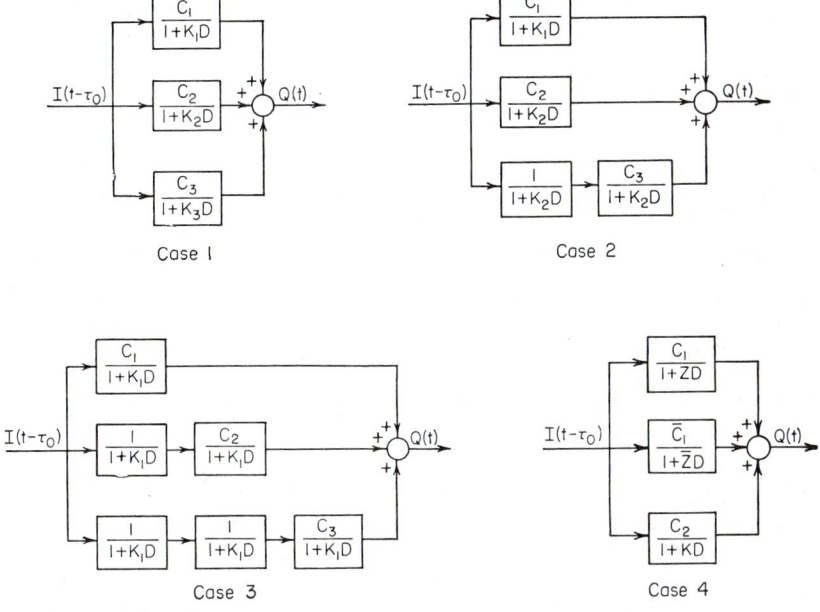

FIG. 14-16. Block diagrams of Kulandaiswamy's models.

reservoir. All blocks are connected in various ways, in series and/or in parallel form, to represent different cases. It should be noted that these mathematical models are slightly different from the other conceptual models described above. Each block does not exactly simulate a physical linear reservoir because, in system analysis, the input and output to an operational block can be made unequal. In Case 1, for example, three functions of equal magnitude of $I(t - \tau_0)$ are inputs to three blocks in parallel. The sum of these outputs from the blocks can be automatically adjusted to equal to the total input instead of three times input; this sum should be equal to the outflow $Q(t)$ which, by law of continuity, must be equal to $I(t - \tau_0)$. The circle, or so-called *junction*, in the block diagrams represents the addition of the block outputs to produce the outflow $Q(t)$, and the positive or negative sign indicates that the inputs to the junction must be multiplied by the appropriate sign before being added together.

The IUH models described above are physical conceptual models as well as mathematical conceptual models. They can be solved and fitted to actual hydrologic data

Table 14-5. Characteristics of Conceptual Models for IUH

Model	Date	Simulation	Parameters	Consideration of nonlinear effects
Zoch [24]	1934	A linear channel in series with a linear reservoir	K, T	No
Clark [63]	1945	A linear channel in series with a linear reservoir	K, T	No
O'Kelly [66]	1955	A linear channel in series with a linear reservoir	K, T	No
Sugawara and Maruyama [37]	1956	Linear reservoirs in series and/or in parallel	K's	No
Nash [67]	1957	Identical linear reservoirs in series	K, n	No
Dooge [68]	1959	A linear channel in series with a linear reservoir for each of the subareas in series	K, n, T	No
Rose et al. [87]	1961	5 linear reservoirs in series	K's	No
Singh [71]	1961	A linear channel and 2 linear reservoirs in series	K_1, K_2, T	Yes
Diskin [74]	1964	2 series of reservoirs in parallel with reservoirs identical in each series	$\alpha, \beta_1, K_1, K_2, n_1, n_2$	Yes
Kulandaiswamy [20]	1964	Linear reservoirs in series and/or in parallel	$a_0, a_1, a_2, b_0, b_1, \tau_0$	Yes

for the purpose of hydrograph analysis. The computations can be simplified by use of electronic computers. Table 14-5 is a summarized list showing the characteristics of the models.

E. Nonlinearity of Runoff Distribution

The conventional procedure of unit-hydrograph analysis uses the principle of superposition by which the direct runoff due to effective rainfall at successive time intervals can be added. Sherman [23] voiced his belief that this is the summation process of nature, and his contemporaries accepted this view without qualifications. Later investigators began to doubt the validity of the principle of superposition as they found that the unit hydrograph so obtained varied with hydrologic conditions. In 1952, Paynter [64] mentioned the possibility of nonlinear effects in flood routing and suggested that the causes may be due to the seasonal variation and magnitude of floods. He indicated that the IUH, which is mathematically the derivative of the admittance function (Subsec. III-F-1), is not an invariable curve but one of the family of curves which vary according to the season and the magnitude of flow being considered.

In 1960, Minshall's work [88] was discussed by Amorocho [89] who demonstrated that the nonlinearity of the rainfall-runoff process may be explained by the variation of the peak ordinate and the time to peak of the unit hydrographs derived from small watersheds up to 290 acres. As a discussion, Amorocho [89] showed that the unit hydrographs obtained by Minshall can fit in the two-parameter equation proposed by Edson [31] and Nash [67], and he derived relationships between the parameters K and n and the rainfall intensity.

In 1961, Amorocho and Orlob [90, 91] proposed a nonlinear analysis of hydrologic systems by means of a generalized functional series originally introduced by Wiener [92] for mathematical analysis of nonlinear systems. This functional series may be viewed as a conceptual model containing a number of operating systems in parallel, all receiving an input function and producing output functions. The output from the first system is obtained by an ordinary convolution integral containing the input and a fixed kernel function. The output from the second system is obtained by a generalized convolution integral containing the input, and a two-dimensional kernel function. The outputs from other systems are obtained in a similar manner. The final output is the sum of all outputs from the systems in parallel. Amorocho and Orlob made a series of experiments on a small, gravel-covered, rectangular model basin and analyzed the results using the above principle.

In 1962, Singh [71] used his model and the models by Nash and Dooge (Subsec. III-G-2) for the study of nonlinearity of the hydrologic records which contain some 24 storms over 6 drainage basins with areas ranging from about 0.5 to 875 sq mi. The parameters of these models were found to vary with an equivalent instantaneous effective rainfall R_e, which leads to the conclusion of nonlinearity. Singh found that the parameters in his model vary systematically with R_e in such a way that both T and K_2 decrease with an increase in R_e while K_1 is kept constant. He also noted an increase in R_e with a decrease in time to peak and an increase in the peak ordinate of the IUH.

In 1964, Diskin [74] completed a study of the nonlinearity of runoff by Laplace transforms (Subsec. III-G-2). Theoretically speaking, perfect linear hydrologic records from a watershed would all be represented by a single point on the plane of Fig. 14-15. Analyzing records of rainfall and runoff for 14 drainage basins with sizes ranging from 30 to 1,420 sq mi, he found that the parameters of the special transfer function derived from the data for any one drainage basin appear to vary systematically. This indicates that the shape of the IUH varies systematically for the storms of the drainage basin. The scatter of the plotted data indicates that there may be a random component that is superimposed on this systematic variation.

At the time of Diskin's study, Kulandaiswamy [20] made an independent investigation of the nonlinearity of runoff by means of mathematical conceptual models in system analysis (Subsec. II-G-2). In this investigation, 6 drainage basins with sizes ranging from about 7 to 1,141 sq mi were analyzed. The results showed that the coefficients a_0, a_1, and a_2 in Eq. (14-50) decrease exponentially with increase in the peak discharge of direct runoff, thus indicating a nonlinear system, while the coefficients b_0 and b_1 in Eq. (14-50) do not vary with any definable trend.

Mathematically speaking, a drainage basin system is linear if the differential equation of the input and output relation [Eq. (14-49)] is linear, and it is nonlinear if the differential equation is nonlinear. For a linear system, the convolution integral expressed by Eq. (14-17) applies; for a nonlinear system, the integral does not apply. For practical purposes, the linearity of the rainfall-runoff relationship may be defined on the basis of the IUH. The rainfall-runoff relationship of a drainage basin is said to be linear if its IUH has a unique unchangeable shape. However, the studies mentioned above indicate that the IUH for most drainage basins varies from storm to storm and from season to season for a given drainage basin, thus indicating a nonlinearity with respect to storm and seasonal variations. There may be other factors that would effect nonlinearity, but they have not yet been fully investigated primarily because their effects were not so outstanding in problems that had been investigated.

IV. SPACE DISTRIBUTION OF RUNOFF

A. Overland Flow

The overland flow was first investigated by Horton [93], producing a semiempirical formula as in Eq. (20-14). This formula was devised for turbulent flow with high discharge. The use of this formula will produce a hydrograph for the overland flow due to a uniform rate of effective rainfall lasting indefinitely. The hydrograph of a finite duration can be derived by the principle of superposition or the S-hydrograph method (Subsec. III-B-2).

If the rational formula is used, the time of concentration t in min for overland flow may be estimated by the following formula proposed by Kerby [94]:

$$t_c^{2.14} = \frac{2}{3} \frac{Ln}{\sqrt{S}} \qquad (14\text{-}55)$$

where L is the length of flow in ft, S is the slope of the surface, and n is the roughness coefficient recommended for use in Eq. (20-14).

In open-channel hydraulics, overland flow is considered as *sheet flow*, a spatially varied unsteady flow. Theoretical investigation was first made by Keulegan [95] and followed with experimental analysis by Izzard [96]. Other investigations were also made by Richey [97], Behlke [98], Woo and Brater [99], and Yu and McNown [100]. For practical hydrologic analysis, Izzard's method has been found to be satisfactory. Izzard found that the form of the rising segment of a hydrograph can be

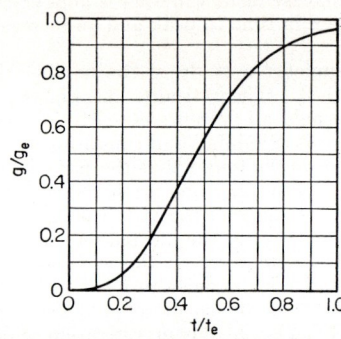

Fig. 14-17. Dimensionless hydrograph of overland flow. (*After Izzard* [96]).

represented by a single dimensionless curve as shown in Fig. 14-17. The notations involved are:

q = discharge of the overland flow, in cfs per ft of width, at time t since the effective rainfall began.

q_e = discharge of the overland flow, in cfs per ft of width, at equilibrium. At the equilibrium condition, the intensity of effective rainfall is equal to the outflow discharge. If I is the effective rainfall intensity in in. per hr and L is the length of the overland flow in ft, then

$$q_e = \frac{IL}{43{,}200} \qquad (14\text{-}56)$$

It should be noted that the equilibrium condition is reached asymptotically.

t = time in min since the effective rainfall began.

t_e = time of equilibrium in min. Since the equilibrium condition is approached asymptotically, the time t_e must be determined arbitrarily as the time when q reaches $0.97 q_e$, or $q/q_e = 0.97$. It was found empirically that the volume of water, represented by D_e in ft³, in the overland flow on a strip of unit width at equilibrium (the area above the curve in Fig. 14-17) is substantially equal to the volume of water that has been discharged in the time required to reach equilibrium (the area below the curve). Thus, the equilibrium time is expressed by

$$t_e = \frac{2D_e}{60 q_e} \qquad (14\text{-}57)$$

D = detention in ft³; that is, the volume of water in overland flow on a strip of unit width at the time t since the effective rainfall began.

SPACE DISTRIBUTION OF RUNOFF   14-37

D_e = detention in ft³ at equilibrium. It was found empirically that this could be expressed in general by

$$D_e = KLq_e^{1/3} \tag{14-58}$$

Actually the exponent was found to vary from about 0.2 for very smooth pavement to nearly 0.4 for turf. The value of K depends on effective rainfall intensity I, the slope of surface S, and a roughness factor c, that is,

$$K = \frac{0.0007I + c}{S^{1/3}} \tag{14-59}$$

The equation was developed for slope not steeper than about 0.04. The value of c was evaluated as follows:

Type of surface	Value of c
Very smooth asphalt pavement	0.0070
Tar and sand pavement	0.0075
Crushed-slate roofing paper	0.0082
Concrete pavement, normal condition	0.0120
Tar-and-gravel pavement	0.0170
Closely clipped sod	0.0460
Dense bluegrass turf	0.0600

When rain ceases, the runoff decreases. The time t_r from the beginning of the recession segment of the hydrograph to the point where $q/q_e = r$ is

$$t_r = \frac{D_0 F(r)}{60 q_e} \tag{14-60}$$

where D_0 is the detention corresponding to D_e after the cease of rainfall, which is the detention when $I = 0$; and where

$$F(r) = 0.5(r^{-2/3} - 1) \tag{14-61}$$

Equation (14-60) was derived mathematically from the finding that detention on the recession curve is proportional to the one-third power of discharge; that is, $D/D_0 = (q/q_e)^{1/3} = r^{1/3}$.

Using the dimensionless hydrograph and the above equations, it is possible to construct a hydrograph for overland flow due to an effective rainfall of given intensity and duration. It is understood that the dimensionless hydrograph and the equations given above were based on experimental conditions under which the flow was laminar at all times. According to Izzard, the method should be limited to cases where the product of I and L is less than 500. However, later applications indicated that the method was satisfactory to turbulent flow as well.

From 1948 to 1954, the Los Angeles District of the U.S. Army Corps of Engineers conducted an extensive experiment in the airfield at Santa Monica, California, in order to study the relationship between simulated rainfall and runoff on paved surfaces as well as on simulated turfed surfaces. The experiments were made on various combinations of rainfall intensity (¼ to 10 in. per hr), slope (0.5, 1, and 2 per cent), length of flow on 500-ft-long and 3-ft-wide surfaces, and surface roughness (concrete, natural grass, and simulated turf). The experimental data, including 601 discharge hydrographs and 153 depth hydrographs, were later analyzed by Yu and McNown [100] in 1963. In the analysis a numerical procedure for solving the equations of continuity and motion was used, using the roughness coefficients determined from the experimental data. It was found that the computed hydrographs agreed with the measured data satisfactorily for practical purposes.

B. Streamflow

The mechanics of flow distribution along stream channels is a specialized subject in open-channel hydraulics [101, 102] and fluvial hydraulics [103], and hence it is beyond the scope of this section. However, some discussions are given in Sections 4-II, 7, 15, 17-II, and 25-II.

In general, the discharge in a river tends to increase in the downstream direction because of the progressively increasing drainage area. However, the discharge in some streams, particularly those in arid areas, decreases downstream. According to the principle of *hydraulic geometry* as studied by Leopold and Maddock [104], when discharges are of equal frequency at different points along a river, that is, equalled or exceeded the same per cent of time, the velocity as well as the width and depth of flow increases with discharge downstream.

In 1945, L'vovich [105] made a study on world distribution of annual runoff. The results of this study shown in Table 14-6 were converted to English units by Langbein and others [106]. From this information, it may be noted that South and North

Table 14-6. World Distribution of Runoff*

Continent (or other area)	Atlantic slope		Pacific slope		Regions of interior drainage		Total land area	
	Area, thousands of sq mi	Runoff, in.	Area, thousands of sq mi	Runoff, in.	Area, thousands of sq mi	Runoff, in.	Area, thousands of sq mi	Runoff, in.
Europe (including Iceland)...	3,073	11.7	......	....	661	4.3	3,734	10.3
Asia, (including Japanese and Philippine Islands)........	4,626	6.4	6,422	11.8	5,273	0.66	16,321	6.7
Africa (including Madagascar).	5,110	14.0	2,109	8.6	4,291	0.54	11,510	8.0
Australia (including Tasmania and New Zealand).........	......	....	1,634	5.5	1,441	0.24	3,075	3.0
South America.............	6,041	18.7	519	17.5	381	2.6	6,941	17.7
North America (including West Indies and Central America).................	5,657	10.8	1,914	19.1	322	0.43	7,893	12.4
Greenland and Canadian Archipelago...............	1,499	7.1	......	....	......	....	1,499	7.1
Malayan Archipelago........	......	....	1,012	63.0	......	....	1,012	63.0
Total or average..........	26,006	12.4	13,610	15.5	12,369	0.82	51,985	10.5

* After L'vovich [105].

America are more favored with water than any of the other continents; also, that the area tributary to the Atlantic Ocean is roughly double that tributary to the Pacific, though the total runoff is only one and a half times as large. Areas of internal drainage total 24 per cent of the land surface. They are dominantly arid and are located largely in Asia, Africa, and Australia. The distribution of average annual runoff in the United States is shown in Fig. 22-4. The distribution of monthly runoff at selected stations in the United States is shown in Fig. 14-18.

Total annual discharge in the rivers of the world is about 8,200 cu mi, representing about one-third of the annual precipitation. Table 14-7 lists some large rivers in the world with their average annual discharges, drainage areas, and lengths [108]. Such information was first investigated by Dr. H. B. Guppy, a British naturalist, in 1880, and has been revised from time to time as new information was obtained. In the United States, similar information for 26 large rivers is given in Table 14-8. The locations of the large rivers in the United States with their average discharges are shown in Fig. 14-19. Of these rivers, only 10 are independent; that is, the rivers discharge directly into oceans, but they carry over 75 per cent of the total 1,800,000 cfs drained

Fig. 14-18. Distribution of monthly runoff in the United States (*Langbein and Wells* [107]).

14-39

Table 14-7. Large Rivers in the World (Subject to Revision)

Name	Average annual discharge, cfs	Drainage area, sq mi	Length, mi
Amazon (S. Am.)	7,200,000	2,772,000	3,900
LaPlata-Paraná (S. Am.)	2,800,000	1,198,000	2,450
Congo (Africa)	2,000,000	1,425,000	2,900
Yangtze (Asia)	770,000	750,000	3,100
Ganges-Brahmaputra (Asia)	707,000	793,000	1,800
Mississippi-Missouri (N. Am.)	620,000	1,243,700	3,892
Yenisei (Asia)	610,000	1,000,000	3,550
Mekong (Asia)	600,000	350,000	2,600
Orinoco (S. Am.)	600,000	570,000	1,600
Mackenzie (N. Am.)	450,000	682,000	2,525
Nile (Africa)	420,000	1,293,000	4,053
St. Lawrence (N. Am.)	400,000	565,000	2,150
Volga (Europe)	350,000	592,000	2,325
Lena (Asia)	325,000	1,169,000	2,860
Ob (Asia)	....	1,000,000	2,800
Danube (Europe)	315,000	347,000	1,725
Zambesi (Africa)	....	513,000	2,200
Indus (Asia)	300,000	372,000	1,700
Amur (Asia)	....	787,000	2,900
Niger (Africa)	....	584,000	2,600
Columbia (N. Am.)	235,000	258,000	1,214
Yukon (N. Am.)	150,000	330,000	2,300
Huang (Asia)	116,000	400,000	2,700
São Francisco (S. Am.)	....	252,000	1,811
Euphrates (Asia)	....	430,000	1,700
Murray-Darling (Australia)	13,000	414,000	2,345

Table 14-8. Large Rivers in the United States*

Name	Average annual discharge, cfs	Drainage area, sq mi	Length, mi
Mississippi	620,000	1,243,700	3,892
St. Lawrence	400,000	565,000	2,150
Ohio (T)	255,000	203,900	1,306
Columbia	235,000	258,200	1,214
Mississippi (T)	91,300	171,600	1,170
Missouri (T)	70,100	529,400	2,714
Tennessee (TT)	63,700	40,600	900
Mobile	59,000	42,300	758
Red (T)	57,300	91,400	1,300
Arkansas (T)	45,200	160,500	1,450
Snake (T)	44,500	109,000	1,038
Susquehanna	35,800	27,570	444
Alabama (T)	31,600	22,600	720
White (T)	31,000	28,000	690
Willamette (T)	30,700	11,250	270
Wabash (TT)	30,400	33,150	475
Cumberland (TT)	27,800	18,080	720
Illinois (T)	27,400	27,900	420
Tombigbee (T)	27,000	19,500	525
Sacramento	26,000	27,100	382
Apalachicola	25,000	19,500	500
Pend Oreille (T)	24,600	25,820	490
Colorado	23,000	246,000	1,450
Hudson	21,500	13,370	306
Allegheny (TT)	19,200	11,700	325
Delaware	19,000	12,300	390

* From Ref. 109.
T, first-order tributary; TT, second-order tributary.

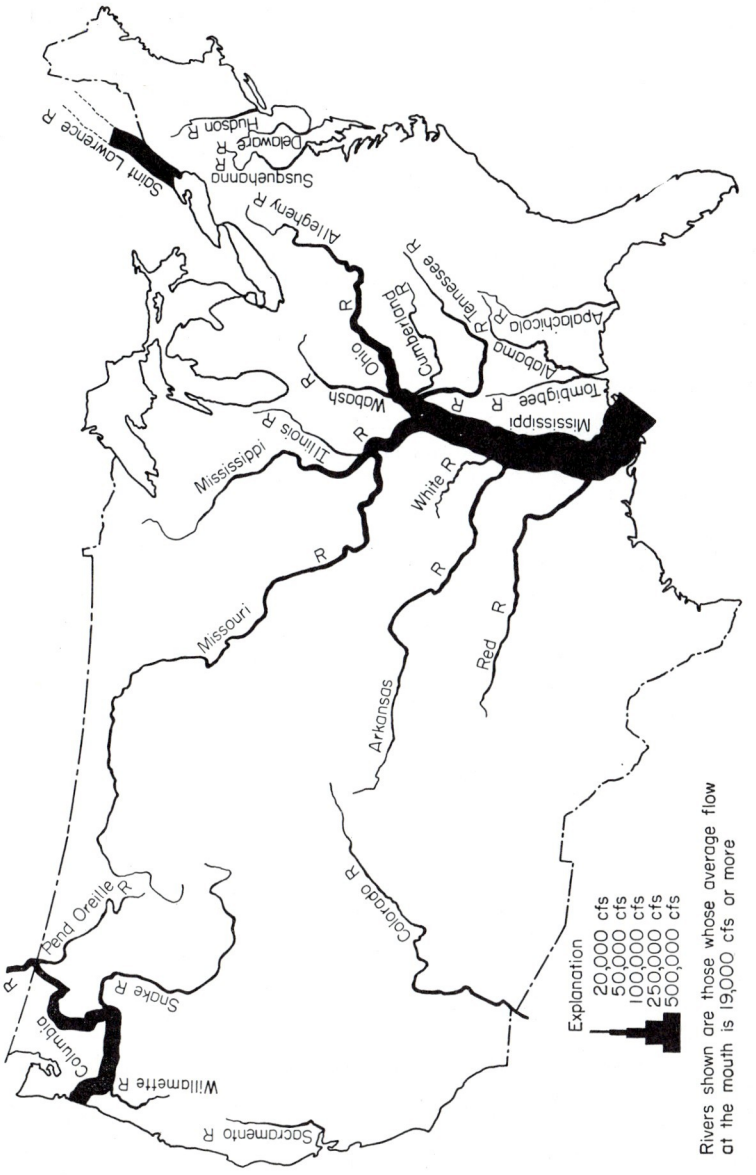

Fig. 14-19. Large rivers in the United States (*U.S. Geological Survey* [109]).

from the continental United States in the river system. Many rivers, mainly in the Southwest, have large drainage areas but relatively low average runoff. The following rivers, arranged in order of drainage area in square miles (shown in parentheses), are believed to discharge less than 10,000 cfs: Rio Grande (171,585), Platte (90,000), Kansas (61,300), Gila (58,100), Brazos (44,500), Green in Utah and Wyoming (44,400), Colorado in Texas (41,500), Pecos (38,300), Canadian (29,700), and Colorado above Green (26,500, giving largest flow in this group).

V. VARIABILITY OF RUNOFF

Variability of runoff can be best studied by statistical analysis, which is discussed in Sections 8-I and 25-I. In this section, only several procedures of analysis which are particularly suitable for the study of runoff variability for the design of storage reservoirs are discussed.

A. Flow-duration Curve

When the values of a hydrologic event are arranged in the order of their descending magnitude, the per cent of time for each magnitude to be equalled or exceeded can be computed. A plotting of the magnitudes as ordinates against the corresponding per cents of time as abscissas results in a so-called *duration curve*. If the magnitude to be plotted is the discharge of a stream, the duration curve is known as a *flow-duration curve* (Fig. 14-20). If the magnitude is the potential power contained in the streamflow, the duration curve is known as the *power-duration curve* which is a very useful

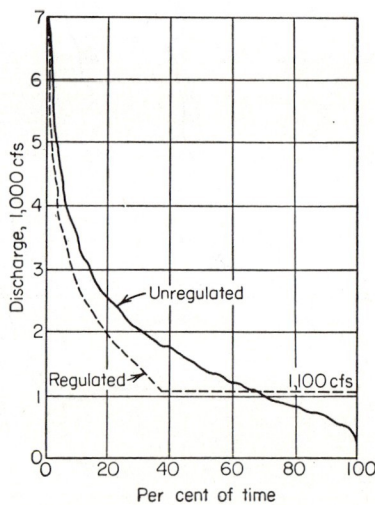

Fig. 14-20. Unregulated and regulated flow-duration curves.

tool in the analysis for the development of water power [110]. In a statistical sense, the duration curve is a cumulative frequency curve of a continuous time series (Sec. 8-I), displaying the relative duration of various magnitudes. The slope of the duration curve depends greatly on the observation period used in the analysis. The mean daily data will yield a much steeper curve than annual data as the latter tend to group and smooth off the variations in the shorter-interval daily data.

Figure 14-20 shows a typical flow-duration curve. Such a curve may be considered to represent the hydrograph of the average year with its flows arranged in order of magnitude. For example, the flow in the average year to be equalled or exceeded

20 per cent of the time is 2,500 cfs. If the flow is to be regulated by a maximum draft of 1,100 cfs all the time, the regulated flows computed on the basis of a given storage function can be used to construct a regulated flow-duration curve as shown in Fig. 14-20. If there were no evaporation losses or leakage of water, the area below the unregulated curve and above the regulated curve should be equal to the area below the regulated curve and above the unregulated curve. It can be seen that there will be excess water during about 37 per cent of the time. In case of an analysis for water-power development, the regulation will produce a potential supply of *firm power* due to 1,100 cfs and the *secondary power* will be supplied by the excess water. In case of a run-of-river power plant with no storage facilities, the unregulated flow-duration curve should be used, and the firm power is customarily assumed on the basis of the flow available 90 to 97 per cent of the time. For many streams, the flow-duration curve may show an abrupt drop somewhere near 97 per cent of time. The per cent of time at the drop may be taken as the basis for an estimate of the firm power.

Table 14-9. Corrections to Variability Index

Terrain	Description	Subtract from 0.60	Add to 0.60
Bedrock..................	Impervious	....	0.1–0.2
	Average	0.0–0.1	0.0–0.1
	Pervious	0.1–0.2	
Soil, deep covering........	Impervious	....	0.1–0.2
	Average	0.0–0.1	0.0–0.1
	Pervious	0.1–0.2	
	Very pervious	0.2–0.3	
	Sand hills	0.3–0.4	
Relief....................	Flat or gently rolling	0.0	0.0
	High in impervious material	0.05–0.1	
	High in average material	0.1–0.2	
	High in pervious material	0.2–0.3	
Moraines.................	Impervious	0.0	0.0
	Average	0.05–0.1	
	Pervious	0.1–0.2	
	Pervious with many kettle holes	0.2–0.4	
Lakes and swamps........	Small	0.05–0.1	
	Moderate	0.1–1.5	
	Extensive	1.5–2.5	

For comparative purposes, flow-duration curves may be plotted in terms of ratios to the average discharge rather than in actual discharge units. The curves may also be constructed on a lognormal probability paper with the flow on logarithmic scale and per cent of time on normal-probability scale. The curve so plotted tends to appear as a straight line, particularly in the middle portion. Lane and Lei [111] made such lognormal probability plottings of the flow records of a large number of streams in the eastern part of the United States. From a flow-duration curve thus plotted, the values of discharge at 10 per cent intervals from 5 to 95 per cent of the time were read off. The logarithms of these discharges were then found and their standard deviation was computed and defined as the *variability index* of the flow which is essentially the standard deviation of the flow statistics [112]. Assuming the index for an average river as 0.60, the corrections due to various basin characteristics were found approximately as listed in Table 14-9. The approximate shape of a duration curve can be obtained by drawing a straight line on logarithmic probability paper with a slope such that the ratio of the discharge exceeded 15.87 per cent of the time to the discharge exceeded 50 per cent of the time is equal to the antilogarithm of the variability index selected with the aid of Table 14-9. It should be noted that the chronological sequence of events is completely masked in a duration curve. In other words, the variability index determined from a flow-duration curve varies with the length of the record owing to the effect of persistence in the hydrologic data (Sec. 8-I).

The variability-index method may be used to construct approximately the flow-duration curves on a basin where no flow records exist. Another approximate method to construct such a curve is to interpolate it from available flow-duration curves on adjacent basins if these basins can be believed to be under similar hydrologic conditions.

The shape of a flow-duration curve may change with the length of record. This property can be used to extend the flow information on a given stream for which short-term records are available and for which simultaneous and long-term records are available on at least one adjacent stream which is believed to be under similar hydrologic conditions. By comparing the flow-duration curves constructed of the short-term record of the given stream and of the corresponding short-period record on the adjacent stream, the flow-duration curve for the long-period record of the adjacent stream can be proportionally adjusted to produce an approximate flow-duration curve for the given stream for the corresponding long-period record [46, pp. 278–283].

Similarly, if a hydrograph is given at a station A, the corresponding hydrograph at an adjacent station B can be estimated by comparing the flow-duration curves of a same period of record at both stations. If there is no flow record at station B, the flow-duration curve at this station can be estimated by the above-mentioned method. From the flow-duration curve for station A, the per cent of time for a discharge on the given hydrograph is first found. For the same per cent of time on the flow-duration curve of station B, the corresponding discharge is estimated. Repeating this procedure for a number of discharges obtained from the given hydrograph for station A, an estimated hydrograph for station B for the corresponding storm can be thus constructed.

The flow-duration curve is believed to have been first used by the American engineers Clemens Herschel and John R. Freeman from early 1880 to 1890 [113]. It is most frequently used for determining water-supply potentials in planning and design of the water-resources projects, particularly the hydropower plants. For detailed discussions on such uses, see Refs. 110, 114–116. For flow-duration analysis of many streamflows, see Refs. 113, 117–122.

B. Flow-mass Curve

A graph of the cumulative values of a hydrologic quantity, generally as ordinate, plotted against time or date as abscissa is known as a *mass curve*. When the discharge of runoff is taken as the hydrologic quantity the mass curve may also be called a *flow-mass curve*. It is the integral curve of the hydrograph which expresses the area under the hydrograph from one time to another. If the flow is the daily discharge expressed in cfs, the area under the hydrograph is the volume of runoff in cfs/day; thus it is the ordinate of the flow-mass curve. As the flow-mass curve represents a summation of flow volume, sometimes it is also called an S-*curve*. It should be noted that the S-curve is different from an S-hydrograph (Subsec. III-B-2) in that the ordinate of the former represents the runoff volume whereas that of the latter represents the runoff rate. Mathematically the flow-mass curve is expressed by

$$V = \int_{t_1}^{t_2} Q_t \, dt \approx \sum_{t=t_1}^{t_2} Q_t \, \Delta t \qquad (14\text{-}62)$$

where V is the volume of runoff, Q_t is the discharge as a function of time t or the ordinate of a hydrograph, and the integration or summation limits are from time t_1 to t_2.

The flow-mass curve is believed to have been first suggested in 1882 by W. Rippl [123], an Austrian engineer, hence its use is also known as the *Rippl Method*. The curve has many useful applications in the design of a storage reservoir, such as the determination of reservoir capacity, operations procedure, and flood routing.

Figure 14-21 shows a typical flow-mass curve. The difference between the ordinates of any two points on the curve is the summation of the flows during the intervening period of time. Thus, if the two points are connected by a straight line, the slope of the line will equal the average flow during that period since it equals the total discharge divided by the corresponding total time interval. In the figure, the slope of the

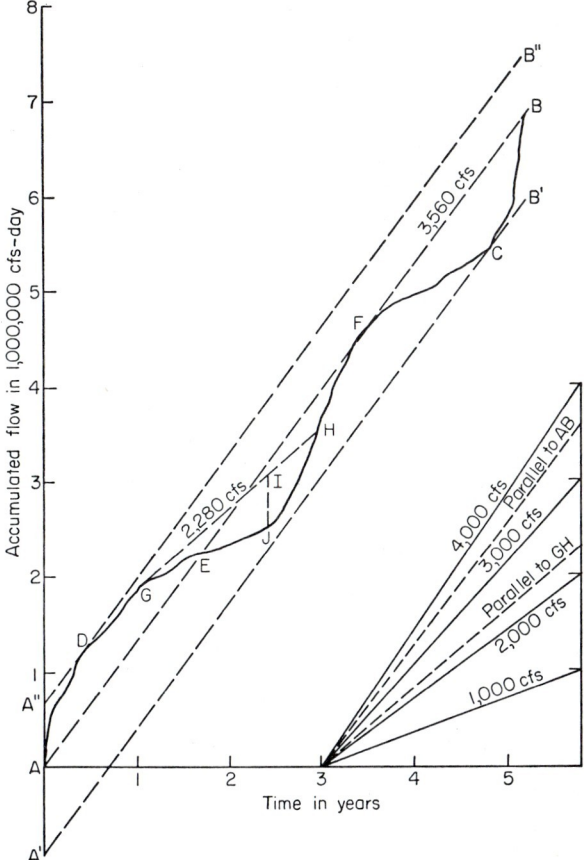

FIG. 14-21. Study of runoff variability by flow-mass curves.

straight line AB joining the end points of the mass curve represents the average discharge over the total period of the plotted record. Two straight lines $A'B'$ and $A''B''$ are drawn parallel to AB and tangent to the mass curve at the lowest tangent point C and the highest tangent point D, respectively. The vertical intercept between these two straight lines represents the storage volume required to permit continuous release of water for the entire period. A reservoir having this capacity would be empty at C and full at D, assuming that it contained a volume equal to AA' at the beginning of the period. If this reservoir were empty in the very beginning, it would be empty again at point E and run empty until a certain point near J, where the tangent is parallel to and at a vertical distance, say d, below AB. It would be empty again at a point between F and C, which is at a vertical distance d below AB. If the reservoir were full in the very beginning, it would spill between A and D and become empty at C.

The *rate* of flow at any point on the mass curve is indicated by the slope of the curve. The rate of *draft*, or *demand*, is the rate at which water is required for use in a power plant, for water supply, or for other requirements. A straight line having a slope equal to this rate is called a *draft line*, or *use line*. Where the mass curve is steeper than the draft line, the natural streamflow is at a higher rate of supply than the rate of draft, and, consequently, some of the streamflow is available for storage or it will spill if the reservoir is already filled. Where the mass curve has a flatter slope than the draft line, water will be withdrawn from storage in order to maintain the required rate of draft.

For a given rate of draft, the draft lines, such as GH in Fig. 14-21, are drawn tangent to the high points of the mass curve. The required storage is given by the maximum intercept, such as IJ, between such lines and the mass curve. On the other hand, if the storage is given, the possible maximum uniform rate of flow can be determined, by reversed procedure, from the maximum slope of all trial draft lines which are drawn tangent to the mass curve and have a maximum departure from it equal to the given storage.

The rate of draft need not necessarily be constant. In fact, the withdrawals from reservoirs in most projects, particularly in irrigation, are variable. In such cases, the draft line is not a straight line but a curve with its slope at different points on the curve determined by the variable rate of draft at the different times. Such a draft line is specifically called a *variable draft line* or *variable use line*, which is essentially a mass curve of the outflow. For an analysis of the problem, the variable draft line may be superimposed on the mass curve with the time scale matching. When the reservoir is assumed full at the beginning of the period under consideration, the draft line and the mass curve should coincide at the beginning time. If the reservoir is assumed empty at the beginning time, the draft line should start at a point vertically above the origin of the mass curve a distance equal to reservoir capacity. When the draft line is above the mass curve, there will be storage, and the maximum vertical intercept between the draft line and the mass curve be the storage capacity required. If the storage of the given reservoir is greater than this required capacity, there will be spilling of water; if it is less, the reservoir will go dry when the required draft becomes greater than the given capacity. A complete diagram of flow-mass curve and variable draft line will show periods and amounts of surplus and deficiency of water. This diagram, usually called a *regulation diagram*, is useful in studying regulation of flow by reservoirs.

From an analysis of storage using the flow-mass curve, the storages required may be plotted as abscissas against corresponding rates of draft as ordinates, resulting in a so-called *storage-draft curve*. This curve shows the relation between the rate of draft of water and the storage capacity required to maintain the draft during the driest season in the future as estimated from past records. Thus the curve enables an immediate determination of the possible dependable rate of draft for any assumed amount of storage in the design.

In practice the recorded flows should be adjusted for evaporation and leakage, which may be expected in a reservoir, before a mass-curve analysis. Thus, it is possible to exhibit a negative slope on the mass curve if the rate of evaporation and leakage exceeds the available rate of flow.

Instead of plotting a mass curve, the departure of the mass curve from normal, such as AB in Fig. 14-21, may be plotted against the time. In other words, the mass curve is plotted about a horizontal axis obtained by rotating the average slope line on the mass curve to the horizontal. Such a plotting is called a *residual mass curve*. The ordinates on the residual mass curve may thus be either positive or negative. The advantages of this plotting are to save additional space needed for plotting a continuously rising mass curve and to accentuate more clearly the peaks and valleys of the cumulative flow records.

C. Time-series Analysis

The mass-curve analysis or its modification is based on a finite period of record within which the sequence of occurrence of the events is assumed definite. In nature,

VARIABILITY OF RUNOFF

however, there is no guarantee that the same sequence will occur again. In fact, this chance is almost nil in the case of hydrologic events. Consequently, the answers obtained by the mass-curve may be deceiving in accuracy. Since future runoff involves probability which is known as stochastic, the stochastic characteristics must be considered in any improved method of analysis. In such methods the hydrologic events may be treated as a time series and the stochastic theory of probability can be applied (Secs. 8-I, 8-III, and 18).

1. Analysis of Ranges. The difference of the maximum and minimum values of a residual mass curve for a given period N is known as the *maximum range* for the period N. According to Hurst [124–127], the maximum range can be considered as (1) the maximum accumulated storage if there is no deficit in outflow (i.e., the outflow from reservoir is equal to the average discharge), (2) the maximum deficit if there is no storage, or (3) the sum of accumulated storage and deficit if there are both storage and deficit. If R is the range of a period of N years, Q is that constant rate of withdrawal, and P is the probability of the range R for given N and Q, the following function can be written:

$$R = f(Q,N,P) \tag{14-63}$$

This relationship can be developed by analyzing the distribution of ranges from actual hydrologic records and should be useful in planning, design, and operation of large reservoirs. An application of the range analysis to the runoff records of the Upper Colorado River at Lees Ferry, Arizona, has been made by Yevdjevich [128, 129]. By investigating the range distribution of many natural and economic phenomena (including runoff, rainfall, temperature, pressure, tree-ring growth, varves, sunspot numbers, and wheat prices), Hurst [124, 127] developed a formula for the determination of the reservoir storage:

$$R = s\left(\frac{N}{2}\right)^{0.73} \tag{14-64}$$

where R is the maximum range or the reservoir storage required to produce a steady discharge equal to the mean over a period of N years, on a stream whose sample standard deviation of annual discharge s is the estimated value of the standard deviation σ of the population.

Equation (14-64) was derived from the data on natural phenomena and was found to be different from the theoretical result based on random normally distributed time series. Leopold [130] believed that the difference was due to persistence in the natural data (Sec. 8-I). Langbein [131] studied the effect of persistence on storage by serial correlation and showed that the serial correlation increases the required storage.

For a random normally distributed time series, Hurst and others [132] derived the following theoretical mean maximum range assuming a binomial approximation for normal distribution:

$$R = 1.25\sigma \sqrt{N} \tag{14-65}$$

Feller [133] later showed that the above equation is for adjusted maximum range. The *adjusted maximum range* is the difference of maximum and minimum values of the residual mass curve, but with the mean changing for every successive period N. He gives the asymptotic form for the mean maximum range as

$$R = 1.596\sigma \sqrt{N} \tag{14-66}$$

the variance of the adjusted maximum range as

$$\sigma_R^2 = 0.0741\sigma^2 N \tag{14-67}$$

and the variance of the maximum range as

$$\sigma_R^2 = 0.2182\sigma^2 N \tag{14-68}$$

The above equations are for large values of N. For any value of N, Anis and Lloyd [134] give the mean maximum range as

$$R = \sqrt{\frac{2}{\pi}} \, \sigma \sum_{i=1}^{N-1} i^{-1/2} \qquad (14\text{-}69)$$

2. Queueing Theory. The hydrologic data series can be treated as a time series by considering it as a *queue*, or *waiting line* [135, 136]. The analogy between queues and storage of runoff was first recognized by Moran [137] and analyzed theoretically by him and others [138–143] to develop the so-called *theory of storage*. The *queueing* (also spelled as *queuing*) *theory* was later extended by Langbein [131] to a general solution of the determination of storage for regulating variable streamflow. The theory has also been applied to economic maximization of water-resources projects involving variable streamflow (see [144, 145] and Sec. 26-II).

A queue involves arriving items or customers that wait to be served at the facility which provides the service they seek. In a queueing system, the customers arrive at the system, wait for service, receive service, and then leave the system. According to the queueing theory, a queueing process can be described by the following factors:

(1) *Arrival rate*. This is generally expressed by a frequency or time distribution of the arrivals or the demands for service.

(2) *Queue discipline*. This is the rule for establishing priority of servicing among those on the queue. Ordinarily, the simplest rule is first come, first served, although special rules can be set up or observed.

(3) *Service function*. This is the rule defining the rate of servicing the items in the queue.

(4) *Attrition rate*. This is the rate of loss from the queues due to *balking* (i.e., customers' reluctance to join the queue any longer) or to *defection* or *reneging* (i.e., customers' leaving the queue because of unsatisfactory service).

(5) *Time of orientation* or *connection delay*. This is the time interval between the demand time and the time of the service engagement; it may be regarded as the time required to identify a free server.

In the case of runoff from a drainage basin or the streamflow through a reservoir, the basin or reservoir can be treated as a queueing system. Thus, the arrival rate is the rate of rainfall over the basin or the inflow to the reservoir. The queue discipline is insignificant on the drainage basin, whereas selective withdrawal of water in the reservoir due to temperature or turbidity requirement may be rigidly enforced. As a service function the physiographic characteristics of a basin control the runoff while the outlet structure of the reservoir regulates the outflow. On the basin or reservoir, the attrition rate is the rate of losing water because of evaporation or other forms of abstractions, and the time of orientation may refer to the time of concentration of the flow.

In the case of a reservoir, for example, the queueing theory can be used to determine the distributions of storage and outflow from a given distribution of inflow through the reservoir of a known storage function.

Considering the flow through a reservoir as a queue, the arrival rate is the frequency distribution function of the inflow P_I, and the service function is the inverse of the storage function represented by $Q = f^{-1}(S)$. For simplicity, no special requirements for withdrawal and no losses for water in the process are considered. Thus, there are no special queue disciplines and no attrition rate.

The storage of the reservoir may be divided into N portions of equal units so that the problem can be considered as a finite Markov chain (Sec. 8-IV). At time t, the storage S_t is taken as n units with $N \geq n \geq 0$. Since $S_t = n$, the outflow at time t is $Q_t = f^{-1}(n)$. At time $t - \Delta t$, the storage was assumed less than m units but more than $m - 1$ units (i.e., $m > S_{t-\Delta t} > m - 1$); if inflow during this interval exceeded outflow by a sufficient amount, then $S_t = n$. Taking an average value of $S_{t-\Delta} = (2m - 1)/2$, the outflow at time $t - \Delta t$ is $Q_{t-\Delta t} = f^{-1}[(2m - 1)/2]$.

By the equation of continuity of flow, the total inflow during the period from

$-\Delta t$ to t is

$$I(m,n) = \sum_{t-\Delta t}^{t} \Delta S + \sum_{t-\Delta t}^{t} Q\,\Delta t$$

$$= n - \frac{2m-1}{2} + \frac{1}{2}\left[f^{-1}(n) + f^{-1}\left(\frac{2m-1}{2}\right)\right] \quad (14\text{-}70)$$

From the inflow frequency distribution function, the probability that the inflow is less than $I(m,n)$ can be found to be $P_I(m,n)$. If $P(m > S_{t-\Delta t} > m-1)$ is the probability that the storage at the beginning of the period is less than m but more than $m-1$, the partial probability of the outflow for the combination of inflow and storage probabilities is $P_I(m,n)P(m > S_{t-\Delta t} > m-1)$. Thus, the total probability that the storage at the end of the period will be n or less is the sum of all partial probabilities, or

$$P(n > S_t > 0) = \sum_{m=1}^{N} P_I(m,n)P(m > S_{t-\Delta t} > m-1) \quad (14\text{-}71)$$

It may be noted that $P(n > S_t > n-1) = P(n > S_t > 0) - P(n-1 > S_t > 0)$. Thus,

$$P(n > S_t > n-1) = \sum_{m=1}^{N} P_I(m,n)P(m > S_{t-\Delta t} > m-1)$$

$$- \sum_{m=1}^{N} P_I(m, n-1)P(m > S_{t-\Delta t} > m-1) \quad (14\text{-}72)$$

For a stationary process, the frequency distribution of storage is independent of time; hence $P(i > S_t > i-1) = P(i > S_{t-\Delta t} > i-1) = P(i > S > i-1)$ for all integers of i with $N \geq i \geq 1$.

Letting $n = 1, 2, \ldots, N$, Eq. (14-72) produces N simultaneous equations which can be written as a matrix equation:

$$[P(n > S > n-1)]_{N \times 1}$$
$$= [P_I(m,n) - P_I(m, n-1)]_{N \times N} \cdot [P(m > S > m-1)]_{N \times 1} \quad (14\text{-}73)$$

where $[P_I(m,n) - P_I(m, n-1)]$ is the transition probability matrix. Equation (14-73) provides a solution of the finite Markov chain. The solution of this equation will give $P(i > S > i-1)$. The method of solving this matrix equation was demonstrated numerically by Langbein [131]. An exact solution was suggested by Moran [143] and another numerical solution based on the principle of stationarity was used by Gani and Moran [138] to verify a Monte Carlo solution of the equation (Sec. 8-IV).

The probability that the storage is less than n is therefore equal to

$$P(n > S > 0) = \sum_{i=1}^{n} P(i > S > i-1) \quad (14\text{-}74)$$

This probability is then plotted against the storage S to produce a curve for the cumulative distribution function of storage. By means of the storage equation, a curve for the frequency distribution function of outflow can be obtained.

When the storage capacity of the reservoir is very large, or N is considered as infinite, the dam creating the reservoir is called by Moran [138] an *infinite dam*. For an infinite dam, Moran [138] and Gani and Prabhu [146] obtained some interesting theoretical solutions.

The above approach of routing flow takes the probability characteristics into account and is called by Langbein [131] *probability routing*. The analysis has been further extended to cases with inflow exhibiting the nature of persistence (Sec. 8-I) by Lloyd [147].

VI. REFERENCES

1. Hoyt, W. G.: An outline of the runoff cycle, *Penn. State Coll. Sch. Eng. Tech. Bull.* 27, pp. 57–69, 1942.
2. Hoyt, W. G.: The runoff cycle, chap. 11D in O. E. Meinzer (ed.), "Physics of the Earth IX," "Hydrology," McGraw-Hill Book Company, Inc., New York, 1942; reprinted by Dover Publications, Inc., New York, 1949, pp. 507–513.
3. Langbein, W. B., and others: Annual runoff in the United States, *U.S. Geol. Surv. Cir.* 52, June, 1949.
4. Linsley, R. K., Jr., and W. C. Ackermann: Method of predicting the runoff from rainfall, *Trans. Am. Soc. Civil Engrs.*, vol. 107, pp. 825–846, 1942.
5. Butler, S. S.: "Engineering Hydrology," Prentice-Hall, Inc., Englewood Cliffs, N.J., 1957, pp. 227–229.
6. Kohler, M. A., and R. K. Linsley, Jr.: Predicting the runoff from storm rainfall, *U.S. Weather Bur. Res. Paper* 34, 1951.
7. Hopkins, C. D., Jr., and D. O. Hackett: Average antecedent temperatures as a factor in predicting runoff from storm rainfall, *J. Geophys. Res.*, vol. 66, no. 10, pp. 3313–3318, 1961.
8. Chow, V. T.: Hydrologic determination of waterway areas for the design of drainage structures in small drainage basins, *Univ. Illinois Eng. Exp. Sta. Bull.* 462, 1962.
9. Kuichling, Emil: The relation between the rainfall and the discharge of sewers in populous districts, *Trans. Am. Soc. Civil Engrs.*, vol. 20, pp. 1–56, 1889.
10. Mulvaney, T. J.: On the use of self-registering rain and flood gauges in making observations of the relations of rainfall and of flood discharges in a given catchment, *Trans. Inst. Civil Engrs. Ir.* (Dublin), vol. 4, pt. 2, p. 18, 1851.
11. Lloyd-Davis, D. E.: The elimination of storm water from sewerage systems, *Min. Proc. Inst. Civil Engrs.* (London), vol. 164, pp. 41–67, 1906.
12. Ramser, C. E.: Runoff from small agricultural areas, *J. Agr. Res.*, vol. 34, no. 9, pp. 797–823, May, 1927.
13. Kirpich, Z. P.: Time of concentration of small agricultural watersheds, *Civil Eng.* (N.Y.), vol. 10, no. 6, p. 362, June, 1940.
14. Design and construction of sanitary and storm sewers, *ASCE Man. Eng. Practice* No. 37 and *WPCF Man. Practice* No. 9, 1960.
15. Krimgold, D. B.: On the hydrology of culverts, *Proc. 26th Ann. Mtg. Highway Res. Bd.*, vol. 26, pp. 214–226, 1946.
16. Barnes, B. S.: Discussion on Analysis of run-off characteristics by O. H. Meyer, *Trans. Am. Soc. Civil Engrs.*, vol. 105, pp. 104–106, 1940.
17. Langbein, W. B.: Some channel storage and unit hydrograph studies, *Trans. Am. Soc. Civil Engrs.*, vol. 21, pp. 620–627, 1940.
18. Linsley, R. K., Jr., M. A. Kohler, and J. L. H. Paulhus: "Hydrology for Engineers," McGraw-Hill Book Company, Inc., New York, 1958, pp. 156–157.
19. Flood-hydrograph analysis and computations, U.S. Army Corps of Engineers, Engineering and Design Manuals, EM 1110-2-1405, 31 August 1959, U.S. Government Printing Office, Washington, D.C.
20. Kulandaiswamy, V. C.: A basic study of the rainfall excess-surface runoff relationship in a basin system, Ph.D. Thesis directed by V. T. Chow, University of Illinois, Urbana, Ill., 1964.
21. Ross, C. N.: The calculation of flood discharge by the use of time contour plan, *Trans. Inst. Engrs.* (Australia), vol. 2, pp. 85–92, 1921.
22. Folse, J. A.: A new method of estimating stream flow based upon a new evaporation formula, part II: a new method of estimating stream flow, *Carnegie Inst. Wash. Pub.* 400, Washington, D.C., 1929.
23. Sherman, L. K.: Stream flow from rainfall by the unit-graph method, *Eng. News-Rec.*, vol. 108, pp. 501–505, Apr. 7, 1932.
24. Zoch, R. T.: On the relation between rainfall and stream flow, *Monthly Weather Rev.*, vol. 62, pp. 315–322, 1934; vol. 64, pp. 105–121, 1936; vol. 65, pp. 135–147, 1937.
25. Bernard, M. M.: An approach to determine stream flow, *Trans. Am. Civil Engrs.*, vol. 100, pp. 347–395, 1935.
26. Snyder, F. F.: Synthetic unit-graphs, *Trans. Am. Geophys. Union*, vol. 19, pp. 447–454, 1938.
27. Turner, H. M., and A. J. Burdoin: The flood hydrograph, *J. Boston Soc. Civil Engrs.*, vol. 28, no. 3, pp. 232–256, July, 1941.
28. Horton, R. E.: Virtual channel-inflow graphs, *Trans. Am. Geophys. Union*, vol. 22, pt. III, pp. 811–820, 1941.

REFERENCES

29. Horner, W. W., and S. W. Jens: Surface runoff determination from rainfall without using coefficients, *Trans. Am. Soc. Civil Engrs.*, vol. 107, pp. 1039–1075, 1942.
30. Commons, G. G.: Flood hydrographs, *Civil Eng.* (N.Y.), vol. 12, pp. 571–572, October, 1942.
31. Edson, C. G.: Parameters for relating unit hydrographs to watershed characteristics, *Trans. Am. Geophys. Union*, vol. 32, pp. 591–596, 1951.
32. Appleby, F. V.: Runoff dynamics, a heat conduction analogue of storage flow in channel work, *Intern. Assoc. Sci. Hydrology, Pub.* 38, vol. 3, pp. 338–348, 1954.
33. Appleby, F. V.: Runoff dynamics, *Civ. Eng.* (London), vol. 51, pp. 772–774, 891–893, 1956.
34. Iwagaki, Yuichi: Fundamental studies on the runoff analysis by characteristics, *Kyoto Univ. Disaster Prevention Res. Inst. Bull.* 10, Kyoto, Japan, December, 1955.
35. Snyder, W. M.: Hydrograph analysis by the method of least squares, *Proc. Am. Soc. Civil Engrs.*, vol. 81, Separate no. 793, September, 1955.
36. Snyder, W. M.: Matrix operations in hydrograph computations, Tennessee Valley Authority, Tributary Area Development, Research Paper 1, Knoxville, Tenn., December, 1961.
37. Sugawara, Masami, and Fumiyuki Maruyama: A method of revision of the river discharge by means of a rainfall model, Symposia Darcy, *Intern. Assoc. Sci. Hydrology, Pub.* 42, vol. 3, pp. 71–76, 1956.
38. Cuénod, M.: Contribution à l'étude des crues. Détermination de la relation dynamique entre les précipitations et le débit des cours d'eau au moyen du calcul à l'aide de suites, principe et application, *La Houille Blanche*, vol. 11, no. 3, pp. 391–403, July–August, 1956.
39. Linsley, R. K., Jr., and N. H. Crawford: Computation of a synthetic streamflow record on a digital computer, *Intern. Assoc. Sci. Hydrology, Pub.* 51, pp. 526–538, 1960.
40. Crawford, N. H., and R. K. Linsley, Jr.: The synthesis of continuous streamflow hydrographs on a digital computer, *Stanford Univ., Dept. Civil Eng. Tech. Rep.* 12, July, 1962.
41. Crawford, N. H., and R. K. Linsley, Jr.: Estimate of the hydrologic results of rainfall augmentation, *J. Appl. Meteor.*, vol. 2, no. 3, pp. 426–427, June, 1963.
42. Holtan, H. N., and Overton, D. E.: Analyses and applications of simple hydrographs, *J. Hydrology*, vol. 1, no. 3, pp. 250–264, 1963.
43. Reich, B. M.: Design hydrographs for very small watersheds from rainfall, *Colorado State Univ., Civil Eng. Sec. Rept.* CER 62BMR41, July, 1962.
44. "Report of the Committee on Floods," *J. Boston Soc. Civil Engrs.*, vol. 17, no. 7, pp. 285–464, September, 1930.
45. Brater, E. F.: The unit hydrograph principle applied to small watersheds, *Trans. Am. Soc. Civil Engrs.*, vol. 105, pp. 1154–1178, 1940.
46. Wisler, C. O., and E. F. Brater: "Hydrology," 2d ed., John Wiley & Sons, Inc., New York, 1959, p. 248.
47. Linsley, R. K., Jr., M. A. Kohler, and J. L. H. Paulhus: "Applied Hydrology," McGraw-Hill Book Company, Inc., New York, 1949, pp. 448–449.
48. Collins, W. T.: Runoff distribution graphs from precipitation occurring in more than one time unit, *Civ. Eng.* (N.Y.), vol. 9, no. 9, pp. 559–561, September, 1939.
49. Johnstone, Don, and W. P. Cross: "Elements of Applied Hydrology," The Ronald Press Company, New York, 1949, pp. 143–146.
50. Morgan, R., and D. W. Hullinghors: Unit hydrographs for gaged and ungaged watersheds (unpublished manuscript), U.S. Engineers Office, Binghamton, New York, July, 1939.
51. Chow, V. T.: Hydrologic studies of urban watersheds, rainfall and runoff of Boneyard Creek, Champaign-Urbana, Ill., *Univ. Illinois Civil Eng. Studies, Hydraulic Eng. Ser. No. 2*, November, 1952.
52. McCarthy, G. T.: The unit hydrograph and flood routing, presented at the Conference of the North Atlantic Division, U.S. Corps of Engineers, June, 1938; rev., U.S. Engineer's Office, Providence, R.I., 1939.
53. Taylor, A. B., and H. E. Schwarz: Unit hydrograph lag and peak flow related to basin characteristics, *Trans. Am. Geophys. Union*, vol. 33, pp. 235–246, 1952.
54. Linsley, R. K., Jr., M. A. Kohler, and J. L. H. Paulhus: "Hydrology for Engineers," McGraw-Hill Book Company, Inc., New York, 1958, p. 207.
55. Williams, H. M.: Discussion on Military airfields: design of drainage facilities, by G. A. Hathaway, *Trans. Am. Soc. Civil Engrs.*, vol. 110, pp. 820–826, 1945.
56. Mitchell, W. D.: Unit hydrographs in Illinois, Illinois Division of Waterways, prepared in cooperation with U.S. Geological Survey, 1948.
57. Hydrology, Suppl. A to Sec. 4 in "Engineering Handbook," U.S. Department of Agriculture, Soil Conservation Service, Washington, D.C., 1957.

58. Hickok, R. B., R. V. Keppel, and B. R. Rafferty: Hydrograph synthesis for small arid land watersheds, *Agr. Eng.*, vol. 40, no. 10, pp. 608–611, 615, October, 1959.
59. Bender, D. L., and J. A. Roberson: The use of a dimensionless unit hydrograph to derive unit hydrographs for some Pacific Northwest basins, *J. Geophys. Res.*, vol. 66, no. 2, pp. 521–527, 1961.
60. Gray, D. M.: Synthetic unit hydrographs for small watersheds, *Proc. Am. Soc. Civil Engrs., J. Hydraulics Div.*, vol. 87, no. HY4, pp. 33–53, 1961.
61. Gray, D. M.: Derivations of hydrographs for small watersheds from measurable physical characteristics, *Iowa State Univ., Agr. Home Econ. Exp. Sta. Res. Bull.* 506, Ames, Iowa, June, 1962.
62. Morgan, P. E., and S. M. Johnson: Analysis of synthetic unit-graph methods, *Proc. Am. Soc. Civil Engrs., J. Hydraulics Div.*, vol. 88, no. HY5, pt. 1, pp. 199–220, September, 1962.
63. Clark, C. O.: Storage and the unit hydrograph, *Trans. Am. Soc. Civil Engrs.*, vol. 110, pp. 1419–1446, 1945.
64. Paynter, H. M.: Methods and results from MIT studies in unsteady flow, *J. Boston Soc. Civil Engrs.*, vol. 39, no. 2, pp. 120–165, April, 1952.
65. Paynter, H. M.: Computer techniques in hydrology: flood routing by admittance method, in "A Palimpsest on the Electronic Analog Art," George A. Philbrick Researches, Inc., Boston, Mass., 1955, 1958, 1960, pp. 239–245.
66. O'Kelly, J. J.: The employment of unit-hydrographs to determine the flows of Irish arterial drainage channels, *Proc. Inst. Civil Engrs.*, vol. 4, pt. 3, pp. 365–412, 1955.
67. Nash, J. E.: The form of the instantaneous unit hydrograph, *Intern. Assoc. Sci. Hydrology, Pub.* 45, vol. 3, pp. 114–121, 1957.
68. Dooge, J. C. I.: A general theory of the unit hydrograph, *J. Geophys. Res.*, vol. 64, no. 1, pp. 241–256, 1959.
69. Boersma, Lyckle: Rainfall-runoff relations for small watersheds, Ph. D. Thesis, Cornell University, June, 1959.
70. O'Donnell, T.: Instantaneous unit hydrograph derivation by harmonic analysis, *Intern. Assoc. Sci. Hydrology, Pub.* 51, pp. 546–557, 1960.
71. Singh, K. P.: A non-linear approach to the instantaneous unit hydrograph, Ph. D. Thesis directed by V. T. Chow, University of Illinois, Urbana, Ill., 1962.
72. Henderson, F. M.: Some properties of the unit hydrograph, *J. Geophys. Res.*, vol. 68, no. 16, pp. 4785–4793, Aug. 15, 1963.
73. Wu, I-Pai: Design hydrographs for small watersheds in Indiana, *Proc. Am. Soc. Civil Engrs., J. Hydraulics Div.*, vol. 89, no. HY6, pt. 1, pp. 35–66, November, 1963.
74. Diskin, M. H.: A basic study of the linearity of the rainfall-runoff process in watersheds, Ph. D. Thesis directed by V. T. Chow, University of Illinois, Urbana, Ill., 1964.
75. Amorocho, J., and G. T. Orlob: Nonlinear analysis of hydrologic systems, *Univ. Calif. Water Resources Center Contribution No.* 40, Berkeley, Calif., November, 1961.
76. Aseltine, J. A.: "Transform Method in Linear System Analysis," McGraw-Hill Book Company, Inc., New York, 1958.
77. Jaeger, J. C.: "An Introduction to Laplace Transformations with Engineering Applications," Methuen & Co., Ltd., London, and John Wiley & Sons, Inc., New York, 1959.
78. LePage, W. R.: "Complex Variables and the Laplace Transform for Engineers," McGraw-Hill Book Company, Inc., New York, 1961.
79. Savant, C. J., Jr.: "Fundamentals of the Laplace Transformation," McGraw-Hill Book Company, Inc., New York, 1962.
80. Rainville, E. D.: "The Laplace Transform: An Introduction," The Macmillan Company, New York, 1963.
81. Nash, J. E.: The form of the instantaneous unit hydrograph, *Intern. Assoc. Sci. Hydrology, Pub.* 45, vol. 3, pp. 114–121, 1957.
82. Nash, J. E.: Determining runoff from rainfall, *Proc. Inst. Civil Engrs.*, vol. 10, pp. 163–184, June, 1958.
83. Nash, J. E.: The effect of flood elimination works on the flood frequency of the River Wandle, *Proc. Inst. Civil Engrs.*, vol. 13, pp. 317–338, July, 1959.
84. Nash, J. E.: Systematic determination of unit hydrograph parameters, *J. Geophys. Res.*, vol. 64, no. 1, pp. 111–115, January, 1959.
85. Nash, J. E.: A note on an investigation into two aspects of the relation between rainfall and storm runoff, *Intern. Assoc. Sci. Hydrology, Pub.* 51, pp. 567–578, 1960.
86. Nash, J. E.: A unit hydrograph study, with particular reference to British catchments, *Proc. Inst. Civil Engrs.*, vol. 17, pp. 249–282, November, 1960.
87. Rosa, J. M., and others: Electronic analog, U.S. Agricultural Research Service, 22 pp., unpublished material, Moscow, Idaho, 1961.
88. Minshall, N. E.: Predicting storm runoff on small experimental watersheds, *Proc. Am. Soc. Civil Engrs., J. Hydraulics Div.*, vol. 86, no. HY8, pp. 17–38, August, 1960.

REFERENCES

89. Amorocho, Jaime: Discussion of Predicting storm runoff on small experimental watersheds by N. E. Minshall, *Proc. Am. Soc. Civil Engrs.*, *J. Hydraulics Div.*, vol. 87, no. HY2, pp. 185–191, March, 1961.
90. Amorocho, J., and G. T. Orlob: Nonlinear analysis of hydrologic systems, *Univ. Calif. Water Resources Center, Contribution No.* 40, November, 1961.
91. Amorocho, J.: Measures of the linearity of hydrologic systems, *J. Geophys. Res.*, vol. 68, no. 8, pp. 2237–2249, 1963.
92. Wiener, Norbert: Response of non-linear devices to noise, *Mass. Inst. Tech. Radiation Lab. Rep.* 129, April, 1942.
93. Horton, R. E.: The interpretation and application of runoff plot experiments with reference to soil erosion problems, *Proc. Soil Sci. Soc. Am.*, vol. 3, pp. 340–349, 1938.
94. Kerby, W. S.: Time of concentration for overland flow, *Civil Eng.* (N.Y.), vol. 29, no. 3, p. 174, March, 1959.
95. Keulegan, G. H.: Spatially variable discharge over a sloping plane, *Trans. Am. Geophys. Union*, vol. 25, pt. VI, pp. 959–968, 1944.
96. Izzard, C. F.: The surface-profile of overland flow, *Trans. Am. Geophys. Union*, vol. 25, pt. VI, pp. 959–968, 1944.
97. Richey, E. P.: The fundamental hydraulics of overland flow, Ph. D. Thesis, Stanford University, October, 1954.
98. Behlke, C. E.: The mechanics of overland flow, Ph. D. Thesis, Stanford University, July, 1957.
99. Woo, D. C., and E. F. Brater: Spatially varied flow from controlled rainfall, *Proc. Am. Soc. Civil Engrs.*, vol. 88, no. HY6, pt. 1, pp. 31–56, November, 1962.
100. Yu, Y. S., and J. S. McNown: Runoff from impervious surfaces, prepared for U.S. Army Engineers Waterways Experiment Station, Corps of Engineers, Feb. 15, 1963.
101. Chow, V. T.: "Open-channel Hydraulics," McGraw-Hill Book Company, Inc., New York, 1959.
102. Chow, V. T.: Open channel flow, Sec. 24 in V. L. Streeter (ed.), "Handbook of Fluid Dynamics," McGraw-Hill Book Company, Inc., New York, 1961.
103. Leliavsky, Serge: "An Introduction to Fluvial Hydraulics," Constable & Co., Ltd., London, 1955.
104. Leopold, L. B., and Thomas Maddock, Jr.: The hydraulic geometry of stream channels and some physiographic implications, *U.S. Geol. Surv. Prof. Paper* 252, 1953.
105. L'vovich, M. I.: Elementy vodnogo rezhima rek zemnogo shara (Elements of water regime in rivers of the terrestial globe), Trudy nauchno-iss ledovatel'skikh uchrezhdeniĭ, Gidroloziia suski (Hydrology of the Land), Ser. 4, no. 13, Gosudarstvennyĭ Gidrologicheskiĭ Institut (State Hydrological Institute), Sverdlovsk-Moskva, Gidrometeoizdat, 1945.
106. Langbein, W. B., and others: Annual runoff in the United States, *U.S. Geol. Surv. Cir.* 52, June, 1949.
107. Langbein, W. B., and J. V. B. Wells: The water in the rivers and creeks, in "Water," Department of Agriculture, Yearbook of Agriculture, 1955.
108. Principal rivers of the world, *The Military Engr.*, vol. 50, no. 337, pp. 386–387, September-October, 1948.
109. Large Rivers of the United States, *U.S. Geol. Surv. Cir.* 44, May, 1949.
110. Doland, J. J.: "Hydro Power Engineering," The Ronald Press Company, New York, 1954, chaps. 3 and 8.
111. Lane, E. W., and Kai Lei: Stream flow variability, *Trans. Am. Soc. Civil Engrs.*, vol. 115, pp. 1084–1134, 1950.
112. Chow, V. T.: Discussion of Stream flow variability by E. W. Lane and Kai Lei, *Trans. Am. Soc. Civil Engrs.*, vol. 115, pp. 1099–1100, 1950.
113. Foster, H. A.: Duration curves, *Trans. Am. Soc. Civil Engrs.*, vol. 99, pp. 1213–1235, 1934.
114. Hickox, G. H., and G. O. Wessenauer: Application of duration curves to hydroelectric studies, *Trans. Am. Soc. Civil Engrs.*, vol. 98, pp. 1276–1290, 1933.
115. Barrows, H. K.: "Water Power Engineering," 3d ed., McGraw-Hill Book Co., Inc., New York, 1943, chap. 3.
116. Creager, W. P., and J. D. Justin (ed.): "Hydroelectric Handbook," 2d ed., John Wiley & Sons, Inc., New York, 1950, chap. 10.
117. Saville, Thorndike, and J. D. Watson: An investigation of flow-duration characteristics of North Carolina streams, *Trans. Am. Geophys. Union*, vol. 14, pp. 406–425, 1933.
118. Morgan, J. H.: Flow-duration characteristics of Illinois streams, *Trans. Am. Geophys. Union*, vol. 17, pp. 419–426, 1936.
119. Hall, C. H., and C. V. Youngquist: Ohio stream drainage areas and flow duration tables, *Ohio State Univ. Exp. Sta. Bull.* 3, vol. 11, no. 3, May, 1942.

120. Cross, W. P., and R. J. Bernhagen: Flow duration, Pt. 1 of Ohio stream-flow characteristics, *Ohio Dept. Nat. Resources Div. Water Bull.* 10, 1949.
121. Mitchell, M. D.: "Water-supply Characteristics of Illinois Streams," Illinois Division of Waterways, prepared in cooperation with U.S. Geological Survey, 1950.
122. Mitchell, W. D.: "Flow-duration of Illinois Streams," Illinois Division of Waterways, prepared in cooperation with U.S. Geological Survey, 1957.
123. Rippl, W.: The capacity of storage reservoirs for water-supply, *Proc. Inst. Civil Engrs.*, vol. 71, pp. 270–278, 1883.
124. Hurst, H. E.: Long-term storage capacity of reservoirs, *Trans. Am. Soc. Civil Engrs.*, vol. 116, pp. 770–799, 1951.
125. Hurst, H. E.: Measurement and utilization of the water resources of the Nile basin, *Proc. Inst. Civil Engrs.*, vol. 3, pt. III, pp. 1–26, April, 1954; discussions, pp. 26–50; correspondence, pp. 580–594.
126. Hurst, H. E.: Methods of using long-term storage in reservoirs, *Proc. Inst. Civil Engrs.*, vol. 5, pt. I, no. 5, pp. 519–543, September, 1956; discussions, pp. 543–590.
127. Hurst, H. E.: A suggested statistical model of some time series which occur in nature, *Nature*, vol. 180, p. 494, Sept. 7, 1957.
128. Yevdjevich, V. M.: Some general aspects of fluctuations of annual runoff in the Upper Colorado River basin, Part III of "Past and Probable Future Variations in Stream Flow in the Upper Colorado River," Civil Engineering Section, Colorado State University, Fort Collins, Colo., CER61VMY54, October, 1961.
129. Yevdjevich, V. M.: Climatic fluctuations studied by using annual flows and effective annual precipitations, in "Changes of Climate," Arid Zone Research, Proceedings of the Rome Symposium, UNESCO and the World Meteorological Organization, United Nations Education, Scientific and Cultural Organization, Place de Fontenoy, Paris, 1963, pp. 183–200.
130. Leopold, L. B.: Probability analysis applied to a water-supply problem, *U.S. Geol. Surv. Cir.* 410, 1959.
131. Langbein, W. B.: Queuing theory and water storage, *Proc. Am. Soc. Civil Engrs., J. Hydraulics Div.*, vol. 84, no. HY5, pt. 1, pp. 1–24, October, 1958.
132. Hurst, H. E., P. Phillips, R. P. Black, and Y. M. Simaika: Supplement to vol. VII of "The Nile Basin," Physical Department, Cairo, Egypt, 8 vols., 1931–1950, Government Press, Cairo, 1949.
133. Feller, William: The asymptotic distribution of the range of series of independent random variables, *Ann. Math. Statist.*, vol. 22, pp. 427–432, 1951.
134. Anis, A. A., and E. H. Lloyd: On the range of partial sums of a finite number of independent normal variates, *Biometrika*, vol. 40, pp. 35–42, 1953.
135. Saaty, T. L.: "Elements of Queueing Theory," McGraw-Hill Book Company, Inc., New York, 1961.
136. Cox, D. R., and W. L. Smith: "Queues," Methuen & Co., Ltd., London, and John Wiley & Sons, Inc., New York, 1961.
137. Moran, P. A. P.: A probability theory of dams and storage systems, *Austral. J. Appl. Sci.*, vol. 5, no. 2, pp. 116–124, 1954.
138. Gani, J., and P. A. P. Moran: The solution of dam equations by Monte Carlo methods, *Austral. J. Appl. Sci.*, vol. 6, no. 3, pp. 267–273, 1955.
139. Moran, P. A. P.: A probability theory of dams and storage systems: modifications of the release rules, *Austral. J. Appl. Sci.*, vol. 6, no. 2, pp. 117–130, 1955.
140. Moran, P. A. P.: A probability theory of a dam with a continuous release, *Quart. J. Math.* (Oxford, 2), vol. 7, pp. 130–137, 1956.
141. Kendall, D. G.: Some problems in the theory of dams, *J. Roy. Statist. Soc.*, ser. B, vol. 19, pp. 207–212, 1957.
142. Moran, P. A. P.: The statistical treatment of flood flows, *Trans. Am. Geophys. Union*, vol. 38, pp. 519–523, 1957.
143. Moran, P. A. P.: "The Theory of Storage," Methuen & Co., Ltd., London, and John Wiley & Sons, Inc., New York, 1959.
144. Bryant, G. T., and H. A. Thomas, Jr.: A stochastic model for flow regulation by an impounding reservoir, paper presented at Annual Convention of the American Society of Civil Engineers, New York, Oct. 16–17, 1961.
145. Fiering, M. B.: Queuing theory and simulation in reservoir design, *Proc. Am. Soc. Civil Engrs., J. Hydraulics Div.*, vol. 87, no. HY6, pp. 39–69, November, 1961.
146. Gani, J., and N. U. Prabhu: Stationary distributions of the negative exponential type for the infinite dam, *J. Roy. Statist. Soc.* ser. B, vol. 19, pp. 342–351, 1957.
147. Lloyd, E. H.: A probability theory of reservoirs with serially correlated inputs, *J. Hydrology*, vol. 1, no. 2, pp. 99–128, September, 1963.

Section 15

STREAMFLOW MEASUREMENT

MARION CLIFFORD BOYER, *Natural Resources Coordinator, Fresno County, California.*

I. Introduction	15-3
A. Quantities and Dimensions	15-3
II. History	15-4
A. Stage Records	15-4
B. Measurement of Discharge	15-4
C. Stage-Discharge Relations	15-5
III. Determination of Stage	15-5
A. Stage Indicators	15-6
1. Staff Gage	15-6
2. Chain, Tape, and Wire Gages	15-6
3. Float Gage	15-6
4. Electric Tape Gage	15-6
5. Pressure Transmitter	15-6
6. Crest-stage Indicator	15-7
B. Automatic Instruments	15-7
1. Water-stage Recorder	15-7
2. Automatic Printer	15-7
3. Long-distance Transmitter	15-8
4. Bubble Gage	15-8
5. Tensometric Recorder	15-9
C. Auxiliary Data	15-9
IV. Methods of Velocity Measurement	15-9
A. Mechanical Devices	15-9
1. Rotating Meters	15-9
2. Dynamometer	15-10
3. Float	15-11
4. The Pitot Tube	15-12
B. Chemical and Electrical Methods	15-12
1. Salt Velocity	15-12
2. Salt Dilution	15-12
3. Radioactive Tracers	15-12
4. Oxygen Polarography	15-12
5. Hot-wire Anemometer	15-13
6. Electromagnetic Flowmeter	15-13
7. Ultrasonic Flowmeter	15-13
V. The Rotating Meter	15-14
A. Calibration	15-14

B. Precision of Operation.. 15-14
     1. Ratings in Still and Moving Water.................. 15-14
     2. Effects of Turbulence............................. 15-14
     3. Effects of Obliquity.............................. 15-15
  C. Field Comparison.. 15-15
VI. Velocity Observation.. 15-15
  A. Distribution in the Vertical............................ 15-15
  B. Setting the Meter....................................... 15-15
  C. Timing the Rotation..................................... 15-16
  D. Meter Circuitry... 15-16
  E. Timing Procedure.. 15-16
  F. Counting and Recording Automats......................... 15-16
VII. Determination of Depth..................................... 15-16
  A. Wading Rod.. 15-17
  B. Cable and Weight.. 15-17
  C. Fathometer.. 15-17
  D. Sources of Error.. 15-17
VIII. Discharge Measurement..................................... 15-17
  A. Types of Measurement.................................... 15-17
     1. Wading.. 15-18
     2. Cableway.. 15-18
     3. Bridge.. 15-18
     4. Boat.. 15-19
     5. Ice... 15-20
  B. Recording the Data...................................... 15-20
  C. Factors Affecting Accuracy.............................. 15-20
     1. Instrumental...................................... 15-20
     2. Methodological.................................... 15-20
     3. Personal.. 15-20
     4. Computational..................................... 15-20
     5. Sampling.. 15-25
  D. Computation Procedure................................... 15-25
  E. A Method for Rapid Computation.......................... 15-25
  F. Correction Factors...................................... 15-25
     1. Angularity.. 15-25
     2. Suspension Coefficient............................ 15-25
     3. Air-line and Wet-line Corrections................. 15-26
IX. The Hydrometric Station..................................... 15-26
  A. Selection of Site....................................... 15-26
     1. Accessibility..................................... 15-26
     2. Adequacy.. 15-26
     3. Stability... 15-27
     4. Permanency.. 15-27
  B. Equipment for Obtaining Stage Records................... 15-28
     1. Nonrecording Gage................................. 15-29
     2. Recording Gage.................................... 15-29
  C. Maintenance of Datum.................................... 15-30
  D. Auxiliary Gage to Measure Fall.......................... 15-30
  E. Artificial Controls..................................... 15-30
     1. Weirs and Critical-depth Flumes................... 15-30
     2. Overflow Dam...................................... 15-31
  F. Calibration of Hydroelectric Stations................... 15-31
X. Indirect Determinations of Peak Discharge.................... 15-31
  A. Channel Reach... 15-32
  B. Contracted Opening...................................... 15-32
  C. Culvert... 15-33
  D. Broad-crested Weir...................................... 15-33
  E. Dam... 15-33

XI Stage-Discharge Relation.................................. 15-34
       A. Conditions Affecting Controls......................... 15-34
          1. Changing Channel.................................. 15-34
          2. Alluvial Channel................................... 15-35
          3. Variable Backwater................................. 15-35
          4. Rapidly Changing Stage............................ 15-36
          5. Variable Channel Storage.......................... 15-37
          6. Aquatic Vegetation................................. 15-37
          7. Ice.. 15-37
       B. Complex Controls..................................... 15-38
       C. High-water Extensions of the Rating.................. 15-38
  XII. Computation of Runoff.................................... 15-39
       A. Daily Gage Heights................................... 15-39
          1. Nonrecording Gage.................................. 15-39
          2. Recording Gage..................................... 15-39
       B. Rating Table... 15-39
       C. Computation Procedure............................... 15-39
          1. Changing Channel.................................. 15-39
          2. Alluvial Channel................................... 15-39
          3. Variable Backwater................................. 15-39
          4. Rapidly Changing Stage............................ 15-40
          5. Variable Channel Storage.......................... 15-40
          6. Aquatic Vegetation................................. 15-40
          7. Ice.. 15-40
       D. Monthly and Yearly Computations..................... 15-40
 XIII. References.. 15-41

I. INTRODUCTION

This section presents basic information on the practices, arts, and techniques of streamflow measurement. It summarizes the present status and offers suggestions for improving the accuracy and efficiency of streamflow computations. The rapidly increasing demands on the surface-water resources of most of the world make it imperative that they be determined with as high a degree of precision as is practicable. Future streamflow determinations will require a greatly expanded network of hydrometric stations and a high degree of efficiency in their operation. Trained hydrographers, competent to make streamflow measurements and to convert them into figures of runoff, are becoming increasingly scarce, and every means must be taken to increase their efficiency by reducing the time required to perform their functions. This can be accomplished by improvements in instrumentation and techniques and a more intensive use of the high-speed computing devices which the space age has made available.

The general methods of streamflow measurement and the computation of runoff are covered elsewhere more comprehensively than the limitation of space here permits. The reader is referred to the works of Hoyt and Grover [1], Grover and Harrington [2], Corbett [3], and Kolupaila [4]. Special reference is made to the "Bibliography of Hydrometry" by Kolupaila [5], which is the most comprehensive work of its kind, presenting a compendium of references in many languages, covering the practice of *hydrometry*[1] throughout most of the civilized world.

A. Quantities and Dimensions

Table 15-1 gives the quantities used in this section, together with their symbols, dimensions, and basic units in the foot-pound-second and centimeter-gram-second systems.

[1] According to Kolupaila [6], hydrometry is the science of water measurements.

Table 15-1. Quantities, Symbols, and Dimensions

Quantity	Symbol	Dimension	Basic units ft-lb-sec	Basic units m-kg-sec (cm-g-sec)
Area	A, a	L^2	ft^2	m^2
Depth	D, d, y	L	ft	m
Discharge	Q	L^3/T	cfs	cm (m^3/sec)
Fall	F	L	ft	m
Head, gage height	h	L	ft	m
Hydraulic radius	R	L	ft	m
Length	L	L	ft	m
Slope	S	L/L	ft/ft	m/m
Velocity	U, V, v	L/T	ft/sec	m/sec
Shear intensity	τ	F/L^2	lb/ft^2	kg/m^2
Mass density	ρ	FT^2/L^4	lb-sec^2/ft^4	kg-sec^2/m^4
Shear velocity	$\sqrt{\tau_0/\rho}$	L/T	ft/sec	m/sec

II. HISTORY

Very early in history man recognized the great importance of water to his existence. Among the oldest monuments are the remains of water-supply structures and irrigation systems in China, India, Babylon, Egypt, the Roman Empire, and Central America. The skillful builders of those great works obviously were quite familiar with water measurement and control.

According to Kolupaila [5], the oldest hydrometric documents are the markings of flood stages of the Nile River carved in its cliffs between Semneh and Kumneh, 255 miles upstream from Aswan. As Kolupaila points out, it was approximately 35 centuries after the first recorded measurements of river stage that a mechanical means for measuring stream velocity was proposed by Domenico Guglielmini (1655–1710).

Early in the history of America the development of water power for use in New England textile mills, the construction of the great navigation canals in the East and Middle West, and a little later the growth of placer mining and then irrigation in the arid West, all brought into sharp focus the need for precise determination of the flow of the nation's rivers and streams. The work of the American pioneers in hydrometry was outstanding, and much of it was widely adapted as standard in other countries.

A. Stage Records

The first records of river stages that have been preserved in the United States [6] were originated by Winthrop Sargent at Natchez, Miss., in 1798. They were of the Mississippi River floods and were continued until 1848 at Natchez and later at Carrollton, La. Continuous daily observations were begun in July, 1846.

The first self-registering gage in America was constructed in 1845 by Saxton and introduced into practice by Hunt in 1853 [6]. It consisted of a wire transmission from a float to a pencil which marked a sheet of paper that was advanced by a pendulum clock. In 1876, A. Fteley placed in operation a recording gage for measuring stages of the Sudbury River in Massachusetts. The rack drive and other principles embodied in this recorder formed the basis for the highly accurate and dependable water-stage recorders now in use.

B. Measurement of Discharge

The role of velocity in the determination of discharge was understood imperfectly until the great Renaissance upturning of interest in science. Almost every scientist of that time paid some attention to flowing water and to the measurement of its velocity.

Guglielmini in 1692 described a method of velocity measurement using a suspended ball deflected by the current. In 1732 Henri Pitot invented the impact tube, known as the *Pitot tube*, or *pitometer*, of great importance to modern fluid mechanics. In 1790 Reinhard Woltman, a German hydraulic engineer, introduced a hydrometric current meter having vanes rotating around a horizontal axis. It proved to be the most suitable instrument for river investigations.

The first American to construct and successfully apply a *current meter* was Daniel Farrand Henry [6], in 1867. He made a significant contribution by developing a "telegraphic current meter," outfitted with an electric circuit for transmitting signals of its rotation. This eliminated the necessity of removing the meter from the water after each operation to read a dial system, as was previously required. Electric transmission provided a means for mechanically recording observations and made possible perfect control of the meter during its operation. He also introduced the wire-and-weight method of meter support, thus permitting soundings in deeper and swifter streams than could be accomplished by rod support. Henry tried runners of two types—a screw in a cylindrical ring and four hemispherical cups on a horizontal axis (actually a rotor from the Robinson anemometer of meteorological fame), mounted transverse to the direction of the current.

In 1870 T. G. Ellis introduced a current meter consisting of a series of cups revolving around a vertical axle. W. G. Price in 1882 redesigned the Ellis meter with simple bearings protected from water and silt and eliminated some of the older meter's other shortcomings. The Price meter has proved eminently satisfactory and in modified form is still used by American engineers.

The concept of discharge as being the product of a cross-sectional area and a velocity normal to the area appears to have been recognized early in the history of hydrometry. The first systematic efforts at discharge determination in America appear to have been started along the lower Mississippi River about 1838. The measurement of discharge of such a large river was extremely difficult because of the simple instruments and imperfect equipment of that time. Soundings were made with a weight on a line, and velocities were observed by timing surface floats (wood chips) over a 1-mile course. The surface velocity was reduced by 10 per cent to obtain the mean velocity.

With the introduction of the telegraphic meter by Henry and the substitution of a weight and line for a rod as its support, the art of gaging large streams advanced rapidly until the present, when even the greatest floods on the largest rivers can be measured with good accuracy.

C. Stage-Discharge Relations

During the eleventh or twelfth century A.D., a large and intricate irrigation system was in operation in the Murghab River in Merv Oasis of Central Asia. Water distribution to the various canals was carefully controlled by lifting the head gate and reading the exact water height on a gage. This is apparently the earliest record of a people having knowledge of a relation between stage and discharge.

The first permanent gaging stations in America were installed in Eaton and Madison Brooks, in Madison County, New York, in 1835. It is inferred that a stage-discharge relation was used in connection with their operation.

Thus step by step were developed the instruments, the methods, and the techniques of streamflow measurement. As the need grew for high precision, the instruments were improved, the methods were reevaluated and changed where necessary, and hydrographers were trained in the techniques. Developments continue, and the flow of streams is measured today with an accuracy not dreamed of only a few decades ago.

III. DETERMINATION OF STAGE

The term *stage*, as used in streamflow measurement, refers to the water-surface elevation at a point along a stream, measured above an arbitrary datum. Stage is determined by indicators of various types, the readings of which are either taken at intervals by an observer or are recorded continuously by automatic instruments.

A. Stage Indicators

Stage indicators are classified according to the method by which they measure the stage and the manner in which they are read. The principal devices are staff gage; chain, tape, and wire gages; pressure transmitters; and crest-stage indicators.

1. Staff Gage. A staff gage is a graduated scale set in a stream by fastening it to a pier, wall, supporting block, or other structure. It is read by observing the level of the water surface in contact with it, with proper allowance for the meniscus. The gage may extend upward in one section to cover the expected range in stage; it may be set vertically in several sections at different locations; or it may be placed as a sloping scale up the stream bank. One of the major problems in installing a staff gage is to protect it from damage by boats, ice, or flood-transported debris and to locate it so that flow disturbances across the scale are at a minimum.

Fig. 15-1. Wire-weight gage. (*Leupold and Stevens Instruments, Inc.*)

2. Chain, Tape, and Wire Gages. Stage is determined with these devices by lowering a weight to the water surface. The chain gage is read from a reference bead moving along a graduated scale. The tape gage is read at a pointer in contact with the tape. The wire gage is read by means of a mechanical counter attached to the reel on which the wire is wound (Fig. 15-1).

These gages have the advantages of ease in installation and freedom from damage. They also can be so installed as to be readily accessible under all conditions of flow.

3. Float Gage. This type is constructed by fastening one end of a tape to a float resting on the water surface, bringing the other end up and over a wheel, and counterweighting it. A pointer is mounted by the wheel so that it rests against the tape. Readings are obtained from the tape at the pointer. This gage is used frequently in connection with water-stage recorder installations. In fact, the tape is often utilized to drive the recorder wheel.

4. Electric Tape Gage. It is not always possible to observe the instant of contact of a gage weight with the water surface. If it is made a part of an electric circuit, with the return through the water, at the instant of contact a galvanometer needle will deflect or a click will be heard in a headset. Thus the water surface is determined with precision.

5. Pressure Transmitter. The stage in a stream may be converted to pressure by means of a cylinder and flexible diaphragm. This pressure is transmitted via a light fluid through a tube to a sensitive gage. The principal objection to the pressure transmitter is that it is inordinately sensitive to temperature changes.

6. Crest-stage Indicator. This indicator is used to delineate the peak stage of a flood at points other than at a hydrometric station. Such data are valuable in the establishment of flood profiles and in determining slope for the investigation of formulas for the computations of flood flows in streams.

A crest-stage indicator consists of a pipe set vertically with an open, screened bottom and a vented top. It contains a graduated wooden staff gage, held in place by a cap on the pipe. A small quantity of powdered cork is introduced into the top of the pipe and washed to the bottom with water, where a fine screen retains it. When a flood passes, the cork is lifted and at the crest is left clinging to the graduated staff. When the gage is visited, the stage is read and the device returned to operating condition by pouring a little water in to wash the cork back down the pipe.

B. Automatic Instruments

The automatic instruments used in streamflow measurement are float- or probe-actuated water-stage recorders or automatic printers, radio and land-line transmitters,

Fig. 15-2. Water-stage recorder. (*Leupold and Stevens Instruments, Inc.*)

a recently developed device called a *bubble gage*, and a tensometric recorder. Each is referred to some type of nonrecording gage.

1. Water-stage Recorder. The most common of the water-stage recorders is the float-actuated instrument. The float is connected to the recorder by a wire or tape passing over a wheel and counterweighted. As the float moves, the wheel is turned and drives a pen on a rack-and-pinion mechanism. Pen travel is continuous. When it reaches one edge of the chart, an automatic reversing arrangement sends it in the opposite direction without loss of motion. Figure 15-2 shows a continuous recorder.

2. Automatic Printer. In 1912 a printing gage was developed [6] which automatically printed the gage height on a paper strip every 15 min. This device met

15–8   STREAMFLOW MEASUREMENT

with little success. At the present time, however, high-speed computers of many types and descriptions have been developed. It is now possible to program one so that a punched tape, with gage heights recorded several times a day, can be fed through and the discharge computed automatically. The U.S. Geological Survey is presently (1963) using digital recorders which are battery operated and record a 4-digit number on a 16-channel paper tape at preselected time intervals. Electronic translators are used to convert the 16-channel punch-tape records to 7-channel paper tape suitable for input into a digital computer. This will allow automatic computation of daily mean gage height and daily mean discharge.

3. Long-distance Transmitter. A long-distance transmitter sends a coded signal which can be translated into stage. Transmission is either by radio or over

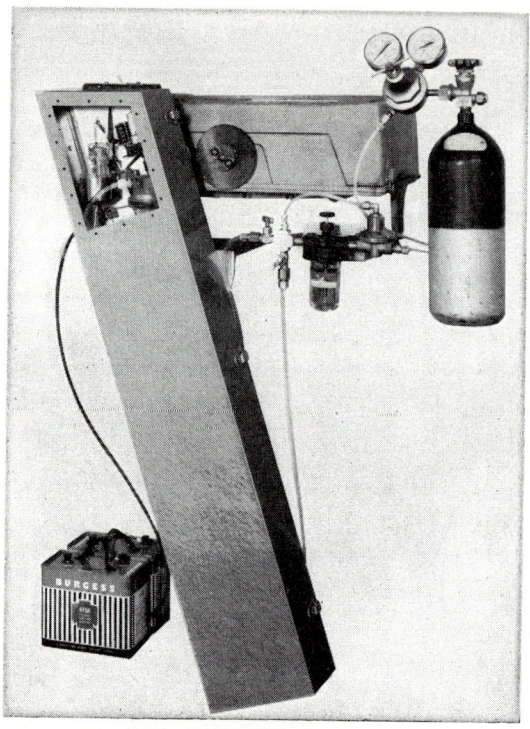

Fig. 15-3. Bubble gage. (*Leupold and Stevens Instruments, Inc.*)

telephone or telegraph circuits. Transmitters are of two types, those which operate at intervals and those which respond to a call. The former may be made fully automatic, in that a receiver will transfer the message to a tape. The latter type is particularly useful for obtaining stages during floods. For example, a network of telephone stage transmitters is located around Indianapolis, Ind. When there is danger of flooding, these are called at intervals, the gage heights noted, graphs plotted, and the time and height of peaks estimated. On the basis of these data, warnings are issued as to the magnitude and time of occurrence of flooding on a river and its principal tributaries. By this precaution lives have been saved and a considerable amount of property damage has been prevented.

4. Bubble Gage. The bubble gage (Fig. 15-3), recently perfected by the U.S. Geological Survey, consists of a specially designed servomanometer, transistor control, gas-purge system, and recorder. It operates on the principle of nitrogen gas being bubbled slowly through a small tube and discharged freely into a liquid through an

METHODS OF VELOCITY MEASUREMENT   15–9

orifice at a fixed elevation. The pressure in the tube corresponds to the head of the liquid over the orifice, except for the minor effect of weight of the gas. The pressure is transferred to a mercury manometer, one side of which is movable. The movable reservoir is equipped with a float and sensitive double-acting switch. If the mercury pool moves because of change in pressure on the fixed side, the switch is closed and the pool is returned to a null position by means of a servomotor. This motor drives a pen on a recorder chart so that a continuous record of stage is obtained.

This device has several advantages. The need for costly wells and recorder structures is eliminated, changes of stage in the order of 100 ft or more can be recorded, and the instrument can be a thousand feet or more distant from the orifice.

5. Tensometric Recorder. An article was translated recently from the Russian language which described an interesting device for measuring the height of waves [7]. It consisted of a diaphragm attached to a flexible metal strip mounted within an air chamber and secured at its ends. Wires were fastened on each side of the strip which have the characteristic of variable electric resistance with tension. These comprise the opposite sides of a Wheatstone bridge. Variations in pressure on the diaphragm change the tension in the wires, which in turn produces a voltage change in the bridge. This is amplified and recorded. The device is practically free of temperature effects. Such an instrument might be adapted to the measurement of river stages.

C. Auxiliary Data

Certain data other than stages are useful in analyzing the stage-discharge relation and the effects of various climatic conditions on streamflow. These include air and water temperature, wind movement and direction, precipitation, and barometric pressure. On an experimental basis water-stage recorders have been adapted also to record some of these phenomena. When these adaptations become standard equipment, the science of potamology will be greatly benefited.

IV. METHODS OF VELOCITY MEASUREMENT

Among the different methods which have been developed for the measurement of stream velocities are mechanical devices which are rotated or deflected by the current, instruments for the conversion of velocity head into potential head, chemical methods, and others. Because of the widely varying conditions of streamflow, ranging from very low to very high and turbulent velocities, extremely rough cross sections, and the like, no single method of velocity measurement is universally applicable. The rotating meter covers the greatest range of conditions, but is not suitable for velocities below about 0.10 ft/sec nor under conditions of extreme turbulence. There is a real need for further development of simple and practical velocity-measuring devices for use under extreme conditions.

A. Mechanical Devices

The three principal mechanical devices are the rotating meter, the dynamometer, and the float. The first converts a part of the stream momentum into angular momentum; the second converts momentum into force; and the third travels as a part of the stream.

1. Rotating Meters. Rotating current meters fall into two general classes, according to the orientation of the revolving axle; that is, the axle may be vertical or may be horizontal and parallel to the direction of flow. Rotation about the vertical axle is accomplished by means of cups or vanes, that about the horizontal axle by means of screw- or propellor-shaped blades.

The Price meter is the most familiar example of the first type. It was a natural evolution from the Robinson anemometer, which used hemispherical cups mounted around a vertical axle. Figure 15-4 shows the small Price meter mounted 1 ft above the bottom of a 100-lb lead weight as used by the U.S. Lake Survey [8] in the determination of turbulence effects on vertical and horizontal axle meters. A pygmy Price meter, approximately one-half the size of the small Price, has been developed for use in shallow streams.

In 1930 M. C. Boyer proposed to the U.S. Geological Survey the substitution of vertical vanes for cups in the small Price meter. The purpose was to eliminate overregistration when measuring from an oscillating cable or when integrating by vertical movement of the meter. It was considered also particularly suitable for use beneath an ice cover, as it could be introduced easily through a hole drilled through the ice. This meter now is undergoing field tests. One of the experimental models is shown in Fig. 15-5.

A patent has been granted to Tice [9] for an S-shaped rotor fastened to a horizontal disk. Savonius [10] has experimented with a bivane wheel of S-shaped cross section.

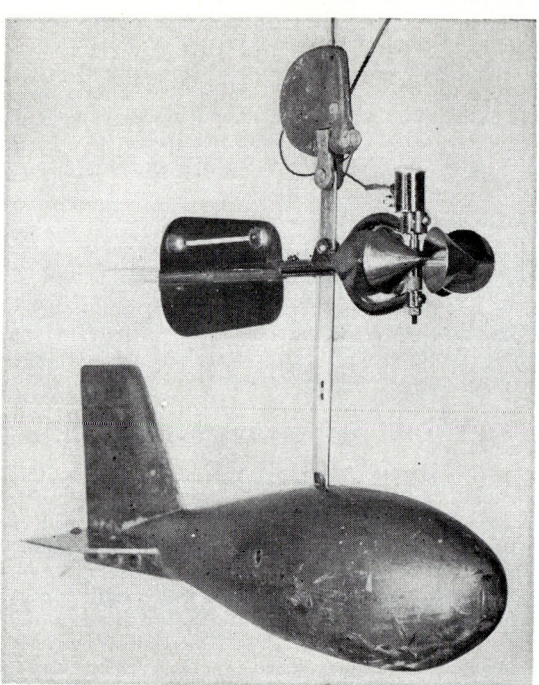

Fig. 15-4. Small Price meter and sounding weight. (*U.S. Lake Survey*.)

Several other hydrographers [5] have considered similar changes in the vertical-axle meter.

The reader is referred to Kolupaila [5] for bibliography describing various types of horizontal-axle meters, which find more favor in Europe than in the United States. Two of the most popular presently in use are the Neyrpic, manufactured in France, and the Ott, produced in the Federal Republic of Germany. Figure 15-6 shows the Neyrpic and Fig. 15-7 shows the Ott meters as they were used by the U.S. Lake Survey [8] in their studies of turbulence effects on meter registration.

2. Dynamometer. The dynamometer translates the momentum force of stream velocity into either deflection or stress. This is measured and calibrated against velocity or discharge. The deflection principle has been incorporated into the Keeler meter [11] and successfully applied to the measurement of outflow from Lake Winnipesaukee in New Hampshire. A hydrometric pendulum, developed at the Hydraulic Laboratories in Delft, has been utilized to measure velocities in the extremely turbulent rivers of Nigeria [12].

METHODS OF VELOCITY MEASUREMENT

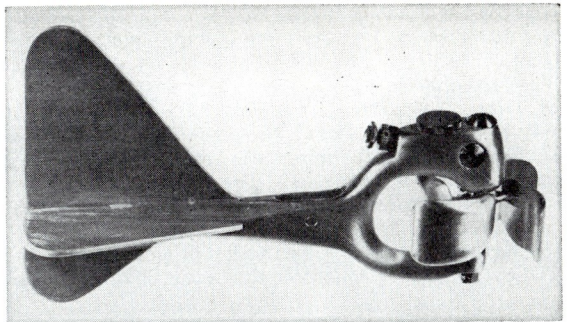

Fig. 15-5. Experimental vertical-vane meter.   (*U.S. Geological Survey.*)

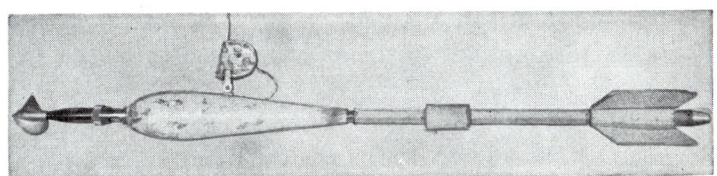

Fig. 15-6. Neyrpic meter and weight.   (*U.S. Lake Survey.*)

Fig. 15-7. Ott meter and sounding weight.   (*U.S. Lake Survey.*)

3. Float.   Approximate determinations of velocity can be made by floats. They should be used only when more precise methods are not available. A float consists of (1) a chip of wood tossed onto the surface of the water, (2) a device with adjustable buoyancy designed to travel at a selected depth below the stream surface and carrying a marker flag, or (3) a long rod floating vertically and reaching nearly to the stream bottom. Long, straight, and uniform channels are required for float measurements.

Observations must be taken along several ranges across the section and within the reach. If surface floats are used, a correction of approximately 0.8 to 0.9 is required to obtain the mean velocity. If a rod is floated with its lower end near the bottom, it presumably indicates the mean velocity.

4. The Pitot Tube. Operation of the Pitot tube is based on the principle that if a rounded body is immersed in a fluid, the pressure at its nose is a maximum, the velocity head $v^2/2g$ being converted into static head. A small hole is drilled at this point and connected to a pressure gage. The latter will indicate the sum of the pressure in the fluid and the pressure created by the conversion of the velocity head. If a second hole is drilled in the flow boundary at 90° with the direction of flow and carefully finished, a pressure gage attached to it will measure only the static pressure in the fluid. These openings are connected to the legs of a differential manometer for measurement.

The Pitot tube is not well adapted to streamflow measurement. Its greatest usefulness is in the measurement of high velocities in chutes, overfalls, or pressure conduits.

B. Chemical and Electrical Methods

Various chemical and electrical methods have been proposed for the measurement of streamflow. Their development has been the result of the need for greater precision, particularly in highly turbulent flow or very low velocities, and for more satisfactory ways of measuring under difficult or unusual conditions. The chemical methods include salt velocity, salt dilution, and the detection of radioactive tracers. The electrical devices include oxygen polarography, the hot-wire anemometer, electromagnetic voltage generation, and supersonic waves.

1. Salt Velocity. The salt-velocity method is based on the principle that salt introduced into the stream will increase its electrical conductivity. A small quantity of concentrated salt solution is quickly introduced from a pipe at several points across a stream section by means of a quick-operating valve. Electrodes are mounted at the ends of a uniform reach beginning a short distance downstream from the injection pipe. They are connected to a recording galvanometer with return circuit through the water. When the salt solution passes the first electrode, it produces a "hump" in the graph, followed by a second hump when the salt reaches the second electrode. The time between the centers of gravity of these humps divided into the length of reach is taken as the average stream velocity in the reach. The product of this velocity and the average cross-sectional area gives the discharge. The method is cumbersome and requires several skilled operators.

2. Salt Dilution. In the salt-dilution method a concentrated salt solution is introduced at a constant and measured rate. At a point some distance downstream, after complete mixing has taken place, samples are drawn and analyzed. The weight of the salt passing the sampling point per second must equal the sum of the weight of salt ordinarily carried by the stream and the weight of the concentrated solution added per second. A simple calculation will give the volume of water per second passing the sampling point. The method can be applied where the flow is so turbulent or the cross section so rough that no other method is feasible. If the stream is large, a high rate of salt introduction is required. This method requires skilled operators.

3. Radioactive Tracers. A variation of the salt-velocity method utilizes radioactive tracers. Some advantage may be gained by a saving in quantity of material required if the apparatus is sufficiently sensitive. The increased cost and the care needed to avoid danger of radiation exposure make the method somewhat disadvantageous.

4. Oxygen Polarography. If a cathode, well insulated except for its tip, is inserted into an electrolyte containing dissolved oxygen, a concentration gradient will be established at the solid-liquid interface, the magnitude of which is a function of the oxygen concentration and the voltage drop at the interface. This will be accompanied by a flow of electric current through the system. If the electrolyte is quiescent and the voltage across the interface is progressively increased, the current flow will increase slowly for a time with a large increase in voltage, then will increase at a much more rapid rate. On the other hand, if the electrolyte is stirred at a constant rate, the current will increase rapidly with increase in voltage to a certain point, whereupon the voltage will increase rapidly with little change in the current flow. This phenomenon, known as *oxygen polarography*, is described by Kolthoff and Lingane [13].

If the electrolyte is stirred at different rates, a series of curves is produced, varying with rate of rotation (Fig. 15-8). Some early experiments with a device based on this phenomenon were reported in 1953 [14].

This effect was found to be highly responsive to low velocities. It is thought that it could be developed to the point where it would be useful in the measurement of velocities below 0.1 ft/sec, a range which is not covered with accuracy by present meters.

Since 1953 some further attention has been paid to the electrical circuitry which could be developed, utilizing the transistor, which is rugged and well adapted to field instrumentation. It should be possible to develop a two-part electrical circuit, one part to follow the voltage-current relation developed by the probe, the other to follow a straight-line relation as shown on Fig. 15-8. Null points would be established which would permit calibration of probe current as a function of velocity.

5. Hot-wire Anemometer. The hot-wire anemometer, or its modern counterpart, the *warm-film* anemometer, consists of a probe whose surface temperature is raised relative to the fluid which it is measuring. Passage of the fluid over the warm surface cools it, which changes the resistance. This is reflected by a change in current or voltage across the probe, which can be calibrated against velocity. The instrumentation for this device is at present rather complicated and not suited to use in the field. The State University of Iowa, Iowa Institute of Hydraulic Research, has developed a warm-film anemometer for laboratory use.

6. Electromagnetic Flowmeter. The electromagnetic flowmeter operates on the principle that an electrical conductor, in this case water, passing through a magnetic field has a current generated within it. The generated current is gathered by two electrodes set with their axes mutually at right angles to the direction of flow and of the magnetic field. Experiments by the U.S. Geological Survey [15] with one of these instruments, furnished by the U.S. Navy, indicate that the device is eminently satisfactory in the measurement of tidal velocities on the St. Johns River at Jacksonville, Fla. Its size and cost preclude its use as a general field instrument, but it may prove valuable in obtaining continuous records of velocity under special conditions.

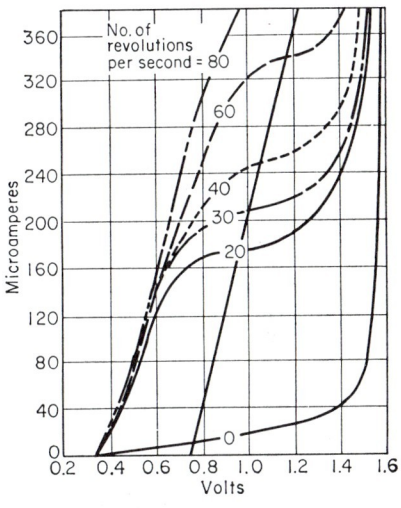

Fig. 15-8. Effect of electrolyte movement on current-voltage relation in oxygen polarography.

7. Ultrasonic Flowmeter. This device depends for its operation upon the Doppler effect on ultrasonic waves passing through water [15]. A transmitter directs a signal toward a receiver some distance upstream. The ultrasonic waves moving upstream are compressed; those returning are attenuated. The magnitude of this effect can be recorded and related to water velocity. Many variables are involved, including temperature effects, thermoclines, random reflections, water-surface interference, and the like.

V. THE ROTATING METER

The rotating meter is the most common device used for the gaging of streams. Each type operates with a high degree of precision. Each responds differently to the effects of turbulence and obliquity of flow.

A. Calibration

The rotating current meter is calibrated by towing at selected speeds in a long tank [3]. A curve of relation is prepared for the individual meter, with revolutions per second along one axis, and velocity, in feet per second, along the other. The points are connected by one or more straight lines. It is the present practice of the U.S. Geological Survey to rate the Price meter for rod suspension only. A coefficient is applied for other suspensions.

B. Precision of Operation

The precision with which a current meter measures stream velocity is a function of (1) the relation between velocity indication by a meter being towed in still water and being held in flowing water, (2) the effect of turbulent eddies on registration of the mean motion, and (3) the effect of the current striking the meter obliquely.

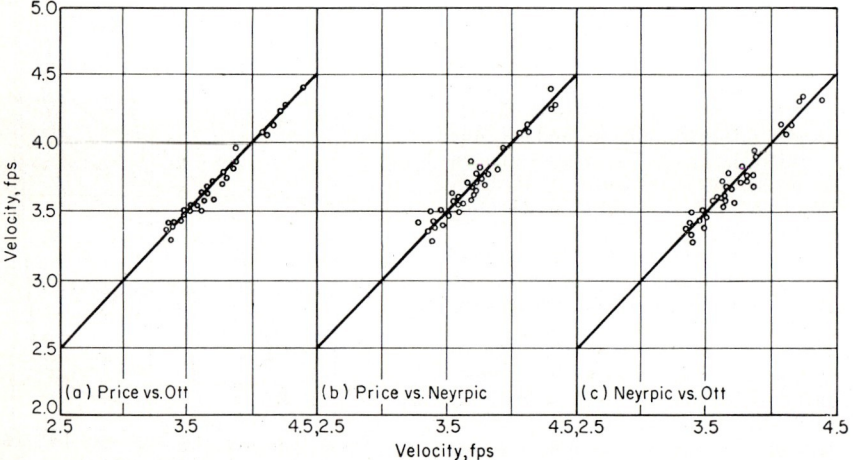

Fig. 15-9. Meter-registration comparisons. (a) Price vs. Ott; (b) Price vs. Neyrpic; (c) Neyrpic vs. Ott. (*U.S. Lake Survey*.)

1. Ratings in Still and Moving Water. It is not feasible to rate a current meter in flowing water, as there is no way to make precise determinations of the velocity past the meter without disturbing the flow. Ratings are therefore made by towing in still water. Comparisons cannot be made between ratings of a meter in still and moving water, except on a volumetric basis.

2. Effects of Turbulence. Turbulent flow affects the recording of mean motion by a rotating current meter. This phenomenon was investigated by Yarnell and Nagler [16]. They concluded that a flowing stream may be so turbulent as to change very noticeably the readings obtained by a current meter; that the inertial characteristics of the wheel are not a factor; and that the variations in meter readings seem to be due to the effect of obliquity of flow. Also, they concluded that if the meter is rigidly supported with its axis parallel to the general direction of streamflow, turbulence causes the vertical-axle cup meter to overregister and the horizontal-axle meter to underregister.

Following their conclusions, these investigators pointed out that a large proportion of actual streamflow measurements are made by cable suspension, in which the meter has considerable freedom for both lateral and vertical motion.

3. Effects of Obliquity. The studies by Yarnell and Nagler showed that substantial errors could result from a meter being held rigidly at an angle to the mean stream motion. The Price meter overregisters, whereas horizontal-axle meters underregister. If the meters were free to move, as on a line-and-weight suspension, they would wander with the flow and the result might be different. The work of Yarnell and Nagler should be extended to cover this condition of meter operation.

C. Field Comparison

The U.S. Lake Survey [8] recently made a careful comparison of velocity indications by the small Price, the Ott, and the Neyrpic meters in the Stella Niagara hydraulic section in the lower Niagara River. The three meters are shown in Figs. 15-4, 15-6, and 15-7. The tests were run at depths up to 50 ft and velocities of approximately 3 ft/sec. The measured discharge was 192,000 cfs. The results of these tests are shown in Fig. 15-9. It will be noted that there is no consistent difference in velocity indication among the meters.

VI. VELOCITY OBSERVATION

A. Distribution in the Vertical

For the condition of turbulent flow over a rough boundary, the following equation [17] may be written:

$$v = 2.5 \sqrt{\frac{\tau_0}{\rho}} \ln \frac{y}{y_0} \tag{15-1}$$

where v is the velocity, $\sqrt{\tau_0/\rho}$ the shear velocity, y the distance above the boundary, and y_0 the distance of the point of zero velocity above the boundary. Integrating over the range from y_0 to D, dividing by D, and considering y_0/D to be negligible,

$$V = 2.5 \sqrt{\frac{\tau_0}{\rho}} \ln \frac{D}{ey_0} \tag{15-2}$$

where V is the mean velocity, D is the depth, and $e = 2.718\ldots$, or the base of natural logarithms. This equation is equivalent to Eq. (15-1) by substituting D/e, or approximately $0.368D$, for y in the latter.

It is the American practice to take single velocity observations at 0.6 the depth, measured from the surface, in shallow streams. This corresponds closely to $0.368D$ in Eq. (15-2). Also, in deep streams observations are taken at 0.2 and 0.8 the depth and averaged. If these depth ratios are substituted in Eq. (15-1) and the velocities averaged, the result is nearly equivalent to substituting $0.368D$ in Eq. (15-1). This compares favorably also with the mean depth as obtained from Eq. (15-2).

B. Setting the Meter

As noted above, it is general practice in the United States to set the meter at 0.6 the depth in shallow streams, or take the average of readings at 0.2 and 0.8 the depth in deeper streams. The minimum depth at which the two-point settings can be made is determined by the meter suspension. If a rod is used, the minimum is 1.5 ft in smooth cross sections. A lesser depth would bring the meter too close to the bottom at the lower setting. If the meter is suspended on a hanger above a weight, the minimum depth for 0.2- and 0.8-depth settings is 5 times the distance from the center of the meter to the bottom of the weight.

Occasionally, observations in very deep streams are made at 0.2, 0.6, and 0.8 depths and the velocity at 0.6 depth averaged with that determined from the average of the observations at 0.2 and 0.8 depth. When measurements are made under ice cover, the meter is sometimes set at 0.5 the depth and a coefficient applied, usually 0.88. Some streams are too swift and deep during floods to permit observation of the velocity at 0.8 depth. Observations are taken just below the surface or at 0.2 the depth. These are corrected by a coefficient determined from complete measurements made at lower stages.

C. Timing the Rotation

A velocity observation is made by timing the number of revolutions which the meter makes during a selected period of time. Ordinarily, it is the practice to time the number of revolutions for about one minute. Periods appreciably less may introduce errors because of stream pulsations, while longer periods do not increase the accuracy commensurate with the additional time required.

D. Meter Circuitry

Modern meters use a make-and-break electrical contact, with the signals made audible in a headset. If the meter is suspended on a rod, current from a battery flows through a wire to the contact chamber of the meter and returns through the rod. If the meter is suspended on a hanger above a weight, the connecting wire is carried to the meter as the insulated core of the cable, and the cable provides the return path. Contact is made at either single- or five-revolution intervals by means of a spring wire and eccentric cam and gear in the contact chamber.

E. Timing Procedure

A stop watch is started at the beginning of a click heard in the receiver. The clicks are counted over a period between 40 and 60 sec. As a practical matter, the number of clicks should be recorded in multiples of 5 or 10, to simplify application of the meter rating table. The rate of counting should not exceed two per second on the single-count circuit since some personal error may be introduced. The circuit is set on the five-revolution counter in measuring swift streams.

F. Counting and Recording Automats

Several devices have been developed for counting and recording the revolutions of a current meter. They are principally laboratory devices, utilizing the make-and-break principle or variations in inductance, capacitance, or generated voltage. There is a definite need for a self-registering counter for field use. It would simplify the work of the hydrographer, reduce the chance of personal error, and provide a means for rapid computation of the velocity for a discharge measurement. With the advent of the transistor, there is now available a simple and rugged device for electronically controlling an electrical circuit. It should be a simple matter to construct an amplifying device so that each contact could be made to actuate a mechanical counter. This counter would be started and stopped with a timer. Because of the low power demand of a transistor circuit, this device could easily be made portable and sufficiently rugged for field use.

VII. DETERMINATION OF DEPTH

Depth is measured by several methods. They include graduations on the wading rod, tags on the cable which supports the meter-and-weight assembly, or a mechanical indicator attached to the reel on which the supporting cable is wound.

A. Wading Rod

The wading rod is graduated in tenths of a foot, with the zero at the bottom of the base plate. The depth is read by estimating the point at which the level surface of the water intersects the rod. Care should be taken to avoid over- or underreading because of the disturbance to the water surface. Observations can usually be made to half-tenths with good accuracy.

B. Cable and Weight

The meter-and-sounding-weight assembly is supported on a steel cable. This is fastened to a rubber-covered two-conductor cable which serves as a hand line for making measurements with 15- and 30-lb sounding weights. If heavier weights are required, the steel cable is wound on a sounding reel. When the hand line is used, the steel cable is tagged at 2-ft intervals from the bottom of the weight, thus facilitating the determination of depth. The sounding reel is equipped with a mechanical counter so that the depth may be read directly.

C. Fathometer

The fathometer has proved successful in the determination of stream depths. A small transducer has been adapted for installation in a sounding weight, and the depth can be read from the fathometer by placing the weight and the transducer just beneath the water surface. A sounding weight is still necessary for the setting of the meter. If the river is too deep or swift to permit setting the meter at the 0.8 depth for an observation, readings may be taken at 0.2 depth and adjusted. With a fathometer, it would not be necessary to depend on a cross section developed at a lower stage for computation of meter setting.

D. Sources of Error

The two principal sources of error are oversounding in alluvial streams from the sounding weight sinking into the stream bottom and apparent oversounding in swift streams caused by drag on the line and the meter-and-weight assembly. Also, if the meter is raised and lowered in the same spot or held within close proximity of the bottom, the scouring action caused by the velocity disturbance tends to cause the formation of a hole beneath it.

A sensitive bottom indicator can be made by placing a trigger mechanism and disk below the sounding weight. When this device contacts the bottom, a signal is heard in the telephone receiver. If the weight is then lowered until the bottom is indicated in the normal manner by slack in the supporting cable, the magnitude of the oversounding will be indicated.

When a meter and weight are dragged downstream in swift currents, the indicated depth is increased both by the bow in the line beneath the water surface and by its angularity in the air. The U.S. Geological Survey has developed a set of tables for making both air-line and submerged-line corrections [3], based on the angle which the cable makes with the vertical. This is measured by means of a protractor attached to the pulley over which the cable passes as it leaves the supporting boom. These corrections are given in Tables 15-3 and 15-4. The air correction can be eliminated by tagging the cable and making the depth determination by observing the tags as they pass beneath the water surface.

VIII. DISCHARGE MEASUREMENT

A. Types of Measurement

A discharge measurement is classified according to the manner in which the stream crossing is made, i.e., by wading, cableway, bridge, boat, or ice.

1. Wading. If a stream is shallow and relatively slow moving, a wading measurement is indicated. The limit is determined by the ability of the hydrographer to cross safely and to stand in position while making an observation. Experience indicates an upper limit at a discharge of 12 cfs/ft of width.

The stream is examined for a cross section in which the flow is as smooth and uniformly distributed as possible. Care should be taken to avoid sections in which the flow may be at varying angles, as this requires the measurement of the effect of angularity and reduces the accuracy of the determination.

A tag line, usually a tape or small-diameter metal cable with beads marking 2- or 5-ft intervals, is stretched across the stream. The engineer notes the width of flow during this procedure, which enables him to determine the spacing of the soundings.

The number of soundings to be taken depends upon the regularity of the cross section and the distribution of the flow. Usually not less than 20 should be made. The soundings are closely spaced through irregular parts of the section or where the velocity is changing rapidly and are spaced more widely in the center of the stream where the flow is uniform. However, the flow through the section between any pair of adjacent soundings should not be more than 5 per cent of the total.

If the small Price meter is used, velocity observations are generally taken at 0.2 and 0.8 depth if the depth is 1.5 ft or greater and at 0.6 depth if it is less than 1.5 ft. If the pygmy Price meter is used, the depth above which 0.2 and 0.8 observations are taken may be as little as 1.0 ft.

In wading measurements the engineer should stand in a position which will least affect the distribution of flow passing the current meter. With the meter rod at the tag line, the engineer will face along the line toward the bank, standing 1 to 3 in. downstream from the tag line and 18 in. or more from the meter rod, supporting the rod with his upstream arm. Care must be taken to keep the rod in a vertical position and to point the meter directly into the flow.

If the flow of the stream is not at right angles to the tag line, a coefficient of adjustment is determined. The U.S. Geological Survey prints a template on the meter-note forms. The edge of the form is aligned with the tag line, a pencil is laid across the form through a point and in line with the direction of the current meter, and the coefficient (the cosine of the angle) is read directly.

2. Cableway. A cableway consists of a cable stretched across the stream, supported at the banks by A-frames and securely anchored at each end. It is placed sufficiently high to permit the operation of a cable car above the elevation of the maximum flood or to clear navigation if the river carries water-borne traffic. The cable is designed to support safely a gaging car, its occupants, and the measuring equipment. A high degree of safety must be provided, for during floods the meter may become entangled in debris and, before it can be cut loose, impart a large strain on the cableway.

The gaging car may be of the sitdown type, in which the hydrographer sits and makes soundings with a hand line operated along the side of the car or by means of a small reel supported on a frame. If large weights are required for soundings, a stand-up car is used, which allows the hydrographer greater freedom of movement and permits the installation of a large reel (Fig. 15-10).

The cable is marked off in intervals by short stripes around its circumference. The spacing between stripes is determined by the width of the stream. They are generally at 2-, 5-, 10-, 25-, or 50-ft intervals, the interval being such that the normal cross section is covered by at least 30 marks. The markings are coded so that stationing can be recognized. The marking is generally begun from the center of the A-frame support on the bank at which the measurement is started.

3. Bridge. Two types of bridge are used for streamflow measurements, those constructed especially for that purpose and those used by highway or rail transportation. The first type serves essentially the same purpose as a cableway. Railroad and highway bridges, frequently the only available crossings from which streamflow measurements can be made, have two serious drawbacks: they are dangerous, and the stream cross sections beneath them are usually disrupted by piers or piles.

Discharge measurements are made from bridges by hand line, hand-operated reels and booms, power-operated reels and booms attached to trucks or trailers, or power-operated reels and booms mounted on prime movers constructed especially for the purpose.

Measurements are generally made from the downstream side. As the meter drags downstream it does not foul against the bridge substructure and its operation can be observed. During periods of floating drift or ice it is necessary to station an observer along the upstream side of the bridge. Distances for soundings are determined by marking the handrail of the bridge in the same manner as a cableway, permission first being obtained from the proper authority.

Soundings are made in the same manner as from cableways, except that more sections are required, particularly around piers and piles. Spacing is selected which will give 30 or more sections if pier interference is not considered, with additional observations near the piers. When a pier is approached, short sections are taken, the

FIG. 15-10. Discharge measurement from cableway and stand-up car.

last being as close to the pier as safety of equipment permits. The principal error is introduced by the turbulence generated by the pier. If the bridge is at an angle to the flow, a cosine correction is made.

4. Boat. Measurement from a boat is a satisfactory way of determining stream discharge if conditions are favorable for its operation. The requirements are that the stream be safe for boats and that a suitable cross section be available. A boat measurement is in a sense a combination of wading and cableway operation, the hydrographer working from just above the water surface but using meter and sounding weight rather than a rod for depth determination.

Cross-section distances are established from a tagged cable stretched across the stream just above the water surface. This line is similar to that used for wading measurements, but much stronger, since it must span a greater distance and is utilized to hold the boat in position against the current.

The sounding equipment is operated by a reel, with the line passing over a boom extending ahead of the bow of the boat. It should extend sufficiently far forward so that the meter will be free of any interference by the boat when the 0.2-depth velocity observation is taken. The sections should number 30 or more, depending on the uniformity of depth and velocity distribution.

The usual precautions of life preservers and other safety devices must be taken. In addition, if the section is in a reach where river traffic passes or is used by sports enthusiasts, a quick-operating release with an operator is provided at one end of the tagged cable, so that the cable may be quickly lowered to the stream bed.

5. Ice. When a stream is frozen over and there are no open-water sections available, it is necessary to make discharge measurements through the ice cover. This is accomplished by cutting a series of holes, sufficiently large to admit a meter mounted on a rod or hanger and weight. If it is necessary to measure through the ice each winter, a site should be selected during the open-water period.

Depths for ice measurements are determined by subtracting the ice thickness from the total depth as observed with the meter equipment. Ice thickness is measured by means of an L-shaped flat steel rod, with the horizontal part at least 6 in. long. This is thrust beneath the ice and hooked to the underside. The thickness is read on graduations of the vertical section. Usually both this reading and the total depth reading are referred to the water surface.

Velocities are taken at 0.2 and 0.8 the depth beneath the ice if practicable, otherwise at 0.5 or 0.6 depth, with a coefficient being applied, usually 0.88 for the first and 0.92 for the second. The meter is suspended on a rod, if the stream is shallow, or on a hanger and weight. A light reel, mounted on a framework over a pair of runners, is an efficient device for use in measurements with sounding weight.

Personal safety is of great importance in ice work. Care should be taken not to go out on ice that is too thin. If the channel is partly open and a measurement must be made, a light boat equipped with reel and boom can be pushed along and used to cross the open water.

Power-driven ice augers have been developed for cutting holes in ice. The vertical-vane current meter, mounted at the end of a rod without tailpiece, is being used experimentally for ice measurements in shallow streams. It is easily inserted in a hole cut by the ice auger.

B. Recording the Data

The data to be recorded include distance from initial point, depth, depths of observation, revolutions of the meter, and time in seconds. In addition, if the direction of flow is not perpendicular to the section, the cosine of the angle of departure is indicated. A typical measurement, recorded on Form 9-275-F of the U.S. Geological Survey is shown in Fig. 15-11.

C. Factors Affecting Accuracy

The factors which affect the accuracy of a discharge measurement include instrumental, methodological, personal, computational, and sampling.

1. Instrumental. Instrumental error is introduced if the meter operation differs from that for which the rating was determined. This may be the result of damage or external interference, such as the accumulation of moss or trash around the shaft or the collection of slush ice on the rotor. This form of inaccuracy can be reduced by care in maintenance and operation of the meter.

2. Methodological. Methodological errors include improper setting of the meter for point observation of velocity and failure to correct apparent depths for the angularity of the sounding line or widths for the angularity of flow with respect to the cross section.

3. Personal. Personal error is introduced if the hydrographer tends to set the meter higher or lower than the computed setting; is inattentive in counting and timing the revolutions; does not frequently check his stop watch for accuracy; or otherwise fails to follow the procedure for making a streamflow measurement.

4. Computational. Computational errors include improper application of the rating table and the simple arithmetic errors of averaging, multiplying, adding, and subtracting.

DISCHARGE MEASUREMENT 15–21

```
9-275-F                UNITED STATES              Meas. No. _____
Jan. 1956        DEPARTMENT OF THE INTERIOR
                      GEOLOGICAL SURVEY           Comp. by _____
                    WATER RESOURCES DIVISION
                 DISCHARGE MEASUREMENT NOTES      Checked by _____
```

Wabash River at Montezuma
Date Mar. 21, 1960 Party Smith and Jones
Width 1153 Area 2229.0 Vel. 3.33 G. H. 24.24 Disch. 74330
Method 2-.8 No. secs. 37 G. H. change +.13 in 2 hrs. Susp. 150 C
Method coef. 1.0 Hor. angle coef. 1.0 Susp. coef. 1.0 Meter No. 2459

GAGE READINGS			
Time	Elec	Recorder	Inside	Outside
7:35	24.15	24.16	24.15	24.17
8:00	start		24.19	
10:10	End			
10:30			24.32	
11:00	24.37	24.38	24.37	24.39
Weighted M. G. H.			24.24	
G. H. correction			0	
Correct M. G. H.				

Date rated 5/25/59 Used rating
for rod ___ susp. Meter 1.0 ft.
above bottom of wt. Tags checked ___
Spin before meas. OK after OK
Meas. plots ___% diff. from ___ rating
Wading, cable, ice, boat, upstr., downstr., side
bridge at feet, mile, above, below
gage, and _____
Check-bar, chain found 44.501
changed to OK at ___
Correct ___
Levels obtained ___

Measurement rated excellent (2%), good (5%), fair (8%), poor (over 8%), based on following
conditions: Cross section _____
Flow Uniform   Weather Good
Other _____   Air 50 °F @
Gage _____   Water 46 °F @
                    Record removed NO   Intake flushed Yes
Observer Visited

Control Clean

Remarks No overflow.

G. H. of zero flow _____ ft.

16—58354-4

Fig. 15-11. Discharge-measurement notes.

.0 .10 .20 .30 .40 .50 .60 .70 .75

River at—

Angle coefficient	Dist. from initial point	Width	Depth	Observation depth	Revolutions	Time in seconds	VELOCITY At point	VELOCITY Mean in vertical	Adjusted for hor. angle or --------	Area	Discharge	
												.80
	10	10	0	Left edge of water							0	
					8:00 AM							.85
	30	25	6.3	.2	20	46	.98	1.08		158	171	
				.8	25	47	1.19					
	60	30	20.8	.2	40	40	2.22	2.50		624	1560	.90
				.8	50	40	2.77					.92
	90	30	27.2	.2	60	42	3.17	3.25		816	2650	.94
				.8	60	40	3.33					
	120	30	33.5	.2	80	45	3.94	4.08		1000	4080	.96
				.8	80	42	4.22					.97
	150	30	37.8	.2	80	49	3.62	4.16		1130	4700	.98
				.8	100	47	4.71					.99
	180	23	34.8	.2	100	45	4.92	3.52		800	2820	
				.8	40	42	2.12					
○	196	8	33.8	.2	100	42	5.27	3.48		270	940	1.00
				.8	40	53	1.68					
			PIER									
	202	9	41.5	.2	100	40	5.53	4.27		374	1600	.99
				.8	80	59	3.01					.98
	220	24	43.8	.2	100	41	5.40	4.02		1050	4220	.97
				.8	50	42	2.64					.96
	250	30	42.1	.2	150	54	6.15	5.59		1260	7040	
		(8:50 AM)		.8	100	44	5.03					.94
	280	30	37.5	.2	150	49	6.78	6.16		1120	6900	.92
				.8	100	40	5.53					.90
	310	30	38.4	.2	150	51	6.51	5.72		1150	6580	
				.8	100	45	4.92					
	340	30	35.5	.2	150	47	7.06	6.34		1060	6720	.85
				.8	150	59	5.63					
	370	29	30.3	.2	150	53	6.26	4.80		879	4220	
				.8	60	40	3.33					.80

.0 .10 20 .30 .40 .50 .60 .70 .75

Fig. 15-11. (*Continued*)

DISCHARGE MEASUREMENT 15–23

.0	.10	.20	.30	.40	.50	.60	.70	.75	
398	14	25.5	.2	100	47	4.71	4.76	357	1700
			.8	100	46	4.81			.80
		PIER							
404	13	32.5	.2	100	42	5.27	5.40	422	2280
			.8	100	40	5.53			.85
430	28	23.7	.2	80	49	3.63	3.36	664	2230
			.8	60	43	3.09			
460	35	14.6	.2	20	40	1.12	1.08	511	552	.90
			.8	20	43	1.04			.92
500	45	11.8	.2	20	41	1.09	1.24	531	658	.94
			.8	25	40	1.40			
550	50	11.0	.2	20	40	1.12	.94	550	517	.96
			.8	15	44	.77			.97
600	25	11.5	.2	7	50	.33	.32	288	92	.98
	(9:25 AM)	.8	7	54	.30				.99
		PIER							
605	27.5	12.0	.2	25	42	1.33	1.33	330	439
			.8	25	42	1.33			1.00
660	47.5	11.2	.2	25	53	1.06	1.09	532	580
			.8	20	40	1.12			
700	45	10.0	.2	20	43	1.04	1.34	450	603	.99
			.8	30	41	1.63			.98
750	51	11.5	.2	50	41	2.71	2.54	586	1490	.97
			.8	50	47	2.36			.96
802	26	11.9	.2	25	40	1.40	1.36	309	420	.94
			.8	30	51	1.31			.92
		PIER							
807	26.5	14.0	.2	60	41	3.24	3.20	371	1190	.90
			.8	60	42	3.17			
860	46.5	16.2	.2	40	49	1.82	1.43	753	1080
			.8	20	43	1.04			.85
900	45	16.7	.2	15	55	.62	.72	752	541
			.8	15	42	.81			.80

Fig. 15-11. (*Continued*)

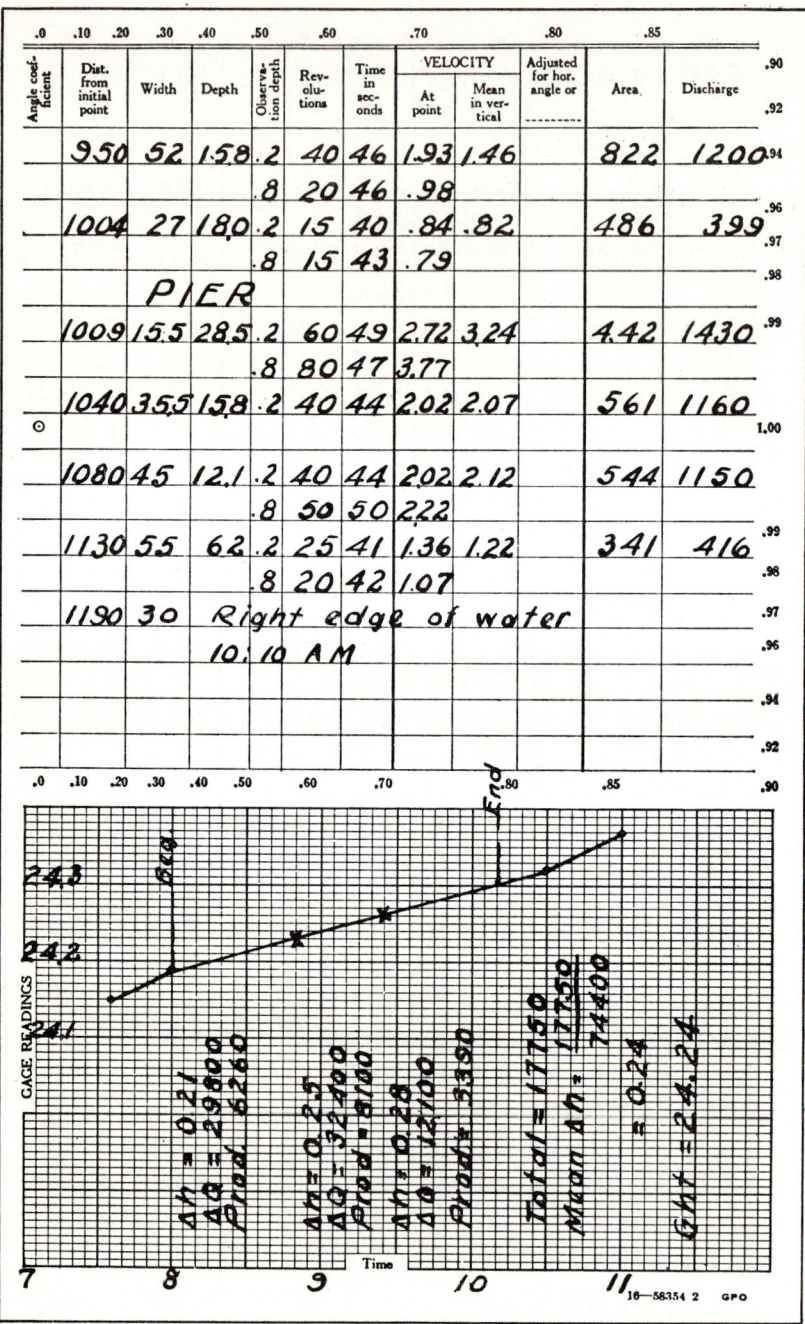

Fig. 15-11. (*Continued*)

5. Sampling. Sampling errors include insufficient or improperly distributed soundings or poor selection of cross section.

All five types of error can be reduced by proper training of the hydrographer by someone long familiar with the procedure of stream gaging. Good judgment plays an important role.

D. Computation Procedure

A discharge measurement is computed by determining the discharge per unit width for each sounding, multiplying it by the width corresponding to one-half the distance to each adjacent sounding to determine the partial discharge. The partial discharges are added to obtain the total.

The gage height for a discharge measurement is computed as a weighted mean in accordance with the equation

$$h = \frac{\Sigma h \, \Delta Q}{Q} \qquad (15\text{-}3)$$

A graph of the gage-height change with time is plotted. Times are noted during the progress of the discharge measurement. The partial discharges ΔQ are individually multiplied by their average gage heights h for the corresponding times. The sum is divided by the total discharge Q.

E. A Method for Rapid Computation

A discharge measurement requires considerable time for its computation and checking. Much of the time is spent in entering the meter rating table to obtain and check the velocities. If each velocity observation were taken over a constant length of time, 50 sec for example, the computation could be made using the revolutions rather than velocity. This can be done with negligible error for the Price meter, as its rating usually consists of two straight lines, meeting at approximately one revolution per second and having only slightly different slopes.

The procedure is as follows:

1. Make the discharge measurement by recording the time nearest to 50 sec and the corresponding revolutions.
2. If the observation is taken at 0.6 depth, multiply the number of revolutions by the partial area and put in the discharge column of the note form. This column should be re-marked "$N \times$ area" to avoid confusion. For each observation where 0.2- and 0.8-depth readings were taken, average the revolutions and multiply by the area.
3. Summate the products of "$N \times$ area."
4. Divide the summation of the product by the total area. This gives the average number of revolutions for the measurement.
5. Average the times.
6. Interpolate in the current-meter rating table for the velocity. If this method were adopted, it would not be necessary to prepare a rating table. The equation between revolutions per second and velocity would suffice.

The discharge measurement shown in Fig. 15-11 was recomputed by adjusting the observations to 50 sec and applying this method. The mean velocity thus obtained was 3.315 ft/sec, as compared with 3.334 ft/sec by the conventional method.

F. Correction Factors

The correction factors to a discharge measurement include adjustments for the effect of angularity, a coefficient for meter suspension, and air-line and wet-line corrections to observed depths in swift streams.

1. Angularity. If the flow is not at right angles to the direction of the cross section, a correction for angularity is applied. This is the cosine of the angle which the flow makes with the cross section. Properly, it should be applied to the width.

2. Suspension Coefficient. A current meter is rated on rod suspension. The average velocity for the measurement is corrected by a coefficient which is a function of the method of suspension. Table 15-2 is a list of the principal methods of suspension and the coefficients to be applied.

Table 15-2. Coefficients for Converting Round-rod Suspension Ratings to Rating with Columbus Weights*

Suspension†	Velocity‡	Coefficient
15 C .5′	0.25 .5–2.9 3.0-up	1.01 .99 .995
30 C .5′	0.25 .5-up	1.020 1.005
50 C .55′ (meter at hole of C-type hanger marked "15")	.5 1.0–4.9 5.0-up	1.005 .995 1.00
50 C .9′ 75 C 1.0′ 100 C 1.0′	.5-up	1.00
150 C 1.0′	.5-up	.995

* A standard design adopted by the U.S. Geological Survey [3] in 1934.
† Denotes distance from horizontal axis of current meter to bottom of weight, as 15 C .5′ is 0.5 ft above bottom of 15-lb Columbus (C) weight.
‡ Coefficients for velocities not shown may be obtained by interpolation.

3. Air-line and Wet-line Corrections. The air-line and wet-line corrections are determined from the angle which the sounding line makes with the vertical and the height of the protractor above the water surface. These are applied to the observed depth to obtain the true depth. In Table 15-3 are given the air-line corrections for angles from 4 to 36° and distances above the water surface from 10 to 100 ft. In Table 15-4 are given the wet-line corrections for the same ranges in angle and depth. These are reproduced from *U.S. Geological Survey Water-Supply Paper* 888 [3].

When making air-line and wet-line corrections, the observed depth is first corrected for the air-line effect to obtain the wet-line observation. This, in turn, is corrected to determine the true depth.

IX. THE HYDROMETRIC STATION

A. Selection of Site

There are four criteria for establishing a hydrometric station: accessibility, adequacy, stability, and permanency.

1. Accessibility. The station should be accessible under all conditions, particularly during floods. In areas where severe winter weather is a problem, it is important to consider methods of access during periods of snow cover. The helicopter is proving a boon in reaching stations which are far from open roads in winter.

2. Adequacy. By adequacy is meant the ability of the station to cover the full range of discharge which may occur. This includes gage equipment which will encompass the full range of stage and structures from which discharge measurements may be made under all conditions. If the stage records are to be obtained by means of a water-stage recorder, the intakes to the well must be below the lowest flow and the shelf of the recorder shelter above the highest flood.

THE HYDROMETRIC STATION

Table 15-3. Air-line Adjustments to Soundings

Vertical length, ft	4°	6°	8°	10°	12°	14°	16°	18°	20°	22°	24°	26°	28°	30°	32°	34°	36°
10	0.02	0.06	0.10	0.15	0.22	0.31	0.40	0.51	0.64	0.79	0.95	1.13	1.33	1.55	1.79	2.06	2.36
12	0.03	0.07	0.12	0.19	0.27	0.37	0.48	0.62	0.77	0.94	1.14	1.35	1.59	1.86	2.15	2.47	2.83
14	0.03	0.08	0.14	0.22	0.31	0.43	0.56	0.72	0.90	1.10	1.32	1.58	1.86	2.17	2.51	2.89	3.30
16	0.04	0.09	0.16	0.25	0.36	0.49	0.64	0.82	1.03	1.26	1.51	1.80	2.12	2.48	2.87	3.30	3.78
18	0.04	0.10	0.18	0.28	0.40	0.55	0.73	0.93	1.16	1.41	1.70	2.03	2.39	2.78	3.23	3.71	4.25
20	0.05	0.11	0.20	0.31	0.45	0.61	0.81	1.03	1.28	1.57	1.89	2.25	2.65	3.09	3.58	4.12	4.72
22	0.05	0.12	0.22	0.34	0.49	0.67	0.89	1.13	1.41	1.73	2.08	2.48	2.92	3.40	3.94	4.54	5.19
24	0.06	0.13	0.24	0.37	0.54	0.73	0.97	1.24	1.54	1.88	2.27	2.70	3.18	3.71	4.30	4.95	5.67
26	0.06	0.14	0.26	0.40	0.58	0.80	1.05	1.34	1.67	2.04	2.46	2.93	3.45	4.02	4.66	5.36	6.14
28	0.07	0.15	0.28	0.43	0.63	0.86	1.13	1.44	1.80	2.20	2.65	3.15	3.71	4.33	5.02	5.77	6.61
30	0.07	0.17	0.29	0.46	0.67	0.92	1.21	1.54	1.93	2.36	2.84	3.38	3.98	4.64	5.38	6.19	7.08
32	0.08	0.18	0.31	0.49	0.71	0.98	1.29	1.65	2.05	2.51	3.03	3.60	4.24	4.95	5.73	6.60	7.55
34	0.08	0.19	0.33	0.52	0.76	1.04	1.37	1.75	2.18	2.67	3.22	3.83	4.51	5.26	6.09	7.01	8.03
36	0.09	0.20	0.35	0.56	0.80	1.10	1.45	1.85	2.31	2.83	3.41	4.05	4.77	5.57	6.45	7.42	8.50
38	0.09	0.21	0.37	0.59	0.85	1.16	1.53	1.96	2.44	2.98	3.60	4.28	5.04	5.88	6.81	7.84	8.97
40	0.10	0.22	0.39	0.62	0.89	1.22	1.61	2.06	2.57	3.14	3.79	4.50	5.30	6.19	7.17	8.25	9.44
42	0.10	0.23	0.41	0.65	0.94	1.29	1.69	2.16	2.70	3.30	3.97	4.73	5.57	5.60	7.53	8.66	9.91
44	0.11	0.24	0.43	0.68	0.98	1.35	1.77	2.26	2.82	3.46	4.16	4.95	5.83	6.81	7.88	9.07	10.39
46	0.11	0.25	0.45	0.71	1.03	1.41	1.85	2.37	2.95	3.61	4.35	5.18	6.10	7.12	8.24	9.49	10.86
48	0.12	0.26	0.47	0.74	1.07	1.47	1.93	2.47	3.08	3.77	4.54	5.40	6.36	7.43	8.60	9.90	11.33
50	0.12	0.28	0.49	0.77	1.12	1.53	2.02	2.57	3.21	3.93	4.73	5.63	6.63	7.74	8.96	10.31	11.80
52	0.13	0.29	0.51	0.80	1.16	1.59	2.10	2.68	3.34	4.08	4.92	5.86	6.89	8.04	9.32	10.72	12.28
54	0.13	0.30	0.53	0.83	1.21	1.65	2.18	2.78	3.47	4.24	5.11	6.08	7.16	8.35	9.68	11.14	12.75
56	0.14	0.31	0.55	0.86	1.25	1.71	2.26	2.88	3.59	4.40	5.30	6.31	7.42	8.66	10.03	11.55	13.22
58	0.14	0.32	0.57	0.89	1.30	1.78	2.34	2.98	3.72	4.55	5.49	6.53	7.69	8.97	10.39	11.96	13.69
60	0.15	0.33	0.59	0.93	1.34	1.84	2.42	3.09	3.85	4.71	5.68	6.76	7.95	9.28	10.75	12.37	14.16
62	0.15	0.34	0.61	0.96	1.39	1.90	2.50	3.19	3.98	4.87	5.87	6.98	8.22	9.59	11.11	12.79	14.64
64	0.16	0.35	0.63	0.99	1.43	1.96	2.58	3.29	4.11	5.03	6.06	7.21	8.48	9.90	11.47	13.20	15.11
66	0.16	0.36	0.65	1.02	1.47	2.02	2.66	3.40	4.24	5.18	6.25	7.43	8.75	10.21	11.83	13.61	15.58
68	0.17	0.37	0.67	1.05	1.52	2.08	2.74	3.50	4.36	5.34	6.44	7.66	9.01	10.52	12.18	14.02	16.05
70	0.17	0.39	0.69	1.08	1.56	2.14	2.82	3.60	4.49	5.50	6.62	7.88	9.28	10.83	12.54	14.44	16.52
72	0.18	0.40	0.71	1.11	1.61	2.20	2.90	3.71	4.62	5.65	6.81	8.11	9.55	11.14	12.90	14.85	17.00
74	0.18	0.41	0.73	1.14	1.65	2.27	2.98	3.81	4.75	5.81	7.00	8.33	9.81	11.45	13.26	15.26	17.47
76	0.19	0.42	0.75	1.17	1.70	2.33	3.06	3.91	4.88	5.97	7.19	8.56	10.08	11.76	13.62	15.67	17.94
78	0.19	0.43	0.77	1.20	1.74	2.39	3.14	4.01	5.01	6.13	7.38	8.78	10.34	12.07	13.98	16.09	18.41
80	0.20	0.44	0.79	1.23	1.79	2.45	3.22	4.12	5.13	6.28	7.57	9.01	10.61	12.38	14.33	16.50	18.89
82	0.20	0.45	0.81	1.27	1.83	2.51	3.30	4.22	5.26	6.44	7.76	9.23	10.87	12.69	14.69	16.91	19.36
84	0.20	0.46	0.83	1.30	1.88	2.57	3.39	4.32	5.39	6.60	7.95	9.46	11.14	12.99	15.05	17.32	19.83
86	0.21	0.47	0.85	1.33	1.92	2.63	3.47	4.43	5.52	6.75	8.14	9.68	11.40	13.30	15.41	17.73	20.30
88	0.21	0.48	0.87	1.36	1.97	2.69	3.55	4.53	5.65	6.91	8.33	9.91	11.67	13.61	15.77	18.15	20.77
90	0.22	0.50	0.88	1.39	2.01	2.75	3.63	4.63	5.78	7.07	8.52	10.13	11.93	13.92	16.13	18.56	21.25
92	0.22	0.51	0.90	1.42	2.06	2.82	3.71	4.73	5.90	7.22	8.71	10.36	12.20	14.23	16.48	18.97	21.72
94	0.23	0.52	0.92	1.45	2.10	2.88	3.79	4.84	6.03	7.38	8.90	10.58	12.46	14.54	16.84	19.38	22.19
96	0.23	0.53	0.94	1.48	2.14	2.94	3.87	4.94	6.16	7.54	9.09	10.81	12.73	14.85	17.20	19.80	22.66
98	0.24	0.54	0.96	1.51	2.19	3.00	3.95	5.04	6.29	7.70	9.27	11.03	12.99	15.16	17.56	20.21	23.13
100	0.24	0.55	0.98	1.54	2.23	3.06	4.03	5.15	6.42	7.85	9.46	11.26	13.26	15.47	17.92	20.62	23.61

3. Stability. The stage-discharge relation should change with time as little as possible. It is important both from the standpoint of cost of operation and accuracy of the runoff computations to choose a site so that the relation between stage and discharge remains reasonably stable.

4. Permanency. The station should be so situated that the installation is not likely to be disturbed. One of the most important features of a streamflow record is its unbroken length. Also, it should be above points of flow diversion.

Table 15-4. Wet-line Adjustments to Soundings

Wet-line depth, ft	4°	6°	8°	10°	12°	14°	16°	18°	20°	22°	24°	26°	28°	30°	32°	34°	36°
10	0.01	0.02	0.03	0.05	0.07	0.10	0.13	0.16	0.20	0.25	0.30	0.35	0.41	0.47	0.54	0.62	0.70
12	0.01	0.02	0.04	0.06	0.09	0.12	0.15	0.20	0.24	0.30	0.36	0.42	0.49	0.57	0.65	0.74	0.84
14	0.01	0.02	0.04	0.07	0.10	0.14	0.18	0.23	0.29	0.35	0.41	0.49	0.57	0.66	0.76	0.87	0.98
16	0.01	0.03	0.05	0.08	0.12	0.16	0.20	0.26	0.33	0.40	0.47	0.56	0.65	0.76	0.87	0.99	1.12
18	0.01	0.03	0.06	0.09	0.13	0.18	0.23	0.30	0.37	0.45	0.53	0.63	0.73	0.85	0.98	1.12	1.26
20	0.01	0.03	0.06	0.10	0.14	0.20	0.26	0.33	0.41	0.50	0.59	0.70	0.82	0.94	1.09	1.24	1.40
22	0.01	0.04	0.07	0.11	0.16	0.22	0.28	0.36	0.45	0.55	0.65	0.77	0.90	1.04	1.20	1.36	1.54
24	0.01	0.04	0.08	0.12	0.17	0.24	0.31	0.39	0.49	0.60	0.71	0.84	0.98	1.13	1.31	1.49	1.68
26	0.02	0.04	0.08	0.13	0.19	0.25	0.33	0.43	0.53	0.64	0.77	0.91	1.06	1.23	1.41	1.61	1.81
28	0.02	0.04	0.09	0.14	0.20	0.27	0.36	0.46	0.57	0.69	0.83	0.98	1.14	1.32	1.52	1.74	1.95
30	0.02	0.05	0.10	0.15	0.22	0.29	0.38	0.49	0.61	0.74	0.89	1.05	1.22	1.42	1.63	1.86	2.09
32	0.02	0.05	0.10	0.16	0.23	0.31	0.41	0.52	0.65	0.79	0.95	1.12	1.31	1.51	1.74	1.98	2.23
34	0.02	0.05	0.11	0.17	0.24	0.33	0.44	0.56	0.69	0.84	1.01	1.19	1.39	1.60	1.85	2.11	2.37
36	0.02	0.06	0.12	0.18	0.26	0.35	0.46	0.59	0.73	0.89	1.07	1.26	1.47	1.70	1.96	2.23	2.51
38	0.02	0.06	0.12	0.19	0.27	0.37	0.49	0.62	0.78	0.94	1.12	1.33	1.55	1.79	2.07	2.36	2.65
40	0.02	0.06	0.13	0.20	0.29	0.39	0.51	0.66	0.82	0.99	1.18	1.40	1.63	1.89	2.18	2.48	2.79
42	0.03	0.07	0.13	0.21	0.30	0.41	0.54	0.69	0.86	1.04	1.24	1.47	1.71	1.98	2.28	2.60	2.93
44	0.03	0.07	0.14	0.22	0.32	0.43	0.56	0.72	0.90	1.09	1.30	1.54	1.80	2.08	2.39	2.73	3.07
46	0.03	0.07	0.15	0.23	0.33	0.45	0.59	0.75	0.94	1.14	1.36	1.61	1.88	2.17	2.50	2.85	3.21
48	0.03	0.08	0.15	0.24	0.35	0.47	0.61	0.79	0.98	1.19	1.42	1.68	1.96	2.27	2.61	2.98	3.35
50	0.03	0.08	0.16	0.25	0.36	0.49	0.64	0.82	1.02	1.24	1.48	1.75	2.04	2.36	2.72	3.10	3.49
52	0.03	0.08	0.17	0.26	0.37	0.51	0.67	0.85	1.06	1.29	1.54	1.82	2.12	2.45	2.83	3.22	3.63
54	0.03	0.09	0.17	0.27	0.39	0.53	0.69	0.89	1.10	1.34	1.60	1.89	2.20	2.55	2.94	3.35	3.77
56	0.03	0.09	0.18	0.28	0.40	0.55	0.72	0.92	1.14	1.39	1.66	1.96	2.28	2.64	3.05	3.47	3.91
58	0.03	0.09	0.19	0.29	0.42	0.57	0.74	0.95	1.18	1.44	1.72	2.03	2.37	2.74	3.16	3.60	4.05
60	0.04	0.10	0.19	0.30	0.43	0.59	0.77	0.98	1.22	1.49	1.78	2.10	2.45	2.83	3.26	3.72	4.19
62	0.04	0.10	0.20	0.31	0.45	0.61	0.79	1.02	1.26	1.54	1.84	2.17	2.53	2.93	3.37	3.84	4.33
64	0.04	0.10	0.20	0.32	0.46	0.63	0.82	1.05	1.31	1.59	1.89	2.24	2.61	3.02	3.48	3.97	4.47
66	0.04	0.11	0.21	0.33	0.48	0.65	0.84	1.08	1.35	1.64	1.95	2.31	2.69	3.12	3.59	4.09	4.61
68	0.04	0.11	0.22	0.34	0.49	0.67	0.87	1.12	1.39	1.69	2.01	2.38	2.77	3.21	3.70	4.22	4.75
70	0.04	0.11	0.22	0.35	0.50	0.69	0.90	1.15	1.43	1.74	2.07	2.45	2.86	3.30	3.81	4.34	4.89
72	0.04	0.12	0.23	0.36	0.52	0.71	0.92	1.18	1.47	1.79	2.13	2.52	2.94	3.40	3.92	4.46	5.03
74	0.04	0.12	0.24	0.37	0.53	0.73	0.95	1.21	1.51	1.84	2.19	2.59	3.02	3.49	4.03	4.59	5.17
76	0.05	0.12	0.24	0.38	0.55	0.74	0.97	1.25	1.55	1.88	2.25	2.66	3.10	3.59	4.13	4.71	5.30
78	0.05	0.12	0.25	0.39	0.56	0.76	1.00	1.28	1.59	1.93	2.31	2.73	3.18	3.68	4.24	4.84	5.44
80	0.05	0.13	0.25	0.40	0.58	0.78	1.02	1.31	1.63	1.98	2.37	2.80	3.26	3.78	4.35	4.96	5.58
82	0.05	0.13	0.26	0.41	0.59	0.80	1.05	1.34	1.67	2.03	2.43	2.87	3.35	3.87	4.46	5.08	5.72
84	0.05	0.13	0.27	0.42	0.60	0.82	1.08	1.38	1.71	2.08	2.49	2.94	3.43	3.96	4.57	5.21	5.86
86	0.05	0.14	0.28	0.43	0.62	0.84	1.10	1.41	1.75	2.13	2.55	3.01	3.51	4.06	4.68	5.33	6.00
88	0.05	0.14	0.28	0.44	0.63	0.86	1.13	1.44	1.80	2.18	2.60	3.08	3.59	4.15	4.79	5.46	6.14
90	0.05	0.14	0.29	0.45	0.65	0.88	1.15	1.48	1.84	2.23	2.66	3.15	3.67	4.25	4.90	5.58	6.28
92	0.06	0.15	0.29	0.46	0.66	0.90	1.18	1.51	1.88	2.28	2.72	3.22	3.75	4.34	5.00	5.70	6.42
94	0.06	0.15	0.30	0.47	0.68	0.92	1.20	1.54	1.92	2.33	2.78	3.29	3.84	4.44	5.11	5.83	6.56
96	0.06	0.15	0.31	0.48	0.69	0.94	1.23	1.57	1.96	2.38	2.84	3.36	3.92	4.53	5.22	5.95	6.70
98	0.06	0.16	0.31	0.49	0.71	0.96	1.25	1.61	2.00	2.43	2.90	3.43	4.00	4.63	5.33	6.08	6.84
100	0.06	0.16	0.32	0.50	0.72	0.98	1.28	1.64	2.04	2.48	2.96	3.50	4.08	4.72	5.44	6.20	6.98

B. Equipment for Obtaining Stage Records

Stage records are obtained either from nonrecording gages, which are read by observers, or from water-stage recorders, which produce *stage-time graphs*. Once- or twice-daily readings on a nonrecording gage generally are sufficient on large streams. More frequent observations are required on small streams, especially during floods. The recording-gage installation is much to be preferred.

THE HYDROMETRIC STATION 15-29

1. **Nonrecording Gage.** Nonrecording gages are of several types. The one chosen depends on the characteristics of the site. The most important consideration in installing a nonrecording gage is to place it so that it can be read at any stage of the stream. Figure 15-12 shows a wire-weight gage on a bridge.

It is recommended that a peak-stage indicator be installed in connection with a nonrecording gage to assure recovery of this very important record after each flood.

2. **Recording Gage.** A recording-gage installation consists of a water-stage recorder in a shelter over a well at the bank of a stream or on a bridge pier. The well is connected to the stream by intake pipes. The recording gage has several advantages, among which are continuity of record, reliability, relative freedom from human error, and smaller operating cost. A recording installation affords the space for radio or land-line transmitting equipment, if required.

a. Gage Well. The gage well should be deep enough to be below the lowest stage and high enough to be above the maximum. It is made large enough to accommodate the float and counterweight, clock weight, and intake-cleaning devices and to provide room for cleaning.

Fig. 15-12. Wire-weight gage on bridge.

b. Recorder Installation. The water-stage recorder is installed on a shelf in the shelter (Fig. 15-13). The float, its counterweight, and the clock weight have free travel to the bottom of the well.

c. Frost Protection. If the water in the gage well is likely to freeze during winter, the float is operated in an oil tube. This tube, somewhat larger than the float, is suspended directly beneath it. The lower end is placed below the expected level of freezing. The tube is set high enough to cover the range in stage anticipated during the winter. It is filled to a depth of several inches with oil, on which the float rests. The recorder is set to an auxiliary electric gage in the shelter. A hole is chopped in the ice in the well when it is read.

d. Intake Flushing Device. The intakes are flushed periodically in order to ensure a connection between the gage well and the stream. A tank is mounted beneath the floor of the shelter and connected to the intakes by a three-way system of valves. Water is pumped into the tank and discharged through the intakes.

e. Reference Gage. One or more gages are installed in the shelter as references for the water-stage recorder. Most recording instruments come equipped with tape-driven float wheels. The tapes are graduated and are set to read above the station datum. In addition, stations are equipped with electrical tape gages. In Fig. 15-13 are shown a graduated float tape on the recorder and an electrical tape gage at the left. Usually an outside reference gage is utilized also.

C. Maintenance of Datum

It is important that all the gages be set to a single datum. This is accomplished by establishing three or more reference marks, firmly positioned. None should be fastened to the gage shelter. The marks are so situated that the least number of setups of the leveling instrument will be required to read elevations from the marks and to "shoot" the gages. All gages are set to the datum as determined from these marks. The datum of the gages should be checked against the reference marks at least once a year.

Fig. 15-13. Water-stage recorder installation.

The datum of the gages should be tied precisely into mean sea level, whenever possible, or to a city or highway datum. This permits reestablishment of the gage to its original datum, in the event it is destroyed. Gage readings may be related also to a common datum in the preparation of flood profiles and in the establishment of flood grades for housing developments, industrial plants, and the like.

D. Auxiliary Gage to Measure Fall

At those hydrometric stations where variable backwater occurs it is necessary to utilize the fall in a reach as a second variable in the determination of discharge. An auxiliary gage is installed at some distance downstream from the principal gage. This is generally a water-stage recorder. The distance depends on the fall of the stream, the availability of a site, and its accessibility. The datum of the auxiliary gage should be the same as that of the principal gage.

E. Artificial Controls

Artificial controls are specially constructed weirs or venturi flumes or overflow dams already on the stream. Figure 15-14 shows a hydrometric station with low-water control and measuring bridge.

1. Weirs and Critical-depth Flumes. The weir acts as an artificial control by introducing a section of critical depth and a fall into the stream; and the venturi flume, by creating a hydraulic jump. The weir has the disadvantage of causing a sizable drop. The *venturi*, or *Parshall, flume*, used extensively in irrigation work in the United States [18, 19], permits the recovery of nearly all the head developed through the hydraulic jump. It will function even when nearly submerged. Two gages are used for calibration, one upstream from the throat and one in the throat of the flume. Recording instruments have been developed which integrate the flow or indicate the discharge rate on a dial.

Fig. 15-14. Hydrometric station with low-water control and measuring bridge. (*U.S. Geological Survey*.)

2. Overflow Dam. An overflow dam is sometimes used as a control for a hydrometric station. It is not entirely satisfactory since the rating is insensitive and is affected by flow regulation and the accumulation of debris on the dam crest. Overflow dams must be used under some conditions, as, for example, for the determination of low flows on the Ohio River. These dams maintain the water levels for locks used in navigating that stream.

F. Calibration of Hydroelectric Stations

Hydroelectric generating stations may be calibrated and used as hydrometric stations. The turbines are rated under varying load conditions, usually by current-meter or salt-velocity measurements. The flows through the controlling gates and over the dam or through the spillway are referred to gages in the pool and rated by current meter, when practicable. Model studies have been used also for this purpose.

X. INDIRECT DETERMINATIONS OF PEAK DISCHARGE

It is often necessary to make determinations of peak discharge by methods other than direct velocity observation. This is true particularly for small ungaged streams, which rise to flood heights and recede too quickly to be reached and measured with a current meter. It should be remembered that no indirect method of discharge determination can be of an accuracy equal to a meter measurement, or even to a series of surface velocity observations made by timing floats or drift.

There are several methods for making indirect determinations of peak discharge, principal among which utilize a channel reach, a contracted opening, a culvert, a broad-crested weir such as a roadway or embankment, or a dam.

A. Channel Reach

The flow in a channel reach is computed by one of the open-channel formulas. The method is known as the *slope-area method*. The most commonly used formula in the method is the Manning equation,

$$Q = \frac{1.49}{n} A R^{2/3} S^{1/2} \tag{15-4}$$

where Q is the discharge, n is the coefficient of roughness, A is the cross-sectional area, R is the hydraulic radius, and S is the slope of the energy gradient. This may be written

$$\frac{Q}{S^{1/2}} = \frac{1.49}{n} A R^{2/3} \tag{15-5}$$

The right side of this equation, which contains only the physical characteristics of the cross section, is referred to as the *conveyance K*.

It is seldom possible to select a channel reach which is completely uniform. Usually, the cross section will consist of a main channel and overflow area. The conveyance and the average velocity will not be the same at each section. The cross section is to be split into several parts, according to the variation in the roughness coefficient n. The conveyance is computed for each part. The mean depth may be substituted for hydraulic radius without significant error.

The friction-head loss h_f in the reach is equal to the fall of the water surface F plus the weighted change in velocity head minus a measure of turbulence loss h_i.

$$h_f = F + \Delta\left(\frac{\alpha V^2}{2g}\right) - h_i \tag{15-6}$$

If the downstream velocity head is less than the upstream, h_i is taken as one-half the change in weighted velocity head. If the downstream velocity head is greater, h_i is negligible.

The correction factor α to the velocity head for total cross section is determined by computing the value $\alpha K^3/a$ for each part of the cross section of area a, adding the values, and dividing the sum by the quantity K^3/A for the total cross section of area A. The slope S as used in the Manning equation is h_f/L, where L is the length of the reach under consideration. Computations are made by successive approximations until Eq. (15-5) is satisfied.

The reader is referred to Chow [19] for a detailed description of the slope-area method and the examples for its application and computation.

The field work for making a slope-area determination includes careful marking of the high-water profile of the flood through the reach, precise levels to determine the elevations of the marks, and careful selection and surveying of the cross sections that will be utilized for the determination. It must be remembered that the surface of a stream during flood is a markedly warped surface and the trace of its passage along the banks provides at best only an approximation of the true slope of the water surface, in so far as a mathematical determination of flow is concerned.

B. Contracted Opening

Flow through a contracted opening, a bridge opening flowing partly full, for example, is computed by means of the maximum drop in the water surface through the structure. This method is known as the *contracted-opening method* [19]. The equation used in this method is

$$Q = CA_c \sqrt{2g\left(\Delta h + \alpha \frac{V_A^2}{2g} - h_f\right)} \qquad (15\text{-}7)$$

where C = a coefficient of discharge
A_c = cross-sectional area at minimum section parallel to constriction between abutments
Δh = difference in elevation of water surface between approach section and minimum section
α = an adjustment factor to determine weighted average velocity head at approach section
V_A = average velocity in approach section
h_f = friction-head loss between approach section and minimum section

The approach section is located above the beginning of drawdown. It is recommended that it be taken at least one bridge-opening width upstream from the constriction. It includes the entire flow width. The contracted section is the section of minimum area between the abutments. The high-water marks for determining the drawdown must be carefully set and should be established as soon after the passage of the flood as possible. They are set along both abutments and on the stream banks both upstream and downstream. The U.S. Geological Survey has prepared a circular [19, 20] giving complete details of this procedure.

C. Culvert

Flow through a culvert can occur under a wide variety of conditions. There are six general types, depending on the location of the control section and the relative heights of headwater and tailwater surfaces. The reader is referred to Chow [19] or a circular [21] of the U.S. Geological Survey for a complete analysis of this problem.

D. Broad-crested Weir

Flow over a broad-crested weir is considered to pass through critical depth at the weir crest. The discharge is given by the equation

$$Q = CL\sqrt{g}\, d^{1.5} \qquad (15\text{-}8)$$

where C is a coefficient of discharge, L is the length of the weir crest, g is the gravitational constant, and d is the depth of flow over the weir. Application of this equation for various types of weirs is discussed in a U.S. Geological Survey circular [22]. Flow over roadways and embankments comes within this category.

E. Dam

Computation of flow over a dam has long been used as a means of determining discharge by indirect methods. The discharge is given by the weir formula

$$Q = CL\left(d + \frac{V^2}{2g}\right)^{3/2} \qquad (15\text{-}9)$$

where C is a coefficient of discharge, L is the length of the overflow section, d is the depth of flow over the dam, and $V^2/2g$ is the velocity head in the approach section.

There are many articles dealing with flow over dams. The reader is referred to one by the U.S. Geological Survey [23] and to two by the U.S. Bureau of Reclamation [24, 25].

XI. STAGE-DISCHARGE RELATION

The stage-discharge relation is defined by the complex interaction of channel characteristics, including cross-sectional area, shape, slope, and roughness. The combination of these effects has been given the designation *control*. A control is permanent if the stage-discharge relation which it defines does not change with time; otherwise it is impermanent. The latter has been designated a *shifting control*.

Figure 15-15 illustrates the stage-discharge relation for a permanent control. The measurements which define this relation were made over a period of 20 years. Data for this and the following graphs were furnished through the courtesy of the U.S. Geological Survey. The gage heights have been reduced in this particular case by 1.0 ft in plotting, in order to establish the zero stage at approximately the elevation of zero flow. Note that the relation is a straight-line function on logarithmic plotting.

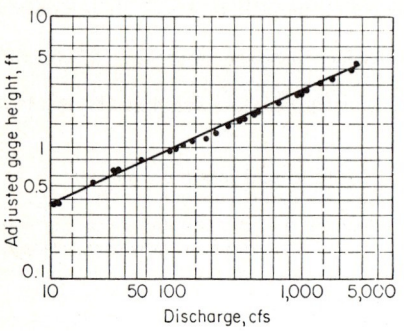

FIG. 15-15. Permanent stage-discharge relation for Little Ossipee River near South Limington, Maine. (*U.S. Geological Survey.*)

The shifting control and its effects on the stage-discharge relation are of great importance in the operation of the hydrometric station and the computation of runoff.

A. Conditions Affecting Controls

Controls may change because of the effects of a changing channel, scour and fill in an alluvial channel, backwater, rapidly changing stage, variable channel storage, aquatic vegetation, and the freezing and breaking of ice.

1. Changing Channel. A channel will undergo change as the result of its modification by dredging, the construction of bridges, or the encroachment of phreatic

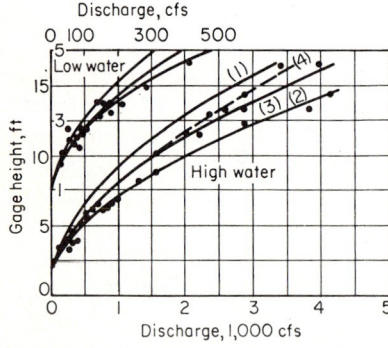

FIG. 15-16. Shifting stage-discharge relation, caused by dredging, for Little Wabash River near Huntington, Ind. (*U.S. Geological Survey.*)

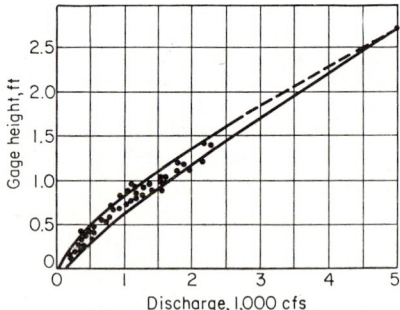

FIG. 15-17. Stage-discharge relation for an alluvial channel for North Platte River at Torrington, Wyo. (*U.S. Geological Survey.*)

vegetation. Figure 15-16 shows an example of the effects of channel dredging. This channel was cleaned in 1949, at which time the stage-discharge relation moved from curve 1 to curve 2. Following this work, the channel began to fill and to become obstructed by the regrowth of vegetation along its banks. The relation has been moving progressively back toward the original condition, having reached curve 3 by 1960.

The dashed curve 4 represents a temporary condition, when ice jams downstream from the hydrometric station caused backwater.

2. Alluvial Channel. An alluvial channel is one whose bed is composed of unconsolidated silts, sands, and gravels. The bed is constantly in motion and highly unstable. Figure 15-17 illustrates the stage-discharge relation for an alluvial stream. Two curves have been drawn to define the limits of the stage-discharge relation as determined by measurements made over a period of approximately 10 years.

The manner in which the stage-discharge relation for an alluvial stream varies is not completely understood. It is assumed that it shifts between two limiting curves which converge at high stages.

3. Variable Backwater. This condition is produced when the normal water-surface slope is decreased as the result of the normal stage being increased at some point downstream. This results from the operation of a dam or an increase in discharge in a downstream tributary or in the stream into which the gaged stream empties.

The effect of backwater is determined by introducing the fall through a reach of channel downstream from the hydrometric station as a third variable. This is graphically incorporated into a three-dimensional *stage-fall-discharge relation*. If the fall is referred to a certain reference value, then a two-dimensional *stage-discharge relation* or *rating* may be developed.

a. Constant-fall Rating. For this rating the fall is taken equal to the arithmetic average of all measured falls or any convenient value of an even foot. All observed discharges may be adjusted to the condition of the constant fall by the following equation:

$$\frac{Q}{Q_0} = \left(\frac{F}{F_0}\right)^p \qquad (15\text{-}10)$$

where Q and F are, respectively, the discharge and fall at any observed condition, p is an exponent close to $\frac{1}{2}$, and Q_0 is the discharge when the fall is equal to a certain constant F_0.

According to Eq. (15-10), values of Q/Q_0 are plotted against F/F_0 and a smooth curve can be fitted. From this curve the adjusted value of Q/Q_0 is read for each given value of F/F_0. The adjusted Q_0 is then computed by dividing the measured Q by Q/Q_0. By further adjustment, if necessary, a relation of Q_0 to F_0 is obtained. This is the *constant-fall rating*.

b. Limiting-fall Rating. The stage-discharge relation is usually not affected by downstream conditions until a *limiting*, or *normal, fall* is reached. Any lesser fall will produce backwater and decrease the discharge for the same stage. The limiting fall is a function of stage.

The stage-discharge relation for a hydrometric station on the Kaskaskia River at New Athens, Ill., is markedly affected at times by backwater from the Mississippi River, into which it empties, 41 miles downstream. Figure 15-18 shows the relation between stage and discharge, and Fig. 15-19 shows the relation between stage and limiting fall F_L. At approximately bankfull stage, the limiting fall decreases markedly, then increases. Presumably this is the result of the sudden change in cross section and roughness.

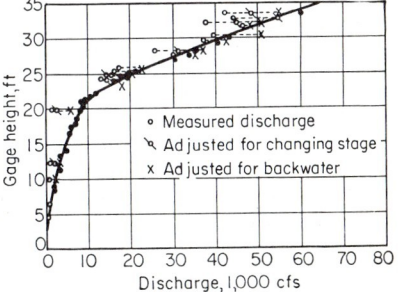

Fig. 15-18. Stage-discharge relation affected by variable backwater and changing stage for Kaskaskia River at New Athens, Ill. (*U.S. Geological Survey.*)

A relation between the ratio of measured to limiting fall F_M/F_L and measured to backwater-free discharge Q_M/Q_R is developed graphically from the discharge measurements. This relation for the Kaskaskia River is shown in Fig. 15-20. It is interesting to note that the discharge ratio is nearly proportional to the square root of the fall ratio. Each discharge measurement is adjusted according to this ratio and used to define the normal stage-discharge relation. The adjusted discharges are shown in Fig. 15-18. The relation thus obtained is known as *limiting-fall*, or *normal-fall, rating*.

4. Rapidly Changing Stage. If the discharge changes rapidly during a measurement, the slope of the water surface will either be greater or less than that for a constant stage S_c, depending on whether the discharge is increasing or decreasing. This effect was recognized many years ago [3]. For the development of a normal stage-discharge relation, the following *Jones formula* [26] can be used:

$$\frac{Q_R}{Q_M} = \sqrt{\frac{S_c}{S_c + (1/U)(dh/dt)}} \quad (15\text{-}11)$$

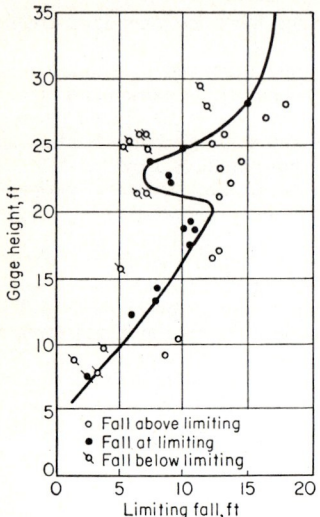

FIG. 15-19. Relation between stage and limiting fall for Kaskaskia River at New Athens, Ill. (*U.S. Geological Survey.*)

where Q_R is the adjusted discharge for defining the rating, Q_M is the measured discharge, S_c is the slope of the water surface for condition of constant discharge, dh/dt is the rate of change of stage, and U is the velocity of the flood wave.

Inverting and simplifying this equation, it becomes

$$\frac{Q_M}{Q_R} = \sqrt{1 + \frac{1}{US_c}\frac{dh}{dt}} \quad (15\text{-}12)$$

The term $1/US_c$ is a function of stage. If a sufficient number of measurements have been made under varying conditions of changing stage, the curve of relation

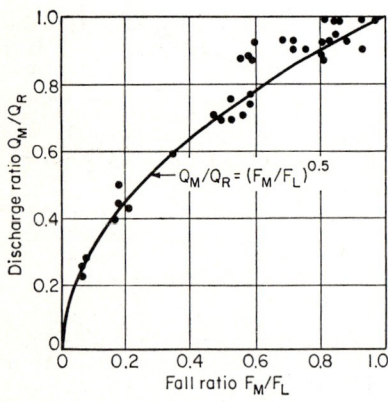

FIG. 15-20. Relation between fall ratio and discharge ratio for Kaskaskia River at New Athens, Ill. (*U.S. Geological Survey.*)

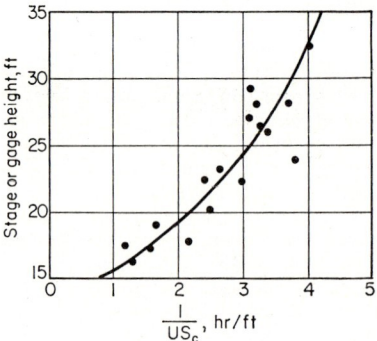

FIG. 15-21. Relation between changing-stage adjustment factor and stage for Kaskaskia River at New Athens, Ill. (*U.S. Geological Survey.*)

between $1/US_c$ and the stage can be established. Figure 15-21 shows this relation for the Kaskaskia River.

Figure 15-22 shows the relation between $\dfrac{1}{US_c}\dfrac{dh}{dt}$ and $(Q_M/Q_R)^2 - 1$, a transposition of Eq. (15-11). The data are somewhat erratic, but since the accuracy of the individual measurement is in the order of 5 per cent, small changing-stage effects might be obscured. Those discharge measurements which were affected by changing stage have been adjusted and replotted on Fig. 15-18. Some were affected by backwater also. In that case, the adjustment for changing stage was applied first, followed by the backwater correction.

5. Variable Channel Storage. Some streams occupy relatively small channels during low flows, but overflow onto wide flood plains during high discharges. On the rising stage the flow away from the stream causes a steeper slope than that for a constant discharge and produces a highly variable discharge with distance along the channel. After passage of the flood crest, the water reenters the stream and again causes an unsteady flow, together with a stream slope less than that for constant discharge. The effect on the stage-discharge relation is to produce what is called a *loop rating* for each flood. This is illustrated by Fig. 15-23.

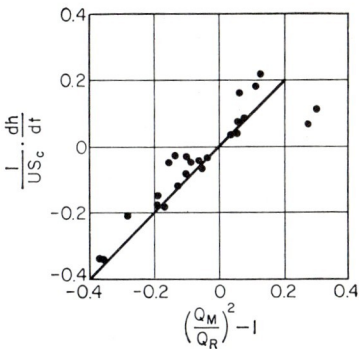

FIG. 15-22. Relation between changing-stage adjustment and discharge ratio for Kaskaskia River at New Athens, Ill. (*U.S. Geological Survey.*)

6. Aquatic Vegetation. The effect of aquatic vegetation is to reduce a channel cross section and increase the roughness, thereby increasing the stage for a given discharge. The stage-discharge relation for this effect is shown in Fig. 15-24. The lower curve is the relation for vegetation-free conditions. The dashed line shows the approximate upper limit of this effect.

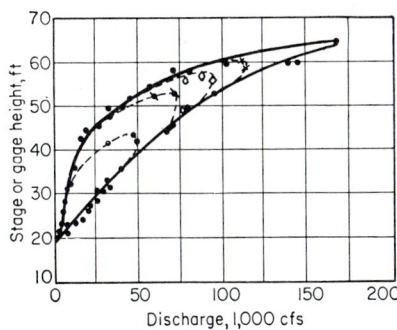

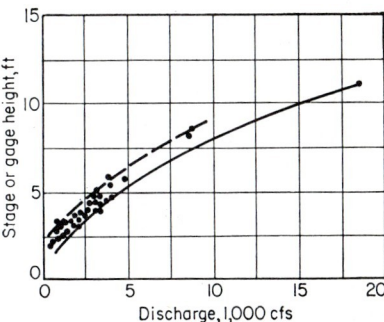

FIG. 15-23. Stage-discharge relation for variable channel storage for Black Warrior River at Tuscaloosa, Ala. (*U.S. Geological Survey.*)

FIG. 15-24. Stage-discharge relation showing effect of aquatic vegetation for Colorado River at Austin, Tex. (*U.S. Geological Survey.*)

7. Ice. The effect of ice on the stage-discharge relation is extremely complex. In areas which experience severe winters, the streams become ice-covered and remain so for considerable periods. The cross-sectional area is reduced, the wetted perimeter is greatly increased, and the resistance to flow is increased. In more temperate zones ice may form and disappear several times during a winter. The U.S. Geological

Survey has made a comprehensive study of the effects of ice on streamflow [27], to which the reader is referred.

B. Complex Controls

A complex control is one for which the stage-discharge relation shows a reversal in direction. That reversal results from a control which functions at low stages but is submerged by effects from the channel downstream at high stages. Figure 15-25

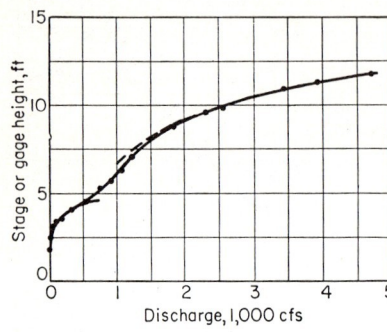

FIG. 15-25. Stage-discharge relation for complex control for Suncook River at North Chichester, N.H. (*U.S. Geological Survey.*)

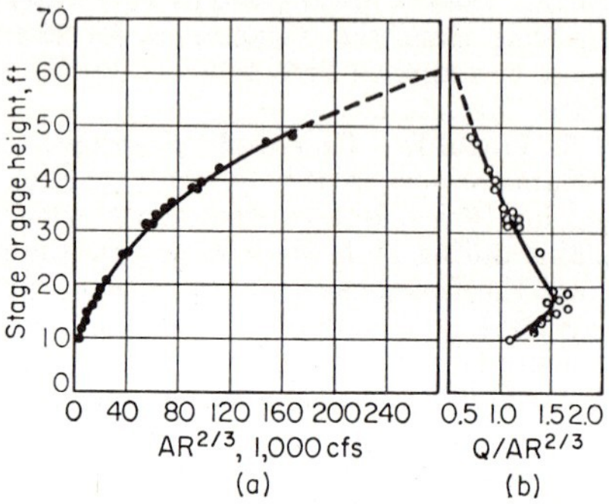

FIG. 15-26. (*a*) Relation between stage and conveyance, and (*b*) relation between stage and slope roughness for Pend Oreille River below Z-Canyon, Wash. (*U.S. Geological Survey.*)

shows the stage-discharge relation for a gaging station which has a concrete low-water control constructed to stabilize the lower end of the rating.

C. High-water Extensions of the Rating

Occasionally, it is necessary to extend a stage-discharge relation for several feet above the highest discharge measurement. An extension is made of the *conveyance factor* $AR^{2/3}$, as determined from discharge measurements and the measurement cross section developed from field data to cover the desired range in stage. Next, the *slope-roughness factor* $1.49 \sqrt{S}/n$, which is equivalent to $Q/AR^{2/3}$, is computed for each discharge measurement, plotted, and extended upward to the desired stage. The product of the geometric and slope-roughness factors gives the discharge. Figure 15-26*a* is a plot of the relation between stage and conveyance factor, and Fig. 15-26*b* of slope roughness and stage, for the Pend Oreille River below Z-Canyon, Washington. The data were plotted to a gage height of 48 ft, the highest for which figures were available. The curves were extended to 60.22 ft, the stage of the highest discharge measurement. The extension gives 162,000 cfs. The measured discharge was 172,000 cfs.

This method is improved by the installation of auxiliary gages upstream and downstream from the station to determine the water-surface slope, thus reducing the slope-roughness relation to one involving roughness only.

There are other methods by which rating curves may be extended, including logarithmic plotting of discharge against stage or of mean depth against average discharge per foot of width, extension of area and velocity curves, and extension of discharge as a function of $A \sqrt{D}$ (*Stevens's method*, where D is equal to cross-sectional area A divided by channel width at the water surface). For details of these methods see Ref. 3.

XII. COMPUTATION OF RUNOFF

The computation of runoff consists of applying a stage-discharge relation to daily gage heights to determine mean daily discharges. Monthly and annual data are then computed.

A. Daily Gage Heights

1. Nonrecording Gage. The average of the observer's daily readings is used unless the stage is changing rapidly. During such periods the gage readings are plotted and a stage graph sketched in, from which the average daily readings are computed.

2. Recording Gage. The graph from a water-stage recorder is converted into average daily gage heights by area balancing above and below a median line. This line, etched on plastic, is placed on the graph and adjusted by inspection.

During periods of rapidly changing stage, part-day averages are computed. The discharge is determined for each, and a weighted discharge computed.

There are certain corrections which must be applied to a recorder graph before the gage heights are computed. These include adjustments for incorrect time, errors in setting, improper reversals, and horizontal and vertical corrections for paper shrinkage or expansion. Time corrections are made by distributing the error in 1-hr intervals through the chart. The other errors are accounted for by adding or subtracting corrections to the recorded average gage height.

B. Rating Table

The rating table is prepared from the stage-discharge relation as determined by the discharge measurements. If the relation is changing with time, a characteristic relation is established and adjustments made to it by applying corrections to the gage height before entering the table. These corrections, known as *shifts*, are determined from the discharge measurements. For each measured discharge the gage height is taken from the rating table. The difference between the rating gage height and the measurement gage height is the shift. If the measurement gage height is greater, the shift is negative; if less, the shift is positive. In other words, a shift is the correction which is applied to the stage of a discharge measurement to bring the measurement to the standard curve. Shifts vary both with time and fluctuations in streamflow. The problem is to distribute these shifts properly between discharge measurements so that the figures of daily discharge will be accurately determined.

C. Computation Procedure

Each type of stage-discharge relation requires a different procedure for the computation of daily discharge. Following is a brief description of each.

1. Changing Channel. A standard stage-discharge relation is established, based on a study of the discharge measurements. Shifts are computed for each measurement and distributed according to time and flow variation. It is to be remembered that the most pronounced changes usually occur during floods, except for the growth of phreatic vegetation, which is a summer phenomenon.

2. Alluvial Channel. The stage-discharge relation for an alluvial channel seems to change without reason. Frequent measurements are necessary to define the shifts. Figure 15-27 shows the daily discharge and the shifts as determined by discharge measurements for an alluvial stream. There appears to be no systematic variation of the shifts with either discharge or time. The stage-discharge relation for this station is shown in Fig. 15-17.

3. Variable Backwater. Computations for backwater are made by the same procedure that is followed in adjusting the discharge measurements. If the effect is changing rapidly, it is necessary to compute the discharge for several instants and plot a discharge-time graph, from which the average daily figures are taken.

15-40 STREAMFLOW MEASUREMENT

4. Rapidly Changing Stage. Computations for rapidly changing stage are made in the same manner as for the adjustment of the discharge measurement. This condition, too, may necessitate the construction of discharge-time graphs for periods of changing effect.

5. Variable Channel Storage. Computations for changing channel storage generally require graphical determination of the discharge from a loop rating for each rise, covering that portion of the curve from the lower to the upper rating, as shown in Fig. 15-23.

6. Aquatic Vegetation. The growth and decay of aquatic vegetation follows the seasons closely. Shifts based on a stage-discharge relation for vegetation-free conditions may be drawn from one measurement to the next. If a flood occurs, the vegetation will be washed out and the stage-discharge relation will return temporarily to the vegetation-free condition. Figure 15-28 illustrates this.

7. Ice. The computation of runoff during periods of ice formation and melting is most difficult. Shifts determined from the open-water relation are too erratic to be

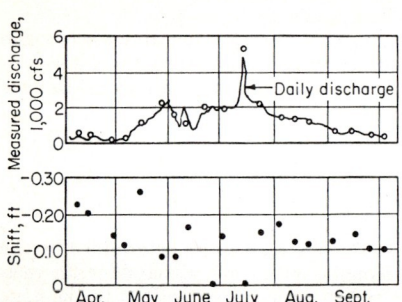
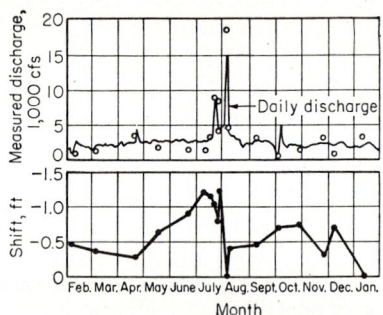

Fig. 15-27. Daily discharge and shift corrections for alluvial stream for North Platte River at Torrington, Wyo. (*U.S. Geological Survey*.)

Fig. 15-28. Daily discharge and shift corrections for effect of aquatic vegetation for Colorado River at Austin, Tex. (*U.S. Geological Survey*.)

used. The practice is to apply the winter gage heights to the open-water rating and to plot an apparent hydrograph with the discharge as the logarithmic ordinate and days as the rectangular abscissa. The discharge for each measurement made during the winter is then plotted on the proper days. Next, an estimated true hydrograph is sketched in, based on these measurements, variations in maximum and minimum temperatures, and whatever reports on ice conditions may be available. A station affected by ice is compared on a drainage-area ratio basis with other stations in the vicinity. This is done by marking each hydrograph with a tick indicating one cubic foot per second per square mile. If two hydrographs are overlaid and matched on this point, they are comparable on a unit-discharge basis.

When estimating the true hydrograph, consideration must be given to variations in temperature and barometric pressure, the occurrence of precipitation, and the trends of the gage-height record [27]. Comprehensive notes on ice, snow, and weather conditions, together with photographs made at the time of each discharge measurement, are extremely helpful in the computation of runoff under ice.

D. Monthly and Yearly Computation

It is the practice in the United States to convert figures of daily discharge into monthly and yearly averages and volumes. For each month is given the total in cfs-days, the maximum and minimum, and the runoff in inches, cubic feet, millions of gallons, or acre-feet, depending upon the section of the country in which the station is located. The same figures are given for the year. The maximum momentary discharge is given, together with its gage height and time of occurrence. Also, other peaks occurring above an established base flow are given. These are very helpful in making flood-frequency studies.

The U.S. Geological Survey has adopted a 12-month period for streamflow computation, known as the *water year*. This year begins on October 1 and ends on September 30. It was chosen for two reasons: (1) to break the record during the low-water period near the end of the summer season, and (2) to avoid breaking the record during the middle of the winter, so as to eliminate computation difficulties during the ice period.

XIII. REFERENCES

1. Hoyt, J. C., and N. C. Grover: "River Discharge," 4th ed., John Wiley & Sons, Inc., New York, 1916.
2. Grover, N. C., and A. W. Harrington: "Stream Flow," John Wiley & Sons, Inc., New York, 1943.
3. Corbett, D. M., and others: Stream-gaging procedure, *U.S. Geol. Surv. Water-Supply Paper* 888, 1943.
4. Kolupaila, Steponas: "Hidrometrija" (Hydrometry), Vytauto Didžiojo Universiteto Technikos Fakultetas, vol. 1, 1939, vol. 2, Kaunas, Lithuania, 1940.
5. Kolupaila, Steponas: "Bibliography of Hydrometry," University of Notre Dame, Notre Dame, Ind., 1960.
6. Kolupaila, Steponas: Early history of hydrometry in the United States, *Proc. Am. Soc. Civil Engrs., J. Hydraulics Div.*, vol. 86, no. HY1, January, 1960.
7. Vilenski, Ya. G., and B. S. Glukhovski: Tensometric shipboard wave recorder GM-16, *Trans. State Organ. Oceanographic Inst.*, issue 47, Moscow, 1959.
8. Townsend, F. W., and F. A. Blust: A comparison of stream velocity meters, *Proc. Am. Soc. Civil Engrs.*, vol. *J. Hydraulics Div.*, 86, no. HY4, pp. 11–20, April, 1960.
9. Tice, C. L.: U.S. Patent 1,709,100, Apr. 16, 1929.
10. Savonius, S. J.: The S-rotor and its applications, *Mech. Eng.*, vol. 53, no. 5, pp. 333–338, May, 1931.
11. Keeler meter, in Hydrographic accessories, *Leupold & Stevens Instrs. Bull.* 18, Portland, Ore.
12. Netherlands Engineering Consultants, the Hague: "River Studies and Recommendations on the Improvement of Niger and Benue," North Holland Publishing Company, Amsterdam, 1959.
13. Kolthoff, I. M., and J. J. Lingane: "Polarography," Interscience Publishers, Inc., New York, 1952, vols. I and II.
14. Boyer, M. C., and E. M. Lonsdale: The measurement of low water velocities by electrolytic means, *Proc. Third Midwestern Conf. on Fluid Mechanics*, University of Minnesota, Institute of Technology, Minneapolis, Minn., June, 1953, pp. 455–462.
15. Barron, E. G.: New instruments of the U.S Geological Survey for the measurement of tidal flow, *Eighth Natl. Conf. Am. Soc. Civil Engrs., Hydraulics Div.*, Colorado State University, Fort Collins, Colo., July 1–3, 1959.
16. Yarnell, D. L., and F. A. Nagler: Effect of turbulence on the registration of current meters, *Trans. Am. Soc. Civil Engrs.*, vol. 95, pp. 766–860, 1931.
17. Rouse, Hunter: "Elementary Mechanics of Fluids," p. 193, John Wiley & Sons, New York, 1946.
18. U.S. Bureau of Reclamation: "Water-measurement Manual," 1953.
19. Chow, V. T.: "Open-channel Hydraulics," McGraw-Hill Book Company, Inc., New York, 1959.
20. Kindsvater, C. E., R. W. Carter, and H. J. Tracy: Computation of peak discharge at contractions, *U.S. Geol. Surv. Circ.* 284, 1953.
21. Carter, R. W.: Computation of peak discharge at culverts, *U.S. Geol. Surv. Circ.* 376, 1957.
22. Tracy, H. J.: Discharge characteristics of broad-crested weirs, *U.S. Geol. Surv. Circ.* 397, 1957.
23. Horton, R. E.: Weir experiments, coefficients, and formulas, *U.S. Geol. Surv. Water-Supply Paper* 200, 1907.
24. U.S. Bureau of Reclamation: Studies of crests for overfall dams, Boulder Canyon Project Final Reports, Part VI, Hydraulic Investigations, *Bull.* 3, 1948.
25. Bradley, J. N.: Discharge coefficients for irregular overfall spillways, *U.S. Bur. Reclamation, Eng. Monographs*, no. 9, March, 1952.
26. Jones, B. E.: A method for correcting river discharge for a changing stage, *U.S. Geol. Surv. Water-Supply Paper* 375(e), pp. 117–130, 1916.
27. Hoyt, W. G.: The effects of ice on stream flow, *U S. Geol. Surv. Water-Supply Paper* 337, 1913.

Section 16

ICE AND GLACIERS

MARK F. MEIER, *Project Hydrologist, U.S. Geological Survey.*

I. Introduction	16-2
II. Distribution of Glaciers and Perennial Ice	16-3
III. Properties and Structure of Ice	16-4
A. Crystalline Structure of Ice	16-4
B. Growth of Ice Crystals	16-5
C. Some Physical Properties of Ice	16-6
1. Mechanical Properties	16-6
2. Geometric Properties	16-6
3. Thermal Properties	16-7
4. Effect of Salt	16-7
IV. Formation of Ice in Water, Snow, and Soil	16-7
A. Ice in Turbulent Streams	16-7
B. Ice Formation in Quiet Water	16-8
C. Conversion of Snow to Firn to Ice	16-9
D. Formation of Ice in the Ground	16-10
V. Glaciers	16-10
A. Definition and Classification of Glaciers	16-10
1. Shape (Morphological) Classification	16-12
2. Temperature (Geophysical) Classification	16-13
3. Activity (Dynamic) Classification	16-14
B. Glacier Nourishment and Wastage	16-14
1. Accumulation	16-15
2. Ablation	16-15
3. The Firn Limit	16-17
4. The Mass Budget	16-17
C. Glacier Runoff	16-19
1. Origin of Glacier Runoff	16-19
2. Distinctive Characteristics	16-20
3. Sediment Content of Glacier Streams	16-21
D. Glacier Flow	16-22
1. The Mechanism of Flow	16-22
2. Velocity Distribution	16-22
E. Response of Glaciers to Climatic Changes	16-25
1. Instability of Glacier Tongues	16-25
2. Theory of Kinematic Waves	16-26
3. Behavior of Typical Glaciers	16-27
4. Effect on Streamflow	16-28
F. Some Special Hydrologic Problems Connected with Glaciers	16-28
1. Reconstructing Past Climates from Glacier Variations	16-28
2. Natural and Artificial Regulation of Glacier Runoff	16-29
3. Jökulhlaups, or Glacier Floods	16-30
VI. References	16-31

I. INTRODUCTION

This section is written to acquaint the practicing hydrologist with the properties and characteristics of ice and glaciers. In this brief introduction to a large field, two objectives are recognized: (1) to present a short but broad introduction to the whole subject, with emphasis on underlying principles rather than description, and (2) to elaborate on those aspects which are now—or which may be in the near future—of direct value to an understanding and effective utilization of this unique form of water resource. Other sections contain specific discussion of the ephemeral snow cover (Sec. 10) and of lake (Sec. 23) and river ice (Sec. 10), which is omitted here.

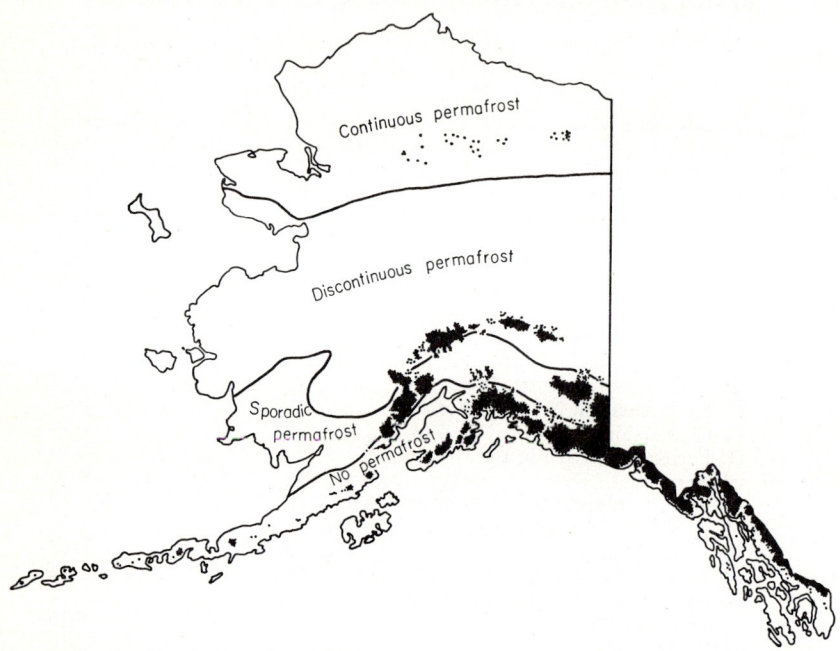

Fig. 16-1a. Map of Alaska showing distribution of permafrost and glaciers. Isolated dots indicate the existence of local glaciers. Black areas indicate where glaciers cover more than 50 per cent of the land area.

Ice is an important water resource, a significant portion of the hydrosphere, and a regular participant in the hydrologic cycle. About 80 per cent of all the fresh water in the world is in the form of glacier ice. Frozen lake, river, and groundwater account for an additional small fraction of a per cent, but the hydrologic significance of this ice is great. One example is permafrost, common in Alaska, which creates hydrologic problems that are as difficult to solve as they are poorly understood. The hydrology of frozen water is a young science, and many fundamental questions and problems are just now being uncovered.

Most of the world's ice remains in storage from year to year, responding only to long-term climatic fluctuations. The amount released seasonally is, however, significant in many civilized lands. Glacier melt, for instance, contributes about 800,000 acre-ft of water to July and August streamflow originating in Washington, a state which normally experiences little or no summer precipitation.

The study of ice in all its aspects is now termed *glaciology* in most European and Eastern countries, and this usage is etymologically correct. Many Americans restrict the term glaciology to the study of glaciers alone, but this practice appears to be diminishing. The term *cryology*, although proposed frequently, has never attained wide usage outside of the field of low-temperature physics.

II. DISTRIBUTION OF GLACIERS AND PERENNIAL ICE

Permanent, or perennial, ice occurs as the great continental ice sheets of Antarctica and Greenland, as the many cirque, valley, and ice-cap glaciers of the high altitudes and high latitudes, as ground ice in the permafrost regions, and as floating ice shelves attached to Antarctica, the Canadian Arctic islands, and Greenland. The amount of water stored in the continental ice sheets is huge; latest results from International Geophysical Year studies in Antarctica indicate that this single body of ice has a volume of about 7×10^6 cu mi (30×10^6 km^3). The water equivalent of this volume is equal to the total present amount of precipitation during about 60 years over the *whole* earth, and if this ice were melted, sea level would be raised by nearly 300 ft.

In Alaska perennial ice occurs in the ground as part of the permafrost layer and on the land in the form of glaciers. As can be seen from Fig. 16-1a, permafrost occurs only in discontinuous patches in central and southern Alaska, but is continuous in the north, where it reaches thicknesses of more than 1,000 ft. This ice consists of small, irregular grains bonding soil or rock particles together or as relatively pure segregations in the form of wedges or layers.

Glaciers cover about 20,000 sq mi (about 3 per cent) of the area of Alaska. This state presents the greatest display of large and varied glaciers outside of the polar regions. The area of greatest glacier development centers around the Fairweather Range and the St. Elias Mountains at the northern end of the Alaskan panhandle (Fig. 16-1a). In this area are valley glaciers up to 90 miles long and piedmont glaciers with areas greater than 1,000 sq mi. The ice cover is also heavy all along the southern

FIG. 16-1b. Map of western United States showing the location of glacierized areas. Black areas indicate the presence of many glaciers. (*After Meier* [1].)

Table 16-1. Distribution of Glaciers by State

State	Total no. of glaciers	No. of glaciers larger than 0.39 sq mi (1 km^2)	Total glacier area, sq mi	Estimated annual streamflow from glaciers, acre-ft
Washington	674	83	153.4	1,430,000
Wyoming	81	12	18.3	107,000
Montana	106	3	10.3	60,000
Oregon	38	7	8.0	64,000
California	80	3	7.3	39,000
Colorado	10	0	0.7	3,400
Nevada	1	0	0.1	400
Total	990	108	198	1,700,000

coast, especially in the Chugach Mountains and the high areas around Prince William Sound. The glacier cover tapers off rapidly inland so that the Wrangell, Talkeetna, and Alaska Ranges have less ice. The Brooks Range in the north supports only tiny glaciers, except near its eastern extremity, where valley glaciers up to 10 miles long occur.

Glaciers, but no permafrost, occur in many other states. In the United States south of Alaska there are nearly 1,000 glaciers covering about 198 sq mi of area. In Washington alone about 40 million acre-feet of water is currently stored as glacier ice. The distribution of existing glaciers in the conterminous states is illustrated in Fig. 16-1b and tabulated in Table 16-1. These data are taken from a compilation by Meier [1].

III. PROPERTIES AND STRUCTURE OF ICE

A. Crystalline Structure of Ice

The water molecule consists of one oxygen atom and two hydrogen atoms. The oxygen atom shares the two electrons of the hydrogen atoms, and the two hydrogen

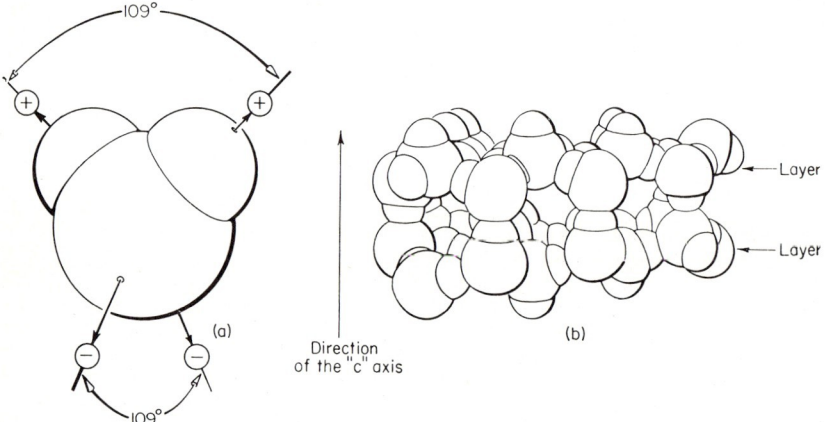

FIG. 16-2. Diagram showing (a) the water molecule and (b) a small portion of an ice crystal. The angle shown for the water molecule is that which is assumed when the molecule fits into an ice-crystal lattice.

atoms are disposed at an angle of 109° with the center of the molecule (Fig. 16-2a).

These molecules can join to form a crystal lattice in which each hydrogen atom lies between two oxygen atoms; this type of attachment is called a *hydrogen bond*. The resultant lattice is the characteristic crystal structure of ice (Fig. 16-2b). This structure, in which each molecule is surrounded by only four immediate neighbors, is a very open structure, and accordingly ice is a substance with an abnormally low density. The lattice is made up of "puckered," or "dimpled," layers of atoms. The direction perpendicular to these layers is the *main crystallographic axis*, the c-axis. Viewed in the direction of the c-axis, the atoms are seen to be arranged in a hexagonal arrangement. The hexagonal symmetry of a snowflake is one manifestation of this internal structure.

Because the lattice is made up of similar layers, it is possible to slide these layers like a deck of cards without destroying the homogeneity of the crystal lattice. This process, which describes how ice crystals deform under stress, is called *intracrystalline gliding*.

When ice melts, the lattice structure is partially destroyed and the water molecules are packed more closely together, causing water to have a higher density than ice.

PROPERTIES AND STRUCTURE OF ICE 16-5

However, many of the hydrogen bonds remain, and aggregates of molecules with the open-lattice structure persist in water at the freezing point. With an increase in temperature, some of the aggregates break up, causing a further increase in density of the liquid. At a temperature of 39°F, the normal expansion due to increase in molecular agitation overcomes this effect, and at higher temperatures water shows the normal decrease in density with rising temperature.

Ice can also exist with other crystalline structures. Bridgman (see [2]) has discovered at least seven different forms of ice which can exist at high pressures; one of these is stable under certain pressure conditions at temperatures as high as 400°F. However, none of these dense, high-pressure forms of ice can exist in nature, at least not in visible quantities, because the pressures existing at the base of the thickest ice sheet are still several times too low. There is a possibility that these other forms may exist as films of absorbed (hygroscopic) water only a few molecules thick.

B. Growth of Ice Crystals

Whenever the temperature of an ice crystal is raised, even infinitesimally, above the melting temperature, the crystal begins to melt. Thus superheating of ice is not generally observed. The reverse, however, is not true; supercooling of liquid water or water vapor is necessary in order to form ice. The formation of ice directly from water vapor requires a nucleus—a foreign particle, or preferably a crystal with a lattice resembling that of ice. The necessary amount of supercooling depends on the quality of available nuclei. If no nuclei are present, ice cannot form from the vapor even at very low temperatures (down to about $-40°F$).

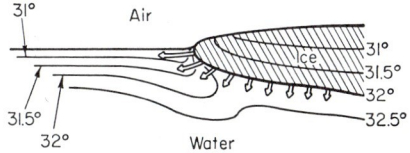

Fig. 16-3. Diagram showing the growth of small ice crystal lying in supercooled water. The arrows indicate the direction and amount of ice-crystal growth. Note that the growth takes place most rapidly along the water surface. (*After Devik* [3].)

The growth of ice crystals in supercooled water is also retarded unless an embryo crystal is present. Small crystals may form spontaneously, but these are not stable unless they exceed a critical size, which is a function of the degree of supercooling. After an embryo crystal has reached the critical size, it enlarges rapidly. Supercooling is still necessary, however, because the heat of solidification must be carried away. This requires that the temperature on one or both sides of the crystal must be lower than the temperature of the crystal surface (32°F). The growth of a small ice crystal which floats on water is shown in Fig. 16-3. The heat of crystallization can be carried away most efficiently where the temperature gradient is steepest, so the ice crystal grows most rapidly in the surface layer. This explains why ice crystals can grow over the surface of still water in an amazingly short time.

The shape of small ice crystals depends on relative growth rates in different crystallographic directions. Growth in the direction of the *c*-axis is slow because the density of atoms on faces perpendicular to this axis is high (Fig. 16-2b). Growth perpendicular to the *c*-axis is more rapid because these surfaces have a lower density of atoms. It is for this reason that the first visible ice crystals in water have the shape of flat disks. If the water is still, these disks usually lie flat on the surface. At first the periphery of the disk is circular, because this form requires the least surface energy. As the disk enlarges, however, the surface-energy consideration is overcome by the problem of disposing of the latent heat of solidification. Then any small irregularities in the disk tend to enlarge as points, or dendrites (branching, treelike forms), because these can dispose of the heat most efficiently. Finally, the points, or dendrites, in certain preferred crystallographic directions grow faster than their neighbors, and the crystal assumes the familiar star, or snowflake, shape. If the embryo disk is not lying parallel to the water surface but is tipped, only those points that are parallel to the water surface will extend rapidly, giving the crystal a needle shape.

C. Some Physical Properties of Ice

1. Mechanical Properties. The strength, elastic properties, and plastic properties of ice depend on many things: grain size, grain shape, impurities, rate of application of stress, temperature, and the orientation of the crystal lattice. Because of this, values of mechanical properties reported in the literature range between wide limits. For further information, see the extensive compilations by Dorsey [2] or Mantis [4].

a. Elastic Properties. *Young's modulus* (resistance to elongation) is about 10^{11} dynes/cm^2 (1.5×10^6 psi), about two-thirds that of lead and one-twentieth that of steel. The *shear modulus* is about 3.4×10^{10} dynes/cm^2 (0.43×10^6 psi). *Poisson's ratio* is approximately 0.36.

b. Plastic Properties. Ice has frequently been analyzed as a viscous material in studies of flow, but the measured values of *viscosity* range from 10^9 to 10^{15} poises, a range of one million times! The reason for this is twofold: (1) the plastic properties are very sensitive to temperature; and (2) at a given temperature the "viscosity" is a function of the applied stress. The relation of shear stress to rate of shear strain (the *Glen flow law* [5]) is given in Fig. 16-4, and may be expressed as

$$\dot{\epsilon} = \left(\frac{\sigma}{B}\right)^n \qquad (16\text{-}1)$$

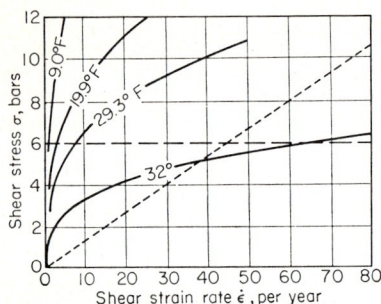

FIG. 16-4. The flow law of ice as measured by Glen [5]. Note the strong dependence of the flow law on temperature. The dashed line represents the behavior of a substance obeying a condition of perfect plasticity. The dotted line represents the behavior of a viscous liquid.

where the shear strain rate $\dot{\epsilon}$ is measured in years^{-1}, and the shear stress σ, in bars (1 bar = 14.5 psi). According to Glen [5], the empirical constants B and n have values equal to about 1.75 and 3.2, respectively, for ice at 32°F and stresses greater than 1 bar. The value of B is very sensitive to temperature and rises to 3.31 at a temperature of 29.3°F and to 7.56 at a temperature of 9°F. A perfectly plastic approximation, also shown in Fig. 16-4, has proved useful in analyzing the flow of temperate glaciers.

c. Ultimate Strength. This property is very sensitive to the rate of application of stress, temperature, orientation of the crystal lattice, and other factors. If the stress is applied very slowly, ice flows plastically and has no strength. Typical values obtained with conventional testing machines are as follows: *crushing tests*, 560 psi; *bending tests*, about 200 psi; *tensile strength*, 135 psi; and *shear strength*, 102 psi.

2. Geometric Properties. Typical measured values of *density*, *porosity*, *air permeability*, and *grain size* for new snow, old snow, firn, and glacier ice are given in Table 16-2.

Table 16-2. Typical Values of Density, Porosity, Air Permeability, and Grain Size for New Snow, Old Snow, Firn, and Glacier Ice

	Density, g/cm^3	Porosity, per cent	Air permeability, g/cm^2/sec	Grain size, mm
New snow	0.01–0.3	99–67	>400–40	0.01–5
Old snow	0.2–0.6	78–35	100–20	0.5–3
Firn	0.4–0.84	56–8	40–0	0.5–5
Glacier ice	0.84–0.917	8–0	0	1–>100

3. Thermal Properties. The *heat of fusion* of water to ice is 79.8 cal/g (144 Btu/lb), and the specific heat of ice is 0.506 (in either cal/°C/g or Btu/°F/lb) at 32°F. The *thermal conductivity* at this temperature is about 54×10^{-4} cal/°C/sec (1.2×10^{-5} Btu/°F/sec). The *linear coefficient of expansion* at the melting temperature is 52.7×10^{-6} per degree Centigrade (29.3×10^{-6} per degree Fahrenheit), and the *bulk coefficient* is 153×10^{-6} per degree Centigrade (85×10^{-6} per degree Fahrenheit). The *depression of the melting point due to hydrostatic pressure* is 0.00752°C/atm (0.92×10^{-4}°F/psi).

4. Effect of Salt. A small amount of salt has a profound effect on the physical and thermal properties of ice. Sea water has no discrete freezing point. Normal sea water with a salinity of 3.25 per cent begins to freeze at a temperature of 28.8°F, but does not become completely solid until it cools below -65°F. The ice which first forms has a salinity less than that of the sea water and contains numerous cells of brine. As freezing progresses, the brine cells become smaller and more of the salt is taken up into the ice. The result of rapid freezing of sea water to -65°F is a material which is 6.3 per cent solid salts and 93.7 per cent pure ice. In the normal freezing process, however, salt tends to be excluded and drains from the ice mush. This continues as the ice ages. Most sea ice has a salinity of 0.3 to 0.5 per cent, and sea ice more than one year old commonly has a salinity of only 0.1 per cent. However, this very small amount of impurity causes gross differences in the properties and structures between fresh and sea ice.

The strength of sea ice depends upon the area of the brine cells and is approximately proportional to the square root of the brine content. Brine cells are normally arranged in layers between plates of ice about 0.5 mm thick. The volume increases in an irregular way as the temperature is lowered. For this reason thermal-contraction cracks in sea ice are not as common as in lake ice, and the stresses imposed within a sheet of sea ice as the temperature changes can be extremely complex.

IV. FORMATION OF ICE IN WATER, SNOW, AND SOIL

A. Ice in Turbulent Streams

Only a small degree of supercooling is necessary to form ice in rapidly flowing streams because the turbulence can carry away the heat of fusion very quickly. Ice formation frequently starts with the development of many tiny disk crystals called *frazil*. In quiet water these crystals would rise to the surface, but the rate of rise is so slow and the temperature of a turbulent stream is so uniform that they form and appear in all depths in a stream. In addition to frazil formed by small embryo disk crystals, dendritic, or star-shaped, crystals may grow on the surface in protected, quiet-water locations. When these fragile crystals are swept into the turbulent water, they are broken up into many jagged and needlelike pieces. Frazil ice particles frequently collect by adhesion or regelation to form larger masses which move along with the current. As the ice content of the water increases, the water becomes oily or milky in appearance and the viscosity of the water increases.

Ice can also form on underwater objects. This material, called *anchor ice*, has been attributed to the loss of heat by subsurface objects due to outgoing radiation. However, it has been shown that even the clearest water is quite opaque to this radiation, and this mechanism of the formation of anchor ice has been questioned. It may be that the only fundamental difference between anchor ice and frazil is that anchor ice is attached to some underwater object. The coating of anchor ice may be several inches thick and may tend to grow more rapidly on sharp corners. This causes the irregularities in the stream bed to be accentuated, and this anchor ice may eventually dam up the stream. Thus it is possible to develop a staircase of a series of small ice dams with some still water trapped behind them. The increased viscosity due to the ice content, the damping of turbulent eddies, and the rise of the river bed due to the formation of anchor ice also cause a slight increase in the river stage.

In addition to the underwater ice, ice forms rapidly on the surface of most flowing rivers. At first this consists of agglomerations of broken surface crystals and frazil ice which unite to form round pans. These pans grow by accretion, and the open water between them becomes smaller in area and freezes rapidly. When the ice cover on the surface is complete, the river regime changes from open-water flow to flow beneath the ice cover. The frictional losses against the rough undersurface of the ice reduce the mean velocity, and the river stage must rise in order to pass the same volume of flow. The rise in stage in turn places a large volume of water in temporary storage and causes a temporary decrease in discharge. After a short time, usually not more than a few days, conditions equalize and the discharge returns to its prefreezing condition. During the winter the ice thickens because of the conduction of heat from the water through the ice to the air and the addition of snow to the upper surface. Overflows of water may saturate the snow or cause the formation of ice layers in or on top of the snow.

In very cold regions rivers may freeze solid. If there is a perennial spring or continuing base flow from the groundwater table, this water may force its way to the surface, causing the accumulation of thick masses of ice called *icings*, or *aufeis*. In northern and western Alaska aufeis accumulations may become so large and thick that they may persist the year around. In more temperate latitudes, ice can be deposited over large areas of a river flood plain because of overflows.

A number of different mechanisms operate to remove the ice cover in the spring and summer. Turbulent water erodes and thaws the underside of the ice cover, forming open leads where the current is especially concentrated. Heat from the sun and air causes the surface ice to melt. The ice formed on the surface frequently has the shape of pencils standing vertically, with the impurities concentrated in the thin films between these crystals. When the ice cover begins to melt, thawing is concentrated along the boundaries between the crystals. This causes the crystals to become loose, and the ice cover as a whole becomes weak and "rubbery." Meltwater causes the river stage to rise in the spring, and this induces tension cracks in the ice cover, further weakening the mass. Finally, the ice cover begins to break at one point and to move with the stream. This initiates a chain reaction, so that a complete ice cover can be removed in a matter of hours.

B. Ice Formation in Quiet Water

The first ice to form in still water takes the form of small disks which lie on the water surface. After reaching a critical size, the ice disks begin to branch out in dendritic shapes. The dendrites may grow across the surface at a very rapid rate until the whole lake or pond is covered by needles, or branching forms of ice. This commonly occurs during one single night. The result is a sheet of ice which is very thin but homogenous, presenting a smooth, unbroken surface. Because the embryo ice disks are parallel to the surface, the crystallographic c-axes are normally perpendicular to the water surface. However, there have been instances reported in which an ice cover was formed with the c-axes predominantly horizontal.

If there is a slight wind and some wave motion in the water, the thin sheet ice may be broken into irregular needlelike fragments, causing a fine mush on the surface. This mush usually agglomerates to form round pans of ice separated by relatively clear water and frazil slush. Eventually these formations unite to cause a complete ice cover. This type of ice cover is not homogenous, but is rough and has an initial thickness of 2 to 4 in. The c-axes of crystals in an ice cover of this type may have random orientation.

As the ice cover thickens, the crystals grow downward in the shape of pencils, or columns, and any impurities or entrapped air bubbles are concentrated along the margins of these pencils. In the case of sea ice, brine pockets are included within the ice crystals themselves. The c-axes of extending sea-ice crystals are most likely to be horizontal, but these crystals are also elongated in a vertical direction. The bottom surface of an ice cover in the process of development usually shows a very jagged, feathery appearance. Snowfall on top of the ice may weight the ice cover, causing it to submerge and draw water into the snow, which subsequently refreezes, forming a complex structure.

The breakup of an ice cover in a pond, lake, or the sea proceeds in much the same

C. Conversion of Snow to Firn to Ice

Newly fallen snow is a light, loose material usually ranging in density from less than 0.1 g/cm³ to about 0.3 g/cm³. The vapor pressure at the points and edges of the snow crystals is higher than in the reentrants. Molecules tend to evaporate from the sharp points and condense in the reentrants, and the crystal becomes rounded. Localized melting, physical crushing, wearing, or compaction also tends to transform the original angular crystals into nearly spherical granules. Vapor transfer within the snowpack responding to local temperature gradients may cause the growth of delicate depth hoar crystals within certain zones, but this phenomenon usually occurs

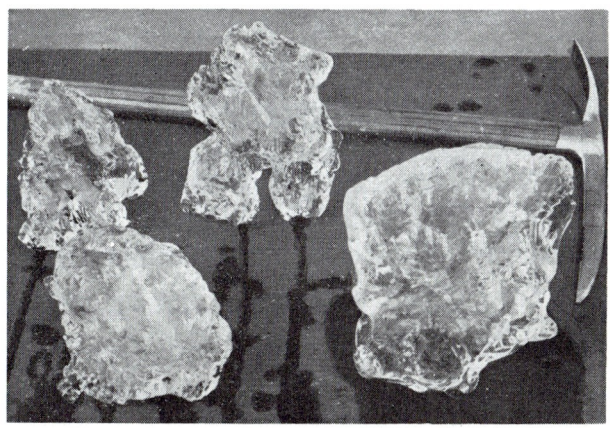

FIG. 16-5. Four single crystals from the terminus of South Cascade Glacier. The head of the ice axe is 1 ft long. (*Photograph by M. F. Meier, 1958.*)

only in relatively thin layers. In general, the density of the snowpack increases with time—rapidly in the first few hours or days and then more slowly. Densities of 0.4 to 0.5 g/cm³ are common in snow layers several months old which were deposited in a mild, temperate environment. The greatest density which can be attained merely by shifting the grains around so that they fit more snugly together is about 0.55 g/cm³.

Further densification occurs at a markedly lower rate, involves deformation, refreezing, recrystallization, and other processes, and produces a compact, dense material called *névé*, or *firn*. Because of the change in the rate of densification at a density of 0.55 g/cm³, some authorities have suggested that this density should mark the dividing line between snow and firn. Other glaciologists, however, note that layers having densities both above and below 0.55 g/cm³ can occur in a given snowpack and believe that firn is better defined on a genetic or time basis. They suggest that the accumulated material changes, by definition, from snow to firn as soon as it survives one complete ablation (melt) season.

Further recrystallization causes the air spaces between grains to become more compressed and restricted. At a density of between 0.82 and 0.84 g/cm³, the air spaces become sealed off and the material becomes impermeable to air and water and is now defined as ice. Further densification takes place; old glacier ice usually has a density of about 0.90, and the density within individual grains approaches very closely the theoretical density of pure ice, 0.917 g/cm³. The change in physical properties during the transition from new snow to glacier ice is given in Table 16-2. Sufficient time and a relaxation of flowage stresses may produce very large, irregular single crystals of ice (Fig. 16-5).

In the polar regions the transition from snow to ice may occur abruptly. The winter snow cover may entirely melt. However, the melt from this snow layer immediately refreezes upon contact with cold ice below, resulting in the formation of a layer of superimposed ice. A thin, ephemeral layer of superimposed ice may also form on temperate glaciers because of below-freezing temperatures in a subsurface layer during the amelioration of the winter-chilled zone.

D. Formation of Ice in the Ground

The growth of ice in the ground is difficult to analyze because of many variables. Essentially, the freezing of ice in the ground is a four-phase process in which air, water, ice, and soil particles all have a mutual influence. The heat-flow characteristics are complicated because of the inhomogeneous materials. The freezing point of water is modified by the presence of impurities in the soil, and especially by the size of the soil particles. Experiments have shown that the freezing point of pure water in capillary tubes is much lower than in an unconfined beaker. In a tube 1.6 mm in diameter the freezing point is depressed by 11.5°F, and in a tube 0.06 mm in diameter, by more than 33°F. This means that water in large masses may be frozen whereas water in small interstices may be liquid at the same temperature. This can cause the movement of water from the small capillary-size interstices into the larger area. If a cylinder of sand or silt with a uniform moisture content is frozen from the top down, the frozen sample will have much more ice at the top than at the bottom. If the material is a fine-grained material such as clay, the ice will be found to be largely segregated into lenses, or layers of pure ice. This result is of great importance to the maintenance of roads and other structures in cold countries.

The type and condition of soil primarily influence the structure of the frost which forms. Porous soils which are not completely saturated develop a honeycomblike ice structure which is permeable. Fine-grained soils may develop a much more compact structure in which the ice grains effectively cement the mass into an impermeable material. Humus appears to have the effect of impeding the cementation process.

Frost forms very slowly in the soil if there is an effective insulating layer on top of the ground. Dry vegetation is especially effective. Frost in the soil can be dissipated rapidly if the insulating layer is absent or if water can move through the soil. Therefore gravelly soils with little vegetation do not retain frost as long as compact vegetated soils. A thick snow cover is also an effective insulator.

If the climate is very rigorous, a layer of frozen ground may be formed which persists from year to year. This is called *permafrost* and is common in Alaska (Fig. 16-1a). The surface layer of the ground (the active layer) normally thaws during the summer and refreezes during the winter, but the permafrost below remains frozen and impermeable. In areas mantled with peat or a dense mat of living vegetation, the active layer is generally thin and permafrost occurs close to the surface. In areas of bare gravel or exposed bedrock, the active layer may be quite thick. Permafrost is absent or lies at great depths between lakes or ponds because these bodies of standing water readily absorb the sun's heat during the summer. The heat loss in winter is largely absorbed in converting water into ice in the ponds, and new frozen ground is not formed. Permafrost is more widely and continuously distributed in lowlands than it is in the mountains in spite of lower temperatures prevailing in the mountains. This is because the lowland vegetation retards thawing in the summer whereas the thick mountain snowpack tends to impede the formation of permafrost in the winter.

Permafrost controls the distribution of groundwater in much of Alaska. The few unfrozen zones found in the permafrost zone are generally good aquifers, but their number and capacity are very limited. In cold areas shallow permafrost forms readily beneath the poorly drained surfaces and tends to seal off deep aquifers. In much of arctic Alaska the only potable water in large quantities is found in deep lakes and a few major perennial streams.

V. GLACIERS

A. Definition and Classification of Glaciers

A *glacier* may be defined as a body of ice originating on land by the recrystallization of snow or other forms of solid precipitation and showing evidence of past or present flow. However, many glaciologists hold to a more stringent definition: in order to qualify as a glacier, an ice mass must have (1) an area where snow or ice usually accumulates in excess of melting and (2) another area where the wastage of snow or ice usually exceeds the accumulation, and there must be (3) a slow transfer of mass by creep from the first region to the second.

FIG. 16-6a. Aerial photograph of a portion of the Juneau Ice Field, Alaska. (*Trimetrogen aerial photograph by U.S. Air Force*, 1942.)

Most recognized glaciers fall within both these definitions, but some tiny masses of ice in high mountains defy precise definition. The difficulty has been circumvented by some authors who term these small ice masses *glacierettes*.

Glaciers exist in a wide variety of forms and characteristics. They range in size from continental ice sheets to tiny masses only a few acres in extent, and in shape from steep narrow ice cascades in high mountains to smooth sheets lying on flat plains. Some are active ice streams in areas of very high precipitation and large amounts of meltwater runoff, whereas others occur in such cold desert environments that yearly nourishment is only a few millimeters and runoff is negligible. These wide differences in appearance, regimen, and hydrologic characteristics can be illustrated and described by referring to the three systems by which glaciers are classified: by shape, by temperature, and by activity.

1. Shape (Morphological) Classification. Glaciers can be separated into two or three groups, depending upon whether they engulf the topography or are confined by it. Ahlmann [6] has divided glaciers into three categories and has noted that each category is characterized by a distinctive area-altitude (hypsometric) graph.

1. Glaciers extending in continuous sheets, the ice moving outward in all directions. This includes *continental glaciers* such as Antarctica and Greenland, *glacier caps* of

Fig. 16-6b. Chocolate Glacier, a valley glacier on Glacier Peak, Washington. The terminus of this glacier advanced 1,300 ft between 1950 and 1955. (*Photograph by A. S. Post*, 1960.)

more limited extent such as the many in the Canadian Arctic Archipelago, and *highland glacier systems* such as the Juneau Ice Field in southeastern Alaska (Fig. 16-6a).

2. Glaciers confined to a more or less marked path, directing its main movement. This group includes both independent glaciers and outlets of ice from glaciers of group 1. Included here are the *valley glaciers* of the Cascade Range (Fig. 16-6b), *cirque glaciers*, which are the most common type of glacier in the American Rocky Mountains (Fig. 16-6c), and many less common types of glaciers (Fig. 16-6d). Among these other types are *transection glaciers*, in which a whole valley system is more or less filled by ice, *wall-sided glaciers, summit glaciers, hanging glaciers, ice aprons, cliff glaciers, crater glaciers, regenerated* or *reconstituted glaciers*, and *glacier tongues afloat*.

3. Glacier ice spreading in large or small cakelike sheets over the level ground at the foot of glaciated regions. No glaciers of this type are independent. This group includes *piedmont glaciers*, such as the huge Malaspina and Bering Glaciers of southeastern Alaska; *foot glaciers*, which are the lower and more extended portions of

group 2 glaciers; and *shelf ice*, such as that which occurs along the northern portion of Ellesmere Island in Canada.

2. Temperature (Geophysical) Classification. Lagally [7] in 1932 and Ahlmann [8] in 1933 independently arrived at a thermal classification of glaciers which is of fundamental importance. The three types of glaciers according to Ahlmann are as follows:

1. *Temperate glaciers.* Throughout these glaciers the temperatures correspond to the melting point of ice, except in the wintertime, when the top layer is frozen to a

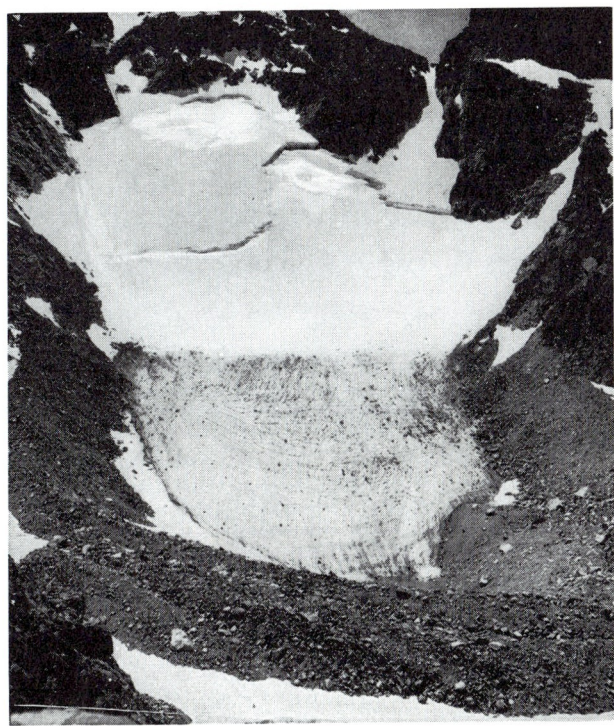

FIG. 16-6c. A small cirque glacier (Heap Steep Glacier) in the Wind River Range, Wyoming. Note the snow-covered accumulation area in the upper part of the glacier, the dirty ablation area in the lower part, and the sharp firn line between the two areas. In the foreground is a bulky compound moraine. (*Photograph by M. F. Meier*, 1950.)

depth not exceeding several tens of feet (Fig. 16-7a). Temperate glaciers can produce copious runoff. Almost all the glaciers of Scandinavia, the Alps, and the United States (excluding northern Alaska) fall into this group.

2. *High-polar glaciers.* In these glaciers the temperature is below the freezing point to a considerable depth, and even in summer the temperature is so low that as a rule there is no melting (Fig. 16-7b). There can be no runoff from high polar glaciers. Most of Antarctica falls under this classification.

3. *Subpolar glaciers.* In these glaciers the summer temperature allows surface melting and the formation of liquid water, but the main mass of ice at depth is below the freezing temperature, so that this liquid water refreezes at lower horizons (Fig. 16-7c). There can be little runoff from this type of glacier except near the margins. The Greenland Icecap at lower elevations, and especially its southern parts, is a subpolar glacier.

FIG. 16-6d. Small glaciers of diverse form in the Northern Cascade Range. In the center is South Cascade Glacier, a small valley glacier. At the left is a summit glacier. Along the right-hand margin of the photograph are several hanging, or wall-sided, glaciers. In the center distance is an ice apron and a valley glacier. (*Photograph by A. S. Post, 1958.*)

3. Activity (Dynamic) Classification. The activity or passivity of a glacier depends upon its depth, speed of flow, and material balance. The rate of movement of an active glacier is generally high because it must transport a large amount of precipitation from one area to another, where the amount of melting is likewise high. A passive or inactive glacier need have only sluggish movement because it occurs in an environment where the accumulation on its surface and the wastage are both small.

In general, active glaciers occur in maritime environments at relatively low latitudes, and inactive, or passive, glaciers occur in high latitudes and in very continental environments. The activity of a glacier has no relation to whether it is advancing or retreating at any instant. The advance or retreat corresponds to the glacier's state of health, whereas the degree of activity corresponds to its metabolism. A quantitative measure of the activity of a glacier will be presented in the next section.

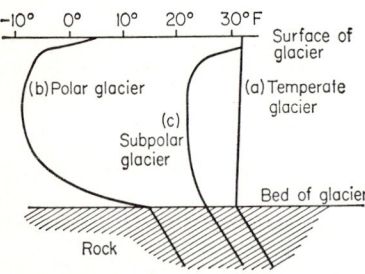

FIG. 16-7. Diagrammatic representation of temperature profiles within temperate, polar, and subpolar glaciers (not drawn to scale).

B. Glacier Nourishment and Wastage

Glaciers grow and decline according to changes in snowfall and melting during the different seasons. In order for a glacier to remain at a constant size, there must be a balance between the accumulation and loss of mass. The balance is referred to as the glacier's regime, or the glacier's mass budget. If income (accumulation) exceeds expenditure (wastage or ablation), the glacier grows, and if expenditure exceeds income, the glacier shrinks. In general, glaciers strive to maintain a balanced budget, expanding to compensate for surplus income and contracting to reduce expenditure during times of net deficit. A complete set of definitions for glacier mass-budget terms has been given by Meier [9].

1. **Accumulation.** Accumulation refers to all processes which increase the mass of ice and snow of the glacier. Accumulation is predominantly due to snow. However, in some environments condensation on the surface in the form of hoarfrost is important, and rime and hail may deliver small amounts of ice to the surface. If any of the glacier surface layers are below freezing in temperature, rain falling on the glacier will refreeze to form ice, adding to the accumulation. In mountainous areas the accumulation of snow on glaciers is often found to be greater than the accumulation of snow on ridges or neighboring slopes, because snowfall is transported by wind and avalanches. This can happen at the time of snow precipitation or at a much later date. Some small glaciers owe their existence to the concentration of accumulation due to these two agencies.

As time passes, the accumulation from a given winter is transformed from snow to firn and finally to ice. This transition results in a general increase in grain size and density and a partial or complete homogenization of the contrasts in grain size and density between individual storm deposits and between layers of individual years. A pit, or core, taken from the accumulation zone of a glacier shows this transition (Fig. 16-8). Annual layers may be recognizable in old, hard firn because of the contrast between midwinter snow (fine grains and dense) and fall or spring snow (coarser-grained and less dense). By the time the density reaches that of ice, the dust blown onto the surface during summer is the only obvious mark of the annual layers. Annual layers from cores obtained at great depth in Greenland and Antarctica can be identified only by use of the mass spectrometer; the oxygen-18 isotope is relatively less abundant in colder precipitation. Thus changes in the O^{18}/O^{16} ratio reflect changes in the temperature of precipitation and can be used to identify summer and winter layers.

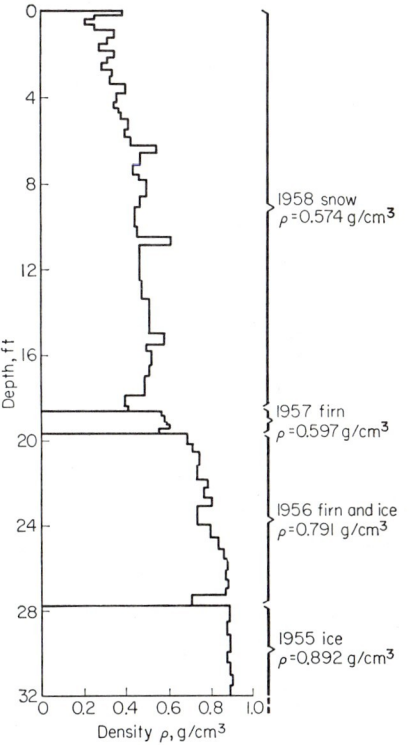

Fig. 16-8. Typical profile of density at various depths for a glacier in a maritime climatic environment. Note that the density for one winter's snowfall is low at the top, reaches a maximum in the middle, and is low at the very bottom. This is because cold midwinter snow is finer-grained and compacts more readily than fall or spring snow. Each previous year's snow is denser than the layer above. The density of ice is reached at a moderate depth of this glacier.

2. **Ablation.** Ablation refers to all processes by which solid material is removed from the glacier. In temperate glaciers melting is the dominant process and normally accounts for all but a few per cent of the ablation. Evaporation is normally quantitatively unimportant, because the heat of vaporization of ice is much higher than the heat of fusion. Glaciers which terminate in lakes or oceans lose mass by the calving

of icebergs. In some high arctic environments, appreciable snow and ice are removed by wind erosion; in some unusual cases, glaciers lose mass by the breaking off of avalanches.

Most ablation occurs on the surface of a glacier. However, the loss of potential energy due to ice flow supplies a small amount of heat which is available to melt ice below the surface. The heat flux from the earth cannot be conducted through a temperate glacier and normally melts a small amount of ice at the base. However, these two heat sources normally melt less than a cubic inch of ice per year per square inch of glacier area. In many temperate glaciers melting at the surface exceeds 10 or 20 ft/year.

During the ablation season the surface level of a glacier drops with respect to a stake emplaced in the surface. This lowering of the surface is not entirely due to ablation, but is partly due to compacting, or densification, of the snow layers beneath. Thus, in order to measure ablation, one must measure the thickness and the density of a surface snow layer at each time of measurement. To be strictly correct, measurement of ablation requires measurement not of the bulk density of the snow or firn, but of the density of the ice phase alone. The bulk density of a snow sample is the sum of the ice density and the water density. Ablation affects only the ice density, and the gain or loss of liquid water is not involved in ablation. The correct measurement of ablation is illustrated in Fig. 16-9.

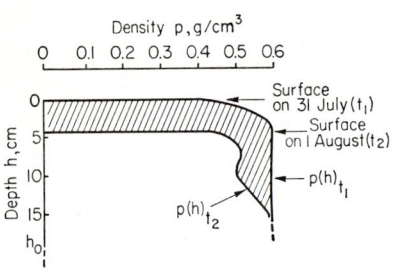

FIG. 16-9. The proper measurement of ablation. Ablation from July 31 to August 1 is equal to the area (shaded) between successive density-depth curves:

$$a_b = \int_0^{h_0} \rho(h)_{t_1} \, dh - \int_0^{h_0} \rho(h)_{t_2} \, dh$$

where h_0 is a depth below which the density does not change from t_1 to t_2. For this specific day, $a_b = 3.0$ g/cm^2 (3.0 cm of water equivalent), of which 2.3 g/cm^2 was due to melting of the surface layer and 0.7 g/cm^2 to subsurface melting. The lowering of the snow surface in this time interval was 4.2 cm. (After LaChapelle [10].)

As a further complication, it must be noted that a stake emplaced in a glacier may be moved either upward or downward because of glacier flow. Thus topographic surveys cannot be used to measure ablation. The proper measurement of ablation is not an easy task.

Ablation of snow, firn, or ice is caused predominantly by radiation intercepted at the surface and by the transfer of heat from the air (convection). Of the two components, radiation is usually the more important. Condensation and evaporation may release or absorb sufficient heat to make significant contributions to the energy budget although their contribution to the mass budget is usually negligible. The relative roles of different sources of ablation energy are illustrated in Table 16-3. These data were obtained from the Blue Glacier, Olympic Mountains, Washington, and appear to be typical of many glaciers in the northwestern United States. The data from 1958 represent a period of settled, warm, clear summer weather, whereas the 1960 data were taken during a typical fall storm with fog, rain, and strong winds. Note that during clear weather solar radiation accounted for the bulk of the ice melt. During the fall storms the rate of melt was high because of large amounts of incoming long-wave radiation (from warm clouds and fog), heat transfer from the air, and heat derived from condensation. Rain contributed almost nothing to the ice melt.

During the course of a year the relative contributions of radiation and convection may change considerably. In western United States solar radiation is greatest in spring and early summer, but the albedo (reflectivity) of the glacier surface is also high. Wind velocities and air temperatures may be fairly low. Thus the rate of melt of ice and snow is generally low. In midsummer the radiation income is slightly diminished but the average albedo of the glacier surface may have decreased markedly, making the radiation a highly effective melting agent. However, even a light fall of snow on the glacier surface will greatly affect the rate of melt because it can abruptly raise the albedo from about 0.3 to about 0.8, thus decreasing the amount of radiation absorbed by a factor of $3\frac{1}{2}$ times. During fall storms the combination of warm air and strong winds causes high rates of melt due largely to the exchange of heat from the air. Radiation from the sun is decreased at this time of year, but long-wave back radiation from warm clouds may contribute significant amounts of energy for ice melt.

Table 16-3. Sources of Energy for Ablation on Blue Glacier, Washington*

	Solar radiation	Long-wave radiation	Eddy conduction	Condensation	Evaporation	Rain	Ablation
July 12 to Aug. 20, 1958:							
Energy income, langleys/day†	+293	−120	+104	+26	−32	0	−272
Per cent of total	69	−28	25	6	−8		−64
Ablation, inches of water melted	1.44	−0.59	0.51	0.13	−0.16	+0.01	1.33
Inches of water sublimated				−0.016	0.020		0.004
Aug. 29–30, 1960:							
Energy income, langleys/day†	+33	+45	+109	+114	0	+3	−304
Per cent of total	11	15	36	38	0	1	−100
Ablation, inches of water melted	0.16	0.22	0.53	0.56	0	0.03	1.49

* Data from LaChapelle [11, 12].
† 1 langley = 1 cal/cm^2.

3. The Firn Limit. The highest level on a glacier to which the winter snow cover retreats during an ablation season is normally called the *firn limit*. The firn limit corresponds rather closely to the elevation at which the accumulation equals the ablation in temperate glaciers. However, in subpolar and polar glaciers there is deposition (freezing) of meltwater below the firn limit. Therefore, for these glaciers, it is necessary to define an *equilibrium limit* as that point below the firn limit where the net budget equals zero and accumulation equals ablation. On a valley glacier exactly in balance with a fixed climatic environment, the firn limit would return to the same position each year. However, in winter, glacier flow would carry the edge of the residual firn blanket some distance downglacier. The surface of this ideal glacier would therefore display a continual succession of older firn layers below the annual firn limit. If this succession of older firn layers below the firn limit is not seen, one can safely assume that the climate has not been constant and that the firn-limit elevation has been changing.

The area above the firn (or equilibrium) limit is called the *accumulation zone*, and the area below is the *ablation zone*. The mass balance of a glacier depends on the ratio of accumulation area to ablation area. Therefore the elevation of the firn limit is a sensitive indicator of a glacier's state of health. Fortunately, the position of the firn limit can be observed readily in the field (Fig. 16-6c).

4. The Mass Budget. The mass budget of a glacier represents the difference between the accumulation and the ablation. If accumulation exceeds ablation, the mass budget is considered positive and the glacier is growing. The mass budget can be defined as a volume change, or a change of mass. Because the density of a glacier varies with time, it is generally preferable to speak in terms of mass budgets instead of volume budgets. Thus one refers to cubic feet of water equivalent when defining the budget terms. One can consider the budget of a glacier as a whole, or one can define the mass budget for a portion of the glacier or at a given point on the glacier. Since accumulation and ablation are predominantly surface phenomena, the mass budget of a glacier as a whole divided by the area of the glacier is equal to the average of the mass budgets per unit area at all points on the surface of a glacier. In defining the average mass budget for the whole glacier or the budget at a specific point, one generally speaks in terms of feet of water equivalent.

One defines the rate of accumulation at any instant or over a short period of time as a_c (dimensions of length per unit time), the rate of ablation as a_b, and the instantaneous mass budget as a ($= a_c - a_b$). These quantities vary markedly with time (Fig. 16-10). The net budget at the end of a year is of interest because this indicates whether the glacier is growing or shrinking. Inspection of Fig. 16-10 shows that a budget year extending from a minimum (or maximum) in the instantaneous net budget of one year to the corresponding minimum (or maximum) of the next year must be defined in order to present a valid measure of the net change of the glacier from year to year. Measurements on fixed dates will not give proper results; in general, a budget year will not last exactly 365 days. If the budget year begins

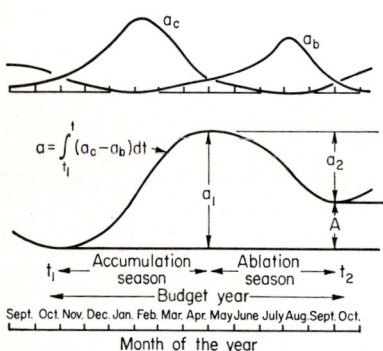

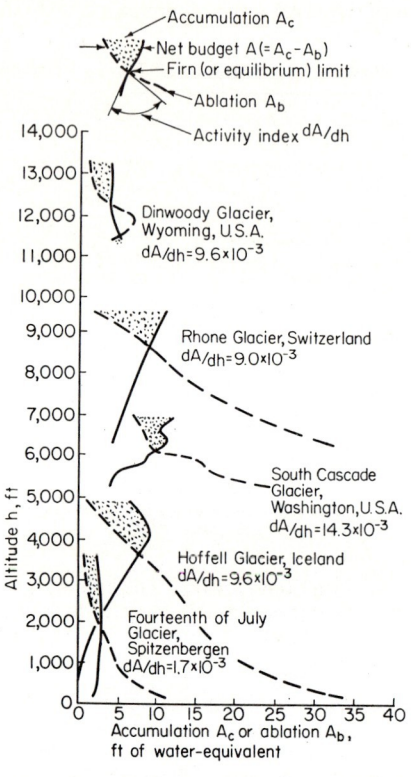

FIG. 16-10. Diagram showing seasonal variation of accumulation, ablation, and net budget. The top figure gives the variation of accumulation and ablation rates, and the bottom figure gives the cumulative-total (integrated) mass budget for one budget year.

FIG. 16-11. Curve showing the variation of accumulation and ablation with altitude and the activity index for five typical glaciers.

at t_1 and ends on t_2, the total accumulation is $A_c = \int_{t_1}^{t_2} a_c \, dt$, and the total ablation A_b is defined similarly. The net budget is

$$A = A_c - A_b = \int_{t_1}^{t_2} a \, dt = \int_{t_1}^{t_2} (a_c - a_b) \, dt \qquad (16\text{-}2)$$

The net budget can be measured several ways. The accumulation and ablation totals A_c and A_b can be measured directly, but measuring snowfall in a glacier environment is a difficult procedure. If A is positive, as depicted in Fig. 16-10, it can be measured directly at t_2. Another common procedure is to measure the net budget at its greatest value in spring (a_1) and then to measure the subsequent *net* ablation (a_2). Then $A = a_1 - a_2$, and the method is valid for either positive or negative **net budgets**.

The mass budget as measured at specific points on the glacier is a function of position and especially of altitude. In general, ablation decreases and accumulation increases with increasing altitude, so the net budget is negative at low elevations and positive at high elevations. In general, the shape of the curve expressing the variation of net budget with altitude (dA/dh) does not change markedly from year to year, but it may be displaced to higher or lower elevations. This may cause great changes in the budget for the glacier as a whole because the amount of area devoted to net accumulation ($A > 0$) may change markedly with respect to that undergoing net ablation ($A < 0$).

A very significant parameter is the *activity index*—the slope of the net budget-altitude curve—evaluated at the firn limit (where $A = 0$). If the activity index is high, the glacier must transfer large quantities of net accumulation at high altitudes to the intense ablation conditions at low altitudes. If, on the other hand, the index has a low value, the glacier need transport only small quantities. Therefore the gradient of net budget is a quantitative measure of the activity of the glacier. Glaciers of high activity index occur in temperate maritime environments, whereas those of lower activity indices occur in high polar and continental environments. Some known values of activity index and the vertical gradients of the mass-budget terms for five typical glaciers are shown in Fig. 16-11.

C. Glacier Runoff

1. Origin of Glacier Runoff. A temperate glacier is essentially a reservoir which gains precipitation in both liquid and solid form, stores a large share of this precipitation, and then releases it with little loss at a later date. However, the hydrologic characteristics of this reservoir are complex because its physical attributes change during a year

In late spring the glacier is covered entirely by a thick snowpack at the melting temperature. Meltwater and liquid precipitation must travel through the snowpack by slow percolation (unsaturated flow) until reaching well-defined meltwater channels in the solid ice below. In summer the snowpack becomes thinner and drainage paths within the snow become more defined. However, perched water tables may exist in local areas. Some bare ice is exposed, and on this there may be surface drainage. Thus meltwater and liquid precipitation can be transmitted through the glacier more rapidly than in the spring. In fall a thin, dense snow layer covers only part of the glacier and bare ice is exposed over the rest of the glacier. Meltwater and liquid precipitation travel very quickly from the surface to the outflow stream. In winter snow accumulates and the surface layer freezes. This completely stops the movement of surface meltwater and precipitation. Any rain which falls on the frozen surface can only refreeze to join the ice reservoir. A small amount of water deep within the glacier slowly drains out during the winter. In early spring the surface begins to

Table 16-4. Seasonal Change in Glacier-runoff Characteristics

Season	Snowpack thickness	Albedo	Diurnal fluctuation in streamflow	Amount of runoff	Characteristics of direct precipitation-runoff
Winter............	Moderate to high	Very high	Nil	Slight	All precipitation stored
Spring............	Highest	High	Slight	Moderate	Subdued, delayed
Summer..........	Moderate	Moderate to low	High	High	Slight delay
Fall (before a snow-pack accumulates)	Low	Low	Moderate	Moderate	No delay, very "flashy"

thaw. Meltwater and rain are effective agents of heat transfer and quickly thaw holes in the winter-chilled layer. Gradually the area between the holes also becomes thawed, and the snowpack reaches a uniform temperature at the melting point. This condition occurs first at the lower elevations of the glacier, and the thawed zone gradually moves to higher altitudes as spring progresses. These changes in the physical characteristics of the ice reservoir result in changes in the characteristics of the stream flow which emerges from the glacier. These are summarized in Table 16-4.

2. Distinctive Characteristics. The daily fluctuation of meltwater discharge from a glacier in midsummer is pronounced. The average daily fluctuation for a small glacier in the Northern Cascades is shown in Fig. 16-12. The maximum daily discharge is about four times the minimum discharge, and the diurnal fluctuation in streamflow is delayed several hours after the fluctuation in melt. Although the daily pattern of ice melt may vary markedly, the shape of the daily streamflow hydrograph is remarkably consistent from day to day, season to season, and glacier to glacier.

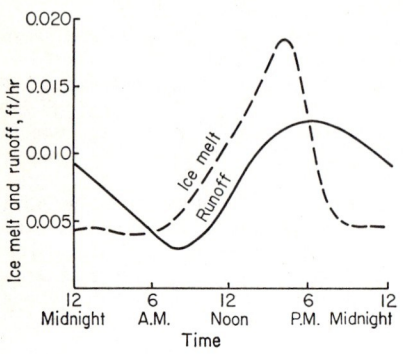

FIG. 16-12. Diurnal variation of ice melt and runoff from South Cascade Glacier during a 14-day period of clear warm weather. (*After Meier and Tangborn* [13].)

Glaciers retain winter precipitation and release it in the summer when the albedo is low and the temperatures are high. Figure 16-13 shows the distribution of temperature, precipitation, and streamflow during a typical year for the South Cascade Glacier drainage basin in the Northern Cascade Range of Washington. The lack of direct relation between precipitation and runoff is evident for all seasons except for late summer.

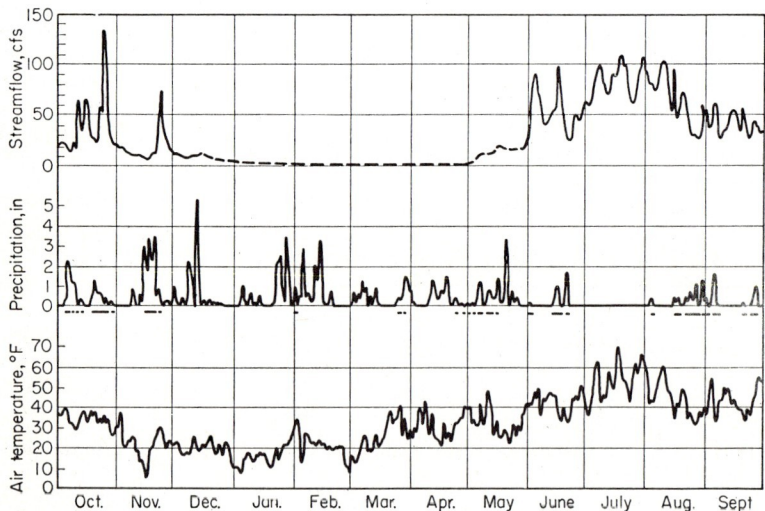

FIG. 16-13. Daily averages of streamflow, precipitation, and temperature from the South Cascade Glacier drainage basin during water year 1960. Precipitation is reported in inches of water. The dots or lines under the precipitation graph indicate that the precipitation occurred in whole or in part as rain during these days. (*After Meier and Tangborn* [13].)

The release of water is closely related to the amount of energy supplied by the sun and the atmosphere. Air temperature, as measured at a typical weather station, is not the direct cause of ice melt. However, the high-radiation conditions which do cause high melt rates are frequently reflected by high air temperatures. Therefore one should expect a fair correlation between air temperature and glacier runoff. Glacier runoff is, however, delayed, and correlation between daily temperatures and daily streamflows is generally unsatisfactory. If one considers the air temperatures for a number of days before the day on which streamflow is measured, the correlation can be greatly improved. Dr. W. H. Mathews of the University of British Columbia studied the streamflow from the Athabasca Glacier, Alberta, Canada, and found that the daily mean streamflow could be given by the relation $Q = e + f(\Sigma T)$, where $T = T_z + KT_{z-1} + K^2T_{z-2} + \cdots + K_sT_{z-s}$. In these expressions Q is the computed mean flow for any day, e and f are coefficients determined by regression analysis, K is a recession coefficient with a value less than 1, which takes into consideration that the flow may be influenced by the previous day's weather, T_z is the mean temperature measured during the day z at a nearby weather station, and s indicates the number of previous days considered. Mathews found that the coefficient K normally had a value of 0.7 to 0.8 and that correlations could not be improved by values of s greater than 6 (about 7 per cent of the discharge was found to be related to air temperatures more than 7 days before).

Glacier-streamflow records sometimes show sudden, unexpected high discharges. These are referred to as glacier floods, outbreaks, bursts, or by the Icelandic term *jökulhlaup*. Jökulhlaups are discussed more fully in Subsec. V-F-3.

The yearly total runoff from a glacierized basin does not necessarily equal the sum of precipitation plus condensation minus evaporation, because of natural changes in storage. These storage changes can be appreciable, as shown by Table 16-5, which presents data of two consecutive years obtained at South Cascade Glacier. It is clearly advantageous to mankind that the changes in storage take place to provide more streamflow during warm-dry years and less during cool-wet years.

Table 16-5. Water Budget for South Fork Cascade River at South Cascade Glacier, Washington, for Two Contrasting Years

Water year	South Fork Cascade River at South Cascade Glacier, Wash.			Newhalem, Wash.	
	Precipitation,* in.	Runoff,† in.	Change in storage,‡ in.	Precipitation departure,‡ in.	Temperature departure,‡ °F
1958	130§	200§	−51	−14.2	+3.0
1959	210	191	+17	+32.5	−0.8

* As measured at Station P-1, 6,160-ft elevation, on South Cascade Glacier.
† Average values for whole drainage basin.
‡ Departures from 1931–1955 means, by U.S. Weather Bureau.
§ Part of the record estimated.

An excellent summary of the water-budget relations in a glacierized basin is given by Kasser [14].

3. Sediment Content of Glacier Streams. Glaciers carry large amounts of debris, as evidenced by their massive moraines. Observation of the stream emerging from a glacier also indicates the active movement of large boulders and all smaller sizes of rock. The distinctive green color of Alpine lakes indicates that the glacier runoff is also characterized by sediment in the fine colloidal sizes.

Few quantitative data exist on the sediment content of glacier streams. What few data are available are not too acceptable for extrapolation because glaciers differ greatly in their activity and the bedrock on which glaciers move differs greatly in resistance. In fact, there is no accepted quantitative theory to explain or predict glacier erosion.

Some studies have been made of the suspended-sediment component in glacier streams. It has been found in central Alaska that as one traces a river back toward its source in a glacier, the suspended-sediment concentration increases. Extrapolating a number of measurements, Borland [15] suggested that the suspended-sediment content of a stream where it emerges from a glacier would be of the order of 65 acre-ft/sq mi/year (1.2 in./year).

The dissolved and suspended material in the meltwater stream issuing from a small glacier in the Northern Brooks Range of Alaska has been studied by Rainwater and Guy [16]. They found a strong diurnal fluctuation in suspended-sediment concentration, with peak concentrations preceding peaks in water discharge by 1 to 2 hr. The suspended-sediment discharge ranged up to 55 tons/day/sq mi of drainage basin, but averaged only about 13 tons/day/sq mi during July and August. The water discharge-weighted mean suspended-sediment concentration for this period was about 500 ppm. The concentration of dissolved solids was only 9 ppm in a sample of the surface ice of the glacier, but ranged from 11 to 32 ppm in the meltwater stream. A nearby nonglacierized tributary stream contained 54 ppm of dissolved solids.

D. Glacier Flow

1. The Mechanism of Flow. Glaciers differ from other bodies of land-borne ice in their ability to flow. This distinguishing behavior has long fascinated scientists, and measurements of surface velocity on glaciers have been made for more than two hundred years. A knowledge of the mechanics of glacier flow is important to understanding how glaciers and glacier runoff react to climatic change.

Deformation of a mass of ice crystals can take place in three different ways: *intergranular adjustment, recrystallization,* and *intragranular gliding.* Intergranular adjustments involve rotation of individual grains like corn in a chute. This mechanism is significant only in surface snow layers because grains in the ice portion of a glacier are generally of irregular shape and are tightly interlocked (Fig. 16-5).

Deformation can take place between grains by recrystallization: local concentrations of stress cause the transfer of atoms from one crystal lattice to another adjacent one in such a way as to relieve the stress concentration. This causes a change or migration of the boundary between grains. The large crystals which exist near the termini of glaciers demonstrate the great efficacy of recrystallization, for they have grown from single snowflakes.

A third means of deforming ice is through intragranular gliding. The layered structure of the crystal lattice (Fig. 16-2) can be deformed without breaking the continuity of the lattice by slip parallel to the layers, as in the sliding of a deck of cards. The very small yield stress of ice indicates that this intragranular gliding takes place principally through the migration of dislocations or small imperfections in the ice. Measurements of crystal axes in glaciers indicate that the glide planes are distinctly oriented in relation to the planes of maximum shear stress, but the exact relation is far from simple. In general, glacier flow must be recognized as taking place largely through recrystallization and intragranular gliding operating simultaneously. Favorably oriented grains are deformed by gliding, and recrystallization tends to restore the nearly equal dimensions of the grains so that none become greatly elongated.

2. Velocity Distribution. *a. Two-dimensional Laminar Flow.* The following simplified treatment of the velocity pattern in glaciers is taken from the work of Nye. For a more refined treatment, the reader is again referred to Nye [17]. The simplest case with which to deal is that of an ice mass which is infinitely wide and is flowing down an inclined plane in such a way that the lines of flow are everywhere parallel to the bed (Fig. 16-14). The shear stress on any layer at depth d measured perpendicular to the surface is

$$\sigma_{xy} = \rho g d \sin \alpha \qquad (16\text{-}3)$$

GLACIERS

where ρ is the density of ice, g is the acceleration of gravity, and α is the slope of the bed or surface of the glacier. It is assumed that ρ is constant throughout this glacier. The strain rate σ_{xy} represents du/dy, where u is the velocity. Inserting the stress equation, Eq. (16-3), in the Glen [5] flow law [Eq. (16-1)] and integrating, the difference between the velocity at the surface u_s and the velocity at depth d is found to be

$$u_s - u_d = \frac{k}{n+1} \sin^n \alpha \, d^{n+1} \qquad (16\text{-}4)$$

where $k = (\rho g/B)^n$.

The discharge (the volume passing through any cross section in unit time for unit thickness perpendicular to the diagram) is given by further integration:

$$q = u_b h + \frac{k}{n+2} \sin^n \alpha \, h^{n+2} \qquad (16\text{-}5)$$

where u_b is the velocity at the bed, and h is the thickness of the glacier.

The theory so far does not allow prediction of how fast the glacier slips on its bed. Therefore these relations cannot be used to predict the depth of a glacier from a knowledge of its surface velocity and slope. However, if the discharge q were known, along with velocity and slope, the depth could be deduced. The velocity of slip on the bed has been measured on very few glaciers, and the results range from negligible sliding to values of more than 90 per cent of the surface velocity. Only one theory has been proposed for predicting the velocity of sliding on the bed. This theory, which is due to Weertman [18], suggests that the sliding velocity is given by

$$u_b = \left(\frac{\sigma_{xy}}{a}\right)^m = \left(\frac{\rho g}{a} \sin \alpha\right)^m h^m \qquad (16\text{-}6)$$

where $m = \frac{1}{2}(n+1)$, σ_{xy} is evaluated at the bed, and a is a constant determined partly by the roughness of the bed. Unfortunately, this formula has not yet been tested by field studies.

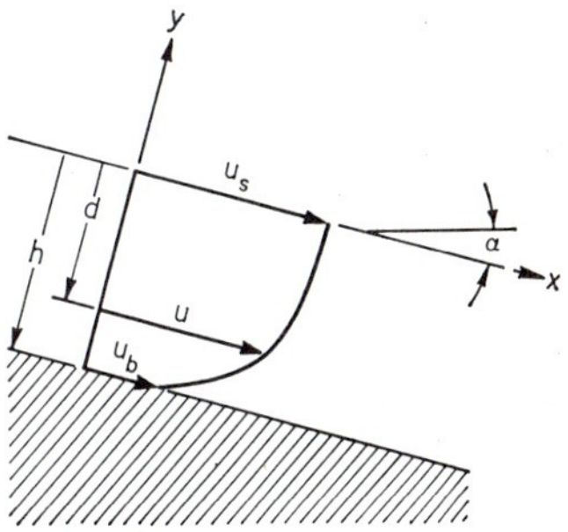

FIG. 16-14. Longitudinal section of a glacier showing laminar flow.

If the velocity of a glacier is entirely due to sliding, the discharge is

$$q = \left(\frac{\rho g}{a}\right)^m h^{m+1} \sin^m \alpha \qquad (16\text{-}7)$$

This approximation is widely used.

b. Effect of Valley Sides. If glacier ice were a Newtonian viscous liquid, it would be relatively easy to evaluate the differential flow in channels having elliptical or other analytical cross sections. However, for a material obeying the Glen flow law, the mathematical difficulties are more formidable. It is possible to analyze the flow in a channel formed by half of a circular cylinder. This problem is radially symmetrical, and the surfaces of maximum shear stress are all half cylinders parallel to the bed. Statical considerations show that the variation of shear stress with depth is still linear, but that the rate of increase is just half as rapid as with a wide valley of the same slope. If R is the maximum depth (the radius of the cylinder), Eqs. (16-4) and (16-5) take the following forms:

$$u_s - u_b = \left(\frac{1}{2}\right)^n \frac{k}{n+1} R^{n+1} \sin^n \alpha \qquad (16\text{-}8)$$

$$q = \tfrac{1}{2}\pi R^2 u_b + \frac{\pi k}{n+3}\left(\frac{1}{2}\right)^{n+1} R^{n+3} \sin^n \alpha \qquad (16\text{-}9)$$

The relations for more complex channel cross sections are not yet exactly determined. For a glacier of arbitrary but constant cross section, the average shear stress on the bed can be determined from simple statics. It is

$$\sigma_{\mathrm{av}} = \rho g \frac{C}{p} \sin \alpha \qquad (16\text{-}10)$$

where C is the area of the cross section (measured perpendicular to the bed), and p is the "iced" perimeter of the cross section. The term C/p is analogous to the *hydraulic radius* of a river channel. One can proceed further by assuming that the linear relation of σ_{xy} with depth always holds and that σ_{xy} at the bed along the center line is equal to σ_{av}. Then Eq. (16-10) can be inserted in the flow law expressed by Eq. (16-1) and integrated to obtain a relation for the velocity at depth analogous to Eqs. (16-4) and (16-8).

c. Extending and Compressing Flow. On real glaciers it is almost impossible to find areas where the simple laminar flow condition exists (Fig. 16-14). In general, there will be a gain or loss of material at the surface, the surface slope will be different from the bed slope and the surface velocity vector will not be parallel to either of these, and the velocity will increase or decrease along the length of the glacier. All these complexities can occur in a steady-state glacier (one which is adjusted to an unvarying climate so that it does not change dimensions with time). Mathematical analysis of the flow when these complications are present is still possible, but is rather involved. The interested reader is again referred to Nye [19], for only a few significant observations can be presented here.

If a longitudinal stress exists at the surface of a glacier, the flow pattern is modified from the simple relations given in the previous section. This longitudinal stress can be caused by a longitudinal curvature of the bed, a transverse constriction or widening of the channel, or the addition or removal of material from the surface. Using Weertman's theory for the sliding of a glacier on its bed [Eq. (16-6)] and the Glen flow law [Eq. (16-1)], Nye has shown that the longitudinal strain rate is

$$r = \frac{mu}{mu_b + u_s} \left(\frac{a}{h} + u_b \frac{\partial \alpha}{\partial x} \cot \alpha - \frac{1}{h} \frac{\partial h}{\partial t} - \frac{\bar{u}}{w} \frac{dw}{dx} \right) \qquad (16\text{-}11)$$

where a is the rate of net accumulation, $\bar{u}$ is the velocity averaged over the complete cross section, and w is the width. The factor $mu/(mu_b + u_s)$ generally ranges between 0.6 and 0.7.

If r is positive, the region is said to be undergoing extending flow; that is, a line marked down the center of the glacier will stretch in time and the velocity increases downglacier. Transverse crevasses indicate this condition.

If r is negative, the region is undergoing compressing flow. A longitudinal line on the glacier will shorten in time, and the velocity decreases downglacier. Transverse structures, such as crevasses, tend to close and become compressed.

If the surface curvature $\partial \alpha/\partial x$, the change in width dw/dx, and the change in thickness of the glacier with time $\partial h/\partial t$ are negligible, the longitudinal strain rate will be determined by the accumulation a. Thus the flow is generally extending in the accumulation zone and compressing in the ablation zone. This can also be seen by considering a glacier in a channel of uniform cross section. The discharge q must increase from zero to a maximum at the firn line (all the net accumulation of the accumulation zone must pass through the firn-line cross section). Below the firn line the discharge must decrease to a very low value at the terminus. If the cross-sectional areas of the glacier are constant, the velocity must increase from the head to the firn line (extending flow) and decrease from the firn line to the terminus (compressing flow).

The decrease of surface velocity from firn line to terminus and from the surface to the bed, as well as the divergence of flow vectors from the surface slope, are shown for a real glacier in Fig. 16-15. The divergence of the velocity vectors in the ablation zone is necessary to compensate for the material lost by ablation. In the case of the glacier pictured, the compensation was not complete; the flux to the surface was 5 to 16 ft/year less than the ablation, so the glacier was thinning at this rate. The transverse variation of surface velocity on this glacier is also depicted in Fig. 16-15.

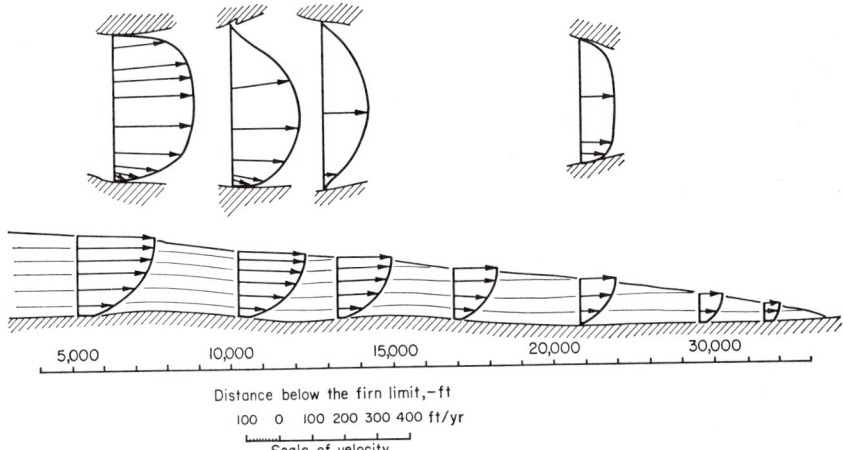

Fig. 16-15. Profiles of velocity of Saskatchewan Glacier. The four upper profiles show the transverse variation in surface velocity. The lower diagram is a vertical section along the center line of the glacier, showing vertical profiles of velocity and streamlines. (*Modified from Meier* [20].)

E. Response of Glaciers to Climatic Changes

1. Instability of Glacier Tongues. The extreme sensitivity of glaciers to climatic change is well known. Prolonged changes in the mean annual or mean summer temperature of less than one degree may instigate glacier advances or retreats amounting to hundreds, or even thousands, of feet. One explanation of this sensitivity is that glaciers exist according to a delicate balance between accumulation and ablation, and a slight disturbance in this balance results in an appreciable reaction. However, the rapid, and sometimes unpredictable, changes in the position of the terminus of some glaciers require additional explanation. The following simple argument, attributed to Nye [19], shows that the thickness of a glacier reach undergoing compressing flow is unstable.

Figure 16-16 shows a small length δx of a glacier in which the average velocity u at section A is greater than the average velocity at section B (compressing flow). It is assumed that a steady-state condition has been reached. For a unit width, the discharge through section A is uh. The portion of this discharge removed by ablation a at the surface is $a\,\delta x$, and that which flows out through section B is $uh - a\,\delta x$.

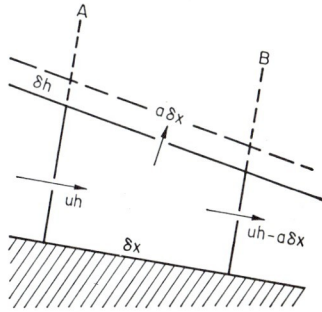

Fig. 16-16. Diagram showing the instability of a glacier surface profile in a region of compressing flow.

Suppose that an extra layer of ice of thickness δh is artificially laid on top of the glacier and that the rate of ablation remains the same. The velocity will increase to $u + \delta u$, and the discharge through section A will increase, approximately, by $u\,\delta h$, on account of the increased thickness and by $h\,\delta u$ on account of the increased velocity. Now if u is proportional to h^m, as shown by Eq. (16-6), $\delta u/u = m\,\delta h/h$, and the increase in discharge due to the velocity increase is $mu\,\delta h$. The total increase in discharge is $(m+1)uh$ through section A. Thus the increase in discharge is proportional to the steady-state velocity. The increase in discharge through section B is less because the steady-state velocity is less. There is therefore an accumulation of material between the two sections, and the level of the surface rises. Thus the small addition of thickness causes an unstable thickening of the glacier.

Similar arguments show that a small reduction in thickness is also unstable in a zone of compressing flow. In a zone of extending flow the glacier thickness is stable. Most glaciers exhibit extending flow in their upper regions and compressing flow in their lower regions. Thus one should expect only the lower regions to show large changes in thickness, and this is, in fact, true.

2. Theory of Kinematic Waves. Throughout most of a glacier the ice density is constant in time and space. This means that an equation of continuity can be written as follows:

$$\frac{\partial q}{\partial x} + \frac{\partial h}{\partial t} = a \qquad (16\text{-}12)$$

where q is the discharge, x is the distance along the longitudinal axis of the glacier, h is the thickness of the glacier, t is the time, and a is the rate of *net* accumulation (accumulation minus ablation) on the surface. This continuity equation follows from simple geometrical considerations. If there is a relation between q and h, such as Eq. (16-5) or (16-7), then it follows directly from Eq. (16-12) that a disturbance in h or q will be propagated down the glacier in the form of a kinematic, or traveling, wave.

The mathematical difficulties are formidable if one tries to analyze large changes in the glacier profile by working directly from Eq. (16-12). When the actual profile of a glacier is only slightly different from its steady-state profile, these difficulties can be partly eliminated by working in terms of the difference (perturbation) between the two profiles. The following development is attributed to Weertman [21] and Nye [22].

It is assumed that the glacier is unlimited in width and flows down a slope in the x-direction and that the thickness $h(x,t)$ is such that $(\partial h/\partial x)_t \ll 1$. The steady-state properties of the glacier are designated by the subscript 0, and the actual properties at any instant are given by the steady-state value plus a small perturbation which carries the subscript 1. Thus $q = q_0 + q_1$, $h = h_0 + h_1$, $\alpha = \alpha_0 + \alpha_1$, and

$$a = a_0 + a_1$$

Then the equation of continuity can be written as

$$\frac{\partial q_1}{\partial x} + \frac{\partial h_1}{\partial t} = a_1 \qquad (16\text{-}13)$$

Because q is a function of x, h, and α, the following can be written for small perturbations:

$$q_1 = c_0 h_1 + D_0 \alpha_1$$

where $c_0 = \partial q/\partial h$ and $D_0 = \partial q/\partial \alpha$, both derivatives being evaluated at the steady-state values of h_0 and α_0. Now $\alpha_1 = \partial h_1/\partial x$, and by substituting in Eq. (16-13), one obtains an equation which describes the non-steady-state changes in the glacier profile

$$\frac{\partial h_1}{\partial t} = a_1 - \frac{dc_0}{dx} h_1 - \left(c_0 - \frac{dD_0}{dx}\right)\frac{\partial h_1}{\partial x} + D_0 \frac{\partial^2 h_1}{\partial x^2} \qquad (16\text{-}14)$$

The four terms on the right-hand side of Eq. (16-14) can be interpreted as follows. The first term indicates that the height perturbation h_1 increases at a rate given by the perturbation in accumulation a_1. The second term represents an exponential change in h_1. The third term indicates that a kinematic wave of constant h_1 travels downglacier at a speed $c_0 - dD_0/dx$. The fourth term represents the diffusion (broadening) of the disturbance h_1 in accordance with the diffusion equation, the diffusivity being D_0.

Several models of glaciers have been discussed by Weertman and Nye. One of the simplest of these is a glacier whose forward motion consists entirely of slippage on the bed, as shown by Eq. (16-7). In this case the wave velocity c_0 and diffusivity D_0 can be evaluated as follows:

$$c_0 = \left(\frac{\partial q}{\partial h}\right)_0 = (m+1)\frac{q_0}{h_0} = (m+1)u_0 \approx 3u_0 \qquad (16\text{-}15)$$

where u_0 is the steady-state velocity, and

$$D_0 = \left(\frac{\partial q}{\partial \alpha}\right)_0 = mq_0 \cot \alpha_0 \approx 2q_0 \cot \alpha_0 \qquad (16\text{-}16)$$

The exponential decay of the disturbance proceeds at a rate determined by the coefficient of $-h_1$ in Eq. (16-14). This is

$$\frac{dc_0}{dx} = (m+1)\frac{du_0}{dx} = (m+1)r_0 \approx 3r_0 \qquad (16\text{-}17)$$

where r_0 is the steady-state rate of longitudinal extension (positive) or compression (negative). If r_0 is negative, the disturbance will grow in an unstable way, as mentioned earlier.

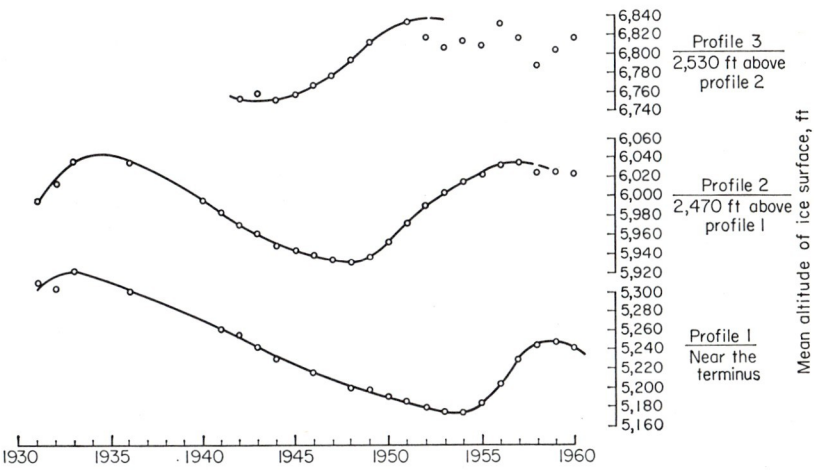

Fig. 16-17. Variation of altitude of the ice surface at three different profiles on Nisqually Glacier. (*After Johnson* [23].)

3. Behavior of Typical Glaciers. The longitudinal profile of a valley glacier is usually concave in the lower reaches of these glaciers, and ablation predominates over accumulation. These effects cause compressing flow, as represented by Eq. (16-11), and therefore an instability to climatic changes. One well-documented typical case is the Rhone Glacier in Switzerland. During the period of 1874 to 1909, the mass budget was slightly negative. This resulted only in a slight thinning over the upper three-fourths of the glacier but an accelerated wasting away of the tongue. The change in the compressing zone of the tongue was more than 11 times larger than the greatest change in the extending zone and caused a retreat of the terminus which accelerated with time.

This instability appears to be terminated by the arrival of a kinematic wave which originates at the point where the flow changes from extending to compressing. Kinematic waves, however, are not obvious to the casual observer, and only a few measurement programs have demonstrated their existence. One of the best of these is the work of Johnson [23] on Nisqually Glacier, Washington. His data clearly show that a thickening in the region of the firn limit (altitude 6,800 ft) was propagated rapidly downglacier (Fig. 16-17). The average ice velocity was about 200 ft/year, but the wave traveled at a speed of about 700 ft/year, as predicted by Eq. (16-15). The ice velocity at profile 2 increased from about 20 to 400 ft/year as the wave passed. The crest of the wave traveled faster than the trough, causing a progressive change in the shape of the wave. As this overrunning effect continues, a *shock wave* may develop; the perturbation may ride out over the older, slower ice as a discrete or separate glacier. Such an effect is frequently observed.

According to theory, all glaciers should show kinematic-wave behavior. However, such waves have not been observed on some well-studied glaciers. On the other hand, some glaciers seem to exhibit remarkable shock-wave effects in lieu of a more continuous adjustment. Post [24] has described several glaciers in Alaska, which, after decades of inactivity, suddenly advanced their termini from 2.5 to 4 miles in 3 to 9 months. The reasons for these divergences in behavior are not completely understood.

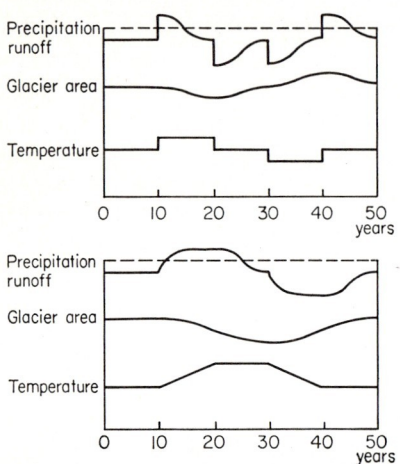

FIG. 16-18. Diagram showing the variation of glacier area and runoff caused by sudden or gradual changes in air temperature.

4. Effect on Streamflow. In general, evaporation losses from a glacier are small, and the runoff is almost equal to the precipitation under steady-state conditions. If the climate changes, the glacier must change its dimensions, and during this change the precipitation-runoff relation may change markedly. Furthermore, some glaciers appear to respond to climatic change in an erratic way, advancing or retreating beyond an equilibrium position. Figure 16-18 shows diagrammatically the effect on glacier runoff of sudden or gradual changes in air temperature for a constant precipitation (snowfall) regime.

F. Some Special Hydrologic Problems Connected with Glaciers

1. Reconstructing Past Climates from Glacier Variations. A recurring problem in hydrology is that of determining where modern water-supply data fit into the long-term pattern of fluctuating water supplies. An even more fundamental problem is the determination of the cause of climatic variation. Historical records, especially in the Western Hemisphere, are short, so that indirect evidence must be obtained from features that respond to and preserve records of climatic changes. Glaciers can provide this evidence because (1) they advance or retreat in response to very small changes in climate, and (2) these advances and retreats leave a stamp on the landscape in the form of moraines, terraces, or changes in vegetation (Fig. 16-19). Frequently these advances or retreats can be precisely dated. One of the best techniques is through the use of botanical evidence; this has been described by Lawrence [25] and Sigafoos and Hendricks [26].

It is difficult to relate glacier changes directly in terms of temperature or precipitation variations. In general, glaciers respond to changes in winter precipitation and/or summer temperature, but the actual relations are much more complicated. Glacier data do, however, show a general correlation with many other types of climatic data because cool climates tend also to be wet and warm climates tend to be dry. These correlations are poor when the sets of data are compared on a year-to-year basis, but the correlations are excellent when time intervals of several decades are considered.

Glacier data correlate well with other climatic data extending back 70,000 years ago, according to Flint and Brandtner [27]. Glacier retreats indicated a warm-dry interval from 3,000 to 7,000 years ago. A major readvance of the ice took place at the beginning of the sixteenth century; most glaciers seen today are products of this "Little Ice Age." This cool-wet period persisted until the beginning of the twentieth century. At least in the Pacific Northwest, the climate then became warmer and drier until about 1945, when it rather abruptly changed back to cool and wet. There is no evidence in the glacier record for the existence of a "normal," or steady-state, climate.

FIG. 16-19. Moraines and trimlines below North Mowich Glacier, Mt. Rainier, Washington. Several moraines indicating different advances of the ice in the last two hundred years are indicated by the symbol m. Advances of this glacier also caused destruction of existing vegetation. After recession of the glacier, the former ice boundary is marked by a sharp change in the age, species, or density of the vegetation. This sharp boundary is called a *trimline*. The time of retreat of the ice can be dated by botanical studies in the vicinity of the trimline. Old (t_1) and young (t_2) trimlines are shown in this photograph. (*Photograph by A. S. Post*, 1958.)

2. Natural and Artificial Regulation of Glacier Runoff. Water supplies for hydroelectric power, irrigation, and other industrial uses are most valuable when the supply of water can be delivered at the time of demand. In many parts of the West most of the precipitation occurs during the winter, when water is in less demand, and during the summer the precipitation is low at the time of highest demand. Furthermore, the supply of precipitation water varies from year to year in an erratic manner. Thus it is necessary to even out flows during the season and from year to year in order to make full use of the water. This is commonly done by channeling the flow into large artificial reservoirs at low altitudes. However, such reservoirs are expensive, cause an appreciable loss of water by evaporation, and have limited capacity to equalize flows over a long period of time.

Water stored in high-altitude snowfields and glaciers can be naturally or artificially regulated, and the mechanics of this regulation avoid some of the disadvantages inherent in normal reservoir operation. Streamflow from glacier ice is naturally regulated so that only slight additional manipulation may be needed. As mentioned earlier in this section, the greatest natural release of water from glaciers takes place in July and August, coinciding with the time of maximum temperatures. During hot-dry years more water is released, and during cool-wet years water is added to storage.

If this natural regulation is insufficient, there are several techniques of artificial regulation which are at least theoretically possible. Melt rates can be increased or decreased by dusting or coating the snow and ice surfaces. If 70 per cent of the melt normally occurs because of radiation and if the albedo of the snow can be lowered from a normal value of 0.6 to a value of 0.1, the total melt will be increased by about 87 per cent. For a typical glacier in the Northern Cascades of Washington this would result in the additional yield of about 8 acre-ft of water per acre of glacier surface over and beyond the natural melt. Actual measurements of the augmentation of snow and ice melt by additives to the surface have not yet been sufficiently conclusive to determine if such a large addition is possible on a practical basis. However, numerous studies are in progress at the present time. It appears that foundry sand, fly ash, or coal dust is an efficient material for coating the surface. It has also been found that the coating must be very thin, not more than 1 mm thick, in order to increase melt rates. A thick coating has the opposite effect of insulating the snow and ice layers.

The application of additives to the glacier surface might be visualized as one way to guarantee firm hydroelectric power, and it is assumed that such an operation need be carried out only once in a decade or so during abnormally low runoff years. The question which then needs to be answered is, how can the glacier reservoir be restored to its former size? This could be done by three different artificial ways: (1) the application of insulating surface coatings such as sawdust, (2) cloud seeding during the winter to increase precipitation in the mountains, and (3) the use of drift fences to drift more snow on glaciers and cause less to be blown into the low valleys. There is also a fourth possibility, that the glacier might tend to regain its former size naturally. LaChapelle has determined that if the Blue Glacier in the Olympic Peninsula were completely removed, it would tend to reform under the present climatic regime.

One important restriction on the artificial dusting of snow or ice surfaces is the fact that many glaciers occur in areas of great scenic beauty and in wilderness preserves. The possible economic benefits would have to be weighed against the potential destruction of wilderness values.

3. Jökulhlaups, or Glacier Floods. Sudden outbursts of water from glaciers or from glacier-dammed lakes are of economic as well as scientific interest. The Icelandic term *jökulhlaup* (pronounced approximately like "yokel-loup" and the last syllable as in "out") is used for this phenomenon. Frequently these outbreaks occur at periodic intervals and generally have no direct relation to meteorologic conditions. Some jökulhlaups are minor outbursts which are recognizable only on streamflow records. Others, such as the outbursts at the Kautz and Nisqually Glaciers on Mount Rainier, are large enough to destroy roads and bridges. However, the releases of water may be truly catastrophic in timing and amount. For instance, Lago Argentino in Argentina discharges from 1.6 to 4.2 million acre-feet of water every year or so, and Grimsvötn in Iceland periodically spills up to 8 million acre-feet of water in a flood in which streamflow discharges have reached nearly 2 million cubic feet per second [28]. In August, 1959, a glacier burst in a remote mountain valley in the Karakorum Range in Kashmir and caused a flood rise of more than 100 ft at a distance of more than 25 miles from the point of outburst. These sporadic outbreaks have exacted a heavy toll in human life and property and caused the destruction of whole villages.

The best known examples of large jökulhlaups in North America occur at Lake George (Knik Glacier) near Anchorage and the Tulsequah Glacier near Juneau, Alaska.

Jökulhlaups may involve several types of impounded water bodies. These may be large visible lakes in lateral stream valleys dammed by ice in the main valley or lakes in main valleys dammed by ice from a tributary valley, or the water may be entirely below the surface of the ice and not visible.

A periodicity in glacier-lake outbursts is frequently observed. This can be due to several causes. In some instances, such as in the dam of the Rio Ploma by the Nevado Glacier in Argentina, the glacier lake is formed by the periodic advance of an unstable glacier tongue. However, in most other cases, the frequency of dumping depends on a certain necessary head of water and the rate of inflow. The period is most commonly about 1 year, but it ranges up to 10 years or more.

The discharge hydrographs from jökulhlaups generally resemble normal storm-unit hydrographs, except that the time scale is reversed. The flow starts out at a low rate

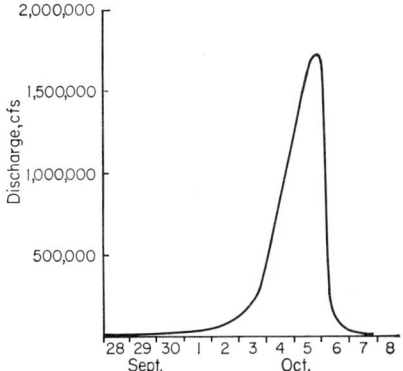

FIG. 16-20. Hydrograph of the jökulhlaup of Grimsvötn, Iceland, in 1922. The total amount of water discharged has been estimated as 1.7 cu mi (5.7 × 10^6 acre-ft). (*After Thorarinsson* [28].)

and increases exponentially until a peak is reached, and then the discharge drops off abruptly (Fig. 16-20).

The cause of the sudden outbreak of water generally appears to be due to the enlargement of a discharge channel by percolating water. Although glacier ice is generally impermeable, there are small channels or cracks in the ice at all times. Furthermore, the flowage of a glacier over a rough channel can cause breaks along the margin or beneath the glacier which permit the escape of small amounts of water. Once a small trickle of water finds a route through the ice dam, it can enlarge the channel at an increasingly rapid rate. Jökulhlaups in Norway have been studied by Liestøl [29], who suggested that the loss of potential energy from the lake level was sufficient to provide enough energy to cause the rapid enlargement of a drainage channel by melting. If the impounded water is open to the atmosphere, it will absorb heat from the sun and the air, and it has been shown that this is sufficient to permit the Lake George drainage channel to enlarge at the observed rate. Unfortunately, little is known of the exact mechanics of outbreak of jökulhlaups. They are such powerful and distinctive phenomena that much more study is indicated.

VI. REFERENCES

1. Meier, M. F.: Distribution and variation of glaciers in the United States exclusive of Alaska, *Helsinki Assembly, Intern. Assoc. Sci. Hydrol. Publ.* 54, pp. 420–429, 1960.
2. Dorsey, N. E.: "Properties of Ordinary Water Substances," Reinhold Publishing Corporation, New York, 1940.
3. Devik, Olaf: Freezing water and supercooling, *J. Glaciol.*, vol. 1, no. 6, pp. 307–309, 1949.

4. Mantis, H. T. (ed.): Review of the properties of snow and ice, *U.S. Army Corps of Engineers, Snow, Ice, and Permafrost Res. Estab., Rept.* 4, 1951.
5. Glen, J. W.: The creep of polycrystalline ice, *Proc. Roy. Soc. London*, ser. A, vol. 228, pp. 519–538, 1955.
6. Ahlmann, H. W.: Glaciological research on the north Atlantic coasts, *Roy. Geograph. Soc. Res. Ser.* 1, 1948.
7. Lagally, Max: Zur Thermodynamik der Gletscher, *Z. Gletscherk.*, vol. 20, pp. 199–214, 1932.
8. Ahlmann, H. W.: Scientific results of the Swedish-Norwegian Arctic Expedition in the summer of 1931, pt. VIII, *Geograf. Ann.*, vol. 15, pp. 161–216, 261–295, 1933.
9. Meier, M. F.: Proposed definitions for glacier mass budget terms, *J. Glaciol.*, vol. 4, pp. 252–261, 1962.
10. LaChapelle, E. R.: Errors in ablation measurements from settlement and subsurface melting, *J. Glaciol.*, vol. 3, no. 26, pp. 458–467, 1959.
11. LaChapelle, E. R.: Annual mass and energy exchange on the Blue Glacier, *J. Geophys. Res.*, vol. 64, no. 4, pp. 443–449, 1959.
12. LaChapelle, E. R.: "The Blue Glacier Project 1959 and 1960," University of Washington, Department of Meteorology and Climatology, Seattle, Wash., 1960.
13. Meier, M. F., and W. V. Tangborn: Distinctive characteristics of glacier runoff, *U.S. Geol. Surv., Profess. Paper* 424, 1961.
14. Kasser, Peter: Der Einfluss von Gletscherrückgang und Gletschervorstoss auf den Wasserhaushalt (The influence of glacier recession and glacier advance on the water budget), *Wasser- und Energiewirtsch.*, Zurich, no. 6, 1959.
15. Borland, W. M.: Sediment transport of glacier-fed streams in Alaska, *J. Geophys. Res.*, vol. 66, no. 10, pp. 3347–3350, 1961.
16. Rainwater, F. H., and H. P. Guy: Some observations on the hydrochemistry and sedimentation of the Chamberlin Glacier area, Alaska, *U.S. Geol. Surv. Profess. Paper* 414-C, 1961.
17. Nye, J. F.: The mechanics of glacier flow, *J. Glaciol.*, vol. 2, no. 12, pp. 82–93, 1952.
18. Weertman, Johannes: On the sliding of glaciers, *J. Glaciol.*, vol. 3, no. 21, pp. 33–38, 1957.
19. Nye, J. F.: The motion of ice sheets and glaciers, *J. Glaciol.*, vol. 3, no. 26, pp. 493–507, 1959.
20. Meier, M. F.: Mode of flow of Saskatchewan Glacier, Alberta, Canada, *U.S. Geol. Surv. Profess. Paper* 351, 1960.
21. Weertman, Johannes: Traveling waves on glaciers, *Symp. Chamonix, Intern. Assoc. Sci. Hydrol. Publ.* 47, pp. 162–168, 1958.
22. Nye, J. F.: The response of glaciers and ice sheets to seasonal and climatic changes, *Proc. Roy. Soc. London*, ser. A, vol. 256, pp. 559–584, 1960.
23. Johnson, Arthur: Variation in surface elevation of the Nisqually Glacier, Mt. Rainier, Wash., *Intern. Assoc. Sci. Hydrol. Bull.* 19, pp. 54–60, 1960.
24. Post, A. S.: The exceptional advances of the Muldrow, Black Rapids and Susitna Glaciers, *J. Geophys. Res.*, vol. 65, no. 11, pp. 3703–3712, 1960.
25. Lawrence, D. B.: Estimating dates of recent glacier advances and recession rates by studying tree growth layers, *Trans. Am. Geophys. Union*, vol. 31, pp. 243–248, 1951.
26. Sigafoos, R. S., and E. L. Hendricks: Botanical evidence of the modern history of Nisqually Glacier, Washington, *U.S. Geol. Surv. Profess. Paper* 387-A, 1961.
27. Flint, R. F., and Friedrich Brandtner: Climatic changes since the last interglacial, *Am. J. Sci.*, vol. 259, pp. 321–328, 1961.
28. Thorarinsson, Sigurdur: Some new aspects of the Grimsvötn problem, *J. Glaciol.*, vol. 2, no. 14, pp. 267–275, 1953.
29. Liestøl, Olav: Glacier dammed lakes in Norway, *Norsk Geograf. Tidsskr.*, vol. 15, pp. 122–149, 1955–1956.

Section 17-I

SEDIMENTATION

PART I. RESERVOIR SEDIMENTATION

LOUIS C. GOTTSCHALK, *Staff Geologist, Engineering Division, U.S. Soil Conservation Service.*

I. Introduction	17-2
II. The Problem	17-3
A. Relationship of Sedimentation to Reservoir Function	17-3
1. Loss of Storage and Services	17-3
2. Deposition at Outlet Gates	17-4
B. Aggradation above Reservoirs	17-5
C. Degradation below Reservoirs	17-5
III. Erosion	17-6
A. Normal Erosion versus Accelerated Erosion	17-6
B. Water Erosion	17-6
1. Kinds of Water Erosion	17-6
2. Sheet Erosion	17-6
3. Musgrave Equation for Predicting Rate of Sheet Erosion	17-7
4. Universal Equation for Predicting Rates of Sheet Erosion	17-7
5. Channel Erosion	17-9
C. Wind Erosion	17-10
1. Empirical Equation for Estimating Rate of Wind Erosion	17-10
D. Ice Erosion	17-10
E. Gravity Erosion	17-10
F. Effect of Human Activities on Erosion	17-11
IV. Movement of Sediment from Watersheds	17-11
A. Sediment Yield	17-11
B. Sediment-production Rate	17-11
C. Sediment-delivery Ratio	17-12
1. Influence of Size of Drainage Area on the Sediment-delivery Ratio	17-12
2. Influence of Topography and Channel Density on Sediment-delivery Ratio	17-13
3. Influence of Relief and Length of Watershed Slopes on Sediment-delivery Ratio	17-13
4. Influence of Precipitation and Runoff on Sediment-delivery Ratio	17-13
V. Sediment Characteristics	17-14
A. Particle Characteristics	17-14
1. Mineralogical Composition	17-14
2. Shape	17-15
3. Size	17-15

B. Bulk Characteristics.................................... 17-15
   1. Grain-size Distribution............................. 17-15
   2. Specific Weight of Sediment........................ 17-15
   3. Effect of Mineral Constituents on Specific Weight....... 17-18
   4. Effect of Thickness of Deposits on Specific Weight...... 17-18
   5. Effect of Reservoir Operation on Specific Weight....... 17-19
C. Influence of Erosion and Transport on Sediment Characteristics.. 17-19
   1. Sheet Erosion...................................... 17-19
   2. Channel Erosion.................................... 17-19
   3. Wind Erosion...................................... 17-19
   4. Transportation..................................... 17-19
D. Characteristics of Sediments Delivered to Reservoirs....... 17-20
VI. Trap Efficiency of Reservoirs............................... 17-21
  A. Factors Influencing Trap Efficiency...................... 17-21
    1. Sediment Characteristics........................... 17-21
    2. Detention-storage Time............................ 17-21
    3. Nature of Outlet................................... 17-21
  B. Computing Trap Efficiency............................. 17-21
VII. Distribution of Sediment in Reservoirs....................... 17-22
  A. Processes of Sediment Deposition........................ 17-22
  B. Delta Deposits... 17-23
  C. Bottom-set Beds....................................... 17-24
  D. Predicting Sediment Distribution........................ 17-24
    1. Area-increment Method............................. 17-24
    2. Empirical Area-reduction Method................... 17-25
    3. Elevation of Sediment-accumulation Method......... 17-25
VIII. Sediment Yields from Watersheds........................... 17-25
  A. Measuring Sediment Yields............................. 17-25
    1. Reservoir-sedimentation Surveys.................... 17-25
    2. Sediment-load Measurements...................... 17-26
  B. Methods of Predicting Sediment Yields from Watersheds... 17-26
    1. Predicting Sediment Yields from Comparative Watersheds 17-26
    2. Installation of Sediment-load Measuring Stations....... 17-26
    3. Estimating Watershed Erosion and the Delivery Ratio of Sediment... 17-27
IX. Rates of Reservoir Sedimentation........................... 17-27
  A. Annual Rate of Deposition.............................. 17-27
  B. Measured Rates of Reservoir Sedimentation.............. 17-27
X. Control of Reservoir Sedimentation.......................... 17-29
  A. Design of Reservoir.................................... 17-29
  B. Venting Sediment..................................... 17-30
  C. Sediment Removal..................................... 17-30
  D. Reduction of Sediment Yield............................ 17-31
    1. Vegetative Screens................................. 17-31
    2. Watershed Structures.............................. 17-31
    3. Watershed Land-treatment Measures............... 17-32
  E. Evaluation of Selected Control Measures................ 17-32
XI. References.. 17-33

I. INTRODUCTION

This section deals with the general subject of erosion and sediment yields from watersheds as related to processes and rates of reservoir sedimentation. The processes of erosion, entrainment, transportation, and deposition of sediment are complex,

Methods have not yet been developed to extrapolate existing results of fundamental research to broad, complex areas, such as watersheds, for prediction of the expected rate or processes of reservoir sedimentation. Although considerable basic data have been assembled and comprehensive research has been initiated in the past quarter century, much yet remains to be done before the prediction of rates and processes of reservoir sedimentation can achieve the degree of accuracy desired. The purpose of this section is to outline results of fundamental research and technological studies related to reservoir sedimentation. Basic considerations are outlined for those areas where research is lacking to aid the technician to recognize specific problems and develop sound solutions. Finally, general guide lines are presented for developing control measures applicable to specific watershed and reservoir conditions.

II. THE PROBLEM

Erosion, transportation, and deposition of sediment are natural processes which have occurred throughout geologic times. The extent of erosion and, consequently,

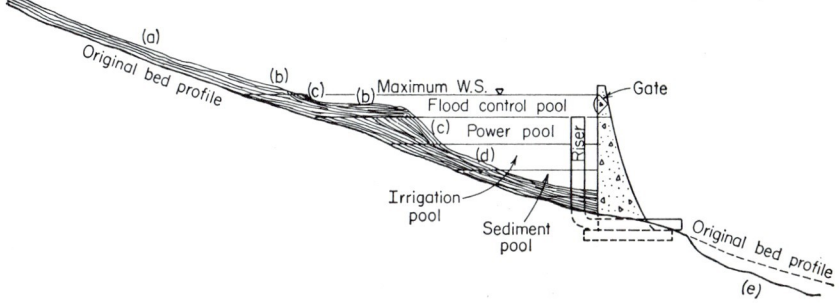

Fig. 17-I-1. Schematic representation of sedimentation processes associated with reservoir construction. (*a*) Area of aggradation; (*b*) top-set beds; (*c*) foreset beds; (*d*) bottom-set beds; (*e*) area of degradation.

the amount of sediment which moves out of a watershed vary greatly from one problem area to another, depending upon geologic, climatic, physical, vegetative, and other conditions. Sedimentation of a reservoir created by a dam constructed on a natural water course is inevitable. The problem of concern is the rapidity of sedimentation and the period of time which will elapse before the usefulness of the storage works is seriously impaired or destroyed. The replacement cost of storage lost to sediment accumulation in American reservoirs amounts to millions of dollars annually.

A dam on a stream channel changes the hydraulic characteristics of flow and the sediment-transport capacity. In the process of adjusting to the equilibrium for the new conditions, additional problems may be created, both upstream and downstream from the structure. Consequently, in the design of dams, the engineer must be alert to these problems and be able to anticipate the extent of changes which might occur outside of the reservoir area, as well as to provide for protection against rapid siltation. Depositional and degradational areas associated with reservoir construction are shown in Fig. 17-I-1.

A. Relationship of Sedimentation to Reservoir Function

1. Loss of Storage and Services. Dams are constructed to create retention or detention water storage or to maintain a static water-surface level. Examples of retention reservoirs are those designed to store water for such purposes as irrigation; municipal, industrial, and domestic water supply; and hydropower. These reservoirs accumulate water during periods of plentiful supply and store it for relatively long periods of time, to be used when the demand exceeds supply. Detention reservoirs are those designed for flood-prevention purposes, groundwater recharge, and other purposes where storage is of more temporary nature, and release rates are regulated to provide downstream discharge rates in accordance with objectives of the project. Dams which do not depend on carryover storage for proper functioning include run-of-the-river dams to create head for power purposes; water-diversion structures; debris dams; and dams to control water levels for other purposes, such as navigation and erosion control.

The maintenance of storage capacity in both detention and retention structures is vital to their proper functioning. The loss of storage space to sediment eliminates such space for further water-storage purposes and directly affects the services dependent upon that water storage.

An example is the case history of Spring Lake, a municipal water-supply reservoir serving Macomb, Ill. A study completed by Stall and others [1] indicates how the useful life of a reservoir can be seriously impaired by sedimentation. Storage began in Spring Lake in 1927. By 1947, about 47 per cent of the water-storage capacity had been lost because of sedimentation. During the period 1935 to 1947 pumpage from the reservoir was increased from 318,000 to 555,000 gpd because of demands of population increase and increase in per capita use. The results of the study indicated that, with further increased water demands and decreased storage due to sedimentation, the city would face a water shortage in event of a serious drought after the reservoir had been in operation only 27 years. By 1947 the reservoir was beginning to experience excessive drawdowns, a condition which becomes evident in all storage reservoirs when demand approaches supply. The city has subsequently found it necessary to dredge the reservoir to restore capacity in order to avoid an eventual water shortage.

To avoid rapid loss in water-storage space and reduction in useful life, sufficient storage space must be provided for accumulation of sediment in the reservoir, or means of reducing sediment inflow or its removal must be provided. For example, if it is determined in project evaluation that the reservoir should function at full efficiency for a given period of years, storage space must be provided in the total reservoir capacity to offset that which will be lost to sedimentation during the same period.

Multiple-purpose reservoirs which combine retention and detention features require an evaluation to determine probable sediment distribution, as well as volume, in each increment of storage for sediment accumulation. The reason for this is that in multiple-purpose reservoirs, storage for different purposes is allocated at different elevations in a reservoir, while sedimentation does not occur at equal rates between these different elevations. Consequently, the average rate of storage loss for the reservoir as a whole cannot be applied to determine the sediment-storage requirements for different increments of storage.

2. Deposition at Outlet Gates. Distribution of sediment is also an important consideration in the location of gates or outlets to avoid areas where rapid sedimentation might occur. The Shandaken Tunnel intake of Schoharie Reservoir, Prattsville, N.Y., is located in the upper one-third of the reservoir. Water from Schoharie Reservoir is diverted through the 18-mile tunnel into Esopus Creek, whence it flows into the Ashokan Storage Reservoir as part of the New York City water-supply system. Excessive drawdowns, aided by a submerged levee used for diverting flow into the tunnel, has resulted in excessive sedimentation in front of the tunnel intake. This requires periodic maintenance despite the fact that the reservoir is silting at the very low rate of 0.07 per cent annually.

Deposition may also occur at gates located at the dam, particularly during periods of heavy drawdown. According to the Annual Reports of the Puerto Rico Water Resources Authority, a flood on May 19, 1940, brought down so much sediment into the Guayabal Irrigation Reservoir, on the south coast of Puerto Rico, that the outlet works were completely buried. Operations were crippled, and it was impossible for several days to deliver water needed for irrigation. The reservoir was subsequently dredged, and the dam raised 16 ft (1950).

B. Aggradation above Reservoirs

The construction of a dam on a water course changes the hydraulic characteristics of flow and resultant sediment-transport capacity, which may create serious problems both upstream and downstream from the reservoir. Deposition in delta and above crest elevations of reservoirs may cause serious *aggradation* upstream from the reservoir. Such aggradation results in reduced capacity of upstream channels to contain flows which create more frequent or permanent flooding; elevating of groundwater tables and swamping of lands in the valleys; and may actually bury improvements under sediment. The extent of upstream aggradation depends upon the stream gradient, the size gradation of sediment, and the degree of fluctuation of the reservoir surface. Upstream aggradation is minimized on steep-gradient streams transporting primarily fine-grained materials. Low-gradient streams which transport heavy loads of sand to large detention reservoirs may result in aggradation extending for many miles upstream from the head of backwater of the reservoir. The effects of highly fluctuating reservoirs are periodic and transitory in the initial stages, with alternating periods of aggradation and degradation, depending upon the inflow in respect to reservoir levels. However, as the reservoir fills with sediment and as given inflows produce higher surface elevations in the reservoir, a net upstream aggradation takes place.

It is reported that Imperial Dam, a comparatively low dam built on the Lower Colorado River in 1938, caused deposits of sediment for a distance of 55 miles upstream. The river bed at Picacho, Calif., upstream from the dam, was raised nearly 10 ft between 1938 and 1948. Aggradation above Elephant Butte Dam, between 1914 and 1954, raised the river valley and channel of the Rio Grande about 13 ft at San Marcial, N.M., which is located just above the head of backwater at crest stage, and about 7 ft at a point about 10 miles upstream. As a result, it became necessary to construct a levee to protect San Marcial from flood damages and from actual burial in the deposits. In addition, an extensive system of drainage works was required to control seepage problems.

C. Degradation below Reservoirs

A change in the hydraulic characteristics of flow with the removal of the sediment load from the outflow of a reservoir establishes a new set of conditions, to which the stream below a dam will eventually adjust in order to achieve equilibrium. Where the outflow has sufficient tractive force to initiate movement of materials in the channel below the structure, channel *degradation* immediately takes place. The lowering of main channel grades results in correlative degradation of tributary channels. This degradation may initiate a new cycle of headward erosion in tributary watersheds. In severe cases it may also result in the lowering of groundwater tables, with resulting damage to agricultural land; in the undermining of bridge piers; and in other damaging effects.

The rate of degradation below reservoirs depends upon the type of material in the channel and the hydraulic characteristics of outflow. Materials which are readily entrained in the clear-water discharge from reservoirs are removed from channels until such time as an equilibrium grade, adjusted to flow conditions and channel materials, can be established.

Numerous and prolonged outflows of clear water from reservoirs into channels with unlimited depths and reaches of transportable materials result in rapid degradation, which may extend for many miles downstream. Where materials in channel beds are more resistant to erosion than those in the channel banks, bank erosion and stream meandering may proceed at a greater pace than degradation to satisfy natural conditions of balance between flow and sediment load.

Minor degradation below dams is sometimes desirable and beneficial, since it increases channel capacity and improves drainage of adjoining lands. In some instances, depending upon the nature of channel materials and scour forces, loss of channel capacity and vegetative encroachment may occur.

Tremendous quantities of materials have been removed from the channel below Hoover Dam (Lake Mead). Measurements reported by Borland and Miller [2] indicate that 151,830,000 yd³ have been removed from the channel for a distance of 92 miles below the dam during the period 1935 to 1951. Below this point, for a distance of about 28 miles to the head of Lake Havasu, a net aggradation of 107,322,000 yd³ took place during the same period. The depth of degradation at Willow Beach, 12 miles downstream from Hoover Dam, amounted to about 14 ft between 1935 and 1949.

III. EROSION

Erosion is defined as the wearing away of the land. Agents of erosion are water, wind, ice, and gravity. To this list might be added human activities, since mining, excavation for buildings, highways, and other activities of man are locally important in the changing of the shape of the landscape. Sediment is the by-product of erosion. The term is usually applied to eroded material which has been transported and deposited by water, but is sometimes used to denote deposits resulting from wind, ice, and other forces. An understanding of the processes of erosion is necessary in order to develop mechanical and vegetative control measures and to predict adequately sediment yields from watersheds.

A. Normal Erosion versus Accelerated Erosion

Normal erosion, often referred to as *geologic norm*, is erosion of land in its natural environment undisturbed by human activity. It includes processes of weathering and the removal of materials by gravity, wind, water, and ice. These processes have been active through geologic time, and tremendous volumes of materials have been translocated on the earth's surface. Geologic erosion occurs today as it did in the past. In some parts of the country, particularly in geologically young areas where materials are not well indurated and where aridity is conducive to sparse vegetation, it may be severe and may account for a large portion of the sediment load transported by streams. Severe geologic erosion is readily discernible in so-called *badland* areas of the United States.

Under natural conditions in the more humid areas, dense vegetation affords protection against removal of materials made available for transportation by weathering processes, and geologic erosion is slow. Disturbance of this cover by man's activity, such as overgrazing of grasslands, removal of timber by logging or burns, and breaking of sod cover by plowing, disturbs the natural conditions, and the rate of erosion becomes greatly *accelerated*.

The removal of natural timber cover from a particular area and subsequent up- and downhill plowing may increase the erosion rate more than one-hundred-fold.

B. Water Erosion

1. Kinds of Water Erosion. Water is the most widespread agent of erosion and accounts for the bulk of sediments transported to reservoirs. Water circulating in the earth's crust is a primary adjunct to chemical weathering of rocks. Precipitation detaches surface particles by raindrop impact, and runoff transports detached particles to points of deposition.

Water erosion may be classified into two general types: sheet erosion and channel erosion. *Sheet erosion* is the detachment of the material from the land surface by raindrop impact and its subsequent removal by prechannel or overland flow. *Channel erosion* is the removal and transport of material by concentrated flow.

2. Sheet Erosion. Ellison [3] has shown experimentally the tremendous kinetic energy exerted by rainfall, which may be defined by the following equation:

$$E = KV^{4.33} d^{1.07} I^{0.65} \qquad (17\text{-I-}1)$$

where E is the soil intercepted in splash samplers, during a 30-min period, in g, V is the velocity of drops in fps, d is the diameter of drops in mm, I is the intensity of rainfall in in./hr, and K is a constant.

The detaching force, for a given rainfall, is the product of the kinetic energy and the duration of rainfall. Research studies indicate that erosion is most closely related to the amount of rainfall occurring during a 30-min period. Forces of resistance include the nature of the soil and cover. Strongly bonded particles are resistant to detachment by raindrop impact, while a vegetative canopy over the soil breaks up the raindrop energy During a single rainstorm on a bare soil, as much as 100 tons of soil per acre may be detached and splashed into the air. Some particles may be cast as high as 2 ft into the air and travel for horizontal distances of 5 ft.

Erosion per se requires transport as well as detachment of particles. On level fields, without appreciable surface flow, the net erosion loss of detached particles is small. As the degree of slope and the length of slope increase, discharge and velocity increase and, correlatively, so does the rate of removal of detached particles for a given soil type.

3. Musgrave Equation for Predicting Rate of Sheet Erosion. Musgrave [4], in 1947, reported on the results of analyses of soil-loss measurements for some 40,000 storms occurring on fractional-acre plots in the United States. The results of this study indicate that soil loss varies in accordance with the following relationship:

$$E = IRS^{1.35}L^{0.35}P_{30}^{1.75} \tag{17-I-2}$$

where E is the soil loss in acre-in., I is the inherent erodibility of the soil in in., R is a cover factor, S is the degree of slope in per cent, L is the length of slope in ft, and P_{30} is the maximum 30-min amount of rainfall, 2-year frequency, in in. Equation (17-I-2) is applicable to long-term average soil losses for broad areas.

4. Universal Equation for Predicting Rates of Sheet Erosion. Further progress has been made since 1947 in both collection and analyses of data, which has resulted in development of new relationships, leading to greater refinement of predictions of short-term rates of soil loss for localized areas. This has resulted in the development of a new so-called *universal equation* by the Agricultural Research Service of the U.S. Department of Agriculture [5]. This equation takes into account the influence of the total rainfall energy for a specific area rather than rainfall amount. Total rainfall energy can be readily computed for localized areas from existing U.S. Weather Bureau data. The universal equation is as follows:

$$A = RKLSCP \tag{17-I-3}$$

where A is the average annual soil loss in tons/acre, R is the rainfall factor, K is a soil-erodibility factor, LS is a slope length and steepness factor, C is a cropping and management factor, and P is the supporting conservation practice, such as terracing, strip cropping, and contouring. In this equation the *rainfall erosion factor* R is determined by the following relationship, as reported by Wischmeier and Smith [6]:

$$R = \frac{\Sigma EI}{100} \tag{17-I-4}$$

where E is the storm energy in ft-tons/acre-in., and I is the maximum 30-in. intensity in in./hr.

Analyses of U.S. Weather Bureau rainfall statistics are in progress at the present time to develop *rainfall-erosion index* values in a form for convenient use. The average annual rainfall-erosion index developed for use in southeastern United States is shown in Fig. 17-I-2.

The *soil-erodibility factor* K defines the inherent erodibility of the soil. It is expressed as the soil loss in tons per acre for each unit of rainfall-erosion index for the locality and for continuous fallow tillage on a 9 per cent slope, 73 ft in length. Standard K values have been established for only a few soil types in the country where actual soil-loss measurements have been made. Research now in progress may eventually provide criteria for estimating erodibility. Numerous factors influence erodibility of cohesive soils, including, but not limited to, texture, grain-size distribution, nature of clay minerals, thickness and permeability of strata, and organic content. Present estimates of soil erodibility must be made on the basis of known erosion characteristics of the soil.

The *slope-length-steepness factor*, or *soil-loss ratio, LS*, is determined by dividing the existing length and steepness by the standard 9 per cent slope, 73 ft in length. The soil-loss ratio from any given slope conditions can be readily determined by means of a set of curves developed by the U.S. Agricultural Research Service [5], shown in Fig. 17-I-3.

The *cropping-management factor* C is a complex factor to evaluate, because of the many different cropping and management combinations which might be used in a given area. This is further complicated by the variable distribution of the rainfall-erosion potential during different periods of canopy provided by the crop during seedbed preparation and growth stages and before and after harvesting. Fertilizing, mulching, crop residues, crop sequence, and other factors also influence the rate of soil loss.

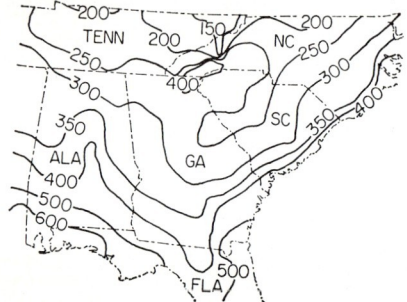

Fig. 17-I-2. Average annual rainfall erosion index for southeastern United States. (*U.S. Department of Agriculture* [5].)

Fig. 17-I-3. Soil-loss ratio for different degrees and lengths of slopes. (*U.S. Department of Agriculture* [5].)

The rainfall-energy distribution curve differs from one area to another. For example, at Memphis, Tenn., 48 per cent of the total annual rainfall energy occurs before June 1, while at Madison, Wis., only 14 per cent occurs by that date. It is quite obvious that the value for the C factor would be different for different areas. Analyses of basic data are in progress by the Agricultural Research Service to evaluate cropping-management conditions in terms of soil loss from clean-cultivated fallow land and to develop ready reference tables for use in specific areas. Some of these have been developed and are available [7].

A base value of $C = 1$ is used for clean-cultivated, fallow, straight-row, and up- and downhill cultivation. All other values of C are less than 1. The calculated average annual C values for a 4-year rotation of wheat, meadow, and 2 years of corn, for example, would be determined as follows:

$$C = \frac{C_w + C_m + C_{c1} + C_{c2}}{T} \qquad (17\text{-}I\text{-}5)$$

where C is the average annual cropping-management factor, C_w is the cropping-management factor for wheat, C_m is the cropping-management factor for meadow, C_{c1} is the cropping-management factor for corn following meadow, C_{c2} is the cropping-management factor for second year of corn following meadow, and T is the time in years.

The calculated C factors for the above cropping practices for the rainfall-energy conditions for central Indiana are reported by Wischmeier [7] to be $C_w = 0.086$, $C_m = 0.016$, $C_{c1} = 0.220$, and $C_{c2} = 0.285$.

The average annual C value for central Indiana for a 4-year rotation of wheat, meadow, and 2 years of corn would therefore be 0.152.

The application of contour practices reduces soil erosion in varying amounts, depending upon the practice and the length and degree of slope. Where no contour practices have been applied and where the cultivation has been straight-row up- and downslope, the *practice factor* $P = 1$. The application of contour practices reduces the factor to less than 1. The practice factor for contouring normally is between 0.5 and 0.6 for slopes of less than 12 per cent; about 0.8 for slopes between 12 and 18 per cent; and 0.9 for slopes from 18 to 24 per cent. Contouring has practically no influence on reducing soil loss for slopes in excess of 24 per cent. Practice factors for strip cropping and terracing range from 0.25 to 0.3 for slopes up to 12 per cent; from 0.3 to 0.4 for slopes between 12 and 18 per cent; and from 0.4 to 0.45 for slopes from 18 to 24 per cent.

5. Channel Erosion. Channel erosion is dependent upon the energy exerted by forces of concentrated water flow. Gully, stream bank, stream bed, and flood-plain scour are examples of erosion caused by concentrated flow of water. Fundamental aspects of particle detachment, entrainment, and transport of sediment by concentrated flow are outlined in Sec. 17-II.

The rate of channel erosion depends upon the hydraulic characteristics of flow and the inherent erodibility of channel materials. For simplicity, channel materials may be classified as *noncohesive* and *cohesive*. Clean, uncemented, coarse-grained materials, such as sand and gravel, which have unstable structure in the dry state, are considered as noncohesive materials. Some materials, such as fine sands or materials containing minor amounts of clay or other materials, may be temporarily stable when compressed in the moist state, but readily lose their stability upon drying out. Cohesive materials are those having particles bonded tightly together by cementation, cohesion, or adhesion and remain stable in the dry state.

The rate of erosion and transport of noncohesive materials may be readily predicted by a number of empirical equations discussed later in this section. Resistance to boundary shear in noncohesive materials is that due solely to the size, shape, and specific gravity of the particle itself and the slope of the bed. In cohesive materials, the bonding of particles must first be broken before entrainment and transport can occur. Thus it is not unusual to find fine-grained materials, such as very stiff clays and hardpans, which resist detachability for tractive forces of 0.4 to 0.6 psf sufficient to dislodge and transport coarse gravels in noncohesive materials.

Though bonding agents may consist of such materials as lime, silica, organic matter, iron, and others, the most prevalent bonding materials in soils are the clay minerals. The clay minerals vary in grain size, shape, and molecular structure. Various types of clay minerals are often mixed in a soil, although one type usually predominates. As a result, wide variations in cohesive and adhesive forces exist, resulting in soils of variable consistency, which, from all outward appearances, may seem to be essentially similar. The strength of materials bonded by clay particles also varies in respect to the moisture content.

The parameters influencing the relative erodibility of cohesive materials are not well understood and have not been evaluated to the extent that reliable quantitative values can be established at this time for properties of soils which influence rates of channel erosion. On the basis of existing studies, it appears that the tractive-force theory (Sec. 17-II) is best adapted for quantitatively expressing boundary shear at the interface of flowing water and cohesive bed materials. Such properties as mean particle size, per cent clay, void ratio, plasticity index, and dispersion ratio and others have been tested by different investigators in attempts to evaluate resistance to shear [8, 9]. The results to date have been variable, but hold promise of producing a reliable method for predicting rates of channel erosion.

Of the properties tested to date, it appears that the plasticity index and the dispersion ratio provide the best direct measurement of cohesiveness in soils. Other factors reflect mainly the grain-size distribution. The *plasticity index* of a soil is the numerical difference between the liquid limit and the plastic limit. The *dispersion ratio* of a soil is the per cent ratio of silt and clay in suspension to the total silt and clay when agitated by a standard procedure [10]. In general, soils with plasticity indices of less than 10 and with 75 per cent of the particles less than 0.25 mm are classified as being highly erosive.

Rates of channel erosion can often be determined readily by field measurements and comparison with earlier channel cross sections and profiles. Channel widening, such as might occur in gully erosion or stream-bank recession, can be readily detected and measured by comparison of position of banks as shown in aerial photographs of different dates.

C. Wind Erosion

Wind is an important agent of erosion, particularly in arid and semiarid areas. The rate of wind erosion is influenced by such factors as wind velocity, reach, moisture content of soil, particle sizes, surface roughness, and vegetative cover and other factors. Fine-grained materials, such as dust and volcanic ashes, may be transported in suspension for thousands of miles by wind currents. Coarser materials travel much slower and for shorter distances. Fine gravels may be transported a few feet, fine sand usually less than a mile, and very fine sand several miles. Wind-blown sand is readily identified in the field by well-rounded grains and pitted surfaces which have a frosted appearance.

Wind-eroded material is an important source of sediment only when materials are deposited by the wind in stream channels, where they can be subsequently entrained and transported by concentrated water flow. Locally, such materials may represent a sizable portion of the total sediment load brought to a reservoir. Generally, wind erosion represents only a minor source of reservoir sediment.

1. Empirical Equation for Estimating Rate of Wind Erosion. As a result of wind-tunnel measurements on farm fields in the Southern Great Plains in the United States, Chepil and Woodruff [11] propose the following equation for estimating rate of wind erosion:

$$X = \frac{491.3 I}{(RK)^{0.835}} \qquad (17\text{-}I\text{-}6)$$

where X is the wind erosion in tons/acre; I is an erodibility index based on per cent of soil fraction greater than 0.84 mm in diameter, R is the crop residue in lb/acre, and K is a ridge-roughness equivalent in in.

The ridge-roughness equivalent is difficult to measure in the field without wind-tunnel equipment. To aid in the selection of the ridge-roughness equivalent, Chepil and Woodruff [11] present a series of 18 photographs showing different degrees of field-roughness conditions with corresponding ridge-roughness equivalents.

D. Ice Erosion

The freezing of water in fractures and crevices aids in the disintegration of rock, while ice masses, such as glaciers, may grind materials into rock flour which is transported as *glacier milk* in melt waters. The disintegration of rock by ice constitutes only a small part of the complex weathering process; thus ice itself is relatively unimportant in reducing rock to particle sizes readily transported by water. The production of rock flour is limited to minor areas of glacier formation. Locally, such as in areas where streamflow is derived from glacial meltwater, ice may represent an important erosion agent in respect to source of sediments delivered to a reservoir.

E. Gravity Erosion

Gravity erosion occurs when forces caused by the weight of material exceed forces of resistance caused by the frictional resistance or shearing strength of the material. Examples of gravity erosion include landslides, bank sloughing, talus movement, and mud flows. Water as an erosive agent may contribute to gravity erosion by removal of toe support of slopes or decreasing shear strength in cohesive materials through increase of moisture content. Locally, gravity erosion may constitute a primary source of sediment delivered to reservoirs when masses of materials are moved to locations where they can be attacked by water, particularly concentrated flow. Although sloughing of wave-cut banks along shore lines of reservoirs appears to be an important source of sediment, actually it usually contributes less than 10 per cent of the total sediment deposited in reservoirs.

F. Effect of Human Activities on Erosion

Human activities, such as disturbance of vegetative cover in land cultivation, excavations for buildings, highways and appurtenances, gravel pits and washing plants, mining, and other activities of man may cause serious disturbance of natural conditions and provide an important source of sediment contributing to reservoirs.

Urbanization of watershed areas has become a matter of much concern in some areas in respect to its influence on erosion and sediment yields. A study of sedimentation of Lake Barcroft, Fairfax County, Virginia, completed by Holeman and Geiger [12], indicated that construction activities in the 14.5-sq-mi watershed above the reservoir greatly influenced erosion conditions, which resulted in an increase in the rate of sedimentation of the reservoir. Construction of houses, roads, parking lots, and shopping centers accelerated about 1942 and reached an optimum about 1951 to 1952. It is estimated that in 1951 alone, 9 per cent of the watershed area was converted from agricultural use, mostly woodland and pasture, to urban development. The sediment yield from the watershed during the period of urbanization was almost triple that which prevailed during the period prior to urbanization.

Gravel pits, mine and slag dumps, strip-mining activities, highway construction, lumbering, and other activities of man may expose materials to serious erosion. Some manufacturing activities, such as copper smelting, may produce fumes which are toxic to protective vegetation, resulting in the devegetation of large areas and establishing excessive erosion conditions.

IV. MOVEMENT OF SEDIMENT FROM WATERSHEDS

A. Sediment Yield

The total amount of on-site sheet and channel erosion in a watershed is known as the *gross erosion*. All eroded materials in a watershed do not get into the stream system. Particles detached on comparatively level areas, with little or no surface runoff, for example, move only short distances and consequently are not transported to downstream points in a watershed. Soil which is eroded from sloping land may become lodged at fence rows and vegetated areas or deposited below breaks in slopes in the form of *colluvium*. Some material may be carried to the stream system, only to be deposited on flood plains as *alluvium* or *channel splays*, and some may be deposited in the form of *bar materials* in the channels themselves. The total amount of eroded material which does complete the journey from source to a downstream control point, such as a reservoir, is known as the *sediment yield*.

B. Sediment-production Rate

The sediment yield for a given problem area varies considerably in accordance with the total size of the contributing drainage area. In order to provide yield data on a comparable basis, it is customary to reduce sediment-yield data to the yield per unit of drainage area. Sediment yield expressed in this manner is normally referred to as the rate of sediment production, or the *sediment-production rate*. The sediment-production rate is computed by dividing the annual sediment yield by the area of the watershed. It may be expressed in terms of tons, or acre-feet, of sediment per square mile of drainage area per year.

C. Sediment-delivery Ratio

The ratio between the amount of sediment yield and the gross erosion in a watershed is called the *sediment-delivery ratio*. Sediment yield and gross erosion are normally expressed in tons. The sediment-delivery ratio is expressed as the per cent of sediment yield to the gross erosion.

Measurements of sediment accumulations in reservoirs and sediment-load records in streams show wide variations in sediment yields from watersheds. An analysis of some 1,100 records of sediment yields in the United States indicate annual sediment-production rates as high as 78 acre-ft to less than 0.01 acre-ft/sq mi. These variations are due to variations in gross erosion and differences in the percentage of eroded material transported to the point of measurement. Measurements show that as little as 5 per cent and as much as nearly 100 per cent of the materials eroded in some watersheds may be delivered to a downstream point.

The efficiency of a stream system in moving eroded materials from their sources to a downstream point of measurement is dependent upon a complex array of hydrophysical conditions. The physical features which appear to be significant are size of drainage area, watershed slopes, and degree of channelization. Hydrologic characteristics include precipitation and runoff characteristics, taking into account peak rates of flow and total amounts of runoff. The science of sedimentology has not progressed to the state where the relative influence of each of the individual watershed and hydrologic factors has been evaluated and their relative influence on the delivery ratio of sediment has been determined to the degree of accuracy desired.

1. Influence of Size of Drainage Area on the Sediment-delivery Ratio. Size of drainage area is a most important consideration in respect to the total yield of sediment from a watershed. Its relative importance in respect to influence on the delivery ratio and sediment production rate has been questioned.

Table 17-I-1 indicates that the arithmetic average of sediment-production rates obtained from about 1,100 existing measurements in the United States declines as the drainage-area size increases. Thus watersheds of less than 10 sq mi in area, on the average, produce more than 7 times as much sediment per unit of drainage than watersheds exceeding 1,000 sq mi in area.

Table 17-I-1. Arithmetic Average of Sediment-production Rates for Various Groups of Drainage Areas in the United States

Watershed-size range, sq mi	No. of measurements	Average annual sediment-production rate, acre-ft/sq mi
Under 10	650	3.80
10– 100	205	1.60
100–1,000	123	1.01
Over 1,000	118	0.50

The tabulated statistics could readily be interpreted to mean that the rate of sediment delivery decreases as the size of drainage area increases and in accord with streamflow. Supporting this theory is the fact that the probability of entrapment and lodgment of a particle being transported downstream increases as the size of drainage area increases. Thus wide flood plains on streams draining large watersheds provide an environment for deposition which does not exist on steeper channels of smaller headwater tributaries. Furthermore, the probability of complete coverage by a single storm event, coincident with widespread erosion and high rates of runoff per unit of drainage area, is much greater for small watersheds than for large ones.

Watershed size may incorporate other factors which in themselves may have a significant influence on the sediment-production rate yet are also related to watershed size. One of these is the greater steepness of the watershed slopes for smaller tributary watersheds in a specific drainage basin. Thus the relationship of watershed size may be purely accidental, and not significant when other parameters are evaluated separately.

Multiple-regression analyses of sediment data for the Missouri Basin loess hills by Gottschalk and Brune [13] and for the Columbia River Basin by Flaxman and Hobba [14] indicate that the sediment-production rate varies as the 0.8 power of the drainage-area size. Other researchers have found sediment production to vary from the 0.6 to 1.1 power of the drainage area.

2. Influence of Topography and Channel Density on Sediment-delivery Ratio. The important topographic features which appear to influence the delivery ratios and sediment-production rates are the degree and length of watershed slopes and the channel density of the watershed. The delivery ratio of sediment is much higher for steep watersheds with well-defined channels than it is for watersheds of low relief and poorly defined channels such as waterways. Thus Gottschalk [15] found that the density of incised channels of the watershed, expressed in feet per acre, was a significant factor influencing sediment-production rates from watersheds in the Pierre-shale area of central South Dakota. Conversely, Stall and Bartelli [16] found that the density of nonincised channels was a significant factor in respect to sediment yield from small watersheds in the Springfield Plain problem area of Illinois.

3. Influence of Relief and Length of Watershed Slopes on Sediment-delivery Ratio. Some investigators have found the watershed slope to be a significant factor influencing the sediment-delivery ratio and sediment yield. Several methods of expressing watershed slopes quantitatively have been developed in recent years. These are discussed in Sec. 4-II. One method, the *relief ratio*, has been tested in several regression analyses and found to be significant. The relief ratio B is readily determined from existing topographic maps by $R = h/L$, where R is the relief ratio, h is the relief of watershed between the minimum and the maximum elevation in ft, and L is the maximum length of watershed in ft.

Maner [17] found that the delivery rate of sediment from 25 watersheds in the Red Hills physiographic area of Texas, Oklahoma, and Kansas was highly correlated with the relief ratio and could be expressed by $\log DR_e = 2.94259 - 0.82363 \operatorname{colog} R$, where DR_e is the estimated sediment delivery rate in per cent of actual gross erosion

Schumm [18] also found a high correlation between sediment production and the relief ratio for a variety of small drainage basins in the Colorado Plateau Province. On the other hand, Stall [16], who tested this factor, did not find it to be significant for the Springfield Plain area of Illinois.

The relief ratio normally increases with decreasing drainage-area sizes of subwatershed areas of a given drainage basin. This may account in part for higher sediment-production rates found for smaller drainage-area sizes in a specific problem area. In mountainous headwater areas the relief ratio may stay essentially the same for drainage areas, irrespective of size, because of relatively uniform watershed slopes of mountain fronts.

4. Influence of Precipitation and Runoff on Sediment-delivery Ratio. Precipitation, in respect to amount, location, and seasonal distribution, influences both the sediment yield and the sediment-delivery ratio. Where high-intensity storms occur on watersheds at a time when cover conditions offer minimum protection against erosion, rates of gross erosion are high. Where maximum rainfall energy occurs during periods when the ground surface is frozen, lower rates of erosion prevail. Where such precipitation occurs in sufficient amounts to create runoff, sediment is transported in varying amounts, depending upon the nature and frequency of the runoff.

The rate of runoff is dependent upon a number of factors, such as the nature of precipitation, infiltration, antecedent moisture conditions, and physical characteristics of the watershed, including topography and shape. All conditions being equal, runoff is greater for watersheds with steeper slopes and greater drainage density. Little work has been done to date to relate precipitation and runoff characteristics to the delivery ratio of sediment.

V. SEDIMENT CHARACTERISTICS

A knowledge of the sediment characteristics, primarily the grain-size distribution and the volume-weight relationship, is necessary to provide for sediment requirements in the design of a reservoir. The grain-size distribution is important in (1) assigning a trap-efficiency value to the reservoir (Subsec. 17-I-VI), (2) predicting the horizontal and vertical distribution of sediment to determine sediment requirements for specific allocated storages (Subsec. 17-I-VII), and (3) predicting the ultimate volume-weight relationship for determining space required for the sediment deposited in the reservoir.

A. Particle Characteristics

1. Mineralogical Composition. A *sediment particle* is defined as a mineral or rock fragment. A coarse-grained sediment particle is one having a diameter equal to or greater than 0.062 mm (Table 17-I-2). Coarse-grained sediment particles may consist of a large variety of sedimentary, metamorphic, or igneous rock types. The finer the particle, the greater the trend toward individual mineral constituents. Thus the clay-size particles (less than 0.004 mm) normally consist of a separate group of

Table 17-I-2. Grade Scale Developed by American Geophysical Union, Subcommittee on Sediment Terminology

Size			Class
Millimeters	Microns	Inches	
4,000–2,000	....	160–80	Very large boulders
2,000–1,000	....	80–40	Large boulders
1,000–500	....	40–20	Medium boulders
500–250	....	20–10	Small boulders
250–130	....	10–5	Large cobbles
130–64	....	5–2.5	Small cobbles
64–32	....	2.5–1.3	Very coarse gravel
32–16	....	1.3–0.6	Coarse gravel
16–8	....	0.6–0.3	Medium gravel
8–4	....	0.3–0.16	Fine gravel
4–2	....	0.16–0.08	Very fine gravel
2.00–1.00	2,000–1,000	....	Very coarse sand
1.00–0.50	1,000–500	....	Coarse sand
0.50–0.25	500–250	....	Medium sand
0.25–0.125	250–125	....	Fine sand
0.125–0.062	125–62	....	Very fine sand
0.062–0.031	62–31	....	Coarse silt
0.031–0.016	31–16	....	Medium silt
0.016–0.008	16–8	....	Fine silt
0.008–0.004	8–4	....	Very fine silt
0.004–0.0020	4–2	....	Coarse clay
0.0020–0.0010	2–1	....	Medium clay
0.0010–0.0005	1–0.5	....	Fine clay
0.0005–0.00024	0.5–0.24	....	Very fine clay

clay minerals with specific crystal structure and chemical composition, usually hydrous-aluminum silicates.

The mineralogy of original source rocks, together with chemical weathering processes and mechanical weathering, including abrasion due to transport, determines the ultimate size, weight, and shape of sediment particles. The mineral nature of sediment deposits is greatly influenced by proximity to source. The mineral content contributes to the sorting phenomena in that the heavy minerals will deposit at higher velocities than lighter minerals of equal size. Hard minerals such as quartz will resist abrasion to a greater degree than soft, unstable minerals such as gypsum or limestone. Some minerals and rocks disintegrate along crystal faces or cleavage planes to form platy particles, while others form equidimensional particles. The nature of the clay minerals in reservoir sediment influences the degree of consolidation and the volume-weight relationship, because of size, shape, and molecular structure.

2. Shape. As indicated in the foregoing paragraph, shape is determined to a large extent by the nature of the parent mineral or rock. Thus shales, interbedded sandstones and shales, and some schists disintegrate into coarse-grained slabs. Freshly fractured rocks, which lack cleavage or fracture planes, break down into angular and subangular particles and eventually, through abrasion by transport, to subrounded and rounded particles. Fine-grained quartz particles, such as rock flour, are equidimensional, while most clay minerals are platy. The critical tractive force required to initiate movement of one shape would be entirely different from that required for another shape of the same volume and weight. Deposits of platy minerals form a metastable structure, which results in slower rates of consolidation.

3. Size. Perhaps the most important single physical factor related to sediment transport and deposition is the size of particle. Broadly, this factor determines how and when a particle will move or deposit with a given set of hydraulic characteristics of flow. Size of particles has an important influence on the amount of sediment trapped and the distribution and ultimate volume of sediment in a reservoir.

Numerous grade scales have been developed to establish the limits of size of each classification. The more important scales in use are the Wentworth, U.S. Bureau of Soils, the Unified Soil Classification System, and the American Geophysical Union (AGU), Subcommittee on Sediment Terminology. The last, shown in Table 17-I-2, is generally the grade scale used in sedimentation studies at the present time in the United States (see also Subsec. 17-II-VI-D).

B. Bulk Characteristics

1. Grain-size Distribution. The bulk characteristics of sediments which are of particular concern in reservoir sedimentation problems are the grain-size distribution and the specific weight of deposited sediment. Sediments with more than 50 per cent of the particles larger than 0.062 mm are hereafter referred to as coarse-grained sediments, while those with more than 50 per cent in grains of less than 0.062 are designated fine-grained sediments.

All size classes of sediment occupy reservoir space upon deposition and therefore are of concern. However, some classes, particularly clay- and silt-size particles, do not always achieve maximum consolidation immediately upon deposition and, consequently, may not reach their ultimate specific weight for many years to come. The gradation of particles influences the distribution of sediment in reservoirs. Figure 17-I-4 shows variations in average grain-size distribution curves for some reservoir deposits.

2. Specific Weight of Sediment. The term *specific weight* is used here to denote the dry weight of sediment particles (solids) of a total, in-place volume of sediment mass. It is expressed in terms of pounds per cubic foot. The term is synonymous with such terms as dry density, or unit dry weight, found in soil-mechanics nomenclature. It is not to be confused with the term density, which is usually reserved to apply to a unit mass of a solid.

The specific weight of sediment must be predicted in order to estimate the storage space which will be displaced by sediment in a given period of time. A given volume of sediment in place is composed of both solids and voids. The voids may be filled with air, gases, or water. If completely saturated, all pore spaces are filled with water. The volume of voids, known as the *porosity* n, varies, depending upon the size distribution of particles and their arrangement in respect to each other. The ratio of the volume of voids to the volume of solids is known as the *voids ratio* e. The voids ratio for a given volume of sediment mass is computed by $e = n/(1 - n)$.

The voids ratio of spheres of uniform size and efficiently packed is theoretically computed as 0.35. The voids ratio of clean uniform sand under the same condition of packing is normally 0.40. In uniform materials the voids ratio increases as the particle size decreases. The voids ratio also decreases for well-graded mixtures of particles where finer materials fill voids between coarser materials.

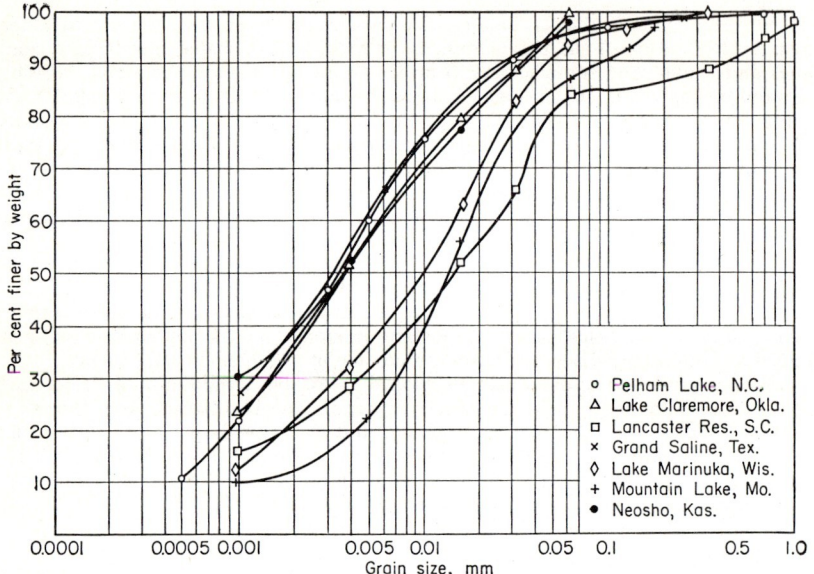

Fig. 17-I-4. Average grain-size distribution of some reservoir sediments.

The arrangement of particles has considerable bearing on the voids ratio, particularly in fine-grained sediments. Close packing of particles reduces the voids ratio. Such sediments are said to be densely compacted. Fine-grained sediments, particularly platy minerals such as micas and clay minerals, bridge across during sedimentation processes to form a metastable structure. Such sediments have high void ratios, in some instances up to 12, and are said to be loosely compacted. This structure breaks down, and consolidation develops in time, as a rearrangement of particles takes place, mainly through vertical stresses imposed by superincumbent deposits.

The specific weight γ_s of a unit mass of sediment, in pounds per cubic foot, is computed by

$$\gamma_s = \frac{(1 - e)GV_w}{V_s} \tag{17-I-7}$$

where e is the voids ratio, G is the specific gravity of sediment, V_w is the unit weight of water equal to 62.5 lb/ft³, and V_s is the volume of sediment in ft³.

SEDIMENT CHARACTERISTICS   17-17

If a cubic foot of sediment is composed of uniform sand grains with a specific gravity of 2.65 and a voids ratio of 0.4, the specific weight will be equal to 99 lb/ft³. The volume of solids is difficult to measure and the voids ratio difficult to predict in fine-grained sediments. Common practice, therefore, is to obtain field measurements and relate them to time and depth of deposits. The specific weight of a sediment deposit may be readily obtained by an undisturbed sample of known volume, drying it under controlled conditions in the laboratory and determining its dry weight.

Sands and gravels achieve maximum consolidation in relatively short periods of time subsequent to deposition, whereas fine-grained sediments require much longer time for maximum consolidation to take place. Lane and Koelzer [19] have presented the following general equation, based on time and the grain-size constituents of sediment, for estimating the unit weight of sediment for design purposes:

$$W = W_1 + K \log T \qquad (17\text{-}I\text{-}8)$$

where W is the unit weight, or density, of sediment after T years of compaction, W_1 is the initial unit weight considered at the end of 1 year, and K is a constant. Table 17-I-3 lists the various values of W_1 and K for different types of materials and different conditions of reservoir operation.

Table 17-I-3. Density Values for Use in Design*

Reservoir operation	Sand		Silt		Clay	
	W_1	K	W_1	K	W_1	K
Reservoir always submerged or nearly submerged...	93	0	65	5.7	30	16.0
Normally a moderate reservoir drawdown..........	93	0	74	2.7	46	10.7
Normally considerable reservoir drawdown........	93	0	79	1.0	60	6.0
Reservoir normally empty.......................	93	0	82	0.0	78	0.0

* After Lane and Koelzer [19].

The relationships given by Lane and Koelzer apply to the unit weight of the first year's deposit after T years. This must be integrated to include subsequent deposits of more recent vintage to obtain an average specific weight of deposits in the reservoir at the end of the design period. Miller [20] further refined the data of Lane and Koelzer to determine average unit weight of deposits for a given period of time. The condensed equations shown in Table 17-I-4 were developed by him for this purpose.

Table 17-I-4. Equations for Converting Lane's and Koelzer's Unit Weight W_1 to Average Weight for Different Periods of Time*
(In years, indicated by subscripts)

$$W_{10} = W_1 + 0.675K$$
$$W_{20} = W_1 + 0.938K$$
$$W_{30} = W_1 + 1.093K$$
$$W_{40} = W_1 + 1.210K$$
$$W_{50} = W_1 + 1.298K$$
$$W_{60} = W_1 + 1.372K$$
$$W_{70} = W_1 + 1.438K$$
$$W_{80} = W_1 + 1.493K$$
$$W_{90} = W_1 + 1.542K$$
$$W_{100} = W_1 + 1.588K$$

* After Miller [20].

Table 17-I-5 shows the relationship of specific weight to grain-size distribution and reservoir operation used by the U.S. Soil Conservation Service, for general design purposes (50-year design period) of floodwater-retarding and multiple-purpose reservoirs.

Table 17-I-5. Ranges in Specific Weight Used by the U.S. Soil Conservation Service for General Design Purposes

Grain size	Permanently submerged, lb/ft³	Aerated, lb/ft³
Clay	40–60	60–80
Silt	55–75	75–85
Clay-silt mixtures (equal parts)	40–65	65–85
Sand-silt mixtures (equal parts)	75–95	95–110
Clay-silt-sand mixtures (equal parts)	50–80	80–100
Sand	85–100	85–100
Gravel	85–125	85–125
Poorly sorted sand and gravel	95–130	95–130

3. Effect of Mineral Constituents on Specific Weight. The relationships in Table 17-I-5 are generally valid for coarse-grained sediments that achieve maximum consolidation soon after deposition. There is a wide variation in the size of clay minerals. Thus the diameter of a particle of the clay mineral kaolinite may be as large as fine silt. Conversely, the clay mineral montmorillonite is so fine grained, it appears as fog under the electronic microscope. The small grain size of montmorillonite clay mineral, plus the fact that montmorillonite has an unstable molecular structure, results in highly expansive deposits, with a high voids ratio and low specific weight on initial deposition. Besides lower initial compaction, it has a greater range of compressibility and requires the longest period of time to achieve ultimate consolidation of any of the clay minerals. The mineralogical character of clays also determines such properties as cohesion, adhesion, and tensile strength, which influence consolidation. Even the same clay mineral, with different absorbed ions, may have different rates of consolidation. Thus a general classification of "clay" based upon grain sizes of less than 0.004 mm falls short for adequate prediction of the ultimate specific weight of sediment.

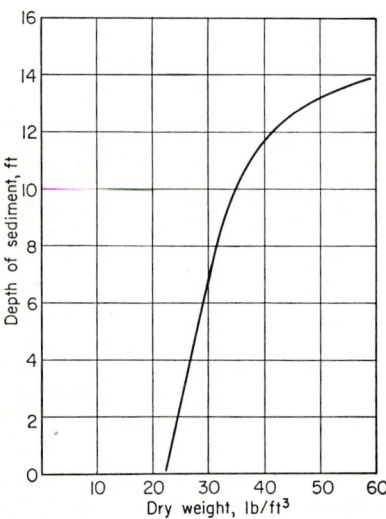

FIG. 17-I-5. Relationship of specific weight to depth, Pelham Lake sediments.

4. Effect of Thickness of Deposits on Specific Weight. Electronic equipment, which is suitable for measuring the specific weight at different depths in submerged sediment of reservoirs, has been developed only recently. This equipment promises to provide more representative measurements of the specific weight of sediment at different depths in deposits. Many analyses of specific weight in the past have been based on samples obtained from the upper strata of deposits. These samples were taken with improvised, thick-wall drive samplers. Consequently, they do not reflect the true average specific weight of sediment where substantial thicknesses of superincumbent strata are involved.

Experimental results of the ultimate volume of given weights of sediment from Pelham Lake, North Carolina, when permitted to settle in a 6-in. lucite tube, reported by Gottschalk [21], show the following relationship with depth:

$$\rho = 22.85 + 1.09d + 0.00129e^{0.695d} \qquad (17\text{-}I\text{-}9)$$

where ρ is the specific weight of sediment in lb/ft^3, d is the depth of sediment in ft, and e is the base of natural logarithms.

The grain-size distribution curve (Fig. 17-I-4) indicates that about 50 per cent of the sediment would be classified as clay, 47 per cent as silt, and 3 per cent sand in the AGU grade scale. Figure 17-I-5 shows the relationship between the observed specific weight with the depth for the Pelham Lake sediment.

5. Effect of Reservoir Operation on Specific Weight. Fine-grained sediment deposits, which are exposed to air drying, develop higher specific weights than do deposits continually submerged. This is an important consideration in allocating space for sediment accumulation for reservoirs which may become dry periodically.

Reservoirs designed for water-supply purposes in humid areas, for example, in their early life stages seldom experience excessive drawdowns through operation or lowered water surfaces due to extended periods of drought. A greater volume per unit of weight must therefore be provided for sediment accumulation under these circumstances. Conversely, deposits in reservoirs designed for single purposes, such as floodwater retarding, may be dry for long periods of time, resulting in high consolidation of deposits and smaller volume per unit weight. Differences of effects of aeration of deposits are reflected in Tables 17-I-3 and 17-I-5.

C. Influence of Erosion and Transport on Sediment Characteristics

1. Sheet Erosion. The mineral composition of sediment is controlled to a large extent by the mineral constituents of rocks from which it is derived. Sediments derived from crystalline rocks contain a great variety of minerals, while sediments derived from offshore marine deposits display the greatest uniformity of particle characteristics. The primary mineral constituents of sediments are quartz, feldspar, mica, and the clay minerals.

The type of erosion has a profound effect on the particle size. Sheet erosion produces primarily fine-grained sediment, since prechannel flow seldom exceeds 2 or 3 fps and is capable of transporting only the finer grains detached by raindrop impact. Ellison [3] reported that, of the sediment removed by prechannel flow from a silt loam in Ohio, 95 per cent had a grain-size diameter of less than 0.05 mm, which would be classified as silt and clay fractions. Only 75 per cent of the grains in the original soil was finer than 0.54 mm.

Stall and others [1] present basic data on the grain-size distribution of watershed soils and reservoir deposits, which further illustrate the process of selective erosion. The principal watershed soils above Spring Lake, Illinois, contain particle sizes ranging from 70 to 80 per cent silt-size particles (2 to 50 microns) and 15 to 23 per cent clay-size particles (less than 2 microns). The bulk of sediment deposited in the reservoir averages 50 per cent silt and 45 per cent clay. The delta deposits are composed of 69 per cent silt and 28 per cent clay. The primary source of sediment transported to the lake is derived from sheet erosion in the watershed.

2. Channel Erosion. Because of higher velocities and greater critical tractive force of concentrated channel flow, such flows may produce either fine-grained or coarse-grained sediments, depending upon the character of bed and bank materials.

3. Wind Erosion. Sediment derived from wind erosion is limited to particle sizes generally less than 0.5 mm in size. Thus dune sands range between 0.5 and 0.25 mm, with the bulk generally 0.3 mm in diameter. Materials transported in the air are much smaller.

4. Transportation. Transportation by water affects the size of particles by (1) sorting and (2) abrasion. The process of sorting takes place through the interaction of hydraulic characteristics of flow, primarily the tractive force, and the size, shape, and specific gravity of particles (Sec. 17-II). When the transport capacity decreases, because of a change of slope, reduction in discharge, or increasing cross-sectional area, the heavier, larger, equidimensional particles are deposited, while the smaller, lighter, and platy particles remain in suspension. The change in hydraulic characteristics of the flow entering a reservoir results in the deposition of coarse-grained sediments in the headwater areas and fine-grained sediments in the lower reaches.

Abrasion takes place as a result of particles rubbing and bouncing against one another, wearing away the larger particles, while the smaller particles are crushed between the larger particles or on the bed of a stream. The net effect is diminution of the average particle size. Abrasion is most rapid for soft rocks, brittle rocks, and rocks with definite fracture or cleavage patterns. The relative rate of abrasion in descending order of some common rocks is as follows: shale, sandstone, marble, limestone, dolomite, granite, and quartz.

It has been reported by numerous investigators that the mean grain size decreases downstream in larger river basins. Large particle sizes in headwater mountain streams are apparent even to the casual observer. Studies by the U.S. Army Corps of Engineers show that the average load of sediment of the Mississippi in the delta area consists of approximately 7 per cent sand, 38 per cent silt, and 55 per cent clay.

D. Characteristics of Sediments Delivered to Reservoirs

With highly complex and variable processes of erosion, entrainment, and transport of the sediment involved, almost any range of sediment characteristics is conceivable which would be applicable to a given set of watershed, hydrologic, and hydraulic conditions. However, the bulk of sediment delivered to most reservoirs is fine-grained, and since fine-grained sediments occupy reservoir space as well as coarse-grained sediments, these become of primary concern in developing sediment requirements. Glymph [22], who compiled data on sediment sources from 113 watersheds, located mostly in the humid agricultural area and ranging in size from less than 0.1 sq mi to 437 sq mi in area, found that the dominant source of sediment was sheet erosion. Of the 113 watersheds studied, sheet erosion accounted for 90 per cent or more of the sediment yield in over 50 per cent of the watersheds. In 73 of the watersheds it accounted for more than 75 per cent of the sediment. Since sheet erosion is con-

Table 17-I-6. Average-grain-size Distribution of Reservoir Deposits

Climatic area	Drainage area, sq mi	Average-grain-size components		
		Clay, %	Silt, %	Sand, %
Humid				
W. Frankfort Reservoir, Ill.	4	19	78	3
Lancaster Reservoir, S.C.	9	28	56	16
Lake Bracken, Ill.	9	52	42	1
Lake Lee, N.C.	51	36	57	7
High Point Reservoir, N.C.	63	45	35	20
Lake Marinuka, Wis.	139	30	64	6
Lake of the Ozarks, Mo.	14,000	46	50	4
Subhumid				
Ardmore Club Lake, Okla.	4	40	24	36
Moran Reservoir, Kans.	5	52	44	4
Mission Lake, Kans.	8	27	17	56
Lake Merritt, Tex.	12	56	41	3
Lake Claremore, Okla.	56	51	46	3
Lake Brownwood, Tex.	1,544	70	28	2
Great Salt Plains Reservoir, Okla.	3,156	51	41	8
Semiarid				
Wellfleet Reservoir, Nebr.	15	9	14	77
Sheridan Reservoir, Kans.	463	85	15	
Tongue Reservoir, Wyo.	1,740	35	46	18
Canton Reservoir, Okla.	12,483	41	39	20
Arid				
Conchos Reservoir, N.M.	7,350	22	52	26
Lake Mead	168,000	28	25	47

ducive to production of fine-grained sediments, it follows that the bulk of sediment transported to reservoirs is, at least in the humid agricultural areas where sheet erosion is the predominant type, in the silt and clay range. This is borne out by grain-size analyses of sediment samples taken from various reservoir deposits. Some of these are shown in Table 17-I-6.

VI. TRAP EFFICIENCY OF RESERVOIRS

Of the sediment brought into a reservoir by stream inflow, only a portion may be trapped and retained in a reservoir, the balance being transported through and carried out of the reservoir by outflow water. The ability of a reservoir to trap and retain sediment is known as the *trap efficiency*, and it is expressed as the per cent of sediment yield (incoming sediment) which is retained in the basin.

A. Factors Influencing Trap Efficiency

1. Sediment Characteristics. The trap efficiency of a reservoir depends upon the sediment characteristics and the rate of flow through the reservoir. From Table 17-I-6 it may be seen that sediment deposits in reservoirs vary greatly in respect to the grain-size distribution of particles. As stream flow enters a reservoir, the cross-sectional area of flow normally is increased, resulting in a reduction of velocity and a corresponding decrease in sediment-transport capacity. The coarse-grained particles are dropped immediately near the head of backwater, while the finer grains remain in suspension, to be transported until such time as they too are deposited or carried out from the reservoir in outflow water. The per cent of total incoming sediment which will be transported out of any reservoir depends primarily upon the fall velocity of particles and the rate at which the particles are transported through the reservoir.

The fall velocity of particles in water depends upon a number of variables, including the size and shape of the particle and the chemical composition and viscosity of water. These are discussed in Sec. 17-II. Electrochemical processes play an important role in determining the fall velocity of fine-grained sediments of less than 10 microns in diameter, such as clays and colloids. The character and extent of dissolved solids in native waters influence the fall velocity and ultimately the trap efficiency. Thus, in some areas, clays and colloids are aggregated into clusters which have settling properties similar to larger grains. Conversely, highly dispersed particles may stay in suspension for long periods of time and be transported out of the reservoir.

2. Detention-storage Time. The rate of flow of water through a reservoir determines the detention-storage time. This is influenced by the volume of inflow in respect to available storage capacity and the rate of outflow. Operation of a reservoir is a controlling factor in respect to detention storage. Thus a storage reservoir for irrigation or water-supply purposes may retain water for many months, resulting in nearly complete desilting. In contrast, a dry flood-control reservoir may empty in a matter of only a few days, with only partial deposition of its incoming sediment load. Such reservoirs would be expected to have a lower trap efficiency than a storage reservoir.

3. Nature of Outlet. The size and location of outlets also influences the trap efficiency. The size of outlet affects the detention-storage time directly. Thus outlets designed to pass a sizable portion of flood inflow in short periods of time trap less sediment than outlets which permit only minor releases over longer periods of time. The location of the outlet in respect to distribution of sediment concentration within the reservoir basin determines the extent of sediment venting which will occur. Bottom outlets are more effective in removing sediment where greater concentrations occur in lower portions of the basin. This is normally the case in most reservoirs.

B. Computing Trap Efficiency

Although many factors may influence trap efficiency, the relative influence of each has not been evaluated to the extent that quantitative values can be assigned to individual factors. The detention-storage time in respect to character of sediment appears to be the most significant controlling factor in most reservoirs. Brune [23] has related the trap efficiency of storage-type reservoirs to the remaining capacity in respect to the annual inflow. He has developed the generalized trap-efficiency envelope curves shown in Fig. 17-I-6. These curves serve a very useful purpose for estimating the trap efficiency for storage-type reservoirs. Their applicability for estimating trap efficiency of reservoirs of greatly differing reservoir shapes, operation, and sediment characteristics remains to be demonstrated.

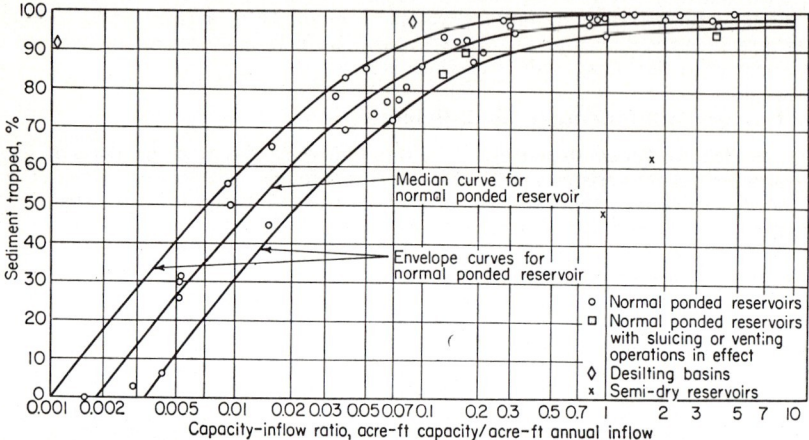

FIG. 17-I-6. Trap efficiency as related to capacity-inflow ratio. (*After Brune* [23].)

VII. DISTRIBUTION OF SEDIMENT IN RESERVOIRS

The character of sediment and processes of deposition influence the distribution of sediment in a reservoir. Such a distribution must be anticipated in the placement of gates and outlets to avoid premature loss of services dependent upon proper functioning of storage increments.

A. Processes of Sediment Deposition

The distribution of sediment in a reservoir is dependent upon several interrelated factors, including nature of sediment, inflow-outflow relations, shape of reservoir, and reservoir operation. Contrary to general belief, sediment deposition is not always concentrated in the lower increments of storage in the reservoir basin.

The forces acting upon a sediment particle brought to a reservoir by stream flow include a horizontal component due to the force of water acting upon the particle in the direction of flow and a vertical component due to force of gravity. A particle remains in suspension and is transported into the reservoir so long as turbulence exists, creating an upward force equal to, or exceeding, that of the force of gravity. When a flow enters a reservoir the increased cross-sectional area and wetted perimeter result in a decrease in velocity and turbulence of the original streamflow. Eventually, these are dissipated over large areas of the reservoir and become ineffective as a transporting medium. The particle then settles to the bottom and is said to be deposited.

Existing evidence indicates that in some reservoirs the sediment-laden inflow, upon entering a reservoir, retains its identity and transports the finer particles the entire length of the reservoir to the dam. When this occurs, the flow apparently follows the thalweg, or prereservoir channel. Because of this phenomenon, a greater percentage of the finer particles is transported to the lower part of the reservoir than would otherwise occur if the inflow were dispersed and dissipated upon immediately entering the reservoir.

Where the incoming sediment load contains appreciable coarse-grained materials or coagulated fine-grained materials, maximum deposition occurs in the headwater area of the reservoir, where the transport capacity of stream inflow is immediately diminished. The sands and gravels are deposited first, progressively finer materials being deposited as the transport capacity is further diminished by the stilling effect of the reservoir.

Clays and colloids which remain in suspension for long periods of time are transported for greater distances into the reservoir. Figure 17-I-7 shows differences of character of sediment in delta and bottom-set beds for several reservoirs. When the detention-storage time is inadequate for such materials to become completely deposited, a portion of the fine-grained sediment will be carried out of the reservoir

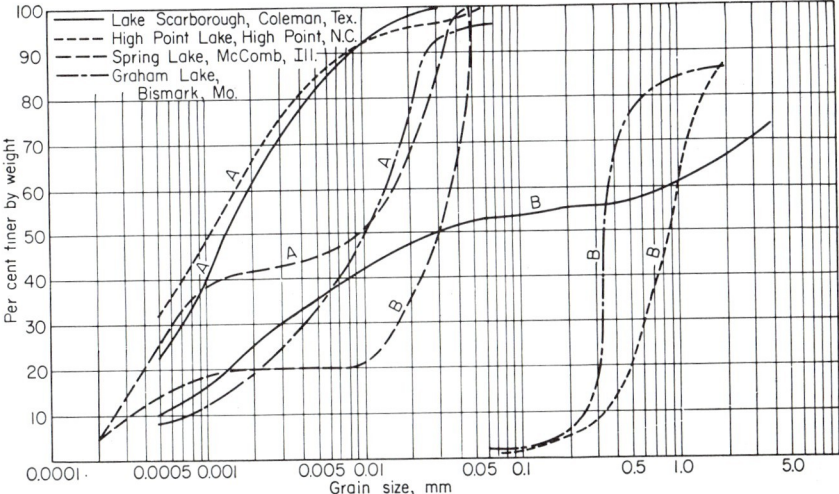

Fig. 17-I-7. Comparative grain-size distribution curves of delta and bottom-set beds of some reservoirs. *A*, lower lake bottom-set deposits; *B*, upper lake delta deposits.

with the outflow. If the incoming sediment load consists primarily of highly dispersed, fine-grained materials, maximum deposition occurs in the lower reaches of the reservoir.

Where maximum initial deposition occurs in the headwater areas, the deposits advance into the reservoir basin with time. Excessive drawdowns aid this process. On such occasions, stream inflow erodes materials previously deposited at higher water stages and transports them to lower elevations in the basin. The distribution of deposits in a reservoir basin may be further augmented by contributions of sediment from tributary watersheds downstream from headwater areas.

B. Delta Deposits

The processes of delta formation have been described by Bondurant [24]. Essentially, inflow maintains its identity as a definite current for an appreciable distance into the reservoir pool. These deposits form a chute-shaped delta ridge. Fine-grained sediments are carried into the pool and deposited on either side of the delta ridge or return to the head of the pool in back eddies.

Bondurant [24] defines two forms of deltas: *ideographic* and *complex*. The ideographic form consists of progressive deposition of coarse-grained sediment downstream with decreases in average velocity of flow, the fine-grained sediments being transported and deposited in an unknown sequence downstream. The ideographic form of delta develops in reservoir backwater areas, where inflow occupies the entire valley section from wall to wall, such as a narrow valley or canyon.

In wider valleys the delta form becomes more complex. Under such conditions the sediment ridge extends into only a portion of the pool. In time, the ridge will extend above normal pool level and develop a reduced slope. Subsequent floods breach the natural levees adjacent to the channels, resulting in development of a new channel and ridge over fines previously deposited in the headwater areas. Most deltas in reservoirs are of the complex type.

C. Bottom-set Beds

The deposition of fines normally extends throughout the reservoir basin but appears to be concentrated in certain locations. For the most part such deposits are aphanitic.

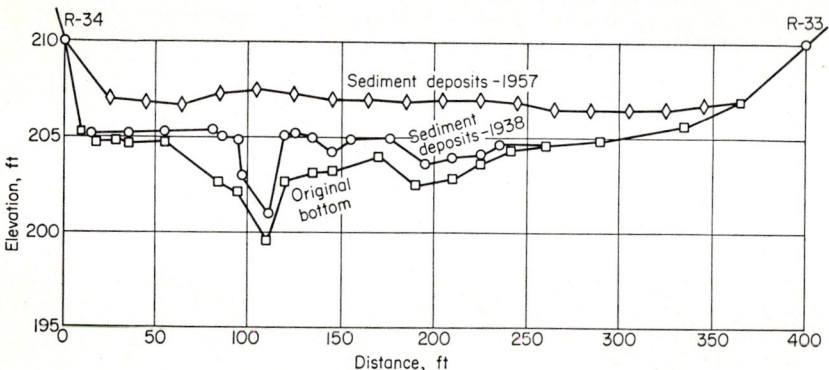

Fig. 17-I-8. Cross sections of sediment deposits, Lake Barcroft, Virginia.

Cross sections of deposits in many reservoirs indicate that maximum depths of deposits occur in the prereservoir channels. Also, in the early stages of deposition, submerged levees occur adjacent to the prereservoir channels, indicating that some identity of flow is maintained in these channels which is capable of transporting fines and influencing deposition for considerable distances in reservoir basins. Deposits on the submerged flood-plain areas have lesser thicknesses than the levee and channel deposits. The flood-plain deposits are strikingly uniform in thickness over the basin floor and faithfully reproduce the undulations of the prereservoir bottom. In the more advanced stages of deposition, the deposits become smoothed out between the valley walls, obliterating all traces of undulations and levee and channel configuration. These processes are illustrated in Fig. 17-I-8.

D. Predicting Sediment Distribution

Several methods have been developed for predicting sediment distribution for design purposes. Borland and Miller [25] have developed two different methods, the area-increment method and the empirical area-reduction method, particularly applicable to large multiple-purpose reservoirs. Heinemann [26] has developed a method for determining elevation of sediment accumulation near the dam for establishing minimum elevation of the principal spillway of small floodwater-retarding structures in the deep loess area of western Iowa.

1. Area-increment Method. The basic equation used for this method is as follows:

$$V_s = A_0(H - h_0) + V_0 \qquad (17\text{-}I\text{-}10)$$

where V_s is the sediment volume in acre-ft to be distributed in the reservoir; A_0 is the area correction factor in acres, which is the original reservoir area at the new zero elevation of the dam; V_0 is the sediment volume in acre-ft below new zero elevation; H is the reservoir depth at the dam in ft, from stream bed to maximum normal water surface; and h_0 is the depth in ft to which the reservoir is completely filled with sediment.

Equation (17-I-10) is a mathematical expression stating that the total volume of sediment V_s consists of a portion, which is distributed uniformly over the height $H - h_0$, plus the portion below new zero elevation of the reservoir. Computation of V_s requires the use of an original area-capacity curve of the reservoir.

2. Empirical Area-reduction Method. This method was developed from analysis of sediment-distribution data obtained from resurveys of reservoirs. A series of curves defining type of sediment deposit for different types of reservoirs, gorge, hill, flood plain, foothill, and lake was developed by Borland and Miller [25] to determine the constants needed for mathematical computation of a design curve. The equation used for this purpose is

$$A_p = C_p{}^m (1 - p)^n \qquad (17\text{-}I\text{-}11)$$

where A_p is a dimensionless relative area at relative distance p above the stream bed, and C, m, and n are dimensionless constants, which are determined by the type of reservoir.

3. Elevation of Sediment-accumulation Method. Regression analyses of sediment accumulation in small retarding-type reservoirs in the deep loess area of western Iowa by Heineman [26] indicated that the minimum elevation of sediment accumulation immediately upstream from the dam could be expressed by

$$Y = 22.6 + 0.886D - 81.2n - 0.175C + 0.494W \qquad (17\text{-}I\text{-}12)$$

where Y is the original reservoir depth filled with sediment, in per cent; D is the total original storage depletion at the end of the design period, in per cent; n is the original n value, or the slope on log-log paper of depth vs. capacity; C is the total storage capacity remaining at the end of the design period, in acre-ft; and W is the specific weight of sediment, in lb/ft^3.

VIII. SEDIMENT YIELDS FROM WATERSHEDS

A. Measuring Sediment Yields

1. Reservoir-sedimentation Surveys. The sediment yield from a watershed may be determined by measuring the accumulation of sediment in a reservoir of known age and adjusting for losses over the spillway, or by periodic sampling of the streamflow.

Methods of conducting reservoir-sedimentation surveys have been outlined by Gottschalk [27]. The volume of sediment accumulated in a reservoir is computed as the difference between the present water capacity of the reservoir and a known water capacity at some prior date, usually the original capacity at date of completion of the dam. The total volume of sediment is converted to dry weight of sediment on the basis of the average specific weight of deposits. The latter is determined by obtaining representative undisturbed samples for laboratory determination of specific weight or by field determination using a calibrated density probe developed for this purpose. The total weight of sediment accumulation in the reservoir plus that which is estimated to have passed through and out of the reservoir, based on the estimated trap efficiency, provides the total estimated sediment yield for the period of record covered by the survey. Reliable long-term records may be readily and economically obtained by this method. Rates of sediment yields, determined by measurements of sediment accumulation in 700 reservoirs and debris basins in the United States, have been summarized and reported by the U.S. Federal Inter-Agency Committee on Water Resources, Subcommittee on Sedimentation [28].

2. Sediment-load Measurements. Sediment yields may be determined by periodic sampling of the streamflow to measure sediment concentration for various water discharges. Sampling equipment and methods are described by the U.S. Federal Inter-Agency River Basin Committee, Subcommittee on Sedimentation [29] (see Sec. 17-II). A sediment-rating curve is developed by plotting sediment concentration against water discharge. Average sediment concentration for various discharges is then related to expected frequencies of the various discharges to estimate long-term suspended load. Methods of analyses of sediment-rating curves have been outlined by Miller [30]. The estimated bed-material load, which is transported by the flow in the zone of flow normally not sampled by suspended-load samplers, must be added to the long-term suspended load to obtain the long-term total sediment yield. Reliable estimates of long-term yield may be obtained by this method, provided that suitable information is obtained to estimate properly the unmeasured load and that the sampling is representative of the variability of discharges to be expected for the long-term period. An inventory of existing sediment-load measurements in the United States has been compiled by the U.S. Federal Inter-Agency River Basin Committee, Subcommittee on Sedimentation [31].

B. Methods of Predicting Sediment Yields from Watersheds

Three methods are in general use where no existing basic data are available for the watershed above a proposed dam. These are (1) use of existing data from comparative watersheds, (2) installation of sediment-load measuring stations, and (3) estimating watershed erosion and the delivery ratio of sediment.

1. Predicting Sediment Yields from Comparative Watersheds. Predicting the expected sediment yield to a proposed reservoir may sometimes be solved by adopting a rate equal to that measured on a nearby similar watershed. This method is reliable if both watersheds are essentially the same in respect to soils, topography, size, cover, and other conditions influencing erosion and movement of sediment. The method is best adapted to small watersheds with single-cover conditions such as forest or rangeland.

Considerable error in estimating may occur when using data from one watershed if the two watersheds involved have mixed cover-slope complexes or have other characteristics which differ in respect to influence on erosion or sediment-delivery ratio. When such conditions exist, data for estimating must be expanded to include ranges in factors as a basis for estimating the yield from a watershed without counterpart. Thus Gottschalk [15] demonstrated that a fairly reliable estimate of the rate of sedimentation of small stock ponds and reservoirs in the Pierre-shale area of central South Dakota could be determined by the relationship

$$S = 0.0522C + 0.0027A + 0.2681T - 1.7974 \qquad (17\text{-}I\text{-}13)$$

where S is the total sediment accumulation in acre-ft, C is the capacity of the pond or reservoir in acre-ft, A is the net drainage area in acres, and T is the age in years.

Equation (17-I-13) was derived from regression analyses of measured sediment accumulation in 18 ponds and reservoirs. In this particular area, soils and land use were similar, but variations in rates of accumulation still existed, depending upon the relationship between watershed area and capacity and the time. In other studies of groups of watersheds, investigators have found different parameters which better define conditions that influence erosion or movement of sediment for the specific problem area involved. These include such parameters as gross erosion, density of incised channels, density of nonincised channels, mean slope of third-order streams, and hydrologic events and others.

2. Installation of Sediment-load Measuring Stations. The sediment yield from a watershed may be obtained at or near the site of a proposed structure by installation of a stream-gaging station and periodic sampling of streamflow prior to the development of design. This method is used for major structures on large watersheds where investigations, prior to design of the structure, permit an adequate period of sampling to obtain measurements representative of expected long-term discharges. The results of such measurements, adjusted for expected long-term sediment yield, are used directly for design purposes.

3. Estimating Watershed Erosion and the Delivery Ratio of Sediment. Sufficient data on sediment yield are not always available to evaluate the relative effects of factors influencing sediment yield. Also, sufficient time may not be adequate to conduct sediment-load measurements prior to design of a structure. In such instances the sediment yield may be predicted by estimating the gross erosion in the watershed and adjusting for the expected delivery ratio of sediment to the structure site. This method is applicable to smaller watersheds and particularly useful where abnormal changes in land use and agricultural practices are anticipated in the watershed during the design life of the structure.

The estimated amount of sheet erosion is calculated by the method outlined in Subsec. 17-I-III-B-3. Basic information on land use, slopes, and other factors needed to estimate sheet erosion can be obtained for most areas through the U.S. Soil Conservation Service. The amount of sediment derived from channel and other types of erosion in the watershed must be added to the amount derived for sheet erosion to estimate gross erosion. Rates of channel erosion are determined in the field, the usual method consisting of comparison of current channel conditions with those shown on older aerial photographs. The delivery ratio of eroded material is then determined on the basis of the size of the watershed, the topography, and the degree of channelization. It is used to estimate the amount of eroded material delivered to the structure.

IX. RATES OF RESERVOIR SEDIMENTATION

A. Annual Rate of Deposition

The annual rate of sedimentation of a reservoir may be expressed by the following equation:

$$S_a = \frac{\Sigma(w_1 A_1 + w_2 A_2 + \cdots + w_n A_n) D R_e T_e}{21.78 \gamma_s} \qquad (17\text{-I-}14)$$

where S_a is the annual rate of deposition in acre-ft, $A_1, A_2, \ldots, A_n$ are the erosion rates in tons/acre; $w_1, w_2, \ldots, w_n$ are the incremental areas of watershed applicable to each erosion rate in acres, DR_e is the delivery ratio in per cent, T_e is the trap efficiency of the reservoir in per cent, and γ_s is the specific weight of sediment in lb/ft³.

Normally, the rate of sedimentation is expressed in terms of per cent of capacity lost annually. This is obtained by $S = 100 S_a/C$, where S is the annual loss of capacity in per cent, S_a is the annual deposition in acre-ft, and C is the capacity of the reservoir in acre-ft.

B. Measured Rates of Reservoir Sedimentation

The factors that influence the rate of sheet and channel erosion, the transportation of eroded materials to a reservoir, and the trapping of sediment within a reservoir are complex and highly variable. The geology of an area, the nature of soils, slopes, rainfall, runoff, hydraulic characteristics, cover, and other conditions vary greatly from one physiographic province to another. Wide variations in rates of erosion and sediment transport may be found even within a small watershed area where different cover, topography, soils, and other conditions prevail. Capacities of reservoirs per unit of drainage also vary, depending upon design requirements for hydrologic conditions and services to be provided. It follows, then, that wide variations in rates of sediment production and reservoir sedimentation exist. It is not possible to select a rate which might be considered as "typical" for any physiographic province, or even a general watershed area.

The U.S. Inter-Agency Committee on Water Resources, Subcommittee on Sedimentation, has assembled and published summary data on about 1,100 reservoir-

Table 17-I-7. Rates of Sedimentation of Selected Reservoirs in the United States

Name and location	Net drainage area, sq mi	Year storage began	Period of record, years	Original capacity, acre-ft	Annual sediment-production rate, tons/sq mi	Loss of storage, % Annual	Loss of storage, % Total
Northeast							
Barcroft, Alexandria, Va.	14.3	1915	42.6	2,092	618	0.38	12.1
Schoharie, Prattsville, N.Y.	312	1926	23.8	63,812	217	0.07	1.75
Byllesby, Byllesby, Va.	1,310	1912	23.7	8,892	238	2.54	60.2
Southeast							
Franklinton, Franklinton, N.C.	1.12	1925	13.3	34.7	743	1.60	21.3
Concord, Kannapolis, N.C.	4.54	1925	10.2	1,201	2,235	0.65	6.58
Roxboro, Roxboro, N.C.	7.52	1924	22.6	531	447	0.69	15.6
Issaqueena, Clemson, N.C.	13.9	1938	11.4	1,836	1,470	1.01	11.5
Harris, Tuscaloosa, Ala.	29.8	1929	24.5	2,241	189	0.16	3.87
High Point, High Point, N.C.	62.3	1928	10.3	4,354	544	0.71	7.26
Spartanburg, Spartanburg, S.C.	90.8	1926	20.9	3,506	423	0.84	17.5
Nolichucky, Greenville, Tenn.	1,182	1913	39.8	21,750	280	1.40	55.6
Norris, Norris, Tenn.	2,823	1936	10.3	2,045,300	450	0.05	0.54
Lay, Clanton, Ala.	9,077	1913	22.3	156,525	116	0.52	11.5
Midwest							
Caldwell, Waberly, Ohio	1.00	1949	12.0	88	331	0.29	3.45
Decker, Piqua, Ohio	2.30	1940	10.0	115	1,032	1.83	18.27
Shepard Mountain, Ironton, Mo.	3.96	1929	10.0	171	471	0.78	7.83
Westville, Alliance, Ohio	8.22	1913	37.0	994	287	0.16	6.10
Upper Pine, Eldora, Iowa	13.8	1934	13.3	660	1,490	2.38	31.5
Carlinville, Carlinville, Ill.	25.8	1939	10.4	1,725	1,020	1.40	14.52
Bloomington, Bloomington, Ill.	60.3	1929	22.7	6,678	514	0.50	11.46
Crab Orchard, Carbondale, Ill.	160	1940	11.2	67,320	1,976	0.45	5.09
Springfield, Springfield, Ill.	258	1934	14.6	61,039	660	0.30	4.36
Taneycomo, Branson, Mo.	4,606	1913	22.4	43,980	256	2.06	46.1
Lake of the Ozarks, Eldon, Mo.	13,900	1931	17.8	2,087,223	598	0.31	5.50
South Central							
Loring, Zwolle, La.	0.95	1928	26.0	663	3,002	0.23	6.03
Grand Saline, Grand Saline, Tex.	2.02	1925	13.2	531	691	0.31	4.14
Ardmore Club, Ardmore, Okla.	3.91	1922	15.5	1,797	2,234	0.55	8.51
Boomer, Stillwater, Okla.	8.67	1925	10.3	2,812	2,522	0.59	6.08
Scarborough, Coleman, Tex.	10.6	1923	17.0	2,153	907	0.40	6.78
Clinton, Canute, Okla.	23.1	1930	19.8	4,415	3,804	1.23	24.28
Eddleman, Graham, Tex.	41.4	1929	25.3	6,538	687	0.40	10.12
Abilene, Abilene, Tex.	97.5	1921	27.0	10,325	274	0.19	5.22
Spavinaw, Spavinaw, Okla.	397	1924	11.0	31,686	353	0.34	3.72
Eagle Mountain, Ft. Worth, Tex.	809	1934	18.0	211,000	2,601	0.69	13.74
Dallas, Denton, Tex.	1,157	1928	10.5	180,759	1,304	0.72	7.57
Altus, Altus, Okla.	2,116	1948	12.6	156,668	778	0.70	8.81
Northern Great Plains							
Bennington, Rago, Kans.	1.40	1929	11.2	75	5,311	5.00	56.0
Kirk, Iola, Kans.	2.36	1897	42.0	111	450	0.91	38.3
Baker, Baker, Mont.	5.01	1908	29.1	756	1,478	1.15	33.60
Mission, Horton, Kans.	7.76	1924	13.0	1,852	3,874	1.20	15.6
Ericson, Ericson, Nebr.	41.0	1915	32.9	1,066	756	1.08	35.4
Sheridan, Quinter, Kans.	463	1948	10.8	436	123	4.06	43.9
Buffalo Bill, Cody, Wyo.	1,460	1910	31.0	455,838	461	0.11	3.5
Guernsey, Guernsey, Wyo.	5,400	1927	26.4	73,810	236	1.49	39.3
Seminoe, Leo, Wyo.	7,317	1939	11.5	1,020,000	153	0.08	0.9
Southwestern							
Camp Marston, Julian, Calif.	1.59	1918	33.0	44	183	0.50	16.3
St. Marys, Walnut Creek, Calif.	2.97	1928	23.0	134	647	1.47	33.9
Gilmore, Bellota, Calif.	4.92	1917	28.0	579	144	0.11	3.1
Upper Crystal Springs, San Francisco, Calif.	12.0	1878	57.8	29,138	1,843	0.06	3.4
Morena, San Diego, Calif.	109	1910	38.3	66,767	2,444	0.31	11.7
Muddy Creek, Caddoa, Colo.	152	1919	20.0	16,918	877	0.48	9.64
Hodges, Escondido, Calif.	301	1919	29.5	36,601	531	0.29	8.5
Cucharas, Walsenburg, Colo.	608	1912	27.0	38,274	1,216	1.47	39.8

Table 17-I-7. Rates of Sedimentation of Selected Reservoirs in the United States (Continued)

Name and location	Net drainage area, sq mi	Year storage began	Period of record, years	Original capacity, acre-ft	Annual sediment-production rate, tons/sq mi	Loss of storage, % Annual	Loss of storage, % Total
Southwestern (Continued)							
Sevier Bridge, Nephi, Utah	1,089	1908	24.0	250,000	776	0.26	6.2
Piute, Marysville, Utah	2,436	1910	28.0	81,200	139	0.32	8.9
Roosevelt, Globe, Ariz.	5,760	1909	36.8	1,522,200	1,112	0.25	9.2
McMillan, Carlsbad, N.M.	12,600	1894	46.1	91,000	147	1.25	57.5
John Martin, Cuddoa, Colo.	17,080	1942	9.5	701,775	396	0.60	5.54
Elephant Butte, Truth or Consequences, N.M.	25,866	1915	32.3	2,634,800	798	0.51	16.6
Mead, Boulder City, Nev.	167,600	1935	13.7	31,250,000	877	0.33	4.6
Northwestern							
Mud Springs, Mountain Home, Idaho	1.06	1939	12.0	12	46	0.25	3.0
High Valley Ranch, Yakima, Wash.	4.10	1939	12.0	9	30	0.93	11.2
Emigrant Gap, Ashland, Ore.	61.2	1924	27.0	8,300	280	0.16	4.3
Cold Springs, Cold Springs, Ore.	186	1908	43.0	49,709	1,070	0.24	10.1
Arrowrock, Boise, Idaho	2,170	1915	32.6	279,250	173	0.09	2.8
Black Canyon, Emmett, Idaho	2,540	1924	12.0	37,659	173	0.89	10.7

sedimentation surveys and resurveys of about 700 reservoirs and debris basins in the United States [28]. These records provide useful information for determining ranges of sediment yields which might be expected for different drainage basins in the United States. Rates of sedimentation of some selected reservoirs in the United States, together with the sediment-production rates, are shown in Table 17-I-7.

X. CONTROL OF RESERVOIR SEDIMENTATION

The sediment yield from the watershed, the trap efficiency of the reservoir, and the ultimate specific weight of deposited sediment fix the volume of sediment which must be considered in the design of a structure to avoid premature loss of storage dedicated to a specific purpose for a given period of time. If economic or physical limitations are such that storage space cannot be incorporated in the reservoir to offset storage volume lost to sediment accumulation, other means of reducing sediment inflow or periodic removal of sediment must be considered to assure proper functioning of the reservoir for the period for which it is designed. This chapter briefly outlines provisions which might be considered in the design and maintenance of structures to assure proper functioning. Details of specific methods have been presented by Brown [32], to which the reader is referred.

A. Design of Reservoir

The basic considerations in developing a reservoir are hydrologic conditions and the service requirements for the reservoir for a given period of time. To prevent premature loss of storage and the services dependent upon it, an additional storage volume is usually incorporated in the reservoir for sediment accumulation. This volume of storage, generally referred to as the *sediment pool*, is equivalent to the volume of sediment expected to deposit in the reservoir during the design life of the structure. This positive method of control is generally adopted for the design of most government-built reservoirs in the United States today.

The sediment pool of many reservoirs built in the past was often allocated in the lower elevations of the reservoir, and in some instances the dead storage served as the sediment pool. More recent studies indicate this procedure to be inadequate for reservoirs where sediment will be distributed at higher elevations. In single-purpose storage reservoirs, the allocation of storage space by elevation is important in determining the elevation of the outlet sill. In multiple-purpose reservoirs sediment storage, to be fully effective, must be allocated within the storage dedicated to a specific purpose rather than in a sediment pool in the bottom of the reservoir.

B. Venting Sediment

Overflow spillways which encourage the discharge of surface water from a reservoir may actually skim off clear water and retain the more turbid water from which the sediment eventually becomes deposited. Occasionally, appreciable amounts of sediment may be vented from reservoirs by use of gated outlets or special curtains to encourage movement of high concentrations of sediment in suspension, through or over the dam. Under certain conditions sediment moves through reservoirs in well-defined stratified flows [33]. Gould [34] reports that 12 conspicuous underflows traveled from the head of Lake Mead to Hoover Dam (distances varying between 70 and 120 miles) during the first 7 years of operation.

The placing of outlets in the lower portion of the dam for the purpose of drawing off the lower strata of water having higher concentrations of sediment has been put to actual practice, but records of the effectiveness of these outlets, in terms of per cent of total sediment yield discharged, are lacking. Separate techniques need to be developed for each reservoir, particularly where problems of water wastage are acute, in order to operate gates during periods of maximum sediment concentrations.

An alternative method of venting, which has been proposed but not yet evaluated, consists of the installation of a curtain immediately upstream from an overflow spillway. The curtain is designed to deflect the reservoir flow beneath the curtain to intercept the lower strata of reservoir water prior to discharge over the spillway.

The effectiveness of venting methods depends upon maintaining the movement of sediment while still in suspension. Deposition of sediment reduces the bottom slope, and the cohesive and adhesive forces on particles in deposited sediment are greater than the tractive forces of flow normally developed by release of water through gated outlets. As a result, erosion and the entrainment of deposited sediment particles are minor, with the exception of a very localized area adjacent to the intake.

C. Sediment Removal

Deposited sediment may be removed from reservoirs periodically by hydraulic or mechanical means. This method has been used successfully to restore the capacity of many small reservoirs. It often represents the only means of restoring capacity where there are physical limitations on the height of a dam or where alternative sites are not available.

The type of equipment used for most economical removal of sediment depends upon the nature of sediment and location of the disposal area. Hydraulic dredging is the most economical for fine-grained, submerged materials where nearby disposal areas are available. Sometimes such materials are pumped over the dam. However, returning sediment in excessive quantities to streamflow below the dam may create downstream damages in excess of those experienced to the reservoir itself. Often disposal sites can be created by leveeing areas for sediment storage adjacent to the reservoir.

Coarse-grained materials require mechanical equipment for removal. Drag-line equipment may be used successfully for the removal of coarse-grained materials. A front-end loader or a power shovel may be used for removal of deposits if not submerged.

Off-site disposal of sediment deposits which requires rehandling of materials and trucking is costly, unless the sediment has commercial value such as for topsoil, land fill, or concrete aggregate.

D. Reduction of Sediment Yield

Sediment-storage requirements for a reservoir may be reduced if sediment can be retained in the watershed by one means or another. Such retention may be affected by development of vegetative screens in the reservoir area and by installation of structures and land-treatment measures in the watershed.

1. Vegetative Screens. Vegetation normally seeds on exposed delta deposits of reservoirs to form natural vegetative screens which reduce inflow velocities and increase the roughness coefficient, encouraging deposition above the reservoir crest elevation within the reservoir basin or immediately upstream. Often the natural seeding processes can be accelerated to develop more effective screens. This may be brought about by the introduction of faster-growing and more effective species of plants, by extending plantings to unseeded areas, and by increasing the plant density of natural screens. The deposition of sediment above crest can often be enhanced by construction of diversions in the delta areas to develop better distribution of incoming floodwaters and deposition in backwater areas.

The effectiveness of vegetative screens in reducing sediment inflow to active storage space depends upon the character of the sediment and the sediment-storage space available at the head of the reservoir. Generally, the additional amount which could be stored above crest level within a reservoir area by this method will not exceed 10 per cent of the total incoming load. However, under certain conditions, such materials may initiate upstream aggradation, resulting in serious seepage and floodwater damages upstream.

2. Watershed Structures. Several types of structures may be built in a watershed for the specific purpose of reducing sediment yield to a reservoir. These include such structures as sedimentation basins to trap sediment below eroding areas and erosion-control structures to halt the production of sediment.

Since the per-unit cost of storage in a specific area normally declines as the size of the reservoir increases, it is generally more economical to incorporate additional capacity in a reservoir itself for sediment accumulation than to develop a separate structure upstream specifically for the purpose of storing sediment which would otherwise be carried to the reservoir. However, this alternative may be justified if physical limitations of the reservoir site prohibit the development of additional capacity needed for sediment storage and if other methods such as sediment removal are neither physically nor economically feasible. If other benefits are associated with the upstream structure, its installation may be justified on the basis of on-site benefits as well as benefits to downstream reservoirs. For example, a gully-control structure may be essential for reduction of land-loss and land-depreciation damages, as well as to reduce sediment production.

A sedimentation basin may be developed to store sediment permanently for the design life of the reservoir or to store sediment for specific storm runoffs for periodic cleanout. The latter type of structure facilitates removal and disposal of sediment which might not be possible in the reservoir area itself. Normally, a sedimentation basin is designed for reducing production of coarse-grained sediments and is not an effective means of reducing movement of fine-grained materials.

The selection of sites for effective sedimentation basins requires evaluation and delineation of the principal sediment-source areas in the watershed. Where sediment sources are widespread in the watershed, the construction of basins on scattered individual subwatersheds will not provide a large degree of reduction of sediment yield to the reservoir. Conversely, a sedimentation basin on a subwatershed which produces the bulk of sediment yield to the reservoir will have a decided effect in reducing total sediment movement to the reservoir.

The reduction of channel erosion normally requires a structure. In a watershed, where the primary source of sediment is derived from channel erosion, installation of control structures may have an appreciable effect in reducing sediment yield to a reservoir. Such structures include drop inlets and chutes for reduction of gully erosion, stream-bank revetment to reduce stream-bank erosion, and sills or drop structures for stream-bed stabilization.

Other types of structures which may also prove expedient under certain specific conditions include bypass canals to divert sediment-laden flows around reservoirs.

3. Watershed Land-treatment Measures. Where the primary source of sediment is sheet erosion, land-treatment measures provide an effective and economical means of reducing erosion and sediment yield. Reduction in the rate of surface runoff and sheet erosion results in increased productivity of soil. Hence various types of conservation measures have been developed for specific soil, cover, slope, and rainfall complexes for the specific purpose of reducing erosion and runoff and increasing productivity. These measures include soil improvement, proper tillage methods, strip cropping, terracing, and crop rotations and others.

The improvement of soil by the addition of organic residues increases the infiltration rate and reduces sheet runoff and sheet erosion. The infiltration rate may also be increased by tillage methods such as basin listing and contour plowing. The installation of terraces not only promotes infiltration, but also reduces the effective length of slope, with corresponding reduction of erosive velocities.

Of all the conservation measures influencing sheet erosion, perhaps the most effective and easiest to manipulate is the cover. Conversion of land use from up- and downhill intertilled crops and fallow to small grains, meadow, and other uses with protective cover results in substantial reduction of rate of sheet erosion.

Reservoir owners normally own only small parcels, if any, of watershed lands and consequently can exercise only minor control over the use and treatment of watershed lands. However, during the last two decades great strides have been made in the promotion of conservation measures and their application to land in the United States. By 1960 nearly 2,900 soil-conservation districts, consisting of 1.8 million owners and operators of over 1.6 billion acres of agricultural land, have been organized under state and Federal laws. According to records of the U.S. Soil Conservation Service, conservation cropping systems have been applied to more than 88 million acres of land, while contour farming has been applied to nearly 38 million acres. More than a million farm ponds have been constructed, and nearly 1.2 million miles of terraces have been built under this program. As this program progresses, noticeable reductions in sediment yields of many watersheds will occur spontaneously, to the benefit of reservoir owners. Reservoir owners often can help in accelerating the installation of conservation measures in a particular watershed by encouraging landowners to adopt such measures for the mutual benefit of both landowners and downstream interests.

E. Evaluation of Selected Control Measures

Because of the many variables which influence sediment yields and the variable characteristics of site conditions, no stereotyped pattern of sediment control can be established to solve the sediment problem of a specific proposed reservoir. In most instances it will be cheaper to incorporate storage in the reservoir than to provide other means of control. This is particularly true where the sediment yield from a watershed is low and where large storage capacities are developed per unit of drainage. In such instances the additional capacity needed for sediment accumulation is only a small fraction of the total and requires a dam which may be only a matter of inches higher than that needed solely for hydrologic reasons or service requirements.

Where high sediment yields prevail, or where the storage capacity per unit of drainage is low, the additional storage and height of dam needed to provide for sediment control may be appreciable. In such instances, additional sites should be investigated, if available, to determine whether watersheds with lower sediment yields prevail in the area.

If it is not physically feasible to construct a dam of suitable height to provide for sediment storage in addition to hydrologic or service requirements, alternative methods of sediment control need to be considered to assure proper functioning of the reservoir. These include the possibility of venting sediment, periodic removal of sediment from the reservoir, development of vegetative screens, and installation of structural or land-treatment measures in the watershed. Each method needs to be evaluated separately to determine its feasibility, effectiveness, and cost, in selecting the most practical method.

XI. REFERENCES

1. Stall, J. B., L. C. Gottschalk, A. A. Klingebiel, E. L. Sauer, and E. E. De Turk: The silt problem at Spring Lake, Macomb, Ill., *Illinois State Water Surv. Div. Rept. Invest.* 4, 1949.
2. Borland, W. M., and C. R. Miller: Sediment problems of the Lower Colorado River, *Proc. Am. Soc. Civil Engrs., J. Hydraulics Div.*, vol. 86, no. HY4, pp. 61–87, April, 1960.
3. Ellison, W. D.: Some effects of raindrops and surface flow on soil erosion and infiltration, *Trans. Am. Geophys. Union*, vol. 26, no. 3, pp. 415–429, 1945.
4. Musgrave, G. W.: The quantitative evaluation of factors in water erosion: a first approximation, *J. Soil Water Conserv.*, vol. 2, no. 3, pp. 133–138, July, 1947.
5. A universal equation for predicting rainfall-erosion losses, *U.S. Agr. Res. Serv., Spec. Rept.* 22–26, March, 1961.
6. Wischmeier, W. H., and D. D. Smith: Rainfall energy and its relationship to soil loss, *Trans. Am. Geophys. Union*, vol. 39, no. 2, pp. 285–291, April, 1958.
7. Wischmeier, W. H.: Cropping-management factor evaluations for a universal soil loss equation, *Proc. Soil Sci. Soc. Am.*, vol. 24, no. 4, pp. 322–326, 1960.
8. Lane, E. W.: Progress report on results of studies on design of stable channels, *U.S. Bur. Reclamation Hydraulic Lab. Rept.* Hyd-352, June, 1952.
9. Smerdon, E. T., and R. P. Beasley: Critical tractive forces in cohesive soils, *Agr. Eng.*, vol. 42, pp. 26–29, January, 1961.
10. Middleton, H. W.: Properties of soils which influence erosion, *U.S. Dept. Agr. Tech. Bull.* 178, March, 1930.
11. Chepil, W. S., and N. P. Woodruff: Estimations of wind erodibility of field surfaces, *J. Soil Water Conserv.*, vol. 9, no. 6, pp. 257–265, 285, 1954.
12. Holeman, J. N., and A. F. Geiger: Sedimentation of Lake Barcroft, Fairfax County, Va., *U.S. Soil Conserv. Serv.* SCS-TP-136, March, 1959.
13. Gottschalk, L. C., and G. M. Brune: Sediment design criteria for the Missouri Basin loess hills, *U.S. Soil Conserv. Serv.* SCS-TP-97, October, 1950.
14. Flaxman, E. M., and R. L. Hobba: Some factors affecting rates of sedimentation in the Columbia River Basin, *Trans. Am. Geophys. Union*, vol. 36, no. 2, pp. 293–303, 1955.
15. Gottschalk, L. C.: Silting of stock ponds in land utilization area, SD-LU-2, Pierre, South Dakota, *U.S. Soil Conserv. Serv. Spec. Rept.* 9, May, 1946.
16. Stall, J. B., and L. J. Bartelli: Correlation of reservoir sedimentation and watershed factors, Springfield Plain, Illinois, *Illinois State Water Surv. Div. Rept. Invest.* 37, 1959.
17. Maner, S. B.: Factors affecting sediment delivery rates in the Red Hills physiographic area, *Trans. Am. Geophys. Union*, vol. 39, no. 4, pp. 669–675, 1958.
18. Schumm, S. A.: The relation of drainage basin relief to sediment loss, *Tenth Gen. Assembly, Intern. Assoc. Sci. Hydrol.*, Rome, 1954, vol. 1, pp. 216–219.
19. Lane, E. W., and V. A. Koelzer: Density of sediments deposited in reservoirs, Rept. 9 in "A Study of Methods Used in the Measurement and Analysis of Sediment Loads in Streams," U.S. Corps of Engineers, St. Paul District Sub-Office, Hydraulic Laboratory, University of Iowa, Iowa City, Iowa, November, 1943.
20. Miller, C. R.: Determination of the unit weight of sediment for use in sediment volume computations, *U.S. Bur. Reclamation Project Planning Div. Mem.*, Feb. 17, 1953.
21. Gottschalk, L. C.: Analysis and use of reservoir sedimentation data, *Proc. Federal Inter-Agency Sedimentation Conf.*, pp. 131–138, Denver, 1948.
22. Glymph, L. M., Jr.: Importance of sheet erosion as a source of sediment, *Trans. Am. Geophys. Union*, vol. 38, no. 6, pp. 903–907, 1957.
23. Brune, G. M.: Trap efficiency of reservoirs, *Trans. Am. Geophys. Union*, vol. 34, no. 3, pp. 407–418, June, 1953.
24. Bondurant, D. C.: Report on reservoir delta reconnaissance, *U.S. Army Corps Engrs. Missouri River Div. Ser.*, no. 6, June, 1955.
25. Borland, W. M., and C. R. Miller: Distribution of sediment in large reservoirs, *Trans. Am. Soc. Civil Engrs.*, vol. 125, pt. I, pp. 166–180, 1960.
26. Heinemann, H. G.: Sediment distribution in small floodwater retarding reservoirs in the Missouri Basin loess hills, *U.S. Agr. Res. Serv.* ARS 41–44, February, 1961.
27. Gottschalk, L. C.: Measurement of sedimentation in small reservoirs, *Trans. Am. Soc. Civil Engrs.*, vol. 117, pp. 59–71, 1952.
28. Summary on reservoir sedimentation surveys made in the United States through 1953, *U.S. Inter-Agency Comm. on Water Resources, Subcomm. on Sedimentation, Sedimentation Bull.* 6, August, 1957.

29. Measurement of the sediment discharge of streams, Rept. 8 in "Measurement and Analysis of Sediment Loads in Streams," U.S. Corps of Engineers, St. Paul District Sub-office, Hydraulic Laboratory, University of Iowa, Iowa City, Iowa, March, 1948.
30. Miller, C. R.: Analysis of flow-duration: sediment-rating curve method of computing sediment yield, *U.S. Bur. Reclamation Project Planning Div.*, April, 1951.
31. Inventory of published and unpublished sediment-load data in the United States, *U.S. Federal Inter-Agency River Basin Comm., Subcomm. on Sedimentation, Sedimentation Bull.* 1, April, 1949, and *Suppl.* 1946–1950, *Sedimentation Bull.* 4, April, 1952.
32. Brown, C. B.: The control of reservoir silting, *U.S. Dept. Agr. Misc. Publ.* 521, 1943, slightly revised, August, 1944.
33. Bell, H. S.: Stratified flow in reservoirs and its use in prevention of silting, *U.S. Dept. Agr. Misc. Publ.* 491, September, 1942.
34. Gould, H. R.: Turbidity currents, in Comprehensive survey of sedimentation in Lake Mead, 1948–49, *U.S. Geol. Surv. Profess. Paper* 295, pp. 201–207, 1960.

Section 17-II

SEDIMENTATION

PART II. RIVER SEDIMENTATION

HANS ALBERT EINSTEIN, *Professor of Hydraulic Engineering, University of California.*

I. Introduction..17-36
   A. General..17-36
   B. Dual Control of Sediment Transport......................17-36
   C. Bed-material Load..17-37
   D. Wash Load..17-37
   E. Effect of Erratic Supply.......................................17-37
   F. Long-time Adjustment of Over- or Undersupply of Sediment 17-38
   G. Segregation of Sediment Sizes..............................17-38
   H. The Test Reach..17-39
II. Basic Theory Governing Sediment Motion.................17-39
   A. Definition of Sediment and Its Size Fractions............17-39
   B. Size Analysis..17-40
   C. Settling Velocity..17-41
   D. Surface Drag on the Alluvial Stream Bed..................17-41
   E. Bar Resistance...17-44
   F. Velocity Distribution on a Sediment Bed...................17-45
   G. Suspension...17-45
   H. Integration of the Sediment Transport by Suspension....17-46
   I. Bed Load...17-49
   J. Auxiliary Functions of the Bed-load Calculation.........17-50
   K. Interrelation between Bed Load and Suspended Load...17-51
   L. The Bed-load Function..17-51
   M. Tractive-force Equations......................................17-52
   N. Similarity..17-53
   O. Theories of Kennedy and Lacey............................17-53
   P. Saltation...17-54
III. Sediment-load Measurement and Sampling................17-54
   A. General..17-54
   B. Measurement of Wash Load..................................17-55
   C. Measurement of the Bed-material Load...................17-56
   D. Bed Sampling..17-57
IV. Calculation of the Bed-material Load........................17-57
   A. The Einstein Procedure..17-57
   B. Tractive-force Formulas......................................17-58

  V. Total-load Determination.................................. 17-58
    A. The Eroding Channel.................................... 17-58
    B. Alluvial Rivers.. 17-58
    C. Depth-integrating Sampling............................. 17-59
    D. The Modified Einstein Method........................... 17-60
  VI. Terminology... 17-60
    A. General Term for Material.............................. 17-61
    B. Types of Sediment...................................... 17-61
    C. Rate of Flow of Material............................... 17-61
    D. Grade Scale of Size.................................... 17-61
    E. Diameters of a Particle................................ 17-62
    F. Concentration of Suspended Material.................... 17-62
    G. Other Terms.. 17-62
  VII. Notation... 17-63
  VIII. References... 17-64

I. INTRODUCTION

A. General

  The hydrologist is basically concerned with the description of natural water courses and with the runoff from various watersheds. An important part of this runoff is the sediment which the flow carries along. The following problem therefore usually arises with respect to sediment: Given a drainage area and a cross section of its stream, how can the rates of sediment load passing through this cross section be predicted and described as a function of certain quantity or quantities?
  It will be seen that not all the sediment passing a cross section moves according to the same laws. This is the most difficult aspect of sediment transport and causes considerable confusion among workers who are not sufficiently familiar with the subject. It is necessary, therefore, to give first a rather qualitative description of the laws governing sediment transport before the individual formulas and measuring procedures can be described in detail. The most baffling aspect of some transportation laws is the fact that they may accurately describe an aspect of transport for a part of the sediment load, but not the corresponding relationship for another part. This may soon become clear in later discussions.

B. Dual Control of Sediment Transport

  Every sediment particle which passes a particular cross section of the stream must satisfy the following two conditions: (1) It must have been eroded somewhere in the watershed above the cross section; (2) it must be transported by the flow from the place of erosion to the cross section.
  Each of these two conditions may limit the sediment rate at the cross section, depending on the relative magnitude of two controls: the availability of the material in the watershed and the transporting ability of the stream. In most streams the finer part of the load, i.e., the part which the flow can easily carry in large quantities, is limited by its availability in the watershed. This part of the load is designated as *wash load*. The coarser part of the load, i.e., the part which is more difficult to move by flowing water, is limited in its rate by the transporting ability of the flow between the source and the section. This part of the load is designated as *bed-material load*. The latter designation will be explained further.
  The basic difference between wash load and bed-material load can best be visualized in a concrete-lined channel. If the flow is large and fast, the flow condition is not in any way affected by adding small amounts of a fine and easily transported material. The added material is the wash load, which would move in suspension

with the flow at the same average velocity without ever settling out. If the flow velocity and discharge are now reduced and/or if the material is increased in size and rate, sediment will begin to deposit on the channel bottom at a certain point and a granular sediment bed will develop. The movement of sediment particles over a bed containing such particles is the bed-material load [1].

C. Bed-material Load

The sediment bed affords a continuous and full availability of its particles in the channel for transportation. Such a channel transports the particles represented in the bed always to its capacity, and this *capacity load*, which depends on the nature of the channel and the sediment, may be predicted as a function of the flow. In any channel with a given sediment bed the bed-material load is strictly a unique function of the flow, called the *bed-load function*.

Values of the bed-load function can be determined by direct measurement of the load. Such load measurements show large variations with time and the location on the bed. These variations, which are comparable with the fluctuations of velocity in a turbulent fluid flow, may be either random or periodic. They are often related to the bars and dunes of the bed and are in general not well understood. The calculated bed-load function is obtained only as the time-average values of the transport. Only these time-average rates can be predicted today. It is important to remember that the fluctuations may have periods comparable with the time which is required for the above-mentioned bars and other bed irregularities to pass a given point. It is thus clear that significant averages are obtained only if the sampling time is sufficiently long compared with the period of fluctuation. Some of these periods are of the order of seconds and are easily averaged by most sampling methods. Other fluctuations, particularly those due to large gravel bars in rivers, may show periods of a full season and create an almost insurmountable obstacle to proper sampling [2].

D. Wash Load

In contrast to the bed-material load, the wash load is represented in the bed only in very small amounts. These small amounts do not constitute a sufficient reserve to maintain a full-capacity transport if the supply of such particles from upstream becomes temporarily insufficient. Consequently, the transport through the reach will reduce to a lower value of supply. The rate of wash load passing through a channel reach is thus unaffected by the transport capacity of the reach.

Whatever the rate of wash-load supply and transport, neither scour nor deposition of wash load may occur in the channel. This fact has two important consequences. First, since no trace is left in the channel by the passing wash load, it is impossible to predict or hindcast the occurrence and rate of wash-load motion in any way from observations in the channel without direct measurement during the period of flow. Second, since no significant values of wash load are either scoured or deposited in the reach, the geometry of the channel section will not be changed. Thus wash load does not have any direct effect on the stability of the movable bed [3].

Part of the wash load contributes to the stability of the banks, however. In some channels of ephemeral streams which do not maintain a water table near the channel, it has been observed that a strong infiltration into banks and bed often occurs during flows. Part of the wash load, mostly silts and some clay, is then deposited on the bank surface as a caked layer which stabilizes and imperviates those banks which otherwise would be subject to strong erosion. Some such bank deposition occurs in perennial streams too, but is mostly counterbalanced by occasional scour.

E. Effect of Erratic Supply

The sediment supply from a watershed is usually not a function of the stream discharge. At the beginning of a storm the water may find much more loose material ready to move than at the end of the storm. A winter storm may find entirely different watershed conditions from those of a summer storm. One tributary may contribute a much higher sediment concentration than another, making the sediment supply dependent on the exact location of the storm center. For these and many other reasons the sediment supply is not a unique function of the stream discharge.

The ability to transport solid material through the stream channel is always satisfactory for transporting wash load, but it follows a unique function for the bed-material load, as was explained above. At the upper end of the alluvial channel, i.e., of the channel in which a sediment bed causes the transport rate to follow a bed-load function, the sediment still arrives at the rates dictated by the supply from the watershed. In that part of the channel one must expect a considerable amount of temporary deposition and scour to occur by which the supply is corrected to follow the bed-load function of the channel. The uppermost part of any alluvial channel must therefore be expected to be temporarily unstable.

Similar conditions of temporary instability occur at other locations where the bed-load function changes. This condition is found downstream at places of significant changes in the channel geometry and below the junction of significant tributaries.

One is not able as yet to predict the cross section which a given stream will assume under given slope and sediment conditions. This information must be obtained from the river itself by prototype observations. It is not advisable, however, to make such observations from reaches which are temporarily unstable.

F. Long-time Adjustment of Over- or Undersupply of Sediment

If there is a permanent over- or undersupply of sediment in the size range of the bed-material load with respect to the transport capacity of the channel, one must expect slow aggradation or degradation to take place. As long as the rates of deposition or scour are not more than fractions of a foot per year, the methods of sediment-load calculation, which were derived for equilibrium flows, still apply to these slowly varying conditions if the particular hydraulic parameters are introduced for the various periods and reaches. A gradual change of the load along the river channel must be introduced, however, in order to satisfy the continuity equation for sediment.

Most rivers which had centuries, even thousands of years, to reach their equilibrium may generally be assumed to be in equilibrium, i.e., neither to aggrade nor to degrade. However, there are a number of rather common causes for nonequilibrium:

1. Where a river discharges into a lake or ocean, it deposits its sediment in the form of a delta, continually extending the length of the river channel. With the lake or ocean at a constant level, the slope in the channel is constantly reduced, causing deposition.

2. In many rivers tectonic elevation changes cause changes of the river slopes, which the river counteracts by deposition or scour.

3. Change of land use in the watershed or change of climate may gradually change the sediment supply. Similar effects stem from forest fires in mountainous areas. All such changes of supply call for the adjustment of the channel slope or section.

4. The construction of reservoirs in the watershed may change both the sediment supply and the hydrology of the stream, calling for readjustment of the bed-load function to new conditions.

5. Improvements of the river channel upstream may prevent local flooding and deposit of sediment, but often cause the sediment load and the flood heights to increase in the downstream.

G. Segregation of Sediment Sizes

The difference between wash-load and bed-material load has already shown that different sediment sizes may behave entirely differently in the same stream. Similar quantitative differences in behavior exist between different sizes of the bed-material load. It has been observed often that the grain-size distributions in the bed and in the transport are quite different even within the size range of the bed-material load.

Thus it is necessary to describe the sediment transport by a weight rate and its size composition.

Within the bed, too, one may find segregation of grain sizes. In gravel-carrying streams it is quite common to find the entire bed surface being covered with gravel during low flows. However, no sand appears. Upon removal of the top layer of the gravel one may find the entire bed underneath to consist of sand, with some gravel mixed in it. Questions then arise: What is the material constituting the effective composition of the bed? Is it the gravel of the top layer or of the mixture underneath? The answer may be that the gravel top layer is the effective bed for the low water, but the mixture underneath is the effective bed for all other flows which are capable of disturbing the gravel layer.

Many streams show a rather heterogeneous bed that has individual bars of very different composition and appearance. Such a bed can be described only statistically.

Another type of segregation may exist in river beds where all the coarse particles are concentrated in lenses, or layers, at greater or lesser depth below the bed surface. Many sand streams show gravel deposits of this type. These gravel deposits are extremely important as an indication of the deepest local scour during flood stage and as stabilizing agents in case of general scour.

H. The Test Reach

A rather important concept for the calculation of the sediment flow and the sediment transport in a river or river system is the *test reach*. This is the portion of a channel which is sufficiently long for determining the slope of the channel with sufficient accuracy. For this purpose it would be most helpful if the cross section were possibly uniform within the channel length and if the channel did not contain any great irregularities, such as extremely sharp bends or rock islands or sills. No important tributaries should join the river within or immediately above the test reach.

The tendency of scour or deposition in a river channel can often be shown by comparison of the sediment load in two test reaches, of which one is at the upstream end and the other at the downstream end of the river reach. Such a comparison must take into account all intermediate sediment sources, such as tributaries, bank erosion, and sediment losses, for instance, by over-bank deposition or artificial removal of sand and gravel.

The concept of the test reach is equally important for the calculation of river conditions as for measurements. The quality of the results in such an investigation may depend on the proper choice of the various test reaches. It is most important that the test reach be characteristic for the river, that it be stable, and that any measuring or gaging section be representative. It is advantageous if the slope through the test reach does not change with the discharge. A most useful test reach is between two gaging stations, of which one may be temporary or easily relocated, because the local slope often changes with discharge.

II. BASIC THEORY GOVERNING SEDIMENT MOTION

A. Definition of Sediment and Its Size Fractions

Sediment has been described generally as solid particles which are being moved or have been moved by a fluid. The following discussion will be limited to particles moved by water; motion by air (wind) may be extremely important under some natural circumstances, but not in direct connection with hydrologic studies. Many sediments, according to the above definition, are solidified by secondary deposition of a binder between the loose grains. This text on hydrology will not discuss these secondary aspects, except when the binder consists of fine particles deposited by the same flow which has moved the larger particles.

Sediments, in general, are classified according to size, specific weight, shape, mineralogical composition, color, and other aspects. With respect to its movement by the water, the grain size is the most important factor since it causes the widest range of mobility. Table 17-II-1 shows the major classifications [4, p. 92] (see also Subsec. VI). It will be seen later that this general classification also divides the sediment into size fractions for characteristic modes of movement. The gravel and boulders may be expected to move predominantly as bed load, while silt and clays move predominantly in suspension. The sand will undergo both types of motion.

Table 17-II-1. General Classification of Sediment Size

Size	Designation	Remark
$D < 0.5 \mu$	Colloids	Always flocculated
$0.5 \mu < D < 5 \mu$	Clay	Sometimes or partially flocculated
$5 \mu < D < 64 \mu$	Silt	Nonflocculating-individual crystals
$64 \mu < D < 2$ mm	Sand	Rock fragments
2 mm $< D$	Gravel, boulders	Rock fragments

The specific gravity of most sand and coarser sediments is between 2.65 and 2.70. This is equivalent to the specific gravity of quartz, feldspar, and calcite combined with small amounts of heavier minerals. Only few sediments have sufficiently different specific gravities to cause a significant change of their property of being moved [5]. The movement of flocculated particles, on the other hand, is governed by the size and bulk density of the flocks, but not of the minerals. The various shapes of sediment grains have little influence on the ability to move. Only for some very flat materials such as mica is their property to be moved by water significantly dependent on their particle shape.

B. Size Analysis

Sediments of sand size and coarser are usually analyzed by sieving. In the United States one uses almost exclusively the *Rotap mechanical shaker* and *standard sieves* [4, p. 96]. The agitation of the Rotap is satisfactory for all sizes below 5 mm. Above that size, hand sieving is more satisfactory. The size of the sample (100 to 200 g sand) and the duration of the sieving period (5 to 10 min) must be standardized for satisfactory duplication of the results. For very accurate work, each set of sieves should be calibrated in order to be comparable with the microscopic analysis. A triple-beam balance reading to $1/100$ g is often used for weighing. Cumulative weighing of consecutive sieve fractions will reduce the effect of many errors.

Sediments of silt size and finer are usually analyzed by settling in a still column of water. Various standard methods give various degrees of accuracy. The fastest and least expensive is the *hydrometer method* [6], which determines in terms of time the total concentration at a distance below the water surface indicated by the instrument. Smallest reading on sensitive hydrometers is 10^{-4} g/cm^3. Slightly better accuracy can be obtained by the *pipette method* [7], by which samples of constant volume are taken at a given elevation below the surface. The samples are analyzed for total solids by evaporation in small beakers. Both methods need correction for dissolved solids. The so-called *bottom-withdrawal tube* [8] permits sampling in terms of time for all particles which have settled to the bottom. All these three methods begin with settling from an originally fully dispersed system. This procedure does not apply to sands larger than the very finest, however; it has proved to be almost impossible to disperse them evenly in such a system.

For a simple and rapid determination of the size-frequency distribution of sand samples containing particles from 62 μ to 2 mm, the *visual-accumulation-tube sand-size analyzer* can be used [9]. This instrument consists of a glass sedimentation tube, a valve mechanism, and a recorder. It records results in terms of the fall velocity of the individual particles of the sample.

BASIC THEORY GOVERNING SEDIMENT MOTION

Samples containing both sands and silts must be analyzed by first wet-sieving the entire sample through the finest standard sieve. In that process the moist sample is placed in the sieve and the sieve is alternately submerged in water in a flat dish and lifted above water surface until all fines are washed out. This method prevents drying of the clays, the property of which may be permanently changed by drying. For dry samples, the samples are first sieved and the fines in the pan analyzed by one of the above three methods.

Special methods must be applied to small samples such as those obtained from suspended-load sampling, because there is usually not sufficient sand in these samples for a sieve analysis, not even if the small 3- or 2-in. sieves are used. Also, the fines are usually not in sufficient quantity for a normal-size analysis.

In these cases the analysis of sand may be made by counting grains under a microscope or with a microprojector after wet-sieving the sample through a 200-mesh sieve. The exact analysis of the fines is usually not necessary, and a breakdown of the fines at a 2- to 5-μ level is usually sufficient to estimate the effect of flocculation on deposition density, for instance. Also, a bottom-withdrawal tube has been developed at the Omaha District Laboratory of the U.S. Army Corps of Engineers. By this device the entire sand sample is introduced at the top end of the tube in the beginning of the measuring period.

All mechanical analyses are rather expensive. Very often great economy can be achieved by analyzing combined samples, without much loss of information. This is particularly true if only data on average composition are needed.

C. Settling Velocity

When a sediment grain moves through the water, it experiences considerable resistance, which is a function of the Reynolds number of this movement. When the particle moves downward, a settling velocity will be reached at which the resistance equals the weight of the grain in water.

For laminar and turbulent flow around the grain, the settling velocities for spherical grains have been shown to be as follows [7]:

Laminar flow: $$v_s = (s - 1) \frac{g}{18\nu} D^2 \qquad (17\text{-II-1})$$

Turbulent flow: $$v_s = \sqrt{(s - 1) \frac{4}{3} \frac{gD}{C_r}} \qquad (17\text{-II-2})$$

where v_s = settling velocity, cm/sec
s = specific gravity of sediment grain with respect to fluid
D = grain diameter, cm
g = acceleration of gravity, or 980 cm/sec^2
ν = kinematic viscosity of fluid, cm^2/sec
C_r = coefficient of resistance, depending on Reynolds number, with a value of about 0.5 for a large range of Reynolds numbers above critical

Rubey [10] has derived the constants for these equations, which apply to natural sediment grains. He also describes a transition between the two ranges as derived from experiments. These are given in Fig. 17-II-1 for grains of $s = 2.65$ at 16°C. The curve will roughly describe the sediment of most streams. For important studies a measurement of the settling velocity for the particular sediment is suggested.

D. Surface Drag on the Alluvial Stream Bed

At the beginning of the century it was already observed that the friction factors of alluvial stream bottoms change significantly with changing transport conditions [11]. Similar observations in sand and gravel streams led to a description of this friction, which may be used for a numerical prediction of the friction on a given bed as a function of the flow [12]. This interpretation is based on the concept of linearity of friction, which postulates that *shape resistance* and *surface drag* on such a surface may be added arithmetically [13]. The shear stress τ_b along the bottom of an open-channel flow is by definition [14]

$$\tau_b = \gamma R_b S_e = \gamma \frac{A_b}{b} S_e \qquad (17\text{-II-}3)$$

where S_e is the energy slope, R_b is the hydraulic radius with respect to the bed, A_b is the part of the cross-sectional area pertaining to the bed, b is the width of the bed, and γ is the unit weight of water. Therefore R_b is the hydraulic radius of that part of the water body which contributes the energy that is dissipated over the unit bed area.

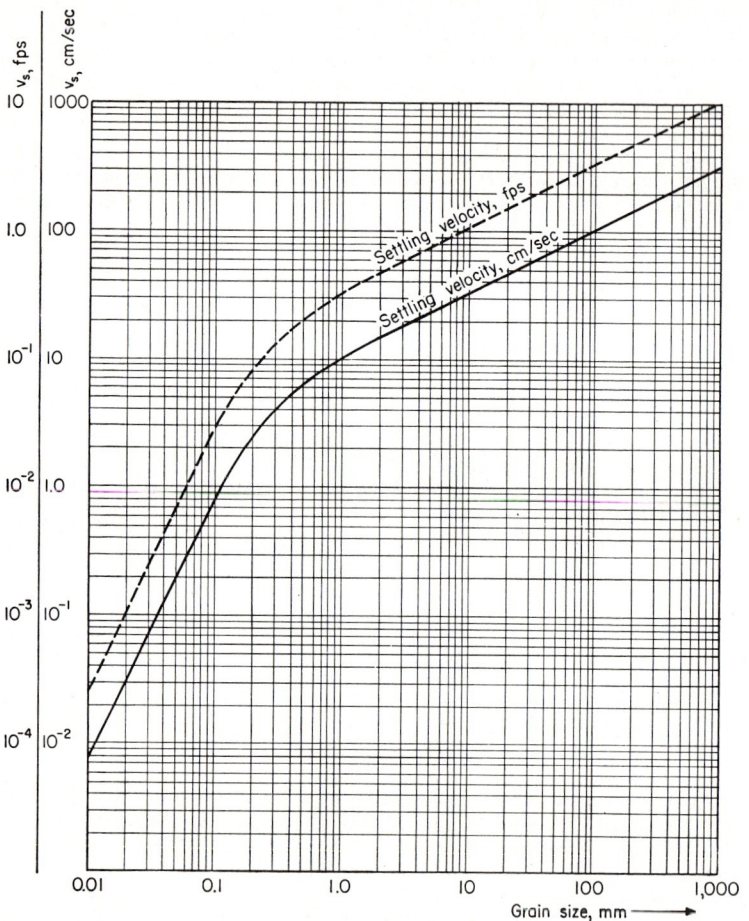

FIG. 17-II-1. Settling velocity v_s for quartz grains of various sizes according to Rubey [10].

Most alluvial beds are hydraulically rough; i.e., the friction along them may be described by formulas such as the Manning equation. The variation of the roughness with sediment transport rate is caused by a change of the shape resistance while the surface drag remains essentially independent of the sediment load. Using Strickler's formula [11] for expressing the Manning roughness coefficient in terms of the grain size, the shear stress τ_b' due to surface drag becomes (in foot-pound-second units)

$$\tau_b' = \gamma R_b' S_e = \frac{\gamma u^2}{44^2} \left(\frac{D}{R_b'}\right)^{1/3} \qquad (17\text{-II-}4)$$

BASIC THEORY GOVERNING SEDIMENT MOTION

Introducing the friction velocity $u_*' = \sqrt{\tau_b'/\rho}$,

$$u_*'^2 = \frac{\tau_b'}{\rho} = gR_b'S_e = \frac{gu^2}{44^2}\left(\frac{D}{R_b'}\right)^{1/3} \tag{17-II-5}$$

where u is the velocity of flow, and D is the representative grain size of the bed in ft, usually taken somewhat coarser than the average size. The size of which 65 per cent of the weight is finer had been successfully used for D, designated by D_{65}. Also, R_b' is the hydraulic radius due to surface drag, ρ is the density of the fluid, and g is the acceleration of gravity. The *prime* used in the notation pertains to surface drag.

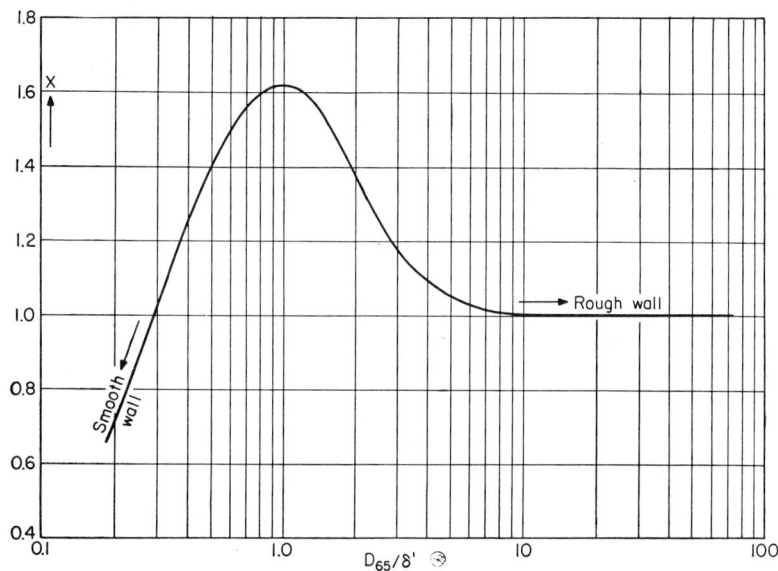

FIG. 17-II-2. Correction factor x in terms of D_{65}/δ for surface drag.

The velocity of flow due to the surface drag may be derived from Nikuradse's experiments for rough channels in the form (the constants are according to Keulegan [15])

$$\frac{u}{u_*'} = 5.75 \log\left(12.27 \frac{R_b'x}{D}\right) \tag{17-II-6}$$

where u is the average velocity of flow, x is a correction factor for the transition from hydraulically rough to hydraulically smooth surface, and $D = D_{65}$. The correction factor x is a function of the ratio D_{65}/δ; that is,

$$\frac{D_{65}}{\delta} = \frac{D_{65}u_*'}{11.6\nu} \tag{17-II-7}$$

where δ is the thickness of the laminar sublayer. The factor x is given in Fig. 17-II-2 as a function of D_{65}/δ.

Flume experiments have shown that at very high rates of bed-load movement the high sediment concentration near the bed caused the entire velocity distribution and the average velocity, u in Eq. (17-II-6), to change. Both flume and river measurements indicated these changes, but no theory exists today which can describe and predict the hydraulic behavior of high bottom concentrations [16].

E. Bar Resistance

To the surface drag of the bed surface one must add the *bar*, or *shape*, *resistance*, which includes all resistance due to bars, dunes, and ripples, as well as other channel irregularities that are caused by uneven distribution of the sediment transport over the bed area. Einstein and Barbarossa [12] proposed to designate this part of the friction by *double prime*, so that the shear stress of this part is

$$\tau_b'' = \gamma R_b'' S_e \qquad (17\text{-}II\text{-}8)$$

In terms of a friction velocity $u_*'' = \sqrt{\tau_b''/\rho}$,

$$u_*''^2 = \frac{\tau_b''}{\rho} = g R_b'' S_e \qquad (17\text{-}II\text{-}9)$$

This bar resistance is not given analytically today, but has been found empirically to be a function of the *intensity of bed-load transport* Φ_* [17], which itself is a unique

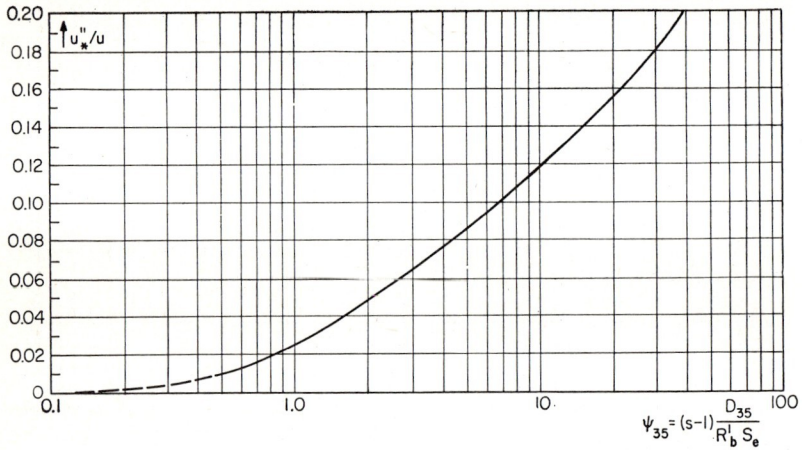

Fig. 17-II-3. Relationship between u_*''/u and Ψ_{35} for bar resistance.

function of a hydraulic parameter Ψ_{35}, defined as

$$\Psi_{35} = (s-1)\frac{D_{35}}{R_b' S_e} \qquad (17\text{-}II\text{-}10)$$

where s is the specific gravity of the sediment, and D_{35} is the grain size, of which 35 per cent of the bed material by weight is finer. In Fig. 17-II-3 this relationship is given as a graph of u_*''/u against Ψ_{35}, where u is the average velocity of flow in the channel section.

The total friction due to surface drag and bar resistance is therefore equal to

$$\tau_b = \tau_b' + \tau_b'' \qquad (17\text{-}II\text{-}11)$$

In terms of friction velocities with the total friction velocity $u_* = \sqrt{\tau_b/\rho}$, Eq. (17-II-11) becomes

$$u_*^2 = u_*'^2 + u_*''^2 \qquad (17\text{-}II\text{-}12)$$

In terms of hydraulic radii,

$$R_b = R_b' + R_b'' \qquad (17\text{-}II\text{-}13)$$

The curve in Fig. 17-II-3 was derived from data of river measurements. The original data show some scatter, but indicate that vegetation in the channel or along the banks tends to give higher values of u_*''/u, and thus the additional friction from these sources should be added separately. The curve, however, includes a normal amount of such additional friction. Rating curves for a wide variety of river sections have been predicted by this method with good accuracy. It may be mentioned that the original curve from which Fig. 17-II-3 was derived was plotted for u/u_*'' rather than for u_*''/u. Also, the bed friction in ordinary experimental flumes is not in all ranges of conditions described by this curve, and deviations are most common for flume experiments with high Ψ_{35} values.

F. Velocity Distribution on a Sediment Bed

On a hydraulically rough bed, as developed by a sediment deposit of coarse grains, the velocity distribution of a turbulent flow is given by Keulegan [15] as

$$\frac{u_y}{u_*} = 5.75 \log \frac{30.2y}{D} \qquad (17\text{-}II\text{-}14)$$

where u_y is the flow velocity at a distance y above the bed at which the friction velocity is u_* and which is composed of grains with a representative size D. For this grain size a value somewhat coarser than the average has proved to be representative. The value D_{65}, the grain size of which 65 per cent of the weight is finer, has proved to be quite satisfactory. Thus D may be taken as D_{65}. For finer sediments the bed may become hydraulically smooth and Eq. (17-II-14) can be corrected by the correction factor x of Fig. 17-II-2:

$$\frac{u_y}{u_*} = 5.75 \log \frac{30.2xy}{D} \qquad (17\text{-}II\text{-}15)$$

It is known today that this distribution undergoes drastic changes when the sediment concentration at the bed increases above 100 g/liter, but no generally accepted better formula can yet be given for this case [16]. Up to this concentration, sediment in suspension as well as the bed-load movement seems to have only insignificant effect upon the velocity distribution.

In the immediate proximity of the bed, viscosity becomes important. Along hydraulically smooth beds one may distinguish a laminar sublayer in which the velocity follows the function

$$u_y = \frac{y u_*^2}{\nu} \qquad (17\text{-}II\text{-}16)$$

The transition between the two velocity distributions as given by Eqs. (17-II-15) and (17-II-16) takes place at a distance δ from the wall equal to

$$\delta = 11.6 \frac{\nu}{u_*} \qquad (17\text{-}II\text{-}17)$$

where δ represents the thickness of the laminar sublayer, which determines whether a grain is influenced by the turbulent flow directly or through the cushioning viscous sublayer [14]. The same parameter δ can also be used to estimate the effect of viscosity along rough boundaries.

G. Suspension

Suspension is a mode of sediment movement in which the particle's weight is supported by the surrounding fluid. Since the particle continuously settles with its settling velocity v_s in relation to the surrounding fluid (Subsec. II-C), a continuous or equilibrium suspension is possible only if the flow provides a countermotion which raises the particles with an equal velocity. This upward motion is provided by the so-called *turbulent exchange*, by which fluid is being continuously exchanged between horizontal layers over finite distances. With the rising fluid originated from layers of higher concentration near the bed and the descending fluid originated from higher layers of lower concentration, a surplus of upward-moving sediment particles over the downward-moving sediment occurs. This surplus provides the upward motion of the particles, which counterbalances the general settling of the sediment. The exchange of fluid between the various layers is also responsible for the exchange of momentum in the flow. If the exchange of sediment and momentum is assumed to take place at the same intensity and over the same distances, the following distribution of sediment concentration in the vertical is obtained [18]:

$$\frac{c_y}{c_a} = \left[\frac{a(d-y)}{y(d-a)}\right]^z \qquad (17\text{-II-}18)$$

where
$$z = \frac{v_s}{0.4 u_*} \qquad (17\text{-II-}19)$$

and c_y = concentration at distance y above bed
c_a = concentration at distance a above bed
d = water depth
y = variable distance from bed
a = a reference distance above bed
v_s = settling velocity of particular grain size
0.4 = von Kármán constant
u_* = friction velocity of flow (at bed)

Equation (17-II-18) has been found to describe the sediment distributions very well, both in the laboratory and in the prototype, even under conditions of high sediment concentration (above 100 g/liter) near the bed. Equation (17-II-19), on the other hand, is not quite satisfactory, since very large deviations have been observed from the predicted values of z. These deviations were tentatively explained by the inadequacy of the theory [19] and by the change of the von Kármán constant [16]. Thus the distribution of the concentration cannot be predicted with certainty for all river conditions.

The distribution curve by Eq. (17-II-18) cannot be applied all the way to the bed, because $y = 0$ will give $c_y = \infty$, which is impossible. It is also physically impossible that suspension operates down to the bed. The linear scale of the turbulence [18], or, in other words, the size of the eddies of which the turbulence is composed, reduces near the bed approximately in proportion to the distance from the bed. A particle which is small compared with the size of the eddies can easily be moved by an eddy if it is in the interior of the eddy. If the eddies become smaller than the particle, the particle cannot be part of such an eddy, and hence cannot be moved by it as a smaller particle would. At a distance of about 2 grain diameters from the bed, suspension becomes impossible. This 2-grain-size layer is called *bed layer*. It may be noted that both the thickness of this bed layer and the function by Eq. (17-II-19) are different for the various grain sizes in a given flow.

H. Integration of the Sediment Transport by Suspension

The total rate of transport q_s over part of a vertical or the entire vertical down to 2 diameters of the representative grain size D from the bed may be obtained by integration of the product of the velocity u and concentration c from y to d for a unit width of the channel, or

$$q_s = \int_y^d uc \, dy \qquad (17\text{-II-}20)$$

where the velocity of flow u may be taken from Eq. (17-II-15) or from Eq. (17-II-14), with u_*' substituted for u_*. The velocity distribution based on u_*' is probably more appropriate in the close vicinity of the bed, where both c and dc/dy are maximum. This is proposed in *Bulletin* 1026 of the U.S. Department of Agriculture [17]. The value u_* is probably more appropriate in the remaining part of the vertical. In deep

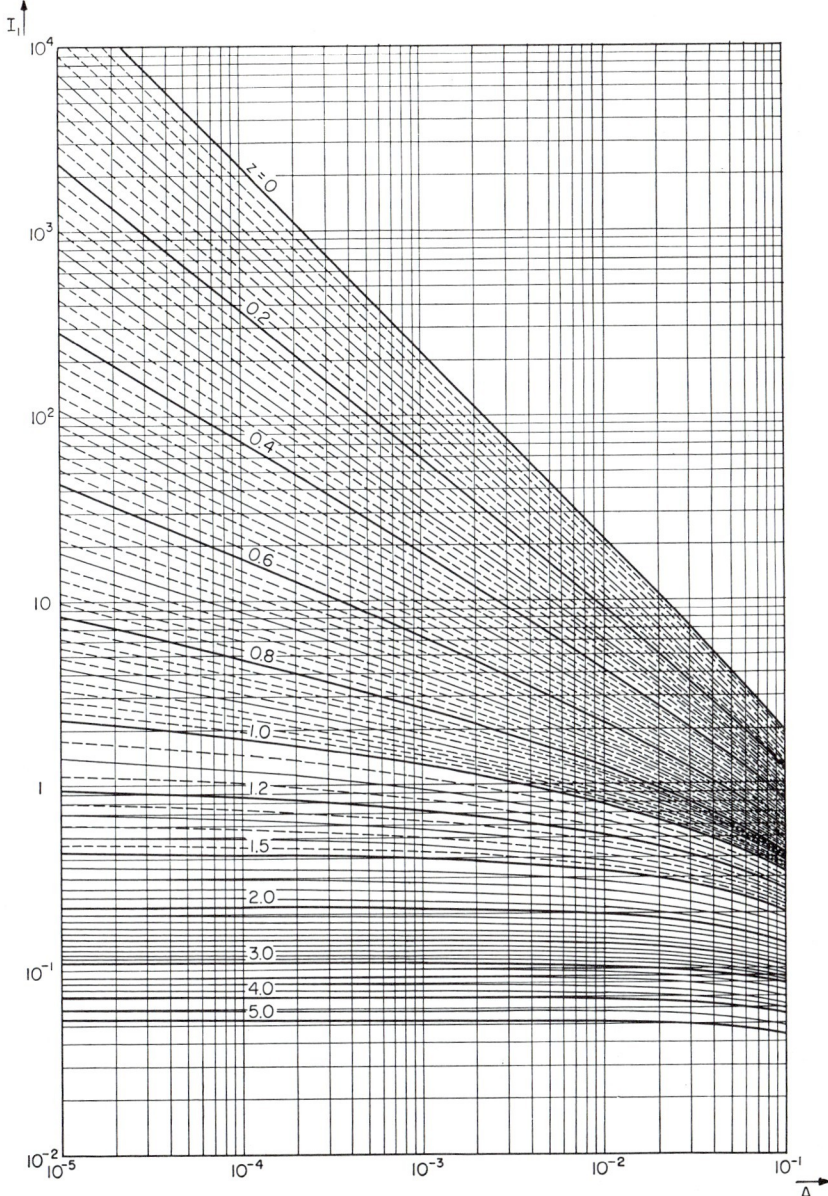

Fig. 17-II-4. Function $I_1 = 0.216 \dfrac{A^{z-1}}{(1-A)^z} \int_A^1 \left(\dfrac{1-y}{y}\right)^z dy$ in terms of A for various values of $z = v_s/0.4u^*$.

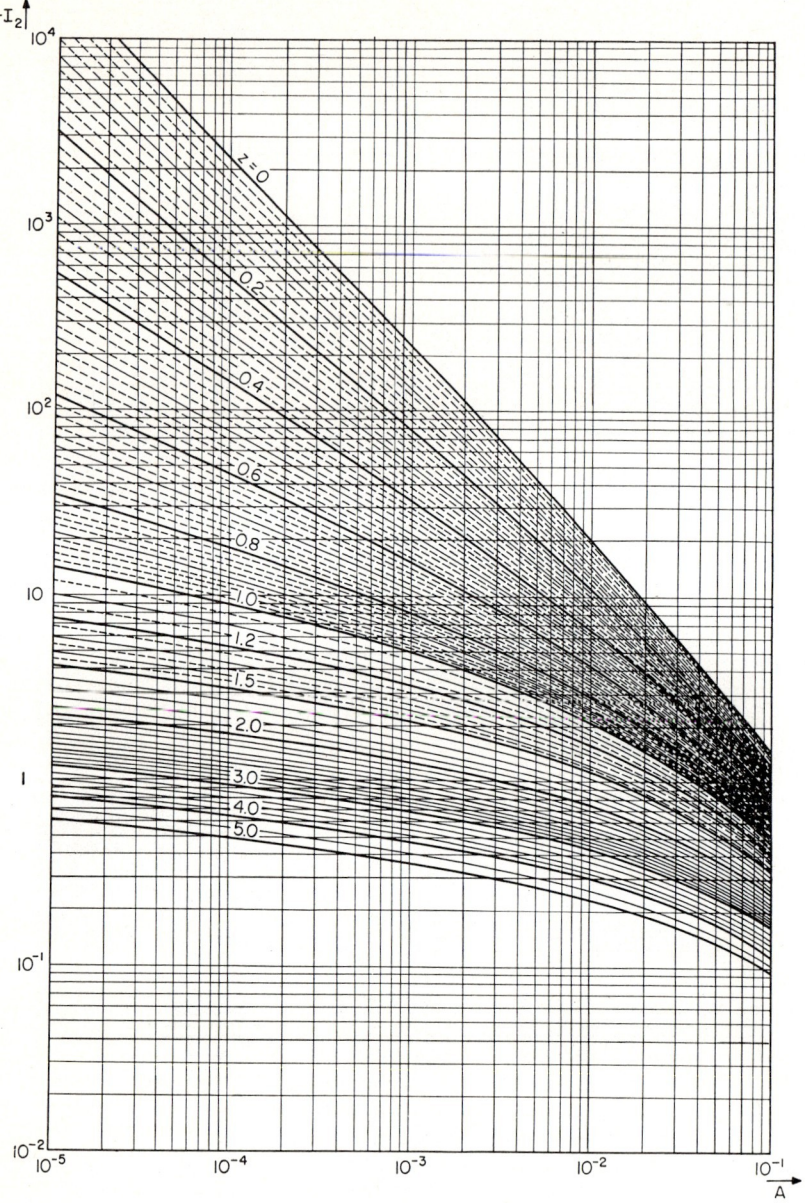

Fig. 17-II-5. Function $I_2 = 0.216 \dfrac{A^{z-1}}{(1-A)^z} \int_A^1 \left(\dfrac{1-y}{y}\right)^z \ln y \, dy$ in terms of A for various values of $z = v_s/0.4u^*$.

sections the latter may give better results. There has been no particular experience, however, by which to compare the two methods explicitly. The concentration c may be taken from Eqs. (17-II-18) and (17-II-19) such that y is the only free variable of the integration. If the lower limit of integration is assumed to be identical with the reference level for the concentration at distance a in Eq. (17-II-18) and the upper limit at the free-water surface at $y = d$, the rate of sediment load q_s per unit time and unit width of channel with the settling velocity v_s and the corresponding value of z is (for derivation see Ref. 17)

$$q_s = \int_a^d c u_y \, dy$$

$$= 5.75 u_* \, d \, c_a \left(\frac{A}{1-A}\right)^z \left[\log \frac{30.2 \, d \, x}{D_{65}} \int_A^1 \left(\frac{1-y}{y}\right)^z dy \right.$$

$$\left. + \int_A^1 \left(\frac{1-y}{y}\right)^z \log y \, dy \right] \quad (17\text{-}II\text{-}21)$$

where $A = a/d$. Numerical values for the two integrals may be derived from the graphs in Figs. 17-II-4 and 17-II-5. This derivation does not include the effect of high sediment concentrations near the bed, which cannot as yet be expressed.

It may be observed that the total rate q_s of suspended load per unit width of channel for a given grain-size range represented by the representative size D is a function of the bottom shear u_* or $u_*{}'$, and of A, which contains the water depth d as well as the reference concentration c_a of the bed layer.

I. Bed Load

For the purpose of this discussion *bed load* may be defined as the load of bed material in the bed layer where suspension is impossible for fluid-dynamic reasons. Sediment grains in the bed layer are not vertically supported by the flow, but rest on the bed almost continuously while sliding, rolling, and jumping along. They move regularly, exchanging places with similar particles in the nonmoving bed. This exchange between moving and resting particles leads to the derivation of

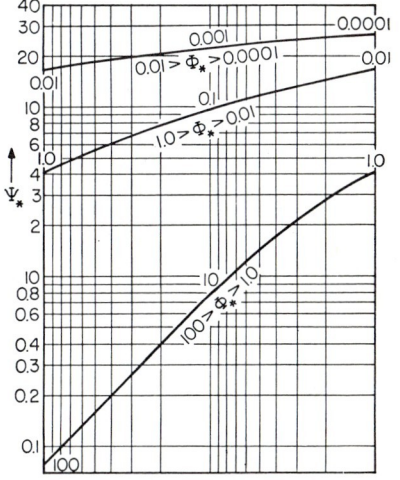

FIG. 17-II-6. Relationship between Φ_* and Ψ_*.

the bed-load equation as Eqs. (17-II-22) and (17-II-23) (see Fig. 17-II-6). The derivation of this equation will not be described here, but the result may be given for the use of practical applications.

The bed-load transport per unit of time and channel width for a particular-size fraction may be given in a dimensionless function known as *intensity of bed-load transport*:

$$\Phi_* = \frac{i_B}{i_b} \left[\frac{q_B}{\rho_s g} \left(\frac{\rho_f}{\rho_s - \rho_f}\right)^{1/2} \left(\frac{1}{g D^3}\right)^{1/2}\right] \quad (17\text{-}II\text{-}22)$$

where i_B = fraction of bed load in particular size
  i_b = fraction of bed material in particular size
  q_B = bed-load rate in weight per unit of time and channel width
  ρ_s = density of sediment
  ρ_f = density of fluid
  g = acceleration due to gravity
  D = grain diameter of particular-size fraction

Φ_* has been found to be a unique function of a hydraulic parameter Ψ_*, known as *flow intensity:*

$$\Psi_* = \xi Y \left[\frac{\log 10.6}{\log (10.6 X x/D_{65})} \right]^2 \frac{\rho_s - \rho_f}{\rho_f} \frac{D}{R_b' S_e} \quad (17\text{-II-}23)$$

where X = a reference grain size for particular bed
R_b' = hydraulic radius of bed for grain roughness or surface drag
S_e = energy slope of flow
x = correction factor of Fig. 17-II-2
ξ = a correction of effective flow for various grains
Y = a correction of lift force in transition between hydraulically rough and smooth beds

The above relationship between Φ_* and Ψ_* is given in graphic form in Fig. 17-II-6 and represents the *bed-load equation.* In other words, the bed-load equation represents the general relationship between bed-load rate, flow condition, and composition of the bed material [18]. Any consistent set of units may be used in Eqs. (17-II-22) and (17-II-23). Several items in the bed-load equation are discussed below.

J. Auxiliary Functions of the Bed-load Calculation

The reference grain size X is the smallest of the grain sizes in a given bed which is fully affected by the turbulent flow. It is defined differently in hydraulically rough and smooth beds.

For rough bed: $\Delta > 1.80\delta'$ $X = 0.77\Delta$
For smooth bed: $\Delta < 1.80\delta'$ $X = 1.39\delta'$ (17-II-24)

where $\Delta = D_{65}/x$, and δ' is the thickness of laminar sublayer δ in Eq. (17-II-17), using u_*' for u_* or R_b' for R_b.

The reference grain size X is used for the determination of the correction ξ, which reduces drastically the transport of the fine particles. In case of the rough bed, the fine particles hide between the larger ones, while on a smooth bed, they are submerged in the laminar sublayer. The two effects seem to

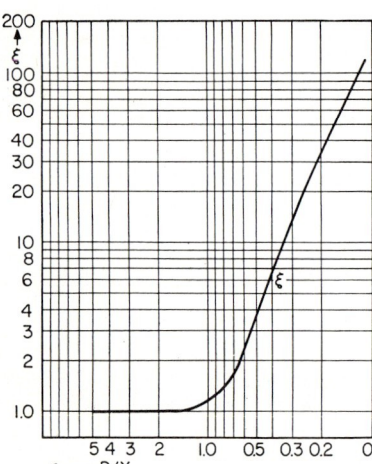
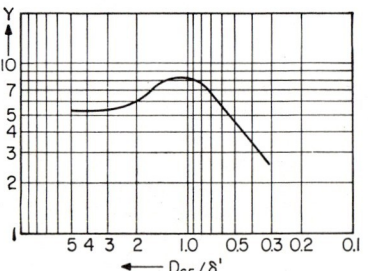

Fig. 17-II-7. Correction factor ξ in terms of D/X.

Fig. 17-II-8. Correction factor Y in terms of D_{65}/δ.

be the same, and are described by Fig. 17-II-7. The value of ξ changes from grain size to grain size in the same bed and flow.

The ξ curve does not extend indefinitely toward higher values of ξ, however. It appears [20] that for any flow the ξ curve reaches its highest point at a value of D/X where the Reynolds number of the particles in the form $u_*'D/\nu$ reaches the value 3. For values of D/X smaller than that at which $u_*'D/\nu = 3$, the factor ξ remains about constant at the value at $u_*'D/\nu = 3$. For uniform grain, ξ becomes unity.

The correction factor Y is given in Fig. 17-II-8 as a function of D_{65}/δ'. It does not vary as much as ξ in the known range. The curve cannot be extrapolated for $D_{65}/\delta' <$ 0.3. The curve may bend up again, but no conclusive measurements exist in the range where $D_{65}/\delta' < 0.3$. For uniform bed material $Y = 1.0$.

K. Interrelation between Bed Load and Suspended Load

If a particle size of the bed moving as bed load has a z value [Eq. (17-II-19)] below 5, it will also go into suspension. Bed load as defined here moves exclusively in the bed layer, i.e., a layer $2D$ thick directly above the bed. Suspension is possible only above this layer, and its prediction requires the knowledge of a reference concentration, preferably at the edge of the bed layer (Subsec. II-G). Assuming a bed-load movement in the bed layer, Einstein [17] derived the reference concentration at $2D$ from the bed as

$$c_a = C \frac{i_B q_B}{2 D u_*} \qquad (17\text{-}II\text{-}25)$$

where C is a constant determined experimentally as $1/11.6$. For a grain-size fraction of the representative size D, the total sediment load per unit width $i_t q_t$ is

$$i_t q_t = i_B q_B (1 + P I_1 + I_2) \qquad (17\text{-}II\text{-}26)$$

where $P = 2.30 \log(30.2 \, dx/D_{65})$.

In these equations $i_B q_B$ is the bed-load rate for the size fraction D, d is the water depth, x is the correction factor in Fig. 17-II-2, D_{65} is the grain diameter, of which 65 per cent of the bed is finer, and I_1 and I_2 are the two integral expressions in Figs. 17-II-4 and 17-II-5, respectively.

Item A of Eq. (17-II-21) assumes the value $2D/d$ and changes its value from one grain size to the next in the same flow.

L. The Bed-load Function

The *bed-load function* [17] is the rate at which various discharges will transport the different grain sizes of the bed material in a given channel. In other words, the bed-load function gives the equilibrium transport of sediment, both in amount and composition, for all sediment discharges for a given channel. This condition may be obtained by the application of the various equations and graphs described above. In a practical application it has proved to be most advantageous to begin the hydraulic computation by assuming various values of R_b'. Table 17-II-2 gives the headings of a table for such a computation. Each row in this table refers to one particular water discharge. The low values of R_b' are best taken at closer intervals than the larger values. The items A_t and P_b are the total cross-sectional area and the wetted perimeter of the bed, respectively, for a given stage. Other symbols have been defined previously. This table gives all the data for a gage rating curve, which can often be compared with observed values.

Table 17-II-2. Headings of the Table for Hydraulic Computation

(1) R_b', ft	(2) u_*', fps	(3) δ, ft	(4) D_{65}/δ	(5) x	(6) D_{65}/x, ft	(7) u, fps	(8) Ψ_{35}	(9) u_*''/u	(10) u_*'', fps	(11) R_b'', ft	(12) R_b, ft	(13) Stage, ft

(14) A_t, ft²	(15) P_b, ft	(16) Q, cfs	(17) X, ft	(18) Y	(19) $\log \dfrac{10.6 X x}{D_{65}}$	(20) $\left[\dfrac{\log 10.6}{\log (10.6 X x / D_{65})}\right]^2$	(21) P

Values from Table 17-II-2 are used in Table 17-II-3 to compute the sediment load. In the latter table each row gives the calculation of load for one grain size range at a given stage. There are as many rows in Table 17-II-3 as the product of the number of stages times the number of grain-size fractions.

Table 17-II-3. Headings of the Table for Sediment Computation

(1) D, ft	(2) i_b	(3) R_b', ft	(4) $\dfrac{D}{X}$	(5) ξ	(6) Ψ_*	(7) Φ_*	(8) $i_B q_B$, lb/ft-sec	(9) $A = \dfrac{2D}{d}$	(10) z	(11) I_1	(12) I_2	(13) $1 + PI_1 + I_2$	(14) $i_t q_t$,→ lb/ft-sec	(15) $i_t Q_t$,→ tons/day

In Table 17-II-3 it may be noted that all I_2 values are negative. Column 15 is derived from column 14 by multiplying the value in the latter by the width of the bed and also by 43.1 to change the units.

The best way to present the results of computation is in the form of cumulative curves for transport of all grains coarser than the various size limits in terms of discharge. This is demonstrated in Fig. 17-II-9 for Big Sand Creek, Mississippi [17].

The entire description of the hydraulics and of the sediment movement given above follows the *method of Einstein* [17], which is today probably the most generally applicable but also the most complicated. This method has been found to give good results, except under conditions where the concentration c_a rises above 100 g/liter [16]. For certain cases many other formulas give excellent results and are much easier to apply. Some of these are given below.

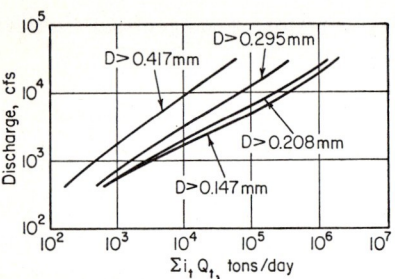

FIG. 17-II-9. Bed load for Big Sand Creek.

M. Tractive-force Equations

In reviewing Eqs. (17-II-18) to (17-II-26), which describe the sediment transport in a general way, one notices that the transport on a given sediment bed depends mainly on the friction velocity u_*, but modified by the shape resistance of the bed, the thickness of the laminar sublayer δ, and the order of magnitude of A. It is not surprising, therefore, that most of the older empirical bed-load formulas give the sediment transport as a function of u_*, usually as a quadratic function of the shear stress τ_0 or the tractive force. A large number of combinations of constants have been given for this function by various authors.

The first such formula was published in 1879 by du Boys [21] in the form

$$q_s = \Psi_{27} \frac{\tau_0}{\gamma} \left(\frac{\tau_0}{\gamma} - \frac{\tau_c}{\gamma} \right) \qquad (17\text{-II-}27)$$

where q_s is the sediment transport in lb/sec/ft of width, τ_0 is the shear stress at the bed in psf, and τ_c is a critical value of τ_0 at which q_s becomes zero. The value γ is the unit weight of fluid in lb/ft³. The parameter Ψ_{27} is supposed to be a constant for a given bed composition. However, there is no connection between this parameter and that of Eq. (17-II-23).

Table 17-II-4. Constants Ψ_{27} and τ_c in Eq. (17-II-27) for Various Grain Sizes

Mean diameter, mm	General classification	Value of Ψ_{27}	Value of τ_c, psf
1/8	Fine sand	523,000	0.0162
1/4	Medium sand	312,000	0.0172
1/2	Coarse sand	187,000	0.0215
1	Very coarse sand	111,000	0.0316
2	Gravel	66,200	0.0513
4	Gravel	39,900	0.089

Equation (17-II-27) was supposedly designed to describe the movement of gravel in the Rhone River of France. This same formula was used in 1935 by Straub [22] to describe the load in sand streams such as the Missouri and its tributaries. For this purpose Table 17-II-4 was developed to define Ψ_{27} and τ_c as a function of the grain size.

The results obtained by this formula were compared with those by the Einstein method described before, but it was impossible to identify the type of the load described by this formula. However, the formula may be used to ascertain the scour or deposition in a river reach by noting the difference in the computed loads in successive reaches.

N. Similarity

The first sign of dissatisfaction with purely empirical expressions came with the introduction of the similarity and dimensional analysis in hydraulic studies. It became clear around 1930 that the transport of coarse sediment (gravel) could be duplicated by model experiments using Froudian law. This indicated that the bed-load equation must conform with this law. The first formula, which was purposely designed to satisfy similarity, is the Meyer-Peter formula [23],

$$\frac{q^{2/3} S_e}{D} = A + B \frac{q_s^{2/3}}{D} \qquad (17\text{-}II\text{-}28)$$

where A and B are constants, q is the flow discharge per unit of time and width, S_e is the energy slope, D is the representative grain size, and q_s is the rate of sediment load per unit of time and width. It can be shown that this formula satisfies the Froude law of similarity. If q and q_s are measured in kilograms per meter-second (kg/m-sec), S_e is expressed as a fraction, and D is expressed in meters, the constants become $A = 17.0$ and $B = 0.40$. For a sediment mixture, D is to be replaced by D_{35} (the grain size of which 35 per cent of the bed material by weight is finer). Equation (17-II-28) gives very similar results, as does the bed-load equation in Fig. 17-II-6. This has been shown by Chien [24]. The equation, however, does not include any suspension in q_s.

Shields [25] postulated in 1936 that the ratio $\tau_0/(\gamma_s - \gamma)D$ is a function of the ratio of D/δ, where τ_0 is the shear stress at the channel bed, γ_s the unit weight of sediment, γ the unit weight of water, D the grain size, and δ the thickness of laminar sublayer. This results in a dimensionless plot which is generally known as the *Shields diagram*. Experimental data indicate that the value of $\tau_0/(\gamma_s - \gamma)D$ is practically constant for uniform sediment. For average conditions the value is in the order of 0.04. The value is higher for nonuniform, sticky, or flocculent materials. For fine irregular sands, the value is about 0.04 when the motion of sediment in smooth channels is incipient, and much higher when ripples and bed undulations formed of the same material are present. Short ripples tend to form at low values of D/δ, and long bars at high values.

A much more comprehensive description of the similarity of sediment transport is given by Einstein and Chien [26] where the Froudian similarity of sediment transport in open-channel flow is extended to cases in which distortion must be applied. This may be necessary in large streams with fine sediment in the bed which goes into suspension during the higher stages. In this case the value D/δ must be taken into account, as well as the relative roughness $A/2$ of the channel; the latter is used for duplicating both the friction and suspension. (Since A is by definition $2D/d$, or 2 grain diameters divided by the water depth, $A/2$, or the grain diameter divided by the water depth, may be interpreted as relative roughness.) It is shown [26] that an accurate similarity is impossible with these distortions, but an acceptable approximation can be achieved.

O. Theories of Kennedy and Lacey

The following formulas have been derived empirically from measurements in irrigation canals in India and are supposed to describe the conditions under which these canals are stable for the existing sediment supply. It is clear that such formulas may be exceedingly valuable for the design of such canals under similar flow and sediment conditions. It has been established that, at different locations, some of the constants must be given different values in order to compensate for the different sizes and rates of the sediment. Today these formulas still have a great appeal to many engineers because of their simplicity, but it would be unfortunate if their use were extended to cover wide ranges of conditions without discretion [27]. (See also Sec. 7.)

Kennedy [28] suggested, in 1895, the following two formulas. The first is the friction formula

$$V = 0.84 d^{0.64} \qquad (17\text{-II-}29)$$

where V is the average flow velocity in fps, and d is the water depth in ft. The numerical constants are for the sandy silt of the upper Bari Doab Canal, India, and should assume different values at other locations. The other formula gives the suspended sediment load q_s as a function of V,

$$q_s = cb V^{2.5} \qquad (17\text{-II-}30)$$

where c is a constant, and b is the bottom width of the canal.

Later, these formulas were found to be oversimplified, and Lacey [29] introduced, in 1929, a *regime theory* as an improvement. Lacey's theory may be described by three equations as follows:

$$\begin{aligned} V &= 16.0 R^{2/3} S_e^{1/3} \\ Qf &= 3.8 V^6 \\ P &= 2.668 Q^{1/2} \end{aligned} \qquad (17\text{-II-}31)$$

where V is the average flow velocity in fps, R is the hydraulic radius in ft, S_e is the energy slope, Q is the discharge in cfs, f is a silt factor, and P is the wetted perimeter, or approximately the width of the channel in ft. The most interesting point of this theory is the fact that three equations are established instead of the usual two. The first is a friction equation, the second a sediment-load equation, and the third a width equation which determines a stable channel width. It appears that this simple form of a width equation can be established only for rather limited channel conditions and therefore represents a rather serious limitation to the applicability of the theory to channels other than irrigation canals.

P. Saltation

Bagnold [30], in his classic paper on the transport of sand by wind, has introduced, defined, and described the type of sediment movement which is called *saltation*. Saltating particles are strongly accelerated near the bed and begin their motion almost vertically upward, being gradually accelerated horizontally in the direction of flow. Their trajectory reaches a height several hundred diameters above the bed, and then gradually curves back to the bed at a great distance downstream from the starting point. Saltation of this kind is impossible in water, as Kalinske [31] has shown very convincingly. Water with its density 1,000 times higher than air causes the saltating particle to reach the velocity of the surrounding fluid after a flight of only a few diameters. Saltation is unimportant as a separate mode of motion in water, but may be part of the bed-load movement, where the rolling and sliding particles sometimes jump at small distances. Its effect on the sediment transport is actually included in the previously described bed-load movement.

III. SEDIMENT-LOAD MEASUREMENT AND SAMPLING

A. General

As a part of the hydrologic investigation of a watershed, it is often necessary to determine the sediment load passing a given stream section. This information is required for the prediction of the useful life span of proposed reservoirs, but may

SEDIMENT-LOAD MEASUREMENT AND SAMPLING 17-55

also be required for the design of other flood-control structures such as the height of levees, the depth of footings of levees, bridge and other foundations, and the proper width of regulated river sections.

Sediment-load sampling is extremely expensive. Only in very few cases is it justified to determine loads as a matter of general interest. In special investigations sampling may be required in order to answer certain important questions in connection with the design and construction of particular river works. In such cases it is possible to design the system of measurements according to the particular need for the information. In the following paragraphs it will be shown how some of such problems can best be solved.

B. Measurement of Wash Load

In most rivers wash load constitutes the predominant bulk of the sediment load. As a rough estimate, it is between 80 to 90 per cent of the total load. Wash-load rates cannot be considered as a function of the river discharge even if the general trend shows such a function. Therefore, before a set of measurements may be used even for a statistical prediction of the future rates, it is necessary to measure wash-load rates for numerous flood hydrographs, both during rising and falling stages, possibly during dry and wet years and for winter and summer floods (where both exist). All this information may then be used to predict future sediment loads by statistical analysis.

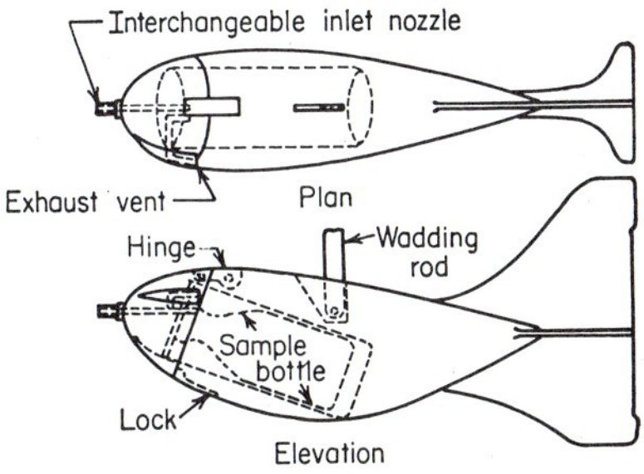

How should one sample wash load? First, one must select the exact sampling location. Since wash load is the finer part of the load, it should not only be expected to be predominantly in suspension, but also to be almost evenly distributed over the entire cross section. The exact location of

FIG. 17-II-10. An improved depth-integrating sampler.

the sampling point is thus not of particular significance. It is important only to sample in the main thread of the flow, not too close to the banks or to "dead corners." Best suited for the sampling are sections in which there is strong local mixing, as caused by high local roughness, in order to assure representative samples. Such conditions may be found at the downstream of small weirs or rapids.

Modern samplers for measuring wash load are *depth-integrating* and *point-integrating* *samplers*. For accurate and reliable field measurement, a good sampler is necessary. Under the sponsorship of several Federal agencies, a series of depth-integrating samplers [32] have been developed. These samplers give an average concentration, not only for a certain time period, but also over almost the entire water depth. Other advantages are minimum disturbance of flow, least influence from short-period fluctuations in sediment concentration, providing results that can be related to velocity measurements, and ease in handling and operation. A typical sampler of the series (Fig. 17-II-10) consists of a streamlined case carrying a standard milk bottle as a sample collector. The exhaust vent allows escape of air when water enters the bottle and also keeps the inlet velocity equal to approximately that of the local stream. Interchangeable inlet nozzles of various sizes are available to adjust the rate of filling of the bottle. The tail vanes are provided only for large samplers for keeping them stable when cable-supported.

At a uniform speed, the sampler is lowered to the bottom of the stream and then raised to the surface. The sample thus collected is an integrated quantity, with the relative portion collected at any depth in proportion to the velocity (or discharge) at that depth. The sampler has one disadvantage. Because of its shape, the nozzle cannot be lowered to within a few inches above the stream bed, where the load cannot therefore be sampled. In shallow streams, the resulting errors may be appreciable

(Subsec. V-C). The collected samples are then filtered, and the sediment dried and analyzed.

The point-integrating sampler is designed to collect continuously at a given point over an interval of time. It is generally used in deep, swift streams, where the depth-integrating sampler is unsuitable.

In making the field measurement, the necessary number of individual samples will vary considerably from river to river. The maximum concentration is often reached in many streams before the peak discharge, and the maximum sediment rate occurs usually between the maxima of discharge and concentration. In a reasonably homogeneous watershed it is usually possible to determine both the concentration and the sediment rate by taking about ten samples each during the rising and receding stages, respectively. The frequency of sampling should be the highest at high stages. Small watersheds may not need that many samples, but complicated composite watersheds may need more samples for an adequate coverage.

Each sample should be analyzed for total suspended solids, with occasional determination of the dissolved load. If the deposition in a reservoir is to be predicted, either all individual samples or some composite samples should be analyzed with respect to sand, silt, and clay sizes. This is important since clay will usually deposit with much smaller densities than coarser particles. An attempt should be made to establish a relationship between discharge and sediment runoff for purposes of sediment-load prediction. This may be more easily done for the silt load alone than for the clay since the latter is apt to be more dependent on seasons and antecedent flows.

C. Measurement of the Bed-material Load

For the bed-material load it is required to measure the rate at which grains of various sizes of the bed move along the channel. This movement is either as bed load only or partially as bed load and partially as suspended load. The bed-material load is a function of the discharge and channel characteristics and can therefore be calculated. Measurement of the bed-material load is rather difficult, but has been attempted in a few cases for the purpose of checking computation methods derived in the laboratory.

Bed-material load in gravel sizes is usually measured by box-type, or basket-type, traps with screen walls. The efficiency of such devices is low, usually about 30 per cent, and must be determined by calibration for a particular flow and sediment condition. The efficiency may be increased by reducing the size of the entrance with respect to the body size [33].

Samplers in general use for bed-material load are of several types: (1) *box*, or *basket, type*, (2) *pan type*, (3) *pressure-difference type*, and (4) *structure type*. The box-type, or basket-type, sampler consists essentially of a perforated container as a catcher. Although it is a simple device, its use requires skill because of the disturbance of local flow and the tendency to dig into the bed. The pan-type sampler is a wedge-shaped box, with the downstream half of the top surface open to catch the sample. The pressure-difference type is a sampler of improved design which provides a larger exit than entrance in order to reduce the flow disturbance at the upstream side and approximate an intake velocity to that of the stream. The structure-type sampler is designed for measurements of long intervals of time. It is a permanent structure built in a small stream or canal for taking intermittent or even continuous records of the total rate of transport. The usual design consists of open or grated depressions on the bottom of the channel at the end of an essentially uniform reach. The materials settled and collected in the depressions are withdrawn for measurement by pumping or sluicing.

Bed-material load in sand sizes moving as bed load may be measured by smaller instruments of similar type for which the sediment efficiency can be increased to 100 per cent. These devices can sample the entire load moving within a bottom layer of a thickness as determined by the height of the entrance opening [34]. Suspended particles above that layer may be measured by using suspended-load samplers. Use of the latter is identical with that for measuring wash load.

For a proper interpretation of the results it is necessary to make a full-size analysis of all samples for the range of grain sizes classified as bed-material load. No extended sets of such measurements are usually needed since they are intended only for checking computations. It is possible to combine the determination of wash load and that of bed-material load in suspension from the same suspended-load samples. Only very few measurements of bed-load rates by means of traps have been made [32-37]. These measurements were used to check the applicability of computation methods to the streams under consideration. The equations given in Subsecs. II-C to II-L have proved to be the most applicable to description of sediment flow and transport in all these rivers, with the stated restrictions.

D. Bed Sampling

The calculation of the bed-material load in a given channel requires knowledge of the bed composition. Strictly speaking, this size composition should be known for various stages of flow, since at higher stages layers in the bed may be uncovered which are not exposed at low stages. Therefore it is difficult not only to sample and describe such a heterogeneous bed, but also to interpret the results. It is also difficult to predict which layers constitute the active bed at various flow stages. Usually, the average composition of all samples at a given depth is considered.

In sampling a dry bed, the only question is in regard to the sample size and location. Usually, a representative composition is desired for a reach of the river. This can be best obtained by sampling with a grid system over the entire bed area, by combining the samples, and by analyzing the combined sample. There is no general rule as to the depth at which the bed is active at various flow stages. There is no upper limit to the size of the sample that should be taken, except the limiting expense of removing and analyzing the material. The minimum size is governed by the condition that the possible error in any one of the size fractions will be a function of the number of particles in this fraction of the sample [38]. If, for instance, a bed material contains a small percentage of gravel which is important for the bed, this fraction may determine the necessary sample size. This is the case, for instance, when possible scour in a river bed is to be predicted. Samples are then best taken in depth, by boring.

Bed sampling under flowing water is very difficult because of the danger of losing the fine particles. Clamshell and similar grabbing devices are often used, but must be carefully inspected for possible leaks by which fine particles may be lost as the sample is lifted from the flowing water. Also, bucket-type samplers have been used, which will sample the bed as they are dragged over it. Again, a loss of fine bed particles is possible.

IV. CALCULATION OF THE BED-MATERIAL LOAD

A. The Einstein Procedure

Following the procedure given by Einstein [17] and the steps given in Tables 17-II-2 and 17-II-3, the formulas in Subsec. II can be used to determine the sediment load as a function of the bed composition and the flow. Table 17-II-2 shows the hydraulic computation by assuming a series of values for R_b'. By assuming R_b' one can avoid all trial-and-error steps in a channel with equal bank and bed roughness. Where the bank has a different roughness, the total depth for any R_b' value must be found by trial and error. Thus it complicates the step from column 12 to column 13. One may therefore introduce a column 12a, in which R_w is given as a function of the slope and the average velocity u (from column 7), using the Manning equation with a known or estimated value for the bank roughness n_w. Equation (17-II-32) indicates that the total cross-sectional area A is divided practically but fictitiously into one part A_b pertaining to the bed and the other part A_w pertaining to the bank, that is,

$$A = A_b + A_w \qquad (17\text{-II-}32)$$

where the individual areas may be expressed in terms of the product of the particular hydraulic radius and its wetted perimeter [13].

In Table 17-II-3 the calculation of the load must be performed separately for each size fraction into which the bed material is divided. For each chosen test reach, one may determine the bed-material load for the sizes of the bed except the 5 to 10 per cent finest particles. For all the particles finer than this limit, calculation of the load is not reliable. However, it is often found by measurement that even the finer particles may follow a well-defined function of the flow. This may be due to the flow in an upstream reach in which the finer particles are in great proportion either in the bed or in the banks.

B. Tractive-force Formulas

Calculations of bed-material load are often performed for channel sections similar to the existing one which is used to study the influence of the proposed channel modifications. In such a case a breakdown of the load into various sizes is not important, while only the bulk rate is needed to predict possible scour or deposition. Then the calculation of Table 17-II-3 may be replaced by the application of formulas such as those given in Subsecs. II-M and II-N. For most reliable results, one should first determine the specific constants for the river in question from measurements in places where the load is predominantly sand and moves partly in suspension. The given constants for most tractive-force-type formulas are for bed-load movement only, but are directly applicable to gravel. It may be noted that the relationship of Fig. 17-II-6 is also of this kind, if a representative uniform grain size is used for the bed, in which case the corrections X and ξ become unity. The size of that 35 per cent of the bed by weight that is finer has been found to be generally representative.

V. TOTAL-LOAD DETERMINATION

A. The Eroding Channel

One may define an *eroding channel* as one in which sufficient amounts of sediment are not available for the river to transport according to a bed-load function. All sediment sizes are moved in such a channel according to their available quantities since the carrying capacity always surpasses the supply. In such a channel the entire load must be measured. The carrying capacity of the channel may be taken as an upper limit of the load.

The method of total-load measurement and the instrumentation depend mostly on the size of the sediment. For particles finer than sand, one may use the same suspended-load samplers [32] that are used for the sampling of suspended wash load in alluvial rivers. Wash load of the pebble and gravel sizes and coarser has been measured only with the construction of retention basins in which the sediment is deposited, measured, and removed [39], while the finer part of the load is measured by sampling with suspended-load samplers in the overflow water. However, most of such load determinations in eroding rivers have been made in existing reservoirs and lakes by lake and delta surveys. Results of such surveys have been systematically collected and are used today to predict by analogy the expected load in unmeasured watersheds. Some attempts have been made to rationalize the results by statistical analysis [40, 41].

Wherever sampling is possible below rapids or drops in which a high degree of turbulence exists, all particle sizes may be evenly distributed over the entire section, so that simple grab samples may give the average concentration for all grain sizes. This may include sizes up to small gravel [42].

One rather important group of rivers shows a well-developed bed at low stages, usually interspersed by rock sills or small rapids. Such rivers often fully erode these deposits at flood stage and act like eroding systems. They must be measured and described as eroding rivers since the low-stage deposits are neither stable nor permanent.

B. Alluvial Rivers

The sizes of grains which constitute the bulk of the bed moving in alluvial rivers is a unique function of the water discharge. This relationship may be established either by sampling or by calculation or by a combination of both. Once the relationship is established, further measurement of these sizes is unnecessary. It may be kept in mind, however, that this function may depend somewhat on the water temperature, particularly for the finer sizes. It may be necessary to establish separate curves of relationship for summer and winter flows.

The grains of sizes finer than the bed material are essentially moving as wash load and generally cannot be expected to move as a function of the water discharge. There are many known cases, however, where grain sizes far down in the wash-load range (according to the bed composition) were still found to follow a unique relationship with the discharge. This was explained by the fact that either somewhere in the bed upstream or in the banks, large deposits of these sizes existed and established an additional range of "bed material" with a corresponding bed-load function. It is very valuable, therefore, to establish in each case, by direct sampling of the load, the actual limit between the wash load and that part of the load which is a function of the discharge. Only the actual wash load must then be sampled over an extended period in order to establish its statistical availability.

The sampling may be done in two different ways. The first is to sample in the very turbulent water of a pool below a natural or artificial drop. Any sample taken in such a pool will give very closely the average concentration and size composition of the entire flow. If such a channel section can be found, the samples taken there should have the average concentration of the flow. The total sediment load is obtained by multiplying this concentration by the flow discharge. No correction of any sort is necessary [42].

When no such highly turbulent section can be found along the stream, another method is to sample at a possibly normal flow section which is preceded by a reach of nearly uniform flow. In this type of section one may expect to find the suspended particles distributed according to the suspended-load theory. When the constants of the distribution can be determined by sampling, it is possible to calculate the total load.

If one uses point-integrating samplers and determines the concentration of the various grain sizes at two or more points of the vertical, one can determine the constants of the distributions using Eqs. (17-II-18) and (17-II-19). These involve the exponent z and the friction velocity u_*, assuming the value 0.4 to be correct. The same value u_* may also be determined from velocity measurements in the vertical using Eq. (17-II-15). It has been found that the two values agree reasonably well except in two cases: (1) if the section has strong secondary currents and (2) if the sediment concentration near the bed is higher than 10 per cent by weight. In both cases strong disagreement of the values occurs, but it is not clear which constants in the two equations are being affected. This phase of the problem is being investigated now at several laboratories.

The sampling may also be obtained by depth integration, which is explained in the next paragraphs.

C. Depth-integrating Sampling

This method is often applied in order to determine the average concentration of the flow or the total suspended load. Assume that a sampler is used which can sample at all elevations from the water surface down to the bed. According to Eq. (17-II-20), the load in a vertical strip of flow may be obtained by such a sampler (1) having an entrance velocity equal to the local flow velocity u and (2) moving at a constant speed v_0 through the vertical during the sampling period. All modern samplers fulfill the first condition, because it has been found that only samplers satisfying this condition will give the true concentration and composition of the sample [32]. The second condition can be satisfied by moving the sampler during the sampling period at constant speed from the surface to the bed and back to the surface. For this procedure the sampler needs no valve in the sampling tube and is for that reason much more reliable. The entrance velocity is equal to u, and the vertical velocity is $v_0 = dy/dt$. Thus $dt\, v_0 = dy$, and

$$\int cu\, dt = \frac{1}{v_0} \int cu\, dy \qquad (17\text{-II-}33)$$

which means that the intake of the sampler is proportional to the sediment transport rate in the vertical covered by the motion of the sampler. Unfortunately, all suspended-load samplers can sample only to a point 3 or more in. above the bed, leaving an "unmeasured" layer near the bed. The load in this layer, which includes the bed layer in which the bed load proper moves, must be determined otherwise and added to the "measured" load. Chien [43] has shown how the unmeasured load may be determined.

D. The Modified Einstein Method

Under this title a computation method was proposed by Colby and Hembree [42, 44, 45], which was used to test the total-load measurements on the Niobrara River, Nebraska. This method follows about the same procedure as that described earlier in Subsec. II-L, with the following modifications:

1. The calculation is based on a measured mean velocity rather than on the slope, and the depth is observed for each velocity.
2. The friction velocity u_* and the corresponding suspended-load exponent z are determined from the observed z value for a dominant grain size. Values of z for other grain sizes are derived from that of the dominant size and are assumed to change with the 0.7 power of the settling velocity.
3. A slightly changed ξ curve against D/X is introduced.
4. The depth d is used to replace R_b' in Eq. (17-II-6).

Modifications 1 and 2 are fully logical deviations from the standard procedure, but do not actually change the equations. The derivation of the z values for other grain sizes from the corresponding value for the dominant grain size is an attempt to compensate for observed deviations of measured z values from those predicted by Eq. (17-II-19) [19].

Modification 3 has been found to be in accord with similar deviations found in flume experiments with sands containing a wide variety of grain sizes [20].

Modification 4 makes the calculation much easier, but may introduce significant errors in cases where R_b'' is large compared with R_b', that is, at low sediment rates. At high sediment rates this deviation may be insignificant.

Colby and Hembree [42] show how total load $i_t q_t$ is obtained from the measured load of a given grain-size range by multiplying it by a correction factor. In the derivation of this correction factor they use two new integral values in addition to those given in Figs. 17-II-4 and 17-II-5. This is not necessary, however, since this correction factor may be expressed by the integral values I_1 and I_2 in the form, already introduced,

$$\frac{i_t q_t}{i_{sm} q_{sm}} = \left(\frac{E}{A}\right)^{z-1} \left(\frac{1-A}{1-E}\right)^z \frac{(1 + PI_1 + I_2)_A}{(PI_1 + I_2)_E} \qquad (17\text{-}II\text{-}34)$$

where $i_t q_t$ is the total load in a given size range of bed-material load, $i_{sm} q_{sm}$ is the measured (depth-integrated) suspended load in the same size range, $A = 2D/d$ is the ratio of the bed-layer thickness of 2 grain sizes to the water depth, E is the ratio of the unmeasured layer thickness at the bed to the water depth, P is the value $2.30 \log(30.2 dx/D_{65})$ as in Eq. (17-II-26), and I_1 and I_2 are the integral values given in Figs. 17-II-4 and 17-II-5. The subscripts A and E indicate whether the integral values must be read for the value A or for E to replace A.

The value of z is determined by a trial-and-error method from the measurement for the predominant grain-size range, while all other z values derived from this method are proportional to the 0.7 power of the particular settling velocity.

The method is described in detail in the *U.S. Geologic Survey Water-Supply Paper 1357* [42], well illustrated by examples. Its more general applicability must be proved by further measurements in streams of different character and size than the Niobrara and the North Loup Rivers, Nebraska, that have been studied.

VI. TERMINOLOGY

A Subcommittee on Sediment Terminology, organized in 1941 by the American Geophysical Union, Section of Hydrology, recommended, in 1947 [46], a list of termi-

TERMINOLOGY

nology for sediments. While these terms and their definitions may not necessarily conform with those used in this Handbook, they should be of general interest to readers of the literature in sediments. The recommended terminology is extracted from the Committee report as follows.

A. General Term for Material

Sediment is fragmental material transported by, suspended in, or deposited by water or air, or accumulated in beds by other natural agents; any detrimental accumulation, such as loess. Ordinarily, this excludes ice or floating organic material.

B. Types of Sediment

Contact load is the material rolled or slid along the bed in substantially continuous contact with the bed.

Saltation load is the material bouncing along the bed, or moved, directly or indirectly, by the impact of the bouncing particles.

Suspended load can be used for the material either (1) moving in suspension in a fluid, being kept by the upward components of the turbulent currents or by colloidal suspension, or (2) collected in or computed from samples collected with a suspended-load sampler. (A *suspended-load sampler* is a sampler which attempts to secure a sample of the water with its sediment load without separating the sediment from the water.) In distinguishing the two meanings, the former may be called the *true suspended load*.

Bed load may be used to designate either (1) coarse material moving on or near the bed, or (2) material collected in or computed from samples collected in a bed-load sampler or trap.

Bed-material load is part of the sediment load of a stream which is composed of particle sizes found in appreciable quantities in the shifting portions of the stream bed.

Wash load is that part of the sediment load of a stream which is composed of particle sizes smaller than those found in appreciable quantities in the shifting portions of the stream bed.

C. Rate of Flow of Material

In distinction from "water discharge," the amount of material moved in unit time should be qualified as, for example, *sediment discharge* or *saltation-load discharge*.

D. Grade Scale of Size

1. Scale of large sizes in metric and English units:

Class name	Metric unit, mm	English unit, in.
Very large boulders	4,096–2,048	160–80
Large boulders	2,048–1,024	80–40
Medium boulders	1,024–512	40–20
Small boulders	512–256	20–10
Large cobbles	256–128	10–5
Small cobbles	128–64	5–2.5
Very coarse gravel	64–32	2.5–1.3
Coarse gravel	32–16	1.3–0.6
Medium gravel	16–8	0.6–0.3
Fine gravel	8–4	0.3–0.16
Very fine gravel	4–2	0.16–0.08

2. Scale of small sizes in metric units:

Class name	Millimeters	Millimeters	Microns
Very coarse sand............	2–1	2.000–1.000	2,000–1,000
Coarse sand................	1–½	1.000–0.500	1,000–500
Medium sand...............	½–¼	0.500–0.250	500–250
Fine sand..................	¼–⅛	0.250–0.125	250–125
Very fine sand.............	⅛–1/16	0.125–0.062	125–62
Coarse silt.................	1/16–1/32	0.062–0.031	62–31
Medium silt................	1/32–1/64	0.031–0.016	31–16
Fine silt...................	1/64–1/128	0.016–0.008	16–8
Very fine silt..............	1/128–1/256	0.008–0.004	8–4
Coarse clay size............	1/256–1/512	0.004–0.0020	4–2
Medium clay size...........	1/512–1/1,024	0.0020–0.0010	2–1
Fine clay size..............	1/1,024–1/2,048	0.0010–0.0005	1–0.5
Very fine clay size.........	1/2,048–1/4,096	0.0005–0.00024	0.5–0.24

E. Diameters of a Particle

Sieve diameter is the size of sieve opening through which the given particle will just pass.

Nominal diameter is the diameter of a sphere of the same volume as the given particle.

Sedimentation diameter is the diameter of a sphere of the same specific gravity and the same terminal uniform settling velocity as the given particle in the same sedimentation fluid.

Classification of wide-range sizes. A sediment covering a wide range of sizes may be classified in one of the major size classes previously discussed as clay, silt, or sand if its geometric mean diameter lies in that range. If the geometric mean is unknown the median diameter may be used instead.

F. Concentration of Suspended Material

Methods of determination: (1) weight of dried sediment divided by weight of sample, including weight of solids, water, and dissolved material, (2) weight of the dried sediment divided by weight of distilled water with a volume equal to that of the sampler, and (3) weight of the dried sediment divided by weight of the water in the sample, including the dissolved material. Either method 2 or 3 is used when concentration is less than 1 per cent.

The *ratio* used in expressing the concentration of suspended sediment should be given in either parts per million (ppm) or per cent.

G. Other Terms

A *sand wave* is a ridge on the bed of a stream, formed by the movement of the bed material, which is usually approximately normal to the direction of flow and has a shape somewhat resembling a water wave.

A *dune* is a sand wave of approximately triangular cross section (in a vertical plane in the direction of flow), with gentle upstream slope and steep downstream slope, which travels downstream by the movement of the sediment slope with the deposition of it on the downstream slope.

An *antidune* is a sand wave, indicated on the water surface by a regular undulating wave, in appearance like that formed behind a stern-wheel steamboat. These ridges move, usually upstream. The surface waves become gradually steeper in their upstream sides, until they break like surf and disappear. These waves are usually in series and often re-form after disappearing.

VII. NOTATION

(All quantities are measured in any consistent set of units.)

- A a/d, also $2D/d$; a constant in the Meyer-Peter formula
- A_b part of channel cross-sectional area due to bed
- A_t total channel cross-sectional area
- A_w part of channel cross-sectional area due to bank
- a a reference distance above channel bed
- B a constant in the Meyer-Peter formula
- b width of channel bed
- C a constant
- C_r a coefficient of resistance
- c a constant in the Kennedy formula; sediment concentration in g/liter
- c_a sediment concentration at distance a above channel bed
- c_y sediment concentration at distance y above channel bed
- D diameter of a particle
- D_{35} particle diameter of which 35 per cent of material by weight is finer
- D_{65} particle diameter of which 65 per cent of material by weight is finer
- d water depth
- E ratio of unmeasured layer thickness at bed to water depth
- f a silt factor in the Lacey formula
- g acceleration of gravity
- I_1 an integral function
- I_2 an integral function
- i_B fraction of bed load in a particular size
- i_b fraction of bed material in a particular size
- $i_{s_m}q_{s_m}$ measured suspended load in a given size fraction
- $i_t Q_t$ total sediment load in tons/day in a given size fraction
- $i_t q_t$ total sediment load per unit width of channel bed in lb/ft-sec in a given size fraction
- n_w bank-roughness coefficient
- P wetted perimeter of a channel cross section; $2.30 \log (30.2 dx/D_{65})$
- P_b wetted perimeter of channel bed
- Q flow discharge
- q flow discharge per unit width of channel
- q_B bed load per unit time and width of channel bed
- q_s rate of sediment load per unit width of channel bed
- R hydraulic radius
- R_b hydraulic radius due to channel bed
- R_b' hydraulic radius due to surface drag
- R_b'' hydraulic radius due to bar resistance
- R_w hydraulic radius due to bank roughness
- S_e energy slope of flow
- s specific gravity of particles with respect to fluid
- u velocity of flow
- u_y velocity of flow at a distance y above channel bed
- u_* friction velocity $= \sqrt{\tau_b/\rho}$
- u_*' friction velocity $= \sqrt{\tau_b'/\rho}$
- u_*'' friction velocity $= \sqrt{\tau_b''/\rho}$
- V average flow velocity
- v_0 vertical velocity of moving a depth-integrating sampler
- v_s settling velocity of particles
- X a reference particle size for a particular bed
- x correction factor

Y a correction of lift force in transition between hydraulically rough and smooth bed
y distance above channel bed
z an exponent $= v_s/0.4u_*$
γ unit weight of water or fluid
γ_s unit weight of sediment
Δ D_{65}/x
δ thickness of laminar sublayer
δ' thickness of laminar layer based on u_*'
ν kinematic viscosity
ξ a correction of effective flow for various particles
ρ density of fluid
ρ_f density of fluid
ρ_s density of sediment
τ_b shear stress
τ_b' shear stress due to surface drag
τ_b'' shear stress due to bar resistance
τ_c critical value of τ_0 at which q_s is zero
τ_0 shear stress at channel bed
Φ_* intensity of bed-load transport, or a function for bed-load transport per unit of time and width of channel bed for a particular fraction of particles
Ψ_{27} a parameter in the du Boys formula
Ψ_{35} a hydraulic parameter to correlate Φ_* with D_{35}
Ψ_* flow intensity, or a function for correlating effect of flow with intensity of sediment transport

VIII. REFERENCES

1. Einstein, H. A., Alvin Anderson, and J. W. Johnson: A distinction between bed load and suspended load in natural streams, *Trans. Am. Geophys. Union*, vol. 21, pp. 628–633, 1940.
2. Einstein, H. A.: Determination of rates of bed-load movement, *Proc. Federal Inter-Agency Sediment Conf.* Denver, Colo., May 6–8, 1947, pp. 75–114, U.S. Bureau of Reclamation, January, 1948.
3. Wright, C. A.: Experimental study of the scour of a sandy river bed by clear and by muddy water, *U.S. Natl. Bur. Std. J. Res.*, vol. 17, RP 907, August, 1936.
4. Dalla Valle, J. M.: "Micromeritics," Pitman Publishing Corporation, New York, 1948.
5. Rittenhouse, Gordon: Transportation and deposition of heavy minerals, *Bull. Geol. Soc. Am.*, vol. 54, pp. 1725–1780, December, 1943.
6. American Society for Testing Materials: Tentative method for grain-size analysis of soils, ASTM D422-54T, in "Procedures for Testing Soils," April, 1958, pp. 83–93.
7. Krumbein, W. C., and F. J. Pettijohn: "Laboratory Manual of Sedimentary Petrology," Appleton-Century-Crofts, Inc., New York, 1938.
8. U.S. Inter-Agency River Basin Committee, Subcommittee on Sedimentation: Accuracy of sediment size analyses made by the bottom withdrawal tube method, *Rept.* 10 in "Measurements and Analysis of Sediment Loads in Streams," St. Anthony Falls Hydraulic Laboratory, Minneapolis, Minn., April, 1953.
9. U.S. Inter-Agency Committee on Water Resources, Subcommittee on Sedimentation: The development and calibration of the visual-accumulation tube, *Rept.* 11 in "Measurement and Analysis of Sediment Loads in Streams," St. Anthony Falls Hydraulic Laboratory, Minneapolis, Minn., 1957.
10. Rubey, W. W.: Settling velocities of gravel, sand and silt particles, *Am. J. Sci.*, vol. 25 no. 148, pp. 325–338, April, 1933.
11. Strickler, A.: Beiträge zur Frage der Geschwindigkeitsformel und der Rauhigkeitszahlen für Ströme, Kanäle und geschlossene Leitungen, *Mitt. Eidgenoess. Amtes Wasserwirtsch.* 16, Bern, Switzerland, 1923.
12. Einstein, H. A., and N. Barbarossa: River channel roughness, *Trans. Am. Soc. Civil Engrs.*, vol. 117, pp. 1121–1132, 1952.
13. Einstein, H. A., and R. B. Banks: Linearity of friction in open channels, *Intern. Assoc. Sci. Hydrol. Publ.* 34, vol. 3, pp. 488–498, 1951.
14. Chow, V. T.: "Open-channel Hydraulics," McGraw-Hill Book Company, Inc., New York, 1959, chap. 8.

15. Keulegan, G. H.: Laws of turbulent flow in open channels, *U.S. Natl. Bur. Std. J Res.*, vol. 21, pp. 707–741, December, 1938.
16. Einstein, H. A., and Ning Chien: Effects of heavy sediment concentration near the bed on veloci y and sediment distribution, *U.S. Corps Engrs. Missouri River Div. Sediment Ser.*, no. 8, Omaha, Nebr., August, 1955.
17. Einstein, H. A.: The bed-load function for sediment transportation in open channel flows, *U.S. Dept. Agr., Soil Conserv. Serv., Tech. Bull.* 1026, September, 1950.
18. Vanoni, V. A.: Transportation of suspended sediment by water, *Trans. Am. Soc. Civil Engrs.*, vol. 111, pp. 67–133, 1946.
19. Einstein, H. A., and Ning Chien: Second approximation to the solution of the suspended load theory, *U.S. Army Corps Engrs. Sediment Ser.*, no. 3, Missouri River Division, Omaha, Nebr., January, 1954.
20. Einstein, H. A., and Ning Chien: Transport of sediment mixtures with large ranges of grain sizes, *U.S. Army Corps Engrs. Sediment Ser.*, no. 2, Missouri River Division, Omaha, Nebr., 1953.
21. Du Boy, Paul: Études du régime du Rhône et l'action exercée par les eaux sur un lit à fond de graviers indéfiniment affouillable, *Ann. Ponts et Chausseés*, ser. 5, vol. 18, pp. 141–195, 1879.
22. Straub, L. G.: Silt investigation on the Missouri River basin, H.D. 238, 73d Congr., 2d Sess., Appendix XV, pp. 1125–1140, 1935.
23. Meyer-Peter, E., H. Favre, and H. A. Einstein: Neuere Versuchsresultäte über den Geschiebetrieb, *Schweiz. Bauzt.*, vol. 103, no. 13, pp. 147–150, March, 1934.
24. Chien, Ning: The Meyer-Peter formula for bed-load transport and Einstein bed-load function, *U.S. Army Corps Engrs., Sediment Ser.*, no. 7, Missouri River Division, Omaha, Nebr. (unpublished).
25. Shields, A.: Anwendung der Aechlichkeitsmechanik und der Turbulenzforschung auf die Geschiebebewegung, *Mitt., Preuss. Versuchsanstalt Wasserbau Schiffbau*, Berlin, 1936.
26. Einstein, H. A., and Ning Chien: Similarity of distorted river models with movable bed, *Trans. Am. Soc. Civil Engrs.*, vol. 121, pp. 440–457, 1956.
27. Inglis, Sir C. C.: Historical note to empirical equations, developed by engineers in India for flow of water and sand in alluvial channels, *Proc. Intern. Assoc. Hydraulic Structures Res.*, Second Meeting, Stockholm, June, 1948, appendix 5.
28. Kennedy, R. G.: The prevention of silting in irrigation canals, *Proc. Inst. Civil Engrs. London*, vol. 119, pp. 281–290, 1895.
29. Lacey, Gerald: Stable channels in alluvium, *Proc. Inst. Civil Engrs. London*, vol. 229, pp. 259–384, 1930.
30. Bagnold, R. A.: The movement of desert sand, *Proc. Roy. Soc. London*, ser. A, no. 892, vol. 157, pp. 594–620, December, 1936.
31. Kalinske, A. A.: Criteria for determining sand-transport by surface creep and saltation, *Trans. Am. Geophys. Union*, vol. 23, pp. 639–643, 1942.
32. Federal Inter-Agency River Basin Committee, Subcommittee on Sedimentation: Measurement of the sediment discharge of streams, Rept. 8 in "Measurement and Analysis of Sediment Loads in Streams," U.S. Corps of Engineers, St. Paul District Sub-Office, Hydraulic Laboratory, University of Iowa, Ames, Iowa, March, 1948.
33. Bogardi, J.: Solid transportation by rivers, with special reference to measurements made in Hungary, *Houille Blanche*, no. 2, pp. 108–131, March-April, 1951.
34. Schaank, E. M. H., and G. Slotboom: Enkele Mededeelingen Betreffende de Zandbeweging opden Neder-Rijn (Sand transport in the Lower Rhine), *Ingenieur*, no. 51, *Bouw- en Waterbouw Kunde* 18, 1937.
35. Untersuchungen in der Natur über Bettbildung, Geschiebe-und Schwebestaffuhrung, *Mitt. Eidgenoess. Amtes Wasserwirtsch.* 33, Bern, Switzerland, 1939.
36. Nesper, Felix: Die Internationale Rheinregulierung. III. Ergebnisse der Messungen über die Geschiebe- und Schlammfuhrung des Rheins au der Brugger Rheinbrucke, *Schweiz. Bauzt.*, vol. 110, no. 12, September, 1937.
37. Einstein, H. A.: Bed-load transportation in mountain creek, *U.S. Dept. Agr., Soil Conserv. Serv., Tech. Paper* SCS-TP-55, August, 1944.
38. Rittenhouse, Gordon, and M. P. Connaughton: Errors of sampling sands for mechanical analysis, *Sedimentary Petrol.*, April, 1944.
39. Ferrell, W. R.: "Report on Debris Reduction Studies for Mountain Watersheds," Los Angeles County Flood Control District, Los Angeles, November, 1959.
40. Brune, G. M.: Rates of sediment production in mid-western United States, *U.S. Dept. Agr., Soil Conserv. Serv.*, SCS-TP-65, Milwaukee, Wis., August, 1948.
41. Gottschalk, L. C., and G. M. Brune: Sediment design criteria for the Missouri Basin loess hills, *U.S. Dept. Agr., Soil Conserv. Serv.*, SCS-TP-97, Milwaukee, Wis., October, 1950.

42. Colby, B. R., and C. H. Hembree: Computations of total sediment discharge Niobrara River near Cody, Nebraska, *U.S. Geol. Surv. Water-Supply Paper* 1357, 1955.
43. Chien, Ning: The efficiency of depth-integrating suspended-sediment sampling, *Trans. Am. Geophys. Union*, vol. 33, no. 5, pp. 693–698, October, 1952.
44. Schroeder, K. B., and C. H. Hembree: Application of the modified Einstein procedure for computation of total sediment load, *Trans. Am. Geophys. Union*, vol. 37, no. 2, pp. 197–212, April, 1956.
45. U.S. Bureau of Reclamation, Sedimentation Section, Hydrology Branch, Project Investigation Division: "Step Method for Computing Total Sediment Load by the Modified Einstein Procedure," rev., July, 1955.
46. Lane, E. W., and others: Report of the Subcommittee on Sediment Terminology, *Trans. Am. Geophys. Union*, vol. 28, no. 6, pp. 936–938, December, 1947.

Other general references:

47. Brown, C. B.: Sediment transportation, chap. 12 in Hunter Rouse (ed.), "Engineering Hydraulics," John Wiley & Sons, Inc., New York, 1950.
48. Leliavsky, Serge: "An Introduction to Fluvial Hydraulics," Constable & Co., Ltd., London, 1955.
49. U.S. Bureau of Reclamation: Sedimentation, in "Reclamation Manual," since 1948.
50. Chien, Ning: The present status of research on sediment transport, *Trans. Am. Soc. Civil Engrs.*, vol. 121, pp. 833–868, 1956.
51. U.S. Inter-Agency Committee on Water Resources, Subcommittee on Sedimentation (formerly, between June 1946 and May 1954, Federal Inter-Agency River Basin Committee): "A Study of Methods Used in Measurement and Analysis of Sediment Loads in Streams," conducted jointly by an interdepartmental committee representing the following agencies: Tennessee Valley Authority, Corps of Engineers, Department of Agriculture (changed to Soil Conservation Service after *Rept.* 9 and to Agricultural Research Service after *Rept.* 10), Geological Survey, Bureau of Reclamation, Indian Service (changed to Office of Indian Affairs after *Rept.* 5 but dropped after *Rept.* 9), Iowa Institute of Hydraulic Research (dropped after *Rept.* 9), Coast and Geodetic Survey (added after *Rept.* 9), Forest Service (added after *Rept.* 9), and Federal Power Commission (added after *Rept.* 10). The reports under the above general heading are:

Rept. 1, Field practice and equipment used in sampling suspended sediment, August, 1940.
Rept. 2, Equipment used for sampling bed-load and bed material, September, 1940.
Rept. 3, Analytical study of methods of sampling suspended sediment, November, 1941.
Rept. 4, Methods of analyzing sediment samples, November, 1941.
Rept. 5, Laboratory investigation of suspended sediment samples, December, 1941.
Rept. 6, The design of improved types of suspended sediment samplers, May, 1952.
Rept. 7, A study of new methods for size analysis of suspended sediment samples, June, 1943.
Rept. 8, Measurement of the sediment discharge of streams, March, 1948.
Rept. 9, Density of sediments deposited in reservoirs, November, 1943.
Rept. 10, Accuracy of sediment size analyses made by the bottom withdrawal tube method, April, 1953.
Rept. 11, The development and calibration of the visual accumulation tube, 1957.
Rept. 12, Some fundamentals of particle-size analysis, December, 1957.
Rept. 13, The single-stage sampler for suspended sediment, 1961.
Rept. AA, Federal Inter-Agency sedimentation instruments and reports, May, 1959.
Rept. A, Preliminary field tests of the U.S. sediment-sampling equipment in the Colorado River basin, April, 1944.
Rept. B, Field conferences on suspended-sediment sampling, September, 1944.
Rept. C, Comparative field tests on suspended-sediment samplers, progress report, December, 1944.
Rept. D, Comparative field tests on suspended-sediment samplers, progress report, as of January, 1946.
Rept. E, Study of methods used in measurement and analysis of sediment loads in streams, July, 1946.
Rept. F, Field tests on suspended-sediment samplers, Colorado River at Bright Angel Creek near Grand Canyon, Ariz., August, 1951.
Rept. G, Preliminary report on U.S. DH-48 (hand) suspended-sediment sampler (superseded by material in *Rept.* 6).

REFERENCES

Rept. H, Investigation of intake characteristics of depth-integrating suspended-sediment samplers at the David Taylor Model Basin, November, 1954.

Rept. I, Operation and maintenance of U.S. P-46 suspended-sediment sampler.

Rept. J, Operating instructions, suspended-sediment hand sampler, U.S. DH-48.

Rept. K, Operator's manual (preliminary), the visual-accumulation-tube method for sedimentation analysis of sands.

Rept. L, Visual-accumulation tube for size analysis of sands, September, 1954.

Rept. M, Operation and maintenance of U.S. BM-54 bed-material sampler, November, 1958.

Rept. N, Intermittent pumping-type sampler, progress report, February, 1960.

Rept. O, Instructions for sampling with U.S. D-49 suspended-sediment sampler, March, 1960.

Rept. P, Investigations of differential-pressure gages for measuring suspended-sediment concentrations, June, 1961.

Section 18

DROUGHTS AND LOW STREAMFLOW

H. E. HUDSON, JR., *and* **RICHARD HAZEN,** *Partners, Hazen and Sawyer, Engineers, New York.*

I. Introduction	18-1
II. Extent and Distribution of Droughts	18-2
III. Long-term Trends	18-6
IV. Hydrologic Relations in Droughts	18-7
V. Drought Severity, Frequency, and Duration	18-10
VI. Storage Calculations	18-15
VII. Evaporation	18-16
VIII. Other Reservoir-design Considerations	18-18
IX. Reservoir Operation	18-20
X. Low-flow Maintenance	18-22
XI. Diversion Works	18-24
XII. References	18-25

I. INTRODUCTION

This section discusses droughts and includes a description of their nature, occurrence, and effects. Processing of hydrologic data for water-resource development of streams and impoundments is covered, using quantitative expressions of the extent, intensity, frequency, and duration of droughts. The figures and data presented illustrate techniques for processing data and do not necessarily represent design values.

The U.S. Weather Bureau has defined a *drought* as a "lack of rainfall so great and long continued as to affect injuriously the plant and animal life of a place and to deplete water supplies both for domestic purposes and for the operation of power plants, especially in those regions where rainfall is normally sufficient for such purposes" [1]. This definition is satisfactory, with the qualification that lack of rainfall in certain areas does not necessarily indicate a drought if the streamflows and groundwaters in the area are derived from rainfall in distant places.

The term "drought" has different connotations in various parts of the world: in Bali, a period of 6 days without rain is a drought; in parts of Libya, droughts are recognized only after 2 years without rain; in Egypt, any year the Nile River does not flood is a drought, regardless of rainfall. In this section, the term "drought" refers to periods of unusually low water supply, irrespective of a demand for water in the specific place. Arid and semiarid parts of the world are covered in Sec. 24. For frequency analysis of droughts and low streamflow, see Sec. 8-I.

II. EXTENT AND DISTRIBUTION OF DROUGHTS

Droughts may be local, confined to a single river basin or area, or they may be widespread, extending over many states and parts of the world. When droughts are widespread, however, they are likely to be very severe in only a limited part of the total area affected. The extent of a number of severe droughts in the United States is shown in Fig. 18-1. Figure 18-2 shows the distribution of drought conditions in the United States for a wet year (1915) and a drought year (1934). The expansion of arid conditions from a small area to a large fraction of the west and the eastward sweep of drier classifications during the dry year are readily apparent.

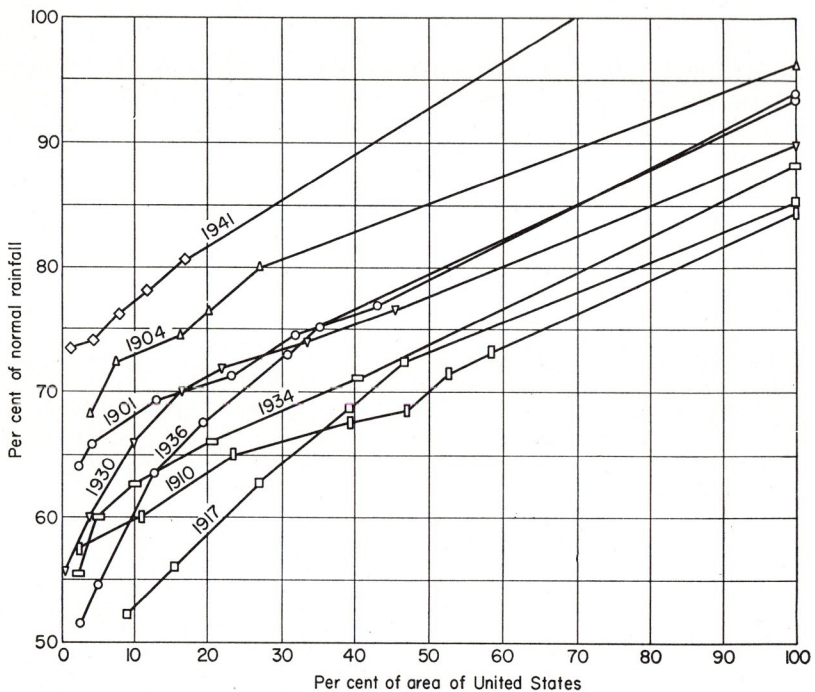

FIG. 18-1. Distribution of severe droughts in the United States. (*After Hunt* [2].)

A discussion of world-wide distribution of arid and semiarid lands is given in Subsec. 24-I-C.

The areas most subject to drought are those in which the variations in annual rainfall are relatively greatest. Hazen's 1916 studies [5] showed that, in the areas in western United States that have frequent droughts, the coefficient of variation of the annual rainfall exceeds 0.35 and that, where droughts are less frequent and less severe in the East, the coefficient of variation of annual rainfall ranged from 0.15 to 0.25. Figure 18-3, showing the average variation in annual rainfall over the world, points to the same conclusion.

Low total precipitation and high variability tend to go hand in hand because, where the total annual precipitation is small, it is generally due to a relatively small number of storms or rainy periods. Since the number of events involved is small, it is natural that the variability will be great.

EXTENT AND DISTRIBUTION OF DROUGHTS 18-3

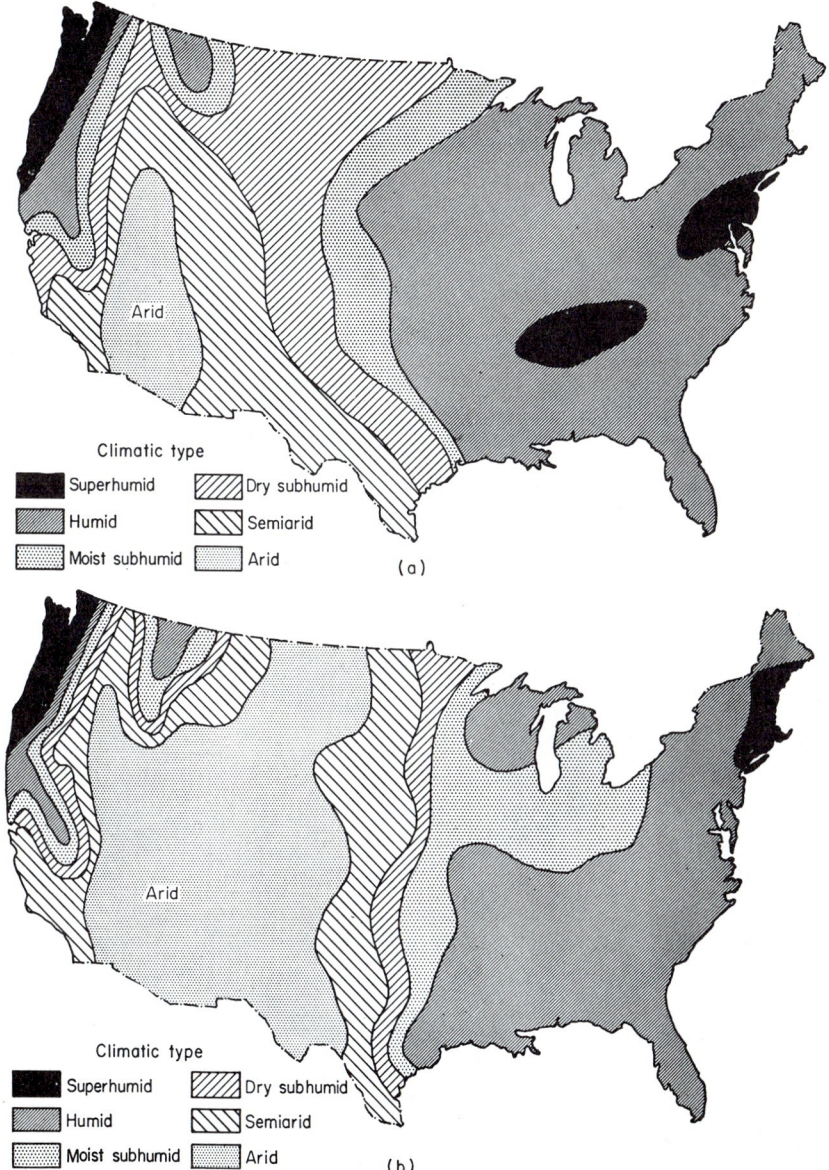

Fig. 18-2. Distribution of climatic types in the United States in (a) the wet year (1915) and (b) the drought year (1934). The expansion of arid climate in 1934 is noteworthy. (*After Tannehill* [3] *and Thornthwaite* [4].)

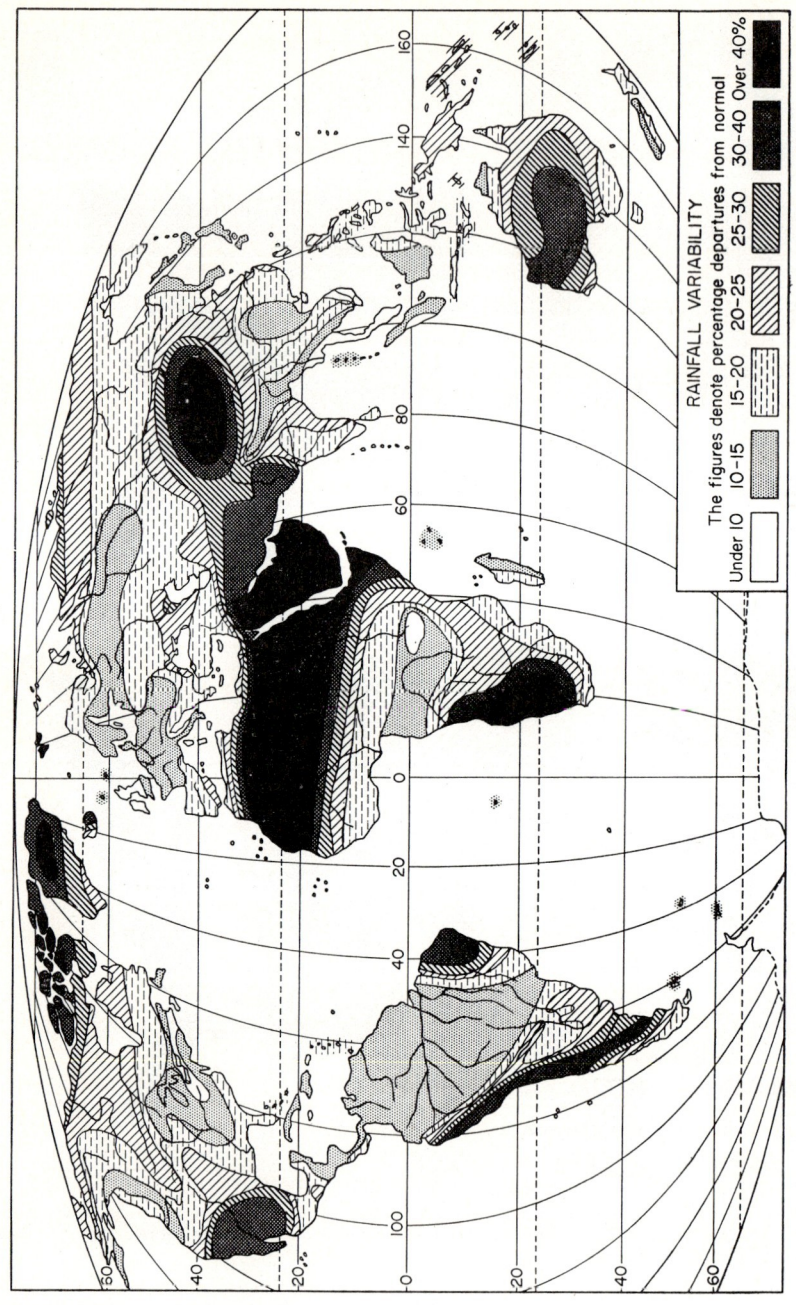

Fig. 18-3. Variability of the annual rainfall. (*After Petterssen* [6] *and Biel* [7].)

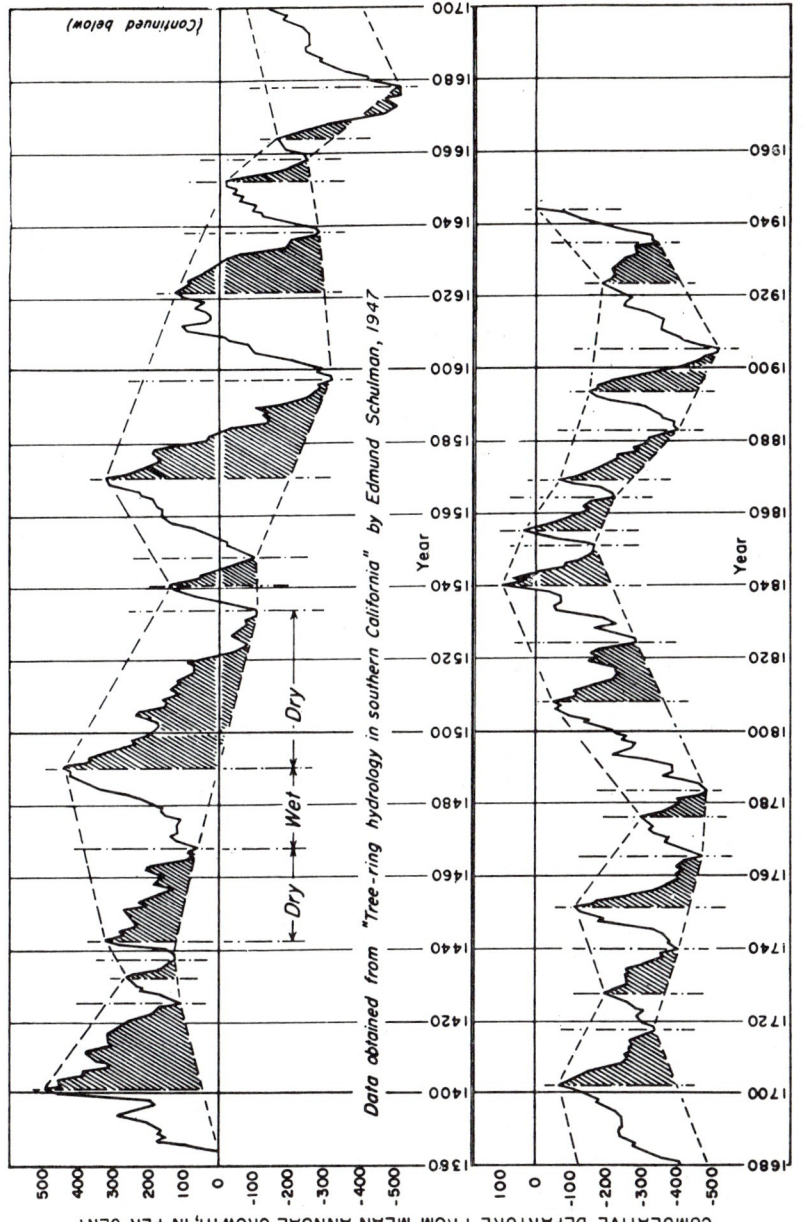

Fig. 18-4. Annual growth of big-cone spruce in southern California mountains. (*After Troxell* [8].)

18-6 DROUGHTS AND LOW STREAMFLOW

III. LONG-TERM TRENDS

Rainfall records do not reveal any appreciable change in the climate of the United States since 1800. However, studies of tree rings, which record long-term trends of rainfall, indicate that there have been definite periods of wet and dry weather. Figure 18-4 shows the variations in growth of big-cone spruce trees in southern California mountains [8] and thus, indirectly, the water supply available for the period 1385 to 1944. The tree-ring data correlate well with nearby rainfall records. There is no clear explanation of the cycles of rainfall, but extended dry periods must be considered in planning water developments if shortages are to be avoided. As may be noted from Fig. 18-4, forecasts based on records between 1780 and 1840 would have been much more optimistic than forecasts based on the period 1840 to 1920. The

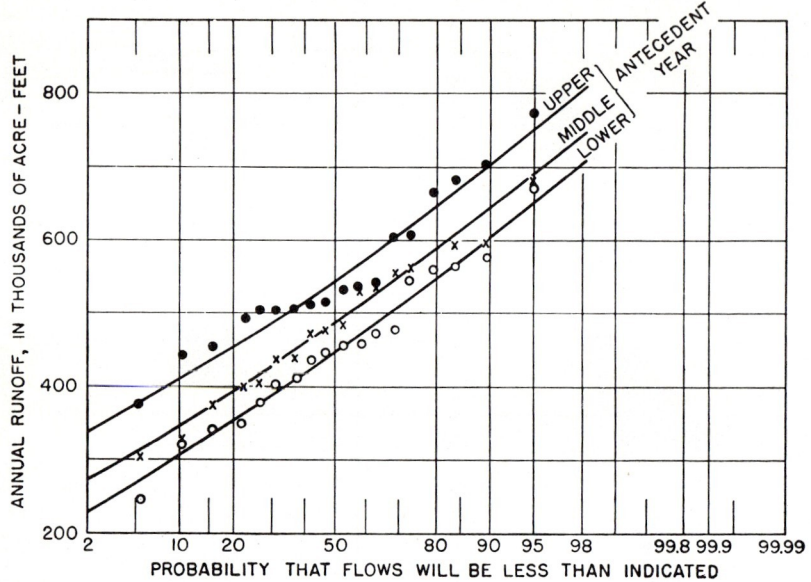

FIG. 18-5. Variation of annual flow of Cedar River, depending on antecedent flow. (*After* Langbein [9].)

inadequacy of rainfall or runoff records of 50 years or less is apparent. Where even shorter records are available, they should be compared with longer records for adjacent and similar watersheds to determine whether the records available are representative (Subsec. 9-V-A).

Analysis of short-term records evidences a tendency for dry years to "bunch" together. This indicates that the sequence of dry years is not random. Langbein has demonstrated the tendency for one dry year to follow another by his analysis of the flow of the Cedar River near Landsberg, Wash. [9]. In these studies, the annual flows were divided into three equal groups: the years with the highest flows of record, the years with the lowest flows of record, and the remainder. Then the annual discharges for the *year following* each of those so grouped were selected and arrayed in three groups. These relationships are shown in Fig. 18-5. The tendency for the flow following a year of low runoff to be lower than the year following a year of high runoff is evident from the curves. Langbein's quantitative evaluation of serial sequential correlation has been confirmed by studies elsewhere [10, 11].

Because of the existence of long-term trends and the presence of serial or sequential correlation it is important to consider *drought duration*, as well as severity.

IV. HYDROLOGIC RELATIONS IN DROUGHTS

During periods of deficient precipitation the deviation from normal conditions is greater for stream runoff than for rainfall. Figure 18-6 shows cumulative departures from normal rainfall for five cities in the southern Illinois drought area in 1952 to 1955. In the worst 30-month period of the midwestern drought of 1952 to 1954, the rainfall was about 75 per cent of normal but the runoff declined to below 10 per

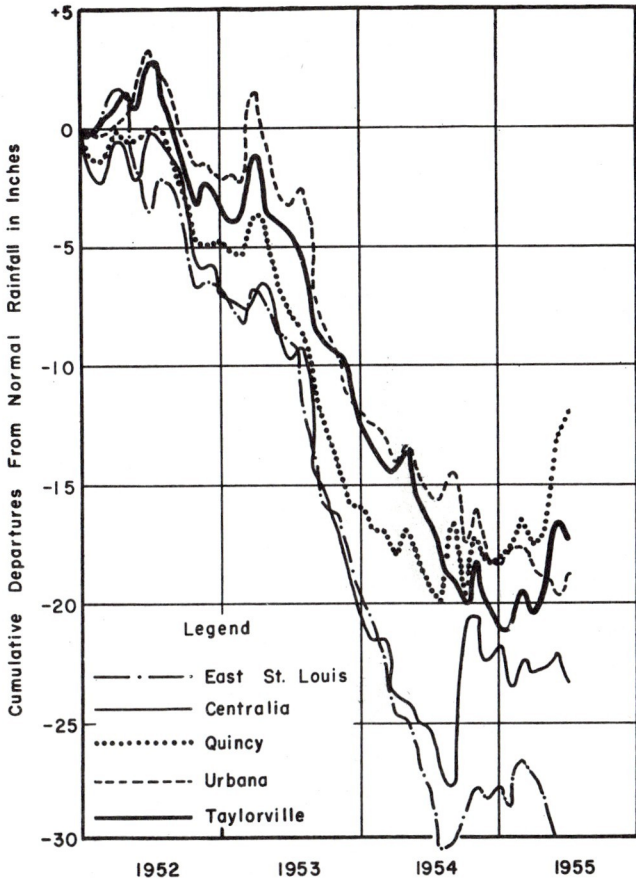

FIG. 18-6. Cumulative rainfall departures for 1952 to 1955. (*After Hudson and Roberts* [12].)

cent of normal. The deficiency in runoff in southern Illinois in December, 1954, is shown on Fig. 18-7.

During droughts the quantity of moisture drawn from storage by transpiration increases, resulting in the exhaustion of soil moisture early in the growing season. This is reflected in lower water levels in shallow wells and in deep wells subject to recharge in the drought area. High temperatures aggravate the situation by increasing the transpiration and evaporation requirements. The magnitude of these demands vs. precipitation in the United States is illustrated in Fig. 18-8.

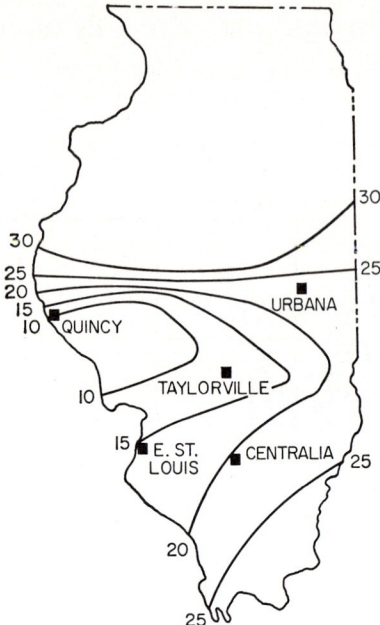

Fig. 18-7. Per cent of normal runoff, 30 months ending December 1954, in Illinois. Rainfalls as per cent of normal were Quincy, 80 per cent; Urbana, 77 per cent; Taylorville, 75 per cent; East St. Louis, 75 per cent; Centralia, 78 per cent. (*After Hudson and Roberts* [12].)

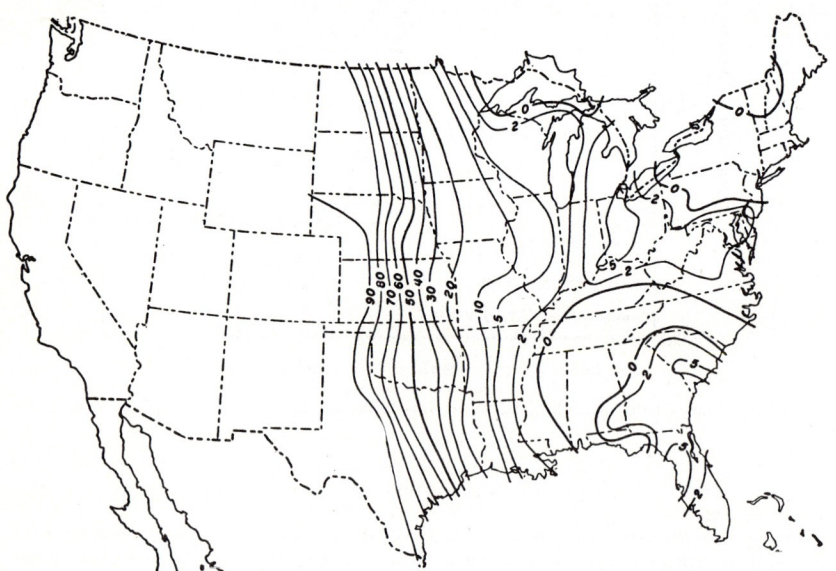

Fig. 18-8. Map of the United States showing percentage of years that annual precipitation has been less than demands for evaporation and transpiration. Throughout the West, except in mountain areas and the Pacific Northwest, annual demands of evaporation and transpiration always, or nearly always, exceed annual precipitation. (*After Hoyt* [13].)

The response of groundwater levels to droughts is illustrated in Fig. 18-9 for the Santa Ana Valley in California. The periods marked "wet" and "dry" were designated from data for the entire southern California area rather than for the Santa Ana Valley. The figure illustrates the correlation between streamflow, groundwater levels, and long-term climatic fluctuation, as well as an apparent progressive overdraft of

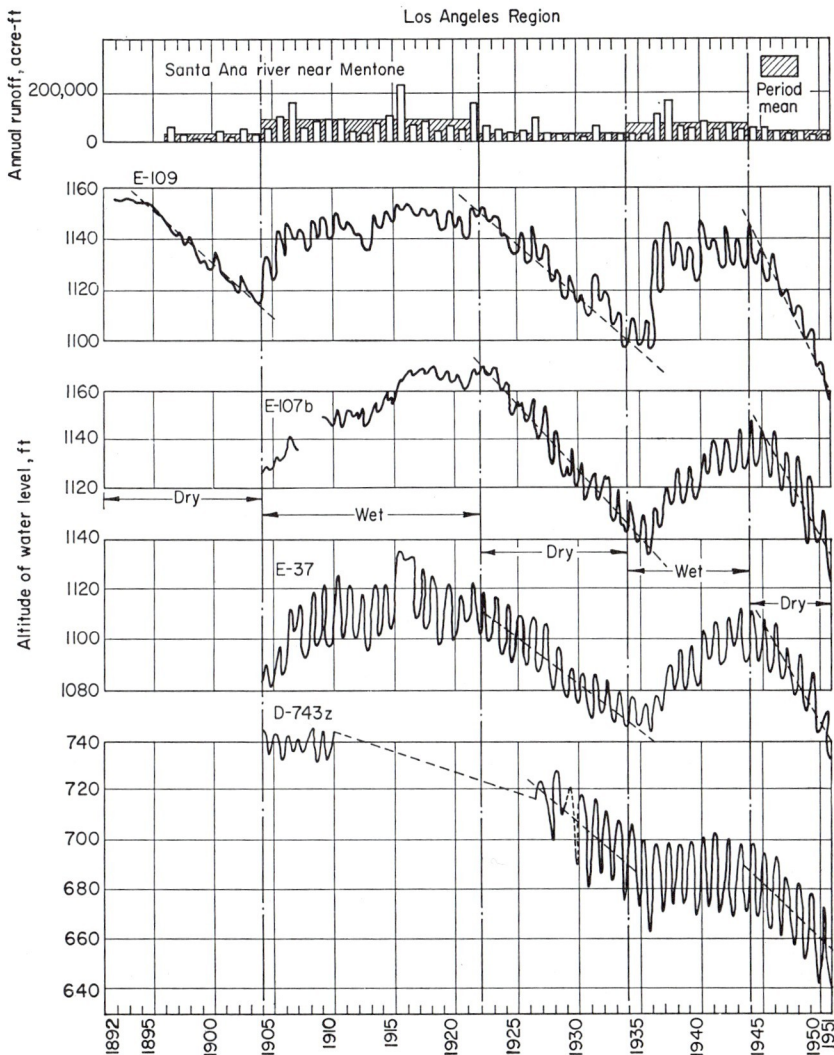

Fig. 18-9. Annual runoff of the Santa Ana River and altitude of ground water at selected wells in the Upper Santa Ana Valley. (*After Troxell* [8].)

water. The water-level variations caused by annual recharge during rainy seasons and by seasonal variations in draft are also shown. Because of the size of underground storage, the depletion of groundwater reservoirs during a single drought is not common. However, depletion may occur when draft rates greatly exceed recharge capabilities or where the groundwater reservoir is relatively shallow.

It is usually assumed that the severity of a drought is independent of the size of the area involved. This amounts to assuming that the low-flow unit discharge (discharge per unit drainage area) will not be affected by changes in area, an assumption that is known to be incorrect for flood flows. While mean flows are proportional to drainage areas, a recent drought-frequency study indicates that unit discharges during drought periods are smaller from small areas when compared with like unit discharges from larger areas [14]. Low-flow studies [15] in Michigan covered 10 basins that had several stream gages and that were free of nonnatural hydrologic variables such as regulation. The data from these basins have been combined in Fig. 18-10. The data indicate that the discharge during a 7-day drought was proportional to the drainage area raised to the 1.25 power. Expressed in another way, the unit discharge varies as the 0.25 power of the drainage area.

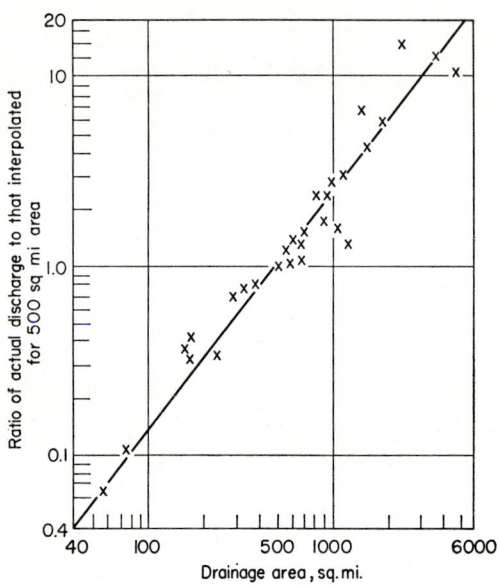

Fig. 18-10. Variance of discharge with drainage area, Michigan streams, 7-day droughts occurring once in 10 years.

Factors such as topography, geology, and land grading influence the severity of droughts. With like amounts of rain, a steep area or an area having soil with low permeability and little underground storage may fail to supply vegetative needs and may yield zero streamflow at critical periods. On the other hand, a highly permeable area with ground storage may support sizable streamflows throughout dry areas, except in the more elevated portions of the area. Man-made changes such as grading, farming, drainage works, and residential improvements aggravate drought conditions by speeding up water runoff and by reducing surface pondage, infiltration, and underground storage.

V. DROUGHT SEVERITY, FREQUENCY, AND DURATION

The severity of droughts may be measured by various parameters: deficiencies in rainfall and runoff, decline in soil moisture, reduction in groundwater levels, and the storage required to meet prescribed drafts or demands.

Because droughts are the result of a cumulative deficiency, records for individual days, months, and in some cases even years are not significant. A cumulative plotting of rainfall such as that in Fig. 18-6, or a mass diagram of runoff, will show the effect of extended dry periods. Such curves may be constructed for the entire record, for the driest period of record, or for several dry periods in order to weigh the drought severity.

Early hydrologic practice depended on analysis of hydrologic records to find the most severe period recorded. Design was then based on this single extreme period. In recent years practice included, in addition to evaluation of extreme drought severity, estimates of the *probability of occurrence* of a drought of given severity and *duration*.

The several methods of analyzing drought frequencies and duration are based on the assumption that meteorological conditions recorded in the past will be repeated. The absence of long records in many places, the long-term variations in rainfall and runoff, and the topographical changes brought about by man militate against precise forecasts. However, the reliability of statistical methods, first applied to forecasting drought frequencies and storage requirements in 1914 [16], has improved with the availability of longer and more widespread flow records. These longer records in some parts of the country reduce the need for forecasting extreme droughts by extrapolation. In many instances, statistical methods are more useful for estimating the probable frequency or return period of a drought of stated severity than for forecasting the worst drought to be expected over a long period of years.

Whether the investigation applies to rainfall, runoff, storage required, or to some other parameter of drought, a statistical analysis begins with an array, or arrangements in order of magnitude, of recorded data from which the probable frequency of a stated condition can be estimated. Data are selected by one of two methods: either one extreme value is chosen for each unit of time, such as the lowest monthly flow each year; or the lowest monthly flows in a period of years are chosen, regardless of when they occurred. With the latter method (the *method of exceedences*), the number of values chosen need not equal the number of units of time.

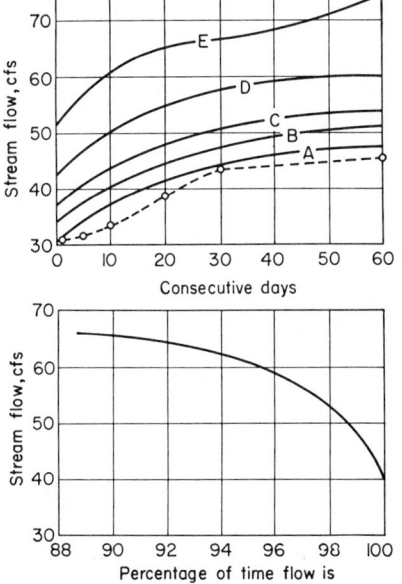

FIG. 18-11. Flow-duration curves for Frankstown Branch, Juniata River, Williamsburg, Pa., 1930–1948. (*After Hazen* [17].)

A duration curve of rainfall or runoff such as that shown for the discharge of the Frankstown Branch of the Juniata River in the lower part of Fig. 18-11 is almost the simplest type of statistical analysis. Its weakness is that it deals only with discrete values of flow and reveals nothing about the sequence of the low flows nor whether the low flows occurred consecutively over a few weeks or were scattered throughout the year.

The analysis can be made more useful by determining the flows over a given period of consecutive days. Such an analysis is shown in the upper part of Fig. 18-11. The dotted line represents the lowest average flow of record over various duration periods. Curves A to E show the flows for various duration periods for frequency or recurrence intervals of 50, 20, 10, 4, and 2 years, respectively. A number of water-resource agencies have published similar analyses of records [14, 15, 18, 19].

A difficulty sometimes encountered in frequency analysis of sequential events is overlapping of data and repeated appearance of extreme values. Thus, in an analysis

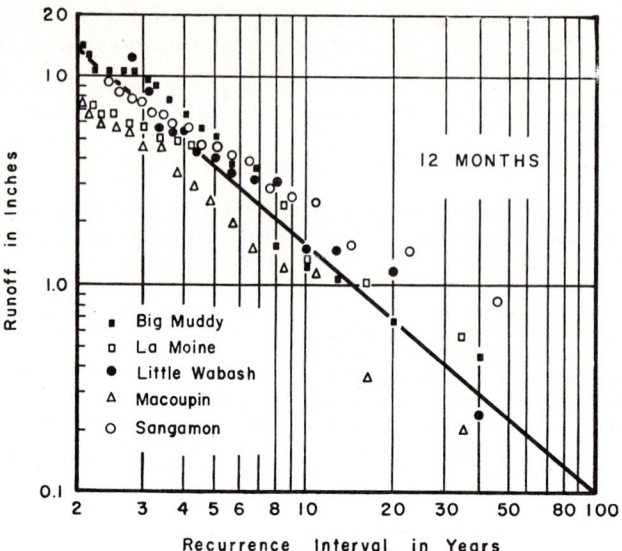

Fig. 18-12. Low-flow frequency data consolidated on basis of flow in inches. (*After* Hudson and Roberts [12]).

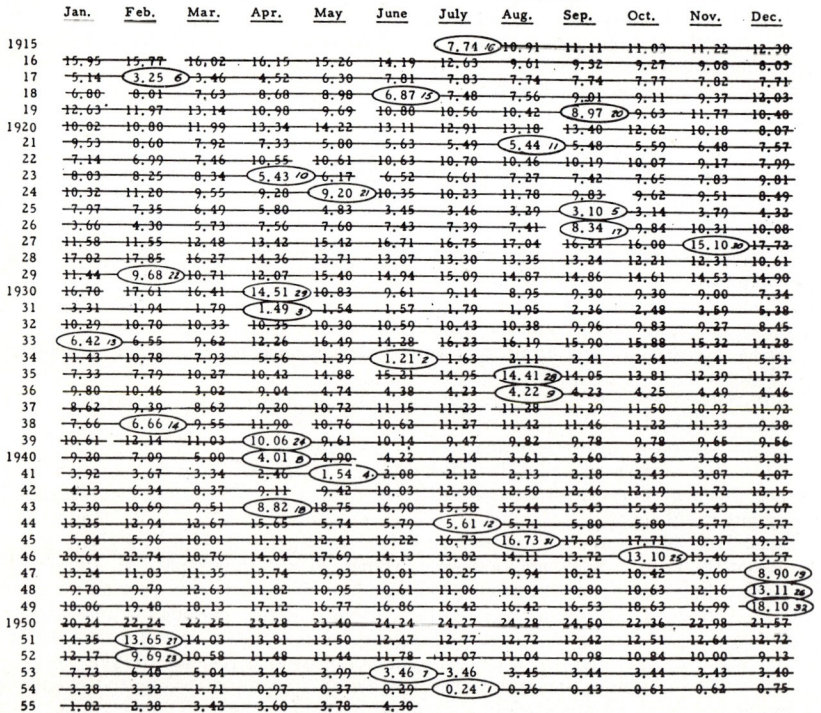

Table 18-1. Twelve Month's Running Totals of Runoff, in Inches
(Little Wabash River at Wilcox, Ill.)

DROUGHT SEVERITY, FREQUENCY, AND DURATION   18-13

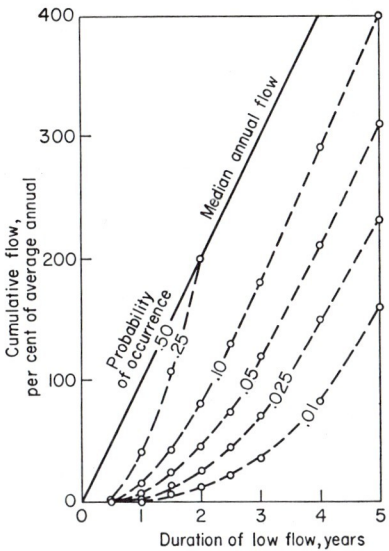

FIG. 18-13. Frequency and duration of low flows for five southern Illinois streams, 1914–1955.

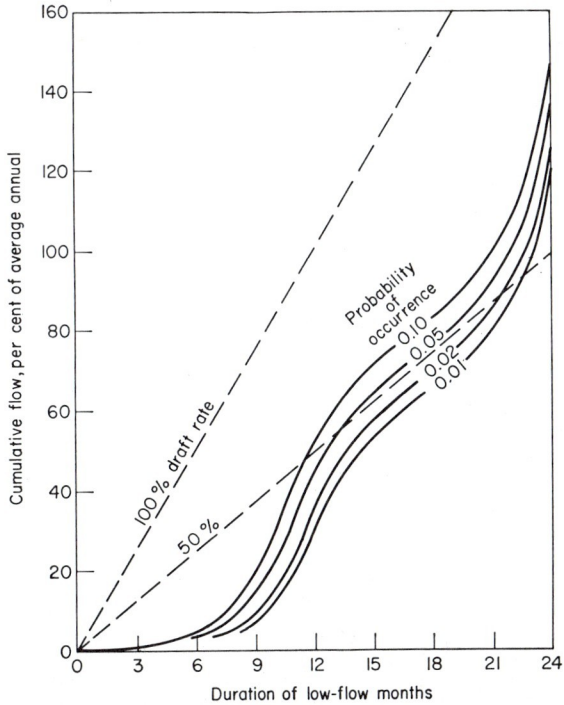

FIG. 18-14. Synthetic mass curves showing seasonal variations (Orange County, New York).

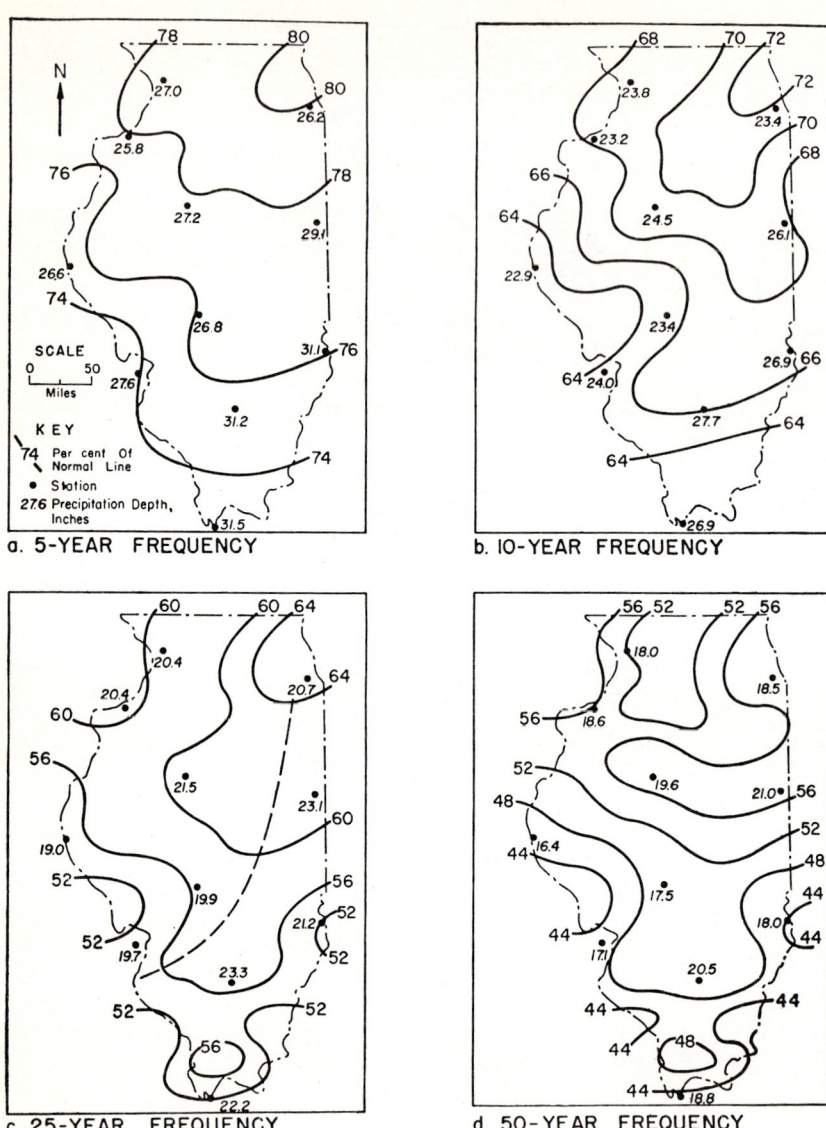

Fig. 18-15. Frequency of 12-month drought periods, expressed as per cent of normal annual precipitation. (*After Huff and Changnon* 2[1].)

of droughts lasting 2 years, certain low-flow months might appear twice. This overlapping can be eliminated by the screening procedure described below. If carried out manually, the procedure is laborious, but it can be executed speedily with the aid of punched-card or taped streamflow records and computing machines [20].

The screening procedure is illustrated by Table 18-1, which contains *running totals* of the runoff of the Little Wabash River at Wilcox, Ill. The figure entered under each month is the total flow of the 12 months, ending that month, expressed in inches.

The lowest total flow in the table is chosen for subsequent arraying. To avoid overlapping of data, the prior and subsequent 11 totals are excluded from further consideration. The next lowest total is then chosen, and the 11 values either side of it are disqualified, etc. These exceedences are arrayed and analyzed with the appropriate frequency function. A logarithmic-probability analysis of droughts lasting 12 months on the Little Wabash River and other nearby streams is shown in Fig. 18-12.

Since there is a tendency for droughts to persist, it is desirable to check runoff characteristics of periods of various durations. Using the *running-totals* technique described above, frequency analyses were prepared for duration periods of 6 months to 5 years on southern Illinois streams. The results, converted from inches to per cent of annual flow, are summarized in Fig. 18-13. This figure shows the probability of occurrence of drought events of stated duration and severity.

The duration periods chosen for Fig. 18-13 were too few to disclose seasonal variations. When smaller time periods are used, seasonal effects are revealed. These *synthetic mass curves* can be used to solve reservoir storage problems. Figure 18-14 was prepared to show a generalized solution of storage problems in small streams in Orange County, New York.

The areal extent of droughts may be described by an extension of sequential-analysis techniques. Huff and Changnon [21] studied drought-duration periods ranging from 12 to 60 months. For each period, they prepared isohyetal maps of rainfall for return periods of 5 to 50 years. Examples of these are given in Fig. 18-15.

VI. STORAGE CALCULATIONS

The basis of storage calculations is the mass-curve technique, which may be applied either with or without consideration of the sequence of low flows. Either graphic or arithmetic procedures may be used, as well as residual-mass techniques [22].

The most common form of this method involves the determination of storage volumes required for all low-flow periods and the array and plotting of these volumes to determine frequencies at which stated storage values are required [16].

The rate of withdrawal of water from the reservoir is called the *draft rate*. Generally, the flow withdrawn for an intended purpose is called the *net draft rate*, and where the reservoir has several purposes, the net draft rate will be the sum of the drafts for the various purposes. There are uncontrollable withdrawals, such as those caused by evaporation and seepage. These should be added to the net draft rate to produce the *gross draft rate*. *Safe yield*, then, must be expressed in terms of net draft rate.

Sequential runoff data, using synthetic mass curves, have been used for storage computations. Figure 18-16 illustrates the resultant draft-storage curves of stated frequency using runoff data from Fig. 18-13. Information on the duration of the recession for various draft, storage, and frequency conditions are also presented in Fig. 18-16.

It is impossible to provide storage sufficient to meet low-flow hydrologic risks of great rarity. The custom, instead, is to design for a stated risk and to add a reserve storage allowance. Extraordinary droughts are met by cutting draft rates. For water-utility practice in the northeastern United States, it is common practice to design for a drought of probability 0.05 and to add a reserve of 25 per cent of the computed storage volume [23]. This empirical allowance provides protection against a drought of probability 0.01 (once in 100 years). An allowance giving similar security for southern Illinois streams requires an additional 100 per cent storage. The coefficients of variation of annual flows are typically 0.25 for the northeastern United States and 0.55 for southern Illinois.

The need for providing adequate storage cannot be overemphasized. During the 1952 to 1955 drought, 23 impounded supplies in Illinois failed at expected draft rates because of lack of capacity and 26 failed because the capacity was insufficient to meet increases in demands [12]. In this same period 62 Kansas cities had to resort to emergency sources and 175 cities in Kansas were short of water [24].

New possibilities for generalized solutions to low-flow storage problems lie in *queuing theory*, as outlined by Langbein [9]. For this approach, mathematical or graphic relations may be set up covering the arrival rate (inflow), queue discipline and service function (management of storage), departure rates (draft rates), and attrition rate (seepage and evaporation losses, etc.). See Subsec. 14-V-C-2.

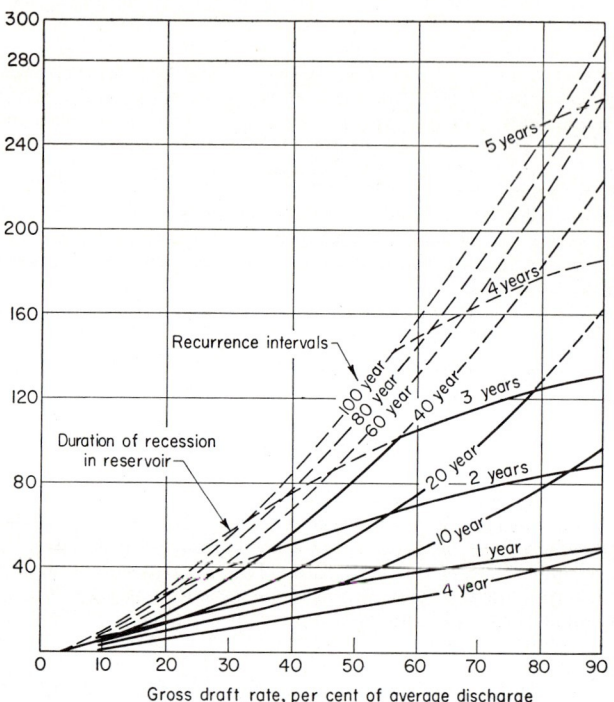

FIG. 18-16. Storage required for maintenance of various draft rates during low-flow periods of various recurrence intervals and durations for southern Illinois streams, 1914–1955. (*After Hudson and Roberts* [12].)

VII. EVAPORATION

Evaporation losses can be substantial in shallow reservoirs or in reservoirs storing more than a year's flow. Evaporation is a minor consideration in deeper reservoirs storing less than one year's flow. Evaporation losses are calculated for the reservoir surface area exposed to evaporation during drawdown. Normally, this is taken as two-thirds of the maximum area of the reservoir. In some areas, evaporation calculations should be applied to as high as 90 per cent of the nominal reservoir area.

Seasonal variations in evaporation rates are important, particularly for irrigation and water-supply developments, since maximum evaporative losses frequently coincide with maximum draft rates. Evaporation data are usually not plentiful enough to permit correlation with climatic conditions, and it is necessary to use average values for computing losses during droughts. In Fig. 18-17, monthly evaporation data have been cumulated. The maximum rates during this 3-year interval corresponded with the summer season. The dashed line tangent to the peaks represents the maximal evaporative rate [12]. Combining appropriate rainfall data with the maximal evaporative losses, net evaporative losses for drought periods of various frequencies and duration in southern Illinois were estimated in Fig. 18-18. Application

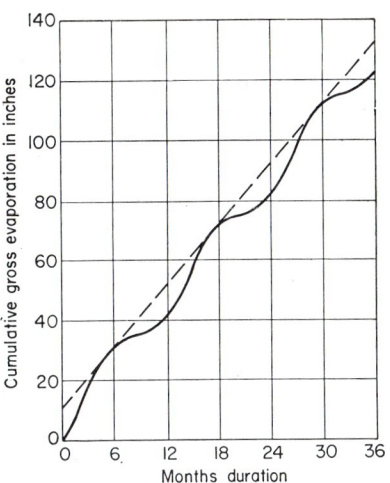

FIG. 18-17. Cumulative evaporation at Vandalia, Ill. (*After Hudson and Roberts* [12].)

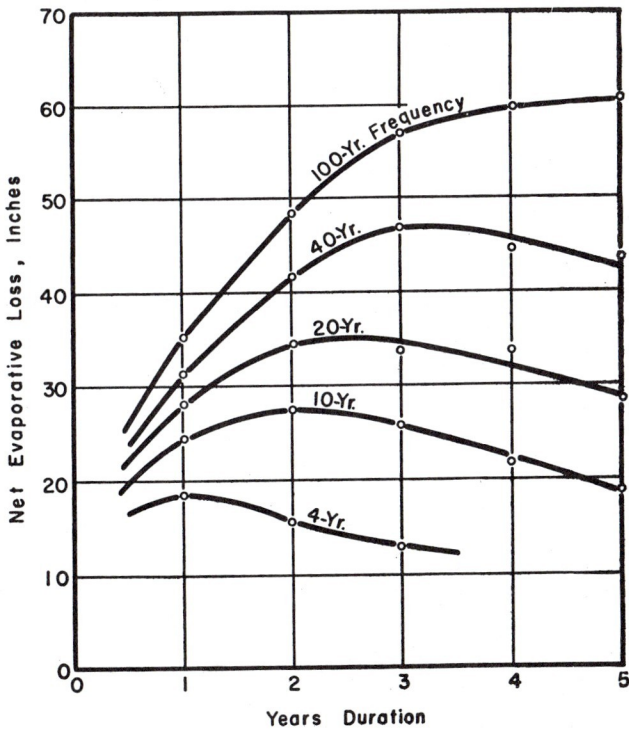

FIG. 18-18. Net evaporative losses in southern Illinois reservoirs. (*After Hudson and Roberts* [12].)

of these data to typical Illinois reservoirs resulted in the preparation of an estimate of net evaporative draft rates caused by evaporation in the southern Illinois region as shown in Fig. 18-19. For the area studied, overall evaporative draft estimates were relatively uniform regardless of drought severity and seldom approached 15 per cent of the average annual discharge into the impoundment [12].

In regions where evaporation is high, the losses may limit the feasible size of a storage development.

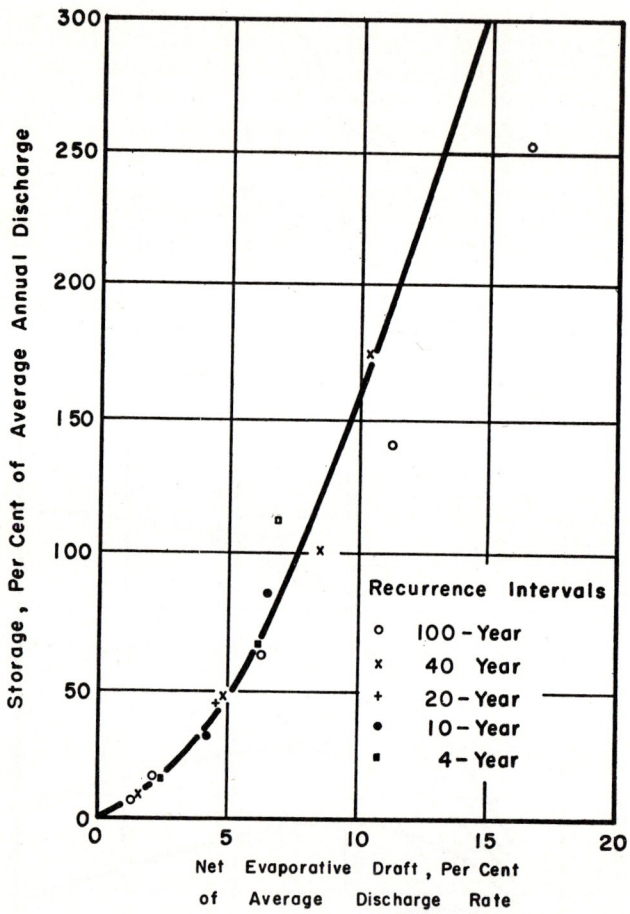

Fig. 18-19. Net evaporative draft from reservoirs in southern Illinois. (*After Hudson and Roberts* [12].)

VIII. OTHER RESERVOIR-DESIGN CONSIDERATIONS

In addition to the basic hydrologic aspects discussed above, factors of importance include practical considerations of sites and economics, effects of silting, and variations in demands. Practical considerations play an important and sometimes dominant role in the storage developments. Fluctuations in interest rates for financing may speed or halt a project. The nature of the dam and reservoir sites may have considerable influence, sometimes making it possible to provide much greater capacity at small cost. Excess reservoir capacity may be provided to obtain gravity flow as well as additional storage capacity. Uncertainty in prediction of future loads may dictate larger storage allowances. Construction costs have been rising steadily for many decades, and it may be a good investment to overbuild if it can be done with relatively low costs. Geomorphological and hydrogeological factors require examination.

Silting of reservoirs often unexpectedly speeds the approach of reservoir inadequacy. When the effects of silting combine with unexpected increases in draft rates, topped off by a drought, the result may be disastrous. Adequate allowance for silt storage is essential (Sec. 17-I), and periodic checks on silt accumulations are important in foreseeing coming deficiencies. Since loss of reservoir capacity due to silting is proportional to drainage area, there is a trend toward the development of smaller drainage areas with larger reservoirs. The use of these smaller watersheds and higher draft rates intensifies the effects of droughts and requires increased care in drought analysis.

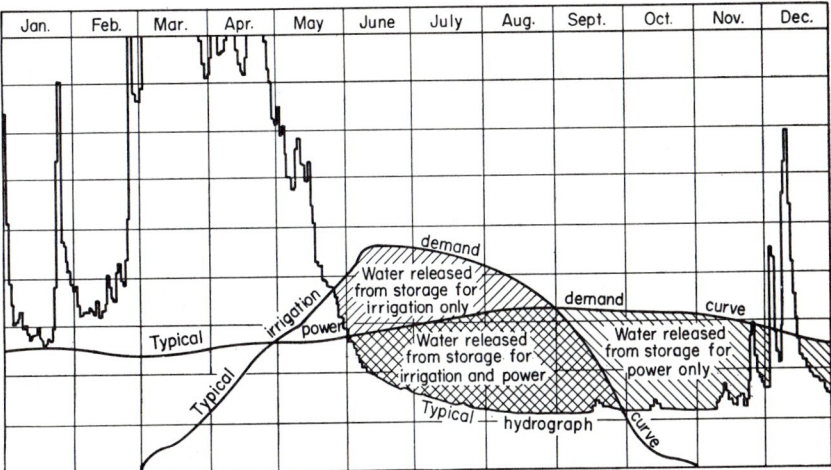

FIG. 18-20. Typical hydrograph of Sierra Nevada River with corresponding typical power and irrigation demand curves. (*After Galloway* [25].)

Seepage losses tend to be lower during droughts than during periods of normal streamflow, since the seepage driving force is reduced by reservoir drawdown. Seepage-loss estimates are difficult and require thorough knowledge of groundwater and geologic conditions. Ordinarily, seepage losses will be small, seldom as great as 10 per cent of the draft rate. Grouting, cutoff walls, bentonite blankets, or other means may be desirable to reduce seepage losses when they are substantial.

Variations in draft rate occur in many applications involving low-flow problems. They are of two types: short-term normal seasonal variances, usually typical of the particular draft purpose, and longer-term changes that may result from growth in need, or from deliberate manipulation of drafts, where this is possible. The short-term variances, often ignored, combine with high evaporative losses and low inflow to cause unexpected difficulty. A simplified approach to this problem is illustrated in Fig. 18-19 for evaporative-draft variances. Where difficulties may be severe, the effect should be checked by a detailed analysis of the critical period for the probable combination of drafts, flows, and losses. A dual-use project demand curve for power and irrigation [25] is shown in Fig. 18-20.

Unnoticed long-term increases in municipal water-supply demand commonly cause unexpected supply depletion. They can be offset by conservation measures such as investigation of transmission and distribution systems and repairs of leaks, restrictions on water use, and emergency use of other sources of supply. Such measures have reduced demands on reservoirs by as much as 50 per cent.

IX. RESERVOIR OPERATION

Once a reservoir is placed in operation, there is opportunity to gain further information on inflow and draft characteristics under conditions of high security. Most reservoirs are sized to meet a design draft rate that will not be reached for several decades; hence the reserve allowance for early years is high. If spillage is gaged, draft and stages recorded, and evaporation measured or estimated, inflow may be calculated. These records should be kept as an extension of the streamflow records used for design and as a basis for reservoir control.

Where streamflow is not extremely variable, analysis of annual low-flow periods, rather than exceedences, may be useful. Probability curves for flows of 1- to 12-months duration for the Croton River, New York, are shown in Fig. 18-21. Since drawdown of the Croton Reservoir usually begins June 1, periods of various lengths ending in June were used to facilitate forecasting the probability of recovery of storage [26, 27].

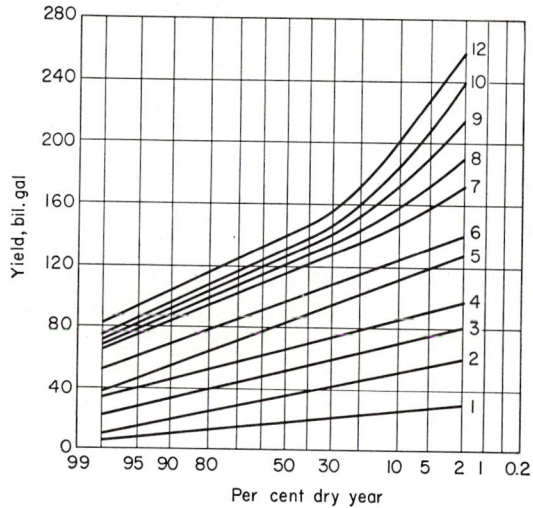

FIG. 18-21. Probability curves, Croton watershed, New York. (*After Clark* [27])

Control curves for New York City reservoirs, based on the conservative assumption that a reservoir must be full on the succeeding June 1, are shown in Fig. 18-22. These are based on the runoff data given in Fig. 18-21. An appropriate adjustment is made for *frozen storage* to take into account snowmelt and groundwater that is frost-locked in winter but released in spring. The draft rate used was aqueduct capacity, which was 375 million gallons per day, as compared with the *safe yield*, estimated at 330 million gallons per day. For each date and reservoir stage, the control curves show the probability that the reservoir will refill by June 1. The heavy lines in Fig. 18-22 show the actual storage situation during 1948 to 1950, when an extreme situation was encountered. Additional control curves were computed for even lower probabilities, and these showed that at one point, the probability of refilling one of the reservoirs by the next June fell below 0.10. The situation was more favorable in other reservoirs, and by shifting drafts, risks were evened out.

New York launched an intensive conservation campaign in December, 1949, which reduced the overall draft rate by approximately 30 per cent. Much of the recovery of reservoir levels was attributable to this reduced draft rate [26].

The effect of reduction in draft rate is illustrated in Fig. 18-23, which is based on the data given in Fig. 18-13. For this example, the control curves are based on duration periods that begin at the start of drawdown rather than on periods ending

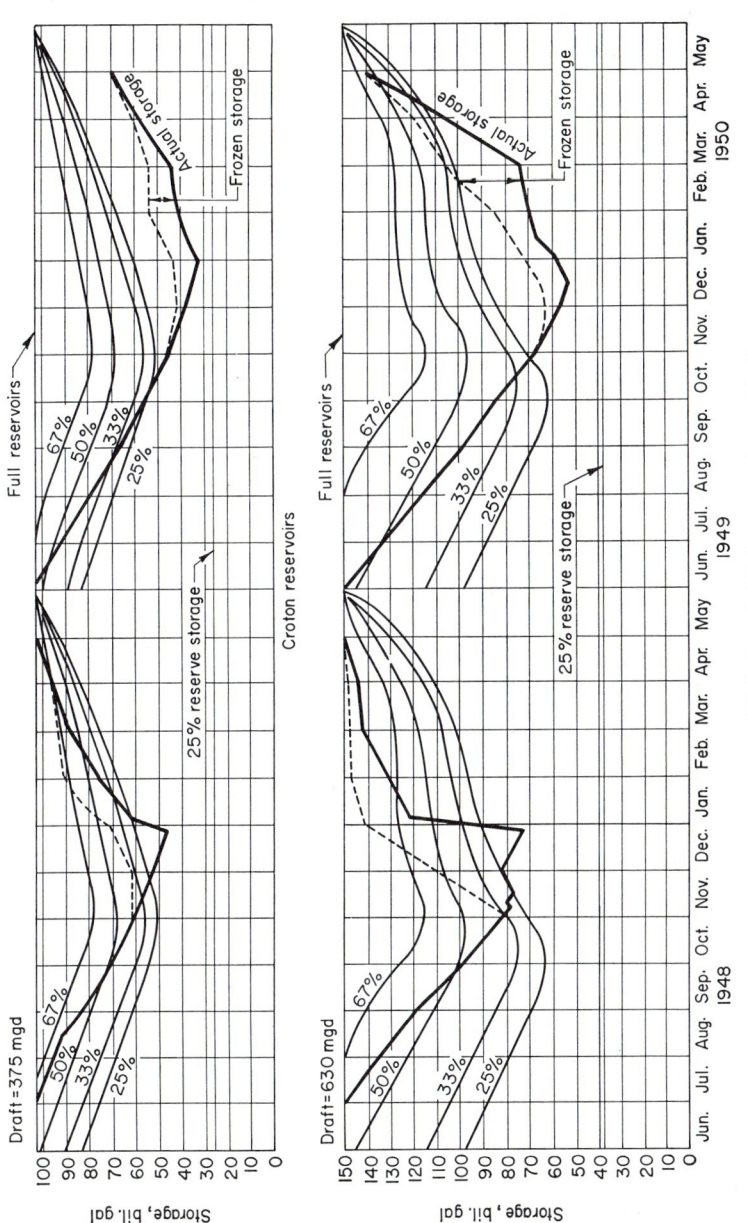

Fig. 18-22. Control chart for Catskill and Croton supplies, New York. (*After Clark* [27].)

18–21

at a certain date. This procedure enables calculations of the probability of ultimate reservoir recovery, rather than the probability of recovery by a stated date. Figure 18-23 shows calculated reservoir drawdowns for various probabilities with gross draft

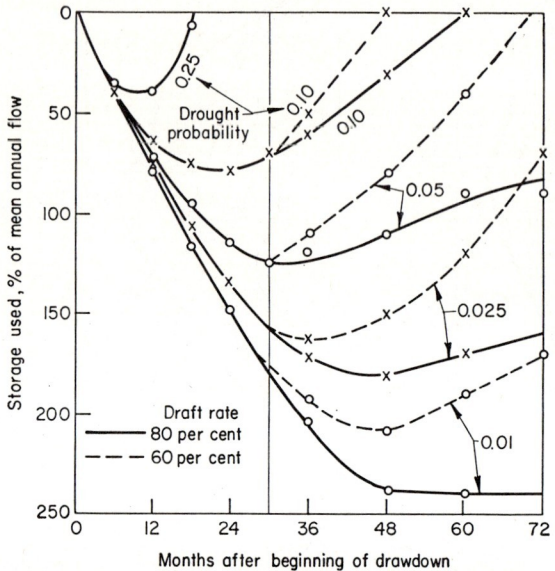

FIG. 18-23. Control curves for reservoirs in southern Illinois.

rate at 80 per cent of mean annual flow. It also shows the calculated effect of cutting the draft rate from 80 to 60 per cent after the drawdown had been going on for 30 months.

X. LOW-FLOW MAINTENANCE

The Wisconsin Valley Improvement Company, chartered in 1907, has built and operated reservoirs to reduce floods and to augment low flows for power generation. Stream-pollution abatement is now recognized as a valuable by-product. The Hudson River Regulating District was established in the state of New York in 1922 for flood control and river regulation. Stream-pollution abatement and the improved quality of raw water throughout the Hudson River are important benefits.

The Savage River reservoir on the headwaters of the north branch of the Potomac River, with a capacity of 6,600 million gallons, is used by the Upper Potomac River Commission primarily to increase dry-weather flows. The Commission is also engaged in waste-treatment plant construction and operation as an activity coordinated with flood control and low-flow regulation.

Low-flow maintenance is involved in storage studies and may take two forms: (1) outflows below a specified limit maintained equal to inflows, or (2) fixed minimum flow maintained. For small projects where downstream water rights are not established or are purchased, low flows may be cut off altogether. Schemes (1) and (2) are fitted to water-rights requirements, and downstream control flows are established. Figure 18-24a shows draft storage curves for a combination of water supply and low-flow maintenance, maintaining downstream flows equal to the natural inflows to the reservoir for the worst case on record (1930 to 1932) for the Hocking River. For comparative purposes, Fig. 18-24b shows additional storage requirements for maintenance of fixed minimum flows [28].

Drought frequency and severity data have direct applicability to pollution problems. From the low-flow data, oxidation capabilities of a stream and the degree of waste treatment required to meet various droughts may be determined. In one such instance, costs of waste treatment to meet droughts of various magnitudes were worked out, together with alternative possibilities that, in certain circumstances,

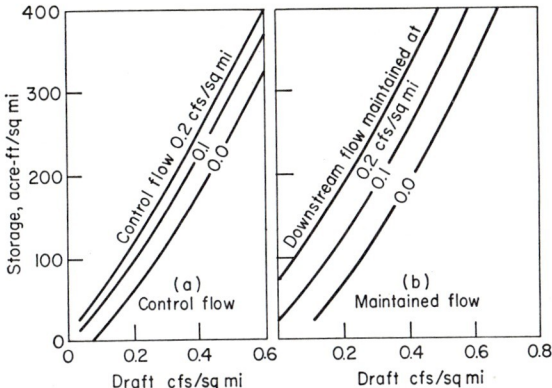

Fig. 18-24. Storage curves for the Hocking River at Athens, Ohio, based on driest period of record. (*After Hazen* [28].)

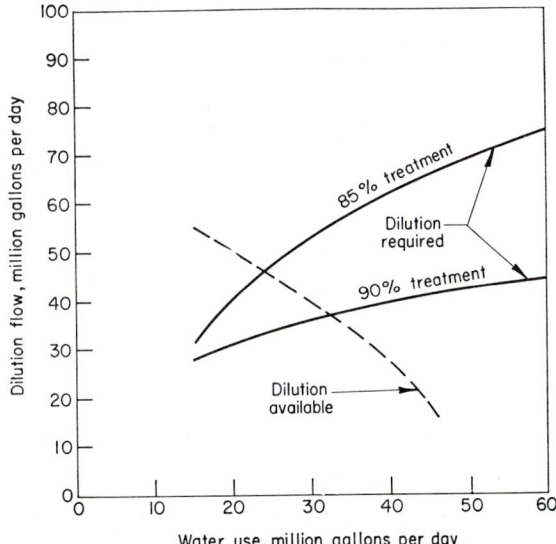

Fig. 18-25. Capability of impoundment for dual water-supply and sewage-dilution purposes at Greensboro, N.C.

reduction in industrial production might be justifiable, rather than costly treatment or low-flow maintenance, in order to meet low-flow oxygenation requirements [29].

Analysis of a dual-purpose project designed for water supply and low-flow maintenance is illustrated in Fig. 18-25. For the particular impoundment illustrated after making reserve allowances and calculating for a 20-year drought, the net yield for water supply was determined for various maintained low flows downstream. For several different sewage flows (taken equal to water-supply drafts), the dilution requirements downstream were determined for two degrees of sewage treatment. The resulting curves in Fig. 18-25 yield a unique solution for the maximum capability of the project for each degree of waste treatment. The intersection of the *dilution available curve* with the *dilution required curve* for sewage treatment effecting 90 per cent purification shows that the maximum feasible development of 69 million gallons per day will allow 33 million gallons per day for water use and require 36 million gallons per day for downstream dilution.

XI. DIVERSION WORKS

Diversion works may be considered for diverting the flow from one river to a reservoir or another drainage basin, or for diversion of flows into a side-channel reservoir by pumping. Ordinarily, it is not practical to divert all flows; required capacities

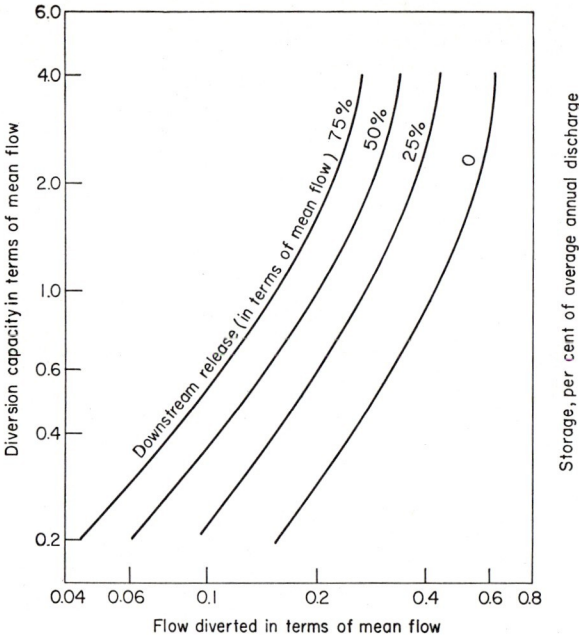

Fig. 18-26. Diversion curves, Wallkill River at Pellet's Island Mountain, New York, with no storage. (*After Hazen* [28].)

would be too great. During low-flow periods the available water may be very much less than the diversion capacity. Therefore, the works must be built to take the water when it is available up to reasonable limits. The capacity to be provided and the quantity of water that can be diverted can be determined by analyzing the daily flow records. In general, the variations from year to year follow a pattern, which is similar for streams in the same general area. If storage is provided at the diversion works, the effect of minor variations in flow can be eliminated.

Where downstream water users must be considered, the quantity of water that can be diverted is reduced. In Fig. 18-26 is shown the diversion capacity required in the 95 per cent year for diverting water from Wallkill River, a tributary of the Hudson River in New York State, based on the streamflow records from 1921 to 1935 [28]. These curves show not only the capacity required if there is no downstream discharge, but also how much additional capacity must be provided to take care of various downstream releases. It will be noted from these curves that increasing the diversion capacity above 2 or 3 times the mean flow does not increase appreciably the amount of water that can be diverted. In general, a diversion capacity 3 times the mean is about the economic limit, unless water is extremely valuable. The effect of releases downstream is very pronounced.

Figure 18-27 shows the effect of providing storage capacity at the diversion works [28]. The benefits of even a small amount of storage at the diversion works are apparent. It will be noted here also that a diversion capacity of about 3 times the mean flow makes it possible to obtain 80 per cent of the water and that further increases in capacity show a much lower return.

In estimating yields from these curves it is assumed that the diversion works would be operated to take all the water available up to their capacity. In other words, gates on a gravity canal would be kept wide open, and pumps would be operated

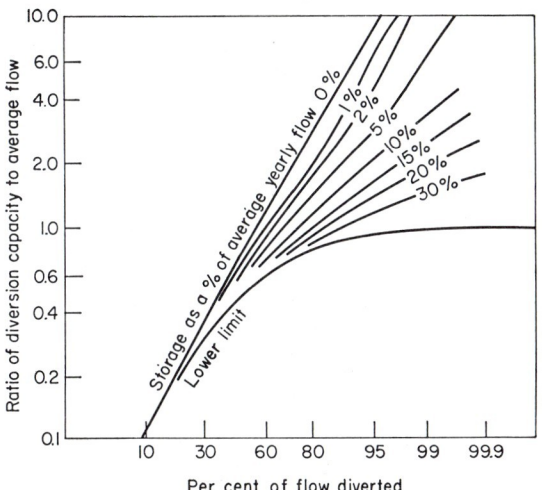

Fig. 18-27. Diversion curves, Manhan River, Massachusetts, 1897–1915. (*After Hazen* [28].)

continuously as long as water was available. It is also assumed that the receiving reservoir could store the water.

In planning diversions, adverse periods need to be investigated in detail. For Monticello, N.Y., for example, because of recreational considerations, diversion was confined to the winter months of October through May. In some situations mineral quality may prohibit diversion during periods of low flow, and diversion computations may have to be limited to periods of flow when acceptable quality is available.

XII. REFERENCES

1. Havens, A. V.: "Drought and Agriculture," *Weatherwise*, vol. 7, pp. 51–55, 68, 1954.
2. Hunt, H. J.: Droughts, chap. 12 in D. W. Mead (ed.), "Hydrology," McGraw-Hill Book Company, Inc., New York, and Mead and Hunt, Inc., Madison, Wis., 1950.
3. Tannehill, I. R.: "Drought: Its Causes and Effects," Princeton University Press, Princeton, N.J., 1947.
4. Thornthwaite, C. W.: Atlas of climatic types in the United States, 1900–1939, *U.S. Dept. Agr., Misc. Publ.* 421, 1941.
5. Hazen, Allen: Variation in annual rainfall, *Eng. News*, vol. 75, no. 1, pp. 4–5, Jan. 6, 1916.
6. Petterssen, Sverre: "Introduction to Meteorology," 2d ed., McGraw-Hill Book Company, Inc., New York, 1958.

7. Biel, E. R.: Veränderlichkeit der Jahresumme der Niederschläge auf der Erde (mit Weltkarte), *Geograf. Jahrb. Oesterr.*, pls. XIV, XV, Vienna, 1929.
8. Troxell, H. C.: Water resources of southern California with special reference to the drought of 1944–51, *U.S. Geol. Surv. Water-Supply Paper* 1366, 1957.
9. Langbein, W. B.: Queuing theory and water storage, *Proc. Am. Soc. Civil Engrs., J. Hydraulics Div.*, vol. 84, no. HY5, pt. 1, pp. 1–24, October, 1958.
10. Matalas, N. C.: "Statistical Analyses of Droughts," Ph.D. thesis, Howard University, Washington, D.C., 1958.
11. Kritsky, S. N., and M. F. Merkel: (a) Methods of quantitative estimation of lingering droughts on rivers, pp. 130–131, and (b) The utilization of water resources of rivers in arid zones, pp. 227–228, *Intern. Assoc. Sci. Hydrol. Publ.* 51, 1960.
12. Hudson, H. E., Jr., and W. J. Roberts: 1952–55 Illinois drought with special reference to impounding reservoir design, *Illinois State Water Surv. Bull.* 43, 1955.
13. Hoyt, W. G.: Droughts, chap. 12 in O. E. Meinzer (ed.), "Hydrology," vol. IX of "Physics of the Earth," Dover Publications, Inc., New York, 1949.
14. Furness, L. W.: Kansas streamflow characteristics, pt. 2, Low flow frequency, *Kansas Water Resources Board Tech. Rept.* 2, June, 1960.
15. Velz, C. J., and J. J. Gannon: "Drought flow characteristics of Michigan streams," Water Resources Commission of the State of Michigan, Lansing, Mich., June, 1960.
16. Hazen, Allen: Storage to be provided in impounding reservoirs for municipal water supply, *Trans. Am. Soc. Civil Engrs.*, vol. 77, pp. 1539–1667, 1914.
17. Hazen, Richard: Economics of stream flow regulation, *J. Am. Water Works Assoc.*, vol. 48, no. 7, pp. 761–767, July, 1956.
18. O'Connor, D. J.: Statistical analysis of drought flows of rivers in New York State, *New York State Water Pollution Control Board Res. Rept.* 1, June, 1959.
19. Schwob, H. A.: Low-flow characteristics of Iowa streams, *Iowa Nat. Resources Council Bull.* 9, 1958.
20. Gooch, R. S.: Application of the digital computer to reservoir yield studies, *Public Works*, no. 8, pp. 89, 91–92, August, 1958.
21. Huff, F. A., and Changnon, S. A.: Drought characteristics in a continental humid climatic region, *Intern. Assoc. Sci. Hydrol. Publ.* 51, pp. 22–33, 1960.
22. Mitchell, W. D.: "Water-supply Characteristics of Illinois Streams," Illinois Division of Waterways and U.S. Geological Survey, 1950.
23. Brush, W. W.: Safe yield from surface storage reservoirs: summation, panel discussions, *J. Am. Water Works Assoc.*, vol. 42, no. 9, p. 838, September, 1950.
24. Metzler, D. F.: Recommended action against effect of severe droughts in Kansas, *J. Am. Water Works Assoc.*, vol. 48, pp. 999–1004, August, 1956.
25. Galloway, J. D.: Discussion of Stream regulation with reference to irrigation and power by J. C. Stevens, *Trans. Am. Soc. Civil Engrs.*, vol. 90, pp. 989–992, 1927.
26. Clark, E. J.: Controlled draft from reservoirs—impounding reservoirs, *J. Am. Water Works Assoc.*, vol. 48, pp. 349–354, April, 1956.
27. Clark, E. J.: Safe reservoir yield—New York control curves: panel discussions, *J. Am. Water Works Assoc.*, vol. 42, pp. 823–827, September, 1950.
28. Hazen, Richard: Analysis and use of surface water data, Proceedings of 1951 Conference on Water Resources, *Illinois State Water Surv. Bull.* 41, 1952.
29. Velz, C. J.: Utilization of natural purification capacity in sewage and industrial waste, *Sewage Ind. Wastes*, vol. 22, pp. 1601–1613, December, 1950.

Section 19

QUALITY OF WATER

SHEPPARD T. POWELL, *Consulting Engineer, Baltimore, Maryland.*

I. Introduction	19-2
II. Constituents and Properties of Water	19-3
A. Cations	19-3
1. Calcium and Magnesium	19-3
2. Sodium	19-4
3. Potassium	19-4
4. Iron and Manganese	19-5
5. Other Cations	19-5
B. Anions	19-5
1. Chloride	19-5
2. Sulfate	19-6
3. Bicarbonate, Carbonate, and Hydroxide	19-6
4. Nitrate	19-7
5. Other Anions	19-7
C. Nonionic Constituents	19-7
1. Oxides (Silica)	19-7
2. Oily Substances	19-8
3. Phenols	19-8
4. Synthetic Detergents	19-8
5. Dissolved Gases	19-8
6. Aquatic Flora and Fauna	19-9
D. Properties	19-9
1. Hardness	19-9
2. Alkalinity	19-10
3. Acidity	19-10
4. Hydrogen-ion Concentration (pH)	19-10
5. Specific Electrical Conductance	19-11
6. Color	19-11
7. Turbidity	19-11
8. Taste and Odor	19-11
9. Oxygen Demand	19-11
10. Per Cent Sodium	19-12
11. Sodium-adsorption Ratio (SAR)	19-12
12. Radioactivity	19-12
13. Density	19-12
14. Temperature	19-13
III. Water-quality Determination	19-13
A. Sampling Techniques	19-13
B. Analytical Procedures	19-14

  C. Units of Expression.................................. 19-15
   1. Weight per Weight............................... 19-15
   2. Weight per Volume.............................. 19-15
   3. Equivalents per Weight........................... 19-15
   4. Calcium Carbonate Equivalents..................... 19-16
   5. Hypothetical Combinations........................ 19-17
   6. Miscellaneous Units.............................. 19-17
 IV. Water-quality Characteristics............................ 19-17
  A. Surface Water..................................... 19-17
   1. Rivers.. 19-18
   2. Lakes and Reservoirs............................. 19-20
   3. Estuaries....................................... 19-21
   4. Oceans... 19-22
  B. Groundwater...................................... 19-23
 V. Water-quality Deterioration............................. 19-23
  A. Natural Pollutants.................................. 19-23
  B. Man-made Pollutants............................... 19-24
   1. Municipal Wastes................................ 19-24
   2. Industrial Wastes................................ 19-25
   3. Agricultural Wastes.............................. 19-25
  C. Waste-water Reclamation and Re-use................. 19-25
  D. Waste-assimilative Capacity......................... 19-25
 VI. Water-quality Requirements for Specific Uses............. 19-27
  A. Domestic and Municipal Water Supply................ 19-27
  B. Industrial Water Supply............................. 19-28
   1. Food and Beverage Industries...................... 19-28
   2. Water for Textile and Paper Manufacture............ 19-28
   3. Pharmaceutical and Antibiotic Requirements......... 19-29
   4. Boiler Feedwater................................ 19-29
   5. Cooling Water.................................. 19-30
   6. Water-quality Requirements for Internal-combustion
    Engines... 19-30
   7. Transporting and Process......................... 19-30
  C. Agricultural Water Supply........................... 19-31
  D. Aquatic Life....................................... 19-31
  E. Recreation.. 19-32
 VII. Water Conditioning for Selective Uses.................... 19-32
  A. Conventional Methods.............................. 19-32
  B. Demineralization of Saline Waters.................... 19-32
VIII. Sources of Water-quality Data............................. 19-35
 IX. References... 19-35

I. INTRODUCTION

 The term "quality" as applied to water embraces the combined physical, chemical, and biological characteristics of the most abundant compound on the surface of the earth. Water is vital to the existence of all life as we know it and is essential, either directly or indirectly, to almost all activities of man. Water quality is a dominant factor in determining the adequacy of any supply to satisfy the requirements of these uses.

 Water is never found in its pure state in nature. Essentially, all water will contain substances derived from the natural environment or from the waste products of man's activities. These constituents are basic criteria in the determination of water quality. In addition, the various properties of water imparted by the constituents serve further to define the quality of water.

Water quality is determined by analytically measuring the concentrations of the various constituents and the effects, or properties, caused by the presence of these substances. In determining the quality of a natural water-supply source, the procedures used in sampling the supply are very important. The units of expression of water quality are significant, too, since there are possibilities of confusion and misinterpretation of water-quality data.

Each major type of water-supply source has certain water-quality characteristics which are valuable in the preliminary and formative phases of the development of a supply. The outstanding characteristics of surface waters, such as rivers, lakes, reservoirs, estuaries, and the oceans, and groundwaters should be recognized.

Extreme water-quality deterioration resulting from pollution is a vital concern of those responsible for the application of water for various purposes. The principal sources of pollution should be recognized. Increased attention is being given to the reclamation and re-use of waste waters and the ability of receiving waters to assimilate wastes. These factors, too, are important, and hydrologists should be cognizant of their implications.

Information about the water-quality requirements for various beneficial uses is important in the evaluation of a water-supply source. The available treatment methods which must be provided to render an unsuitable water supply satisfactory for the intended purpose should be known to the worker in the field of hydrology. An outline of the various sources of water-quality data can be a helpful guide.

The information presented here is a guide to the overall evaluation of water quality. In many cases it is not feasible to consider the topic in detail and there is no alternative but recourse to the specific references cited.

II. CONSTITUENTS AND PROPERTIES OF WATER

Water quality embraces the individual and combined effects of the substances present. The utility of a supply to serve various uses is determined, to a large extent, by the constituents found in the water, as well as its properties.

It is not feasible in a text of this type to discuss all the constituents and properties of water. A great many constituents are associated only with those waters which pass through very unusual geological formations or receive wastes from various industrial processes. In addition, a number of properties have little or no effect on water quality. Only those constituents and properties which significantly affect the application of water to beneficial uses are considered. Additional information on these and other water-quality factors is available in the literature [1–6].

A. Cations

The cations (positively charged ions) usually determined in routine analyses of natural waters include calcium, magnesium, sodium, and potassium. These substances are generally present in all natural waters and in many instances make up almost all the cations. Because of their importance, each will be considered subsequently in more detail.

Other cations, usually present in small or in trace concentrations, include:

Aluminum	Nickel	Cobalt	Strontium
Iron	Copper	Arsenic	Barium
Manganese	Tin	Selenium	Beryllium
Titanium	Lead	Cadmium	Lithium
Chromium	Zinc	Antimony	

1. Calcium and Magnesium. These two cations are similar in many respects and are both dissolved freely from many rocks and soils. Calcium is an abundant constituent in waters which are associated with calcite, dolomite, and gypsum. Sources of magnesium include dolomite, magnesite, and micas.

Concentrations of calcium and magnesium found in natural waters vary over a wide range, depending on the history of the water sample with respect to geological formations and the geochemical characteristics of the water. In general, concentrations of calcium in fresh waters are somewhat greater than those of magnesium, because of the greater abundance of calcium in the earth's crust. Magnesium concentrations, however, may often be greater than the calcium concentrations, especially in oceanic or estuarine waters.

Hardness, a property of water historically characterized by the formation of insoluble salts of the fatty acids found in soaps, is attributable principally to the presence of calcium and magnesium ions. Scales and incrustations associated with hardness in water are primarily caused by the formation of salts of calcium and magnesium upon evaporation or heating.

Some of the effects of calcium and magnesium as they relate to hardness are considered later. Calcium and magnesium are both essential elements in man's diet. In addition, both of these substances are generally desirable in limited quantities for most beneficial uses of water. The U.S. Public Health Service recommends that magnesium not exceed 50 parts per million (ppm) in water to be used for drinking. This is based primarily on the laxative effects of high magnesium concentrations on persons not physiologically adapted to the supply.

2. Sodium. Sodium is the most important and abundant of the alkali metals in natural waters. Nearly all sodium compounds are readily soluble, and since it does not take part in significant precipitation reactions, sodium generally remains in solution.[1]

The principal source of sodium in water is the evaporate sediments, although sewage, industrial wastes, and oil-field drainage also contribute significant amounts.

Concentrations of sodium vary widely, depending upon the origin of the water. Essentially all waters contain some sodium, ranging from a few parts per million where leaching has removed most soluble rock minerals up to 100,000 ppm where water is found in contact with evaporate beds. In sea water, where sodium is the most abundant cation, its concentration is about 10,000 ppm.

Sodium salts are abundant in nature and are commonly used in many industrial processes. Some of these compounds, in excessive concentrations, are harmful to man and may be injurious to fish and other aquatic life. Probably the most significant effect of high sodium concentrations in water is related to irrigation. Such water, when used to irrigate soils containing exchangeable calcium and magnesium ions, will result in base-exchange reactions in which the sodium replaces these cations. This leads to the creation of alkali soils which impair the value for agricultural uses.

In many analyses, a computed value for sodium is given. It is assumed that the difference, in equivalents per million, between the determined anions and cations is equal to the amount of sodium and potassium present. This computed value for the two cations is then expressed as an equivalent amount of sodium. The possibilities of gross errors in the computed value are great, especially when the dissolved solids concentration of water is high. The proper analytical techniques should be used to secure reliable information.

3. Potassium. Potassium, another of the alkali-metals group, is similar in many respects to sodium, although it differs significantly in several factors which relate to its presence in water and its effects on other uses. While sodium is readily available for solution, rocks containing significant amounts of potassium are generally resistant to weathering, and evaporates containing large amounts of potassium are uncommon. In addition, several processes tend selectively to return potassium to the solid phase by refixing it in new minerals which resist attack. Thus potassium is normally present in water in smaller amounts than sodium. Concentrations of potassium in most natural waters are generally less than 10 ppm.

Potassium is an essential nutrient for both plant and animal life. Traces of this element are desirable for plant life, but excessive concentrations are harmful, and it is recognized that this constituent should be balanced with the other mineral require-

---

[1] Sodium salts can be completely removed by passage of water through special types of ion-exchange resins.

ments. Potassium is usually of little significance with regard to the creation of alkali soils, since the concentrations are relatively minor.

As pointed out in the prior discussion of sodium, potassium is commonly given as a calculated value expressed in terms of an equivalent amount of sodium. Analytical methods for determining potassium are available and should be used when possible.

4. Iron and Manganese. Although both iron and manganese are usually present in only minor concentrations, these two constituents are of major significance with respect to water-quality criteria for domestic and industrial supplies. Iron is much more abundant in rocks than is manganese and therefore is usually found in greater concentrations in natural waters.

Both iron and manganese normally occur in water at two levels of oxidation. Iron may be present as the bivalent (Fe^{++}) ferrous iron or the trivalent (Fe^{3+}) ferric iron. Manganese generally occurs in the bivalent (Mn^{++}) manganous form or the quadrivalent (Mn^{4+}) manganic form. Both are readily oxidized to the ferric and manganic forms, usually hydroxides, which are relatively insoluble. In the oxidized forms, the solubility of the metals is largely controlled by the pH[1] of the solution.

Surface waters, which are usually alkaline[1] and nearly saturated with dissolved oxygen, contain much lower concentrations of dissolved iron and manganese than those containing little or no oxygen. Groundwater frequently contains significant amounts of these metals in solution. Acid-mine drainage contains relatively high concentrations of iron in both the ferrous and the ferric state. Iron and manganese are often dissolved in the bottom waters of deep, stratified lakes and reservoirs. Microorganisms are known to be active in the solubility relationships of these metals, especially iron. The U.S. Geological Survey [7] has published detailed works on the chemistry of iron in natural waters.

Both metals are objectionable in domestic and industrial water supplies, and desirable threshold values for most uses have been well established. The U.S. Public Health Service Drinking Water Standards [8] recommend maximum concentrations for iron and manganese of 0.3 and 0.05 ppm, respectively.

Iron and manganese are generally reported as parts per million of the ions. Unless the water sample is quite acid, concentrations are not reported as chemical equivalents since much of hydroxide may be in the colloidal state. It is therefore not dissociated and does not enter into the cation-anion balance.

5. Other Cations. There are several cations which are significant with respect to water. Most of these are found in waters receiving wastes from various industrial establishments, and also in nature at some locations.

Chromium, selenium, arsenic, barium, cadmium, and lead have been given mandatory limits in drinking water by the 1962 U.S. Public Health Service Drinking Water Standards [8]. In addition, recommended limits have been established for copper and zinc. Most of these specific constituents may be harmful to crops, stock, wildlife, fish, and other aquatic life, as well as to human beings and in industrial use, if present in sufficiently high concentrations.

B. Anions

The negatively charged ions, or anions, usually determined in routine water analyses are chloride, sulfate, bicarbonate, carbonate, hydroxide, and nitrate. In most natural waters these constituents account for essentially all the anions present.

Other anions, usually present in small, or trace, concentrations, include:

    Fluoride   Boron
    Nitrite   Iodide
    Bromide   Cyanide
    Phosphate

1. Chloride. Chlorides are present in all natural waters. Sources of this anion are numerous and include many sedimentary rocks, particularly the evaporates, salt

---

[1] Terms such as pH and alkaline are discussed in subsequent portions of this section.

"seeps," oil-field drainage, domestic and industrial contamination, and, to some degree, air-borne matter resulting from ocean spray. High chlorides in both surface and groundwater are often due to contamination from ocean water and many brackish supplies.

Concentrations of chloride in natural waters vary widely. During high and normal flow periods, the rivers of humid regions will generally have concentrations less than a few parts per million. In some arid regions the chloride concentration may be several hundred parts per million of chloride. Well water may contain chloride concentrations of a similar range. Significant increases are common in water obtained from coastal wells which are overdrawn, since this lowers the water table and allows encroachment of the high-chloride sea water. This occurs also in many tidal streams during drought periods. The concentration of chloride in sea water is about 19,000 ppm.

Excessive chloride concentrations are undesirable for many uses of water. The U.S. Public Health Service [8] recommends that chloride not exceed 250 ppm in drinking water. Although chloride may be harmful to some persons, limitations are generally based on palatability considerations and corrosive effects on various domestic and municipal equipment. Many industries limit chloride concentrations to prevent damage to manufacturing equipment and various industrial processes. Very high concentrations of chloride are harmful in irrigation water applied to less resistant crops and in water fed to stock.

2. Sulfate. In most natural waters, sulfate is found in smaller concentrations than chloride. The sulfate ion is the fully oxidized form of sulfur and is derived principally from evaporate sediments such as gypsum and anhydrite. Sulfate may also result from the oxidation of the reduced forms of sulfur, such as sulfide and sulfite, or from the wastes of various industries.

Many sulfates are readily soluble in water, and once dissociated, the sulfate ion is relatively stable. Certain cations, namely, barium, strontium, and lead, form fairly insoluble sulfate salts, and significant concentrations of sulfate are unlikely when these cations are present. Sulfates may be very high in some well waters and in surface waters of arid regions where sulfate minerals are present.

There are some uses of water where moderately high sulfate concentrations (200 to 300 ppm) are highly objectionable. The U.S. Public Health Service [8] recommends that sulfates in drinking water not exceed 250 ppm, although public water supplies having concentrations in excess of this value are not uncommon. Some beverage industries and the ice-producing industry limit the sulfate concentration in their water supplies. Sulfates are often directly related to concrete corrosion, but there are several other factors which influence the corrosive action of sulfate-bearing waters on concrete structures.

3. Bicarbonate, Carbonate, and Hydroxide. These three anions may be conveniently grouped for purposes of discussion. Together they contribute essentially all the alkalinity, or acid-neutralizing power, of water. Other constituents, namely, borate, phosphate, silicate, and anionic organic matter, contribute alkalinity, but these are usually of little significance. In practice, it generally is assumed that bicarbonate, carbonate, and hydroxide contribute to the alkalinity of water, and the amounts of each are determined from the amounts of acid required to lower the pH value to characteristic end points (see alkalinity, Subsec. II-D-2).

Natural water seldom contains hydroxide, although it may be present in waters receiving highly alkaline wastes from various industries or in waters associated with fresh concrete. Water softening by the excess lime-soda process will result in hydroxide in the finished water.

Most waters contain bicarbonate and carbonate, except those which are strongly acid (pH below about 4.5). The relative amounts of these two anions depend on the pH of the water and other factors, with bicarbonates increasing as the pH decreases. Most natural waters (pH of 7.0 to 8.0) will contain much more bicarbonate than carbonate.

Bicarbonate, carbonate, and hydroxide are not, in themselves, generally regarded as undesirable for municipal supplies, but are highly objectionable for many industrial supplies. The detrimental characteristics of bicarbonate, carbonate, and hydroxide are usually considered in light of their association with alkalinity and hardness.

4. Nitrate. The principal form of nitrogen in water is in the completely oxidized state as nitrate. Other forms of nitrogen are important in sanitary analyses of waters containing organic wastes.

The primary source of nitrates in water is the end product of the aerobic stabilization of substances containing organic nitrogen. The use of fertilizers may add nitrates to the drainage from the treated areas. In surface waters, the nitrates are quickly returned to the organic nitrogen state by the photosynthetic action of aquatic plants, both phytoplankton[1] and rooted plants. The presence of nitrates in excess of about 5 ppm in surface water may be indicative of organic pollution. In the evaluation of such a situation, some of the other forms of nitrogen are also determined to aid in the interpretation of the nitrate concentrations.

Groundwaters may contain very little nitrate or, in some cases, as much as several hundred parts per million. Here again, organic pollution may be the primary source of nitrates. In some instances, the continued use of fertilizers will contribute significant amounts of nitrate to surface water or groundwater. Since photosynthetic activity does not occur in groundwaters, the nitrates are not returned to the organic nitrogen phases.

Nitrates in relatively high concentrations are generally undesirable in drinking water and in many industrial water supplies. The U.S. Public Health Service Drinking Water Standards [8] recommend that nitrate concentrations not exceed 45 ppm. Limited concentrations may be beneficial for fish propagation by improving the growth of food organisms. Excessive amounts may lead to prolific and objectionable growth of aquatic plants. Irrigation water containing nitrates generally stimulates plant growth because of the fertilizer value imparted by this constituent.

5. Other Anions. Few of the other anions found in water are of great significance. Some, such as bromide, nitrite, phosphate, and cyanide, may be found in waters receiving municipal and industrial wastes. Bromide is not common, although it may be found in wastes from the pharmaceutical industry and in groundwater which is influenced by oceanic or other highly saline waters. Nitrite in surface water is quickly oxidized to the nitrate form, which is then utilized in the photosynthetic processes. Phosphates, like nitrates, are utilized by plants and are usually not found in significant concentrations in surface waters.

Boron is a very important constituent in irrigation water supplies. Concentrations of only a few parts per million, or less for some crops, may cause significant damage. Extensive studies of the effects of boron in irrigation waters have been made by various groups, notably the U.S. Department of Agriculture [9].

Another significant anion sometimes found in minor concentrations is fluoride. In some groundwaters, and occasionally in surface waters, fluoride concentrations may be as high as 15 ppm or more. Generally, however, fluoride concentrations are less than 1 ppm.

Fluorides are regarded by many as a beneficial constituent in drinking water, in concentrations which do not exceed about 1.0 ppm, since evidence indicates that this anion aids in the reduction of dental decay. Excessive concentrations, however, may be quite harmful in drinking water, and the U.S. Public Health Service has prescribed limits on the fluoride concentration [8].

C. Nonionic Constituents

Not all the constituents in water are present in dissociated form as ions. Some of the nonionic substances present in water supplies significantly affect the general quality of a supply.

1. Oxides (Silica). Silicon dioxide (SiO_2), or silica, is an oxide of silicon commonly found in natural waters. Silica, although quite insoluble in natural water, may be fairly readily dissolved or occur as finely divided colloidal matter originating from silicate rocks. Waters passing through volcanic deposits may have silica concentra-

---
[1] Free-drifting plant life, mainly algae.

tions on the order of 100 ppm or higher,[1] although most natural waters have concentrations less than 40 ppm.

From the standpoint of potability and general water quality for domestic and municipal uses, silica is not a significant constituent. It is, however, undesirable in many industrial supplies, especially in boiler feedwater. It forms very hard deposits on boiler tubes and, at high concentrations, tends to "carry over" with the steam and deposit on the turbine blading. As the operating pressure of the boiler increases, the allowable silica concentration in the feedwater decreases.

Silica is generally reported as the oxide (SiO_2) in concentration units. Since it is not in ionic form, it should not be reported in equivalent weight units.

2. Oily Substances. In general, oily substances, which may include oil, fat, grease, wax, and other similar materials, are not present in natural, unpolluted waters. They are present, however, in many specific areas where municipal or industrial wastes are discharged to the watercourses. Many groundwater supplies in oil-field operation are frequently polluted by oil.

Oily substances may be obtained from activities involving the production, transportation, and processing of oils, as well as municipal and industrial effluents which are not adequately treated. The California Water Pollution Control Board [10] has published an excellent report on the presence of oily substances in water, which should be consulted for additional information.

3. Phenols. Natural, unpolluted waters rarely contain phenol or phenolic compounds. Such material may be present in water which receives industrial wastes. Chemical plants, gas works, oil refineries, and coke ovens are typical sources of phenolic material.

Phenols are readily soluble in water to high concentrations. Since such materials are highly objectionable in water, pollution-control agencies usually place strict limitations on the discharge of wastes containing phenols.

The strong, characteristic tastes and odors associated with phenols make waters containing much of these substances unsuitable for many food and beverage industries. Chlorinated drinking water is especially distasteful when phenols are present. The 1962 U.S. Public Health Service Drinking Water Standards [8] recommend a limit for phenolic compounds of 0.001 ppm (one part per billion) as phenol.

4. Synthetic Detergents. The rapid rise in the use of synthetic detergents in the past decade has resulted in significant amounts of these surfactant[2] substances in waste-receiving streams. There are many varieties of synthetic detergents in use today for general domestic cleaning and some industrial applications. One of the most commonly used is alkyl benzene sulfonate (ABS).

Waters receiving municipal and some industrial wastes may be expected to contain these materials. Although the self-purification powers of streams will gradually reduce the concentrations of synthetic detergents, they have a tendency to persist and may be found well downstream of the waste source. Some groundwaters may also be contaminated with synthetic detergents. The U.S. Public Health Service recommends that synthetic detergents, as ABS, not exceed 0.5 ppm in drinking water.

5. Dissolved Gases. The most significant dissolved gases present in natural waters are oxygen and carbon dioxide. Nitrogen is present in most waters, but it has little significance with respect to general water quality. Other gases sometimes found include ammonia compounds, methane, hydrogen sulfide, and minute quantities of the rare gases. Most of these less significant gases are not commonly found in natural, unpolluted waters, although methane and hydrogen sulfide may be present in swamp water and in deep-well supplies which contain no oxygen.

Oxygen in water is derived principally from the atmosphere. Another significant source of oxygen in surface waters may be the photosynthetic processes of aquatic plants. Prolific algal growths are responsible for the creation of oxygen concentrations in excess of the normal solubility (supersaturation).

Like all gases, the solubility of oxygen varies directly with pressure and inversely

---

[1] High concentrations of soluble silica are found in glacier water in Iceland, Chile, and other areas.

[2] Surface active agent.

with temperature. At normal atmospheric pressure and a temperature of 70°F, fresh water is capable of dissolving about 9.0 ppm of oxygen. The solubility decreases slightly with increased concentrations of dissolved solids and with increase in temperature. It should be pointed out that there is, at present, an unresolved question regarding oxygen-solubility–temperature relationships. Recent work in England [11] indicates saturation concentrations to be as much as 4 per cent lower than the commonly accepted values [4] shown in Fig. 19-1. Current research in several countries should provide data to resolve the differences.

Groundwater, particularly from deep wells, will usually contain little or no dissolved oxygen. Upon exposure to air, oxygen is readily absorbed.

Dissolved oxygen is vital to fish and other aquatic life and is generally not objectionable for municipal uses and some industrial uses. However, the presence of oxygen in water greatly accelerates the rate of corrosion of metals. The corrosive action of dissolved oxygen in water depends upon a number of factors, among which are pH and temperature. Great care must be taken to remove oxygen from boiler feedwater by conventional equipment, such as deaerating heaters and cold-water vacuum deaerators, with supplemental removal by deoxygenating chemicals.

The presence of free carbon dioxide in water is related to the pH and alkalinity of the solution. Surface waters generally contain minor concentrations of carbon dioxide, while occasionally highly mineralized groundwaters may contain several hundred parts per million.

6. Aquatic Flora and Fauna. A complete discussion of the plants and animals which may be found in water is well beyond the scope of this section. Some mention of them is essential, however, since they influence directly and indirectly the quality of water. Aquatic flora and fauna affect water quality by their very presence, in addition to the influence of the by-products of their photosynthetic and biological processes.

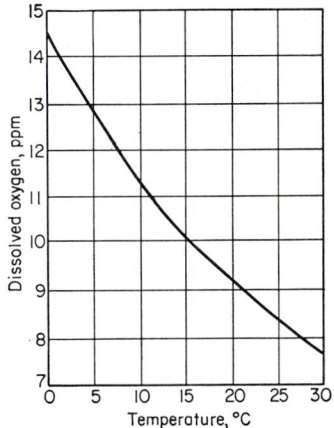

Fig. 19-1. Solubility of oxygen in fresh water.

Many beneficial uses of water are significantly affected by the presence of certain specific plants and animals or prolific growths of common aquatic life. Any evaluation of a water-supply source should include an investigation to determine the potential effects of the aquatic life in the supply on the uses of the water.

D. Properties

Many of the constituents of water, both ionic and nonionic, impart certain quality characteristics which may be termed *properties*. The most significant properties of water which relate to quality are considered here.

1. Hardness. One of the most important properties of water is *hardness*. Historically, hardness has been characterized by the formation of insoluble salts of the fatty acids found in soaps and by the deposition of scale on heated surfaces. For years, the standard analytical method for determining hardness was by the addition of a standard soap solution to a sample until a persistent lather was obtained. This method is not now generally used for precise water measurement, but is widely used for water-softening control procedures. Hardness is due to the presence of calcium and magnesium in the water, although other cations, such as iron, manganese, aluminum, zinc, and strontium, also react as hardness.

Hardness of water generally indicates the sum of the calcium and magnesium expressed as an equivalent amount of calcium carbonate ($CaCO_3$). If unusual amounts of the other hardness-producing cations are present, they should also be included.

Hardness may be calculated from the results of a complete mineral analysis, or it may be determined by a compleximetric titration using EDTA (ethylenediaminetetraacetic acid and its sodium salts). This procedure is described in Standard Methods for the Examination of Water and Wastewater [4]. The method used should be cited, and if a calculated value is given, the cations used in the calculation should be noted.

The total hardness is often divided into carbonate (temporary) and noncarbonate (permanent) hardnesses. When the carbonate and bicarbonate alkalinity is equal to, or greater than, the total hardness, all the hardness is estimated as *carbonate hardness*. If the total hardness exceeds the carbonate and bicarbonate alkalinity, the excess is considered *noncarbonate hardness*. The principal anions associated with noncarbonate hardness are sulfate, chloride, and nitrate. The terms *temporary* and *permanent* hardness are the terms which were used in the past, but have been largely abandoned in modern usage and are now identified as carbonate and noncarbonate hardness.

Water supplies vary widely in their hardness characteristics. The U.S. Geological Survey has summarized hardness data for the United States [12]. Because of the wide ranges of hardness encountered, terms such as "hard," "soft," and "moderately hard" should not be used, except for a specific location or when the terms have been quantitatively defined.

Although a limited amount of hardness is not objectionable for domestic and municipal uses, and some industrial uses, excessive amounts are undesirable. Hardness limits for municipal uses vary widely, and the objectionable characteristics are related to nuisances and often costly maintenance problems. Limits are usually based on public acceptance and the general quality of the supplies in the region. Industries have generally established hardness limitations, most of which are considerably below that which would be acceptable for municipal use. Many industries and some municipalities soften their water supplies to prevent objectionable effects and high cost of commercial or industrial operations.

2. Alkalinity. The ability of water to neutralize acid is termed *alkalinity* and is due primarily to the presence of hydroxide, bicarbonate, and carbonate. The presence of borate, phosphate, silicate, and other ionic constituents imparts additional alkalinity to water, but these are usually of little significance except in certain areas. Like hardness, alkalinity is usually expressed as an equivalent amount of calcium carbonate ($CaCO_3$).

In the titration of alkalinity, the pH of the sample is reduced to characteristic end points of 8.2 and 4.5, which signify the conversion of the carbonate to bicarbonate ions and, in turn, the bicarbonate ions to carbonic acid. The amounts of acid, expressed as calcium carbonate, required to neutralize the alkaline salts and reach the end points indicate the amounts of hydroxide, carbonate, and bicarbonate which were present in the sample [1, 4]. A pH value of 7.0, which indicates the neutrality point between acidic and basic solutions, has no special significance in the determination or expression of alkalinity.

Excessive alkalinity is undesirable in water for some uses, mainly because of its association with waters having excessive hardness or high concentrations of sodium salts. If the alkalinity is equal to, or greater than, the hardness (both expressed as calcium carbonate equivalents), all the hardness will be present as *carbonate hardness*. When the alkalinity is less than the hardness, the excess will be in the form of *noncarbonate hardness*.

3. Acidity. Water which has a pH below about 4.5 has a property termed *acidity*. Acidity is caused primarily by the presence of free mineral acids and carbonic acid.

Surface waters seldom contain any acidity except where acid-mine drainage flows to the stream or industrial wastes are discharged without proper neutralization. Groundwaters may be acidic because of high concentrations of uncombined carbon dioxide.

Like alkalinity, acidity is expressed as an equivalent amount of calcium carbonate.

4. Hydrogen-ion Concentration (pH). The negative logarithm to the base 10 of the hydrogen-ion concentration of an aqueous solution is termed the *pH value*. A pH value of 7.0 denotes a neutral water, or one in which there is a balance between dissociated hydrogen and hydroxyl ions. An excess of hydrogen ions indicates an acid solution with a corresponding pH value less than 7.0. Conversely, an excess of hydroxyl ions indicates a basic solution which has a pH value greater than 7.0. The full pH scale is from 0 to 14.

Most natural waters are within one or two units of neutrality, with groundwaters usually on the acid side (pH less than 7) and surface waters usually on the basic side (pH greater than 7). In addition, most natural waters are buffered solutions which resist changes in pH upon the addition of acids or bases.

5. Specific Electrical Conductance. Essentially, all water is capable of conducting an electric current. The standard measure of this capability is generally termed *specific electrical conductance*, although the terms *volume conductivity, conductivity per centimeter*, and *electrical conductivity* are sometimes used. In each instance the term refers to the conductance of a cube of the solution one centimeter on a side.

Specific electrical conductance is temperature-dependent and should be reported for a standard temperature, usually 25°C.

Conductance depends on the concentration of ionized mineral salts in solution, and, to a limited extent, there are simple relationships between the two. As the conductivity values increase, the relationships between conductivity and dissolved solids concentrations become more uncertain. The validity of the relationships generally becomes doubtful for conductivities higher than 50,000 micromhos.

Specific electrical conductance is measured with a conductivity cell balanced with a Wheatstone bridge. The resistance of a standardized potassium chloride solution is measured to determine the characteristics of the conductivity cell.

6. Color. *Color* in water is attributable to materials in solution. These materials are primarily organic compounds leached from decaying vegetation and inorganic colored compounds found in industrial-waste effluents. Some metallic substances, such as iron compounds, impart color to water.

Surface waters originating in swamps or marshlands may have color of several hundred units on the cobalt-platinum scale. Groundwaters, too, may become very colored if they are associated with peat or lignite deposits. The U.S. Public Health Service [8] limits color for aesthetic quality to 15 units.

The color of water is measured after suspended matter is removed, by comparison with an arbitrary yellow standard of cobalt chloride and potassium chloroplatinate.

7. Turbidity. The *turbidity* of a water sample is a measure of the ability of suspended and colloidal materials to diminish the penetration of light through the sample. Although turbidity is associated with the amounts of these materials present, attempts to correlate the two are not practical, since the ability of suspended and colloidal matter to diffuse and diminish light is dependent upon many factors.

The standard instrument for the determination of turbidity is the *Jackson candle turbidimeter* [4]. In general practice, however, turbidity is measured by comparison of the water sample with suspensions which have been previously standardized against the Jackson candle turbidimeter. The U.S. Geological Survey [6] uses the *Hellige turbidimeter* and other instruments as standard turbidity-measuring devices.

8. Taste and Odor. *Taste* and *odor* in water are properties which are related to sensations experienced by man through sensitive areas on the tongue, mucous membranes, and the olfactory tissue in the nose. It is difficult to differentiate between the two, although some nonvolatile substances can cause tastes without odors.

Tastes and odors may be caused by any number of substances. Some attempts to adopt descriptive terms for odors have been made, but these are generally inadequate.

Results of taste and odor tests are generally reported in terms of *threshold concentrations*, which indicate the lowest concentration which gives a perceptible taste or odor. The variability of the sensory response of man must always be considered in the evaluation of taste and odor data. Many testing programs employ a large group of persons to make such tests.

9. Oxygen Demand. The ability of substances to utilize, either directly or indirectly, the dissolved oxygen in water for their eventual stabilization is termed *oxygen demand*. The two major types are chemical oxygen demand (COD) and biochemical oxygen demand (BOD). These and other types of oxygen demands do not actually indicate water quality. Instead, they reveal a potential for removing dissolved oxygen from water and indicate the "strength" of sewage and industrial wastes or other oxidizable compounds found naturally in many waters.

Chemical oxygen demand is a measure of the materials present in water that may be readily oxidized. Other terms, such as *oxygen consumed* and *dichromate oxygen demand*, have been used synonymously with COD. The value obtained is helpful in ascertaining the amount of organic and reducing material present in streams or industrial wastes. The exact method of analysis used to determine COD should be specified to prevent misinterpretation of the data.

Organic matter that is susceptible to biological decomposition is used by aquatic microorganisms in their metabolic processes. The amount of oxygen removed from the aquatic environment during these processes is termed *biochemical oxygen demand*. Factors which influence the activity of aquatic organisms will affect the BOD values. Most biochemical oxygen-demand values are reported as 5-day, 20°C, BOD. This standard value indicates the amount of oxygen utilized by the microorganisms in 5 days when incubated at a temperature of 20°C (68°F).

10. Per Cent Sodium. An important property of irrigation water is the amount of sodium present with respect to the cationic concentration. The *sodium percentage* is calculated from the following relationship:

$$\text{Per cent sodium} = \frac{Na^+ \times 100}{Na^+ + Ca^{++} + Mg^{++} + K^+}$$

where ion concentrations are given as equivalents per million (epm). Values of per cent sodium are commonly given to indicate the quality of water for irrigation.

Excessive sodium percentages in irrigation waters will result in base-exchange reactions with soils whereby the calcium and magnesium in the soil are replaced by sodium. Approximate predictions of this exchange reaction are possible [13], and the effect does not become important until the per cent sodium is greater than 50 (see following discussion of sodium-adsorption ratio, SAR).

11. Sodium-adsorption Ratio (SAR). This property of water is calculated from the ionic concentrations of sodium, calcium, and magnesium according to the following relationship:

$$SAR = \frac{Na^+}{\sqrt{\frac{Ca^{++} + Mg^{++}}{2}}}$$

where ion concentrations are given as equivalents per million. This empirical ratio has been proposed [14] as a means of evaluating the sodium-adsorption potential of soils which are irrigated with waters containing significant amounts of sodium. It is generally more significant than per cent sodium, since it deals with potential base-exchange relationships in which calcium and magnesium in the soil are replaced by the sodium in the water.

Sodium-adsorption ratios have been related to the overall sodium, or alkali, hazard of irrigation water. Dividing points between low, medium, high, and very high sodium hazards are SAR values of 10, 18, and 26.

12. Radioactivity. The increasing application of nuclear technology is making studies of the radioactivity in water most important. Comprehensive programs are being developed to monitor the radioactivity of water supplies where there is a possibility that such pollutants could enter water supplies. A great deal of study on radioactivity has been conducted by the Atomic Energy Commission, the National Academy of Sciences, the U.S. Public Health Service, and affiliated research groups. The review of radioactivity as a potential pollutant by the California State Water Pollution Control Board [2, 3] is an excellent initial source of additional information on this property of water.

13. Density. Although density does not directly relate to water quality, this property is very important in many respects to various phenomena which significantly affect it. Density stratification in large rivers, lakes and reservoirs, estuaries, and the ocean results in important changes in water quality.

It is significant to point out that water has its maximum density at 4°C (39.2°F). This unusual property is responsible for thermal stratification of deep-water bodies during the summer when warmer, less dense water is on the surface and during the winter (in some regions) when *colder*, less dense water is on the surface. It is important, too, to note that the relative difference in density increases as the temperature increases or decreases from 4.0°C, the point of maximum density.

14. Temperature. A very important property of water, sometimes disregarded, is temperature. Many important physical, chemical, and biological processes are significantly influenced by the temperature of the water used. Temperature is especially important in determining the value of water for cooling purposes. It is a significant factor in other industrial uses and also for some municipal purposes.

Surface-water temperatures usually approximate the average seasonal air temperatures. Groundwater is uniformly cool and usually has a temperature near the long-term average air temperature for the region.

III. WATER-QUALITY DETERMINATION

Three primary factors in determining the quality of water control the usefulness of the data obtained. These are sampling techniques, analytical procedures, and units of expression. Confusion and misinterpretation of data is possible when these factors are not clearly understood.

A. Sampling Techniques

The basic consideration in determining the water-quality characteristics of a supply is obtaining a sample, or series of samples, which is representative of the whole. This is a simple matter for a homogeneous water body which does not change with time. However, few waters are completely homogeneous, and essentially all will vary in overall quality from time to time. For example, lakes and reservoirs typically exhibit vertical stratification during warm months, and river-water quality varies widely, depending on runoff and other conditions. Limited or incomplete data should be used only with recognition of these and other expected variations in water quality.

In developing a water-quality sampling program, the techniques to be used depend largely on the intended application of the water. Where water is stored in reservoirs, for example, the quality extremes typical of a river are significantly damped. The average river-water quality, weighted on the basis of streamflow, is indicative of the overall quality of the water when impounded. If the water is not to be stored, the sampling program must encompass a wide range of water-quality conditions to determine the expected variations.

The extent of such a program depends on the specific use of the water. Where water quality is a vital factor to the specific application and even a very short period of poor water quality would not be permissible, the sampling program must be very detailed. Automatic sampling devices and analytical equipment are used in some investigations. However, if the intended use will not be seriously impaired by brief periods of poor-quality water, or where adequate storage is provided, the sampling program need not be elaborate.

Variations in groundwater quality are generally less rapid and more uniform than surface supplies. Significant changes due to local contamination or saline-water encroachment may occur, and these possibilities should be recognized. Seasonal or annual samples should be frequent enough to indicate water-quality changes in deeper wells. A monthly schedule might be more satisfactory for shallow wells, which could be influenced more rapidly by contaminated infiltration.

Care should be taken to obtain stable conditions of flow from groundwater aquifers prior to taking a water sample. New wells will generally show significant changes in water quality during initial operation. The samples should be taken after the well has been pumped for some time at normal operating rates. Established wells which have not been used for long periods should also be pumped prior to obtaining samples for water-quality analyses.

Many factors are important in the collection and transmission of water samples. Some of the constituents and properties should be determined immediately, since they are subject to change. The allowable time lapse between sampling and analysis is small for determinations of dissolved gases, temperature, pH, and similar water-quality characteristics. The type of sample container is also important, and care should be taken to ensure that the sample is not contaminated. The proper sampling techniques are discussed in the various manuals describing standard procedures for water-quality determinations.

B. Analytical Procedures

The actual determinations of the specific water-quality characteristics are based on procedures and techniques developed through the years. The methods currently in use are too numerous to be discussed or even outlined here. Reference should be made to manuals which prescribe standard procedures for the physical, chemical, and biological examination of water. The most used manuals and the publishing groups are listed below:

Standard Methods for the Examination of Water and Wastewater [4], jointly published by American Public Health Association, American Water Works Association, and Water Pollution Control Federation

Manual on Industrial Water and Industrial Waste Water [15], American Society for Testing Materials

Official Methods of Analysis of the A.O.A.C. [16], Association of Official Agricultural Chemists

Methods for Collection and Analysis of Water Samples [6], U.S. Department of the Interior, Geological Survey

Most of the methods proposed by each group are essentially uniform. Analyses and the related field and laboratory procedures should conform to the standard methods prescribed by the appropriate group from each professional area.

The major analytical techniques include volumetric, gravimetric, colorimetric, and flame-photometric procedures. Also important are the bacteriological, biological, and bioassay procedures necessary in some determinations of water quality and its effects.

Volumetric analyses are those where a test reagent is added to the prepared sample until an end-point reaction occurs. The amount of reagent necessary to obtain the desired reaction indicates the amount of the specific constituent which is present in the sample. The end point may be shown by a color change or by a potentiometric, amperometric, or polarographic instrument. Special techniques are necessary when substances in the sample interfere with the reaction between the test solution and the primary substance to be measured.

Some substances may be determined by both volumetric and gravimetric methods. The *gravimetric techniques* are generally less precise and more time-consuming for some determinations. They require long and tedious procedures, involving precipitation, filtration, washing, ignition, and weighing.

Colorimetric methods are among those most frequently used in water analysis. A sample is suitably prepared, and specific reagents are added to develop a characteristic color, the intensity of which indicates the amount of the specific substance present. The equipment used to measure the color developed varies from simple tube comparators to more exact color photometers and very sensitive spectrophotometers.

Flame-photometric methods are used in the determination of sodium, potassium, lithium, and strontium. Other methods of analysis include *spectroscopic* and *polarographic techniques*.

Bacteriological analyses involve the culture of the bacteria in such a manner that their presence may be detected or visually observed. For a detailed procedure, reference should be made to "Standard Methods for the Examination of Water and Wastewater" [4]. *Biological analyses* include the sampling identification and enumeration of the macro- and microorganisms present in the water. The effects of water quality on the survival and propagation of aquatic organisms are determined by *bioassay procedures*. Standardized methods for such analyses have been proposed [17].

C. Units of Expression

Various units are used to report water-quality data. The proper interpretation and correlation of data from several sources requires the knowledge of units of expression commonly employed and the relationships between them.

Most inorganic salts in aqueous solution are present in dissociated form as charged ions. Although not all constituents and properties may be conveniently reported as ions, those substances which are known to be dissociated in solution are generally expressed as concentrations of the individual ions. Ionic concentrations may be expressed in several different ways.

1. Weight per Weight. One of the most common expressions of concentration is *parts per million* (ppm), or one pound of compound, or ions, per million pounds of water, or one milligram per kilogram of solution.

Oceanographers commonly express the high concentrations of some sea-water constituents in *parts per thousand*, which are equivalent to 1,000 times parts per million. Percentage units are equivalent to 10 times parts per thousand, or 10,000 times parts per million. Concentrations of trace constituents are often reported in terms of parts per billion. A *part per billion* is equal to one microgram per kilogram, or one-thousandth of a part per million.

All such units are based on the weight of the constituent with respect to the weight of the solution. The various units avoid cumbersome decimals for the range of concentrations generally encountered.

2. Weight per Volume. The convenience of volumetric measurements of water, as opposed to gravimetric measurements, leads to many expressions of concentrations in terms of the weight of dissolved substance with respect to the solution volume. One of the more common expressions of this type is *milligrams per liter* (mg/liter), which is essentially equal to parts per million. If it can be assumed that a liter of solution weighs one kilogram, results in terms of milligrams per liter are equivalent to those in parts per million. Corrections for density effects due to dissolved minerals and temperature are necessary to make the two units strictly comparable. Results in milligrams per liter, divided by the specific gravity of the solution, are equal to parts per million by weight. However, unless the dissolved solids concentration is on the order of 7,000 ppm, the assumption of unit density introduces little error. Temperature effects on density are normally insignificant with respect to the equality of the two units.

A weight-per-volume unit frequently used by water-treatment-plant operators is *grains per gallon*.[1] One grain per gallon is equivalent to 17.12 milligrams per liter (mg/liter). The quality of water used for irrigation is sometimes expressed in terms of *tons*[2] *per acre-foot*. One ton per acre-foot is equal to 0.735 mg/liter. The concentrations of materials in flowing streams are often given as *pounds* or *tons per day*. The relationships between these units and weight-per-weight units also depend on density considerations.

3. Equivalents per Weight. Interpretation of analytical data is often facilitated if the constituents are reported in terms of the chemical equivalence of the ions present. When analytical data are presented in terms of equivalent weights, the sum of all anions should essentially equal the sum of all cations. This provides a check on the accuracy and completeness of the analysis.

The *equivalent weight* of an ion is equal to its formula weight divided by the ionic charge. The concentration of an ion can be expressed as the number of equivalent weights in a million parts by weight of the solution, such as milligram equivalents per kilogram. This is commonly expressed as *equivalents per million* (epm). Con-

---
[1] As used throughout this section, *gallon* refers to the U.S. gallon, equal to 0.833 imperial gallon.
[2] As used in this section, *ton* means a short ton, equal to 2,000 lb, or 907.185 kg.

version from parts per million to equivalents per million can be made from the following relationships:

$$\text{epm} = \frac{\text{ppm}}{\text{equivalent weight}}$$

The equivalent weights of most ions of concern in water analysis are given in Table 19-1.

Table 19-1. Formula Weights and Equivalent Weights of Ions Found in Water*

Ion	Formula weight	Equivalent weight	Ion	Formula weight	Equivalent weight
Al^{3+}	27.0	9.00	Fe^{3+}	55.8	18.6
Ba^{++}	137.	68.7	Pb^{++}	207.	104.
HCO_3^-	61.0	61.0	Li^+	6.94	6.94
Br^-	79.9	79.9	Mg^{++}	24.3	12.2
Ca^{++}	40.1	20.0	Mn^{++}	54.9	27.5
CO_3^{--}	60.0	30.0	Mn^{4+}	54.9	13.7
Cl^-	35.5	35.5	NO_3^-	62.0	62.0
Cr^{6+}	52.0	8.67	PO_4^{3-}	95.0	31.7
Cu^{++}	63.6	31.8	K^+	39.1	39.1
F^-	19.0	19.0	Na^+	23.0	23.0
H^+	1.01	1.01	Sr^{++}	87.6	43.8
OH^-	17.0	17.0	SO_4^{--}	96.1	48.0
I^-	127.	127.	S^{--}	32.1	16.0
Fe^{++}	55.8	27.9	Zn^{++}	65.4	32.7

* Weights expressed to three significant figures.

Neglecting the correction for specific gravity, the equivalents-per-weight units, epm, can be equated to the equivalents-per-volume units, *milliequivalents per liter* (meq/liter).

Many analyses are reported both in equivalents-per-weight units (epm) and in weight-per-weight units (ppm). Graphical presentations of analytical data are sometimes more effective if they show concentrations in terms of equivalents.

Nonionizing constituents in solution are not normally reported in terms of equivalent weights, but are generally expressed in parts per million or some other weight units. Silica and iron are generally not reported as equivalents, although there is some indication that these constituents may be somewhat ionized and enter into the cation-anion balance.

4. Calcium Carbonate Equivalents. Some properties of water, namely, hardness and alkalinity, are conventionally expressed in terms of an equivalent quantity of calcium carbonate. Hardness and alkalinity both represent characteristics of water which result from the presence of several ions in solution. Reporting these ions as calcium carbonate is a matter of convenience, and it should not be assumed that the concentrations given represent actual concentrations of calcium carbonate.

Hardness is often calculated from the ionic concentrations of calcium, magnesium, and certain other cations such as iron, aluminum, manganese, and zinc. *Calcium carbonate equivalents* can be obtained by multiplying the ion concentrations by appropriate factors as given in Table 19-2.

Table 19-2. Hardness Conversion Factors

Ion	Factor
Calcium (Ca^{++})	2.497
Magnesium (Mg^{++})	4.116
Iron (Fe^{3+})	1.792
Aluminum (Al^{3+})	5.567
Zinc (Zn^{++})	1.531
Manganese (Mn^{++})	1.822

Ionic concentrations are frequently calculated to their calcium carbonate equivalents in the determination of water-treatment requirements.

5. Hypothetical Combinations. Water-quality data may be expressed in terms of the concentrations of combined salts which would most likely be formed or precipitated from solution. This method of reporting data is misleading, since the substances are present as individual ions rather than as combined salts. However, there are some uses of water analyses which may be served by reports of *hypothetical combinations*. This form of reporting is not now used except for certain specific applications. In studies of boiler feedwater or cooling water, it is important to have an estimate of the nature of the various deposits that might be formed on heat transfer and other surfaces. An arrangement of the ionic constituents in solution according to certain hypotheses [15] will indicate the deposits that will form.

If hypothetical combinations are reported, the water analysis in terms of ionic concentrations should always be given to prevent misleading conclusions.

6. Miscellaneous Units

a. *Electrical Conductance.* The ability of a water solution to conduct an electric current is measured in terms of reciprocal ohms, or mhos. Most waters, with the exception of very strong brines, will have conductance values much less than one mho and the values are more conveniently expressed as micromhos, or millionths of a mho.

b. *Hydrogen-ion Concentration (pH).* The pH of a solution refers to the negative logarithm to the base 10 of the concentration of hydrogen ions in moles per liter.

c. *Color.* The color of a water sample is measured by comparison with an arbitrary *cobalt-platinum scale* which has a maximum value of 500 units. A unit on this color scale is equivalent to one milligram per liter of platinum in the form of the chloroplatinate ion.

d. *Turbidity.* Turbidity is reported in terms of an arbitrary scale based on the Jackson candle turbidimeter. References to the *silica scale* and *parts per million of turbidity* are no longer common practice [4], although some groups have retained these units [6]. Turbidity units are equal numerically to the silica-scale units formerly used.

e. *Temperature.* Although most laboratories report temperature on the centigrade scale, many temperature measurements are reported in Fahrenheit degrees. The following formula may be used to convert temperature in degrees Fahrenheit (°F) to degrees centigrade (°C):

$$°C = 5/9(°F - 32)$$

f. *Radioactivity.* There are two basic units of radioactivity measurement. The quantity of radioactive material present is measured by the disintegration rate in terms of curies (c). A *curie* is equal to 37 billion disintegrations per second. Other terms frequently encountered are the *millicurie* (mc), which is one-thousandth of a curie, and the *microcurie* (μc), which is one-millionth of a curie.

The amount of radioenergy produced by the radioactive substances is measured in terms of the *roentgen* (r). It is based on the amount of X, or gamma, radiation that produces a specific amount of ionization in a standard volume of air. Results are often reported in terms of the *milliroentgen* (mr), which is one-thousandth of a roentgen.

IV. WATER-QUALITY CHARACTERISTICS

The source of a water supply is usually indicative of its major water-quality characteristics. Often a great deal of information can be postulated with knowledge of the nature of the source, whether surface or ground, and certain data concerning climatic, soil, geologic, and other conditions within the watershed. Such an evaluation of a water-supply source is not a substitute for actual water-quality data. Rather, it is an aid to formative planning in the primary phases of development. In addition, the same concepts may be used to develop programs for obtaining water-quality data and to evaluate the completeness of existing, but limited, data.

A. Surface Water

Most of the water used is obtained from surface supplies, namely, (1) rivers, (2) lakes and reservoirs, (3) estuaries, and (4) oceans. There are certain definite water-quality characteristics which may be described for each class. In addition, there are several phenomena which cause significant variations in water quality within a given category.

1. Rivers. A great deal of information on river water quality may be evaluated from the climatic and geologic conditions in the river basin. These two factors play a primary role in the quantity of water that will be available.

In most rivers the normal, or "dry-weather," flow is made up primarily of water which seeps from the ground. This is not always true, of course, where the river is fed by snowmelt or where regulated water releases are provided by storage impoundments. Most of the flow of a river is contributed during the high runoff or flood periods. The amount of water contributed during the dry-weather periods is only a small part of the total.

During the periods of high runoff, most rivers exhibit their most favorable chemical water-quality characteristics. Although the water may contain extremely large amounts of suspended matter, the concentrations of dissolved substances are low, often only a fraction of that found during dry-weather periods. This is characteristic of most rivers in all parts of the country. Figure 19-2 shows the relation-

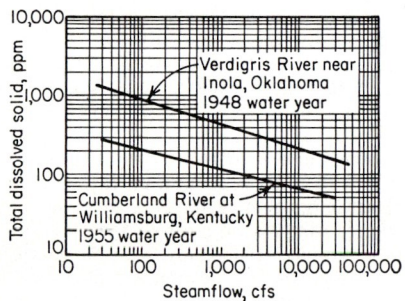
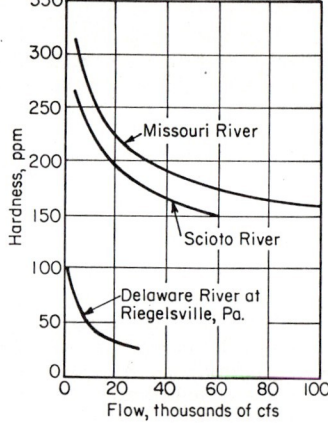

Fig. 19-2. Typical relationship between streamflow and dissolved solids.

Fig. 19-3. Variation of hardness in water in comparison with river flows.

ship between streamflow and total dissolved solids concentrations for several rivers. Variations in hardness with streamflow are shown on Fig. 19-3.

There are some instances where high runoff may cause deterioration in water quality. If precipitation falls selectively on the watershed of a tributary which contributes poor-quality water to a comparatively good-quality river system, the water contributed will cause a transitory deterioration of the water quality in the system. Conditions such as these are not uncommon in many rivers throughout the country. A similar example may be cited where high flows are capable of flushing marshy bogs of their high-colored water without providing adequate dilution.

An important consideration in the evaluation of geologic influence on water quality is the history of an area with respect to its general hydrology. The geology of an area determines the minerals which will be present, but the hydrologic history largely determines the availability of these minerals for solution during periods of high and low flows. In areas where runoff has been relatively abundant and uniform, the previous leaching action of the water has removed those minerals which are easily taken into solution.

Where precipitation is characteristically sparse and poorly distributed in time, the rivers will generally have poor water-quality characteristics for a major part of the time. In such areas, the flood flows of the rivers, which constitute a very large per cent of the total flow, are of relatively good quality, in that they are not capable of leaching large amounts of minerals. During periods of moderate or low flow, evapotranspiration consumes a large part of the runoff, resulting in increases in dissolved solids.

The predominant cations in river water are calcium, magnesium, sodium, and potassium, generally in that order. Some rivers, especially in the west and southwest parts of the United States, contain high proportions of sodium. Iron and manganese are not generally significant in most river waters. Concentrations of metallic salts increase if oxygen is absent or if the water is acidic. Other cations are present in river water in minor or trace concentrations, if at all, except where sewage or industrial wastes are discharged to the stream.

Chloride, sulfate, and bicarbonate are generally the most concentrated anions in river water. Some streams contain some carbonate, and even minor amounts of hydroxide. Nitrate concentrations are usually small, except where organic pollutants drain to the river. Only minor amounts of the other anions may be expected in most natural rivers.

River-water temperature generally follows the long-term average air temperature of the area. Shallow, rapid-flowing streams respond quickly to changes in air temperatures, while deep, sluggish streams are less affected.

Essentially, all rivers contain various amounts of several gases in solution. In rivers unpolluted by oxygen-demanding wastes, the dissolved-oxygen concentration will be near saturation. Concentrations will generally vary throughout the day, being highest just before dark, often greater than 100 per cent saturation, and lowest just before sunrise. These variations are the result of the photosynthetic activity of algae and other aquatic life which utilize oxygen during periods of darkness and produce oxygen when there is light. Thermal stratification of deep rivers may result in decreasing concentrations of dissolved oxygen with increasing depths. Dissolved-oxygen concentrations below waste outfalls may be quite low when the pollution imposes a heavy oxygen demand on the stream. Low concentrations of dissolved oxygen may also be found during certain periods of the year below reservoirs which discharge water from the bottom of the reservoir pool.

Carbon dioxide is usually present in river water, often in concentrations of 10 ppm or more. Since carbon dioxide is a by-product of the metabolism of aquatic life, the concentration generally varies throughout the day, with the highest concentrations usually occurring just before sunrise. Hydrogen sulfide, methane, sulfur dioxide, and ammonia are not usually found in river water, except possibly below outfalls which discharge sewage or industrial wastes. Some springs may discharge water which contains such gases, but they are quickly given off to the atmosphere or oxidized to other forms of less significance.

Normal, unpolluted river water will generally exhibit a 5-day, 20°C, biochemical oxygen demand (BOD) of 1 to 2 ppm. Below waste outfalls this value may be a great deal higher, depending upon the waste characteristics and the degree of waste treatment provided.

The pH value of river water is usually 7 or slightly higher, although much lower values are not to be unexpected. Very low pH values, 5 or less, are typical of rivers which drain coal-mining and similar areas. Organic acids resulting from decaying vegetation also cause lower pH values. Many industrial wastes may cause the pH value of river water to be greater or less than normal values.

Color in river water is commonly below 50 ppm. Much higher concentrations are usually associated with drainage from swamps or with industrial-waste pollution. Tastes and odors are not generally significant in river water except where swamps drain to the watercourse or there are industrial wastes present.

As previously noted, turbidity will be very high during periods of high runoff. This property of river water is extremely variable. Like turbidity, the suspended solids concentration of river water depends on runoff and soil conditions in the watershed and is highly variable. The amounts of suspended matter carried in river water can be surprisingly great, and consideration should be given to this factor in the design of reservoirs and other river-control structures.

River water does not exhibit large amounts of radioactivity except in areas where radioactive waste products are introduced to the river. Present evidence indicates that nuclear explosions have been reflected in the pattern of radioactivity in river water as a result of the deposition of radioactive particles from the atmosphere.

2. Lakes and Reservoirs. The water-quality characteristics of lakes and reservoirs are highly variable and depend on many factors, most of which are too complex to be discussed here. Limnologists have conducted extensive studies of lakes, and detailed discussions of their water-quality characteristics may be found in the classical texts of this field [18, 19]. Many limnological phenomena which affect water quality in lakes are significant, too, in determining the quality of water that will exist in large water-storage reservoirs.

Water storage in reservoirs modifies the wide fluctuations in water quality which are characteristic of the rivers flowing into the impoundment. In general, the quality of water available from an impoundment will be better than the average quality of the influent river water. This effect depends on the amount of storage provided and the water-quality-streamflow relationship for the river.

Bacterial concentrations in reservoirs are reduced as a result of the unfavorable bacteriological environment, coupled with long detention periods. Turbidity is decreased by sedimentation of suspended particles in the relative quiescence of the impoundment. The bleaching action of sunlight is responsible for removing color from water retained in reservoirs.

Not all the effects of impoundments on water quality are good. During certain periods of the year, water stored in reservoirs may exhibit objectionable tastes and odors, high color, objectionable algal growth, low concentrations of dissolved oxygen, and high concentrations of iron, manganese, and hydrogen sulfide. All these undesirable effects on water quality are due largely to thermal stratification of reservoirs during certain periods of the year.

Deep reservoirs can be expected to develop thermal stratification during the warm period of the year. In the spring, warming of the surface waters and the inflowing water from tributaries creates thermal-density differences between the deep water which has been stored through the preceding winter and the warmer, low-density water near the surface. This temperature gradient results in a stable stratification which resists mixing by wind and other forces. Normally, an intermediate layer of maximum temperature gradient, called the *thermocline*, forms between an upper layer called the *epilimnion* and the deep layer called the *hypolimnion*. The thermocline in natural lakes is often observed at depths of 25 to 40 ft below the surface, but may be shallower or deeper in artificial impoundments, depending on withdrawal rates, depth of water outlets, and other factors.

Water above the thermocline in the epilimnion is stirred by wind and wave action, which combines with photosynthesis of algae to maintain dissolved oxygen in the water at near-saturation levels. Below the thermocline, however, little mixing occurs and sunlight, necessary for photosynthesis, does not penetrate to the hypolimnion. The stagnant water, deprived of reaeration, cannot replenish the dissolved oxygen which is consumed by biochemical processes. Thus these processes gradually reduce the dissolved oxygen to levels considerably below saturation values and frequently exhaust it completely. The same biological processes which utilize the oxygen produce carbon dioxide, resulting in the production of acidic conditions.

The quality of water in the hypolimnion becomes progressively worse until late summer or early fall. Cooler weather lowers the temperature of the surface water, increasing its density, so that it sinks and mixes with the deeper water in the hypolimnion. Eventually all the water in a reservoir becomes of relatively uniform temperature and density.

In the southern latitudes, where average winter temperatures are not near freezing, reservoir waters will remain mixed from top to bottom and well aerated during the winter period. In northern latitudes, however, deep reservoirs exhibit winter stratification resulting from cooling of the surface waters to temperatures below about 4°C, the point of maximum density. Colder, but less dense, water will remain on the surface of the reservoir, with bottom-water temperatures at about 4°C. The winter stratification is weak because the density differences are small and wind action may cause a high degree of mixing. When an ice cover forms, the wind disturbance is stopped and the reservoir becomes quite stagnant. The period of spring overturn, brought on by warming air temperatures, starts the stratification cycle again.

Figure 19-4 shows the salient limnological features of a stratified and nonstratified reservoir. The variations in water quality caused by thermal stratification of lakes and reservoirs should not be overlooked in an evaluation of this type of supply. Many of the problems associated with the variations of water quality with depth may be overcome by the selective withdrawal of water by multiple-level intakes.

3. **Estuaries.** An *estuary* may be defined as that reach of a river where river water mixes with and measurably dilutes sea water. The general water-quality characteristics of an estuary depend, therefore, on the relative amounts of the two waters present in the mixture. In addition, there are several factors which affect significant variations in the quality of water at any given location in the estuary.

The upper limit of an estuary is usually not at a fixed station, but moves according to river inflow, tidal actions, and wind forces. In the upper "area" of an estuary the water generally exhibits the quality characteristics of the inflowing river, although

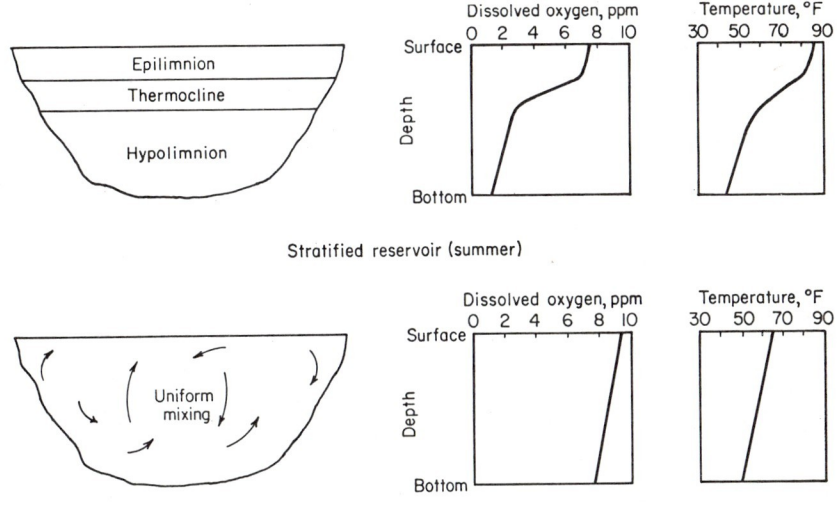

Fig. 19-4. Limnological characteristics of reservoir stratification.

the increased salinity of the water is apparent. In addition, a change may be noted in the relative concentrations of the various constituents found in sea water. In the lower portions of the estuary the water-quality characteristics strongly resemble those of sea water, although the total dissolved solids concentrations are significantly less.

The salinity of the upper portion of an estuary is subject to wide variations. Heavy river inflows will displace significant volumes of saline water seaward, while lower drought flows will permit the salinity of this area to increase. Tidal actions will cause short-lived, but sometimes significant, variations in salinity in the upper reaches. Flood tides carry saline waters upstream, while ebb tides flush them seaward. This condition is shown in Fig. 19-5. Strong winds of long duration, and acting generally along the axis of the estuary, can supplement or oppose these variations caused by river flows and tides.

The lower stretches of an estuary are generally less significantly affected by the actions described previously. Variations are normal, but the stability of the greater volume of water reduces the magnitude of the short-term variations. Long periods of high or low inflow to the estuary can cause significant variations in salinity in the lower reaches of the estuary.

It is not uncommon for estuaries to stratify as the result of density differences which are caused by temperature, salinity, or a combination of the two. Often the less dense surface water flows seaward, while the cooler saline water at the bottom moves in an opposite direction. Stratification in deep sheltered estuaries is usually quite stable during the summer months. In the more shallow portions of an estuary the mixing forces of strong tidal current tend to destroy the stratification and mix the waters of different density.

In northern latitudes, where cold winters are experienced, deeper estuaries will exhibit spring and fall overturns, similar in many respects to those which occur in lakes and reservoirs. The fall overturn is generally more pronounced than that which occurs in the spring. During and following these phenomena, the quality of water in the estuary may be affected by materials, usually found only in the bottom waters, which are distributed throughout the depth of the estuary.

In general, the overall water-quality characteristics of an estuary are not unlike those of larger rivers. Temperatures are usually close to the long-term average air temperature of the area. In deep estuaries, however, significant vertical temperature gradients are not uncommon, with water temperatures decreasing with depth during the summer and, in some locations, increasing with depth during the winter.

The pH value of estuarine water is generally greater than 7 and usually higher than the pH of the river water flowing into the estuary. Estuarine water is highly buffered and resists changes in pH on the addition of acids or bases. The salinity characteristics and the factors causing variations have been previously described. Most other water-quality characteristics are similar to those described for river water.

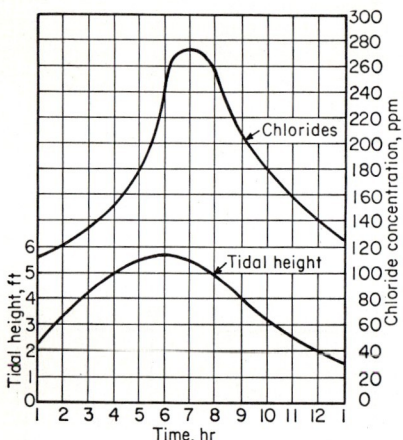

Fig. 19-5. Variations in the salinity of an estuary caused by tidal flows.

4. Oceans. The water-quality characteristics of the oceans are a major interest of oceanographers. The quality of sea water is important for many reasons, such as fisheries management and effects on ocean-going vessels and off-shore structures, and it provides a means of studying the dynamics of oceanic circulation. In addition, recent advances in the conversion of saline water have focused attention on oceanic water as a possible source of potable water supplies in some areas.

It has been found that the relative chemical composition of sea water is essentially constant, although the actual concentration of the various substances may vary. In certain oceanic waters, and in regions near the mouths of large rivers, the relative composition may be changed somewhat. These areas are of minor significance to the oceans as a whole.

The greatest ionic concentrations in sea water are chloride, sodium, sulfate, magnesium, calcium, and potassium, in that order. Many other ionic constituents are present, but in smaller or trace concentrations. The total dissolved solids content of sea water averages about 34,500 ppm. Of this total, chloride accounts for about 55 per cent (about 19,000 ppm), and sodium, 30 per cent (about 10,500 ppm). Complete lists of the constituents of sea water are given in many textbooks and references [20, 21, 1]. (See Subsec. 2-II.)

Oceanographic considerations of water quality are not closely related to the problems encountered by hydrologists, limnologists, and sanitary engineers. The salient water-quality characteristics of interest to oceanographers are primarily salinity and chlorinity.

The *salinity of sea water* has been defined as the grams of solid material in one kilogram of sea water when the organic matter has been oxidized, the carbonate converted to oxide, and the bromine and iodine replaced by chlorine. *Chlorinity* is the value expressed as the grams of "atomic-weight silver" necessary to precipitate the halides (chloride, bromide, and iodide) in 0.3285233 kg of sea water.

Salinity, which is seldom actually measured, is calculated from chlorinity by the following relationship:

$$\text{Salinity} = 0.03 + 1.805 \times \text{chlorinity}$$

where both characteristics are in parts per thousand (grams per kilogram).

These are two of the most important water-quality characteristics considered in oceanographic studies. The overall evaluation of sea-water quality is most complex, and the classical texts of the field should be consulted for detailed reference [20–23] (see also Subsec. 2-II).

B. Groundwater

The quality of water available from underground aquifers may be far superior, in some respects, to the quality of surface water. There are, however, some objectionable quality characteristics of groundwater which should also be recognized. There is no recourse but to obtain representative samples of the specific supply for analysis and evaluation.

Groundwaters are generally free of suspended solids and objectionable color. In some instances, however, the water may be associated with underground beds of peat and decaying vegetation which contribute substantial amounts of color-producing organic matter to the supply. There is usually no contamination of groundwater by sewage or industrial wastes except in fissured aquifers where surface pollution may be easily transmitted to the groundwater reservoir. Contamination may also result from the use of reclaimed wastes to recharge groundwater supplies.

The temperature of underground water is a highly desirable characteristic of such supplies for cooling and other purposes. Groundwater is uniformly cool and usually has a temperature near the long-term average air temperature for the region. It is common for deeper groundwaters to show increasing temperature with increasing depth.

Groundwater generally contains higher dissolved solids concentrations than surface waters of the same locality. Most of the minerals present in greater amounts are those which contribute to hardness (calcium and magnesium) and alkalinity (bicarbonate, carbonate, and hydroxide). This is due largely to the increased amounts of carbon dioxide in the groundwater.

The decomposing organic materials in the soil remove dissolved oxygen and contribute carbon dioxide to underground water. These conditions are favorable to the solution of iron and manganese, which are not uncommon in groundwater supplies. Hydrogen sulfide may also be present in some well waters.

It should be recognized that overdraft on the groundwater aquifer may lead to contamination of the supply. This is not uncommon in coastal areas.

V. WATER-QUALITY DETERIORATION

A. Natural Pollutants

Not all pollution is caused directly by man's activities. Natural water pollution may be considered as changes in the constituents and properties of water in the absence of man's activities which are significantly greater than the normally expected variations. There is no distinct division between the normal quality extremes and natural pollution. Each specific situation must be judged on the basis of the various factors involved.

Drainage from certain watersheds may contribute water of very poor quality to a river system which has good water-quality characteristics in general. This type of natural pollution may be caused by evaporate mineral deposits, salt seeps, drainage from abandoned oil and coal fields, and similar sources of exceptionally poor quality water. Such a situation may sometimes be aggravated by rainfall, which selectively occurs in the watersheds contributing the water of poorer quality (Subsec. IV-A-1). Drainage from oil and coal fields is sometimes considered to be industrial pollution, although the brine and acid wastes from long-abandoned fields are commonly regarded as natural pollutants.

Federal measures are being considered for the control of natural pollution in some of the rivers of the West and Southwest [24]. Increasing attention is being given to such problems, and there will no doubt be several feasible control methods developed in the near future.[1]

Excessive turbidity and suspended solids concentrations during periods of high runoff may be considered as natural pollution, although this condition is normally expected. Loss of vegetation, either by agricultural development, forest fires, and the like, aggravates this condition. Reforestation and other soil-conservation measures are necessary to prevent extremely high turbidities and suspended solids during periods of high surface runoff.

Tastes, odors, and high concentrations of color-producing substances may be expected in waters derived from swamps, marshes, and peat bogs. Some types of aquatic life, principally the algae, may also contribute these objectionable characteristics to water. Prolific algal growths interfere with many beneficial uses of water to a large degree and, in some instances, are regarded as a very serious pollutant of natural origin.

Thermal stratification of hydroelectric impoundments has resulted in an interesting and unique condition of natural pollution in rivers below the impoundments during the warm period of the year. Water discharged through the turbines is usually drawn from the hypolimnion, or lower level of the impoundment, where there is little or no dissolved oxygen in the water. In addition, this water may contain dissolved iron and manganese, excessive carbon dioxide, and occasionally some hydrogen sulfide. A unique installation of a "submerged weir" has been carried out to withdraw surface waters selectively through the turbines and thus overcome this type of water-quality deterioration in the Lower Roanoke River basin [25].

Groundwaters are sometimes polluted naturally by the encroachment of oceanic or coastal waters or adjacent groundwater which is highly mineralized. Excessive pumping from the aquifer lowers the water table and allows poorer-quality water to move into the area. Various schemes have been suggested to prevent such salt-water encroachment (Sec. 13).

B. Man-made Pollutants

Much of the pollution of water is the direct result of man's activities, and a great deal of study has been devoted to the problems involved. A discussion of the pollutants which may be introduced to waters is greatly beyond the scope of this section. Because water pollution is vital to any discussion of water quality, however, the polluting substances produced by man's activities will be broadly outlined. Additional information is available from comprehensive reviews found in the sanitary-engineering literature [2, 3, 25, 26].

1. Municipal Wastes. Man's normal domestic and municipal activities result in the creation of wastes which may be broadly termed *sewage*. Municipal sewage includes, in addition to normal domestic sewage, wastes from commercial and some industrial establishments, hospitals, hotels, institutions, and, in some cases, surface drainage from paved areas. The constituents added to the water during its use will include oils and grease, grit, paper, ground garbage, synthetic detergents, dissolved and suspended organic and inorganic matter, bacteria, and various other substances, depending on the commercial and industrial complex which discharges to the sewers.

The amounts of these materials which will be present in the river receiving municipal sewage depend on the treatment provided. *Primary* treatment usually removes 40 to 70 per cent of the suspended solids and 20 to 40 per cent of the biochemical

---

[1] Potential methods of control include injection of brines into suitable aquifers, storage and evaporation of highly saline waters, diversion of concentrated brines into oceanic waters, and others.

oxygen demand (BOD). Secondary, or *complete*, treatment will generally remove about 75 to 95 per cent of the suspended solids and 85 to 95 per cent of the BOD. The effluent is commonly chlorinated to destroy about 99 per cent of the bacteria.

There will usually be a reduction in the dissolved-oxygen concentration in a stream below a waste outfall contributing substantial amounts of oxygen-demanding sewage. Chlorides, nitrates, and phosphates are present in increased concentrations. Suspended matter and other turbidity-producing substances are also contributed to streams.

2. Industrial Wastes. The types of wastes produced by industries are extremely complex and variable. They may contain organic matter, acid or alkaline substances, chemicals, suspended matter, poisonous substances, radioactive materials, heat, color, tastes and odors, oils, and a wide variety of other substances. The amounts of these materials which are discharged to streams and underground waters are also highly variable. In some industries, adequate treatment methods have been developed and used to render the wastes relatively harmless.

Any study of the water quality of a specific area should include an evaluation of the existing or potential effects of industrial pollutants. The sanitary-engineering literature contains a great deal of information which will guide such an evaluation [2, 3, 27].

3. Agricultural Wastes. The development of agricultural lands and the increasing use of water for irrigation usually result in significant changes in water quality. Increased suspended solids during periods of high surface runoff are commonly associated with agricultural development where soil cover has been removed and adequate soil-conservation measures are not employed.

The increased concentration of dissolved minerals in the return flow from irrigated areas has long been recognized. In arid regions the high rates of evapotranspiration result in large consumptive losses of water, and therefore in increased concentrations of the dissolved constituents in the surface supplies.

In some areas, the return flows from irrigated lands contain increased amounts of hardness. Base-exchange reactions may decrease high sodium concentrations but intensify the concentrations of calcium and magnesium.

Increased attention has recently been given to the effects of insecticides, pesticides, and herbicides which are applied to croplands. Streams draining areas treated with such materials frequently contain significant concentrations of the chemicals reaching rivers and lakes. This problem is receiving much attention since the toxic effects of the materials used are highly detrimental to many uses of water.

C. Waste-water Reclamation and Re-use

Serious water shortages in some areas have given impetus to programs which reclaim waste water and directly re-use it for certain purposes. The effluent from the waste-treatment plant is generally given some additional treatment before the water is used. Thus such "water" supplies will have the water-quality characteristics of the wastes, along with any changes resulting from the additional "water" treatment.

Most of the reclaimed waste water is used in industry, irrigation, or groundwater recharge. Reclaimed sewage and industrial effluents are used primarily to augment industrial water supplies. However, attention is being given to such supplies to solve water shortages for domestic and municipal purposes.

Any program intended to utilize waters directly should be preceded by a thorough evaluation of the physical, chemical, and biological characteristics of the proposed supply and the potential effects of using it. Many of the characteristics of such supplies would be highly objectionable for some agricultural and industrial purposes, and each situation should be carefully studied. Several comprehensive reports [28–33] are available as initial reference sources to guide the evaluation program.

D. Waste-assimilative Capacity

Because of an inherent capacity to assimilate wastes, surface waters and groundwaters are capable of natural purification. There are limits to the degree of purification that can be effected by nature, however, and excessive water-quality deterioration may result if the amounts of wastes discharged are too great. Federal, state, and interstate pollution-control agencies regulate the discharge of waste products to prevent water-quality degradation which affects the other beneficial uses of water. The greatest attention has generally been given to the ability of surface waters to assimilate wastes. Some study has been devoted to the assimilative capacity of groundwaters, however, and a brief mention of this should be made here.

Wastes may enter groundwaters in two basic ways. Seepage from waste lagoons, tile-drainage fields, and recharge of aquifers from rivers may result in contamination of groundwater, but this is usually not significant. Recharging groundwater basins with reclaimed sewage or industrial waste can, on the other hand, cause important changes in the physical, chemical, and biological characteristics of the groundwater supply. Reports concerning studies of the effects of groundwater recharge and the travel of pollution have been published by the California State Water Pollution Control Board [28–31, 33] and provide a basic reference source for detailed information on this subject.

In general, bacteria are quickly removed from groundwaters, depending on the type of aquifer. Organic matter, both suspended and dissolved, is similarly removed in short distances by filtering action and biochemical stabilization in the zone where bacteria are present. Persistent substances include soluble minerals and other constituents such as synthetic detergents, phenols, and the like. Some change in the hardness of the groundwater supply is possible due to the base-exchange reactions of sodium in the recharge water and calcium and magnesium in the aquifer.

The *waste-assimilative capacity* of surface waters is generally considered to be the amount of waste which will not cause water-quality deterioration beyond the limits required for other beneficial uses of the supply. Because oxygen-demanding wastes are most frequently encountered, this capacity is commonly given in terms of the biochemical oxygen demand (BOD) of the waste. Other wastes are very important in determining the assimilative capacity, however, and BOD should not be assumed the only criterion.

The greatest single factor controlling the overall waste-assimilative capacity of surface waters is the amount of dilution that is provided. With fixed limits on the degree of water-quality deterioration that may be tolerated, the waste-assimilative capacity increases as the amount of dilution water increases. It cannot be assumed, however, that all the available water will serve to dilute the polluting substances. In rivers it is not uncommon for waste effluents to remain segregated along a shore line for many miles below the outfall. This effect is in addition to the normal variations in streamflow. The dilution in tidal estuaries is exceedingly complex, and special techniques must be applied to determine the degree of dispersion which will actually take place [34]. Current studies are necessary to determine the dilution which will occur in lakes and oceans [35]. In all studies of assimilative capacity it must be remembered that the amount of dilution water available at any time is a primary concern.

The waste-assimilative capacity in terms of polluting substances which exert high oxygen demands depends not only on the available dilution, but also on the factors which control the oxygen resources in the water body. The determination of the BOD assimilative capacity is based on mathematical formulations of the factors controlling the *oxygen sag* which were developed by Streeter and Phelps [36] and expanded by Fair [37].

There have been many investigations of the factors involved in the oxygen-sag relationships, most of which are contained in the report of a seminar on this topic sponsored by the U.S. Public Health Service [38].

The primary factors controlling the BOD assimilative capacity of surface waters are dilution, temperature, and the relative rates of biochemical deoxygenation and atmospheric reoxygenation. Important too is the minimum dissolved oxygen concentration that may be tolerated. The influence of photosynthetically produced oxygen may sometimes be a very significant factor in the oxygen balance of surface waters. Toxic materials retard the metabolic activities of the microorganisms, thereby decreasing the BOD rate.

Often the waste-assimilative capacity of surface waters is limited by substances other than those which exert an oxygen demand. Toxic materials and those which cause excessive tastes and odors or color are typical substances which limit the assimilative capacity. The controlling factors in such cases are the available dilution and tolerable limits dictated by the pollution-control agency.

VI. WATER-QUALITY REQUIREMENTS FOR SPECIFIC USES

The value of an ample water supply to serve any beneficial use depends on two basic considerations. First, it depends on the water-quality requirements for the particular use and, secondly, on the feasibility and costs of treating the raw water to meet these requirements.

It should be recognized at the outset that the limits outlined here and detailed in the more comprehensive discussions of the general literature are not necessarily rigid. In many situations where the volumes of water used are great, the "limits" are really no more than guides that are useful primarily in comparing several different sources of supply. In some instances, where the amount of water used is small, or water quality is a vital factor in the application, the limits become more rigid and must be met. The application of "standards" must be tempered with judgment, based on all the factors concerning the intended use of the water.

A. Domestic and Municipal Water Supply

Water supplied for domestic and municipal purposes should satisfy the physical, chemical, and bacterial criteria which indicate the safety of water for ingestion, culinary, and sanitary purposes. In addition, the water-quality characteristics of a community supply should fulfill two other important criteria. Water delivered to the consumer must be wholesome and palatable, or many people will reject the general supply and turn to other supplies which are more appealing, but may be less safe.

Furthermore, some recognition of the water-quality requirements of other municipal-water users is necessary. Selection of a water-supply source or the degree of treatment to be provided should take into consideration the general quality requirements of the community, as well as the specific requirements for drinking water. For example, a community economically dependent on tourist trade might find it advisable to provide high-quality water that would be entirely acceptable in all respects to the transient population.

The water-quality requirements of most municipal supplies in the United States are related to the U.S. Public Health Drinking Water Standards of 1962 [8]. Although these Standards officially apply only to "carriers and others subject to Federal Quarantine regulations," they are generally accepted by most state and local health agencies. They have been endorsed by the American Water Works Association [8, 39]. An interesting review of such standards since their inception in 1914 was presented in a symposium given by the American Water Works Association [40].

The 1962 Drinking Water Standards establish water-quality limits on the bacterial, physical, radiological, and chemical characteristics of the supply. Bacterial standards specify the sampling frequency to be used and the allowable presence of coliform bacteria which indicate pollution. Physical standards limit turbidity to 5 units, color to 15 units, and threshold-odor number to 3 units.

The chemical requirements of the 1962 Drinking Water Standards are divided into mandatory and recommended criteria. The mandatory limits are:

Parts per million

Arsenic	0.05
Fluoride	3.4*
Lead	0.05
Selenium	0.01
Silver	0.05
Barium	1.0
Cadmium	0.01
Chromium	0.05
Cyanide	0.2

* Subject to temperature characteristics of the water supply.

Recommended limits are:

	Parts per million
Iron	0.3
Manganese	0.05
Nitrate	45
Detergent (ABS)	0.5
Copper	1.0
Magnesium	50
Zinc	5
Chloride	250
Sulfate	250
Chloroform-soluble extract	0.2
Phenol	0.001
Total dissolved solids	500

Radiological water-quality criteria limit radioactivity in drinking water to:

Radium (226)	3 $\mu\mu$c/liter
Strontium (90)	10 $\mu\mu$c/liter
Gross beta activity	1,000 $\mu\mu$c/liter

Other factors are important with respect to the application of the radiological limitations, and the 1962 Standards should be consulted.

The Drinking Water Standards are not universally applicable, and the regulations of the specific health agency should be consulted to determine the detailed requirements for these and other water-quality characteristics.

B. Industrial Water Supply

Industrial water-quality requirements are highly dissimilar. There are so many uses for water in manufacturing that a complete list of such requirements cannot readily be catalogued. It would be impossible to present here specific quality standards in the great number of industries where water is essential for processing and other uses. Discussions will be limited to certain industries where the quality of water is an all-important consideration. The operation of such industries, as related to their water supplies, is given merely to illustrate the problems, rather than to cover the entire field in this respect.

1. Food and Beverage Industries. Water used in the food and beverage industries should satisfy the general requirements for drinking water, in addition to those for the other process industries. Canning, brewing, ice, and similar industries require water that does not contain objectionable turbidity, color, or tastes and odors. High concentrations of sodium salts are objectionable in the manufacture of ice. Iron and manganese are particularly undesirable. Some hardness is desirable for certain brewing operations, while excessive hardness is highly objectionable in the processing of some vegetables for canning or freezing. The presence of saprophytic (nonpathogenic) organisms is detrimental for use in the food and beverage industries.

Ordinarily, water from underground sources is preferred to surface-water supplies which are subject to contamination, but the very high demand for water in these industries frequently precludes the exclusive use of groundwater because of the limited supply generally available from most water-bearing aquifers. In the larger plants, therefore, water from surface sources must be employed, and since it is seldom that these supplies are suitable in their raw state, treatment is required.

2. Water for Textile and Paper Manufacture. The most important requirements with regard to water quality in the paper and textile industries are freedom from color, turbidity, and suspended matter, and there are also limitations on the content of hardness, manganese, and iron. Control of slime formations, algae growths, and various microorganisms is commonly necessary in paper mills, especially when the raw water is heavily polluted.

Water softening and the removal of color, suspended matter, iron, and managanese are readily effected by standard treatment and filtration facilities. High concentrations of soluble organic matter greatly retard coagulation of color or turbidity as well as the softening reactions of the lime-and-soda process. Difficulty has also been experienced in the removal of iron and manganese in the presence of certain types of organic material.

If reasonably soft water is available for use in the textile industry, softening treatment may not be essential, but this is seldom the case. If hard water is used, the alkali solution (usually soda) results in the formation of calcium and magnesium salts, which are highly objectionable. It is particularly difficult to mercerize fabrics where hard water is used, because of the failure of adequate penetration of the dyes which are subsequently applied.

Difficulty has been experienced at some locations in the textile and silk industries when hard-water supplies have been substituted for a relatively soft public water supply which had been treated to prevent corrosion by maintaining a high pH value. The hard water caused precipitation of salts during penetration of the weighting material into the fabric. Difficulty was also experienced in the woolen industry because of the detrimental effect of the hard water in some steps in the processing of such material.

3. Pharmaceutical and Antibiotic Requirements. Water entering pharmaceutical and biological products requires a higher degree of purification than is necessary in any other industry. Water for these purposes cannot be used satisfactorily as received from any of the ordinary sources available. Even with the most meticulous water-conditioning operation involving clarification and softening, the water produced is still unsatisfactory for some of the more refined requirements in this industry.

Where water is to be used for the manufacture of serum, it must be pyrogen-free. The term pyrogen is not widely recognized in the waterworks field, but is used to define specific requirements in the pharmaceutical industry. The definition of pyrogen is given in Stedman's Medical Dictionary as follows: "An agent which causes a rise in temperature." It is probable that the pyrogens are complex bodies induced by certain types of microscopic organisms or possibly the products of bacterial metabolism. Although the nature and character of pyrogens are still in the "twilight zone" of exact knowledge, much study and research are being conducted to establish the character of these products.

It has been demonstrated that simple distillation does not destroy pyrogens and that they are carried over with the stream or vapor and reappear in the condensate. In solution they withstand the ordinary sterilizing procedures, and although they may be present only in infinitesimal quantities, they are not detected by the usual forms of analysis. Even when present in minute quantities, pyrogens can cause considerable physiological upset when they enter the bloodstream.

To assure pyrogen-free water, serum manufacturers depend on triple distillation, and in some cases complete demineralization of the water by means of ion-exchange equipment is practiced prior to distillation.

4. Boiler Feedwater. Water used for make-up to steam-generating units must satisfy very complex and exacting water-quality requirements. Steam quality and boiler deposits are directly related to the quality of the feedwater. Natural water supplies are unsatisfactory for use as feedwater sources without treatment, even for the low-pressure boilers. The quality of water for use in high-pressure boilers (above 2,000 psi) is the most exacting for any use. Water for such purposes must be treated to procure a purity of 1 ppm, or less, of total dissolved solids.

Water quality which induces or directly causes corrosion of the steam-generating systems and their auxiliary components is universally important. The pH of the feedwater and the amounts of dissolved oxygen and carbon dioxide present are the most significant factors controlling corrosion. Other gases and certain dissolved and suspended solids are also factors in the mechanism of corrosion of ferrous and non-ferrous metals.

The most important solids which must be reduced to very low concentrations are silica, those which contribute to hardness and alkalinity, and metals, such as iron copper, and nickel. Silica can be entrained with steam and subsequently deposited in steam turbines, requiring costly maintenance and repair.

5. Cooling Water. Cooling water should not cause corrosion, form scales, or promote the growth of slimes. The quality requirements for cooling water depend on the type of system used. Water used in once-through cooling systems need not be as good in quality as that used in recirculating systems. Recirculating systems require water of somewhat better quality, since the dissolved solids concentrations increase and must be controlled.

The more important impurities in water which affect its utility for cooling purposes are scale-forming constituents (hardness), suspended matter, dissolved corrosive gases, acids, oil or other organic matter, and slime-forming organisms.

Surface-water supplies used once through for cooling or water recirculated over cooling towers may introduce problems related to all undesirable impurities. Suspended solids in the cooling water are especially objectionable when they lodge on heat-exchanger surfaces. Such materials, when accompanied by bacterial slimes and corrosion products, cause marked loss in cooling efficiency and accelerated corrosive attack on both ferrous and nonferrous metals.

6. Water-quality Requirements for Internal-combustion Engines. Diesel and gasoline engines require water for cooling purposes only. Water is circulated around the combustion chambers to remove the heat produced there, which otherwise would cause excessively high temperatures and damage to the metal of the machinery. For this purpose, a fairly large flow of circulating water is necessary. This water should be free of suspended matter and scale-forming solids in solution. It should also be noncorrosive, unless the water system is constructed of corrosion-resistant alloys.

With a closed primary cooling system, a high-grade water can be used economically, as the losses are negligible. Such water can be produced by softening and the addition of corrosion-inhibiting chemicals.

The heat absorbed by the primary coolant in the engine can be removed by means of a heat exchanger using either air or water as the secondary coolant. Water for this purpose can be of poor quality provided the heat exchanger is constructed of corrosion-resistant material. Salt water is used successfully in many installations.

If the secondary coolant contains a large amount of scale-forming solids, the water has to be treated to avoid heavy encrustations on the heat-exchanger equipment. If the water is discharged to waste from a once-through system, the cost of treatment cannot be justified. However, if the secondary cooling water is conserved by circulating in a cooling tower, some form of treatment may be employed economically.

7. Transporting and Process. Waters that are used to transport or process raw materials must not interfere with the physical, chemical, or biological reactions of the process, deteriorate the quality of the product, or necessitate excessive repair and maintenance of the equipment used. The actual requirements vary widely.

Color is objectionable in many such uses of water, especially in the production of fine-quality papers and textiles. Iron and manganese are similarly objectionable since they may cause staining. High turbidities and excessive tastes and odors are undesirable where they might impair the quality of the product.

Limitations on hardness vary widely, and excessive amounts are undesirable in supplies for the pulp and paper, textile, photography, and other industries. High concentrations of dissolved solids are objectionable for most uses, and the presence of certain ions may interfere with the normal process reactions.

Other important criteria for water supplies for transportation and processing are corrosiveness and the tendency to create and maintain slime growths. The pH value, the amounts of dissolved oxygen and carbon dioxide present, and the alkalinity of the water are important in evaluating the corrosive nature of the supply. Aquatic organisms may produce slime growths which interfere with industrial operations in general.

The water-quality requirements for various industrial uses is a very complex subject and cannot be covered completely here. The foregoing discussion is not intended as a comprehensive and exhaustive discussion of water-quality requirements for various purposes. Rather, it is intended merely to indicate the general importance of water quality and to assist in appraising the value of a potential water supply.

C. Agricultural Water Supply

Agricultural water supplies are used primarily for stock watering and irrigation. The quality of water used for these purposes is a very important factor in determining the acceptability of the supply.

Stock generally require water similar to that which would be suitable for human consumption, although many animals will tolerate water of somewhat poorer quality. Limits on dissolved solids concentrations range up to about 10,000 ppm, depending on the type and age of the stock, the period of use, climatic conditions, and the relative concentrations of specific ions. High proportions of sodium, magnesium, and sulfate are undesirable.

Water consumed by stock should not contain excessive amounts of toxic substances such as selenium, arsenic, lead, zinc, nitrates, fluorides, and other metals. Bacterial contamination is a potential source of infectious diseases. High concentrations of blue-green algae in water consumed by stock have been reported as the cause of fatal poisoning in many instances.

The quality of water used for irrigation is well recognized as an important factor in productivity and quality of the irrigated crops. The most important constituents or properties of irrigation water are boron, per cent sodium, total dissolved solids, chlorides, and sulfates. Specific conductivity is often used as a measure of the suitability of water for irrigation.

It is difficult, if not impossible, to specify limiting water-quality characteristics which are generally applicable. In addition to the various tolerance levels of the different crops, other factors are important in the evaluation of water quality and its effects on crops. Drainage of the irrigated lands is usually a major consideration and may sometimes be more significant than the quality of the irrigation water. Soil types, climatic conditions, and irrigation practices are all important factors. In addition, the relative composition of the water may have a significant effect.

Specific tolerance levels for the various quality characteristics of irrigation water have been reported in some detail [1-3, 9, 13, 14]. These values should be used only with proper recognition of the factors mentioned above.

D. Aquatic Life

The effects of water quality on the survival and successful propagation of the various forms of aquatic life may be broadly classed as (1) direct toxic effects, (2) physiological effects, and (3) environmental effects. These classes are not clearly defined, and there are complex interrelationships between each.

Many substances are directly toxic to aquatic life and, when present in sufficient concentrations, can be fatal. The toxicity of the various substances depends not only on the concentration in the water, but also on the synergistic[1] and antagonistic effects of the other water-quality characteristics. The time-concentration relationships are also important.

Detailed information on lethal concentrations of toxic substances for fish and other aquatic life is abundant in the literature [2, 3, 41-43], but all data must be used with extreme caution.

Concentrations of dissolved oxygen, carbon dioxide, and ammonia are important factors controlling the physiological effects of water quality on aquatic life. In addition, temperature, pH, dissolved solids, and turbidity affect the feeding and spawning activities of most aquatic life. Essential nutrients should be present, along with a balanced and unbroken food chain.

Factors which affect the general environment may also influence the propagation of aquatic life. Silt, clay, and other suspended materials may significantly decrease the penetration of light into the water, thereby decreasing or stopping photosynthetic activity. This may have important effects on the availability of food organisms.

---

[1] *Synergistic* refers to the condition where the total effect is greater than the sum of the independent effects taken separately. The opposite to synergistic is *antagonistic*.

Settleable solids may destroy natural habitats and smother marine shellfish. Physiological and environmental water-quality characteristics have been considered in detail by the Ohio River Valley Water Sanitation Commission and many other groups [44–46].

E. Recreation

Waters that are used for swimming, boating, and aesthetic enjoyment should be free from obnoxious floating or suspended substances and objectionable colors or odors. In addition, bathing waters should not contain substances or organisms which are harmful or irritating when ingested or in contact with the skin. The presence of high concentrations of pathogenic bacteria is undesirable, although the general requirements are less strict than those for drinking water. Health agencies have established standards for recreational waters where swimming is permitted.

VII. WATER CONDITIONING FOR SELECTIVE USES

A. Conventional Methods

Long experience has clearly demonstrated that the quality of water in underground and surface supplies generally does not meet definite specifications for water quality in industrial or domestic uses. Trends in industrial expansion have demonstrated need for water of the highest quality. Public water supplies generally do not meet the specifications, even though the supply may be entirely satisfactory for general municipal uses. Refined processing for special uses is a function of private management. Under conditions where industry depends on private water sources, analogous treatment will occur, the only difference being in the relative quality of the raw material. Beyond this point the refinement of water quality needed will be dependent on the specific product manufactured.

Because of the wide variety of water-quality specifications to meet man's requirements, there are a great variety of treatment processes from which selection can be made. These include plain sedimentation in the impoundments, sedimentation aided by coagulants to speed up the rate of sedimentation, filtration through sand, coal, or other materials, softening either by chemicals or by the passage through ion-exchange resins, distillation, and a variety of selection and arrangement of the foregoing individual processes. Detailed descriptions of these processes and their application do not come within the scope of this text. The reader should refer to the extensive literature[1] on water conditioning and the specific application of processes to meet the quality of specifications required [47–50].

B. Demineralization of Saline Waters

The conversion of sea or brackish waters[2] into usable fresh water has a strong appeal to people in arid regions. In fact, it is being considered as a means to increase supplies in many areas having reasonable amounts of rainfall. The degree of interest and hope of accomplishment are, in general, proportional to local needs and to the short supply of available fresh water.

Most of the known procedures for desalting have been classified broadly into three major groups according to the type of process used, namely, physical, chemical, and electrical. Within each of these general classifications are many specific methods of water conditioning. They include vaporization, crystallization, sublimation, adsorption, ultrasonics, osmosis, ion exchange, electric-ion migration, and numerous other processes and phenomena.[3]

[1] Valuable information on water-treatment methods is given in the *Journal of the American Water Works Association*.

[2] Typical chemical analyses of brackish surface and groundwater supplies are given in Tables 19-3 and 19-4. The average composition of sea water is given in Table 19-5.

[3] For detailed descriptions of these processes, the reader should refer to Saline Water Conversion Reports, issued by U.S. Department of the Interior, Office of Saline Water, 1952–1960.

WATER CONDITIONING FOR SELECTIVE USES

Table 19-3. Typical Analysis of Well Water in Brackish-water Areas
(Results in parts per million—weighted averages)

Location	Dissolved solids	Chloride	Hardness	
			Total	Noncarbonate
Minot, N.D.............	1,280	240	334	0
Hutchinson, Kans........	853	230	352	144
Amphitheater, Ariz......	930	288	25	0
Roswell, N.M...........	1,160	185	664	484
Alice, Tex..............	1,290	448	202	0
Lubbock, Tex...........	1,200	126	589	242
Midland, Tex...........	2,790	698	1,250	1,040
Ogden, Utah............	1,120	595	435	302
Bakersfield, Calif........	1,010	411	628	

Of interest at present is the method of electro-ion migration utilizing membranes which selectively remove cations or anions of the dissolved salts in saline waters. This process has been the subject of much experimentation, and a number of fairly large scale installations are now operating in the United States and in many other countries. New synthetic membranes being developed give hope of increasing the efficiency of this method.

Table 19-4. Typical Analysis of Surface Water Subject to Demineralization Processes
(Results in parts per million—weighted averages)

River and location	Dissolved solids	Chloride	Hardness	
			Calcium magnesium	Non-carbonate
Saline River at Tescott, Kans.........	1,790	622	472	243
Crooked Creek near Nye, Kans........	1,300	674	330	148
Virgin River at Littlefield, Ariz........	1,870	281	1,030	795
Gila River below Gillespie Dam, Ariz...	1,700	620	534	354
Pecos River near Artesia, N.M........	2,580	562	1,280	1,160
Carlsbad Main Canal at Head near Carlsbad, N.M...................	4,020	941	1,990	1,890
Pecos Railroad east of Malaga, N.M....	1,680	392	828	714
Pecos River near Acme, N.M.........	1,820	218	1,110	1,010
Brazos River at Possum Kingdom Dam near Graford, Tex................	1,370	445	476	388
Pecos River below Red Bluff Dam near Orla, Tex........................	3,350	1,150	1,140	1,030
Salt Fork Brazos River near Aspermont, Tex.......................	3,220	1,360	752	656
Double Mt. Fork Brazos River near Rotan, Tex......................	1,300	270	554	455
Salt Fork Brazos River near Peacock, Tex.............................	2,610	1,160	583	470
Arkansas River at Ralson, Okla.......	1,230	541	264	145
Cimarron River at Perkins, Okla......	2,070	1,000	323	202
Cimarron River at Mannford, Okla....	2,420	1,230	435	310
Dolores River near Cisco, Utah.......	1,050	316	371	260

Generalization of any world problem may be grossly misleading and unreliable. This is true of the separation of fresh water from saline supplies. The physical, sociological, economic, and political status of peoples and the meteorological and geographical environments in widely separated regions are so varied that there can be no single solution to the problem. These and other factors must be considered before any decision is reached as to the desalting process most suitable in a particular area.

The crux of realistic accomplishment in demineralization of saline water anywhere rests largely on permissible cost. Determination of the cost of desalted water requires a comprehensive survey of the possibilities of selective water use. Statistical data are needed for appraisal of the economic practicability of partial or total use of desalted water for miscellaneous regional needs.

Table 19-5. Percentage Composition of Dissolved Solids in Sea Water and Amount of Various Ions for "Average" Sea Water

Ion	Symbol	Per cent composition	Amount in "average" sea water, ppm
Chloride...........	Cl^-	55.04	18,980
Sodium............	Na^+	30.61	10,556
Sulfate............	$SO_4^=$	7.68	2,649
Magnesium.........	Mg^{++}	3.69	1,272
Calcium...........	Ca^{++}	1.16	400
Potassium.........	K^+	1.10	380
Bicarbonate.......	HCO_3^-	0.41	140
Bromide...........	Br^-	0.19	65
Boric acid.........	H_3BO_3	0.07	26
Strontium.........	Sr^{++}	0.04	13
Others............	......	0.01	2
Total salts........	......	100.00	34,483

The permissible cost of separating fresh water from saline supplies anywhere depends on the urgency of the existing needs, whether they be for irrigation, industrial, municipal, or other uses. Local conditions must govern the acceptable cost of using all types of saline supply. Comparison of treatment cost in different areas is misleading unless allowance is made for local influences. In areas where no fresh water is available, the acceptable cost of demineralization bears little relation to that of areas having relatively abundant natural fresh-water supplies.

In regions where saline water only is available and all fresh water must be obtained by importation or conversion, strict conservation of all fresh water used must be enforced. Scarcity of fresh water promotes greater tolerance of lower quality in water for many uses which in nonarid areas would be considered more demanding. The economics of any particular situation will always govern the choice between fresh and saline supplies. Untreated or partially desalted saline water must be utilized where and when possible to limit consumption of costly demineralized water, thus minimizing the overall water cost to consumers.

There are many arid and semiarid regions near the sea or other saline-water sources where demineralization systems could be developed. Augmentation of fresh-water supplies in many of these locations is necessary because of industrial growth. The establishment of industries in areas where existing water supplies are inadequate to meet the civil, agricultural, and industrial needs also depends on an additional source of fresh water.

The need for saline-water conversion is not limited to coastal areas. Many inland areas which do not have adequate fresh water have available saline sources. Many inland water sources usually considered to be fresh water are, in reality, fresh only seasonally, and for parts of each year should be classified as saline.

VIII. SOURCES OF WATER-QUALITY DATA

A problem frequently encountered in hydrologic studies requiring information on water quality is that of locating and securing the necessary data. There are many sources of such data, some of which are readily apparent and others not generally recognized. The data sources mentioned here are intended as general guides in obtaining information on water quality.

The basic references to sources of water-quality data for the United States are three publications of the Federal Inter-Agency Committee on Water Resources, Subcommittee on Hydrology. These publications are:

Bulletin 2—Inventory of published and unpublished chemical analyses of surface waters in western United States (October, 1948).

Bulletin 6—Inventory of published and unpublished chemical analyses of surface waters in western United States (February, 1954).

Bulletin 9—Inventory of published and unpublished chemical analyses of surface waters in eastern United States, 1947–55 (September, 1956).[1]

The largest quantity of data on water quality is readily available in *Water-Supply Papers*, published by the U.S. Geological Survey. Many of these data are contained in two series of annual reports, "Quality of Surface Waters of the United States" and "Quality of Surface Waters for Irrigation, Western United States." Additional comprehensive information on the quality of natural waters in specific areas and the quality of municipal water supplies is included in other publications of the Geological Survey.[2]

Another excellent source of data is the National Water Quality Network, coordinated by the U.S. Department of Health, Education, and Welfare, Public Health Service, with the cooperation of state and local agencies. The data collected include radioactivity measurements, plankton and coliform organisms, organic chemicals, and conventional measurements of other constituents and properties. Annual reports of these data are published. Additional data are frequently available as the result of other activities of the U.S. Public Health Service.

Additional information may be available from other United States Federal agencies such as Corps of Engineers, Fish and Wildlife Service, Tennessee Valley Authority, U.S. Department of Agriculture, and others.

Various interstate, state, and local agencies collect and disseminate water-quality data. Interstate agencies are usually those formed to guide river-basin activities such as Ohio River Valley Water Sanitation Commission, Interstate Commission on the Delaware River Basin, Interstate Commission on the Potomac River Basin, and several others. State agencies engaged in water-resources development, pollution control, wildlife management, and similar activities are potential sources of water-quality data. Certain local groups, including universities, engineering offices, water departments, and some promotional agencies, maintain records of water quality for the local region.

IX. REFERENCES

1. Hem, J. D.: Study and interpretation of the chemical characteristics of natural water, *U.S. Geol. Surv., Water-Supply Paper* 1473, 1959.
2. Water quality criteria, *California Water Pollution Control Board Publ.* 3, 1952.
3. Water quality criteria, addendum no. 1, *California Water Pollution Control Board Publ.* 3 (*Addendum* 1), 1954.
4. "Standard Methods for the Examination of Water and Wastewater," American Public Health Association, American Water Works Association, and Water Pollution Control Federation, 11th ed., 1960.
5. "Water Quality and Treatment," 2d ed., American Water Works Association, 1950.
6. Rainwater, F. H., and L. L. Thatcher: Methods for collection and analysis of water samples, *U.S. Geol. Surv. Water-Supply Paper* 1454, 1960.

[1] This is a supplement to *Bulletin* 2.

[2] See complete list in "Publications of the Geological Survey," available from the U.S. Department of the Interior, Geological Survey.

7. U.S. Geological Survey: Chemistry of iron in natural water, pts. A–G, *Water-Supply Paper* 1459, 1959–1960.
8. Public Health Service Drinking Water Standards, 1962. *Public Health Serv. Publ.* 956, 1962.
9. Scofield, C. S., and L. V. Wilcox: Boron in irrigation water, *U.S. Dept. Agr. Tech. Bull.* 264, 1931.
10. Report on oily substances and their effects on the beneficial uses of water, *California Pollution Control Board Publ.* 16, 1956.
11. Truesdale, G. A., A. L. Downing, and G. F. Lowden: The solubility of oxygen in pure water and sea-water, *J. Appl. Chem.*, vol. 5, pp. 53–62, February, 1955.
12. Lohr, E. W., and S. K. Love: The industrial utility of public water supplies in the United States, 1952, pt. 1, States east of the Mississippi River, *U.S. Geol. Surv. Water-Supply Paper* 1299, pp. 13–28, 1954.
13. Wilcox, L. V.: The quality of water for irrigation use, *U.S. Dept. Agr. Tech. Bull.* 962, 1948.
14. Salinity Laboratory Staff: "Diagnosis and Improvement of Saline and Alkaline Soils," U.S. Department of Agriculture Handbook 60, 1954.
15. Manual on industrial water and industrial waste water, *Am. Soc. Testing Mater. Spec. Tech. Publ.* 148-D, 1959.
16. "Official Methods of Analysis of the A.O.A.C.," 8th ed., Association of Official Agricultural Chemists, 1955.
17. Doudoroff, Peter, and others: Bio-assay methods for the evaluation of acute toxicity of industrial wastes to fish, *Sewage Ind. Wastes*, vol. 23, no. 11, pp. 1380–1397, November, 1951.
18. Hutchinson, G. E.: "A Treatise on Limnology," vol. 1, "Geography, Physics, and Chemistry," John Wiley & Sons, Inc., New York, 1957.
19. Welch, P. S.: "Limnology," 2d ed., McGraw-Hill Book Company, Inc., New York, 1952.
20. Sverdrup, H. U., M. W. Johnson, and R. H. Fleming: "The Oceans: Their Physics, Chemistry, and General Biology," Prentice-Hall, Inc., Englewood Cliffs, N.J., 1942.
21. Johnstone, James: "An Introduction to Oceanography," University Press, Liverpool, 1928.
22. Coker, R. E.: "This Great and Wide Sea," University of North Carolina Press, Chapel Hill, N.C., 1947.
23. Marmer, H. A.: "The Sea," Appleton-Century-Crofts, New York, 1930.
24. Beckman, H. C., and others: Water quality control considerations in the Red River compact, *J. Water Pollution Control Federation*, vol. 32, no. 7, pp. 761–774, July, 1960.
25. Senate Select Committee on National Water Resources: Water resources activities in the United States, pollution abatement," 86th Congr., 2d Sess., Comm. Print no. 9, pp. 30–31, 1960.
26. *Proc. Natl. Conf. on Water Pollution*, Dec. 12–14, 1960, U.S. Public Health Service.
27. Rudolfs, Willem (ed.): "Industrial Wastes: Their Disposal and Treatment," Reinhold Publishing Corporation, New York, 1953.
28. Studies of waste water reclamation and utilization, *California Water Pollution Control Board Publ.* 9, 1954.
29. Report on the investigation of travel of pollution, *California Water Pollution Control Board Publ.* 11, 1954.
30. A survey of direct utilization of waste waters, *California Water Pollution Control Board Publ.* 12, 1955.
31. Third report on the study of waste water reclamation and utilization, *California Water Pollution Control Board Publ.* 18, 1957.
32. Senate Select Committee on National Water Resources: Water resources activities in the United States, present and prospective means for improved reuse of water, 86th Congr., 2d Sess., Comm. Print no. 30, 1960.
33. Report on continued study of waste water reclamation and utilization, *California Water Pollution Control Board Publ.* 15, 1956.
34. Diachishin, A. N., S. G. Hess, and W. T. Ingram: Sewage disposal in tidal estuaries, *Proc. Am. Soc. Civil Engrs.*, vol. 79, separate no. 167, 1953.
35. Kersnar, F. J., and D. H. Caldwell: Ocean outfall studies at San Diego, *Sewage Ind. Wastes*, vol. 25, no. 11, pp. 1336–1343, November, 1953.
36. Streeter, H. W., and E. B. Phelps: A study of the pollution and natural purification of the Ohio River. III. Factors concerned in the phenomena of oxidation and reaeration, *Public Health Bull.* 146, February, 1925 (reprinted by Public Health Service, 1958).
37. Fair, G. M.: The dissolved oxygen sag: an analysis, *Sewage Works J.*, vol. 11, no. 3, pp. 445–461, May, 1939.

REFERENCES

38. Proceedings of Seminar on oxygen relationships in streams, *U.S. Public Health Ser. Tech. Rept.* W58-2, Cincinnati, Ohio, Oct. 30–Nov. 1, 1957.
39. American Water Works Association: AWWA resolution endorses USPHS Standards as 'minimum,' *Willing Water*, vol. 5, no. 11, pp. 1–2, October, 1961.
40. Derby, R. L., and others: Water quality standards, *J. Am. Water Works Assoc.*, vol. 52, no. 9, pp. 1159–1188, September, 1960.
41. Ellis, M. M.: Detection and measurement of stream pollution, *U.S. Bur. Fisheries Bull.* 22, 1937.
42. Doudoroff, P., and Max Katz: Critical review of literature on the toxicity of industrial wastes and their components to fish. I. Alkalies, acids, and inorganic gases, *Sewage Ind. Wastes*, vol. 22, no. 11, pp. 1432–1458, November, 1950.
43. Doudoroff, P., and Max Katz: Critical review of literature on the toxicity of industrial wastes and their components to fish. II. The metals, as salts, *Sewage Ind. Wastes*, vol. 25, no. 7, pp. 802–839, July, 1953.
44. Ohio River Valley Sanitation Commission, Aquatic Life Advisory Committee: Aquatic life water quality criteria, First Progress Report, *Sewage Ind. Wastes*, vol. 27, no. 3, pp. 321–331, March, 1955.
45. Ohio River Valley Sanitation Commission, Aquatic Life Advisory Committee: Aquatic life water quality criteria, Second Progress Report, *Sewage Ind. Wastes*, vol. 28, no. 5, pp. 678–690, May, 1956.
46. Ohio River Valley Sanitation Commission, Aquatic Life Advisory Committee: Aquatic life water quality criteria, Third Progress Report, *J. Water Pollution Control Federation*, vol. 32, no. 1, pp. 65–82, January, 1960.
47. Powell, S. T.: "Water Conditioning for Industry," McGraw-Hill Book Company, Inc., New York, 1954.
48. Nordel, Eskel: "Water Treatment for Industrial and Other Uses," Reinhold Publishing Corporation, New York, 1951.
49. Baylis, J. R.: "Elimination of Taste and Odor in Water," McGraw-Hill Book Company, Inc., New York, 1935.
50. Fair, G. M., and J. C. Geyer: "Water Supply and Waste-water Disposal," John Wiley & Sons, Inc., New York, 1954.

Section 20

HYDROLOGY OF URBAN AREAS

STIFEL W. JENS, *Consulting Engineer, St. Louis, Missouri.*
M. B. McPHERSON, *Professor of Hydraulic Engineering, University of Illinois.*

I. Introduction	20-2
II. Qualitative Description of Urban Stormwater Runoff	20-4
A. Precipitation	20-4
B. Interception	20-5
C. Infiltration	20-5
D. Depression Storage and Detention	20-5
E. Overland Flow	20-5
F. Gutter Storage	20-6
G. Conduit Storage	20-7
III. Quantitative Determination of Urban Stormwater Runoff	20-7
A. Empirical Formulas	20-7
B. The Rational Method	20-8
C. Correlation Studies of Rainfall and Runoff	20-9
D. The Hydrograph Method	20-10
1. General	20-10
2. Los Angeles Hydrograph Method	20-10
3. Chicago Hydrograph Method	20-16
4. Synthetic Storm Pattern and Areal Distribution of Rainfall	20-21
E. The Inlet Method	20-22
1. General	20-22
2. Peak Inlet Flows	20-22
3. Inlet-flow Routing	20-22
4. Simplified Procedure	20-26
5. Design Example	20-27
6. Limitations	20-29
7. Comparison of Measured and Computed Runoff	20-29
F. Gutters and Inlets	20-30
1. General	20-30
2. Gutter Capacities	20-30
3. Inlets	20-30

The authors are indebted for constructive comments to Dr. John C. Geyer and Dr. Warren Viessman of The Johns Hopkins University, to Dr. Paul Bock of Travelers Research Center, to Mr. A. L. Tholin and Mr. Clint J. Keifer of the City of Chicago, and to Mr. Lyall A. Pardee of the City of Los Angeles.

   G. Manholes and Junction Chambers................... 20-31
   H. Open-channel Storm Drainage...................... 20-32
 IV. Floods in Urban Areas................................. 20-33
 V. Water Supply.. 20-33
 VI. Water Pollution...................................... 20-33
 VII. Airport Hydrology.................................... 20-34
   A. General.. 20-34
   B. Objectives..................................... 20-34
   C. Hydrologic Data................................ 20-34
   D. Subsurface Drainage............................ 20-35
   E. Surface Drainage............................... 20-35
     1. General................................... 20-35
     2. Grading................................... 20-35
     3. Ponding................................... 20-35
 VIII. Hydrology of Urban Expressways........................ 20-41
   A. General.. 20-41
   B. Design Frequency............................... 20-41
   C. Time of Concentration.......................... 20-41
   D. Gutters and Inlets............................. 20-42
   E. Pumping Stations at Grade Separations........... 20-42
   F. Use of Pondage at Interchanges.................. 20-42
 IX. References... 20-43

I. INTRODUCTION

In 1950 there were 168 metropolitan complexes in the continental United States containing cities with populations exceeding 50,000. These largest urban centers housed 56 per cent of the population on only 7 per cent of the land area [1]. In 1960 the number of continental metropolitan complexes had increased to over 200, with about 63 per cent of the population on 9 per cent of the land area.

It is generally accepted that the contemporary trend toward more intensive urbanization which exists in the United States and in nearly all other nations should continue through the remainder of the century. As a consequence, urban problems associated with the hydrologic aspects of water management should become increasingly more acute. The hydrology of urban areas is already quite complex.

Savini and Kammerer [2] have reported on a review, classification, and preliminary evaluation of the significance of the effects of urbanization on the hydrologic regimen. They stated:

The continuing growth and concentration of population and industry in urban and suburban areas in recent decades has caused a complex merging of social, economic and physical problems. The interrelationships of man and his use and development of the land and water resources is a particularly significant aspect of urbanization, but there has been relatively little study to date of the effect of urban man upon natural hydrologic conditions.

Their analysis of the hydrologic effects during a selected sequence of changes in land and water use associated with urbanization are summarized in Table 20-1. They classify the generic hydrologic effects of urbanization according to

(1) the sequence of usual occurrence, (2) changes separately associated with man's use of water and man's use of the land, (3) type of hydrologic process affected, or (4) changes affecting quantity of water on the one hand and quality of water on the other.

Specific effects resulting from urbanization on man's use of water include:

 1. Increase in both total use and per capita use.
 2. Increasing development of new water-supply sources that may require transportation over great distances.

INTRODUCTION

Table 20-1. Hydrologic Effects during a Selected Sequence of Changes in Land and Water Use Associated with Urbanization*

Change in land or water use	Possible hydrologic effect
Transition from preurban to early-urban stage:	
Removal of trees or vegetation	Decrease in transpiration and increase in storm flow. Increased sedimentation of streams.
Construction of scattered city-type houses and limited water and sewage facilities	
Drilling of wells	Some lowering of water table.
Construction of septic tanks and sanitary drains	Some increase in soil moisture and perhaps a rise in water table. Perhaps some waterlogging of land and contamination of nearby wells or streams from overloaded sanitary drain system.
Transition from early-urban to middle-urban stage:	
Bulldozing of land for mass housing, some topsoil removed, farm ponds filled in	Accelerated land erosion and stream sedimentation and aggradation. Increased flood flows. Elimination of smallest streams.
Mass construction of houses, paving of streets, building of culverts	Decreased infiltration, resulting in increased flood flows and lowered groundwater levels. Occasional flooding at channel constrictions (culverts) on remaining small streams. Occasional overtopping or undermining of banks of artificial channels on small streams.
Discontinued use and abandonment of some shallow wells	Rise in water table.
Diversion of nearby streams for public water supply	Decrease in runoff between points of diversion and disposal.
Untreated or inadequately treated sewage discharged into streams or disposal wells	Pollution of stream or wells. Death of fish and other aquatic life. Inferior quality of water available for supply and recreation at downstream populated areas.
Transition from middle-urban to late-urban stage:	
Urbanization of area completed by addition of more houses and streets and of public, commercial, and industrial buildings	Reduced infiltration and lowered water table. Streets and gutters act as storm drains, creating higher flood peaks and lower base flow of local streams.
Larger quantities of untreated waste discharged into local streams	Increased pollution of streams and concurrent increased loss of aquatic life. Additional degradation of water available to downstream users.
Abandonment of remaining shallow wells because of pollution	Rise in water table.
Increase in population requires establishment of new water-supply and distribution systems, construction of distant reservoirs diverting water from upstream sources within or outside basin	Increase in local streamflow if supply is from outside basin.
Channels of streams restricted at least in part to artificial channels and tunnels	Increased flood damage (higher stage for a given flow). Changes in channel geometry and sediment load. Aggradation.
Construction of sanitary drainage system and treatment plant for sewage	Removal of additional water from the area, further reducing infiltration and recharge of aquifer.
Improvement of storm drainage system	A definite effect is alleviation or elimination of flooding of basements, streets, and yards, with consequent reduction in damages, particularly with respect to frequency of flooding.†

Table 20-1. Hydrologic Effects during a Selected Sequence of Changes in Land and Water Use Associated with Urbanization (Continued)

Change in land or water use	Possible hydrologic effect
Drilling of deeper, large-capacity industrial wells	Lowered water-pressure surface of artesian aquifer; perhaps some local overdrafts (withdrawal from storage) and land subsidence. Overdraft of aquifer may result in salt-water encroachment in coastal areas and in pollution or contamination by inferior or brackish waters.
Increased use of water for air conditioning	Overloading of sewers and other drainage facilities. Possibly some recharge to water table, due to leakage of disposal lines.
Drilling of recharge wells	Raising of water-pressure surface.
Waste-water reclamation and utilization	Recharge to groundwater aquifers. More efficient use of water resources.

* From Savini and Kammerer [2].
† Added by authors.

3. Increasingly frequent conflicts wherein two or more types of water users seek the same supply.
4. Diminished streamflow as a result of diversions of water.
5. Declining water levels and pressure in ground water reservoirs. [Also causing pollution of groundwater by leakage from sanitary sewers.]
6. Increasing number of artificial recharge projects, for purposes of water supply and flood control.
7. Increase in amount of wastes disposed to streams and possible increase in pollution when wastes are inadequately treated.
8. Increased re-use of waste water in agriculture and industry.
9. Land subsidence.

Many of the above effects are so interrelated as a consequence of diminishing quality of supply and increasing demands for quantity, that assignment of relative importance to them is not practical. The primary significance probably is that they are interrelated.

This section will outline current practice in the use of hydrologic data and methods in the solution of urban water problems and needs. Stormwater drainage is given major emphasis, not only because of its considerable economic significance, but also because of the growing evidence of dissatisfaction with established methods of runoff determination and the consequent attempts to develop more realistic and accurate, yet practical, engineering designs. In addition, brief mention is made of the utilization of urban hydrology in connection with designs dealing with floods, water supply, pollution, airports, and expressways.

II. QUALITATIVE DESCRIPTION OF URBAN STORMWATER RUNOFF

The engineer designing facilities for the collection and disposal of stormwater can exercise better judgment if he understands what actually occurs from the time a runoff-producing storm starts until the storm and runoff cease. The principal phases of this part of the hydrologic cycle are as follows.

A. Precipitation

Flow in storm sewer systems is principally by gravity. Like natural drainage basins, smaller sewer branches unite with larger branches, and so on, until a main sewer is reached. The smallest catchment area, of the order of an acre in size, is that tributary to an inlet. For most smaller areas in the upper reaches of an urban drainage system the time required to reach peak runoff after the beginning of a storm is a matter of minutes. Hence high-intensity, short-duration rainfall is normally the main, if not sole, type of precipitation contributing to critical runoff rates. This type of rainfall is usually associated with thunderstorms. Intensity-duration-frequency data of such rainfall in the United States were first analyzed by Yarnell, but are now being modified and improved with additional data by the U.S. Weather Bureau (Subsec. 9-VI-A).

B. Interception

Quantitatively, rainfall interception by vegetation is rarely of importance in connection with urban storm drainage and may properly be ignored in design (Secs. 6 and 22).

C. Infiltration

The phenomena associated with infiltration (Sec. 12) indicate that most field infiltration-capacity curves approach a steady, minimum rate after one or two hours. Relative minimum infiltration capacities for three broad soil groups are as follows [3]:

Soil group	Infiltration capacity, in./hr
Sandy, open-structured	0.50–1.00
Loam	0.10–0.50
Clay, dense-structured	0.01–0.10

The great influence of vegetal cover is evidenced by the fact that bare-soil infiltration capacity can be increased 3 to 7.5 times with good permanent forest or grass cover, but little or no increase results with poor row crops.

Soil infiltration capacity is also affected by antecedent precipitation, such as high-intensity rains of short duration coming after a dry period. However, very few quantitative data are available for evaluating this factor.

D. Depression Storage and Detention

Of the precipitation which reaches roofs, pavements, and pervious surfaces, some is trapped in the many shallow depressions of varying size and depth present on practically all surfaces. The specific magnitude of depression storage has not been measured in the field because of obvious difficulties in obtaining meaningful data. On pervious surfaces where interception is negligible or absent (as in the urban environment), approximate depression storage can be evaluated for rains which start abruptly with intensities in excess of infiltration capacity. The accumulated rainfall less the mass infiltration, in the interval between the beginning of rainfall excess and the start of direct runoff, would equal depression storage plus the amount of detention required to initiate runoff.

The term *detention* refers to the storage effect due to overland flow in transit. Horton [4] stated that this initial detention "commonly ranges from $\frac{1}{8}$ to $\frac{3}{4}$ inch for flat areas and $\frac{1}{2}$ to 1.5 inches for cultivated fields and for natural grass lands or forests." On moderate or gentle slopes he estimated that pervious surface depressions "can commonly hold the equivalent of $\frac{1}{4}$ to $\frac{1}{2}$ inch depth of water and even more on natural meadow and forest land."

Based upon analysis of periods of high rates of rainfall and runoff, Hicks [5] has used depression storage losses of 0.20, 0.15, and 0.10 in. for sand, loam, and clay, respectively. In determining overland flow supply, Tholin and Keifer [6] assumed for one series of analyses an overall total depression storage of $\frac{1}{4}$ in. on pervious areas with a range of depths of specific depressions up to $\frac{1}{2}$ in. and $\frac{1}{16}$ in., on paved areas with a range of depths up to $\frac{1}{8}$ in.; for a second series these depths were doubled.

E. Overland Flow

Flow across a sloping plane surface at unsteady state was studied by Horton [7] and Izzard [8] (see Sec. 14). Horton proposed an equation for overland flow considered suitable for turbulent flow with high discharge on natural surfaces, whereas Izzard developed a dimensionless hydrograph for surface flow which is presumed largely applicable to laminar flow on developed surfaces. There is limited experimental support for the Horton overland-flow equation. The Izzard dimensionless hydrograph was verified in laboratory tests and gave computed overland-flow hydrographs agreeing closely enough for all practical purposes with the measured hydrographs. Reasonable agreement was even achieved for turbulent sheet flow across very wide airport aprons.

Izzard's experiments also considered interruptions, decreases, and increases in uniform supply rates of rainfall, as well as the simulation of runoff from pavements draining across turf [8, p. 137]. The experiments were performed with combinations limited to $iL < 500$, where i is the rainfall intensity in inches per hour, and L is the length in feet of overland flow. Practical guides were developed for the approximation of overland-flow hydrographs under similar circumstances. Izzard developed both the rising and recession portions of a dimensionless overland-flow hydrograph. Inasmuch as the Horton equation applies to a uniform rainfall supply of unlimited duration, a complete hydrograph for a specified duration can be generated by using the principle of superposition (Sec. 14). However, there is no experimental verification of the Horton equation for the recession portion. Because of its comparative simplicity and adaptability to nonuniform supply, the Izzard dimensionless hydrograph is a more useful design tool.

F. Gutter Storage

Routing the overland-flow hydrograph through storage in the gutter or channel leading to an inlet requires an evaluation of the instantaneous storage under the water-surface profile for various rates of flow at the inlet. The overland flow entering the gutter is zero at the upstream end, increasing progressively downstream. The flow in the gutter is spatially varied, and its longitudinal water-surface profile is very complex but has been discussed in detail elsewhere [6, 8–10]. Gutter storage generally has a greater peak-reducing influence than the surface detention of overland flow and requires a longer time to achieve equilibrium outflow. Long gutters sometimes provide a surplus of storage above that required to accommodate the rainfall excess, which results in a maximum gutter outflow rate at the inlet less than the equilibrium rate.

In any gutter reach, at a given rate of outflow there is a related volume of storage, but on the rising side of the hydrograph the storage is greater than for the identical outflow rate on the falling phase. From this fact, Izzard [8] was able to develop a short method for the determination of peak gutter outflow, and equations were derived to determine time to reach equilibrium flow in the gutter, gutter storage at equilibrium, lag time (from the rate of discharge at the end of rain to the same discharge rate on the recession), and the gutter-flow recession curve. All these relationships were predicated upon a uniform rainfall rate.

In an urban runoff study for the City of Los Angeles, Hicks [5] developed equations relating gutter storage and time of flow to supply rate, gutter length, and slope (Subsec. III-D-2-i). He stated [8] that Los Angeles flume experiments with uniformly increasing lateral inflow verified the general shape of analytical rate-length curves obtained via Izzard's equations.

The following approximate, modified Manning equation has also been presented [8] for computing uniform flow in shallow, wide, triangular channels such as swales and gutters:

$$Q = 0.56 \left(\frac{z}{n}\right) S^{1/2} y^{8/3} \qquad (20\text{-}1)$$

where Q is the discharge in cfs, y is the maximum depth of water in ft, z is the ratio of water-surface width to y, n is the Manning's coefficient of roughness consistent with the constants in the equation, and S is the longitudinal slope of the channel. A nomograph for this equation is available [11, fig. 9; 12, fig. 12-27].

G. Conduit Storage

The volume of detention in a conduit can effect a reduction in the peak rate of flow of a hydrograph in the same basic way as any detention storage attenuates the height of an inflow hydrograph. Storage routing can be applied if satisfactory discharge-storage relationships are available. This necessitates the computation of instantaneous backwater curves. Since only the rate of change in storage is necessary to solve the storage equation, it is considered expedient to assume a uniform flow condition for each discharge rate and compute the conduit volume occupied by the flow. This requires a knowledge of actual or assumed conduit cross sections.

In Los Angeles, Hicks [5] has developed percentage-of-peak-reduction factors (Table 20-2) for various times of concentration, based on the premise that increments of conduit detention are independent of the conduit slope or velocity. The total occupied volume in cubic feet for any specific discharge q in cubic feet per second is therefore equal to $60t_f q$, with t_f the time of flow in the reach in minutes.

In Chicago, Tholin and Keifer [6] used a time-offset method for conduit routing because of its simplicity compared with other routing methods (Subsec. III-D-a-Steps 6 & 7) and because more refined routing procedures had not led to significantly different results.

III. QUANTITATIVE DETERMINATION OF URBAN STORMWATER RUNOFF

The financing, installation, and maintenance of urban drainage facilities is usually a tax-supported public service under the responsibility of municipal governments. Human life is seldom threatened by the flooding of these facilities. The principal detrimental effects of flooding are damage to the below-ground sections of buildings and hindrance of traffic. The consequences of flooding range from clearly assessable property destruction to annoying inconvenience. It follows that provision of complete protection from flooding can only rarely be justified. Instead, facilities are designed which will be overtaxed infrequently. There is an obligation to the public, in so far as it is practicable, to equate the cost of a given design to the probable protection and service which it will afford. Otherwise, an adequate criterion for the "most economical design" does not exist.

Some characteristics of precipitation data can be arrayed and frequencies determined with satisfactory exactitude, such as the frequency distribution of maximum average point rainfall of a given duration. Definition of storm-pattern frequency is considerably more elusive. Because there are so many physiographic and physical variables that affect the hydrologic and hydraulic aspects of urban stormwater runoff, there is a tendency to regard the frequency of a runoff rate as the frequency of an associated rainfall event, by default. Until recent years exceedingly little attention was given to urban drainage research. There exists a serious need for much more extensive investigation, consistent with the huge public expenditures in these facilities. Despite the difficulties, a reasonably accurate prediction of average runoff frequency is clearly a desirable fundamental design responsibility.

A. Empirical Formulas

Late in the last century engineers developed empirical formulas to determine design discharges for storm drains. Many such formulas took the general form

$$Q = CAI \left(\frac{S}{A}\right)^x \qquad (20\text{-}2)$$

where Q is the peak discharge in cfs, C is a coefficient, depending on climatic and physiographic conditions of the watershed, A is the drainage area in acres, I is the average rainfall intensity in in./hr, S is the slope of the drainage basin in ft per 1,000 ft, and x is an exponent.

For the famous Bürkli-Ziegler formula, $x = 0.25$. For the McMath formula, $x = 0.5$. The C value of both formulas varies from 0.20 for pervious rural areas to 0.75 for highly impervious built-up areas. About 100 empirical formulas have been collected by Chow [13]. Because of the development of other methods for runoff determination, use of empirical formulas has become almost obsolete in modern engineering design practice.

B. The Rational Method

The rational method, currently used by many design engineers [9], is usually expressed in terms of the following equation:

$$Q = CIA \tag{20-3}$$

where Q is the peak discharge in cfs, C is the runoff coefficient, depending on the characteristics of the drainage area, I is the uniform rate of rainfall intensity in in./hr for a duration equal to the time of concentration, and A is the drainage area in acres.

Average values of the coefficient C commonly used in current practice are detailed in Ref. 9.

The time of concentration is defined as the time which would be required for the surface runoff from the remotes part of the drainage area to reach the point being considered. This time will vary, generally depending on the slope and character of surfaces. Where the drainage area served by an inlet is entirely paved, the time of concentration is assumed to vary from about 5 to 10 min as the length of flow to the inlet varies from 100 to about 500 ft. For turfed areas the time is usually considered to vary from about 10 min for lengths of flow less than 100 ft to about 30 min for 400 to 500 ft. For bare ground the time might be taken somewhere between the values for paved and turfed areas, decreasing with the expected smoothness of the surface.

When the drainage area consists of several different types of surface, the time of concentration is determined by adding the respective times estimated for flow over the lengths of the several surfaces along the path from the remotest point to the inlet. The remotest point should be considered as that from which the time of flow is longest. However, this will not necessarily be the most remote point from the standpoint of measured distance.

Detailed consideration of the several components constituting inlet concentration times is often circumvented through establishment of a fixed *inlet time* for particular types of highly developed urban areas, with 5 to 15 min in common use [9].

The use of a uniform rainfall intensity for a duration equal to the time of concentration is a simplifying assumption since rainfall does not truly persist at a uniform intensity for even so short a time as 5 min. Small watersheds (such as most urban drainage areas) possess neither the overall detention storage nor long times of concentration and other peak-flow-reducing characteristics of large watersheds. Indeed, the brief interval between the occurrence of short, intense rainfall and succeeding peak runoff has a significant effect on the magnitude of the peak rate. The influence of this effect becomes less as the size of the watershed increases.

The choice of a value of C is the most intangible aspect in the use of the *rational method*. Taken literally, it represents the multiplier of a 100 per cent runoff peak (with no infiltration or storage) required to obtain the design peak. This coefficient has to account for the various climatic conditions and physiographic characteristics of the watershed. The judgment required in estimating the value of C is therefore considerable.

For a refinement of the rational method, some designers use values of C varying with the length of time prior to a thorough wetting of the soil. In this procedure it is essential that modification of the coefficient be started at the beginning of the time of concentration, which occurs usually after the beginning of the storm. Other designers use the *zone principle* [14] with or without joint application of the first procedure. By the zone principle, or *time-contour analysis*, the drainage area is divided into zones by contours, each connecting the points from which water will flow to the outlet in an equal time. Each zone is assigned an appropriate value of runoff coefficient, depending on the imperviousness of the area, which is reduced progressively with increasing outlet concentration times. The total discharge is taken as the summation of the discharges from the various zones. Thus an average runoff coefficient for the drainage area can be computed. Proper application of either procedure requires selection of a design pattern of rainfall distribution.

Further refinements of the rational method have been suggested. In 1932 Gregory and Arnold [15] developed a *general rational formula* by taking into account such factors as watershed shape and slope, the pattern of the stream system, and the elements of channel flow. This general rational formula may take the form of Eq. (20-3), in which

$$I = \frac{KT^x}{t^e} \qquad (20\text{-}4)$$

where K = a coefficient depending on geographic location
T = recurrence interval in years for a rainfall intensity of I in in./hr for a duration of t min to be reached or exceeded
x, e = exponents depending on geographic location
t = duration of rainfall intensity in min equal to time of concentration t_c expressed as follows:

$$t_c = \frac{J^{1/e}}{(CAK)^g T^{xg}} \qquad (20\text{-}5)$$

where $g = 1/(4 - e)$
J = a watershed factor = $1/PFS$, where P is a shape factor, F a channel hydraulic factor, and S a slope factor

In 1938 Bernard [16] modified the runoff coefficient in the general rational formula by

$$C = C_{\max} \left(\frac{T}{100}\right)^x \qquad (20\text{-}6)$$

where $C_{\max}$ is a limiting runoff coefficient corresponding to a recurrence interval of 100 years. He also prepared charts for values of $C_{\max}$, e, K, and x for the humid central and eastern half of the United States.

C. Correlation Studies of Rainfall and Runoff

The difficulties and expense attending simultaneous measurement of rainfall intensities and flow rates in stormwater inlets and sewers no doubt account for the relatively few such data available.

Horner and Flynt [17] in 1936 reported an analysis of measurements of intense rainfall and runoff collected over the period 1914 to 1933 for three small urban areas in St. Louis. Numerous correlations were studied in an attempt to find an orderly, relatively simple, and usable relationship between rainfall and runoff. The study included an examination of the variation of P, the ratio of accumulated actual runoff at any particular time after the beginning of the storm to the accumulated computed 100 per cent runoff. The instantaneous values of actual and computed 100 per cent runoff were also compared. It was found that there is appreciable difference between P and the corresponding instantaneous ratios.

For any given drainage area the values of all runoff-rainfall ratios varied between wide limits. No evidence was found of a general correlation between rainfall characteristics of storms and final values of the ratio of accumulated measured runoff to accumulated rainfall. The ratio appears to be affected by antecedent precipitation and seasonal and general climatic conditions.

A close correlation was found between the ratio of maximum measured runoff to maximum computed 100 per cent runoff taken over a 5-min duration period and the ratio of total measured volume of runoff to total measured volume of rainfall for the entire period of the storm.

As a second major part of these St. Louis studies, frequency analyses were made of the measured runoff and rainfall rates considered as independent phenomena. Curves were developed for rainfall and runoff of equivalent probabilities of occurrence.

D. The Hydrograph Method

1. General. Recognition of actual rainfall and runoff phenomena has led to an approach termed the *hydrograph method*, which is based on the following conditions: (See also works by Eagleson, Viessman and Geyer, and Willeke under "other references" in Subsec. IX.)

1. For urban areas, runoff is the residual of rainfall after deducting that which is trapped in surface depressions and infiltrated into the soil.

2. Net rainfall or runoff from the time it begins to flow over an urban lot area is further modified by detention.

3. Detention is the time-varying volume of runoff water which must be built up to induce flow over lots, in gutters, and in conduits.

2. Los Angeles Hydrograph Method. *a. General.* Since 1939, the City of Los Angeles and many of its satellite communities have been using a hydrograph method

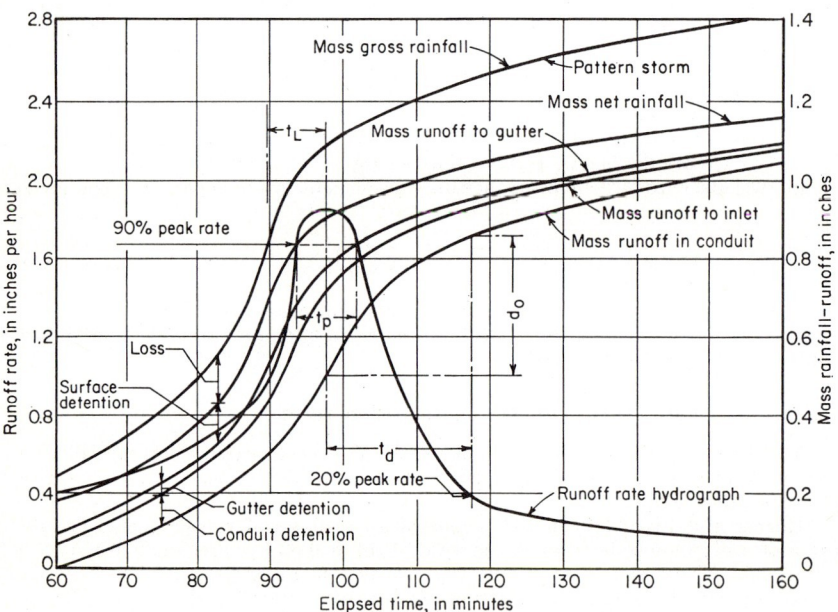

FIG. 20-1. Elements of the hydrograph. (*After Hicks* [5].)

first reported in 1944 [5]. To develop the necessary design charts and graphs in this method, the following data were used: rainfall records from automatic recording and standard rain gages maintained by the U.S. Weather Bureau and other agencies; infiltration capacities for various local soils from experimental data; experimental data for characteristics of overland flow over various types of soils at different slopes; runoff records from gaged urban areas using stilling-well automatic stage records, striped maximum-peak-runoff gages, and stage recorders in flumes; classification of drainage areas based upon zoning requirements, soil maps, and field-survey measurements; and runoff factors F_{RO}, developed from experimental data.

b. *Rainfall.* Since the mixed terrain of mountains, valleys, and coastal plains greatly affects rainfall distribution, a regional index map was prepared showing several isohyetals for the 1-hr 50-year rainfall [5]. Corresponding values for other frequencies and durations were expressed in terms of the values shown on this map.

A design storm pattern was determined by plotting the mass curves for the major storms of the Los Angeles Weather Bureau records with the center of the most intense 5-min rainfall period at a common point. Illustrated in Fig. 20-1 as the mass gross rainfall, this design storm (referred to as a 10-year storm) is between a "medium" and "delayed" pattern.

Areal distribution of rainfall intensity was obtained for various times of concentration [5] by applying the method developed by Marston [18] to a large area centering in Pasadena, Calif., where records were available from a large number of gages. Allowance for areal reduction was made for areas larger than 100 acres.

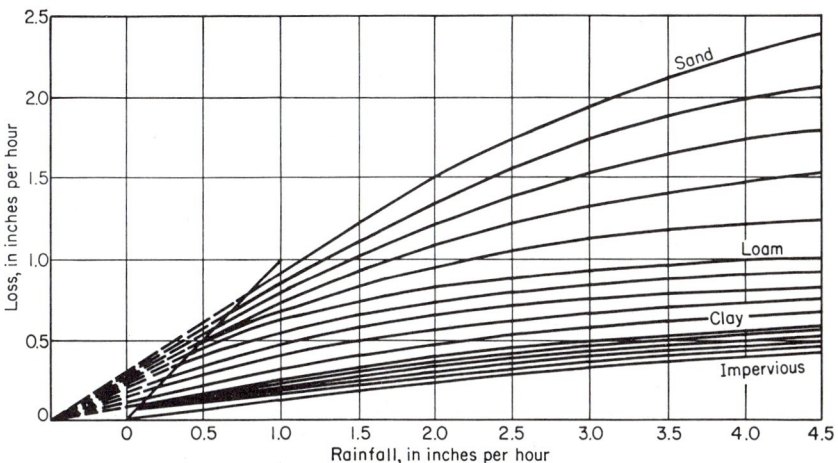

FIG. 20-2. Rainfall vs. loss curves. (*After Hicks* [5].)

For determining antecedent rainfall conditions, use was made of a weighted *moisture factor M*, which relates to the 60-day antecedent seasonal evaporation and weighted 60-day antecedent precipitation.

c. *Infiltration.* The rainfall-vs.-loss curves of Fig. 20-2 were developed from the data for infiltration tests on small soil plots using artificial precipitation, supplemented by analyses of rainfall-runoff data from large drainage areas. These curves are for a moisture factor M of 1, which represents a well-saturated soil, characteristic of the lower capacities usually found in the flat portion of the infiltration-capacity curves. These lower values are used in design for the Los Angeles area because of the relatively high antecedent precipitation of the delayed-peak design storm pattern and of low evaporation rates during the winter rainy season. A measure of depression storage was also determined from some of the large-area rainfall-runoff records and incorporated in the design procedure.

d. *Overland Flow.* In the experiments on overland flow at Los Angeles [5, 19] artificial rainfall was applied to surfaces of various materials in a long flat-bottomed trough with variable slope and length. The following general formulas were derived:

$$\delta_a = \frac{C_1 l^m \sigma^n}{S^p} \tag{20-7}$$

and

$$t_c = \frac{C_2 l^x}{\sigma^y S^z} \tag{20-8}$$

where δ_a is the average depth of flow in in. representing detention effect, t_c is the time of overland flow in min measured from the beginning of supply rate of rainfall excess to the time of full runoff, l is the length of overland flow in ft, σ is the supply rate of rainfall excess in in./hr, and S is the slope of the surface in percentage. The coefficients and exponents in the formulas depend on the surface material as follows:

Surface	C_1	m	n	p	C_2	x	y	z
Tar and sand.......	0.0136	0.318	0.535	0.4615	1.3	0.323	0.64	0.448
Tar and gravel......	0.0257	0.384	0.351	0.367	2.23	0.373	0.684	0.366
Clipped sod.........	0.078	0.322	0.325	0.281	9.34	0.298	0.785	0.302

The time t_c also represents the time difference between the mass curves of rainfall supply and runoff. For an approximation, the time for overland flow may be computed by $t_c = 60\delta_a/\sigma$. The results obtained from these experiments were later found to be in close agreement with those obtained by Izzard [8].

e. Time of Concentration. In urban drainage design, the time of concentration t_c is taken as the sum of the following: time of lot runoff; time of flow in gutters, open swales, or channels to an inlet; and time of flow in the storm drain. From inlet-time charts such as Fig. 20-3, the time of concentration is selected for the initial block for the various types of soil and surface conditions. These charts were derived from analyses of experimental data.

Peak runoff rates for the design storm pattern, corresponding to the 10-year rainfall curve proportioned from the 50-year, 1-hr isohyetal of 1.33 in. passing through the location of the Los Angeles Weather Bureau Station and for 100 per cent imperviousness were computed for various values of t_c and are given in Table 20-2 (column A).

Because the basic peak runoff rates (Table 20-2) are particularly time-sensitive for small t_c, the inlet time should be based on an average for the first few blocks at the head of each trunk and lateral. If the uppermost area has low runoff rates with prolonged flow concentration (such as parks, cemeteries, etc.) and a major part of the lower area is fast-concentrating and highly productive of runoff, the inlet time of concentration should be based upon a weighted average for each of the areas.

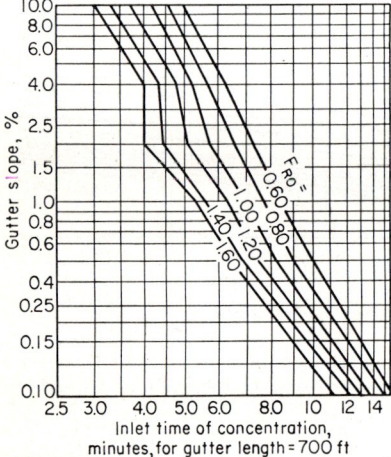

Fig. 20-3. Inlet time of concentration for soil type and classification: 100; 70-1, 70-2, 70-3; 60-1, 60-2, 60-3. (*From City of Los Angeles.*)

f. Soil Types. Los Angeles soils are placed in three principal imperviousness classifications: clay, loam, and sand, designated, respectively, by 1, 2, and 3 for ready reference.

g. Zoning Classifications. Surface improvements are grouped to conform with six broad classifications of 1960 zoning for areal improvements. Surveys have been made to determine the pervious and impervious percentage and the amount of impervious area directly tributary to the gutter without flow over pervious areas.

h. Factor of Runoff F_{RO}. Inlet hydrographs were developed for various combinations of surface and soil. Since these were all similar in shape, a basic inlet hydrograph for 100 per cent imperviousness and 10-year rainfall frequency proportioned from the

1.33 isohyetal of the 50-year, 1-hr rainfall was routed for t_c of 5 to 120 min, yielding the basic peak runoff rates in Table 20-2. Design peak runoff rates are obtained by multiplying the basic peak rate by the appropriate runoff factor F_{RO}, taken from charts, such as Fig. 20-4, which were developed from experimental data for various land-use classifications and soil combinations. The figures on the chart, such as

Table 20-2. Basic Runoff Rates and Reduction Factors for Conduit Detention in Los Angeles*

(Column A: peak runoff rates in cfs/acre; column B: peak-reduction factor due to conduit detention in %/min of flow)

t_c, min	A	B	t_c, min	A	B
5	3.40†	12.00	32	1.51	1.62
6	3.13	10.00	34	1.47	1.51
7	2.93	8.60	36	1.43	1.42
8	2.76	7.50	38	1.40	1.32
9	2.63	6.60	40	1.37	1.24
10	2.50†	5.90	42	1.34	1.18
11	2.41	5.40	44	1.31	1.12
12	2.32	4.85	46	1.29	1.05
13	2.24	4.45	48	1.27	1.00
14	2.17	4.10	50	1.24	0.96
15	2.10	3.80	52	1.22	0.92
16	2.05	3.50	54	1.20	0.88
17	1.99	3.30	56	1.18	0.84
18	1.94	3.10	58	1.17	0.80
19	1.90	2.90	60	1.15	0.77
20	1.85†	2.75	70	1.07	0.64
22	1.78	2.45	80	1.01	0.55
24	1.71	2.25	90	0.96	0.47
26	1.65	2.04	100	0.92	0.42
28	1.60	1.88	110	0.88	0.37
30	1.55	1.75	120	0.85	0.34

* From Hicks [5].
† Illustrated in Hicks [5, p. 1241, fig. 15].

70-1, indicate the surface-area imperviousness in per cent and the permeability in terms of soil type, such as 70 per cent and no. 1 imperviousness for clay. The "isohyetal" refers to 50-year, 1-hr rainfall values.

i. Gutter Flow. Analytical and experimental studies [19] of the effect of gutter storage on hydrographs resulted in formulas similar to Eqs. (20-7) and (20-8), with $C_1 = 0.00044$, $m = 0.697$, $n = 0.794$, $p = 0.417$, $C_2 = 0.0443$, $x = 0.732$, $y = 0.252$, and $z = 0.39$. However, the value of t_c thus computed is the time of flow with fully developed gutter storage; for t_c as defined in Eq. (20-8), multiply by 1.33.

These studies were made on a 30-ft roadway with Manning's $n = 0.012$ and a 4.5-in. crown and zero cross slope, representing average conditions. Equal increments of flow were introduced at regular intervals along the length of a V-shaped experimental trough. The hydrograph of inflow to the inlet was obtained by routing the gutter-inflow hydrograph through the gutter storage.

j. Conduit Storage. To develop a satisfactory method for reflecting the effects of conduit storage, peak-reduction factors (Table 20-2, column B) for conduit detention were computed [5, pp. 1237–1241] and are expressed in percentage per minute of flow (in reach of conduit under design) for various values of t_c, the time of concentration.

HYDROLOGY OF URBAN AREAS

The percentage of peak-rate reduction for a given t_c is irrespective of the slope of the conduit, since the related volume of storage is independent of either the slope or the velocity in the conduit.

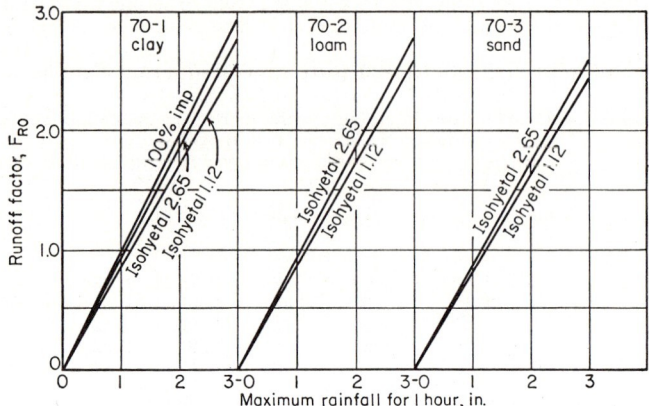

FIG. 20-4. Runoff factors F_{RO}. (*After Hicks* [5].)

k. Development of the Runoff Hydrograph. The mass curve of gross rainfall (in inches) is plotted against clock time, which is the actual time in minutes of the most intense part of the storm. This time is renumbered so that zero time is equal to 1,062 min after start of the particular storm shown in Fig. 20-1, because the design storm pattern has a 24-hr (1,440-min) duration. Beginning of the time of concentration occurs at 1,062 min, which is taken as the initial starting time for the design of storm drains. By doing so, the storage effect is considered and complete saturation of the soil in the drainage area is assumed. Any of the mass curves thus plotted can be translated to rate curves, which are comparable for plotting with the mass and rate curves for the runoff in conduits.

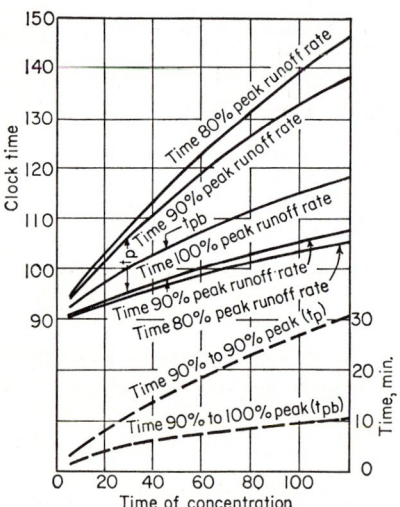

FIG. 20-5. Time elements of the hydrograph. The time of drawdown (100 to 20 per cent peak runoff rate) equals the time of concentration t_c for any t_c exceeding 20 min. (*After Hicks* [5].)

l. Time Elements of the Hydrograph. Certain time elements of the hydrograph (Figs. 20-1 and 20-5) are defined as follows:

Time lag t_L is the time interval in minutes between the centroid of the effective rainfall and the peak runoff rate.

Peak time t_p is the time interval in minutes between the two runoff rates equal to 90 per cent of the peak rate on each side of the runoff peak.

Time of drawdown t_d is the recession time in minutes from 100 to 20 per cent of the peak runoff rate.

From a study of the time elements, the following equation for peak runoff was derived in terms of t_d above:

$$q_c = \frac{K d_0}{t_d} \qquad (20\text{-}9)$$

where q_c is the computed peak runoff rate in in./hr, d_0 is the volume of runoff on the recession side of the hydrograph between 100 and 20 per cent peak runoff rate, in in. of depth on the drainage area, and K is a coefficient varying from 96 for a 5-min time of concentration to 108 for 60 min.

m. *Computation of Runoff.* There are two methods of computing runoff: the *peak-rate method* and the *method of summing hydrographs.* The former is more easily applied. The latter is used when the former is inapplicable.

PEAK-RATE METHOD. By this method the initial time of concentration to the uppermost inlet is obtained from charts such as Figs. 20-3 and 20-4, after evaluating the slope of the gutter, the land-use classification, and the soil type. The basic peak-runoff rate for this initial time (Table 20-2) is multiplied by the factor of runoff (F_{RO} from chart similar to Fig. 20-4) to give the unit runoff in cubic feet per second per acre. This quantity times the acreage of the initial drainage area gives the peak runoff rate in cubic feet per second. The procedure of computation is as follows:

1. List the areas at their respective points of concentration.
2. Subclassify the areas according to zoning classification (Subsec. III-D-2-g).
3. Select the 1-hr rainfall rate from a storm isohyetal map, considering the isohyetal and frequency (Subsec. III-D-2-b).
4. When the drainage area exceeds 100 acres, adjust the 1-hr rate by the areal distribution factor (Subsec. III-D-2-b).
5. Determine the runoff factors (Fig. 20-4) for the adjusted 1-hr rainfall rate and apply them to the basic runoff rate for the time of concentration in Table 20-2 to derive runoff rate per acre by the rational method.
6. Follow through with the routine tabling computation.
7. When a lateral is added, the time of concentration is to be adjusted accordingly.

Experience has caused Los Angeles to modify some of the assumptions and details of methodology since publication of the original method [5]. It has been found that in most instances the cost to construct storm drains to the most remote block of a drainage area is prohibitive. Runoff reaches a storm sewer only after flowing overland for several blocks, usually concentrating in the streets. Because of this, upper watershed flow entering a street intersection may leave by more than one outflow street. From a number of experiments made from 1952 to 1953 (necessarily covering a limited number of conditions), a factor termed *outflow variation* was developed which can be used to modify the theoretical flow distribution for actual field conditions.

Also, the design practice reported in 1944 [5] assumed that storm drains would flow at the most efficient free-flow depth. The design policy in 1960, however, provided for flow with the storm drain slightly under pressure to ensure carrying the design runoff even from a storm greater than the design storm.

Below the junction of a lateral and trunk storm drain, the hydrograph may be quite distorted from the typical hydrograph. The hydrograph for the lateral is taken at the junction, whereas that for the trunk drain is taken at the upstream design point preceding the junction. The time of concentration at such a junction of drains A and B can be adjusted by the following:

$$t_{cx} = \frac{t_{ca}q_a + t_{cb}q_b}{q_a + q_b} \qquad (20\text{-}10)$$

where t_{cx} is the adjusted time of concentration in min, t_{ca} is the time of concentration of drain A in min, t_{cb} is the time of concentration of drain B in min, q_a is the flow in drain A, and q_b is the flow in drain B. The times of concentration in this formula are at the junction of the two drains, regardless of which is the lateral or main drain.

For the main drain, the hydrograph is drawn for the peak flow and concentration time at the location preceding the junction. The peak flow will be unchanged from this location to the junction, but at the junction it will be reduced because of the increase in time of concentration. It will be further reduced by the increase in conduit detention (and the rainfall distribution factor if applicable) because of addition of the lateral and main drain areas.

Open-channel or free-flowing storm drains require an allowance for the velocity of the flood wave, which was found to be about 1.5 times the mean velocity computed for the peak-flow depth. The time of flow so calculated (two-thirds of the time for mean free-flow velocities) for the conduit or channel reach immediately above the junction would thus be added to the time of concentration at the design point preceding the junction.

For pressure drains, the entire time of flow is added to the previous time of concentration, based on the average flowing full velocity.

The traveled hydrograph of the main line and that immediately upstream in the main line should necessarily have the same volumes, and the former must be adjusted until the volumes become the same. The adjusted traveled hydrograph and the lateral hydrograph are summed. The result is usually a distorted hydrograph, which must be corrected to the proper shape between the 80 per cent peak runoff values. This is done with the aid of the time-element relationships of Fig. 20-5. From the corrected combined hydrograph, the peak runoff rate and time of concentration are used for continuation of design computation.

The peak-rate method has been programmed for an electronic computer. This computer program (unpublished) is in use by the City of Los Angeles to determine the surface flows which will reach a proposed storm drain or a freeway in cut or fill and to develop a master plan of the storm drain system.

METHOD OF SUMMING HYDROGRAPHS. The procedure for this method involves the development of a hydrograph from a drainage area at point A on a drain, routing of this hydrograph through a reach of the conduit to point B, where it is combined with all hydrographs from other sources tributary to point B, routing the summed hydrograph to another junction point C, and so on through the system. This method is used when the peak-rate method is inapplicable, usually under the following conditions:

1. When t_{ca} (Eq. 20-10) exceeds 60 min, the method of summing hydrographs is used for the continuation of the trunk design, and the laterals are designed by the peak-rate method.

2. When a retarding reservoir is part of a system, the outfall below the reservoir is designed by the summation of hydrographs, after the inflow to the reservoir has been routed through the reservoir by the routing formula (outflow = inflow ± rate of storage change).

3. Other limitations may be imposed by the judgment of the designer.

n. Comparison of Measured and Computed Runoff. Figure 20-6 shows that the majority of peak runoff rates computed by the Los Angeles hydrograph method are within a tolerance of 20 per cent plus or minus of the actual measured rates (i.e., the two lines). It is also believed that the peak-rate method is a more scientific approach than any other method for the mixed mountainous and alluvial-plain areas and the rainfall pattern in the vicinity of Los Angeles.

3. Chicago Hydrograph Method. *a. The Procedure.* The City of Chicago since the end of World War II has made extensive comprehensive studies of the rainfall-runoff relationship [6, 9, 20]. The hydrograph method thus developed can be summarized in the following 10 steps:

STEP 1: STREET LAYOUT AND LAND USE. A system of parallel main drains at half-mile intervals is assumed, with sets of laterals, each serving two 5-acre blocks (330 by 660 ft) on each side of the main sewer. More than 10,000 such lateral sewers are in the City of Chicago. The most common type of land use, designated by "type 5," is represented by a 10-acre unit with 46.3 per cent imperviousness (36.0 per cent for area of directly connected house roof to sewer and pavement to inlet and 10.3 per cent for areas such as garage roofs and walks discharging onto pervious areas). The following assumptions are introduced: (1) Sidewalks draining onto pervious areas are considered a part of the pervious area; (2) building roofs are converted into a uniform strip of equivalent area, and their flow is assumed to enter the lateral sewer through the house drains; and (3) garage roofs (type 5, Land Use) discharge water onto backyard lawns and are considered a part of the pervious areas. For the other three types of land use studied, part or all of the flow from garage roofs is added directly to the alley gutter.

STEP 2: DESIGN STORM PATTERN. To develop the net supply chronologically, there must be a rainfall hyetograph of specific shape and size from which can be subtracted the appropriate infiltration and depression-storage values. For a specific duration, the most important factors affecting the peak runoff rate are:
1. Volume of water falling during the maximum period
2. Amount of antecedent rainfall
3. Time location of peak rainfall intensity

The first factor can be obtained from an intensity-duration-frequency curve of rainfall. The 5-year frequency curve was converted into an equivalent synthetic hyetograph.

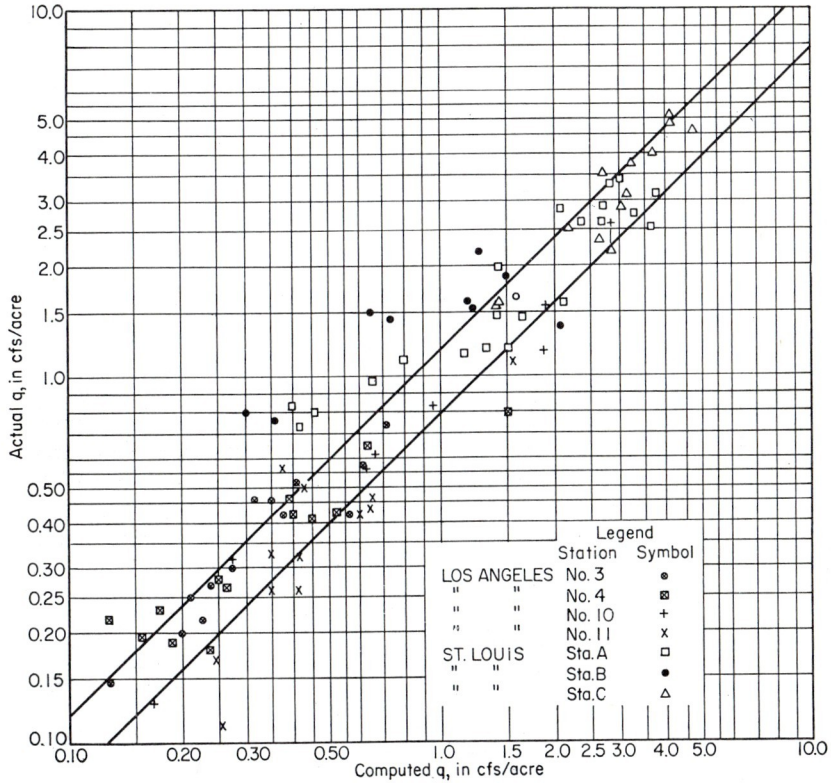

FIG. 20-6. Computed and measured runoff. (*After Hicks* [5].)

A study of storms in Chicago for 25 years (1934 to 1959) indicated that the summer thunderstorm is the critical type.

The majority of actual storm patterns show an advanced peak located at about the three-eighths point of the total storm duration [9, p. 60]. Eighty-three station rainfalls were used in determining mean values of antecedent rainfall and location of the peak for durations of 15, 30, 60, and 120 min. Mass antecedent rainfall computed from the adopted synthetic rainfall pattern closely agreed with the mass antecedent rainfall for the station rainfalls.

A total duration of 3 hr was adopted for the synthetic storm pattern, equivalent to the longest time of concentration of any Chicago drainage system to be designed. Because this storm duration is quite long for an urban-storm-drain design, it was assumed that any additional small amount of rain preceding this 3-hr period would not be significant, and no antecedent rain was assumed prior to the beginning of the 3-hr synthetic storm.

Since the three-eighths advanced pattern was determined from actual Chicago data, it might not be suitable for other localities.

STEP 3: ABSTRACTIONS FROM RAINFALL. *Infiltration* [6]: Whenever the rainfall intensity is less than the infiltration capacity the actual absorption by the soil is limited to the actual mass rainfall. The same assumption used in Ref. 21 was adopted for Chicago; i.e., the time of beginning of excess rainfall occurs when the antecedent mass of rainfall equals the antecedent mass of infiltration. In application, the time of offset of the infiltration-capacity curve is obtained by shifting the mass-infiltration curve to the right until it is tangent to the mass-design-storm curve. Subtracting the shifted mass-infiltration-curve ordinates from the mass-rainfall values yields the mass curve of overland flow and depression-storage supply.

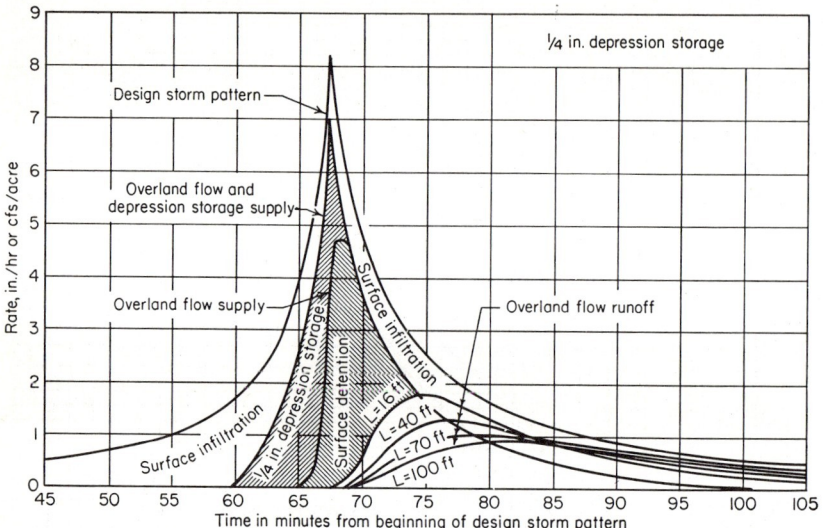

FIG. 20-7. Overland flow on pervious areas. Bases: (1) standard infiltration, (2) depression storage as indicated above, (3) overland-flow factors (equation from Izzard): slope = 0.01, roughness coefficient = 0.06, flow distances as indicated above. (*After Tholin and Keifer* [6].)

Surface depression storage: Analytical approaches were employed to determine the probable depth-distribution relationship of depressions of a given mean depth at any random location, and a normal distribution was selected. A $\frac{1}{4}$-in. overall volume of depression storage was assumed for flat, pervious areas. For pavements a volume of $\frac{1}{16}$-in. overall depression storage was adopted. For comparison, volumes of $\frac{1}{2}$ and $\frac{1}{8}$ in., respectively, were used in an additional computation series [6, pp. 1340–1341].

STEP 4: OVERLAND FLOW. Figure 20-7 shows the overland-flow supply curve for pervious areas after deducting the assumed depression storage and infiltration. The basic relationships and procedures developed by Izzard [8] were adapted to a storage-routing procedure to obtain the hydrographs of overland-flow runoff from the overland-flow supply curves. Figure 20-7 (and [9, p. 62]) illustrates the results of routing the overland-flow supply through detention storage for several lengths of overland flow over turf surfaces. For impervious surfaces, procedures are the same but the supply curve will reflect less depression storage and considerably less detention storage.

STEP 5: GUTTER ROUTING. The mixed flow from the pavement and lawns is routed through the street-gutter detention to determine the hydrograph of flow entering the inlet or catch basin. Unit runoff rates from the overland-flow hydrographs of the elemental strips of pavement and lawn are multiplied by the respective tributary area of each, and the products added together to give the gutter inflow hydrograph. The inflow hydrograph is assumed to be uniformly contributed at the curb along the (110-ft) length of each street-gutter section. The gutter supply thus starts at zero at the upstream end of the gutter section and aggregates the instantaneous rate of the inflow hydrograph at the downstream end of the gutter section, at the inlet.

Assuming a typical or representative gutter cross section, the gutter-storage volume can be determined as a function of inflow and outflow. This function, together with the equation of continuity, is used for the gutter routing.

STEP 6: ROUTING THROUGH LATERAL SEWER. Examination of several methods of routing through the lateral sewer, including storage-routing procedures, led to the use

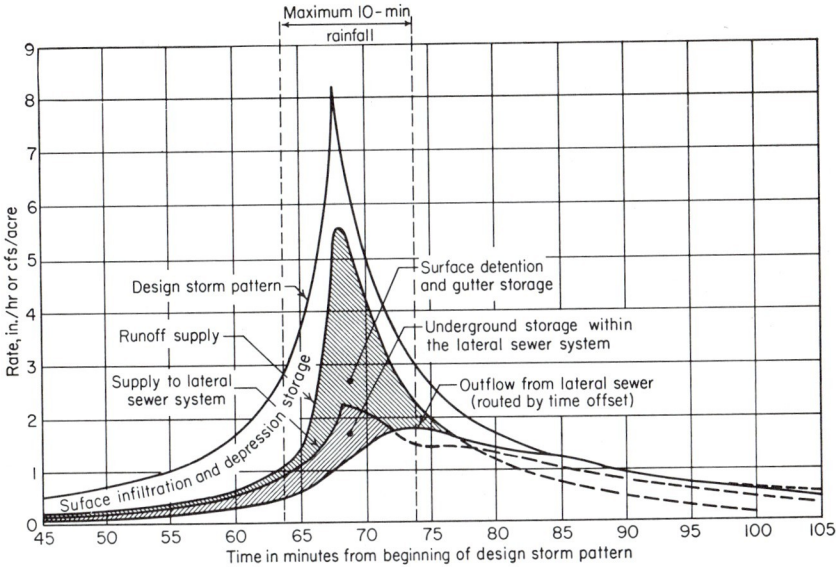

FIG. 20-8. Development of "10-acre" outflow hydrograph. Bases: (1) type 5 land use (36 per cent directly connected impervious area); (2) $\frac{1}{4}$-in. depression storage on pervious areas, $\frac{1}{16}$-in. depression storage on impervious areas. (*After Hicks* [9].)

of a *time-offset method* because of its simplicity. In addition, tests of accuracy showed that the simplified time-offset method (programmed for computer) gave slightly higher peaks at somewhat later times than a more exact method in which storage effects were considered. In the time-offset method, each hydrograph of inflow is shifted or offset the amount of time required for water to travel from its point of entry in the drain to the point under design.

To reduce the labor of combining the many roof-drain and gutter hydrographs, ordinates of all inflow hydrographs are added and this total inflow is divided equally into a number of smaller hydrographs assumed to enter the lateral at uniform intervals. Offsets are distributed by times of flow from the point of entry to the design point.

Figure 20-8 shows a two-block 10-acre outflow hydrograph from the lateral sewer.

STEP 7: ROUTING THROUGH MAIN SEWER. With the assumption that the average hydrograph of all lateral hydrographs enters the main sewer at uniform intervals, the same time-offset routing method was used as outlined for the lateral sewer.

With varying land use or nonuniform spacing of laterals, the inflow hydrographs entering one junction are summed and shifted by travel time to the next junction, where the local inflow hydrographs are added, and so on until the design point is reached.

STEP 8: DESIGN CHARTS. From routings of lateral hydrographs through main sewers, sets of hydrographs for various times of travel were developed. Plotting peak rates of runoff from these sets against the per cent of directly connected imperviousness for each of four land-use types gives a family of curves (such as Fig. 20-9) for various times of travel in the main sewer (to each is added a 10-min time of concentration for travel to the lateral outlet). Peak rates for 100 per cent directly connected impervious areas are extrapolations of the values for the four land-use types, based on the assumption that the runoff would be 90 per cent of the average rainfall rate for completely impervious areas.

STEP 9: HYDROGRAPH SHAPE. The previous step gives the peak rate of flow, but occasionally it is desirable to know the total hydrograph shape such as required for the

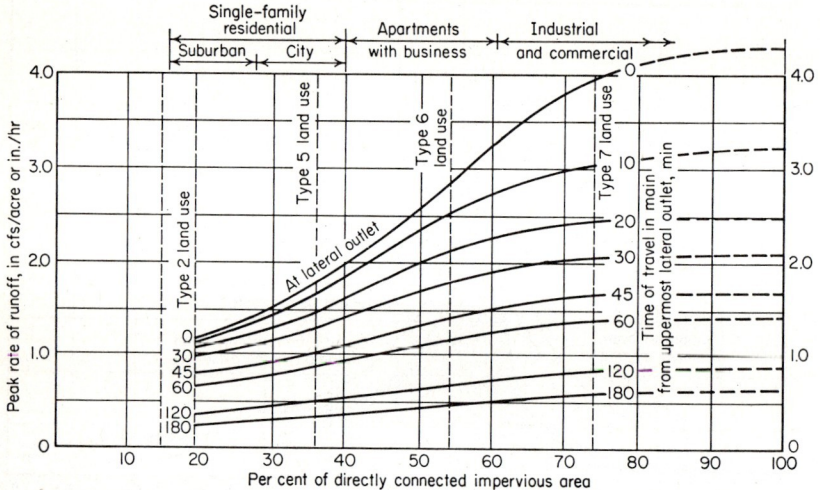

FIG. 20-9. Runoff rates in sewer system vs. travel time and per cent of directly connected impervious areas. Average ground slope 1 per cent. Bases: (1) hydrograph method using "5-year" design storm pattern; (2) standard infiltration curve for types 2, 5, and 6 land use, reduced infiltration curve for type 7 land use; (3) $\frac{1}{4}$-in. depression storage on pervious areas, $\frac{1}{16}$-in. depression storage on impervious areas. (*After Tholin and Keifer* [6].)

design of pumping stations, branch sewers, etc. For such cases mass curves for lateral outflow from uniformly developed areas were constructed for use in determining a complete design hydrograph [6, fig. 19].

STEP 10: AREAL VARIATION OF RAINFALL. The average rainfall within a drainage area decreases as the area increases. Therefore a reduction factor should be applied to the runoff rate determined from the previous steps. A study of the storms in Chicago, adjusted for the shape of drainage areas and percentage of infiltration, resulted in a set of reduction factors which can be used for the City of Chicago. No correction is applied for areas smaller than 640 acres.

b. Comparison of Measured and Computed Runoff. A tipping-bucket rain gage and a parabolic flume were installed (in 1959) to measure the rainfall and runoff on a city area of 13.9 acres in Chicago. Rainfall and runoff records are transmitted by telemeter to offices of the Chicago Bureau of Engineering. Figure 20-10 shows the gaged rainfall and runoff and the computed hydrographs of the storms of July 1 and 29, 1959, indicating a fairly close conformity between the measured and computed runoff. In the computation, the routing time in laterals was assumed to be 8 min, and half of the area of impervious surfaces which drain across pervious areas was treated as directly connected impervious area. The test area will be continuously gaged for verification of the principal assumptions used in the method.

Both the Chicago hydrograph method and the rational method have been programmed for electronic computers such that complicated existing storm drain systems can be analyzed easily or new systems designed. It was found that only a few of the many factors involved in the hydrograph method significantly affect the peak rate. A series of charts similar to Fig. 20-9 could therefore be prepared covering the maximum probable range of each important factor and any combination thereof. A designer could then choose the charts best fitting the rainfall, land use, and physical characteristics of the drainage area under design.

4. Synthetic Storm Pattern and Areal Distribution of Rainfall. Employment of a hydrograph method of analysis requires the derivation or assumption of a design storm pattern. The recurrence interval of maximum average rainfall intensities of specified durations can be determined with reasonable precision from long-term point rainfall records. However, assessment of the probable recurrence interval of a design storm as a whole is extremely difficult and almost wholly empirical. Despite attainment of a successful cross correlation between the constituent intensities of storms as recorded at a discrete point, some question would remain as to whether or not the derived storm pattern would have the same recurrence interval at other locations in the same urban area.

Instead of synthesizing a single storm pattern, a series of analyses might be performed for a set of typical physical conditions, using all major recorded storms. Recurrence intervals could be assigned to the array of computed peak flows for each set of typical conditions to establish a design base. This procedure is reversible because flows could be computed from later rainfall records for an existing drainage system, and their probable frequency determined from the design base. In particular,

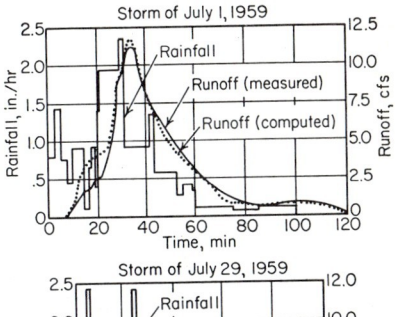

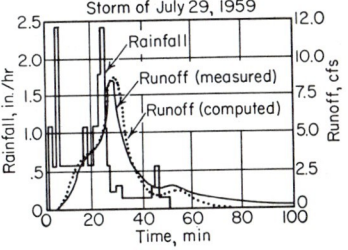

FIG. 20-10. Comparison of measured and computed runoff. Basis:
Total area drained, 13.9 acres
Directly connected impervious area, 5.7 acres
Impervious surfaces discharging into pervious areas, 1.2 acres
Rainfall gaged one-half block from center of drainage area
Runoff measured at lateral outlet by parabolic flume
Standard infiltration curve
$\frac{1}{4}$-in. depression storage on pervious areas
$\frac{1}{16}$-in. depression storage on impervious areas
Roof hydrographs directly connected to lateral
Time-offset routing through lateral sewer of 8 min
(*After Tholin and Keifer* [6].)

a fairly realistic estimate could be made of the probable frequency of flooding for a given design, from which cost-benefit relations could be defined.

Synthesis of design storm patterns has been achieved by the cities of Chicago [6, 9, 20] and Los Angeles [5] from local rainfall data. Several other approaches, or modifications, have been suggested [21–24].

Seldom is a synthetic design storm applied to the analysis of areas larger than about one square mile without some form of intensity-reduction allowance to account for the nonuniform areal distribution of intense rainfall. Reduction factors have been developed from local data for Los Angeles [5, 18] and Chicago [6]. Similar relationships have been developed for specific regions of the United States [25], with reduction factors in terms of distance from the storm center and duration of the intense part of the storm.

The frequency of a synthetic storm pattern which has been amended by an areal reduction factor is clearly indeterminant.

E. The Inlet Method

1. General. The Johns Hopkins University [26] has been conducting extensive research to develop relationships between rainfall and runoff for application to urban-storm-sewer design. A continuing program initiated in 1951 has included measurements of rainfall and runoff on inlet areas and larger composite areas in Baltimore, Md., described in Table 20-3. (See also work by Kaltenbach under "other references" in Subsec. IX.)

The principal influence of hydrologic factors is on the aboveground concentration of runoff, terminating at inlets. The underground collection and transportation of flow from inlets is controlled almost exclusively by the hydraulic characteristics of the sewer system. It was reasoned that advantage could be taken of this approximate separation by evolving suitable simplified inflow hydrographs from actual inlet gagings. From gagings on larger areas an auxiliary relation could be developed to appraise the attenuation of each individual inlet hydrograph (reduction of peak and concomitant extension of the base) on arrival at the collecting-system point under design. The peak flow at the design point could then be obtained by summation of the transposed inlet hydrographs.

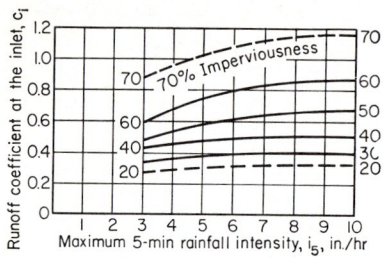

Fig. 20-11. Smoothed plot of c_i versus i_5 versus I. (*After Bock and Viessman* [26b].)

2. Peak Inlet Flows. Because of limitations in the quality and number of measurements, the initially derived relationships [26b] are in need of further verification. Strong but approximate correlations were found between the peak inlet runoff rate (q_i in cubic feet per second) and the maximum 5-min rainfall intensity (i_5, in inches per hour), and between the latter and a dimensionless runoff coefficient c_i for various percentages of imperviousness (Fig. 20-11). The peak runoff rate in cubic feet per second at the inlet was expressed as

$$q_i = c_i i_5 A_i \tag{20-11}$$

where A_i is the inlet area in acres.

Values of c_i greater than 1.0 (Fig. 20-11) may be due to higher intensities within the 5-min maximum. For example, records show that the maximum 2-min intensity can be 1.5 times the 5-min maximum. Initial developments were restricted to i_5 in order that correlations could be established with the long-term U.S. Weather Bureau data for Baltimore (shortest period maximums reported are for 5 min).

Inlet gaging records reveal that peak runoff rates occur about 1 min after the peak rainfall rate and that at neighboring inlets (e.g., at a street intersection) the peak runoff rates occur almost simultaneously. The coincidence of peak flows from adjacent inlet areas which happened to have dissimilar physical features suggested that rainfall is the most important factor in establishing the time when peak runoff will occur at an inlet.

3. Inlet-flow Routing. For routing purposes, the inlet hydrograph is assumed to be of triangular shape with q_i in cubic feet per second as the peak and $2T$ as base (Fig. 20-12a). The term T, in minutes, is defined as the time from the beginning of intense rainfall to the end of the period of maximum rainfall intensity. The actual time of rise of the inlet hydrograph is related in some manner to T. The routing relation used was proposed by Shrank [27, 28] as

$$q_o = q_i \left(\frac{2T}{2T + 0.8L/V} \right) \qquad (20\text{-}12)$$

where q_o is the related downstream peak flow in cfs, L is the length of the drain in feet from the inlet to the design point, and V is the mean velocity of flow in the drain in fpm.

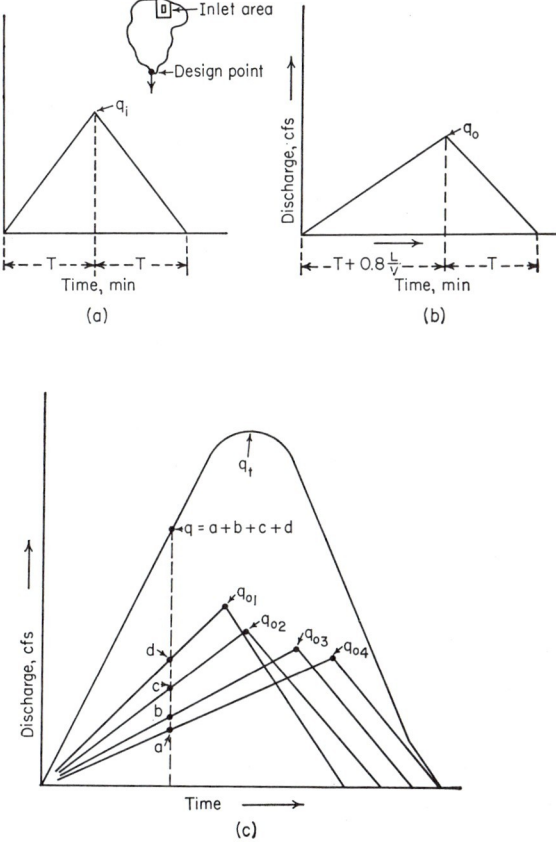

FIG. 20-12. Inlet hydrograph and its synthesis. (a) Triangular hydrograph at the inlet; (b) triangular hydrograph at the design point; (c) synthesis of subarea hydrographs. Total outflow hydrograph produced by summing ordinates of q_0 hydrographs. (*After Bock and Viessman* [26b, c].)

The routed inlet hydrograph at the design point has a triangular shape of reduced peak and lengthened base (Fig. 20-12b). The base is lengthened on the rising side by an amount equal to the time of wave travel in the drain, assumed for simplicity to be $0.8L/V$. Thus the term in parentheses becomes the reduction or attenuation factor applied to the peak flow for a given inlet to obtain its corresponding peak downstream. The total peak flow at a downstream point is obtained by summing the routed-inlet hydrograph ordinates to obtain the maximum q_t (Fig. 20-12c). For this purpose a common time origin is used for all the routed-inlet hydrographs.

Table 20-3. Baltimore Drainage Areas

Drainage area	Inlet no.	Area, acres	Impervious area, %	Average gutter slope, ft/100 ft	Crown slope	Gutter length, ft	Street, ft	Comment
Midwood Group housing; residential; roof water over lawns	1 2 3 4 5	0.522 0.208 0.188 0.641 0.635	54.6 65.1 60.0 55.0 56.0	5.5 6.5 1.4 6.1 6.1	1:16.6 1:21.0 1:16.0 1:14.7 1:18.7	200–400	34 and 30 wide, (concrete with curbs and gutters)	
Enfield Residential, detached houses on large lots; roofs directly to gutter	1 2 3 4 5 6 7 8 9	0.225 0.105 0.194 0.736 0.722 3.178 1.522 0.306	33.3 61.0 72.6 34.0 38.4 25.9 33.5 49.4	0.33 0.53 4.3 4.3 2.0 2.0 6.2 6.2	1:38.8 1:18.0 1:22 1:21.1 1:17.7 1:27.3 1:26.1 1:20			In the case of inlets 1, 2, and 6, gaging difficulties made it impossible to analyze data. Gaging discontinued in 1957.
Uplands Residential, commercial, garden apartments		30.13	52.4					Upper part, slopes less than 1%; middle area, slopes to 2%; lower half, slopes to 6%.
Walker Residential 6,000-sq-ft detached houses; essentially uniform; small undeveloped sections in east and west		153.4	33.0					Average slope about 2%. Gutter flow lengths in many cases 1000′ or more. Flow over pervious surfaces for 200+ ft not uncommon. Well-defined drainage limits.
Yorkwood Garden apartments, group homes		11.05	41.0					Average slope about 4%. Well-defined drainage limits.

20–24

Table 20-3. Baltimore Drainage Areas (Continued)

Drainage area	Inlet no.	Area, acres	Impervious area, %	Average gutter slope, ft/100 ft	Crown slope	Gutter length, ft	Street, ft	Comment
Swansea Group homes, 2,000 to 3,000-sq-ft lots		47.09	44.0					Average slope about 3%. Well-defined drainage limits. Essentially uniform development (measuring flume). Installation: 1959.
South parking lot	1	0.395±	100.0	1.71		280		
	2	0.469±	100.0	2.16		332		
Northwood Group houses and highly developed commercial area		47.41±	68.0					Average grades about 3%. Well-defined drainage limits (measuring flume). Grades generally less than 1%.
Gray Haven		23.29	51.7					
Hamilton Hills Individual homes, ¼-acre lots; roofs directly to gutter	1	0.71	55.6	0.83	1:50	472	34 (concrete, with curbs and gutters)	Installation: 1961. Well-defined drainage limits.
	2	0.97	20.1	0.98	.....	505		
	3	1.83	36.4	0.85	1:42	583		
	4	0.22	96.3	0.86	1:154	583		
	5	1.71	31.8	2.10	1:65	360		
Montebello Group homes and open grassed area; roof directly to gutter	1	0.21	64.6	1.77	1:33	485	24	Five inlet areas. Installation: 1961. Drainage limits for inlet 1 not well defined. Drainage limits for remaining areas well defined.
	2	1.51	8.7	1.73	1:37	470	24	
	3	0.45	57.1	0.81	1:32	153	15	
	4	0.54	64.8	0.79	1:34	352	29	
	5	0.52	65.9	0.85	1:27	352	29	

While there is a degree of judgment involved in reading T from mass-rainfall curves, reasonable interpretive differences will not affect the attenuation factor greatly unless the term $0.8L/V$ has a large value compared with $2T$.

No correlation was found between i_5 and T [26c]. Values of T for the 15 largest values of i_5 from U.S. Weather Bureau data are given in the following table [29]:

Rank of highest i_5 over period of record	Baltimore (1894–1935)		Chicago (1913–1935)		Philadelphia (1894–1955)	
	i_5, in./hr	T, min	i_5, in./hr	T, min	i_5, in./hr	T', min
1	8.9	15	6.6	5	7.7	15
2	7.2	5	6.4	5	7.2	30
3	6.8	5	6.1	15	6.6	15
4	6.7	10	6.0	15	6.0	40
5	6.5	20	5.8*	5	6.0	20
6	6.4	15	5.4	15	6.0	10
7	6.2	5	5.0	10	6.0	10
8	6.1*	10	4.9	10	5.9	25
9	5.9	10	4.9	20	5.9	25
10	5.9	15	4.9	10	5.9	15
11	5.9	25	4.7	20	5.6	10
12	5.6	10	4.7	20	5.5*	15
13	5.5	15	4.7	5	5.5	20
14	5.5	10	4.6	10	5.5	10
15	5.5	10	4.6	10	5.4	5

* Approximate 5-year recurrence interval.

Despite the independence between T and i_5, a flow frequency could be ascertained for a given design point. This would require calculation of the peak flow at the design point for each significant i_5 of record (e.g., values greater than the 1-year frequency value) using its associated T. From an array by rank of the computed flows, a design flow of an assignable frequency could be selected. After all, it is the flow frequency, rather than a rainfall frequency, which is the ultimate objective.

4. Simplified Procedure. To eliminate the laborious detail required to draw and summate individual subarea transposed hydrographs, a simplification of the inlet method [26b], called the *factor procedure*, was developed, using the following equation:

$$q_t = F_t \Sigma q_i \tag{20-13}$$

where q_t is the maximum total flow at the design point as before (Fig. 20-12c), F_t is an attenuation factor, and Σq_i is the sum of the individual subarea values of q_i by Eq. (20-11). An analysis of synthetic-inlet-area complexes resulted in a uniformly consistent solution for F_t (Fig. 20-13). The term L refers to the maximum drain length in feet from the most remote inlet to the design point, and V is the average drain velocity over the same route. Reservations were expressed concerning the generality of F_t, but analysis of the data for 46 storms on 10 different Baltimore drainage areas indicated that q_t values computed by the factor procedure were equally as valid as those obtained by the original method. The maximum and average deviations from the original-method values were 10.7 and 2 per cent, respectively. Peak flows calculated by means of the factor procedure were within an average of 12.2 per cent of the gaged and differed less in the extreme than the original method. Because the maximum L and minimum V for the gagings were about 5,000 ft and 4 fps, respectively, the validity of F_t beyond these limits is uncertain.

The City of Philadelphia has found that results obtained by using Eq. (20-13) can be closely approximated by using $F_t' = 2T/(2T + 0.8L/V)$ for F_t in Eq. (20-13), where L and V are as defined above for F_t [26b, p. 240].

5. Design Example. The application of the inlet method to a hypothetical design problem is illustrated using the Courtleigh drainage area of 67.1 acres in Baltimore (Fig. 20-14). The objective will be to determine the peak flow in the outfall pipe Z for $i_5 = 7.1$ in./hr and $T = 30$ min.

Table 20-4 gives procedural details. From a map showing streets, impervious surfaces, drainage-area limits, inlet and drain locations, etc., subareas should be selected to combine the tributary areas of groups of inlets discharging into street sewers at about the same location, such as the three inlets draining subarea G at a street intersection. Compute the acreage and imperviousness of each subarea (Table 20-4, columns 2 and 4). In column 5 enter the values of c_i obtained from Fig. 20-11. Multiply the values in columns 2, 3, and 5, and enter the product in column 6. Enter the length of flow through the drainage system between each subarea inlet location and the design point in column 7. In the simplified method the controlling L would be 1,350 ft, the maximum. For comparison, a V of both 5 and 10 fps is given in column 9. In practice, an average V for the 1,350 ft from subarea A would be used, in accordance with the hydraulic characteristics of the drains involved. The F_t in column 11 is obtained from Fig. 20-13 for the L/V in minutes of column 10. The q_t of column 12

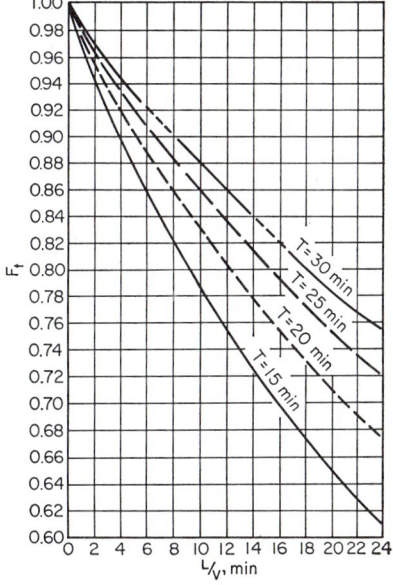

Fig. 20-13. L/V versus F_t curves for T values of 15, 20, 25, and 30 min. (*After Bock and Viessman* [26b].)

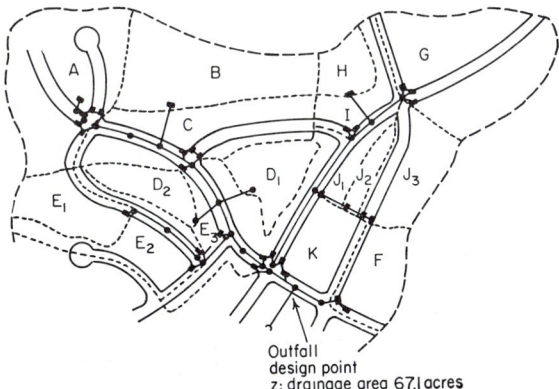

Fig. 20-14. Courtleigh drainage area, Baltimore. (*After Bock and Viessman* [26b].)

is from Eq. (20-13). The values for F_t' are given in column 13 and corresponding q_t in column 14, for comparison. Also shown are values for $T = 15$ min for contrast. Referring to the footnote [26b], for $T = 30$ min and $V = 5$ fps, the two simplified procedures yield results almost identical with the original method. In this example the

Table 20-4. Computations of q_t by the Inlet Method for the Courtleigh Drainage Area*

(1) Subarea	(2) Acres	(3) i_5	(4) Impervious-ness	(5) c_i	(6) q_i	(7) L	(8) T	(9) V	(10) L/V	(11) F_t	(12) $q_t = F_t \Sigma q_i$	(13) $F_t' = \dfrac{2T}{2T + 0.8L/V}$	(14) $q_t = F_t' \Sigma q_i$
A	8.85	7.1	0.50	0.64	40.3	1,350	30	5	4.5	0.939	261†	0.943	262†
B	7.30	7.1	0.30	0.39	20.2	1,190	30	10	2.25	0.966	268	0.970	270
C	6.81	7.1	0.49	0.63	30.5	830	15	5	4.5	0.889	247	0.893	248
D_1	2.56	7.1	0.30	0.39	7.1	740	15	10	2.25	0.937	260	0.943	262
D_2	1.12	7.1	0.30	0.39	3.1	740							
E_1	2.90	7.1	0.42	0.52	10.7	1,060							
E_2	4.19	7.1	0.45	0.56	16.6	650							
E_3	0.93	7.1	0.50	0.64	4.2	440							
G	10.60	7.1	0.50	0.64	48.2	1,260							
H	3.60	7.1	0.25	0.35	8.9	1,230							
I	2.95	7.1	0.50	0.64	13.4	950							
J_3	3.74	7.1	0.51	0.65	17.3	920							
J_2	0.73	7.1	0.55	0.73	3.8	800							
J_1	1.01	7.1	0.50	0.64	4.6	620							
K	6.68	7.1	0.55	0.73	34.6	170							
F	3.11	7.1	0.51	0.65	14.4	150							

$\Sigma A_i = 67.08$ $\Sigma q_i = 277.9$, or 278

* From Bock and Viessman [26b, p. 245].
† Versus 260 cfs [20b] by summing 16 transposed inlet hydrographs by Fig. 20-12c.

various combinations of T and V lead to q_t differing at most by 9 per cent. For larger L's and similar V's the effect of T on the magnitude of q_t would be more pronounced.

6. Limitations. In the design of urban storm drainage, the inlet method (simplified by the factor procedure) can give improved results over those obtained by the rational method within the following limits of application [26d]:

1. Urban residential-type areas with paved streets and gutters and adequate capacity inlets; no extensive unimproved areas such as woods and parks.
2. Inlet areas up to 3 acres.
3. Drainage areas up to about 1 sq mi.
4. Imperviousness, 30 to 60 per cent.
5. Gutter grades of 2 to 7 per cent.
6. Generally steep roofs and surface grades not so flat as to support significant ponding or surface storage (in the areas gaged in Baltimore, roofs are connected to gutters or to lawn-surrounded catch basins or yard drains).
7. Pervious areas underlain generally with clay subsoil strata.
8. Maximum 5-min rainfall intensity greater than 3 in./hr, the approximate lower limit used in the gaging studies (the 5-min 1-year frequency intensity for Baltimore is 4.5 in./hr).

7. Comparison of Measured and Computed Runoff. The inlet method, as well as the rational method, was applied to five areas in Baltimore and to St. Louis [17], Los Angeles [5], and Oxhey Road at Hertfordshire, England [30]. The results

Table 20-5. Summary of Results Using Inlet Method and Rational Method

Area	Total no. of estimates	Number of estimates falling outside of 20% limits	Mean (absolute) deviation, %
By inlet method			
Uplands............................	25	3	8.4
Walker Ave.........................	19	7	21.2
Northwood..........................	4	4	168.3
Yorkewood South....................	7	2	15.2
Swansea............................	5	1	12.7
St. Louis (Clarendon)..............	3	0	6.5
Los Angeles (Station 3)............	1	0	14.8
Oxhey Road.........................	3	1	15.8
Total.............................	67	18	32.9* 13.5†
By rational method			
Uplands............................	25	17	26.9
Walker Ave.........................	19	6	16.6
Northwood..........................	4	3	54.9
Yorkewood South....................	7	5	33.9
Swansea............................	4	2	18.6
St. Louis (Clarendon)..............	3	3	79.0
Los Angeles (Station 3)............	1	1	23.0
Oxhey Road.........................	3	1	22.5
Total.............................	66	38	34.4* 31.5†

* Average deviation.
† Omitting Northwood.

are shown in Table 20-5. The high deviations between the computed and measured peak flows for Northwood may be due to use of overly conservative estimates of c_i for fully impervious areas (a shopping center occupies over one-third of the Northwood area). Discounting the Northwood estimates, the total deviation for the inlet method is less than half that for the rational method. The peak flows computed by the rational method were found to be consistently low.

The more recent studies [26d, e] have been concentrated on completely impervious areas.

A verification study of the inlet method is under way (1963) in the City of Philadelphia, where, as in Chicago, roof drains are almost all directly connected to sewers. Gagings of four drainage areas of 30 to 288 acres and average imperviousness of 48 to 67 per cent for 41 station-storms have been reported [29]. Using the simplified inlet method, fairly good correlation with gaged flows was obtained when nine station-storms with no intense rain immediately preceding i_5 were eliminated (calculated peak flow was about twice the gaged). A subsequent study indicated the possibility of segregating various levels of c_i in terms of antecedent rainfall conditions. Since several subareas are almost completely impervious, improved correlation is expected when more extensive inlet data for 100 per cent imperviousness become available.

F. Gutters and Inlets

1. General. Provision for adequate interception of storm water and its subsequent introduction to a system of drains requires an understanding of the hydraulics of inlets and gutter flow. Capacity of inlets must be matched with conduit capacity for suitable economic handling of stormwater. Untold millions of dollars are invested in underground storm drains which are not working because the surface runoff cannot get into them. In Baltimore it was found [31] that severe flooding of more than 20 "problem" areas was relieved by provision of adequate additional inlet capacity. Considerable research on gutters and inlets has been reported [11, 31–37].

2. Gutter Capacities. The capacity of all types of inlets is closely related to the depth of water in the gutter adjacent to the inlet. Channel characteristics affecting the depth of flow include cross section, grade, and roughness of gutter and pavement surfaces. For triangular gutters and for composite sections with two or more rates of cross slope, Eq. (20-1) can be used to compute uniform-flow depths. Limited field data indicate an n value in the equation equal to 0.021 for a belted or broomed finish on a concrete pavement and 0.018 for a smooth, trowel-finish concrete gutter. Appreciable quantities of sediment raise these coefficients of roughness markedly.

The length of gutter sufficient to establish a uniform-flow depth may be of the order of 50 ft or more. Flow may be far from uniform in the vicinity of inlets because of drawdown or ponding. Thus, for short gutters and ponded inlets, Eq. (20-1) should not be used to estimate depths of gutter flow.

3. Inlets. *a. General Considerations.* In urban storm drainage systems, inlets are usually provided immediately upstream from crosswalks and street intersections and at such intermediate locations as design-flow quantities or topography may require.

The inlet spacing selected should be adequate to limit the spread of water on the pavement and prevent interference with traffic at the design flow (for about a 15- to 20-min duration and the adopted frequency). In the absence of other governing considerations, allowance for a 3-ft encroachment should be reasonable. The design should next be checked for a higher-frequency storm, and the spacing revised where occurrence of excessive flooding is indicated.

Flow bypassing an inlet on a continuous grade must be included in the total gutter flow at the next downstream inlet. All the flow reaching an inlet at the low point of a sag vertical curve must enter the inlet without overtopping the curb or pavement crown.

Good practice suggests provision of three inlets in each sag vertical curve: one at the low point, and one on each side, where the gutter grade is about 0.2 ft higher. The latter two auxiliary inlets minimize the spreading of pondage, thus reducing the deposit of sediment on the pavement and the amount of trash and debris reaching the low-point inlet.

Inlets should be provided at points where an ice slick might otherwise occur (on bridge slabs, on warped transitions between superelevated and normal sections, etc.).

All inlets not at low points are more effective if spaced and designed to intercept somewhat less than the entire gutter flow. Grate inlets should remove 75 to 85 per cent and curb-opening inlets 85 to 90 per cent of the flow. Larson [34] found in some instances that grate capacity can be almost doubled if a small amount of water is allowed to go past the inlet.

b. Grate Inlets. Grate inlets on a continuous grade with efficient openings will intercept all the water flowing in that part of the gutter cross section directly above the grating, plus that which enters laterally at the inlet. For grate lengths less than 2 ft, side inflow is very small. On a uniform, continuous grade, uniformly spaced inlets in series tend to intercept the total runoff by the time the third or fourth inlet is reached, if the carry-over from the first inlets does not exceed about 25 per cent.

The most efficient grates have bars parallel to the direction of flow, with openings totaling at least 50 per cent of the total grate width. Clear length of the openings should be long enough to allow the water entering the grate to fall clear of the downstream end of the slot, with 18 in. the minimum recommended [36]. Efficient grates with smooth top bars generally have the best self-cleansing or nonclogging characteristics.

c. Curb-opening Inlets. The capacity of a curb-opening inlet on a continuous grade varies directly with depth of flow at inlet entrance and length of clear opening. Gutter depth in the vicinity of the inlet for a given discharge varies directly with cross slope of pavement adjacent to the curb, amount and shape of depression or warping of the gutter along the throat of the inlet, and roughness of the gutter surfaces, but inversely with longitudinal gutter slope.

The capacity of a curb-opening inlet located at a low point or sump varies principally with the length of opening and depth of flow at the entrance.

Depressions have been successfully used in some cities to increase curb-opening and/or grate-inlet efficiency. Traffic requirements limit or prohibit their use in some instances. In general, capacity of inlets increases with length, width, and depth of depression. A long, shallow depression can be as effective as a shorter, deeper one. St. Louis experience [33] has indicated that the flow into the curb opening can be increased by decreasing the downstream width of the depression.

d. Combination Inlets. An efficient grate immediately in front of a curb opening has a capacity only slightly more than with the grate alone. As a consequence, only the grate capacity should be depended upon in design.

A curb opening located upstream from a grate will tend to intercept debris and thereby reduce clogging of the grate. Capacity of the curb opening is not affected by the grate, but the downstream grate inlet should be designed to handle the carry-over from the curb opening.

Sumps or low points should be provided with combination inlets, since the curb opening will provide some relief from flooding if the grate becomes clogged.

G. Manholes and Junction Chambers

When storm flow passes through a manhole or other junction, losses occur which must be considered in design. There are virtually no data on which estimates of such losses can be based, other than those from the recent University of Missouri experiments [38]. That work involved systems with all pipes flowing full, and the results apply only to that condition. Pressure changes at manholes and junctions were measured, and experimental coefficients derived for design use. The investigation included both circular and rectangular junctions, with the majority of the studies confined to the latter.

A simple design example [38a] was studied involving eleven inlet manholes and two 48-in.-diameter junction manholes. In the 1,669-ft length of principal mains, the calculated losses at inlets and manholes were found to be 37 per cent of the overall pipe friction loss. This increase is much greater than an estimate made by ordinary design criteria. Most practice has been to assign arbitrary nominal or constant losses at manholes, inlets, and similar junctions. Sometimes the design roughness coefficient of drains is increased by judgment to compensate for these losses. Additional studies on junction losses are needed.

H. Open-channel Storm Drainage

In a major part of the mushrooming urban areas around the old central cities, open channels are largely relied upon for storm drainage. Land development usually leads to the straightening, widening and deepening, and in many instances the relocation of existing natural drainage channels.

Straightening shortens the length and consequently steepens the hydraulic gradient of a winding natural channel. The improved channel with higher velocities has greater capacity to transport sediment, and consequently deposition will occur at downstream reaches with flatter gradients. Because erosion increases with straightening, provision of a protective lining for the improved channel may be necessary.

If the plan of land development permits, the established regimen of a natural stream should be left undisturbed in so far as is practicable. All the costs of various drainage schemes must be estimated in order to choose the most satisfactory, but least costly, plan. The greater land cost for a wide drainage easement to accommodate a natural channel must be compared with the costs of lining and maintaining a straightened channel in a narrower easement.

The major problems encountered with open channels in urban areas are probable erosion of bottom and sides, with consequent potential widening and encroachment on private property; numerous culvert intersections with their inlets and outlets; and the introduction of flows from lateral pipes, branch channels, or ditches.

The first of these problems requires selection of maximum permissible velocities and a decision whether the channel should be lined with vegetation or artificial erosion-resistant lining. Grass-lined channels and swales can be used to advantage where flow quantities are intermittent and not large, grades are not too flat, and the soil can support a good stand of grass. For the design of grassed channels, see Refs. 10 and 39.

Nonerodible linings other than grass can be concrete, riprap, asphaltic concrete, soil-cement, etc. Portland-cement concrete can be cast-in-place, precast in units, or pneumatically placed. Asphaltic concrete [40] can be cast-in-place or preformed as membranes.

Each erosion-resistant lining has advantages under certain circumstances, and costs of the various kinds or types vary widely. The particular method and materials used are dependent upon such local related conditions as the general types of soil encountered, available aggregates, climatological conditions as they affect the growth of grass, and severity of freezing and thawing. Other important considerations are established local construction methods and availability of highly specialized equipment (such as slip-form, grading, and paving equipment) for both construction and maintenance.

The U.S. Bureau of Reclamation has had the leading experience with ditch linings for irrigation canals. Reduction of seepage loss is most frequently the principal reason for lining. However, the areas traversed are not usually residential. Virtually no experience with urban storm channels of different sizes in various types of soils and with various types of erosion-resistant linings is reported in the literature.

Intercepting grass-lined swales adjacent to and parallel with the finished top of channel banks represent a current workable solution for removal of overbank runoff. Their flows should be delivered to drop inlets with pipe outlets on flat slopes, terminating near the channel bottom. All pipe outlets for collected surface runoff should be carefully designed to ensure that their flow will join the main-channel flow with the least possible hydraulic disturbance. Outlet structures in unlined channels should be designed so as to minimize erosion of the main channel and undermining of the lateral outlet.

For a thorough treatment of the hydraulics of open channels see Ref. 10.

IV. FLOODS IN URBAN AREAS

The protection afforded by most sizable urban flood-regulation projects is predicated upon a much rarer frequency of occurrence than for agricultural, forest, or recreational areas. Spillways of reservoirs upstream from a populous area should be designed for the peak runoff from the maximum probable storm. This is the policy of the U.S. Army Corps of Engineers. Levees, flood walls, and channel improvements should be designed to pass a flood from at least a 25-year storm. The U.S. Army Corps of Engineers designs flood-regulation facilities for urban areas, where upstream impoundments are present or planned, to pass safely a flood from a storm of about one-half the severity of the maximum probable storm. Evaluation of costs and benefits of such projects are based upon stage-damage and stage-frequency relationships.

Stage-frequency and other hydrologic determinations required in flood-regulation studies for urban areas are discussed in other sections, notably Secs. 8 and 25.

There is a growing interest in flood-plain regulation or zoning as reasonable alternatives to the construction of extensive local flood retention or control works. Section 25-V covers this topic and flood insurance.

V. WATER SUPPLY

With its population increasing at a rate of 3 million persons per year, the United States should pass the 200 million mark before 1970. By 1980, from 29 to 43 billion gallons per day will be furnished by public supplies. Industrial withdrawals (which began to exceed irrigation withdrawals in 1955) aggregated about 124 billion gallons per day in 1957, and it has been estimated that, by about 1975, this rate will be doubled. Added to this will be the need for considerable quantities of water in atomic-energy activities. Fortunately, most industrial water withdrawn is not consumed, but is ultimately returned to aquifers, streams, and lakes (94 per cent is used for cooling). However, more extensive thermal pollution may occur as use increases.

Many factories and plants depend upon municipal systems, but most large plants with a considerable need for water develop their own supplies. The hydrology applied in the development of industrial and municipal water supplies is similar. Both quantity and quality are basic considerations. Water supply for urban use may be from surface streams or groundwater reservoirs. Hydrology of surface water and groundwater is treated in other sections of this handbook. However, it must be emphasized that the hydrology of urban water supply is frequently regional rather than local in scope.

A characteristic of urban areas is that the concentration of large numbers of people results in water demand exceeding locally generated or accessible supply of potable water. . . .

The hydrologic impact of an urban area . . . may extend far beyond its own borders. . . . . . . 1 out of 8 persons having access to public water supply in the North American continent takes water from a system which transports it from a source 75 miles or more away. . . .

The re-use of water is especially common in heavily populated or highly industrialized regions where most of the municipal and self-supplied water supplies are obtained from sources within the basin. The principal factors limiting re-use of water are consumptive use and contamination or pollution [2].

VI. WATER POLLUTION

Inadequately treated wastes discharged into a stream, lake, reservoir, or ocean can add pollutants which make the surface water unusable (or economically untreatable) for some or all of the more common uses: domestic and industrial supplies, thermal-power-plant cooling water, agriculture, wildlife, recreation, and navigation. "Waste water now constitutes an ever increasing portion of the Nation's water supply" [2]. The Federal Water Pollution Control Act of 1956, supplemented by many state laws, authorizes the U.S. Public Health Service to undertake pollution surveys on all interstate streams and to institute antipollution procedures [41]. The sanitary significance of pollution is discussed in Ref. 42.

Contamination of groundwater can and does occur in urban areas where wells are close to sanitary or combined sewers. Poor pipe joints and leaky manholes can introduce bacteria into the soil, and with sufficient transmissibility the bacteria can be carried to a well. If a well was carelessly built or the casing has corroded, pollution can readily occur. Little is known about the range of travel of bacteria through geological formations; virtually no information is available on the travel of viruses. In sandy formations, bacterial travel in aquifers is limited to about 250 ft [43]. In coarse formations, the distance is longer. In dry, porous, homogeneous materials, bacterial travel is vertical. Viruses may travel farther, but it is thought that "all other pathogenic organisms probably travel shorter distances than bacteria" [43, p. 649].

Septic-tank systems, cesspools, and oxidation ponds introduce synthetic detergents into groundwater [44]. Other troublesome contaminants are salt water (brines) and some of the toxic chemicals. Difficulties in decontamination are exemplified in a paper [45] about the accidental groundwater contamination of a small well supply (12,000 population).

Waste-water reclamation can be achieved by groundwater recharge through direct injection or water spreading, lawn watering, crop irrigation, and treatment for industrial use. The sanitary engineering problems incident to these re-uses have been reviewed in Ref. 43. The control of radioactive pollution of a small Colorado stream is discussed in Ref. 46.

VII. AIRPORT HYDROLOGY

A. General

The application of hydrology to the design of airport drainage is in most respects similar to its use in the design of urban drainage facilities. Differences arise because of the extremely large and relatively flat surfaces of an airport and the operational requirements of smooth, firm, stable pavement subgrades and turfed areas. Open channels or natural watercourses can be tolerated only at the perimeter, well removed from the landing strips and traffic areas.

B. Objectives

An airport drainage system should (1) intercept and divert both surface and groundwater flows originating outside the airfield, (2) remove surface runoff originating on the airport itself, and (3) reduce moisture (or gravitational water) in pavement subgrades and in the surface soils to stabilize and ensure the firmness of the landing and terminal areas. The drainage system should function with a minimum of maintenance difficulties and expense and be adaptable to future expansion.

C. Hydrologic Data

The desirable degree of achievement of the foregoing objectives will vary with each airport. Design should be based upon a careful and thorough understanding of the probable operating procedure and studies of soils and climatological data, including precipitation, location of water table, average frost penetration, and normal annual, maximum, and average depth of snowfall.

Careful and realistically sound engineering of storm drainage for airports is exceedingly important. Underdesign can result in hazardous operational conditions, while overdesign means unnecessary and wasteful investments in excessively large conduits, etc.

D. Subsurface Drainage

In site selection thorough consideration should be given to the need for subsurface drainage [47–49]. Provision of complete subsurface drainage over the entire landing and terminal areas is rarely justified. Consequently, it will generally be necessary to lower the groundwater table only where such a need is clearly evidenced.

Subdrainage should lower the gravitational water table sufficiently to leave no free or capillary water at or above the normal frost depth, particularly under pavements. Seasonal or longer-term cyclic fluctuations of the water table must be carefully considered.

E. Surface Drainage

1. General. Surface drainage for civil airports of any importance is generally designed for the runoffs occurring from rainfalls with an average recurrence interval of 5 years and checked for the extent of damage, if any, that might result from a 10-year rainfall. Of course, the class and volume of traffic to be accommodated, the necessity for uninterrupted service, and other similar factors will exert a major influence on the desirable degree of drainage protection for a particular site. In all instances, minimum requirements must be adequate to avoid serious operational hazards. The difference in cost between adequate and inadequate drainage might well be less than the tangible losses due to one severe accident.

The extensive flat-turf areas of airports are conducive to the accumulation of appreciable surface-detention volumes, with consequent retardation of overland flow. These seldom-used interpavement areas can be utilized to further delay runoff and reduce peak flows by means of forced pondage. Despite attainable economies, forced pondage cannot be used at all airports. However, it is not objectionable to utilize forced pondage at most major civil airports, since landings on unpaved areas are made only in emergencies. Use of forced pondage is also permitted on some military airfields.

2. Grading. Coordination of grading and drainage designs is imperative to secure the most effective surface drainage and to minimize or eliminate the need for subsurface drainage. To satisfy civil and military requirements, transverse grades in paved and shoulder areas are usually set at a maximum of 1.5 per cent. Similarly, longitudinal grades of paved and shoulder areas should not exceed 1.5 per cent for light planes and 1 per cent for heavy (transport) planes, although 0.5 per cent is the preferred maximum for all major airfields. There are also rigorous criteria defining maximum permissible longitudinal pavement and shoulder grade changes for both civil and military airfields.

With a drain pipe approximately parallel to the runway, a 2 per cent longitudinal interrunway grade (along the line of drain inlets) is permitted for military airfields, with a maximum grade change of 3 per cent. Where this prevents pondage, a small transverse ridge not higher than 6 in. above the inlet elevation may be constructed 100 ft downstream from the inlet. The U.S. Civil Aeronautics Administration [47] has similar provisions for civil airports.

3. Ponding. Ponding areas should be provided wherever practicable. To avoid injurious saturation of the ground next to the pavement, the pond edges under maximum design conditions should be at least 75 ft from the pavements. For the preferred 1.5 per cent turf slope, the design water surface would be about 1.5 ft below the elevation of the pavement edges. Often carefully worked out pondage can reduce pipe sizes and overall costs of drainage. To illustrate the methods used in designs involving ponding, a design of a simple drainage system for a small training field is given below (Fig. 20-15 and Table 20-6). The method used was originally developed by Hathaway of the U.S. Army Corps of Engineers [51], based upon the work of Horner [50] for Washington National Airport.

a. Design Storm Criteria. Since this was a World War II military installation, the procedure reported in Ref. 51 was followed in preparing Table 20-6. The design is

Table 20-6. Computation for Drain-inlet Capacities Required to Limit Ponding to Permissible Volumes for Interrunway Areas*

Sheet 1 of 1
Supply curve nos.:
Paved areas: 1.72 } 2 years
Turfed areas: 1.32 }
Paved areas: 2.23 } 5 years
Turfed areas: 1.83 }

Project: Training Airfield (Fig. 20-15)
Location: Anytown, U.S.A.

Inlet no.	Drainage area (DA), acres					Permissible ponding				Average roughness factor n	Average slope S, per cent	Length L, ft			Standard supply curve no. × DA				Weighted supply curve (col. 18 ÷ col. 5)	Drain-inlet capacity				Critical contribution to system		
	Paved $n=0.02$	Unpaved		Total	Depth at inlet, ft	Pond area, thousands ft²	Volume V, thousands ft³	Volume, ft³/acre DA				Actual or effective length, ft	Equivalent L for $n=0.40$ and $S=1.0\%$	L adopted for selecting diagrams	Paved areas	Unpaved areas		Total		t_{cr}, min	q_d, cfs/acre DA	Q_d, cfs (col. 5) × col. 21		T_{cr}, min	q_d, cfs/acre DA	Q_d, cfs (col. 24 × col. 5)
		Bare $n=0.10$	Turf $n=0.40$													Bare	Turf									
(1)	(2)	(3)	(4)	(5)	(6)†	(7)	(8)	(9)†	(10)	(11)	(12)	(13)	(14)	(15)	(16)	(17)	(18)	(19)	(20)	(21)	(22)	(23)	(24)	(25)

Supply curves 1.72 and 1.32 (2-year rainfall)

1	2.80		19.20	22.0	1.83	260	158.6	7200	0.35	0.8	320	320	300	4.81	25.35	30.16	1.37	Not required as appreciable ponding (col. 9) is permissible	0.46†	10.1	Same as shown in col. 21 as appreciable ponding is permissible
2	0.66		17.17	17.83	1.53	98.4	60.2	3330	0.39	0.9	360	390	400	1.13	22.67	23.80	1.33		0.44†	7.9	
3	2.43		13.19	15.62	1.00	154.8	61.9	3960	0.34	0.7	270	280	300	4.18	17.40	21.58	1.38		0.47†	7.4	
4	2.05		21.50	23.55	1.40	133.2	74.6	3170	0.37	0.8	370	390	400	3.53	28.40	31.93	1.36		0.46†	10.8	
5	2.04		21.81	23.85	1.40	117.2	65.7	2760	0.35	0.8	340	330	300	3.51	28.80	32.31	1.35		0.45†	10.7	
6	5.86		15.94	21.80	1.60	131.2	84.0	3850	0.30	0.8	325	280	300	10.08	21.05	31.13	1.43		0.48†	10.5	

Supply curves 2.23 and 1.83 (5-year rainfall)

1	2.80		19.20	22.00	1.83	260	158.6	7200	0.35	0.8	320	320	300	6.24	35.15	37.39	1.70	Not required as appreciable ponding (col. 9) is permissible	0.59†	13.0	Same as shown in col. 21 as appreciable ponding is permissible
2	0.66		17.17	17.83	1.53	98.4	60.2	3330	0.39	0.9	360	390	400	1.47	31.40	32.87	1.84		0.63†	11.2	
3	2.43		13.19	15.62	1.00	154.8	61.9	3960	0.34	0.7	270	280	300	5.42	24.15	29.57	1.89		0.66†	10.3	
4	2.05		21.50	23.55	1.40	133.2	74.6	3170	0.37	0.8	370	390	400	4.57	39.35	43.92	1.87		0.65†	15.3	
5	2.04		21.81	23.85	1.40	117.2	65.7	2760	0.35	0.8	340	330	300	4.55	39.90	44.45	1.86		0.65†	15.5	
6	5.86		15.94	21.80	1.60	131.2	84.0	3850	0.30	0.8	325	280	300	13.06	28.20	41.26	1.89		0.66†	14.3	

Drain-inlet capacities required to remove 2-year design-storm runoff without temporary ponding

1	2.80		19.20	22.00				0	0.35	0.8	610						1.37	33	1.1	24.2				
2	0.66		17.17	17.83				0	0.39	0.9	530						1.33	29	1.2	21.4				
3	2.43		13.19	15.62				0	0.34	0.7	490						1.38	29	1.3	20.3				
4	2.05		21.50	23.55				0	0.37	0.8	580						1.36	31	1.2	28.3				
5	2.04		21.81	23.85				0	0.35	0.8	530						1.35	30	1.2	31.0				
6	5.86		15.94	21.80				0	0.30	0.8	510						1.43	30	1.3	28.3				

Computed by: Date:
Checked by: Date:

* After Hathaway [51].
† Values in col. 21 were determined by minimum-capacity criteria; actual volumes and depths will be somewhat less than values in cols. 6 and 9 during design storm.

based on a 2-year frequency rainfall with a 60-min intensity of 1.72 in./hr. A check analysis was made for a 5-year rainfall with a 60-min rate of 2.23 in./hr.

  b. Infiltration. Study of the soil reports suggested that a reasonable average infiltration-capacity value would be 0.4 in./hr after a 1-hr rainfall, and this was used for all turf areas.

  In the U.S. Army Corps of Engineers' method of airport drainage design, it is assumed (for simplicity) that rainfall is uniform and infiltration capacity is constant over the design duration, with 60-min values used as indices.

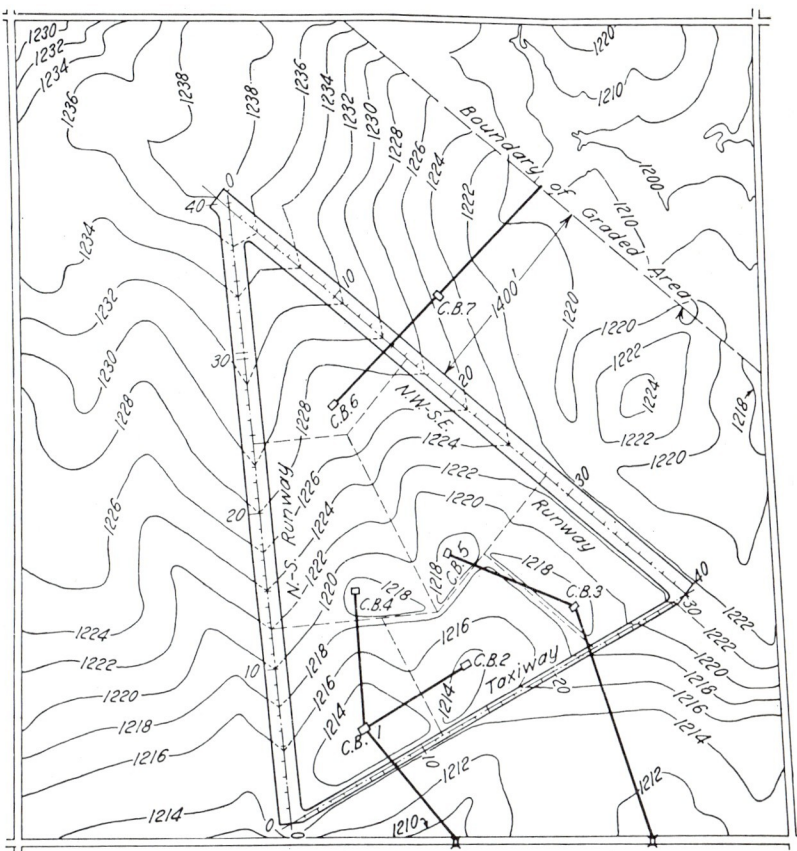

Fig. 20-15. Training airfield, showing drainage system. (*After Horner* [50].)

  For turfed areas it is usually assumed that the 1-hr infiltration rate is approximately 0.5 in./hr, with an upper limit of about 1.0 in./hr. The best estimates would be those based upon infiltration tests on the specific areas, simulating an hour design storm as well as antecedent rainfall that may reasonably be expected prior to the design storm. Jenkins has suggested the values given in Table 20-7 as guides. The values for dry antecedent conditions are for regions where high intensities occur only in summer thunderstorms, and it is therefore unlikely that wet antecedent conditions will prevail at the time of the design rainfall.

  c. Supply Rate. Generalized intensity-duration-frequency curves are adopted for the time distribution of design rainfall. From a study of maximum-average-rainfall data for the United States, correlations were found between 1-hr intensities and the

Table 20-7. One-hour Infiltration Capacities, in Inches per Hour*

Character of entire soil† profile	Dry antecedent conditions		Moist antecedent conditions	
	Dense grass	Poor grass‡ or bare	Dense grass	Poor grass‡ or bare
Tight plastic clays and silty clays....	0.4–0.6	0.2–0.4	0.05–0.2	0.02–0.1
Porous loams and fine sandy soils....	0.8–1.2	0.5–0.7	0.4–0.8	0.2–0.4

* After D. S. Jenkins [51], p. 745.
† Gravels and coarse sands not included. For pavements use zero.
‡ To a small degree, these values for bare soil will vary inversely with intensities. This is not true for dense turf.

intensities of other durations with the same recurrence intervals. From these correlations *standard rainfall intensity-duration curves* were prepared and each curve given a *curve number* equal to its 1-hr value. Yarnell's 1-hr maps are used to determine the 1-hr index rainfall for a given return period at the design site. The difference between the index 1-hr rainfall rate and the index 1-hr infiltration capacity is called the *standard supply rate*. Because infiltration capacity is assumed to parallel the time distribution of rainfall intensity, the standard rainfall intensity-duration curves serve also as *standard supply curves*, with the curve number then becoming a *supply-rate number*. All points on the same curve, for either rainfall or supply, are assumed to have the same average frequency of occurrence.

For the design example, the supply rates for turf areas are 1.32 in./hr for the 1-hr, 2-year frequency and 1.83 in./hr for the 1-hr, 5-year frequency; and for paved areas, 1.72 in./hr for the 1-hr, 2-year frequency and 2.23 in./hr for the 1-hr, 5 year frequency (Table 20-6).

d. *Overall Drainage Considerations.* The objective in this example is to develop as much pondage as possible within the triangle formed by pavement in Fig. 20-15 while satisfying required grading criteria. Runoff from all other areas within the field boundaries would be removed by surface drainage through broad swales. Therefore only the ponded inlet drains will be discussed here. Sizes of the tributary turf and paved areas and the pertinent physical characteristics of the six ponded inlet areas are given in columns 2 to 9 in Table 20-6. In only a few instances is a 2 per cent transverse turf grade required, most of the others falling below 1½ per cent.

e. *Drain-inlet Capacities with Temporary Ponding.* The overland flow for various uniform supply rates for various durations and flow lengths is computed by the Horton equation [7, 10]:[1]

$$q = \sigma \tanh^2 \left[0.922t \left(\frac{\sigma}{nL} \right)^{0.5} S^{0.25} \right] \qquad (20\text{-}14)$$

where q is the rate of overland flow at the lower end of an elemental strip of turfed, bare, or paved surface in cfs/acre of the drainage area or in in./hr, σ is the supply rate in in./hr, n is the retardance coefficient representing the surface roughness, L is the effective length of flow in ft, S is the average surface slope in per cent, and t is the time or duration in min since the supply began.

For $n = 0.40$ and $S = 1$ per cent, the Horton equation is used to compute a set of curves for q, taking σ and t as parameters. This set of curves can also be used for

---

[1] For overland-flow computation Izzard's dimensionless hydrograph will give more realistic results, not only for the rising side, but also for the recession portion of the hydrograph. However, the Corps of Engineers continues to use design aids based on the Horton equation, and therefore it has been used in this example.

AIRPORT HYDROLOGY   20–39

other values of n and S, provided L is replaced by an equivalent length equal to $nL/0.4\sqrt{S}$. Recommended values for n are as follows:

Smooth pavements................................... 0.02
Bare packed soil free of stone....................... 0.10
Poor grass cover or moderately rough bare surface.......... 0.30
Average grass cover................................. 0.40
Dense grass cover................................... 0.80

Column 11 in Table 20-6 gives an estimate of the average surface slope in the ponding area as given on the grading plan.

Column 12 gives the effective length or the weighted overland-flow length from the ridges or watershed boundaries to the inlet when no pondage is to be developed; or to the surface-water line corresponding to a depth at the inlet equal to two-thirds the design depth with ponding. The latter is considered representative of the actual length as the pond surface varies with depth during the design storm. Because water from paved surfaces usually passes over turf prior to reaching ponded inlets, it can be assumed that the pavement runoff is subject to the same retardance as the turf. Since most pavement runoff reaches its peak in a very short time (5 to 10 min) it has been suggested that the length of the paved area can be disregarded or given very little weight in estimating the effective length for composite areas.

Column 13 gives the effective length of overland flow.

Column 14 gives the length to the nearest 100 ft, so that a chart constructed for Eq. (20-14) with $n = 0.40$ and $S = 1$ per cent can be used directly to choose the required drain-inlet capacity in column 21. Of course, the value of q_d can be computed directly by Eq. (20-14), instead of the procedure involving the use of the equivalent length and a set of design curves or charts.

When ponding is provided, however, it is not advisable to pond the water so long or so frequently as to create unstable soil conditions in the ponded area. Therefore a minimum inlet capacity should be required in the design. In the example, the minimum inlet capacities in cubic feet per second per acre of drainage area are considered equal to the supply rates corresponding to a duration of 4 hr on standard supply curves. If a drain inlet is of minimum required capacity, it may be expected that some storage in the ponding basin will result during all storms less than 4 hr in duration that produce supply rates corresponding to the given standard supply curve. In Table 20-6 the drain-inlet capacities in column 21 for all six areas are determined by minimum-capacity criteria. The minimum inlet capacities are greater than actually needed, and therefore permissible pondage both as to volume and depth is in excess of the storage requirements of the design storm.

Figure 20-16 illustrates the results of a hydrograph storage computation for an inlet area with $L = 400$ ft, $n = 0.40$, $S = 1.0$ per cent, and a uniform supply rate of 4 in./hr for a 20-min duration. Hydrograph 3 represents a constant rate of outflow assumed as 1.25 cfs/acre of drainage area corresponding to the mean inflow rate of hydrograph 2. Curve 4 shows the storage volume as it accumulates while inflow exceeds outflow and as it subsequently diminishes. From curve 4 it may be seen that the pond must store a maximum of 1,350 ft³/acre of drainage area; the accumulation period ends about 43 min after the start of runoff; and the pond is drained at about 72 min. Because pond surfaces are usually very large compared with the depths at inlets, it is reasonable to assume that outflow through a drain inlet will be nearly constant when the rate of runoff and the accumulated pondage are sufficient to maintain the full discharge capacity of the inlet. A further assumption is made that all the inflow to the pond will flow out at the same rates until the full capacity of the inlet is reached.

f. Drain-inlet Capacities without Ponding. When no ponding is provided, the computation for inlet capacities becomes simpler, as self-explained in Table 20-6.

By plotting the duration of supply against the rate of runoff computed by the previous design procedure for a given supply curve and effective length of overland flow, it may be found that the greatest discharge occurs at a time t_{cr}. This time is the critical duration of overland-flow supply for runoff from an area unaffected by pondage. Experience suggests adoption of a minimum t_{cr} value of 20 min for turf and 10 min for paved areas. For mixed areas, the value should be weighted.

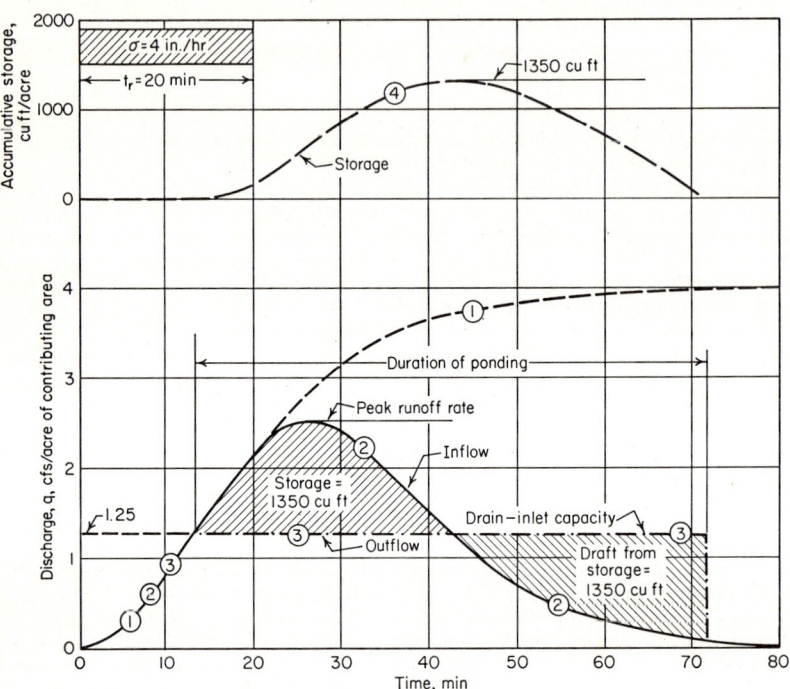

FIG. 20-16. Airfield drainage. Sample computations. Storage required with selected drain-inlet capacity to provide for runoff from 1 acre of turf under assumptions designated.
Notes:
$L = 400$ ft; $S = 1$ per cent; $n = 0.40$.
Length L was chosen to represent weighted mean distance from limits of drainage area to average limits of pond.
Hydrograph 1 represents runoff that would have occurred if given rate of supply of 4 in./hr had continued indefinitely.
Hydrograph 2 corresponds to runoff from supply of 4 in./hr for duration of 20 min.
Hydrograph 3 represents rate of outflow, assuming inlet capacity as designated and constant head-discharge relation. (*After Hathaway* [51].)

To determine the maximum rate of flow at a given point in the drainage system, the assumed uniform intensity of supply must last a duration t_{cr}' given by

$$t_{cr}' = t_{cr} + t_d \qquad (20\text{-}15)$$

where t_{cr} is the duration of supply that would produce the maximum design storm runoff at the critical inlet (usually the most distant inlet upstream from the point under design), and t_d is the time of flow from the critical inlet to the design point. For simplification, values of t_{cr}' are taken to the nearest 5 min.

g. Drain Sizes Required. The usual hydraulic computations were made for the design flows of Table 20-6, and nominal pipe sizes were chosen to fit the grading and outlet conditions. The outlet hydraulic gradients were assumed at the intrados of the pipes, and hydraulic grades worked back from there in the usual manner.

For the case with ponding, check computations for the pipe sizes of the 2-year design storm were made using the greater quantities of inflow which would result from a 5-year design storm. It was found that at inlets 1, 3, and 4 the 5-year hydraulic gradient would be satisfactory, but at inlets 2, 5, and 6 it would exceed the maximum pond levels adopted for the 2-year design. Under the 5-year conditions, sheet flow would take place over the lower portions of the taxiway, but in no place would pondage be closer than about 150 ft from the edge of any runway.

The pipe sizes for the system given in Fig. 20-15, with and without pondage, are

Line	Pipe sizes	
	With ponding, in.	Without ponding, in.
6 to 7	24	33
7 to outlet	18	24
5 to 3	24	36
3 to outlet	30	42
4 to 1	24	30
2 to 1	24	30
1 to outlet	36	48

VIII. HYDROLOGY OF URBAN EXPRESSWAYS

A. General

In the design of urban expressways, provision must be made for more complete storm-runoff removal than is customarily considered desirable for most city streets. A high degree of freedom from traffic interruption and a minimization of hazard to the faster-moving vehicles are necessary.

B. Design Frequency

Storm drains for most urban expressways are designed to handle the runoff from 5- to 10-year-frequency rainfall intensities when flowing full. They should be capable of handling 25-year rainfall intensities without raising the hydraulic gradient of the drains to a height which would diminish the capacities of inlets.

It may not be possible to satisfy both these design criteria because of inadequacies of local storm drains or channels into which the expressway drainage is to be discharged. The absence of satisfactory outlet facilities nearby may make it desirable and practicable to design the expressway drainage as a separate system with its own outfall completely independent of local facilities.

C. Time of Concentration

The rapid movement of overland flow over the almost completely impervious surfaces of expressways (even cut slopes with turf cover will have very short times of equilibrium flow) results in inlet times of 5 min or less. Hence 5 min may be the realistic minimum to use in determining the required design capacity of individual inlets.

Both the rational method and various modifications of it and the overland-flow procedure are used in the design of urban expressways.

Charts based on overland-flow data can be used to estimate the time of concentration of homogeneous surfaces [52].

D. Gutters and Inlets

The hydraulics of gutters and inlets in city streets is directly applicable to urban expressways. In general, the gutters of city streets are hydraulically more efficient because of the higher crowns and greater cross slopes. Expressways have flat cross slopes, and the gutter aprons adjacent to pavement edges rarely have a fall toward the curb steeper than $\frac{1}{2}$ to 1 in./ft. However, a 2-ft-wide gutter on a 1-in./ft transverse slope has been used in many 12-ft traffic-lane expressways.

The following criteria have been used in designing expressway drainage: (1) Depth of flow in the gutter section at the pavement edge is limited to a maximum of 1 in.; (2) no grating shall extend into the running slab; (3) except in emergencies, wheels are not expected to run in the gutter; and (4) any depression below the continuous gutter flow line is undesirable.

Criterion (1) would ensure infrequent short-duration flooding of the running slab, and combined with the design runoff it would lead to the required spacing of grated inlets. Criterion 2 leads to a "carry-over" of at least that part of the longitudinal pavement flow not intercepted at each inlet, and hence complete interception is impossible under this criterion except at low points. Criterion 4 recognizes that on all high-speed urban expressways there will be occasions when vehicle wheels will ride in the gutter.

Spacing of inlets on a continuous grade is determined by evaluating the distance to the point where the computed runoff results in the maximum permissible gutter-flow depth. This evaluation must be reconciled with the capacity and corresponding carry-over of each of the inlets in the series.

Because the longitudinal slopes of most urban expressways and accessory ramps are quite variable with distance, the resulting irregular inlet spacing must be determined by trial and error.

Izzard [11] has developed graphs to facilitate the spacing of curb-opening inlets on continuous grades. Sample computations were given to illustrate the use of the graphs.

Deflectors on curb-opening inlets offer a practical opportunity to reduce inlet lengths where packing of the deflector notches by traffic is unlikely or where maintenance (such as should be in practice on heavily traveled expressways) can be assured after each storm. In laboratory investigations [31] it was found that the deflectors became increasingly efficient with greater longitudinal grades, which is the exact reverse of the relationship for all other types of inlets. As opposed to the depression in front of a curb-opening inlet, deflector notches would not present a traffic hazard. Experience is needed to assure their suitability with respect to self-cleansing or maintenance characteristics, but they have a promising potential.

E. Pumping Stations at Grade Separations

Pumping stations may be required at low points in grade separations because of the impracticality of gravity disposal by means of special outfalls or because design hydraulic gradients of existing storm channels or drains are at elevations too high to service gravity outlets from underpasses. The importance and character of expressways dictate that inlet facilities for pumping stations be conservatively designed. It is suggested that their installed capacities be about one-third larger than the capacity required to accommodate design flows. Pump capacities should likewise be conservative. The installed capacity should be capable of handling the flows resulting from rainfall intensities with an average recurrence interval of 25 to 50 years [9].

F. Use of Pondage at Interchanges

The sweeping access roads of interchanges cover large expanses. Intervening turf areas can in some instances be designed for or adapted to pondage [53]. Design procedures and the economies afforded are similar to those described for airport drainage.

IX. REFERENCES

1. U.S. Census of Population, 1950, U.S. Bureau of the Census, 1952, vol. 1.
2. Savini, John, and J. C. Kammerer: Urban growth and the water regimen, *U.S. Geol. Surv. Water-Supply Paper* 1591-A, 1961.
3. "Hydrology Handbook," American Society of Civil Engineers Manuals of Engineering Practice, no. 28, 1949.
4. Horton, R. E.: Surface runoff phenomena; Part I, Analysis of the hydrograph, *Horton Hydrol. Lab. Pub.* 101, Edwards Brothers, Inc., Ann Arbor, Mich. 1935.
5. Hicks, W. I.: A method of computing urban runoff, *Trans. Am. Soc. Civil Engrs.*, vol. 109, pp. 1217–1253, 1944.
6. Tholin, A. L., and C. J. Keifer: The hydrology of urban runoff, *Trans. Am. Soc. Civil Engrs.*, vol. 125, pp. 1308–1379, 1960.
7. Horton, R. E.: The interpretation and application of runoff plot experiments, with reference to soil erosion problems, *Proc. Soil Sci. Soc. Am.*, vol. 3, pp. 340–349, 1938.
8. Izzard, C. F.: Hydraulics of runoff from developed surfaces, *Proc. 26th Annual Meeting Highway Research Board*, vol. 26, pp. 129–146, 1946.
9. "Design and Construction of Sanitary and Storm Sewers," American Society of Civil Engineers, Manuals of Engineering Practice, no. 37, or Water Pollution Control Federation Manual of Practice, no. 9, 1960.
10. Chow, V. T.: "Open-channel Hydraulics," McGraw-Hill Book Company, Inc., New York, 1959.
11. Izzard, C. F.: Tentative results on capacity of curb opening inlets, in "Surface drainage," *Highway Research Board Res. Rept.* 11-B, December, 1950.
12. West, E. M.: Drainage for highways and airports, sec. 12 in K. B. Woods (ed.), "Highway Engineering Handbook," McGraw-Hill Book Company, Inc., New York, 1960.
13. Chow, V. T.: Hydrologic determination of waterway areas for the design of drainage structures in small drainage basins, *Univ. Ill. Eng. Expt. Sta. Bull.* 462, 1962.
14. Metcalf, Leonard, and H. P. Eddy: "Sewerage and Sewage Disposal," McGraw-Hill Book Company, Inc., 1930, pp. 94–103.
15. Gregory, R. L., and C. E. Arnold: Runoff-rational runoff formulas, *Trans. Am. Soc. Civil Engrs.*, vol. 96, pp. 1038–1099, 1932.
16. Bernard, Merrill: Modified rational method of estimating flood flows, Appendix A, in "Low Dams," National Resources Committee, Washington, D.C., 1938.
17. Horner, W. W., and F. L. Flynt: Relation between rainfall and runoff from small urban areas, *Trans. Am. Soc. Civil Engrs.*, vol. 101, pp. 140–183, 1936.
18. Marston, F. A.: The distribution of intense rainfall and some other factors in the design of storm-water drains, *Trans. Am. Soc. Civil Engrs.*, vol. 87, pp. 535–562, 1924.
19. Hicks, W. I.: Discussion on Surface runoff determination from rainfall without using coefficients by W. W. Horner and S. W. Jens, *Trans. Am. Soc. Civil Engrs.*, vol. 107, pp. 1097–1102, 1942.
20. Keifer, C. J., and H. H. Chu: Synthetic storm pattern for drainage design, *Proc. Am. Soc. Civil Engrs., J. Hydraulics Div.*, vol. 83, no. HY4, Paper 1332, August, 1957; Discussion by M. B. McPherson in vol. 84, no. HY1, Paper 1558, pp. 49–57, February, 1958.
21. Jens, S. W.: Drainage of airport surfaces: some basic design considerations, *Trans. Am. Soc. Civil Engrs.*, vol. 113, pp. 785–809, 1948.
22. Williams, H. M.: Discussion on Drainage of airport surfaces: some basic design considerations by S. W. Jens, *Trans. Am. Soc. Civil Engrs.*, vol. 113, pp. 810–813, 1948.
23. Williams, G. R.: Hydrology, chap. 4 in Hunter Rouse (ed.), "Engineering Hydraulics," John Wiley & Sons, Inc., New York, 1950.
24. Williams, G. R.: Drainage of leveed areas in mountainous valleys, *Trans. Am. Soc. Civil Engrs.*, vol. 108, pp. 83–96, 1943.
25. Hershfield, D. M.: Rainfall frequency atlas of the United States, *U.S. Weather Bur. Tech. Paper* 40, May, 1961.
26. The Johns Hopkins University, Department of Sanitary Engineering and Water Resources: Storm Drainage Research Project Progress Reports: (a) By Paul Bock, July 1, 1955 to June 30, 1956; (b) by Paul Bock and Warren Viessman, Jr., July 1, 1956 to June 30, 1958; (c) by Warren Viessman, Jr., July 1, 1958, to June 30, 1959; (d) by Warren Viessman, Jr., July 1, 1959, to June 30, 1960; (e) by Warren Viessman, Jr., and J. H. McKay, Jr., July 1, 1960 to June 30, 1961; (f) by J. C. Schaake, Jr., July 1, 1961 to June 30, 1962; (g) by J. C. Schaake, Jr., July 1, 1962 to June 30, 1963. Report (f) is on instrumentation and later outlined in Measuring rainfall and runoff at stormwater inlets, by J. W. Knapp, J. C. Schaake, Jr., and Warren Viessman, Jr., *Proc. Am. Soc. Civil Engrs., J. Hydraulics Div.*, vol. 89, no. HY5, Paper 3644, pp. 99–115, September, 1963.

27. Shrank, F.: Kritsche Bemerkungen zu den neuen Verzögerungsberechnungsmethoden, die zur Bestimmung massgebender Abflussziffern für Städeentwässerungen dienen, *Gesundh.-Ing.*, vol. 37, no. 21, pp. 415–417, 1914.
28. Reinhold, F.: Bemerkungen über die Abflachung der Flautwelle in Abflussleitungen, *Gesundh.-Ing.*, vol. 66, no. 15, pp. 166–167, 1943.
29. Willis, M. J.: Collection and analysis of field gaging data, part F of "Review of Storm Drainage Design," Philadelphia Water Department Progress Report, December, 1960 (available at The Franklin Institute Library, Philadelphia).
30. Watkins, L. H.: Surface water drainage research, Progress Report to April, 1959, *Dept. Sci. Ind. Res., Road Res. Lab., Res. Note* RN/3515/LHW, London, June, 1959; and also, The design of urban sewer systems, *Road Res. Tech. Paper* 55, London, 1962.
31. "The Design of Storm-water Inlets," The Johns Hopkins University, Department of Sanitary Engineering and Water Resources, Storm Drainage Research Project, 1956.
32. Larson, C. L.: Experiments on flow through inlet gratings for street gutters, *Highway Res. Board Res. Rept.* 6-B, December, 1948.
33. Horner, W. W.: More engineering on sewer inlets, *Municipal and County Eng.*, Indianapolis, Ind., 1919.
34. Larson, C. L.: Grate inlets for surface drainage of streets and highways, *St. Anthony Falls Hydraulic Lab. Bull.* 2, University of Minnesota, Minneapolis, 1949.
35. Airfield drainage structure investigation, *U.S. Corps Engrs. Hydraulic Lab. Rept.* 54, St. Paul District Sub-office, Iowa City, April, 1949.
36. Guillou, J. C.: The use and efficiency of some gutter inlet grates, *Univ. Illinois Eng. Expt. Sta. Bull.* 450, July, 1959.
37. Wasley, R. J.: Hydrodynamics of flow into curb-opening inlets, *Proc. Am. Soc. Civil Engrs., J. Hydraulics Div.*, vol. 87, no. HY4, *Paper* 2880, August, 1961.
38. Sangster, W. M., H. W. Wood, E. T. Smerdon, and H. G. Bossy: (a) Pressure changes at storm drain junctions, *Univ. Missouri Eng. Expt. Sta. Bull.*, ser. 41, Oct. 15, 1958; (b) Pressure changes at open junctions in conduits, *Proc. Am. Soc. Civil Engrs., J. Hydraulics Div.*, vol. 85, no. HY6, *Paper* 2235, June, 1959. Discussions: vol. 85, no. HY10, *Paper* 2235, October, 1959, and no. HY11, *Paper* 2269, November, 1959. Closure: vol. 86, no. HY5, *Paper* 2489, October, 1960.
39. Handbook of channel design for soil and water conservation, *U.S. Soil Conserv. Serv. Stillwater Outdoor Hydraulic Lab.*, SCS-TP-61, March, 1947, rev. June, 1954.
40. Proceedings of the First Western Conference on Asphalt in Hydraulics, Oct. 19 and 20, 1955, *Univ. Utah Eng. Expt. Sta. Bull.* 78, June, 1956.
41. Hollis, M. D.: Water resources and needs for pollution control, *J. Water Pollution Control Federation*, vol. 32, pp. 225–231, March, 1960.
42. Fair, G. M., and J. C. Geyer: "Water Supply and Waste Water Disposal," John Wiley & Sons, Inc., New York, 1954.
43. Ongerth, H. J., and J. A. Harmon: Sanitary engineering appraisal of waste water reuse, *J. Am. Water Works Assoc.*, vol. 51, no. 5, pp. 647–658, May, 1959.
44. Effects of synthetic detergents on water supplies, AWWA Task Group 2661P Report, *J. Am. Water Works Assoc.*, vol. 51, no. 10, pp. 1251–1254, October, 1959.
45. Parks, W. W.: Decontamination of ground water at Indian Hill, *J. Am. Water Works Assoc.*, vol. 51, no. 5, pp. 644–646, May, 1959.
46. Tsivoglou, E. C., Murray Stein, and W. W. Towne: Control of radioactive pollution of the Animas River, *J. Water Pollution Control Federation*, vol. 32, pp. 262–287, March, 1960.
47. "Airport Drainage," Federal Aviation Agency, Airports Division, Bureau of Facilities and Materiel, 1960.
48. (a) Surface and (b) Subsurface drainage facilities for airfields, pt. 13, chaps. 1 and 2 in Engineering Manual, U.S. Army Corps of Engineers, June, 1955.
49. Subsurface drainage of highways and airports, *Highway Research Board Bull.* 209, 1959.
50. Horner, W. W.: The drainage of airports, *Univ. Illinois Eng. Expt. Sta. Circ.* 49, Nov. 21, 1949.
51. Hathaway, G. A.: Military airfields: design of drainage facilities, *Trans. Am. Soc. Civil Engrs.*, vol. 110, pp. 679–733; discussions, pp. 734–848, 1945.
52. Kerby, W. S.: Time of concentration for overland flow, *Civil Eng.*, vol. 29, p. 60, March, 1959.
53. Forest, E., and H. G. Aronson: (a) Highway ramp areas become flood control reservoirs, *Civil Eng.*, vol. 29, p. 35, February, 1959; (b) Storm water control for a shopping center, *ibid.*, vol. 30, p. 63, February, 1960.

REFERENCES

Other References

54. Bauer, W. J.: Economics of urban drainage design, *Proc. Am. Soc. Civil Engrs., J. Hydraulics Div.*, vol. 88, no. HY6, Paper 3321, pp. 93–114, November, 1962.
55. Kaltenbach, A. B.: Storm sewer design by the inlet method, *Public Works*, vol. 94, no. 1, pp. 86–89, January, 1963.
56. Eagleson, P. S.: Unit hydrographs for sewered areas, *Proc. Am. Soc. Civil Engrs., J. Hydraulics Div.*, vol. 88, no. HY2, Paper 3069, pp. 1–25, March, 1962. Discussions: no. HY4, Paper 3209, July; no. HY5, Paper 3290, September; Paper 3338, no. HY6, November, 1962. Closure: vol. 95, no. HY4, Paper 3580, July, 1963.
57. Viessman, Warren, Jr., and J. C. Geyer: Characteristics of the inlet hydrograph, *Proc. Am. Soc. Civil Engrs., J. Hydraulics Div.*, vol. 88, no. HY5, pt. 1, Paper 3285, pp. 245–268, September, 1962. Discussion: vol. 89, no. HY2, Paper 3468, March, 1963. Closure: no. HY4, Paper 3580, July, 1963.
58. Willeke, G. E.: The prediction of runoff hydrographs for urban watersheds from precipitation data and watershed characteristics, abstract of paper presented at 43rd Annual Meeting of American Geophysical Union, *J. Geophys. Res.*, vol. 67, no. 9, p. 3610, August, 1962.

Section 21

HYDROLOGY OF AGRICULTURAL LANDS

HAROLD O. OGROSKY, *Chief, Hydrology Branch, U.S. Soil Conservation Service.*

VICTOR MOCKUS, *Hydraulic Engineer, U.S. Soil Conservation Service.*

I. Introduction	21-3
II. Typical Problems	21-3
A. Farms	21-3
1. Water Control	21-3
2. General Design Conditions	21-3
3. Extent of Hydrologic Investigations	21-3
B. Small Watersheds	21-4
1. Agricultural Watershed Management	21-4
2. General Design Conditions	21-4
3. Extent of Hydrologic Investigations	21-4
III. Data for Hydrologic Analyses	21-4
A. Precipitation	21-4
1. Rainfall Data	21-4
2. Snow	21-5
3. Interception	21-6
B. Evapotranspiration	21-6
1. The Blaney-Criddle Method	21-6
2. The Penman Method	21-8
C. Watershed Characteristics	21-10
1. Size	21-10
2. Shape	21-10
3. Watershed Slope	21-10
4. Time of Concentration	21-10
5. Hydrologic Soil-cover Complexes	21-11
6. Erosion	21-27
D. Runoff	21-27
1. Data	21-27
IV. Determination of Runoff from Precipitation	21-28
A. Storm Runoff	21-28
1. Antecedent Moisture Condition	21-29
2. Estimate of Direct Runoff by Type 1 Approach	21-29
3. Estimate of Direct Runoff by Type 2 Approach	21-32

    B. Snowmelt Runoff.. 21-32
        1. General.. 21-32
        2. Degree-day Method.. 21-32
    C. Annual and Seasonal Runoff..................................... 21-34
        1. Data Transposition... 21-34
        2. Supplementary Stations.................................... 21-37
        3. Determination of Frequency............................... 21-37
        4. Seasonal Variations.. 21-37
V. Determination of Peak Rates of Runoff............................... 21-37
    A. Runoff from Rainfall... 21-37
        1. The Rational Method...................................... 21-37
        2. The Cook Method... 21-39
    B. Snowmelt Runoff.. 21-41
    C. Regional Analysis... 21-41
VI. Hydrographs.. 21-41
    A. Synthetic Hydrographs... 21-41
        1. Dimensionless Hydrographs................................ 21-41
        2. Triangular Hydrographs................................... 21-44
        3. Composite Hydrographs................................... 21-45
VII. Field Applications.. 21-46
    A. Terraces and Diversions... 21-46
        1. Construction Types.. 21-46
        2. Criteria... 21-46
        3. Hydraulic Considerations.................................. 21-48
        4. Hydrologic Considerations................................. 21-49
    B. Grassed Waterways.. 21-54
        1. Construction Types.. 21-54
        2. Criteria... 21-54
        3. Hydraulic Considerations.................................. 21-55
        4. Hydrologic Considerations................................. 21-58
    C. Grade-stabilization Structures................................... 21-61
        1. Construction Types.. 21-62
        2. Criteria... 21-62
        3. Hydraulic Considerations.................................. 21-62
        4. Hydrologic Considerations................................. 21-65
    D. Farm and Ranch Ponds... 21-65
        1. Construction Types.. 21-65
        2. Criteria... 21-66
        3. Hydrologic Considerations................................. 21-68
    E. Structures for Temporary Floodwater Storage............ 21-71
        1. Criteria... 21-71
        2. Hydrologic Considerations................................. 21-75
    F. Channel Works... 21-76
        1. Construction Types.. 21-76
        2. Criteria... 21-77
        3. Hydrologic Considerations................................. 21-77
    G. Irrigation... 21-78
        1. General... 21-78
        2. Hydrologic Considerations................................. 21-79
        3. Hydraulic Considerations.................................. 21-87
    H. Drainage.. 21-88
        1. Types... 21-88
        2. Legal Aspects.. 21-88
        3. Planning.. 21-89
        4. Surface Drainage... 21-89
        5. Subsurface Drainage....................................... 21-91
        6. Pumping for Drainage..................................... 21-95
VIII. References... 21-95

I. INTRODUCTION

The time is past when fertile land can be agriculturally "mined," then abandoned in a move to new fertile land. In 1960 only about 20 per cent of the agricultural land (including woodland) in the United States was being used properly [1]. The remainder requires treatment or improvement, much of which involves the control of water.

In one year (July, 1958, to July, 1959) over 500 million cubic yards of earth were moved for farm improvements such as terraces, diversions, farm ponds, waterways, land leveling, and drainage ditches. More earth moving can be expected as the application of land-treatment practices becomes more widespread. Although individual jobs or projects are small, the total volume of work is great. The hydrologic aspects of this work require special emphasis.

II. TYPICAL PROBLEMS

A. Farms

1. Water Control. The efficient control and use of water on farms and ranches requires the application of various hydrologic techniques. Although cropland vegetation provides some protection against soil erosion, it is often necessary to increase the protection against runoff from high-intensity rains by strip cropping, contouring, or terracing. A terrace system may be required to convey surface runoff at non-erosive velocities to a disposal system; or diversions, grassed waterways, and grade-stabilization structures may be needed. When soils have high infiltration rates, a terrace system can be designed to store the surface runoff until it can be disposed of by percolation to the soil substrata. On flat or waterlogged soils, artificial drainage by means of ditches or tile drains may be required before the land is cultivated. Where rain is infrequent or poorly distributed during the growing season, the supply and demand of water for small irrigation developments must be determined. Farm ponds may be required to store water for livestock, irrigation, fire protection, or recreation or to stabilize the grade of a waterway. Although most projects of this type are small, success requires adequate consideration of the hydrologic and hydraulic features of design.

2. General Design Conditions. The *level of protection* generally desired for water control on farms is one that will maintain a certain level of soil productivity for each soil type. Permissible soil losses on deep loessial soils are about 5 tons/acre (about 0.035-in. depth over an acre) per year, but they may be much less on other soils.

Standard designs are normally used for terraces, diversions, and grassed waterways; they are usually planned to control runoff from storms of 5- to 10-year frequency. Designs for farm drainage facilities require individual consideration of the availability of outlet locations and of the soil types and water-table depths of each area to be treated.

Irrigation systems may be planned so that the water supply is adequate for a fixed acreage in 8 out of 10 years, or so that the entire supply is used on a varying acreage each year. Similarly, design of farm ponds involves the study of water supply, evaporation, seepage, and spillway requirements. Small pipe spillways are used to carry persistent small flows, and emergency spillways of earth are designed to accommodate larger flows (normally 10- to 50-year floods). In each case, hydrologic design should be commensurate with the cost or importance of the project. On-farm structures usually will be designed for flows of less than the 50-year frequency.

3. Extent of Hydrologic Investigations. Low-cost on-farm water control does not justify extensive and costly hydrologic investigations. Usually, available maps of rainfall, runoff, soils, and topography can be used and standard designs modified as necessary to meet specific needs. When high-cost structures are necessary, field investigations are justified to achieve economy in design and construction.

B. Small Watersheds

1. Agricultural Watershed Management. Management of a small watershed to conserve soil and water requires that the land be "used within its capabilities and treated according to its needs." The objectives are to protect the land against all forms of soil deterioration; to rebuild eroded and depleted soils; to build up soil fertility; to stabilize critical runoff- and sediment-producing areas; to improve grasslands, woodlands, and wildlife lands; to conserve water for beneficial use; to provide needed drainage and irrigation; and to reduce flood and sediment damage. These objectives can be attained by the application of land-treatment practices and water-control structures on individual farms and upstream drainageways through both individual and group action. Complete conservation management of a small watershed may be as complicated as for a major river basin.

2. General Design Conditions. The cost and risk involved in each watershed improvement determine the nature of the hydrologic design. Channels, levees, and floodwater-retarding structures may be designed to provide agricultural lands with full protection against floods greater than a 5-year frequency. Urban or industrial areas normally will be protected against 100-year-frequency floods, with provisions for passing the probable maximum flood around storage dams. The benefits of the project will determine the permissible cost and the level of protection to be provided. However, the benefits of the project should not determine the risk, which should be reduced to a minimum through conservative design and good construction.

3. Extent of Hydrologic Investigations. Watershed programs use a wide range of control measures, and the extent of hydrologic investigation necessary varies accordingly. High-cost or high-risk projects require special hydrologic studies, including field investigations, to attain economy of design and to avoid unnecessary risk of failure.

III. DATA FOR HYDROLOGIC ANALYSES

The type of hydrologic analysis made on a small watershed will depend largely on the availability of hydrologic data. The following types will be considered.

Type 1. The determination of hydrologic elements is made for a rainfall or runoff event of a specific date or time, such as the determination of the hydrograph for a watershed for a given storm or the determination of the total water yield for a given calendar year. Determinations of this type require local hydrologic data of relatively high accuracy.

Type 2. The determination of hydrologic elements is made on the basis of probability, such as the determination of the peak flow of a certain frequency or the determination of the annual water yield of a certain per cent chance from a given watershed. This type of determination is based on frequency studies available in reports or on use of a long record of a nearby hydrologic station.

A. Precipitation

1. Rainfall Data. Local rainfall data consisting of readings of small-bore tube gages can sometimes be obtained from farmers and ranchers. Gages of this type are often read at irregular intervals, and dates should be verified wherever possible, using the nearest U.S. Weather Bureau gages as references. Data for gages in the U.S. Weather Bureau network can be obtained from the published "Climatological Data" for each state. In some localities, data from denser networks of rain gages on small experimental watersheds may be found in publications of the U.S. Agricultural Research Service [2–4]. The chief modern source of generalized rainfall information for a type 2 analysis is the Rainfall Frequency Atlas of the United States [5, 6].

a. Areal Adjustments. Conversion of point rainfall to areal rainfall can be made using Fig. 9-59, which applies to storms of all types. It is not considered necessary to convert from point to areal rainfall for watersheds of less than about 2 sq mi.

b. Distance Adjustments. Figure 21-1 (derived from Fig. 9-59) can be used to estimate the probable rainfall at a point or area some distance away from a rain gage, provided the center of the rainstorm is at the gage.

c. Accuracy. The accuracy of estimates made by using Fig. 21-1 can be determined by using Fig. 21-2, which is adapted from a more detailed analysis [7]. Conservative estimates of rainfall or derived runoff can be made using the indicated range of the estimate. The range will also assist in determining if an isohyetal map of sufficient accuracy can be made from the data of a given network. Figure 21-2 is also useful in cases where distance adjustments are necessary in transposing data.

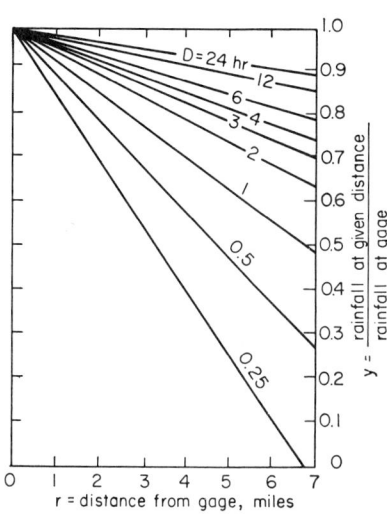

FIG. 21-1. Curves for estimating probable rainfall at a distance from a gage or an assumed center of a storm.

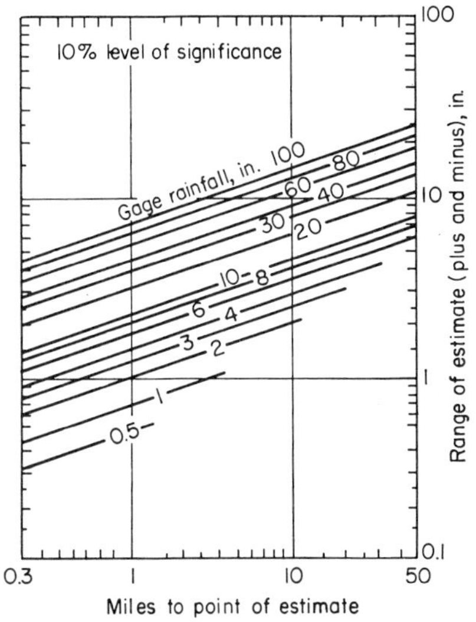

FIG. 21-2. Expected range of error in estimating rainfall at a distance from a gage. (*Modified from Huff and Neill* [7].)

$$y = 1 - \frac{r}{13.4 \sqrt{D}} \quad (r \leq 7 \text{ miles})$$

Example 21-1. Find the probable rainfall at a point 3.5 miles from a rain gage that caught 4.1 in. of rain in a storm of 2.5-hr duration, and determine the range of the estimate.

1. Enter Fig. 21-1 with a distance of 3.5 miles, and (by interpolation) read an adjustment factor of 0.83. The probable rainfall is therefore $0.83 \times 4.1 = 3.4$ in.
2. Enter Fig. 21-2 with the distance of 3.5 miles, interpolate to find the gage rainfall 3.4 in., and read the limit of 1.8 in. on the vertical scale. The probable range in terms of the station catch is $3.4 + 1.8$ to $3.4 - 1.8$, or 5.2 to 1.6 in.

d. Data Transposition. This is the practice of using data from a distant gage or network as though it had been obtained on the problem watershed. In flat areas, such as the Plains states, adjustments may be made using isohyetal-frequency maps [5] as a guide. In hilly or mountainous areas caution must be used in the transposition of data since adjustments for differences in topography will probably be necessary.

2. Snow. Data on the depth, areal distribution, and water content of snow on small watersheds are rarely available. The transposition of data from another area is very likely to lead to erroneous estimates, since local topography and ground cover will greatly affect the drifting and distribution of snow. The type 2 analysis will therefore be most often used when precipitation is in the form of snow.

3. Interception. Rainfall interception by low-lying vegetation has been measured in the field [8] and the laboratory [9], but because of seasonal changes and local variations, only approximate values can be obtained for small-watershed studies. Average interception can be taken into account by including it as part of the "initial abstraction" in methods for estimating runoff. In dense tall vegetation, such as forests, interception may account for as much as 10 per cent of the rainfall [10]. Snow interception by low-lying vegetation is usually ignored in hydrologic studies, but where a large percentage of a small watershed is in forest cover, snow interception may be an important factor in estimating runoff.

B. Evapotranspiration

Evapotranspiration data are used primarily for estimating water requirements for crop production (see also Sec. 11). It is seldom possible to make extensive investigations on small projects, and the water requirement (or *consumptive use*) is estimated by one of several methods [11], two of which are described below. Measured consumptive uses for various crops for selected locations have been published [12, 13].

1. The Blaney-Criddle Method. Seasonal consumptive use for a given crop can be estimated by the Blaney-Criddle relation [11]

$$U = k_s B \qquad (21\text{-}1)$$

where U is the consumptive use of water in in. for the growing season, k_s is an empirical seasonal consumptive-use coefficient for a given crop, and B is the sum of monthly consumptive-use factors for the given season, or $B = \Sigma(tp/100)$, where t is the average monthly temperature in °F and p is the monthly daytime hours as per cent of the year.

When values of the consumptive-use coefficient k for individual months of the season are available, the consumptive use may be estimated by the month:

$$u = \frac{ktp}{100} \qquad (21\text{-}2)$$

where u is the monthly consumptive use in in., and k, t, and p are values for the particular month. Values of k for selected conditions and locations are given in Tables 21-1 and 21-2. Values of p for latitudes 24 to 50°N are given in Table 21-3.

Table 21-1. Consumptive-use Coefficients for Selected Crops*

Crop	Length of growing season or period	Seasonal consumptive-use coefficients k_s	Maximum monthly values of k†
Alfalfa	Frost-free	0.80–0.85	0.95–1.25
Beans	3 months	0.60–0.70	0.75–0.85
Corn (maize)	4 months	0.75–0.85	0.80–1.20
Cotton	7 months	0.65–0.75	0.75–1.10
Citrus orchard	7 months	0.50–0.65	0.65–0.75
Deciduous orchard	Frost-free	0.60–0.70	0.70–0.95
Pasture, grass, hay, annuals	Frost-free	0.75	0.85–1.15
Potatoes	3 months	0.65–0.75	0.85–1.00
Rice	3 to 4 months	1.00–1.20	1.10–1.30
Small grains	3 months	0.75–0.85	0.85–1.00
Sorghum	5 months	0.70	0.85–1.10
Sugar beets	5.5 months	0.65–0.75	0.85–1.00

* From Criddle [11] and Blaney [13].
† Dependent on average monthly temperature and crop stage of growth.

DATA FOR HYDROLOGIC ANALYSES   21-7

Table 21-2. Monthly Consumptive-use Coefficients k for Use in
Blaney-Criddle Method*

Crop	Location	Mar.	Apr.	May	June	July	Aug.	Sept.	Oct.	Nov.
Alfalfa.......	California, coastal	0.60	0.65	0.70	0.80	0.85	0.85	0.80	0.70	0.60
	California, interior	0.65	0.70	0.80	0.90	1.10	1.00	0.85	0.80	0.70
	North Dakota	....	0.84	0.89	1.00	0.86	0.78	0.72		
	Utah, St. George	....	0.88	1.15	1.24	0.97	0.87	0.81		
Corn (maize)..	North Dakota	....	....	0.47	0.63	0.78	0.79	0.70		
Cotton.......	Arizona	....	0.27	0.30	0.49	0.86	1.04	1.03	0.81	
	Texas	0.24	0.22	0.61	0.42	0.50				
Orchard, citrus	Arizona	0.57	0.60	0.60	0.64	0.64	0.68	0.68	0.65	0.62
	California, coastal	....	0.40	0.42	0.52	0.55	0.55	0.55	0.50	0.45
Pasture.......	California, Murrieta	....	....	0.84	0.84	0.77	0.82	1.09	0.70	
Potatoes......	North Dakota	....	....	0.45	0.74	0.87	0.75	0.54		
	South Dakota	....	....	0.69	0.60	0.80	0.89	0.39		
Small grain...	North Dakota	....	0.19	0.55	1.13	0.77	0.30			
Wheat........	Texas	0.64	1.16	1.26	0.87					
Sorghum.....	Arizona	....	....	....	....	0.34	0.72	0.97	0.62	0.60
	Kansas	....	....	....	0.80	0.94	1.17	0.86	0.47	
	Texas	....	....	....	0.26	0.73	1.20	0.85	0.49	
Soybeans.....	Arizona	....	....	....	0.26	0.58	0.92	0.92	0.55	
Sugar beets...	California, coastal	....	0.39	0.38	0.36	0.37	0.35	0.38		
	California, interior	....	0.30	0.60	0.86	0.96	0.91	0.41		
	Montana	....	....	....	....	0.83	1.05	1.02		
Truck crops...	California, interior	0.19	0.26	0.38	0.55	0.71	0.82	0.69	0.37	0.35

* From Blaney [12, 13].

Example 21-2. Determine the monthly consumptive use for a citrus orchard at Santa Ana, Calif. (latitude 34°06′), for the season April to October (7 months). The computations for the following items are shown in Table 21-4.
1. Tabulate average monthly temperatures (column 2).
2. Tabulate daytime hours as a per cent of the year for the location (column 3), using Table 21-3.
3. Tabulate k values (column 4) taken from Table 21-2 for "Orchard, citrus, California, coastal."
4. Using Eq. (21-2), compute average monthly consumptive uses (column 5).

Table 21-3. Monthly Percentage of Daytime Hours of the Year for Latitudes 24 to 50° North of Equator*

Month	Latitudes, degrees north of equator												
	24	26	28	30	32	34	36	38	40	42	44	46	48	50
Jan......	7.58	7.49	7.40	7.30	7.20	7.10	6.99	6.87	6.76	6.62	6.49	6.33	6.17	5.98
Feb......	7.17	7.12	7.07	7.03	6.97	6.91	6.86	6.79	6.73	6.65	6.58	6.50	6.42	6.32
Mar......	8.40	8.40	8.39	8.38	8.37	8.36	8.35	8.34	8.33	8.31	8.30	8.29	8.27	8.25
Apr......	8.60	8.64	8.68	8.72	8.75	8.80	8.85	8.90	9.00	9.05	9.12	9.18	9.25
May......	9.30	9.38	9.46	9.53	9.63	9.72	9.81	9.92	10.02	10.14	10.26	10.39	10.53	10.69
June.....	9.20	9.30	9.38	9.49	9.60	9.70	9.83	9.95	10.08	10.21	10.38	10.54	10.71	10.93
July.....	9.41	9.49	9.58	9.67	9.77	9.88	9.99	10.10	10.22	10.35	10.49	10.64	10.80	10.99
August...	9.05	9.10	9.16	9.22	9.28	9.33	9.40	9.47	9.54	9.62	9.70	9.79	9.89	10.00
Sept.....	8.31	8.31	8.32	8.34	8.34	8.36	8.36	8.38	8.38	8.40	8.41	8.42	8.44	8.44
Oct......	8.09	8.06	8.02	7.99	7.93	7.90	7.85	7.80	7.75	7.70	7.63	7.58	7.51	7.43
Nov.....	7.43	7.36	7.27	7.19	7.11	7.02	6.92	6.82	6.72	6.62	6.49	6.36	6.22	6.07
Dec......	7.46	7.35	7.27	7.14	7.05	6.92	6.79	6.66	6.52	6.38	6.22	6.04	5.86	5.65

* From Criddle [11]. See also Table 11-4.

Table 21-4. Computations of Average Monthly Consumptive Use for a Citrus Orchard, Santa Ana, Calif.

(1) Month	(2) Average monthly temperature t, °F	(3) Average daytime hours, % of the year, p	(4) Average monthly value of k	(5) Average monthly consumptive use u, in.
April	59.9	8.80	0.40	2.11
May	63.5	9.72	0.42	2.59
June	67.1	9.70	0.52	3.38
July	71.4	9.88	0.55	3.88
August	71.9	9.33	0.55	3.69
September	69.5	8.36	0.55	3.20
October	64.7	7.90	0.50	2.56
Seasonal total	....	....	....	21.41

2. The Penman Method. Figure 21-3 shows the van Bavel nomograph [14], which permits a rapid estimate of consumptive use by the Penman method [15] without much loss of accuracy. The sunshine percentage in the nomograph can be determined using the possible sunshine hours listed in Table 21-5, together with local data on observed hours, or more roughly, using the observed seasonal averages given in "Climate and Man" [16]. For conservative (high) estimates of consumptive use for design purposes, such as for an estimate of the maximum probable monthly demand from storage, a high percentage of sunshine can be assumed. The extra-terrestrial radiation for use with the van Bavel nomograph is given in Table 21-6. It should be noted that the type of crop is not considered in this method.

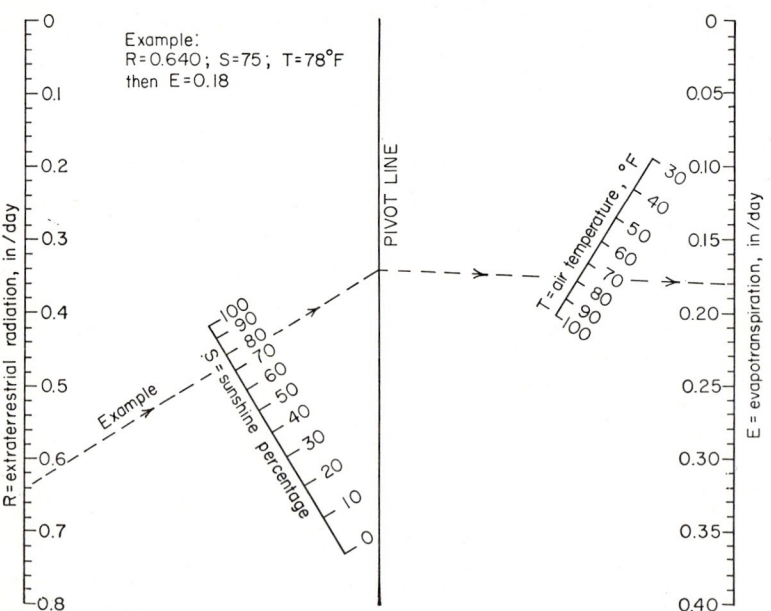

FIG. 21-3. Nomograph for a simplified solution of the Penman evapotranspiration equation. (*After van Bavel* [14].)

DATA FOR HYDROLOGIC ANALYSES

Table 21-5. Possible Hours of Sunshine*

Month	Latitude, degrees north of equator											
	26°	28°	30°	32°	34°	36°	38°	40°	42°	44°	46°	48°
Jan..........	10.7	10.5	10.4	10.2	10.1	10.0	9.8	9.6	9.4	9.3	9.1	8.9
Feb..........	11.2	11.2	11.1	11.0	10.9	10.8	10.7	10.6	10.5	10.4	10.3	10.2
Mar..........	12.0	12.0	12.0	12.0	12.0	12.0	12.0	12.0	12.0	11.9	11.9	11.9
Apr..........	12.7	12.8	12.8	12.9	13.0	13.0	13.1	13.2	13.3	13.4	13.5	13.7
May..........	13.3	13.4	13.6	13.8	13.9	14.0	14.2	14.4	14.5	14.7	14.9	15.1
June.........	13.8	14.0	14.2	14.4	14.5	14.7	14.8	15.0	15.2	15.5	15.7	15.9
July.........	13.6	13.7	13.9	14.1	14.2	14.4	14.6	14.8	15.0	15.2	15.4	15.7
Aug..........	13.1	13.1	13.3	13.4	13.5	13.6	13.7	13.8	13.9	14.0	14.2	14.3
Sept.........	12.3	12.3	12.3	12.3	12.3	12.3	12.4	12.4	12.4	12.4	12.5	12.5
Oct..........	11.6	11.5	11.4	11.4	11.3	11.2	11.1	11.1	11.0	10.9	10.8	10.8
Nov..........	10.8	10.7	10.6	10.5	10.4	10.2	10.1	10.0	9.9	9.8	9.6	9.3
Dec..........	10.4	10.2	10.1	9.9	9.7	9.6	9.4	9.3	9.1	9.0	8.7	8.5

* From van Bavel [14], who used the American Ephemeris and Nautical Almanac for the year 1957. Values are for approximately the 15th of each month.

Example 21-3. Solve the problem of Example 21-2, using the Penman method. The values determined in each of the following items are given in Table 21-7.
 1. Determine the average monthly radiation from Table 21-6 for the location, and list it as shown in column 2.
 2. Estimate the monthly percentage of possible sunshine hours (column 3). For this example it is assumed that the monthly values are not available and that the observed seasonal average of 70 per cent, estimated using maps in "Climate and Man" [16], must be used throughout.
 3. Tabulate average monthly temperatures (column 4).
 4. Estimate the average daily consumptive use (column 5), using the nomograph (Fig. 21-3).
 5. Compute the average monthly consumptive use (column 6), using column 5 and the number of days per month.

Table 21-6. Extraterrestrial Radiation*
(In inches of water per day, evaporation equivalent)

Month	Latitude, degrees north of equator											
	26°	28°	30°	32°	34°	36°	38°	40°	42°	44°	46°	48°
Jan..........	.364	.354	.335	.317	.300	.279	.260	.241	.221	.203	.184	.167
Feb..........	.485	.478	.462	.447	.433	.414	.397	.379	.360	.340	.320	.298
Mar..........	.516	.512	.501	.492	.480	.467	.454	.441	.427	.412	.397	.381
Apr..........	.600	.598	.594	.588	.582	.578	.571	.564	.556	.548	.540	.530
May..........	.620	.620	.622	.623	.623	.622	.622	.620	.619	.617	.614	.610
June.........	.652	.654	.656	.660	.662	.664	.666	.667	.668	.668	.668	.667
July.........	.620	.622	.626	.627	.628	.629	.629	.628	.628	.627	.626	.622
Aug..........	.588	.587	.586	.582	.580	.574	.570	.564	.558	.552	.544	.536
Sept.........	.553	.546	.538	.530	.519	.507	.496	.483	.469	.456	.441	.426
Oct..........	.462	448	.435	.421	.406	.390	.373	.356	.338	.320	.300	.283
Nov..........	.398	.380	.362	.345	.327	.307	.289	.270	.250	.230	.211	.191
Dec..........	.347	.329	.309	.292	.272	.253	.233	.213	.195	.176	.157	.140

* From van Bavel [14], who used Angot's tables as given in David Brunt, "Physical and Dynamical Meteorology," 2d ed., Cambridge University Press, London, 1939, p. 112.

Table 21-7. Data and Results, Monthly Consumptive Use at Santa Ana, Calif.

Month	Average monthly radiation, in./day	Average monthly per cent sunshine	Average monthly temperature, °F	Average consumptive use	
				Daily	Monthly
(1)	(2)	(3)	(4)	(5)	(6)
April...............	0.582	70	59.9	0.115	3.45
May................	0.623	70	63.5	0.131	4.48
June...............	0.662	70	67.1	0.149	4.62
July................	0.628	70	71.4	0.150	4.65
August.............	0.580	70	71.9	0.140	4.34
September..........	0.519	70	69.5	0.120	3.60
October............	0.406	70	64.7	0.086	2.66
Seasonal total......	.....	...	....	.....	27.80

C. Watershed Characteristics

1. Size. Soil maps, aerial photographs, or county maps are used to determine the size of a small watershed. Occasionally, a field reconnaissance is necessary to define noncontributing areas, which must be omitted from the area used in estimating surface runoff and quick-return flow.

2. Shape. The shape of a watershed may be described by a shape index:

$$S_w = \frac{L}{W} = \frac{L^2}{A} \qquad (21\text{-}3)$$

where L is the length of the watershed along the main stream from the watershed outlet to the most distant ridge, W is the average width of the watershed, or A/L, and A is the watershed size. With L in miles, A must be in square miles. The shape index of watersheds is generally found to be more consistent if the valley length is used instead of the length of the low-flow meandering channel.

3. Watershed Slope. The average watershed slope in per cent may be determined from a topographic map by the following formula:

$$S = \frac{MN}{A} \times 100 \qquad (21\text{-}4)$$

where M is the total length of contours within the watershed, in ft, N is the contour interval in ft, and A is the size of the watershed in ft². For very small watersheds the average slope can be taken as the ratio of the difference in elevation between the watershed outlet and the most distant ridge to the approximate average length of the watershed.

4. Time of Concentration. This is the time it takes for water to travel from the most distant point of a watershed to the watershed outlet or to some other downstream point of reference. Figure 21-4 provides an estimate of the time of concentration, assuming that average values of Manning's n and hydraulic radius prevail. This figure is based on the equation

$$T_c = \frac{L^{1.15}}{7{,}700 H^{0.38}} \qquad (21\text{-}5)$$

where T_c is the estimated time of concentration, or the time for water to flow from the most distant point in the watershed to the watershed outlet, in hr; L is the length

DATA FOR HYDROLOGIC ANALYSES   21-11

of the watershed along the main stream from the watershed outlet to the most distant ridge in ft; and H is the difference in elevation between the watershed outlet and the most distant ridge in ft. However, differences in elevation due to overfalls, rapids, or other sudden drops should be subtracted from the value of H before using Fig. 21-4.

5. Hydrologic Soil-cover Complexes. The soils and vegetative covers of a watershed are generally classified separately. A combination of a specific soil and a specific cover is referred to as a *soil-cover complex* [19], and a measure of this complex can be used as a watershed parameter in estimating runoff.

a. Soils. The hydrologic properties of a soil or a group of soils are an essential factor in the hydrologic analysis of watershed data. Soils can be classified according to their hydrologic properties if considered independently of watershed slope and cover. Four major soil groups are recognized for the primary classification of watershed soils:

Group A (low runoff potential). Soils having high infiltration rates even when thoroughly wetted, consisting chiefly of sands or gravel that are deep and well to excessively drained. These soils have a *high* rate of water transmission.

Group B. Soils having moderate infiltration rates when thoroughly wetted, chiefly moderately deep to deep, moderately well to well drained, with moderately fine to moderately coarse textures. These soils have a *moderate* rate of water transmission.

Group C. Soils having slow infiltration rates when thoroughly wetted, chiefly with a layer that impedes the downward movement of water, or of moderately fine to fine texture and a slow infiltration rate. These soils have a *slow* rate of water transmission.

Group D (high runoff potential). Soils having very slow infiltration rates when thoroughly wetted, chiefly clay soils with a high swelling potential; soils with a high permanent water table; soils with a clay pan or clay layer at or near the surface; and shallow soils over nearly impervious materials. These soils have a *very slow* rate of water transmission.

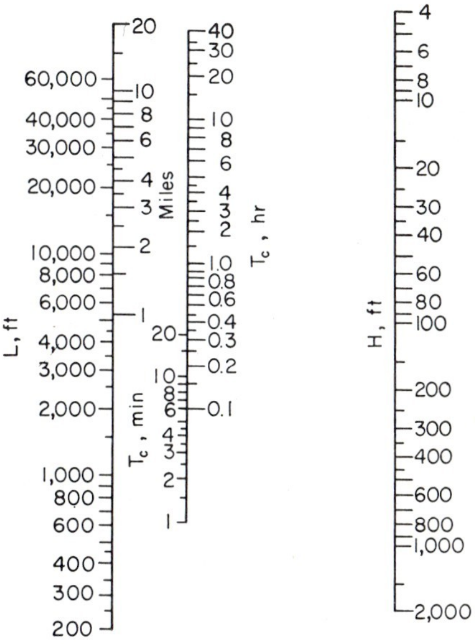

FIG. 21-4. Nomograph for estimating time of concentration. (*Modified from Kirpich* [18].)

Detailed definitions of depth and soil-drainage classes may be found in the "Soil Survey Manual" [20]. Major soils of the United States and Puerto Rico classified by hydrologic soil groups are listed in Table 21-8.

b. Cover. Essentially, cover is any material (but usually vegetative) covering the soil and providing protection from the impact of rainfall. Under ordinary conditions, detailed information about the cover, such as plant density and height, root density and depth, extent of plant cover, and extent and amount of litter, is seldom available. It is therefore necessary to rely largely on the land use as an index of cover conditions in the hydrologic analysis of watersheds. The following definitions describe the cover and land uses listed in Table 21-12, and they apply to field conditions that may be estimated visually.

Crop rotation: The cyclic sequence of agricultural crops on a farm field. Each cycle extends over a period of 2 to 7 or more years. The rotation should be evaluated on the basis of its hydrologic effect, rather than on crop yields. It should be recognized, however, that generally the larger the crop yields, the greater the effect of cover on surface runoff. Hydrologically speaking, rotations range from "Poor" to

Table 21-8. Classification of Soils by Hydrologic Soil Groups*

Aastad B	Allis C	Arredondo A	Barneston A
Aberdeen D	Allison C	Artesia C	Barnett D
Abilene C	Allouez B	Arvada D	Barney A
Abington B	Alma C	Arveson D	Barnhardt A
Acadia D	Almena C	Arzell D	Barnstead B
Acme B	Almirante D	Asa B	Barrancas D
Acton B	Almo D	Asbury B	Barron C
Adair D	Almy B	Ascalon B	Barronett C
Adams A	Alonso C	Ashby C	Barth B
Adamsville C	Alps C	Ashe B	Bartle D
Ade A	Alsen C	Ashkum C	Basher B
Adel B	Altamaha D	Ashley A	Bass B
Adelanto B	Altamont C	Ash Springs C	Bastrop B
Adelphia B	Altavista B	Ashton B	Batavia B
Adler C	Alto C	Ashuelot C	Bates B
Adolph D	Alton B	Ashwood D	Bath C
Afton D	Altoona C	Asotin B	Baudette B
Agar B	Altura B	Assumption B	Baugh C
Agate C	Altvan B	Astoria B	Baxter B
Agawam B	Alvin B	Athelwold B	Bayamon B
Agency C	Alvira C	Athena B	Bayard A
Agnew B	Amalu D	Atherton D	Bayboro D
Aguadilla A	Amarillo B	Athol B†	Bayside C
Aguilita C	Amboy B	Atkins D	Beadle C
Aguirre D	Amelia C	Atterberry B	Bearden C
Ahmeek B	Amenia D	Atwood B	Beardstown C
Ahnberg C	Americus A	Auburn C	Bear Lake D
Ahtanum C	Ames C	Auburndale D	Bear Prairie B
Aiken B	Amite B	Au Gres C	Beatty B
Airmont B	Amity C	Augusta C	Beaucoup C
Akaka B	Amsterdam B	Aurora C	Beauford D
Akan C	Andover D	Austin B	Beaumont D
Akaska B	Andres B	Ava C	Beauregard C
Alachua B	Angie C	Avalanche B	Beaver B
Alaeloa B	Angola C	Avery B	Beaverhead B
Alama C	Ankeny A	Avon B	Beaverton B
Alamance C	Annandale B	Avonburg D	Becket C
Alamosa C	Anoka A	Axtell D	Beckton D
Albaton D	Anselmo B	Ayr B	Beckwith C
Albemarle B	Anthony B		Bedford C
Albertville C	Antigo B		Bedington B
Albia C	Apache B	Babylon A	Beecher C
Albion B	Apakuie A	Baca B	Beechy D
Alcester B	Apishapa C	Bagnell D	Belfast D
Alcoa B	Apison C	Bainville B	Belfore C
Alden D	Applegate C	Baker C	Belgrade B
Alderwood B	Appling B	Balch A	Belknap C
Aldino D	Arch B	Baldock B	Bell D
Alexandria B	Archer C	Baldwin C	Belle B
Alexis B	Arenzville B	Balfour B	Bellingham D
Alford B	Argyle B	Balm A	Belmont B
Algarrobo D	Ark C	Balmorhea D	Beltrami B
Algiers D	Arkport B	Bangor B	Beltsville C
Alicel B	Arland B	Banks A	Belvoir C
Allard B	Armagh D	Bannerville C	Benevola B
Allegheny B	Armer C	Baraboo B	Benfield C
Allen B	Armour B	Barbour B	Benld D
Allendale C	Armuchee C	Barbourville B	Bennington C
Allenwood B	Arnold A	Barclay C	Benoit D
Alligator D	Arnot C	Barnard C	Benson C‡
		Barnes B	

* From U.S. Soil Conservation Service [19].
† The hydrologic grouping for the rocky phases of these soils should be reduced one group.
‡ When these soils are classified as having class 3 or 4 rockiness, the hydrologic grouping should be reduced one group.

DATA FOR HYDROLOGIC ANALYSES   21-13

Table 21-8. Classification of Soils by Hydrologic Soil Groups* (Continued)

Bentonville C	Bloomington B	Bremer B	Burton B
Beotia B	Blount C	Bremo C	Buse B
Berg C	Blue Earth D	Brennan B	Butler D
Bergland D	Bluffton D	Brenner C	Butlertown C
Berkeley C	Bluford D	Brenton B	Butte B
Berks C	Bobtail C	Bresser B	Buxin D
Berkshire B	Bodine B	Brewer D	Buxton C
Bermudian B	Bogota C	Brewster C	Byars D
Bernard D	Bohemian B	Brickton C	Byrds D
Bernardston C	Bold B	Bridgehampton B	
Berrien B	Bolivia B	Bridgeport B	Cabinet C
Berthoud B	Bolton B	Bridger B	Cabo Rojo C
Bertie C	Bombay C	Bridgeville B	Cabot C
Bertolotti A	Bonaccord D	Briggs A	Cacapon B
Bertrand B	Bonaparte A	Briggsdale C	Caddo C
Berwyn C	Bonham B	Brill B	Cagey D
Bethany C	Bonilla B	Brimfield C‡	Caguas D
Bethel D	Bonita D	Brimley B	Cahaba B
Beulah B	Bonner B	Brinkerton D	Cajon A
Beverly A	Bonneville A	Briscoe B	Calais B
Bewleyville B	Bonnie D	Brittain C	Caldwell B
Bibb D	Bono D	Broadbook C	Calhoun D
Bickleton B	Bonpas B	Broadview C	Califon C
Biddeford D	Boomer C	Brockport C	Calloway C
Bienville B	Boone A	Brooke C	Calvert D
Biggs A	Bordeaux B	Brookfield B	Calverton C
Biggsville B	Bosket B	Brookings B	Calvin C
Big Horn C	Boswell D	Brooklyn D	Camaguey D
Billett A	Bow D	Brookston B	Camas A
Billings D	Bowdoin D	Brooksville B	Cambridge C
Binnsville C	Bowdre C	Broughton D	Camden B
Bippus B	Bowie B	Broward C	Cameron C
Birds C	Bowmansville D	Brownfield A	Camillus C
Birdsall D	Boyd D	Browning B	Camp B
Birdsboro B	Boyer B	Brownlee C	Campo C
Birkbeck B	Boynton C	Bruno A	Camroden C
Biscay D	Bozarth C	Bryce D	Canaan C‡
Bitterroot C	Bozeman B	Bub C	Canadian B
Blacklock D	Braceville B	Buchanan C	Canadice C
Blackwater D	Bracken D	Buckingham C	Canandaigua C
Bladen D	Brackett C	Buckland C	Canaseraga B
Blago D	Braddock B	Buckley D	Cane B
Blaine C	Braden C	Buckner A	Caneadea C
Blair C	Bradenton C	Bucks B	Caneyville C
Blairton C	Bradley B	Bucoda D	Canfield B
Blakeland A	Brady B	Bude C	Canoncito D
Blakely B	Braham B	Buncombe A	Canyon C
Blanchard A	Brallier A	Bunkerville D	Cape Fear D
Blanco B	Brandon B	Burchard B	Capron B
Bland C	Brandywine C	Burdett C	Capshaw C
Blandford C	Branford B	Burgess B	Captina C
Blanding B	Brashear C	Burgin D	Capulin B
Blanket D	Bratton B	Burke B	Carbo C
Blanton A	Braxton B	Burkhardt B	Cardiff B
Blencoe C	Brayton C	Burleson D	Cardington C
Blichton C	Brazito A	Burnham C	Carey B
Blockton C	Brazos B	Burnside D	Caribou B
Blodgett B	Brecknock C	Burnsville B	Carlisle D
Blomford B	Breece B	Burnt Fork B	Carlsborg A
Bloomfield A	Breese D	Burrell C	Carlton B

* From U.S. Soil Conservation Service [19].
† The hydrologic grouping for the rocky phases of these soils should be reduced one group.
‡ When these soils are classified as having class 3 or 4 rockiness, the hydrologic grouping should be reduced one group.

Table 21-8. Classification of Soils by Hydrologic Soil Groups* (Continued)

Carmi B	Charlton B	Clarksburg C	Colp D
Carnegie B	Chaseburg B	Clarksdale C	Colrain B
Carnero C	Chastain D	Clarkson C	Colton A
Caroline C	Chatfield C‡	Clarksville B	Colts Neck B
Carrington B	Chatsworth D	Clary B	Columbia B
Carrizo A	Chattahoochee B	Clatsop D	Colville B
Carroll D	Chauncey C	Claverack B	Colwood D
Carson D	Chehalis B	Clawson C	Colyer C
Carstairs C	Chelsea A	Clayton C	Comfrey C
Carver A	Chemawa B	Cleaver D	Comly C
Carytown D	Chanango B	Cleburne B	Commerce C
Casa B	Cheney B	Cle Elum B	Comoro A
Casa Grande C	Chenoweth B	Cleman B	Compton B
Cascade C	Cherette A	Clement C	Conant B
Casco A	Cherokee D	Cleora B	Conasauga C
Casey C	Cherry B	Clermont D	Concord B
Cashmere A	Cherryhill B	Clifton B	Condit D
Cashton D	Cheshire B	Climax D	Condon B
Cass A	Chester B	Clinton B	Conestoga B
Cassville B	Chesterfield B	Clio C	Conesus C
Castana B	Chetek A	Clipper C	Congaree B
Castle D	Chewacla C	Clodine B	Conley C
Castner C	Chewelah B	Cloquallum C	Conotton B
Catalina C	Cheyenne B	Cloquet B	Conover B
Catalpa C	Chickasha B	Cloud D	Conowingo C
Catano A	Chiefland A	Clover Creek C	Constable A
Cataula C	Chigley C	Clovis B	Continental C
Catheart C	Chilgren C	Clyde C	Conway C
Catlett C	Chilhowie D	Clymer B	Cook D
Catlin B	Chili B	Coamo C	Cookeville B
Catoctin C	Chillisquaque C	Cobb B	Cookport C
Catron D	Chillum C	Cochise B	Coolidge B
Catskill C	Chilmark C	Cocoa B	Coolville C
Cattaraugus B	Chilo D	Cody A	Cooney C
Cave C	Chipeta D	Coeburn C	Cooper B
Cavode C	Chippewa D	Cogswell C	Copake B
Cavot B	Chiricahua D	Cokedale C	Copalis C
Cavour D	Choctaw B	Coker D	Copas C
Cayagua C	Choptank A	Cokesbury D	Copeland C
Caylor B	Choteau C	Colbert D	Coplay D
Cayucos D	Christian B	Colby B	Coral B
Cayuga C	Christiana C	Colden D	Corcega D
Cazenovia C	Christianburg D	Coldwater D	Corduroy B
Cecil B	Churchill D	Colebrook Loamy	Corkindale C
Celina B	Ciales B	Fine Sand A	Corley C
Center C	Cialitos C	Colebrook Fine	Cornutt D
Centerton B	Cicero C	Sandy Loam B	Corvallis B
Central A	Cincinnati B	Coleman B	Corwin B
Chagrin B	Cinebar A	Colemantown D	Corydon C
Chalfont D	Cintrona D	Colfax C	Cossayuna C
Chama B	Cisne D	Colinas C	Cotaco C
Chamber D	Cispus A	Collamer B	Coto C
Chamberino C	Clackamas C	Collington B	Cottonwood C
Chamokane B	Claiborne B	Collins C	Cougar C
Chandler C	Clallam C	Collinsville C	Couparle D
Channahon B	Clarence D	Colo C	Coupeville B
Chariton C	Clareville C	Coloma A	Courtland B
Charleston C	Clarinda D	Colonie A	Courtney D
Charlos B	Clarion B	Coloso D	Couse C
Charlotte B	Clark Fork B	Colosse A	Cove D

* From U.S. Soil Conservation Service [19].
† The hydrologic grouping for the rocky phases of these soils should be reduced one group.
‡ When these soils are classified as having class 3 or 4 rockiness, the hydrologic grouping should be reduced one group.

DATA FOR HYDROLOGIC ANALYSES 21-15

Table 21-8. Classification of Soils by Hydrologic Soil Groups* (Continued)

Coveland C	Darling C	Dominic B	Dyke B
Coveytown C	Darnell C	Donerail C	
Covington D	Darret B	Donlonton C	Eakin B
Cowden D	Darwin D	Dorchester B	Easton C
Cowiche B	Davidson B	Dorsey D	Eastonville A
Cowling B	Davie D	Dos Cabezas B	Ebbert D
Coxville D	Dawes C	Doty B	Ebbs B
Crago B	Dayton D	Dougherty B	Ebeys B
Craig C	Deary C	Douglas B	Eckman B
Crandon B	Decatur B	Dover C	Ector C
Crane B	Deckerville D	Dowellton D	Edalgo D
Craven C	Decorra B	Dowling D	Eddy C
Crawford D	Defiance D	Downs B	Eden C
Creal D	Dekalb B†	Doylestown D	Edenton C
Creedmoor C	Delanco C	Dragston C	Edge D
Cresbard C	Delfina C	Drake B	Edgeley C
Crescent B	Dell C	Dresden B	Edgemont B
Crestmore C	Dellrose B	Dripping Springs D	Edgewick A
Crete D	Delphi B	Drummer B	Edgington C
Crevasse A	Delpine D	Drummond D	Edina D
Crider B	Delray D	Drury B	Edinburg C
Crockett D	Del Rey C	Dryad C	Edisto C
Crofton B	Demers C	Duane B	Edith A
Croghan B	Denham D	Dubbs B	Edmonds D
Croom C	Dennis C	Dubois C	Edna D
Crosby C	Denny D	Dubuque	Edneyville B
Crossville B	Denrock D	Deep Phase B	Edom C
Croton D	Denson C	Dubuque B	Edwards D
Crow C	Denton C	Dubuque	Eel C
Crowder D	Depew C	Shallow Phase C	Efland C
Crowley D	Derby A	Duffield B	Egam C
Crown B	Descalabrado B	Duffy C	Egeland B
Crystal B	Deschutes A	Dukes A	Eifort C
Culleoka B	De Soto C	Dulac C	Elbert D
Cullo C	Detour C	Dunbar C	Elburn B
Culpeper B	Detroit B	Duncan C	Elco C
Culvers C	Dewart B	Duncannon B	Eld B
Cumberland B	Dewey B	Duncom D	Eldon C
Curran C	Dexter B	Dundas D	Eldorado C
Curtis B	Diablo D	Dundee C	Elfrida C
Cushman B	Dick A	Dune Sand A	Elioak B
Custer D	Dickey A	Dunellen B	Elk B
Cut Bank B	Dickinson Fine	Dungeness B	Elkins D
Cuthbert C	Sandy Loam A	Dunham B	Elkinsville B
Cypremort C	Dickinson Loam B	Dunkirk B	Elkton D
	Dickson C	Dunlap C	Ellery D
Dade A	Dill B	Dunmore C	Elliber A
Daggett B	Dilldown B	Dunning D	Ellington C
Dakota B	Dillinger B	Du Page B	Elliott C
Dalbo B	Dillon D	Duplin B	Ellis D
Dale C	Dilman B	Dupo C	Ellison B
Dalhart B	Dimmick D	Dupont D	Ellsberry C
Dalton C	Disco B	Durant D	Ellsworth D
Dana B	Dixmont B	Durham B	Elmo C
Dandridge C	Dodgeville	Durkee C	Elmore C
Daniels B	Deep Phase B	Dutchess B†	Elmwood C
Danley C	Dodgeville	Dutson C	Elsinboro B
Dannemora C	Shallow Phase C	Duval B	Elsmere A
Danvers C	Doland B	Dwight D	Elwha C
Darien C	Dominguito D	Dwyer A	Emmert A

* From U.S. Soil Conservation Service [19].
† The hydrologic grouping for the rocky phases of these soils should be reduced one group.
‡ When these soils are classified as having class 3 or 4 rockiness, the hydrologic grouping should be reduced one group.

Table 21-8. Classification of Soils by Hydrologic Soil Groups* (Continued)

Emmet	B	Fargo	D	Freer	C	Glendale	C
Emory	B	Farland	B	Fremont	C	Glendive	A
Empey	C	Farmington	C‡	Frenchtown	D	Glenelg	B
Empeyville	C	Farnum	C	Freneau	D	Glenfield	C
Enders	C	Farragut	C	Frio	B	Glenford	C
Enfield	B	Faunce	A	Frost	D	Glenoma	B
Englund	D	Fauquier	B	Fruita	B	Glenville	C
Ennis	B	Fawcett	C	Frye	C	Gloucester	B
Enon	C	Faxon	D	Fullerton	B	Godwin	B
Ensenada	C	Fayette	B	Fulton	C	Goessel	D
Ensley	D	Fe	D			Gogebic	B
Enstrom	B	Felda	D	Gage	D	Goldridge	A
Enterprise	A	Felida	B	Gainesville	A	Goldsboro	B
Enumclaw	B	Fellowship	C	Gale	B	Goldston	C
Ephrata	A	Fergus	C	Galen	B	Goldvein	C
Epping	D	Fiander	C	Galestown	A	Goliad	C
Era	B	Fidalgo	C	Gallatin	D	Gooch	C
Eram	D	Fillmore	D	Gallion	B	Gore	D
Erie	C	Fincastle	C	Galveston	A	Gorus	B
Ernest	C	Fitch	A	Galvin	B	Goshen	B
Escondido	C	Fitchville	C	Gann	B	Gosport	C
Espinosa	B	Fitzhugh	B	Gannett	D	Gothard	D
Esquatzel	B	Flamingo	D	Gara	C	Gowen	C
Essex	C	Flanagan	B	Gardnerville	C	Grady	D
Estacion	B	Flandreau	B	Garfield	D	Graham	C
Estelline	B	Flasher	A	Garner	D	Grail	C
Estevan	C	Flathead	B	Garrison	A	Granby	D
Estherville	B	Fleetwood	B	Garwin	D	Grande Ronde	C
Esto	C	Fletcher	C	Gasconade	D	Grant	B
Etowah	B	Flint	C	Gaviota	A	Grantsburg	C
Ettrick	D	Flora	D	Gayville	D	Grantsdale	B
Eubanks	B	Florence	C	Gearhart	A	Granville	B
Eufaula	A	Florsheim	D	Geary	B	Grayling	A
Eulonia	C	Floyd	B	Geer	C	Great Bend	B
Eustis	A	Fluvanna	C	Geiger	D	Greeley	B
Eutaw	D	Foard	D	Gem	C	Green Bluff	B
Evans	B	Foley	C	Genesee	B	Greenbush	B
Evendale	C	Folsom	C	Genoa	D	Greendale	B
Everett	A	Fonda	D	Genola	B	Greenfield	B
Everson	C	Fordney	A	Georgetown	C	Greenport	C
Ewing	A	Fordville	B	Georgeville	B	Green River	B
Exline	D	Fore	D	Gerald	D	Greensboro	B
Eylar	D	Forestdale	D	Germania	B	Greenville	B
		Forrest	D	Geronimo	B	Greenwater	A
Faceville	B	Fort Collins	B	Gila	B	Greer	C
Fahey	B	Fort Lyon	B	Gilcrest	B	Grenada	C
Fairfax	B	Fort Meade	A	Gilead	C	Grenville	B
Fairhaven	B	Fort Pierce	B	Giles	A	Gresham	C
Fairhope	C	Fortuna	D	Gilford	D	Greybull	C
Fairmount	D	Fox	B	Gilligan	B	Greys	B
Fajardo	C	Foxhome	B	Gilman	B	Griffin	C
Falaya	C	Frankfort	D	Gilpin	C	Grimstad	C
Falcon	B	Franklinton	C	Gilson	B	Grosclose	C
Falkner	D	Frankstown	B	Gilt Edge	D	Groton	A
Fall	B	Fraternidad	D	Ginat	D	Grove	A
Fallbrook	B	Frederick	B	Gird	B	Groveland	B
Fallsburg	C	Fredon	C	Givin	C	Grover	B
Fallsington	D	Freehold	B	Glasgow	C	Groveton	B
Falun	B	Freeland	C	Glenbar	C	Grundy	C
Fannin	B	Freeon	B	Glencoe	D	Guadalupe	B

* From U.S. Soil Conservation Service [19].
† The hydrologic grouping for the rocky phases of these soils should be reduced one group.
‡ When these soils are classified as having class 3 or 4 rockiness, the hydrologic grouping should be reduced one group.

DATA FOR HYDROLOGIC ANALYSES   21-17

Table 21-8. Classification of Soils by Hydrologic Soil Groups* (Continued)

Guanica D	Hartleton B	Hillsdale B	Hoyleton C
Guayabo C	Hartsburg B	Hilo B	Hoypus A
Guayama C	Hartsells B	Hilton C	Hoytville D
Guckeen C	Harwood B	Hinckley A	Hubbard A
Gudrid B	Haskill A	Hinman D	Huckabee A
Guelph B	Hassel D	Hiwassee B	Huckleberry C
Guernsey C	Hastings C	Hiwood A	Hudson C
Guin A	Hatchie C	Hixton B	Huey D
Guthrie D	Haven B	Hobble B	Huff B
	Havre B	Hobbs B	Huffine B
Habersham B	Haxtun A	Hockley C	Huggins C
Haccke D	Hayden B	Hoffman C	Hugo B
Hackers B	Hayesville B	Hogansburg B	Huikau A
Hackettstown B	Haymond B	Hoko C	Humacao C
Hadley B	Haynie B	Holbrook D	Humbarger B
Hagener A	Hayter B	Holcomb D	Humeston C
Hagerstown B	Hazel C	Holdrege B	Humphreys B
Haig C	Hazen B	Holland B	Hunt D
Haiku C	Heath B	Hollinger C	Hunters B
Haines B	Hebo D	Hollis C‡	Huntington B
Halawa C	Hecla B	Hollister C	Huntsville B
Haleakala A	Hector B	Holloway B	Hurley D
Halewood B	Hedville C	Holly C	Hurst D
Half Moon C	Heisler B	Hollywood D	Hutchinson C
Halfway C	Heitt C	Holmdel B	Hyattsville B
Halii C	Helena C	Holston B	Hyde D
Haliimaile B	Hemmi C	Holt B	Hymon C
Hall B	Hempstead B	Holyoke C‡	
Halsey D	Henderson D	Homer C	Iao B
Hamburg B	Hennepin B	Hondo C	Iberia D
Hammerly C	Henry D	Honeoye B	Ida B
Hamilton B	Henshaw C	Honokaa B	Idana C
Hamlin B	Herbert B	Honolua B	Ihlen D
Hammond D	Herkimer B	Honomanu C	Ilion C
Hampshire C	Hermiston B	Honouliuli C	Illiopolis B
Hampton B	Hermitage B	Hood B	Ima B
Hanalei C	Hermon B	Hoodsport A	Immokalee C
Hanceville B	Hermosa C	Hooker B	Imperial D
Hand B	Hernando B	Hoosic B	Ina C
Hanford B	Herndon C	Hopewell C	Independence A
Hanipoe B	Hero B	Hopper B	Inglefield C
Hannahatchee B	Herrick C	Hoquiam B	Ingomar D
Hanover B	Hershal B	Hord B	Inman C
Hanska C	Hesch B	Hornell C	Inola D
Hanson B	Hesseltine A	Hortman C	Io B
Harbin B	Hesson C	Horton B	Iola A
Harbourton C	Hialeah D	Hosmer C	Iona B
Harlem B	Hiawatha A	Houdek B	Ipava B
Harley C	Hibbard C	Houghton D	Iredell D
Harlingen D	Hibbing B	Houlka D	Irion D
Harmon D	Hickory C	Houlton C	Irish D
Harmony C	Hicks B	Housatonic C	Iron River B
Harpster B	Hidalgo B	Houseville C	Irurena D
Harriet D	Hidewood C	Houston D	Irving D
Harris D	Highfield B	Houston Black D	Irvington C
Harrisburg C	Higley B	Hovde B	Irwin D
Harrison C	Hiko Springs D	Hoven D	Isanti D
Harstine A	Hilger B	Howard B	Islote B
Hartford B	Hilliard B	Howell B	Isom B
Hartland B	Hillsboro B	Hoye C	Issaquah B

* From U.S. Soil Conservation Service [19].
† The hydrologic grouping for the rocky phases of these soils should be reduced one group.
‡ When these soils are classified as having class 3 or 4 rockiness, the hydrologic grouping should be reduced one group.

Table 21-3. Classification of Soils by Hydrologic Soil Groups* (Continued)

Istokpoga D	Keaau D	Kokokahi D	Landisburg C
Iuka C	Kealakekua B	Kokomo D	Lane C
Ivanhoe D	Keansburg D	Kolekole C	Langford C
Ives B	Keating C	Konokti A	Langley B
Izagora C	Kedron C	Koolau D	Langrell B
	Keene C	Kopiah D	Lanham D
Jacana C	Keith B	Kosmos D	Lansdale B
Jackson B	Kelly D	Koster C	Lansdowne C
Jacob D	Kelso C	Kranzburg B	Lansing B
Jaffrey A	Kelton A	Krause B	Lantz D
Jaucas A	Kempsville B	Kreamer C	La Palma B
Jayuya C	Kempton B	Kresson C	Lapine A
Jeanerette C	Kenansville B	Krum C	Lapon D
Jefferson B	Kendaia C	Kukaiau B	Laporte D
Jerauld D	Kendall B	Kunia B	La Prairie B
Jerome D	Kennebec B	Kutztown C	Laredo B
Jessup C	Kennedy B		Lares C
Joe Creek B	Kenney A	La Belle C	Largent C
Johnston D	Kenspur B	Labette C	Largo C
Joliet C	Kent D	Labounty C	Larimer B
Jonesville A	Keomah C	La Brier C	Larkin B
Joplin B	Kerby B	Lacamas D	La Rose B
Josefa D	Keri C	La Casa C	Larry D
Josephine C	Kerrtown B	Lackawanna B	Las Animas A
Joy B	Kershaw A	Ladd C	Lashley B
Juana Diaz B	Kettle B	La Delle B	Las Lucas D
Judith B	Kettleman C	Ladoga C	Las Piedras D
Judson B	Keyesport C	Ladysmith D	Lassen C
Jules B	Keyport C	Lafe D	Las Vegas D
Juliaetta A	Keystone A	Lagonda D	Latah C
Juncos C	Kibbie C	La Grande C	Lauderdale C
Juniata B	Kickerville B	La Hogue B	Laurel B
Junius C	Kilauea A	Lahontan D	Lauren A
Juno A	Kilbourne A	Laidig C	Laveen B
	Kimbrough D	Laidlaw A	La Verkin B
Kaena D	Kinghurst B	Laie D	Lawhorn D
Kahana B	Kings C	Lairdsville B	Lawrence C
Kalamazoo B	Kinross D	Lajas C	Lawrenceville C
Kalihi D	Kipling D	Lake Charles D	Lawson B
Kalispell B	Kipp B	Lake Creek B	Lawton B
Kalkaska A	Kipson D	Lakehurst A	Lax C
Kalmia B	Kirkland D	Lakeland A	Lea B
Kaloko D	Kirvin C	Lakemont C	Leadvale C
Kamananui C	Kistler B	Lakeville	Leaf D
Kanab C	Kitsap C	Sandy Loam A	Leal B
Kanapaha B	Kittitas B	Lakeville Loam B	Leavenworth A
Kaneohe B	Kittson B	Lakewood A	Leavitt B
Kapapala B	Kiwanis B	Lakin A	Leavittville B
Karnak D	Klaberg C	La Lande B	Lebanon D
Karnes B	Klamath C	Lamington D	LeBar B
Karro B	Klaus A	Lamont Fine	Leck Kill B
Kars B	Klej B	Sandy Loam A	Lee D
Kasota C	Kline A	Lamont Loam B	Leeds B
Kasson C	Klinesville C	Lamonta C	Leeper D
Katemcy D	Knappa B	Lamoure C	Leetonia B
Kato B	Knight C	Lamson C	Legore C
Katy D	Knox B	Lanark B	Lehigh C
Kaufman D	Koch C	Lancaster B	Leicester C
Kawaihae C	Koehler B	Land C	Lela D
Kawaihapai A	Kohala B	Landes B	Lempster D

* From U.S. Soil Conservation Service [19].
† The hydrologic grouping for the rocky phases of these soils should be reduced one group.
‡ When these soils are classified as having class 3 or 4 rockiness, the hydrologic grouping should be reduced one group.

Table 21-8. Classification of Soils by Hydrologic Soil Groups* (Continued)

Lena D	Logan C	Madras C	Mason B
Lenoir Fine Sandy Loam B	Logandale C	Madrid B	Massena C
	Lolo B	Maginnis C	Massillon B
Lenoir Silt Loam C	Lomax B	Magnolia B	Matanzas C
	Lonepine B	Mahaska B	Matapeake B
Lenox C	Lone Rock B	Mahnomen B	Matawan C
Leon C	Longford C	Mahoning D	Matlock D
Leona D	Longrie C	Maiden C	Matmon C
Leonardtown D	Lonoke B	Maile B	Matney B
Leota C	Lookout C	Makalapa D	Mattapex C
Leshara B	Loon A	Makawao B	Maumee D
Lester B	Loradale B	Makena B	Maunabo D
LeSueur B	Lorain C	Malaga B	Maury B
Letcher D	Lordstown C	Malaya C	Maverick D
Letort B	Lorella C	Maleza B	May B
Levan A	Lorenzo A	Mamala C	Mayhew D
Lewisberry B	Loring B	Manalapan D	Maynard Lake B
Lewiston C	Los Guineos D	Manana C	Mayo B
Lewisville B	Los Osos C	Manassa B	Mayodan B
Lexington B	Loudon C	Manassas B	Maytown C
Liberty C	Loudonville C	Manastash C	Mazeppa B
Lick B	Louisa B	Manatee D	McAfee B
Lick Creek B	Louisburg B	Manchester A	McAllister C
Lickdale D	Loup D	Mangus C	McBride B
Lightning D	Lowell C	Manhattan B	McDonald C
Lignum C	Loy D	Manheim C	McDowell D
Lihen A	Loysville C	Manlius C	McEwen C
Likes A	Lualualei D	Manor B	McGary C
Lima C	Lubbock C	Mansfield D	McKamie C
Limerick C	Lucas C	Mansic C	McKay D
Lincoln A	Lucien C	Mansker B	McKenna B
Lincroft A	Ludlow C	Mantachie C	McKenzie D
Lindley C	Lufkin D	Manteo D	McLain C
Lindsborg D	Lukin C	Manvel B	McMurray A
Lindside C	Lummi C	Maple D	McNeal C
Lindstrom B	Lun C	Mapleton C‡	McPaul B
Linganore C	Lunt C	Marble A	McPherson D
Link B	Lupton D	Marcus D	Meadin A
Linker B	Lura D	Marcy D	Meadowville B
Linneus B	Luray C	Mardin C	Meadville B
Lino C	Luton D	Marengo C	Mecklenburg C
Lintonia B	Luverne B	Mariana C	Meda B
Lisbon B	Luzena D	Marias D	Medary C
Lismas D	Lycoming C	Marietta B	Medford C
Lismore B	Lyman C†	Marina A	Medina B
Littlefield D	Lynchburg C	Marion D	Medio C
Little Horn B	Lynden A	Marissa C	Meeteetse C
Littleton B	Lynndyl C	Markland C	Mehlhorn C
Litz C	Lyons D	Marksboro B	Meigs C
Livingston D	Lystair A	Marlboro B	Melbourne C
Llave C		Marlow C	Mellenthin B
Lloyd B	Mabi D	Marlton C	Melrose C
Lobdell B	Machete C	Marna D	Melvern C
Lobelville C	Mack B	Marquette B	Melvin D
Lockhard C	Macomber C†	Marshall B	Memphis B
Lockhart B	Macon D	Martha C	Menahga A
Lockport C	Madalin C	Martin Pena D	Mench B
Locust C	Maddock A	Martinsdale B	Menfro B
Lodi B	Maddox C	Martinton C	Menlo D
Lofton C	Madison B	Masada B	Mentor B

* From U.S. Soil Conservation Service [19].

† The hydrologic grouping for the rocky phases of these soils should be reduced one group.

‡ When these soils are classified as having class 3 or 4 rockiness, the hydrologic grouping should be reduced one group.

Table 21-8. Classification of Soils by Hydrologic Soil Groups* (Continued)

Mercedita D	Montara D	Nassau C‡	Norfolk B
Mercer C	Montell D	Natalie C	Norge C
Mereta D	Monteola D	Natchez B	Norma C
Meridian B	Montesano B	Natchitoches D	Northport C
Meros A	Montevallo C	National A	North Powder C
Merrimac B	Montgomery D	Navajo D	Northumberland C
Mertz B	Monticello B	Navasota D	Northville D
Mesa B	Montoya D	Navesink B	Norton B
Meskill D	Moody B	Naylor C	Norwich D
Metea B	Moreau D	Nebish B	Norwood B
Methow B	Morley C	Neble A	Nosbig D
Metolius A	Mormon Mesa D	Needmore C	Nuby C
Mexico D	Moro Bay D	Negley B	Nuckolls C
Mhoon C	Morocco C	Nehalem B	Nueces A
Miami B	Moro Cojo A	Nellis B	Nunda C
Middlebury B	Morrill B	Neosho D	Nunn C
Midland D	Morris C	Neptune A	Nutley D
Midway D	Morrison B	Nereson B	Nymore A
Mifflinburg B	Morrow C	Neshaminy B	
Miguel D	Morse D	Nesika B	Oahe B
Milaca B	Morton B	Nester D	Oakford B
Milam B	Moscow B	Neubert B	Oakland B
Miles B	Moshannon B	Nevada D	Oasis F
Milford C	Mossyrock A	Neville B	Ochlockonee B
Millbrook B	Mottsville A	Newark C	Ochopee D
Mill Creek B	Mount Carroll B	Newart B	Ockley B
Miller D	Mount Lucas C	Newberg B	Oconee C
Millington B	Mountview B	Newberry C	Odessa C
Millsdale B	Mucara B	New Cambria C	Odin C
Milo D	Muir B	Newfane B	O'Fallon D
Milroy D	Muirkirk B	Newkirk B	Ogemaw B
Mimosa C	Mukilteo A	Newport B	Okaw D
Minatare D	Mullins D	Newton D	Okeechobee D
Minco B	Munising B	Newtonia B	Okeelanta D
Mineola B	Munuscong D	Nicholson B	Okemah C
Miner D	Murrill B	Nicholville B	Okenee D
Minnequa B	Muscatine B	Nickel D	Okoboji B
Minora C	Muse B	Nicollet B	Oktibbeha D
Minvale B	Muskingum C	Niles C	Olaa B
Mires A	Muskogee C	Nimrod C	Olequa C
Mission C	Musselshell B	Ninigret B	Olinda B
Mitchell B	Myatt D	Niota D	Olivier C
Moca D	Myersville B	Nipe B	Olmitz B
Modale C		Nisqually A	Olmsted C
Moenkopie C	Naalehu B	Nixa C	Olympic B
Moffat B	Naches B	Nixon B	Omaha B
Mohave C	Nacimiento C	Nixonton B	Omega A
Mohawk B	Nacogdoches C	Noble B	Ona C
Moiese B	Naiwa B	Nobscot A	Onalaska B
Moira C	Nakelele B	Nodaway B	Onamia B
Mokena C	Nantucket C	Nogales C	Onarga B
Molena A	Nanum B	Nohili D	Onaway B
Moline D	Napa D	Nolan C	Ondawa B
Monarda C	Napier B	Nolichucky B	Oneida C
Monee D	Nappanee D	Nolo C	O'Neill B
Monmouth C	Naranjito C	Nonopahu C	Onslow B
Monona B	Narcisse B	Nookachamps D	Ontario B
Monongahela C	Narragansett C	Nooksack C	Ontonagon C
Monroeville C	Nasel C	Nora B	Onyx B
Montalto B	Nason C	Norden B	Ookala B

* From U.S. Soil Conservation Service [19].

† The hydrologic grouping for the rocky phases of these soils should be reduced one group.

‡ When these soils are classified as having class 3 or 4 rockiness, the hydrologic grouping should be reduced one group.

Table 21-8. Classification of Soils by Hydrologic Soil Groups* (Continued)

Oquaga C‡	Palmdale C	Petoskey A	Post D
Oquawka A	Palmyra B	Petrolia D	Potamo D
Ora C	Palouse B	Pettis B	Potter C
Oracle B	Pana B	Pheba C	Pottsville D
Orange D	Pandura B	Phelps B	Poultney B
Orangeburg B	Panton D	Phillips D	Poverty C
Orcas A	Papago B	Philo C	Powder B
Orchard B	Papakating D	Picacho D	Powell C
Ordway D	Papineau C	Pickaway C	Poygan D
Orelia D	Parishville C	Pickford D	Pozo Blanco C
Orella D	Parkdale A	Pickwick B	Prather C
Orient B	Parke B	Pierce B	Pratt A
Orienta B	Parker B	Pierre D	Prentiss C
Orio C	Parkwood C	Piihonua B	Prescott D
Orion C	Parnell D	Pilchuck A	Presque Isle B
Orlando	Parr B	Pilot B	Preston A
High Phase A	Parsons D	Pilot Rock C	Prewitt C
Orlando	Pasco B	Pima C	Prieta B
Low Phase B	Paso Seco C	Pinal D	Princeton B
Orman D	Pasquotank D	Pinckney C	Pring B
Orrville C	Patent C	Pinones D	Proctor B
Ortello A	Patit Creek B	Pinson B	Progresso B
Orting C	Patoutville C	Pintura A	Promise D
Osage D	Patrick B	Pisgah B	Prospect B
Osceola D	Patton C	Pittsfield C‡	Prosser B
Oshawa D	Paulding D	Pittstown C	Providence C
Oshtemo B	Pauwela C	Pittwood B	Provo B
Osmund B	Pawlet B	Placentia D	Prowers B
Oso A	Pawnee D	Plainfield A	Ptarmigan B
Ostrander B	Paxton C	Plano B	Puget B
Otero B	Paymaster B	Plata C	Puhi B
Othello D	Payne D	Platea C	Pulaski B
Otisville A	Peace River D	Platner C	Pulehu A
Otsego C	Peacham D	Plattsmouth B	Pullman D
Ottawa A	Pearman C	Plattville B	Purdy D
Otter D	Pearson C	Pledger D	Purgatory D
Otterholt B	Pecatonica B	Plummer D	Puu Oo B
Ottokee A	Pedernales C	Plymouth B	Puu Pa B
Otway D	Pekin C	Pocomoke D	Puyallup B
Ovid C	Pelan B	Podunk B	
Owaneco D	Pella B	Poinsett B	Quamba D
Owen Creek C	Pembroke B	Poland B	Quandahl B
Owens D	Pence B	Polson C	Quay C
Ozona C	Penn C	Pomello A	Quicksell C
	Pennington B	Pompano D	Quincy A
Paaloa B	Penoyer C	Pomroy B	Quinlan B
Paauhau B	Penrose C	Poncena D	Quonset A
Pace B	Penwood A	Pond Creek C	
Paden C	Peoh C	Pontotoc B	Raber C
Page D	Peone C	Pope B	Rabun B
Pahranagat C	Peotone C	Poppleton A	Racine B
Pahroc D	Pequea C	Poquonock C	Racoon D
Paia B	Perkinsville B	Port D	Radford B
Painesville B	Perks A	Portales B	Radnor D
Paiso C	Perrine D	Port Byron B	Ragnar A
Paiute B	Perry D	Porters B	Rago C
Palatine C	Persayo D	Portland D	Rainbow C
Palestine B	Pershing C	Portsmouth D	Rains D
Palmas Altas D	Peru C	Portugues D	Ralston B
Palm Beach A	Peshastin A	Poskin C	Ramona C

* From U.S. Soil Conservation Service [19].

† The hydrologic grouping for the rocky phases of these soils should be reduced one group.

‡ When these soils are classified as having class 3 or 4 rockiness, the hydrologic grouping should be reduced one group.

Table 21-8. Classification of Soils by Hydrologic Soil Groups* (Continued)

Ramsey	B	Ringling	C	Rucker	B	Saugatuck	C
Randall	D	Ringold	B	Rudyard	C	Sauk	B
Ranger	C	Rio Arriba	D	Rumford	B	Savage	C
Rankin	C	Rio Canas	C	Rumney	C	Savannah	C
Rantoul	D	Rio Lajas	A	Rupert	A	Sawmill	C
Rapidan	C	Rio Piedras	D	Rushtown	A	Sawtooth	C
Rarden	C	Ritchey	B	Rushville	D	Sawyer	C
Raritan	B	Rittman	C	Ruskin	C	Saybrook	B
Raub	B	Ritzville	B	Russell	B	Scandia	B
Rauville	D	Riverside	A	Russellville	C	Scantic	C
Ravalli	D	Riverton	C	Ruston	B	Scarboro	D
Ravenna	C	Roane	C	Rutlege	D	Schapville	C
Ravola	B	Roanoke	D	Ryder	C	Schoharie	C
Ray	B	Robbs	D			Schooley	D
Rayne	B	Robertsville	D	Sabana	C	Schumacher	C
Reagan	C	Robinsonville	B	Sabana Seca	D	Schuylkill	B
Reaton	B	Roby	C	Saco	D	Scio	B
Reaville	C	Rockaway	B	Saffell	B	Sciotoville	C
Rebuck	C	Rockbridge	B	Sage	D	Scipio	D
Red Bay	B	Rockdale	D	Sagemoor	C	Scituate	C
Redfield	B	Rockmart	C	St. Albans	B	Scobey	B
Red Hook	C	Rockport	B	St. Charles	B	Scott	D
Redington	C	Rockton	B	St. Clair	D	Scott Lake	B
Redlands	B	Rockwood	B	St. Helens	A	Scowlale	B
Redmond	B	Rocky Ford	B	St. Joe	B	Scranton	C
Reed	D	Rodman	A	St. Johns	D	Searing	B
Reeser	C	Roe	B	St. Lucie	A	Seaton	B
Reeves	C	Roebuck	D	St. Marys	C	Sebeka	D
Regent	C	Rogers	D	St. Paul	C	Sebewa	D
Regnier	D	Rohrersville	D	Salal	B	Sebring	D
Reinach	B	Rokeby	D	Salem	B	Sedan	C
Reliance	C	Rolfe	C	Salemsburg	B	Sediu	C
Renfrow	D	Romeo	C	Salisbury	D	Segal	D
Reno	D	Romulus	D	Salix	B	Segno	B
Renohill	C	Rosachi	D	Salkum	D	Selah	C
Renova	B	Rosario	C	Salmon	B	Selkirk	D
Renshaw	B	Roscoe	C	Salol	D	Selle	A
Renslow	B	Roscommon	D	Saltillo	C	Selma	B
Rentide	C	Rosebud	B	Saluvia	C	Semiahmoo	B
Reparada	D	Rosedell	D	Salvisa	D	Senecaville	C
Retsof	C	Roselms	D	Samish	D	Sequatchie	B
Rex	C	Rosemount	B	Sammamish	B	Sequoia	C
Reynolds	B	Roseville	B	Sams	B	Serrano	D
Rhinebeck	C	Rositas	A	San Anton	C	Sexton	D
Rhoades	D	Roslyn	C	San Antonio	D	Seymour	C
Richfield	B	Ross	B	San German	C	Shannon	B
Richland	C	Rossmoyne	C	Sango	C	Shapleigh	C‡
Richview	C	Round Butte	D	San Joaquin	D	Sharkey	D
Richwood	B	Routon	D	San Jose	B	Sharon	B
Ridgebury	C	Rowe	D	San Juan	B	Sharpsburg	B
Ridgely	B	Rowland	C	San Saba	D	Shavano	B
Ridgeville	B	Rowley	C	Santa	C	Shelburne	C
Riesel	D	Rox	B	Santa Clara	D	Shelby	C
Riffe	A	Roy	B	Santa Isabel	D	Shelbyville	C
Riga	C	Royalton	C	Santa Lucia	C	Shelmadine	C
Riggs	D	Roza	D	Santiago	B	Shelocta	B
Riley	B	Rozetta	B	Sargeant	D	Shelton	B
Rillito	B	Ruark	C	Sarpy	A	Sheppard	A
Rimer	C	Rubicon	A	Sassafras	B	Sheridan	B
Rinard	D	Rubio	C	Sauble	A	Sherman	B

* From U.S. Soil Conservation Service [19].

† The hydrologic grouping for the rocky phases of these soils should be reduced one group.

‡ When these soils are classified as having class 3 or 4 rockiness, the hydrologic grouping should be reduced one group.

Table 21-8. Classification of Soils by Hydrologic Soil Groups* (Continued)

Shiloh C	Spring Creek D	Sunnyside B	Teton B
Shoals C	Springer A	Sunrise B	Tetonka C
Shook B	Springfield D	Sunsweet C	Thackery B
Shoshone B	Springtown B	Superstition A	Thatuna C
Shouns B	Spur B	Surry B	Thayer B
Shrewsbury D	Staatsburg C	Susquehanna D	Thomasville C
Shubuta C	Stambaugh B	Sutherlin C	Thompson A
Shuwah C	Stamford D	Sutphen D	Thorndike C‡
Sicily B	Stanfield C	Sutton B	Thornton D
Sidell B	Stanton D	Swaim D	Thornwood A
Sierra C	Starks C	Swanton C	Thoroughfare A
Sifton B	Starr B	Swantown A	Thorp C
Signal C	Staser B	Swartswood C	Thurman A
Siler A	State B	Sweden B	Thurmont B
Silerton B	Stecum D	Sweeney C	Thurston B
Silver Creek D	Steekee C	Sweetwater D	Tiburones D
Simcoe B	Steinauer B	Swims B	Tice C
Simla B	Steinsburg B	Switzer D	Tickfaw D
Sims D	Stendal C	Swygert C	Tieton B
Sinai C	Stephensburg C	Sylvan B	Tifton B
Sinclair B	Stephenville B	Symerton B	Tijeras B
Singsaas B	Stetson B		Tilden C
Sioux B	Stevenson B	Tabernash B	Tillman B
Sipple B	Stewart D	Tabler D	Tilsit C
Siskiyou B	Stidham B	Tabor D	Timmer B
Sites C	Stimson B	Taft C	Timmerman A
Skaggs C	Stissing C	Tahoe C	Timpahute D
Skagit B	Stockbridge C	Talante D	Timula B
Skalkaho B	Stockland B	Talbott D	Tinton B
Skamania B	Stockton B	Talcot D	Tiocano D
Skames C	Stoneham B	Talihina D	Tioga B
Skerry C	Stonington B	Talladega C	Tippah D
Skiyou B	Stono C	Tallula B	Tippecanoe B
Skokomish A	Stookey B	Tally B	Tipperary A
Skyberg C	Storden B	Taloka D	Tipton B
Skykomish A	Story C	Tama B	Tirzah B
Sleeth C	Stough C	Tamms C	Tisbury B
Sloan D	Stoy D	Tanama C	Tisch C
Slocum D	Strasburg C	Tanberg D	Tishomingo C
Smoky Butte C	Strauss C	Taneum C	Titusville C
Smolan C	Strawn B	Tanwax A	Tivoli A
Snow B	Stronghurst B	Taos B	Toa C
Soda Lake B	Stryker D	Tarrant D	Tobin B
Sodus C	Stukel C	Tate B	Tobosa D
Sogn C	Stumpp D	Tatum B	Todd B
Soller D	Sudbury B	Taylor C	Toddville B
Solomon D	Suffield C	Teague D	Tokul A
Somers B	Sula B	Teas C	Toledo D
Somerset B	Sulphura C	Tebo B	Tolley B
Sonoita B	Sultan B	Tedrow A	Tolo B
Sontag D	Sumas C	Teja C	Toltec B
Souva D	Summerville C	Tell B	Tombigbee A
Sparta A	Summit C	Teller B	Tonawanda C
Spearfish B	Sumner A	Tellico B	Tongue River C
Spencer B	Sumter D	Tenino A	Tonopah B
Sperry C	Sun D	Tepee D	Toppenish C
Spilo D	Sunbury B	Teresa D	Topton B
Spooner C	Suncook A	Terril B	Torres C
Spottswood B	Sunderland C‡	Terry B	Tortugas D
Spring D	Sunniland C	Tescott B	Tours C

* From U.S. Soil Conservation Service [19].
† The hydrologic grouping for the rocky phases of these soils should be reduced one group.
‡ When these soils are classified as having class 3 or 4 rockiness, the hydrologic grouping should be reduced one group.

Table 21-8. Classification of Soils by Hydrologic Soil Groups* (Continued)

Toutle A	Uinta B	Vira B	Warrenton A
Tovey B	Ulen B	Virden C	Warrior C
Tower D	Ulm B	Virgil B	Warsaw B
Townsbury B	Ulupalakua B	Virgin River D	Wartrace C
Townsend B	Ulysses B	Virtue C	Warwick B
Toxaway D	Umapine B	Vista B	Washburn D
Toyah B	Una D	Vives C	Washington B
Traer C	Unadilla B	Vivi B	Washoe C
Transylvania B	Uncompahgre B	Vlasaty C	Washougal C
Trapper B	Ungers B	Volga D	Washtenaw C
Travessilla D	Union C	Volin B	Wassaic C
Travis C	Unison B	Volke C	Wassuk D
Treadway D	Unity A	Volney B	Watauga B
Trego C	Upshur C	Volperie C	Watchaug B
Trempealeau B	Urbana B	Volusia C	Waterboro D
Trenary B	Urbo D	Vona A	Waterloo C
Trent B	Usine D	Vrooman B	Waterville C
Trexler C	Ursula D		Watseka A
Trinity D	Utica B		Watson C
Tripp B	Utuado B	Wacousta C	Watsonville D
Tromp B	Uvalde B	Wade D	Watt B
Trout River A		Wadell D	Watton C
Trowbridge B		Wadena B	Waubay B
Troxel B	Vader B	Wadesboro B	Waugh C
Troy C	Vaiden D	Wadsworth D	Waukegan B
Truman B	Vale B	Wagner D	Waukesha B
Trumbull D	Valentine A	Waha C	Waukon B
Tubac D	Valera C	Wahee C	Waumbek B
Tucker D	Vallecitos C	Wahiawa B	Wauseon D
Tucumcari B	Valois C	Wahtum C	Waverly D
Tuffit B	Vance C	Waialua B	Wayland C
Tughill D	Vandalia C	Waikaloa B	Wayne B
Tujunga A	Vanderville C	Waikapu B	Waynesboro B
Tuller D	Vanoss B	Wailea B	Wea B
Tully C	Varna C	Waimanalo D	Weaver C
Tumacacori B	Vaucluse C	Waimea B	Webb C
Tumbez D	Vayas D	Waipahu C	Webster C
Tumwater A	Vebar B	Waiska B	Weeksville B
Tunica D	Vega Alta C	Waits B	Wehadkee D
Tunkhannock B	Vega Baja D	Wakeland B	Weikert C
Tupelo D	Vekol C	Wakonda C	Weinbach C
Turbotville C	Velma B	Wallace B	Weir D
Turin C	Venango D	Walla Walla B	Weld C
Turkey Creek B	Venedy D	Waller C	Weller D
Turnbow D	Verdel D	Wallington C	Wellington D
Turner B	Verdigris B	Wallkill C	Wellman C
Turnerville C	Verdun D	Wallpack B	Wellsboro C
Tuscan B	Vergennes D	Walpole C	Wellston B
Tuscarora C	Verhalen D	Walsh D	Wemple B
Tuscola B	Vernon D	Walters D	Wenas C
Tuscumbia D	Verona B	Walton C	Wenatchee B
Tusquitee B	Veyo D	Wampsville B	Wesley C
Tuxedo D	Via C	Wann A	Wessington B
Twin Creek B	Vicksburg B	Wapato C	Westbury C
Twin Lakes B	Victor B	Wapping B	Westfall C
Two Dot B	Victoria D	Ward D	Westland D
Tyler C	Vienna B	Warden B	Westminster C‡
	Vilas A	Warman D	Westmoreland C
	Vinton A	Warne D	Weston C
Udolpho C	Viola D	Warners C	Westphalia B

* From U.S. Soil Conservation Service [19].
† The hydrologic grouping for the rocky phases of these soils should be reduced one group.
‡ When these soils are classified as having class 3 or 4 rockiness, the hydrologic grouping should be reduced one group.

Table 21-8. Classification of Soils by Hydrologic Soil Groups* (Continued)

West Point D	Wickham B	Winslow B	Xenia B
Westport A	Wickiup A	Winston A	
Westville C	Wilbraham C†	Winterset C	Yabucoa D
Wethersfield C	Wilcox D	Witt B	Yadkin B
Weymouth C	Wildwood D	Wolcottsburg C	Yahola B
Whalan B	Wilkes C	Woldale D	Yakima A
Wharton C	Wilkeson B	Wolf B	Yale C
Whatcom D	Will D	Wolftever C	Yauco D
Whately D	Willamette B	Woodbridge C	Yeoman B
Wheeling B	Willard C	Woodglen D	Yoder B
Whidbey A	Williams B	Woodinville B	Yonaba C
Whippany C	Williamsburg C	Woodlyn C	Yordy B
Whitefish B	Williamson B	Wood River D	York C
Whiteford C	Willoughby C	Woodson D	Yunes B
White House C	Willow Creek B	Woodstown C	
Whitelaw C	Wilson D	Woodward B	Zaca D
Whitesburg C	Winchester A	Wooster B	Zahl B
Whiteson D	Windom B	Woostern B	Zaleski B
White Store D	Wind River A	Worland B	Zaneis C
White Swan C	Windsor A	Worsham D	Zapata D
Whitetail B	Windthorst C	Worth C	Zell B
Whitlock A	Winema B	Worthen B	Zimmerman A
Whitman D	Winfield C	Worthington B	Zion C
Whitson D	Wingville B	Wortman C	Zipp C
Whitwell C	Winifred D	Wrightsville D	Zita B
Wibaux C	Winlock D	Wurtsboro C	Zook D
Wichita C	Winnett D	Wykoff B	Zuber B
Wickersham C	Winooski B	Wynoose D	Zwingle D

* From U.S. Soil Conservation Service [19].
† The hydrologic grouping for the rocky phases of these soils should be reduced one group.
‡ When these soils are classified as having class 3 or 4 rockiness, the hydrologic grouping should be reduced one group.

"Good" in proportion to the amount of dense vegetation in the rotation. Poor rotations from a hydrologic standpoint usually contain row crops, small grains, and fallow in various combinations. Good rotations contain a high proportion of alfalfa or other close-seeded legumes or grasses that will improve tilth and increase infiltration. The effect of such crops will carry over into the second or third year.

Table 21-9. Classification of Native Pasture or Range*

Hydrologic condition	Vegetative condition
Poor............	Heavily grazed, no mulch, or having plant cover on less than about 50 per cent of the area
Fair.............	Moderately grazed; between about 50 and 75 per cent of the area with plant cover
Good............	Lightly grazed; more than about 75 per cent of the area with plant cover

* From U.S. Soil Conservation Service [19].

Native pasture or range: Usually divided into three condition classes on the basis of their hydrologic effects (Table 21-9). A watershed consisting primarily of native pasture or range may require a more detailed classification, such as that given in Table 21-10. The weights are determined by field sampling, but this need not be extensive if samples are selected to represent large portions of the watershed. The plus signs in Table 21-10 indicate that the runoff curve number taken from Table 21-12 should be interpolated between that for the class indicated and the one more favorable. For example, a watershed in soil group C having a "Fair +" cover would have a runoff curve number of $(79 + 74)/2 = 76.5$.

Permanent meadow: Ungrazed, native grassland with 100 per cent cover. It represents the upper limit of watershed grass cover.

HYDROLOGY OF AGRICULTURAL LANDS

Woodland: Table 21-11 gives the classification of woodland based on hydrologic effects. It must be kept in mind that these classifications are for hydrologic purposes and are not based on timber production.

  c. Land Treatment. This refers to the cultural or tillage practices used in farming. The following definitions refer to the land treatments or practices listed in Table 21-12.

  Straight-row farming: A type of farming where plowing, planting, cultivating, and other farm operations are carried on without regard to the slope of the land.

  Contour farming: A type of farming where farm operations are carried on by following the general contour of the land. The furrows developed during tillage reduce surface runoff and erosion. The size of the furrows will vary according to the crop and the farm equipment used. The furrows developed in planting small grains and

Table 21-10. Classification of Native Pasture or Range by Density and Weight Sampling*

Areal density, per cent	Air-dry weight of plant and litter, tons/acre		
	Less than 0.5	0.5 to 1.5	Over 1.5
Less than 50	Poor	Poor +	Fair
50 to 75	Poor +	Fair	Fair +
Over 75	Fair	Fair +	Good

* From U.S. Soil Conservation Service [19].

Table 21-11. Classification of Woodlands*

Hydrologic condition	Vegetative condition
Poor	Heavily grazed or regularly burned so that litter, small trees, and brush are destroyed
Fair	Grazed but not burned; there may be some litter, but these woodlands are not fully protected from grazing
Good	Protected from grazing so that litter and shrubs cover the soil

* From U.S. Soil Conservation Service [19].

legumes are small (about 4 in. wide, 4 in. deep, and spaced about 8 to 10 in. apart) and will disappear under the action of rainfall. Furrows developed by planting or tilling row crops are generally larger (12 in. wide, 6 in. deep, and spaced about 40 in. apart). The storage capacity of furrows decreases as watershed slope increases. Although contour farming provides some protection against erosion, the use of alternate strips of dense vegetation (strip cropping) provides greater protection. Usually, strips of hay alternate with strips of less dense cover, such as small-grain or row crops. Occasionally, contour furrows are used on native pasture or rangeland. Their dimensions should vary with climate and topography, decreasing as annual rainfall and land slope decrease. In general, contour furrows on native pasture or range last longer than furrows on cropland, but their permanence varies with the soil type, the intensity of grazing, and the density of the vegetation.

  Terracing: The practice of constructing dikes or dike-ditch combinations to control runoff from farmland. Terraces may be graded, open-end level, or closed-end level. Open-end level terraces under cultivation will usually become graded as a result of erosion and sediment deposition. In Table 21-12, only graded terraces and their grassed waterway outlets are considered. The storage capacity of closed-end level terraces can be determined from Table 21-29. Terraces are seldom used on native pasture or range. If areas of pasture or range are to be terraced, the area should be classified as cultivated terraces for the period of time it takes to revegetate the area after terrace construction.

Table 21-12. Runoff Curve Numbers for Hydrologic Soil-cover Complexes*
(For watershed condition II and $I_a = 0.2S$)

(1) Land use or cover	(2) Treatment or practice	(3) Hydrologic condition	(4) Hydrologic soil group			
			A	B	C	D
Fallow................	Straight row	Poor	77	86	91	94
Row crops............	Straight row	Poor	72	81	88	91
	Straight row	Good	67	78	85	89
	Contoured	Poor	70	79	84	88
	Contoured	Good	65	75	82	86
	Contoured and terraced	Poor	66	74	80	82
	Contoured and terraced	Good	62	71	78	81
Small grain...........	Straight row	Poor	65	76	84	88
	Straight row	Good	63	75	83	87
	Contoured	Poor	63	74	82	85
	Contoured	Good	61	73	81	84
	Contoured and terraced	Poor	61	72	79	82
	Contoured and terraced	Good	59	70	78	81
Close-seeded legumes† or rotation meadow	Straight row	Poor	66	77	85	89
	Straight row	Good	58	72	81	85
	Contoured	Poor	64	75	83	85
	Contoured	Good	55	69	78	83
	Contoured and terraced	Poor	63	73	80	83
	Contoured and terraced	Good	51	67	76	80
Pasture or range.......	....	Poor	68	79	86	89
		Fair	49	69	79	84
		Good	39	61	74	80
	Contoured	Poor	47	67	81	88
	Contoured	Fair	25	59	75	83
	Contoured	Good	6	35	70	79
Meadow (permanent).......	....	Good	30	58	71	78
Woodlands (farm woodlots)...	....	Poor	45	66	77	83
		Fair	36	60	73	79
		Good	25	55	70	77
Farmsteads............	....	.....	59	74	82	86
Roads, dirt‡..........	....	.....	72	82	87	89
Roads, hard-surface‡......	....	.....	74	84	90	92

* From U.S. Soil Conservation Service [19].
† Close-drilled or broadcast.
‡ Including right-of-way.

d. Soil-cover Complexes. Table 21-12 gives runoff curve numbers for various combinations of soils and covers. For special conditions, numbers can be estimated by interpolation or by weighting the numbers of the given complexes.

6. Erosion. Typical soil-erosion conditions on large watersheds can be learned from soil maps [21]. Erosion conditions on a small watershed must usually be determined from a detailed soil map or a field survey by a soil scientist. This information is necessary in estimating upland erosion for sedimentation studies of water-storage sites.

D. Runoff

1. Data. Annual, monthly, and storm runoff data from selected experimental small watersheds in the United States are available in publications of the U.S. Department of Agriculture [2-4]. Runoff data for many small watersheds are available in publications of state agricultural experiment stations and in the *U.S. Geological Survey Water-Supply Papers.*

a. Transposition of Data. The type of runoff being considered has a bearing on the adjustments required in transposing data from a gaged to an ungaged watershed. Runoff from experimental plots and very small watersheds normally will be composed entirely of surface runoff. Runoff from larger watersheds may include surface runoff, quick-return flow, and base flow. Care must be taken, therefore, to determine the runoff characteristics of each watershed and to make the necessary adjustments to offset differences between the watersheds.

b. Supplementary Gaging Stations. A temporary stream gage installed on a watershed to be studied can be used to improve runoff estimates. Depending on the data required, the gage may be either a recorder or a staff type. It should be installed upstream from a control, such as a culvert [22]. Data from the temporary station are related to data for the same period from a nearby permanent station with a long record. A relation between the two stations may be found by regression analysis, and the record at the permanent station used to make a long-term estimate for the watershed under consideration. The length of record needed at the temporary station will depend on the size and number of flows that occur during the period of measurement and on the accuracy required.

IV. DETERMINATION OF RUNOFF FROM PRECIPITATION

A. Storm Runoff

Figure 21-5 shows schematic curves of accumulated storm rainfall P, runoff Q, and infiltration F, plus initial abstraction I_a. For convenience in estimating runoff, initial abstraction I_a consists of all the storm rainfall occurring before surface runoff starts. Let us assume, as in Ref. 19,

$$\frac{F}{S} = \frac{Q}{P_e} \quad (21\text{-}6)$$

where F is the actual infiltration excluding the initial abstraction in inches, S is the potential infiltration in inches, Q is the actual direct runoff in inches, and P_e is the potential runoff or effective storm rainfall, i.e., storm rainfall minus the initial abstraction, in inches.

With $F = P_e - Q$, Eq. (21-6) can be written as

$$Q = \frac{P_e^2}{P_e + S} \quad (21\text{-}7)$$

The initial abstraction I_a, in inches, estimated from an empirical relation based on data from small watersheds, is

$$I_a = 0.2S \quad (21\text{-}8)$$

Thus

$$P_e = P - I_a = P - 0.2S \quad (21\text{-}9)$$

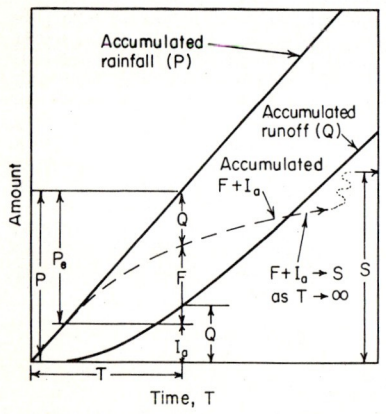

FIG. 21-5. Schematic curves of accumulated P, Q, and $F + I_a$, also showing the relation expressed by Eq. (21-10).

where P is the total storm rainfall in inches. Substituting Eq. (21-9) in Eq. (21-7),

$$Q = \frac{(P - 0.2S)^2}{P + 0.8S} \quad (21\text{-}10)$$

The *runoff curve numbers*, designated as CN, for the hydrologic soil-cover complexes (Table 21-12) are functionally related to S as

$$\text{CN} = \frac{1{,}000}{S + 10} \quad (21\text{-}11)$$

DETERMINATION OF RUNOFF FROM PRECIPITATION

Eliminating S between Eqs. (21-10) and (21-11), the relation between Q, P, and CN can be shown by the curves in Fig. 21-6.

1. Antecedent Moisture Condition. The CN values in Table 21-12 are for average watershed soil-cover and soil-moisture conditions. When it is necessary to make a runoff estimate for below-average (dry) or above-average (wet) moisture

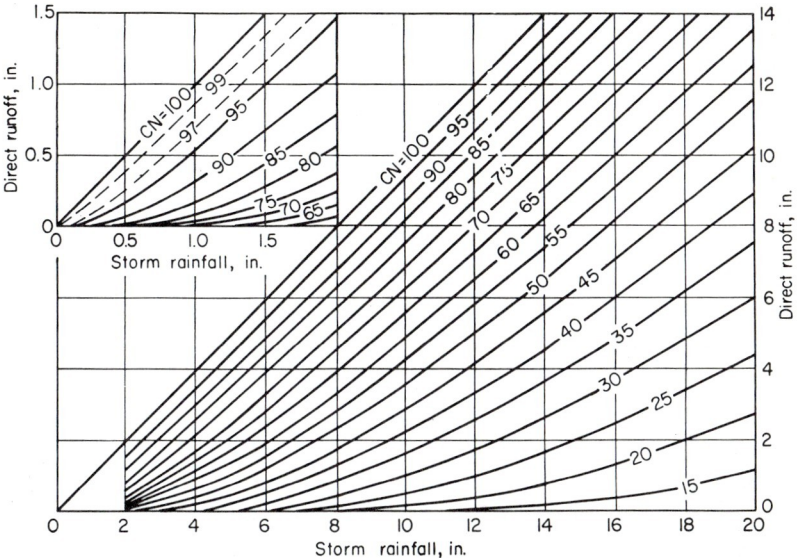

Fig. 21-6. Chart for estimating direct runoff or solution of Eqs. (21-10) and (21-11). (*From U.S. Soil Conservation Service* [19].)

conditions, representative CN values can be approximated using Tables 21-13 and 21-14.

2. Estimate of Direct Runoff by Type 1 Approach. The use of Table 21-12 and Fig. 21-6 will be illustrated with a simple example, after which a detailed type 1 estimate will be made.

Example 21-4. A small watershed having a group C soil and fair pasture cover had a rainfall of 3.23 in. Assuming that soil, cover, and antecedent moisture were in average conditions, estimate the direct runoff.

Table 21-13. Rainfall Limits for Estimating Antecedent Moisture Conditions*

Antecedent moisture condition class	5-day total antecedent rainfall, in.	
	Dormant season	Growing season
I	Less than 0.5	Less than 1.4
II	0.5 to 1.1	1.4 to 2.1
III	Over 1.1	Over 2.1

* From U.S. Soil Conservation Service [19].

1. Using Table 21-12, under soil group C for fair pasture, find CN = 79.
2. Entering Fig. 21-6 with the storm rainfall of 3.23 in. and CN = 79, by interpolation, read $Q = 1.38$ in.

Table 21-14. Runoff Curve Number (CN), Conversions and Constants*

CN for condition II	CN for AMC		S values,† in.	Curve† starts where $P =$ (in.)
	I	III		
(1)	(2)	(3)	(4)	(5)
100	100	100	0.000	0.00
98	94	99	0.204	0.04
96	89	99	0.417	0.08
94	85	98	0.638	0.13
92	81	97	0.870	0.17
90	78	96	1.11	0.22
88	75	95	1.36	0.27
86	72	94	1.63	0.33
84	68	93	1.90	0.38
82	66	92	2.20	0.44
80	63	91	2.50	0.50
78	60	90	2.82	0.56
76	58	89	3.16	0.63
74	55	88	3.51	0.70
72	53	86	3.89	0.78
70	51	85	4.28	0.86
68	48	84	4.70	0.94
66	46	82	5.15	1.03
64	44	81	5.62	1.12
62	42	79	6.13	1.23
60	40	78	6.67	1.33
58	38	76	7.24	1.45
56	36	75	7.86	1.57
54	34	73	8.52	1.70
52	32	71	9.23	1.85
50	31	70	10.0	2.00
48	29	68	10.8	2.16
46	27	66	11.7	2.34
44	25	64	12.7	2.54
42	24	62	13.8	2.76
40	22	60	15.0	3.00
38	21	58	16.3	3.26
36	19	56	17.8	3.56
34	18	54	19.4	3.88
32	16	52	21.2	4.24
30	15	50	23.3	4.66
25	12	43	30.0	6.00
20	9	37	40.0	8.00
15	6	30	56.7	11.34
10	4	22	90.0	18.00
5	2	13	190.0	38.00
0	0	0	Infinity	Infinity

* From U.S. Soil Conservation Service [19].
† For CN in column 1.

DETERMINATION OF RUNOFF FROM PRECIPITATION   21-31

Example 21-5. On June 6, rainfall averaging 4.3 in. fell on a 2,335-acre watershed with the following acreages of cover: corn, straight-row, good rotation, 1,160 acres; clover, straight-row, good rotation, 540 acres; pasture, fair condition, 210 acres; woodland, good condition, 343 acres; and roads, hard surface, 82 acres. Watershed soils are in group C. The antecedent rainfalls were June 1, 0.07 in.; June 2, 0.24 in.; June 3, 0 in.; June 4, 0 in.; and June 5, 0.28 in. Estimate the direct runoff for this storm.

1. Determine the antecedent moisture condition. The total rainfall for the 5 days before the storm is $0.07 + 0.24 + 0 + 0 + 0.28 = 0.59$ in. In Table 21-13, under "growing season," a value of 0.59 in. is found to fall in antecedent moisture condition class I.

2. Prepare a working table (Table 21-15), and list the land use and treatment for each acreage (columns 1 and 2).

3. Determine the CN for antecedent moisture condition II (AMC-II) for each soil-cover complex, using Table 21-12, and list the value of CN in column 4.

4. Determine the value of CN for antecedent moisture condition I (AMC-I). Entering column 1 of Table 21-14 with the values for step 3, read the CN for AMC-I in column 2. Tabulate these values of CN in Table 21-15, column 4.

5. Estimate the direct runoff for individual soil-cover complexes. Enter Fig. 21-6 with $P = 4.3$ in., and read the direct runoffs for the required CN values of Q in column 5.

6. Compute the weighted runoff. The sum of column 6 divided by the sum of column 2 gives the weighted estimate as 1.27 in.

Table 21-15. Computations for Estimate of Direct Runoff, Storm of June 6, 1960

| Land use and treatment | Acres | Curve number | | Q | Acres $\times Q$ |
| | | AMC-II | AMC-I | | |
(1)	(2)	(3)	(4)	(5)	(6)
Corn, straight-row, good rotation..	1,160	85	70	1.54	1,786
Clover, straight-row, good rotation.	540	81	64	1.17	632
Pasture, fair condition............	210	79	62	1.00	210
Woodlands, good condition........	343	70	51	0.48	165
Roads, hard-surface..............	82	90	78	2.15	176
Total......................	2,335	...	...	....	2,969

a. Alternative Method of Weighting Soil-cover Complexes. The detailed weighting method of Example 21-5 should be used where wide differences in cover or soils are found or where the watershed contains large areas of impervious surfaces. In most cases, a shorter method of weighting the CN values can be used, and Fig. 21-6 is used only once with the weighted CN. For the data in Example 21-5, the weighted CN is 65.4. Entering Fig. 21-6 with $P = 4.3$ in. and a CN of 65.4, by interpolation read $Q = 1.22$ in. This alternative method will usually give a smaller total runoff.

b. Storm Duration. Figure 21-6 is for use with storms of a 1-day duration or less. The time of day during which the runoff occurred must be found in detailed rainfall records. When a storm occurs over several days, the runoff estimate may be made on a daily basis, changing the antecedent moisture condition accordingly.

Example 21-6. Given a watershed of 720 acres, with soils in group D, and a cover of fair pasture. Estimate the daily and total runoff for a storm occurring as follows: May 3, 1.65 in.; May 4, 3.27 in.; May 5, 1.02 in.; May 6, 6.23 in. There was no rainfall in the 5-day period preceding May 3.

1. Determine the CN for antecedent moisture conditions I, II, and III. For AMC-II, Table 21-12 gives a CN value of 84 for fair pasture in soil group D. Entering Table 21-14 with CN = 84 for AMC-II, read CN = 68 for AMC-I and CN = 93 for AMC-III.

2. Tabulate the rainfalls and CN values on a working table (Table 21-16), and estimate the direct runoff from Fig. 21-6 as shown in column 5.

The total runoff can also be estimated directly, using the value of CN applicable for the first day of the storm and the total storm rainfall of 12.17 in. Using CN = 68 for AMC-I, $Q = 7.92$ in. In some cases such an estimate will differ even more widely from the one made on a day-by-day basis.

3. Estimate of Direct Runoff by Type 2 Approach. In a frequency analysis, the specific date of a storm is unknown but a specific duration is usually given and the estimate of runoff is for that duration. The average antecedent moisture condition (AMC-II) is used in the type 2 approach.

Example 21-7. Given a watershed of 185 acres, of which 30 acres is in fair woodland and the remainder in fair pasture. All soils are in soil group B. The 6-hr-duration rainfalls for the 2-, 10-, and 100-year frequencies are 2.04, 3.40, and 5.11 in., respectively. Determine the runoff for the three frequencies.

Table 21-16. Computation of Direct Runoff for a Multiple-day Storm

(1) Date	(2) Rainfall, in.	(3) AMC	(4) CN	(5) Q, in.
May 3........	1.65	I	68	0.10
May 4........	3.27	II	84	1.74
May 5........	1.02	III	93	0.47
May 6........	6.23	III	93	5.41
Total........	12.17	...	...	7.72

1. Determine the weighted CN = $(30 \times 60 + 155 \times 69)/185 = 67.54$. Use 68.
2. Entering Fig. 21-6 with a 2-year rainfall of 2.04 in. and a CN = 68, read $Q = 0.21$ in. Similarly, for the 10- and 100-year rainfalls of 3.40 and 5.11 in., find $Q = 0.84$ and 1.96 in., respectively. These runoffs are assumed to have the same frequencies as the rainfalls from which they are derived. The three estimates of runoff can be plotted on probability paper and a frequency line drawn, from which estimates of runoff for other frequencies can be made.

B. Snowmelt Runoff

1. General. The runoff curve numbers for the hydrologic soil-cover complexes of Table 21-12 do not apply for conditions of frozen ground or snowmelt. When the ground is frozen, differences between soils can usually be ignored and cover is at a minimum except in forested areas. Estimates of snowmelt runoff usually cannot be made accurately by the type I approach since the depth, water content, and areal distribution of snow are so variable on a small watershed. Estimates by the type 2 approach can be made using published data on the amount and frequency of snow depth and water content [23, 24]. Except where all runoff is from melting snow, peak rates of flow used for design purposes will generally be higher for the growing season than for winter. Snowmelt, however, may be greater in volume than rainfall and may be important in the design of storage reservoirs. (See also Sec. 10.)

2. Degree-day Method. This method provides a means of estimating the potential snowmelt on a daily basis. Local or transposed temperature and snowfall data are necessary.

a. Degree-day Equation. In its usual form, this equation is

$$M = K(T - 32°) \tag{21-12}$$

where M is the potential daily snowmelt in in., K is a constant representing the watershed condition, and T is the average daily air temperature in °F.

The constant 32° is the base air temperature in degrees Fahrenheit at which snowmelt is assumed to begin. Equation (21-12) can be modified for local conditions by changing the base temperature or by using some combination of maximum and minimum daily temperatures for T, instead of the arithmetic mean. The difference, $T - 32°$, gives the number of degree-days per day. For example, with an average

DETERMINATION OF RUNOFF FROM PRECIPITATION 21-33

daily temperature of 47°F, the number of degree-days for that day is 15. Since M is the potential snowmelt for the day, the actual snowmelt cannot be larger than the water equivalent available for the day. Table 21-17 gives values of K, which include the effects of watershed infiltration.

b. Water Equivalent. The density of freshly fallen snow varies greatly. Water equivalent is the average depth of water, in inches, obtained by melting the snow catch. New snow may have a water content of from 5 to 25 per cent of snow depth. A value of 10 per cent, or 1 in. of water for 10 in. of snow depth, is often used for making general estimates. The density of snow increases with the age of the snowpack at an approximately uniform rate. It is greatest just before the snowmelt season begins, and a water content of about 60 per cent has been measured at such times.

Table 21-17. Values of the Constant K*

Watershed condition	K
Low runoff potential	0.02
Average heavily forested areas; north-facing slopes of open country	0.04–0.06
Average runoff potential	0.06
South-facing slopes of forested areas; average open country	0.06–0.08
High runoff potential	0.30

* From U.S. Soil Conservation Service [19].

c. Temperature Adjustment for Altitude. Air temperature varies with altitude, decreasing about 4°F for a rise of 1,000 ft in elevation. If the average daily air temperature at a station in a valley at 1,200 ft elevation is 51°F on a given day, the estimate for a nearby small watershed at an elevation of 2,300 ft would be 46.6°F.

d. Estimating Snowmelt Runoff. Determinations of snowmelt runoff are usually made on a daily basis, using a bookkeeping system to avoid possible oversights.

Example 21-8. A small watershed on a north-facing slope of open country has snow cover with an initial water equivalent of 1.31 in. A record of air temperature and snowfall is available at a nearby station at about the same elevation. Estimate the daily snowmelt for the period of April 2 to 12.

1. Prepare a working table (Table 21-18), listing the date, precipitation, initial water equivalent on a watershed (first day), and average daily temperature. Note that snow fell on dates as shown in columns 1 and 2.
2. Select K from Table 21-17. For this example, $K = 0.04$.

Table 21-18. Computation of Snowmelt by Degree-day Method

(1)	(2)	(3)	(4)	(5)	(6)	(7)	(8)
Date	Water equivalent carry-over, in.	Precipitation, in.	Total water equivalent available, in.	Average daily temperature, °F	Potential snowmelt, in.	Snowmelt estimate, in.	Remaining water equivalent, in.
Apr. 2	1.31	0	1.31	28	0	0	1.31
Apr. 3	1.31	0	1.31	35	0.12	0.12	1.19
Apr. 4	1.19	0	1.19	47	0.60	0.60	0.59
Apr. 5	0.59	0	0.59	48	0.64	0.59	0
Apr. 6	0	0	0	41	0.36	0	0
Apr. 7	0	0	0	31	0	0	0
Apr. 8	0	2.43*	0.24	28	0	0	0.24
Apr. 9	0.24	1.51*	0.39	30	0	0	0.39
Apr. 10	0.39	0	0.39	33	0.04	0.04	0.35
Apr. 11	0.35	0	0.35	38	0.24	0.24	0.11
Apr. 12	0.11	0	0.11	44	0.48	0.11	0

* Snow.

3. Compute the potential snowmelt, M, by days (column 6), using Eq. (21-12) and data of column 5.
4. Complete the table for the other days.

Each item in column 2 is carried over from column 8 of the preceding day. Items in column 2 plus water equivalents in column 3 give items in column 4, using a general water content of 10 per cent. Results of step 3 are listed in column 6. An item in column 7 is taken from column 4 or column 6, whichever gives the smaller value. The entry in column 8 for the first day minus the entry in column 7 for the next day equals the entry in column 8 for the next day.

C. Annual and Seasonal Runoff

Maps of average or median annual runoff based on records of large streams and rivers [25, 26] are of little value in estimating the water yield of small watersheds. In humid climates, annual yields of small watersheds may be much smaller than those of rivers because the latter contain high base flow. In arid and semiarid climates, annual yields of small watersheds may be greater than those of large streams and rivers that have channel-transmission losses and no base flow. General adjustments for channel losses or gains can usually be made for large areas, but the error of estimate for a specific watershed may be very large.

1. Data Transposition. Data from 70 small watersheds [2] are given in Table 21-19 as a guide for estimating water yield. At Coshocton, Ohio, in a humid climate, the yield increases with the watershed size because of contributions from subsurface sources. The semiarid watersheds in New Mexico show a decreasing water yield as watershed size increases. At Beltsville, Md., variation in soils may account for the range in yields, while at Hastings, Nebr., where the soils are alike, the differences probably result from differences in cover.

Sometimes two or more small watersheds appear identical in all physical factors normally considered as affecting yield, yet the differences in yield are great. Detailed investigations for detecting the causes of these differences are seldom justified for small-project development. The use of a temporary streamflow station will give assurance that a yield estimate is of the right order of magnitude.

a. Transposition Method. Data from Table 21-19 will be used to illustrate a method of transposing data.

Example 21-9. A watershed having a drainage area of 1,200 acres and located near Hastings, Nebr., has soils in group C and cover consisting of 120 acres of meadow and 1,080 acres of cultivated land. Annual precipitation is about 22 in. Transmission losses occur. Determine the average annual water yield.

1. Make a first approximation for the type of cover, using the yields from the two single-cover watersheds listed for Hastings, Nebr., in Table 21-19, by weighting according to acreage:

Cover	Yield, in.	Acres	Yield × acres
Meadow	0.2	120	24
Cultivated	3.5	1,080	3,780
Total	...	1,200	3,804

The weighted yield is 3,804/1,200 = 3.17 in.

2. Adjust the weighted yield for transmission losses. Using linear interpolation with the data from the two larger watersheds at Hastings (481 and 2,086 acres), the modified yield is 2.90 in.

Where a difference in average annual precipitation enters the problem, this should also be taken into consideration. However, large differences (more than about 25 per cent) may lead to serious errors.

Table 21-19. Annual Runoff from Small Agricultural Areas*

Location	Years of record	Land use	Hydrologic soil group, %				Drainage area size, acres	Average† precipitation, in.	Average† runoff, in.
			A	B	C	D			
Arizona, Safford.........	15	80 to 90 % bare	...	7	...	93	519.0	7.2	0.4
	14	75 to 85 % bare	...	10	90	...	764.0	8.1	0.2
Arkansas, Bentonville.....	8	Pasture	...	100	...	...	14.2	49.1	1.2
	8	Cultivated	...	100	...	...	19.4	50.4	6.3
California, Watsonville....	3	Pasture	...	...	...	‡	27.4	35.3	14.4
	3	Brush	...	...	...	‡	10.1	32.8	7.1
Colorado, Colorado Springs.	7	Cultivated	...	...	...	‡	10.6	12.2	1.4
	7	Pasture	...	...	...	‡	35.4	12.7	0.4
Georgia, Watkinsville......	16	Cultivated §	...	70	30	...	19.2	46.3	4.3
Idaho, Moscow............	5	Cultivated	...	62	31	7	146.8	21.9	4.6
Illinois, Edwardsville......	17	Cultivated	...	1	99	...	27.2	36.5	8.5
Indiana, Lafayette.........	11	Cultivated	...	...	100	...	2.0	38.7	0.9
	11	Cultivated	...	...	66	34	2.1	39.2	4.8
Iowa, Clarinda............	8	Cultivated	...	82	18	...	2.0	28.2	2.6
	8	Cultivated §	...	74	26	...	2.0	28.2	1.1
	9	Cultivated §	...	75	25	...	3.1	27.7	0.4
Iowa, Shenandoah.........	5	Cultivated	...	61	39	...	67,200.0	26.9	2.6
Kansas, Hays.............	16	Grass	...	...	100	...	1.6	22.1	1.0
Maryland, College Park....	15	Cultivated	...	10	90	...	8.2	44.2	8.4
	15	Cultivated	...	60	40	...	7.4	44.2	3.0
	15	Cultivated	...	100	...	...	5.6	41.6	1.9
	14	Woods	...	78	22	...	12.0	43.6	0.9
Michigan, East Lansing....	14	Cultivated	...	100	...	...	2.0	30.7	3.6
	14	Woods	...	100	...	...	1.6	32.0	0.4
Missouri, Bethany.........	8	Cultivated	...	...	72	28	7.5	28.7	3.9
	10	Cultivated	...	...	92	8	4.5	29.4	5.0
Missouri, McCredie........	5	Mixed	...	...	22	78	44.3	30.9	3.1
	15	Mixed	...	...	11	89	153.0	36.7	9.0
Nebraska, Hastings........	16	Meadow	...	...	100	...	3.6	22.0	0.2
	16	Cultivated	...	...	100	...	4.2	22.0	3.5
	17	Mixed	...	25	75	...	481.0	21.6	3.0
	17	Mixed	...	31	69	...	2,086.0	21.6	2.4
New Mexico, Albuquerque.	16	77 % bare	...	20	‡	42	97.2	7.8	0.4
	15	75 % bare	...	...	‡	36	183.0	7.5	0.3
New Mexico, Santa Fe....	7	68 % bare	...	...	‡	...	790.0	13.3	0.1
N. Carolina, High Point...	27	Mixed	...	46	54	...	21,100.0	43.7	13.3
Ohio, Coshocton...........	16	Cultivated	...	...	100	...	2.0	35.7	2.0
	15	Cultivated	...	...	100	...	7.2	36.6	4.8
	16	Mixed (grass 48 %)	...	...	100	...	29.0	36.3	7.0
	16	Woods	...	...	100	...	43.6	36.4	12.1
	18	Mixed	...	...	100	...	303.0	37.4	14.3
	17	Mixed	...	...	100	...	920.0	37.0	12.2
	17	Mixed	...	...	100	...	1,520.0	37.0	11.8
	19	Mixed	...	...	100	...	4,580.0	37.4	12.8
	19	Mixed	...	...	100	...	17,500.0	37.4	13.4
Ohio, Zanesville...........	12	Cultivated	...	...	100	...	2.6	37.8	7.7
	12	Pasture	...	...	100	...	3.6	38.2	5.6
	12	Woods	...	...	100	...	2.2	37.4	1.1
Oklahoma, Cherokee......	14	Cultivated	...	100	...	...	4.8	24.1	3.7
	14	Cultivated	...	100	...	...	7.8	24.1	2.8
Oklahoma, Guthrie........	17	Grass and weeds	...	...	100	...	5.3	29.9	2.7
	17	Good grass	...	100	...	...	2.5	30.2	0.8
	12	Good grass	...	100	...	...	9.1	30.5	4.0
Oklahoma, Muskogee......	6	Cultivated	...	...	75	25	65.4	46.2	14.6
Texas, Spur...............	19	Cultivated	...	50	50	...	9.4	20.1	2.7
	16	Cult. (contoured)	...	50	50	...	8.4	20.7	0.4
Texas, Tyler..............	9	Woods	...	...	...	100	7.9	42.4	1.3
	9	Cultivated	...	...	...	100	1.6	41.6	8.0
Texas, Riesel.............	13	Grass	...	...	...	100	3.0	29.8	1.9
	13	Mixed	...	...	...	100	42.3	31.1	3.5
	18	Mixed	...	...	...	100	130.0	32.0	5.8
Virginia, Blacksburg.......	11	Cultivated	...	6	94	...	5.4	38.5	0.7
Virginia, Chatham........	9	Cultivated	...	100	...	...	13.3	40.9	3.3
	9	Cultivated	...	100	...	...	16.1	42.4	6.4
Virginia, Staunton........	6	Mixed	...	100	...	...	2,430.0	35.9	0.8
	6	Mixed	...	100	...	...	6,144.0	35.9	2.8
Wisconsin, Fennimore.....	17	Mixed	...	100	...	...	330.0	32.7	4.6
	17	Mixed	...	100	...	...	22.8	32.9	1.6
Wisconsin, LaCrosse.......	20	Mixed	...	100	...	...	2.7	32.2	1.9
Wisconsin, Coon Valley...	5	Mixed	10	68	‡	...	49,344.0	29.8	7.5

* From U.S. Agricultural Research Service [2].
† Period of record.
‡ Complete soil classification unknown.
§ Terraced.

Table 21-20. Monthly Distribution (Per Cent of Total) of the Average Annual Yield for Selected Watersheds*

Watershed location	Area, acres	Years of record	Jan.	Feb.	Mar.	Apr.	May	June	July	Aug.	Sept.	Oct.	Nov.	Dec.	Remarks
Arizona, Safford........	723	14	0	0	0	0	0	0	46	38	12	4	0	0	Range
Arkansas, Bentonville...	9	8	0	4	6	13	19	25	2	2	8	14	5	2	Pasture
California, Placerville..	41	8	26	32	27	7	0	0	0	0	0	0	2	6	Orchard
Colorado, Colorado Springs.	40	7	0	0	0	0	0	8	46	42	4	0	0	0	Plains, range
Georgia, Watkinsville...	19	16	16	10	15	12	3	3	9	11	1	2	10	8	Piedmont, cultivated
Idaho, Moscow..........	178	5	17	25	41	5	1	0	0	0	0	0	1	10	Plateau, prairie
Illinois, Edwardsville...	290	17	9	11	14	15	10	11	8	9	1	3	5	4	50% cultivated
Nebraska, Hastings.....	481	17	0	1	3	3	14	37	18	5	13	5	1	0	Plains, cultivated
Ohio, Coshocton........	44	16	12	12	22	18	14	7	2	1	2	1	3	6	Woods and brush
Oregon, Newburg.......	13	4	17	26	10	5	3	2	1	1	1	1	8	25	Orchard, 1,500' elevation
Puerto Rico, Mayagüez..	¹⁄₄₀†	9	2	0	0	0	4	5	11	21	26	15	12	4	Grass
Texas, Tyler...........	8	9	10	9	12	13	28	4	1	0	0	1	12	10	Wooded
Virginia, Staunton......	390	6	8	9	9	16	14	11	8	6	3	13	3	0	Mixed cover (cultivated 26%)
Wisconsin, Fennimore...	330	17	8	11	16	6	6	12	12	10	6	5	4	4	Mixed cover

* From Mockus [28].
† Experimental plot with surface runoff only. For watersheds with base flow the per cents will be more uniform.

Example 21-10. Assume that the annual precipitation for the problem watershed of Example 21-9 is 24.5 in., which is about 11 per cent greater than the precipitation at Hastings (22 in.). Determine the water yield.
1. Estimate the yield as shown in Example 21-9. This is 2.90 in.
2. Modify the estimate using a ratio of the precipitations: $2.90 \times 24.5/22.0 = 3.22$ in.

2. Supplementary Stations. A temporary gaging station can be established on the ungaged watershed for use in estimating long-term yields. Data from the temporary station are related to the data obtained at a nearby long-term station by methods of regression analysis and adjusted on that basis. If the watershed is very small, the gaging installation may be a precalibrated weir or Parshall flume (Sec 7 and [27]) to avoid the necessity of rating the station.

3. Determination of Frequency. In some cases, as for irrigation projects, a more dependable yield, smaller and occurring more frequently than the average annual yield, will be selected. If sufficient data are not available for a frequency analysis, Fig. 21-7 may be used to estimate more frequent yields [28].

Example 21-11. The average annual yield for a watershed is 6.5 in. Determine the expected 80 per cent chance yield.
Enter Fig. 21-7 with the given average annual yield on the left-side scale, follow the guide lines as shown by the dashed line, and at the 80 per cent chance read the vertical scale and find $Q = 3.2$ in.

4. Seasonal Variations. The distribution pattern of the volume of flow from small watersheds usually differs widely from that for nearby large streams and rivers. Available data [2] can be used to estimate the seasonal variation of small watersheds. Table 21-20 shows monthly distribution, in per cent, of annual flow for selected stations.

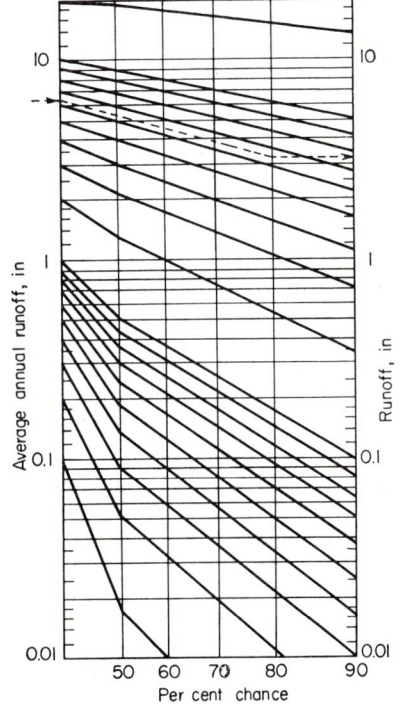

FIG. 21-7. Chart for estimating annual runoff volumes occurring more often than the average annual runoff. (*From Mockus* [28].)

V. DETERMINATION OF PEAK RATES OF RUNOFF

A. Runoff from Rainfall

Two methods are presented for approximating peak rates of runoff from small watersheds when a detailed study of the problem is not justified. More accurate estimates would be expected from methods based on the unit-hydrograph theory, which is discussed later.

1. The Rational Method. The basis for this method is the rational formula

$$q_p = Cia \qquad (21\text{-}13)$$

where q_p is the peak rate of flow in cfs for a given frequency of rainfall intensity, C is a constant ranging from 0 to 1 according to watershed conditions, i is the rainfall rate in in./hr for the given frequency, and a is the drainage area of the watershed in acres.

The rainfall data of the required frequencies can be obtained from published maps or charts [5, 6, 29]. Table 21-21 gives approximate values of C for selected soil and cover conditions. Watershed slope is not directly considered in the selection of C, but the effects of slope are included, since rainfall duration is taken as equal to the time of concentration.

Table 21-21. Values of C for Use in Rational Formula*

Soil type	Watershed cover		
	Cultivated	Pasture	Woodlands
With above-average infiltration rates; usually sandy or gravelly...	0.20	0.15	0.10
With average infiltration rates; no clay pans; loams and similar soils..................................	0.40	0.35	0.30
With below-average infiltration rates; heavy clay soils or soils with a clay pan near the surface; shallow soils above impervious rock....................	0.50	0.45	0.40

* Table modified from Bernard [30].

Example 21-12. A watershed near Coshocton, Ohio, has a drainage area of 74.2 acres, a length of main channel of 3,400 ft, a difference in elevation (watershed outlet to the most distant ridge) of 220 ft, and a mixed cover consisting of approximately 14 per cent woodland, 50 per cent pasture, and 36 per cent cultivated. The soils are about average. Determine the peak rate of runoff of a 10-year frequency.

Table 21-22. Incremental W Values for Use in Cook's Method*

Watershed characteristic	Extent or degree	W
Relief.............	Steep rugged terrain with average slopes generally above 30%	40
	Hilly, with average slopes 10 to 30%	30
	Rolling, with average slopes 5 to 10%	20
	Relatively flat land, slopes 0 to 5%	10
Infiltration (I)......	No effective cover; either rock or thin soil mantle of negligible infiltration capacity	20
	Slow to take up water; clay or other soil of low infiltration capacity	15
	Deep loams with infiltration about that of typical prairie soils	10
	Deep sand or other soil that takes up water readily and rapidly	5
Vegetal cover (C)..	No effective plant cover or equivalent	20
	Poor to fair cover; clean cultivated crops or poor natural cover; less than 10% of the watershed in good cover	15
	About 50% of watershed in good cover	10
	About 90% of watershed in good cover, such as grass, woodlands, or equivalent	5
Surface storage......	Negligible; few surface depressions	20
	Well-defined system of small drainage	15
	Considerable depression storage with not more than 2% in lakes, swamps, or ponds	10
	Surface-depression storage high; drainage system poorly defined; large number of lakes, swamps, or ponds	5

* From U.S. Soil Conservation Service [32].

DETERMINATION OF PEAK RATES OF RUNOFF

Table 21-23. Frequency Factors for Use with Cook's Method*

$(I + C)$†	Average annual precipitation, in.					
	10	20	30	40	60	80
Ratio: 25-year/50-year						
5	0.31	0.38	0.41	0.44	0.48	0.51
10	0.41	0.50	0.55	0.58	0.63	0.66
15	0.50	0.59	0.64	0.68	0.73	0.77
20	0.55	0.65	0.71	0.76	0.82	0.87
25	0.60	0.71	0.78	0.83	0.90	0.92
30	0.64	0.76	0.83	0.89	0.92	0.92
35	0.67	0.81	0.89	0.92	0.92	0.92
40	0.71	0.85	0.92	0.92	0.92	0.92
Ratio: 10-year/50-year						
5	0.05	0.08	0.10	0.12	0.15	0.17
10	0.10	0.16	0.21	0.24	0.30	0.34
15	0.16	0.25	0.31	0.37	0.45	0.51
20	0.21	0.33	0.42	0.49	0.60	0.68
25	0.26	0.41	0.52	0.61	0.75	0.80
30	0.31	0.49	0.62	0.74	0.80	0.80
35	0.36	0.58	0.73	0.80	0.80	0.80
40	0.42	0.66	0.80	0.80	0.80	0.80

* From U.S. Soil Conservation Service [33].
† From Table 21-22.

1. Determine the coefficient C. Using the values for soil and cover found in Table 21-21 and the per cent of each land use, obtain the weighted value of $C = 0.36$.

2. Determine the time of concentration of the watershed. Entering Fig. 21-4 with the channel length of 3,400 ft and the difference in elevation of 220 ft, the time of concentration is found to be 0.19 hr.

3. Determine the rainfall intensity of a 10-year frequency and duration of 0.19 hr. The rainfall for this location is 0.95 in. [5]. Thus the intensity is 5.00 in./hr.

4. Use Eq. (21-13) to find the 10-year-frequency peak rate of flow: $q_p = 0.36 \times 5.00 \times 74.2 = 134$ cfs.

2. The Cook Method. This is the method originally developed by H. L. Cook [31] and later modified by M. M. Culp and others [32]. The method uses an empirical relationship between drainage area and peak flow with modifications for climate, relief, infiltration, vegetal cover, and surface storage. The summation of applicable values of W given in Table 21-22 reflects the hydrologic condition of the watershed. Figure 21-8 shows the relationship between drainage area and peak flow for various hydrologic conditions. Values obtained from Fig. 21-8 are modified for climatic influence by the factors in Fig. 21-9, and values for other frequencies are obtained using the ratios shown in Table 21-23.

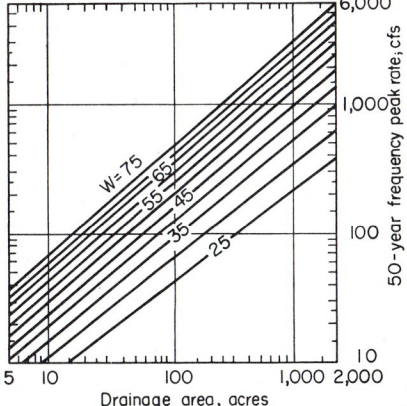

FIG. 21-8. Chart for estimating 50-year-frequency peak rates of flow. (*Modification from Hamilton and Jepson* [31, *fig.* 27].)

Example 21-13. Given the watershed of Example 21-12, with the additional information that the watershed is in hilly country (average slope of watershed, 16 per cent) and the average annual precipitation is 36 in. Determine the peak rate of flow of 10-year frequency.

1. Determine the coefficient W. Using Table 21-22 and interpolating when necessary, the component values of W are:

Relief—hilly	30
Infiltration—interpolating for average condition	12
Cover—64 per cent good; by interpolation	9
Surface storage—well drained	15
$W =$	66

2. Entering Fig. 21-8 with the watershed size of 74.2 acres at curve $W = 66$ (by interpolation), read a discharge of 285 cfs for a 50-year frequency rate.

FIG. 21-9. Distribution of rainfall factors used with the modified Cook's method. (*From Hamilton and Jepson* [31].)

3. Determine the rainfall factor from Fig. 21-9 for this location. For central Ohio, this is 0.75.

4. Determine the frequency factor from Table 22-23. Entering the table with the sum of W for the given conditions of infiltration and cover ($I + C = 12 + 9 = 21$) and the annual precipitation of 36 in., the 10-year frequency ratio is found by interpolation to be 0.48.

5. Compute the 10-year-frequency peak flow: $q_p = 285 \times 0.75 \times 0.48 = 103$ cfs.

a. Discussion. The watershed used in Examples 21-12 and 21-13 is W-183 at Coshocton, Ohio, for which topographic, precipitation, and runoff data are available [2, 4]. For the period of record, 1938 to 1956, the 10-year-frequency peak rate of flow is 105 cfs, determined using the log-normal distribution and the moment method. The period of record for watershed W-183 is too short to provide a firm estimate of the 10-year frequency. Therefore the similarity between the results obtained in the examples and those determined from observatons must be considered accidental. The rational and Cook methods cannot be expected to provide exact answers, since some factors that affect runoff are not considered. However, they are quite useful in making quick estimates on projects where detailed hydrologic studies are not justified.

B. Snowmelt Runoff

The rational and Cook methods are not suitable for estimating peak rates of flow caused by snowmelt. Generally, the largest annual peak rate of flow from a small watershed is caused by rainfall during the growing season. Snowmelt is important mainly in the consideration of the volume of flow. Daily volumes of melt may be determined by the method used in Example 21-8. Methods for developing a hydrograph and determining the peak rate of flow will be discussed later.

C. Regional Analysis

The analysis of hydrologic data from relatively homogeneous areas to obtain general relationships between various hydrologic factors for the region covering the areas is sometimes referred to as *regional analysis*. The shortage of small-watershed data makes this type of analysis difficult to apply. Records of large streams and rivers vary so much that their use in a regional analysis of small watersheds may lead to erroneous conclusions.

Where data are available to develop flood-frequency curves for small watersheds on a regional basis, the peaks for selected frequencies (usually 2-, 10-, and 100-year) can be plotted versus A/T_p, where A is the drainage area and T_p is the time from the beginning of rise to the peak rate of flow. The resulting plotting will show a rather consistent relationship if the soil-cover complexes of the watershed are similar. Where T_p cannot be determined, it is often possible to use the ratio A/T_c in lieu of A/T_p. In some homogeneous areas where T_c is a simple function of area, the peak rates will vary directly with some power of A, usually about 0.5. The U.S. Geological Survey publishes circulars (on a state basis) which give the results of regional analysis of larger watersheds in specific areas. Some of these analyses are also applicable to small watersheds (see also Sec. 25-I).

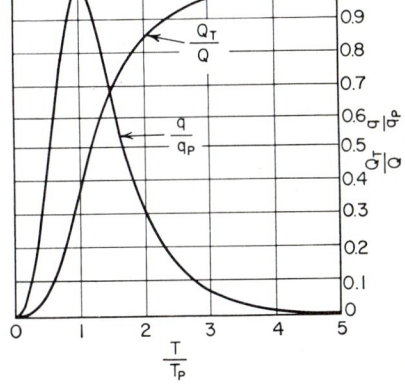

FIG. 21-10. Dimensionless unit hydrograph and mass curve. Q = total direct runoff, Q_T = accumulated direct runoff at time T, q = rate of flow, q_p = peak rate of flow, T = time from beginning of hydrograph rise, and $T_p = T$ at peak rate. (*From Mockus* [34].)

VI. HYDROGRAPHS

The unit-hydrograph procedure can be applied to data from small watersheds. However, the relatively small time units required (sometimes as small as 2 or 3 min) make the derivation of unit hydrographs a difficult task, and as a result, a synthetic hydrograph approach is more often used. See also Sec. 14.

A. Synthetic Hydrographs

A synthetic hydrograph is prepared using the data from a number of watersheds to develop a dimensionless unit hydrograph applicable to ungaged watersheds [19].

1. Dimensionless Hydrographs. The curvilinear hydrograph shown in Fig. 21-10 is a dimensionless unit hydrograph prepared from the unit hydrographs of a variety of watersheds [34]. The recession limb plots as a straight line on semilogarithmic paper. Various relationships between the hydrograph components can be determined, the most useful of which is the equation for the peak rate of flow, derived as shown on Fig. 21-11:

$$q_p = \frac{484AQ}{0.5D + 0.6T_c} \qquad (21\text{-}14)$$

where q_p is the peak rate of flow in cfs, 484 is the constant K obtained as shown on Fig. 21-11, A is the drainage area in sq mi, Q is the volume of runoff in in., D is the

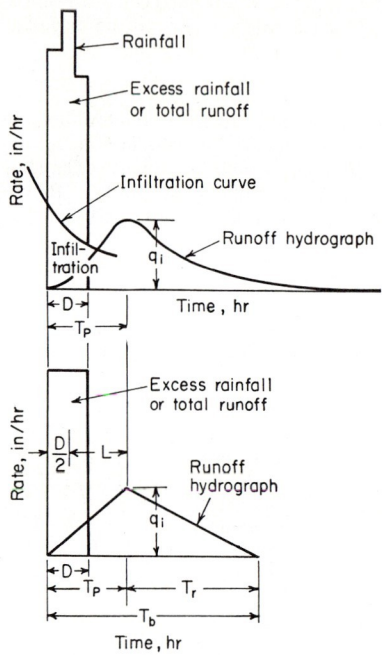

FIG. 21-11. Development of a triangular unit hydrograph. *Symbols:* Q = total runoff in in., q_i = peak rate in in./hr, T_p = time in hr from start of rise to peak rate, T_r = time in hr from peak rate to end of triangle, T_b = time base of hydrograph, D = rainfall excess period in hr, L = lag, time from center of excess rainfall to time of peak, in hr.

Development of peak-rate equation: From the triangle,

$$Q = \frac{q_i T_p}{2} + \frac{q_i T_r}{2}$$

and

$$q_i = \frac{2Q}{T_p + T_r}$$

If

$$T_r = HT_p$$

then

$$q_i = \frac{2Q}{(1+H)T_p}$$

Converting from inches per hour to cubic feet per second by introducing drainage area A in square miles (1 in./hr = 645.3 cfs/sq mi) gives the peak rate q_p:

$$q_p = \frac{KAQ}{T_p} \quad \text{cfs}$$

where $K = 2(645.3)/(1 + H)$ and $T_p = (D/2) + L$. When an empirical value of $H = 1.67$ is used, $K = 484$. (*From Mockus* [34]).

HYDROGRAPHS

duration in hr of the excessive rainfall period of the storm, and $0.6T_c$ is an empirical estimate of lag, based on the time of concentration T_c in hr.

The denominator of Eq. (21-14) is the time from the beginning of the rise of the hydrograph to the peak rate T_p, since it is the sum of one-half the duration of excessive rainfall, $0.5D$, and the lag, $0.6T_c$, as shown on Fig. 21-11.

A study of unit hydrographs of more than 500 large and small watersheds [33] indicates that the expected unit-hydrograph storm duration is approximated by

$$D_e = 2\sqrt{T_c} \tag{21-15}$$

where D_e is the unit-hydrograph storm duration in hr, and T_c is the time of concentration in hr. Therefore, in the development of unit hydrographs, the term $2\sqrt{T_c}$ may be used for D in Eq. (21-14) and written as

$$q_u = \frac{484AQ}{\sqrt{T_c} + 0.6T_c} \tag{21-16}$$

where q_u is the peak rate of flow of the unit hydrograph in cfs. Equation (21-14) will be found useful in constructing hydrographs when D is known or assumed, while Eq. (21-16) will be useful in constructing unit hydrographs when D is unknown. It should be noted that Eqs. (21-14) and (21-16) are empirical variations of the basic equation for the peak rate of a hydrograph [34], which is

$$q_p = \frac{KAQ}{T_p} \tag{21-17}$$

where K is a constant dependent on the shape of the hydrograph.

a. Construction of Simple Hydrographs. Single-peaked hydrographs can be constructed when any three of the four variables in Eq. (21-17) are known.

Example 21-14. A watershed with a drainage area of 1.2 sq mi has a peak rate of flow of 386 cfs and a volume of runoff of 1.4 in. Construct a single-peaked hydrograph using Fig. 21-10 and Eq. (21-17) with $K = 484$.

1. Compute the time to peak T_p. From Eq. (21-17), with $K = 484$, $T_p = 484AQ/q_p = 484 \times 1.2 \times 1.4/386 = 2.11$ hr.
2. Prepare a working table as shown in Table 21-24, listing selected time ratios T/T_p in column 1 and the corresponding ratios of q/q_p from Fig. 21-10 in column 2. T is the time measured from the beginning of the hydrograph rise.
3. Compute the values for column 3 by multiplying the ratios in column 1 by the T_p of 2.11 hr.
4. Compute the values for column 4 by multiplying the ratios in column 2 by the q_p of 386 cfs.
5. Plot the values in column 3 vs. the values in column 4 to obtain the hydrograph.

b. Unit Hydrographs. The unit-hydrograph principle, that the peak rate varies directly with the volume of runoff for a unit storm duration, can be used to convert any hydrograph of unit duration to a unit hydrograph. The hydrograph of Example 21-15 could be converted to a unit hydrograph. Since the peak rate of 386 cfs resulted from a runoff volume of 1.4 in., the peak rate for a unit hydrograph would be $386/1.4 = 276$ cfs. The unit hydrograph is developed by using the unit peak in Example 21-14, step 4.

Equation (21-14) can also be used to convert a given hydrograph into a hydrograph of another duration.

Example 21-15. A watershed with a drainage area of 10.2 sq mi has a 1-hr unit hydrograph with a peak rate of 1,928 cfs. Determine the time to peak T_p and the peak rate of flow q_p for a 3-hr hydrograph.

1. Compute T_p for the 1-hr hydrograph. Using Eq. (21-17) with $K = 484$, $T_p = 484 \times 10.2 \times 1/1,928 = 2.56$ hr.

2. Compute T_p for the 3-hr hydrograph. Since the denominator of Eq. (21-14) is the time to peak, $T_p = 0.5D + 0.5T_c$. If D is increased in this relationship, T_p will be increased by one-half of this amount. Therefore T_p for the 3-hr hydrograph is $2.56 + 0.5(3 - 1) = 3.56$ hr.

3. Compute q_p for the 3-hr hydrograph. When the area A and the volume Q are unchanged, $q_{p1}T_{p1} = q_{p2}T_{p2}$ and the peak rate q_p for the 3-hr unit hydrograph is $1{,}928 \times 2.56/3.56 = 1{,}386$ cfs. The entire hydrograph may be constructed by the method used in Example 21-14.

Table 21-24. Computations for a Simple Hydrograph*

(1) Time ratio T/T_p	(2) Discharge ratio q/q_p	(3) Time,† hr	(4) Discharge,‡ cfs
0	0	0	0
0.1	0.015	0.21	6
0.2	0.075	0.42	29
0.3	0.16	0.63	62
0.4	0.28	0.84	108
0.5	0.43	1.05	166
0.6	0.60	1.26	231
0.7	0.77	1.48	297
0.8	0.89	1.69	343
0.9	0.97	1.90	374
1.0	1.00	2.11	386
1.1	0.98	2.32	378
1.2	0.92	2.53	355
1.3	0.84	2.74	324
1.4	0.75	2.95	289
1.5	0.66	3.16	255
1.6	0.56	3.37	216
1.8	0.42	3.80	162
2.0	0.32	4.22	124
2.2	0.24	4.63	93
2.4	0.18	5.06	70
2.6	0.13	5.48	50
2.8	0.098	5.91	38
3.0	0.075	6.33	29
3.5	0.036	7.38	14
4.0	0.018	8.44	7
4.5	0.009	9.49	3.5
5.0	0.004	10.55	1.5
Infinity	0	.....	0

* From U.S. Soil Conservation Service [19].
† T_p of 2.11 hr times column 1.
‡ q_p of 386 cfs times column 2.

2. Triangular Hydrographs. The curvilinear hydrograph of Fig. 21-10 can often be replaced by an equivalent triangular hydrograph which is more easily constructed and, for routing through reservoirs or stream channels, gives results about as accurate as those obtained using the curvilinear hydrograph. Where the tail of the hydrograph is likely to affect design, the curvilinear hydrograph should be used. Figure 21-11 shows a triangular hydrograph, based on the curvilinear hydrograph of Fig. 21-10, where

$$T_b = T_p + T_r \qquad (21\text{-}18)$$

HYDROGRAPHS

Since $T_r = 1.67 T_p$, by construction, then

$$T_b = 2.67 T_p \tag{21-19}$$

Example 21-16. Compute the peak rate and plot a triangular unit hydrograph for a watershed of 0.87 sq mi with $T_c = 0.20$ hr and D equal to T_c.

1. Compute T_p. If T_p is represented by the denominator of Eq. (21-14) and if the unit duration is taken equal to the time of concentration, then $T_p = 1.1 T_c$, or $T_p = 1.1 \times 0.20 = 0.22$ hr.
2. Compute T_b. Using Eq. (21-19), $T_b = 2.67 \times 0.22 = 0.587$ hr.
3. Compute the peak rate q_p. Using Eq. (21-17) with $K = 484$ and $Q = 1$, $q_p = 484 \times 0.87 \times 1/0.22 = 1,914$ cfs. Note that this rather high peak is due to the occurrence of 1 in. of runoff in the relatively short time of 0.20 hr.
4. Plot the peak rate (1,914 cfs) at time T_p (0.22 hr) and the end of flow ($q = 0$) at time T_b (0.587 hr). Starting with the beginning of rise ($T = 0$, $q = 0$), draw straight lines between the plotted points.

3. Composite Hydrographs. In type I studies, or for other purposes such as spillway design, the hydrograph for a long-duration storm may be required. This type of hydrograph can be constructed from triangular hydrographs without significant loss of accuracy [19].

Table 21-25. Computation of Subhydrograph Peak Rates

Time, hr	Accumulated P, in.	Accumulated Q,* in.	ΔQ, in.	900(ΔQ), cfs
0	0	0		
0.5	0.43	0	0	0
1.0	0.93	0.09	0.09	81
1.5	1.60	0.40	0.31	279
2.0	2.79	1.21	0.81	729
2.5	7.27	5.17	3.96	3564
3.0	8.60	6.43	1.26	1134
3.5	9.47	7.27	0.84	756
4.0	10.18	7.95	0.68	612
4.5	10.79	8.54	0.59	531
5.0	11.30	9.04	0.50	450
5.5	11.80	9.52	0.48	432
6.0	12.20	9.92	0.40	360

* Estimated using CN = 82 on Fig. 21-6.

Example 21-17. Given the 6-hr spillway design storm in columns 1 and 2 of Table 21-25, a drainage area A of 1.86 sq mi, a time of concentration T_c of 1.25 hr, and a watershed runoff curve number of 82. Construct a composite design hydrograph for the watershed.

1. Choose a uniform increment of the duration. In this case the incremental time interval ΔD is taken as 0.5 hr, which is smaller than T_c.
2. Tabulate the accumulated time and rainfall as shown in columns 1 and 2 of Table 21-25.
3. Enter Fig. 21-6 with the rainfall listed in column 2, and, for CN = 82, read and tabulate the accumulated runoff in column 3.
4. Compute and tabulate the incremental runoff in column 4.
5. Compute the peak rate of flow for a storm of 0.5-hr duration with $Q = 1$ in. Using Eq. (21-14), $q_p = 900$ cfs.
6. Compute and tabulate in column 5 the peak rates for the subhydrographs, using the values of Q in column 4 and the peak rate of 900 cfs determined in step 5.
7. Compute T_p and T_b for the subhydrographs, using a ΔD of 0.5 hr. Since the denominator of Eq. (21-14) is T_p, the value of $T_p = 1$ is obtained from the denominator of the equation in step 5. Using Eq. (21-19), $T_b = 2.67$ hr.

8. Plot the subhydrographs as shown on Fig. 21-12. Each subhydrograph starts, peaks, and ends ΔD (= 0.5 hr) later than the preceding one. The peak rates of flow are obtained from column 5 of Table 21-25.

9. Add the ordinates of the subhydrographs, and plot the composite hydrograph as shown in Fig. 21-12. Note there is no subhydrograph for the first time increment where $Q = 0$.

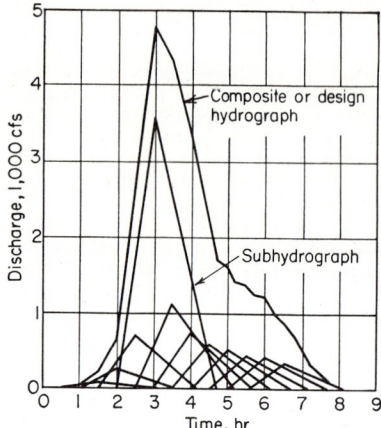

FIG. 21-12. Use of subhydrographs in construction of a composite hydrograph for a design storm. (*U.S. Soil Conservation Service* [19].)

Although it involves more work, the curvilinear hydrograph of Fig. 21-10 can also be used to make a composite hydrograph. The procedure used in Example 21-17 can also be applied for storms, whatever their time or areal distribution of precipitation. Where a watershed is large and runoff is not uniform over the area, hydrographs for the subwatersheds can be made, lagged according to their location in the stream system, and plotted, and consequently a composite hydrograph can be developed.

VII. FIELD APPLICATIONS

A. Terraces and Diversions

1. Construction Types. A terrace is a ridge or channel (or both) constructed in earth to hold back surface runoff or to convey it safely off a field. Terraces are constructed to interfere as little as possible with normal farming operations. Where topography permits, they are made parallel to reduce tillage problems.

Graded terraces intercept surface runoff from farm fields and convey it at nonerosive velocities to a suitable outlet, such as a natural grassed waterway. A waterholding or absorptive-type terrace has no grade and is designed to store surface runoff. This type is used only where soils are highly permeable, so that impounded water seeps into the soil before damaging the crop.

Other terracelike structures, such as contour furrows, ridges, or trenches, are smaller in cross section than terraces and are spaced close enough to provide storage for expected runoff on steep slopes. They are normally used on pasture or rangeland to conserve water or in mountains for erosion control.

Diversions are similar to graded terraces, but are usually larger and have steeper side slopes. They are used primarily to intercept runoff above cultivated fields and convey it to safe outlets or to divert water to water-spreading systems on rangeland.

2. Criteria. Standard design criteria for terraces, developed for local conditions, are available in local offices of the U.S. Soil Conservation Service and Extension Service and at state agricultural colleges. Terraces are usually designed to handle runoff from a 10-year-frequency storm. The design of a terrace consists primarily of determining peak rates of flow at critical locations in the terrace system and selecting cross sections which will carry the design flow and permit normal farm operations. The capacities of upper sections of a terrace could be less than the downstream, but for ease of construction and farm operations a nearly uniform cross section is normally used. The spacing, or lateral distance between terraces, depends on the climate and

Table 21-26. Dimensions for Level and Graded Terraces*

Design depth d, ft	Dimension†	Land slope, per cent											
		1	2	3	4	5	6	7	8	10	12	14	16
1.0	w	20.0	20.0	20.0	20.0	20.0	20.0	20.0	20.0	20.0	20.0	20.0	20.0
	a	5.45	5.83	6.15	6.43	6.67	6.88	7.06	7.22	7.50	7.73	7.92	8.08
	j	0.54	0.58	0.62	0.64	0.67	0.69	0.71	0.72	0.75	0.77	0.79	0.81
	b	9.09	8.33	7.69	7.14	6.67	6.25	5.88	5.56	5.00	4.54	4.17	3.85
	x	10.1	10.8	11.4	11.9	12.3	12.7	13.1	13.4	13.9	14.3	14.7	15.0
1.5	w	30.0	30.0	30.0	30.0	30.0	30.0	30.0	30.0	30.0	30.0	30.0	30.0
	a	8.20	8.75	9.23	9.64	10.0	10.3	10.6	10.8	11.2	11.6	11.9	12.1
	j	0.82	0.88	0.92	0.96	1.00	1.03	1.06	1.08	1.12	1.16	1.19	1.21
	b	13.6	12.5	11.5	10.7	10.0	9.38	8.82	8.33	7.50	6.82	6.25	5.77
	x	22.7	24.3	25.6	26.8	27.8	28.6	29.4	30.1	31.2	32.2	33.0	33.7
2.0	w	40.0	40.0	40.0	40.0	40.0	40.0	40.0	40.0	40.0	40.0	40.0	40.0
	a	10.9	11.7	12.3	12.9	13.3	13.8	14.1	14.4	15.0	15.5	15.8	16.2
	j	1.09	1.17	1.23	1.29	1.33	1.38	1.41	1.44	1.50	1.55	1.58	1.62
	b	18.2	16.7	15.4	14.3	13.3	12.5	11.8	11.1	10.0	9.09	8.33	7.69
	x	40.4	43.2	45.6	47.6	49.4	50.9	52.3	53.5	55.6	57.2	58.6	59.8
2.5	w	50.0	50.0	50.0	50.0	50.0	50.0	50.0	50.0	50.0	50.0	50.0	50.0
	a	13.6	14.6	15.4	16.1	16.7	17.2	17.6	18.1	18.8	19.3	19.8	20.2
	j	1.36	1.46	1.54	1.61	1.67	1.72	1.76	1.81	1.88	1.93	1.98	2.02
	b	22.7	20.8	19.2	17.9	16.7	15.6	14.7	13.9	12.5	11.4	10.4	9.62
	x	63.1	67.5	71.2	74.4	77.2	79.6	81.7	83.6	86.8	89.4	91.6	93.5
3.0	w	60.0	60.0	60.0	60.0	60.0	60.0	60.0	60.0	60.0	60.0	60.0	60.0
	a	16.4	17.5	18.5	19.3	20.0	20.6	21.2	21.7	22.5	23.2	23.8	24.2
	j	1.64	1.75	1.85	1.93	2.00	2.06	2.12	2.17	2.25	2.32	2.38	2.42
	b	27.3	25.0	23.1	21.4	20.0	18.8	17.6	16.7	15.0	13.6	12.5	11.5
	x	90.9	97.2	102.6	107.1	111.1	114.6	117.6	120.4	125.0	128.8	131.9	134.6

* From U.S. Soil Conservation Service [33].
† Dimensions w, a, j, and b are in feet and are shown on Fig. 21-13. Dimension x is the cubic yards of excavation per 100 lineal ft of terrace.

on the erodibility of the soil; it usually ranges from 35 to 250 ft. A general spacing equation [35] is

$$L = \frac{200}{S} + 50 \qquad (21\text{-}20)$$

where L is the horizontal distance between terraces in ft, and S is the land slope in per cent.

In the southeastern United States, where rainfall is greater, terraces need to be about 25 ft closer together [36]; the correct distance can be obtained by using 25 instead of 50 as the constant in Eq. (21-20).

Typical cross sections for one type of terrace are shown in Fig. 21-13. Dimensions and excavation quantities are given in Table 21-26, and design factors are given in Tables 21-27 and 21-28. Other standard cross sections and tables can be developed for any locality, using methods described in this section.

Design criteria for closed-end level terraces have been developed in states having large areas of highly permeable soils. The cross sections and dimensions shown in

Fig. 21-13 and Table 21-26 can also be used for level terraces, and Table 21-29 gives the corresponding amounts of storage in terms of inches of runoff from the terraced area. Contour furrows or trenches are similar to closed-end level terraces in purpose and design [37], but their dimensions and cross sections depend partly on the limitations in the use of construction equipment on steep mountain slopes.

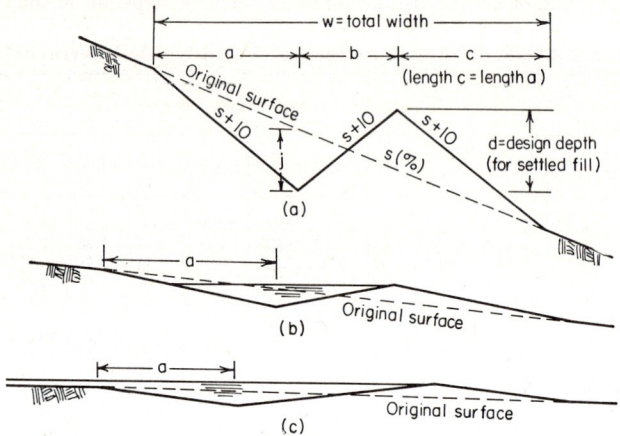

Fig. 21-13. (a) Definition sketch of a terrace cross section; vertical scale exaggerated. Sketches to scale showing (b) terrace on steep slope with flow area confined to excavated area, land slope = 10 per cent, design depth d = 3 ft; and (c) terrace on flat slope with flow area exceeding excavated area, land slope = 2 per cent, design depth d = 3 ft. Both sections are shown before aging.

Diversions also can be designed according to standard criteria [38], but the greater variation in conditions under which they are used generally makes it advisable to design each one individually.

3. Hydraulic Considerations. Detailed hydraulic studies are not necessary for closed-end level terraces or contour furrows and trenches, but it is important that

Table 21-27. Values of K in the Formula $q = Kar^{2/3}$ for Various Values of Channel Slope and Roughness Coefficient n*

Channel slope		Manning's n						
%	ft/ft	0.02	0.03	0.04	0.05	0.06	0.07	0.08
0.1	0.001	2.3	1.6	1.2	0.94	0.78	0.67	0.59
0.2	0.002	3.3	2.2	1.6	1.3	1.1	0.94	0.85
0.3	0.003	4.1	2.7	2.0	1.6	1.4	1.2	1.0
0.4	0.004	4.7	3.2	2.3	1.9	1.6	1.3	1.2
0.5	0.005	5.2	3.5	2.6	2.1	1.8	1.5	1.3
0.6	0.006	5.7	3.7	2.8	2.3	1.9	1.6	1.4
0.7	0.007	6.2	4.2	3.1	2.5	2.1	1.8	1.6
0.8	0.008	6.6	4.5	3.3	2.7	2.2	1.9	1.7
0.9	0.009	7.1	4.7	3.5	2.8	2.4	2.0	1.8
1.0	0.010	7.4	5.0	3.7	3.0	2.5	2.1	1.9

* From U.S. Soil Conservation Service [33].

NOTE: In the formula $q = Kar^{2/3}$, q is the discharge in cfs; $K = 1.49\sqrt{s}/n$, where s is the channel slope in ft/ft and n is the Manning roughness coefficient; a is the channel cross-sectional area in sq ft; and r is the hydraulic radius in ft.

Table 21-28. Velocities in Terrace Channels*

Design depth d, ft	Values of K from Table 21-27								
	0.6	0.8	1.0	1.5	2.0	3.0	4.0	5.0	6.0	7.0
	Average velocity, fps								
1.0	0.4	0.5	0.6	0.9	1.3	1.9	2.5	3.2	3.8	4.4
1.5	0.5	0.7	0.8	1.2	1.6	2.5	3.3	4.1	5.0	5.8
2.0	0.6	0.8	1.0	1.5	2.0	3.0	4.0	5.0	6.0	7.0
2.5	0.7	0.9	1.2	1.7	2.3	3.5	4.6	5.8	7.0	8.1
3.0	0.8	1.0	1.3	2.0	2.6	3.9	5.2	6.6	7.9	9.2

* From U.S. Soil Conservation Service [33].

nonerosive velocities be maintained in graded terraces and diversions. Diversion design may require the development of detailed water-surface profiles. Smaller diversions can usually be designed from tables such as Table 21-30.

4. Hydrologic Considerations. *a. Level Terraces.* The hydrologic design of level terraces usually will involve a quick check to ascertain that the proposed terrace system will contain the volume of runoff of the specified frequency. Table 21-29 can be used to determine the storage in terms of inches of runoff for various slope and ridge heights based on the cross sections shown in Fig. 21-13.

Table 21-29. Storage Above Closed-end Level Terraces, in Inches Depth of Runoff from the Area Between Terraces*

Average land slope, %	Average distance between terraces,† ft	Maximum depth of storage, or design depth d (Fig. 21-13), for settled fill without freeboard, ft				
		1	1.5	2	2.5	3
		Storage in inches				
1	250	0.79	1.77	3.14	4.91	7.07
2	150	0.82	1.84	3.27	5.10	7.35
3	117	0.84	1.89	3.37	5.26	7.57
4	100	0.87	1.95	3.47	5.42	7.81
5	90	0.89	2.00	3.56	5.56	8.00
6	83	0.90	2.03	3.61	5.65	8.13
7	79	0.89	2.01	3.57	5.58	8.04
8	75	0.89	2.00	3.56	5.56	8.00
10	70	0.86	1.93	3.43	5.36	7.71
12	67	0.81	1.83	3.26	5.09	7.33
14	64	0.78	1.76	3.13	4.88	7.03
16	62	0.74	1.67	2.98	4.65	6.70

* From U.S. Soil Conservation Service [33].
† Computed using Eq. (21-20). With smaller average distances the storage is greater. For example, with a 2 per cent land slope and a 1.5-ft design depth d, the storage for a terrace spacing of 115 ft is $1.84 \times 159/115 = 2.40$ in

21-50 HYDROLOGY OF AGRICULTURAL LANDS

Example 21-18. A system of closed-end level terraces is proposed for a field with a fairly uniform slope of 3 per cent and soils having a fairly high infiltration rate (the soil is estimated to be midway between hydrologic soil groups A and B). A good rotation is planned, with the poorest condition from the hydrologic standpoint occurring when the field is in *contoured row crop*. Determine the height of terrace ridge (d) required to contain the runoff from a 25-year-frequency storm of 24-hr duration.

1. Determine the runoff curve number (CN) for "Row crop, contoured, good condition," using Table 21-12. By interpolation between values for soil groups A and B, the value is found to be 70.

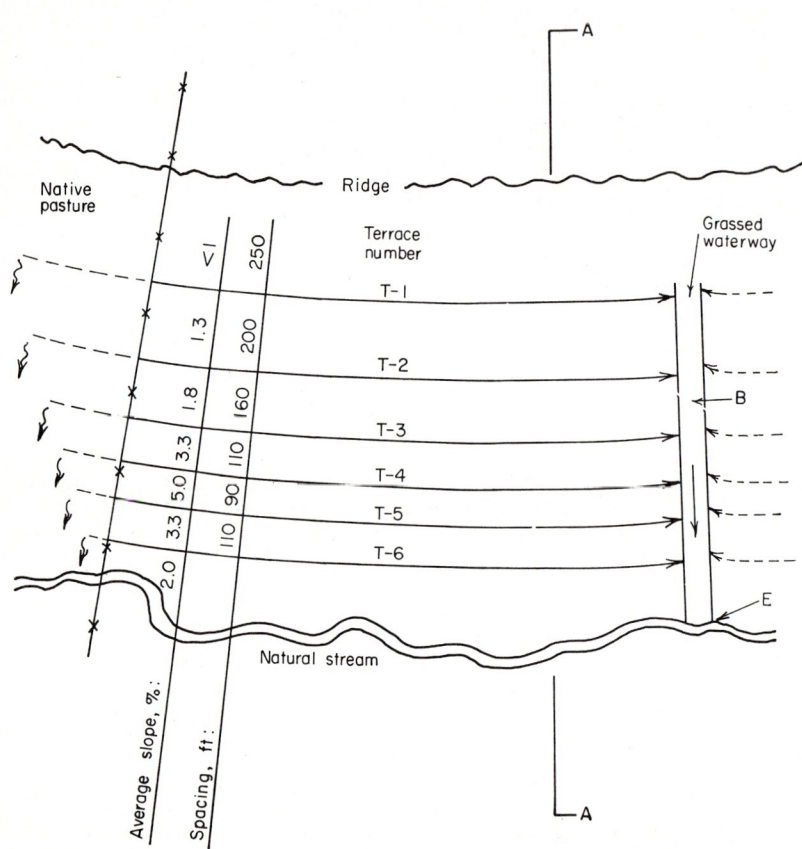

FIG. 21-14. Typical graded-terrace layout with alternative possibilities for water disposal by use of available grassed area (native pasture) or by use of grassed waterway into which one or two terrace layouts may discharge flows.

2. Determine the rainfall for a 25-year-frequency storm of 24-hr duration for this location. From available maps [5] this is found to be 4.88 in.

3. Estimate the surface runoff. Enter Fig. 21-6 with $P = 4.88$ in. and CN = 70 and read $Q = 1.95$ in.

4. Estimate the total runoff. Assuming that the ponding surface during the rain will be approximately one-fourth of the area between the terraces and that this area will produce 100 per cent runoff, the total runoff for the area would be $\frac{3}{4} \times 1.95 + \frac{1}{4} \times 4.88 = 2.68$ in.

5. Determine the required ridge height. Enter Table 21-29 with a land slope of 3 per cent and runoff of 2.68 in., and find that a ridge height of nearly 2 ft is required. More

precisely, a required height of 1.8 ft is determined by logarithmic interpolation. A freeboard of about 0.5 ft would normally be added.

b. Graded Terraces. When standard dimensions such as those of Table 21-26 are used, the hydrologic design will be based on determination of the peak rates of flow at critical points along the proposed terrace system and the choice of a cross section that will carry the maximum flow. Flood routing ordinarily will not be necessary.

Example 21-19. Given the cultivated field shown in Fig. 21-14, design a graded-terrace system to control a 10-year-frequency peak rate of runoff.

1. Make a field reconnaissance to determine land slopes and to select tentative locations of the terraces and outlet. A sketch map or aerial photograph of the field will be sufficient for making a tentative layout and for determining drainage areas. Land slopes are measured with a hand level, and distances are determined from the aerial photo or by rough chaining or pacing. The terrace outlet is located in a natural depression to save construction costs. When a natural outlet is not available, the outlet is located near a field boundary for convenience in farming operations.

2. Determine the horizontal spacing of the terraces. Where terracing has been used in an area, local experience should be considered in determining the horizontal spacing. In this case, spacing was determined by Eq. (21-20) and is shown in Fig. 21-14. Note that it is possible to use parallel terraces because of the uniformity of the slope of the field. In any case, terraces should be kept as nearly parallel as possible, even if it is necessary to change terrace grades or to make some cuts and fills [36].

3. Determine the drainage area of each terrace. For this example they are:

Terrace	T-1	T-2	T-3	T-4	T-5	T-6
Drainage area, acres	9.2	7.4	6.0	4.2	3.5	4.4

4. Determine the design flow at the terrace outlets. Using the method illustrated in Example 21-13, the design flow (10-year-frequency peak rate) for the poorest hydrologic condition is 14 cfs at the end of terrace T-1. The peak rates for the remaining terraces will be proportional to the drainage areas; for this given range of areas (Fig. 21-8) the design rates are:

Terrace	T-1	T-2	T-3	T-4	T-5	T-6
Design peak, cfs	14.0	11.3	9.1	6.4	5.3	6.7

5. Estimate the probable range in Manning's n and the maximum nonerosive velocity. The channel velocity will be greatest when the field is relatively bare. Under these conditions, Manning's n for the terrace channel would probably be about 0.03. During the preharvest period for grain crops, Manning's n would probably increase to about 0.06. The maximum nonerosive velocity for the soil type is estimated to be about 2.5 fps.

6. Determine the design depth (d in Fig. 21-13). The size of the channel is determined by the capacity required when Manning's n reaches its maximum. Use the terrace with the largest ratio of design discharge to land slope (q/s), and determine one design depth for the entire system. This will usually be the top terrace, as it is in this case. Entering Table 21-27 with the maximum value of n (= 0.06), select a terrace grade and determine the value of $K = 1.49 \sqrt{s}/n$. Assuming a grade of 0.5 per cent for the first try, K is found to be 1.8. Next, determine the ratio q/K, which for terrace T-1 is $14.0/1.8 = 7.8$. Enter Fig. 21-15 with this ratio and the land slope (1 per cent) for T-1, and read a design depth d of just over 1 ft (use $d = 1$ ft).

7. Determine if the permissible maximum velocity (2.5 fps) will be exceeded with a design depth of 1 ft. Determine the value of K for the minimum expected value of Manning's n (= 0.03) and the terrace grade of 0.05 per cent. Using Table 21-27, find $K = 3.5$. Enter Table 21-28 with $d = 1$ and $K = 3.5$, and find the velocity, by interpolation, to be

Table 21-30. Discharge Factors, Flow Areas, and Flow Depths for Diversions with Side Slope $z = 1.5$ and 4*

(q = discharge, cfs; n = Manning's n; s = channel slope, ft/ft; a = flow area, ft²; D = design flow depth, ft; b = bottom width, ft)

qn	$s = 0.001$				$s = 0.003$				$s = 0.005$			
	a†	D in ft for $b =$			a†	D in ft for $b =$			a†	D in ft for $b =$		
		0	6	12		0	6	12		0	6	12
Side slope $z = 1.5$												
0.1	3.02	1.4	...	...	2.00	1.2	...	...	1.65	1.1		
0.3	6.90	2.1	1.0	...	4.57	1.7	...	...	3.78	1.6		
0.5	10.1	2.6	1.3	...	6.70	2.1	0.9	...	5.53	1.9		
1	17.0	3.4	1.9	...	11.3	2.7	1.4	...	9.30	2.5	1.2	
1.5	23.0	3.9	2.4	...	15.3	3.2	1.8	...	12.6	2.9	1.5	
2	28.6	4.4	2.8	2.0	19.0	3.6	2.1	...	15.6	3.2	1.8	
3	38.8	5.1	3.5	2.6	25.7	4.1	2.6	1.8	21.2	3.8	2.3	
4	48.1	5.7	4.0	3.0	31.9	4.6	3.0	2.1	26.3	4.2	2.7	1.8
5	56.8	6.2	4.5	3.3	37.7	5.0	3.4	2.4	31.1	4.6	3.0	2.1
6	65.3	6.6	4.9	3.7	43.2	5.4	3.7	2.7	35.7	4.9	3.3	2.4
8	81.0	7.4	5.6	4.4	53.6	6.0	4.3	3.2	44.3	5.4	3.8	2.8
10	95.8	8.0	6.2	4.9	63.4	6.5	4.8	3.6	52.4	5.9	4.3	3.1
12	110	8.6	6.8	5.5	72.7	7.0	5.2	4.0	59.9	6.3	4.6	3.5
14	123	9.1	7.3	5.9	81.7	7.4	5.6	4.4	67.3	6.7	5.0	3.8
16	136	9.5	7.7	6.3	90.3	7.8	6.0	4.8	74.3	7.0	5.3	4.1
18	149	10.0	8.2	6.8	98.7	8.1	6.3	5.1	81.3	7.4	5.6	4.4
20	161	10.4	8.5	7.1	106	8.4	6.7	5.3	88.0	7.7	5.9	4.7
Side slope $z = 4$												
0.1	3.58	0.9	...	...	2.37	0.8	...	...	1.96	0.7		
0.3	8.13	1.4	...	...	5.38	1.2	...	...	4.46	1.1		
0.5	12.0	1.7	1.2	...	7.93	1.4	...	...	6.55	1.3		
1	20.1	2.2	1.6	...	13.3	1.8	1.2	...	11.0	1.7	1.1	
1.5	27.2	2.6	2.0	...	18.1	2.1	1.5	...	14.9	1.9	1.3	
2	33.8	2.9	2.3	1.8	22.4	2.4	1.7	...	18.5	2.2	1.5	
3	45.8	3.4	2.7	2.2	30.4	2.8	2.1	...	25.1	2.5	1.9	
4	56.9	3.8	3.1	2.6	37.7	3.1	2.4	1.9	31.1	2.8	2.2	
5	67.2	4.1	3.4	2.9	44.6	3.3	2.7	2.2	36.8	3.0	2.4	1.9
6	77.1	4.4	3.7	3.2	51.0	3.6	2.9	2.4	42.1	3.2	2.6	2.1
8	95.6	4.9	4.2	3.7	63.3	4.0	3.3	2.8	52.1	3.6	3.0	2.5
10	113	5.3	4.6	4.1	74.9	4.3	3.7	3.1	61.8	3.9	3.3	2.7
12	130	5.7	5.0	4.4	85.9	4.6	4.0	3.4	70.9	4.2	3.5	3.0
14	146	6.0	5.3	4.7	96.4	4.9	4.2	3.7	79.6	4.5	3.8	3.2
16	161	6.3	5.8	5.0	107	5.2	4.5	4.0	88.1	4.7	4.0	3.4
18	175	6.6	5.9	5.3	116	5.4	4.7	4.1	96.2	4.9	4.2	3.7
20	190	6.9	6.2	5.6	126	5.6	4.9	4.3	104	5.1	4.4	3.8

* From U.S. Soil Conservation Service [33].

† Cross-sectional areas are for the triangular channel ($b = 0$) but can be used with the trapezoidal channels, in the range shown, without significant error in mean velocities. For example, in the table for $z = 1.5$, under $s = 0.001$ at $qn = 10$, the given area is 95.8 ft² while the actual area for that channel with $b = 12$ ft is 97.5 ft².

2.2 fps. Since the velocity is below the maximum permissible, the selected grade of 0.5 per cent and depth of 1 ft may be used.

8. Determine the other terrace dimensions using Table 21-26.

9. Complete the design of the terrace system by determining the dimensions of the terrace outlet, using the methods of Example 21-21.

c. Diversions. Standard dimension tables for diversions are usable for a variety of diversion design problems [38]. Table 21-30 is a condensed design table to be used as in the following example.

Example 21-20. Design a diversion to protect the irrigated land shown in Fig. 21-16. The climate is semiarid (annual rainfall is 12 in.), and the design is to provide protection from a 10-year-frequency peak rate of flow.

1. Make a field survey or use available maps or aerial photographs to delineate the drainage area above the land to be protected. Determine the approximate slope of the

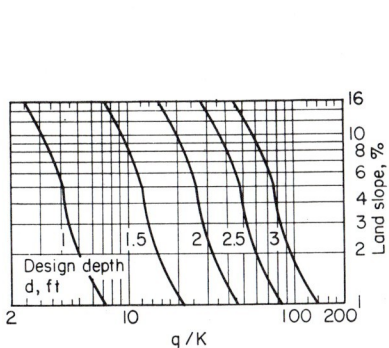

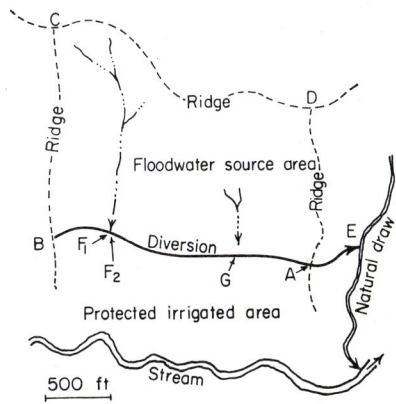

FIG. 21-15. Design chart for graded terraces. (*U.S. Conservation Service* [33].)

FIG. 21-16. Use of a diversion to protect agricultural land from upland runoff.

drainage area, select the approximate location of the diversion, make a profile of the selected location, and take cross sections at changes in land slope along the profile. In this case, point A was selected as far down the slope as possible and a grade channel slope of 0.001 was assumed for tentatively locating the diversion in order to determine drainage areas.

2. Estimate the 10-year-frequency peak rates of flow at critical points along the diversion. In this case, critical points will be at F_1, F_2, A, and E. Drainage areas at these points are 4, 51, 155, and 166 acres, respectively. Using the method shown in Example 21-13 with $W = 75$ for all areas, a rainfall factor of 0.50 from Fig. 21-9, a frequency factor of 0.35 from Table 21-23, and peak rates taken from Fig. 21-8, the 10-year-frequency peak rates at F_1, F_2, A, and E are 5.3, 48, 126, and 135 cfs, respectively.

3. Estimate a value of Manning's n for the diversion channel. In this case, the channel can be expected to be sparsely vegetated and a conservative (high) estimate of n would be 0.040.

4. Estimate the maximum nonerosive velocity for the channel soil and vegetative conditions. The final design velocity should be reasonably close to the maximum nonerosive velocity, to avoid both scour and silting. For this example, 3.5 fps will be used.

5. Select channel slopes and dimensions. For this example, the side slope z will be 1.5. Where it is necessary to cross the diversion with farm machinery, a value of $z = 4$ is more appropriate. Compute qn and a for each critical point on the diversion. For example, the design discharge at A is 126 cfs; $n = 0.040$; therefore $qn = 5.0$. The nonerosive velocity is 3.5 fps, and $a = q/v$; therefore $a = 36.0$ ft². Entering Table 21-30 where $z = 1.5$ and on the line where $qn = 5.0$, a value of $a = 37.7$ ft² is found for a slope of 0.003. Since the value of $a = 37.7$ is very near the required $a = 36.0$, use the slope of 0.003 at this point. A bottom width of $b = 6.0$ ft is used in this example. Computations for all

critical points along the diversion are listed below:

Item	At F_1	At F_2	At A	At E
10-year-frequency design flow = q, cfs	5.3	48	126	135
Manning's n	0.040	0.040	0.040	0.040
Discharge factor = qn	0.2	1.9	5.0	5.4
Nonerosive velocity, fps	3.5	3.5	3.5	3.5
Minimum flow area, ft^2	1.5	13.7	36.0	38.6
Selected upstream channel slope, ft/ft	0.005	0.005	0.003	0.003
Design flow area, ft^2	2.8	15.0	37.7	40.4
Design velocity, fps	1.9	3.2	3.3	3.3
Design bottom width, ft	0	0	6	6
Design flow depth, ft	1.4	3.1	3.4	3.5

6. Check to determine if the selected channel slopes give suitable elevations at B and E, Fig. 21-16. If the diversion goes too far uphill or downhill at B, it may be desirable to repeat the calculations of step 5.

Item	B	F_2	A	E
Distance, ft	450	1,580	440	
Slope, ft/ft	0.005	0.003	0.003	
Rise, ft	2.25	4.74	−1.32	
Elevation, ft	106.99	104.74	100.0	98.68

7. Add freeboard. This will normally be 0.5 ft, but more may be needed at points F and G, where flow concentrates.

B. Grassed Waterways

Grassed waterways are the most economical and most widely used means of conveying surface runoff from farm fields at nonerosive velocities. They provide protection against soil erosion, and the vegetation in them (usually grasses or grass-legume mixtures) can be grazed or mowed for hay.

1. Construction Types. A grassed natural draw is the most desirable waterway, since it is already available and the time and cost of shaping the channel and establishing vegetation are eliminated. However, when additional flow is added, as from a diversion or terrace system, its capacity should be checked. Natural waterways are generally parabolic in section; constructed waterways may be parabolic, trapezoidal, or triangular, with shallow depths of flow. When design discharges are large, it is often necessary to divide the flow and use two or more waterways to avoid excessive velocities. Where soils remain wet, it may also be necessary to use tile drains to maintain satisfactory vegetative cover.

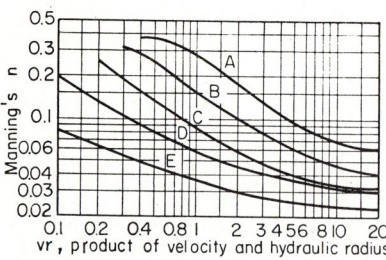

Fig. 21-17. Relation between Manning's roughness coefficient and the product of velocity and hydraulic radius. The curves A to E represent various degrees of vegetal retardance, A for very high, B for high, C for moderate, D for low, and E for very low vegetal retardance.

2. Criteria. Grassed waterways are usually designed to convey a 10-year-frequency peak rate of flow. They are planned without the use of flood routing. The depth of the design flow in the waterway should exceed critical depth. Nor-

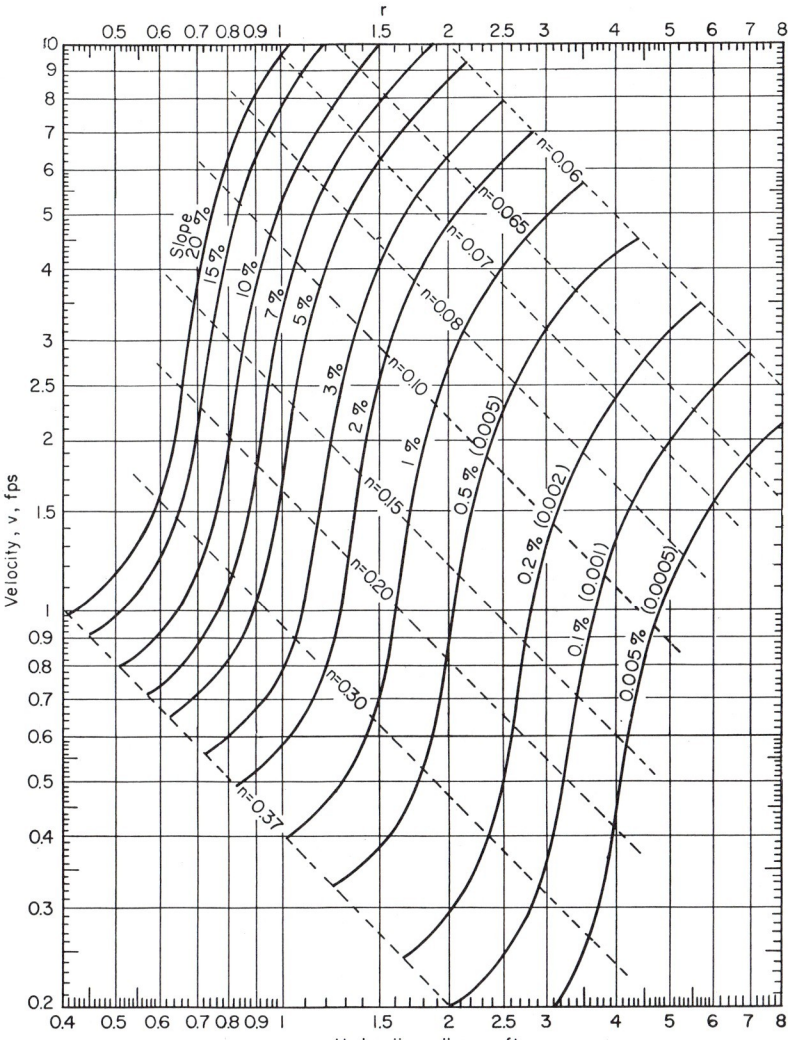

Fig. 21-18. Solution of the Manning equation for retardance A (very high vegetal retardance). (*U.S. Soil Conservation Service* [39].)

mally, the top widths of constructed waterways will be kept between 10 and 100 ft and the design depth in the center of the waterway will be at least 0.5 ft.

3. Hydraulic Considerations. Graphs and charts are available for the hydraulic design of constructed waterways [39]. Some of them are reproduced as Figs. 21-17 to 21-22. The hydraulic features of parabolic channels [40, 44] can be used to estimate the necessary top widths of natural parabolic waterways. If one top width and corresponding center depth are known for a parabolic channel, the top width for any other depth may be estimated by the following relationship:

$$T = T_m \sqrt{\frac{D}{D_m}} \qquad (21\text{-}21)$$

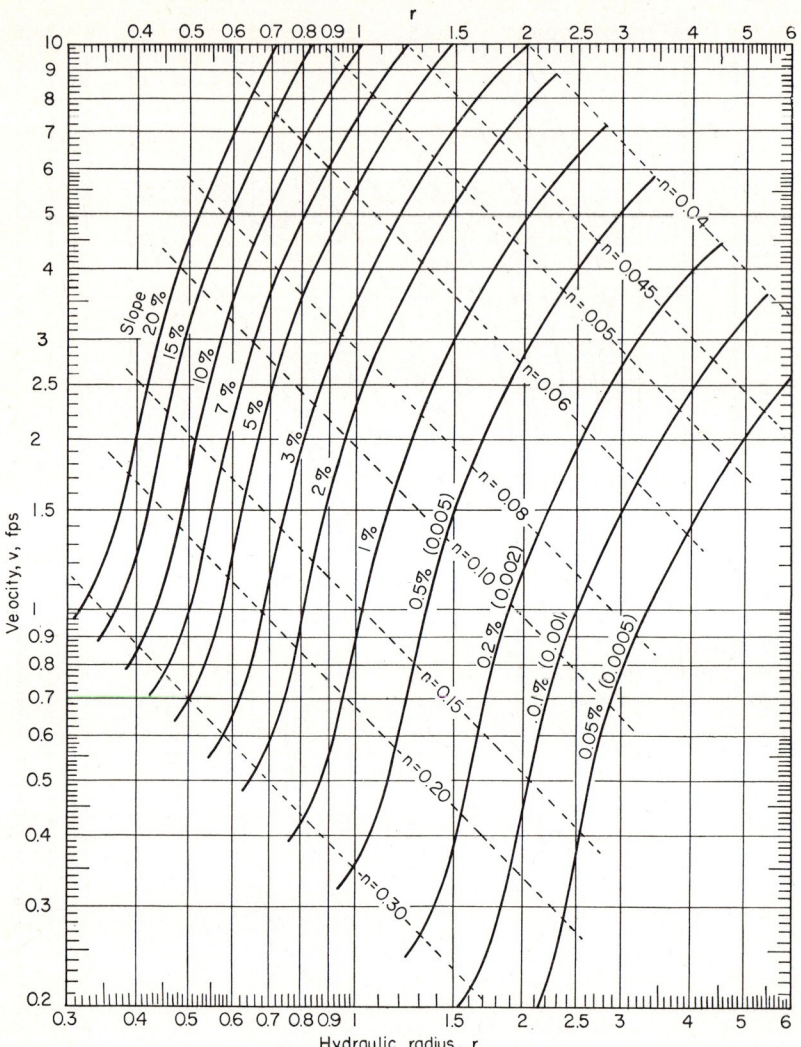

FIG. 21-19. Solution of the Manning equation for retardance B (high vegetal retardance). (*U.S. Soil Conservation Service* [39].)

where T is the top width for depth of flow D in ft, D is the depth of flow in the center of the channel in ft, T_m is the measured top width at depth of flow D_m in ft, and D_m is the measured depth of flow in the center of the channel in ft.

The hydraulic radius for a shallow parabolic channel can be closely approximated by

$$r = \tfrac{2}{3}D \tag{21-22}$$

where r is the hydraulic radius in ft, and D is the depth of flow in the center of the channel in ft.

The area of a parabolic channel ($a = \tfrac{2}{3}TD$) may therefore be closely approximated

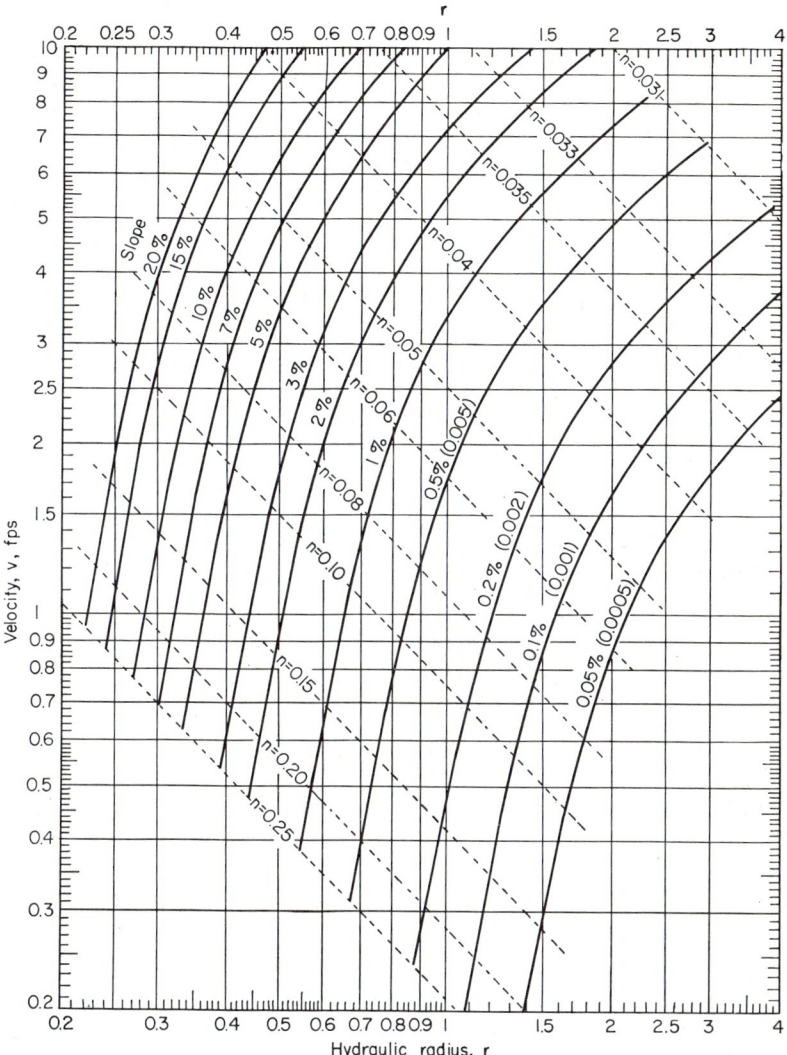

Fig. 21-20. Solution of the Manning equation for retardance C (moderate vegetal retardance). (*U.S. Soil Conservation Service* [39].)

by

$$a = rT \qquad (21\text{-}23)$$

and since

$$q = av \qquad (21\text{-}24)$$

where q is the discharge in cfs, a is the cross-sectional area in ft², and v is the average velocity in fps, the discharge for shallow parabolic channels may be closely approximated by

$$q = \frac{2vDT}{3} \qquad (21\text{-}25)$$

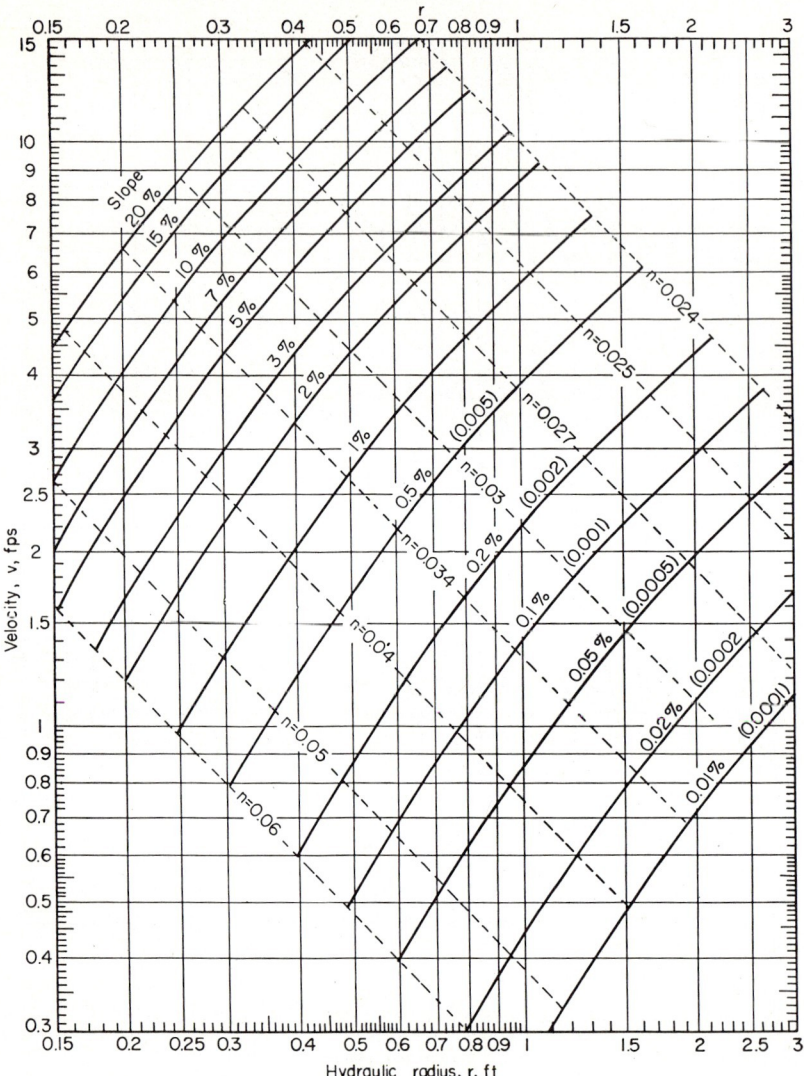

Fig. 21-21. Solution of the Manning equation for retardance D (low vegetal retardance). (*U.S. Soil Conservation Service* [39].)

The critical velocity-depth relationships for use in the design of shallow parabolic channels, shown in Table 21-31, are based on Eq. (21-22) and the relation $D_c = 0.0466v_c^2$ where D_c is the critical depth in feet and v_c is the critical velocity in feet per second.

4. Hydrologic Considerations. Grassed waterways are designed to convey peak rates of flow, disregarding effects of channel storage. These peak rates must be determined for application in the design.

Example 21-21. Design a grassed waterway as an outlet for the terrace system shown in Fig. 21-14. The waterway will be located at B, on easily eroded soil, varying in slope

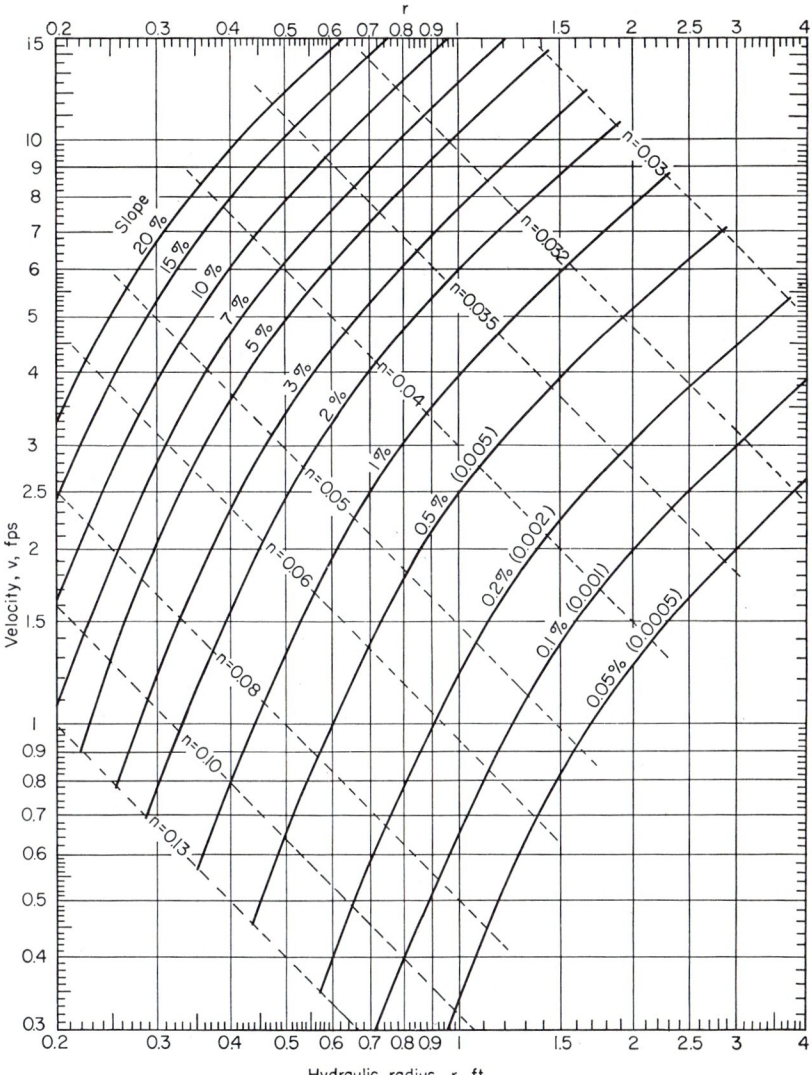

Fig. 21-22. Solution of the Manning equation for retardance E (very low vegetal retardance). (*U.S. Soil Conservation Service* [39].)

as shown in Fig. 21-23. The plan is to seed the waterway with a brome-alfalfa mixture and to keep it mowed when the vegetation is established. For details on the design of the terrace, see Example 21-19.

1. Determine the design rates of flow for critical points along the waterway. The drainage areas are shown in Table 21-34, line 1. The peak rates of flow, obtained by the method illustrated in Example 21-13, are listed in line 2. Note that accumulated drainage areas are used in determining these peak rates. Slightly higher peak rates would be obtained if it were assumed that peak flows for individual terraces (see Example 21-13, step 4) would coincide in time.

2. Determine the slope of the waterway. Minor irregularities will be removed during

HYDROLOGY OF AGRICULTURAL LANDS

Table 21-31. Critical Velocity-depth Relations for Parabolic Channels

v_c, fps	D_c, ft	r, ft	v_c, fps	D_c, ft	r, ft
1.0	0.047	0.03	5.0	1.16	0.78
1.5	0.105	0.07	5.5	1.41	0.94
2.0	0.186	0.12	6.0	1.68	1.12
2.5	0.291	0.19	6.5	1.97	1.31
3.0	0.419	0.28	7.0	2.28	1.52
3.5	0.571	0.38	7.5	2.62	1.75
4.0	0.746	0.50	8.0	2.98	1.99
4.5	0.944	0.63			

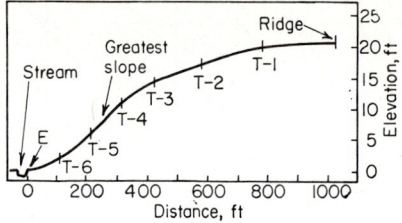

FIG. 21-23. Profile of the terrace outlet used in Example 21-22.

Table 21-32. Permissible Velocities for Channels Lined with Vegetation*

Cover	Slope range, %	Permissible velocity, fps	
		Erosion-resistant soils	Easily eroded soils
Bermuda grass..............	Up to 5 5 to 10 Over 10	8 7 6	6 5 4
Buffalo grass Kentucky bluegrass Smooth brome Blue grama	Up to 5 5 to 10 Over 10	7 6 5	5 4 3
Lespedeza sericea Weeping lovegrass Yellow bluestem Kudzu Alfalfa Crabgrass	Up to 5†	3.5	2.5
Common lespedeza‡ Sudan grass‡	Up to 5§	3.5	2.5

* From U.S. Soil Conservation Service [39]. Values apply to average, uniform stands of cover. Velocities exceeding 5 fps are to be used only where good covers and proper maintenance can be obtained.

† Not to be used on slopes steeper than 5 per cent except for side slopes in a combination channel.

‡ Annuals—used on mild slopes or as temporary cover until permanent covers are established.

§ Use on slopes steeper than 5 per cent is not recommended.

construction; however, the general slope of the land will be followed, and the values in per cent are listed in Table 21-34, line 3. Note that maximum slope is between T-4 and T-5; since this will be a high-velocity section in the waterway, it is used as a starting point in design.

3. Determine the "permissible" velocity. Using Table 21-32, a velocity of 5 fps is found to be satisfactory for easily eroded soils on a slope of up to 5 per cent. A cover of brome is used since it will dominate the mixture.

4. Determine the vegetal retardance. Table 21-33 gives a vegetal retardance of D for a good stand of vegetation from 2 to 6 in. (mowed) in height.

5. Determine whether or not critical flow will occur. Enter Fig. 21-21 (D retardance) with $v = 5$ fps and $s = 5$ per cent, and read $r = 0.51$. This value of r is less than that

Table 21-33. Guide to Selection of Vegetal Retardance*

Average height of vegetation, in.	Degree of retardance	
	Good stand	Fair stand
More than 30	A	B
11 to 24	B	C
6 to 10	C	D
2 to 6	D	D
Less than 2	E	E

* From U.S. Soil Conservation Service [39].

Table 21-34. Computations for Design of Grassed Waterway

Item	Location						
	T-1	T-2	T-3	T-4	T-5	T-6	E
1. Accumulated area, acres	9.2	16.6	22.6	26.8	30.3	34.7	34.7
2. Peak, cfs	14	23	30	35	40	45	45
3. Slope, %	1.1	1.6	2.6	4.2	4.2	2.7	1.4
4. r, ft	0.64	0.58	0.50	0.42	0.42	0.49	0.60
5. v, fps	2.20	2.42	2.80	3.34	3.34	2.86	2.34
6. a, ft²	6.4	9.5	10.7	10.5	12.0	15.7	19.2
7. T, ft	10	16	21	25	29	32	32
8. D, ft	0.96	0.87	0.75	0.63	0.63	0.74	0.90

shown for $v = 5$ fps in Table 21-31; therefore critical flow will occur, and it is necessary to select a lower design velocity. A suitable value is $v = 3.5$ fps, with $r = 0.40$, $n = 0.051$, and $a = q/v = 40/3.5 = 11.4$ ft².

6. Determine the channel shape. For a parabolic section, the design center depth D is found to be 0.60 ft by using Eq. (21-22). The top width T is found to be 28.3 ft by using Eqs. (21-22) and (21-23), or $T = 3a/2D$. Note that this value is used for the section between T-4 and T-5.

7. Determine top widths and center depths for other locations in the waterway by repeating steps 5 and 6, and list the results in Table 21-34.

8. Use Eq. (21-21) to compute the widths for two or three other depths in each section to obtain depths and widths for construction.

C. Grade-stabilization Structures

When the velocity of flow exceeds allowable limits for vegetative waterways, then structures of masonry, concrete, or metal can be used to reduce grades.

1. Construction Types. Chute spillways, drop spillways, and pipe spillways with hood or drop inlets are the principal types of structures used in grade stabilization of waterways (Figs. 21-24 and 21-25). *Chute spillways* are used where it is necessary to make large drops in channel elevation, as in reservoir spillways or in the stabilization of deep gullies. *Drop spillways* are used where a small drop in channel elevation is required and it is necessary to handle large rates of flow; a box inlet is often used to increase the capacity of the structure. *Pipe spillways* are often used with earth dams. They have relatively low capacities, and water is stored behind the dam to reduce the peak rates of outflow. Inlets of the drop or hood type are used to reduce the priming head, and antivortex devices and trash racks are used to maintain the hydraulic capacity [41].

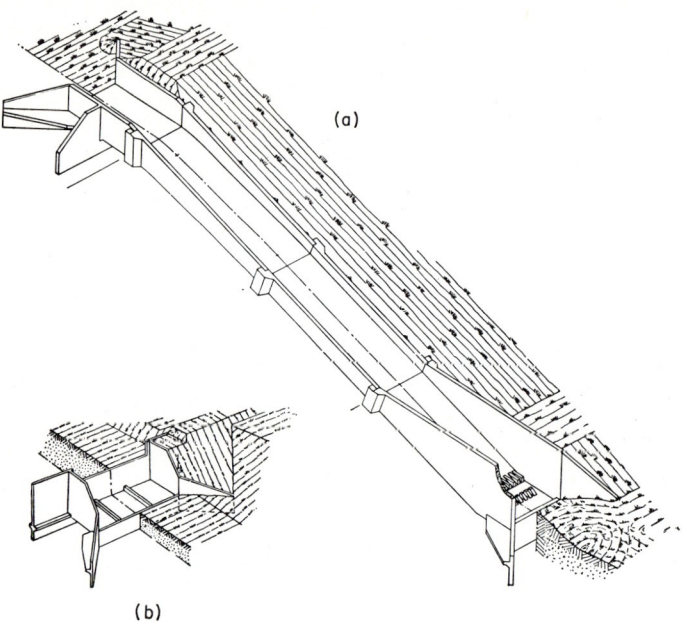

Fig. 21-24. (a) Typical chute spillway with one wing wall cut away to show baffles. (b) Typical drop spillway.

2. Criteria. Small drop spillways in waterways are generally designed for the same flow as the waterway. Larger and more costly grade-stabilization structures are designed to control at least a 25-year-frequency flow, and a freeboard of 1 ft or more is used with chute and drop spillways. Freeboard for pipe spillways is provided in the design of the emergency spillway for the dam.

3. Hydraulic Considerations. Spillway entrance capacities will be discussed briefly since they will be used in hydrologic computations.

a. Chute and Drop Spillways. Inflow capacities can be estimated using

$$q = 3.1LH^{3/2} \qquad (21\text{-}26)$$

where q is the discharge in cfs, L is the length of weir in ft, and H is the depth of flow above the crest of the weir in ft.

The length of the weir of a box inlet is twice the length of the box plus the width; if necessary, this figure is modified to allow for the approach channel width. The depth of flow H is determined at a point a short distance upstream from the crest of the weir. When the velocity of approach is great, H is the energy head (depth plus velocity head). Proportioning the hydraulic capacities of various parts of structures follows the requirements of nonuniform flow [42].

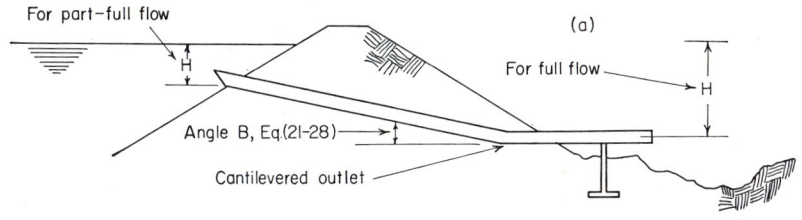

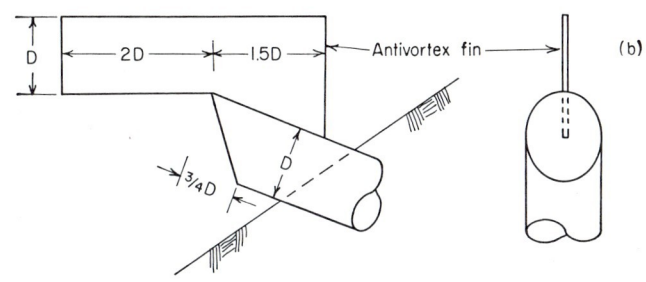

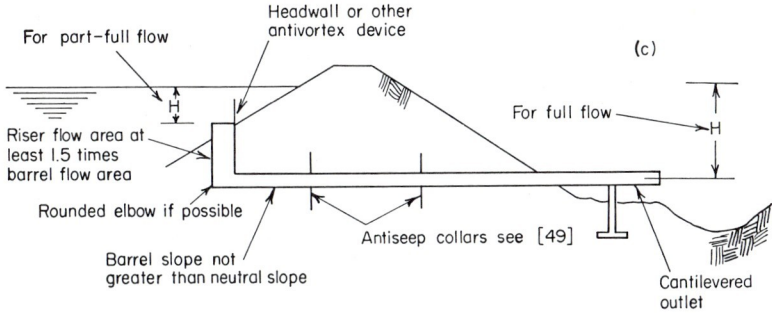

Fig. 21-25. Definition sketches for small dam with a pipe spillway: (a) Use of pipe spillway with hood inlet, (b) details of hood inlet showing antivortex fin, and (c) use of pipe spillway with drop inlet.

b. *Pipe Spillways.* The inflow capacity of a pipe spillway at full flow is estimated by

$$q = a \sqrt{\frac{2gH}{1 + K_e + K_b + K_p L}} \qquad (21\text{-}27)$$

where q = discharge, cfs
a = cross-sectional area of pipe, ft^2
g = 32.2 ft/sec^2
H = total head, ft
K_e = coefficient for entrance loss
K_b = coefficient for bend loss
K_p = coefficient for pipe friction loss
L = length of pipe, ft

Table 21-35. K_b for Selected Values of n and B for Pipe Spillways

	Concrete pipe ($n = 0.015$)	Corrugated-metal pipe ($n = 0.025$)
Straight............	0	0
Angle = 30°.........	0.15	0.25
Angle = 60°.........	0.30	0.50
Angle = 90°.........	0.45	0.75

Table 21-36. Values of K_p for Selected Values of n and D for Pipe Spillways*

Pipe diameter, in.	Manning's n			
	0.010	0.015	0.020	0.025
6	0.0467	0.105	0.187	0.292
12	0.0185	0.0417	0.0741	0.116
18	0.0108	0.0243	0.0431	0.0674
24	0.00735	0.0165	0.0294	0.0459
30	0.00546	0.0123	0.0218	0.0341
36	0.00428	0.00963	0.0171	0.0267
42	0.00348	0.00784	0.0139	0.0218
48	0.00292	0.00656	0.0117	0.0182
54	0.00249	0.00561	0.00997	0.0156
60	0.00217	0.00487	0.00866	0.0135

* From Culp et al. [43].

Table 21-37. Upstream Head and Discharge Ratios for Pipes Flowing Partly Full*

h/D	0.2	0.4	0.6	0.8	1.0	1.2
$q/D^{5/2}$	0.16	0.46	0.88	1.56	2.20	2.80

* From Culp et al. [43].

A value of $K_e = 0.5$ may be taken for the commonly used sharp-edged pipe. K_b will be equal to zero if the pipe is straight throughout; however, if there is a bend, the coefficient is [43]

$$K_b = \frac{nB}{3} \quad (21\text{-}28)$$

where n is Manning's n for the pipe, and B is the deflection angle (Fig. 21-25) in deg. Selected values of K_b are given in Table 21-35. Values of K_p for circular pipe [43] may be determined by

$$K_p = \frac{5{,}100 n^2}{D^{4/3}} \quad (21\text{-}29)$$

where D is the inside diameter of the pipe in inches. Table 21-36 gives values of K_p for selected diameters.

At low heads the pipe flows partly full, with the inlet acting as a weir. Under such conditions, the ratios of Table 21-37 may be used with h equal to depth of water in feet above the pipe entrance invert.

FIELD APPLICATIONS

c. *Submergence.* The type of pipe flow and the capacity are affected by submergence of the outlet. The entrance capacity of a chute spillway is not affected, but drop structures of some types may have a reduced discharge when tailwater is high. The effects of submergence can be determined by methods given in King's handbook [44].

4. Hydrologic Considerations. A "full-flow" design is used where the storage behind the structure is negligible and the spillway must carry the entire flow at its peak rate. The hydrologic studies consist of determining the peak rate of flow of the design storm. The methods used in Examples 21-12 and 21-13 are generally applied in the design of small or inexpensive structures; for the larger and more costly structures the methods of Examples 21-14 to 21-17 are used. Where upstream channel storage must be considered, Fig. 21-26 can be used to make a quick estimate of the effects of storage [45] on the design peak rate for weir-type spillways. The peak rate, volume of runoff, and available storage must be known.

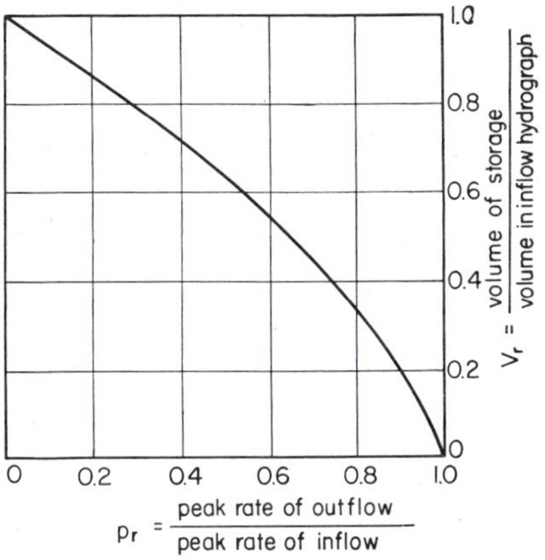

Example 21-22. A drop spillway is to be used for channel stabilization of the watershed of Example 21-13. A 25-year-frequency peak rate of flow is to be used as the emergency-spillway design discharge. The storage between the crest of the weir and the proposed 3-ft depth of flow over the weir is 0.32 in. Determine the length of weir L required.

1. Determine the 50-year-frequency peak rate of flow. This is given in Example 21-13, steps 1 and 2, and is 285 cfs.
2. Determine the time of concentration. This is given in Example 21-12, step 2, as $T_c = 0.19$ hr.
3. Estimate the volume of storm runoff using Eq. (21-16): $Q = 2.80$ in.
4. Determine the effect of channel storage on the design rate of flow. Compute the ratio $V_r = 0.32/2.80 = 0.114$. Enter Fig. 21-26 with V_r and find $p_r = 0.91$. Compute the reduced peak rate, $q_u = 285 \times 0.91 = 259$ cfs.
5. Compute the length of weir. Using Eq. (21-26) with $H = 3$ ft and $q = 259$ cfs gives $L = 16.1$ ft.

FIG. 21-26. Dimensionless diagram for estimation of effect of reservoir storage on peak discharge through spillways of the weir type. (*From Hartman and Wilke* [45].)

$$p_r = \frac{\text{peak rate of outflow}}{\text{peak rate of inflow}}$$

$$V_r = \frac{\text{volume of storage}}{\text{volume in inflow hydrograph}}$$

D. Farm and Ranch Ponds

The relatively low cost of construction has made the small farm or ranch pond increasingly popular. It is expected that over a million small ponds will be built in the United States in the next forty years [1]. They provide water for livestock, domestic use, irrigation, fire protection, fish propagation, and recreation.

1. Construction Types. Where topography does not provide a good site at the desired location, a pond can be built off stream and water supplied by a diversion from one or more streams. Two main types of farm and ranch ponds are used.

a. *Dam.* The most common layout of a pond is illustrated in Fig. 21-27. An earth dam is built across a natural waterway, the pond area is fenced to prevent pollution by livestock, and water for livestock is piped from the pond to a supply tank outside the fence. Bank slopes at the reservoir's edge should be steep to reduce evaporation and seepage and to aid in mosquito control.

b. *Dugout.* In flat country a dugout is used (Fig. 21-28). Excavated earth is either wasted or placed to construct a dike to increase the area draining into the reservoir. In northern states, it may be placed so that drifting snow will contribute to the water supply. Sides of the dugout are steep, and the reservoir area is usually fenced. If it is not fenced, one side of the dugout is made with a 4:1 side slope to allow access for cattle. Dugouts may be cross-shaped or rectangular, whichever happens to be easier to construct with the available machinery.

2. Criteria. Economic studies are not needed for these small structures, and dimensional criteria are based on experience, preferably local.

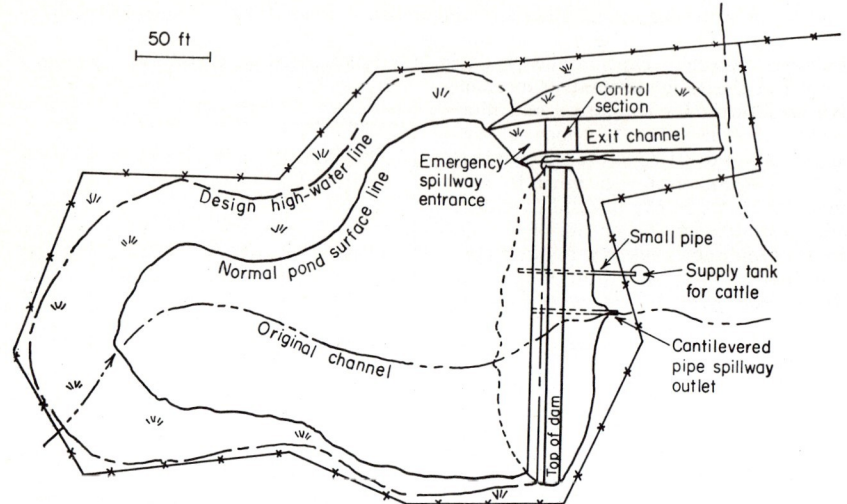

FIG. 21-27. Typical layout of a farm pond.

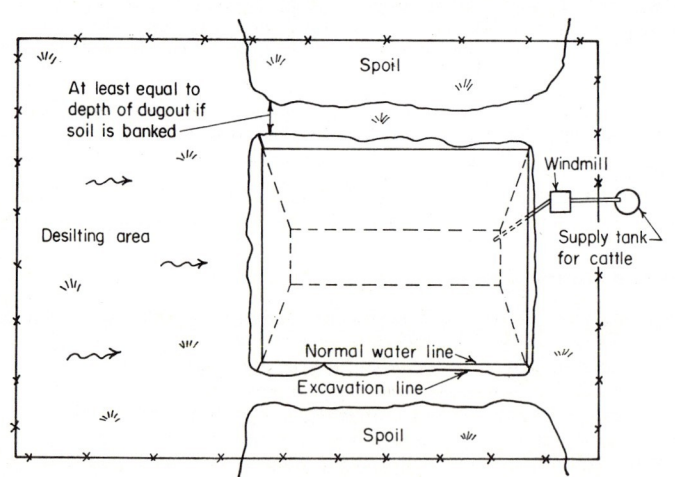

FIG. 21-28. Typical layout for a dugout type of farm pond.

a. Drainage Area. The area contributing runoff to the pond should be selected to avoid runoff from barnyards, feeding lots, and similar sources of pollution.

b. Reservoir Features. Small-surfaced, deep, and steep-sided reservoirs are the most desirable to keep evaporation losses low and to control insects. When seepage losses are high, special treatment may be necessary [46].

c. Storage. The average annual runoff of an area is commonly used in determining a reasonable storage for the drainage area. Instead of making a special study, an estimate of watershed yield can be made using Fig. 21-29, which allows for normal evaporation and seepage [31].

d. Low-flow Spillway. A small pipe spillway is used to maintain the reservoir level below the emergency-spillway crest and thereby eliminate small persistent flows that could erode the grassed spillway. The capacity of such a pipe spillway should exceed the capacity of base flow, if such flow occurs. The inlet should be from 1 to 3 ft below the crest of the emergency spillway, and suitable trash racks should be provided to prevent plugging of the inlet.

e. Emergency-spillway Location. Either one or two spillways can be used, depending on the topography. In some valleys, natural swales along the sides of the reservoir can be used as emergency spillways. In such cases, fill for the dam is usually

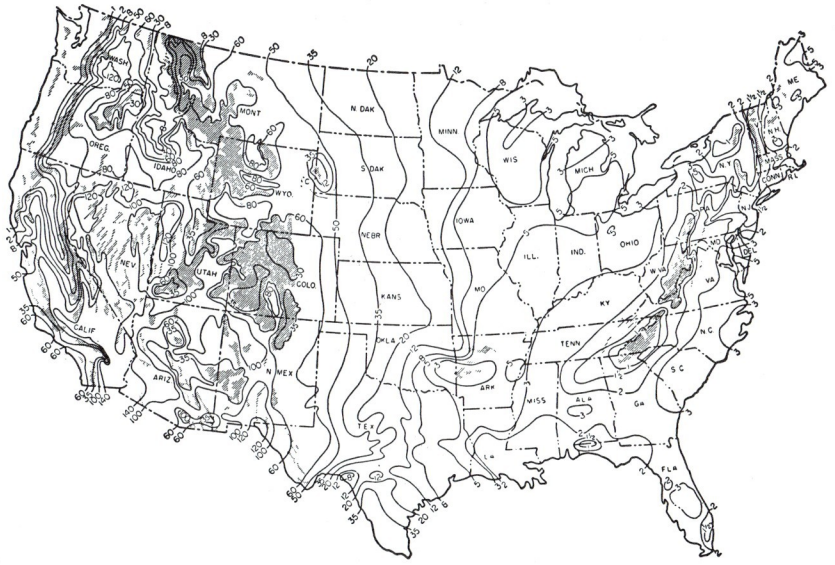

Fig. 21-29. A general guide for estimating drainage area required for farm ponds. Numbers on the map are number of acres of drainage area required per acre-foot of storage. *Note:* Mountainous areas have been crosshatched. The numbers may not apply to these areas since rainfall in them is very spotty and varies sharply. (*Hamilton and Jepson* [31].)

taken from the reservoir area; however, the depth of excavation must be limited if geologic features make a soil blanket necessary to prevent excessive seepage. In any case, emergency spillways should be located to avoid discharging water on the toe of the dam.

f. Emergency-spillway Capacity. It is common practice to design the emergency spillway of a farm pond without flood routing. Ordinarily, it is designed to pass the peak rate of a 25-year-frequency flow, but if the pond is exceptionally large or the dam costly, it is desirable to construct the spillway to pass at least a 100-year-frequency flood. Table 21-40 can be used to make quick estimates of spillway dimensions (Fig. 21-30) for the ordinary farm pond. It should be remembered, however, that the safety of a structure depends on good construction and a good foundation, as well as on adequate spillway capacity. Ponds used for fish propagation require a wide shallow spillway to reduce the loss of fish. Instead of a normal design depth of about 2 ft, the spillway of a fish pond should be less than 0.5 ft deep. Freeboard of 1 or 2 ft is usually added to the design depth. In all cases, state laws should be checked to see if a specific freeboard is required. Critical flow is used in the design of the grassed exit channel, since the spillway vegetation is expected to recover between infrequent uses.

g. Sedimentation. When the area draining into the pond has primarily grass or forest cover, the sediment inflow can usually be ignored in design. If the area is cultivated, soil-conservation measures can be taken to reduce sedimentation, and allowance should be made for the expected sediment inflow by increasing the capacity of the pond. Fenced filter strips of grass 100 ft or more wide upstream from the pond are effective in removing some of the sediment.

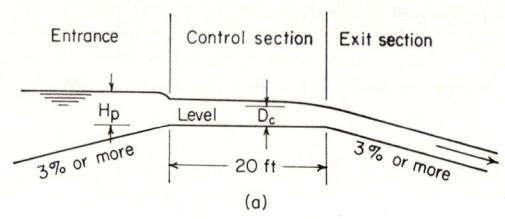

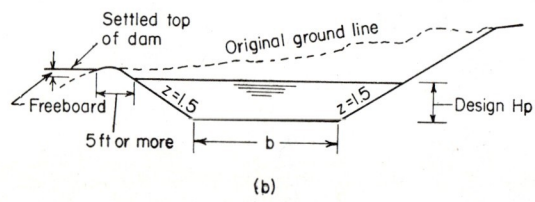

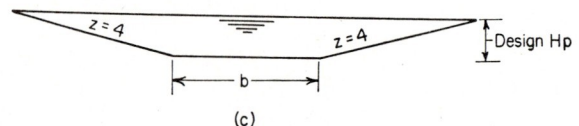

FIG. 21-30. (a) Definition sketch for the profile of spillway of Table 21-40; (b) typical cross-sectional layout for a spillway with side slopes of 1.5:1 ($z = 1.5$); (c) typical cross section with side slopes of 4:1 ($z = 4$).

3. Hydrologic Considerations. Rainfall, runoff, evaporation, and seepage data are generally not available for specific farm and ranch pond sites; even if they are, the cost of an extensive hydrologic investigation is seldom justified. General guides have been developed which take into consideration the principal factor affecting pond design. Seepage usually must be estimated for each site, and in karst or gypsum areas it may be so high that pond construction is not practical. Seepage varies greatly, but it is wise to allow for a loss of at least 3 ft of water per year when estimating storage. Typical pond depths by climatic areas are given in Table 21-38, and a generalized approach to the design of a farm pond is given in the following example.

Example 21-23. Determine the hydrologic factors affecting the design of a pond, and show how they are applied in the design. The pond is to be constructed in the semiarid climate of eastern Colorado, near the Kansas state line. The topography is rolling, and an earth dam will be used to create a reservoir for recreation and for stock water supply.

1. Estimate the required storage in acre-feet. Large livestock (horses, cattle) will require about 1 acre-ft/year per 75 head, and small stock (sheep, hogs) will require 1

FIELD APPLICATIONS 21-69

acre-ft/year per 750 head [31]. Recreational needs will vary with the desires of the owner and for this example it is assumed that 15 acre-ft will satisfy all needs.

Table 21-38. Typical Average Pond Depths*

Climate	Average depth, ft
Wet	5
Humid	6–7
Moist subhumid	7–8
Dry subhumid	8–10
Semiarid	10–12
Arid	12–14

* Hamilton and Jepson [31].

2. Estimate the required drainage area. Using Fig. 21-29, it is determined that approximately 60 acres of drainage area are required at this location for each acre-foot of required storage. In this case, a 900-acre drainage area should provide an adequate water supply. This estimate includes allowances for average evaporation and seepage for the locality [31]. Local experience may indicate the need for a larger drainage area, but smaller areas are seldom satisfactory.
3. Select a desirable average pond depth. From Table 21-38, a depth of 10 to 12 ft is recommended for this climatic area. For this example, a 12-ft depth will be used.
4. Obtain a topographic map or aerial photographs of the locality, and locate possible pond sites with drainage areas of 900 acres or more.
5. Make a field investigation of the various sites. The most desirable site would be easily accessible, would require a short dam below a broad pond area with steep sides, would require the removal of few trees, would have a suitable foundation for the dam, would show no evidence of excessive seepage through geologic formations, would have suitable fill material near the site, and would have a natural swale suitable for use as an emergency spillway. Few sites would have all these features, and the best one is selected by considering its practicability from the standpoint of use and potential construction costs.
6. Estimate the pond and dam dimensions for the selected site to make sure that the necessary storage and depth can be attained. Determine the desirable maximum pond elevation, which may include additional depth for sedimentation or excessive seepage. This elevation will be used to determine the crest of the pipe spillway. The crest of the emergency spillway will normally be 1 to 3 ft above the crest of the pipe spillway. Estab-

Table 21-39. Approximate Discharges, in Acre-feet per Day, for Pipe Spillways on Steep Slopes, with Entrance Control*

Head† ft	Inside pipe diameter, in.						
	6	9	12	15	18	21	24
0.5	0.8	1.0	1.3	1.6	1.8	2.3	2.7
1	1.7	3.1	4.4	5.4	5.8	6.3	7.4
1.5	2.1	4.6	7.3	9.8	11	14	16
2	2.5	5.5	9.4	14	18	22	25
2.5	2.7	6.2	11	16	22	28	34
3	3.0	6.8	12	19	26	34	41
4	...	7.9	14	22	31	42	53
6	...	...	17	27	39	52	68
8	...	...	....	....	44	61	79
10	...	...	....	....	....	68	88

* U.S. Soil Conservation Service [33].
† Head is water depth, in feet, above invert of pipe inlet. When head exceeds about 3 pipe diameters, the pipe may flow full, with discharge depending on pipe length and roughness, tailwater height, etc.

lish the elevation of the top of the settled dam by adding 2 ft plus freeboard to the elevation of the crest of the emergency spillway. In some states, a minimum freeboard is specified by law. When good fill material is available, an upstream slope of 3:1 and a downstream slope of 2:1 are generally used. The top width of the dam is normally deter-

Table 21-40. Bottom Widths, in Feet, for Standard Emergency Spillways (Fig. 21-30)*

Design discharge, cfs	Control section, side slopes = 1.5:1, H_p =					Control section, side slopes = 4:1, H_p =				
	1	2	3	4	5	1	2	3	4	5
20	7	...	...	...	...	5				
30	11	...	...	...	...	9				
40	15	4	...	...	...	13				
50	19	5	...	...	...	17				
60	23	6	...	...	...	21				
70	27	7	...	...	...	25	4			
80	31	9	...	...	...	29	5			
90	35	10	...	...	...	33	6			
100	39	11	4	...	...	37	7			
120	47	14	5	...	...	45	10			
140	55	16	7	...	...	53	13			
160	63	19	8	...	...	61	16	4		
180	71	22	9	4	...	69	18	5		
200	79	24	11	5	...	77	21	6		
220	87	27	12	5	...	85	24	7		
240	95	30	14	6	...	93	26	9		
260	103	32	15	7	...	101	29	10		
280	111	35	16	8	...	109	31	11		
300	119	38	18	9	4	117	34	13		
350	139	44	21	11	6	137	41	16	4	
400	159	51	25	13	7	157	47	20	6	
450	179	57	28	15	9	177	54	23	9	
500	199	64	32	17	10	197	60	27	11	
600	239	77	38	22	13	237	73	33	15	5
700	279	90	45	26	16	277	87	40	19	8
800	319	103	52	30	19	317	100	47	23	11
900	...	116	59	34	22	...	113	54	28	14
1,000	...	130	66	39	25	...	127	61	32	17
1,200	...	156	80	47	31	...	153	75	40	23
1,400	...	182	94	56	37	...	179	89	50	29
1,600	...	209	107	64	43	...	206	102	58	35
1,800	...	235	121	73	49	...	232	116	66	41
2,000	...	261	135	81	55	...	258	130	74	47

* From U.S. Soil Conservation Service [33].

mined using $W = 2\sqrt{Y} + 2$, where Y is the settled height of the dam in feet and W is the top width in feet.

7. Determine the size of the pipe spillway. To avoid prolonged use of the emergency spillway, the pipe spillway should be large enough to carry base flow. Table 21-39 can be used to determine the minimum pipe size required. Note that an average flow of 1 cfs for 1 day is equal to approximately 2 acre-ft.

8. Determine the size of the emergency spillway. The difference in elevation between the spillway crest and the maximum pond surface is H_p and should be about 2 ft during the spillway design flow. The design peak is used without routing the flow through the structure. In this case, where it has been decided to design for a 25-year-frequency flow, the methods illustrated in Examples 21-12 and 21-13 give a design discharge of 1,100 cfs. Entering Table 21-40 with this discharge and for $H_p = 2$ ft, by interpolation find the bottom width for the selected side slopes. Using side slopes of 1.5:1, the bottom width is about 143 ft. If this width is not practical for the site, the bottom width for two small spillways can be determined similarly.

Using these dimensions, the dam and pond area can be staked for construction.

E. Structures for Temporary Floodwater Storage

Properly located temporary storage for floodwater will reduce downstream flood damage. Earth dams are generally used to provide temporary storage, and often some permanent storage for irrigation, water supply, or recreation is incorporated in the plan to develop the full potential of the dam site.

1. Criteria. The design of a dam and reservoir to provide temporary floodwater storage involves a balancing of the reservoir storage and the low-flow release rate and provision of an emergency spillway to discharge unusually heavy flood flows without endangering the structure.

Table 21-41. Typical Minimum Floodwater Storage, Emergency-spillway and Freeboard Design-storm Criteria*

Type of structure	Design-storm precipitation for determining:		
	Temporary storage	Emergency-spillway capacity	Freeboard capacity
Type 1—low cost and low downstream hazard in the event of failure	25-year frequency	¼ probable maximum 6-hr precipitation	⅜ probable maximum 6-hr precipitation
Type 2—moderate cost with no downstream hazard in the event of failure; or low cost with moderate downstream hazard (serious economic loss)	50-year frequency	⅜ probable maximum 6-hr precipitation	⅝ probable maximum 6-hr precipitation
Type 3—high cost; or low or moderate cost with high downstream hazard in the event of failure (may involve loss of life)	100-year frequency	½ probable maximum 6-hr precipitation	Probable maximum 6-hr precipitation

* From U.S. Soil Conservation Service [33].

a. Design Storms. Three different design storms are used in proportioning the structure. A storage design storm is drawn up to determine the greatest volume of floodwater that needs to be handled during varying periods. Another design storm is used to determine the capacity of the emergency spillway. A third is used to determine freeboard. In all cases, state laws should be checked, since they often establish spillway requirements. Design storms may be selected on the basis of frequency or in terms of the "probable maximum precipitation" for the locality. The duration used for the emergency spillway and freeboard design storms will depend on the time of concentration of the watershed; however, some convenient standard duration, such as 6 hr, will suffice for most small watersheds. Table 21-41 shows typical minimum design-storm criteria in terms of frequency and "probable maximum precipitation." Figures 21-31 to 21-35 show the minimum emergency-

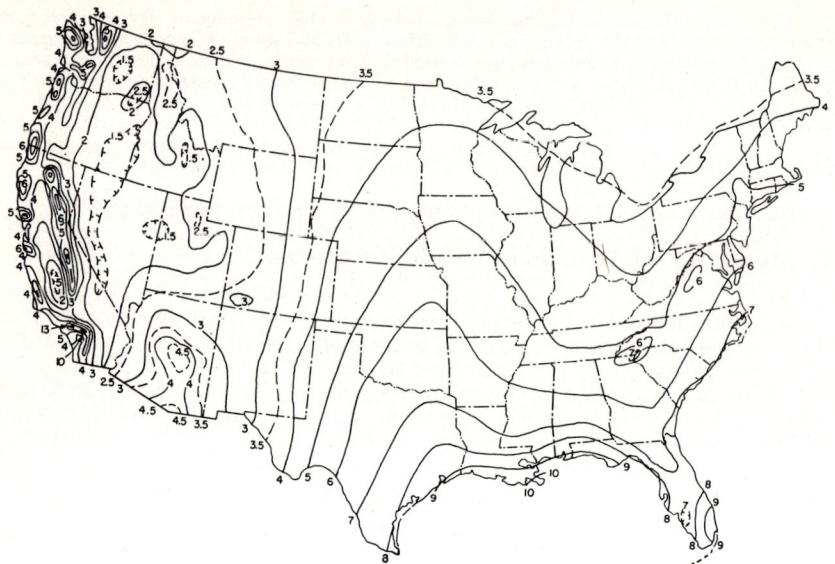

Fig. 21-31. Typical emergency-spillway design storm for type 1 structures, based on 6-hr precipitation (inches), used in construction of the emergency-spillway design hydrograph. (*U.S. Weather Bureau* [5]).

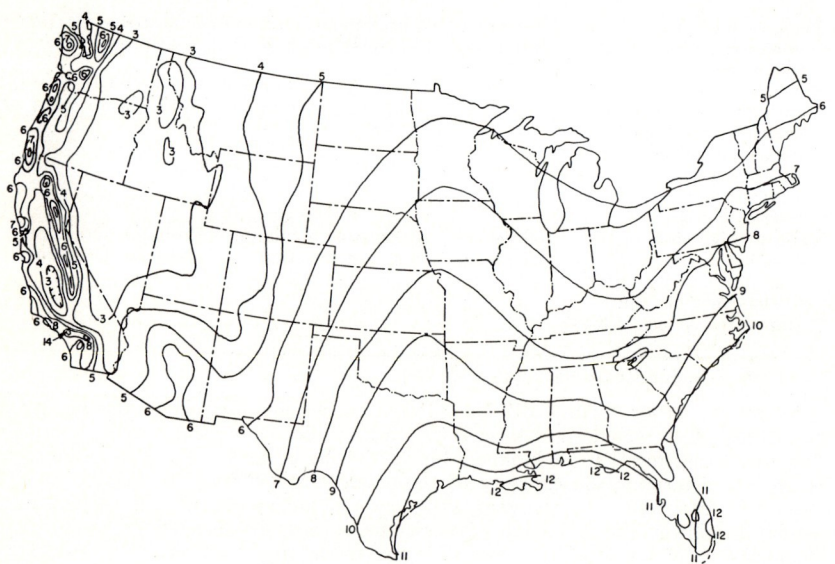

Fig. 21-32. Typical emergency-spillway design storm for a type 2 structure or typical freeboard design storm for a type 1 structure. Values in terms of inches of precipitation represent the minimum 6-hr precipitation used in developing design hydrographs. (*U.S. Weather Bureau* [5]).

FIELD APPLICATIONS 21–73

Fig. 21-33. Typical freeboard design storm for a type 2 structure, based on 6-hr precipitation (inches), used in construction of the freeboard design hydrograph. (*U.S. Weather Bureau* [5].)

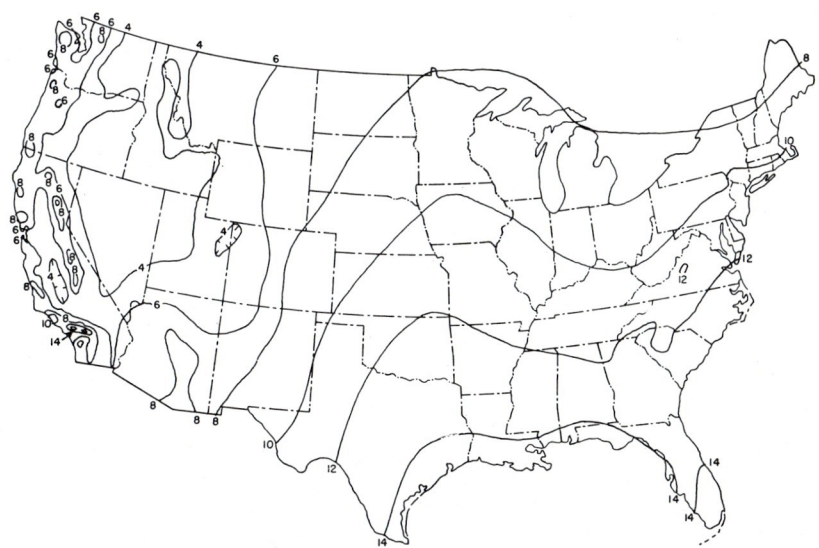

Fig. 21-34. Typical emergency-spillway design storm for a type 3 structure, based on 6-hr precipitation (inches), used in construction of the emergency-spillway design hydrograph. (*U.S. Weather Bureau* [5].)

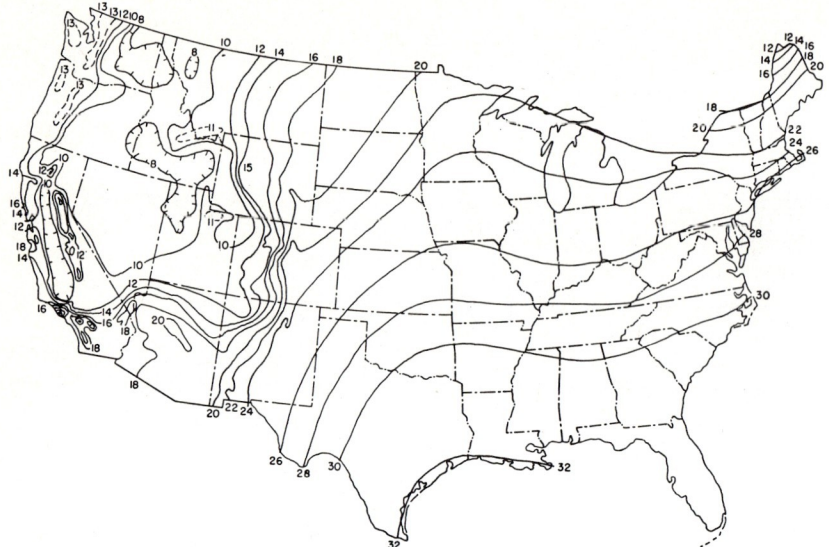

Fig. 21-35. Typical freeboard design storm for a type 3 structure. This is the probable maximum 6-hr precipitation (inches) and is used as the precipitation in constructing the freeboard design hydrograph. (*U.S. Weather Bureau* [5].)

spillway and freeboard design storms of 6-hr duration for the three types of small floodwater storage structures used by the U.S. Soil Conservation Service.

b. Storage. The storage requirement is dependent on the rate at which the water can be released. Site conditions sometimes are such that it is cheaper to build a dam higher (create more storage) and take advantage of a natural emergency-spillway site than to construct for minimum design requirements. Since bankful capacities in terms of cubic feet per second per square mile generally decrease downstream, channel capacities should be checked at various points. When several structures are

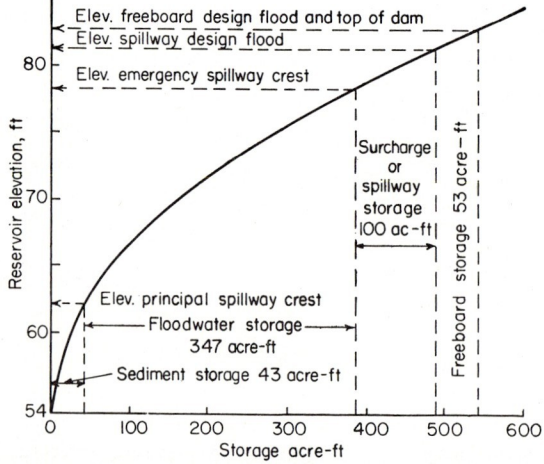

Fig. 21-36. Graph of storage vs. elevation, used in design of a floodwater-retarding reservoir.

FIELD APPLICATIONS

involved, flood routings are used to assist in determining suitable low-flow release rates. For details relating to the design of the emergency spillway see Ref. 48.

2. Hydrologic Considerations. The watershed of Example 21-17 will be used to illustrate the design principles, but details will be given only for the hydrologic design. A floodwater-retarding dam (providing temporary storage) is to be constructed at the watershed outlet to give protection from a 50-year-frequency flood to the agricultural land in the next mile downstream. Bankful channel capacity at the end of the protection zone is 20 cfs/sq mi, and the drainage area at that point is 5.3 sq mi. Farm-machinery buildings along the channel and a county road crossing call for a type 2 structure, according to the conditions shown in Table 21-41. The following paragraphs briefly cover the various phases of design.

a. Site Investigations. A field survey is made of the area to determine cross sections at the dam site, topography of the reservoir, and possible emergency-spillway location. Geologic investigations are also made [49].

b. Reservoir Capacity. Field survey data are used to make an elevation-storage curve (Fig. 21-36).

c. Sediment Storage. An estimate is made of the volume of sediment likely to be deposited in the reservoir during the life of the structure, which is generally assumed to be 50 years or more. The expected sediment in this case is 43 acre-ft, as shown in Fig. 21-36.

d. Floodwater Storage. Whenever available, streamflow records are used to determine the storage requirement. Usually, however, such records are not available on small watersheds, and storage requirement is determined from rainfall records as illustrated in the following example.

Example 21-24. Determine the storage required for the conditions given above.

1. Determine the 50-year-frequency rainfall amounts for durations of 4, 6, 8, 12, and 24 hr, using published data [5].
2. Plot the rainfall amounts of step 1 vs. duration on logarithmic paper (Fig. 21-37a). Draw a straight line through the plotted points and extend it to longer durations, usually 4 or 5 days, or even as much as 10 days in some locations of high rainfall.

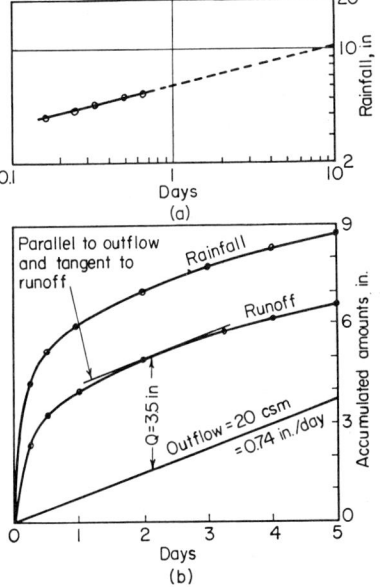

FIG. 21-37. Method of determining required storage for a given frequency of emergency-spillway flow: (a) extension of limited rainfall data, (b) graphical solution for required storage.

3. Determine the runoff curve number for the watershed. This is given as 82 in Example 21-17.
4. Determine the runoff amounts for selected rainfalls plotted in step 2, using Fig. 21-6 to estimate runoff.
5. Plot accumulated runoff vs. time (Fig. 21-37b).
6. Compute the average release rate of the low-flow spillway. In this case the selected release rate is 20 cfs/sq mi because of downstream limitations. At the dam, where the drainage area is 1.86 sq mi, the release rate would average 20 × 1.86 = 37 cfs. Since an average release rate of 1 cfs/sq mi will discharge 0.0372 in. runoff from the drainage area, the rate of 20 cfs/sq mi will release 20 × 0.0372 = 0.74 in./day.
7. Plot the accumulated release on Fig. 21-37b. This will be a straight line as shown.
8. Find the maximum difference between the accumulated inflow of flood runoff and the accumulated release. Set a drafting triangle to the slope of the release line, slide it upward to the inflow line, and where the slope is tangent to the inflow, find the maximum difference as shown on Fig. 21-37b. This is the required 50-year-frequency floodwater storage.

e. Elevation of Emergency Spillway. Convert the required floodwater storage to acre-feet, and enter the elevation-storage curve of Fig. 21-36 to find the required elevation of the emergency-spillway crest.

f. Emergency-spillway Dimensions. These are determined by flood-routing the spillway design flood to get the bottom width at the control section, using given side slopes (usually 3:1) and routing the freeboard design hydrograph to get the freeboard. Table 21-40 can be used to get a preliminary idea of the spillway size. Detailed spillway dimensions are determined as follows:

1. Compute a rating curve for the selected spillway.
2. Plot a discharge storage curve, starting the values at the proposed emergency-spillway crest, and develop flood-routing curves.
3. Select the emergency-spillway design flood. The probable maximum 6-hr rainfall for this location is 24.4 in. (Fig. 21-35). Following the criteria of Table 21-41, the design storm rainfall is three-eighths of this, or 9.15 in. The procedure of Example 21-17 is used to develop the design hydrograph.
4. Route the emergency-spillway design flood, assuming the reservoir is partly still (a 4- to 10-day drawdown of the full pool is usually assumed). The maximum allowable velocity for the given vegetative conditions of the exit channel is selected from Table 21-32. Velocities exceeding the critical are permissible because of the infrequency of spillway usage. The routing determines the depth of flow at the control section of the emergency spillway. Several routings may be needed to find the desired combination of depth of flow and control-section width.
5. Select the freeboard design flood. From step 3 and Table 21-41, the freeboard design storm rainfall is $\frac{2}{3} \times 24.4 = 16.27$ in. The design-flood hydrograph is developed by the method of Example 21-17.
6. Route the freeboard design hydrograph. The spillway dimensions established in step 4 are used, and the routing gives the maximum depth of flow expected in the control and upstream sections of the emergency spillway. In this case, limiting velocities are not used as a design criterion because of the remote possibility of such an event. The elevation of the top of the dam is based on the depth in the upstream section or reservoir.

F. Channel Works

Low initial cost makes stream channel works a popular flood-prevention measure. However, there may be a high maintenance cost, especially where floods occur annually. Temporary upstream floodwater storage can be used to make channel works longer-lived and more effective. Since channel works may increase downstream flood peaks, some upstream storage is often essential to avoid increased damage below the project.

1. Construction Types

a. Levees. Levees (or *earth dikes*) are usually made of random earth fill and are used to confine streamflow within a specified area along the stream or to prevent flooding due to waves or tides. *Floodwalls* serve the same purpose but are often built of reinforced concrete for the protection of valuable properties. Levees are relatively inexpensive but, like floodwalls, they may fail through overtopping or through leaks or seepage. Lowlands protected by levees usually need drainage ditches to collect local runoff, which is passed through the levee by a conduit having a flap gate on the stream side. The local runoff drains after the flood drops below the flap gate; or if there is danger of great damage, it is pumped. Since levees on both sides of a stream prevent the use of natural valley storage during floods, the flood peak is increased between the levees and may persist for some distance downstream. In Kansas, levee failures "were traceable to numerous causes, among which were poor construction, grades not in conformity with flood profiles, inadequate maintenance, as well as failure to provide sufficient floodway" [50].

b. Floodways. Large-capacity channels constructed to divert flows from damageable areas are usually referred to as *floodways*. They are usually additions to existing stream systems.

c. *Channel Improvements.* There are several kinds of channel improvements with the common purpose of increasing the capacity of an existing channel. Those most frequently used will be discussed briefly.

Cutoffs are channels constructed to bypass large bends, or oxbows, and thereby relieve an area normally subjected to flooding or channel erosion.

Pilot channels consist of a series of cutoffs for converting a meandering stream into a straight channel of greater slope. Pilot channels are built only large enough to start flow along the new course, with the expectation that erosion will occur during floods and create channels of adequate capacity. Channels upstream from pilot channels will usually degrade unless bedrock or some other grade control intervenes. In extreme cases, degradation may continue in tributary streams and may even create a serious gully problem in the entire watershed. A straight, steep channel may also result in higher flood peaks and greater sediment loads downstream, thereby offsetting benefits at the site. This type of channel improvement is not recommended unless drop structures or other means of grade control are used to control tributary degradation.

Clearing and snagging is the improvement of a channel by clearing brush and trees and removing fallen trees and other channel obstructions. The channel capacity is increased by a reduction in Manning's n. This type of channel work also must be used with caution, since it may cause bank erosion and increase downstream peak flows. Annual maintenance of this improvement is generally necessary.

Channel lining with durable materials is usually too expensive for the protection of agricultural lands and is used mainly in cities or in other high-value-property areas. Well-constructed channel linings have a relatively long life and low maintenance cost. They can double or triple the channel capacity, but in so doing may increase downstream flood peaks.

2. Criteria

a. *Levees.* The 10-year-frequency flood is a common level of design for levees protecting agricultural land. Top widths of important levees should be of roadway width (8 ft or more) to permit the use of vehicles for inspection and maintenance. Side slopes should be at least 4:1 if levees are to be grazed.

b. *Floodways.* Criteria for floodways apply principally to appurtenances such as diversion dams and boundary levees and to the design of a stable channel. They will not be discussed here.

c. *Channel Improvements.* The level of protection of agricultural lands by channel improvement seldom exceeds control of a 10-year-frequency flood. Where cutoffs or pilot channels are used, criteria do not apply, since there is no control over the resulting channel development. In general, criteria used in all channel improvements should be directed toward avoiding significant increases in downstream peaks and degradation of the upstream channel system.

3. Hydrologic Considerations. Hydrology strictly for design purposes can be relatively simple, but when downstream property owners might suffer increased flood damage because of upstream channel works, elaborate hydrologic studies are usually necessary. In such cases, detailed topographic maps are needed for construction of stage-storage curves; water-surface profiles are needed for conditions before and after improvement; and a considerable number of detailed flood routings must be made. Even the smallest channel project may require much data and analysis if an accurate answer is sought.

a. *Levees.* Figure 21-38, based on a levee analysis [50], can be used for the design of levees on small watersheds and for the preliminary design on large watersheds. Note that when the capacity of the channel is increased by enlargement or cleaning, the valley capacity is also increased.

Example 21-25. Levees on both sides of a channel are proposed in a valley 2,500 ft wide. Estimate the average height of levees required to contain the 10-year-frequency flood when the floodway width between the levees is 300 ft.

1. Determine the 10- and 100-year-frequency flood peaks. For this example, these are 9,200 and 16,800 cfs, respectively.

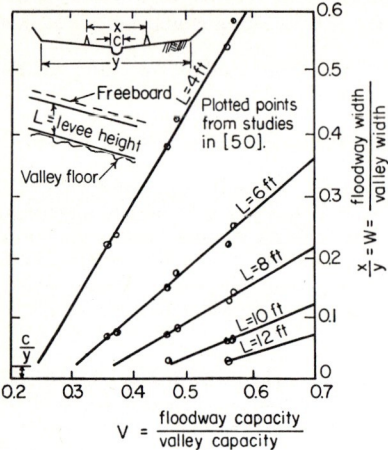

FIG. 21-38. Graph for making preliminary estimates of required levee height. *Note:* Use 100-year-frequency flood width and discharge when valley boundaries are indefinite. (*From U.S. Soil Conservation Service* [33].)

2. Compute ratios of capacity and width. $V = 9{,}200/16{,}800 = 0.55$; $W = 300/2{,}500 = 0.12$.

3. Estimate the levee height. Enter Fig. 21-38 with V and W and find $L =$ about 8.1 ft. Freeboard (additional height) may be used for possible fill or foundation settlement, wave action, or other contingencies.

4. Check floodway velocities and downstream effects. The hydrograph for the design flood can be routed through the floodway. Figure 21-38 can be used for more complicated problems, such as the determination of the most economic floodway width and levee height. For such cases, data on crop-production costs and benefits, earth-moving costs, and flood frequencies will be needed. Since it is easier to raise the levee height than to widen the floodway, the floodway is generally constructed as wide as possible at the start. The narrower the floodway or the greater the height of the levees, the greater the flood damage when the levees are breached. This potential damage should enter into the economic study if one is made.

b. *Floodways.* A design flood is routed through the proposed floodway. Where detailed studies are not justified, as with very small projects, the use of Manning's formula may be sufficient for estimating the floodway capacity.

c. *Channel Improvements.* Since pilot channels are intended to be enlarged by erosion, no hydrology is involved in the channel design. If it is necessary to make an estimate of the eventual effects of the enlarged channel on flood damages downstream, routings of past floods will be sufficient. Flow-duration curves may be needed for studies of upstream degradation due to change in channel slope, although it can generally be assumed that the pilot channel will eventually return to the grade of the original channel unless bedrock interferes. Induced erosion upstream may tend to give the new channel an increased but stable slope temporarily.

Example 21-26. A meandering stream has a slope of 4 ft/mile ($s = 0.00076$). A length of 2.2 miles of the stream is shortened to 1.3 miles by means of a pilot channel. Determine the depth of overfall that can be expected to develop at the head of the pilot channel.

Compute the expected overfall depth using the relation

$$O_e = G(C_o - C_n) \qquad (21\text{-}30)$$

where O_e is the expected overfall depth in ft, G is the present channel grade in ft/mile, C_o is the old channel length in miles, and C_n is the new channel length in miles. In this example, $O_e = 3.6$ ft.

Minor channel works such as clearing and snagging can be evaluated using Manning's formula to determine channel velocities. The reduced travel time through the reach can be used in routing the design hydrograph to determine downstream effects. The triangular hydrograph of Fig. 21-11 is suitable for such routings.

G. Irrigation

1. General. The hydrologic considerations in irrigation design vary with the size and type of project. Large-scale irrigation developments [51] may involve engineering, economic, social, legal, and political problems concerning water and its use; sometimes they may be regional or international in scope. Engineering and legal

FIELD APPLICATIONS   21-79

problems are common to small-scale projects, and economics may be a critical factor with marginal projects or in the selection of the type of irrigation.

a. Methods of Irrigating. The method or type of irrigation varies with topography, soils, crops, available water supply, and often with the irrigator's personal preference. The most general types are as follows:

(I) Surface irrigation [52]
    (A) Flood irrigation
        (1) Graded-border irrigation. A field is normally divided into strips 30 to 60 ft wide and 300 to 1,300 ft long, separated by low dikes; water is applied at the high end of the strip and advances as a sheet.
        (2) Level-border or basin irrigation. Level plots are surrounded by dikes; water is applied rapidly and held until it soaks into the soil.
        (3) Contour-ditch irrigation. Controlled flooding from ditches running along the approximate contour allows the water to flow down the land slope; ditches are from 25 to 200 ft apart, depending on topography, slope, and soils.
    (B) Furrow irrigation
        (1) Furrows. Water flows down the furrows between crop rows, which may be on approximate contours or down the slope.
        (2) Corrugations. Shallow furrows 15 to 30 in. apart are run downhill from head ditches; they are used for irrigating hay and small grain.
(II) Subsurface irrigation. An artificial water table is created and maintained at a depth within reach of plant roots. This is used with very permeable soils, such as peats, where topography must be level and smooth [53].
(III) Sprinkler irrigation. Water is pumped through portable pipes and applied by overhead sprinklers. This method is usable with most soils, crops, and land slopes, but water must be clean and free of debris. It is also usable for frost control [54].

b. Hydrologic Aspects. Irrigation project design is a complex field, and specialists, or irrigation engineers, are needed for major projects. These specialists use hydrology in determining the source, amount, quality, and seasonal distribution of water supply; in estimating consumptive use; in planning and designing storage dams and major canals; in planning and designing protection of the irrigation system from cross drainage (flood flows from natural waterways crossing the canals and laterals); in determining infiltration rates and water-storage capacity of the soil; in estimating erosion rates in field distribution systems; and in controlling seepage and excess surface-water drainage.

c. Legal Aspects. Rights to the use of surface water and groundwater are generally subject to the constitutions, statutes, compacts, treaties, and court decisions of the states and countries in which the water exists [55]. These rights vary from state to state. The prospective irrigator should have, or be sure he is able to get, necessary water rights before a detailed planning of the irrigation system begins. (See Sec. 27.)

2. Hydrologic Considerations. Useful information on the engineering design and layout of all types of on-farm irrigation systems is given in several sources [56–58]. Methods are available [59] for the evaluation of old irrigation systems and can also be used in the design of new systems. Additional hydrologic design will generally consist of determining the quantity of water available and the capacity of structures needed to control cross drainage Cross-drainage problems can be greatly reduced (but not completely eliminated) by the use of inverted siphons to convey irrigation water past cross-flowing streams.

a. Small-irrigation-project Design. The major steps in designing a project usually follow the general order shown below, but some may be taken together or may be reversed in different localities. Large projects require much more extensive studies [51, 60].

(1) Water Supply. It is futile to start detailed planning unless necessary water rights are available. A preliminary estimate of the needed supply can be made for a

given acreage, or the acreage that can be irrigated with a given supply can be determined, using the methods described here. Detailed studies can proceed after the water right is available. In states where rights are based on priority, late rights will usually apply only to high flood stages or seepage water and are the least desirable; full or early rights will permit better design. Snowmelt runoff is a desirable source of supply, especially when it occurs just before and during the irrigation season and when little or no storage is needed. Groundwater (artesian or pumped) is another very desirable source. If the rate of supply is low, surface storage can be used to accumulate enough water for each irrigation. A lake or a base flow in a stream or river is a desirable source. Surface runoff from rainfall is the least desirable, but often the only available source; it will be highly variable, and hence large storage may be needed. Combinations of various types of water supply can also be used to develop an efficient project.

(2) *Location of Irrigable Land.* The irrigable land should be near the source of water in order to reduce water losses and conveyance costs. Level or gently sloping land (preferably not over 2 per cent slope) is desirable for all types of irrigation, but not necessary for sprinkler irrigation. Land leveling [61] will make uneven land more suitable; it will increase initial project cost but will reduce annual operating cost. Furrow irrigation on the contour can be used on steeper land but requires more operating care; generally, sprinkler irrigation is used on such land. For maximum efficiency, the irrigated area should be regular and compact in shape.

(3) *Quality of Soil and Water.* A soil survey will aid in selecting suitable irrigation sites. Deep, medium-textured soils are preferable for irrigation. Clay-pan soils, shallow soils, very highly permeable soils with little soil-water storage, and saline or alkaline soils will require extra preparation and management. Irrigation water should be such that it will not cause soil deterioration by increasing salinity. Moderately saline waters, however, can be so managed as to avoid soil damage [62]. Some crops tolerate more salinity than others [63], and crop losses can be minimized by wise selection of crops and proper management practices.

(4) *Supply Design.* A well, lake, or stream can be used as the water supply with a pump or diversion dam to withdraw the water. A supply canal or other means of conveyance, such as pipes or flumes with special structures for crossing streams and gullies, brings the water to the farm. Local flood runoff that might destroy these conveyances must be controlled by dams or diversions or avoided by use of inverted siphons.

(5) *On-farm Design.* Amounts of consumptive use by crops, together with operating losses, must be estimated. Soil intake rates must be determined for sprinkler irrigation, and the most efficient pipe layout must be designed. Ditches, drops, controls, and layouts should be inexpensive and such that they can be maintained by the farmer [57]. Waste control may also be necessary.

b. Water Supply. The precision with which the available water supply and seasonal distribution must be determined depends partly on which of the following two plans is used:

Plan 1. The water supply is determined for a specified per cent chance, usually 80 per cent, which means that the project will have sufficient water for 4 out of 5 years and a deficiency of 1 out of 5, and a considerable quantity may go unused (Fig. 21-39). The controlling factor is the deficiency that can be tolerated. A fixed acreage of irrigated land is used. This is a suitable plan where the irrigation layout can have little flexibility.

Plan 2. The irrigated acreage or the number of irrigations per season varies from year to year, and all available water is used. This plan is generally used where water is scarce.

The total water supply required for an irrigation season is the sum of the uses and losses, less the amount of rain contributing directly to soil moisture. The following are the usual items considered:

1. Seasonal consumptive use. The seasonal total is generally computed for the most demanding crop or combination of crops.

2. *Effective rainfall.* This is the rain falling on the irrigated land and adding to soil moisture.

3. *Farm losses.* Annual on-farm losses vary with the type of irrigation and other factors such as seepage and evaporation.

4. *Leaching requirements.* Estimation of this need is made only where required by the quality of water or soils.

5. *Canal seepage losses.*

6. *Reservoir losses* if a reservoir is used.

7. *Other losses.* Such losses are due mainly to waste through accidents of weather and operation.

c. Seasonal Distribution. Storage is often necessary to assure a supply of water when it is needed. Seasonal distribution ordinarily is of little concern where wells are used; then storage is needed only if delivery from the well is so small that a supply must be accumulated before irrigating. A supply from natural springs may decrease as the season progresses, and thus storage may be needed. Supply from a lake will generally be sufficient if pumps and pipes have sufficient capacity. Snowmelt supply begins in spring and usually continues into the middle of summer; the seasonal pattern is fairly similar each year, although daily fluctuations may be great.

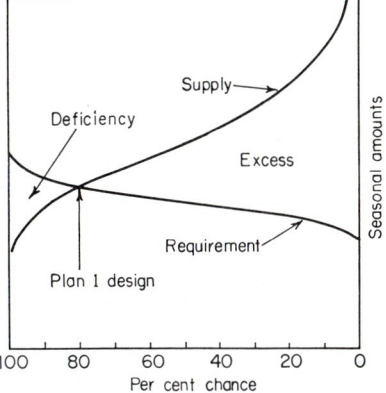

Fig. 21-39. Frequency curves of irrigation supply and requirement, redrawn on arithmetic coordinate paper. Example with negative correlation between supply and requirement.

Snow surveys help to estimate the supply (but not necessarily the distribution) in advance. Base flow and return flow from upstream irrigation may have a consistent seasonal pattern, which can be determined from streamflow records or from observation over several years. Supplies from storm surface runoff are the most variable, both in distribution and total amount. In drier climates total runoff may be several hundred per cent greater than the average in one year and almost zero the next. Table 21-20 shows some typical seasonal distributions on small watersheds. Storage dams are usually needed for surface-runoff control. The required storage is estimated from historical records; if they are not available, runoff and distribution must be estimated.

d. Consumptive Use. The various amounts of water required by different crops are determined by methods illustrated in Examples 21-2 and 21-3. The time to apply irrigation water is usually estimated by means of a table such as Table 21-42. The irrigation should start when the soil-moisture deficiency is about 50 per cent, since part of the field will have even greater deficiencies by the time water can be supplied. The amount of each application varies with soil type and crop. Table 21-44 shows some typical depths of irrigation. The time it takes to irrigate a field depends on the amount of water supplied, the depth of irrigation, and the intake of the soil.

Example 21-27. A cotton field on medium-textured soil in a humid climate has a soil-moisture deficiency of 50 per cent. Determine a maximum application amount of irrigation water, in inches of depth over the field.

1. In Table 21-43 find a maximum of 2.3 in. of moisture-holding capacity per foot of soil depth for medium-textured soil. In Table 21-44, find a maximum irrigation depth of 36 in. for cotton in a humid area.

2. Compute the maximum irrigation-water application. The moisture-holding capacity (inches per foot) times the deficiency (divided by 100 to make a decimal) times the irrigation depth for the crop (divided by 12 to convert from inches to feet) gives the required application of water in inches: $2.3 \times (50/100) \times (36/12) = 3.45$ in. If a 75 per cent efficiency in application is expected, the total application must be $3.45 \times (100/75) = 4.60$ in.

Table 21-42. Soil Moisture and Appearance Relationship Chart*

Moisture deficiency, in./ft	Soil-texture classification				Moisture deficiency, in./ft
	Coarse (loamy sand)	Light (sandy loam)	Medium (loam)	Fine (clay loam)	
	Field capacity				
0.0	Leaves wet outline on hand when squeezed	Appears very dark; leaves a wet outline on hand; makes a short ribbon	Appears very dark; leaves a wet outline on hand; will ribbon out about 1 in.	Appears very dark; leaves slight moisture on hand when squeezed; will ribbon out about 2 in.	0.0
0.2	Appears moist; makes a weak ball	Quite dark color; makes a hard ball	Dark color; forms a plastic ball; slicks when rubbed	Dark color; will slick and make ribbon easily	0.2
0.4	Appears slightly moist; sticks together slightly	Fairly dark color; makes a good ball	Quite dark; forms a hard ball	Quite dark, will make a thick ribbon; may slick when rubbed	0.4
0.6	Very dry, loose; flows through fingers (wilting point)	Slightly dark color; makes a weak ball	Fairly dark; forms a good ball	Fairly dark; makes a good ball	0.6
0.8		Lightly colored by moisture; will not make ball	Slightly dark; forms a weak ball	Will make ball; small clods will flatten out rather than crumble	0.8
1.0					1.0
1.2		Very slight color due to moisture (wilting point)	Lightly colored; small clods crumble fairly easily	Slightly dark; clods crumble	1.2
1.4					1.4
1.6			Slight color due to moisture, small clods are hard (wilting point)	Some darkness due to unavailable moisture; clods are hard, cracked (wilting point)	1.6
1.8					1.8
2.0					2.0

*From Merriam [64].

Table 21-43. Moisture-holding Capacities of Soils*

Soil texture or type	Inches water per foot of soil
Very coarse texture; very coarse sands	0.40–0.75
Coarse texture; coarse sands, fine sands, and loamy sands	0.75–1.00
Moderately coarse textures; sandy loams, loams, and silt loams	1.50–2.30
Moderately fine textures; clay loams, silty clay loams, and sandy clay loams	1.75–2.50
Fine textures; sandy clays, silty clays, and clays	1.60–2.50
Peats and mucks	2.00–3.00

* From Quackenbush et al. [57].

e. Effective Rainfall. Water supply for consumptive use by crops consists of the effective rainfall on the irrigated area and the water brought in by conveyances. *Effective rainfall* is the total rainfall falling during the growing season minus that occurring immediately after an irrigation when the soil is already filled to capacity, so that additional moisture either goes to deep storage beyond the root zone or is lost

Table 21-44. Crop Irrigation Depths
(Soil depth in inches)

Crop	Humid areas*	Semiarid to arid areas†
Alfalfa	36–42	60–120
Beans	....	36–48
Beets (sugar)	....	48–72
Broccoli	....	24
Cabbage	....	24
Clover (ladino)	....	24
Corn (maize)	24–36	48–60
Cotton	24–36	48–72
Grapes	24–30	48–72
Orchards:		
Citrus	....	48–72
Deciduous	36–60	72–96
Pasture	18–36	36–48
Peas	....	36–48
Potatoes (white)	12–24	26–48
Small grain	18–30	48
Sorghum	20–30	
Soybeans	18–36	
Tobacco	15–24	
Tomatoes	....	72–120
Truck crops:		
Shallow-rooted	9–12	
Medium-rooted	12–24	
Deep-rooted	24–30	

* From Quackenbush et al. [57].
† From U.S. Soil Conservation Service [65]. Larger figure applies to arid areas.

as runoff. Unpublished work by Renfro [66] was used to develop Table 21-45 and Eq. (21-31), which can be used for general estimates:

$$\text{Effective rainfall} = ER_g + A \quad (21\text{-}31)$$

where E is the ratio from Table 21-45, R_g is the growing season rainfall in inches, and A is the average irrigation application in inches.

Example 21-28. Given an average seasonal rainfall at Lincoln, Nebr., of 19.1 in., a crop consumptive use of 32.9 in., and an average irrigation application of 3 in., find the total effective rainfall.

1. Compute the ratio of consumptive use U in inches to rainfall during the growing season R_g in inches. This is $32.9/19.1 = 1.72$.
2. In Table 21-45, for $U/R_g = 1.72$, find the ratio E, which is 0.59 by interpolation.
3. Compute the effective rainfall using Eq. (21-31): $0.59 \times 19.1 + 3 = 14.3$ in.

Table 21-45. Ratios for Use in Estimating Effective Rainfall*

U/R_g	E	U/R_g	E
0	0	2.4	0.72
0.2	0.10	2.6	0.75
0.4	0.19	2.8	0.77
0.6	0.27	3.0	0.80
0.8	0.35	3.5	0.84
1.0	0.41	4.0	0.88
1.2	0.47	4.5	0.91
1.4	0.52	5.0	0.93
1.6	0.57	6.0	0.96
1.8	0.61	7.0	0.98
2.0	0.65	9.0	0.99
2.2	0.69		

* From U.S. Soil Conservation Service [33].
NOTATION: U = consumptive use in in., R_g = rainfall during the growing season in in., and E = ratio for finding total effective rainfall.

f. Field Efficiency. The per cent of the total volume of water delivered to the field and finally consumed by evapotranspiration is referred to as *field efficiency* [57]. Efficiency varies with many factors, including the experience of the irrigation operator. Assuming proper planning and operation of the field irrigation system, the ranges of efficiency to be expected are given in Table 21-46.

Table 21-46. Irrigation Field Efficiencies*

Method of irrigation	Range of efficiency, %
Graded borders	60 to 75
Basins and level borders	60 to 80
Contour ditch	50 to 55
Furrows	55 to 70
Corrugations	50 to 70
Subsurface	Up to 80
Sprinklers	65 to 75

* From Quackenbush et al. [57] and U.S. Soil Conservation Service [58].

Example 21-29. If the expected efficiency is 60 per cent, the consumptive use for the irrigation season is 2.4 acre-ft (28.8 in.) of water per acre, and effective rainfall is 11 in., the amount to be delivered to the farm is $(28.8 - 11) \times {}^{10}\!\%_{60} = 29.7$ in., or 2.5 acre-ft/acre.

g. Leaching Requirement. When soils and water are of good quality, large applications of irrigation water will leach out plant nutrients and will require additional drainage and will therefore be wasteful. However, with saline or alkaline conditions, which may result either from poor soil or poor water, it is generally necessary to apply extra water to leach the salts out of the root zone and prevent movement of salts to the surface, where they will further injure crops. Drainage of the leaching water is necessary and is usually accomplished with deep ditches. Level borders or basins rather than furrow irrigation should be used under these conditions. The amount of water required for leaching can be estimated [62, 63], but usually local experience will govern.

h. Delivery Losses. Canal or ditch seepage is the principal loss between the water source and the delivery point. This not only reduces supply, but may also cause damage to land adjacent to the canal or ditch. Seepage losses may vary from 10 to 70 per cent of the amount entering a canal [68]. Additional losses occur from transpiration by vegetation along a canal or ditch. Where the water supply is limited, control of this vegetation by cutting or spraying with chemicals may be necessary [67]. Evaporation from a canal water surface is ordinarily not considered. In general, where water losses are likely to be large, pipe, chutes, flumes, or lined canals or ditches can be used to convey the water [69, 70]. Table 21-47 can be used as a guide to the average upper limits of losses to be expected. An equation for estimating delivery losses is

$$q_d = q_0 - \frac{SW}{86,400} \tag{21-32}$$

where q_d is the discharge at the delivery point in cfs, q_0 is the inflow to the ditch in cfs, S is the seepage rate in ft^3/ft^2 of the wetted area, from Table 21-47, and W is the wetted area in ft^2. The figure 86,400 is the conversion factor to convert days to seconds.

Example 21-30. An irrigation delivery ditch is 1.7 miles long, is cut through sand and sandy loam, and has a 1-ft bottom width with 1.5:1 side slopes. If 4.1 cfs enters the ditch at the design depth of 0.7 ft, estimate the discharge that can be expected at the farm after steady flow is established.

1. Select the seepage rate from Table 21-47. Using the fourth item (sand and sandy loam), this is 3.4 ft^3/ft^2 of wetted area per day.
2. Compute the total wetted area. The wetted perimeter times the length of ditch in feet, or 3.52 × 1.7 × 5,280, is 31,596 ft^2 (use 31,600 ft^2).

Table 21-47. Average Maximum Seepage Rates for Canals*

Canal soil material	Seepage, ft^3/ft^2 of wetted area per day
Sandy loam	8.2
Gravelly loam	5.3
Fine sandy loam and adobe	3.8
Sand and sandy loam	3.4
Loam and sandy loam	3.3
Adobe	3.0
Fine sandy loam	2.1
Loam and adobe	1.4
Loam	1.1
Silty clay	0.9
Sand and silty clay	0.4
Sand and clay	0.1
Loam and gravelly loam	0.1

* From Rohwer and Stout [68].

3. Compute the rate at the delivery point: 4.1 − (3.4 × 31,600)/86,400 = 2.86 cfs, or about 2.9 cfs after steady flow is established.

i. Storage Losses. These are evaporation and seepage losses in the reservoir and are to some extent controllable [49, 69, 70].

j. Other Losses. An estimate of operational losses, due to accidents or other uncontrolled losses, is generally used as a safety factor in supply design.

k. Recovery of Losses. Seepage and waste water (return flow) are often recovered, in streams or underground, and used for further irrigation. In some localities water is used many times before it becomes unsuitable for further use. Where states permit rights to seepage flows, it may be illegal to reduce seepage or waste if doing so violates someone's water right [69].

l. Estimation of Required Reservoir Storage. A summary of the gains and losses can be made as shown in the following example.

Example 21-31. Estimate the required irrigation storage, using a seasonal distribution of surface runoff for a location in the midwestern United States where corn and beans are to be irrigated.

1. Determine the total seasonal irrigation-water requirement. Prepare a table such as Table 21-48, which shows the various uses and their quantities. Item 1 is a total of

Table 21-48. Computation of Total Required Supply

Item	Acre-feet	Accumulated acre-feet
1. Consumptive use, 80 acres:		
Beans (3 months), 14.4 in............................	96	
Corn (4 months), 21.1 in.............................	142	142*
2. Effective rainfall: average 4-month irrigation-season rainfall, 16.3 in.; average application, 4 in.; total effective rainfall = 0.50 × 16.3 + 4 = 12.2 in.	−81	61
3. Field efficiency: furrow irrigation, $61 \times {}^{100}/_{55}$...........	....	111
4. Leaching requirement...................................	0	111
5. Other losses: estimated at 10 per cent.................	11	122
6. Canal seepage losses for a total of 5 runs (6 acre-ft per run)...	30	152
7. Reservoir losses:		
Evaporation, 8-acre surface...........................	43	195
Seepage...	24	219

* Use the larger of the two values in middle column.

monthly increments computed using the methods illustrated in Examples 21-2 and 21-3. Two crops are to be grown, but only one occupies the entire acreage in any one year; therefore the greater of the two consumptive uses is taken. Item 2 is computed using Eq. (21-31), with the selected consumptive use of item 1 (21.1 in.) in the ratio U/R_g. Item 3 is computed using the method of Example 21-29 and an expected field efficiency of 55 per cent. Item 4 is zero since no leaching is required. Item 5 is based on local experience for this case. Item 6 is estimated using the method of Example 21-30. Item 7 is an estimate of reservoir evaporation and seepage losses, assuming the reservoir to be two-thirds full. The total use of 219 acre-ft is equal to the design supply. If it is greater than the amount available, the acreage may be reduced, or the losses determined more closely, or more runoff must be diverted into the reservoir.

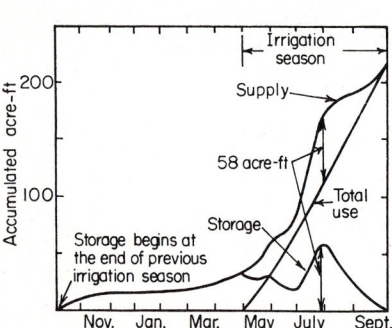

Fig. 21-40. Graphical solution for required irrigation storage, assuming all losses occur during irrigation season, proportional to consumptive use.

2. Determine the amount and distribution of supply, using local streamflow data if available, or using Table 21-20, and plot the accumulated average (or 80 per cent chance, if used) supply as shown on Fig. 21-40.

3. Plot an accumulated total-use curve as shown on Fig. 21-40, assuming that the total use and losses occur during the irrigation season, with losses proportional to the monthly consumptive use.

4. Find the maximum difference between the supply and total use (Fig. 21-40), or plot the storage curve as shown and find the maximum ordinate. The required storage for this example is 58 acre-ft. The total use should not rise above the supply. If it does, a more refined analysis can be made, with losses used as they occur, or the irrigated acreage can be reduced. Otherwise, some method of reducing losses is needed.

After determining storage requirements, it is necessary to estimate the maximum required capacities of the conveyance system, using the probable largest single irrigation run as a design level.

FIELD APPLICATIONS 21-87

3. Hydraulic Considerations. The hydraulics involved in the design of storage or diversion dams and water conveyances will not be discussed here. However, special applications in irrigation are discussed below.

a. Surface Irrigation. When furrow, corrugation, or border irrigation is used, small siphon tubes made of light metal, rubber, or plastic may be used to convey water from the supply ditch to the field. Table 21-49 gives siphon-tube capacities [57].

Table 21-49. Discharge, in Gallons per Minute, from Siphons Operating under Different Heads*

Diameter of siphon, in.	Head, in.					
	2	3	4	5	6	9
½	1.3	1.6	1.8	2	2.1	2.7
¾	3	4	5	5.5	6	7
1	4	5	7	8	9	11
1¼	8	10	12	13	15	18
1½	13	16	18	21	24	28
1¾	17	21	25	28	32	38
2	21	27	32	36	40	

* From Quackenbush et al. [57]. Head is the difference in height of water in supply ditch and at discharge end of tube.

Gated pipes are more efficient than supply ditches. When they are used no siphon tubes are needed. Other types of field distribution systems, such as flexible hose of various materials or buried pipelines with risers, are sometimes used. The hydraulics of these systems must be specially computed, or the suitable discharge found by trial during irrigation. Erosion in furrows or corrugations is minimized by using Criddle's empirical relation [52],

$$q_m = \frac{10}{S} \quad (21\text{-}33)$$

where q_m is the permissible maximum discharge in gpm, and S is the furrow slope in per cent.

Determination of desirable maximum length of run involves consideration of land slope and soil-infiltration capacity. If a run is too long, water soaks in too deep at the head of the furrow by the time the stream reaches the lower end. If the run is too short, extra supply ditches are needed and irrigation labor increases. Erosion from runoff due to rainfall must also be considered. Typical lengths of run are shown in Table 21-50.

b. Subsurface Irrigation. Water is usually applied through open ditches, but underground means, such as moles or tile drains, can be used. The supply ditches are also used for lowering the water level. They contain check structures having gates or flashboards spaced at vertical intervals of about 6 in. The flashboards are in place when irrigation water is being added, and out when excess water is being removed. The ditch system does not require extensive hydraulic computations; the use of Manning's formula is adequate. The design must provide for the proper arrangement of ditches to maintain a uniform depth of water table and the location of a suitable outlet for drainage. Since the soils suitable for subsurface irrigation are highly permeable, there will be no runoff from the irrigated area except from unusually high intensity rains.

c. Sprinkler Irrigation. Permanent or portable sprinkler-irrigation systems include a pump at the water source (with debris screens if lake or stream water is used), a main supply pipe, and lateral pipelines on which sprinkler nozzles are mounted at a level suitable to the crop. Often a fertilizer solution is added to the irrigation water in the line between the water source and the pump. The hydraulics of the pipelines has been reduced to a minimum of design effort [71, 72].

d. Measurement of Flow. When irrigation water is purchased, or when water rights limit the supply within a certain range of discharge, the flow is measured with weirs or Parshall flumes, for which discharge tables and construction information are available [27].

Table 21-50. Lengths of Run for Furrow or Corrugation Irrigation, in Feet

Furrow grade, %	Fine-textured soils			Medium-textured soils			Moderately coarse textured soils		Coarse-textured soils	
	Irrigation application, in.									
	2	4	6	2	4	6	2	4	2	4
Eastern (humid) United States*										
0.1	800	800	....	800	800	....	660	920	300	425
0.2	1,000	1,000	....	920	1,000	....	560	800	260	360
0.3	950	1,000	....	720	1,000	....	450	640	205	290
0.4	800	900	....	620	880	....	380	540	175	250
0.5	550	550	....	550	620	....	340	480	150	220
Western (semiarid) United States†										
0.25	1,050	1,500	1,800	825	1,150	1,400	500	700	225	325
0.50	725	1,000	1,250	550	775	950	325	475	150	225
0.75	575	800	975	425	625	750	275	375	125	175
1.0	475	675	850	375	525	650	225	325	100	150
1.5	375	550	675	300	425	525	175	250	...	125
2.0	325	475	575	250	350	425	150	225	...	100
2.5	300	425	500	225	325	400	125	200		
3.0	275	375	450	200	300	350	125	175		
4.0	225	325	400	175	250	300	100	150		

* From Quackenbush et al. [57].
† U.S. Soil Conservation Service [65].

H. Drainage

Agricultural drainage is the removal of surplus gravitational water from the surface or subsurface of farmland to improve soil conditions for plant growth.

1. Types. *Surface drainage* is accomplished by grading and smoothing the land to remove barriers and to fill in depressions, by digging ditches to remove water, or by diverting runoff from adjacent source areas to ditches or natural waterways.

Subsurface drainage removes excess water from within the soil by means of tile or mole drains, lowering the water table to below the root zone.

Levees or dikes are used in conjunction with drainage works to exclude flood or sea water. Pumps are used to remove excess local water from the protected area.

2. Legal Aspects. Drainage by individual farmers is practically a free enterprise, with only the common-sense limitation of preventing injury to neighboring lands. Most drainage systems, however, must pass through several farms, and since

1790, drainage laws have established procedures and safeguards for developing large drainage projects [73]. (See Sec. 27.)

a. Drainage Districts. A *drainage district* is a legal subdivision of government to provide drainage of land for the benefit of farmers within the district. It is directed by a group of elected or appointed commissioners. The district develops plans for water disposal, provides the organization and controls funds, obtains easements and rights-of-way, and distributes construction and operation costs among its members. It has the power to tax within the district and usually to sell bonds for financing its work.

b. Drainage Associations. **Drainage associations** are unincorporated groups of voluntary members operating under articles of association. Unless restricted by state law, the association may attain official status by recording the articles of association with the county or parish clerk. Organization of the association is simple, and officials are elected to operate it. Funds must be raised by voluntary subscription, since there are no provisions for taxation.

c. Informal Groups. A few landowners may create an informal group by signing a group agreement. The entire operation of the group is based on voluntary agreement.

3. Planning. History shows many failures of drainage projects. Difficulties in many cases were due to poor planning, piecemeal methods of drainage, or lack of maintenance. Lack of proper machinery for large construction and simple maintenance was often the cause of failure before the general use of the dragline excavator began in 1906. Modern equipment, such as trenching machines, dragline excavators, and more efficient pumps, has reduced the physical problem of installing drainage systems. The size of a drainage project will determine the intensity of the planning, but in general the method of approach is the same for all projects. Planning consists for the most part of answering these questions [74]: Does the excess water come from rainfall, tides, irrigation, seepage, or artesian pressure? Can the excess water be removed from the soil? How much water must be drained or excluded? Is there an outlet for drained water? What system will give the best results? Even small-scale planning requires a review of available data for the problem area, such as geologic reports, previous surveys or plans, and state engineering publications. Field reconnaissance is needed to form an idea of further needed investigation, and topographic maps, soil surveys, and water-table surveys are usually needed.

a. Causes of Drainage Problems. Successful drainage of wet lands may require removal of both surface and subsurface water, but generally only one type is involved. Surface-drainage problems are caused by ponding due to lack of slope, by the small capacity of natural or constructed channels, or by the lack of proper outlets. Often outlets cannot be made low enough to drain the wet land because of high elevations of lakes, tidewaters, or highway culverts. Subsurface-drainage problems may be due to a high water table, to subsurface barriers of low permeability such as clay lenses, to seepage from irrigation canals and levees, or to other causes.

4. Surface Drainage. A surface-drainage system is composed of *field drains*, which collect the excess water from the land surface and conduct it to *lateral ditches*, in which it flows to a *main ditch* and then to the *drainage outlet*, usually a natural waterway.

a. Field Drains. These are shallow ditches with flat side slopes, which farm machinery can cross. A field drain is normally 9 to 12 in. (sometimes as much as 18 in.) deep with side slopes of 6:1 or flatter. Horizontal spacing of field drains depends on the soils, the topography, and the amount of drainage expected. Where the drains are parallel, they are usually not more than 650 ft apart on sandy soils, 200 ft apart on organic soils, and about 300 ft apart on other soils. Various layouts of field drains are used [35]. They may be in a cross-slope parallel system, a random system, or a bedding system.

b. Lateral Ditches. These are deeper than field drains, usually with a depth of a foot or more, and often with flat side slopes that farm machinery can cross.

c. Main Ditch. The main ditch is normally constructed with a dragline and requires most of the planning and design in a drainage project. Main ditches generally run along property lines or roads, although sometimes a small natural channel may be used (after enlargement) in its original location. The capacity of main ditches often needs to be greater than needed to carry water from the laterals because of runoff from uplands.

d. Hydrologic Considerations. Guides and tables for the design and layout of field drains and farm lateral ditches are available [35, 75]. Generally, the chief problem is the determination of the water and sediment capacity of the main ditch. The customary hydrologic approach is difficult to apply on very flat land with indefinite drainage boundaries; in such cases the field experience of drainage engineers becomes the principal basis of design. *Drainage curves* (Fig. 21-41) are normally used except in areas where drainage is needed because of irrigation water. Since the curves of Fig. 21-41 are generalized over large areas, no specific frequency applies for any particular curve. In localities needing drainage, natural channels are usually physiographically undeveloped and the land is so flat that normal floodwaters do not flow to them. Thus drainage ditches are used also to provide some function of flood control, and the division between functions is often impossible to make. Areas of application of the curves of Fig. 21-41 are as follows:

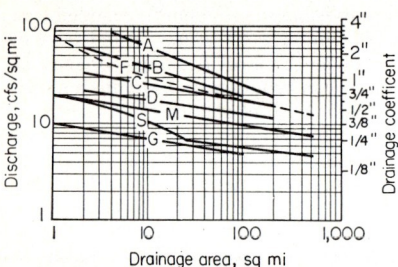

Fig. 21-41. Drainage-ditch design curves. (*U.S. Soil Conservation Service* [75, 76].)

Curve A. North-central and northeastern United States, for good protection from overflows; Gulf states, maximum for hill areas [75].

Curve B. North-central and northeastern United States, for excellent drainage; Gulf states, minimum for hill areas [75].

Curve C. North-central and northeastern United States, for good drainage and a general curve for grain crops; Mississippi Delta, a general curve; western plains of Oklahoma and Texas [75].

Curve D. North-central and northeastern United States, for fair drainage and a general curve for improved pastures; Gulf states, for improved pastures and riceland [75].

Curve F. Florida Everglades [76].

Curve G. Gulf and Atlantic Coast "flatwood" areas [75].

Curve M. Red River Valley, Minnesota, and North Dakota, for areas requiring better drainage [75].

Curve S. Red River Valley, Minnesota, and North Dakota, for ordinary drainage [75].

Obviously, there is no upper limit to such curves, but in practice it is seldom necessary to drain more than 3 in./day from an area. The scale on the right of Fig. 21-41 gives the *drainage coefficients*, which are the water depths in inches drained from an area in one day. These coefficients enable the designer to compare drainage methods by ditches, tile lines, and pumps.

e. Hydraulic Considerations. Once the capacity of a main ditch is selected, the ditch design must overcome the limitations due to the following factors:

1. Small range of possible channel slopes. Elevation of the outlet will often control the ditch slope. Where the ditch flows into a river, backwater effects can be expected at times when the ditch should be draining land.

2. Practical difficulties of maintaining low values of Manning's n. Ditch maintenance is seldom a regular practice; controlling vegetation every two or three years is desirable. Generally, the average value of n will be between 0.03 and 0.06 [77].

3. Hydraulic radius. Deep ditches can be constructed, but their effective depth will be controlled by elevations of existing road culverts and bridges and by effective elevation of the outlet.

4. Sediment inflows. The sediment problem in drainage ditches is usually caused by erosion on upstream areas rather than on the land being drained. Upstream sediment can be controlled to some extent by using conservation measures such as

strip cropping, contouring, and terracing on the uplands. In downstream areas sediment traps can be built, or the sediment-laden floodwaters can sometimes be diverted. In areas such as the Missouri River bottoms of Iowa, large sediment inflows are used to raise the general elevation of the bottomland. Runoff from a stream carrying sediment from the uplands is diverted to a section of land reserved for settling out the soil material. When this section has been built up, another section is used for the settling basin while the first is deep-plowed, ditched, and used for crop production. If the sediments are sterile, the raised lands must be made fit for crop production by the use of fertilizers and soil-improving crops.

The hydraulic determinations in ditch design include (1) determining the most desirable hydraulic grade line, (2) assuming desirable ditch cross sections to match required depths, and (3) developing profiles of ditch bottoms for the hydraulic grade line selected, using a backwater method such as Leach's [40, 44]. Effects of combined flow from laterals are also considered in ditch design [35, 75].

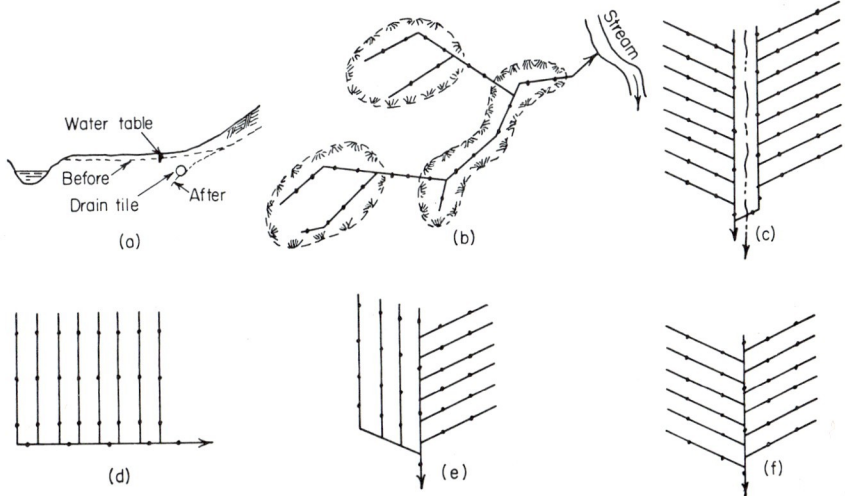

FIG. 21-42. Tile-drain layouts. (a) Interception method; (b) random; (c) double main; (d) parallel; (e) gridiron; (f) herringbone.

5. Subsurface Drainage. Beauchamp [78] presents a comprehensive discussion of subsurface drainage by use of tile drains. In tile drainage, free water from the soil enters the tile line through openings at the joints and flows to an outlet. The depth and spacing of tile lines depend on the quantity of water to be removed per 24 hr (the drainage coefficient), the permeability of the soil, and the depth of the root zone. Permeability can be estimated by one of several methods [74, 75]; the requirement in appraising an area to be tiled is adequate sampling and correct interpretation of the measurements.

a. Layouts. The distribution of the wet land to be drained determines the pattern of the tile layout. Where seepage is caused by a perched water table or an irrigation canal on a hillside, it can be intercepted by a tile line above the wet land (Fig. 21-42a). If the wet land is in the form of potholes or scattered small areas, a random system of tile lines can be used (Fig. 21-42b). A large area of wet land can be drained by any one of a number of layout patterns designated by such terms as "parallel," "gridiron," "herringbone," or "double-main" (Fig. 21-42).

b. Lateral Spacing. Permeability measurements can be used to determine lateral spacing, but usually the spacings shown in Table 21-51 are used. Lateral tile lines can always be added to the tile system if the main lines have adequate capacities.

c. *Depth of Tile Lines.* The usual depth of tile lines in mineral soils varies from 3 to 4 ft in humid areas and from 6 to 8 ft in western irrigated lands. Shallow depths (under 30 in.) are used only when absolutely necessary. When lines are less than 2 ft deep in mineral soils they are likely to be broken by the passage of farm machinery. Maximum depth is limited by the quality of tile [78]. In organic soils tile needs to be at least 4 ft deep, since the soils will subside through oxidation and compaction, and wind erosion may further lower the surface [79].

Table 21-51. Horizontal Spacing of Tile Lines*

Soil type	Permeability	Spacing, ft
Clay, clay loam	Very slow	30– 70
Silt, silty clay loam	Slow to moderately slow	60–100
Sandy loam	Moderately slow to rapid	100–300
Mucks, peat		50–200

* From Beauchamp [78].

Table 21-52. Drainage Coefficients for Tile Lines*

Soil	Drainage coefficient, in.	
	Field crops	Truck crops
No surface water admitted to the tiles and complete surface drainage provided otherwise		
Mineral	3/8–1/2	1/2–3/4
Organic	1/2–3/4	3/4–1 1/2
Surface water admitted to tile lines through blind inlets		
Mineral	1/2–3/4	3/4–1
Organic	3/4–1	1 1/2–2
Surface water admitted to tile lines through open inlets		
Mineral	1/2–1	1–1 1/2
Organic	1–1 1/2	2–4

* From Beauchamp [78].

d. *Tile-line Slope.* The slope (feet per foot) of any tile line should not be less than 0.0007 for 5-in. tile or 0.0005 for larger tile. Flatter slopes permit the lines to fill with sediment. The maximum permissible slope of a tile line varies with soil type, but in general it should not be more than 0.01. Slopes of as much as 0.02 are feasible with some soils, but steep slopes often have undesirable effects. When a steep line is not running full, it may run at less than critical depth, with resulting surges that cause inflows of soil at the joints, and thus may cause shifting and blocking of the tile line. If it runs full and under pressure, the line may shift and break. Special precautions need to be taken when high velocities or pressure may develop in a section of tile line [78]. Under such conditions tile joints can be covered with durable materials such as tar-impregnated paper and the joints should be made as small as possible. No inflow of local water should be permitted at such joints. Watertight lines are

FIELD APPLICATIONS

used where methods described above are not safe. Such installations provide little drainage but serve only as transmission lines through steep reaches of the system. Slopes of tile lines can usually be kept low enough by use of one of the patterns shown in Fig. 21-42, although it may be necessary to use a main line with some steep sections.

e. Surface-water Inlets. Special inlets are used when it is necessary to get surface waters directly into the tile lines. An *open* inlet is a section of pipe rising from the line to the surface and screened to prevent entry of debris. A *blind* inlet is a section

Table 21-53. Comparable Tile or Pipe Diameters for Equal Discharges*

	Diameter† in inches when Manning's n =			
0.011	0.013	0.015	0.017	0.025
5	5.3	5.6	5.9	6.8
6	6.4	6.7	7.1	8.2
8	8.5	9.0	9.4	10.9
10	10.6	11.2	11.8	13.6
12	12.8	13.5	14.1	16.3
14	14.9	15.7	16.5	19.1
16	17.0	18.0	18.8	21.8
18	19.2	20.2	21.2	24.5
20	21.3	22.5	23.6	27.2
22	23.4	24.7	25.9	29.9
24	25.5	27.0	28.3	32.7
26	27.7	29.2	30.6	35.4
28	29.8	31.4	33.0	38.1
30	31.9	33.7	35.3	40.8
32	34.0	35.9	37.7	43.5
34	36.2	38.2	40.0	46.3
36	38.3	40.4	42.4	49.0
38	40.4	42.7	44.8	51.7
40	42.6	44.9	47.1	54.4
42	44.7	47.2	49.5	57.1
44	46.8	49.4	51.8	59.8
46	48.9	51.7	54.2	62.6
48	51.1	53.9	56.5	65.3

* From U.S. Soil Conservation Service [33].
† Diameter for $n = 0.011$ is the base; diameters for other values of n are to the nearest tenth of an inch to aid in selection of a standard size.

of line backfilled with crushed rock, etc., graded from coarse to fine, the top foot being coarse sand or porous soil [78].

f. Drainage Coefficients. Table 21-52 gives values of drainage coefficients based on the use of Fig. 21-43 [78].

g. Tile Capacities. Figure 21-43 gives capacities and other information for tile drains flowing full but not under pressure; Manning's n is 0.011 for this chart. Table 21-53 shows comparable sizes for some larger n values.

h. Tile-line-capacity Design. It is not desirable to have a tile line flow under pressure, but it will be difficult to keep a line from running full except by overdesign of the tile diameter. In general, it should be controlled through the velocity. Velocities greater than 3 fps will nearly always cause trouble, and although soil texture partly determines the safe velocity, the optimum will usually be about 1.5 fps. When

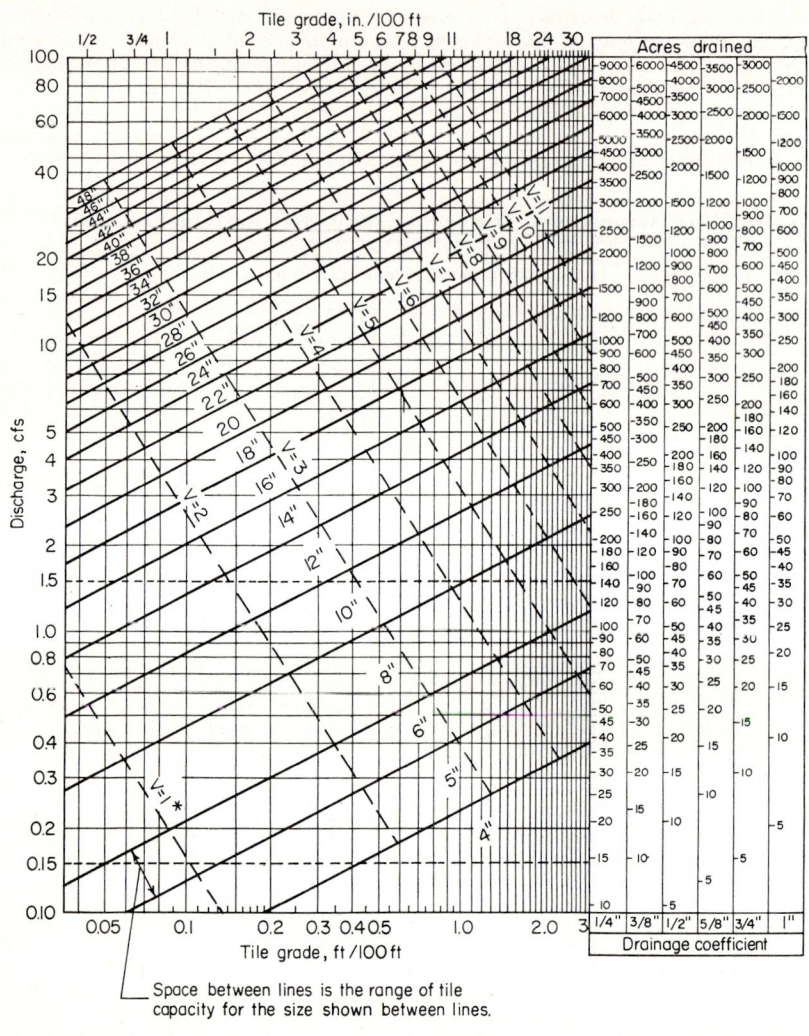

*V = velocity, fps

Fig. 21-43. Drain-tile design chart. The tile size and capacity are determined by extending a horizontal line from the area drained to intersect the proper slope line. The chart is based on the Yarnell-Woodward formula $V = 138 r^{\frac{2}{3}} s^{\frac{1}{2}}$, where V = velocity in fps, r = hydraulic radius in ft, and s = slope in ft/ft.

it is necessary to use steep slopes, the methods described by Beauchamp [78] should be used.

i. Tile-line Outlet. The end of the tile line is usually a section of corrugated metal pipe cantilevered into a waterway, with the opening of the pipe outlet screened to keep out small animals.

j. Mole Drains. These are unlined, rounded or egg-shaped channels, 20 to 24 in. under the surface of highly cohesive or fibrous soils. They are formed by means of a *moling plow*. Mole drains have an effective life of 2 to 5 years, although some drainage effect may last twice that long. They are used to supplement ditch and tile drains where soil conditions would require very close spacing. Spacings of less than 10 ft are often used with mole drains. Criteria and installation practices are given by Stephens and others [75, 79].

6. Pumping for Drainage. Sutton [76] gves information on pumping for drainage. Pump size and operation costs are reduced by use of storage. The drainage coefficient for this combination reaches a maximum of 3 in. for some areas along the Gulf Coast.

VIII. REFERENCES

1. Estimated water requirements for agricultural purposes and their effects on water supplies, United States Senate, Select Committee on National Water Resources, Committee Print no. 13, 1960, table 4.
2. Monthly precipitation and runoff for small agricultural watersheds in the United States, U.S. Agricultural Research Service, 1957.
3. Annual maximum flows from small agricultural watersheds in the United States, U.S. Agricultural Research Service, 1958.
4. Selected runoff events for small agricultural watersheds in the United States, U.S. Agricultural Research Service, 1960.
5. Rainfall frequency atlas of the United States, *U.S. Weather Bur. Tech. Paper* 40, 1961.
6. Rainfall intensity-frequency regime, *U.S. Weather Bur. Tech. Paper* 29, pts. I to V, 1957–1960. See also Refs. 68–73 of Sec. 9.
7. Huff, F. A., and J. C. Neill: Rainfall relations on small areas in Illinois, *Illinois State Water Surv. Bull.* 44, 1957.
8. Musgrave, G. W., and R. A. Norton: Soil and water investigations, Clarinda, Iowa, *U.S. Soil Conserv. Serv. Tech. Bull.* 558, 1937.
9. Clark, O. R.: Interception of rainfall by prairie grasses, weeds, and certain crop plants, *Ecol. Monographs*, vol. 10, no. 2, pp. 243–277, April, 1940.
10. Kittredge, Joseph: "Forest Influences," McGraw-Hill Book Company, Inc., New York, 1948.
11. Criddle, W. D.: Methods of computing consumptive use of water, *Proc. Am. Soc. Civil Engrs., J. Irrigation and Drainage Div.*, vol. 84, no. IR1, pp. 1–27, January, 1958.
12. Blaney, H. F., H. R. Haise, and M. E. Jensen: Monthly consumptive use by irrigated crops in western United States, *Provisional Suppl. to SCS-TP-96*, U.S. Soil Conservation Service, 1960.
13. Blaney, H. F.: Monthly consumptive use requirements for irrigated crops, *Proc. Am. Soc. Civil Engrs., J. Irrigation and Drainage Div.*, vol. 85, no. IR1, pp. 1–12, March, 1959.
14. van Bavel, C. H. M.: Estimating soil moisture conditions and time for irrigation with the evapotranspiration method, *U.S. Agr. Res. Serv. Paper* ARS 41-111, 1956.
15. Penman, H. L.: Estimating evaporation, *Trans. Am. Geophys. Union*, vol. 37, no. 1, pp. 43–50, 1956.
16. "Climate and Man," Yearbook of Agriculture, 1941, U.S. Department of Agriculture, pp. 740–741.
17. Langbein, W. B., and others: Topographic characteristics of drainage basins, *U.S. Geol. Surv. Water-Supply Paper* 968-C, 1947.
18. Kirpich, Z. P.: Time of concentration of small agricultural watersheds, *Civil Eng.*, vol. 10, no. 6, p. 362, June, 1940.
19. "National Engineering Handbook," sec. 4, supplement A, Hydrology, U.S. Soil Conservation Service, 1957 (3d ed. in press, 1963).
20. "Soil Survey Manual," U.S. Department of Agriculture Handbook 18, 1951.
21. Soil survey reports for individual counties in the United States, U.S. Soil Conservation Service, published as available.
22. Carter, R. W.: Computation of peak discharge at culverts, *U.S. Geol. Surv. Circ.* 376, 1957.
23. Wilson, W. T., and R. D. Tarble: Estimated frequencies and extreme values of snowpack water equivalent at major cities in the United States, *Trans. Am. Geophys. Union*, vol. 33, no. 6, pp. 871–880, December, 1952.
24. Myers, V. A.: Frequency variation of snow depths in the Missouri and Upper Mississippi Basins, *Monthly Weather Rev.*, vol. 81, no. 6, p. 162, June, 1953.
25. Langbein, W. B., and others: Annual runoff in the United States, *U.S. Geol. Surv. Circ.* 52, 1949.
26. Long-duration runoff volumes, *U.S. Army, Corps of Engrs., Tech. Bull.* 5, Sacramento, Calif., 1958.

27. Parshall, R. L.: Measuring water, *U.S. Soil Conserv. Serv. Circ.* 843, 1950.
28. Mockus, Victor: Estimation of annual water yields from ungaged watersheds of 10 to 2000 acres in size, U.S. Soil Conservation Service, standard drawing ES-1014, 1958.
29. Rainfall intensities for local drainage design in western United States, *U.S. Weather Bur. Tech. Paper* 28, 1956.
30. Bernard, Merrill: Discussion of Runoff: rational runoff formulas, *Trans. Am. Soc. Civil Engrs.*, vol. 96, p. 1161, 1932.
31. Hamilton, C. L., and H. G. Jepson: Stock-water developments: wells, springs, and ponds, *U.S. Dept. Agr. Farmers' Bull.* 1859, 1940 (Cook's method, p. 39).
32. "Engineering Handbook," U.S. Soil Conservation Service, Milwaukee, Wis., 1942 (out of print).
33. U.S. Soil Conservation Service, unpublished material.
34. Mockus, Victor: Use of storm and watershed characteristics in synthetic hydrograph analysis and application, U.S. Soil Conservation Service, 1957.
35. "Engineering Handbook for Soil Conservationists in the Corn Belt," U.S. Soil Conservation Service, Agriculture Handbook 135, Milwaukee, Wis., 1958, p. 8-3.
36. Blakely, D. B., J. J. Coyle, and J. G. Steele: Erosion on cultivated land, in "Soil," Yearbook of Agriculture, 1957, U.S. Department of Agriculture, pp. 290–307.
37. Craddock, G. W.: Floods controlled on Davis County watersheds, *J. Forestry*, vol. 58, no. 4, pp. 291–293, April, 1960.
38. "Engineering Handbook for Work Unit Staffs," U.S. Soil Conservation Service, Alexandria, La., 1956.
39. Handbook of channel design for soil and water conservation, *U.S. Soil Conserva. Serv. Stillwater Outdoor Hydraulic Lab. Tech. Paper* SCS-TP-61, rev., 1954.
40. Chow, V. T.: "Open-channel Hydraulics," McGraw-Hill Book Company, Inc., New York, 1959.
41. Culp, M. M.: U.S. Soil Conservation Service, unpublished material.
42. Doubt, P. D.: Chute spillways, sec. 14 in "National Engineering Handbook," U.S. Soil Conservation Service, 1955.
43. Culp, M. M., and others: Hood inlets for culvert spillways, *U.S. Soil Conserv. Serv. Tech. Release* 3, 1956.
44. King, H. W., and E. F. Brater, "Handbook of Hydraulics," 5th ed., McGraw-Hill Book Company, Inc., New York, 1963.
45. Hartman, M. A., and R. W. Wilke: Downstream effects of land treatment and upstream floodwater-retarding structures, *U.S. Soil Conserv. Serv. Tech. Paper* SCS-TP-130, 1956, fig. 31.
46. Pond sealing with polyphosphates, U.S. Soil Conservation Service, Upper Darby, Pa., 1958.
47. Culp, M. M., and C. A. Reese: Drop spillways, in "National Engineering Handbook," sec. 11, U.S. Soil Conservation Service, 1954.
48. Culp, M. M.: Earth spillways, *U.S. Soil Conserv. Serv. Tech. Release* 2, 1956.
49. "Design of Small Dams," U.S. Bureau of Reclamation, 1960.
50. The Neosho river basin plan, Report of the Kansas State Board of Agriculture, Division of Water Resources, Topeka, Kans., August, 1947.
51. Riter, J. R., and Charles LeMoyne, Jr.: Planning a large irrigation project, in "Water," Yearbook of Agriculture, 1955, U.S. Department of Agriculture, pp. 328–333.
52. Phelan, J. T., and W. D. Criddle: Surface irrigation methods, in "Water," Yearbook of Agriculture, 1955, U.S. Department of Agriculture, pp. 258–266.
53. Renfro, G. M., Jr.: Applying water under the surface of the ground, in "Water," Yearbook of Agriculture, 1955, U.S. Department of Agriculture, pp. 273–278.
54. Quackenbush, T. H., and D. G. Shockley: The use of sprinklers for irrigation, in "Water," Yearbook of Agriculture, 1955, U.S. Department of Agriculture, pp. 267–273.
55. Hutchins, W. A.: Selected problems in the law of water rights in the West, *U.S. Dept. Agr. Misc. Publ.* 418, 1942.
56. "National Engineering Handbook," sec. 15, Irrigation, U.S. Soil Conservation Service, 1959. Chapters of this handbook are published separately by the U.S. Government Printing Office, see Refs. 61, 71, and 72.
57. Quackenbush, T. H., and others: "Conservation Irrigation in Humid Areas," U.S. Soil Conservation Agriculture Handbook 107, 1957.
58. Irrigation on western farms, *Agr. Inform. Bull.* 199, U.S. Soil Conservation Service and U.S. Bureau of Reclamation, 1959.
59. Criddle, W. D., and others: "Methods for Evaluating Irrigation Systems," U.S. Soil Conservation Service Agricultural Handbook 82, 1956.
60. Quackenbush, T. H.: Developing storage for irrigation water in humid areas of the United States, *Proc. Am. Soc. Civil Engrs., J. Irrigation and Drainage Div.*, vol. 85, no. IR3, pp. 41–47, September, 1959.

REFERENCES

61. Land leveling, chap. 12 in "National Engineering Handbook," sec. 15, Irrigation, U.S. Soil Conservation Service, 1959.
62. Eaton, F. M.: Formulas for estimating leaching and gypsum requirements of irrigation waters, *Texas Agr. Expt. Sta. Misc. Publ.* 111, College Station, Tex., 1954.
63. Richards, L. A. (ed.): Diagnosis and improvement of saline and alkali soils, U.S. Department of Agriculture Handbook 60, 1954.
64. Merriam, J. L.: Field method of approximating soil moisture for irrigation, *Trans. Am. Soc. Agr. Engrs.*, Special Soil and Water Edition, vol. 3, no. 1, pp. 31–32, 1960.
65. "Instructions and Criteria for Preparation of Irrigation Guides," U.S. Soil Conservation Service, Portland, Ore., 1957.
66. Renfro, G. M., Jr.: U.S. Soil Conservation Service, Spartanburg, S.C., unpublished material, 1959.
67. "Control of weeds on irrigation systems," U.S. Bureau of Reclamation, 1949.
68. Rohwer, Carl, and O. V. P. Stout: Seepage losses from irrigation canals, *Colo. Agr. Expt. Sta. Tech. Bull.* 38, Fort Collins, Colo., 1948.
69. Lauritzen, C. W.: Ways to control losses from seepage, in "Water," Yearbook of Agriculture, 1955, U.S. Department of Agriculture, pp. 311–320.
70. Evaporation reduction and seepage control, United States Senate, Select Committee on National Water Resources, Committee Print no. 23, 1960.
71. Irrigation pumping plants, chap. 8 in "National Engineering Handbook," sec. 15, Irrigation, U.S. Soil Conservation Service, 1959.
72. Sprinkler irrigation, chap. 11 in "National Engineering Handbook," sec. 15, Irrigation, U.S. Soil Conservation Service, 1960.
73. Wooten, H. H., and L. A. Jones: The history of our drainage enterprises, in "Water," Yearbook of Agriculture, 1955, U.S. Department of Agriculture, pp. 478–491.
74. Donnan, W. W., and G. B. Bradshaw: Drainage investigation methods for irrigated areas in western United States, *U.S. Soil Conserv. Serv. Tech. Bull.* 1065, 1952.
75. Drainage, sec. 16, in "National Engineering Handbook," U.S. Soil Conservation Service, 1958.
76. Sutton, J. G.: Design and operation of drainage pumping plants, *U.S. Soil Conserv. Serv. Tech. Bull.* 1008, 1950.
77. Ramser, C. E.: Flow of water in drainage channels, *U.S. Dept. Agr. Tech. Bull.* 129, 1929.
78. Beauchamp, K. H.: Tile drainage: its installation and upkeep, in "Water," Yearbook of Agriculture, 1955, U.S. Department of Agriculture, pp. 508–520.
79. Stephens, J. C.: Drainage of peat and muck lands, in "Water," Yearbook of Agriculture, 1955, U.S. Department of Agriculture, pp. 539–557.

Section 22

HYDROLOGY OF FOREST LANDS AND RANGELANDS

HERBERT C. STOREY, *Director, Division of Watershed Management and Recreation Research, U.S. Forest Service.*

ROBERT L. HOBBA, *Hydrologist, U.S. Forest Service.*

J. MARVIN ROSA, *Hydraulic Engineer, U.S. Agricultural Research Service.*

I. Introduction	22-1
II. Features of Hydrologic Significance	22-4
A. Organic Matter	22-7
B. Plant Roots	22-8
C. Plant and Animal Life	22-8
D. Sheltering	22-9
III. Hydrologic Processes Affected	22-9
A. Storage and Drainage	22-9
B. Overland Flow	22-10
C. Snowmelt	22-10
D. Erosion and Sedimentation	22-11
IV. Hydrologic Evaluation of Land Treatment	22-12
A. Evaluation by Infiltration Procedure	22-12
B. Evaluation by Snowmelt Analysis	22-19
C. Evaluation by Multiple Regression	22-25
1. Dependent Variables	22-28
2. Independent Variables	22-28
D. Evaluation by Regional Analysis	22-31
1. Determining Effect of Fire on Peak Discharges	22-31
2. Determining Effect of Fire on Erosion Rates	22-38
E. Evaluation by Hydrograph Analysis	22-44
F. Evaluation by Runoff-curve-number Procedure	22-47
V. Glossary	22-51
VI. References	22-52

I. INTRODUCTION

This section has been prepared for land managers, engineers, hydrologists, and others who need information on how forest and range cover affects the hydrologic functions of watersheds. Emphasis is placed upon explanation of the hydrologic processes undergone by water in places where forest and range vegetation can affect it. It is intended that this section overlap as little as possible the closely related subjects which have been treated in other sections of this handbook.

With a view to determining the relative importance of forest and range vegetation in the regulation of streamflow, in water supplies, in erosion control, or in influencing other watershed values, maps are presented that show the forest and range areas of the United States and the average annual precipitation and streamflow in the United States (Figs. 22-1 to 22-4). Examination of these maps shows that forested areas generally occur in regions of relatively high precipitation and correspondingly high water yield. This points up the importance of a thorough understanding of the hydrology of such areas. Various methods are also provided for appraising the effects of given land conditions, treatments, and uses on streamflow behavior.

Included at the end of this section is a list of definitions of terms used in this section that may not be familiar to certain readers of this handbook.

Public recognition of the close relationships between vegetation and water began in Europe many centuries ago. Kittredge [1, p. 6] traces the history of forest influences from the thirteenth century, and the development of man's thinking on the subject can be used as a guide to the evolution of the specific field of forest and range hydrology.

As early as 1215, Louis VI of France proclaimed an ordinance entitled "The Decree of Waters and Forests." In 1342, a community in Switzerland reserved a forest for protection against avalanches, and between 1535 and 1777 protection forests were proclaimed in 322 instances.

The study of forest influences stemmed from observations in Europe on the flow of springs and streams and their connection with forest destruction. An area east of Venice, on the Adriatic coast, was converted to a "desert," because of extreme changes in the microclimate after forest removal, loss of humus, and higher temperatures.

The Landes area in southwest France provides a classic example of the effects of deforestation and the stabilization resulting from reforestation. After the original forest was destroyed, sand dunes were successfully restabilized with forest cover.

Considerable attention has been given to torrent control in the European Alps, and the Forest Department of France had to undertake torrent control as early as 1860.

Interest in forest influences in the United States started early in the history of this country.

The movement of and damage by shifting sand resulted in measures for stabilization as early as 1739. In that year the citizens of Truro, Mass., passed an act providing a penalty for grazing of the meadows near the beach. This was necessary because cutting of the woods and grazing just back from the beach had resulted in shifting sands burying meadows near the shore. Later in the same year a similar act was passed for Plum Island near Ipswich, Mass., which in its preamble recognized fires as well as cutting of the trees as a cause of the movement of sand.

In 1849, a report of the U.S. Patent Office discussed the destruction of forests and the resulting influence on water flows. In 1858, R. U. Piper, in "Trees of America," mentioned melting of snow and quoted W. C. Bryant as follows: "Streams are drying up and from the same cause, the destruction of our forest." In the period 1891 to 1893, 17,500,000 acres of public domain lands in California was reserved primarily for watershed protection.

Kittredge [1, p. 15] sets out the following five stages of development in the history of forest influences in the United States:

1. The earliest interest resulted from the necessity of preventing damage by shifting sand.

2. Supposed effects of deforestation on climate were early sources of concern.

3. The third stage emphasized the unfavorable effect of timber cutting upon the regulation of streamflow and the maintenance of navigation.

4. The propaganda stage followed when many of the alleged benefits of forest influences were put forward in the movement to "sell" forestry.

5. The recent and present stage of scientific inquiry and support for the effects of

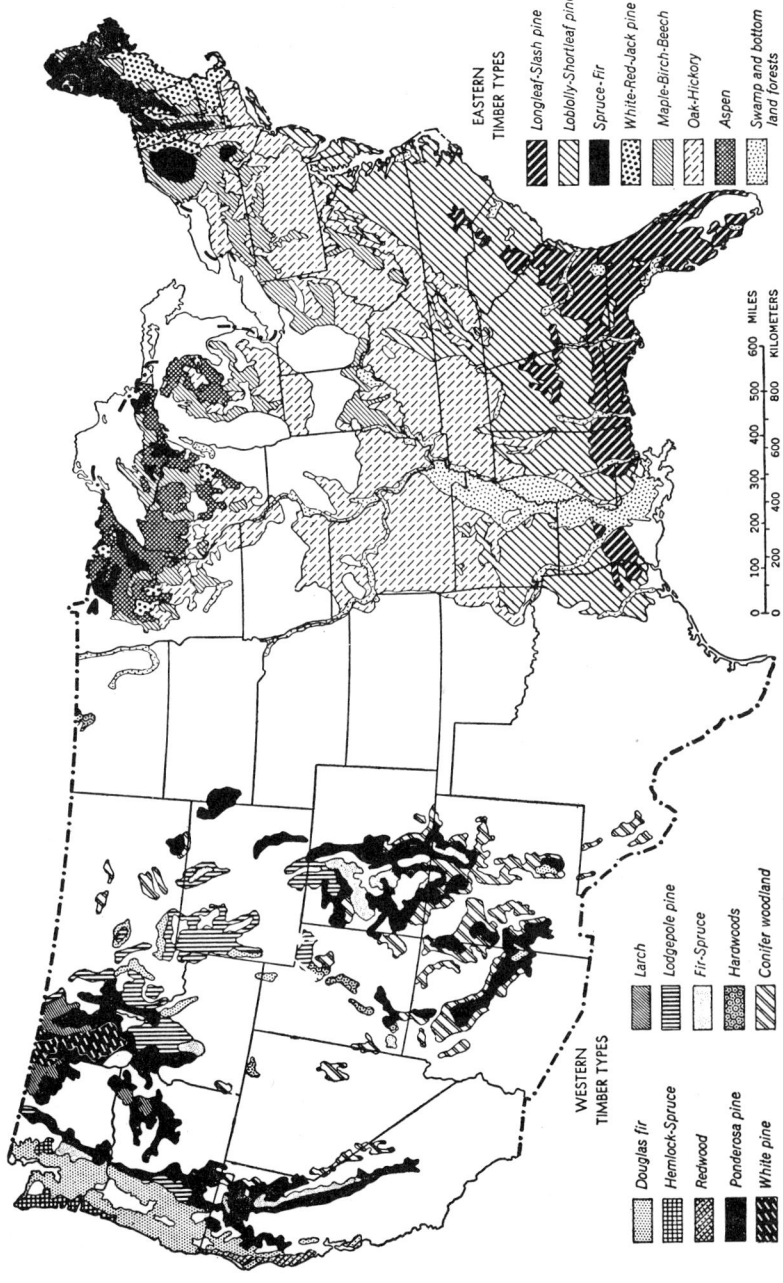

Fig. 22-1. Major forest types in the United States. Areas not typed may have some commercial timber, usually covering less than 10 per cent of the land. Extensive areas of chaparral and other noncommercial forests are not shown. (*Based on map: Area Characterized by Major Forest Types in the United States, U.S. Department of Agriculture, Forest Service, 1949. From the National Survey of Forest Resources.*)

forests on climate, soil, and water, which began in Europe about 1870 and has been continued, at least intermittently, in Germany, Austria, France, Russia, Sweden, Switzerland, Japan, and China, began in the United States about 1908, with the

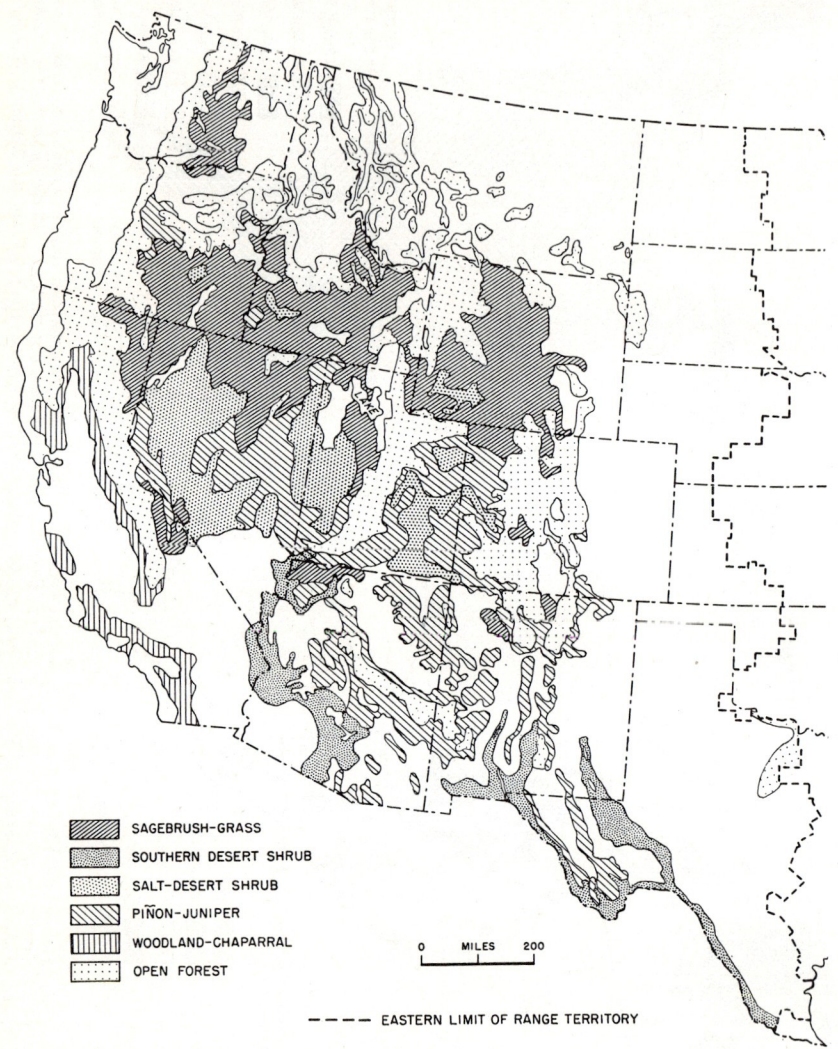

Fig. 22-2. Major range types in western United States.

initiation of the collection of forest meteorological data at the forest experiment stations. More recently the emphasis has been focused on erosion, soil stabilization, and flood control.

II. FEATURES OF HYDROLOGIC SIGNIFICANCE

Information on the hydrologic behavior of forest lands and rangelands and the relationship of land use to water yield and streamflow has been accumulated through

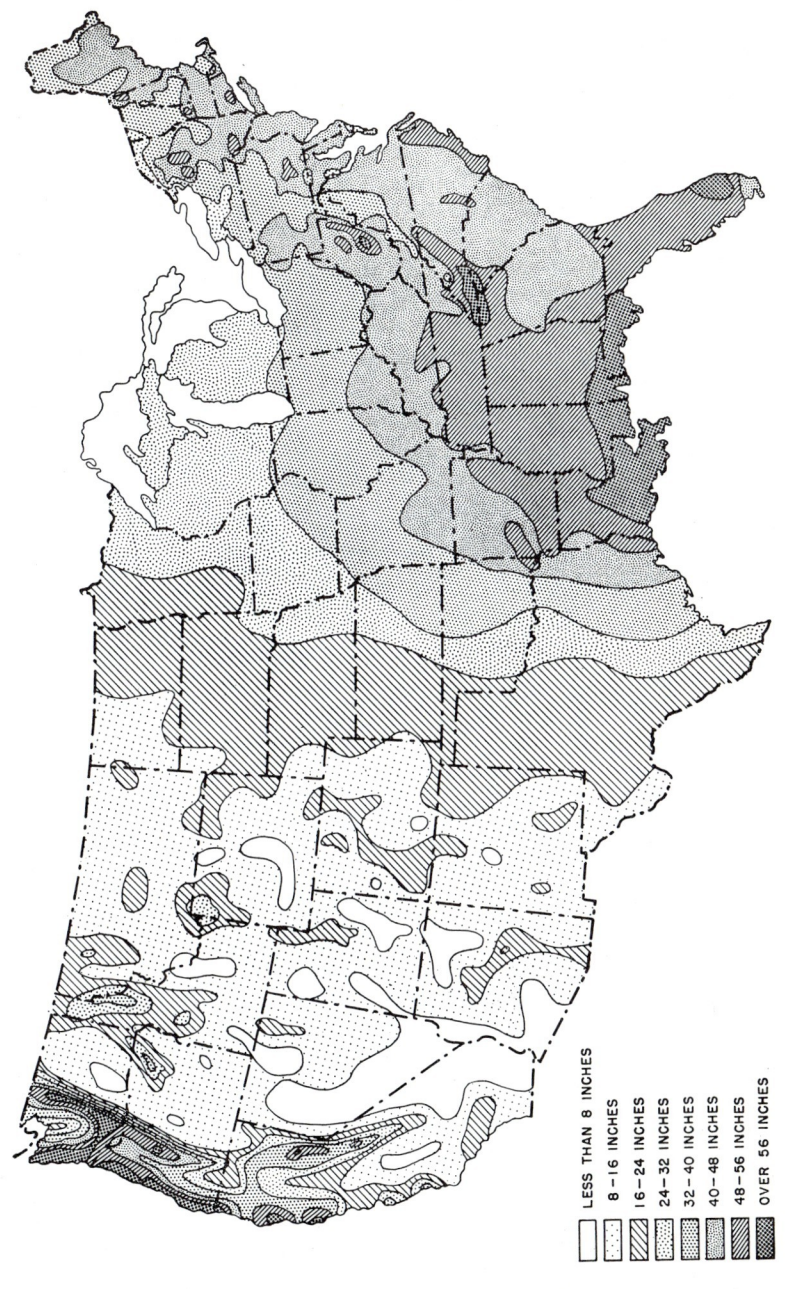

Fig. 22-3. Normal annual amount of precipitation in the United States.

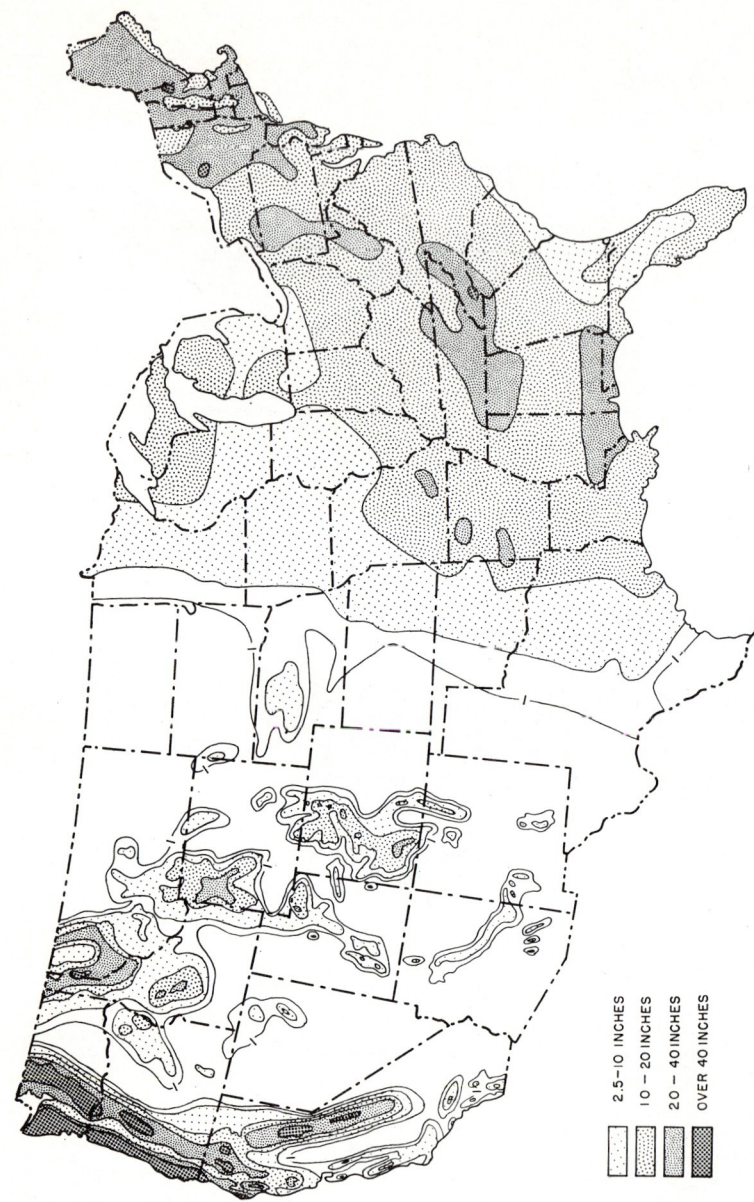

Fig. 22-4. Average annual runoff in the United States.

the studies, experiments, and observations of competent observers over a long period of years. These studies have gradually led to an understanding of some of the basic relationships between land use and runoff, debris deposition, the shoaling of stream channels, silting of reservoirs, storm flows, and other phenomena that have followed excessive logging, cultivating, burning, and grazing of forest lands and rangelands. They have revealed something of the nature of the change in water behavior that follows modifications in the land cover with man's use of the land. Such information is important to the hydrologist, for many hydrologic problems are solved by determining the disposition of all available water under various watershed conditions.

Among the hydrologic functions of vegetative cover are breaking the impact of rainfall, direct interception of a part of the precipitation by the aerial portions of the plants, dissipation of soil moisture by transpiration, reduction in the loss of soil moisture by evaporation, binding the soil against erosion, and holding some moisture by the "blotter" effect of the litter. The beneficial effects of vegetal cover are the result of the following:

1. Building up and maintaining the organic content of the soil, thus developing a more open soil structure and thereby greatly increasing both the infiltration capacity and the storage capacity of the soil layer within the root zone.

2. Establishing and maintaining a partial or complete cover of undecomposed or partly decomposed organic matter at or near the surface of the soil, which tends to prevent surface water from picking up fine soil particles and thereby protects the pores and interstices of the soil from being clogged or closed to the passage of water. The net result is to keep the water passageways through the surface soil open to continued infiltration of water and movement to lower soil levels.

3. Keeping the water spread out over the surface of the land and mechanically retarding or hindering runoff, so that water moves more slowly and thereby affords much more time for absorption.

4. Increasing infiltration and reducing overland flow, resulting in less erosion of the soil and less formation of gullies.

5. Shading the ground and minimizing wind movement, reducing snow-melting rates. This lessens rapid runoff and promotes infiltration.

A. Organic Matter

The influence of vegetation upon infiltration and soil water storage is due particularly to the effect of organic matter on and in the soil and to plant roots. Repeated measurements have shown a positive correlation between the quantity of organic matter present in a soil and its water-holding capacity.

The physical condition of forest soils influences their water-absorbing capacity [2]. Samples of the upper 9 in. of soil under several old-growth stands in oak-hickory and other hardwood types in the Ohio Valley were found to be 13 per cent lighter at ovendryness than equal volumes of soil from adjacent cultivated fields and heavily used pastures, indicating more pore spaces in the forest soils. At a 3-in. depth, 14 times as much water was absorbed per minute under the forest as by the field soil, and at a 1-in. depth, over 50 times as much.

Increased soil porosity and water-absorbing capacity have been found to follow forest planting in fields formerly cultivated. At 1-in. depth, the average rate of water absorption in the soil under a 17-year-old forest plantation was 107 cm^3/min, as contrasted with only 8 cm^3 in an open adjacent field.

The results of a series of measurements of the absorptive capacity of cherty and sandy soils in northern Arkansas and on yellow silt loam soils in southern Illinois are given in Table 22-1.

The data show that in the Illinois silt loam soils a great decrease in water-absorptive capacity followed fires on forest land and heavy use and trampling of pastured soils. In the case of the undisturbed woods, the rate of absorption for each successive liter of water applied remained relatively constant throughout the study, whereas in the case of burned woods and pastured soils the rate of absorption diminished with each application. For both burned woods and pastured soils, the rate of absorption of the

Table 22-1. Rate of Water Absorption per Second per Square Foot of Soil for Three Soil Types under Different Site Conditions*

Soil type and locality	Site conditions	Volume of water, cm³, absorbed per second on application of:†			
		1st liter	2d liter	3d liter	4th liter
Yellow silt loam, Illinois	Undisturbed oak woods	21.83	23.36	22.78	21.23
	Burned oak woods	7.60	4.63	3.40	2.64
	Poorly managed pasture	2.52	1.34	1.01	0.86
Cherty silt loam, Arkansas	Undisturbed oak woods	55.87	44.87	38.76	32.05
	Burned oak woods	14.25	9.78	6.12	5.10
	Poorly managed pasture	17.73	10.47	6.16	4.74
Sandy soil, Arkansas	Old-field pine woods	53.19	35.21	21.10	14.71
	Poorly managed pasture	12.32	7.66	8.04	6.37
	Undisturbed oak woods	64.10	46.08	40.00	30.50
	Poorly managed pasture	24.33	16.84	14.35	12.92

* From U.S. Department of Agriculture [3].
† Four successive applications of water were made.

fourth liter of water was only about 35 per cent that of the first. This decrease in water absorption is the result of compaction of the bare surface soil due to soil puddling and sealing of the soil pores during rainfall. The permeability of the forest soil, which is preserved by the protective covering of the litter, is greatly diminished when the litter is destroyed by fire or by grazing and soil aggregates are broken down.

B. Plant Roots

Channels left by decayed roots also perform an important function in percolation and storage of water. These roots ramify through the soil in an intricate network, the density of which depends on the type and density of the vegetation. Near the surface this network is particularly close. Below 2 ft it is somewhat less dense. While the roots are alive, their growing tips force a way into minute cracks in the soil granules and through small passageways between soil grains, expand and enlarge the opening, or break the granules into still finer particles. When the roots die, they soon decay, leaving channels through which water may pass through the soil.

C. Plant and Animal Life

The soil under a relatively undisturbed forest and range cover is the home of much animal life. Many animals, including most of the rodents and insectivores, dwell or burrow in the soil. Some, like the mole and pocket gopher, spend most of their life in the ground. The soil also teems with animal and plant life too small to be seen without a hand lens or microscope. In forests of the Appalachian region, up to 1,000 individual microarthropods (such as spiders, springtails, centipedes, etc.) may inhabit each square foot of forest litter. These small creatures, which feed upon the litter and other organic material, aid greatly in incorporating organic residues into the soil. Earthworms, which inhabit the soil, feed on fallen leaves and use them as litter in their burrows. Even at a low density of three worms per square foot, these animals may rework as much as half a ton of leaves per acre per season. In the process of nutrition the worms pass great quantities of soil and organic debris through their bodies, thereby, together with bacterial action, promoting humification and the incorporation of organic matter with mineral soil.

Soils under forest and good grass cover contain fungus and bacterial growths. Fungus mycelia grow downward along the cracks in the soil and thus increase the intensity of the lines of cleavage. Such growths may take place to a depth of several feet. Fungi also attack dead-plant material and help to break down and incorporate it into the soil. All such life, plant and animal, influences the moisture intake and moisture-holding capacity of the soil, either directly or indirectly. Although the influence of such organisms has never been determined quantitatively, it unquestionably is a highly important factor affecting infiltration rates. The larger forms provide openings through which water can readily pass; the smaller forms help to maintain an open soil structure by their constant working of the soil and incorporation of organic material. These activities are conducive to development and maintenance of a relatively high water-absorptive capacity.

Any modification of the plant cover and surface soil by cultivation, burning, or overgrazing induces conditions unfavorable to the optimum development of these soil fauna and flora and results in a reduction in the capacity of the soil to take up water.

D. Sheltering

Vegetation shades the ground and minimizes wind movement. The effects tend to reduce evaporation rates and snow-melting rates. It has been found that evaporation of water from snow in forest areas may be only about one-third as fast as on open areas. Also, snow-melting rates have been observed to be as much as one-third more rapid in the open than in the forest. It is not uncommon for snow to remain on forest slopes from 1 to 6 weeks after the disappearance of initially greater quantities of snow on nonforest areas.

III. HYDROLOGIC PROCESSES AFFECTED

In previous sections some of the variables affecting runoff have been examined. Such variables as precipitation, climate, geology, and soil cannot be controlled directly by man's activities. There is, however, another set of watershed influences which can be manipulated, supplying a degree of control over rates and amounts of runoff, erosion, and sedimentation. These are the influences of vegetation.

In managing watershed lands, man has learned that his use or misuse of vegetation on the land may have a very great effect on water production and purity and on floods and erosion. It is widely recognized that vegetation increases the amount of large pore space in the soil, thus minimizing surface runoff and encouraging the storage of water in the soil; that it may augment available soil storage space by removing water through transpiration; and that it stabilizes the soil by physical means.

To better understand the effect of vegetation on runoff, it is desirable to examine the hydrologic processes that are affected by forest and range cover. Section 6 of this handbook presents a discussion of interception, infiltration, and evapotranspiration and the vegetation effects on them. In addition to these processes, there are others that are of major importance in understanding the complete hydrologic relations for forest and range watersheds. A discussion of them follows.

A. Storage and Drainage

Water occupies the soil in three forms: hygroscopic, capillary, and gravitational.

Hygroscopic water is of little hydrologic interest because it is tightly held as a thin film around the soil particles and cannot be removed by gravity or capillary force.

Capillary water is held in the small pore spaces of the soil. The volume of these spaces determines the retention-storage capacity of the soil. Capillary water is gradually removed as the soil surface is dried by evaporation, and the water supply at the surface is replenished by upward movement of water from below. Plants supplement this process by drawing water from the soil and passing it to the air in the process of transpiration. In this manner vegetation removes soil water and enhances the flood-control capacity of the soil at the expense of water yield. This effect tends to be offset, however, by the cooling effect of vegetation on the soil surface and the reduction of wind, thereby decreasing evaporational losses.

Gravitational water, as the name implies, is drained from the larger pore spaces of the soil by the force of gravity. This drainage may be vertical or laterally downslope, depending upon the uniformity of soil texture and the presence or absence of hardpan layers within the soil profile. Vegetation may improve the drainage by root penetration of the hardpan layers and by improving the granular structure of the soil. The temporary detention of the gravitational water in the soil reduces flood runoff without affecting water yield.

B. Overland Flow

Precipitation reaching the soil surface, which does not infiltrate or pond in small depressions, moves downhill over the soil surface as *overland flow*. Rarely does water move over the land as a sheet. Because of surface irregularities, areas of even a few square feet exhibit microdrainage patterns from which water flows downhill, merging with other microchannels which merge into rills, gullies, or streams that provide drainage facilities for runoff water. Overland flow is of hydrologic importance for several reasons: it moves quickly to stream channels, thereby causing the flashiest flood peaks; it is only slightly subject to evaporation because of its short time in transit, and sometimes a greater proportion of overland flow than of subsurface flow contributes to streamflow; and it has the capacity, by virtue of its velocity, to detach soil particles and is therefore an important agent in eroding soil and impairing water quality in streams by increased turbidity. The velocity of overland flow largely determines its flood-and-erosion potential. Water flows more rapidly down steep slopes, over smooth surfaces, and in concentrated channels. Vegetation increases surface roughness, and litter particles on the soil surface form small dams and obstructions which slow down the velocity of overland flow and discourage concentration in rills and gullies.

C. Snowmelt

The rate of snowmelt at any location is determined to a great extent by the vegetative influences upon the various sources of heat. Many observations show that snow disappears more quickly in the open than under a forest cover. The means by which the forest cover retards snowmelt, however, are many and varied. The six noteworthy sources of heat producing snowmelt and the manner in which they are affected by a forest cover are as follows (see also Sec. 10):

1. *Heat stored in underlying soil.* The ground litter acts as an insulating blanket to protect the snowpack from soil heat. The effect of soil heat on snowmelt is apparently relatively small, and the effect of cover on this source of heat is consequently unimportant.

2. *Heat in surrounding air.* Heat is transferred to the snowpack by conduction from the surrounding air. The air in a dense forest cover is cooler during the time of snowmelt than in adjacent open areas. This source of heat from still air is relatively small and unimportant.

3. *Heat content of rain falling on snowpack.* Some of the worst spring floods have occurred following a heavy rainfall on a melting snowpack. Warm rains are accompanied by relatively high temperature and often by moderate winds, which are conducive to rapid snowmelt. When the soil surface is saturated or frozen, a high percentage of the total rainfall runs off as overland flow. A driving rain also acts to break down the snow structure and speed up the release of stored moisture. The rapid runoff may wash unmelted snow into stream channels. These factors combine to produce extremely high rates of runoff. The quantity of snow actually melted by the heat of rainwater is relatively small. Vegetation has a deterrent effect on this type of runoff through its influence on local climate, interception of rainfall, prevention of solid-ice formation at the soil surface, improvement of infiltration capacity, and the water-storage capacity of the litter.

4. *Solar radiation.* Direct solar radiation is one of the largest sources of heat that produces snowmelt. Radiation increases with elevation, and on high mountain ranges it may be the principal factor in the rate of snowmelt. The function of the forest as an

energy converter responsive to the entire spectrum is perhaps responsible for its influence on snowmelt [4, p. 74].

5. *Turbulent exchange of heat from air.* While the heat transfer from still air is relatively small, a turbulent air mass is capable of transferring large quantities of heat and water vapor to the snowpack. Vegetation influences this convective heat transfer by reducing wind velocity and turbulence.

6. *Condensation of atmospheric moisture.* The transfer of air moisture by turbulent exchange follows the same laws as the transfer of heat from the air. The heat produced by condensation of water vapor on the snow is an important source of heat for snowmelt. Vegetation influences this transfer also, by controlling air movement.

D. Erosion and Sedimentation

Erosion is the process by which soil particles are detached from the soil surface. Transportation is the process of moving these detached soil particles to another location. Deposition is the process of dropping these soil particles at a new location. Erosion, transportation, and deposition are natural processes which have sculptured many of the topographic features of the present landscape. These processes have been functioning throughout geologic history. This type of normal, or geologic, erosion is generally beneficial, because it proceeds at a rate slower than that at which new soil is being formed and it helps to keep the soil in a productive condition. This natural balance is easily destroyed by a change in the density or character of the native cover, which may result in accelerated erosion.

Erosion is of two general types: *sheet erosion* and *channel erosion*. The latter includes rill and gully erosion as well as stream-channel erosion.

Sheet erosion is the removal of soil more or less uniformly over the entire soil surface. This type of erosion is not as spectacular as channel erosion and is often not readily apparent to the casual observer. Evidences of sheet erosion are plants standing with part of the root system exposed, indicating that the soil has been washed away around it; soil pedestal; the invasion of plant species that are indicators of a deteriorated soil; and a silt load in streams in the absence of channel erosion. While not as obvious as channel erosion, sheet erosion is probably responsible for damaging a greater land area.

Rill erosion results from scouring of microchannels in which water concentrates as it runs downslope. This type of erosion is more readily recognizable, and the rills merge to form gullies which are almost impossible to overlook. Channel erosion may take place also in streams and rivers as evidenced by bank cutting and changes in channel alignment.

The relative quantities of soil movement resulting from sheet and channel erosion are highly variable. Generally, sheet erosion is higher in areas of high rainfall, while channel erosion is greater in more arid regions.

The rate of erosion depends upon the length and degree of slope, rainfall intensity, type and density of vegetal cover, and the inherent erodibility of the soil.

Erodibility—the tendency of soil to be detached and carried away—varies with the soil texture and the amount of organic matter and colloids in the soil. Both the particle size and the degree of aggregation, or binding together of soil particles, are important. Clay, for example, has very small particles which are easily transported by water, but are not easily detached because of the high aggregation. Sand, on the other hand, is very easily detached, but is not easily transported because of the larger particle size.

Soil particles may be detached either by the impact of raindrops falling on the bare soil or by water flowing over the soil surface. Although interest is given here primarily in erosion caused by water, it is appropriate to mention that erosion can be caused by wind, which may detach, transport, and deposit soil particles in much the same way as water does, and by gravitational forces, which may cause particles or large areas of a hillside to move downslope.

Raindrops striking the bare soil act like miniature bombs to break up soil aggregates and spatter soil particles as much as 2 ft into the air. This spattering effect of rainfall is evidenced by soil deposits on foilage and along the sides of building foundations.

Some conception of the striking force of rainfall is adduced from the fact that raindrops strike the ground at velocities of about 30 fps, and 1 in. of water over an acre of ground weighs more than 110 tons. Raindrop splash can be an important factor in erosion. This type of erosion can be prevented or reduced by maintenance of a dense ground cover.

Water flowing over the soil surface also possesses energy by virtue of its mass and velocity. This energy can be expended upon obstacles in its path. Depending upon how big and well anchored these obstacles are, the water may be slowed down and diverted or the obstacle may be moved or even picked up and carried downslope. In this way soil particles may be moved or picked up and carried by the flowing water.

The velocity of flowing water is an important factor since the kinetic energy varies as the square of the velocity. If the velocity is doubled, the cutting power is increased 4 times, the quantity of material of a given size that can be carried is increased 32 times, and the volume size of particles that can be carried increases 64 times. Slope is an important factor since velocity varies as the square root of the slope. Water will flow down a 40 per cent slope with twice the velocity of that on a 10 per cent slope. Vegetation acts to reduce erosion by slowing down the velocity of water flowing over the soil surface. Since erosion is caused by overland flow, the effects of vegetation in reducing and retarding overland flow also operate to reduce erosion.

IV. HYDROLOGIC EVALUATION OF LAND TREATMENT

In the preceding discussion attention has been called to hydrologic processes, the behavior of water on the land, and the qualitative aspects of how vegetation affects this behavior. The following discussion will deal with the extent to which there has been a destruction or modification of the vegetative cover brought about by such actions as fire, grazing, or logging and will present methods of hydrologic analyses to evaluate land-treatment (cover-change) effects on streamflow.

The hydrologic evaluation of vegetative changes requires an understanding of the particular hydrologic processes that are operating in the watershed under study. Physical watershed characteristics or meteorological processes, which affect streamflow, vary by sections of the country and even by seasons of the year. Therefore procedures for making the hydrologic evaluation can vary according to these conditions. For example, if the influencing factor of flood runoff is high-intensity rainfall, a method to determine the difference between floods that will occur without and with certain land treatment (cover-condition changes) is by evaluation of the effect of the measures upon the infiltration and storage capacities of the soil. If the influencing factor is snowmelt, the determination of cover effect is by an evaluation of the effect of the measures upon snow distribution, rates of snowmelt, and storage capacities of the soil.

Examples of the evaluation of land-treatment programs are found in typical flood-control-survey reports of the Department of Agriculture to the Congress of the United States. Soil Conservation Service and Forest Service hydrology handbooks [5, 6] also describe in detail evaluation methods that can be used under different hydrologic situations which may be encountered. In general, all methods use a procedure which may be briefly outlined as follows:

1. The soils, vegetation, use, and condition of the watershed are surveyed to determine the area of all existing complexes of soil, cover, and condition of cover.

2. Estimates are made of the area of the complexes in the future without and with a land-treatment program in effect.

3. Estimates of streamflow from the watershed are made for vegetative conditions with and without the proposed land-treatment program. The difference between the streamflow for the two conditions is a measure of the effect of the land-treatment measures.

A. Evaluation by Infiltration Procedure

Several methods of analysis, which utilize infiltration rates for given soil-cover complexes, have been developed. Differences in the methods lie mainly in the determination of infiltration rates and the accounting for groundwater storage. The method described here has been proposed by Whelan, Miller, and Cavallero [7].

The first step in this method consists of dividing the storm rainfall into a selected number of rainfall-depth classes. Average total storm rainfall is then determined for each class. Rainfall intensities for each rainfall-depth class are also determined as based on intensities of a nearby recording rain gage.

For comparison purposes and when available, streamflow records are next analyzed to determine the total flood runoff, the surface runoff, and the subsurface runoff. The difference between the total storm rainfall and the surface runoff gives the amount of infiltrated water for the storm.

Since the total amount of water that infiltrates into a soil profile during a storm depends upon the percolation rate, transmission velocity, and retention- and detention-storage values of each horizon in the soil profile, a soil and land-use inventory must be made to determine the areal extent of each soil-cover complex and the average depth of the soil horizons. For open land the topsoil and the B- and C-horizons are measured, and for forest land the humus, including the A_1-horizon, and the lower A-, the B-, and the C-horizons are measured. The forest land is classified by humus depth and type and by the presence or absence of grazing. All areas are classified by surface-soil texture and degree of soil drainage.

In connection with the land-use inventory, a study of soil-water relationships must also be made in the field [8]. Undisturbed soil samples by horizons are used to obtain percolation rates, transmission velocities, and retention- and detention-storage values for each horizon. Average values are determined for the individual horizons of each soil-cover complex. These values are then applied to the respective areas of the soil-cover complexes as established by the soil and land-use inventory.

Surface-detention storage occurs whenever the rainfall intensity exceeds the infiltration capacity or the percolation rate of the uppermost soil horizon. When the rainfall intensity is high and sustained enough to saturate the uppermost soil horizon, the percolation rate of the surface soil then is limited to that of the next horizon that is not saturated. In computations of surface runoff by this method, surface detention-storage space is considered instantly available for use whenever rainfall excess occurs. When the rainfall excess exceeds this storage, the difference becomes surface runoff. In a severe storm of long duration, surface detention storage may be utilized over and over again because of the variation in rainfall intensities.

Retention storage is the amount of water the soil can hold against the pull of gravity. Detention storage is the difference between saturation and retention storage. Using antecedent storm conditions and retention-storage values as obtained from analysis of the undisturbed soil samples, an estimation of the retention-storage space available is made at the beginning of the storm. It is assumed that any deficiency in retention storage will be used first as the wet front moves downward.

Percolation rates as measured in the field laboratory are adjusted for natural conditions by trial-and-error methods until they give a volume of surface runoff commensurate with the surface runoff as estimated from the gaged hydrograph.

Transmission velocity can be computed from the equation $V = Q/A$, where Q is the percolation rate, A is the percentage of pore space in detention storage, and V is the transmission velocity. When the supply of water equals or exceeds the percolation rate and detention storage and the percolation rate remains constant at increased soil depths, the transmission velocity of the wet front is constant. Any change in either detention storage or percolation rate as the wet front moves to a lower soil horizon, however, will be reflected by changes in the transmission velocity. When the rate of supply is less than the percolation rate, the transmission velocity varies directly as the rate of supply. In the computations the time of transmission through each horizon is used; it is equal to the depth of the horizon divided by the transmission velocity.

Subsurface runoff is considered on a watershed basis rather than by separate soil-cover complexes. Lateral drainage of the soil profile by subsurface flow makes it possible for additional water to be infiltrated into the soil profile. The effect of subsurface flow can be compensated for by increasing the percolation rates by the rate of contribution to lateral flow. This compensation is included as part of the adjustment

Table 22-2. Computations Used in "Routing" Infiltrated Water through the Soil to Determine Surface Runoff*
(For grazed woodland with mull humus and medium-textured, imperfectly drained soil)

Rainfall				Infiltration															
				Surface			Humus			Lower A			Upper B			Lower B			
Time increment, hr	Rainfall intensity, in./hr	Amount of rainfall, in.	Inflow, in.	Storage, in.	Outflow, in.	Accumulated time, hr	Storage, in.	Outflow, in.	Accumulated time, hr	Storage, in.	Outflow, in.	Accumulated time, hr	Storage, in.	Outflow, in.	Accumulated time, hr	Storage, in.	Outflow, in.	C Storage, in.	Total rainfall infiltrated, in.	Runoff, in.
0.167	4.63	0.772	.772	....	.772	0.167	.106	.666	.144	.568	.098	.041	.098	....	....	....	....	0.585	0.772	
0.167	2.32	0.387	.387	....	.387	....	.053	.440	....	.607	.410	.208	.499	....	....	....	....	.585	1.159	
0.500	0.91	0.450	.450	....	.450	....	.021	.482	....	.377	.712	.708	.754	.457	0.403	.426	.031	.616	1.609	
0.167	0.31	0.052	.052	....	.052	....	.007	.066	....	.393	.050	....	.754	.050	....	.426	.050	.666	1.661	
0.167	0.23	0.039	.039	....	.039	....	.005	.041	....	.384	.050	....	.754	.050	....	.426	.050	.716	1.700	
0.083	1.69	0.142	.142	....	.142	....	.039	.108	....	.467	.025	....	.754	.025	....	.426	.025	.741	1.842	
0.583	0.27	0.154	.154	....	.154	....	.006	.187	....	.479	.175	....	.754	.175	....	.426	.175	.916	1.996	
0.333	0	0	....	....	....	....	....	.006	....	.385	.100	....	.754	.100	....	.426	.100	1.016	1.996	
0.167	1.01	0.167	.167	....	.167	....	.023	.144	....	.479	.050	....	.754	.050	....	.426	.050	1.066	2.163	
0.417	0.06	0.026	.026	....	.026	....	.001	.048	....	.402	.125	....	.754	.125	....	.426	.125	1.191	2.189	
0.250	0.52	0.129	.129	....	.129	....	.012	.118	....	.445	.075	....	.754	.075	....	.426	.075	1.266	2.318	
0.083	1.54	0.129	.129	....	.129	....	.035	.106	....	.526	.025	....	.754	.025	....	.426	.025	1.291	2.447	
0.083	3.60	0.296	.296	....	.296	....	.092	.239	....	.740	.025	....	.754	.025	....	.426	.025	1.316	2.743	
0.167	0.47	0.078	.078	....	.078	....	.120	.050	....	.740	.050	....	.754	.050	....	.426	.050	1.366	2.821	
0.333	0.17	0.013	.013	....	.013	....	.033	.100	....	.740	.100	....	.754	.100	....	.426	.100	1.466	2.834	
0.167	1.54	0.257	.257	....	.257	....	.240	.050	....	.740	.050	....	.754	.050	....	.426	.050	1.516	3.091	
0.167	8.26	1.375	.288	.100	.188	....	.378	.050	....	.740	.050	....	.754	.050	....	.426	.050	1.566	3.379	1.087
0.167	0	0	....	.050	.050	....	.378	.050	....	.740	.050	....	.754	.050	....	.426	.050	1.616	3.379	
0.033	12.75	0.426	.060	.100	.010	....	.378	.010	....	.740	.010	....	.754	.010	....	.426	.010	1.626	3.439	0.366
0.216	1.25	0.270	.065	.100	.065	....	.378	.065	....	.740	.065	....	.754	.065	....	.426	.065	1.691	3.504	0.205
0.250	1.75	0.438	.075	.100	.075	....	.378	.075	....	.740	.075	....	.754	.075	....	.426	.075	1.766	3.579	0.363
0.333	0.04	0.013	.013	.013	.100	....	.378	.100	....	.740	.100	....	.754	.100	....	.426	.100	1.866	3.592	
0.250	0.05	0.013	.013	....	.026	....	.329	.075	....	.740	.075	....	.754	.075	....	.426	.075	1.941	3.605	
0.083	0.47	0.039	.039	....	.039	....	.343	.025	....	.740	.025	....	.754	.025	....	.426	.025	1.966	3.644	
1.667	0	0	....	....	....	....	....	.343	....	.583	.500	....	.754	.500	....	.426	.500	2.466	3.644	

1.000	0.05	0.052	0.052		0.052		0.001	0.051		0.334	0.300		0.754		0.300		0.426	0.300	2.766	3.696
1.000	0	0						0.001		0.035	0.300		0.754		0.300		0.426	0.300	3.066	3.696
1.000	0.01	0.013	0.013		0.013		0.001	0.013		0.001	0.047		0.501		0.300		0.426	0.300	3.366	3.709
1.000	0.04	0.039	0.039		0.039			0.038		0.004	0.035		0.236		0.300		0.426	0.300	3.666	3.748
0.333	0.08	0.026	0.026		0.026		0.002	0.025		0.008	0.021		0.157		0.100		0.426	0.100	3.766	3.774
0.250	0.52	0.129	0.129		0.129		0.012	0.119		0.054	0.073		0.155		0.075		0.426	0.075	3.841	3.903
0.167	1.54	0.257	0.257		0.257		0.035	0.234		0.157	0.131		0.236		0.050		0.426	0.050	3.891	4.160
0.083	3.09	0.257	0.257		0.257		0.071	0.221		0.249	0.129		0.340		0.025		0.426	0.025	3.916	4.417
0.167	0.78	0.129	0.129		0.129		0.018	0.182		0.080	0.351		0.641		0.050		0.426	0.050	3.966	4.546
0.500	0.10	0.052	0.052		0.052		0.002	0.068		0.010	0.138		0.629		0.150		0.426	0.150	4.116	4.598
0.167	2.00	0.335	0.335		0.335		0.046	0.291		0.206	0.095		0.674		0.050		0.426	0.050	4.166	4.933
0.167	0.62	0.103	0.103		0.103		0.014	0.135		0.211	0.130		0.754		0.050		0.426	0.050	4.216	5.036
0.167	3.87	0.643	0.643		0.643		0.089	0.568		0.729	0.050		0.754		0.050		0.426	0.050	4.266	5.679
0.250	2.16	0.541	0.475	0.100	0.375		0.378	0.086		0.740	0.075		0.754		0.075		0.426	0.075	4.341	6.154
0.167	0	0		0.050	0.050		0.378	0.050		0.740	0.050		0.754		0.050		0.426	0.050	4.391	6.154
0.083	2.78	0.232	0.075	0.100	0.025		0.378	0.025		0.740	0.025		0.754		0.025		0.426	0.025	4.416	6.229
0.083	0.31	0.026	0.025	0.100	0.025		0.378	0.025		0.740	0.025		0.754		0.025		0.426	0.025	4.441	6.254
1.333	0	0			0.100		0.078	0.400		0.740	0.400		0.754		0.400		0.426	0.400	4.841	6.254
0.083	0.31	0.026	0.026		0.026		0.079	0.025		0.740	0.025	5.69	0.754		0.025		0.426	0.025	4.866	6.280 0.066
16.000	0	0				1.000		0.079			0.819			0.874	1.573			1.999	8.865	6.280
1.000	0.01	0.013	0.013		0.013			0.013		0.002	0.011		0.004		0.007		0.003	0.004	6.869	6.293 0.157
1.000	0.01	0.013	0.013		0.013			0.013		0.002	0.013		0.004		0.013		0.003	0.013	6.882	6.306 0.001
1.000	0.01	0.013	0.013		0.013			0.013		0.002	0.013		0.004		0.013		0.003	0.013	6.895	6.319
1.000	0.01	0.013	0.013		0.013			0.013		0.002	0.013		0.004		0.013		0.003	0.013	6.908	6.332
10.083	0	0							0.977		0.002				0.006			0.009	6.917	6.332
0.083	0.31	0.026	0.026		0.026		0.007	0.019		0.019	0.011								6.917	6.358
1.583	0	0						0.007	0.060		0.026				0.026			0.026	6.943	6.358
0.250	0.62	0.154	0.154		0.154		0.014	0.140	0.227	0.064	0.076	0.124	0.076		0.146	0.236	0.146		6.943	6.512
0.417	0.05	0.026	0.026		0.026		0.001	0.039		0.005	0.098	0.541	0.028		0.146	0.319	0.150		6.943	6.538
0.083	1.24	0.103	0.103		0.103	0.250	0.029	0.074	0.060	0.075	0.004		0.022		0.010			0.006	6.949	6.641
57.167		8.886																		6.705 2.245

*From Whelan, Miller, and Cavallero [7, p. 8].

factor, where percolation rates as measured in the laboratory are made to agree with the flood hydrograph analysis.

In the routing procedure of this analysis, the infiltrated water is routed through the soil by carefully accounting for the time stages in the movement of the wet front and the changes in water storage in each soil horizon. The computations are facilitated by assuming detention storage, percolation rate, and time of transmission to have uniform values throughout each soil horizon.

As an example, the details of the routing procedure are shown (Table 22-2) for grazed woodland with mull humus and a medium-textured, imperfectly drained soil. The maximum storage capacity (retention plus detention storage), percolation rate, and time of transmission for the different soil horizons are as follows:

Horizon	Retention storage, in.	Detention storage, in.	Percolation rate, in./hr	Time of transmission, hr
Surface..........		0.100		
Humus...........	0.870	0.378	15.10	0.023
Lower A.........	1.404	0.740	6.50	0.103
Upper B..........	1.490	0.754	2.40	0.305
Lower B..........	1.630	0.426	1.40	0.298
C...............			0.30	

In the tabular computations the first line of Table 22-2 shows that, during the first period of uniform rainfall intensity (0.167 hr), 0.772 in. of rain fell at an intensity of 4.63 in./hr. Since this intensity is less than the percolation rates of the humus (15.10 in./hr) and the lower A-horizon (6.50 in./hr), this rain is infiltrated into those upper horizons as fast as it falls. But the upper B-horizon has a percolation rate of 2.40 in./hr, so that the water can infiltrate into this horizon no faster than the percolation rate. At the end of this time increment, the humus has stored 0.106 in. of water. This is calculated as transmission time of the soil horizon (0.023 hr) times the rainfall intensity. The rest of the rainfall for this period flows down into the lower A-horizon.

Water flows into the lower A-horizon for a period of 0.144 hr. This is calculated as rainfall period (0.167 hr) minus transmission time in the humus (0.023 hr). Water flows out of the lower A-horizon for 0.041 hr, that is, 0.144 hr minus the horizon's transmission time, 0.103 hr.

Since the percolation rate of the upper B-horizon is less than the rainfall intensity, this rate controls the inflow to the upper B-horizon. The percolation rate (2.40 in./hr) multiplied by the length of time water flows into the horizon (0.041 hr) gives the amount of water that infiltrates into this horizon: 0.098 in., all of which is stored in this horizon. The storage in the lower A-horizon is computed as the difference between inflow (0.666 in.) and outflow (0.098 in.), or 0.568 in.

This procedure is carried out for each period of uniform rainfall intensity during the storm. As the amount of rainfall increases and the soil horizons become saturated, the percolation rates determine the rate at which water infiltrates through the various horizons. The amount of surface runoff is computed by subtracting the amount of water infiltrated from the total rainfall. In the example 8.886 in. of rain resulted in 2.245 in. of surface runoff.

The routing procedure is shown graphically in Fig. 22-5. The uppermost graph in this figure shows amounts of rainfall by successive periods of uniform intensity. The unshaded portions of the bars show infiltrated water, and the shaded portions show surface runoff. The lower graphs show the relationship of storage to storm duration for each soil horizon.

During the first part of the storm the soil is able to absorb 3.4 in. of rainfall in 3 hr and 40 min without any surface runoff occurring. However, at the end of this period the lower A- and the B-horizons are saturated and can transmit water only at the

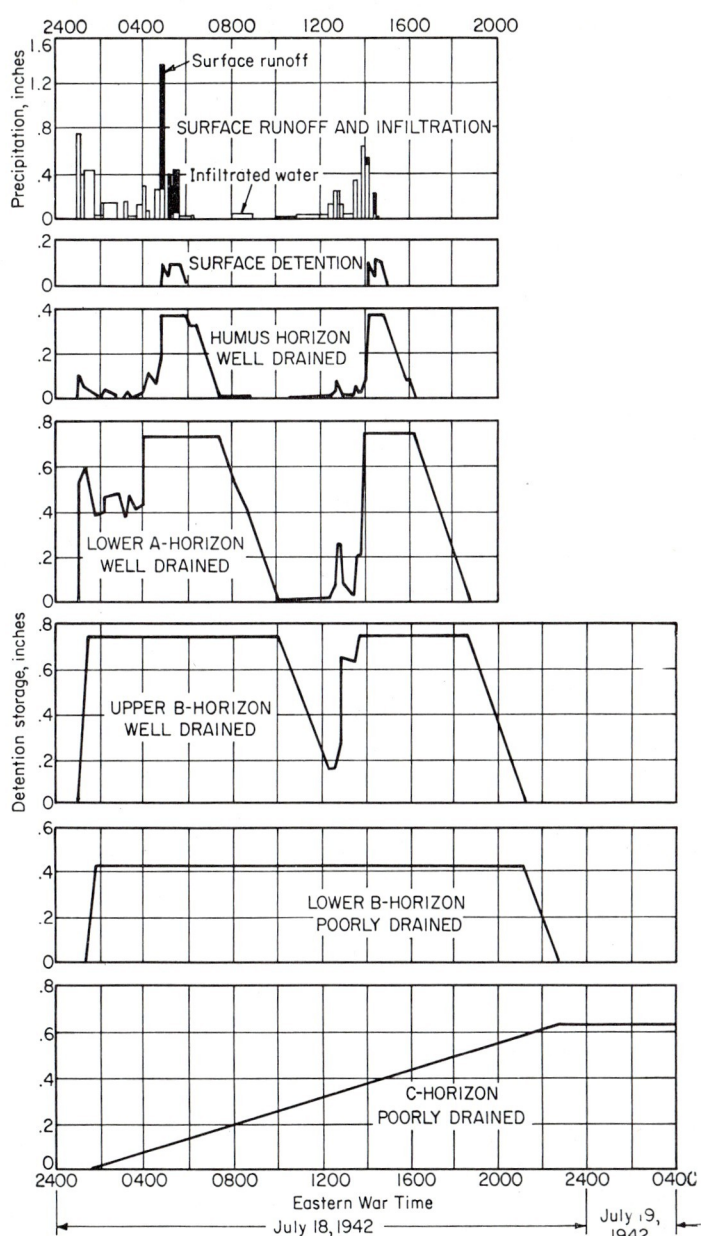

Fig. 22-5. Sample "routing" of infiltrated water through the soil horizons of a grazed forest with mull humus and medium-textured, imperfectly drained soil. (*After Whelan, Miller, and Cavallero* [7].)

percolation rate of the C-horizon. The next two bursts of rain, lasting 20 min, completely utilize the remaining storage space in the humus horizon and on the surface, which results in more than 1 in. of runoff. Since the humus can now absorb water only at the percolation rate of the C-horizon, 1.1 in. of rainfall in the next 40 min results in about 0.9 in. of surface runoff.

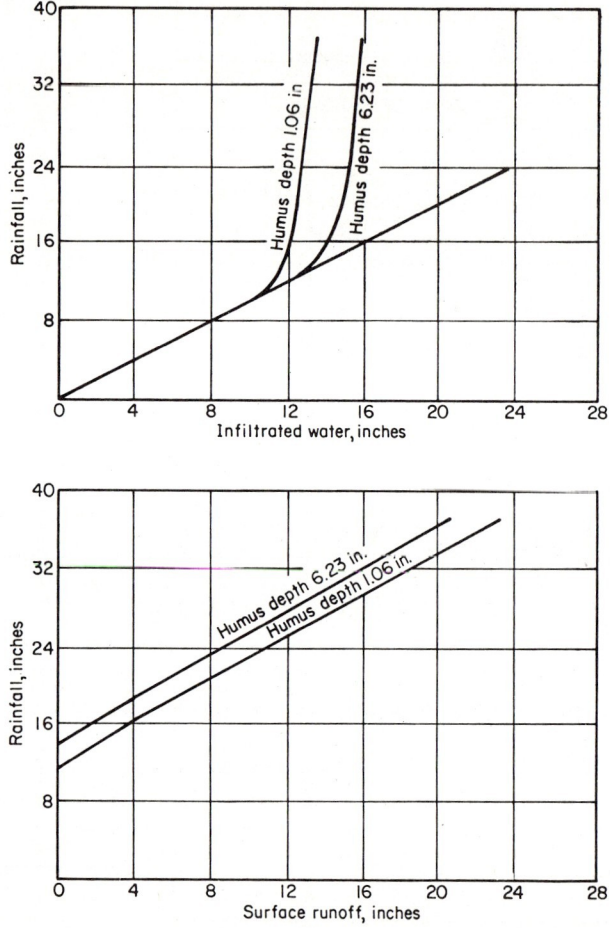

Fig. 22-6. Summary of "routing" studies in the Allegheny River watershed, showing how increased depth of humus reduces surface runoff. (*After Whelan, Miller, and Cavallero* [7].)

During the next 8 hr a large part of the water, which is temporarily stored in the humus horizon and lower A- and well-drained B-horizons, drains into the C-horizon. This makes additional storage space available in the upper horizons. Several additional bursts of rain now cause these horizons to be saturated again and result in about 0.2 in. of surface runoff.

For any given storm and its antecedent condition, there is a definite limit to the total amount of water that can infiltrate into a soil. The maximum limit for a given storm occurs when the humus horizon and lower A- and B-horizons are saturated for the duration of the storm and supply water to the C-horizon at a constant rate. This condition probably seldom occurs, because intervals of little or no rainfall generally cause a slackening in the supply of water at which time the humus and lower A-horizons have time to drain.

The recovery of detention-storage space in the upper soil horizons—as in the example—enables the soil to absorb a series of rainfall bursts. Thus the aggregate gain in infiltrated water is far more than the detention-storage capacities of the humus horizon and lower A-horizon.

Figure 22-6 shows the results of applying the routing procedure on the Allegheny River watershed in western New York and Pennsylvania. The upper graph shows how the amount of infiltrated water for a given total storm rainfall increases with humus depth. The lower graph shows how surface runoff for a given total storm rainfall decreases with humus depth.

B. Evaluation by Snowmelt Analysis

The effects of forest cover on snow distribution and snowmelt might be summarized as follows:

1. A part of the snowfall is intercepted by the crowns of the trees and evaporated back into the air.
2. The wind velocity is reduced by the trees, tending to cause deposition of the snow in the forest, particularly in the small openings between the trees.
3. The forest canopy reduces the intensity of solar radiation and the maximum temperatures at the ground surface, and hence retards the rate of melting and the date of disappearance of the snow.

Is the melt rate greater in forested or open areas? Sometimes it is greater in the forest and sometimes in the open, depending on the size of clearings and other factors. Usually the greater deposition is in the open. Small openings in the forest about one to two tree heights in diameter usually collect the maximum snowpack [9, p. 49]; larger openings may be scoured of snow by wind. Melt rates are generally considerably greater in large clearings than in forests. The higher wind velocities over the open result in greater melts due to convection and condensation. Melt due to absorbed solar radiation is greater in the open.

It has been observed that the last snow patches to disappear in the spring are usually found in small clearings in the forest, having a diameter about the same as the heights of the surrounding trees. Melt rates on large, open areas are affected by the existence of surrounding forests or bare ground, which constitutes a source of sensible heat. Large, nonforested, snow-covered areas would not have as high melt rates as would a large clearing in an otherwise forested area.

Upper-air temperatures and humidities do not serve as good indices of snowmelt since the air may be warmed and its humidity increased in passing over trees and bare ground. Only surface temperatures and humidities tend to better reflect this heat transfer. Since more of the radiant energy is manifest in the air temperature and humidity in forested areas than in barren areas, temperature and humidity indices should be expected to increase in accuracy with the degree of forest cover.

There is no universal index for describing accurately the snowmelt-runoff regime for all areas. Index coefficients vary with local conditions of weather, time, snow condition, vegetation, and terrain. They are limited in applicability to the specific area, time of year, and weather conditions for which they are derived.

The main assumption which permits index analysis of equations is that loss is equal to evapotranspiration and can be estimated by the same indices that described melt. This assumption for the active melt period requires that all initial losses for conditioning the snowpack for runoff and satisfying soil-moisture deficits have been met. All transitory storage in the soil and underlying rocks is accounted for by use of flow-recession curves. During periods of uniform climate on limited areas, there appears to be a fixed proportion of available energy used for transpiration and melt in forested areas. Evapotranspiration loss under this condition is equal to about 10 per cent of the melt.

In densely forested areas, direct observations of solar radiation are not essential since 80 to 90 per cent of the radiant energy is absorbed by the forest. The energy stored in the forest is released to the snow by long-wave radiation, convection, and condensation. Diurnal temperature range is a fair index of solar radiation in heavily forested areas. Long-wave radiation is important to the snowmelt process for any density of forest cover. Since the temperature of the forest canopy is generally above freezing, the forest canopy radiates more energy to the snow than the snow emits. The use of air temperature as a long-wave index is ideally suited to densely forested areas but is inadequate for open areas.

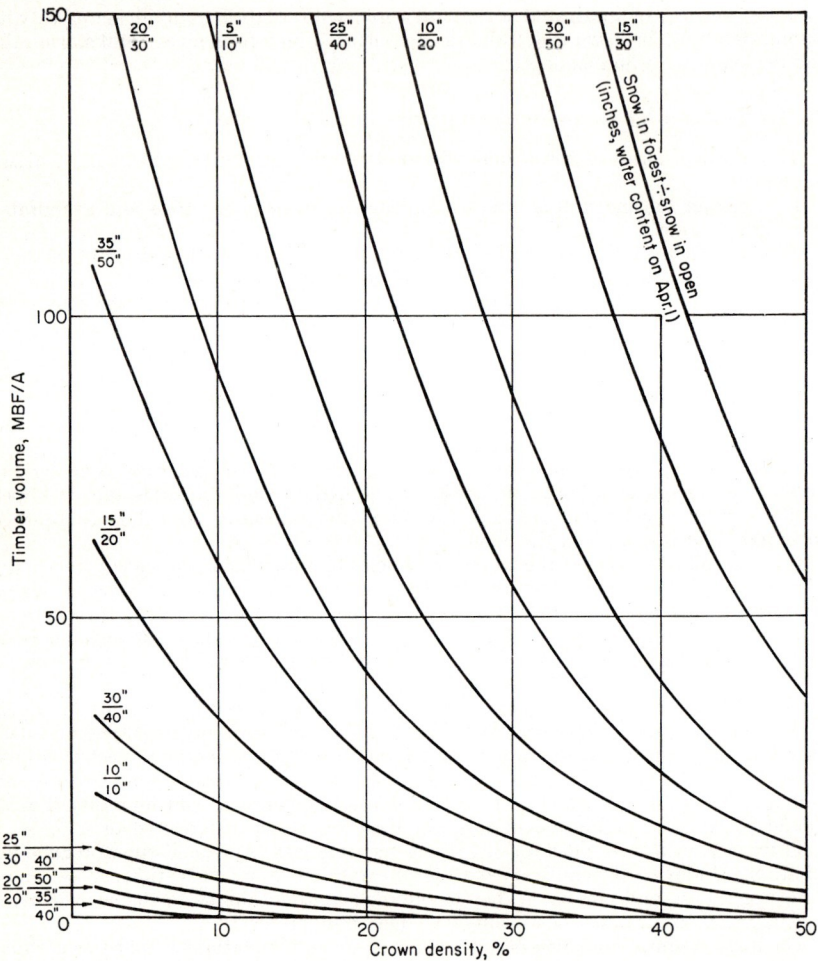

FIG. 22-7. Snow storage in forest and open, Idaho and Utah, 1949–1952.

A method has been proposed by Rosa [10] whereby hydrographs from melting snow can be computed from the following physical factors and relations: (1) snow storage and melting, related to climatic, topographic, and forest characteristics; (2) losses from evapotranspiration, varying with climatic factors; (3) rainfall contribution, depending on the effective area producing runoff; (4) soil-moisture storage, determined by plant-soil types and antecedent conditions; (5) groundwater discharge, related to storage derived from streamflow depletion; and (6) observed streamflow, consisting of groundwater plus routed channel inflows from other sources.

All these factors and relations have long been recognized as influencing streamflow;

however, the charts and illustrated procedures now make possible a practical means of estimating the probable effect of forests on spring floods. If the watershed is largely open or only sparsely forested, radiation equations should be used to compute snowmelt. However, for heavily forested watersheds, snowmelt can be computed from temperature by the degree-day method.

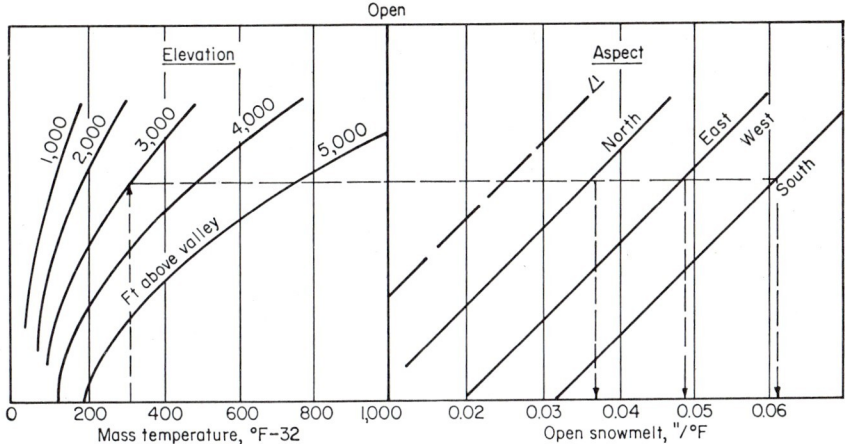

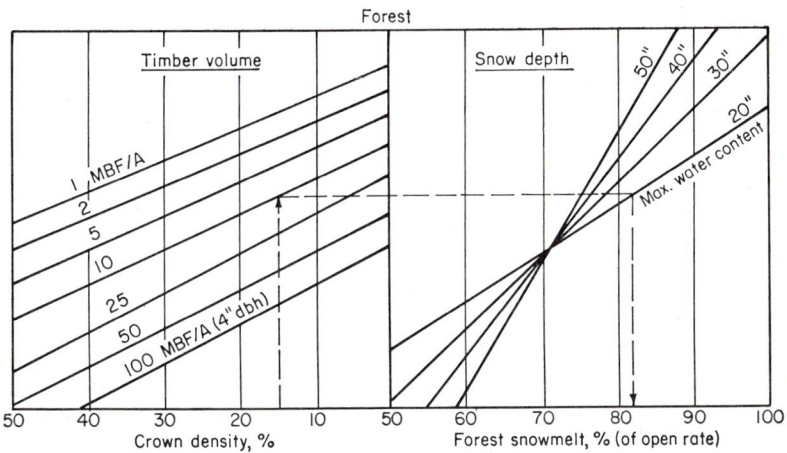

FIG. 22-8. Snowmelt in forest and open, Idaho, Wyoming, and Utah, 1949–1952.

Snowmelt is computed for forested areas from mean daily temperatures, using graphs showing the separate effects of aspects, elevations, and seasonal mass temperatures (Table 22-3).

Rates of melting are determined as if the area were open and then adjusted for conditions within forests. The distribution of the snow cover under forest canopies and in small openings is estimated from such variables as timber volume, crown density, and total snowfall (Fig. 22-7). For any daily temperature, then, the method is to read basic open melting rates from Fig. 22-8 for each 500-ft elevation zone and aspect.

Table 22-3. Example of Snowmelt Hydrograph Computations, Idaho

Date	Temperature, °F −32			5000–5500 ft			5500–8000 ft 5 elevation zones	8000–8500 ft			Snow-melt total, in. × mi	Rain,* in. Rain at Deadwood Dam, elevation 5300 ft	Loss,† in. × mi Evapotranspiration loss (Fig. 22-9)	Runoff-routed		
	Mean daily Temperature at Deadwood Dam, elevation 5300 ft	Total mass Cumulative temperature above 32°, after Jan. 1		North slope Snowmelt on north-facing slopes (Fig. 22-8)	South slope Snowmelt on south-facing slopes (Fig. 22-8)	East-west Snowmelt on other slopes (Fig. 22-8)	Description of procedures	North slope	South slope	East-west				Inflow, cfs Total snowmelt runoff less losses	Storage, cfs × days Groundwater storage (Fig. 22-10)	Outflow, cfs Discharge vs. storage (Fig. 22-10)
							(1) Snowmelt rates (in./°F − 32) by aspect and elevation (Fig. 22-8)									
							(2) Rate (in./°F − 32) × temperature (°F − 32) × area (sq mi)‡									
							(3) For snowmelt rates in forests									
							(4) Snowmelt (in.) × area of commercial timber, (sq mi)‡									
							(5) Snowmelt (in.) × area of lodgepole pine (sq mi)‡									
								Computed snowmelt								
May 1	....	146		0.055	0.070	0.060	(1) Open (see above)	0.020	0.050	0.035	....	....	....	....	....	§
May 2	2.5	....		....	0.1	0.5	(2) Open (see above)	0.2	....	0.1	20.3	0.26	0.1	240	3400	170
May 3	2.0	....		0.1	....	0.4	(3) Forest (see above)	....	....	....	....	....	....	210	3470	170
May 4	....	154		0.5	....	1.5	(4) Forest (see above)	....	....	....	14.0	0.40	0.1	450	3510	175
May 5	4.0						(5) Forest (see above)						0.2		3780	190
May 6	15.5	....		....	....	0.057	(1) Open (see above)	0.022	0.052	0.037	....	0.02	0.9	1780	5330	265
May 7	10.5	....		0.052	0.067	2.1	(2) Open (see above)	0.7	....	0.4	90.5	0.20	0.7	1200	6240	310
May 8	1.0	....		0.3	....	....	(3) Forest (see above)	....	....	....	....	0.15	....	110	6040	300
May 9	2.0	....		0.6	....	1.9	(4) Forest (see above)	....	....	....	....	0.05	....	240	5980	300
May 10	6.0	189		2.1	....	6.4	(5) Forest (see above)	....	....	....	62.3	0.90	....	720	6390	320

* Effective rainfall on snowmelt area estimated from temperature; remaining precipitation added to snowpack.
† Includes soil-moisture deficit, satisfied by fall rains in this case.
‡ Areas by elevations and aspects from air photos; areas of timber types from timber surveys distributed by elevations and aspects.
§ Snowmelt and runoff during April not shown.

22–22

HYDROLOGIC EVALUATION OF LAND TREATMENT 22–23

These rates apply as if the area were in large openings. Melting rates within forests (expressed as per cent of the rate in the open), classified for aspects and elevation zones, are read from Fig. 22-8 at the intersection of timber volume and crown density, with an interpolation for depth of snowpack. These percentages divided by 100 and multiplied by melt rate in the open are multiplied by the areas within each aspect-elevation class (Table 22-4).

The total snow cover in open areas is obtained from snow surveys and rain gages in the vicinity. These measurements of snow-water content are related to elevation in order to estimate winter precipitation throughout the tributary watersheds (Fig. 22-9). Rainfall during the preceding October to December has to be added to the soil storage. In some abnormal years this antecedent rainfall is enough to satisfy soil-moisture deficits and set the stage for a major flood.

Evaporative losses produce a difference between streamflow and the apparent snowmelt as observed by periodic snow surveys. Evaporation from the snow surfaces must be estimated and deducted from the observed rate of snow depletion. Temperature and wind movement are the main variables in evaporation from open surfaces. In

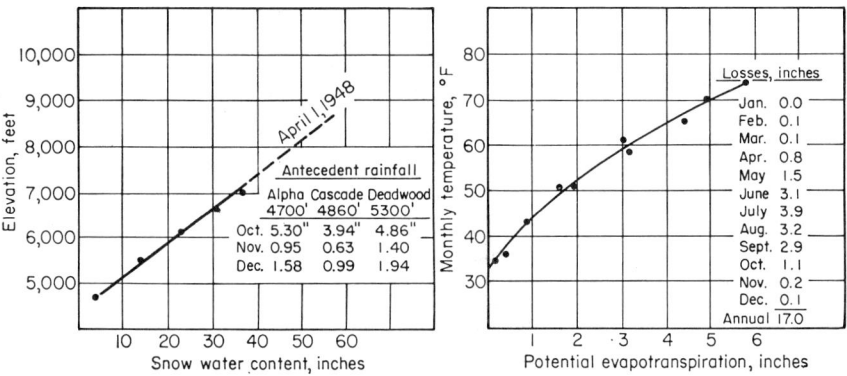

FIG. 22-9. Snow cover and potential losses, South Fork, Salmon River, Idaho, 1948.

forest the insolation from the sun and reduction in wind velocity are undoubtedly related to the character of the forest type. The total loss depends upon the elevation and the extent of forest and open areas on various exposures.

Transpiration losses, especially later in the season, also must be included in depletion of soil moisture; these losses depend on variables as described above. The combined losses of evapotranspiration have been determined at experimental locations throughout the country and related to climatic factors of temperature and daylight hours. Fairly uniform losses have been computed for the same season at many locations over wide areas and therefore can be applied throughout the region. It seems best to combine all evapotranspiration losses into about 17 in. annually in the West for the forested mountains (Fig. 22-9) and estimate daily variations from climatic factors.

Precipitation occurring after the last snow survey near the time of maximum snow accumulation must be added to the snowpack at high elevations. If temperatures indicate that rain was falling, this moisture may go immediately to recharge the groundwater storage on that portion of the watershed near the snow line where snow is ripe from rapid melting or where the soil is saturated and percolation rates are high enough to allow water movement as fast as it is supplied. Streams may sometimes receive inflow from rainfall as surface runoff in those parts of the watershed occupied by exposed bedrock, roadways, or water surfaces. Generally, rainfall does not come in such large storms as to require computation of the effect of warm rain in accelerating snowmelt, although formulas are available for this phenomenon.

Soil-moisture storage capacity depends on soil texture and depth and amount of litter and humus. To obtain information on watershed storage a series of well-dis-

Table 22-4. Distribution of Forest and Open by Elevations and Aspects, Idaho
(In square miles)

Tributary aspect	4000–4500 Open	4000–4500 Forest	4500–5000 Open	4500–5000 Forest	5000–5500 Open	5000–5500 Forest	5500–6000 Open	5500–6000 Forest	6000–6500 Open	6000–6500 Forest	6500–7000 Open	6500–7000 Forest	7000–7500 Open	7000–7500 Forest	7500–8000 Open	7500–8000 Forest	8000–8500 Open	8000–8500 Forest	8500–9000 Open	8500–9000 Forest	9000–9500 Open	9000–9500 Forest	Total Open	Total Forest
SouthFork near Knox:																								
North	…	…	0.6	1.0	…	4.5	0.3	7.5	…	7.6	1.5	4.6	1.3	2.1	0.5	0.4	…	…	…	…				
South	…	…	1.7	0.5	1.4	…	4.2	0.6	6.4	…	2.4	0.2	0.8	0.8	…	…	…	…						
East	…	…	…	2.3	2.1	3.6	1.1	2.0	4.2	0.3	3.6	0.8	0.5	0.4	0.1	…								
West	…	…	2.4	…	2.6	0.9	2.4	2.3	3.3	1.8	2.6	0.7	1.6	…	0.6	0.2	…	…						
Total	…	…	4.7	3.8	6.1	9.0	8.0	12.4	13.9	9.7	10.1	6.3	4.2	2.5	0.6	0.6	…	…	…	…	47.7	44.3
Lake Fork near McCall:																								
North					0.7	0.9	…	1.6	0.9	0.9	0.2	0.8	0.3	1.2	0.8	0.6	0.8	0.6	1.0	0.3	0.2	0.1		
South					…	…	0.8	0.6	1.0	0.4	0.2	0.2	0.8	0.3	0.4	0.3	2.4	0.4	0.4	…	0.1	…		
East					0.5	0.5	1.5	1.0	1.0	0.4	1.9	0.6	1.7	1.1	1.7	0.5	2.6	0.5	0.3	0.3	…	0.1		
West					…	…	2.8	0.9	2.0	0.4	2.9	0.8	2.3	0.7	2.3	0.7	1.6	2.0	0.9	0.6	0.2	0.2		
Total					1.2	1.4	5.1	4.1	4.9	2.3	4.7	2.4	5.1	2.6	5.2	2.7	7.4	2.0	2.6	0.6	0.5	0.2	36.7*	18.3
Mud Creek near Tamarack:																								
North	…	…	0.5	…	1.1	…	…	0.1	…	0.1														
South	0.9	0.8	0.5	…	0.5	0.5	0.5	0.1	…	0.3														
East	0.9	0.8	1.9	2.2	0.7	0.1	0.1	0.1	0.3	…														
West	…	…	1.5	1.7	…	…	0.2	…	…	…														
Total	1.8	1.6	4.4	3.9	2.3	0.6	0.8	0.2	0.3	0.1	…	0.1											9.6	6.4

* Includes large area of exposed bedrock.

tributed ¼-acre plots are established. Soil-profile descriptions on the plots include depth of litter, humus, soil texture by horizons, and total depth to bedrock. Total watershed storage is obtained by expanding the plot information through delineation of rock outcrop, cropland, open areas, and forest types on aerial photographs.

Long records of streamflow from tributary watersheds are used to determine the rates of depletion of groundwater storage after having been recharged in the spring. The daily drop in runoff is related to the discharge to build a depletion curve as shown in Fig. 22-10. In the example area, streams generally drop only about 5 per cent/day at lower stages. By adding up the volume of such depletion curves, a storage-vs.-discharge relation can be drawn for each watershed (Fig. 22-10). By entering this figure for a given increase in storage, the increased flow in the stream can be read. The change is very slight at low stages of later summer, but becomes quite pronounced as groundwater storage builds up in the spring.

For larger watersheds the channel inflow must be routed downstream to the gage in order to reproduce hydrographs. However, the smaller the area, the less the time lag required to move the water from the melt zone down to the stream gage. When

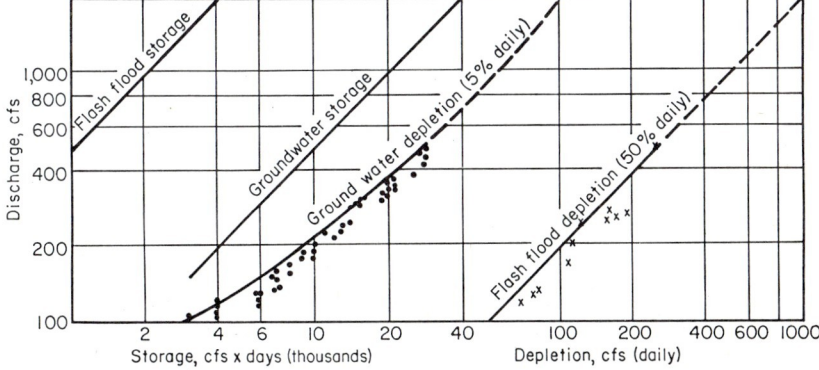

FIG. 22-10. Storage and depletion curves, South Fork, Salmon River, Idaho, 1929–1948.

channel sections are typically uniform and only a few miles of channel is involved, this channel-storage factor can be handled by shifting the time scale slightly during periods when the flow is as fairly steady as snowmelt runoff.

The general shape of the computed hydrographs from snowmelt should be in agreement with the groundwater or base flow, ignoring the minor fluctuations due to rainfall. This is a check to indicate that all the basic relationships of climatic, topographic, groundwater, and cover influences are in the proper order of magnitude and variation as far as the melting of snow is concerned. However, the flashy hydrographs from rainfall superimposed on the base flow may not conform to overall groundwater characteristics.

A preliminary comparison is made between the present runoff and the expected future runoff with different forest conditions—all hydrographs computed by the same procedures. The conditions assumed for the example are that the good forest represents complete protection from destructive cutting and fires for a long period of time, while the poor condition occurs following unduly heavy cutting in the commercial timber, but without eroding the soil or reducing its moisture-holding capacity (Fig. 22-11).

C. Evaluation by Multiple Regression

To estimate the effect of land-treatment measures on flood runoff or peak discharge, a multiple-regression analysis first draws on past experiences, observations, and assumptions. Of first concern is a test to determine whether the experiences apply to the area, whether the observations may be given quantitative measurement, and whether or not the assumptions are true. For example, destruction of vegetative cover has been hypothesized as contributing to flood magnitudes. By measuring the variations in flood magnitudes associated with variations in cover in the past, a basis is obtained for estimating what further reductions in cover or increase in cover will do to flood magnitudes.

Flood discharge is the result of a number of variables with respect to time and place. Among the major factors are the meteorological events antecedent to the flood and the physiographic and geologic characteristics of the watershed. The variations due to

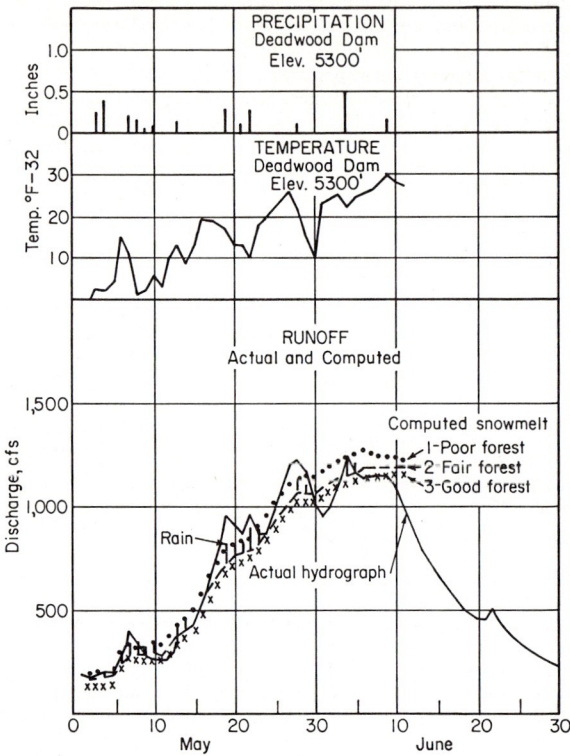

FIG. 22-11. Computed snowmelt hydrographs for forest conditions, South Fork, Salmon River, Knox, Idaho, 1948.

the vegetative differences will become apparent and measurable when the variation in flood discharge attributable to the other causes are measured and their effects isolated.

Numerical values can be assigned to most of the factors associated with flood runoff. With numerical values as an index to certain hydrologic processes, statistical methods can predict the general laws which apply most of the time. No method can hope to reveal the laws of nature to such a degree as to permit perfect results to be determined in all cases. In studying the trend of data with the passage of time, it is most important to be able to calculate that which is normal for the event. Multiple regression is especially adapted to tell, without bias, what value can usually be expected with a given set of variables. Any residual difference between actual and computed values must therefore be associated with some factors yet untested. By progressively testing all known watershed and meteorological data, it should be possible to write an equation which would predict the natural phenomena and later to associate induced changes on the watershed with the unexplained differences.

To test blindly all data on the watershed and find certain factors significant cannot in itself establish definite cause-and-effect relationship. The interaction of certain variables can accidentally give apparent significance to variables that may have no physical importance. In all the watershed research of past years and in the practical experience of hydrologists of all agencies, there is a tremendous background for the selection of factors which could be expected to influence the hydrology of a watershed. Therefore the statistical approach should be guided as much as possible by such information. There is no reason to believe that the factors tested are the only variables involved. Additional research and experience will reveal factors that should be included to obtain greater precision in delineation of relations.

Any method of analysis of rainfall-runoff relations generally starts with commonly accepted procedures in the preliminary stages of computing basic data; that is, rainfall records must be arranged by some system, hydrographs must be separated into flow components, etc. In this phase, detailed storm studies and similar analyses are completed only to the stage where reasonably accurate concepts are obtained of the factors involved in the runoff relation. With statistical methods in mind for handling a great number of variables in the end, the physical approach can stop when basic data have been determined and need not be carried to such refinements as subdividing drainages, synthesizing hydrographs, or flood routing. The point here is that practically the same basic data on most factors are required by any method and, further, that the adequacy of available information predetermines the final result. To carry the physical approach to great detail often confuses the problem by reason of the amount of personal judgment involved in refining the computations. By contrast, statistical methods restrain the personal bias of individuals once the basic data have been established by comparable methods.

The method of multiple regression combines least-square fitting of the best equation to the data and gives a measure of the significance of each of the effects, the probable range of error in evaluation of the individual effects, and the error in the total effect. The tests of significance provide a criterion to judge whether a given variable should be retained or dropped from the analysis. The individual effects obtained are true partial effects, and may be used to estimate the effect of changing a single variable. The overall results are expressed in the form of an equation.

For an explanation of the mathematical principle of multiple regression as used in hydrologic problems see Sec. 8-11 and Ref. 11.

Available hydrologic records are reviewed with the aim of selecting a series of typical watersheds for analysis. If possible, these should be watersheds with 20 years or more of streamflow records. The areas selected should contain a good range in sizes, conditions, cover types, damage areas, and flood and land-use history. The number of watersheds to be used depends, in part, on the number of records available. If possible, the minimum number should be at least 7 or 8 and the maximum, 20 to 25. The hydrologic analysis will be improved by having a larger number of watersheds, up to this maximum.

If floods in the area are due mainly to snowmelt, to rain, or to combinations of both, the analyst should select the watersheds with this point in mind. If, in any given area, snow and rain cause distinctly different groups or types of floods, it is desirable to sort the watersheds and the flood data so as to provide separate analyses.

For each sample watershed a flood-recurrence-interval chart using standard plotting methods is constructed. If frequency charts have been made and are obtainable for the watersheds, considerable time may be saved by using them.

The flood-frequency curves are examined for the various sample watersheds to eliminate any obvious inconsistencies. This can be accomplished by plotting peak discharges in cubic feet per second on a single sheet of semilogarithmic paper, making any necessary adjustments in curved slopes so as to keep the lines in reasonable harmony. Also, any necessary extensions should be made in the length of the recurrence-interval curves for the watersheds with the shorter records.

Divide the combined chart into three or more "strata" of recurrence intervals, the number of strata k depending on the number of watersheds studied. The objective is to make the product of k times n (the number of watersheds) larger than 60, and preferably about 80. If there are 15 sample watersheds, for example, 5 strata would be suggested so as to give a total of 75 flood observations. The strata may be graduated in width so as to make the number of floods in each of the strata more nearly the same. An ideal arrangement would be to have the same number of flood events in each stratum. When the stratification has been laid out, one flood event is drawn at random from each of the strata for each watershed; it is necessary to make sure that data on the necessary correlated variables, such as precipitation, can be set up for each flood selected.

The above steps will give a sample set of flood data, 60 to 80 in number, for further analysis. For each sample flood event, a series of descriptive data is developed, expressing flood magnitude and watershed characteristics.

In this method of analysis and in developing these variables, it is worth commenting that the analyst should orient his thinking away from the idea of synthesizing actual hydrographs from the several variables involved. Considering all the errors and biases inherent in the available data on watershed and storm characteristics, such a synthesis is obviously impractical. Instead, the aim should be to set up a series of variables that are highly correlated with flood discharge; although they cannot be directly combined to give the observed discharge for each flood, their relation to it can be estimated in an efficient and unbiased manner by multiple-regression analysis.

From this viewpoint it is neither desirable nor necessary to attempt a synthesis of a flood hydrograph from infiltration data. Instead, these data are used in accepted procedures to supply an estimate of surface runoff from any given storm. This estimate is used as a dimensionless index of cover-soil, land-use condition and inserted as another independent variable into the regression analysis.

Information helpful in characterizing a variable is (1) a descriptive title, (2) a symbol which stands for the variable, (3) the exact operational definition of the variable, that is, how a value for the variable is obtained in practice, (4) units in which the variable is expressed, (5) what the variable is intended to characterize or represent, and (6) variations in the variable, its units, its descriptive definition, etc.

The variables may be divided into two large categories: dependent and independent variables. Within the independent group, crude separation into three groups might be indicated: (1) meteorological variables, (2) watershed constants or land variables which are not subject to rapid and substantial fluctuations in time, and (3) watershed-management variables or land variables which are subject to relatively rapid fluctuations due to management.

1. Dependent Variables

Peak discharge from a watershed, symbol, Qp: The highest discharge recorded or reported during the interval of a defined event, expressed in cubic feet per second (cfs). It is intended to express the relative magnitude of a flood as to the effects it can produce, that is, impact damages or sediment production.

Unit peak discharge, Qpm: The highest discharge for a flood event, expressed in cfs or cfs/sq m.

Estimated peak discharge, $Q'p$: The highest discharge for an interval as estimated for some watershed condition where means other than the actual measurement of the discharge are used, expressed in cfs.

Storm runoff from a watershed, Qsr: The total discharge as a result of a storm of known rainfall and duration, expressed in inches depth or acre-feet.

Discharge component, $Q'pi$, which is any one of the segments that the peak discharge Qp could be separated into, expressed in cfs. From as few as two to as many as eight segments have been used; the operational definitions should be furnished by the user.

Total streamflow for a particular period, Qri, where i represents the period used. This variable, expressed in inches, is often used in analyses of water yield where Qri is the streamflow for a year or for several months of the year, such as the summer season.

2. Independent Variables

a. Meteorological

Storm precipitation, Pi: The precipitation associated with the defined event, where i indicates the period of time over which the precipitation is averaged; for example, Pmh would be the maximum hourly precipitation and Pmd would indicate the maximum daily precipitation. Since watershed precipitation is in itself an abstraction, the operational definition should be outlined by the analyst. This variable is expressed in inches.

Storm temperature, Ti: The temperature characteristic which characterizes the snowmelt potential and/or the rain-snow characteristics of the precipitation in the defined event. For example, Tm might be defined to be the highest minimum daily temperature during the defined event estimated for the mean elevation of the watershed. Tw might be the mean winter temperature (November to March), preceding the flood event. The operational definition should be furnished by the user. An alternative of this variable is the area of the watershed where rain falls instead of snow, RA.

Ripe snow area, RSA: The area of the watershed where snow is in a condition such that addition of heat by storm temperature or rain will cause snow to melt and lose water. This is expressed in per cent of watershed area or square miles. Operational definition of variables is to be furnished by the user.

Antecedent precipitation, aPi: The precipitation occurring during some period antecedent to the defined event or some part of the defined event, where i is the period in months, days, or hours, during which the precipitation is totaled or averaged. This is intended to express watershed wetness and/or snow storage, in inches.

Antecedent temperature, aTi: The temperature occurring during some period prior to the defined event, where i is the period in months, days, or hours, during which the temperature is totaled or averaged.

Snow-storage area, SA: The area of the watershed on which snow cover exists at the start of the defined event, expressed in per cent or square miles.

Snowmelt, SM: Maximum daily snowmelt during the defined event, expressed in area-inches.

b. Watershed Constants

Watershed area, A: The total area of the watershed above the discharge-measuring points, expressed in square miles.

Standard precipitation excess, xPs: The component of precipitation excess obtained when infiltration for some standard land-use condition is subtracted from maximum precipitation for a defined event; thus xPs is the component of precipitation excess attributed to soil characteristics of the watershed. For areas where infiltration is limiting, this variable is intended to express the effect of the soil-geologic characteristics on flood production.

If soil-water storage is not a limiting factor, infiltration capacity should determine the magnitude of the peak discharge. If, on the other hand, soil storage is limiting, the figure obtained by subtracting available soil storage from the precipitation estimate is likely to be larger than that obtained by using infiltration capacity.

Geological and/or soil effectiveness, GE or SE: The relative effectiveness of the soil-geologic types in contributing to flood production. Variations: percentage of geologic types, percentages in SCS hydrologic soil group (Sec. 21), or percentage of soil depth-texture groups, etc.

Elevation, Ei, where i represents the moment which is being used to characterize the distribution of elevations in the watershed. Variations: mean elevation or area of watershed of low elevation (below the mean elevation of the area).

Physiographic effectiveness, PhE: An index of the watershed characteristic that indicates the effectiveness of physiography on flood runoff. One or more of the following variables may be used: simple parameters of orientation, equivalent slope, per cent of north-facing slope, characteristics of the area-elevation curve, or the stream physiography variables of Horton [12].

c. Management

Precipitation excess, xPu: That component of precipitation excess attributed to the deviation of land use from a standard condition, obtained by subtracting the standard precipitation excess xPs from the precipitation excess obtained from the difference between the maximum rainfall and the actual infiltration capacities of the soil and land-use types on the area during the discharge event.

For areas where infiltration is limiting, this variable is intended to express the effect of land-use characteristics on flood production, independent of those soil characteristics which are not subject to management.

The suggested steps for obtaining this variable are:

1. For the same depth-texture classes as used for determining standard precipitation excess, make up a table of infiltration capacities for each of the series of cover types and land uses. Subdivide each of these land uses and cover classes into hydrologic condition classes (as Good, Fair, and Poor).

Infiltration data are obtained from all possible sources. In some areas standard infiltration capacities have been built up for soil classes in the area. (If no other data are available, it may be possible to obtain reasonable results by employing the values shown in Ref. 13, p. 48, table 9.)

2. Calculate a series of soil-storage indices, and then make up a table of precipitation-excess indices corresponding to the standard soil indices used in determining the standard precipitation excess.

3. From each of these indices subtract the standard index for each soil to give a series of net cover indices.

4. Calculate a weighted-average precipitation excess for each storm and for each watershed. Use the results as a cover variable.

Forest age-stocking effectiveness, ASE: The effect on floods of variation in forest age and stocking from watershed to watershed. This variable is intended to be an expression of the forest's effects on floods due to the complex effects on: (1) surface storage, interception, surface detention, litter storage, and snow accumulation and melt; (2) soil and rock storage, infiltration, soil porosity, evaporation plus transpiration, soil freezing; and (3) time-of-concentration factors such as roughness, depth, and velocity of flow. The variable is computed as the product of a weighted age-effectiveness and of stocking for individual units of the forests in watersheds, expressed in per cent. *Age-effectiveness* is taken proportional to foliage increase with age as follows: 0 for age zero, 0.3 for 10 years, 0.7 for 20 years, and 1.0 for 30 years and greater [1, pp. 31–37]. Thus, a forest stand with 55 per cent stocking and 20 years of age has $ASE = 0.55 \times 0.7 = 39$ per cent [14, p. 36].

Area bare-cultivated, BC: The area of the watershed which was in bare cultivation at the time of the defined event producing the discharge, expressed in per cent of total watershed area, square miles, or acres.

Denuded area, DA: The area of the watershed which was burned, clear-cut, or otherwise denuded at the time of the defined event, expressed in per cent of total watershed area, square miles, or acres. This is intended to express residual effects of denudation which are not fully eliminated by recovery of the vegetation.

Cover density, Ci: This defines the cover component, that is, canopy density, herbaceous density, litter density, or total density, expressed in per cent of full cover.

An example of a multiple-regression analysis for land-treatment evaluation is furnished by Anderson [15]. The study was made to obtain a hydrologic basis for evaluation of upstream-flood-prevention programs of the U.S. Department of Agriculture in southern California.

The relation of peak discharge to storm and watershed variables was determined by analysis of 127 sets of storm-watershed data involving 29 watersheds and 23 storms. Storms were selected to give full representation of three size groups: small, medium, and large on dry, moderately wet, and wet watersheds. Replication was obtained whenever possible in storm size group. The watersheds were selected to give as wide a representation in space as possible, including only those watersheds for which there were adequate precipitation measurements.

The resulting regression equations and the definition of the variables are given below:

$$\log Q = 3.245 + 1.084 \log A + 0.569 \log P_a + 1.226 \log P_{24} + 0.785 \log P_i \\ + 3.084 \log T_m - 0.991 \log C \quad (22\text{-}1)$$

where Q is the momentary peak discharge in cfs, A is the drainage area in sq mi, P_a is the antecedent precipitation in 21 day-in., P_{24} is the maximum 24-hr precipitation in in., P_i is the maximum 1-hr precipitation in in., T_m is the minimum daily temperature in °F, and C is the cover density in per cent.

By analysis of variance and t tests the variables were significant in their effects on flood-peak discharge. The multiple-correlation coefficient of the equation is 0.954.

The equation and the partial regression coefficients are used to establish a hydrologic base against which the land-treatment effects can be compared. The base usually used is the expected frequency of flood flows under present watershed conditions. This is obtained by:

1. Adjusting the existing records of discharges from a watershed as measured under the past variable watershed conditions to their expected values under a constant watershed condition (the present), using the regression coefficient for the cover variable C in Eq. (22-1).

2. Extending the discharge record to a uniform long-term basis by:

   (a) Correlation of adjusted discharges between watersheds.
   (b) Correlation of adjusted discharges with recorded meteorological events by using the partial regression coefficients found for precipitation and temperature in Eq. (22-1). The assumption is made that events of the past, when adjusted and extended to a long-term basis, will be representative of expected events in the future (should the constant conditions be maintained).

The adjusted and extended discharges representing present conditions are next converted to their expected magnitudes under future watershed conditions without and with a land-treatment program. This is done by:

1. Expressing the program in terms of the changes which (and time) will bring about in the cover parameter used in the equation
2. Using these changes to obtain from the equation the effects on peak discharges

D. Evaluation by Regional Analysis

Experience has shown that the removal of vegetative cover by fire may change the runoff and erosion characteristics of watersheds. The evaluation of fire effect on storm runoff and sediment yield is essential if an adequate appraisal of the total damage due to wildfire is to be obtained. This information is needed to determine the level of fire protection justified.

Rowe, Countryman, and Storey [16] describe an analysis which was made to determine the effect of fire on storm runoff and erosion rates in southern California. The basic purpose of the analysis was to provide the following information:

1. The most probable frequency of various-size storm flows under conditions of normal vegetative cover
2. The effect of fire on the size of these flows
3. The residual effects of past fires on runoff flows
4. The normal annual erosion rates
5. The effect of fire on erosion rates
6. The residual effect of past fires on erosion rates

1. Determining Effect of Fire on Peak Discharge. Preliminary analyses by the authors indicated that peak discharge is well correlated with flood damages and is particularly sensitive to fire effects. Peak discharge was therefore used in the study as the basic measure of watershed discharge.

Storm frequencies were developed to supplement the shorter-time streamflow data used in determining peak-flow frequencies. By determining the relation between storm precipitation and peak discharge it was possible to compute discharge frequencies using the longer-time base of the precipitation records. To facilitate determination of storm frequencies, watershed units were grouped into five storm zones. Each zone consisted of a series of adjacent watershed units in which there were approximately equal number and reasonably uniform occurrences of storms.

A key watershed was selected in each storm zone to establish the frequency of normal storm peak discharges of various sizes. Each key watershed was selected as being generally representative of the watersheds in the particular storm zone within which it occurs. A long record of precipitation and streamflow and a long unburned vegetation cover condition were also essential requirements of the key watershed.

Peak discharges of the key watershed were grouped into size classes, expressed in cubic feet per second per square mile, and the average number of storm peak discharges per year in each size class was determined. The average number of storms per year in each discharge class was also determined for each storm-frequency class.

The number of observed peak discharges in each storm-frequency class was then checked against the corresponding number of storms given on the storm-frequency curve. When these differed, the observed number was adjusted to conform to the number given by the longer period of record from the frequency curve. The adjustments were made by computing the ratio of the number of storms in each size class of the storm-frequency curve to the observed number for the period of the discharge record. The adjusted data were tabulated by classes and plotted on logarithmic paper. Irregularities in the number of events between size classes were eliminated by fitting a curve to the plotted data (curve A in Fig. 22-12). The number of events per discharge class given by this curve was used as a base for establishing the peak-discharge frequencies for all watershed units within the storm zone.

Frequencies of normal discharge for gaged watersheds, other than key watersheds, were also developed to determine the average size of storm peak discharges for these units and as an aid in developing a system of watershed ratings for units for which streamflow data were lacking. These frequencies were based on the relation of the observed peak discharges of the individual watersheds to the corresponding discharges of the key watersheds. A curve fitted to the plotted data as shown in Fig. 22-12, curve B, establishes the frequency curve of normal peak discharges for these gaged watersheds.

In the southern California study, discharge measurements adequate for determining frequencies were available for only about one-third of the watersheds analyzed. Estimates of normal peak discharges for the ungaged watersheds were based on numerical ratings of the effect of various watershed physiographic factors on peak discharge. The ratings are derived from analyses of discharge data and the physiography of watershed units with streamflow records. The ratings were determined by storm zones and are an expression of the effects upon peak discharge of local differences in watershed topography, soil, geology, and precipitation.

In the determination and application of the watershed ratings in the described study the following conditions existed:

1. Peak discharge was expressed in terms of unit area (cubic feet per second per square mile), thus permitting direct comparison of the effects of differences in watershed characteristics on the discharges.

2. Although rates of peak discharge varied with local differences in the watersheds, the general pattern of hydrologic behavior was similar, particularly for watersheds within the same storm zone.

3. Records of streamflow were available for nearly all the larger, more important watersheds for which peak discharge was the most variable and difficult to estimate.

4. Watersheds for which discharge records were available were rather evenly distributed throughout each storm zone, thus providing a reasonably complete sample of conditions existing in the watersheds included in the study.

Each of the watershed factors to be rated was subdivided into classes. Class division of some factors, such as watershed size, was based on actual areas of the units involved. Many of the factors, such as watershed shape, could only be classified qualitatively. The classes of these factors were assigned numbers, the class numbers indicating the relative effect of the class on peak discharge. Thus class I would indicate effects tending to produce the highest discharges; class II, effects tending to produce next highest, etc. The number of classes in each breakdown, except those in which the actual dimensions were used, was contingent upon the range of conditions encountered and their estimated effects on normal peak discharge. A representative watershed of each class was selected as a model to facilitate the classification of other watershed units.

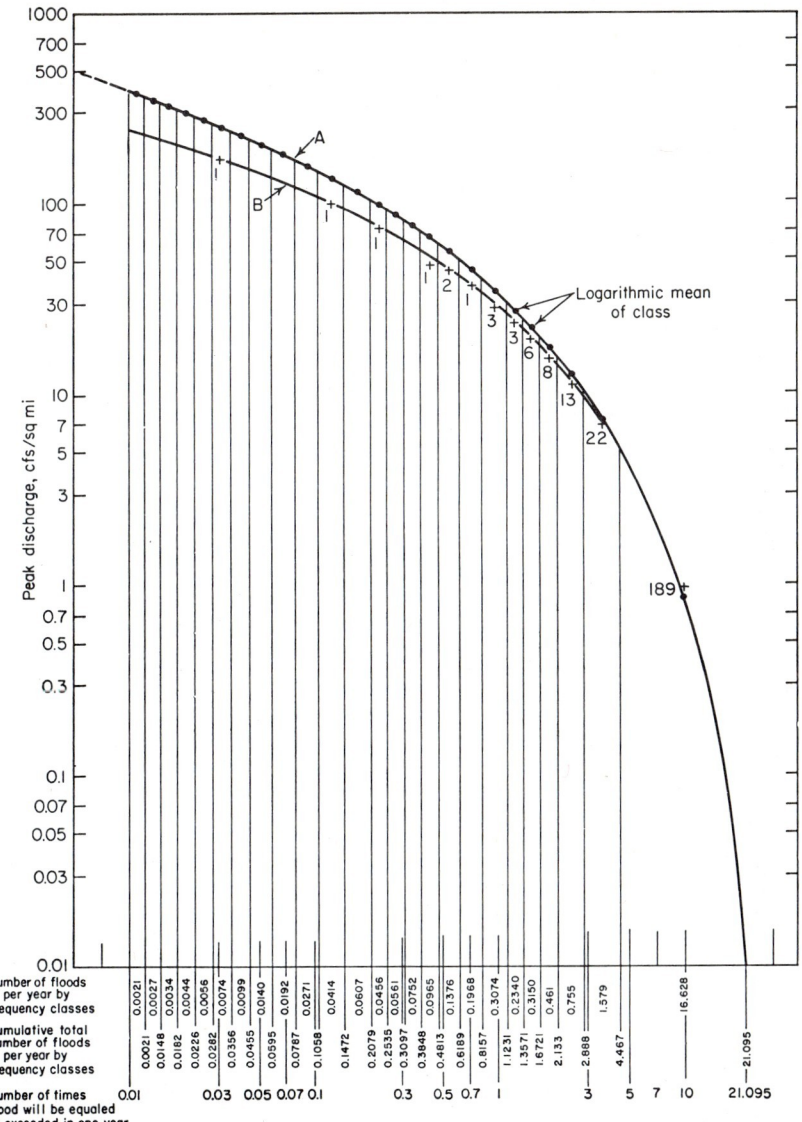

Fig. 22-12. Determination of peak-discharge frequencies for watersheds having streamflow records. *A*, frequency curve of key watershed; *B*, frequency curve of typical watershed developed by comparison of recorded discharges with corresponding discharges of key watershed; 3+, observed discharges with number of observations.

Topographic factors considered in the watershed ratings included (1) size in square miles, (2) shape (fan, rectangular, etc.), (3) density of surface-water sources (number and length of channels per unit area), (4) average gradient of side slopes, (5) flow characteristics of principal stream channels (shape, roughness, etc.), and (6) average gradient of principal stream channels. The effects of some factors like length of stream channels were excluded from individual consideration because they were reflected in other ratings, such as those of watershed size and shape.

Soil factors considered in the watershed ratings included (1) relative infiltration capacity and (2) water-storage capacity (wilting point to field capacity plus one-quarter storage between field capacity and field saturation).

The limits of the infiltration and water-storage classes of the different soil formations were established on the basis of soil textures, depths, and the results of past infiltration and soil-moisture sampling.

Geologic factors included (1) relative permeability, (2) maximum total water storage, and (3) amount of rock outcrop. Other differences in the soil and geologic factors, such as drainage characteristics, also influenced peak discharge, but these effects were small or were reflected in other ratings.

Relative permeability of the different geologic (rock) formations was determined by comparing the amounts of penetration of storm precipitation in watersheds containing rock substratum representative of the different formations but with other factors such as slope, topography, cover, etc., approximately the same. Only major storms with high amounts and rates of rainfall were used for this purpose. Rainfall penetration into the substratum during a storm was determined from the following:

$$F_s = P_s - (M_i + I_s + Q_s) \tag{22-2}$$

where F_s is the rainfall penetration into the substratum during a storm, P_s is the storm rainfall, M_i is the increase in soil-water storage, I_s is the storm interception loss, and Q_s is the storm runoff, all in in. The results of a series of such calculations for the different geologic formations permitted determination of their relative permeability and establishment of permeability classes.

Both type and density of the vegetation were considered important factors in watershed performance. However, there proved to be a close correlation between type and density of the cover. Preliminary vegetation rating classes, therefore, were based solely on the mean areal density of the unburned watershed cover.

Precipitation classes were expressed in terms of the average annual rainfall, in inches depth, of the respective watersheds. Reasons for the use of the actual rainfall data were explained in the succeeding discussion of the watershed ratings.

After the watershed rating factors for each storm zone were established and divided into standardized rating classes, the ratings for each unit were assigned and tabulated as illustrated for a sample watershed in Table 22-5. The numerical significance of each class rating was determined by comparing normal peak discharges of a watershed with other watersheds that were similar in all hydrologic characteristics except the one being evaluated. These numerical ratings were expressed as ratios to the corresponding peak discharge of the key watershed. The ratings thus determined indicate the effect of differences in each factor on peak discharge, but do not constitute a measure of the total effect of all differences on peak discharge.

With the rating factors for the different watershed characteristics evaluated in terms of peak discharge, the determination of normal peak discharges for the watersheds lacking measurements of streamflow was possible. Numerical ratings were determined and tabulated for each watershed by selected peak-discharge classes, as illustrated for the sample watershed in Table 22-5, columns 4 to 7. The numerical ratings expressed the effect of each factor on peak discharge when all the other factors were similar to those of the key watershed. To get the total effect of all the factors that varied from those of the key watershed, it was necessary to compute the algebraic sum of all the corrections. For the size discharge represented in column 4 of Table 22-5, the effect of size of the sample watershed would be to reduce the peak discharge to 0.83 of the key watershed, or a reduction of 0.17. Similarly, the reduction for watershed shape would be 0.44; for main-channel flow characteristics, 0.10; and for precipitation differences, 0.09. This would give a total reduction of 0.80. The peak discharge per square mile of the sample watershed would thus be only 0.20 that of the key watershed.

This final figure may be more readily computed by finding the sum of the individual ratings of the discharge class and subtracting one less than the total number of ratings.

Table 22-5. Use of Watershed Ratings in Computing Peak Discharge

Item	Rating class		Rating for sample watershed[a]			
	Key watershed	Sample watershed	Very small discharge[b]	Small discharge[c]	Medium discharge[d]	Large discharge[e]
(1)	(2)	(3)	(4)	(5)	(6)	(7)
Topographic factors:						
Size	10.6[f]	26.7[f]	0.83	0.88	0.92	0.94
Shape	I−	IV−	0.56	0.68	0.78	0.84
Density of surface watercourses	II	II	1.00	1.00	1.00	1.00
Average gradient of side slopes	II	II	1.00	1.00	1.00	1.00
Flow characteristics of main channel	II−	III−	0.90	0.94	0.97	0.97
Average gradient of main channel	III+	III−	1.00	0.99	0.99	0.99
Soil-geologic factors:						
Soil infiltration capacities	IV+	IV	1.00	1.00	0.98	0.96
Soil water-storage capacities	II	II−	1.00	0.99	0.96	0.94
Rock permeability	III+	IV+	1.00	0.99	0.94	0.92
Rock water-storage capacities	IV−	V	1.00	0.99	0.96	0.94
Rock outcrop	III	II	1.00	1.00	1.04	1.06
Precipitation factor	36.6[g]	32.6[g]	0.91	0.91	0.91	0.91
Integrated topographic, soil-geologic, and precipitation ratings			0.20	0.37	0.45	0.47
Peak discharge of key watersheds, cfs/sq mi			0.81	33.5	152.0	389.0
Computed peak discharge of sample watershed, cfs/sq mi			0.16	12.4	68.0	183.0

[a] Ratings expressed as ratios of peak discharge of sample watershed to corresponding discharges of key watershed.
[b] Peak discharge of size equaled or exceeded an average of ten times each year.
[c] Peak discharge of size equaled or exceeded an average of once each year.
[d] Peak discharge of size equaled or exceeded an average of once each 10 years.
[e] Peak discharge of size equaled or exceeded an average of once each 100 years.
[f] Area in square miles.
[g] Precipitation in inches depth.

For column 4 of Table 22-5 this would be 11.20 − (12 − 1), or 0.20. This method was used to compute the "integrated" ratings for the selected discharge classes as shown in Table 22-5.

The peak discharge for each of these selected discharge classes was then computed by multiplying the discharge of the key watershed by the integrated rating. These computed discharges were plotted on the same frequency scale as the key watershed (Fig. 22-13). A smooth curve drawn through the points then established the discharge-frequency curve for the watershed (curve B).

Determination of the effects of fire on peak discharge was made in order to predict the most probable peak discharges of individual watersheds by years, from time of burning until complete recovery of vegetation. Since burning has no effect on storm frequency, the only difference in the discharge-frequency curve for a burned unit before and after burning is in the size of the average discharge for each frequency class. The average discharge per class can be expected to change from year to year as the watershed recovers from the fire.

The effect of complete burning of watershed cover was determined by (1) comparing peak discharge rates of burned watersheds with those of similar but unburned watersheds for the same storm, and (2) comparing peak discharge rates from similar storms on the same watershed before and after burning. Watersheds used in the determinations were restricted to those in which measurements of streamflow were available for the burned and comparable unburned watersheds and in which the normal vegetative cover consisted of chaparral associations of average density.

These comparisons were made storm by storm by years after the last fire. Similarity of watersheds was judged by comparison of normal discharge-frequency curves and watershed ratings. Comparative watersheds should be close to or adjacent to the burned units.

The most probable size of each peak-discharge event of the burned watershed, had the unit remained unburned, was first determined. The frequency of each discharge peak from the burned watershed was assumed to be the same as the frequency of the corresponding peak from the key and other nearby unburned watersheds. Using this frequency, an estimated peak for the burned unit in an unburned condition was read directly from its normal frequency curve. The ratio of the observed peak of the burned unit to the computed peak for unburned conditions was then calculated to obtain the *fire-effect ratio*. This procedure of computing peak discharges following burning automatically corrects for differences in the physiographic characteristics of the watersheds, for, as previously shown, these are reflected in the normal curves.

Similar fire-effect ratios were also developed by comparing peak discharges occurring from watersheds before complete burning with those occurring in the same watersheds after burning. The similarity of storms is established on the basis of uniformity in amounts of maximum 24-hr and total-storm precipitation, occurrence of the storms in relation to antecedent precipitation and time of year, and similarity in size of storm peak discharges in the key and in adjacent unburned watersheds of the storm zone.

The fire-effect ratios were plotted over their corresponding frequencies on logarithmic paper for all years after burn for which data were adequate. Variations in the ratios within years were eliminated by fitting a smooth curve to the plotted points. Thus a series of curves giving average fire-effect ratios by discharge frequency classes and years after burn was developed.

To smooth out differences in rates of change in the ratios between years and to provide ratios for years for which data are inadequate, ratios for each frequency class were read from the curves and plotted over time in years. Smooth curves were then drawn through these data. Ratios read from these curves were then replotted in their original form, and the recovery curves redrawn as illustrated in Fig. 22-14, for selected years following burning. The fire-effect ratios as given by these final curves were used as the basis for computing the average, or most probable, peak discharge for all watersheds within the storm zone.

In the southern California study a period of 70 years was used as a standard recovery-from-fire period for all watersheds. This was done to simplify both the peak-flow and damage-rate computations. Although it was recognized that the recovery period

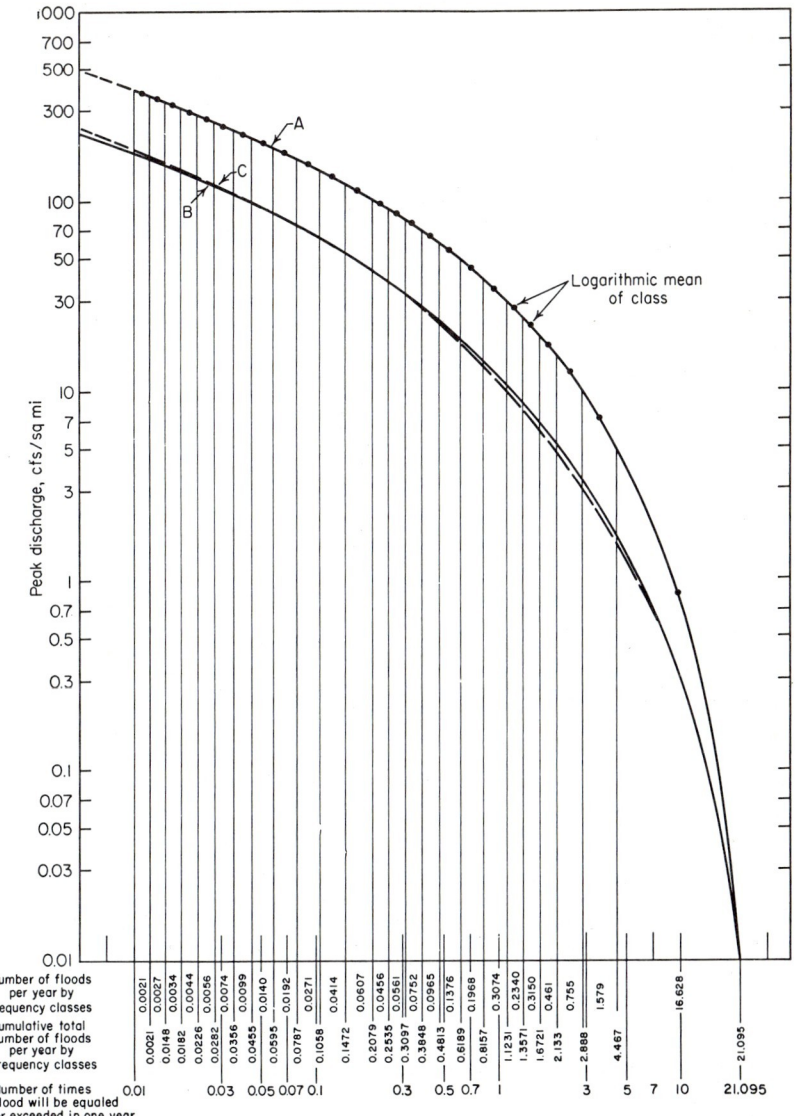

Fig. 22-13. Determination of peak-discharge frequencies for watersheds not having streamflow records. Frequency curves of normal peak discharges for A, key watershed, Santa Anita Canyon, Los Angeles, storm zone; B, sample watershed computed from ratings; C, same watershed as B but computed from observed-discharge data as illustrated by Fig. 22-12.

would vary between units, the comparatively small fire effect after the first few years and greatly increased computational work did not appear to justify the refinement of using varying recovery periods.

2. Determining Effects of Fire on Erosion Rates.[1] Data used in determining annual rates of normal erosion consisted of measurements of total siltation in reservoirs situated in watersheds with normal vegetation cover. Periods for which erosion records were available ranged from 1 to 15 years. Because of the short duration of the erosion records and because the siltation measurements were at irregular intervals, including various numbers and types of storms, it was necessary to determine relations between the recorded peak discharges and corresponding erosion rates. These relations were then used to compute normal erosion rates of the individual watersheds on the basis of the normal peak-discharge frequencies.

Details of this procedure are explained in the following discussion by use of data from the Santa Anita Canyon and from certain other watersheds on the south slope of the San Gabriel Mountains.

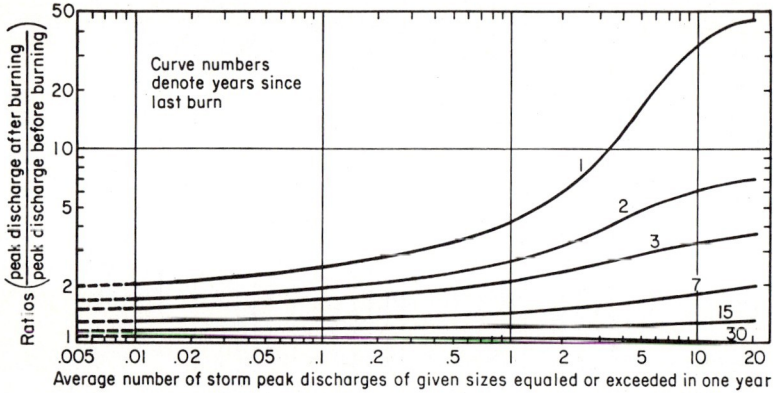

FIG. 22-14. Average effect of fire on peak discharge by frequency classes and years following burning at Los Angeles storm zone.

The first step in determining relations between erosion and normal peak discharge was to prorate measured erosion to the individual discharges that produced it. To accomplish this, a representative cross section of the stream channel just upstream from the reservoir was selected for determining velocities by peak-discharge sizes. The shape, cross-sectional dimensions, and average slope of this section were determined, and a roughness coefficient 0.05 was assumed. Using Scobey's graphical solution of Kutter's formula [17], a series of velocities was computed for various depths. A velocity-flow graph was then plotted showing velocity in feet per second by flow in cubic feet per second. Velocities for all discharge peaks during the period of siltation measurements were determined directly from this graph, and the total eroded material was distributed to individual peak discharges in proportion to the fifth power of the velocity [18], as illustrated in Table 22-6.

This type of computation was repeated for each period of record. The erosion rates (Table 22-6, column 6) were next plotted against the corresponding units' peak discharges (column 7) on logarithmic paper as shown in Fig. 22-15. A smooth curve drawn through the mean of these data was taken as representing the average relation between peak discharge and erosion rates of the watershed.

Peak-discharge–erosion-rate curves similar to that for Santa Anita were developed for four additional watersheds in the same storm zone for which these data were available. Although these curves represented a wide sample of south-slope watersheds,

[1] *Erosion rates* as used here is the volume of eroded material (soil and rock) discharged from a watershed by specified flood flows or in given units of time.

Table 22-6. Computation of Relations of Erosion Rates and Normal Peak Discharge at Santa Anita Canyon, Los Angeles, Storm Zone

(Period, 1940–1941 rainy season; total siltation for period, 31 acre-ft)

(1) Peak discharge, cfs	(2) V, velocity, fps	(3) V^5, fifth power of velocity	(4) V^5, proportion of total, %	(5) Erosion, acre-ft	(6) Erosion rate, cu yd/sq mi	(7) Peak discharge, cfs/sq mi
58	6.0	7,703	1.49	0.46	68.7	5.5
121	7.3	21,000	3.90	1.21	180.7	11.5
63	6.1	8,521	1.67	0.52	77.7	6.0
79	6.5	11,500	2.23	0.69	103.0	7.5
266	9.1	62,900	11.71	3.63	542.1	25.3
80	6.5	11,500	2.23	0.69	103.0	7.6
235	8.8	51,500	9.66	2.99	446.6	22.4
342	9.8	89,300	16.54	5.13	766.2	32.6
193	8.3	40,100	7.43	2.30	343.5	18.3
123	7.4	22,100	4.09	1.27	189.7	11.7
88	6.7	13,400	2.42	0.75	112.0	8.4
465	10.6	134,000	24.91	7.72	1,153.0	44.3
153	7.8	28,300	5.20	1.61	240.4	14.6
175	8.1	34,600	6.51	2.02	301.7	16.7
Total......	....	536,424	99.99	30.99		

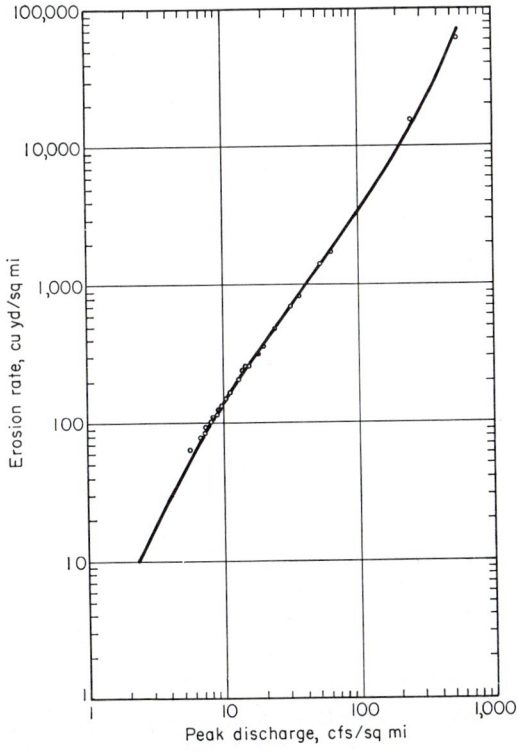

FIG. 22-15. Relation of erosion rates to normal peak discharges in Santa Anita Canyon, Los Angeles, storm zone.

they showed very little variation when superimposed. The five curves were therefore used to develop a single mean curve to represent the average relations between normal peak discharge and erosion rates for all the watersheds of the zone. This mean curve varied only slightly from that shown for Santa Anita Canyon in Fig. 22-15.

Using the peak-discharge–erosion-rate curve as a base, normal erosion for each individual watershed was computed from its normal-peak-discharge frequency curve. The procedure followed in these computations is illustrated in Table 22-7. The data

Table 22-7. Computation of Normal Annual Erosion Rate at Santa Anita Canyon, Los Angeles, Storm Zone

(1) Mean peak discharge, cfs/sq mi	(2) No. of events per year	(3) Weighted peak discharge, cfs/sq mi	(4) Erosion rate, cu yd/sq mi	(5) Weighted erosion rate, cu yd/sq mi
0.849	16.628	14.12	0	0
7.10	1.579	11.21	87	137
12.25	0.755	9.25	189	143
17.30	0.461	7.98	309	142
22.35	0.315	7.36	460	145
27.40	0.234	6.41	570	133
34.56	0.3074	10.62	780	240
44.64	0.1968	8.78	1,110	218
54.83	0.1376	7.54	1,470	202
65.31	0.0965	6.30	1,880	181
75.75	0.0752	5.70	2,260	170
86.20	0.0561	4.84	2,720	153
96.92	0.0456	4.42	3,210	146
113.40	0.0607	6.88	4,000	243
135.3	0.0414	5.60	5,100	211
157.5	0.0271	4.27	6,400	173
180.2	0.0192	3.46	7,900	152
203.3	0.0140	2.85	9,500	133
226.8	0.0099	2.24	11,300	112
250.8	0.0074	1.86	13,300	98
274.9	0.0056	1.54	15,700	88
300.0	0.0044	1.32	18,400	81
325.0	0.0034	1.10	21,700	74
350.0	0.0027	0.94	25,200	68
375.0	0.0021	0.79	29,200	61
389.0	0.0100	3.89	32,000	320
Total............	21.0951	141.27	......	3,824

in columns 1 and 2 are taken directly from the Santa Anita normal-peak-discharge frequency curve. Erosion rates (column 4), corresponding to each discharge peak, were read from the peak-discharge–erosion-rate curve for the zone. The "weighted peak discharge" of column 3 is the product of columns 1 and 2, and the "weighted erosion rate" of column 5 is the product of columns 2 and 4. The sum of column 5 is the most probable, or normal, annual erosion rate of the watershed.

Annual rates of normal erosion were computed in this way for the five watersheds referred to earlier. When plotted on logarithmic paper over the sum of their corresponding weighted peak discharges, these normal rates formed a straight line, as illustrated by Fig. 22-16. The sums of the weighted peak discharges were then computed for each of the watershed units of the zone. Based on this datum the annual rate of normal erosion for each unit was read directly from the erosion-rate curve of Fig. 22-16, thus eliminating the need for computing the weighted erosion rates for individual watersheds.

Determination of the effects of fire upon erosion was based on comparison of erosion rates of completely burned watersheds with those of unburned watersheds. Details of the procedure are here illustrated by use of data from the Los Angeles storm zone.

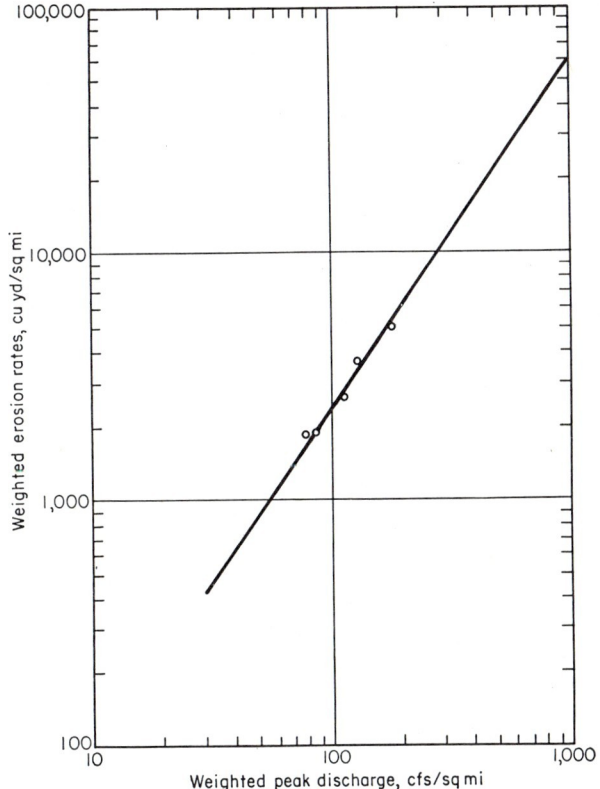

FIG. 22-16. Relation of annual erosion rates to weighted peak discharge at Los Angeles storm zone.

Records of siltation from debris basins situated in watersheds completely burned over during the 1930s were used to establish average erosion rates by years from time of burning until return to normal. Siltation records from Santa Anita Canyon, the key watershed of the zone, were used to establish erosion rates for unburned conditions during the same time.

Streamflow records were not available for the burned watersheds. In order to establish the relation between erosion in these watersheds and peak discharge, it was necessary to correlate the erosion with peak discharges of the unburned key watershed. This was accomplished by first computing ratios of erosion from the burned watersheds to erosion for corresponding periods from the key watershed. These ratios became practically constant in 9 to 10 years, indicating the establishment of relatively stable watershed conditions and normal erosion rates

Using the ratios thus established, it was then possible to compute the most probable rates that would result if the key watershed were burned. This computation was made as follows. The first year after the fire in one of the completely burned watersheds, erosion totaled about 75,000 cu yd/sq mi. The ratios described above indicate that the key watershed, if burned at the same time, would have had an erosion rate for the same period of about 110,000 cu yd/sq mi. During this period of siltation six storms occurred with peak discharges that would produce erosion. The estimated erosion (110,000 cu yd/sq mi) was prorated to each of these peak discharges in the same manner as described for computing normal rates. The computed erosion rate

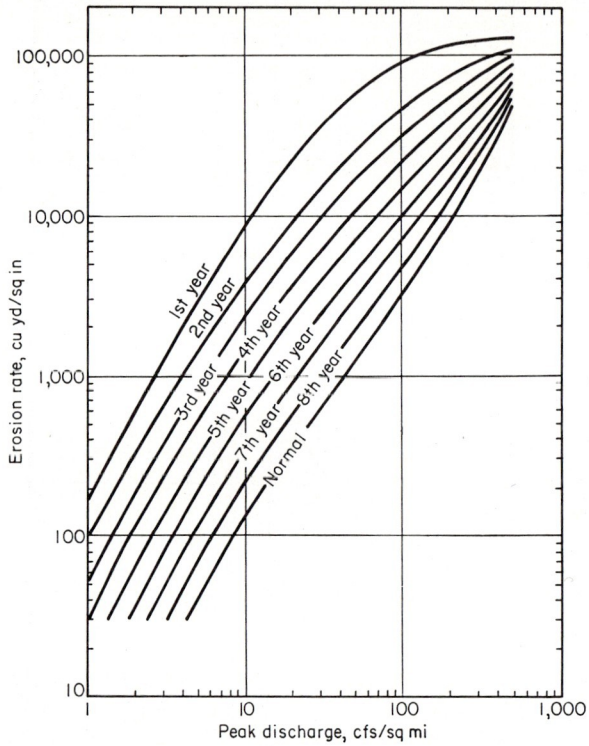

Fig. 22-17. Relation of erosion rates to storm peak discharge following burning at Los Angeles storm zone.

allotted to each peak discharge was next plotted on logarithmic paper over the recorded peak discharge. Similar erosion rates computed by use of data from other burned watersheds were also used. A smooth curve was then fitted to these data. The same procedure was used for other years after burning, and a series of curves was established showing the average relation between normal peak discharge and erosion by years after burning. These curves are illustrated in Fig. 22-17.

Probable average annual erosion rates were computed for the key watershed for each year following a complete burn, using the same method as described for computing the normal rate. These rates were then plotted over time in years since burning, and irregularities between years eliminated by drawing a smooth curve through the plotted points. The average annual rates were read from this curve and plotted over the weighted normal peak discharge of the key watershed. The relation between weighted normal peak discharge and annual erosion rate for individual watersheds was established by drawing a curve through the plotted point for each year after burn and parallel to the normal-erosion-rate curve for the watersheds from Fig. 22-16. This resulted in the series of curves shown in Fig. 22-18.

Estimates of the most probable annual erosion rates by years following fire for watersheds of the Los Angeles storm zone were read directly from the curves of Fig. 22-18. Entry to these curves was made through either the normal weighted peak discharge or normal annual erosion rate of the individual watershed. The estimated

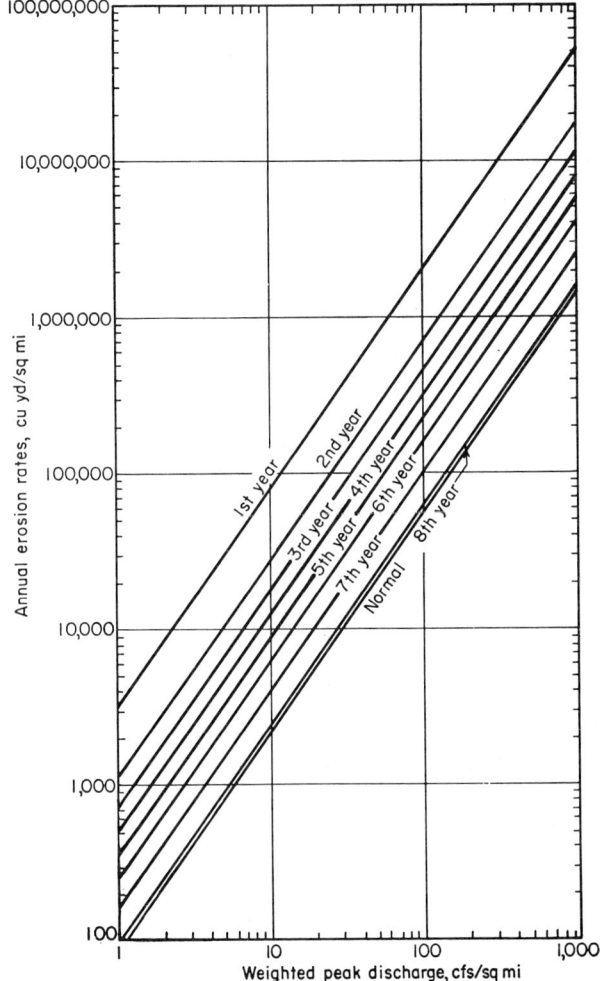

Fig. 22-18. Relation of annual erosion rates to weighted peak discharge following burning at Los Angeles storm zone.

watershed erosion rates following burning were thus related directly to frequencies of normal peak discharge. As indicated in previous discussions, frequencies of peak discharge varied with storm zones. However, the relations between normal erosion rates and erosion rates following fire were reasonably constant from storm zone to storm zone. Thus these relations, developed for the Los Angeles storm zone (Fig. 22-18), were also used to estimate the effects of fire on erosion rates in all zones.

E. Evaluation by Hydrograph Analysis

Basin constants are widely used to represent watershed characteristics. For example, the unit hydrograph has been accepted as a practical method of integrating the effects of size of drainage area, shape, stream pattern, channel capacities, stream gradients, and land slopes. Basically, the *unit hydrograph* is defined as the discharge hydrograph resulting from one inch of direct runoff generated uniformly over the tributary area at a uniform rate during a specified time ([13, p. 105] or Sec. 14). It

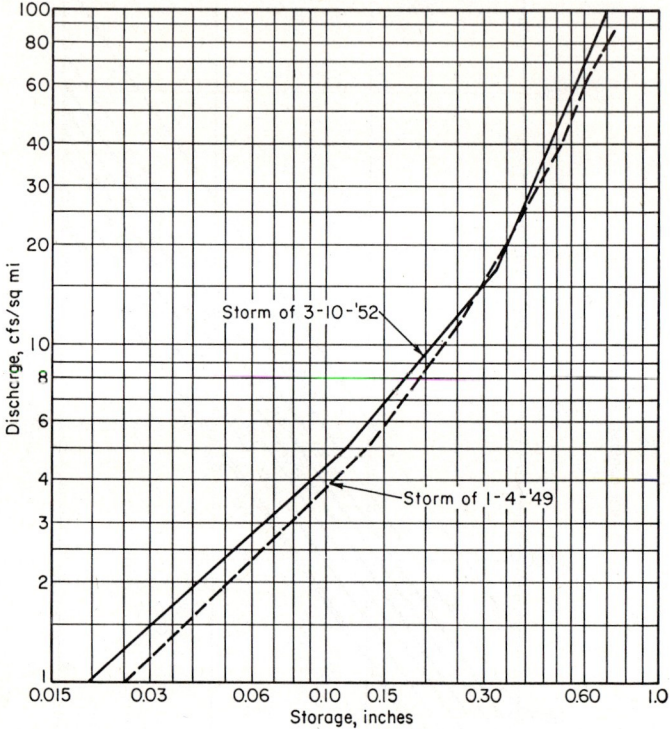

FIG. 22-19. Storage-discharge relationship. Coweeta experimental forest watershed 8, with 1,877 acres.

might be implied that the flow is largely surface runoff and that the soil-cover conditions also approach "constants." Originally, the effect of cover conditions was not widely recognized as fundamentally changing the shape of the hydrograph. However, distribution graphs from forested experimental watersheds show radical differences of 1 to 2 hr in base time and one-third the peak rates for adjacent barren or abandoned farmlands [19, p. 1154]. Nevertheless, such observations have not been utilized extensively in synthetic unit-hydrograph derivations.

The time lag for runoff from ungaged areas is commonly determined by channel lengths and slopes or other surface characteristics of the drainage pattern. The typical shape of unit hydrographs is then applied to the direct runoff as if it were surface flow. Runoff which reaches the channel by subsurface routes must have a much flatter hydrograph because of considerable groundwater storage [20, p. 828]. Hydrograph analyses should recognize that soil-cover conditions change with time and land use on different tributaries. Appropriate hydrograph shapes for the different propor-

Table 22-8. Hydrograph Computations
(In Coweeta Experimental Forest for storm of Jan. 4, 1949, on watershed 8 with drainage area = 1,877 acres)

Precipitation			Subsurface routing					Groundwater routing			Hydrograph	
Date, 1949	Time, hr	Accumulated, in.	Inflow accumulated, in.	Inflow increment, in.	Storage, in.	Outflow, cfs/sq mi	Outflow, cfs	Inflow, cfs-hr	Storage, cfs-hr	Outflow, cfs	Total runoff, computed, cfs	Total runoff, actual, cfs
(1)	(2)	(3)	(4)	(5)	(6)	(7)	(8)	(9)	(10)	(11)	(12)	(13)
Jan. 4	0300	0.12	...	...	...	...	...	...	...	15.4	16	16
	0600	0.50	...	...	...	...	...	...	...	16.0	19	20
	1100	0.65	...	...	...	...	...	...	...	18.5	20	20
	1500	0.65	...	...	...	...	...	...	...	19.0	19	18
	1800	0.80	...	...	...	...	...	...	...	19.0	19	19
	2000	1.25	0.02	0.02	0.020	1.0	3	225	4,100	20.0	23	26
	2200	1.70	0.06	0.04	0.056	2.0	6	715	4,280	22.5	28	26
	2400		0.11	0.05	0.100	4.0	12	280	4,915	25.0	37	42
Jan. 5	0100	1.93	0.15	0.04	0.134	5.0	15		5,125	26.5	42	49
	0200	2.16	0.19	0.04	0.166	6.5	19	245	5,065	27.5	47	54
	0300	2.32	0.22	0.03	0.186	7.5	22	770	5,270	28.5	50	55
	0400	2.48	0.25	0.03	0.204	8.5	25	750	6,000	29.5	55	56
	0500	2.69	0.29	0.04	0.231	10.0	30		6,705	30.5	60	57
	0600	2.90	0.33	0.04	0.255	11.5	34	360	7,040	31.5	65	61
	0700	3.12	0.37	0.04	0.277	13.5	40	360	7,375	32.5	72	66
	0800	3.30	0.41	0.04	0.296	15.0	44	245	7,595	33.5	77	70
	0900	3.48	0.45	0.04	0.313	17.0	50	245	7,810	34.5	84	72
	1000	3.67	0.50	0.05	0.336	19.0	56	320	8,100	35.5	91	74
	1100	3.96	0.57	0.07	0.376	23.0	68	320	8,390	36.5	104	89
	1200	4.25	0.65	0.08	0.420	30.0	88	340	8,700	38.0	126	105
	1300	4.54	0.75	0.10	0.473	33.0	97	265	8,930	39.0	136	122
	1400	4.84	0.85	0.10	0.522	40.0	117	265	9,160	40.5	157	140
					0.458	32.0	94	265	9,390	40.0	134	130
					0.408	26.5	78	415	9,770	40.0	118	113
					0.367	21.5	63	395	10,130	40.0	103	105
					0.333	18.5	54	360	10,450	40.0	94	97
					0.304	16.0	47	375	10,785	40.0	87	89
					0.279	14.0	41			40.0	81	81
					0.257	12.0	35			40.0	75	77
					0.238	10.5	31			39.0	74	74
					0.222	9.5	28			39.0	70	70
					0.207	8.5	25			38.5	64	66

22–45

tions of overland-vs.-subsurface flow might improve the reproduction of recorded floods.

Rather than measuring or indexing factors to reproduce storm hydrographs, an alternative method might be to let streamflow integrate basin characteristics. Using the depletion rates of recorded runoff, from small watersheds with uniform soil-cover conditions, storage-discharge curves can be derived for rainless periods. These can be separated into segments representing the subsurface flow immediately after storms and the groundwater flow many days later (Fig. 22-19). These storage curves can be entered with inflow from precipitation after deducting a percolation or loss rate

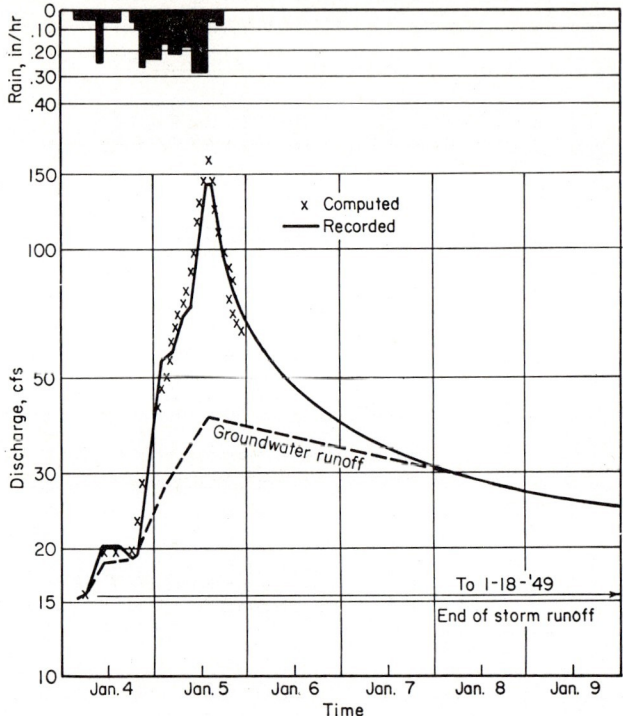

Fig. 22-20. Computed and gaged hydrographs. Coweeta experimental forest watershed 8, storm of January, 1949: Precipitation = 5.08 in., Accumulated precipitation = 6.58 (30-day antecedent rainfall).

(Table 22-8). Computations are then similar to flood routing down a channel by the storage-indication method. Simply stated, the computation required is to find the best relation of storage to discharge which, in combination with a loss rate to separate types of flow, will give a reasonable reproduction of actual hydrographs. By successive trials, the time relation can be found which will best reproduce hydrographs with this routing procedure.

As an illustration of the method, the computations for the columns of Table 22-8 and Fig. 22-20 are accomplished as follows:

1. The time and the accumulated precipitation depth for each time period were tabulated in columns 2 and 3, respectively.
2. After successive trials for different values of S, accumulated subsurface inflow Q_i was computed in column 4 from the equation $Q_i = P^2/(P + S)$, using an S value of 23 (see Sec. 21 for equation derivation and discussion of S values).

3. Increments of subsurface runoff from column 4 were determined and tabulated in column 5.

4. Subsurface runoff was then routed in columns 6 and 7, using the storage-discharge relation as developed from the recorded runoff of several streams (Fig. 22-19).

5. In column 8, the outflow expressed in cfs/sq mi was converted to cfs.

6. Increments of precipitation computed from column 3 minus increments of subsurface inflow from column 5 were converted to cfs-hr and tabulated in column 9.

7. Starting with 15.4 cfs antecedent base flow, groundwater inflow was routed using a base-flow depletion rate of about 10 per cent daily as determined from records for rainless periods. These values were tabulated in columns 10 and 11.

8. Subsurface runoff (column 8) and groundwater runoff (column 11) were added and tabulated in column 12.

9. The gaged runoff was tabulated in column 13. For comparison, see Fig. 22-20.

F. Evaluation by Runoff-curve-number Procedure

This procedure is explained in Sec. 21. The *runoff curve numbers*, as developed by this method, show the relative value of the hydrologic-soil-cover complexes in producing direct runoff. Table 22-9 shows curve numbers used by the U.S. Department of Agriculture for forest lands and rangelands.

Table 22-9. Runoff Curve Numbers for Hydrologic Soil-cover Complexes*
A. Forest land for antecedent moisture condition II, where $I_a = 0.2S$

Hydrologic condition class	Hydrologic soil group			
	A	B	C	D
I. Poorest (1.0–1.9)	56	75	86	91
II. Poor (2.0–2.9)	46	68	78	84
III. Medium (3.0–3.9)	36	60	70	76
IV. Good (4.0–4.9)	26	52	62	69
V. Best (5.0–5.9)	15	44	54	61

B. Forest-range areas in western United States for antecedent moisture condition III, where $I_a = 0.2S$

Cover	Hydrologic condition class	Hydrologic soil group			
		A	B	C	D
Herbaceous	Poor	...	90	94	97
	Fair	...	84	92	95
	Good	...	77	86	93
Sagebrush	Poor	...	81	90	
	Fair	...	66	83	
	Good	...	55	66	
Oak-aspen	Poor	...	80	86	
	Fair	...	60	73	
	Good	...	50	60	
Juniper	Poor	...	87	93	
	Fair	...	73	85	
	Good	...	60	77	

* From U.S. Soil Conservation Service [5, p. 3.9-4].

The hydrologic condition class of forest cover can be estimated by several methods. Figure 22-21 gives an alignment chart by which the hydrologic condition class can be estimated. The following definitions are used with Fig. 22-21:

Humus depth: Humus is the organic layer occurring immediately beneath the litter layer from which it is derived. It may be composed of (1) *mulls,* consisting of the A_1-horizon of an intimate mixture of organic matter and mineral soil, or (2) *mors,* consisting of the H layer, or A_{02}-horizon, which is practically pure organic matter, unrecognizable as to origin, lying on the mineral soil.

On forested areas the depth of humus increases with age of stand until an equilibrium is reached between the processes that build up the humus and those that break it down. Although as much as 12 in., or even more, may be produced under very favorable conditions, a depth of 5 to 6 in. is considered to be the maximum attainable

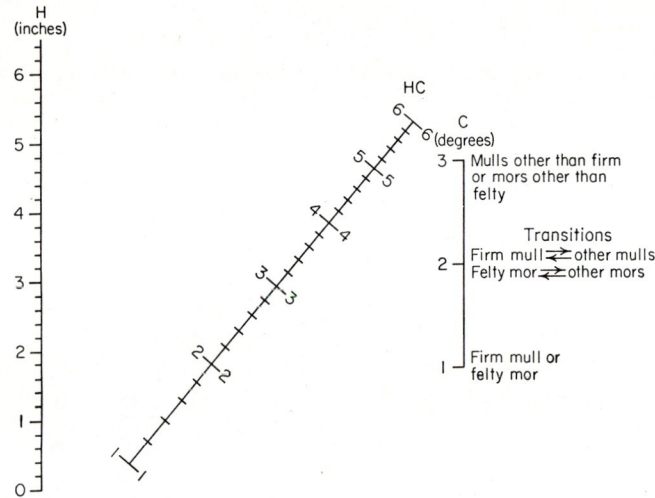

FIG. 22-21. Chart for determining hydrologic condition of forest and woodland. *H,* humus depth; *HC,* hydrologic condition class; *C,* compactness factor. (*From U.S. Forest Service* [6].)

under average conditions. Since as much as ½ in. of water can be held in detention storage by each inch of humus, depth of humus is an important criterion of hydrologic condition.

Humus type: Humus is very sensitive to land use and treatment. Under good use and treatment, it is porous and has high infiltration and detention-storage capacities, but poor management practices such as burning, overcutting, and overgrazing convert porous humus to a compact one which impedes the absorption of water. Since compaction is the principal factor affecting infiltration and because the degree of compaction is closely related to humus type, humus type is used as the criterion for compaction.

Three degrees of compaction, as expressed by *compactness factors,* are used in determining hydrologic condition in Fig. 22-21:

1. *Compact.* Compactness factor 1, characterized by firm mull or felty mor.

Firm mull: A_1-horizon, an intimate mixture of mineral soil with a gradual transition to underlying horizon and with generally less than 5 per cent organic matter by weight, massive and firm.

Felty mor: H layer, with practically no mixing of organic matter with mineral soil, abrupt transition from surface organic matter to underlying horizon, feels and looks felty because of the presence of fungal hyphae and/or plant residues but not living roots.

2. *Moderately Compact.* Compactness factor 2, characterized by transitions between (a) firm mull and other mulls and mors other than felty mors, and (b) felty mors and other mors, and mulls other than firm mull.

3. *Loose, Not Compact, Friable.* Compactness factor 3, characterized by mulls other than firm mulls, and mors other than felty mor.

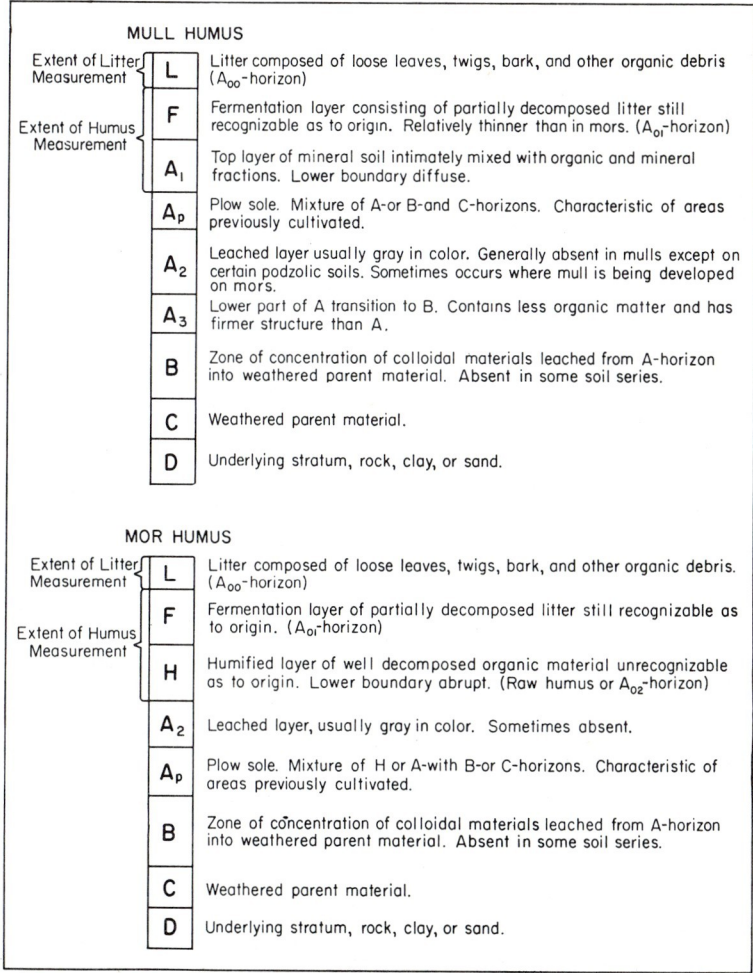

Fig. 22-22. Relative positions of organic layers in soil profiles of forested areas. (*U.S Forest Service* [6].)

The relative positions of organic layers and the extent of litter and humus measurements are indicated on Fig. 22-22. At each sampling point, litter depth is measured to the nearest $\frac{1}{4}$ in., humus type is identified, and humus depth is measured to the nearest $\frac{1}{10}$ in. Because of the variation in humus depth, an average of three measurements should be taken as the depth at each sampling point.

Figure 22-21 is used by placing a straight edge on scale H, at the point indicating depth of humus layer, and on scale C, indicating degree of compactness of the humus layer. The hydrologic condition class is then read on scale HC to the one-tenth class. The forest floor should be covered by at least ¼ in. of litter in the late summer and ½ in. at other seasons as protection to the humus layer. If this minimum amount of litter is not present, the hydrologic condition class is reduced by five-tenths of a class, but not lower than class 1. For example, if the adjusted hydrologic condition falls below 1.0, as 0.9 or 0.5, the hydrologic condition class would be still reported as 1.0.

Another method of estimating hydrologic condition class makes use of rated disturbance factors instead of compaction factors. The *disturbance factors* are rated at each observation point on the basis of disturbance from fire, grazing, and logging, according to the following table, which contains numerical values that can be added to give a relative total disturbance factor:

Degree of disturbance	Cause of disturbance		
	Fire	Grazing	Logging
Negligible	1	2	1
Moderate	3	5	3
Severe	4	7	4

When this total disturbance factor is 5 or less, the degree of disturbance for the observation point is classed as *negligible;* when the total is 6 to 8, the degree is *moderate;* and when the total is 9 and over, the degree is *severe*. These classes should agree rather well with the compactness factors of 3, 2, and 1, respectively, of the previous

Table 22-10. Determination of Hydrologic Condition Class

Litter $(L + F)$, depth in in.	Humus (H or A_1), depth in in.	Hydrologic condition class		
		TDF* 3–6	TDF 7–8	TDF 9+
0+	0–0.49	II	I	I
0–0.99	0.5–0.99	II	II	I
1+	0.5–0.99	III	II	I
0–0.99	1.0–1.49	III	II	I
1+	1.0–1.49	III	III	I
0–0.99	1.5–1.99	III	III	II
1+	1.5–1.99	IV	III	II
0+†	2.0–4.99	IV	III	III
0+†	5.0+	V	IV	III

* TDF = total disturbance factor.
† When less than 0.2 in., hydrologic condition is lowered one condition class.

method. By using Table 22-10 the average depth of humus and litter can be matched with the degree of disturbance to determine the hydrologic condition class.

In the more arid rangeland sections of the country, surveys to obtain information on hydrologic condition can be based on a classification of the vegetation by a measurement of ground-cover density.

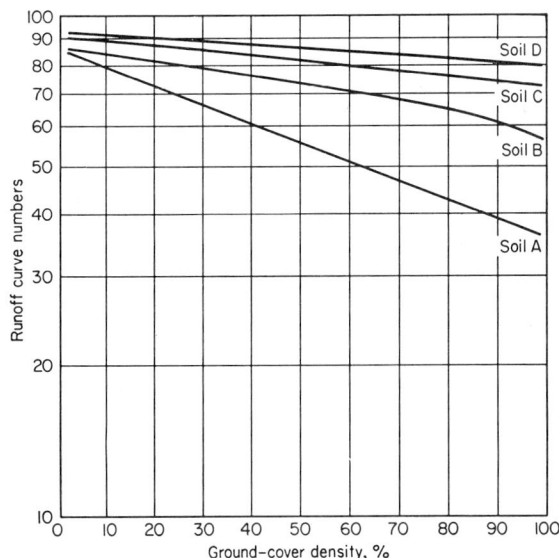

FIG. 22-23. Runoff curve numbers for hydrologic soil-cover–density complexes on rangelands with antecedent moisture condition II.

Runoff curve numbers can then be determined from Fig. 22-23, which gives a relationship of the curve numbers to ground-cover density by hydrologic soil groups.

V. GLOSSARY

Complex, soil cover: A group of similar areas having comparable physical conditions of soil, slope, litter, and cover.
Conifer: A tree belonging to the order Coniferae, usually evergreen with cones and needle-shaped leaves, and producing wood known commercially as "softwood."
Crown, density: The compactness of the crown cover of the forest, depending upon (1) the distance apart and (2) the compactness of the individual crowns.
Density, ground: The per cent of ground surface which appears to be completely covered by vegetation and litter as viewed from a vertical position.
Forest, influences: All effects upon water supply, soil, climate, and health resulting from the presence of forests.
Forest, type: A descriptive term used to group stands of similar character with regard to composition and development due to certain ecological factors, by which they may be differentiated from other groups of stands.
Hardwood: Generally, one of the botanical groups of trees that have broad leaves, in contrast to the conifers.
Humification: The process of humus formation.
Litter: The uppermost layer of the organic debris, composed of freshly fallen or slightly decomposed organic materials.
Percolation, rate: The quantity of water, in inches, that passes a given point in the soil profile in a given time.
Stand, old-field: A stand which has taken possession of land once used for agricultural crops or for pasture.
Stand, old-growth: A stand consisting mainly of mature and/or overmature trees.
Stocking: An indication of the number of trees in a stand as compared with the desirable number for best growth and management, such as well-stocked, overstocked, partially stocked.
Timber, volume: The amount of wood in a stand according to some unit of measurement (board feet, cubic feet, etc.).
Transmission, rate: The vertical distance, in inches, that water travels through the soil in a given time.

VI. REFERENCES

1. Kittredge, Joseph: "Forest Influences," McGraw-Hill Book Company, Inc., New York, 1948.
2. Auten, J. T.: Porosity and water absorption of forest soils, *J. Agr. Res.*, vol. 46, pp. 997–1014, 1933.
3. Relationship of vegetal cover and other watershed treatment to floods and associated problems, Provisional Report to Subcommittee on Vegetal Cover and Floods of the Water Resources Committee, April, 1937, U.S. Department of Agriculture, p. 47.
4. Miller, D. H.: Transmission of insolation through pine forest canopy, as it affects the melting of snow, Festschrift zum Siebzigsten Geburtstag von Prof. Dr. H. Burger, *Swiss Inst. Forestry Res. Publ.*, vol. 35, no. 1, 1959.
5. Hydrology, sec. 4 in "Engineering Handbook," supplement A, U.S. Soil Conservation Service, 1957.
6. "Forest and Range Handbook 2518," U.S. Department of Agriculture, Forest Service, 1959.
7. Whelan, D. E., L. E. Miller, and J. B. Cavallero: A method of determining surface runoff by routing infiltrated water through the soil profiles, *U.S. Forest Serv. Northeast. Forest Expt. Sta. Paper* 54, 1952.
8. Trimble, G. R., Jr.: A method of measuring increase in soil depth and water-storage capacity due to forest management, *U.S. Forest Serv. Northeast. Forest Expt. Sta. Paper* 47, 1952.
9. Anderson, H. W., R. M. Rice, and A. J. West: Snow in forest openings and forests stands, *Proc. Soc. Am. Forestry*, pp. 46–50, 1958.
10. Rosa, J. M.: Forest snowmelt and spring floods, *J. Forestry*, vol. 54, pp. 231–235, April, 1956.
11. Ford, P. M.: Multiple correlation in forecasting seasonal runoff, *U.S. Bur. Reclamation Eng. Monograph* 2, 1949.
12. Horton, R. E.: Erosional development of streams and their drainage basins; hydrophysical approach to quantitative morphology, *Geol. Soc. Am. Bull.*, vol. 56, pp. 275–370, March, 1945.
13. "Hydrology Handbook," American Society of Civil Engineers, Manuals of Engineering Practice, no. 28, 1949.
14. Anderson, H. W., and R. L. Hobba: Forests and floods in the northwestern United States, Symposium Hannoversch-Münden, *Intern. Assoc. Sci. Hydrology Publ.* 48, pp. 30–39, 1959.
15. Anderson, H. W.: Flood frequencies and sedimentation from forest watersheds, *Trans. Am. Geophys. Union*, vol. 30, no. 4, pp. 567–586, August, 1949.
16. Rowe, P. B., C. M. Countryman, and H. C. Storey: Hydrologic analysis used to determine effects of fire on peak discharges and erosion rates, U.S. Forest Service, California Forest and Range Experiment Station, February, 1954 (mimeo.).
17. Gibson, A. H.: "Hydraulics and Its Applications," D. Van Nostrand Company, Inc., Princeton, N.J., 1930, pp. 288 and 289.
18. Ayres, Q. C.: "Soil Erosion and Its Control," McGraw-Hill Book Company, Inc., New York, 1936, p. 22.
19. Brater, E. F.: The unit hydrograph principle applied to small watersheds, *Trans. Am. Soc. Civil Engrs.*, vol. 105, pp. 1154–1178, 1940.
20. Linsley, R. K., and W. C. Ackermann: A method of predicting the runoff from rainfall, *Trans. Am. Soc. Civil Engrs.*, vol. 107, pp. 825–835, 1942.

Section 23

HYDROLOGY OF LAKES AND SWAMPS

JAMES H. ZUMBERGE, *President, Grand Valley State College, Allendale, Michigan.*

JOHN C. AYERS, *Oceanographer, Great Lakes Research Division, Institute of Science and Technology, and Professor of Oceanography, University of Michigan.*

I. Introduction	23-2
II. A Classification of Lakes	23-3
A. Lake Defined	23-3
B. Lake Classification	23-3
1. Tectonic Lake Basins	23-4
2. Lake Basins Produced by Volcanic Activity	23-4
3. Lake Basins Produced by Landsliding	23-4
4. Lake Basins Produced by the Action of Glaciers	23-4
5. Solution Lakes	23-5
6. Lake Basins Formed by Fluviatile Action	23-5
7. Lake Basins Formed by Wind Action	23-5
8. Lake Basins Produced by Shoreline Processes	23-6
9. Lake Basins Formed by Organic Accumulation	23-6
10. Lake Basins Formed by the Activity of Higher Organisms	23-6
11. Lake Basins Formed by the Impact of Meteorites	23-7
III. Lake Morphometry and Morphology	23-7
A. Morphometry	23-7
1. Hydrographic and Bathymetric Charts	23-7
2. Morphometric Parameters	23-8
B. Morphology	23-10
IV. Lakes in the Hydrologic Cycle	23-11
A. Lakes as Natural Reservoirs	23-11
B. Lake Extinction	23-11
C. Fluctuations in Lake Levels	23-11
D. The Great Lakes	23-12
1. Nature, Magnitude, and Timing of Lake-level Fluctuations in the Great Lakes	23-13
2. Factors Controlling Water-level Fluctuations in the Great Lakes	23-13
V. Thermal Properties of Lake Water	23-15
A. The Temperature-Density Relationship	23-15
B. Liquid Nature	23-16
C. Thermal Capacity	23-16
D. Latent Heats	23-16

E. Thermal Expansion.................................... 23-17
  F. Thermal Conductivity................................. 23-17
  G. Thermal Stratification............................... 23-17
    1. The Turnovers..................................... 23-17
    2. Summer Stratification............................. 23-18
  H. Heat Budgets... 23-18
    1. Birgean Heat Budgets.............................. 23-19
    2. Analytical Heat Budgets........................... 23-19
VI. Lake Ice.. 23-20
  A. Structure, Classification, and Thickness............. 23-20
    1. Sheet Ice... 23-21
    2. Agglomeritic Ice.................................. 23-21
    3. Ice Thickness..................................... 23-21
    4. Empirical Prediction of Ice Thickness.............. 23-21
    5. Ice Breakup....................................... 23-22
  B. Vehicular Traffic and Aircraft Landings on Lake Ice...... 23-23
  C. Effect of Lake Ice on Shore Installations............ 23-24
VII. Hydromechanics of Lakes.............................. 23-24
  A. Surface Waves....................................... 23-24
    1. Open-water Waves.................................. 23-24
    2. Changes in Waves Approaching Shore................ 23-25
  B. Seiches... 23-26
    1. Surface Seiches................................... 23-26
    2. Internal Seiches.................................. 23-27
  C. Currents.. 23-27
    1. Alongshore, or Littoral, Currents................. 23-28
    2. Open-lake Currents................................ 23-29
VIII. Swamps and Bogs..................................... 23-30
  A. Origin and Definitions.............................. 23-30
    1. Swamps and Bogs Differentiated.................... 23-30
IX. References... 23-31

I. INTRODUCTION

Lakes and swamps are transitory features of the earth's surface, and each has a birth, life, and death, related to certain geologic and biologic processes. Both lakes and swamps involve standing surface water, but beyond this one common point, they are quite different and will be treated separately. Lake origins and various lake processes will be considered first, after which a separate account of swamps, marshes, and bogs will be given.

The lakes of the world are basically a natural resource available for use by man. Throughout the history of civilization lakes have supplied such basic needs as food, primary water supply, and transportation routes. More recently, in addition to the above, lakes have provided energy to drive water-propelled turbines and have made possible recreational activities of great variety.

Only in rare instances have lakes been destroyed or severely modified by man because they obstructed the exploitation of a more valuable resource. Rabbit Lake in Minnesota and Steep Rock Lake in Ontario were drained to obtain high-grade iron-ore bodies which they covered. Man generally finds it to his advantage not only to preserve existing lakes, but also, because of the many benefits accruing from them, to build artificial lakes where nature has not provided them.

Before present lake conditions can be fully understood and appreciated, it is often necessary to view the modern lake against the background of its origin and developmental history. Lakes are dynamic systems, some of which have been in operation for thousands of years, whereas others are quite ephemeral. In either case, a given lake has a previous history which may reach far back into geological time and which may be very complex. The modern dynamic system is not only the result of contemporary meteorological and physiographic processes, but also in part a reflection of the previous history of the lake.

The history of a lake begins with the origin of the basin which contains it. The following pages summarize the various processes which produce lakes, and a classification based thereon is reviewed.

II. A CLASSIFICATION OF LAKES

A. Lake Defined

The matter of a precise definition of a lake has received insufficient attention. From the geological point of view, a lake consists of two distinct parts, the *basin* and the *water body*. It is obvious that the latter could not exist without the former, and both should be taken into account in any workable definition.

Zumberge [1] defined a *lake* as an inland basin filled with water. A further restriction of size seems necessary, as pointed out by Welch [2], who made a distinction between lakes and ponds. The former, he suggested, should have "an area of open, relatively deep water sufficiently large to produce somewhere on its periphery a barren wave-swept shore." In contradistinction are *ponds*, which Welch defines as "very small, very shallow bodies of standing water in which quiet water and extensive occupancy by higher aquatic plants are common characteristics."

Obviously, very small basins such as stream-cut potholes or tiny solution pits in limestone must be excluded from the definition of a lake, but exactly where the cutoff in size comes has never been defined accurately. A really quantitative definition may well require some recognition of Welch's stipulation of a wave-swept shore, possibly expressed in terms of some numerical index containing maximum wave size.

As used in this section, a lake is an inland basin filled or partially filled by a water body whose surface dimensions are sufficiently large to sustain waves capable of producing a barren wave-swept shore.

B. Lake Classification

The diverse origins of lake basins have been recognized since the late nineteenth century. In 1882, Davis [3] grouped the basin-forming processes into three geologic categories, *constructive*, *destructive*, and *obstructive*. The emphasis here is on the gross process rather than the specific agent and tends to obscure the fact that lakes of similar geologic origin are concentrated in certain geographical regions of the earth.

In 1895, Russell [5], recognizing the deficiencies in Davis's scheme, published a system of lake classification "based on the natural agencies which produce depressions in the earth's surface." Ten major natural agencies, nine geologic and one organic, were defined by Russell; these have served as the basis for every genetic classification that has appeared since. The most recent and complete is that of Hutchinson [4]. He recognizes eleven major genetic processes, which produce a total of 76 different lake types. His examples are selected from all the continents of the earth and provide the best international outlook on lake classification available.

The classification used in this section in general follows Hutchinson's arrangement, which is based on the work of Russell. Restrictions of space preclude discussion or listing of all Hutchinson's 76 types, but the important categories are introduced and elaborated where necessary. The interested reader should consult Ref. 4. (For desert lakes, see Sec. 24.)

The following categories theoretically should accommodate lakes of every conceivable origin, natural or artificial:
1. Tectonic lake basins
2. Lake basins produced by volcanic activity
3. Lake basins produced by landsliding
4. Lake basins produced by the action of glaciers

5. Solution lakes
6. Lake basins formed by fluviatile action
7. Lake basins formed by wind action
8. Lake basins produced by shoreline processes
9. Lake basins formed by organic accumulation
10. Lake basins formed by the activity of higher organisms
11. Lake basins formed by the impact of meteorites

1. Tectonic Lake Basins. Tectonic forces, although of deep-seated origin, are capable of producing profound changes at the earth's surface. Submarine structural basins, or depressions, formed by differential marine sedimentation may become lakes when uplifted above sea level, or broad upwarping of the earth's crust can impound the waters of whole drainage systems. Faulting is a major result of tectonic activity and is responsible for the origin of primary basins, of which the graben type are the best known. Lake Baikal, the deepest lake in the world, lies in a multiple graben that may have been in existence 60 million years since late Mesozoic or early Tertiary time [4].

2. Lake Basins Produced by Volcanic Activity. Hutchinson lists 13 different kinds of basins related to volcanic activity, but several are variations of a single type, and we need consider only three kinds here. The first is the nearly circular lake which occupies volcanic craters or calderas. *Craters* are inverted conical depressions at the crest of a volcanic cone; *calderas* are much larger depressions resulting from collapse of the central part of a volcano after quantities of lava have been discharged from the underlying magmatic reservoir [6]. Crater Lake, Oregon, is a classical caldera, although one could hardly deduce this fact from its name [7].

Other volcano-related lakes originate when a lava flow dams a valley, or when growth of volcanic cones obstructs a preexisting drainage system.

Another type of lake basin results when the solidified crust of a new lava flow collapses after the still-fluid lava beneath has drained away.

3. Lake Basins Produced by Landsliding. Landsliding is a general term applied to all surficial movements of earth materials under the direct influence of gravity. Slides may be initiated by events such as wave action, earthquakes, heavy rains, artificial excavations, etc. When a landslide occurs, it results in the sudden translocation of earth materials from a higher to a lower elevation. If the final resting place of landslide debris is across a stream valley, the impounded waters form an elongate lake. The lake may drain suddenly when the water spills over the slide dam, producing disastrous floods downstream. A landslide lake will tend to fill rapidly with sediment because the inflowing stream has its sediment-carrying capacity lowered as it enters the newly formed lake.

Smaller lakes of landslide origin may form between the slide and the scar or on the irregular surface of the landslide debris.

4. Lake Basins Produced by the Action of Glaciers. Glaciers, especially those of Pleistocene age, have produced more lake basins than any other single agency. Glacially produced basins range in size from the smallest lakes to the five Great Lakes.

Modern glaciers also are lake-producing agencies. Lakes lying entirely on glacier ice are formed by differential melting and are rare. More common are lakes formed by the damming action of glacier tongues across valleys. Another type is formed by the damming action of a moraine; such basins may persist long after the glacier which formed the moraine has disappeared.

Both valley glaciers and continental ice sheets are able to erode the land surface over which they move and to scour or quarry deep basins in rock. These basins may lie in deep rock valleys (the so-called *fjord lakes*), or they may have the amphitheater form of cirque lakes, which are so common in mountainous regions formerly occupied by alpine glaciers. Many of the lake basins in bedrock formerly covered by ice sheets are the result of differential glacial quarrying rather than concentrated abrasion by debris-laden basal ice [9].

Glacial deposition accounts for a variety of lake basins, most of which are in one way or another related to the collapse of outwash material surrounding masses of stagnant ice, though some are formed by differential deposition of glacial till. Other lakes related to glacial deposition are those formed by dams of till or outwash.

Hutchinson [4] includes in this category the *thermokarst* lakes produced by collapse of the ground resulting from the melting of ice or frozen ground in regions underlain by permafrost [10].

5. Solution Lakes. Geological provinces characterized by carbonate rocks or evaporite rocks are susceptible to solution by groundwater, which results eventually in the formation of rock basins called *sinks*. Some sinks, such as those found in Florida [11], are funnel-shaped in cross section and circular in plan view. Others are complex in form because they represent coalesced solution pits. All solution lakes are connected to subsurface drainage channels during some phase of their history; not only is a constant supply of water needed to foster the solution process, but it is also necessary to remove the dissolved materials.

Solution lakes characteristically exhibit rapid fluctuations in water level because of their close relationship to groundwater. Some even drain suddenly through a subsurface conduit when a natural underground siphon action is initiated by abnormally high water levels, or when sediments blocking a sublacustrine outlet are suddenly displaced.

It is possible that, in their later stages, solution lakes may become sealed off from the surrounding groundwater by sediment, much of which is insoluble residue originally present in the bedrock.

6. Lake Basins Formed by Fluviatile Action. Running water is a common geologic agent capable of producing closed basins by erosion or deposition or a combination of both. The greatest hydraulic action of streams occurs at the base of escarpments over which the streams plunge as waterfalls. The turbulent action of the water commonly excavates a basin, which later may contain a lake, such as Falls and Castle Lakes in the Grand Coulee of Washington [12].

Far more common are lakes formed by stream deposition at the confluence of a tributary and the main channel. In some cases the main channel is dammed by excess sediment brought in by a more competent tributary. In other instances the tributary is dammed by progressive alluviation in the main channel. Either one of these situations can usually be traced to some previous episode in the history of the tributary or the trunk stream wherein a proportionally larger discharge was suddenly reduced, thereby reducing the stream's capacity to transport its bed load. The Lake Pepin–St. Croix system on the upper Mississippi River [1] is an excellent example of lakes produced by change in stream regime.

Several kinds of lakes occur on floodplains of mature rivers. One of the best known is the *oxbow lake*, which is an abandoned meander loop of a river. Another type is produced during flood stages when natural levees form a dam between the main channel and the floodplain, establishing a very shallow saucer-shaped basin. Other small elongate lakes on floodplains are merely channels of the master stream which are abandoned as it shifts its course in any manner other than by a breakthrough at the narrow neck of a meander loop. Such lakes are common along meandering rivers and are known as *meander scrolls*.

River-mouth deltas contain closed basins formed by the growth of bars and levees along distributaries. Deltaic levee lakes [4] (Lake Pontchartrain on the Mississippi Delta is an example) occur on most of the world's major river deltas [5].

7. Lake Basins Formed by Wind Action. Wind, as a geological agent, is capable of producing closed depressions by the erosive process of deflation, by deposition of wind-borne sediment, or by combination of both. Most basins originating by wind action are very shallow, and many contain water only during certain seasons.

Basins of this sort were formed during arid phases of the Pleistocene in nonglaciated regions and have undergone several episodes of alternating wet and dry climates. Judson [13] showed that the deflation basins of eastern New Mexico were formed during a dry period of the Pleistocene, after the calcareous cement of the local sandstone was removed by leaching during a previous wetter period.

The *pans* of South Africa are regarded by some observers as being purely deflationary in origin, but others have emphasized the role of large herds of hoofed animals that carry away prodigious amounts of mud on their feet when visiting small depressions containing ephemeral water. Hutchinson [4, pp. 130–134] gives a full account of pans.

When drifting sand in the form of migrating dunes obstructs natural drainage lines, **a lake is formed.** Such a lake is Moses Lake in Washington, described by Russell [14].

In areas of active sand dunes the continual interplay between shifting dunes and deflation produces closed basins of various shapes and sizes, which will become lakes if climatic change permits stabilization of the ground surface and a rise in the water table.

8. Lake Basins Produced by Shoreline Processes. Wave and current action along seacoasts or shores of inland lakes transports clastic sediment and redistributes it in such a manner as to produce isolated basins by the formation of bars across the openings of embayments in the shore. Numerous lakes of this origin (once connected to the sea) occur on the New England coast. Many lakes of this origin are known from Minnesota, Wisconsin, and Michigan, where larger lakes served as the parent water bodies.

A special case of a coastal lake formed by the building of coastal deposits occurs when an offshore island is joined to the mainland by two bars called *tombolos*. Johnson [15] cites Lake Stagno di Orbetello on the west coast of Italy as an example.

Inland lake basins may undergo bisection into two separate basins when spits or points of accretionary origin join through lakeward growth from opposite shores.

9. Lake Basins Formed by Organic Accumulation. This category contains few lakes in actual numbers, but involves distinct basin-forming processes associated with dead or living organisms. Basins of this type include those formed when large masses of plant debris dam a stream, as well as lakes produced by the accumulation of coral fragments in tropical and subtopical oceanic regions.

Russell [5] and Murray [16] believed that some of the tundra lakes of arctic regions were formed by the growth of vegetation around lingering snowbanks. Although such an origin may be possible, it is now known that most tundra lakes are produced through the melting of permafrost, with resulting local subsidence [10, 17].

Typical lagoons in coral atolls of the Pacific Ocean possess several outlets to the sea, but a few, such as the one on Washington Island, have coral rims which completely isolate them from the surrounding ocean [18]. Some of these coral lakes retain underground connections with the sea and therefore rise and fall with the tide, but others are sealed off completely by fine-grained sediments of organic origin [19].

10. Lake Basins Formed by the Activity of Higher Organisms. Man has acquired considerable skill in building large artificial lakes by the construction of dams across river valleys. This practice may have started as early as 2000 B.C. in Egypt [4], and engineers of ancient Syrian civilizations made lakes to serve as reservoirs for water supply [20]. Dam design and construction have improved greatly since the days of the ancients, but the nature of the impounded water bodies remains identical because they all lie in river valleys. Except in deep gorges, artificially dammed lakes are dendritic in shape and are invariably deeper at the dam end than farther upstream.

As soon as sufficient water had collected behind the dam, wind-induced waves and currents begin to modify the new shoreline by erosion and deposition. The lake also begins to fill with sediment brought in by the retarded stream even before the reservoir has reached the level proposed by its designers. Some artificial reservoirs (such as Lake Mead formed by Hoover Dam on the Colorado River) have estimated lives of several hundred years, but all are doomed to extinction by the accumulation of streamborne sediments, a process discussed elsewhere (Sec. 17-1).

Although man's massive concrete-and-steel dams and immense earth-fill dams are spectacular feats of engineering, one can hardly be less impressed by the diligent work of the beaver, who instinctively apprehends some of the finer points of dam construction. From sticks, logs, and mud, the American and European beavers build dams that are not only structurally sound, but also large enough to impound several acres of standing water. One beaver dam near Three Forks, Mont., was found to be more than 2,000 ft long [21], and another on Grand Island in Lake Superior is nearly 1,500 ft long [4]. Many other less spectacular examples are known from the forested regions of North America, Scandinavia, and northern U.S.S.R.

Not to be neglected in this category are the quarries, pits, holes, and other excavations in the earth which man has abandoned over the centuries. When abandoned, these excavations normally fill with water and remain for decades before they are obliterated by natural or artificial means. Some of these old excavations become useful recreation sites, while others become dumping areas for various domestic and industrial wastes. This latter practice is hardly to be condoned, under most circumstances, because of the groundwater contamination that may result.

11. Lake Basins Formed by the Impact of Meteorites. The earth's surface is occasionally bombarded by meteorites of varying sizes. The impact of such a body produces an explosion, the blast of which creates a crater in a manner similar to that of an artillery shell. In both cases, the trajectory of the missile is not reflected in the shape of the crater it produces. These explosion craters are usually circular in plan view. Chubb Crater in Ungava, Quebec, the largest confirmed meteorite crater on record, has a diameter of about 2 miles and contains a lake of 800-ft depth [22].

Probably the most controversial group of lakes in so far as their origin is concerned are the Carolina Bays on the Atlantic coastal plain of the southeastern United States. The various hypotheses, including meteorite impact, devised to explain the origin of these remarkable basins are discussed by Hutchinson [4]. The interested reader should also consult the original works on the subject [23–27].

III. LAKE MORPHOMETRY AND MORPHOLOGY

A. Morphometry

The three-dimensional form of a lake basin and several aspects of the nature of the lake within it depend in part upon the kind of topography in which the lake was formed, in part upon the physical means by which the lake was brought into being, and in part upon conditions and events in the lake and in the drainage basin since the lake was formed. Lakes formed in mountainous terrain have different form characteristics from those of lakes in lowlands. Rift lakes formed by faulting are different in morphology from oxbow lakes abandoned by river meanders. Lakes, after being formed, are subject to filling in at different rates and with different materials, depending upon the tributary gradients and the rock types, erosive agents, biological activity, and land-use practices occurring in their drainage basins.

The measurement of the form characteristics of lakes and lake basins is termed *morphometry*. The determination and use of standard morphometric parameters have advantages in enabling the quantitative expression of aspects of lake form and the meaningful comparison of one lake with another.

This discussion follows Welch [28] and Hutchinson [4]. These authors have followed Birge and Juday [29], who in turn have drawn from Penck [30, 31], Halbfass [32], and Gravelius [33].

1. Hydrographic and Bathymetric Charts. The basic device in morphometry is a *hydrographic chart* of suitable size and construction. Other things being equal, the larger the chart the better the results, but mere enlargement of existing maps is usually not sufficient, for cartographers in general pay insufficient attention to accurate depiction of lake outlines. Probably the best delineation of lake outline for a small lake is obtained from well-centered aerial photographs of known scale. For larger lakes a mosaic of aerial photographs at known scale provides an accurate shore outline, gives the shapes and locations of islands, and indicates the sand spits, bars, and shoals of the submerged topography.

Hydrographic maps of high accuracy can be made by applying the methods of ordinary surveying at times when the lake is covered with ice. This method provides detailed depiction of shoreline and also allows depth soundings to be made from a stable platform.

Under open-water conditions accurate hydrographic charts may be made by combining ordinary surveying, for shore outline, with lines of depth soundings made from a boat. In the latter case, the position of each sounding may be located by paired transit angles to the boat taken by transits at the ends of a base line ashore or paired sextant angles taken from the boat on carefully located landmarks ashore. Accepted procedures for soundings from boats include (1) the running of parallel range lines or compass-course lines of soundings across the lake, with location of periodic soundings by transit or sextant angles, (2) running of numerous compass-course lines or range lines perpendicular to shore and extending to a point beyond the middle of the lake. Range lines involve range markers of high visibility so set on shore that the boat moves normal to shore when the boatman keeps the lower front range marker directly in front of the higher back marker. The normal-to-shore method has the advantage of yielding a higher density of soundings in mid-lake and usually will be more successful in locating the point of maximum depth. Periodic sounding locations are determined by transits ashore or sextant angles shot from the boat.

Excellent expositions of charting procedures, of the care of instruments and gear, and of pitfalls to be avoided are given by Welch [28].

For large lakes, or in cases where especially detailed depth soundings are required, considerable time and labor may be saved by using a portable recording fathometer. An excellent instrument, battery-powered and suitable for use in boats of all sizes, is made by the Raytheon Company, Waltham, Mass. There may be others of which the authors are unaware.

The recording fathometer draws a profile of the bottom traversed; it allows the boat to travel at any reasonable speed; distance control is given by means of a marking signal permanently affixed to the record at the instant of each of the periodic transit or sextant location fixes; and the record is accessible for hand labeling at any time during the sounding operation. Care must be taken to increase all recorded depths by the amount that the sounding head was under water.

The development of an accurate hydrographic chart of the lake is, regardless of the methods used, the first requirement of morphometry. It consists of (1) an accurate depiction of the shoreline, (2) a sufficient covering of the water area, with accurate depth soundings accurately located, and (3) depictions of islands, bars, shoals, etc., with a degree of accuracy suitable to the needs of the project in hand. A hydrographic chart with depth contours drawn at regular intervals becomes a *bathymetric chart*. The depth interval chosen for contouring should be as small as feasible.

From the bathymetric chart can be obtained a number of morphometric parameters which quantitatively express aspects of the lake-form characteristics and which allow quantitative comparisons between lakes.

2. Morphometric Parameters. While different authors have used sets of parameters that have varied in some respects, there is at least implicit agreement upon certain ones. Those most commonly used, along with some of the variations in terminology, are given by Welch [28] and Hutchinson [4].

Maximum length, Lm: The length of a line along the water's surface between the most widely separated extremities of the lake. This line may vary in position and need not be straight. It should be so located as to express as accurately as possible the true open-water length; this line may cross islands, but no other land.

In cases where an island near one end of the lake produces a sheltered portion, or where a long shallow embayment contributes to the maximum length but provides an environment distinctly more sheltered than the rest of the lake, it is desirable to designate another length parameter.

Maximum effective length, Lme: This is the *straight-line* length separating the most distant points between which the actions of wind and waves can occur without interruption by land.

Maximum width, Wm: The straight-line distance between the most separated shores, at approximately right angles to the maximum length.

Direction of major axes: Compass directions of the length and width axes.

Area, A or A_0: Surface area of water, determined from the hydrographic or bathymetric chart, usually determined by planimetry. This is the area enclosed within the zero-depth contour. In volume computation, and for other purposes, it is necessary to determine the area enclosed by subsurface contours. Such areas are designated by A, with an appropriate depth subscript.

Mean width, $\bar{W}$: Surface area divided by maximum length.

Maximum depth, d_m: The greatest depth sounded.

Volume, V: Volume of a lake may be determined by either of two methods, both of which require the measurement of the areas enclosed by the several depth contours. The areas enclosed by the depth contours may be plotted against depth, and the area under the curve so obtained may be planimetered or otherwise measured. In another method, the areas enclosed by successive pairs of depth contours are averaged and multiplied by the contour interval to yield a series of volume elements which are summed.

The formula

$$\frac{h}{3}(A_1 + A_2 + \sqrt{A_1 A_2}) = V_{A_1 A_2} \tag{23-1}$$

is also used to find the volume between two adjacent depth contours h distance apart and of which A_1 is the area enclosed by the upper and A_2 that enclosed by the lower. Summation of the results of repeated successive applications of Eq. (23-1) will also yield lake volume.

Mean depth, $\bar{d}$: Volume of a lake divided by the surface area.

Development of volume, D_V: Development of volume is a comparison of the volume of the lake to that of a cone, with a basal area equal to the lake's surface area and a height equal to the lake's maximum depth. It is expressed as the ratio

$$D_V = \frac{3\bar{d}}{d_m} \tag{23-2}$$

The ratio approaches unity when the lake basin approaches conicity. It is less than unity when the basin's sides are essentially convex toward the water, and more than unity in lakes whose basin walls are essentially concave toward the water.

Shoreline, L: The length of shoreline circumference. It may be measured with a map measurer (rotometer or chartometer) or stepped off in small length segments with dividers.

Development of shoreline, D_L: The ratio of shoreline length to the circumference of a circle whose area is equal to the surface area of the lake.

$$D_L = \frac{L}{2\sqrt{\pi A}} \tag{23-3}$$

This ratio cannot be less than unity. It may be taken as a measure of the building effects of near-shore processes which increase the value of L by constructing shoreline irregularities.

Hypsographic curves: Curves of area or volume plotted against depth are useful in depicting certain structural characteristics of the lake basin. They also permit the determination of area or volume at any desired depth. Actual areas or volumes may be plotted, or the plot may be in terms of per cent of surface area or per cent of total volume.

Slope of basin: Need may arise for a general expression for the slope of the basin between adjacent depth contours or for the mean slope of the entire lake basin. Slope between adjacent contours can be defined by

$$S = \tfrac{1}{2}(C_1 + C_2)\frac{I}{A_B} \tag{23-4}$$

where C_1 and C_2 are the lengths of the contours, I is the contour interval, and A_b is the area of the bottom included between the contours.

The mean slope of the lake basin as a whole may be obtained in per cent from

$$S = \frac{(\tfrac{1}{2}C_0 + C_1 + C_2 + \cdots + C_{n-1} + \tfrac{1}{2}C_n)d_m}{nA_0} \tag{23-5}$$

where C_0, C_1, etc., are lengths of the contours, n is the number of contours, d_m is maximum depth, and A_0 is surface area of the lake.

Presentations of morphometric data should include the identification of the basic map or chart used, the scale of that map or chart, and the datum upon which that map or chart is based.

B. Morphology

Although lakes have pronounced tendencies to individuality and to nonconformity to classification systems, there are some general characteristics that are of value, communicatively and for comparison purposes, in giving lake descriptions.

Hutchinson [4] has evolved a system of classifying lakes according to their mode of origin. This has been summarized in Subsec. II.

A number of fairly characteristic shoreline shapes have been recognized and are of descriptive value, though there are complete series of intergradations between the different shape types.

1. Circular
2. Subcircular
3. Elliptical
4. Subrectangular-elongate
5. Dendritic (branching)
6. Lunate (crescentic)
7. Triangular
8. Irregular (joined basins)

Morphology is determined initially and primarily by the terrain in which lakes are formed and by the mode of their formation. These factors determine the initial forms of the submerged basin and of the shoreline. After the lake has formed, it undergoes a characteristic series of events, which are visibly expressed in varying degrees, depending upon the nature of the basin materials. Somewhat oversimplified, the subsequent history of a lake may be as follows.

A newly formed lake has a shoreline of submergence in which the shore is essentially only a submerging portion of the adjacent terrain. Such a lake has an irregular shoreline and does not exhibit wave-cut notches or cliffs (of any size), nor does it show the wave-deposited underwater shelf which terminates in a typical dropoff to the deep basin.

Erosional action of waves begins to cut a beach and to develop an underwater shelf as soon as the lake is formed and filled. Alongshore transport of sediments by littoral currents, usually created by waves striking the shore obliquely, begins to smooth out indentations of the shoreline soon after the lake fills. Erosion of headlands provides the major portion of the littoral sediments, though beach building all around the lake perimeter also contributes sediments.

In what may be loosely termed the lake's "maturity," or mid-life, it exhibits a wave-cut cliff, a beach, and an underwater shelf terminating in a dropoff. The headlands at this time are eroded to nearly the maximum permitted by the materials of which they are composed. Indentations of the shoreline at this time tend to fill in or to be cut off by bars across their mouths. The shoreline of a mature lake begins to approach a configuration along which the littoral currents can run with a minimum of energy loss due to shoreline irregularity. During maturity, further, the smoothing of the shoreline is aided by downcutting of outlets, which lowers the lake level and makes at least some of the sediments of the underwater shelf available for reworking by waves and currents.

In a lake's late maturity or early old age, it typically demonstrates, as a result of further outlet erosion, a shoreline of emergence with old raised wave-cut cliffs and beaches, at least partially evident. The underwater shelf in this stage typically is wide and sufficiently shallow so that waves will break well out from shore, with the result that the bottom near shore is able to accumulate fine sediments, and land and aquatic plants begin to invade from the shores. Sediments of the deep basin at this stage begin to show significant amounts of organic material, and filling of the deep basin accelerates.

A lake in old age is typically shallow, weedy, and bordered by a band of swamp of greater or less extent. Biologically, a lake in old age is characterized by high rates of production of new plant and animal material. Filling of the basin goes on at increasing rates, with organic materials derived from the shores as the dominant filling material. The shoreline generally becomes increasingly smooth and regular as embayments become filled and taken over by swamp.

The death of a lake occurs when it has become predominately a swamp, usually with a stream running through the originally deepest parts. Outlets have by this stage been eroded down to grade or down to some stable sediment. Lakes whose outlets become filled during the aging process end as bogs (Subsec. VIII).

The majority of lakes have a ratio of mean depth to maximum depth which exceeds 0.33, approaching the value that a conical depression would have. With increasing age, this ratio tends to increase as maximum depths are filled in. Lakes in easily erodible rocks generally have d/d_m ratios between 0.33 and 0.5. In block-fault (graben) lakes, values of this ratio may exceed 0.5. Minimum values of the ratio occur in lakes with deep holes. Neumann [34] believes that a lake basin is best described by an elliptic sinusoid with a d/d_m ratio of 0.467.

IV. LAKES IN THE HYDROLOGIC CYCLE

A. Lakes as Natural Reservoirs

The hydrologic cycle is driven by solar energy and gravity. Most of the precipitation which falls on the land surface is derived from oceanic evaporation [35]. Water falling on land is eventually returned to the sea, though some of it reaches the oceanic reservoir by complex and devious routes. Lakes are natural reservoirs in which water is temporarily stored during its passage to the sea.

Lakes and swamps are supplied by meteoric water and are sensitive to variations in the net rate of supply. This sensitivity is registered in rise or fall in lake level, reflecting volumetric water changes in the lake basins. Lake-level fluctuations not only record changes in water gain, but also changes in water losses, both of which are a function of climatological variations with time.

B. Lake Extinction

Shallow lakes may dry up completely during droughts, only to regain their former levels during subsequent wetter periods. It is a matter of observation that lake extinction can be caused by climatic change. The extinction of a lake through loss of water only is termed *temporary extinction* [1]. So long as the closed basin remains intact, it always has the potential of becoming a lake again.

If the basin itself is destroyed by one or a combination of processes, it is then *permanently extinct* [1]. This condition may be attained catastrophically very soon after the basin is formed (as in the case of a breached landslide across a river) or may take place only after episodes of geologic time have passed. In the latter case the lake may have experienced temporary extinction many times and undergone severe modification of its shoreline.

Destruction of a lake basin involves many different geological processes. These include not only erosive processes, whereby the rim of the basin is breached, but also depositional processes, which reduce or destroy the basin's volume by filling with mineral and organic sediments. Although the latter are generally long-term processes in natural lakes, they are of immediate importance to the hydrologist in the case of artificial lakes where reservoir sedimentation is a major consideration in predicting useful life (Sec. 17-I).

C. Fluctuations in Lake Levels

The water level of a lake is a function of the volume contained in the lake basin. The rate of change of water volume is controlled by the rate at which water enters the basin from all sources minus the rate at which water is lost by evaporation from its surface and discharged by surface and subsurface effluents.

The magnitude of any positive or negative element in the water-balance equation is determined by the climatic and ecological setting within which the lake occurs. Some lakes with small drainage basins and no influents or outlets have a very simple water balance in that the major source of income is by direct precipitation on the lake surface and the chief loss is by evaporation. Other lakes directly connected to the water table with surface inlets and outlets have a more complex relationship between water gain and water loss. Water income from melting snow or glaciers, or water loss by the transpiration of higher aquatic plants around the shore, may be extremely difficult to measure. From the engineering point of view it is important to consider the equation of water balance because any use to which the lake or its shoreline is put will inevitably require some knowledge of lake-level fluctuations.

Hutchinson considers a theoretical model in which rain falling on the drainage basin reaches the lake in a short time. The ideal case would be one which considers the mean annual level of a lake lying in a region of short and well-defined rainy seasons. If the area of the lake is small compared with the area of the drainage basin including the lake, the following equation may be written [4, p. 236]:

$$A\dot{z} = A(P_r - E_v) + (A' - A)(P_r' - E_v') - AE_f \qquad (23\text{-}6)$$

where P_r = mean precipitation over lake surface
P_r' = mean precipitation over rest of drainage basin
E_v = mean evaporation from lake surface
E_v' = mean evaporation from rest of drainage basin
E_f = mean discharge by any channel per unit of lake area
$\dot{z}$ = mean rate of increase of depth
A' = area of drainage basin including lake
A = area of lake

This equation serves more as a tool for evaluating the response of lake level to the various parameters than for the forecasting of future levels. Both A and E_f are dependent on depth, with a sensitivity that is determined by the shape of the basin and by the variation in discharge of the outlet as the hydraulic head changes. Steep-sided lakes reduce the dependency of area on depth, but in saucer-shaped basins the relationship A/z is extremely sensitive.

Hutchinson [4, p. 237] sums up the theoretical considerations as follows:

In general, the presence of an effluent, particularly one which is highly sensitive to increases in z, a high ratio of $\bar{A}/A'$, and a gentle slope, will make the lake more sensitive to short-period changes in meteorological variables, whereas the smallness or absence of an effluent, a small ratio $\bar{A}/A'$, and a steep-sloping lake basin will make the lake less sensitive to minor variations and perhaps permit its variation in level to pick out selectively long-period variations in climate. ($\bar{A}$ = mean area of lake over a range of values of depth z.)

The validity of this theoretical treatment rests on the assumption that the time involved in the transfer of water from watershed to lake is negligible compared with the time intervals used in the gradation of the data. Thus actual cases amenable to rigorous analysis according to Eq. (23-6) are relatively few in nature, since by far the majority of the lakes of the world lie in forested areas—a circumstance which generally serves to produce a considerable delay between storm peaks and high discharge rates in stream channels. Only in semiarid regions where lakes have no outlets is the theoretical model most likely to be realized (Sec. 24).

D. The Great Lakes

A brief account of the fluctuations of water levels in the five Great Lakes and of the effect of these fluctuations on the various uses to which these lakes are put will demonstrate the importance of water-level changes to such diverse interests as navigation, hydroelectric-power development, water diversion, and shore-property utilization.

The Great Lakes and their connecting waterways constitute the largest inland system of fresh water in the world. They extend from the Gulf of St. Lawrence to the headwaters of the St. Louis River in Minnesota, a distance of about 2,000 miles. The drainage basin in which they lie has an area of approximately 325,000 sq mi, of which somewhat less than a third is open-water surface.

Lake Superior, the largest (31,820 sq mi) and highest (603 ft above sea level), discharges into Lake Huron via the St. Marys River at an average rate of 73,000 cfs. Lakes Huron (23,000 sq mi) and Michigan (22,400 sq mi) both stand at 580 ft by virtue of their broad and deep connection at the Straits of Mackinac. The Straits contain a submerged river channel which joined the two basins at a period in former geologic time when both lakes stood at much lower elevations [36].

The effluent waters of Lakes Superior, Michigan, and Huron all pass out the St. Clair–Detroit River outlet of Lake Huron. Powers and Ayers [37] indicate that the summertime relationship between the lake effluents and the outflow is about

Superior, 72 : Michigan, 56 : Huron, 69 : outflow, 197

The St. Clair–Detroit River drainage flows into Lake Erie (572 ft above sea level and 9,940 sq mi) at an average rate of 175,000 cfs. Erie is the shallowest of the Great Lakes and the only one whose bottom does not extend below sea level. Erie discharges 194,000 cfs into Lake Ontario (7,540 sq mi) via the Niagara River, which drops 326 ft (172 ft at Niagara Falls) in its 36-mile course. The waters from all the lakes eventually discharge into the Atlantic Ocean through the St. Lawrence River at an average rate of 231,000 cfs.

1. Nature, Magnitude, and Timing of Lake-level Fluctuations in the Great Lakes. Records published by the U.S. Lake Survey Office in Detroit show that since 1860 the lake levels have fluctuated from a little over 4 ft on Lake Superior to over 6.5 ft on Lake Ontario. Superimposed on these long-range fluctuations are seasonal variations which average from 1 ft on Lakes Michigan and Huron to 1.8 ft on Lake Ontario. Peak levels are usually reached in Lakes Erie and Ontario in June, in Michigan and Huron in July or August, and in Superior in September.

2. Factors Controlling Water-level Fluctuations in the Great Lakes. The factors which affect seasonal and yearly fluctuations of the Great Lakes can be separated into two categories, natural and artificial, the former being much more important than the latter. The natural factors include precipitation on the watershed, evaporation from the lake surfaces, flow in the connecting waterways, and crustal movements. Artificial factors influencing lake levels are controlling locks and dams on connecting channels, dredging of channels, and water diversions into and out of the lake system.

a. Natural Factors. During spring and early summer the lakes normally experience a rising stage because the net inflow from snowmelt and precipitation on the watershed exceeds the losses by outflow and evaporation. After the summertime highs are reached, the lakes decline gradually to their low winter stages because of the reduction in precipitation and the general increase in evaporation during late summer and early fall. Peak lows of winter involve temporary entrapment of frozen precipitation within the watersheds.

This seasonal pattern is the "normal" situation for the lakes. However, when several years of low or high annual precipitation follow one another, the lake levels respond accordingly. The most recent example was the abnormally high stage which occurred on all five lakes in 1951 to 1952 [38]. At that time Lake Ontario and Lake Erie reached all-time highs, and the other three lakes experienced highs that were exceeded only a few times in the previous 90 years. The high levels were caused by heavier-than-average precipitation during the previous 10 years (when precipitation over the watershed was 2.23 in./year above the normal annual average based on the 1900–1950 record). In 1950 and 1951, the two years prior to the attainment of the high levels, precipitation was almost 6 in./year above average [39]. This sequence of wet years not only caused the summer highs of each year to exceed the peak of the former summer, but also reduced the amount of the autumn and winter decline.

Although attempts have been made to relate the trends in lake levels to cyclic causes such as sunspot maxima and minima, no clear-cut relationship to this or any other cyclic phenomenon has yet been found [38, 39]. On the other hand, the effects of wet years and dry years on lake levels have prompted the development of moderately successful means of forecasting the levels of the Great Lakes [40, 41].

Precipitation over the watershed of the Great Lakes is measured by a system of meteorological stations in the United States and Canada, but evaporation from open-water surfaces is a much more elusive parameter to evaluate. Formulas for determining evaporation from a lake surface have been devised, but McDonald [39] claims that no precise means has yet been devised. He does give, however, approximate estimates of 1.5 ft for Lake Superior and 3 ft for Lake Erie. The subject is by no means dead (Secs. 11 and 24).

The discharge from one lake to another is a function of the hydraulic head and channel characteristics; if either of these is altered naturally or by artificial means, there will be corresponding changes in discharge, which in turn affect the water balance in the lakes at either end of the connecting channel.

The final natural factor relevant to changes in levels of the Great Lakes is movement of the earth's crust. Crustal movements derive from postglacial warping due to removal of glacier ice which once covered the area. Such movements have not only had a pronounced effect on the whole geologic history of the Great Lakes [8], but are believed to be still in progress. McDonald [39] reports that the coasts of Lakes Erie and Ontario, most of the Lake Michigan coast, and the southern shore of Lake Superior are sinking in relation to their respective outlets, causing slowly rising water levels in those sectors. On the other hand, the coast of most of Lake Huron, the northeastern shore of Lake Michigan, and the north shore of Lake Superior are rising with respect to their outlets; hence levels are slowly declining. Determinations by the U.S. Lake Survey indicate that the crustal downwarping is producing a rising water level of 1.1 ft per century on the west end of Lake Ontario and a falling water level of 0.7 ft per century at French River in Georgian Bay on Lake Huron. Though these movements are slow when measured on the time scale of human activity, they cannot be ignored in long-range planning for the Great Lakes.

b. Artificial Factors. When human beings try to improve on nature's own regulatory mechanisms, some benefit may, in fact, be derived; but not infrequently, along with artificial manipulation, come a host of other problems, some of which are political or economical, while others are purely technical. In man's attempt to influence the Great Lakes, the main target is a man-made system of lake-level regulation, plus the development and maintenance of deep navigational channels in the connecting waterways. The opinion on water levels in the Great Lakes is divided between the navigational and hydroelectric-plant interests that favor higher levels and the riparian ownership interests that favor stabilized lower levels. Engineers are thus confronted with a very complex problem from the outset, and many of them feel that a solution which will be fair and equitable to all legitimate interests belongs in the realm of wishful thinking. That this pessimistic viewpoint is not shared by all, however, is exemplified by the undertaking and completion of the St. Lawrence Seaway. But even before this forward step in lake and harbor engineering was on the drawing boards, the Great Lakes were subjected to man-made regulatory installations, some of which are still in operation.

A system of controlling gates in the lock and dam on the St. Marys River was inaugurated in 1911 to reduce the natural outflow from Lake Superior during periods of drought and maintain water depths in such important harbors as Duluth.

Another aid to navigation is the dredging of the connecting waterways to provide maximum water depths for deep-draft ships carrying important bulk cargoes such as iron ore, limestone, coal, grain, and petroleum products. The St. Clair–Detroit River system between Lake Huron and Lake Erie has undergone systematic dredging and channel improvement since 1870. This has facilitated the passage of deep-draft ships through the otherwise too-shallow river system and, according to Hoad [38], has reduced the original 9-ft drop between Huron and Erie to about 7.5 ft. Hoad further speculates that this permanent lowering may account for the fact that the 1952 summer high-water levels broke all records back to 1860 in Erie and Ontario, while in

Lakes Huron and Michigan the 1952 summertime peak was below three others (1871, 1876, and 1886) which occurred prior to the completion of the dredging operations.

Another means of influencing water levels in the Great Lakes is by diversion of water into the lakes from adjoining watersheds or out of the lakes through canals. Two diversions into Lake Superior are now in operation. The waters of Long Lake and the Ogoki River of the Hudson Bay watershed were diverted into Lake Superior in the 1940s. The maximum flow of these two diversions is 5,000 cfs, a small amount compared with the average discharge of 194,000 cfs through the Niagara River. The purpose of the diversion was to permit the Ontario Provincial Power Commission to withdraw an equal volume of water from the Niagara River for power purposes, without impairing the beauty of Niagara Falls. The addition of a 5,000-cfs flow through the Great Lakes may be expected to raise their levels by 2 or 3 in. [38].

On the other side of the Great Lakes water balance sheet is the small negative effect on water levels brought about by the diversion of an average of 6,795 cfs (1900–1924 basis) down the Chicago Sanitary and Ship Canal to the Mississippi River. This outflow costs Lakes Michigan and Huron about 5.5 in. of water for the whole period of time in question [38]. In the late 1950s the Chicago Sanitation District asked to increase the diversion, a proposal which generated much opposition from public and private organizations on both side of the international border. Hoad [38] is of the opinion that any additional diversion by Chicago up to 3,000 or 4,000 cfs would have only a very small effect on the water levels of Lakes Michigan and Huron and might, moreover, even improve the salubrity of the water in the southern end of Lake Michigan, from which several heavily populated communities, including Chicago, obtain their entire water supply.

Clearly, the task of artificial "control" of levels in the Great Lakes is a formidable one, but it is more likely to be solved by sound research and technology than by debates before courts or international commissions, where the participants lose sight of the fact that the net effect on lake levels of all the existing diversions and control devices is on the order of 2 in., an insignificant amount compared with the 4 to 6 ft involved in long-term fluctuations, or even the 1 to 2 ft resulting from seasonal variations [39].

V. THERMAL PROPERTIES OF LAKE WATER

A. The Temperature-Density Relationship

Water has a number of unique physical characteristics. Henderson [42] in "The Fitness of the Environment" summarized the impressiveness of the suitability of these properties for the internal and external environments of life and life processes.

Water's anomalous density decrease below 4°C entered scientific literature at least as early as 1833 and 1834, when Whewell [43] and Prout [44] independently took emphatic notice of it. Röntgen [45] first brought out the idea that the anomalous properties of water resulted from variations in the degree of association of water molecules. Dorsey [46] reviewed the growth of concepts of water structure to his day; Hutchinson [4] traces the development of concepts since that time, and the authors have followed his presentation.

According to present beliefs, the anomalous density characteristics of water are related to the retention of varying degrees of the crystalline structure of ice in liquid water at ordinary temperatures. Ice exhibits a crystalline structure similar to that of the mineral *tridymite*. In ice's crystalline structure the water molecules (their molecular movements slowed and their interatomic distances decreased by removal of heat) in major part occupy a tetrahedral latticework. In this latticework each oxygen atom is associated with four others so arranged that they form the apices of a tetrahedron, with the reference atom in the center and with their hydrogen atoms forming hydrogen bonds, or bridges, between the five oxygen atoms.

Each apical oxygen atom of the tetrahedron is simultaneously associated with three others. These, along with the original reference atom (now serving as an apical atom), form another tetrahedron, interlocked with the first. The four apical atoms of the first tetrahedron are thus also the central atoms of four tetrahedrons interlocked with the first, all of which use the original reference atom as an apical atom. Repetitiously, each member of the four sets of secondarily associated oxygens is associated with three others and serves as the central atom of still another tetrahedron whose tip is an apical atom of the initial tetrahedron. In this fashion a latticework of interlocked tetrahedrons is formed. In ice the maximum numbers of water molecules are involved in the latticework of tetrahedrons, and interstitial space within the latticework is at its maximum; the reduced density of ice (sp. gr. 0.92) is a reflection of maximal interstitial space.

The addition of heat to change ice at 32°F to water at 32°F causes rupture of about 15 per cent of the hydrogen bonds. As a result, portions of the latticework detach and occupy part of the interstitial space of the lattice, with associated increase in density (ice water sp. gr. 0.99).

Further additions of heat rupture more of the hydrogen bonds, and the interstitial space of the lattice becomes increasingly filled with lattice fragments. Interstitial space in the latticework becomes minimal at 3.94°C (usually approximated as 4°C, or 39.2°F), and water achieves its maximum density (sp. gr. 1.00) at this temperature. In the temperature interval between 0 and 4°, it appears that introduced heat energy goes primarily into the production of hydrogen-bond ruptures. Above 4°C, heat additions appear to produce increased interatomic distances as well as bond ruptures, and water density decreases steadily after 4° is exceeded.

B. Liquid Nature

In addition to its unusual temperature-density relationship, water is unique in being one of the very few substances which exists in the liquid state at the ranges of temperature and pressure encountered on the earth's surface. Aside from the aqueous liquids, the other natural liquids are native mercury and the several liquid forms of petroleum. Hutchinson [4] reasons from the chemical compositions and boiling points of other dihydrides that water would boil at −80°C (−112°F) if it had a normal set of characteristics.

C. Thermal Capacity

Water has the highest specific heat and thermal capacity of all solids and liquids except liquid ammonia. At 0°C the specific heat of water is 1.01 cal/g, or 1 Btu/lb/°F. Thermal capacity is specific heat multiplied by mass. The high thermal capacity of water allows it to take up or give off large quantities of heat with minimal temperature change. This quality bears importantly upon the climate on the downwind sides of large lakes. In the oceans, the permanent surface currents flowing poleward from the tropics transport great quantities of heat to the temperate and subpolar regions.

D. Latent Heats

The *latent heat of evaporation* of water is the highest of all substances: 539.55 cal/g, or 972 Btu/lb; these values are for boiling temperatures. When water evaporates at lower temperatures the latent heat is somewhat higher: at 0°C it is about 600 cal/g (1080 Btu/lb).

The high latent heat of evaporation of water is of importance in nature primarily in that it regulates the evaporative phase of the hydrologic cycle. Without their high heat of evaporation, natural waters would suffer more severe evaporative losses than they do.

Water's *latent heat of fusion* is the highest of all substances except liquid ammonia. To convert one gram of ice at freezing temperature to water at the same temperature, 79.67 cal of heat must be added. To melt one pound of ice without temperature change requires 143.4 Btu.

High latent heat of fusion gives lake waters a material thermostatic effect at freezing temperatures. This effect, coupled with the lower density of ice and the 4°C temperature of maximum density, means that subsurface temperatures in ice-covered lakes do not fall below freezing.

The *latent heat of sublimation* of ice is 679 cal/g, or 1222 Btu/lb.

E. Thermal Expansion

Having its maximum density at about 4°C, water undergoes a normal thermal expansion (decreasing density with increasing temperature) from that temperature upward, and an inverse thermal expansion (decreasing density with decreasing temperature) from that point downward. The values in Table 23-1 are from the Smithsonian Tables and illustrate the point.

Table 23-1. Density of Water as a Function of Temperature, at 1 Atm

Temperature, °C (°F)	Density, g/ml
−10 (14.0)	0.99815
0 (32.0)	0.99987
4 (39.2)	1.00000
10 (50.0)	0.99973
20 (68.0)	0.99823
30 (86.0)	0.99567
100 (212.0)	0.95833

F. Thermal Conductivity

Thermal-conductivity values for various substances are commonly tabulated as the flow of heat per second through sheets of the material 1 cm or 1 in. thick, of 1 cm^2 or 1 ft^2 face area, and under a thermal gradient of 1°C or 1°F between faces. The figures of Table 23-2 are for water under confined conditions and with molecular conductivity as the active agent. Unfortunately, confined water or water in laminar flow (with its concomitant implication of molecular conductivity) does not exist to any significant extent in nature. Natural movements of surface water, and natural conditions in it, almost always involve turbulent flow and eddy phenomena. The limited numbers of determinations under these conditions have given values of the coefficient of eddy conductivity that range from 2 to 300,000 times those of molecular conductivity obtained in the laboratory.

Table 23-2. Thermal Conductivity of Water

Temperature, °C (°F)	Cal/sec/cm^2 under 1°C/cm	Btu/sec/ft^2 under 1°F/in.
0 (32)	0.00139	0.00111
4 (39.2)	0.00138	0.00110
15 (59.0)	0.00144	0.00115
20 (68.0)	0.00143	0.00114

G. Thermal Stratification

1. The Turnovers. The fact that water has its temperature of maximum density above the freezing point brings about, in the typical lake, twice-yearly periods of uniform temperature when vertical circulation occurs.

During autumn, the shortening days, decreasing altitude of the sun, and declining air temperatures cool the surface waters. The cooled surface waters, being denser, settle away from the surface and are replaced by warmer water from below. As cooling approaches 4°C, the cooled surface waters become denser than any lying beneath, and they sink through to the bottom. In this period, the *fall turnover*, the water becomes uniform in temperature as a result of this convective circulation. The whole water body at this time receives renewed supplies of dissolved atmospheric gases, and gaseous materials accumulated in the subsurface water beneath the temperature and density stratification of summer are given off. The fall turnover usually begins shortly before surface temperatures of 4°C are reached and usually goes on until temperatures slightly lower are attained, for at this time the density differences from top to bottom are so slight that turbulence from wind action can provide mechanical mixing reaching to the bottom.

Continued autumnal cooling lowers surface-water temperatures below 4°, and an inverse temperature structure develops, with water at about 4° on the bottom and colder, less dense water above. The fall turnover may cease, at any time after 4° and prior to freeze-up, if the density of the surface water becomes sufficiently reduced to prevent the local winds from mixing it down and dispersing it.

The formation of ice cover provides protection from wind action, and water temperatures during winter typically go from 0°C (32°F) against the ice at the surface to about 4°C (39.2°F) at the bottom.

Spring warming results in meltoff of the ice cover and in rising surface-water temperatures. When the surface temperature approaches 4°, convective sinking, aided by the mechanical stirring of the wind, again produces a period of uniform vertical temperature and of vertical convective circulation. This *spring turnover* comes to an end when the surface water becomes sufficiently warm and less dense to prevent wind mixing from destroying the vertical density gradient.

2. Summer Stratification. From spring until the height of summer, heat enters the surface water and is mixed downward for a limited distance by wind action. The result is the formation of an upper layer of water, of varying thickness, in which temperatures are relatively uniform and higher than those in the rest of the water. This relatively uniform mixed upper layer is the *epilimnion*.

Beneath the epilimnion is a water layer in which the temperature decreases rapidly (and the density *increases* rapidly) with depth. This is the *thermocline*, which is the major density discontinuity and which limits the downward extent of the epilimnion. The thermocline is poorly developed in spring, but becomes increasingly thin and sharp as summer goes on. At its full development, the thermocline commonly exhibits temperature drops of more than 1°F/ft.

Below the thermocline lies the rest of the subsurface water, the *hypolimnion*. Its temperatures typically are low and decrease only a small amount from the bottom of the thermocline to the lake bottom. The hypolimnion is largely cut off from the atmosphere and from wind action by the epilimnion and thermocline. The content of dissolved gases and chemicals in the hypolimnion may remain all summer very much as it was left by the spring turnover if the lake is clean and deep; it may suffer depletion of oxygen and accumulation of H_2S and other chemicals from the bottom if the lake is dirty and/or shallow. Summer stratification gradually breaks down during the fall, and is destroyed at or just before the time of the fall turnover.

Very large lakes, such as the Great Lakes, which do not develop complete ice cover in winter, have a modified form of annual temperature cycle. In these lakes the fall turnover is continued throughout the winter by strong wind action. These lakes may attain water temperatures less than 4° from surface to bottom in severe winters; Lake Michigan was observed by Church [47] to attain vertically uniform temperatures below 2.5°C (36.5°F) in the winter of 1941–1942. In the presence of high winds and lacking ice cover, such lakes continue their fall turnover through the time of the typical spring turnover. They enter into summer stratification in late spring.

H. Heat Budgets

Heat budgets of several kinds have been devised as means of expressing aspects of the thermal regimes of lakes. In essence, heat budgets attempt to assess the internal distribution of the annual heat input received by the lake. The older, more classic type of heat budget is that devised by Birge, and consequently called the *Birgean heat budget*. Birgean heat budgets are primarily concerned with the division of annual

THERMAL PROPERTIES OF LAKE WATER   23-19

heat input between winter and summer. More informative, and much more difficult, are the so-called *analytical heat budgets*. These attempt, first, to determine the amounts of heat received from the sun, sky, atmosphere, environment, and influent waters, and second, to follow the attrition of this heat by the several sources of heat loss until it is possible to arrive at the heat storage in the water and sediments of the lake.

1. Birgean Heat Budgets. In its common usage the Birgean heat budget considers the summer and winter heat incomes. *Summer heat income* is defined as the amount of heat required to raise the lake from a uniform temperature of 4°C to its highest observed summer heat content. *Winter heat income* is the heat required to raise the lake from its minimum winter heat content to a uniform temperature of 4°C. The sum of summer and winter heat incomes gives the annual heat budget. Computation of a Birgean heat budget requires an accurate bathymetric chart, with depth contours relatively closely spaced and water-temperature values in the height of summer and the dead of winter at the same depths as the depth contours. The areas enclosed by the several depth contours are measured, and the heat content of the lake at a uniform 4°C is computed. Finding the summer heat income involves repeated computations of heat content during summer until a maximum has been reached. Subtraction of the heat content at 4° from the maximum heat content gives the summer heat income. Determination of winter heat income involves finding the minimum heat content of the lake during winter and subtracting it from the content at uniform 4°C.

Heat content may be determined by plotting against depth d the product of area at that depth and the difference of temperature (from 4°) at depth d, determining the area under the resulting curve, and dividing by the surface area of the lake. Alternatively, heat content can be obtained by summing the products of volume and average temperature difference (from 4°) for a series of layers bounded by adjacent depth contours and dividing the sum by the surface area of the lake. Results from either method are expressed in calories per square centimeter.

From the heat contents of the Birgean heat budget can be derived the mean time rate of heat-content change for either the summer or the winter heat incomes, provided the dates of the summer and winter peaks and of the lake's spring passage through 4°C are determined. Also, from the mean time rate of heat-content change, the mean time rate of change of mean lake temperature can be obtained by dividing by the lake volume.

2. Analytical Heat Budgets. Birgean heat budgets are of value for limited purposes, but much more detailed understanding of the thermal regime is obtained when the analytical heat budget is used. The analytical budget treats the rates of transfer of the several forms of radiant and thermal energy; all its terms are relative to unit surface area; and the budget aims at obtaining the storage of heat (Q_t) in the lake and its sediments.

The sources of energy are solar radiation (Q_S), radiation from the sky (skylight) (Q_H), long-wave (heat) radiation from the atmosphere (Q_A), similar heat radiation from mountains adjoining the lake (Q_M), thermal energy supplied by condensation of water vapor on the lake surface (Q_c), conduction of sensible heat from the air when it is warmer than the water (Q_s), and heat carried in by influent water (Q_i). Terms which are losses are reflection from the water surface (Q_R), scattering of radiation in the water, with loss as scattered energy reemerges from the water (Q_u), long-wave back radiation from the water (Q_W), energy used in the evaporation process (Q_E), energy transported away as the thermal content of evaporated water (Q_e), transfer of sensible heat from water to air (Q_s, may be positive or negative), and heat carried out by effluent water (Q_i, may be positive or negative).

Certain of the terms are customarily grouped:

$$Q_F = Q_S + Q_H - Q_R - Q_u \qquad (23\text{-}7)$$

Q_F is referred to as the net radiation flux. Also

$$Q_B = Q_F + (Q_A \pm Q_M - Q_W) \qquad (23\text{-}8)$$

Q_B being designated the net radiation surplus.

The equation for total energy then becomes

$$Q_t = Q_B \pm Q_s \pm Q_i - Q_E - Q_e + Q_c \tag{23-9}$$

Ordinarily, the terms Q_M, Q_i, and Q_e are negligible for practical work, but in mountainous country, or for lakes with large inlets and outlets, or in semiarid territory, the appropriate one probably should be included. Q_c, the energy brought in by condensation onto the water surface, can be ignored unless foggy weather is a dominant feature of the local climate.

The solution of an analytical heat budget requires data on the radiation flux, sufficiently detailed temperature series within the lake and in the air above the lake, and temperatures of inlets and outlets if flow-through is significant. Details of the evaluation of the several terms are given in the Lake Hefner studies [48]. Most difficult to evaluate are the evaporative and the sensible-heat terms. A common circumvention of this difficulty, *when the other terms are known*, is the employment of the *Bowen ratio* R_b,

$$R_b = \frac{Q_s}{Q_E} = 61 \times 10^{-5} P_h \frac{T_w - T_a}{P_s - P_w} \tag{23-10}$$

where P_h is the barometric pressure at the altitude of the lake surface in the same units as P_s and P_w, which are, respectively, the saturation pressure of water vapor at water-surface temperature T_w and the observed pressure of water vapor in the air at the same standard height as the observed air temperature T_a.

Estimates of the evaporative term Q_E and of net sensible-heat transfer can be obtained by other means if wind measurements over the lake are available.

$$Q_E = 0.18L(P_s - P_w)W^{0.8} \tag{23-11}$$
and
$$Q_s = 4.4(T_w - T_a)W^{0.8} \tag{23-12}$$

where W is the wind velocity in cm/sec measured at 6 m above water, L is the latent heat of vaporization of water at surface temperature T_w, P_w and T_a are as above but measured at 2 m above the lake surface, and P_s is saturation pressure at T_w.

VI. LAKE ICE

Before mechanical refrigeration devices became widespread, the chief value of lake ice was in its use as a coolant in the preservation of perishable foodstuffs. Interest in ice as a natural structure received great stimulus from practical problems encountered during World War II and has continued at a high level.

Whether of only seasonal duration on lakes in temperate latitudes or as a semipermanent feature of lakes in arctic and antarctic regions, lake ice is now employed as roadways for vehicles and as landing strips for aircraft. Ice also acts as a potent destructive agent on shore installations, both as a result of thermal contraction and expansion during the ice season and by shoreward movement under wind pressure during spring breakup. Ice is also a hazard to shipping in navigable waters such as the Great Lakes.

Before considering lake ice in any of these roles, it will be profitable to look into the origin, classification, and structure of this natural crystalline material (see also Sec. 16).

A. Structure, Classification, and Thickness

The ice cover of a lake begins to form after the water reaches a more or less isothermal condition a few degrees below the 4°C (39.2°F) temperature of maximum density; then, on a still cold night, freezing begins in the surface.

The following discussion of the origin and classification of lake ice is taken for the most part from the work of Wilson, Zumberge, and Marshall [49]. Two factors are involved in the formation of the ice cover. The first is the freezing of the upper water layer itself, producing a smooth homogeneous ice called *sheet ice*. The second is the fusion of individual ice masses produced by the breakup and refreezing of an ice sheet in its early stages of development, or by the introduction of snow into the surface water of a lake during the initial stages of freezing. The ice formed when either of these second conditions applies is known as *agglomeritic ice* and generally can be distinguished from sheet ice by its rough surface and nontransparency.

1. Sheet Ice. Sheet ice is the purest and strongest form of lake ice. It is produced by rapid freezing of the surface-water film when there is no wind or snow. The ice crystals of the original skim grow in such a manner as to form a columnar aggregate in which the crystallographic axes are vertical in orientation. Individual crystals in a single mass of sheet ice tend to be uniform in size through a given horizontal plane, but the average crystal diameter generally increases with depth, thereby reducing the number of crystals per unit area. Most individual crystals in sheet ice range in size from less than 0.1 in. to about 1 in., but larger crystals ranging from 6 in. to 1 ft have been observed in some of the near-shore areas of protected bays in the Great Lakes. The presence of single large crystals in an otherwise uniformly textured ice has also been observed, but the occurrence is considered rare.

2. Agglomeritic Ice. Agglomeritic ice has a more complex history or origin than sheet ice. Its main characteristic is that it contains older ice or snow masses bonded together by frozen lake water. One of the most common ways in which agglomeritic ice is formed is by the breakup of relatively thin sheet ice. The individual ice floes are moved against each other by wind and wave action so that they become abraded and broken before being fused together during the next freeze-up.

Another type of agglomeritic ice is produced when a heavy snow falls on a lake surface at a time when the water is so cold that it cannot melt the snow. Wave action breaks the water-saturated snow blanket into *slush balls*, or *slush pans*, that are incorporated into the ice cover when final freeze-up occurs.

Alternating freeze-up and breakup occur most frequently during the early part of the ice season, when the ice cover is relatively thin. Depending on the number of these cycles and the addition of snow, the resulting winter's ice layer may have a very complex developmental history. It is not unusual for a single lake to exhibit the same ice structure year after year in the same part of the lake, especially in bays and protected areas, where shore configuration is a dominant factor in the exposure of that particular lake segment to wind stress.

3. Ice Thickness. Once the ice cover is established for the duration of the winter, it thickens by accretion from below at a rate governed by the temperature gradient through the ice. Snow cover exerts a strong influence on lake ice thickness in two ways. First, it acts as an insulator between the ice surface and the air, thereby reducing the temperature gradient. Second, it depresses the ice by its own weight, causing water from below to leak through cracks in the ice and partially flood the ice surface. When this surficial layer of water freezes with the snow, it forms a whitish névé-like ice generally known as *snow ice*.

A vertical section through lake ice bearing a heavy cover of snow in midwinter might show conditions similar to the following, which are typical of the lakes around Marquette, Mich.: dry snow of variable density, 18 to 20 in.; crust of snow ice, 2 to 3 in.; unfrozen slush, 4 to 6 in.; snow ice, 4 to 6 in.; and sheet ice, 10 to 12 in.

The crust of snow ice in the case just cited is an irritating inconvenience when ice studies or surveys are being conducted, because it is often too weak to bear a man unless he is on skis or snowshoes. Even if the crust possesses sufficient bearing strength to hold a man, it is normally not strong enough to support a vehicle of any kind, making vehicular traffic extremely difficult, if not preventing it entirely.

4. Empirical Prediction of Ice Thickness. In operations in which natural lake-ice cover is involved, some indication of ice thickness during any part of the ice season is desirable. When the ice cannot be measured by direct means, one must resort to predictive methods such as that given by Assur [50]. He presents a simple empirical equation relating sheet-ice thickness to accumulated degree-days below freezing,

$$h_i = \alpha(1.06\sqrt{S}) \qquad (23\text{-}13)$$

where h_i is the ice thickness in in., α is the coefficient of snow cover and local conditions

Table 23-3. Values of α for Eq. (23-13) and β for Eq. (23-14)*

α	β	Local conditions (mainly snow cover)
1.00	3.6	Theoretical maximum (never reached under natural conditions)
0.9–0.85	4.4–5.0	Maximum for ice not covered with snow
0.75–0.65	6.7–9.0	Medium-size lakes with moderate snow cover

* From Assur [50].

(see Table 23-3), and S is the accumulated degree-days since freeze-up, based on °F below freezing.

The values in Table 23-3 are based on average conditions and do not take into account variations in snow cover. Equation (23-13), therefore, should be used only as a rough approximation and is not intended to replace detailed ice surveys.

If there are data on climatological conditions in the area where the lake is situated, the time rate of ice thickening, after the ice cover is established, can be forecast according to another expression given by Assur [50],

$$\Delta t = \frac{\beta(1 + h_i)}{32 - F} \quad (23\text{-}14)$$

where Δt is the time, in days, required for ice to increase in thickness by 2 in., h_i is the measured or estimated ice thickness in in., F is the average expected mean air temperature in °F, and β is the coefficient of snow cover and local conditions, from Table 23-3.

The chief value of Eq. (23-14) is in predicting the number of days necessary for an established ice cover to increase from one thickness to another. Calculations using this equation are based on 2-in. increments of additional ice thickness, which are added algebraically to arrive at the total additional time required for the ice to reach the desired thickness. Assur gives the following example:

A certain aircraft requires 22 in. of lake ice for safe landing operations. An advance party measures an ice thickness of 16 in. on a certain date and evaluates the snow cover. The expected mean air temperature for the next month is 12°F. One wishes to know how long it will require this ice layer to attain the desired thickness of 22 in. From Table 23-3, β is estimated at 8.0, and Eq. (23-14) is applied as follows:

$$\text{From 16 to 18 in. } \Delta t = \frac{8(1 + 16)}{32 - 12} = 6.8 \text{ days}$$

$$\text{From 18 to 20 in. } \Delta t = \frac{8(1 + 18)}{20} = 7.6 \text{ days}$$

$$\text{From 20 to 22 in. } \Delta t = \frac{8(1 + 20)}{20} = 8.4 \text{ days}$$

$$\text{Total} \cong 23 \text{ days}$$

5. Ice Breakup. The dissipation of lake ice in the spring is accomplished by melting of the upper part first, producing a *rotted* appearance. Melting along grain boundaries renders sheet ice into a loosely bound crystal aggregate, which is especially pronounced in ice having a columnar structure. Rotted columnar lake ice is known as *candled ice;* it will disintegrate into long columnar crystals at the slightest blow from pickax or ice ax. Beneath the upper candled ice the lake ice may still retain considerable bearing strength, although the rate of melting varies so much from one place to another on a single lake that it is wise to suspend all operations soon after the melt season begins.

On many lakes the greatest melting rate occurs near the shore, where solar radiation is absorbed by the darker materials right at the shoreline. By thawing around the shore it is not uncommon for a coherent ice cover to be entirely freed, so that it can be attacked by wind and waves along the edge, a condition which accelerates the breakup.

LAKE ICE

The ice around surface or subsurface influents also melts faster than the rest of the ice cover because of additional heat in the incoming surface water or groundwater. The ice cover may be very thin, or absent entirely, during the whole winter in places where an influent enters a temperate-zone lake. These areas should be determined by ice surveys and avoided in so far as operations with vehicles or aircraft are concerned.

B. Vehicular Traffic and Aircraft Landings on Lake Ice

It is well known that automobiles, trucks, tractors, and even trains can safely negotiate lake ice during the winter months in some parts of the world. In most circumstances where such vehicular traffic occurs, the safety of the ice is a matter of experience rather than of calculated bearing strength.

Military operations on ice during World War II and in Korea pointed up the need for a thorough study of moving loads on lake and sea ice. Two important contributions were made, one by Assur [50] and the other by Wilson [51].

Assur pointed out that the bearing capacity of ice is based on its resistance to bending when loaded. The curvature of bending is dependent on such parameters as Young's modulus, Poisson's ratio, ice thickness, pressure of the water against the ice layer when deflection occurs, and the concentration of the load. It is beyond the purpose of this work to consider all phases of the results obtained by either Assur or Wilson; the reader desiring details should go to the original papers. At least one significant result of Wilson's work should be reported here. It concerns the deflection of lake ice under a moving load.

Wilson investigated the deflection of lake and sea ice produced by moving trucks or landing airplanes and conducted field experiments in order to test a theory on the coupling or resonance produced in the ice when moving loads attained certain critical velocities.

For example, a 6,000-lb weapons carrier traveling on Lake Mille Lacs, Minnesota, produced the deflections given in Table 23-4.

Table 23-4. Ice Deflections Produced by a 6,000-lb Vehicle Traveling at Different Velocities over Ice 2 Ft Thick and a Water Depth of 10 Ft*

Velocity, mph	Maximum deflection, in.
6	0.11
10.5	0.16
16.5	0.25
19	0.09
40	0.04

* From Wilson [51].

The experimental values of Table 23-4 clearly show that a critical velocity is reached around 16 mph under the conditions which existed at the time of the tests in February, 1955. At the critical velocity, the deflection of the ice was about 2.5 times that at slow speeds. Wilson's work indicates that in water less than 30 ft deep, the critical velocity depends only on the thickness of the ice as demonstrated by Tables 23-5 and 23-6 (taken from Assur's paper but based on Wilson's work).

Table 23-5. Critical Velocities of Moving Loads on Lake Ice over Shallow Waters of a Given Depth*

Water depth, ft	Critical velocity, mph†
4	9
6	11
8	12
10	14
15	17
20	19
30	22

* From Wilson [51].
† Approximate values only. Exact values depend on ice thickness.

Table 23-6. Critical Velocities of Moving Loads over Ice on Deep Water*

Feet of ice	Critical velocity, mph	Minimum water depth which can be considered deep, ft
0.5	19	40
1	23	60
2	31	100
3	37	140
4	41	180
6	48	290
8	53	410

* From Wilson [51].

No over-ice operations should be undertaken without preliminary surveys because the ice cover may not be homogeneous everywhere on the lake, and water depths may vary considerably from place to place. Open cracks also lead to additional complications; if they are wider than 0.5 in. and open to the water, they should be avoided.

C. Effect of Lake Ice on Shore Installations

Lake ice may have considerable effect on the lake shore, not only during the spring breakup, when slabs of ice are driven ashore by winds, but also throughout the winter, when thermal expansion of the ice cover produces thrust against the shore and against exposed installations thereon. Studies by Zumberge and Wilson [52] on Wamplers Lake in southern Michigan showed that a temperature rise of 1°F/hr, prolonged over a period of 12 hr on an 8-in. ice sheet, was sufficient to cause pronounced thrust on the shore of a lake 1 mile long and 1½ miles wide. The study also showed that the direction of the thrust was not everywhere orthogonal to the shore because of the elongate shape of the lake.

Thermal ice push attains maximum development as follows. When the ice cover has become more or less isothermal from top to bottom (that is, 32°F at the ice-water interface and near 32°F at the ice-air interface), a drop in air temperature will cause contraction cracks to develop. Water rises in these cracks and freezes, increasing the total mass of the ice cover. When the ice surface is warmed by a subsequent rise in air temperature, the cover expands and exerts a thrust at the ice-shore contact. Docks, retaining walls, and any other type of shore structure will be subject to severe damage when this sequence of temperature changes occurs.

The thrust exerted on dams by expanding ice is of special interest to engineers concerned with the design of such structures [53–55]. As a precautionary measure against ice thrust, the levels of some reservoirs are materially reduced during the winter months; in others, some evaluation of the force of the ice thrust is considered in the design of the dam [56].

VII. HYDROMECHANICS OF LAKES

A. Surface Waves

1. Open-water Waves. Ordinary surface waves on water are the result of movement of water particles. The wave shape progresses visibly, but individual water particles move in vertical circular orbits and make only minor net forward progress. In open water the vertical orbital movements are of maximum size at the surface, but their diameters decrease rapidly with depth and they practically disappear at depths approximating one-half a wavelength.

The simplest water waves are a train of alternate crests and troughs which appear to progress across the water surface. This is the so-called *progressive wave*. In some harbors, over shoals, or at the junctions of currents, there sometimes occur two or more progressive waves running in opposite directions (frequently as a result of the reflection of one progressive wave back through itself). Under these conditions there occurs a *standing wave* whose visible characteristics are wave crests and troughs which succeed each other in the same location, with no visible forward motion.

The simple progressive wave is not commonly well seen in nature. The general condition is that of the simultaneous existence of numerous waves of various wavelengths, heights, and directions—in other words, a wave spectrum. Under these conditions the dominant wave may show its direction and part of its shape, but the typical long crest is disrupted by the other waves, and short isolated waves more or less in line with each other are formed.

Waves originate under the impetus of wind, but the mechanism of origin is incompletely known. The initial stage is a series of ripples whose mode of formation is not clear. Once ripples are present, waves grow under wind pressure on the upwind sides of the crests, aided by reduced atmospheric pressure in the wind eddy behind each crest. Growing waves increase in size and velocity of progress as long as the causative wind is steady or increasing. In this growth they pass from the ripple stage, with its sine-wave outline, to the typical wave shape, which is essentially trochoidal in cross section.

For a given wind velocity, Sverdrup and Munk [57] derive a relationship for the maximum possible wave height in meters regardless of fetch (over-water distance of wind action) or duration of wind,

$$h_m = \frac{0.26 W^2}{g} \qquad (23\text{-}15)$$

where W is wind velocity in m/sec, and g is 9.80 m/sec². Stevenson [58] concluded from observation that under strongest winds the height of the highest waves is proportional to the square root of the fetch,

$$h_m = \tfrac{1}{3} \sqrt{F} \qquad (23\text{-}16)$$

where h_m is wave height in m, and F is the fetch in km.

The largest waves in lakes have a wavelength to wave-height ratio of about 10:1, though the theoretical ratio is 7:1. In most lakes the fetch may be expected to be insufficient for maximum waves to develop. Even in small lakes whitecaps form, but the best evidence is that they are the tops of high waves that are blown into instability before they reach maximum-wave size.

2. Changes in Waves Approaching Shore. In open water where depth exceeds one-half a wavelength, waves whose amplitude (displacement from mean water level, or one-half height) is small compared with wavelength have definite and useful relationships between wavelength L, wave period T, and wave velocity C:

$$C = \frac{L}{T} = \sqrt{\frac{gL}{2\pi}} = \frac{gT}{2\pi} \qquad (23\text{-}17)$$

$$L = CT = \frac{gT^2}{2\pi} = \frac{2\pi C^2}{g} \qquad (23\text{-}18)$$

$$T = \frac{L}{C} = \sqrt{\frac{2\pi L}{g}} = \frac{2\pi C}{g} \qquad (23\text{-}19)$$

When waves travel into shoaling water the wave period remains unchanged but the wave velocity decreases and wavelength decreases. The effect of shoaling begins when the depth d decreases to about half the wavelength. When d/L is less than $\tfrac{1}{2}$, wave velocity is

$$C = \sqrt{gd} \qquad (23\text{-}20)$$

At $d/L = \frac{1}{2}$, the cone-shaped arrangement of decreasing water-particle orbits beneath each wave reaches the bottom. Decreasing depth forces more and more of the lower orbital motions into ellipses and, against the bottom, into straight-line oscillations. Deformation of the orbital motions produces disturbances in the internal-energy distribution of the wave, with the result that wave height increases and the sinusoidal or trochoidal form is replaced by an asymmetry toward shore.

Waves approaching shore break in depths approximately equal to the wave height (theoretically at $d/h = 1.28$). On beaches with slopes between 1:20 and 1:100, only about 10 per cent of the wave energy is lost in frictional effects prior to breaking; about 90 per cent is delivered in the breaking process.

The energy delivered to shore by waves is proportional to the square of their average height. The matter of energy delivery is an essential of shore-erosion studies and of the engineering of marine installations. Both these matters are presumed to be normal components of engineering competence and will not be taken up here.

B. Seiches

1. Surface Seiches. A *seiche* is an oscillation of a water surface and may be of almost any period and height. Ideally, each lake basin has a natural series of longitudinal and transverse seiches, of diminishing periods and amplitudes, which are resonance phenomena of the basin and are determined by the dimensions of the basin. Also present in lakes are forced seiches, whose periods and amplitudes are directly determined by the periodicity and energy of the seiche-producing force. Forced seiches are essentially foreign to the lake basin and, for lack of resonance effects, are damped out quickly after the causative agent ceases.

Wind and barometric pressure are the two most common seiche-producing forces. Wind-produced seiches follow cessation or shift of wind after a period of relatively steady wind direction. Surface water blown against the downwind shore (*wind setup*) is released by the change of wind, and the lake surface undergoes a seesaw oscillation. When observed from a single shore point, this oscillation consists of a series of rises and falls of water level superficially similar to the sea tide; hence the term *wind tide*, which often is applied to the seiches. Wind tide is also applied to wind setup.

The simple seiche may oscillate on one or more nodal lines, but the uninodal seiche is the most common. If the initial wind setup is released by wind shift, instead of wind cessation, energy input in the new wind direction may add a lateral component, and an oscillation rotating around a nodal point may result.

The barometric seiche, or storm surge, is set off by the passage of a sharp barometric high or low across the lake. While such passages are common, there is little seiche produced unless the rate of passage is such as to reinforce one of the natural seiche periods of the basin. In the latter case, the natural seiche is abruptly reinforced and a single high surge (or short series of surges) results. These are the dramatic, and frequently destructive, seiches which come to public attention via newspapers. The seiche which drowned several persons at Chicago on June 26, 1954, was of this kind.

Occasionally a forced seiche is spectacular enough to receive public attention. Many of the lakes of northwestern Europe are recorded as having had pronounced forced seiches in 1755, when Lisbon was devastated by an earthquake. The artificial harbors at Capetown, South Africa, have received much engineering attention as a result of seiches within them which were forced by sea waves at the entrances. In these cases moored ships moved sufficiently to interfere with cargo transfers, and damage to ships occurred.

For most practical work, the periods of a lake's seiches may be calculated with sufficient accuracy from

$$T_n = \frac{1}{n} \frac{2L}{\sqrt{gd}} \tag{23-21}$$

where T_n is the period (in the time units used for g) of the seiche with n number of nodes, L is the mean length of the basin, d is the mean depth, and g is the acceleration of gravity. For precise work Eq. (23-21) is not sufficient; it treats the lake as a rectangular basin of uniform depth. Hutchinson [4] presents a discussion of the more specific formulas needed when shape of the lake basin must be considered in seiche computations.

Theoretically, each shape of lake basin has a typical series of seiche periodicities, but in practice, these vary around the series typical of a rectangular basin:

$$T_1:T_2:T_3:T_4:T_5 = 100:50:33.3:25:20 \tag{23-22}$$

In this formula, T is the period of the seiche, with the number of nodes that are indicated by the subscripts, and the right side of the proportion indicates the ratios of the various periods to each other.

Theoretically, also, each basin shape has a typical series of locations for the nodes of its seiches. For the simple rectangular basin the ideal locations of nodes are at the mid-length for T_1; at $\frac{1}{3}$ and $\frac{2}{3}$ of the length for T_2; at $\frac{1}{4}$, $\frac{1}{2}$, and $\frac{3}{4}$ of L for T_3; at each $\frac{1}{5}$ of L for T_4; and at each $\frac{1}{6}$ of L for T_5. In practice, however, the typical series of node locations is of little value, for shoal areas and constricted areas attract nodes while deep areas appear to repel them.

2. Internal Seiches. Not only can a lake oscillate as a whole (considered to be the case in surface seiches), but if it is stratified, the layers of various densities can oscillate relative to one another. The limited data indicate that very small density differences are sufficient to produce layers in which internal seiches can occur. In practical lake work the most important internal seiche is that involving oscillation of the thermocline, and a two-layer theory is usually adequate.

The internal seiche of a stratified lake appears to be activated by the same factors that produce the surface seiche. When wind setup occurs at one end of the lake, the thermocline at that end is depressed, and throughout the lake the thermocline is raised or depressed oppositely to the elevations of the surface. When the causative force releases and the surface seiche begins to operate, the internal seiche is also set in motion. Being confined between overlying water and the bottom, the internal seiche is subject to greater frictional effects than is the surface seiche and has a considerably longer period. In addition to its longer period, the internal seiche has a much greater amplitude than the surface seiche. This may be a frequency-amplitude relationship similar to that in a harmonic oscillator under constant-energy conditions.

The period of the internal seiche for the simple case of a rectangular basin of length L with two layers of thickness d_e and d_h and densities of ρ_e and ρ_h is given by

$$T_i = \frac{2L}{\sqrt{\dfrac{g(\rho_h - \rho_e)}{1/d_h + 1/d_e}}} \tag{23-23}$$

where the subscripts e and h refer to the epilimnion and hypolimnion (defined under Thermal Stratification in Subsec. IV).

The primary effect of an internal seiche is movement of the confined waters of the hypolimnion. This movement of these colder waters frequently brings water of unusually low temperature to water intakes.

The reader is referred to Hutchinson [4] for further discussion and for contact with the literature on seiches.

C. Currents

Since lakes almost always have inlets and outlets, there commonly is a current-producing tendency resulting from flow-through; this is the so-called *gradient*, or *slope current*. However, in most lakes, the width of the basin is materially greater than that of the inlets and outlets, and the gradient current is negligible in comparison with currents produced by external forces such as wind.

Lake currents are separable into *periodic* and *nonperiodic* currents. Periodic currents are water flows associated in various ways with seiches. Because such currents are usually reversing, they may have important local ventilative or mixing effects and may cause reversals of flow in tributary streams, but they do not usually play clearly discernible parts in the lake-wide patterns of water movement by which influent tributary waters pass through the lake, or by which dissolved or suspended materials become widely distributed. The latter are results of the nonperiodic currents, which are irregularly variable in direction and velocity but have a time-mean tendency toward unidirectionality that is related to the similar time-mean directionality of the local prevailing wind.

Nonperiodic currents form as the result of transfer of wind-stress energy into the surface water. While there is lack of agreement as to details, there is general unity in the belief that stress-energy input varies with some power of wind velocity. Most investigations have put this power relationship as dependent upon exponents ranging from 1.8 to 3; the majority of investigators incline to the W^2 relationship. A commonly used expression for wind stress is

$$T = 3.2 \times 10^{-6} W^2 \qquad (23\text{-}24)$$

where T is the wind stress in g/cm/sec^2, and W is the wind velocity in cm/sec. Hutchinson [4] reviews the several methods of stress computation.

Stress energy entering the water produces acceleration of the water particles. Part of the acceleration produces waves, and part results in water current. Surface current produced by wind is usually found to have velocities between 1 and 3 per cent of the wind velocity, and 2 per cent is commonly used.

Wind-produced surface currents flow in directions essentially downwind, but they also have components of direction due to the rotation of the earth. Theoretically, such currents in open deep lakes of the Northern Hemisphere should move at about 45° to the right of wind movement. Theoretically, this right deflection should decrease with decreasing depth of water, and also with decreasing latitude, but the few studies in shallow water indicate that consideration of the Coriolis deflection is necessary to careful work in such environments. For many, but not all, practical purposes the Coriolis right deflection in shallow waters is small enough to be neglected.

Residual current effects of previous winds and windage effects on current-measuring devices are the primary reasons for the common failure to observe the Coriolis deflection in shallow water. The most accurate current tracers are discolorations of the water produced by dyes, water-base paints, clay, and the like.

1. Alongshore, or Littoral, Currents. For convenience, the wind-driven surface currents of a lake can be divided into alongshore and open-lake currents, the division being made at that distance from shore where effects of local shore and bottom topography become unimportant.

The alongshore currents are primarily the result of the interaction of local topography and local winds; the effects of the earth's rotation are demonstrable only in occasional special aspects of the alongshore currents. Because each local situation has its own topographic and wind peculiarities, it is imperative that their relationship to alongshore currents be studied when problems involve these currents. Only a few generalized suggestions relative to alongshore currents can be made in a discussion such as this.

Barring exceptional topographic conditions, the direction of the alongshore current will probably be downwind so long as the wind is within 45° of being parallel to the shore. For most practical purposes, winds blowing from off the land set up negligible alongshore currents. It is with winds blowing onto shore from the offshore quadrant that shore installations encounter their most severe problems. It is also under these winds that effects of the Coriolis parameter are apt to be demonstrable in alongshore currents.

Onshore winds tend to result in wind setup against the shore (with the possibility of problems from high water level). Onshore winds bring waves of maximum size and potential destructiveness, whose action reaches farther landward as water level rises with setup. Onshore winds, traveling over deep water before coming to shore, are most apt to have developed a visible Coriolis deflection in the currents they produce.

If there are no unusual topographic factors, a developed Coriolis component in the currents will show up in unexpected relationships between wind direction, setup, and direction of alongshore current.

On a west shore running north-south, onshore winds would be from the quadrant between northeast and southeast. If a Coriolis deflection of about 40° right is present in the wind-driven water, a wind from the east would move water shoreward in about a west-northwest direction, and setup against shore probably would occur. Similarly, a wind from the northeast would move water almost directly onto shore, with establishment of setup. A southeast wind, however, would move water in an almost north direction nearly parallel to shore, and setup would not be expected.

In this example, a southeast wind blowing onto shore would be expected to produce northward alongshore current. Winds from northeast and east, however, might be expected to produce setup against the shore. On the sloping face of the setup, *southward* alongshore current might be expected as a secondary current running parallel to the face of the slope in balance between Coriolis force directed up the slope and the downslope component of gravity. Similar situations have long been known in the coastal currents of the oceans [59, p. 501].

2. Open-lake Currents. In open waters away from shore, the energy of wind stress is apparently transferred downward into the surface water through a series of alternately clockwise and counterclockwise horizontal helices of water-particle motions. The helices have been observed and studied, but their relationship to wind-driven currents is as yet unclear.

The actual transformation of wind energy into water current can vary greatly, depending upon the depth, conformation, and size of the lake. Possibly simplest is the case of the small circular shallow lake, whose current pattern is usually expected to be a central downwind movement with associated upwind return currents around the sides of the lake.

In lakes of moderate-to-large size and reasonable depth, the wind-induced current pattern usually takes the form of a main current accompanied by interlocked large and small eddies whose precise pattern is determined by the interaction of flow-through, earth-rotation effects, and local incidents of topography and hydrodynamics.

In this size range of lakes there is a tendency for the field of water density to come into equilibrium with the wind-forced current. In the equilibrium condition in the Northern Hemisphere, stability can be achieved only when the less dense water lies on the current's right and the more dense water on the current's left. The current runs on the sloping surface between more dense and less dense waters; it runs parallel to the contours of the slope, and is held on the slope by being in balance between Coriolis force (directed up the slope) and the downslope component of gravity.

The primary relationship of wind and current is that in which the current and the axis of the slope of the water surface are both oriented to the right of the wind movement. Secondary relationships of wind and current occur in the presence of physical or hydrodynamic barriers, against which less dense water involved in the initial right-deflected transport can pile up. On the sloping face of the pile-up (setup, when a shore is the barrier) a secondary current develops and flows, in geostrophic balance betweeen Coriolis force and the downslope component of gravity, with the less dense water on its right, and in a direction determined by the orientation of the barrier. Shores, eddy centers, and the sides of other currents are the common barriers.

It may be expected that in every lake there is some maximum fetch of open water where the stress-energy input of the prevailing wind is greatest. A primary relationship of water current to wind may be expected to develop in this part of the lake, the remainder of the current pattern consisting of secondary currents and of interrelated eddies, which for the most part rotate like intermeshed gears. To winds from directions other than the prevailing, maximal fetches oriented in other directions are available, and the current pattern in a lake may be expected to be different for each wind direction (provided that the wind is of sufficient strength and duration to develop it).

Currents have momentum, and upon the cessation of a wind, the current pattern will persist (with losses due to friction) until another wind enforces another current pattern. Effects of momentum bear heavily upon the events that follow a change of wind.

The simplest case is that of a wind which, having ceased, begins to blow again from the same direction as before. The energy input of the new wind would go into velocity accelerations of the persisting current. When wind shift occurs, the new energy input produces accelerations which modify the existing currents, and a new current pattern in equilibrium with the new wind cannot be developed until modification of the old current pattern is complete.

Some of the relationships between new wind energy and old current momentum are qualitatively evident. The time lag between application of the new stress and modification of the currents appears to be inversely dependent upon the ratio of new stress magnitude to existing current momentum and directly dependent upon the direction difference (up to 180°) between the new transport direction and the existing current direction. When less water moves, there is less momentum, and new currents form more quickly; in water 1 to 6 ft deep new currents will develop in less than 2 hr after a wind shift, as compared with about a day's lag in the deep waters of one of the Great Lakes.

In a lake of moderate or large size the current pattern present at a given time is predominantly that produced by the last wind. In Lakes Michigan and Huron, Ayers et al. [60] obtained evidence that energy from the winds of the preceding 10 days was present in exponentially diminishing amounts, though wind of the preceding day was the dominant factor in the observed currents.

VIII. SWAMPS AND BOGS

A. Origin and Definitions

1. Swamps and Bogs Differentiated. The terms *swamp* and *bog* are commonly used synonymously, but in a strict ecological sense, as well as in their physical and chemical characteristics, they are two different phenomena. If for any practical reason a hydrologist is confronted with either of these natural dynamic systems, it will be to his advantage to know something about their contrasting characteristics.

Generally speaking, it might be said that a bog is the end stage in the life history of a lake (though not all lakes end as bogs). Its chief distinctive characteristic is a floating vegetal mat attached to the shore [61]. A swamp, on the other hand, is a vegetated land area saturated with water. Dansereau and Segadis-Vianna [62] summarized the distinction between a bog and a fresh-water swamp by describing the former as a large cushion (a mantle of vegetation high in mosses and other spongy plants) and the latter as a wet prairie (characterized by a grassy or prairielike vegetation). The same authors consider the terms swamp and marsh to be synonymous, a usage adopted in this section.

a. Bogs. True bogs occur most commonly in the areas where lake basins produced by Pleistocene glaciation are abundant, although they are known from lower latitudes. The regions dominated by a cold climate and constantly high relative humidity, such as the glaciated regions of northern Europe and North America, contain abundant bogs. In tropical regions, bogs occur only in mountainous areas, where climatic conditions of the more temperate latitudes are duplicated.

A typical bog contains a floating mat of vegetation which may cover the entire water surface in late stages of development. The mat may be several feet thick; when it is about 2 ft thick it can hold the weight of a man. Peat, an organic accumulation of fibrous or woody material, is the organic sediment resulting from the dead and partially decayed vegetation of the bog mat. A boring through a peat bed will penetrate woody peat near the surface and will encounter the fibrous variety consisting of remains of rushes, sedge, and sphagnum moss plant at depth.

The presence of a continuous bog mat on a lake prohibits the deposition of mineral sediments, so that, characteristically, the bottom of a bog lake consists wholly of organic material, which eventually becomes peat. Much of the organic sediment is in colloidal form and remains in suspension for a long time. The surface of the lake bottom is not sharply defined, but grades from a firm bottom upward to a colloidal sludge, or muck, commonly known as a *false bottom*. Because of the false-bottom condition in bog lakes, depths measured by ordinary sounding line will show considerable variation at the same place when made by different individuals.

Chemically, bog lakes are generally strongly acid, with a low percentage of oxygen saturation, a condition which explains the scarcity of aquatic animal life in bogs [62]. Plants growing on the mat increase the water loss from the bog lake because of their water loss through transpiration. (For a fuller treatment of bog vegetation, see Dansereau and Segadis-Vianna [62].)

  b. *Swamps or Marshes.* Whereas a bog represents the end stage in the life cycle of a lake, swamps need not have any relationship to lake basins, although many commonly do. Aside from lake-basin origins, swamps originate on floodplains of rivers, on deltas at mouths of rivers entering fresh-water lakes, along lake margins, or in the vicinity of springs. Salt-water marshes, or swamps, develop under a wide variety of conditions ranging from those of a lagoonal environment along the Atlantic Coast of North America [63–65] to the mangrove swamps of southern Florida along the Gulf Coast [66]. They are extensive in marine deltas such as the delta of the Mississippi River and are found also in association with tidal flats.

Because of their wide geographic distribution and different ecological and physiographic environment, no consistent general statements may be made which apply to all kinds of swamps, other than that they all are dominated by water and heavy vegetation. The subject of salt marshes is so complex and the literature so voluminous as to make a full presentation in this section impossible. Fresh-water swamps, too, have been studied extensively, and because of their frequent confusion with bogs, it may be profitable to dwell briefly on some of the basic features of swamps.

Swamps lack the characteristic floating mat of vegetation which makes them different from bogs. Swamp sediments, although high in organic content, may also contain a considerable amount of inorganic material. Some floodplain swamps receive a heavy influx of mineral sediment each time the main stream overflows its banks. Swamps do not generally contain false bottoms, although, where finely disseminated organic sediment and oozes predominate, the swamp bottom may be very soft and possess very low bearing strength.

Fresh-water swamps in the glaciated regions of North America exhibit a systematic zonation of vegetation from the swamp "shore" toward the wetter parts of the swamp. Swamps marginal to shallow lakes ideally contain gradational zones of different plant associations, each of which reflects the prevailing local moisture conditions. The various vegetal zones are known collectively as a *succession* [67]. The succession represents a series of plants invading a new habitat. The initial plants that gain a foothold on the shallow lake waters are the *pioneers*, which are in turn followed by species with less tolerance for moisture than the pioneers. A plant succession in which the moisture conditions are improved by the pioneer species so that the more water-intolerant species can follow is a *hydrarch succession* [68].

A *bog mat* is a special kind of hydrarch succession, but in both swamps and bogs the invasion of an aquatic environment by successive plant communities, each with less aquatic affinities than the one next adjacent to lakeward, is a basic tenet of plant ecologists.

IX. REFERENCES

1. Zumberge, J. H.: The lakes of Minnesota: their origin and classification, *Minn. Geol. Surv. Bull.* 35, 1952.
2. Welch, P. S.: "Limnology," 2d ed., McGraw-Hill Book Company, Inc., New York, 1952.
3. Davis, W. M.: On the classification of lake basins, *Proc. Boston Soc. Nat. Hist.*, vol. 21, pp. 315–381, 1882.
4. Hutchinson, G. E.: "A Treatise on Limnology," John Wiley & Sons, Inc., New York, 1957.
5. Russell, I. C.: "Lakes of North America," Ginn and Company, Boston, 1895.
6. Williams, H.: Calderas and their origin, *Univ. Calif. Dept. Geol. Bull.*, vol. 25, pp. 239–346, 1941.
7. Williams, H.: The geology of Crater Lake, Oregon, *Carnegie Inst. Wash. Bull.*, vol. 540, 1942.

8. Hough, J. L.: "Geology of the Great Lakes," University of Illinois Press, Urbana, Ill., 1958.
9. Zumberge, J. H.: Glacial erosion in tilted rock layers, *J. Geol.*, vol. 63, pp. 149–158, 1955.
10. Wallace, R. E.: Cave-in lakes in the Nabesna, Chisana, and Tanana River Valleys, Eastern Alaska, *J. Geol.*, vol. 56, pp. 171–181, 1948.
11. Cooke, C. W.: Scenery of Florida, *Florida Geol. Surv. Bull.*, vol. 17, 1939.
12. Bretz, J H.: The Grand Coulee, *Am. Geograph. Soc. Spec. Publ.*, 15, 1932.
13. Judson, S.: Depressions of the northern portion of the southern high plains of New Mexico, *Bull. Geol. Soc. Am.*, vol. 61, pp. 253–274, 1950.
14. Russell, I. C.: Geological reconnaissance in central Washington, *U.S. Geol. Surv. Bull.* 108, 1893.
15. Johnson, D. W.: "Shore Processes and Shoreline Development," John Wiley & Sons, Inc., New York, 1919.
16. Murray, Sir J.: The characteristics of lakes in general, and their distribution over the surface of the globe, in J. Murray and L. Puller (eds.), "Bathymetric Survey of the Scottish Fresh-water Lochs," Challenger Office, Edinburgh, 1910.
17. Hopkins, D. M.: Thaw lakes and sinks in the Imuruk Lake area, Seward Peninsula, Alaska, *J. Geol.*, vol. 57, pp. 119–131, 1949.
18. Wentworth, C. K.: Geology of the Pacific equatorial islands, *Occas. Papers Bishop Mus.*, vol. 8, no. 15, 1931.
19. Dixon, W. A.: Notes on the meteorology and natural history of a guano island, *J. Roy. Soc. New South Wales*, vol. 11, pp. 165–175, 1878.
20. Drower, H. S.: Water supply, irrigation and agriculture, in M. S. Singer, E. J. Holmyard, and A. R. Hall (eds.), "A History of Technology," Oxford University Press, London, 1954.
21. Mills, E. A.: "In Beaver World," Houghton Mifflin Company, Boston, 1913.
22. Meen, V. B.: Chubb Crater, Ungava, Quebec, *Bull. Geol. Soc. Am.*, vol. 61, p. 1485, 1950.
23. Cooke, C. W.: Elliptical bays in South Carolina and the shape of eddies, *J. Geol.*, vol. 48, pp. 205–211, 1940.
24. Johnson, D. W.: "Origin of the Carolina Bays," Columbia University Press, New York, 1942.
25. Johnson, D. W.: Mysterious craters of the Carolina coast, *Am. Scientist*, vol. 32, pp. 1–22, 1944.
26. Prouty, W. F.: Carolina bays and their origin, *Bull. Geol. Soc. Am.*, vol. 63, pp. 167–224, 1952.
27. Frey, D. G.: Stages in the ontogeny of the Carolina bays, *Verhandl. Intern. Ver. Limnol.*, vol. 12, pp. 660–668, 1955.
28. Welch, P. S.: "Limnological Methods," McGraw-Hill Book Company, Inc., New York, 1948.
29. Birge, E. A., and Chancey Juday: The inland lakes of Wisconsin: the hydrography and morphometry of the lakes, *Wisconsin Geol. Nat. Hist. Surv. Bull.* 27, *Sci. Ser.*, no. 9, Madison, Wis., 1914.
30. Penck, Albrecht: Morphometrie des Bodensees, *Jahresber, Geograph. Gesell. München*, 1894.
31. Penck, Albrecht: "Morphologie der Erdoberfläche," 2 vols., Engelhorn, Stuttgart, 1894.
32. Halbfass, W.: Morphometrie des Genfer Sees, *Z. Gesell. Erdkunde zu Berlin*, vol. 32, 1897.
33. Gravelius, Harry: *Z. Gewasserk.*, vol. 9, 1910. (Reference in Birge and Juday [29].)
34. Neumann, J.: Maximum depth and average depth of lakes, *J. Fisheries Res. Board Can.*, vol. 16, pp. 923–927, 1959.
35. Benton, G. S., R. T. Blackburn, and V. O. Snead: The role of the atmosphere in the hydrologic cycle, *Trans. Am. Geophys. Union*, vol. 31, pp. 61–73, 1950.
36. Stanley, G. M.: The submerged valley through Mackinac Straits, *J. Geol.*, vol. 46, pp. 966–974, 1938.
37. Powers, C. F., and J. C. Ayers: Water transport studies in the straits of Mackinac region of Lake Huron, *Limnol. Oceanogr.*, vol. 5, pp. 81–85, 1960.
38. Hoad, W. C.: High water in the Great Lakes, *Mich. Alumnus Quart. Rev.*, vol. 59, no. 10, pp. 37–45, 1952.
39. McDonald, W. E.: Variations in Great Lake levels in relation to engineering problems, in *Proc. Fourth Conf. on Coastal Eng.*, Council on Wave Research, Berkeley, Calif., 1954.
40. Pierce, D. M., and F. E. Vogt: Method for predicting Michigan-Huron lake level fluctuations, *J. Am. Water Works Assoc.*, vol. 45, pp. 502–520, 1953.
41. Brunk, I. W.: Precipitation and the levels of Lakes Michigan and Huron, *J. Geophys. Res.*, vol. 64, pp. 1591–1595, 1959.

REFERENCES

42. Henderson, L. J.: "The Fitness of the Environment," The Macmillan Company, New York, 1913.
43. Whewell, W.: "Astronomy and General Physics Considered with Reference to Natural Theology: Bridgewater Treatises on the Power, Wisdom, and Goodness of God as Manifested in the Creation, Treatise III," Carey, Lea, & Blanchard, Philadelphia, 1833.
44. Prout, W.: "On Chemistry, Meteorology and the Function of Digestion, Considered with Reference to Natural Theology: Bridgewater Treatises on the Power, Wisdom, and Goodness of God as Manifested in the Creation, Treatise VIII," Carey, Lea, & Blanchard, Philadelphia, 1834.
45. Röntgen, W. C.: Über die Constitution des flüssigen Wassers, *Ann. Phys. Leipzig*, new series, vol. 45, pp. 91–97, 1892.
46. Dorsey, N. E.: "Properties of Ordinary Water-substance in All Its Phases: Water-vapor, Water, and All the Ices," American Chemical Society Monograph Series, Reinhold Publishing Corp., New York, 1940.
47. Church, P. E.: The annual temperature cycle of Lake Michigan. I. Cooling from late autumn to the terminal point, 1941–42, *Univ. Chicago, Inst. Meteorol., Misc. Repts.*, no. 4, Chicago, 1942.
48. U.S. Geological Survey: Water-loss investigations: Lake Hefner studies, technical report, *U.S. Geol. Surv. Profess. Paper* 269, 1954.
49. Wilson, J. E., J. H. Zumberge, and E. W. Marshall: A study of ice on an inland lake, *U.S. Army Corps Engrs., Snow, Ice, Permafrost Res. Estab., Rept.* 5, pt. 1, Wilmette, Ill., 1954.
50. Assur, A.: Airfields on floating ice sheets for routine and emergency operations, *U.S. Army Corps Engrs., Snow, Ice, Permafrost Res. Estab., Rept.* 36 (prelim. ed.), Wilmette, Ill., 1956.
51. Wilson, J. T.: Coupling between moving loads and flexural waves in floating ice sheets, *U.S. Army Corps Engrs., Snow, Ice, Permafrost Res. Estab., Rept.* 34, Wilmette, Ill., 1955.
52. Zumberge, J. H., and J. T. Wilson: Quantitative studies on thermal expansion and contraction of lake ice, *J. Geol.*, vol. 61, pp. 374–383, 1953.
53. Brown, Ernest, B. R. McGrath, and R. E. Kennedy: Discussion of Thrust exerted by expanding ice sheet by Edwin Rose, *Trans. Am. Soc. Civil Eng.*, vol. 112, pp. 886–895, 1947.
54. Rose, Edwin: Thrust exerted by expanding ice sheet, *Trans. Am. Soc. Civil Eng.*, vol. 112, pp. 871–885, 1947.
55. Hill, H. M.: Field measurements of ice pressure at Hastings locks and dam, *Military Engr.*, pp. 119–122, March–April, 1935.
56. Brown, Ernest, and G. C. Clark: Ice thrust in connection with hydro-electric plant design, *Eng. J.*, pp. 18–25, January, 1932.
57. Sverdrup, H. U., and W. H. Munk: Wind, sea and swell: theory of relations for forecasting, *U.S. Navy Hydrographic Office Publ.* 601, 1947.
58. Stevenson, Thomas: Observations on the relation between the height of waves and their distance from the windward shore, *Edinburgh New (Jameson's) Phil. J.*, vol. 53, pp. 358–359, 1852.
59. Sverdrup, H. U., M. W. Johnson, and R. H. Fleming: "The Oceans," Prentice-Hall, Inc., Englewood Cliffs, N.J., 1942.
60. Ayers, J. C., D. C. Chandler, G. H. Lauff, C. F. Powers, and E. B. Henson: Currents and water masses of Lake Michigan, *Great Lakes Res. Inst. Publ.* 3, University of Michigan, Ann Arbor, Mich., 1958.
61. Weaver, J. E., and F. E. Clements: "Plant Ecology," McGraw-Hill Book Company, Inc., New York, 1938.
62. Dansereau, P., and F. Segadas-Vianna: Ecological study of the peat bogs of eastern North America, *Can. J. Botany*, vol. 30, pp. 490–520, 1952.
63. Shaler, N. S.: Beaches and tidal marshes of the Atlantic coast, *Natl. Geograph. Monograph* 5, 1895.
64. Johnson, D. W.: "The New England–Acadian Shore Line," John Wiley & Sons, Inc., New York, 1925.
65. Steers, J. A., and H. D. Thomas: Vegetation and sedimentation as illustrated in the region of the Norfolk salt marshes, *Proc. Geologists' Assoc. (Engl.)*, vol. 40, pp. 341–352, 1930.
66. Vaughan, T. W.: The geological work on mangroves in southern Florida, *Smithsonian Inst. Misc. Collections*, vol. 52, pp. 461–466, 1909.
67. Oosting, H. J.: "The Study of Plant Communities," W. H. Freeman and Company, San Francisco, 1948.
68. Cooper, W. S.: The climax forest of Isle Royale, Lake Superior and its development, *Botan. Gazette* vol. 55, pp. 1–44, 1913.

Section 24

HYDROLOGY OF ARID AND SEMIARID REGIONS

R. O. SLATYER, *Climatologist, Division of Land Research and Regional Survey, Commonwealth Scientific and Industrial Research Organization, Australia.*

J. A. MABBUTT, *Geomorphologist, Division of Land Research and Regional Survey, Commonwealth Scientific and Industrial Research Organization, Australia.*

I. Introduction... 24-2
   A. Scope... 24-2
   B. Characteristics of Arid and Semiarid Regions............ 24-2
   C. World Distribution of Arid and Semiarid Climates........ 24-8
II. Precipitation in Arid and Semiarid Regions................ 24-9
   A. General Characteristics............................... 24-9
   B. Long-term Fluctuations............................... 24-10
   C. Geographical Variability.............................. 24-12
   D. Significance of Dew.................................. 24-13
   E. Modification of Precipitation.......................... 24-15
      1. Cloud Seeding................................... 24-15
      2. Interception of Precipitation by Vegetation........ 24-15
III. Evaporation and Transpiration in Arid and Semiarid Regions... 24-16
   A. Evaporation from Open-water Surfaces.................. 24-16
   B. Evaporation from Bare Soils........................... 24-17
   C. Evapotranspiration from Arid-zone Plant Communities.... 24-18
   D. Modification of Evapotranspiration..................... 24-20
IV. Surface Drainage in Arid and Semiarid Regions.............. 24-22
   A. General Drainage Characteristics....................... 24-22
   B. Weathering Processes and Landforms................... 24-22
      1. Weathering Processes and Vegetative Cover........ 24-22
      2. Desert Landforms................................ 24-23
   C. Runoff.. 24-23
   D. Streamflow.. 24-24
      1. Minor Channels.................................. 24-24
      2. Larger Channels................................. 24-24
   E. Sheet Flow.. 24-25
   F. Desert Lakes.. 24-26
      1. Tectonic Lakes.................................. 24-26
      2. Lakes Formed by the Disorganization of Drainage.... 24-26

The authors wish to acknowledge the assistance of Mr. C. S. Christian and Dr. T. G. Chapman, both of CSIRO, in the general compilation and presentation.

        3. Lakes at Drainage Terminals.......................... 24-26
        4. Lakes Formed in Uneven Surfaces of Aeolian Deposition.. 24-27
        5. Deflation Lakes.. 24-27
        6. Water Holes in Drainage Channels..................... 24-27
    G. Modification of Surface Drainage.......................... 24-27
V. Groundwater in Arid and Semiarid Regions................... 24-27
    A. General Considerations..................................... 24-27
    B. Recharge... 24-29
    C. Phreatic, or Unconfined, Groundwater...................... 24-30
        1. Rock-floored Alluvial Basins in River Valleys........... 24-30
        2. Stratigraphic Traps in Which Groundwater Accumulation Is Primarily Conditioned by the Form of Deposition.... 24-30
        3. Traps Due to Intrusive and Volcanic Rocks............ 24-31
        4. Structural Traps.. 24-31
    D. Artesian Water... 24-32
    E. Modification of Groundwater Storage..................... 24-32
VI. Irrigation in Arid Regions—Some Special Features............ 24-34
    A. Some Adverse Effects of Arid Climatic Conditions on Physiological Responses of Plants............................. 24-34
        1. Relation between Plant Growth and Water Requirements. 24-34
        2. The Influence of Local Advection...................... 24-36
    B. Salinity Problems in Irrigation Agriculture................. 24-39
        1. Origin of Saline and Alkali Soils....................... 24-39
        2. Characteristics of Saline and Alkali Soils............... 24-39
        3. Salinity Effects on Plants.............................. 24-41
VII. References... 24-42

I. INTRODUCTION

A. Scope

Fundamental aspects of hydrology have previously been considered in sections devoted to the various aspects of the hydrologic cycle. This section is specifically concerned with the manner in which the various components of the hydrologic cycle are of significance in the arid and semiarid regions. Accordingly, the subject matter is devoted, first, to definition and description of arid and semiarid zones, second, to the special features of the arid hydrologic cycle; and finally, to some features and problems of irrigation in arid regions.

The water available for use in arid regions falls into two main categories: first, that which is regional in the sense that it is derived from the precipitation actually occurring within the region and which may reappear as streamflow or groundwater; and second, that which is derived exogenously from more humid areas and which enters the arid zone as streamflow, such as the perennial streams which traverse many of the major arid zones of the world or as extensive groundwater reserves which are charged from aquifers that extend beyond the arid region.

In the first case, some irrigation developments may occur, but the areas as a whole retain an arid-zone economic basis, and the irrigation economy is linked closely with the arid-type productivity characteristic of the region. In the second case, utilization of the available water may render the area no longer arid in the sense that its economic basis is changed to that of a high-production non-water-limiting area.

These two instances refer to extremes, and in reality there is a wide range of intermediate situations. However, for the purposes of this section, emphasis is placed on the first, since it is characteristic of what might be termed the *natural* arid zones.

B. Characteristics of Arid and Semiarid Regions

Arid means dry, or parched, and the primary determinant of aridity in most areas is lack of rainfall. In consequence, most arid regions can be accurately delineated by

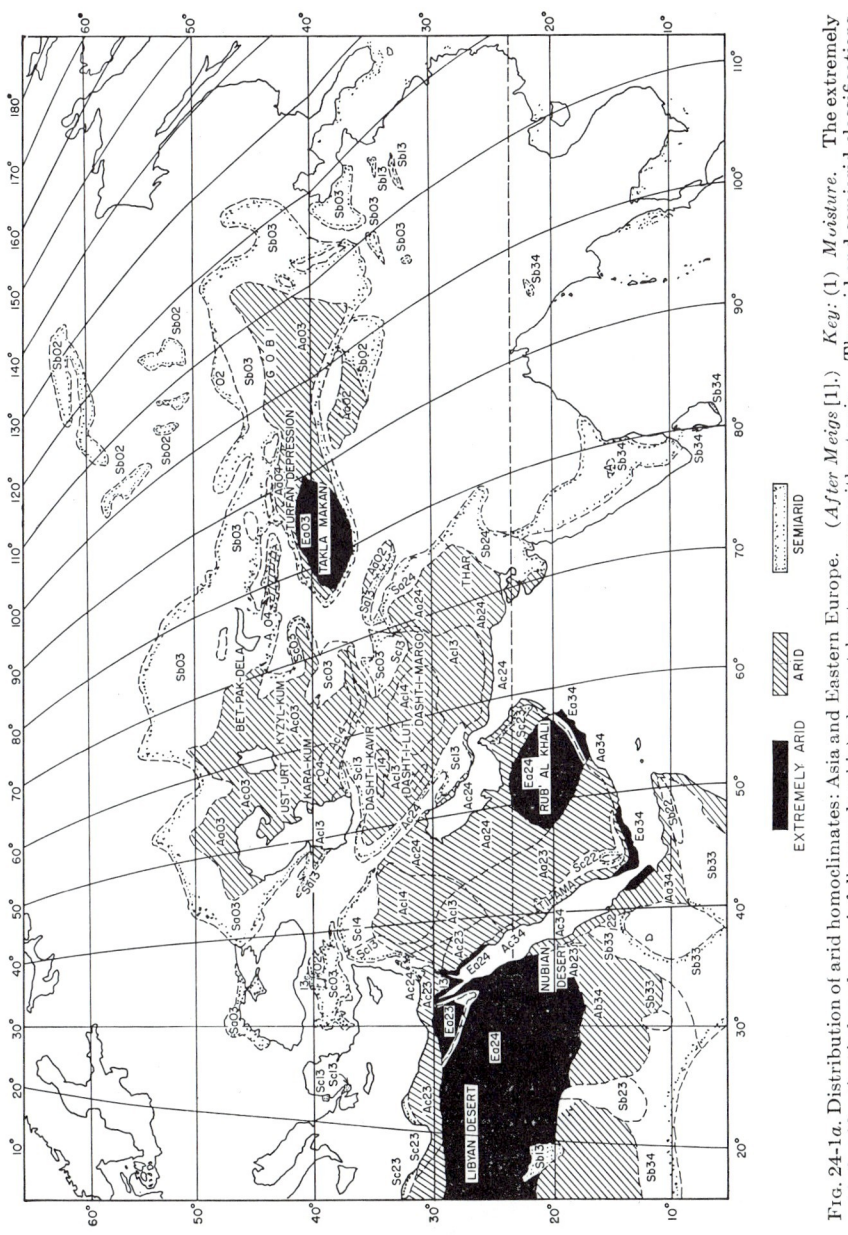

FIG. 24-1a. Distribution of arid homoclimates: Asia and Eastern Europe. (*After Meigs* [1].) *Key:* (1) *Moisture.* The extremely arid classification is based on rainfall records which show at least one year without rain. The arid and semiarid classifications are based on the deficit of precipitation in relation to potential evapotranspiration, using the index described by Thornthwaite in 1948: E, extremely arid; A, arid; S, semiarid. (2) *Season of precipitation.* a, no distinct season; b, summer precipitation; c, winter precipitation. (3) *Temperature.* In the climatic classification symbol used on the map, for example, Sb24, the first digit represents the coldest month and the second digit the warmest month based on mean monthly temperatures as follows: 0, below 0°C; 1, 0 to 10°; 2, 10 to 20°; 3, 20 to 30°; 4, above 30°. Thus hot is represented by (24, 33, 34); mild, by (22, 23); cool winter, by (12, 13, 14); and cold winter, by (02, 03, 04).

24-4 HYDROLOGY OF ARID AND SEMIARID REGIONS

a knowledge of climatic characteristics. However, other features of the land surface may also be used to delimit arid zones, since the geomorphology, soils, and vegetation have their own distinctive characteristics.

Of the many classifications based on climate, the most recent, by Meigs [1], has had wide international acceptance and has been recognized by the World Meteorological

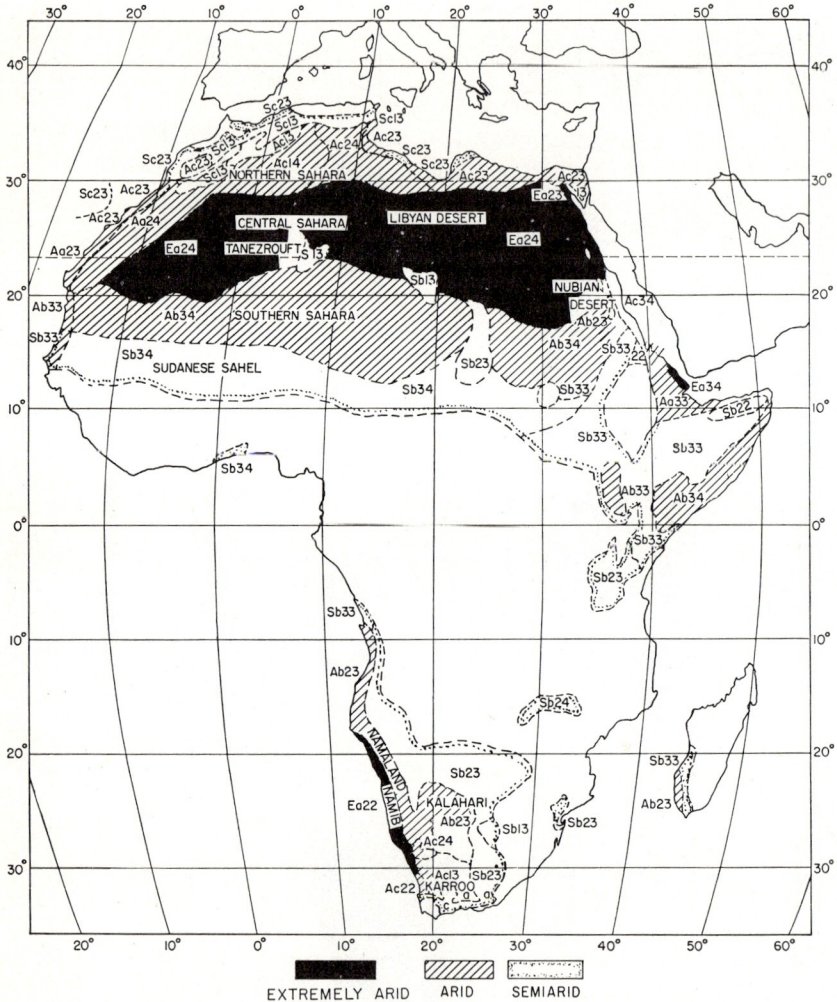

FIG. 24-1b. Distribution of arid homoclimates: Africa and Southern Europe. (*After Meigs* [1].) *Key:* see Fig. 24-1a.

Organization as providing valid comparative data for the arid and semiarid regions of different continents, even though its detailed application and interpretation must necessarily be limited. His maps prepared for UNESCO are reproduced in five parts in Fig. 24-1a to e.

Meigs defined the arid areas as those in which the rainfall on a given piece of land is

INTRODUCTION

not adequate for regular crop production, and the semiarid areas as those in which the rainfall is sufficient for short-season crops and where grass is an important element of the natural vegetation. The basis for the division was the system developed by Thornthwaite [2], which used an index based upon the adequacy of precipitation in relation to the needs of plants. Where the precipitation analyzed month by month is just adequate to supply all the water that would be needed for maximum evaporation and transpiration in the course of a year, the *moisture index* is considered to be 0. Climates with an index between 0 and -20 are regarded by Thornthwaite as *subhumid;* between -20 and -40, *semiarid;* below -40, *arid*. The potential evapotranspiration and the amount of water deficiency or surplus are worked out for each month with the

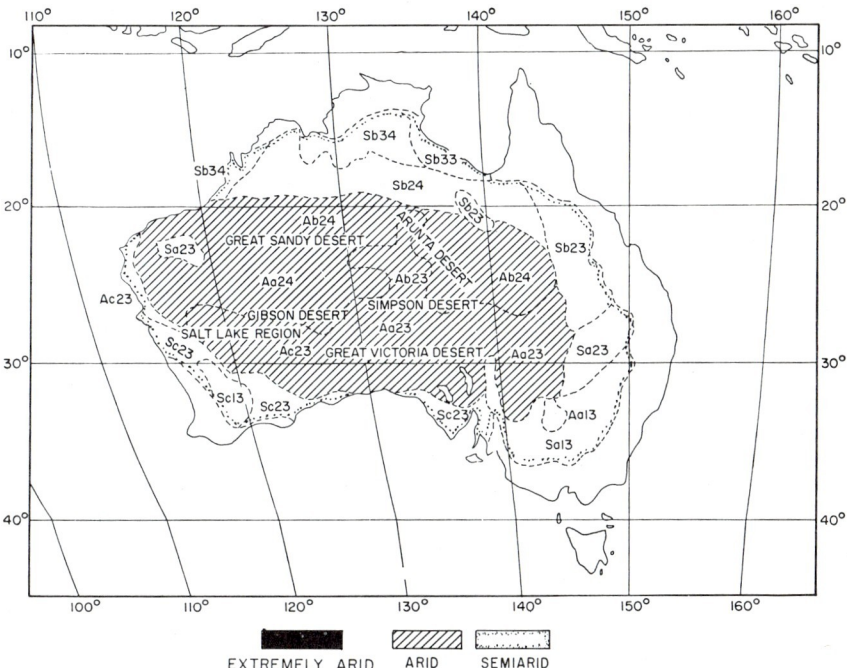

Fig. 24-1c. Distribution of arid homoclimates: Australia and Southern Europe. (*After* Meigs [1].) *Key:* see Fig. 24-1a.

aid of a series of tables and nomograms, involving the use of precipitation and temperature data, with adjustments for length of month, length of day in relation to latitude and season, and water-holding capacity of an average soil. In Fig. 24-1 the arid regions are represented by the symbol A and the semiarid by S.

A third symbol, E, is used to represent *extremely arid* regions. This is mapped on the basis of the definition used by L. Emberger [1], which defines the *extreme desert* as an area in which there is not a regular seasonal rhythm of rainfall and at least 12 consecutive months without rainfall have been recorded.

The season of precipitation is indicated on the maps by the use of the letters a for no distinct seasonality of precipitation, b for summer concentration, and c for winter concentration, and four temperature divisions have been introduced by Meigs to make the climatically defined regions more valid.

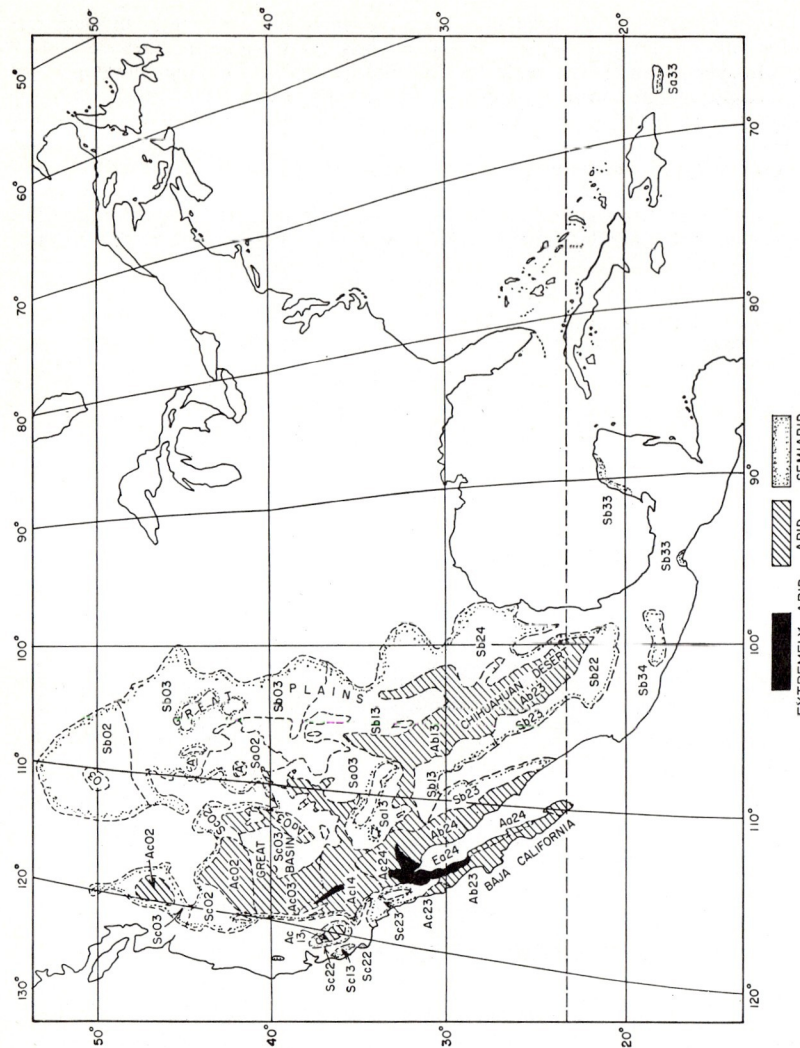

Fig. 24-1d. Distribution of arid homoclimates: North America and Southern Europe. (*After Meigs* [1].) *Key:* see Fig. 24-1a.

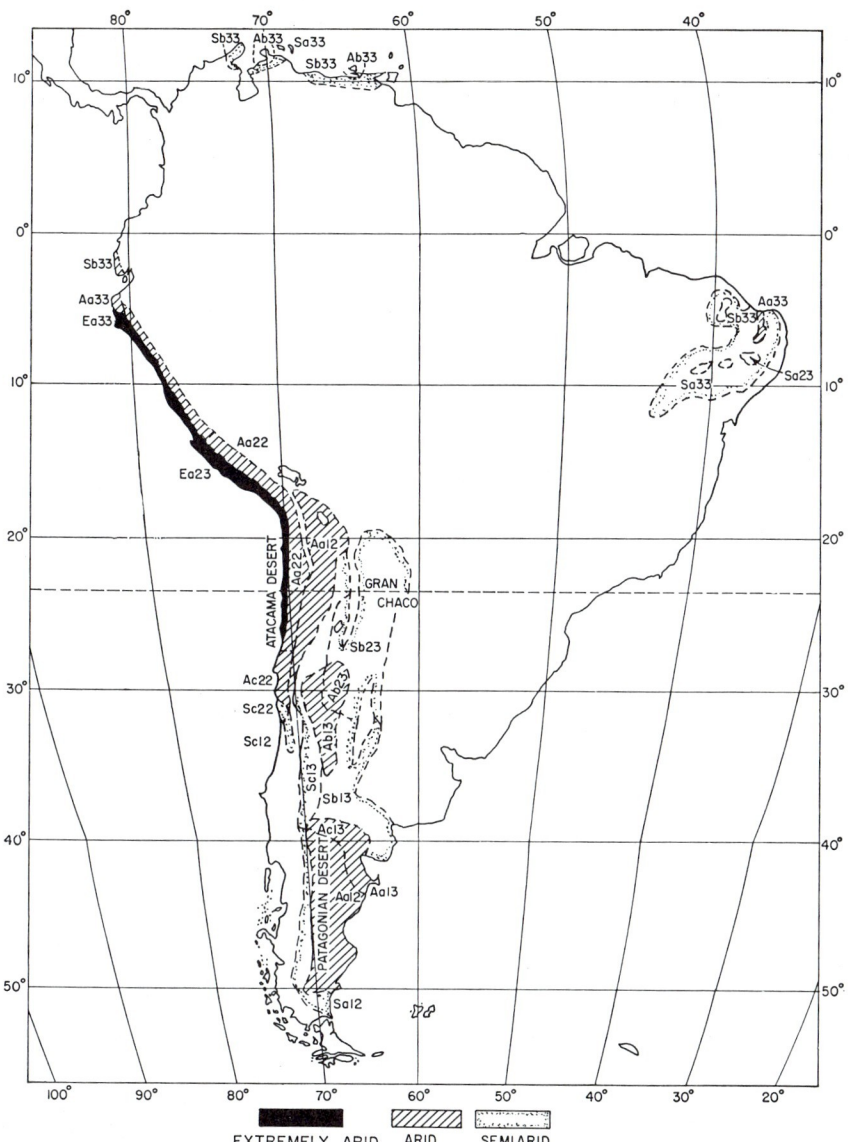

Fig. 24-1e. Distribution of arid homoclimates: South America and Southern Europe. (*After Meigs* [1].) *Key:* see Fig. 24-1a.

The mean temperatures of the coldest and warmest months, respectively, are indicated on the UNESCO maps by a pair of numbers denoting the following ranges:

0 = mean temperature below 0°C (32°F)
1 = mean temperature between 0 and 10°C (32 and 50°F)
2 = mean temperature between 10 and 20°C (50 and 68°F)
3 = mean temperature between 20 and 30°C (68 and 86°F)
4 = mean temperature above 30°C (86°F)

An area marked on the maps as Sb24, for example, has a semiarid climate, with summer rainfall concentration, a coldest month with mean temperature between 10 and 20°C, and a hottest month above 30°C.

Where water is available for irrigation either from groundwater, rivers, or reservoirs, temperature becomes a more important factor than precipitation. On the original maps, therefore, patterns of stippling are used to denote four broad temperature categories: *hot* (24, 33), *mild* (22, 23), *cool winter* (12, 13, 14), *cold winter* (02, 03, 04). For the semiarid areas, vertical shading has been applied over the stippling and the arid areas are crosshatched. In Fig. 24-1, however, only three types of shading are shown. The combination 01 does not appear on the maps. Such areas, with temperatures of the warmest month below 10°C, are too cold for agriculture, whether or not water is available, and for the purposes of this classification they are regarded as *arctic* or *high alpine*.

C. World Distribution of Arid and Semiarid Climates

From Fig. 24-1 the arid and semiarid regions are seen to occur in five great provinces, separated from one another by oceans or by wet equatorial zones. Meigs has described each of these five provinces as a core of desert, partly surrounded by semiarid lands, bordering the west coasts of the continents, chiefly from about 15 to 35° latitude, and extending inland and poleward as far as 55° latitude. Toward the equator from the axis of each dry province the high temperatures and summer rainfall of the tropics prevail (Ab, Sb, 02, 03, 04). Highlands introduce irregularities in the form of lower temperatures.

The *North Africa–Eurasia dry province* is larger than all the remaining dry areas of the world combined. It includes the world's largest desert, the Sahara, and a series of other hot deserts and semiarid areas, continuing eastward through the Arabian Peninsula and along the Persian Gulf to Pakistan and India. Along the northern fringes lie the mild or cool winter dry areas of the Mediterranean coast and Iran, and still farther northward and eastward lie the vast deserts and steppes of the U.S.S.R., Chinese Turkestan, and Mongolia, with subfreezing winters and warm or hot summers.

The *North American dry province* resembles the North Africa–Eurasia province in variety of subdivision types, though the subdivisions are much smaller in America. Dry upland areas analogous to those of Iran, Turkestan, and upland Arabia make up much of the province in the United States and Mexico, leaving only a small area bordering the Gulf of California and its northward extension in California and Arizona to compare with the hot Saharan climate. The Great Plains of the United States and Canada find their climatic analogs in the Russian steppes.

The *South African dry province* consists chiefly of the narrow elongated coastal desert of the Namib and Luanda and the Karoo and Kalahari desert and steppe uplands. Dry outliers, omitted from most world climatic maps, are located in southern Madagascar and Mozambique. Dry climates with cold winters (02, 03, 04) are not represented in any continents south of the equator.

The *Australian dry province* occupies the entire continent except for a small fringe. Hot climates prevail in the northern half of the province, mild climates in the south, with cool winters appearing in southern uplands.

The *South American dry province* is confined to an attenuated strip resulting from the form of the continent and the existence of the Andean barrier along the west coast. The world's driest desert borders the west coast almost to the equator. In the south, the eastern side of the continent, Argentine Patagonia, is dry. In the central Andean plateaux a chilly upland desert region forms a link between the western and eastern dry lands. Along the northern coast of Venezuela and the neighboring islands of Curaçao and in the *caatinga* area of eastern Brazil, there are small semiarid and arid tropical areas in about the latitude of Somaliland.

Meigs makes special mention of the narrow west and east coastal deserts that fringe all five of the provinces. Both coastal types are characterized by humid air and low daily ranges of temperature. The west-coast type, which is well developed in North and South Africa, North and South America, and Australia, is cool for its latitude and in many areas is marked by frequent fogs from the sea. The east-coast type, along the Red Sea, Gulf of Aden, Persian Gulf, and Gulf of Oman, has very high humidities along with intense heat.

In order to place the major continental land masses in perspective from the point of view of continental rainfall, it is interesting to observe that the mean annual rainfall for the land areas of the world has been given by Nimmo [3] as 26 in. For South America the figure is 53.1 in.; for Africa, 28.0 in.; Asia, 25.4 in.; Europe, 24.3 in.; North America (including the West Indies), 23.7 in.; and Australia, 16.5 in. The extremely high proportion of Australia occupied by arid and semiarid lands indicated in the homoclimatic maps can well be appreciated from these figures. A further indication of the relative aridity of Australia can be obtained from runoff data. Nimmo [3] has estimated total annual runoff from Australia as only 200 million acre-feet, representing an average depth of 1.3 in., which may be compared with a figure of 9.7 in. for all the land areas of the world. Australia's largest river, the Murray, has an average annual discharge (assuming no diversions) of 12 million acre feet. On the same basis the Mississippi discharges the same amount in little more than one week.

II. PRECIPITATION IN ARID AND SEMIARID REGIONS

The primary feature of precipitation in arid areas is the high variability of the small amount received. In fact, it is not uncommon for the standard deviation of the mean annual rainfall to exceed the mean value. In most arid regions precipitation characteristics follow somewhat similar patterns, reflecting a high order of variability in time and space of individual storms, of seasonal rainfall, and of annual and cyclical totals. These features of arid-zone precipitation are best illustrated by specific reference to various arid regions in different continents. The reader may refer to White [4], Wisler and Brater [5], and UNESCO [6] for more comprehensive information.

A. General Characteristics

For the purpose of illustrating the general characteristics of arid-zone precipitation, some examples are given for Alice Springs, Australia. This area is situated on the Tropic of Capricorn, in the center of the vast Australian arid and semiarid region which constitutes four-fifths of that continent. Average annual rainfall is 10 in., with a standard deviation of 5.1 in., and approximately three-quarters of the total is received, on the average, in the summer six months, October to March.

The general features of the distribution of annual rainfall for the 85-year period 1874 to 1958 are given in Fig. 24-2. The highest total during this period has been 28.57 in., and the lowest 2.37 in. The high order of variability and the marked skewness of the frequency distribution are immediately apparent, both features being very characteristic of arid-rainfall regimes. In Fig. 24-3a a frequency diagram of monthly rainfall is provided, reflecting an even higher order of variability.

The distribution of daily rainfall throughout the year is given in Fig. 24-3b. It can be seen that the concept of sporadic and torrential rain so often thought to be characteristic of arid regions is not supported by this analysis. At Alice Springs over 80 per cent of the daily falls are less than 0.50 in., and these falls constitute nearly half of the annual rainfall. In fact, in most arid regions, rain frequently falls, particularly in winter, in groups of several wet days associated with regional influences. In summer, prolonged rainfall can follow the intrusion of moist air into arid zones, but thunderstorm activity is more usual, and seldom results in rain falling on two or more consecutive days. These features, common to most arid regions, are reflected in Fig. 24-3c, where the frequency of occurrence of wet periods (rainy periods in which not more than two consecutive dry days occur), with rainfall exceeding specified amounts, is shown throughout the year. The marked changes in rainfall frequency between seasons and the associated increased rainfall expectancy in summer are very apparent, as is the absence of frequent periods of high rainfall amounts.

Although Alice Springs might be regarded as one of the "wetter" arid localities, the general rainfall features in summer and winter are characteristic of many of the arid regions of the world and reflect the high degree of variability of annual and seasonal rainfall and the nature of the daily rainfall composition.

B. Long-term Fluctuations

Long-term fluctuations are a feature of regional precipitation, but in arid and semiarid regions they are of vital importance to the maintenance of plants and animals, since the periods of dry years set the lower limits for persistence. Wallén [7] has made an interesting analysis of Mexican rainfall, which provides a good example of the long-term variability encountered. Fluctuations in annual precipitation were smoothed by calculating 10-year overlapping, or moving, means, and these are presented for the period 1870 to 1950 for the semiarid station of Tacubaya, D.F. (Fig. 24-4). Although Tacubaya is not arid, the general curve indicates that there

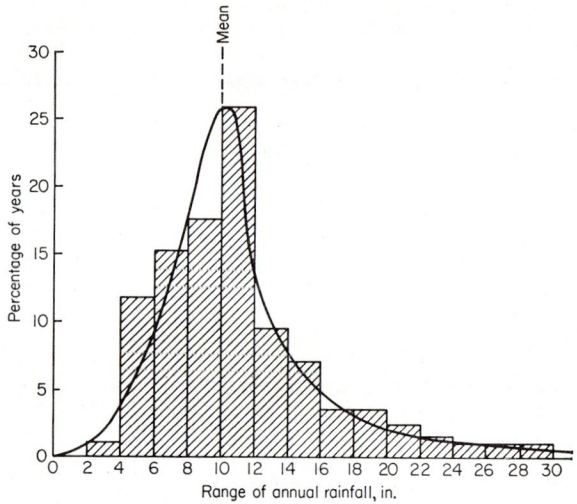

Fig. 24-2. Annual rainfall distribution, 1874–1959, at Alice Springs, Central Australia.

was a decrease of rainfall in central Mexico from the period 1878 to 1887 to the period 1892 to 1901 and followed by an increase up to around 1935 and a subsequent decline to the end of the period of analysis. The fluctuations are very wide, the 10-year overlapping means ranging from 600 to 800 mm and the individual annual totals from 440 mm in 1894 to 958 mm in 1925.

Studies of the seasonal fluctuations within these periods revealed that the increase during the first 20 years of this century was caused by an increase of both summer and winter precipitation, whereas the increase from 1920 to 1935 was associated with an increase of summer and autumn precipitation. The subsequent decrease has been associated with decreased rainfall in the same seasons. Wallén attributed the considerable increase of summer rainfall to a strengthening of the summer easterlies caused by a general northward movement of the subtropical high-pressure cells. These synoptic features are commonly associated with summer rainfall in the area. The decrease in precipitation in 1935 is thought to have been due to a gradual retreat of the subtropical high-pressure cells, a feature suggested by other evidence from northern latitudes [7].

Because of the extremely short geological period for which rainfall observations have been made, fluctuations of rainfall over extended periods have been examined geochronologically to reveal and date climatic changes, the length of particular

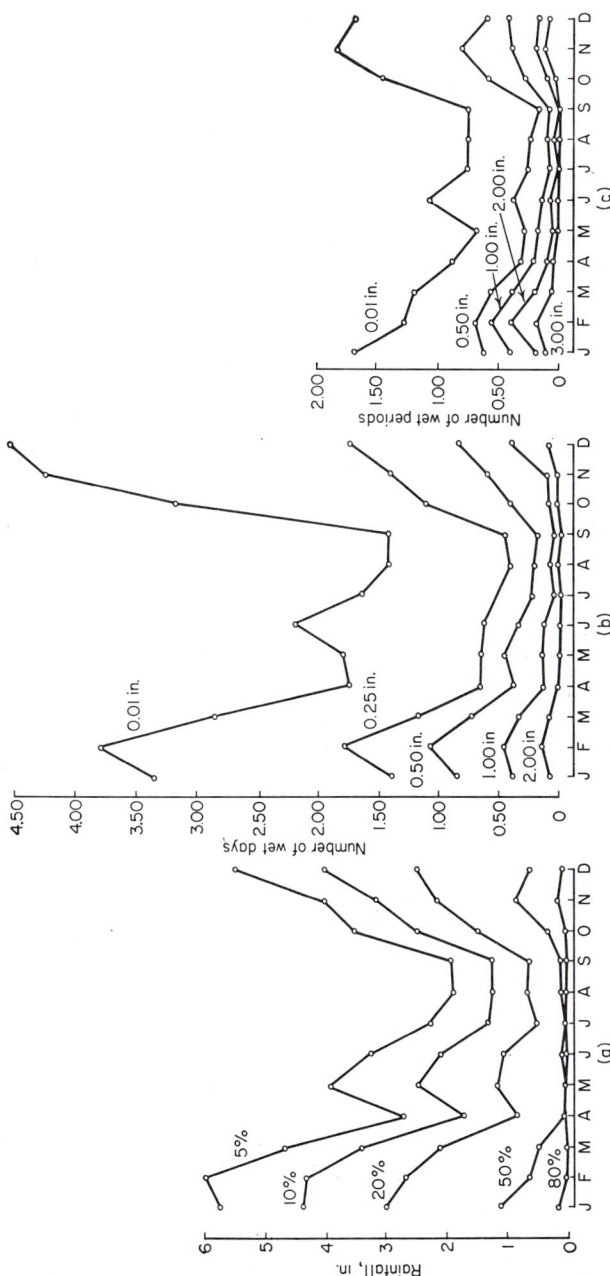

Fig. 24-3. Rainfall at Alice Springs, Australia. (a) Percentage of years in which rainfall in individual months exceeds indicated amounts; (b) frequency of occurrence of rainy days per month in which rain can be expected in excess of indicated amounts; (c) frequency of occurrence of rainy periods in which rain can be expected in excess of indicated amounts.

climatic patterns, and the recurrence of climatic cycles. Although Smiley [8] considers that no evidence of what might be termed a true cycle has been described, even with refined dating techniques, he has provided very interesting data for the southwestern United States. These data indicate that definite periods of below-average rainfall occurred between about the fourth and twelfth centuries and again in the sixteenth century, and periods of above-average rainfall occurred from the first to the third century, during the fourteenth and fifteenth centuries, and since the beginning of the seventeenth century. Smiley's data are supported by geobotanical information, which, for instance, indicates that areas now covered by *Pinus ponderosa* were, in the sixth century, supporting *Pinus edulis*, a species characteristic of more arid habitats. Likewise, the presence of *Prosopis* in the Sonoran desert can be traced back only for the last 300 or 400 years.

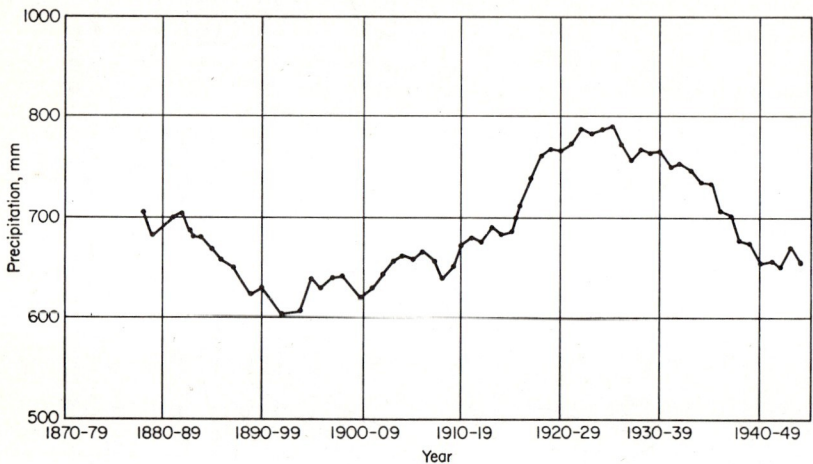

FIG. 24-4. Long-term fluctuations at Tacubaya, D.F., Mexico. (*After Wallén* [7].)

C. Geographical Variability

In regions where the topography is of major importance in controlling precipitation, the pattern of precipitation from individual storms is often quite similar to that of the annual geographical distribution. The latter is, in consequence, used as a model for determining the distribution pattern of the former. For rainfall during short periods the results are expressed on an isopercental basis. This shows the relationship between the two patterns since it is based on the ratio of individual storm precipitation to annual or seasonal precipitation at all recording stations in the locality under study. This method of studying local rainfall variability has been used extensively in mountainous regions, and a characteristic example has been provided for the Central Valley of California by Christian and Francis [9]. It can be appreciated that, if the storm follows exactly the normal annual isohyetal pattern, all the ratios will be the same and the most probable storm isohyetal map would appear as a scaled-down version of the normal annual map. Alternatively, if the storm characteristics change across the region in a different manner from the normal annual isohyetal pattern, this tendency is revealed by drawing isopercental lines through the points of measurement. The pattern of the storm depicted in Fig. 24-5 is clearly revealed in this way.

As distinct from local rainfall variability just described, regional variability of rainfall can be better described by methods such as those employed by Wallén [7] for the central plateau of Mexico. Using a probability analysis devised by Landsberg [10], Wallén has provided a special risk diagram (Fig. 24-6), which shows the probability of having individual annual rainfall amounts as a function of mean annual rainfall.

PRECIPITATION IN ARID AND SEMIARID REGIONS 24-13

This diagram is obviously of value to hydrologists and agriculturalists. It is of interest to consider some examples of rainfall probabilities expressed in this way because they indicate the order of variability in annual rainfall of this arid and semi-arid region. It is seen that the risk of receiving individual annual totals of less than 500 mm is 40 per cent in areas of 530 mm mean annual rainfall and 20 per cent in areas receiving 65 mm mean rainfall. As the rainfall requirements decrease, variability increases and a relatively higher mean annual rainfall is needed. Thus, should a crop

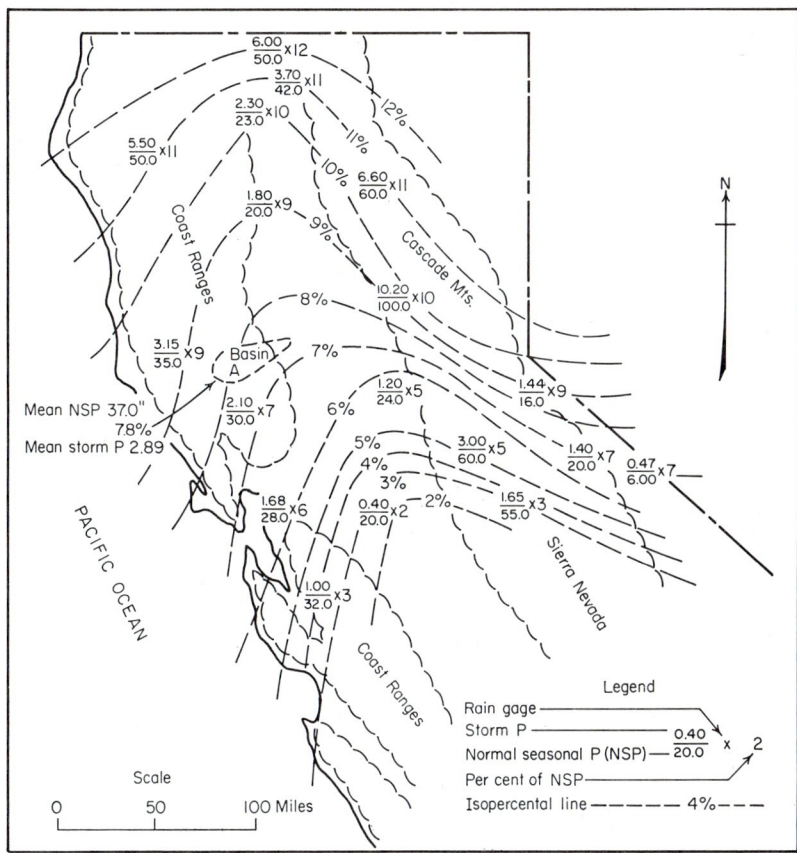

FIG. 24-5. Isopercental map of a storm in Central Valley, California. (*After Christian and Francis* [9].)

require 250 mm of rain, annual totals of less than 250 mm can be expected in only 20 per cent of the years in areas with mean annual values of 400 mm. Such areas, in consequence, should be moderately safe for cultivation.

D. Significance of Dew

Much has been written about the significance of dew as a water source in arid zones, and comprehensive reviews are available on this subject from the viewpoint of the physics of dewfall and the possible physiological benefits of dew to plants by Monteith [11], Stone [12], Milthorpe [13], and Slatyer [14]. Because the importance of dew has possibly been overestimated, it is of value to mention some of the primary features of dew deposition and their significance in arid and semiarid regions.

Visible dew—the appearance of condensed water on soil, plant, or other exposed surfaces—can be divided into three separate categories: *dewfall*, which represents condensation associated with the downward flux of atmospheric water vapor; *distillation*, which represents condensation associated with the upward flux of soil water vapor; and *guttation*, which is the physiological plant process resulting in exudation of water from specialized cells of the plant epidermis. Although this last phenomenon is frequently the source of most of the visible water on plant surfaces, it is nonphysical in origin and hence need not be considered here. It is also readily distinguishable, because of the form and distribution of the water droplets, from condensed surface water.

Hofmann [15] and Monteith [11] have provided valuable accounts of the physics of dew formation, from which several features are revealed. In the first place significant dewfall is dependent on turbulent transfer of water vapor from an appreciable layer of air. With wind speeds of less than 0.5 m/sec, evaporation always occurred. Monteith regards as ideal conditions for dewfall a clear sky for rapid cooling, a relative humidity (at screen height) of at least 75 per cent at sunset, and a wind speed of 1 to 3 m/sec maintained throughout the night. With regard to distillation, the flux of water vapor to the soil surface is determined by the gradients of soil temperature and humidity, by the surface radiation balance, and by the sensible and latent heat fluxes in the atmosphere. In England, Monteith has shown that distillation is far more common than dewfall and is the major source of dew in that country.

In arid regions both dewfall and distillation can be expected to be less than in humid localities because of the much lower levels of atmospheric and soil humidity. After rain, when this factor is removed, heavy dews may be expected because other meteorological factors, particularly the clear skies necessary for rapid cooling, are usually favorable. However, at such times the extra contribution to total precipitation is of reduced value. The contribution made by dew to the total precipitation in arid regions is confused to some extent, because distillation and dewfall have not usually been separated and the former process does not represent any net gain of water by the earth's surface.

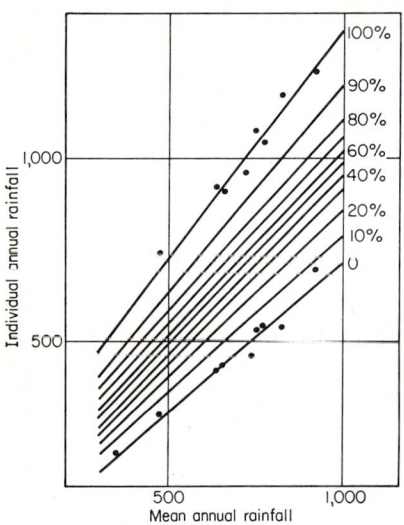

FIG. 24-6. Risk diagram of annual rainfall in the central plateau of Mexico. (*After Wallén* [7].)

The highest values reported in arid regions are those of Duvdevani [16], who reported total amounts in Israel of 30 mm over 200 dew nights throughout the year and maximum daily amounts of 0.2 mm. If advective effects are ignored, it is most unlikely that the total dew condensed could exceed a value represented by the amount of back radiation, which would seldom exceed a value equivalent to 1.0 mm per night. Masson [17] has set 0.9 mm as the extreme limit for condensation and considers heavy dews as being of the order of 0.2 mm. Maximum values of 0.1 mm were found by Arvidsson [18] in Egypt.

Advective influences are undoubtedly responsible in part for the heavy dews observed particularly in west-coast desert areas such as those of California, Israel, and Western Australia, where there is advective inflow of humid air into areas where meteorological conditions are optimal for dewfall. Also, microadvective effects are of considerable importance, not only with regard to the dew deposited on any one surface, but also because of the consequent difficulty of interpretation of data from exposed dew gages. Projecting objects such as isolated shrubs are frequently referred to as "attracting" dew. This phenomenom reflects the additional condensation of water from vapor horizontally passing the exposed surface, as well as the additional dewfall and distillation which follow the exposure of the surface to more rapid heat loss.

In summary, assuming a maximum value of 0.2 mm per night for dewfall and a figure of, say, 5 mm for daily evaporation, the total hydrologic contribution of dew can be seen to be insignificant and of negligible importance in terms of the normal hydrologic components. The primary benefits of dew in arid regions appear to be through the physiological effects on the vegetation, a feature which is discussed in more detail elsewhere by Stone [12] and Slatyer [14].

A practical application of the physical principles underlying dew deposition is seen in Middle Eastern desert areas, where stone mounds are used to collect dew. These mounds provide very favorable radiation surfaces for dew deposition and are suitably exposed to advective dewfall. Moreover, the impermeability of the individual stones results in water accumulation at the base of the mound when it is protected from evaporation. Such mounds are used as water collectors for human use and for plant growth.

E. Modification of Precipitation

Two main types of precipitation modification may be considered here: cloud seeding, which seeks to increase total precipitation, and interception by vegetation, which modifies the pattern of rainfall actually reaching the soil surface.

1. Cloud Seeding. A voluminous literature has been accumulated on artificial rainmaking and has been reported in detail in Secs. 3 and 9. In arid regions useful accounts of cloud-seeding experiments have been prepared by Bowen [19], Brier [20], Schaefer [21], and Fournier d'Albe [22], and an investigation of the possibilities of achieving increased precipitation through this technique was conducted by the World Meteorological Organization (WMO) [23], which concluded: (1) Operations which have so far been carried out have produced results that could be termed, at best, as inconclusive, neither the complete failure of the methods employed nor the certainty of getting substantial increases of rainfall having been demonstrated. (2) The most favorable meteorological conditions for the artificial inducement of precipitation are to be sought in regions and during seasons when natural precipitation is most likely. (3) Present-day techniques, either cold or warm cloud seeding, have very little value in augmenting the precipitation in areas of very low rainfall or during dry periods in regions of medium rainfall. For the conclusions reached by the U.S. Advisory Committee on Weather Control, see Subsec. 9-III.

Although significant improvements in technique have been made since the WMO report was issued, the fact that favorable meteorological conditions for cloud seeding seldom occur in arid regions means that the probability of economically successful cloud seeding remains very low.

2. Interception of Precipitation by Vegetation (see also Sec. 6). The interception of precipitation by vegetation results in the incoming precipitation being partitioned into three components, since part is retained on the foliage, part is channeled down the branches and stems, and part falls through unimpeded or drips from the branches. Runoff is also modified because of the altered intensity and distribution characteristics and the improved soil permeability caused by the presence of the vegetation. This particularly applies to the portion of the precipitation which is channeled down the trunks of the trees since root density and soil permeability are highest in this zone. In arid regions this is of special importance since it concentrates water in the most suitable biological situation and also raises the chances of deep infiltration.

Measurements of interception of precipitation have usually been made with trees or other woody plants because of the difficulties in arranging adequate interception devices on more ephemeral or smaller plants. The amount of precipitation intercepted depends on the density of the vegetation, its morphological structure, and the duration and intensity of the precipitation. As an example of the degree of interception which can take place with arid-zone plant communities, the studies of Hamilton and Rowe [24] may be cited. These investigations were conducted at research centers in central and southern California to determine the loss of rainfall as the result of its interception by shrub vegetation. Two main vegetation types were examined, one being a partially deciduous chaparral and the other an evergreen vegetation. Gross rainfall, throughfall (the quantity of rain actually falling to the ground), and stemflow (the rain which reached the ground as flow down the shrub stems) were measured, and interception loss was computed from these measurements.

Five per cent of the gross rainfall was lost annually in the semideciduous vegetation. The interception loss under autumn-winter conditions, when the deciduous shrubs were bare, was slightly more than 4 per cent, but under spring-summer conditions, when the deciduous shrubs were in leaf, the loss was about 14 per cent. The highest spring-summer loss, although due largely to the increased density of the vegetative cover when in full foliage, was also due in part to smaller average size of the spring-summer storms. By comparison, 8 per cent was lost annually in the evergreen-vegetation experiment.

Stemflow was influenced to a considerable extent by the branching habit and character of the bark of the various shrub species. The deciduous-shrub vegetation, which was composed largely of smooth-bark and upright-stem species, yielded an average of over 11 in. of stemflow out of an average annual precipitation of 38 in. On the other hand, the evergreen vegetation yielded an average of only a little more than 2 in. of stemflow out of an average rainfall of 27 in. This latter community was comprised of species having comparatively rough bark and a spreading branch habit. Throughfall, stemflow, and interception losses were generally directly proportional to storm size. For small storms the amount of interception loss was as much as 50 to 75 per cent of the gross rainfall, and for large storms as little as 3 to 6 per cent. The relation between precipitation and interception loss was curvilinear for small storms of less than 0.3 in. and linear for heavier storms.

Most studies from other arid regions provide figures of the same general magnitude as those given by Hamilton and Rowe. In Australia, Slatyer [25], in the study of *Acacia* shrub vegetation, found that the interception by individual trees of *Acacia aneura* was almost 40 per cent of the total rainfall incident on the horizontal projection of the tree canopy. This figure was maintained unless the rainfall intensity was greater than 1.0 in./hr. In general, Slatyer found that stemflow did not commence till about 0.1 in. of rain had fallen, this amount of water being needed to wet the plant surface completely. Even so, the yield from a fall of rain of 0.60 to 0.80 in. was usually in the order of 20 gal for a normal adult tree 12 ft in diameter, so the contribution to the soil water storage adjacent to the bole of the tree can be appreciated.

In some circumstances vegetation may increase precipitation; this happens in areas where ground fogs are frequent or on mountains high enough to be frequently in cloud, particularly if there is wind. The fog or cloud droplets are collected by leaves and twigs and reach the ground either by dripping or by stemflow. The advective inflow of moist air, mentioned earlier in connection with dew, is frequently of influence in this regard. Early investigations of this phenomenon have been well summarized by Geiger [26].

III. EVAPORATION AND TRANSPIRATION IN ARID AND SEMIARID REGIONS

A. Evaporation from Open-water Surfaces

Evaporation from free-water surfaces, ranging from small farm ponds to irrigation channels and large reservoirs, is a major factor in reducing water storage in arid and semiarid regions. Perhaps the most comprehensive studies of free-water evaporation and the relationship between evaporation from standard evaporation pans and tanks and larger water surfaces are those conducted at Lakes Hefner and Mead in the western United States [27, 28] (see also Secs. 6 and 11).

Lake Hefner is a reservoir of about 4 sq mi surface area, near Oklahoma City, in central Oklahoma. The climate is subhumid, with a mean annual rainfall of 31 in., most of which is received in the summer six months. Lake Mead, on the other hand, is over 200 sq mi in surface area and is located on the boundary of northwestern Arizona and southwestern Nevada, in an arid area of about 5 in. total rainfall. At Lake Hefner lake evaporation for the year which was studied intensively totaled about 55 in., and that from a land-based Class A U.S. Weather Bureau evaporimeter (raised pan, 4 ft in diameter and 10 in. high) was about 80 in., providing a pan coefficient of 0.69. At Lake Mead, lake evaporation averaged about 70 in. for the period 1936 to 1949, compared with land-based pan values of 120 in., indicating a pan coefficient of about 0.59.

The higher evaporation at Lake Mead, even though the exposed water surface was much more extensive, is indicative of the higher radiation load and greater aridity. The pan values demonstrate the same point, and the significantly lower pan coefficient in the more arid area demonstrates the degree to which advected energy from the more arid surround at Lake Mead affected the relative losses from evaporimeter and lake.

Follansbee [29] has given a list of evaporation values for a number of reservoirs in the United States and some examples from other countries. These values suggest typical figures for most reservoirs in arid and semiarid regions of the southwestern United States of about 60 to 70 in./year. The highest reservoir value quoted is 123 in./year from Atbara in the Sudan. Although this figure may be questioned, it provides some idea of the values which free-water evaporation can reach in extremely arid regions located at low latitudes and altitudes.

B. Evaporation from Bare Soils

In arid regions an important feature of the water balance is the high proportion of incoming water which is returned to the atmosphere by evaporation from the soil surface. In many areas of coarse-textured soils on more or less level land surfaces, drainage patterns are not apparent on aerial photographs and it is assumed that surface runoff is of little significance. Deep drainage and infiltration to groundwater are also of rare occurrence, and direct evaporation and transpiration are the main operative hydrologic processes. Because of the low plant density, direct evaporation of water from the soil is of enhanced importance, and frequently as much as one-half of the annual rainfall can be lost in this manner (see also Secs. 6 and 11).

Philip [30, 31] has given a good account of the factors affecting evaporation, although restricted to isothermal soil conditions, and a typical evaporation-time relationship, assuming initially saturated soil and constant meteorological conditions, is given in Fig. 24-7a. In Fig. 24-7b actual experimental data from a water-balance study conducted near Alice Springs in arid central Australia [25] is shown to demonstrate the close relationship between the actual and idealized situations. These data indicate the rapid drop of evaporation when the soil surface becomes dry, because a vapor barrier develops and evaporation is limited by the rate of transfer of water vapor to the surface. In this phase it is virtually independent of atmospheric conditions. However, the fact that under arid conditions, falls of rain or rainy periods do not usually wet the soil deeply and are frequently followed by long periods of dry weather means that much of the rain received in light falls is lost by direct evaporation before the vapor barrier develops.

As well as direct evaporation losses, the steep temperature gradients in the surface-soil horizon are of considerable influence in modifying the pattern of soil-water distribution. On a diurnal basis this is reflected in a marked movement of water to the soil surface at night, where it may condense and contribute to the total visible "dew" appearing before sunrise, as has been mentioned in Subsec. II-D. Most of this water is evaporated early on the following morning, reducing soil water storage further than would be expected under isothermal conditions. However, during the morning, when the heat flux is downward into the soil, water movement occurs under this influence, tending to protect the soil water from evaporation. Philip [30] has shown that under some conditions the downward flux can inhibit evaporation more than the upward flux increases it. Seasonally similar situations occur, evaporation tending to be reduced in spring and early summer, when the downward heat flux is greatest, and increased in autumn and early winter, when the upward flux is greatest. This provides the paradox that water loss from the soil may be least when atmospheric conditions appear most favorable for evaporation, and vice versa.

The possible magnitude of water loss by the upward movement of water to the soil surface may be gaged from the experiments of Hilgeman [32], who noted that water loss from bare soil in Arizona, over a period of 22 months, totaled 9.8 in., representing 47 per cent of the water available for plant growth.

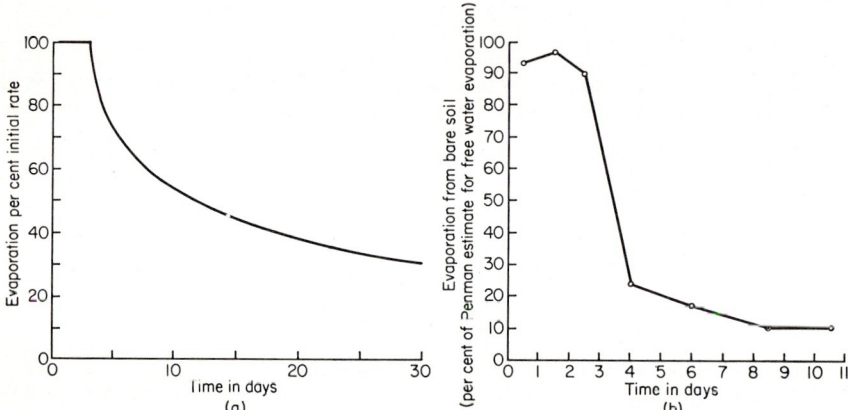

Fig. 24-7. Evaporation from initially saturated bare soil. (a) Theoretical relationship (*after Philip* [30]); (b) actual data from Alice Springs, Australia (*after Slatyer* [25]).

C. Evapotranspiration from Arid-zone Plant Communities

The evaporation component of evapotranspiration conforms, in general, with the expectancy indicated for bare-soil evaporation (see also Secs. 6 and 11). Modifying factors are introduced by the physical presence of the vegetation, through shading, windbreak effects, and the influence of water extraction by plant roots. The effect of shading is to reduce evaporation rates by reducing incident radiation on the soil surface and simultaneously decreasing the heat flux in and out of the soil, thus reducing both upward and downward water movement induced by temperature gradients. Although shading tends to reduce evaporation rate, it frequently means that the soil surface remains wet for considerably longer periods, and the total loss of water may be greater than from exposed soil surfaces, in which the vapor barrier develops sooner.

Transpiration, while basically a physical process in that the energy for the evaporation of water from plants is derived from solar radiation, is also affected by the physiological and anatomical characteristics of the plants themselves.

A great deal is already known about the evapotranspiration of plant communities when water is nonlimiting. In arid regions, however, this is of little consequence, since all plants subsisting on natural rainfall alone are under severe soil-water stress for prolonged periods. During such periods evapotranspiration drops rapidly as soil-water stress increases.

In systems such as vegetation or dry soil, the evaporating surface is located within the aerodynamic surface. Vapor transport between these two surfaces is by molecular diffusion, and its rate of movement is therefore influenced by the effective length of the interposed path. With reference to transpiration, Milthorpe [33] has pointed out that water evaporates and diffuses by two separate pathways in the leaf. First, the vapor escaping from the liquid surface located within the cell walls abutting on air spaces diffuses through the surface of the cell walls, the substomatal cavities, and the stomatal pores; second, the water evaporating within the epidermal cell walls diffuses through the cuticle. Although all these sources of resistances to vapor movement contribute to the total resistance in the diffusion path of the leaf, Milthorpe has found that the only major sources of resistance are located in the cuticle and the stomatal pores.

In arid-zone plants these anatomical features are strongly developed. In the first place, because of the extremely xeromorphic nature of arid-zone plants, the cuticular conductances are extremely low. In other words, when the stomata are closed, transpiration is virtually zero. Second, because stomatal closure can be caused by internal plant water deficits, it can be anticipated that maximum stomatal opening will be of rare occurrence during most of the year in arid zones. In fact, it appears that, on any one day, limited stomatal opening may be expected to occur for not more than one or two hours each morning.

A good deal of controversy has arisen as to the effect of soil-water stress on transpiration because of conflicting evidence from the experiments of different investigators. Veihmeyer and Hendrickson [34] have claimed that transpiration is unaffected by soil-water stress until the permanent-wilting percentage is reached. Similar results have been obtained by other workers [35, 36], but most studies have indicated that reductions in transpiration rate commence at much lower suction [37–39]. Since some reduction in transpiration rate must occur in conjunction with stomatal closure, it seems that initial reductions in rate should be observed at stress values well below those at permanent wilting. Veihmeyer and Hendrickson [40, 41] have also claimed that under field conditions absorption from any soil zone ceases when the soil water content is reduced to the permanent-wilting percentage. There seems no valid reason why this should occur, and results from other field studies,

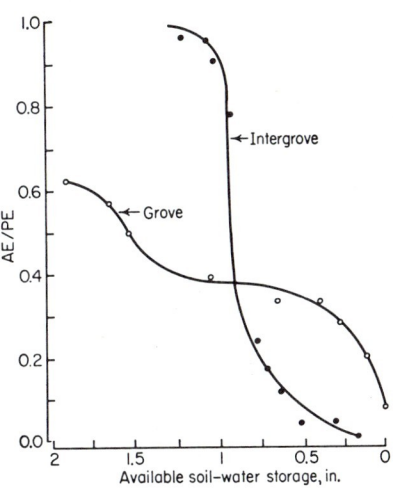

Fig. 24-8. Evapotranspiration-decline curves for a typical arid plant community (*Acacia aneura*) near Alice Springs, Australia.

particularly in arid regions, demonstrate continued absorption beyond the permanent-wilting percentage, extending sometimes after the death of the plant [42–45].

In general, it may be expected that the influence of soil-moisture stress on transpiration will vary considerably from plant to plant, depending on evaporating conditions, on effectiveness and sensitivity of stomatal closure, and on the availability of water for absorption (dependent in turn upon the extent of the root system, on rate of flow of soil water to the root surface, on water stress at the root surface, and on the rate of root growth into the soil mass). Resistance to the passage of water into and through plants caused by reduced internal permeability usually appears to be small compared with the resistance at the leaf-air surface [14].

Because of the low plant density and relative physiological inactivity characteristic of arid regions, rates of evapotranspiration, even when soil water is freely available, are usually well below those characteristic of physiologically active, dense communities, which may reach levels close to those for potential evapotranspiration. Characteristic evapotranspiration-decline curves for an *Acacia* shrub community near Alice Springs in Central Australia are given in Fig. 24-8. This plant community is of special interest in that the shrubs are grouped to long groves which are separated by grassy intergroves. In the groves plant cover is of the order of 40 per cent, whereas in the intergroves, the cover is of the order of 15 per cent. The former situation thus provides a shadier environment with a higher proportion of permanent vegetation. This is reflected in the diagram, where curves for grove and intergrove are shown. Actual evapotranspiration is plotted as a function of potential evapotranspiration (after Penman [103]) against available soil water storage.

Initial rates, when the soil surface is wet, are seen to approximate potential evapotranspiration levels in the exposed intergroves, but are only of the order of 0.6 of the potential rate in the shaded location. Surface drying occurred more rapidly in the former location, and the decline in rate was also more rapid. In the latter location evapotranspiration was reduced, on drying of the surface soil, to a value equivalent to about 0.4 of the potential rate, at which level it was maintained until only about 0.5 in. of available soil storage remained.

Although evapotranspiration by typical arid plant communities accounts for most of the rainfall in arid regions, loss of water by phreatophytes tapping streamflow or groundwater is frequently of greater hydrologic significance because it can drastically influence the limited water yield.

Phreatophytes are characteristic of streamline vegetation and areas of shallow water table throughout the world's arid regions, but most attention has been paid to them in the arid regions of the United States. Robinson [46] has estimated that phreatophytes cover about 15 million acres in the 17 western states and that they may transpire 20 to 25 million acre-feet of water. It is thought that the water used by these plants might represent the largest source of reclaimable water in that area. Because of the "oasis" form of distribution, evapotranspiration from phreatophytic communities usually contains a large advective component, and there is good evidence that it may exceed free-water evaporation.

Most studies of evapotranspiration by phreatophytes have been made with lysimeters or with streamflow gages. The former technique involves difficulties in simulating a uniform surround of similar vegetation; the latter involves difficulties in interpretation if there is inadequate knowledge of some of the local hydrologic features. Taylor [47] has presented data collected with the latter technique which provide good evidence of the magnitude of phreatophytic losses. They refer to the flow of water past two bedrock control points in a canyon near San Bernadino in central California. A marked diurnal fluctuation in the upper and lower gagings is evident and reflects the diurnal evapotranspiration regime. For the 6 days depicted, the mean loss per day was 0.32 in., almost equaling the loss from a nearby shallow black-pan evaporimeter (0.33 in./day). On one day when a hot advective wind was blowing down the canyon and air temperature reached 111°F (44°C), Taylor reported a value of 0.52 in., exceeding the pan value of 0.50 in., and no doubt including a large advective component.

The largest phreatophytic area in the world is that in the Nile valley, where the waters from the central African lakes are dispersed in the great swamps of the Sudd region of the southern Sudan. At present it is estimated that one-half of the Nile flow is lost by evaporation in these swamps [48], and construction of the Jonglei Canal is planned to eliminate this tremendous wastage.

D. Modification of Evapotranspiration

Modification of evapotranspiration can only follow some modification in the energy balance of the evaporating surface since the incoming and advective components of the total radiation flux must be dissipated by one of the normal heat-transfer mechanisms. Since evapotranspiration constitutes the latent-heat-flux component of the energy balance, modification of evapotranspiration rates can be expected by controlling the net amount of energy used for evaporation, and this is normally achieved by altering the physical nature of the evaporating surface.

In dry soils and plants this happens naturally since the evaporating surface is located within the physical surface, but no such barrier exists in a natural water surface. In consequence, most methods of reservoir evaporation control have involved reduction of the exposed reservoir surface or moderation of the conditions which affect evaporation, or both. The usual procedure adopted has been to reduce the energy load on the evaporating surface rather than to introduce a resistance between the evaporating surface itself and the free circulating air. In most cases this has been achieved by shading or reducing surface area relative to depth, and details of these and other techniques have been described by Beadle and Cruse [49]. Windbreaks have also been used, and are normally quite effective in arid areas, where they reduce the important advective component of the surface-energy balance.

In recent years considerable interest has been focused on the use of monomolecular films to reduce evaporation, the primary mode of action presumably being to introduce a diffusion barrier between the evaporating surface and the free air. Even though the film is only one molecule thick, significant reductions in evaporation have been observed [50, 51]. See also Secs. 6 and 11.

The degree of reduction is greatest on small storages, but Mansfield [51] has observed reductions of the order of 20 per cent over a long period in a 2,000-acre reservoir near Broken Hill in western New South Wales. Difficulties in using the technique have been caused mainly by the difficulty in maintaining the film over extended water surfaces. Two of the major problems which still remain to be solved are the actions of microorganisms and wind, but it is apparent that the method has considerable potential for reducing evaporation, and hence enabling more effective water storage. The method shows particular promise in arid regions, where even small reductions in total evaporation appear to be economically sound.

With reference to evapotranspiration, the energy balance can be modified most effectively by reducing either the total incoming radiation reaching the soil surface, the albedo of the surface, or the proportion of the net radiation used for sensible and latent-heat transfer. The incoming radiation reaching the soil surface can be changed by the introduction of a mechanical factor between the sun and the ground surface, such as a straw mulch. In hot arid zones trees have a significant effect in this respect, although their influence in competing with other plants may be too detrimental to enable their use. Surface soil mulches have a similar effect and also increase the length of the diffusion path between the evaporating soil surface and the free air.

The albedo of soil and vegetation communities ranges between 5 and 30 per cent and is higher if the color of the vegetation changes from green to yellow and the wavelength increases from the middle of the visible spectrum. If species which have high albedos can be developed genetically, this feature may offer a promising method of reducing net radiation in hot, dry climates. Likewise, lighter soils can be expected to have higher albedos than darker ones.

Modification of the proportion of water used as evaporation and transpiration can be achieved by regulating the surface-runoff factor of the water balance and inducing water accumulation in selected areas. This subject will be discussed later with reference to surface runoff and recharge of aquifers, but it also has implications with regard to evaporation. The advantages of using natural runoff in this manner or of artificially inducing and regulating runoff are obvious; for instance, two adjacent areas might be envisaged, receiving, say, a fall of 1 in. of rain. Under normal conditions 0.25 in. of this might be expected to evaporate, leaving 0.75 in. If, however, 0.25 in. runoff is induced from one area to the other, the net storage in one area after evaporation becomes 1 in. and on the other, 0.5 in. In this way a greater component of the total water is available for transpiration, although the total evapotranspiration over a period of time may not be affected unless water penetration proceeded to an extent sufficient to contribute to groundwater.

In the case of phreatophytes the primary objective is to destroy the vegetation and so conserve water through elimination of undesirable transpiration. Both chemical and mechanical means have been used to achieve control, and the savings in water loss are of the magnitude indicated previously. Because of the "oasis" character of phreatophyte distribution and the influence of advection on transpiration, complete control should be sought in all cases. Otherwise quite severe reductions of phreatophyte density may not markedly reduce total transpiration, since the additional exposure of the remaining plants may result in a relative increase in transpiration per plant which would compensate in part for the reduced plant numbers.

IV. SURFACE DRAINAGE IN ARID AND SEMIARID REGIONS

A. General Drainage Characteristics

A major hydrologic feature of arid and semiarid regions is the extent of interior drainage [52]. This is partly explained by the persistence of structural basins as drainage units in deserts and partly by the disorganization of former outgoing drainage systems.

It is often possible to distinguish between the *theoretical* extent of an arid drainage basin, as given by its hypsometric unity, and the *effective* extent subject to present surface drainage [53]. The latter may be restricted to a peripheral zone which may cover less than 25 per cent of the theoretical basin, suggesting that there has been an upslope retraction of drainage which originally graded the basin and which formed the alluvial deposits in its lower parts. Many desert drainage nets are known to predate present aridity. The *wadi systems* of the northern Sahara, for instance, probably originated under moister conditions in the Tertiary era, as did much of the drainage of the Australian arid zone. Subsequent desiccation has resulted in the obliteration or disorganization of the lower sectors of such drainage and in a change from perennial or intermittent to ephemeral regimes in the upper sectors, with consequent effects on the balance of erosion, transport, and deposition and on channel form and grade. In this way, inherited systems of drainage have become adjusted to the desert environment.

River regimes are mainly ephemeral, although many semiarid areas of the United States are typified by seasonal, snow-fed streams [5]. An overriding characteristic is extreme variability in both frequency and volume of discharge.

Despite variations due to differences of structure and of geomorphic history, surface drainage in arid and semiarid areas shows many characteristic features. It includes upland sectors of relatively organized, incised channels, plain tracts across which larger channels persist but which are otherwise characterized by poorly defined or diffuse local drainage, and terminal lowlands which may contain salt lakes but which are commonly riverless and often have extensive sand cover. Drainage activity is primarily erosional in the uplands and becomes increasingly aggradational in the lower zones.

There are large differences in the sizes of the drainage units and their climatic and base-level controls. They range downward from the Australian Lake Eyre basin, 480,000 sq mi in extent, to watersheds where relief and structure patterns are much smaller. Their sources may lie outside the limits of the arid zone, as in western North America, or may have rainfall characteristics similar to the areic (i.e., without definite surface-drainage patterns) desert plains, as in Central Australia. They may be enclosed basins of interior drainage or may drain to the ocean, locally with exogenous perennial channels.

Useful general accounts of surface drainage in arid regions are given by UNESCO [6], Dixey [54], and Christian, Jennings, and Twidale [55].

B. Weathering Processes and Landforms

1. Weathering Processes and Vegetative Cover. Weathering in deserts is slow and somewhat superficial, the secondary comminution of rock fragments being particularly retarded. Weathering mantles are accordingly coarse-textured and thin, and soils are generally absent from slopes steeper than 5 per cent. At the same time, rock fragments are very persistent and most surfaces bear a stony veneer.

The low density of desert vegetation is an important factor in landform evolution. First, vegetation plays a reduced part in weathering attack on desert surfaces; second it is relatively ineffective in binding detrital mantles or in preserving minor channel forms; third, it provides incomplete protection against raindrop impact or against scour by diffuse runoff; fourth, it is inefficient in storing rainfall, and hence regulating streamflow.

2. Desert Landforms. A major feature of desert relief is the simple contrast of upland and plain. Because of the thinness of slope mantles and the lack of rounding by mass movement as in humid areas, hill slopes are largely controlled by the nature of the rock. Slopes on harder, massive strata are controlled by bedding and jointing planes, with deep reentrants restricted to planes of structural weakness. Slopes on softer rocks may be determined by the angle of repose of loosened rock material and are frequently dissected by shallow gullies. All hill slopes tend to be steeper and more rectilinear than their humid counterparts because their lower sections are relatively free from slope wash, grading, and colluvial mantling.

The desert plains generally have gentler gradients than those of humid areas because of ready transport on sparsely vegetated surfaces.

They consist partly of erosional surfaces or pediments, thinly veneered with rock waste, which adjoin the hill foot and which may ingress as erosional embayments into the uplands. Aggradational surfaces are more extensive and may adjoin the upland drainage outlets or may be restricted to lower tracts. The profiles of both erosional and aggradational types of surfaces are somewhat similar, being concave slopes with gradients approaching, but rarely exceeding, 5 per cent in the upper parts.

The junction of hill and plain is characteristically abrupt. This is partly due to the extremely selective response of arid landforms to differences in lithology. It also results from the contrasting nature of the controls active on the two surfaces. The hill slopes are gravity-controlled, whereas the plains are graded surfaces of sediment transport [56]. Because of an absence of human influence, these two slope types are left clearly demarcated.

The rapid transition from hill to plain results in a limitation of hill and valley zones. Where soft rocks have been protected by flattish hard cappings, dissected marginal or foothill zones of some extent may be formed, but elsewhere primary gullies open out rapidly on the piedmont plains.

C. Runoff

Much of the rain falling on to desert surfaces is absorbed by the porous mantle of soil and rock waste or is lost by evaporation. Since desert rainfall is naturally infrequent and short-lived, these initial losses are proportionately greater than in humid regions, so that, in general, percentage runoff tends to decrease with annual rainfall and is less than 10 per cent in most arid zones.

The intensity of the rainfall may be more important than the total amount, although runoff will tend to increase with the duration of the fall. Dubief [53] notes that in the Ahaggar, summer falls as low as 0.20 in. will yield runoff if the intensity approaches 1.0 in./hr, whereas no flow may result from larger amounts of less intensive winter rain.

Because of high evaporation losses, the degree of slope exercises a stronger control on the amount of runoff than in humid areas. On hill slopes, percentage runoff may be as high as in humid uplands, the initial losses being compensated by steeper gradients, bare rocky surfaces, and low wastage through vegetation, but rain falling on gently sloping plains will be less effective in producing runoff than in humid areas.

In the absence of dense vegetation, the nature of the surface is an important factor in controlling the amount and character of runoff. On desert hill slopes there is little retention of rainfall by the slope mantle and runoff is very rapid. There is also a deficiency of fine material on such slopes, so that surface runoff is underloaded and the length of overland flow is reduced. As a result, a fine meshwork of slope rills is formed, and these link up into a channel network incised to varying degree, depending on rock resistance.

In contrast, the plains are usually mantled with fine detritus, which initially absorbs much of the precipitation. From the outset, runoff tends to be diffuse, and its subsequent concentration into linear flow is opposed by lack of surface storage and by the abundant load. Even where runoff enters such surfaces as linear flow, it tends to become dispersed by heavy load.

The contrast between hill slope and plain is thus reflected in the basic distinction of two types of desert drainage, namely, streamflow and sheet flow.

Aeolian sand surfaces represent a third type of surface, yielding little or no runoff. Localized surface flow and effluent seepage into dune swales or sand-plain depressions may take place, but it is normally insufficient to give rise to organized drainage.

Because of the thinness of soils and the sparsity of vegetation, these surfaces offer little storage for rainfall; such runoff as does occur will ensue rapidly, and surface flow will be less restricted than on similar slopes in humid areas. For the same reasons, there can be little feedback from the surface layer, so that runoff will rapidly cease when the rain stops. This leads to a concentration of runoff into short flood phases.

D. Streamflow

Despite their ephemeral character, drainage channels in arid regions show the same tendency to adjust in equilibrium with hydraulic factors as do perennial streams [57], and in fact, such channel adjustments are facilitated by the paucity of restraining vegetation (see also Sec. 17-II).

The main contrast between the perennial and ephemeral channel is that the load-discharge ratio increases down valley in the latter. This is mainly because of loss by percolation into the sandy bed which is not compensated by base flow from groundwater. Thus seepage of 2 in./hr into the bed of the Todd River at Alice Springs, Australia, has been observed during the short periods of flooding. A secondary factor is the tendency to pick up load from the river bed. In consequence, erosional activity in headwater areas tends to change to aggradation down valley. Eventually, unless receiving flow from other sources, the channel loses itself in a terminal zone of deposition.

1. Minor Channels. The headwaters are typically shallow gullies, only a few feet wide, with gradients similar to those of the hill slopes on which they are developed. Concentration of runoff into higher-order channels allows vertical incision and a very rapid flattening of the profile down valley, which parallels the abrupt change from hill slope to plain described above. The valley form varies with rock type, being gorge-like in harder rocks and widening in softer strata, but the greater effectiveness of channel erosion relative to slope grading normally results in narrower, steeper-sided valleys than occur in humid areas. Drainage density in these upland sectors normally exceeds that in humid areas because of the shorter distance of overland flow on these freely drained desert slopes [58].

At an early stage the channels show the trenchlike form characteristic of ephemeral drainage, usually with a shallow fill of coarse alluvium. The channel may occupy the whole floor of gorge tracts, or it may be entrenched into floodplain terraces. The irregularity of streamflow and the unsorted nature of the available load result in irregular profiles, with rapid alternation of erosional and depositional sectors along the valley and a lack of accordance between main and tributary channels.

In arid Central Australia, about 0.5 in. of rain of moderate intensity will cause flow in these minor channels, lasting for an hour or two only. Such flow may occur as many as five times each year. However, about 2 in. is needed for bankful flow, and such falls can be expected less frequently than once per year.

2. Larger Channels. The headwater sectors of desert drainage differ little in profile or in channel form from those in humid areas, but there is an increasing divergence down valley in response to the increasing load-discharge ratio. In response to heavy-load conditions, there is a greater increase in velocity down valley than in the perennial river, caused by the maintenance of steeper longitudinal gradients through aggradation or by the much slower increase in depth down valley in relation to discharge, resulting in a shallow, trenchlike channel.

Shoaling is an adjustment to a plentiful, coarse bed load, which demands high cross-sectional velocity and shear stress. Leopold and Miller [57] have described the rolling of large cobbles and small boulders in flash floods with depths no greater than half the diameter of the object moved. In Central Australia, although the larger channels may attain ½ mile in width, their depths rarely exceed 15 ft. This channel form persists into the lower sectors, since there is less longitudinal sorting of alluvium than in the perennial stream, and the down-valley decrease in the caliber of the bed load is slow. In the terminal zones of finer alluvial deposition, however, the main channel may narrow, deepen, and adopt the semicircular perimeter of the perennial stream prior to splitting into smaller tributaries.

Channel width increases down valley as the square root of discharge, as in the perennial stream. It is greatest in piedmont sectors, but tends to decrease progressively below the entry of the main tributaries. Channel widening results from bank collapse following wetting rather than from undercutting by floods [57], and the channel is mainly flooded with an alluvial fill which increases in thickness down valley.

The bed is generally flat or gently sloping in cross section, with banks and low islands of roughly sorted gravels and sands. The bed contours result from the pattern of deposition during the declining phase of the last major flood, and they in turn will determine the advance of the next flow. The long profile shows considerable irregularities, particularly in upper sectors, where localized scouring may take place below rock bars.

The channel course varies from slightly sinuous and braiding to meandering or strongly braiding. These latter extremes represent adjustments of channel form to different conditions of discharge, amount and caliber of load, and slope. In general, for a given discharge, braiding is associated with steeper river gradients than meanders and with coarser load. On a given gradient, braiding is associated with a higher bankful or dominant discharge. The braiding channel has a greater width-depth ratio suited to the movement of coarse bed load and is particularly characteristic of ephemeral drainage.

In piedmont sectors, flow may occur more than once a year, following rain of 1 to 2 in. For example, an average rainfall of 1.76 in. over a watershed of 230 sq mi gave a total flow of 3.7 million cubic feet, lasting for 60 hr in the Todd River at Alice Springs, Australia. The advance of a small flash flood in an arroyo channel tributary to the Rio Grande in New Mexico is described by Leopold and Miller [57]. Peak flood was reached in less than 10 min by a succession of small surges, each a few inches in height. High flood lasted only 10 min, and flow decreased to an insignificant amount in less than 2 hr. Only part of the channel was occupied by this flood. Down valley, floods naturally occur at much longer intervals.

E. Sheet Flow

All *interfluves*, that is, up to 95 per cent of surfaces yielding runoff, must be subject to diffuse flow of some form, in both humid and arid regions. On loose desert surfaces, with discontinuous vegetation, the effectiveness of such flow is increased, as evidenced by the low gradients on wash slopes. The loose detrital mantle results in high-load-discharge conditions, which, combined with rapid evacuation of runoff, are opposed to the selective incision of drainage channels on low gradients and which may even succeed in dispersing channeled runoff.

Sheet flow may be erosional or depositional; the former activity is seen in the regrading of pediments and erosional lowlands; the latter results in the formation of alluvial fans and plains.

Joly [59] distinguishes two types of sheet flow, that deriving from adjacent hill slopes, either as hill wash or as flow in small hill-slope gullies which flood out at the hill foot, and that derived locally from rainfall on the lowlands. The former is naturally more rapid and coordinated.

Because of the short duration of *sheet floods* there are few observational records. Joly has given a valuable description of a sheet flood in the northwest Sahara, derived from adjacent hill slopes. It followed heavy rain lasting 1 hr and occurred on a slope of 5 per cent which was already wet from an earlier storm. He records the following stages: (1) The soil surface becomes saturated. (2) Puddles form and are linked by shifting rills of water. (3) Connected flow begins in very small preexisting gutters. (4) The gutters overflow and unite to form a sheet of water with a maximum depth of $1\frac{1}{2}$ in. The water is heavily charged with fine sand, which is redeposited as small deltas and banks in the lee of obstacles. Very small cobbles are occasionally moved. This phase lasted 10 min. (5) Flow declines, coarser load is first deposited, and the water sheet splits into very shallow ponds which drain by a newly formed system of small rills into very shallow channels. As flow declines further, these rills incise themselves slightly into the thin detrital cover already formed, and tongues of finer material are then deposited in these flow lines. (6) The soil dries rapidly.

Davis [60] has distinguished between sheet floods and stream floods. The latter have the spasmodic nature of sheet floods, but are confined to definite shallow channel zones. The earlier and later stages of runoff are commonly of the stream-flood type, as in the sequence described above. The sheet-flood stage is then of relatively short duration. Davis also suggested that sheet flooding is associated with more smoothly graded surfaces, while Blissenbach [61] maintains that it is relatively more important in arid than in semiarid areas.

On open plains less spectacular forms of sheet runoff occur. Central Australia contains many thousands of square miles of gently sloping alluvial or erosional plains traversed by shallow alluvial drainageways generally less than 10 ft below the adjacent interfluves. Gradients are mainly between 0.5 and 0.1 per cent. In such areas, after initial losses there is a slow movement of very thin, discontinuous sheets of water down slope, where it collects and covers extensive areas to a depth generally less than 1 ft. Flow along these shallow drainage depressions is extremely slow, even after 1-in. fall of rain. It is possible that the connected graded drainage systems of which they form part may be inherited from an earlier, moister period.

Alluvial fans or aprons are formed in piedmont sectors where heavily loaded hill wash or channel flow is checked at the change of gradient between hill slope and plain. The upper sectors of the larger fans may have shallow, braiding channels. These may be replaced in intermediate sectors by shifting zones of more diffuse flow, commonly seen on air photographs as networks of ill-defined broad flow lines. They may pass downslope into riverless areas, in which case the lowermost sectors show little evidence of surface drainage. Elsewhere, the fan drainage may persist and connect with larger drainage channels downslope, and many fans of this type are subject to dissection in their lower tracts [61].

F. Desert Lakes

The large number of ephemeral *desert lakes*, *pans*, or *playas* is reflective of the disorganized drainage of such areas. The main types of desert-lake basin and their hydrologic significance are given briefly below.

1. Tectonic Lakes. These are generally the largest and occupy the centers of structural depressions. Lake Eyre in Central Australia exemplifies this type. They tend to be the focus of interior drainage systems, which may show various stages of disorganization. Such lakes commonly date from earlier periods of heavier rainfall, and although structurally emplaced, their reduced outlines may be determined by deflation, sand movement, or alluvial deposition by tributary drainage. Despite the rarity of surface flooding and the high evaporation losses, the water table is normally close to the surface. Gautier [62] has shown that the *shotts* of Tunisia act as evaporative surfaces in a hydraulic system which includes both shallow and deep groundwater as well as surface flow. Such lakes may contain thick salt crusts.

2. Lakes Formed by the Disorganization of Drainage. These commonly occupy parts of former valleys which have become discontinuous as the result of climatic deterioration. They may retain former tributary systems, or they may be isolated from surface drainage. They are a form of barrier lake, dammed by alluviation in the former valley or by the encroachment of aeolian sands. They may be fed by surface flow or by shallow groundwater from infiltration from the margins of the basins. They do not necessarily form terminals of subsurface drainage.

3. Lakes at Drainage Terminals. These are normally shallow basins of alluvial deposition, terminal or lateral to major drainage channels, fed by floodwater arms or by distributary channels. They may be fresh or brackish, and are fed entirely by surface flow.

4. **Lakes Formed in Uneven Surfaces of Aeolian Deposition.** They include shallow pans in dune swales or in hollows in sand plain, and they may be fed by local runoff and seepage or by the flooding out of exogenous drainage.

5. **Deflation Lakes.** These include *lunettes* [63], which are part erosional, part barrier lakes, and also shallow clay pans formed by the deflation of alluvial topsoils [64].

6. **Water Holes in Drainage Channels.** These are situated in locally scoured tracts in active and abandoned ephemeral channels, and retain water after flooding.

G. Modification of Surface Drainage

Several techniques have been used to modify surface drainage, but the most significant in arid and semiarid regions is probably that commonly referred to as *water spreading*, implying the diversion of channel flow over riparian tracts. Subject to water rights down valley, it aims to bring about complete infiltration of the spasmodic flood flow of semiarid channels. It may be employed for flood irrigation of pastures or crops or for artificial recharge of groundwater (discussed later).

Water spreading to improve native pastures has been fully described by Stokes, Larson, and Pearse [65]. Apart from a suitable drainage basin, the primary requirement is a gently sloping surface with soil cover. On larger projects a retention dam for temporary water storage may be built, combined with a system of dykes to convey water to spreading grounds. Elsewhere, a series of smaller diversion barrages may be constructed across the channel, or small diversion ditches may be carried from the main channel without storage works.

The two main methods of spreading from diversion dykes are the *ponding* and *wild-flooding* systems. The former is suited to gentler slopes, preferably less than 1 per cent. It consists of a series of ponding dykes, so constructed that the water moves slowly from one dyke to the next, covering the area between them with a sheet of water. On steeper slopes the wild-flooding type of spreading is used. This consists of releasing water at numerous points at the head of a slope and allowing it to flow freely, with secondary spreader dykes to keep the flow diffused. Such spreading schemes are particularly suited to secondary drainage outlets in piedmont zones, where flood discharge is unlikely to exceed that which can be contained by the earth-fill type of storage dam.

A simpler form of water spreading from diversion barrages across the drainage channel has been successfully employed in an alluvial valley, longitudinal gradient 1 in 500, tributary to the Fitzroy River in the West Kimberley area of Western Australia. Earth-fill dams were constructed at intervals of about 1 mile along the main channel, which is up to 50 ft wide and entrenched up to 10 ft in its fine-textured alluvium. Diversion wings were extended up to 50 yd from the channel, employing the natural slope of the discontinuous levee to carry water to the backplain. The water then passed back to the creek via shallow floodplain channels.

Water spreading as applied to crop irrigation is exemplified by the *saaidam* (sowing dam) system employed in shallow alluvial valleys in the northwest of the Cape Province of South Africa. This is an area receiving an average of 5 in. of rain annually, where the rivers generally come down in flood on one or two occasions each year. When the river runs, the water is diverted at favorable points into shallow basins formed by mud embankments up to 3 ft high. The basins are interconnected by sluice gates, so that when the floor of one basin is thoroughly wetted, the water can be allowed to pass down valley. Wheat is then sown in the mud floor and may come to maturity without further watering.

V. GROUNDWATER IN ARID AND SEMIARID REGIONS

A. General Considerations

Since the origin of most groundwater is meteoric, the primary control of supply is the amount of infiltration in the intake zone and, where the intake zone is sited in a low-rainfall area, groundwater supply will be correspondingly small. However, many intake zones lie outside the arid zone in which the water is exploited. In North Africa, for instance, the Atlas Mountains are intake zones for the adjoining Saharan lowlands, while much of arid Central Australia draws its subsurface water supplies from the eastern highlands.

Many of the basic controls are geological. There must be a suitable aquifer with outcrop in the intake zone, and the migrating water must be contained or trapped to make it available.

Relief is also important since the groundwater table tends to follow the ground surface, but with more subdued relief. Given uniform geological conditions, it is deeper below interfluves, which are areas of runoff, and shallower below valley zones which receive runoff. It will of necessity be more strongly inclined in mountainous areas, with consequent effect on supply. In areas of complex geological structure it may be at considerable depth. Springs form where the ground surface intersects the water table, as in incised valleys or piedmont zones. In lowland areas, the water table is generally gently inclined, parallel with the surface, and at shallow depth.

Relief also affects groundwater indirectly by its effect on recharge. This is particularly true of the arid zone, where preliminary concentration by runoff is a normal condition of infiltration. The steep, rocky hill slopes of desert uplands give a relatively high, rapid runoff response, whereas the low gradients of desert plains lead to more sluggish or dispersed runoff and higher evaporation loss. Accordingly, intake is much more pronounced in upland zones.

Subsurface drainage tends to parallel surface drainage, although with more generalized patterns, and to share the same base-level controls. In arid areas this is less likely to be sea level than a local base level of an interior drainage basin. Under natural conditions, most large central lakes or pans, such as the intermont shotts of Algeria, also form subsurface drainage foci and act as seepage evaporative zones within a hydrologic system.

Surface-drainage capture may thus cause significant changes in subsurface drainage and groundwater levels, and Lòpez de Llergo [66] has described the draining of lakes and the depletion of groundwater resources in the Tertiary lacustrine fill of the Mexican altiplano through capture by headward regression of streams in the adjoining Sierra Madre ranges.

Drainage also exercises a more localized control on groundwater levels through its effect on recharge. The general pattern of arid drainage is one of effective concentration of runoff in close valley networks in the uplands and of dispersal and loss by percolation in the lowlands. This tends to cause alluvial deposition on low ground, which is of great importance to the distribution of shallow groundwater reserves. Much of the percolation of surface drainage occurs in the piedmont zones, particularly where they consist of coarse materials. Where large channels persist across desert lowlands, they may control the level of the water table locally. For instance, at times of flood the Murray River in Australia may raise the water table in a tract extending up to 7 miles from the channel, while at low river level there is return seepage [67] and the effect of the Nile flood is still more widespread. It is a feature of the small ephemeral channels of truly desert origin, however, that base flow from groundwater does not occur to any extent, and when the channel deposits are particularly impermeable, the river may flow on a perched water table independent of regional groundwater.

The spasmodic, ephemeral flow of desert drainage is reflected in episodic recharge of underground water supplies and corresponding irregularity in the level of the water table. However, since subsurface flow is slow compared with that of surface drainage and since groundwater supplies are not subject to the same high losses as surface water in arid areas, the fluctuations of rainfall and runoff are reflected in water-table movements only in reduced measure. This equalization of supply is perhaps the most valuable attribute of groundwater in arid regions.

An idealized scheme of relief and groundwater in a semiarid area of range and basin fill has been described by Loehnberg [68]. Folded mountain ranges and the piedmont fringe of coarse alluvial materials form the intake zones, generally with deeper groundwater, which may, however, emerge in intermont and piedmont springs. The lower part of the alluvial fill is regarded as consisting of less permeable, fine-textured sediments, and the water table here approaches the surface closely, eventually emerging in a broad seepage zone in a central playa.

Loehnberg's scheme is applicable to much of the western United States and the Sahara, but is only locally relevant to the Southern Hemisphere deserts, where the lowlands may comprise broad structural basins of sedimentary rocks or plains eroded on the crystalline basement.

Groundwater exists in two forms: unconfined, or phreatic, groundwater, governed solely by hydraulic gradients, and artesian, or piestic, groundwater, confined by a cover rock or aquiclude and under pressure such that it will attain a piezometric level above the ground surface.

B. Recharge

Recharge of groundwater may take place directly from rainfall, from surface drain age following concentration by runoff, or from an aquifer with intake outside the area. The last case is not under arid-climatic control and will not be treated here.

Recharge from rainfall is relatively less important in arid than in humid areas, mainly because of high evaporation losses. Most estimates place it below 5 per cent of the rainfall and less than 10 per cent of the total recharge. The most important factor controlling recharge from rainfall is the nature of the ground surface and the subsurface layer, since rapid infiltration is essential.

On highly permeable sand surfaces with little or no surface runoff, any local recharge must clearly come from rainfall. However, observations from the Kalahari Sands [67] and the Australian sand plain agree that this water seldom reaches the water table. The widespread occurrence of soluble surface deposits, such as surface limestones, also indicates that moisture penetration is normally restricted. On the other hand, Schoeller [69] has claimed that rainfall can be the only source of the remarkably fresh groundwater at Beni-Abbès in the Western Grand Erg of the Algerian Sahara, and Ward [70] states that infiltration of about 5 per cent of a total rainfall of 15.5 in. occurs in coastal dunes of South Australia. It seems likely that such recharge may be limited to years of exceptional rainfall.

Fissured rock surfaces with low field capacity are probably most favorable to rainfall recharge. Structure may be a secondary factor here, for where the dip is into the slope, infiltration is favored at the expense of runoff.

Any factor affecting runoff, in particular degree of slope, will clearly influence the proportion of rainfall which enters the soil. On gently sloping interfluves, for instance, rainfall will obviously form a larger part of total recharge, although this may be small in amount.

South Australia provides an interesting example of the influence of vegetation cover and transpiration loss on infiltration from rainfall [70]. On sandy plains with little surface drainage, the supplies of useful groundwater are found under grassland belts, adjoining areas of mallee (*Eucalyptus*) scrub yielding only small supplies.

The nature of the rainfall itself will also control the amount of infiltration. The most favorable conditions are long-continued steady rainfall with low temperatures and evaporation. These are obviously rarely attained in deserts, but Schoeller [71] has stressed that the high-latitude Northern Hemisphere deserts receive most of their rain in winter and that this may explain a greater degree of infiltration from rainfall in such areas.

Generally, aquifers fed directly by rainfall have high-quality water, but are commonly associated with a low ratio of storage to flow, and hence with considerable water-table fluctuations.

In areas with organized surface drainage, by far the greatest contribution to groundwater is via the drainage, whereby a greater head of water is applied to limited areas for longer periods of time [53].

In upland areas, recharge from streams will penetrate the underlying rock through suitable fissures, the outstanding example being in limestone areas, where whole sectors of the drainage may pass underground. The sandy floors of ephemeral drainage channels generally allow rapid infiltration of flow. Babcock and Cushing [72] have described infiltration into the channel of a piedmont arroyo in Arizona. More than 50 per cent of discharge was lost within 20 miles of the mountain front by infiltration into the coarse alluvium in which the channel was entrenched. Infiltration rates ranged from 2 to 25 in./day. They were smaller in the violent silt-laden summer floods than in the winter flows, and there was a progressive decrease in seepage rates down valley with increasing fineness of channel-bed material. Major floods from larger channels, covering larger areas and probably remaining for some time in natural depressions, provide greater recharge at longer intervals.

Recharge into alluvial fill is particularly important in piedmont zones, where it is favored by the greater frequency and volume of flow and by the coarseness of the alluvium. Examples of this are given by Miles [73] for the streams entering the Adelaide Plains from the Mount Lofty Ranges in South Australia and by Linsley [74] for the similar intake areas fringing the Santa Clara Valley in southern California.

Whether direct from rainfall or via drainage, recharge in desert areas will naturally be irregular and discontinuous, reflecting the variable nature of rainfall and runoff. Discontinuous recharge is normally opposed by deep and permeable zones of intermittent saturation between the ground surface and the water table.

C. Phreatic, or Unconfined, Groundwater

A useful review of the geology of the main types of aquifers in arid regions has been given by Picard [75]. He stresses the importance of connected fissuring rather than porosity as determining the capacity and yield of aquifers, for which reason limestones may be more important than sandstones. While this is true of Israel and the Moroccan Atlas, porous sandstones are elsewhere of the greatest importance as aquifers, particularly the Mesozoic sandstones of the Algerian and Tunisian Atlas and of the Great Australian Basin. Unconsolidated alluvium or marine sands of younger geological age form another important type of aquifer. The main retaining strata are argillaceous rocks, dense limestones, and impervious lava sheets.

In addition to a suitable aquifer, some form of trap is necessary to contain and store groundwater. The following outline classification is based mainly on Picard [75].

1. Rock-floored Alluvial Basins in River Valleys. These occur upslope from rock barriers, mainly in intermont zones, as a result of uneven channel scour followed by alluviation. The basin in the Todd Valley at Alice Springs, Australia, has an area of 3 sq mi and a depth of alluvium between 20 and 60 ft and is underlain by metamorphic rocks. Actual recharge is in excess of current pumping of 15 million cubic feet per year, as there is no sign of regional depletion.

In many arid areas underlain by nonporous crystalline rocks there is no continuous water table but a general rest level [67]. The groundwater moves to find storage in fissures or weathering pockets, some of which may be inherited from earlier, moister periods.

In such small basins recharge is very irregular and there is a considerable variation in the amount and quality of supply.

2. Stratigraphic Traps in Which Groundwater Accumulation Is Primarily Conditioned by the Form of Deposition. This class includes various forms of unconsolidated "fill," which is very extensive in arid and semiarid lowlands.

a. Underflow in a Stream Channel. Influent seepage into porous stream-channel deposits is important in arid areas, and in fact accounts for many of the characteristic features of the ephemeral channel regime. Water is generally available at shallow depths in the larger channels for varying periods following stream floods. Flow may occur above rock floors in upland tracts or on an impervious alluvium in areas of deeper fill. In such cases underflow will be short-lived and supply and quality will vary. Underflow may connect with regional groundwater, the water table generally descending away from the channel, as in the case of the Murray River, referred to above.

Schoeller [71] has remarked that very few of the Saharan wells are sited outside wadi floors, and this close connection between shallow groundwater supplies and drainage lines holds good for most arid areas. Lowdermilk [76] has described how underflow in arroyo channels in Arizona is intercepted by subsurface dams, or *tapoons*, which give increased recharge to shallow wells sunk in the channel nearby.

b. Basin and Valley Fill. These range from geologically youthful intermont fills of the Basin and Range Province of the western United States to extensive coastal plains, as in Israel, and to shallow alluvial fills in old broad valleys on high-level erosional plains, which occur extensively in interior Australia and Africa.

In basins with deep fill and strong marginal relief, the main intake zones are the piedmont tracts of coarser alluvium at the outlets of upland drainage. The water table may deepen and pass below the level of small-scale exploitation away from the upland border. In the central parts of such basins the water table may again approach the ground surface toward the central playa. Since such central areas may consist of fine-textured alluvium, yields may be low, while quality is often poor.

Storage-flow ratios are normally medium to high. Picard [75], no doubt influenced by the experience of the Galilean coastal plain, states that yields tend to be medium or low. The permeability and porosity of the fill materials influence other yields which vary considerably. Particularly high yields may come from sand or gravel lenses, as in the "deep leads" of the Victorian and Western Australian goldfields and in the deep-lying sandbeds of the Murrumbidgee Plains [77]. Layers of fine alluvium form perched water tables. In the lower parts of intermont basins, and particularly in coastal tracts, fresh water may rest on salt water, and problems of salt-water ingress may arise in exploitation.

The valley fills of erosional desert plains are generally shallow and are fed by extensive watersheds of gentle relief. This type of terrain is well exemplified by the interior plateau of Western Australia, where mature drainageways with gradients between 1 in 250 and 1 in 1,000 have been alluviated and where such depositional plains occupy more than 30 per cent of the surface. The fill consists mainly of medium- and fine-textured alluvium with storage capacity generally less than 5 per cent, and subsurface flow may be less than 15 ft/day [78]. Although recharge may be less than 5 per cent of the total rainfall, extensive tributary watersheds may result in significant storage in such valleys. Quality is particularly good in the upper, marginal sectors and remains suitable for stock in all but the lowest parts of the trunk valleys near salt-lake terminals.

c. Sand Dunes. Sand dunes commonly contain small reserves fed directly from rainfall. In South Australia, supplies from inland dunes have been used for the transcontinental railway near Ooldea, while coastal dunes provide small supplies at many points. In the coastal dunes the fresh water may rest on a salt-water table.

Larger supplies are obtained where dunes have obstructed surface drainage. At Walvis Bay in South-West Africa, the outlet of the Kuiseb River is blocked by coast dunes and the ephemeral floods rarely break through to the sea. The dunes supply good water in adequate amounts for this small port.

3. Traps Due to Intrusive and Volcanic Rocks. These fall into two classes:

a. Dyke Barriers. In South Africa, dolerite dykes act as barriers to migrating groundwater in the Mesozoic Karroo sandstones. Although individually small, water storages of this type probably account for more than half of all groundwater supplies in interior South Africa.

b. Lava Sheets and Fills. Impervious lava sheets and fills interstratified with permeable sediments and tuffs have widespread occurrence in the Middle East, in particular in Yemen. Best yields are obtained from wells drilled to the impervious sedimentary basement on the downdip side [75], but supplies are very variable.

4. Structural Traps. The largest groundwater reserves are formed where earth movements have caused the uplift of permeable rocks in mountain intake areas and have disposed these aquifers to form groundwater reservoirs.

a. Fold Structures. Close folding, involving compaction, thrusting, and fracturing of sedimentary rocks, is unfavorable both to groundwater storage and exploitation. On the other hand, open folding may provide abundant groundwater under pressure in synclinal tracts. The high intermont plateau of Algeria can be regarded as a synclinorium of regional size in which the shotts are artesian seepages, with bores providing up to 3.5 ft^3/hr.

b. Basin Structures. These are extensive in the basin and swell patterns of the

Southern Hemisphere continents. Since they form the most important areas of artesian water, they are discussed below.

c. Fault Structures. Faulting may form a structural trap by bringing an aquifer into contact with impervious rock. The displacement and tilting of fault-block segments may also cause outcrop or shallowing of the piezometric surface and the ponding of groundwater supply, as exemplified in the Galilean uplands, where the Cenomanian limestones yield an average of 200 m³/hr in bores in fault-scarp zones [75].

D. Artesian Water

Artesian waters occur where the piezometric level is above the ground surface, i.e., where there is sufficient pressure to cause outflow at the surface. Pressure of flow is due mainly to the hydraulic head, or height below the intake zone, but may also result from elastic deformation of the aquifer. Storage-to-flow ratios of artesian water are typically very high.

Artesian water is mainly associated with basin structures in fairly consolidated sedimentary rocks. The requirements are one or more porous aquifers with outcrop in a relatively humid intake area at high level, aquicludes above and below the groundwater reservoir, and sufficient amplitude of structure and relief to bring the ground surface below the piezometric level.

Australia provides many examples of artesian basins, and useful summaries are given by Ward [70] and Hills [79]. The typical features are illustrated by the Great Australian Basin, the largest in Australia, with an area of 580,000 sq mi. It is a complex basin, with three main aquifers consisting of sandstones ranging from Triassic to lower Cretaceous in age and with Cretaceous shales forming the main aquiclude. The main intake area is in the east of the basin, where the sandstones outcrop over an area of 40,000 sq mi between 1,000 and 2,000 ft above sea level and with an average rainfall of 25 in. River floodouts provide a smaller intake in the northwest and west of the basin, where the aquifer is at the surface. The deepest part of the structure is in the southwest, where the basement rocks are 7,000 ft below sea level in Lake Eyre. Bores are more than 4,000 ft deep in this area. The surface form of the basin corresponds largely to its structure, so that the direction of subsurface flow accords closely with the surface drainage. There is a northern component tributary to the Gulf of Carpentaria, while the remainder is directed southwestward to Lake Eyre or south-southwestward along the line of the Darling River. The piezometric level is never far above the topographic surface. The natural outlets of the artesian waters are the calcareous mound springs which occur in the lowest parts, as in Lake Eyre, and also at rises in the level of the basement rock.

Yields generally exceed 1,500 ft³/day. Salinity increases with distance from the intake areas, but the waters of the Great Australian Basin are notable for their low proportion of calcium and magnesium carbonates and of chlorides. It is possible to distinguish the carbonated water of the eastern intake from the sulfated water originating from the minor intake areas in the west. The temperature of the water increases with depth, temperature gradients being abnormally high in this basin. The highest temperature recorded is 230°F (110°C) from a depth of 5,700 ft at Springleigh in western Queensland.

Following an early period of diminishing pressure and loss of artesian flow locally, associated with unregulated exploitation, it would appear that some areas near the eastern intake are already approaching the steady state. Legislation has been introduced to control the rate of diminution of pressure and flow by the throttling down or shutting of bores and by control of the development of new bores.

E. Modification of Groundwater Storage

Water storage as groundwater is not restricted to arid and semiarid regions, but has particular advantages in these areas. In the first place, the heavy evaporation losses from surface reservoirs are avoided. Second, underground storage capacity exceeds that of most surface structures. This is important because of the highly variable character of desert streamflow, particularly in semiarid regions of intermittent snow-fed streams, where high streamflow and agricultural water needs may not coincide seasonally, or where seasons of high surface flow may alternate with drought years. Underground storage is relatively cheap. It avoids the difficulties of major constructional work in tectonically unstable areas such as California. Silting difficulties, due to the characteristically high load carried by desert floods, are largely avoided. Aquifer recharge fits well into rational schemes of flood control. Lastly, recharge serves to purify water supplies for drinking purposes. An annotated bibliography of this subject has been prepared by Todd [80].

There are two main methods of recharge. In the case of closed aquifers, infiltration may be carried out from wells sunk through the impervious cover into the aquifer, preferably to the water table. Infiltration from such wells may be aided by galleries driven into the aquifer. This method is also suitable in areas where land values are high. The water should be clear, while blockage of the well intake by silting must be avoided by screening and surging.

In open aquifers, recharge may be carried out by infiltration from the stream channel or by water spreading over porous intake beds. This latter method is particularly suited to piedmont fans, where channels of moderate size open on to suitably sloping alluvial terrain built of very porous, coarse deposits. The piedmont zone on the east of the Californian Central Valley, with its snow-fed streams from the Sierra Nevada, is one of the main areas of development.

Check dams are built near the valley outlet as diversion or occasionally as storage structures, from which water spreading may be carried out. Recharge may be from excavated basins or impoundments of drainage channels or from a furrow or ditch reticulation. Basins are more costly to develop and maintain, but bring a higher percentage of groundwater into contact with the intake surface.

Recharge is carried out at as high a level as possible in the watershed, thus allowing maximum distribution of stored groundwater. For instance, distributary ditches are run close to the heads of alluvial fans, from which the natural slope is utilized.

The main factors controlling recharge are as follows;

1. *Nature of the catchment.* Minor upland catchments are best suited to water spreading, since flooding in large channels is difficult to control.

2. *Slope.* Where gradients are steep, check banks and diversion banks are necessary to reduce rates of flow. The radial slopes of alluvial cones provide very favorable dispersal forms.

3. *Nature of the intake surface.* The surface must consist of porous deposits such as unconsolidated sands and gravels. Infiltration rates cited by Todd [80] mainly range from 0.5 to 10 ft/day. Infiltration rates have been found to be higher on vegetated than on bare surfaces, which tend to seal more rapidly.

4. *Nature of the water.* A major problem affecting infiltration is silting, since streams of the type employed for water spreading are commonly heavily silt-charged. In extreme cases, settling ponds may be used before reticulation and spreading. Ditches can be kept clear of silt by maintaining sufficient grade and velocity to prevent the deposition of suspended load, while flood scour may be utilized to prevent the sealing of channel-basin systems. Basin floors are often scarified at fairly frequent intervals, while ridging of the floors also restricts the area of silt blockage.

Sodium-rich water tends to reduce infiltration by causing dispersal of clay material, but hard waters are quite suitable for recharge.

5. *Head of water.* Maximum recharge is obtained by spreading a thin sheet of water over a large area. Infiltration basins are generally shallow, depths being maintained less than 1 ft.

6. *Groundwater levels.* The water table should be low or falling, and there should be no obstruction by perched water tables.

The most important area of water spreading in piedmont zones of high relief is that of southern California, where spreading basins extend over more than 12,200 acres with an infiltration capacity of more than 5,000 ft^3/sec. Clyde [81] estimates the average recharge in this area per year as 75,000 acre-ft.

As an example of aquifer recharge in lowland areas, Lowdermilk [82] describes a method employed in alluvial valleys on the crystalline plateau of Southern Rhodesia. Earth-filled dams are constructed across these shallow drainageways, exposing the coarse weathered regolith in barrow pits on the upslope side. These pits act as shallow reservoirs and intake basins, which may be filled by surface flow on two or three occasions each year. Leakage occurs beneath the dams and replenishes shallow wells sunk in the alluvium on the down-valley side.

VI. IRRIGATION IN ARID REGIONS—SOME SPECIAL FEATURES

A. Some Adverse Effects of Arid Climatic Conditions on Physiological Responses of Plants

1. Relationship between Plant Growth and Water Requirements. Extreme radiation conditions and low atmospheric humidity are probably the major climatic features which may adversely influence plant growth under irrigation in arid regions.

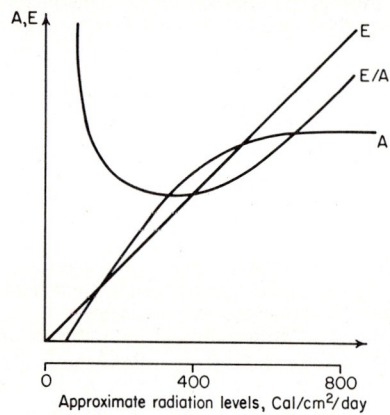

FIG. 24-9. Typical interrelationships between transpiration, assimilation, and radiation. (*After de Wit* [84].)

When water is nonlimiting, transpiration is primarily controlled by the prevailing weather. Under conditions where an extended surface of uniform, physiologically active plant material is available, transpiration can be quantitatively assessed from weather elements. Growth, in terms of net assimilation or dry-matter increment, is influenced by the same weather elements, but whereas evaporation and transpiration are roughly proportional to the total incoming radiation, net assimilation reaches a maximum rate which depends on the *saturation radiation intensity* and on *leaf area index* (the ratio of total leaf area to total ground surface). A number of research workers have investigated the relationship between transpiration and growth, and the "water-requirement," or "transpiration-ratio," experiments of Briggs and Shantz [83] are classical in this regard. While the transpiration-ratio concept is open to specific criticisms in application, it is of value to examine it from the viewpoint of efficient utilization of available water, the primary objective of arid-zone plant production.

Recently de Wit [84] has shown the typical relationship between transpiration and assimilation to be as in Fig. 24-9. This indicates that while transpiration E is proportional to radiation, positive assimilation A does not commence until radiation reaches a certain minimum intensity. It then increases more or less proportionally with radiation until moderate intensities are reached, beyond which little, if any, increase in assimilation occurs. Thus, at high radiation intensities, the ratio E/A varies more or less proportionally with radiation, at low-moderate intensities it is almost constant, and at very low intensities it increases again. This relationship is affected by temperature since photosynthesis is less influenced by temperature than respiration, but this influence is only of importance when respiration is large compared with photosynthesis. Since, at normal light and temperature conditions, the ratio of these processes is about 1:7 [85], the effect of respiration is likely to be significant only at low light intensities and high temperatures. It is apparent that these conditions do not frequently apply in arid regions.

Although Fig. 24-9 suggests that it is desirable to grow plants under conditions which favor a constant and low E/A ratio, most arid regions are located between latitudes 20 to 40°, and characteristic values of total incoming radiation range in such regions from 400 cal/cm²/day in winter to about 800 in summer. From Fig. 24-9 it is apparent that the E/A ratio commences to increase once again at values close to the former figure, so that for only short periods each year is plant production likely to be as efficient in many arid regions as in areas receiving less radiation. Apart from this important feature, several other factors affect the E/A ratio and are of special influence in arid regions. These include internal water deficits, plant species, and density.

There is now general agreement that the availability of soil water for plant growth decreases progressively as soil-moisture stress increases [86] and that the first evidence of decreased growth occurs at quite low stress levels. Evidence to the contrary [87] is based mostly on field experimentation, the proper interpretation of which is rendered difficult because of the variation which exists between different experiments with respect to root distribution, total depth of the root zone, and other experimental factors. Reductions in growth appear to be caused primarily by decreasing hydration of the plants, since the degree of water stress in the plant is limited by the soil-water stress, and with increasing stress there is essentially decreasing turgor in the plant tissues. This is reflected in a progressive decrease in growth.

It is to be expected that growth should cease at or before the turgor pressure in the cells of the active tissues falls to zero since it is turgor pressure in any one plant cell which causes cell and tissue expansion. This contention is supported by most studies that bear directly on this point [88, 89], as well as by direct evidence from experiments concerned with effects of water stress on elongation and photosynthesis. Photosynthesis is directly affected by water stress as a result of decreased tissue hydration and disturbed metabolism, and indirectly as a result of stomatal closure and its influence on CO_2 exchange. In general, apparent photosynthesis commences to decline at low water-stress levels and to decrease progressively until it reaches zero at a stress level fairly close to zero turgor pressure, but depending on the plant and tissue involved [90–92].

The effect of decreased turgor on transpiration is not as direct as on essentially physiological functions. Transpiration is basically a passive process and, as such, is determined primarily by the energy available for evaporation, the vapor-pressure gradient from leaf to air, and the rate of water supply to the roots. Since the effect of low leaf water content in itself is not likely to affect transpiration until severe wilting occurs, the main regulating influence of the plant lies in the effectiveness of stomatal closure. Stomatal closure affects the diffusion of CO_2 and water vapor similarly, and it is to be expected that the effect of water deficits on the efficiency of assimilation will be small. In general, this view is also supported by evidence in the literature, most studies demonstrating that as water stress increases, transpiration and apparent photosynthesis, or growth, decrease more or less proportionally [93, 94]. Beyond the point of zero turgor pressure, further increase in water stress usually results in complete inhibition of photosynthesis, but cuticular transpiration may continue for a significant period. Thus in this stage the transpiration-assimilation ratio increases steeply and the efficiency of assimilation declines.

The early studies of Briggs and Shantz [83] indicated that under the same climatic conditions the transpiration ratio of different species varied considerably. The interpretation of these investigations conducted in pots in a greenhouse requires some clarification, but it is still apparent that some species are more efficient producers of dry matter than others, for the same expenditure of water. Watson [95] has shown that this difference may be largely attributable to different morphological characteristics, and in sugar beet, for instance, it appears probable that the higher yields characteristic of the more modern varieties are due primarily to a different leaf arrangement, which enables more efficient light utilization, and hence a higher rate of photosynthesis per plant.

Under field conditions transpiration may reach maximum rates in a developing crop before photosynthesis is at a maximum, since leaf area index values below the optimum for the latter process may be adequate for peak transpiration rates. Thus transpiration may be increased relative to assimilation (even though an absolute increase in assimilation may also occur because of better illumination) and the efficiency of plant production reduced. This may provide the explanation for the more efficient transpiration-growth ratios of evergreen trees compared with deciduous forests, or of forests generally compared with agricultural crops. In regions where leaf area index values are seldom optimal for assimilation, it can be expected that the efficiency of production will always be below maximum values.

In summary, it appears possible to state, without misleading oversimplification, that the efficiency of water utilization (i.e., the relationship between transpiration and

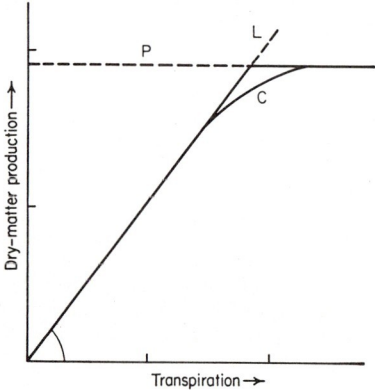

FIG. 24-10. Typical interrelationships between transpiration, dry-matter production, and plant characteristics.

growth) is relatively unaffected by availability of soil water, unless this factor reaches extreme limits, and that radiation exerts a profound influence on the efficiency of assimilation as it rises above moderate levels. A simple model (Fig. 24-10) may be constructed to describe this relationship in the form suggested by de Wit [84], in which the slope of the line L varies with plant species, density, and horizontal extension of the plant surface. It provides an estimate of the relationship between transpiration and production when yields are limited only by shortage of water and is a characteristic for any one plant community, whether natural vegetation or an irrigated crop. The line P represents the expected yields when water is nonlimiting; that is, after a certain stage is reached, further water increments will not increase yields. Theoretically, the minimum amount of water necessary to obtain maximum production is given by the intersect of lines L and P. Under practical conditions the particular characteristics of the plant-soil surface may impose modifications and the line C may approximate reality. Even so, this generalized relationship has special significance to an understanding of the physiological and agronomic background to irrigation agriculture.

2. The Influence of Local Advection. The efficiency of plant production in arid irrigation areas may also be affected by advection, which tends to increase evapotranspiration relative to assimilation. Problems of advection are of particular importance in arid regions where irrigated fields adjoin deserts.

Comparing the situation in an irrigated area with that of the adjacent dry land, it is obvious that the extra water available causes more energy to be consumed in evaporation and less in heating the air and the soil. The temperature of the lowest layers of air is therefore reduced by irrigation, while the humidity of these layers is increased. In addition, the lower surface temperature leads to a reduction of the emission of long-wave radiation from the ground. The degree of modification of the energy balance depends mainly on the irrigation rate, on the dryness of the dry land, and on the rate of advection of warm and dry air into the wet region.

Most attempts at quantitative studies in micrometeorology have been based on the assumption of horizontal homogeneity and equilibrium profiles. The possibility of advection has been recognized mainly in a negative way in that advection is sometimes invoked to explain otherwise inexplicable observations.

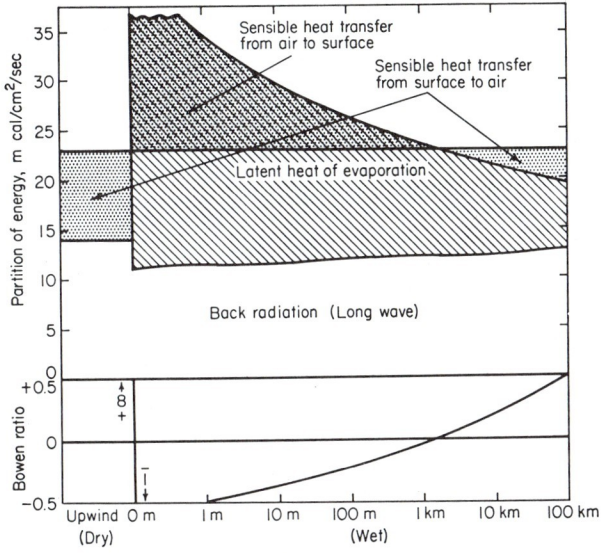

FIG. 24-11. Energy partition in a typical irrigation area downwind of a dry surface. (*After Philip* [102].)

Until recently little was known quantitatively about these effects, but recently a number of studies have been carried out [96–101]. To give an example of the magnitude of advective factors in arid regions, a situation is quoted from Philip [102] for a typical irrigation area located in an arid zone. Wind speed at 2 m is assumed to be 4 m/sec. Upwind conditions are assumed to be dry, with zero evaporation, and the energy available for long-wave back radiation and sensible heat transfer is 23 mcal/cm^2/sec. Absolute humidity is 10 g/m^3, and air temperature at 2 m is 30.2°C. Downwind in the irrigated area, the surface is assumed to be completely wet, with the albedo the same as that of the dry area and likewise the soil heat flux. Philip has drawn up a typical energy balance for this situation, and the partition of energy is shown in Fig. 24-11, in which the energy available to the surface varies between the dry and wet areas and with distance downwind into the wet area. The figure gives a graphic picture of how the advective effect operates. One immediately obvious point is that the change in radiation balance is quite small. This is obviously a second-order effect; the major effect is of violent change in the partition of the remainder of the energy between the latent heat of evaporation and sensible heat transfer to the air. This is especially important immediately downwind of the leading edge, but the region of advective inversion extends for over 1 km, so that at this point the heat flux is still from air to surface.

Also, in this region of advective inversion the energy used as latent heat of evaporation may greatly exceed that available from the radiation balance. The excess is supplied by heat transfer from the air. These changes are reflected in the spatial variation of the *Bowen ratio*, the ratio of the sensible heat transfer to the latent heat transfer (Subsecs. 11-II and 23-V). In the upwind dry area this ratio is $+\infty$; immediately downwind of the leading edge it is -1. It remains negative for over 1 km; then it becomes positive and continues to increase slowly. The magnitude of the advection effect is at once obvious.

The variation of evaporation rate with position in the irrigation area is replotted in Fig. 24-12, together with estimates of the potential evapotranspiration of the area computed on the basis of one-dimensional considerations. The upper estimate is based on calculations which assume that air temperature and humidity at 2 m are unaffected by irrigation; the lower estimate, on the assumption that the net radiant energy less soil heat flux is available for evaporation. The former is more or less the estimate which would be made by the Penman [103] approach, while the latter would be obtained by the energy-balance approximations suggested by House, Rider, and

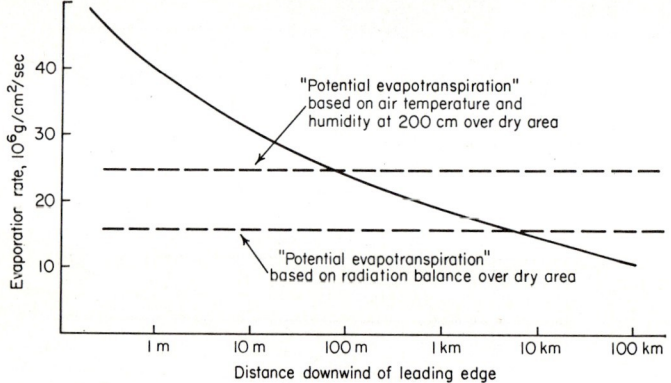

Fig. 24-12. Variation of evapotranspiration rate with position in irrigation area (using the situation described in Fig. 24-11), compared with estimates of evapotranspiration based on one-dimensional considerations. (*After Philip* [102].)

Tugwell [104], which give reasonable results over the diurnal cycle in extensive humid areas. (It might be mentioned that Penman did not intend that his equation or formulation should be used in what might be termed "oasis" areas, but it provides an estimate of the type of calculation available from a one-dimensional approach.)

It is obvious that these approaches are not well suited to the estimates of evapotranspiration in arid regions, and further, that any system of lysimeters or small test areas with or without guard areas is likely to give misleading predictions of the water requirements of areas which differ much from them in size. It is also of considerable interest in this diagram to note that the value computed for potential evapotranspiration based on the first method used is not obtained until 100 m downwind, and the value estimated by radiation-balance data is not obtained until 10 km downwind.

de Vries [100] has given similar examples based on an investigation in an arid area of southern Australia, and it can be appreciated that similar conditions will obtain in most other arid regions. Likewise, de Vries has shown the type of change in climate of the irrigation area itself, particularly the increase in humidity, which increases with distance downwind from the leading edge of the irrigated area.

It follows from this discussion that, from the viewpoint of water economy, it is advantageous to have a few large irrigated areas instead of many scattered small ones. This applies notably to conditions such as those in the Australian Riverina, where irrigated pastures often have linear dimensions of the order of only 100 m.

B. Salinity Problems in Irrigation Agriculture

1. Origin of Saline and Alkali Soils. Salinity problems usually arise from two sources, either through the use of saline waters for irrigation or through over-irrigation and a consequent rise of a saline water table to the root zone. This question is so vital to arid-zone irrigation that the question of whether irrigation agriculture can persist permanently is very largely a question of eliminating salinity effects. Since all water used for irrigation contains some soluble material, accumulation of these solutes in the root zone will ultimately affect plant growth unless some leaching occurs.

A useful account of the origin of saline water has been given by Chebotarev [105]. In the first instance, the original sources of salt constituents are the primary minerals found in the soils and in the exposed rocks of the earth's crust. As a result of chemical decomposition and physical weathering, the soluble constituents are gradually released from the minerals. In humid areas the salts are carried downward by rain into the groundwater and are ultimately transported by streams to the ocean. In arid regions, however, leaching is usually local in nature and the soluble salts may not be transported far. This is not only because there is less rainfall to leach and transport the salts, but also because the high evaporation rates characteristic of arid climates tend to concentrate the salts in groundwater, soils, and dry lakes.

In arid regions surface-drainage channels are usually poorly developed and drainage basins may have no outlet to permanent streams. The salt-bearing waters drain from the surrounding higher lands of the basin to the lower lands and may temporarily flood the soil surface or form permanent salty lakes. This occurs under characteristic arid conditions, and the subsequent mobility of the soluble salts depends to a considerable extent on changes in the hydrology of the areas themselves; for instance, increasing salinity in nonirrigated areas has been noticed following land clearing and a change in the hydrologic cycle resulting from the different magnitude of the various components. In some arid areas, rather than local salt deposits, extensive areas of country are affected by salinity to a slight extent, and this problem becomes acute only when a change in the hydrologic cycle, due to removal of vegetation or introduction of irrigation, tends to concentrate the soluble salts in the root zone of the soil.

In addition, salts of oceanic origin are deposited by rain, the amount in any one locality depending on the nature and amount of the rainfall, distance from the coast, and local orographic conditions. This has occurred extensively in southern Australia, and it is generally considered [106] that most of the salt now present in the soils of many semiarid and arid areas is the result of accumulations of small annual amounts brought in by rain over a long period during recent geological time. It has also been shown [107] that in some arid areas some of the salt in the rainwater is almost certainly of (local) terrestrial origin. Typical accessions of salt from oceanic origin range from 8 to 30 lb/acre/year of sodium chloride in Victorian regions having 20 to 30 in. of annual rainfall, and in areas within 50 miles of the coast the amounts can be 60 to 100 lb/acre/year.

Saline and alkali soil problems most often develop as a result of the irrigation of level valley lands which may be nonsaline and well drained under natural conditions, but may have drainage facilities inadequate to take care of the additional groundwater resulting from irrigation practice. In such situations the groundwater level may be raised to the root zone in a relatively short time. Water then moves to the soil surface as a result of evaporation, increasing the salt content of the surface soil and of the soil water and causing the development of saline and alkali soil conditions. Saline and alkali problems also arise if drainage facilities are adequate, but insufficient irrigation water is applied to provide for the necessary leaching of excess salts, or from the direct effect of irrigation with water which is too saline or has undesirable quality features.

2. Characteristics of Saline and Alkali Soils. Richards [108] has divided saline and alkali soils into three groups, and his definitions have formed the basis for the following notes.

Saline is used to describe soils in which the conductivity of the saturation extract is more than 4 μmhos/cm at 25°C and the exchangeable sodium percentage is less than

15. Ordinarily the pH is less than 8.5. In such soils excess soluble salts may be removed by leaching and again become normal soils. These soils are often recognized by the presence of white crusts of salts on the surface. The chemical characteristics of soils classed as saline are mainly determined by the kind and amount of salts present. Sodium seldom constitutes more than one-half of the soluble cations, and hence is not adsorbed on the exchange complex to any significant extent. The relative amounts of calcium and magnesium present in the soil solution and on the exchange complex may vary considerably. The chief anions are chloride, sulfate, and sometimes nitrate. Small amounts of bicarbonate may occur, but soluble carbonates are almost invariably absent. In addition to the readily soluble salts, saline soils may contain salts of low solubility, such as calcium sulfate and calcium and magnesium carbonates. Because of the presence of excess salts and the absence of significant amounts of exchangeable sodium, saline soils are generally flocculated, and as a consequence, the permeability is equal to, or higher than, that of similar nonsaline soils. These are not serious problem soils.

The term *saline alkali*, or *saline sodic*, is applied to soils in which the conductivity of the saturation extract is greater than 4 μmhos/cm at 25°C and the exchangeable sodium percentage is greater than 15. These soils form as a result of the combined processes of salinization and alkalization. As long as excess salts are present, the appearance and properties of these soils are generally similar to those of saline soils. Under conditions of excess salts the pH reading is seldom higher than 8.5 and the particles remain flocculated. If the excess soluble salts are leached downward, the properties of these soils may change markedly and become similar to those of non saline sodic soils. As the concentration of salts in the soil solution is lowered, some of the exchangeable sodium hydrolizes and forms sodium hydroxide. This may change to sodium carbonate upon reaction with carbon dioxide absorbed from the atmosphere. In any event, upon leaching, the soil may become strongly alkaline, with pH readings above 8.5. The soil particles may then disperse and the soil become unfavorable for the entry and movement of water and for tillage. Reintroduction of soluble salts may lower the pH reading and restore the particles to a flocculated condition, but until the excess salts and exchangeable sodium are removed from the root zone and a favorable physical condition of the soil is reestablished, the soils are difficult to manage. Saline sodic soils sometimes contain gypsum. When such soils are leached, calcium dissolves and the replacement of exchangeable sodium by calcium takes place concurrently with the removal of the excess salts, leading to development of a soil with quite desirable cation-exchange properties.

Nonsaline alkali, or *nonsaline sodic*, describes soils in which the exchangeable sodium percentage is greater than 15 and the conductivity of the saturation extract is less than 4 μmhos/cm at 25°C. pH readings usually range between 8.5 and 10. These are the *black alkali* soils which frequently occur in semiarid and arid regions. Except when gypsum is present in the soil or irrigation water, the drainage and leaching of saline sodic soils lead to the formation of nonsaline sodic soils. As mentioned above, the removal of excess salts in the former groups of soils tends to increase the rate of hydrolysis of the exchangeable sodium and often causes a rise of pH. Dispersed and dissolved organic matter present in the soil solution of highly alkaline soils may be deposited on the soil surface by evaporation, causing darkening and giving rise to the term black alkali. If allowed sufficient time, nonsaline sodic soils develop characteristic morphological features, because the partially sodium saturated clays are highly dispersed and may be transported downward through the soil and accumulate at lower levels. As a result, a few inches of the surface soil may become relatively coarse in texture and friable, but below where the clay accumulates, the soil may develop a dense layer of low permeability and of columnar or prismatic structure. Usually, however, alkali conditions develop in such soils as a result of irrigation. In such cases, sufficient time has usually not elapsed for the development of the typical columnar structure, but the soil has low permeability and is extremely difficult to till. The exchangeable sodium present in nonsaline sodic soils can have a marked influence on the physical and chemical properties. As the proportion of exchangeable sodium increases, the soil tends to become more dispersed. The pH reading may increase to

10, and the soil solution of such soils, although relatively low in soluble salts, has a composition which differs considerably from that of normal or saline soil. While the anions present consist mainly of chloride, sulfate, and bicarbonate, small amounts of carbonate often occur. At high pH readings and in the presence of carbonate ions, calcium and magnesium are precipitated; hence the soil solutions of nonsaline sodic soils usually contain only small amounts of these cations, sodium being predominant. These soils are extremely difficult to manage because the change in morphology is much more permanent. Because of the high sodium concentration, the colloids swell when wet and prevent water transmission. Other types of clays, including calcium clays, do not show this degree of colloidal swelling. One way to improve such soils is therefore to add calcium, which will replace the sodium in the clay complex and shift that element into solution. Massive applications of gypsum may achieve this quite effectively, but these have frequently been too difficult to apply, from both economic and technological angles. Recently, in Australia, an effective method of introducing the calcium ion has been developed by Davidson and Quirk [109] on the basis of dissolving in the irrigation water the minimum amount of electrolyte needed to maintain flocculation of the particular soil. In practice, gypsum is dissolved in the irrigation water instead of being merely suspended. The concept of maintaining flocculation by applying an electrolyte in this matter has not previously been applied to irrigation agriculture. It should not be confused with the addition of small amounts of gypsum to rather saline irrigation water in California so as to maintain a favorable calcium-sodium balance and avoid further adverse chemical changes in the clay. The amount of electrolyte required was 10 meq of calcium per liter in the irrigation water to prevent deflocculation.

3. Salinity Effects on Plants. In recent years a number of excellent reviews of the effect of salinity on soils and plants have been prepared, and the reader may be referred to them for more details [110–115]. Several of these papers give detailed responses of specific crops to salinity effects and the relative sensitivity of these crops to specific ion effects. For this reason only general aspects of salt tolerance will be considered here. It is commonly recognized that plant growth bears a definite relationship to the osmotic pressure of the root medium and that different species, and sometimes even different varieties, exhibit different sensitivities to increase in osmotic pressure.

In order to provide a dynamic basis for understanding osmotic effects on plant growth, Wadleigh and Ayers [88] proposed the concept of *total soil-moisture stress*, which they defined as a summation of the osmotic pressure of the soil solution and the soil-moisture tension. Accordingly, Wadleigh [116] developed a mathematical method to integrate the variables affecting moisture availability in the root zone. The concept of total soil-moisture stress has since been used extensively and has provided a firmer basis for irrigation technology where salinity is a problem.

The physiological basis for salt tolerance can be assessed in several different ways. Hayward [111] used three criteria of salt tolerance on the basis of the proposals made by Hayward and Wadleigh [117]. First, *salt tolerance* may be defined as the capacity of a plant to persist under conditions of increasing salinity. This criterion is used by the plant ecologist in evaluating halophytic environments, since the species most capable of persisting in such habitats constitute the climax vegetation of that area. Second, salt tolerance may be regarded on the basis of the productive capacity of a plant at a given level of salinity. This method of appraising salt tolerance may result in a different evaluation, since the capacity of a given variety to produce well at moderate levels of salinity does not always indicate its ability to survive at the higher levels. The agronomist finds this method of evaluation particularly useful in comparing the performance of strains and varieties of a given crop. The third criterion is the relative performance of a crop at a given level of soil salinity as compared with its behavior on a comparable nonsaline soil. The advantage of this method is that comparisons between species can be made more readily than by the two preceding methods. An evaluation of salt tolerance on this basis provides a more useful basis for selecting crops to be grown on a moderately saline soil.

In assessing the effect of saline and alkali soils on plant growth, Hayward [111] considered that it was related to (1) the increased osmotic pressure of the soil solution, which results in an accompanying decrease in the physiological availability of water to the plant, and (2) the accumulation of toxic quantities of various ions within the plant as a result of the increase of those ions in the more concentrated soil solution. Reviewing experimental evidence, Hayward concluded that accumulation of salts in the substrate inhibits plant growth primarily because of the increase in the osmotic pressure of the soil solution and the consequent decrease of water available to the plant. In a number of experiments, linear reductions in growth have been found with increasing osmotic concentrations of the substrates, and the reduction was largely independent of whether the added salts were, for example, chloride or sulfate, if they were supplied on an isoosmotic basis. The mode of action of the osmotic factor is still obscure, primarily because any simple osmometer concept which might imply that the salts are not diffusible in the plant system is obviously invalid. Bernstein [114] considers that the osmotic basis for growth depression may be associated with some osmotic subunit of the plant cell which cannot adjust, or that the adjustment is osmotic pressure of the cell sap is attained only at the cost of some retardation in growth. While this remains to be clarified, Hayward considered that the well-defined relationship between growth and osmotic pressure appeared to be more than coincidental. Associated with this reduction in plant growth is a reduction in transpiration, and good examples of this phenomenon are reported by Strogonov [115]. However, growth responds more sensitively and more profoundly to osmotic pressure than to transpiration, so osmotic substrates cannot be used to improve the efficiency of water utilization.

VII. REFERENCES

1. Meigs, Peveril: World distribution of arid and semi-arid homoclimates, *Rev. Res. on Arid Zone Hydrol.*, UNESCO, Paris, pp. 203–210, 1953.
2. Thornthwaite, C. W.: An approach toward a rational classification of climate, *Geograph. Rev.*, vol. 38, pp. 55–94, 1948.
3. Nimmo, W. H. R.: The world's water supply and Australia's portion of it, *J. Australian Inst. Eng.*, vol. 21, pp. 29–34, 1949.
4. White, G. F. (ed.): "The Future of Arid Lands," *Publ.* 43, American Association for the Advancement of Science, 1956.
5. Wisler, C. O., and E. F. Brater: "Hydrology," 2d ed., John Wiley & Sons, Inc., New York, 1959.
6. A series of *Rev. Res.* (with yellow cover) and *Proc. Symp.* (with grey cover) on UNESCO Arid Zone Research:

    I. Reviews of research on arid zone hydrology, 1953.
    II. Proceedings of the Ankara Symposium on arid zone hydrology, 1953.
    III. Directory of institutions engaged in arid zone research, 1953.
    IV. Utilization of saline water. Reviews of research, 1954.
    V. Plant ecology. Proceedings of the Montpellier Symposium, 1955.
    VI. Plant ecology. Reviews of research, 1955.
    VII. Wind and solar energy. Proceedings of the New Delhi Symposium, 1956.
    VIII. Human and animal ecology. Reviews of research, 1957.
    IX. Guide book to research data on arid zone development, 1957.
    X. Climatology. Reviews of research, 1958.
    XI. Climatology and microclimatology. Proceedings of the Canberra Symposium, 1958.
    XII. Arid zone hydrology. Reviews of research, 1959.
    XIII. Medicinal plants of the arid zones, 1960.
    XIV. Salinity problems in the arid zones. Proceedings of the Teheran Symposium, 1961.
    XV. Plant-water relationships in arid and semi-arid conditions. Reviews of research, 1960.
    XVI. Plant-water relationships in arid and semi-arid conditions. Proceedings of the Madrid Symposium, 1961.
    XVII. A history of land use in arid regions, 1961.
    XVIII. The problems of the arid zone. Proceedings of the Paris Symposium, 1962.

REFERENCES

XIX. Nomades et nomadisme au Sahara, 1962.
XX. Changes of climate. Proceedings of the Rome Symposium organized by UNESCO and WMO, 1963.
XXI. Bioclimatic map of the Mediterranean zone. Explanatory notes, 1963.

7. Wallén, C. C.: Fluctuations and variability in Mexican rainfall, in G. F. White (ed.), "The Future of Arid Lands," *Publ.* 43, American Association for the Advancement of Science, Washington, D.C., 1956, pp. 141–155.
8. Smiley, T. L.: "Geochronology as an aid to the study of arid lands," in G. F. White (ed.), "The Future of Arid Lands," *Publ.* 43, American Association for the Advancement of Science, Washington, D.C., 1956, pp. 161–171.
9. Christian, F. G., and W. J. Francis: Semiarid Regions, in C. O. Wisler and E. F. Brater (eds.), "Hydrology," 2d ed., John Wiley & Sons, Inc., New York, 1959, pp. 286–300.
10. Landsberg, Helmut: Statistical investigations into the climatology of rainfall on Oahu (T.H.), *Hawaiian Rainfall Contri. Meteorol. Monographs*, American Meteorological Society, vol. 1, no. 3, pp. 7–23, 1951.
11. Monteith, J. L.: Dew, *Quart. J. Roy. Meteorol. Soc.*, vol. 83, pp. 322–341, 1957.
12. Stone, E. C.: Dew as an ecological factor. I. A review of the literature, *Ecology*, vol. 38, pp. 407–413, 1957.
13. Milthorpe, F. L.: The income and loss of water in arid and semi-arid zones, *UNESCO, Rev. Res., on Arid Zone Plant-Water Relationships in Arid and Semi-arid Conditions*, Paris, pp. 9–36, 1960.
14. Slatyer, R. O.: Absorption of water by plants, *Botan. Rev.*, vol. 26, pp. 332–392, 1960.
15. Hofmann, G.: Die Thermodynamik der Taubildung, *Ber. Deut. Wetterdienst*, vol. 3, no. 18, 1955.
16. Duvdevani, S.: Dew gradients in relation to climate, soil and topography, in "Desert Research," *Spec. Publ.* 2, Research Council of Israel, Jerusalem, pp. 136–152, 1953.
17. Masson, H.: La rosée et les possibilités de son utilisation, *Ann. Ecole Super. Sci. Inst. Hautes Etudes Dakar*, vol. 1, 1954.
18. Arvidsson, I.: Austrocknungs- und Dürreresistenzverhältnisse einiger Repräsentanten öländischer Pflanzenvereine nebst einigen Bemerkungen über die Wasserabsorption durch oberirdische Organe, *Oikos*, 1951, suppl. 1.
19. Bowen, E. G.: Induced precipitation, in G. F. White (ed.), "The Future of Arid Lands," *Publ.* 43, American Association for the Advancement of Science, Washington, D.C., 1956, pp. 291–299.
20. Brier, G. W.: Some problems in utilizing water resources of the air, in G. F. White (ed.), "The Future of Arid Lands," *Publ.* 43, American Association for the Advancement of Science, Washington, D.C., 1956, pp. 314–319.
21. Schaefer, V. J.: Some relationships of experimental meteorology to arid land water sources, in G. F. White (ed.), "The Future of Arid Lands," *Publ.* 43, American Association for the Advancement of Science, Washington, D.C., 1956, pp. 300–313.
22. Fournier d'Albe, E. M.: The modification of microclimates, *UNESCO, Rev. Res. on Climatol.*, Paris, pp. 126–146, 1958.
23. Preliminary report on artificial inducement of precipitation with special reference to the arid and semi-arid regions of the world, World Meteorological Organization, Rome, 1954.
24. Hamilton, E. L., and P. B. Rowe: Rainfall interception by chaparral in California, California Forest and Range Experiment Station, 1949.
25. Slatyer, R. O.: Methodology of a water balance study conducted on a desert woodland (*Acacia aneura* F. Muell.) community in central Australia, *UNESCO Proc. Madrid Symp. on Plant-Water Relations in Arid and Semi-arid Conditions*, pp. 15–26, 1961.
26. Geiger, Rudolf: "The Climate near the Ground," transl. by M. N. Stewart, Harvard University Press, Cambridge, Mass., 1957.
27. Water-loss investigations: Lake Hefner studies, technical report, *U.S. Geol. Surv. Profess. Paper* 269, and *U.S. Geol. Surv. Circ.* 229, 1954.
28. "Lake Mead Comprehensive Survey of 1948–49," vols. I–III, U.S. Department of the Interior, 1954.
29. Follansbee, Robert: Evaporation from reservoir surfaces, *Trans. Am. Soc. Civil Engrs.*, vol. 99, pp. 704–715, 1934.
30. Philip, J. R.: Evaporation, and moisture and heat fields in the soil, *J. Meteorol.*, vol. 14, pp. 354–366, 1957.
31. Philip, J. R.: Evaporation from soil, *UNESCO Proc. Canberra Symp. on Climatol. and Microclimatol.*, pp. 117–122, 1958.
32. Hilgeman, R. H.: Changes in soil moisture in the top eight feet of bare soil during 22 months after wetting, *J. Am. Soc. Agron.*, vol. 40, pp. 919–925, 1948.
33. Milthorpe, F. L.: Plant factors involved in transpiration, *UNESCO Proc. Madrid*

Symp. on Plant-Water Relationships in Arid and Semi-arid Conditions, pp. 107–115, 1961.
34. Veihmeyer, F. J., and A. H. Hendrickson: Does transpiration decrease as the soil moisture decreases? *Trans. Am. Geophys. Union*, vol. 36, pp. 425–448, 1955.
35. Furr, J. R., and C. A. Taylor: Growth of lemon fruits in relation to moisture content of the soil, *U.S. Dept. Agr. Tech. Bull.* 640, pp. 1–72, 1939.
36. Wadsworth, H. A.: Soil moisture and the sugar cane plant, *Hawaiian Planters' Record*, vol. 38, pp. 111–119, 1934.
37. Chung, C. H.: A study of certain aspects of the phenomenon of transpiration periodicity, Ph.D. thesis, Ohio State University, Columbus, Ohio, 1935.
38. Martin, E. V.: Effect of soil moisture on growth and transpiration in *Helianthus annuus*, *Plant Physiol.*, vol. 15, pp. 449–466, 1940.
39. Slatyer, R. O.: Evapotranspiration in relation to soil moisture, *Neth. J. Agr. Sci.*, vol. 4, pp. 73–76, 1956.
40. Veihmeyer, F. J., and A. H. Hendrickson: Soil moisture in relation to plant growth, *Ann. Rev. Plant Physiol.*, vol. 1, pp. 285–304, 1950.
41. Hendrickson, A. H., and F. J. Veihmeyer: Permanent wilting percentages of soils obtained from field and laboratory trials, *Plant Physiol.*, vol. 20, pp. 517–539, 1945.
42. Alway, F. J.: Studies on the relation of the non-available water of the soil to the hygroscopic coefficient, *Nebr. Agr. Expt. Sta. Bull.* 3, 1913.
43. Alway, F. J., G. R. McDole, and R. S. Trumbull: Relation of minimum moisture content of subsoil of prairies to hygroscopic coefficient, *Botan. Gaz.*, vol. 67, pp. 185–207, 1919.
44. Batchelor, L. D., and H. S. Reed: The seasonal variation of the soil moisture in a walnut grove in relation to the hygroscopic coefficient, *Calif. Agr. Expt. Sta. Tech. Paper* 10, 1923.
45. Haise, H. R., H. J. Haas, and L. R. Jensen: Soil moisture studies of some great plains soils. II. Field capacity as related to $\frac{1}{3}$ atmosphere percentage and "minimum point" as related to 15- and 26-atmosphere percentages, *Proc. Soil Sci. Soc. Am.*, vol. 10, pp. 20–25, 1955.
46. Robinson, T. W.: Phreatophytes and their relation to water in western United States, *Trans. Am. Geophys. Union*, vol. 33, pp. 57–61, 1952.
47. Taylor, C. A.: Transpiration and evaporation losses from areas of native vegetation, *Trans. Am. Geophys. Union*, vol. 15, pp. 554–559, 1934.
48. Simaika, Y. M.: Alternative uses of limited water supplies in the Egyptian region of the United Arab Republic, *UNESCO Proc. Paris Symp. on Problems of the Arid Zone*, pp. 381–393, 1962.
49. Beadle, B. W., and R. R. Cruse: Evaporation control as a means of water conservation. *39th Texas Water Sewage Works Assoc. Short School*, College Station, Tex., 1957.
50. Mansfield, W. W.: Influence of monolayers on the natural rate of evaporation of water, *Nature*, vol. 175, p. 247, 1955.
51. Mansfield, W. W.: The influence of monolayers on evaporation from water storages. II. Evaporation and seepage from water storages, *Australian J. Appl. Sci.*, vol. 10, pp. 65–72, 1959.
52. de Martonne, Emmanual: Regions of interior-basin drainage, *Geograph. Rev.*, vol. 17, pp. 397–414, 1927.
53. Dubief, J.: Essai sur l'hydrographie superficielle au Sahara, vol. 1, Direction du Service de la Colonisation et de l'Hydraulique, Direction des Études Scientifiques, Birmandreis, Algiers, vol. 1, 1953.
54. Dixey, Frank: Geology and geomorphology, and ground water hydrology, *UNESCO Proc. Paris Symp. on Problems of the Arid Zone*, pp. 23–52, 1962.
55. Christian, C. S., J. N. Jennings, and C. R. Twidale: Geomorphology in "UNESCO Guide Book to Research Data for Arid Zone Development," by B. T. Dickson (ed.), Paris, 1957, pp. 51–65.
56. Holmes, C. D.: Geomorphic development in humid and arid regions: a synthesis, *Am. J. Sci.*, vol. 253, pp. 377–390, 1955.
57. Leopold, L. B., and J. P. Miller: Ephemeral streams: hydraulic factors and their relation to the drainage net, *U.S. Geol. Surv. Profess. Paper* 282-A, 1956.
58. Melton, M. A.: An analysis of the relations among climate, surface properties, and geomorphology, *U.S. Office Naval Res. Project* NR 389-042, Tech. Rept. 11 Columbia University, Department of Geology, New York, 1957.
59. Joly, F.: Quelques phénomènes d'écoulement sur la bordure du Sahara dans les confins algéro-marocains et leurs conséquences morphologiques, *Compt. Rend. XIX Congrès Géol. Intern.*, Algiers, 1952, pt. VII, Déserts actuels et anciens, pp. 135–146, 1953.
60. Davis, W. M.: Sheetfloods and streamfloods, *Bull. Geol. Soc. Am.*, vol. 49, pp. 1337–1416, 1938.

61. Blissenbach, Erich: Geology of alluvial fans in semi-arid regions, *Bull. Geol. Soc. Am.*, vol. 65, pp. 175–190, 1954.
62. Gautier, M.: Les chotts, machines évaporatoires complexes, in Actions éoliennes, phénomènes d'évaporation et d'hydrologie superficielle dans les régions arides, *Colloq. Intern. Centre Natl. Rech. Sci. (Paris)*, no. 35, pp. 317–324, 1953.
63. Hills, E. S.: The lunette: a new land form of aeolian origin, *Australian Geograph.*, vol. 3, pp. 15–21, 1939.
64. Jaeger, F.: Die Trockenseen der Erde, *Petermanns Mitt.*, suppl. 236, 1939.
65. Stokes, C. M., F. D. Larson, and C. K. Pearse: "Range Improvement through Waterspreading," U.S. Government Printing Office, 1954.
66. Lòpez de Llergo, Rita: The importance of the phenomenon of capture in changes in the character of hydrological basins and in the growth of desert and semi-desert areas, *UNESCO Proc. Ankara Symp. on Arid Zone Hydrol.*, Paris, pp. 151–157, 1953.
67. Dixey, Frank: Some recent studies in groundwater problems, *UNESCO Proc. Ankara Symp. on Arid Zone Hydrol.*, Paris, pp. 43–59, 1953.
68. Loehnberg, Alfred: Water supply and drainage in semi-arid countries, *Trans. Am. Geophys. Union*, vol. 38, pp. 501–510, 1957.
69. Schoeller, Henri: L'hydrogéologie d'une partie de la vallée de la Saoura et du Grand Erg occidental, *Bull. Soc. Geol. Franç.*, vol. 15, pp. 563–585, 1945.
70. Ward, L. K.: "Underground Water in Australia," Tait Publishing Co., Melbourne, 1951.
71. Schoeller, Henri: "Arid Zone Hydrology: Recent Developments," UNESCO, Paris, 1959.
72. Babcock, H. M., and E. M. Cushing: Recharge to groundwater from floods in a typical desert wash, Penal Country, Arizona, *Trans. Am. Geophys Union*, vol. 23, pp. 49–55, 1942.
73. Miles, K. R.: Origin and salinity distribution of artesian water in the Adelaide plains, South Australia, *Econ. Geol.*, vol. 46, pp. 193–207, 1951.
74. Linsley, R. K.: Some aspects of precipitation and surface stream flow in groundwater recharge, *UNESCO Proc. Ankara Symp. on Arid Zone Hydrol.*, Paris, pp. 140–149, 1953.
75. Picard, Leo: Outline of ground-water geology in arid regions, *UNESCO Proc. Ankara Symp. on Arid Zone Hydrol.*, Paris, pp. 165–175, 1953.
76. Lowdermilk, W. C.: Floods in deserts, in "Desert Research," *Spec. Publ.* 2, Research Council of Israel, Jerusalem, pp. 365–371, 1953.
77. Langford-Smith, T.: The dead river systems of the Murrumbidgee, *Geograph. Rev.*, vol. 50, pp. 368–389, 1960.
78. Chapman, T. G.: Hydrology survey at Wiluna, W. A., abstract of findings and recommendations, *CSIRO Div. Land Res. Reg. Surv. Tech. Mem.* 60/4, 1960.
79. Hills, E. S.: Regional geomorphic patterns in relation to climatic types in dry areas of Australia, in "Desert Research," *Spec. Publ.* 2, Research Council of Israel, Jerusalem, pp. 355–364, 1953.
80. Todd, D. K.: Annotated bibliography on artificial recharge of groundwater through 1954, *U.S. Geol. Surv. Water-Supply Paper* 1477, 1959.
81. Clyde, G. D.: Utilization of natural underground water storage reservoirs, *J. Soil Water Conserv.*, vol. 6, pp. 15–19, 1951.
82. Lowdermilk, W. C.: Some problems of hydrology and geology in artificial recharge of underground aquifers, *UNESCO Proc. Ankara Symp. on Arid Zone Hydrol.*, Paris, pp. 158–161, 1953.
83. Briggs, L. J., and H. L. Shantz: The water requirements of plants. II. A review of the literature, *Bull. U.S. Bur. Plant Ind.*, no. 285, 1913.
84. de Wit, C. T.: Transpiration and crop yields, *Verslag Landbouwk. Onderzoek*, no. 64.6, pp. 1–88, 1958.
85. Thomas, M. D.: Effect of ecological factors on photosynthesis, *Ann. Rev. Plant Physiol.*, vol. 6, pp. 135–156, 1955.
86. Richards, L. A., and C. H. Wadleigh: Soil water and plant growth, in Soil physical conditions and plant growth, *Agronomy*, vol. 2, pp. 73–251, 1952.
87. Veihmeyer, F. J.: Soil moisture in W. Ruhland (ed.), "Encyclopaedia of Plant Physiology," vol. 3, pp. 64–123, Springer-Verlag OHG, Berlin, 1956.
88. Wadleigh, C. H., and A. D. Ayers: Growth and biochemical composition of bean plants as conditioned by soil moisture tension and salt concentration, *Plant Physiol.*, vol. 20, pp. 106–132, 1945.
89. Slatyer, R. O.: The influence of progressive increases in total soil moisture stress on transpiration, growth and internal water relationships of plants, *Australian J. Biol. Sci.*, vol. 10, pp. 320–336, 1957.
90. Schneider, G. W., and N. F. Childers: Influence of soil moisture on photosynthesis, respiration and transpiration of apple leaves, *Plant Physiol.*, vol. 16, pp. 565–584, 1941.

91. Loustalot, A. J.: Influence of soil moisture conditions on apparent photosynthesis and transpiration of pecan leaves, *J. Agr. Res.*, vol. 71, pp. 519–532, 1945.
92. Bordeau, P. S.: Oak seedling ecology determining segregation of species in piedmont oak-hickory forests, *Ecol. Monographs*, vol. 24, pp. 297–320, 1954.
93. Scofield, C. S.: The water requirement of alfalfa, *U.S. Dept. Agr. Circ.* 735, 1945.
94. Haynes, J. L.: The effect of availability of soil moisture upon vegetative growth and water use in corn, *J. Am. Soc. Agron.*, vol. 40, pp. 385–395, 1948.
95. Watson, D. J.: The physiological basis of variation in yield, *Advanc. Agron.*, vol. 4, pp. 105–145, 1952.
96. Dzerdzeevskii, B. L.: Puti preobrazovaniya klimaticheskikh uslovii Prikaspiya, *Izv. Akad. Nauk SSRS, Ser. Geograf.*, no. 1, pp. 3–13, 1952.
97. Lemon, E. R., A. H. Glaser, and L. E. Satterwhite: Some aspects of the relationship of soil, plant, and meteorological factors to evapotranspiration, *Proc. Soil Sci. Soc. Am.*, vol. 21, pp. 464–468, 1957.
98. Halstead, M. H., and W. Covey: Some meteorological aspects of evapotranspiration, *Proc. Soil Sci. Soc. Am.*, vol. 21, pp. 461–464, 1957.
99. Tanner, C. B.: Factors affecting evaporation from plants and soils, *J. Soil Water Conserv.*, vol. 12, pp. 221–227, 1957.
100. de Vries, D. A.: The influence of irrigation on the energy balance and the climate near the ground, *J. Meteorol.*, vol. 16, pp. 256–270, 1959.
101. Philip, J. R.: The theory of local advection, I, *J. Meteorol.*, vol. 16, pp. 535–547, 1959.
102. Philip, J. R.: Advection in the arid zone: theoretical, *Proc. Australian Arid Zone Tech. Conf.*, 1960.
103. Penman, H. L.: Natural evaporation from open water, bare soil and grass, *Proc. Roy. Soc. (London)*, ser. A, vol. 193, pp. 120–146, 1948.
104. House, G. J., N. E. Rider, and C. P. Tugwell: A surface energy-balance computer, *Quart. J. Meteorol. Soc.*, vol. 86, pp. 215–231, 1960.
105. Chebotarev, I. I.: The salinity problem in the arid regions, *Water and Water Eng.*, pp. 10–68, January, 1955.
106. Anderson, V. G.: The origin of the dissolved inorganic solids in natural waters with special reference to the O'Shannassy River catchment, Victoria, *J. Australian Chem. Inst.*, vol. 8, pp. 130–148, 1941.
107. Hutton, J. T., and T. I. Leslie: Accession of non-nitrogenous ions dissolved in rain-water to soils in Victoria, *Australian J. Agr. Res.*, vol. 9, pp. 492–507, 1958.
108. Richards, L. A.: "Diagnosis and Improvement of Saline and Alkali Soils," U.S. Department of Agriculture Handbook 60, 1954.
109. Davidson, J. L., and J. P. Quirk: Influence of dissolved gypsum on pasture establishment on irrigated sodic clays, *Australian J. Agr. Res.*, vol. 12, pp. 100–110, 1961.
110. Grillot, Georges: The biological and agricultural problems presented by plants tolerant of saline or brackish water and the employment of such water for irrigation, *UNESCO, Rev. Res. on Utilization of Saline Water*, Paris, pp. 9–35, 1954.
111. Hayward, H. E.: Plant growth under saline conditions, *UNESCO, Rev. Res. on Utilization of Saline Water*, Paris, pp. 37–71, 1954.
112. Hayward, H. E., and Leon Bernstein: Plant-growth relationships on salt-affected soils, *Botan. Rev.*, vol. 24, pp. 584–635, 1958.
113. Bernstein, Leon, and H. E. Hayward: Physiology of salt tolerance, *Ann. Rev. Plant Physiol.*, vol. 9, pp. 25–46, 1958.
114. Bernstein, Leon: Salt-affected soils and plants, *UNESCO Proc. Paris Symp. on Problems of the Arid Zone*, pp. 139–174, 1962.
115. Strogonov, B. P.: The water regime of plants on saline soils, *Rev. Res. on Arid Zone Plant-Water Relations*, UNESCO, Paris, 1959.
116. Wadleigh, C. H.: The integrated soil moisture stress upon a root system in a large container of saline soil, *Soil Sci.*, vol. 61, pp. 225–238, 1946.
117. Hayward, H. E., and C. H. Wadleigh: Plant growth on saline and alkali soils, *Advan. Agron.*, vol. 1, Academic Press Inc., New York, pp. 1–38, 1949.

Section 25-I

HYDROLOGY OF FLOW CONTROL

PART I. FLOOD CHARACTERISTICS AND FLOW DETERMINATION

TATE DALRYMPLE, *Hydraulic Engineer, U.S. Geological Survey.*

I. Introduction	25-2
II. Measurable Features of Flood	25-2
A. Flood Elevation	25-3
B. Flood Discharge	25-4
C. Flood Volume	25-4
D. Duration of Floods	25-5
III. Peak Discharge	25-5
A. Flood Formulas	25-5
B. Floods of Selected Frequency	25-5
1. Discharge-frequency Curves	25-6
2. Stage-frequency Curves	25-6
C. Maximum of Experience	25-7
1. Myers-Jarvis Enveloping Curves	25-7
2. Flood Regions	25-8
D. Records of Floods	25-8
1. Peak Discharges of Record	25-15
E. Maximum Probable Flood	25-15
F. Flood Volumes	25-16
G. Floods from Small Rural Basins	25-16
1. The U.S. Bureau of Public Roads Method	25-17
2. The California Method	25-20
3. The Cook Method	25-20
4. The Chow Method	25-22
IV. Design Flood	25-25
A. General	25-25
B. Standard Project Flood	25-26
1. Statistical Analyses of Streamflow Records	25-26
2. Standard-project-flood Estimates	25-26
3. Maximum-probable- (or Maximum-possible-) flood Estimates	25-26
C. SPS Criteria for Small Drainage Basins	25-27
1. Generalized Estimates of Maximum Possible Precipitation	25-27
2. SPS Index Rainfall	25-27

       3. Tabulation and Conversion of Rainfall Depths.......... 25-27
       4. Generalized Depth-Area Curves....................... 25-29
       5. SPS Depth-Area-Duration Curves..................... 25-29
       6. Time Distribution of 24-hr SPS Rainfall................ 25-29
       7. Time Distribution of Maximum 6-hr SPS Rainfall....... 25-29
       8. SPS Isohyetal Pattern................................ 25-30
   D. SPF Estimates for Small Drainage Basins.................. 25-30
   E. SPF as Percentage of Maximum Probable Flood........... 25-31
   F. SPF Estimates for Large Drainage Basins.................. 25-31
   G. Projects for Which SPF Estimates Are Required........... 25-32
V. References... 25-32

I. INTRODUCTION

A flood is any relatively high flow that overtops the natural or artificial banks in any reach of a stream. When banks are overtopped, water spreads over the floodplain and generally comes into conflict with man. Since the floodplain is a desirable location for man and his activities, it is important that floods be controlled so that the damage done does not exceed an acceptable amount. Many millions of dollars have been spent in the past to control floods, and certainly many more millions will be spent in the future. The flood problem has been discussed at length in publications by Leopold and Maddock [1] and by Hoyt and Langbein [2].

Man must acquaint himself with the characteristics of floods if he is to control them. Floods vary as the weather—from month to month and year to year. The first step in becoming better acquainted with floods is to measure them. When this has been done and a record is available for a long period of time, analytical studies will provide a better understanding of the phenomenon.

Floods for design purposes may be measured hydraulically or determined hydrologically. Records of measured discharge usually are scarce or do not exist for small drainage basins, say, less than 10 sq mi, and for urban areas. They are not available, either, for floods of maximum probability, which perhaps are greater than any that have occurred in the past. In these cases, flood discharges may be determined by hydrologic methods.

Floods have been systematically measured in the United States for less than a hundred years, and it has been only in the last fifty years that a wide coverage of observation stations has been obtained. Many Federal, state, municipal, and private agencies have observed flood heights and made measurements or estimates of flood discharges. A need for one agency to devote full time to this job early became obvious, and thus led to the U.S. Geological Survey becoming the predominant agency in the field of streamflow measurement. Today (1964), over 7,000 gaging stations are being operated by the Survey, in cooperation with other Federal and state agencies. The records obtained are public information, and are used by many people. It is these records that have furnished a basis for studying floods and analyzing them in a way that will be useful in the design of projects concerned with the control and use of floodwaters.

The problem of obtaining basic data, analyzing them best for every need, and getting the information to those who must make the decisions is the theme of a recent book by Langbein and Hoyt [3]. Various discussions of floods are given in Secs. 8-I, 10, 14, 15, 20 to 22, 25-II to 25-V, and 26-I. In this section of the handbook, the purpose is to describe the elements of floods to be measured, the estimation of peak discharges of flood, and the determination of design floods.

II. MEASURABLE FEATURES OF FLOOD

Floods may be measured as to height, area inundated, peak discharge, and volume of flow. The height of a flood is of interest to those planning to build structures along or across streams; the area inundated is of interest to those planning to occupy in any manner the floodplains adjacent to a stream; the peak discharge is of interest to those designing spillways, bridges, culverts, and flood channels; and the volume of flow is of interest to those designing storage works for irrigation, water supply, and flood control.

The height of a flood may be measured at a point, as at a gage in a fixed location, or in a reach, as is defined by a profile along one or both banks. The area inundated is measured by outlining the edge of water on a map; often this area is defined by developing flood profiles along each bank. The peak discharge is measured in terms of time rate, as cubic feet per second. The volume of flow as commonly used is measured in acre-feet or in cubic feet per second-days.

A. Flood Elevation

The elevation of a flood peak is the most often used fact of all flood data. Care should be used in recording peak elevations; it is not as simple a measurement as it at first may appear. A natural stream in flood does not produce a smooth surface. Waves and surges may cause fluctuations of a foot or more. These fluctuations must be considered when recording a flood height.

Gaging stations measure a flood at a point. The gage may be of a recording type, operated by a float in a stilling well or by gas bubbling through a tube fastened in the stream bed, or of a nonrecording type, as a staff mounted on a tree or a post or as a weighted tape lowered from a bridge.

Recording gages are affected by the velocities past the end of the intake pipe; high velocities may result in recorded elevations being one, two, or even more feet lower than the actual water-surface elevation in the stream. This difference, or *drawdown*, may be minor in the process of computing discharge, but it can be a major factor in the design of embankment heights, bridge waterways, and other engineering structures. Fortunately, velocity and drawdown are low in streams in flat topography where a small difference in elevation would be critical, and also in steep-sloped, mountainous streams, where drawdown may be high but a large difference in elevation may not be critical.

The maximum flood height at a staff gage may be read by an observer at the time of flood peak or may be indicated on the gage or on the object to which the gage is fastened by a line of mud, leaves, oil, or other marking substance left. A weight-type gage must be read at the time of flood peak.

Peak stages often can be obtained from debris lines as marks on the ground, on trees and posts, and on and in buildings. The most clearly defined marks may not always be the best. Best high-water marks are those made where velocities are low, or for high velocities, where there is little opportunity for a velocity-head recovery to raise the water surface locally above the general level. High-water marks on the upstream side of a building located in an area of high velocity should be used with understanding.

Caution should be used in combining high-water elevations from a recording gage with those obtained from outside marks, since the tendency of one may register elevations lower and the other higher than the true height. Floodwater rising inside a tight building may trap air, and not reach the same height as in the stream outside. A critical appraisal of high-water elevations should be made before using them for the design of structures where a small change in elevation may be critical.

A flood profile may be defined by plotting high-water elevations as determined at many points vs. distance along streams. The profile should be drawn on the basis of the plotted points, but judgment must be exercised since often some inconsistencies will be evident. Profiles are more realistic when drawn through the well-defined points, rather than when a straight line is drawn to average them.

Many high-water points must be determined to define a flood profile through a developed area. Points should be obtained up- and downsteam from all bridges and other contractions. In open areas, high-water marks along banks parallel to the flow are preferred. More care should be used in obtaining high-water elevations in flat-sloped streams than in steep-sloped areas.

B. Flood Discharge

The measurement of flood discharges can be made most satisfactorily by the current-meter method. The making of current-meter measurements involves no great difficulty within itself; the difficulty is usually in ability to travel to the gaging site and in availability of a suitable structure from which to work. If either of these necessary conditions is lacking, the discharge must be determined by some other method.

Frequently, the peak discharge is needed at a nongaged site. When it is important to know the peak discharge of a flood where a current-meter measurement was not made, indirect methods may be used. The most commonly used types of indirect measurements are by slope area, contraction, culvert, and flow over dams and embankments. These may be called *hydraulic methods*. Such methods are discussed in Secs. 7 and 15.

Flood discharges may also be determined by analyses of hydrologic data or by formulas developed empirically. These methods may consider precipitation, land use, the size, shape, and slope of basin, and other parameters believed of importance by the designer of the method. These may be called *hydrologic methods*. Such methods are also discussed in Secs. 10 and 20 to 22.

Hydrologic methods for computing discharges are widely used for obtaining data for the design of spillways, reservoirs, urban drainage systems, culverts, and other structures draining large as well as small basins. Few small basins are gaged, and by the application of hydrologic methods, design discharges can be computed.

C. Flood Volume

A measure of the volume of flow can easily be obtained at a reservoir, especially if the reservoir capacity is large in relation to the volume. The volume of a flood flow is given by the difference in storage at the beginning and end of the flood plus volume of flow passing the dam during the same time interval.

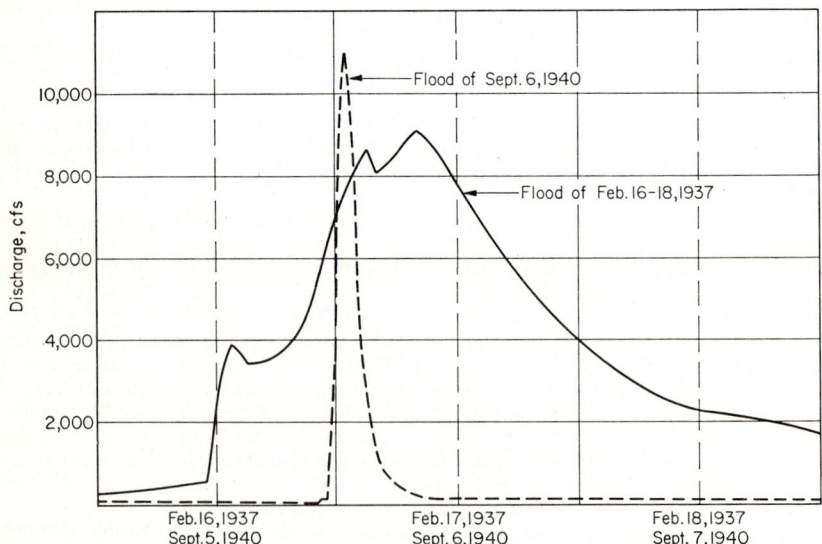

Fig. 25-I-1. Discharge hydrographs, Gila River, near Virdon, N.Mex. (*After Smith and Heckler* [4].)

PEAK DISCHARGE   **25-5**

The volume of flow during a flood period can be computed from a record of daily discharges, such as is obtained at thousands of gaging stations in the United States. An average discharge of 1 cfs for 1 day is about 2 acre-ft of water, or 1 cfs-day = 1.983471, or approximately 2 acre-ft. Thus the volume of flood flow in acre-feet may be obtained by adding the daily mean discharges in cubic feet per second for the flood period and multiplying the sum by 2.

D. Duration of Floods

The time that a stream remains at flood stage is important in many instances. A high peak of short duration usually has a relatively small volume of flow and thus may be completely controlled by a reservoir, while a lower peak of long duration and large volume will not be controlled by a reservoir of the same size. Many roads are built across floodplains on low embankments that will be overtopped by the higher floods. It is therefore important to know how long the road will be flooded.

The design of flood walls and embankments must consider only the height of the flood peak. The design of storage reservoirs must also consider the volume of flood flow.

The duration of flood peaks varies widely; some rise and recede within hours, while others remain at a high stage for days. An example of the two kinds of peaks is given by the hydrographs for a New Mexico stream [4] (Fig. 25-I-1). A study of duration of flood peaks can best be made by means of hydrographs obtained from gaging stations and the unit hydrographs and distribution graphs derived from them (Sec. 14), since these graphs will show clearly the time distribution of flood flows.

III. PEAK DISCHARGE

The peak discharge that has been, or may be, experienced is a pertinent item in the design of structures along or across a stream. The peak discharge may be computed from a formula, from a frequency concept (Sec. 8-I), or from flood records.

A. Flood Formulas

Many formulas have been derived for computing flood discharges [5, 6], but most are considered inadequate for engineering design. Studies based on flood records are usually preferred, but flood records are scarce for small drainage areas, and often use must be made of some formula.

The most common hydrologic method for computing peak discharge is the so-called *rational formula*. While this formula is sometimes used to compute discharge for drainage areas of many square miles, its use should be limited to areas of less than 100 acres, possibly 200 acres at most. The formula is given and thoroughly discussed in Secs. 14, 20, and 21.

B. Floods of Selected Frequency

Knowledge of the magnitude and probable frequency of recurrence of floods is necessary to the proper design and location of many structures. Either overdesign or underdesign of the structure involves excessive costs on a long-time basis. The problem is an economic one, involving computation of the total annual cost of building and maintaining a structure of a given design compared with the cost for other designs. Knowledge of flood frequency is necessary also to flood insurance and flood zoning, activities which are now being considered on a broad scale (Sec. 25-V).

The U.S. Geological Survey has published many flood-frequency reports, usually on a state-wide basis, from which the magnitude and probable recurrence interval of floods may be determined at any place on a stream, within limits of the basic data, whether or not a gage is located nearby. Several methods are in use for deriving flood-frequency curves. The Survey method has been applied extensively, and is described in the literature [7]. (See also Sec. 8-I.)

25-6 FLOOD CHARACTERISTICS AND FLOW DETERMINATION

1. Discharge-frequency Curves. The Survey method of analysis provides two curves. The first expresses the flood discharge-time relation, showing variation of peak discharge, expressed as a ratio to the mean annual flood, with recurrence interval. The second relates the mean annual flood to the size of area alone, or to the size of area and other significant basin characteristics. Each of these curves is applicable to a region, and a frequency curve may be defined for any place in the region by use of these two curves.

An example of the method is given by the two curves developed for application to the Youghiogheny and Kiskiminetas River basins, in Pennsylvania and Maryland. The curve of Fig. 25-I-2 shows relation of discharge, in ratio to the mean annual flood, to the recurrence interval in years. To be useful, the mean annual flood must be known. This can be read from the curve of Fig. 25-I-3 that shows the relation of the mean annual flood to the size of drainage area.

To define a flood-frequency curve at any place within the basins, the size of the drainage area must be determined, and then the mean annual flood is read from Fig. 25-I-3. This discharge, multiplied by the flood ratios for selected recurrence intervals, taken from Fig. 25-I-2, will give points to plot the complete frequency curve.

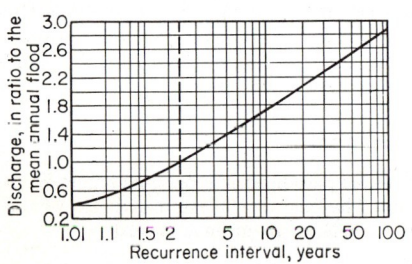
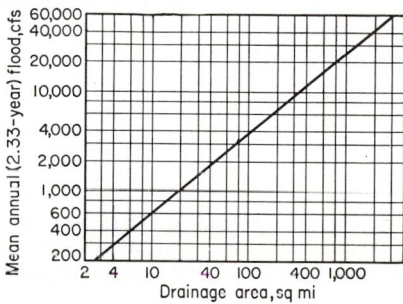

Fig. 25-I-2. Regional flood-frequency curve.

Fig. 25-I-3. Variation of mean annual flood with drainage area.

Two definitions will be helpful: (1) the *mean annual flood* is defined as the flood having a recurrence interval of 2.33 years; and (2) *recurrence interval* is the average interval of time within which the given flood will be equaled or exceeded once. A flood having a recurrence interval of 10 years is one that has a 10 per cent chance of recurring in any year. Likewise, a 50-year flood has a 2 per cent chance, and a 100-year flood has a 1 per cent chance, of recurring in any year.

Flood-frequency reports by the U.S. Geological Survey authors have been published for Alabama [8], Arkansas [9], Columbia River Basin [10], Connecticut [11], Delaware River Basin [12], Florida [13], Georgia [14], Illinois [15], Indiana [16], Iowa [17], Kansas [18], Kentucky [19], Louisiana [20], Maryland [21], Minnesota [22], Mississippi [23], Missouri [24], Montana (eastern part) [25], Nebraska [26, 27], New York [28], North Carolina [29], North and South Dakota [30], Ohio [31], Pennsylvania [32], Tennessee [33], Utah [34], Washington [35], and Wisconsin [36]. Similar reports are in preparation for other areas as a part of a program eventually to cover the entire United States. Regional rather than state-wide reports will also be published for regions corresponding to the "parts" now used for publication of streamflow records by the U.S. Geological Survey as shown on Fig. 25-I-4.

2. Stage-frequency Curves. When it is necessary to compute frequencies of stage occurrence, thought must be given to the stage-discharge relation. If the stage-discharge relation has remained stable throughout the period of record, frequencies can be first computed in terms of discharge, then transformed to stages by means of the stage-discharge relation. If a stream has frequent random shifts, a stage-frequency study can be made based on stages alone. Caution should be used in defining

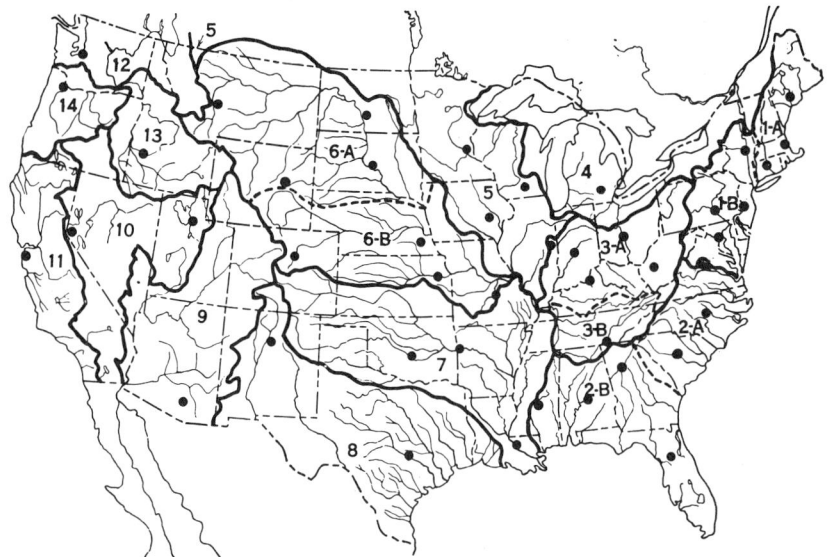

Fig. 25-I-4. Areas included in the "parts" for publication of annual *Water-Supply Papers* and location of major field offices of the U.S. Geological Survey.

a stage-frequency curve at its upper, or high-stage, end, since a short extrapolation may result in inferring an impossible stage-discharge relation (Sec. 15).

C. Maximum of Experience

Efforts to measure outstanding floods have been increased in the past thirty years, until today most noteworthy floods are measured, including those occurring both at gaging stations and at miscellaneous points. This activity has resulted in a mass of flood information that, when properly analyzed, can provide a firm basis for design.

Many formulas have been derived and methods originated for computing a "maximum" flood to be expected at a given site. Various flood-estimating methods are discussed in *U.S. Geological Survey Water-Supply Paper* 771 [5], by Linsley, Kohler, and Paulhus [37], and by others. Most of these methods make use of the known maximum floods at various points without regard to the frequency of the floods. When the region over which floods are compared is so large that it includes areas of dissimilar hydrologic characteristics, a gross error may be committed by such methods in selecting a design discharge.

1. Myers-Jarvis Enveloping Curves. Probably the most widely used flood formula is that developed by Major E. D. T. Myers. As modified by Jarvis [38, p. 531], this formula is

$$Q = 100p \sqrt{M} \qquad (25\text{-}I\text{-}1)$$

where Q is the discharge in cfs, p is the numerical percentage rating on the Myers scale, and M is the size of drainage area in sq mi.

In ordinary use, peak discharges Q in cubic feet per second per square mile are plotted on logarithmic paper against the drainage area M in square miles. Then a straight line, called the *Myers curve*, with a slope of 1 in vertical to 2 in horizontal, is drawn as an envelope through the upper points. This curve is intended to give an estimate of flood peaks that could occur anywhere in the region.

25-8 FLOOD CHARACTERISTICS AND FLOW DETERMINATION

The Myers-Jarvis formula makes two questionable assumptions: (1) that flood peaks vary as the 0.5 power of the drainage area, and (2) that flood-producing characteristics of all streams in the region under consideration follow the same law as expressed by the formula. Experience, however, does not support these assumptions.

2. Flood Regions. Preparation of regional flood-frequency reports has led to the division of states, or other large areas, into *flood-frequency regions* (having like flood-frequency characteristics) and *hydrologic areas* (having like mean annual flood characteristics). For instance, four flood-frequency regions and eight hydrologic areas have been defined in Missouri, resulting in 14 combinations of the two divisions. Each of these 14 combinations has different flood-producing potential, and hence flood peaks experienced in one area should be used in other areas with caution.

As an example, Fig. 25-I-5 shows a plot of peak discharges, in cubic feet per second per square mile vs. drainage area in square mile, for Missouri streams. Point A plots well above the others, 2.5 times above the line defining the 50-year flood, and would be considered as a very high flood. However, all points except A are in a region D-8 of low flood potential. Point A is in a region B-5 of high flood potential, and Fig. 25-I-6 gives a plot for this region, showing the 50-year-flood line and the position of point A.

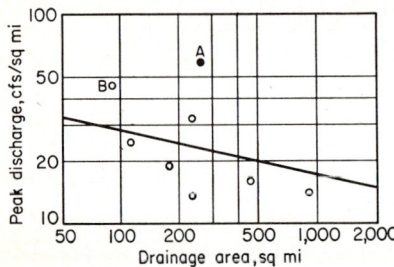

Fig. 25-I-5. Relation of maximum to 50-year flood in region D-8.

Fig. 25-I-6. Relation of maximum to 50-year flood in region B-5.

From Fig. 25-I-6, it is apparent that the flood represented by point A is not very high, but really is very low, being only 0.3 of the 50-year flood.

It is interesting to note from Fig. 25-I-5 that the highest flood in region D-8 (point B), having a Myers rating of 4.3 per cent, is a much greater flood than flood A, which has a Myers rating of 9.6 per cent, and the first flood (point B) is 1.6 times the 50-year flood, while the second (point A) is only 0.3 the 50-year event.

D. Records of Floods

The main source of streamflow information in the United States is the *Water-Supply Papers* of the U.S. Geological Survey. The U.S. Weather Bureau annually publishes daily stages of many streams in *Daily River Stages*. The U.S. Soil Conservation Service, the U.S. Army Corps of Engineers, and the U.S. Bureau of Reclamation obtain records of some streams. Many state agencies also operate gaging stations. Records can usually be obtained directly from the local offices of these agencies.

The U.S. Geological Survey publishes annually a series of *Water-Supply Papers*, each containing streamflow records for a particular section of the United States. Areas included in each of the sections, called "parts," are shown on the map of Fig. 25-I-4. *Water-Supply Papers* may be purchased from the Superintendent of Documents, Washington 25, D.C. [39]. Sets of *Water-Supply Papers* may be consulted in libraries of the principal cities in the United States and in the offices of the Water Resources Division of the U.S. Geological Survey. These offices are located in nearly every state, usually at the state capital.

A series of compilation reports have been published also as *Water-Supply Papers*, one volume for each "part." These reports contain, among other information,

Table 25-I-1. Selected Peak Discharges in the United States
(C, creek; R, river; F, fork; B, brook)

Stream and location	Drainage area, sq mi	Date	Flood peak, cfs	cfs/sq mi
Alabama				
Norton C., Saraland	4.15	Apr. 13, 1955	6,030	1,450
Pigeon C., Gosport	22.6	July 8, 1956	18,800	832
Jackson C., Winn	42.7	July 8, 1956	34,000	796
Chickasaw C., Whistler	123	Apr. 13, 1955	42,000	341
Boguechitto C., Orrville	276	Dec. 29, 1942	47,000	170
Conecuh R., Andalusia	1,300	Mar. 15, 1929	154,000	118
Escambia R., Century	3,700	March, 1929	315,000	85.1
Tennessee R., Florence	30,810	Mar. 19, 1897	444,000	14.4
Arizona				
Wash, Gunters Ranch	3.8	Sept. 26, 1948	6,700	1,760
Picacho Wash, Yuma	41.5	Sept. 5, 1939	37,000	892
Agua Fria R., Lake Pleasant Dam	1,460	Jan. 28, 1916	105,000	71.9
Gila R., Sentinel	51,600	February, 1891	250,000	4.83
Colorado R., Lees Ferry	107,900	July 7, 1884	300,000	2.78
Colorado R., Yuma	242,900	Jan. 22, 1916	250,000	1.03
Arkansas				
Potash Sulphur C., Hot Springs	1.25	July 16, 1963	2,440	1,950
Tigre C., Hot Springs	8.3	July 16, 1963	12,800	1,540
Cove C., Lee Creek	36.9	May 5, 1960	33,600	911
Rolling Fork, Dequeen	181	Aug. 27, 1947	110,000	608
Little Missouri R., Murfreesboro	380	Mar. 30, 1945	120,000	316
Buffalo R., Rush	1,091	Aug. 19, 1915	164,000	150
White R., Calico Rock	9,965	Jan. 31, 1916	350,000	35.1
Red R., Fulton	46,444	Feb. 24, 1938	338,000	7.28
Arkansas R., Little Rock	135,960	Apr. 20, 1927	813,000	5.98
Mississippi R., Arkansas City	1,130,700	Apr. 20, 1927	2,472,000	2.19
California				
San Antonio C., Claremont	16.9	Mar. 2, 1938	21,400	1,270
E. F. San Gabriel R., Camp Bonita	88.2	Mar. 2, 1938	46,000	522
San Gabriel R., Azusa	202	Mar. 2, 1938	90,000	446
Smith R., Cresent City	613	Dec. 22, 1955	165,000	269
Eel R., Scotia	3,113	Dec. 22, 1955	541,000	174
Klamath R., Klamath	12,100	Dec. 22, 1955	425,000	35.1
Colorado				
Tributary to Cold Spring Gulch	0.63	Sept. 2, 1938	2,050	3,250
Cold Spring Gulch, Morrison	4.48	Sept. 2, 1938	9,000	2,010
Cameron Arroyo, Pueblo	7.3	June 3, 1921	13,900	1,900
Rock C., Pueblo	59	June 3, 1921	53,900	913
S. F. Republican R., Idalia	1,300	May 31, 1935	103,000	79.2
Connecticut				
Valley B., West Hartland	7.2	Aug. 19, 1955	8,260	1,150
E. B. Salmon B., North Granby	13.2	Aug. 19, 1955	14,300	1,080
Naugatuck R., Waterbury	138	Aug. 19, 1955	75,900	550
Farmington R., Collinsville	360	Aug. 19, 1955	140,000	389
Connecticut R., Thompsonville	9,661	Mar. 20, 1936	282,000	29.2

Table 25-I-1. Selected Peak Discharges in the United States (Continued)

Stream and location	Drainage area, sq mi	Date	Flood peak, cfs	cfs/sq mi
Florida				
Pine Barren C., Barth.........	75.3	Apr. 14, 1955	24,800	329
Choctawhatchee R., Caryville	3,499	Mar. 17, 1929	206,000	58.9
Suwannee R., Ellaville........	6,580	Apr. 7, 1948	95,300	14.5
Apalachicola R., Chattahoochee.....................	17,100	Mar. 20, 1929	293,000	17.1
Georgia				
Little Tired C., Cairo.........	9.5	April, 1948	6,540	688
Talona C., Whitestone........	26	Apr. 7, 1938	20,400	785
Tired C., Cairo...............	60	Apr. 1, 1948	28,100	468
Savannah R., Augusta........	7,508	1796	360,000	47.9
Idaho				
Highland Valley Gulch, Boise.	0.39	Aug. 20, 1959	2,100	5,380
Squaw C., Boise..............	1.47	Aug. 20, 1959	7,320	4,980
Maynard Gulch, Boise........	2.25	Aug. 20, 1959	9,540	4,240
N. F. Payette R., McCall....	144	June 10, 1933	4,260	29.6
Weiser R., Weiser.............	1,460	Dec. 23, 1955	19,900	13.6
Payette R., Emmett..........	2,680	May 1, 1938	22,800	8.5
Snake R., Weiser.............	69,200	Mar. 30, 1910	120,000	1.7
Illinois				
Muddy C., Wheeler...........	8.1	May 24, 1943	3,600	444
Seminary C., Whitehall.......	16.8	May, 1943	7,200	429
Otter C., Palmyra............	61.5	May, 1943	15,000	244
Horse C., Cotton Hill........	127	May, 1943	26,000	205
Shoal C., Breese..............	760	May 19, 1943	52,000	68.4
Kankakee R., Wilmington....	5,250	May 21, 1943	42,000	8.0
Indiana				
Owl C., trib., Morton.........	2.64	June 28, 1957	2,870	1,090
Little Raccoon C., Guion.....	30.7	June 28, 1957	23,700	772
Raccoon C., Mansfield........	240	June 28, 1957	38,400	160
E. F. White R., Bedford.....	3,870	March, 1913	155,000	40.1
Wabash R., Vincennes........	13,700	Mar. 29, 1913	255,000	18.6
Wabash R., Mt. Carmel......	28,600	Mar. 30, 1913	428,000	15.0
Iowa				
Union Park C., Dubuque.....	1.0	July 9, 1919	3,000	3,000
Waymon C., Garber..........	6.98	May 31, 1958	15,500	2,220
Catfish C., Dubuque..........	40	Aug. 16, 1918	28,000	700
Devils C., Santa Fe Br, Viele.	143	June 10, 1905	80,000	559
Salt C., Elberon..............	200	June 17, 1944	34,000	170
Middle R., Indianola..........	502	June 13, 1947	34,000	67.7
Maquoketa R., Maquoketa...	1,550	June 27, 1944	48,000	31.0
Iowa R., Iowa City...........	3,230	June, 1851	70,000	21.7
Des Moines R., Tracy........	12,400	June 14, 1947	155,000	12.5
Kansas				
Rock C., Burlington..........	8.8	July, 1951	9,560	1,090
Mill C., Alta Vista...........	18.7	July, 1951	19,800	1,060
Big Bull C., Hillsdale........	147	July 11, 1951	45,200	307
Neosho R., Iola...............	3,818	July 13, 1951	436,000	114
Neosho R., Parsons...........	4,905	July 14, 1951	410,000	83.6
Kansas R., Bonner Springs...	54,190	July 13, 1951	510,000	9.4

PEAK DISCHARGE 25-11

Table 25-I-1. Selected Peak Discharges in the United States (Continued)

Stream and location	Drainage area, sq mi	Date	Flood peak, cfs	cfs/sq mi
Kentucky				
Big Run, Brookville	4.52	May 13, 1955	11,600	2,570
Rockhouse C., Matthews	13.0	July 30, 1961	11,500	885
Elm Fork, Lenox	59.4	July 30, 1961	22,200	374
S. F. Cumberland R., Nevelsville	1,271	Mar. 23, 1929	130,000	102
Big Sandy R., Louisa	3,892	Mar. 2, 1955	89,400	23.0
Ohio R., Louisville	91,170	Jan. 26, 1937	1,110,000	12.2
Louisiana				
Garrett C., Jonesboro	2.14	Apr. 24, 1953	1,670	780
Hemphill C., Hot Wells	18.0	Apr. 29, 1953	8,320	462
Big C., Pollock	51	Apr. 24, 1953	23,500	461
Tenmile C., Elizabeth	92.4	May 18, 1953	31,900	346
Whiskey Chitto C., Oberlin	510	May 18, 1953	144,000	282
Calcasieu R., Kinder	1,700	May 19, 1953	182,000	107
Atchafalaya R., Krotz Springs	. . . .	Mar. 6, 1950	624,000	
Maryland				
Trib. to Broad C., Pylesville	3.96	July 15, 1951	4,410	1,110
Chaptico C., Chaptico	10.7	Sept. 10, 1950	7,800	729
Big Elk C., Elk Mills	52.6	July 5, 1937	10,600	202
Octoraro C., Rising Sun	193	Aug. 9, 1942	35,000	181
N. B. Potomac R., Cumberland	875	June 1, 1889	89,000	102
Potomac R., Hancock	4,073	Mar. 18, 1936	340,000	83.5
Potomac R., Pt. of Rocks	9,651	Mar. 19, 1936	480,000	49.7
Massachusetts				
Powermill Brook, Westfield	2.50	Aug. 19, 1955	5,740	2,300
Lamberton B., West Brookfield	4.47	Aug. 19, 1955	4,140	926
M. B. Westfield R., Goss Heights	52.6	Sept. 21, 1938	19,900	378
Westfield R., Knightville	162	Sept. 21, 1938	37,900	234
Chicopee R., Indian Orchard	688	Sept. 21, 1938	45,200	65.7
Merrimack R., Lowell	4,424	Mar. 20, 1936	173,000	39.1
Connecticut R., Montague City	7,840	Mar. 20, 1936	236,000	30.1
Michigan				
N. B. Ammond C., Rose City	0.646	May 20, 1959	1,210	1,870
S. B. Ammond C., Rose City	1.14	May 20, 1959	1,710	1,500
Shepards C., Selkirk	4.51	May 20, 1959	727	161
Escanaba R., Cornell	870	May 7, 1960	10,500	12.1
Tittabawassee R., Midland	2,400	Mar. 21, 1948	34,000	14.2
Grand R., Grand Rapids	4,900	Mar. 28, 1904	54,000	11.0
Minnesota				
Willow C., Rochester	17.6	June 4, 1958	6,240	355
Yellow Medicine R., Granite Falls	653	June 18, 1957	11,800	18.1
Minnesota R., Mankato	14,900	Apr. 26, 1881	90,000	6.0
Mississippi R., Anoka	19,100	Apr. 14, 1952	75,900	4.0
Mississippi R., St. Paul	36,800	Apr. 16, 1952	125,000	3.4

FLOOD CHARACTERISTICS AND FLOW DETERMINATION

Table 25-I-1. Selected Peak Discharges in the United States (Continued)

Stream and location	Drainage area, sq mi	Date	Flood peak, cfs	cfs/sq mi
Mississippi				
Caney C., Eureka Springs....	4.85	Mar. 27, 1954	14,700	3,030
Long C., Eureka Springs.....	12.8	May 27, 1954	19,500	1,520
Long C., Pope...............	30.8	May 27, 1954	31,900	1,040
W. F. Tombigbee R., Nettleton...................	617	Mar. 22, 1955	151,000	245
Tombigbee R., Columbus.....	4,490	Apr. 8, 1892	268,000	59.7
Missouri				
Green Acre B., Rolla.........	0.62	June 9, 1950	1,900	3,060
Clear C., trib., Holt..........	6.52	July 22, 1947	14,000	2,147
Little Gravois C., Bagnell....	24.1	Aug. 2, 1944	31,000	1,286
Fishing R., Kearney.........	39.4	June 22, 1947	30,000	761
Cuivre R., Troy..............	903	Oct. 5, 1941	120,000	133
Meramec R., Eureka.........	3,788	Aug. 22, 1915	175,000	46.3
Osage R., Bagnell............	14,000	May 19, 1943	220,000	15.7
Missouri R., Hermann.......	528,200	June, 1844	892,000	1.7
Montana				
Little Beaver C., trib., Baker.	0.50	Sept. 5, 1954	1,680	3,360
Fred Burr C., Victor.........	18.6	May, 1948	23,000	1,240
Custer C., Miles City........	155	June 19, 1938	21,000	135
Flathead R., Columbia Falls..	4,464	June, 1894	135,000	30.2
Clark Fork, Heron...........	22,006	June, 1894	195,000	8.9
Yellowstone R., Sidney.......	68,812	June 21, 1921	159,000	2.3
Nebraska				
Unnamed C., York...........	6.9	July 9, 1950	23,000	3,330
Dry C., Curtis...............	17.0	June 21, 1947	25,900	1,520
Little Nemaha R., Syracuse...	218	May 9, 1950	225,000	1,030
North Loup R., St. Paul......	1,270	June 6, 1896	90,000	70.9
Republican R., Cambridge....	8,870	June 1, 1935	280,000	31.6
Nevada				
Martin C., Paradise Valley ...	172	Jan. 21, 1943	9,000	52.3
E. F. Carson R., Gardnerville	344	Dec. 23, 1955	17,600	51.2
Carson R., Carson City.......	876	Dec. 23, 1955	30,000	34.2
Truckee R., Reno............	1,067	Dec. 23, 1955	20,800	19.5
Humboldt R., Argenta.......	7,490	February, 1962	6,000	0.80
Humboldt R., Imlay.........	15,700	May 9, 1952	6,080	0.39
New Jersey				
Cakepaulin C., Lansdowne....	13.7	Aug. 18, 1955	7,230	528
Neshanic R., Reaville........	25.7	Aug. 19, 1955	8,830	344
Flat B., Flatbrookville.......	65.1	Aug. 19, 1955	9,560	147
N. B. Raritan R., Raritan...	190	Aug. 19, 1955	20,700	109
Delaware R., Belvidere.......	4,535	Aug. 19, 1955	273,000	60.2
Delaware R., Trenton........	6,780	Aug. 20, 1955	329,000	48.5
New Mexico				
El Rancho Arroyo, Pojoaque..	6.7	Aug. 22, 1952	44,600	6,670
Trujillo Arroyo, Hillsboro....	22	June 10, 1914	20,800	945
Sapello C., Watrous..........	284	Sept. 29, 1904	62,900	222
Canadian R., French.........	1,480	September, 1904	156,000	105
Canadian R., Logan..........	10,031	Sept. 30, 1904	278,000	27.7

PEAK DISCHARGE

Table 25-I-1. Selected Peak Discharges in the United States (Continued)

Stream and location	Drainage area, sq mi	Date	Flood peak, cfs	cfs/sq mi
New York				
Depot C., Sidney Center.....	0.60	June 10, 1954	1,660	2,767
Glen C., Townsend..........	2.91	July, 1935	7,330	2,520
Trumansburg C., Trumansburg....................	11.5	July, 1935	17,800	1,550
Taughannock C., Halseyville..	56.7	July, 1935	42,100	743
Esopus C., Coldbrook........	192	Mar. 30, 1951	59,600	310
Schoharie C., Prattsville......	236	Oct. 16, 1955	55,200	234
Delaware R., Port Jervis.....	3,076	Aug. 19, 1955	233,000	75.7
North Carolina				
Big B., Sunburst............	0.4	Aug. 30, 1940	4,500	11,200
Big C., Burnett Siding.......	1.32	Aug. 30, 1940	12,900	9,770
Elk C., Elkville.............	50.9	Aug. 13, 1940	71,500	1,400
Yadkin R., Wilkesboro.......	493	Aug. 14, 1940	160,000	325
Cape Fear R., Lillington......	3,440	Sept. 19, 1945	150,000	43.6
Roanoke R., Scotland Neck...	8,700	Aug. 19, 1940	260,000	29.9
North Dakota				
M. F. Crooked C., Korinen...	3.9	July 28, 1951	5,700	1,460
Oak C., trib. no. 4, Bottineau.	7.7	July 6, 1955	2,330	303
Government C., Richardton..	30.5	Apr. 16, 1950	4,300	141
Tongue R., Akra............	148	Apr. 18, 1950	11,800	79.7
Turtle R., Manvel...........	525	Apr. 19, 1950	28,000	53.3
Cannonball R., Breien........	4,100	Apr. 19, 1950	94,800	23.1
Red R., of the North, Drayton	27,600	May 12, 1950	86,500	3.13
Missouri R., Williston........	164,500	Apr. 4, 1930	231,000	1.40
Ohio				
Duck C., Pleasant Valley.....	0.78	Aug. 7, 1947	2,980	3,820
E. F. Honey C., New Carlisle.	6.7	July, 1918	14,800	2,210
Brush C., West Milton.......	16.0	July, 1939	13,500	844
Lost C., Troy...............	52	March, 1913	29,700	571
Mad R., Springfield..........	485	Mar. 25, 1913	55,400	114
Miami R., Hamilton.........	3,630	Mar. 26, 1913	352,000	97.0
Oklahoma				
Hudson C., Narcissa.........	13.4	May 18, 1943	15,000	1,120
Sallisaw C., Sallisaw.........	182	Apr. 15, 1945	110,000	604
Illinois R., Tahlequah........	959	May 10, 1950	150,000	156
Spring R., Quapaw...........	2,510	May 19, 1943	190,000	75.7
Neosho R., Langley..........	10,335	May 20, 1943	300,000	29.0
Arkansas R., Muscogee.......	84,133	May 21, 1943	700,000	8.32
Oregon				
Butter C., trib., Echo........	1.42	June 9, 1948	5,260	3,700
Willow C., Hepner...........	87	June 14, 1903	36,000	414
Santiam R., Jefferson........	1,790	Nov. 21, 1921	202,000	113
Willamette R., Salem........	7,280	Dec. 4, 1861	500,000	68.1
Columbia R., The Dalles.....	237,000	June 6, 1894	1,240,000	5.23

Table 25-I-1. Selected Peak Discharges in the United States (Continued)

Stream and location	Drainage area, sq mi	Date	Flood peak, cfs	cfs/sq mi
Pennsylvania				
Twomile C., trib., Port Allegany	0.053	July 18, 1942	640	12,100
Lillibridge, Port Allegany	6.73	July 18, 1942	16,000	2,380
Annin C., Turtlepoint	11.4	July 18, 1942	24,000	2,100
E. B. Wallenpaupack C., Greentown	33.9	Aug. 19, 1955	33,000	982
Brodhead C., Analomink	124	Aug. 18, 1955	72,200	582
First F., Sinnemahoning C., Sinnemahoning	245	July 18, 1942	80,000	327
Kiskiminetas R., Vandergrift	1,825	Mar. 18, 1936	185,000	101
South Dakota				
Castle C., trib., no. 2, Rochford	0.019	July 28, 1955	98.9	5,200
Iron C., Rochford	1.25	July 28, 1955	2,410	1,930
Camel C., Ladner	5.2	July 28, 1951	7,500	1,440
Mush C., Pierre	14.6	Aug. 10, 1956	3,620	248
N. F., Grand R., Lodgepole	1,060	Apr. 7, 1952	21,300	20.1
Bad R., Fort Pierre	3,107	April, 1927	50,000	16.1
Cheyenne R., Hot Springs	8,710	May 12, 1920	114,000	13.1
Tennessee				
E. F. Globe C., McKenzie School	6.6	June 18, 1939	16,300	2,470
Richland C., Pulaski	366	March, 1902	100,000	273
Caney F., Rock Island	1,640	Mar. 23, 1929	210,000	128
Duck C., Paint Rock Bridge	3,500	February, 1948	220,000	62.9
Tennessee R., Knoxville	8,934	Mar. 10, 1867	270,000	30.2
Cumberland R., Clarksville	16,000	Jan. 24, 1937	290,000	18.1
Tennessee R., Savannah	33,140	Mar. 21, 1897	450,000	13.6
Texas				
Bunton C., Kyle	4.13	June 30, 1936	13,800	3,350
Little Red Bluff C., Carta Valley	10.3	June 24, 1948	30,000	2,910
Mailtrail C., Loma Alta	75.3	June 24, 1948	170,000	2,260
W. Nueces R., Kickapoo Springs	402	June 14, 1935	580,000	1,440
W. Nueces R., Cline	880	June 14, 1935	536,000	609
Nueces R., Uvalde	1,930	June 14, 1935	616,000	319
Llano R., Castell	3,747	June 14, 1935	388,000	104
Little R., Cameron	7,000	Sept. 10, 1921	647,000	92.4
Pecos R., Comstock	35,293	June 28, 1954	948,000	26.9
Utah				
South Coal Fork, Mt. Pleasant	1.2	Aug. 25, 1961	3,310	2,760
Farley Canyon, Hite	12.5	Sept. 8, 1961	7,500	600
Saleratus Wash, Green River	120	Sept. 21, 1962	19,500	162
Virgin R., Virgin	934	Mar. 3, 1938	13,500	14.5
Colorado R., Cisco	24,100	July 4, 1884	125,000	5.2
Virginia				
Coal Run, Stokesville	2.4	June 17, 1949	4,420	1,840
Little R., Stokesville	25	June 17, 1949	32,900	1,320
North R., Stokesville	65	June 17, 1949	69,000	1,060
Rappahannock R., Remington	616	Oct. 16, 1942	90,000	146
S. F. Shenandoah R., Front Royal	1,638	Oct. 16, 1942	130,000	79.4
James R., Scottsville	4,571	November, 1877	160,000	35.0

PEAK DISCHARGE

Table 25-I-1. Selected Peak Discharges in the United States (Continued)

Stream and location	Drainage area, sq mi	Date	Flood peak, cfs	cfs/sq mi
Washington				
S. F. Pine Canyon C., Waterville	5.4	May, 1948	25,000	4,630
N. F. Skokomish R., Hoodsport	57.2	Nov. 5, 1934	27,000	472
Queets R., Clearwater	445	Nov. 3, 1956	118,000	265
Sauk R., Sauk	714	Nov. 27, 1949	82,400	115
Cowlitz R., Castle Rock	2,238	Dec. 23, 1933	139,000	62.1
Columbia R., Int. Boundary	59,700	June, 1894	680,000	11.4
West Virginia				
N. F. Yellow C., Big Spring	1.51	Aug. 5, 1943	4,700	3,100
Little Wheeling C., Elm Grove	62.0	July 10, 1937	25,900	417
Potomac R., Hancock	4,073	Mar. 18, 1936	340,000	83.4
Ohio R., Parkersburg	35,600	Mar. 29, 1913	593,000	16.7
Wisconsin				
Bishops B., Viroqua	7.1	Aug. 26, 1959	5,820	820
Big Eau Pleine R., Stratford	224	Sept. 9, 1938	41,000	183
Black R., Neillsville	756	Sept. 10, 1938	48,800	65
Red Cedar R., Colfax	1,100	Apr. 3, 1934	21,900	19.9
Wisconsin R., Merrill	2,780	Aug. 31, 1941	49,400	17.8
Chippewa R., Chippewa Falls	5,600	Sept. 1, 1941	102,000	18.2
Wisconsin R., Muscoda	10,300	Sept. 16, 1938	80,800	7.8
Mississippi R., Prescott	44,800	Apr. 17, 1952	155,000	3.5
Wyoming				
Sand Coulee, Dubois	13	July 24, 1934	6,700	515
Little Red C., Circle	18	July 24, 1934	6,450	358
Salt C. (Sec. 36, T.41N., R.79W.)	520	Sept. 27, 1923	32,000	61.5
Big Horn R., Thermopolis	8,080	July 24, 1923	29,800	3.7

tabulations of the momentary maximum discharge for each year from the beginning of record through 1950. Records are included for each gaging station that has been operated by the Survey prior to 1950. A list of these reports also may be obtained from the Survey [39]. In addition to these reports, special reports of outstanding flood events have been published. In 1950 and subsequent years, a report has been published annually containing data for all outstanding floods that occurred during the year. A list of flood reports published as *Water-Supply Papers* may be obtained from the Survey.

1. **Peak Discharges of Record.** Peak discharges at selected points in the United States are listed in Table 25-I-1. These discharges, listed by states, are about the highest of record for each state. A line joining the plotted points of these discharges will generally be above all points that may be plotted for other discharges. Many peak discharges have been observed in recent years on small streams, but are not included in the table because the size of drainage area is not known. No attempt has been made to locate the flood peaks listed in their appropriate flood regions, since these regions have not been defined for most of the country.

E. Maximum Probable Flood

The *maximum probable flood* has been defined [1, p. 112] as the largest flood for which there is any reasonable expectancy in this climatic era. This is a very large flood indeed, and seldom would be used in design except for reservoir spillways, where failure could lead to great damage and loss of life. The determination of the maximum probable flood involves a detailed study of storm patterns, transposition of storms to a position that will give maximum runoff, and computation of the maximum flood by the unit-hydrograph method.

The maximum probable flood is of importance in the design of large dams and will be discussed in Subsec. IV-B. Computation of this flood in the United States is done chiefly by the U.S. Army Corps of Engineers and the U.S. Bureau of Reclamation. Reference should be made to procedures developed by these agencies for details of computation.

F. Flood Volumes

The volume of floodwater flowing past a point is of interest to those having responsibility for the design of storage works for domestic and industrial water supply, irrigation, and flood control. A duration curve of daily discharges will show what per cent of time the flow is less than a given amount, but flow for a per cent of time does not take into account the number of consecutive days for which the flow may be less. It is important to know consecutive days of deficient flow to design a storage project properly (Sec. 18).

Frequency curves may be prepared for the maximum runoff of 1 day, 3 days, 7 days, etc. The technique for defining such frequency curves is similar to that for defining frequency curves of momentary peak discharges. Regional curves can be prepared for each runoff period, from which estimates of the volume of flow can be made at ungaged places.

Little work has been done in the past on computing volume frequencies, partly because of the laborious tasks of compiling the basic data. Streamflow records are now being put on tape, from which volume figures can be quickly tabulated by machine. The U.S. Geological Survey is now computing the highest mean discharge, for some gaging stations, for 1, 3, 7, 15, 30, 60, 90, 120, 150, 183, and 274 consecutive days. Flood-volume frequency curves may be defined from these tabulations by a method similar to that used for peak-discharge frequency curves.

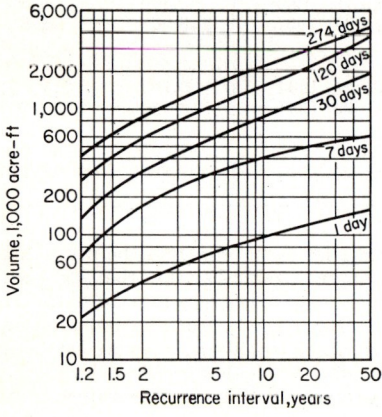

FIG. 25-I-7. Annual flood volume-frequency curves.

Processed records for Big Blue River at Randolph, Kans., list the highest mean discharge for each water year from 1921 to 1959 for the 11 periods mentioned above. For each period, data were plotted to define a frequency curve; the curves for 1, 7, 30, 120, and 274 days are presented in Fig. 25-I-7. As an example, the curves show that the 1-day volume of flow will equal or exceed 96,000 acre-ft and the 274-day volume of flow will equal or exceed 2,180,000 acre-ft, on the average of once every 10 years. Other values may be read from appropriate curves.

G. Floods from Small Rural Basins

The small watershed and its characteristics have been described by the Committee on Runoff of the American Geophysical Union [40, p. 379] as follows:

From the hydrologic point of view, a distinct characteristic of the small watershed is that the effect of overland flow rather than the effect of channel flow is a dominating factor affecting the peak runoff. Consequently, a small watershed is very sensitive to high-intensity rainfalls of short durations, and to land use. On larger watersheds, the effect of channel flow or the basin storage effect becomes very pronounced so that such sensitivities are greatly suppressed. Therefore, a small watershed may be defined as one that is so small that its sensitivities to high-intensity rainfalls of short durations and to land use are not suppressed by the channel-storage characteristics. By this definition, the size of a small watershed may be found from few acres to 1000 acres, or even up to 50 sq mi. The upper limit of the area depends on the condition at which the above-mentioned sensitivities become practically lost due to the channel-storage effect.

Current methods for estimating peak runoff from small watersheds were outlined as follows:

1. Empirical or semi-empirical methods developed from experience and judgment, involving the use of rules of thumb and empirical formulas, curves, and tables.
2. Statistical analysis of flood records on regional basis, involving probable flood frequency and drainage basin characteristics which are applied through a family of index curves to areas which involve no appreciable change in natural runoff characteristics.
3. Rational method.
4. Synthetic flood hydrographs based on the application of unitgraphs to design storm patterns.
5. Correlation analysis for developing formulas or charts showing the effect on peak runoff by various hydrologic factors such as rainfall intensities, land use and slope, frequency of occurrence, and watershed dimensions.

Empirical methods for computing flood discharges have been, and are, widely used. Over 100 such formulas have been proposed [6]. Most, if not all, of these are inadequate in evaluating the hydrologic factors involved. The increasing accumulation of records of discharge of streams draining small basins has decreased the need for these empirical formulas, and they are being supplanted by more logical methods.

Results of statistical analyses of flood records, involving computation of regional flood-frequency curves, are being used increasingly as such information becomes available. This is a logical procedure, since it makes use of the large amount of discharge data that is being collected at many gaging stations. This method is preferable to the others mentioned above, but its use for small drainage basins is handicapped by the deficiency of data.

The rational method, described in Secs. 14 and 20, is generally applicable to urban areas of not over 200 acres.

Synthetic flood hydrographs are difficult to compute for small basins because of lack of rainfall and runoff records with which to compute unit hydrographs and because of the widely varied physical nature of small basins. The obstacles to use of this method probably will be surmounted, but it will require more data than are generally available at the present time.

Correlation analyses for developing formulas showing effect of various hydrologic factors and basin dimensions hold great promise for future development of a better tool than is now available. But this tool must wait until more study has been given to the problem.

Several well-developed methods for the determination of design peak discharge from small rural drainage basins are given below:

1. The U.S. Bureau of Public Roads Method (BPR Method). The BPR method for estimating peak rates of runoff from small drainage basins makes use of a *topographic index* and a *precipitation index*. These indices vary from place to place, resulting in a series of relationships, expressed as curves, for different parts of the United States.

The development of the BPR method was based on runoff records of natural streams excluding (1) watersheds with man-made controls such as diversion or storage reservoirs, (2) watersheds with 1 per cent or more of the drainage area in lakes, swamps, or excessive floodplain storage, (3) watersheds with 20 per cent or more of the area in urban development, and (4) experimental watersheds with controlled land use when such land use varied throughout the period of runoff record or was different from the prevailing land use ascribed to mixed-cover watersheds in the vicinity.

25-18 FLOOD CHARACTERISTICS AND FLOW DETERMINATION

A complete description of the development of the BPR method and directions for its use may be found in Ref. 41. From this publication directions for its use have been prepared for individual states. The following description was taken from such a report [42] prepared for application in Ohio. Other reports have been, or are, in preparation for other states. To estimate the peak rate of runoff at a proposed stream crossing for any desired recurrence interval, the procedure is as below:

1. Measure the basin area in acres (say, 500 acres).
2. Compute the topographic index T as follows:
   a. Measure the length L in miles of the principal stream from the site of the proposed crossing to the headwater.
   b. Divide this length into two reaches, the upper reach being 0.3 of the total length, or $0.3L$, and the lower reach the remaining $0.7L$.
   c. Determine the elevations of the stream channel at the upper and lower limits of each reach.
   d. Compute the average slope of each reach as the fall of the stream channel within the reach, in feet, divided by the length of the reach, in miles. The slopes so computed are designated as S_1 and S_2 for the upper and lower reach, respectively.
   e. Divide the length of the stream channel for each reach, in miles, by the square root of the corresponding slope, and add the quotients. This is the topographic index (say, 0.10), or

$$T = \frac{0.3L}{\sqrt{S_1}} + \frac{0.7L}{\sqrt{S_2}} \tag{25-I-2}$$

3. Determine the precipitation index P from Fig. 25-I-8 as follows:
   a. Locate the proposed crossing on the map of Fig. 25-I-8.
   b. Read the value of P to the nearest 0.1 corresponding to the location (say, 2.2).
   c. Note the drainage area above the proposed crossing as being in zone I, zone II, or zone III. If all or a large portion lies outside the zone boundaries shown, the estimates of runoff obtained cannot be considered as reliable as for areas within the zone boundaries.

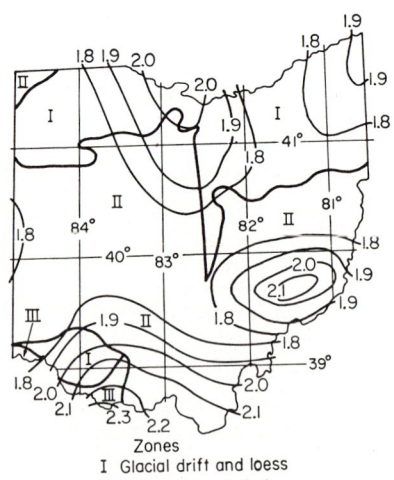

Zones
I Glacial drift and loess
II Sandstone and shale
III Limestone

FIG. 25-I-8. Zones and precipitation index P for Ohio. (*U.S. Bureau of Public Roads* [41].)

4. Estimate a 10-year flood, Q_{10}. (For purpose of this example, consider the site being located in zone III.) The procedure is as follows:
   a. Enter Fig. 25-I-9 with the drainage-basin area, and move vertically up the graph to the measured value of T.
   b. Move horizontally across the graph to the value determined for P.
   c. Move vertically up the graph and read the estimated value of Q_{10} ($= 1,000$ cfs). The accuracy of this first estimate will depend on the degree to which the drainage characteristics of the basin are similar to those of the basins on which the graphs of Fig. 25-I-9 were based. These drainage characteristics are compared as shown below.

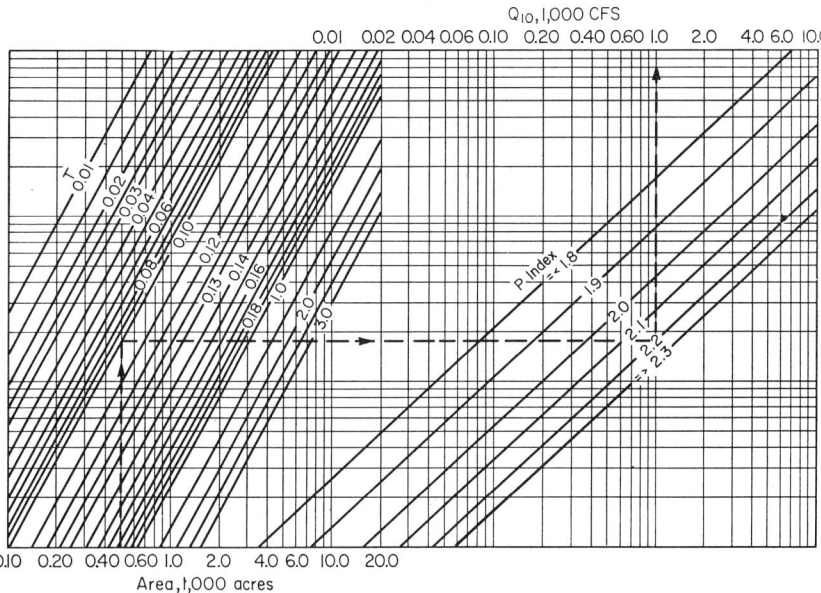

FIG. 25-I-9. Curves for estimating Q_{10} for zone III, Ohio. (*U.S. Bureau of Public Roads* [41].)

   d. For basins located in zone III, enter Fig. 25-I-10 with the drainage area in acres, and move vertically up the graph to the value of P. Then move horizontally across the graph to the ordinate scale and read the estimated T' ($= 0.2$).
   e. Express the difference between T' and T as a per cent of the estimated value, or

$$\Delta = \frac{T' - T}{T'} \times 100 \tag{25-I-3}$$

   (In the present example, $\Delta = 50$.)
   f. If the absolute value of Δ is less than 30 per cent, the value of the first estimate of Q_{10} (peak discharge with 10-year recurrence interval) needs no modification and becomes the final estimate.
   g. If the absolute value of Δ is equal to or greater than 30 per cent, the value of the first estimate must be modified by multiplying it by a coefficient C as shown in the next step.
5. Determine C as follows:
   a. Divide the measured value of T ($= 0.10$) by the estimated value ($= 0.20$). Enter Fig. 25-I-11 with this ratio ($= 0.5$) and move vertically up the graph

to the curve. Then move horizontally across the graph and read the value of coefficient C (= 0.47).

  b. Multiply the first value of Q_{10} (= 1,000 cfs) by coefficient C to obtain the final estimate of Q_{10} (= 470 cfs).

6. To obtain an estimate of Q_{25} (peak discharge with 25-year recurrence interval), Q_{50} (peak discharge with 50-year recurrence interval), or Q_{100} (peak discharge with 100-year recurrence interval), multiply Q_{10} by 1.3, 1.5, or 1.9, respectively.
7. To obtain estimates of Q for other recurrence intervals, plot the values for discharges determined in step 6, draw a complete frequency curve, and interpolate the estimates from the curve.

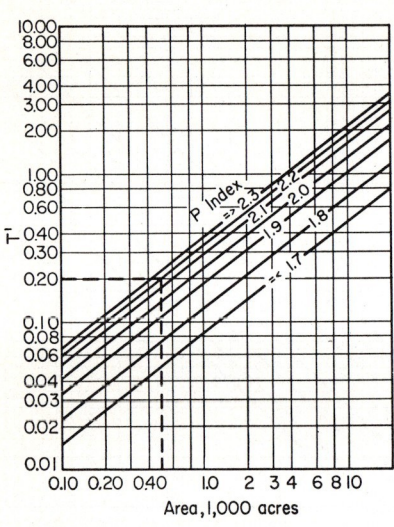

Fig. 25-I-10. Curves for estimating topographic index T' for zone III, Ohio. (*U.S. Bureau of Public Roads* [41].)

Fig. 25-I-11. Curves for determining coefficient C for zones I, II, and III, Ohio. (*U.S. Bureau of Public Roads* [41].)

8. If the drainage area of the stream crossing being considered is less than 100 acres' determine runoff estimates for three areas larger than 100 acres, and plot the three estimates to logarithmic scales. Draw a straight line through the three points so plotted and extrapolate downward to the area being considered.
9. It is suggested that estimates made for urban areas be increased by 1.5 to 2 times the discharge given by this procedure. If diversion or storage reservoirs are effective or the area contains a large percentage of natural storage, estimates made by this procedure will be high and a reduction of estimated peaks can sometimes be justified. The reduction in this latter case depends on the amount of storage.

  2. The California Method. The California Division of Highways [43] has developed a method for the calculation of design discharges for culvert design. The method is based on the rational formula, and for practical application, it has been expressed by a nomograph as shown in Fig. 25-I-12, which is self-explanatory. The design discharge obtained by this method is known as the *limit design discharge*, which is the momentary peak of flow with a recurrence interval of 100 years.

  3. The Cook Method. This method was developed by the engineers of the U.S. Soil Conservation Service. A detailed description of the method is given in Sec. 21.

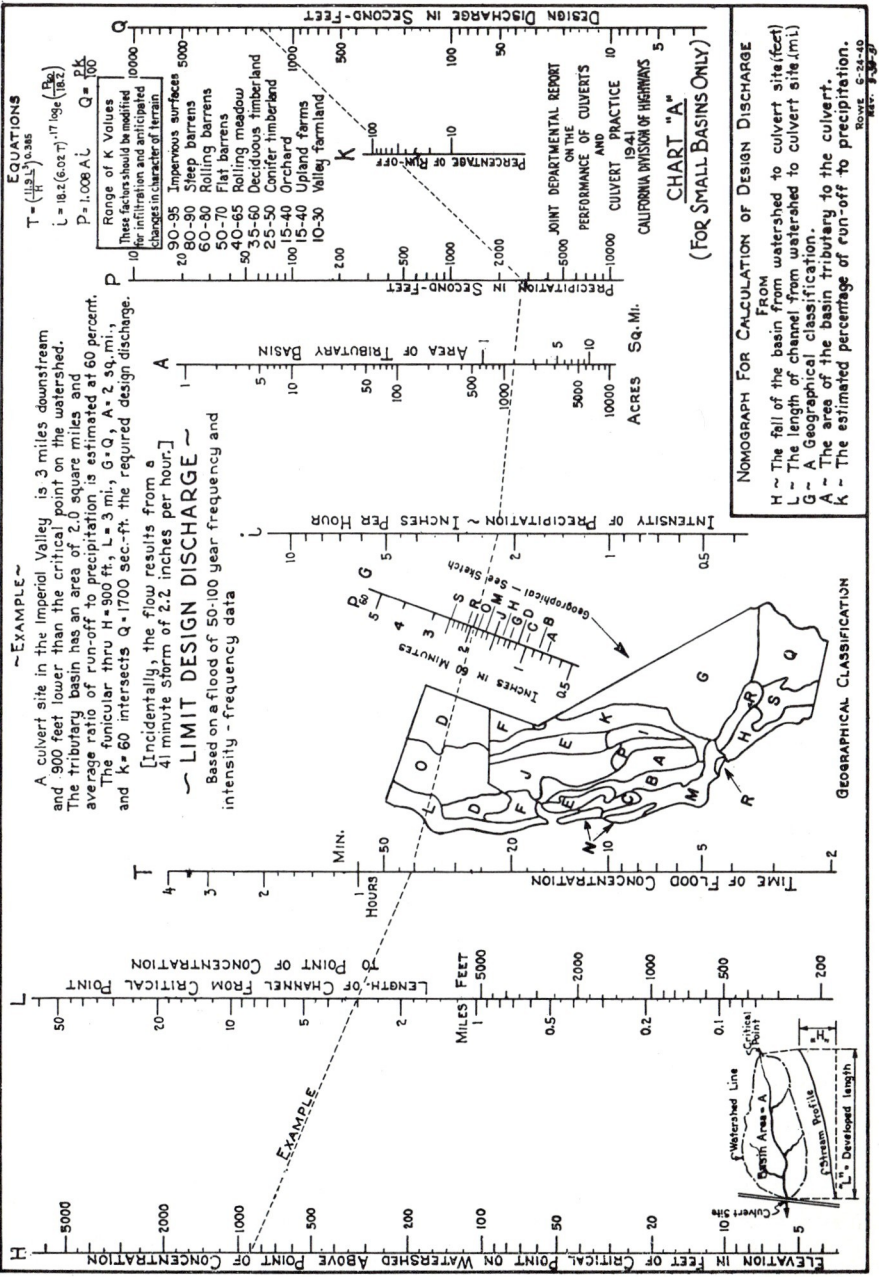

Fig. 25-I-12. Nomograph for the calculation of design discharge for culverts by the California method. (*State of California Division of Highways* [43].)

25-22 FLOOD CHARACTERISTICS AND FLOW DETERMINATION

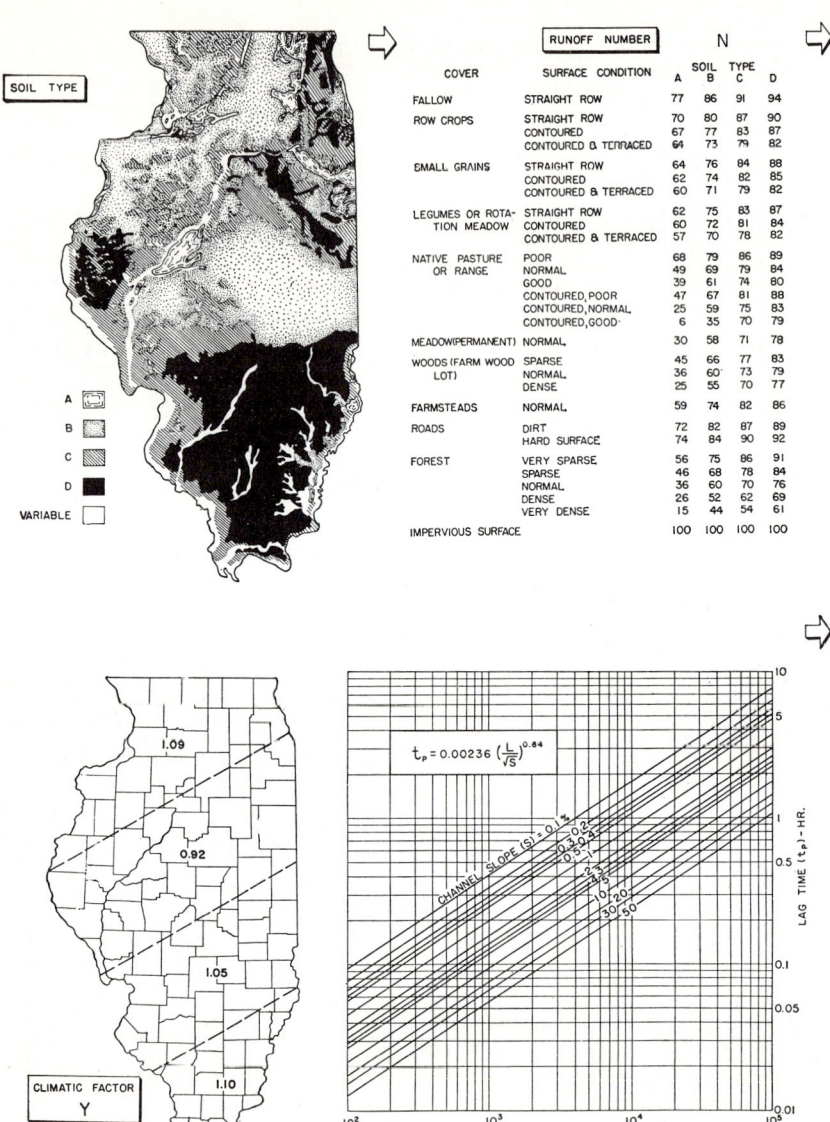

FIG. 25-I-13. Chart for the determination of design discharge

4. The Chow Method. Chow [6] has developed a practical method for the determination of peak discharge from small rural basins of less than 6,000 acres for the design of waterway openings of culverts and other small drainage structures. By this method the direct peak discharge from a drainage basin is computed as a product of the rainfall excess and the peak discharge of a unit hydrograph, or

$$Q = R_e P \qquad (25\text{-}I\text{-}4)$$

where R_e is the rainfall excess in in. for a given duration of t hr at the location investi-

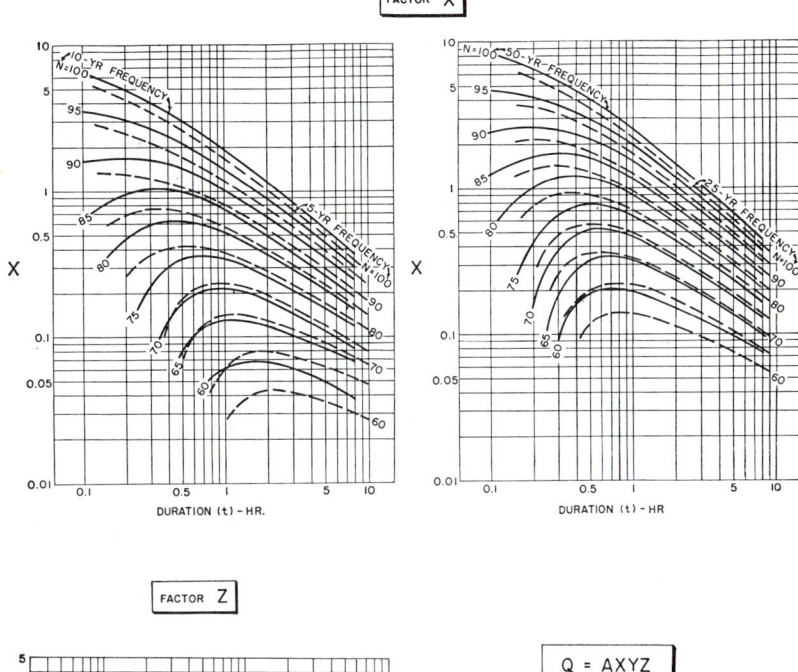

for small basins by the Chow method. (*After Chow* [6].)

gated, and P is the unit hydrograph in cfs/in. of direct runoff for the duration of t hr of rainfall excess.

Considering a continuous rainfall excess of a rate of 1 in. per t hr, the equilibrium direct discharge from a drainage basin of A acres is equal to $1.008A/t$ cfs. The peak discharge of the unit hydrograph for a duration of t hr may be expressed as a fraction Z of the equilibrium discharge, or

$$P = \frac{1.008AZ}{t} \qquad (25\text{-I-}5)$$

where Z is called a *peak-reduction factor*. Substituting Eq. (25-I-5) in Eq. (25-I-4),

$$Q = \frac{1.008 R_e A Z}{t} \qquad (25\text{-}I\text{-}6)$$

which may be written as

$$Q = AXYZ \qquad (25\text{-}I\text{-}7)$$

where X is a *runoff factor*, or

$$X = \frac{R_{e0}}{t} \qquad (25\text{-}I\text{-}8)$$

where R_{e0} is the rainfall excess in in. at a given base location, being adjusted for the effect of variable rainfall distribution in the duration of t hr, and Y is a *climatic factor*. Assuming $R_e/R_{e0} = R/R_0$, this climatic factor is

$$Y = \frac{1.008 R}{R_0} \qquad (25\text{-}I\text{-}9)$$

where R_0 is the rainfall in in. at the base location in duration t hr, and R is the rainfall in in. at the given location under investigation in duration t hr.

Equation (25-I-7) gives the design peak discharge. If the base flow at the time of the peak discharge is Q_b, then the design peak discharge is

$$Q_d = Q + Q_b \qquad (25\text{-}I\text{-}10)$$

In applying Eq. (25-I-7) to the state of Illinois, available hydrologic data were analyzed and a design chart (Fig. 25-I-13) prepared accordingly. The design discharge is defined as the maximum discharge that would occur under the average physiographic condition of the watershed due to rainfall of a given frequency and of various durations and due to the base flow. By means of the design chart, the procedure for the determination of a design discharge is as follows:

1. From the soil map in the design chart, determine the soil type. The soil map was developed from an extensive study of the drainage properties of soils in Illinois.
2. From the runoff-number table, determine the *runoff number* for the soil type and the given cover and surface conditions. If the watershed has composite soil types and cover and surface conditions, a weighted runoff number should be computed. The runoff number is a modified *hydrologic-soil-cover complex number* developed by the U.S. Soil Conservation Service (Sec. 21).
3. Assign a duration t.
4. From the curves for the runoff factor, determine the value of X for the assigned duration, the given frequency, and the selected runoff number. The families of curves for X were computed by Eq. (25-I-8), using the rainfall depth-duration-frequency curves at Urbana, Ill., as the base location and the average rainfall and rainfall-excess relationship developed by the U.S. Soil Conservation Service (Sec. 21).
5. From the chart for climatic factors, determine the value of Y. The climatic factors in the four regions in the map were computed by Eq. (25-I-9), using rainfall data supplied by the U.S. Weather Bureau and the Illinois State Water Survey.
6. From the chart for *lag time* determine the value of t_p for the given drainage area and the given length and slope of the stream. The lag time t_p is the time interval in hours between the center of mass of rainfall excess and the resulting runoff peak. The chart was developed from a correlation analysis of 60 runoff peaks from 20 small drainage basins in the Midwestern area (2.79 to 4,580 acres).
7. Compute the ratio t/t_p.
8. From the curve for peak-reduction factor, determine the value of Z for the computed t/t_p. The curve was developed from data of 60 runoff peaks on 20 small watersheds. For t/t_p greater than 4.0, the value of Z is equal to 1.0.
9. Compute the discharge by Eq. (25-I-7).
10. Repeat the steps for other assigned durations.

11. Plot the computed discharges against durations. By definition the largest discharge is the design discharge.

12. If the stream is perennial, a proper base flow should be estimated and then added to the discharge determined in the previous step.

The Chow method applies also to other areas than Illinois, provided adequate hydrologic data are available to determine the various factors. The method is advantageous in minimizing the guesswork and thus producing practically the same result when used by different individuals. On the other hand, the method is based on a given frequency of rainfall, instead of runoff, which is a shortcoming. This shortcoming is due to lack of sufficient runoff-frequency data for small watersheds, but it may be overcome in the future when such data become available and the method is modified accordingly.

IV. DESIGN FLOOD

A. General

A *design flood* is the flood adopted for the design of a structure after consideration of economic and hydrologic factors. It is seldom economically practicable to design for the maximum probable flood (Subsec. IV-B) and often not for the maximum of record. The design flood is usually selected by exercise of engineering judgment after consideration of the pertinent facts. Pertinent facts are represented by streamflow records which are either computed from precipitation records or by application of hydrologic principles to measured physical factors or directly observed and analyzed to best apply to the particular situation.

Where an area is thickly inhabited or developed industrially and the failure of protective works will result in loss of life and great property damage, a design on the basis of the maximum probable flood may be justified. In agricultural areas where failure would result only in flooding of crops, a design for a much smaller degree of protection would be justified. Varying conditions lie between these extremes, and varying design discharges would be called for in providing protective works.

The maximum probable flood may be used as the design flood in some instances, but the design of most structures is based upon a flood less than this, and perhaps one less than the maximum of experience.

Selection of the design flood requires a hydroeconomic analysis of the problem. As the size of design floods selected for study increases, the capital cost of the structure will also increase. At the same time, the probability of damage will decrease. By taking both of these elements on an annual cost basis and adding the two, a total annual cost curve will result. The most economical structure can be found from this curve. Figure 25-I-14 shows a schematic total annual cost curve as it might be defined for a bridge. In this illustration, the most economical bridge would be one designed to pass the 20-year flood, or 35,000 cfs. It will be noted that the lower part of the total cost curve is rather flat, indicating that to design for a larger flood than 35,000 cfs would cost little more.

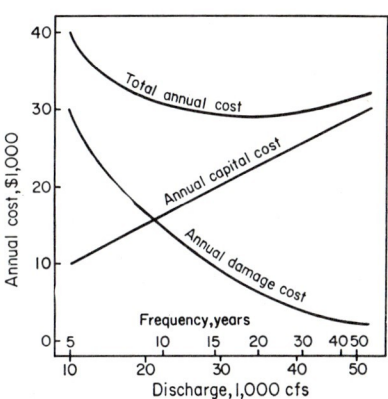

Fig. 25-I-14. Hydroeconomic analysis for bridge design.

Structures of major size will justify an economic study, but for minor structures such a study is not feasible. Generally, streamflow records are available to help in the design of works along the larger streams, but when considering problems involving the smaller streams, say, those that drain 100 sq mi or less, it usually becomes necessary to compute discharges by hydrologic methods.

B. Standard Project Flood[1]

In the planning and design of flood-control and multiple-purpose projects, the U.S. Army Corps of Engineers makes detailed studies of hydrologic information to aid in selection of the design flood. The term *design flood* refers to the flood-hydrograph or peak-discharge value finally adopted as the basis for design of a particular project or section thereof after full consideration has been given to flood characteristics, frequencies, and potentialities and the economic and other practical considerations entering into selection of the design discharge criteria.

In the design of flood-control projects it would of course be desirable to provide protection against the maximum probable flood, if this were feasible within acceptable limits of cost. However, it is seldom practicable to provide absolute flood protection by means of local protection projects or reservoirs since the costs are usually too high. As a rule, some risk must be accepted in the selection of design flood discharges. A decision as to how much risk should be accepted in each case is of utmost importance and should be based on careful consideration of flood characteristics and potentialities in the basin, the class of area to be protected, and economic limitations.

Three classes of basic flood estimates are required in general flood-control planning and design investigations:

1. Statistical Analyses of Streamflow Records. These include flood-frequency estimates (preferably on a regional basis) and various correlations of flood characteristics and hydrologic features of the drainage basin. Flood-frequency determinations are used primarily as a basis for estimating the mean annual benefits that may be expected from the control or reduction of floods of relatively common occurrence.

2. Standard-project-flood (SPF) Estimates. These represent flood discharges that may be expected from the most severe combination of meteorologic and hydrologic conditions that are considered reasonably characteristic of the geographical region involved, excluding extremely rare combinations.

3. Maximum-probable- (or Maximum-possible-) flood (MPF) Estimates. These represent flood discharges that may be expected from the most severe combination of critical meteorologic and hydrologic conditions that are reasonably possible in the region. Applications of such estimates are usually confined to the determination of spillway requirements for high dams, but in unusual cases may constitute the design flood for local protection works where an exceptionally high degree of protection is advisable and economically obtainable (Sec. 9).

The SPF estimate must reflect a generalized analysis of flood potentialities in a region, as contrasted with an analysis of flood records at the specific locality, which may be misleading because of the inadequacies of records or abnormal sequences of hydrologic events during the period of streamflow observation. The SPF represents the flood discharge that should be selected as the design flood for the project, or approached as nearly as practicable in consideration of economic or other governing limitations, where some small degree of risk can be accepted but an unusually high degree of protection is justified by hazards to life and high property values within the area to be protected. Estimates completed to date indicate that SPF discharges are generally equal to 40 to 60 per cent of maximum probable floods for the same basins.

In general terms, the SPF may be defined as a hydrograph representing runoff from the *standard project storm* (SPS). The SPS estimate for a particular drainage area and season of year in which snowmelt is not a major consideration should represent the most severe flood-producing rainfall depth-area-duration relationship and isohyetal pattern of any storm that is considered reasonably characteristic of the region in which the drainage basin is located, giving consideration to the runoff characteristics and existence of water-regulation structures in the basin. In deriving SPS rainfall estimates applicable to seasons and areas in which melting snow may contribute a substantial volume of runoff to the SPF hydrograph, appropriate allowances for snowmelt are included with and considered as a part of the SPS rainfall quantities in computing the

---

[1] The material for the discussion of this topic has been taken from the U.S. Army Corps of Engineers, *CWE Bulletin* 52-8 [44].

SPF hydrograph. Where floods are predominantly the result of melting snow, the SPF estimate is based on estimates of the most critical combinations of snow, temperature, and water losses considered reasonably characteristic of the region.

C. SPS Criteria for Small Drainage Basins

For drainage areas less than approximately 1,000 sq mi, located east of 105° longitude, the generalized SPS criteria are as follows:

1. Generalized Estimates of Maximum Possible Precipitation (MPP) (Sec. 9). Figure 25-I-15 shows such estimates for 24-hr rainfall over a 200-sq-mi area, as estimated by the Hydrometeorological Section of the U.S. Weather Bureau and explained in Ref. 45.

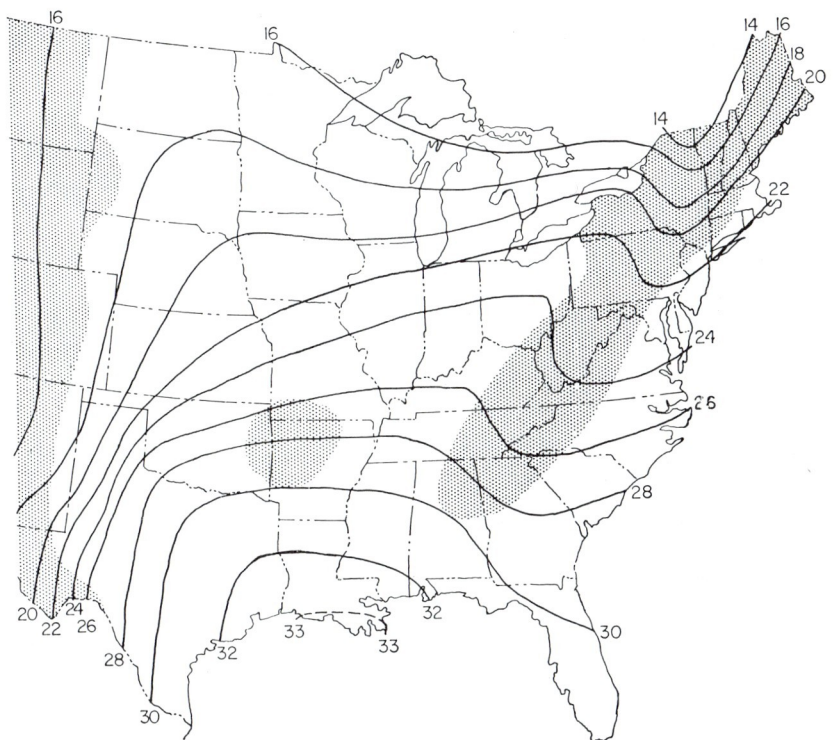

Fig. 25-I-15. Generalized estimates of maximum possible precipitation. (*U.S. Army Corps of Engineers* [44].)

2. SPS Index Rainfall. This represents the maximum average depth of rainfall during the SPS and in general is equal to approximately 40 to 60 per cent of the MPP in step 1, with a general average of about 50 per cent. Figure 25-I-16 shows such SPS index rainfall isohyets for 24-hr rainfall over a 200-sq-mi area. They were prepared by reducing isohyets of MPP in step 1 to 50 per cent of original values and reshaping the isohyets in certain regions by moderate amounts to conform with supplementary studies of rainfall characteristics in those regions.

3. Tabulation and Conversion of Rainfall Depths. The average depths of rainfall in recorded major storms over selected areas in various periods of time (obtainable from "Storm Rainfall in the Eastern United States" published by the U.S. Army Corps of Engineers) are tabulated and converted to per cent of SPS index rainfall.

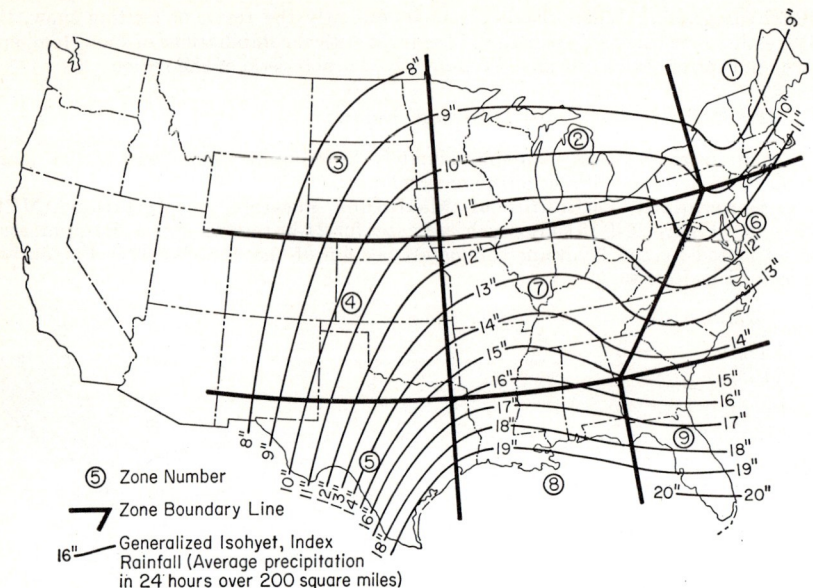

Fig. 25-I-16. SPS index rainfall isohyets. (*U.S. Army Corps of Engineers* [44].)

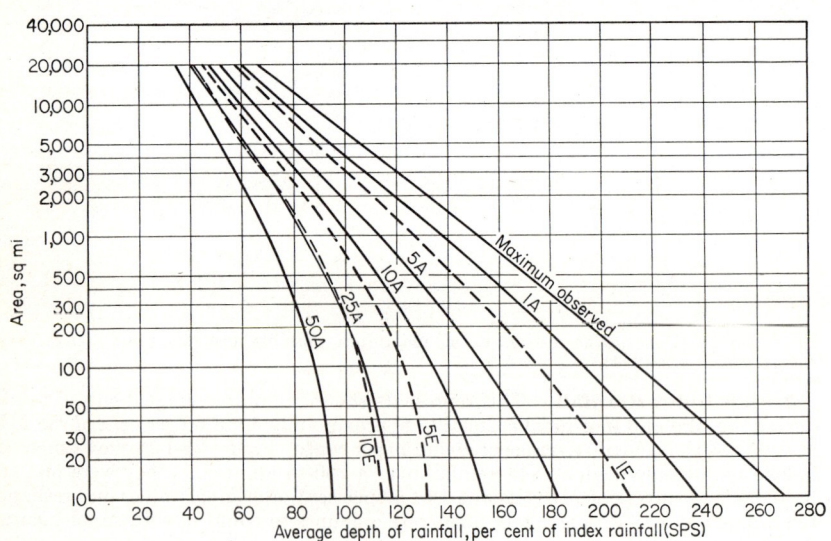

Fig. 25-I-17. Generalized depth-area curves for 24-hr rainfall. 5A, average of largest 5 per cent of storm-study values; 5E, amounts exceeded by 5 per cent of storm-study values. Total number of storm-study values for each area size is 400. (*U.S. Army Corps of Engineers* [44].)

4. Generalized Depth-Area Curves. Considering all storms tabulated in step 3, the 24-hr average rainfall depths in per cent of SPS index rainfall are arranged in order of diminishing magnitude for each of the given areas. The number of rainfall depths exceeding various values is determined and converted to per cent of the total number of storms studied. By plotting for each selected area the rainfall depths equaled or exceeded by 1, 5, and 10 per cent of the total number of values, a set of curves numbered $1E$, $5E$, and $10E$, respectively, is obtained as shown in Fig. 25-I-17. The curve for maximum observed is an envelope of the highest values of rainfall depths. The curves for $1A$, $5A$, $10A$, $25A$, and $50A$ are obtained by plotting the mean of the highest 1, 5, 10, 25, and 50 per cent, respectively, of the rainfall depths for the selected areas.

Similarly, generalized depth-area curves for 48- and 96-hr storm periods are prepared.

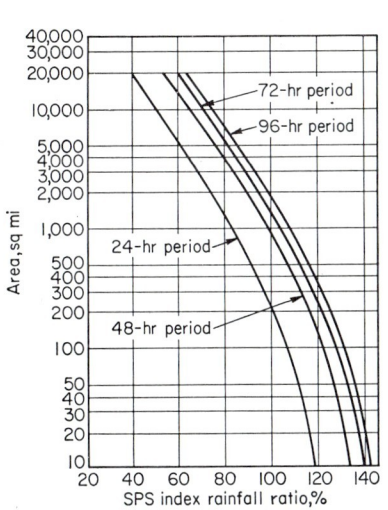

FIG. 25-I-18. SPS depth-area-duration relationships by 24-hr storm increments. (*U.S. Army Corps of Engineers* [44].)

FIG. 25-I-19. Time distribution of 24-hr SPS rainfall. (*a*) SPS 24-hr precipitation over 200 sq mi vs. per cent in 6-hr periods; (*b*) typical arrangement of 6-hr rainfall quantities in SPS. (*U.S. Army Corps of Engineers* [44].)

5. SPS Depth-Area-Duration Curves. The depth-area curves designated by $25A$ for 24-, 48-, and 96-hr periods are selected as indices to the volume of precipitation that should be assumed as occurring in the corresponding periods of standard project storms. The curves for $25A$ can be replotted to produce the depth-area-duration as shown in Fig. 25-I-18.

6. Time Distribution of 24-hr SPS Rainfall. Actual storms representing the maximum rainfall data in 6 and 12 hr over selected areas less than 1,000 sq mi are used to construct curves in Fig. 25-I-19*a*. By an extensive study of the time-distribution pattern of actual storms, the sequence of 6-hr rainfall increments as shown in Fig. 25-I-19*b* is adopted on the basis that such a sequence would produce critical runoff from most basins.

7. Time Distribution of Maximum 6-hr SPS Rainfall. In order to assure safe estimates of peak discharges to be expected from SPS rainfall over drainage areas less than approximately 300 sq mi on the average, the maximum 6-hr rainfall of the SPS is broken down into shorter unit periods, and higher intensities are assumed for the shorter intervals according to the following criteria:

Time Distribution of Maximum 6-hr SPS Rainfall, in Per Cent of Total 6-hr Rainfall

Rainfall period (subdivision of 6-hr period)	Selected unit rainfall duration				
	6 hr	3 hr	2 hr	1 hr	0.5 hr
1st	100	20	19	6	3
2d		80	69	8	3
3d			12	14	4
4th				55	5
5th				11	6
6th				6	12
7th					43
8th					8
9th					6
10th					4
11th					3
12th					3
Total.........	100	100	100	100	100

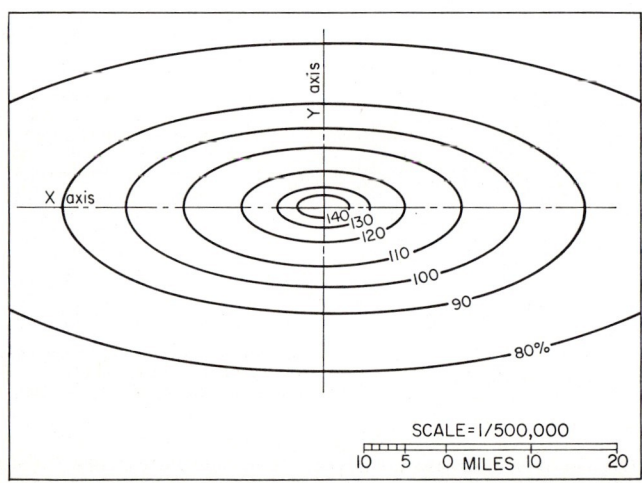

FIG. 25-I-20. SPS isohyetal pattern. (*U.S. Army Corps of Engineers* [44].)

The "selected unit rainfall duration" is determined roughly from the synthetic-unit-hydrograph equation, $t_r = t_p/5.5$, where t_p is the lag time from mid-point of unit rainfall duration t_r to peak of unit hydrograph in hr [46, p. 11]. Rounded-off values are used in the table.

8. SPS Isohyetal Pattern. By an extensive study of actual storms, a typical isohyetal pattern as shown in Fig. 25-I-20 is recommended for SPS computations. This pattern, expressed in per cent of index rainfall corresponding to depth-area relation for a 96-hr storm period, may be oriented in any direction.

D. SPF Estimates for Small Drainage Basins

Utilizing the generalized criteria for SPS, the preparation of an SPF estimate for a small drainage basin (less than 1,000 sq mi) involves the following steps:

DESIGN FLOOD   25-31

1. From Fig. 25-I-16 interpolate the SPS index rainfall corresponding to the given location of the project basin.

2. From Fig. 25-I-18 and with the interpolated SPS index rainfall, obtain the SPS index rainfall ratios for 24-, 48-, 72-, and 96-hr periods and an area equal to the total area of the project basin. Multiply these ratios by the SPS index rainfall to obtain average depths of rainfall within an isohyetal area equal to the project basin area and determine incremental 24-hr values by subtraction.

3. Arrange the 24-hr SPS rainfall values determined in the previous step in a sequence favorable to production of critical runoff.

4. Prepare an overlay of the isohyetal map shown in Fig. 25-I-20, converting the isohyet values to inches of 96-hr SPS rainfall by multiplying the percentage values on the pattern by the SPS index rainfall value determined in step 1.

5. Superimpose the project basin outline over the map prepared in step 4, and planimeter to determine the average depth of total storm rainfall over the basin and each subdivision thereof that is to be considered in estimating the SPS rainfall excess and runoff rates.

6. Subdivide the total storm SPS rainfall values obtained in step 5 into 6-hr values in accordance with criteria described in Subsec. IV-C-6. The same sequence and 6-hourly percentage distribution of rainfall are assumed for each day of the SPS.

7. If the computed value of t_p (lag time from mid-point of unit rainfall duration to peak of unit hydrograph) is less than 16 hr, subdivide the maximum 6-hr SPS rainfall value of the maximum 24-hr rainfall into shorter unit durations in accordance with criteria given in Subsec. IV-C-7.

8. Subtract estimated infiltration losses from SPS rainfall values obtained in steps 6 and 7 to obtain rainfall excess for computing SPF discharges by unit hydrographs.

E. SPF as Percentage of Maximum Probable Flood

As stated previously, estimates completed to date indicate that SPF discharges based on detailed studies usually equal 40 to 60 per cent of the MPF for the same basin; a ratio of 50 per cent is considered representative of average conditions. Inasmuch as computation of MPF estimates are normally required as the basis of design of spillways for high dams, it is convenient to estimate the SPF for reservoir projects as equal to 50 per cent of the MPF hydrograph to avoid the preparation of a separate SPF estimate. Accordingly, this conversion is acceptable for reservoir projects in general. The rule may also be applied in estimating SPF hydrographs for basins outside of the region and range of areas covered by generalized charts presented herein, where MPF estimates based on detailed hydrometeorological investigations have been completed. Where snowmelt and extreme ranges in topography are major factors to be taken into consideration, it is appropriate to estimate the MPF hydrograph for the basin by considering optimum combinations of critical flood-producing factors and assuming the SPF hydrographs as equal to 50 per cent of the maximum probable discharges. This approximation is based on the conclusion that critical conditions can be determined from analyses of meteorological and topographic influences, while a substantial period of hydrometeorological records is required to determine appropriate combinations of flood-producing factors meeting SPF specifications.

F. SPF Estimates for Large Drainage Basins

The basic principles involved in the preparation of SPS and SPF estimates for large drainage basins are the same as those applicable to basins of less than 1,000 sq mi in area, which have been discussed. However, generalizations of criteria become more difficult as the size of the area increases. While SPF discharge estimates for small areas are usually governed largely by the maximum 6- or 12-hr rainfall associated with a severe thunderstorm situation, floods of SPF category on large drainage basins are generally the result of a succession of relatively distinct rainfall events. Although the intensity and quantities of rainfall are important factors in the production of a flood in a large drainage basin, the location of successive increments of rainfall in the basin and the synchronization of intense bursts of rainfall with progression of runoff are of equal or greater importance in many cases than quantity of total precipitation. For example, the total rainfall over the Kansas River basin during the period May 25 to 31, 1903, which produced an estimated peak discharge of 260,000 cfs on the Kansas River at Kansas City, was almost identical with the total precipitation that occurred over the basin on July 9 to 12, 1951, to produce a peak discharge of 510,000 cfs. Other examples of a similar nature might be cited. Accordingly, selection of an SPS for a large basin cannot be predicted on a statistical analysis of precipitation quantities alone, but must be based on a review of hydrometeorological data for several outstanding storms of record in the basin and adjacent regions in relation to hydrologic characteristics of the basin under study. Consideration of major floods of record and historical account should also play an important part in the selection of SPF criteria and final estimate.

G. Projects for Which SPF Estimates Are Required

In some cases the SPF estimate may have a major bearing on selection of the design flood for a particular project; in other cases the estimate may serve only as an indication of the partial degree of protection proposed for the project. SPF estimates are useful in connection with practically all flood-control investigations, but to reduce the work required in project studies, preparation of SPF estimates is required only for those projects in which the proposed design flood is greater than one of estimated 25-year frequency. To reduce the work involved further, approximate estimates are acceptable where it is apparent that the design flood is strictly limited by economic considerations and is substantially less than the SPF. The basis of SPF estimates should be clearly stated in each case.

V. REFERENCES

1. Leopold, L. B., and Thomas Maddock, Jr.: "The Flood Control Controversy," The Ronald Press Company, New York, 1954.
2. Hoyt, W. G., and W. B. Langbein: "Floods," Princeton University Press, Princeton, N.J., 1955.
3. Langbein, W. B., and W. G. Hoyt: "Water Facts for the Nation's Future," The Ronald Press Company, New York, 1959.
4. Smith, Winchell, and W. L. Heckler: "Compilation of Flood Data in Arizona, 1862–1953," U.S. Geological Survey, open-file report, 1955.
5. Jarvis, C. S., and others: Floods in the United States, *U.S. Geol. Surv. Water-Supply Paper* 771, 1936.
6. Chow, V. T.: Hydrologic determination of waterway areas for the design of drainage structures in small drainage basins, *Univ. Ill. Eng. Expt. Sta. Bull.* 462, 1962.
7. Dalrymple, Tate: Flood-frequency analyses (in "Manual of Hydrology," pt. 3, Flood-flow Techniques), *U.S. Geol. Surv. Water-Supply Paper* 1543-A, 1960.
8. Peirce, L. B.: Floods in Alabama, magnitude and frequency, *U.S. Geol. Surv. Circ.* 342, 1954.
9. Patterson, J. L.: "Floods in Arkansas, Magnitude and Frequency," U.S. Geological Survey, in cooperation with Arkansas State Highway Commission, open-file report, 1961.
10. Rantz, S. E., and H. C. Riggs: Magnitude and frequency of floods in the Columbia River Basin, *U.S. Geol. Surv. Water-Supply Paper* 1080, pp. 317–469, 1949.
11. Bigwood, B. L., and M. P. Thomas: A flood-flow formula for Connecticut, *U.S. Geol. Surv. Circ.* 365, 1955.
12. Tice, R. H.: "Delaware River Basin Flood Frequency," U.S. Geological Survey, open-file report, 1958.
13. Pride, R. W.: "Floods in Florida, Magnitude and Frequency," U.S. Geological Survey, open-file report, 1958.
14. Bunch, C. M., and McGlone Price: "Floods in Georgia, Magnitude and Frequency," U.S. Geological Survey, open-file report, 1962.
15. Mitchell, W. D.: "Floods in Illinois, Magnitude and Frequency," State of Illinois, Department of Public Works and Buildings, Division of Waterways, prepared in cooperation with U.S. Geological Survey, 1954.

REFERENCES

16. Green, A. R., and R. E. Hoggatt: "Floods in Indiana, Magnitude and Frequency," U.S. Geological Survey, open-file report, 1960.
17. Schwob, H. H.: Iowa floods, magnitude and frequency, *Iowa Highway Res. Board Bull.* 1, 1953.
18. Ellis, D. W., and G. W. Edelen, Jr.: Kansas streamflow characteristics, pt. 3, Flood frequency, *Kansas Water Resources Board Tech. Rept.* 3, 1960.
19. McCabe, J. A.: Floods in Kentucky, magnitude and frequency, *Kentucky Geol. Surv. Inform. Circ.* 9, ser. X, 1962.
20. Cragwall, J. S., Jr.: "Floods in Louisiana, Magnitude and Frequency," State of Louisiana, Department of Highways, 1952.
21. Darling, J. M.: "Floods in Maryland, Magnitude and Frequency," U.S. Geological Survey, open-file report, 1959.
22. Prior, C. H.: Magnitude and frequency of floods in Minnesota, *Minn. Dept. Conserv. Bull.* 12, 1961.
23. Wilson, K. V., and I. L. Trotter, Jr.: Floods in Mississippi, magnitude and frequency, *Mississippi State Highway Dept. Bull.*, 1961.
24. Searcy, J. K.: Floods in Missouri, magnitude and frequency, *U.S. Geol. Surv. Circ.* 370, 1955.
25. Berwick, V. K.: "Floods in Eastern Montana, Magnitude and Frequency," U.S. Geological Survey, open-file report, 1958.
26. Furness, L. W.: "Floods in Nebraska, Magnitude and Frequency," State of Nebraska, Department of Roads and Irrigation, 1955.
27. Beckman, E. W., and N. E. Hutchinson: Floods in Nebraska on small drainage areas, magnitude and frequency, *U.S. Geol. Surv. Circ.* 458, 1962.
28. Robison, F. L.: Floods in New York, magnitude and frequency, *U.S. Geol. Surv. Circ.* 454, 1961.
29. Forrest, W. E., and P. R. Speer: "Floods in North Carolina, Magnitude and Frequency," U.S. Geological Survey, open-file report, 1961.
30. McCabe, J. A.: "Floods in North and South Dakota, Frequency and Magnitude," U.S. Geological Survey, open-file report, 1957.
31. Cross, W. P., and E. E. Webber: Floods in Ohio, magnitude and frequency, *Ohio Dept. Nat. Resources Bull.* 32, 1959.
32. Busch, W. F., and L. C. Shaw: "Floods in Pennsylvania, Frequency and Magnitude," U.S. Geological Survey, open-file report, 1960.
33. Jenkins, C. T.: "Floods in Tennessee, Magnitude and Frequency," State of Tennessee, Department of Highways, 1960.
34. Berwick, V. K.: Floods in Utah, magnitude and frequency, *U.S. Geol. Surv. Circ.* 457, 1962.
35. Bodhaine, G. L., and D. M. Thomas: "Floods in Washington, Magnitude and Frequency," U.S. Geological Survey, open-file report, 1960.
36. Ericson, D. W.: "Floods in Wisconsin, Magnitude and Frequency," U.S. Geological Survey, in cooperation with the State Highway Commission of Wisconsin, open-file report, 1961.
37. Linsley, R. K., Jr., M. A. Kohler, and J. L. H. Paulhus: "Applied Hydrology," McGraw-Hill Book Company, Inc., New York, 1949.
38. Jarvis, C. S.: Floods, chap. XI-G, in O. E. Meinzer (ed.), "Hydrology," McGraw-Hill Book Company, Inc., New York, 1942.
39. "Publications of the Geological Survey," U.S. Department of the Interior, May, 1958. Also see monthly supplement.
40. Chow, V. T., as chairman, and others: Report of the Committee on Runoff, 1955–56, *Trans. Am. Geophys. Union*, vol. 38, no. 3, pp. 379–384, June, 1957.
41. Potter, W. D.: Peak rates of runoff from small watersheds, *U.S. Bur. Public Roads, Hydraulic Design Ser.*, no. 2, April, 1961.
42. Estimating peak rates of runoff from small watersheds in Ohio, *U.S. Bur. Public Roads, Hydraulic Eng. Circ.* 4, April, 1961.
43. "California Culvert Practice," 2d ed., State of California, Department of Public Works, Division of Highways, 1953.
44. Standard project flood determinations, *U.S. Dept. of the Army Civil Works Eng. Bull.* 52–8, Mar. 26, 1952.
45. Generalized estimates of maximum possible precipitation over the United States east of the 105th meridian, for areas of 10, 200, and 500 square miles, *U.S. Weather Bur. and U.S. Corps of Engineers Hydrometeorol. Rept.* 23, 1947.
46. "Hydrologic and Hydraulic Analyses, Flood-hydrograph Analyses and Computations," U.S. Army Corps of Engineers, Civil Works Construction, Engineering Manual, March, 1948, pt. CXIV, chap. 5.

Section 25-II

HYDROLOGY OF FLOW CONTROL

PART II. FLOOD ROUTING

EDWARD A. LAWLER, *Chief, Hydrology and Reservoir Regulation Section, Ohio River Division, Corps of Engineers, U.S. Army.*

I. Introduction: Definition and Scope	25-34
II. Basic Data	25-35
III. Mathematics of Flood Routing	25-37
IV. Routing Methods	25-38
A. Invariable Discharge-Storage Relationship	25-38
1. The Puls Method	25-38
2. The Coefficient Method	25-40
B. Variable Discharge-Storage Relationship	25-40
1. The Muskingum Method	25-40
2. The Working-value Method	25-44
C. Lag Methods	25-46
1. The Successive Average-lag Method	25-46
2. The Progressive Average-lag Method	25-47
D. Graphical Methods	25-47
1. The Simplified Muskingum Method	25-49
2. The Working-value Method	25-49
E. Routing at Junctions	25-50
1. The Working-value Method	25-50
F. Routing of Reservoir Modifications	25-52
G. Mechanical Routers	25-53
H. Electronic Computers	25-54
1. The Analog Computer	25-54
2. The Differential Analyzer	25-55
3. The Digital Computers	25-58
V. Stage Routing	25-58
VI. Flow-line Computation	25-58
VII. References	25-58

I. INTRODUCTION: DEFINITION AND SCOPE

Many communities owe much of their prosperity to the advantages offered by adjacent or nearby streams, the more important being adequate commercial and municipal water supplies, navigation, power development, and recreation. Adverse effects, however, are experienced when erratic flows occur, particularly those causing damaging floods. When economically feasible, measures must be taken to eliminate or control these erratic occurrences. Such measures may include construction of flood-control reservoirs, levees, and flood walls; channel improvements; etc. Procedures for evaluating these measures both in the design and operation phases and methods of predicting flood crest are therefore economically very important. These procedures and methods generally come under the subject of flood routing.

Flood routing may be defined as the procedure whereby the time and magnitude of a flood wave at a point on a stream is determined from the known or assumed data at one or more points upstream. Flood routing may be considered under two broad but somewhat related types, namely, *reservoir routing* and *open-channel routing*. The former type provides methods for evaluating the modifying effects on a flood wave passing through a reservoir. In design and planning it applies to the determination of the location and capacity of reservoirs, of the size of outlet structures and spillways, etc. Reservoir routing and the procedures are discussed in Sec. 25-III.

Open-channel routings are used to determine the time and magnitude of flood waves in rivers, to develop design elevations for flood walls and levees, to estimate benefits from completed or proposed reservoirs, etc. This section attempts to discuss the most suitable and acceptable procedures in general use.

The theoretical analysis of the movement of flood waves is quite complex, and the methods that attempt a strict mathematical approach are generally impracticable. Even with the assistance of high-speed electronic computers, approximate numerical techniques and simplifying assumptions are necessary. The derivation of equations, development of involved mathematical procedures, and theoretical discussions can be found in the listed references.

It should be recognized that the differences between various routing procedures arise to a considerable extent from minor variations in algebraic manipulation or graphical presentation or from refinements in the basic assumptions. The choice of a procedure depends on many factors, including the nature of available data and personal preference.

II. BASIC DATA

All flood-routing methods are based on some knowledge of the river reach under consideration. The knowledge may be as meager as only maps defining topographic features, with no record of floods, or it may include a rather complete history of floods, profiles, cross sections, etc. Generally, routing procedures are based on the relationship between stage and storage or discharge and storage. Two methods are available for developing this relationship. The first involves the determination of volumes of storage at different levels from valley cross sections or detailed topographic maps. The second method, more commonly used, determines the storage volumes by the analysis of floods of record, assuming that the relationships established by the record will be valid for future floods. Data required for this analysis are the flow records at the upstream, downstream, and tributary stations, as well as the precipitation records for the ungaged areas. Flows at gaged points are to be determined from rating curves (Sec. 15), while flows from the ungaged areas are to be estimated from precipitation records, using unit-hydrograph computations (Sec. 14). The hydrograph of the total flow into a reach, known as the *inflow hydrograph*, is compiled by adding the flows from the upstream station, the gaged tributaries, and the ungaged areas. Figure 25-II-1 illustrates a typical reach AB, where stations A and B are, respectively, the upstream and downstream stations on the main stream and stations C and D are on tributaries entering the main stream. The dotted line embraces the ungaged area, called the *local area*. In Fig. 25-II-2 the curve $abdeghi$ represents the total inflow hydrograph into the reach for a specific flood. The area under the curve for each time period is the volume of flow during that period, and the total area under the curve is the volume of flow for the entire flood wave. Flow (actually, the rate of flow) is expressed in cubic feet per second (cfs), and volume in cubic feet per second per day, or day-second-feet (dsf). Curve *acdfghi* represents the flow at the downstream end of the reach, known as *outflow*, and must equal the total inflow in volume. During the first portion of the flood wave, periods 1 to 8, inflow exceeds outflow, and thus water is being stored in the reach. The area *abdca*, or the difference between the inflow and outflow hydrographs, represents the volume stored. During the time periods 9 to 17, outflow exceeds inflow, and thus water is being drawn from the storage.

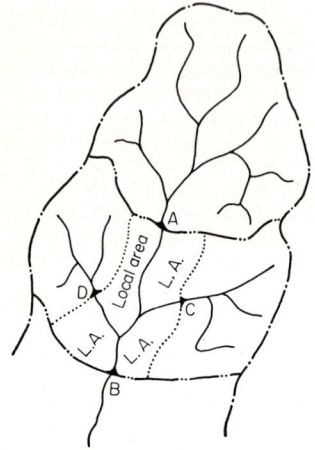

FIG. 25-II-1. A typical drainage area.

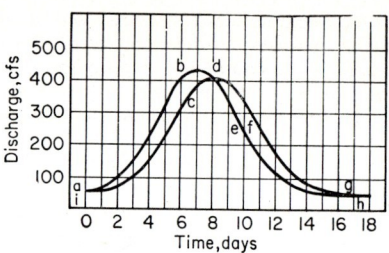

FIG. 25-II-2. Typical inflow and outflow hydrographs.

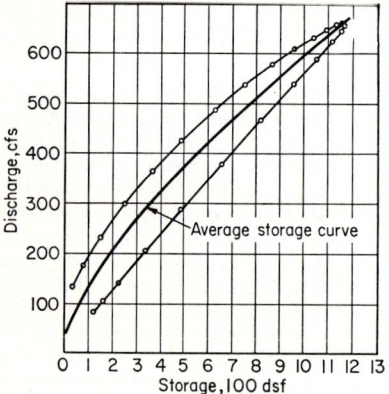

FIG. 25-II-3. Discharge-storage curve.

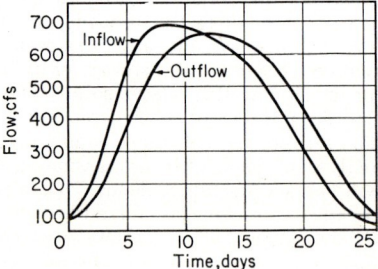

FIG. 25-II-4. Inflow and outflow hydrographs.

The volume of storage depleted is represented by area *dfged*, which equals *abdca*. A method of tabulating these flows is shown in Table 25-II-1. At any time period, the difference between the sum of inflows and the sum of outflows is the volume of storage in the reach. Figure 25-II-3 is a plotting of the storage volume vs. outflow discharge for the periods shown in Table 25-II-1. The mean curve of the points, expressing the average relationship between outflow and storage, is the average *storage curve*. The total inflow and outflow hydrographs for the data are plotted in Fig. 25-II-4.

MATHEMATICS OF FLOOD ROUTING

Table 25-II-1. Tabulation of Flows

(1)	(2)									(3)	(4)	(5)	(6)	(7)	(8)	(9)
	Ungaged area															
Routing period, days	Run- off, cfs	Distribution, per cent							Base flow, cfs	Total flow, cfs	Station C, cfs	Station D, cfs	Station A, cfs	Total in- flow, cfs	Station B, cfs	Storage incre- ment, dsf	Cumu- lative stor- age, dsf
		5	15	25	25	15	10	5									
1	...	..	..	..	..	..	..	..	2	2	10	1	80	93	85	8	118
2	140	7	..	..	..	..	..	..	2	9	18	3	107	137	102	35	153
3	40	2	21	..	..	..	..	..	2	25	32	5	146	208	141	67	220
4	65	3	6	35	..	..	..	..	2	46	50	12	212	320	205	115	335
5	12	1	10	10	35	..	..	..	2	58	68	24	292	442	290	152	487
6	23	1	2	16	10	21	..	..	2	52	82	37	375	546	380	166	653
7	6	0	3	3	16	6	14	..	2	44	90	48	448	630	470	160	813
8	...	..	..	1	6	3	10	4	2	33	89	46	510	678	539	138	951
9	...	..	..	..	2	6	2	7	2	21	79	33	558	691	591	100	1,051
10	...	..	..	..	..	2	4	1	2	12	64	25	591	692	627	65	1,116
11	...	..	..	..	..	..	1	2	2	5	50	17	612	684	648	36	1,152
12	...	..	..	..	..	..	..	0	2	3	35	11	622	671	660	11	1,163
13	...	..	..	..	..	..	..	..	2	2	24	8	623	657	664	−7	1,156
14	...	..	..	..	..	..	..	..	2	2	17	6	613	638	660	−22	1,134
15	...	..	..	..	..	..	..	..	2	2	11	4	592	609	650	−41	1,093
16	...	..	..	..	..	..	..	..	2	2	9	3	563	577	635	−58	1,035
17	...	..	..	..	..	..	..	..	2	2	8	2	522	534	610	−76	959
18	...	..	..	..	..	..	..	..	2	2	7	1	474	484	58−	−96	863
19	...	..	..	..	..	..	..	..	2	2	6	1	417	426	540	−116	747
20	...	..	..	..	..	..	..	..	2	2	6	1	357	366	488	−122	625
21	...	..	..	..	..	..	..	..	2	2	5	1	290	298	430	−132	493
22	...	..	..	..	..	..	..	..	2	2	5	1	227	235	365	−130	363
23	...	..	..	..	..	..	..	..	2	2	4	1	176	183	300	−117	246
24	...	..	..	..	..	..	..	..	2	2	4	1	130	137	233	−96	150
25	...	..	..	..	..	..	..	..	2	2	3	1	97	103	178	−75	75
26	...	..	..	..	..	..	..	..	2	2	3	1	75	81	132	−51	24
27	...	..	..	..	..	..	..	..	2	2	2	1	70	75	100	−25	−1

III. MATHEMATICS OF FLOOD ROUTING

The fundamental law of unsteady flow is based on the principles of conservation of energy and conservation of matter. It may be expressed by the following two partial differential equations:

$$\frac{\partial H}{\partial x} + \frac{V}{g}\frac{\partial V}{\partial x} + \frac{1}{g}\frac{\partial V}{\partial t} + \frac{V^2}{C^2 R} = 0 \qquad (25\text{-II-}1)$$

$$A\frac{\partial V}{\partial x} + V\frac{\partial A}{\partial x} + B\frac{\partial H}{\partial t} = 0 \qquad (25\text{-II-}2)$$

where H denotes the elevation of water surface, V is the velocity, A is the cross-sectional area of the flow in the channel, B is the surface width of the flow, g is the acceleration of gravity, C is Chézy's resistance factor, R is the hydraulic radius, x is the distance along the channel, and t is time.

In Eq. (25-II-1), the terms represent, in order, the depth taper, the velocity head, the acceleration head, and the friction slope. In Eq. (25-II-2), the terms represent in order the prism storage, the wedge storage, and the rate of rise. By replacing the derivatives by finite differences, a numerical method of solution of these equations can be developed [1–3]. Neglecting the energy relationship represented by Eq.

(25-II-1), many procedures of *storage routing* based on Eq. (25-II-2) have been developed. They have general applications and give relatively acceptable results. By introducing Eq. (25-II-1), more theoretically acceptable methods are possible, but they suffer the handicap of greater complexity. Other methods, called *lag methods*, based on the time displacement of average inflow values, have been used successfully. Some other methods, empirically derived from stage relationships, are used commonly for forecasting purposes. Mechanical and electronic machines, utilizing various mathematical equations or relationships, have also been designed and used satisfactorily. Development of routing equations and their theoretical discussions can be found in Refs. 2 to 5.

Routing methods based on complicated mathematics and hydraulic principles have also been developed, such as the *method of characteristics* and the *method of diffusion analogy* [3]. In general, these methods assume a uniform channel and therefore become cumbersome or impracticable when applied to irregular natural channels. With simplified assumptions and the aid of digital computers, their application to practical problems may be possible. However, a discussion of these methods is usually considered in hydraulics, and hence it is beyond the present scope.

IV. ROUTING METHODS

Of initial importance prior to the selection of a routing method is the selection of a proper *routing period*. This is the time interval at which the ordinates of a hydrograph used in the routing are represented. The period must be sufficiently short to define the hydrograph adequately. Theoretically, it should be equal to, or somewhat shorter than, the travel time of the flow through the reach. Also, the period must be short enough so that the hydrograph during the period approximates a straight line.

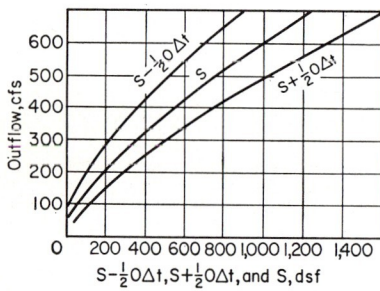

FIG. 25-II-5. Curves for the Puls method. $\Delta t = 1$ day.

A. Invariable Discharge-Storage Relationship

1. The Puls Method. This method assumes invariable discharge-storage relationships and neglects the variable slope occurring during the passage of a flood wave. Although used quite satisfactorily for reservoir routing, it gives a poor approximation for open-channel routing. The method is presented here as a means of introducing a basis from which more complete methods have been developed.

In a given time interval the difference between inflow and outflow is equal to the change in storage:

$$I - O = \Delta S \qquad (25\text{-II-}3)$$

or if expressed in finite time intervals,

$$\tfrac{1}{2}(I_1 + I_2)\,\Delta t - \tfrac{1}{2}(O_1 + O_2)\,\Delta t = S_2 - S_1 \qquad (25\text{-II-}4)$$

where the subscripts indicate the routing periods, and I, O, and S are instantaneous values of inflow, outflow, and storage, respectively, at the beginning of the routing periods indicated. Arranging the equation so that all known values are on the left, the expression becomes

$$\tfrac{1}{2}(I_1 + I_2)\,\Delta t + S_1 - \tfrac{1}{2}O_1\,\Delta t = S_2 + \tfrac{1}{2}O_2\,\Delta t \qquad (25\text{-II-}5)$$

Routing is accomplished by substituting the known values in the above equation to obtain $S_2 + \tfrac{1}{2}O_2\,\Delta t$. Then O_2 is obtained from the relationship between O_2 and $S_2 + \tfrac{1}{2}O_2\,\Delta t$. Figure 25-II-5 shows the storage curve obtained from the procedure of either of the two methods described in Subsec. II. The figure also shows the $S - \frac{1}{2}O\,\Delta t$ and $S + \frac{1}{2}O\,\Delta t$ curves, which are obtained, respectively, by subtracting from and adding to the abscissa of the storage curve one-half of the value of $O\,\Delta t$. At the beginning of a routing period the known values are the inflows for periods 1 and 2 and the outflow for period 1. It is required to determine the outflow for period 2. Referring to the routing computation in Table 25-II-2, the routing steps are:

Step 1. Compute $\frac{1}{2}(I_1 + I_2)$ in col. 3.

Step 2. From the $S - \frac{1}{2}O\,\Delta t$ curve, read the value of $S_1 - \frac{1}{2}O_1\,\Delta t$ corresponding to a given O_1 value. For example, $S_1 - \frac{1}{2}O_1\,\Delta t = 0$ in col. 5 for $O_1 = 85$ in col. 4.

Step 3. Compute $S_2 + \frac{1}{2}O_2\,\Delta t$ in col. 6 by adding the value in col. 3 to that in col. 5. For example, $115 + 0 = 115$.

Step 4. From the $S + \frac{1}{2}O\,\Delta t$ curve, read the value of O_2 corresponding to that of $S_2 + \frac{1}{2}O_2\,\Delta t$. For example, $O_2 = 103$ for $S_2 + \frac{1}{2}O_2\,\Delta t = 115$.

Step 5. Determine $S_1 - \frac{1}{2}O_1\,\Delta t$, or actually $S_2 - \frac{1}{2}O_2\,\Delta t$, for the next routing period by subtracting O_2 from $S_2 + \frac{1}{2}O_2\,\Delta t$ or by reading it from the $S - \frac{1}{2}O\,\Delta t$ curve for an O_2 value. For example, $S_2 - \frac{1}{2}O_2\,\Delta t = 12$ for $O_2 = 103$, or $115 - 103 = 12$.

Table 25-II-2. Computation for the Puls Method

(1)	(2)	(3)	(4)	(5)	(6)
Routing period, days	I, cfs	$(I_1 + I_2)/2$, cfs	O, cfs	$S - \frac{1}{2}O\,\Delta t$, dsf	$S + \frac{1}{2}O\,\Delta t$, dsf
1	93		85	0	115
2	137	115	103	12	184
3	208	172	143	41	305
4	320	264	206	99	480
5	442	381	288	192	686
		494			
6	546	588	373	313	901
7	630	654	456	445	1,099
8	678	684	527	572	1,256
9	691	692	582	674	1,366
10	692	688	621	745	1,433
11	684	678	644	789	1,467
12	671	664	656	811	1,475
13	657	648	658	817	1,465
14	638	624	655	810	1,434
15	609	594	645	789	1,383
16	577	556	626	757	1,313
17	534	509	602	711	1,220
18	484	455	569	651	1,106
19	426	396	529	577	973
20	366	332	482	491	823
21	298	266	427	396	662
22	235	209	364	298	507
23	183	160	299	208	368
24	137	120	237	131	251
25	103	92	179	72	164
26	81		131		

To obtain O_3 for period 3, repeat steps 1 to 5, and so on.

The above procedure is for the original *Puls method*, which was developed by L. G. Puls [6] of the U.S. Army Corps of Engineers, Chattanooga District. A so-called *modified Puls method* [7] requires the construction of only two curves: S curve and $S + \frac{1}{2}O\,\Delta t$ curve. For an initial outflow O_1, the storage S_1 is obtained from the S

curve and the quantity $S_1 - \frac{1}{2}O_1 \Delta t$ can be computed. In accordance with Eq. (25-II-5) the average inflow plus the quantity $S_1 + \frac{1}{2}O_1 \Delta t$ gives the quantity $S_2 + \frac{1}{2}O_2 \Delta t$. Thus the outflow O_2 corresponding to $S_2 + \frac{1}{2}O_2 \Delta t$ can be obtained from the $S + \frac{1}{2}O \Delta t$ curve.

2. The Coefficient Method. This method assumes that storage is directly proportional to outflow, or

$$S_2 - S_1 = K(O_2 - O_1) \tag{25-II-6}$$

From Eq. (25-II-4),

$$\frac{1}{2}(I_1 + I_2) \Delta t - \frac{1}{2}(O_1 + O_2) \Delta t = K(O_2 - O_1) \tag{25-II-7}$$

or

$$O_2 = O_1 + C(I_1 - O_1) + \frac{1}{2}C(I_2 - I_1) \tag{25-II-8}$$

where

$$C = \frac{\Delta t}{K + 0.5 \Delta t} \tag{25-II-9}$$

and K is a proportional constant equal to the reciprocal of the slope of the storage curve, used either as a constant or as a variable function of the outflow. When K is used as a variable, derived curves for C and $\frac{1}{2}C$ as functions of the outflow are plotted. The appropriate values of C and $\frac{1}{2}C$ are then read from the curves for each routing step corresponding to the value of outflow at the time under consideration. The computational procedure using Eq. (25-II-8) is apparent.

B. Variable Discharge-Storage Relationship

1. The Muskingum Method. This method was developed by G. T. McCarthy and others in connection with studies of the Muskingum Conservancy District Flood Control Project of the U.S. Army Corps of Engineers in 1934–1935 [8–9]. The method involves the concept of wedge and prism storages (Fig. 25-II-6).

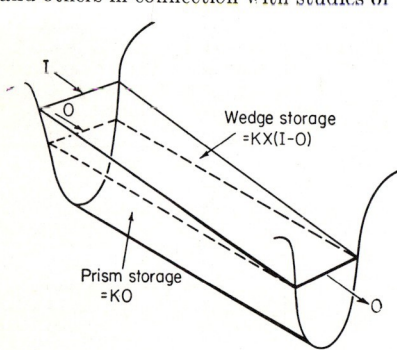

FIG. 25-II-6. Prism and wedge storages in a river channel.

Storage volume can be correctly related to outflow with a simple linear function only when inflow and outflow are equal, that is, when a steady flow exists. During the advance of a flood wave, however, inflow always exceeds outflow, thus producing a wedge of storage, called *wedge storage*. Conversely, during the recession, outflow exceeds inflow, resulting in a negative wedge storage. The wedge can be related to the difference between the instantaneous values of inflow and outflow. In Fig. 25-II-6, the wedge storage is represented by $KX(I - O)$. In addition, there is a storage of prism, or *prism storage*, as represented by KO. In these expressions, K is a coefficient and X a parameter. The total storage is therefore

$$S = KO + KX(I - O) \tag{25-II-10}$$

This is known as the *Muskingum equation*.

Using symbols defined previously, Eq. (25-II-10) may be written as

$$S_2 - S_1 = K[X(I_2 - I_1) + (1 - X)(O_2 - O_1)] \tag{25-II-11}$$

Combining this equation with Eq. (25-II-4) and simplifying,

$$O_2 = C_1'I_2 + C_2'I_1 + C_3'O_1 \tag{25-II-12}$$

where
$$C_1' = \frac{\Delta t - 2KX}{2K(1 - X) + \Delta t} \quad \text{(25-II-13)}$$

$$C_2' = \frac{\Delta t + 2KX}{2K(1 - X) + \Delta t} \quad \text{(25-II-14)}$$

and
$$C_3' = \frac{2K(1 - X) - \Delta t}{2K(1 - X) + \Delta t} \quad \text{(25-II-15)}$$

By an algebraic modification, this equation can be written as

$$O_2 = O_1 + C_1(I_1 - O_1) + C_2(I_2 - I_1) \quad \text{(25-II-16)}$$

where
$$C_1 = \frac{\Delta t}{K(1 - X) + 0.5 \Delta t} \quad \text{(25-II-17)}$$

and
$$C_2 = \frac{0.5 \Delta t - KX}{K(1 - X) + 0.5 \Delta t} \quad \text{(25-II-18)}$$

For ease in routing, curves of C_1 and C_2 (as functions of outflow) can be constructed for use, with K and X taken as parameters.

Two approaches are available for the determination of X. The first, providing a simultaneous determination of K and X, is as follows. Combining Eq. (25-II-10) with Eq. (25-II-4) and solving for K,

$$K = \frac{0.5 \Delta t[(I_2 + I_1) - (O_2 + O_1)]}{X(I_2 - I_1) + (1 - X)(O_2 - O_1)} \quad \text{(25-II-19)}$$

Successive values of the numerator (representing storage increment) and the denominator (representing weighted-flow increment) are computed for a flood with known values of inflow and outflow, assuming various parametric values of X. The computed values of the accumulated numerator and denominator are then plotted, usually producing curves in the form of loops. The assumed value of X that resulted in a loop closest to a single line is accepted as the correct value. The reciprocal of the slope of the single line gives the value of K. Figure 25-II-7 illustrates a group of loops deter-

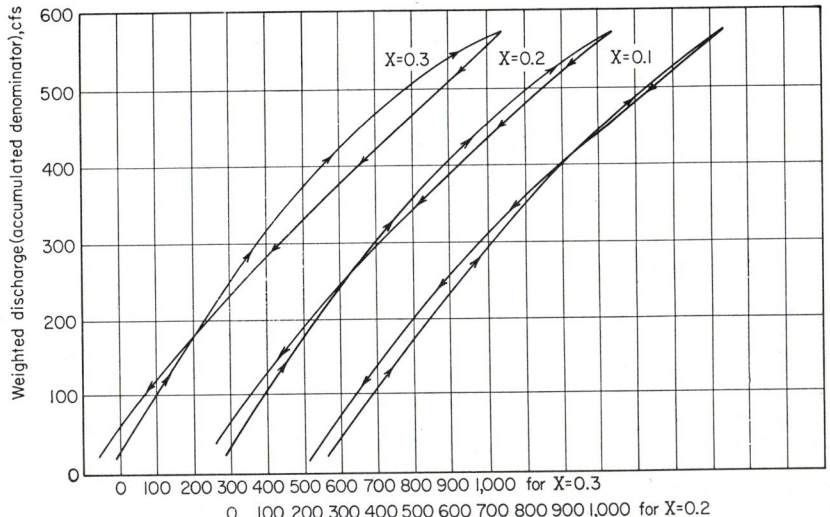

FIG. 25-II-7. Storage loops. $\Delta t = 1$ day.

mined from the data in Table 25-II-1. In this example a value between 0.1 and 0.2 appears to fit the data best. A value of $X = 0.15$ is therefore assumed as correct, and the corresponding $K = 2.3$, approximately.

The second approach assumes a value for K or various values for K (as a function of outflow) determined from the slope of the storage curve obtained by the procedure described in Subsec. II. Several floods with known inflows and outflows are then routed, using Eq. (25-II-12) or (25-II-16) and assigning various values of X. The value of X which produces the routed outflow hydrographs most nearly reproducing the

Table 25-II-3. Computation for Muskingum Method with $C_1 = 0.4$ and $C_2 = 0.1$

(1)	(2)	(3)	(4)	(5)	(6)	(7)	(8)
Routing period, days	I, cfs	$I_2 - I_1$, cfs	$C_2(I_2 - I_1)$, cfs	$I_1 - O_1$, cfs	$C_1(I_1 - O_1)$, cfs	(4) + (6), cfs	O, cfs
1	93	44	4.4	8	3.2	7.6	85
2	137	71	7.1	44	17.6	24.7	93
3	208	112	11.2	90	36.0	47.2	118
4	320	122	12.2	155	62.0	74.2	165
5	442	104	10.4	203	81.2	91.6	239
6	546	84	8.4	215	86.0	94.4	331
7	630	48	4.8	205	82.0	86.8	425
8	678	13	1.3	166	66.4	67.7	512
9	691	1	0.1	111	44.4	44.5	580
10	692	−8	−0.8	68	27.6	26.8	624
11	684	−13	−1.3	33	13.2	11.9	651
12	671	−14	−1.4	8	3.2	1.8	663
13	657	−19	−1.9	−11	−4.4	−6.3	668
14	638	−29	−2.9	−24	−9.6	−12.5	662
15	609	−32	−3.2	−40	−16.0	−19.2	649
16	577	−43	−4.3	−53	−21.3	−25.6	630
17	534	−50	−5.0	−70	−28.0	−33.0	604
18	484	−58	−5.8	−87	−34.8	−40.6	571
19	426	−60	−6.0	−104	−41.6	−47.6	530
20	366	−58	−5.8	−116	−46.4	−52.2	482
21	298	−63	−6.3	−132	−52.8	−59.1	430
22	235	−52	−5.2	−136	−54.4	−59.6	371
23	183	−46	−4.6	−131	−52.4	−57.0	314

actual hydrographs is assumed to be the best. Adjustment of K may also be required to give good results. Effects of modifying K and X and their relationship upon each other will become apparent as various trial computations are made.

A further consideration in this method is the routing period being used. If the ordinates of hydrographs are too far apart to define the hydrograph adequately, the storage curve will be in error, which will affect the values of X and K thus derived. The best results are obtained when the routing period is not less than $2KX$ or more than K. In some cases routing periods exceeding K or less than $2KX$ may be used without serious error. However, it is best to keep routing periods within these limits.

Table 25-II-3 illustrates the method of routing using Eq. (25-II-16). In the example both K and X are assumed to be constant; therefore C_1 and C_2 are constant throughout the procedure. Known values at the beginning of the routing are the total inflows for all routing periods and the outflow for the first routing period. Routing is accomplished in the following steps:

Step 1. Compute $I_2 - I_1$ in col. 3.
Step 2. Compute $C_2(I_2 - I_1)$ in col. 4.
Step 3. Compute $I_1 - O_1$ in col. 5.
Step 4. Compute $C_1(I_1 - O_1)$ in col. 6.
Step 5. Add col. 4 and col. 6, resulting in col. 7.
Step 6. Compute O_2 in col. 8 (= col. 7 + O_1 in col. 8).
Repeat steps 1 to 6 for subsequent routing periods.

Plotting of storage curves such as the one in Fig. 25-II-3 has shown that K or their slope at a point on the curve varies with the outflow. When this variation is substantial, the derived coefficients in Eqs. (25-II-12) and (25-II-16) will vary accordingly. Figure 25-II-8 shows a plotting of the K values for $X = 0.15$ derived from the storage curves of Fig. 25-II-7. Curves for C_1 and C_2 determined from this K curve, using an X of 0.15, are also shown. In use, values of C_1 and C_2 for each routing period are read from the curves corresponding to the outflow O. Table 25-II-4 illustrates the routing computation.

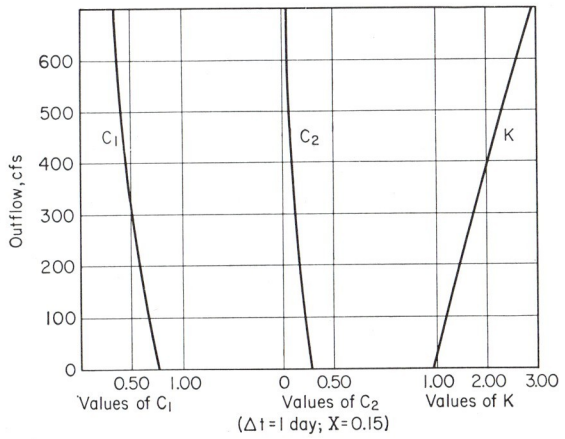

FIG. 25-II-8. Muskingum routing coefficients.

The value of X used is generally assumed to be a constant. Although there is evidence that X may vary, any attempt to consider X as a variable is a refinement infrequently necessary. Moreover, there appears to be no analytical method for the determination of the variation of X.

If a major tributary enters the main stream within a routing reach, the tributary stream may be influenced by flows in the main stream to a varying degree. For those cases where the influence is great, the procedure described in Subsec. IV-E should be used. Where the tributary stream is relatively unaffected by the main stream, routing can be accomplished by introducing an independent variable in Eq. (25-II-16), or

$$O_2 = O_1 + C_1(I_1 - O_1) + C_2(I_2 - I_1) + C_3(F_2 - F_1) \quad (25\text{-II-}20)$$

where F_2 and F_1 are tributary flows for the periods indicated by the subscripts, and C_3 is a coefficient derived for the tributary from the expression

$$C = \frac{Z}{K(1 - X) + 0.5\,\Delta t} \quad (25\text{-II-}21)$$

where Z represents the influence of storage in the tributary above the confluence produced by the variation in the tributary flow.

Table 25-II-4. Computation for Muskingum Method with Variable Coefficients

(1) Routing period, days	(2) I, cfs	(3) $I_2 - I_1$, cfs	(4) C_2	(5) $C_2(I_2 - I_1)$, cfs	(6) O, cfs	(7) $I_1 - O_1$, cfs	(8) C_1	(9) $C_1(I_1 - O_1)$, cfs	(10) $O_2 - O_1$, cfs
1	93	44	.23	10.1	85	8	.67	5.4	15.5
2	137	71	.22	15.6	101	36	.66	23.8	39.4
3	208	112	.20	22.4	140	68	.63	42.8	65.2
4	320	122	.16	19.5	205	115	.58	66.7	86.2
5	442	104	.12	12.5	291	151	.52	78.5	91.0
6	546	84	.09	7.6	382	164	.46	75.4	83.0
7	630	48	.08	3.8	465	165	.42	69.3	73.1
8	678	13	.07	0.9	538	140	.38	53.2	54.1
9	691	1	.05	0	592	99	.37	36.7	36.7
10	692	−8	.05	−0.4	629	63	.35	22.0	21.6
11	684	−13	.05	−0.7	651	33	.34	11.2	10.5
12	671	−14	.05	−0.7	661	10	.34	3.4	2.7
13	657	−19	.04	−0.8	664	−7	.34	−2.4	−3.2
14	638	−27	.05	−1.4	661	−23	.34	−7.8	−9.2
15	609	−32	.05	−1.6	652	−43	.34	−14.6	−16.2
16	577	−43	.05	−2.2	636	−59	.35	−20.6	−22.8
17	534	−50	.05	−2.5	613	−79	.36	−28.5	−31.0
18	484	−58	.06	−3.5	582	−98	.37	−36.2	−39.7
19	426	−60	.07	−4.2	542	−116	.38	−44.1	−48.3
20	366	−68	.07	−4.8	494	−128	.41	−52.5	−57.3
21	298	−63	.08	−5.0	437	−139	.43	−59.7	−64.7
22	235	−52	.10	−5.2	372	−137	.47	−64.4	−69.6
23	183	−46	.12	−5.5	302	−119	.51	−60.7	−66.2
24	137	−34	.14	−4.8	236	−99	.55	−54.4	−59.2
25	103	−22	.17	−3.7	177	−74	.59	−43.6	−47.3

If the routing period is taken equal to $2KX$, Eq. (25-II-16) becomes

$$O_2 = O_1 + \frac{\Delta t}{K}(I_1 - O_1) \qquad (25\text{-}II\text{-}22)$$

This equation is theoretically valid only when Δt, K, and X are related as specified, implying constant K and X. However, this expression has been used with some satisfaction by considering a variable K when the deviation from the initial assumption is not too great. Acceptable limits of this deviation can be evaluated by routing specific floods.

The Muskingum method is by no means exact inasmuch as only a few terms in Eqs. (25-II-1) and (25-II-2) are considered in the derivation. However, the method is adequate for many purposes, particularly in the planning and designing stages of flood-control or multiple-purpose projects. It may be used not only for routing discharging hydrographs, but also for routing holdouts, i.e., reductions in natural flows resulting from the control of floods by reservoirs.

2. The Working-value Method. A concept which is useful in the evaluation of wedge storage involves a virtual *working discharge*, which represents a steady flow that would produce a storage equal to that produced by the actual inflow I and outflow

O. By reference to Fig. 25-II-9, in which the working discharge D is shown, it can be seen that the wedge storage previously represented by $KX(I - O)$ is also equal to $K(D - O)$. Thus

$$D = XI + (1 - X)O \qquad (25\text{-}II\text{-}23)$$

For outflows of routing periods 1 and 2, the above equation gives

$$O_1 = D_1 - \frac{X}{1-X}(I_1 - D_1) \qquad (25\text{-}II\text{-}24)$$

$$O_2 = D_2 - \frac{X}{1-X}(I_2 - D_2) \qquad (25\text{-}II\text{-}25)$$

Combining the above two expressions with Eq. (25-II-4),

$$S_2 - S_1 = \tfrac{1}{2}\Delta t \frac{1}{1-X}(I_1 + I_2) - \tfrac{1}{2}\Delta t \frac{1}{1-X}(D_1 + D_2) \qquad (25\text{-}II\text{-}26)$$

from which

$$S_1(1-X) + \tfrac{1}{2}\Delta t(I_1 + I_2) - \tfrac{1}{2}\Delta t\, D_1 = S_2(1-X) + \tfrac{1}{2}\Delta t\, D_2 \qquad (25\text{-}II\text{-}27)$$

Let
$$S_1(1-X) + \tfrac{1}{2}\Delta t\, D_1 = R_1 \qquad (25\text{-}II\text{-}28)$$
and
$$S_2(1-X) + \tfrac{1}{2}\Delta t\, D_2 = R_2 \qquad (25\text{-}II\text{-}29)$$
Then
$$R_2 = R_1 + \tfrac{1}{2}\Delta t(I_1 + I_2) - \Delta t\, D_1 \qquad (25\text{-}II\text{-}30)$$

where R_1 and R_2 are called *working values*, representing indices of storage.

Combining Eq. (25-II-23) with Eq. (25-II-10) and solving for S,

$$S = KD \qquad (25\text{-}II\text{-}31)$$

For routing periods 1 and 2, this equation gives

$$S_1 = K_1 D_1 \qquad (25\text{-}II\text{-}32)$$
$$S_2 = K_2 D_2 \qquad (25\text{-}II\text{-}33)$$

The determination of the routing period, K, and X is the same as that described in the Muskingum method (Subsec. IV-B-1). If K is a constant, then $K_1 = K_2 = K$. If K is a variable, then the value of K_1 or K_2 can be determined from its relationship with the outflow such as shown in Fig. 25-II-8.

Knowing I_1, I_2, and O_1, routing can be accomplished mathematically by the above equations. First, by Eq. (25-II-23), compute the value of D_1 for the given I_1, O_1, and X. Then, by Eq. (25-II-32), compute S_1, using K_1 corresponding to O_1 if K_1 is a variable. By Eq. (25-II-28), compute R_1. By Eq. (25-II-30), compute R_2. By Eqs. (25-II-29) and (25-II-33), D_2 can be solved, using K_2 corresponding to an assumed O_2

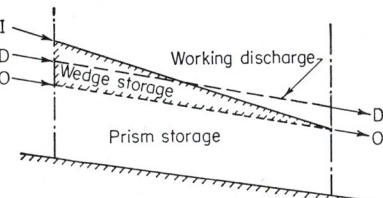

Fig. 25-II-9. Schematic representation of prism and wedge storages and working discharge.

if K_2 is a variable. Finally, by Eq. (25-II-25), compute O_2, which should check the assumed O_2. If necessary, a new value of O_2 should be assumed and the previous steps repeated until the value is checked agreeably.

However, the routing procedure can be much simplified by using a graphical approach with a sliding mechanical device (Subsecs. IV-D-2 and IV-E) or by using a specifically constructed slide rule [5, 10–11].

The working-value method is considered more advantageous than the Muskingum method if an independent variable, such as tributary inflow (Subsec. IV-D-2) or controlled discharge through a gated low dam, is involved.

C. Lag Methods

Methods of routing by time displacement of average inflow in some cases yield adequate approximations of flood-discharge hydrographs or crests. They are generally not based on mathematical relationships of motion or channel storage, but developed rather from intuitive or empirical processes. In practice, they have been applied to long reaches of slowly fluctuating streams. The short cuts in computations thus obtained constitute the principal advantage over the Muskingum method. Since the methods are more or less empirical, their reliability and applicability are limited. The two most commonly used such so-called lag methods are the *successive average-lag method* and the *progressive average-lag method* [10].

1. The Successive Average-lag Method. This method, developed by F. E. Tatum [12] of the U.S. Army Corps of Engineers, Rock Island District, is based upon the following assumptions: (1) the discharge-storage relation, and hence the shape of the flood hydrograph, tends to vary uniformly along the channel; (2) the average of inflows at routing periods 1 and 2 at a point in the channel, or $\frac{1}{2}(I_1 + I_2)$, will be experienced at some point downstream at period 2, and this relation between discharges at the two points applies to all routing periods of an equal time interval; and (3) the change in shape of the hydrograph between the two points reflects the cumulative effect of all storage characteristics of the channel reach. Therefore the routing process may be repeated for as many subreaches as desired in order to determine the change in shape of the hydrograph as a result of routing through channel storage. Probably the successive hydrographs obtained by this procedure are equally spaced in time, but not necessarily in distance along the channel. For some subreaches the velocity of flood-wave movement may be greater or less than for others because of local irregularities.

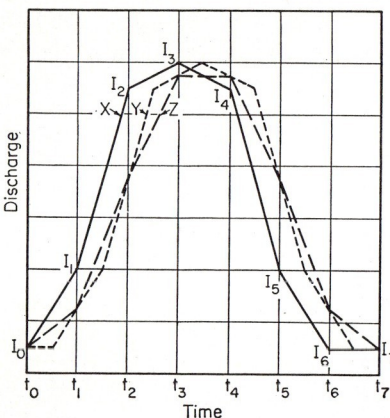

FIG. 25-II-10. Schematic diagram for successive average-lag method.

In this method routing periods are selected such that the inflow hydrograph during the interval is essentially a straight line. The basis for the method becomes apparent by referring to Fig. 25-II-10, where the hydrograph X at point A is defined by inflows $I_0, I_1, I_2, I_3, \ldots, I_n$, respectively, at routing periods $t_0, t_1, t_2, t_3, \ldots, t_n$. Hydrograph Y is the same hydrograph translated one-half routing period without modification. The flows for hydrograph Y at routing periods t_1, t_2, t_3, etc., are therefore $\frac{1}{2}(I_0 + I_1)$, $\frac{1}{2}(I_1 + I_2)$, $\frac{1}{2}(I_2 + I_3)$, etc. Constructing straight lines between these produces hydrograph Z, representing the first-step hydrograph of the method. By continuing and repeating the process, the wave form flattens in its downstream procession, and the resulting hydrograph approximates the hydrograph at the end of the last subreach. By a proper selection of routing periods and the number of subreaches, an observed hydrograph at the downstream point may therefore be expressed in terms of the upstream hydrograph. Thus, for one subreach, the ordinates are

$$O_2 = \frac{1}{2}(I_1 + I_2)$$

For two subreaches, the ordinates are

$$O_3 = \frac{1}{2}\left(\frac{I_1 + I_2}{2} + \frac{I_2 + I_3}{2}\right) = \frac{1}{4}(I_1 + 2I_2 + I_3)$$

If the inflow ordinates are $I_1, I_2, I_3, \ldots, I_{n+1}$, and n is the number of subreaches, the ordinates are

$$O_{n+1} = C_1 I_1 + C_2 I_2 + C_3 I_3 + \cdots + C_{n+1} I_{n+1} \qquad (25\text{-II-}34)$$

where
$$C_1 = \frac{1}{2^n}$$

$$C_2 = \frac{n}{2^n}$$

$$C_3 = \frac{n(n-1)}{2^n 2!}$$

$$\cdots \cdots \cdots$$

$$C_n = \frac{n(n-1)(n-2)\cdots(2)}{2^n(n-1)!}$$

and
$$C_{n+1} = \frac{n!}{2^n n!} = \frac{1}{2^n}$$

Routing constants for various numbers of routing subreaches are given in Table 25-II-5. An example of routing for four subreaches is shown in Table 25-II-6, which is self-explanatory.

Table 25-II-5. Routing Constants for Successive Average-lag Method

Coefficients	Number of subreaches									
	1	2	3	4	5	6	7	8	9	10
C_1	.5000	.2500	.1250	.0625	.0313	.0156	.0078	.0039	.0020	.0010
C_2	.5000	.5000	.3750	.2500	.1562	.0937	.0547	.0313	.0176	.0098
C_3	.....	.2500	.3750	.3750	.3125	.2344	.1641	.1094	.0703	.0440
C_4	.....	.....	.1250	.2500	.3125	.3126	.2734	.2187	.1641	.1172
C_5	.....	.....	.....	.0625	.1562	.2344	.2734	.2734	.2460	.2050
C_6	.....	.....	.....	.....	.0313	.0937	.1641	.2187	.2460	.2460
C_7	.....	.....	.....	.....	.....	.0156	.0547	.1094	.1641	.2050
C_8	.....	.....	.....	.....	.....	.....	.0078	.0313	.0703	.1172
C_9	.....	.....	.....	.....	.....	.....	.....	.0039	.0176	.0440
C_{10}	.....	.....	.....	.....	.....	.....	.....	.....	.0020	.0098
C_{11}	.....	.....	.....	.....	.....	.....	.....	.....	.....	.0010

2. The Progressive Average-lag Method. This method has been used by the U.S. Army Corps of Engineers, Tulsa and Kansas City Districts [13]. By this method, two or more inflows are averaged and the mean value is lagged by the time of travel of the flood wave to produce the discharge and time of occurrence of one value of the outflow hydrograph. The process is repeated for other outflow values until the outflow hydrograph is determined. In this procedure, as different from the successive average-lag method, equal weight is given each inflow value in deriving an outflow, and the length of period for which inflow values are averaged to obtain an outflow does not necessarily relate to the flood-wave travel time. In application, the length of the inflow period is determined by trial until a satisfactory agreement is obtained between the computed and observed peak outflows. This length is usually found to be three-quarters to twice the travel time.

D. Graphical Methods

Graphical methods of flood routing are many. Two important ones are discussed below.

Table 25-II-6. Computation for Successive Average-lag Method

(1) Routing period, days	(2) I, cfs	(3) IC_1 .0625	(4) IC_2 .2500	(5) IC_3 .3750	(6) IC_4 .2500	(7) IC_5 .0625	(8) Outflow, cfs
		6.25					
		6.26	25.00				
	100	6.25	25.00	37.50			
	100	6.25	25.00	37.50	25.00		
1	100	6.25	25.00	37.50	25.00	6.25	100.00
2	200	12.50	25.00	37.50	25.00	6.25	106.25
3	400	25.00	50.00	37.50	25.00	6.25	143.75
4	700	43.75	100.00	75.00	25.00	6.25	250.00
5	1100	68.75	175.00	150.00	50.00	6.25	450.00
6	1400	87.50	275.00	262.50	100.00	12.50	737.50
7	1600	100.00	350.00	412.50	175.00	25.00	1,062.50
8	1700	106.25	400.00	525.00	275.00	43.75	1,350.00
9	1600	100.00	425.00	600.00	350.00	68.75	1,543.75
10	1400	87.50	400.00	637.50	400.00	87.50	1,612.50
11	1100	68.75	350.00	600.00	425.00	100.00	1,543.75
12	700	43.75	275.00	525.00	400.00	106.25	1,350.00
13	400	25.00	175.00	412.50	350.00	100.00	1,062.50
14	200	12.50	100.00	262.50	275.00	87.50	737.50
15	100	6.25	50.00	150.00	175.00	68.75	450.00
16	100	6.25	25.00	75.00	100.00	43.75	250.00
17	100	6.25	25.00	37.50	50.00	25.00	143.75
18	100	6.25	25.00	37.50	25.00	12.50	106.25
19	100	6.25	25.00	37.50	25.00	6.25	100.00

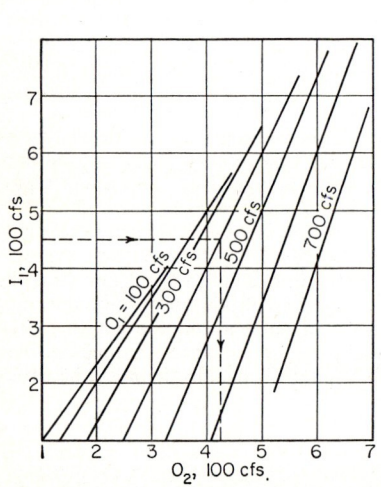

Fig. 25-II-11. Graphical routing for the simplified Muskingum method. $\Delta t = 1$ day.

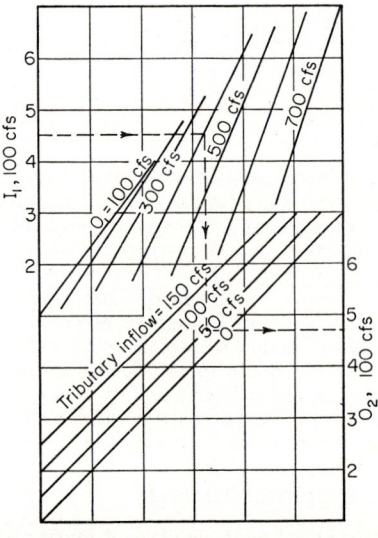

Fig. 25-II-12. Graphical routing with tributary inflow. $\Delta t = 1$ day.

1. **The Simplified Muskingum Method.** Some of the numerical methods described previously may lend themselves to graphical solutions. The preference of one method over the other may, however, be determined by the individual. For example, Eq. (25-II-22) can be solved quite readily through the use of the diagram given in Fig. 25-II-11. The diagram is developed from the variable K values given in Fig. 25-II-8. Routing is accomplished by entering the diagram with the total inflow for the beginning of period 1, moving horizontally to the parametric value of outflow for period 1, and then dropping vertically to read the value of outflow for the beginning of period 2. The example shows an inflow of 450 cfs and an outflow of 400 cfs for period 1, producing an outflow of 426 cfs for period 2. This type of diagram can be modified further by the addition of one or more parameters representing tributary contribution, local ungaged contribution, etc. Figure 25-II-12 illustrates the graphical solution of Fig. 25-II-11 modified by the addition of a tributary entering the reach. It is assumed that the influence of the tributary or the flow at the lower end of the reach has been determined empirically. The example illustrates an inflow

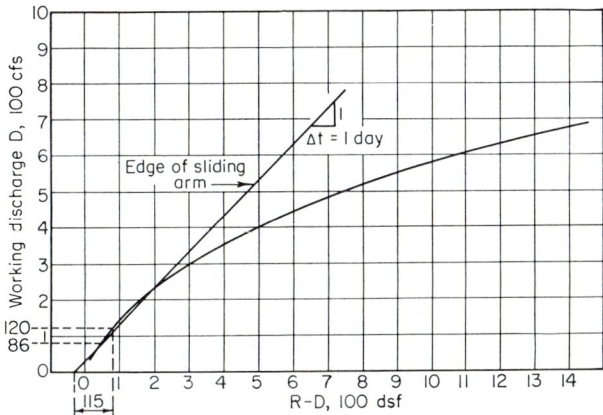

FIG. 25-II-13. R-D versus D curve.

of 450 cfs and outflow of 400 cfs for period 1, with a contribution of 50 cfs from the tributary for period 1, producing an outflow for period 2 of 480 cfs.

2. **The Working-value Method.** Equations (25-II-28) and (25-II-29), employing the working value R, may be solved graphically with the use of a sliding arm, set to a slope equal to $1/\Delta t$. A working curve obtained by plotting R-D versus D should then be prepared from computations, using the following relationships:

$$D = IX + (1 - X)O \qquad (25\text{-II-}23)$$
$$S = KD \qquad (25\text{-II-}31)$$
and
$$R = S(1 - X) + \tfrac{1}{2} \Delta t\, D \qquad (25\text{-II-}35)$$

which is a general expression of Eqs. (25-II-28) and (25-II-29).

Figure 25-II-13 is a plotting of the R-D curve for the variable K values with a constant value of 0.10 for X. Table 25-II-7 illustrates the routing procedure as follows:

Step 1. Compute the initial value of D (col. 4) from known values of I and O for period 1. For example, by Eq. (25-II-23), $D = 93(0.1) + 85(0.9) = 86$ cfs.
Step 2. Compute the average inflow (col. 3) from the known values in col. 2.
Step 3. Set the slope of a sliding arm equal to $1/\Delta t$.
Step 4. Set the sliding arm to intersect the curve at the computed value of $D = 86$ cfs.
Step 5. From the intersection of the sloping arm with the abscissa axis, measure off a distance equal to the average inflow of 115 cfs.

Step 6. A vertical from this point to the curve gives the ordinate $D = 120$ cfs as shown in col. 4.
Step 7. Compute the values in cols. 5 and 6.
Step 8. Compute the outflow ($=$ col. 4 $-$ col. 6, or $120 - 2 = 118$ cfs).

Table 25-II-7. Computation for Graphical Routing by Working-value Method
$X = 0.10 \quad \Delta t = 1$ day

(1)	(2)	(3)	(4)	(5)	(6)	(7)
Routing period, days	I, cfs	$\dfrac{I_1 + I_2}{2}$, cfs	D, cfs	$I - D$, cfs	$\dfrac{X(I - D)}{1 - X}$, cfs	O, cfs
1	93	115	86	...	..	85
2	137	172	120	17	2	118
3	208	264	175	33	4	171
4	320	381	244	76	8	236
5	442	494	327	115	13	314
6	546	588	415	131	15	390
7	630	654	481	150	17	464
8	678	689	538	140	16	522
9	691	691	578	113	13	555
10	692	688	606	86	10	596
11	684	678	625	59	7	618
12	671	664	636	35	4	632
13	657	648	640	17	2	638
14	638	623	642	-4	0	642
15	609		637	-28	-3	640

To continue the computation for the subsequent routing periods, repeat steps 4 to 8, and so on.

E. Routing at Junctions

The simplest case of routing at junctions is where streams are of steep slope or tributaries entering along the main stream are relatively small so that the discharge from one stream has little effect upon stages upstream from the junction in the other stream. This can readily be solved by routing the branches separately as previously described and then combining the routed discharges. However, at the junction of two streams where changes in flow in one branch materially affect the stage in the other branch, complications will arise in attempting to evaluate the effect of one stream upon the other. There appears to be no practical direct method available for this evaluation. Although several procedures have been advanced, they may lead to serious errors. One such method, for instance, involves the use of an average storage curve which is developed from the combined inflows and the outflow below the junction. Serious errors in this method may result in certain cases where the percentage of contribution from each branch varies widely; thus the storage curve derived for the average condition cannot be taken as a true representation.

1. The Working-value Method. When the flows in two branches of a river mutually affect the flow conditions, a trial-and-error procedure, using a working discharge D and a working value R, may be used adequately. This method recognizes the variation of flow distributions through the use of working curves. Development of the working curves requires the determination of storage in each reach under steady-flow conditions for the combined flow as well as for various distributions of flow between the two branches. Storage values may be determined by backwater computations from valley cross sections, or from observed flow profiles and discharges. The working curves for each branch represent the relationship between a working discharge D and a working value R. The working discharge D is subject to corrections

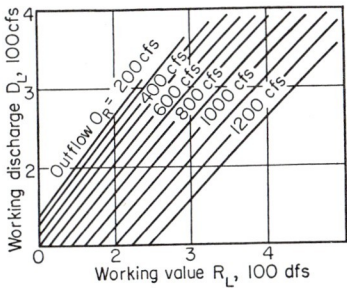
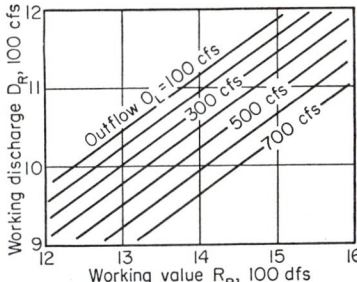

Fig. 25-II-14. Junction routing curve for left branch. $\Delta t = 1$ day.

Fig. 25-II-15. Junction routing curve for right branch. $\Delta t = 1$ day.

in order to obtain the actual outflow, as influenced by the wedge storage. The computation for each routing period is accomplished by a process of successive trials. Table 25-II-8 and Figs. 25-II-14 and 25-II-15 illustrate the procedure.

Table 25-II-8. Junction Routing by Working-value Method

Routing period, days	Left branch ($X = 0.15$)					Right branch ($X = 0.20$)					Total O, cfs
	I, cfs	$(I_1 + I_2)/2$, cfs	R, cfs	D, cfs	O, cfs	I, cfs	$(I_1 + I_2)/2$, cfs	R, cfs	D, cfs	O, cfs	
1	250		292	230	226	1,050		1,273	1,000	988	1,214
2	300	275	337	280	276	1,100	1,075	1,348			
				270	265			1,348	1,045	1,031	1,296
3	350	325	392	322	317	1,150	1,125	1,428			
					300			1,428	1,094	1,080	
					305			1,428	1,090	1,075	
				315	309			1,428	1,090	1,075	1,384
4	400	375	452	375	371	1,200	1,175	1,513			
				363	356			1,513	1,140	1,125	1,481

Step 1. Compute the working discharge D for each branch using Eq. (25-II-23) for the given data at the beginning of the flood wave:

$$D_{1L} = 226(0.85) + 250(0.15) = 230 \text{ cfs}$$
$$D_{1R} = 988(0.80) + 1{,}050(0.20) = 1{,}000 \text{ cfs}$$

Step 2. From the working curves determine R_1 for each reach:

$R_{1L} = 292$ cfs for $D_{1L} = 230$ cfs and $O_{1R} = 988$ cfs from Fig. 25-II-14
$R_{1R} = 1{,}273$ cfs for $D_{1R} = 1{,}000$ cfs and $O_{1L} = 226$ cfs from Fig. 25-II-15

Step 3. Compute R_2 for each branch using Eq. (25-II-30):

$$R_{2L} = 292 + 275 - 230 = 337 \text{ cfs}$$
$$R_{2R} = 1{,}273 + 1{,}075 - 1{,}000 = 1{,}348 \text{ cfs}$$

Step 4. Assuming no change in flow in the right branch during the period, D_2 in the left branch is determined from the working curve as

$$D_{2L} = 280 \text{ cfs} \quad \text{for } R_{2L} = 337 \text{ cfs and } O_{2R} = O_{1R} = 988 \text{ cfs}$$

Step 5. Compute O_2 for the left branch using Eq. (25-II-25):

$$O_{2L} = 280 - 0.176(300 - 280) = 276 \text{ cfs}$$

Step 6. It is apparent that the outflow from the left branch will be a value between 226 and 276 cfs. Therefore a value of 265 cfs is assumed.

Step 7. From the working curve determine D_2 for the right branch:

$$D_{2R} = 1{,}045 \text{ cfs} \quad \text{for } R_{2R} = 1{,}348 \text{ cfs and } O_{2L} = 265 \text{ cfs}$$

Step 8. Compute O_2 for the right branch using Eq. (25-II-25):

$$O_{2R} = 1{,}045 - 0.25(1{,}100 - 1{,}045) = 1{,}031 \text{ cfs}$$

Step 9. Using the new outflow from the right branch, determine D_2 for the left branch from the working curve:

$$D_{2L} = 270 \text{ cfs} \quad \text{for } R_{2L} = 337 \text{ cfs and } O_{2R} = 1{,}031 \text{ cfs}$$

Step 10. Compute O_2 for the left branch using Eq. (25-II-25):

$$O_{2L} = 270 - 0.176(300 - 270) = 265 \text{ cfs}$$

This value checks the assumed value at step 6. If the computed value does not agree with the assumed value, a new value should be assumed and steps 7 to 10 repeated.

Step 11. Total outflow at the junction is the sum of the outflows from both branches:

$$\text{Total } O_2 = 265 + 1{,}031 = 1{,}296 \text{ cfs}$$

To obtain outflow for the next period repeat steps 3 to 11.

F. Routing of Reservoir Modifications

In order to determine the benefits derived from operating flood-control reservoirs and the effects of proposed reservoirs, it is necessary to route the reservoir modifications. This may be accomplished in two ways: (1) to route the natural and the modified hydrographs and compare their results at the routing points under consideration; (2) to route the differences, or *reservoir holdouts*, between the natural and the modified hydrographs. In general, routing the differences is simpler than routing the modified hydrograph, with which tributary inflow must be combined. By routing the differences, it is not necessary to collect and use the data from the main stream, tributaries, and ungaged areas for all the reaches. The routed differences may be simply applied to the observed flows at the points under consideration. Any inaccuracies inherent in the routing method used are minimized since only the routed differences are applied to the observed hydrographs at the points of interest. Since only the modifications at the time of crest discharges at the routing points are of interest, the routing may be limited to the crest periods only.

The method for the routing of reservoir modifications can be a choice of one of the following. For routing the natural and modified hydrographs or either of them, the methods generally used are the Puls method, the coefficient method, the Muskingum method, and the working-value method. For routing the differences between the natural and modified hydrographs, the methods recommended are the progressive average-lag method and the coefficient method.

When routing methods involving coefficients varying with the magnitude of outflow are used, constant values for the coefficients corresponding to the maximum flow may be assumed without introducing serious errors, because only the flows near maximum condition are usually of interest.

G. Mechanical Routers

Mechanical devices for flood routing are many, but most of them are designed only for reservoir routing. Two types of such devices have been developed by the U.S. Army Corps of Engineers. They are an integrating machine designed by J. F. Tarpley, Jr. [14] and a rolling-type flood router by F. B. Harkness [15].

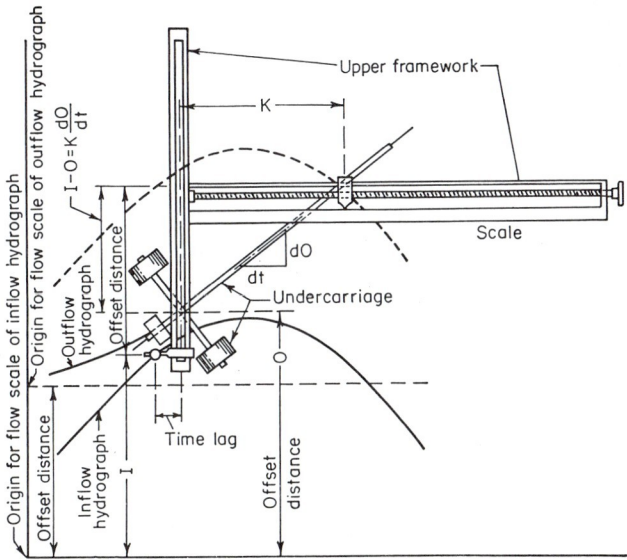

Fig. 25-II-16. Harkness flood router.

The *Harkness flood router* was designed to solve Eq. (25-II-22) in differential form, or

$$\frac{dO}{dt} = \frac{I - O}{K} \qquad (25\text{-II-}36)$$

It can therefore be used for open-channel routing where the Muskingum method with the routing period equal to $2KX$ is applicable. Figure 25-II-16 is a simplified sketch of the device. It is comprised of two principal parts: a rigid T-shaped frame and an undercarriage whose movement is guided by the T frame. The undercarriage consists of a pair of freely rotating wheels, mounted on an axle, and a shaft extending perpendicular to the axle from a point midway between the wheels. The T frame consists of a slotted vertical member; a horizontal guide bar, on which is mounted a K scale; an indicator attached to a sliding collar on the guide bar; an adjusting screw for setting the indicator; and a tracing pointer for following the inflow hydrograph. The tongue of the undercarriage shaft is free to move along the guide bar under the sliding collar, and the wheel assembly is allowed to move vertically in the slotted member of the T frame. The tracing pointer is offset vertically to allow the free movement of the undercarriage. Provision is also made to offset the pointer horizontally to introduce

25-54 FLOOD ROUTING

a time lag. A scriber, to trace the routed-outflow hydrograph, is mounted on the underside of the carriage at the lower end of the shaft.

During use, the slotted member of the T frame is always perpendicular to the base line of the inflow hydrograph. As the inflow hydrograph is traced, the movement of the undercarriage is such that the angular relationship of the shaft and the horizontal member of the T frame makes the tangent of the angle equal to $(I_1 - O_1)/K$, or dO/dt as given in Eq. (25-II-36). The scriber will then draw the outflow hydrograph. The Harkness router was later improved for high-speed operations. In the original design the T frame of the device was attached to a drafting machine and made to move to the left in following the inflow hydrograph. Subsequent improvements have made the device so as to mount more rigidly on a bridge, permitting only vertical movement of the T frame. Motor-driven rolls are used to move the plotted hydrograph under the router, thus providing easy and fast tracing.

H. Electronic Computers

Electronic computers have been used to a varying degree in solving flood-routing problems. Now becoming more available, they provide a very useful tool when

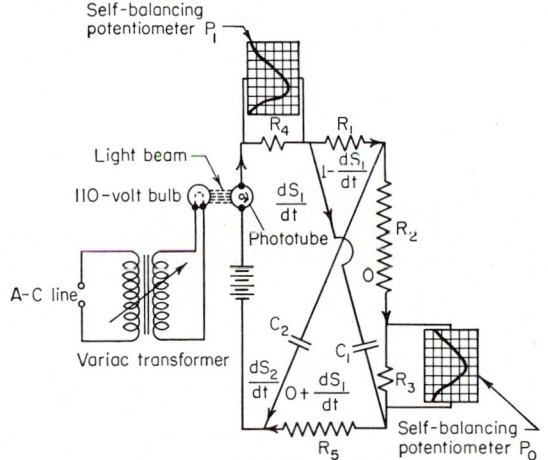

FIG. 25-II-17. Circuit for U.S. Weather Bureau electronic router.

repeated routings are required, making possible more routings in less time and permitting the use of methods derived from complicated principles. The electronic computer may be an analog or a digital computer. See Sec. 29.

1. The Analog Computer. The principle of the electronic analog is to utilize the analogy between the equation for an electric circuit and the equation for routing [16, 17]. The U.S. Weather Bureau has developed a simple electronic analog on the basis of an analogy in which the storage elements satisfy Eq. (25-II-10). The storage S corresponds to the charge on the condensers, and the inflow I and outflow O to the electric current.

Figure 25-II-17 shows the circuit diagram for the Weather Bureau analog. With a given inflow current corresponding to I on the chart of potentiometer P_I and an outflow current corresponding to O on potentiometer P_O, the currents in the circuit are shown in the diagram. By varying the light intensity on the photocell, the inflow current can be controlled so that the pen of potentiometer P_I follows the inflow hydrograph plotted on the chart. With fixed condensers, values of K and X are determined by changing the resistances R_1, R_2, and R_3, until the outflow hydrograph agrees with the given hydrograph. The resistances R_4 and R_5 are small enough to be neglected. When proper values of K and X are obtained and fixed, a new inflow hydrograph may be routed by introducing it to P_I and reading the routed outflow at P_O. The circuit allows only constant K values, but, by stopping the machine during a run and resetting the resistance dials, step variation of K is possible.

There are certain *special-purpose analogs* which are built for the whole of a specific basin. Such an analog is used to study the river and reservoir elevations in the basin as affected by a proposed reservoir-operation schedule. The machine can solve the problem repeatedly and fast for various proposed schedules, by several times a second. The results obtained at any point of interest in the basin can be examined on an oscilloscope. Thus any proposed operation schedule can be evaluated rapidly and then revised, if necessary, to obtain the most beneficial effects.

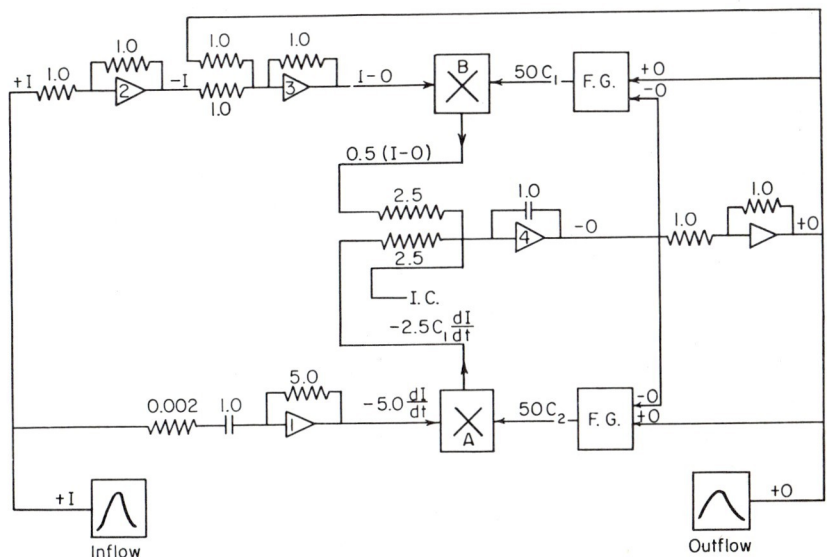

Fig. 25-II-18. Patch diagram for a single-reach routing.

2. The Differential Analyzer. This is a general-purpose analog [18]. All variables are expressed in terms of voltage, and the equations expressing the interrelationship between the variables are solved by the various components. These components perform the mathematical operations of addition, subtraction, multiplication, differentiation, integration, arbitrary function generation, etc. The equations, usually in differential form, are solved in a continuous manner at a convenient time scale. The number of reaches used in the problem and the complexity of the equation are limited only by the number of components available in the machine.

Expressed in a differential form, Eq. (25-II-16) may be written as

$$\frac{dO}{dt} = C_1(I - O) - C_2\frac{dI}{dt} \qquad (25\text{-II-}34)$$

The patch diagram, showing how the various computer components are interconnected to satisfy this equation for a single reach, is shown in Fig. 25-II-18. Components marked FG are diode function generators which can continuously develop the variable coefficients C_1 and C_2 from continuously computed values of outflow O. The initial value of outflow is introduced as an initial condition. The computer continuously computes the rate of change of outflow and, when integrated and recorded, results in a continuous hydrograph of outflow. The inflow is introduced into the circuit as a variable voltage, developed by a curve follower which automatically tracks the plotted

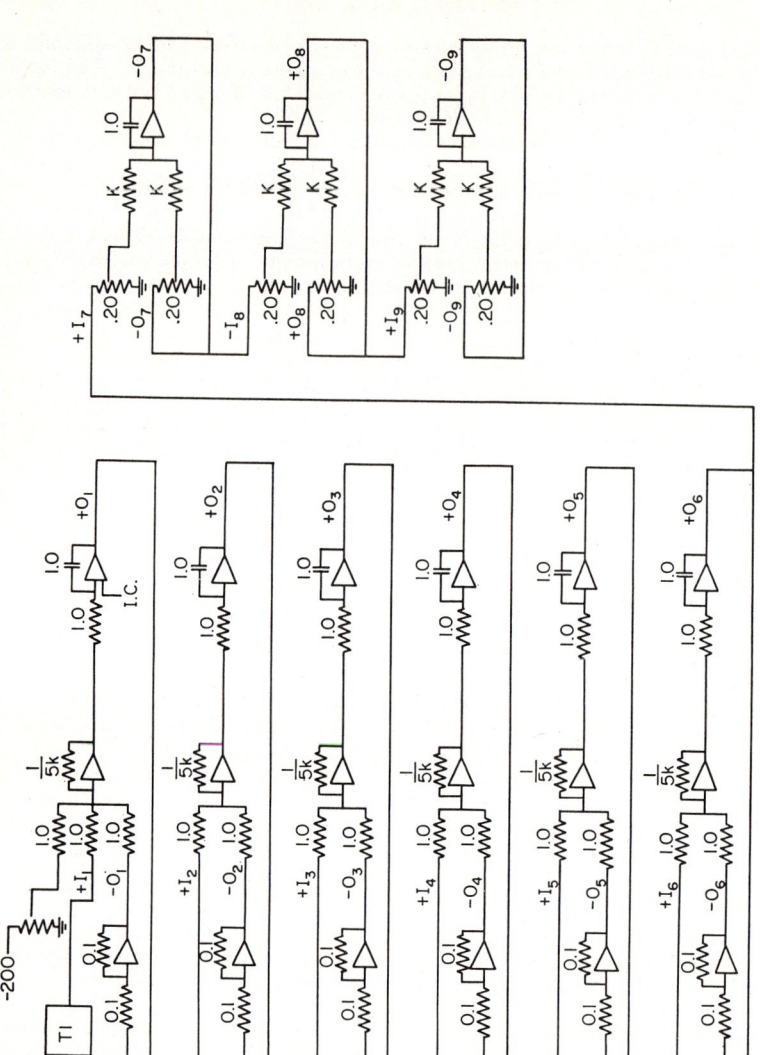

Fig. 25-II-19. Patch diagram for a nine-reach routing.

inflow hydrograph. The rate of change of this voltage is determined by the differentiating amplifier 1 and fed to multiplier A, where it is multiplied by the coefficient C_2 to obtain the product $C_2\, dI/dt$. The voltage from the input device is also fed to a sign changing amplifier 2 and then to a summing amplifier 3, along with the value of outflow to obtain $I - O$. This in turn is fed to multiplier B to obtain the product $C_1(I - O)$. The two products are summed algebraically and integrated in amplifier 4 to produce the value of outflow O, which is recorded on a plotter.

Multiple-reach routing can be accomplished by duplicating the circuit, the outflow from the first reach thus becoming part of the inflow to the second reach. Additional inflows, such as flows from tributaries and ungaged areas, are introduced by another curve follower. A convenient time scale for this problem is 5 machine-seconds equal to 1 day; thus a 20-day flood would be run in 100 sec. The time scale is limited only by the physical ability of the curve follower to trace or follow the plotted inflow hydrograph. A time scale of 5 sec equal to 20 days has been tried with satisfactory results, but a slower time scale is recommended.

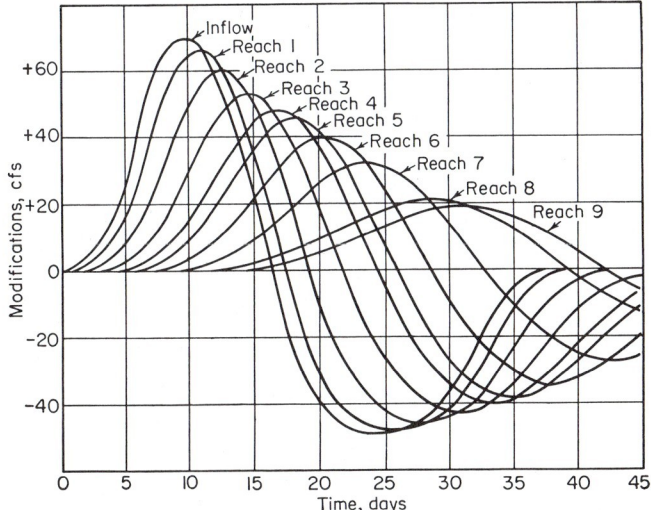

Fig. 25-II-20. Routing of reservoir modifications.

The choice of the scale for the variables usually limits the solution within the operating range of ± 100 volts, the general voltage limitation of the computer. However, a judicious choice of scales will give results consistent with the accuracy of the basic data, particularly for high-quality computers. Generally speaking, computers of this type are more accurate than special-purpose machines, because of the better quality of the components. Also, the higher cost of general-purpose analyzers can be justified because of their ability to solve other problems, thus having greater flexibility in its utilization.

A patch diagram showing the circuits for the routing of reservoir modifications through nine reaches of a stream is given in Fig. 25-II-19. This represents the circuit for routing the effects of the Pittsburgh District reservoirs from Wheeling, W.Va., to Metropolis, Ill., near the mouth of the Ohio, a distance of almost 900 miles. The routed modifications, both positive and negative, at the nine points of interest, as well as the initial inflow, are shown in Fig. 25-II-20. The machine operating time for this 45-day period was accomplished in 100 sec. By conservative estimate, the time required for equivalent manual computations would be 8 hr.

The patch diagrams shown in the figures were used for solution by a Goodyear electronic differential analyzer (GEDA), but the circuits would be similar when applied to any general-purpose computer.

3. The Digital Computers. Digital computers perform a routing procedure in finite-time periods, following the mathematical operations by manual computations. The speed of these operations, however, is many times faster than that required by manual computations, and thus reduces greatly the time of solution of routing problems. Consequently, more complex equations and more comprehensive routings become practical when solved by the machine. Programs for various problems have been written and are available for all the commercially available computers, from the small desk-size computers to the giant computers now being manufactured [18–20].

Representative of the desk-size computers is the Burroughs E-101, a simply programmed computer having limited memory capacity. Time ratios of 20:1 or better are available for routing and solving other hydrologic problems by the machine. The program instructions for a flood-routing problem have provisions for tributary or local flow in each reach and for routing any predetermined number of reaches. Data are introduced by either a keyboard or by punched paper tape, and the results are printed in tabular form by the computer. The only peripheral equipment required is the paper-tape punch for punching the input data. For keyboard entry even this is not necessary. In this case data for one routing period for all reaches are fed into the computer. All the computations for that period are made by the machine, and the results thus obtained are printed before the data for the next routing period are entered.

Some other digital computers can store the program as well as all the data before computations begin. The results can also be stored until the solution of the problem is completed, at which time all the results are registered on some type of output media, such as paper tape, magnetic tape, or punch cards. Printing and editing are performed by some peripheral device.

V. STAGE ROUTING

For the purpose of flood forecasting or control and operation of multiple-purpose river projects, the stage of flow is usually a major concern. By means of a stage-discharge relationship or a discharge rating curve, the routed discharge can be converted to stage. Since river forecasts often require immediate issue of news in terms of stage for flood warning, a direct stage-routing procedure will eliminate the extra time need for conversion of stage to discharge.

Stage routing usually relates the upstream and downstream stages by means of a correlation chart. A predicted stage at the downstream can be readily obtained from the chart when the upstream stage or stages are given. For such purposes, Lane [21, 22] has proposed a method using multiple-line charts.

VI. FLOW-LINE COMPUTATION

Closely related to flood routing is the flow-line computation. The term *flow line* means a longitudinal profile along a stream showing at all points the maximum height attained by a flood. The computation of flow line is important in establishing level grades and flood-wall elevations and in general in evaluating the modification of channel and overbank cross sections brought about by natural or man-made changes.

The flow-line computation involves essentially the determination of water-surface profiles at points along a channel. It requires the backwater-profile computation, which is a subject in hydraulics and beyond the scope of this section [3].

VII. REFERENCES

1. Isaacson, E. J., J. J. Stoker, and B. A. Troesch: Numerical solution of flood prediction and river regulation problems, New York University, Institute of Mathematical Sciences, Rept. I, October, 1953, Rept. II, January, 1954, and Rept. III, October, 1956.

REFERENCES

2. Stoker, J. J.: Water waves, *Pure Appl. Math.*, Interscience Publishers, Inc., New York, vol. 4, 1957.
3. Chow, V. T.: "Open-channel Hydraulics," McGraw-Hill Book Company, Inc., New York, 1959.
4. Chow, V. T.: Open channel flow, sec. 24 in V. L. Streeter (ed.), "Handbook of Fluid Dynamics," McGraw-Hill Book Company, Inc., New York, 1961.
5. Gilcrest, B. R.: Flood routing, chap. 10 in Hunter Rouse (ed.), "Engineering Hydraulics," John Wiley & Sons, Inc., New York, 1950.
6. Puls, L. G.: Flood regulation of the Tennessee River, 70th Congr., 1st Sess., H.D. 185, pt. 2, appendix B, 1928.
7. Flood routing, chap. 6.10, in Flood Hydrology, pt. 6, in Water Studies, vol. IV of U.S. Bureau of Reclamation Manual, 1949.
8. Method of flood-routing, Report on Survey for Flood Control, Connecticut River Valley, vol. 1, sec. 1, appendix, U.S. Army Corps of Engineers, Providence District, 1936.
9. Carter, R. W., and R. G. Godfrey: Storage and flood routing, in "Manual of Hydrology," pt. 3, Flood-flow Techniques, *U.S. Geol. Surv. Water-Supply Paper* 1543-B, 1960.
10. "Routing of Floods through River Channels," Engineering and Design Manual, EM 1110-2-1408, U.S. Corps of Engineers, Mar. 1, 1960.
11. Steinberg, I. H.: A flood-routing device, *Trans. Am. Geophys. Union*, vol. 28, no. 2, pp. 247–254, April, 1947.
12. Tatum, F. E.: A simplified method of routing flood flows through natural valley storage, unpublished memorandum, U.S. Army Corps of Engineers, Rock Island District, May 29, 1940.
13. Missouri River, 73d Congr., 2d Sess., H.D. 238, p. 105, par. 166, 1935.
14. Tarpley, J. F., Jr.: A new integrating machine, *Military Engr.*, vol. 32, no. 181, pp. 39–43, 1939.
15. Harkness, F. B.: Harkness flood router: specification of construction and operation, U.S. Patent no. 2,550,692, May 1, 1951.
16. Linsley, R. K., Jr., M. A. Kohler, and J. L. H. Paulhus: "Applied Hydrology," McGraw-Hill Book Company, Inc., New York, 1949.
17. Paynter, H. M.: Flood routing by admittances, in "A Palimpsest on the Electronic Analog Art," Geo. A. Philbrick Researches, Inc., Boston, 1955, pp. 239–246.
18. Lawler, E. A., and F. V. Druml: Hydraulic problem solution on electronic computers, *Proc. Am. Soc. Civil Engrs., J. Waterways and Harbors Div.*, vol. 84, no. WW1, pp. 1–38, January, 1958.
19. Rockwood, D. M.: Columbia basin streamflow routing by computer, *Proc. Am. Soc. Civil Engrs., J. Waterways and Harbors Div.*, vol. 84, no. WW5, pp. 1–15, December, 1958.
20. Morrice, H. A. W., and W. N. Allen: Planning for the ultimate hydraulic development of the Nile valley, *Proc. Inst. Civil Engrs.*, vol. 14, pp. 101–156, October, 1959.
21. Lane, E. W.: Predicting stages for the Lower Mississippi, *Civil Eng. (N.Y.)*, vol. 7, pp. 122–125, February, 1937.
22. King, R. E.: Stage predictions for flood control operations, *Trans. Am. Soc. Civil Engrs.*, vol. 117, pp. 690–698, 1952.
23. Yevdjevich, V. M.: Flood routing methods, discussion and bibliography, *U.S. Geol. Surv. and Soil Conserv. Serv. Project* 4, 1960.

Section 25-III

HYDROLOGY OF FLOW CONTROL

PART III. RESERVOIR REGULATION

EDWARD J. RUTTER, *Chief (retired), Flood Control Branch, Tennessee Valley Authority.*

LE ROY ENGSTROM, *Late Chief, River Control Branch, Tennessee Valley Authority.*

I. Introduction	25-61
II. Purposes of Reservoir Regulation	25-61
A. Storage of Excess Water	25-61
B. Release of Stored Water for Beneficial Uses	25-61
III. Classification of Reservoir Regulations	25-62
A. Single-purpose Reservoirs	25-62
B. Multiple-purpose Reservoirs	25-62
IV. Reservoir-design Studies	25-64
A. Physical and Economic Factors	25-64
B. Hydrologic Studies	25-65
C. Storage Requirements	25-68
D. Design Floods	25-72
E. Operation Planning	25-74
F. Reservoir Routing	25-77
1. Level Storage	25-79
2. Slope Storage	25-80
3. Hydraulic Routing	25-82
V. Operating Schedules and Guides	25-85
A. Rigid Schedules	25-86
B. Semirigid Schedules	25-87
C. Long-range-planning Schedules	25-87
D. Examples of Reservoir Operation	25-87
VI. Operating Organization	25-95
VII. Hydrologic Network	25-96
VIII. References	25-96

Mr. Engstrom died on Dec. 23, 1960, after many years of brilliant service in engineering—six years with the U.S. Geological Survey and twenty-five years with the Tennessee Valley Authority. His contributions to the science and engineering of hydrology in general, and to this Handbook in particular, will be long remembered.—Editor-in-chief.

PURPOSES OF RESERVOIR REGULATION

I. INTRODUCTION

This section discusses the purposes of reservoir regulation, separating them generally into the storage of excess water, on one hand, and the release of stored water for beneficial uses, on the other. Regulations are classified according to whether they are for a single purpose or for several purposes. Next, some of the many studies that are required for adequate design of a project or a system of projects are reviewed, including studies of physical and economic factors, storage requirements, hydrology and design floods, seasonal and day-by-day operation, and reservoir routing. Types of operating schedules are then discussed, and examples of operating plans of several multiple-purpose projects are presented to show how the principles and objectives are attained for the several purposes. The organization needed for operation of reservoirs is then described, and finally the necessity of an adequate network for reporting hydrologic data is emphasized.

The role of hydrology in reservoir regulation begins with the early design of a project and continues throughout its actual operation. The design of a reservoir requires a knowledge of the quantity of stream flow and its occurrence with respect to area and time, and the operation of a reservoir requires the analysis of streamflow based on prereservoir flow records and on current streamflow and precipitation estimates.

II. PURPOSES OF RESERVOIR REGULATION

A. Storage of Excess Water

The purpose of reservoir regulation is to smooth out the peaks and valleys of streamflow so as to obtain greater beneficial use of water resources. Water is stored at times of excess flow either for the purpose of reducing downstream flood damage or to conserve water for later use at times of low flow. In those reservoirs built with flood control as a major purpose it is essential that the reservoir capacity reserved for storage of flood water be emptied as soon as practicable after a flood. In some cases, because of a definite seasonal pattern of floods, stored floodwater may also be retained, at least partially, for later conservation uses. In those reservoirs built for conservation uses, streamflow in excess of current requirements may be stored in the storage zones reserved for those purposes and not released until needed later. The concept of flood-control storage is empty space; that of conservation storage is stored water for later use.

B. Release of Stored Water for Beneficial Uses

The release of stored water may be for a variety of uses—power generation, irrigation, industrial and public water supply, fish and wildlife preservation, condensing water for steam-electric power plants, maintenance of navigation depths, prevention of salt-water intrusion, dilution for sanitary purposes, and others. Some of these are the major uses for which a reservoir project is designed; others are of minor significance; and still others result incidental to the planning for major purposes. They all are usually called conservation uses since the excess water is conserved or stored for them. A reservoir may also be used for nonconservation purposes—for recreation such as fishing, boating, and swimming.

There is often geographical conflict or competition for the same use; for example, owners of irrigable land in one location may object to those in another location receiving irrigation water. On the other hand, there may be competition between different uses, as, for example, between public water supply and recreation or between flood control and power. Such conflicts and competition must be compromised in project design.

A decision as to the purpose or purposes of a reservoir must be made in advance of design so as to permit proper and adequate consideration of basic principles related to those purposes. Other purposes, however, may be made a part of a project at a later date. For example, a single-purpose flood-control reservoir with an ungated low-level conduit for controlling the flow could later have the inlet level raised so as to maintain a permanent pool for recreation or to provide a specific reservation for a conservation use. Such a change would also require raising the dam to provide the same amount of flood storage.

III. CLASSIFICATION OF RESERVOIR REGULATIONS

Reservoirs and their operations are generally classified according to whether they are for a single purpose or for more than one purpose. Single-purpose reservoirs are usually simpler in design and operation, but operation may become complex if they are part of a large integrated system. Multiple-purpose reservoirs are more complex in both design and operation. Figure 25-III-1 shows diagrammatically types of dams and reservoirs used for single and multiple purposes.

All reservoirs, for any purpose, must have adequate outlet capacity to care for an appropriate extreme flood, or succession of large floods, taking into account the effect of dependable storage.

A. Single-purpose Reservoirs

The essential feature of single-purpose reservoirs for recreation or for fish and wildlife preservation is a constant or nearly constant pool level. Such a reservoir is shown diagrammatically in Fig. 25-III-1a. No fluctuation of the pool will occur except during floods or because evaporation exceeds water supply. Since the spillway is ungated, the outflow control is built into the project by the length or size of the spillway, and the outflow varies with the three-halves power of the head on the spillway.

Single-purpose reservoirs for flood detention (Fig. 25-III-1b) are held empty, except during floods, by a low-level ungated outlet, the size of which is predetermined for a specified outflow rate. The outflow varies with the one-half power of the head on the outlet after the water level is above the entrance plus the discharge over the spillway, which varies with the three-halves power of the head on the crest. The operation is automatic, requiring no attendant, and the reservoirs will reduce all flood peaks to some degree immediately downstream. Flood-detention reservoirs of the Miami Conservancy District (Ohio) are of this type. On the other hand, in a large watershed the combined discharge of several detention reservoirs on tributaries may add significantly to the peak of a flood farther downstream. Such reservoirs also do not make economical or efficient use of the storage space; that is, they may store when storage is not needed, as in the early part of a flood when flows are relatively low.

Single-purpose reservoirs (Fig. 25-III-1c) for conservation uses, such as public water supply, irrigation, power, and downstream wildlife preservation, require a gated outlet to control the release of water and reservoir capacity to store excess flow in sufficient amounts to supply the demand during periods of deficient flow. In the case of a single-purpose power project, the turbine gates control the outflow when the water level is below the spillway. The total outflow capacity is similar to that of Fig. 25-III-1b, but because of the gate on the low-level outlet the discharge can be held to zero until the spillway level is reached. The storage capacity is governed by the water supply available in the stream or by physical features of the site.

B. Multiple-purpose Reservoirs

Any number of uses compatible with the physical limitations of the reservoir site and the available water supply may be combined in a multiple-purpose reservoir. Figure 25-III-1a and c to h shows types of dams used for multiple-purpose reservoirs. The combination of uses may be in effect a number of separate uses added together at one location, or there may be joint use of some of the available water or available storage space. The best combined beneficial use probably would be a compromise between different uses or different geographical locations.

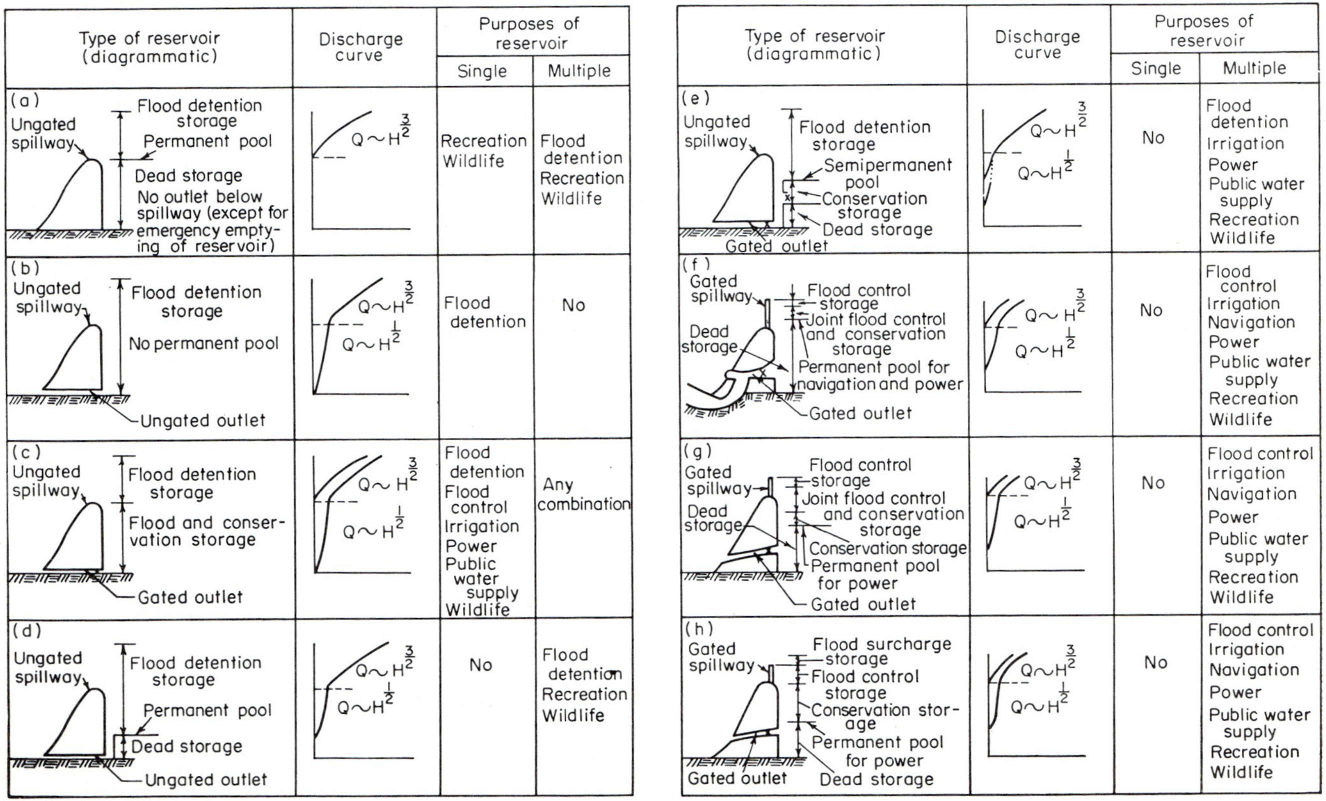

Type of reservoir (diagrammatic)	Discharge curve	Purposes of reservoir — Single	Purposes of reservoir — Multiple
(a) Ungated spillway; Flood detention storage; Permanent pool; Dead storage; No outlet below spillway (except for emergency emptying of reservoir)	$Q \sim H^{3/2}$	Recreation Wildlife	Flood detention Recreation Wildlife
(b) Ungated spillway; Flood detention storage; No permanent pool; Ungated outlet	$Q \sim H^{3/2}$; $Q \sim H^{1/2}$	Flood detention	No
(c) Ungated spillway; Flood detention storage; Flood and conservation storage; Gated outlet	$Q \sim H^{3/2}$; $Q \sim H^{1/2}$	Flood detention Flood control Irrigation Power Public water supply Wildlife	Any combination
(d) Ungated spillway; Flood detention storage; Permanent pool; Dead storage; Ungated outlet	$Q \sim H^{3/2}$; $Q \sim H^{1/2}$	No	Flood detention Recreation Wildlife
(e) Ungated spillway; Flood detention storage; Semipermanent pool; Conservation storage; Dead storage; Gated outlet	$Q \sim H^{3/2}$; $Q \sim H^{1/2}$	No	Flood detention Irrigation Power Public water supply Recreation Wildlife
(f) Gated spillway; Flood control storage; Joint flood control and conservation storage; Permanent pool for navigation and power; Dead storage; Gated outlet	$Q \sim H^{3/2}$; $Q \sim H^{1/2}$	No	Flood control Irrigation Navigation Power Public water supply Recreation Wildlife
(g) Gated spillway; Flood control storage; Joint flood control and conservation storage; Conservation storage; Permanent pool for power; Dead storage; Gated outlet	$Q \sim H^{3/2}$; $Q \sim H^{1/2}$	No	Flood control Irrigation Navigation Power Public water supply Recreation Wildlife
(h) Gated spillway; Flood surcharge storage; Flood control storage; Conservation storage; Permanent pool for power; Dead storage; Gated outlet	$Q \sim H^{3/2}$; $Q \sim H^{1/2}$	No	Flood control Irrigation Navigation Power Public water supply Recreation Wildlife

Fig. 25-III-1. Classification of reservoirs.

Storage reservations for conservation uses may include (1) fixed year-round allocations solely for specific purposes or (2) variable allocations, depending on the expected need during specified seasons or months of the year or on the anticipated runoff where reliable forecasts can be made.

Such uses as recreation, fish and wildlife preservation, and flood detention may be combined with dams of the types in Fig. 25-III-1a and d, the latter being the type usually built by the U.S. Soil Conservation Service in their agricultural flood-damage-prevention program. The addition of storage as in Fig. 25-III-1e for conservation uses—irrigation, power, public water supply—requires a higher dam than for flood detention alone, a gated outlet, and a semipermanent pool.

Multiple-purpose reservoirs built to maintain a permanent pool for navigation and power head and to provide joint and single-purpose conservation and flood-control storage; thus flood-control and flood-detention reservoirs become extremely complex both in their design and operation. Figure 25-III-1f to h shows diagrammatic examples of several combinations of such uses. Outflows usually are completely gate-controlled, although an emergency free-overfall spillway may be necessary in extreme floods. In some reservoirs extreme floods will rise above the maximum controlled level at the dam.

IV. RESERVOIR-DESIGN STUDIES

Reservoir-design studies as they relate to hydrology include the determination of the amount of storage needed for conservation or flood-control purposes; pool levels for recreation and navigation; discharge capacity of spillways and other outlets to take care of maximum probable floods; discharge capacity to provide for release of conservation storage; and operation studies during critical high- and low-flow periods. Structural-design studies and social problems arising from population adjustments in a reservoir are not directly related to hydrology.

A. Physical and Economic Factors

A number of physical factors must be considered in the design and operation of reservoirs. The first consideration is whether the topography of the stream valley provides a feasible dam and reservoir site with adequate storage capacity to satisfy the flood-control and conservation-storage requirements. Second, if feasible sites are available but the storage capacity is inadequate to satisfy all needs completely, then an allocation of the storage to the various purposes must be made as equitably as possible, the adopted allocation being a compromise between the various uses. (See Sec. 26 for a more complete discussion of both physical and economic factors.)

Even if feasible sites are available, other considerations may lead to the selection of less desirable alternative sites. For example, the presence of a large city along a river bank may prevent the adoption of the best and cheapest dam site in the vicinity and thus force the selection of a less desirable site. In other locations maximum pool levels and useful storage may be restricted to less than fully adequate amounts because of low-lying urban centers. For example, urban and industrial development in a river valley can limit the upper pool level for flood storage, while minimum depth for navigation and minimum levels for power can limit the lower flood-storage level. Thus the flood-storage capacity may be limited even though the topography of the river valley would permit a fully adequate amount. The presence of highways, railroads, bridges, and dams also may cause similar limitations. In such cases the cost of replacement of the development probably would be more than the benefits that would be obtained. Other factors, such as historical landmarks, which cannot be replaced at any cost, can limit the selection of reservoir sites and operating levels.

Economic factors affecting the design and operation of reservoirs are capital costs of construction and land; annual costs of amortization, interest, operation, and maintenance; lump-sum benefits for increased values; and annual benefits due to storage of floodwater and release of conservation storage. Standard Federal practice for justification of a project has been that annual benefits exceed annual costs; that is, the benefit-cost ratio must be greater than 1.

To determine the most economic project of a number of projects or the most economic degree of development of a single project, consideration should be given to the net annual benefits—annual benefits less annual costs. The most economic project or degree of development would be in the range of both the greatest benefit-cost ratio and the greatest net annual benefits.

In comparing a number of alternative feasible reservoir sites, each producing the same benefits, the most economic site usually is that with the lowest capital cost.

B. Hydrologic Studies

The two essential kinds of basic data needed for reservoir-design studies are adequate topographic maps and hydrologic records. Watershed divides can be outlined on maps, and the size of the drainage area above a reservoir site determined with a

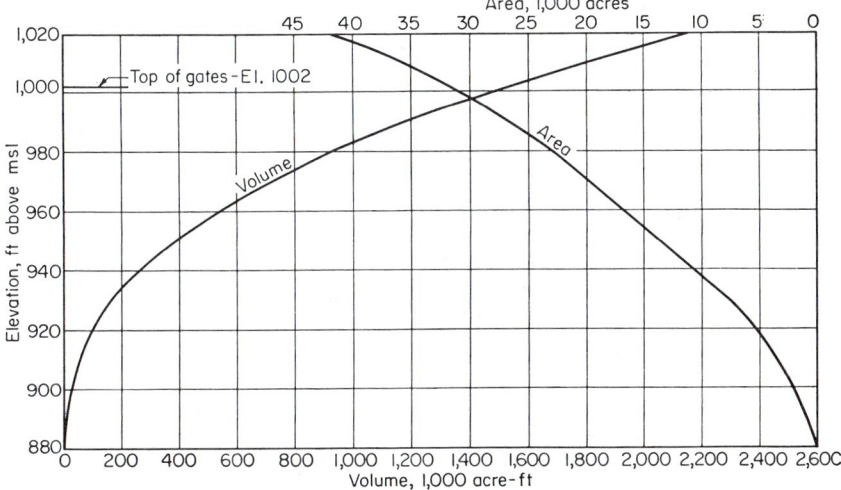

Fig. 25-III-2. Level reservoir area and volume curves (Douglas Reservoir, TVA).

planimeter. Reservoir volumes are computed from reservoir areas which are determined by planimetering the area enclosed above a dam site for successive elevation contours on the maps. The mean of two successive contour areas multiplied by the contour interval gives the interval volume, and the summation of interval volumes gives the total volume below any elevation. The scale of the maps and the contour interval affect the accuracy of results, but modern-day maps at a scale of 1:24,000 and contour intervals of 10, 20, or 40 ft give satisfactory results for all but the smallest areas. Figure 25-III-2 shows typical area and volume curves for a level reservoir.

In reservoirs having a significant slope in the water surface, reservoir volumes are determined for an assumed range of flow conditions by dividing the reservoir into small reaches, computing a level-volume curve for each as described above, and then summing the volumes read from the reach curves for the elevations corresponding to the range of flow conditions. Figure 25-III-3 shows volume curves for a range of steady flows and headwater elevations where there is a significant rise in the water surface. Figure 25-III-4 shows volume curves for a range of inflows, outflows, and headwater elevations. Because of variations in flows and elevations within long reservoirs during different floods, the conditions assumed for constructing the curves may not represent the correct relation between inflow, outflow, elevation, and storage, and adjustments may have to be made in the use of such volume curves.

Records of streamflow are essential for determining the amount of water available for conservation purposes. These records reveal flood peaks and volumes which are used to determine the amount of storage needed to control floods and to design spillways and other outlets (see Sec. 15 for methods of measuring streamflow). Streamflow is usually depicted in graphic form as a hydrograph of rate of discharge against time, as in Fig. 25-III-5. Curves derived from hydrographs are the flow-duration curve of discharge rate in order of magnitude against percentage of time and the mass curve of

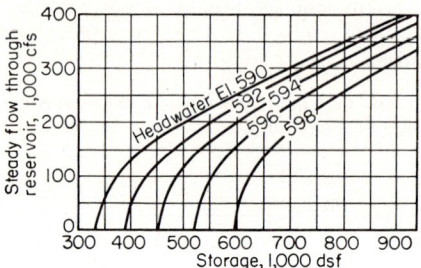

Fig. 25-III-3. Volume curves for steady flows in a reservoir (Guntersville Reservoir, TVA).

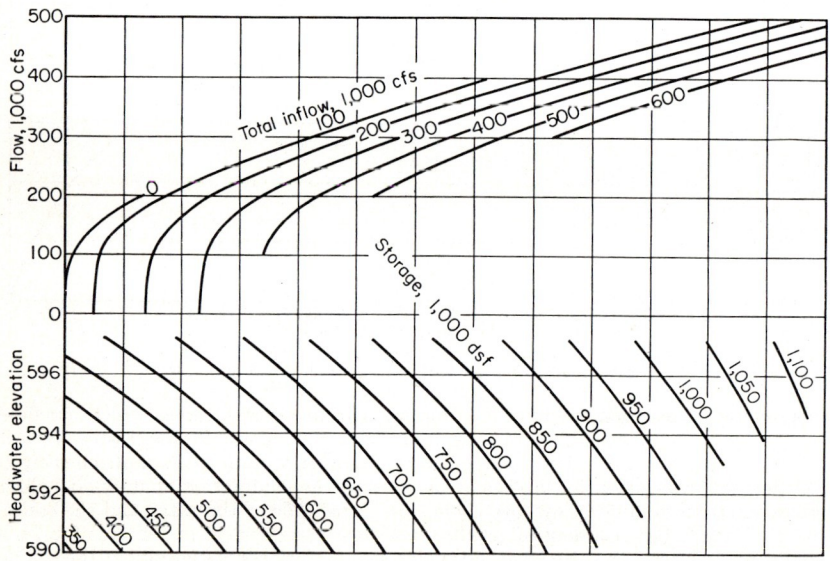

Fig. 25-III-4. Volume curves for a range of inflows, outflows, and headwater elevations (Guntersville Reservoir, TVA).

accumulated discharge against time, also shown in Fig. 25-III-5. These latter curves are used to facilitate storage computations.

In addition to streamflow records at gaging stations, high-water elevations of historic floods should be located to aid in determining the height and discharge of a flood at any point in those reservoirs having a significant rise in the water surface. From this information and valley cross sections, hydraulic roughness coefficients, such as n in the Manning or Kutter formulas, can be computed for successive reaches. From these coefficients flow profiles for a range of conditions, as in Fig. 25-III-6, are computed, and then reservoir volumes, as in Figs. 25-III-3 and 25-III-4.

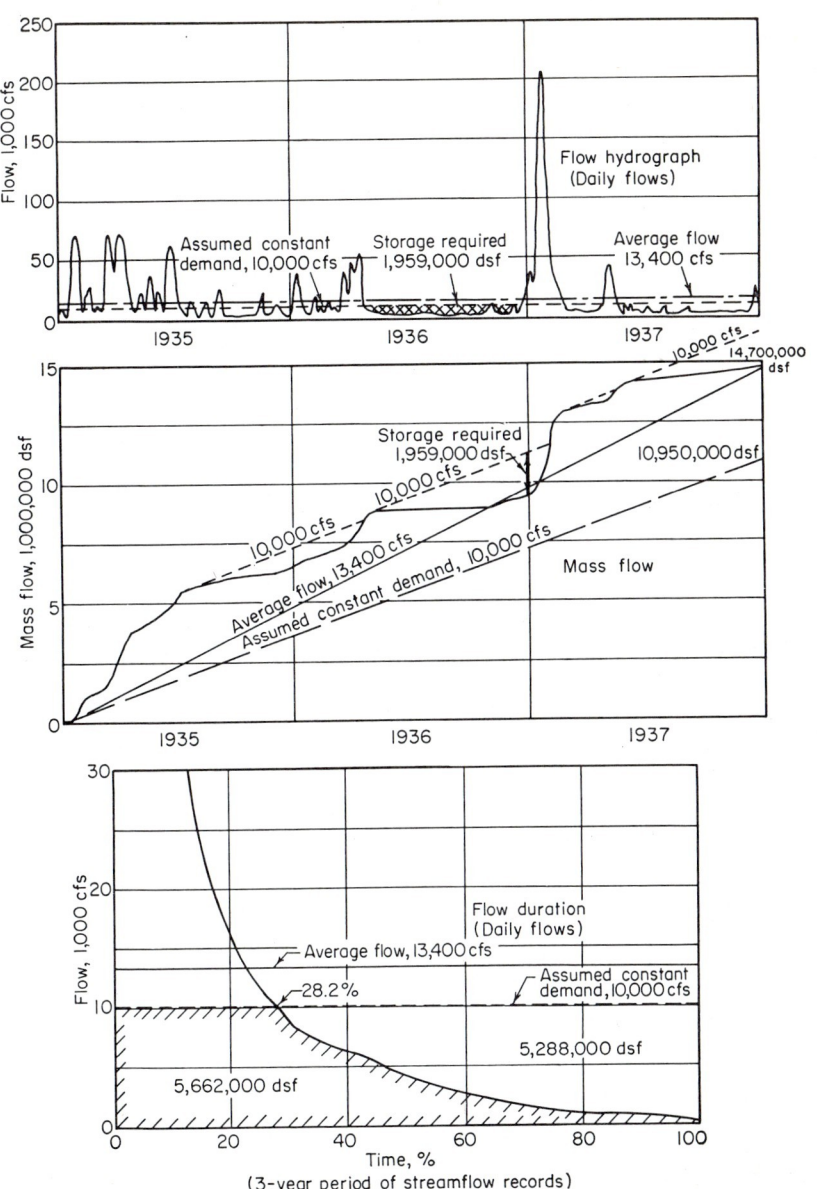

Fig. 25-III-5. Flow hydrograph, mass curve, and flow duration.

In the design and operation of flood-storage reservoirs, the seasonal occurrence of floods may be an important factor. If floods occur at any time, storage to control those floods must be available at all times. But if there is a definite seasonal pattern shown by long-term dependable records, then the full amount of flood storage need be reserved only during the indicated flood season. Figure 25-III-7 is a plot of flood occurrence for a stream on which floods have occurred in any season, and Fig. 25-III-8 is one for a stream having a definite flood pattern confined to the season from December to early April. The flood record by years above the record by days shows the length of the record and other pertinent data.

Records of rainfall are used to supplement streamflow records or as a basis for computing streamflow where there are no flow records. Storm rainfall records both for the

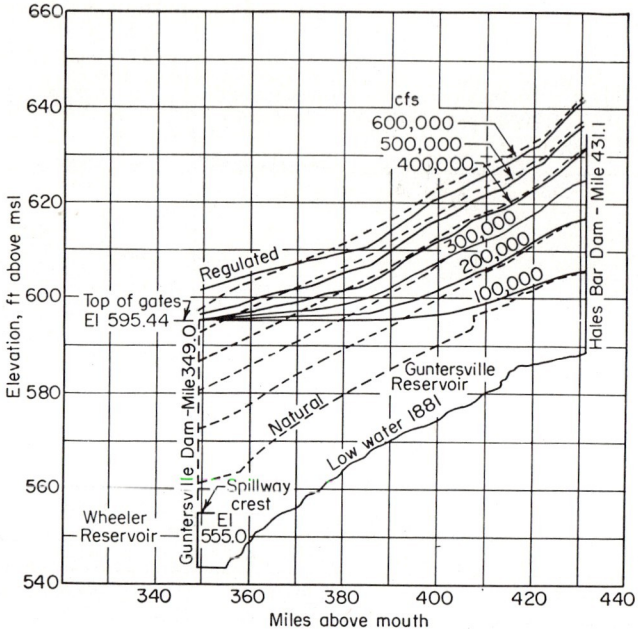

FIG. 25-III-6. Reservoir flow profiles (Guntersville Reservoir, TVA).

area being studied and for other areas whose streams could be transposed are particularly useful in computing hypothetical maximum floods for reservoir or spillway design (Sec. 9).

C. Storage Requirements

The complete development of the reservoir storage potential of a watershed may not provide as much storage as is needed for any one or all potential uses. In this case supplemental means must be found to provide protection if flood control is involved and the available storage must be allocated to the various uses.

The amount of storage to be provided in a flood-control reservoir depends on the degree of protection needed and on the nondamaging capacity of the stream channel. The latter may be increased by building protective works, such as levees or walls, or by channel improvements. The increased channel capacity thus permits a smaller reservoir capacity, and an economic balance between reservoir control and the other means of protection must be made. For example, it may be physically possible to reduce a design flood by reservoir storage to the nondamaging stage, but it may also be physically possible to build levees to a height substantially above this stage. The proper balance between the two depends on economic factors, as well as possible inconvenience of high levees in a city.

The desired degree of protection determines the magnitude of the flood adopted as a basis for reservoir design. Complete protection against the maximum probable flood should be the goal where loss of life or disastrous property damage is involved. To

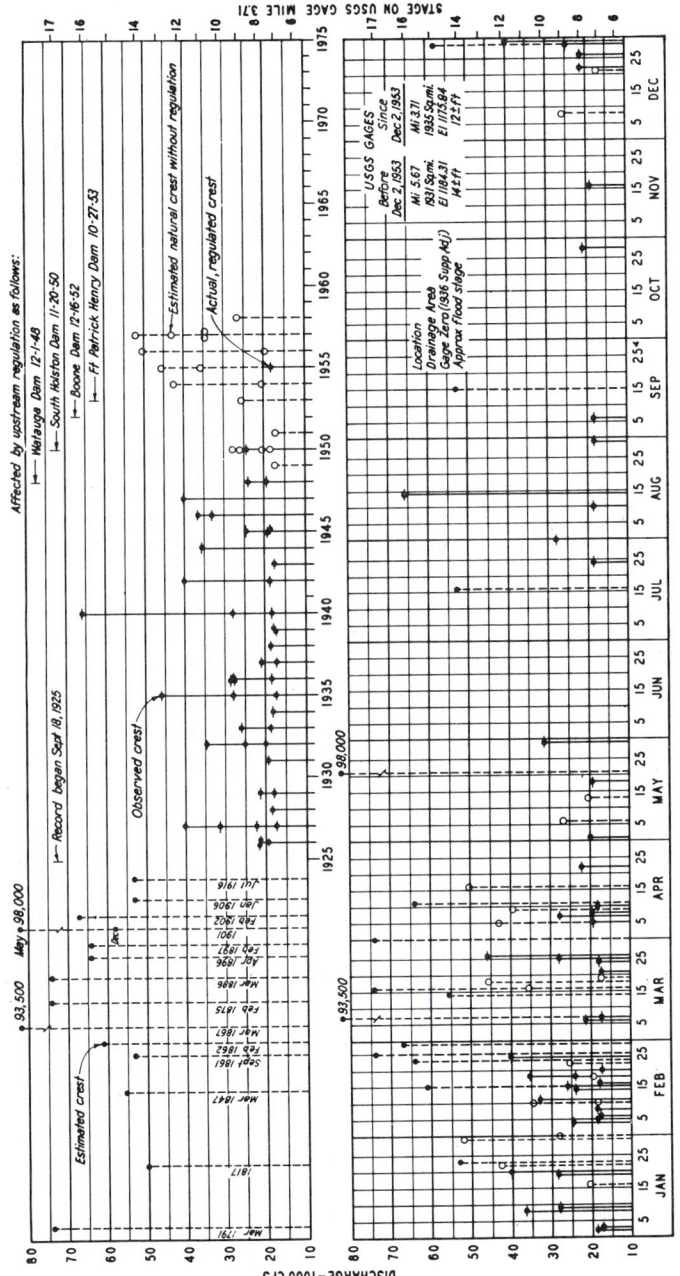

FIG. 25-III-7. Flood record showing occurrences at any time of the year (South Fork Holston River at Kingsport, Tenn.).

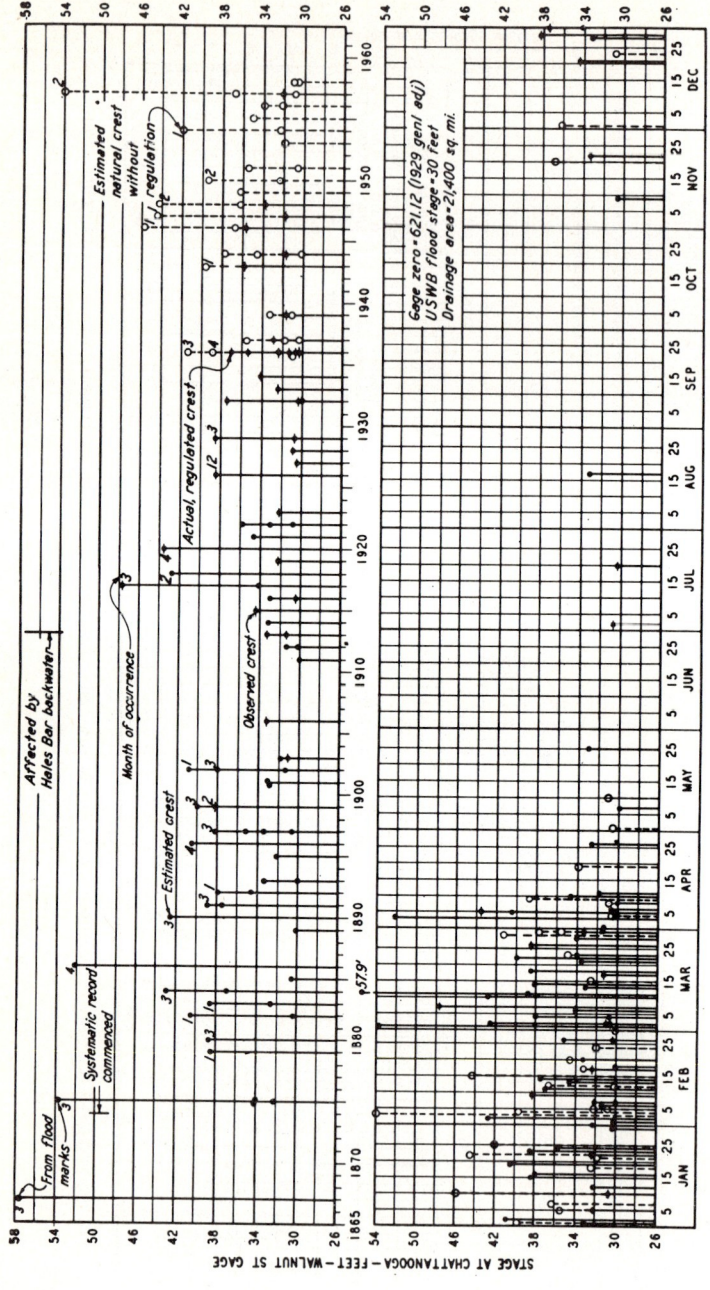

FIG. 25-III-8. Flood record showing seasonal pattern of occurrence (Tennessee River at Chattanooga, Tenn.). Affected by reservoir operation as follows:

Norris, 3-4-36
Chickamauga, 1-15-40
Hiwassee, 2-8-40
Cherokee, 12-5-41

Watts Bar, 1-1-42
Nottely, 1-24-42
Chatuge, 2-12-42

Douglas, 2-19-43
Ft. Loudoun, 8-2-43
Fontana, 11-7-44

Watauga, 12-1-48
South Holston, 11-20-50
Boone, 12-16-52

accomplish complete protection it is necessary to store only the flow above the non-damaging stage. The amount of this flow to be stored is represented in Fig. 25-III-5 by the shaded area between the horizontal line at that flow and the hydrograph. The operation of a reservoir or a system of reservoirs in such an ideal manner, especially if it is located several days away in time of water travel, cannot be depended upon in practice, and a greater capacity is therefore needed to allow for less than ideal operation. The necessary excess capacity can be determined only after a study of several assumed reservoir operations based on practical rules that can be followed in actual operation. Moreover, a strict adherence to mathematical calculations of required storage may result in inadequate capacity.

In a system of reservoirs the distribution of storage on the different tributary areas should be based largely on the average contribution of those tributaries to the critical flood flows at the protection point. Greater amounts of storage may be provided in those reservoirs where storage is relatively cheap.

Where loss of life or disastrous property damage is not involved, protection against a flood smaller than the maximum probable flood may be adequate. For example, the amount of storage needed for the protection of agricultural land from crop loss or erosion is often based on moderate-size floods, those occurring on the average about once in 10 to 15 years during the crop season.

Conservation-storage requirements vary widely, depending on the amount and variation of natural streamflow and on the demand of the particular use. For example, the demand for irrigation water depends on the type of irrigating system, the kind of crop, and the irrigable acreage. The demand for municipal and industrial water is influenced largely by present population and estimated future growth and present and anticipated use by large industries. Water to provide navigation depths depends on the flow required to sustain the required depth at critical locations in the navigation channel. In the determination of any conservation-storage requirement, appropriate allowance should be made for losses such as evaporation from the reservoir surface.

In a simple case the amount of storage needed to supply a constant demand can be determined from a hydrograph of streamflow, a mass curve, or a duration curve. On the flow hydrograph of Fig. 25-III-5, the amount of storage required to maintain a constant flow of 10,000 cfs is represented by the area between the horizontal line at that flow and the hydrograph and also by the greatest vertical distance between a line showing the mass discharge for the demand rate and the mass curve of natural flow. Since the storage required depends on the severity of the period of deficient flow, it is important to have a long record of streamflow to assure that the driest period is included. For example, the storage required to maintain a specified minimum discharge over a 100-year period was used in studies for the development of the Nile River to its fullest extent for irrigation (see work by Hurst in the References, Subsec. VIII). Such determinations of long-term storage are unusual and are possible only where flow records, rainfall records, or records of other natural phenomena are available.

The amount of storage needed to supply a constant demand can be determined from a hydrograph of streamflow, a mass curve, or a duration curve. Such curves are shown in Fig. 25-III-5 for a 3-year period of streamflow records for a river having a wide fluctuation in rate of flow. A record of only 3 years is too short for storage determinations. It is shown only to explain the use of the curves. The average flow for the 3-year period is 13,400 cfs, as indicated by the horizontal lines on the flow hydrograph and duration curves and by the sloped line connecting the end points of the mass curve. The sloped line is a mass curve for a uniform rate of flow. The streamflow rate is less than the average flow when the slope of the mass curve is less than that of the straight line. The average flow during any period is represented on the mass curve by the slope of a line connecting the end points of that period. The storage needed to supply a uniform demand rate in a dry period is represented by the greatest vertical distance of the mass demand curve above the mass curve of streamflow. For example, assuming the reservoir would be full on Apr. 24, 1936, a storage of 1,959,000 day-sec-ft (dsf) would be needed to maintain a flow of 10,000 cfs up to Dec. 31, 1936, after which the reservoir would refill and would be full again on Jan. 30, 1937. The dry period of 1935–1936 does not require as much storage to maintain

10,000 cfs, and storage required for the dry period of 1937-1938 cannot be definitely determined because streamflow is not shown after Dec. 31, 1937.

On the flow-duration curve the area below the assumed demand flow of 10,000 cfs and the duration curve represents the total natural flow up to that rate available in the 3-year period. This is equal to 5,662,000 dsf. The full rate of 10,000 cfs is available 28 per cent of the time, and less than that rate is available for the remainder of the time. The area below the assumed flow of 10,000 cfs but above the duration curve represents the amount of flow required from storage, equal to 5,288,000 dsf. This latter amount is not the required capacity of a reservoir because the streamflow may fluctuate above and below the demand rate and the reservoir may rise and fall several times during the 3-year period. The maximum required reservoir storage is determined from study of the mass curve or the hydrograph. Reservoir storage requirements to supply nonuniform demand rates may also be determined from a study of flow hydrographs and mass curves, but in a system of several reservoirs the determination may best be made by accounting for daily flows and changes in storage.

D. Design Floods

The design of any water-regulating structure requires the adoption of a design-flood flow. The risks involved for the kind of property affected and the type of structure under consideration—whether a concrete dam or an earth dam—will influence the designer's judgment in the selection of the *design-flood* magnitude. The magnitude of the adopted flow in turn affects the proportions of the structure and, consequently, its economic feasibility. See also Sec. 25-II.

Two types of extreme floods given consideration in the design of dams and reservoirs are the maximum probable flood and the maximum possible flood. The *maximum probable flood* is the greatest flood that may reasonably be expected, taking into account all pertinent conditions of location, meteorology, hydrology, and terrain. In some areas such a flood has already been observed, and there is a reasonably good chance of its occurrence in any area and in any year. The frequency of its occurrence is not susceptible of determination. It is used for designing spillways and flood-protection works where loss of life or disastrous property damage is involved. The magnitude of such a flood may be from one-half to three-fourths of that of the maximum possible flood. The *standard project flood* used by the U.S. Army Corps of Engineers in the design of reservoir projects is in this category.

The *maximum possible flood* is the greatest flood to be expected assuming complete coincidence of all factors that would produce heaviest rainfall and maximum runoff. It would result from the maximum possible rainfall as determined from maximum observed rainfall adjusted for differences between observed and potential moisture charges. The frequency of this flood is not susceptible of determination, but its occurrence would be highly improbable. It is used to test the effectiveness of spillways and other outlet works in preventing the failure of a dam.

The principal use of the maximum probable flood is to determine the reservoir storage to be provided, and hence the height of the dam and also the discharge capacity, and hence the physical dimensions of spillways and other outlet works, with appropriate consideration of the available flood-storage capacity at the site or in upstream reservoirs. Where flood storage is not available, only the peak discharge rate of the flood need be determined. Where storage is available, the volume as well as the peak rate must be known. The three factors—outlet capacity, storage, and height of dam—are closely interrelated, and variations in one will affect the other two.

The freeboard between the maximum level of the maximum probable flood and the top of the dam depends on the type of structure and the seriousness of its failure at downstream locations. With high earth dams the freeboard allowance should be from 10 to 15 ft, and with concrete dams, from 5 to 7 ft. A smaller freeboard may be provided at low dams protecting uninhabited agricultural land.

A large dam that will be depended upon for the protection of life and valuable property, though designed for the maximum probable flood, should also be investigated for the effect of the maximum possible flood. For example, the maximum water level, the outlet works, and freeboard may be determined on the basis of the maximum probable flood, but it should also be determined that the maximum possible flood will not overtop the structure and cause failure.

In the case of flood control for agricultural land only, a design flood smaller than either the maximum probable or maximum possible flood may be adopted. An occasional flooding of farmland, or the loss of a crop, would not be a catastrophe. In some areas such flooding might even be beneficial. The benefits from providing protection, therefore, may not justify the cost of protection against extremely large floods, and smaller floods, as determined by the economics of the situation, would be used in the design of dams and reservoirs.

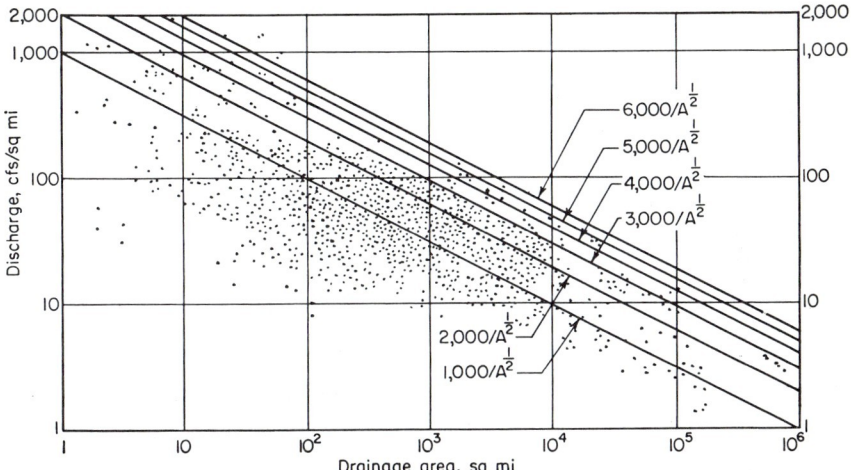

Fig. 25-III-9. Extreme flood discharges to 1940, eastern United States. The points on the diagram represent extreme flood rates reported on streams east of the Rocky Mountains, excluding the smaller rivers draining directly to the Gulf of Mexico or through the southern states to the Atlantic Ocean. Since the diagram was prepared primarily for use on drainage areas of 500 sq mi or more, no attempt was made to include all reported flood rates on small streams. The points shown represent drainage basins of widely varying characteristics. The significance of some of the most extreme rates for any particular problem should be considered in the individual case. Arid or semiarid areas were deducted from the large drainage areas in determining the position of points showing flood rates below these arid regions. The inclined lines on the diagram represent a variation in extreme flood rates inversely proportional to the square root of the drainage area.

Three basic characteristics of a flood-discharge hydrograph are (1) peak rate, (2) total runoff volume, and (3) successive variations in the rate of flow. Another important factor is the time of occurrence.

In the determination of the peak-discharge rate of the maximum probable flood, whether by the transposed storm and unit-hydrograph technique or by other methods, the reasonableness of the adopted flow should be checked and compared with maximum observed floods on streams in the same broad meteorological and physiographic region. The theory is that on some streams in those broad regions the maximum flood of which those streams are capable has occurred within the period of knowledge and that the maximum capabilities of each stream will be equal to the maximum flood of the entire group of streams. Figure 25-III-9 shows a logarithmic plotting of flood peak discharges per square mile of drainage area against the drainage area in square miles for watersheds generally east of the Rocky Mountains. The slope of the mass of the points on the chart shows in a general way that the flood rate varies inversely with the square root of the drainage area. The study for which Fig. 25-III-9 was prepared in 1940 was concerned principally with watersheds of more than 500 sq mi. Great floods occurring since that date have exceeded some of the highest rates shown. A line having a slope of $\frac{1}{2}$ drawn through the uppermost points would determine the reasonableness of maximum flood rates determined by other methods for any area of comparable meteorological and physiographic characteristics.

Design-flood hydrographs for watersheds up to several thousand square miles in area are developed from storm rainfall that is transposable within broad regions but adjusted for differences in moisture charge which depend on meteorological factors such as dewpoint and barrier altitude. The adjusted rain is converted to runoff by subtracting infiltration and other losses, and the flow hydrograph is then computed by the unit-hydrograph technique (Sec. 14), by flood routing in natural streams, or a combination of the two methods. Appropriate snowmelt contribution should be added to the hydrograph.

For large watersheds of major river systems draining areas of diverse meteorological and hydrologic patterns, design-flood hydrographs are best determined by combining maximum observed floods on the several major tributaries, with appropriate shifting with respect to time, as indicated to be possible meteorologically, so as to concentrate the flow at the critical location.

E. Operation Planning

Planning for the operation of reservoirs is an essential feature of project design. As soon as the location of a reservoir has been established and the amount of storage for each purpose has been determined, an intensive study of methods to attain the best operation should follow. Reservoirs which are automatic in their operation (Fig. 25-III-1a, b, and d) have their operating plan built into them. Such reservoirs, for example, provide a pool level having a specified area for recreation, or in the case of some flood-control reservoirs, the spillway elevation and length and the size of other outlets are designed to provide a predetermined degree of flood protection.

If the only purpose of the reservoir is the release of conservation storage and it is essential that the regulated streamflow is never less than a specified minimum amount, as may be the case for municipal water supply, irrigation, or navigation purposes, then the operating plan is simple. The required minimum flow is determined by the anticipated demand, and the storage necessary for maintaining that flow is determined by analysis of streamflow or other hydrologic records. During periods when the natural flow is less than the required rate, sufficient water is released from storage to maintain the required rate.

Reservoirs having gate-controlled spillways or other types of gated outlets usually require attendants for operating the gates so as to regulate the water levels, and these reservoirs require a predetermined plan of operation. Such a plan is especially necessary in a large system of reservoirs involving several uses, some of which may be conflicting. The plan specifies the limits of storage that may be used for each purpose. The limits may vary, depending on the season of the year or the current hydrologic runoff conditions. While it is important to have a plan, it should not be considered unchangeable; demands for specific uses may change, or it may be found that a project was initially underdesigned or overdesigned because of insufficient data.

As an example of an annual operation plan, Fig. 25-III-10 shows the schedule for a tributary reservoir providing flood-control storage and conservation storage for power and navigation. The reservoir is a unit of the TVA system. The storage reservation for flood control on March 15 was determined as the amount necessary, in conjunction with other reservoirs and levees, for controlling the maximum probable flood at a critical downstream location. The greater flood-storage reservation on January 1 gives assurance that the March 15 reservation will be available in case a series of floods make it difficult to draw down the reservoir to the March 15 level. The lesser March 31 reservation makes allowance for the decreased chance of floods near the end of the valley-wide flood season, as indicated by the flood record on Fig. 25-III-8. Flood storage above elevation 1034 is also available for reducing extreme floods, the estimated maximum probable flood reaching elevation 1044. After March 31 the danger of major valley-wide floods is past and the reservoir is filled as high as elevation 1020, the maximum elevation reached depending on streamflow and the demand for conservation-storage use. Storage above elevation 1020 is reserved for flood control through the summer months. The normal minimum drawdown level

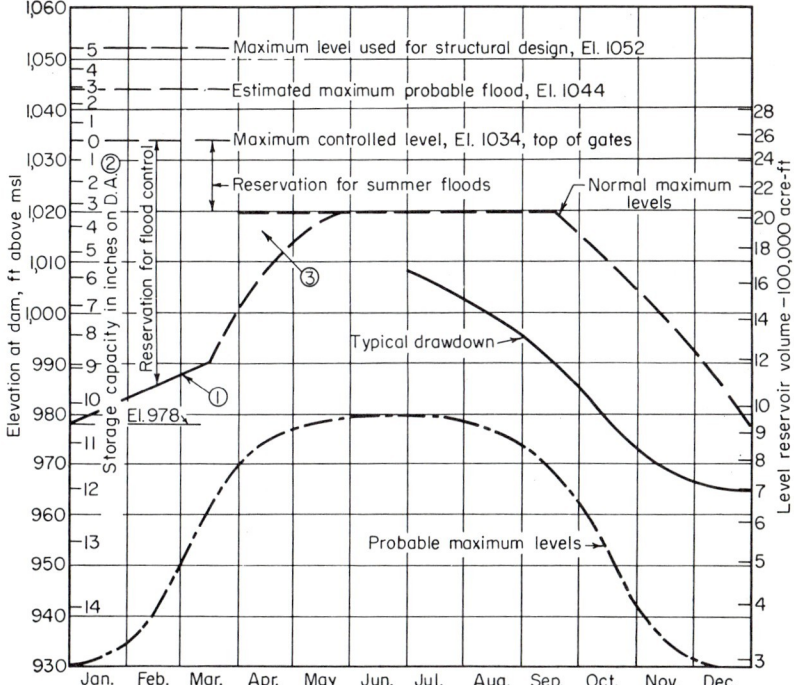

Fig. 25-III-10. Annual plan of operation, multiple-purpose tributary reservoir (Norris Reservoir, TVA). (1) Maximum multipurpose levels during flood period to be exceeded only during flood-control operations; (2) based upon drainage area 2,912 sq mi; (3) after March 31, filling depends on current hydrologic conditions.

for conservation storage is elevation 930. Probable minimum and maximum levels shown in Fig. 25-III-10 are the expected range of operation except during a flood. The allocation of storage is as follows:

Storage	Elevation	Acre-feet
Below normal minimum conservation pool............	826–930*	286,000
Conservation...	930–978	646,000
Joint conservation and flood control..................	978–1,020	1,115,000
Flood control...	1,020–1,034	520,000
Flood surcharge...	1,034–1,044	425,000
Total to El 1044.......................................	....	2,992,000

* Drawdown below El 930 may be made under extreme conditions.

An example of an annual operation plan of a multiple-purpose main-river reservoir which also provides flood-control and conservation storage for power and navigation is shown in Fig. 25-III-11. In addition to conservation storage it provides a permanent pool for navigation. The minimum pool, elevation 675, was determined by the specified navigation depth at critical points in the reservoirs, and the maximum pool, elevation 685.44, was determined by reservoir limitations and the location of the next upstream dam site. Flood-control or conservation storage therefore was limited to the zone between these two levels, but during the usual valley-wide flood season the full amount was reserved for flood control, except for minor fluctuations

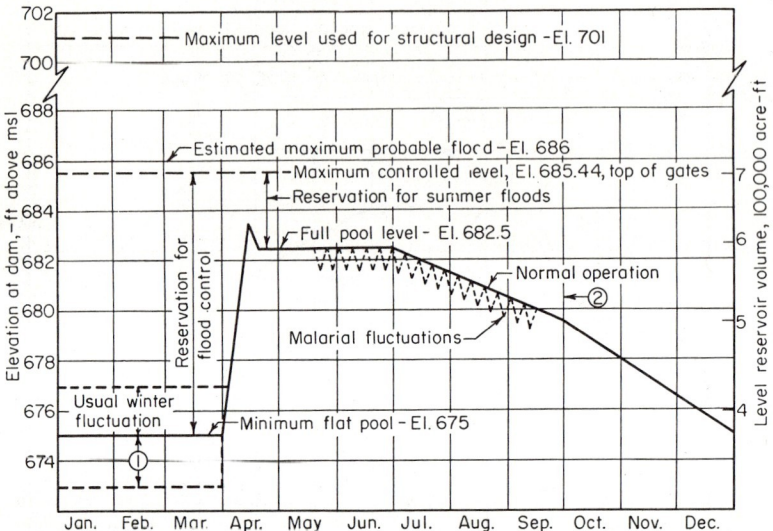

Fig. 25-III-11. Annual plan of operation, multiple-purpose main-river reservoir (Chickamauga Reservoir, TVA). (1) Drawdown zone for maintaining flat-pool volume prior to flood crest may extend to El 673 at dam. (2) The pool may be raised as high as El 682.5 after October 1 for power storage, but after December 1 it must be kept within the winter fluctuating range.

due to turbine operation. In order to retain storage capacity for flood control, drawdown to elevation 673 at the dam is permitted provided navigation depths are maintained throughout the reservoir. After March 31 the reservoir is filled to elevation 682.5, and the zone between elevation 682.5 and 685.44 is the minimum reservation for flood storage. The fluctuating dashed lines show weekly 1-ft changes in level for malaria control, and the 1-ft rise above 682.5 about the middle of April is for the same purpose. Substantial water-surface slope occurs in this reservoir during periods of high flow. The allocation of level-pool storage is as follows:

Storage	Elevation	Acre-feet
Below navigation pool	628–675	375,900
Joint conservation and flood control	675–682.5	220,900
Flood control	682.5–685.44	108,500
Total level-pool storage to El 685.44	....	705,300

In addition to seasonal operating guides, rigid or fixed rules of operation of gate-controlled reservoirs during flood periods are necessary in making studies of the effect of reservoir operations on downstream discharges and stages. The rules should be designed to give comparable results in the study of different floods and to approximate the results expected in actual operations. They also should be designed so that the available storage is utilized advantageously in damaging floods. Obviously, a rule designed to give optimum regulation in a flood as great as the maximum probable flood probably would not give optimum regulation in small floods and, in fact, might give higher peak discharges than would have occurred without the regulation. Fixed rules of operation can best be determined after ideal operations to determine the maximum needed reduction have been made for the expected range of floods. In one type of rule the rate of discharge that should prevail during a flood is dependent on the headwater elevation—the higher the reservoir, the greater the discharge. In another type the discharge is related to the rate of rise of headwater elevation. Use of the rules would begin when it is recognized that a flood is developing. No foreknowledge of the streamflow is necessary. The use of the rules results in an operation similar to that of a detention reservoir but retains the benefits of gate-controlled outlet works for quick drawdown after the flood. An example of a fixed rule is shown in Fig. 25-III-13, and the results of its application in a flood are shown in Fig. 25-III-14.

Under natural flood conditions water is stored temporarily as the stage rises. This storage, often referred to as valley storage, reduces the peak inflow. Under reservoir conditions the space previously available for natural valley storage may already be filled, and because of the increased depth the arrival of upstream inflow to the dam is accelerated. For example, if a reservoir having control gates at the top of the dam is at the level of the top of gates when a flood occurs and no rise in water level can be made, resulting outflows will exceed those which would have occurred under natural conditions. A negative flood-control benefit might result. To counteract this loss of storage in those cases where the change would be undesirable, provision should be made to permit a rise in headwater elevation during floods. This may be done by reserving flood storage below the normal maximum level or by permitting storage above that level. Since spillway designs, freeboard safety, and flowage rights usually provide for stages higher than gate-top levels, use of induced storage can be made within gate-design limits to avoid negative flood-control benefits. Induced surcharge is accomplished by partial openings of spillway gates to force storage to higher levels. The maximum levels of induced surcharge are usually limited by gate design. Use of induced surcharge storage is discussed in the U.S. Army Corps of Engineers Manual EM-1110-2-3600 (see References). An example of an emergency schedule with provision for induced surcharge is illustrated in Fig. 25-III-22 for the Wolf Creek project, Kentucky.

F. Reservoir Routing

The method used to route water through a reservoir will depend on whether the reservoir is level or whether it has substantial slope, as under natural river conditions, and on the operation of control gates in the dam. If the reservoir is level, as is usual with deep reservoirs on relatively steep streams, and if there is no change in the position of control gates in the dam, the outflow is a function only of the storage volume during the period of routing, and routing is a simple procedure. If the reservoir has substantial slope or if the gates are operated during the routing period, then the routing process becomes complex (see also Sec. 25-II).

The use of reservoir volume curves for flood routing which give the total volume between the profile as defined by inflow, outflow, and headwater elevation and the former natural low-water profile (Figs. 25-III-4 and 25-III-6), together with an estimated inflow, assures full accounting of the fact that reservoirs occupy some of the channel and overflow area formerly filled by a natural flood. Under natural conditions the valley storage is being filled without artificial control from the beginning of the rise up to the peak. Under reservoir conditions, as the flood increases, the headwater at the dam can be held down by increasing the discharge over the spillway, but when the peak flow arrives, the headwater elevation is allowed to rise by regulating the outflow. Thus, even though the total storage volume filled from the beginning of the rise to the peak may be less under reservoir conditions than was filled naturally, the effective storage for peak reduction is greater with the reservoir.

The fundamental relation between inflow, outflow (discharge), and storage is almost universally used in reservoir routing. This simple relation, stated briefly, is: over a given period of time, inflow minus outflow is equal to the change in storage. In equation form this may be written

$$\frac{I_1 + I_2}{2} t - \frac{O_1 + O_2}{2} t = S_2 - S_1 \qquad (25\text{-III-}1)$$

where I is the inflow rate, O is the outflow rate, S is the volume in storage, t is the length of the time interval, and the subscripts 1 and 2 refer to the beginning and end of the time interval, respectively. If the unit of flow is cubic feet per second (cfs) and

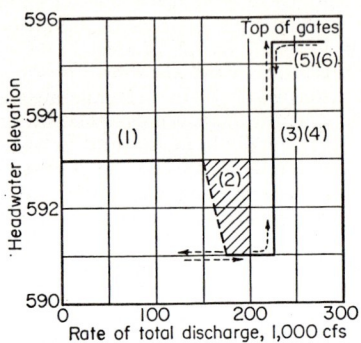

FIG. 25-III-12. Fixed rule for operation during floods (Guntersville Reservoir, TVA). Normal flood-season level is El 593. In case of flood, follow instructions 1, 2, and 3. Should headwater already be above El 593, lower to that level immediately. In emptying the reservoir after a flood crest, follow appropriate instruction 4, 5, or 6. Consideration must also be given to flood conditions on the lower Tennessee River.
 1. Hold El 593 until discharge equals 150,000 cfs.
 2. Lower headwater to El 591 by increasing discharge to as much as 200,000 cfs. If after reaching El 591 the flood does not develop, return headwater to normal. Otherwise hold El 591 until discharge increases to 225,000 cfs.
 3. Hold 225,000 cfs until gate-top level is reached or until headwater starts to fall.
 4. For floods in which the highest elevation reached is below gate-top level, lower headwater to normal by continuing discharge of 225,000 cfs.
 5. For floods in which gate-top level is reached, hold that elevation until discharge recedes to 225,000 cfs. Then lower headwater to normal by continuing 225,000 cfs.
 6. For floods in which the spillway capacity at gate-top level is exceeded and the headwater rises above that level, continue discharging at capacity until headwater returns to gate-top level. Hold that elevation until discharge recedes to 225,000 cfs. Then lower headwater to normal by continuing 225,000 cfs.

the time period is 1 day, the two left-hand terms give the total inflow and outflow in day-second-feet (dsf), a common unit of storage equal to 1.983, or nearly 2 acre-ft. Clearing the equation of fractions, it becomes $I_1 + I_2 - O_1 - O_2 = 2(S_2 - S_1)$, with inflow and outflow in cfs units and storage in dsf units. Transposing so that the known terms are on the left, the equation may be written

$$I_1 + I_2 + 2S_1 - O_1 = O_2 + 2S_2 \qquad (25\text{-III-}2)$$

Appropriate changes must be made if other time periods are used.
The routing process requires a knowledge of the following: (1) total inflow, which may be the sum of several components such as the discharge at a dam at the upper end of the reservoir, the discharge of a contributing river within the reservoir, and the discharge of all small creeks entering the reservoir (local inflow); and (2) the relationship between storage and other variables such as inflow, outflow, and headwater

elevation. With these factors known, an outflow hydrograph can be obtained by the repetitive solution of the storage equation.

When applying Eq. (25-III-2), the outflow at the beginning of the first period, O_1, must be estimated unless it is specified by initial conditions. For any initial storage and with the inflow known, the equation can be solved for $O_2 + 2S_2$ for the first period. From a known relationship between the sum of $O + 2S$ for the full range of their values, O_2 can be determined. By subtraction, $2S_2$ can be determined. Because O_2 and $2S_2$ are equivalent to O_1 and $2S_1$ of the next period, the whole process may be repeated to give succeeding outflows.

1. Level Storage. An example of preparing the discharge-storage relationship for a routing period of 1 day for the simple case of an uncontrolled spillway and level-pool reservoir follows. The routing curve for a level reservoir is prepared from the elevation-storage curve and from the elevation-outflow relationship, assuming a free-overfall spillway, as shown in Fig. 25-III-12 and Table 25-III-1. For a range of elevations the appropriate outflows are added to twice the corresponding storage values. These sums are then plotted against the outflow to construct the routing curve shown in Fig. 25-III-12.

Table 25-III-1. Computation of Routing Curve for Level Reservoir

Elevation	Outflow O, cfs	Storage S, dsf	$2S$, dsf	$O + 2S$
950	0	189	378	378
970	0	358	716	716
971	1	368	736	737
973	7	388	776	783
975	15	409	818	833
980	46	465	930	976
985	88	524	1,048	1,136
990	140	588	1,176	1,316
1,000	270	732	1,464	1,734

In the convenient form of Table 25-III-2 for repetitive calculation, the initial reservoir elevation was specified to be 965.0. The initial storage was then read from the elevation-storage curve and doubled to give the initial $2S_1$. The initial outflow O_1 was read from the outflow-elevation curve.

Table 25-III-2. Calculation of Outflow Using Level Storage

Items	Period, days								
	1	2	3	4	5	6	7	8	9
I_1	5	25	102	190	197	155	75	30	10
I_2	25	102	190	197	155	75	30	10	
$2S_1$	620	650	768	987	1,172	1,223	1,157	1,045	950
Sum	650	777	1,060	1,374	1,524	1,453	1,262	1,085	
O_1	0	0	9	64	138	163	133	84	51
$O_2 + 2S_2$	650	777	1,051	1,310	1,386	1,290	1,129	1,001	
HW, El$_1$	965.0	966.8	972.7	982.3	990.0	991.9	989.1	984.7	980.8

Inflow at the beginning of each period, I_1, was specified in tabular or plotted form (Fig. 25-III-12). I_2 was tabulated by shifting I_1 one period earlier. To begin routing, add I_1, I_2, and $2S_1$ of the first period and subtract O_1, giving $O_2 + 2S_2$ of period 1. Enter the routing curve and read O_2 as zero. Enter this as O_1 of period 2, subtract it from $O_2 + 2S_2$ of period 1, and enter this difference as $2S_1$ of period 2. Repeat the entire process of $I_1 + I_2 + 2S_1 - O_1 = O_2 + 2S_2$ for each succeeding period. Elevations are determined by entering the elevation-storage curve with one-half each $2S_1$ value and reading the corresponding elevation. Computed outflows and elevations are shown in Fig. 25-III-12.

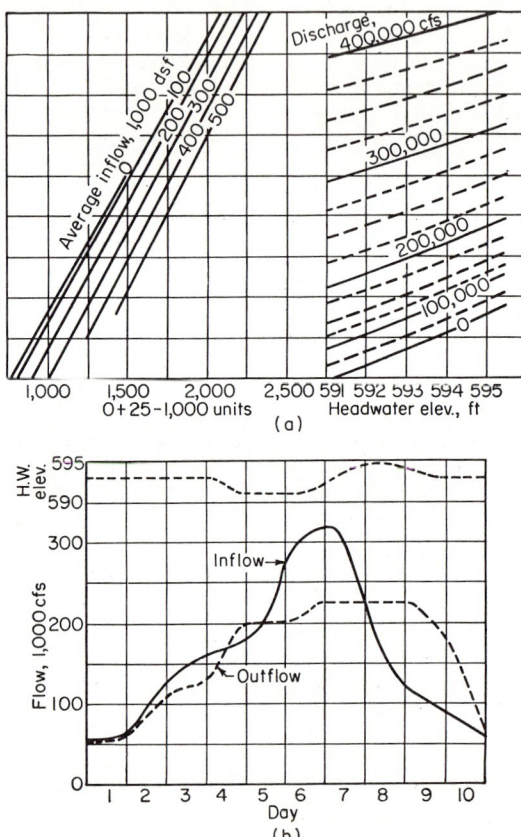

FIG. 25-III-13. Sloped storage routing curves and example of routing (Guntersville Reservoir, TVA). (a) Routing curve 1-day time period; (b) hydrographs.

2. Slope Storage. Reservoir routing using slope storage involves the use of a fourth variable, headwater elevation, in addition to those of inflow, outflow, and storage. An example of such a routing using a fixed rule of operation is shown below. The rule being followed is given in Fig. 25-III-13. The four-variable routing curve is shown in Fig. 25-III-14.

The routing curve is entered with average inflow and $O_2 + 2S_2$ to determine a vertical location on the diagram. Then a horizontal guideline through this location is followed until it intersects with either the desired headwater elevation to determine discharge or with the desired discharge to determine headwater elevation.

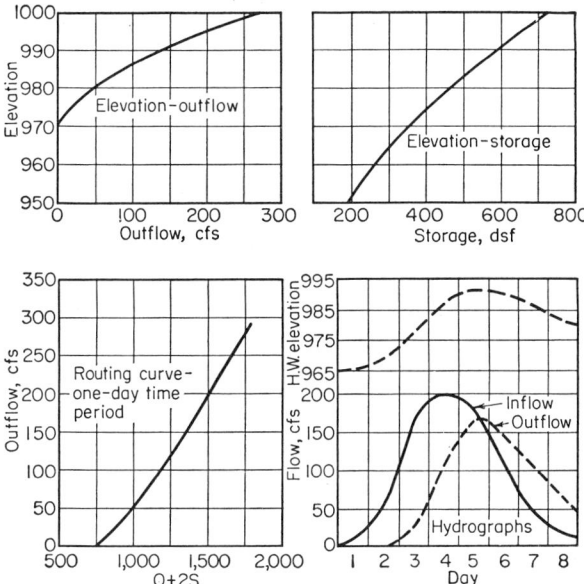

Fig. 25-III-14. Level-reservoir routing.

Table 25-III-3. Calculation of Outflow or Headwater Elevation Using Slope Storage and Fixed Operating Rule
(All flow and storage units in 1,000's)

Items	Period, days										
	1	2	3	4	5	6	7	8	9	10	11
Mean I, cfs........	60	95	142	170	230	300	270	170	105	75	
I_1, cfs.............	55	65	125	160	180	280	320	220	120	90	60
I_2, cfs.............		125	160	180	280	320	220	120	90	60	
$2S_1$, dsf...........		898	919	966	980	1,041	1,214	1,304	1,194	998	901
Sum...............		1,088	1,204	1,306	1,440	1,641	1,754	1,644	1,404	1,148	
O_1, cfs.............		60	109	129	197	202	225	225	225	181	66
$O_2 + 2S_2$..........	958	1,028	1,095	1,177	1,243	1,439	1,529	1,419	1,179	967	
HW, El_1..........	593.0	593.0	593.0	593.0	591.0	591.0	592.4	594.7	594.2	593.0	593.0

In detail the day-by-day description of the fixed-rule operation follows:

Period 1. This corresponds to initial conditions.

Period 2. To begin, the reservoir was assumed to be at normal level, elevation 593.0. The discharge O_1 on period 2 was made equal to the average inflow of the previous day, 60,000 cfs. The routing curve was entered with these three known conditions to determine $O_2 + 2S_2$ of day 1 (= $O_1 + 2S_1$ of day 2). $2S_1$ of day 2 was computed by subtracting the initial discharge. $O_2 + 2S_2$ for day 2 was then computed.

Periods 3 and 4. Headwater elevation was continued at 593.0, according to the rule, and the necessary discharge was computed.

Period 5. The discharge would have had to exceed 150,000 cfs to maintain the headwater elevation at 593.0. The fixed rule dictated an increase of discharge to as much as 200,000 cfs. A discharge of 197,000 cfs caused the headwater to fall to the minimum level, elevation 591.0.

Period 6. Minimum headwater elevation 591.0 was maintained. This required a discharge of 202,000 cfs.

Periods 7 and 8. Discharge was limited to 225,000 cfs. The headwater rose to a peak elevation of 594.7.

Period 9. Discharge was held at 225,000 cfs to begin to reduce the headwater elevation to normal.

Periods 10 and 11. Normal headwater elevation 593.0 was reached and held, resulting in discharges of 181,000 and 66,000 cfs.

Computed outflows and elevations are shown on Fig. 25-III-14.

3. Hydraulic Routing. If the correct relationship between inflow, outflow, and storage can be determined reasonably accurately and defined by practical curves, then the foregoing method of reservoir routing involving slope storage will give satisfactory results for computing discharge at a dam and headwater and tailwater elevations. The method makes no direct use of the basic hydraulics of unsteady flow in open channels. Flows and elevations within a reservoir are not directly determined and must be estimated from flow profiles and rating curves.

Hydraulic routing is based on the solution of the basic differential equations for unsteady nonuniform flow in open channels. A strict solution of these equations for flood routing is extremely complicated and difficult. Many authors have made outstanding contributions to the solution of equations for unsteady flow, but until recent years the amount of manual numerical work required for solution was too time-consuming for practical purposes. With the development of high-speed electronic computing equipment suitable for carrying out the numerical calculations and with the development of appropriate numerical procedures, the solution of the relevant differential equations is now practical.

Although a knowledge of the theory of characteristics is necessary for a complete understanding of the method of finite differences described hereafter, the method can be used without this complete understanding. Using finite differences, the derivatives in the differential equations are replaced by difference quotients, and the approximate solutions are found by solving linear equations for values of the desired quantities, water-surface elevation, and flow velocity.

The basic differential equations governing nonsteady flow in open channels are

$$B\frac{\partial H}{\partial t} + \frac{\partial (AV)}{\partial x} = q \qquad (25\text{-III-}3)$$

$$\frac{\partial V}{\partial t} + V\frac{\partial V}{\partial x} + \frac{V}{A}q = gS_0 - gS_f - g\frac{\partial y}{\partial x} \qquad (25\text{-III-}4)$$

where B = width in ft of channel at water surface
 H = elevation in ft of water surface above a datum
 A = cross-sectional area in sq ft
 q = volume of local inflow in cu ft per unit length and time into channel over river banks and through tributaries
 V = velocity of flow in fps
 S_0 = slope of channel bottom
 S_f = friction slope
 g = acceleration due to gravity in ft/sec^2
 y = depth of flow in ft
 t = time in sec
 x = distance along channel in ft

Equation (25-III-3) is the continuity equation for unsteady flow in open channels and represents the law of conservation of mass. Equation (25-III-4) is the general dynamic equation for gradually varied unsteady flow and represents the law of conservation of energy.

Using finite differences or difference quotients in place of derivatives for a series of net points as shown in Fig. 25-III-15a and substituting in Eq. (25-III-3) gives

$$H_P = H_M + \frac{1}{B_M}\left[\frac{\Delta t}{\Delta x}(A_L V_L - A_R V_R) + \frac{2\,\Delta t\, q}{B_M}\right] \quad (25\text{-III-5})$$

Applying the same difference and substitution procedure as outlined above to Eq. (25-III-4) for the same net points as shown in Fig. 25-III-15a, using the Manning formula to define the friction slope, and solving for the velocity at point P produces

$$V_P = V_M + \frac{\Delta t}{\Delta x}\left[\frac{V_L^2 - V_R^2}{2} + g(H_L - H_R)\right] - \left[\frac{2\,\Delta t\, q_{LR} V_M}{A_M} + \frac{2\,\Delta t\, g V_M |V_M|}{(1.49/n)^2 R^{4/3}}\right] \quad (25\text{-III-6})$$

where $|V_M|$ is the absolute value of V_M disregarding sign, n is the Manning roughness coefficient, and R is the hydraulic radius in ft.

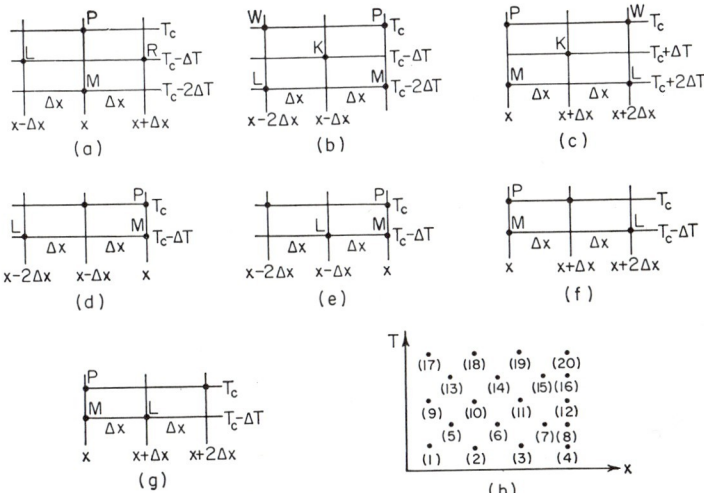

Fig. 25-III-15. Net points used in finite-difference schemes.

Assuming that appropriate quantities at points L, R, and M are known or can be determined, the solutions of Eqs. (25-III-5) and (25-III-6) give the water-surface elevation and flow velocity at point P. It is obvious from Fig. 25-III-15a that these two equations can be used only if values at $x - \Delta x$ and $x + \Delta x$ are available. In other words, values to the left and right of line x must be known before values at point P can be computed. Because of this limitation, these equations are used to compute only *interior points*. Should line x happen to be a computational boundary such as a dam and values along line $x - \Delta x$ not be appropriate because of being outside the boundary, then different equations must be used. These different equations which follow are used to compute quantities at all boundary points.

Boundary points are locations where certain conditions are prescribed. For flood routing in a reservoir, typical prescribed boundary conditions could be the change in headwater elevation with respect to time or a change in discharge with respect to time. Prescribed conditions at the upstream limits of the reservoir might be release or discharge from another dam or desired tailwater elevation changes. A boundary condition must be prescribed at each end of the routing reach.

Considering first boundary points where the discharge Q is prescribed, applying finite differences to a series of net points as shown in Fig. 25-III-15b, and substituting in Eq. (25-III-15a) for *right boundary Q prescribed* give

$$H_P = H_M + H_L - H_W + \frac{1}{B_K}\left[4\,\Delta t\,q_K + \frac{\Delta t}{\Delta x}(A_W V_W + A_L V_L - Q_P - Q_M)\right] \quad (25\text{-III-7})$$

Assuming $Q = VA$ at any point, then

$$V_P = \frac{Q_P}{A_P} \quad (25\text{-III-8})$$

For the left boundary Q prescribed as represented in Fig. 25-III-15c, using the same procedure as was used with the right boundary gives for *left boundary Q prescribed*

$$H_P = H_M + H_L - H_W + \frac{1}{B_K}\left[4\,\Delta t\,q_K - \frac{\Delta t}{\Delta x}(A_W V_W + A_L V_L - Q_P - Q_M)\right] \quad (25\text{-III-9})$$

and

$$V_P = \frac{Q_P}{A_P} \quad (25\text{-III-10})$$

At boundary points where the elevation H is prescribed, computational procedure is different from interior points and Q prescribed boundaries in that, for H prescribed, a value of velocity V must be computed at each Δt interval along the x line. A review of Fig. 25-III-15a to c will indicate that values for these conditions are computed along any x line at $2\,\Delta t$ intervals. Because of this unusual condition when H is prescribed, two equations for V are needed if the right boundary H is prescribed, and similarly two equations for V are needed if the left boundary H is prescribed.

For the boundary condition as shown in Fig. 25-III-15d, which is an even-to-odd-line calculation and with H prescribed, solution of Eq. (25-III-4) for V gives for the *right boundary even-to-odd line, with H prescribed condition,*

$$V_P = V_M + \beta_M B_M (H_M - H_P) + \frac{1}{2}\left\{\left[\frac{V_L^2 - V_M^2}{2} + g(H_L - H_M) + \beta_M(Q_L - Q_M)\right]\right.$$
$$\left. \cdot \frac{\Delta t}{\Delta x} - \frac{2\,\Delta t\,gV_M|V_M|}{(1.49/n)^2 R^{4/3}} + q_{LM}\left(\beta_M - \frac{V_M}{A_M}\right)\right\} \quad (25\text{-III-11})$$

where $\beta_M = \sqrt{g/A_M B_M}$.

For the *right-boundary odd-to-even line with H prescribed condition,* as shown by Fig. 25-III-15e, use

Eq. (25-III-11) with $\frac{\Delta t}{\Delta x}$ replaced by $2\frac{\Delta t}{\Delta x}$ \quad (25-III-12)

At the *left-boundary even-to-odd line with H prescribed condition,* as shown by Fig. 25-III-15f, use

Eq. (25-III-11) with β_M replaced by $-\beta_M$ and $\frac{\Delta t}{\Delta x}$ replaced by $-\frac{\Delta t}{\Delta x}$ \quad (25-III-13)

Finally, at the *left-boundary odd-to-even line with H prescribed condition,* as shown by Fig. 25-III-15g, use

Eq. (25-III-11) with β_M replaced by $-\beta_M$ and $\frac{\Delta t}{\Delta x}$ replaced by $-2\frac{\Delta t}{\Delta x}$ \quad (25-III-14)

Summarizing, *Eqs. (25-III-5) and (25-III-6) are used for all interior points, while Eqs. (25-III-7) to (25-III-14) are used for boundary conditions.* The proper boundary equation depends on the boundary quantity prescribed.

OPERATING SCHEDULES AND GUIDES 25–85

Because of the characteristics of the two basic differential equations used to derive Eqs. (25-III-5) to (25-III-14), there exists a maximum permissible $\Delta t/\Delta x$ ratio which can be used to advance the solution from time t to $t + \Delta t$. This maximum permissible ratio is fixed by the inequality

$$\frac{\Delta t}{\Delta x} \leq \frac{1}{V + c} \qquad (25\text{-III-}15)$$

where V is the velocity, and c is the propagation velocity of small wavelets given by the formula

$$c = \sqrt{gy_m} \qquad (25\text{-III-}16)$$

with y_m as the mean depth of the river. For example, the maximum $\Delta t/\Delta x$ ratio would be 0.024 sec/ft for a depth y_m of 40 ft and a maximum value of V of 5 fps.

Equation (25-III-15) gives only the ratio of Δt to Δx. Obviously, an infinite number of Δt's and Δx's exist which will satisfy the equation; therefore Δt and Δx must be set by other means. Generally, the accuracy of results is improved by use of smaller Δt's and shorter Δx's, but this increases the computational time required for routing. Desired accuracy must be weighted against computational time.

In order to begin the computational procedure, it is best to start with steady or nearly steady conditions. For these conditions the discharge Q and elevation H must be available or assumed for all net points along two consecutive t lines. As an example, in Fig. 25-III-15h, values of Q and H must be obtainable at numbered net points 1 to 8 before numerical solution of the equation can begin. With Q and H values at these net points and assuming $V = Q/A$, values of V can be computed at points 1 to 8. The method for determining A, B, R, n, and q will be discussed later

In order to explain the computational procedure, assume that the right boundary H is prescribed and the left boundary Q is prescribed. Based on values at points 2, 5, and 6, values for H and V at point 10 can be computed using Eqs. (25-III-5) and (25-III-6). In like manner, values for H and V at point 11 can be computed based on values at 3, 6, and 7. This procedure can be continued for an indefinite number of interior points along the x axis for a particular t line.

At the left boundary where Q is prescribed in Fig. 25-III-15h and using values at points 1, 2, 5, and 10, a value of H at point 9 can be computed using Eq. (25-III-9). The quantity V at point 9 can be computed by dividing the prescribed Q by the area at point 9.

At the right boundary where H is prescribed and based on values at points 7 and 8, a value for V at point 12 can be computed using the odd-to-even-line equation (25-III-12). Then, using the even-to-odd-line equation (25-III-11) and values at points 11 and 12, a value for V at point 16 can be computed.

Following the same repetitive procedure as described above, values of H and V at points on successive time t lines can be computed.

For any given river it is necessary to have data available capable of yielding the cross-sectional area A, hydraulic radius R, channel width B, and roughness coefficient n. The first three quantities are geometrical and can be determined from topographic maps or field surveys. Cross sections to be used for determining the area and hydraulic radius should be selected so as not to include ineffective or slack-water regions. The channel width, in turn, must include slack regions because it is used in reality to determine changes in storage. The product of the width, Δx, and ΔH, gives changes in storage for the conservation-of-mass equation. The roughness coefficient can be best determined from a flow profile for steady conditions. Local inflow can best be determined from stream gages on tributaries and from rainfall, using unit hydrographs.

V. OPERATING SCHEDULES AND GUIDES

Schedules and guides for reservoir operation should be developed in a preliminary form in the operation planning stage and used to determine in advance the most effective use of reservoir storage. Later refinements are usually necessary, based on further study and on actual operating experience. Schedules may vary from rigid or fixed rules to be followed during floods by nontechnical operators at dams to general seasonal guides and long-range plans for the storage and release of water for conservation purposes. They may be in the form of graphs, tabulations, or narrative or a combination of the three.

A. Rigid Schedules

The most rigid schedules are those built into the physical structures of single-purpose, ungated flood-control projects, such as most U.S. Soil Conservation Service projects and the detention reservoirs of the Miami Conservancy District (Fig. 25-III-16). An example of the results of such automatic operation is shown for the Miami projects in Figs. 25-III-17 and 25-III-18.

Rigid schedules for flood-control operation may be necessary for use by the non-technical operating personnel at gated structures in case communication with the operating center fails. Such schedules can also serve as guides to operating-center personnel during extreme floods, particularly if communication with the reporting hydrologic network is lost. Results of their use in regulating floods of record and maximum probable or project floods are known from previous study. Schedules are usually based on some combination of reservoir elevation, stage at a downstream control point, and reservoir inflow or rate of change in reservoir elevation. Figures

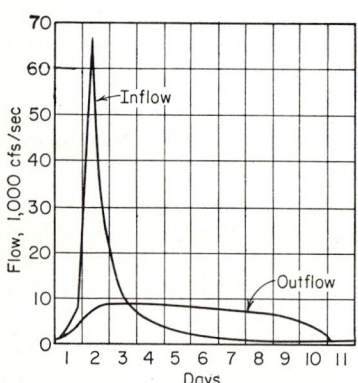

Fig. 25-III-17. Regulating effect of Germantown detention reservoir on flood like that of March, 1913.

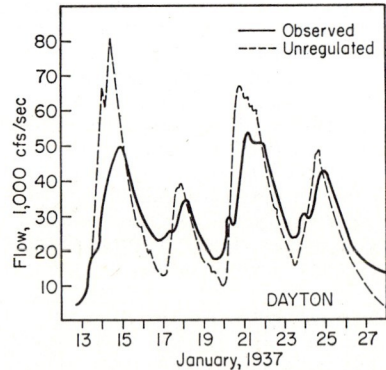

Fig. 25-III-18. Effect of the four upstream detention reservoirs on flows at Dayton during the flood of 1937, which was the highest since that of 1913.

Fig. 25-III-16. Map of Miami River showing location of detention reservoirs and cities and towns where river improvement was provided.

25-III-22 and 25-III-23 for the Wolf Creek project illustrate one type of such schedule and the results of its use in regulating a project flood. A simpler type of fixed-rule operation, based on headwater elevation only, is shown in Fig. 25-III-13.

B. Semirigid Schedules

The day-to-day operation of most gated reservoirs and reservoir systems is based on current forecasts of streamflow, with such adjustments as may be prudent, based on the current precipitation outlook. Such operation can be more effective in an individual flood event. This does not mean that maximum or ideal regulation should be attempted for each flood. Weather forecasting is not sufficiently advanced to ensure that additional flood-producing rains will not develop in many cases during the predicted regulation period. In other cases, the weather outlook may be definite enough so that the entire hydrograph of the flood can be forecast with assurance in advance. Examples of such schedules cannot be given because they involve day-to-day decisions based on judgment but supported by the knowledge gained from studies of past floods. Should conditions become more serious and the flood develop into a maximum probable flood, operations can be shifted to the rigid schedules whose effect on extreme floods is known from previous planning study.

C. Long-range-planning Schedules

Long-range-planning schedules apply principally to the use of water for conservation purposes and to reservoirs and systems where storage is large compared with annual streamflow. Where dual use of storage is made for seasonal flood-control and conservation purposes, schedules should contain rigid safeguards based on current conditions. This is illustrated by the variations in filling in the planned operation of the Missouri River Main Stem Reservoir System, illustrated in Fig. 25-III-29. In this case current filling is dependent upon existing snow cover and its runoff potential, either for producing later flood events or supplying water for later conservation use. Long-range planning and scheduling involve a distribution of the storage and use of water against the long-term pattern of streamflow. In a given year an equitable distribution must be scheduled for such diverse uses as irrigation, navigation releases, power generation, and storage changes, either plus or minus, to best carry out the long-range potential of the system. Excess flows in a given year should not be wasted, but should be stored for later use to augment the flows during the deficient streamflow years which the long-term historical record indicates will occur.

D. Examples of Reservoir Operation

The following examples of reservoir operation are for (1) a single-purpose flood-control system; (2) a multiple-purpose flood-control and power system; (3) a multiple-purpose navigation, flood-control, and power system; and (4) a multiple-purpose flood-control, irrigation, sanitation-and-water-supply, navigation, and power system.

FLOOD CONTROL: MIAMI FLOOD–CONTROL PROJECT— MIAMI CONSERVANCY DISTRICT—MIAMI RIVER AND TRIBUTARIES, OHIO

The Miami project shown in Fig. 25-III-16 consists of five detention reservoirs, channel improvements, and levees to provide protection to towns and cities on the Miami River from Piqua to Hamilton. It is designed to provide protection against all anticipated floods, allowing for a flood 40 per cent greater than the disastrous one of March, 1913. An emergency spillway at each detention reservoir provides structure safety should a greater flood occur. The reservoirs are empty except during floods, and normal flows are passed through uncontrolled conduits acting as open channels. During floods, these conduits act as orifices designed in connection with downstream river improvements to limit discharges to safe channel capacities. The effect of the system on major Ohio River floods is dependent upon the overall distribution of flood flows. Figure 25-III-17 shows the regulating effect at one of the reservoirs (Germantown), and Fig. 25-III-18 shows the effect at Dayton of the four upstream reservoirs.

Drainage areas:
  Dayton —2,598 sq mi
  Hamilton—3,670 sq mi
Controlled by detention basins:
  Above Piqua   —   255 sq mi
  Above Dayton   —2,455 sq mi
  Above Hamilton —2,725 sq mi
Official-plan flood:
  9.5 to 10.0 in. runoff in 3 days

FLOOD CONTROL AND POWER: WOLF CREEK PROJECT— U.S. ARMY CORPS OF ENGINEERS— CUMBERLAND RIVER, KENTUCKY

Wolf Creek (Fig. 25-III-19) is one of three multiple-purpose flood-control and power reservoirs on the upper Cumberland River and tributaries, constructed and operated by the U.S. Army Corps of Engineers, Nashville, Tenn., District. The

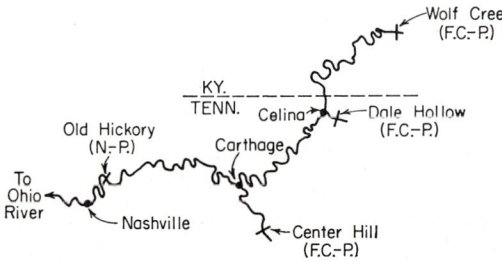

Fig. 25-III-19. Map of the Cumberland River. (F.C. = flood control, N. = navigation, P. = power.)

flood-control objective is to minimize downstream agricultural and urban damage. Operation is coordinated with that at the other two flood-control–power projects, Dale Hollow and Center Hill. Power operation at these projects and at the downstream navigation and power projects, Old Hickory and Cheatham, is integrated with, and power distributed through, the adjoining TVA system. Statistics on the Wolf Creek project are as follows:

Drainage area: 5,789 sq mi
Average annual runoff: 22 in.
Storage space:
  Power—El 673–723, 6.94 in. runoff
  Flood control—El 723–760, 6.78 in. runoff

Operation for power is seasonal in character. It is estimated that one year in three the headwater level would be above or below the normal operation zone shown on the annual guide (Fig. 25-III-20), depending upon streamflow, power demand, and combined available water in TVA and Cumberland Basin reservoirs.

Storage reservation for flood control is constant throughout the year, but additional space might be available because of previous drawdown for power. Under normal flood-control operation, as shown in Fig. 25-III-21, releases are limited, in conjunction with those from Dale Hollow and Center Hill, to avoid exceeding the following non-damaging rates downstream.

	Noncrop season, Dec. 15–Apr. 15, cfs	Crop season, Apr. 15–Dec. 15, cfs
Celina	40,000	30,000
Carthage	72,000	45,000
Nashville	100,000	54,000

Power releases are cut off if necessary. Studies show that floods of record can be controlled to these rates except when downstream uncontrolled local runoff exceeds them.

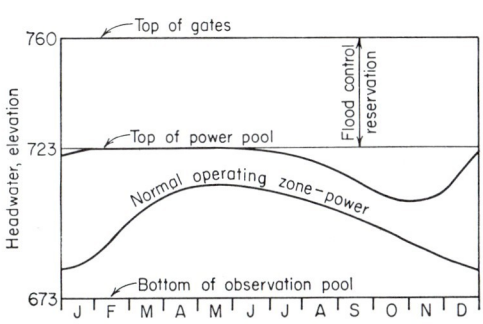

Fig. 25-III-20. Annual operating guide (Wolf Creek Reservoir, Cumberland River).

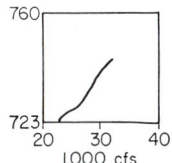

Fig. 25-III-21. Normal flood-control release schedule unless downstream control rates require smaller outflow (Wolf Creek Reservoir, Cumberland River).

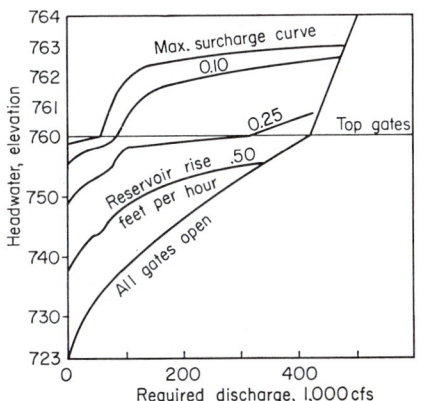

Fig. 25-III-22. Emergency operation schedule (Wolf Creek Reservoir, Cumberland River).

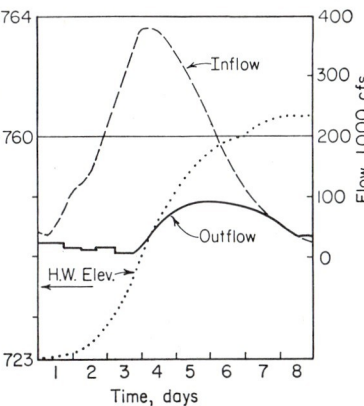

Fig. 25-III-23. Operation in standard project flood (Wolf Creek Reservoir, Cumberland River).

If forecasts of inflow show that limiting to normal releases will fill the reservoir to above the gate-top level or if communications are interrupted, special operation is scheduled or the emergency schedule is followed. The emergency schedule provides for an induced surcharge of as much as 3 ft above normal gate-top level, if necessary (limited by elevation of gate mechanism). Under the emergency schedule, as shown in Fig. 25-III-22 (top left), the standard project flood would be reduced from a maxi-

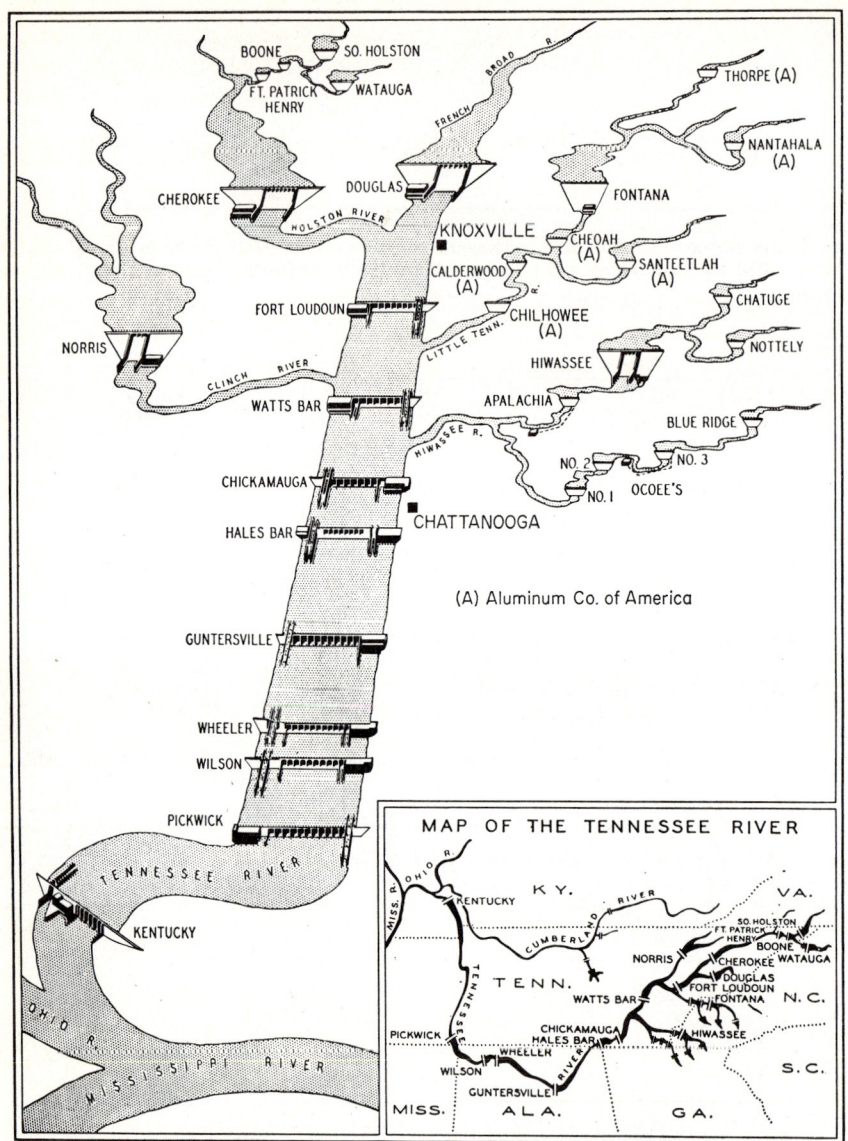

Fig. 25-III-24. TVA water-control system.

mum inflow rate of 380,000 cfs to a maximum outflow rate of 90,000 cfs, resulting in an induced surcharge of about 0.7 ft (Fig. 25-III-23).

Special provision is made for limiting flood-control releases which would arrive on flood crests on the Ohio and Mississippi Rivers. Completion of Barkley Dam near the mouth of the Cumberland River will provide substantial additional regulation for lower Ohio and Mississippi River floods.

NAVIGATION, FLOOD CONTROL, POWER: TVA SYSTEM—TENNESSEE VALLEY AUTHORITY, TENNESSEE RIVER, AND TRIBUTARIES

The TVA water-control system in the Tennessee Valley, shown in Fig. 25-III-24, consists of 31 major projects—19 multiple-purpose and 12 single-purpose power projects. Nine of the multiple-purpose projects are on the main Tennessee River, and 10 are on the major tributary streams above Chattanooga. Six of the single-purpose power projects are owned by the Aluminum Company of America but are operated by TVA under agreement. Navigation and flood-control purposes have priority over power.

The 9 multiple-purpose projects on the main stream provide the 650-mile navigation channel from the Ohio River to Knoxville. Regulation for navigation on the Tennessee River is accomplished by observing minimum drawdown levels of main-stream projects. Withdrawal of storage from the reservoir system in the summer and fall increases navigation depths on the lower Ohio and Mississippi Rivers, reducing the amount of annual dredging.

Drainage area:
  Chattanooga—21,400 sq mi
  Kentucky Dam—40,200 sq mi
Flood-control storage reservation (Jan. 1)—11,800,000 acre-feet = 5.5 in. runoff
Average annual rainfall—52 in.
Average annual runoff—22 in.

Regulation for power is seasonal, utilizing much of the reserved flood-control storage space after the major valley-wide flood season is over. Power operation of the 31 hydroelectric plants is integrated with 12 large TVA steam plants and 5 Corps of Engineers and 1 TVA hydro plant in the adjacent Cumberland River Basin.

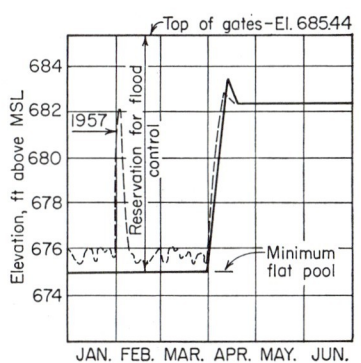

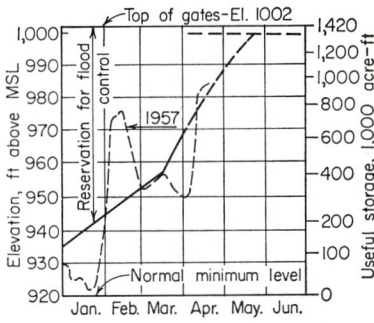

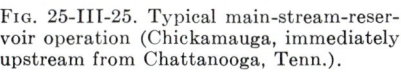

FIG. 25-III-25. Typical main-stream-reservoir operation (Chickamauga, immediately upstream from Chattanooga, Tenn.).

FIG. 25-III-26. Typical tributary-storage-reservoir operation (Douglas on the French Broad River, Tennessee).

Flood-control objective is the reduction of flood damage within the valley and along the lower Ohio–Mississippi Rivers. The greatest flood-damage potential within the valley is at Chattanooga, Tenn., where levees are needed to provide complete protection. Flood damage on the lower Ohio–Mississippi Rivers is to lands not protected by the levee system. Major valley-wide floods are limited to the period from mid-December to early April (Fig. 25-III-8). Flood-control reservation, reflecting this seasonal pattern, is shown for two typical reservoirs in Figs. 25-III-10 and 25-III-11.

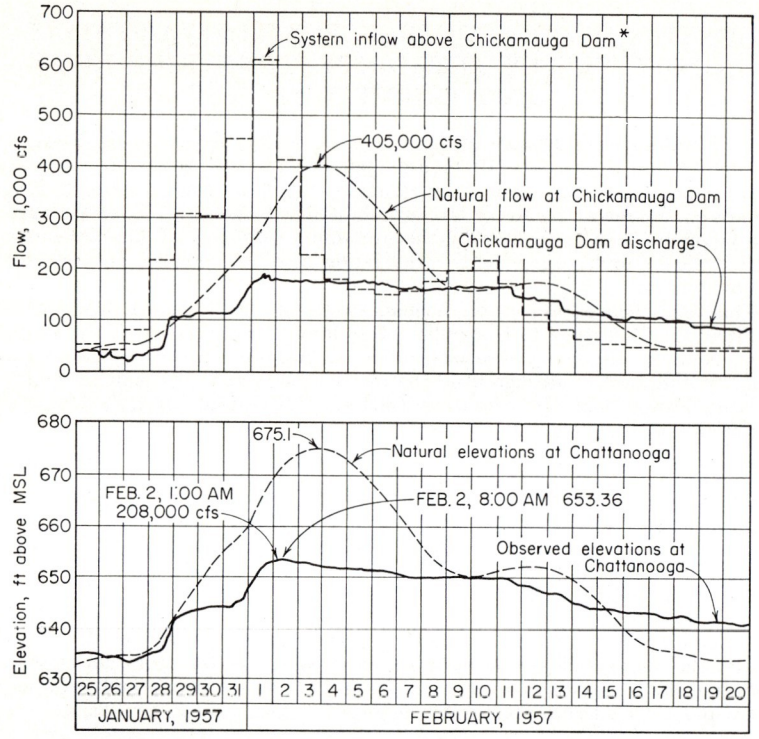

Fig. 25-III-27. Hydrographs at Chickamauga Dam and Chattanooga, Tenn. System inflow is summation of simultaneous inflows to all upstream reservoirs.

Flood stage at Chattanooga—El 651.12.
Estimated damage at observed crest of 653.36—20,000.
Estimated damage at natural crest of 675.1—would have been—66,000,000.

Operation for flood control within the valley is illustrated in Figs. 25-III-25 to 25-III-27, which show actual operation for the January-February, 1957, flood. Without regulation this would have been the second highest flood of record (since 1867) at Chattanooga, Tenn.

FLOOD CONTROL, IRRIGATION, SANITATION AND WATER SUPPLY, NAVIGATION, POWER: MISSOURI RIVER— MAIN STEM RESERVOIR SYSTEM— U.S. CORPS OF ENGINEERS

The Missouri River drains an area of about 530,000 sq mi, of which nearly 280,000 sq mi is controlled by the Main Stem Reservoir System. Although the river originates in the high elevations of the Rocky Mountains, the greatest portion of the drainage area is a flat or gently rolling plains region. Average annual streamflow at the mouth is about 57 million acre-feet, while at Sioux City, Iowa, it averages 24 million acre-feet. Flows at Sioux City have ranged from about 11 million to 37 million acre-feet. Principal sources of runoff above Sioux City are the plains-area snowmelt occurring in March or April and the mountain snowmelt augmented by rains during the May-to-July period. During the remainder of the year flows above this point are generally low. Below Sioux City flows originate primarily as a result of warm-season rainfall and occasional winter-snow accumulations.

Upon completion, the Missouri River Main Stem Reservoir System, shown in Fig. 25-III-28, will consist of six multiple-purpose projects constructed by the Corps of Engineers for the purpose of flood control and conservation uses. The six-reservoir

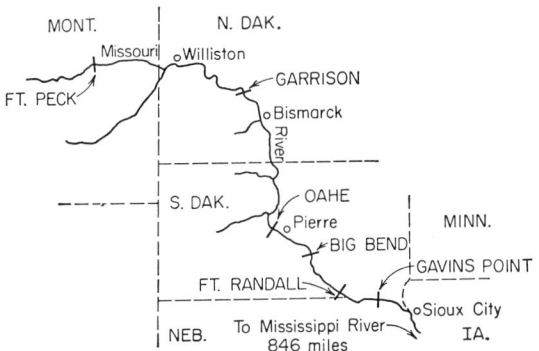

Fig. 25-III-28. Missouri River Main Stem Reservoir System.

system provides flexibility in distribution of storage for efficient operation, while the seasonal nature of flood events makes joint use of part of the flood-control storage space possible. The gross storage capacity of 76 million acre-feet is equivalent to over 3 years' average streamflow at Sioux City.

Tentative Initial Allocation of Storage Space (1,000 Acre-ft)

	Inactive	Conservation carryover, multiple-use	Annual flood control and multiple use	Exclusive flood control	Total
Fort Peck........	4,500	11,200	2,700	1,000	19,400
Garrison.........	4,900	13,600	4,400	1,600	24,500
Oahe*...........	5,500	13,800	3,200	1,100	23,600
Big Bend*.......	1,465	260	0	175	1,900
Fort Randall.....	1,400	2,400	1,400	900	6,100
Gavins Point.....	156	218	103	64	541
Total†.........	17,921	41,478	11,803	4,839	76,041

* Under construction 1960.
† In addition, as of 1959 there were some 59 tributary projects completed or under construction upstream from Gavins Point, with a total storage capacity of 8,430,000 acre-feet.

Principal conservation functions of the system are as described below:

1. *Irrigation.* It is contemplated that 1¾ million acres of land will be irrigated by direct diversions from the reservoir system. Additionally, a portion of the power revenues from these projects will be used to defray the expense of other irrigation facilities in the basin.

2. *Water supply and stream sanitation.* The reservoirs will provide a firm water supply to meet needs of downstream municipal water supply and of sanitation.

3. *Navigation.* Reservoir releases, in combination with downstream bank stabilization and river training works, will provide a 9-ft navigation channel from Sioux City to the mouth of the Missouri River for up to 8 months of each year.

4. *Hydroelectric power.* The main-stem power installations, when completed, will have an installed capacity of 2,048,000 kw. Annual energy generation will average about 9 billion kilowatthours. The power is marketed by the Bureau of Reclamation through its transmission network.

The use of water for these conservation purposes is based on annual plans prepared by the Reservoir Control Center of the Corps of Engineers in Omaha, Nebr., and approved by a coordinating committee consisting of representatives from the Federal agencies and Basin states involved. Daily regulation for these purposes, and for flood control, is directed by the Reservoir Control Center on the basis of current and forecast conditions.

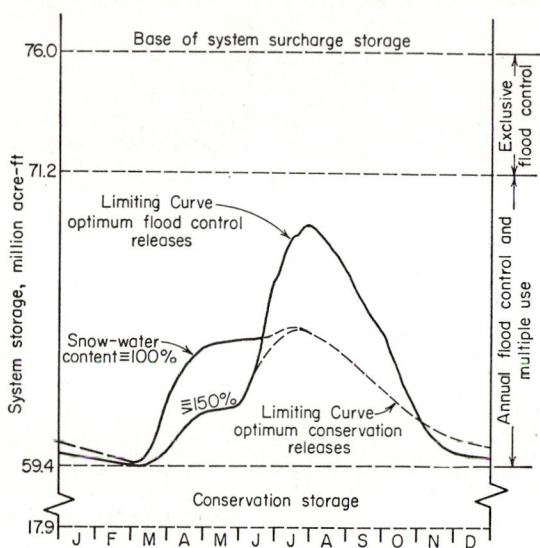

Fig. 25-III-29. Preliminary system operating curves (Missouri River Main Stem Reservoir System).

The Main Stem Reservoir System, in combination with existing and proposed downstream levees and tributary reservoirs, will afford a high degree of flood protection through the lower basin. The flood-control objective is to prevent system releases from contributing significantly to downstream flood flows. This is accomplished by release schedules based primarily on system storage and downstream hydrologic conditions. Studies indicate that all floods of record, as well as floods of standard-project magnitude, can be regulated to less than 100,000 cfs at Sioux City by utilization of the available flood-storage space.

Preliminary system operating curves which give the required drawdown and seasonal-fill limitations of the joint-use race during nonflood periods are shown in Fig. 25-III-29. Upstream snow-water content, as compared with normal, affects the amount of allowable early filling. With system storage at a level above the optimum flood-control release curve, releases at the maximum rate allowable by the existing downstream flood potential are necessary reasonably to ensure evacuation of the flood-control space prior to the next flood season. With system storage at a level below the optimum conservation release curve, a careful use of the available water for conservation purposes is necessary in order to prevent excessive drawdown into the carryover storage space. Storage levels between these two curves result in intermediate operations.

In addition to the regulating curves shown, other curves have been developed to guide usage of the exclusive flood-control space. The graph in the upper right of

Fig. 25-III-30 shows the use of system storage in regulating a synthetic project flood, which exceeds any flood of past record, by these techniques, while the graph to the right of Fig. 25-III-31 shows the resulting regulation at Sioux City for this flood. Emergency regulation curves have also been developed to provide for control of

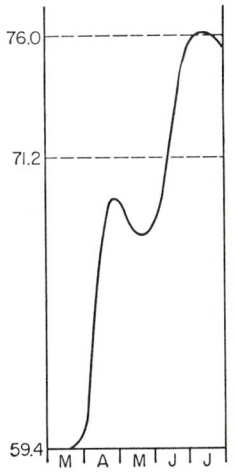
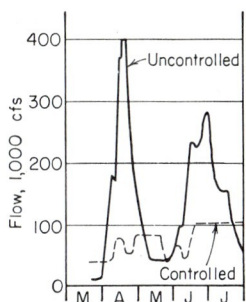

Fig. 25-III-30. System storage in synthetic project flood.

Fig. 25-III-31. Sioux City, Iowa, hydrographs of synthetic project flood.

floods approaching spillway-design magnitude through the use of surcharge storage space and spillway releases.

VI. OPERATING ORGANIZATION

Operating-organization requirements vary considerably with the purposes and complexity of the regulating system, and no fixed pattern can be established. Flood-control-detention basins are automatic in their operation. Gated flood-control reservoirs may need detailed attention only during flood periods. Normal conservation releases need only routine attention. Regulation for multiple uses, however, particularly where hydroelectric power is involved, may require detailed daily attention. The size of the staff may be determined by the current forecasts of flow needed for daily power operation rather than by the less frequent needs for flood control.

Most flood-control operating procedures are based on current forecasts of inflows to reservoirs, usually combined with forecasts of unregulated streamflow at downstream control points. Depending upon the dynamics of the river, decisions may be necessary on a 24-hr basis to vary upstream releases to provide nondamaging stages downstream. If storage is not adequate for complete protection, which is usually the case, competent personnel must be available to determine when regulation should be shifted from complete protection to that of minimizing damage, or in extreme floods, to ensure safety of the regulating structure.

Forecasts of streamflow are equally essential to the economical use of stored water for power, both on a daily and long-range basis. Increases in streamflow which would have no significance for flood control may be extremely significant for power operations. Unless power releases from upstream storage reservoirs are reduced in time, the combined flow may exceed turbine capacities at downstream run-of-river plants. Integration of a hydro power system with a thermal generating system increases the complexities.

The day-to-day operation of most gated reservoirs and reservoir systems is based on current forecasts of streamflow, with such judgment adjustments as may be prudent in view of the current precipitation outlook. Such operation can be more effective in an individual flood event. This does not mean that maximum or ideal regulation should be attempted for each flood. Weather forecasting is not sufficiently advanced to ensure that additional flood-producing rains will not develop in many cases during the predicted regulation period. In other cases the weather outlook may be definite enough so that at a given time the remainder of the flood hydrograph may be forecast with assurance in advance. Day-to-day judgments, however, must be backed up by a knowledge of the results of preplanned operation schedules for extreme floods.

In TVA's large multiple-purpose reservoir system of more than 30 reservoirs, the number of personnel required for daily water-control operations varies with streamflow conditions. Inasmuch as there is a seasonal pattern in the runoff characteristics of the Tennessee River Basin, more personnel are usually involved in the period from early winter to mid-spring than in the remaining portion of the year. The factor governing the number of personnel is the requirement that the hydrologic data must be received and processed into streamflow predictions and water-control operations must be established within a period of about 4 hr each morning. These operations involve prediction and routing of streamflow and determination of discharges and elevations at each of the dams and at other critical places, as well as special operations for other uses and issuance of bulletins and warnings. The forecasting assignments during the flood season can usually be handled by a staff of 13 engineers and aides. During the remainder of the year a staff of 11 can usually handle the work. The use of electronic computers is being increased in various phases of the detailed computations to expedite the work, thereby reducing the time from receipt of raw data to the final issuance of water-control operations.

VII. HYDROLOGIC NETWORK

To make streamflow forecasts, current reports must be available from an adequate network of rainfall and river-stage stations. Adequacy is dependent upon the character of the drainage basin and the critical need of the regulating system. For the operation of the more than 30 dams in the TVA system in the Tennessee River Basin, daily rainfall reports are received from 190 stations, a basin-wide average of one station per 200 sq mi. The average ranges from about one per 100 sq mi in the mountainous eastern third of the basin to one per 300 to 400 sq mi in the rolling terrain of the western third. Daily reports of river stages are received from 40 stations. During critical flood periods, more frequent reports of both precipitation and river stages may be obtained. Hourly reports from each of the major dams provide information on their inflow and releases.

Dependable communications are extremely important, both from the regulating center to the dams and from the hydrologic reporting network to the center. If telephone service is not adequate, it should be supplemented by radio.

VIII. REFERENCES

Carter, R. W., and R. G. Godfrey: Storage and flood routing, in "Manual of Hydrology," pt. 3, Flow-flow Techniques, Methods and Practices of the Geological Survey, *U.S. Geol. Surv. Water-Supply Paper* 1543-13, 1960.

Chow, V. T.: "Open-channel Hydraulics," McGraw-Hill Book Company, Inc., New York, 1959.

Churchill, M. A.: Effect of storage improvements on water quality, *Trans. Am. Soc. Civil Engrs.*, vol. 123, pp. 419–464, 1958.

Floods and flood control, Tennessee Valley Authority, Technical Report 26, 1961.

Hurst, H. E.: Long-term capacity of reservoirs, *Trans. Am. Soc. Civil Engrs.*, vol. 116, pp. 770–799, 1951.

Isaacson, E. J., J. J. Stoker, and B. A. Troesch: Numerical solution of flood prediction and river regulation problems, New York University, Institute of Mathematical Sciences, Reports 2 and 3, 1954 and 1956.

Johnson, W. E.: Missouri River basin plan in operation, *Trans. Am. Soc. Civil Engrs.*, vol. 122, pp. 654–665, 1957.

REFERENCES

Johnstone, Don, and W. P. Cross: "Elements of Applied Hydrology," The Ronald Press Company, New York, 1949.

Knappen, T. T., J. H. Stratton, and C. V. Davis: River regulation by reservoirs, sec. 1 in C. V. Davis (ed.), "Handbook of Applied Hydraulics," 2d ed., McGraw-Hill Book Company, Inc., New York, 1952, pp. 1–21.

Koelzer, V. A.: The use of statistics in reservoir operations, *Trans. Am. Soc. Civil Engrs.*, vol. 122, pp. 1187–1201, 1957.

Morgan, A. E.: "The Miami Conservancy District," McGraw-Hill Book Company, Inc., New York, 1951.

Multiple-purpose reservoirs: a symposium, *Trans. Am. Soc. Civil Engrs.*, vol. 115, pp. 789–892, 1950. This symposium consists of the following papers: Malcolm, Elliott: Their relation to flood control and navigation, pp. 792–796; Debler, E. B.: Development of policy by the Bureau of Reclamation, pp. 797–802; Bowden, N. W.: General problems of design and operation, pp. 803–817; Cochran, A. L.: Their use for flood control, pp. 818–824; Pafford, R. J.: Their use for navigation, pp. 825–832; Nelson, W. R.: Application of general policies when used for irrigation, pp. 833–843; Thomas, J. B.: Coordination with the electric utility industry, pp. 844–859; De Luccia, E. R.: Influence of Federal Power Commission on design and operation, pp. 860–865; Dieffenbach, Rudolph: Their relation to fish and wildlife, pp. 866–870; Wirth, C. L.: Planning for the recreational use of reservoirs, pp. 871–876; McBride, Don: State and Federal government participation, pp. 877–887; Hill, R. A.: Summary and review of principles, pp. 888–892.

Pafford, R. J.: Operation of Missouri River main-stem reservoirs, *Trans. Am. Soc. Civil Engrs.*, vol. 124, pp. 381–394, 1959.

Project operation studies, pt. 5 of Water Studies, vol. IV of U.S. Bureau of Reclamation Manual, 1950.

Reservoir regulation, U.S. Army Corps of Engineers Manual EM 1110-2-3600, "Engineering and Design," May 25, 1959.

Reservoirs, chap. 6 in "Flood Control," The Engineer School, Fort Belvoir, Va., 1940, pp. 181–224.

Riesbol, H. S.: Snow hydrology for multiple-purpose reservoirs, *Trans. Am. Soc. Civil Engrs.*, vol. 119, pp. 595–613, 1954.

Rutter, E. J.: Flood control operation of Tennessee Valley Authority reservoirs, *Trans. Am. Soc. Civil Engrs.*, vol. 116, pp. 671–707, 1951.

Stoker, J. J.: Water waves, *Pure Appl. Math.*, Interscience Publishers, Inc., New York, vol. 4, 1957.

Woodward, S. M.: Hydraulics of the Miami Flood Control Project, The Miami Conservancy District Technical Report, pt. VII, Dayton, Ohio, 1920.

Yevdjevich, V. M.: Flood routing methods, discussion and bibliography, *U.S. Geol. Surv* and *Soil Conserv. Serv. Project* 4, 1960.

Section 25-IV

HYDROLOGY OF FLOW CONTROL

PART IV. RIVER FORECASTING

T. J. NORDENSON, *Chief, Hydrologic Investigations Section, U.S. Weather Bureau.*

M. M. RICHARDS, *Special Assistant, Hydrologic Services Division, U.S. Weather Bureau.*

I. Introduction	25-99
A. Need for River Forecasts	25-99
B. General Problems in Connection with River Forecasting	25-99
II. Basic Data Requirements	25-99
A. Data for Procedure Development	25-99
B. Reporting Network for River Forecasting	25-101
III. Basic River-forecasting Procedures	25-101
A. Rainfall-Runoff Relations	25-102
B. Unit Hydrographs	25-103
C. Streamflow Routing	25-103
IV. River-forecasting Example	25-104
A. Computation of Runoff	25-104
B. Forecast for Headwater Point	25-106
C. Forecast for Downstream Point	25-107
D. Remarks	25-107
V. Complicating Factors	25-107
A. Areas Where Unit Hydrographs Are Inadequate	25-107
B. Routing	25-109
C. Snow	25-109
D. Use of Forecast Precipitation	25-109
E. Stage-Discharge Relations	25-109
F. Average Precipitation Where Topography Is a Factor	25-109
VI. Specialized Forecasts	25-110
A. Flash-flood Warning	25-110
B. Water-supply Forecasts	25-110
C. Low-flow Forecasts	25-110
VII. References	25-111

I. INTRODUCTION

The river-forecasting procedures described in this section are for the most part those used by the U.S. Weather Bureau, the Federal agency responsible for river and flood forecasting and the dissemination of public warnings in the United States [1, 2]. Other United States Federal agencies, such as the Bureau of Reclamation, Army Corps of Engineers, Tennessee Valley Authority, International Boundary and Water Commission, and Bonneville Power Administration, and private power companies also engage in river forecasting as required for reservoir operations. Forecasting techniques vary considerably, and there are many unusual problems that require special solutions. Only the basic river-forecasting principles and some of the more common operational problems will be discussed.

A. Need for River Forecasts

Water-control structures such as dams, levees, etc., offer a positive method of reducing or eliminating the damages caused by flooding. In numerous situations, however, topographic (lack of potential dam sites) and/or economic factors make the control of floods impractical or unjustifiable. In these situations river forecasting provides an alternative means of reducing flood damage and loss of life [1, 3]. Advance warning of an approaching flood permits evacuation of people, livestock, and equipment. The warning time available determines how much evacuation is possible. River forecasts are required for estimating inflow to reservoirs in order to permit the most efficient operation for flood control or other purposes. In addition, there is an increasing demand for day-to-day forecasts of river stages and discharges by those interested in navigation, water supply, stream pollution, and many other related fields.

B. General Problems in Connection with River Forecasting

The tools of the river forecaster include rainfall-runoff relations, unit hydrographs, routing methods, recession curves, and stage-discharge relations. Because of the importance of the time factor, great stress must be placed on the development of forecast procedures that will enable flood warnings to be issued at the earliest possible time. Methods for collection and handling of basic data, preparation of forecasts, and dissemination of these forecasts must be carefully organized in order to speed the operation and minimize human and mechanical errors. A warning received too late to permit evacuation of people and removal of property from the threatened area is of no value.

II. BASIC DATA REQUIREMENTS

Any forecasting service is dependent on adequate data. The development of the river-forecasting procedures requires historical hydrologic data, while the preparation of operational forecasts requires sufficient current information.

A. Data for Procedure Development

Available historical hydrologic data consist of streamflow records collected by the U.S. Geological Survey and river-stage data and daily and hourly rainfall values collected by the U.S. Weather Bureau. As a generalization, it can be said that it is necessary to have a minimum of 10 years of basic hydrologic data available in order to develop adequate river-forecasting procedures. The main requirement, however, is that the period of record should contain a representative range of peak flows. Short records with a limited range of peak flows make it necessary to extrapolate the relations, with a probable loss of accuracy.

Rainfall records should be adequate to provide reasonable estimates of the average precipitation over the area under study. The density of rainfall reports required

25-100   RIVER FORECASTING

WB Form 612-18   Station   Date
(Rev. 9-57)   U.S. DEPARTMENT OF COMMERCE--WEATHER BUREAU

REPORTING INSTRUCTIONS
(River District Offices will cross out items not applicable)

RIVER AND RAINFALL STATION

TIMES OF OBSERVATION

1. Your regular daily observation of river stage and precipitation should be taken at 7 a.m. each day. (EMPTY THE NON-RECORDING RAIN GAGE AFTER EACH 7 A.M. OBSERVATION).

2. Special observations when made should be taken at 1 p.m., 7 p.m., and 1 a.m. These special observations should be taken ONLY when a report is required in accordance with instructions (see below).

WHEN TO REPORT

BASED ON RIVER STAGE:
Your station is designated as a DAILY REPORTING station, send a report immediately after each 7 a.m. observation. Also send in extra reports at 1 p.m., 7 p.m., and 1 a.m., when the stage is above _____ feet.
Your station is designated as an OCCASIONAL REPORTING station, make your first report whenever the stage has reached _____ feet. Continue to report daily at 7 a.m. until the stage goes below this limit. If the stage goes above _____ feet, make extra reports at 1 p.m., 7 p.m., and 1 a.m., until the stage goes below this limit. NOTE: Always report precipitation and river stage in same message. If no precipitation has occurred, report "None" for amount.

BASED ON PRECIPITATION:
a. Make an initial report at 7 a.m., 1 p.m., 7 p.m., or 1 a.m., whenever 0.50 inch or more of precipitation has accumulated in the rain gage.
b. After the first report of precipitation has been made, CONTINUE REPORTING at each observation time (1 p.m., 7 p.m., 1 a.m., 7 a.m.) as long as any additional precipitation has occurred since your previous report.
c. If you have made a final report, but it begins to rain again in less than 24 hours, start reporting again, just as though you had not stopped. That is, you should not consider the storm to be over until there has been no precipitation for 24 hours.

WHAT TO REPORT

Your report should include the following information in the order listed (Numbers refer to WB Form 612-24):
(1) Time of observation (hour).
(2) Amount of precipitation in gage at time of observation, in figures (inches and hundredths).
(3) Character of precipitation as it fell (rain, snow, sleet, etc.).
(4) Amount of precipitation measured at the PREVIOUS 7 A.M. OBSERVATION, in figures (inches and hundredths). This information should be sent ONLY in your first report of a series of reports. The amount, when sent, should always be preceded by "Previous 7 a.m." In subsequent reports omit this section entirely.
(5) Weather at time of observation (clear, cloudy, raining, snowing, etc.).
(6) Depth of snow or ice on ground, in figures (nearest inch). The figure showing depth should always be followed by the word "Inch" or "Inches". If there is no snow on the ground, omit this section entirely.
(7) River stage at time of observation (feet and tenths), in figures.
(8) Tendency of river stage at time of observation

(rising, falling, or stationary).
(9) River stage observed at PREVIOUS 7 A.M. OBSERVATION (feet and tenths), in figures. This information should be sent ONLY in your FIRST report of a series of reports and need not be sent by stations which telephone or telegraph daily reports regularly. The stage when sent should be preceded by "Previous 7 a.m." In subsequent reports omit this section entirely.
(10) Special effort should be made to obtain a reading at the crest. This reading should be included in the next report.
(11) Give the approximate time of occurrence of crest.
(12) Remarks. Any special comments which you feel would be of real value to the forecaster, such as: If snow is melting, state whether slowly or rapidly. If thunderstorm or unusually heavy shower occurred within a short period of time give time of beginning or ending. If there is ice in the river give any changes, such as "ice breaking up", "ice jam forming below station", etc. If instructed, include temperature readings.
(13) Last name of observer.

PREPARATION OF REPORT

1. The special River Rainfall Report card (WB Form 612-24) furnished will assist you in arranging your report in the proper order. This form has numbered blocks for each of the items to be reported by river and rainfall observers.
2. You should enter the designated information in all blocks. Each report must be complete. Your report will then be ready for transmission in message form as follows (Indicate whether report has been telephoned or telegraphed).

Sample Messages:
(First of a series--"7 P.M. 0.65 SNOW PREVIOUS 7 A.M. 0.30 CLOUDY 6 INCHES 10.2 RISING PREVIOUS 7 A.M. 4.8 SNOW MELTING RAPIDLY JONES"
(Subsequent report)-"7 A.M. 1.85 SNOW AND RAIN CLEAR 3 INCHES 15.2 FALLING CREST 17.8 AT 3 A.M.   JONES"

SENDING THE REPORT

Telephone number to call: _____
_____
_____

Telegraphic address: _____
_____
_____

1. All messages should be sent COLLECT.
2. If you customarily report by telephone and the lines are out of order report by telegram, if possible.

3. If you customarily report by telegram, use telephone when telegraph office is closed.
4. In an emergency, when all land lines of communication are out, contact your local or state police who may be able to transmit your report by police radio, or local "HAM" radio operator.
5. If you have difficulty in getting your telephone call through to the Weather Bureau during period of heavy rain or flood, tell your telephone operator that it is an EMERGENCY WEATHER REPORT.

NOTES

1. Promptly after each observation, mail the River Rainfall Report card which you have filled out, to _____
....
2. When additional supplies are needed, notify ....
....
3. SPECIAL INSTRUCTIONS: ....
....

USCOMM-WB-DC

FIG. 25-IV-1. Example of instructions to observers.

varies with basin topography and meteorological factors. Areas where precipitation is extremely spotty (e.g., where showery-type precipitation predominates) require a greater density than areas where the precipitation is of a more uniform nature.

If possible, rated gaging stations operated by the U.S. Geological Survey and others are selected as the forecast points. Occasionally, it is necessary to issue forecasts for a gage which is not rated.

B. Reporting Network for River Forecasting

The primary data required operationally are precipitation (rain or snow), snow on the ground (water equivalent, if possible), air temperature, and river stage or discharge. The number of reporting stations depends upon hydrologic need and availability of observers and communications. Criteria for reporting are standardized as much as possible, but may vary somewhat from one area to another. Sample instructions to observers appear in Fig. 25-IV-1.

The frequency of reports is a function of basin characteristics. In some areas once-daily reports of rainfall and river stages may be adequate. Forecasts for small basins with rapid concentration times may require reports at intervals of 6 hr, or even less, during high-water situations.

It would be desirable to have observers report daily, but economic considerations usually dictate that the observer report only on certain predetermined criteria of precipitation amount or river stage.

In recent years, there has been a significant advance in the hydrologic applications of radar [4, 5]. Information obtained from the radar scope can be used to estimate storm rainfall with a measure of success. Radar indicates the existence of centers of high-intensity rainfall and aids in interpreting the time and areal distribution of rainfall over the basin. Such information is of particular value in dealing with floods over very small watersheds and analyzing thunderstorm-type rainfall. A radar-beacon precipitation gage has been developed which makes it possible to obtain reports from inaccessible areas where there is a lack of observers and communication facilities [6].

III. BASIC RIVER-FORECASTING PROCEDURES

Where adequate data are available and forecasts of the complete hydrograph are required, a reasonably standardized approach to river forecasting has been developed. Rainfall-runoff relations (Fig. 25-IV-2) are used to estimate the amount of water

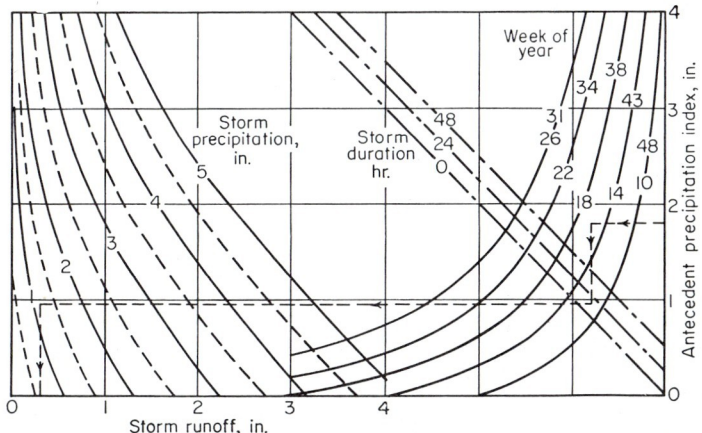

Fig. 25-IV-2. Rainfall-runoff relation.

expected to appear in the streams, while unit hydrographs (Fig. 25-IV-3) and streamflow-routing procedures (Fig. 25-IV-4), in one form or another, are utilized to determine the time distribution of this water at a forecast point. Stage-discharge relations (Fig. 25-IV-5) are then utilized to convert these flows to stages. The basic forecast procedures required are discussed only briefly since they have been described in detail in other sections of the handbook. (See Sec. 25-II.)

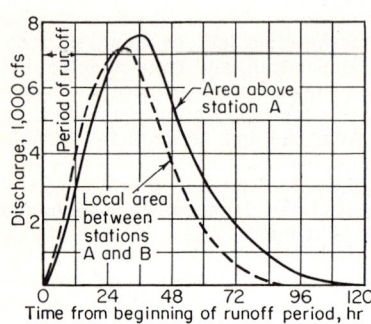

FIG. 25-IV-3. Twelve-hour unit hydrographs.

FIG. 25-IV-4. Muskingum routing diagram. $K = 18$ hr; $X = 0$; routing period $\Delta t = 12$ hr.

A. Rainfall-Runoff Relations (see Sec. 14)

The rainfall-runoff relation correlates storm rainfall, antecedent basin conditions, storm duration, and the resulting storm runoff (usually expressed as an average depth, in inches, over the basin). The basic technique in use by the U.S. Weather Bureau is the *coaxial graphical method* [7–9] (see also Sec. 8-II). An example is shown in Fig. 25-IV-2. Such a relation is developed using data from one or more headwater areas in the basin for which forecasts are required. Studies must be limited to areas for which the runoff can be evaluated (from the hydrograph) for each individual storm event. In larger basins where more than one area can be analyzed (e.g., the drainage areas above stations A and C in Fig. 25-IV-6), it is necessary to determine which relation is applicable to the downstream areas where detailed studies are usually not practical. Storm runoff can be estimated for the local inflow areas, such as A to B and B and C to D, and tested in the relation. Factors such as soil type, land use, ground cover, etc., are also considered.

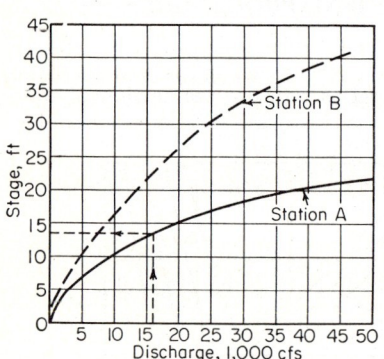

FIG. 25-IV-5. Stage-discharge relations.

In this rainfall-runoff relation the antecedent basin conditions are represented by two variables. The first is an *antecedent precipitation index* (API), which is essentially the summation of the precipitation amounts occurring prior to the storm weighted according to time of occurrence. The API for today is equal to k times the API for yesterday plus the average basin precipitation observed for the intervening day. The value of k used by most Weather Bureau River Forecast Centers is 0.90. An example of the computations is shown in Table 25-IV-1. The second variable is week of the year in which the storm occurs (e.g., the first week in January being 1, etc.). Week of

BASIC RIVER-FORECASTING PROCEDURES

the year introduces the average interception and evapotranspiration characteristics of each season, which, when combined with the antecedent precipitation index, provides an index of antecedent soil conditions.

Table 25-IV-1. Computation of Antecedent Precipitation Index

		Month April Year Date	10	11	12	13	14	15	16	17	18	19	20	21
Area above A	1	Yesterday's API × 0.9	1.93	1.74	1.97	1.77	1.59	1.43	2.01	1.81	1.63	2.55	4.90	
	2	Average basin precipitation		0.45				0.80			1.20	2.90		
	3	API = [(1) + (2)]	1.93	2.19	1.97	1.77	1.59	2.23	2.01	1.81	2.83	5.45		
Area A to B	4	Yesterday's API × 0.9	1.71	1.54	1.82	1.64	1.48	1.33	1.87	1.68	1.51	2.17	4.74	
	5	Average basin precipitation		0.48				0.75			0.90	3.10		
	6	API = [(4) + (5)]	1.71	2.02	1.82	1.64	1.48	2.08	1.87	1.68	2.41	5.27		

The value of storm duration used in the runoff relation is not critical and can be adequately derived from 6-hourly precipitation records. One method defines the duration as the sum of those 6-hourly periods with more than 0.2 in. of rain plus one-half the periods with less than 0.2 in. (e.g., four periods each with more than 0.2 in. and two periods with less than 0.2 in. would be considered a storm duration of $4 \times 6 + 2 \times 3$, or 30 hr).

The storm precipitation is the average over the basin. If a sufficient number of precipitation stations are available, an arithmetic mean is usually sufficient, although the Thiessen weighting method or isohyetal maps can be used [7, 8].

The storm runoff in most river-forecasting relations is direct runoff. Direct runoff is assumed to be the water which reaches the stream by traveling over the soil surface and through the upper soil horizons and has a rapid time of concentration. It is composed of surface runoff, channel precipitation, and interflow. The groundwater flow is discharged to the stream over a much longer period of time. Any of several methods of hydrograph analysis may be employed, but care must be taken to use the same method operationally as was used in development.

B. Unit Hydrographs (see Sec. 14)

The rainfall-runoff relation provides an estimate of the volume of water which will run off for a given storm situation. It is then necessary to determine the distribution of this water with respect to time at the forecast point. The unit hydrograph is a simple and generally effective method for accomplishing this [10]. In order to deal effectively with uneven distribution of runoff in time, unit hydrographs for short periods are used, very often for 6- or 12-hr durations. The increment of runoff is estimated for each time period, with the contributions from each interval superimposed upon the previous contributions.

C. Streamflow Routing (see Sec. 25-II)

The next basic problem is to predict the movement and change in shape of a flood wave as it moves downstream. Specifically, the river forecaster is interested in determining the shape of the flood wave from station A as it arrives at station B after being modified by lag and storage in the reach from A to B (Fig. 25-IV-6). Numerous routing methods are available, ranging from very complex storage functions to simple lagging procedures (Sec. 25-II). The Muskingum type of routing was selected for the forecast example [11]. In preparing a forecast for station B it is also necessary to determine the contribution of flow from the local drainage area between A and B.

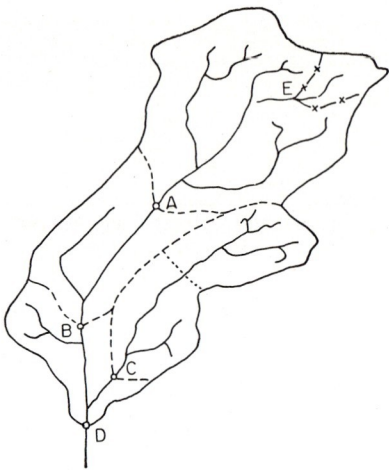

FIG. 25-IV-6. River-basin map.

The procedure for handling the local drainage area is similar to that for a headwater area, i.e., estimate the runoff from the local area and distribute by means of a unit hydrograph.

IV. RIVER-FORECASTING EXAMPLE

An example of a basic river forecast will be described in detail. A hypothetical river basin (Fig. 25-IV-6) has been selected in order to illustrate some of the special forecasting problems (discussed under Subsec. V). The rainfall-runoff relation, unit hydrographs, and routing method are the operational procedures for an actual river basin. However, because of use of a hypothetical basin, the forecast points will be designated as stations A and B.

It is assumed that the storm began about 7:00 P.M. on April 17, and a forecast is being made on the basis of rainfall reported up to 7:00 A.M. on April 19.

A. Computation of Runoff

The computation of storm runoff is shown in Table 25-IV-2. The antecedent precipitation index (API) selected is the value prior to the storm. The week of the year is determined by the date of the beginning of the storm, April 17, which falls in the sixteenth calendar week. The average rainfall amounts above station A and between stations A and B for 12-hr increments are entered on lines 3 and 10.

Dashed lines on the runoff relation (Fig. 25-IV-2) indicate the computation of runoff for station A for 7:00 A.M. on April 18. Enter the relation with API, move left to the week of year, vertically to storm duration, left to storm precipitation, and down to obtain storm runoff. Enter this value on line 6. This process is repeated at the end of each 12-hr period, using precipitation accumulated to that time. The 12-hr increments of runoff (line 7) are determined by subtracting each total storm-runoff value from the previous one and are entered on lines 1 and 12 of the forecast computation sheet (Table 25-IV-3).

Table 25-IV-2. Computation of Storm Runoff
(Using runoff relation in Fig. 25-IV-2)

Month April Year _____

			17		18		19	
			7 A.M.	7 P.M.	7 A.M.	7 P.M.	7 A.M.	7 P.M.
Drainage area above station A	1	Antecedent precipitation index	1.81					
	2	Week of year	16					
	3	12-hr precipitation increment, in.			1.20	0.80	2.10	
	4	Total storm precipitation, in.			1.20	2.00	4.10	
	5	Duration of storm, hr			12	24	36	
	6	Total storm runoff, in.			0.30	0.70	2.25	
	7	12-hr runoff increment, in.			0.30	0.40	1.55	
Drainage area between stations A and B	8	Antecedent precipitation index	1.68					
	9	Week of year	16					
	10	12-hr precipitation increment, in.			0.90	1.05	2.05	
	11	Total storm precipitation, in.			0.90	1.95	4.00	
	12	Duration of storm, hr			12	24	36	
	13	Total storm runoff, in.			0.15	0.65	2.10	
	14	12-hr runoff increment, in.			0.15	0.50	1.45	

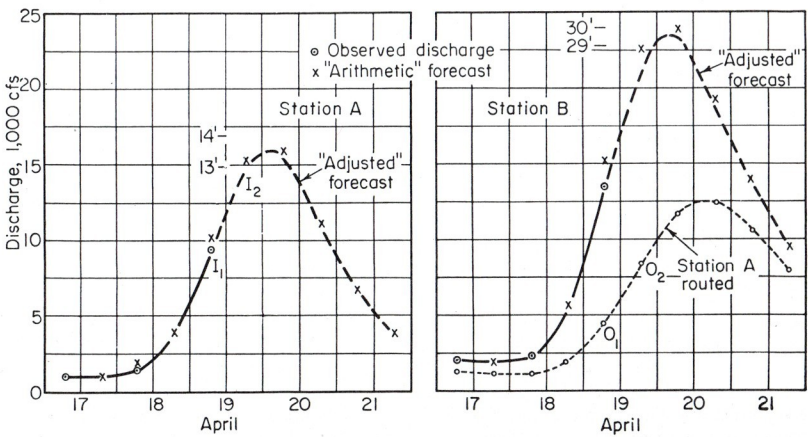

Fig. 25-IV-7. Forecast hydrographs.

Table 25-IV-3. Forecast Computation Sheet
(All discharge values in units of 1,000 cfs)

		Month April	17		18		19		20		21		22		23	
		Year	7 A.M.	7 P.M.	7 A.M.	7 P.M.	7 A.M.	7 P.M.	7 A.M.	7 P.M.	7 A.M.	7 P.M.	7 A.M.	7 P.M.	7 A.M.	7 P.M.
Station A	1	Forecast 12-hr RO, in.			0.30	0.40	1.55									
	2	Distribution of RO			0.9	1.9	2.3	1.7	1.0	0.5	0.3	0.1				
	3	" "				1.2	2.5	3.0	2.2	1.3	0.7	0.4	0.1			
	4	" "					4.6	9.8	11.8	8.7	5.1	2.8	1.4	0.5	0.2	
	5	" "														
	6	Total			0.9	3.1	9.4	14.5	15.0	10.5	6.1	3.3	1.5	0.5	0.2	
	7	Base flow	1.0	1.0	0.9	0.9	0.8	0.8	0.7	0.7	0.7	0.7				
	8	Arithmetic forecast	1.0	1.0	1.8	4.0	10.2	15.3	15.7	11.2	6.8	4.0				
	9	Adjusted forecast (I)	1.0	1.0	1.5	4.0	9.5	14.7	15.5	10.9	6.8	4.0				
Routing	10	$I_1 + I_2$	2.2	2.0	2.5	5.5	13.5	24.2	30.2	26.4	17.7	10.8				
	11	A routed to B (O)	1.3	1.2	1.2	2.0	4.4	8.3	11.7	12.4	10.6	8.0				
Station B	12	Forecast 12-hr RO, in.			0.15	0.50	1.45									
	13	Distribution of RO			0.6	1.0	1.0	0.6	0.3	0.1						
	14	" "				2.0	3.4	3.3	1.8	0.8	0.3	0.1				
	15	" "					5.8	9.9	9.6	5.4	2.5	0.9	0.1			
	16	" "														
	17	Total			0.6	3.0	10.2	13.8	11.7	6.3	2.8	1.0	0.1			
	18	Base flow	0.8	0.8	0.7	0.7	0.6	0.6	0.6	0.6	0.6	0.6				
	19	Arithmetic forecast	2.1	2.0	2.5	5.7	15.2	22.7	24.0	19.3	14.0	9.6				
	20	Adjusted forecast	2.1	2.0	2.3	5.5	13.5	21.0	23.3	18.7	13.6	9.6				

B. Forecast for Headwater Point (Station A)

The 12-hr runoff increments are converted to discharge, using the 12-hr unit hydrograph for station A (Fig. 25-IV-3). Each 12-hr ordinate of the unit hydrograph is multiplied by the first runoff (RO) increment (0.30 in.) and entered in line 2 (Table 25-IV-3) with the first value in the same column as the runoff increment (this is the ending time of the 12-hr period when the runoff occurred). This process is repeated on lines 3 and 4 for the other increments of runoff, and the total for each time entered on line 6. Base flow (line 7) includes all flow from events preceding the storm.

The *arithmetic* forecast, the sum of total runoff (line 6) and base flow (line 7), is entered on line 8 and plotted on the hydrograph (Fig. 25-IV-7). This arithmetic forecast is the unadjusted result of the forecasting procedures, and the forecaster must then draw an *adjusted* forecast, reconciling the arithmetic forecast with available observed data. The adjusted forecast is shown as a solid line when based on observed data and as a dashed line in the forecast period. The adjusted values are entered on line 9 for routing to station B.

The final step in preparing the forecast is the conversion of forecast discharge to stage using the stage-discharge relation (Fig. 25-IV-5). The forecast for station A could be stated as "crest of 13.5 ft at 2:00 A.M. on April 20" or as "crest of 13 to 14 ft early on April 20." Quoting a specific figure, such as 13.5 ft, might give the impression to the recipient of the forecast that it is likely to verify within tenths of a foot, which may not be the case.

C. Forecast for Downstream Point (Station B)

The adjusted flows for station A (line 9) are routed to station B, using the routing diagram (Fig. 25-IV-4). Successive pairs of inflows (line 9) are added to obtain the $I_1 + I_2$ values (line 10). The computation of the routed value for 7:00 P.M. on the 19th (8.3) is indicated by dashed lines on the routing diagram.

The forecast of flow from the local area is made in the same way as for station A. The arithmetic forecast is the sum of the routed value (line 11), the total runoff (line 17), and the base flow (line 18). These values are plotted on the hydrograph and adjusted on the basis of observed data.

The forecast for station B might be given as "crest of 29.5 ft at 4:00 A.M. on April 20" or as "crest of 29–30 ft early on the morning of April 20." It is a good practice to enter these forecasts on a tabulation sheet (Table 25-IV-4) as soon as completed to minimize the possibility of mistakes in transmitting the forecast to the user.

D. Remarks

It should be clearly understood that the above example demonstrates only one of many ways for deriving forecasts for stations A and B. Different methods could be used for estimating runoff, distributing runoff, and routing streamflow. The forecaster might also prefer to perform all or part of these computations on the hydrograph. A variety of forecasting techniques are required to handle most effectively the different river conditions encountered in the United States.

V. COMPLICATING FACTORS

The example given describes the basic techniques needed to handle most river-forecasting situations. Operationally there are often some complicating factors; a few of the most common ones will be discussed briefly.

A. Areas Where Unit Hydrographs Are Inadequate

The unit-hydrograph theory assumes uniform areal distribution of runoff. This is rarely the case, but in a fan-shaped basin, as above station A (Fig. 25-IV-6), it is usually not critical. In long, narrow basins as that above station C, the distribution of runoff may be very important. One solution is the development of special unit hydrographs based on various areal concentrations such as upstream, uniform, and downstream. Another solution is the division of the area into two zones, as indicated by a dotted line in Fig. 25-IV-6, and developing synthetic unit hydrographs for each of the subareas. The unit hydrograph for the upstream area can be prerouted to the forecast point C [7]. This approach provides flexibility in the handling of nonuniform areal distributions, but does appreciably increase the time required to prepare the forecast. It is also possible to divide the basin into zones based on estimated travel times and develop a channel inflow which can be routed to the forecast point [8].

In some basins it has been necessary to use a different unit hydrograph, usually cresting earlier and higher, for extreme floods from that for moderate floods [8, 12]. Another solution is to derive a unit hydrograph from moderate floods and develop a correction graph relating the computed peak discharge, using this unit hydrograph against observed peak flow for a number of storms of record [8]. The volume of the hydrograph should be maintained in adjusting the peak flow.

Table 25-IV-4. Forecast Tabulation Sheet

No.	Forecast for:	Forecast				Time forecast issued				Latest stage available when forecast prepared					Based on precip. up to:			Remarks
		Crest, ft	Hour	Date	By:	Hour	Date	By:	Stage	Hour	Date		Hour	Date	
1	Station A	13–14	2 A.M.	20	Mr.	9 A.M.	19	Mr.	9.9	7 A.M.	19		7 A.M.	19	
2	Station B	29–30	4 A.M.	20	Mr.	9 A.M.	19	Mr.	20.1	7 A.M.	19		7 A.M.	19	
3															

25–108

B. Routing

Often the inflow to be routed consists of the flow from two or more upstream stations (e.g., stations B and C to D in Fig. 25-IV-6). One solution is to add the flows at the upstream points, lagging one an appropriate time if necessary. This lag, based on the travel time from each station to the forecast point, synchronizes the two flows. When the stream channels through which two upstream inflows travel are significantly different, it may be necessary to route each inflow separately.

When the local area between two points is so shaped that most of its flow travels through the main channel, the unit hydrograph for the local area may be expressed in terms of its distribution as it reaches the main channel. This contribution is added to the upstream flows, and this total inflow routed to the outflow point (local inflow between B, C, and D might well be added to the flows at B and C and routed to D).

C. Snow (see Sec. 10)

Precipitation in the form of snow and snow on the ground complicates the forecasting problem. In mountainous areas it is necessary to estimate the elevation at which the precipitation changes from snow to rainfall. This level determines the portion of the basin contributing to rainfall runoff.

When rainfall occurs with snow on the ground, snowmelt and rainfall are added and this total used in computing runoff. The snowmelt is usually computed on the basis of degree-day factor, but studies are being made to use an energy-budget approach.

D. Use of Forecast Precipitation

In some extreme headwater areas with a flash-flood problem, the time to peak is so rapid that forecasts based on observed rainfall have practically no time advantage. In such areas the warning must be based on forecast precipitation.

At the present time practically all forecasts, except as noted above, are based on observed precipitation. The U.S. Weather Bureau has established a special Quantitative Precipitation Forecast Program, and it may be practical at some future time to utilize the forecast rainfall in making river predictions.

E. Stage-Discharge Relations (see Sec. 15)

One of the final products of the forecaster is the stage to which the river will rise. When the forecast point is a rated station and the stage-discharge relation is defined by a single curve as shown in Fig. 25-IV-5, there is no problem in converting discharge to stage. If the gage is not rated, then it is necessary to develop a synthetic rating using concurrent records at the unrated gage and the nearest rated station. In a few situations the forecast procedure may be developed in terms of stage.

In many cases the relationship of discharge to stage is complicated by the effects of slope, backwater, and scour. Another problem is the extension of rating curves beyond the maximum observed discharge so that forecasts of record-breaking flows can be converted to stage.

F. Average Precipitation Where Topography Is a Factor (see Sec. 9)

In mountainous areas where topographic features affect the precipitation, average basin-precipitation values determined by arithmetic means or Thiessen weights can be considered only as an index to the actual amounts. One solution is to have the operational network of reporting stations used in forecasting reasonably consistent with that used in procedure development. Another solution is the use of the *per cent normal method* [13]. Storm precipitation in mountainous areas tends to conform to the normal annual isohyetal pattern. Storm-precipitation values at each station can be expressed in per cent of its annual normal, and these percentage values averaged for the basin. The basin normal annual precipitation multiplied by this storm per cent value provides an average storm precipitation. Use of this per cent normal method reduces the need for a consistent reporting network.

This technique also provides a method for estimating average precipitation when only a portion of the basin is contributing to runoff. From the normal annual isohyetal map, the normal annual precipitation below various elevations can be determined. The contributing area is related to elevation. The normal annual precipitation below this elevation would be multiplied by storm per cent value to obtain the average precipitation over the contributing area.

VI. SPECIALIZED FORECASTS

A. Flash-flood Warning

When the time factor is so short that forecasts must be prepared locally (e.g., station E in Fig. 25-IV-6), procedures of a rather specialized form have been developed. The flash-flood-warning procedures are derived from conventional rainfall-runoff relations and unit hydrographs. An example is shown in Table 25-IV-5.

This relation is given to a local flood-warning representative, who is regularly furnished current index values (based on API and the week of the year) by the responsible river-forecasting office. The local representative has a special reporting network, and these rainfall data are used to prepare the flood warnings.

B. Water-supply Forecasts

In the western states the seasonal accumulation of snow is the principal source of streamflow, and the delay between accumulation and melt permits advance forecasting of the volume of runoff. The forecast is based upon correlations of runoff to either observed precipitation at regular climatological stations or water equivalent of the snowpack as determined by snow surveys or a combination of the two.

C. Low-flow Forecasts

Interest in navigation, water supply, stream pollution, etc., has increased the demand for forecasts of low flows. Special procedures are required to forecast ground-water runoff in addition to direct runoff.

Table 25-IV-5. Flash-flood-warning Procedure
(Rainfall, in.)

Duration of rain, 6 hr; time to crest after end of heavy rain, 7 hr				Index*	Duration of rain, 12 hr; time to crest after end of heavy rain, 5 hr			
Crest stage, ft					Crest stage, ft			
12	14	16	18		12	14	16	18
2.80	3.40	4.20	4.80	1	3.30	4.00	4.50	5.30
2.90	3.60	4.30	5.00	2	3.50	4.10	4.70	5.50
3.00	3.70	4.40	5.20	3	3.60	4.20	4.80	5.70
3.20	3.90	4.60	5.40	4	3.70	4.40	5.00	5.90
3.40	4.00	4.80	5.60	5	3.90	4.50	5.20	6.10
3.60	4.20	5.00	5.80	6	4.10	4.70	5.40	6.30

* Provided by the responsible river-forecast office.

VII. REFERENCES

1. River forecasting and hydrometeorological analysis, in water resources activities in the United States, U.S. Senate, 86th Cong., 1st Sess., Committee Print 25, Select Committee on National Water Resources, November, 1959.
2. Bernard, Merrill: Flood forecasts that reduce losses, *Eng. News-Rec.*, vol. 141, pp. 64–66, Nov. 25, 1948.
3. Hoyt, W. G., and W. B. Langbein: "Floods," Princeton University Press, Princeton, N.J., 1955.
4. Bigler, S. G., and R. D. Tarble: Applications of radar weather observations to hydrology, Final report under Weather Bureau Contract CWB-9090, Texas A & M Research Foundation, Texas A & M College, College Station, Tex., November, 1957.
5. Tarble, R. D.: The use of radar in detecting flood potential precipitation and its application to the field of hydrology, M.S. Thesis, Department of Oceanography and Meteorology, Texas A & M College, College Station, Texas, May, 1957.
6. Soltow, D. R., and R. D. Tarble: The use of a radar beacon for telemetering precipitation data, *J. Geophys. Res.*, vol. 64, no. 11, pp. 1863–1866, 1959.
7. Linsley, R. K., Jr., M. A. Kohler, and J. L. H. Paulhus: "Applied Hydrology," McGraw-Hill Book Company, Inc., New York, 1949.
8. Linsley, R. K., Jr., M. A. Kohler, and J. L. H. Paulhus: "Hydrology for Engineers," McGraw-Hill Book Company, Inc., New York, 1958.
9. Kohler, M. A., and R. K. Linsley, Jr.: Predicting the runoff from storm rainfall, *U.S. Weather Bur. Res. Paper* 34, September, 1951.
10. Sherman, L. K.: Streamflow from rainfall by the unit-graph method, *Eng. News-Record*, vol. 108, pp. 501–505, Apr. 7, 1932.
11. McCarthy, G. T.: The unit hydrograph and flood routing, presented at conference of U.S. Corps of Engineers, North Atlantic Division, June, 1938.
12. Flood-hydrograph analyses and computations, in "Engineering and Design," Engineering and Design Manual EM 1110-2-1405, U.S. Corps of Engineers, pp. 14, 15, Aug. 31, 1959.
13. Linsley, R. K., Jr.: Frequency and seasonal distribution of precipitation over large areas, *Trans. Am. Geophys. Union*, vol. 28, no. 3, pp. 445–450, June, 1947.

Section 25-V

HYDROLOGY OF FLOW CONTROL

PART V. FLOODPLAIN ADJUSTMENTS AND REGULATIONS

GILBERT F. WHITE, *Professor of Geography, The University of Chicago.*

```
    I. Introduction............................................. 25-113
       A. Results of Controlling Floods....................... 25-113
       B. Alternatives to Controlling Floods.................. 25-113
       C. Combinations........................................ 25-113
   II. Range of Possible Adjustments........................... 25-113
       A. Bearing the Loss.................................... 25-114
       B. Flood Abatement..................................... 25-114
       C. Flood Control....................................... 25-114
       D. Land Elevation...................................... 25-114
       E. Emergency Evacuation and Rescheduling............... 25-114
       F. Structural Adjustment............................... 25-114
       G. Land-use Change..................................... 25-114
       H. Insurance........................................... 25-117
  III. Practical Range of Choice............................... 25-118
   IV. Critical Characteristics of Flood Hazard................ 25-118
       A. Depth............................................... 25-118
       B. Duration............................................ 25-119
       C. Velocity............................................ 25-119
       D. Rate of Rise........................................ 25-119
       E. Recurrence.......................................... 25-119
       F. Seasonality......................................... 25-119
    V. Classification of Floodplain Land....................... 25-120
       A. Floodway............................................ 25-120
       B. Pondage.............................................  25-120
       C. Shifting Boundaries................................. 25-120
       D. Essential Encroachments............................. 25-120
   VI. Regulation of Floodplain Use............................ 25-120
       A. Encroachment Lines.................................. 25-120
       B. Zoning Ordinances................................... 25-122
       C. Subdivision Regulations............................. 25-123
```

The author is indebted for helpful criticism to Tate Dalrymple, George W. Edelen, Robert W. Kates, Joseph I. Perrey, John R. Sheaffer, and especially to James R. Goddard.

D. Building Codes.. 25-123
E. Miscellaneous Ordinances............................. 25-124
F. Acquisition... 25-124
VII. Dissemination of Information............................. 25-124
VIII. References.. 25-124

I. INTRODUCTION

Measures to control flood flows are only one set of adjustments that can be made to flood hazard. The appraisal of such engineering measures, if they are to be viewed in relation to their full consequences, therefore requires analysis of other possible adjustments. In this section the possible range of human adjustments to floods is outlined, attention is drawn to critical characteristics of flood events in making the various adjustments, the basis for classifying floodplain lands is discussed, and forms of regulating floodplain use are described.

A. Results of Controlling Floods

Experience shows that the construction and operation of engineering works to control flood flows may reduce or curb flows without necessarily reducing the threat to the health, safety, and welfare to the community or the flood-damage potential [11]. This may be because (1) while protection is given to life and property within a levee, new development takes place in unprotected areas; (2) while flood frequency and magnitude are reduced along the floodplain by channel improvements or reservoir operation upstream, new encroachments at lower levels cause greater threats to the public and greater losses, even though some floods are lower than formerly; or (3) the losses which occur when a levee is overtopped or channel capacity is exceeded by a flow greater than the design flood are catastrophic and outweigh the benefits gained in other years. A full analysis of an engineering project for controlling flood flows takes into account these possible consequences, as well as the changes anticipated in computing benefits (Sec. 26-I) and the possible effects upon channel and hydrologic conditions at and downstream from the site.

B. Alternatives to Controlling Floods

Where flood control is impracticable upon technical, social, or economic grounds, it is important to the people of the area concerned that the engineering studies suggest other adjustments which might be made. This involves stating the possible alternatives and compiling as much information as may be relevant to appraising each one.

C. Combinations

It may be that the most effective social solution to a flood problem is not one single type of adjustment but rather a combination of several. For example, a levee in one part of town may be helpfully supplemented by land-use adjustment in an unprotected floodway area and by structural adjustments in a sparsely built-up sector; or flood control by reservoirs may be combined with land-use regulations.

II. RANGE OF POSSIBLE ADJUSTMENTS

In any situation where there is human occupancy of a floodplain, the theoretically possible adjustments which might be made to the flood hazard fall into eight classes, none of which is necessarily exclusive.

A. Bearing the Loss

The occupants may bear the occasional loss, allowing the financial burden to fall upon households or business firms which may or may not have set aside reserves against this contingency. Some large firms do so explicitly. Others prefer to treat flood losses as unpredictable events covered under a general contingency account. Still others make no provision for flood losses.

The willingness and ability of households to bear flood losses are influenced in part by the availability of grants from relief agencies such as the American Red Cross. Likewise, business interests have often received special assistance after a flood disaster, such as public loans at low interest rates and income-tax advantages.

B. Flood Abatement

In some areas it may be practicable to reduce the frequency or magnitude of the smaller and more frequent floods by improved land practices and associated measures in the watershed (Secs. 21 and 22).

C. Flood Control

This is accomplished by constructing conventional channel improvement, levees, cutoffs, and storage or detention reservoirs (Sec. 25-III).

D. Land Elevation

By land fill it may be practicable in some situations to raise land surface above the level of floodwaters. This may or may not encroach upon the flowage area of the valley cross section.

E. Emergency Evacuation and Rescheduling

Property may be removed from the reach of floods, and property movement and production operations may be rescheduled so as to avoid or minimize losses due to interruption. This requires a relatively accurate system of flood forecasting (Sec. 25-IV) and a plan for emergency action to be taken when the critical forecast is received. The relative susceptibility of different types of flood losses to emergency action is given in Table 25-V-1.

F. Structural Adjustment

Well in advance of a flood warning, structures may be altered so as to prevent or reduce flood losses when the water rises. These structural changes may include such permanent alterations as the bricking in of low openings, installment of cutoff valves on sewers, and rearrangement of electric circuits so as to place them above the reach of water.

Many temporary structural adjustments are dependent for their full success upon the receipt of an adequate flood warning. Emergency evacuation and structural adjustment are closely linked, and their combination for modifying a given property is called *floodproofing* [5]. The chief forms of floodproofing are listed in Table 25-V-2, with indications in each case as to the type of material protected, the permanency of the measure, and the prerequisites to its effective operation [5].

G. Land-use Change

The use of floodplain land may be changed so as to introduce a practice that is less susceptible to flood loss. This may range from transfer of an entire town from a riverine to an upland site to a modification of a factory layout so that parking lots are in the flood zone and vulnerable buildings are on higher land. In urban areas the direction of change without public regulation is historically toward more vulnerable uses.

Table 25-V-1. Relative Susceptibility of Flood Losses to Reduction by Emergency Measures Based on Flood Forecasting

Class of loss*	Degree to which loss may be reduced by emergency measures based on timely and accurate flood forecasts			Emergency measures
	Large	Medium	Small or none	
Agricultural				
Crops				
Unharvested mature crops	–	x	–	Rescheduling—early or more rapid harvest
Decrease in yield	–	–	x	
Reseeding perennial crops	–	–	x	
Crops not planted	–	–	x	
Replanting of crops	–	–	x	Rescheduling—delay in planting
Stored crops	x	–	–	Removal
Orchard	–	–	x	
Farm timber	–	–	x	
Livestock and livestock products	x	–	–	Removal
Residence—see Urban residential				
Furnishings	x	–	–	Removal
Personal belongings	x	–	–	Removal
Other farm buildings	–	–	x	
Farm machinery and equipment	x	–	–	Removal, protection
Automobiles, trucks, wagons, boats	x	–	–	Removal
Fences, roads, and outdoor improvements	–	–	x	
Drainage and irrigation works	–	–	x	
Land	–	–	x	
Nonfarm income	–	–	x	
Evacuation and reoccupation	–	–	x	Rescheduling
Urban residential				
Residence				
Foundation	–	–	x	
Superstructure	–	–	x	
Improvements (fixed)	–	x	–	Protection
Decorations	–	–	x	
Furnishings	x	–	–	Removal
Personal belongings	x	–	–	Removal
Garage and other buildings	–	–	x	
Automobiles, wagons, trucks	x	–	–	Removal
Grounds and outdoor improvements	–	–	x	
Loss of property income	–	–	x	
Evacuation and reoccupation	–	–	x	
Retail and wholesale commercial				
Building—see Urban residential				
Furnishings	x	–	–	Removal
Equipment	x	–	–	Removal, protection
Stock of merchandise	x	–	–	Removal
Minor buildings	–	–	x	
Automobiles, wagons, trucks, etc.	x	–	–	Removal
Grounds and improvements	–	–	x	

Table 25-V-1. Relative Susceptibility of Flood Losses to Reduction by Emergency Measures Based on Flood Forecasting (Continued)

Class of loss*	Degree to which loss may be reduced by emergency measures based on timely and accurate flood forecasts			Emergency measures
	Large	Medium	Small or none	
Business interruption				
Production of goods and services	–	x	–	Rescheduling
Productive equipment and supplies	–	x	–	Protection, removal, rescheduling
Excess cost of delayed sales	–	–	x	
Evacuation and reoccupation	–	–	x	
Manufacturing				
Buildings—see Urban residential				
Office furnishings and records	x	–	–	Removal
Plant machinery	–	x	–	Removal, protection
Stock of raw materials or finished goods	x	–	–	Removal, protection
Minor buildings	–	–	x	
Automobiles, wagons, trucks, etc.	x	–	–	Removal
Grounds and improvements	–	–	x	
Business interruption				
Production of goods	–	x	–	
Productive equipment and supplies	–	x	–	
Excess cost of delayed production	–	–	x	
Evacuation and reoccupation	–	–	x	
Public buildings and grounds				
Building—see Urban residential				
Furnishings	x	–	–	Removal
Equipment	–	x	–	Removal, protection
Stocks of supplies	x	–	–	Removal
Public records, books, or other valuables	x	–	–	Removal
Minor buildings	–	–	x	
Grounds	–	–	x	
Automobiles, wagons, trucks, etc.	x	–	–	Removal
Evacuation or reoccupation	–	–	x	
Public services				
Indoor equipment, housing and grounds	–	x	–	Protection
Outdoor stationary equipment	–	x	–	Protection
Outdoor mobile equipment	x	–	–	Removal
Underground facilities	–	x	–	
Lines, mains, tracks, poles, etc.	–	–	x	
Emergency service	x	–	–	Rescheduling
Evacuation and reoccupation	–	–	x	Rescheduling
Railroads				
Minor buildings	–	–	x	
Outdoor stationary equipment	–	x	–	Protection
Roadway	–	–	x	
Rolling stock	x	–	–	Removal
Goods in transit	x	–	–	Rescheduling, removal
Emergency service	–	x	–	Rescheduling
Evacuation and reoccupation	–	–	x	Rescheduling

Table 25-V-1. Relative Susceptibility of Flood Losses to Reduction by Emergency Measures Based on Flood Forecasting (Continued)

Class of loss*	Degree to which loss may be reduced by emergency measures based on timely and accurate flood forecasts			Emergency measures
	Large	Medium	Small or none	
City streets and highways				
Roadway	–	–	x	
Cost of rerouting traffic	–	x	–	Rescheduling
Cost of fighting flood	–	x	–	Rescheduling
Cleanup	–	–	x	
Bridges				
	–	x	–	Protection
Piers and abutments	–	x	–	Protection
Superstructure	–	–	x	
Approaches	–	–	x	
Utilities				
Miscellaneous river structures—dams, revetments, levees, etc.	–	–	x	
Relief expenditures				
Evacuation and rescue work	–	x	–	Rescheduling
Emergency supplies	–	x	–	Rescheduling
Administration on rescue camps	–	x	–	Rescheduling
Care of sick and injured	–	x	–	Rescheduling
Policing	–	x	–	Rescheduling
Flood fighting	–	x	–	Rescheduling
Cleanup (public)	–	–	x	
Public health				
Sickness and injury	–	x	–	Rescheduling
Emergency public health activities	–	x	–	Rescheduling
Loss of life	–	x	–	Removal, rescue work, improved public health work

* Classification from National Resources Committee, Report of the Subcommittee on Flood Damage Data, Mar. 15, 1939, mimeographed.

H. Insurance

Although insurance against flood losses is not generally available in the United States, there is an inactive program for Federal-state subsidized insurance under the Flood Insurance Act of 1956. In a few instances firms have been able to obtain insurance coverage by private companies. Such coverage typically is available where there is clear evidence as to the physical characteristics of flood hazard and where the firm has been ready to make structural adjustments which would minimize losses from the more frequent floods, as in the Golden Triangle of Pittsburgh.

Table 25-V-2. Floodproofing Measures

Measure	Material protected	Class of measure	Prerequisites	
			Structural	Hydrologic
Seepage control............	St-Co	P-C	Well constructed	None
Sewer adjustment.........	St-Co	P-C	None	H-W
Permanent closure.......	St-Co	P	Impervious walls	H-S
Openings protected.......	St-Co	C-E	Impervious walls	H-S-W
Interiors protected........	St	P-C	None	S-W
Protective coverings......	St-Co	P-C-E	None	H-W-F
Fire protection............	St-Co	P	None	None
Appliance protection......	Co	E	None	W
Utilities service............	Co	P-C-E	None	S-W-V
Roadbed protection.......	St	P-E	Sound structure	H-W-V-D
Elevation..................	St-Co	P-C-E	Sound structure	S-W-V-F
Temporary removal.......	Co	E	None	W-F
Rescheduling..............	Co	E	Alternatives	W
Proper salvage............	Co		None	None
Watertight caps...........	Co	P-C	None	W
Proper anchorage.........	St-Co	P-C	Sound structure	S-W-V-D
Underpinning.............	St	P	Sound structure	V
Timber treatment.........	St	P	None	None
Deliberate flooding........	St-Co	E	None	None
Structural design..........	St-Co	P	Design	H-S
Reorganized use..........	Co	P	Alternatives	None

St = structure   P = permanent   H = hydrostatic pressure   F = flood-to-peak interval
Co = content   C = contingent   S = stage of flood   V = velocity of flow
   E = emergency   W = warning   D = duration of flood

III. PRACTICAL RANGE OF CHOICE

Although each of these adjustments may be theoretically available to an individual, firm, or public agency responsible for the management of floodplain land, in practice it may not be a viable choice. For example, physical conditions may make flood abatement impossible, or the flash character of the flood rise may render emergency evacuation ineffective, or the cost of the adjustment may be far in excess of prospective returns. One or more of the adjustments may be eliminated early in the analysis as infeasible on technical or economic grounds.

Moreover, the political guides which affect water management at any given time may rule out certain types of adjustments. For example, if the Federal flood-control policy is to pay for total cost of reservoir projects but to make no contribution toward floodproofing or changes in land use, the local agencies may find it impracticable to consider floodproofing even though the latter measure may be less costly.

IV. CRITICAL CHARACTERISTICS OF FLOOD HAZARD

When flood hazard is examined in its relation to possible practical adjustments, it is found that certain physical characteristics play critical roles in limiting the choice.

A. Depth

The maximum height which waters reach above the land surface may restrict the type of adjustment for particular land uses. It is not practical to seal structures against water depths which render them buoyant; most floodproofing measures for fixed property are ineffective where depths exceed 10 ft; and some crops which can survive 1 ft of water will be killed by greater depths.

B. Duration

The length of time an area is under water will vary with the size, shape, and slope of the watershed. This is important, particularly if the floods affect water supplies or sewage disposal or interrupt industry and commercial enterprises.

C. Velocity

Velocities of the water are affected by the slope and the roughness coefficient of the stream. Structures that may be subjected to high velocities should be designed to withstand those forces. There is a real hazard to life in swiftly flowing water. Combinations of or exceeding 3 ft of depth and 3 fps velocity are quite hazardous.

D. Rate of Rise

The length of time required for a stream to rise to flood stage sets limits within which emergency evacuation and rescheduling can take place. The time of rise from flood stage to peak is another limiting factor. A flood-to-peak interval of 3 hr seems a minimum for most warning systems to be effective, and 12 hr or more probably is required for full operation of large-scale removal [6].

E. Recurrence

The most common measure of flood recurrence is the perimeter of the largest known flood, with or without an estimate of recurrence interval (Secs. 8-I, 14, 20, and 25-I). Quite aside from the frequency estimates for purposes of designing engineering works, at least four other measures of flood recurrence have significance in the choice among possible adjustments [12].

The frequently recurring flood is one which may be expected on the average once in 5 to 10 years. It is often used, for example, as the basis for planning agricultural land use since greater floods cause only small additional crop damage, and it is used in setting some encroachment lines.

The maximum experienced flood is the flood of maximum height or crest flow within the memory of active managers of floodplain property. Generally, 25 years seems to be the limiting time. Being the largest flood with which local property managers are familiar, it is often considered to be the greatest they should expect, but engineering studies reveal that greater floods will occur and must be considered for reasonable safety.

The regional flood, as determined by methods used by the Tennessee Valley Authority [8], is a flood which may reasonably be expected at any time within the period of useful life of modern-day structures. Such a flood is suitable for establishing limits below which residences and buildings that suffer heavy damage should be prohibited.

The maximum probable flood is the greatest flood that may reasonably be expected (Sec. 25-I), taking into account all pertinent conditions of location, meteorology, hydrology, and terrain. It is an extremely large flood, likely to occur at rare intervals, but it may occur in any year. It is suited for the safe design of major structures on a stream. It is generally identical with the design flood, but only when it is should it be used in computing benefits from engineering measures. Its probability nevertheless will be a factor in assessing the need for other measures, such as evacuation plans or land-use changes, which might then be of major importance [9].

F. Seasonality

In agricultural land use the season of probable occurrence (and the probability of recurrence within a given season) appears to have a relation to choice of crop and of planting time, making some crops, such as soybeans, preferable to others in particular circumstances of climate and flood seasonality [2]. Certain seasonal industries and commercial establishments are subject to seasonal floods.

V. CLASSIFICATION OF FLOODPLAIN LAND

In addition to defining the flood hazard in terms of flood characteristics, it is important in arriving at choices among adjustments to classify the land of the floodplain into at least two classes, on the basis of capacity at peak flow. The distinction, as will be shown to be important for floodplain regulation, is between (1) floodway, or flowage area, necessary to carry the peak flow and (2) pondage area [8].

A. Floodway

Floodway, or flowage land, is that land on which the elevating of land surface (filling) or the construction of structures would cause a significant increase in water elevation at or above the site. It is the principal flow-carrying part of the natural cross section of the stream at a given stage of erosional development, and encroachment upon it will increase flood heights.

B. Pondage

Pondage land is that land on which water is stored as dead water as the flood rises. These dead areas do not contribute to the downstream passage of flow. Hence, if the areas are filled or built upon, there would be no significant increase in water elevation there or above the site. This is designated *flood fringe* in the Tennessee Valley Authority studies (Fig. 25-V-1).

C. Shifting Boundaries

The line between floodway and pondage lands obviously shifts according to the magnitude of the flood for which the hydraulic capacity of the channel is estimated. Portions of the pondage area for a small flood may be essential flowage area for a larger flood. The distinction, therefore, must be shown in terms of a stated magnitude or recurrence interval, rather than in absolute terms, if its meaning is to be clear.

D. Essential Encroachments

In classifying floodplains, it is important to recognize certain structures or fills which, while reducing the hydraulic capacity of the floodway, are essential to particular uses. Examples are highway fills, bridge piers and abutments, dams, thermal-electric power plants, and waterway terminal facilities. It may be necessary to take such encroachments into account in computing floodway capacity and pondage areas.

VI. REGULATION OF FLOODPLAIN USE

A major use for descriptions of flood characteristics and of floodplain land is in designing regulations of land use. These regulations are to be established for the protection of the health, safety, and general welfare of the community. In accomplishing this purpose, Murphy [4] states, they may (1) prevent encroachment upon the floodway cross section, (2) prevent carelessness in maintenance of channels, (3) prevent structures which, if floated in time of flood, would destroy bridges and other communications vital to the community and cause hazard to other property, (4) restrict uses which would be hazards to health and welfare, (5) protect land owners from being victimized, and (6) restrict uses which would lead to undue claims upon the public treasury for remedy. (For the last two points see Ref. 1.)

A. Encroachment Lines

These lines, as set by some states and local governments, establish zones within which further encroachment is not permitted. The criteria for setting the line range

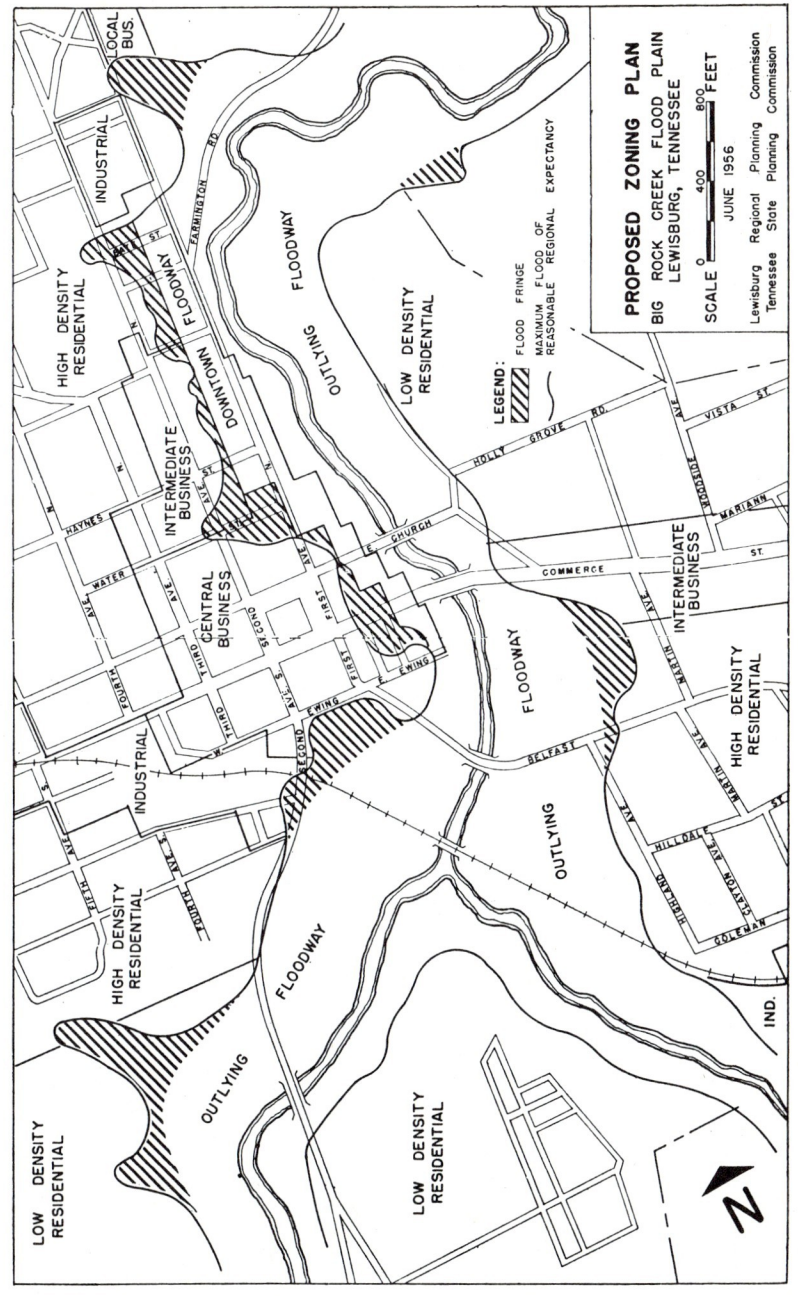

Fig. 25-V-1. Sample map of floodplain showing floodway and pondage areas. (*After Tennessee Valley Authority* [8].)

25-122 FLOODPLAIN ADJUSTMENTS AND REGULATIONS

from the annual flood to seven times the mean annual flood, or in a few local instances, to floods with estimated frequencies of as little as once in 100 years. This is illustrated by the diagram of state lines in Fig. 25-V-2.

B. Zoning Ordinances

Municipal, county, state, and conservancy ordinances may under general zoning classifications or in special enactments provide for areas having a particular designa-

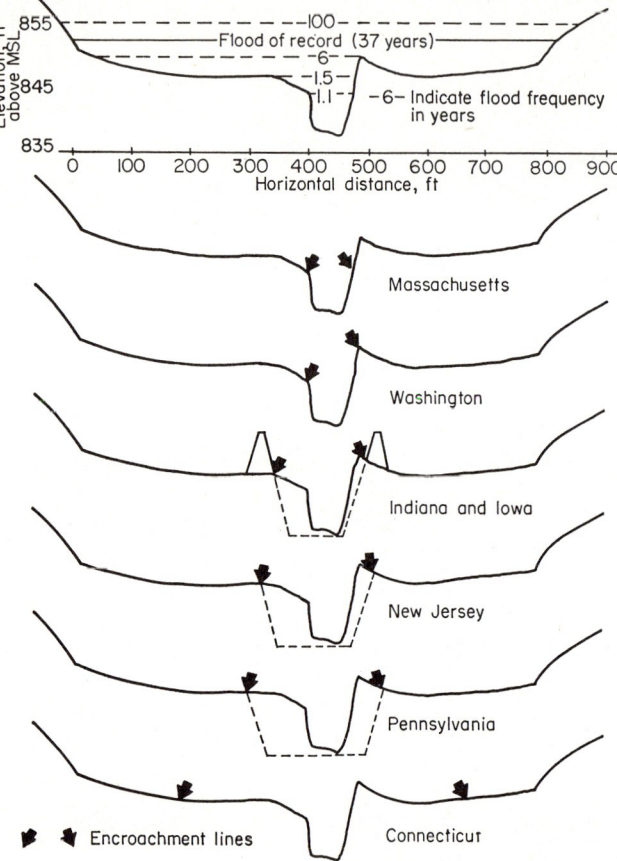

Fig. 25-V-2. Impression of methods of determining state channel-encroachment lines. (*After Murphy* [4].)

tion as "flood," "floodway," "agriculture," or "floodplain" areas, in which development is zoned for specified uses. These may include recreational open-space, agricultural, open industrial, and similar uses. The criteria for setting the floodplain may range from the zone of annual flooding to the maximum known flood, the regional flood, or the maximum probable flood.

It is important to recognize that zoning is not restriction against all uses, but it does involve a description of uses to be encouraged in harmony with one or more of the six aims stated above. Floodplain zoning is not intended to keep people out of the floodplain; it is intended to promote beneficial use of the floodplain, and it may properly allow certain uses which would be subject to flood loss but which nevertheless would be considered beneficial. The determination of flood lines for zoning ordinances is more than a hydraulic exercise and involves a plan of development for the entire area covered by the ordinance, taking into account community aims and the suitability and availability of land in the light of other criteria as well as flood hazard [1, 3, 7].

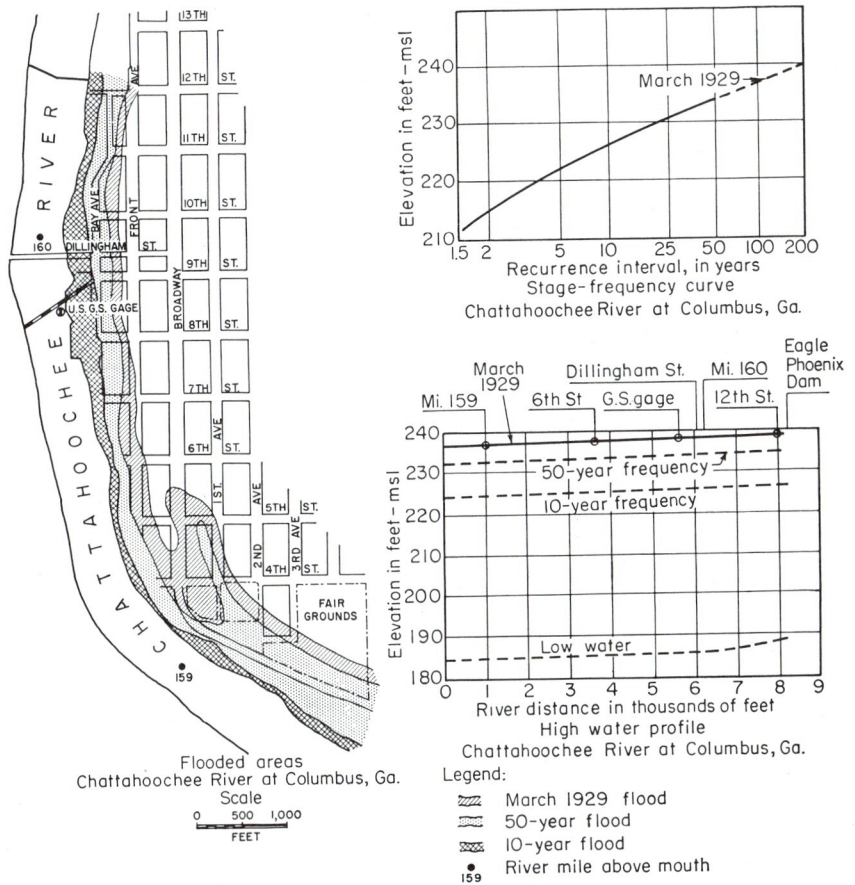

FIG. 25-V-3. Flood-risk map for Chattahoochee River at Columbus, Ga. (*After Murphy and U.S. Geological Survey* [4].)

C. Subdivision Regulations

Wherever land proposed for subdivision is subject to overflow or may be flooded as a result of structures downstream or from increased runoff caused by prospective land improvements upstream, there is opportunity to specify elevation of lands below which structures may not be built. The regulations may specify that the flood problem must be corrected or the land itself be subdivided in such a way as to provide flood-free building sites. They also may designate areas reserved as floodways.

D. Building Codes

In protecting "people from physical or financial injury resulting from negligence of property owners," building codes may specify minimum structural-design criteria, minimum elevations for footings or floors, minimum specifications for foundations or walls, and anchoring of buildings and tanks and may prohibit basements.

E. Miscellaneous Ordinances

In addition to building codes, special provisions for structures in flood-hazard areas are found in some housing codes and fire-prevention codes.

F. Acquisition

Titles or easements may be acquired by public agencies as a means of regulating development in a floodplain. Here, as in the case of zoning ordinances, the acquisition is a tool of planned development. It may include the permanent evacuation of an entire town, the maintenance of flowage rights below and around a reservoir, or the extension of a public park system along a valley bottom. One of the more systematic applications of land acquisition may be in urban renewal areas where floodplain land may be designated for redevelopment or conservation according to plans making allowance for flood hazard.

VII. DISSEMINATION OF INFORMATION

It is important to make available in an understandable form information as to the characteristics of flood hazard and floodplain lands as indicated under Subsecs. VI-D and VI-E. The information is necessary in order to develop sound public support of the regulatory measures. It has a bearing, although often a very minor one, upon the ways in which individuals, firms, and public agencies manage floodplain property. In the hands of financial agencies it may have a major effect upon negotiations for loans and mortgages.

The principal methods of making the hydraulic and hydrologic information available to property managers include (1) marking the high-water lines of floods, (2) erecting warning signs, (3) issuing maps showing the extent of stated floods [10], and (4) issuing special hazard reports [8, 12].

VIII. REFERENCES

1. Dunham, Allison: Flood control via the police power, *Univ. Penn. Law Rev.*, vol. 107, pp. 1098–1132, 1959.
2. Kates, R. W.: Seasonality, *Univ. Chicago, Dept. Geography Res. Paper* 70, 1961.
3. Moore, J. A.: Planning for flood damage prevention, *Georgia Inst. Tech. Eng. Expt. Sta. Spec. Rept.* 35, 1958.
4. Murphy, F. C.: Regulating flood-plain development, *Univ. Chicago, Dept. Geography, Res. Paper* 56, 1958.
5. Sheaffer, J. R.: Flood-proofing: an element in a flood damage reduction program, *Univ. Chicago, Dept. Geography, Res. Paper* 65, 1960.
6. Sheaffer, J. R.: Flood-to-peak interval, *Univ. Chicago, Dept. Geography Res. Paper* 70, 1961.
7. Siler, R. W., Jr.: "Flood Problems and Their Solution through Urban Planning Programs," Tennessee State Planning Commission, Nashville, Tenn., 1955.
8. A program for reducing the national flood damage potential, Tennessee Valley Authority, U.S. 86th Cong., 1st Sess., Senate Committee on Public Works, 1959.
9. Floods on Tennessee River, Chattanooga and Dry Creeks and Stringers Branch, *Tenn. Valley Authority Rept.* 0-5865, 1959.
10. Map of flood hazard at Topeka, Kansas, U.S. Geological Survey, 1959.
11. White, G. F., W. C. Calef, J. W. Hudson, H. M. Mayer, J. R. Sheaffer, and D. J. Volk: Changes in urban occupance of flood plains in the United States, *Univ. Chicago, Dept. Geography, Res. Paper* 57, 1958.
12. Wiitala, S. W., K. R. Jetter, and A. J. Sommerville: "Hydraulic and Hydrologic Aspects of Flood-plain Planning," *U.S. Geol. Sur. Water-Supply Paper* 1526, 1961.

Section 26-I

WATER RESOURCES

PART I. PLANNING AND DEVELOPMENT

J. W. DIXON, *Water Resources Specialist, U.S. Area Redevelopment Administration.*

I. Introduction	26-1
II. Multiple-purpose Projects	26-3
III. Items to Be Considered in Planning a Multiple-purpose Project	26-4
A. Physical and Related Items	26-4
B. Economic Aspects of Project Formulation	26-6
IV. Hydrologic Appraisal of Water Resources	26-6
A. Collection of Hydrologic Data	26-7
1. Kinds of Data	26-7
2. Hydrologic-data Networks	26-7
3. Sources of Hydrologic Data	26-10
B. Utilization of Hydrologic Data	26-13
V. Project Formulation	26-13
A. Project Investigation and Planning	26-13
1. Reconnaissance	26-13
2. Detailed Investigation	26-13
3. Definite Planning	26-15
B. Economic Aspects of Project Formulation	26-15
1. Definitions	26-17
2. Benefit Study	26-18
3. Cost Study	26-19
4. Benefit-Cost Analysis	26-20
5. Project Composition	26-23
VI. References	26-25

I. INTRODUCTION

The wealth of technical data on the subject of water-resources planning and development is so extensive and it has been prepared in such detail by its authors during the past thirty years, that to attempt to synthesize it here would result in the consumption of more than the allowable space. It would also produce a galaxy of formulas, tables, charts, graphs, and lengthy discussions on philosophies and practices.

Table 26-I-1. Purposes of a Water-resources Project*

Serial no.	Purpose	Description	Type of works and measures
1	Flood control	Flood-damage prevention or reduction, protection of economic development, conservation storage, river regulation, recharging of groundwater, water supply, development of power, protection of life	Dams, storage reservoirs, levees, floodwalls, channel improvements, floodways, pumping stations, floodplain zoning, flood forecasting
2	Irrigation	Agricultural production	Dams, reservoirs, wells, canals, pumps and pumping plants, weed-control and desilting works, distribution systems, drainage facilities, farmland grading
3	Hydroelectric	Provision of power for economic development and improved living standards	Dams, reservoirs, penstocks, power plants, transmission lines
4	Navigation	Transportation of goods and passengers	Dams, reservoirs, canals, locks, open-channel improvements, harbor improvements
5	Domestic and industrial water supply	Provision of water for domestic, industrial, commercial, municipal, and other uses	Dams, reservoirs, wells, conduits, pumping plants, treatment plants, saline-water conversion, distribution systems
6	Watershed management	Conservation and improvement of the soil, sediment abatement, runoff retardation, forests and grassland improvement, and protection of water supply	Soil-conservation practices, forest and range management practices, headwater-control structures, debris-detention dams, small reservoirs, and farm ponds
7	Recreational use of water	Increased well-being and health of the people	Reservoirs, facilities for recreational use, works for pollution control, reservation of scenic and wilderness areas
8	Fish and wildlife	Improvement of habitat for fish and wildlife, reduction or prevention of fish or wildlife losses associated with man's works, enhancement of sports opportunities, provision for expansion of commercial fishing	Wildlife refuges, fish hatcheries, fish ladders and screens, reservoir storage, regulation of streamflows, stocking of streams and reservoirs with fish, pollution control, and land management
9	Pollution abatement	Protection or improvement of water supplies for municipal, domestic, industrial, and agricultural use and for aquatic life and recreation	Treatment facilities, reservoir storage for augmenting low flows, sewage-collection systems, legal control measures
10	Insect control	Public health, protection of recreational values, protection of forests and crops	Proper design and operation of reservoirs and associated works, drainage, and extermination measures
11	Drainage	Agricultural production, urban development, and protection of public health	Ditches, tile drains, levees, pumping stations, soil treatment
12	Sediment control	Reduction or control of silt load in streams and protection of reservoirs	Soil conservation, sound forest practices, proper highway and railroad construction, desilting works, channel and revetment works, bank stabilization, special dam construction and reservoir operations

Table 26-I-1. Purposes of a Water-resources Project* (Continued)

Serial no.	Purpose	Description	Type of works and measures
13	Salinity control	Abatement or prevention of salt-water contamination of agricultural, industrial, and municipal water supplies	Reservoirs for augmenting low streamflow, barriers, ground-water recharge, coastal jetties
14	Artificial precipitation	Control of precipitation within meteorological limits	Portable cloud-seeding equipment, ground generators
15	Employment	Stimulation of employment and sources for increased income in depressed areas of unemployment and under-development	Area Redevelopment Act [24] and Area Redevelopment Administration [25]
16	Public works acceleration	Acceleration of Federal, state, and local constructions of public works on cost-sharing basis	Public Law 87-658 [26]
17	New water-resources policies	New policies to be used by Federal agencies, according to S. Doc. no. 97 [27], approved by the President May 15, 1962, affecting the economics of project justification as well as project formulation and composition	Senate Document no. 97 [27]

* Adapted in part from Report of President's Water Resources Policy Commission [2, p. 47] and from ECAFE [10, p. 2].

This section therefore chooses to present an outline of the important considerations that are related to hydrology. It defines some basic terminology used in water-resources studies; lists major items to be considered in planning a multiple-purpose water project; discusses hydrologic appraisal of water resources; and describes project formulation, including a working example for economic analysis of water projects. For design, legal, and policy aspects of water-resources projects, detailed discussions are presented by other authors in Secs. 26-II, 27, and 28. For general references, see Refs. 1 to 60 in Subsec. VI.

II. MULTIPLE-PURPOSE PROJECTS

Modern water-resources projects are mostly planned for more than one of the many purposes such as those listed in Table 26-I-1, because such *multiple-purpose projects* usually have a greatly improved economic justification that is made either purposely or incidentally.

There are, basically, two types of multiple-purpose projects: the *multiple-purpose single-development project* and the *multiple-purpose areal-development project*, which is a part of a basin-wide or regional type of plan. The former is designed to serve a limited geographical area in at least some of its major functions. The latter is designed and operated to serve as an element in a plan of much larger scope. It is even possible that the structures are identical, and their differences of function result solely from the different methods by which they are operated. There is, of course, also the *single-purpose project*, which is designed and operated basically to achieve one objective, but which may, purely incidentally, produce some other benefits on occasion, but not on an assured basis.

Thus a single-development multiple-purpose project may be built as the first unit in an ultimate basin-wide plan and be operated accordingly. As more dams are added to the system, the method of operation of this first dam may be altered to complement the operation of the whole series of dams. For this same purpose, single-purpose dams may be added to the system.

It is no wonder, therefore, that confusion has arisen in the minds of some people about just what a multiple-purpose project is. A multiple-purpose project can be defined as one that can be operated on an assured basis to provide more than one purpose and whose design is such that its method of operation can be altered from time to time, if desirable, to change the emphasis of its services so that it can always contribute most beneficially, whether it is to serve as a single-development project or as a productive element in a larger system of dams, rendering under either circumstance the optimum of economic benefits that are desired with each service at a cost that is warranted. Thus the following features become evident: (1) If a reservoir of a given capacity is to serve several purposes such as power generation, municipal water supply, irrigation, navigation, and flood control, its value for any one purpose cannot be the maximum possible for that purpose, for the reservoir operating plan must include operations for all purposes, thus probably reducing the maximum possible benefits for each service somewhat in order to provide for inclusion of the other purposes. (2) The sum of all the benefits from the services will equal the optimum and together will exceed the maximum benefit from any one function. (3) Meticulous studies of benefits and costs for different capacities of reservoirs, different types of dams, and the inclusion or exclusion of each of the several services must be made in order to obtain this optimum economic and physical balance.

A multiple-purpose areal project will usually include a number of separate units or elements, such as dams for various purposes. Any individual unit or element may, in itself, serve either multiple purposes or only a single purpose, but the composite of all units must serve, as best possible, the overall needs of the area.

III. ITEMS TO BE CONSIDERED IN PLANNING A MULTIPLE-PURPOSE PROJECT

There are nearly as many outlines as there have been authors on this general subject. Some outlines are long; some are short. The following condensed outline is by no means complete, but serves primarily as an illustrative example and also as a checklist of major factors to be considered in an area investigation for the project planning and development.

A. Physical and Related Items

1. Project area
    a. Physical geography: location and size; physiography; climate; soils
    b. Settlement: history; population; cultural background, both rural and urban
    c. Development: industry; transportation; communication; commerce; power; land uses; water uses; minerals; undeveloped resources
    d. Economic conditions: general; relief problems; community needs; national needs
    e. Investigations and reports: previous investigations; history; scope
2. Hydrologic data
    a. Hydrologic records and networks: gaging and observation stations; data-collecting agencies
    b. Hydrometeorological data: precipitation; evaporation and evapotranspiration
    c. Surface water: low flows; normal flows; maximum floods; "design flood"; drought; quality
    d. Groundwater: aquifers; recharge; quality
3. Supply of water
    a. Sources of supply: surface-water supply; groundwater supply; reservoirs

b. Variation of supply: variability; consumptive use; regulation; diversion requirements; return flow; evapotranspiration losses; seepage losses or gains
   c. Quality of water: physical, chemical, biological, and radioactive qualities; quality requirements; pollution
   d. Legal rights: water rights; development of project rights; operation rights
4. General considerations for design and planning
   a. Geology: explorations; geological formations; foundation problems
   b. Design problems: design criteria; methods of analysis; project operation and maintenance
   c. Construction problems: accessibility to project site; rights of way and relocation; construction materials; construction period; flow diversion; manpower; equipment; accessibility
   d. Alternative plans: comparison of alternative plans; supplementary plans; possible alternative plans; relationships to areas to be served
   e. Estimates of costs
   f. Intrastate, interstate, and international problems
   g. Organizations involved: public and/or private; technical, social, and political
5. Flood control
   a. Flood characteristics of the project area: historical floods; flood magnitude and frequency
   b. Design criteria: project design storms and floods; degree of protection
   c. Damage: survey of flooding areas and things affected by floods, nearby or quite a distance away, including commerce, good will, dates of delivery of goods, etc.
   d. Measure of control: reduction of peak flow by reservoirs; confinement of flow by levees, floodwalls, or a closed conduit; reduction of peak stage by channel improvement; diversion of floodwater through bypasses or floodways; floodplain zoning and evacuation; floodproofing and flood insurance of specific properties; reduction of flood runoff by watershed management
   e. Existing remedial works
6. Agricultural use of water (irrigation and related drainage)
   a. Factors for land classification: soil texture; depth to sand, gravel, shale, raw soil, or penetrable lime zone; alkalinity; salinity; slopes; surface cover and profile; drainage; waterlogging
   b. Present and anticipated development: crops; livestock; financial resources; improvements; organizations; development period
   c. Water requirements, if any: total crop requirement; irrigation-water demand; farm-delivery losses; diversion amounts
   d. Available water: sources; quality; quantity; distribution
   e. Irrigation methods: flooding; furrow irrigation; sprinkling; subirrigation; supplemental irrigation
   f. Structural works: storage reservoirs; dams; spillways; diversion works; canals and distribution systems
7. Hydroelectric power
   a. Development: sources; present potential and future capacities
   b. Alternative sources of power: stream; oil; gas; nuclear power; interties
   c. Types of power plants: run-of-river; storage; pumped storage
   d. Structural components: dams; canals; tunnels; penstock; forebay; powerhouse; tailrace
   e. Power problems: load demand and distribution; interties (interconnections with other power transmission systems)
   f. Markets; revenues; costs
8. Navigation
   a. Water traffic: present and future needs and savings in shipping costs, if any, on the basis of which the justifications are primarily judged at the present time
   b. Alternative means of transportation: air; land
   c. Navigation requirements: Depth, width, and alignment of channels; locking time; current velocity; terminal facilities

　　　　　d. Methods of improving navigation: channel improvement by reservoir regulation; contraction works; bank stabilization, straightening, and snag removal; lock-and-dam construction; canalization; dredging
　9. Domestic and industrial water supply
　　　　　a. Sources of supply: surface and/or groundwater; location and capacity; desalinization
　　　　　b. Water demand: climate; population characteristics; industry and commerce; water rates and metering; size of project area; fluctuation
　　　　　c. Water requirements: quantity; pressure; quality (tastes, odors, color, turbidity, bacteria content, chemicals, temperature, etc.)
　　　　　d. Methods of purification: plain sedimentation; chemical sedimentation or coagulation; filtration; disinfection; aeration; water softening
　　　　　e. Treatment plant: location; design; purpose or purposes
　　　　　f. Distribution systems: reservoirs; pumping stations; elevated storage; layout and size of pipe systems; location of fire hydrants
　　　　　g. Waterworks organizations: maintenance and operation of supply, distribution, and treatment facilities
　10. Recreational use of water
　　　　　a. Population tributary (population near enough to the project area to use it for recreational purposes)
　　　　　b. Facilities: boating; fishing; swimming; etc.
　　　　　c. Water requirements: Depth of water; area of water surface; sanitation
　11. Fish and wildlife
　　　　　a. Biological data: species; habits
　　　　　b. Facilities: reservoirs; fish ladders
　　　　　c. Water requirements: temperature; current velocity; biological qualities
　12. Drainage
　　　　　a. Existing projects
　　　　　b. Drainage conditions: rainfall excess; soil condition; topography; disposal of water
　　　　　c. Drainage system: urban; farmland
　13. Water-quality control
　　　　　a. Problems involved: sources; nature and degree of pollution; sediment; salinity; temperature; oxygen content; radioactive contamination
　　　　　b. Hydrologic information and measurement
　　　　　c. Methods of control

B. Economic Aspects of Project Formulation

　1. Benefits and damages: identification and evaluation
　2. Costs: identification and estimation
　3. Financial feasibility
　4. Allocation of costs
　5. Reimbursement requirements and sharing of allocated costs
　6. Methods and costs of financing the project, whether Federal, state, or local, bringing all benefits and all costs to an annual basis and recognizing interest on the investment not only during construction, but throughout the entire proposed "life of the project"
　7. Benefit-cost-ratio analysis: alternative plans

IV. HYDROLOGIC APPRAISAL OF WATER RESOURCES

　The hydrologic appraisal of water resources is the basic requirement for planning, designing, constructing, and operating water-resources projects. Its objective is to determine the source, extent, and dependability of supply and the character of water on which an evaluation of control and utilization is to be based. The appraisal involves, among many other factors, the collection and utilization of hydrologic information and data as one of the important factors.

HYDROLOGIC APPRAISAL OF WATER RESOURCES 26–7

A. Collection of Hydrologic Data

1. Kinds of Data. Hydrologic data required for the planning of water-resources development include precipitation, river stage, river discharge, sediment transportation, yield and storage of groundwater, quality of water, and hydrometeorological data, as well as other related meteorological data such as temperature, total snowmelt and rates thereof, and rainfall-runoff relationship. The principal kinds of hydrologic data required for various purposes of water-resources projects are as follows:

Flood control. Precipitation (depth of storm precipitation, intensity, duration, areal distribution and path; maximum probable precipitation; snow surveys); river stage (peak stages, stage hydrograph during floods, flood-wave profiles along the stream and tributaries); discharge (peak rates, frequencies, hydrographs); sediment (rate of suspended and bed-load transportation).

Irrigation. Precipitation (amount during crop season, annual variation, minimum amount, snow surveys); evapotranspiration; river stage (stage hydrograph at intake and canal outfall); discharge (hydrographs at intake site for wet, average, and dry years; minimum rate, duration, and frequency); groundwater (storage and yield; groundwater levels and quality).

Hydroelectric power. Precipitation (amount for design period, snow surveys); evaporation in reservoirs; river stage (stage hydrographs at intake and tailrace); discharge (hydrograph for the design period; minimum rate, duration, and frequency; flood flows, duration, and frequency); sediment (rate and types of suspended and bed-load transportation near intake or inflow into a reservoir, composition and compaction of sediment as it may ultimately affect the usable capacity of a reservoir).

Navigation. River stage (stage hydrograph; flood-wave profile; low-water duration and frequency); discharge (flood discharge and current velocity); sediment (bed-load movement over shoals; scouring and deposition on river bed); groundwater (groundwater level along artificial canals, seepage flow from and into the canals).

Domestic and industrial water supply, recreational use of water, fish and wildlife conservation, water-quality control. Precipitation (amount, duration, and frequency; snow surveys); evapotranspiration (amount in reservoirs); river or reservoir stage (stage hydrograph at intake and tail water); discharge (hydrograph at intake, minimum discharge, duration, and frequency); sediment (suspended-load concentration and composition); groundwater (groundwater level, yield, and storage); quality of water (physical, chemical, and biological characters).

Drainage. Precipitation (storm-rainfall amount, intensity, duration, and frequency); river stage (stage hydrograph and volumetric quantity of water, in particular during floods); groundwater (groundwater level and infiltration).

2. Hydrologic-data Networks. The establishment of stations and networks for the collection of hydrologic data is not always a product of scientific planning, but as nearly so as practicable. Because of rapid expansion and progress in water-resources development, there is need for further intelligent *network design* in order to produce the maximum hydrologic information for a given investment of time and money.

Langbein [16, 28] has given a number of examples to illustrate the principle of network design. One hypothetical and, depending upon the character and size of the river basin, generalized example involves the planning of a reservoir survey which is to estimate the total storage capacity in a river basin containing, perhaps, a great many reservoirs, large and small.

It is of course essential to know approximately the capacity of any proposed reservoir before any really intelligent plan can be prepared for its use and before judgment can be rendered as a part of its justification. In the case of very large reservoirs, this is an expensive and time-consuming task. Aerial mapping is perhaps the best for overall maps and can be checked very accurately by successive aerial photographs as the reservoir fills. But a good working knowledge of its elevation-capacity relationship is necessary for operation studies, spillway designs, capacities to meet various purposes, etc. Because of limited time and funds in some cases, the reservoirs or other waterway improvements cannot all be surveyed. A report on any such reservoir should have minimum use and be marked clearly as "preliminary" or "reconnaissance," and the report should state clearly that the reservoir capacity used and the estimates of costs and operation are of judgment quality only. The problem is therefore to determine how to make the estimate most efficiently. An intelligent approach in this case would be to survey the large reservoirs down to a certain capacity and to leave sufficient funds for surveying the small reservoirs on a random basis. For a given amount of money for the survey, the cutoff point can be readily defined on the basis of statistical principles so that the overall error in capacity is at a minimum. Let N be the number of reservoirs having capacities less than the cutoff value and having a standard deviation of s. From the N reservoirs, n reservoirs are to be surveyed. The error involved in this sampling would be $Ns/\sqrt{n}$. Plot this computed error against variously assumed cutoff capacities. A cutoff point for the minimum error can be obtained to serve as a basis for planning the survey intelligently.

Another example describes a river-measurement program. The two controlling factors in this case are the number of streams and the probable period of operation. The factors are in a sense in competition for the available funds. A desirable plan would be to build the hydrologic network around base stations which are operated permanently in order to record the time variations in streamflow in a locality and around subsidiary stations which are operated on a stream just long enough to "sample" particularly the medium and low-flow character of the stream. Some of the most disastrous floods, so far as local areas are concerned, have been on relatively small tributaries. Hence one must not be misled that this short time record supports more than a token understanding of the stream. A hydrometeorological study would be required to obtain a really workable "sample." For ordinary uses, the short record obtained from the subsidiary stations can be extended through correlation with records from the nearest base stations. The problem is therefore to find the optimum division of effort between base and subsidiary stations that will produce the optimum number of stations operated for a given investment of time and money. The problem can likewise be solved by statistical techniques.

The desirable density of hydrologic networks should reflect the hydrologic regimen of an area and the ultimate use to which the data are to be directed. A general survey of water resources may be served by a relatively scanty network, while the development of a specific important water project may require a much higher density of numbers and of types of stations. Because of a large number of factors involved in this problem, the range of adequate network requirements is broad. It is particularly difficult to specify the required density of hydrologic stations on a unit-area basis, except perhaps in the pool and tail-water gages at dams. Investigation of the problem indicates that population density may be one of the major factors affecting the density of hydrologic stations. Large population creates problems of water supply, flood damage, and pollution and requires adequate flood forecasting and warning, and therefore indirectly gives rise to a need for hydrologic stations with adequate communication facilities and interpretative services and publicity. Furthermore, the state of economic development and capital and operation costs in population centers varies with the magnitude of the water problems, and hence influences the need for hydrologic information. The same is true for large water-resource developments in sparsely populated areas such as the upper Missouri River system and the Colorado River system. Hydrologic and operating forecasting systems are always necessary and should be "tailor-made" to suit the circumstances.

Figure 26-I-1 shows the data on density-of-stream gaging stations in various countries. The lines labeled "relative density" represent the percentage of the various countries that have densities less than indicated. The "reasonable minimum objective" lines represent a compromise between uniform adequacy in regard to population and that desirable on a purely geographic basis. Similarly, Fig. 26-I-2 shows the data on density-of-precipitation gages in various countries. These data should be considered very approximate since many other influencing factors are ignored.

According to Langbein [28], the minimum objectives for a balanced distribution of hydrometeorological networks are as follows:

   a. Density of precipitation stations should be at least as indicated in Fig. 26-I-2. Approximately 1 station in 5 should be equipped with a recording precipitation gage.

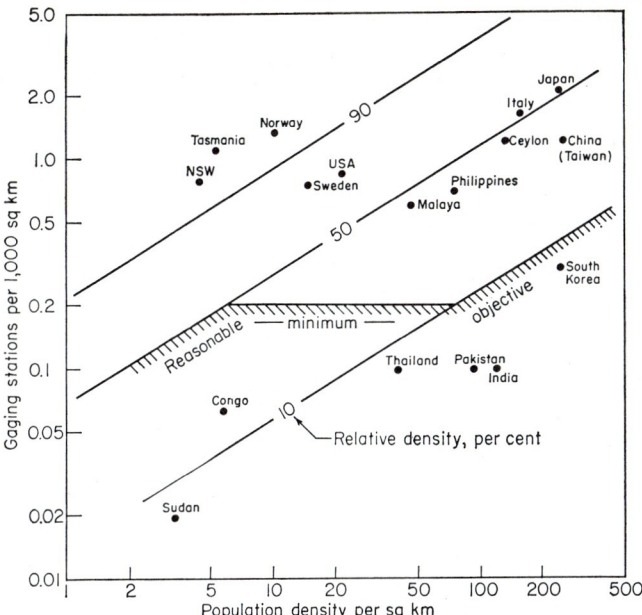

Fig. 26-I-1. Comparative areal densities of streamflow gaging stations. (*After Langbein, United Nations ECAFE* [28].)

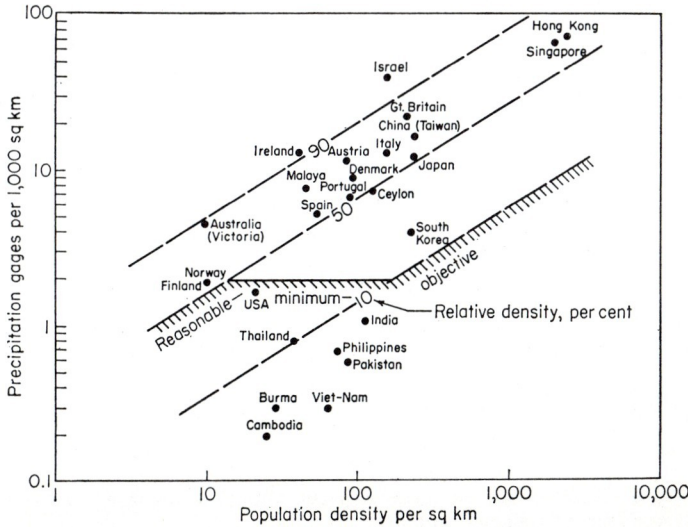

Fig. 26-I-2. Comparative densities of national networks of precipitation gages. (*After Langbein, United Nations ECAFE* [28].)

where daily observations are impossible, recording gages or storage gages should be used, depending on the conditions to be met.

b. A density of first-order hydrometeorological stations of about 1 per 15,000 km^2. Station location should be selected to sample the major climatic areas within the region.

c. Second-order hydrometeorological stations approximately equal in number to the first-order stations and distributed more or less uniformly within the first-order network.

d. Approximately 1 of every 10 hydrometeorological stations should be equipped as an evaporation station, with pan, anemometer, water temperature, and humidity observations.

e. Snow survey stations as required.

3. Sources of Hydrologic Data. Principal hydrologic data are usually published by private and public agencies. The professional hydrologists, however, usually obtain much information outside the formally published data. The informally recorded information may be found in newspapers, diaries of pioneers, and files of state and local historical societies. Less reliable, but vital, information may be obtained from field observations, such as flood marks on trees and buildings, and from people who have lived for a long time in the project area.

Table 26-I-2 gives a checklist of basic hydrologic data collected by Federal agencies in the United States. Further information on sources of hydrologic data in the United States may be found in Refs. 16 and 29 to 31 and from the following publications.

U.S. Army Corps of Engineers: *Annual Report of Chief of Engineers; Water Resources Development of the Corps of Engineers* (report for each state); division and district *Reports* and *Technical Bulletins;* Waterways Experiment Station publications; *Snow, Ice, and Permafrost Research Establishment* (SIPRE) *Reports;* most other reports published as Congressional documents.

U.S. Bureau of Reclamation: *Technical Reports; Planning Reports.*

U.S. Bureau of the Census: *Census of Agriculture* (decennial, *Irrigation of Agricultural Lands, Drainage of Agricultural Lands*); *Areas of the United States* (including areas of inland waters; by counties); *Survey of Manufactures* (including water use).

U.S. Department of Agriculture: *Yearbook of Agriculture; Technical Bulletins; Miscellaneous Bulletins; Journal of Agricultural Research;* publications of the Soil Conservation Service—*Snow Surveys and Water Supply Forecasts, Technical Paper Series, Progress Reports, Physical Land Surveys,* and *Hydrologic Bulletins;* publications of the Agricultural Research Service—*Monthly Precipitation and Runoff for Small Agricultural Watersheds in the United States, Annual Maximum Flows from Small Agricultural Watersheds in the United States,* and *Selected Runoff Events for Small Agricultural Watersheds in the United States.*

U.S. Federal Power Commission: Statistics on hydroelectric-power generation, installation, and resources; several series on installed generating capacity of hydro- and steam-power plants.

U.S. Geological Survey: *Professional Papers; Water-Supply Papers; Circulars; Hydrologic Investigations Atlases; Water Resources Review;* publications of the Geological Survey, topographic maps; and many water-resources reports of the Survey published by cooperating state agencies.

U.S. Inter-Agency Committee on Water Resources: *Bulletins; Subcommittee on Sedimentation Reports.*

U.S. International Boundary and Water Commission (United States and Mexico): *Water Bulletins.*

U.S. Mississippi River Commission: *Annual Report, Stages and Discharges—Mississippi River and Its Outlets and Tributaries.*

U.S. Public Health Service: *Water Pollution Series;* inventories of public water supplies and municipal water systems; inventories of municipal sewerage works.

U.S. Tennessee Valley Authority: *Precipitation in the Tennessee River Basin* (monthly reports, annual summaries); *Geologic Bulletins; Flood Reports; Technical Reports; Operation of TVA Reservoirs* (monthly reports, annual summaries); *TVA Reservoir Elevations and Storage Volumes* (annual).

U.S. Weather Bureau: *Climatological Data; Hourly Precipitation Data; Local Clima-*

Table 26-I-2. Checklist of Basic Hydrologic Data Collected by the United States Federal Agencies*

Item	Agricultural Research Service	Bureau of the Census	Coast and Geodetic Survey	Department of Commerce†	Corps of Engineers	Fish and Wildlife Service	Forest Service	Geological Survey	Department of the Interior†	International Boundary and Water Commission	Mississippi River Commission	Naval Observatory	Public Health Service	Bureau of Reclamation	Smithsonian Institution	Soil Conservation Service	Tennessee Valley Authority	Weather Bureau
Consumptive use	×						×	×						×		×	×	
Density currents					×			×		×				×			×	
Drought					×		×	×		×				×		×	×	×
Evaporation	×				×			×						×			×	×
Floods					×		×	×									×	×
Groundwater								×						×		×	×	
Humidity																	×	×
Ice	×						×									×	×	×
Infiltration							×							×		×	×	
Interception	×						×							×		×	×	
Lysimeters	×						×							×			×	
Precipitation					×		×	×		×						×	×	×
Chemical quality of water										×			×				×	
River, lake, reservoir stages	×				×					×				×			×	
Runoff	×				×		×	×		×				×		×	×	
Sedimentation:																		
Suspended load					×			×						×		×	×	
Reservoir sedimentation					×		×	×						×		×	×	
Snow	×				×			×						×		×	×	×
Soil moisture	×				×		×	×		×				×		×	×	×
Solar radiation												×			×		×	×
Storms							×	×			×					×	×	×
Streamflow	×				×		×	×			×			×		×	×	
Air temperature	×						×										×	×
Soil temperature	×				×		×									×	×	×

26-11

Table 26-I-2. Checklist of Basic Hydrologic Data Collected by the United States Federal Agencies* (Continued)

Item	Agricultural Research Service	Bureau of the Census	Coast and Geodetic Survey	Department of Commerce†	Corps of Engineers	Fish and Wildlife Service	Forest Service	Geological Survey	Department of the Interior‡	International Boundary and Water Commission	Mississippi River Commission	Naval Observatory	Public Health Service	Bureau of Reclamation	Smithsonian Institution	Soil Conservation Service	Tennessee Valley Authority	Weather Bureau
Water temperature			x		x	x		x x						x			x	x
Tides		x x		x x														
Water use and waste disposal:																		
Drainage													x					
Industrial use					x x		x x		x				x					
Municipal use													x					
Sewage and waste disposal								x		x							x x	
Wind	x																	x
Experimental watersheds	x																	

* Adapted from Langbein and Hoyt [16].
† Agencies other than the Census Bureau, Coast and Geodetic Survey, and Weather Bureau (very few, special installations).
‡ Agencies other than the Fish and Wildlife Service, Geological Survey, and Bureau of Reclamation (very few, special installations).

tological Data; Weekly Weather and Crop Bulletin; Monthly Weather Review; Daily River Stages; Climatography of the United States (temperature and precipitation); Substation History; Technical Papers; Research Papers; Daily Weather Map; Water Supply Forecasts for the Western United States (monthly, January 1 to May 1); Average Monthly Weather Resumé and Outlook (bimonthly); Hydrologic Bulletins (1940–1948; Hydrometeorological Reports (with U.S. Army Corps of Engineers). All weather data are kept in the U.S. Weather Bureau National Weather Records Center, Arcade Building, Asheville, N.C.

B. Utilization of Hydrologic Data

Hydrologic data are a basic ingredient of the planning and development of water-resources projects. Better data are required for better project decisions. Unfortunately, they are rarely adequate for all needful purposes. Deficient hydrologic data may result in the danger of under- or overdesigning, the handicap of insufficient operation, and unnecessary delays and expense in initiating projects [32]. Langbein and Hoyt [16] classified the deficiencies in three fundamental categories: (1) distribution of water information over the country is not most effective; (2) the basic data programs have emphasized data collection to the neglect of advancement of basic principles; and (3) distribution of water facts among those who can use this information is inadequate.

Despite deficiencies, adequate analysis and intelligent utilization of available data are necessary for planning and developing water projects. Methods of analyzing hydrologic data are variously described in other sections of this handbook. Many tools are available for hydrologic analysis, including recorder graphs, hydrographs, duration curves, mass curves, double-mass curves, probability curves, sediment-concentration rating curves, and various methods of frequency and correlation analysis. When hydrologic data are unavailable at the project site, indirect determinations may be employed. For example, runoff data may be determined indirectly by the following methods: extension of a short-term runoff record by using a long-term precipitation record from the same station; extension of runoff data from one or several stations in a river basin to another station in the same basin; comparison of precipitation and runoff relations in a given basin with those of an adjacent or nearby basin; and the use of empirical formulas [10, pp. 17–18].

V. PROJECT FORMULATION

Project formulation, or *plan formulation*, is the procedure to evaluate all pertinent information and to prepare a plan of water-resources-project development. It deals with project investigation, planning, and economic analysis and the method of financing recommended to be used. It should be noted that the principles of project formulation are almost universal, but practice varies considerably from agency to agency and from country to country.

A. Project Investigation and Planning

1. Reconnaissance. The purpose of this step is to determine the feasibility and extent of a water-development project by the simplest means and within the shortest time possible. Usually, a reconnaissance party is formed, composed of an experienced planning engineer and a few assistants. This party collects information and data already available from any reasonably reliable sources and conducts field examinations. The findings thus obtained are covered in a *reconnaissance report*. The main purpose of this report is to ascertain the desirability of continuing the investigation on a more refined basis.

2. Detailed Investigation. This step usually follows reconnaissance if the latter has demonstrated justification for further refined investigation. Major items for this detailed investigation would include various types of surveys, such as control survey, topographic survey, route survey, etc.; the purpose or purposes of the contemplated project; hydrologic studies; geologic studies; power and other potential markets; agricultural investigation; structural study; transportation facilities; alternative proposals; and socioeconomic study.

a. Selection of a Tentative Plan. In the process of detailed investigation, a tentative plan or alternative plans for effective development of the available water resources are sometimes selected with respect to regional or basin-wide needs. The purposes of the plan are often expressed in physical and economic terms, such as achieving a certain degree of flood control, supplying a specific amount of water or power to cities and industries, recreation, fish and wildlife conservation, and drainage of a given acreage of land. Engineering studies are then made in order to develop a plan that will most nearly satisfy the needs at least cost. Such studies include mainly elements of the plan and the types and controlling dimensions of the structures, with consideration of the hydrologic and economic studies.

The hydrologic studies are made to identify available yields or products for each chosen purpose or service. The operations of reservoirs and other structures involved in the proposed plans are studied on the basis of the historical hydrologic data, always taking into account the *design flood*, which must be determined conservatively, i.e., "on the safe side." Such data, whether modified or not, must simulate the actual and reasonably prospective hydrologic behavior in the region or basin in the design period of the plan.

The economic studies include the estimate of capital costs and of the benefits created by the plans. Such studies will lead to the determination of the benefit-cost ratio (Subsec. V-B-4). The capital costs of all engineering works and their operation, maintenance, and repairing during the design period should be evaluated. The benefit for each purpose, including, among other things, the direct and indirect and induced employment, must be estimated and the overall benefit of the plan computed. From the computed costs and benefits, the benefit-cost ratio can be computed to serve as a basis of determining justification of the required investment.

b. Evolvement of the Optimal Plan. The tentative plan or plans may be improved by *incremental*, or *marginal, analysis* [1]. The procedure is to modify the tentative plan by addition or subtraction from it of substantial segments of purpose, levels of output, or sizes of structure and to determine through operation studies the resulting physical effects on the plan as a whole and on the consequent benefits and costs. The differences between the benfits and costs of the project before and after modification represent, respectively, the incremental gross benefits and the incremental costs corresponding to each incremental modification. The incremental analysis of the plan continues until all feasible increments have been tried and no further growth of the total net benefits is obtained. The analysis therefore results in a series of alternative plans from which the optimal plan is evolved, taking into account, of course, the anticipated changes in the economy of the area over a period of, say, 15 to 25 years. The optimal plan is the one that includes all increments to which higher benefits than costs are attached and that excludes all increments which act to the contrary. Usually, the resulting optimal plan differs from the originally conceived plan or plans because of refinements and alternatives incorporated as the study progressed.

c. Preparation of Project Report. All information obtained from the detailed investigation should be presented in a project report. The purpose of this report is to present the facts with a sufficient degree of exactitude to permit the superior administrator to make a positive recommendation for or against the project in its proposed form, scope, conclusions, and probable results. The report should cover the size and character of the project; the type of lands to be served and their location; the cost of the project; the services to be rendered, such as flood control, irrigation, navigation, or any other; allocation of cost to each of the services; the sharing of the allocated costs among the purposes to be served; the manner and extent for each service; the benefits and costs, including methods of financing by various means; and all other germane facts. The investigations which underlie the report are carried to a degree of finality that will assure the proper decision on the merit and effects of the proposal. Preliminary designs are made, with all controlling dimensions and capacities, for it is during the preparation of this report that practically all the original, conceptual, and creative planning is done, but they do not as a rule go into final designs. Such planning reports are used primarily in connection with the authorization of Federal projects or for decisions by private groups who are interested and who may have paid for all or a part of the investigations.

The project report may be prepared in two parts: a general report and a detailed technical report. The *general report* is composed in clear language, with necessary appendixes to support the conclusions and to supplement the data. It should be complete in itself, but it must be concise and informative to both technical and nontechnical readers. As a guide, the following headings are suggested for such a report: general description, present use of water resources, need for development, plan of development, economic analysis, financing, related investigations by others, conclusions, and recommendations. The *technical report* should present a discussion of subjects necessary to a thorough understanding of the findings, conclusions, and recommendations made in the general report. The items listed in Subsec. III may be suggested as the headings in such a report. The contents of the report should contain not only the detailed results of investigation, but also the basis of analyses and computations and the reference to the methods used.

3. Definite Planning. This step follows the detailed investigation covered by the project report, after it is approved. The plan covers a complete layout and the dimensions of all structures, with specificity. It is reported in a *definite-plan report*. This report is a refinement of the project report and serves as a basis for the preparation of the actual design drawings and specifications. It is during this stage, particularly, that many design savings can be recognized and incorporated and that any operational hydraulic or structural-model studies are usually made.

When the development of water resources aims at a large river basin, a *basin-plan report* is necessary. This report assembles and relates reconnaissance and project reports for all projects in the river basin and correlates them into an overall basin plan. Its purpose is to relate the specific project proposals to the conditions and requirements of the basin and to set up an overall, but intentionally fluid, pattern for ultimate basin-wide development.

A basin plan is not necessarily the summation of the optimum plans for its separate projects, for in a basin plan a project may be converted into a single-purpose project, while some of its other possible attributes are carried further in other projects of the overall plan. The basin plan is therefore to achieve the optimum benefits at minimum costs *and still* meet the special needs of immediate areas, as well as those of the basin as a whole.

B. Economic Aspects of Project Formulation

Although many Federal agencies and practically no state, local, and private interests are required to compute and publish the relationship between benefits and costs (the benefit-cost ratio, or B/C), it is a requirement of the leading Federal water-resources agencies. Because of the simplicity of its final figure presentation, the benefit-cost ratio is being used with increasing regularity, even by those who are not required to do so. Its use is probably applied to half to perhaps two-thirds of the costs of all the dollars spent annually on water-resources types of projects at all governmental levels.

It is discussed here as a preamble to *project formulation* because, in 1961 and 1962, certain Acts of the United States Congress (Public Law 87-27 [24] and Public Law 87-658 [26], together with Senate Document no. 97 [27], a new Federal water-resources policy approved by the President on May 15, 1962) have substantially altered the previous criteria for project justification, in terms of the benefit-cost ratio, and therefore in terms of project formulation. Official Federal interagency committees have been at work on the general benefit-cost-ratio subject continuously since 1946, but the following features constitute the most significant single set of changes yet adopted, though many elements of the earlier studies are naturally reflected in the new documents.

Participating funds, in terms of long-term loans at low rates of interest plus grants (some repayable in whole or in part from the project, if it moves on to completion) are provided. Also, funds have been made available for two general types of research: first, research seeking the fundamental causes of unemployment or underemployment, and second, research leading toward stimulation of existing or potential industries in depressed areas. This latter type of research ranges from market research to processing and manufacture research and to the creation or adaptation of new types of industries in areas of unhealthy economic atmospheres.

In addition, Senate Document no. 97 [27] alters and relaxes many of the basic concepts of *Budget Bureau Circular* A-47, which has been the controlling Federal water-resources-policy documant for at least a decade.

Some of these new elements influence very substantially the benefit-cost ratios in depressed areas only (as defined in the public laws of Congress, since the statuses of the depressed areas may change periodically), while other provisions affect Federal projects throughout the country generally, wherever they may be. An example of the latter is the fact that the project benefits to recreation and to the enhancement of fish and wildlife are now made generally nonreimbursable benefits in the same sense that flood-control and navigation benefits are nonreimbursable. Of course, some cost sharing may be applicable under certain circumstances, particularly for very localized benefits, and become a part of the cost side of the benefit-cost formula, but the full benefit may be included in the benefit side of the formula. The very fact that the full benefits have now been so classified alters the whole matter of project formulation. Already pressures are beginning to be felt by some states and local communities to utilize this approach. If history repeats, as it has done so often in these matters, project formulation will, within a matter of years, take such procedures for granted and use them as normal, just as in the cases in the recent past where large dams, built by non-Federal funds, have secured from Congress a nonreimbursable allocation of cost to flood control, provided that the dams are operated in such ways as to produce a reduction of floods.

This type of transitional change forces the hydrologist into a much more exacting position in regard to this use of hydrologic data and analyses, particularly the operations of a certain dam to achieve predecided flood-control benefits as a part of a system of dams. These have to be done in accordance with established water rights, water policies of the several agencies, water laws of the separate states, and, in so far as they may exist, interstate water compacts and treaties, such as those with Canada and Mexico.

The foregoing are the simpler types of changes in project formulation brought about by these documents. In the "depressed areas" as roughly defined by the Congressional formulas (a surprisingly large percentage of the total area of the 50 states, varying from time to time from one-fourth to one-third), there are other items that may now be classed as "benefits." These are specifically designed to promote stabilization or improvement in the economies of these depressed areas. For example, in those areas, the following criteria are being seriously considered in some informal interagency quarters as constituting an acceptable basis for measurement of such benefits:

1. Some proportion of the labor costs of construction.
2. Some proportion of the costs of other resources required for project construction, including, particularly, machinery and manufactured equipment permanently built into the project.
3. Part of the cost of operation and maintenance of each completed and separately usable portion of a project, which involves study of what changes in the economy of an area may be expected to occur with and without the project.
4. A portion of the direct labor costs at new or enlarged industrial facilities for a reasonable period of time, based upon the same type of analysis as discussed in item 3.
5. The benefits resulting from induced, or secondary labor (related services and commerce), which differ from project to project, but on the average may approximate half of the above four benefits.

Detailed economic analyses of the effects of these items are proceeding at the time of this writing (1963), and it cannot be said exactly what the precise results will be. It can be said, however, that in the depressed areas, these now permissible benefits will result in a change in both the manner and substance of project formulation.

Since all these reflect the work of the hydrologist fundamentally, as well as of the engineer, economist, administrator and financier (public or private), and businessman, one thing is certain: any historic formulas or definitions for project formulation are now in a state of reexamination and the new working procedures are not yet clearly available. They are in a state of flux, both in the public and, so far to a lesser degree, in the private fields of project formulation.

Further, there is an entirely different tool being developed for use in project formulation. To date, it is strictly a tool and has nothing to do with changes resulting from altered policies or legislation, but its ultimate potentials must not be overlooked. This tool is the electronic computer. The electronic-computer techniques may some day reach a point where special "programming" may permit them to indicate direct, or at least relative, decisions about project formulation. Today they are only a useful tool to expedite the many computations that are necessary, and they do not have imagination or judgment of their own, since their "answers" reflect only the combinations of the raw materials put into them. They give only the mathematical answers to the formulas that are introduced through "programs" produced by the imagination and judgment of the hydrologists, engineers, economists, and others, and put into the machines. Thus electronic computers are a tool for rapid computation, but so far are limited to the operational parts of project formulation.

Some of the effects of the laws and purposes of Senate Document no. 97 [27] are clear, such as the facts that recreation and fish and wildlife conservation are already declared, as a matter of policy, to be benefits. Work on the precedures is already well under way in both the executive and legislative branches of the United States government, but accepted techniques for implementing those and the other types of benefits are not yet final.

Therefore, in order to proceed, the author has chosen to utilize, for illustrative purposes, the heretofore generally accepted methods of project formulation. However, in using the following procedures, the reporter should ascertain the applicability of any or all of the anticipated changes described above and incorporate them, if feasible, in his final report. The source of his information should be the President's Water Resources Council as prescribed in Senate Document no. 97 [27], or the same source if it is established by the pending legislation, Senate Bill no. 1111 [33], or in the absence of both, the Secretary of the Interior, who served as chairman of the Council that prepared Senate Document no. 97 [27].

Thus the heretofore acceptable procedures are presented below. These may properly be used preliminary to explication of the effects of the changes discussed above and should be modified as rapidly as applications of the new laws and policies to project formulation are made clear.

1. **Definitions.** Terminology commonly used in the economic analysis of water-resources projects may be defined as follows:

Project costs are the value of goods and services used to establish, maintain, and operate a project.

Associated costs are the value of goods and services needed, beyond project cost, to make the output of the project available for use or sale.

Indirect costs, also known as *secondary costs*, are the value of goods and services other than project and associated costs, used as a result of the project, including the costs of processing the immediate products of the project.

Intangible costs are costs which cannot be estimated in monetary terms.

Direct benefits, also known as *primary benefits*, are the value of the immediate products and services resulting from the measures for which project and associated costs are incurred.

Indirect benefits, also known as secondary benefits, are the value in addition to direct benefits as a result of activities stemming from or induced by the project.

Intangible benefits are benefits that cannot be expressed in monetary terms.

Project benefits are the direct benefits attributable to the project plus any net indirect benefits, that is, the net value of the goods and services produced by the project and by activities stemming from or induced by the project after deducting all nonproject (associated and indirect) costs involved. Such benefits must be net of all costs other than those designated as project costs.

Separable costs are costs which are clearly chargeable to a single project function, such as the cost of the powerhouse, navigation locks, or flood-control dams. The separable costs for a single function are usually estimated by computing the cost of the project with that function omitted. In other words, the separable cost of the function is the total project cost less the estimated cost with that function omitted.

Joint costs are the total cost less the sum of the separable costs.

Benefit-cost ratio is the ratio of total annual benefit (direct plus indirect) to total annual cost (project, associated, and indirect costs).

Marginal cost is the increase in cost per unit increase in output of the project.

Marginal benefit is the increase in benefit per unit increase in output of the project.

Marginal benefit-cost ratio, also known as *incremental benefit-cost ratio*, is the ratio of marginal benefit to marginal cost. In other words, it is the ratio of increments of benefits to corresponding increments of costs.

2. Benefit Study. In the process of economic analysis, benefits and costs must be first identified and then expressed in monetary terms as far as possible. The total benefit is equal to the measure of economic improvement received "with development" less that "without development." The total benefits and costs must be converted to an annual basis. Because of the time factor, all benefits must be discounted to the time when the project will start to operate.

Examples of various benefits associated with different purposes of a water-resources project are as follows.

a. Flood Control

Direct benefits: reduction or prevention of direct physical damage to properties, such as land and agricultural crops; change of lands to higher or more intensive uses.

Indirect benefits: reduction or prevention of indirect damages, such as loss of income, wages, good will; interruption of services, utilities, and transportation; disruption of markets; temporary rentals; unusual expenditures.

Intangible benefits: prevention of loss of life, personal injury, and sickness; maintenance of public morale.

b. Irrigation

Direct benefits: increase in farm products; increase in cash income.

Indirect benefits: increased net income of local and nonlocal handling, processing, and marketing of farm goods, and of enterprises supplying goods and services to farmers; increased land value of local residential property.

Intangible benefits: greater stability and welfare of community, strengthened security; better diet and health; new opportunities for settlement, employment, and investment.

c. Navigation

Direct benefits: increase of undeveloped or new traffic; lower costs, attracting existing traffic to improved transport; elimination or reduction of damages, delays, and hazards associated with existing traffic and resulting from improvement of an existing waterway; recreational value for small-boat traffic.

Indirect benefits: increase in net income due to new or increased economic activity engendered by the project.

Intangible benefits: flexibility of total transportation; availability for use in national emergencies; stimulation of business activity.

d. Hydroelectric Power

Direct benefits: gross power revenue adjusted for gains or losses at downstream plants.

Indirect benefits: share of returns to distributors of project power; saving to consumers from lower power rates; benefits attributable to power in final production of goods and services.

Intangible benefits: increased comfort and convenience; improved living conditions; increased industrial development; national security; conservation of nonrenewable fuels.

e. Domestic and Industrial Water Supply
Direct benefits: measured by alternative cost, or water revenues.
Indirect benefits: usually described but not measured.
Intangible benefits: increased industrial development; greater municipal and national security; higher living standards; increased potentialities for municipal growth.

f. Fish and Wildlife and Recreation
Direct benefits: commercial value of fish and game caught measured in terms of expected market prices; value of increased recreational use of hunting, fishing, and outdoor recreation.
Indirect benefits: increased land value of local residential property.
Intangible benefits: personal enjoyment; better health and outlook; better work and production.

g. Water-quality Control
Direct benefits: preservation of aquatic life; decreased cost of treatment of domestic and industrial water supply and waste material; value of damage prevented.
Indirect benefits: increased uses of water.
Intangible benefits: prevention of smells and obnoxious nuisance conditions; better recreation facilities; safer water supply; improved scenic values.

For different purposes of water-resources projects, there are various methods of measuring their benefits [34–38]. Discussion of this subject is beyond the scope of this treatment.

3. Cost Study. The capital costs of water-resources projects should include all tangible costs, directly or indirectly identifiable, incurred from the start of preliminary investigation to the point of beneficial use and occupancy of the project. Estimates of these costs should be based on price levels prevailing at the time of the estimate and then reviewed later whenever necessary.

In establishing benefit-cost ratios, annual costs should be computed on the basis of averages over the period of amortization of the project. The major items to be included in computing annual costs are (1) direct operation and maintenance expenses, (2) administration and general expenses, (3) interest, amortization, and interim repayment, and (4) taxes of various kinds.

In computing item 3, it is necessary to know the probable life of each of the major parts of the project. Unfortunately, there are no universally accepted standard figures for these "lives." Although recommended figures can be found in the literature [10], they should be adopted with necessary modifications with respect to the circumstances under consideration.

The probable life of a project for economic analysis is specifically known as *economic life*. The "life" is affected by such factors as physical depreciation, obsolescence, changing requirements for project services, and time discount and allowances for risk and uncertainty. It is determined by the point in time at which the effect of these factors causes the costs of continuing the project to exceed the additional benefits to be expected from continuation. As so used, the economic life is generally shorter than the physical life of a project, and never longer than the estimated physical life. According to the U.S. Federal Inter-Agency River Basin Committee [1], it is recommended that the maximum period of economic analysis for river-basin projects be the expected economic life of the project or 100 years, whichever is shorter. See also Ref. 27.

Because of the utilization of one structure to provide the services of different natures and the requirements of laws to make some services reimbursable, others nonreimbursable, some interest-bearing, and others non-interest-bearing, government agencies usually find it necessary to allocate the total project cost among the several functions of the project [39]. However, the cost allocation is not a part of the benefit-cost-ratio analysis for economic justification of the project.

Although a number of laws provide for allocations of cost, no law or adopted Federal policy yet provides how such allocations should be made. This is being given further detailed study under the conditions of approval by the President of Senate Document no. 97 [27]. And, in addition, he has provided for the establishment of cost-sharing procedures within the structure of the cost-allocation procedure. Many methods of cost allocation have been proposed. The only sound basis, however, is equitable consideration of all purposes.

There are many methods for allocating costs. According to the U.S. Federal Inter-Agency River Basin Committee [1], the *separable costs–remaining benefits method* is recommended for general use in allocating costs on Federal multiple-purpose river-basin projects. By this method, the allocation to each purpose is equal to the separable cost assigned to the purpose plus a share of the joint costs distributed in proportion to the remaining benefits of the purpose. The so-called remaining benefit is the difference between the separable cost and the estimated benefit of the purpose. However, the estimated benefit should be less than the cost of an alternative single-purpose project. Otherwise the latter would be used in lieu of the estimated benefit in the computation.

4. Benefit-Cost Analysis. The procedure of the benefit-cost analysis may be illustrated by a working example. Necessarily, this example is an abbreviated presentation of the procedure, but displays the techniques and the underlying philosophies.

It must be appreciated that to apply this procedure in any country other than the United States of America will necessitate consideration of the resources-development policies of that country. Modification of the procedure to fit such policy will be essential. It must also be kept in mind that assigning monetary values to most of the intangible benefits is difficult to impossible, and this subject must be approached with caution. The following working example is given, notwithstanding the fact that it may be controversial in certain quarters.

a. Definitions for the Working Example

General basis: Wherever possible, benefits and costs are expressed in monetary terms; other effects are described as nonmonetary quantities or in narrative. Federal costs of installing and operating the project are compared with increased incomes of beneficiaries less non-Federal costs.

Time basis: All benefits and costs are converted to equivalent average annual amounts for a selected period of analysis. This period is the estimated economic life of principal project works or 100 years, whichever is less.

Interest rate: A rate of $2\frac{7}{8}$ per cent (1963) is used for interest during the period of construction and for calculating annual equivalent Federal investment costs, annual equivalent replacement costs, non-Federal public investment cost (e.g., municipal or state), and annual equivalent benefits. A rate of 4 per cent is used for annual cost of private investments.

Price levels: Current prices are used for installation costs, and estimated average future prices, for benefits and deferred costs.

Irrigation benefits: These include (1) direct benefits from increase in family living (home-grown food, rental value of dwelling, etc.), increase in net cash farm income, and accumulation of equity; (2) indirect benefits from increased profits of local and non-local handling, processing, and marketing of farm products and of enterprises supplying goods and services to farmers and increased land value of local residential property; and (3) public benefits from provision of new opportunities for settlement, employment, and investment, improved community facilities and services, and stabilization of local or regional economy.

Power benefits: Benefits are based on the cost of the most economical alternative source of power (equivalent to power revenues plus savings to consumers from lower power rates).

Domestic, municipal, and industrial water-supply benefits: Benefits are based on the cost of the most economical alternative source of water supply (equivalent to revenues plus savings to consumers from lower water rates).

Flood-control benefits: Direct benefits are based on reductions in flood damage to land and other properties, in terms of restoration cost or reduction in value, and on reduced damage to crops in terms of market value adjusted by replanting possibilities and production costs not incurred. Indirect benefits are based on higher-grade use of land formerly flooded in terms of increased earnings and on reduced interruption of business, industry, and commerce in terms of net loss of income or added operating costs.

Navigation benefits: Direct benefits are based on savings to shippers compared with

PROJECT FORMULATION   **26–21**

the alternative cost, savings in time and operating cost from improvement of an existing waterway and estimated recreational value of harbors and waterways to small-boat traffic. In some cases, indirect benefits from stimulation of business activity are considered.

Pollution-abatement benefits: Benefits are based on the cost of the most economical alternative method of waste treatment or disposal or on reduction in operating costs of existing facilities.

Fish- and wildlife-conservation benefits: Benefits are based on increased value of annual yields for hunting and fishing in terms of expenditures by sportsmen and gross market value of commercial fish and fur.

Salinity-control benefits: Benefits are based on the value of damage prevented, reduction in costs of some types of sediment removal, increased value of services of reservoirs and channels, and value of extended life of facilities.

Recreation benefits: Benefits are based on judgment evaluation.

Project cost: This includes construction cost (excluding the cost of investigation prior to authorization of the project) plus interest at $2\frac{7}{8}$ per cent on half of the cost for the period of construction, less the present worth of salvage value remaining at the end of the period of analysis. The project cost represents the net project investment.

Annual cost: This is the annual equivalent of the net project investment plus the average annual cost of operation, maintenance, and replacement. The annual cost represents the average equivalent project cost.

b. The Working Example

Time basis: The useful life of the major structure, a concrete dam, is estimated at 150 years. The period of analysis is set at 100 years from the date of initial operation of the project.

Interest rate: For 100 years at $2\frac{7}{8}$ per cent interest, $1 today is worth $0.0306 of annuity, $1 of annuity over 100 years is worth $32.722 today, and $1 at the one-hundredth year is worth $0.0589 today.

Price levels: Future agricultural prices are estimated at 215 per cent of the prices in the years 1910 to 1914. Future operating and maintenance costs are estimated at 180 per cent of the *Engineering News-Record* base-period prices for the year 1939. Installation costs are estimated at prices current at the time of investigation, e.g., 230 per cent of the 1939-base-period prices. These figures should be updated by use of ENR indices to the date when any project report is prepared.

Irrigation benefits: A summary of the farm budgets in terms of annual values is as follows:

Item	With irrigation	Without irrigation	Difference	Benefit
Type of farm	Irrigated	Dry		
Number of farms	100	10	90	
Acres per farm	160	1,600		
Direct benefits:				
Farm products sold	$1,350,000	$100,000	$1,250,000	
Home consumption	50,000	3,000	47,000	
Rental value of home	25,000	2,000	23,000	
Gross farm income	$1,425,000	$105,000	$1,320,000	
Production expenses	900,000	60,000	840,000	
Farm living expenses	112,500	22,500	90,000	
Net cash income	$ 412,500	$ 22,500	$ 390,000	$390,000
Farm investment	4,000,000	400,000	3,600,000	36,000*
Total				$426,000

* Increase in equity over 100 years by straight-line method.

Indirect benefits:
Farm products sold	$1,350,000	$100,000	$1,250,000	$200,000†
Farm purchases	960,000	80,000	880,000	150,000†
Total				$350,000

Intangible benefits:
Stabilization of economy (expansion, if inherently practicable)	74,000†
Total	$ 74,000
Total irrigation benefits	$850,000

† Statistical projections.

Power benefits:
Installed hydroelectric capacity		50,000 kw
Equivalent steam-electric capacity		45,000 kw
Investment for steam at $160 per kilowatt		$ 7,200,000
Average annual hydroelectric generation	337,000,000 kwhr	
Equivalent annual steam-electric generation	234,300,000 kwhr	
Annual cost of steam generation at 3.81 mills/kwhr (3.81 × 234,300,000 × 0.001)	$892,700	
Annual hydroelectric expenses at 0.7475 mill/kwhr	251,900	
Increase in annual steam expenses over hydro	$640,800	
Present value of increased annual expenses (640,800 × 32.722)		$20,968,000
Investment in power transmission to hydro site due to shifting of load center		5,492,000
Total alternative steam-plant expenses		33,660,000
Annual power benefits computed over 100 years (0.0306 × 33,660,000)		$ 1,030,000

Domestic, municipal, and industrial water benefits: Evaluation is similar in principle to that for power benefits, but calculations are related to quantity of water and water rates, adjusted for qualitative differences such as purification treatment. In this example, such benefits are assumed nil.

Flood-control benefits: These benefits are flood damages reduced by the project. The computation of annual values of flood damages is summarized as follows:

Item	Without project	With project	Preventable damage
Direct crop losses	$ 75,000	$ 40,000	$ 35,000
Direct damage to irrigation works	65,000	20,000	45,000
Indirect crop losses due to failure of irrigation works	600,000	250,000	350,000
Damage to urban property	50,000	0	50,000
Damage to city water works	20,000	0	20,000
Damage to railroad	10,000	0	10,000
Damage to rural areas	60,000	20,000	40,000
Total	$880,000	$330,000	$550,000
Annual flood-control benefits			$550,000

Navigation benefits: The minimum navigable depth of the river is 7 ft without the project, and it would be 8 ft when the project is established. Accordingly, the estimated savings in annual maintenance of the channel due to the project is $100,000.

Pollution-abatement benefits: The present minimum streamflow near the pollution-affected city is 450 cfs in summer and 55 cfs in winter. Under project conditions, the minimum streamflow would be 325 cfs in summer and 200 cfs in winter. Accordingly, both primary and secondary treatments of sewage are needed at the present, whereas only the primary treatment would be necessary under the project conditions. The savings in cost of the secondary treatment due to the project is estimated as $55,000 per year. This is the annual pollution-abatement benefits.

Fish- and wildlife-conservation benefits: Computation of these benefits can be summarized as follows (the numbers for waterfowl and fishery are measured by number of days in a year used by hunting and fishing, and other numbers are annual harvest):

Item	With project Number	With project Value	Without project Number	Without project Value	Difference, value
Big game, deer	76	$ 7,600	1	$100	$ 7,500
Upland game:					
Quail	100	700	20	140	560
Rabbits	620	620	80	80	540
Fur animals:					
Skunk	95	95	5	5	90
Mink	16	160			160
Beaver	50	1,000			1,000
Muskrat	150	150			150
Waterfowl:					
Ducks	10,000,000	50,000			50,000
Geese	300,000	3,000			3,000
Fishery	60,000	210,000			210,000
Total		$273,325		$325	$273,000
Annual fish and wildlife benefits					$273,000

Salinity-control benefits: These benefits are estimated similarly to flood-control benefits, except that the damages are attributable to saline water rather than to floods. In this example there are no such benefits.

Sediment-control benefits: These benefits are estimated similarly to flood-control and salinity-control benefits, except that the damages are attributable to sediment rather than to floods or saline water. They are usually included with flood-control benefits in cases where sediment damage occurs primarily during floods. In this example such benefits are nil.

Recreation benefits: The estimated maximum attendance of visitors per day is 300 persons. The total attendance during the year is thus estimated as 10,000 persons. Assuming an admission fee of $1 per person, the judgment estimate of the annual recreational benefits is therefore $10,000.

Total benefits and justifiable investments: These are shown in Table 26-I-3.

5. Project Composition. By applying the benefit-cost analysis to each separate purpose or service, it is possible, at least to some degree, to ascertain which of the purposes should be given paramount consideration in formulating the plan of development. The problem of choice of the purpose and of its degree of service to be rendered is always present. Too often the choice is made arbitrarily, or by "cloakroom compromise."

Whether a project is privately or federally developed, it undoubtedly has direct and intangible benefits and costs. The problem is therefore to determine, in a practical way, how to compare the benefits and the costs. It would be possible, but impracticable, for example, to measure the benefits of a sugar-beet-producing irrigation project by determining the benefits, first, to the farmer, second, to the refiner, third, to the packer and trucker, fourth, to the shipper, fifth, to the wholesale grocer, sixth, to the merchandiser, and seventh, to the consumer of the sugar, who uses it in his coffee 2,000 miles away. By the same token, it would be impossible or impractical to assign costs realistically to each of these steps. A practical cutoff line must therefore be determined both for benefits and for costs, and this cutoff line must be an impartial one, which gives equal emphasis to the benefits and to the costs. One of the most intensive studies of benefits and costs is described in Ref. 1. However, such studies are still in progress since there have been differences of opinion on some matters even among informed personnel, and a few limited but controversial and highly complicated matters remain for further study.

A small chart can be used to illustrate the use of benefit-cost analysis in the determination of the size and characteristics a project should have or the degree of service that any one function of a multiple-purpose project might best be planned to render. To prepare such a chart, the benefit-cost ratio is computed separately for each particular service to be rendered. This can be done for a number of projects of different sizes. The points are then plotted for each service, and a curve is drawn through them. These individual *service curves* can be used to ascertain the best degrees of development

Table 26-I-3. Total Benefits and Justifiable Investments

Purpose	Annual benefits	Annual operation and maintenance	Annual net project benefit	Capitalized net benefit
Irrigation	$ 850,000	$120,000	$ 730,000	$23,887,000
Power	1,030,000	250,000	780,000	25,523,000
Flood control	550,000	50,000	500,000	16,361,000
Navigation	100,000	10,000	90,000	2,945,000
Pollution abatement	55,000	1,000	54,000	1,767,000
Fish and wildlife	273,000	32,000	241,000	7,886,000
Recreation	10,000	5,000	5,000	164,000
Total	$2,868,000	$468,000	$2,400,000	$78,533,000

Project-construction cost:
Separable costs:
Irrigation	$15,000,000
Power	16,000,000
Flood control	1,500,000
Recreation	54,000
Joint costs (in reservoir)	25,000,000
Total cost*	$57,554,000
Capitalized net benefits	78,533,000
Savings from the joint development	$20,979,000

Project investment:
Construction cost	$57,554,000
Previous investigation cost	54,000
Cost of future work	$57,500,000
Interest during 8 years of construction	6,612,000
Cost at beginning of period of analysis	64,112,000
Present worth of terminal salvage value	1,712,000
Net project investment	$62,400,000

Annual equivalent project cost:
Annual equivalent of net project investment ($0.0306 \times 62,400,000$)	1,909,000
Annual operation, maintenance, and replacement	468,000
Total annual costs	2,377,000

Benefit-cost ratio:
Total annual benefits	$ 2,868,000
Total annual costs	2,377,000
Annual excess of benefits over costs	$ 491,000
Benefit-cost ratio	1.2:1.0

* Depending upon the outcome of decisions and procedures yet to be established. See discussion at beginning of Subsec. V-B, as other costs such as recreation, fish and wildlife, and area redevelopment for depressed areas should be included here.

for each particular service in relation to other possible project services. Having determined the relative degrees of services to be provided, a composite curve can be drawn for sizes of completely formulated multiple-purpose projects. Such a curve is shown in Fig. 26-I-3. For the multiple-purpose project shown on the graph, the following may be noted:

1. The project size must be at least as great as that at point A, for anything smaller than that would not produce a project with a benefit-cost ratio of at least 1:1.
2. The project which will produce the maximum benefit-cost ratio is at point B. One obtains the greatest return per dollar invested with a project built for this size.
3. The project having the maximum net benefits, that is, the last added increment has a benefit-cost ratio of 1:1, is shown at point C.
4. The largest project that can be built and still have an overall benefit-cost ratio of at least 1:1 is shown at point D.

It is recognized, of course, that any project built to the size between C and D has an incremental benefit-cost ratio of less than 1:1 for the increment or portion of its size which is between points C and D. In other words, one can still justify the total project to size D if sheer size is an objective, even though one would have had a more economic project at any other point between A and D.

It is believed that, in most cases, public projects built to the size at point C are best, for they will make maximum use of a reservoir site or other resource, compared with money invested in the project. For private, *risk capital* investment, where the profit motive ordinarily controls, projects should usually be built to the size at point B, unless some governmental arrangement is made to furnish the incremental cost to enlarge the project from size B to size C, in return for assurances that the project will be so operated as to produce the added public benefits between those two points.

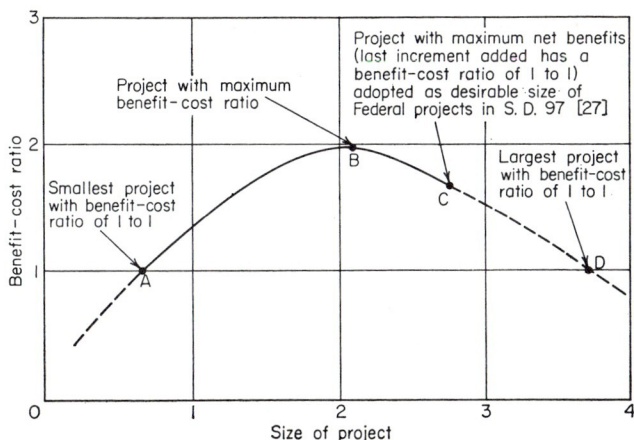

Fig. 26-I-3. Selection of project on the basis of benefit-cost ratio.

VI. REFERENCES

1. "Proposed Practices for the Economic Analysis of River Basin Projects," Report of the Subcommittee on Benefits and Costs, Federal Inter-Agency River Basin Committee, May, 1950, revised by Subcommittee on Evaluation Standards, May, 1958.
2. "A Water Policy for the American People," vol. 1 (of 3 vols.), Report of the President's Water Resources Policy Commission, 1950.
3. Multiple-purpose reservoirs: a symposium, *Trans. Am. Soc. Civil Engrs.*, vol. 115, pp. 789–908, 1950.
4. "Water Resources," vol. IV, Proceedings of the United Nations Scientific Conference on the Conservation and Utilization of Resources, Aug. 17–Sept. 6, 1949, Lake Success, N.Y., United Nations Department of Economic Affairs, New York, 1951.
5. Huffman, R. E.: "Irrigation Development and Public Water Policy," The Ronald Press Company, New York, 1953.
6. Bowman, J. S.: Multipurpose river developments, *Trans. Am. Soc. Civil Engrs.*, vol. CT, pp. 1125–1131, 1953.
7. Leopold, L. B., and Thomas Maddock, Jr.: "The Flood Control Controversy: Big Dams, Little Dams, and Land Management," The Ronald Press Company, New York, 1954.
8. "Water," Yearbook of Agriculture, 1955, U.S. Department of Agriculture, 1955.
9. Hoyt, W. G., and W. B. Langbein: "Floods," Princeton University Press, Princeton, N.J., 1955.
10. Multiple-purpose river basin development, pt. 1, Manual of River Basin Planning, no. 7, Flood Control Series, United Nations Economic Commission for Asia and the Far East, *United Nations Publ.* ST/ECAFE/SER.F/7, Jan. 24, 1955.
11. "Water Resources and Power," 2 vols., A Report to the Congress by the Commission on Organization of the Executive Branch of the Government, June, 1955.
12. "Report on Water Resources and Power," 3 vols., prepared for the Commission on Organization of the Executive Branch of the Government by the Task Force on Water

Resources and Power, 1955. A study made by the Task Force for the Commission's use, not to be confused with the Commission's own report to the Congress [11].
13. Krutilla, J. V., and Otto Eckstein: "Multiple Purpose River Development: Studies in Applied Economic Analysis," The Johns Hopkins Press, Baltimore, 1958.
14. Eckstein, Otto: "Water-resource Development: The Economics of Project Evaluation," Harvard University Press, Cambridge, Mass., 1958.
15. McKean, Roland: "Efficiency in Government through Systems Analysis with Emphasis on Water Resources Development," John Wiley & Sons, Inc., New York, 1958.
16. Langbein, W. B., and W. G. Hoyt: "Water Facts for the Nation's Future: Uses and Benefits of Hydrologic Data Programs," The Ronald Press Company, New York, 1959.
17. Ackerman, E. A., and G. O. G. Löf: "Technology in American Water Development," The Johns Hopkins Press, Baltimore, 1959.
18. Shih, Yang-Chéng: "American Water Resources Administration," 2 vols., The Bookman Associates, Inc., New York, 1959.
19. Hirshleifer, Jack, J. C. De Haven, and J. W. Milliman: "Water Supply: Economics, Technology, and Policy," The University of Chicago Press, Chicago, 1960.
20. "Water Resources Activities in the United States," made pursuant to S. Res. 48, 86th Congress, Senate 1959–1960, Select Committee on National Water Resources, Government Printing Office, 1961. Contains 32 Committee prints in three volumes:

Vol. 1—Background

Print 1. Water facts and problems, by Luna Leopold, U.S. Geological Survey, August, 1959.
Print 2. Reviews of national water resources during the past fifty years, by Barbara Jibrin, Library of Congress, Legislative Reference Service, October, 1959.
Print 3. National water resources and problems, by U.S. Geological Survey, January, 1960.
Print 4. Surface water resources of the United States, by U.S. Geological Survey, January, 1960.
Print 5. Population projections and economic assumptions, by the Committee Staff, U.S. Census Bureau and Resources for the Future Staff, Inc., Washington, D.C., March, 1960.
Print 6. Views and comments of the states, a compilation of reports by state officials, January, 1960.

Vol. 2—Future Needs

Print 7. Future water requirements for municipal use, by Department of Health, Education, and Welfare, Public Health Service, January, 1960.
Print 8. Future water requirements of principal water-using industries, separate reports by Department of the Interior, Department of Commerce and the Bureau of Mines, April, 1960.
Print 9. Pollution abatement, by U.S. Public Health Service, January, 1960.
Print 10. Electric power in relation to the nation's water resources, separate reports by Federal Power Commission, Edison Electric Institute, American Public Power Association, Rural Electrification Administration, Department of Agriculture, and Brace C. Netschert, Resources for the Future, Inc., January, 1960.
Print 11. Future needs for navigation, by U.S. Army Corps of Engineers, May, 1960.
Print 12. Land and water potentials and future requirements for water, by Department of Agriculture, December, 1959.
Print 13. Estimated water requirements for agricultural purposes and their effects on water supplies, by Department of Agriculture, December, 1959.
Print 14. Future needs for reclamation in the United States, by Department of the Interior, Bureau of Reclamation, April, 1960.
Print 15. Floods and flood control, by U.S. Army Corps of Engineers, July, 1960.
Print 16. Flood problems and management in the Tennessee River Basin, by the Tennessee Valley Authority, December, 1959.
Print 17. Water recreation needs in the United States, 1960–2000, by Department of the Interior, National Park Service, May, 1960.
Print 18. Fish and wildlife and water resources, by Department of the Interior, Fish and Wildlife Service, April, 1960.
Print 19. Water resources of Alaska, by Department of the Interior, January, 1960.
Print 20. Water resources of Hawaii, by Department of the Interior, January, 1960.

Vol. 3—Meeting Needs

Print 21. Evapotranspiration reduction, two separate reports: pt. I, Phreatophytic and hydrophytic plants along western streams, by Department of the Interior, and pt. II, Vegetative management and water yields in the 17 western states, by Department of Agriculture, February, 1960.

Print 22. Weather modification, two separate reports: pt. I, Progress and possibilities in weather modification, by Department of Commerce, Weather Bureau, and pt. II, Weather modification as a new technique to increase water supplies, by A. M. Eberle, staff member, Senate Committee on Public Works, January, 1960.

Print 23. Evaporation reduction and seepage control, by Department of the Interior, Bureau of Reclamation, December, 1959.

Print 24. Water quality management, by Department of Health, Education, and Welfare, Public Health Service, February, 1960.

Print 25. River forecasting and hydrometeorological analysis, by Department of Commerce, Weather Bureau, November, 1959.

Print 26. Saline water conversion, by Department of the Interior, Office of Saline Water, November, 1959.

Print 27. Application and effects of nuclear energy, by U.S. Atomic Energy Commission, December, 1959.

Print 28. Water resources research needs, by Department of Agriculture, February, 1960.

Print 29. Water requirements for pollution abatement, by George W. Reid, University of Oklahoma, Norman, Okla., July, 1960.

Print 30. Present and prospective means for improved reuse of water, by Abel Wolman Associates, Baltimore, Md., March, 1960.

Print 31. The impact of new techniques on integrated multiple-purpose water development, by Edward A. Ackerman, Carnegie Institution of Washington, Washington, D.C., March, 1960.

Print 32. A preliminary report on the supply of and demand for water in the United States as estimated for 1980 and 2000, by Nathaniel Wollman, Resources for the Future, Inc., Washington, D.C., August, 1960.

21. Golzé, A. R.: "Reclamation in the United States," McGraw-Hill Book Company, Inc., New York, 1952; 2d ed., The Caxton Printers, Ltd., Caldwell, Idaho, 1962.
22. Linsley, R. K., Jr., and J. B. Franzini: "Water-resources Engineering," McGraw-Hill Book Company, Inc., New York, 1963.
23. Fry, A. S., and others: Basic considerations in water resources planning, Progress Report of the Committee on Water Resources Planning, *Proc. Am. Soc. Civil Engrs., J. Hydraulics Div.*, vol. 88, no. HY5, pt. 1, pp. 23–55, September, 1962.
24. An act to establish an effective program to alleviate conditions of substantial and persistent unemployment and underemployment in certain economically distressed areas, U.S. Public Law 87-27, 87th Cong., 1st Sess., May 1, 1961 (75 *Stat.* 47 to 63). This act may be cited as the Area Redevelopment Act.
25. Planning for new growth—new jobs, U.S. Department of Commerce Area Redevelopment Administration, Area Redevelopment Bookshelf of Community Aids, OEDP (Overall Economic Development Program), *ARA Publ.* 62-A, 1962.
26. An act to provide authority to accelerate public works programs by the Federal government and state and local bodies, U.S. Public Law 87-658, 87th Cong., 2d Sess., Sept. 14, 1962 (76 *Stat.* 541). This act may be cited as the Public Works Acceleration Act.
27. Policies, standards, and procedures in the formulation, evaluation, and review of plans for use and development of water and related land resources, 87th Cong., 2d Sess., S. Doc. no. 97, 1962.
28. Langbein, W. B.: Hydrologic data networks and methods of extrapolating or extending available hydrologic data, in Hydrologic Networks and Methods, Flood Control Series, no. 15, United Nations Economic Commission for Asia and the Far East and World Meteorological Organization, 1960.
29. Principal federal sources of hydrologic data, Water Resources Committee, Special Advisory Committee on Hydrologic Data, *Natl. Resources Planning Bd. Tech. Paper* 10, May 1943.
30. Linsley, R. K., Jr., M. A. Kohler, and J. L. H. Paulhus: Sources of hydrologic and meteorological data, appendix B in "Applied Hydrology," McGraw-Hill Book Company, Inc., New York, 1949.
31. Miller, D. W., J. J. Geraghty, and R. S. Collins: "Water Atlas of the United States," Water Information Center, Inc., Port Washington, N.Y., 1962.
32. "Deficiencies in Basic Hydrologic Data," Report of the Special Advisory Committee on Standards and Specifications for Hydrologic Data, National Resources Committee, Water Resources Committee, 1936.
33. A bill to provide for the optimum development of the nation's natural resources through

the coordinated planning of water and related land resources, through the establishment of a water resources council and river basin commissions, and by providing financial assistance to the states in order to increase state participation in such planning, S. Bill S. 1111, 88th Cong., 1st Sess., Mar. 15, 1963. This bill may be cited as a bill for the Water Resources Planning Act of 1963.
34. Measurement aspects of benefit-cost analysis, Federal Inter-Agency River Basin Committee, Subcommittee on Benefits and Costs, November, 1948. Reprinted as Appendix II in [1].
35. Selby, H. E.: Indirect benefits from irrigation development, *J. Land and Public Utility Economics*, vol. 20, no. 1, pp. 45-51, February, 1944.
36. Foster, E. E.: Evaluation of flood losses and benefits, *Trans. Am. Soc. Civil Engrs.*, vol. 107, pp. 871-894, 1932.
37. Landenberger, E. W.: "Multiple-purpose Reservoirs and Pollution Control Benefits," Ohio River Valley Water Sanitation Commission, Cincinnati, Ohio, January, 1953.
38. Renshaw, E. F.: Measurement of benefits from navigation projects, *Am. Econ. Rev.*, vol. 47, no. 5, pp. 652-662, September, 1957.
39. Bennett, N. B.: Cost allocation for multiple-purpose water projects, *Trans. Am. Soc. Civil Engrs.*, vol. 123, pp. 85-92, 1958.
40. Dixon, J. W.: Justification of projects, selection of a project from among alternatives, a discussion of methods of reimbursement and a procedure for benefit-cost analyses, *Proceedings of the Regional Technical Conference on Water Resources Development in Asia and the Far East, Bangkok, 1956*, United Nations Economic Commission for Asia and the Far East, Flood Control Series no. 9, ST/ECAFE/SER.F/9, pp. 81-85, September, 1956.
41. Dixon, J. W.: Planning an irrigation project today, *Centennial Trans. Am. Soc. Civil Engrs.*, vol. CT, pp. 357-387, 1953.
42. River basin development, *Law and Contemporary Problems*, School of Law, Duke University, vol. 22, no. 2, pp. 155-322, spring, 1957. This issue contains: M. G. Shimm, Foreward, pp. 155-156; G. F. White, A perspective of river basin development, pp. 157-187; A. T. Lenz, Some engineering aspects of river basin development, pp. 188-204; W. E. Folz, The economic dynamics of river basin development, pp. 205-235; C. F. Kraenzel, The social consequences of river basin development, pp. 221-236; Hubert Marshall, The evaluation of river basin development, pp. 237-257; Norman Wengert, The politics of river basin development, pp. 258-275; W. S. Hutchins and H. A. Steele, Basic water rights doctrines and their implications for river basin development, pp. 276-300; F. J. Trelease, A model state water code for river basin development, pp. 301-322.
43. Water resources, *Law and Contemporary Problems*, School of Law, Duke University, vol. 22, no. 3, pp. 323-537, summer, 1957. This issue contains: M. G. Shimm, Foreward, pp. 323-324; E. A. Engelbert, Federation and water resources development, pp. 325-350; R. C. Martin, The Tennessee Valley Authority: a study of federal control, pp. 351-377; L. E. Craine, The Muskingum Watershed Conservancy District: a study of local control, pp. 378-404; R. J. Morgan, The small watershed program, pp. 405-432; R. E. Huffman, The role of private enterprise in water resources development, pp. 433-443; J. W. Fester, National water resources administration, pp. 444-471; I. K. Fox, National water resources policy issues, pp. 472-509; H. C. Hart, Crisis, community, and consent in water politics, pp. 510-537.
44. Manual on the planning of small water projects, outlining considerations that should be included in planning small projects for water supply, sewage disposal and sanitation, flood protection, irrigation development, land drainage, hydroelectric power, recreation and wildlife, and conservation and water flow retardation, prepared by National Resources Planning Board, *House Document* 408, 82nd Congress, 2d Session, U.S. Government Printing Office, 1952.
45. Thomas, H. E.: "The Conservation of Ground Water," McGraw-Hill Book Company, Inc., 1951.
46. "Low Dams," Subcommittee on Small Water Storage Projects, Water Resources Committee, National Resources Committee, Washington, D.C., 1938.
47. Ford, P. M.: Multiple correlation in forecasting seasonal runoff, *U.S. Bur. Reclamation Eng. Monograph* 2, Denver, Colo., 2d ed., rev., June, 1959.
48. Briggs, L. J., and H. L. Shantz: Daily transpiration during the normal growth period and its correlation with the weather, *J. Agr. Res.*, vol. 7, no. 4, pp. 155-212, 1916.
49. Sherman, L. K.: Streamflow from rainfall by the unit-graph method, *Eng. News-Record*, vol. 108, pp. 501-505, 1932.
50. Barnes, B. S.: The structure of discharge-recession curves, *Trans. Am. Geophys. Un.*, vol. 20, pt. 4, pp. 721-725, 1939.

REFERENCES

51. Van't Hul, A. W.: A progress report on the disposition of sediment in reservoirs, U.S. Bureau of Reclamation, Sedimentation Section, Hydrology Division, January, 1950.
52. "Drainage," Student Reference E. 011, U.S. Army Engineer School, Fort Belvoir, Virginia, August, 1961.
53. Douglass, A. E.: Tree rings and their relation to solar variations and chronology, *Report of Smiths. Inst.* no. 3152, pp. 304–312, 1931.
54. Glock, W. S.: "Principles and Methods of Tree-ring Analysis," Carnegie Institution of Washington, Washington, D.C., 1937.
55. Straus, M. W.: Global hydroeconomics—a world survey—a report on 21 nations, a paper for International Commission on Large Dams, 1948. (Copies available from U.S. Department of Interior, Bureau of Reclamation, Washington 25, D.C.)
56. Fox, I. K.: Trends in river basin development, *Proc. Nat. Watershed Cong. in Columbus, Ohio, May 7–9, 1962,* Resources for the Future, Inc., *Reprint* no. 37, Washington, D.C., 1962.
57. Dexheimer, W. A.: International water problems and progress made through treaties, compacts and agreements, a paper presented at the Sectional Meeting of the World Power Committee at Rio de Janeiro, July 25–31, 1954.
58. "Seminar on River Basin Planning, Ft. Belvoir, Virginia, 27–31, May, 1963," U.S. Department of the Army, Office of the Chief of Engineers, Washington 25, D.C., 1963.
59. Hockensmith, R. D. (ed.): "Water and Agriculture," *Pub.* 62, American Association for the Advancement of Science, Washington, D.C., 1960.
60. Thorne, Wynne (ed.): "Land and Water Use," *Pub.* 73, American Association for the Advancement of Science, Washington, D.C., 1963.

Section 26-II

WATER RESOURCES

PART II. SYSTEM DESIGN BY OPERATIONS RESEARCH

VEN TE CHOW, *Professor of Hydraulic Engineering, University of Illinois.*

I. Introduction 26-30
II. Design Criteria for Optimization 26-31
    A. Objectives of Water-resources Development 26-31
        1. Objectives as Operational Criteria 26-31
        2. Constraints 26-32
        3. Objective Functions 26-32
    B. Economic Concepts 26-33
III. Design by Simulation 26-35
    A. Simulation Analysis 26-35
    B. Sampling Techniques 26-36
        1. Random Sampling 26-37
        2. Systematic Sampling 26-37
    C. Operating Procedures 26-38
    D. Sequential Generation of Hydrologic Data 26-38
IV. Design by Mathematical Models 26-38
    A. Nature of Models 26-38
    B. Deterministic Models 26-39
        1. Linear Programming 26-39
        2. Dynamic Programming 26-41
        3. Other Models 26-43
    C. Stochastic Models 26-43
        1. Stochastic Linear Programming 26-43
        2. Stochastic Dynamic Programming 26-44
        3. Other Models 26-44
V. References 26-44

I. INTRODUCTION

Operations research as a science was developed as early as the late 1800s, but its name was originated only during World War II. The term has been defined in many ways.

The author is indebted to many colleagues, especially Dr. Abel Wolman and Dr. J. C. Geyer, for their valuable comments.

In a general sense, it may be defined as the application of scientific methods and tools "to problems involving the operations of systems so as to provide those in control of the operations with optimum solutions to the problems" [1]. These techniques usually call for quantification of data, creation of an analytical model to describe the operations, and finally, study of the model in light of the objectives of the investigator.

The plan, or arrangement, of a water-resources project may be considered as a *system* [1]. The project formulation of the system may be called *system design*. Modern water-resources projects often constitute very complex systems which may be created through different combinations of system units (reservoirs, power plants, canals, etc.), levels of output, and allocation of capacity of the units to various purposes (flood control, irrigation, hydroelectric power, etc.) at different times. The objective of system design is to select the combination of these variables that maximizes net benefits in accordance with the requirements of the design criteria. The design so achieved is known as the *optimal design*. The maximization, or optimization, is subject to the requirements of the design criteria, or *constraints*, that are imposed. The constraints may be technical, budgetary, social, or political, and the benefits may be real or implied. Hence the optimal design is subject to technical as well as economic and sociopolitical limitations.

Because of the unlimited, almost infinite number of combinations that can be arranged in a multiunit-multipurpose water-resources system, the optimal design cannot possibly be obtained by the conventional approach of incremental analysis (Subsec. 26-I-V-A-2b; [2]). By the techniques of operations research [3], however, it is possible to consider simultaneously a large number of alternative system designs and thereby isolate the optimal design.

The use of operations research in water-resources planning and development was developed largely during the last decade. Special contributions to the knowledge are particularly due to a team of research workers participating in the Harvard Water Program (1955–1960) [4–6]. In this program, the river-basin system is analyzed, using a model system as the test vehicle. Two approaches of analysis are employed. One is to simulate the system on a high-speed digital computer and thereby select the best combination of variables by observing the response of the simulated system to various alternative combinations. The other is to use simplified mathematical models which can be solved directly for the optimal design for relatively simple problems. The materials in Subsecs. II and III are largely summarized and adapted from relevant chapters of Ref. 6. However, the differences, additions, and modifications in the materials are entirely due to author's own views. The author is indebted to the Harvard University Press for permission to use the original materials. Because the use of operations research in water-resources system design is just beginning and no extensive applications or verifications have yet been made in practical problems, this section offers only a brief outline of the principles involved, not detailed procedures. The purpose is merely to introduce the concept of such techniques, but by no means to recommend any methodology to be followed. There is no doubt that the methodology will change and be greatly improved in time to come.

II. DESIGN CRITERIA FOR OPTIMIZATION

A. Objectives of Water-resources Development[1]

1. Objectives as Operational Criteria. The prime objective of a public water-resources development is generally considered as maximizing the national welfare or the regional welfare, as the case may be. However, this main objective can be interpreted in many ways, for example, as follows:

To optimize *economic efficiency;* that is, to maximize the increase in real national income resulting from a national or regional investment in a water-resources system.

To generate *income redistribution;* that is, to maximize redistribution of income generated by the water-resources system among classes of people or regions served by the system.

---

[1] See Ref. 6, chap. 2.

To approach *full employment*

To promote and sustain *economic growth*

To achieve certain *intangible objectives* such as national defense and preserving wilderness areas

All these objectives may in fact be achieved by means of water-resources development and may be translated into operational criteria for system design. However, in general they conflict and also compromise with each other to various degrees. Of all the objectives, economic efficiency and income redistribution are found to be far more tractable than others.

2. Constraints. When a system design is developed for a selected design criterion to perform best in terms of one objective, it may be subject to a specified level of performance in terms of another objective. This condition is called a *constraint*. There are various kinds of constraints in water-resources system design, such as technical, budgetary, institutional, and legal. The technical constraint is a physical limitation, for example, of a hydrologic or geographical nature. The budgetary constraint is to set a limit on expenditures, which may be an objective of public policy to keep public water-resources developments within a desired scope. The institutional constraint is to satisfy the specification in an agreement such as an interstate compact or international agreement which divides the water of a river between neighboring states or countries. The legal constraint is to specify the use of water according to law.

3. Objective Functions. When an objective is translated into a design criterion, it may be written in the form of a mathematical expression, known as the *objective function*. An objective function may or may not be subject to constraints. Some examples are as follows [5]:

When no constraints are present, the objective function for economic efficiency may be expressed as

$$\max \sum_{j=1}^{n} \sum_{t=1}^{T} \frac{B_{pjt} - M_{jt}}{(1+i)^t} - \sum_{j=1}^{n} K_j \qquad (26\text{-II-1})$$

where B_{pjt} is the gross efficiency benefits of the jth unit in the tth year, and the unit may also be a project increment; M_{jt} is the operation and maintenance costs of the jth unit in the tth year; K_j is the capital cost of the jth unit; i is the discount rate; n is the number of units; and T is the life in years of the longest-lived unit.

When an income-redistribution constraint is applied, the objective function for economic efficiency may be written as

$$\max \sum_{j=1}^{n} \sum_{t=1}^{T} \frac{B_{pjt} - M_{jt}}{(1+i)^t} - \sum_{j=1}^{n} K_j \qquad (26\text{-II-2})$$

subject to

$$\sum_{j=1}^{n} \sum_{t=1}^{T} \frac{(B_p + B_s)_{jt}}{(1+i_r)^t} \geq B_r$$

where B_p is the primary system benefits (efficiency), as B_{pjt}; B_s is the secondary system benefits, accruing to the intended beneficiaries; B_r is a specified level in the region of development; and i_r is the discount rate for income redistribution.

When an efficiency constraint is applied to the income redistribution as a design criterion, the objective function is

$$\max \sum_{j=1}^{n} \sum_{t=1}^{T} \frac{(B_p + B_s)_{jt} - M_{jt}}{(1+i_r)^t} - \sum_{j=1}^{n} K_j \qquad (26\text{-II-3})$$

subject to

$$\sum_{j=1}^{n} \sum_{t=1}^{T} \frac{B_{pjt} - M_{jt}}{(1+i)^t} - \sum_{j=1}^{n} K_j \geq 0$$

where the symbols are explained as above.

It should be noted that the income-redistribution objective may be applied to all regions or to one or more specific regions, depending on the planning policy.

There is much to be discussed with respect to the selection of objective functions in water-resources system design relating to economic and sociopolitical matters. Such discussion, dealing with economic feasibility and public policy on water-resources development, must be considered as beyond the scope of this section (but see [6–9] and Subsecs. 28-II-D and -E).

B. Economic Concepts[1]

For the optimal design, i.e., one to achieve maximum benefits, the condition in economic analysis requires that the system is to be developed to the level at which the *marginal benefit* of producing output is equal to the *marginal cost* of producing it. Marginal cost and marginal benefit are, respectively, the first derivative of the cost and the first derivative of the benefit with respect to the output. In other words, they are, respectively, the cost increment and the benefit increment per unit incremental output at the determined level.

For optimization of a complex multiunit-multipurpose water-resources system, the above condition for one purpose may be written as

$$\frac{MP_1}{MC_1} = \frac{MP_2}{MC_2} = \cdots = \frac{MP_j}{MC_j} = 1 \qquad (26\text{-II-}4)$$

or

$$\frac{\partial P_1}{\partial C_1} = \frac{\partial P_2}{\partial C_2} = \cdots = \frac{\partial P_j}{\partial C_j} = 1 \qquad (26\text{-II-}5)$$

where MP_j is the marginal gross benefit, or productivity, MC_j is the marginal cost, $\partial P_j/\partial C_j = MP_j/MC_j$, and $j = 1, 2, \ldots$, denoting the different units. For n purposes, there will be n such equations as the above.

It should be noted that the preceding equations are applicable to no budgetary constraints. If such constraints exist, the system cannot be developed to the point of satisfying the above conditions. In this case, the marginal cost of producing the last increment of benefit must be equal for all purposes in order to attain the maximum efficiency [6, p. 36]. This can be expressed as

$$\frac{\partial P_1}{\partial C_1} = \frac{\partial P_2}{\partial C_2} = \cdots = \frac{\partial P_n}{\partial C_n} > 1 \qquad (26\text{-II-}6)$$

In analyzing a water-resources system for the above-stated economic conditions, several types of economic and technological relationships should be considered, namely, the input-cost function, the production function, the annual-cost function, and the annual-benefit function.

The *input-cost function* is the multidimensional relationship between the total cost of the project and the magnitudes of the project variables. The *production function* is an input-output function which shows the relationship between the magnitudes of the project variables and their feasible combinations of outputs. From the input-cost function and production function, the *annual-cost function* can be derived to show the relationship between the annual cost and the outputs of the project variables. Finally, the *annual-benefit function* is the relationship between the average annual gross benefits and the outputs of the project variables.

The principle involved may be illustrated by a simple example of the development of a reservoir for flood control (adapted from Ref. 6, chap. 5). In this case, the input-cost function can be represented by the relationship between the total cost in dollars to be invested in the project and the reservoir capacity in acre-feet, or the height of the dam in feet. The production function is the relationship between the reservoir capacity, or the dam height, and the project output. The output of a flood-control reservoir can be measured by the peak flood flows in cubic feet per second for which complete protection is provided. Since flood flows are stochastic in nature, the

---

[1] See Ref. 6, chaps. 2 and 3.

reservoir output must be evaluated by probability and hydrologic analyses of the flood data at the project site. Finally, an annual-cost function can be derived as shown in Fig. 26-II-1.

The benefits derived from a flood-control reservoir are measured by the reduction in

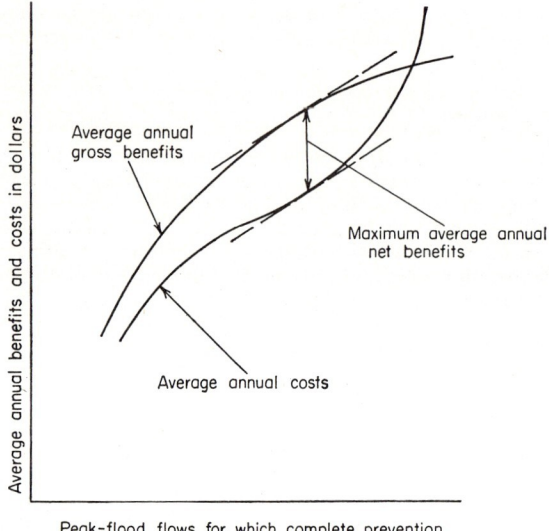

Fig. 26-II-1. Determination of optimal scale of development for a flood-control reservoir.

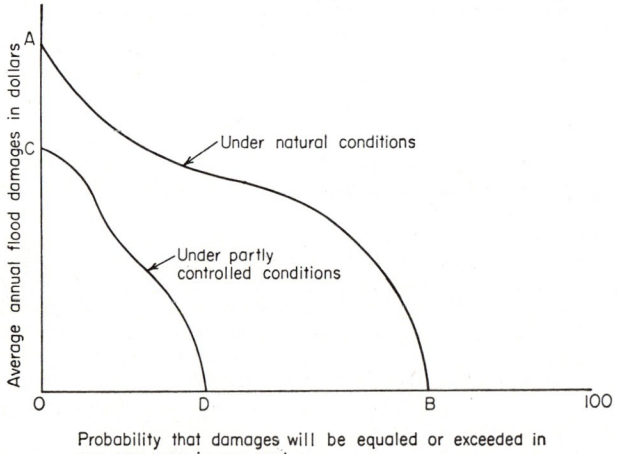

Fig. 26-II-2. Flood-damage frequency curves.

flood damage by the project. The flood damage can be evaluated by a flood survey. Flood occurrence being stochastic in nature, a probability analysis of the flood flow is necessary. The curves in Fig. 26-II-2 are the results of such a survey and analysis. Curve *AB* shows the flood-damage probability function under natural conditions.

Curve CD shows the residual flood-damage probability function when floods of a given probability, say, 30 per cent, and smaller floods are completely controlled. Thus the area $ACBD$ represents the average annual gross benefits of the project. From a flood-frequency curve, the peak flood flow for the given probability can be obtained. Hence the annual benefits can be plotted against peak flood flows for given probabilities as shown in Fig. 26-II-1.

The marginal benefit can be represented by the slope of a tangent to the annual-benefit curve. The marginal cost is represented by the slope of a tangent to the annual-cost curve. At the optimal scale of development, the two slopes should be equal or the two tangents must be parallel. This condition is indicated in Fig. 26-II-1. According to Eq. (26-II-4), $MP = MC$, or

$$\frac{d(\text{benefit})}{d(\text{output})} = \frac{d(\text{cost})}{d(\text{output})} \qquad (26\text{-II-}7)$$

Mathematically, the above condition may lead to either a maximum or a minimum. In order to assure the optimization being a maximum, the following secondary condition must hold:

$$\frac{d^2(\text{benefit})}{d(\text{output})^2} < \frac{d^2(\text{cost})}{d(\text{output})^2} \qquad (26\text{-II-}8)$$

The above analysis can be extended to more reservoirs or structural units. When the problem is too complex, however, such analyses may become very difficult, and simulation techniques and mathematical models will then be applicable.

In making economic decisions by theoretical analysis, the result cannot always be dependable, because the hydrologic and economic data are subject to a wide range of uncertainty. Therefore such decisions must be modified by experienced judgment in accordance with the intangible factors involved. (See Subsecs. 28-II-D and -E.)

III. DESIGN BY SIMULATION

A. Simulation Analysis [1]

As stated at the outset, a modern water-resources system may be created through almost infinite combinations of a large number of system variables. Conventional methods of analysis (Sec. 26-I-V) are practically unable to determine the optimal design of the system, because they use only desk calculators and thus can consider only selected parts of the system, generally using historical hydrologic data of a limited period of record. By means of high-speed digital computers (Sec. 29), however, it is possible to simulate by simplified systems the behavior of relatively complex water-resources systems for periods of any desired length; to perform numerous and repetitive computations needed for many combinations of the system variables, which cannot be done by desk calculators; and finally, to evolve an optimal or near-optimal design of the system.

The use of simulation analysis of water-resources systems was begun in 1953 by the U.S. Army Corps of Engineers on the Missouri River [10]. In this analysis, the operation of six reservoirs on the Missouri River was simulated on the Univac I computer to maximize power generation subject to constraints for navigation, flood control, and irrigation specifications. In the following year, a two-reservoir system on the Rio Grande was simulated on an IBM 701 computer by the International Boundary and Water Commission, United States and Mexico. In 1955, Morrice and Allan [11, 12] simulated the Nile Valley plan on an IBM 650 computer to determine the particular combination of reservoirs, control works, and operating procedures that would maximize the use of irrigation water. Later, simulation analysis was also made of the Columbia River Basin on an IBM 650 to analyze the power outputs of a large number of alternative system designs [13, 14]. In all these analyses, no economic optimization was involved.

[1] Adapted from Ref. 6, chaps. 6, 9, and 10.

In the Harvard Water Program [6], simulation analysis was applied for the economic optimization of water-resources system designs. To illustrate such analyses, simulation was made of a simplified river-basin system consisting of four reservoirs having three purposes. This system is purely hypothetical, but it would create an adequate number of interrelationships to typify the complexities of an actual system of substantial proportions. The hydrologic data used in the simulation analysis of this system were the U.S. Geological Survey runoff records on the Clearwater River Basin in Idaho. By sampling techniques, this analysis was able to evolve a design that yielded net benefits of $811 million against $724 million for the best design by conventional methods.

Other simulation analyses were made by Brittan [15] and Fiering [16]. Brittan dealt with the integration of an energy-producing Glen Canyon Dam into the already-existing power system on the Colorado River in order to approximate maximum return. Brittan also evaluated various operating criteria to attempt an optimum allocation of the available water. Fiering proposed a method for the optimal design of a single multiple-purpose reservoir by computer simulation studies of a simple coded model.

In essence, the simulation analysis is intended to reproduce the behavior or performance of a river-basin system on the computer in every important aspect of the system variables, such as system units, inputs, outputs, and ranges of scale [6, p. 250]. The simulation represents all the inherent characteristics and probable responses of the system to control by a model that is largely arithmetic and algebraic in nature but includes also some nonmathematical logical processes. The following procedure was used in the Harvard Water Program [6, pp. 326–328].

1. Assemble and arrange the basic data for the system (including hydrologic data) in a form easily handled by the computer.

2. Program or formulate an operating procedure to serve as the fundamental control for the simulation.

3. Code the basic system data and the operating procedure for use on the computer.

The results of the analysis are expressed in economic terms by including cost and benefit functions in the simulation program and by evaluating deficits and surpluses in outputs as well as target outputs themselves. These results are then converted into present values of net benefits at a number of alternative interest rates.

Application of simulation analysis to the design of water-resources systems has limitations. The major ones are as follows [6, p. 252]:

1. It does not yield an immediate optimal answer. Each answer by the analysis pertains merely to the operation study of the selected combination of system variables, while the possible combinations are numerous.

2. It is not flexible in handling various operating procedures of the system. The computer can be instructed to follow only one operating procedure at a time. Thus each operating procedure has to be analyzed separately, and this is time-consuming.

3. It does not work satisfactorily with most historical hydrologic data. The simulation analysis requires data of good representation of the possible occurrences of the hydrologic events, but most hydrologic data have short records and may constitute poor representations of the possibilities.

These limitations can be overcome partially by sampling techniques, operating procedures, and sequential generation of hydrologic data, which are discussed below.

B. Sampling Techniques*

The net benefit of a simplified river-basin system in response to various simulating combinations of many system variables, say, n in number, may be conceived to represent an n-dimensional surface. This surface may be called the *net-benefit response surface*. On this surface, there may be several points at which the net benefits are maximum. Among these points, there is at least one point which locates the maximum of all maxima, or the greatest net benefits.

Because of the enormous number of possible combinations of the system variables,

---

*See also Ref. 6, chap. 10, except the classification of random methods.

the point of greatest net benefits cannot be easily determined even with the use of high-speed computers. Many combinations and points of maximum net benefits, however, can be eliminated from consideration on account of certain side conditions and limitations in the range of and due to the nature of the variables. The only time-saving practical method of locating the point of optimality, or points of local maxima, is to sample the variables and thereby explore the response surface and eliminate undesirable combinations from the computation.

There are two broad categories of sampling methods: random sampling and systematic sampling [17–21]. All sampling methods may be used either separately or in combination. The selection of methods depends mainly on the nature of the response surface under investigation.

1. Random Sampling. In *random sampling*, the values (levels or magnitudes) of the system variables for each combination are selected purely by chance from an appropriate population of values of the variables. These selected values should be within their respective ranges and subject to whatever side conditions or limitations may be associated with the different variables. The maximum benefit for each combination is then determined. The combination with the greatest benefit will be very nearly the maximum of maxima. Two types of random-sampling methods, stratified random method and creeping random method, may be briefly described as follows.

In the *stratified random method*, the multidimensional response surface is arbitrarily divided into a number of subregions. A combination of system variables is picked at random from each subregion, and the trial computation is made. The procedure is repeated until local maxima, and then the optimum combination, are located.

In the *creeping random method*, an initial guess of the optimal combination is made. The populations of the variables are assumed to have, for example, a normal distribution. When such distributions are established from the initial guess, random variables are selected from them to form new combinations. If a new combination yields better benefits than the initial combination, the variables forming this combination are used as the next initial guess, and the process is repeated. If the variance of the normal distribution is reduced each time, the local maxima, and finally the optimal combination, can be found by spot checks.

2. Systematic Sampling. In *systematic sampling*, the values of system variables are selected in accordance with some ordering principle, and the entire range of combinations of the variables is systematically examined. The methods of systematic sampling particularly applicable to river-basin system designs are the uniform-grid method, steepest-ascent method, single-factor method, and incremental-analysis method.

In the *uniform-grid method*, also known as the *factorial method*, the values of variables are taken at uniform intervals, or grids, over the entire range of each variable. Naturally, the finer the grid, the larger will be the number of combinations of variables to be considered. Hence this method may usually be applied in two steps. In the first step, a coarse grid is used to determine the region of local maxima, or summits. This is followed by the second step, in which a finer grid is used to analyze the benefit functions around the local summits. From the local summits, the optimal one can be found. In this method, a greater number of grid nodes is used only for more important variables. Thus the entire response surface of the system can be mapped with a small number of variables.

In the *steepest-ascent method*, an iterative process is used to direct sequential sampling from lower toward higher elevations on the response surface along the path of steepest ascent. The initial combinations of variables may be determined approximately by a coarse grid or a random sampling. When a starting point or initial combination is at hand, the magnitude of each variable is changed separately by an arbitrary preselected amount, while the values of other variables are held constant. These corresponding changes in the net-benefit function per unit change of the variables represent the approximate partial derivatives of the net-benefit function with respect to each variable. By knowing these partial derivatives, the direction of the steepest ascent can be determined and followed until the summit of the net-benefit hill on the response surface is reached. In this procedure, it must be made sure that the benefit hill considered is the optimal, but not other local hills on the response surface. This can usually be verified by spot checks. When the number of variables exceeds 10, this method may become very cumbersome.

In the *single-factor method*, a sequential procedure similar to the steepest-ascent method is used except that only the values of a selected single variable are changed step by step, while other variables are held constant, until the greatest benefit response of the system to this variable has been registered. The process is then repeated with each of the other variables in succession until there is no further improvement in the design as a whole.

In the *incremental-analysis*, or *marginal-analysis, method*, a sequential procedure similar to the single-factor method is used, except that two variables are altered at a time, with other variables held constant.

C. Operating Procedures

The flexibility of considering the operating procedure in simulation analysis can be improved by formulating a relatively small number of separate operating procedures, each with its own logic and each to be programmed for the computer. When a specific operating procedure is developed and programmed, a set of studies is run to determine the result of using this procedure over the range of combinations of specified inputs and outputs. Then a second operating procedure is formulated and tested by the same set of studies. This process is repeated for other suggested operating procedures, until an operating procedure which gives better results in terms of net benefits than all other procedures tested is found. This method was used satisfactorily in the Harvard Water Program [6].

There have been other suggestions for solving this operating-procedure problem, such as developing a generalized master operating procedure within which the detailed logical rules can be varied within certain limits and developing an optimum operating procedure as an integral part of the computer program. However, such suggestions have yet to be studied and developed.

D. Sequential Generation of Hydrologic Data

In simulation analysis a large quantity of hydrologic data is required in order to test the broad range of response of the proposed water-resources systems. Historical hydrologic data are usually limited in their length of record, but they can be used to generate a hypothetical series of hydrologic data that may have a sequence of any desired length suitable for simulation analysis. Detailed discussion on this subject is presented in Sec. 8-IV.

IV. DESIGN BY MATHEMATICAL MODELS

A. Nature of Models

Complex water-resources systems may also be analyzed for optimal design, water allocation, and project operation by suitable mathematical techniques known as *mathematical models*. The model adopted may be much simpler than the actual system, but it retains the essential properties of the latter. A simple model can be solved easily by mathematical methods, but it must be noted that the results so obtained should be considered approximate. On the other hand, a relatively complicated model may delineate the system better and yield more exact results, but it may be difficult to solve. Usually the model becomes complicated because of the difficulties in hydrology and the multiplicity in system units and purposes. Consequently, a compromise has to be made between the hydrology and the system so that the model can be readily amenable to mathematical analysis. For complicated systems, the hydrologic data used are generally simplified or approximated.

Mathematical models used in water-resources design are usually identified by the mathematical techniques employed to solve the problem. They are generally the techniques of operations research for making optimal decisions. The models may be categorized according to the nature of hydrologic data. When the probability of hydrologic data is ignored, the mathematical model is known as a *deterministic model*. When the hydrologic uncertainty is considered, the model is called a *stochastic model*.

The use of mathematical models will give the designer an insight into the behavior of the system under various conditions of planning and operation and therefore will enable him to base his decision on the expected behavior of the system and thus to reduce his dependence on experience and intuition.

B. Deterministic Models

1. Linear Programming. *Linear programming* had its inception within the field of military logistics by George B. Dantzig in 1947 to assist in overall planning of the multitudinous activities of the United States Air Force. This is a mathematical technique to optimize an objective if the latter can be approximated by a linear function and the constraints can be expressed as linear equalities or inequalities [22]. The method of linear programming has been applied to water-resources design by Massé and Gibrat [23, 24], Lee [25], Castle [26], Heady [27], and Dorfman [28]. The principle of such applications can be best illustrated by a simple example as follows:[1]

A single multiple-purpose reservoir is subjected to analysis by linear programming for its beneficial use of water. The hydrologic data used for inflow to the reservoir are the inflow hydrograph corrected for estimated evaporation and leakage. The initial reservoir level is given or so chosen that the reservoir is full, empty, or at optimal condition of operation.

The duration for analysis is assumed to be one year. It can be divided into a number of equal intervals, say, 12 months. Let $Q_1, Q_2, \ldots, Q_{12}$ be the respective volumes of monthly inflow, V_0 be the initial storage, and $V_1, V_2, \ldots, V_{12}$ be the volumes of water planned to be used in the respective months. The twelve volumes of water used monthly should add up to the total volume of outflow from the reservoir, including any unavoidable spills in an average year. Since the total outflow up to the nth month cannot exceed the sum of inflow volumes up to the nth month plus the initial storage, the following inequality can be written

$$\sum_{i=1}^{n} V_i \leq \sum_{i=1}^{n} Q_i + V_0 \qquad (26\text{-II-}9)$$

where $n < 12$. Also, the total volume of water in storage at any time cannot exceed the maximum useful storage capacity V_m of the reservoir. Thus

$$V_0 + \sum_{i=1}^{n} Q_i - \sum_{i=1}^{n} V_i \leq V_m \qquad (26\text{-II-}10)$$

where $n < 12$. When $n = 12$, the above inequalities become the equation

$$\sum_{i=1}^{12} V_i = \sum_{i=1}^{12} Q_i \qquad (26\text{-II-}11)$$

Here, although it may be unrealistic, it is assumed that the total outflow in every year is exactly equal to the total inflow and that there is no carry-over from year to year. If the storage V_0' at end of the year is different from V_0, then

$$V_0 + \sum_{i=1}^{12} Q_i - \sum_{i=1}^{12} V_i = V_0' \qquad (26\text{-II-}12)$$

---

[1] This example was originally prepared by Sundaresa Ramaseshan, a doctorate candidate at the University of Illinois, under the author's direction.

In this case, the total inflow is not equal to the total outflow. For simplicity of analysis, however, no carry-over is assumed and the initial and final reservoir storages are assumed either empty or full, that is, either $V_0 = V_0' = 0$ or $V_0 = V_0' = V_m$.

Let $P_1, P_2, \ldots, P_{12}$ be the net-benefit values per unit volume of water in the 12 monthly intervals. The total annual benefit is therefore equal to

$$B = \sum_{i=1}^{12} P_i V_i \qquad (26\text{-II-}13)$$

The object of this problem is to maximize this benefit function subject to constraints of the inequalities (26-II-9) and (26-II-10) and Eq. (26-II-11).

In linear programming, the inequalities (26-II-9) and (26-II-10) may be changed to equations by introduction of so-called *slack variables* of no benefit value, $S_1, S_2, \ldots, S_{11}$ and $S_1', S_2', \ldots, S_{11}'$, respectively, in (26-II-9) and (26-II-10). Thus

$$\sum_{i=1}^{n} V_i + S_n = \sum_{i=1}^{n} Q_i + V_0 \qquad (26\text{-II-}14)$$

and

$$V_0 + \sum_{i=1}^{n} Q_i - \sum_{i=1}^{n} V_i + S_n' = V_m \qquad (26\text{-II-}15)$$

where $n < 12$.

Subject to the linear equations (26-II-11), (26-II-14), and (26-II-15), the linear benefit function B given by Eq. (26-II-13) can be maximized by linear programming.

There are many methods of solution in linear programming. One of several iterative methods, known as the *simplex method*, can be used to solve the above equations. This method is based on a fundamental theorem [29] which states that, *if a linear programming problem with m constraints and m + n total variables has an optimal solution, then there is a solution in which at least n of the variables are equal to zero.* For the given problem, $n + m = 34$, $m = 23$, and thus $n = 11$. (The variables are 12 monthly flows and 22 slack variables. The constraints are given by Eqs. (26-II-14), (26-II-15), and (26-II-11) as 11, 11, and 1 in number respectively.) Hence the solution has at most 23 real variables, and the 11 others are zero. The steps of the solution are:

1. Obtain a basic solution for the linear constraints. The basic solution will include only m of the $m + n$ variables, while the remaining n variables are set equal to zero.
2. Substitute this solution into the linear function B, and determine the value of B.
3. Test whether the value of B so obtained is the absolute maximum value possible for the linear function. This is done by considering the change in benefit function due to a unit change in each of the variables and subject to the constraints. When the change is negative for the change of every variable, the benefit is maximum.
4. If the change is nonnegative, replace one of the m variables by one of the n variables. The procedure is repeated until B becomes maximum.

There are computer solutions for linear-programming problems. Thus time for repeated computations can be saved by computers. In the above problem, it is therefore possible to repeat the computation for different values of V_0 so that the optimum initial reservoir storage can be obtained. It is also possible to repeat computations for different values of V_m. Knowing the maximum benefit for each V_m as B_m and the corresponding cost of the dam, C_m, the net optimal benefit is given by $B_n = B_m - C_m$. The design reservoir capacity is the one for which B_n is maximum.

The actual situation for the design of reservoirs is more complicated. Generally, reservoirs serve multiple purposes, and each purpose has maximum and minimum limitations. Also, there are wasteful spills. Further, the net unit benefit for any purpose is not a constant, but usually follows the law of diminishing returns. These complications, however, can be resolved by modification of the procedure. For example, the total flow through the reservoir can be divided into a number of components, each assigned to a given purpose. Unintentional overflow may be considered as a purpose with zero benefit. The maximum and minimum limitations can be taken care of as constraints. The net-benefit vs. stage-of-development curve, if not a straight line, can be subdivided into a number of segments, and each segment approximated by a straight line which represents constant unit benefit for the segment. The linear-programming solution can thus be applied to each segment. This method is known as the *quasilinear programming*, which has also been suggested for use to approximate nonlinear objective functions or cost functions in a linear-programming model [5, 6, 30].

2. Dynamic Programming. *Dynamic programming*, originated and developed largely by Richard Bellman [31, 32], deals with the mathematical theory of multistage decision processes and thereby provides a sequential optimization procedure in which the optimization is done in steps. The method is applicable to those problems in which time plays an essential role and in which the sequence of decision is vital. In general, a *decision* may be defined as the choice of one of two or more alternatives. *Multistage decision process* is a process for making a sequence of decisions. Most multistage decision processes are "programming" problems, and since time is almost always important in such processes, the programming is "dynamic" in nature.

A sequence of decisions is defined as a *policy*. Those policies which are most desirable according to some predetermined criterion are called *optimal policies*. In making optimal policies, the dynamic programming is based on a so-called *principle of optimality: An optimal policy has the property that, whatever the initial state and the initial decision, the remaining decisions must constitute an optimal policy with regard to the state resulting from the first decision.* By this principle, the dynamic programming is formulated as follows.

Through a number of n stages in a development, it is necessary, at the beginning of the first stage, to divide the amount x_1 of an effort or quantity into two parts, y_1 and $x_1 - y_1$, with beneficial returns from y_1 given by $g(y_1)$ and from $x_1 - y_1$ given by $h(x_1 - y_1)$. It may be noted that $0 \leq y_1 \leq x_1$. Also, the benefit functions $g(y_1)$ and $h(x_1 - y_1)$ are usually of convex type (i.e., represented by curves convexed upward).

After a part of the original amount x_1 has been expended in the first stage, there remains an amount of effort equal to $ay_1 + b(x_1 - y_1)$, where ay_1 is the amount salvaged from y_1 and $b(x_1 - y_1)$ from $x_1 - y_1$, with $0 \leq a \leq 1$ and $0 \leq b \leq 1$. With the same benefit functions, the remaining amount is again divided into two parts. In this manner the process is continued in several stages. The problem is therefore to determine the allocation of effort at each stage so that the total benefit is maximized for n stages.

In the above problem, the total beneficial return from successive allocations may be written as

$$B_n(x_1) = \sum_{i=1}^{n} [g(y_i) + h(x_i - y_i)] \qquad (26\text{-II-}16)$$

where x_i and y_i at stage i correspond, respectively, to x_1 and y_1 at stage 1. Any choice of y_i $(i = 1, \ldots, n)$ constitutes a policy, and a choice which maximizes $B_n(x_1)$ is an optimal policy. For an optimal policy, $B_n(x_1)$ depends only on the original effort x_1 and on the number of stages n.

Let $f_n(x_1)$ be the total optimal beneficial return from n stages, using an optimal policy and starting with an initial amount x_1. Thus, from Eq. (26-II-16),

$$f_n(x_1) = \max \sum_{i=1}^{n} [g(y_i) + h(x_i - y_i)] \qquad (26\text{-II-}17)$$

or

$$f_n(x_1) = \max [g(y_1) + h(x_1 - y_1) + f_{n-1}(x_2)] \qquad (26\text{-II-}18)$$

subject to $0 \leq y_1 \leq x_1$. In the above equation, $f_{n-1}(x_2) = f_{n-1}[ay_1 + b(x_1 - y_1)]$, representing the total optimal beneficial return from the remaining $n - 1$ stages.

For an intermediate stage i, Eq. (26-II-18) may be written as

$$f_{n-i+1}(x_i) = \max [g(y_i) + h(x_i - y_i) + f_{n-i}(x_{i+1})] \qquad (26\text{-II-}19)$$

with $0 \leq y_i \leq x_i$ and $x_{i+1} = ay_i + b(x_i - y_i)$. This is the basic functional equation of dynamic programming. It is a recursion equation which is subject to solution by methods of recurrence in mathematics. This equation represents the total optimal beneficial return for a total allocation x_i to the $n - i + 1$ stages from stage i to n as the maximum of the beneficial return from an allocation y_i to the stage i and the optimal allocation of $x_i - y_i$ to the remaining $n - i$ stages from stage $i + 1$ to n. A number of simplified approximate solutions of this equation have also been developed by Bellman and his school.

Hall and Buras [33, 34] have applied the technique of dynamic programming to problems of optimizing the development of water resources. This is done through a series of levels of similar development problems, beginning at the highest level and branching downward to the various final suballocations for beneficial returns. In case of reservoir designs, they assumed that three levels of decisions are involved. The highest level is the optimal development of storage volume x_i of each reservoir at plausible sites $i = 1, \ldots, n$. The second level is the optimal use x_{ij} of each storage volume x_i for purposes $j = 1, \ldots, m$, such as irrigation, flood control, and hydroelectric power. The third level is the optimal allocation x_{ijk} of the use x_{ij} for each purpose j to geographical regions $k = 1, \ldots, p$.

At the highest level of decision, let $f_{1i}(x_i)$ be the optimal net beneficial return from the reservoir site i as a function of x_i, let q_i be the level of development required at site i, and let Q be an upper limit beyond which there would be no water for storage even in the long range. The problem is then to determine x_i for maximizing the total benefit from all possible sites.

By the general theory of dynamic programming and the notations used previously, it can be seen that in this problem $x_i = q_i, y_i = x_i, x_i - y_i = q_i - x_i, g(y_i) = f_{1i}(x_i)$, $h(x_i - y_i) = 0, a = 0$, and $b = 1$. From Eq. (26-II-17), the total beneficial return is

$$f_n(Q) = \max \left[\sum_{i=1}^{n} f_{1i}(x_i) \right] \qquad (26\text{-II-}20)$$

subject to $0 \leq x_i$ and $\sum_{i=1}^{n} x_i \leq Q$. For $n - i + 1$ additional sites remaining for development, one of them is taken as the first site to be developed, leaving $n - i$ sites yet to be developed. By Eq. (26-II-19), the recursion equation is

$$f_{n-i+1}(q_i) = \max \left[f_{1i}(x_i) + f_{n-i}(q_i - x_i) \right] \qquad (26\text{-II-}21)$$

with $0 \leq x_i \leq q_i$ and $0 \leq q_i \leq Q$. Beginning with Eq. (26-II-20) for $i = 1$, the optimal policy for a particular site may be determined by Eq. (26-II-21) only after all levels and all n possible sites have been investigated. However, the optimal functions can be determined for each site i starting from the optimal function for the site $i - 1$.

Similarly, at the second level of decision, the optimal net beneficial return $f_{1i}(x_i)$ in the first level, depending on x_i or the volume allocated to site i, is obtained by maximizing the net benefits $f_{2ij}(x_{ij})$ from uses x_{ij} to purposes j as follows:

$$f_{1i}(x_i) = \max \left[\sum_{j=1}^{m} f_{2ij}(x_{ij}) \right] \qquad (26\text{-II-}22)$$

with $0 \leq x_{ij}$ and $\sum_{j=1}^{m} x_{ij} \leq x_i$. The corresponding recursion equation is

$$f_{i(m-j+1)}(q_{ij}) = \max \left[f_{2ij}(x_{ij}) + f_{i(m-j)}(q_{ij} - x_{ij}) \right] \qquad (26\text{-II-}23)$$

with $0 \leq x_{ij} \leq q_{ij}$ and $0 \leq q_{ij} \leq q_i$, where q_{ij} is the level of use required for each purpose j at site i.

At the third level, the optimal net beneficial return $f_{2ij}(x_{ij})$ in the second level, depending on x_{ij} or the volume of water allocated to purpose j, is obtained by maximizing the total benefit from all p regions, or

$$f_{2ij}(x_{ij}) = \max \left[\sum_{k=1}^{p} f_{3ijk}(x_{ijk}) \right] \qquad (26\text{-II-}24)$$

with $0 \leq x_{ijk}$ and $\sum_{k=1}^{p} x_{ijk} \leq x_{ij}$. The recursion equation is

$$f_{ij(p-k+1)}(q_{ijk}) = \max \left[f_{3ijk}(x_{ijk}) + f_{ij(p-k)}(q_{ijk} - x_{ijk}) \right] \qquad (26\text{-II-}25)$$

with $0 \leq x_{ijk} \leq q_{ijk}$ and $0 \leq q_{ijk} \leq q_{ij}$, where q_{ijk} is the level of use at each geographical region k for purpose j from site i. The optimal beneficial return $f_{3ijk}(x_{ijk})$, depending on x_{ijk}, or the volume of water allocated to the region k, can be determined either by subsequent optimizing at the fourth level or by a best estimate thereof.

The problem is solved by optimizing first at the lowest level, or the third level in the above example, for each x_{ij}, with $i = 1, \ldots, n$ and $j = 1, \ldots, m$. The results from the third-level optimization are used for the second-level optimization. The optimization at the second level is then carried for each x_i, with $i = 1, \ldots, n$. Finally, the highest, or first-level, optimization is done to yield the optimal solution.

In general, the optimization is started at the lowest level. The results from each level of optimization are used for optimization on the next-higher level. This procedure is repeated until the optimal solution is obtained. In a dynamic programming, it is also possible to take into account the various types of constraints that may be present.

Hall [35] also applied the dynamic-programming technique to the optimal design of aqueduct capacity to serve a number of geographic districts located in sequence along the supply line of water. Many others [36–41] applied the technique to economic operation problems of hydrothermal power systems.

3. Other Models. There are other miscellaneous deterministic models that have been proposed for optimal designs of water-resources systems. Some of them may be shown similar to either linear or dynamic programming. Some require linear or dynamic programming for their solutions.

A *deterministic nonlinear programming* has been proposed by Cypser [42, 43] for minimizing annual cost in a hydrothermal electrical system. This model was solved by the gradient method in calculus of variations. A simplied nonlinear-programming model was also formulated by Bernholtz [44].

The *theory of inventory* [45, 46] and the *theory of storage* [47] are often used to determine optimal policies for hydroelectric operations. Although the two theories are called differently by different authors, they are essentially the same type of mathematical models. In France, Giguet [48] and Massé [24, 49, vol. 1] used such theories in formulating a deterministic hydroelectric model to minimize the cost of supplying electrical energy in the problem of scheduling the use of stored water. In the United States, Karlin [46] and Koopmans [50] have applied the inventory theory to determine water-storage policies in hydroelectric systems. Gessford [51–53] and Karlin described a deterministic model based on the formulation by Massé [49]. Koopmans proposed a model of a hydroelectric system that combined one storage reservoir and power plant with a thermal plant of unlimited capacity operating at increasing incremental cost. He assumed that the streamflow and the demand for power during a future period of time are completely known or certain.

C. Stochastic Models

1. Stochastic Linear Programming. Based on inventory control, Manne [54] formulated a typical stochastic model and optimized it by means of linear programming. In water-resources applications, he later developed a single-reservoir, three-period model in which both current inflow and initial storage are assumed known [55]. Thomas and Watermeyer [56] expanded Manne's theory to reservoir systems and formulated a *stochastic sequential model*. In this model, both inflow and storage were assumed variable in defining the initial stage of the reservoir system. The consistency of the linear-programming solution was also verified by the queuing theory.

2. **Stochastic Dynamic Programming.** Little [57] used the functional-equation approach of the dynamic programming to inventory problems and thus to formulate a stochastic dynamic-programming model for determining the optimal water-storage policy for an electric generating system consisting of one hydroelectric unit (reservoir and power plant) and one or more supplementary thermal plants. This system provided an approximate model of the Grand Coulee plant on the Columbia River. The results of this study indicate that streamflow during a time period cannot be assumed to be an independent random variable with the same probability distribution from time period to time period. Little's work was later simplified to a deterministic model by Koopmans [50], as mentioned previously.

Other stochastic models using inventory theory and solved by dynamic programming were proposed by Gessford and Karlin [51–53]. Buras [58–60] set up stochastic models for watershed-management and surface-reservoir-and-aquifer problems to which a solution by dynamic-programming methods was applied. Burt [61] applied dynamic programming to water-inventory problems involved with a stochastic supply of water. Hall and Howell [62] optimized the size of a single-purpose reservoir by applying dynamic programming to sequentially generated flow data.

3. **Other Models.** Labaye [63] and J. G. Polifka [64] formulated simple stochastic economic models for reservoir problems. By means of an inventory model, Massé [49, vol. 2] described a stochastic model for reservoir regulation, and Avi-Itzhak and Ben-Tuvia [65] gave a solution to optimize a reservoir system. Based on the queuing theory, Bryant and Thomas [66] presented a stochastic model for flow regulation by reservoirs.

V. REFERENCES

1. Churchman, C. W., R. L. Ackoff, and E. L. Arnoff: "Introduction to Operations Research," John Wiley & Sons, Inc., New York, 1957.
2. "Proposed Practices for the Economic Analysis of River Basin Projects," Report of the Subcommittee on Benefits and Costs, Federal Inter-Agency River Basin Committee, May, 1950, revised by Subcommittee on Evaluation Standards, May, 1958, Government Printing Office, 1959.
3. McKean, R. N.: "Efficiency in Government through Systems Analysis: With Emphasis on Water Resources Development," Publications in Operations Research, no. 3, sponsored by the Operations Research Society of America, John Wiley & Sons, Inc., New York, 1958.
4. Maass, Arthur, and M. M. Hufschmidt: In search of new methods for river system planning, *J. Boston Soc. Civil Engrs.*, vol. 46, no. 2, pp. 99–117, April, 1959.
5. Maass, Arthur, and M. M. Hufschmidt: Report on the Harvard program of research in water resources development, in "Resources Development: Frontiers for Research," Western Resources Conference, 1959, University of Colorado Press, Boulder, Colo., 1960, pp. 133–179. Excerpts are published in appendix 4 of The impact of new techniques on integrated multiple-purpose water development, United States Senate, Select Committee on National Water Resources, 88th Cong., 2d Sess., pursuant to S. Res. 48, March, 1960, Committee Print no. 31, 1960, pp. 64–77. Further studies concerning water qualities are reported in "Operations Research in Water Quality Measurement," U.S. Public Health Service Contract PH86-62-140 with Harvard University, February 15, 1963.
6. Maass, Arthur, M. M. Hufschmidt, Robert Dorfman, H. A. Thomas, Jr., S. A. Marglin, and G. M. Fair: "Design of Water-resource Systems: New Techniques for Relating Economic Objectives, Engineering Analysis, and Governmental Planning," Harvard University Press, Cambridge, Mass., 1962.
7. Eckstein, Otto: "Water-resource Development: The Economics of Project Evaluation," Harvard University Press, Cambridge, Mass., 1958.
8. Krutilla, J. V., and Otto Eckstein: "Multiple Purpose River Development: Studies in Applied Economic Analysis," The Johns Hopkins Press, Baltimore, 1958.

REFERENCES

9. Hirshleifer, Jack, J. C. De Haven, and J. W. Milliman: "Water Supply: Economics, Technology, and Policy," The University of Chicago Press, Chicago, 1960.
10. "Report of Use of Electronic Computers for Integrating Reservoir Operations," vol. 1, DATAmatic Corporation, prepared in cooperation with Raytheon Manufacturing Company for the U.S. Army Corps of Engineers, Missouri River Division, January, 1957.
11. Morrice, H. A. W.: The use of electronic computing machines to plan the Nile Valley as a whole, General Assembly of Toronto, Sept. 3–14, 1957, *Intern. Assoc. Sci. Hydrology Publ.* 45, vol. 3, pp. 95–105, 1958.
12. Morrice, H. A. W., and W. N. Allan: Planning for the ultimate hydraulic development of the Nile Valley, *Proc. Inst. Civil Engrs.*, vol. 14, pp. 101–156, 1959.
13. Brown, F. S.: Water resource development: Columbia River Basin, in Report of Meeting of Columbia Basin Inter-Agency Committee, Portland, Oregon, December, 1958.
14. Lewis, D. J., and L. A. Shoemaker: Hydro system power analysis by digital computer, *Proc. Am. Soc. Civil Engrs., J. Hydraulics Div.*, vol. 88, no. HY3, pt. 1, pp. 113–130, May, 1962.
15. Brittan, M. R.: A probability model for integration of Glen Canyon Dam into the Colorado River System, a doctorate dissertation at University of Colorado, Bureau of Economic Research, Boulder, Colo., August, 1960.
16. Fiering, M. B.: Queuing theory and simulation in reservoir design, *Trans. Am. Soc. Civil Engrs.*, vol. 127, pt. I, pp. 1114–1144, 1962.
17. Brooks, S. H.: A discussion of random methods for seeking maxima, *J. Operations Res.*, vol. 6, no. 2, pp. 244–251, March–April, 1958.
18. Brooks, S. H.: A comparison of maximum-seeking methods, *J. Operation Res.*, vol. 7, no. 4, pp. 430–457, July–August, 1959.
19. Cochran, W. G.: "Sampling Techniques," John Wiley & Sons, Inc., New York, 1953.
20. Cochran, W. G., and G. M. Cox: "Experimental Designs," 2d ed., John Wiley & Sons, Inc., New York, 1957.
21. Harling, John: Simulation techniques in operations research: a review, *J. Operations Res.*, vol. 6, no. 3, pp. 307–319, 591, 1958.
22. Garvin, W. W.: "Introduction to Linear Programming," McGraw-Hill Book Company, Inc., New York, 1960.
23. Massé, P., and R. Gibrat: Applications of linear programming to investments in the electric power industry, *Management Sci.*, vol. 3, no. 2, pp. 149–166, January, 1957.
24. Massé, Pierre: "Optimal Investment Decisions; Rules for Action and Criteria for Choice," transl. from the French by Scripta Technica, Inc., Prentice-Hall, Inc., Englewood Cliffs, N.J., 1962.
25. Lee, I. M.: Optimum water resources development, *Univ. Calif. Giannini Foundation Rept.* 206, Berkeley, Calif., 1958.
26. Castle, E. N.: Programming structures in watershed development, chap. 12, in G. S. Tolley and F. E. Riggs (eds.), "Economics of Watershed Planning," The Iowa State University Press, Ames, Iowa, 1961, pp. 167–178.
27. Heady, E. O.: Mathematical analysis: models for quantitative application in watershed planning, chap. 14-A, in G. S. Tolley and F. E. Riggs (eds.), "Economics of Watershed Planning," The Iowa State University Press, Ames, Iowa, 1961, pp. 197–216.
28. Dorfman, Robert: Mathematical analysis: design of the Simple Valley project, chap. 14-B, in G. S. Tolley and F. E. Riggs (eds.), "Economics of Watershed Planning," The Iowa State University Press, Ames, Iowa, 1961, pp. 217–229.
29. Vazsonyi, Andrew: "Scientific Programming in Business and Industry," John Wiley & Sons, Inc., New York, 1958.
30. Ramaseshan, S.: Discussion of aqueduct capacity under an optimum benefit policy by Warren A. Hall, *Proc. Am. Soc. Civil Engrs., J. Irrigation and Drainage Div.*, vol. 88, no. IR2, pp. 97–100, June, 1962.
31. Bellman, Richard: An introduction to the theory of dynamic programming, *RAND Corp., Rept.* R-245, Santa Monica, Calif., 1953.
32. Bellman, Richard: "Dynamic Programming," Princeton University Press, Princeton, N.J., 1957.
33. Hall, W. A., and Nathan Buras: The dynamic programming approach to water-resources development, *J. Geophys. Res.*, vol. 66, no. 2, pp. 517–520, February, 1961.
34. Hall, W. A., and Nathan Buras: Optimum irrigated practice under conditions of deficient water supply, *Trans. Am. Soc. Agr. Engrs.*, gen. ed., vol. 4, no. 1, pp. 131–134, 1961.
35. Hall, W. A.: Aqueduct capacity under an optimum benefit policy, *Proc. Am. Soc. Civil Engrs., J. Irrigation and Drainage Div.*, vol. 87, no. IR3, pp. 1–11, September, 1961.

36. Bernholtz, B., W. Shelson, and O. Kesner: A method of scheduling optimum operation of Ontario Hydro's Sir Adam Beck–Niagara generating station, *Trans. Am. Inst. Elec. Engrs., Power Apparatus and Systems*, vol. 77, pt. 3, no. 39, pp. 981–991, December, 1958.
37. Bernholtz, B., and L. J. Graham: Hydrothermal economic scheduling, pt. I, Solution by incremental dynamic programming, *Trans. Am. Inst. Elec. Engrs., Power Apparatus and Systems*, vol. 79, pt. III, pp. 921–929, December, 1960.
38. Fukao, T., T. Yamazaki, and S. Kimura: An application of dynamic programming to economic operation problems of a power system, *J. Inst. Elec. Engrs. (Tokyo)*, vol. 79, p. 158, February, 1959 (in Japanese); also in *Electrotech. J. Japan*, vol. 5, no. 2, 1959 (in English).
39. Fukao, T., and T. Yamazaki: A solution method of economic operation of hydrothermal power systems including flow-interconnected hydro power plants (dynamic linear programming method), *J. Inst. Elec. Engrs. (Tokyo)*, vol. 79, p. 601, May, 1959 (in Japanese).
40. Kirnura, M., and others: Numerical solution by dynamic programming of the problem of economic load allocation in a hydrothermal electric system, *J. Inst. Elec. Engrs. (Tokyo)*, March, 1959 (in Japanese).
41. Tzvetkov, E. V.: Economic operation of hydrogenerating stations operating in parallel with steam generating stations, *Elektr. Stantsii (Moscow)*, vol. 5, pp. 38–43, May, 1959.
42. Cypser, R. J.: The optimum use of water storage in hydrothermal electric systems, Sc.D. thesis, Massachusetts Institute of Technology, Mass., January, 1953.
43. Cypser, R. J.: Computer search for economical operations of a hydrothermal electric system, *Trans. Am. Inst. Elec. Engrs.*, vol. 73, pp. 1260–1266, discussions on pp. 1266–1267, October, 1954.
44. Bernholtz, B.: Optimum allocation of units in a hydroelectric generating station, *SIAM Rev.*, Society for Industrial and Applied Mathematics, Philadelphia, 1960.
45. Arrow, K. J., Samuel Karlin, and Herbert Scarf: "Studies in the Mathematical Theory of Inventory and Production," Stanford Mathematical Studies in the Social Sciences, no. I, Stanford University Press, Stanford, Calif., 1958.
46. Karlin, Samuel: The mathematical theory of inventory processes, chap. 10 in E. F. Beckenbach (ed.), "Modern Mathematics for the Engineer," McGraw-Hill Book Company, Inc., pp. 228–258, 1961.
47. Moran, P. A. P.: "The Theory of Storage," Methuen's Monographs on Applied Probability and Statistics, Methuen & Co., Ltd., London, 1959.
48. Giguet, R.: Étude de l'exploitation optimum d'un barrage en régime connu, *Mém. Ann. Ponts Chaussées*, vol. 115, no. 2, pp. 145–172, March–April, 1945.
49. Massé, Pierre: "Les réserves et la regulation de l'avenir dans la vie économique," vol. 1, "Avenir déterminé," and vol. 2, "Avenir aléatoire," Actualités Scientifiques et Industrielles, nos. 1007 and 1008, Hermann & Cie, Paris, 1946.
50. Koopmans, T. C.: Water storage policy in a simplified hydroelectric system, *Cowles Foundation Discussion Paper* 26, Yale University, New Haven, Conn., Mar. 15, 1957. This paper is not offered to the general public, but it was later published in Proceedings of the First International Conference on Operational Research, Oxford, 1957, Operations Research Society of America, Baltimore, and printed by John Wright and Sons Ltd., Stonebridge Press, Bristol, England, 1957, pp. 191–227.
51. Gessford, John: The use of reservoir water for hydroelectric power generation, dissertation, Stanford University, Stanford, Calif., 1957.
52. Gessford, John, and Samuel Kralin: Optimal policy for hydroelectric operations, chap. 11, pp. 179–200, in [45].
53. Gessford, John: Scheduling the use of water power, *Management Sci.*, vol. 5, no. 2, pp. 179–191, January, 1959.
54. Manne, A. S.: Linear programming and sequential decisions, *Management Sci.*, vol. 6, no. 3, pp. 259–267, April, 1960.
55. Manne, A. S.: Product-mix alternatives: flood control, electric power, and irrigation, *Cowles Foundation Discussion Paper* 95, Yale University, New Haven, Conn., Oct. 14, 1960.
56. Thomas, H. A., Jr., and Peter Watermeyer: Mathematical models: a stochastic sequential approach, chap. 14, pp. 540–561, in [6].
57. Little, J. D. C.: The use of storage water in a hydroelectric system, *J. Operations Res.*, vol. 3, no. 2, pp. 187–197, May, 1958.
58. Buras, Nathan: Dynamic programming methods applied to watershed management problems, *Trans. Am. Soc. Agr. Engrs.*, vol. 5, no. 1, pp. 3–5, 1962.
59. Buras, Nathan: Conjunctive operation of dams and aquifers, *Proc. Am. Soc. Civil Engrs., J. Hydraulics Div.*, vol. 89, no. HY6, pt. 1, pp. 111–131, November, 1963.

REFERENCES

60. Buras, Nathan: Conjunctive operation of a surface reservoir and a ground water aquifer, *Intern. Assoc. Sci. Hydrology*, Pub. 63, pp. 492–501, 1963.
61. Burt, O. R.: The economics of conjunctive use of ground and surface water, Ph.D. thesis, University of California, Berkeley, Calif., 1962.
62. Hall, W. A., and D. T. Howell: The optimization of single-purpose reservoir design with the application of dynamic programming to synthetic hydrology samples, *J. Hydrology*, vol. 1, no. 4, pp. 355–363, 1963.
63. Labaye, G.: Le problème des évacuateurs de crues de Serre-Ponçon: essai de détermination d'un optimum économique, in "L'Utilisation de la statistique dans les problèmes d'hydraulique et l'hydrologie," Proceedings of the Conference of the S.H.F. (Société Hydrotechnique de France), Comité Technique, Mar. 15, 1956, pp. 47–64.
64. Hufschmidt, M. M.: Application of basic concepts: graphic techniques, chap. 5, pp. 226–244, in [6].
65. Avi-Itzhak, B., and S. Ben-Tuvia: A problem of optimizing a collecting reservoir system, *Operations Res.*, vol. 11, no. 1, pp. 122–136, January–February, 1963.
66. Bryant, G. T., and H. A. Thomas, Jr.: A stochastic model for flow regulation by an impounding reservoir, paper presented at the Annual Convention of the American Society of Civil Engineers, New York, Oct. 16–17, 1961.

Section 27

WATER LAW

FRANK J. TRELEASE, *Dean and Professor of Law, College of Law, The University of Wyoming.*

I. Introduction	27-2
A. The Function of Law	27-2
B. Sources of Law	27-3
II. Surface Streams	27-4
A. Riparian Rights	27-4
1. General Statement	27-4
2. Riparian Land	27-4
3. Riparian Uses	27-5
4. Prescriptive Rights	27-8
5. Loss of Riparian Rights	27-9
6. Extent of Riparian Doctrines	27-10
7. Statutory Modifications	27-10
B. Prior Appropriation	27-11
1. General Statement	27-11
2. Elements of an Appropriation	27-11
3. Appropriators	27-12
4. Beneficial Use	27-12
5. Quantity of Water	27-13
6. Place of Use	27-14
7. Priority	27-14
8. Preferences	27-15
9. Procedures	27-16
10. Changes in Appropriations	27-17
11. Transfers of Appropriations	27-18
12. Loss of Appropriations	27-18
13. Extent of Appropriation Doctrines	27-18
C. Obstruction and Pollution	27-19
1. Obstruction of Streams	27-19
2. Changes in Watercourses	27-19
3. Flowage	27-20
4. Pollution	27-20
D. Interstate Streams	27-21
1. Equitable Apportionment	27-21
2. Interstate Compacts	27-21
E. Federal Water Law	27-22
1. General Statement	27-22
2. Commerce Power	27-22

        3. Other Federal Powers.................................. 27-23
        4. Navigation and Flood Control........................ 27-23
        5. Federal Power Licenses............................... 27-24
        6. Reclamation and Irrigation........................... 27-24
        7. Department of Agriculture Programs................... 27-25
   III. Other Sources of Water..................................... 27-26
        A. Classification of Waters................................ 27-26
           1. General Statement................................. 27-26
           2. Watercourses...................................... 27-26
           3. Lakes and Ponds................................... 27-27
           4. Springs... 27-27
           5. Surface and Floodwaters........................... 27-27
           6. Return Flows and Waste Water...................... 27-27
           7. Foreign and Added Waters.......................... 27-28
           8. Groundwaters...................................... 27-28
           9. Atmospheric Water................................. 27-29
        B. Surface-water Law..................................... 27-30
           1. Drainage.. 27-30
           2. Rights to Capture................................. 27-30
        C. Groundwater Law...................................... 27-31
           1. Ownership Rule.................................... 27-31
           2. Reasonable-use Rule............................... 27-31
           3. Correlative Rights................................ 27-31
           4. Conservation Statutes............................. 27-31
           5. Appropriation..................................... 27-32
    IV. Conclusion... 27-32
        A. Law and Science....................................... 27-32
        B. The Law in Action..................................... 27-33
     V. References... 27-34

I. INTRODUCTION

A. The Function of Law

The function of law is to regulate the relations between men or groups of men. In playing this part, the law serves essentially a dual purpose; it provides a mechanism for the solution of conflicts after they have arisen, and it furnishes a guide for the ordering of future conduct. In general, it may be said that the major purposes of law are to adjust the relations and control the activities of men so that the greatest good redounds to the greatest number of people. Much law does not literally regulate conduct in the sense of requiring or forbidding certain action; it instead provides an area of free choice, setting outside limits within which a person may act as he chooses. Many of these laws, such as those relating to property and contracts, unobtrusively form the basic framework of society.

It is true that men have not always agreed upon the form that the law should take to accomplish these objectives and that some may find their desires blocked by laws that to them seem outmoded or unsuitable. But by and large the law at any particular time and place represents the will of the majority for encouraging action deemed desirable and for discouraging or forbidding action considered to be in conflict with some broad public interest.

Law is a complex subject. Often a final legal decision can be reached only after an extensive lawsuit and an appeal, requiring the participation of trained attorneys and judges with years of experience. Yet law is also a part of our everyday lives. As a person drives an automobile down the street, or pays his bills at the end of the month, he is governed by law and is applying law he knows to the common situations of his private life. The purpose of this summary of water law is to acquaint the hydrologist or engineer with the principal doctrines of law that deal with the subject of his professional activities. It is hoped that this exposition will provide some guidelines which he may follow in the planning and construction of works and enable him to avoid risks of legal liability, and even risks of lawsuits, when the law or the facts are uncertain. Physical solutions to legal problems may exist in that design features of a project may eliminate objections and prevent future lawsuits. Although it is not possible in this short treatment to make the reader an expert in water law, it is possible to point out factors that must be considered, procedures that must be followed, and danger spots where competent legal advice may be required before action is taken. In addition, a knowledge of fundamental legal principles may be of help to the scientist if he is called as an expert witness to describe or explain the factual background of lawsuits over water. These legal principles may also be of aid to him in determining the nature and type of physical data to be collected, where there are possibilities that such data may become important in future controversies.

B. Sources of Law

Water law is primarily a division of property law. Most of it is concerned with water rights, rights to divert and use water or to use it in place. Some water law is included in the general field of torts, wrongs done to other persons and property for which the wrongdoer must pay or may be enjoined from perpetrating. Some acts in relation to water have been declared to be crimes. Doctrines of constitutional law determine the power of governments to engage in water-control activities or to regulate individual conduct relating to water. Most water law is state law and stems from *common law* declared by judges or from statutes passed by legislatures. While the decisions and statutes may vary from state to state, the principles behind them are often similar, or may be grouped into categories on the basis of similar characteristics.

American law is primarily a system of common law governed by the doctrine of precedent. As disputes arise, they are settled in lawsuits, which not only result in a decision of the instant case, but also provide an indication of how courts will handle similar cases that may arise in the future. The rule a court announces in making a decision becomes a precedent, which it and other courts will follow. Most precedents are found in the reports of decisions of higher courts in appealed cases, for few trial courts write opinions. When the supreme court of a state has once announced a rule of law, it and all trial judges of that state will follow that rule in later cases. But trial and appellate judges in other states may choose another rule. When the same type of case has been decided by the courts of several states, and some choose one rule and some another, there is a "split of authority" and a majority and a minority rule. A court will sometimes overrule its prior decisions, if convinced that they are wrong, or may refuse to follow a precedent set by another court. While Federal courts have the last word on the Federal Constitution and statutes, they must follow state law on matters of local property and tort law.

Legislation is the other principle source of law. If a rule of common law is found undesirable, or if there is uncertainty as to what rule will be applied, a statute may change the rule or adopt a particular rule. Often a statute will engraft a particular exception on the common law. Often legislation will supersede it almost entirely on a particular subject, by the adoption of a comprehensive regulatory code. A statute must be followed by a court, but courts may determine its applicability to a particular case and may interpret its language. Thus a considerable body of case law about statutes has come into being, and where states have similar statutes, the decisions of one court are often accepted as precedent by another.

Constitutions are a special form of legislation that bind not only courts but legislatures. Some state constitutions have provisions laying the basis for water-right laws; others are silent. But every state has general sovereignty to enact laws affecting property within its borders and also has the "police power" to regulate the use of property and the conduct of its citizens in the interests of the public health, safety, and welfare. Both Federal and state constitutions may authorize governmental water activities, though this is not always done in specific terms. Negatively, constitutions may prohibit certain types of laws, such as special or local laws, or may place limits on some types of public financing, and all contain restrictions against laws that unreasonably deprive persons of liberty or property without due process of law.

Administrative law comes into being when legislatures regulating a complete subject delegate power to administer the law to an executive agency with a technically trained staff or to a particular official, usually an expert. These administrators will frequently adopt subsidiary rules and regulations and, in the day-to-day administration of the law, will make decisions and adopt practices that come to govern future cases much as do judicial precedents. The courts will review the decisions of administrative agencies and officials, to determine whether they acted within their jurisdiction and delegated powers and whether they correctly interpreted the law. But in many cases courts will refuse to review administrative decisions on questions that call for expert technical judgment, especially on fact questions.

Local governments such as cities and counties may adopt laws, usually called *ordinances*, that may affect water use and control. These must be authorized by legislation and must be reasonable, for a local purpose and not inconsistent with state-wide policies or legislation.

In the following discussion many decisions and statutes are cited and may be consulted in respect to specific problems. But a comprehensive treatment is impossible within the space allotted, and these citations are only to illustrate typical cases and laws. Although some references are made to more detailed discussions of law, the necessity for legal assistance must again be stressed, where the particular form of applicable law is important, or where the answer to conflicting views of the facts may require the application of one rule of law rather than another.

II. SURFACE STREAMS

A. Riparian Rights

1. General Statement. The major feature of the *doctrine of riparian rights* is that under it the law gives equal rights to the use of water to the owners of land which borders upon or touches a stream or watercourse. Another important principle is that a riparian right to the use of water exists whether the use is made or not; hence a riparian owner can initiate a use at any time and insist that his rights be accommodated with other uses or that a share of the water be allotted to him. Developed in the humid climates of England and eastern United States, riparian law seems to be based upon an unspoken premise that if rights to use are restricted to those persons that have access to the water through the ownership of the banks, and if those persons will restrict their demands on the water to reasonable uses, there will be enough for all.

But the law of riparian rights exists not only in eastern United States. Although the pioneers that settled the western states developed the law of prior appropriation when they used the streams of the public domain,[1] some of the courts of these states applied riparian principles to disputes involving private property, apparently without any conscious choice of one type of law over the other. Later, as water rights based upon the different doctrines came into conflict, these states made various judicial and legislative adjustments and modifications which permitted both bodies of law to remain in effect at the same time or which subordinated riparian rights to appropriation law.[2]

2. Riparian Land. Riparian rights are a form of real property, a part of land law. The rights are "appurtenant" to the land; that is, they are attached to it in the sense that a person who purchases or inherits riparian land automatically acquires the water right, although it is not specifically mentioned. The nature of the right is "usufructuary"; the riparian does not own the water, but owns only the right to use it on his riparian land and to have it flow to his land so that it may be used.

In modern water law, the determination of whether land is riparian or nonriparian

---

[1] See Subsec. II-B-1.
[2] See Subsecs. II-A-6 and II-B-13.

has two significant aspects. First, the ownership of riparian land gives the owner rights to the use of the waters of the stream and the corresponding power to resist uses by nonriparians or excessive uses by other riparians. Second, since uses of water are restricted to riparian lands, at least where there are conflicting demands upon the stream, the extent of the riparian land measures the extent to which irrigation may be practiced as an exercise of the riparian right.

The first consideration in determining the riparian character of land is always whether or not it touches the stream. Thus land, to be riparian, must have the stream flowing over it or along its border. Where a stream has both high banks and low banks, ownership of land touching the high banks, or of the bottomlands, may give rights to use the stream if regularly occurring high water, which can be regarded as the natural flow of the stream, touches or runs over the land.[3] Land is contiguous to the stream if it is on a pool or slough in contact with the stream, even though there is no current past the land.[4]

A cardinal rule of riparian law is that land, to be riparian, must be within the watershed of the stream, and it is immaterial that the land beyond the watershed is a part of a single tract which extends to the stream.[5] The determination of the extent of a watershed is primarily a question of physical fact, since the watershed is the drainage area or catchment basin whose surface runoff feeds the stream, but sometimes legal problems arise as to the extent of a watershed where tributaries break up the area into subwatersheds.[6]

Where a riparian tract has been split up among several owners, or where several tracts have been acquired by a single owner, the courts of different states have not agreed as to whether all the land is riparian. Several states have adopted the rule that riparian rights cannot be extended to lands that were not within the original tract to which riparian rights first attached.[7] This is known as the *source-of-title test*, and under this rule a tract not riparian when acquired cannot be made such by purchase of the land lying between it and the stream, nor can riparian lands be enlarged by the acquisition of title to lands contiguous to the riparian land but not to the stream. Similarly, although all the lands were once a part of a single riparian tract, once the title has been split so that a part has become nonriparian, the reunion of title in a single person cannot restore its riparian character.

Other states have rejected the doctrine of source of title, and instead adopted the *test of unity of title*, by which all land in a single ownership is regarded as a single tract whose riparian character is to be determined by its contiguity to the stream.[8] Thus the fact that the owner procured the tract washed by the stream at one time and subsequently purchased the land adjoining it is not a valid objection to the use of the water on the land last acquired.

Some modification of these rules is possible by agreement between the parties expressed in the deed transferring the land. When the owner of a riparian tract conveys away a portion of the land that is not contiguous to the stream, by a deed that is silent as to water rights, the conveyed parcel loses its riparian status.[9] But where appropriate words show the intention of the grantor to preserve the riparian character of the land conveyed, the deed passes title to the same rights that the grantor enjoyed to use the water on the noncontiguous land.[10]

3. Riparian Uses. In considering the uses to which the waters of streams may be put under the doctrine of riparian rights, it is necessary to explore the fundamental

---

[3] *Ventura Land & Power Co. v. Meiners* (1902) 136 Cal. 284, 68 P. 818, 89 A.S.R. 128.
[4] *Turner v. James Canal Co.* (1909) 155 Cal. 82, 99 P. 520, 22 L.R.A. (n.s.) 401, 132 A.S.R. 59.
[5] *Anaheim Union Water Co. v. Fuller* (1907) 150 Cal. 327, 88 P. 978, 11 L.R.A. (n.s.) 1062.
[6] *Crane v. Stevinson* (1936) 5 Cal.2d 387, 54 P.2d 1100.
[7] *Boehmer v. Big Rock Irrigation District* (1897) 117 Cal. 19, 48 P. 908; *Crawford Co. v. Hathaway* (1903) 67 Neb. 325, 93 N.W. 781.
[8] *Jones v. Conn* (1901) 39 Ore. 30, 64 P. 855, 65 P. 1068; *Clark v. Allaman* (1905) 71 Kan. 206, 80, P. 571.
[9] *Anaheim Union Water Co. v. Fuller* (1907) 150 Cal. 327, 88 P. 978, 11 L.R.A. (n.s.) 1062.
[10] *Strong v. Baldwin* (1908) 154 Cal. 150, 97 P. 178, 129 A.S.R. 149.

nature of the right of the riparian proprietor. The courts have not been in complete accord as to the exact nature of the fundamental underlying concept, and two distinct theories have arisen which have a material bearing upon the legality of any use of the water. These have been designated as the *natural-flow theory* and the *reasonable-use theory*.

The natural-flow theory was first evolved. According to it, the fundamental right of a riparian is to have the streamflow as it was accustomed to flow in nature, unimpaired in quality and undiminished in quantity. Of course, the enforcement of such a theory to the ultimate would prevent almost all use of the water, but even in its strictest form it has been held to permit some use by riparian proprietors. Although it has been criticized as nonutilitarian, the rule was first adopted in a mill-dam economy and worked very well to preserve the streams for the most important demands made upon them: the furnishing of power to grist mills, sawmills, and factories. When applied to more modern situations, however, the rule might prohibit many beneficial uses of water, although those uses might cause no one any harm and although the water would run to waste if not used.

Today the great majority of American jurisdictions apply the theory that the fundamental right of the proprietor is to the reasonable use of the water of the stream and to be free from unreasonable interferences with his use. This view of riparian rights is today held to some extent in most of the eastern states[11] and in all the western states that give any recognition to riparian doctrines.[12] However, courts that once adopted the natural-flow theory have not been absolutely consistent in modifying it with the reasonable-use rule, and some inconsistencies and illogical results have followed. The most frequent variation of the rule of reasonable use is to hold that any nonriparian use of the water is unreasonable as a matter of law, regardless of its reasonableness in fact or whether or not actual harm is being done.[13] But in most states today, even in controversies between riparians and nonriparians, the riparian can complain only when nonriparian appropriations cause an actual loss or injury to his use of water under reasonable methods of use and diversion.[14]

In some early cases the courts divided riparian uses into *natural* and *artificial*. The former were those absolutely necessary for man's survival, such as domestic and stock-water purposes; the latter were those that increase his comfort and prosperity, such as irrigation and manufacturing.[15] The importance of this classification was that the upper riparian on a small stream might consume all the water for his natural wants. This was originally adopted as a significant exception to the natural-flow doctrine, and with the ascendancy of the reasonable-use rule, the classification has fallen somewhat into disuse. Nevertheless, a preference is still given to the domestic demands of riparians over the use of water for other purposes.[16]

A number of uses have received judicial approval, and their limits have been defined to some extent. Domestic use includes water for drinking, cooking, laundry and sanitation, and other household purposes. In some courts it has been held that domestic use includes the watering of small household gardens and barnyard stock.[17] Water for any number of livestock is a proper riparian use if reasonable in relation to the needs of others, but water for large herds of stock may not be included with domestic use so as to entitle the user to a preference.[18] A substantial quantity of water may be necessary to fulfill domestic uses where people are gathered in hotels,

---

[11] *Ulbricht v. Uefaula Water Co.* (1889) 86 Ala. 587, 6 So. 78; *Sanborn v. Peoples Ice Co.* (1900) 82 Minn. 43, 84 N.W. 641.

[12] *Peabody v. Vallejo* (1935) 2 Cal.2d 351, 40 P.2d 486; *McDonough v. Russell-Miller Milling Co.* (1917) 38 N.D. 465, 165 N.W. 504; *Watkins Land Co. v. Clements* (1905) 98 Tex. 578, 86 S.W. 733, 70 L.R.A. 964, 107 A.S.R. 653.

[13] *Anaheim Union Water Co. v. Fuller* (1907) 150 Cal. 327, 88 P. 978, 11 L.R.A. (n.s.) 1062; *Harvey Realty Co. v. Wallingford* (1930) 111 Conn. 352, 150 A. 60.

[14] *Watkins Land Co. v. Clements* (1905) 98 Tex. 578, 86 S.W. 733, 70 L.R.A. 964, 107 A.S.R. 653; *Brown v. Chase* (1923) 125 Wash. 542, 217 P. 23.

[15] *Evans v. Merriweather* (1842) 3 Scam. (Ill.) 492, 18 A.D. 106.

[16] *Church v. Barnes* (1933) 175 Wash. 327, 27 P.2d 690.

[17] *Hough v. Porter* (1909) 51 Ore. 318, 98 P. 1083.

[18] *Cowell v. Armstrong* (1930) 210 Cal. 218, 290 P. 1036.

apartment houses, or resorts or where state institutions such as penitentiaries and insane asylums, or even military camps, are given the privilege of taking water for domestic use.[19] But domestic use does not include municipal uses in the nonriparian areas of cities. A city situated on the banks of a stream is not a riparian proprietor in any sense that would permit it to divert water and sell it to inhabitants who live on land not adjacent to the stream.[20] The city must procure its municipal supplies in some other manner.[21]

In the arid and semiarid West, where irrigation is necessary to the successful cultivation of the soil, there is no doubt that irrigation is a proper riparian use, if the quantity taken is reasonable in relation to the needs of others. It has also been held to be a reasonable use in the East.[22]

One of the earliest and most clearly established uses of streams is for the generation of water power for the operation of mills and factories. Today the most important application of water power is its conversion into hydroelectric power by means of turbines and generators, and the reasonable use of the waters of streams and rivers for this purpose is recognized as an exercise of riparian rights.

There has been little uniformity in the holdings of courts in regard to the right of a riparian proprietor to store the waters of a stream for future use. The detention of waters is a use, or means of accomplishing a use, that is permissible if reasonable, and reasonableness is a question of fact. But it has been held that the seasonal storage of water for power purposes in such quantities as would amount to a complete obstruction of the usual flow and the withdrawal of the waters from the lands of other riparians is not a proper riparian use.[23] On the other hand, a temporary detention by a dam that does not impound a large quantity of water but simply creates a head and passes the water through a powerhouse is a proper riparian use.[24] Clear distinctions between these types of dams have not been drawn and may be difficult. The right to store the waters of spring flows, for irrigation later in the season, is permissible when it can be done consistently with the rights of the lower proprietors,[25] but not when it would deprive other owners of the use and service of the stream.[26] In recent years many small detention dams, gulley plugs, dikes, and stock-water reservoirs have been constructed in order to check runoff, create natural subirrigation, aid in flood protection, and conserve the soil. Where these are built upon watercourses, they raise problems of riparian rights. The courts have recognized the value of such soil- and water-conservation measures and ruled that it is a reasonable use to erect small dams on a small intermittent stream for such purposes where the flow of the stream was not materially reduced.[27]

These are the most common riparian uses, but there is no fixed category, and almost any application of water that fulfills a need or desire of man can be considered a proper use as long as it is reasonably exercised with due regard to equal rights of other riparians. The courts have upheld many other uses of water that serve some commercial or industrial need, such as the harvesting of ice, the use in railroad engines, the use of water for cooling in steam power plants, and the use in oil-drilling operations.[28] This does not mean that a commercial or industrial application is necessary to a proper

[19] *Prather v. Hoberg* (1944) 24 Cal.2d 549, 150 P.2d 405; *United States v. Fallbrook Public Utility District* (S.D. Cal. 1951) 101 F. Supp. 298, (1952) 108 F. Supp. 72, 109 F. Supp. 28.
[20] *Emporia v. Soden* (1881) 25 Kan. 588, 37 A.R. 265.
[21] See Subsec. III-A-5.
[22] *Anderson v. Cincinnati Southern Ry.* (1887) 86 Ky. 44, 5 S.W. 49; *Meyers v. Lafayette Club* (1936) 197 Minn. 241, 266 N.W. 861.
[23] *Herminghaus v. Southern California Edison Co.* (1926) 200 Cal. 81, 252 P. 607.
[24] *Mentone Irrigation Co. v. Redlands Electric Light & Power Co.* (1909) 155 Cal. 323, 100 P. 1082, 22 L.R.A. (n.s.) 382, 17 Am. Ann. Cas. 1222.
[25] *Stacy v. Delery* (1909) 57 Tex. Civ. App. 242, 122 S.W. 301.
[26] *Still v. Palouse Irrigation & Power Co.* (1911) 64 Wash. 606, 117 P. 466.
[27] *Heise v. Schultz* (1949) 167 Kan. 34, 204 P.2d 706.
[28] *McDonough v. Russell-Miller Milling Co.* (1917) 38 N.D. 465, 165 N.W. 504; *Atchison, Topeka & Santa Fe Ry. Co. v. Shriver* (1917) 101 Kan. 257, 166 P. 519; *Fairbury v. Fairbury Mill & Elevator Co.* (1932) 123 Neb. 558, 243 N.W. 774; *Smith v. Stanolind Oil & Gas Co.* (1946) 197 Okla. 499, 172 P.2d 1002.

riparian use; the water may be used merely to satisfy a desire for pleasure or aesthetic enjoyment. It may be used in a swimming pool, in connection with a park or resort, or in place for boating, bathing, and fishing.[29] On lakes where the chief value of the surrounding land is for resort purposes and summer homes, the water level may not be lowered to the point where the lake is turned into a mud flat and these site values are destroyed.[30]

The reasonableness of a particular use of water by a riparian is a question of fact, and each case must be determined with reference to its own facts and circumstances. The use of water by one riparian that causes substantial harm to another can generally be said to be unreasonable unless the utility of the use outweighs the gravity of the harm. Reasonableness is not determined solely by reference to the needs of the user nor solely from the standpoint of the person harmed; the mere fact of substantial inconvenience or damage to other riparians does not itself make the use unreasonable.[31] Wasteful uses, or a wasteful method of use, may be unreasonable.[32]

Even a nonriparian use may be reasonable under certain circumstances. Although it was formerly the law that the use of water on nonriparian lands was unlawful, today it is the general rule that if the use is by a riparian, the fact that the use is on nonriparian land is merely one factor to be considered in its reasonableness.[33] But in some states, if the use is by a nonriparian, a riparian has an absolute right to legal protection for his use, without regard to the reasonableness of the nonriparian use.

It follows where this is the rule that a riparian owner cannot grant the right to use the waters to persons who are not themselves riparian proprietors, and thus give them rights that would be good against downstream riparians.[34] But the grant of such a use would be binding between the parties, and such grants may therefore enable a nonriparian user to buy out some of the larger riparian demands and thus ensure a sufficient supply of water.[35]

Although it is said that all riparian proprietors on a stream have equal rights to the use of the waters, this generalization is not to be applied literally to the quantity of water to which the proprietors are entitled. Though the rights may be equal, the relative reasonableness of various uses will frequently result in an unequal division of the water. Generally, the riparian owner who desires to use the water for a consumptive use such as irrigation is not entitled to any specific quality of water measured either in total amount or in rate of flow; he is entitled to only a reasonable quantity of water consistent with the demands of others for like purposes. Theoretically, what is reasonable will vary from year to year and from season to season, not only because of the varying volume of water flowing in the stream, but also because of the varying extent of the use of water by others.[36] In most jurisdictions there is no mathematical rule or formula for determining the quantity of water to which a riparian will be entitled, but in a few cases a specific fraction or percentage of the flow has been allowed.[37]

4. Prescriptive Rights. A riparian may lose the right to complain of a nonriparian or excessive use if that use has been made continuously and adversely for the period of the statute of limitations (which requires lawsuits to be brought within a certain time), and conversely, *prescriptive rights* to the use of water may be acquired by such adverse use. The use of water is adverse to a riparian proprietor if it is hostile, open, and notorious and is an invasion of his rights. If a use is made with the permisssion or license of the riparian owner, no prescriptive right can be obtained regardless of the length of time the use continues.

[29] *In re* Clinton Water District (1950) 136 Wash. 284, 218 P.2d 209; *Broady v. Furray* (1933) 163 Okla. 204, 21 P.2d 770.
[30] *Los Angeles v. Aitken* (1935) 10 Cal. App. 460, 52 P.2d 585.
[31] *Turner v. James Canal Co.* (1909) 155 Cal. 82, 99 P. 520, 22 L.R.A. (n.s.) 401, 132 A.S.R. 59.
[32] *Campbell v. Grimes* (1901) 62 Kan. 503, 64 P. 62.
[33] *Elliot v. The Fitchburg Railroad Co.* (1852) 10 Cush. (Mass.) 191.
[34] *Gould v. Eaton* (1897) 117 Cal. 539, 49 P. 577.
[35] *Duckworth v. Watsonville Water & Light Co.* (1907) 150 Cal. 520, 89 P. 338.
[36] *Meng v. Coffey* (1903) 67 Neb. 500, 93 N.W. 713.
[37] *Harris v. Harrison* (1892) 93 Cal. 676, 29 P. 325; *Hunter Land Co. v. Laugenour* (1926) 140 Wash. 588, 250 P. 41.

SURFACE STREAMS 27-9

Since the use must be wrongful as against the riparian owner, the statute of limitations begins to run only if the lower riparian has a cause of action against the user. Thus, where the natural-flow theory of riparian rights is applied, the statute will begin to run from the moment the nonriparian use is begun, for any sensible depletion of the natural flow of water is actionable even though it interferes with no use of the riparian proprietor and causes him no damage. On the other hand, where the reasonable-use theory is applied, the statute will not begin to run until actual damage results from an unreasonable interference with the uses of the riparian.[38]

The period of the statute of limitations is set by state law, and varies from 5 years to 20 years. If the natural-flow theory is used and the period is short, prescriptive rights to nonriparian uses can be obtained with ease, for people will not often begin a lawsuit if they are not actually damaged. But where the period of limitations is long, and especially where the rule of reasonable use is applied in these circumstances, prescriptive rights are difficult to obtain, for if actual and substantial damage is done to a lower riparian, he is very likely to go to court.

In an ordinary case, no prescriptive right can be obtained by a water user against riparians upstream from his point of diversion, regardless of the length of time the use continues.[39] The rule is sometimes expressed in the expression: "Prescription does not run upstream." It is derived from the rule stated above, that a use, to be adverse, must be an invasion of the rights of the riparian owner sufficient to give him a cause of action, and in most cases a riparian is not harmed by anything that is done below him. However, it is possible for a prescriptive right to be obtained against an upstream owner where the downstream user goes upstream to construct a ditch across the upper owner's land and has continued to use and maintain the ditch for the limitation period.[40]

A prescriptive right is not itself a riparian right, but may be more accurately described as a power to take the water without reference to the rights of riparian owners who have been affected by the adverse use. The right obtained by prescription is absolute; there are no correlative rights between the riparian and the prescriptive user. The extent of such a right is measured by the extent of the adverse use.[41]

5. Loss of Riparian Rights. Generally, a riparian right cannot be lost by abandonment, simply by nonuse of the water. Since use does not create the right, disuse cannot destroy it.[42] However, in some states which have statutory proceedings for claiming and adjudicating all water rights on a stream system, the failure to take part in such proceedings and claim riparian rights may result in their loss.[43]

A riparian may also lose his right to object to an upstream use by the operation of the legal doctrine known as *estoppel*. A riparian who has permitted a nonriparian to construct a dam and ditch on his land at great expense is "estopped" (prevented) from revoking the license and destroying the value of the irrigated nonriparian land.[44] And silent acquiescence in an upstream use for which large sums of money have been spent for the public benefit may estop the riparian from procuring an injunction that would destroy the public use, though he may still have his legal remedy for damages to compensate him for the rights he has lost.[45]

The most frequent way in which riparian rights are lost is by their being taken by condemnation for the fulfillment of a public purpose. When a public or semipublic agency needs water, it has the power to take it, as long as it pays just compensation for the loss it causes. Any governmental authority has this *right of eminent domain*, and corporations organized for public purposes, such as public-utilities or irrigation-water distribution agencies, may be given a similar power by grant from the state that creates them.

---

[38] *Pabst v. Finmand* (1922) 190 Cal. 124, 211 P. 11.
[39] *Cory v. Smith* (1929) 206 Cal. 508, 274 P. 969.
[40] *Dontanello v. Gust* (1915) 86 Wash. 268, 150 P. 420.
[41] *Larsen v. Apollonio* (1936) 5 Cal.2d 440, 55 P.2d 196.
[42] *Redwater Land & Canal Co. v. Reed* (1910) 26 S.D. 466, 128 N.W. 702.
[43] *Wilson v. Angelo* (1934) 176 Wash. 157, 28 P.2d 276.
[44] *Motl v. Boyd* (1926) 116 Tex. 82, 286 S.W. 458.
[45] *Miller & Lux v. Enterprise Canal & Land Co.* (1915) 169 Cal. 415, 147 P. 567.

Riparian rights on navigable waters over which the United States has jurisdiction may be lost, without compensation, by the exercise of Federal powers. Such rights are subject to the *navigation servitude*, and the right itself is defined as not good against the government. Therefore the United States may refuse a Federal Power Commission license to exercise riparian rights by building a power dam, it may improve navigation by destroying a power dam, or it may construct a dam on a site held by a power company,[46] without paying anything for the loss of water rights.

6. Extent of Riparian Doctrines. Originally, the law of riparian rights was the sole law applicable to surface streams in the 31 eastern states of the nation, including that tier of states running northward from Louisiana to Minnesota. It is still the major law in all these states except Iowa and Mississippi, which have recently adopted appropriation laws somewhat similar to those in effect in the West.

In what may be called the Rocky Mountain states (Arizona, Colorado, Idaho, Montana, Nevada, New Mexico, Utah, and Wyoming) and in Alaska, riparian rights to the use of water have never been recognized and the law of prior appropriation is the sole basis of rights to the use of water. In the early history of the states on the west coast (Washington, Oregon, and California) and on the Great Plains boundary between the humid and arid areas (the Dakotas, Nebraska, Kansas, Oklahoma, and Texas), riparian rights as well as appropriation rights were recognized. In these states many rights to use water are based on early exercises of the riparian rights of the landowners along the streams, but today the use of riparian rights as the foundation of a new claim to the use of water is possible only in California, Texas, North Dakota, and perhaps Washington.[47] Nevertheless, the existence, extent, and protection of a large portion of the rights to use water in the west coast and Great Plains states depend upon the principles of riparian law, and the further development of the water resources of those states through the appropriation of water will be conditioned and limited by the riparian rights that have heretofore been vested. The nature and extent of the adjustments to reconcile these two inconsistent doctrines of law and to coordinate the water rights of persons under them will be described in the section on prior appropriation.

In Hawaii, the doctrine of riparian rights has been superimposed upon an ancient system of water rights based upon that state's unique system of land titles—*ahupuaas, ilis,* and *kuleanas*—derived from the old customs of the original Hawaiian people. The overlord of an *ahupuaa*, a large wedge of land stretching from the sea to the mountain tops, had rights to use and dispose of all the normal flow of streams within his land unit that he had not previously assigned to tenants and holders of *kuleanas*, small tracts of cultivated lands. The owner of an *ili* had a similar right within his tract, usually a large area carved out of an *ahupuaa*. Within these units the holders of lands to which water rights were appurtenant had equal claims to the waters, but not strictly riparian claims, since there was no requirement that the holding be adjacent to the stream. Nor was the right appurtenant to the entire *ahupuaa* riparian; the water might be applied to land outside of it. But when a stream flows between or through two *ahupuaas* or *ilis*, the modern courts have applied riparian principles to divide surplus freshet waters between the holders of the lands.[48]

7. Statutory Modifications. In many of the eastern states the basic riparian law has been changed by statutes which regulate exercises of certain types of riparian rights or create exceptions and permit the acquisition of some water rights outside the riparian pattern. The earliest of these were the *Mill Dam Acts*, which encouraged the development of water power by giving privileges to overflow upstream lands upon payment of damages and by giving the protection of priority to the person that first located his mill, so that he could not be overflowed from below by a subsequently con-

---

[46] *United States v. Appalachian Electric Power Co.* (1940) 311 U.S. 377, 61 SCt. 291, 85 Led. 243; *United States v. Chandler Dunbar Water Power Co.* (1913) 229 U.S. 53, 33 SCt. 667, 57 Led. 1063; *United States v. Twin City Power Co.* (1956) 350 U.S. 222, 4 Led. 2d 1188, 80 SCt. 1136.

[47] *Wallace v. Weitman* (1958) 52 Wash.2d 585, 328 P.2d 157; see *Brown v. Chase* (1923) 125 Wash. 542, 217 P. 345.

[48] *Carter v. Territory of Hawaii* (1917) 24 Haw. 47; *Territory of Hawaii v. Gay* (1930) 31 Haw. 376.

structed mill pond.[49] Legislatures have granted many charters to companies organized to construct canals and to improve navigable and nonnavigable waters and to develop water power. Other special privileges have been given to favored industries such as cranberry growers and mining companies.[50]

Many states today regulate the building of dams for the development of hydroelectric power and other purposes,[51] and some require riparians to obtain permits and licenses from the state.[52] The owners of such reservoirs may be permitted to do things outside the riparian pattern, such as holding back a stated proportion of the flow,[53] or diverting a river across an oxbow to develop a head for power.[54] In some states public districts or private persons or companies that capture and store flood or excess waters may use the water for any purpose, including nonriparian uses.[55]

In a number of states riparians located on lakes, or on rivers flowing in and out of lakes, are restricted by statutes permitting administrative agencies to set and maintain proper lake levels.[56]

Several eastern states have adopted a permit system for the diversion and use of waters, usually surplus waters, for irrigation.[57] These look not unlike the permits required in most western states to initiate appropriations, but they have quite generally been administratively construed to be regulations of riparian rights. Thus only riparians may obtain such permits and only riparian land may be irrigated. Irrigation permits are granted only if there will be no injury to riparian rights or public rights of fishing and recreation and may be terminated if such injury appears. However, similar permits are granted to favored mining industries in Wisconsin and Minnesota that are not limited to riparian land and that exist for a specific period.[58] These have many aspects of appropriations.

B. Prior Appropriation

1. General Statement. The two cardinal principles of the *doctrine of prior appropriation* are that beneficial use of water, not land ownership, gives the basis of the right to use water and that priority of use, not equality of right, is the basis of the division of water between appropriators when there is not enough for all.

Prior appropriation originated as an example of American common law, arising from the customs of the people and the necessities of the climate. The originators of the doctrine were the gold miners who came to California in 1849 and organized *mining districts* to create some semblance of order on the then ungoverned public domain. They established the same rule for ownership of mining claims and the right to use water to wash the gold from the gravel. This rule was known as prior appropriation—the law of the first taker. The courts adopted this rule and applied it to the ranchers and farmers who followed the miners into the West, recognizing that the first person to irrigate land had a better right to the water than a later claimant. Today all the western states have adopted statutes codifying these rules, so that prior appropriation is very largely a matter of state legislation. However, these statutes, with similar backgrounds, often copied from other states, are generally interpreted in the light of similar laws, so that the substance of the law of prior appropriation is fairly homogeneous.

2. Elements of an Appropriation. An appropriation is the right to use a specific quantity of the water of a public source of supply for a beneficial purpose, if that quantity is available free from the claims of prior appropriators. Most definitions that

---

[49] *Mass. Ann. Laws* c. 253 Secs. 1, 2; *Wis. Stat.* Secs. 31.31, 31.32.
[50] *Wis. Stat.* Secs. 94.26, 107.05; *Minn. Stat.* Sec. 105.64.
[51] *Ky. Rev. Stat.* Sec. 182.010; *Ala. Code* Tit. 10 Sec. 178, Tit. 38 Sec. 116.
[52] *Wis. Stat.* Sec. 30.01.
[53] *Wis. Stat.* Sec. 30.34.
[54] *Mo. Rev. Stat.* Sec. 10313.
[55] *Ind. Stats.* Sec. 27-1403; *Fla. Laws* 1957 c. 57-380 Sec. 8.
[56] *Ind. Stats.* Sec. 27-627; *Mich. Comp. L.* Sec. 218.01.
[57] *Minn. Stat.* Sec. 105.38; *N.C. Stat.* Sec. 113-8.1; *Wis. Stat.* Sec. 30.18.
[58] *Wis. Stat.* Sec. 107.05; *Minn. Stat.* Sec. 105.64.

have been stated by the courts are not true definitions, but are rather descriptions of typical appropriations or of typical methods used to acquire an appropriation. A synthesis of these is that an appropriation requires (1) the diversion of water from a stream or other source, (2) the intent to appropriate, (3) notice of the appropriation to others, (4) compliance with state procedural requirements, and (5) the application of the water to a beneficial use. However, not all these are literally required in each case. The intent to appropriate is obvious in most cases and is clarified and crystallized where a permit to appropriate must be obtained. Notice of the appropriation, at first given in the form of a posted notice or by the physical demonstration of construction of the works, is now supplied by the filing of an application for a permit in most states.[59] While it is frequently asserted that an actual diversion of the water is an element of an appropriation, it is also possible to appropriate water by intercepting it before it reaches the stream,[60] by taking advantage of natural irrigation by overflow (eventually installing ditches and headgates),[61] and by simply pasturing large herds of cattle and sheep near the source.[62] The application of water to a beneficial use is the final step in a completed appropriation, but a valid right may exist though the water has not yet been put to a beneficial use, if the works have not yet been finished,[63] or because the water is held for sale.[64]

3. Appropriators. In general, since a water right is regarded as real property, any natural or artificial person or governmental agency with the capacity to hold and deal with interests in land may be an appropriator. Corporations, Indians, aliens, married women, and infants may appropriate water,[65] except in a very few states that have adopted some restrictions. In most states an appropriator for irrigation purposes must be in possession of the land, but the extent of his title is immaterial; he need not own it, lease it, or hold any particular form of interest in it. However, in Arizona and Washington an appropriator for irrigation purposes must either own the land on which the water is to be used or be in possession of the land with the present intent and legal ability to acquire ownership, as a lessee with option to purchase or as a bona fide settler under Federal public-land laws.[66]

Some states have rules that certain types of distributing organizations cannot be appropriators for irrigation purposes, but that the consumer who applies the water to the land, rather than the organization that diverts it, is the true appropriator.[67] This problem of who owns the water right may be significant in certain aspects of the relation of the distributor to the consumer or to the public, but at least such an agency is generally regarded as having the capacity to initiate the appropriation and to hold and defend it for the benefit of those who will eventually consume the water.

4. Beneficial Use. A number of uses of water have been approved as beneficial by courts and legislatures. Domestic use is everywhere recognized as such, since water is necessary to sustain the life and health of man. Domestic use that will support an appropriation is similar to the same concept in riparian law[68] and covers drinking and cooking, general household purposes, and in some places the irrigation of lawns and gardens and the watering of domestic and barnyard animals. The use of water for herds of stock and all types of domestic animals is a beneficial purpose.[69] Cities and towns may appropriate water for municipal purposes, to supply the municipality and inhabitants with water for domestic uses, for irrigation of lawns and gardens, for sanitary purposes, and for use in shops, business establishments, and factories.[70] A

[59] See Subsec. II-B-9.
[60] *Clausen v. Armington* (1949) 123 Mont. 1, 212 P.2d 440.
[61] *In re Silvies River* (1925) 115 Ore. 27, 237 P. 322.
[62] *Steptoe Live Stock Co. v. Gulley* (1931) 53 Nev. 163, 295 P. 772.
[63] *United States v. Big Bend Transit Co.* (D. Wash. 1941) 42 F. Supp. 459.
[64] *Bailey v. Tintinger* (1912) 45 Mont. 154, 122 P. 575.
[65] *Santa Paula Water Works Co. v. Peralta* (1896) 113 Cal. 38, 45 P. 168.
[66] *Tattersfield v. Putnam* (1935) 45 Ariz. 156, 41 P.2d 228; *Avery v. Johnson* (1910) 59 Wash. 332, 109 P. 1028.
[67] *Slosser v. Salt River Valley Canal Co.* (1901) 7 Ariz. 376, 65 P. 332.
[68] *Montrose Canal Co. v. Loutsenhizer Ditch Co.* (1896) 23 Colo. 233, 48 P. 532.
[69] *First State Bank v. McNew* (1928) 33 N.M. 414, 269 P. 56.
[70] *Denver v. Brown* (1913) 56 Colo. 216, 138 P. 44.

city may appropriate more water than it presently needs in órder to provide for future growth.[71]

Water is used everywhere in the West to supply the deficiencies of natural rainfall, and it has never been questioned that irrigation in general is a beneficial use. The method of application, by flooding, channeling, or sprinkling, is immaterial,[72] and the crop grown is immaterial, whether pasture grass or cultivated crops of high value.[73]

Mining was the earliest of beneficial uses, and has been extended in modern times to drilling for oil and gas. The uses for sawmills and ore-reduction mills were also very early purposes for which appropriations were allowed, and today water may be appropriated for any form of manufacturing. The use of water for the production of electricity or of direct mechanical power is everywhere recognized as useful and beneficial.

Water has been put to many other uses, such as railway use, production of steam, refrigeration, cooling, the manufacture of ice, and for fish hatcheries.

In modern times a new beneficial use, recreation, has come to the forefront. Diversions or impoundments of water have been approved for beautifying parks and resorts[74] and for forming pools and lakes for swimming, boating, fishing, and hunting.[75] What amounts to an appropriation of water in place for such purposes has been made in some states by the adoption of laws which prohibit the appropriation of some beautiful lakes and streams.[76] It seems to be generally recognized that public recreational facilities are beneficial in a broad sense to a large segment of the population and, when operated for profit, are an important source of wealth.

In rather recent times the concept that an appropriation will not be permitted for a wasteful use has been expanded into a new rule, that a particular use must not only be embraced within the general class of beneficial uses, but it must also be a reasonable and economic use of the water in view of other present and future demands upon the source of supply. Thus the use of large quantities of water to clean debris from a reservoir during the irrigation season was held not to be a beneficial use, despite the values of such use to a power company.[77] Today, as supplies of unappropriated water dwindle, it may be said that each new use must be reasonable and economic in the light of other demands for the little water remaining to be allocated.

5. Quantity of Water. An appropriation is always stated in terms of the right to take a definite amount of water. Direct flow rights are stated in terms of the maximum current or flow that may be diverted from the stream; storage rights are expressed in terms of the total volume of water that may be stored. When the first appropriations were made on the public domain, the capacity of the ditch was the measure of the right, and today it still may be an outside limit on the appropriation. The amount of water claimed may be a similar limit. But in general the amount of water that an appropriator is entitled to divert is measured by the beneficial use to be served, by the need for sufficient water to accomplish the object of the appropriation. If the water is to be used for irrigation, the appropriator will be allowed the quantity necessary to irrigate the particular tract of land properly. In some states this may be determined in each case, considering the climate, altitude, location and slope of land, composition of soil, type of crop for which the land is suitable, and other similar factors.[78] In others, there are statutes which place a definite limit upon the quantity which an appropriator can receive.[79]

An appropriation acquired by building a reservoir and storing water in it is measured by the storage capacity of the reservoir, what it will hold as a result of a single filling

---

[71] *Denver v. Sheriff* (1939) 105 Colo. 193, 96 P.2d 836.
[72] *Charnock v. Higuerra* (1896) 111 Cal. 473, 44 P. 171, 32 L.R.A. 190, 52 Am. St. 195.
[73] *Wyoming v. Colorado* (1922) 259 U.S. 419, 42 SCt. 552, 66 Led. 999.
[74] *Cascade Town Co. v. Empire Water & Power Co.* (D. Colo. 1910) 181 Fed. 1011, rev'd on other grounds, 9th Cir. (1913) 205 Fed. 123.
[75] *State v. Red River Valley Co.* (1945) 51 N.M. 207, 182 P.2d 421.
[76] *Ore. Rev. Stat.* Sec. 538.110; *Idaho Code* Sec. 67-4301.
[77] *In re* Deschutes River (1930) 134 Ore. 623, 286 P. 563, 294 P. 1049.
[78] *Farmers Highline Canal & Reservoir Co. v. Golden* (1954) 129 Colo. 575, 272 P.2d 629.
[79] *Wyo. Stat.* Sec. 41-181; *Idaho Code* Sec. 42-202; *Nev. Comp. Laws* Sec. 7899.

each year.[80] An off-channel reservoir to which the water is carried from the stream by a ditch or pipe is governed by the same rule plus the additional limitation of the capacity of the diversion works.[81] The rights of a reservoir appropriator are also limited to the amount that can be applied to the beneficial use to be served, but he may be allowed carry-over storage, so that water saved during wet years can provide a supply during dry years.[82]

6. Place of Use. With few exceptions, an appropriation can be made in order to use the water at any place where it is needed. Diversions out of the watershed have been permitted and protected from the beginnings of the doctrine of prior appropriation.[83] Such diversions from one watershed to another are common in the West, but in Nebraska water cannot be transported out of major watersheds,[84] in Colorado a district cannot take water across the continental divide to the eastern slope unless it protects present and future uses on the western slope,[85] and in California water must be reserved for future uses in the county or watershed in which it originates.[86]

Originally, the fact that the water would be used in another state was immaterial. Many diversions were made on interstate streams where the water was diverted in one state, transported across the state line, and used in another.[87] Today a number of states modify the rule by statute. The most extreme law is that of Colorado, which absolutely prohibits the transportation of water into any other state.[88] More typical statutes permit appropriations for out-of-state use if the state in which the use is to be made grants reciprocal privileges, or if the legislature gives specific approval of each diversion.[89]

7. Priority. On a typical western stream where there are many irrigators with water rights initiated at different times, there may be water for all if the stream is high. As the quantity of water decreases, the diversion works of the appropriators are shut off in inverse order of priority. The last ditch is the first closed, and the earliest is never closed. As the stream rises, water users are allowed to take water in order of seniority. The burden of shortage thus falls on those with the later rights; there is no proration in times of scarcity. The amount that each appropriator is entitled to receive remains fixed, if there is sufficient flow in the stream to supply it; otherwise the senior rights are supplied in full, while the junior rights are shut off completely.

The right of the senior appropriator extends both upstream and downstream. He may take water needed by a junior appropriator below him, while the junior appropriator upstream must permit the water to go past his point of diversion when it is needed to supply the senior right. A junior may have to leave more in the stream than the technical amount of the senior appropriation if a portion of the flow will be lost in transit.[90] A senior appropriator from a stream is prior in right to junior appropriators whose diversions are from tributaries entering the stream above the senior's point of use.[91]

Junior appropriators who take water from a source that has already been partially appropriated get the right to use such water as is not needed to satisfy the rights of prior appropriators. The downstream junior appropriator is entitled to insist that the senior takes no more than his appropriation allows.[92] An upstream junior may construct diversion works for his own use so long as he releases the quantity of water

---

[80] *Windsor Reservoir & Canal Co. v. Lake Supply Ditch Co.* (1908) 44 Colo. 214, 98 P. 729.
[81] *Big Wood Canal Co. v. Chapman* (1927) 45 Idaho 380, 236 P. 45.
[82] *Federal Land Bank v. Morris* (1941) 112 Mont. 445, 116 P.2d 1006.
[83] *Coffin v. Left Hand Ditch Co.* (1882) 6 Colo. 443.
[84] *Osterman v. Central Nebraska Public Power & Irrigation District* (1936) 131 Neb. 356, 268 N.W. 334.
[85] Colo. Rev. Stat. Sec. 149-6-13.
[86] Cal. Water Code Secs. 10505, 11460, 11463.
[87] *Willey v. Decker* (1903) 11 Wyo. 496, 73 P. 210.
[88] Colo. Rev. Stat. Sec. 147-1-1.
[89] Utah Code Sec. 73-2-8; Wyo. Comp. Stat. Sec. 41-151.
[90] *Raymond v. Wimsette* (1892) 12 Mont. 551, 31 P. 537.
[91] *Strickler v. Colorado Springs* (1891) 16 Colo. 61, 26 P. 313, 25 A.S.R. 245.
[92] *Clausen v. Armington* (1949) 123 Mont. 1, 212 P.2d 440.

needed by the senior[93] and may make substantial changes in the regimen of the stream so long as he observes the senior's fundamental right to receive water. For instance, he may use the water to produce power or in some other nonconsumptive manner, and he may even take the direct flow of the stream to which the senior is entitled if he replaces it with stored water or substitutes water imported from another source.[94]

A water right obtained under the doctrine of prior appropriation must be defined in terms of a specific date, in order that its place in the schedule of priorities may be fixed. The date of the appropriation is generally the date on which the project was initiated. Before the inception of statutory regulation of appropriations, the date of the appropriation was that of the first act consisting of an open and notorious physical demonstration of a purpose to acquire the water right. This was usually the date of the start of construction, or of the survey, or of the posting of a notice or the filing of a claim. At present nearly all states provide a procedure for obtaining permits from state water officials before making an appropriation, and in these states the date of appropriation is the date of filing the application for the permit.

Although an appropriation is generally not complete until the water has been applied to a beneficial use, the date of the beginning, rather than the date of completion, is chosen as the better measure of priority, provided the appropriator has used reasonable diligence in completing the work. Generally, whether or not reasonable diligence has been used is a question of fact, considering the magnitude of the undertaking, the natural obstacles encountered, and financial and legal difficulties.[95] In many states where permits must be obtained for appropriations, the rules of diligence have been crystallized into specific statutory requirements of the times within which the work must be begun or completed.

Rotation is a system of distributing water among users by periods of time, permitting each user in turn to take the total share of all for a short interval. This is not an exception to the doctrine of priority, but only a variation in its administration. Rotation is generally practiced by agreement among irrigators when the water supply is sufficient for all, but may be more efficiently used if large withdrawals by each in turn are taken instead of small simultaneous withdrawals by all. Over the rotation period each appropriator gets as much water as he would have had, had he taken his small share continuously. Occasionally courts have decreed a rotation scheme among claimants to a small stream or to low-water summer flows.[96]

An appropriation may entitle the owner to take water only during certain periods or seasons, so that persons who initiated their rights at later dates may have the better claim to water during other times of the year. Such seasonal or periodic priorities can arise because the claim by which the appropriation was initiated specified certain periods for its use. This may be required by statute.[97] Where the appropriator has not so limited himself, some courts have inferred time limits on appropriations from the early practices of the appropriator and have said that all subsequent uses of the water must be made within the same times as were the original uses.[98] The better rule is that unless the claim upon which the right is founded imposes time limits on its use, the appropriator can take the water at any time he can put it to the beneficial use for which it was appropriated.[99]

8. Preferences. Preferences are exceptions to the rule of priority. A preference allocates the water to what has been legislatively deemed to be a higher and better use regardless of the time of initiation of the use. There is wide variation as to what uses shall be preferred. There is general agreement only that man's personal needs come first, so that domestic and municipal purposes head every list, and power and navigation are generally found near the bottom. But irrigation, manufacturing, mining, and railroad transportation are given some preferred status in various states.

[93] *Knutson v. Huggins* (1941) 62 Idaho 662, 115 P.2d 421.
[94] *Dry Gulch Ditch Co. v. Hutton* (1943) 170 Ore. 656, 133 P.2d 601.
[95] *In re* Hood River (1924) 114 Ore. 112, 227 P. 1065.
[96] *Albion-Idaho Land Co. v. Naf Irrigation Co.* (10th Cir. 1938) 97 F.2d 439.
[97] Tex. Civ. Stat. Art. 7467c.
[98] *Oliver v. Skinner* (1951) 190 Ore. 423, 226 P.2d 507.
[99] *Harkey v. Smith* (1926) 31 N.M. 521, 247 P. 550.

A true preference exists when a junior right to a preferred use is placed at the top of the priority list, so that in times when water is short, senior nonpreferred rights are shut off while the preferred user still draws water. Stated another way, a true preference exists when the preferred use may be initiated without regard to the fact that the supply is already fully appropriated for other purposes. There are very few true preferences of this nature in western water law. One exists in Texas, where, since 1931, all appropriations, except of the Rio Grande, are granted subject to future appropriations for municipal purposes,[100] and in California some power appropriations have been granted subject to future appropriations of the water for agricultural or municipal purposes.[101] Some interstate compacts have protected the future growth of upstream states by declaring that the impounding and use of water for power shall be subservient to use and consumption of water for agricultural and domestic purposes.[102] Federal law gives upstream consumptive uses a preference over the use of water downstream to maintain a flowing navigation channel.[103]

Most so-called preferences in western water law merely give the preferred user the power to condemn water supplies from senior users, upon the payment of compensation.[104] The exercise of this form of preference results in a permanent change of the water right from the inferior use to the preferred. In all states, cities and towns are given the power to acquire water rights in this fashion.

In many of the states that require an application for a permit to appropriate water, the water administrators are instructed to prefer some uses over others when several applications are pending and the available water is insufficient for all. Typically, these preferences are that the water should go first for domestic and municipal purposes, then to agriculture, then to power.[105]

9. Procedures. Early in the development of the priority doctrine, appropriations were made simply by applying the water to beneficial use, and procedural requirements for recording them were minimal or did not have to be complied with. Court procedures for bringing all interested parties into one proceeding in which all their claims could be adjudicated and definitely placed on record were developed in several of the western states. These judicial procedures are still used in the states of Montana, Colorado, and Texas, but most states have found them costly and cumbersome and have replaced them with administrative proceedings. Typically, the water officials in control of the waters of the state initiate proceedings in which each claimant to water on the stream system in question is required to appear and present his claim. The facts of the claim are checked by the administrators; they may be contested by other appropriators, and the officials make a determination as to the existence and date of the appropriation, the purpose for which the water is used, the land on which it is used, and the quantity of water appropriated. In some states the administrative decision is final, but dissatisfied persons may appeal to the courts. In others, the administrative findings are submitted to a court, which makes the final determination. The result is much the same under either system. These adjudication procedures settle all claims to past rights; a person who fails to file a claim loses his water right. In some states, when lawsuits between individuals involve technical questions of water administration and use, the courts may call upon the officials for assistance by referring the case to them to make findings and recommendations.[106]

At first there were no procedures for initiating an appropriation, although it became customary in some localities to post notices of the claim on the stream. Early statutes codified this requirement of posting and also required recording of the claim in the

---

[100] *Tex. Civ. Stat.* Art. 7472, 7472a.
[101] *East Bay Municipal Utility District v. Department of Public Works* (1934) 1 Cal.2d 476, 35 P.2d 1027.
[102] *Colorado River Compact* Art. IV(b), 63 *Stat.* 31.
[103] 33 U.S.C. 701-lb.
[104] *Sterling v. Pawnee Ditch Extension Co.* (1908) 42 Colo. 421, 94 P. 339, 15 L.R.A. (n.s.) 238; *Wyo. Comp. Stat.* Sec. 41-4.
[105] *Ariz. Rev. Stat.* Sec. 45-147; *Cal. Water Code* Sec. 1254.
[106] *Laramie Rivers Co. v. LeVasseur* (1949) 65 Wyo. 414, 202 P.2d 680; *Tulare Irrigation District v. Lindsay-Strathmore Irrigation District* (1935) 3 Cal.2d 489, 45 P.2d 972.

county courthouses. However, there were no ways of discovering which claims had been actually followed by a true appropriation, and a person who actually put water to beneficial use was regarded as having a valid appropriation though he had not followed such procedures. Colorado adopted a requirement for central filing of maps and statements of claim in the office of an official called the state engineer, but again no penalty attached to the failure to follow these procedures.[107] To remedy this situation, the permit system was invented. Under it a person desiring to make an appropriation must file an application for a permit with the state water officials, accompanied by proper maps and descriptive statements. The permit requires the work to be begun within a certain period of time and to be completed within another period. Typically, the appropriator is required to report back and request a certificate of appropriation, which he obtains upon a showing that he has completed the work in compliance with the terms of the permit and has applied the water to beneficial use. His certificate is then added to the list of water rights adjudicated in the proceedings described, above thus keeping the list up to date. All the appropriation states except Alaska, Colorado, and Montana now have this permit system. In all the states that have it, except one, obtaining a permit is an absolute prerequisite to obtaining a water right,[108] but in Idaho the only penalty for failure to request a permit is that the water right will date from the time it was actually applied to a beneficial use instead of from the date of the application.[109]

In all these states, again except for Idaho, the state water agency is empowered to deny the permit on various grounds. Most common of these is that the proposed use of the water would not be in the public interest. If the proposed appropriation would injure other appropriators, retard the development of the state, result in an uneconomic project, interfere with the economic feasibility of a much larger and more desirable project, or produce similar undesirable results, the permit may be denied or be granted subject to certain conditions which will ensure the use of the water within the public interest.[110]

10. Changes in Appropriations. A water right is private property, and in most states it can be sold or used by its owner at any place to which he can transport the water. Changes can be made, not only of the place of use, but also in the point of diversion, type of use, time of use, or place of storage. But the privilege of making such changes is subject to the rule that a change must not injure the vested water rights of other appropriators. The agencies and courts that regulate the appropriation and distribution of water are given the power to approve or forbid changes on this ground, after proceedings at which all interested parties are represented. Some states have adopted further restrictions on change of use or place of use, which range from prohibition of all transfers (except to preferred uses)[111] to forbidding changes of irrigation water to other lands unless it is impracticable to use the water beneficially or economically on the original land.[112]

The restriction on changes that cause damage is not merely an application of the rule of priority; it is applicable if any person, senior or junior, will suffer as a result of the change. A change from a nonconsumptive use to a consumptive one will obviously injure downstream appropriators. The loss of benefits from return flows is the most common type of damage that will prevent a change, but the appropriator may be permitted to change the place of use of the amount of his consumptive use, though not of his total diversion,[113] and other conditions may be imposed to permit a change to as great an extent possible, yet prevent the infliction of damage.[114]

[107] *DeHaas v. Benesch* (1947) 116 Colo. 344, 181 P.2d 453.
[108] *Wyoming Hereford Ranch v. Hammond Packing Co.* (1925) 33 Wyo. 14, 236 P. 764.
[109] *Nielson v. Parker* (1911) 19 Idaho 727, 115 P. 488.
[110] *Young & Norton v. Hinderlider* (1910) 15 N.M. 666, 110 P. 1045; *Tanner v. Bacon* (1943) 103 Utah 494, 136 P.2d 957; *Kirk v. State Board of Irrigation* (1912) 90 Neb. 627, 134 N.W. 167.
[111] *Wyo. Comp. Stat.* Sec. 41-2.
[112] *Nev. Comp. Laws* Secs. 7893, 7944; *N.D. Code* Sec. 61-1404; *Okla. Stat.* Tit. 82 Sec. 34; *S.D. Code* Sec. 61.0141.
[113] *Vogel v. Minnesota Canal & Reservoir Co.* (1910) 47 Colo. 534, 107 P. 1108.
[114] *Farmers Highline Canal & Reservoir Co. v. Golden* (1954) 129 Colo. 575, 272 P.2d 629.

11. Transfers of Appropriations. An appropriation is regarded as real property, and where it can be sold to a person who will use it at a different place or for a different use, the transfer is ordinarily made by a deed.[115] However, transfers made by assignments of permits and certificates of appropriation have been recognized administratively. Water rights for the irrigation of land are generally regarded as appurtenant to the land; hence a sale of the land will carry the water right with it, although the water right was not specifically mentioned in the deed.[116]

12. Loss of Appropriations. An appropriation is a property right, and its ownership, like that of land, is held in perpetuity, although some may be granted for a limited period.[117] However, it may be terminated if it is not used.

From the earliest times it has been recognized that the nonuse of water, coupled with an intent not to resume the use, amounts to an *abandonment* that terminates the water right and makes the water available for use and appropriation by others. No particular period of time is required for an abandonment, but long unexplained nonuse will often cause a court to say that the right is abandoned, although there is no direct evidence of the intent of the appropriator.[118]

Most of the appropriation-doctrine states have enacted statutes which embody an entirely different concept, that of *forfeiture* for nonuse. If the appropriator does not use his water for a period, which varies in different states from 3 to 10 years, the water right is forfeited if the nonuse is voluntary and is not forced upon the appropriator by circumstances over which he has no control.[119] In some states the forfeiture is not complete until special proceedings are brought to terminate the right;[120] in most, the forfeiture can be asserted at any time and in any legal proceeding in which the question arises.

13. Extent of Appropriation Doctrines. Prior appropriation is the only method of obtaining a right to water use in nine states: Alaska, Arizona, Colorado, Idaho, Montana, Nevada, New Mexico, Utah, and Wyoming.[121] These follow what is known as the *Colorado doctrine*.

The so-called *California doctrine* is a general term describing several forms of combining both riparian and appropriation law in a single state. In Kansas, Nebraska, Oregon, and South Dakota, riparian rights were once recognized, but since the enactment of modern water codes, no new water uses can be initiated by virtue of riparian rights, although riparian uses that existed prior to the codes are confirmed as vested rights.[122] All new uses of water must be initiated by obtaining a permit for an appropriation. A similar result has quite recently been reached in the eastern states of Iowa and Mississippi, which have adopted substantially the western appropriation law with some modifications.[123] In California, North Dakota, Oklahoma, and to some extent in Washington,[124] new riparian uses can be started that are superior to prior appropriations from the stream, unless the appropriation

---

[115] *Bates v. Hall* (1908) 44 Colo. 360, 98 P. 3.
[116] *Frank v. Hicks* (1894) 4 Wyo. 502, 35 P. 475.
[117] S.D. Code Sec. 61.0152.
[118] *Green Valley Ditch Co. v. Frantz* (1913) 54 Colo. 226, 129 P. 1006.
[119] *Ramsay v. Gottsche* (1937) 51 Wyo. 516, 69 P.2d 535.
[120] *Horse Creek Conservation District v. Lincoln Land Co.* (1939) 54 Wyo. 320, 92 P.2d 572.
[121] *Van Dyke v. Midnight Sun Mining & Ditch Co.* (9th Cir. 1910) 177 F. 85 (Alaska); *Clough v. Wing* (1888) 2 Ariz. 317, 17 P. 453; *Coffin v. Left Hand Ditch Co.* (1882) 6 Colo. 443; *Drake v. Earhart* (1890) 2 Idaho 750, 23 P. 541; *Mettler v. Ames Realty Co.* (1921) 61 Mont. 152, 201 P. 702; *Jones v. Adams* (1885) 19 Nev. 78, 6 P. 442; *Albuquerque Land & Irrigation Co. v. Gutierrez* (1900) 10 N.M. 117, 61 P. 357; *Stowell v. Johnson* (1891) 7 Utah 215, 26 P. 290.
[122] *State ex rel Emery v. Knapp* (1949) 167 Kan. 546, 207 P.2d 440; *McCook Irrigation & Water Power Co. v. Crews* (1905) 70 Nev. 109, 96 N.W. 996, 102 N.W. 249; *in re* Hood River (1924) 114 Ore. 112, 227 P. 1065; S.D. Code Sec. 61.0101.
[123] *Iowa Code* Sec. 455A (minimum flows preserved, 10-year-cancellable permits); *Miss. Code* Sec. 5906-04 (terminable appropriations).
[124] *Meridian, Ltd. v. San Francisco* (1939) 13 Cal.2d 424, 90 P.2d 537; *Sturr v. Beck* (1890) 133 U.S. 541, 10 SCt. 350, 33 Led. 761; *Smith v. Stanolind Oil & Gas Co.* (1946) 197 Okla. 499, 172 P.2d 1002; *Wallace v. Weitman* (1958) Wash. 2d 585, 328 P.2d 157.

was made on public land before the riparian lands were settled or the appropriator has a prescriptive right. Appropriations that do not interfere with reasonable riparian uses are permitted, and some appropriators such as public districts may condemn riparian rights, paying compensation to the owners. In Texas, riparians have superior rights to waters below the line of highest ordinary flow, while appropriators may take flood and surplus water.[125]

In some eastern states elements of priority have been engrafted on the riparian system by statutes giving special privileges that amount to the protection of priority to certain favored industries such as mill dams, cranberry growers, and certain mining enterprises.[126]

C. Obstruction and Pollution

1. Obstruction of Streams. The placing of dams, bridges, or other structures in a water course in such a manner as to deprive other persons of rights or privileges to use it may be a violation of riparian rights or of an appropriation, in accordance with principles already noted. It may also be wrongful from other standpoints.

Navigable streams have long been regarded as public highways open to all persons, who may use them for commercial highways and general transportation. In some states these public rights have been held to include individual privileges to boat, hunt, fish, swim, and engage in other forms of water recreation.[127] Any obstruction of a stream that interferes with the exercise of these privileges is as much a wrong as a fence across a highway would be, and the person who builds such an obstruction may be forced to remove it at the suit of public authorities. If the stream is one navigable by the Federal test, that is, suitable for commercial navigation, the United States may require the removal of the structure.[128] The states may also control obstructions to such streams, and in many states a much more liberal interpretation of navigability gives them authority over very small streams, capable of floating a saw log or even a light skiff or canoe.[129]

Such obstructions may not be absolutely prohibited, but may be licensed by the state or Federal authority.[130] Such a license is necessary even though the obstruction is made in the course of exercising a riparian right or making an appropriation. Other state powers over streams are exercised for the protection of fish and wildlife habitat. Licenses are frequently conditioned upon the licensee furnishing facilities for locks and fish ladders that will allow the structure to exist without destruction of the public rights.[131]

In the West, where state constitutions and statutes declare that all streams are the property of the public or of the state, some states have similar rules against obstruction of any stream, regardless of its navigability, without first obtaining a permit.[132]

2. Changes in Watercourses. It is unlawful to dig a new channel for a watercourse and divert the entire stream away from the land of a person who has rights in it.[133] Similarly, the destruction of the navigable capacity of water by drainage operations is an injury to public rights.[134]

Flood-control structures and operations have often been the subject of litigation. The release of waters from a flood-control dam in such quantities as to surcharge the stream so that lower lands are flooded or eroded may lead to liability.[135] Liability

---

[125] *Motl v. Boyd* (1926) 116 Tex. 82, 286 S.W. 458.
[126] See footnotes 49 and 50.
[127] *Muench v. Public Service Commission* (1952) 261 Wis. 492, 53 N.W.2d 514.
[128] *Union Bridge Co. v. U.S.* (1908) 204 U.S. 364.
[129] *Muench v. Public Service Commission* (1952) 261 Wis. 492, 53 N.W.2d 514.
[130] 16 U.S.C. 797(e); *Wis. Stat.* Sec. 31.05.
[131] *Wis. Stat.* Sec. 31.18.
[132] *Wyo. Comp. Stat.* Sec. 41-26.
[133] *Bigelow v. Draper* (1896) 6 N.D. 152, 69 N.W. 570.
[134] *In re* Crawford County Drainage Dist. (1924) 182 Wis. 404, 196 N.W. 874.
[135] *Bemmerly v. Lake County* (1942) 55 Cal. App.2d 829, 132 P.2d 249.

is also imposed upon landowners who protect their own river banks by levees, dikes, or ripraps that throw the water against a neighbor's land and thereby cause more damage than would result from the normal streamflow.[136]

3. Flowage. A person who backs up the water of a watercourse by a dam so that it overflows the land of his upper riparian neighbor commits a trespass on the upper land. He is liable for damage done and may be enjoined from committing such damage in the future.[137] A similar liability exists when the raising of the level of the watercourse interferes with drainage works.[138] Flowage rights may be obtained by purchase, and are sometimes obtained by prescription when the owner of the overflowed land fails to complain or to assert his rights in court for the period of the statute of limitations.[139] Statutory privileges to overflow upper land upon the payment of damages have been given to some dam builders,[140] and a Federal, state, or local public agency may condemn the land (or flowage privileges if the land is not to be permanently under water) upon the payment of just compensation. Similar privileges of eminent domain are often given to public utilities such as electric power companies.

4. Pollution. The early law regulating pollution was enforced almost entirely through the process of individual suits for what was termed a private nuisance. A private nuisance results from acts of one person that unreasonably interfere with the interests of another person in the use or enjoyment of land, and the fouling of a stream to the extent that unpleasant odors render the lower premises unpleasant for living, or normal uses of the water are prevented, is a nuisance. The reasonableness of the interference is determined by weighing the gravity of the harm to plaintiff against the utility of defendant's conduct. It can be seen that under such a rule much actual pollution was possible, and much occurred. To a certain extent a stream is itself an efficient disposal and treatment plant, and to the extent that the level of pollutants is not deleterious, there is no ground for complaint. But a person who adds polluting materials to those already in the stream and raises the level of foreign material above the point of tolerance can be held guilty of creating a nuisance.[141]

Pollution can result from any activity that renders the water unusable by or obnoxious to other persons. The watercourse may be clogged with a solid material. The stream may be made unfit for use by animal wastes, by chemical wastes, or by heating.[142] Pollution may be caused incidentally, by changing a stream so that part of its water is turned into a stagnant pool,[143] or by withdrawing so much water that the level of pollution is raised beyond tolerable limits.[144]

The concept of public nuisance has also been used to some degree to control pollution. A public nuisance is an act which causes inconvenience or damage to the public, as distinguished from one or a few individuals, and includes any interference with the public health, safety, or convenience. Thus the pollution of a stream which merely inconveniences several riparian owners is a private nuisance only, but may become a public one if it kills the fish or creates a menace to the health of a community. A public nuisance is subject to abatement at the suit of state officials. It may also constitute a crime.

Today, pollution is generally regulated by statutes giving administrative agencies the power to set standards to be met for the elimination of pollution before an industry starts operations and to require industries and communities to clean up existing sources of pollution. Even though these standards result in the permission of some pollution and may be in form a license or permit,[145] such permission or license creates no rights

---

[136] *Ladd v. Redle* (1903) 12 Wyo. 362, 75 P. 691.
[137] *Bobo v. Young* (1952) 258 Ala. 222, 61 So. 2d 814.
[138] *U.S. v. Kansas City Life Insurance Co.* (1950) 339 U.S. 799, 70 SCt. 885, 94 Led. 1277.
[139] *Weed v. Keenan* (1888) 60 Vt. 74, 13 A. 840.
[140] *McMillan v. Noyes* (1909) 75 N.H. 258, 72 A. 759.
[141] *West Muncie Strawboard Co. v. Slack* (1904) 164 Ind. 21, 72 N.E. 879.
[142] *Sandusky Portland Cement Co. v. Dixon Pure Ice Co.* (7th Cir. 1915) 221 F. 200.
[143] *Crum v. Mt. Shasta Power Corporation* (1934) 220 Cal. 295, 30 P.2d 30.
[144] *Meridian, Ltd. v. San Francisco* (1939) 13 Cal.12d 424, 90 P.2d 537.
[145] *Ohio Rev. Code* Sec. 6111.03; *Minn. Stat.* Sec. 144.372.

to pollute as against a third person and gives no guarantee that future standards will not be more stringent and the permission withdrawn. A violation of the permit, or of the orders of the agency, is a public nuisance and may be punished criminally or be enjoined at the suit of the state.

The Federal Water Pollution Control Act authorizes the Secretary of Health, Education and Welfare to develop, in cooperation with state and interstate agencies, comprehensive programs for eliminating or reducing pollution, and compacts and uniform state laws for the control of pollution, to encourage programs of research, investigation and training in methods of pollution control, and to grant Federal funds to the states for aid in the construction of such programs. The Act makes the pollution of interstate waters subject to abatement in a Federal court, after a lengthy process of conferences, recommendations, and public hearings.[145a]

D. Interstate Streams

1. Equitable Apportionment. Where rivers flow across state lines or form state boundaries, disputes have frequently arisen between individuals in these states or between the states themselves over the proper utilization and control of the water. The Constitution of the United States gives jurisdiction to the United States Supreme Court over controversies between states.[146]

In such suits the major principle of decision is that each state bordering on the river is entitled to an equitable apportionment of the benefits resulting from the flow of the river. In the cases from the West, which have principally involved irrigation, an apportionment of benefits has for the most part come to mean a literal division or partition of the waters. In the eastern states the benefits for which protection has been sought are the preservation of navigability, of fisheries and oyster beds, of flow for power plants, and of flows for carrying sewage. The *principle of equitable apportionment* was first announced in a case between a state recognizing only appropriative rights and a state which gave effect to riparian rights,[147] and obviously neither doctrine could be applied without doing violence to the internal law of one or the other state. When a relatively simple dispute arose between two states which each adhered to prior appropriation, the court said that to apply the doctrine of priority in such a case would be the equitable method of apportionment.[148] But in a more complex situation, where the application of priority would disrupt an economy built on junior appropriations and operated for years without objection, the court said that protection of established uses would be more equitable than strict priority.[149] Similarly, in the East, when proposals for large withdrawals of water to supply metropolitan areas precipitated similar litigation, the court said that the fact that both states follow the riparian doctrine does not dictate its use as the basis for settlement of such an interstate controversy.[150]

Congress has the power to apportion the water of a navigable, interstate stream among the states through which it flows, and to direct the Secretary of the Interior to divide any surplus or portion among these states in such a way as to carry out Federal purposes.[150a]

2. Interstate Compacts. The states having interests in a single river may, with the consent of Congress, settle their differences by *compact* instead of resorting to litigation.[151] Such a pact will operate as a restriction upon private rights held under state law but inconsistent with the compact. Where compacts apportion the water between the states, the apportionment is regarded as an equitable division of the waters similar to that made by a decree of the United States Supreme Court in an

---

[145a] 63 U.S.C. Sec. 466–466j.
[146] *U.S. Const.* Art. III, Sec. 2.
[147] *Kansas v. Colorado* (1907) 206 U.S. 46, 27 SCt. 655, 51 Led. 956.
[148] *Wyoming v. Colorado* (1922) 259 U.S. 419, 42 SCt. 552, 66 Led. 999.
[149] *Nebraska v. Wyoming* (1945) 325 U.S. 589, 65 SCt. 1332, 89 Led. 1815.
[150] *Connecticut v. Massachusetts* (1931) 282 U.S. 660, 51 SCt. 286, 75 Led. 602.
[150a] *Arizona v. California* (U.S. 1963) 88 SCt. 1468.
[151] *U.S. Const.* Art. I, Sec. 10.

interstate lawsuit,[152] and in negotiating the compacts, the states are governed by the principles of equitable apportionment, and the resulting compact may be regarded as a settlement in advance of such a potential lawsuit. Typically, compacts between western states have included such subjects as protection of existing rights, apportionment of unappropriated water, principles of river regulation by existing and proposed works, and establishment of an order of preferential uses and have set up agencies for planning and advising the state and Federal governments as to future development of the stream. In the East the most usual provisions have related to flood control, cooperation for future beneficial uses, maintenance of minimum flows, and pollution control, and a number of interstate planning and advisory agencies have been established.

E. Federal Water Law

1. General Statement. Unlike the states, which have general governmental powers over waters and water rights, the Federal government is one of delegated powers and may engage in water regulation or resource development only to the extent that a specific authorization is found in the Federal Constitution. The national interests that are served by Federal water-resource programs and laws are those inherent in the word "nation": the use of the country's waters for the free flow of trade and travel between its different sections, for the strengthening of the country both internally and in its relations with foreign nations, and for the conduct of its national business.

2. Commerce Power. The most important source of Federal jurisdiction over water is in relation to navigable waters. This depends upon a rather attenuated construction of article 1, section 8, of the Constitution, giving Congress the power "to regulate commerce . . . among the several states." It was early decided that "commerce" includes "transportation," which in turn includes "navigation,"[153] and later the power to regulate navigation was held to comprehend the control of navigable waters.[154]

Navigable waters of the United States are those which are navigable in fact, susceptible of use as highways for commerce, over which trade and travel may be conducted.[155] But actual use is not necessary, and a stream may be navigable if artificial aids and improvements will make it suitable for use.[156] The power to control these waters includes the power to destroy the navigable capacity of the waters and prevent navigation.[157] It also includes the power to protect the navigable capacity by preventing diversions of the river or its tributaries,[158] or by preventing obstructions by bridges or dams,[159] and by constructing flood-control structures on the river, its nonnavigable tributaries, or even on the watersheds.[160] The power to prevent obstruction in turn leads to powers to license obstruction,[161] and the power to obstruct leads to the power to generate electricity from the dammed water.[162] Using this somewhat flimsy-looking, but by no means shaky, structure for a foundation, Congress has built a huge program of river regulation and water control. Huge multipurpose projects combine features of navigation improvement, flood prevention, power production, irrigation, and recreation and involve in many cases the development of entire river basins.

[152] *Hinderlider v. LaPlata River & Cherry Creek Ditch Co.* (1938) 304 U.S. 92, 58 SCt. 803, 82 Led. 1202.
[153] *Gibbons v. Ogden* (1824) 9 Wheat. 1.
[154] *Gilman v. Philadelphia* (1865) 3 Wall. 713.
[155] The Daniel Ball (1870) 10 Wall. 557.
[156] *United States v. Appalachian Electric Power Co.* (1940) 311 U.S. 377, 61 SCt. 291, 85 Led. 243.
[157] *Oklahoma v. Guy F. Atkinson Co.* (1941) 313 U.S. 508, 61 SCt. 1050, 85 Led. 1487.
[158] *United States v. Rio Grande Dam & Irrigation Co.* (1899) 174 U.S. 690, 19 SCt. 770, 43 Led. 1136.
[159] *Economy Power & Light Co. v. United States* (1921) 256 U.S. 113.
[160] *Oklahoma v. Guy F. Atkinson Co.* (1941) 313 U.S. 508, 61 SCt. 1050, 85 Led. 1487.
[161] *United States v. Appalachian Electric Power Co.* (1940) 311 U.S. 377, 61 SCt. 291, 85 Led. 243.
[162] *Ashwander v. T.V.A.* (1936) 297 U.S. 288.

3. Other Federal Powers. Other powers are occasionally called upon to justify Federal action. One of these is the *proprietary power* arising out of the authority given to Congress to make all and needful rules and regulations respecting the territory or other property belonging to the United States.[163] This is one of the bases for the Reclamation Act,[164] under which the public domain of the United States has been improved by irrigation for settlement. It also gives the government the authority to license power projects on nonnavigable streams flowing through reserved public lands, free from the control of the states.[165]

The power to declare war and levy taxes and appropriate money to provide for the common defense has once been used to justify a Federal water-resource development for the generation of power and the production of nitrates for the manufacture of munitions.[166] A number of treaties, conventions, and protocols have been negotiated with Canada and Mexico that limit the activities of the United States or its citizens on boundary waters and waters that flow into boundary waters or across the boundaries.[167] Obligations of the United States established by such treaties give to the nation an additional basis for authorizing works of improvement on international waterways.[168]

Still another power is found in the "general welfare clause" giving Congress the power to levy taxes and appropriate funds to provide for the general welfare of the United States.[169] This has been held to give a specific power to the national government and has been used to substantiate large-scale projects for reclamation, irrigation, and other internal improvements.[170]

4. Navigation and Flood Control. There are few laws relating generally to the improvement of navigable waters, but there are several hundreds of specific acts relating to particular projects for improving navigation. Since 1824, when Congress first authorized the President to employ officers of the U.S. Army Corps of Engineers to make surveys, plans, and estimates of nationally important roads and canals, the Corps has been put in charge of river and harbor improvements under each specific act, and in 1935 Congress provided generally that investigations and improvements of rivers, harbors, and other waterways shall be under the jurisdiction of and be prosecuted by the Corps of Engineers.[171]

Assumption of Federal responsibility for flood control is a comparatively recent development. The first large appropriation for flood control, on the Mississippi and Sacramento rivers, was made in 1917, and the first major plan for the control of a river, the Mississippi, was authorized in 1928. In 1936 the *Flood Control Act*[172] authorized numerous projects throughout the nation for the purpose of curbing destructive floods that upset orderly processes and caused loss of life and property and impaired and obstructed navigation and in general constituted a menace to national welfare. In that Act Congress provided that no money should be expended on construction of any project until the state or local political subdivision gave assurance that they will (1) provide all necessary lands and property rights, (2) protect the United States from damage claims, and (3) maintain and operate the works after completion. In 1938 it was provided that these conditions would not apply to main-stem dams or channel-improvement projects, but they still apply to local works, unless waived in particular cases by Congress. No other contribution or reimbursement is required from beneficiaries of navigation or flood-control projects.

Federal works for the primary purpose of improving navigation or controlling floods may be multipurpose projects with many subsidiary features. When a dam is erected for navigation purposes and a surplus of water is produced, the government has the

---

[163] *U.S. Const.* Art. IV, Sec. 3.
[164] As amended, 43 U.S.C. Secs. 371–615.
[165] *Federal Power Commission v. Oregon* (1955) 349 U.S. 435, 75 SCt. 832, 99 Led. 1215.
[166] *Ashwander v. T.V.A.* (1936) 297 U.S. 288.
[167] Documents on the Use and Control of the Waters of Interstate and International Streams (G.P.O. 1956).
[168] *Arizona v. California* (1931) 283 U.S. 423.
[169] *U.S. Const.* Art. I, Secs. 8 and 9.
[170] *U.S. v. Gerlach Livestock Co.* (1950) 339 U.S. 725.
[171] 33 U.S.C. Sec. 540.
[172] 33 U.S.C. Sec. 701.

right to control or dispose of such water as an incident to the right to make the improvement. The water power resulting from impounding navigable waters is property belonging to the United States, and the electric power produced is likewise government property, which it may sell and dispose of as it sees fit.[173] Additional storage capacity may be added to flood-control reservoirs for domestic water supply or other conservation storage, when the cost of the increased storage is contributed by local agencies.[174] The 1944 Flood Control Act[175] authorized the Secretary to make contracts with states, municipalities, private concerns, or individuals for providing surplus water for domestic and industrial use. This Act also authorizes the Secretary of the Interior to construct, operate, and maintain, under reclamation law, additional works in connection with Corps of Engineers projects, when the project may be utilized for irrigation purposes specifically authorized by Congress.

Measures for the protection and enhancement of fish and wildlife habitat and recreational opportunities may also be included as subsidiary features of a navigation or flood-control project.[176] The features are jointly planned by the Department of the Interior's Fish and Wild Life Service and the state department in charge of fish and game, and generally the completed facilities and recreational areas are turned over to the state agencies for supervision and administration.

5. Federal Power Licenses. The Federal Power Commission is an independent agency whose duties are to make broad investigations of water resources and power potentialities and to license hydroelectric dams and structures on navigable waters.

No agency, private, state, or municipal, may construct a dam or other works for the development, transmission, or utilization of power from waters over which Congress has jurisdiction without a license from the Commission issued upon a finding that the project is desirable and justified in the public interest.[177] Plans for structures affecting the navigable capacity of any waters of the United States must be approved by the Chief of Engineers, and if a project is on a stream which is nonnavigable but affects navigation or commerce, the proposed developer must file a declaration of intention with the Commission, which may, if it finds that the interest of commerce would be affected, require that a license be obtained. Preference is given in the issuance of licenses to states, municipalities, and public-power districts, and if the Commission feels that the development of the water resources should be undertaken by the United States, it may deny a license and submit a report to Congress with its recommendations.

Licenses are issued for a period not exceeding 50 years, after which the United States may, on 2 years' notice, take over and maintain the project, paying the net investment of the licensee in the project plus reasonable damages. If the United States does not take over the project, the Commission may issue a new license or a year-to-year annual license.

All licenses are subject to numerous conditions. The project must be adapted to a comprehensive plan for improving the waterways for commerce, for the improvement and utilization of water-power development, and for other beneficial public uses, including recreation, and the Commission may require the modification of any project to secure such a plan. If the project is on navigable waters, the Commission may also require the licensee to construct, without expense to the United States, locks, booms, and sluices for navigation.

6. Reclamation and Irrigation. The United States first initiated substantial Federal aid to irrigation in 1902, with the passage of the *Reclamation Act*.[178] This Act established a fund with moneys received from the sale of lands in the semiarid western states, to be used by the Secretary of the Interior for irrigation projects in those states. Public lands susceptible of irrigation by such projects might be settled by homesteaders, in tracts of 40 to 160 acres, as might be reasonably required for the

---

[173] *Ashwander v. T.V.A.* (1936) 297 U.S. 288.
[174] 33 U.S.C. Sec. 701h.
[175] 33 U.S.C. Sec. 708.
[176] 16 U.S.C. Sec. 662.
[177] 16 U.S.C. Secs. 791a *et seq.*
[178] As amended, 43 U.S.C. Secs. 371–615.

support of a family. Rights to the use of water on private lands might be sold for a tract not exceeding 160 acres, or 320 acres to husband and wife. Charges were assessed against each acre of land irrigated, which were to be repaid into the reclamation fund in 10 annual installments without interest. The Secretary was authorized to acquire necessary rights or property by purchase or condemnation, and in acquiring water rights he was required to proceed in conformity with state laws. The principles of prior appropriation were made the basis of water rights on reclamation projects.

Most of the foregoing provisions have been amended, supplemented, or superseded by a long series of amendatory acts, some general in nature, some referring to specific projects, but the basic theory of reclamation remains. The *Reclamation Project Act* of 1939[179] authorized multipurpose projects for irrigation, hydroelectric power, flood control, navigation improvement, municipal and industrial water supply, recreation, and fish and wildlife preservation. The reclamation fund has long since been exhausted, and today projects are built with general funds. Repayment periods have been stretched out, and now much of the reimbursement to the government comes from power revenues. The 160-acre limitation has been relaxed in a number of cases, being waived in some areas and increased in some high-altitude valleys where 160 acres will not support a family.

On special projects, Congress may in its authorizing legislation eliminate the requirement that state laws be observed, and instead provide that the Secretary of the Interior may enter into contracts with water users and distribute the water in accordance with these contracts, without regard to priority of appropriation and without regard to state law in choosing between users.[179a]

Under the *Small Reclamation Projects Act* of 1956,[180] the United States may make grants or loans to states, conservation and irrigation districts, and water users' associations for new projects or for rehabilitation of old projects that supply irrigation, power, and domestic and industrial water. Grants are made to the extent of nonreimbursable benefits such as flood control, navigation, and recreation values, and loans are made up to $5 million, repayable out of the irrigation, power, and water-supply features.

7. Department of Agriculture Programs. The Department of Agriculture operates primarily on the principle of direct relations with farmers and landowners, and is concerned in the area of the small streams and the watershed rather than with the large and navigable main streams. The original Soil Conservation Service Program,[181] directed toward control of soil erosion and flood protection, was augmented by the Agricultural Conservation Program,[182] which added the objectives of water conservation and the beneficial use of water on individual farms, including measures to prevent runoff, the building of check dams and ponds, and facilities for applying water to land. The Service conducts surveys, investigations, and research and provides financial and technical assistance to the landowners through soil-conservation districts created under state law.

The 1936 Flood Control Act[183] provided that the Department of Agriculture should investigate watersheds and measures for runoff and water-flow retardation and soil-erosion prevention, and in 1938 the Department was authorized to encourage works of improvements along these lines.[184] This program has also been prosecuted through the mechanism of benefits to soil-conservation districts. The *Water Facilities Act*[185] makes loans available to farmers for the construction of ponds, reservoirs, and other facilities for water storage and utilization. All these programs for the most part envisage small individual projects and works on the farmer's land.

[179] 43 U.S.C. Sec. 485h.
[179a] *Arizona v. California* (U.S. 1963) 88 SCt. 1468.
[180] 43 U.S.C. Sec. 422a.
[181] 16 U.S.C. Sec. 807.
[182] 16 U.S.C. Sec. 590h.
[183] 33 U.S.C. Sec. 701b.
[184] 33 U.S.C. Sec. 701b-1.
[185] 16 U.S.C. Secs. 590r-x.

In 1954 the *Small Watersheds Act*[186] was enacted to fill the gap between these programs, which emphasize land treatment and very small individual structures, and the program of the Corps of Engineers, with its emphasis on large dams on the main stems. As amended, this Act provides for financial and technical assistance to state and local agencies in the construction of multipurpose projects for flood control and the conservation, development, utilization, and disposal of water in watersheds up to 250,000 acres. The Federal government assumes up to $250,000 of the cost of flood-prevention features and loans the local agency the balance, up to $5 million. Dams may be constructed that will impound 25,000 acre-ft, including 5,000 ft of floodwater detention, and even larger projects are possible with the approval of Congress and other interested agencies.

III. OTHER SOURCES OF WATER

A. Classification of Waters

1. General Statement. Although scientists regard the hydrologic cycle as a unity, the law has usually applied different rules to different segments of the cycle. The legal problems of the past arose principally between persons competing for water at the same point in the cycle. Disputes between competitors for groundwater or between users or drainers of diffused surface waters were not seen to require rules identical with those developed to settle conflicts over the use of streamflows, and scientific knowledge of the nature of water occurrence and the interrelationships of different classes of water was lacking. Thus each type of case was handled separately, and different bodies of law grew up, concerned with different types of water occurrence. It is therefore important to determine the legal classification of an occurrence of water in order to discover the rules which the courts will apply to it.

2. Watercourses. Although the law of streams or watercourses has already been discussed in detail, consideration of what constitutes a watercourse has been postponed to this point. When the definition is important, the conflict is usually whether a particular source of water is a watercourse or belongs in some other class of water. A comparison between the classes is therefore necessary, and definitions of all the classes are here brought together.

A *watercourse* is usually defined as a current, or stream of water, flowing in a channel with a well-defined bed and banks. This is more descriptive than definitive, and sometimes the description is embroidered with additions requiring that a watercourse must have a regular source, that it must discharge into another body of water, or that it must flow in a particular direction.[187] Yet apparently no one of these elements is essential. The flow of water need not exist; the stream may be intermittent.[188] Flows of water have been held to be watercourses despite the fact that in places the stream is not strong enough to create for itself bed and banks.[189] Occasional rains and thunderstorms have been held to be a sufficient source.[190] A slough connecting two rivers that reverses its flow when one is higher than the other has been held to be a watercourse.[191] A stream that loses itself in sands may be a watercourse.[192]

This seeming paradox could be avoided by adding many descriptive exceptions, or more properly, by a redefinition of a watercourse as water so confined and sufficiently persistent or recurring as to offer the advantages of typical forms of water use.[193] The usual question in these cases is whether the owner of the land on which the water is

---

[186] 16 U.S.C. Secs. 1001–1008.
[187] *State v. Hiber* (1935) 48 Wyo. 172, 44 P.2d 1005.
[188] *Hoefs v. Short* (1925) 114 Tex. 501, 273 S.W. 785, 40 A.L.R. 833.
[189] *Miller & Lux v. Madera Canal & Irrigation Co.* (1907) 155 Cal. 59, 99 P. 502, 22 L.R.A. (n.s.) 391.
[190] *Jaquez Ditch Co. v. Garcia* (1912) 17 N.M. 160, 124 P. 891.
[191] *Turner v. James Canal Co.* (1909) 155 Cal. 82, 99 P. 520, 22 L.R.A. (n.s.) 401, 132 A.S.R. 59.
[192] *Medano Ditch Co. v. Adams* (1902) 29 Colo. 317, 68 P. 431.
[193] *Hoefs v. Short* (1925) 114 Tex. 501, 273 S.W. 785, 40 A.L.R. 833.

found must recognize the rights of others, as he must if it is a stream, or may capture or dispose of these waters without reference to such rights, as he may if it is mere surface water. The definition is flexible, but there is need for flexibility in these cases. *Dry arroyos* and *washes* that only occasionally carry the water of a thunderstorm are the principal sources of water in parts of the Southwest, where they are held to be watercourses, but the same need might not exist in more humid areas. There may be need for different rules in cases where the parties are seeking to get rid of the water and in cases where both parties are seeking to use the water.

Artificial watercourses have been created by man's activities, and all rights applicable to natural watercourses attach to them. An artificial watercourse may be a new stream channel into which a river or stream is turned, a drainage ditch intended as a permanent structure, or a channel formed by collected return flows from irrigation.[194]

3. Lakes and Ponds. *Lakes* and *ponds* that form the source of a watercourse, or through which a watercourse flows, are a part of the watercourse.[195] The use of water from such a lake or pond may be obtained under riparian or appropriation law, whichever is applicable, and the rights of landowners upon or water users from the lake must be correlated with the water rights held in the connecting stream. Such rights may also exist in bodies of water that have neither a visible source of supply nor an outlet.[196] *Sloughs* and *marshes* through which a stream flows are also parts of a watercourse, as well as *bays, inlets,* and *pools* in contact with the stream whose waters mingle with the waters of the stream though no current flows through them.[197] If an intermittent stream leaves pools when the water ceases to flow, the water in them is subject to similar water rights.[198]

4. Springs. A *spring* that forms the source of and supplies the watercourse is a part of the stream and is subject to the riparian or appropriative rights of all those who have interests in the stream itself. On the other hand, springs that do not flow in sufficient quantity to form a stream and leave the land upon which the waters arise are generally held by courts or declared by statutes to belong to the owner of the land,[199] although in some states they have been made subject to appropriation by others.[200] A person who "develops" a spring, causing it to flow or to flow in greater amounts, is regarded as taking percolating waters.[201]

5. Surface and Floodwaters. *Surface waters* are all forms of water on the surface of the earth that are diffused across broad areas and are not, or have not yet, collected in a watercourse. They may be in motion, when their source is rainfall, melting snow, springs, or seepage. They may be still waters, in which motion is practically imperceptible, such as water standing in swamps, bogs, or marshes. *Floodwaters* that have overflowed the banks of a watercourse are also generally classified as surface waters.

6. Return Flows and Waste Water. When water is spread over land to irrigate crops, some of it is incorporated in the plant structure, some is lost through evaporation from the ground or transpiration from plants, and some finds its way to aquifers not directly connected to streams. This is *consumptive use*. The remainder, flowing on the surface to a nearby stream or seeping underground toward the stream, is known as *waste water,* or *return flow.* Water seeping from a ditch or reservoir is also described as waste water. Return flow at or near the surface is treated as surface water, and the laws of drainage are applied to it when the landowners seek to get rid of it. But when efforts are made by irrigators to capture and re-use this water, a host of legal complications arise.

When return flows from irrigation form a part of the stream in a jurisdiction that

---

[194] *Jack v. Teegarden* (1949) 151 Neb. 309, 37 N.W.2d 387; *Binning v. Miller* (1940) 55 Wyo. 451, 102 P.2d 54.
[195] *Duckworth v. Watsonville Water & Light Co.* (1907) 150 Cal. 520, 89 P. 338.
[196] *Proctor v. Sim* (1925) 134 Wash. 606, 236 P. 114.
[197] *Turner v. James Canal Co.* (1909) 155 Cal. 82, 99 P. 520, 22 L.R.A. (n.s.) 401, 132 A.S.R. 59.
[198] *Humphreys-Mexia Co. v. Arseneaux* (1927) 116 Tex. 603, 297 S.W. 225.
[199] *Deseret Live Stock Co. v. Hooppiania* (1925) 66 Utah 25, 239 P. 479.
[200] *Parker v. McIntyre* (1936) 47 Ariz. 484, 56 P.2d 1337.
[201] *Bullock v. Tracy* (1956) 4 Utah 2d 370, 294 P.2d 707.

recognizes riparian law and riparian rights attached to such waters, so that any change that would deprive the lower riparian of the return flow is a violation of his rights.[202]

Most controversies over this type of water have arisen between appropriators, and the law of appropriation of return flows is quite confused. Controversies usually arise when the first user of the water seeks to recapture it and re-use it. The person opposing the re-use may be senior or junior to the first user. The attempted recapture may be made before any second use has been made or long after the second users have come to rely upon the presence of the water. The first user may be genuinely attempting to reduce waste and to use his appropriation more efficiently, or he may be trying to extend his right beyond its original bounds. The attempted recapture may be made while the water is still diffused or percolating or after it reaches a natural stream. Statutes enacting generalities about "waste and seepage" have been applied indiscriminately to these varying situations, and language from cases deciding one type of controversy has been applied without discretion to other and quite different disputes.

This tangle of laws cannot be completely unraveled here. But some general principles may be stated and some general observations may be made. An appropriator of the natural flow of the stream is regarded as beneficially using all that he diverts even though he consumes only a part of it and the rest returns to the stream. If he can increase his benefits by increasing his consumption, he is privileged to recapture waste water still on his land and re-use it on the land for which it was originally appropriated.[203] In most states he can postpone this recapture indefinitely. However, if the water has reentered the natural stream, most courts have said that he has lost control over it.[204] If the first user attempts to recapture the water and use it on new lands other than those to which the original right was appurtenant, he cannot do so if the interests of secondary users have intervened.[205]

Looking at the matter from the standpoint of the person attempting to capture return flows originating from the water rights and lands of others, he may obtain a valid appropriation of these return flows as against third persons also seeking to use them, though he may not have the right to expect the continuance of the return flow if the original appropriator exercises the rights of recapture outlined above or should cease to irrigate.[206]

7. Foreign and Added Waters. *Foreign waters* are those imported into a basin from another watershed. If these are added to the new stream, the stream is being used as a part of a ditch system, and the person adding the water has the right to redivert it from the stream at the place where it is needed for use. Once it has served the purpose for which it was added to the stream and returned to the stream as return flow, it is subject to the rules stated above applicable to all return flow.[207] Some courts have permitted importers of water to recapture return flow from foreign water even after it has reached a natural stream, especially when the foreign water has been imported with the original intent of getting the complete benefit from it by using it, recapturing it, and using it again.[208]

Other cases in which water is added to a stream are governed by the same rules as are applicable to foreign water. This added water may include water stored in a reservoir and transported to the place of use by releasing it into a natural stream,[209] or *developed water*, that is added by activities such as unwatering a mine or draining land.[210]

8. Groundwaters. The courts have divided groundwaters into two major classes, *percolating water* and *underground streams*. Percolating waters were once defined as

---

[202] *Southern California Investment Co. v. Wilshire* (1904) 144 Cal. 68, 77 P. 767.
[203] *Binning v. Miller* (1940) 55 Wyo. 451, 102 P.2d 54.
[204] *Lasson v. Seely* (1951) 120 Utah 679, 238 P.2d 418.
[205] *Comstock v. Ramsey* (1913) 55 Colo. 244, 133 P. 1107.
[206] *Bower v. Big Horn Canal Association* (1957) 77 Wyo. 80, 307 P.2d 593.
[207] *Coryell v. Robinson* (1948) 118 Colo. 225, 194 P.2d 342.
[208] *Miller v. Wheeler* (1909) 54 Wash. 429, 103 P. 641.
[209] *United States v. Haga* (D. Idaho 1921) 276 F. 41.
[210] *Ripley v. Park Center Land & Water Co.* (1907) 40 Colo. 129, 90 P. 75; *Platte Valley Irrigation Co. v. Buckers Irrigation, Milling & Improvement Co.* (1898) 25 Colo. 77, 53 P. 334.

those that ooze, seep, or filter through the soil without a definite channel in a course that is unknown or not discoverable, while an underground stream flows as a definite current in a well-defined channel, discoverable by men without special scientific training.[211] One form of an underground stream is the subflow of a surface stream, percolating from the banks and supporting the stream, so that drawing off the subsurface water would appreciably and directly diminish the surface flow.[212]

Scientifically, this is nonsense, but legally, the distinction is important and has some justification. Percolating water could be captured by the landowner under most early rules, but takers of water from underground streams were subject to the law applicable to surface watercourses (riparian or appropriative) and had to respect the rights of others. The courts applied these concepts flexibly to achieve generally desirable results. A mine owner who, in unwatering his works, dried up a neighbor's spring or well was not liable if he had no foreknowledge of the result; the court said water was percolating.[213] Where a stream flowed in part underground and in part on the surface of a narrow valley bounded by canyon walls and it was clearly desirable that the rights of all parties that withdrew waters from the stream or from wells be correlated, the groundwater was designated an underground stream.[214] Even a broad aquifer filling an entire valley floor was called an underground lake to protect a city's water supply from depletion by drillers.[215] Wells drilled so close to a stream as to affect its flow were held to be taking the underflow in order to protect prior rights to the stream.[216]

Today, these concepts might be redefined so as to give them some scientific content. Rules of percolating water are applicable to most aquifers, whether water-table or artesian conditions exist. But a narrow confined aquifer may be treated as a stream when the court believes that surface-water rules present a more desirable solution to a conflict than do the rules of percolating water. One court has said that the presence of such a stream may be proved by scientific as well as by lay evidence.[217] When the pumping of groundwater has a substantial and demonstrable effect on an influent stream, the court is likely to say that the water table near the stream is the subflow, in order to be able to correlate withdrawals from what is in fact a single source of supply.

Under some modern statutes, the distinction between these classes of water has ceased to exist, where all ground and surface waters are subject to the same rule of law and water users must respect the rights of prior users.[218]

9. Atmospheric Water. There has been much speculation on the law of weather control, but few cases about it. Some states have required persons engaged in weather-modification activities to register and obtain licenses and to report their operations.[219] Only two reported cases, neither decisive, have been litigated. A New York lower court was asked by a resort owner to enjoin a city from experimenting with "rain making" aimed at filling municipal reservoirs. The injunction was denied on the ground that the possibility of inconvenience to plaintiff's guests was vastly outweighed by the desirability of supplying the inhabitants of the city with water.[220] The Texas courts currently have before them the other phase of weather modification, hail suppression. A temporary injunction has been issued against cloud seeding directly over the land of a person who claimed that, as a result, he failed to receive the benefits of rain on his rangeland. The injunction merely preserves the *status quo*

---

[211] *Clinchfield Coal Corp. v. Compton* (1927) 148 Va. 437, 139 S.E. 308.
[212] *Maricopa County Municipal Water Conservation District v. Southwest Cotton Co.* (1931) 39 Ariz. 65, 4 P.2d 369.
[213] *Wheatley v. Baugh* (1885) 25 Pa. St. 528.
[214] *Verdugo Canon Water Co. v. Verdugo* (1908) 152 Cal. 655, 93 P. 1021.
[215] *Los Angeles v. Hunter* (1909) 156 Cal. 603, 105 P. 755.
[216] *Montecito Valley Water Co. v. Santa Barbara* (1904) 144 Cal. 578, 77 P. 1113.
[217] *Maricopa County Municipal Water Conservation District v. Southwest Cotton Co.* (1931) 39 Ariz. 65, 4 P.2d 369.
[218] *Wyo. Comp. Stat.* Sec. 41-133; *Templeton v. Pecos Valley Artesian Conservancy District* (1958) 65 N.M. 59, 332 P.2d 465.
[219] *Cal. Water Code* Secs. 400–415; *Wyo. Comp. Stat.* Secs. 9-267 to 9-276.
[220] *Slutsky v. New York* (1950) 97 N.Y.S.2d 238.

pending a full hearing on the case.[221] Neither case decides who "owns" the clouds, or whether landowners have a right to rain that would naturally fall, or a right not to have "artificial" rain fall on their land.

B. Surface-water Law

1. Drainage. In the law of drainage two rules have competed for acceptance by the courts and a third has developed. Some courts follow the *common-enemy rule*, under which the possessor of land may protect himself against surface water as best he can, building dikes to keep the water off his land or drains to cast it down onto lower lands.[222] Others apply the *civil-law rule*, borrowed from European countries, which subjects each piece of land to a servitude for the natural flow of water across it, so that a landowner cannot prevent water from coming to his land, nor may he collect it so it flows from his land in unusual quantities or change the direction of the natural drainage.[223]

Each of these rules has obvious disadvantages, and both have been modified in some stated by statute or decision. Thus some courts applying the civil-law rule have held that the natural flow may be changed if damage is slight or if the landowner is protecting himself from extraordinary floodwater.[224] Some courts have imposed liability despite the common-enemy rule, where the upper owner has discharged unusually large flows on the lower.[225] A few courts have blended such modifications into a third rule that "reasonable" interference with surface water is permissible when the utility of the drainage outweighs the gravity of the harm.[226] Statutes requiring railroads to maintain ditches along, and openings in, road beds are common.[227]

Most drainage is performed today by *drainage districts*, which are public bodies established to build common works to drain a large area of land. They are formed with the approval of a court or county board when the majority of the landowners in the area desire the improvement. They are financed by assessments on all improved lands, and these assessments can be levied and enforced regardless of whether a minority landowner desires to participate. The actions of such public districts are not limited by the laws of private drainage, and if damage is done to others, the district may nevertheless proceed, making compensation payments where necessary and adding them to the costs of the project.

2. Rights to Capture. Practically all courts have held that the landowner "owns" surface water diffused over his land and may capture it and use it on his land or sell its use to others.[228] Lower landowners cannot complain, even though they may have initiated a use of the water at a prior time.[229]

Although surface runoff is an important source of natural streams, there seems to be no case in which stream users have complained of an interference with their rights by a capture of surface water. A California statute declares that soil-conservation practices are not an interference with water rights.[230] It is doubtful that a court would prohibit a landowner from protecting his farm soil or making rangeland fit for its best use by building stock-water dams on draws that are not watercourses. One court has held that, while the owner of boggy lands may drain them, he cannot lead the

---

[221] *Southwest Weather Research Inc. v. Duncan* (Tex. 1959) 327 S.W.2d 417.

[222] *Young v. Moore* (Mo. 1951) 236 S.W.2d 740; *Stacy v. Walker* (1953) 222 Ark. 819, 262 S.W.2d 889.

[223] *Robinson v. Belanger* (1952) 332 Mich. 657, 52 N.W.2d 538; *Gott v. Franklin* (1948) 307 Ky. 466, 211 S.W.2d 680.

[224] *Bolinger v. Murray* (1931) 18 La. App. 158, 137 So. 761; *McManus v. Otis* (1943) 61 Cal. App.2d 432, 143 P.2d 380.

[225] *Clark v. Springfield* (Mo. 1951) 241 S.W.2d 100.

[226] *Stouder v. Dashmer* (1951) 242 Iowa 1340, 49 N.W.2d 859; *Bush v. Rochester* (1934) 191 Minn. 591, 255 N.W. 256.

[227] *Peterson v. Northern Pacific R.R.* (1916) 132 Minn. 265, 156 N.W. 121.

[228] *Riggs Oil Co. v. Gray* (1934) 46 Wyo. 504, 30 P.2d 145.

[229] *State v. Hiber* (1935) 48 Wyo. 172, 44 P.2d 1005.

[230] *Cal. Water Code* Sec. 1252.1.

water to another watershed and cut off a flow of water upon which an appropriator has come to rely.[231]

C. Groundwater Law

1. Ownership Rule. In the first cases involving disputes over groundwater, the courts applied a rule, developed in other types of cases, that the owner of the surface of land owns everything beneath the land as well and held that a landowner could treat percolating water found beneath the surface as his absolute property.[232] He might capture it and use it or dispose of it, without regard to the adverse affects his activities might have on his neighbors. In some later decisions the courts adhering to this rule have modified it by saying that liability may follow if the acts of the landowner are excessively wasteful or if the damage is maliciously caused.[233]

2. Reasonable-use Rule. A number of courts, dissatisfied with the obvious limitations of the rule of absolute ownership, have modified it by saying that the use of the water by the landowner must be reasonable.[234] Reasonableness in these cases is tested with relation to the use of the water on the overlying land and includes withdrawal of water necessary or desirable for the purposes of manufacturing, mining, or agriculture, despite the fact of damage to neighboring landowners.[235] Unreasonable uses have been held to include malicious or wasteful extractions that interrupt a neighbor's use and the transportation of water away from the land for sale at distant places.[236]

3. Correlative Rights. In California, the basic law of percolating water bears a close relation to that state's law of surface streams. The overlying landowners have correlative rights in the underground source; each may use it reasonably on his land; but if the supply is insufficient for all, it must be apportioned fairly and justly between them. If there is a surplus over that needed for the overlying land, it may be appropriated for distant use, and priority in time gives the better right as between such appropriators.[237] The total quantity taken should be limited to the annual recharge.[238] If overdrafts by all consumers exceed the safe yield of an aquifer, each is committing a wrong against the others, but if this condition persists for the period of the statute of limitations (5 years), each has a prescriptive right, and a proportionate reduction in pumping can be enforced against all.[239]

4. Conservation Statutes. Artesian waters have been the subject of statutory regulation in many states. These statutes generally require that artesian wells be controlled and be capped when not in use, so as to prevent a loss of pressure.[240] Some forbid drillers and owners to permit artesian water to escape into pervious strata before reaching the surface.[241]

Well drilling is regulated in many states, where state boards are given power to control the drilling, casing, or deepening of wells.[242]

Licensing of new wells is required in several states where local overdrafts have become a serious problem.[243] Small domestic or replacement wells are usually exempted. A statute which authorizes a state agency to deny permits for new wells

---

[231] *Black v. Taylor* (1953) 128 Colo. 449, 264 P.2d 502.
[232] *Acton v. Blundell* (1842) 12 M&W 324, 152 Eng. Rep. 1223; *Huber v. Merkel* (1903) 177 Wis. 355, 94 N.W. 354.
[233] *Rose v. Socony-Vacuum Corp.* (1934) 54 R.I. 411, 173 A. 627.
[234] *Forbell v. City of N.Y.* (1900) 184 N.Y. 522, 58 N.E. 644.
[235] *Sloss-Sheffield Co. v. Wilker* (1936) 231 Ala. 511, 165 So. 764.
[236] *Barclay v. Abraham* (1903) 121 Iowa 619, 96 N.W. 1080; *Canada v. Shawnee* (1937) 179 Okla. 53, 64 P.2d 694.
[237] *Katz v. Walkinshaw* (1903) 141 Cal. 116, 70 P. 663, 99 A.S.R. 35.
[238] *Burr v. Maclay Water Co.* (1908) 154 Cal. 428, 98 P. 260.
[239] *Pasadena v. Alhambra* (1949) 33 Cal.2d 908, 207 P.2d 17.
[240] *Mont. Rev. Code* Secs. 89-2902, 89-2904; *S.D. Code* Sec. 61.0403.
[241] *Cal. Water Code* Sec. 305.
[242] *Colo. Rev. Stat.* Sec. 147-19-10; *S.D. Code* Sec. 61-0409.
[243] *N.Y. Stat.* Sec. 821a; *N.J. Stats.* Sec. 588:4A-1.

for the irrigation of previously uncultivated land in critical areas has been held to be constitutional.[244] In Texas, the ownership rule is considerably modified where *conservation districts* have been established in critical areas. The districts may impose extensive regulation of drilling and pumping to conserve water and prevent waste.[245]

5. Appropriation. In nine western states, Idaho, Kansas, Nevada, New Mexico, Oklahoma, Oregon, Utah, Washington, and Wyoming, the principles of appropriation have been applied to underground water as well as to surface streams. Beneficial use, not ownership of overlying land, is the basis of a right to groundwater in these states.[246] Under this rule a permit to appropriate is required. The statutes generally exempt domestic or stock-water wells. They cover all groundwaters in most states, but only underground streams, channels, and basins with reasonably ascertainable boundaries, in others.[247] Strict regulation of drilling and withdrawals is applicable only in critical areas in some.[248]

The major feature of these appropriation laws is the power of the state to deny permits for new wells when there is no unappropriated water available. Well-spacing regulations may also be imposed. If wisely administered, these provisions will result in effect in there being no junior appropriators, since all pumping will be within the limits of the safe yield, but some statutes make provisions for apportionment of water according to priority in cases of overdraft.[249] In areas of groundwater mining, the safe yield can be construed as the quantity that will permit pumping for a period sufficient to amortize investments in facilities and land and to realize the maximum benefits from the water.

Colorado has the rule that underground water tributary to a surface stream is subject to appropriation and regulation under the laws applicable to streams.[250] Little regulation has been accomplished under this rule. Well permits are issued automatically, but local districts may cooperate with a state commission in establishing pumping regulations and restrictions.[251]

IV. CONCLUSION

A. Law and Science

Major doctrines of law have always borne some relation, however crude, to the science of hydrology. It has already been pointed out that where streamflows were abundant, the riparian law was adopted to make water available to those with easy access to it and to provide a flexible mechanism, the lawsuit, to settle the few disputes that arose. Where water was available only in limited quantities, prior appropriation was invented to permit its most beneficial use. Groundwater rules were based on the best knowledge available or on suppositions that appeared reasonable at the time. The rule of ownership of groundwater was based, more than a century ago, on a theory of *vagrant drops* that ooze or percolate in all directions. This scientific theory was an adequate reason for saying that the overlying landowner might do with it as he pleased, since no other rule seemed possible of enforcement. Even today the rule may do no harm; it may be a perfectly adequate rule to encourage development by agricultural landowners overlying an unconsolidated alluvial aquifer in an area where perennial rains give a large amount of recharge. As large withdrawals by cities became a major threat to the security of agricultural water uses, the courts evolved the reasonable-use rule, which in effect made cities pay for the damage caused. When serious problems of overdevelopment arose, legislatures and courts met the problems by a variety of methods: the appropriation doctrine that strictly defines rights and

---

[244] *Southwest Engineering Co. v. Ernst* (1955) 79 Ariz. 404, 291 P.2d 764.
[245] *Texas Laws* 1949 c. 306.
[246] *Fairfield Irrigation Co. v. Carson* (1952) 122 Utah 225, 247 P.2d 1004.
[247] *N.M. Stat. Ann.* Sec. 75-11-1; *Wash. Rev. Code* Sec. 9.440.010.
[248] *Idaho Code* Sec. 42-233a; *Wyo. Comp. Stat.* Sec. 41-139.
[249] *Ore. Rev. Stat.* Sec. 537.735; *Wyo. Comp. Stat.* Sec. 41-132.
[250] *Colorado Springs v. Bender* (1961) 148 Colo. 458, 366 P.2d 552.
[251] *Colo. Rev. Stat.* Secs. 147-19-3, 4.

limits withdrawals to the safe yield, the freezing of correlative rights or reasonable uses, and the licensing of new wells.

As scientific knowledge has increased, the law has kept pace, though sometimes a step behind. While some courts have adhered to old rules and in controversies between users of streams and of related groundwaters have permitted the person with superior access to take the water without regard to the other person's rights,[252] one state legislature has declared that interrelated sources shall be treated in law as a single supply,[253] and some courts have done the same in the absence of statute.[254]

B. The Law in Action

A statement of the basic law can be misleading if applied too literally to a situation. If a rule of law is found to have serious disadvantages, the people concerned may often make corrective adjustments in their legal relations. When it is said, for instance, that nonriparian uses violate lower riparian rights, this does not mean that a particular proposed nonriparian use is completely forestalled. Adjustments may be made; the person with the superior right may be willing to relinquish his right by contract or deed if payment is made to him or if some other advantage will accrue to him. Indeed, no objection may be made at all. Much nonriparian development, particularly for public water supply, has taken place in the East without objection, for the simple reason that withdrawals were insignificant in relation to the abundance of water.

The district is a solution to other problems. If the common-enemy rule of surface drainage will not work, the landowners, instead of fighting each other in court, may arrange for cooperative effort through a drainage district. A junior water right on a fluctuating western stream may seem insecure. Yet in many areas of the West irrigators receive water under arrangements which at first blush seem to bear little relation to the doctrine of priority. Where the irrigator is a consumer of water supplied by a mutual water company or an irrigation district, he is in most cases the holder of a right to receive a fractional part of the common supply without regard to priority in the sense of the time when he put the water to use upon his land. The water-distributing company or district has an appropriation to which the doctrine of priority is applicable, but if it has stored a season's supply or has supplemented insecure junior appropriations by importation of water, the water user's share becomes a firm right in a firm supply and is not subject to diminution or stoppage because of a late priority. In the East, conservancy districts and water-management districts eliminate objections to nonriparian use and eliminate the instability of fluctuating riparian rights by storing floodwaters and equating the supply.

Law has been the foundation of water development. United States laws have served the people and the nation well. As the basis of property rights, the law has encouraged private development, yet when private action has threatened the public interest, property rights have been modified or restricted. The law has enabled governments to undertake public development beyond the capacity of private enterprise.

The law grows and changes with the needs of the people. If a law places a serious barrier to desirable development and the barrier cannot be surmounted by economic or cooperative action, that law is ripe for judicial or legislative change. Almost every doctrine shows this characteristic of growth. The common law of riparian rights was not a static thing, applied by courts without reference to economic conditions. Under the limited technology of the last century, most uses of water were uses in place, and the natural-flow doctrine passed the water on to each user in turn. As depletive uses became more common, the reasonable-use rule was evolved. Where appropriations gave the best promise of development, as in California, a narrow definition of riparian lands was adopted that freed much water from riparian claims. Where it appeared that riparian law was entrenched, a broad definition of such lands gave development

---

[252] *Pecos County Water Control District v. Williams* (Tex. Civ. App. 1954) 271 S.W.2d 503.
[253] *Wyo. Comp. Stat.* Sec. 41-133.
[254] *Templeton v. Pecos Valley Artesian Conservancy District* (1958) 65 N.M. 59, 332 P.2d 465.

a wider scope. When the West was first settled and it was seen that water was the limiting factor on development, the courts immediately adopted a doctrine of water rights suitable to conditions in the pioneering community, and the quick development of the mineral and land resources of the West was fostered by giving almost unrestricted opportunity to appropriate water as private property and to move water to where it was most needed. Public regulation came later, as a second stage, when free appropriation began to show defects that allowed individuals to act in a manner contrary to public interests.

The law is a human institution; it reflects the needs and desires of man. It changes when a sufficient number of a community or group accept an idea and give it the public support that will guarantee its acceptance by the courts or its enactment by a legislature. The future course of the law, where it fails to accord with scientific principles or seems to hinder development, is perhaps as much in the hands of the scientist and the engineer as of anybody. The exposition and dissemination to courts, legislatures, and the community of scientific information and of the desirability of development will enable the people and the lawmakers to adopt suitable legal institutions intelligently and rationally.

V. REFERENCES

I. General

Hutchins, W. A.: Selected problems in the law of water rights in the West, *U.S. Dept. Agr. Misc. Publ.* 418, 1942.

Lasky, Moses: From prior appropriation to economic distribution of water by the state—via irrigation administration, *Rocky Mt. L. Rev.*, vol. 1, pp. 161–216, vol. 2, pp. 35–58, 1929.

Martz, C. O.: Seepage rights in foreign waters, *Rocky Mt. L. Rev.*, vol. 22, pp. 407–421, 1950.

Shih, Y. C.: "American Water Resources Administration," Bookman Associates, New York, 1956.

Stark, D. D.: Weather modification: water—three cents per acre-foot, *Calif. L. Rev.*, vol. 45, no. 5, pp. 698–711, 1957.

"State Water Law in the Development of the West," Report Submitted to the Water Resources Committee by its Subcommittee on State Water Law, National Resources Planning Board, U.S. Government Printing Office, 1943.

Ziegler, W. T.: Statutory regulation of water resources, in "Water Resources and the Law," University of Michigan Law School, Ann Arbor, Mich., 1958, pp. 89–129.

———: Water use under common law doctrines, in "Water Resources and the Law," University of Michigan Law School, Ann Arbor, Mich., 1958, pp. 49–86.

II. Surface Streams

Busby, C. E.: American water rights law: a brief synopsis of its origin and some of its broad trends with special reference to the beneficial use of water resources, *S.C.L.Q.*, vol. 5, no. 2-A, pp. 106–129, 1952.

Ellis, H. H.: Development and elements of the riparian doctrine: with reference to the eastern states, *Proc. Water Rights Conf.*, Michigan State University, East Lansing, Mich., 1960, pp. 19–33.

Haber, David, and S. W. Bergen (eds.): "The Law of Water Allocation in the Eastern United States," The Ronald Press Company, New York, 1958.

Hutchins, W. A., and H. A. Steele: Basic water rights doctrines and their implications for river basin development, *Law & Contemp. Prob.*, vol. 22, no. 2, pp. 276–300, 1957.

Kinney, C. S.: "Treatise on the Law of Irrigation," 2d ed., Bender-Moss Co., San Francisco, 1912.

Trelease, F. J.: Coordination of riparian and appropriation rights to the use of water, *Texas L. Rev.*, vol. 33, pp. 24–69, 1954.

———: Preferences to the use of water, *Rocky Mt. L. Rev.*, vol. 27, pp. 133–160, 1955.

———: The concept of reasonable beneficial use in the law of surface streams, *Wyo. L. J.*, vol. 12, pp. 1–21, 1957.

———: A model state water code for river basin development, *Law & Contemp. Prob.*, vol. 22, no. 2, pp. 301–322, 1957.

Wagner, S. C.: Statutory stream pollution control, *U. Pa. L. Rev.*, vol. 100, pp. 225–241, 1951.

REFERENCES

"Water Resources and the Law," University of Michigan Law School, Ann Arbor, Mich., 1958.

Wiel, S. C.: "Water Rights in the Western States," 3d ed., Bancroft Whitney Co., San Francisco, 1911.

III. Groundwater

Kirkwood, M. R.: Appropriation of percolating water, *Stan. L. Rev.*, vol. 1, pp. 1–22, 1948.

McGuinness, C. L.: Water law with special reference to ground water, *U.S. Geol. Surv. Circ.* 117, 1951.

McHendrie, A. W.: Law of underground water, *Rocky Mt. L. Rev.*, vol. 31, pp. 1–19, 1940.

Piper, A. M.: Interpretation and current status of ground water rights, *Proc. Water Rights Conf.*, Michigan State University, East Lansing, Mich., 1960, pp. 68–76.

—————— and H. E. Thomas: Hydrology and water law: what is their future common ground? in "Water Resources and the Law," University of Michigan Law School, Ann Arbor, Mich., 1958, pp. 7–24.

Thomas, H. E.: Water rights in areas of ground-water mining, *U.S. Geol. Surv. Circ.* 347, 1955.

——————: Hydrology vs. water allocation in the eastern United States, in "The Law of Water Allocation in the Eastern United States," The Ronald Press Company, New York, 1958, pp. 165–180.

Uelmen, D. L.: Law of underground water: a half-century of *Huber v. Merkel*, *Wis. L. Rev.*, vol. 1953, pp. 491–515, 1953.

IV. Federal and International

Berber, F. J.: "Rivers in International Law," (transl. by R. K. Batstone) Stevens & Son, Ltd., London, and Oceana Publications Inc., New York, 1959.

Corker, C. E.: Water rights and federalism: the western water rights settlement bill of 1957, *Calif. L. Rev.*, vol. 45, no. 5, pp. 604–637, 1957.

Gisvold, Per: "A Survey of the Law of Water in Alberta, Saskatchewan and Manitoba," The Queen's Printer, Ottawa, Ont., 1959.

King, D. B.: Interstate water compacts, in "Water Resources and the Law," University of Michigan Law School, Ann Arbor, Mich., 1958, pp. 353–422.

Martz, C. O.: The role of the federal government in state water law, *Kan. L. Rev.*, vol. 5, 626–648, 1957.

Trelease, F. J.: Federal jurisdiction over water in the eastern U.S., *Proc. Water Rights Conf.*, Michigan State University, East Lansing, Mich., 1960, pp. 1–17.

——————: Reclamation water rights, *Rocky Mt. L. Rev.*, vol. 32, pp. 464–501, 1960.

"Water Resources Law," vol. 3 of Report of the President's Water Resources Policy Commission, U.S. Government Printing Office, 1950.

Witmer, T. R. (ed.): "Documents on the Use and Control of the Waters of Interstate and International Streams," U.S. Government Printing Office, 1956.

V. State Water Law

State water-rights laws and related subjects: a bibliography, *U.S. Dept. Agr. Econ. Res. Div. Farm Econ. Div. Misc. Publ.* 921, 1962.

Arkansas
  Aspects of legal control of surface water, Arkansas Legislative Council, Research Department, Research Memorandum 1 on Proposal 6, 1954.

California
  Hutchins, W. A.: "The California Law of Water Rights," State of California, Sacramento, 1956.
  Legal problems in water rights (symposium), *Calif. L. Rev.*, vol. 45, no. 5, 1957.

Colorado
  Breitenstein, J. S.: Some elements of Colorado water law, *Rocky Mt. L. Rev.*, vol. 22, pp. 343–355, 1950.
  Danielson, P. A.: Water administration in Colorado: higher-ority or priority? *Rocky Mt. L. Rev.*, vol. 30, no. 3, pp. 293–314, 1958.

Connecticut
  Fisher, C. O., Jr.: Connecticut law of water rights, appendix A in "Water Resources of Connecticut," Report to General Assembly by the Water Resources Commission, Hartford, Conn., 1956.

Delaware
  Ellis, H. H., and R. O. Bausman: Some legal aspects of water use in Delaware, *Delaware Agr. Expt. Sta. Tech. Bull.* 314, 1955.
Florida
  Maloney, F. E.: Florida's new water resources law, *U. Fla. L. Rev.*, vol. 10, no. 2, pp. 119–152, 1957.
Georgia
  Agnor, W. H.: Riparian rights in Georgia, *Ga. Bar J.*, vol. 18, pp. 401–414, 1956.
Hawaii
  Hutchins, W. A.: "The Hawaiian System of Water Rights," City and County of Honolulu, Board of Water Supply, Honolulu, 1946.
Idaho
  Hutchins, W. A.: "The Idaho Law of Water Rights," Idaho State Department of Reclamation, Boise, Idaho, 1956.
Illinois
  Cribbet, J. E.: Illinois water rights law and what should be done about it, Illinois State Chamber of Commerce, Water Resources Committee, Chicago, 1958.
  Drablos, C. J. W., and B. A. Jones, Jr.: Illinois highway and agricultural drainage laws, *Univ. Ill. Eng. Expt. Sta. Cir.* 76, Urbana, Ill., 1963.
  Hannah, H. W.: Illinois farm drainage law, *Univ. Ill. Coll. Agr. Ext. Ser. Agr. Home Econ. Cir.* 751, Urbana, Ill., 1960 ed.
  "Smith-Hurd Illinois Annotated Statutes—The Drainage Code, Effective January 1, 1956," Burdette Smith Co., Chicago, Ill. and West Publishing Co., St. Paul, Minn., 1956.
  "Summary of Illinois Laws Relating to Drainage and Flood Control," All-University Committee on Community Problems, University of Illinois, Urbana, Ill., September, 1959.
Iowa
  O'Connell, Jeffrey: Iowa's new water law, *Proc. Water Rights Conf.*, Michigan State University, East Lansing, Mich., 1960, pp. 54–67.
Kansas
  Hutchins, W. A.: "The Kansas Law of Water Rights," Kansas State Board of Agriculture, Division of Water Resources, Topeka, Kan., 1957.
  Symposium on water law, *Kan. L. Rev.*, vol. 5, no. 4, 1957.
Kentucky
  Gregory, J. A., Jr.: Riparian rights: analysis of new statutory provisions, *Ky. L. J.*, vol. 43, pp. 407–415, 1955.
Louisiana
  Borton, M. E., and H. H. Ellis: Some legal aspects of water use in Louisiana, *Louisiana State Univ. Agr. Expt. Sta. Bull.* 537, 1960.
Maryland
  Bohanan, L. B.: Trends and developments in irrigation in Maryland, *Univ. Maryland Agr. Expt. Sta. Misc. Publ.* 244, College Park, Md., 1955.
Massachusetts
  Haar, C. M., and Barbara Gordon: Legislative change of water law in Massachusetts, "The Law of Water Allocation in the Eastern United States," The Ronald Press Company, New York, 1958, pp. 1–47.
Michigan
  Lauer, T. E., D. B. King, and W. L. Ziegler: Water law in Michigan, in "Water Resources and the Law," University of Michigan Law School, Ann Arbor, Mich., 1958, pp. 423–531.
Montana
  Hutchins, W. A.: "The Montana Law of Water Rights," *Montana Agr. Expt. Sta. Bull.* 545, Bozeman, Mont., 1958.
Nebraska
  Doyle, J. A.: Water rights in Nebraska, *Neb. L. Rev.*, vol. 20, pp. 1–22, 1941, *Neb. L. Rev.*, vol. 29, pp. 385–415, 1950.
Nevada
  Hutchins, W. A.: "The Nevada Law of Water Rights," Nevada State Engineer, Carson City, Nev., 1955.
New Mexico
  Hutchins, W. A.: "The New Mexico Law of Water Rights," *N.Mex. State Eng. Tech. Rept.* 4, Santa Fe, N.Mex., 1955.
New York
  Martin, R. C.: Water resources law, chap. 4 in "Water for New York," Syracuse University Press, Syracuse, N.Y., 1960.

North Carolina
  Ellis, H. H.: Some legal aspects of water use in North Carolina, in "The Law of Water Allocation in the Eastern United States," The Ronald Press Company, New York, pp. 1958, 189–370.
Ohio
  Callahan, C. C.: Principles of water rights law in Ohio, State of Ohio Department of Natural Resources, Division of Water, Columbus, Ohio, 1957.
Oklahoma
  Hutchins, W. A.: "The Oklahoma Law of Water Rights," Oklahoma Planning and Resources Board, Division of Water Resources, Oklahoma City, 1955.
Oregon
  Hutchins, W. A.: The common-law riparian doctrine in Oregon: legislative and judicial modifications, *Ore. L. Rev.*, vol. 36, pp. 192–220, 1957.
Pennsylvania
  Anderson, W. C., and F. H. Cook: Summary of legal principles governing the use of water for irrigation in Pennsylvania, *Penn. State Coll. Sch. Agr. Prog. Rept.* 31, State College, Pa., 1950.
Texas
  Proc. Water Law Conf., University of Texas School of Law, Austin, Tex., 1952, 1954, 1956.
Washington
  Morris, A. A.: Washington water rights: a sketch, *Wash. L. Rev.*, vol. 31, pp. 243–260, 1956.
West Virginia
  Lugac, M. E.: "Water Rights Law in West Virginia," Joint Committee on Government and Finance, Charleston, W.Va., 1957.
Wisconsin
  Beuscher, J. H.: Wisconsin's law of water use, *Wis. Bar Bull.* 31, pp. 30–50, 1958.
  Modjeska, L. M.: Wisconsin's water diversion law: a study of administrative case law, *Wis. L. Rev.*, vol. 1959, pp. 279–311, 1959.

Section 28

WATER POLICY

S. V. CIRIACY-WANTRUP, *Professor of Agricultural Economics, University of California.*[1]

I. Meaning of Water Policy	28-2
A. Definition of Terms	28-2
B. Water Policy, Water Projects, and the Significance of Self-supply	28-2
C. Self-supply and the Significance of Groundwater	28-4
D. Water Policy and the Water Market	28-4
E. Implications of This Section	28-5
II. Objectives and Implementation of Water Policy	28-6
A. Water Policy and Economic Optima	28-6
B. The Problem of Valuation	28-6
C. The Problem of Institutional Constraints	28-7
D. The Problem of Uncertainty	28-8
E. Implementation of Objectives of Water Policy	28-9
F. Water-allocation Policy	28-10
G. Water-development Policy	28-12
III. Economic Projections in Water Policy	28-14
A. Significance of Water-requirement Projections	28-14
B. Water Requirements and Water Demand	28-15
C. Validity of Economic Projections	28-17
D. Implications for Water Policy	28-19
IV. Organization and Coordination of Water Policy in the United States	28-20
A. The Federal Government	28-20
1. Navigation	28-20
2. Flood Control	28-20
3. Irrigation	28-21
4. Municipal and Industrial Water	28-22
5. River-basin Development	28-22
6. Coordination in the Federal Executive Branch	28-22
7. Coordination in the Federal Legislative Branch	28-23
B. State Governments	28-23
1. Proposals for Better Coordination within State Governments	28-23
2. Public Districts	28-24
3. Interstate Cooperation	28-24
4. State-Federal Cooperation	28-25

[1] Also Agricultural Economist in the Agricultural Experiment Station and in the Giannini Foundation, University of California, Berkeley, Calif.

I. MEANING OF WATER POLICY

A. Definition of Terms

The term *policy* is used with many different meanings, both in scientific and in popular language. For the present purpose, it is helpful to restrict its meaning to interrelated actions (*action systems*), real or hypothetical, of organized *publics* such as Federal, state, and local governments, including public districts. This restriction of meaning is in accordance with the etymological origin of the term. Any private individuals, firms, or associations may have opinions, attitudes, and proposals pertaining to policy; they may aid or hinder the formation and implementation of policy, and they are always affected by it. The term itself, however, will be used here in the restricted sense of *public* policy. Accordingly, the term *water policy* refers to actions of governments at various levels and in various branches (legislative, judicial, executive), affecting the development and allocation of water resources.

The unifying principle that transforms a number of individual government actions into a system termed "policy" is supplied, first of all, by the viewpoint of the scientific observer interested in the many kinds of relations that exist between individual actions in objectives, planning, execution, and effects. In this sense, water policy is both a conceptual tool of analysis—a construct—and a field of scientific inquiry.

Ideally, a second unifying principle is supplied by the purposes of the governments undertaking or considering actions: ideally, individual actions constitute a segment of a purposefully coordinated system. In actuality, however, the purposes of actions by different governments and different branches of the same government are frequently not coordinated. Furthermore, objectives of an individual action may be disconnected with or contrary to those of others by the same government agency. Such lack of coordination and multiplicity of objectives is an important subject in the study of water policy.

Reference was just made to the terms *development* and *allocation* of water resources. In political reality, actions in these two spheres are closely related. In economic analysis, it is useful to separate them. The term *water-development policy* refers to actions affecting the increase of quantities of water available for distribution and use. The term *water-allocation policy* refers to actions affecting the distribution of given quantities of water among different uses—such as domestic, industrial, agricultural, and recreational—and users—such as farms, industrial firms, and households. This terminology follows common usage that generally, although sometimes rather vaguely, employs the terms "development" and "allocation" in connection with natural resources. In technical terms, water-development policy refers to actions affecting the length of the water-use vector; water-allocation policy refers to actions affecting the direction of this vector.[2]

B. Water Policy, Water Projects, and the Significance of Self-supply

It is implied in the definition of terms just given that water policy includes, but is not limited to, public investment in water projects. The study of water policy is a far more comprehensive field of scientific inquiry than the study of public water projects. The economics of public water projects—such as benefit-cost analyses, formal programming, and other techniques of evaluating such projects and the whole problem of efficiency in government investment—constitutes only a segment, and sometimes only a small segment, of water policy.

Reasons for this statement are not far to seek. In the United States—and this is true also for most countries of Western society—water is largely developed and allocated through decentralized decision making of self-supplying farms, industrial corporations, and nonprofit water organizations. These water *firms* are the operating subsectors in the water economy. Individual Federal and state projects may be

---

[2] For an explanation of these terms, see S. V. Ciriacy-Wantrup: Economics of joint costs in agriculture, *J. Farm Economics*, vol. 23, no. 4, pp. 771–818, November, 1941.

regarded as subsectors in this sense, subject to *rules of the game*, not greatly different from those applying to other subsectors. These rules of the game and their modifications are the domain of water policy.

The significance of self-supply is so great for the economic and political issues of water policy that a few quantitative data on the relative importance of various types of subsectors in the above sense may be helpful.

According to the 1950 Census, 47 per cent of the irrigated acreage in the United States (58 per cent in California) was supplied by single-farm irrigation enterprises.[3] Of the industrial use of water in the United States, 97 per cent (in California, likewise, 97 per cent) was supplied by individual company systems.[4]

The second largest part of aggregate water use was supplied by water users themselves cooperatively, through nonprofit water organizations such as mutual water companies and public districts. In the United States, 28 per cent (in California, 12 per cent) of irrigated acreage was supplied by mutual water companies.[5] The corresponding figure for public districts is 18 per cent (in California, 25 per cent).[6] Regarding domestic water use, 87 per cent of a population of 79 million in communities of more than 25,000, covered by a survey of the U.S. Public Health Service in 1957, was supplied by water systems owned by municipalities or municipal water districts.[7] A comparable figure for California is 89 per cent.[8] All these water organizations have in common that in their formation, operation, and growth, water consumers have a direct and significant influence that is outside the demand-supply mechanism of a market. In many respects, the factors affecting decision making in these organizations are similar to those affecting self-supplying firms.

Only a small part of the aggregate water supply is produced for sale by profit-seeking firms. Most of these, in turn, are regulated by state public utility commissions. In the United States, only 3 per cent of the irrigated acreage is supplied by such firms. In California, the corresponding figure is 4 per cent. Of industrial water use, only 5 per cent is supplied by such firms in the United States and 3 per cent in California. Of municipal water use, the data in the above-mentioned survey indicate that only 13 per cent of the population surveyed is supplied by privately owned systems in the United States and 11 per cent in California.

[3] Developed from data given by the U.S. Bureau of the Census, U.S. Census of Agriculture: 1950, Irrigation of agricultural lands, The United States, 1952, vol. III, p. 58, table 16; U.S. Bureau of the Census and U.S. Agricultural Research Service, U.S. Census of Agriculture: 1954, Irrigation in humid areas, a cooperative report, Special Report, 1956, vol. III, part 6, p. 86, table 14.

[4] This information is developed from data given in U.S. Bureau of the Census, Industrial water use, U.S. Census of Manufactures: 1954, Bulletin MC209 (suppl.), 1955, pp. 209–221 and 209–213, table 1; pp. 209–214 and 209–215, table 2; pp. 209–218 and 209–219, table 6; pp. 209–220, 209–221, 209–226, and 209–227, table 7; pp. 209–228 and 209–229, table 8; pp. 209–230 and 209–231, table 9. Water supplied includes fresh, brackish, and mine water, but excludes the quantity of water necessary if there were no recirculation. An insignificant proportion of this supply is provided by combination systems and sources not specified. Steam electric plants account for 80 per cent of the industrial water use in California and 67 per cent in the United States. Steam electric plants use 87 per cent of the total brackish water used in industry in California and 77 per cent in the United States. Brackish water in the tables cited is not differentiated on the basis of supplying systems. In the present computations, brackish water is counted as supplied by company systems. If steam electric plants are excluded from industrial water use, 85 per cent was supplied by company systems in California and 86 per cent in the United States (see also footnote 26).

[5] U.S. Bureau of the Census, U.S. Census of Agriculture: 1950, . . . , and U.S. Bureau of the Census and U.S. Agricultural Research Service, U.S. Census of Agriculture: 1954, . . . (see footnote 3).

[6] *Ibid.*

[7] John R. Thoman and Kenneth H. Jenkins: Inventory of 1956 water supply facilities in communities of 25,000 and over, *J. Am. Water Works Assoc.*, vol. 50, p. 1078, table 3, August, 1958.

[8] Data summarized from U.S. Department of Health, Education and Welfare, Public Health Service, Division of Sanitary Engineering Services, Municipal water facilities, communities of 25,000 population and over, Continental United States and Territorial Possessions, as of January 1, 1958, *Public Health Serv. Publ.* 661, 1959, pp. 10–15.

C. Self-supply and the Significance of Groundwater

The significance of self-supply in the water economy is related to that of groundwater. Groundwater is largely developed through investment by self-supplying water firms, and not by state and Federal investment. Furthermore, in the development of groundwater, private investment is more significant than the investment by public districts.

The significance of groundwater is especially great for agricultural use. For industrial and municipal use, in the United States as a whole, surface water is quantitatively more important than groundwater. For industry, the locational attraction to surface water is partly based on the adequacy of "free" brackish and sea water for steam-electric plants—the largest industrial users (see footnote 4). Other locational attractions to surface water for industry and urban communities alike are the effects of navigable water on costs of transportation and the suitability of large bodies of water for waste disposal. In the arid and semiarid basins of the West, on the other hand, groundwater is no less significant for industrial and domestic use than for agricultural use.

The significance of groundwater for agricultural use may be illustrated with a few figures. Groundwater development was responsible for 67 per cent of total public and private irrigation development in the 17 western states since 1940. If Federal water development is excluded, this figure rises to 89 per cent. The share of groundwater development by decades is also interesting. Before 1900, groundwater was responsible for only 1 per cent of total public and private irrigation development in the 17 western states. Between 1900 and 1909, this percentage was 8 per cent; from 1910 to 1919, 27 per cent; from 1920 to 1929, 52 per cent; from 1930 to 1939, 42 per cent; from 1940 to 1949, 63 per cent; from 1950 to 1958, 69 per cent.[9]

The increasing significance of groundwater is one of the reasons why the differentiation between public water policy and public water projects was emphasized above. Groundwater development, being based largely on private investment, is highly sensitive to changes in general economic conditions for investment. For example, during the depression decade 1930 to 1939, groundwater development decreased absolutely—and even more, relatively to surface-water development. During this decade, public investment in water development, consisting largely of surface-water development, showed a strong increase.

The significance of groundwater development and the increasing need for integrating groundwater and surface-water development poses important issues for water policy that are different from those of public investment. Recent economic literature focusing exclusively on the efficiency of public investment in water projects must be diagnosed as myopia that overlooks some of the most important issues of water policy.

D. Water Policy and the Water Market

Present-day economics in Western countries is largely concerned with the analysis of markets and market prices. Accordingly, economic policy is to a large extent price and market policy.

The preceding subsections showed that decisions about production and use of water are largely internal within self-supplying water firms. In other words, such decisions are not expressed through the firms' behavior in a water market. Neither are they coordinated through such a market. Hence market-oriented economic concepts have more limited analytical significance for explaining and evaluating the behavior of water-producing and water-consuming firms than is true for many other fields of economic inquiry.

In water economics, the price per unit—for example, per acre-foot—is only a part of the total payment complex. The economic, institutional, and technological factors

---

[9] Computed from data given in U.S. Congress, Senate, Select Committee on National Water Resources, Water resources activities in the United States: future needs for reclamation in the Western States, Committee Print no. 14, 86th Cong., 1st Sess., 1960, S. Res. 48.

responsible for this have recently been investigated.[10] Some of them will be touched upon later. Costs of water development are covered through various forms of taxes, special assessments, and fees. While there is an upward trend in the portion covered by prices, and some economic argument for continuation of this trend, a major portion of the water bill will continue to be met in other ways than through prices. Furthermore, as already implied, water prices are always under strong institutional influence, such as the rules of a public district, contracts with the U.S. Bureau of Reclamation and other public agencies, or the regulation of a public utility commission.

Seasonal and permanent transfers of water between water firms occur. In special cases—for example, if water consumers are owners of mutual water companies—seasonal water transfers show market characteristics.[11] Permanent transfer is generally in terms of water rights and is governed by water law. Water exchanges (water for water, differentiated in terms of time and location) are not uncommon. They are individual transactions, usually not involving pecuniary considerations. But they are potentially important for increasing efficiency of water allocation with respect to time, location, and uses. More research is needed on the social performance of these transfer mechanisms. But it is already fairly clear that there is little meaning in speaking of a water market in which water-supplying and water-demanding industries meet.

This situation raises a question with respect to the analytical contribution of the term *water industry* that has become popular recently. In economic theory, this term is employed as a useful fiction. Such usefulness is contingent upon the identification of reasonably homogeneous products and services and of aggregate demand-and-supply functions of groups of profit-seeking firms consuming or producing such products and services. The term water industry may mislead the unwary to believe that the structure, functioning, and performance of water institutions—to use the old-fashioned but by no means obsolete term—can be analyzed in terms of conventional market concepts. Such concepts form the basis for the field of industrial organization that is a useful frame of reference for the study of other natural-resource industries, for example, the oil industry. In the oil industry, self-supply is an insignificant portion of the market supply.

E. Implications of This Section

It is evident from what has been said so far that, as compared with other economic policies, water policy is less concerned with markets and prices and more with the laws, regulations, and administrative structures under which self-supplying individual firms and nonprofit organizations make decisions. This situation poses a challenge to scientific inquiry that must be faced squarely: institutional influences are so diverse, so pervasive, so widely distributed over time, so difficult to isolate and quantify, so resistant to controlled experiment, and so closely related to the social conditioning of the political preferences and the emotions of the investigator, that the temptation is great to remain on the descriptive level instead of proceeding toward analysis.

Over many years, the descriptive approach to water institutions has yielded much valuable material, contributed largely by noneconomists. The works of Elwood Mead, Frank Adams, and many others from the engineering profession fall in this class. This material is now available to the social sciences for analysis focusing on the *structure*, the *functioning*, and the *performance* of water institutions. Water policy as a field of scientific inquiry is analytically oriented institutional economics. In such an economics, theoretical constructs and their testing are no less needed than in the economics of the marketplace.

[10] M. F. Brewer: Water-pricing and allocation with particular reference to California irrigation districts, *Univ. Calif., Giannini Foundation Rept.* 235, Berkeley, Calif., October, 1960. Mimeographed.
[11] Raymond L. Anderson: Operation of the irrigation water rental market in the South Platte Basin, *J. Farm Economics*, vol. 42, no. 5, pp. 1501–1502, December, 1960.

II. OBJECTIVES AND IMPLEMENTATION OF WATER POLICY

A. Water Policy and Economic Optima

In identifying the objectives of water policy, one encounters what might be called the problem of unity of social objectives and criteria. Objectives of water policy cannot, in principle, be divorced from the objectives of other policies. Such objectives are interrelated, and social-welfare criteria are no different in water policy than in other fields.

Since Pareto,[12] economists have taken a special interest in optimizing social welfare and in the criteria of such optimizing. Space does not permit discussing the pros and cons of this literature. It must be noted, however, that optima of social welfare and formal criteria for optimizing are constructs in the sense of useful scientific fictions.[13] Such constructs are not operational policy objectives. These fictional constructs are useful as organizing principles for the great number of variables and kinds of relations that must be considered in welfare economics—to decide which ones to bring into the analysis explicitly, which to neglect, which to combine with others, and which to take into account as constraints.

Information about the most essential variables and their relations is too vague for projecting quantitatively an optimum expansion path of social welfare over time in a dynamic framework. Such a framework means not merely that time is taken into account by "dating" and by projected economic "growth," but that economic change and its uncertainty—especially uncertain changes of technology, of consumer preferences, and of social institutions—are explicitly taken into account. The operational objective of policy decisions in such a framework is the modest one of making successive incremental improvements in welfare, considering a limited number of alternatives.

For policy decisions of a more limited scope—for example, for making choices between alternative water projects—incremental improvements can be determined cardinally in pecuniary terms. Benefit-cost analysis and formal programming can be employed for this purpose. For policy decisions of a broader scope, incremental improvements in social welfare can be determined only ordinally in terms of the direction of changes, the relative speed of changes, and their sequence in time. This discussion is concerned especially with policy decisions of the latter kind.

Operational implementation of such decisions will be considered presently. First, however, there is need for further clarification of the reasons why, in public policy, economic optima are scientific fictions rather than quantitatively definable operational objectives. These reasons may be discussed as three interrelated problem areas which will be called here (1) the problem of valuation, (2) the problem of institutional constraints, and (3) the problem of uncertainty.

The importance of these three problems is not confined to water policy, but applies to other policy decisions as well. On the other hand, the conceptual and operational difficulties related to these problems are less for quantitative optimizing when employed in decision making by subsectors—in the above sense—than if used as an aid to policy decisions.[14]

B. The Problem of Valuation

Quantitative optimizing in economics requires that commensurate indicators of value attach to all physical inputs and outputs of resource use, at least in relative terms. Market prices are taken as starting points and bench marks in valuation.

---

[12] Vilfredo Pareto: "Cours d'économique politique," F. Route, Libraire-Éditeur, Lausanne, 1897. An excellent bibliography of welfare economics is appended to E. J. Mishan: A survey of welfare economics, 1939–1959, *Econ. J.*, vol. 70, pp. 197–265, June, 1960.

[13] The scientific use of fictions is discussed in S. V. Ciriacy-Wantrup: Policy considerations in farm management research in the decade ahead, *J. Farm Economics*, vol. 38, no. 5, pp. 1301–1311, December, 1956.

[14] While the conceptual limitations of quantitative optimizing for subsectors are less than for policy decisions, the problems posed by the availability of data for smaller statistical aggregates are frequently greater.

While one may be in favor of judiciously expanding the area of pecuniary evaluation for purposes of policy decision of a more limited scope,[15] one should not forget that there will always remain a large portion of social benefits and cost for which such evaluation would be meaningless. Neglect of such "extramarket" benefits and costs introduces a systematic bias into quantitative optimizing.

Resource economists have stressed the significance of specific extramarket benefits and costs for some time.[16] In 1958, John Kenneth Galbraith called attention to them with special reference to affluent societies, like that of the United States.[17] Emphasis is given to the increasing lack of what is called "social balance" between the products supplied by the market economy and products such as educational and recreational facilities, which are publicly supplied and financed by general taxation rather than sale. The supply of extramarket goods has an inherent tendency to lag behind the supply of market goods because modern advertising and emulation, which are largely responsible for demand shifts in affluent societies, operate exclusively in favor of market goods.

Extramarket goods, however, are only a part of the valuation problem. No less significant is the question, to what extent are present and projected market prices, where they exist, valid and relevant indicators for public policy? To what extent, for example, are the prices of agricultural products under extensive direct and indirect government support relevant for programming public investment in water resources when compared with public investment in health, education, and housing?

At this point, the author does not refer to the price and income effects of the public investment on external economies and diseconomies and on economic fluctuations. Solution of these problems is often difficult, but practical approximations can be found. More basic difficulties are created by the fact that the functioning and the results of the price system itself are profoundly affected by income distribution, market structure, and many public policies inside and outside agriculture. Quantitative optimizing looks at these influences as institutional conditions which, together with the technological ones, are introduced into the optimizing calculus as constraints. Only too often investigators are not aware of the severe limitations which this procedure imposes on the validity and relevance of quantitative optimizing if the results are to serve as a basis for decision making in public policy. The implications are so important that two of them need to be explored further.

C. The Problem of Institutional Constraints

First, when social institutions are used as constraints, they become conceptually indistinguishable from social objectives. In this respect, they are different from technological constraints. As explained elsewhere,[18] in natural-resource policy, changes of social institutions are among the most significant controllable variables and relations. In other words, in natural-resource policy, social institutions must frequently be regarded as means rather than ends. Hence the distinction in econometrics between the part of the model that constitutes the *objective function* to be maximized or minimized and the part that constitutes the constraints describing the structure of the operation and the relations between variables becomes misleading if the conceptual difference between technological and institutional constraints is not sufficiently recognized.

Second, when social institutions are used as constraints in a quantitative optimizing

---

[15] S. V. Ciriacy-Wantrup: Benefit-cost analysis and public resource development, *J. Farm Economics*, vol. 37, no. 4, November, 1955, paper presented before the Annual Meeting of the American Association for the Advancement of Science, Joint Session, Section K (Economics) and M (Engineering), Dec. 27, 1954.

[16] For example, S. V. Ciriacy-Wantrup: "Resource Conservation, Economics and Policies," University of California Press, Berkeley, Calif., 1952; 2d rev. ed., University of California, Division of Agricultural Sciences, Agricultural Experiment Station, Berkeley, Calif., 1963.

[17] J. K. Galbraith: "The Affluent Society," Houghton, Mifflin Company, Boston, 1958.

[18] S. V. Ciriacy-Wantrup: "Resource Conservation, Economics and Policies," especially chaps. 16 to 21. See footnote 16.

calculus, a new optimum must be calculated for each combination of constraints that is considered. The optima calculated for different sets of constraints are then compared. Recently, a whole literature has grown up around this approach, known as *the theory of second best*.[19] This term merely indicates that there is at least one constraint additional to the ones existing in the *Pareto optimum* mentioned above.

The exponents of this theory claim that the major contribution is a negative one: If a deviation from one of the Pareto optimum conditions prevails, the best course of action is not an attempt to attack this deviation and keep all others intact. On the contrary, a second-best solution is usually obtained only by departing from all other Pareto conditions. To apply only a part of the Pareto conditions would move the economy away from, rather than toward, a second-best position. In consequence, the exponents of this theory direct their criticism against what they call *piecemeal welfare economics*.

If this criticism is valid—the author believes it has some merit—does it not point to a basic weakness in the logic of quantitative economic optimizing itself? If one tries to avoid the futility of piecemeal welfare economics and strives for bold changes in the combination of constraints, can one be sure that quantitative optima are comparable in a meaningful way? Is it not unavoidable that such bold changes affect some structural elements of the optimizing calculus, among them especially preferences, technology, and the motivation of human agents in their various functions in the economy? Are we not confronted with a problem of identification, in the econometric sense, on a grand scale?

D. The Problem of Uncertainty

The third set of problems may now be considered. Uncertainty is especially important in those areas of economic decision making where time is a significant element in differentiating alternatives. This is especially true for water economics.

The most important uncertainties—those created by changes of technology, of preferences, and of institutions—increase with time. The probability of such changes is not amenable to precise quantitative measurement. At best, the direction, the relative speed, and the range of such changes can be vaguely projected (Subsec. III). Other uncertainties—for example, the occurrence of drought, floods, and hail storms—do not increase with time, and their probability can be measured quantitatively. Economists frequently refer to these uncertainties as *risk*, in order to differentiate them from the former. Techniques to allow for uncertainties of the latter kind are being developed. The former kind of uncertainty imposes severe limitations on the relevance of quantitative optimizing for policy decisions.

Practitioners of quantitative optimizing who are also competent economists are aware of these limitations.[20] On the other hand, one may wonder whether the enthusiasm of many less critical optimizers might not delay acceptance of formulations that would allow for uncertainty.

In natural-resource economics, allowance for uncertainty can often be made through formulation of the policy objective itself. Such a formulation the author has called *the safe minimum standard of conservation*. In water policy, it can be effectively employed, for example, in the areas of water-quality management and groundwater development.

A detailed explanation of the theoretical development of this concept and its operational application is found elsewhere.[21] Space limitations permit merely stating that one aspect of the economic rationale of the safe minimum standard of conservation as a policy objective is to minimize maximum possible social losses connected with avoid-

---

[19] R. G. Lipsey and R. K. Lancaster: The general theory of second best, *Rev. Econ. Studies*, vol. 24 (1), no. 63, pp. 11–32, 1956–1957. The earlier literature is cited in this article.

[20] Robert Dorfman, for example, in a recent appraisal of operations research, states: "Another important limitation, in which less progress has been made, is that linear programming formulations do not allow for uncertainty" in Operations research, *Am. Econ. Rev.*, vol. 50, no. 4, pp. 575–623, September, 1960.

[21] S. V. Ciriacy-Wantrup: "Resource Conservation Economics and Policies," especially chaps. 17 and 18. See footnote 16.

able irreversibilities. In this respect, the safe minimum standard of conservation may be regarded as a conceptual relative of the well-known min-max solution, or *saddle point*, in a two-person, strictly determined game.[22] This is not to suggest, however, that all aspects of the economics of conservation should be forced into the framework of modern game theory—as "man playing against nature," in an almost literal sense.

E. Implementation of Objectives of Water Policy

It may be concluded from this sketch of the three important problems that quantitative optimizing is most useful for the implementation of policy objectives when prices can serve as relevant indicators of value, when the treatment of institutions as constraints is logical, and when the influence of time and uncertainty is small. In water policy, such usefulness is limited on all three counts.

It is interesting to observe that the more sophisticated practitioners of quantitative optimizing design their models in such a way that the three limitations are involved as little as possible. The resulting models, however, are engineering rather than economic ones. Such models throw light on only a limited segment of policy issues. Examples in water policy are Dorfman's "Simple Valley" and Tolley's "optimal water allocation."[23] The former model is purely hypothetical, designed to illustrate techniques. The latter model deals *ex post facto* with an actual water allocation. Under assumptions most favorable to linear programming, allocative efficiency is improved by no more than 5 per cent, as compared with that accomplished by existing institutional arrangements. Careful comparative studies of this kind are only too rare.

On the positive side, the three problems just reviewed underline the conclusion reached above at the end of Subsec. I: The rules of the game with which water policy is concerned should not be used one by one or set by set and introduced into a quantitative optimizing calculus as constraints. Rather, these rules constitute structured systems that function as wholes, each with particular patterns of change over time. These systems can be studied in structure, functioning, performance, and change over time. They are created by men and can be modified through the legislative, the judiciary, and the executive branches of government, each with a different range over which such modification can be accomplished.

The purpose of these systems is *not* to obtain quantitative optima of welfare at given points in time under given conditions projected for these points in time. Instead, their purpose is to maintain and to increase welfare continuously under constantly changing conditions that at any point in time can be projected only vaguely and are always uncertain with respect to actual occurrence. The implications of this for the type of projection that is useful in water policy will be explained further below (Subsec. III).

It follows from their purpose that to appraise the performance of these systems by introducing arbitrary temporal cross sections of them, either actual or hypothetical, as alternative combinations of constraints in a quantitative optimizing calculus is inadequate. Performance can be appraised only by criteria applied to a whole system as it functions over time. Such criteria need not be pecuniary. For the system that is of special interest for discussion in this section, namely, water law, it has been shown elsewhere that nonpecuniary criteria, such as security against legal, physical, and tenure uncertainties and flexibility in its various legal and economic categories, can effectively be employed.[24]

[22] Johann von Neumann and Oskar Morganstern: "Theory of Games and Economic Behavior," Princeton University Press, Princeton, N.J., 1944. See also Robert Dorfman, Paul A. Samuelson, and Robert M. Solow: "Linear Programming and Economic Analysis," McGraw-Hill Book Company, Inc., New York, 1958, chaps. 15 and 16.

[23] Robert Dorfman, Simple Valley, in G. S. Tolley and F. E. Riggs (eds.): "Economics of Watershed Planning," Iowa State University Press, Ames, Iowa, 1960. G. S. Tolley and V. S. Hastings: Optimal water allocation: the North Platte River, *Quart. J. Economics*, vol. 74, no. 2, pp. 279–295, May, 1960.

[24] S. V. Ciriacy-Wantrup: Concepts used as economic criteria for a system of water rights, *Land Economics*, vol. 32, no. 4, pp. 295–312, November, 1956. Also published in D. Haber and S. W. Bergen (eds.): "The Law of Water Allocation in the Eastern United States," The Ronald Press Company, New York, 1958, pp. 531–552.

To put the foregoing in somewhat different terms, implementation of the objectives of water policy involves programming of water institutions rather than programming (in the technical sense of linear programming) of water resources themselves. Resources programming must be pragmatic; that is, it must regard institutions as means (tools) or as ends (objectives), depending on the purposes of the analysis.[25] In the analysis of water policy, water institutions must be regarded as means.

F. Water-allocation Policy

For analyzing further how the objectives of water policy are implemented through water institutions, water allocation will be dealt with first, and then water development. The meanings of these two terms were defined previously (Subsec. I-A).

In the arid and semiarid regions in the United States and elsewhere, water allocation among uses and users has always been a policy problem under greatly changing conditions affecting the aggregate quantity of water and the quantitative relations between uses. In the beginning of water development—for example, in the gold-rush days in the mother-lode country of California—water allocation was such a problem when all uses, including agricultural, were small, but when industrial use, namely, hydraulic mining, was the dominant one. In present-day California, water allocation is such a problem, even though other uses are a fraction of a quantitatively dominant agricultural use.[26] Since water allocation is always vital for societies in arid and semiarid regions, effective institutional arrangements have been developed which govern it. The result is an allocative system which cannot be understood in its structure, functioning, and performance by taking its legal provisions one by one.

If one wants to undertake an economic appraisal of the allocative performance of this system, one cannot be content with appraising quantitative allocation prevailing at a particular point of time. The purpose of this system is not to optimize water allocation in particular instants. What needs to be appraised is the direction and speed of reallocation. As indicated earlier, incremental improvements in these respects are the main policy objectives. The first step toward such improvements is an understanding of the existing system and of the process of its change. Each state is a laboratory in which this system has developed and is still developing. When individual provisions are modified, such changes must be fitted into the whole system. If the system as a whole is judged inadequate, a better substitute must be offered. Only too often, criticism of water allocation at a particular point of time is voiced by economists without regard to the nature of the decision problems that water allocation poses for policy. Optimizing as a fictional construct is confused with an actual policy objective.

As an illustration, focus may be placed on four facts already mentioned: (1) that water law in the West has developed with and around the growth of agriculture, (2) that agricultural water use is now quantitatively dominant, (3) that nonagricultural

---

[25] On the schism between "orthodox" and "pragmatic" attitudes toward institutions, see F. O. Sargent: A methodological schism in agricultural economics, *Can. J. Agric. Economics*, vol. 8, no. 2, pp. 45–52, 1960.

[26] For comparing different uses quantitatively, two factors are frequently not sufficiently considered: (1) whether conveyance losses are included or excluded for agricultural use and (2) whether water use by steam electric plants—the quantitatively most significant one among industrial uses—is included or excluded for industrial use. In California, for example, agricultural use is 87 per cent, industrial use, 5 per cent, and domestic use, 8 per cent of total use, if conveyance losses are included and steam electric plants are excluded. Agricultural use is 67 per cent, industrial use, 25 per cent, and domestic use, 8 per cent, if conveyance losses are excluded and steam electric plants are included. In terms of water consumption, the former is a more appropriate comparison, provided that double counting is avoided. Some conveyance losses are used via groundwater and counted then. On the other hand, there is considerable interfirm re-use of water both in agricultural and in industrial use. The quantitative extent of such re-use is not known. Steam electric plants use, largely, cooling water that is not usable for other purposes and, in any event, is not consumed. Potentially, of course, most domestic use can be made nonconsumptive (see also footnote 4).

water uses are increasing at a rate greater than agricultural use, and (4) that water is used in agriculture with relatively low average value productivity. Do these four facts indicate that western water law misallocates water to the advantage of agriculture, as is commonly alleged? The author believes the answer must be negative or, in more guarded terms, "not necessarily."[27]

As to the past, only a fraction of present water development would exist if agricultural use had not become dominant. In other words, no large quantities of developed water would be available now for reallocation. As to the present, the relevant criterion, as stated above, is not whether misallocation exists at the moment, but whether continuous reallocation is too slow. As to relative-value productivities of water in different uses, it is the marginal and not the average-value productivity that is the proper basis for continuous reallocation. The figures that are presented in the literature refer to average values frequently aggregating over highly dissimilar situations. If one wanted to be facetious, one could say that the average-value productivity of water in nonagricultural uses would be negative if the whole or a large portion of agricultural water were to be reallocated to nonagricultural uses. Moreover, care must be observed that marginal productivities are taken at comparable stages of water distribution and refinement. Agriculture uses water wholesale and largely unrefined. Domestic use, on the other extreme, is retail and frequently refined. Costs of water distribution and refinement are by far the largest items in the retail water bill.

The indictment of misallocation then boils down to an allegation that the rate of water reallocation from agricultural to nonagricultural uses is too slow. Invariably, this allegation is based on two structural elements of western water law: (1) preference of agricultural over industrial use in most states and (2) priority in time of agricultural use that becomes relevant in those states operating under water laws based on the appropriation doctrine.

It is true that the statutory preference given to agricultural over industrial use is obsolete. This has been pointed out on previous occasions.[28] It was suggested then that statutory preference be eliminated altogether and that it be left to the courts or special water-rights boards to determine which is the higher use in each situation of conflict. On the other hand, the economic significance of agricultural preference and priority is already limited by other provisions in western water law. There are no less than seven of these that are relevant.

First, municipal use has preference over agricultural use. A large part of municipal water use is for commercial and industrial purposes, although it is difficult to differentiate this part statistically from domestic use.

Second, under several state laws, municipal water use enjoys the right of water reservation. This means that municipalities can hold water rights for future rather than present need, without being subject to the due-diligence clause that is such an important part of the appropriation doctrine.

Third, municipalities can acquire agricultural water rights through eminent-domain proceedings. Frequently, the mere threat of such proceedings is sufficient. Owens Valley in California is an example.

Fourth, in most states, many industrial self-suppliers rely on riparian and correlative groundwater rights rather than appropriative rights. They are therefore not affected by the preference and priority clauses of the appropriation law.

Fifth, water organizations such as irrigation districts and the U.S. Bureau of Reclamation that originally developed water under agricultural preferences and priorities now deliver water and hydroelectric power on a large scale for industrial and municipal purposes. Irrigation-district laws have been adapted to permit such deliveries. Often industrial and municipal use takes place on the same acreage where irrigation agriculture has been replaced by urban development. Per-acre requirements of irrigation agriculture are more than sufficient to cover the needs of these "higher" uses. This is an example of what was suggested above, namely, that water

---

[27] On this point, see also Stephen C. Smith: Legal and institutional controls in water allocation, *J. Farm Economics*, vol. 42, no. 5, pp. 1345–1358, December, 1960.

[28] S. V. Ciriacy-Wantrup: Some economic issues in water rights, *J. Farm Economics*, vol. 27, no. 5, pp. 875–885, December, 1955.

development historically undertaken mainly for agriculture is now of direct benefit for other uses.

Sixth, it has been shown that water development itself tends to reduce the economic significance of the superiority of a senior over a junior right under appropriation.[29] This superiority is based mainly on greater security against *physical uncertainty*, as distinct from *tenure uncertainty*, that is, against variability over time of the quantity of water usable under the right due to seasonal or annual variability of natural runoff and groundwater recharge. Storage above and below ground is the major technical possibility of reducing physical uncertainty. After storage capacity has been provided and is managed with a view to reducing physical uncertainty, the relative economic status among senior and junior rights changes without changes in their relative legal status. Priority in time, in conjunction with the quantitative definition of appropriative rights, limits the number of rights that can be served with the regulated flow. Rights exceeding this flow become, economically speaking, less meaningful the better the flow is regulated. For rights that can be served with the regulated flow, the new situation is not greatly different from that prevailing under a water-delivery contract where a limited number of users are equal in right, although the quantities to which their contracts entitle them may differ.

This situation is related to the seventh, and last, point. Water rights are increasingly vested in water organizations such as districts, Federal bureaus, and state water departments. Contracts between water users and these organizations rather than private water rights become the operationally important aspect of water allocation. These organizations do not serve agriculture alone and can reallocate water over time under the terms of the contracts by following appropriate statutory procedures.[30]

Space does not permit more than a sketch of these seven points. Enough has been said to illustrate the proposition that economists should carefully study the actual functioning and performance of water law before this institution is criticized for failing to perform effectively as the major tool of water-allocation policy.

This does not imply, of course, that a given temporal cross section of water law in a given place (state) is perfect and that, over time, water law is adapting to economic change at an adequate rate. Economists should be continuously alert for possible improvements in water law. While economics cannot define quantitative optima of water allocation which the law—as "social engineering"—should aim to realize, economics can explain whether and why the reallocative performance of water law is too slow. This area of water policy is a promising field for cooperative research between the economist and the student of law.

G. Water-development Policy

Water-development policy was differentiated previously from water-development projects. The economics of the latter has been discussed elsewhere.[31]

Water-development policy, like water-allocation policy, becomes operative mainly through water law in a broad sense. But different aspects of the law are involved. Among these several may be mentioned. First, there is the blend in state water laws between riparian and appropriation doctrines; this blend is significant because the riparian doctrine is relatively less favorable to water development, with one important exception to be discussed presently. There are, second, the laws concerning water reservations for future development by particular uses and regions; municipal reservations and the area-of-origin legislation in California are examples. Third, there are the antipollution laws that affect the broad field of water-quality management; this field becomes increasingly significant as water development is intensified and

---

[29] S. V. Ciriacy-Wantrup: Concepts used as economic criteria for a system of water rights. See footnote 24.

[30] Stephen C. Smith: Resource policies and the changing West, *Land Economics*, vol. 36, no. 1, pp. 26–34, February, 1960.

[31] S. V. Ciriacy-Wantrup: Cost allocation in relation to western water policies, *J. Farm Economics*, vol. 36, no. 1, pp. 108–128, February, 1954. Benefit-cost analysis and public resource development, pp. 676–689. Philosophy and objectives of watershed development, *Land Economics*, vol. 35, no. 3, pp. 211–221, August, 1959.

natural purification processes are overloaded. Fourth, there are the laws establishing and regulating water organizations; the problems of coordinating these organizations through "superdistricts" has become especially acute for water development. These problems are related to a fifth aspect, namely, the need for integrating groundwater development with that of surface water.

In view of the impossibility of discussing all these aspects in the space assigned and of the increasing significance of groundwater development indicated above (Subsec. I), the last two aspects may be selected here for more detailed discussion. This discussion is intended to be merely illustrative of the type of problem that is faced in water-development policy.

Groundwater law has been an important factor in permitting overdevelopment (in the physical sense of overdraft) if it is based on riparian ideology such as the correlative-rights doctrine, developed mainly in California. This is the exception to the rule that the riparian doctrine is less favorable to water development than the appropriation doctrine. On the other hand, after adjudication of a groundwater basin, correlative rights acquire the economic characteristics of appropriation rights. They become quantitatively defined, transferable, and secure against tenure uncertainty. Adjudication can be used to "freeze" groundwater development by limiting it to the safe yield. Appropriation is used in several western states for the same purpose.

Adjudication, however, is not a sufficient condition for maintaining the economic superstructure based on many years of groundwater overdraft. Such maintenance is impossible without importation of surface water. Rather, adjudication may be regarded as a necessary condition for such a purpose. This is important because, at present, adjudication is an expensive and time-consuming process even under the legal simplification of the *court reference procedure*. This process is well illustrated by the two important reference cases in Los Angeles County, Calif.—"Raymond Basin" and "West Coast"—in which the court reference procedure was tested.[32]

At first glance, adjudication of correlative rights is not a necessary condition for integration of ground- and surface-water development. All groundwater rights could be transferred without adjudication to the water organization that imports the surface water. This water organization would then allocate quantities of ground or surface water or combinations of the two to individual users. Where pumping was done by individual users, they would be appropriately compensated, since overdraft during extended periods of time is an essential part of integrated development of groundwater with surface water. The most important resource of a groundwater basin, namely, its storage capacity, could then be utilized to counteract seasonal and cyclical variability of precipitation without a complex system of price and other inducements that would become necessary if all rights to ground and surface water were not held by the same water organization.

There is some question whether private groundwater rights will be surrendered voluntarily in this way without prior adjudication. Groundwater rights are valuable private-property rights because local groundwater is generally much cheaper than imported surface water. Owners therefore will not be willing to surrender these rights without quantitative definition and adequate compensation. Such definition and compensation also become necessary if private groundwater rights are taken by the water organization through eminent-domain proceedings. If, for these reasons, adjudication of groundwater rights is looked upon as a necessary condition for integrated ground- and surface-water development—with or without surrendering them to

---

[32] California, Report of Referee, Department of Public Works, Division of Water Resources, Sacramento, 1943. In the Superior Court of the State of California in and for the County of Los Angeles, no. 1323. City of Pasadena, a municipal corporation, Plaintiff, v. City of Alhambra, a municipal corporation, *et al.*, Defendants. For the decision in this case, see *Pasadena v. Alhambra*, 33 Cal.2d 908, 207 Pac.2d 17 (1949); certiorari denied, *California-Michigan Land & Water Co. v. Pasadena*, 339 U.S. 397 (1950). California, Draft of Report of Referee, Department of Public Works, Division of Water Resources, Sacramento, February, 1952. In the Superior Court of the State of California in and for the County of Los Angeles, no. 506806. California Water Service Company, a corporation, *et al.*, Plaintiffs, v. City of Compton, *et al.*, Defendants; California Water Service Company, a corporation, *et al.*, Plaintiffs, v. Alexander Abercromby, *et al.*, Defendants.

a water organization—it would seem worthwhile to make even greater efforts to simplify existing procedures with a view to making them less expensive and time-consuming.

The problem of private groundwater rights is only one aspect of a water-development policy aimed at integration of ground and surface water. Another aspect is the establishment, regulation, and supervision of the type of water organization that actually does the integrating. It is fairly clear that water organizations of sufficient size are needed to cover the groundwater basin that is to be managed and to import and distribute adequate quantities of surface water. If present water organizations are not of sufficient size, the question arises whether they should amalgamate or whether they should federate and form a superdistrict within which each organization would still maintain a degree of independence. The trend is toward superdistricts. The most outstanding example in terms of size is the Metropolitan Water District of Southern California, in which the municipal water organization of Los Angeles is the most important member. There are others of this kind in California and elsewhere.

What is the implication of such superdistricts for the state that establishes, regulates, and supervises them? Some students believe that water development in the West is an undertaking too big for even the largest superdistricts and that it should be undertaken by the state. Others believe that superdistricts have become so big financially and politically that the state should step into direct water development through planning, constructing, and operating projects in order to preserve balance in the rate of water development between the various regions of the state. Still others believe that state-wide water development should be left to the interaction of water organizations, including superdistricts; that the state should stay out of direct water development; and that in such a less involved position it could more effectively play its important role as the locus of water-development policy, in the sense in which the word is used in this discussion.

These and other views on appropriate institutional arrangements for integrated ground- and surface-water development are in conflict. This conflict is acute in California. It is mentioned, not in order to take sides, but to suggest that some of the crucial issues of water development are in this area. With water-development policy, as with water-allocation policy, the functioning, performance, and change over time of an institutional system are involved, rather than quantitative optimizing at given points in time under projected conditions.

Research in this area need not return to the descriptive approach to water institutions. As stated earlier (Subsec. I), there is now considerable historical evidence on water institutions that can be approached analytically by the social sciences. An analysis of the structure, functioning, performance, and change over time of superdistricts is an example. Such studies can be extended to other countries where interesting material is available, for example, for the Metropolitan Water District of the Ruhr, constituting Germany's industrial heartland. This district has been in successful operation since 1913. Some of the same problems now faced in the United States, such as the relation between superdistricts and state, have been solved there. It would be unfortunate if the current emphasis in economic research on public investment in water-development projects should lead to the neglect of the crucial issues of water-development policy.

III. ECONOMIC PROJECTIONS IN WATER POLICY

A. Significance of Water-requirement Projections

Objectives and implementation of water policy always consider *future* conditions. Most attempts to provide a quantitative basis for water policy emphasize projections of future water requirements. The President's Water Resources Policy and Materials Policy Commissions have made such studies.[33] More recently, the Select Committee

---

[33] U.S. President's Water Resources Policy Commission, "A Water Policy for the American People," 1950, vol. 1. U.S. President's Materials Policy Commission, "Resources for Freedom," 1952, vol. 5 (Selected Reports to the Commission).

on National Water Resources of the U.S. Senate has made a thorough investigation of this problem.[34] Numerous individual authors have used such projections with or without modification.[35] For individual states, there are detailed regional projections of water requirements.[36] Every benefit-cost analysis of water-development projects contains a section on future water requirements.

Some of these projections are extrapolations of past trends in water use. Others focus on projections of the consumption of agricultural products. Water requirements are then determined on the basis of trends in average yields and water duties per acre. Projections of consumption of agricultural products are often interesting statistically.[37] Usually, a regression to gross national product or national income is employed. These, in turn, are projected through trends in population, labor force, and average productivity of labor, capital, or other inputs.

The meaning of water requirements in these projections is rarely identical with that of water demand in the professional language of the economist. The first aim, therefore, is to clarify the relation between these two concepts and to consider their advantages and disadvantages: how significant each is in providing needed information and what difficulties are encountered in applying them in water economics.

The second aim is to review the logical validity of projecting water requirements into the more distant future, say, for a decade or more. The intent is not to discredit current projections of water requirements or to suggest that one might dispense with long-run economic projections in water policy. Rather, we shall inquire into possible differences in the logical validity of different kinds of projections. If it can be shown that there are such differences, greater weight in policy decisions and greater statistical effort should be given to those kinds of projections the logical validity of which appears relatively greater.

The third aim is to consider implications for water policy. Superiority of a projection in logical validity does not necessarily also mean superiority in usefulness for policy. In view of the objectives and implementations of water policy discussed earlier (Subsecs. I and II), what can one conclude about the types of projections that are relevant for such policy?

B. Water Requirements and Water Demand

As just stated, existing projections of water requirements are neither conceptually nor empirically identical with projections of water demand. In the literature, besides the word "requirement," similar words with a physical-engineering connotation are

[34] U.S. Congress, Senate, Select Committee on National Water Resources, Water resources activities in the United States: land and water potentials and future requirements for water, Committee Print no. 12, 86th Cong., 1st Sess., 1960, S. Res. 48. U.S. Congress, Senate, Select Committee on National Water Resources, Water resources activities in the United States: estimated water requirements for agricultural purposes and their effects on water supplies, Committee Print no. 13, 86th Cong., 2d Sess. (i.e., 1st Sess.), 1960, S. Res. 48. U.S. Congress, Senate, Select Committee on National Water Resources, Water resources activities in the United States: future needs for reclamation in the Western States, Committee Print no. 14, 86th Cong., 1st Sess., 1960, S. Res. 48. See also Ref. 20 in Sec. 26-I.

[35] Sherman E. Johnson: Prospects and requirements for increased output, *J. Farm Economics*, vol. 34, no. 5, pp. 682–694, December, 1952. Carl P. Heisig: Long-range production prospects and problems, *J. Farm Economics*, vol. 35, no. 5, pp. 744–753, December, 1953. Colin Clark: Afterthoughts on Paley, *Rev. Economics and Statistics*, vol. 36, no. 3, pp. 267–273, August, 1954. Edward S. Mason: Afterthoughts on Paley: a comment, *Rev. Economics and Statistics*, vol. 36, no. 3, pp. 273–278, August, 1954. John D. Black: Resources needed in American agriculture, *J. Farm Economics*, vol. 39, no. 5, pp. 1074–1086, December, 1957. Edward A. Ackerman: Water resource planning and development in agriculture, *J. Soil Water Conserv.*, vol. 14, no. 3, pp. 112–117, May, 1959.

[36] California, State Water Resources Board, Water utilization and requirements of California, *State Water Resources Board Bull.* 2, June, 1955.

[37] National Bureau of Economic Research, Conference on Research in Income and Wealth: "Long-range Economic Projection: Studies in Income and Wealth," vol. 16, Princeton University Press, Princeton, N.J., 1954.

employed. Such words are "consumption," "utilization," "needs," "use," "potential use," and "ultimate use." Commonly, however, the word "demand" is employed interchangeably, and the identity of projected demand with projected "requirements," "use," "consumption," and so on, is clearly implied. In some cases, such identity is explicitly stated. Nevertheless, projections of water requirements have two major characteristics which differentiate them from demand projections.

In the first place, requirement projections do not separate demand and supply conceptually or statistically. This lack of separation has operational advantages because, as shown above (Subsec. I), such separation is more difficult in water economics than in the economic analysis of most commodities. The conceptual defect is not too great if supply can be regarded as fixed or "ultimately" fixed and if demand is not regarded as the "end" toward which policy is oriented, as, for example, in military logistics, where orientation comes from outside of economics. On the other hand, if demand is to serve as a principle of orientation for water policy, that is, to help in planning on the supply side, problems of demand and supply need to be separated conceptually and, in empirical investigation, variables pertaining to demand ("ends") must be differentiated from those pertaining to supply ("means"). To illustrate: the great increase of water use in the 17 western states in the last few decades was based, as it is known (Subsec. I), on groundwater development. For this development, changes on the supply side—changes in technology and resulting decreases in costs per foot of lift of well drilling and pumping—were largely responsible. Overdraft on groundwater, on the other hand, has led to increased lift and costs per unit of water pumped. Furthermore, in many basins, groundwater development is identical with groundwater mining. Extrapolating water requirements on the basis of past and present groundwater use is therefore especially hazardous.

In the second place, projections of water requirements do not consider sufficiently, if at all, the functional relation—and its changes over time—between prices (unit values) of water, on one side, and physical quantities, on the other. This relation is the foundation of demand and supply concepts. On theoretical grounds, supported by somewhat meager evidence, it may be assumed that the price elasticity of the demand for water is small within the relevant range for each major water use such as domestic, industrial, agricultural, and recreational. In spite of this small price elasticity of water demand, insufficient consideration of the price-to-quantity relation leads to an upward bias in projections of water requirements. The explanation is connected with the elasticity concept itself. Two characteristics of the concept are pertinent here.

The elasticity concept is a proportional one. There is little doubt that *proportionally* the upward changes of water prices that can be expected in most parts of this country and the world will be large. In the past, water has been used without charge or with only a nominal charge per unit. It has already been mentioned (Subsec. I) that there is an increasing trend in that portion of water costs covered by prices in relation to the portion that is covered through various forms of taxes, special assessments, and fees. But this trend starts at low—sometimes zero—levels. Increases will be large proportionally. With such large proportional changes of prices, even small elasticities of water demand, say, around -0.10, lead to considerable absolute changes in quantities.

Further, the elasticity concept is an instantaneous one. Price elasticity may be quite different under "long-run" and under "short-run" assumptions. Given a sufficient period of time, price elasticity of demand for an aggregate of water uses may be considerable if quantitatively important uses are priced out of the market. For example, rising water prices may force a curtailment of irrigated agriculture in favor of domestic, industrial, or recreational uses. In agriculture, large quantities of water are used with relatively low average value productivities. In other words, projections of water requirements already imply policy decisions with respect to water pricing and water allocation among major uses. Since water-requirement projections in most studies are to serve as a basis for policy decisions, this point is significant.

The foregoing does not imply that separation of demand and supply in water economics is a simple matter, conceptually or empirically. Again, the demand for groundwater may serve as an illustration. Water in agriculture is an input, and demand for water is derived demand. A meaningful demand function for one input requires reference to prices and quantities of complementary and competing inputs. Electric power is an input that is used for many purposes besides groundwater pumping. If diesel or gasoline power is used for pumping, it frequently comes from power take-off of tractors which are used for many other purposes. Technology in water application is closely related to the technology of applying other inputs, for example, fertilizer. In all these cases, changes of prices and quantities of other inputs affect both water supply and water demand at the same time.

The difficulties of separating water supply and demand are especially great if one considers aggregates. The explanation was given when the significance of self-supply and the nature of the water "market" were discussed (Subsec. I). A question was raised about the identification of a "water industry" to which aggregate demand and supply concepts could be applied.

Can separation of water demand and supply be simplified by regarding overall regional supply as fixed or ultimately fixed? This approach is frequently adopted for the input "land." One may also note that the concept *ultimate use* occurs with increasing frequency in regional-water-requirement projections.[38]

Simplification through assuming regional water supply as fixed or ultimately fixed does not appear helpful. Water is a mobile input. Modern technology tends to increase the economic significance of physical mobility. The history of water institutions indicates considerable adaptability to the physical facts and the economic potentialities of regional water mobility.

Modern technology tends to make the assumption of ultimately fixed regional water supply questionable on several other grounds. Purification of mineralized waters is already economical under special conditions; it is only a question of time until existing knowledge makes more general application economical. The implications for domestic and industrial uses in regions with economical access to the ocean—for example, for southern California—are potentially great.[39] Modification of weather conditions with the objective of increasing precipitation and of reducing evaporation can no longer be regarded merely as a scientific dream. Management of the vegetative cover of watersheds to influence evapotranspiration, infiltration, and snow storage affects the volume, quality, and timing of water yield. Technological developments in geological exploration, well drilling, and pumping are constantly increasing groundwater supplies. Control over new sources of energy, at present through fission of heavy elements and in the future through fusion of light elements, will accelerate all these developments—and greatly increase the problems of waste disposal.

Must one conclude, then, that the conceptual and empirical difficulties in identifying water demand are so great that one is forced, after all, to rely on water-requirement projections? Before this question can be answered the logical validity of economic projections and its implications for water policy must be considered.

C. Validity of Economic Projections

Projection over time is a special problem of inductive inference. In such projections, all undetermined cases of a hypothesis or a system of hypotheses—a theory—are future cases. A projection is then called *prediction*.

Predictions differ with respect to the degree of articulation in the formulation of hypotheses and with respect to the degree of quantification. On the basis of these differences, projection is sometimes differentiated from *forecast* and *estimate*. Without endorsing such differentiation, it may be noted here that projections, forecasts, and estimates must be classed as predictions as far as the criteria for their logical validity are concerned. On the other hand, a prediction is not a prophecy. The only criterion that can be applied to a prophecy is the eventual outcome. In contrast, the eventual outcome of an individual case is not, in itself, a sufficient criterion of validity for a

---

[38] California, State Water Resources Board, *op. cit.*
[39] Present estimates of costs of purification of around $120 per acre-foot (not including costs of distribution) would exclude nearly all irrigation uses.

prediction. A prediction, in order to be valid, requires tested theories in the sense of lawlike generalizations.

Criteria to determine whether a generalization deserves or does not deserve the designation lawlike, and therefore can be used for prediction, have occupied formal logic for a long time. Recently, Nelson Goodman has suggested a criterion which has attracted some attention in the philosophy of science.[40] His contribution is of interest for the present purpose.

In short, for establishing a criterion for lawlike, Goodman proposes not only to examine the predictive record of a theory, but also to investigate how well this record is *entrenched* in scientific language. Like Hume, Goodman appeals to past recurrences, but to recurrences in the explicit use of terms as well as to recurrent features of what is observed.

Such an approach may be called *semantic* because language becomes the medium through which predictive theories are separated from those that are not. Such separation does not, of course, imply that theories with predictive power will be confirmed by all or any future observations. As just emphasized, facts that are as yet unobserved cannot be used as criteria for predictability.

A semantic criterion for predictability may appear weak. Basically, however, Goodman's concept of lawlike is similar to the concept of a law in modern physics and in other natural sciences. Both concepts of a law are merely stochastic.

Goodman is not specific about the meaning of the terms "entrenched" and "scientific language." Apparently, he is thinking largely about the natural sciences, where agreement among "competent" men on the meaning of these terms may not be too difficult. In the social sciences, on the other hand, there are few theories that could be called entrenched through agreement of competent men. This difference between the natural and the social sciences does not render the Goodman criterion unhelpful for the present purpose. Rather, this difference indicates that the term *theory* does not have the same meaning in the natural and social sciences. It is increasingly recognized—among economists, at any rate—that most theories in the social sciences are, strictly speaking, models.[41]

In studying the linguistic terms of lawlike generalizations, it becomes apparent that a high degree of quantification is not a requirement. On the contrary, most lawlike generalizations are phrased on a rather low quantitative level. For example, lawlike generalizations about the relations between variables refer to less quantified characteristics—such as general direction of change (increase, decrease), ordinal characteristics of magnitude of changes (greater, smaller, equal, proportional, and so on), ordinal characteristics of temporal distribution of changes (earlier, later, simultaneous), and tendencies toward correction (equilibrium) or cumulation (disequilibrium)—rather than to cardinal characterization of parameters. On the basis of the Goodman criterion, it may be concluded that predictive power of a theory and degree of quantification are not correlated positively.[42]

---

[40] Nelson Goodman: "Fact, Fiction, and Forecast," Harvard University Press, Cambridge, Mass., 1955, especially pp. 63–120. R. Carnap: On the application of inductive logic, *Philosophy Phenomenol. Res.*, vol. 8, no. 1, pp. 133–147, September, 1947. For the classical views, see John Stewart Mill: "A System of Logic," Longmans, Green & Co., Ltd., London, 1843, new impression, 1947, especially book 3, chap. 3.

[41] H. A. Simon: "Models of Man," John Wiley & Sons, Inc., New York, 1957. H. G. Papandreou: "Economics as a Science," J. B. Lippincott Company, Philadelphia, 1958.

[42] There is, of course, no implication here that measurement is not necessary in science in order to ascertain "reasonable agreement" between theory and observation. Measurement, however, is commonly thought of as significant, not merely for testing, but also for the discovery of theories. There is some doubt whether, in the discovery of new theories, that is, in the creative act of innovation, degree of quantification is so important as is generally supposed in contemporary literature. Support for this doubt has recently come from a historian of modern physical science. Thomas S. Kuhn: The function of measurement in modern physical science, Conference on the History of Quantification in the Sciences, Nov. 20–21, 1959, New York, sponsored by the Joint Committee on the History of Science, National Research Council, and Social Science Research Council. Processed.

D. Implications for Water Policy

What are the implications of this conclusion for the issue under discussion, namely, whether demand or requirement projections should be emphasized in water policy?

If by demand theory one means the broad generalizations in the Marshall-Henderson-Hicks formulation, they will pass Goodman's predictivity test.[43] By the authors themselves and by most economists, these generalizations are referred to as "the" demand laws. Yet their language is couched in terms that refer only to direction of change, that is, increase or decrease of prices and quantities. Elasticities, if mentioned at all, are stated in terms of ordinal characteristics.

On the other hand, there is little possibility of passing the Goodman predictivity test if, by demand theory, one means a demand function with a quantitative characterization of parameters that would allow demand projections comparable, in numerical precision, with existing projections of water requirements.

Projections of water requirements may be regarded as a species of economic model. Even the best of this particular species is incomplete in the sense that the most significant dynamic variables, namely, changes of technology, preferences, and institutions, are not included, or only to a small extent. Besides, models are not substitutes for theories. Models are designed for better understanding of individual cases, past or future, and may be used for testing theories in the process of validation. Models, however, carry no predictive power.

To question the validity of water-requirement projections in terms of criteria for predictability implicitly raises a question of relevance. Projections based on lawlike generalizations—for example, on the demand laws—would be on a quantification level far lower than that of existing water-requirement projections. It may be asked, therefore, what quantitative level of projections is relevant for water policy. The answer to this question was implied in the previous discussion of the objectives and the implementation of water policy.

It was emphasized (Subsec. II-E) that the purpose of water institutions (through which water policy is largely implemented) is *not* to obtain quantitative optima of welfare at given points in time under given conditions projected for these points in time. Rather, their purpose is to maintain and to increase welfare continuously under conditions that at any point in time can be projected only vaguely and are always uncertain with respect to actual occurrence. Thus demand projections, expressed in terms of general direction of demand changes (increases, decreases), ordinal characteristics of increases and decreases of demand for different water uses (greater, smaller, etc.), and temporal distribution of demand changes (earlier, later, simultaneous) for different uses, would still be useful for water-allocation and water-development policy.

In water-allocation policy much can be learned from observing historically the allocative performance of water institutions in relation to changes of demand for different uses. The result of this comparison can be used for institutional change. Changes in the allocative system can also be made in anticipation of the above general characteristics of demand changes that can be projected. On the other hand, precise quantitative projections of water requirements for different uses are not relevant for water-allocation policy. Such projections merely beg the question.

In water-development policy, as in allocation policy, the performance of water institutions in response to demand changes can be appraised for the past, and the results of this appraisal used for institutional change. Here again, changes in the institutional system can be made in anticipation of those general characteristics of demand changes that can be projected. On the other hand, to use temporal cross sections of such institutions as constraints in a model projecting water requirements quantitatively for the year 2000 and beyond is likely to be misleading for water-development policy.

---

[43] Alfred Marshall: "Principles of Economics," 8th ed., Macmillan & Co., Ltd., London, 1930, especially book 5. H. D. Henderson: "Supply and Demand," Harcourt, Brace & World, Inc., 1922, chap. 2. J. R. Hicks: "A Revision of Demand Theory," Clarendon Press, Oxford, 1956.

IV. ORGANIZATION AND COORDINATION OF WATER POLICY IN THE UNITED STATES[44]

A common theme pervading the vast literature on the organization and coordination of water policy in the United States is the need for agency reorganization and better coordination between agencies. In accordance with the present emphasis on the dynamic character of water institutions in relation to economic change, it will be helpful to discuss the evolution of government responsibility in water policy. This discussion is of necessity historical and descriptive, rather than analytical, as in the preceding discussion.

A. The Federal Government

Any discussion of organization and coordination of water policy in the United States must be realistic in terms of the constitutional framework within which government is organized. Two aspects of this framework are especially relevant here.

First, the Constitution divides Federal responsibility in the water field, as in most others, between the legislative, the executive, and the judicial branches. State constitutions have similar provisions.

Second, the Constitution divides government responsibility between Federal and state authorities. Under the Constitution, a major portion of the responsibility for water policy, as interpreted previously, is reserved to the states. The Federal government is the locus of water policy mainly for actions in which international and interstate relations are involved. For water policy, the power to regulate interstate commerce is the most important of those reserved to the Federal government.[45] In the main, Federal water policy has been and is connected with Federal investment in water projects undertaken under a gradually broadening interpretation of the commerce clause. These projects loom so large in the water economy, however, that Federal policies with respect to them have wide repercussions.

1. Navigation. Until the turn of the century, Federal agencies limited their activity to the improvement of navigable streams.[46] The U.S. Army Corps of Engineers continues to be the primary agency responsible for planning, constructing, and operating these projects. However, the growing importance of multiple-purpose river development and the interrelation between the use of water for transport and other purposes have inevitably drawn into this area other agencies, such as the U.S. Bureau of Reclamation, the U.S. Coast Guard, and the U.S. Geological Survey.

2. Flood Control. Until comparatively recently, the Constitution has been interpreted in such a way as to restrict Federal activity in the undertaking of "internal improvements,"[47] with the exception of navigation improvements justified by the previously mentioned commerce clause. The cumulative increase in flood damage, both in terms of lives lost and personal property damaged, was a major factor which brought about a gradual broadening in constitutional interpretation. Thus, in 1916, when California and the states bordering the lower Mississippi were subjected to major floods, Congress responded by authorizing the nation's first Federal flood-control

---

[44] In assembling this material, help from Mr. Gardner Brown, Jr., Research Assistant, Giannini Foundation of Agricultural Economics, is acknowledged.

[45] United States, Constitution, art. I, sec. 8, par. 3.

[46] Up to 1910, inclusive, treasury expenditures for navigation projects amounted to one-half billion dollars. In the following 44 years, appropriations for river and harbor improvement and maintenance and for canals were increased ninefold. These figures serve to show the relative difference of Federal financial involvement in the two periods. See U.S. Commission on Organization of the Executive Branch of the Government, "Water Resources and Power: A Report to the Congress, 1955," vol. 1, p. 77.

[47] U.S. Congress, Senate, Select Committee on National Water Resources, Water resources activities in the United States: flood and flood control, Committee Print no. 15, 86th Cong., 1st Sess., 1960, S. Res. 48, p. 10. This source has a brief summary of the historical legislative developments with respect to flood control, pp. 10–15.

projects.⁴⁸ Congressional appropriations for the works were contingent upon local contributions amounting, for example, to 50 per cent for the Sacramento and 25 per cent for the Mississippi.

Congress responded to the devastating floods of 1927 with the Flood Control Act of 1928. Financial responsibility was changed, because local levee districts along the Mississippi were unable to contribute their allocated share of construction costs. For these reasons, the Federal government assumed a higher degree of financial responsibility.⁴⁹

The floods of the mid-thirties, the need for public works projects during the Great Depression, and the change in the attitude of Congress as to the Federal role in this realm are three elements which led to the Flood Control Act of 1936. In it, Federal policy was made fairly explicit: the Federal government should participate with states and other local legal bodies in flood-control projects if the benefits "to whomsoever they may accrue" exceed costs. The U.S. Army Corps of Engineers was to be the principal agency responsible for preparing and carrying out flood-control schemes on the major rivers, the Department of Agriculture limiting its scope to that of retarding water flow and preventing soil erosion in the upper reaches of streams by land-treatment measures.

The Flood Control Act of 1944 expanded the government's policy, particularly in the area of resolving interagency and Federal-state conflicts. In this Act, the Corps was given responsibility for operating flood-control storage behind federally constructed structures. The Secretary of the Interior became responsible for marketing power produced at Corps projects, and all plans were to be reviewed by states and all interested agencies that would be affected.

The purview of the U.S. Department of Agriculture was expanded by the enactment of the Watershed Protection and Flood Control Act of 1954 (Public Law 566). The Act permitted this agency to construct engineering works for controlling floods on watersheds of less than 250,000 acres. In so far as these Public Law 566 projects are single-purpose, cost-sharing responsibilities, they hardly differ from the larger Corps projects.

3. Irrigation. Federal interest in irrigation dates back to 1902, when Congress enacted the Reclamation Act, giving the U.S. Bureau of Reclamation responsibility for undertaking irrigation development. The main objective of this legislation was to stimulate settlement in the arid lands of the West. Expectations were that Federal irrigation programs, by inducing the formation of secondary and tertiary industry, would form the basis for regional development. Federal policy with respect to irrigation projects has become a potent factor affecting state water policies. The most important aspects of Federal policy in this respect are related to the allocation and the repayment of costs.

Statutes and executive orders have made the allocation of total construction costs to various project purposes the basis of repayment. According to the same institutions, costs allocated to some purposes (for example, flood control) need not be repaid at all; some purposes (for example, irrigation) need not pay the interest portion; other purposes (for example, municipal water) must repay all costs allocated to them, including interest; and for still other purposes (for example, power), allocated costs include costs which other purposes cannot repay.⁵⁰

Generally, it is in the interest of both equity and economic efficiency for beneficiaries from public resource development to pay for the benefits received, provided such benefits are practically assessable and provided that enough incentive is left for beneficiaries to participate in resource development.⁵¹ Payments under these provisions are the

⁴⁸ Previous to 1916, there had been authorization for projects providing flood control, but these were called "navigation improvements."

⁴⁹ However, as the Act was originally conceived, these projects were to be carried out under the spirit of Federal-local "partnership" policy, with the local benefactors paying at least 50 per cent of the project costs.

⁵⁰ In the Act of 1937 authorizing the Central Valley Project of California, power is designated as "a means of financially aiding and assisting other functions." This role of power goes back to the Reclamation Act of 1906.

⁵¹ S. V. Ciriacy-Wantrup: Cost allocation in relation to Western water policies, *op. cit.*

best guarantee that the determination of economic feasibility will receive scrutiny by state and local governments and by private groups of beneficiaries.

Such a principle would also be an alternative to policies designed to reduce speculation and "unearned increment" of private income. The major Federal policy in this respect is the so-called *160-acre limitation*. Actually, the quantitative definition of acreage limitation in irrigation projects is, to a large extent, left to the discretion of the Secretary of the Interior. By statute, 160 acres is mentioned as a maximum. It can be—and has been—reduced by the Secretary to as low as 40 acres for some projects. On the other hand, the limitation has been interpreted in such a way that man and wife may operate 320 acres. Sometimes the pressure of economic change forces a change in the rulings of the Secretary. The Orland Reclamation Project in California, for example, was operated under a limitation of 160 acres from 1907 to 1916, 40 acres from 1916 to 1953, and 160 acres since 1953.

Another problem which in Federal policy is closely linked to cost allocation is that of pricing water and power. Rates for both products are either directly or indirectly connected to cost allocation by legislation. Two factors are largely responsible. First, net revenues obtained from the sale of water and power are by far the most important—and thus far have usually been the only—financial source of repayment. Second, in view of the statutory differentiation between reimbursable and nonreimbursable costs, all reports on public multiple-purpose projects contain implications or direct suggestions about rates "necessary" to support cost allocations to water and power or to make these purposes "self-supporting."

4. Municipal and Industrial Water. The provision of municipal and industrial water supplies from Federal projects is thus far primarily a byproduct of multiple-purpose development. Federal activity in this area is limited to carrying out investigations of future needs and to formulating project plans incorporating municipal and industrial water-requirement projections.

The quality aspects of water have been attacked through the Water Pollution Control Act of 1948. The U.S. Department of Health, Education, and Welfare is the Federal agency charged with responsibility for pollution that passes state lines. However, agreement is necessary between the state where pollution originates and the state that brings the complaint. The Act is likely to stimulate state compacts (Subsec. IV-B-3). In 1956, the Act was amended to include Federal grants to municipalities of $50 million per year on a matching basis for building sewage facilities. In the future, this area of Federal activity is likely to become of far greater significance than it has been in the past. Proposals have been made to extend Federal authority over pollution to all navigable waters.

5. River-basin Development. In 1907, President Theodore Roosevelt made the often-quoted statement that "a river is, from its headwaters to its mouth, a single unit and should be developed with this concept in mind." Many proposals have been made to carry out this idea by setting up, through Federal legislation, unified planning-administrative-construction agencies for the major river basins of the country, modeled after the Tennessee Valley Authority. In view of existing weakness in intra-Federal, state-Federal, and interstate coordination, the TVA constitutes a valuable experiment. Moreover, the existence of the TVA and proposals for its extension have prompted some Federal agencies to cooperate with each other and with the states.

Whether the TVA is a model that should be used in other river basins would seem to depend on whether there are no better alternatives by which coordination in water policy could be achieved. At best, the alternatives offer solution for individual geographic areas only. They do not solve the essential problems involved in the coordination of water policy for the country as a whole. On the other hand, if these problems could be solved, separate authorities for each river valley would be unnecessary.

6. Coordination in the Federal Executive Branch. With the expansion of Federal interest and participation in water-resources development, no less than 21 commissions have been appointed to study the status of United States water resources. All these commissions have made recommendations for better coordination to eliminate waste and duplication.

Proposals were generally one of two types: (1) placing all activities related to water in one agency and (2) establishing a board to review all projects. These proposals have been analyzed in detail and improvement offered, in an earlier publication.[52]

7. Coordination in the Federal Legislative Branch. Because of the constitutional division of responsibility referred to earlier, coordination within the executive branch of the government, even if effectively realized, appears insufficient to achieve a well-coordinated water policy. Since it seems probable that water policy will continue to be formulated through individual acts, legislative as well as executive coordination is needed. For this purpose, it would seem feasible to develop still further a tendency toward consolidation of legislative committees, a tendency already manifest in Congress. A single committee for flow resources, including water, in each house of Congress could be a strong force for coordination of resource policies.

A further need in Congress is more adequate professional staffing of committees, preferably with specialists drawn from a permanent pool like the Legislative Reference Service. While the activities of such a staff might involve some overlapping with the planning agency of the executive branch, a certain amount of overlapping seems neither avoidable nor regrettable in the functioning of a democratic government.

B. State Governments

State governments, as indicated earlier, are constitutionally the primary locus of water policy. As emphasized in previous discussions in this section, the most important area of state policy is water law. This includes legislation, administrative regulations, and judicial decisions. Furthermore, local governments, including public districts, are established, regulated, and supervised by the states. For example, zoning and similar regulations are the domain of local governments based on state enabling legislation. So far, the states have only to a very limited extent undertaken the construction of water projects. This may be different in the future. California is a state in which a break with the past is under way.

1. Proposals for Better Coordination within State Governments. Generally, coordination of water policy within state governments is weak. This results from the single-purpose approach to interrelated resource problems and from fragmentation of administrative functions into separate agencies. As an illustration, in earlier years, problems of water pollution were directly related to public health; thus authority to regulate pollution rested in the public health departments. As concern for the pollution problem grew, pollution abatement in some cases was fractured into its component causes and effects, the fish and game departments regulating pollution affecting fish, the agencies concerned with mining regulating pollution as a result of these activities, the highway departments regulating pollution caused by misuse of state highway drains, and the agricultural agencies controlling pollution injurious to livestock.

Three recent examples illustrate possible ways in which greater coordination within a state might be achieved. The Department of Water Resources in California was established after considerable study. It was argued that a separate department is best if water resources are the most important aspect of the total resource program. However, in 1961 the State Legislature created a comprehensive Resources Agency dealing with all natural resources. In Arizona, recommendations were made that consolidation of agencies would best be achieved by establishing a new Department of Natural Resources. The study commission in Utah, on the other hand, advocated maintaining the *status quo* (the separation of the State Engineer's Office and the Water Power Board) on the premise that "functions of regulation and distribution ought to be administered apart from those relating to development and utilization." It is not immediately apparent why this separation should be made, for effective administration will continue to require the coordination of these two agencies.

It would seem desirable to consolidate executive agencies in the states in the same way as suggested for Federal agencies—with water and all other flow resources administered by a single agency—and for the same reasons. Similarly, in the legislative

---

[52] S. V. Ciriacy-Wantrup: "Resource Conservation, Economics and Policies," chap. 21. See footnote 16.

branch, consolidation of committees dealing with water and other flow resources into a single committee and adequate professional staffing would seem as essential for effective coordination of water policy within each state as within the Federal government. Furthermore, an overall planning agency would also seem desirable for each state.

2. Public Districts. The public district is a legally authorized agency representing a community of interest and generally organized for realizing a specific objective. Public districts possess several characteristics which make them suitable for attending to a wide range of water-management problems such as sewage, irrigation, water conservation, reclamation, and drainage.

Water problems are not necessarily confined to local political boundaries. Thus, when a problem exhibiting such geographic characteristics arises, a public district can be formed in such a way that all and exclusively those individuals are included for whom the problem has relevance. Public districts also reflect intertemporal flexibility, since these districts are locally created under state enabling legislation. As conditions change, the district's powers can be altered by appropriate legislation after receiving community consent.[53]

The district's ability to focus attention on one problem can be another positive attribute where local interests feel that their problem will not receive sufficient attention if existing agencies with many responsibilities undertake to solve it. Moreover, existing agencies may lack legal authority to perform certain desired functions.

Finally, public districts which have powers to assess and float bond issues may be utilized to tackle water problems where existing units of government are at their legal tax and debt limits or there may be doubts about obtaining voter approval for raising the additional revenues which an additional service would require.[54]

While public districts have many advantages, they also have an important disadvantage. It was just noted that these units facilitate the grouping of special interests, but this may lead to excessive fragmentation of functions, which in turn prevents effective utilization of water resources. In the Santa Clara Valley, California, for example, there are the South Santa Clara Valley Water Conservation District, the Santa Clara Valley Water Conservation District, the Santa Clara County Parks and Recreation Commission, and the Santa Clara County Flood Control and Water Conservation District. The result of this fragmentation has been interagency conflict, and it has been difficult to formulate an integrated plan for operating reservoirs and other facilities.[55]

In several states, public districts have been created in order to achieve integrated management of surface and groundwater. The program adopted has been artificial groundwater recharge.[56] In California, the district form of organization has been adopted because present groundwater rights (under the correlative-rights doctrine) has forced responsibility for groundwater management upon local water users rather than upon a state agency.[57]

Soil-conservation districts and county flood-control and water-conservation districts are the principal local governmental units which participate in the small-watershed-protection program.

3. Interstate Cooperation. Turning now to the problem of better interstate cooperation in water policy, some past attempts to set up suitable institutional machinery may be noted: the Interstate Commissions of the Delaware and Potomac Basins, the Arkansas-Oklahoma Interstate Water Resources Committee, and agencies

---

[53] Stephen C. Smith: Problems in the use of the public district for ground water management, *Land Economics*, vol. 32, no. 3, p. 264, August, 1956.

[54] For a rather extensive analysis of public districts, in particular their characteristics and their many reasons for having been created, see John C. Bollens: "Special District Governments in the United States," University of California Press, Berkeley, Calif., 1957.

[55] Stephen C. Smith: Public districts in the management of California's ground water, paper prepared for the Conference on Ground Water, University of Colorado, Boulder, Colo., Aug. 24–25, 1960, p. 4. Processed.

[56] *Ibid.*, p. 2.

[57] Stephen C. Smith: Problems in the use of public district for ground water management, p. 261. See footnote 53.

set up by state compacts. Many such agencies have been initiated or aided by the Council of State Governments. The Council might well consider whether interstate cooperation in the planning and execution of water and other resource policies should not constitute one of its major objectives. Since, as has been repeatedly emphasized, regulation of resource use is constitutionally left to the states, and since many resource problems concern more than one state, these problems can be attacked most effectively through cooperative action by all states concerned.

Although state compacts are not a necessary condition for better interstate cooperation in water policy, they are generally helpful. The experience gained in negotiating, concluding, and administering state compacts has resulted in a better integration of state policies, even for resources not directly affected.

4. State-Federal Cooperation. As mentioned earlier, the Flood Control Act of 1944 provides that states be consulted during the planning stages of federally sponsored projects and that they have the opportunity to review the final project proposals. In practice, however, consultation during these stages may be perfunctory, sometimes virtually nonexistent.

States' rights in the realm of natural resources may best be defended if the states themselves shoulder the duties with which these rights are constitutionally associated. More progressive states realize this. Several states have effective policies of water development. Federal agencies could encourage state participation in responsibility by consulting state agencies regarding formulation and implementation of policies before submitting executive proposals to Congress, rather than, as sometimes happens, regarding the states merely as tools for the implementation of policy. More active state participation might be facilitated by better intrastate coordination of water policy and by better institutional machinery for coordinating resource policies between the states and the Federal government. The Pacific Coast Board of Intergovernmental Relations, although it did not deal with resource policies, and although its conclusions were merely advisory, points the way in which appropriate institutional machinery might be formed.

One of the barriers to effective Federal-state cooperation in water policy is the multiplicity of Federal agencies, often having conflicting policies, with which the states must now deal. Consolidation of Federal agencies along the lines sketched above would do much to improve Federal-state relations. At the same time, it would reduce the problem of states playing one Federal agency against the other, a game which does not always result in the emergence of the soundest policy. Adoption of a regional organization within Federal agencies, to a greater extent than heretofore, and possibly with somewhat smaller units, would further facilitate coordination of Federal and state water policies.

Section 29

APPLICATION OF ELECTRONIC COMPUTERS IN HYDROLOGY

VEN TE CHOW, *Professor of Hydraulic Engineering, University of Illinois.*

I. Introduction	29-1
A. Significance of Electronic Computers in Hydrology	29-1
B. Analog Computer versus Digital Computer	29-2
II. Analog Computers	29-3
A. Fundamentals of Analog Computers	29-3
1. Types of Analog Computers	29-3
2. The Differential Analyzer	29-4
B. Application in Hydrology	29-7
1. Surface Water	29-7
2. Groundwater	29-11
III. Digital Computers	29-14
A. Fundamentals of Digital Computers	29-14
1. Types of Digital Computers	29-14
2. Computer Capabilities	29-16
3. Computer System	29-16
4. Numerical System	29-17
5. Solution of a Problem	29-18
6. Programming	29-18
B. Application in Hydrology	29-21
1. Surface Water	29-21
2. Groundwater	29-23
IV. References	29-23
A. On Both Analog and Digital Computers	29-23
B. On Analog Computers	29-23
C. On Digital Computers	29-24
D. On Analog-computer Application in Surface-water Hydrology	29-25
E. On Analog-computer Application in Goundwater Hydrology	29-26
F. On Digital-computer Application in Hydrology	29-28
G. On Digital-computer Application in Surface-water Hydrology	29-28
H. On Digital-computer Application in Ground-water Hydrology	29-30

I. INTRODUCTION

A. Significance of Electronic Computers in Hydrology

The electronic computer has become one of the most important tools of modern research and practicing hydrologists. The major reasons for this are that hydrologic analysis and design necessitate processing of a large amount of quantitative data which have been accumulating at a rapid rate (Subsec. 8-I-I), and that theoretical approaches have been gainfully introduced into modern quantitative hydrology and such approaches involve complicated mathematical procedures and models which can be solved practically only by high-speed computers (Subsec. 1-II-H). The history of the application of electronic computers to the solution of hydrologic problems has been brief, but it has gained such great momentum that it will have far-reaching implications in the future development of hydrology.

In general, possible applications of electronic computers in hydrology lie in three overlapping areas: solution of specified mathematical equations describing hydrologic laws and phenomena, simulation of hydrologic systems and subsystems, and control of hydrologic instrumentation and experimentation. The first two areas have been developed considerably, whereas the third area has been explored only very recently particularly in the field of automatic instrumentation for hydrologic measurement (Subsec. 15-III-B-2) and in laboratory hydrologic experimentation. Under the sponsorship of the U.S. National Science Foundation the author has initiated at the University of Illinois in 1963 a research project on basic investigation of watershed hydraulics. In planning this project, electronic computers were introduced to control the laboratory operation of an instrumented rainfall-runoff system designed for the experimental study of flow over artificial drainage basins.

Owing to the rapid rate of technological progress in the field of electronic computers and their utilization, specific details and equipment descriptions would have only temporary meaning. For such information the reader should refer to the current technical literature of the various computer manufacturers. Furthermore, the principles and methods in computer design and programming change so fast that new ideas and approaches are constantly replacing the current ones. Therefore, the material presented in this section is intended to be very brief, covering a short introduction to computer types and characteristics and a brief survey of the computer application in hydrology supplemented with a list of references. For comprehensive information on electronic computers and their use and applications, see Refs. 1–54.

B. Analog Computer versus Digital Computer

Electronic computers may be classified into two basic categories: *analog* and *digital*. The analog and digital characteristics may be combined, however, to produce a machine known as a *hybrid* computer.

The analog computer operates on continuous variables which are represented by physical or mathematical analogies between the computer variables and the variables of a given problem to be solved by the computer. The digital computer, on the other hand, operates on discrete variables represented by digits of numbers and does arithmetic by manipulating the digits and executing the basic arithmetic operations in a manner similar to a human arithmetician.

A fair comparison between the two types of computers is not always easy, because not only does it depend on their own characteristic features but also on many interrelated factors that are involved in the particular problem or problems to be solved by the computer. Only in very general terms, the following comparative characteristics may be mentioned:

Size. The analog computer can be of any size, whereas the digital computer is generally large since the required special circuits usually fix a minimum size.

Speed. The analog computer operates continuously and rapidly, while the digital computer possesses great inherent speed. For certain types of problems, an analog computer can generally reach a solution more quickly than a comparable digital machine, but in practice it is almost always possible to construct a digital computer which can reach a solution sufficiently rapidly.

Accuracy. Because of various circuit losses the analog computer has limited precision, generally of 0.1 to 1.0 per cent in error. Theoretically, the digital computer can be built to any desired precision depending mainly on its storage capacity and word length and it has no intrinsic source of error other than the cumulative round-off error.

Function. The analog computer can be built with circuits designed for almost any operations. On the digital computer, all functions must be performed in terms of four basic operations: addition, subtraction, comparing, and storing.

Operation. The analog computer can be set up and operated by persons with very little training; whereas the use of digital computer requires a thorough knowledge of programming either in machine language or in automatic programming language. Since programming may be very involved, it is less easy to introduce unforeseen modifications into a problem or its solution in a digital computer than in an analog computer.

Flexibility. Once the circuit is completed, the use of an analog computer is usually limited to the problem under consideration, but it can easily be adapted to problems within its range or by reconnecting the circuit. The digital computer can be adapted to any type of problem simply by changes in the program.

Cost. The cost of an analog computer is approximately proportional to its size, whereas the digital computer is likely to have a large minimum cost.

II. ANALOG COMPUTERS

A. Fundamentals of Analog Computers

1. Types of Analog Computers. There are many types of electronic analog computers. According to the purpose of application, they may be classified into two categories: *special-purpose* and *general-*, or *multiple-, purpose*. According to the manner of representation, they may be grouped into two kinds: *direct*, or *physical*, and *indirect*, or *mathematical*. The direct computer is also known as a *simulator*, and the indirect computer simply as an *analog computer*.

The special-purpose analog computer is a combination of basic computing components that are selected to perform one special application. The general-purpose analog computer, on the other hand, contains more basic components than may be required for accomplishing any one particular application and thus it can solve a wide variety of problems.

The direct analog computer is so characterized by its design that problem variables and problem parameters are represented directly by variables and parameters on the computer. Its examples are equivalent circuits and network analyzers. The *equivalent circuit* is the direct analogy of an electrical circuit for a problem system, such as the U.S. Weather Bureau electronic flood router (Fig. 25-II-17). The *network analyzer* is a general-purpose computer with its operation based upon the analogy of the mechanical behavior of the problem elements to the corresponding electrical behavior of the circuit components, such as the McIlroy network analyzer (Subsec. 7-VII-H). Because of the relatively high cost and less perfect simulation, the use of network analyzers is generally limited to the analysis of electrical power-distribution systems as well as to the analysis of aircraft structural problems.

The indirect analog computer is so characterized by its design that it can carry out or assist in the solution of algebraic or differential equations. The most commonly used indirect electronic analog computer today is the *differential analyzer* (see also Subsec. 25-II-IV-H-2).

While the power network analyzer was developed in the 1920s, the differential analyzer came into being relatively late only immediately after World War II. During World War II, one basic computer component, the operational amplifier, was introduced by C. A. Lovell and D. B. Parkinson of Bell Telephone Laboratories. Because of the difference in operation time, two methods of operating the differential analyzer can generally be identified. The so-called *long-time, slow* or *one-shot* differential analyzer makes use of integrators which have a relatively long time constant, so that typical problem solutions may take several seconds to minutes and thereby the results can be recorded on strip-chart recorders or servo-driven plotters. The so-called *short-time, high-speed*, or *repetitive* differential analyzer, on the other hand, has the integrator time constants so designed that solutions can be completed in several milliseconds and then repeated many times per second; thus the results can be displayed on an oscilloscope. In the one-shot computer field, examples of the well-known models are REAC of the Reeves Instruments Inc., Electronic Associates' PACE, Berkeley's EASE, and Goodyear's GEDA. The high-speed computer becomes possible because of the use of modern repetitive amplifiers developed by G. A. Philbrick [18]. The most widely used model today is Philbrick's GAP/R.

2. The Differential Analyzer. Because of the advantage in amplification by voltage amplifiers built with vacuum tubes, voltage is generally selected as the *computer variable* in a differential analyzer. With the advent of transistors, however, it may become suitable to select current as the computer variable because transistors approximate true current amplifiers. While either voltage or current is selected as the computer variable, it is always a function of the time taken as the independent variable. Since voltage is selected as the computer variable in most computers today, the following will be discussed on this basis.

a. Computing Elements. A differential analyzer is essentially the assemblage of a number of *computing elements* which are connected together for the solution of problems. In general, the computing elements may be classified into two groups: *operating elements* and *supplementing elements*. The operating elements are employed to effect linear or nonlinear computing operation. For nonlinear operation, they are required for the solution of nonlinear algebraic and differential equations. The supplementing elements do not take a direct part in the problem solution but they are the required auxilliary equipment, for example, to permit repetitive operation and to produce suitable display of the results. Such elements include oscilloscope, plotter, voltage generator, switching system, power supplies, etc.

Table 29-1 lists a number of basic computing elements and their representative symbols. The circuit symbol shows the input and feedback elements in detail and is useful when studying the operation of specific computer units. The block symbol is a compact notation generally used in schematic presentation of circuits for the analog solution of entire problems. The input voltage e_i and the output voltage e_0 are represented with the ungrounded terminals only, since the other terminals are grounded and usually are not shown.

The *high-gain d-c amplifier* performs the operation of amplification of input (grid) voltage by a gain of $-A$. As the gain is a function of frequency having magnitude and phase, it is an approximation. This approximation is not sufficiently accurate for the high-gain amplifier used alone, but it is suitable for most network approximations involving the amplifier as an element such as in the case of analog computers.

The *constant multiplier* is a modification of the high-gain amplifier by the addition of suitable large resistors R_1 and R_2. In order to approximate the multiplication under consideration, the gain A must be very large, say 10^8. Also the input voltage is usually scaled so that the output will not exceed the linear range of the amplifier.

The *summing amplifier* or *summer* is an extension of the constant multiplier by the addition of several input resistors so as to permit (negative) summation of several inputs.

The *integrating amplifier* or *integrator* is a modification of the constant amplifier in which a low-loss capacitor C is used as the feedback impedance in place of the resistor connected across the high-gain amplifier. This unit can perform the operation of integration. In the circuit, K is the initial charge on the capacitor. By adding more input resistors in parallel, the integrator can perform the operations of summation as well as integration.

The *operational amplifier* is a modification of the constant amplifier in which the input impedance z_i and feedback impedance z_f are used in place of the resistors R_1 and R_2 respectively. Again, the gain A must be sufficiently high in order to attain good approximation of the operation.

The *potentiometer* is a multiplier which is used most frequently as the input to an operational amplifier. However, the potentiometer can perform only multiplication of a voltage by a constant a less than 1, and this constant is only approximate and does not exactly equal the potentiometer reading. Thus, the potentiometer must usually be set to the desired accuracy, to about three places, either manually or by a servomechanism.

The *function multiplier* has various designs and it can perform the multiplication of two input functions e_1 and e_2. The constant a is usually positive and is about 0.01 in order of magnitude. When multipliers are employed in feedback circuits, they can

Table 29-1. Some Basic Computing Elements

Element	Operation	Input-output relation	Circuit symbols	Block symbols
High-gain d-c amplifier	High-gain amplification	$e_o = -Ae_i$		
Constant multiplier	Constant multiplication	$e_o = -\frac{R_2}{R_1} e_i$		
Summing amplifier, or summer	Summation	$e_o = -\left(\frac{R_3}{R_1} e_1 + \frac{R_3}{R_2} e_2 \right.$ $\left. + \ldots + \frac{R_3}{R_n} e_n\right)$		
Integrating amplifier, or integrator	Integration	$e_o = -\frac{1}{RC} \int_0^t e_i \, dt + K$		
Integrating summer	Integration and summation	$e_o = -\int_0^t \left(\sum_{i=1}^n \frac{1}{R_i C} e_i\right) dt + K$		
Operational amplifier	Complex transfer function	$e_o = -\frac{z_f}{z_i} e_i$		
Potentiometer	Constant multiplication by $a<1$	$e_o = a e_i$		
Function multiplier	Multiplication	$e_o = -a e_1 e_2$		
Function generator	Function generation	$e = f(e_i)$		

be made to form the reciprocal and the square root of a function and to perform other nonlinear operations.

The *function generator* has a variety of designs. It can establish a functional relation $f(e_i)$ between a variable e_i and a function f. Special generators, for example, which convert rectangular coordinates to polar coordinates and vice versa, are known as *resolvers*.

Differentiation can be performed by using an inductor in the feedback network while the input impedance remains resistive, or by interchanging the capacitor and resistor of the integrating circuit. However, this operation is seldom used in the solution of problems on analog computers because it will produce noise amplification which is undesirable. Unless the noise can be kept at an acceptable low level, it is preferable to rewrite the differential equations so that they can be solved by other mathematical operations.

b. Computer Setup and Operation. The use of a differential analyzer for the solution of the set of differential equations representing a given problem is simply to vary the voltage in the computer with time in a manner prescribed by the equations. In the computer setup the computer establishes electronic circuit relations between voltages, or *computer variables*, to simulate the mathematical relations between the original *problem variables*.

The first step in a computer setup is to establish the relations between the computer variables X's and the problem variables x's by *transformation equations*. In all ordinary cases these equations are of the form

$$X = a_x x \qquad (29\text{-}1)$$

where a_x is a dimensional coefficient known as a *scale factor*. A scale factor should be chosen so that it is as large as possible in order to minimize percentage errors due to stray voltages, but, on the other hand, the resulting computer variables must not exceed ± 100 volts (one *computer unit*) so that all vacuum tubes operate within the linear range of their characteristics. These two requirements are conflicting and hence some compromise must be made. In general, a scale factor can be chosen as a round number somewhat less than that given by the following rule:

$$a_x = \frac{1}{\text{maximum expected value of } x} \text{ [computer units/problem units]} \qquad (29\text{-}2)$$

This rule ensures that the maximum expected excursion of the computer variable X will never exceed one computer unit. It may be noted that there are other minor considerations which may affect the choice of the scale factor, and also that there may be several scale factors for various problem variables as determined usually by one transformation equation.

Similarly, the computer time τ is related to the real time t in the problem by a transformation equation

$$\tau = \alpha_t t \qquad (29\text{-}3)$$

where α_t is a dimensionless *time-scale factor*. The time-scale factor should be chosen so that it is small in order to minimize computer errors which are developed mostly because of increase in time, and, on the other hand, so that it is large enough to ensure reliability of computer servos if they are used but is not too large to cause instability and overloading of the circuit. In general, the time scale must be selected on the basis of a compromise of the above conditions and by taking into consideration the nature of the particular problem in question.

After the scale factor and the time-scale factor are chosen, the corresponding transformation equations are substituted in the mathematical equations of the given problem. The resulting mathematical equations show the problem relations in terms of the computer variables and are called the *computer equations*. Each of these computer equations is then represented by a *block*, or *patch, diagram* which outlines the arrangement of computing elements that will establish the relationships between the computer variables. The block diagram describes the relation between computer variables (voltages) and computing elements (computer operators) just as the mathematical equations describe the relation between problem variables and operators.

The computing elements described by the block diagram can therefore be connected to form an electric circuit for the solution of the problem. The computer variables (voltages) are set to correct the initial conditions prescribed by the problem. The number of initial conditions usually equal the number of integrators involved in the problem and each integrator output must be set to the corresponding initial-condition voltage. The computing elements are then made operative and direct the voltages in the computer to vary in a manner defined by the given differential equations. The output voltage variations with time are recorded and produce the solutions of the problem. After the solutions are obtained, the computer is stopped at a time chosen by the operator. The computer is at once reset to its initial condition and is ready for the next run with changed coefficients, initial conditions, etc. By means of repetitive computers, the steps of the solution of a problem can be repeated automatically at a rapid speed (10 to 60 cps).

B. Application in Hydrology

1. Surface Water. The earliest use of electronic analog computers in surface-water hydrology was to study flood flows in rivers. As reported in 1947 [55], an analog computer was developed by Prof. C. E. Warren of the Ohio State University at the suggestion of the U.S. Army Corps of Engineers to route floods in the Ohio River. This study was further extended at the Ohio River Division Office of the Corps. In 1954, a GEDA was employed to route floods by the Muskingum equation (Figs. 25-II-18 and 19) [5, p. 6–41; 56], to compute surges in sewers, and to analyze siphon air vents and the fluctuation of hydropower surge chambers [57–58]. At the North Pacific Division of the Corps at Portland, Oregon, a small battery-powered computer consisting of two stages of resistors and capacitors was built in 1956 for flood routing [59], but it was found of limited use since digital computers were made available for the computation at about the same time.

At the U.S. Weather Bureau, an equivalent-circuit computer was first reported in 1948 [60] for use of flood routing by the Muskingum method (Fig. 25-II-17). Further reports on this analog computer are given in Refs. 61–64.

In 1953, Messerle [65] reported that a "pondage simulator" had been developed at the University of Melbourne, Australia, for the solution of the pondage equation relating the flood flow and stage in a reservoir. He further used a differential analyzer at the University of Sydney, Australia, to determine the optimum design features of hydroelectric systems on Tooma River [66]. Since 1954, Ishihara and his colleagues [67–76] at Disaster Prevention Research Institute of Kyoto University, Japan, have been reporting the development and applications of an electronic nonlinear flood router which was designed on the basis of Hayami's flood theory of diffusion analogy [77, pp. 601–603].

The first application of electronic analogs to unsteady flow in open-channel systems is believed to have been made by Glover, Herbert, and Daum of the U.S. Bureau of Reclamation. In 1953 [78], they reported the use of a direct analog based on the required square law of resistance for the determination of tidal flow distribution in the channel network of the Delta area of California at the east of Suisan Bay. Further development of a nonlinear direct analog for flow in channel systems was reported by Harder and others [79–83] of the University of California. This analog, developed in a study first supported by the University of California Water Resources Center and later by the Kansas City District Office of the U.S. Army Corps of Engineers, was used to simulate flood control systems in the Kansas River basin and Sacramento River tidal estuary.

The use of the unit-hydrograph principle (Subsecs. 14-III-B, C, and D) in analog computer design for runoff simulation was essentially discussed in 1952 by Paynter [84] who suggested the use of admittance, the integral of IUH, for flood routing. Other developments along this approach have been reported in Refs. 59, 85–96. Dziatlik [87] used the equations for the long-distance electric transmission line with variable parameters to simulate the unsteady flow in rivers. Halek [88] discussed the use of electrical models for solving hydraulic problems at the Hydraulic Research Institute in Brno. A portable multiple-stage storage routing analog for converting effective rainfall to direct runoff has been developed at the University of Idaho in cooperation with the North Branch, Soil and Water Conservation Research Division of the U.S. Agricultural Research Service at Moscow, Idaho [90–94]. Also, a quasilinear analog was developed at the U.S. Geological Survey [95–96] in Phoenix, Arizona, for simulating the runoff-producing characteristics of a drainage system and for synthesizing flood-frequency distribution.

Falk [97] demonstrated that an analog computer is applicable to the solution of differential equations in the mathematical models of various problems encountered in water pollution control. For the study of hydropower engineering problems electronic analog techniques were applied to surge and water hammer by Paynter [98] and to penstock and governor systems by Koenig and Knudtson [99].

The application of analog computers to surface-water hydrology can be illustrated by the computer simulation of the conceptual models of drainage basins described in Subsection 14-III-D. The components of various models can be represented by different analog circuits.

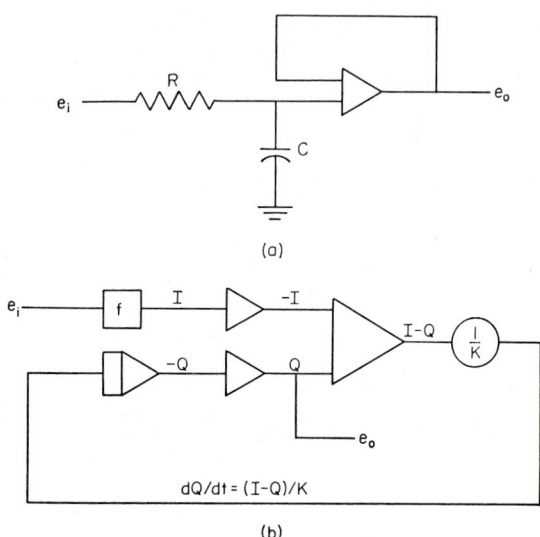

FIG. 29-1. Analog circuits for a linear reservoir. (a) Direct analog; (b) indirect analog.

Figure 29-1a represents the simple RC circuit of a direct analog for a linear reservoir which has the following storage function:

$$S = KQ \qquad (14\text{-}28)$$

where S is the storage, K is the storage coefficient, and Q is the discharge. Substituting this function in the hydrologic continuity equation

$$I - Q = \frac{dS}{dt} \qquad (14\text{-}29)$$

gives

$$I - Q = K\frac{dQ}{dt} \qquad (29\text{-}4)$$

This equation is analogous to the circuit equation of the direct analog in Fig. 29-1a; i.e.,

$$e_i - e_0 = RC\frac{de_0}{dt} \qquad (29\text{-}5)$$

where R is the resistor, C is the capacitor, e_i is equivalent to I, e_0 is equivalent to Q, and RC is equivalent to K. The operational amplifier in the circuit is a voltage-

ANALOG COMPUTERS

transferring device which provides a means of interconnecting a series of reservoirs. This device serves as a buffer and can ensure positive direction of flow of current through series connections although it is not absolutely necessary. An indirect analog circuit for the linear reservoir is shown in Fig. 29-1b in which suitable computing elements described in Table 29-1 are connected to complete the circuit.

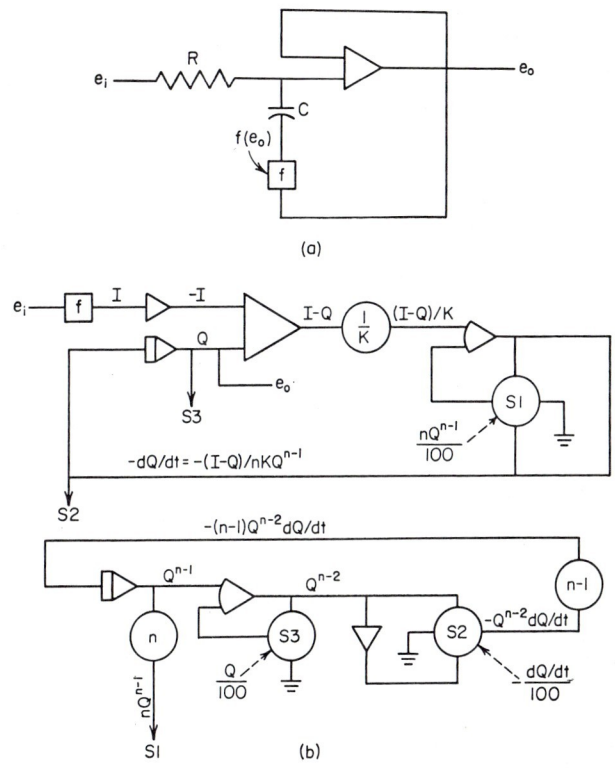

FIG. 29-2. Analog circuits for a nonlinear exponential-type reservoir. (a) Direct analog; (b) indirect analog.

For a nonlinear reservoir of exponential type, the storage function may be written

$$S = KQ^n \tag{29-6}$$

where K is the storage coefficient and n is an exponent. Combining this equation with Eq. (14-29), the hydrologic continuity equation of the reservoir may be written

$$I - Q = nKQ^{n-1}\frac{dQ}{dt} \tag{29-7}$$

This equation can be simulated by the direct analog circuit in Fig. 29-2a which, according to Shen [96], has the following circuit equation:

$$e_i - e_0 = RCe_0^{n-1}\frac{de_0}{dt} \tag{29-8}$$

assuming

$$f(e_0) = e_0 - \frac{e_0^n}{n} \tag{29-9}$$

29-10 APPLICATION OF ELECTRONIC COMPUTERS IN HYDROLOGY

The function generator in the circuit is to generate the nonlinear function $f(e_0)$ represented by Eq. (29-9). This device may be a nonlinear voltage amplifier which consists of an operational amplifier and a group of diodes, each having a series resistance that conducts at a specific voltage level.

Figure 29-2b shows an indirect analog circuit for the exponential reservoir. In the circuit, the symbol of a circle with an S inside it represents a servomechanism which, when combined with a high-gain amplifier or an integrating amplifier, can perform function division or multiplication and is known as a *servo divider* or a *servo multiplier*, respectively.

A linear channel (Subsec. 14-III-D) has the property of

$$Q = f(t - T) = \text{constant} \tag{29-10}$$

where t is the time and T is a translation time lag. This may be simulated electrically by a *delay line* or a *phase-shifting* circuit that may be commercially available.

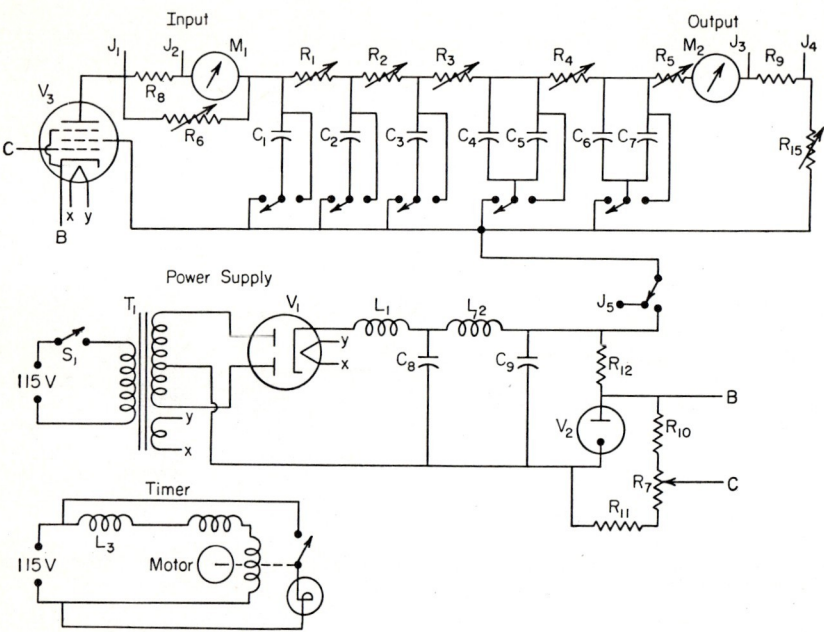

FIG. 29-3. Circuit of a multiple-stage storage routing analog. (*Rosa* [91].)

By connecting the analog circuits for various combinations of linear or nonlinear reservoirs and linear channels, the conceptual models for drainage basins described in Subsection 14-III-D can be simulated electrically. Figure 29-3 shows a multiple-stage storage routing analog that was built by L. A. Beyers of the University of Idaho and tested by J. M. Rosa of the U.S. Agricultural Research Service in 1961 [90, Fig. 2, p. 16; 91, Fig. 8, p. 26]. This is a direct analog which simulates a series of five linear reservoirs with variable storage coefficients. Table 29-2 is a list of parts of the analog which is portable and can be built easily [91, Table I, p. 21]. Beyers [92] also developed an indirect analog to simulate the snowmelt. This snowmelt analog and an analog to simulate effective rainfall, if any, can be connected in parallel as input to the multiple-stage storage routing analog to complete a snowmelt-runoff analog computer which can compute the runoff hydrograph due to snowmelt and effective rainfall. Robinson and Beyers [93] further devised an analog to simulate Eq. (21-7), which can compute the effective rainfall increments of a hyetograph from average rainfall rates,

The effective rainfall increments so computed can be routed through the multiple-stage routing analog to produce the runoff hydrograph at the drainage-basin outlet.

2. Groundwater. The first use of electrical analog techniques in groundwater hydrology is believed to have been proposed by Pavlovskiĭ [100] in 1918 for the analysis of seepage under dams. After Pavlovskiĭ, similar but somewhat more elaborate analogs were devised by many others [101–121]. The principle of such

Table 29-2. List of Parts for Analog Circuit Shown in Fig. 29-3

V_1—6X4 vacuum tube..	$ 1.65
V_2—VR 105 gas regulator tube...	1.25
V_3—6BA6 vacuum tube...	2.00
T_1—Pri.-115 v, sec.-6.3 v, 350 V.C.T., 50 ma transformer................	7.35
M—Synchronous clock motor...	3.00
M_1—0-1-ma milliammeter..	10.40
M_2—0-1-ma milliammeter..	10.40
S_1—Dpdt switch..	0.40
S—Spdt switch (4)...	1.00
L_1—48-mh inductance...	1.25
L_2—9.6-mh inductance..	1.00
L_3—1.6-h inductance...	1.35
C_1, C_5—200-mf, 250 WVDC capacitor..................................	8.00
C_2—300-mf, 200 WVDC capacitor.......................................	2.80
C_3, C_4, C_6, C_7—500-mf, 200 WVDC capacitor......................	43.20
C_8, C_9—40-mf, 600 WVDC capacitor...................................	2.50
R_1, R_2—25-kilohm potentiometer.......................................	2.50
R_3, R_4, R_5—10-kilohm potentiometer................................	3.75
R_6—5-kilohm potentiometer..	1.25
R_7—100-kilohm potentiometer..	1.25
R_8, R_9—50-kilohm resistor..	0.30
R_{10}—20-kilohm resistor..	0.15
R_{11}—180-kilohm resistor...	0.15
R_{12}—10-kilohm, 2-watt resistor...	0.85
Chassis, brackets, clamps, knobs, jacks, miscellaneous hardware...............	15.00
	$122.75

analogs is based on the analogy between Darcy's law of seepage flow and Ohm's law of flow of electric current. Darcy's law is

$$v = K \frac{\partial h}{\partial s} \qquad (13\text{-}20)$$

where v is the velocity of flow, K is the coefficient of permeability, h is the total head, and s is the distance along the average direction of flow. Ohm's law of an electric current can be expressed by

$$I = \sigma \frac{\partial V}{\partial x} \qquad (29\text{-}11)$$

where I is the electric current, σ is the conductivity, V is the voltage, and x is the length of conductor. By comparing Eqs. (13-20) and (29-11), it can be seen that I is equivalent to v, σ to K, V to h, and x to s. For equipotential lines, V is constant and so h is constant. For an insulated boundary, $\partial V/\partial x = 0$; and so, for an impervious boundary, $\partial h/\partial s = 0$. For both seepage flow and flow of electric current, the Laplace equation is satisfied; that is $\nabla^2 h = 0$ and $\nabla^2 V = 0$. Therefore, an electric analog model can be built to reproduce groundwater boundaries, and the electric field produced by a voltage applied to the model will simulate the groundwater flow in the prototype. The equipotential lines and the lines of current flow to be traced in the model will be equivalent to the equipotential lines and flow lines, respectively, of the groundwater flow net. The material used to construct the conductor in the analog model may be solid, liquid, or gelatin. Solid conductors may be constructed from thin metal sheets, carbon paper [103], and graphite [109] (Fig. 29-4). Liquid conductors may be made of an electrolyte of low conductivity such as a dilute copper sulfate solution [106]. Gelatin conductors can be formed by adding sodium chloride or copper

29-12 APPLICATION OF ELECTRONIC COMPUTERS IN HYDROLOGY

sulfate to a hot gelatin and pouring it into a mold, representing the groundwater boundaries, where it solidifies [115]. Color tracer electrical models used in the study of movement of injected fluid in an oil reservoir [122–124] can also be adapted to groundwater studies. The above analog models are usually made for two-dimensional representations. However, three-dimensional representations can be made by conductors of variable thickness [106, 111].

Another type of electric analog for groundwater study involves the use of *resistance network* [125–140] which can be used to analyze unsteady seepage flow and more complicated groundwater problems. The resistance network has been applied to seepage

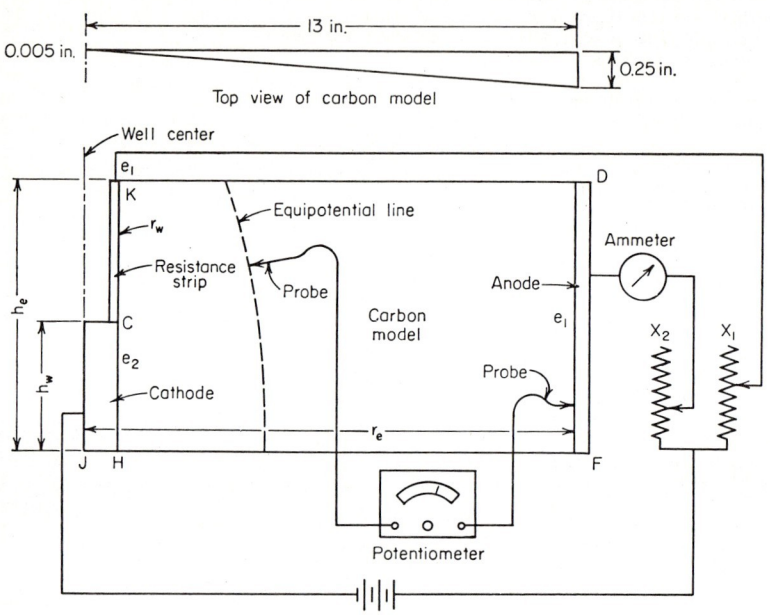

FIG. 29-4. Graphite model showing electrical circuit and connections to study unconfined radial flow to a well. e_1 = potential at anode, e_2 = potential at cathode, h_e = thickness of water-bearing medium measured vertically between the bottom of the well and the undisturbed groundwater surface, h_w = depth of water in the well, r_e = radius of circle of influence, r_w = radius of pumped well, X_1 and X_2 = variable resistances. (*Babbitt and Caldwell* [109]*.*)

problems by Hanks and Bowers [126] of the Western Soil and Water Management Research Branch, U.S. Agricultural Research Service, Manhattan, Kansas, and by Bouwer and others [127–130] of the U.S. Water Conservation Laboratory, Tempe, Arizona. It has been further applied to both steady and unsteady flow through either homogeneous or nonhomogeneous aquifers by Stallman [131–134] and Skibitzke [135–137] of the U.S. Geological Survey, and by others [138–140].

The resistance network consists mainly of *junctions* which are formed by resistors and capacitors and are connected to form a finite-difference grid representing the finite-difference form of the following partial differential equation of unsteady potential flow obtained from Eq. (13-29) with T replacing Kb:

$$\frac{\partial^2 h}{\partial x^2} + \frac{\partial^2 h}{\partial y^2} + \frac{\partial^2 h}{\partial z^2} = \frac{S}{T}\frac{\partial h}{\partial t} \tag{29-12}$$

where T is the coefficient of transmissibility and the other notations have been defined for Eq. (13-29). The finite-difference approximation of the above equation for a three-dimensional grid system of equal grid intervals of $\Delta x = \Delta y = \Delta z = \Delta s$ (Fig. 29-5a) may be written as

$$\sum_{i=2}^{7} h_i - 6h_1 = (\Delta s)^2 \frac{S}{T} \frac{\Delta h}{\Delta t} \tag{29-13}$$

where h_1 is the head at node 1, and h_i denotes the heads at nodes $i = 2, 3, 4, 5, 6,$ and 7.

The junction of the resistance network to simulate Eq. (29-13) is shown in Fig. 29-5b. It consists of six resistors and one capacitor connected to a common terminal, with the

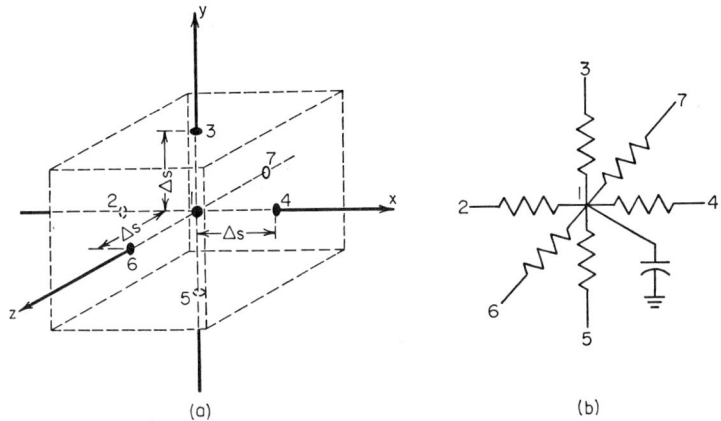

FIG. 29-5. Electronic simulation of a three-dimensional grid system of groundwater flow. (a) Grid element of the groundwater system; (b) simulated junction of resistance network.

other terminal of the capacitor connected to ground. If the resistors are of equal magnitude, the electrical potentials for the junction can be expressed as

$$\sum_{i=2}^{7} V_i - 6V_1 = RC \frac{\Delta V}{\Delta t} \tag{29-14}$$

where V_1 is the electrical potential at the common terminal 1, V_i denotes the electrical potentials at terminals $i = 2, 3, 4, 5, 6,$ and 7, R is the resistance, and C is the capacitance. Comparison of Eqs. (29-13) and (29-14) shows the similarity of the two equations. Thus, V_1 is analogous to h_1, V_i to h_i, and RC to $(\Delta s)^2 S/T$, or R to $1/T$ and C to $(\Delta s)^2 S$. Therefore, the junction shown by Fig. 29-5b represents an electrical analogy of a finite-difference grid element of Fig. 29-5a which approximates the unsteady three-dimensional flow condition of groundwater through an element within a homogeneous, isotropic, infinite aquifer. In other words, the interior of such an aquifer can be represented by a resistance network formed by a number of junctions like that shown in Fig. 29-5b. The volume of the grid element $(\Delta s)^3$ is small compared with the volume of the aquifer. The accuracy of the analogy, of course, depends on the selected value of Δs. The relations between the problem variables and the computer variables are

$$q = a_1 \check{Q} \tag{29-15a}$$
$$h = a_2 V \tag{29-15b}$$
$$Q = a_3 I \tag{29-15c}$$
$$t_d = \alpha_t t_s \tag{29-15d}$$

where q is the quantity of flow in gal, $\check{Q}$ is the electric charge in coulombs, h is the head in ft, V is the voltage in volts, Q is the discharge in gpd, I is the electrical current in amp, t_d is the time in days in the problem, t_s is the time in sec in the analog, a_1, a_2, and a_3 are scale factors, and α_t is the time-scale factor. The relations between the scale factors and the time-scale factor are

$$\alpha_t = \frac{a_1}{a_3} \qquad (29\text{-}16)$$

$$RT = \frac{a_3}{a_2} \qquad (29\text{-}17)$$

$$\frac{C}{7.48(\Delta s)^2 S} = \frac{a_2}{a_1} \qquad (29\text{-}18)$$

Equation (29-17) is obtained from Eq. (29-15c) through the analogy between Ohm's law and Darcy's law. In Eq. (29-18), C is the capacitance in farads, Δs is in ft, and S is dimensionless. The magnitude of the resistors and capacitors in the interior part of the network can be determined from Eqs. (29-15) to (29-18). The scale factors and the time-scale factor may be so chosen that use may be made of standard tolerance, low-wattage fixed carbon resistors (10^2 to 10^7 ohms) and of low-voltage rating capacitors (10 to 10^5 $\mu\mu$f). They may also be chosen for convenience in reading voltages and in making resistors and capacitors to be of simple magnitudes.

The junction shown in Fig. 29-5b may be modified to simulate various boundary conditions and nonhomogeneity. For an impervious barrier at the node i, the resistor R_i may be removed. For a recharge boundary at the node i, the other terminal of the resistor R_i in a junction may be connected to ground. For a nonhomogeneous aquifer, the resistors and capacitors may be of different magnitudes. For a two-dimensional aquifer, the two resistors having outer terminals 6 and 7 may be removed.

The pumping of a well in the aquifer is simulated by a function generator which may consist of a waveform generator to produce sawtooth pulses and control the repetition rate of computation, and a pulse generator to produce rectangular pulses of various durations and magnitudes on command of the waveform generator. The duration of the rectangular pulse is equal to t_s in sec. By applying Ohm's law in Eq. (29-15c), the pumping rate in gpd is expressed by

$$Q = \frac{\Delta V}{R_i} a_3 \qquad (29\text{-}19)$$

where ΔV is the voltage drop required to effect Q in volts across the resistor R_i in ohms. The pulse generator is connected to the junction representing the pumping well. For monitoring purposes, an oscilloscope may be connected to any junction to exhibit a time-voltage graph, analogous to a time-drawdown graph, at the junction.

III. DIGITAL COMPUTERS

A. Fundamentals of Digital Computers

1. Types of Digital Computers. Digital computers may be classified in many ways. According to the intended application, they may be classified as *general-purpose* and *special-purpose* computers. In this section, only general-purpose computers are discussed because they are used most frequently in solving hydrologic problems. Considering the size, digital computers vary in cost by a factor of a hundred and in speed by a factor of a thousand, and also they vary over a very broad range of capacity and flexibility. Although it is customary to class computers as *small-*, *medium-*, and *large-scale* machines, such classification really has no precise meaning but rather is a vague indication of the overall magnitude.

The tree shown in Fig. 29-6 gives a graphical portrayal of the digital-computer heritage. It illustrates the accelerated evolution of electronic digital computers which has taken place since the world's first electronic digital computer ENIAC appeared in 1945. The right limb of the tree shows the history of high-speed parallel "scientific" computers, represented by ORDVAC. The left limb shows the development of

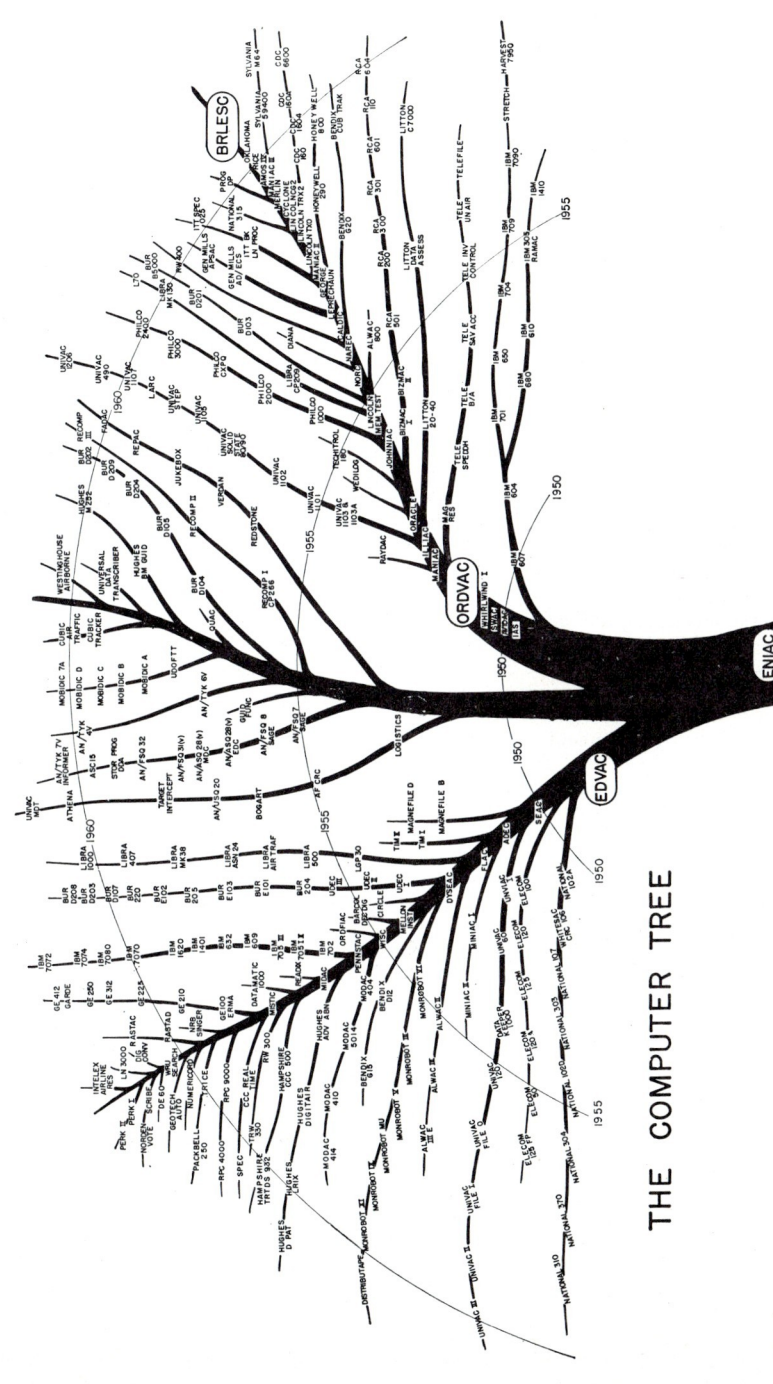

Fig. 29-6. The computer tree. (Courtesy of U.S. Army Ballistic Research Laboratories, Aberdeen Proving Ground, Maryland.)

slower but less expensive serial "business" computers, represented by EDVAC. The center limb shows the computers that were developed specifically to meet military needs. Manufacturers have entered the electronic computer field at different times shown by different branches; while university and government sponsored computers are shown along the limbs. The radial distance from the ENIAC is an approximate indication of the year each computer was either developed, constructed, or placed in operation. The ENIAC, designed and constructed at the University of Pennsylvania, EDVAC, ORDVAC, and BRLESC were sponsored or manufactured by the Ballistic Research Laboratories at the U.S. Army Aberdeen Proving Ground, Maryland. Wilcox [54] of the U.S. Office of Naval Research proposed to add a pot to this tree to represent Howard Aiken's Harvard Mark I, a relay machine developed partly under U.S. Navy sponsorship in 1944, on which the fundamental logical operation involved in electronic digital computers were first demonstrated.

2. Computer Capabilities. The following are three basic capabilities of the electronic digital computer:

a. Stored Program. The computer can automatically follow a set of stored instructions expressed in its own *machine language*. This is possible because the machine has a built-in *memory* or *storage* unit within which the *program*, i.e., a set of instructions and the problem data, can be stored. In order to command the computer, the program must be prepared in complete detail. One instruction may dictate the operation to perform and the location of the number to be involved in the operation; another will dictate the further use of the result so obtained. These instructions, called the *stored program*, are stored in the computer's control unit in the proper sequence required to solve a given problem. The various operations covered in these instructions are usually expressed by an *alphameric* (a generic term for numeric digits, alphabetic characters, and special characters) code.

b. Logical Decision. The computer can make simple logical decisions and automatically alter the sequence of operations. This is made possible by building such logical decisions into the program. In general, any problem which can be basically reduced to a series of rules, each with only two possible outcomes, say "yes" or "no," can be handled by the machine.

c. Arithmetic Operations. The computer can perform simple arithmetic operations, such as add, subtract, multiply, and divide, at extremely high speed. Although the computer cannot directly perform advanced mathematical operations such as integration, differentiation, solving a trigonometric function, and taking a square root, such operations can be reduced to a series of basic arithmetic operations by the techniques of *numerical analysis* which approximate integration by summation, differentiation by finite differences, trigonometric function by power series, and square root by the iteration process.

3. Computer System. A computer is formed by five separate components or units as shown below:

$$\begin{array}{c} \text{Storage Unit} \\ \Updownarrow \\ \text{Input Unit} \rightarrow \text{Control Unit} \rightarrow \text{Output Unit} \\ \Updownarrow \\ \text{Arithmetic-logic Unit} \end{array}$$

a. Input Unit. This unit supplies the information given in a problem in terms of numbers, letters, and symbols which the computer can accept. The information is usually fed into the system from punched cards, punched paper tape, or magnetic tape, or inserted manually from a keyboard or switches, or transmitted automatically from analog-to-digital or machine-to-digital converters.

b. Control Unit. This unit directs the sequence of operations, interprets the coded instructions, and initiates commands for execution of the instructions. The unit contains a *program register* which can temporarily store current instructions and control the operation during the execution of instructions.

c. Arithmetic-logic Unit. This unit performs arithmetic and logical operations such as adding, subtracting, multiplying, and comparing two numbers at a time. It contains an *accumulator* which can store information temporarily and can form algebraic sums or make other arithmetic-logic operations. In *serial computers*, addition is performed serially by adding two digits of a pair of numbers at a time. In *parallel computers*, this is done in parallel by adding all digits of a pair of numbers simultaneously. Since other arithmetic operations are usually performed by repeated additions, the speed of a computer is mainly influenced by the arithmetic mode. The arithmetic operation time varies from microseconds in high-speed machines, to milliseconds in medium-speed machines, and to seconds in low-speed machines. This time often includes the storage *access time* which is the time it takes to deliver the information from storage when called upon for use.

The arithmetic-logic unit can also make logical decisions; distinguish positive, negative, and zero values; and transfer this information to other units of the computer.

d. Storage Unit. This unit is constructed mainly of electromechanical, magnetic, or electric devices in the form of cores, drums, or tapes which can store data and instructions internally until needed. Each storage location is identified by an individual location number called an *address*.

A storage unit may be called *external, internal,* or *buffer* depending on whether the storage facilities are independent, dependent, or synchronizing two different forms of storage (usually external and internal) of the computer.

The storage location to hold one computer word is usually designated by a specific address. The speed of storage is measured in terms of access time. The time required to transfer one computer word from a designated register is the *read time* and that required to transfer one computer word from a register to a designated storage location is the *write time*. The storage may be *serial* or *parallel* depending on whether the words are located one at a time or all at the same time, respectively. The time required to transfer one computer word from one storage device to another is the *word time*. The capacity of a storage unit may vary from 2 to over 64,000 words.

e. Output Unit. This unit is to present the results obtained from the computer on punched cards or tapes, magnetic tape, and oscilloscope, or in printed form by electrical typewriters.

4. Numerical System. *a. Number System.* The digital computer works with the digits of numbers. The familiar number systems used in computers are *binary* (using a base of 2) and *decimal* (using a base of 10). The decimal system uses ten symbols, while the binary system uses only two: 0 and 1. A *b*inary dig*it* is frequently called a *b*it. Because binary machines are simpler to build than decimal machines, except for problems involving large amounts of input and output, the binary system is nearly always used for the internal machine representation of numbers. The actual conversion of the decimal to binary notation and vice versa is performed automatically by the machine.

b. Word Length. A *word* or *machine word* is a set of digits occupying one storage location and treated by the computer as a unit. The word can be either a number or an instruction. The number of digits in a word is termed the *word length*. A common word length is 10 decimal digits and sign (plus or minus) for decimal computers and 36 to 40 binary digits plus sign for binary computers. Some machines have a variable word length.

c. Location of Decimal Point. There are two systems of locating the decimal point in arithmetic operations; namely the *fixed point* and the *floating point*. In fixed-point systems, the calculation is done usually with integers only and without any decimal point or decimal portions. In the floating-point system, the computer itself is used to determine the decimal point and a number is specified in two parts, a mantissa containing the significant digits and an exponent denoting the power of the base by which the mantissa is to be multiplied. A floating-point machine is more expensive because of its more complicated design but it makes the problem easier to program for a computation.

d. Instruction Words. In some computers each instruction is one word long; in others two or more instructions can be put in a single word. An instruction is therefore composed of a number of digits; every digit has its particular functions. Some of the digits indicate the arithmetic and step operations to be performed and are called the *operation code*. Others indicate the *data address*, or the *memory address* from or to which numbers are to be taken. Still others indicate the *instruction address*, or the execution of the instruction stored in the location. For example, an instruction represented by 19 1348 0254 means: multiply (represented by the operation code 19) the number in the arithmetic unit by the number stored in location 1348 (data address) and then execute the instruction stored in location 0254 (instruction address).

5. Solution of a Problem. The solution of a problem by a digital computer consists of several basic steps: problem formulation, mathematical analysis, programming, coding, code checking, program documentation, and production. Distinction between the second, third, and fourth steps is not always clear, depending largely on the language used to communicate with the computer. The step for program documentation is needed only when dealing with production problems where the programmer is not the sole user.

a. Problem Formulation. This step consists mainly of the preparation of a written statement defining the problem. This should include all necessary sketches and notations to make the problem clear, detailed input and output information, ranges and precision of the numbers involved, and specifications or restrictions of the problem, if any.

b. Mathematical Analysis. This step is to provide the necessary equations and numerical techniques for solving the problem. It usually involves the derivation of the equations and the expression of the equations by numerical analysis for computer solution.

c. Programming. This step is to prepare the complete set of instructions necessary to solve the given problem. In a broad sense, it is often used to include the complete plan for the solution of a problem, including the numerical analysis, the coded instructions, and specifications for input and output data. Further details about programming will be discussed later.

d. Coding. This step involves the preparation in proper sequence of the detailed set of instructions. The instructions must be coded by basic symbols so that they can be accepted by the computer. Such a set of coded instructions is called a *routine*. A routine which can be easily incorporated into a larger problem and which will perform some special computations, such as the solution of a system of linear equations, or evaluation of a trigonometric function, is known as *subroutine*. Details of coding depend entirely on the particular type of computer to be used and are therefore beyond the present discussion.

e. Code Checking. Since coding usually involves a large number of detailed steps, it is a major source of human errors. *Code checking* or *debugging* (to find the "bugs" or errors) is necessary and usually done by various methods. A complete code checking may be made by a hand-calculated test solution which provides a set of correct answers to compare with the computer answers.

f. Program Documentation. When a correct program has been developed, it may be fully documented for future use by other programmers. This step should describe the nature of the program, related technical and operating instructions, and any other information which other users of the program should know in order to use it correctly in the future. Generally a complete program write-up is prepared and deposited in a computer program library.

g. Production. This step is to execute the program by the computer and to provide the results by a plotter or other device for output as described before.

6. Programming. In a broad sense, programming may include several or all steps in solving a problem on a computer. Here, however, the subject is discussed in a rather limited scope with respect only to flow diagramming and automatic programming.

a. Flow Diagramming. A *flow diagram* or *flow chart* is a group of symbols which are used to describe the procedure for the solution of a problem. Various symbols are used by different authors. Table 29-3 shows some typical symbols and their examples. The *direction* of flow showing the relationship of one symbol to another is represented by arrows. The example shows that from the start of the program, A is executed, then B is executed, and finally the program is ended. The *connector* represented by a circle may be a start or a stop. The stop sign indicates the end of the program.

DIGITAL COMPUTERS 29-19

If there are several ways to end the program, there may be several stop symbols. The *input/output* symbol represented by a rectangle with a cutoff corner refers to any operation that involves an input or output device. The *operation* shown by a rectangular box represents any steps in the program that are not represented by special symbols. The *decision* shown by a rhomboid defines the logic of the program. Fundamentally, the logical choice is made by giving a yes-or-no answer to a single question, or by providing a greater, equal, or lesser answer to the question. If the answer is one of the two or three answers, the computation goes to that particular place in the program for the next series of operations.

Several typical operations or procedures that may be involved in a program can be shown by flow diagrams. The *arithmetic operation* is characterized by a linear flow diagram as shown in Table 29-3 by the example for the direction of flow. In general,

Table 29-3. Typical Symbols for Flow Diagrams

Symbol	Interpretation	Example
→, ←	Direction	Start → A → B → Stop
○	Connector	Start Stop
▱	Input/output	Read a card Print Y
▭	Operation	Compute $A = B+C$ Increase A by 2 Find average rainfall
◇	Decision	Yes ← Another case? ↓ No ; $X>Y$ ← ◇ → $X<Y$ ↓ $X=Y$; Is $X=0$? → Yes ↓ No

this diagram is the simplest to program because the program is essentially the same as the mathematical equation to be evaluated.

The *iterative procedure* can be performed by much more efficient use of the computer memory and logic, for which a small number of instructions are used repeatedly to process different sets of data. Such procedures are typified by the flow diagram in Fig. 29-7 which contains a closed *loop*. Loops may involve a fixed number of steps or may be terminated when a given criterion is reached. Generally, there are four essential steps in writing a loop: initialization, execution, modification, and testing. *Initialization* includes the preliminary steps necessary to restore the initial conditions before entering a loop. *Execution* includes a series of sequential steps required to determine the incremental and cumulative values each time through the loop. *Modification* includes the steps necessary to change the variables and also the address in preparation for the next cycle through the cycle. *Testing* is the step which determines whether the loop should be repeated again or the computation should leave the loop and proceed elsewhere in the program. In complicated programs, a series of loops may be used within loops. In a normal iterative procedure, the same instruction operates on different sets of data. In a special case, the instruction may operate on successive approximation of a single final answer.

There is a large group of problems where the computation is relatively simple, but the answer is to be obtained through a large number of simple *logical decisions* of the yes-no or greater-equal-less type. For such processes there are no formal rules for flow diagramming but they all consist of repeated application of the decision symbol.

b. Automatic Programming. A computer is constructed basically to respond only to the basic codes or *machine language*. These codes or the language are the information recorded in a form which may be used directly by the computer without prior translation. In order to communicate with the computer, the user must be able to provide his program in terms of the machine language. Naturally, the machine language is developed for the convenience of the computer design and operation but not from the user's point of view. For the user, the machine language is often too complicated to learn easily, and its use is too time-consuming to solve complex problems. Further, the machine language is different for different types of

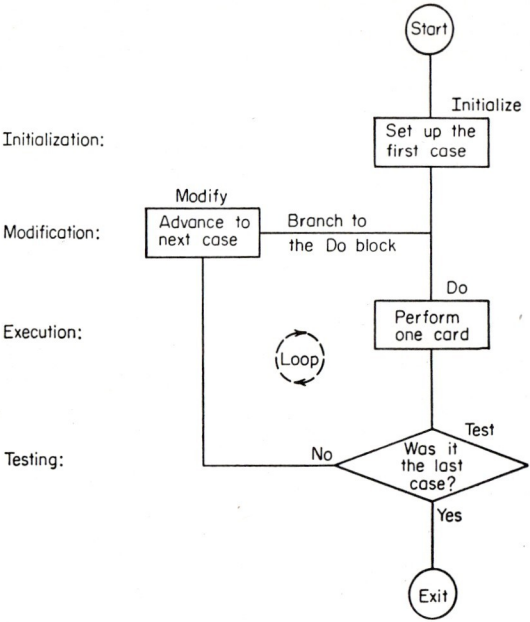

Fig. 29-7. Example and parts of a loop.

computers. Thus, the user has to learn a new language every time he works with a new machine. All such difficulties are now being gradually relieved by means of *automatic programming*. By automatic programming, the user may learn how to write a program without having to learn the intricacies of the computer and its language.

Automatic programming is now commonly achieved by the use of routines called *compilers* and *assemblers*. These programs automatically assemble or compile several programs containing symbolic notations or language into a complete program that is expressed in absolute machine language and is ready to be used by the computer. For this type of programming, there is a need for a *processor* which is a program for the computer that tells it how to translate the program written in the symbolic language into a program written in machine language. The program which defines the operations for the computer and which is written by the programmer in the symbolic language is called the *source program*. The source program is then put into the computer along with the processor. The computer, following the instructions from the processor, converts the source program into a program in machine language, called the *object program*, ready to be executed on the machine. The symbolic language is largely independent of the computer but the processor is dependent on the object machine for its proper execution. Thus, each object machine must have its own processor.

One widely used automatic programming system is the FORTRAN (for FORmula TRANslation) which has been developed and improved mainly by the International Business Machine Corporation. The FORTRAN language is a mathematical-type language which is composed of the individual commands or statements of a program, operators such as + or −, and expressions such as $A + B - C$. The FORTRAN processor is a program for the computer which tells it how to translate the program written in FORTRAN language into a program expressed in machine language. The program written in FORTRAN is subsequently translated by the computer, under the control of the processor, into machine language for execution. FORTRAN has been made available for the IBM 650, 1620, 705, 6070, 704, 709, 7090, and 7094 and it is also constantly being developed for use in other types of computers.

A further development in automatic programming is the adoption of *algorithmic language*. This is an arithmetic language designed for combinatorial analysis by which numerical procedures may be precisely and logically presented to a computer in a standard form. The language is intended not only as a means of directly presenting any numerical procedure to any suitable computer for which a compiler is available, but also as a means of communicating numerical procedures among individuals. A number of ALGOL (for ALGOrithmic Language) have been developed for scientific and engineering computation. Similar to ALGOL, COBOL (for COmmon Business Oriented Language) has also been devised to serve as a means of communicating computational algorithms particularly those relating to business data processing. Both developments were sponsored by the Association of Computing Machinery and the U.S. Department of Defense. Bugliarello [146] proposed a content-oriented computer language for hydrology and hydraulic engineering, which he called HYDROL.

B. Application in Hydrology

Electronic digital computers have been used extensively in hydrology [141–152]. A number of published studies on such applications are to be reported below.

1. Surface Water. *a. Open-channel Flow.* One of the earliest applications of digital computers to hydrologic problems is the computation of backwater curves for the design of the St. Lawrence Seaway Project in 1956 [153]. The computation was made on the FERUT computer at the University of Toronto Physics Department for the Ontario Hydro-Electric Power Commission. It was reported that the machine produced results of the analysis within a year which otherwise would have taken 50 man-years to accomplish. In connection with this project, an IBM 602A computer was also employed by the Commission to calculate surges in hydropower-plant tunnels.

Computation of unsteady flow by computers was first attempted by the Institute of Mathematical Sciences of New York University in cooperation with the U.S. Army Corps of Engineers Ohio River Division for the analysis of floods in the Ohio River [154–155]. In this analysis, Eqs. (25-II-1) and (25-II-2) were solved on Remington-Rand UNIVAC I (for UNIVersal Automatic Computer I). Similar analyses were also made later such as on IBM 650 [156], on IBM 704 and 7090 [157], and on other computers [158].

Use of digital computers was further considered for the calculation of river tides [159]. A FORTRAN program has been prepared for the solution of storm surges in Louisiana [160].

b. Hydrologic Routing. Assuming a series of linear reservoirs, Eq. (14-31) was solved on the IBM 650 for streamflow routing in the Columbia River Basin by the U.S. Army Corps of Engineers North Pacific Division at Portland, Oregon [161–162]. Also, the Muskingum method with variable coefficients and a level-pool-reservoir routing (Sec. 25-II) were used for the development of hydrographs for a complex network of subareas in the analysis of floods in the Kansas River basin. This analysis was made on an IBM 650 computer by the U.S. Army Corps of Engineers Missouri River Division at Omaha, Nebraska [163]. It was also expanded to include the determination of releases of water in a system of reservoirs on an IBM 704 computer. The entire program was also prepared for the IBM 1620.

The routing of flow from rainfall to runoff has been done on the Burroughs 220 and IBM 7090 for developing synthetic streamflow records on small drainage areas [164–165] (Subsec. 14-III-A-5). Many other studies of hydrographs were also performed on digital computers such as those described in Refs. 166–170.

c. *River Hydraulics.* A number of problems in river hydraulics have been programmed for computer solution. The modified Einstein method for the analysis of total sediment load (Subsec. 17-II-V-D) was programmed on an IBM 650 by the U.S. Bureau of Reclamation [171–172]. The computations on grain-sorting in rivers and river models were executed on the Stantec-Zebra computer of the Institute of Applied Mathematics of the Technological University of Delft [173]. The analysis of critical tractive force for cohesive soils was made by a multiple linear correlation programmed on an IBM 650 [174]. An electronic computer was also used in the investigation of a bifurcation in the Lower Rhine River [175].

d. *Water Supply.* In the water-supply field, digital computers have been employed to study various problems [176] such as reservoir yield and water quality [177–179].

e. *Water Resources Planning.* Because of the complexity of water-resources problems, the digital computer is well suited for the analysis, design, and planning of water-resources projects. Several topics of flood routing, reservoir operation, etc., relating to water resources have already been discussed above. The U.S. Army Corps of Engineers investigated the use of computers for integrating reservoir operations as early as 1957 [180].

The use of digital computers in simulation of a whole river basin plan was first made in the mid-1950s on planning of the Nile River basin by the advisors in the Colonial Office of Great Britain and other European engineers for the Ministry of Irrigation and Hydro-Electric Power, Republic of Sudan. The plan was to control the Nile and its tributaries in order to provide the largest possible amount of irrigation water for the Sudan and Egypt. According to Morrice and Allen [181–182], the plan included five existing major dams and many other proposed structures (10 dams, two equalizing reservoirs, one canal, one tunnel, and one barrage). Monthly inflows were calculated to decimal places in billions of cubic meters for the 48-year record (January, 1905 to December, 1952) under analysis. The plan was to maximize the sum of Egypt's and the Sudan's share of flow at Aswan. For the development of this plan, continuity equations were written for various reaches of the river, for certain reservoirs, and for the main confluences. Also, control equations were written for river discharges at 16 reservoir locations. The inputs and the solution of the continuity and location equations were represented by 14 flow diagrams to be programmed on the IBM 650 computer. The flow diagrams consisted essentially of the continuity and control equations interspersed with checks in the form of questions or criteria. A compressed overall flow diagram for the computation was later given by Fitzpatrick [183]. The use of the computer enabled the engineers to try out computations for 300 project combinations. The Nile system, with an extensive number of control points and reservoirs, has proved the economy of the use of digital computers. It was estimated that a normal machine run covering one 48-year period took from 25 to 50 min, whereas at least 1,200 to 1,500 man-hours of hand calculations would have been required.

Simulation of river basins on computers was carried further in other studies. Subsequent to the analysis of the Nile River plan, the Tigris River basin in Iraq was similarly investigated by computer simulation. In the Harvard Water Program (Subsec. 26-II-I), simulation of a prototype river basin system was programmed for FORTRAN II performed on the IBM 704 [184–185].

The U.S. Bureau of Reclamation used the IBM 650 to determine the effect of various controlled amounts of fresh water inflow on the salinity of a series of lakes including Devils and Stump Lakes, North Dakota [186], and a Burroughs 205 Datatron to determine the firm power output from the steamflow record for Canyon Ferry and eight other power plants for forecast and flood-control operation with various uses of water for irrigation [141]. Other uses of digital computers for hydropower system analysis were also variously reported [187–189].

2. Groundwater. The digital computer has been employed in groundwater hydrology to solve the Theis equation, Eq. (13-43). Foley [190] reported that this equation has been programmed for solution on the LGP-30 and UNIVAC computers. The problems that were involved in the solution included calculation of drawdown, evaluation of the coefficients of storage and transmissibility, determination of economical well spacing, and computation of the collective capacity of a group of individual wells. Walton and Neill [191] later programmed the Theis equation on the University of Illinois ILLIAC computer to analyze groundwater supply of a deeply buried artesian aquifer in the Chicago region. In this investigation, the hydrogeologic boundaries were simulated for a mathematical model consisting of real and image wells. The computer required about 30 min to calculate 600 drawdowns at a rate of $0.35 per min. It took about one man-day to solve the whole problem as compared with 20 man-days for solution by conventional methods.

Fayers and Sheldon [192] have formulated a three-dimensional model of a geologic basin through the use of the Laplace, Poisson, and diffusion equations. This model was then solved by successive overrelaxation on the IBM 704 computer. Approximately 2 min of the computer time were required to perform 20 iterations for this problem. The objective of this study was to evaluate quantitatively the magnitude and direction of steady flow in aquifer systems and geologic basins and thus to enable prediction of the migration of hydrocarbons and also estimation of water supply and water conservation requirements.

IV. REFERENCES

A. On Both Analog and Digital Computers

1. Namyet, S.: Analog and digital computers in civil engineering, *J. Boston Soc. Civil Engrs.*, vol. 44, no. 1, pp. 36–56, January, 1957.
2. Scott, N. R.: "Analog and Digital Computer Technology," McGraw-Hill Book Company, Inc., New York, 1960.
3. Advani, R. M.: Importance of electronic computers and similar other devices in the field of hydraulic engineering, *Proc. Intern. Assoc. Hyd. Res. 9th Convention, Dubrovnik, Yugoslavia*, 1961, pp. 622–625.
4. Handel, Paul: "Electronic Computers: Fundamentals, Systems, and Applications," Prentice-Hall, Inc., Englewood Cliffs, N.J., 1961.
5. Huskey, H. D., and G. A. Korn: "Computer Handbook," McGraw-Hill Book Company, Inc., New York, 1961.
6. "Computer Basics," 6 vols., Howard W. Sams and Company, Inc., Indianapolis, Ind.; and The Bobbs-Merrill Company, Inc., New York, 1962.
7. Kitov, A. I., and N. A. Krinitskiĭ: "Electronic Computers," translated from the Russian by R. P. Froom, Pergamon Press, New York, 1962.
8. Automatic data processing glossary, U.S. Bureau of the Budget, U.S. Government Printing Office, Washington, D.C., December, 1962.

B. On Analog Computers

9. Wass, C. A. A.: "Introduction to Electronic Analogue Computers," Pergamon Press, London, 1955.
10. Korn, G. A., and T. M. Korn: "Electronic Analog and Hybrid Computers," McGraw-Hill Book Company, Inc., New York, 1964.
11. Noon, W.: Application of analog computer to engineering problems, *Gen. Motors Eng. J.*, vol. 4, no. 2, pp. 34–37, April–May–June, 1957.
12. Karplus, W. J.: "Analog Simulation," McGraw-Hill Book Company, Inc., New York, 1958.
13. Karplus, W. J., and W. W. Soroka: "Analog Methods; Computation and Simulation," 2d ed., McGraw-Hill Book Company, Inc., New York, 1959.
14. Smith, G. W., and R. C. Wood: "Principles of Analog Computation," McGraw-Hill Book Company, Inc., New York, 1959.
15. Warfield, J. N.: "Introduction to Electronic Analog Computers," Prentice-Hall, Inc., Englewood Cliffs, N.J., 1959.

16. Éterman, I. I.: "Analogue Computers," translated from the Russian by G. Segal, Pergamon Press, New York, 1960.
17. Jackson, A. S.: "Analog Computation," McGraw-Hill Book Company, Inc., New York, 1960.
18. Paynter, H. M. (ed.): "A Palimpsest on the Electronic Analog Art," George A. Philbrick Researches, Inc., Boston, Mass., 1955, 1958, 1960.
19. Rogers, A. E., and T. W. Connolly: "Analog Computation in Engineering Design," McGraw-Hill Book Company, Inc., New York, 1960.
20. Truitt, T. D., and A. E. Rogers: "Basics of Analog Computers," J. F. Rider, New York, 1960.
21. Fifer, Stanley: "Analogue Computation: Theory, Techniques and Applications," 4 vols., McGraw-Hill Book Company, Inc., New York, 1961.
22. Williams, R. W.: "Analogue Computation: Techniques and Components," Academic Press, Inc., New York, 1961.
23. MacKay, D. M., and M. E. Fisher: "Analogue Computing at Ultra-high Speed," Chapman and Hall, Ltd., London, 1962.
24. Tomovic, Rajko, and W. J. Karplus: "High-speed Analog Computers," John Wiley & Sons, Inc., New York, 1962.
25. Williams, R. W.: "Analogue Computation," Academic Press Inc., New York, 1962.
26. Johnson, C. L.: "Analog Computer Techniques," 2d ed., McGraw-Hill Book Company, Inc., New York, 1963.
27. Murphy, Glenn, D. J. Shippy, and H. L. Luo: "Engineering Analogies," Iowa State University Press, Ames, Iowa, 1963.

C. On Digital Computers

28. Livesley, R. K.: "An Introduction to Automatic Digital Computers," Cambridge University Press, London, 1957.
29. McCraken, D. D.: "Digital Computer Programming," John Wiley & Sons, Inc., New York, 1957.
30. Alt, F. L.: "Electronic Digital Computers," Academic Press Inc., New York, 1958.
31. Gotlieb, C. C., and J. N. P. Hume: "High-speed Data Processing," McGraw-Hill Book Company, Inc., New York, 1958.
32. Mandl, Matthew: "Fundamentals of Digital Computers," Prentice-Hall, Inc., Englewood Cliffs, N.J., 1958.
33. Susskind, A. K.: "Notes on Analog-Digital Conversion Techniques," John Wiley & Sons, Inc., New York, 1958.
34. Jeenel, Joachim: "Programming for Digital Computers," McGraw-Hill Book Company, Inc., New York, 1959.
35. McCormick, E. M.: "Digital Computer Primer," McGraw-Hill Book Company, Inc., New York, 1959.
36. Williams, S. B.: "Digital Computing Systems," McGraw-Hill Book Company, Inc., New York, 1959.
37. Wrubel, M. H.: "A Primer of Programming for Digital Computers," McGraw-Hill Book Company, Inc., New York, 1959.
38. Bartee, T. C.: "Digital Computer Fundamentals," McGraw-Hill Book Company, Inc., New York, 1960.
39. Irwin, W. C.: "Digital Computer Principles," Van Nostrand, Princeton, N.J., 1960.
40. Ledley, R. S.: "Digital Computer and Control Engineering," McGraw-Hill Book Company, Inc., New York, 1960.
41. Evans, G. W., II, and C. L. Perry: "Programming and Coding for Automatic Digital Computers," McGraw-Hill Book Company, Inc., New York, 1961.
42. Leeds, H. D.: "Computer Programming Fundamentals," McGraw-Hill Book Company, Inc., New York, 1961
43. Leeds, H. D., and G. M. Weinberg: "Computer Programming Fundamentals," McGraw-Hill Book Company, Inc., New York, 1961.
44. McCraken, D. D.: "A Guide to Fortran Programming," John Wiley & Sons, Inc., New York, 1961.
45. Nathan, Robert, "Computer Programming Handbook: A Guide for Beginners," Prentice-Hall, Inc., Englewood Cliffs, N.J., 1961.
46. Siegel, Paul: "Understanding Digital Computers," John Wiley & Sons, Inc., New York, 1961.
47. Ledley, R. S.: "Programming and Utilizing Digital Computers," McGraw-Hill Book Company, Inc., New York, 1962.
48. Leeson, D. N., and D. L. Dimitry: "Basic Programming Concepts and the IBM 1620 Computer," Holt, Rinehart and Winston, New York, 1962.

REFERENCES

49. Staff of Technical Training Department, Burroughs Corporation: "Digital Computer Principles," McGraw-Hill Book Company, Inc., New York, 1962.
50. Welch, H. J.: The use of computers in civil engineering education, College of Engineering, The University of Michigan, June 1, 1962.
51. Arden, B. W.: "An Introduction to Digital Computing," Addison-Wesley Publishing Company, Reading, Mass., 1963.
52. Green, B. F.: "Digital Computers in Research," McGraw-Hill Book Company Inc., New York, 1963.
53. Streeter, V. L.: Use of digital computers in teaching hydraulic design, *J. Eng. Ed.*, vol. 53, no. 5, pp. 284–288, January, 1963.
54. Wilcox, R. H.: Unconventional computers of the future, *Naval Res. Rev.*, vol. 17, no. 1, pp. 1–8, January, 1964.

D. On Analog-computer Application in Surface-water Hydrology

55. Electrical computer aids flood studies, *Civil Eng.*, vol. 17, no. 3, p. 116, March, 1947.
56. GEDA application to flood control and water distribution problems, Goodyear Aircraft Corporation, GER-6053, May, 1954.
57. Lawler, E. A.: The use of the analog computer in flood routing and other hydraulic problems, in "Analog Computers, Their Industrial Applications," Midwest Research Institute, Kansas City, Mo., April, 1956, pp. 76–108.
58. Lawler, E. A., and F. V. Druml: Hydraulic problem solution on electronic computers, *Proc. Am. Soc. Civil Engrs., J. Waterways and Harbors Div.*, vol. 84, no. WWI, pp. 1–38, January, 1958.
59. Rockwood, D. M., and C. E. Hildebrand: An electronic analog for multiple-stage reservoir-type storage routing, *Tech. Bull.* 18, U.S. Army Corps of Engineers, Civil Works Investigation, Project CW-171, March, 1956, pp. 1–12.
60. Linsley, R. K., Jr., L. W. Foskett, and M. A. Kohler: Electronic device speeds flood forecasting, *Eng. News-Rec.*, vol. 141, no. 26, pp. 64–66, December 23, 1948.
61. Linsley, R. K., Jr., L. W. Foskett, and M. A. Kohler: Use of electronical analogy in flood wave analysis, *Intern. Assoc. Sci. Hydrology Pub.* 29, pp. 221–227, 1948.
62. Kohler, M. A.: Application of electronic flow routing analog, *Proc. Am. Soc. Civil Engrs.*, vol. 78, separate no. 135, pp. 1–11, June, 1952.
63. Light, Phillip: Electronic streamflow routing, Upper Mississippi River, U.S. Weather Bureau River Forecast Center, St. Louis, Mo., August, 1953.
64. Kohler, M. A.: Application to stream-flow routing, in electrical analogies and electronic computers: a symposium, *Trans. Am. Soc. Civil Engrs.*, vol. 118, pp. 1028–1038, 1953.
65. Messerle, H. K.: Electronic high speed simulation of hydraulic problems, *J. Inst. Engrs.* (Australia), vol. 25, no. 3, pp. 35–41, March, 1953.
66. Messerle, H. K.: Differential analyser solution of hydroelectric systems, *Houille Blanche*, no. 6, pp. 813–836, 1956.
67. Ishihara, Tojiro, Shoitiro Hayami, and Shigenori Hayami: On the electronic analog computer for flood routing, *Proc. Japan Acad.* (Tokyo), vol. 31, no. 9, pp. 891–895, 1954.
68. Ishihara, Tojiro, and Yasuo Ishihara: On the electronic analog computer for flood routing (in Japanese), *Trans. Japan Soc. Civil Engrs.* (Tokyo), no. 24, pp. 44–57, April, 1955.
69. Ishihara, Tojiro, and Yasuo Ishihara: Electronic analog computer for flood flows in the Yodo River (in Japanese), *Proc. Japan Soc. Civil Engrs.*, vol. 41, no. 8, pp. 21–24, 1956.
70. Ishihara, Tojiro, and Yasuo Ishihara: On the electronic analog computer for reservoir routing (in Japanese), *Trans. Japan Soc. Civil Engrs.*, vol. 41, no. 2, pp. 58–63, February, 1956.
71. Ishihara, Tojiro, Shoitiro Hayami, and Shigenori Hayami: Electronic analog computer for flood flows, *Proc. Region. Tech. Conf. Water Resources Develop. Asia and Far East*, U.N. Econ. Com. Asia and Far East, Flood Control Ser. 9, Bangkok, 1956, pp. 170–174.
72. Ishihara, Tojiro: Application of electronic analog computer for flood routing to actual rivers (in Japanese), *Trans. Japan Soc. Civil Engrs.* (Tokyo), no. 43, pp. 43–47, February, 1957.
73. Ishihara, Yasuo: On the applieation of an electronic analog computer for flood routing to actual rivers (in Japanese), *Trans. Japan Soc. Civil Engrs.*, no. 43, pp. 43–47, 1957.
74. Ishihara, Yasuo: Electronic analog computer for runoff, *Trans. Japan Soc. Civil Engrs.*, no. 60, pp. 37–53, January, 1959.
75. Ishihara, Yasuo: On an electronic analog computer for runoff (in Japanese), *Proc. Japan Soc. Civil Engrs.*, no. 60, pp. 37–45, 1959.

76. Ishihara, Tojiro, and Yasuo Ishihara: Runoff analysis by analog computer, *Proc. Intern. Assoc. Hyd. Res. 9th Convention, Dubrovnik, Yugoslavia,* 1961, pp. 749–759.
77. Chow, V. T.: "Open-channel Hydraulics," McGraw-Hill Book Company, Inc., New York, 1959.
78. Glover, R. E., D. J. Herbert, and C. R. Daum: Application to an hydraulic problem, in electrical analogies and electronic computers: a symposium, *Trans. Am. Soc. Civil Engrs.,* vol. 118, pp. 1010–1016, 1953.
79. Harder, J. A.: The theoretical basis for non-linear electric analogs for open channel flow, *Univ. of Calif. Inst. Eng. Res.,* ser. 107, no. 1, June, 1957.
80. Harder, J. A., L. Mockros, and R. Nishizaki: Flood control analogs, *Univ. of Calif. Water Resources Center Contribution* 24, Berkeley, Calif., 1960.
81. Einstein, H. A., and J. A. Harder: An electric analog model of a tidal estuary, *Trans. Am. Soc. Civil Engrs.,* vol. 126, pt. IV, pp. 855–868, 1961.
82. Harder, J. A., and F. D. Masch: Non-linear tidal flows and electric analogs, *Proc. Am. Soc. Civil Engrs., J. Waterways and Harbors Div.,* vol. 87, no. WW4, pp. 27–39, November, 1961.
83. Harder, J. A.: Analog models for flood control systems, *Trans. Am. Soc. Civil Engrs.,* vol. 128, pt. I, pp. 993–1004, 1963.
84. Paynter, H. M.: Methods and results from MIT studies in unsteady flow, *J. Boston Soc. Civil Engrs.,* vol. 39, no. 2, pp. 120–165, April, 1952.
85. Paynter, H. M.: Computer techniques in hydrology: Flood routing by admittance methods, in "A Palimpsest on the Electronic Analog Art," George A. Philbrick Researches, Inc., Boston, Mass., 1955, 1958, 1960, pp. 239–245.
86. Paynter, H. M.: Hydraulics by analog, part I, *J. Boston Soc. Civil Engrs.,* vol. 46, no. 3, pp. 197–219, July, 1959.
87. Dziatlik, H.: Model elektryczny rzeki (Electrical model for a river), *Przeglad Elektrotechniczny* (Poland), pp. 447–455, July 21, 1955.
88. Halek, V.: Hydraulic problems using electrical analogy (in Czech), *Vodni Hospodarstvi,* no. 10, pp. 297–305, 1958.
89. Kalinin, G. P., and A. G. Levin: Use of an electronic analog computer in the forecasting of rain floods, *Meterologiia i Gidrologiia,* no. 12, pp. 14–18, December, 1960. Translation from the Russian by U.S. Joint Publications Research Service, Washington, D.C.
90. Rosa, J. M., and others: Electronic analog, U.S. Agricultural Research Service, unpublished material, Moscow, Idaho, 1961.
91. Rosa, J. M.: The electronic analogy to a watershed, M. S. Thesis, University of Idaho, Moscow, Idaho, 1963.
92. Beyers, L. A.: The electronic analogy to spring snow melt on a watershed, M. S. Thesis, University of Idaho, Moscow, Idaho, 1962.
93. Robinson, A. D., and L. A. Beyers: An analog computer for calculating runoff increments from average rainfall rates, *Res. Rept.* 353, U.S. Agricultural Research Service, Soil and Water Conservation Research Division, in cooperation with the Idaho Agricultural Experiment Station, Moscow, Idaho, 1962.
94. Rosa, J. M.: The use of analog models in the solution of hydrologic problems, paper presented at the Soil Conservation Service's Hydraulics' Meeting, New York City, August 12–16, 1963.
95. Shen, John: An analog solution of the turbulent-diffusion equation for open-channel flow, *U.S. Geol. Surv. Profess. Paper* 450-E, pp. 169–171, 1963.
96. Shen, John: Use of hydrologic models in the analysis of flood runoff, *U.S. Geol. Survey Surface Water Br., Res. Sec. Prel. Rept.,* March, 1963.
97. Falk, L. L.: Analog computer—a modern tool in water pollution control, *Proc. Am. Soc. Civil Engrs., J. Sanitary Eng. Div.,* vol. 88, no. SA6, pp. 31–59, November, 1962.
98. Paynter, H. M.: Surge and water hammer problems, *Trans. Am. Soc. Civil Engrs.,* vol. 118, pp. 962–989, 1953; also in Ref. 18, pp. 217–223.
99. Koenig, E. C., and H. A. Knudtson: Computer studies of penstock and governor systems, *Proc. Am. Soc. Civil Engrs., J. Power Div.,* vol. 85, no. P05, pp. 1–18, December, 1957.

E. On Analog-computer Application in Groundwater Hydrology

100. Pavlovsky (Pavlovskiĭ), N. N.: Motion of water under dams, *Trans. 1st Intern. Cong. on Large Dams,* Stockholm, vol. 4, pp. 179–192, 1933.
101. Lane, E. W., F. B. Campbell, and W. H. Price: The flow net and the electric analogy, *Civil Eng.,* vol. 4, pp. 510–514, October, 1934.
102. Harza, L. F.: Uplift and seepage under dams on sand, *Trans. Am. Soc. Civil Engrs.,* vol. 100, pp. 1352–1385, 1935.

103. Wyckoff, R. D., and D. W. Reed: Electrical conduction models for the solution of water seepage problems, *J. Appl. Phys.*, vol. 6, pp. 395–401, 1935.
104. Khosla, R. B. A. N., N. K. Bose, and E. McK. Taylor: Design of weirs on permeable foundations, *Pub.* 12, Central Board of Irrigation, Simla, India, September, 1936.
105. Rel'tov, B. F.: Electrical analogy applied to three dimensional study of percolation under dams built on previous heterogeneous foundations, *Trans. 2nd Cong. on Large Dams*, Washington, D.C., pp. 73–85, 1936.
106. Vreedenburgh, C. G. F., and O. Stevens: Electric investigation of underground water flow nets, *Proc. Conf. Soil Mech. and Foundation Eng.*, vol. 1, Harvard University, Cambridge, Mass., pp. 219–222, 1936.
107. Stevens, O.: Electrical determination of the line of seepage and flow net of a groundwater flow through joint regions with different anisotropy, *De Ingenieur in Ned. Indie*, vol. 9, pp. 205–212, 1938.
108. Selim, M. A.: Dams on porous media, *Trans. Am. Soc. Civil Engrs.*, vol. 112, pp. 488–505, 1947.
109. Babbitt, H. E., and D. H. Caldwell: The free surface around, and the interference between, gravity wells, *Univ. of Illinois Eng. Exp. Sta. Bull.* 374, Urbana, Ill., 1948.
110. Muskat, Morris: The theory of potentiometric models, *Trans. Am. Inst. Min. and Metal Engrs.*, vol. 179, pp. 216–221, 1949.
111. Zangar, C. N.: Theory and problems of water percolation, *U.S. Bur. Reclam. Eng. Mono.* 8, Denver, Colo., April, 1953.
112. Felius, G. P.: Recherches hydrologiques par des modèles électriques (Hydrologic studies by electric models), *Intern. Assoc. Sci. Hydrology Pub.* 37, vol. 2, pp. 162–169, 1954.
113. Luthra, S. D. L., and G. Ram: Electrical analogy applied to study seepage into drain tubes in stratified soil, *J. Central Board Irrig. and Power* (India), vol. 11, pp. 398–405, 1954.
114. Johnson, H. A.: Seepage forces in a gravity dam by electrical analogy, *Proc. Am. Soc. Civil Engrs.*, vol. 81, separate 757, pp. 1–16, 1955.
115. Opsal, F. W.: Analysis of two- and three-dimensional ground-water flow by electrical analogy, *The Trend in Eng. at the Univ. of Washington*, vol. 7, no. 2, Seattle, pp. 15–20, 32, 1955.
116. Mack, L. E.: Evaluation of a conducting-paper analog field plotter as an aid in solving ground-water problems, *Kans. Geol. Surv. Bull.* 127, pt. 2, 1957.
117. Zee, C. H., D. F. Peterson, Jr., and R. O. Bock: Flow into a well by electric and membrance analogy, *Trans. Am. Soc. Civil Engrs.*, vol. 122, pp. 1088–1105, 1957.
118. Monke, E. J.: A study of water flow patterns near subsurface drains, Ph.D. thesis, University of Illinois, Urbana, Ill., 1959.
119. Todd, D. K., and Jacob Bear: River seepage investigation, *Univ. of Calif. Water Resources Center Contribution* 20, September 1, 1959.
120. Bullen, R. O.: An electrical analogue investigation of the standpipe field permeability test, *Proc. Intern. Assoc. Hyd. Res. 9th Convention, Dubrovnik, Yugoslavia*, pp. 287–298, 1961.
121. Sherwood, C. B., and H. Klein: Use of analog plotter in water-control problems, *Ground Water*, vol. 1, no. 1, pp. 8–10, 15, January, 1963.
122. Wyckoff, R. D., H. G. Botset, and Morris Muskat: Flow of liquids through porous media under the action of gravity, *J. Appl. Phys.*, vol. 3, pp. 90–113, 1932.
123. Wyckoff, R. D., and H. G. Botset: An experimental study of the motion of particles in systems of complex potential distribution, *J. Appl. Phys.*, vol. 5, pp. 265–275, 1934.
124. Botset, H. G.: The electrolytic model and its application to the study of recovery problems, *Trans. Am. Inst. Min. Metal. Engrs.*, vol. 165, pp. 15–25, 1946.
125. Luthin, J. N.: An electrical resistance network solving drainage problems, *Soil Sci.*, vol. 75, pp. 259–274, 1953.
126. Hanks, R. J., and S. A. Bowers: Non-steady-state moisture, temperature, and soil air pressure approximation with an electric simulator, *Soil Sci. Am. Proc.*, vol. 24, no. 4, pp. 247–252, July–August, 1960.
127. Bouwer, Herman, and W. C. Little: A unifying numerical solution for two-dimensional steady flow problems in porous media with an electrical resistance network, *Proc. Soil Sci. Soc. Am.*, vol. 23, pp. 91–96, 1959.
128. Bouwer, Herman: A study of final infiltration rates from cylinder infiltrometers and irrigation furrows with an electrical resistance network, *Trans. 7th Cong. Intern. Soc. Soil Sci.*, vol. 1, pp. 448–457, Midson, Wisconsin, 1960.
129. Bouwer, Herman: Variable head technique for seepage meters, *Proc. Am. Soc. Civil Engrs., J. Sanitary Eng.*, vol. 87, no. IR1, pp. 31–44, March, 1961.
130. Bouwer, Herman: Analyzing ground-water mounds by resistance network, *Proc. Am.*

Soc. Civil Engrs., J. Irrig. and Drainage Div., vol. 88, no. IR3, pp. 15–36, September, 1962.
131. Stallman, R. W.: From geologic data to aquifer analog models, *Am. Geol. Inst.*, vol. 5, no. 7, pp. 8–11, April, 1961.
132. Stallman, R. W.: Electric analog of three-dimensional flow to wells and its application to unconfined aquifers, *U.S. Geol. Surv. Open-file Rept.*, 1961.
133. Stallman, R. W.: Electric analog of three-demensional flow to wells and its application to unconfined aquifers, *U.S. Geol. Surv. Water-Supply Paper* 1536-H, 1963.
134. Stallman, R. W.: Calculation of resistance and error in an electric analog of steady flow through nonhomogeneous aquifers: General ground-water techniques, *U.S. Geol. Surv. Water-Supply Paper* 1544-G, 1963.
135. Skibitzke, H. E.: Electronic computers as an aid to the analysis of hydrologic problems, *Intern. Assoc. Sci. Hydrology Pub.* 52, pp. 347–358, 1961.
136. Skibitzke, H. E.: Hydrologic models of ground-water movements, *Proc. 6th Arizona Watershed Symp.*, pp. 17–19, Phoenix, Ariz., 1962.
137. Skibitzke, H. E.: The use of analog computers for studies in ground-water hydrology, *J. Inst. Water Engrs.*, vol. 17, no. 3, pp. 216–230, London, 1963.
138. Brown, R. H.: Progress in ground water studies with the electronic analog model, *J. Am. Water Works Assoc.*, vol. 54, no. 8, pp. 943–958, 1962.
139. Walton, W. C., and T. A. Prickett: Hydrogeologic electric analog computer, *Proc. Am. Soc. Civil Engrs., J. Hydraulics Div.*, vol. 89, no. HY6, pp. 67–91, November, 1963.
140. Robinson, G. M.: The evaluation of water resources development by analog techniques, *Intern. Assoc. Sci. Hydrology Pub.* 64, pp. 442–454, 1963.

F. On Digital-computer Application in Hydrology

141. Abstracts of electronic computer programs developed by Bureau of Reclamation, U.S. Bureau of Reclamation, Electronic Computer Program Abstract Issue no. 1, October, 1959.
142. Swain, F. E., and H. S. Riesbol: Electronic computers used for hydrologic problems, *Proc. Am. Soc. Civil Engrs., J. Hydraulics Div.*, vol. 85, no. HY11, pp. 21–29, November, 1959.
143. Beck, H. A., and D. W. Webber: Electronic computer applications in the Bureau of Reclamation, a paper presented at the Second Annual Digital Computer Applications Symposium, Denver Research Institute, University of Denver, Denver, Colo., November 5–6, 1959.
144. Harbeck, G. E., Jr., and W. L. Isherwood, Jr.: Digital computers for water resources investigations, *Proc. Am. Soc. Civil Engrs., J. Hydraulics Div.*, vol. 85, no. HY1, pp. 31–38, November, 1959.
145. Snyder, W. M.: Hydrologic studies by electronic computers in TVA, *Proc. Am. Soc. Civil Engrs., J. Hydraulics Div.*, vol. 86, no. HY2, pp. 1–10, February, 1960.
146. Bugliarello, George: Toward a computer language for hydraulic engineering, *Proc. Intern. Assoc. Hyd. Res. 9th Convention, Dubrovnik, Yugoslavia*, 1961, pp. 714–729.
147. Normand, M.: L'électronique au service de l'hydrologie (Use of electronics in hydrology), *Proc. Intern. Assoc. Hyd. Res. 9th Convention, Dubrovnik, Yugoslavia*, 1961, pp. 665–672.
148. Welch, H. J.: The use of computers in civil engineering education, Department of Civil Engineering, The University of Michigan, June 1, 1962.
149. Streeter, V. L.: Use of digital computers in teaching hydraulic design, *J. Eng. Ed.*, vol. 53, no. 5, pp. 284–288, January, 1963.
150. TVA computer programs for hydrologic analyses, *Res. Paper* 3, Tennessee Valley Authority, Office of Tributary Area Development, Knoxville, Tenn., June, 1963.

G. On Digital-computer Application in Surface-water Hydrology

151. Kennedy, E. J.: Streamflow records by digital computers, *Proc. Am. Soc. Civil Engrs., J. Irrigation and Drainage*, vol. 89, no. IR3, pp. 29–36, September, 1963.
152. Kelly, L. L.: Electronic computers in watershed research, *Proc. Am. Soc. Civil Engrs., J. Irrigation and Drainage*, vol. 89, no. IR3, pp. 37–43, September, 1963.
153. Backwater curves computed by machine, *Eng. News-Rec.*, vol. 156, no. 19, pp. 62 and 64, May 10, 1956.
154. Isaacson, E. J., J. J. Stoker, and B. A. Treosch: Numerical solution of flood prediction and river regulation problems, New York University, Institute of Mathematical Sciences, Rept. I, October, 1953, Rept. II, January, 1954, and Rept. III, October, 1956.

REFERENCES

155. Isaacson, E., J. J. Stoker, and A. Troesch: Numerical solution of flow problems in rivers, *Proc. Am. Soc. Civil Engrs., J. Hydraulics Div.*, vol. 84, no. HY5, pt. 1, pp. 1–18, October, 1958.
156. Guyot, M. T., J. Nougaro, Cl. Thirriot: Étude numérique des régimes transitories dans les canaux (Numerical study of unsteady flow in open channels), *Proc. Intern. Assoc. Hyd. Res. 9th Convention, Dubrovnik, Yugoslavia*, 1961, pp. 820–831.
157. Faure, J., and N. Nahas: Deux problèmes de mouvements non permanents à surface libre résolus sur ordinateurs électroniques. (Two problems of unsteady flow with free-surface solved on electronic computers), *Proc. Intern. Assoc. Hyd. Res. 9th Convention, Dubrovnik, Yugoslavia*, 1961, pp. 854–869.
158. Preissmann, A., and J. Cunge: Calcul des intumescences sur machines électroniques (Computing water-wave propagation by electronic computers), *Proc. Intern. Assoc. Hyd. Res. 9th Convention, Dubrovnik, Yugoslavia*, 1961, pp. 656–664.
159. Canisius, P., and F. K. Rubbert: The application of electronic computers to the predictive calculation of river tides, *Proc. Intern. Assoc. Hyd. Res. 9th Convention, Dubrovnik, Yugoslavia*, 1961, pp. 790–792.
160. Davis, H. R.: Computer solution of storm surges, *Proc. Am. Soc. Civil Engrs., J. Waterways and Harbors Div.*, vol. 88, no. WW3, pp. 117–123, August, 1962.
161. Rockwood, D. M.: Columbia basin streamflow routing by computer, *Civil Eng.*, vol. 28, no. 5, pp. 348–351, May, 1958.
162. Rockwood, D. M.: Columbia basin streamflow routing by computer, *Proc. Am. Soc. Civil Engrs., J. Waterways and Harbors Div.*, vol. 84, no. WW5, pt. 1, pp. 1–15, December, 1958.
163. Northrop, W. L., and C. W. Timberman: Use of computers for Kansas River flood studies, *Proc. Am. Soc. Civil Engrs., J. Hydraulics Div.*, vol. 87, no. HY4, pt. 1, pp. 113–150, July, 1961.
164. Linsley, R. K., Jr., and N. H. Crawford: Computation of a synthetic streamflow record on a digital computer, *Intern. Assoc. Sci. Hydrology Pub.* 51, pp. 526–538, 1960.
165. Crawford, N. H., and R. K. Linsley, Jr.: Synthesis of continuous streamflow hydrographs on a digital computer, *Stanford Univ. Dept. Civil Eng. Tech. Rept.* 12, 1962.
166. Laurenson, E. M.: Hydrograph synthesis by runoff routing, *Univ. of New South Wales, Water Res. Lab. Rept.* 66, December, 1962.
167. Singh, K. P.: A non-linear approach to the instantaneous unit-hydrograph theory, Ph.D. thesis directed by V. T. Chow, University of Illinois, Urbana, Ill., 1962.
168. Diskin, M. H.: A basic study of the linearity of the rainfall-runoff process in watersheds, Ph.D. thesis directed by V. T. Chow, University of Illinois, Urbana, Ill., 1964.
169. Kulandaiswamy, V. C.: A basic study of the rainfall excess-surface runoff relationship in a basin system, Ph.D. thesis directed by V. T. Chow, University of Illinois, Urbana, Ill., 1964.
170. Ramaseshan, Sundaresa: A stochastic analysis of rainfall and runoff characteristics by sequential generation and simulation, Ph.D. thesis directed by V. T. Chow, University of Illinois, Urbana, Ill., 1964.
171. Step method for computing total sediment load by the modified Einstein procedure, U.S. Bureau of Reclamation, Project Investigations Division, July, 1955.
172. Determination of the total sediment load in a stream by the modified Einstein procedure, U.S. Bureau of Reclamation, Electronic Computer Program Description No. HY-100, September, 1959.
173. de Vries, M.: Computations on grain-sorting in rivers and river models, *Proc. Intern. Assoc. Hyd. Res. 9th Convention, Dubrovnik, Yugoslavia*, 1961, pp. 876–880.
174. Thomas, C. W., and P. F. Enger: Use of an electronic computer to analyse data from studies of critical tractive forces for cohesive soils, *Proc. Intern. Assoc. Hyd. Res. 9th Convention, Dubrovnik, Yugoslavia*, 1961, pp. 760–771.
175. Sijbesma, R. P., and M. de Vries: Computations of boundary conditions of a river model with movable bed, *Proc. Intern. Assoc. Hyd. Res. 9th Convention, Dubrovnik, Yugoslavia*, 1961, pp. 870–875.
176. Gooch, R. S.: Electronic computers in the water supply field, *J. Am. Water Works Assoc.*, vol. 52, no. 3, pp. 311–314, March, 1960.
177. Gooch, R. S.: Application of digital computer to reservoir yield studies, *Pub. Works*, vol. 89, no. 8, pp. 91–92, August, 1958.
178. Jackson, S., E. L. MacLemon, and R. E. Speece: Estimating reservoir yields on digital computer, *J. Am. Water Works Assoc.*, vol. 51, no. 1, pp. 51–54, January, 1959.
179. "Operations Research in Water Quality Measurement," U.S. Public Health Service Contract PH 86-62-140 with Harvard University, February 15, 1963.
180. "Report of use of electronic computers for integrating reservoir operations," vol. 1, DATAmatic Corporation, prepared in cooperation with Raytheon Manufacturing

Company for the U.S. Army Corps of Engineers, Missouri River Division, January, 1957.
181. Morrice, H. A. W.: The use of electronic computing machines to plan the Nile valley as a whole, *Intern. Assoc. Sci. Hydrology Pub.* 45, vol. 3, pp. 95–105, 1958.
182. Morrice, H. A. W., and W. N. Allen: Planning for the ultimate hydraulic development of the Nile valley, *Proc. Inst. Civil Engrs.*, vol. 14, pp. 101–156, 1959.
183. Fitzpatrick, A. C.: Computer simulation of the Nile River system, *Civil Eng.* (London), vol. 55, no. 651, pp. 1291–1292, October, 1960.
184. Maass, Arthur, and M. M. Hufschmidt: Report on the Harvard program of research in water resources development, in "Resources Development: Frontiers for Research," Western Resources Conference, 1959, University of Colorado Press, Boulder, Colo., 1960, pp. 133–179.
185. Manzer, D. F., and M. P. Barnett: Analysis by simulation: programming techniques for a high-speed digital computer, in Arthur Maass and others: "Design of Water-resources Systems," Harvard University Press, Cambridge, Mass., 1962, pp. 324–390.
186. Restoration of Devils and Stump Lakes, U.S. Bureau of Reclamation, Electronic Computer Program Description No. HY 101, April, 1961.
187. Pasquali, Romolo, and Giorgio Rossi: Résolution de problèmes hydrologiques par le calculateur électronique (Solution of hydrologic problems by electronic computers), *Proc. Intern. Assoc. Hyd. Res.* 9th Convention, Dubrovnik, Yugoslavia, 1961, pp. 947–955.
188. Grimes, P. R., and G. H. Von Gunten: Electronic computer use in scoping power projects, *Proc. Am. Soc. Civil Engrs., J. Power Div.*, vol. 87, no. P01, pt. 1, pp. 1–6, January, 1961.
189. Lewis, D. J., and L. A. Shoemaker: Hydro system power analysis by digital computer, *Trans. Am. Soc. Civil Engrs.*, vol. 128, pt. I, pp. 1074–1091, 1963.

H. On Digital-computer Application in Groundwater Hydrology

190. Foley, Joseph: Computer application in groundwater hydrology, *Proc. Am. Soc. Civil Engrs., J. Irrig. and Drainage*, vol. 86, no. IR3, pt. 1, pp. 83–99, September, 1960.
191. Walton, W. C., and J. C. Neill: Analyzing groundwater problems with mathematical models and a digital computer, *Intern. Assoc. Sci. Hydrology Pub.* 52, pp. 336–346, 1960.
192. Fayers, F. J., and J. W. Sheldon: The use of a high-speed digital computer in the study of the hydrodynamics of geologic basins, *J. Geophys. Res.*, vol. 67, pp. 2421–2431, 1962.

INDEX

A profile, **7**-39
Abandonment, **27**-18
Ablation, **16**-15
Ablation zone, **16**-17
Absolute zero, **7**-6
Abstractions, effective, **14**-2
    initial, **12**-29
Access time, **29**-17
Accumulation zone, **16**-17
Accumulator, **29**-17
Acidity, **19**-10
Ackerpodsole, **5**-3
Activity index, **16**-19
Ad Hoc Committee on International Programs in Atmospheric Sciences and Hydrology (CIPASH), **1**-10
Ad Hoc Panel on Hydrology, **1**-2, **1**-12
Address, **29**-17
    data, **29**-18
    instruction, **29**-18
    memory, **29**-18
Adhesion, **7**-4
Adiabatic cooling, **3**-6, **3**-7
Adiabatic process, **3**-7, **7**-3
Administrative law, **27**-4
Admittance function, **14**-25
Advances in Hydroscience, **1**-17
Advection, **11**-5
    effects on plant production, **24**-36
Advection fog, **2**-22
Adverse slope, **7**-39
AEC (U.S. Atomic Energy Commission), **1**-14
Aeronomy, **3**-2
Aerosol, **9**-11
Age-effectiveness, **22**-30
Agglomeritic ice, **23**-21
Aggradation, **17**-5
Aggregate, **5**-22
Agricultural Conservation Program, **27**-25
Agricultural land, flood losses on, **25**-115
    hydrology of, **21**-3
Agricultural Research Center (*see* U.S. Agricultural Research Center)
Agronomy, **1**-5
AGU (*see* American Geophysical Union)
Agulhas Current, **2**-7
A-horizon, **4**-7
Ahupuaa, **27**-10
Air, **3**-2
    composition of, **3**-3
    conduction of, **3**-14

Air, dry, **3**-2
    freezing index of, **10**-47
    in soil, **5**-19
    (*See also* Atmosphere)
Air-line correction, **15**-26
Air mass, **3**-29
    precipitation model, **9**-14
Air-mass showers, **9**-22
Air-mass-source regions, **3**-29
Air permeability, of ice and snow, **16**-6
    of soil, **5**-20
Airport, drainage, **20**-35
    hydrology of, **20**-34
Albedo, **3**-11, **10**-29
Alfisol, **5**-8
ALGOL (ALGOrithmic Language), **29**-21
Alice Springs, Australia, evapotranspiration at, **24**-17 to **24**-19
    rainfall at, **24**-10, **24**-11
Alkali soil, **24**-39
    black, **24**-40
Alkalinity, **19**-10
All-Russian Hydrologic Congress, **1**-9
Allegheny River, length versus basin area, **4**-49
    routing of infiltrated water on watershed, **22**-18
Alluvial channel, **7**-26, **15**-35, **15**-39
Alluvial river, **17**-58
Alluvial-slope spring, **4**-35
Alluvial soil, **5**-5
Alluvium, **5**-3
Alluvium splay, **17**-11
Alongshore current, **23**-28
Alphameric code, **29**-16
Alpine meadow soil, **5**-5
Alternate depths, **7**-37
AMC (antecedent moisture condition), **14**-6, **21**-29
American Geophysical Union (AGU), **1**-14
    founding of, **1**-9
    hydrology committee in, **1**-10
    Infiltration, Committee on, **12**-2
    publications of, **1**-17
    Sediment Terminology, Subcommittee on, **17**-14, **17**-15, **17**-60 to **17**-63
American Meteorological Society, **1**-15
    on cloud seeding, **9**-15
American Public Health Association, **19**-14
American Society of Agricultural Engineers, **1**-15

INDEX

American Society of Agronomy, **1**-15
American Society of Civil Engineers, **1**-14, **14**-7
American Society of Limnology and Oceanography, **1**-15
American Society for Testing Materials, **19**-14
American Water Works Association, **1**-15, **19**-14, **19**-27, **19**-32
Amplifier, summing, **29**-4
Amplitude, **8**-11
Analog computer (*see* Computer, analog)
Analysis of variance, **8**-69, **8**-72
Anchor ice (*see* Ice, anchor)
Anderson Bill (*see* Water Resources Research Act)
Ando soil, **5**-6
Anemometer, **15**-13
  hot-wire, **15**-13
  Robinson, **15**-9
  warm-film, **15**-13
Anion, **19**-5
Annual-benefit function, **26**-33
Annual-cost function, **26**-33
Annual exceedance series, **8**-20
Annual maximum series, **8**-20
Annual minimum series, **8**-20
Annual series, **8**-20, **9**-49
Annual yield, frequency determination of, **21**-37
Antagonistic effect, **19**-31
Antarctic convergence, **2**-8
Antecedent condition, of moisture, **14**-6, **21**-29
  of snowmelt, **10**-32
Antecedent moisture condition (AMC), **14**-6, **21**-29
Antecedent precipitation, **22**-29
Antecedent precipitation index (API), **14**-6, **25**-102
Antecedent soil moisture, **12**-28
Antecedent temperature, **22**-29
Antecedent temperature index (ATI), **14**-6
Anticlinal spring, **4**-35
Anticyclone, **3**-30
Antidune, **7**-26
  defined, **17**-62
Antilles Current, **2**-7
API (antecedent precipitation index), **14**-6, **25**-102
Appalachian Highlands, **14**-23
Approach segment, limb, or curve, **14**-8
Appropriation, change in, **27**-17
  elements of, **27**-11
  of groundwater, **27**-32
  loss of, **27**-18
  transfer of, **27**-18
Appropriator, **27**-12
Aquatic life, **19**-31
  fauna, **19**-9
  flora, **19**-9
  vegetation, **15**-37
    effect of discharge computation, **15**-40
Aquiclude, **4**-10

Aquifer, **4**-9, **13**-3
  in arid regions, **24**-30
  artesian, **13**-4
  composed, of aggregates, **4**-21
    of carbonate rocks, **4**-21
    of clastic rocks, **4**-21
  confined, **13**-4 to **13**-6, **13**-16, **13**-18
  leaky, **13**-21
  occurrence in U.S., **4**-11, **4**-12, **4**-14 to **4**-20
  unconfined, **13**-5, **13**-6, **13**-17
Aquifer boundary, location of, **13**-23
Aquifer loss, **13**-27
Aquifuge, **4**-10
Aquitard, **4**-10
Arapahoe National Forest, avalanche, **10**-42
Archimedes principle, **7**-8
Arctic alpine, **24**-8
Arctic convergence, **2**-8
Arctic front, **3**-29
Area bare-cultivated, **22**-30
Area-depth curves for rainfall, **9**-57
Area-increment method, **17**-24
Area Redevelopment Act, **26**-3, **26**-15, **26**-27
Area-reduction method, **17**-25
Arid area, defined, **24**-4
Arid homoclimate, distribution of, **24**-3 to **24**-9
Arid region, **24**-2
  artesian aquifer in, **13**-4
  irrigation in, **24**-34
Arid and semiarid regions, evaporation in, **24**-16
  groundwater in, **24**-27
  hydrology of, **24**-2
  legal aspects in, **27**-7
  precipitation in, **24**-9
  runoff from, **24**-23
  streamflow in, **24**-24
  surface drainage in, **24**-22
  transpiration in, **24**-16
  world distribution of, **24**-8
Arid zone, **24**-2
Arid Zone Program, UNESCO, **1**-10
Arid Zone Research, **1**-12
Aridisol, **5**-8
Arithmetic-logic unit, **29**-16
Arithmetic mean, **8**-6
Arithmetic operation, **29**-16
Arizona Department of Natural Resources, **28**-23
Arkansas River Basin, water-supply forecast, **10**-26
Arkansas-Oklahoma Interstate Water Resources Committee, **28**-24
Arrival rate, **14**-48
ARS (*see* U.S. Agricultural Research Service)
Artesian aquifer, **13**-4
Artesian spring, **4**-35
Artesian water, **24**-32
Artesian well, **13**-15
Arthur D. Little, Inc., **1**-14

INDEX

Artificial flow control, **15**-30
Artificial precipitation, **26**-3
Artificial recharge of groundwater, **13**-41
Ashokan Storage Reservoir, N.Y., **17**-4
Assembler, **29**-20
Assimilation, **24**-34
Associated cost, **26**-17
Association of Official Agricultural Chemists, **19**-14
Assur's equations, **23**-21, **23**-22
Asymmetry, **8**-8
ATI (antecedent temperature index), **14**-6
Atlantic Ocean, **14**-38
Atlantic States, **14**-23
Atmometer, **11**-9
  Bellani, **11**-10
  Livingston, **11**-10
  Piche, **11**-10
Atmosphere, **1**-2, **3**-2, **3**-6
  adiabatic cooling of, **3**-6, **3**-7
  moisture condensation in, **22**-11
  pressure of (*see* Atmospheric pressure)
  stability of, **3**-8
  standard, **3**-9
  water in, **27**-29
  (*See also* Air)
Atmospheric moisture, condensation of, **22**-11
Atmospheric pressure, **7**-6, **11**-5
  affecting evaporation, **11**-5
  effects of groundwater, **13**-36
Atmospheric sciences, international programs, **1**-10
Atmospheric water, defined legally, **27**-29
Atomic Energy Commission (*see* U.S. Atomic Energy Commission)
Attrition rate, **14**-48
Aufeis, **16**-8
Automatic programming, **29**-20
Autoregression process, **8**-85
Auxiliary gage, **15**-30
Avalanches, slab, **10**-41
  snow, **10**-41
Average, **8**-6
  moving, **8**-81

Backwater, legal aspects, **27**-20
  variable, **15**-35, **15**-39
Backwater curve, **7**-38
Bacteriological analysis, **19**-14
Badland, **17**-6
Balking, **14**-48
Baltic Hydrologic Conference, **1**-9
Baltimore drainage areas, **20**-24, **20**-25
Bank storage, **13**-34
Bar materials, **17**-11
Bar resistance, **17**-44
Bari Doab Canal, India, **17**-54
Barkley Dam, **25**-90
Barograph, **7**-7
Barometer, mercury, **7**-8
  recording, **7**-7
Barometric efficiency, **13**-36
Barometry, **3**-9

Barrier height, **9**-31
Barrier spring, **4**-35
Base flow, **4**-3, **14**-2
Base-flow separation, **14**-11
Base runoff, **4**-3, **14**-2
Base value, **8**-20
Basin constant for snowmelt, **10**-39
Basin fill, **24**-31
Basin index, **10**-34
Basin method for groundwater recharge, **13**-43
Basin-plan report, **26**-15
Basin slope, coefficient for, **14**-23
Basin snowmelt, **10**-39
Basin system, nonlinearity of, **14**-13
Basins, collinear, **4**-42
  homothetic, **4**-42
  of lake, **23**-3
  (*See also* Drainage basin)
Bathymetric chart, **23**-8
Bay, defined legally, **27**-27
Beauchamp, K. H., on subsurface drainage, **21**-91
Bed layer, **17**-46
Bed load, **7**-31, **17**-49, **17**-61
Bed-load calculation, **17**-50
Bed-load equation, **17**-50
Bed-load formulas, **7**-31
Bed-load function, **17**-5, **17**-31, **17**-37
Bed-load transport, intensity of, **17**-44, **17**-49
Bed-material load, **7**-29, **17**-36, **17**-37, **17**-56, **17**-57, **17**-61
  calculation of, **17**-57
  channel capacity for, **17**-37
  measurement of, **17**-56
  samplers for, **17**-56
Bed sampling, **17**-57
Beds, mortar, **4**-8
  plane, **7**-26
Bell Telephone Laboratories, **29**-3
Bellani atmometer, **11**-10
Bellman, Richard, on dynamic programming, **26**-41
Bend-loss coefficient, **21**-64
Benefit-cost analysis, **26**-24, **28**-2
  working example, **26**-20 to **26**-23
Benefit-cost ratio, **26**-14, **26**-15, **26**-18, **26**-25
  analysis of, **26**-19
  on choice of plotting-position formula, **8**-29
  marginal, **26**-18
  for reservoir design, **25**-65
Benefits, direct, **26**-17
  indirect, **26**-17
  intangible, **26**-17
  marginal, **26**-33
  primary, **26**-17
  project, **26**-17
  remaining, **26**-20
  terminology on, **26**-17
  for various purposes, **26**-18, **26**-19
Benquela Current, **2**-7
Bergeron-Findeisen condition, **9**-15

INDEX

Bering Glacier, Alaska, **16**-12
Bernard, Merrill, on hydrometeorology, **1**-9
Bernoulli's theorem, **1**-7
Best estimate, **8**-51
Beta coefficient, **8**-64
Beta function, **8**-71
B-horizon, **4**-7
Bicarbonate, **19**-6
Bifurcation ratio, **4**-44
Big Blue River, Rondolph, Kan., flood records, **25**-16
Big Muddy River, Ill., low-flow frequency, **18**-12
Big Sand Creek, Miss., bed load in, **17**-52
Binary system, **29**-17
Binomial distribution, **8**-13
Bioassay procedures, **19**-15
Biochemical oxygen demand (BOD), **19**-11
Biohydrology, **1**-5
Biological analysis, **19**-14
Birgean heat budget, **23**-18
Bit, **29**-17
Bivariate distribution, **8**-50
Bjerknes cyclone model, **3**-32
Bjerknes' theorem, **2**-11
Black Warrior River, Tuscaloosa, Ala., stage-discharge relation, **15**-37
Blaney-Criddle's equation, **11**-25
Blaney-Criddle's method, **21**-6
Blaney-Morin's equation, **11**-25
Blench's equations, **7**-27
Block diagram, **29**-6
Block-fault lake, **23**-11
Block method for infiltration analysis, **12**-17
Blue Glacier, Olympic Mountains, Wash., **16**-16, **16**-17, **16**-30
Blue River, Dillon, Ohio, water-supply forecast, **10**-33
Board of Engineers for Rivers and Harbors (*see* U.S. Board of Engineers for Rivers and Harbors)
BOD (biochemical oxygen demand), **19**-11
Bog mat, **23**-31
Bog soil, **5**-5
Bogs, **23**-30
Boiling point, **7**-4
Boltzmann constant, **11**-29
Boltzmann transformation, **5**-19
Boneyard Creek, Champaign-Urbana, Ill., hydrograph analysis, **14**-18
Borda tube, **1**-7
Bose's equation, **7**-27
Boston Society of Civil Engineers, **14**-13, **14**-25
Bottom-set beds, **17**-24
Bottom-water layer, **2**-6
Bottom-withdrawal tube, **17**-40
Boundary layer, **7**-15
  laminar, **7**-15
  turbulent, **7**-15
Boundary spring, **4**-35
Bourdon gage, **7**-7

Bowen's ratio, **11**-11, **23**-20, **24**-38
Boyle's law, **7**-3
BPR (*see* U.S. Bureau of Public Roads)
Brackish water, demineralization of, **19**-33
Braiding, **7**-29
Braunerde, **5**-5
Brazil Current, **2**-7
Brazos River, **14**-42
BRLESC, **29**-16
Brown forest soil, **5**-5
Brown podzolic soil, **5**-5
Brown soil, **5**-5
Brownian movement, **5**-4
Brunizem, **5**-6
Bubble gage, **15**-7, **15**-8
Buckingham's pi theorem, **7**-5
Buffer unit, **29**-17
Building codes for flood regulation, **25**-123
Bulk coefficient, **16**-7
Bulk density, **5**-4, **5**-15
Buoyancy, **7**-8
  center of, **7**-8
Bürkli-Ziegler formula, **20**-8
Burroughs E-101, **25**-58
Burroughs 220, **29**-22

C profile, **7**-39
Caatinga area, **24**-8
Cable-tool method, **13**-28
Calcium, **19**-3
Calcium carbonate equivalents, **19**-16
Calculated risk, **9**-59
Caldera, **23**-4
Caliche, **4**-8
California Current, **2**-7
California Debris Commission, **1**-13
California Department of Water Resources, **28**-23
California, Division of Highways, **25**-20
California method, for culvert discharge, **25**-21
  for plotting positions, **8**-28
California Research Corporation, **1**-14
California State Department of Water Resources, snow survey, **10**-21
California State Water Pollution Control Board, **19**-8, **19**-12, **19**-26
California streams, flow frequency of, **8**-3
Canadian River, **14**-42
Canaries Current, **2**-7
Candled ice, **23**-22
Canyon Ferry, **29**-22
Capacity, specific, **4**-13, **13**-27
  thermal, of water, **23**-16
  waste-assimilative, **19**-26
Capacity-inflow ratio, **17**-22
Capacity load, **17**-37
Cape Horn Current, **2**-7
Capillarity, **5**-9
Capillary conductivity, **5**-15, **5**-19
Capillary potential, **5**-19
Capillary rise, **5**-10, **5**-11
Capillary water, **22**-9
Caprock, **4**-9

INDEX

Carbon dioxide, 3-4
Carbonate, **19**-6
Carbonate hardness, **19**-10
Castle Lake, Grand Coulee, Wash., **23**-5
Catchment, **14**-2
Catena, **5**-6
Cation, **19**-3
Cauchy-type distribution, **8**-16
Cedar River, Landsberg, Wash., variation of annual flow, **18**-6
Celerity, **2**-16, **7**-32
Center of buoyancy, **7**-8
Center of pressure, **7**-8
Central Snow Conference, **10**-22
Central Technical Unit, ARS, **1**-12
Central Valley Project, Calif., **28**-21
Cetyl alcohol, **11**-14
CGS (U.S. Coast and Geodetic Survey), **1**-13
Chain gage, **15**-6
Channel, alluvial, **7**-26, **15**-35, **15**-39
  changing, **15**-39
  cross-section geometry of, **4**-61
  eroding, **17**-58
  linear, **14**-27
  maintenance, constant of, **4**-54
  natural, **7**-25
  open (*see* Open channel)
  parabolic, **21**-60
  pilot, **21**-77
  (*See also* Streams)
Channel density, **17**-13
Channel encroachment, **25**-120, **25**-122
Channel erosion, **17**-6, **17**-9, **17**-19
Channel improvement, **21**-77
Channel lining, **21**-60, **21**-77
Channel precipitation, **14**-2
Channel profile, law of declivities, **4**-57
  upconcavity, **4**-57
Channel regime equations, **7**-27, **7**-28
Channel regime theory, **7**-26, **17**-54
Channel slope, versus ground slope, **4**-62
  valleyside, **4**-62
Channel splay, **17**-11
Channel spring, **4**-35
Channel storage, **15**-37, **15**-40
Characteristic hydrograph, **14**-13, **14**-25
Characteristics, method of, **14**-13, **25**-38, **25**-82
Chattahoochee River, Columbus, Ga., flood-risk map, **25**-123
Chegodayev formula, **8**-29
Chemical oxygen demand (COD), **19**-11
Chemosphere, **3**-10
Chernozen, **5**-5
Chestnut soil, **5**-5
Chézy's discharge coefficient, **7**-23
Chézy's equation, **7**-23
Chézy's formula, **1**-7
Chi-square, **8**-69
Chi-square distribution, **8**-69
Chi-square parameter, **8**-47
Chi-square test, **8**-47
Chicago Bureau of Engineering, **20**-20
Chicago hydrograph method, **20**-16

Chicago Sanitary and Ship Canal, **23**-15
Chickamauga Dam, hydrograph at, **25**-92
Chickamauga Reservoir, operation of, **25**-76, **25**-91
Chief of Engineers, Office of the, **1**-13
Chloride, **19**-5
Chlorinity, **2**-3, **19**-23
Chocolate Glacier, Glacier Park, Wash., **16**-12
C-horizon, **4**-7, **5**-3
Chow, V. T., *Advances in Hydroscience*, **1**-17
Chow's equation for hydrologic frequency, **8**-23
Chow's method, for aquifer coefficients, **13**-18
  for peak discharges, **25**-22
Christian's Creek, Va., length and basin area relation, **4**-50
Chute spillway, **21**-62
CIPASH (Ad Hoc Committee on International Programs in Atmospheric Sciences and Hydrology), **1**-10
Circularity ratio, **4**-51
Circulation, global, **3**-24
Circumpolar Current, **2**-7
Civil-law rule, **27**-30
Clapotis, **2**-19
Clays, **5**-3, **5**-7
Clear-cut, definition, **6**-27
Clear-cutting, **6**-24
Clearwater River, Idaho, snow gages, **10**-21
Clearwater River, base flow at Spalding, Idaho, **10**-32, **10**-33
Clearwater River Basin, Idaho, simulation analysis, **26**-36
Cliff spring, **4**-35
Climate, of arid and semiarid regions, **24**-8
  homoclimatic distribution, **24**-3 to **24**-9
  reconstructing past, **16**-28
  type distribution in U.S., **18**-3
Climatic cycle, **8**-12
Climatic factors affecting runoff, **14**-5
Climatology, **1**-5
Climax, **6**-27
Clouds, **3**-20 to **3**-22
  altocumulus, **3**-21
  altostratus, **3**-21
  cirrocumulus, **3**-20
  cirrostratus, **3**-20
  classification of, **3**-20
  convective, **3**-30
  cumulonimbus, **3**-10, **3**-22
  cumulus, **3**-10, **3**-20, **3**-21
  cumulus congestus, **3**-21
  cumulus humulis, **3**-21
  droplet growth in, **3**-18
  heap-shaped, **3**-20
  layer, **3**-20
  mackerel, **3**-21
  mesoscale system of, **3**-10
  nimbostratus, **3**-21
  ocean, **2**-22
  seeding of, **9**-15, **24**-15
    legal aspects, **27**-30

INDEX

Clouds, stratocumulus, **3**-21
    stratus, **3**-20, **3**-22
CN (*see* Runoff curve number)
Coastal lake, **23**-6
Coaxial correlation, **9**-31, **14**-6
Coaxial graphical method, **8**-65, **25**-102
Cobalt-platinum scale, **19**-17
COBOL (COmmon Business Oriented Language), **29**-21
COD (chemical oxygen demand), **19**-11
Code, alphameric, **29**-16
    building, for flood regulation, **25**-123
    operation, **29**-18
Code checking, **29**-18
Coding, **29**-18
Coefficient, of basin slope, **14**-23
    bend-loss, **21**-64
    beta, **8**-64
    bulk, **16**-7
    correlation (*see* Correlation coefficient)
    consumptive-use, **21**-6
    of curvature, **5**-7
    of determination, **8**-54, **8**-62
    discharge, Chézy's, **7**-23
    drag, **7**-43, **7**-44
    drainage, **21**-90
    energy-flux correction factor, **7**-12, **15**-32
    entrance-loss, **21**-64
    form-loss, **7**-19, **7**-20
    of linear expansion, **16**-7
    Manning's roughness, **7**-24, **21**-54
    Markov-chain, **8**-93
    momentum-flux correction factor, **7**-10
    of permeability, **4**-13, **13**-8
        field, **4**-13, **13**-8
        laboratory, **13**-8
    pipe-friction-loss, **21**-64
    regression, **8**-51, **8**-56
        multiple, **8**-60
    resistance, **7**-16
    retardance, **20**-38
    of roughness, Manning's, **7**-24, **21**-54
    runoff, **14**-8, **20**-8, **21**-38
    of skewness, **8**-8
    storage, **4**-13, **13**-4, **14**-27, **29**-8
    surface-tension, **5**-9
    translation, **14**-28
    of transmissibility, **4**-13, **13**-8
    of uniformity, **5**-7
    of variation, **8**-7
    wind, **2**-9
Coefficient method, **25**-40
Cold front, **3**-29
Cold Regions Research and Engineering Laboratory, **10**-50
Cold spring, **4**-35
Colebrook-White equation, **7**-16
Colloidal stability, **9**-11
Colluvium, **17**-11
Color, **19**-11, **19**-17
Colorado doctrine, **27**-18
Colorado River, **23**-6
    annual runoff of, **14**-42
    at Lees Ferry, Ariz., **8**-92 to **8**-94, **14**-47

Colorado River, runoff at Cameo, **10**-32
    shift corrections, at Austin, Tex., **15**-40
    simulation analysis of, **26**-36
    stage-discharge relation, at Austin, Tex., **15**-37
Colorado River System, hydrologic network of, **26**-8
Colorado sunken pan, **11**-7, **11**-8
Colorimetric method, **19**-14
Columbia River base flow at Casttegar, B.C., **10**-32
    flow above Kootenay, **10**-33
Columbia River Basin, sediment data, **17**-13
    simulation analysis of, **26**-35
Combination inlet, **20**-31
Commerce power, **27**-22
Commercial district, flood losses in, **25**-115
Commission for Hydrometeorology (Hydrological Meteorology), **1**-12
Commission on Organization of the Executive Branch of the Government, **1**-10
Committee on Water Resources Research, **1**-12
Common-enemy rule, **27**-30
Common law, **27**-3
Commonwealth Scientific and Industrial Research Organization, **11**-14
Compact, **27**-21
    interstate, **27**-21
Compaction, effect on infiltration, **6**-14
Compactness factor, **22**-48
Compiler, **29**-20
Complete-duration series, **8**-19
Complex, **5**-6
Complex control, **15**-38
Complex delta, **17**-23
Composite hydrograph, **21**-45
Compressibility, **7**-3
Compressing flow, **16**-24
Computer, **25**-54, **29**-1 to **29**-30
    analog, **29**-2
        application in groundwater, **29**-11
        application in surface water, **29**-7
        classification of, **29**-3
        direct, **29**-3
        for flood routing, **25**-54
        general-purpose, **29**-3
        indirect, **29**-3
        for linear reservoir, **14**-30
        mathematical, **29**-3
        multiple-purpose, **29**-3
        physical, **29**-3
        special-purpose, **29**-3
    differential analyzer (*see* Differential analyzer)
    digital, **29**-2
        application in groundwater, **29**-23
        application in surface water, **29**-21
        classification of, **29**-14
        for flood routing, **25**-58
        general-purpose, **29**-14
        for multiple regression, **8**-61

Computer, digital, parallel, **29**-17
    serial, **29**-17
        special-purpose, **29**-14
    hybrid, **29**-2
Computer equation, **29**-6
Computer system, **29**-16
Computer tree, **29**-15
Computer unit, **29**-6
Computer variable, **29**-4, **29**-6
Computing elements, **29**-4, **29**-5
Concentration, hydrogen-ion (*see* pH value)
    threshold, **19**-11
    time of (*see* Time, of concentration)
Concentration segment, limb, or curve, **14**-8
Concentric rings, infiltrometer, **12**-7
Conceptual models of IUH, **14**-32
Condensation, **3**-18, **7**-4, **9**-10
Condensation nuclei, **3**-18, **9**-10
Condition of instability, **9**-18
Conductance, electrical, **19**-11, **19**-17
Conduction of air, **3**-14
Conductive capacity, **3**-13
Conductivity, **16**-7
    capillary, **5**-15, **5**-19
    per centimeter, **19**-11
    electrical, **19**-11, **19**-17
    hydraulic, **5**-16, **13**-8
    specific, **6**-5, **11**-17
    thermal, **5**-20, **16**-7, **23**-17
    volume, **19**-11
Conduit, flow in closed, **7**-15
    forces on, **7**-9
Conduit storage, **20**-7, **20**-13
Cone of depression, **13**-16
Conference of American Hydrologists, **1**-10
Confidence band, **8**-31
Confidence interval, **8**-50
Confidence limits, **8**-31, **8**-50, **8**-83
Confined aquifer, **13**-4 to **13**-6, **13**-16, **13**-18
Confining bed, **4**-10
Conifer, **22**-51
Connection delay, **14**-48
Connector, **29**-18
Conservation district, **27**-32
Conservation statute, **27**-31
Constant of channel maintenance, **4**-54
Constant-fall rating, **15**-35
Constant multiplier, **29**-4
Constitutions, **27**-3
Constraints, **26**-31, **26**-32
Consumptive use, **11**-2, **21**-6, **21**-81
    defined legally, **27**-27
Consumptive-use coefficient, **21**-6
Contact load, **17**-61
Contact spring, **4**-35
Continuity equation, **1**-5, **7**-10
Contour farming, **21**-26
Contracted opening, **7**-48
Contracted-opening method, **15**-32
Contraction, end, **7**-45

Control, flood (*see* Flood control)
    of flow, **15**-34
        complex, **15**-38
        conditions affecting, **15**-34
        (*See also* Flow control)
    insect, **26**-2
    malarial, **25**-76
    salinity, **26**-3
    sediment, **26**-2
    shifting, **15**-34
    water-quality (*see* Water-quality control)
Control curves, **8**-31
Control unit, **29**-16
Convective process, **2**-20
Convective rainfall, **9**-21
Convergence, **9**-13
    Antarctic, **2**-8
    Arctic, **2**-8
    subtropical, **2**-8
Conveyance, **15**-32
Conveyance factor, **15**-38
Convolution integral, **14**-24
Cooke Commission, **1**-10
Cook's method, **21**-39, **25**-20
Cooling, **9**-10
    adiabatic, **3**-6, **3**-7
Coral lakes, **23**-6
Coriolis' effect, **23**-28
Coriolis' force, **2**-9, **3**-23
Correction factors (*see* Energy-flux correction factor coefficient; Momentum-flux correction factor coefficient)
Correlation, **8**-45
    coaxial, **8**-65, **9**-31, **14**-6, **25**-102
    curvilinear, **8**-65
    multiple, **8**-48
    multiple linear, **8**-59
    rainfall and runoff, **20**-9
    serial, **8**-79, **8**-85
    simple, **8**-48
    simple linear, **8**-50
Correlation analysis, for periodicity, **8**-12
    for snowmelt, **10**-38
Correlation coefficient, **8**-52, **8**-55
    multiple, **8**-62
    partial, **8**-62
    serial, **8**-79, **8**-85
Correlation index, **8**-59
Correlation and regression analyses, **8**-86
Correlative right, **27**-31
Correlogram, **8**-85
Cortex, **6**-4
Cost, allocation, **26**-20
    associated, **26**-17
    direct, **26**-17
    indirect, **26**-17
    intangible, **26**-17
    joint, **26**-18
    marginal, **26**-18
    project, **26**-17
    secondary, **26**-17
    separable, **26**-18
    terminology on, **26**-17
    for various purposes, **26**-19
Countercurrents, equatorial, **2**-7

INDEX

Court reference procedure, **28**-13
Courtleigh drainage area, Baltimore, Md., **20**-27
Covariance, **8**-69
  analysis of, **8**-75, **8**-76
Cover density of watershed, **22**-30
Cow Creek, Death Valley, Calif., annual evaporation, **11**-8
Coweeta experimental forest watershed, **22**-44, **22**-46
Coweeta Hydrologic Laboratory, **6**-24, **6**-30
Crater, **23**-4
Crater Lake, Oregon, a classical caldera, **23**-4
Creeping random method, **26**-37
Crest-gage indicator, **15**-7
Crest segment, **14**-9
Criddle's erosion equation, **21**-87
Critical-depth flume, **15**-30
Critical flow, **7**-23
Critical slope, **7**-23
Critical velocity, **21**-60
  of loads on ice, **23**-23, **23**-24
Crop rotation, **21**-11
Cropping-management factor, **17**-8
Cross-Doland method, **7**-22
Croton Aqueduct, **1**-8
Croton Reservoir, **18**-20
Croton River, N.Y., probability curve of, **18**-20
Crown, **6**-27
  density, **22**-51
Cryology, **1**-5, **16**-3
Crystallographic axis, main, **16**-4
Cuesta spring, **4**-35
Culvert, **15**-33
Cumberland River, flood control and power of, **25**-88
  hydropower of, **25**-91
  map of, **25**-88
Cummings radiation integrator, **11**-13
Curb-opening inlet, **20**-31
Curie, **19**-17
Current, Agulhas, **2**-7
  Antarctic, **2**-8
  Antarctic bottom, **2**-8
  Antarctic circumpolar, **2**-8
  Antilles, **2**-7
  Arctic, **2**-8
  Arctic bottom, **2**-8
  Arctic circumpolar, **2**-8
  Benguela, **2**-7
  Brazil, **2**-7
  Californian, **2**-7
  Canaries, **2**-7
  Cape Horn, **2**-7
  Circumpolar, **2**-7
  East Australian, **2**-7
  East Greenland, **2**-7
  Equatorial Counter-, **2**-7
  Falkland, **2**-7
  Guiana, **2**-7
  Kuroshio, **2**-7
  Labrador, **2**-7
  North Atlantic, **2**-7

Current, North Equatorial, **2**-7
  North Pacific, **2**-7
  North-east Monsoon, **2**-7
  Oyashio, **2**-7
  Peru, **2**-7
  Somali, **2**-7
  South Equatorial, **2**-7
  South-west Monsoon, **2**-7
  Subantarctic intermediate, **2**-8
  Subarctic intermediate, **2**-8
  West Australian, **2**-7
  (*See also* Currents)
Current meter, **2**-13, **15**-5, **15**-9
  Ellis, **1**-8
  horizontal-axle, **15**-10, **15**-11
  Neyrpic, **15**-10, **15**-11
  Ott, **15**-10, **15**-11
  Price, **1**-8, **15**-5, **15**-9
  pygmy Price, **15**-9
  rotating, **15**-9
  vertical-axle, **15**-10, **15**-11
  Woltman's, **1**-7, **15**-5
Currents, alongshore, **23**-28
  compensation, **2**-6
  convective, **3**-30
  counter-, **2**-8
  deep-water, **2**-8
  density, **2**-6, **2**-11 to **2**-13
  drift, **2**-6, **2**-9, **2**-10
  Ekman's theory, **2**-9, **2**-10
  in estuaries, **2**-12, **2**-13
  geostrophic, **2**-13
  gradient, **23**-27
  lake, **23**-27
  littoral, **23**-28
  measurement of, **2**-13, **2**-14
  nonperiodic, **23**-27
  in Northern Hemisphere, **2**-7
  ocean, **2**-6
    hydrodynamics of, **2**-8
  open-lake, **23**-29
  periodic, **23**-27
  slope, **2**-6, **2**-11, **23**-27
  subtropical deeper, **2**-8
  tidal, **2**-6
  types of, **2**-6 to **2**-8
  wind, **2**-6
  (*See also* Current)
Curvature, coefficient of, **5**-7
Curve fitting, **8**-46
  for frequency data, **8**-29
  by polynomial, **8**-81
  for regression and correlation, **8**-45
Cutoff, **21**-77
Cuttings, clear, **6**-24
  liberation, **6**-26
Cycle, climatic, **8**-12
  in precipitation, **9**-46
  sunspot, **8**-12
Cyclone, **3**-10, **3**-29, **3**-30
  Bjerknes' model, **3**-32
  extratropical, **3**-29 to **3**-32
  frontal model, **9**-19
  tropical, **3**-30

DAD analysis (depth-area-duration analysis), 9-32
D'Alembert's principle, 1-7
Dalton, John, on evaporation, 1-8
Dalton's evaporation equation, 11-4
Dalton's law, 2-22, 11-3
Dam, 15-33
  aggradation effect of, 17-5
  degradation effect of, 17-6
  for farm and ranch ponds, 21-65
  infinite, 14-49
  overflow, 15-31
Dancing of power lines, 10-45
Dantzig, G. B., on linear programming, 26-39
Darcy, 13-9
Darcy-Weisbach equation, 7-16
Darcy's law, 1-8, 5-16, 11-17, 13-7, 29-11
Data, experimental, 8-2
  historical, 8-2
  hydrologic (see Hydrologic data)
  point, 8-36
  rainfall (see Rainfall data)
Data address, 29-18
Daytime hours, 11-27, 21-7
Debugging, 29-18
Decimal system, 29-17
Decision, defined, 26-41
Decision symbol, 29-19
Deep percolation, 14-2
Deep-water layer, 2-6
Defection, 14-48
Definite plan report, 26-15
Deflation lake, 24-27
Degradation, 17-5
  below reservoir, 17-5
Degraded chernozem, 5-5
Degree-day, 10-33
Degree-day equation, 21-32
Degree-day method, 21-32
Degree of freedom, 8-69
Delay line, 29-10
Delft Hydraulic Laboratory, 15-10
Delivery loss, 21-85
Delivery ratio, 17-27
Delta, 23-5
  complex, 17-23
  ideographic, 17-23
Delta deposit, 17-23
Deltaic levee lake, 23-5
Demineralization, 19-32, 19-33
Density, 7-2
  crown, 22-51
  ground, 22-51
  probability, 8-6
  of sediment, 17-17
  of snow and ice, 16-6
  of water, 19-12
Density current, in estuaries, 2-11 to 2-13
Denuded watershed area, 22-30
Depletion curve, 14-8
Depression, cone of, 13-16
  wake, 3-33
Depression storage, 20-5, 20-18

Depth-area curves, 25-28, 25-29
Depth-area-duration (DAD) analysis, 9-32
Depth-area-duration pattern, 9-60
Depth of flow, determination of, 15-16
Depth of frictional resistance, 2-10
Depth-integrating sampler, 17-55
Depth-integrating sampling, 17-59
Depth profile, 4-29
Deschutes River Valley, Ore., 4-3, 4-4
Desert, extreme, 24-5
Desert lake, 24-26
Desert landforms, 24-23
Desert soil, 5-5
Design discharge, 25-24
  limit, 25-20
  peak, 25-24
Design flood, 25-25, 25-72, 26-14
Design period, 8-34
Design storm, for airport, 20-35
  for cities, 20-21
  for emergency spillways, 21-72 to 21-74
Detention, 12-9, 20-5
Detention-flow-relationship method, for infiltration analysis, 12-14
Detention storage, 22-13
Detergents, synthetic, 19-8
Determination, coefficient of, 8-54
  multiple, coefficient of, 8-62
Deterministic model, 8-9, 26-39
Deterministic nonlinear programming, 26-43
Deterministic process, 8-9
Deterministic system, 8-10
Developed water, defined legally, 27-28
Deviation, mean, 8-7
  standard (see Standard deviation)
Devils Lake, N.D., 29-22
Dew, 3-23, 6-7, 24-13
Dew point, 3-5
Dewfall, 24-14
DF theory (Dupuit-Forchheimer theory), 5-17
D-horizon, 5-3
Diagram, block, 29-6
  flow, 29-18
  freezing-point, 10-13
  hydrophase, 14-10, 14-11
  Moody, 7-16, 7-17
  Muskingum routing, 25-102
  patch, 29-6
  regulation, 14-46
  Shields, 17-53
  specific-head, 7-37
  temperature-salinity, 2-4
  time-area, 14-9, 14-28
  time-area-concentration, 14-28
Diameter-limit cut, 6-25, 6-27
Diamond dust, 3-22
Dichromate oxygen demand, 19-12
Differential analyzer, 29-3 to 29-7
  for flood routing, 25-55
  high-speed, 29-3
  long-time, 29-3
  one-shot, 29-3

Differential analyzer, repetitive, **29**-3
  short-time, **29**-3
  slow, **29**-3
Diffusion, **5**-20
  molecular, **9**-10
Diffusion analogy, **29**-7
  method of, **25**-38
Diffusion-pressure deficit, **6**-4
Diffusivity, **5**-19, **5**-20
  thermal, **5**-21
Digital computer (*see* Computer, digital)
Dike, **21**-88
Dilution available curve, **18**-24
Dilution required curve, **18**-24
Dimensional analysis, **7**-5
  of drainage basins, **4**-41
Dimensionless hydrograph, **14**-13, **14**-24, **14**-36, **20**-6, **21**-41
  Izzard's, **20**-38
  for overland flow, **20**-6
Dimensionless unit hydrograph, **21**-41
Dimple spring, **4**-35
Dirac-delta function, **14**-28
Direct cost, **26**-17
Direct runoff, **14**-2, **21**-29
Direct-runoff hydrograph (DRH), **14**-12
Direct surface runoff, **14**-2
Disaster Prevention Research Institute, **29**-7
Discharge, **7**-9
  Chézy's coefficient of, **7**-24
  determination of peak, **15**-31
  working, **25**-44
Discharge curves, for reservoirs, **25**-63
Discharge diagram, **7**-37, **7**-38
Discharge formulas, **20**-7
Discharge-frequency curve, **25**-6
Discharge hydrograph, **14**-8
Discharge measurement, **15**-17
  accuracy of, **15**-20
  corrections of, **15**-25
  over ice, **15**-20
  in streams, **15**-4
Dispersion ratio, **17**-9
Distillation, **24**-14
Distribution graph, **14**-13, **14**-22
Distribution, of drainage basin surface-slope, **4**-63
  of drought, **18**-2
  frequency, **8**-5
  probability, **8**-6
  of sediment, **17**-22, **17**-24
  statistical, **8**-6
    binomial, **8**-13
    bivariate, **8**-50
    Cauchy type, **8**-16
    chi-square, **8**-69
    exponential, **8**-35
    extremal (*see* Extremal distribution)
    F, **8**-70
    gamma, **8**-14
    Gumbel, **8**-16
    linearization of, **8**-5
    logarithmically transformed, **8**-17
    logextremal, **8**-17, **8**-35

Distribution, statistical, lognormal, **8**-17, **8**-25, **8**-35
    truncated, **8**-17
  marginal, **8**-48
  multivariate, **8**-50
  normal, **8**-14, **8**-24
  partly bounded, **8**-17
  Pearson (*see* Pearson distributions)
  Poisson, **8**-14
  rectangular, **8**-13, **8**-14
  Slade, **8**-17
  totally bounded, **8**-17
  truncated lognormal, **8**-17
  Weibull, **8**-16
Ditch, **21**-89
Ditch method for groundwater recharge, **13**-43
Diversion works, **18**-24, **21**-46, **21**-53
Doctrine, of prior appropriation, **27**-11
  of riparian rights, **27**-4
Domain, eminent, right of, **27**-9
Double-mass curve, **8**-76, **9**-26
Douglas Reservoir, area and volume curves of, **25**-65
  operation of, **25**-91
Draft, **14**-46
Draft line, **14**-46
  variable, **14**-46
Draft rate, **18**-15
  gross, **18**-15
  net, **18**-15
Drag, **7**-14, **7**-42
  on alluvial stream bed, **17**-41
  shear, **7**-14, **7**-42
  on sphere, **7**-44
  surface, **17**-41
Drag coefficient, **7**-43, **7**-44
Drag equation, **7**-14
Drain, field, **21**-89
  layouts of, **21**-91
  mole, **21**-94
  tile, **21**-91 to **21**-95
Drain-spacing, equation for, **5**-17
  nomograph for, **5**-18
Drain-tile design chart, **21**-94
Drainage, **21**-88, **26**-2
  airport, **20**-35
  in arid and semiarid regions, **24**-22
  highway, **20**-41
  hydrologic data for, **26**-7
  legal aspects of, **21**-88, **27**-30
  pumping for, **21**-95
  storm, by open channels, **20**-32
  subsurface, **21**-91
    in airport, **20**-35
  surface, **21**-88, **21**-89
    (*See also* Surface drainage)
Drainage area, effect on sediment delivery, **17**-12
  (*See also* Drainage basin; Watershed)
Drainage association, **21**-89
Drainage basin, **14**-2
  areal aspects of, **4**-48 to **4**-56
  basin geometry, climatic factors on, **4**-72
    relation to streamflow, **4**-72

INDEX

Drainage basin, channel gradient, **4**-56
 circulatory ratio, **4**-51
 dimensional analysis, **4**-41
 discharge, area relation to, **4**-50
 drainage density, **4**-52
 elongation ratio, **4**-51
 frequency distribution of area, **4**-48
 isotangent slope map of, **4**-63, **4**-64
 large, **14**-5
 lemniscate ratio of, **4**-51
 length, area relation to, **4**-49
 maximum relief of, **4**-66
 morphometric analysis of, **4**-43
 relief of, **4**-65
 relief aspect of, **4**-56
 relief ratio of, **4**-66
 shape of, **4**-51
 small, **14**-5
 SPF estimates for large and small, **25**-30, **25**-31
 statistical analysis of, **4**-42
 steady state, **4**-71
 surface-slope distribution of, **4**-63
 theory of dynamics of, **4**-69
 (*See also* Watershed)
Drainage coefficient, **21**-90
Drainage curves, **21**-90
Drainage density, **4**-52
 factors controlling, **4**-70
Drainage district, **21**-89, **27**-30
Drainage law, **27**-30
Drainage outlet, **21**-89
Drainage pattern, dendritic, **4**-3
 trellis, **4**-2
Drawdown, in flood elevation, **25**-3
 in wells, **13**-16
DRH (direct-runoff hydrograph), **14**-12
Drinking Water Standards (1962), **19**-27
Drizzle, **3**-22
Drop spillway, **21**-62
Drought, **18**-1
 distribution of, **18**-2
 extent of, **18**-2
 hydrologic relations of, **18**-7
 minimum, **8**-35
 probability of occurrence of, **18**-11
 in U.S., **18**-2
Drought characteristic, **8**-35
Drought duration, **18**-6, **18**-10
Drought frequency, **18**-10
 analysis of, **8**-4
Drought severity, **18**-10
Dry arroyos, defined legally, **27**-27
Dry provinces of the world, **24**-8
Dry-weather flow, **19**-18
Dugout for farm and ranch ponds, **21**-65
Duhamel integral, **14**-24
Dune, **7**-26
 defined, **17**-62
 sand, **24**-31
 washout, **7**-26
Dupuit-Forchheimer (DF) theory, **5**-17
Dupuit's well equation, **13**-17
Dupuit-Thiem's well formula, **1**-8

Duration curve, **14**-42
 (*See also* Flow-duration curve)
Duty of water, **11**-2
Dyke barrier, **24**-31
Dynamic programming, **26**-41
Dynamic sections, method of, **2**-14
Dynamic system, **8**-10
Dynamic viscosity, **7**-4
Dynamometer, **15**-10

E/A (Transpiration-assimilation ratio), **24**-34
Earth dike, **21**-76
Earthquake Lake, tectonic movement, **4**-3
EASE, **29**-4
East Australian Current, **2**-7
East Greenland Current, **2**-7
Easterly waves, **3**-33
Eastern Snow Conference, **10**-23
Eaton Brook, N.Y., gaging station, **15**-5
ECAFE (Economic Commission for Asia and the Far East), **1**-12
Ecological succession, **6**-2
Ecology, plant, **1**-5, **6**-2, **6**-27
Economic analysis, **26**-6, **26**-15
 benefit-cost (*see* Benefit-cost analysis)
 incremental, **26**-14, **26**-38
 interest rate, **26**-20
 marginal, **26**-14, **26**-38
 price level, **26**-20
 separable costs–remaining benefits method, **26**-20
Economic Commission for Asia and the Far East (ECAFE), **1**-12
Economic efficiency, **26**-31
 objective function for, **26**-32
Economic growth, **26**-32
Economic life, **26**-19
Economic optima and water policy, **28**-6
Economic prediction, **28**-17
Economic projection, validity of, **28**-17
 in water policy, **28**-14
Economics of water projects, **28**-2
Eddy viscosity, **7**-13
EDTA (ethylenediaminetetra-acetic acid), **19**-10
EDVAC, **29**-16
Effective abstractions, **14**-2
Effective length of record, **8**-85
Effective precipitation, **14**-2
Effective rainfall, **21**-83
Effective-rainfall hyetograph (ERH), **14**-12
Effluent stream, **14**-4
Einstein, A. H., on bed-load function, **1**-9
Einstein method, **17**-57
 modified, **17**-60
 computer solution, **29**-22
Ekman spiral, **2**-9
Electric tape gage, **15**-6
Electrical conductance, **19**-11, **19**-17
Electrical conductivity, **19**-11, **19**-17
Electromagnetic flowmeter, **15**-13
Electromagnetic method, **2**-14
Electronic analog (*see* Computer, analog)

INDEX

Electronic computer (*see* Computer)
Elephant Butte Dam, **17**-5
Elevation head, **7**-12
Ellis current meter, **1**-8, **15**-5
Elongation ratio, **4**-51
Emberger, L., on extreme desert, **24**-5
Emergency spillway, **21**-67
   design storm for, **21**-72 to **21**-74
Eminent domain, right of, **27**-9
Employment, **26**-3
   full, **26**-31
Encroachment, channel, **25**-120, **25**-122
   salt water, **13**-46
End contraction, **7**-45
Energetics of soil moisture, **11**-16
Energy, free, **11**-16
   specific, **7**-36
   surface, **5**-9, **7**-4, **16**-5
Energy-balance method, **2**-21, **11**-11, **11**-25
Energy-budget method, **11**-11
Energy equation, **7**-12
Energy-flux correction factor coefficient, **7**-12, **15**-32
Engineers Joint Council, **1**-10
ENIAC, **29**-14 to **29**-16
Entisol, **5**-8
Entrance-loss coefficient, **21**-64
Epidermis, **11**-20
Epilimnion, **19**-20, **23**-18
Equation of state, **7**-3
Equatorial countercurrent, **2**-7
Equilibrium equation, **13**-16
Equilibrium limit, **16**-17
Equitable apportionment, principle of, **27**-21
Equivalent circuit, **29**-3
Equivalent weight, **19**-15
ERH (effective-rainfall hyetograph), **14**-12
Erie Canal, **1**-8
Erodibility, **22**-11
Eroding channel, **17**-58
Erosion, **17**-6, **21**-27, **22**-11
   channel, **17**-6, **17**-19
   control structures for, **17**-31
   effect of fire on, **22**-38
   gravity, **17**-10
   gross, **17**-11
   human effects on, **17**-11
   by ice, **17**-10
   normal, **17**-6
   rill, **22**-11
   sheet, **17**-6, **17**-19, **22**-11
   by water, **17**-6
   by wind, **17**-10, **17**-19
Erosion proportionality factor, **4**-69
Erosion rate, **22**-38
Errors, accidental, **8**-18
   Gaussian law of, **8**-3, **8**-14
   of inconsistency, **8**-45
   normal law of, **8**-3
   observation, **8**-18
   random, **8**-44
   sampling, **26**-8
   standard, **8**-7
   systematic, **8**-18, **8**-45

Esopus Creek, N.Y., **17**-4
Estoppel, **27**-9
Estuary, water quality in, **19**-21
Eulerian method, **2**-13
Euler's constant, **8**-16
Evaporation, **6**-18, **7**-4, **11**-2
   in arid and semiarid regions, **24**-16
   from base soils, **24**-16
   control from soil, **11**-19
   effect of atmospheric pressure, **11**-5
   effect of thinning on, **6**-25
   factors affecting, **11**-4
   from free-water surfaces, **11**-3
   latent, **11**-31
   maps for U.S., **11**-8, **11**-9
   measurement of, **11**-6
   from open-water surfaces, **24**-16
   opportunity of, **4**-2
   in reservoirs, **18**-16
   from saturated soils, **11**-19
   on sea surface, **2**-21
   from soil surface, **11**-15
   versus transpiration, **11**-19
Evaporation equations, **11**-4, **11**-10
Evaporation pan, **11**-6
Evaporation reduction, **11**-14
Evapotranspiration, **6**-2, **6**-17, **11**-2, **11**-23, **21**-6
   from arid-zone plants, **24**-18
   ecological aspects of, **6**-23
   effect of harvesting on, **6**-25
   effect of groundwater, **13**-35
   by forest, **6**-21
   by herbaceous vegetation, **6**-21
   modification of, **24**-20
   potential, **11**-23
   silvicultural aspects, **6**-23
Evapotranspiration equations, **11**-25 to **11**-30
Exceedances, method of, **18**-1
Excessive precipitation, **9**-35
Execution, **29**-19
Exosphere, **3**-11
Expansion, coefficient of, **16**-7
Experimental data, **8**-2
Exponential distribution, **8**-35
Extending flow, **16**-24
External unit, **29**-17
Extraterrestrial radiation, **11**-29, **21**-9
Extremal distribution, **8**-3, **8**-16, **8**-25
   justification for, **8**-35
   probability paper for, **8**-28
   Type I, **8**-16, **8**-17, **8**-25, **8**-28, **8**-30, **8**-35
   Type II, **8**-16, **8**-17
   Type III, **8**-16, **8**-17, **8**-28, **8**-35
Extremal probability paper, **8**-28
Extreme desert, **24**-5
Extreme-value distribution, **9**-49
   (*See also* Extremal distribution)
Extreme-value probability, **8**-3, **8**-4
Extreme-value series, **8**-20

F distribution, **8**-70
F test, **8**-73, **8**-78

INDEX

Factor of outflow variation, **20**-15
Factor procedure, **20**-26
Factorial method, **26**-37
Fair-weather runoff, **4**-3, **14**-2
Falkland Current, **2**-7
Fall velocity, **7**-43
Falling segment, limb, or curve, **14**-8
Falls Lake, Grand Coulee, Wash., **23**-5
False bottom, **23**-31
FAO (Food and Agriculture Organization), **1**-10, **1**-11
Farm, water control on, **21**-3
Farm pond, **21**-65
  typical layout, **21**-66
Farming, contour, **21**-26
  land treatment for, **21**-26
  straight-row, **21**-26
Fathometer, **15**-17
Federal Constitutions, **27**-22
Federal Council for Science and Technology, **1**-2, **1**-10, **1**-12
Federal Housing Administration, **5**-6
Federal Inter-Agency Committee on Water Resources (*see* U.S. Inter-Agency Committee on Water Resources)
Federal Inter-Agency River Basin Committee, **17**-66, **26**-19, **26**-20
  (*See also* U.S. Inter-Agency Committee on Water Resources)
Federal Power Commission (*see* U.S. Federal Power Commission)
Federal power licenses, **27**-24
Federal-State-Private Cooperative Snow Survey System, **10**-21
Fernow Experimental Forest, W.Va., **6**-24
Ferrito zone, **4**-9
FERUT, **29**-21
Fetch, **2**-18
Field capacity, **5**-15, **11**-17
Field drain, **21**-89
Field efficiency, **21**-84
Field-moisture deficiency, **14**-6
Fine-sediment load, **7**-29
Finite-differences, method of, **25**-83
Fire effect, on erosion, **22**-38
  on runoff, **22**-31
Fire-effect ratio, **22**-36
Firm power, **14**-43
Firn, **16**-9
Firn limit, **16**-17
Fish and wildlife, **26**-2
  benefits, **26**-19
  economic analysis, example for, **26**-20 to **26**-23
  hydrologic data, **26**-7
Fisher-Tippett theory of extreme values, **8**-3
Fisher's transformation, **8**-56
Fitting parameter, **8**-47
Fitzgerald's evaporation equation, **11**-4
Fitzroy River, Australia, **24**-27
Fixed point, **29**-17
Fjord lake, **23**-4
Flame-photometric method, **19**-14
Flash-flood warning, **25**-110

F-layer, **6**-27
Float for velocity determination, **15**-11
Float gage, **15**-6
Floating point, **29**-17
Flocculating agent, **5**-4
Flood(s), **25**-2
  control of (*see* Flood control)
  for design, **25**-25, **25**-72, **26**-14
  duration of, **25**-5
  frequently occurring, **25**-119
  glacier, **16**-30
  maximum experienced, **25**-119
  maximum possible, **25**-72
  maximum probable, **25**-72, **25**-119
  mean annual, **8**-36, **25**-6
  measurable features of, **25**-2
  project, standard, **9**-65, **25**-26, **25**-72
  synthetic, **25**-95
  records of, **25**-8
  recurrence of, **25**-119
  regional, **25**-119
  regression and correlation of, **8**-61 to **8**-63
  sheet, **24**-25
  from small rural basins, **25**-16
  stream, **24**-26
  in urban areas, **20**-33
Flood abatement, **25**-114
Flood control, **26**-2
  alternatives to, **25**-113
  benefits, **26**-18
  economic analysis, example for, **26**-20 to **26**-23
  hydrologic data for, **26**-7
  legal aspects, **27**-23
  in Miami River, **25**-87
  in Missouri River, **25**-92
  results of, **25**-113
  in TVA system, **25**-91
  water policy on, **28**-20
  Wolf Creek Project, **25**-88
Flood Control Acts, **27**-23 to **27**-25, **28**-21, **28**-25
Flood damage, **25**-118
  frequency curves for, **26**-24
Flood discharge, **25**-4
  east of Rocky Mountains, **25**-73
  in eastern U.S., **25**-73
Flood elevation, **25**-3
Flood formulas, **25**-5
Flood frequency, **25**-5
  analysis, **8**-3
Flood-frequency curve, **25**-6
Flood-frequency regions, **25**-8
Flood fringe, **25**-120
Flood hydrograph, characteristics of, **25**-73
  for reservoir design, **25**-67
Flood insurance, **25**-117
Flood Insurance Act, **25**-117
Flood irrigation, **21**-79
Flood losses, **25**-114, **25**-115
  in industrial areas, **25**-116
  in public areas, **25**-116
  on railroads, **25**-116
  in urban areas, **25**-115

INDEX

Flood profile, **25**-3
Flood records, South Fork Holston River,
    Kingsport, Tenn., **25**-69
  Tennessee River, Chattanooga, Tenn.,
    25-70
Flood regions, **25**-8
Flood regulation, building codes for, **25**-123
Flood-risk map, **25**-123
Flood routing, **25**-35
  characteristics, method of, **25**-38
  coefficient method, **25**-40
  by computer, **25**-58, **29**-7 to **29**-11,
    29-21, **29**-22
  diffusion analogy, method of, **25**-38
  graphical methods, **25**-47
  Harkness flood router, **25**-53
  at junctions, **25**-50
  lag methods, **25**-38, **25**-46
  mathematics of, **25**-37
  Muskingum method, **25**-40
    simplified, **25**-49
  Puls method, **25**-38
    modified, **25**-39
  for reservoir modification, **25**-52, **25**-57
  stage routing, **25**-58
  U.S. Weather Bureau electronic router,
    25-54
  working-value method, **25**-44, **25**-49,
    25-50
Flood volume, **25**-4, **25**-16
Flood warning, **25**-110
Flooding method for groundwater
    recharge, **13**-44
Floodplain, **25**-2
  classification, **25**-120
  regulation, **25**-120
  zoning, **25**-121 to **25**-124
Floodproofing, **25**-114
  measures of, **25**-118
Floodwall, **21**-76
Floodwaters, defined legally, **27**-27
Floodway, **21**-76, **25**-120
Flotation, **7**-8
Flotation forces, **7**-9
Flow, base, **4**-2, **14**-3
    separation of, **14**-11
  in closed conduits, **7**-15
  compressing, **16**-24
  continuity equation of, **7**-10
  critical, **7**-23
  depth determination of, **15**-16
  dry-weather, **19**-18
  extending, **16**-24
  of glaciers, **16**-22
  gradually varied, **7**-10, **7**-38
  groundwater, **14**-2
  gutter routing of, **20**-18
  in gutters, **20**-6, **20**-13, **20**-18
  inlet, **20**-22
  irrotational, **7**-10
  laminar, **7**-12
  mass, **5**-20
  measurement of, **7**-44
  nonuniform, **7**-10, **7**-23
  in open channels, **7**-22

Flow, overland (*see* Overland flow)
  potential, **7**-10
  radial, **13**-16 to **13**-22
  rapid, **7**-23
  regional, **25**-119
  resistance to, **7**-14
  return, **27**-27
  rotational, **7**-10
  routing of (*see* Routing)
  separation of, **7**-10
  shear of, **7**-12
  sheet, **14**-36, **24**-25
  steady, **7**-9
    of groundwater, **13**-13
    nonuniform, **7**-19
    in open channels, **7**-23
    radial, **13**-16
    uniform, **7**-16
  subsurface, **14**-2
  subsurface storm, **14**-2
  tranquil, **7**-23
  turbulent, **7**-13, **7**-23
  uniform, **7**-10, **7**-23
  unsteady, **7**-9
    of groundwater, **13**-14
    in open channels, **25**-82
    radial, **13**-18
  varied, **7**-23
  virgin, **14**-2
Flow chart, **29**-18
Flow control, artificial, **15**-30
  complex, **15**-38
  (*See also* Control, of flow)
Flow diagram, **29**-18
Flow-duration curve, **8**-3, **8**-9, **14**-42
  for reservoir design, **25**-67
Flow line, **25**-58
Flow-line analysis, for groundwater, **13**-14
Flow-line computation, **25**-58
Flow-mass curve, **14**-44
Flow net, **7**-10, **13**-14
Flow profiles, **7**-38 to **7**-42
  for reservoirs, **25**-68
Flowage, **27**-20
Flowmeter, **15**-13
  ultrasonic, **15**-13
Fluid, characteristics in soil, **12**-4
  dynamics, **7**-9
  pressure, **7**-6
  statics, **7**-6
  viscosity, **7**-12
Fogs, **2**-22, **3**-22, **6**-7
  advection, **2**-22
  ocean, **2**-22
  radiation, **2**-22
  steam, **2**-22
Food and Agriculture Organization (FAO),
    1-10, **1**-11
Forecasting, precipitation, **9**-25
  rainfall, **9**-25
  river, **25**-99
  of runoff for river regulation, **10**-27
  of seasonal water yield, **10**-23
Forecasts, low-flow, **25**-110
  water-supply, **25**-110

INDEX

Foreign waters, defined legally, **27**-28
Forest, hydrologic features of, **22**-4
   influences of, **22**-51
      development in U.S., **22**-2
      major types in U.S., **22**-3
   terminology for, **6**-27, **22**-51
   type, **22**-51
Forest age-stocking effectiveness, **22**-30
Forest hydrology, **22**-1, **22**-2
Forest and ranges, animal life in, **22**-8
Forfeiture, **27**-18
Form-loss coefficient, **7**-19, **7**-20
Form losses, **7**-20
FORTRAN, **20**-21, **29**-22
Fourier's law, **5**-20
FPC (*see* U.S. Federal Power Commission)
Francis' weir formula, **1**-8, **7**-45
Franklin County Conservancy District, **1**-9, **1**-14
Frankstown Branch, Juniata River, Williamsburg, Pa., flow-duration curve, **18**-11
Frazil, **16**-7
Free energy, **11**-16
Free surface, **7**-22
Freeboard, **21**-71
   of flood-control dams, **25**-72
Freezing index, **10**-47
Freezing nuclei, **3**-20, **9**-13
Freezing-point diagrams for solutions, **10**-13
French Broad River Basin, Tenn., rain-gage network, **9**-8
French River, Georgian Bay, Lake Huron, water level in, **23**-14
Frequency, **8**-5, **8**-22
   of annual yield, **21**-37
   design, values for urban expressways, **20**-41
   of drought, **18**-10
   formulas for rainfall, **9**-60, **9**-61
   stream, **4**-55
Frequency analysis, **8**-3
   for drought and low streamflow, **8**-4
   for flood and streamflow, **8**-3
   procedure of, **8**-17
   for rainfall, **8**-4, **9**-49
   of water quality, **8**-4
   of water waves, **8**-4
Frequency distribution, **8**-5
Frequency factor, **8**-23
   for extremal distribution, **8**-25
   for lognormal distribution, **8**-26
Frequency synthesis of rainfall regimes, **9**-60
Freshwater Biological Association, England, **1**-11
Friction velocity, **17**-43
Friction-head loss, **15**-32
Frictional resistance, depth of, **2**-10
Front, **3**-29, **9**-18
   Arctic, **3**-29
   cold, **3**-29
   Mediterranean, **3**-29
   polar, **3**-29
   warm, **3**-29

Frontal model of cyclone, **9**-19
Frost, effect on infiltration, **6**-15
   hoar, **3**-23
   in soil, **10**-47
Frost action, **10**-48
Frost heave, **10**-48
Frost zone, annual, **10**-47
Froude number, **7**-5
Frozen storage, **18**-20
Full employment, **26**-32
Fuller's formula, **8**-23
Function, annual-benefit, **26**-33
   annual-cost, **26**-33
   beta, **8**-71
   gamma, **8**-69
      incomplete, **8**-14
   impulse, **14**-28
   input-cost, **26**-33
   kernel, **14**-24
   Laplace-transform, **14**-26
   objective, **26**-32
   production, **26**-33
   pulse, **14**-28
   service, **6**-48
   special transfer, **14**-31
   well, **13**-18
Function generator, **29**-5
Function multiplier, **29**-5
Furrow irrigation, **21**-79
Furrow method for groundwater recharge **13**-43

Gage, auxiliary, **15**-30
   Bourdon, **7**-7
   bubble, **15**-7, **15**-8
   chain, **15**-6
   crest indicator, **15**-7
   electric tape, **15**-6
   float, **15**-6
   radioisotope for snow, **10**-20
   reference, **15**-29
   Roda, **8**-12
   staff, **15**-6
   stream, auxiliary, **15**-30
      nonrecording, **15**-39
      recording, **15**-29, **15**-39
   tape, **15**-6
   wire, **15**-6
Gage density, **26**-9
Gage pressure, **7**-6
Gaging station, density of, **26**-9
   supplementary, **21**-28, **21**-37
Galloping of power lines, **10**-45
Galton's law, **8**-3, **8**-17
Gamma distribution, **8**-14
Gamma function, **8**-14, **8**-69
   incomplete, **8**-14
Ganguillet-Kutter's formula, **1**-8
GAP/R, **29**-4
Gardiner River, Mammoth, Wyo., seasonal runoff, **10**-34
Gate, **7**-46
   deposition at outlet, **17**-4
   sluice, **7**-46

Gaussian law of errors, **8**-3, **8**-14
GEDA, **25**-58, **29**-4, **29**-7
GEK (geomagnetic electrokinetograph), **2**-14
General project report, **26**-15
Generating process, **8**-84
Geohydrologic unit, **4**-10
Geohydrology, **1**-5
Geologic norm, **17**-6
Geological effectiveness, **22**-29
Geological Society of America, **1**-15
Geomagnetic electrokinetograph (GEK), **2**-14
Geometric mean, **8**-6
Geometry number, **4**-68
Geomorphology, **1**-5
Geophysical Research Board, National Academy of Sciences, **1**-10
Geostrophic method, **2**-13
Geostrophic motion, **2**-13
Germantown detection reservoir, **25**-86
GGI (State Hydrologic Institute, Leningrad), **1**-15
GGO (Main Geophysical Observatory, Leningrad), **1**-15
Ghyben-Herzberg concept, **13**-48
Ghyben-Herzberg principle, **1**-8
Ghyben-Herzberg relation, **13**-48
Gila River, **14**-42
  near Virdon, N.Mex., discharge hydrographs of, **25**-4
Glacier caps, **16**-12
Glacier floods, **16**-21, **16**-30
Glacier flow, **16**-22
Glacier milk, **17**-10
Glacier runoff, **16**-19
  regulation, **16**-29
Glacier streams, **16**-21
Glacier tongues, **16**-12, **16**-25
Glacier variation for reconstructing past climate, **16**-28
Glacierettes, **16**-11
Glaciers, **16**-2, **16**-10
  ablation of, **16**-15
  ablation zone, **16**-17
  accumulation, **16**-15
  accumulation zone, **16**-17
  activity of, **16**-14
  activity index, **16**-19
  in Alaska, **16**-2
  behavior, **16**-27
  cirque, **16**-12
  classification, **16**-10
  cliff, **16**-12
  continental, **16**-12
  crater, **16**-12
  density profile of, **16**-15
  distribution by state, **16**-3
  foot, **16**-12
  hanging, **16**-12
  high-polar, **16**-13
  highland, **16**-12
  hydrologic problems connected with, **16**-28
  piedmont, **16**-12

Glaciers, qualifications of, **16**-11
  reconstituted, **16**-12
  regenerated, **16**-12
  response to climatic changes, **16**-25
  subpolar, **16**-13
  summit, **16**-12
  temperate, **16**-13
  temperature profiles in, **16**-14
  transection, **16**-12
  valley, **16**-12
  wall-sided, **16**-12
  in western U.S., **16**-3
Glaciology, **1**-5, **16**-2
Glaze, **3**-22
Glei, **5**-6
Glen Canyon Dam, **26**-36
Glen's flow law, **16**-6
Goodman's predictivity test, **28**-19
Goodness of fit, **8**-47
Goose Lake, Tex., subsidence of, **4**-25
Graben lakes, **23**-11
Grade stabilization, **21**-61
Gradient current, **23**-27
Grading in airport drainage, **20**-35
Gradually varied flow, **7**-38
Grand Coulee Power Plant, system model of, **26**-44
Grassed waterways, **21**-54
Grassland Research Institute, **1**-11
Grate inlet, **20**-31
Gravels, **5**-7
Gravimetric method, **19**-14
Gravitational water, **5**-4, **5**-15, **20**-10
Gravity, specific, **7**-3
Gravity erosion, **17**-10
Gravity spring, **4**-35
Gravity wave, **2**-16, **2**-17
  small, **7**-32
Gray-brown podzolic soil, **5**-5
Gray wooded soil, **5**-5
Great Australian Basin, **24**-32
Great Lakes, **23**-12 to **23**-15
  currents in, **23**-30
  diversion of water from, **23**-15
  glacial action on, **23**-4
  ice, in, **23**-20
  live roof loads, **10**-43
  snowstorms, **9**-22
  stratification in, **23**-18
  U.S. Lake Survey, **1**-13
  water level in, **23**-13 to **23**-15
Green River, Utah and Wyo., **14**-42
Greenland Icecap, **16**-13
Grid-square method, **4**-63
Ground density, **22**-51
Ground Water journal, **1**-17
Groundwater, appropriation of, **27**-32
  in arid and semiarid regions, **24**-27
  atmospheric pressure effects on, **13**-36
  computer application in, analog, **29**-10
    digital, **29**-23
  conjunctive use of, **13**-41
  dating, **13**-13
  depletion curve of, **14**-8
  development law for, **28**-13

Groundwater, distribution of use of, **13**-7
  evapotranspiration effects on, **13**-35
  exploration of (see Groundwater exploration)
  flow equations of, **13**-13
  flow-line analysis of, **13**-14
  fluctuations of, **13**-33
    indicating evapotranspiration, **11**-25
  in hydrologic cycle, **13**-3
  legal aspects of, **27**-28, **27**-37
  nuclear explosion for, **13**-7
  occurrence of, **13**-3
  overdraft of, **13**-38
  phreatic, **24**-30
  qualities of, **19**-23
  radioactive tracer for, **13**-12
  recession curve of, **14**-8, **14**-9
  recharge of (see Groundwater recharge)
  references on laws, **27**-34
  refraction at permeable boundaries, **13**-15
  regarding water policy, **28**-4
  safe yield of, **13**-38
  salt-water intrusion in, **13**-46
  secondary cementation on, **4**-8
  secular and seasonal effects, on, **13**-33
  steady flow of, **13**-13
    radial, **13**-16
  storage of (see Groundwater storage)
  streamflow effects on, **13**-34
  tidal effects on, **13**-37
  unconfined, **24**-30
  in U.S., **4**-27
  unsteady flow of, **13**-14
Groundwater cement, **4**-8
Groundwater exploration, **4**-26 to **4**-33
  method of, cable-tool and rotary, **4**-32
    electric, **4**-32
    geochemical, **4**-31
    gravity, **4**-30
    radiation, **4**-33
    reflection, **4**-30
    refraction, **4**-30
    resistivity, **4**-28
    seismic, **4**-30
    sonic, **4**-33
    subsurface geochemical, **4**-33
    subsurface geologic, **4**-31
    subsurface geophysical, **4**-32
    surficial geophysical, **4**-28
Groundwater flow, **14**-2
Groundwater laterite soil, **5**-5
Groundwater law, **27**-31
Groundwater management, **13**-38
Groundwater movement, **4**-8, **4**-22 to **4**-26, **13**-7
  tracing of, **13**-12
Groundwater packed soil, **5**-5
Groundwater recharge, artificial, **13**-41
  method of, basin, **13**-43
    ditch, **13**-43
    flooding, **13**-44
    furrow, **13**-43
    induced, **13**-46
    modified-stream-bed, **13**-43

Groundwater recharge, method of, by pits, **13**-44
  by wells, **13**-44
  in U.S., **13**-41
Groundwater reservoir by nuclear explosion, **13**-7
Groundwater runoff, **14**-2
Groundwater storage, **4**-22 to **4**-26
  modification of, **24**-32
Groundwater use in U.S., **13**-6
Grumusol, **5**-6
Guard cells, **11**-20
Guayabal Irrigation Reservoir, Puerto Rico, sedimentation, **17**-4
GUGMS (Main Administration of the Hydrometeorological Service, U.S.S.R.), **1**-15
Guiana Current, **2**-7
Gulf of Mexico, **14**-23
Gulf Stream, **2**-7
Gumbel, E. J., on extreme-value theory, **1**-9
Gumbel distribution, **8**-16
Gumbel-Powell probability paper, **8**-28
Guntersville Reservoir, flow profiles of, **25**-68
  operation of, **25**-78
  routing through, **25**-80
  volume curves of, **25**-66
Guttation, **11**-20, **24**-14
Gutter, **20**-30, **20**-42
Gutter flow, **20**-6, **20**-13
  routing of, **20**-18
Gutter storage, **20**-6

H profile, **7**-39
Hagen-Poiseuille's equation, **1**-8
Hail, **3**-23
  small, **3**-23
  soft, **3**-23
Halkias-Veihmeyer-Hendrickson's evapotranspiration equation, **11**-26
Hantush-Jacob's method, **13**-21
Hardness, **19**-9
  carbonate, **19**-10
  conversion factors of, **19**-16
  noncarbonate, **19**-10
  permanent, **19**-10
  temporary, **19**-10
Hardpan, **4**-8, **4**-9
  ironstone, **4**-9
Hardwood, **22**-51
Hargreave's equation, **11**-26, **11**-28
Harkness flood router, **25**-53
Harmonic analysis, **8**-11
Harmonic mean, **8**-6
Harmonics, tidal, **2**-18
Harvard Mark I, **29**-16
Harvard Water Program, **26**-31, **26**-36, **26**-38
  computer simulation, **29**-22
Harvesting system, **6**-24
Hawaii, water law, **27**-10, **27**-36

Hayami's flood theory of diffusion analogy, **29**-7
Haze, **3**-22
Hazen, Allen, on hydrologic statistics, **1**-9
Hazen-Williams' formula, **7**-18
Hazen's formula, **8**-28
Head, **7**-6
  elevation of, **7**-12
  piezometric, **7**-12
  on pipe spillways, **21**-69
  pressure, **7**-12
  specific, **7**-36
Heap Steep Glacier, Wind River Range, Wyo., **16**-13
Heat, latent (*see* Latent heat)
  specific, **10**-5
Heat balance, annual, **3**-12
Heat budget, analytical, **23**-19
  Birgean, **23**-8, **23**-19
  in lakes, **23**-18
Heat-budget method, **2**-21
Heat exchange, **10**-30
Heat of fusion, **16**-7
Heat income, **23**-19
Heat index, **11**-28
Heat lows, **3**-30
Hedke's equation, **11**-25
Hellige turbidimeter, **19**-11
Hexadeconal, **11**-14
High alpine, **24**-8
High-gain d-c amplifier, **29**-4
Highway drainage, **20**-41
Highway interchanges, **20**-42
Highways, flood losses on, **25**-116
Historical data, **8**-2
Histosol, **5**-8
H-layer, **6**-27
Hocking River, Athens, Ohio, storage curves of, **18**-23
Homogeneity, meteorological, **9**-28
  space, **8**-13
  statistical, **8**-10
  time, **8**-10
Homogeneity of sample, **8**-18
Homogeneity test, **8**-36
Homothetic basins, **4**-42
Hoop tension, **7**-9
Hoover Commission, Second, **1**-10
Hoover Dam, **17**-6, **23**-6
Horizons, soil, **12**-3
Horton, R. E., on infiltration theory, **1**-9
Horton's equation, **11**-4, **12**-11, **20**-38
Horton's number, **4**-70
Hot spring, **4**-35, **4**-36
Hot-wire anemometer, **15**-13
Hudson River Regulating District, **18**-22
Humic-glei soil, **5**-5
Humic-gley soil, **4**-8
Humid Tropics Research, **1**-12
Humidity, relative, **3**-4
  specific, **3**-5
Humification, **22**-51
Humphreys and Abbot's streamflow measurement, **1**-8

Humus, **22**-48
Hunting, **14**-19
Hurricane, **3**-30, **3**-34
  tracks of, **3**-35
Hurst, H. E., on Nile River, **25**-71
Hurst's formula for reservoir storage, **14**-47
Hydathodes, **11**-20
Hydrarch succession, **23**-31
Hydraulic conductivity, **5**-16, **13**-8
Hydraulic geometry, **14**-38
Hydraulic jump, **7**-34
Hydraulic probability paper, **8**-3
Hydraulic radius, **7**-23, **16**-24, **21**-56
Hydraulic rotary method, **13**-28
Hydraulic routing, **25**-82
Hydraulics of wells, **13**-15
Hydrobiology, **1**-5
Hydrodynamics, of lakes, **23**-24
  of rain, **9**-9
Hydroeconomic analysis, **25**-25
Hydroeconomics, **1**-5
Hydroelectric station, calibration of, **15**-31
Hydrogen bond, **16**-4
Hydrogen-ion concentration (*see* pH value)
Hydrogeology, **1**-5, **4**-2
Hydrograph, **21**-41
  analysis of (*see* Hydrograph analysis)
  characteristic, **14**-13, **14**-25
  at Chickamauga Dam, **25**-92
  complex, **14**-8
  components of, **14**-8, **14**-10
  composite, **21**-45
  defined, **14**-8
  dimensionless (*see* Dimensionless hydrograph)
  direct-runoff, **14**-12
  discharge, **14**-8
  flood (*see* Flood hydrograph)
  of glacier floods, **16**-31
  point of rise, on, **14**-8
  points of inflection on, **14**-8
  for river forecasting, **25**-105
  S-, **14**-16
  simple, **14**-8
  at Sioux City, Iowa, **25**-95
  for snowmelt, **10**-37
  stage, **14**-8
  synthetic, **21**-41
  triangular, **20**-23, **21**-42, **21**-44
  unit (*see* Unit hydrograph)
Hydrograph analysis, Chicago method **20**-16
  for infiltration, **12**-13
  from infiltrometer record, **12**-8
  for land treatment evaluation, **22**-44
  methods of, **14**-13
Hydrograph method, **20**-10
Hydrographer, **1**-2
Hydrographic chart, **23**-7
Hydrographic Department, British Admiralty, **1**-2
Hydrography, **1**-2
HYDROL, **29**-21
Hydrologic areas, **25**-8

INDEX

Hydrologic cycle, **1**-2 to **1**-4
  early concepts, **1**-7
  groundwater in, **13**-3
  for lakes, **23**-11
  (*See also* Water cycle)
Hydrologic data, in airport hydrology, **20**-34
  collection, by Federal agencies, **26**-11, **26**-12
    by U.S. Geological Survey, **25**-99
    by U.S. Weather Bureau, **25**-99
  continuous series, **8**-44
  discontinuous series, **8**-44
  inherent defectiveness in, **8**-18
  observation errors in, **8**-18
  sampling of, **8**-18
  sequential generation of, **26**-38
  series of classification of, **8**-20
    selection of, **8**-19
  sources of, **26**-10
  transposition of, **21**-5, **21**-28
  treatment of raw, **8**-18
  utilization of, **26**-13
  for water resources appraisal, **26**-7
Hydrologic education, **1**-17
Hydrologic equation, **1**-4, **13**-39
Hydrologic model, **8**-9
Hydrologic network, **25**-96
  design of, **26**-7
Hydrologic organizations, **1**-11 to **1**-16
Hydrologic process, **8**-9
  classification of, **8**-10
  deterministic, **8**-9
  non-pure-random, **8**-9
  nonstationary, **8**-9
  probabilistic, **8**-9
  pure-random, **8**-9
  stationary, **8**-9
  stochastic, **8**-9
Hydrologic publications, **1**-16, **1**-17
Hydrologic range, **9**-5
Hydrologic routing, **29**-21
Hydrologic series, **8**-44
  continuous, **8**-44
  discontinuous, **8**-44
Hydrologic soil-cover complex, **21**-11
Hydrologic soil-cover complex number, **25**-24
Hydrologic soil groups, **21**-12 to **21**-25
Hydrologic system, **8**-9
Hydrologic year (*see* Water year)
Hydrology, Ad Hoc Panel on, **1**-2, **1**-12
  advanced degrees in, first, **1**-17
  of agricultural lands, **21**-3
  airport, **20**-34
  applied, **1**-2
  of arid and semiarid regions, **24**-2
  book on, first American, **1**-16
    first British, **1**-16
  defined, **1**-1 to **1**-2
  engineering, **1**-2
  of forest lands and rangelands, **22**-1
  historical development of, **1**-7 to **1**-10
  international programs on, **1**-10
  journal of, **1**-16

Hydrology, of lakes and swamps, **23**-2
  medical, **1**-1, **1**-11
  parametric, **8**-91
  periodicals on, **1**-16
  scientific, **1**-2
  scope, **1**-5
  stochastic, **8**-91
  synthetic, **8**-91
  U.S.S.R. connotation of, **1**-15
  urban, **20**-2 to **20**-24
    of urban expressways, **20**-41
Hydromechanics of lakes, **23**-24
Hydrometeorological network, **26**-8
Hydrometeorology, **1**-5
Hydrometer, **7**-9
Hydrometer method, **17**-40
Hydrometric documents, oldest, **15**-4
Hydrometric pendulum, **15**-10
Hydrometric station, **15**-26
Hydrometry, **15**-3
  defined, **1**-2
Hydrophase diagram, **14**-10, **14**-11
Hydrophobic property, **11**-14
Hydrophyte, **6**-5, **6**-27
Hydropower, **26**-2
  benefits of, **26**-18
  economic analysis, example of, **26**-20 to **26**-23
  in Great Lakes, **23**-12
  hydrologic data of, **26**-7
  legal aspects of, **27**-7, **27**-11
  licenses, **27**-24
  TVA system, **26**-91
  Wolf Creek Project, **25**-88
Hydroscience, Advances in, **1**-17
Hydrosphere, **1**-2
Hydrostatic pressure, **7**-6
Hydroxide, **19**-6
Hyetograph, analysis of, **14**-17
  effective-rainfall, **14**-12
Hygroscopic nuclei, **9**-10
Hygroscopic water, **16**-5, **22**-9
Hypolimnion, **19**-20, **23**-18
Hypothetical combinations, **19**-17
Hypsographic curve, **2**-2, **23**-9
Hypsometric analysis, **4**-68
Hysteresis, **5**-14

IAEA (*see* International Atomic Energy Agency)
IAHR (International Association for Hydraulic Research), **1**-11, **1**-17
IAL (International Association of Theoretical and Applied Limnology), **1**-11
IAMAP (International Association of Meteorology and Atmospheric Physics), **1**-10
IAPO (International Association of Physical Oceanography), **1**-9, **1**-11
IASH (*see* International Association of Scientific Hydrology)
ICAE (International Commission of Agricultural Engineering, **1**-11

INDEX

Ice, **16**-2
  agglomeritic, **23**-21
  air permeability of, **16**-6
  anchor, **10**-46, **16**-7
  aufeis, **16**-8
  breakup of, **23**-22
  candled, **23**-22
  conversion from firn and snow, **16**-9
  critical velocity of loads on, **23**-23, **23**-24
  crystals, growth of, **16**-5
  density of, **16**-6
  discharge measurement over, **15**-20
  drift, **2**-6
  effect, on discharge computation, **15**-40
    on erosion, **17**-10
    on stage-discharge relation, **15**-37
  erosion by, **17**-10
  fast, **2**-6
  formation of, **16**-7
  frazil, **10**-46, **16**-7
  in ground, **16**-10
  lake, **23**-20
  latent heat of sublimation, **23**-17
  pack, **2**-6
  Poisson's ratio of, **16**-6
  porosity of, **16**-6
  properties, **16**-4, **16**-6
  in quiet water, **16**-8
  sea, **2**-6, **16**-7
  sheet, **23**-20
  shelf, **16**-13
  snow, **23**-21
  in streams, **10**-46
  strength of, **16**-6
  structure of, **16**-4, **23**-15
  thickness of, **23**-21
  in turbulent streams, **16**-7
  viscosity of, **16**-6
  Young's modulus of, **16**-6
Ice aprons, **16**-12
Ice-crystal effect on rain drops, **9**-12
Ice-forming factor, **10**-34
Ice mass, **16**-11
Ice needles, **3**-22
Ice thickening, time rate of, **23**-22
Ice-thickness equation, **23**-21
Iceberg, **2**-6
ICID (International Commission on Irrigation and Drainage), **1**-11
Icings, **16**-8
ICOLD (International Commission on Large Dams), **1**-9, **1**-11
Ideal-gas equation, **7**-3
Ideal solution, **11**-6
Ideographic delta, **17**-23
IGC (International Grassland Congress), **1**-11
IHB (International Hydrographic Bureau), **1**-11
IHD (International Hydrologic Decade), **1**-10
IILC (International Institute for Land Reclamation and Improvement), **1**-11

Ili, **27**-10
ILLIAC, **29**-23
Illinois River, at Peoria, Ill., gravel pits for groundwater recharge, **13**-44
Illinois State Geophysical Survey, **1**-14
Illinois State Water Survey, **1**-14, **25**-24
  rain-gage network, **9**-8
Image well, **13**-24
Images, method of, **13**-22
Imperial Dam, **17**-5
Impulse function, **14**-28
Inceptisol, **5**-8
Income redistribution, **26**-31
Incremental analysis, **26**-14, **26**-38
Index, activity, **16**-19
  antecedent precipitation, **14**-6, **25**-102
  basin, **10**-34
  correlation, **8**-59
  freezing, for air, **10**-47
  heat, **11**-28
  infiltration, **12**-28
  leaf area, **24**-34
  moisture, **24**-5
  plasticity, **5**-7, **17**-9
  precipitation, **25**-17
  precipitation-evaporation, **11**-33
  rainfall-erosion, **17**-7
  topographic, **25**-17
  variability, **8**-3, **14**-43
Indian Ocean, **2**-7
Indirect cost, **26**-17
Induced groundwater recharge, **13**-46
Inference, statistical, **8**-55 to **8**-58
  in regression coefficients, **8**-64
  in regression and correlation, **8**-50, **8**-59
Infiltration, **6**-2, **6**-13, **12**-2
  affected by rainfall distribution, **12**-29
  in airport drainage, **20**-37
  in Chicago hydrograph method, **20**-18
  detention-flow relationship of, **12**-14
  ecological aspects of, **6**-15
  effects on land treatments, **22**-12
  factors affecting, **6**-13 to **6**-15, **12**-2
  Horton's equation for, **12**-11
  in Los Angeles hydrograph method, **20**-11
  mass curve of, **12**-11
  measurement of, **12**-6
  rainfall distribution affecting, **12**-29
  relation to runoff, **12**-19
  in runoff computations, **12**-22
  silvicultural aspects of, **6**-16
  in urban hydrology, **20**-5
  water depth affecting, **12**-6
Infiltration analysis, **12**-8 to **12**-19
  average-value method, **12**-17
  block method, **12**-17
  time-condensation method, **12**-16
Infiltration-capacity curve, **12**-11
Infiltration gallery, **13**-31
Infiltration index, **12**-28
Infiltration-rate curves, standard, **12**-27
Infiltration rate of soils, **12**-26
Infiltration recovery, **12**-23

Infiltration routing, **22**-12 to **22**-19
   through soil, **22**-14
Infiltrometer, **12**-6
Infinite dam, **14**-49
Influent stream, **14**-4
Inglis' equations, **7**-27
Initial abstractions, **12**-29
Initial depth, **7**-11, **7**-34
Initialization, **29**-19
Inlet, **20**-30, **20**-42
   capacities of, with ponding, **20**-38
      without ponding, **20**-39
   combination, **20**-31
   curb-opening, **20**-31
   defined legally, **27**-27
   grate, **20**-31
Inlet flow, routing of, **20**-22
Inlet method, **20**-22
Inlet time, **20**-8
Input, **8**-10
Input-cost function, **26**-33
Input unit, **29**-16
Insect control, **26**-2
Instability, condition of, **9**-18
   of glaciers, **16**-25
Instability line, **3**-29, **9**-19
Instability showers, **9**-22
Instantaneous unit hydrograph (IUH), **14**-24
   conceptual models of, **14**-27 to **14**-34
   determination of, **14**-25
   Laplace transform of, **14**-26
   special, **14**-31
Institutional constraints in water economics, **28**-7
Instruction address, **29**-18
Instruction word, **29**-17
Intangible benefit, **26**-17
Intangible cost, **26**-17
Integrating amplifier, **29**-4
Integration method for evapotranspiration, **11**-25
Integrator, **29**-4
Interbasin areas, **4**-48
Interbasin length, **4**-47
Interception, **6**-2, **6**-6, **24**-15
   ecological aspects of, **6**-11
   effect, on rainfall intensity, **6**-7
      on transpiration, **6**-9
   effect of thinning on, **6**-25
   factors influencing, **6**-8, **6**-9
   by forest, **6**-10
   by herbaceous vegetation, **6**-9
   for hydrologic analyses of agricultural lands, **21**-6
   percentages for different species, **6**-9
   silvicultural aspects of, **6**-11
   snow, **6**-8
   in urban hydrology, **20**-5
   for various forests, **6**-11
Interception loss, **6**-7
Interest rate, **26**-20
Interface, **7**-22
   of salt-water intrusion, **13**-49 to **13**-52
Interflow, **14**-2
Interfluve, **24**-25

Intergranular adjustment, **16**-22
Intermittent spring, **4**-35
Internal seiche, **23**-27
Internal unit, **29**-17
International Association for Hydraulic Research (IAHR), **1**-11
   publications of, **1**-17
International Association of Hydrogeologists, **1**-11
International Association of Meteorology and Atmospheric Physics (IAMAP), **1**-10
International Association of Physical Oceanography (IAPO), **1**-9, **1**-11
International Association of Scientific Hydrology (IASH), **1**-9, **1**-11
   Commission on Snow and Ice, **10**-2
   founding of, **1**-9
   on IHD, **1**-10, **1**-11
   publications of, **1**-16
International Association of Theoretical and Applied Limnology (IAL), **1**-11
International Atomic Energy Agency (IAEA), **1**-10, **1**-11
International Boundary and Water Commission, United States and Mexico, **1**-14
   on rainfall data, **9**-9
   simulation analysis by, **26**-35
International Commission of Agricultural Engineering (ICAE), **1**-11
International Commission on Irrigation and Drainage (ICID), **1**-11
International Congress on Large Dams (ICOLD), **1**-9, **1**-11
International Critical Tables, **5**-21
International Grassland Congress (IGC), **1**-11
International Hydrographic Bureau (IHB), **1**-11
International Hydrologic Decade (IHD), **1**-10
International Hydrologic Seminars, **1**-12
International Institute for Land Reclamation and Improvement (IILC), **1**-11
International Programs in Atmospheric Sciences and Hydrology, **1**-10
International Snow Classification, **10**-2
International Society of Medical Hydrology and Climatology, **1**-11
International Society of Soil Science, **1**-9
International Union of Geodesy and Geophysics (IUGG), **1**-9, **1**-11
International Water Supply Association (IWSA), **1**-12
Interstate Commission on the Delaware River Basin, **19**-35, **28**-24
Interstate Commission on the Potomac River Basin, **19**-35, **28**-24
Interstate Water Resources Committee of Arkansas-Oklahoma, **28**-24
Intolerance, **6**-27
Intracrystalline gliding, 16-4

Intragranular gliding, **16**-22
Intrinsic permeability, **13**-9
Ionosphere, **3**-11
Iowa Institute of Hydraulic Research, **15**-13
Iron, **19**-5
Iron zone, **4**-9
Ironstone hardpans, **4**-9
Irrigation, in arid regions, **24**-34, **24**-36
  benefits of, **26**-18
  economic analysis, example for, **26**-20 to **26**-23
  effect of advection on, **24**-36
  flood, **21**-79
  on frequency of annual yield, **21**-37
  frequency curves of water supply for, **21**-81
  furrow, **21**-79
  hydraulic considerations of, **21**-78
  hydrologic data for, **26**-7
  legal aspects of, **27**-11, **27**-79
  methods of, **21**-79
  in Missouri River System, **25**-92
  requirements of, **11**-2
  salinity problems in, **24**-39
  small project design for, **21**-79
  subsurface, **21**-79, **21**-87
  surface, **21**-87
  water distribution in furrow, **11**-18
  water policy on, **28**-21
  water qualities for, **19**-31
Irrigation depths for various crops, **21**-83
Irrigation district, **28**-11
Irrotational flow, **7**-10
Isochrones, **14**-30
Isohyet, **9**-28
Isohyetal method, **9**-28
Isopercental map of storm, **24**-13
Isopercentual method, **9**-29, **9**-31
Isopycnals, **2**-4
Isothermal process, **7**-3
Isotherm, **3**-35
Iterative procedure, **29**-19
IUGG (International Union of Geodesy and Geophysics), **1**-9, **1**-11
IUH (*see* Instantaneous unit hydrograph)
IWSA (International Water Supply Association), **1**-12
Izzard's dimensionless hydrograph, **20**-6, **20**-38

Jackson candle turbidimeter, **19**-11
Jackson Lake, Wyo., snow-water equivalent, **10**-8
Jacob's method for determining aquifer coefficients, **13**-18
Jet stream, **9**-19
Joint cost, **26**-18
Jökulhlaup, **16**-21, **16**-30
Jones formula, **15**-36
Jonglei Canal, **24**-20
Junction, **14**-32, **29**-12
Junction chamber, **20**-31
Juneau Ice Field, Alaska, 16-11, 16-12

Kansas River, **14**-42
  flood control, **29**-7
  flood routing, **29**-21
Karst, **4**-4
Kaskaskia River, New Athens, Ill., stage-discharge relation, **15**-35 to **15**-37
Kautz Glacier, Mt. Rainier, Wash., **16**-30
Keeler's meter, **15**-10
Kennedy's equation, **7**-27
Kennedy's formula, **17**-54
Kennedy's method, **11**-12
Kennedy's theory, **17**-53
Kentucky Dam, **25**-91
Kernel function, **14**-24
Kerr Committee, **1**-10
Kinematic viscosity, **7**-4
Kinematic wave, **16**-26
King's River, Calif., snow gage, **10**-20, **10**-21
Kirchhoff's law, **3**-12
Kiskiminetas River, Pa. and Md., flood frequency of, **25**-6
Knik Glacier, near Anchorage, Alaska, **16**-30
Knudsen's formula, **2**-3
K-T curve, **8**-23
Kuiseb River, Walvis Bay, Africa, **24**-31
Kuleana, **27**-10
Kuroshio Current, **2**-7
Kutter's formula, **22**-38

La Compagnie Générale le Géophysique, **4**-30
La Moine River, Ill., low-flow frequency of, **18**-12
Labrador Current, **2**-7
Lacey's equations, **7**-27
Lacey's theory, **17**-53
Lag methods, **25**-38, **25**-46
Lag time, **14**-25, **20**-14, **25**-24
  defined, **14**-9
Lagoon, **23**-6
Lake Baikal, world's deepest, **23**-4
Lake Barcroft, Va., sedimentation, **17**-11
Lake basins, **23**-3 to **23**-7
Lake current, **23**-27
Lake Erie, water level in, **23**-13, **23**-14
Lake extinction, **23**-11
Lake Eyre, Australia, **24**-26, **24**-32
Lake George, near Anchorage, Alaska, **16**-30
Lake Hefner, Oklahoma City, Okla., **24**-17
  evaporation study, **11**-11, **11**-13
Lake Hefner evaporation equation, **11**-4
Lake Huron, **23**-13
  currents in, **23**-30
  water diversion from, **23**-15
  water level in, **23**-14
Lake ice, **23**-20
Lake Mead, Ariz. and Nev., **23**-6, **24**-17
  evaporation study, **11**-14
  sedimentation, **17**-6
Lake Mead evaporation equation, **11**-4

Lake Michigan, **23**-13, **23**-24
  current in, **23**-30
  water diversion from, **23**-15
  water level in, **23**-14
Lake Mille Lacs, Minn., loads on ice, **23**-23
Lake Ontario, **23**-13
  water level in, **23**-13, **23**-14
Lake Pepin-St. Croix system, **23**-5
Lake Pontchatrain, **23**-5
Lake Stagno di Orbetello, Italy, **23**-6
Lake Superior, **23**-6, **23**-13, **23**-15
  outflow reduction in, **23**-14
Lake water, thermal properties, **23**-15
Lake Winnipesaukee, N.H., outflow measurements, **15**-10
Lakes, artificial, **23**-6
  block-fault, **23**-11
  caldera, **23**-4
  classification of, **23**-3
  coastal, **23**-6
  coral, **23**-6
  craters, **23**-4, **23**-7
  defined, **23**-3
    legally, **27**-27
  deflation, **24**-27
  deltaic levee, **23**-5
  desert, **24**-26
  fjord, **23**-4
  fluctuation of level in, **23**-11
  graben, **23**-11
  hydromechanics of, **23**-24
  lagoons, **23**-6
  landings on ice in, **23**-23
  landslide, **23**-4
  lunettes, **24**-27
  meander scrolls, **23**-5
  oxbow, **23**-5
  pans, **23**-5
  playas, **24**-26
  shotts, **24**-26
  sinks, **23**-5
  solution, **23**-5
  tectonic, **23**-4, **24**-26
  thermokarst, **23**-5
  tombolos, **23**-6
  tundra, **23**-6
  water qualities in, **19**-20
Laminar boundary layer, **7**-15
Laminar flow, **7**-12, **7**-13, **7**-23
  of glaciers, **16**-22
Laminar sublayer, **7**-15
Land subsidence, **13**-34
Land treatment, **17**-32, **21**-26
  effects on runoff, **22**-15 to **22**-31
  hydrologic evaluation of, **22**-12
Land use, **21**-11
  according to infiltration, **12**-28
  changes on floodplain, **25**-114
  (*See also* Land treatment)
Land-use array, **12**-27
Landing, **6**-27
Landings on lake ice, **23**-23
Landsliding, **23**-4
Lane-Kalinske's method, **7**-31

Langrangian method, **2**-13
Laplace transform, **14**-26
Laplace-transform function (LTF), **14**-26
Laplace's equation, **5**-17
Lapse rate, **3**-7
Larchmont Engineering Company, **10**-45
Large watersheds, versus small watersheds, **25**-16
  SPF estimates for, **25**-31
Latent evaporation, **11**-31
Latent heat, of evaporation, **23**-16
  of fusion, **10**-5, **23**-16
  of sublimation, **23**-17
Lateral ditch, **21**-89
Laterites, **12**-3
Latosol, **5**-8
Lava fills, **24**-31
Lava sheets, **24**-31
Law, in action, **27**-33
  administrative, **27**-4
  civil-law rule, **27**-30
  common, **27**-3
  common-enemy rule, **27**-30
  compact, **27**-21
  constitutions, **27**-3
  correlative right, **27**-3
  Dalton's (*see* Dalton's law)
  Darcy's (*see* Darcy's law)
  of declivities, **4**-57
  of errors, **8**-3
  Fourier's, **5**-20
  function of, **27**-2
  Glen's flow, **16**-6
  on groundwater, **27**-31, **28**-13
  Kirchhoff's, **3**-12
  legislation, **27**-3
  logarithmic, **7**-15
  one-seventh power, **7**-15
  outflow, **5**-15
  Planck's, **10**-28
  Rault's, **11**-6
  and science, **27**-32
  of small numbers, **8**-14
  sources of, **27**-3
  of stream areas, **4**-48
  of stream lengths, **4**-46
  of stream numbers, **4**-44
  of stream slopes, **4**-60
  of streams, **4**-54
  (*See also* Water law)
Leaching, **5**-19, **21**-84
Leaf area index, **24**-34
Leaky aquifer, **13**-21
Least-squares method, **8**-31, **8**-46
Legal aspects, in arid and semiarid regions, **27**-7
  for drainage, **21**-88
  in irrigation, **21**-79
Legislation, **27**-3
Lemniscate ratio, **4**-51
Leonardo da Vinci on hydrologic cycle, **1**-7
Levee, **21**-76, **21**-88
Level of confidence, **8**-50
LGP-30, **29**-23
Light's equation, **10**-38

Limit design discharge, **25**-20
Limiting fall, **15**-35
Limiting-fall rating, **15**-35
Limnology, **1**-5
Lindley's equation, **7**-27
Linear channel, **14**-27
Linear programming, **26**-39
  stochastic, **26**-43
Linear reservoir, **14**-27
  electronic analog, **14**-30, **29**-8
Linearity, principle of, **14**-14
  of regression, **8**-74
  test for, **8**-75
Linearization of statistical distribution, **8**-5
Lining of channels, **21**-60, **21**-77
Liquid limit, **5**-6
Lithologic factors, **4**-2
Lithosol, **4**-8, **5**-4
Lithosphere, **1**-2
Litter, **6**-27, **22**-51
Little, Arthur D., Inc., **1**-14
Little Ice Age, **16**-29
Little Ossipee River, Limington, Maine, stage-discharge relation, **15**-34
Little Wabash River, Huntington, Ind., stage-discharge relation, **15**-34
Little Wabash River, Wilcox, Ill., low-flow frequency of, **18**-12
Littoral current, **23**-28
Livingston atmometer, **11**-10
Lloyd-Davis' method, **14**-6
Load, bed (*see* Bed load)
  bed-material (*see* Bed-material load)
  channel capacity for, **17**-37
  contact, **17**-61
  fine-sediment, **7**-29
  snow, **10**-42
  suspended, **7**-31, **17**-61
  wash, **7**-29, **17**-36, **17**-37, **17**-61
    measurement of, **17**-55
Loess, **5**-3
Logarithmic law, **7**-15
Logarithmically transformed distributions, **8**-17
Logextremal distributions, **8**-17, **8**-35
Logextremal probability paper, **8**-28
Logical decision, **29**-16
Lognormal distribution, **8**-17, **8**-25, **8**-35
  frequency factor, **8**-26
  truncated, **8**-17
Lognormal probability law, **8**-3, **8**-4
Lognormal probability paper, **8**-28, **14**-43
Long Lake, **23**-15
Loop, **29**-19
Loop rating, **15**-37
Los Angeles County Flood Control District, **1**-9, **1**-14
Los Angeles hydrograph method, **20**-10
Los Angeles storm zone, **22**-38 to **22**-43
Losses, **14**-2
  aquifer, **13**-27
  bend, **21**-64
  delivery, **21**-85
  entrance, **21**-64

Losses, due to floods (*see* Flood losses)
  form, **7**-20
  friction-head, **15**-32
  interception, **6**-7
  major, **7**-20
  minor, **7**-20
Lower Colorado River, sedimentation, **17**-5
Low-flow forecasts, **25**-110
Low-flow maintenance, **18**-22
Low-flow spillway, **21**-67
Lowering segment, limb, or curve, **14**-8
Lowry-Johnson's equation, **11**-25
LTF (Laplace-transform function), **14**-26
Lunette, **24**-27
Lysimeter, **6**-27, **11**-24

M profiles, **7**-39
McCarthy, G. T., on Muskingum method **25**-40
Machine language, **29**-16, **29**-20
Machine word, **29**-17
McIlroy network analyzer, **7**-22, **29**-3
McMath formula, **20**-8
Macoupin River, Ill., low-flow frequency of, **18**-12
Madison Brook, N.Y., gaging station, **15**-5
Madison River, Montana, **4**-3
Magnesium, **19**-3
Malarial control, **25**-76
Malaspina Glacier, Alaska, **16**-12
Malhotra's equation, **7**-27
Manganese, **19**-5
Manhan River, Mass., diversion curves of, **18**-25
Manhole, **20**-31
Manning's equation, **7**-24
  modified for gutter flow, **20**-6
Manning's formula, **1**-8, **15**-32
  applied to overland flow, **12**-15
Manning's roughness coefficient, **7**-24
  for grassed waterways, **21**-54
Manometer, differential, **7**-8
  micro-, **7**-8
  open, **7**-8
  single-tube, **7**-8
Mansfield, W. W., on hexadeconal, **11**-14
Marginal analysis, **26**-14, **26**-38
Marginal benefit, **26**-18, **26**-33
Marginal benefit-cost ratio, **26**-18
Marginal cost, **26**-18
Marginal distribution, **8**-48
Mariotte, Edmé, on flow measurement, **1**-7
Markov chain, **8**-93, **14**-49
Markov-chain coefficient, **8**-93
Markov-chain model, **8**-92, **8**-94
Markov process, **8**-86, **8**-89, **8**-93
Marshall-Henderson-Hicks formulation, **28**-19
Marshes, **23**-31
  defined legally, **27**-27
Mass, air, **3**-29
  water, **2**-4
Mass budget of glacier, **16**-17

INDEX

Mass curve, **14**-44
 for infiltration, **12**-11
 for reservoir design, **25**-67
 residual, **14**-46
Mass flow, **5**-20
Mass-transfer method, **11**-13
Mathematic models, **26**-38
Matric suction, **5**-15
Maximum experienced flood, **25**-119
Maximum likelihood, method of, **8**-31
Maximum possible flood, **25**-72
 (*See also* Maximum probable flood)
Maximum possible precipitation (MPP), **25**-27
 (*See also* Probable maximum precipitation)
Maximum probable flood (MPF), **25**-15, **25**-26, **25**-72, **25**-119
Maximum range, **14**-47
 adjusted, **14**-47
 mean, **14**-47, **14**-48
Mead, D. W., on first course in hydrology, **1**-17
Meadow, permanent, **21**-25
Mean, **8**-6
 arithmetic, **8**-6
 geometric, **8**-6
 harmonic, **8**-6
Mean annual flood, **8**-25, **8**-36, **25**-6
Mean deviation, **8**-7
Mean range, **2**-18
Meander scroll, **23**-5
Meandering, **7**-26
Median, **8**-7
Median cross, **8**-83
Median-cross test, **8**-83
Mediterranean front, **3**-29
Mediterranean Sea, **1**-7
Meigs' maps of arid and semiarid regions, **24**-3 to **24**-9
Melting point, **10**-9
 depression of, **16**-7
Memory address, **29**-18
Memory unit, **29**-16
Mesa spring, **4**-35
Mesophyll, **6**-4, **11**-20
Mesophyte, **6**-6, **6**-27
Mesoscale phenomena, **3**-33
Mesoscale systems, **3**-10
Mesosphere, **3**-10
Metacenter, **7**-8
Meteorological homogeneity, **9**-28
Meteorology, **1**-5
 defined, **3**-2
 synoptic, **9**-18
Method of characteristics, **14**-13, **25**-38, **25**-82
 of finite-differences, **25**-83
 of images, **13**-22
 of least squares, **8**-31, **8**-46
 of maximum likelihood, **8**-31
 of moments, **8**-30
 of successive approximation, **14**-15
 of summing hydrographs, **20**-15, **20**-16

Meyer-Peter's formula, **17**-53
Meyer's evaporation equation, **11**-4
Mexican Spring, N.Mex., rain-gage network, **9**-8
Miami Conservancy District, **1**-9, **1**-14
 flood-control project, **25**-87
 operating reservoirs, **25**-86
 rainfall frequency, **9**-49
 rainfall frequency data, **8**-4
 single-purpose reservoir, **25**-62
 soil absorption, **12**-2
 storm pattern, **9**-60
Miami Flood-Control Project, **25**-87
Miami River, flood control of, **25**-87
 map of, **25**-86
 soil absorption, **12**-2
Michigan streams, low flow frequency of, **8**-4
Micromanometer, **7**-8
Middle Branch Westfield River, near Goss Heights, Mass., serial correlation of, **8**-79
 serial dependence of, **8**-83
 trend analysis of, **8**-82
Mill Dam Acts, **27**-10
Minikin's formulas, **2**-20
Minimum drought, **8**-35
Mining district, **27**-11
Min-max solution, **28**-9
Mississippi River, **4**-3, **23**-5, **23**-15
 flood control in, **25**-91
 flood-control policy for, **28**-20, **28**-21
 flood records, **15**-4
 floods in, **25**-90
 flow measurement on, **1**-8
 gaging, **15**-5
 legal aspects on, **27**-23
 navigation in, **25**-91
 at St. Louis, correlation analysis of, **8**-58
 swamps in, **23**-31
Mississippi River Commission, **1**-13
 founding of, **1**-8
Mississippi Valley Committee, Public Works Commission, **1**-9
Missouri River, flood control in, **25**-92
 irrigation by, **25**-92
 map of, **25**-92
 pollution control in, **25**-92
 reservoir system of, **25**-87
 sand load, **17**-53
 upper basin river ice, **10**-34
 water supply from, **25**-92
Missouri River Basin, sediment data, **17**-13
Missouri River Reservoir System, **25**-92 to **25**-95
 hydrologic network, **26**-8
 operating curves, **25**-94
 simulation analysis, **26**-35
Mixing ratio, **3**-5
Mode, **8**-7
Model, deterministic, **8**-9
 hydrologic, **8**-9
 mathematical, **26**-38

Model, probabilistic, **8**-9
 stochastic, **8**-9, **26**-39, **26**-43
 stochastic sequential, **26**-44
Modification, **29**-19
Modified Einstein method, **17**-60
Modified-stream-bed method for groundwater recharge, **13**-43
Mohn-Sandström-Helland-Hansen formula, **2**-14
Moisture-adjustment factor, **9**-62
Moisture condensation, **22**-11
Moisture condition, antecedent, **14**-6, **21**-29
Moisture content, **5**-15
Moisture deficiency, **21**-82
Moisture index, **24**-5
Moisture percentage, **5**-15
Molar viscosity, **7**-13
Mole drain, **21**-94
Molecular diffusion, **9**-10
Molecular viscosity, **7**-13
Mollisol, **5**-8
Moments, method of, **8**-30
 statistical, **8**-8
Momentum equation, **7**-10
Momentum-flux correction factor coefficient, **7**-10
Monadnock phase, **4**-69
Monocacy River at Jug Bridge, Md., basin-recharge correlation for, **8**-66
Monoclinal spring, **4**-35
Monofilm, **11**-15
Monomolecular film, **24**-21
Monsoons, **3**-27
Monte Carlo method, **8**-91
Monte Carlo solution, **14**-49
Moody diagram, **7**-16, **7**-17
Morphology of lakes, **23**-7, **23**-10
Morphometric analysis, **4**-43
Morphometric parameters, **23**-8
Morphometry of lakes, **23**-7
Mortar beds, **4**-8
Mosby's formula, **11**-12
Moving average, **8**-81
Moving-average process, **8**-84
MPF (*see* Maximum probable flood)
MPP (maximum possible precipitation), **25**-27
 (*See also* Possible maximum precipitation)
Mud Lake Basin, Idaho, aquifer test in, **4**-22
Mulch, **11**-19
Multiple-purpose projects, **26**-3
 for areal development, **26**-3
 planning items, **26**-4 to **26**-6
 for single development, **26**-3
Multiple-purpose reservoir, ASCE symposium, **25**-97
 linear programming for, **26**-39
Multiple regression, **8**-48
 on land-treatment effects, **22**-15 to **22**-31
 for sediment data analysis, **17**-13
Multistage decision process, **26**-41
Multivariate analysis, **8**-66
Multivariate distribution, **8**-50

Murghab River, Merv Oasis, Central Asia, irrigation system, **15**-5
Murray River, Australia, **24**-28, **24**-30
Musgrave's equation, **17**-7
Muskingum Basin, Ohio, rain-gage network, **9**-8, **9**-30, **9**-31
Muskingum equation, **25**-40
Muskingum method, **25**-40
 simplified, **25**-49
Muskingum routing diagram, **25**-102
Muskingum Valley Conservancy District, **1**-14, **25**-40
 founding of, **1**-9
Myers' curve, **25**-7
Myers' formula, **25**-7
Myers' rating, **25**-8
Myers-Jarvis' enveloping curve, **25**-7
Myers-Jarvis' formula, **25**-7

National Academy of Sciences, **1**-9, **1**-10
 Geophysical Research Board, **1**-10
National Archives on rainfall records, **9**-9
National Bureau of Standards, publications of, **1**-17
National Oceanographic Data Center, **1**-13
National Reactor Test Site, **4**-22
National Research Council, **1**-9
National Resources Board, **1**-9
National Resources Committee, **1**-9
National Science Foundation (NSF), **1**-14, **29**-2
 study on radioactivity, **19**-12
National Water Quality Network, **19**-35
National Water Well Association, **1**-15
 journal of *Ground Water*, **1**-17
Natural-flow theory, **27**-6
Navigation, **26**-2
 benefits of, **26**-18
 economic analysis for, example of, **26**-20 to **26**-23
 hydrologic data for, **26**-7
 legal aspects of, **27**-23
 in TVA system, **25**-91
Navigation improvements, **28**-21
Navigation servitude, **27**-10
NBS (U.S. National Bureau of Standards), **1**-12
Net-benefit response surface, **26**-36
Network, synoptic, **9**-7
Network analyzer, **29**-3
Neutron soil-moisture meter, **5**-16
Névé, **16**-9
Neyrpic meter, **15**-11, **15**-14
Niagara River, waterpower of, **23**-15
Nikuradse's experiment, **17**-43
Nile River, drought in, **18**-1
 flood gage, **15**-4
 reservoir storage requirement, **25**-71
 Roda gage, **8**-12
Nile River basin, computer simulation of, **29**-22
Nile Valley plan, simulation analysis of, **26**-35

INDEX

Niobrara River, Nebr., total load measurement, **17**-60
Nisqually Glacier, Mt. Rainier, Wash., kinematic wave, **16**-28, **16**-30
Nitrate, **19**-7
Nominal diameter, **17**-62
Noncalcic brown soil, **5**-5
Nonconformity spring, **4**-35
Nonequilibrium equation, **13**-18
  limitations of, **13**-20
Nonlinear reservoir, **29**-9
Nonlinear system, **14**-34
Nonlinearity, of basin system, **14**-13
  of runoff distribution, **14**-34
Nonperiodic current, **23**-27
Nonrandom element, **8**-79
Nonrecording stream gage, **15**-29, **15**-39
Nonsaline alkali soil, **24**-40
Nonsaline sodic soil, **24**-40
Nonstationary process, **8**-9
Nonthermal spring, **4**-35
Nonuniform flow, **7**-10
  in open channels, **7**-23
Normal depth, **7**-23
Normal distribution, **8**-14, **8**-24
Normal erosion, **17**-6
Normal fall, **15**-35
Normal-fall rating, **15**-36
Normal law of errors, **8**-3
Normal-ratio method, **9**-28, **9**-32
Norris Reservoir, operation of, **25**-75
North Atlantic Current, **2**-7
North-east Monsoon Current, **2**-7
North Equatorial Current, **2**-7
North Loup River, Nebr., total load measurement, **17**-60
North Mowich Glacier, Mt. Rainier, Wash., **16**-29
North Pacific Current, **2**-7
North Platte River, annual runoff of, **4**-5
  shift corrections at Torrington, Wyo., **15**-40
  stage-discharge relation at Torrington, Wyo., **15**-34
Nozzle, **7**-44
NSF (National Science Foundation), **1**-14, **29**-2
Nuclear explosion for goundwater, **13**-7
Nuclei, hygroscopic, **9**-10
Nuclei condensation, **9**-10
Nuclei diffusion, **9**-10
Nuclei freezing, **9**-13
Nuclei sublimation, **9**-13
Number system, **29**-17
Numerical analysis, **29**-16

Object program, **29**-21
Objective function, **26**-32, **28**-7
Occlude, **3**-29
Ocean clouds, **2**-22
Ocean currents, **2**-6
  hydrodynamics of, **2**-8, **2**-9
Ocean fogs, **2**-22

Oceanography, defined, **2**-2
  physical, **2**-2
Oceanology, **1**-5, **2**-2
Oceans, **2**-2
  characteristics of, **2**-2
  heat budget of, **2**-21
  interaction with atmosphere, **2**-20 to **2**-22
  water qualities in, **19**-22
Odor, **19**-11
Off-control data, **8**-33
Office of Science and Technology, Executive Office of the President, **1**-12
Offset S-hydrograph, **14**-16
Ogoki River, **23**-15
Ohio River, flood control in, **25**-91
  flood routing by computer in, **29**-7
  floods in, **25**-90
  navigation in, **25**-91
  overflow dam on, **15**-31
Ohio River Valley Water Sanitation Commission, **19**-32
  on water-quality data, **19**-35
Ohm's law, **29**-11
160-acre limitation, **28**-22
One-seventh power law, **7**-15
1,000-year flood, **8**-3, **8**-92
Onion Creek, South Yuba River, Calif., snow course, **10**-25
Ontario Hydro-Electric Power Commission, **29**-21
Open channel, storm drainage by, **20**-32
  transitions in, **7**-36
  unsteady-flow equations, **25**-82
Open-channel flow, **7**-22
  computer application of, **29**-21
  profiles of, **7**-39 to **7**-42
Open-channel routing, **25**-35
Open-lake current, **23**-29
Open-water wave, **23**-24
Operating elements, **29**-4
Operating procedure in simulation analysis, **26**-38
Operation code, **29**-18
Operation symbol, **29**-19
Operational amplifier, **29**-4
Operational block, **14**-33
Operations research, **26**-30
Optimal design, **26**-31
Optimal policy, **26**-41
Optimum water allocation, Tolley's, **28**-9
Ordinance, **27**-4
ORDVAC, **29**-14 to **29**-16
Organic layers in forested areas, **22**-49
Organic matter on soil, **22**-7
Organic soils, **5**-7
Orifice, **7**-44
Orifice meter, **7**-44
Orland Reclamation Project, California, **28**-22
Orographic rainfall, **9**-24
Osmotic forces, **5**-15
Ott meter, **15**-11, **15**-14
Outflow law, **5**-15
Outflow variation, factor of, **20**-15

Outlet, submergence, of, **21**-65
Output, **8**-10
Output unit, **29**-16, **29**-17
Overdraft of groundwater, **13**-38
Overfall, **21**-78
Overflow dam, **15**-31
Overland flow, **14**-2, **14**-35, **20**-5, **20**-10, **20**-11, **20**-18
  application of Manning's formula, **12**-15
  detention-flow relationships of, **12**-10
  in forests and ranges, **22**-10
  Horton's equation, **20**-38
  interbasin length of, **4**-47
  length of, **4**-47
  time of concentration for, **14**-36
Owens Valley, California, **28**-11
Ownership rule, **27**-31
Oxbow lake, **23**-5
Oxides, **19**-7
Oxisol, **5**-8
Oxygen, solubility in water, **19**-9
Oxygen consumed, **19**-12
Oxygen polarography, **15**-12
Oxygen sag, **19**-26
Oyashio Current, **2**-7
Ozone, **3**-3

PACE, **29**-4
Pack ice, **2**-6
Packing factor, **13**-10
Palisy, Bernard, on hydrologic cycle, **1**-7
Pan coefficient, **11**-7
Pans, **23**-5, **24**-26
  evaporation, **11**-7
  slush, **23**-21
  theoretical, **11**-7
Panther Creek, El Paso, Ill., hydrograph of, **14**-9 to **14**-11
Parameter, statistical, **8**-6
Parametric hydrology, **8**-91
Pareto optimum, **28**-8
Parshall flume, **7**-47, **15**-30
Parsons' method, **9**-29
Partial-duration series, **8**-19, **9**-49
  theoretical justification, **8**-35
Partial series, **8**-20
Particle diameters, **17**-62
Pasture, classification of, **21**-25
Patch diagram, **29**-6
PE index (precipitation-evaporation index), **11**-33
PE-index method (*see* PE index)
P/E ratio (precipitation-evaporation ratio), **11**-32
Peak discharge, **25**-5
  effect of fire on, **22**-31
  unit, **22**-28
  in U.S., **25**-9 to **25**-15
Peak point, **14**-8
Peak-rate method, **20**-15
Peak-reduction factor, **25**-24
Peak segment, **14**-9
Peak time, **20**-14

Pearson distributions, **8**-3, **8**-14, **8**-25
  K-T curves for, **8**-24
  Type I, **8**-15, **8**-24, **8**-25
  Type III, **8**-15, **8**-24, **8**-25
Pearson frequency distributions, **8**-3
Pearson's skewness, **8**-8
Peat, **5**-7
Pecos River, **14**-42
Pedalfer, **4**-8, **5**-6
Pedocal, **4**-7, **5**-6
Pedological system, **5**-4 to **5**-6
Pehham Lake, N.C., sediment weight, **17**-18
Pend Oreille River, below Z-Canyon, Wash., stage-conveyance-roughness relations, **15**-38
Penman's equation, **11**-26, **11**-28
Penman's method, **21**-8
Per cent normal method, **25**-109
Percentage-of-mean-annual method, **9**-29
Perched spring, **4**-35
Percolating water, defined legally, **27**-28
Percolation, deep, **14**-2
Percolation rate, **22**-51
Perennial spring, **4**-35
Periodic current, **23**-27
Periodicity, **8**-11
  correlation analysis for, **8**-12
Permafrost, **10**-48, **16**-3, **16**-10
  in Alaska, **16**-2
Permafrost line, **10**-48
Permafrost table, **10**-48
Permanent hardness, **19**-10
Permanent International Association of Navigation Congress (PIANC), **1**-12
Permanent-wilting percentage, **6**-27, **11**-17, **24**-19
Permanent-wilting point, **5**-15, **11**-17
Permeability, **4**-12 to **4**-13, **5**-16
  air, in ice, **16**-6
  in snow, **16**-6
  in soil, **5**-20
  coefficient of, **13**-8
    field, **4**-13, **13**-8
    laboratory, **13**-8
  intrinsic, **13**-9
  measurement of, **13**-10
    field, **13**-10
    laboratory, **13**-10
  of soil, **5**-16
  soil scale for, **13**-10
  specific, **13**-9
  units and conversions, **13**-9
Permeability formulas, **13**-10
Permeameter, **13**-10
Permissible velocity, **7**-26, **21**-60
Perrault, Pierre, on hydrologic concepts, **1**-7
Persistence, **8**-12
  in probability routing, **14**-49
Peru Current, **2**-7
pH value, **19**-10, **19**-17
Phase-shifting circuit, **29**-10
Phenols, **19**-8

INDEX

Φ index, **12**-29
Phreatic groundwater, **24**-30
Phreatophyte, **6**-27, **24**-21
Physiographic effectiveness, **22**-29
Physiographic factors affecting runoff, **14**-5
Phytometer, **11**-23
Pi theorem, **4**-70, **7**-5
PIANC (Permanent International Association of Navigation Congress), **1**-12
Piche atmometer, **11**-10
Piezoelectric cell, **7**-8
Piezometer, **7**-8
Piezometric head, **7**-12
Pilot channel, **21**-77
Pioneer species, **6**-27
Pioneers of swamps, **23**-31
Pipe-friction-loss coefficient, **21**-64
Pipe spillway, **21**-62
Pipes, branching, **7**-21
    compound, **7**-20
    forces on, **7**-9
    looping, **7**-21
    network of, **7**-22
Pipette method, **17**-40
Pitometer, **15**-5
Pitot tube, **1**-7, **15**-5, **15**-12
Pittsburgh Flood Commission, **1**-9, **1**-14
Plan formulation, **26**-13
    (*See also* Project formulation)
Planck's law, **10**-28
Plane bed, **7**-26
Planosol, **5**-6
Plant ecology, **1**-5, **6**-2, **6**-27
Plant growth, water requirement for, **24**-34
Plant life, in forests and ranges, **22**-8
Plant roots, **22**-8
Plastic limit, **5**-7
Plasticity index, **5**-7, **17**-9
Platform, continental, **2**-2
    oceanic, **2**-2
Platte River, **14**-42
Playa, **24**-26
Playfair's law of streams, **4**-54
Plotting-position formulas, **8**-29
Plotting positions, **8**-28
PMP (probable maximum precipitation), **9**-62, **21**-71
PMP generalized charts, **9**-63, **9**-64
Pocket spring, **4**-35
Podsol (podzol), **5**-4
Point data, **8**-36
Point-integrating sampler, **17**-55
Point of rise on hydrograph, **14**-8
Points of inflection on hydrographs, **14**-8
Poisson distribution, **8**-14
Poisson's ratio of ice, **16**-6
Polar front, **3**-29
Polarographic technique, **19**-14
Polarography, oxygen, **15**-12
Pole, **6**-27
Policy, defined, **26**-41, **28**-2
Pollutant, **19**-23 to **19**-27

Pollution, **20**-33
    abatement, **26**-2
    example of economic analysis, **26**-20 to **26**-23
    legal aspects of, **27**-19 to **27**-21
Pollution control in Missouri River, **25**-92
Pondage, **25**-120
    in airports, **20**-35
    at highway interchanges, **20**-42
Pondage simulator, **29**-7
Ponding, **24**-27
    in airport drainage, **20**-35
Ponds, **23**-3
    defined legally, **27**-27
    on farms and ranches, **21**-65
Pools, defined legally, **27**-27
Population, **8**-4
Porosity, **4**-12, **13**-4
    effective, **13**-4
    of sediment, **17**-16
    in sedimentary materials, **13**-4
    of snow and ice, **16**-6
Possible maximum precipitation, **9**-62
Potamology, **1**-5
Potassium, **19**-4
Potential, thermodynamic, **11**-16
Potential flow, **7**-10
Potentiometer, **29**-4
Potomac River, **18**-22
Potomac River basin, **4**-50
Potometer, **11**-23
Power-duration curve, **14**-42
Power lines, dancing and galloping, **10**-45
Power spectrum, **8**-93
Practice factor, **17**-9
Prairie soil, **5**-5
Prandtl-Kármán logarithmic velocity distribution, **7**-31
Precipitation, antecedent, **22**-29
    in arid and semiarid regions, **24**-9
    artificial, **26**-3
        production of, **9**-14
    average in mountains, **25**-109
    channel, **14**-2
    classification of, **3**-22
    correlation with runoff, **14**-5
    cycles in, **9**-46
    distribution in U.S., **1**-6
    effect on sediment delivery, **17**-13
    effective, **14**-2
    excessive, **9**-35
    forecasting of, **9**-25, **25**-109
    mean annual in U.S., **9**-40
    modification of, **24**-15
    monthly variation in U.S., **9**-38
    normal annual values of, **9**-28
        in U.S., **22**-5
    possible maximum, **9**-62
    probable maximum, **9**-60, **9**-62, **21**-71
    processes of, **3**-19
    regression and correlation analysis of, **8**-53
    solid, **10**-3
    storm, **22**-28
    trend in, **9**-46

Precipitation, in urban hydrology, **20**-5
  various values in U.S., **9**-41, **9**-42, **9**-45, **9**-46
Precipitation-evaporation index (PE), **11**-33
Precipitation-evaporation (P/E) ratio, **11**-32
Precipitation excess, **14**-2, **22**-29
  standard, **22**-29
Precipitation fluctuation at Tacubaya D. F., Mexico, **24**-12
Precipitation gages, density of, **26**-9
  valley stations, **10**-26
  (*See also* Rain gages)
Precipitation index, **25**-17
Precipitation normal, **8**-7
Preclimax, **6**-27
Prediction, economic, **28**-17
Prediction reliability, **8**-34
Predictivity test, Goodman's, **28**-19
Prefrontal showers, **9**-22
Prescriptive rights, **27**-8
Presidential Advisory Committee on Water Resources, **1**-10
President's Committee on Water Flow, **1**-9
President's Water Resources Policy Commission, **1**-10, **28**-14
President's Water Resources Policy and Materials Policy Commissions, **28**-14
Pressure, **7**-6
  atmospheric (*see* Atmospheric pressure)
  center of, **7**-8
  external, **7**-9
  gage, **7**-6
  hydrostatic, **7**-6
  measurement of, **7**-7
  saturation, **7**-4
  vapor, **7**-4, **11**-3
  variation with elevation, **7**-7
Pressure cell, **7**-8
Pressure charts, **3**-24 to **3**-28
Pressure drag, **7**-14, **7**-42
Pressure head, **7**-6, **7**-12
Pressure resistance, **7**-14
Pressure-ridge method, **13**-51
Pressure transmitter, **15**-6
Price current meter, **1**-8
Price level, **26**-20
Price meter, **15**-5, **15**-9
Principle, Archimedes', **7**-8
  D'Alembert's, **1**-7
  of equitable apportionment, **27**-21
  of linearity, **14**-14
  of optimality, **26**-41
  of proportionality, **14**-14
  of superposition, **14**-14
  of time invariance, **14**-15
Prior appropriation, doctrine of, **27**-11
Prism storage, **25**-40
Probabilistic model, **8**-9
Probabilistic process, **8**-9
Probability, **8**-2
  cumulative, **8**-6
  of drought occurrence, **18**-11

Probability, extreme-value, **8**-3, **8**-4
  Galton's law of, **8**-3
  Gaussian law of, **8**-3
  lognormal law of, **8**-3, **8**-4
Probability density, **8**-6
Probability distribution, **8**-6
Probability paper, **8**-27
  extremal, **8**-28
  Gumbel-Powell, **8**-28
  hydraulic, **8**-3
  logextremal, **8**-28
  lognormal, **8**-28, **14**-43
  skew-frequency, **8**-3
  Weibull, **8**-28
Probability routing, **8**-9, **14**-49
Probable maximum precipitation (PMP), **9**-60, **9**-62, **21**-71
  (*See also* Maximum possible precipitation)
Problem formulation, **29**-18
Problem variable, **29**-6
Process, **8**-9
  adiabatic, **3**-7, **7**-3
  advective, **2**-20
  autoregression, **8**-85
  convective, **2**-20
  deterministic, **8**-9
  generating, **8**-84
  hydrologic (*see* Hydrologic process)
  isothermal, **7**-3
  moving-average, **8**-84
  multiple-decision, **26**-41
  nonstationary, **8**-9
  probabilistic, **8**-9
  stationary, **8**-9
  stochastic, **8**-9
  sum-of-harmonics, **8**-84
Processor, **29**-20
Production function, **26**-33
Program, **29**-16
  object, **29**-21
  source, **29**-20
  stored, **29**-16
Program documentation, **29**-18
Program register, **29**-16
Programming, **29**-18
  automatic, **29**-20
  deterministic nonlinear, **26**-43
  dynamic, **26**-41
  linear, **26**-39
  nonlinear, **26**-43
  quasilinear, **26**-41
  stochastic dynamic, **26**-44
  stochastic linear, **26**-43
Progressive average-lag method, **25**-46
Progressive wave, **23**-24
Project benefit, **26**-17
Project composition, **26**-24
Project cost, **26**-17
Project flood, synthetic, **25**-95
Project formulation, **26**-13, **26**-31
  economic aspects, **26**-6, **26**-15
  investigation and planning, **26**-13 to **26**-15
Project report, **26**-14

Proportionality, principle of, **14**-14
Proprietary power, **27**-23
Public areas, flood losses in, **25**-116
Public districts, **28**-24
Public health, flood losses on, **25**-117
Public Works Acceleration Act, **26**-3, **26**-15, **26**-27
Puls' method, **25**-38
  modified, **25**-39
Pulsations, **8**-12
Pulse function, **14**-28
Pumping for drainage, **21**-95
Pumping station, **20**-42
Pumping tests, **13**-24
  assumptions in formulas for, **14**-23
Puerto Rico Water Resources Authority, **17**-4
Pumping-trough method, **13**-51
Pyrogen, **19**-29

Quality of water (*see* Water quality)
Quasilinear programming, **26**-41
Queue, **14**-48
Queue discipline, **14**-48
Queueing theory (or quuing theory), **14**-48, **18**-16

Rabbit Lake, Minn., **23**-2
Radar beacon, **9**-8
Radar, Weather Bureau, WSR-57, **9**-5
Radial flow, **13**-16 to **13**-22
Radiation, **3**-13, **10**-28
  affecting transpiration, **11**-22
  extraterrestrial, **11**-29, **21**-9
  net, **11**-13
  saturation intensity of, **24**-34
  solar, **22**-10
Radiation fog, **2**-22
Radioactive tracer, **15**-12
Radioactivity, **19**-12, **19**-17
Radioisotope as groundwater tracer, **13**-12
Radioisotope snow gage, **10**-20
Radiometer, **11**-12
Rain, **3**-22
  drop sizes of, **9**-11
  formation of, **9**-9
  freezing, **3**-22
  hydrodynamics of, **9**-9
  physics of, **9**-9
Rain drops, **3**-19
Rain gages, **9**-2 to **9**-5
  density of, **9**-30, **9**-31
  float-type, **9**-5
  networks of, **9**-7
  nonrecording, **9**-4
  radio, **9**-8
  recording, **9**-5
  tipping-bucket, **9**-4, **9**-5
  weighing type, **9**-4, **9**-5
Rain shadow, **9**-24
Rainfall, areal reduction of, **21**-5
  areal variation in Chicago, **20**-20
  by artificial stimulation, **3**-20
  average depth of, **9**-28

Rainfall, bucket surveys of, **9**-8
  convective, **9**-21
  depth-area relationships, **9**-57, **25**-29
  depth-duration-frequency interpolation, **9**-57
  effective, **14**-2, **21**-83
  with extratropical systems, **9**-18
  forecasting of, **9**-25
  frequency analysis of, **8**-4, **9**-49
  frequency formulas for, **9**-61
  gross, **6**-7
  history of recording, **9**-9
  in Los Angeles hydrograph method, **20**-11
  measurement of, **9**-2 to **9**-9
    by radar, **9**-5
  net, **6**-7
  orographic, **9**-24
  point rate-duration-frequency of, **9**-50
  seasonal probability of, **9**-58
  space-time characteristics of, **9**-26
  standard intensity-duration curves, **20**-38
  terminal velocity of, **3**-19
  variability of annual, **18**-4
Rainfall data, **9**-9
  average around world, **3**-36
  frequencies, **9**-50 to **9**-56
  maximum in U.S., **9**-48
  by Miami Conservancy District, **8**-4
  records, **9**-8, **9**-9
    interpolation of, **9**-28
  transposition of, **21**-28
  world's record point values, **9**-47
  Yarnell's, **8**-4, **9**-49
Rainfall erosion factor, **17**-7
Rainfall-erosion index, **17**-7
Rainfall excess, **14**-2
Rainfall frequency atlas, **9**-49
  of U.S., **21**-4
Rainfall-runoff relation, **25**-101
Rainfall simulator, **12**-6
Ranch ponds, **21**-65
Rand Corporation, **1**-14, **8**-93
Random-coordinate method, **4**-63
Random element, **8**-79
Random errors, **8**-44
Random number, **8**-92
  pseudo-, **8**-93
Random sample, **8**-18
Random sampling, **26**-37
Random variable, **8**-4, **8**-47
Range, **8**-7
  adjusted maximum, **14**-47
  analysis of, **14**-47
  maximum, **14**-47
  mean, **2**-18
  mean maximum, **14**-47, **14**-48
  tide, **2**-19
Ranges, classification of, **21**-25
  hydrologic features of, **22**-4
  major types in U.S., **22**-4
Rapid flow, **7**-23
Rating table, **15**-39

INDEX

Rational formula, **14**-6, **20**-8, **25**-5
  assumptions involved, **14**-7
  general, **20**-9
  history of, **14**-6
  runoff coefficients for, **14**-8
Rational method, **20**-8, **21**-37
Rault's law, **11**-6
Raymond Basin, Los Angeles County, **28**-13
REAC, **29**-4
Real time, **29**-17
Reasonable-use rule, **27**-31
Reasonable-use theory, **27**-6
Recession analysis for snowmelt, **10**-35
Recession constant, **14**-6, **14**-9
Recession curve, analysis for detention, **12**-9
  groundwater, **14**-8, **14**-9
Recession segment, limb, or curve, **14**-8
Recharge of groundwater (*see* Groundwater recharge)
Reclamation, legal aspects of, **27**-24
Reclamation Act, **27**-23, **27**-24, **28**-21
Reclamation Engineering Center, **1**-14
Reclamation Project Act, **27**-25
Reconnaissance report, **26**-13
Recording barometer, **7**-7
Recording stream gage, **15**-29, **15**-39
Recreation, benefits of, **26**-19
  example of economic analysis, **26**-20 to **26**-23
  hydrologic data for, **26**-7
  use of water for, **26**-2
  water quality for, **19**-32
Recrystallization, **16**-22
Rectangular distribution, **8**-13, **8**-14
Recurrence interval, **8**-22, **9**-59, **25**-6
Red desert soil, **5**-5
Red-yellow podzolic soil, **5**-5
Reddish-brown lateritic soil, **5**-5
Reddish-brown soil, **5**-5
Reddish chestnut soil, **5**-5
Reddish prairie soil, **5**-5
Reelfoot Lake, regarding tectonic movement, **4**-3
Reference gage, **15**-29
Regelation, **10**-9
Regime equations, **7**-27
Regime theory, **7**-26, **17**-54
Regional analysis, **8**-13, **8**-36, **21**-41
  for evaluation of land treatment, **22**-31
Regional flood, **25**-119
Regional Technical Conferences on Water Resources Development, **1**-12
Regosol, **5**-4
Regression, curvilinear, **8**-58, **8**-65
  linearity of, **8**-74
  multiple, **8**-48
  simple, **8**-48
  simple linear, **8**-50
Regression coefficient, **8**-51, **8**-56
  multiple, **8**-60
  statistical inference of, **8**-64

Regression and correlation, coaxial graphical method for, **8**-65
  curvilinear, **8**-65
  defined, **8**-45
  inference of, **8**-59
  models for analysis by, **8**-47
  multiple, **8**-65
  multiple curvilinear, **8**-65
  multiple linear, **8**-59
  simple curvilinear, **8**-58
  simple linear, **8**-50
Regression and correlation analyses, **8**-86
  general models for, **8**-47
  statistical inference of, **8**-50
Regression line, **8**-45, **8**-50
Regulation diagram, **14**-46
Regur, **5**-6
Relative roughness, **7**-14
Relief aspect, **4**-56
Relief ratio, **17**-13
Remaining benefit, **26**-20
Rendzina soil, **5**-6
Reneging, **14**-48
Renfro's effective-rainfall equation, **21**-83
Republic of Sudan, Ministry of Irrigation and Hydro-Electric Power, **29**-22
Reservoir holdouts, **25**-52
Reservoir regulation, classification of, **25**-62
  purposes of, **25**-61
Reservoir routing, **25**-35
Reservoir sedimentation, **17**-3
Reservoir storage, Hurst's formula for, **14**-47
Reservoirs, aggradation above, **17**-5
  area curves of, **25**-65, **25**-66
  capacity-inflow ratio of, **17**-22
  degradation below, **17**-5
  design of, **25**-64
  design considerations on, **18**-18
  economic factors involved in, **25**-64
  evaporation from in U.S., **24**-17
  evaporation losses in, **18**-16
  flow profiles in, **25**-68
  hydrologic studies on, **25**-65
  level-routing through, **25**-81
  linear, **14**-27
  multiple-purpose, **25**-62
  nonlinear, **29**-9
  operating organization for, **25**-95
  operating schedules and guides for, **25**-85
  operation of, **18**-20
  operation examples for, **25**-87 to **25**-95
  operation planning for, **25**-74
  optimal development for flood-control, **26**-34
  physical factors involved in, **25**-64
  routing through, **25**-35, **25**-77
  sediment design for, **17**-29
  sediment distribution in, **17**-22
  sediment sizes in, **17**-20
  sedimentation in, **17**-3
  sedimentation control in, **17**-29
  sedimentation rates in, **17**-27
    in U.S., **17**-28, **17**-29

Reservoirs, sedimentation in small, **21**-68
  sedimentation survey of, **17**-25
  silting in, **18**-19
  single-purpose, **25**-62
  storage, **14**-47, **18**-15, **21**-65, **25**-68
  storage calculations of, **14**-47, **18**-15
  storage relation to discharge, **21**-65
  storage requirements, **25**-68
  system analysis, by dynamic programming, **26**-41
    by linear programming, **26**-39
  types of, **25**-63
  volume curves of, **25**-65, **25**-66
  water qualities in, **19**-20
Residual mass curve, **14**-46
Residuals, **8**-54
Resistance, bar, **17**-44
  coefficient of, **7**-16
    Colebrook-White equation for, **7**-16
    Moody diagram for, **7**-16, **7**-17
  pressure, **7**-14
  shape, **17**-41
  shear, **7**-14
Resistance coefficient, **7**-16
Resistance to flow, **7**-14
Resistance network, **29**-12
Resistivity, apparent, **4**-28
Resistivity methods, **4**-28
Resolver, **29**-5
Response surface, **26**-36
Retardance coefficient, **20**-38
Retardance of grassed waterways, **21**-54 to **21**-59
Retention, specific, **13**-4
Retention forces, **5**-14
Retention storage, **22**-13
Return flow, defined legally, **27**-27
Return period, **8**-22, **9**-59
Reverse rotary method, **13**-30
Reynolds number, **7**-5
Rhine River, flow measurement in, **1**-8
Rhone Glacier, Switzerland, **16**-27
Rhone River, France, gravel movement, **17**-53
Right of eminent domain, **27**-9
Rights to capture, **27**-30
Rill erosion, **22**-11
Rime, **3**-23
Rio Grande, **14**-42
  aggradation, **17**-5
  flash flood in, **24**-25
  legal aspects of, **27**-16
  simulation analysis of reservoir system in, **26**-35
Rio Grande Basin, on rainfall data, **9**-9
  water-supply forecast, **10**-26
Riparian doctrine, **27**-4
  extent of, **27**-10
Riparian land, **27**-4
Riparian rights, doctrine of, **27**-4
  loss of, **27**-9
Riparian uses, **27**-5
Ripe snow area, **22**-29
Ripening, **10**-9
Ripple, **7**-26

Ripple's method, **14**-44
Rising segment, limb, or curve, **14**-8
Risk, **28**-8
  calculated, **9**-59
River-basin development, policy of, **28**-22
River basin map, **25**-104
River forecasting, **25**-99
  complicating factors in, **25**-107 to **25**-110
  data requirements for, **25**-99
  examples of, **25**-104 to **25**-107
  general problems in, **25**-99
  need for, **25**-99
  rainfall-runoff relations for, **25**-101
  specialized, **25**-110
River hydraulics, computer application in, **29**-22
River stage records, **15**-4
River-measurement program, design of, **26**-8
Rivers, alluvial, **17**-58
  in U.S., **14**-40, **14**-41
  water characteristics in, **19**-18
  in world, **14**-40
  (*See also* Channel; Streams)
Rock interstices, **4**-13
Roda gage, **8**-12
Roentgen, **19**-17
Rohwer's evaporation equation, **11**-4
Rotap mechanical shaker, **17**-40
Rotary-percussion method, **13**-30
Rotating meter, **15**-9, **15**-14
Rotational flow, **7**-10
Roughness, of channel bed, **7**-30
  Manning's coefficient of, **7**-24, **21**-54
  relative, **7**-14
Roughness factor for overland flow, **14**-37
Routine, **29**-18
Routing, of floods, **25**-35
  of gutter flow, **20**-18
  hydraulic, **25**-82
  hydrologic, computer application in, **29**-21
  infiltration, **22**-12 to **22**-19
  of inlet flow, **20**-22
  level-reservoir, **25**-81
  open-channel, **25**-35
  period of, **25**-38
  probability, **8**-9, **14**-49
  reservoir, **25**-35, **25**-77
  for river forecasting, **25**-109
  through sewers, **20**-19
  stage, **25**-58
  storage, **25**-38
  streamflow, **25**-104
  (*See also* Flood routing)
Routing equation, Shrank's, **20**-23
Routing period, **25**-38
Ruggedness ratio, **4**-67
Running totals, **18**-14
Runoff, annual, **21**-34
    average in U.S., **22**-6
  from arid and semiarid regions, **24**-23
  base, **4**-3, **14**-2
  coefficient of, **14**-8, **20**-8, **21**-38

INDEX

Runoff, computation of, **15**-39, **25**-104
  correlation with precipitation, **14**-5
  defined, **14**-2
  direct, **14**-2, **21**-29
  direct surface, **14**-2
  distribution of, nonlinearity in, **14**-34
    in U.S., **14**-39
    in world, **14**-38
  effect of fire on, **22**-31
  effect on sediment delivery, **17**-13
  factor of, **20**-12
  factors affecting, **14**-4
  fair-weather, **4**-3, **14**-2
  from glaciers, **16**-19
  groundwater, **14**-2
  infiltration in computations of, **12**-22
  peak rates of, **21**-37
  relation, to infiltration, **12**-19
    to rainfall, **12**-5
  seasonal, **21**-34
  from small watersheds, **21**-35
  snowmelt, **21**-32, **21**-41
  storm, **14**-2, **21**-28
  subsurface, **14**-2
  surface, **14**-2
  sustained, **4**-3, **14**-2
  time distribution of, **14**-8
  total, **14**-2
  transposition of, **21**-34
  variability of, **14**-42
Runoff coefficient, **14**-8, **20**-8, **21**-38
Runoff curve number (CN), **21**-27, **21**-28, **22**-47, **22**-51
Runoff cycle, **14**-3
Runoff data, on small watersheds, **21**-28
  transposition of, **21**-28
Runoff dynamics, **14**-13
Runoff factor, **25**-24
Runoff number, **25**-24
Runoff phenomena, **14**-3
Runoff process, **14**-3

S profile, **7**-39
Sacramento method of orographic precipitation, **9**-29
Sacramento River, legal aspects of, **27**-23
  tidal estuary, **29**-7
Saddle point of min-max solution, **28**-9
Safe minimum standard of conservation, **28**-8
Safe yield, **13**-38
  determination of, **13**-40
  of reservoir storage, **18**-15
  of storage reservoir, **18**-20
Sainflou theory, **2**-19
St. Clair-Detroit River, **23**-3, **23**-14
  water level in, **23**-14
St. John River, Jacksonville, Fla., tidal velocity measurement, **15**-13
St. Lawrence River, Canada, ice engineering, **10**-5
St. Lawrence River at Ogdensburg, correlation analysis for, **8**-58
St. Lawrence Seaway, **23**-14

St. Lawrence Seaway Project, **29**-21
St. Louis Creek, Fraser, Colo., snowmelt discharge, **10**-35 to **10**-39
St. Louis River, Minn., **23**-13
St. Mary's River, lock and dam on, **23**-14
St. Maurice River Basin, degree-days for, **10**-33
Saline alkali soil, **24**-40
Saline sodic soil, **24**-40
Saline soil, **24**-39
Saline water, demineralization of, **19**-32
Saline Water Conversion Reports, **19**-32
Salinity, **2**-3, **19**-23
  effects on plants, **24**-41
  in estuaries, **19**-22
  in irrigation, **24**-39
  of sea water, **19**-22
Salinity control, **26**-3
Salmon River Basin, snowmelt flood, **10**-33
Salt dilution, **15**-12
Salt effect on ice, **16**-7
Salt tolerance, **24**-41
Salt velocity, **15**-12
Salt-water intrusion, **13**-46
  control of, **13**-51
  interface of, **13**-49 to **13**-52
  in U.S., **13**-47
Salt water sources, **13**-46
Saltation, **7**-29, **17**-54
Saltation load, **17**-61
Saltation-load discharge, **17**-61
Sample, homogeneity of, **8**-18
  random, **8**-18
  space dependence of, **8**-18
  spot, **8**-18
  stratified, **8**-18
  time dependence of, **8**-18
  unbiased, **8**-18
Sample space, **8**-4
Sampler, for bed-material load, **17**-56
  depth-integrating, **17**-55
  point-integrating, **17**-55
  suspended-load, **17**-61
Sampling, of bed, **17**-57
  random, **26**-37
  for simulation analysis, **26**-36
  systematic, **26**-37
Sampling error, **26**-8
Sampling reliability, **8**-31
San Dimas Experimental Forest, rain-gage network, **9**-8
Sand dune, **24**-31
Sand-shape factor, **13**-10
Sand-size analyzer, visual-accumulation-tube, **17**-40
Sand wave, **17**-62
Sangamon River, Ill., low-flow frequency of, **18**-12
Santa Ana River, Calif., annual runoff of, **18**-9
Santa Clara County Flood Control and Water Conservation District, **28**-24
Santa Clara Valley, Calif., **28**-24
Santa Clara Valley Water Conservation District, **28**-24

INDEX

Sapling, **6**-27
SAR (sodium-adsorption ratio), **19**-12
Saskatchewan Glacier, velocity profile, **16**-25
Saturated soils, evaporation from, **11**-19
Saturation, **3**-4, **11**-3
Saturation pressure, **7**-4
Saturation radiation intensity, **24**-34
Savage River reservoir, **18**-22
Sawlog, **6**-27
Scale factor, **29**-6
Schlumberger configuration, **4**-29
Schlumberger Well Surveying Corporation, **1**-14
Schmidt's method, **2**-21
Schoharie Reservoir, Plattsville, N.Y., sedimentation at gates, **17**-4
Schuster test, **8**-11
Scrolls, meander, **23**-5
SCS (*see* U.S. Soil Conservation Service)
S-curve, **14**-44
Sea, **2**-18
  defined, **2**-18
  state of, **2**-18
Sea breeze, **3**-28
Sea ice, **2**-6, **16**-7
Sea level, defined, **2**-14
  mean, **2**-14
  special distribution of, **2**-15
Sea state, **2**-18
Sea water, **2**-2 to **2**-6
  composition of, **19**-34
  demineralization of, **19**-32
  salinity of, **19**-22
Second Hoover Commission, **1**-10
Secondary cost, **26**-17
Secondary power, **14**-43
Secondary succession, **6**-27
Sediment, characteristics of, **17**-14
  defined, **17**-39, **17**-61
  delivered to reservoirs, **17**-20
  delivery ratio of, **17**-12, **17**-27
  deposition processes of, **17**-22
  discharge of, **17**-61
  in glacier streams, **16**-21
  movement from watersheds, **17**-11
  production rate of, **17**-11
  removal of, **17**-30
  supply of, **17**-37, **17**-38
  suspension of, **17**-45
  terminology for, **17**-60
  types of, **17**-61
  venting of, **17**-30
Sediment-accumulation method, **17**-25
Sediment control, **17**-32, **26**-2
Sediment delivery, effect of channel density on, **17**-13
Sediment distribution, prediction of, **17**-24
  in reservoirs, **17**-22
Sediment load, measurement of, **17**-26, **17**-54
  measuring stations of, **17**-26
  sampling of, **17**-54
  total, **7**-31

Sediment particles, defined, **17**-14
  grain-size distribution of, **17**-15
  settling velocity of, **17**-41
  shape of, **17**-15
  sizes, **17**-15, **17**-61, **17**-62
    analysis of, **17**-40
    fractions of, **17**-39
    segregation of, **17**-38
  specific weight of, **17**-15
Sediment pool, **17**-29
Sediment transport, **7**-29
  control in rivers, **17**-36
  similarity of, **17**-53
  by suspension, **17**-46
  by water, **17**-19
Sediment yield, **17**-11
  measurement of, **17**-25
  reduction in, **17**-31
  from watersheds, **17**-26
Sedimentation, **22**-11
  basin, **17**-31
  in reservoirs, **17**-3
    control of, **17**-29
    rate of, **17**-29
    survey of, **17**-25
  in small reservoirs, **21**-68
Sedimentation diameter, **17**-62
Seed-tree cut, **6**-27
Seed-tree method, **6**-24
Seeding cloud (*see* Clouds, seeding of)
Seepage, under dams, **13**-15
  in reservoirs, **18**-19
  storm, **14**-2
Seiches, **23**-26
  internal, **23**-26
  surface, **23**-27
Seine River, Paris, **1**-7
Selection cut, **6**-27
Self-supply, **28**-2 to **28**-4
Semiarid zones or regions, **24**-2
Separable cost, **26**-18
Separable costs–remaining benefits method, **26**-20
Separation, **7**-10
Sequent depth, **7**-11, **7**-34
Sequential generation of hydrologic data, **26**-38
  of hydrologic information, **8**-91
Sequential system, **8**-10
Serial correlation, **8**-79, **8**-85
Service function, **14**-48
Servo divider, **29**-10
Servo multiplier, **29**-10
Sesquioxides, **5**-6
Settling velocity, **17**-41
Sewage, **19**-24
Sewers, routing through, **20**-19
Shantung, **5**-5
Shape resistance, **17**-41
Shear drag, **7**-14, **7**-42
Shear in flow, **7**-12
Shear resistance, **7**-14
Shear stress, **7**-42
Sheet erosion, **17**-6, **17**-19, **22**-11
Sheet-erosion equation, **17**-7

INDEX

Sheet floods, 24-25
Sheet flow, 14-36, 24-25
Sheet ice, 23-20
Shelf ice, 16-13
Shelter belt, 6-27
Shelterwood cut, 6-24, 6-27
Sherman, L. K., on unit hydrograph, 1-9
Shields diagram, 17-53
Shifting control, 15-34
Shifts, 15-39
Shock wave, 16-28
Shore installation, effect of ice on, 23-24
Shoreline, 23-9
Shotts, 24-26
Showers, air-mass, 9-22
  instability, 9-22
  prefrontal, 9-22
Shrank's routing equation, 20-23
S-hydrograph, 14-16
  offset, 14-16
Sierozem, 5-5
Sierra Nevada River, 18-19
Sieve diameter, 17-62
Silica, 19-7
Silica scale, 19-17
Silt, 5-3, 5-7
Silting of reservoirs, 18-19
Silviculture, 1-5, 6-2, 6-27
Similarity of sediment transport, 17-53
Simple correlation, 8-48
  linear, 8-50
Simple Valley, Dorfman's, 26-39, 28-9
Simplex method, 26-40
Simulation analysis, 26-35
  operating procedure, 26-38
Simulator, 29-3
Single-factor method, 26-38
Single-purpose project, 26-3
Sinks, 23-5
SIPRE (Snow, Ice, and Permafrost Research Establishment), 10-49
  publications, 26-10
Skew-frequency probability paper, 8-3
Skewness, 8-8
  coefficient of, 8-8
  measures of, 8-8
  Pearson's 8-8
Slab avalanches, 10-4
Slack variables, 26-40
Slade distributions, 8-17
Sleet, 3-22
Slope, channel, 4-57
  coefficient for basin, 14-23
  critical, 7-23
  of lake basins, 23-9
  mild, 7-23
  valleyside, 4-62
Slope-area method, 15-32
Slope current, 23-27
Slope-length-steepness factor, 17-8
Slope-roughness factor, 15-38
Slough, defined legally, 27-27
Slush balls, 23-21
Slush pans, 23-21
Slutzky-Yule effect, 8-82

Small Reclamation Projects Act, 27-25
Small watersheds, 21-4
  versus large watersheds, 25-16
  SPE estimates for, 25-30
Small Watersheds Act, 27-26
Smeaton's scale models, 1-7
Smith, William, on hydrogeology, 1-8
Smithsonian Institution, meteorological tables, 3-5
  on rainfall records, 9-9
Smog, 3-22
Snake River, near Minidoka, Idaho, aquifer test in, 4-22
Snake River, Wyo., snow-water equivalent, 10-8
Snow, 3-22, 9-65, 10-2
  air permeability of, 16-6
  artificial production of, 16-45
  classification of, 10-2
    field, 10-4
  conversion, to firn to ice, 16-9
  design floods due to, 10-27
  distribution in U.S., 10-7
  granular, 3-23
  in hydrologic analyses of agricultural lands, 21-5
  induced melting, 10-49
  interception by, 6-8
  international classification, 10-2
  loose, 10-41
  measurement of, 10-9
  observation from air, 10-19
  ripening of, 10-8
  for river forecasting, 25-109
  for water-yield forecasting, 10-23
Snow avalanches, 10-41
Snow boards, 10-13
Snow course, 10-14
Snow conferences, 10-22, 10-23
Snow gages, 10-12
  radioisotope, 10-20
Snow, Ice, and Permafrost Research Establishment (SIPRE), 10-49
Snow ice, 23-21
  publications, 26-10
Snow interception, 6-8
Snow loads, 10-42
Snow sampling, Federal sampler for, 10-15
  kit for, 10-17
Snow stake, 10-14
Snow-storage area, 22-29
Snow survey, data, 10-21
  record form, 10-18, 10-19
  safety guide, 10-14
Snow surveying, 10-14
  on ground, 10-17
Snowfall recorded extremes, 10-6
Snowmelt, 22-10, 22-29
  analog for, 29-10
  antecedent conditions of, 10-32
  correlation analysis for, 10-38
  effects of rainfall on, 10-33
  energy sources for, 10-27
  evaluation of analysis of, 22-19

INDEX

Snowmelt, hydrograph of, **10**-37
  hydrograph computation for, **22**-20
  induced, **10**-49
  recession analysis of, **10**-35
  site conditions for, **10**-31
Snowmelt equation, Light's, **10**-38
  U.S. Army Corps of Engineers, **10**-39
Snowmelt runoff, **21**-32, **21**-41
  factors affecting, **10**-27
Snowpack, characteristics of, **10**-30
  estimated weight in U.S., **10**-44
Sodium, **19**-4
Sodium-adsorption ratio (SAR), **19**-12
Sodium percentage, **19**-12
Sodium salts, **19**-4
Soil, **4**-6, **5**-3
  ackerpodsole, **5**-3
  air permeability of, **5**-20
  alfisol, **5**-8
  alkali, **24**-39
  alluvial, **5**-5
  Alpine meadow, **5**-5
  ando, **5**-6
  aridisol, **5**-8
  azonal, **4**-8, **5**-4
  black alkali, **24**-40
  bog, **5**-5
  bottomland forest, **5**-9
  Braunerde, **5**-5
  brown, **5**-5
  brown forest, **5**-5
  brown podzolic, **5**-5
  brunizen, **5**-6
  caliches, **4**-8
  catena, **5**-6
  chernozem, **5**-5
  chestnut, **5**-5
  clay, **5**-3, **5**-7
  coarse-grained, **5**-6
  complex, **5**-6
  degraded chernozem, **5**-5
  desert, **5**-5
  entisol, **5**-8
  fine-grained, **5**-6
  fines, **5**-6
  formation of, **5**-3
  frost in, **10**-47
  geologic factors in development, **4**-6
  glei, **5**-6
  gravel, **5**-7
  gray-brown podzolic, **5**-5
  gray wooded, **5**-5
  groundwater laterite, **5**-5
  groundwater packed, **5**-5
  grumusol, **5**-6
  histosol, **5**-8
  horizons of, **4**-7
  humic-glei, **5**-5
  humic-gley, **4**-8
  hydromorphic, **4**-8
  inceptisol, **5**-8
  infiltration rates of, **12**-13, **12**-26
  intrazonal, **4**-8, **5**-4
  lateritic, **5**-5
  latosol, **5**-8
Soil, lithosols, **4**-8, **5**-4
  mechanical composition of, **5**-3, **5**-4
  mollisol, **5**-8
  noncalcic brown, **5**-5
  nonconformity, **4**-35
  nonlinear sodic, **24**-40
  nonsaline alkali, **24**-40
  nonsaline sodic, **24**-40
  normal profile of, **4**-7
  organic, **5**-7
  oxisol, **5**-8
  oxygen diffusion in, **5**-20
  parent material of, **5**-3
  peat, **5**-7
  pedalfer, **4**-8, **5**-6
  pedocal, **4**-7, **5**-6
  planosol, **5**-6
  podsol (podzol), **5**-4
  podzolized, **5**-4
  prairie, **5**-5
  red desert, **5**-5
  red-yellow podzolic, **5**-5
  reddish-brown, **5**-5
  reddish-brown lateritic, **5**-5
  reddish chestnut, **5**-5
  reddish prairie, **5**-5
  regosol, **5**-4
  regur, **5**-6
  rendzina, **5**-6
  saline and alkali, **24**-39
  saline alkali, **24**-40
  saline sodic, **24**-40
  shantung, **5**-5
  sierozem, **5**-5
  silt, **5**-3, **5**-7
  solonchak, **5**-5
  solonetz, **5**-5
  soloth, **5**-5
  solum, **5**-3
  spodosol, **5**-8
  storage capacity of, **12**-3
  surface entry of water, **12**-2
  tundra, **5**-5
  ultisol, **5**-8
  unsaturated, **5**-19
  upland forest, **5**-9
  vertisol, **5**-8
  water retention in, **5**-9
  water transmission through, **12**-2
  wiesenboden, **5**-5
  yellowish-brown lateritic, **5**-5
  zonal, **5**-4, **5**-5
    profile of, **4**-7
Soil aeration, **5**-20
Soil air, **5**-19, **5**-20
Soil array, **12**-27
Soil classification, **5**-4, to **5**-9
  AASHO system, **5**-6
  comprehensive pedological system, **5**-8
  by hydrologic soil groups, **21**-12 to **21**-25
  land capacity system, **5**-8
  unified system, **5**-6
Soil compaction, **5**-22
Soil Conservation Service Program, **27**-25
Soil-cover complex, **21**-11, **21**-27, **21**-51

INDEX

Soil effectiveness, **22**-29
Soil effects on surface water, **4**-3
Soil-erodibility factor, **17**-7
Soil-loss ratio, **17**-8
Soil moisture, affecting transpiration, **11**-22
    antecedent, **12**-28
    to appearance relationship, **21**-82
    energetics of, **11**-16
    movement of, **11**-17
    neutron meter for, **5**-16
    profiles of, **5**-16
    sampling of, **11**-23
Soil-moisture content, **6**-14
Soil-moisture stress, total, **24**-41
Soil-moisture tension, **5**-13, **5**-14
Soil permeability, **5**-16
Soil phases, **5**-6
Soil profiles, **4**-7, **5**-3
    of forested areas, **22**-49
Soil scale, **13**-10
Soil series, **5**-6
Soil structure, **5**-21, **6**-13
    measurement of, **5**-22
Soil suction, **5**-15
"Soil Survey Manual," **5**-6
Soil surveys, **5**-6
Soil temperature, **5**-20
Soil texture, **6**-13
Soil texture classification, **5**-2, **5**-3, **21**-82
Soil thermal constants, **5**-20
Soil types, **5**-6
    in Los Angeles, **20**-12
Soil water, **5**-9
Solar radiation, **22**-10
Solonchak, **5**-5
Solonetz, **5**-5
Soloth, **5**-5
Solum, **5**-3
Solute suction, **5**-15
Solution, ideal, **11**-6
Solution lake, **23**-5
Somali Current, **2**-7
Sounding, **15**-19, **15**-27
Source program, **29**-20
Source-of-title test, **27**-5
South Cascade Glacier, **16**-9, **16**-14, **16**-20, **16**-21
South Coastal Basin, Calif., grain-sizes, **13**-4, **13**-5
South Equatorial Current, **2**-7
South Fork Holston River, Kingsport, Tenn., flood record of, **25**-69
South Fork, Salmon River, Knox, Idaho, snow cover and losses, **22**-23
    snowmelt hydrographs, **22**-26
    storage and depletion curves, **22**-25
South Santa Clara Valley Water Conservation District, **28**-24
South-west Monsoon Current, **2**-7
Southwest Research Institute, **1**-14
Space dependence, **8**-18
Space homogeneity, **8**-13
Special IUH, **14**-31
Special-purpose analogs, **29**-3
    for flood routing, **25**-55

Special transfer function, **14**-31
Specific capacity, **4**-3, **13**-27
Specific conductivity, **6**-5, **11**-17
Specific electrical conductance, **19**-11
Specific energy, **7**-36
Specific gravity, **7**-3
Specific head, **7**-36
Specific-head diagram, **7**-37
Specific heat, **10**-5
Specific permeability, **13**-9
Specific retention, **13**-4
Specific weight, **7**-3
    design values of, **17**-18
    effect of deposit thickness on, **17**-18
    effect of reservoir operation on, **17**-19
    of sediment, **17**-15
Specific yield, **13**-4
Spectroscopic technique, **19**-14
Sperry Rand Research Center, Inc., **1**-14
SPF (*see* Standard project flood)
SPF estimates, for large drainage basins, **25**-31
    for small drainage basins, **25**-30
Spillway, chute, **21**-62
    design storm for emergency, **21**-72 to **21**-74
    drop, **21**-62
    emergency, **21**-67
    low-flow, **21**-67
    ogee, **7**-47
    overflow, **7**-47
    pipe, **21**-62
Splay, alluvium, **17**-11
    channel, **17**-11
Spodosol, **5**-8
Spreen method, **9**-29
Spring Lake, Macomb, Ill., sedimentation, **17**-4
Springs, **4**-33 to **4**-36
    alluvial-slope, **4**-35
    anticlinal, **4**-35
    artesian, **4**-35
    barrier, **4**-35
    boundary, **4**-35
    channel, **4**-35
    classifications of, **4**-35
    cliff, **4**-35
    cold, **4**-35
    contact, **4**-35
    cuesta, **4**-35
    defined legally, **27**-27
    dimple, **4**-35
    gravity, **4**-35
    hot, **4**-35, **4**-36
    intermittent, **4**-35
    mesa, **4**-35
    monoclinal, **4**-35
    nonthermal, **4**-35
    perched, **4**-35
    perennial, **4**-35
    pocket, **4**-35
    synclinal, **4**-35
    talus, **4**-35
    thermal, **4**-35

Springs, valley, **4**-35
  warm, **4**-35
  water-table, **4**-35
Sprinkler irrigation, **21**-79, **21**-87
SPS (*see* Standard project storm)
SPS depth-area-duration relationships, **25**-29
SPS index rainfall, **25**-27, **25**-28
SPS isohyetal pattern, **25**-30
SPS rainfall, time distribution of, **25**-29
SPS rainfall distribution, **25**-29
Squall line, **3**-29, **9**-19
Stability, colloidal, **9**-11
  in fluid, **7**-8
Staff gage, **15**-6
Stage, determination of, **15**-5
Stage-discharge rating, **15**-35
Stage-discharge relation, **15**-5, **15**-34, **15**-35
  for river forecasting, **25**-102, **25**-109
Stage-fall-discharge relation, **15**-35
Stage-frequency curve, **25**-6
Stage hydrograph, **14**-8
Stage indicator, **15**-6
Stage routing, **25**-58
Stage-time graph, **14**-8, **15**-28
Stand, **6**-27, **22**-51
Standard atmosphere, **3**-9
Standard deviation, **8**-7, **8**-50
  of correlation coefficient, **8**-55
  of residuals, **8**-54, **8**-61
Standard error, **8**-7
Standard Methods for the Examination of Water and Wastewater, **19**-10, **19**-14, **19**-35
Standard project flood (SPF), **9**-65, **25**-26, **25**-72
  for operation of Wolf Creek Reservoir, **25**-89
  (*See also* SPF estimates)
Standard project storm (SPS), **9**-64, **25**-26 to **25**-30
Standard sieves, **17**-40
Standard supply curves, **20**-38
Standard supply rate, **20**-38
Standardized variable, **8**-56
Stanford Watershed Models, **14**-13
Stanislaus River, Calif., runoff per degree-day, **10**-34
Stantec-Zebra computer, **29**-22
State water laws, references on, **27**-35 to **27**-37
Station-year method, **8**-36, **9**-27
Statistical distribution, **8**-6
  (*See also* Distribution, statistical)
Statistical homogeneity, **8**-10
Statistical inference (*see* Inference, statistical)
Statistical moments, **8**-8
Statistical parameter, **8**-6
Statistical variables, **8**-4
Statistically homogeneous area, **8**-13
Statistics, **8**-2
Stationary process, **8**-9
Statutory modifications, **27**-10

Steady flow, **7**-9
  of groundwater, **13**-13
  nonuniform, **7**-19
  in open channels, **7**-23
  radial, **13**-16
  uniform, **7**-16
Steam fog, **2**-22
Steep Rock Lake, Ontario, **23**-2
Steepest-ascent method, **26**-37
Stefan-Boltzmann constant, **11**-13
Stefan's law, **3**-12, **10**-28
Stemflow, **6**-7
Step method, **7**-42
Steven's method, **15**-38
Stochastic dynamic programming, **26**-44
Stochastic hydrology, **8**-91
Stochastic linear programming, **26**-43
Stochastic model, **8**-9, **26**-39, **26**-43
Stochastic process, **8**-9
Stochastic sequential model, **26**-44
Stochastic system, **8**-10
Stocking, **22**-51
Stock-water well, legal aspects of, **27**-32
Stomata, **6**-27, **11**-20
Storage, in banks, **13**-34
  channel, variable, **15**-37, **15**-40
  in conduits, **20**-7, **20**-13
  depression, **20**-5, **20**-18
  detention, **22**-13
  frozen, **18**-20
  of groundwater (*see* Groundwater storage)
  gutter, **20**-6
  induced surcharge, **25**-77
  level, **25**-79
  parallel, **29**-17
  prism, **25**-40
  relation to reservoir discharges, **21**-65
  retention, **22**-13
  serial, **29**-17
  slope, **25**-80
  for temporary floodwater, **21**-71
  theory of, **14**-48, **26**-43
  wedge, **25**-40
Storage capacity in soil, **12**-3
Storage coefficient, **4**-13, **13**-4, **14**-27, **29**-8
Storage curve, **25**-36
Storage-draft curve, **14**-46
Storage equation, **1**-5
Storage loop, **25**-41
Storage reservoir, **14**-47, **18**-15, **21**-65, **25**-68
Storage routing, **25**-38
Storage unit, **29**-16, **29**-17
Stored program, **29**-16
Storm, design (*see* Design storm)
  standard project, **9**-64
  unit, **14**-14
Storm drainage by open channels, **20**-32
Storm precipitation, **22**-28
Storm region, **9**-65
Storm runoff, **14**-2, **21**-28
Storm seepage, **14**-2
Storm temperature, **22**-29

INDEX

Storms, synthetic patterns in cities, **20**-21
   transposition of, **9**-60, **9**-62
      tropical, **3**-34, **9**-23
         revolving, **3**-30
Straight-row farming, **21**-26
Stratified random method, **26**-37
Stratosphere, **3**-10
Stream floods, **24**-26
Stream frequency, **4**-55
Stream length, **4**-45
Stream orders, **4**-43
Streamflow, **14**-2, **14**-37
   in arid and semiarid regions, **24**-24
   climatic effect on, **16**-28
   effects on groundwater, **13**-34
   gaging station density for, **26**-9
   history of measurement, **15**-4
   routing for river forecasting, **25**-104
Streamline, **7**-9
Streams, effluent, **14**-4
   influent, **14**-4
   interstate, legal aspects of, **27**-21
   jet, **9**-19
   law of slopes, **4**-60
   profile analyses of, **4**-56 to **4**-61
   profile segmentation of, **4**-60
   slope factor of, **4**-61
   surface, legal aspects of, **27**-4, **27**-34
   underground, **27**-28
   (*See also* Channel; Rivers)
Streamtube, **7**-9
Strickler's formula, **17**-42
Structural factors affecting surface water, **4**-2
Stump Lake, N.D., **29**-22
Sublimation, **10**-29, **11**-3
   latent heat of, **23**-17
Sublimation nuclei, **9**-13
Submerged objects, **7**-42
Submerged surfaces, forces on, **7**-8
Submergence of outlet, **21**-65
Subroutine, **29**-18
Subsurface barrier, **13**-51
Subsurface drainage, **21**-91
   in airports, **20**-35
Subsurface flow, **14**-2
Subsurface irrigation, **21**-79, **21**-87
Subsurface runoff, **14**-2
Subsurface storm flow, **14**-2
Subsurface water, **13**-5
Subtropical convergence, **2**-8
Succession, ecological, **6**-2
   hydrarch, **23**-31
   of swamps, **23**-31
Successive approximation, method of, **14**-15
Successive average-lag method, **25**-46
Sudbury River, Mass., recording gage, **15**-4
Sulfate, **19**-6
Sum-of-harmonics process, **8**-84
Summer, **29**-4
Summing amplifier, **29**-4
Summing hydrographs, method of, **20**-15, **20**-16
Suncook River, North Chichester, N.H., stage-discharge relation, **15**-38

Sunshine table, **11**-27, **21**-9
Sunspot cycle, **8**-12
Superposition, principle of, **14**-14
Supplementing elements, **29**-4
Supply curves, standard, **20**-38
Supply rate number, **20**-38
Suprapermafrost, **10**-48
Surcharge storage, induced, **25**-77
Surf beat, **2**-18
Surface active agent, **19**-8
Surface drag, **17**-41
Surface drainage, **21**-88, **21**-89
   in airports, **20**-35
   in arid and semiarid regions, **24**-22
   modification of, **24**-27
Surface energy, **5**-9, **7**-4, **16**-5
Surface irrigation, **21**-87
Surface layer, **2**-6
Surface runoff, **14**-2
   direct, **14**-2
Surface seiche, **23**-26
Surface streams, legal aspects of, **27**-4
   references on laws, **27**-34
Surface tension, **5**-9, **7**-4
Surface-tension coefficient, **5**-9
Surface water, characteristics of, **19**-17
   defined legally, **27**-27
   geologic factors affecting, **4**-2
   lithologic factors of, **4**-2
   soil effects on, **4**-3
   structural factors of, **4**-2
   topographic effects on, **4**-3 to **4**-6
   (*See also* Runoff)
Surface-water law, **27**-30
Surface wave, **2**-15 to **2**-17, **23**-24
Surfaces, submerged, forces on, **7**-8
Surfactant, **19**-8
Surges, **7**-32 to **7**-34
   moving, **7**-35
   standing, **7**-34
Suspended load, **7**-31, **17**-61
   true, **17**-61
Suspended-load sampler, **17**-61
Suspended material, concentration determination of, **17**-62
Suspension, **7**-29, **17**-45
Sustained runoff, **4**-3, **14**-2
Swamps, **23**-30
   defined, **23**-31
Swan River, Mont., temperature effect on snow, **10**-32
Swell, **2**-17
Synclinal spring, **4**-35
Synergistic effects, **19**-31
Synoptic meteorology, **9**-18
Synoptic network, **9**-7
Synthetic detergent, **19**-8
Synthetic hydrograph, **21**-41
Synthetic hydrology, **8**-91
Synthetic project flood, **25**-95
Synthetic unit hydrograph, **14**-13, **14**-22
System, **8**-10, **26**-31
   basin, nonlinearity of, **14**-13
   binary, **29**-17
   deterministic, **8**-10

System, dynamic, **8**-10
 harvesting, **6**-24
 hydrologic, **8**-9
 mesoscale, **3**-10
 nonlinear, **14**-34
 number, **29**-17
 pedological, **5**-4 to **5**-6
 sequential, **8**-10
 stochastic, **8**-10
 Wadi, **24**-22
System analysis, **14**-31 to **14**-33
System design, **26**-31
Systematic errors, **8**-18, **8**-45
Systematic sampling, **26**-37

t test, **8**-57, **8**-86
Tagg's method, **4**-28
Talus spring, **4**-35
Tape gage, **15**-6
Task Force on Water Resources and Power, **1**-10
Taste, **19**-11
Technical report, **26**-15
Tectonic lake, **23**-4, **24**-26
Temperature, absolute zero, **7**-6
 affecting evaporation, **11**-4
 affecting transpiration, **11**-21
 antecedent, **22**-29
 boiling point, **7**-4
 dew-point, **3**-5
 freezing point, **10**-13
 frost-point, **3**-5
 isobaric equivalent, **3**-5
 melting point, **10**-9
  depression of, **16**-7
 potential, **3**-8
 soil, **5**-20
 storm, **22**-29
 virtual, **3**-9
 of water, **19**-13
 wet-bulb, **3**-5
Temperature scales, **19**-17
Temperature variation, with altitude, **3**-7, **3**-11
 annual, **3**-15, **3**-16
 diurnal, **3**-17
 with vapor pressure, **3**-4
Temperature-salinity (TS) diagram, **2**-4
Temporary hardness, **19**-10
Tennessee River, **25**-91
 at Chattanooga, Tenn., flood record, **25**-70
Tennessee River Basin, **1**-14
Tennessee Valley Authority (*see* TVA)
Tensiometer, **5**-11
Tension, hoop, **7**-9
 surface, **5**-9, **7**-4
Tensometric recorder, **15**-9
Terrace, **21**-46
 design, **21**-46 to **21**-53
 graded, **21**-51
 level, **21**-49
Terracing, **21**-26

Test, of significance, **8**-55, **8**-82, **8**-83
 for correlation, **8**-86
 for serial dependence, nonparametric, **8**-83
  parametric, **8**-82
 of unity of title, **27**-5
Test reach, **17**-39
Testing, **29**-19
Theis, C. V., on nonequilibrium theory, **1**-9
Theis' equation, **13**-18
Theis' formula, assumptions of, **4**-24
Theoretical pan, **11**-7
Theory, of drainage-basin dynamics, **4**-69
 of extreme values, **8**-3
 of inventory, **26**-43
 of natural flow, **27**-6
 of reasonable use, **27**-6
 of second best, **28**-8
 of storage, **14**-48, **26**-43
 of unit hydrographs, **14**-13, **14**-14
 of vagrant drops, **27**-32
Thermal capacity of water, **23**-16
Thermal conductivity, **5**-20, **16**-7
 of water, **23**-17
Thermal diffusivity, **5**-21
Thermal expansion, of water, **23**-17
Thermal spring, **4**-35
Thermal stratification, **23**-17
Thermocline, **19**-20, **23**-18
 seasonal, **2**-6
Thermocline layer, **2**-6
Thermodynamic potential, **11**-16
Thermokarst lake, **23**-5
Thiem's equation, **13**-16
Thiessen method, **9**-28
Thornthwaite-Holzman equation, **11**-14
Thornthwaite precipitation-effectiveness index, **4**-53
Thornthwaite's climate classification, **24**-5
Thornthwaite's equation, **11**-27
Thornthwaite's PE index, **4**-53, **4**-54
Three-point method, **8**-46
Threshold concentration, **19**-11
Throughfall, **6**-7
Throughput, **8**-10
Thunder, **9**-22
Thunderstorm high, **3**-33
Thunderstorms, **9**-21, **9**-22
Tidal efficiency, **13**-38
Tidal harmonics, **2**-18
Tide level, mean, **2**-14
Tide species, **2**-18
Tides, **2**-17 to **2**-19
 astronomical, **2**-18
 diurnal, **2**-18
 effects on groundwater, **13**-37
 meteorological, **2**-18
 mixed, **2**-18
 neap, **2**-18
 partial, **2**-18, **2**-19
 range, **2**-18, **2**-19
 spring, **2**-18
 tropic, **2**-18
Tigris River basin, **29**-22

Tile drain, **21**-91 to **21**-95
Till, **5**-3
Timber, volume, **22**-51
Time, access, **29**-17
  of concentration, **14**-7, **20**-8, **20**-12, **20**-41, **21**-10
    Kirpich's formula for, **14**-7
    nomograph for, **21**-11
    for overland flow, **14**-36
  of drawdown, **20**-14
  of equilibrium, **14**-7
  lag, **14**-9, **14**-25, **25**-24
  of orientation, **14**-48
  of peak flow, **14**-9, **14**-11
  real, **29**-17
  of rise, **14**-9
Time-area-concentration diagram, **14**-28
Time-area diagram, **14**-9, **14**-28
Time-condensation method, **12**-16
Time-contour analysis, **20**-9
Time-contour plan, **14**-13
Time dependence, **8**-18
Time homogeneity, **8**-10
Time invariance, principle of, **14**-15
Time lag, **20**-14
Time-offset method, **20**-19
Time-scale factor, **29**-6
Time series, **8**-9, **8**-78
  characteristics of, **8**-79
  circular, **8**-82
  cyclical, **8**-80
  nonstationary, **8**-78
  stationary, **8**-78
Time-series analysis, **14**-46
Todd River, Australia, **24**-25
Tolerance, **6**-27
Tombolos, **23**-6
Tooma River, **29**-7
Topographic effect, on sediment delivery, **17**-13
  on surface water, **4**-3 to **4**-6
Topographic index, **25**-17
Tornado, **3**-30, **9**-21
Total runoff, **14**-2
Tracer, for groundwater movement, **13**-12
  radioactive, **15**-12
Tractive force, **7**-26, **17**-52
Tractive-force equations, **17**-52
Tractive-force formulas, **17**-58
Tranquil flow, **7**-23
Transformation equation, **29**-6
Transition probability matrix, **14**-49
Translation coefficient, **14**-28
Transmissibility, coefficient of, **4**-13, **13**-8
Transmission rate, **22**-51
Transpiration, **6**-17, **11**-2, **11**-20, **12**-2
  in arid and semiarid regions, **24**-16
  cuticular, **11**-20
  determination of, **11**-22
  effect of thinning on, **6**-25
  factors affecting, **6**-19, **11**-21
  stomatal, **11**-20
Transpiration-assimilation ratio (E/A), **24**-34
Transpiration ratio, **11**-23, **24**-34

Transport, momentum, **2**-9
  vorticity, **2**-9
Transposition, of rainfall data, **21**-21
  of runoff data, **21**-28, **21**-34
Transposition method for runoff, **21**-34
Trap efficiency, **17**-21
  computation of, **17**-21
Travelers Research Center, Inc., **1**-14
Trenches, **2**-2
Trend, **8**-10, **8**-79
  analysis of, **8**-81
  cyclic, **8**-57
  in precipitation, **9**-46
Trials, **8**-4
Triangular hydrograph, **20**-23, **21**-42, **21**-44
Triple point, **10**-8
Tropical depressions, **3**-34
Tropopause, **3**-10
Troposphere, **3**-10
TS-diagram (temperature-salinity diagram), **2**-4
TsIP (Central Forecasting Institute, Moscow), **1**-15
Tsunami, **2**-18
Tubes, **7**-44
  infiltrometer, **12**-7
Tulsequah Glacier, near Juneau, Alaska, **16**-30
Tundra lake, **23**-6
Tundra soil, **5**-5
Turbidimeter, **19**-11
Turbidity, **19**-11, **19**-17
Turbulence, **7**-12
Turbulent boundary layer, **7**-15
Turbulent exchange, **17**-46, **22**-11
Turbulent flow, **7**-13, **7**-23
Turning-point test, **8**-83
Turnover, **23**-17
Tuslumne River, Calif., runoff per degree-day, **10**-34
TVA (Tennessee Valley Authority), **1**-14, **28**-22
  Chickamauga Reservoir, **25**-76, **25**-91
  Douglas Reservoir, **25**-65, **25**-91
  flood control in, **25**-91
  flood fringe in, **25**-120
  founding of, **1**-9
  Guntersville Reservoir, **25**-66, **25**-68, **25**-78, **25**-80
  hydrologic data publications of, **26**-10
  hydrologic network in, **25**-96
  hydropower in, **25**-91
  navigation in, **25**-91
  Norris Reservoir, **25**-75
  on rainfall data, **9**-9
  reservoir system in, **25**-96
  water-control system in, **25**-90
  on water-quality data, **19**-35
TVA system, **25**-91
Twin Falls Soil Conservation District, soil moisture under snowpack, **10**-25
Twin Lakes, Colo., avalanche, **10**-42
Type curves, **13**-18
  for leaky aquifer, **13**-22
Typhoons, **3**-30

INDEX

UCOH (Universities Council on Hydrology), **1**-10, **1**-17
UCOWR (Universities Council on Water Resources), **1**-17
Ultimate use of water, **28**-17
Ultisol, **5**-8
Ultrasonic flowmeter, **15**-13
Uncertainty, physical and tenure, **28**-12
   on water economics, **28**-8
Unconfined aquifer, **13**-5, **13**-6, **13**-17
Unconfined groundwater, **24**-30
Underground streams, defined legally, **27**-28
UNESCO (United Nations Education, Scientific, and Cultural Organization), **1**-10, **1**-11, **1**-12
   on IHD, **1**-10
UNESCO Arid Zone Program, **1**-10
UNESCO Arid Zone Research publications, **24**-42
UNESCO maps of arid and semiarid regions, **24**-3 to **24**-9
Uniform flow, **7**-10
   in open channels, **7**-23
Uniform-flow equations, **7**-23
Uniform-grid method, **26**-37
Uniformity, coefficient of, **5**-7
U.S.S.R. organizations relating to hydrology, **1**-15
U.S.S.R. State Hydrologic Institute (GGI), **1**-15, **1**-16
Unit-graph, **14**-13
Unit hydrograph, **14**-13, **22**-44
   applied to small drainage area, **14**-14
   base length of, **14**-24
   defined, **14**-13
   derivation for other durations, **14**-19
   dimensionless, **21**-41
   duration of, **14**-13
   inadequate for forecasting, **25**-107
   instantaneous (*see* Instantaneous unit hydrograph)
   linear theory of, **14**-14
   nonlinear theory of, **14**-14
   for river forecasting, **25**-102, **25**-103
   synthetic, **14**-13, **14**-22
   theory of, **14**-13
   widths of, **14**-24
Unit-hydrograph analysis, **14**-15 to **14**-22
Unit-hydrograph synthesis, **14**-22
Unit peak discharge, **22**-28
Unit storm, **14**-14
United Nations, **1**-10
   (*See also* ECAFE; FAO; UNESCO; WHO; WMO; WRDC)
United Nations Bureau of Technical Assistance and Operations, **1**-12
United Nations Educational, Scientific and Cultural Organization (*see* UNESCO)
U.S. Advisory Committee on Weather Control, **9**-15, **24**-15
U.S. Agricultural Research Center, **1**-12
U.S. Agricultural Research Service (ARS), **1**-12
   Central Technical Unit, **1**-12

U.S. Agricultural Research Service (ARS), flood routing analog, **29**-7, **29**-10
   resistance network, **29**-12
   sheet erosion, **17**-7, **17**-8
   small watershed publications, **21**-4
U.S. Air Force Wright Development Center, aircraft icing, **10**-5
U.S. Army, Aberdeen Proving Ground, **29**-15, **29**-16
   Ballistic Research Laboratories, **29**-15, **29**-16
   Coastal Engineering Research Board, **1**-13
   Coastal Engineering Research Center, **1**-13
   Cold Regions Research Engineering Laboratory, **1**-13
   Corps of Engineers, **1**-13, **14**-12, **14**-24
   on airport drainage design, **20**-35, **20**-37, **20**-38
   on authorities in water resources, **27**-23, **27**-24, **27**-26
   Chattanooga District, on flood routing, **25**-39
   Cold Regions Research and Engineering Laboratory, **10**-50
   on computers in water resources planning, **29**-22
   on flood-control policy, **28**-21
   on flood protection in urban areas, **20**-53
   founding of, **1**-8
   freezing index, **10**-47
   on hydrologic data publications, **26**-10
   ice bibliographies, **10**-6
   Kansas City District, on flood analog, **29**-7
   Los Angeles District, **14**-37
   Manual EM-1110-2-3600, **25**-77
   on maximum probable flood, **25**-16
   on maximum probable snowmelt flood, **10**-33
   on mechanical flood router, **25**-53
   Missouri River Division, on flood analysis, **29**-22
   Missouri River Reservoir System, **25**-92
   on Muskingum method, **25**-40
   on navigation water policy, **28**-20
   North Pacific Division, on flood routing by computer, **29**-7
   on streamflow routing, **29**-21
   Ohio River Division, on flood routing, **29**-21
   by computer, **29**-7
   Omaha District Laboratory, **17**-41
   radioisotope snow gage, **10**-20
   rainfall records, **9**-46
   Reservoir Control Center, Omaha, Nebr., **25**-94
   Rock Island District, on flood routing, **25**-46
   Sacramento District, precipitation estimates on, **9**-29

INDEX

U.S. Army, Corps of Engineers, sediment load in Mississippi delta, **17**-20
  on simulation analysis, **26**-35
  SIPRE, **10**-49
  snow classification, **10**-4
  snow observation, **10**-3
  snowmelt equations, **10**-34, **10**-39
  on standard project flood, **25**-26, **25**-72
  standard project storm, **9**-64
  storm pattern, **9**-60
  "Storm Rainfall in the Eastern United States," **25**-27
  on streamflow records, **25**-8
  Tulsa and Kansas City Districts, on flood routing, **25**-47
  on water-quality data, **19**-35
  Waterways Experiment Station (WES), **1**-13, **7**-29
    founding of, **1**-9
  on Wolf Creek Project, **25**-88
  Willamette Basin Snow Laboratory, **10**-30
Signal Corps, on rainfall records, **9**-9
U.S. Army Weasel and M-7 over snow vehicles, **10**-17
U.S. Atomic Energy Commission (AEC), **1**-14
  Plowshare program, **19**-7
  on radioactivity, **19**-12
U.S. Beach Erosion Board, **1**-13
U.S. Board of Engineers for Rivers and Harbors, **1**-13
U.S. Bureau of the Budget, Budget Bureau Circular A-47, **26**-16
U.S. Bureau of the Census, on hydrologic data publications, **26**-10
  on water uses, **28**-3
U.S. Bureau of Plant Industry, sunken evaporation pan, **11**-7
U.S. Bureau of Public Roads (BPR), **1**-13
  founding of, **1**-9
  freezing index, **10**-47
  method, **25**-17
U.S. Bureau of Reclamation (USBR), **1**-14
  on authorities in water resources, **27**-24
  on channel lining, **20**-32
  on computer application, **29**-22
  on flow over dams, **15**-33
  founding of, **1**-9
  on hydrologic data publications, **26**-10
  irrigation districts of, **28**-11
  on irrigation policy, **28**-21
  on maximum probable flood, **25**-16
  on navigation policy, **28**-20
  on rainfall records, **9**-46
  on snow gage system, **10**-13
  on storm pattern, **9**-60
  on streamflow records, **25**-8
  on unsteady flow by computer, **29**-7
  on water price, **28**-5
U.S. Bureau of Soils, soil classification, **17**-15
U.S. Bureau of State Services, **1**-13
U.S. Civil Aeronautics Administration, **20**-35

U.S. Coast and Geodetic Survey (CGS), **1**-13
U.S. Coast Guard, on navigation policy, **28**-20
U.S. Department of Agriculture, **1**-12
  on authorities in water resources, **27**-25
  Bulletin 1026, **17**-46
  on hydrologic data publications, **26**-10
  on irrigation water quality, **19**-7
  on runoff curve numbers, **22**-47
  on runoff data, **21**-27
  on soil survey, **5**-6, **5**-7
  on upstream-flood-prevention program in California, **22**-30
  on water-quality data, **19**-35
U.S. Department of the Army, **1**-13
U.S. Department of Commerce, **1**-12
U.S. Department of Defense, **29**-21
U.S. Department of Health, Education, and Welfare, **1**-13
  on pollution policy, **28**-22
  on water-quality data, **19**-35
U.S. Department of the Interior, **1**-14
U.S. Department of the Navy (*see* U.S. Navy)
U.S. Federal Power Commission, **1**-14, **27**-24
  on hydrologic data publications, **26**-10
U.S. Fish and Wildlife Service, **27**-24
  on water-quality data, **19**-35
U.S. Forest Service, **1**-12
  avalanche, **10**-42
  Coweeta Hydrologic Laboratory, N.C., **6**-24
  "Forest and Range Handbook," **22**-12
  founding of, **1**-9
U.S. Geological Survey (USGS), **1**-14
  on annual flood mean, **8**-25
  on bubble gage, **15**-8
  on corrections on streamflow measurements, **15**-17
  on electromagnetic flowmeter, **15**-13
  on electronic analog, **29**-8
  on floating pan, **11**-7, **11**-8
  on flood-frequency reports, **25**-5, **25**-6
  founding of, **1**-8
  on geologic information, **4**-27
  on groundwater fluctuation, **11**-25
  on hydrologic data, **25**-99
    publications, **26**-10
  ice effect on streamflow, **15**-38
  on indirect flow measurement, **15**-33
  major field offices of, **25**-7
  on navigation policy, **28**-20
  on regional analysis, **8**-36
  on regional flood-frequency, **21**-41
  on resistance network, **29**-12
  on runoff records on Clearwater River, **26**-36
  on salt-water-flow system, **13**-50
  on small Price meter, **15**-10
  on snow conferences, **10**-23
  on stage-discharge relation, **15**-34
  on streamflow measurement, **15**-18, **15**-20 to **15**-24, **25**-2

INDEX

U.S. Geological Survey (USGS), on water hardness, **19**-10
  on water-quality analysis, **19**-14
  Water Supply Papers, **25**-8
  on water turbidity, **19**-11
  on water year, **8**-12, **15**-41
U.S. Housing and Home Finance Agency, snowpack maps, **10**-43
U.S. Hydrographic Office, **1**-2
U.S. Inter-Agency Committee on Water Resources, on hydrologic data publications, **26**-10
  Subcommittee, on Hydrology, **1**-14, **19**-35
  on Sedimentation, **17**-25 to **17**-27, **17**-66
  (*See also* Federal Inter-Agency River Basin Committee)
U.S. International Boundary and Water Commission, on hydrologic data publications, **26**-10
U.S. Lake Survey, **1**-13, **23**-14
  Price meters, **15**-9 to **15**-11
U.S. Mississippi River Commission, on hydrologic data publications, **26**-10
U.S. National Bureau of Standards, **1**-12
U.S. National Hydraulics Laboratory, **1**-9, **1**-12
U.S. National Institutes of Health, **1**-13
U.S. Navy, **29**-16
  Bureau of Naval Weapons, **1**-13
  Bureau of Ships, **1**-13
  Bureau of Yards and Docks, **1**-13
  on electromagnetic flowmeter, **15**-13
  Hydrographic Office, **1**-13, **2**-6
  Office of the Chief of Naval Operation, **1**-13
  Office of Naval Research, **1**-13
  on snow compaction, **10**-41
U.S. Office of Saline Water, **1**-14, **19**-32
U.S. Patent Office, on forest destruction, **22**-2
U.S. Public Health Service, **1**-13
  on Drinking Water Standards, **19**-5, **19**-7, **19**-8, **19**-27, **19**-28
  on hydrologic data publications, **26**-10
  on pollution survey, **20**-34
  on survey of water use, **28**-3
  on water-quality criteria, **19**-4 to **19**-8, **19**-11, **19**-12, **19**-26, **19**-35
U.S. Public law 87-27 (*see* Area Redevelopment Act)
U.S. Public Law 87-658 (*see* Public Works Acceleration Act)
U.S. Public Law 566 (*see* Watershed Protection and Flood Control Act)
U.S. Senate Bill 1111 (*see* Water Resources Planning Act)
U.S. Senate Bill S-2 (*see* Water Resources Research Act)
U.S. Senate Document No. 97, **26**-3, **26**-15 to **26**-17, **26**-19
U.S. Senate Select Committee on National Water Resources, **1**-10, **28**-15
  Committee prints, **26**-26, **26**-27

U.S. Soil Conservation Service (SCS), **1**-12, **14**-24, **25**-24
  on authorities in water resources, **27**-25
  on Cook's method, **25**-20
  on cropping systems, **17**-32
  on design criteria for terrace and diversions, **21**-46
  "Engineering Handbook," **22**-12
  founding of, **1**-9
  on multiple-purpose reservoirs, **25**-64
  on operating flood-control projects, **25**-86
  on sediment size, **17**-17
  on snow survey, **10**-21, **10**-22
  on snow-survey form, **10**-17
  on soil classification, **5**-8, **21**-2 to **21**-25
  soil moisture under snowpack, **10**-25
  on storm pattern, **9**-60
  on streamflow records, **25**-8
  water-yield forecast, **10**-26
U.S. Weather Bureau (USWB), **1**-13, **25**-24
  a *Circular B,* **10**-9, **10**-13
  on Class A evaporimeter, **24**-17
  Class A Land Pan, **11**-6 to **11**-8
  on coaxial graphical method, **25**-102
  Daily River Stages, **25**-8
  on defining drought, **18**-1
  on electronic flood router, **25**-54, **29**-3, **29**-7
  on evaporation record, **11**-5
  on excessive precipitation, **9**-35, **9**-39
  founding of, **1**-8, **1**-9
  on hydrologic data, **25**-99
  publication of, **26**-10
  National Weather Record Center, **9**-9, **26**-13
  on precipitation forecasting, **9**-25
  on precipitation normal, **8**-7
  Quantitative Precipitation Forecast Program, **25**-109
  on radioisotope snow gage, **10**-20, **10**-21
  on rain gages at City of Los Angeles, **20**-10, **20**-11
  on rain-gage network, **9**-8
  on rainfall computation, **9**-28
  on rainfall data, **21**-4
    at Baltimore, Chicago, and Philadelphia, **20**-26
  on rainfall energy and erosion, **17**-7
  rainfall frequency atlas, **9**-49, **9**-60
  on rainfall frequency data, **8**-4, **20**-5
  on recording rain gage, **9**-5
  River Forecast Centers, **25**-102
  on river forecasting instructions, **25**-100
  on snow data, **10**-6, **10**-22
  on snowpack maps, **10**-43
  on storage precipitation gage, **10**-12
  on sunshine table, **11**-27
  water-supply forecast, **10**-25, **10**-26, **10**-32
UNIVAC, **29**-21, **29**-23
Universal sheet-erosion equation, **17**-7
Universe, **8**-4, **8**-50
Universities Council on Hydrology (UCOH), **1**-10, **1**-17

Universities Council on Water Resources (UCOWR), **1**-17
Unsteady flow, **7**-9
  of groundwater, **13**-14
Unsteady radial flow, **13**-18
Upconcavity of channel profile, **4**-57
Upland forest soil, **5**-9
Upper Colorado River, at Lees Ferry, Ariz., **8**-92 to **8**-94, **14**-47
Upper Potomac River Commission, **18**-22
Urban areas, flood losses in, **25**-115
  floods in, **20**-33
  water supply in, **20**-33
Urban expressway, **20**-41
Urban hydrology, **20**-2 to **20**-4
  design example for, **20**-27
  rainfall-runoff correlation for, **20**-9
Urbanization, hydrologic effects on, **20**-2 to **20**-4
USBR (*see* U.S. Bureau of Reclamation)
Use line, **14**-46
  variable, **14**-46
USGS (*see* U.S. Geological Survey)
USWB (*see* U.S. Weather Bureau)
Utah State Engineer's Office and Water Power Board, **28**-23

Vacuum, **7**-6
Vagrant drops, theory of, **27**-32
Valley fill, **24**-31
Valley spring, **4**-35
Valuation in water economics, **28**-6
van Bavel nomograph, **21**-8
Vapor density, **3**-5
Vapor pressure, **3**-4, **7**-4, **11**-3
Vapor transfer, **11**-25
Vapor transport, **9**-15
Variability, of a spring, **4**-35
  of mean streamflow, **8**-13
  measures of, **8**-7
Variability index, **8**-3, **14**-43
Variable, **8**-44
  autocorrelated, **8**-47
  computer, **29**-4, **29**-6
  continuous, **8**-5
  dependent, externally, **8**-48
    internally, **8**-47
  discrete, **8**-5
  independent, externally, **8**-48
    internally, **8**-47
  problem, **29**-6
  random, **8**-4, **8**-47
  serially correlated, **8**-47
  slack, **26**-40
  standardized, **8**-56
  statistical, **8**-4
  transformation, **8**-48
Variance, **8**-7
  analysis of, **8**-69, **8**-72
Variate, **8**-4
Variation, coefficient of, **8**-7
  secular, **8**-12
Varied flow, gradually, **7**-38
  in open channels, **7**-23

Vegetal cover, benefits of, **22**-7
Vegetal retardance, **21**-61
Vegetation, effect on infiltration, **6**-14
  effects of water, **6**-3
Vegetative screens, **17**-31
Velocity, **7**-9
  critical (*see* Critical velocity)
  distribution of, **7**-15
    in glaciers, **16**-22
    on sediment bed, **17**-45
    in streams, **15**-15
  fall, **7**-43
  friction, **17**-43
  measurement of, **15**-9
    by float, **15**-11
  permissible, **7**-26, **21**-60
  salt, **15**-12
  settling, **17**-41
  terminal, **3**-19
    of rain, **9**-12
    of waves, **23**-25
Venturi flume, **7**-47, **15**-30
Venturi meter, **7**-47
Vertisol, **5**-8
Virgin flow, **14**-2
Virtual channel-flow graph, **14**-13
Viscosity, **7**-3
  dynamic, **7**-4
  eddy, **7**-13
    dynamic, **7**-13
    kinematic, **7**-13
  fluid, **7**-12
  of ice, **16**-6
  kinematic, **7**-4
  molar, **7**-13
  molecular, **7**-13
Visual-accumulation-tube sand-size analyzer, **17**-40
Vitruvius, Marcus, on hydrologic cycle, **1**-7
Voids ratio, **17**-16
Volume conductivity, **19**-11
Volume weight, **5**-4
Volumetric analysis, **19**-14
von Kármán's constant, **17**-46
von Kármán's universal constant, **7**-31

W index, **12**-29
W_{min} index, **12**-29
Wadi system, **24**-22
Wading rod, **15**-16
Waiting line, **14**-48
Wake depression, **3**-33
Walker test, **8**-11
Wallkill River, Pellet's Island Mountain, N.Y., diversion curves of, **18**-24
Warm-film anemometer, **15**-13
Warm front, **3**-29
Warm spring, **4**-35
Wash load, **7**-29, **17**-36, **17**-37, **17**-61
  measurement of, **17**-55
Washes, defined legally, **27**-27
Washington National Airport, **20**-35
Washout dune, **7**-26

INDEX

Waste-assimilative capacity, **19**-26
Waste water, defined legally, **27**-27, **29**-24
   reclamation of, **19**-25
Wastes, **19**-24 to **19**-27
   assimilative capacity of, **19**-26
Water, for agricultural uses, **19**-31
   for aquatic life, **19**-31
   artesian, **24**-32
   atmospheric, **27**-29
   available, **5**-15
   beneficial use of, **27**-12
   for boilers, **19**-29
   brackish, **19**-33
   capillary, **22**-9
   color of, **19**-11, **19**-17
   constituents in, **19**-3
   for cooling, **19**-30
   developed, **27**-28
   domestic and municipal uses, **19**-27
   duty of, **11**-2
   effects on erosion, **17**-6
   effects on vegetation, **6**-3
   for food and beverage industries, **19**-28
   foreign, **27**-28
   gravitational, **5**-4, **5**-15, **20**-10
   hygroscopic, **16**-5, **22**-9
   for internal-combustion engines, **19**-30
   in lakes, **23**-15
   latent heat of fusion of, **10**-5
   maximum density of, **19**-13
   oxygen solubility in, **19**-9
   percolating, **27**-28
   for pharmaceutical uses, **19**-29
   precipitable, **3**-6
   properties of, **19**-3
   quality of (*see* Water quality)
   reclamation of, **19**-25
   re-use of, **19**-25
   salinity in sea, **19**-22
   sampling techniques for, **19**-13
   in sea (*see* Sea water)
   soil, **5**-9
   specific heat of, **10**-5
   spreading of, **13**-42
   standards for drinking, **19**-27
   subsurface, **13**-5
   surface, **27**-27
      (*See also* Surface water)
   for textile and paper manufacture, **19**-28
   thermal capacity of, **23**-16
   thermal conductivity of, **23**-17
   thermal expansion of, **23**-17
   for transporting and processing raw materials, **19**-30
   ultimate use of, **28**-17
   vadose, **4**-6 to **4**-9
   waste (*see* Waste water)
Water-allocation policy, **28**-2, **28**-10
Water-balance method, **11**-11
Water body, **23**-3
Water-budget method, **11**-11
Water conditioning, **19**-32
Water control on farm, **21**-3
Water cycle, **3**-15
   (*See also* Hydrologic cycle)

Water demand, **28**-15
Water-development policy, **28**-2, **28**-12
Water diversion in Great Lakes, **23**-10
Water economics, **28**-6 to **28**-9
Water equivalent, **21**-33
Water Facilities Act, **27**-25
Water firms, **28**-2
Water hole, **24**-27
Water law, **27**-2
   Colorado doctrine, **27**-18
   conservation statute, **27**-31
   doctrine, of prior appropriation, **27**-11
      of riparian rights, **27**-4
   on drainage, **27**-30
   Federal, **27**-22
   on reasonable use, **27**-31
   references on Federal and international, **27**-35
   for various states, **27**-35 to **27**-37
   (*See also* Law)
Water market, **28**-4
Water mass, **2**-4
Water movement, in plants, **6**-4
   in unsaturated soils, **5**-19
Water policy, on allocation, **28**-2
   defined, **28**-2
   on development, **28**-12
   economic projection in, **28**-14
   on groundwater, **28**-4
   implementation of objectives, **28**-9
   implications for, **28**-19
   organization and coordination in U.S. **28**-20 to **28**-23
   as project purpose, **26**-3
Water Policy Panel, Engineers Joint Council, **1**-10
Water pollution, **20**-33
Water Pollution Control Act, **20**-34, **27**-21, **28**-22
Water Pollution Control Federation, **14**-7, **19**-14
Water power, analysis by duration curve, **14**-42
Water quality, **19**-2
   characteristics of, **19**-17
   control of (*see* Water-quality control)
   data, **19**-35
   deterioration of, **19**-23
   determination of, **19**-13
   frequency analysis of, **8**-4
   law for management of, **28**-12
   legal aspects about, **27**-5
   National Water Quality Network, **19**-35
   in U.S., **19**-35
Water-quality control, benefits of, **26**-19
   on hydrologic data, **26**-7
Water quantity, legal aspects about, **27**-13
Water requirement, **11**-2, **21**-6, **24**-34, **28**-15
Water resources, allocation policy on, **28**-2
   design criteria for, **26**-31
   development objects of, **26**-31
   development policy of, **28**-2
   hydrologic appraisal of, **26**-6

INDEX

Water resources, multiple-purpose project (*see* Multiple-purpose projects)
   operations research for, **26**-30
   project formulation (*see* Project formulation)
   project selection by benefit-cost ratio, **26**-25
   single-purpose project, **26**-3
   system design of, **26**-31
Water Resources Development Center, **1**-10, **1**-12
Water resources planning, computer application in, **29**-22
Water Resources Planning Act, **26**-17, **26**-28
Water-resources project, purposes of, **26**-2, **26**-3
Water Resources Research Act, **1**-10
Water-resources system, **26**-31
   mathematical models, **26**-38
Water rights, **27**-14
Water spreading, **24**-27
Water-stage recorder, **15**-7
Water supply, **19**-27
   benefits of, **26**-19
   domestic and industrial, **26**-2
   economic analysis, example for, **26**-20 to **26**-23
   forecasting, **25**-110
   hydrologic data for, **26**-7
   for irrigation, **21**-79
   legal aspects of, **27**-6
   (*See also* Water law)
   of Missouri River, **25**-92
   policy on, **28**-22
   in urban areas, **20**-33
Water Supply Paper, **15**-26, **17**-60, **19**-35, **25**-7, **25**-8, **25**-15, **26**-10
Water table, in arid and humid zones, **4**-26
   perched, **5**-19
Water-table cement, **4**-8
Water-table spring, **4**-35
Water-table well, **13**-15
Water vapor, **3**-4
   transport of, **9**-15
Water year, **8**-12, **15**-41
Water yield, forcasting of, **10**-23
Watercourse, defined legally, **27**-26
   legal aspects on changes, **27**-19
Watershed, **14**-2
   annual yields of, **21**-36
   characteristics of, **21**-10
   constant for condition of, **21**-33
   cover of, **21**-11
   cover density of, **22**-30
   denuded area on, **22**-30
   ratings of, **22**-32 to **22**-38
   runoff from small, **21**-36
   shape of, **21**-10
   size of, **21**-10
   slope of, **21**-10
   small, **21**-4
   small versus large, **25**-16
   (*See also* Drainage area; Drainage basin)
Watershed constants, **22**-29

Watershed management, **21**-4, **26**-2
Watershed 97, Coshocton, Ohio, hydrograph analysis of, **14**-12
Watershed Protection and Flood Control Act, **28**-21
Watershed rating factor, **22**-32 to **22**-38
Waterways, grassed, **21**-54
Wave height, by wind, **23**-25
Wave period, **23**-25
Wave pressure, **2**-19
Wave velocity, **23**-25
Wavelength, **2**-16, **23**-25
Waves, **7**-32
   breaking, **2**-20
   capillary, **2**-16
   celerity of, **2**-16, **2**-17, **7**-32
   clapotis, **2**-19
   deep-water, **2**-17, **7**-33
   easterly, **3**-33
   energy, **2**-17
   frequency analysis of, **8**-4
   gravity, **2**-16, **2**-17
      small, **7**-32
   infragravity, **2**-16
   internal, **2**-19
   kinematic, **16**-26
   long-period, **2**-16
   ocean, **2**-14 to **2**-20
   open-water, **23**-24
   ordinary gravity, **2**-16
   period of, **2**-16
   progressive, **23**-24
   sand, **17**-62
   shallow-water, **2**-17, **7**-33
   shock, **16**-28
   standing, **7**-26
   stationary, **7**-33
   surface, **2**-15 to **2**-17, **23**-24
   tidal, **2**-18
   trans-tidal, **2**-17
   ultra-gravity, **2**-16
   wind, **2**-17
Waving, **14**-19
Weather, **3**-10
   convective, **3**-30, **3**-32, **3**-33
   extremes around the world, **10**-11
Weather modification, legal aspects, **27**-29
Weather regimes, **3**-29, **3**-35 to **3**-38
Weather systems, **3**-29
Weathering process, **24**-22
Weber number, **7**-5
Wedge storage, **25**-40
Weibull distribution, **8**-16
Weibull formula, **8**-29
Weibull probability paper, **8**-28
Weight, specific (*see* Specific weight)
   volume, **5**-4
Weir, **7**-45, **15**-30
   broad-crested, **15**-33
   Cipolletti, **7**-46
   sharp- and broad-crested, **7**-45
   suppressed, **7**-45
   trapedzoidal, **7**-46
   triangular, **7**-46
   uncontracted, **7**-45

INDEX

Weir formula, **15**-33
Welfare economics, piecemeal, **28**-8
Well function, **13**-18
Well interference, **13**-25
Well losses, **13**-27
Well systems, **13**-25
Wells, **4**-33, **4**-34
   artesian, **13**-15
   capacity of, specific, **4**-3, **13**-27
   collector, **13**-31
   completion of, **13**-30
   deep, **13**-28
   development of, **13**-30
   drilling of, **13**-28
   flow to, near boundaries, **13**-22
     near stream, **13**-22
   gravel-packed, **13**-31
   hydraulics of, **13**-15
   image, **13**-24
   maintenance and repair of, **13**-31
   penetrating, partially, **13**-26
   sanitary protection on, **13**-31
   shallow, **13**-28
   testing of, **13**-30
   in uniform flow, **13**-17
   water-table, **13**-15
Wenner electrode configuration, **4**-28, **4**-29
WES (U.S. Army Waterways Experiment Station), **1**-9, **1**-13
West Australian Current, **2**-7
West Coast, Los Angeles County, **28**-13
West Coastal Basin, Los Angeles, Calif., control of salt-water intrusion, **13**-51
Western Snow Conference, **10**-22
Wet-line corrections, **15**-26
Wetting angle, **5**-10
White caps, **2**-17
White Hollow watershed, Tenn., **6**-25
White House Office, Executive Office of the President, **1**-12
White's equation, **7**-27
WHO (World Health Organization), **1**-10, **1**-12
Wiesenboden, **5**-5
Wild-flooding, **24**-27
Willamette Basin Snow Laboratory, **10**-30
Wilting point, **5**-15
   permanent, **5**-15, **11**-17
Wind coefficient, **2**-9
Wind erosion, **17**-10, **17**-19
Wind factor, **11**-5
Wind stress formula, **23**-28
Wind systems, **3**-23
Winds, affecting evaporation, **11**-5
   affecting transpiration, **11**-22
   average zonal speed around world, **3**-24
   charts of, **3**-24 to **3**-28
   effect on erosion, **17**-10, **17**-19

Winds, geostrophic, **3**-24
   mountain and valley, **3**-28
   thermal, **3**-24
Wire gage, **15**-6
Wisconsin Valley Improvement Company, **18**-22
WMO (*see* World Meteorological Organization)
Wolf Creek Project, **25**-86, **25**-88
Wolf Creek Reservoir, operating schedule and guide of, **25**-89
Woltman's current meter, **1**-7, **15**-5
Woodland, **21**-26
Word, **29**-17
Word length, **29**-17
Word time, **29**-17
Working discharge, **25**-44
Working-value method, **25**-44, **25**-49, **25**-50
Working values, **25**-44
World Health Organization (WHO), **1**-10, **1**-12
World Meteorological Organization (WMO), **1**-10, **1**-12
   on climate classification, **24**-4
   on cloud seeding, **9**-15
   on precipitation modification, **24**-15
World Power Conference, **1**-9, **1**-11
World's rainfall records, **9**-47
World's weather extremes, **10**-11
WPC (*see* World Power Conference)
WRDC (Water Resources Development Center), **1**-10, **1**-12
Write time, **29**-17

Xerophyte, **6**-5, **6**-28
Xylem, **6**-4, **6**-28

Yarnell rainfall frequency data, **8**-4, **9**-49
Yarnell-Woodward's formula, **21**-94
Yellowish-brown lateritic soil, **5**-5
Yellowstone River, Mont., **4**-3, **4**-4
Yellowstone River Basin, Wyo., seasonal runoff, **10**-34
Yield, specific, **13**-4
Youghiogheny River, Pa. and Md., flood frequency of, **25**-6
Young's modulus of ice, **16**-6

Zonal soils, **5**-4, **5**-5
Zone, ablation, **16**-17
   accumulation, **16**-17
   ferrito, **4**-9
Zone ordinances, **25**-121
Zone principle, **20**-8
Zoning, floodplain, **25**-121
   in urban hydrology, **20**-12